caput- *head* decapitate, remove the head

carcin- *cancer* carcinogen, a cancer-causing agent

cardi, cardio- *heart* cardiotoxic, harmful to the heart

carneo- *flesh* trabeculae carneae, ridges of muscle in the ventricles of the heart

carot- *1) carrot, 2) stupor* 1) carotene, an orange pigment; 2) carotid arteries in the neck, blockage causes fainting

cata- *down* catabolism, chemical breakdown

caud- *tail* caudal (directional term)

cec- *blind* cecum of large intestine, a blind-ended pouch

cele- *abdominal* celiac artery, in the abdomen

cephal- *head* cephalometer, an instrument for measuring the head

cerebro- *brain, especially the cerebrum* cerebrospinal, pertaining to the brain and spinal cord

cervic-, cervix *neck* cervix of the uterus

chiasm- *crossing* optic chiasma, where optic nerves cross

chole- *bile* cholesterol; cholecystokinin, a bile-secreting hormone

chondr- *cartilage* chondrogenic, giving rise to cartilage

chrom- *colored* chromosome, so named because they stain darkly

cili- *small hair* ciliated epithelium

circum- *around* circumnuclear, surrounding the nucleus

clavic- *key* clavicle, a "skeleton key"

co-, con- *together* concentric, common center, together in the center

coccy- *cuckoo* coccyx, which is beak-shaped

cochlea *snail shell* the cochlea of the inner ear, which is coiled like a snail shell

coel- *hollow* coelom, the ventral body cavity

commis- *united* gray commissure of the spinal cord connects the two columns of gray matter

concha *shell* nasal conchae, coiled shelves of bone in the nasal cavity

contra- *against* contraceptive, agent preventing conception

corn-, cornu- *horn* stratum corneum, outer layer of the skin composed of (horny) cells

corona *crown* coronal suture of the skull

corp- *body* corpse, corpus luteum, hormone-secreting body in the ovary

cort- *bark* cortex, the outer layer of the brain, kidney, adrenal glands, and lymph nodes

cost- *rib* intercostal, between the ribs

crani- *skull* craniotomy, a skull operation

crypt- *hidden* cryptomenorrhea, a condition in which menstrual symptoms are experienced but no external loss of blood occurs

cusp- *pointed* bicuspid, tricuspid valves of the heart

cutic- *skin* cuticle of the nail

cyan- *blue* cyanosis, blue color of the skin due to lack of oxygen

cyst- *sac, bladder* cystitis, inflammation of the urinary bladder

cyt- *cell* cytology, the study of cells

de- *undoing, reversal, loss, removal* deactivation, becoming inactive

decid- *falling off* deciduous (milk) teeth

delta *triangular* deltoid muscle, roughly triangular in shape

den-, dent- *tooth* dentin of the tooth

dendr- *tree, branch* dendrites, telodendria, both branches of a neuron

derm- *skin* dermis, deep layer of the skin

desm- *bond* desmosome, which binds adjacent epithelial cells

di- *twice, double* dimorphism, having two forms

dia- *through, between* diaphragm, the wall through or between two areas

dialys- *separate, break apart* kidney dialysis, in which waste products are removed from the blood

diastol- *stand apart* cardiac diastole, between successive contractions of the heart

diure- *urinate* diuretic, a drug that increases urine output

dors- *the back* dorsal; dorsum; dorsiflexion

duc-, duct *lead, draw* ductus deferens which carries sperm from the epididymis into the urethra during ejaculation

dura *hard* dura mater, tough outer meninx

dys- *difficult, faulty, painful* dyspepsia, disturbed digestion

ec-, ex-, ecto- *out, outside, away from* excrete, to remove materials from the body

ectop- *displaced* ectopic pregnancy; ectopic focus for initiation of heart contraction

edem- *swelling* edema, accumulation of water in body tissues

ef- *away* efferent nerve fibers, which carry impulses away from the central nervous system

ejac- *to shoot forth* ejaculation of semen

embol- *wedge* embolus, an obstructive object traveling in the bloodstream

en-, em- *in, inside* encysted, enclosed in a cyst or capsule

enceph- *brain* encephalitis, inflammation of the brain

endo- *within, inner* endocytosis, taking particles into a cell

entero- *intestine* enterologist, one who specializes in the study of intestinal disorders

epi- *over, above* epidermis, outer layer of skin

erythr- *red* erythema, redness of the skin; erythrocyte, red blood cell

eso- *within* esophagus

eu- *well* enesthesia a normal state of the senses

excret *separate* excretory system

exo- *outside, outer layer* exophthalmos, an abnormal protrusion of the eye from the orbit

extra- *outside, beyond* extracellular, outside the body cells of an organism

extrins- *from the outside* extrinsic regulation of the heart

fasci-, fascia- *bundle, band* superficial and deep fascia

fenestr- *window* fenestrae of the inner ear; fenestrated capillaries

ferr- *iron* transferrin, ferritin, both iron-storage proteins

flagell- *whip* flagellum, the tail of a sperm cell

flat- *blow, blown* flatulence

folli- *bag, bellows* hair follicle

fontan- *fountain* fontanels of the fetal skull

foram- *opening* foramen magnum of the skull

foss- *ditch* fossa ovalis of the heart; mandibular fossa of the skull

gam-, gamet- *married, spouse* gametes, the sex cells

gangli- *swelling, or knot* dorsal root ganglia of the spinal nerves

gastr- *stomach* gastrin, a hormone that influences gastric acid secretion

gene *beginning, origin* genetics

germin- *grow* germinal epithelium of the gonads

gero-, geront- *old man* gerontology, the study of aging

gest- *carried* gestation, the period from conception to birth

glauc- *gray* glaucoma, which causes gradual blindness

glom- *ball* glomeruli, clusters of capillaries in the kidneys

glosso- *tongue* glossopathy, any disease of the tongue

gluco-, glyco- *gluconeogenesis, the production of glucose from non-carbohydrate molecules

glute- *buttock* gluteus maximus, largest muscle of the buttock

gnost- *knowing* the gnostic sense, a sense of awareness of self

gompho- *nail* gomphosis, the term applied to the joint between tooth and jaw

gon-, gono- *seed, offspring* gonads, the sex organs

gust- *taste* gustatory sense, the sense of taste

hapt- *fasten, grasp* hapten, a partial antigen

hema-, hemato-, hemo- *blood* hematocyst, a cyst containing blood

hemi- *half* hemiglossal, pertaining to one-half of the tongue

hepat- *liver* hepatitis, inflammation of the liver

hetero- *different or other* heterosexuality, sexual desire for a person of the opposite sex

hiat- *gap* the hiatus of the diaphragm, the opening through which the esophagus passes

hippo- *horse* hippocampus of the brain, shaped like a seahorse

hirsut- *hairy* hirsutism, excessive body hair

hist- *tissue* histology, the study of tissues

holo- *whole* holocrine glands, whose secretions are whole cells

hom-, homo- *same* homeoplasia, formation of tissue similar to normal tissue; homocentric, having the same center

hormon- *to excite* hormones

humor- *a fluid* humoral immunity, which involves antibodies circulating in the blood

hyal- *clear* hyaline cartilage, which has no visible fibers

hydr-, hydro- *water* dehydration, loss of body water

hyper- *excess* hypertension, excessive tension

hypno- *sleep* hypnosis, a sleeplike state

hypo- *below, deficient* hypodermic, beneath the skin; hypokalemia, deficiency of potassium

hyster-, hystero- *uterus or womb* hysterectomy, removal of the uterus; hysterodynia, pain in the womb

ile- *intestine* ileum, the last portion of the small intestine

im- *not* impermeable, not permitting passage, not permeable

inter- *between* intercellular, between the cells

intercal- *insert* intercalated discs, the end membranes between adjacent cardiac muscle cells

intra- *within, inside* intracellular, inside the cell

iso- *equal, same* isothermal, equal, or same, temperature

jugul- *throat* jugular veins, prominent vessels in the neck

juxta- *near, close to* juxtaglomerular apparatus, a cell cluster next to the glomeruli in the kidneys

karyo- *kernal nucleus* karyotype, the assemblage of the nuclear chromosomes

kera- *horn* keratin, the water-repellent protein of the skin

kilo- *thousand* kilocalories, equal to one thousand calories

kin-, kines- *move* kinetic energy, the energy of motion

labi-, labri- *lip* labial frenulum, the membrane which joins the lip to the gum

lact- *milk* lactose, milk sugar

lacun- *space, cavity, lake* lacunae, the spaces occupied by cells of cartilage and bone tissue

lamell- *small plate* concentric lamellae, rings of bone matrix in compact bone

lamina *layer, sheet* basal lamina, part of the epithelial basement membrane

lat- *wide* latissimus dorsi, a broad muscle of the back

laten- *hidden* latent period of a muscle twitch

later- *side* lateral (directional term)

leuko- *white* leukocyte, white blood cell

leva- *raise, elevate* levator labii superioris, muscle that elevates upper lip

lingua- *tongue* lingual tonsil, adjacent to the tongue

lip-, lipo- *fat, lipid* lipophage, a cell that has taken up fat in its cytoplasm

lith- *stone* cholelithiasis, gallstones

luci- *clear* stratum lucidum, clear layer of the epidermis

lumen *light* lumen, center of a hollow structure

lut- *yellow* corpus luteum, a yellow, hormone-secreting structure in the ovary

lymph *water* lymphatic circulation, return of clear fluid to the bloodstream

macro- *large* macromolecule, large molecule

macula *spot* macula lutea, yellow spot on the retina

magn- *large* foramen magnum, largest opening of the skull

mal- *bad, abnormal* malfunction, abnormal functioning of an organ

mamm- *breast* mammary gland, breast

mast- *breast* mastectomy, removal of a mammary gland

mater *mother* dura mater, pia mater, membranes that envelop the brain

meat- *passage* external auditory meatus, the ear canal

medi- *middle* medial (directional term)

medull- *marrow* medulla, the middle portion of the kidney, adrenal gland, and lymph node

mega- *large* megakaryocyte, large precursor cell of platelets

meio- *less* meiosis, nuclear division that halves the chromosome number

melan- *black* melanocytes, which secrete the black pigment melanin

men-, menstru- *month* menses, the cyclic menstrual flow

meningo- *membrane* meningitis, inflammation of the membranes of the brain

mer-, mero-, *a part* merocrine glands, the secretions of which do not include the cell

meso- *middle* mesoderm, middle germ layer

meta- *beyond, between, transition* metatarsus, the part of the foot between the tarsus and the phalanges

metro- *uterus* metroscope, instrument for examining the uterus

micro- *small* microscope, an instrument used to make small objects appear larger

mictur- *urinate* micturition, the act of voiding the bladder

mito- *thread, filament* mitochondria, small, filamentlike structures located in cells

mnem- *memory* amnesia

mono- *single* monospasm, spasm of a single limb

morpho- *form* morphology, the study of form and structure or organisms

multi- *many* multinuclear, having several nuclei

mural *wall* intramural ganglion, a nerve junction within an organ

muta- *change* mutation, change in the base sequence of DNA

myelo- *spinal cord, marrow* myeloblasts, cells of the bone marrow

myo- *muscle* myocardium, heart muscle

nano- *dwarf* nanometer, one billionth of a meter

narco- *numbness* narcotic, a drug producing stupor or numbed sensations

natri- *sodium* atrial natriuretic factor, a sodium-regulating hormone

necro- *death* necrosis, tissue death

neo- *new* neoplasm, an abnormal growth

nephro- *kidney* nephritis, inflammation of the kidney

neuro- *nerve* neurophysiology, the physiology of the nervous system

noci- *harmful* nociceptors, receptors for pain

nom- *name* innominate artery; innominate bone

noto- *back* notochord, the embryonic structure that precedes the vertebral column

nucle- *pit, kernel, little nut* nucleus

nutri- *feed, nourish* nutrition

ob- *before, against* obstruction, impeding or blocking up

oculo- *eye* monocular, pertaining to one eye

odonto- *teeth* orthodontist, one who specializes in proper positioning of the teeth in relation to each other

olfact- *smell* olfactory nerves

oligo- *few* oligodendrocytes, neuroglial cells with few branches

onco- *a mass* oncology, study of cancer

oo- *egg* ocyte, precursor of female gamete

ophthalmo- *eye* ophthalmology, the study of the eyes and related disease

orb- *circular* orbicularis oculi, muscle that encircles the eye

orchi- *testis* cryptorchidism, failure of the testes to descend into the scrotum

org- *living* organism

ortho- *straight, direct* orthopedic, correction of deformities of the musculoskeletal system

osm- *smell* anosmia, loss of sense of smell

osmo- *pushing* osmosis

osteo- *bone* osteodermia, bony formations in the skin

oto- *ear* otoscope, a device for examining the ear

ov-, ovi- *egg* ovum, oviduct

oxy- *oxygen* oxygenation, the saturation of a substance with oxygen

pan- *all, universal* panacea, a cure-all

papill- *nipple* dermal papillae, projections of the dermis into the epidermal area

para- *beside, near* paraphrenitis, inflammation of tissues adjacent to the diaphragm

pect-, pectus *breast* pectoralis major, a large chest muscle

pelv- *a basin* pelvic girdle, which cradles the pelvic organs

peni- *a tail* penis; penile urethra

penna- *a wing* unipennate, bipennate muscles, whose fascicles have a feathered appearance

pent- *five* pentose, a 5-carbon sugar

continues

Free Student Aid.

Log on.

Tune in.

Succeed.

To help you succeed in anatomy and physiology, your professor has arranged for you to enjoy access to great media resources, including an interactive CD-ROM and The Anatomy & Physiology Place web site. You'll find that these resources that accompany your textbook will enhance your course materials.

What your system needs to use these media resources:

WINDOWS™
- 250 MHz
- Windows-98/NT/2000/XP
- 32 MB RAM installed, 64 preferred
- 800x600 screen resolution
- Thousands of colors
- Browsers: Internet Explorer 5.0 and higher; Netscape 4.7, 7.0
- Plug-Ins: Flash player, QuickTime

NOTE: Use of Netscape 6.0 and 6.1 are not recommended due to a known compatibility issue between Netscape 6.0 and 6.1 and the Flash and Shockwave plug-ins.

MACINTOSH™
- 233 MHz PowerPC
- OS 9.2 or higher
- 32 MB RAM minimum
- 800x600 screen resolution
- Thousands of colors.
- Browsers: Internet Explorer 5.0; Netscape 4.7
- Plug-ins: Flash player, QuickTime

NOTE: Use of Netscape 6.0 and 6.1 are not recommended due to a known compatibility issue between Netscape 6.0 and 6.1 and the Flash and Shockwave plug-ins.

Got technical questions?
For technical support, please visit www.aw-bc.com/techsupport and complete the appropriate online form. Technical support is available Monday-Friday, 9 a.m. to 6 p.m. Eastern TIme (US and Canada).

Here's your personal ticket to success:

How to log on to www.anatomyandphysiology.com:

1. Go to www.anatomyandphysiology.com.
2. Click *Anatomy & Physiology*, Second Edition.
3. Click "Register."
4. Scratch off the gold foil coating below to reveal your pre-assigned access code.
5. Enter your pre-assigned access code exactly as it appears below.
6. Complete the online registration form to create your own personal user Login Name and Password.
7. Once your personal Login Name and Password are confirmed by email, go back to www.anatomyandphysiology.com, type in your new Login Name and Password, and click "Log In."

Your Access Code is:

If there is no gold foil covering the access code above, the code may no longer be valid. In that case, you need to either:
- Purchase a new student access kit at your campus bookstore.
- Purchase access online using a major credit card. Go to www.anatomyandphysiology.com, click on *Anatomy & Physiology*, **Second Edition**, and then click **Buy Now**.

Important: Please read the License Agreement, located on the launch screen, before using The Anatomy & Physiology Place. By using the web site, you indicate that you have read, understood, and accepted the terms of this agreement.

Anatomy & Physiology

Second Edition

Elaine N. Marieb, R.N., Ph.D.
Holyoke Community College

PEARSON

Benjamin Cummings

San Francisco Boston New York
Cape Town Hong Kong London Madrid Mexico City
Montreal Munich Paris Singapore Sydney Tokyo Toronto

Publisher: Daryl Fox
Sponsoring Editor: Serina Beauparlant
Development Manager: Claire Alexander
Project Editor: Mary Ann Murray
Contributing Editor: Paige Farmer
Managing Editor: Wendy Earl
Production Supervisor: Janet Vail
Media Producer: Ziki Dekel
Editorial Assistant: Sarah Kaminker
Cover Image: Kenneth X. Probst
Text and Cover Designer: tani hasagawa
Compositor: GTS Graphics
Proofreader: Martha Ghent
Indexer: Karen Hollister
Manufacturing Buyer: Stacey Weinberger
Executive Marketing Manager: Lauren Harp

Photo and illustration credits follow the Glossary.

Library of Congress Cataloging-in-Publication Data

Marieb, Elaine Nicpon
    Anatomy & physiology / Elaine N. Marieb.—2nd ed.
        p. cm.
    Includes index.
    ISBN 0-8053-5913-3
    1. Human physiology. 2. Human anatomy. I. Title: Anatomy and physiology. II. Title.

QP34.5.M454 2005
612—dc22

2003063201

Pearson Benjamin Cummings™ is a trademark of Pearson Education, Inc.

ISBN 0-8053-5913-3 (For sale copy)
ISBN 0-8053-5912-5 (Complete package)
ISBN 0-8053-5916-8 (Professional copy with CD-ROM)
4 5 6 7 8 9 10—VHP—07 06 05

www.aw-bc.com

About the Author

I dedicate this work to my students present and past, who have always inspired me to "push the envelope."

For Elaine N. Marieb, taking the student's perspective into account has always been an integral part of her teaching style. Dr. Marieb began her teaching career at Springfield College where she taught anatomy and physiology to physical education majors. She then joined the faculty of the Biological Science Division of Holyoke Community College in 1969 after receiving her Ph.D. in zoology from the University of Massachusetts at Amherst. While teaching at Holyoke Community College, where many of her students were pursuing nursing degrees, she developed a desire to better understand the relationship between the scientific study of the human body and the clinical aspects of the nursing practice. To that end, while continuing to teach full time, Dr. Marieb pursued her nursing education, which culminated in a Master of Science degree with a clinical specialization in gerontology from the University of Massachusetts. It is this experience, along with stories from the field—including those of former students now in health careers—that has informed the development of the unique perspective and accessibility for which her texts and laboratory manuals are known.

In her ongoing commitment to students and her realization of the challenges they face, Dr. Marieb has given generously to provide opportunities for students to further their education. She contributes to the New Directions, New Careers Program at Holyoke Community College by providing several full-tuition scholarships each year for women who are returning to college after a hiatus or attending college for the first time and would be unable to continue with their studies without financial support. She funds the E. N. Marieb Science Research Awards at Mount Holyoke College, which promotes research by undergraduate science majors, and is a contributor to the University of Massachusetts at Amherst where she generously provided funding for reconstruction and instrumentation of a cutting-edge cytology research laboratory that bears her name.

In 1994, Dr. Marieb received the Benefactor Award from the National Council for Resource Development, American Association of Community Colleges, which recognizes her ongoing sponsorship of student scholarships, faculty teaching awards, and other academic contributions to Holyoke Community College. In May 2000, the science building at Holyoke Community College was named in her honor.

Additionally, while actively engaged as an author, Dr. Marieb serves as a consultant for the Benjamin Cummings/A.D.A.M.® *InterActive Physiology* CD-ROM series, and is an active member of the Human Anatomy and Physiology Society (HAPS). This text—*Anatomy & Physiology*, Second Edition—is the latest expression of her commitment to student needs in their pursuit of the study of A&P.

Preface to the Instructor

The first edition of *Anatomy & Physiology*—"Slim" to those at Benjamin Cummings—has proven the saying, "Give the people what they want and they will buy it." Although Slim is shorter and cheaper than the original two-semester *Human Anatomy & Physiology* text, I, as author, was sure that no one would seriously consider adopting an anatomy and physiology text that pared down pedagogical perks and omitted topics of pregnancy, developmental aspects (embryonic and fetal development and aging), and heredity; I have been proven to be dead wrong. So we are off to see the wizard again with this second edition of *Anatomy & Physiology*.

As before, all the major organ systems are covered in this text version. As always, I continue to use the pyramid approach in presenting the material, building a broad base to support all that comes later. Unifying themes are emphasized, and explanations are reinforced with comfortable analogies and familiar examples that focus on mechanisms and cause and effect relationships. Thus, students can gain an integrated and continuously enhanced understanding of the workings of the human body.

Unifying Themes

Three integrating themes organize, unify, and set the tone of this text:

• **Interrelationships of body organ systems:** The fact that nearly all regulatory mechanisms require interaction of several organ systems is continually emphasized. For example, Chapter 24, which deals with the structure and function of the urinary system, discusses the vital importance of the kidneys not only in maintaining adequate blood volume to ensure normal blood circulation, but also in continually adjusting the chemical composition of blood so that all body cells remain healthy.

• **Homeostasis:** The normal and most desirable condition of body functioning is homeostasis. Its loss or destruction always leads to some type of pathology—temporary or permanent. Pathological conditions are integrated with the text to clarify and illuminate normal functioning, not as an end in and of themselves. For example, Chapter 18, which deals with the structure and function of blood vessels, explains how the ability of healthy arteries to expand and recoil ensures continuous blood flow and proper circulation. The chapter goes on to discuss the effects on homeostasis when arteries lose their elasticity: high blood pressure and all of its attendant problems. These homeostatic imbalances are indicated visually by an orange symbol with an H in its center. Whenever students see the imbalance symbol in text, the concept of disease as a loss of homeostasis is reinforced.

• **Complementarity of structure and function:** Students are encouraged to understand the structure of an organ, a tissue, or a cell as a prerequisite to comprehending its function. Concepts of physiology are explained and related to structural characteristics that promote or allow the various functions to occur. For example, the lungs can act as a gas exchange site because the walls of their air sacs present an incredibly thin barrier between blood and air.

Pedagogical Features Retained

Pedagogical features that have proven their worth and are still with us include the following:

• Chapter outlines and student objectives preface each chapter and guide students to concepts that are most important to know.

• Illustrated tables summarize complex information and serve as "one-stop shopping" study tools.

• Key questions accompany selected figures and prod the student to interpret the concepts or illustrated processes, or to make predictions about "what's next." Answers can be found upside down on the same page.

- Homeostatic imbalance icons alert the students to consequences when the body is not functioning optimally.
- Review questions help students to evaluate their progress in the subject.
- The award-winning *InterActive Physiology®* 8-System Suite now comes packaged with each new copy of the text and provides students with multimedia backup for the topics covered.

New to the Second Edition

- **Shorter, more concise text with enhanced readability.** The entire manuscript was perused by a developmental editor whose ministrations tightened the text without sacrificing important content or the analogies and synonyms that enhance student learning.
- **Factual material updates.** A major undertaking of this edition was to update the factual material in all subject areas—a monumental task that demands painstaking selectivity. Certain areas are updated in every edition—immunity, cell signaling, and the offerings of genetic engineering—and this edition is no exception. However, other subjects of study are also attended to as appropriate for the changing literature. Specific examples from Chapter 3 are sufficient to illustrate this trend: Information on membrane transport, particularly facilitated diffusion and vesicular transport, is updated, and the explanation of how the membrane potential is generated and maintained is made more explicit. Chapter 3 has improved: more detailed descriptions of DNA and protein synthesis, accompanied by new schematic illustrations that clearly depict these events.
- **Reorganization of specific chapters and topics.** Chapter 9 on muscle tissue has been reorganized so that the discussion of the physiology of skeletal muscle fibers now encompasses muscle fiber contraction, which in turn is described in its temporal sequence in the process of excitation and contraction. The chapter also includes a new discussion of the relationship between stimulus intensity and muscle tension that recognizes the order in which motor units of various sizes and muscle cell types are recruited.
- **Revised art program.** Once again, every piece of art has been touched and the color palette has been revised to provide art that has richer, more organic color. Many pieces have been touched up to enhance their detail and many new views have been added,

especially in the nervous system chapters. Selected physiology figures have been rerendered to reflect recent research on functional processes. All the tissue figures in Chapter 4 as well as flowcharts throughout the book have been redone for consistency and clarity. As before, the art has a consistent color protocol to encourage automatic learning. Every time you see an ATP starburst it is bright yellow; oxygen is red; cellular cytoplasm is routinely beige; and so on.

Additionally, there are many more photomicrographs and clinical photos in this edition and many of those retained from earlier editions have been upsized. As pointed out by a conscientious reviewer, "If you can't see it clearly, then it's not worth having."

- **Multimedia offerings enhanced.** The *InterActive Physiology®* 8-System Suite is now packaged free with each new copy of the text. The Anatomy & Physiology Place has been updated, and innovative media new to the Second Edition includes the *Instructor's Resource* CD-ROM: *Images, Animations, and Lecture Presentation*, which is explained in greater detail in the following discussion of the supplements that accompany the text.

A Comprehensive Teaching and Learning Package

The supplements to the Second Edition of *Anatomy & Physiology* have been developed to be more innovative and thorough than ever before—in print, on CD-ROM, and on the Web—for instructors and students alike.

The Anatomy & Physiology Place (www.anatomyandphysiology.com)

The A&P Place has been updated to accompany the Second Edition, and includes learning activities, web exercises, clinical case studies, quizzing, and more. For instructors, a useful Syllabus Manager is available to help organize the course.

Course Management Systems, Including WebCT and Blackboard

Useful for online course management and distance learning courses, access codes for preloaded content in WebCT and Blackboard are available alone or packaged with the text. Content includes selections from The A&P Place and the entire computerized Test Bank.

InterActive Physiology® 8-System Suite

Now packaged free with each new copy of the text, this award-winning tutorial series uses detailed animations and engaging quizzes to help students advance beyond simple memorization to a genuine understanding of the most difficult topics in A&P. This CD-ROM includes a long-awaited eighth module on the endocrine system.

Instructor's Resource CD-ROM

This powerful classroom presentation tool offers all of the illustrations, photos, and tables from the text—approximately 750 images—in both jpeg and PowerPoint format, as well as fully customizable lecture slides in PowerPoint with art and text for each chapter. Also included are *InterActive Physiology®* animations in a ready-to-use format so instructors can quickly pull together relevant classroom presentations.

Instructor's Resource Guide

This fully revised instructor's guide provides useful information for both new and seasoned instructors. Detailed objectives and lecture outlines provide guidelines for topics that are covered in the text, and lecture and discussion suggestions offer helpful hints for how to present the material. The suggested media and reading lists have been fully updated. *InterActive Physiology®* review worksheets help to reinforce physiological concepts that are presented on the CD-ROM, and can be used for homework, lab assignments, review, or quzzing. New to this edition is a Visual Resource Guide with thumbnails of all the illustrations from the text.

Test Bank

With more than 3,000 test questions, this Test Bank has been updated with new and revised questions that cover all major topics at a range of difficulty levels. Question types include figure, fill-in-the-blank, matching, short answer, true/false, multiple-choice, essay, and clinical questions. All questions in the printed Test Bank are available in electronic format through TestGen-EQ software. This cross-platform CD-ROM allows instructors to easily generate tests via a user-friendly interface in which they can view, edit, sort, and add questions.

Transparency Acetates

This expanded package includes all illustrations, photos, and illustrated tables from the text—approx-

imately 750 images—with labels that have been enlarged for easy viewing in the classroom or lecture hall.

A Brief Atlas of the Human Body

This four-color atlas is available bundled with the text upon request and includes 107 bone and 47 soft tissue photographs with easy-to-read labels. Featuring photos taken by renowned biomedical photographer Ralph Hutchings, this high-quality photographic atlas makes an excellent resource for the classroom and laboratory, and is referenced in appropriate figure legends throughout the text.

Student Study Guide

Revised to accompany the Second Edition of *Anatomy & Physiology*, the study guide offers a wide variety of exercises that address different learning styles. The three major sections, Building the Framework, Challenging Yourself, and Covering All Your Bases, help students build a base of knowledge using recall, reasoning, and imagination that can be applied to solving problems in both clinical and nonclinical situations.

Laboratory Manual for *Anatomy & Physiology*

The Second Edition of this four-color lab manual is designed for instructors who are teaching a two-semester allied health-related course in anatomy and physiology, but who need fewer laboratory exercises. This laboratory manual features 27 complete exercises, along with review sheets, plus a histology atlas and a complete exercise on the use and care of the microscope in the Appendix.

Human Anatomy & Physiology Laboratory Manuals

Elaine N. Marieb's three widely used and acclaimed laboratory manuals are designed to meet the varying needs of most laboratory courses: *Human Anatomy & Physiology Laboratory Manual*: *Main Version*, Sixth Edition Update; *Human Anatomy & Physiology Laboratory Manual*: *Cat Version*, Seventh Edition Update; *Human Anatomy & Physiology Laboratory Manual*: *Pig Version*, Seventh Edition Update. Shrink-wrapped with each laboratory manual is the *PhysioEx*™ 4.0 CD-ROM, which features ten easy-to-use computerized physiology experiments, including an entirely new laboratory simulation on Acid/Base Balance as well as a new Histology Tutorial. The *PhysioEx*™ labs are accompanied with written exercises and review sheets, and can be accessed on the CD-ROM that accompanies the lab manuals or via the Web at www. physioex.com.

The Tutor Center

Free tutoring is available to students when instructors bundle the Tutor Center with *Anatomy & Physiology,* Second Edition. The Tutor Center is staffed with qualified A&P instructors who can tutor students on all the material covered in the text, including the art and multiple choice/matching questions at the end of each chapter. Tutoring content is restricted to the text material, and tutors will discuss only answers that are provided in the textbook. After an instructor authorizes student access to this service, students can contact the Tutor Center by phone, fax, email, or the Internet; the Tutor Center is open five days a week. Instructors can contact their local Benjamin Cummings sales representative for information about bundling the Tutor Center with their text. For more information, visit www.aw.com/tutorcenter.

Additional Supplements Available from Benjamin Cummings

A.D.A.M.® *Interactive Anatomy Student Lab Guide,* Second Edition
By Mark Lafferty and Samuel Panella

A.D.A.M.® *Interactive Anatomy Student Package*

A.D.A.M.® *Anatomy Practice*

Human Cadaver Dissection Videos
By Rose Leigh Vines, et al.

Student Video Series for Human Anatomy & Physiology, Volume 1

Student Video Series for Human Anatomy & Physiology, Volume 2

Anatomy Flashcards
By Glenn Bastian

Anatomy & Physiology Coloring Workbook: A Complete Study Guide, Seventh Edition
By Elaine N. Marieb

The Physiology Coloring Book, Second Edition
By Wynn Kapit, Robert I. Macey, Esmail Meisami

The Anatomy Coloring Book, Third Edition
By Wynn Kapit and Lawrence M. Elson

A Color Atlas of Histology
By Dennis Strete

Histology for the Life Sciences
By Allen Bell and Victor Eroshenko

Contact your Benjamin Cummings sales representative or your campus bookstore for more information.

Preface to the Student

Human anatomy and physiology is more than just interesting—it is fascinating. To help get you involved in the study of this exciting subject, a number of special features are incorporated throughout the book.

The tone of *Anatomy & Physiology* is intentionally informal and unintimidating. There is absolutely no reason that you can't enjoy your learning task. This book is meant to be a guide to the understanding of your own body, not an encyclopedia of human anatomy and physiology. I have tried to be selective about the information included and have chosen facts that stress essential concepts. Physiological concepts are explained thoroughly; abundant analogies are used and examples are taken from familiar events whenever possible.

The illustrations and tables are designed with your learning needs in mind. The tables are summaries of important information in the text and should be valuable resources for reviewing for an exam. In all cases, the figures are referenced where their viewing would be most helpful in understanding the text. Those depicting physiological mechanisms often use the flowchart format so that you can see where you've been and where you're going. Additionally, several key figures come equipped with questions that will help you to interpret the figures or to apply and integrate what you have learned. (Answers are provided at the bottom of the same page.)

Each chapter begins with an outline of its major topics (and their page references), with specific learning objectives listed for each topic. Important terms within the chapter text are bold-faced (in dark type), and phonetic spellings are provided for terms that are likely to be unfamiliar to you. To use these phonetic aids, you will need to remember the following rules:

1. Syllables are separated by dashes.

2. Accent marks follow stressed syllables. Primary stress is shown by ' and secondary stress by ".

3. Unless otherwise noted, assume that vowels at the ends of syllables are long. Assume that all other vowels are short unless a bar, indicating a long vowel sound, appears above them.

For example, the phonetic spelling of "thrombophlebitis" is throm"bo-fleh-bi'tis. The fourth syllable (bi') receives the greatest stress, and the first syllable (throm") gets the secondary stress. The "o" in the second syllable and "i" in the fourth syllable are long. Further explanation of the pronunciation system can be found at the beginning of the glossary.

Any exam causes anxiety. To help you prepare for an exam or comprehend the material you have just read, thorough summaries complete with page references are found at the ends of the chapters, as are review questions that use a combination of testing techniques (multiple choice, matching, short answer essay, critical thinking, and clinical applications). Also remember that key figure questions are sprinkled throughout the chapter.

I hope that you enjoy *Anatomy & Physiology* and that this book makes learning about the body's structures and functions an exciting and rewarding process. Perhaps the best bit of advice I can give you is that memory depends on understanding. Thus, if you strive for understanding instead of rote memorization, your memory will not fail you often.

I would appreciate hearing from you about your experiences with this textbook or suggestions for improvements in future editions.

Elaine N. Marieb
Anatomy and Physiology
Benjamin Cummings
1301 Sansome Street
San Francisco, CA 94111

Acknowledgments

As noted in the First Edition, the kernel from which this text germinated was cast by Daryl Fox, VP at Benjamin Cummings. Although I hate to have to tell him he was right about the demand for a text like "Slim," if I must, I must. Serina Beauparlant, the sponsoring editor of this edition, was ever vigilant about solving problems. Claire Alexander, Development Manager, and Mary Ann Murray, Project Editor, oversaw the project while Paige Farmer, the contributing editor, was involved in scrutinizing the manuscript and delivering it to production. Sarah Kaminker, Editorial Assistant, provided support every step of the way, and Stacey Weinberger's expertise as manufacturing buyer has served us well.

I also wish to thank the designer, tani hasagawa, for her text and cover design, Kenneth X. Probst for the cover image, and the entire staff at Wendy Earl Productions, including Wendy Earl, Managing Editor, the production supervisor, Janet Vail, and proofreader, Martha Ghent.

Finally, I wish to express my appreciation to Lauren Harp, the Executive Marketing Manager for Applied Sciences at Benjamin Cummings.

I owe a huge debt of gratitude to my reviewers and colleagues from the United States and abroad who, over the years, have made helpful suggestions and offered constructive criticisms to my work. A list of reviewers follows.

Please—if you are so inclined—let me know what you think of this edition. Your suggestions and comments provide much of the impetus for embarking on the next edition, and I would appreciate hearing from you about ways in which this book might be improved still further in future editions.

Elaine N. Marieb
Anatomy and Physiology
Benjamin Cummings
1301 Sansome Street
San Francisco, CA 94111

Reviewers of *Anatomy & Physiology*, Second Edition

Terry Ayers
James A. Rhodes State College

Cyndy Beck
George Mason University

Bert Blumenkranz
North Shore Community College

Ed Bradel
University of Cincinnati

Frank Breazeale
TriCounty Technical College

Doug Chandler
Arizona State University

Marnie Chapman
Alaska University

Sonya Conway
Northern Illinois University

Veronica Dougherty
Cecil Community College

Rick Douglas
Montana Tech

Henry Fabian
Merritt College

Maria Florez-Duquet
California Polytechnic State University

Eric Genz-Mould
Shoreline Community College

Dan Gibson
Worcester Polytechnic Institute

Susan Gilmore
Cayuga County Community College

Christopher Harendza
Montgomery County Community College

Greg Hartman
McNeese State University

Richard Hengst
Purdue University North Central

Marty Hicks
Community College of Southern Nevada

Jeff Hochbaum
Middlesex County College

John King
Arizona Western College

Sandra Kollar
Mount Union College

Clinton Krager
Broome Community College

Doug Light
Ripon College

David Lott
Clarion University of Pennsylvania

Ann Markel
Eastern Maine Technical College

Diane Mauldin
Bellevue Community College

Richard McCullough
Saddleback College

Ken McDowell
Sinclair Community College

Robb Moats
University of North Florida

Brett Niedens
West Hills Community College

Keith Nissen
Hibbing Community College

Heidi Peterson
Indian Hills Community College

David Pierotti
Northern Arizona University

Martha Pippitt
Rockford College

Thoniot Prabhakaran
Southwest Texas State

Marge Przygoda
Middlesex County College

Stacy Pugh-Towe
Crowder College

Tom Rachow
Missouri Western State College

Nancy Rauch
Merritt College

Stephen Reinbold
Longview Community College

Laura Ritt
Burlington County Community College

Heiko Schoenfuss
St. Cloud State

Tim Scott
Texas A&M

Steve Smith
Graceland University

Stephanie Spooner
Cypress College

Tim Stabler
Indiana University Northwest

John Straneva
SUNY Cortland

Melody Sweet
Syracuse University

Jan Webber
Huron University

Harold Wilkinson
Millikin University

Matthew Williamson
Georgia Southern University

Lynda Wilson
Valencia Community College West Campus

Willard Woodward
Parkland College

Jean Zorko
Stark State College

Contents

Chapter 18 continues

UNIT FIVE Continuity

Chapter 26 continues

1

THE HUMAN BODY: AN ORIENTATION

An Overview of Anatomy and Physiology (pp. 2–3)

1. Define anatomy and physiology and describe their subdivisions.
2. Explain the principle of complementarity.

Levels of Structural Organization (pp. 3–4)

3. Name the different levels of structural organization that make up the human body, and explain their relationships.
4. List the 11 organ systems of the body, identify their components, and briefly explain the major function(s) of each system.

Maintaining Life (pp. 4–9)

5. List the functional characteristics necessary to maintain life in humans.
6. List the survival needs of the body.

Homeostasis (pp. 9–12)

7. Define homeostasis and explain its significance.
8. Describe how negative and positive feedback maintain body homeostasis.
9. Describe the relationship between homeostatic imbalance and disease.

The Language of Anatomy (pp. 12–20)

10. Describe the anatomical position.
11. Use correct anatomical terms to describe body directions, regions, and body planes or sections.
12. Locate and name the major body cavities and their subdivisions, and list the major organs contained within them.
13. Name the serous membranes and indicate their common function.
14. Name the nine regions or four quadrants of the abdominopelvic cavity and list the organs they contain.

As you make your way through this book, you will be learning about one of the most fascinating subjects possible—your own body. Such a study is not only highly personal, but timely as well. The current information blizzard brings news of some medical advance almost daily. If you are to appreciate emerging discoveries in genetic engineering, to understand new techniques for detecting and treating disease, and to make use of published facts on how to stay healthy, it is important to learn about the workings of your body. For those preparing for a career in the health sciences, the study of anatomy and physiology has added rewards because it provides the foundation needed to support your clinical experiences.

In this chapter we define and contrast anatomy and physiology and discuss how the human body is organized. Then we review needs and functional processes common to all living organisms. Three essential concepts—*the complementarity of structure and function, the hierarchy of structural organization*, and *homeostasis*—will unify and form the bedrock for your study of the human body. The final section of the chapter deals with the language of anatomy—terminology that anatomists use when they describe the body or its parts.

An Overview of Anatomy and Physiology

Two complementary branches of science—anatomy and physiology—provide the concepts that help us to understand the human body. **Anatomy** studies the *structure* of body parts and their relationships to one another. Anatomy has a certain appeal because it is concrete. Body structures can be seen, felt, and examined closely; it is not necessary to imagine what they look like. **Physiology** concerns the *function* of the body, in other words, how all the body parts work and carry out their life-sustaining activities. When all is said and done, physiology is explainable only in terms of the underlying anatomy.

Topics of Anatomy

Anatomy is a broad field with many subdivisions, each providing enough information to be a course in itself. **Gross, or macroscopic, anatomy** is the study of large body structures visible to the naked eye, such as the heart, lungs, and kidneys. Indeed, the term *anatomy* (derived from the Greek words meaning "to cut apart") relates most closely to gross anatomy because in such studies preserved animals or their organs are dissected (cut up) to be examined. Gross

anatomy can be approached in different ways. In **regional anatomy,** all the structures (muscles, bones, blood vessels, nerves, etc.) in a particular region of the body, such as the abdomen or leg, are examined at the same time. In **systemic anatomy** (sis-tem'ik)*, the anatomy of the body is studied system by system. For example, when studying the cardiovascular system, you would examine the heart and the blood vessels of the entire body. Another subdivision of gross anatomy is **surface anatomy,** the study of internal structures as they relate to the overlying skin surface. You use surface anatomy when you identify the bulging muscles beneath a bodybuilder's skin, and clinicians use it to locate appropriate blood vessels in which to feel pulses and draw blood.

Microscopic anatomy concerns structures too small to be seen with the naked eye. For most such studies, exceedingly thin slices of body tissues are stained and mounted on slides to be examined under the microscope. Subdivisions of microscopic anatomy include **cytology** (si-tol'o-je), which considers the cells of the body, and **histology** (his-tol'o-je), the study of tissues.

Developmental anatomy traces structural changes that occur in the body throughout the life span. **Embryology** (em"bre-ol'o-je), a subdivision of developmental anatomy, concerns developmental changes that occur before birth.

Some highly specialized branches of anatomy are used primarily for medical diagnosis and scientific research. For example, *pathological anatomy* studies structural changes caused by disease. *Radiographic anatomy* studies internal structures as visualized by X-ray images or specialized scanning procedures. In *molecular biology*, the structure of biological molecules (chemical substances) is investigated. Molecular biology is actually a separate branch of biology, but it falls under the anatomy umbrella when we push anatomical studies to the subcellular level. As you can see, subjects of interest to anatomists range from easily seen structures down to the smallest molecule.

One of the most important tools for studying anatomy is a mastery of anatomical terminology. Others are observation, manipulation, and, in a living person, palpation (feeling organs with your hands) and auscultation (listening to organ sounds with a stethoscope). A simple example will illustrate how some of these tools work together in an anatomical study. Let's assume that your topic is freely movable joints of the body. In the laboratory, you will be able to *observe* an animal joint,

---

*For the pronunciation guide rules, see the Preface to the Student.

noting how its parts fit together. You can work the joint (*manipulate* it) to determine its range of motion. Using *anatomical terminology*, you can name its parts and describe how they are related so that other students (and your instructor) will have no trouble understanding you. The list of word roots (on the inside of the book cover) and the glossary will help you with this special vocabulary.

Although most of your observations will be made with the naked eye or with the help of a microscope, medical technology has developed a number of sophisticated tools that can peer into the body without disrupting it.

Topics of Physiology

Like anatomy, physiology has many subdivisions, most of which consider the operation of specific organ systems. For example, **renal physiology** concerns kidney function and urine production; **neurophysiology** explains the workings of the nervous system; and **cardiovascular physiology** examines the operation of the heart and blood vessels. While anatomy provides us with a static image of the body's architecture, physiology reveals the body's dynamic nature.

Physiology often focuses on events at the cellular or molecular level. This is because the body's abilities depend on those of its individual cells, and cells' abilities ultimately depend on the chemical reactions that go on within them. An understanding of physiology also rests on principles of physics, which help to explain electrical currents, blood pressure, and the way muscles use bones to cause body movements among other things. Basic chemical and physical principles are presented in Chapter 2 and throughout the book as needed to explain physiological topics.

Complementarity of Structure and Function

Although it is possible to study anatomy and physiology individually, they are really inseparable because function always reflects structure. That is, what a structure can do depends on its specific form. This is called the **principle of complementarity of structure and function.** For example, bones can support and protect body organs because they contain hard mineral deposits; blood flows in one direction through the heart because the heart has valves that prevent backflow; and the lungs can serve as a site for gas exchange because the walls of their air sacs are extremely thin. Throughout this book, a description of a structure's anatomy is accompanied by an explanation of its function, and structural

characteristics contributing to that function are emphasized.

Levels of Structural Organization

The human body has many levels of structural organization (Figure 1.1). The simplest level of the structural hierarchy is the **chemical level,** which we study in Chapter 2. At this level, *atoms*, tiny building blocks of matter, combine to form *molecules* such as water and proteins. Molecules, in turn, associate in specific ways to form *organelles*, basic components of the microscopic cells. *Cells* are the smallest units of living things. The **cellular level** is examined in Chapter 3. All cells have some common functions, but individual cells vary widely in size and shape, reflecting their unique functions in the body.

The simplest living creatures are composed of single cells, but in complex organisms such as human beings, the hierarchy continues on to the **tissue level.** *Tissues* are groups of similar cells that have a common function. The four basic tissue types in the human body are epithelium, muscle, connective tissue, and nervous tissue. Each tissue type has a characteristic role in the body, which we explore in Chapter 4. Briefly, epithelium covers the body surface and lines its cavities; muscle provides movement; connective tissue supports and protects body organs; and nervous tissue provides a means of rapid internal communication by transmitting electrical impulses.

An *organ* is a discrete structure composed of at least two tissue types (four is more commonplace) that performs a specific function for the body. At the **organ level,** extremely complex functions become possible. Let's take the stomach for an example. Its lining is an epithelium that produces digestive juices; the bulk of its wall is muscle, which churns and mixes stomach contents (food); its connective tissue reinforces the soft muscular walls; and its nerve fibers increase digestive activity by stimulating the muscle to contract more vigorously and the glands to secrete more digestive juices. The liver, the brain, and a blood vessel are very different from the stomach, but they are organs as well. You can think of each organ of the body as a specialized functional center responsible for a necessary activity that no other organ can perform.

The next level of organization is the **organ system.** Organs that work together to accomplish a common purpose make up an *organ system.* For example, the heart and blood vessels of the cardiovascular system see to it that blood circulates

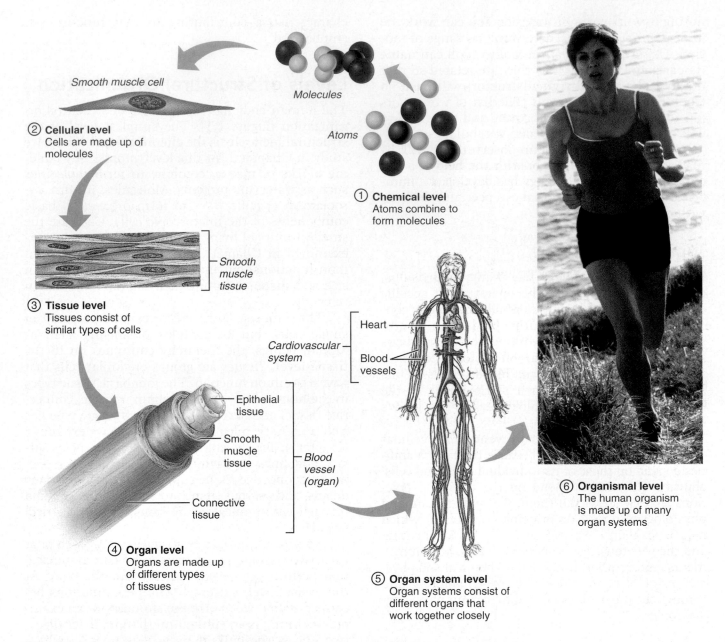

Smooth muscle cell

② **Cellular level**
Cells are made up of
molecules

Molecules

Atoms

① **Chemical level**
Atoms combine to
form molecules

Smooth
muscle
tissue

③ **Tissue level**
Tissues consist of
similar types of cells

Epithelial
tissue

Smooth
muscle
tissue

Connective
tissue

④ **Organ level**
Organs are made up
of different types
of tissues

*Cardiovascular
system*

Heart

Blood
vessels

*Blood
vessel
(organ)*

⑤ **Organ system level**
Organ systems consist of
different organs that
work together closely

⑥ **Organismal level**
The human organism
is made up of many
organ systems

FIGURE 1.1 Levels of structural organization. Components of the
cardiovascular system illustrate the levels of structural organization in a human being.

continuously to carry oxygen and nutrients to all
body cells. Besides the cardiovascular system, the
other organ systems of the body are the integumentary, skeletal, muscular, nervous, endocrine, respiratory, digestive, lymphatic, urinary, and reproductive
systems. Figure 1.3 on pp. 6 and 7 provides a brief
overview of the 11 organ systems, which we study in
more detail in Units 2–5.

The highest level of organization is the *organism*, the living human being. The **organismal level**
represents the sum total of all structural levels working together to promote life.

Maintaining Life
Necessary Life Functions

Now that you know the structural levels composing
the human body, the question that naturally follows
is: What does this highly organized human body do?
Like all complex animals, human beings maintain
their boundaries, move, respond to environmental
changes, take in and digest nutrients, carry out metabolism, dispose of wastes, reproduce themselves,
and grow. We will discuss each of these necessary life
functions briefly here and in more detail in later
chapters.

It cannot be emphasized too strongly that the multicellular state and parceling out of vital body functions to several *different* organ systems result in interdependence of all body cells. Organ systems do not work in isolation; they work cooperatively to promote the well-being of the entire body. This theme is emphasized throughout the book, and some of the organ systems making major contributions to selected functional processes are identified in Figure 1.2. Also, as you read this section you will want to refer to the more detailed descriptions of the organ systems in Figure 1.3.

Maintaining Boundaries Every living organism must **maintain its boundaries** so that its internal environment (inside) remains distinct from the external environment surrounding it (outside). In single-celled organisms, the external boundary is a limiting membrane that encloses its contents and admits needed substances while restricting entry of potentially damaging or unnecessary substances. Similarly, all the cells of our body are surrounded by a selectively permeable membrane. Additionally, the body as a whole is enclosed and protected by the integumentary system or skin, which protects our internal organs from drying out (a fatal change), bacteria, and the damaging effects of heat, sunlight, and an unbelievable number of chemicals in the external environment.

Movement **Movement** includes the activities promoted by the muscular system, such as propelling ourselves from one place to another by running or swimming, and manipulating the external environment with our nimble fingers. The skeletal system provides the bony framework that the muscles pull on as they work. Movement also occurs when substances such as blood, foodstuffs, and urine are propelled through internal organs of the cardiovascular, digestive, and urinary systems, respectively. On the cellular level, the muscle cell's ability to move by shortening is more precisely called **contractility.**

Responsiveness **Responsiveness,** or **irritability,** is the ability to sense changes (stimuli) in the environment and then respond to them. For example, if you cut your hand on broken glass, a withdrawal reflex occurs—you involuntarily pull your hand away from the painful stimulus (the broken glass). It is not necessary to think about it; it just happens! Likewise, when carbon dioxide in your blood rises to dangerously high levels, chemical sensors respond by sending messages to brain centers controlling respiration, and your breathing rate speeds up.

Because nerve cells are highly irritable and communicate rapidly with each other via electrical im-

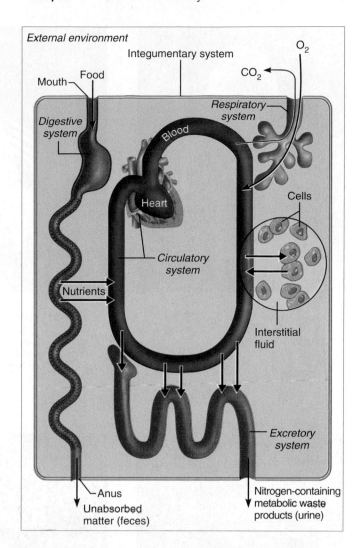

FIGURE 1.2 **Examples of selected interrelationships among body organ systems.** The integumentary system protects the body as a whole from the external environment. The digestive and respiratory systems, in contact with the external environment, take in nutrients and oxygen, respectively, which are then distributed by the blood to all body cells. Elimination of metabolic wastes is accomplished by the urinary and respiratory systems.

pulses, the nervous system is most involved with responsiveness. However, all body cells are irritable to some extent.

Digestion **Digestion** is the breaking down of ingested foodstuffs to simple molecules that can be absorbed into the blood. The nutrient-rich blood is then distributed to all body cells by the cardiovascular system. In a simple, one-celled organism such as an amoeba, the cell itself is the "digestion factory," but in the multicellular human body, the digestive system performs this function for the entire body.

Metabolism **Metabolism** (mĕ-tab′o-lizm; "a state of change") is a broad term that includes all chemical

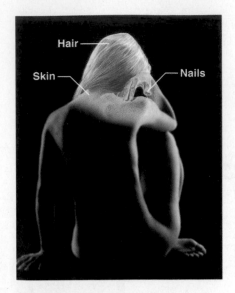

(a) Integumentary System
Forms the external body covering;
protects deeper tissues from injury;
synthesizes vitamin D; site of cutaneous
(pain, pressure, etc.) receptors, and
sweat and oil glands.

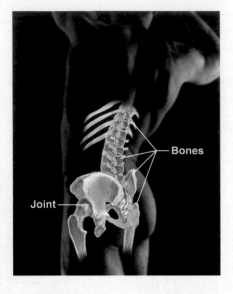

(b) Skeletal System
Protects and supports body organs;
provides a framework the muscles use
to cause movement; blood cells are
formed within bones; stores minerals.

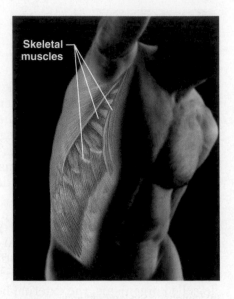

(c) Muscular System
Allows manipulation of the environment,
locomotion, and facial expression; maintains posture; produces heat.

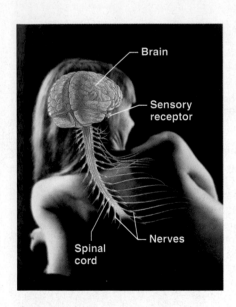

(d) Nervous System
Fast-acting control system of the body;
responds to internal and external
changes by activating appropriate
muscles and glands.

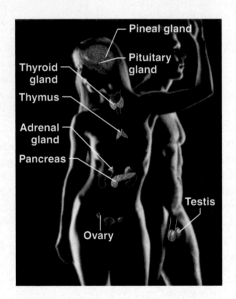

(e) Endocrine System
Glands secrete hormones that regulate
processes such as growth, reproduction
and nutrient use (metabolism) by body
cells.

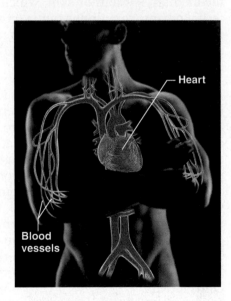

(f) Cardiovascular System
Blood vessels transport blood, which
carries oxygen, carbon dioxide,
nutrients, wastes, etc.; the heart pumps
blood.

FIGURE 1.3 **Summary of the body's organ systems.** The structural
components of each system are illustrated, and its major functions are listed below.

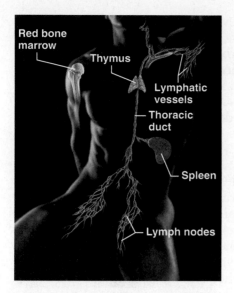

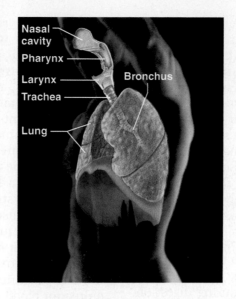

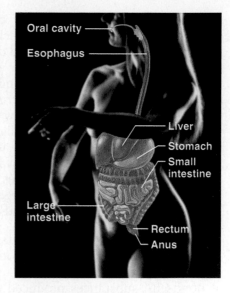

(g) Lymphatic System/Immunity
Picks up fluid leaked from blood vessels and returns it to blood; disposes of debris in the lymphatic stream; houses white blood cells (lymphocytes) involved in Immunity. The immune response mounts the attack against foreign substances within the body.

(h) Respiratory System
Keeps blood constantly supplied with oxygen and removes carbon dioxide; the gaseous exchanges occur through the walls of the air sacs of the lungs.

(i) Digestive System
Breaks down food into absorbable units that enter the blood for distribution to body cells; indigestible foodstuffs are eliminated as feces.

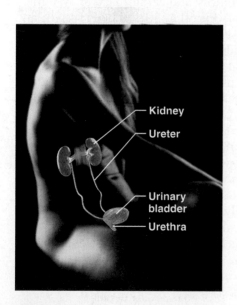

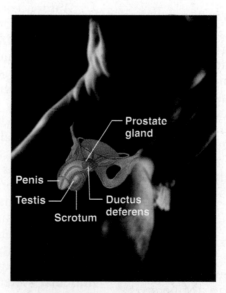

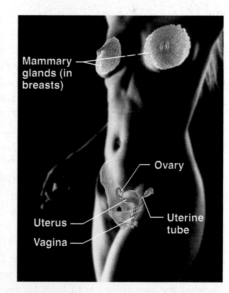

(j) Urinary System
Eliminates nitrogenous wastes from the body; regulates water, electrolyte and acid-base balance of the blood.

(k) Male Reproductive System
Overall function is production of offspring. Testes produce sperm and male sex hormone; ducts and glands aid in delivery of sperm to the female reproductive tract. Ovaries produce eggs and female sex hormones; remaining structures serve as sites for fertilization and development of the fetus. Mammary glands of female breasts produce milk to nourish the newborn.

(l) Female Reproductive System

reactions that occur within body cells. It includes breaking down substances into their simpler building blocks (more specifically called *catabolism*), synthesizing more complex cellular structures from simpler substances *(anabolism)*, and using nutrients and oxygen to produce (via *cellular respiration*) ATP, the energy-rich molecules that power cellular activities. Metabolism depends on the digestive and respiratory systems to make nutrients and oxygen available to the blood and on the cardiovascular system to distribute these needed substances throughout the body. Metabolism is regulated largely by hormones secreted by endocrine system glands.

Excretion   Excretion is the process of removing *excreta* (ek-skre'tah), or wastes, from the body. If the body is to operate as we expect it to, it must get rid of nonuseful substances produced during digestion and metabolism. Several organ systems participate in excretion. For example, the digestive system rids the body of indigestible food residues in feces, and the urinary system disposes of nitrogen-containing metabolic wastes, such as urea, in urine. Carbon dioxide, a by-product of cellular respiration, is carried in the blood to the lungs, where it leaves the body in exhaled air.

Reproduction   Reproduction can occur at the cellular or organismal level. In cellular reproduction the original cell divides, producing two identical daughter cells that may then be used for body growth or repair. Reproduction of the human organism, or making a whole new person, is the major task of the reproductive system. When a sperm unites with an egg, a fertilized egg forms, which then develops into a baby within the mother's body. The reproductive system is directly responsible for producing offspring, but its function is exquisitely regulated by hormones of the endocrine system.

Because males produce sperm and females produce eggs (ova), there is a division of labor in the reproductive process, and the reproductive organs of males and females are different (see Figure 1.3). Additionally, the female's reproductive structures provide the site for fertilization of eggs by sperm, then protect and nurture the developing fetus until birth.

Growth   Growth is an increase in size of a body part or the organism. It is usually accomplished by increasing the number of cells. However, individual cells also increase in size when not dividing. For true growth to occur, constructive activities must occur at a faster rate than destructive ones.

Survival Needs

The ultimate goal of all body systems is to maintain life. However, life is extraordinarily fragile and requires that several factors be present. These factors, which we will call *survival needs*, include nutrients (food), oxygen, water, and appropriate temperature and atmospheric pressure.

Nutrients, taken in via the diet, contain the chemical substances used for energy and cell building. Most plant-derived foods are rich in carbohydrates, vitamins, and minerals, whereas most animal foods are richer in proteins and fats. Carbohydrates are the major energy fuel for body cells. Proteins, and to a lesser extent fats, are essential for building cell structures. Fats also cushion body organs, form insulating layers, and provide a reserve of energy-rich fuel. Selected minerals and vitamins are required for the chemical reactions that go on in cells and for oxygen transport in the blood. The mineral calcium helps to make bones hard and is required for blood clotting.

All the nutrients in the world are useless unless **oxygen** is also available. Because the chemical reactions that release energy from foods are *oxidative* reactions that require oxygen, human cells can survive for only a few minutes without oxygen. Approximately 20% of the air we breathe is oxygen. It is made available to the blood and body cells by the cooperative efforts of the respiratory and cardiovascular systems.

Water accounts for 60 to 80% of body weight and is the single most abundant chemical substance in the body. It provides the watery environment necessary for chemical reactions and the fluid base for body secretions and excretions. Water is obtained chiefly from ingested foods or liquids and is lost from the body by evaporation from the lungs and skin and in body excretions.

If chemical reactions are to continue at life-sustaining rates, **normal body temperature** must be maintained. As body temperature drops below 37°C (98°F), metabolic reactions become slower and slower, and finally stop. When body temperature is too high, chemical reactions occur at a frantic pace and body proteins lose their characteristic shape and stop functioning. At either extreme, death occurs. Most body heat is generated by the activity of the muscular system.

Atmospheric pressure is the force that air exerts on the surface of the body. Breathing and gas exchange in the lungs depend on *appropriate* atmospheric pressure. At high altitudes, where atmospheric pressure is lower and the air is thin, gas exchange may be inadequate to support cellular metabolism.

The mere presence of these survival factors is not sufficient to sustain life. They must be present in *appropriate* amounts; excesses and deficits may be equally harmful. For example, oxygen is essential, but excessive amounts are toxic to body cells. Similarly, the food we eat must be of high quality and in proper amounts; otherwise, nutritional disease,

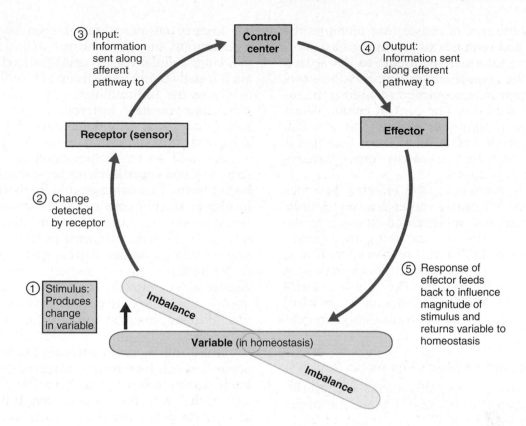

③ Input:
Information
sent along
afferent
pathway to

Control center

④ Output:
Information sent
along efferent
pathway to

Receptor (sensor)

Effector

② Change
detected
by receptor

⑤ Response of
effector feeds
back to influence
magnitude of
stimulus and
returns variable to
homeostasis

① Stimulus:
Produces
change
in variable

Imbalance

Variable (in homeostasis)

Imbalance

FIGURE 1.4 **The elements of a control system.** Communication between the receptor, control center, and effector is essential for normal operation of the system.

obesity, or starvation is likely. Also, while the needs listed above are the most crucial, they do not even begin to encompass all of the body's needs. For example, we can live without gravity if we must, but the quality of life suffers.

Homeostasis

When you think about the fact that your body contains trillions of cells in nearly constant activity, and that remarkably little usually goes wrong with it, you begin to appreciate what a marvelous machine your body is. Walter Cannon, an American physiologist of the early twentieth century, spoke of the "wisdom of the body," and he coined the word **homeostasis** (ho"me-o-sta'sis) to describe its ability to maintain relatively stable internal conditions even though the outside world changes continuously. Although the literal translation of homeostasis is "unchanging," the term does not really mean a static, or unchanging, state. Rather, it indicates a *dynamic* state of equilibrium, or a balance, in which internal conditions vary, but always within relatively narrow limits. In general, the body is in homeostasis when its needs are adequately met and it is functioning smoothly.

Maintaining homeostasis is more complicated than it appears at first glance. Virtually every organ system plays a role in maintaining the constancy of the internal environment. Adequate blood levels of vital nutrients must be continuously present, and heart activity and blood pressure must be constantly monitored and adjusted so that the blood is propelled to all body tissues. Also, wastes must not be allowed to accumulate, and body temperature must be precisely controlled. A wide variety of chemical, thermal, and neural factors act and interact in complex ways—sometimes helping and sometimes hindering the body as it works to maintain its "steady rudder."

Homeostatic Control Mechanisms

Communication within the body is essential for homeostasis. Communication is accomplished chiefly by the nervous and endocrine systems, which use electrical impulses delivered by nerves or blood-borne hormones respectively as information carriers. The details of how these two great regulating systems operate are covered in later chapters, but the basic characteristics of control systems that promote homeostasis are explained here.

Regardless of the factor or event being regulated—this is called the **variable**—all homeostatic control mechanisms have at least three interdependent components (Figure 1.4). The first component, the receptor, is some type of sensor that monitors the environment and responds to changes, called *stimuli*, by sending information (input) to the second component, the *control center*. Information flows from the receptor to the control center along the so-called *afferent pathway*. The **control center**, which determines the *set point* (the level or range at which a variable is to be maintained), analyzes the input it receives and then determines the appropriate response or course of action.

The third component, the **effector**, provides the means for the control center's response (output) to the stimulus. Information flows from the control center to the effector along the *efferent pathway*. The results of the response then *feed back* to influence the stimulus, either depressing it (negative feedback) so that the whole control mechanism is shut off or enhancing it (positive feedback) so that the reaction continues at an even faster rate.

Negative Feedback Mechanisms

Most homeostatic control mechanisms are **negative feedback mechanisms.** In these systems, the output shuts off the original stimulus or reduces its intensity. These mechanisms cause the variable to change in a direction *opposite* to that of the initial change, returning it to its "ideal" value; thus the name "negative" feedback mechanisms.

A good example of a nonbiological negative feedback system is a home heating system connected to a temperature-sensing thermostat. The thermostat houses both the receptor and the control center. If the thermostat is set at 20°C (68°F), the heating system (effector) is triggered ON when the house temperature drops below that setting. As the furnace produces heat and warms the air, the temperature rises, and when it reaches 20°C or slightly higher, the thermostat triggers the furnace OFF. This process results in a cycling of "furnace-ON" and "furnace-OFF" so that the temperature in the house stays very near the desired temperature of 20°C. Your body "thermostat," located in a part of your brain called the hypothalamus, operates in a similar fashion.

Regulation of body temperature is only one of the many ways the nervous system maintains the constancy of the internal environment. Another type of neural control mechanism is seen in the *withdrawal reflex* referred to earlier, in which the hand is jerked away from a painful stimulus such as broken glass. The endocrine system is equally important in maintaining homeostasis, however, and a good example of a hormonal negative feedback mechanism is the control of blood glucose levels by pancreatic hormones (Figure 1.5).

To carry out normal metabolism, body cells need a continuous supply of glucose, their major fuel for producing cellular energy, or ATP. Blood sugar levels are normally maintained around 90 milligrams (mg) of glucose per 100 milliliters (ml) of blood. Let's assume that you have lost your willpower and have just downed four jelly doughnuts. The doughnuts are quickly broken down to sugars by your digestive system, and as the sugars flood into the bloodstream, blood sugar levels spike upward, disrupting homeostasis. The rising glucose levels stimulate the insulin-producing cells of the pancreas, which respond by secreting insulin into the blood. Insulin accelerates the uptake of glucose by most body cells. It also encourages storage of excess glucose as glycogen in the liver and muscles so that a reserve supply of glucose is "put into the larder," so to speak. Consequently, blood sugar levels ebb back toward the normal set point, and the stimulus for insulin release diminishes.

Glucagon, another pancreatic hormone, has the opposite effect. Its release is triggered as blood sugar levels decline below the set point. Suppose, for example, that you have skipped lunch and it is now about 2 PM. Since your blood sugar level is running low, glucagon secretion is stimulated. Glucagon targets the liver, causing it to release its glucose reserves from glycogen into the blood. Consequently, blood sugar levels increase back into the homeostatic range.

The body's ability to regulate its internal environment is fundamental, and all negative feedback mechanisms have the same goal: preventing sudden severe changes within the body. Body temperature and blood glucose levels are only two of the variables that need to be regulated. There are hundreds! Other negative feedback mechanisms regulate heart rate, blood pressure, the rate and depth of breathing, and blood levels of oxygen, carbon dioxide, and minerals. Now, let's take a look at the other type of feedback control mechanism—positive feedback.

Positive Feedback Mechanisms

In **positive feedback mechanisms,** the result or response enhances the original stimulus so that the activity (output) is accelerated. This feedback mechanism is "positive" because the change that occurs proceeds in the *same* direction as the initial disturbance, causing the variable to deviate further and further from its original value or range. In contrast to negative feedback controls, which maintain some physiological function or keep blood chemicals within narrow ranges, positive feedback mechanisms usually control infrequent events that do not require continuous adjustments. Typically, they set off a series of events that may be self-perpetuating

? (1) *What is the control center in this example?* **(2)** *What is the stimulus for insulin release?* **(3)** *What is the response to insulin release?*

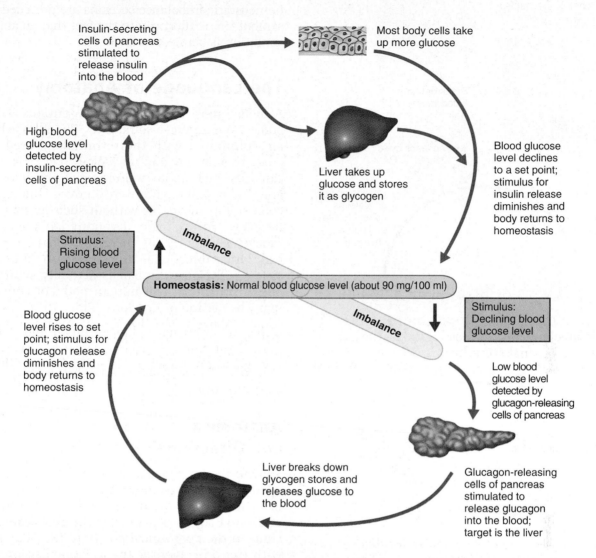

Insulin-secreting cells of pancreas stimulated to release insulin into the blood

Most body cells take up more glucose

High blood glucose level detected by insulin-secreting cells of pancreas

Liver takes up glucose and stores it as glycogen

Blood glucose level declines to a set point; stimulus for insulin release diminishes and body returns to homeostasis

Stimulus: Rising blood glucose level

Imbalance

Homeostasis: Normal blood glucose level (about 90 mg/100 ml)

Imbalance

Stimulus: Declining blood glucose level

Blood glucose level rises to set point; stimulus for glucagon release diminishes and body returns to homeostasis

Low blood glucose level detected by glucagon-releasing cells of pancreas

Liver breaks down glycogen stores and releases glucose to the blood

Glucagon-releasing cells of pancreas stimulated to release glucagon into the blood; target is the liver

FIGURE 1.5 **Regulation of blood glucose levels by a negative feedback mechanism involving pancreatic hormones.**

and that, once initiated, have an amplifying or waterfall effect. Positive feedback mechanisms are often referred to as *cascades* (from the Italian word meaning "to fall") because of these characteristics. Positive feedback mechanisms are likely to race out of control, and so they are rarely used to promote the moment-to-moment well-being of the body. However, there are at least two familiar examples of their use as homeostatic mechanisms—enhancement of labor contractions during birth and blood clotting.

Blood clotting is a normal response to a break in the lining of a blood vessel and is an excellent example of an important body function controlled by positive feedback. Basically, once vessel damage has occurred (1), blood elements called platelets immediately begin to cling to the injured site (2) and release chemicals that attract more platelets (3). This rapidly growing pileup of platelets initiates the sequence of events that finally forms a clot (4) (Figure 1.6).

Homeostatic Imbalance

Homeostasis is so important that most disease is regarded as a result of its disturbance, a condition

■ (1) *The pancreas.* (2) *Rising blood glucose levels.* (3) *Blood glucose levels fall as body cells take it up from the blood.*

(1) *Why is this control mechanism called a "positive feedback mechanism"?* **(2)** *What event ends the cascade or chain reaction seen in this positive feedback control mechanism?*

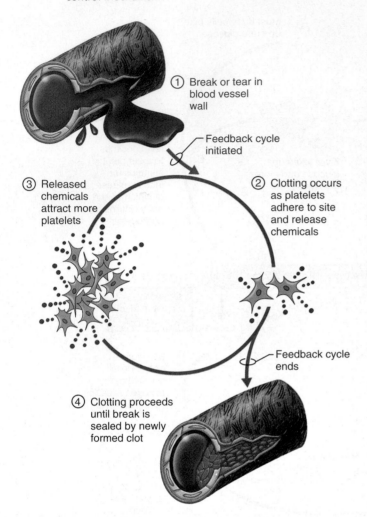

① Break or tear in blood vessel wall

Feedback cycle initiated

③ Released chemicals attract more platelets

② Clotting occurs as platelets adhere to site and release chemicals

Feedback cycle ends

④ Clotting proceeds until break is sealed by newly formed clot

FIGURE 1.6   **Summary of the positive feedback mechanism regulating blood clotting.**

called **homeostatic imbalance.** As we age, our body's control systems become less efficient, and our internal environment becomes less and less stable. These events increase our risk for illness and produce the changes we associate with aging.

Another important source of homeostatic imbalance occurs when the usual negative feedback mechanisms are overwhelmed and destructive positive feedback mechanisms take over. Some instances of heart failure reflect this phenomenon.

(1) Because the response leads to an even greater response (i.e., platelets cling to damaged vessels and release chemicals that attract more platelets, which release more chemicals, etc.) rather than shutting off the stimulus. (2) When the clot seals the break in the vessel, the cascade ends. ■

Examples of homeostatic imbalance are provided throughout this book to enhance your understanding of normal physiological mechanisms. These homeostatic imbalance sections are preceded by the symbol ❧ to alert you to the fact that an abnormal condition is being described.

The Language of Anatomy

Although most of us are naturally curious about our bodies, our interest sometimes dwindles when we are confronted with the terminology used in the study of anatomy and physiology. Let's face it. You can't just pick up an anatomy and physiology book and read it as though it were a novel. Unfortunately, confusion is inevitable without such specialized terminology. For example, if you are looking at a ball, "above" always means the area over the top of the ball. Other directional terms can also be used consistently because the ball is a totally symmetrical object, a sphere. The human body, of course, has many bends and protrusions. Thus, the question becomes: Above what? To prevent misunderstanding, anatomists use universally accepted terms to identify body structures precisely and with a minimum of words. This language of anatomy is presented and explained next.

Anatomical Position and Directional Terms

To describe body parts and position accurately, we need an initial reference point and must indicate direction. The anatomical reference point is a standard body position called the **anatomical position.** In the anatomical position, the body is erect with feet only slightly apart. This position is easy to remember because it resembles "standing at attention," except that the palms face forward and the thumbs point away from the body. You can see the anatomical position in Table 1.1 (top) and Figure 1.7a. It is essential to understand the anatomical position because most of the directional terms used in this book refer to the body *as if it were in this position, regardless of its actual position.* Another point to remember is that the terms "right" and "left" refer to those sides of the person or the cadaver being viewed—not those of the observer.

Directional terms allow us to explain where one body structure is in relation to another. For example, we could describe the relationship between the ears and the nose informally by stating, "The ears are located on each side of the head to the right and left of the nose." Using anatomical terminology, this condenses to, "The ears are lateral to the nose."

TABLE 1.1	Orientation and Directional Terms		
Term	**Definition**	**Example**	
Superior (cranial)	Toward the head end or upper part of a structure or the body; above		The head is superior to the abdomen
Inferior (caudal)	Away from the head end or toward the lower part of a structure or the body; below		The navel is inferior to the chin
Anterior (ventral)*	Toward or at the front of the body; in front of		The breastbone is anterior to the spine
Posterior (dorsal)*	Toward or at the back of the body; behind		The heart is posterior to the breastbone
Medial	Toward or at the midline of the body; on the inner side of		The heart is medial to the arm
Lateral	Away from the midline of the body; on the outer side of		The arms are lateral to the chest
Intermediate	Between a more medial and a more lateral structure		The collarbone is intermediate between the breastbone and shoulder
Proximal	Closer to the origin of the body part or the point of attachment of a limb to the body trunk		The elbow is proximal to the wrist
Distal	Farther from the origin of a body part or the point of attachment of a limb to the body trunk		The knee is distal to the thigh
Superficial (external)	Toward or at the body surface		The skin is superficial to the skeletal muscles
Deep (internal)	Away from the body surface; more internal		The lungs are deep to the skin

*Whereas the terms *ventral* and *anterior* are synonymous in humans, this is not the case in four-legged animals. *Ventral* specifically refers to the "belly" of a vertebrate animal and thus is the inferior surface of four-legged animals. Likewise, although the dorsal and posterior surfaces are the same in humans, the term *dorsal* specifically refers to an animal's back. Thus, the dorsal surface of four-legged animals is their superior surface.

As you study this figure, determine precisely where you would be injured if you
(1) pulled a muscle in your inguinal region, **(2)** cracked a bone in your olecranal region,
and **(3)** received a cut on the fibular region.

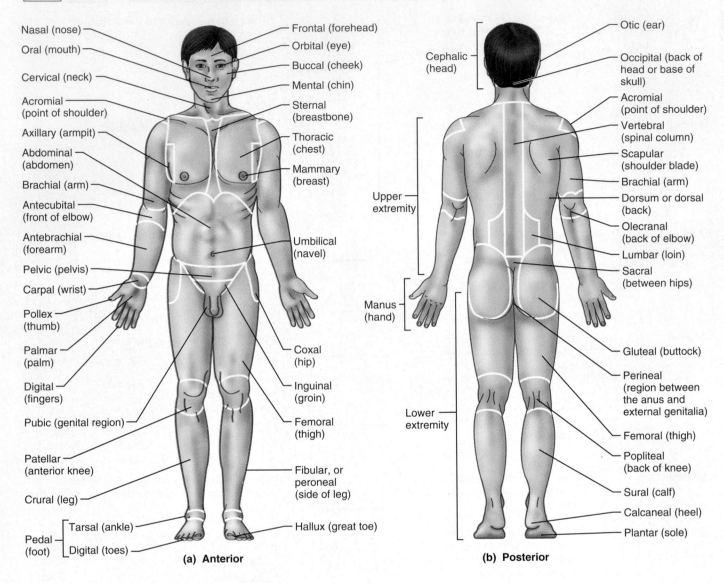

FIGURE 1.7 Regional terms used to designate specific body areas. (a) The
anatomical position. **(b)** The heels are raised slightly to show the plantar surface of
the foot.

Clearly, using anatomical terms saves words and is less ambiguous. Commonly used orientation and directional terms are defined and illustrated in Table 1.1. Many of these terms are also used in everyday conversation, but keep in mind that their anatomical meanings are very precise.

Regional Terms

The two fundamental divisions of our body are its *axial* and *appendicular* (ap″en-dik′u-lar) parts. The **axial part,** which makes up the main *axis* of our body, includes the head, neck, and trunk. The **appendicular part** consists of the *appendages,* or *limbs,* which are attached to the body's axis. **Regional terms** used to designate specific areas within the major body divisions are indicated in Figure 1.7. The common term for each of these body regions is also provided (in parentheses).

(1) Your groin. **(2)** Posterior aspect of the elbow. **(3)** The side of your leg. ■

Anatomical Variability

Although we use common directional and regional terms to refer to all human bodies, we know from observing the faces and body shapes of people around us that we humans differ in our external anatomy. The same kind of variability holds for our internal organs as well. In some bodies, for example, a nerve or blood vessel may be somewhat out of place, or a small muscle may be missing. However, well over 90% of all structures present in any human body match the textbook descriptions. Extreme anatomical variations are seldom seen because they are incompatible with life.

Body Planes and Sections

For anatomical studies, the body is often *sectioned* (cut) along a flat surface called a plane. The most frequently used body planes are sagittal, frontal, and transverse planes, which lie at right angles to one another (Figure 1.8). A section is named for the plane along which it is cut. Thus, a cut along a sagittal plane produces a sagittal section.

A **sagittal plane** (saj'ĭ-tal; "arrow") is a vertical plane that divides the body into right and left parts. A sagittal plane that lies exactly in the midline is the **median plane,** or **midsagittal plane** (Figure 1.8c). All other sagittal planes, offset from the midline, are **parasagittal planes** (*para* = near).

Frontal planes, like sagittal planes, lie vertically. Frontal planes, however, divide the body into anterior and posterior parts (Figure 1.8a). A frontal plane is also called a **coronal plane** (kŏ-ro'nal; "crown").

A **transverse,** or **horizontal, plane** runs horizontally from right to left, dividing the body into superior and inferior parts (Figure 1.8b). Of course, many different transverse planes exist, at every possible level from head to foot. A transverse section is also called a **cross section.** Cuts made diagonally between the horizontal and the vertical planes are called **oblique sections.** Such sections are often confusing and difficult to interpret and are not used much.

The ability to interpret sections made through the body, especially transverse sections, is important in the clinical sciences. New medical imaging devices produce sectional images rather than three-dimensional images. It can be difficult to decipher an object's overall shape from sectioned material. A cross section of a banana, for example, looks like a circle and gives no indication of the whole banana's crescent shape. Likewise, sectioning the body or an organ along different planes often results in very different views. For example, a transverse section of the body trunk at the level of the kidneys would show kidney structure in cross section very nicely; a frontal section of the body trunk would show a different view of kidney anatomy; and a midsagittal section would miss the kidneys completely. With practice, you will gradually learn to relate two-dimensional sections to three-dimensional shapes.

Body Cavities and Membranes

Within the axial portion of the body are two large cavities called the dorsal and ventral body cavities. These cavities are closed to the outside and each contains internal organs.

Dorsal Body Cavity

The **dorsal body cavity,** which protects the fragile nervous system organs (Figure 1.9), has two subdivisions. The **cranial cavity,** in the skull, encases the brain. The **vertebral,** or **spinal cavity,** which runs within the bony vertebral column, encloses the delicate spinal cord. Because the spinal cord is essentially a continuation of the brain, the cranial and spinal cavities are continuous with one another.

Ventral Body Cavity

The more anterior and larger of the closed body cavities is the **ventral body cavity** (see Figure 1.9). Like the dorsal cavity, it has two major subdivisions, the *thoracic cavity* and the *abdominopelvic cavity.* The ventral body cavity houses internal organs collectively called the **viscera** (vis'er-ah; *viscus* = an organ in a body cavity), or **visceral organs.**

The superior subdivision, the **thoracic cavity** (tho-ras'ik), is surrounded by the ribs and muscles of the chest. The thoracic cavity is further subdivided into lateral **pleural cavities** (ploo'ral), each housing a lung, and the medial **mediastinum** (me"de-ah-sti'num). The mediastinum contains the **pericardial cavity** (per"ĭ-kar'de-al), which encloses the heart, and it also surrounds the remaining thoracic organs (esophagus, trachea, and others).

The thoracic cavity is separated from the more inferior **abdominopelvic cavity** (ab-dom'ĭ-no-pel'vic) by the diaphragm, a dome-shaped muscle important in breathing. The abdominopelvic cavity, as its name suggests, has two parts. However, these regions are not physically separated by a muscular or membrane wall. Its superior portion, the **abdominal cavity,** contains the stomach, intestines, spleen, liver, and other organs. The

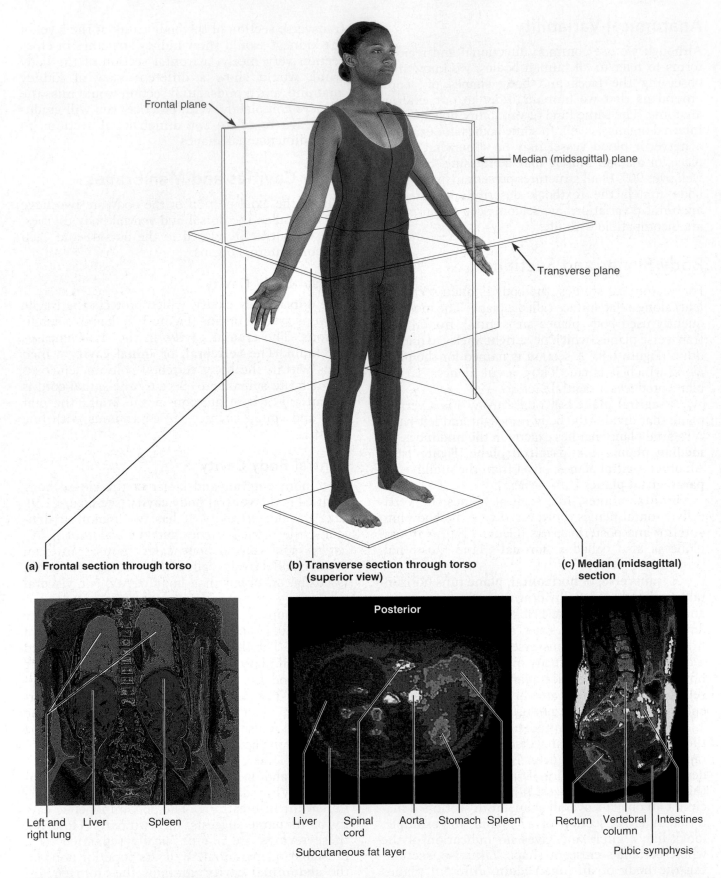

Frontal plane

Median (midsagittal) plane

Transverse plane

(a) Frontal section through torso

(b) Transverse section through torso (superior view)

(c) Median (midsagittal) section

Posterior

Left and right lung   Liver   Spleen

Liver   Spinal cord   Aorta   Stomach   Spleen

Subcutaneous fat layer

Rectum   Vertebral column   Intestines

Pubic symphysis

FIGURE 1.8   **Planes of the body—frontal, transverse, and median (midsagittal) with corresponding magnetic resonance imaging (MRI) scans.**

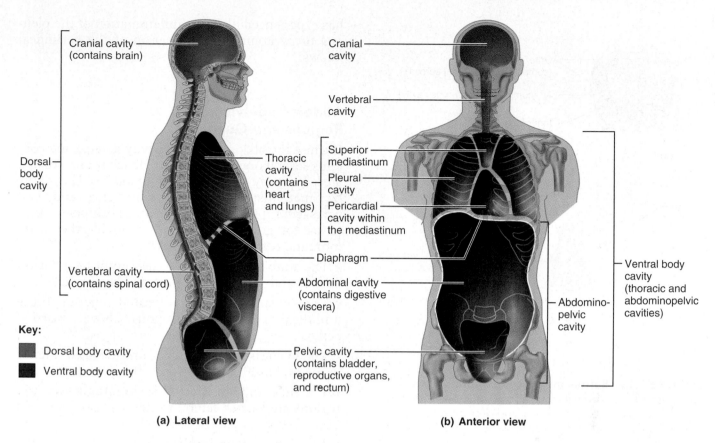

Key:

■ Dorsal body cavity

■ Ventral body cavity

(a) Lateral view

(b) Anterior view

FIGURE 1.9 **Dorsal and ventral body cavities and their subdivisions.**

inferior part, the **pelvic cavity,** lies in the bony pelvis and contains the bladder, some reproductive organs, and the rectum. The abdominal and pelvic cavities are not aligned with each other. Instead, the bowl-shaped pelvis tips away from the perpendicular.

⬤H HOMEOSTATIC IMBALANCE

When the body is subjected to physical trauma (as often happens in an automobile accident), the abdominopelvic organs are most vulnerable. This is because the walls of the abdominal cavity are formed only by trunk muscles and are not reinforced by bone. The pelvic organs receive a somewhat greater degree of protection from the bony pelvis. ●

Membranes in the Ventral Body Cavity

The walls of the ventral body cavity and the outer surfaces of the organs it contains are covered by a thin, double-layered membrane, the **serosa** (se-ro'sah), or **serous membrane.** The part of the membrane lining the cavity walls is called the **parietal serosa** (pah-ri'ĕ-tal; *parie* = wall). It folds in on itself to form the **visceral serosa,** covering the organs in the cavity.

You can visualize the relationship between the serosal layers by pushing your fist into a limp balloon (Figure 1.10a). The part of the balloon that clings to your fist can be compared to the visceral serosa clinging to the organ's external surface. The outer wall of the balloon then represents the parietal serosa that lines the walls of the cavity. (However, unlike the balloon, it is never exposed but is always fused to the cavity wall.) In the body, the serous membranes are separated not by air but by a thin layer of lubricating fluid, called **serous fluid,** which is secreted by both membranes. Although there is a potential space between the two membranes, the slitlike cavity is filled with serous fluid.

The slippery serous fluid allows the organs to slide without friction across the cavity walls and one another as they carry out their routine functions. This freedom of movement is especially important for mobile organs such as the pumping heart and the churning stomach.

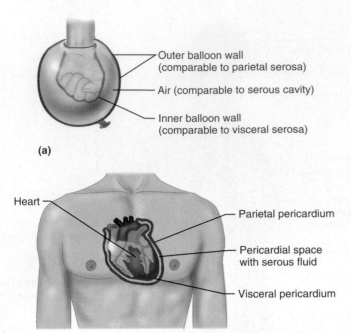

(a)

(b)

FIGURE 1.10   Serous membrane relationships.
(a) A fist is thrust into a flaccid balloon to demonstrate the relationship between the parietal and visceral serous membrane layers. **(b)** The serosae associated with the heart. The parietal pericardium is the outer layer lining the pericardial cavity; the visceral pericardium clings to the external heart surface.

The serous membranes are named for the specific cavity and organs with which they are associated. For example, as shown in Figure 1.10b, the *parietal pericardium* lines the pericardial cavity, while the *visceral pericardium* covers the heart within that cavity. Likewise, the *parietal pleura* (ploo′rah) lines the walls of the thoracic cavity, and the *visceral pleura* covers the lungs; and the *parietal peritoneum* (per″ĭ-to-ne′um) is associated with the walls of the abdominopelvic cavity, while the *visceral peritoneum* covers most of the organs within that cavity. (The pleural and peritoneal serosae are illustrated in Figure 4.9c on p. 127.)

ⓗ *HOMEOSTATIC IMBALANCE*

When serous membranes are inflamed, they typically produce less lubricating serous fluid. This leads to excruciating pain as the organs stick together and drag across one another, as anyone who has experienced *pleurisy* (inflammation of the pleurae) or *peritonitis* (inflammation of the peritonea) knows. ●

Abdominopelvic Regions and Quadrants

Because the abdominopelvic cavity is large and contains several organs, it helps to divide it into smaller areas for study. One division method, used primarily by anatomists, uses two transverse and two parasagittal planes. These planes, positioned like a tic-tac-toe grid on the abdomen, divide the cavity into nine **regions** (Figure 1.11):

■ The **umbilical region** is the centermost region deep to and surrounding the umbilicus (navel).

■ The **epigastric region** is located superior to the umbilical region (*epi* = upon, above; *gastri* = belly).

■ The **hypogastric (pubic) region** is located inferior to the umbilical region (*hypo* = below).

■ The **right** and **left iliac**, or **inguinal** (ing′gwĭ-nal), **regions** are located lateral to the hypogastric region (*iliac* = superior part of the hip bone).

■ The **right** and **left lumbar regions** lie lateral to the umbilical region (*lumbus* = loin).

■ The **right** and **left hypochondriac regions** flank the epigastric region laterally (*chondro* = cartilage).

Medical personnel usually use a simpler scheme to localize the abdominopelvic cavity organs (Figure 1.12). In this scheme, one transverse and one median sagittal plane pass through the umbilicus at right angles. The resulting **quadrants** are named according to their positions from the subject's point of view: the **right upper quadrant (RUQ), left upper quadrant (LUQ), right lower quadrant (RLQ),** and **left lower quadrant (LLQ).**

Other Body Cavities

In addition to the large closed body cavities, there are several smaller body cavities (Figure 1.13). Most of these are in the head and most open to the body exterior:

1. Oral and digestive cavities. The oral cavity, commonly called the mouth, contains the teeth and tongue. This cavity is part of and continuous with the cavity of the digestive organs, which opens to the exterior at the anus.

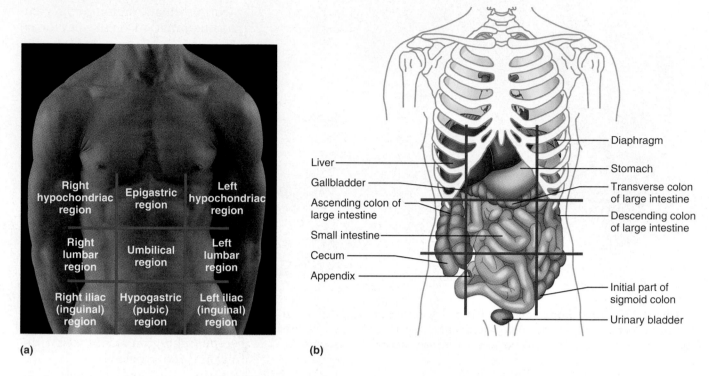

(a)

(b)

FIGURE 1.11 **The nine abdominopelvic regions. (a)** Division of the abdominopelvic cavity into nine regions delineated by four planes. The superior transverse plane is just inferior to the ribs; the inferior transverse plane is just superior to the hip bones; and the parasagittal planes lie just medial to the nipples. **(b)** Anterior view of the abdominopelvic cavity showing the superficial organs.

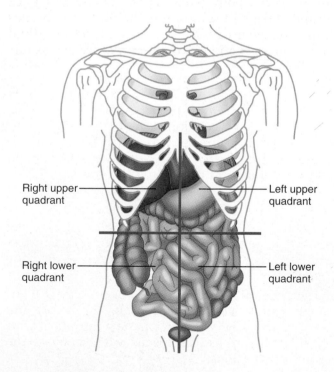

FIGURE 1.12 **The four abdominopelvic quadrants.** In this scheme, the abdominopelvic cavity is divided into four quadrants by two planes. The figure shows superficial organs within each quadrant.

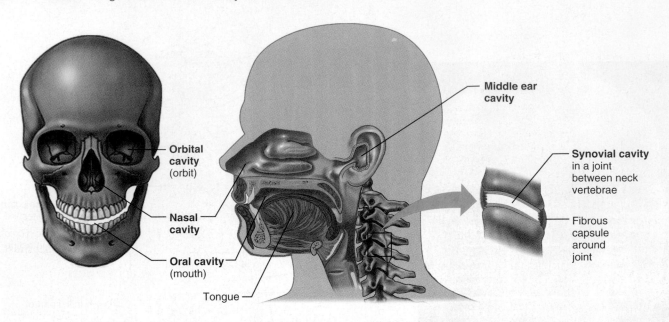

FIGURE 1.13 **Other body cavities.** The oral, nasal, orbital, and middle ear cavities are located in the head and open to the body exterior. Synovial cavities are found in the joints between many bones such as the vertebrae of the spine.

2. Nasal cavity. Located within and posterior to the nose, the nasal cavity is part of the respiratory system passageways.

3. Orbital cavities. The orbital cavities (orbits) in the skull house the eyes and present them in an anterior position.

4. Middle ear cavities. The middle ear cavities carved into the skull lie just medial to the eardrums. These cavities contain tiny bones that transmit sound vibrations to the hearing receptors in the inner ears.

5. Synovial (sĭ-no′ve-al) **cavities.** Synovial cavities are joint cavities. They are enclosed within fibrous capsules that surround freely movable joints of the body (such as the elbow and knee joints). Like the serous membranes, membranes lining synovial cavities secrete a lubricating fluid that reduces friction as the bones move across one another.

Review Questions

Multiple Choice/Matching

(Some questions have more than one correct answer. Select the best answer or answers from the choices given.)

1. The correct sequence of levels forming the structural hierarchy is (a) organ, organ system, cellular, chemical, tissue, organismal; (b) chemical, cellular, tissue, organismal, organ, organ system; (c) chemical, cellular, tissue, organ, organ system, organismal; (d) organismal, organ system, organ, tissue, cellular, chemical.

2. The structural and functional unit of life is (a) a cell, (b) an organ, (c) the organism, (d) a molecule.

3. Which of the following is a *major* functional characteristic of all organisms? (a) movement, (b) growth, (c) metabolism, (d) responsiveness, (e) all of these.

4. Two of these organ systems bear the *major* responsibility for ensuring homeostasis of the internal environment. Which two? (a) nervous system, (b) digestive system, (c) cardiovascular system, (d) endocrine system, (e) reproductive system.

5. In (a)–(e), a directional term [e.g., distal in (a)] is followed by terms indicating different body structures or locations (e.g., the elbow/the wrist). In each case, choose the structure or organ that matches the given directional term.

   (a) distal: the elbow/the wrist
   (b) lateral: the hip bone/the umbilicus
   (c) superior: the nose/the chin
   (d) anterior: the toes/the heel
   (e) superficial: the scalp/the skull

6. Assume that the body has been sectioned along three planes: (1) a median sagittal plane, (2) a frontal plane, and (3) a transverse plane made at the level of each of the organs listed below. Which organs would not be visible in all three cases? (a) urinary bladder, (b) brain, (c) lungs, (d) kidneys, (e) small intestine, (f) heart.

7. Relate each of the following conditions or statements to either the dorsal body cavity or the ventral body cavity.

   (a) surrounded by the bony skull and the vertebral column
   (b) includes the thoracic and abdominopelvic cavities
   (c) contains the brain and spinal cord
   (d) contains the heart, lungs, and digestive organs

8. Which of the following relationships is *incorrect*?
   (a) visceral peritoneum/outer surface of small intestine
   (b) parietal pericardium/outer surface of heart
   (c) parietal pleura/wall of thoracic cavity

9. Which ventral cavity subdivision has no bony protection? (a) thoracic cavity, (b) abdominal cavity, (c) pelvic cavity.

Short Answer Essay Questions

10. According to the principle of complementarity, how does anatomy relate to physiology?

11. Construct a table that lists the 11 systems of the body, name two organs of each system (if appropriate), and describe the overall or major function of each system.

12. List and describe briefly five external factors that must be present or provided to sustain life.

13. Define homeostasis.

14. Compare and contrast the operation of negative and positive feedback mechanisms in maintaining homeostasis. Provide two examples of variables controlled by negative feedback mechanisms and one example of a process regulated by a positive feedback mechanism.

15. Describe and assume the anatomical position. Why is an understanding of this position important? What is the importance of directional terms?

16. Define plane and section.

17. Provide the anatomical term that correctly names each of the following body regions: (a) arm, (b) thigh, (c) chest, (d) fingers and toes, (e) anterior aspect of the knee.

18. (a) Make a diagram showing the nine abdominopelvic regions, and name each region. Name two organs (or parts of organs) that could be located in each of the named regions. (b) Make a similar sketch illustrating how the abdominopelvic cavity may be divided into quadrants, and name each quadrant.

2

CHEMISTRY COMES ALIVE

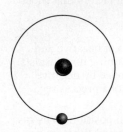

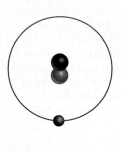

PART 1: BASIC CHEMISTRY

Definition of Concepts: Matter and Energy (pp. 23–24)

1. Differentiate between matter and energy and between potential energy and kinetic energy.
2. Describe the major energy forms.

Composition of Matter: Atoms and Elements (pp. 24–28)

3. Define chemical element and list the four elements that form the bulk of body matter.
4. Define atom. List the subatomic particles; describe their relative masses, charges, and positions in the atom.
5. Define atomic number, atomic mass, atomic weight, isotope, and radioisotope.

How Matter Is Combined: Molecules and Mixtures (pp. 28–29)

6. Distinguish between a compound and a mixture. Define molecule.
7. Compare solutions, colloids, and suspensions.

Chemical Bonds (pp. 29–33)

8. Explain the role of electrons in chemical bonding and in relation to the octet rule.
9. Differentiate among ionic, covalent, and hydrogen bonds.
10. Compare and contrast polar and nonpolar compounds.

Chemical Reactions (pp. 34–37)

11. Define the three major types of chemical reactions: synthesis, decomposition, and exchange. Comment on the nature of oxidation-reduction reactions and their importance.
12. Explain why chemical reactions in the body are often irreversible.
13. Describe factors that affect chemical reaction rates.

PART 2: BIOCHEMISTRY

Inorganic Compounds (pp. 37–41)

14. Explain the importance of water and salts to body homeostasis.
15. Define acid and base, and explain the concept of pH.

Organic Compounds (pp. 41–55)

16. Describe and compare the building blocks, general structures, and biological functions of carbohydrates, lipids, proteins, and nucleic acids.
17. Explain the role of dehydration synthesis and hydrolysis in the formation and breakdown of organic molecules.
18. Describe the four levels of protein structure.
19. Describe the general mechanism of enzyme activity.
20. Describe the function of molecular chaperones.
21. Compare and contrast DNA and RNA.
22. Explain the role of ATP in cell metabolism.

Why study chemistry in an anatomy and physiology course? The answer is simple. Your entire body is made up of chemicals, thousands of them, continuously interacting with one another at an incredible pace. Although it is possible to study anatomy without much reference to chemistry, chemical reactions underlie all physiological processes—movement, digestion, the pumping of your heart, and even your thoughts. This chapter presents the basic chemistry and biochemistry (the chemistry of living material) needed to understand body functions.

PART 1: BASIC CHEMISTRY

Definition of Concepts: Matter and Energy

Matter

Matter is the "stuff" of the universe. More precisely, matter is anything that occupies space and has mass. With some exceptions, it can be seen, smelled, and felt. For all practical purposes, we can consider mass to be the same as weight. However, this usage is not quite accurate. The *mass* of an object, which is equal to the actual amount of matter in the object, remains constant wherever the object is. In contrast, weight varies with gravity. So while your mass is the same at sea level and on a mountaintop, you weigh just slightly less on that mountaintop. The science of chemistry studies the nature of matter, especially how its building blocks are put together and interact.

States of Matter

Matter exists in *solid, liquid,* and *gaseous states*. Examples of each state are found in the human body. Solids, like bones and teeth, have a definite shape and volume. Liquids such as blood plasma have a definite volume, but they conform to the shape of their container. Gases have neither a definite shape nor a definite volume. The air we breathe is a gas.

Energy

Compared with matter, **energy** is less tangible. Energy has no mass and does not take up space. It can be measured only by its effects on matter. Energy is defined as the capacity to do work, or to put matter into motion. The greater the work done, the more energy is used doing it. A baseball player who has just hit the ball over the fence uses much more energy than a batter who just bunts the ball back to the pitcher.

Kinetic Versus Potential Energy

Energy exists in two forms, or work capacities, each transformable to the other. **Kinetic energy** (ki-net′ik) is energy in action. Kinetic energy is seen in the constant movement of the tiniest particles of matter (atoms) as well as in larger objects (a bouncing ball). It does work by moving objects, which in turn can do work by moving or pushing on other objects. For example, a push on a swinging door sets it into motion.

Potential energy is stored energy, that is, inactive energy that has the *potential*, or capability, to do work but is not presently doing so. The batteries in an unused toy have potential energy, as do your leg muscles when you sit still on the couch. When potential energy is released, it becomes kinetic energy and so is capable of doing work. For example, dammed water becomes a rushing torrent when the dam is opened, and that rushing torrent can move a turbine at a hydroelectric plant, or charge a battery.

Actually, energy is a topic of physics, but matter and energy are inseparable. Matter is the substance, and energy is the mover of the substance. All living things are composed of matter and they all require energy to grow and function. It is the release and use of energy by living systems that gives us the elusive quality called life. Thus, it is worth taking a brief detour to introduce the forms of energy used by the body as it does its work.

Forms of Energy

▪ **Chemical energy** is the form stored in the bonds of chemical substances. When chemical reactions occur that rearrange the atoms of the chemicals in a certain way, the potential energy is unleashed and becomes kinetic energy, or energy in action.

For example, some of the energy in the foods you eat is eventually converted into the kinetic energy of your moving arm. However, food fuels cannot be used to energize body activities directly. Instead, some of the food energy is captured temporarily in the bonds of a chemical called **adenosine triphosphate** (ah-den′o-sēn tri″fos′fāt), or **ATP.** Later, ATP's bonds are broken and the stored energy is released as needed to do cellular work. Chemical energy in the form of ATP is the most useful form of energy in living systems because it is used to run all functional processes.

▪ **Electrical energy** results from the movement of charged particles. In your home, electrical energy is found in the flow of electrons along the household wiring. In your body, electrical currents are generated when charged particles called *ions* move along or across cell membranes. The nervous system uses electrical currents, called *nerve impulses*, to transmit messages from one part of the body to another.

Electrical currents traveling across the heart stimulate it to contract (beat) and pump blood. (This is why a strong electrical shock, which interferes with such currents, can cause death.)

- **Mechanical energy** is energy *directly* involved in moving matter. When you ride a bicycle, your legs provide the mechanical energy that moves the pedals.

- **Radiant energy,** or **electromagnetic energy** (e-lek″ tro-mag-net'ik), is energy that travels in waves. These waves, which vary in length, are collectively called the *electromagnetic spectrum* and include visible light, infrared waves, radio waves, ultraviolet waves, and X rays. Light energy, which stimulates the retinas of our eyes, is important in vision. Ultraviolet waves cause sunburn, but they also stimulate our body to make vitamin D.

Energy Form Conversions

With few exceptions, energy is easily converted from one form to another. For example, chemical energy (gasoline) that powers the motor of a speedboat is converted into the mechanical energy of the whirling propeller that allows the boat to skim across the water. Energy conversions are quite inefficient; some of the initial energy supply is always "lost" to the environment as heat. (It is not really lost because energy cannot be created or destroyed, but that part given off as heat is at least partly *unusable*.) It is easy to demonstrate this principle. Electrical energy is converted into light energy in a lightbulb. But if you touch a lit bulb, you will soon discover that some of the electrical energy is producing heat instead. Likewise, all energy conversions that occur in the body liberate heat. This heat contributes to our relatively high body temperature, which has an important influence on body functioning. For example, when matter is heated, the kinetic energy of its particles increases and they begin to move more quickly. The higher the temperature, the faster the body's chemical reactions occur. We will learn more about this later.

Composition of Matter: Atoms and Elements

All matter is composed of **elements,** unique substances that cannot be broken down into simpler substances by ordinary chemical methods. Among the well-known elements are oxygen, carbon, gold, silver, copper, and iron. At present, 112 elements are known with certainty (and numbers 114 and 116 are alleged). Of these, 92 occur in nature. The rest are made artificially in particle accelerator devices.

Four elements—carbon, oxygen, hydrogen, and nitrogen—make up about 96% of body weight; 20 others are present in the body, some in trace amounts. The elements contributing to body mass and their importance are given in Table 2.1.

Each element is composed of more or less identical particles or building blocks, called **atoms.** The smallest atoms are less than 0.1 nanometer (nm) in diameter, and the largest only about five times as large. (1 nm = 0.0000001 [or 10^{-7}] centimeter [cm], or 40 billionths of an inch!)

Every element's atoms differ from those of all other elements and give the element its unique physical and chemical properties. *Physical properties* are those we can detect with our senses (such as color and texture) or measure (such as boiling point and freezing point). *Chemical properties* pertain to the way atoms interact with other atoms (bonding behavior) and account for the facts that iron rusts, animals can digest their food, and so on.

Each element is designated by a one- or two-letter chemical shorthand called an **atomic symbol,** usually the first letter(s) of the element's name. For example, C stands for carbon, O for oxygen, and Ca for calcium. In a few cases, the atomic symbol is taken from the Latin name for the element. For example, sodium is indicated by Na, from the Latin word *natrium.*

Atomic Structure

The word *atom* comes from the Greek word meaning "indivisible." However, we now know that atoms are clusters of even smaller particles called protons, neutrons, and electrons and that even those subatomic particles can be subdivided with high-technology tools. Even so, the old idea of atomic indivisibility is still helpful because an atom loses the unique properties of its element when it is split into its subatomic particles.

An atom's subatomic particles differ in mass, electrical charge, and position in the atom. An atom has a central **nucleus** containing protons and neutrons tightly bound together. The nucleus, in turn, is surrounded by orbiting electrons (Figure 2.1). **Protons** (p^+) bear a positive electrical charge, and **neutrons** (n^0) are neutral. Thus, the nucleus is positively charged overall. Protons and neutrons are heavy particles and have approximately the same mass, arbitrarily designated as 1 **atomic mass unit** (1 amu). Since all of the heavy subatomic particles are concentrated in the nucleus, the nucleus is fantastically dense and accounts for nearly the entire mass (99.9%) of the atom. The tiny **electrons** (e^-) bear a negative charge equal in strength to the positive charge of the proton. However, an electron has only about 1/2000 the mass

TABLE 2.1 Common Elements Composing the Human Body

Element	Atomic Symbol	Approx. % Body Mass*	Functions
Major (96.1%)			
Oxygen	O	65.0	A major component of both organic (carbon-containing) and inorganic (non-carbon-containing) molecules; as a gas, it is needed for the production of cellular energy (ATP)
Carbon	C	18.5	A primary component of all organic molecules, which include carbohydrates, lipids (fats), proteins, and nucleic acids
Hydrogen	H	9.5	A component of all organic molecules; as an ion (proton), it influences the pH of body fluids
Nitrogen	N	3.2	A component of proteins and nucleic acids (genetic material)
Lesser (3.9%)			
Calcium	Ca	1.5	Found as a salt in bones and teeth; its ionic (Ca^{2+}) form is required for muscle contraction, conduction of nerve impulses, and blood clotting
Phosphorus	P	1.0	Part of calcium phosphate salts in bones and teeth; also present in nucleic acids; part of ATP
Potassium	K	0.4	Its ion (K^+) is the major positive ion (cation) in cells; necessary for conduction of nerve impulses and muscle contraction
Sulfur	S	0.3	Component of proteins, particularly muscle proteins
Sodium	Na	0.2	As an ion (Na^+), sodium is the major positive ion found in extracellular fluids (fluids outside of cells); important for water balance, conduction of nerve impulses, and muscle contraction
Chlorine	Cl	0.2	Ionic chlorine (Cl^-) is the most abundant negative ion (anion) in extracellular fluids
Magnesium	Mg	0.1	Present in bone; also an important cofactor in a number of metabolic reactions
Iodine	I	0.1	Needed to make functional thyroid hormones
Iron	Fe	0.1	Component of hemoglobin (which transports oxygen within red blood cells) and some enzymes

Trace (less than 0.01%)

Chromium (Cr); cobalt (Co); copper (Cu); fluorine (F); manganese (Mn); molybdenum (Mo); selenium (Se); silicon (Si); tin (Sn); vanadium (V); zinc (Zn)

These elements are referred to as *trace elements* because they are required in very minute amounts; many are found as part of enzymes or are required for enzyme activation.

*Percentage of "wet" body mass; includes water.

of a proton, and the mass of an electron is usually designated as 0 amu.

All atoms are electrically neutral because the number of protons in an atom is precisely balanced by its number of electrons (the + and − charges will then cancel the effect of each other). Thus, hydrogen has one proton and one electron, and iron has 26 protons and 26 electrons. For any atom, the number of protons and electrons is always equal.

The **planetary model,** illustrated in Figure 2.1a, is a simplified (and now outdated) model of atomic structure. As you can see, it depicts electrons moving around the nucleus in fixed, generally circular orbits. But we can never determine the exact location of electrons at a particular time because they jump around following unknown trajectories. So, instead of speaking of specific orbits, chemists talk about **orbitals**—regions around the nucleus in which a given electron or electron pair is likely to be found most of the time. This more modern model of atomic structure, called the **orbital model,** is more useful for predicting the chemical behavior of atoms. As illustrated in Figure 2.1b, the orbital model depicts *probable* regions of greatest electron density by

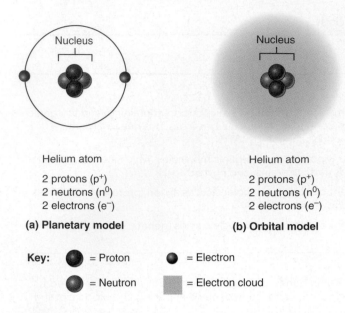

Helium atom

2 protons (p⁺)
2 neutrons (n⁰)
2 electrons (e⁻)

(a) Planetary model

Helium atom

2 protons (p⁺)
2 neutrons (n⁰)
2 electrons (e⁻)

(b) Orbital model

Key: = Proton   = Electron

 = Neutron   = Electron cloud

FIGURE 2.1   **The structure of an atom.** The dense central nucleus contains the protons and neutrons. **(a)** In the planetary model of atomic structure, the electrons move around the nucleus in fixed orbits. **(b)** In the orbital model, electrons are shown as a cloud of negative charge (electron cloud).

denser shading (this haze is called the *electron cloud*). However, because of the simplicity of the planetary model, most descriptions of atomic structure in this text will use that model.

Hydrogen, with just one proton and one electron, is the simplest atom. You can visualize the spatial relationships in the hydrogen atom by imagining it enlarged until its diameter equals the length of a football field. In that case, the nucleus could be represented by a lead ball the size of a gumdrop in the exact center of the sphere and its lone electron pictured as a fly buzzing about unpredictably within the sphere. Though not completely accurate, this mental picture should serve as a reminder that most of the volume of an atom is empty space, and nearly all of its mass is concentrated in the central nucleus.

Identifying Elements

All protons are alike, regardless of the atom considered. The same is true of all neutrons and all electrons. So what determines the unique properties of each element? The answer is that atoms of different elements are composed of *different numbers* of protons, neutrons, and electrons.

The simplest and smallest atom, hydrogen, has one proton, one electron, and no neutrons (Figure 2.2). Next in size is the helium atom, with two protons, two neutrons, and two orbiting electrons. Lithium follows with three protons, four neutrons, and three electrons. If we continued this step-by-step progression, we would get a graded series of atoms containing from 1 to 112 protons, an equal number of electrons, and a slightly larger number of neutrons at each step. However, all we really need to know to identify a particular element are its atomic number, mass number, and atomic weight. Taken together, these provide a fairly complete picture of each element.

Atomic Number

The **atomic number** of any atom is equal to the number of protons in its nucleus and is written as a subscript to the left of its atomic symbol. Hydrogen, with one proton, has an atomic number of 1 ($_1$H); helium, with two protons, has an atomic number of 2 ($_2$He); and so on. The number of protons is always equal to the number of electrons in an atom; so the atomic number *indirectly* tells us the number of electrons in the atom as well. As explained shortly, this is important information indeed, because electrons determine the chemical behavior of atoms.

Mass Number and Isotopes

The **mass number** of an atom is the sum of the masses of its protons and neutrons. (The mass of the electrons is so small that it is ignored.) Hydrogen has only one proton in its nucleus, so its atomic and

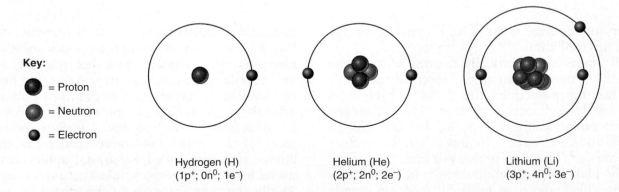

Key:

 = Proton

 = Neutron

 = Electron

Hydrogen (H)
(1p⁺; 0n⁰; 1e⁻)

Helium (He)
(2p⁺; 2n⁰; 2e⁻)

Lithium (Li)
(3p⁺; 4n⁰; 3e⁻)

FIGURE 2.2   **Atomic structure of the three smallest atoms.**

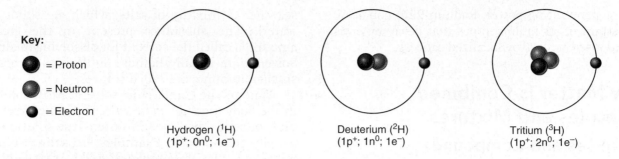

Key:

⚫ = Proton

🔘 = Neutron

● = Electron

Hydrogen (^{1}H)
($1p^+$; $0n^0$; $1e^-$)

Deuterium (^{2}H)
($1p^+$; $1n^0$; $1e^-$)

Tritium (^{3}H)
($1p^+$; $2n^0$; $1e^-$)

FIGURE 2.3 **Isotopes of hydrogen.**

mass numbers are the same: 1. Helium, with two protons and two neutrons, has a mass number of 4. The mass number is usually indicated by a superscript to the left of the atomic symbol. Thus, helium is $^4_2$He. This simple notation allows us to deduce the total number and kinds of subatomic particles in any atom because it indicates the number of protons (the atomic number), the number of electrons (equal to the atomic number), and the number of neutrons (mass number minus atomic number).

From what we have said so far, it may appear as if each element has one, and only one, type of atom representing it. This is not the case. Nearly all known elements have two or more structural variations called **isotopes** (i′so-tōps), which have the same number of protons (and electrons), but differ in the number of neutrons they contain. Earlier, when we said that hydrogen has a mass number of 1, we were speaking of ^{1}H, its most abundant isotope. Some hydrogen atoms have a mass of 2 or 3 amu (atomic mass units), which means that they have one proton and, respectively, one or two neutrons (Figure 2.3). Carbon has several isotopic forms. The most abundant of these are ^{12}C, ^{13}C, and ^{14}C. Each of the carbon isotopes has six protons (otherwise it would not be carbon), but ^{12}C has six neutrons, ^{13}C has seven, and ^{14}C has eight. Isotopes are also written with the mass number following the symbol: C-14, for example.

Atomic Weight

It would seem that atomic weight should be the same as atomic mass, and this would be so if atomic weight referred to the weight of a single atom. However, **atomic weight** is an average of the relative weights (mass numbers) of *all* the isotopes of an element, taking into account their relative abundance in nature. As a rule, the atomic weight of an element is approximately equal to the mass number of its most abundant isotope. For example, the atomic weight of hydrogen is 1.008, which reveals that its lightest isotope (^{1}H) is present in much greater amounts in our world than its ^{2}H or ^{3}H forms.

Radioisotopes

The heavier isotopes of many elements are unstable, and their atoms decompose spontaneously into more stable forms. This process of atomic decay is called *radioactivity*, and isotopes that exhibit this behavior are called **radioisotopes** (ra″de-o-i′so-tōps). The disintegration of a radioactive nucleus may be compared to a tiny explosion. It occurs when subatomic *alpha (α) particles* (packets of 2p + 2n), *beta (β) particles* (electronlike negative particles), or *gamma (γ) rays* (electromagnetic energy) are ejected from the atomic nucleus. Why this happens is complex, and you only need to know that the dense nuclear particles are composed of even smaller particles called *quarks* that associate in one way to form protons and in another way to form neutrons. Apparently, the "glue" that holds these nuclear particles together is weaker in the heavier isotopes. When disintegration occurs, the element may transform to a different element.

Radioisotopes gradually lose their radioactive behavior. The time required for a radioisotope to lose one-half of its activity is called its *half-life*. The half-lives of radioisotopes vary dramatically from hours to thousands of years.

Because radioactivity can be detected with scanners, and radioactive isotopes share the same chemistry as their more stable isotopes, radioisotopes are valuable tools for biological research and medicine. Most radioisotopes used in the clinical setting are used for diagnosis, that is, to localize and illuminate damaged or cancerous tissues. For example, iodine-131 is used to determine the size and activity of the thyroid gland and to detect thyroid cancer, and the sophisticated PET scans use radioisotopes to probe the workings of molecules deep within our bodies. All types of radioactivity, regardless of the purpose for which they are used, damage living tissue.

Alpha emission has the lowest penetrating power and is least damaging to living tissue. Nonetheless, inhaled alpha particles from decaying radon are second only to smoking as a cause of lung cancer. (Radon results naturally from decay of uranium in the ground.) Gamma emission has the greatest penetrating power. Radium-226, cobalt-60, and certain other radioisotopes that decay by emission are used to destroy localized cancers.

How Matter Is Combined: Molecules and Mixtures

Molecules and Compounds

Most atoms do not exist in the free state, but instead are chemically combined with other atoms. Such a combination of two or more atoms held together by chemical bonds is called a **molecule.**

If two or more atoms of the *same* element combine, the resulting substance is called a *molecule of that element.* When two hydrogen atoms bond, the product is a molecule of hydrogen gas and is written as H_2. Similarly, when two oxygen atoms combine, a molecule of oxygen gas (O_2) is formed. Sulfur atoms commonly combine to form sulfur molecules containing eight sulfur atoms (S_8).

When two or more *different* kinds of atoms bind, they form molecules of a **compound.** Two hydrogen atoms combine with one oxygen atom to form the compound water (H_2O), and four hydrogen atoms combine with one carbon atom to form the compound methane (CH_4). Notice again that molecules of methane and water are compounds, but molecules of hydrogen gas are not, because compounds always contain atoms of at least two different elements.

Compounds are chemically pure, and all of their molecules are identical. Thus, just as an atom is the smallest particle of an element that still exhibits the properties of the element, a molecule is the smallest particle of a compound that still displays the specific characteristics of the compound. This is an important concept because the properties of compounds are usually very different from those of the atoms they contain. Indeed, it is next to impossible to tell what atoms are in a compound without analyzing it chemically.

Mixtures

Mixtures are substances composed of two or more components *physically intermixed.* Although most matter in nature exists in the form of mixtures, there are only three basic types: *solutions, colloids,* and *suspensions.*

Solutions

Solutions are *homogeneous* mixtures of components that may be gases, liquids, or solids. Examples include the air we breathe (a mixture of gases) and seawater (a mixture of salts, which are solids, and water). The substance present in the greatest amount is called the **solvent** (or dissolving medium). Solvents are usually liquids. Substances present in smaller amounts are called **solutes.**

Water is the body's chief solvent. Most solutions in the body are *true solutions* containing gases, liquids, or solids dissolved in water. True solutions are usually transparent. Examples are saline solution (table salt [NaCl] and water) and a mixture of glucose and water. The solutes of true solutions are minute, usually in the form of individual atoms and molecules. Consequently, they are not visible to the naked eye, do not settle out, and do not scatter light. If a beam of light is passed through a true solution, you will not see the path of light.

Concentration of Solutions   True solutions are described in terms of their *concentration,* which may be indicated in various ways. Solutions used in a college laboratory or a hospital are often described in terms of the **percent** (parts per 100 parts) of the solute in the solution. This designation always refers to the solute percentage, and unless otherwise noted, water is assumed to be the solvent.

Another way to express the concentration of a solution is in terms of its **molarity** (mo-lar′ĭ-te), or moles per liter, indicated by *M*. This method is more complicated but much more useful. To understand molarity, you must first understand what a mole is. A **mole** of any element or compound is equal to its atomic weight or **molecular weight** (sum of the atomic weights) weighed out in grams. This concept is easier than it seems, as illustrated by the following example.

Glucose is $C_6H_{12}O_6$, which indicates that it has 6 carbon atoms, 12 hydrogen atoms, and 6 oxygen atoms. To compute the molecular weight of glucose, you would look up the atomic weight of each of its atoms in the periodic table and compute its molecular weight as follows:

Atom	Number of atoms		Atomic weight		Total atomic weight
C	6	×	12.011	=	72.066
H	12	×	1.008	=	12.096
O	6	×	15.999	=	95.994
					180.156

Then, to make a *one-molar* solution of glucose, you would weigh out 180.156 grams (g), called a gram molecular weight, of glucose and add enough water to make 1 liter (L) of solution. Thus, a one-molar solution (abbreviated 1.0 *M*) of a chemical substance is one gram molecular weight of the substance (or one gram atomic weight in the case of elemental substances) in 1 L (1000 ml) of solution.

The beauty of using the mole as the basis of preparing solutions is its precision. One mole of any substance always contains exactly the same number of solute particles, that is, 6.02×10^{23}. This number is called **Avogadro's number** (av″o-gad′rōz). So whether you weigh out 1 mole of glucose (180 g) or 1 mole of water (18 g) or 1 mole of methane (16 g), in each case you will have 6.02×10^{23} molecules of that substance.* This allows almost mind-boggling precision to be achieved.

Colloids

Colloids (kol′oidz), also called *emulsions,* are *heterogeneous* mixtures that often appear translucent or milky. Although the solute particles are larger than those in true solutions, they still do not settle out. However, they do scatter light, and so the path of a light beam shining through a colloidal mixture is visible.

Colloids have many unique properties, including the ability of some to undergo **sol-gel transformations,** that is, to change reversibly from a fluid (sol) state to a more solid (gel) state. Jell-O, or any gelatin product, is a familiar example of a nonliving colloid that changes from a sol to a gel when refrigerated (and that will liquefy again if placed in the sun). Cytosol, the semifluid material in living cells, is also a colloid, and its sol-gel changes underlie many important cell activities, such as cell division.

Suspensions

Suspensions are *heterogeneous* mixtures with large, often visible solutes that tend to settle out. An example of a suspension is a mixture of sand and water. So is blood, in which the living blood cells are suspended in the fluid portion of blood (blood plasma). If left to stand, the suspended cells will settle out unless some means—mixing, shaking, or in the body, circulation—is used to keep them in suspension.

As you can see, all three types of mixtures are found in both living and nonliving systems. In fact, living material is the most complex mixture of all, since it contains all three kinds of mixtures interacting with one another.

---

* The important exception to this rule concerns molecules that ionize and break up into charged particles (ions) in water, such as salts, acids, and bases (see p. 38). For example, simple table salt (sodium chloride) breaks up into two types of charged particles; therefore, in a 1.0-*M* solution of sodium chloride, there are actually *2 moles* of solute particles in solution.

Distinguishing Mixtures from Compounds

Now we are ready to zero in on how to distinguish mixtures and compounds from one another. Mixtures differ from compounds in several important ways:

1. The chief difference between mixtures and compounds is that no chemical bonding occurs between the components of a mixture. The properties of atoms and molecules are not changed when they become part of a mixture. Remember they are only physically intermixed.

2. Depending on the mixture, its components can be separated by physical means—straining, filtering, evaporation, and so on. Compounds, by contrast, can be separated into their constituent atoms only by chemical means (breaking bonds).

3. Some mixtures are homogeneous, whereas others are heterogeneous. To say that a substance is *homogeneous* means that a sample taken from any part of the substance has exactly the same composition (in terms of the atoms or molecules it contains) as any other sample. A bar of 100% pure (elemental) iron is homogeneous, as are all compounds. *Heterogeneous* substances vary in their makeup from place to place. For example, iron ore is a heterogeneous mixture that contains iron and many other elements.

Chemical Bonds

As noted earlier, when atoms combine with other atoms, they are held together by **chemical bonds.** A chemical bond is not a physical structure like a pair of handcuffs linking two people together. Instead, it is an energy relationship between the electrons of the reacting atoms, and it is made or broken in less than a trillionth of a second.

The Role of Electrons in Chemical Bonding

Electrons forming the electron cloud around the nucleus of an atom occupy regions of space called **electron shells** that consecutively surround the atomic nucleus. The atoms known so far can have electrons in seven shells (numbered 1 to 7 from the nucleus outward), but the actual number of electron shells occupied in a given atom depends on the number of electrons that atom has. Each electron shell contains one or more orbitals.

It is important to understand that each electron shell represents a different **energy level,** because this

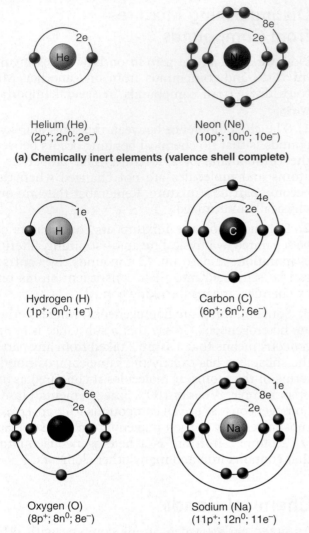

Helium (He)
($2p^+$; $2n^0$; $2e^-$)

Neon (Ne)
($10p^+$; $10n^0$; $10e^-$)

(a) Chemically inert elements (valence shell complete)

Hydrogen (H)
($1p^+$; $0n^0$; $1e^-$)

Carbon (C)
($6p^+$; $6n^0$; $6e^-$)

Oxygen (O)
($8p^+$; $8n^0$; $8e^-$)

Sodium (Na)
($11p^+$; $12n^0$; $11e^-$)

(b) Chemically active elements (valence shell incomplete)

FIGURE 2.4   Chemically inert and reactive elements.
(a) Helium and neon are chemically inert because in each case the outermost energy level (valence shell) is fully occupied by electrons. **(b)** Elements in which the valence shell is incomplete are chemically reactive. Such atoms tend to react with other atoms to gain, lose, or share electrons to fill their valence shells. (*Note:* For simplicity, each atomic nucleus is shown as a sphere with the atom's symbol; individual protons and neutrons are not shown.)

prompts you to think of electrons as particles with a certain amount of potential energy. In general, the terms *electron shell* and *energy level* are used interchangeably.

The amount of potential energy an electron has depends on the energy level it occupies, because the attraction between the positively charged nucleus and negatively charged electrons is greatest closest to the nucleus and falls off with increasing distance. This statement explains why electrons farthest from the nucleus (1) have the greatest potential energy (it takes more energy to overcome the nuclear attrac-

tion and reach the more distant energy levels), and (2) are most likely to interact chemically with other atoms (they are the least tightly held by their own atomic nucleus and the most easily influenced by other atoms and molecules).

Each electron shell can hold a specific number of electrons. Shell 1, the shell immediately surrounding the nucleus, accommodates only 2 electrons. Shell 2 holds a maximum of 8; shell 3 has room for 18. Subsequent shells hold larger and larger numbers of electrons, and the shells tend to be filled with electrons consecutively. For example, shell 1 is completely filled before any electrons appear in shell 2.

When considering bonding behavior, the only electrons that are important are those in the atom's outermost energy level. Inner electrons usually do not take part in bonding because they are more tightly held by the atomic nucleus. When the outermost energy level of an atom is filled to capacity or contains eight electrons, the atom is stable. Such atoms are *chemically inert,* that is, unreactive. A group of elements called the *noble gases,* which include helium and neon, typify this condition (Figure 2.4a). On the other hand, atoms in which the outermost energy level contains fewer than eight electrons (Figure 2.4b) tend to gain, lose, or share electrons with other atoms to achieve stability.

A possible point of confusion must be addressed here. In atoms that have more than 20 electrons, the energy levels beyond shell 2 can contain *more* than eight electrons. However, the number of electrons that can participate in bonding is still limited to a total of eight. The term **valence shell** (va'lens) is used specifically to indicate an atom's outermost energy level *or that portion of it* containing the electrons that are chemically reactive. Hence, the key to chemical reactivity is the **octet rule** (ok-tet'), or **rule of eights.** Except for shell 1, which is full when it has two electrons, atoms tend to interact in such a way that they have eight electrons in their valence shell.

Types of Chemical Bonds

Three major types of chemical bonds—*ionic, covalent,* and *hydrogen bonds*—result from attractive forces between atoms.

Ionic Bonds

Atoms are electrically neutral. However, electrons can be transferred from one atom to another, and when this happens, the precise balance of + and − charges is lost and charged particles called **ions** are formed. An **ionic bond** (i-on'ik) is a chemical bond

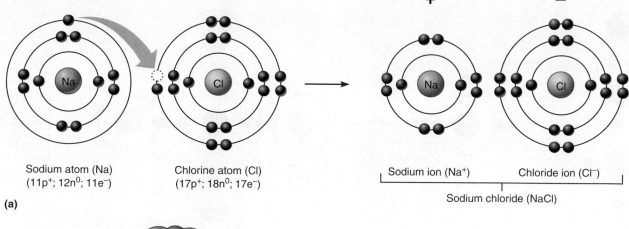

Sodium atom (Na)
($11p^+$; $12n^0$; $11e^-$)

Chlorine atom (Cl)
($17p^+$; $18n^0$; $17e^-$)

(a)

Sodium ion (Na$^+$)   Chloride ion (Cl$^-$)

Sodium chloride (NaCl)

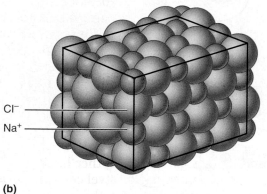

Cl$^-$
Na$^+$

(b)

FIGURE 2.5 Formation of an ionic bond. (a) Both sodium and chlorine atoms are chemically reactive because their valence shells are incompletely filled. Sodium gains stability by losing one electron, whereas chlorine becomes stable by gaining one electron. After electron transfer, sodium becomes a sodium ion (Na$^+$), and chlorine becomes a chloride ion (Cl$^-$). The oppositely charged ions attract each other. **(b)** Large numbers of Na$^+$ and Cl$^-$ ions associate to form salt (NaCl) crystals.

between atoms formed by the transfer of one or more electrons from one atom to the other. The atom that gains one or more electrons, the *electron acceptor*, acquires a net negative charge and is called an **anion** (an'i-on). The atom that loses electrons (the *electron donor*) acquires a net positive charge and is called a **cation** (kat'i-on). (It might help to think of the "t" in "cation" as a + sign.) Both anions and cations are formed whenever electron transfer between atoms occurs. Because opposite charges attract, these ions tend to stay close together, resulting in an ionic bond.

One example of ionic bonding is the formation of sodium chloride (NaCl), by interaction of sodium and chlorine atoms (Figure 2.5). Sodium, with an atomic number of 11, has only one valence shell electron, and it would be very difficult to attempt to fill this shell by adding seven more. However, if this single electron is lost, shell 2 with eight electrons becomes the valence shell (outermost energy level containing electrons) and is full. Thus, by losing the lone electron in its third energy level, sodium achieves stability and becomes a cation (Na$^+$). On the other hand, chlorine, atomic number 17, needs only one electron to fill its valence shell. By accepting an electron, chlorine becomes an anion and achieves stability. When these two atoms interact, this is exactly what happens.

Sodium donates an electron to chlorine, and the ions created in this exchange attract each other, forming sodium chloride. Ionic bonds are commonly formed between atoms with one or two valence shell electrons (the metallic elements, such as sodium, calcium, and potassium) and atoms with seven valence shell electrons (such as chlorine, fluorine, and iodine).

Most ionic compounds fall in the chemical category called *salts*. In the dry state, salts such as sodium chloride do not exist as individual molecules. Instead, they form **crystals,** large arrays of cations and anions held together by ionic bonds (see Figure 2.5b).

Sodium chloride is an excellent example of the difference in properties between a compound and its constituent atoms. Sodium is a silvery white metal, and chlorine in its molecular state is a poisonous green gas used to make bleach. However, sodium chloride is a white crystalline solid that we sprinkle on our food.

Covalent Bonds

Electrons do not have to be completely transferred for atoms to achieve stability. Instead, they may be *shared* so that each atom is able to fill its outer electron shell at least part of the time. Electron sharing produces molecules in which the shared electrons

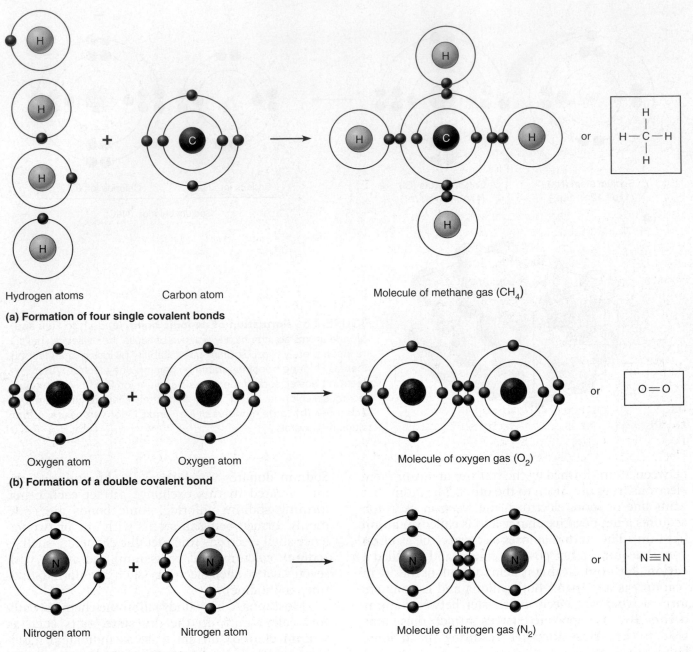

Hydrogen atoms   Carbon atom   Molecule of methane gas (CH$_4$)

(a) Formation of four single covalent bonds

Oxygen atom   Oxygen atom   Molecule of oxygen gas (O$_2$)

(b) Formation of a double covalent bond

Nitrogen atom   Nitrogen atom   Molecule of nitrogen gas (N$_2$)

(c) Formation of a triple covalent bond

FIGURE 2.6   Formation of covalent bonds. Left, the electron-sharing patterns of interacting atoms; center, the resulting molecules; right, the structural formula shows each covalent bond as a single line between the atoms that share an electron pair.

occupy a single orbital common to both atoms and constitute **covalent bonds** (ko-va′lent).

Hydrogen with its single electron can fill its only shell (shell 1) by sharing a pair of electrons with another atom. When it shares with another hydrogen atom, a molecule of hydrogen gas is formed. The shared electron pair orbits around the molecule as a whole, satisfying the stability needs of each atom. Hydrogen can also share an electron pair with different kinds of atoms to form a compound (Figure 2.6a).

Carbon has four electrons in its outermost shell, but needs eight to achieve stability, whereas hydrogen has one electron, but needs two. When a methane molecule (CH$_4$) is formed, carbon shares four pairs of electrons with four hydrogen atoms (one pair with each hydrogen). Again, the shared electrons orbit and "belong to" the whole molecule, ensuring the stability of each atom. When two atoms share one pair of electrons, a *single covalent bond* is formed (indicated by a single line connecting the atoms, such as H—H). In some cases, atoms share two or three electron pairs (Figure 2.6b and c), resulting in *double* or *triple covalent bonds* (indicated by double or triple connecting lines such as O=O or N≡N).

Polar and Nonpolar Molecules In the covalent bonds discussed thus far, the shared electrons are shared equally between the atoms of the molecule. The molecules formed are electrically balanced and are called **nonpolar molecules** (because they do not have separate + and − poles of charge). This is not always the case. When covalent bonds are formed, the resulting molecule always has a specific three-dimensional shape, with the bonds formed at definite angles. A molecule's shape helps determine what other molecules or atoms it can interact with; it may also result in unequal electron pair sharing and **polar molecules.** This is especially true in nonsymmetrical molecules that have atoms with different electron-attracting abilities. In general, *small* atoms with six or seven valence shell electrons, such as oxygen, nitrogen, and chlorine, are electron-hungry and attract electrons very strongly, a capability called **electronegativity.** Conversely, most atoms with only one or two valence shell electrons tend to be **electropositive;** that is, their electron-attracting ability is so low that they usually lose *their* valence shell electrons to other atoms. Potassium and sodium, each with one valence shell electron, are good examples of electropositive atoms.

Carbon dioxide and water illustrate how molecular shape and the relative electron-attracting abilities of atoms determine whether a covalently bonded molecule is nonpolar or polar. In carbon dioxide (CO_2), carbon shares four electron pairs with two oxygen atoms (two pairs are shared with each oxygen). Oxygen is very electronegative and so attracts the shared electrons much more strongly than does carbon. However, because the carbon dioxide molecule is linear and symmetrical (Figure 2.7a), the electron-pulling ability of one oxygen atom is offset by that of the other, like a standoff between equally strong teams in a game of tug-of-war. As a result, the shared electrons orbit the entire molecule and carbon dioxide is a nonpolar compound.

On the other hand, a water molecule (H_2O) is V-shaped (Figure 2.7b). The two hydrogen atoms are located at the same end of the molecule, and oxygen is at the opposite end. This arrangement allows oxygen to pull the shared electrons toward itself and away from the two hydrogen atoms. In this case, the electron pairs are *not* shared equally, but spend more time in the vicinity of oxygen. Because electrons are negatively charged, the oxygen end of the molecule is slightly more negative (indicated with a delta and minus as δ^-) and the hydrogen end slightly more positive

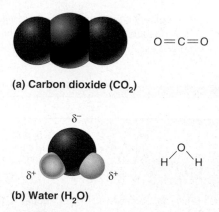

(a) Carbon dioxide (CO_2)

(b) Water (H_2O)

FIGURE 2.7 **Molecular models of the three-dimensional structure of carbon dioxide and water molecules.**

(indicated by δ^+). Because water has two poles of charge, it is a *polar molecule*, or **dipole** (di′pōl). Polar molecules orient themselves toward other dipoles or toward charged particles (such as ions and some proteins), and they play essential roles in chemical reactions in body cells. The polarity of water is particularly significant, as you will see later in this chapter.

Different molecules exhibit different degrees of polarity, and we can see a gradual change from ionic to nonpolar covalent bonding as summarized in Figure 2.8. Ionic bonds (complete electron transfer) and nonpolar covalent bonds (equal electron sharing) are the extremes of a continuum, with various degrees of unequal electron sharing in between.

Hydrogen Bonds

Unlike the stronger ionic and covalent bonds, hydrogen bonds are more like attractions than true bonds. Hydrogen bonds form when a hydrogen atom, already covalently linked to one electronegative atom (usually nitrogen or oxygen), is attracted by another electron-hungry atom, and forms a "bridge" between them. Hydrogen bonding is common between dipoles such as water molecules, because the slightly negative oxygen atoms of one molecule attract the slightly positive hydrogens of other molecules (Figure 2.9). The tendency of water molecules to cling together and form films, referred to as *surface tension*, helps explain why water beads up into spheres when it sits on a hard surface.

Although hydrogen bonds are too weak to bind atoms together to form molecules, they are important as *intramolecular bonds*, which bind different parts of a single large molecule together into a specific three-dimensional shape. Some large biological molecules, such as proteins and DNA, have numerous hydrogen bonds that help maintain and stabilize their structures.

Assume imaginary compound XY has a covalent polar bond. How does its charge distribution differ from that of XX molecules?

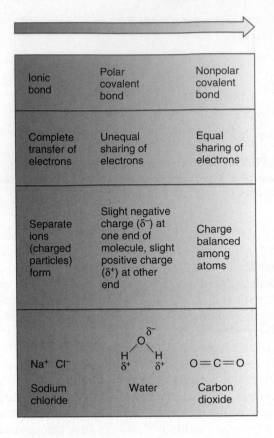

Ionic bond	Polar covalent bond	Nonpolar covalent bond
Complete transfer of electrons	Unequal sharing of electrons	Equal sharing of electrons
Separate ions (charged particles) form	Slight negative charge (δ^-) at one end of molecule, slight positive charge (δ^+) at other end	Charge balanced among atoms
Na$^+$ Cl$^-$	$\begin{array}{c} \delta^- \\ O \\ H \quad\quad H \\ \delta^+ \quad\quad \delta^+ \end{array}$	O=C=O
Sodium chloride	Water	Carbon dioxide

FIGURE 2.8 Comparison of ionic, polar covalent, and nonpolar covalent bonds. This display compares the status of electrons involved in bonding and the electrical charge distribution in the molecules formed.

Chemical Reactions

As noted earlier, all particles of matter are in constant motion because of their kinetic energy. Movement of atoms or molecules in a solid is usually limited to vibration because the particles are united by fairly rigid bonds. But in liquids or gases, particles dart about randomly, sometimes colliding with one another and interacting to undergo chemical reactions. A **chemical reaction** occurs whenever chemical bonds are formed, rearranged, or broken.

Because X and Y are different atoms, one is bound to be more electronegative than the other. Consequently, there will be a separation of charge in XY, which is a dipole. In XX, the two atoms are the same and there will be equal e⁻ sharing; no dipole is formed. ■

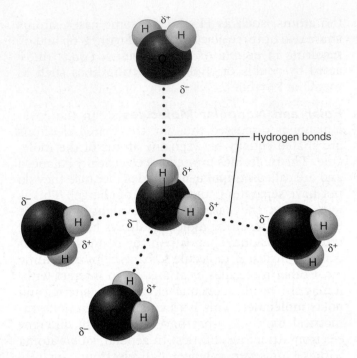

FIGURE 2.9 Hydrogen bonding between polar water molecules. The slightly positive ends (indicated by δ^+) of the water molecules become aligned with the slightly negative ends (indicated by δ^-) of other water molecules.

Chemical Equations

Chemical reactions can be written in symbolic form as **chemical equations.** For example, joining two hydrogen atoms to form hydrogen gas is indicated as

$$\underset{\text{(reactants)}}{H + H} \rightarrow \underset{\text{(product)}}{H_2} \text{ (hydrogen gas)}$$

and combining four hydrogen atoms and one carbon atom to form methane is written

$$4H + C \rightarrow CH_4 \text{ (methane)}$$

Notice that a number written as a *subscript* indicates that the atoms are joined by chemical bonds. But a number written as a *prefix* denotes the number of *unjoined* atoms or molecules. Hence, CH_4 reveals that four hydrogen atoms are bonded together with carbon to form the methane molecule, but 4H signifies four unjoined hydrogen atoms.

A chemical equation is like a sentence describing what happens in a reaction. It contains the following information: the number and kinds of reacting substances, or **reactants;** the chemical composition of the **product(s);** and in balanced equations, the relative proportion of each reactant and product. In the examples used above, the reactants are atoms, as indicated by their atomic symbols (H, C). The product in each case is a molecule, as represented by its **molecular formula** (H_2, CH_4). The equation for the formation of methane may be read as *either* "four hydrogen atoms plus one carbon atom yield one molecule of methane" *or* "four moles of hydrogen atoms plus one mole of carbon yield one mole of methane." Using moles is more practical because it is impossible to measure out one atom or one molecule of anything!

Patterns of Chemical Reactions

Most chemical reactions exhibit one of three recognizable patterns: they are either *synthesis, decomposition,* or *exchange reactions.*

When atoms or molecules combine to form a larger, more complex molecule, the process is a **synthesis,** or **combination, reaction.** A synthesis reaction, which always involves bond formation, can be represented (using arbitrary letters) as

$$A + B \rightarrow AB$$

Synthesis reactions are the basis of constructive, or **anabolic,** activities in body cells, such as joining small molecules called amino acids into large protein molecules (Figure 2.10a). Synthesis reactions are conspicuous in rapidly growing tissues.

A **decomposition reaction** occurs when a molecule is broken down into smaller molecules or its constituent atoms:

$$AB \rightarrow A + B$$

Essentially, decomposition reactions are reverse synthesis reactions; bonds are broken. Decomposition reactions underlie all degradative, or **catabolic,** processes that occur in body cells. For example, the bonds of glycogen molecules are broken to release simpler molecules of glucose sugar (Figure 2.10b).

Exchange, or **displacement, reactions** involve both synthesis and decomposition; bonds are both made and broken. In an exchange reaction, parts of the reactant molecules change partners, so to speak, producing different product molecules:

$$AB + C \rightarrow AC + B \quad \text{and} \quad AB + CD \rightarrow AD + CB$$

An exchange reaction occurs when ATP reacts with glucose and transfers its end phosphate group (indicated by a circled P in Figure 2.10c) to glucose, forming glucose-phosphate. At the same time, the ATP becomes ADP. This important reaction occurs whenever glucose enters a body cell and it effectively traps the glucose fuel molecule inside the cell.

Another group of very important chemical reactions in living systems are **oxidation-reduction reactions—redox reactions** for short. Oxidation-reduction reactions are decomposition reactions in that they are the basis of all reactions in which food fuels are catabolized for energy (that is, in which ATP is produced). They are also a special type of exchange

? *Which of the reaction types depicted occurs when fats are digested in your small intestine?*

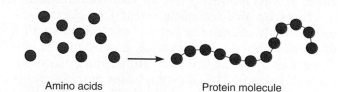

Amino acids → Protein molecule

(a) Example of a synthesis reaction: amino acids are joined to form a protein molecule

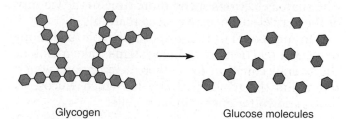

Glycogen → Glucose molecules

(b) Example of a decomposition reaction: breakdown of glycogen to release glucose units

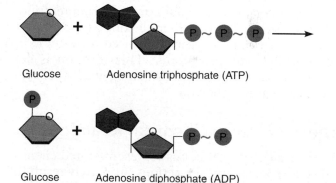

Glucose + Adenosine triphosphate (ATP)

Glucose phosphate + Adenosine diphosphate (ADP)

(c) Example of an exchange reaction: ATP transfers its terminal phosphate group to glucose to form glucose-phosphate

FIGURE 2.10 Patterns of chemical reactions. (a) In synthesis reactions, smaller particles are bonded together to form larger, more complex molecules. **(b)** Decomposition reactions involve bond breaking. **(c)** In exchange, or displacement, reactions, bonds are both made and broken.

reaction because electrons are exchanged between the reactants. The reactant losing the electrons is referred to as the *electron donor* and is said to be **oxidized;** the reactant taking up the transferred electrons is called the *electron acceptor* and is said to become **reduced.**

Redox reactions occur when ionic compounds are formed. Recall that in the formation of NaCl

Digestion of fats (lipids) to their fatty acid and glycerol building blocks is an example of (b) a decomposition reaction. ■

(see Figure 2.5), sodium loses an electron to chlorine. Consequently, sodium is oxidized and becomes a sodium ion, and chlorine is reduced and becomes a chloride ion. However, not all oxidation-reduction reactions involve *complete transfer* of electrons—some simply change the pattern of electron sharing in covalent bonds. For example, a substance is oxidized both by losing hydrogen atoms and by combining with oxygen. The common factor in these events is that electrons that formerly "belonged" to the reactant molecule are lost, either entirely (as hydrogen is removed and takes its electron with it) or relatively (the shared electrons spend more time in the vicinity of the very electronegative oxygen atom).

To understand the importance of oxidation-reduction reactions in living systems, take a look at the overall equation for *cellular respiration*, which represents the major pathway by which glucose is broken down for energy in body cells:

$$C_6H_{12}O_6 + 6O_2 \rightarrow 6CO_2 + 6H_2O + ATP$$

glucose   oxygen   carbon   water   cellular
                              dioxide   energy

As you can see, it is an oxidation-reduction reaction. Glucose is oxidized to carbon dioxide as it loses hydrogen atoms, and oxygen is reduced to water as it accepts the hydrogen atoms. This reaction is described in detail in Chapter 23, along with other topics of cellular metabolism.

Energy Flow in Chemical Reactions

Because all chemical bonds represent stored chemical energy, all chemical reactions ultimately result in net absorption or release of energy. Reactions that release energy are called **exergonic reactions.** These reactions yield products that have less energy than the initial reactants, but they also provide energy that can be harvested for other uses. With a few exceptions, catabolic and oxidative reactions are exergonic. In contrast, the products of energy-absorbing, or **endergonic,** reactions contain more potential energy in their chemical bonds than did the reactants. Anabolic reactions are typically energy-absorbing endergonic reactions. Essentially this is a case of "one hand washing the other"—the energy released as fuel molecules are broken down (oxidized), is captured in ATP molecules and then used to synthesize the complex biological molecules the body needs to sustain life.

Reversibility of Chemical Reactions

All chemical reactions are theoretically reversible. If chemical bonds can be made, they can be broken, and vice versa. Reversibility is indicated by a double arrow. When the arrows differ in length, the longer arrow indicates the major direction in which the reaction proceeds:

$$A + B \rightleftharpoons AB$$

In this example, the forward reaction (reaction going to the right) predominates. Over time, the product (AB) accumulates and the reactants (A and B) decrease in amount.

When the arrows are of equal length, as in

$$A + B \rightleftharpoons AB$$

neither the forward reaction nor the reverse reaction is dominant; that is, for each molecule of product (AB) formed, one product molecule breaks down, releasing the reactants A and B. Such a chemical reaction is said to be in a state of **chemical equilibrium.** Once chemical equilibrium is reached, there is no further *net change* in the amounts of reactants and products. Product molecules are still formed and broken down, but the situation already established when equilibrium is reached (such as greater numbers of product molecules) remains unchanged. This is analogous to the admission scheme used by many large museums in which tickets are sold according to time of entry. If 300 tickets are issued for the 9 AM admission, 300 people will be admitted when the doors open. Thereafter, when 6 people leave, 6 are admitted, and when another 15 people leave, 15 more are allowed in. Although there is a continual turnover, the museum contains about 300 art lovers throughout the day.

Although all chemical reactions are reversible, many show so little tendency to go in the reverse direction that they are irreversible for all practical purposes. This is true of many biological reactions. Chemical reactions that release energy when going in one direction will not go in the opposite direction unless energy is put back into the system. For example, when our cells break down glucose via the reactions of cellular respiration to yield carbon dioxide and water, some of the energy released is trapped in the bonds of ATP. Because the cells need ATP's energy for various functions (and more glucose will be along with the next meal), this particular reaction is never reversed in our cells. Furthermore, if a product of a reaction is continuously removed from the reaction site, it is unavailable to take part in the reverse reaction. This situation occurs when the carbon dioxide that is released during glucose breakdown leaves the cell, enters the blood, and is eventually removed from the body by the lungs.

Factors Influencing the Rate of Chemical Reactions

For atoms and molecules to react chemically, they must *collide* with enough force to overcome the repulsion between their electrons. Interactions between valence shell electrons—the basis of bond making and breaking—cannot occur long distance. The force of collisions depends on how fast the particles are moving. Solid, forceful collisions between rapidly moving particles in which valence shells overlap are much more likely to cause reactions than are those in which the particles graze each other lightly.

Temperature

Increasing the temperature of a substance increases the kinetic energy of its particles and the force of their collisions. Therefore, chemical reactions proceed quicker at higher temperatures.

Concentration

Chemical reactions progress most rapidly when the reacting particles are present in high numbers, because the chance of successful collisions is greater. As the concentration of the reactants declines, chemical equilibrium eventually occurs unless additional reactants are added or products are removed from the reaction site.

Particle Size

Smaller particles move faster than larger ones (at the same temperature) and tend to collide more frequently and more forcefully. Hence, the smaller the reacting particles, the faster a chemical reaction goes at a given temperature and concentration.

Catalysts

Although many chemical reactions in nonliving systems can be speeded up simply by heating, drastic increases in body temperature are life threatening because important biological molecules are destroyed. Still, at normal body temperatures, most chemical reactions would proceed far too slowly to maintain life were it not for the presence of catalysts. **Catalysts** (kat'ah-lists) are substances that increase the rate of chemical reactions without themselves becoming chemically changed or part of the product. Biological catalysts are called **enzymes** (en'zīmz). The mode of action of enzymes is described later in this chapter.

PART 2: BIOCHEMISTRY

Biochemistry is the study of the chemical composition and reactions of living matter. All chemicals in the body fall into one of two major classes: organic or inorganic compounds. **Organic compounds** contain carbon. All organic compounds are covalently bonded molecules, and many are large.

All other chemicals in the body are considered **inorganic compounds.** These include water, salts, and many acids and bases. Organic and inorganic compounds are equally essential for life. Trying to decide which is more valuable is like trying to decide whether the ignition system or the engine is more essential to run your car!

Inorganic Compounds

Water

Water is the most abundant and important inorganic compound in living material. It makes up 60 to 80% of the volume of most living cells. Among the properties that make water vital are its:

1. **High heat capacity.** Water has a high heat capacity; that is, it absorbs and releases large amounts of heat before changing appreciably in temperature itself. This prevents sudden changes in temperature caused by external factors, such as sun or wind exposure, or by internal conditions that release heat rapidly, such as vigorous muscle activity. As part of blood, water redistributes heat among body tissues, ensuring temperature homeostasis.

2. **High heat of vaporization.** When water evaporates, or vaporizes, it changes from a liquid to a gas (water vapor). This transformation requires that large amounts of heat be absorbed to break the hydrogen bonds that hold water molecules together. This property is extremely beneficial when we sweat. As perspiration (mostly water) evaporates from our skin, large amounts of heat are removed from the body, providing an efficient cooling mechanism.

3. **Polar solvent properties.** Water is an unparalleled solvent; indeed, it is often called the **universal solvent.** Biochemistry is "wet chemistry." Biological molecules do not react chemically unless they are in solution, and virtually all chemical reactions occurring in the body depend on water's solvent properties.

Because of their polar nature, water molecules orient themselves with their slightly negative ends toward the positive ends of the solutes, and vice versa, first attracting them, then surrounding them. This property of water explains why ionic compounds and other small reactive molecules (such as acids and bases) *dissociate* in water, their ions separating from each other and becoming evenly scattered in the water, forming true solutions (Figure 2.11). Water also forms **hydration layers** (layers of water molecules) around large charged molecules such as proteins, shielding them from the effects of other charged substances in the vicinity and preventing them from settling out of solution. Such protein-water mixtures are *biological colloids.* Cerebrospinal fluid and blood are examples of colloids.

Water is the body's major transport medium because it is such an excellent solvent. Nutrients,

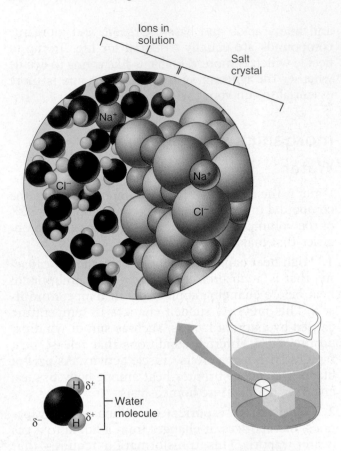

Ions in solution

Salt crystal

Na⁺

Cl⁻

Na⁺

Cl⁻

H δ^+

δ^-

H δ^+

Water molecule

FIGURE 2.11   Dissociation of a salt in water. The slightly negative ends of the water molecules (δ^-) are attracted to Na^+, whereas the slightly positive ends of water molecules (δ^+) orient toward Cl^-, causing the ions to be pulled off the crystal lattice.

respiratory gases, and metabolic wastes carried throughout the body are dissolved in blood plasma, and many metabolic wastes are excreted from the body in urine, another watery fluid. Specialized molecules that lubricate the body (e.g., mucus) also use water as their dissolving medium.

4. Reactivity. Water is an important *reactant* in many chemical reactions. For example, foods are digested to their building blocks by adding a water molecule to each bond to be broken. Such decomposition reactions are more specifically called **hydrolysis reactions** (hi-drol'ĭ-sis; "water splitting"). Conversely, when large carbohydrate or protein molecules are synthesized from smaller molecules, a water molecule is removed for every bond formed, a reaction called **dehydration synthesis.**

5. Cushioning. By forming a resilient cushion around certain body organs, water helps protect them from physical trauma. The cerebrospinal fluid surrounding the brain exemplifies water's cushioning role.

Salts

A **salt** is an ionic compound containing cations other than H^+ and anions other than the hydroxyl ion (OH^-). As already noted, when salts are dissolved in water, they dissociate into their component ions (Figure 2.11). For example, sodium sulfate (Na_2SO_4) dissociates into two Na^+ ions and one SO_4^{2-} ion. This occurs easily because the ions are already formed. All that remains is for water to overcome the attraction between the oppositely charged ions. All ions are **electrolytes** (e-lek'tro-līts), substances that conduct an electrical current in solution. (Note that groups of atoms that bear an overall charge, such as sulfate, are called *polyatomic ions.*)

Salts commonly found in the body include NaCl, Ca_2CO_3 (calcium carbonate), and KCl (potassium chloride). However, the most plentiful salts are the calcium phosphates that make bones and teeth hard. In their ionized form, salts play vital roles in body function. For instance, the electrolyte properties of sodium and potassium ions are essential for nerve impulse transmission and muscle contraction. Ionic iron forms part of the hemoglobin molecules that transport oxygen within red blood cells, and zinc and copper ions are important to the activity of some enzymes. Other important functions of the elements found in body salts are summarized in Table 2.1.

Ⓗ *HOMEOSTATIC IMBALANCE*

Maintaining proper ionic balance in our body fluids is one of the most crucial homeostatic roles of the kidneys. When this balance is severely disturbed, virtually nothing in the body works. All the physiological activities listed above and thousands of others are disrupted and grind to a stop. ●

Acids and Bases

Like salts, acids and bases are electrolytes; that is, they ionize and dissociate in water and can then conduct an electrical current.

Acids

Acids have a sour taste, can react with (dissolve) many metals, and "burn" a hole in your rug. But the most useful definition of an acid for our purposes is that it is a substance that releases **hydrogen ions** (H^+) in detectable amounts. Because a hydrogen ion is just a hydrogen nucleus, or "naked" proton, acids are also defined as **proton donors.**

When acids dissolve in water, they release hydrogen ions (protons) and anions. It is the concentration of protons that determines the acidity of a solution. The anions have little or no effect on acidity. For example, hydrochloric acid (HCl), an acid produced by stomach cells that aids digestion, dissociates into a proton and a chloride ion:

$$HCl \rightarrow H^+ \quad Cl^-$$
$$\text{proton} \quad \text{anion}$$

Other acids found or produced in the body include acetic acid ($HC_2H_3O_2$, commonly abbreviated as HAc), the acidic portion of vinegar; and carbonic acid (H_2CO_3). The molecular formula for an acid is easy to recognize because the hydrogen is written first.

Bases

Bases have a bitter taste, feel slippery, and are **proton acceptors**—that is, they take up hydrogen ions (H^+) in detectable amounts. Common inorganic bases include the *hydroxides* (hi-drok'sīds), such as magnesium hydroxide (milk of magnesia) and sodium hydroxide (lye). Like acids, hydroxides dissociate when dissolved in water, but in this case **hydroxyl ions (OH^-)** (hi-drok'sil) and cations are liberated. For example, ionization of sodium hydroxide (NaOH) produces a hydroxyl ion and a sodium ion; the hydroxyl ion then binds to (accepts) a proton present in the solution. This produces water and simultaneously reduces the acidity (hydrogen ion concentration) of the solution:

$$NaOH \rightarrow Na^+ + OH^-$$
$$\text{cation} \quad \text{hydroxyl ion}$$

and then

$$OH^- + H^+ \rightarrow H_2O$$
$$\text{water}$$

Bicarbonate ion ($HCO3^-$), an important base in the body, is particularly abundant in blood. **Ammonia (NH_3)**, a common waste product of protein breakdown in the body, is also a base. It has one pair of unshared electrons that strongly attracts protons. By accepting a proton, ammonia becomes an ammonium ion:

$$NH_3 + H^+ \rightarrow NH_4^+$$
$$\text{ammonium}$$
$$\text{ion}$$

pH: Acid-Base Concentration

The more hydrogen ions in a solution, the more acidic the solution is. Conversely, the greater the concentration of hydroxyl ions (the lower the concentration of H^+), the more basic, or *alkaline* (al'kuh-līn), the solution becomes. The relative concentration of hydrogen ions in various body fluids is measured in concentration units called **pH units** (pe-āch').

The idea for a pH scale was devised by a Danish biochemist and part-time beer brewer named Sören Sörensen in 1909. He was searching for a convenient means of checking the acidity of his alcoholic product to prevent its spoilage by bacterial action. (Many bacteria are inhibited by acidic conditions.) The pH scale that resulted is based on the concentration of hydrogen ions in a solution, expressed in terms of moles per liter, or molarity. The pH scale runs from 0 to 14 and is logarithmic; that is, each successive change of one pH unit represents a tenfold change in hydrogen ion concentration (Figure 2.12). The pH of a solution is thus defined as the negative logarithm of the hydrogen ion concentration [H^+] in moles per liter or $-\log[H^+]$.

At a pH of 7 (at which [H^+] is 10^{-7} M), the number of hydrogen ions exactly equals the number of hydroxyl ions (pH = pOH), and the solution is *neutral*—neither acidic nor basic. Absolutely pure (distilled) water has a pH of 7. Solutions with a pH below 7 are acidic; the hydrogen ions outnumber the hydroxyl ions. The lower the pH, the more acidic the solution. A solution with a pH of 6 has ten times as many hydrogen ions as a solution with a pH of 7.

Solutions with a pH higher than 7 are alkaline, and the relative concentration of hydrogen ions decreases by a factor of 10 with each higher pH unit. Thus, solutions with pH values of 8 and 12 have, respectively, 1/10 and 1/100,000 (1/10 × 1/10 × 1/10 × 1/10 × 1/10) as many hydrogen ions as a solution of pH 7. Notice that as the hydrogen ion concentration decreases, the hydroxyl ion concentration rises, and vice versa. The approximate pH of several body fluids and of a number of common substances also appears in Figure 2.12.

Neutralization

When acids and bases are mixed, they react with each other in displacement reactions to form water and a salt. For example, when hydrochloric acid and sodium hydroxide interact, sodium chloride (a salt) and water are formed.

$$HCl + NaOH \rightarrow NaCl + H_2O$$
$$\text{acid} \quad \text{base} \quad \text{salt} \quad \text{water}$$

This type of reaction is called a **neutralization reaction,** because the joining of H^+ and OH^- to form water neutralizes the solution. Although the salt produced is written in molecular form (NaCl), remember that it actually exists as dissociated sodium and chloride ions when dissolved in water.

Buffers

Living cells are extraordinarily sensitive to even slight changes in the pH of the environment. In high concentrations, acids and bases are extremely damaging

Concentration in moles/liter

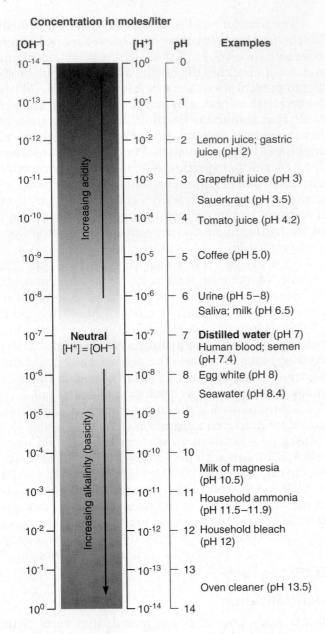

FIGURE 2.12 **The pH scale and pH values of representative substances.** The pH scale is based on the number of hydrogen ions in solution. The actual concentration of hydrogen ions ($[H^+]$) expressed in moles per liter and the corresponding hydroxyl concentration ($[OH^-]$) are indicated for each pH value noted. At a pH of 7, the concentrations of hydrogen and hydroxyl ions are equal and the solution is neutral.

to living tissue. (Imagine what would happen to all those hydrogen bonds in biological molecules with large numbers of free H^+ running around.) Homeostasis of acid-base balance is carefully regulated by the kidneys and lungs and by chemical systems (proteins and other types of molecules) called **buffers.** Buffers resist abrupt and large swings in the pH of body fluids by releasing hydrogen ions (acting as acids) when the pH begins to rise and by binding hydrogen ions (acting as bases) when the pH drops. Because blood comes into close contact with nearly every body cell, regulation of its pH is particularly critical. Normally, blood pH varies within a very narrow range (7.35 to 7.45). If blood pH varies from these limits by more than a few tenths of a unit, it may be fatal.

To comprehend how chemical buffer systems operate, you must thoroughly understand strong and weak acids and bases. The first important concept is that the acidity of a solution reflects *only* the free hydrogen ions, not those still bound to anions. Consequently, acids that dissociate completely and irreversibly in water are called **strong acids,** because they can dramatically change the pH of a solution. Examples are hydrochloric acid and sulfuric acid. If we could count out 100 hydrochloric acid molecules and place them in 1 ml of water, we could expect to end up with 100 H^+, 100 Cl^-, and no undissociated hydrochloric acid molecules in that solution.

Acids that do not dissociate completely, like carbonic acid (H_2CO_3) and acetic acid (HAc), are **weak acids.** If we were to place 100 acetic acid molecules in 1 ml of water, the reaction would be something like this:

$$100 \text{ HAc} \rightarrow 90 \text{ HAc} + 10 \text{ H}^+ + 10 \text{ Ac}^-$$

Because undissociated acids do not affect pH, the acetic acid solution is much less acidic than the HCl solution. Weak acids dissociate in a predictable way, and molecules of the intact acid are in dynamic equilibrium with the dissociated ions. Consequently, the dissociation of acetic acid may also be written as

$$\text{HAc} \rightleftharpoons \text{H}^+ + \text{Ac}^-$$

This viewpoint allows us to see that if H^+ (released by a strong acid) is added to the acetic acid solution, the equilibrium will shift to the left and some H^+ and Ac^- will recombine to form HAc. On the other hand, if a strong base is added and the pH begins to rise, the equilibrium shifts to the right and more HAc molecules dissociate to release H^+. This characteristic of weak acids allows them to play extremely important roles in the chemical buffer systems of the body.

The concept of strong and weak bases is more easily explained. Remember that bases are proton acceptors. Thus, **strong bases** are those, like hydroxides, that dissociate easily in water and quickly tie up H^+. On the other hand, sodium bicarbonate (commonly known as baking soda) ionizes incompletely and reversibly. Because it accepts relatively few protons, its released bicarbonate ion is considered a **weak base.**

Now let's examine how one buffer system helps to maintain pH homeostasis of the blood. Although there are other chemical blood buffers, the **carbonic acid-bicarbonate system** is a very important one.

Carbonic acid (H_2CO_3) dissociates reversibly, releasing bicarbonate ions (HCO_3^-) and protons (H^+):

$$H_2CO_3 \underset{\text{Response to drop in pH}}{\overset{\text{Response to rise in pH}}{\rightleftharpoons}} HCO_3^- + H^+$$

H^+ donor (weak acid) H^+ acceptor (weak base)   proton

The chemical equilibrium between carbonic acid (a weak acid) and bicarbonate ion (a weak base) resists changes in blood pH by shifting to the right or left as H^+ ions are added to or removed from the blood. As blood pH rises (i.e., becomes more alkaline due to the addition of a strong base), the equilibrium shifts to the right, forcing more carbonic acid to dissociate. Similarly, as blood pH begins to drop (i.e., becomes more acidic due to the addition of a strong acid), the equilibrium shifts to the left as more bicarbonate ions begin to bind with protons. As you can see, strong bases are replaced by a weak base (bicarbonate ion) and protons released by strong acids are tied up in a weak one (carbonic acid). In either case, the blood pH changes much less than it would in the absence of the buffering system. Acid-base balance and buffers are discussed in more detail in Chapter 25.

Organic Compounds

Molecules unique to living systems—proteins, carbohydrates, lipids (fats), and nucleic acids—all contain carbon and hence are organic compounds. Although organic compounds are distinguished by the fact that they contain carbon and inorganic compounds are defined as compounds that lack carbon, there are a few irrational exceptions to this generalization that you should be aware of. Carbon dioxide, carbon monoxide, and carbides, for example, all contain carbon but are considered inorganic compounds.

What makes carbon so special that "living" chemistry depends on its presence? To begin with, no other *small* atom is so precisely **electroneutral.** The consequence of its electroneutrality is that carbon never loses or gains electrons; it always shares them. Furthermore, with four valence shell electrons, carbon forms four covalent bonds with other elements, as well as with other carbon atoms. As a result, carbon is found in long, chainlike molecules (common in fats), ring structures (typical of carbohydrates and steroids), and many other structures that are uniquely suited for specific roles in the body.

Carbohydrates

Carbohydrates, a group of molecules that includes sugars and starches, represent 1 to 2% of cell mass. Carbohydrates contain carbon, hydrogen, and oxygen, and generally the hydrogen and oxygen atoms occur in the same 2:1 ratio as in water. This ratio is reflected in the word *carbohydrate* (meaning "hydrated carbon").

A carbohydrate can be classified according to size and solubility as a monosaccharide ("one sugar"), disaccharide ("two sugars"), or polysaccharide ("many sugars"). Monosaccharides are the structural units, or building blocks, of the other carbohydrates. In general, the larger the carbohydrate molecule, the less soluble it is in water.

Monosaccharides

Monosaccharides (mon"o-sak'ah-rīdz), or *simple sugars,* are single-chain or single-ring structures containing from three to seven carbon atoms (Figure 2.13a). Usually the carbon, hydrogen, and oxygen atoms occur in the ratio 1:2:1, so a general formula for a monosaccharide is $(CH_2O)n$, where n is the number of carbons in the sugar. Glucose, for example, has six carbon atoms, and its molecular formula is $C_6H_{12}O_6$. Ribose, with five carbons, is $C_5H_{10}O_5$.

Monosaccharides are named generically according to the number of carbon atoms they contain. Most important in the body are the pentose (five-carbon) and hexose (six-carbon) sugars. For example, the pentose *deoxyribose* (de-ok"sĭ-ri'bōs) is part of DNA, and *glucose,* a hexose, is blood sugar. Two other hexoses, *galactose* and *fructose,* are **isomers** (i'so-mers) of glucose. That is, they have the same molecular formula ($C_6H_{12}O_6$), but their atoms are arranged differently, giving them different chemical properties.

Disaccharides

A **disaccharide** (di-sak'ah-rīd), or *double sugar,* is formed when two monosaccharides are joined by **dehydration synthesis** (Figure 2.13b). In this synthesis reaction, a water molecule is lost as the bond is made, as illustrated by the synthesis of sucrose (soo'krōs):

$$2C_6H_{12}O_6 \rightarrow C_{12}H_{22}O_{11} + H_2O$$

glucose + fructose   sucrose   water

Notice that the molecular formula for sucrose contains two hydrogen atoms and one oxygen atom less than the total number of hydrogen and oxygen atoms in glucose and fructose, because a water molecule is released during bond formation.

Important disaccharides in the diet are *sucrose* (glucose + fructose), which is cane or table sugar; *lactose* (glucose + galactose), found in milk; and *maltose* (glucose + glucose), also called malt sugar (see Figure 2.13b). Disaccharides are too large to pass through cell membranes, so they must be digested to their simple sugar units to be absorbed

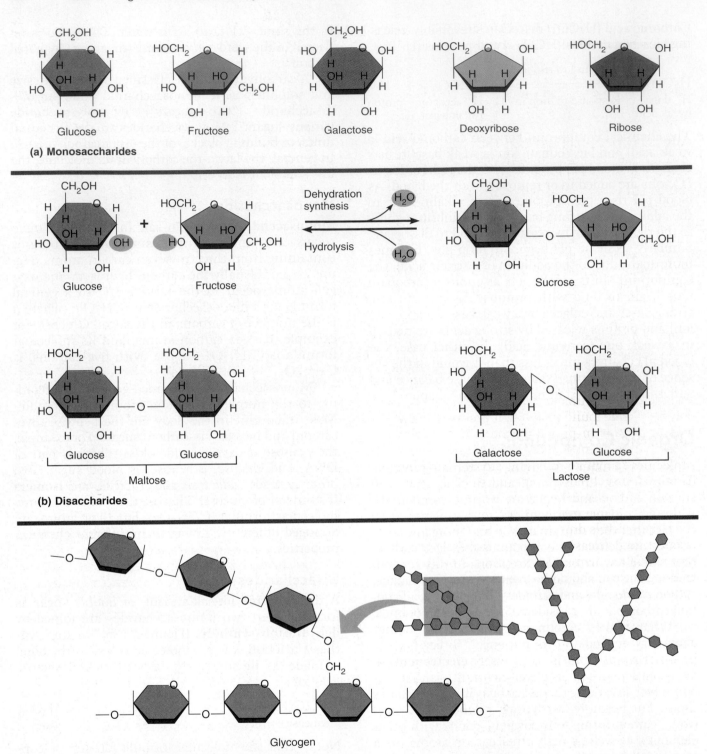

(a) Monosaccharides

Dehydration synthesis

Hydrolysis

Glucose   Fructose   Sucrose

(b) Disaccharides

Glucose   Glucose   Galactose   Glucose

Maltose   Lactose

(c) Portion of a polysaccharide molecule (glycogen)

Glycogen

FIGURE 2.13 **Carbohydrate molecules.*** **(a)** Monosaccharides important to the body. The hexose sugars glucose, fructose, and galactose are isomers: They have the same molecular formula ($C_6H_{12}O_6$), but the arrangement of their atoms differs, as shown here. Deoxyribose and ribose are pentose sugars. **(b)** Disaccharides are two monosaccharides linked together by dehydration synthesis. Formation of sucrose ($C_{12}H_{22}O_{11}$) from glucose and fructose molecules is illustrated. In the reverse reaction (hydrolysis), sucrose is broken down to glucose and fructose. Other important disaccharides are maltose and lactose (both $C_{12}H_{22}O_{11}$). **(c)** Simplified representation of part of a glycogen molecule, a polysaccharide formed from glucose units.

*See footnote on facing page concerning the complete structure of these sugars.

from the digestive tract into the blood. This decomposition process, called **hydrolysis,** is essentially the reverse of dehydration synthesis. A water molecule is added to each bond, breaking the bonds and releasing the simple sugar units.

Polysaccharides

Polysaccharides (pol"e-sak'ah-rīdz) are long chains of simple sugars linked together by dehydration synthesis. Such long, chainlike molecules made of many similar units are called **polymers.** Because polysaccharides are large, fairly insoluble molecules, they are ideal storage products. Another consequence of their large size is that they lack the sweetness of the simple and double sugars. Only two polysaccharides are of major importance to the body: starch and glycogen. Both are polymers of glucose. Only their degree of branching differs.

Starch is the storage carbohydrate formed by plants. The number of glucose units composing a starch molecule is high and variable. When we eat starchy foods such as grain products and potatoes, the starch must be digested for its glucose units to be absorbed. We are unable to digest *cellulose*, another polysaccharide found in all plant products. However, it is important in providing the *bulk* [one form of fiber] that helps move feces through the colon.

Glycogen (gli'ko-jen), the storage carbohydrate of animal tissues, is stored primarily in skeletal muscle and liver cells. Like starch, it is highly branched and is a very large molecule (see Figure 2.13c). When blood sugar levels drop sharply, liver cells break down glycogen and release its glucose units to the blood. Since there are many branch endings from which glucose can be released simultaneously, body cells have almost instant access to glucose fuel.

Carbohydrate Functions

The major function of carbohydrates in the body is to provide a ready, easily used source of cellular fuel. Most cells can use only a few simple sugars, and glucose is at the top of the "cellular menu." As described in our earlier discussion of oxidation-reduction reactions (p. 35), glucose is broken down and oxidized within cells. During these chemical reactions, electrons are transferred. This relocation of electrons releases the bond energy stored in glucose, and this energy is used to synthesize ATP. When ATP supplies are sufficient, dietary carbohydrates are converted to glycogen or fat and stored. Those of us who have gained weight from eating too many carbohydrate-rich snacks have personal experience with this conversion process!

Only small amounts of carbohydrates are used for structural purposes. For example, some sugars are found in our genes. Others are attached to the external surfaces of cells where they act as "road signs" to guide cellular interactions.

Lipids

Lipids are insoluble in water but dissolve readily in other lipids and in organic solvents such as alcohol and ether. Like carbohydrates, all lipids contain carbon, hydrogen, and oxygen, but the proportion of oxygen in lipids is much lower. In addition, phosphorus is found in some of the more complex lipids. Lipids include *neutral fats, phospholipids* (fos"fo-lip'idz), *steroids* (stĕ'roidz), and a number of other lipoid substances. Table 2.2 gives the locations and functions of some lipids found in the body.

Neutral Fats (Triglycerides)

The **neutral fats** are commonly known as *fats* when solid or *oils* when liquid. A neutral fat is composed of two types of building blocks, **fatty acids** and **glycerol** (glis'er-ol) (Figure 2.14a). Fatty acids are linear chains of carbon and hydrogen atoms (hydrocarbon chains) with an organic acid group (—COOH) at one end. Glycerol is a modified simple sugar (a sugar alcohol). Fat synthesis involves attaching three fatty acid chains to a single glycerol molecule by dehydration synthesis; the result is an E-shaped molecule. Because of the 3:1 fatty acid to glycerol ratio, the neutral fats are also called **triglycerides** (tri-glis'er-īdz) or **triacylglycerols** (tri-as"il-glis'er-olz; *acyl* = acid). The glycerol backbone is the same in all neutral fats, but the fatty acid chains vary, resulting in different kinds of neutral fats. Neutral fats are large molecules, often consisting of hundreds of atoms, and ingested fats and oils must be broken down to their building blocks before they can be absorbed. Neutral fats provide the body's most efficient and compact form for storing usable energy fuel, and when they are oxidized they yield large amounts of energy.

The hydrocarbon chains make neutral fats nonpolar molecules. Because polar and nonpolar molecules do not interact, oil (or fats) and water do not mix. Consequently, neutral fats are well suited for storing energy fuel in the body. Deposits of neutral fats are found mainly beneath the skin, where they

---

* Notice that in Figure 2.13 the carbon (C) atoms present at each angle of the carbohydrate ring structures are not illustrated. For example, the illustrations below show the full structure of glucose on the left and the shorthand structure on the right. This type of shorthand is used for nearly all organic ringlike structures illustrated in this chapter.

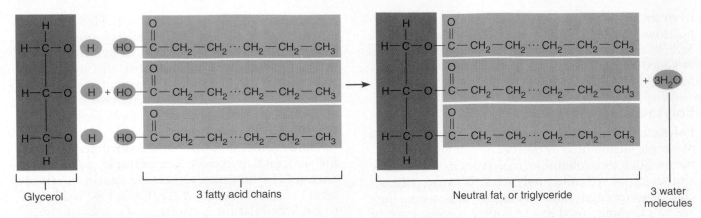

(a) Formation of a triglyceride

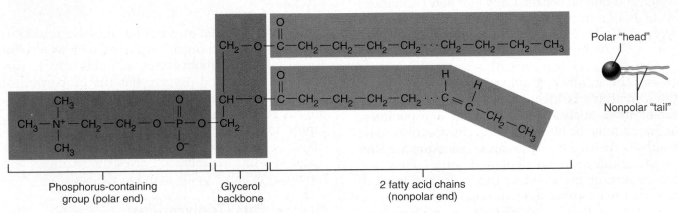

(b) Phospholipid molecule (phosphatidylcholine)

(c) Cholesterol

FIGURE 2.14   **Lipids. (a)** The neutral fats, or triglycerides, are synthesized by dehydration synthesis. **(b)** Structure of a typical phospholipid molecule. Two fatty acid chains and a phosphorus-containing group are attached to the glycerol backbone. In the commonly used schematic to the right, the polar end ("head") is depicted as a sphere, and the nonpolar end ("tail") is depicted as two wavy lines. **(c)** The generalized structure of cholesterol. Cholesterol is the basis for all steroids formed in the body.

insulate the deeper body tissues from heat loss and protect them from mechanical trauma. Women are usually more successful English Channel swimmers than men. This is due partly to their thicker subcutaneous fatty layer, which helps insulate them from the bitterly cold water of the channel.

The length of a neutral fat's fatty acid chains and their degree of saturation with H atoms determine how solid a neutral fat is at a given temperature. Fatty acid chains with only single covalent bonds between carbon atoms are referred to as **saturated.** Fatty acids that contain one or more double bonds between carbon atoms are said to be **unsaturated** (**monounsaturated** and **polyunsaturated** respectively). Neutral fats with short fatty acid chains or unsaturated fatty acids are oils (liquid at room temperature) and are typical of plant lipids. Examples

include olive and peanut oils (rich in monounsaturated fats) and corn, soybean, and safflower oils, which contain a high percentage of polyunsaturated fatty acids. Longer fatty acid chains and more saturated fatty acids are common in animal fats such as butter fat and the fat of meats, which are solid at room temperature.

Phospholipids

Phospholipids are modified triglycerides, that is, they are diglycerides with a phosphorus-containing group and two, rather than three, fatty acid chains (Figure 2.14b). The phosphorus-containing group gives phospholipids their distinctive chemical properties. Although the hydrocarbon portion (the "tail") of the molecule is nonpolar and interacts only with nonpolar molecules, the phosphorus-containing part

TABLE 2.2   Representative Lipids Found in the Body

Lipid Type	Location/Function
Neutral fats (triglycerides)	In fat deposits (subcutaneous tissue and around organs); protect and insulate body organs; the major source of *stored* energy in the body
Phospholipids (phosphatidylcholine; cephalin; others)	Chief component of cell membranes; may participate in the transport of lipids in plasma; prevalent in nervous tissue
Steroids	
Cholesterol	The structural basis for manufacture of all body steroids
Bile salts	Breakdown products of cholesterol; released by the liver into the digestive tract, where they aid fat digestion and absorption
Vitamin D	A fat-soluble vitamin produced in the skin on exposure to UV radiation; necessary for normal bone growth and function
Sex hormones	Estrogen and progesterone (female hormones) and testosterone (a male hormone) are produced in the gonads and are necessary for normal reproductive function
Adrenal cortical hormones	Cortisol, a glucocorticoid, is a metabolic hormone necessary for maintaining normal blood glucose levels; aldosterone helps to regulate salt and water balance of the body by targeting the kidneys
Other lipoid substances	
Fat-soluble vitamins:	
A	Found in orange-pigmented vegetables and fruits; converted in the retina to retinal, a part of the photoreceptor pigment involved in vision
E	Found in plant products such as wheat germ and green leafy vegetables; claims have been made (but not proved in humans) that it promotes wound healing and contributes to fertility; may help to neutralize highly reactive particles called free radicals believed to be involved in triggering some types of cancer
K	Made available to humans largely by the action of intestinal bacteria; also prevalent in a wide variety of foods; necessary for proper clotting of blood
Eicosanoids (prostaglandins; leukotrienes; thromboxanes)	Group of molecules derived from fatty acids found in all cell membranes; the potent prostaglandins have diverse effects, including stimulation of uterine contractions, regulation of blood pressure, control of gastrointestinal tract motility, and secretory activity; both prostaglandins and leukotrienes are involved in inflammation; thromboxanes are powerful vasoconstrictors
Lipoproteins	Lipoid and protein-based substances that transport fatty acids and cholesterol in the bloodstream; major varieties are high-density lipoproteins (HDLs) and low-density lipoproteins (LDLs)

(the "head") is polar and attracts other polar or charged particles, such as water or ions. Molecules that have both polar and nonpolar regions are *amphipathic* (am-fe-path'ik). As you will see in Chapter 3, cells use this unique characteristic of phospholipids in building their membranes. Some biologically important phospholipids and their functions are listed in Table 2.2.

Steroids

Structurally, steroids differ quite a bit from fats. **Steroids** are basically flat molecules made of four interlocking hydrocarbon rings. Like neutral fats, steroids are fat soluble and contain little oxygen. The single most important molecule in our steroid chemistry is *cholesterol* (ko-les'ter-ol) (Figure 2.14c). We ingest cholesterol in animal products such as eggs, meat, and cheese, and our liver produces a certain amount.

Cholesterol has earned bad press because of its role in arteriosclerosis, but it is absolutely essential for human life. Cholesterol is found in cell membranes and is the raw material of vitamin D, steroid hormones, and bile salts. Although steroid hormones are present in the body in only small quantities, they are vital to homeostasis. Without sex hormones, reproduction would be impossible, and a total lack of the corticosteroids produced by the adrenal gland is fatal.

Eicosanoids

The **eicosanoids** (i-ko'sah-noyds) are diverse lipids chiefly derived from a 20-carbon fatty acid (arachidonic acid) found in all cell membranes. Most important of these are the *prostaglandins* and their relatives, which play roles in various body processes (see Table 2.2) including blood clotting, inflammation, and labor contractions.

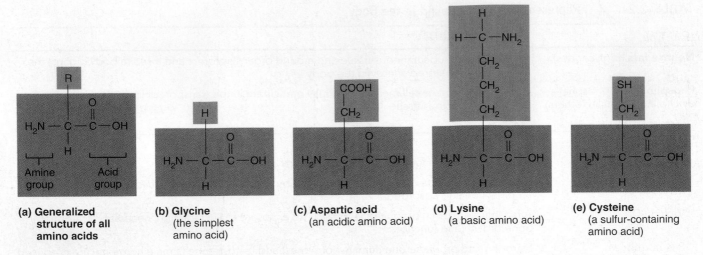

(a) Generalized structure of all amino acids

(b) Glycine (the simplest amino acid)

(c) Aspartic acid (an acidic amino acid)

(d) Lysine (a basic amino acid)

(e) Cysteine (a sulfur-containing amino acid)

FIGURE 2.15   Amino acid structures. (a) Generalized structure of amino acids. All amino acids have both an amine (—NH_2) group and an acid (—COOH) group; they differ only in the atomic makeup of their R groups (green). **(b–e)** Specific structures of four amino acids. **(b)** The simplest (glycine) has an R group consisting of a single hydrogen atom. **(c)** An acid group in the R group, as in aspartic acid, makes the amino acid more acidic. **(d)** An amine group in the R group, as in lysine, makes the amino acid more basic. **(e)** The presence of a sulfhydryl (—SH) group in the R group of cysteine suggests that this amino acid is likely to participate in intramolecular bonding.

Proteins

Protein composes 10 to 30% of cell mass and is the basic structural material of the body. However, not all proteins are construction materials; many play vital roles in cell function. Proteins, which include enzymes (biological catalysts), hemoglobin of the blood, and contractile proteins of muscle, have the most varied functions of any molecules in the body. All proteins contain carbon, oxygen, hydrogen, and nitrogen, and many contain sulfur and phosphorus as well.

Amino Acids and Peptide Bonds

The building blocks of proteins are molecules called **amino acids,** of which there are 20 common types (see Appendix B). All amino acids have two important functional groups: a basic group called an *amine* (ah'mēn) *group* (—NH_2), and an organic *acid group* (—COOH). An amino acid may therefore act either as a base (proton acceptor) or an acid (proton donor). In fact, all amino acids are identical except for a single group of atoms called their *R group*. Differences in the R group make each amino acid chemically unique (Figure 2.15).

Proteins are long chains of amino acids joined together by dehydration synthesis, with the amine end of one amino acid linked to the acid end of the next. The resulting bond produces a characteristic arrangement of linked atoms called a **peptide bond** (Figure 2.16). Two united amino acids form a *dipeptide*, three a *tripeptide*, and ten or more a *polypeptide*. Although polypeptides containing more than 50 amino acids are called proteins, most proteins are **macromolecules,** large, complex molecules containing from 100 to over 10,000 amino acids.

Because each type of amino acid has distinct properties, the sequence in which they are bound together produces proteins that vary widely in both structure and function. We can think of the 20 amino acids as a 20-letter "alphabet" used in specific combinations to form "words" (proteins). Just as a change in one letter can produce a word with an entirely different meaning (flour → floor) or that is nonsensical (flour → fllur), changes in the kinds or

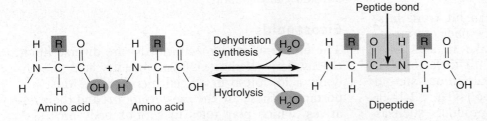

Peptide bond

Amino acid   +   Amino acid   →   Dipeptide

Dehydration synthesis H_2O

Hydrolysis H_2O

FIGURE 2.16   Amino acids are linked together by dehydration synthesis. The acid group of one amino acid is bonded to the amine group of the next, with loss of a water molecule. The bond formed is called a peptide bond. Peptide bonds are broken when water is added to the bond (i.e., during hydrolysis).

positions of amino acids can yield proteins with different functions or proteins that are nonfunctional. Nevertheless, there are thousands of different proteins in the body, each with distinct functional properties, and all constructed from different combinations of the 20 common amino acids.

Structural Levels of Proteins

Proteins can be described in terms of four structural levels. The linear sequence of amino acids composing the polypeptide chain is called the *primary structure* of a protein. This structure, which resembles a strand of amino acid "beads," is the backbone of the protein molecule (Figure 2.17a).

Proteins do not normally exist as simple, linear chains of amino acids. Instead, they twist or bend upon themselves to form a more complex *secondary structure*. The most common type of secondary structure is the **alpha (α) helix,** which resembles a Slinky toy or the coils of a telephone cord (Figure 2.17b). The α helix, formed by coiling of the primary chain, is stabilized by hydrogen bonds formed between NH and CO groups in amino acids in the primary chain which are approximately four amino acids apart. Hydrogen bonds in α helices always link different parts of the *same* chain together. In another type of secondary structure, the **beta (β)-pleated sheet,** the primary polypeptide chains do not coil, but are linked side by side by hydrogen bonds to form a pleated, ribbonlike structure (Figure 2.17c) that resembles an accordion. Notice that in this type of secondary structure, the hydrogen bonds may link together *different polypeptide* chains as well as *different parts* of the same chain that has folded back on itself. A single polypeptide chain may exhibit both types of secondary structure at various places along its length.

Many proteins have *tertiary* (ter'she-a"re) *structure*, the next higher level of complexity, which is superimposed on secondary structure. Tertiary structure is achieved when α-helical or β-pleated regions of the polypeptide chain fold upon one another to produce a compact ball-like, or globular, molecule. The tertiary structure (and underlying secondary structure) of a beta polypeptide chain of hemoglobin, the oxygen-binding protein found in red blood cells, is shown in Figure 2.17d. This unique structure is maintained by both covalent and hydrogen bonds between amino acids that are often far apart in the primary chain. When two or more polypeptide chains aggregate in a regular manner to form a complex protein, the protein has *quaternary* (kwah'ter-na"re) *structure*. Hemoglobin (Figure 2.17e) exhibits this structural level.

Although a protein with tertiary or quaternary structure looks a bit like a clump of congealed pasta, the ultimate overall structure of any protein is very specific and is dictated by its primary structure. This reflects the fact that the types and relative positions of amino acids in the protein backbone determine where bonds can form to produce the complex coiled or folded structures that keep water-loving amino acids near the surface and water-fleeing amino acids buried in the protein's core.

Fibrous and Globular Proteins

The overall structure of a protein determines its biological function. In general, proteins are classified according to their overall appearance and shape as either fibrous or globular.

Fibrous proteins are extended and strandlike. Most exhibit only secondary structure, but some have quaternary structure as well. For example, *collagen* (kol'ah-jen) is a composite of the helical tropocollagen molecules packed together side by side to form a strong ropelike structure. Fibrous proteins are insoluble in water, and very stable—qualities ideal for providing mechanical support and tensile strength to the body's tissues. Besides collagen, which is the single most abundant protein in the body, the fibrous proteins include keratin, elastin, and certain contractile proteins of muscle (Table 2.3). Because fibrous proteins are the chief building materials of the body, they are also known as **structural proteins.**

Globular proteins are compact, spherical proteins that have at least tertiary structure; some also exhibit quaternary structure. The globular proteins are water soluble, chemically active molecules, and they play crucial roles in virtually all biological processes. Consequently, some refer to this group as **functional proteins.** Some of these proteins (antibodies) help to provide immunity, others (protein-based hormones) regulate growth and development, and still others (enzymes) are catalysts that oversee just about every chemical reaction in the body. The roles of these and selected other proteins found in the body are summarized in Table 2.3.

Protein Denaturation

Fibrous proteins are stable, but globular proteins are quite the opposite. The activity of a protein depends on its specific three-dimensional structure, and intramolecular bonds, particularly hydrogen bonds, are important in maintaining that structure. However, hydrogen bonds are fragile and easily broken by many chemical and physical factors, such as excessive acidity or heat. Although individual proteins vary in their sensitivity to environmental conditions, hydrogen bonds begin to break when the pH drops or the temperature rises above normal (physiological) levels, causing proteins to unfold and lose their

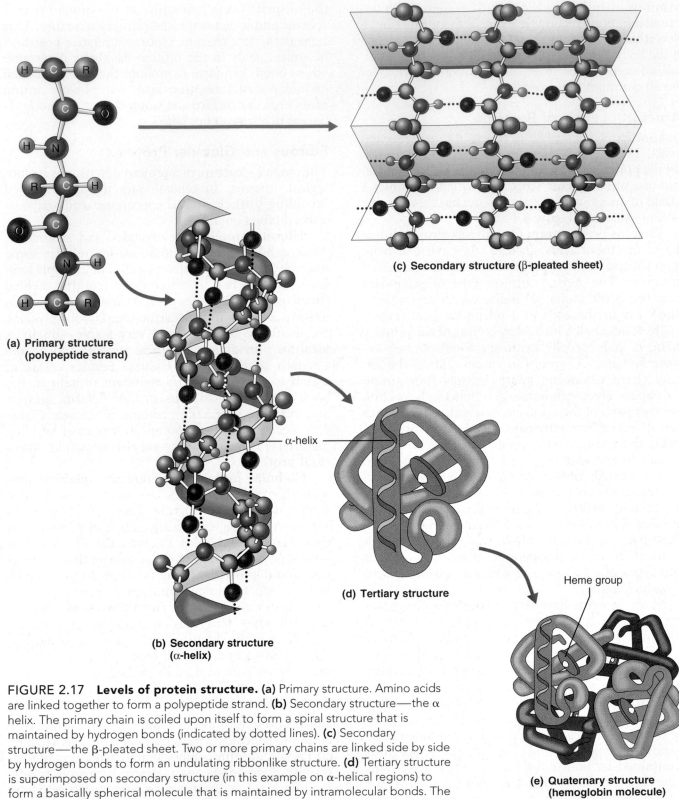

(a) **Primary structure (polypeptide strand)**

(b) **Secondary structure (α-helix)**

(c) **Secondary structure (β-pleated sheet)**

α-helix

(d) **Tertiary structure**

Heme group

(e) **Quaternary structure (hemoglobin molecule)**

FIGURE 2.17 **Levels of protein structure. (a)** Primary structure. Amino acids are linked together to form a polypeptide strand. **(b)** Secondary structure—the α helix. The primary chain is coiled upon itself to form a spiral structure that is maintained by hydrogen bonds (indicated by dotted lines). **(c)** Secondary structure—the β-pleated sheet. Two or more primary chains are linked side by side by hydrogen bonds to form an undulating ribbonlike structure. **(d)** Tertiary structure is superimposed on secondary structure (in this example on α-helical regions) to form a basically spherical molecule that is maintained by intramolecular bonds. The molecule shown is a beta chain, one of the four polypeptide chains of hemoglobin. **(e)** Quaternary structure of hemoglobin. Hemoglobin is formed from four polypeptide chains that are linked together in a specific manner.

TABLE 2.3 Representative Types of Proteins in the Body

Classification According To

Overall Structure	General Function	Examples from the Body
Fibrous	Structural framework/ mechanical support	*Collagen*, found in all connective tissues, is the single most abundant protein in the body. It is responsible for the tensile strength of bones, tendons, and ligaments.
		Keratin is the structural protein of hair and nails and a waterproofing material of skin.
		Elastin is found, along with collagen, where durability and flexibility are needed, such as in the ligaments that bind bones together.
		Spectrin internally reinforces and stabilizes the surface membrane of some cells, particularly red blood cells. *Dystrophin* reinforces and stabilizes the surface membrane of muscle cells. *Titin* helps organize the intracellular structure of muscle cells and accounts for the elasticity of skeletal muscles.
	Movement	*Actin and myosin*, contractile proteins, are found in substantial amounts in muscle cells, where they cause muscle cell shortening (contraction); they also function in cell division in all cell types. Actin is important in intracellular transport, particularly in nerve cells.
Globular	Catalysis	*Protein enzymes* are essential to virtually every biochemical reaction in the body; they increase the rates of chemical reactions by at least a millionfold. Examples include salivary amylase (in saliva), which catalyzes the breakdown of starch, and oxidase enzymes, which act to oxidize food fuels.
	Transport	*Hemoglobin* transports oxygen in blood, and *lipoproteins* transport lipids and cholesterol. Other transport proteins in the blood carry iron, hormones, or other substances.
	Regulation of pH	Many plasma proteins, such as *albumin*, function reversibly as acids or bases, thus acting as buffers to prevent wide swings in blood pH.
	Regulation of metabolism	*Peptide* and *protein hormones* help to regulate metabolic activity, growth, and development. For example, *growth hormone* is an anabolic hormone necessary for optimal growth; *insulin* helps regulate blood sugar levels.
	Body defense	*Antibodies* (immunoglobulins) are specialized proteins released by immune cells that recognize and inactivate foreign substances (bacteria, toxins, some viruses). *Complement proteins*, which circulate in blood, enhance both immune and inflammatory responses. *Molecular chaperones* aid folding of new proteins in both healthy and damaged cells and transport of metal ions into and within the cell.

specific three-dimensional shape. In this condition, a protein is said to be **denatured.** Fortunately, the disruption is reversible in most cases, and the "scrambled" protein regains its native structure when desirable conditions are restored. However, if the temperature or pH change is so extreme that protein structure is damaged beyond repair, the protein is *irreversibly denatured.* The coagulation of egg white (primarily albumin protein) that occurs when you boil or fry an egg is an example of irreversible protein denaturation. There is no way to restore the white, rubbery protein to its original translucent form.

When globular proteins are denatured, they can no longer perform their physiological roles because their function depends on the presence of specific arrangements of atoms, called **active sites,** on their surfaces. The active sites are regions that fit and interact chemically with other molecules of complementary shape and charge. Because atoms contributing to an active site may actually be very far apart in the primary chain, disruption of intramolecular bonds separates them and destroys the active site (Figure 2.18). Hemoglobin becomes totally unable to bind and transport oxygen when blood pH is too acidic, because the structure needed for its function has been destroyed.

Most body proteins are described in conjunction with the organ systems or functional processes to which they are closely related. However, two groups of proteins—*molecular chaperones* and *enzymes*—are intimately involved in the normal functioning of all cells, and so these incredibly complex molecules are considered here.

What event, which is possible in situation (a), can no longer occur in situation (b)?

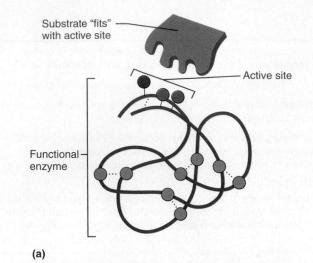

Substrate "fits" with active site

Active site

Functional enzyme

(a)

Substrate unable to bind

Denatured enzyme

(b)

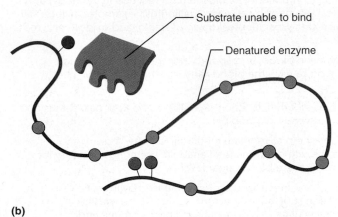

FIGURE 2.18   Denaturation of a globular protein such as an enzyme. (a) The molecule's globular structure is maintained by intramolecular bonds. Atoms composing the active site of the enzyme are shown as stalked particles. The substrate, or interacting molecule, has a corresponding binding site, and the two sites fit together precisely. **(b)** Breaking the intramolecular bonds that maintain the secondary and tertiary structure of the enzyme results in a linear molecule, with atoms of the former active site widely separated. Enzyme-substrate binding can no longer occur.

Molecular Chaperones

In addition to the ubiquitous enzymes, all cells contain a class of unrelated globular proteins called **molecular chaperones**, or **chaperonins**, which, among

The substrate can no longer bind to the enzyme's active site in (b) because the atoms composing the active site are no longer in their functional relationship to one another. ∎

other things, help proteins to achieve their functional three-dimensional structure. Although its amino acid sequence determines the precise way a protein folds, the folding process also requires the help of molecular chaperones to ensure that the folding is quick and accurate. Apparently, molecular chaperones have numerous protein-related roles to play. For example, specific chaperonins

- Prevent accidental, premature, or incorrect folding of polypeptide chains or their association with other polypeptides
- Aid the desired folding and association process
- Help to translocate proteins and certain metal ions (copper, iron, zinc) across cell membranes
- Promote the breakdown of damaged or denatured proteins

The first such proteins discovered were called *heat shock proteins (hsp)* because they seemed to protect cells from the destructive effects of heat. It was later found that these proteins are produced in response to a variety of traumatizing stimuli—for example, in the oxygen-deprived cells of a heart attack patient—and the name *stress proteins* replaced hsp for that particular group of chaperonins. It is now clear that chaperonins are vitally important to cell function in all types of stressful circumstances.

Enzymes and Enzyme Activity

Characteristics of Enzymes   **Enzymes** are globular proteins that act as biological catalysts. *Catalysts* are substances that regulate and accelerate the rate of biochemical reactions but are not used up or changed in those reactions. More specifically, enzymes can be thought of as chemical traffic cops that keep our metabolic pathways flowing. Enzymes cannot force chemical reactions to occur between molecules that would not otherwise react; they can only increase the speed of reaction. Without enzymes, biochemical reactions proceed so slowly that for practical purposes they do not occur at all.

Some enzymes are purely protein. In other cases, the functional enzyme consists of two parts, collectively called a **holoenzyme**—an **apoenzyme** (the protein portion) and a **cofactor.** Depending on the enzyme, the cofactor may be an ion of a metal element such as copper or iron, or an organic molecule needed to assist the reaction in some particular way. Most organic cofactors are derived from vitamins (especially the B complex vitamins); this type of cofactor is more precisely called a **coenzyme.**

Each enzyme is chemically specific. Some enzymes control only a single chemical reaction. Others exhibit a broader specificity in that they can bind with similar (but not identical) molecules and thus regulate a small group of related reactions. The presence of specific enzymes thus determines not only which reactions will be speeded up, but also which reactions will occur—no enzyme, no reaction. This also means that unwanted or unnecessary chemical reactions do not occur. Most enzymes are named for the type of reaction they catalyze. *Hydrolases* (hi'druh-lās-es) add water during hydrolysis reactions; *oxidases* (ok'sĭ-dās-es) add oxygen; and so on. Most enzyme names can be recognized by the suffix *-ase.*

In many cases, enzymes are part of cellular membranes in a bucket-brigade type of arrangment; that is, the product of one enzyme-catalyzed reaction becomes the substrate of the abutting enzyme, and so on. Some enzymes are produced in an inactive form and must be activated in some way before they can function. For example, digestive enzymes produced in the pancreas are activated in the small intestine, where they actually do their work. If they were produced in active form, the pancreas would digest itself. Sometimes, enzymes are inactivated immediately after they have performed their catalytic function. This is true of enzymes that promote blood clot formation when the wall of a blood vessel is damaged. Once clotting is triggered, those enzymes are inactivated. Otherwise, you would have blood vessels full of solid blood instead of one protective blood clot.

Mechanism of Enzyme Activity How do enzymes perform their catalytic role? Every chemical reaction requires that a certain amount of energy, called **activation energy,** be absorbed to prime the reaction. The activation energy pushes the reactants to an energy level where their random collisions are forceful enough to ensure interaction. This is true regardless of whether the overall reaction is ultimately energy absorbing or energy releasing.

One obvious way to increase molecular energy is to increase the temperature, but in living systems this would denature proteins. (This is why a high fever can be a serious event.) Enzymes allow reactions to occur at normal body temperature by decreasing the amount of activation energy required (Figure 2.19). Exactly how enzymes accomplish this remarkable feat is not fully understood. However, we know that they decrease the randomness of reactions by binding the reacting molecules temporarily to the enzyme surface and presenting them to each other in the proper position for chemical interaction to occur.

Three basic steps appear to be involved in the mechanism of enzyme action (Figure 2.20).

① The enzyme's active site must first bind with the substance(s) on which it acts. These substances are called the **substrates** of the enzyme. Substrate binding causes the active site to change shape so that the substrate and the active site fit together precisely.

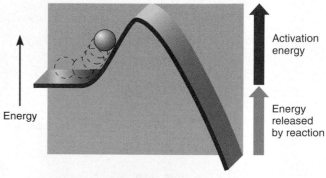

(a) Noncatalyzed reaction

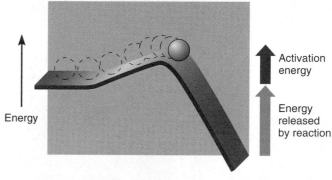

(b) Enzyme-catalyzed reaction

FIGURE 2.19 Enzymes lower the energy barrier for a reaction. (a) The reacting particles, indicated by the ball, must achieve a certain energy level before they can interact. The amount of energy that must be absorbed to reach that state is the activation energy. **(b)** The energy barrier and thus the activation energy is lower in the same reaction catalyzed by an enzyme.

Although enzymes are specific for particular substrates, other (nonsubstrate) molecules may act as *enzyme inhibitors* if their structure is similar enough to occupy or block the enzyme's active site.

② The enzyme-substrate complex undergoes internal rearrangements that form the product.

③ The enzyme releases the product of the reaction. This step shows the catalytic role of an enzyme: If the enzyme became part of the product, it would be a reactant and not a catalyst.

Because the unaltered enzymes can act again and again, cells need only small amounts of each enzyme. Catalysis occurs with incredible speed; most enzymes can catalyze millions of reactions per minute.

Nucleic Acids (DNA and RNA)

The **nucleic acids** (nu-kle'ic), composed of carbon, oxygen, hydrogen, nitrogen, and phosphorus, are the largest molecules in the body. Their structural units,

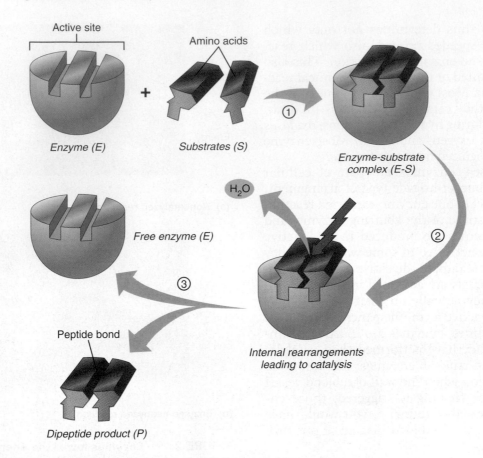

FIGURE 2.20   Mechanism of enzyme action. Each enzyme is highly specific in terms of the reaction(s) it can catalyze and bonds properly to only one or a few substrates. In this example, the enzyme catalyzes the formation of a dipeptide from specific amino acids.

① The enzyme-substrate complex (E-S) is formed.

② Internal rearrangements occur. In this case, energy is absorbed (indicated by the yellow arrow) as a water molecule is removed and a peptide bond is formed.

③ The enzyme releases the product (P) of the reaction, the dipeptide. The free enzyme has not changed in the course of the reaction and is now available to catalyze another such reaction.

Summary: E + S → E-S → P + E

called **nucleotides,** are quite complex. Each nucleotide consists of three components (Figure 2.21a): a nitrogen-containing base, a pentose sugar, and a phosphate group. Five major varieties of nitrogen-containing bases can contribute to nucleotide structure: **adenine** (ad′ĕ-nēn), abbreviated A; **guanine** (gwan′ēn), G; **cytosine** (si′to-sēn), C; **thymine** (thi′mēn), T; and **uracil** (u′rah-sil), U. Adenine and guanine are large, two-ring bases (called purines), whereas cytosine, thymine, and uracil are smaller, single-ring bases (called pyrimidines).

The stepwise synthesis of a nucleotide involves the attachment of a base to the pentose sugar to form first a *nucleoside,* named for the nitrogenous base it contains. The nucleotide is formed when a phosphate group is bonded to the sugar of the nucleoside.

The nucleic acids include two major classes of molecules, **deoxyribonucleic acid (DNA)** (de-ok″sĭ-ri″bo-nu-kle′ik) and **ribonucleic acid (RNA).** Although DNA and RNA are both composed of nucleotides, they differ in many respects. Typically, DNA is found in the nucleus (control center) of the cell, where it constitutes the *genetic material,* or *genes.* DNA has two fundamental roles: It replicates (reproduces) itself before a cell divides, ensuring that the genetic information in the descendant cells is identical, and it provides instructions for building every protein in the body. By providing the information for protein synthesis, DNA determines what

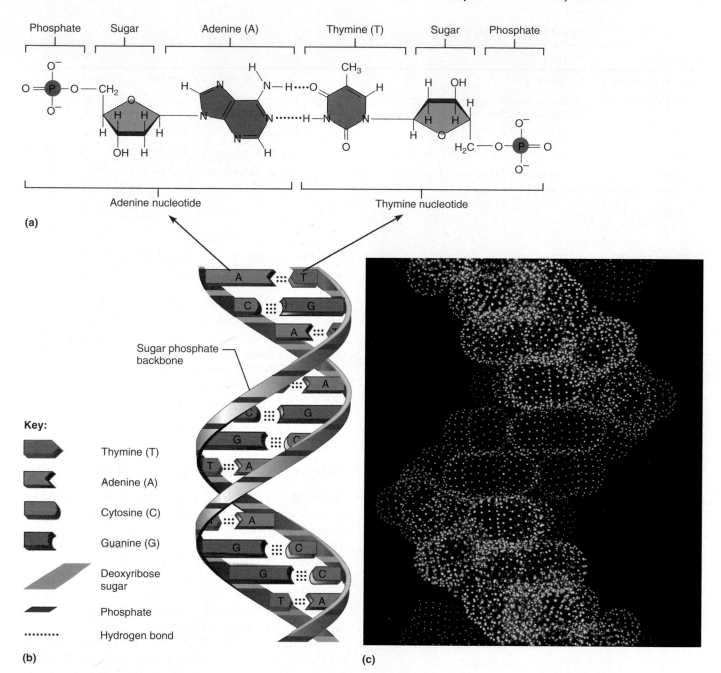

FIGURE 2.21 **Structure of DNA.**
(a) The unit of DNA is the nucleotide, which is composed of a deoxyribose sugar molecule linked to a phosphate group, with a base attached to the sugar. Two nucleotides, linked by hydrogen bonds between their complementary bases, are illustrated. **(b)** DNA is a coiled double polymer of nucleotides (a double helix). The backbones of the ladderlike molecule are formed by alternating sugar and phosphate units. The rungs are formed by the binding together of complementary bases (A-T and G-C) by hydrogen bonds (shown by dotted lines). **(c)** Computer-generated image of a DNA molecule.

type of organism you will be—frog, human, oak tree—and directs your growth and development. Although we have said that enzymes govern all chemical reactions, remember that enzymes, too, are proteins formed at the direction of DNA. RNA is located chiefly outside the nucleus and can be considered a "molecular slave" of DNA. That is,

RNA carries out the orders for protein synthesis issued by DNA. (Viruses in which RNA [rather than DNA] is the genetic material are an exception to this generalization.)

DNA is a long, double-stranded polymer—a double chain of nucleotides (Figure 2.21b and c). The bases in DNA are A, G, C, and T, and its

TABLE 2.4	Comparison of DNA and RNA	
Characteristic	DNA	RNA
Major cellular site	Nucleus	Cytoplasm (cell area outside the nucleus)
Major functions	Is the genetic material; directs protein synthesis; replicates itself before cell division	Carries out the genetic instructions for protein synthesis
Sugar	Deoxyribose	Ribose
Bases	Adenine, guanine, cytosine, thymine	Adenine, guanine, cytosine, uracil
Structure	Double strand coiled into a double helix	Single strand, straight or folded

pentose sugar is *deoxyribose* (as reflected in its name). Its two nucleotide chains are held together by hydrogen bonds between the bases, so that a ladderlike molecule is formed. Alternating sugar and phosphate molecules of each chain form the *backbones* or "uprights" of the "ladder," and the joined bases form the "rungs." The whole molecule is coiled into a spiral staircase–like structure called a **double helix.** Bonding of the bases is very specific: A always bonds to T, and G always bonds to C. A and T are therefore called **complementary bases,** as are C and G. Hence, ATGA on one nu-

cleotide strand would necessarily be bonded to TACT (a complementary base sequence) on the other strand.

RNA molecules are single strands of nucleotides. RNA bases include A, G, C, and U (U replaces the T found in DNA), and its sugar is *ribose* instead of deoxyribose (Table 2.4). The three major varieties of RNA are distinguished by their relative size and shape, and each has a specific role to play in carrying out DNA's instructions. We discuss DNA replication and the relative roles of DNA and RNA in protein synthesis in Chapter 3.

Adenosine Triphosphate (ATP)

Although glucose is the most important cellular fuel, none of the chemical energy contained in its bonds is used directly to power cellular work. Instead, energy released during glucose catabolism is coupled to **adenosine triphosphate (ATP)** synthesis—that is, it is captured and stored as small packets of energy in the bonds of ATP. ATP, in turn, acts as a chemical "drive shaft" to provide a form of energy that is immediately usable by all body cells.

Structurally, ATP is an adenine-containing RNA nucleotide to which two additional phosphate groups have been added (Figure 2.22). Chemically, the triphosphate tail of ATP can be compared to a tightly coiled spring ready to uncoil with tremendous energy when the catch is released. Actually, ATP is a very unstable energy-storing molecule because its three negatively charged phosphate groups are closely packed and repel each other. When its terminal high-energy phosphate bonds are broken (hydrolyzed), the chemical "spring" relaxes and the molecule as a whole becomes more stable. Cells tap ATP's bond energy during coupled reactions by using enzymes to transfer the terminal phosphate groups from ATP to other compounds. The newly *phosphorylated* molecules are said to be "primed" and temporarily become more energetic and capable of performing some type of cellular work. In the process of doing

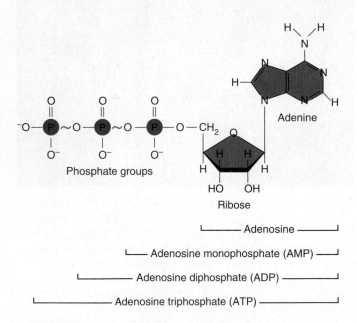

FIGURE 2.22 Structure of ATP (adenosine triphosphate). ATP is an adenine nucleotide to which two additional phosphate groups have been attached during breakdown of food fuels. (The phosphate bonds that can be hydrolyzed to liberate energy are indicated by the wavy lines.) When the terminal phosphate group is cleaved off, energy is released to do useful work and ADP (adenosine diphosphate) is formed. When the terminal phosphate group is cleaved off ADP, a similar amount of energy is released and AMP (adenosine monophosphate) is formed.

their work, they lose the phosphate group. The amount of energy released and transferred during ATP hydrolysis corresponds closely to that needed to drive most biochemical reactions. As a result, cells are protected from excessive energy release that might be damaging, and energy squandering is kept to a minimum.

Cleaving the terminal phosphate bond of ATP yields a molecule with two phosphate groups—*adenosine diphosphate (ADP)*—and an inorganic phosphate group, indicated by P_i, accompanied by a transfer of energy:

$$\mathrm{ATP} \underset{H_2O}{\overset{H_2O}{\rightleftharpoons}} \mathrm{ADP} + P_i + \text{energy}$$

As ATP is hydrolyzed to provide energy for cellular needs, ADP accumulates. Cleavage of the terminal phosphate bond of ADP liberates a similar amount of energy and produces adenosine monophosphate (AMP). ATP supplies are replenished as glucose and other fuel molecules are oxidized and their bond energy is released. The same amount of energy that is liberated when ATP's terminal phosphates are cleaved off must be captured and used to reverse the reaction to reattach phosphates and re-form the energy-transferring phosphate bonds. Without ATP, molecules cannot be made or degraded, cells cannot transport substances across their membrane boundaries, muscles cannot shorten to tug on other structures, and life processes cease (Figure 2.23).

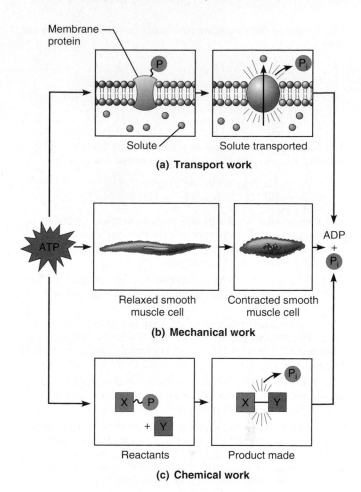

FIGURE 2.23   **Three examples of how ATP drives cellular work.** The high-energy bonds of ATP are like coiled springs that release energy for use by the cell when they are broken. **(a)** ATP drives the transport of certain solutes (amino acids, for example) across cell membranes. **(b)** ATP activates contractile proteins in muscle cells so that the cells can shorten and perform mechanical work. **(c)** ATP provides the energy to drive endergonic (energy-absorbing) chemical reactions.

Review Questions

Multiple Choice/Matching

(Some questions have more than one correct answer. Select the best answer or answers from the choices given.)

1. Which of the following forms of energy are in use during vision? (a) chemical, (b) electrical, (c) mechanical, (d) radiant.

2. All of the following are examples of the four major elements contributing to body mass except (a) hydrogen, (b) carbon, (c) nitrogen, (d) sodium, (e) oxygen.

3. The mass number of an atom is (a) equal to the number of protons it contains, (b) the sum of its protons and neutrons, (c) the sum of all of its subatomic particles, (d) the average of the mass numbers of all of its isotopes.

4. A deficiency in this element can be expected to reduce the hemoglobin content of blood: (a) Fe, (b) I, (c) F, (d) Ca, (e) K.

5. Which set of terms best describes a proton? (a) negative charge, massless, in the orbital; (b) positive charge, 1 amu, in the nucleus; (c) uncharged, 1 amu, in the nucleus.

6. The subatomic particles responsible for the chemical behavior of atoms are (a) electrons, (b) ions, (c) neutrons, (d) protons.

7. In the body, carbohydrates are stored in the form of (a) glycogen, (b) starch, (c) cholesterol, (d) polypeptides.

8. Which of the following does *not* describe a mixture? (a) properties of its components are retained, (b) chemical bonds are formed, (c) components can be separated physically, (d) includes both heterogeneous and homogeneous examples.

9. In a beaker of water, the water-water bonds can properly be called (a) ionic bonds, (b) polar covalent bonds, (c) nonpolar covalent bonds, (d) hydrogen bonds.

10. When a pair of electrons is shared between two atoms, the bond formed is called (a) a single covalent bond, (b) a double covalent bond, (c) a triple covalent bond, (d) an ionic bond.

11. Molecules formed when electrons are shared unequally are (a) salts, (b) polar molecules, (c) nonpolar molecules.

12. Which of the following covalently bonded molecules are polar?

(a) H—Cl   (b) H—C—H   (c) Cl—C—Cl   (d) N≡N

with H above and H below for (b); H above and Cl below for (c)

13. Identify each reaction as (a) a synthesis reaction, (b) a decomposition reaction, or (c) an exchange reaction.

(1) $2Hg + O_2 \rightarrow 2HgO$
(2) $HCl + NaOH \rightarrow NaCl + H_2O$

14. Factors that accelerate the rate of chemical reactions include all but (a) the presence of catalysts, (b) increasing the temperature, (c) increasing the particle size, (d) increasing the concentration of the reactants.

15. Which of the following molecules is an inorganic molecule? (a) sucrose, (b) cholesterol, (c) collagen, (d) sodium chloride.

16. Water's importance to living systems reflects (a) its polarity and solvent properties, (b) its high heat capacity, (c) its high heat of vaporization, (d) its chemical reactivity, (e) all of these.

17. Acids (a) release hydroxyl ions when dissolved in water, (b) are proton acceptors, (c) cause the pH of a solution to rise, (d) release protons when dissolved in water.

18. A chemist, during the course of an analysis, runs across a chemical composed of carbon, hydrogen, and oxygen in the proportion 1:2:1 and having a six-sided molecular shape. It is probably (a) a pentose, (b) an amino acid, (c) a fatty acid, (d) a monosaccharide, (e) a nucleic acid.

19. A neutral fat consists of (a) glycerol plus up to three fatty acids, (b) a sugar-phosphate backbone to which two amino groups are attached, (c) two to several hexoses, (d) amino acids that have been thoroughly saturated with hydrogen.

20. A chemical has an amine group and an organic acid group. It does not, however, have any peptide bonds. It is (a) a monosaccharide, (b) an amino acid, (c) a protein, (d) a fat.

21. The lipid(s) used as the basis of vitamin D, sex hormones, and bile salts is/are (a) neutral fats, (b) cholesterol, (c) phospholipids, (d) prostaglandin.

22. Enzymes are organic catalysts that (a) alter the direction in which a chemical reaction proceeds, (b) determine the nature of the products of a reaction, (c) increase the speed of a chemical reaction, (d) are essential raw materials for a chemical reaction that are converted into some of its products.

Short Answer Essay Questions

23. Define or describe energy, and explain the relationship between potential and kinetic energy.

24. Some energy is lost in every energy conversion. Explain the meaning of this statement. (Direct your response to answering the question, Is it really lost? If not, what then?)

25. Provide the atomic symbol for each of the following elements: (a) calcium, (b) carbon, (c) hydrogen, (d) iron, (e) nitrogen, (f) oxygen, (g) potassium, (h) sodium.

26. Consider the following information about three atoms:

$$^{12}_{6}C \qquad ^{13}_{6}C \qquad ^{14}_{6}C$$

(a) How are they similar to one another? (b) How do they differ from one another? (c) What are the members of such a group of atoms called? (d) Using the planetary model, draw the atomic configuration of $^{12}_{6}C$ showing the relative position and numbers of its subatomic particles.

27. How many moles of aspirin, $C_9H_8O_4$, are in a bottle containing 450 g by weight? (*Note:* The approximate atomic weights of its atoms are C = 12, H = 1, and O = 16.)

28. Given the following types of atoms, decide which type of bonding, ionic or covalent, is most likely to occur: (a) two oxygen atoms; (b) four hydrogen atoms and one carbon atom; (c) a potassium atom ($^{39}_{19}K$) and a fluorine atom ($^{19}_{9}F$).

29. What are hydrogen bonds and how are they important in the body?

30. The following equation, which represents the oxidative breakdown of glucose by body cells, is a reversible reaction.

glucose + oxygen → carbon dioxide + water + ATP

(a) How can you indicate that the reaction is reversible? (b) How can you indicate that the reaction is in chemical equilibrium? (c) Define chemical equilibrium.

31. (a) Differentiate clearly between primary, secondary, and tertiary protein structure. (b) What level of structure do most fibrous proteins achieve? (c) How do globular proteins relate to the structural levels?

32. Dehydration and hydrolysis reactions are essentially opposite reactions. How are they related to the synthesis and degradation (breakdown) of biological molecules?

33. Describe the mechanism of enzyme activity. In your discussion, explain how enzymes decrease activation energy requirements.

34. Explain the importance of molecular chaperones.

3

CELLS: THE LIVING UNITS

Overview of the Cellular Basis of Life (pp. 58–59)

1. Define cell.
2. List the three major regions of a generalized cell and indicate the function of each.

The Plasma Membrane: Structure (pp. 59–63)

3. Describe the chemical composition of the plasma membrane and relate it to membrane functions.
4. Compare the structure and function of tight junctions, desmosomes, and gap junctions.

The Plasma Membrane: Functions (pp. 63–79)

5. Relate plasma membrane structure to active and passive transport mechanisms. Differentiate between these transport processes relative to energy source, substances transported, direction, and mechanism.
6. Define membrane potential and explain how the resting membrane potential is maintained.
7. Describe the role of the glycocalyx when cells interact with their environment.
8. List several roles of membrane receptors.

The Cytoplasm (pp. 79–88)

9. Describe the composition of the cytosol; define inclusions and list several types.
10. Discuss the structure and function of mitochondria.

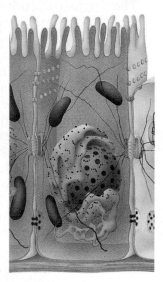

11. Discuss the structure and function of ribosomes, the endoplasmic reticulum, and the Golgi apparatus including functional interrelationships among these organelles.
12. Compare the functions of lysosomes and peroxisomes.
13. Name and describe the structure and function of cytoskeletal elements.
14. Describe the roles of centrioles in mitosis and in formation of cilia and flagella.

The Nucleus (pp. 88–91)

15. Outline the structure and function of the nuclear envelope, nucleolus, and chromatin.

Cell Growth and Reproduction (pp. 91–103)

16. List the phases of the cell life cycle and describe the key events of each phase.
17. Describe the process of DNA replication.
18. Define gene and genetic code and explain the function of genes.
19. Name the two phases of protein synthesis and describe the roles of DNA, mRNA, tRNA, and rRNA in each phase. Contrast triplets, codons, and anticodons.
20. Describe the importance of ubiquitin degradation of soluble proteins.

Extracellular Materials (p. 103)

21. Name and describe the composition of extracellular materials.

Just as bricks and timbers are the structural units of a house, **cells** are the structural units of all living things, from one-celled "generalists" like amoebas to complex multicellular organisms such as humans, dogs, and trees. The human body has 50 to 100 trillion of these tiny building blocks.

This chapter focuses on structures and functions shared by all cells. Specialized cells and their unique functions are addressed in later chapters.

Overview of the Cellular Basis of Life

The English scientist Robert Hooke first observed plant cells with a crude microscope in the late 1600s. However, it was not until the 1830s that two German scientists, Matthias Schleiden and Theodor Schwann, were bold enough to insist that all living things are composed of cells. The German pathologist Rudolf Virchow extended this idea by contending that cells arise only from other cells. Virchow's proclamation was revolutionary because it openly challenged the widely accepted *theory of spontaneous generation,* which held that organisms arise spontaneously from garbage or other nonliving matter. Since the late 1800s, cell research has been exceptionally fruitful and provided us with four concepts collectively known as the **cell theory:**

1. A cell is the basic structural and functional unit of living organisms. So when you define cell properties you are in fact defining the properties of life.

2. The activity of an organism depends on both the individual and the collective activities of its cells.

3. According to the **principle of complementarity,** the biochemical activities of cells are dictated by their specific subcellular structures.

4. Continuity of life has a cellular basis.

All of these concepts will be expanded as we progress. Let us begin with the idea that the cell is the smallest living unit. Whatever its form, however it behaves, the cell is the microscopic package that contains all the parts necessary to survive in an ever-changing world. It follows then that loss of cellular homeostasis underlies virtually every disease.

The trillions of cells in the human body include some 200 different cell types that vary greatly in shape, size, and function (Figure 3.1). The spherical fat cells, disc-shaped red blood cells, branching nerve cells, and cubelike cells of kidney tubules are just a few examples of the shapes cells take. Depending on type, cells also vary greatly in length—ranging from 2 micrometers (1/12,000 of an inch) in the smallest cells to over a meter in the nerve cells that cause you to wiggle your toes. A cell's shape reflects its func-

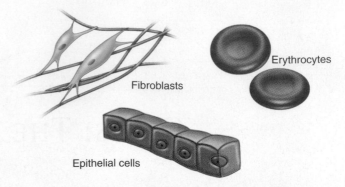

(a) Cells that connect body parts or cover and line organs

(b) Cells that move organs and body parts

(c) Cell that stores nutrients   **(d) Cell that fights disease**

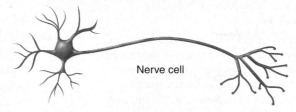

(e) Cell that gathers information and controls body functions

(f) Cell of reproduction

FIGURE 3.1   **Cell diversity.**

tion. For example, the flat, tilelike epithelial cells that line the inside of your cheek fit closely together, forming a living barrier that protects underlying tissues from bacterial invasion.

Regardless of type, all cells are composed chiefly of carbon, hydrogen, nitrogen, oxygen, and trace amounts of several other elements. In addition, all

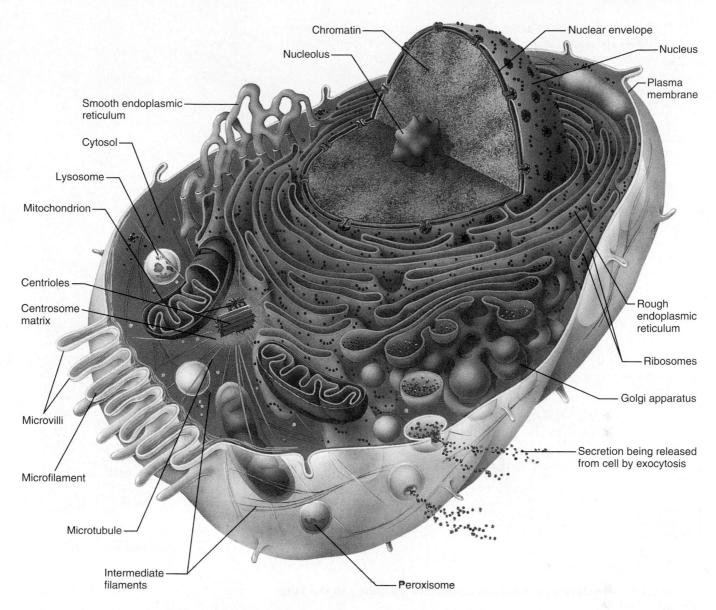

Chromatin
Nucleolus
Nuclear envelope
Nucleus
Plasma membrane
Smooth endoplasmic reticulum
Cytosol
Lysosome
Mitochondrion
Centrioles
Centrosome matrix
Microvilli
Microfilament
Microtubule
Intermediate filaments
Peroxisome
Rough endoplasmic reticulum
Ribosomes
Golgi apparatus
Secretion being released from cell by exocytosis

FIGURE 3.2 Structure of the generalized cell. No cell is exactly like this one, but this composite illustrates features common to many human cells. Note that not all of the organelles are drawn to the same scale in this illustration.

cells have the same basic parts and some common functions. Thus, it is possible to speak of a **generalized,** or **composite, cell** (Figure 3.2). Human cells have three main parts: the plasma membrane, the cytoplasm, and the nucleus. The *plasma membrane,* a fragile barrier, is the outer boundary of the cell. Internal to this membrane is the *cytoplasm* (si'to-plazm), the intracellular fluid that is packed with organelles, small structures that perform specific cell functions. The *nucleus* (nu'kle-us) controls cellular activities and lies near the cell's center. These cell structures are summarized in Table 3.1 (pp. 64–65) and are described next.

The Plasma Membrane: Structure

The flexible **plasma membrane** defines the extent of a cell, thereby separating two of the body's major fluid compartments—the *intracellular* fluid within cells and the *extracellular* fluid outside cells. The term *cell membrane* is commonly used as a synonym for plasma membrane, but because nearly all cellular organelles are enclosed in a membrane, we will always refer to the cell's surface or outer limiting membrane as the plasma membrane in this book. The plasma membrane is much more than a passive envelope. As you will see, its unique structure allows it to play a dynamic role in many cellular activities.

Some membrane proteins float freely in the lipid phase of the membrane. Others are secured in specific locations. According to this diagram, what would serve as the anchoring structures in the latter case?

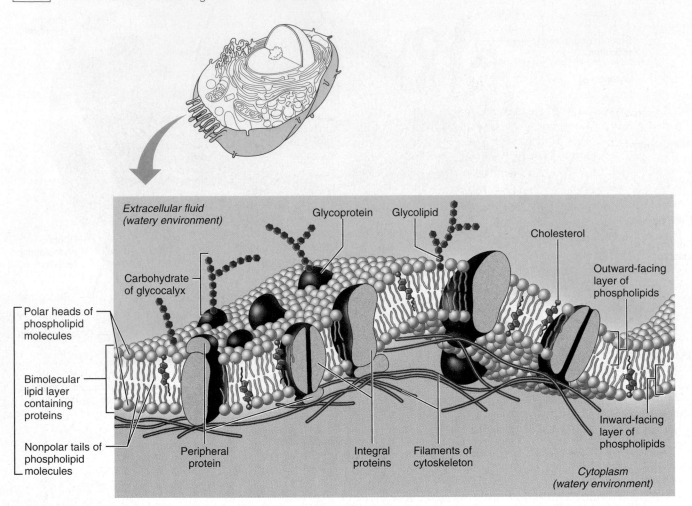

FIGURE 3.3   Structure of the plasma membrane according to the fluid mosaic model.

The Fluid Mosaic Model

The **fluid mosaic model** of membrane structure (Figure 3.3) depicts the plasma membrane as an exceedingly thin (7–10 nm) structure composed of a double layer, or bilayer, of lipid molecules with protein molecules dispersed in it. The proteins, many of which float in the fluid lipid *bilayer*, form a constantly changing mosaic pattern; hence the name of the model.

The lipid bilayer, which forms the basic "fabric" of the membrane, is constructed largely of *phospholipids*, with smaller amounts of *cholesterol* and *glycolipids*. Each lollipop-shaped phospholipid molecule has a polar "head" that is charged and **hydrophilic** (*hydro* = water, *philic* = loving), and an uncharged, nonpolar "tail" that is made of two fatty acid chains and is **hydrophobic** (*phobia* = hating). The polar heads are attracted to water—the main constituent of both the intracellular and extracellular fluids—and so they lie on both the inner and outer surfaces of the membrane. The nonpolar tails, being hydrophobic, avoid water and line up in the center of the membrane. The result is that all biological membranes share a common sandwich-like structure: They are composed of two parallel sheets of phospholipid molecules lying tail to tail, with their polar heads exposed to water inside and outside the cell. This self-orienting property of phospholipids encourages biological membranes to self-assemble into closed, generally spherical, structures and to reseal themselves quickly when torn. It also enables scientists to suck out and replace a cell's nucleus in *cloning experiments*, a hot scientific area today.

■ *Filaments of the cytoskeleton.*

The inward-facing and outward-facing surfaces of the plasma membrane differ in the kinds and amounts of lipids they contain. The majority of membrane phospholipids are unsaturated (like phosphatidyl choline), a condition which kinks their tails (increasing the space between them) and increases membrane fluidity. **Glycolipids** (gli″ko-lip′idz), phospholipids with attached sugar groups, are found only on the outer plasma membrane surface and account for about 5% of the total membrane lipid. Their sugar groups, like the phosphate-containing groups of phospholipids, make that end of the glycolipid molecule polar, whereas the fatty acid tails are nonpolar. Some 20% of membrane lipid is cholesterol, which stabilizes the lipid membrane by wedging its platelike hydrocarbon rings between the phospholipid tails and restraining movement of the phospholipids.

About 20% of the outer membrane surface contains **lipid rafts,** dynamic assemblies of saturated phospholipids (which pack together tightly) associated with unique lipids called sphingolipids and lots of cholesterol. These quiltlike patches are more stable and orderly and less fluid than the rest of the membrane, and can include or exclude specific proteins to various extents. Because of these qualities, lipid rafts are assumed to be concentrating platforms for molecules needed for cell signaling.

There are two distinct populations of membrane proteins, integral and peripheral (see Figure 3.3). Proteins make up about half of the plasma membrane by mass and are responsible for most of the specialized membrane functions (Figure 3.4). **Integral proteins** are firmly inserted into the lipid bilayer. Some protrude from one membrane face only, but most are *transmembrane proteins* that span the entire width of the membrane and protrude on both sides. Whether transmembrane or not, all integral proteins have both hydrophobic and hydrophilic regions. This structural feature allows them to interact both with the nonpolar lipid tails buried in the membrane and with water inside and outside the cell. Transmembrane proteins are mainly involved in transport. Some cluster together to form *channels*, or pores, through which small, water-soluble molecules or ions can move, thus bypassing the lipid part of the membrane. Others act as *carriers* that bind to a substance and then move it through the membrane. Still others are receptors for hormones or other chemical messengers and relay messages to the cell interior (a process called *signal transduction*).

FIGURE 3.4 Some functions of membrane proteins. A single protein may perform some combination of these tasks.

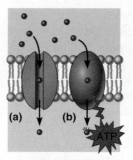

Transport
(a) A protein that spans the membrane may provide a hydrophilic channel across the membrane that is selective for a particular solute. **(b)** Some transport proteins hydrolyze ATP as an energy source to actively pump substances across the membrane.

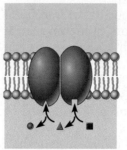

Enzymatic activity
A protein built into the membrane may be an enzyme with its active site exposed to substances in the adjacent solution. In some cases, several enzymes in a membrane act as a team that catalyzes sequential steps of a metabolic pathway as indicated (right to left) here.

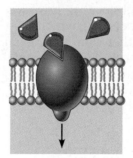

Receptors for signal transduction
A membrane protein exposed to the outside of the cell may have a binding site with a specific shape that fits the shape of a chemical messenger, such as a hormone. The external signal may cause a conformational change in the protein that initiates a chain of chemical reactions in the cell.

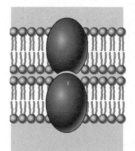

Intercellular joining
Membrane proteins of adjacent cells may be hooked together in various kinds of intercellular junctions. Some membrane proteins (CAMs) of this group provide temporary binding sites that guide cell migration and other cell-to-cell interactions.

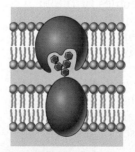

Cell-cell recognition
Some glycoproteins (proteins bonded to short chains of sugars) serve as identification tags that are specifically recognized by other cells.

Attachment to the cytoskeleton and extracellular matrix (ECM)
Elements of the cytoskeleton (cell's internal supports) and the extracellular matrix (ECM) may be anchored to membrane proteins, which help maintain cell shape and fix the location of certain membrane proteins. Others play a role in cell movement or bind adjacent cells together.

Peripheral proteins, in contrast, are not embedded in the lipid. Instead, they attach rather loosely to integral proteins or membrane lipids and are easily removed without disrupting the membrane. Peripheral proteins include a network of filaments that helps support the membrane from its cytoplasmic side. Some peripheral proteins are enzymes. Others are involved in mechanical functions, such as changing cell shape during cell division and muscle cell contraction, or linking cells together.

Many of the proteins that abut the extracellular space are glycoproteins that sport branching sugar groups. The term **glycocalyx** (gli″ko-kal′iks; "sugar covering") is used to describe the fuzzy, sticky carbohydrate-rich area at the cell surface. You can think of your cells as sugar-coated. The glycocalyx is enriched both by glycolipids and by glycoproteins secreted by the cells that cling to its surface.

Because every cell type has a different pattern of sugars in its glycocalyx, the glycocalyx provides highly specific biological markers by which approaching cells recognize each other. For example, a sperm recognizes an ovum (egg cell) by the ovum's unique glycocalyx, and cells of the immune system identify a bacterium by binding to certain membrane glycoproteins in the bacterial glycocalyx.

(H) HOMEOSTATIC IMBALANCE

Definite changes in the glycocalyx occur in a cell that is becoming cancerous. In fact, a cancer cell's glycocalyx may change almost continuously, allowing it to keep ahead of immune system recognition mechanisms and avoid destruction. ●

The plasma membrane is a dynamic fluid structure that is in constant flux. Its consistency is like that of olive oil. The lipid molecules of the bilayer move freely from side to side, parallel to the membrane surface, but their polar-nonpolar interactions prevent them from flip-flopping or moving from one leaflet (half of the bilayer) to the other. Some of the proteins float freely. Others, particularly the peripheral proteins, are restricted in their movements because they are "tethered" to intercellular structures that make up the *cytoskeleton* (see Figure 3.4).

Specializations of the Plasma Membrane

Microvilli

Microvilli (mi″kro-vil′i; "little shaggy hairs") are minute, fingerlike extensions of the plasma membrane that project from a free, or exposed, cell surface (Figures 3.2 and 3.5). They increase the plasma membrane surface area tremendously and are most often found on the surface of absorptive cells such as intestinal and kidney tubule cells. Microvilli have a core of actin filaments. Actin is a contractile protein, but in microvilli it appears to function as a mechanical "stiffener."

Membrane Junctions

Although certain cell types—blood cells, sperm cells, and some phagocytic cells—are "footloose" in the body, many other types, particularly epithelial cells, are knit into tight communities. Typically, three factors act to bind cells together:

1. Glycoproteins in the glycocalyx act as an adhesive.

2. Wavy contours of the membranes of adjacent cells fit together in a tongue-and-groove fashion.

3. Special membrane junctions are formed (Figure 3.5). Because this last factor is the most important let us look more closely at the various types of junctions.

Tight Junctions   In a **tight junction,** a series of integral protein molecules in the plasma membranes of adjacent cells fuse together, forming an *impermeable junction* that encircles the cell (Figure 3.5). Tight junctions help prevent molecules from passing through the extracellular space between adjacent cells. For example, tight junctions between epithelial cells lining the digestive tract keep digestive enzymes and microorganisms in the intestine from seeping into the bloodstream. (Although called "impermeable" junctions, some tight junctions are somewhat leaky and may allow certain types of ions to pass.)

Desmosomes   Desmosomes (des′mo-sōmz; "binding bodies") are *anchoring junctions*—mechanical couplings scattered like rivets along the sides of abutting cells that prevent their separation. On the cytoplasmic face of each plasma membrane is a buttonlike thickening called a *plaque.* Adjacent cells are held together by thin linker protein filaments (cadherins) that extend from the plaques and interdigitate like the teeth of a zipper in the intercellular space. Thicker protein filaments (intermediate filaments, which form part of the cytoskeleton) extend from the cytoplasmic side of the plaque across the width of the cell to anchor to the plaque on the cell's opposite side. Thus, desmosomes not only bind neighboring cells together, they also contribute to a continuous internal network of strong "guy-wires." This arrangement distributes tension throughout a cellular sheet and reduces the chance of tearing when it is subjected to pulling forces. Desmosomes are abundant in tissues subjected to great mechanical stress, such as skin and heart muscle.

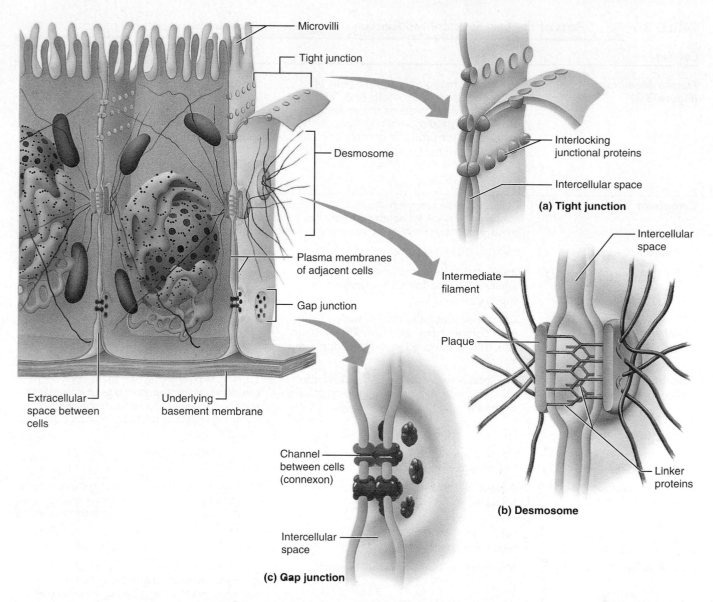

FIGURE 3.5 **Cell junctions.** An epithelial cell is shown joined to adjacent cells by three common types of cell junctions: tight junctions, desmosomes, and gap junctions. (Note: Except for epithelia, it is unlikely that a single cell will have all three junction types.)

Gap Junctions   A **gap junction**, or *nexus* (nek'-sus; "bond"), is a communicating junction that allows chemical substances to pass between adjacent cells. At gap junctions the adjacent plasma membranes are very close, and the cells are connected by hollow cylinders called *connexons* (kŏ-nek'sonz), composed of transmembrane proteins. The many different types of connexon proteins vary the selectivity of the gap junction channels. Ions, simple sugars, and other small molecules pass through these water-filled channels from one cell to the next (Figure 3.5). Gap junctions are present in electrically excitable tissues, such as the heart and smooth muscle, where ion passage from cell to cell helps synchronize their electrical activity and contraction.

The Plasma Membrane: Functions
Membrane Transport

Our cells are bathed in an extracellular fluid called **interstitial fluid** (in"ter-stish'al) that is derived from the blood. Interstitial fluid is like a rich, nutritious "soup." It contains thousands of ingredients, including amino acids, sugars, fatty acids, vitamins, regulatory substances such as hormones and neurotransmitters, salts, and waste products. To remain healthy, each cell must extract from this mix the exact amounts of the substances it needs at specific times.

TABLE 3.1 **Parts of the Cell: Structure and Function**

Cell Part	Structure	Functions
Plasma Membrane *(Figure 3.3)*	Membrane made of a double layer of lipids (phospholipids, cholesterol, and so on) within which proteins are embedded; proteins may extend entirely through the lipid bilayer or protrude on only one face; externally facing proteins and some lipids have attached sugar groups	Serves as an external cell barrier; acts in transport of substances into or out of the cell; maintains a resting potential that is essential for functioning of excitable cells; externally facing proteins act as receptors (for hormones, neurotransmitters, and so on) and in cell-to-cell recognition
Cytoplasm	Cellular region between the nuclear and plasma membranes; consists of fluid **cytosol**, containing dissolved solutes, **organelles** (the metabolic machinery of the cytoplasm), and **inclusions** (stored nutrients, secretory products, pigment granules)	
Cytoplasmic organelles		
▪ Mitochondria (Figure 3.17)	Rodlike, double-membrane structures; inner membrane folded into projections called cristae	Site of ATP synthesis; powerhouse of the cell
▪ Ribosomes (Figures 3.18, 3.19)	Dense particles consisting of two subunits, each composed of ribosomal RNA and protein; free or attached to rough endoplasmic reticulum	The sites of protein synthesis
▪ Rough endoplasmic reticulum (Figures 3.18, 3.19)	Membrane system enclosing a cavity, the cisterna, and coiling through the cytoplasm; externally studded with ribosomes	Sugar groups are attached to proteins within the cisternae; proteins are bound in vesicles for transport to the Golgi apparatus and other sites; external face synthesizes phospholipids and cholesterol
▪ Smooth endoplasmic reticulum (Figure 3.18)	Membranous system of sacs and tubules; free of ribosomes	Site of lipid and steroid synthesis, lipid metabolism, and drug detoxification
▪ Golgi apparatus (Figures 3.20, 3.21)	A stack of smooth membrane sacs and associated vesicles close to the nucleus	Packages, modifies, and segregates proteins for secretion from the cell, inclusion in lysosomes, and incorporation into the plasma membrane
▪ Lysosomes (Figure 3.22)	Membranous sacs containing acid hydrolases	Sites of intracellular digestion
▪ Peroxisomes (Figure 3.2)	Membranous sacs of oxidase enzymes	The enzymes detoxify a number of toxic substances; the most important enzyme, catalase, breaks down hydrogen peroxide
▪ Microtubules (Figures 3.24–3.27)	Cylindrical structures made of tubulin proteins	Support the cell and give it shape; involved in intracellular and cellular movements; form centrioles
▪ Microfilaments (Figures 3.24, 3.25)	Fine filaments of the contractile protein actin	Involved in muscle contraction and other types of intracellular movement; help form the cell's cytoskeleton, cilia, and flagella, if present
▪ Intermediate filaments (Figure 3.24)	Protein fibers; composition varies	The stable cytoskeletal elements; resist mechanical forces acting on the cell
▪ Centrioles (Figure 3.26)	Paired cylindrical bodies, each composed of nine triplets of microtubules	Organize a microtubule network during mitosis to form the spindle and asters; form the bases of cilia and flagella
Cellular Extensions		
▪ Cilia (Figure 3.27)	Short, cell surface projections; each cilium composed of nine pairs of microtubules surrounding a central pair	Move in unison, creating a unidirectional current that propels substances across cell surfaces
▪ Flagella	Like cilium, but longer; only example in humans is the sperm tail	Propels the cell

TABLE 3.1 *(continued)*

Cell Part	Structure	Functions
Inclusions	Varied; includes stored nutrients such as lipid droplets and glycogen granules, protein crystals, pigment granules	Storage for nutrients, wastes, and cell products
Nucleus (Figure 3.28)	Largest organelle; surrounded by the nuclear envelope; contains fluid nucleoplasm, nucleoli, and chromatin	Control center of the cell; responsible for transmitting genetic information and providing the instructions for protein synthesis
▪ Nuclear envelope (Figure 3.28)	Double-membrane structure; pierced by the pores; outer membrane continuous with the endoplasmic reticulum	Separates the nucleoplasm from the cytoplasm and regulates passage of substances to and from the nucleus
▪ Nucleoli (Figure 3.28)	Dense spherical (non-membrane-bounded) bodies; composed of ribosomal RNA and proteins	Site of ribosome subunit manufacture
▪ Chromatin (Figures 3.28, 3.29)	Granular, threadlike material composed of DNA and histone proteins	DNA constitutes the genes

Although there is continuous traffic across the plasma membrane, it is a **selectively,** or **differentially, permeable** barrier, meaning that it allows some substances to pass while excluding others. Thus, it allows nutrients to enter the cell, but keeps many undesirable substances out. At the same time, it keeps valuable cell proteins and other substances in the cell, but allows wastes to exit.

🄗 HOMEOSTATIC IMBALANCE

Selective permeability is a characteristic of healthy, intact cells. When a cell (or its plasma membrane) is severely damaged, the membrane becomes permeable to virtually everything, and substances flow into and out of the cell freely. This phenomenon is evident when someone has been severely burned. Precious fluids, proteins, and ions "weep" from the dead and damaged cells. ●

Substances move through the plasma membrane in essentially two ways—passively or actively. In **passive processes,** substances cross the membrane without any energy input from the cell. In **active processes,** the cell provides the metabolic energy (ATP) needed to move substances across the membrane. The various transport processes that occur in cells are summarized in Table 3.2 on pages 75–76. Let's examine each of these types of membrane transport.

Passive Processes

The two main types of passive transport in cells are *diffusion* (di-fu'zhun) and *filtration*. Diffusion is an important means of passive membrane transport for every cell of the body. By contrast, filtration generally occurs only across capillary walls.

Diffusion Diffusion is the tendency of molecules or ions to scatter evenly throughout the environment (Figure 3.6). Recall that all molecules possess kinetic energy and are in constant motion (see p. 23). As molecules move about randomly at high speeds, they collide and ricochet off one another, changing direction with each collision. The overall effect of this erratic movement is that molecules move away from areas where they are in higher concentration to areas where their concentration is lower, so we say that molecules diffuse *along*, or

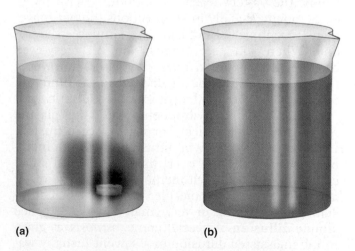

(a)   **(b)**

FIGURE 3.6 **Diffusion.** Molecules in solution move continuously and collide constantly with other molecules. As a result, molecules tend to move away from areas of their highest concentration and become evenly distributed. **(a)** Diagram of dye molecules diffusing from a dye pellet into the surrounding water. **(b)** Dye molecules from pellet have diffused evenly throughout the water.

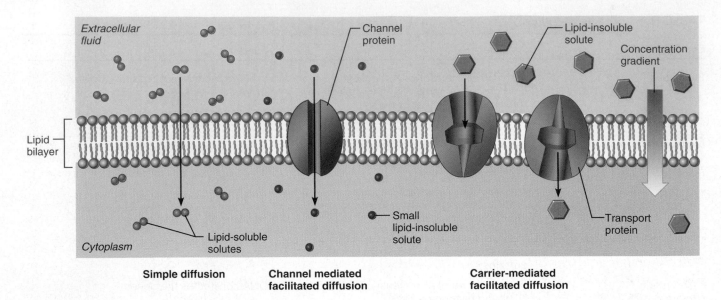

FIGURE 3.7   **Diffusion through the plasma membrane.** In simple diffusion, depicted on the left, fat-soluble molecules diffuse directly through the lipid bilayer of the plasma membrane, in which they can dissolve. In facilitated diffusion via membrane channels, shown next, small polar or charged particles (water molecules or small ions) diffuse through membrane channels constructed by channel proteins. In facilitated diffusion using protein carriers, shown on the right, large, lipid-insoluble molecules (e.g., glucose) are moved across the membrane.

down, their **concentration gradient.** The greater the difference in concentration between the two areas, the faster the net diffusion of the particles.

Because the driving force for diffusion is the kinetic energy of the molecules themselves, the speed of diffusion is influenced by molecular *size* (the smaller, the faster) and by *temperature* (the warmer, the faster). In a closed container, diffusion eventually produces a uniform mixture of molecules. In other words, the system reaches equilibrium, with molecules moving equally in all directions (no *net* movement). Examples of diffusion are familiar to everyone. When you peel onions, you get teary-eyed because the cut onion releases volatile substances that diffuse through the air, dissolve in the fluid film covering your eyes, and form irritating sulfuric acid.

Because of its hydrophobic core, the plasma membrane is a physical barrier to free diffusion. However, a molecule *will* diffuse through the membrane if the molecule is (1) lipid soluble, (2) small enough to pass through membrane channels, or (3) assisted by a carrier molecule. The unassisted diffusion of lipid-soluble or very small particles is called **simple diffusion.** A special name, *osmosis*, is given to the unassisted diffusion of a solvent (usually water) through a membrane. Assisted diffusion is known as *facilitated diffusion.*

1. Simple diffusion. Nonpolar and lipid-soluble substances diffuse directly through the lipid bilayer (Figure 3.7, far left). Such substances include oxygen, carbon dioxide, fat-soluble vitamins, and alcohol. Because oxygen concentration is always higher in the blood than in tissue cells, oxygen continuously diffuses from the blood into the cells, whereas carbon dioxide (in higher concentration within the cells) diffuses from tissue cells into the blood.

2. Facilitated diffusion. Certain molecules, notably glucose and other sugars, amino acids, and ions are transported passively even though they are unable to pass through the lipid bilayer. Instead they move through the membrane by a passive transport process called **facilitated diffusion** in which the transported substance either (1) binds to protein carriers in the membrane and is ferried across or (2) moves through water-filled protein channels.

▪ *Carriers.* A carrier is a transmembrane integral protein (sometimes called a **permease**) that shows specificity for molecules of a certain polar substance or class of substances that are too large to pass through membrane channels, such as sugars and amino acids. Although it was initially believed that the integral proteins that act as carriers either flip-flopped or physically crossed the membrane like ferryboats, the most popular model for this process (Figure 3.7, far right) indicates that changes in the shape of the carrier allow it to first envelop and then release the transported substance, shielding it en route from the nonpolar regions of the membrane. Essentially, the binding site is moved from one face of the membrane to the other by changes in the conformation of the carrier protein.

Just as in any simple diffusion process, a substance transported by carrier-mediated facilitated diffusion, like glucose, moves down its concentration gradient. Glucose is normally in higher concentrations in the blood than in the cells, where it is rapidly used for ATP synthesis. So, within the body glucose transport is *typically* unidirectional—into the cells. However, carrier-mediated transport is *limited* by the number of receptors present. For example, when all the glucose carriers are "engaged," they are said to be *saturated,* and glucose transport is occurring at its maximum rate.

■ *Channels.* Channels are transmembrane proteins that serve to transport substances, usually ions or water, through aqueous channels from one side of the membrane to the other (Figure 3.7). Some channels are always open and simply allow ion or water fluxes according to concentration gradients. In other cases, binding or association sites exist within the channel and the channel is selective due to pore size and the charges of the amino acids lining the channel. Additionally, some channels are always open (the so-called leakage channels), and others are gated and controlled (opened or closed) by various chemical or electrical signals.

Like carriers, many channels can be inhibited by certain molecules, show saturation, and tend to be specific. They also honor the concentration gradient (always moving down the gradient). When a substance crosses the membrane by simple diffusion, the rate of diffusion is not controllable because the lipid solubility of the membrane is not immediately changeable. By contrast, the rate of facilitated diffusion *is* controllable because the permeability of the membrane can be altered by regulating the activity or number of individual carriers or channels.

Oxygen, water, glucose, and various ions are vitally important to cellular homeostasis. Thus, their passive transport by diffusion (either simple or facilitated) represents a tremendous saving of cellular energy. Indeed, if these substances had to be transported actively, cell expenditures of ATP would increase exponentially!

3. Osmosis The diffusion of a solvent, such as water, through a selectively permeable membrane is **osmosis** (oz-mo'sis; *osmos* = pushing). Even though water is highly polar, it passes via osmosis through the lipid bilayer. This is surprising because you'd expect it to be repelled by the hydrophobic lipid tails. Although still hypothetical, one explanation is that random movements of the membrane lipids open small gaps between their wiggling tails, allowing water to slip and slide its way through the membrane by moving from gap to gap. Water also moves freely and reversibly through water-specific channels constructed by transmembrane proteins called **aquaporins (AQP).** Although aquaporins are believed to be present in all cell types, they are particularly abundant in red blood cells and in cells involved in body-wide or organ water balance such as kidney tubule cells.

Osmosis occurs whenever the water concentration differs on the two sides of a membrane. If only distilled water is present on both sides of a selectively permeable membrane, no *net* osmosis occurs, even though water molecules move in both directions through the membrane. If, however, the solute concentration on the two sides of the membrane differs, water concentration differs as well (as solute concentration increases, water concentration decreases). The extent to which water's concentration is decreased by solutes depends on the *number,* not the *type,* of solute particles, because one molecule or one ion of solute (theoretically) displaces one water molecule. The total concentration of all solute particles in a solution is referred to as the solution's **osmolarity** (oz"mo-lar'ĭ-te). When equal volumes of aqueous solutions of different osmolarity are separated by a membrane that is *permeable to all molecules* in the system, net diffusion of both solute and water occurs, each moving down its own concentration gradient (Figure 3.8a). Eventually, equilibrium is reached when the water concentration on the left equals that on the right, and the solute concentration on both sides is the same. If we consider the same system, but make the membrane *impermeable to solute molecules,* we see quite a different result (Figure 3.8b). Water quickly diffuses from the left to the right compartment and continues to do so until its concentration is the same on the two sides of the membrane. Notice that in this case equilibrium results from the movement of water alone (the solutes are prevented from moving). Notice also that the movement of water leads to dramatic changes in the volumes of the two compartments.

The last example mimics osmosis across plasma membranes of living cells, with one major difference. In our examples, the volumes of the compartments are infinitely expandable and the effect of pressure exerted by the added weight of the higher fluid column is not considered. In living plant cells, which have rigid cell walls external to their plasma membranes, this is not the case. As water diffuses into the cell, the point is finally reached where the **hydrostatic pressure** (the back pressure exerted by water against the membrane) within the cell is equal to its **osmotic pressure**—the cell's tendency to resist further (net) water entry. As a rule, the higher the amount of nondiffusible (or nonpenetrating) solutes in a cell, the higher the osmotic pressure and the greater the hydrostatic pressure that must be developed to resist further net water entry.

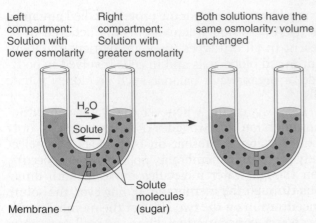

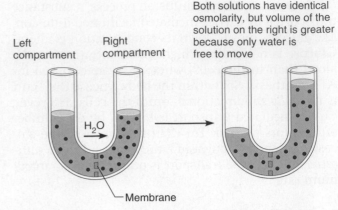

Left compartment: Solution with lower osmolarity

Right compartment: Solution with greater osmolarity

Both solutions have the same osmolarity: volume unchanged

(a) **Membrane permeable to both solute molecules and water**

Left compartment

Right compartment

Both solutions have identical osmolarity, but volume of the solution on the right is greater because only water is free to move

(b) **Membrane impermeable to solute molecules, permeable to water**

FIGURE 3.8 **Influence of membrane permeability on diffusion and osmosis. (a)** In this system, the membrane is permeable to both water and solute (sugar) molecules. Water moves from the solution with lower osmolarity (left compartment) to the solution with greater osmolarity (right compartment). The solute moves along its own concentration gradient in the opposite direction. When the system comes to equilibrium (right), the solutions have the same osmolarity and volume. **(b)** This system is identical to that in (a) except that the membrane is impermeable to the solute. Water moves by osmosis from the left to the right compartment, until its concentration and that of the solutions are identical. Because the solute is prevented from moving, the volume of the solution in the right compartment increases.

However, such major changes in hydrostatic (and osmotic) pressures do not occur in living animal cells, which lack rigid cell walls. Osmotic imbalances cause animal cells to swell or shrink (due to net water gain or loss) until either the solute concentration is the same on both sides of the plasma membrane, or the membrane is stretched to its breaking point.

This leads us to the important concept of *tonicity* (to-nis′ĭ-te). As noted, many molecules, particularly intracellular proteins and selected ions, are prevented from diffusing through the plasma membrane. Consequently, any change in their concentration alters the water concentration on the two sides of the membrane and results in a net loss or gain of water by the cell.

The ability of a solution to change the shape or tone of cells by altering their internal water volume is called **tonicity** (*tono* = tension). Solutions with the same concentrations of nonpenetrating solutes as those found in cells (0.9% saline or 5% glucose) are **isotonic** ("the same tonicity"). Cells exposed to such solutions retain their normal shape, and exhibit no net loss or gain of water (Figure 3.9a). As you might expect, the body's extracellular fluids and most intravenous solutions (solutions infused into the body via a vein) are isotonic. Solutions with a higher concentration of nonpenetrating solutes than is seen in the cell (for example, a strong saline solution) are **hypertonic.** Cells immersed in hypertonic solutions lose water and shrink, or *crenate* (kre′nāt) (Figure 3.9b). Solutions that are more dilute (contain a lower concentration of nonpenetrating solutes) than cells are called **hypotonic.** Cells placed in a hypotonic solution plump up rapidly as water rushes into them (Figure 3.9c). Distilled water represents the most extreme example of hypotonicity. Because it contains *no* solutes, water continues to enter cells until they finally burst or *lyse.*

Notice that osmolarity and tonicity are not the same thing. A solution's osmolarity is based solely on its total solute concentration; its tonicity is based on how the solution affects cell volume, which depends on (1) solute concentration and (2) solute permeability of the plasma membrane. Osmolarity is expressed as osmoles per liter (osmol/L) where 1 osmol is equal to 1 mole of nonionizing molecules.* A 0.3-osmol/L solution of NaCl is isotonic because sodium ions are usually prevented from diffusing through the plasma membrane. But if the cell is immersed in a 0.3-osmol/L solution of a penetrating solute, the solute will enter the cell and water will follow. The cell will swell and burst, just as if it had been placed in pure water.

---

* Osmolarity (Osm) is determined by multiplying molarity (moles per liter, or *M*) by the number of particles resulting from ionization. For example, since NaCl ionizes to $Na^+ + Cl^-$, a 1-*M* solution of NaCl is a 2-Osm solution. For substances that do not ionize (e.g., glucose), molarity and osmolarity are the same.

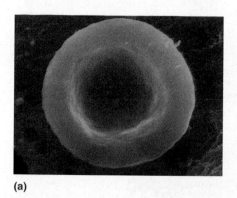
(a)

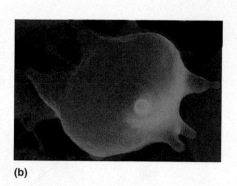
(b)

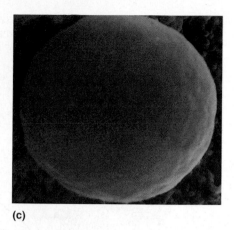
(c)

FIGURE 3.9 **The effect of solutions of varying tonicities on living red blood cells. (a)** In isotonic solutions (same solute/water concentrations as inside cells), cells retain their normal size and shape. **(b)** In a hypertonic solution (containing more solutes than are present inside the cells), the cells lose water and shrink (crenate). **(c)** In a hypotonic solution (containing fewer solutes than are present inside the cell), the cells take in water by osmosis until they become bloated and burst (lyse).

Osmosis is extremely important in determining distribution of water in the various fluid-containing compartments of the body (in cells, in blood, and so on). In general, osmosis continues until osmotic and hydrostatic pressures acting at the membrane are equal. For example, water is forced out of capillary blood by the hydrostatic pressure of the blood against the capillary wall, but the presence in blood of solutes that are too large to cross the capillary membrane draws water back into the bloodstream. As a result, very little net loss of plasma fluid occurs.

Simple diffusion and osmosis are not selective processes. In those processes, whether a molecule can pass through the membrane depends chiefly on its size or its solubility in lipid, not on its unique structure. Facilitated diffusion, on the other hand, *is* often highly selective. The carrier for glucose, for example, combines specifically with glucose, in much the same way an enzyme binds to its specific substrate and ion channels allow only selected ions to pass.

Ⓗ HOMEOSTATIC IMBALANCE

Hypertonic solutions are sometimes infused intravenously into the bloodstream of edematous patients (those swollen because water is retained in their tissues) to draw excess water out of the extracellular space and move it into the bloodstream so that it can be eliminated by the kidneys. Hypotonic solutions may be used (with care) to rehydrate the tissues of extremely dehydrated patients. In less extreme cases of dehydration, drinking hypotonic fluids (colas, apple juice, and sports drinks) usually does the trick. ●

Filtration **Filtration** is the process that forces water and solutes through a membrane or capillary wall by *fluid,* or *hydrostatic,* pressure. Like diffusion, filtration is a passive transport process and involves a gradient. However, the gradient for filtration is a **pressure gradient** that pushes solute-containing fluid (filtrate) from a higher-pressure area to a lower-pressure area. As already mentioned, in the body hydrostatic pressure exerted by the blood forces fluid out of the capillaries, and this fluid contains solutes that are vital to the tissues. Filtration also provides the fluid ultimately excreted by the kidneys as urine. Filtration is not selective; only blood cells and protein molecules too large to pass through membrane pores or the paracellular path (between cells) are held back.

Active Processes

Whenever a cell uses the bond energy of ATP to move solutes across the membrane, the process is referred to as *active.* Substances moved actively across the plasma membrane are usually unable to pass in the necessary direction by any passive transport processes. The substance may be too large to pass through the channels, incapable of dissolving in the lipid bilayer, or unable to move down its concentration gradient. There are two major mechanisms of active membrane transport: active transport and vesicular transport.

Active Transport **Active transport** is similar to facilitated diffusion in that both require carrier proteins that combine *specifically* and *reversibly* with the transported substances. However, facilitated diffusion always honors concentration gradients

(1) Which ions are being pumped against their concentration gradient?
(2) Which ions are being pumped against their electrical gradient?

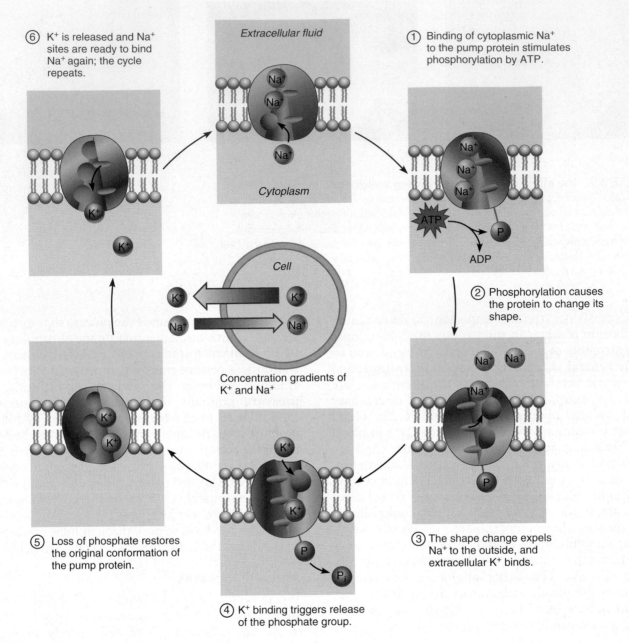

⑥ K⁺ is released and Na⁺ sites are ready to bind Na⁺ again; the cycle repeats.

Extracellular fluid

Cytoplasm

① Binding of cytoplasmic Na⁺ to the pump protein stimulates phosphorylation by ATP.

ATP

ADP

② Phosphorylation causes the protein to change its shape.

Cell

Concentration gradients of K⁺ and Na⁺

③ The shape change expels Na⁺ to the outside, and extracellular K⁺ binds.

⑤ Loss of phosphate restores the original conformation of the pump protein.

④ K⁺ binding triggers release of the phosphate group.

FIGURE 3.10 **Operation of the sodium-potassium pump, an antiport pump (Na⁺-K⁺ ATPase).** Hydrolysis of a molecule of ATP provides energy for an antiport "pump" protein to move three sodium ions out of the cell and two potassium ions into the cell. Both ions are moved against their concentration gradients, indicated in the center diagram by colored arrows moving through the membrane (yellow arrow = Na⁺ gradient; green arrow = K⁺ gradient).

(1) Both kinds of ions are being pumped against their concentration gradient. (2) Sodium ions are being pumped against their electrical gradient. ∎

because its driving force is kinetic energy. In contrast, the active transporters or **solute pumps** move solutes, most importantly ions (such as Na^+, K^+, and Ca^{2+}), "uphill" *against* a concentration gradient. To do this work, cells must expend the energy of ATP.

Active transport processes are distinguished according to their source of energy. In *primary active transport*, the energy to do work comes directly from hydrolysis of ATP. In *secondary active transport*, transport is driven indirectly by energy stored in ionic gradients created by operation of primary active transport pumps. Secondary active transport systems are all *coupled systems*; that is, they move more than one substance at a time. If the two transported substances are moved in the same direction, the system is a **symport system** (*sym* = same). If the transported substances "wave to each other" as they cross the membrane in opposite directions, the system is an **antiport system** (*anti* = opposite, against). Let's examine these processes more carefully.

1. Primary active transport. In primary active transport, hydrolysis of ATP results in the phosphorylation of the transport protein, a step that causes the protein to change its conformation in such a manner that it "pumps" the bound solute across the membrane.

The most investigated example of a primary active transport system is the operation of the **sodium-potassium pump** (Figure 3.10), for which the carrier is an enzyme called **Na^+-K^+ ATPase.** The concentration of K^+ inside the cell is 30–50 times higher than that outside, and the reverse is true of Na^+. These ionic concentration differences are essential for excitable cells like muscle and nerve cells to function normally and for all body cells to maintain their normal fluid volume. Because Na^+ and K^+ leak slowly but continuously through channels in the plasma membrane along their concentration gradient (and cross more rapidly in stimulated muscle and nerve cells), the Na^+-K^+ pump operates more or less continuously as an antiport to simultaneously drive Na^+ out of the cell against a steep concentration gradient and pump K^+ back in. Other examples of primary active transport include the calcium pumps that actively sequester ionic calcium from the intracellular fluid into specific organelles or eject it from the cell.

2. Secondary active transport. A single ATP-powered pump, such as the Na^+-K^+ pump, can indirectly drive the secondary active transport of several other solutes (Figure 3.11). By moving sodium across the plasma membrane against its concentration gradient, the pump stores energy (in the ion gradient). Then, just as water pumped uphill can do work as it flows back down (turn a turbine or water

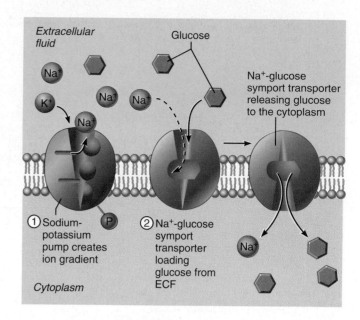

FIGURE 3.11 Secondary active transport. (Sequence of events moves from left to right.) In this example, ① the ATP-driven Na^+-K^+ pump stores energy by creating a steep concentration gradient for Na^+ entry into the cell. ② As Na^+ diffuses back across the membrane through a membrane cotransporter protein, it drives glucose against its concentration gradient into the cell.

wheel), a substance pumped across a membrane can do work as it leaks back, propelled downhill along its concentration gradient. Thus, as sodium moves back into the cell with the help of a carrier protein (facilitated diffusion), other substances are "dragged along" or cotransported by a common carrier protein. For example, some sugars, amino acids, and many ions are cotransported in this way into cells lining the small intestine. Though the cotransported substances both move passively, Na^+ has to be pumped back into the lumen of the intestine to maintain its diffusion gradient. Ion gradients can also be used to drive antiport systems such as those that help to regulate intracellular pH by using the sodium gradient to expel hydrogen ions.

Regardless of whether the energy is provided directly (primary active transport) or indirectly (secondary active transport), each membrane pump or cotransporter transports only specific substances. Hence, active transport systems provide a way for the cell to be very selective in cases where substances cannot pass by diffusion. (No pump—no transport.)

Vesicular Transport Large particles, macromolecules, and fluids are transported across plasma and intracellular membranes by **vesicular transport.** Vesicular transport is the mechanism used for: **exocytosis** (ek"so-si-to'sis; "out of the cell"), moving substances from the cell interior to the extracellular

Phagocytic cells gather in the air sacs of the lungs, especially in the lungs of smokers. What is the connection?

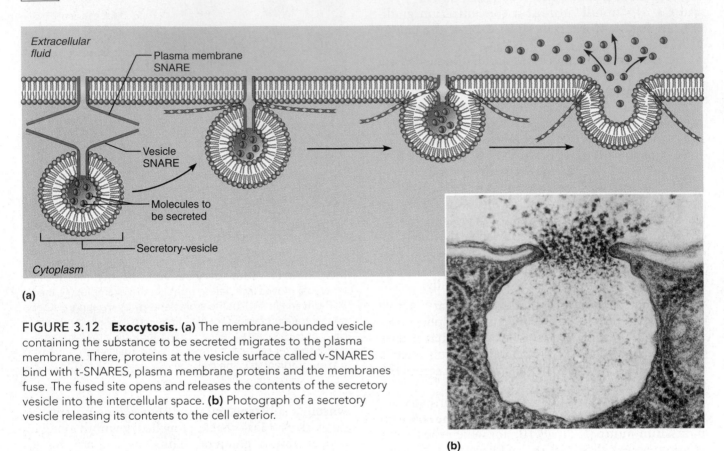

(a)

FIGURE 3.12   Exocytosis. (a) The membrane-bounded vesicle containing the substance to be secreted migrates to the plasma membrane. There, proteins at the vesicle surface called v-SNARES bind with t-SNARES, plasma membrane proteins and the membranes fuse. The fused site opens and releases the contents of the secretory vesicle into the intercellular space. **(b)** Photograph of a secretory vesicle releasing its contents to the cell exterior.

(b)

space; and **endocytosis** (en"do-si-to'sis; "within the cell"), moving substances across the plasma membrane into the cell from the extracellular environment.

Vesicular transport is also used for combination processes such as *transcytosis,* moving substances into, across, and then out of the cell; and *substance, or vesicular trafficking,* moving substances from one area (or organelle) in the cell to another.

Like solute pumping, vesicular transport processes are energized by ATP [or in some cases *GTP* (guanosine triphosphate, another energy-rich compound)].

Lungs, open to the external environment, collect dust and other airborne debris. In the lungs of smokers, carbon particles are added to this debris. ∎

1. Exocytosis. This mechanism, typically stimulated by a cell-surface signal such as binding of a hormone to a membrane receptor, accounts for hormone secretion, neurotransmitter release, mucus secretion, and in some cases, ejection of wastes. The substance to be removed from the cell is first enclosed in a membranous sac called a **vesicle.** The vesicle migrates to the plasma membrane, fuses with it, and then ruptures, spilling the sac contents out of the cell (Figure 3.12). The mechanism involves a "docking" process in which transmembrane proteins on the vesicles, fancifully called v-SNAREs (*v* for vesicle), recognize certain plasma membrane proteins, called t-SNAREs (*t* for target), and bind with them. This causes the lipid layers to "corkscrew" together, rearranging the lipid monolayers without mixing them. As described shortly, membrane material added by exocytosis is removed by endocytosis—the reverse process.

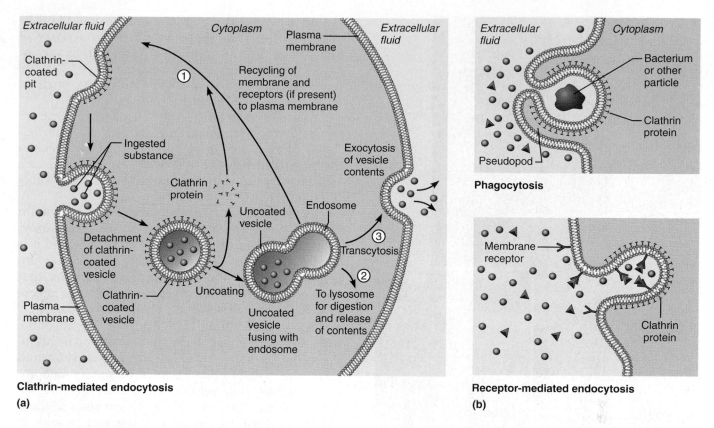

Clathrin-mediated endocytosis
(a)

Receptor-mediated endocytosis
(b)

FIGURE 3.13 **Clathrin-mediated endocytosis. (a)** Events of clathrin-mediated endocytosis. **(b)** Phagocytosis and receptor-mediated endocytosis.

2. Endocytosis, transcytosis, and vesicular trafficking. Virtually all the remaining forms of vesicular transport involve the use of an assortment of protein-coated vesicles of three types and, with some exceptions, all are mediated by membrane receptors. Let's look at the specialties of each type of coated vesicular transport.

■ *Clathrin-coated vesicles.* Clathrin-coated vesicles provide the main route for endocytosis and transcytosis of bulk solids, most macromolecules, and fluids. Basically, the substance to be taken into the cell by endocytosis is progressively enclosed by an infolding portion of the plasma membrane called a *coated pit,* a term that refers to the bristlelike **clathrin** (kla′thrin; "lattice clad") protein coating on the cytoplasmic face of the vesicle (Figure 3.13a). The clathrin coat acts both to deform the membrane to produce the vesicle and in cargo selection.

Once inside, the vesicle loses its fuzzy coat and then typically fuses with a processing and sorting vesicle called an *endosome.* As shown in Figure 3.13a, contents of the endosome may be ① recycled back to the plasma membrane (applies to many receptors and membrane components), ② combined with a *lysosome* (li′so-some), a specialized cell structure containing digestive enzymes, where the ingested substance is degraded or released (if iron or cholesterol), or ③ transported completely across the cell and released by exocytosis on the opposite side (transcytosis). Transcytosis is common in the endothelial cells lining blood vessels because it provides a quick means to get substances from the blood to the interstitial fluid. Based on the nature and quantity of material taken up and the mechanism of uptake, three types of endocytosis that use clathrin-coated vesicles are recognized: phagocytosis, pinocytosis, and receptor-mediated endocytosis.

Phagocytosis (fag″o-si-to′sis; "cell eating") is the type of endocytosis in which some relatively large or solid material, such as a clump of bacteria or cell debris, is engulfed by the cell. When a particle binds to the cell's surface, cytoplasmic extensions called pseudopods (soo′do-pahdz; *pseudo* = false, *pod* = foot) form and flow around the particle and engulf it (Figure 3.13b). The endocytotic vesicle thus formed is called a **phagosome** (fag′o-sōm; "eaten body"). In most cases, the phagosome then fuses with a lysosome and its contents are digested.

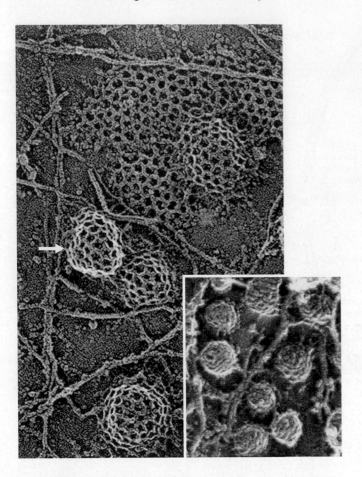

FIGURE 3.14 Clathrin-coated versus caveolin-coated vesicles. Structural differences between the two vesicle types are shown in electron micrographs of cells that have been sheared open by freeze-fracturing. The larger photo shows the inner surface of the plasma membrane, upon which many polygonal lattices are visible. Some lattices are relatively flat (these are the clathrin-coated pits pointed out by the yellow arrow); others curve around a vesicle (clathrin-coated vesicle indicated by the white arrow). These polygonal lattices of clathrin are easily distinguished from the striped coats of caveolin protein that characterize caveolae (seen in the inset).

In the human body only macrophages and certain white blood cells are "experts" at phagocytosis. They help police and protect the body by ingesting and disposing of bacteria, other foreign substances, and dead tissue cells. Most phagocytes move about by **amoeboid motion** (ah-me′boyd; "changing shape"); that is, the flowing of their cytoplasm into temporary pseudopods allows them to creep along.

In **pinocytosis** ("cell drinking"), also called **fluid-phase endocytosis,** a bit of infolding plasma membrane surrounds a very small volume of extracellular fluid containing dissolved molecules. This droplet enters the cell in an endosome. Unlike phagocytosis, pinocytosis is a routine activity of most cells, affording them a nonselective way of sampling the extracellular fluid. It is particularly important in cells that absorb nutrients, such as cells that line the intestines.

As mentioned, bits of the plasma membrane are removed when the membranous sacs are internalized. However, these membranes are recycled back to the plasma membrane by exocytosis, so the surface area of the plasma membrane remains remarkably constant.

Receptor-mediated endocytosis is the main mechanism for the specific endocytosis and transcytosis of most macromolecules by body cells, and it is exquisitely selective (Figure 3.13b). The receptors for this process are plasma membrane proteins that bind only with certain substances. Both the receptors and attached molecules are internalized in a clathrin–coated pit and then dealt with in one of the ways discussed above. Substances taken up by receptor-mediated endocytosis include enzymes, insulin (and some other hormones), low-density lipoproteins (such as cholesterol attached to a transport protein), and iron. Unfortunately, flu viruses and diphtheria toxin use this route to enter and attack our cells.

▪ ***Non-clathrin-coated vesicles.*** Caveolae (ka″ve-o′le; "little caves"), tubular or flask-shaped inpocketings of the plasma membrane seen in many cell types, also seem to be involved in a kind of receptor-mediated endocytosis. Like clathrin-coated pits, they capture specific molecules (folate, tetanus toxin) from the extracellular fluid in coated vesicles and participate in some forms of transcytosis. However, caveolae are smaller than clathrin-coated vesicles. Additionally their coat is thinner and composed of a different protein called **caveolin** (Figure 3.14). Because of the close association of caveoli with lipid rafts that are platforms for a variety of signaling molecules such as G proteins, various types of hormone receptors (e.g., insulin), and enzymes known to be involved in cell regulation, these vesicles may provide important sites for cell signaling and cross talk between the same or different signaling pathways. Their precise role in the cell is still being worked out.

Virtually all intracellular vesicular trafficking, in which vesicles pinch off from organelles and travel to other organelles to unload their contents [for example, proteins made at the endoplasmic reticulum are sent to the Golgi apparatus (see Figure 3.20, p. 82)], uses vesicles coated with coatomer (COP1 and COP2) proteins.

Table 3.2 summarizes the membrane transport processes.

TABLE 3.2   **Membrane Transport Processes**

Process	Energy Source	Description	Examples
Passive Processes			
Diffusion			
Simple diffusion	Kinetic energy	Net movement of particles (ions, molecules, etc.) from an area of their higher concentration to an area of their lower concentration, that is, along their concentration gradient	Movement of fats, oxygen, carbon dioxide through the lipid portion of the membrane
Facilitated diffusion	Kinetic energy	Same as simple diffusion, but the diffusing substance is attached to a lipid-soluble membrane carrier protein or moves through a membrane channel	Movement of glucose and some ions into cells
Osmosis	Kinetic energy	Simple diffusion of water through a selectively permeable membrane	Movement of water into and out of cells directly through the lipid phase of the membrane or via membrane pores (aquaporins)
Filtration	Hydrostatic pressure	Movement of water and solutes through a semipermeable membrane from a region of higher hydrostatic pressure to a region of lower hydrostatic pressure, that is, along a pressure gradient	Movement of water, nutrients, and gases through a capillary wall; formation of kidney filtrate
Active Processes			
Active transport			
Primary active transport	ATP	Transport of substances against a concentration (or electrochemical) gradient; across the plasma membrane by a solute pump; directly uses energy of ATP hydrolysis	Ions (Na^+, K^+, H^+, Ca^{2+}, and others)
Secondary active transport	ATP	Cotransport (coupled transport) of two solutes across the membrane; energy supplied by the ion gradient created by a primary active solute pump (indirectly); symporters move the transported substances in the same direction; antiporters move transported substances in opposite directions across the membrane	Movement of polar or charged solutes, e.g., amino acids (into cell by symporters); Ca^{2+}, H^+ (out of cells via antiporters)
Vesicular transport			
Exocytosis	ATP	Secretion or ejection of substances from a cell; the substance is enclosed in a membranous vesicle, which fuses with the plasma membrane and ruptures, releasing the substance to the exterior	Secretion of neurotransmitters, hormones, mucus, etc.; ejection of cell wastes
Endocytosis • Via clathrin-coated vesicles	ATP		
Phagocytosis	ATP	"Cell eating": A large external particle (proteins, bacteria, dead cell debris) is surrounded by a "seizing foot" and becomes enclosed in a clathrin-coated vesicle	In the human body, occurs primarily in protective phagocytes (some white blood cells and macrophages)
Pinocytosis (fluid phase endocytosis)	ATP	Plasma membrane sinks beneath an external fluid droplet containing small solutes; membrane edges fuse, forming a fluid-filled vesicle; clathrin-coated vesicles formed	Occurs in most cells; important for taking in dissolved solutes by absorptive cells of the kidney and intestine

▶

TABLE 3.2   Membrane Transport Processes *(continued)*

Process	Energy Source	Description	Examples
Endocytosis *(continued)*			
Receptor-mediated endocytosis	ATP	Selective endocytosis and transcytosis; external substance binds to membrane receptors, and clathrin-coated pits are formed	Means of intake of some hormones, cholesterol, iron, and most macro-molecules
• Via caveolin-coated vesicles (caveoli)	ATP	Selective endocytosis (and transcytosis); external substance binds to membrane receptors (often associated with lipid rafts); caveolin-coated vesicles formed	Roles not fully known; proposed roles include cholesterol regulation and trafficking; platforms for signal trans-duction
• Via coatomer-coated vesicles	ATP	Vesicles coated with coatomer proteins pinch off from organelles and travel to other organelles to deliver their cargo	Accounts for nearly all intracellular trafficking of molecules

Generating and Maintaining a Resting Membrane Potential

As you're now aware, the selective permeability of the plasma membrane can lead to dramatic osmotic flows, but that is not its only consequence. An equally important result is the generation of a **membrane potential,** or voltage, across the membrane. A *voltage* is electrical potential energy resulting from the separation of oppositely charged particles. In cells, the oppositely charged particles are ions, and the barrier that keeps them apart is the plasma membrane.

In their resting state, all body cells exhibit a **resting membrane potential** that typically ranges from -5 to -100 millivolts (mV), depending on cell type. Hence, all cells are said to be **polarized.** The minus sign before the voltage indicates that the *inside* of the cell is negative compared to its outside. This voltage (or charge separation) exists only at the membrane. If we were to add up all the negative and positive charges in the cytoplasm, we would find that the cell interior is electrically neutral. Likewise, the positive and negative charges in the extracellular fluid balance each other exactly.

So how does the resting membrane potential come about, and how is it maintained? Although many kinds of ions are found both inside cells and in the extracellular fluid, the resting membrane potential is determined mainly by the concentration gradient of K^+, and by the differential permeability of the plasma membrane to K^+ and other ions. As mentioned, K^+ and protein anions predominate inside body cells and the extracellular fluid contains relatively more Na^+, which is largely balanced by Cl^-. The unstimulated plasma membrane is somewhat permeable to K^+ because of leakage channels, but impermeable to the protein anions. Potassium therefore diffuses out of the cell along its concentration gradient but the protein anions are unable to follow, so loss of positive charges makes the membrane interior more negative (Figure 3.15). As more and more K^+ leaves the cell, the negativity of the inner membrane face becomes great enough to attract K^+ back toward and even into the cell. At the point where potassium's concentration gradient is balanced by the membrane potential $(-70$ mV), one K^+ enters the cell as one leaves, and the resting potential is established.

Other ions do contribute to the resting membrane potential, but only minimally. Sodium is strongly attracted to the cell interior by its concentration gradient, but because the membrane is nearly impermeable to sodium, K^+ outflow is not balanced by Na^+ inflow and Cl^- entry is resisted by the negative charge of the interior, even though the membrane is permeable to Cl^-. Although it is tempting to believe that massive flows of K^+ ions are needed to generate the resting potential, this is not the case. Surprisingly, the number of ions producing the membrane potential is so small that it does not change ion concentrations in any significant way.

In a cell at rest, very few ions cross its plasma membrane. However, Na^+ and K^+ are not at equilibrium and there is some net movement of K^+ out of the cell and of Na^+ into the cell because of its strong pull into the cell by both its concentration gradient and the interior negative charge. If only passive forces were at work, these ion concentrations would eventually become equal inside and outside the cell.

Instead, the cell exhibits a *steady state* in which diffusion causes ionic imbalances that polarize the membrane, and active transport processes *maintain* that membrane potential. The rate of active transport is equal to, and depends on, the rate of Na^+ diffusion into the cell. If more Na^+ enters, more is pumped out. (This is like being in a leaky boat. The more water that comes in, the faster you bail!) The Na^+-K^+ pump couples sodium and potassium transport and, on average, each "turn" of the pump ejects

3Na⁺ out of the cell and carries 2K⁺ back in (see Figure 3.15). Because the membrane is always 50 to 100 times more permeable to K⁺, the ATP-dependent Na⁺-K⁺ pump maintains both the membrane potential and the osmotic balance. Indeed, were Na⁺ not continuously removed from cells, in time so much would accumulate intracellularly that the osmotic gradient would draw water into the cells, causing them to burst.

Now that we've introduced the membrane potential, we can add a detail or two to our discussion of diffusion. Earlier we said that solutes diffuse down their concentration gradient. This is true for uncharged solutes, but only partially true for ions. The negatively and positively charged faces of the plasma membrane can help or hinder diffusion of ions driven by a concentration gradient. It is more correct to say that ions diffuse according to **electrochemical gradients,** thereby recognizing the effect of both electrical and concentration (chemical) forces. Consequently, although diffusion of K⁺ across the plasma membrane is aided by the membrane's greater permeability to it and by the ion's concentration gradient, its diffusion is resisted somewhat by the positive charge on the cell exterior. On the other hand, Na⁺ is drawn into the cell by a steep electrochemical gradient; the limiting factor is the membrane's relative impermeability to it. As we will describe in detail in later chapters, "upsetting" the resting membrane potential by transient opening of Na⁺ and K⁺ channels in the plasma membrane is a normal means of activating neurons and muscle cells.

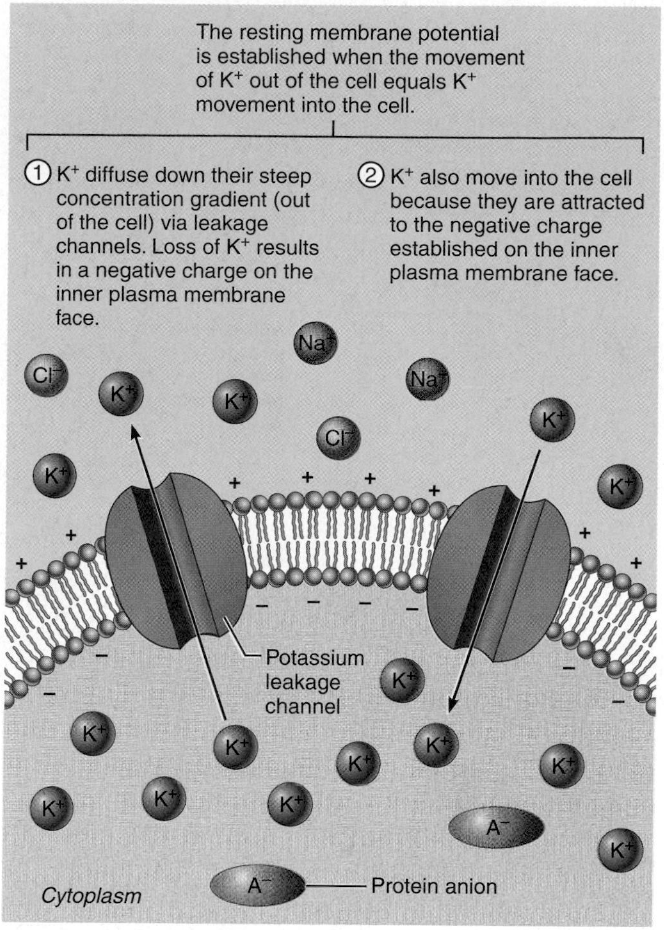

Cell-Environment Interactions

Cells are biological minifactories and, like other factories, they receive and send orders from and to the outside community. But *how* does a cell interact with its environment, and what activates it to carry out its homeostatic functions?

Although cells sometimes interact directly with other cells, this is not always the case. In many cases cells respond to extracellular chemicals, such as hormones and neurotransmitters distributed in body fluids. Cells also interact with molecules in the substratum that act as signposts to guide cell migration during development and repair.

Whether cells interact directly or indirectly, however, the glycocalyx is always involved. The best understood of the participating glycocalyx molecules fall into two large families—cell adhesion molecules and plasma membrane receptors (see Figure 3.4).

Roles of Cell Adhesion Molecules

Thousands of **cell adhesion molecules (CAMs)** are found on almost every cell in the body. They play key

FIGURE 3.15   **Summary of forces that generate and maintain membrane potentials.** The ionic imbalances that produce the membrane potential are due to the differential membrane permeability to the passive diffusion of Na⁺ and K⁺. The potassium ions move through membrane leakage channels, but the membrane is nearly impermeable to Na⁺. The net effect is that the outside membrane face becomes more electrically positive (more positive ions build up) than the inside, which is relatively more negative. The active transport of sodium and potassium ions (in a ratio of 3:2) by the Na⁺-K⁺ pump maintains these conditions.

roles in embryonic development and wound repair (situations where cell mobility is important) and in immunity. These sticky glycoproteins (*cadherins* and *integrins*) act as

1. The molecular "Velcro" cells use to anchor themselves to molecules in the extracellular space and to each other (see desmosome discussion on p. 62)

2. The "arms" that migrating cells use to haul themselves past one another

3. SOS signals sticking out from the blood vessel lining that rally protective white blood cells to a nearby infected or injured area

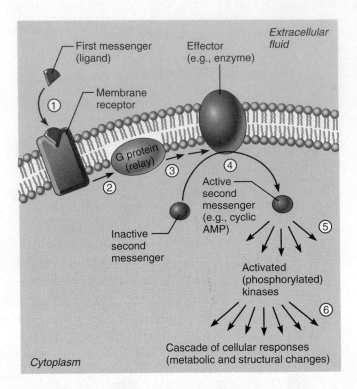

FIGURE 3.16 Model of the operation of a G protein–linked receptor. In this simplified diagram: ① an extracellular molecule (ligand) is a first messenger that binds to a specific receptor protein. ② The receptor activates a G protein that acts as a relay to ③ stimulate an effector protein. ④ The effector is an enzyme that produces a second messenger inside the cell. ⑤ The second messenger, cyclic AMP in this model, activates kinase enzymes. ⑥ The kinase enzymes can activate a whole series of enzymes that trigger the various responses of the cell.

4. Mechanical sensors that respond to local tension at the cell surface by stimulating synthesis or degradation of adhesive membrane junctions

Roles of Membrane Receptors

A huge and diverse group of integral proteins and glycoproteins that serve as binding sites are collectively known as **membrane receptors.** Some function in contact signaling, others in electrical signaling, and still others in chemical signaling. Let's take a look.

Contact Signaling   *Contact signaling,* the actual coming together and touching of cells, is the means by which cells recognize one another. It is particularly important for normal development and immunity. Some bacteria and other infectious agents use contact signaling to identify "preferred" target tissues or organs.

Electrical Signaling   In the process known as *electrical signaling,* certain plasma membrane receptors are channel proteins that respond to changes in membrane potential by opening or closing the "gates" associated with an ion channel. Such voltage-regulated channels are common in excitable tissues like neural and muscle tissues, and indispensable to their normal functioning.

Chemical Signaling   Most plasma membrane receptors are involved in *chemical signaling,* and this group will receive the bulk of our attention. Signaling chemicals that bind specifically to plasma membrane receptors are called **ligands.** Among these ligands are most *neurotransmitters* (nervous system signals), *hormones* (endocrine system signals), and *paracrines* (chemicals that act locally and are rapidly destroyed). Different cells respond in different ways to the same ligand. Acetylcholine, for instance, stimulates skeletal muscle cells to contract, but inhibits heart muscle. Thus, a target cell's response depends on the internal machinery that the receptor is linked to, not the specific ligand that binds to it.

Though cell responses to receptor binding vary widely, there is a fundamental similarity. When a ligand binds to a membrane receptor, the receptor's structure changes, and cell proteins are altered in some way; for example, muscle proteins change shape to generate force. Some of these membrane receptor proteins are *catalytic proteins* that function as enzymes. Others, such as the *chemically gated channel-linked receptors* common in muscle and nerve cells, respond to ligands by transiently opening or closing ion gates or channels, which in turn changes the excitability of the cell. Still other receptors are coupled to enzymes or ion channels by a regulatory molecule called a G protein. Because nearly every cell in the body displays at least some of these receptors, we will spend a bit more time with them. The lipid rafts mentioned earlier sequester many receptor-mediated elements facilitating signaling processes.

G protein–linked receptors exert their effect indirectly through a **G protein,** which acts as a middleman or relay to activate (or inactivate) a membrane-bound enzyme or ion channel (Figure 3.16). As a result, one or more intracellular chemical signals, commonly called **second messengers,** are generated and connect plasma membrane events to the internal metabolic machinery of the cell. Two very important second messengers are **cyclic AMP** and ionic calcium, both of which typically activate protein kinase enzymes. By transferring phosphate groups from ATP to other proteins, the protein kinases can activate a whole series of enzymes that bring about the desired cellular activity. Because a single enzyme can catalyze hundreds of reactions, the amplification effect of such a chain of events is tremendous. These and other receptor systems are described in greater detail in Chapter 15.

One more signaling molecule must be mentioned, even though it doesn't fit any of the mechanisms described. *Nitric oxide (NO)*, made of a single atom of nitrogen and one of oxygen, is one of nature's simplest molecules, an environmental pollutant, and the first gas known to act as a biological messenger. Because of its tiny size, it slips into and out of cells easily. Its one unpaired electron makes it a highly reactive free radical that reacts with head-spinning speed with other key molecules to spur cells into a broad array of activities. You will be hearing more about NO later (in the neural, cardiovascular, and immune system chapters).

The Cytoplasm

Cytoplasm ("cell forming material") is the cellular material between the plasma membrane and the nucleus. It is the site where most cellular activities are accomplished. Although early microscopists thought that the cytoplasm was a structureless gel, the electron microscope has revealed that it consists of three major elements: the cytosol, organelles, and inclusions.

The **cytosol** (si'to-sol) is the viscous, semitransparent fluid in which the other cytoplasmic elements are suspended. It is a complex mixture with properties of both a colloid and a true solution. Dissolved in the cytosol, which is largely water, are proteins, salts, sugars, and a variety of other solutes.

The **cytoplasmic organelles** are the metabolic machinery of the cell. Each type of organelle is "engineered" to carry out a specific function for the cell—some synthesize proteins, others package those proteins, and so on.

Inclusions are chemical substances that may or may not be present, depending on cell type. Examples include stored nutrients, such as the glycogen granules abundant in liver and muscle cells; lipid droplets common in fat cells; pigment (melanin) granules seen in certain cells of skin and hair; and crystals of various types.

Cytoplasmic Organelles

The cytoplasmic organelles ("little organs") are specialized cellular compartments, each performing its own job to maintain the life of the cell. Some organelles, the *nonmembranous organelles*, lack membranes. Examples are the cytoskeleton, centrioles, and ribosomes. However, most organelles are bounded by a membrane similar in composition to the plasma membrane (minus the glycocalyx), and

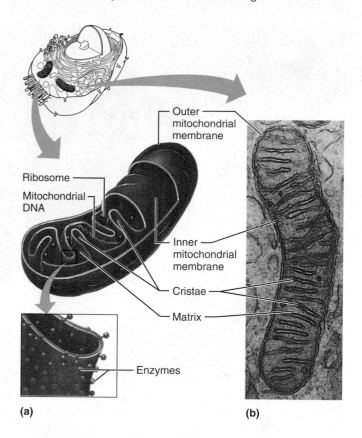

FIGURE 3.17 **Mitochondrion. (a)** Diagrammatic view of a longitudinally sectioned mitochondrion. **(b)** Electron micrograph of a mitochondrion (28,400×).

this membrane enables *membranous organelles* to maintain an internal environment different from that of the surrounding cytosol. This compartmentalization is crucial to cell functioning. Without it, thousands of enzymes would be randomly mixed and biochemical activity would be chaotic. Besides providing "splendid isolation," an organelle's membrane unites it with the rest of an interactive intracellular system called the *endomembrane system* (see p. 84). The system's membranous organelles include the mitochondria, peroxisomes, lysosomes, endoplasmic reticulum, and Golgi apparatus. Now, let us consider what goes on in each of the workshops of our cellular factory.

Mitochondria

Mitochondria (mi"to-kon'dre-ah) are threadlike (*mitos* = thread) or sausage-shaped membranous organelles (Figure 3.17). In living cells they squirm, elongate, and change shape almost continuously. They are the power plants of a cell, providing most of its ATP supply. The density of mitochondria in a particular cell reflects that cell's energy requirements, and mitochondria are generally clustered where the action is. Busy cells like kidney and liver

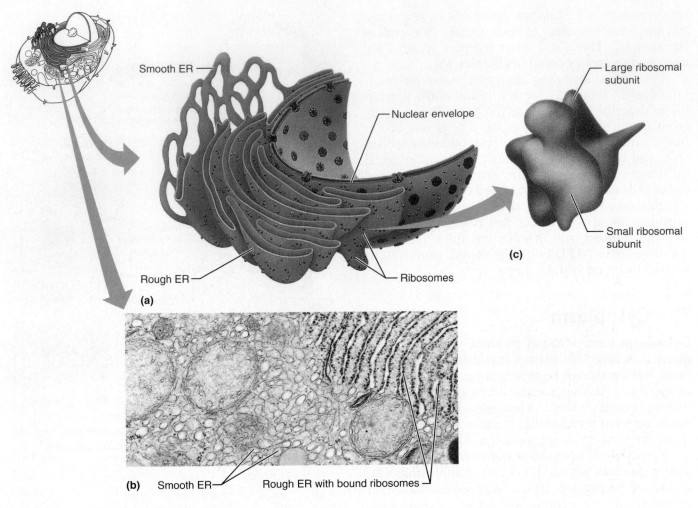

Smooth ER

Nuclear envelope

Large ribosomal subunit

Rough ER

Ribosomes

Small ribosomal subunit

(a)

(c)

(b)   Smooth ER   Rough ER with bound ribosomes

FIGURE 3.18   **The endoplasmic reticulum. (a)** Diagram of the rough ER in a liver cell; its connections to smooth ER areas are also illustrated. **(b)** Electron micrograph of smooth and rough endoplasmic reticulum (15,000×). **(c)** Diagram of a ribosome showing the large and small subunits.

cells have hundreds of mitochondria, whereas relatively inactive cells (such as unchallenged lymphocytes) have just a few.

Mitochondria are enclosed by *two* membranes, each with the general structure of the plasma membrane. The outer membrane is smooth and featureless, but the inner membrane folds inward, forming shelflike **cristae** (krĭ'ste; "crests") that protrude into the *matrix*, the gel-like substance within the mitochondrion. Intermediate products of food fuels (glucose and others) are broken down to water and carbon dioxide by teams of enzymes, some dissolved in the mitochondrial matrix and others forming part of the crista membrane.

As the metabolites are broken down and oxidized, some of the energy released is captured and used to attach phosphate groups to ADP molecules to form ATP. This multistep mitochondrial process

(described in Chapter 23) is generally referred to as *aerobic* (a-er-o'bik) *cellular respiration* because it requires oxygen.

Mitochondria are complex organelles: They contain their own DNA and RNA and are able to reproduce themselves. Although mitochondrial genes (some 37 of them) direct the synthesis of some of the proteins required for mitochondrial function, the DNA of the cell's nucleus encodes the remaining proteins needed to carry out cellular respiration. When cellular requirements for ATP increase, the mitochondria synthesize more cristae or simply pinch in half (a process called *fission*) to increase their number, then grow to their former size.

Intriguingly, mitochondria are similar to a specific group of bacteria (purple bacteria phylum), and mitochondrial DNA is similar to that found in bacterial cells. It is now widely believed that mitochondria arose from bacteria that invaded the ancient ancestors of plant and animal cells.

Ribosomes

Ribosomes (ri'bo-sōmz) are small, dark-staining granules composed of proteins and a variety of RNA called *ribosomal RNA*. Each ribosome has two globular subunits that fit together like the body and cap of an acorn (Figure 3.18c). Ribosomes are sites of protein synthesis, a function we discuss in detail later in this chapter.

Some ribosomes float freely in the cytoplasm; others are attached to membranes, forming a complex called the *rough endoplasmic reticulum*. These two ribosomal populations appear to divide the chore of protein synthesis. **Free ribosomes** make soluble proteins that function in the cytosol. **Membrane-bound ribosomes** synthesize proteins destined either for incorporation into cell membranes or for export from the cell. Ribosomes can switch back and forth between these two functions, attaching to and detaching from the membranes of the endoplasmic reticulum, according to the type of protein they are making at a given time.

Endoplasmic Reticulum

The **endoplasmic reticulum** (en"do-plaz'mik re-tik'u-lum; "network within the cytoplasm"), or **ER,** is an extensive system of interconnected tubes and parallel membranes enclosing fluid-filled cavities, or **cisternae** (sis-ter'ne), that coils and twists through the cytosol. The ER is continuous with the nuclear membrane and accounts for about half of the cell's membranes. There are two distinct varieties of ER: rough ER and smooth ER.

Rough Endoplasmic Reticulum   The external surface of the **rough ER** is studded with ribosomes (Figure 3.18a). Proteins assembled on these ribosomes thread their way into the fluid-filled interior of the ER cisternae, where various fates await them (as described shortly). The rough ER has several functions. Its ribosomes manufacture all proteins secreted from cells. Thus, the rough ER is particularly abundant and well developed in most secretory cells, antibody-producing plasma cells, and liver cells, which produce most blood proteins. It is also the cell's "membrane factory" because integral proteins and phospholipids that form part of all cellular membranes are manufactured there. The enzymes needed to catalyze lipid synthesis have their active sites on the external (cytosolic) face of the ER membrane, where the needed substrates are readily available.

As mentioned, ribosomes attach to and disattach from the rough ER. When a short "leader" pep-

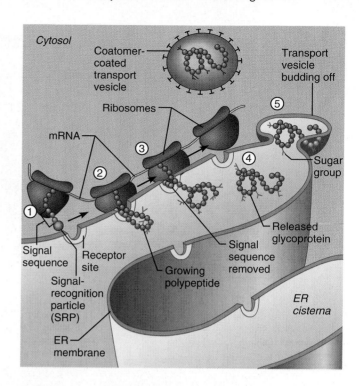

FIGURE 3.19 **The signal mechanism targets ribosomes to the ER for protein synthesis.** Enlarged view of a portion of the rough ER. According to the signal mechanism of protein synthesis: ① Presence of a short signal sequence on a newly forming protein causes the mRNA-ribosome complex to be directed to the rough ER by a signal recognition particle (SRP), which binds to a receptor site that includes a pore and an enzyme to clip the signal sequence. ② Once attached to the ER receptor site, the SRP is released and the growing polypeptide snakes through the ER membrane into the cisterna. ③ The signal sequence initially remains attached to the receptor but shortly it is clipped off by an enzyme. As protein synthesis continues, sugar groups may be added to the protein. ④ In this example, the completed protein is released from the ribosome and folds into its 3-D conformation, a process aided by molecular chaperones (see p. 53). Transmembrane proteins are only partially translocated and remain embedded in the membrane. ⑤ The protein is enclosed within a coatomer—coated transport vesicle that pinches off the ER. The transport vesicles make their way to the Golgi apparatus, where further processing of the proteins occurs (see Figure 3.20).

tide called a **signal sequence** is present in a protein being synthesized, the associated ribosome attaches to the membrane of the rough ER. This signal sequence [with its attached "cargo" of ribosome and messenger RNA (mRNA)] is guided to the appropriate receptor sites on the ER membrane by a **signal-recognition particle (SRP)**, which cycles between the ER and the cytosol. The subsequent events occurring at the ER are detailed in Figure 3.19.

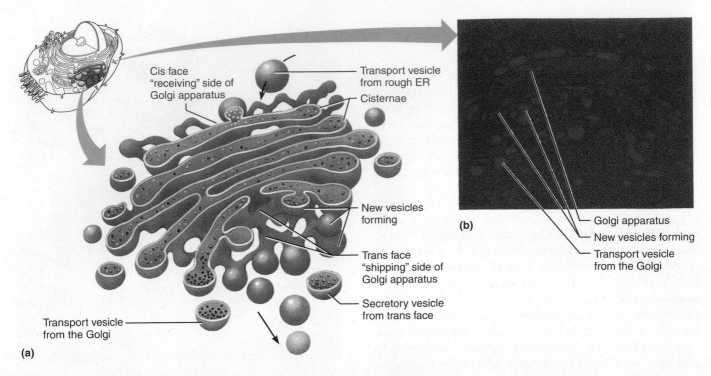

FIGURE 3.20   Golgi apparatus. (a) Three-dimensional diagrammatic view of the Golgi apparatus. **(b)** Electron micrograph of the Golgi apparatus (28,000×). Notice the vesicles in the process of pinching off from the membranous Golgi apparatus. (The various and abundant vesicles shown would have coatomer proteins on their external surfaces; these proteins have been omitted for simplicity.)

Smooth Endoplasmic Reticulum   The **smooth ER** (see Figures 3.2 and 3.18) is a continuation of the rough ER and consists of tubules arranged in a looping network. Its enzymes (all integral proteins forming part of its membranes) play no role in protein synthesis. Instead, they catalyze reactions involved with the following processes:

1. Lipid metabolism, cholesterol synthesis, and synthesis of the lipid components of lipoproteins (in liver cells)

2. Synthesis of steroid-based hormones such as sex hormones (testosterone-synthesizing cells of the testes are full of smooth ER)

3. Absorption, synthesis, and transport of fats (in intestinal cells)

4. Detoxification of drugs, certain pesticides, and carcinogens (in liver and kidneys)

5. Breakdown of stored glycogen to form free glucose (in liver cells especially)

Additionally, skeletal and cardiac muscle cells have an elaborate smooth ER (called the sarcoplasmic reticulum) that plays an important role in calcium ion storage and release during muscle contrac-tion. Except for the examples given above, most body cells contain little, if any, true smooth ER.

Golgi Apparatus

The **Golgi apparatus** (gol'je) consists of stacked and flattened membranous sacs, shaped like hollow dinner plates, associated with swarms of tiny membranous vesicles (Figure 3.20). The Golgi apparatus is the principal "traffic director" for cellular proteins. Its major function is to modify, concentrate, and package the proteins and lipids made at the rough ER. The transport vesicles that bud off from the rough ER move to and fuse with the membranes at its convex *cis face,* the "receiving" side of the Golgi apparatus. Inside the apparatus, the proteins are modified: Some sugar groups are trimmed while others are added, and in some cases, phosphate groups are added. The various proteins are "tagged" for delivery to a specific address, sorted, and packaged in at least three types of vesicles that bud from the concave *trans face* (the "shipping" side) of the Golgi stack.

Vesicles containing proteins destined for export pinch off from the trans face as **secretory vesicles,** or

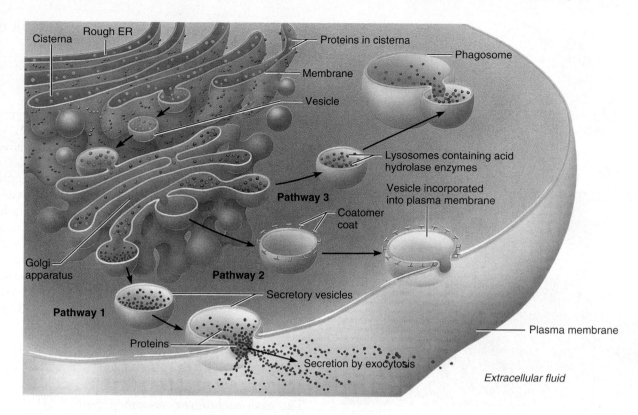

FIGURE 3.21 **Role of the Golgi apparatus in packaging proteins for cellular use and for secretion.** The sequence of events from protein synthesis on the rough ER to the final distribution of those proteins. Protein-containing vesicles pinch off the rough ER and migrate to fuse with the membranes of the Golgi apparatus. Within the Golgi compartments, the proteins are modified; they are then packaged within different Golgi vesicle types, depending on their ultimate destination (pathways 1–3 as shown).

granules, which migrate to the plasma membrane and discharge their contents from the cell by exocytosis (pathway 1, Figure 3.21). Specialized secretory cells, such as the enzyme-producing cells of the pancreas, have a very prominent Golgi apparatus. Besides its packaging-for-release function, the Golgi apparatus pinches off vesicles containing lipids and transmembrane proteins destined for a "home" in the plasma membrane (pathway 2, Figure 3.21) or other membranous organelles. It also packages digestive enzymes into membranous lysosomes that remain in the cell (pathway 3 in Figure 3.21, and discussed next).

Lysosomes

Lysosomes ("disintegrator bodies") are spherical membranous organelles containing digestive en- zymes (Figure 3.22). As you might guess, lysosomes are large and abundant in phagocytes, the cells that dispose of invading bacteria and cell debris. Lysosomal enzymes can digest almost all kinds of biological molecules. They work best in acidic conditions and thus are called *acid hydrolases* (hi"drah-la'siz). The lysosomal membrane is adapted to serve lysosomal functions in two important ways: (1) It contains H^+ (proton) "pumps", ATPases that gather hydrogen ions from the surrounding cytosol to maintain the organelle's acidic pH, and (2) it retains the dangerous acid hydrolases while permitting the final products of digestion to escape so that they can be used by the cell or excreted. Hence, lysosomes provide sites where digestion can proceed *safely* within a cell.

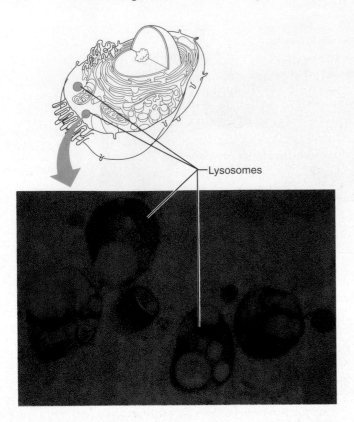

Lysosomes

FIGURE 3.22 Lysosomes. Electron micrograph of a cell containing lysosomes (90,000×).

Lysosomes function as a cell's "demolition crew" by

■ Digesting particles taken in by endocytosis, particularly ingested bacteria, viruses, and toxins

■ Degrading worn-out or nonfunctional organelles

■ Performing metabolic functions, such as glycogen breakdown and release

■ Breaking down nonuseful tissues, such as the webs between the fingers and toes of a developing fetus and the uterine lining during menstruation

■ Breaking down bone to release calcium ions into the blood

Some unique lysosomes called *secretory lysosomes* are found mainly in white blood cells, immune cells, and melanocytes (pigment-producing cells). They appear to have the ability to store and then secrete by exocytosis newly synthesized proteins.

The lysosomal membrane is ordinarily quite stable, but it becomes fragile when the cell is injured or deprived of oxygen and when excessive amounts of vitamin A are present. Lysosomal rupture results in self-digestion of the cell, a process called **autolysis** (aw"tol'ĭ-sis). Autolysis is the basis

for desirable destruction processes (see fourth bullet point above).

HOMEOSTATIC IMBALANCE

Glycogen and certain lipids in the brain are degraded by lysosomes at a relatively constant rate. In *Tay-Sachs disease,* an inherited condition seen mostly in Jews from Central Europe, the lysosomes lack an enzyme needed to break down a glycolipid abundant in nerve cell membranes. As a result, the nerve cell lysosomes swell with the undigested lipids, which interfere with nervous system functioning. Affected infants typically have doll-like features and pink translucent skin. At 3 to 6 months of age, the first signs of disease appear (listlessness, motor weakness). These progress to mental retardation, seizures, blindness, and ultimately death within 18 months. ●

Summary of Interactions in the Endomembrane System

The **endomembrane system** (Figure 3.23) is a system of organelles (most described above) that work together mainly (1) to produce, store, and export biological molecules, and (2) to degrade potentially harmful substances. It includes the ER, Golgi apparatus, secretory vesicles, and lysosomes, as well as the nuclear membrane—that is, all of the membranous organelles or elements that are either structurally continuous or arise via forming or fusing transport vesicles. There are continuities between the nuclear envelope (itself an extension of the rough ER) and the rough and smooth ER. The plasma membrane, though not actually an *endo* membrane, is also functionally part of this system. Besides these direct structural relationships, a wide variety of indirect interactions (indicated by arrows in Figure 3.23) occur among the members of the system. Some of the vesicles "born" in the ER migrate to and fuse with the Golgi apparatus or the plasma membrane, and vesicles arising from the Golgi apparatus can become part of the plasma membrane, secretory vesicles, or lysosomes.

Peroxisomes

Peroxisomes (pĕ-roks'ĭ-sōmz; "peroxide bodies") are membranous sacs containing a variety of powerful enzymes, the most important of which are oxidases and catalases. Oxidases use molecular oxygen (O_2) to detoxify harmful substances, including alcohol and formaldehyde. However, their most important function is to neutralize danger-

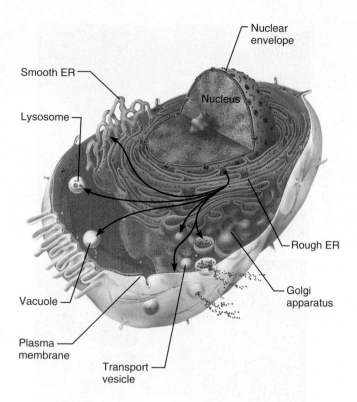

Smooth ER

Lysosome

Nucleus

Nuclear envelope

Rough ER

Golgi apparatus

Vacuole

Plasma membrane

Transport vesicle

FIGURE 3.23 The endomembrane system.

ous **free radicals,** highly reactive chemicals with unpaired electrons that can scramble the structure of biological molecules. Oxidases convert free radicals to hydrogen peroxide, which is also reactive and dangerous but is quickly converted to water by catalase enzymes. Free radicals and hydrogen peroxide are normal by-products of cellular metabolism, but they have devastating effects on cells if allowed to accumulate. Peroxisomes are especially numerous in liver and kidney cells, which are very active in detoxification.

Although peroxisomes look like small lysosomes (see Figure 3.2), they are self-replicating organelles formed by a simple pinching in half of preexisting peroxisomes. Unlike lysosomes, they do not arise by budding from the Golgi apparatus.

Cytoskeleton

The **cytoskeleton,** literally, "cell skeleton," is an elaborate series of rods running through the cytosol. This network acts as a cell's "bones," "muscles," and "ligaments" by supporting cellular structures and providing the machinery to generate various cell movements. The three types of rods in the cytoskeleton are *microtubules, microfilaments,* and *intermediate filaments.* None of these is membrane covered.

Microtubules (mi″kro-tu′būlz), the elements with the largest diameter, are hollow tubes made of spherical protein subunits called *tubulins* (Figure 3.24). Most microtubules radiate from a small region of cytoplasm near the nucleus called the *centrosome* (see Figure 3.26). These stiff but bendable microtubules determine the overall shape of the cell, as well as the distribution of cellular organelles. Mitochondria, lysosomes, and secretory granules attach to the microtubules like ornaments hanging from the limbs of a Christmas tree. These attached organelles are continually pulled along the microtubules and repositioned by **motor proteins** (*kinesins, dyneins,* and others) that act like train engines on the microtubular "railroad tracks" (see Figure 3.25a). Microtubules are remarkably dynamic organelles, constantly growing out from the centrosome, disassembling, and then reassembling.

Microfilaments (mi″kro-fil′ah-ments), the thinnest elements of the cytoskeleton, are strands of the protein *actin* ("ray"). Each cell has its own unique arrangement of microfilaments; thus no two cells are alike. However, nearly all cells have a fairly dense cross-linked network of microfilaments (Figure 3.24) attached to the cytoplasmic side of their plasma membrane that strengthens the cell surface. Most microfilaments are involved in cell motility or changes in cell shape. You could say that cells move when they get their act(in) together. For example, actin microfilaments interact with another protein, **myosin** (mi′o-sin), to generate contractile forces in a muscle cell (Figure 3.25b) and to form the cleavage furrow that pinches a cell in two during cell division. Microfilaments attached to CAMs (see Figure 3.4) are responsible for the crawling movements of amoeboid motion, and for the membrane changes that accompany endocytosis and exocytosis. Except in muscle cells, where they are highly developed and long-lived, microfilaments are constantly breaking down and re-forming from smaller subunits whenever and wherever their services are needed.

Intermediate filaments are tough, insoluble protein fibers that have a diameter between those of microfilaments and microtubules (Figure 3.24). Constructed like woven ropes, intermediate filaments are the most stable and permanent of the cytoskeletal elements and have high tensile strength. They act as internal guy wires to resist pulling forces exerted on the cell, and they attach to desmosomes. Because the protein composition of intermediate filaments varies in different cell types, there are numerous names for these cytoskeletal elements—for example, they are called neurofilaments in nerve cells and keratin filaments in epithelial cells.

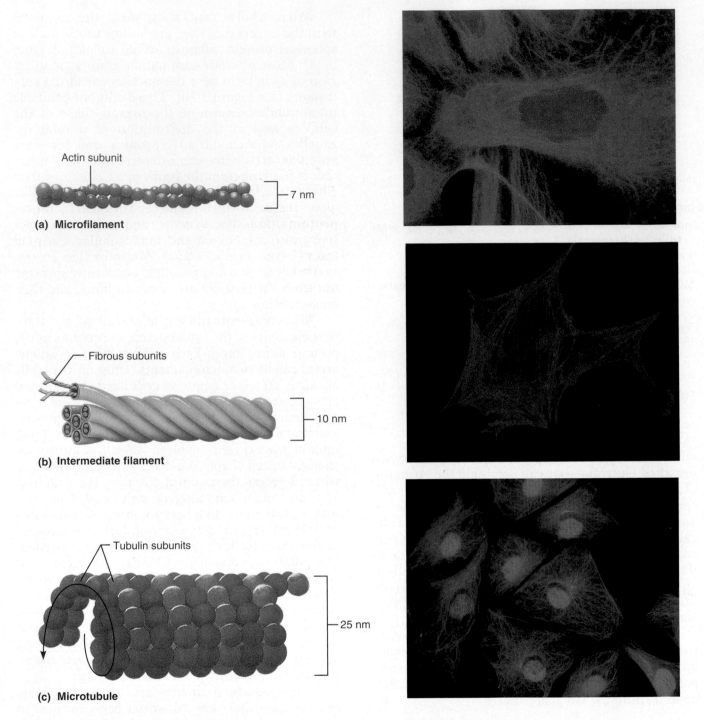

Actin subunit

— 7 nm

(a) Microfilament

Fibrous subunits

— 10 nm

(b) Intermediate filament

Tubulin subunits

— 25 nm

(c) Microtubule

FIGURE 3.24 Cytoskeleton. Diagrammatic views (left) and photos (right) of the cytoskeletal elements. **(a)** Microfilaments show up as green strands. **(b)** Intermediate filaments form a red bat-like network in this view. **(c)** Microtubules appear as green networks surrounding the cells' blue nuclei.

Centrosome and Centrioles

As mentioned, microtubules are anchored at one end in a region near the nucleus called the **centrosome.** The centrosome, which acts as a *microtubule organizing center*, has few distinguishing marks other than a granular-looking *matrix* that contains paired **centrioles,** small, barrel-shaped organelles oriented at right angles to each other (Figure 3.26). The centrosome matrix is best known for its generation of microtubules and its role of organizing the mitotic spindle (see Figure 3.32 on pp. 94–95) in cell division. Each centriole consists of a pinwheel array of nine *triplets* of microtubules, arranged to form a hollow tube. Centrioles also form the bases of cilia and flagella.

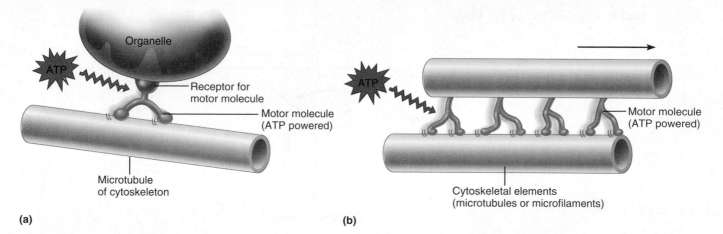

(a)

(b)

FIGURE 3.25 **Interaction of motor molecules (motor proteins) with cytoskeletal elements.** Microtubules and microfilaments function in motility by interacting with protein complexes called motor molecules. The various types of motor molecules, all powered by ATP, work by changing their shapes, moving back and forth something like microscopic legs. With each cycle of shape changes, the motor molecule releases at its free end and grips at a site farther along the microtubule or microfilament. **(a)** Motor molecules can attach to receptors on organelles, such as mitochondria or ribosomes, and enable the organelles to "walk" along the microtubules of the cytoskeleton. **(b)** In some types of cell motility, motor molecules attached to one element of the cytoskeleton can cause it to slide over another element. For example, the sliding of one set of microfilaments past another underlies muscle contraction, and the sliding of neighboring microtubules moves cilia.

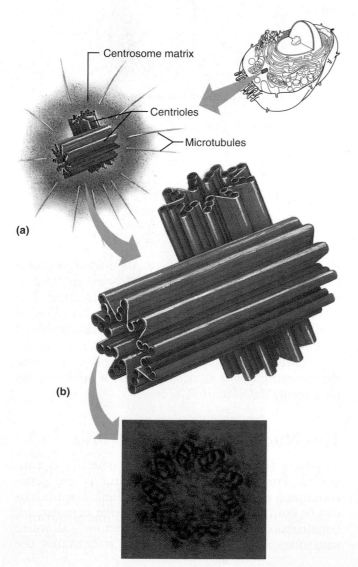

(a)

(b)

(c)

Cellular Extensions

Cilia and Flagella

Cilia (sil'e-ah; "eyelashes") are whiplike, motile cellular extensions that occur, typically in large numbers, on the exposed surfaces of certain cells. Ciliary action is important in moving substances in one direction across cell surfaces. For example, ciliated cells that line the respiratory tract propel mucus laden with dust particles and bacteria upward away from the lungs.

When a cell is about to form cilia, the centrioles multiply and line up beneath the plasma membrane at the free cell surface. Microtubules then begin to "sprout" from each centriole, forming the ciliary projections by exerting pressure on the plasma membrane. When the projections formed by centrioles are substantially longer, they are called **flagella** (flah-jel'ah). The single example of a flagellated cell in the human body is a sperm, which has one propulsive flagellum, commonly called a tail. Notice that cilia *propel other substances* across a cell's surface, whereas a flagellum *propels the cell itself.*

Centrioles forming the bases of cilia and flagella are commonly referred to as **basal bodies** (ba'sal)

FIGURE 3.26 **Centrioles. (a)** and **(b)** Three-dimensional views of a centriole pair oriented at right angles, as they are usually seen in the cell. The centrioles are located in an unconspicuous region to one side of the nucleus called the centrosome, or cell center. **(c)** An electron micrograph showing a cross section of a centriole (180,000✕). Notice that it is composed of nine microtubule triplets.

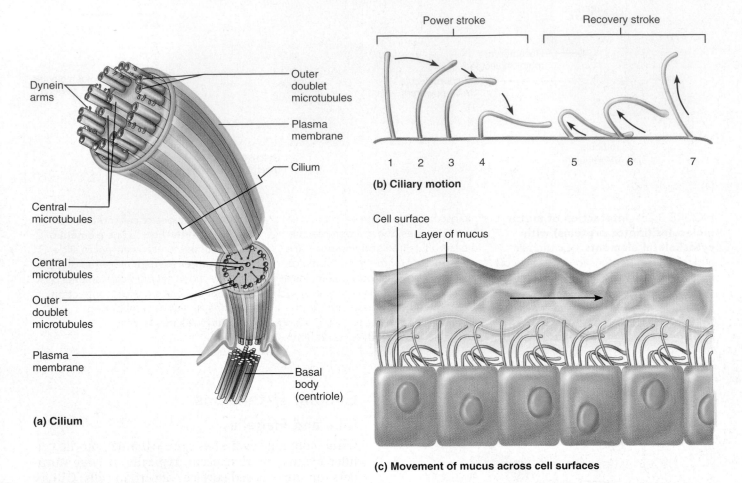

Power stroke   Recovery stroke

1   2   3   4   5   6   7

(b) Ciliary motion

Cell surface
Layer of mucus

(c) Movement of mucus across cell surfaces

Dynein arms
Outer doublet microtubules
Plasma membrane
Cilium
Central microtubules
Central microtubules
Outer doublet microtubules
Plasma membrane
Basal body (centriole)

(a) Cilium

FIGURE 3.27   Cilia structure and function. (a) Three-dimensional diagram of a cross section through a cilium, showing its nine pairs of peripheral microtubules and one central microtubule pair. **(b)** Diagram of the phases of ciliary motion. 1–4 are part of the power (propulsive) stroke; 5–7 are phases of the recovery (nonpropulsive) stroke when the cilium is resuming its initial position. **(c)** A representation of the traveling wave created by the activity of many cilia acting together to propel mucus across the cell surfaces.

(Figure 3.27a) because they were thought to be different from the structures seen in the centrosome. We now know that centrioles and basal bodies are identical. However, the pattern of microtubules in the cilium or flagellum itself (nine *doublets*, or pairs, of microtubules encircling one central pair) differs slightly from that of a centriole (nine microtubule *triplets*).

Just how cilia activity is coordinated is not fully understood, but microtubules are definitely involved. Extending from the microtubule doublets are arms composed of the motor protein *dynein*. Cilia move when the dynein side arms grip adjacent doublets and start to crawl along their length, much like cats dig in their claws to climb a tree (see Figure 3.25b). The collective bending action of all the doublets causes the cilium to bend.

As a cilium moves, it alternates rhythmically between a propulsive *power stroke*, when it is nearly straight and moves in an arc, and a *recovery stroke*,

when it bends and returns to its initial position (Figure 3.27b). With these two strokes, the cilium produces a pushing motion in a single direction. The activity of cilia in a particular region is coordinated so that the bending of one cilium *is* quickly followed by the bending of the next and then the next, creating a current at the cell surface that brings to mind the traveling waves that pass across a field of grass on a windy day (Figure 3.27c).

The Nucleus

Anything that works, works best when it is controlled. For cells, the control center is the gene-containing **nucleus** (*nucle* = pit, kernel). The nucleus can be compared to a computer, design department, construction boss, and board of directors—all rolled into one. As the genetic library, it contains the

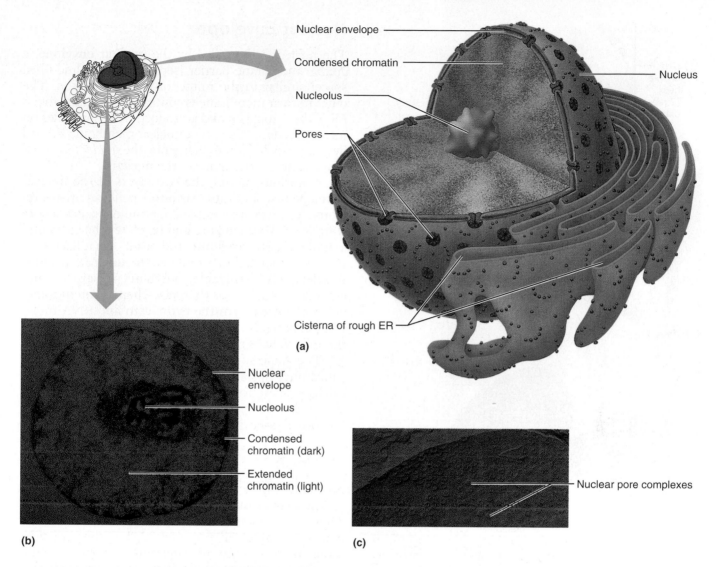

Nuclear envelope

Condensed chromatin

Nucleolus

Pores

Nucleus

Cisterna of rough ER

(a)

Nuclear envelope

Nucleolus

Condensed chromatin (dark)

Extended chromatin (light)

(b)

Nuclear pore complexes

(c)

FIGURE 3.28 The nucleus. (a) Three-dimensional diagrammatic view of the nucleus, showing the continuity of its double membrane with the ER. **(b)** Freeze-fracture electron micrograph (5,200×) of the nucleus showing nuclear envelope, nuclear pores, a nucleolus, and condensed chromatin regions. **(c)** Freeze-fracture micrograph (16,000×) of the external leaflet of the nuclear envelope. Notice the raised nuclear pore (NP) complexes.

instructions needed to build nearly all the body's proteins. Additionally, it dictates the kinds and amounts of proteins to be synthesized at any one time in response to signals acting on the cell.

Most cells have only one nucleus, but some, including skeletal muscle cells, bone destruction cells, and some liver cells, are **multinucleate** (mul″tĭ-nu′kle-āt); that is, they have many nuclei. The presence of more than one nucleus usually signifies that a larger-than-usual cytoplasmic mass must be regulated. With one exception, all of our body cells are nucleated. The exception is mature red blood cells, whose nuclei are ejected before the cells enter the bloodstream. These **anucleate** (a-nu′kle-āt; *a* = without) cells cannot reproduce and therefore live in the bloodstream for only three to four months before

they begin to deteriorate. Without a nucleus, a cell cannot produce mRNA to make proteins, and when its enzymes and cell structures start to break down (as all eventually do), they cannot be replaced.

The nucleus, averaging 5 μm in diameter, is larger than any of the cytoplasmic organelles. Although most often spherical or oval, its shape usually conforms to the shape of the cell. The nucleus has three recognizable regions or structures: the *nuclear envelope (membrane)*, *nucleoli*, and *chromatin* (Figure 3.28). Besides these structures, there are several distinct compartments (a splicing factor compartment and others) rich in specific sets of proteins. These are not limited by membranes and are in a constant state of dynamic flux. Much remains to be learned about them.

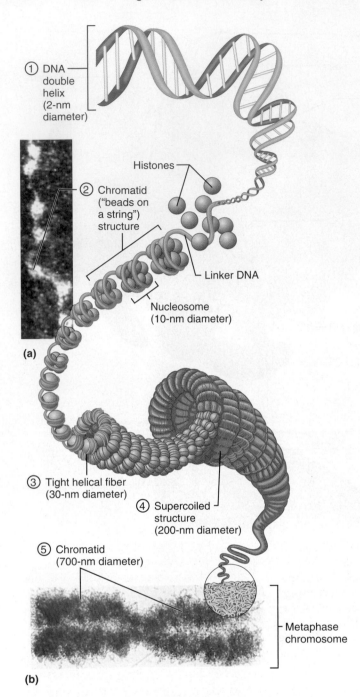

① DNA double helix (2-nm diameter)

Histones

② Chromatid ("beads on a string") structure

Linker DNA

Nucleosome (10-nm diameter)

(a)

③ Tight helical fiber (30-nm diameter)

④ Supercoiled structure (200-nm diameter)

⑤ Chromatid (700-nm diameter)

Metaphase chromosome

(b)

FIGURE 3.29 Chromatin and chromosome structure.
(a) Electron micrograph of chromatin fibers (10,000×).
(b) DNA packing in a chromosome. The levels of increasing structural complexity (coiling) from the DNA helix to the metaphase chromosome are indicated in numerical order. (Metaphase is a stage of nuclear division that occurs before the genetic material is distributed to the daughter cells.) Notice the structure of the nucleosomes, the fundamental units of chromatin that give it the "beads on a string" appearance. Each nucleosome is composed of eight histone proteins wrapped by two winds of the DNA helix.

Nuclear Envelope

The nucleus is bounded by the **nuclear envelope,** a *double* membrane barrier (separated by a fluid-filled space) similar to the mitochondrial membrane. The outer nuclear membrane is continuous with the rough ER of the cytoplasm and is studded with ribosomes on its external face. The inner nuclear membrane is lined by a network of protein filaments (the *nuclear lamina*) that maintains the shape of the nucleus.

At various points, the two layers of the nuclear envelope fuse, and **nuclear pores** penetrate these regions. An intricate complex of proteins, called a *pore complex,* lines each pore and regulates the entry and exit of large particles into and out of the nucleus.

Like other cell membranes, the nuclear envelope is selectively permeable, but here passage of substances is much freer than elsewhere. Protein molecules imported from the cytoplasm and RNA molecules exported from the nucleus pass easily through the relatively large pores.

The nuclear envelope encloses a jellylike fluid called nucleoplasm (nu'kle-o-plazm) in which other nuclear elements are suspended. Like the cytosol, the nucleoplasm contains dissolved salts, nutrients, and other essential solutes.

Nucleoli

Nucleoli (nu-kle'o-li; "little nuclei") are the dark-staining spherical bodies found within the nucleus. They are not membrane bounded. Typically, there are one or two nucleoli per nucleus, but there may be more. Because they are sites where ribosome subunits are assembled, nucleoli are usually very large in growing cells that are making large amounts of tissue proteins. Nucleoli are associated with *nucleolar organizer regions,* which contain the DNA that issues genetic instructions for synthesizing ribosomal RNA (rRNA). As molecules of rRNA are synthesized, they are combined with proteins to form the two kinds of ribosomal subunits. (The proteins are manufactured on ribosomes in the cytoplasm and "imported" into the nucleus.) Most of these subunits leave the nucleus through the nuclear pores and enter the cytoplasm, where they join to form functional ribosomes.

Chromatin

Seen through a light microscope, **chromatin** (kro'-mah-tin) appears as a fine, unevenly stained network, but special techniques reveal it as a system of bumpy threads weaving their way through the nucleoplasm (Figure 3.29a). Chromatin is composed of approximately equal amounts of **DNA,** which constitutes our genetic material, and globular **histone proteins** (his'tōn). The fundamental units of chromatin, **nucleosomes** (nu′kle-o-sōmz; "nuclear bodies"), consist of discus-shaped cores or clusters of eight histone proteins connected like beads on a string by a DNA molecule. The DNA winds (like a ribbon of Velcro) around each nucleosome and continues on to the next cluster via *linker DNA* segments (Figure 3.29a). In addition to providing a physical means for packing the very long DNA molecules in a compact, orderly way, the histones play an important role in gene regulation. In a nondividing cell, for example, addition of phosphate or methyl groups to histone exposes different DNA segments, or genes, so that they can dictate the specifications for protein synthesis. Such active chromatin segments, referred to as *extended chromatin*, are not usually visible under the light microscope. The generally inactive *condensed chromatin* segments are darker staining and so are more easily detected (see Figure 3.28b). Understandably, the most active body cells have much larger amounts of extended chromatin.

When a cell is preparing to divide, the chromatin threads coil and condense enormously to form short, barlike bodies called **chromosomes** ("colored bodies") (Figure 3.29b). Chromosome compactness avoids entanglement and breakage of the delicate chromatin strands during the movements that occur during cell division. The functions of DNA and the events of cell division are described in the next section.

Cell Growth and Reproduction

The Cell Life Cycle

The **cell life cycle,** the series of changes a cell goes through from the time it is formed until it reproduces, encompasses two major periods: *interphase,* in which the cell grows and carries on its usual activities, and *cell division,* or the *mitotic phase,* during which it divides into two cells (Figure 3.30).

Interphase

Interphase is the period from cell formation to cell division. Early cytologists, unaware of the constant molecular activity in cells and impressed by the obvious movements of cell division, called interphase the resting phase of the cell cycle. (The term *interphase* reflects this idea of a stage *between* cell divisions.) However, this image is grossly misleading because during interphase a cell is carrying out all its routine activities and is "resting" only from dividing. Perhaps a more accurate name for this phase would be *metabolic phase* or *growth phase.*

Subphases   In addition to carrying on its life-sustaining reactions, an interphase cell prepares for

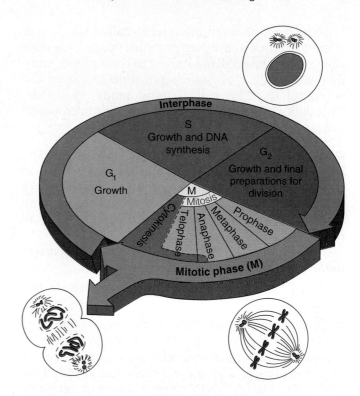

FIGURE 3.30   The cell cycle. During G_1, cells grow rapidly and carry out their routine functions; the centrioles begin to replicate as this phase ends. The S phase begins when DNA synthesis starts and ends when the DNA is replicated. In G_2, materials needed for cell division are synthesized and growth continues. During the M phase (cell division), mitosis and cytokinesis occur, producing two daughter cells.

the next cell division. Interphase is divided into G_1, S, and G_2 subphases (the Gs stand for gaps before and after the S phase; S is for synthetic). In all three subphases, the cell grows by producing proteins and organelles; however, chromatin is reproduced only during the S subphase. During **G_1 (gap 1),** the cell is metabolically active, synthesizing proteins rapidly and growing vigorously. This is the most variable phase in terms of length. In cells that divide rapidly, G_1 typically lasts several minutes to hours; in those that divide slowly, it may last for days or even years. Cells that permanently cease dividing are said to be in the **G_0 phase.** For most of G_1, virtually no activities directly related to cell division occur. However, as G_1 ends, the centrioles start to replicate in preparation for cell division.

During the **S** phase, DNA replicates itself, ensuring that the two future cells being created will receive identical copies of the genetic material. New histones are made and assembled into chromatin. One thing is sure: Without a proper S phase, there can be no correct mitotic phase. (We will describe DNA replication next.)

The final phase of interphase, called G_2, is very brief. Enzymes and other proteins needed for division are synthesized and moved to their proper sites. By the end of G_2, centriole replication (begun in G_1) is complete. The cell is now ready to divide. Throughout S and G_2, the cell continues to grow and carries on with business as usual.

DNA Replication   Before a cell can divide, its DNA must be replicated exactly, so that identical copies of the cell's genes can be passed on to each of its offspring. Replication begins simultaneously on several chromatin threads and continues until all the DNA has been replicated.

This process, still being worked out, appears to involve the following events:

1. The DNA helices begin unwinding from the nucleosomes.

2. A *helicase* enzyme untwists the double helix and gradually separates the DNA molecule into two complementary nucleotide chains, exposing the nitrogenous bases. The site of separation is called the *replication bubble;* the Y-shaped region at each end of the replication bubble is called the *replication fork* (see Figure 3.31b).

3. Each nucleotide strand then serves as a *template,* or set of instructions, for building a new complementary nucleotide strand from free DNA precursors dissolved in the nucleoplasm.

Recall from Chapter 2 that nucleotide base pairing is always complementary: *adenine (A)* bonds to *thymine (T),* and *guanine (G)* bonds to *cytosine (C).* Hence the order of the nucleotides on the template strand determines the order on the strand being built. For example, a TACTGC sequence on a template strand would bond to new nucleotides with the order ATGACG.

4. At sites where DNA synthesis is to occur, the needed "machinery" gradually accumulates until several different proteins (mostly enzymes) are present in a large complex called a **replisome.** The actual initiation of DNA synthesis requires formation of short (about ten bases long) **RNA primers** by primase enzymes which are part of the replisome (Figure 3.31a). These primers are eventually replaced by DNA nucleotides.

5. Once the primer is in place, **DNA polymerase III** comes into the picture. Continuing from the primer, it positions complementary nucleotides along the template strand and then covalently links them together. DNA polymerase works only in one direction. Consequently, one strand, the *leading strand,* is synthesized continuously following the movement of the replication fork. The other strand, called the *lagging strand,* is constructed in segments in the opposite direction and requires that a primer initiate replication of each segment.

6. The short segments of DNA are then spliced together by **DNA ligase.** The end result is that two DNA molecules are formed from the original DNA helix and are identical to it. Each new molecule consists of one old and one new nucleotide strand. This mechanism of DNA replication is referred to as **semiconservative replication** (Figure 3.31b). Replication also involves the generation of two new telomeres (*tel* = end; *mer* = piece), snugly fitting nucleoprotein caps that prevent degradation of the ends of the chromatin strands.

7. As soon as replication ends, histones (synthesized in the cytoplasm and imported into the nucleus) associate with the DNA, completing the formation of two new chromatin strands. The chromatin strands, united by a buttonlike centromere (believed to be a stretch of repetitive DNA), condense to form **chromatids** (see Figure 3.29b). The chromatids remain attached, held together by the centromere and a protein complex called *cohesin,* until the cell has entered the anaphase stage of mitotic cell division. They are then distributed to the daughter cells, as described next, ensuring that each has identical genetic information.

Cell Division

Cell division is essential for body growth and tissue repair. Cells that continually wear away, such as cells of the skin and intestinal lining, reproduce themselves almost continuously. Others, such as liver cells, divide more slowly (to maintain the particular size of the organ they compose) but retain the ability to reproduce quickly if the organ is damaged. Most cells of nervous tissue, skeletal muscle, and heart muscle lose their ability to divide when they are fully mature, and repairs are made with scar tissue (a fibrous type of connective tissue).

Events of Cell Division   In most body cells, cell division, which is called the **M (mitotic) phase** of the cell life cycle (see Figure 3.30), involves two distinct events: **mitosis** (mi-to'sis; *mit* = thread; *osis* = process), or division of the nucleus, and **cytokinesis** (si-to-ki-ne'sis; *kines* = movement), or division of the cytoplasm. A somewhat different process of nuclear division called *meiosis* (mi-o'sis) produces sex cells (ova and sperm) with only half the number of genes found in other body cells. The details of meiosis are discussed in Chapter 26. Here we concentrate on mitotic cell division.

Mitosis   Mitosis is the series of events that parcel out the replicated DNA of the mother cell to two daughter cells (Figure 3.32). It is described in terms

Text continues on page 96.

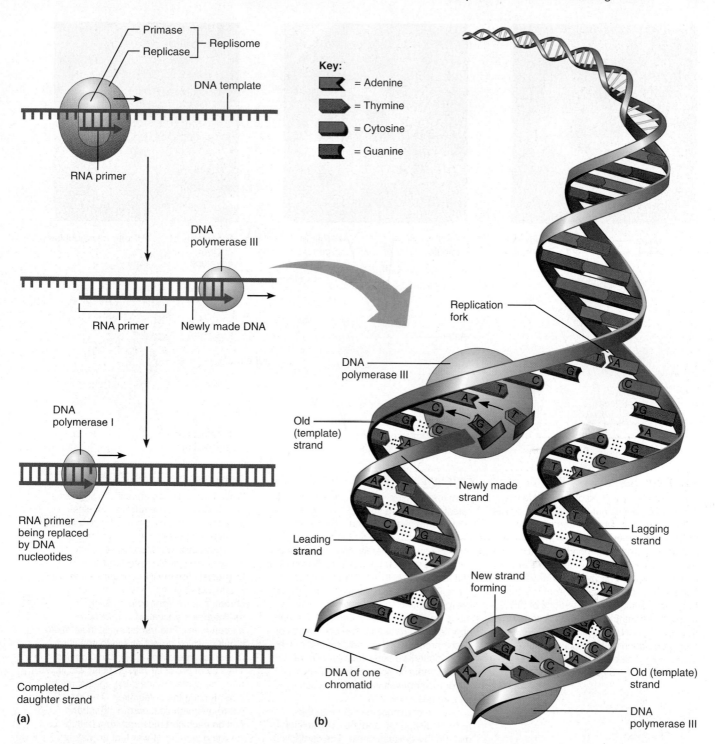

Key:
- = Adenine
- = Thymine
- = Cytosine
- = Guanine

FIGURE 3.31 **Replication of DNA.** **(a)** Role of the RNA primer. The RNA primer initiates DNA synthesis. The primer is synthesized by primase, an RNA polymerase, that is part of a protein complex called a replisome and that uses a single strand of DNA as its template. When a short stretch of RNA is available, DNA polymerase III is able to initiate DNA synthesis. The primer is eventually removed by DNA polymerase I, which replaces the ribonucleotides with deoxynucleotides. **(b)** Sequential events of DNA synthesis. The DNA helix uncoils, and the hydrogen bonds between its base pairs are broken. Then each nucleotide strand of the DNA acts as a template for the construction of a complementary nucleotide strand, as illustrated in the bottom portion of the diagram. Because the DNA polymerases work in one direction only, the two new strands (leading and lagging) are synthesized in opposite directions. Each DNA molecule formed consists of one old (template) strand and one newly assembled strand and constitutes a chromatid of a chromosome.

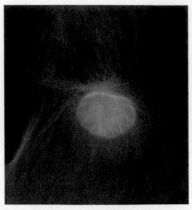

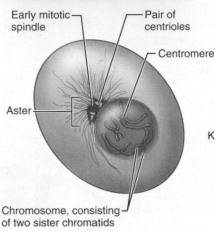

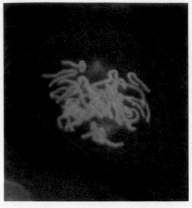

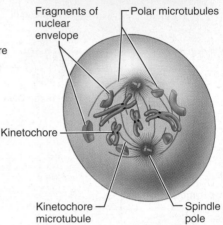

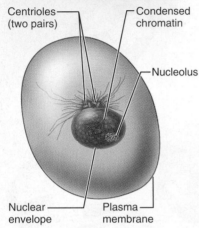

Centrioles (two pairs) — — Condensed chromatin

— Nucleolus

Nuclear envelope — — Plasma membrane

Early mitotic spindle — — Pair of centrioles

— Centromere

Aster —

Chromosome, consisting of two sister chromatids

Fragments of nuclear envelope — — Polar microtubules

Kinetochore —

Kinetochore microtubule — — Spindle pole

Interphase ➤

Interphase is the period of a cell's life when it is carrying out its normal metabolic activities and growing. During interphase, the chromosomal material is seen in the form of extended and condensed chromatin, and the nuclear envelope and nucleolus are intact and visible. Microtubule arrays (asters) are seen extending from the centrosomes. During various periods of this phase, the centrioles begin replicating (G_1 through G_2), DNA is replicated (S), and the final preparations for mitosis are completed (G_2). The centriole pair finishes replicating into two pairs during G_2.

Early prophase ➤

As mitosis begins, microtubule arrays called *asters* ("stars") are seen extending from the centrosome matrix around the centrioles. Early in *prophase,* the first and longest phase of mitosis, the chromatin threads coil and condense, forming barlike *chromosomes* that are visible with a light microscope. Since DNA replication has occurred during interphase, each chromosome is actually made up of two identical chromatin threads, now called *chromatids.* The chromatids of each chromosome are held together by a small, buttonlike body called a *centromere* and a protein complex called cohesin. (After the chromatids separate, each is considered a new chromosome.)

As the chromosomes appear, the nucleoli disappear, and the cytoskeletal microtubules disassemble. The centriole pairs separate from one another. The centrioles act as focal points for growth of a new assembly of microtubules called the **mitotic spindle.** As these microtubules lengthen, they push the centrioles farther and farther apart, propelling them toward opposite ends (poles) of the cell.

Late prophase ➤

While the centrioles are still moving away from each other, the nuclear envelope fragments, allowing the spindle to occupy the center of the cell and to interact with the chromosomes. Meanwhile, some of the growing spindle microtubules attach to special protein-DNA complexes, called *kinetochores* (ki-ne′to-korz), at each chromosome's centromere. Such microtubules are called *kinetochore microtubules.* The remaining spindle microtubules, which do not attach to any chromosomes, are called *polar microtubules.* The tips of the polar microtubules are linked near the center; these push against each other forcing the poles apart. The kinetochore microtubules, on the other hand, pull on each chromosome from both poles, resulting in a tug-of-war that ultimately draws the chromosomes to the middle of the cell.

FIGURE 3.32 The stages of mitosis. The light micrographs show dividing lung cells from a newt. The chromosomes appear blue and the microtubules green. (The red fibers are intermediate filaments.) The schematic drawings show details not visible in the micrographs. For simplicity, only four chromosomes are drawn.

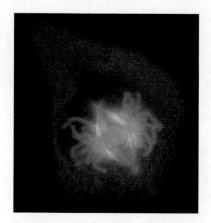

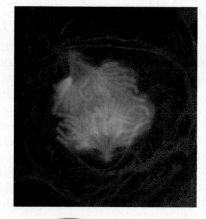

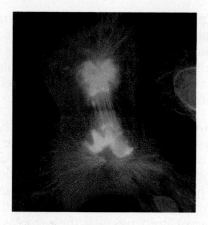

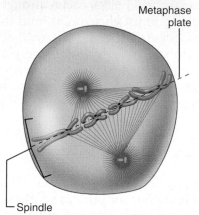

Metaphase
plate

Spindle

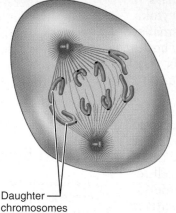

Daughter
chromosomes

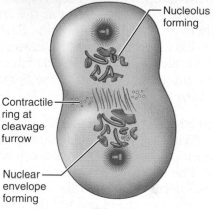

Nucleolus
forming

Contractile
ring at
cleavage
furrow

Nuclear
envelope
forming

Metaphase

Metaphase is the second phase of mitosis. The chromosomes cluster at the middle of the cell, with their centromeres precisely aligned at the exact center, or *equator,* of the spindle. This arrangement of the chromosomes along a plane midway between the poles is called the *metaphase* plate.

An enzyme called *separase* cleaves cohesin, triggering separation of the chromatids at the metaphase–anaphase transition.

Anaphase

Anaphase, the third phase of mitosis, begins abruptly as the centromeres of the chromosomes split, and each chromatid now becomes a chromosome in its own right. The kinetochore fibers, moved along by motor proteins in the kinetochores, rapidly disassemble at their kinetochore ends by removing tubulin subunits, and gradually pull each chromosome toward the pole it faces. By contrast the polar microtubules slide past each other and lengthen (a process presumed to be driven by kinesin motor molecules), and push the two poles of the cell apart, causing the cell to elongate. Anaphase is easy to recognize because the moving chromosomes look V-shaped. The centromeres, which are attached to the kinetochore microtubules, lead the way, and the chromosomal "arms" dangle behind them. Anaphase is the shortest stage of mitosis; it typically lasts only a few minutes.

This process of moving and separating the chromosomes is helped by the fact that the chromosomes are short, compact bodies. Diffuse threads of extended chromatin would tangle, trail, and break, which would damage the genetic material and result in its imprecise "parceling out" to the daughter cells.

Telophase and cytokinesis

Telophase begins as soon as chromosomal movement stops. This final phase is like prophase in reverse. The identical sets of chromosomes at the opposite poles of the cell uncoil and resume their threadlike extended-chromatin form. A new nuclear envelope, derived from the rough ER, re-forms around each chromatin mass. Nucleoli reappear within the nuclei, and the spindle breaks down and disappears. Mitosis is now ended. The cell, for just a brief period, is binucleate (has two nuclei) and each new nucleus is identical to the original mother nucleus.

As a rule, as mitosis draws to a close, *cytokinesis* completes the division of the cell into two daughter cells. Cytokinesis occurs as a contractile ring of peripheral microfilaments forms at the *cleavage furrow* and squeezes the cells apart. Cytokinesis actually begins during late anaphase and continues through and beyond telophase.

of four phases: **prophase, metaphase, anaphase,** and **telophase,** but it is actually a continuous process, with one phase merging smoothly into the next. Its duration varies according to cell type, but it typically lasts about an hour or less in human cells from start to finish.

Cytokinesis   Cytokinesis, or the division of the cytoplasm, begins during late anaphase and is completed after mitosis ends. The plasma membrane over the center of the cell (the spindle equator) is drawn inward to form a **cleavage furrow** by the activity of a *contractile ring* made of actin filaments. The furrow deepens until the cytoplasmic mass is pinched into two parts, so that at the end of cytokinesis there are two daughter cells. Each is smaller and has less cytoplasm than the mother cell, but is *genetically* identical to it. The daughter cells then enter the interphase portion of the life cycle until it is their turn to divide.

Control of Cell Division   The signals that prod cells to divide are poorly understood, but we know that the ratio of cell surface area to cell volume is important. The amount of nutrients a growing cell requires is directly related to its volume. Volume increases with the cube of cell radius, whereas surface area increases with the square of the radius. Thus, for example, a 64-fold (4^3) increase in cell volume is accompanied by only a 16-fold (4^2) increase in surface area. Consequently, the surface area of the plasma membrane becomes inadequate for nutrient and waste exchange when a cell reaches a certain critical size. Cell division solves this problem because the smaller daughter cells have a favorable surface-area-to-volume ratio. These surface-volume relationships help explain why most cells are microscopic in size.

Two other mechanisms that influence when cells divide are chemical signals (growth factors, hormones, and others) released by other cells and the availability of space. Normal cells stop proliferating when they begin touching, a phenomenon called *contact inhibition.* Cancer cells lack many of the normal controls and therefore divide wildly, which makes them dangerous to their host.

The cell life cycle is controlled by a system that has been compared to an automatic washer's timer control. Like that timer, the cell life cycle control system is driven by a built-in clock. However, just as the washer's cycle is subject to adjustments (by regulating the flow from the faucet, say, or by an internal water-level sensor) the cell cycle is regulated by both internal and external factors.

Although triggers for cell division are still being researched, we do know that a number of "switches" and checkpoints are involved. Thus far, two groups of proteins appear crucial to the ability of a cell to accomplish the S phase and enter mitosis—**cyclins** (regulatory proteins whose levels rise and fall during each life cycle) and **Cdks** (cyclin-dependent kinases), which are present in a constant concentration in the cell and are activated by binding to particular cyclins. In response to specific signals, a new batch of cyclins accumulates during each interphase (Figure 3.33a). Subsequent joining of specific Cdk and cyclin proteins initiates enzymatic cascades that phosphorylate histones and other proteins needed for the various stages of cell division. At the end of mitosis, the cyclins are abruptly destroyed by proteasomes.

A number of crucial checkpoints occur throughout interphase. One important checkpoint, and the first to be understood, occurs late in G_2 (Figure 3.33) when a threshold amount of a protein complex called **MPF** (M-phase promoting factor) is required to give the okay signal to pass the G_2 checkpoint and enter the M phase. Later in M phase, MPF is inactivated.

Protein Synthesis

In addition to directing its own replication, DNA serves as the master blueprint for protein synthesis. Although cells also make lipids and carbohydrates, DNA does not dictate their structure. DNA specifies *only* the structure of protein molecules, including the enzymes that catalyze the synthesis of all classes of biological molecules. Most of the metabolic machinery of the cell is concerned in some way with protein synthesis. This is not surprising, seeing as structural proteins constitute most of the dry cell material, and functional proteins direct almost all cellular activities. Essentially, cells are miniature protein factories that synthesize the huge variety of proteins that determine the chemical and physical nature of cells—and therefore of the whole body.

Proteins, you will recall from Chapter 2, are composed of polypeptide chains, which in turn are made up of amino acids. For purposes of this discussion, we can define a **gene** as a segment of a DNA molecule that carries instructions for creating one polypeptide chain (however, some genes specify the structure of certain varieties of RNA as their final product).

The four nucleotide bases (A, G, T, and C) are the "letters" of the genetic dictionary, and the information of DNA is found in the sequence of these bases. Each sequence of three bases, called a **triplet,** can be thought of as a "word" that specifies a particular amino acid. For example, the triplet AAA calls for the amino acid phenylalanine, and CCT calls for glycine. The sequence of triplets in

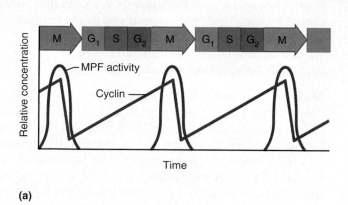

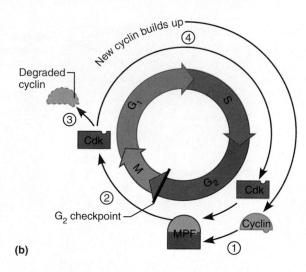

(a)

(b)

FIGURE 3.33 **Cell-cycle control at the G₂ checkpoint.** The stepwise processes of the cell cycle are timed by rhythmic fluctuations in the activity of protein kinases called cyclin-dependent kinases (Cdks), because they are active only when bound to a cyclin, a protein whose concentration varies cyclically. Here we focus on a Cdk-cyclin complex called MPF, which acts at the G₂ checkpoint to trigger mitosis. **(a)** The graph shows how MPF activity fluctuates with the level of cyclin in the cell. The cyclin level rises throughout interphase (G₁, S, and G₂ phases), then falls abruptly during mitosis (M phase). The peaks of MPF activity and cyclin concentration correspond. The Cdk itself is present at a constant level (not shown). **(b)** ① By the G₂ checkpoint (black bar), enough cyclin is available to produce many molecules of MPF. ② MPF promotes mitosis by phosphorylating various proteins, including other enzymes. ③ One effect of MPF is the initiation of a sequence of events leading to the breakdown of its own cyclin. ④ The Cdk component of MPF is recycled. Its kinase activity will be restored by association with new cyclin that accumulates during interphase.

each gene forms a "sentence" that tells exactly how a particular polypeptide is to be made; it specifies the number, kinds, and order of amino acids needed to build a particular polypeptide. Variations in the arrangement of A, T, C, and G allow our cells to make all the different kinds of proteins needed. Even an unusually "small" gene has an estimated 210 base pairs in sequence. Because the ratio between DNA bases in the gene and amino acids in the polypeptide is 3:1, the polypeptide specified by such a gene would be expected to contain 70 amino acids. Additionally, most genes of higher organisms contain **exons**, amino acid–specifying informational sequences, separated by introns. **Introns** are noncoding segments that range from 60 to 100,000 nucleotides. Hence, most genes are much larger than you might expect.

The Role of RNA

By itself, DNA is like a strip of magnetic recording tape: The information it contains cannot be used without a decoding mechanism. Furthermore, most polypeptides are manufactured at ribosomes in the cytoplasm, but in interphase cells, DNA never leaves the nucleus. So, DNA requires not only a decoder, but a messenger as well. The decoding and messenger functions are carried out by RNA, the second type of nucleic acid.

As you learned in Chapter 2, RNA differs from DNA in being single stranded and in having the sugar ribose instead of deoxyribose and the base uracil (U) instead of thymine (T). Three forms of RNA act together to carry out DNA's instructions for polypeptide synthesis:

1. **Transfer RNA (tRNA),** small, roughly cloverleaf-shaped molecules

2. **Ribosomal RNA (rRNA),** part of the ribosomes

3. **Messenger RNA (mRNA),** relatively long nucleotide strands resembling "half-DNA" molecules, or one of the two strands of a DNA molecule

All three types of RNA are formed on the DNA in the nucleus in much the same way as DNA replicates itself: The DNA helix separates and one of its strands serves as a template for synthesizing a complementary RNA strand. Once formed, the RNA molecule is released from the DNA template and migrates into the cytoplasm. Its job done, the DNA simply recoils into its helical, inactive form. Most of the nuclear DNA codes for the synthesis of short-lived mRNA, so named because it carries, from gene to ribosome, the "message" containing the instructions for building a polypeptide. Thus, the end product of most genes is a polypeptide. Only small amounts of the DNA code for the synthesis of rRNA (the nucleolar organizer regions mentioned

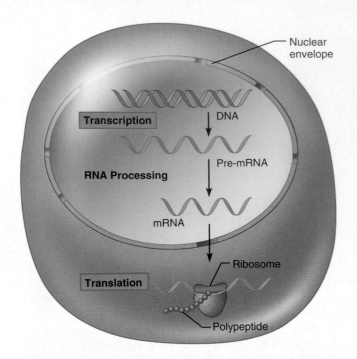

FIGURE 3.34   Simplified scheme of information flow from the DNA gene to protein structure. Information flows from DNA to mRNA during transcription. (The RNA is first synthesized as pre-mRNA, which is processed by enzymes before leaving the nucleus.) During translation, information flows from mRNA to synthesize a protein as amino acids are joined together in the designated order.

previously) and tRNA, both of which are long-lived and stable. Because rRNA and tRNA do not transport codes for synthesizing other molecules, they are the final products of the genes that code for them. Ribosomal RNA and tRNA act together to "translate" the message carried by mRNA.

The rules by which the base sequence of a gene is translated into an amino acid sequence are called the **genetic code.** Essentially, polypeptide synthesis involves two major steps: (1) *transcription,* in which DNA's information is encoded in mRNA, and (2) *translation,* in which the information carried by mRNA is decoded and used to assemble polypeptides. These steps are shown in simplified form in Figure 3.34 and described in more detail next.

Transcription

The word *transcription* often refers to one of the jobs done by a secretary—converting notes taken in shorthand into a typed or electronic copy. In other words, the same information is transformed, or transferred, from one form or format to another. In cells, **transcription** involves the transfer of information from a DNA gene's base sequence to the complementary base sequence of an mRNA molecule.

The form is different, but the same information is being conveyed. Once the mRNA molecule is made, it detaches and leaves the nucleus via a nuclear pore. Only DNA and mRNA are involved in the transcription process.

Let us follow the transcription process when a particular polypeptide is to be made.

Making the mRNA Complement   Transcription cannot begin until a gene-activating chemical called a *transcription factor* stimulates loosening of the histones at the site-to-be of gene transcription and then binds to the promoter. The **promoter** is a special DNA sequence adjacent to the *startpoint* of the structural gene that specifies where mRNA synthesis starts and which DNA strand is going to serve as the *template strand.* It typically contains an area called the *tata box* in addition to several other factors (Figure 3.35a). Once gene activation has occurred, transcription is ready to begin.

The transcription factor mediates the binding of **RNA polymerase,** the enzyme that oversees the synthesis of mRNA. Once bound, the polymerase unwinds 16 to 18 base pairs of the DNA helix at a time (Figure 3.35b). Then, using incoming ribonucleoside triphosphates as substrates, it aligns them with complementary DNA bases on the template strand and joins the RNA nucleotides together. For example, if a particular DNA triplet is AGC, the mRNA sequence synthesized at that site will be UCG. [The uncoiled DNA strand not used as a template is called the *coding strand* because the mRNA to be built will have the same (coded) sequence except for the use of U instead of T in mRNA.] As RNA polymerase elongates the mRNA strand one base at a time, it unwinds the DNA helix in front of it, and rewinds the helix behind it so that only a small region—the *DNA-RNA hybrid* (Figure 3.35b)—of the template DNA and mRNA is united at any one time. When the polymerase codes a special sequence called a *termination signal,* transcription ends and the newly formed messenger pulls off the DNA template.

FIGURE 3.35   An overview of transcription.
(a) Transcription of DNA occurs in these stages: binding of RNA polymerase to a DNA promoter; initiation of transcription on the template DNA strand; elongation of the mRNA chain; and termination of transcription accompanied by the release of RNA polymerase and the completed mRNA from the DNA template. **(b)** Elongation of mRNA. During mRNA elongation, RNA polymerase binds to about 30 base pairs of DNA. At any given moment, 16 to 18 base pairs of DNA are unwound, and the most recently synthesized mRNA is still hydrogen-bonded to the DNA, forming a short RNA-DNA hybrid. This hybrid is about 12 base pairs long or shorter.

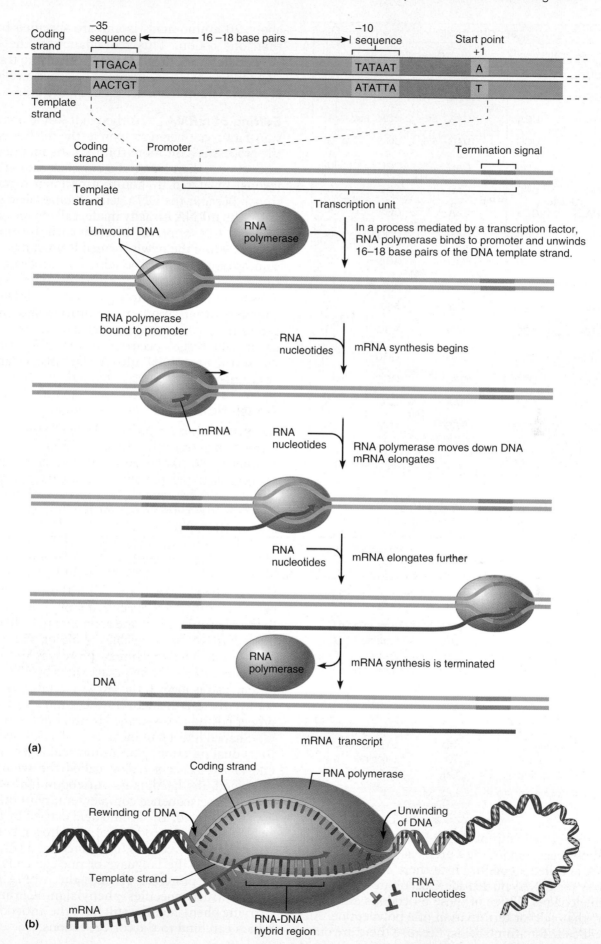

(a)

(b)

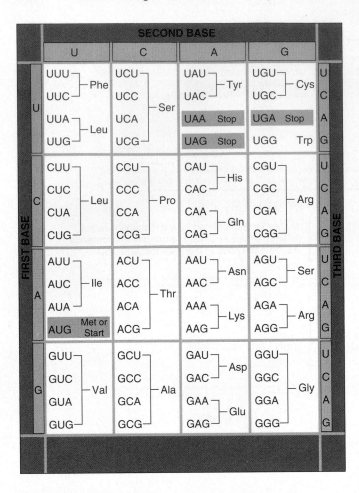

FIGURE 3.36   The genetic code. The three bases in an mRNA codon are designated as the first, second, and third, starting at the 5′ end. Each set of three specifies a particular amino acid, represented here by a three-letter abbreviation (see list below). The codon AUG (which specifies the amino acid methionine) is the usual start signal for protein synthesis. The word *stop* indicates the codons that serve as signals to terminate protein synthesis.

Abb.*	Amino acid	Abb.*	Amino acid
Ala	alanine	Leu	leucine
Arg	arginine	Lys	lysine
Asn	asparagine	Met	methionine
Asp	aspartic acid	Phe	phenylalanine
Cys	cysteine	Pro	proline
Glu	glutamic acid	Ser	serine
Gln	glutamine	Thr	threonine
Gly	glycine	Trp	tryptophan
His	histidine	Tyr	tyrosine
Ile	isoleucine	Val	valine

* Abbreviation for the amino acid

For each triplet, or three-base sequence, on DNA, the corresponding three-base sequence on mRNA is called a **codon.** Since there are four kinds of RNA (or DNA) nucleotides, there are 4^3, or 64, possible codons. Three of these 64 codons are "stop signs" that call for termination of a polypeptide. All the rest code for amino acids. Because there are only about 20 amino acids, some are specified by more than one codon. This redundancy in the genetic code helps protect against problems due to transcription (and translation) errors. The genetic code and a complete codon list is found in Figure 3.36.

Editing of mRNA   Although it would appear that translation can begin as soon as the mRNA is made, the process is a bit more complex. As mentioned before, mammalian DNA like ours has coding regions (exons) separated by noninformational regions (introns). Because the DNA gene is transcribed sequentially, the mRNA initially made, called *pre-mRNA* or *primary transcript*, is littered with intron "nonsense." Before the newly formed RNA can be used as a messenger, it must be edited or processed—sections corresponding to introns must be removed. This job is done by *spliceosomes*, large RNA-protein complexes that snip out the introns and splice together the remaining exon-coded sections in the order in which they occurred in the DNA gene, producing the functional mRNA that directs translation at the ribosome.

Translation

A translator takes a message in one language and restates it in another. In the **translation** step of protein synthesis, the language of nucleic acids (base sequence) is translated into the language of proteins (amino acid sequence). The process of translation occurs in the cytoplasm and involves all three varieties of RNA (Figure 3.37).

When it reaches the cytoplasm, the mRNA molecule carrying instructions for a particular protein binds to a small ribosomal subunit by base pairing to rRNA. Then tRNA comes into the picture, its job being to *transfer* amino acids, dissolved in the cytosol, to the ribosome. There are approximately 20 different types of tRNA, each capable of binding with a specific amino acid. The attachment process is controlled by a synthetase enzyme and is activated by ATP. Once its amino acid is loaded, the tRNA migrates to the ribosome, where it maneuvers the amino acid into the proper position, as specified by the mRNA codons.

Shaped like a handheld drill, tRNA is well suited to its dual function. The amino acid is bound to one end of tRNA, at a region called the stem. At the other end, the head, is its **anticodon** (an″ti-ko′don), a three-base sequence complementary to the mRNA codon calling for the amino acid carried by that particular tRNA. Anticodons form hydrogen bonds with complementary codons, meaning that a tRNA is the link between the language of nucleic acids and the language of proteins. Thus, if the mRNA codon is UUU, which specifies phenylalanine, the tRNAs carrying phenylalanine will have the anticodon AAA, which can bind to the correct codons.

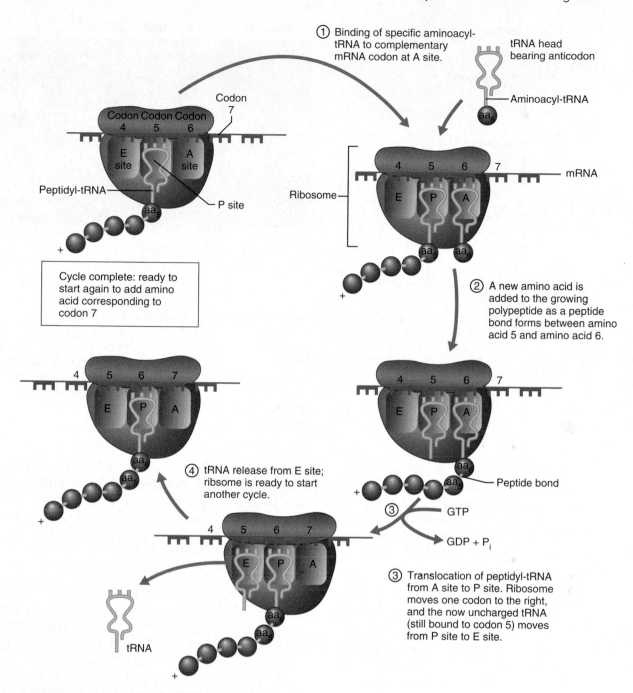

FIGURE 3.37 Polypeptide chain elongation in translation. This process, shown as a cycle, depicts the subsequent events when the polypeptide chain is five amino acids long. Following translocation (step 3) and tRNA release, the ribosome is ready to accept the next aminoacyl-tRNA.

The ribosome is more than just a passive attachment site for mRNA and tRNA. Besides its binding site for mRNA, it has three binding sites for tRNA: an A (aminoacyl) site for incoming tRNA, a P (peptidyl) site for the tRNA holding the growing polypeptide chain, and an E (exit) site for outgoing tRNA. Like a vise, the ribosome holds the tRNA and mRNA close together to coordinate the coupling of codons and anticodons, and positions the next (in-coming) amino acid for addition to the growing polypeptide chain.

Initiation of translation requires that a number of events occur. The mRNA attaches to the small ribosomal subunit by a *leader sequence* of bases that functions only to make this attachment. As a methionine-charged **initiator tRNA** also binds, a large ribosomal subunit attaches to the smaller one, forming a functional ribosome, with the

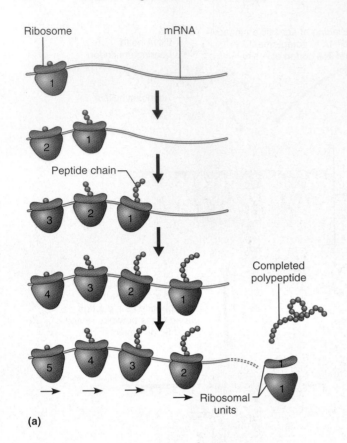

(a)

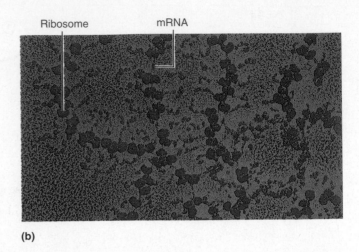

(b)

FIGURE 3.38 **Polyribosomes.** Each polyribosome consists of one strand of mRNA that is being read by several ribosomes simultaneously. **(a)** In this diagram, the mRNA is moving through the ribosome to the left and the "oldest" functional ribosome is farthest to the right. Polyribosome arrays allow a single strand of mRNA to produce hundreds of the same polypeptide molecules in a short time. **(b)** Micrograph (120,000×).

mRNA positioned in the "groove" between the two ribosomal particles. The ribosome scans along the leader sequence until it reaches the "start" codon (AUG) which is recognized and bound by the anticodon (UAC) of the initiator tRNA. This initiation process requires the help of a number of initiation factors and is energized by GTP.

When the initiator tRNA binds to the mRNA, it occupies the P site on the ribosome and the A and E sites are vacant. Now the ribosome slides the mRNA strand along, bringing the next codon into position to be "read" by an aminoacyl-tRNA coming into the A site (see ① in Figure 3.37). It is at this point that the ribosome does a little "proofreading" to make sure of the codon-anticodon match. That accomplished, ② an enzyme in the large ribosomal particle peptide-bonds the amino acid of the initiator tRNA to that of the tRNA at the A site. ③ The ribosome then translocates the tRNA that is now carrying two amino acids to the P site, and ratchets the initiator tRNA to the E site, ④ from which it moves away from the ribosome. This kind of "musical chairs" situation continues: the peptidyl-tRNAs transferring their polypeptide cargo to the aminoacyl-tRNAs, and then the P site to E site, and A site to P site movements of the tRNAs.

As mRNA is progressively read, its initial portion passes through the ribosome and may become attached successively to several other ribosomes, all reading the same message simultaneously and sequentially. Such a multiple ribosome-mRNA complex is called a *polyribosome* (Figure 3.38), and it provides an efficient system for producing many copies of the same protein. The mRNA strand continues to be read sequentially until its last codon, the *stop codon* (one of UGA, UAA, or UAG) enters the ribosomal groove. The stop codon is the "period" at the end of the mRNA sentence that tells the ribosome that translation of that mRNA is finished. The completed polypeptide chain is then released from the ribosome (Figures 3.37 and 3.38), and the ribosome separates into its two subunits.

The genetic information of a cell is translated into the production of proteins via a sequence of information transfer that is completely directed by complementary base pairing (Figure 3.39). Thus, the transfer of information goes from DNA base sequence (triplets) to the complementary base sequence of mRNA (codons) and back to the DNA base sequence (anticodons).

Cytosolic Protein Degradation

All cellular proteins are eventually degraded. Organelle proteins are digested by lysosomes, but lysosomal enzymes do not have access to damaged or unneeded soluble proteins in the cytosol that need to be disposed of. To prevent such proteins from accumulating while preventing wholesale destruction

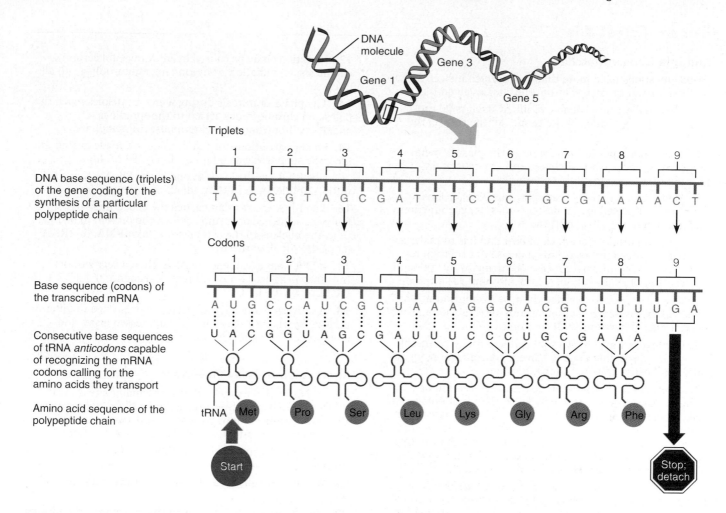

FIGURE 3.39 **Information transfer from DNA to RNA.** Information is transferred from the DNA of the gene to the complementary messenger RNA molecule, whose codons are then "read" by transfer RNA anticodons. Notice that the "reading" of the mRNA by tRNA anticodons reestablishes the base (triplet) sequence of the DNA genetic code (except that T is replaced by U).

of virtually all soluble proteins by the cytosolic enzymes, doomed proteins are marked for attack (proteolysis) by attachment of proteins called *ubiquitin* (u-bĭ'kwĭ-tin) ligases in an ATP-dependent reaction. The tagged proteins are hydrolyzed to small peptides by soluble enzymes or by giant complexes of protein-digesting enzymes called *proteasomes*. (Structurally, proteasomes are remarkably similar to one variety of molecular chaperone and may in fact be members of that protein group.)

Extracellular Materials

Extracellular materials are any substances contributing to body mass that are found outside the cells. One class of extracellular materials is *body fluids*, mainly interstitial fluid, blood plasma, and cerebrospinal fluid. These fluids are important transport and dissolving media. *Cellular secretions*, also extracellular materials, include substances that aid in digestion (intestinal and gastric fluids) and those that act as lubricants (saliva, mucus, and serous fluids).

By far the most abundant extracellular material is the *extracellular matrix*. Most body cells are in contact with a jellylike substance composed of proteins and polysaccharides. These molecules are secreted by the cells and self-assemble into an organized mesh in the extracellular space, where they serve as a universal "cell glue" that helps to hold body cells together. The extracellular matrix is particularly abundant in connective tissues—so much so, in some cases, that it (rather than living cells) accounts for the bulk of that tissue type. Depending on the structure to be formed, the extracellular matrix in connective tissue ranges from soft to rock-hard.

Review Questions

Multiple Choice/Matching

(Some questions have more than one correct answer. Select the best answer or answers from the choices given.)

1. The smallest unit capable of life by itself is (a) the organ, (b) the organelle, (c) the tissue, (d) the cell, (e) the nucleus.

2. The major types of lipid found in the plasma membranes are (choose two) (a) cholesterol, (b) neutral fats, (c) phospholipids, (d) fat-soluble vitamins.

3. Membrane junctions that allow nutrients or ions to flow from cell to cell are (a) desmosomes, (b) gap junctions, (c) tight junctions, (d) all of these.

4. A person drinks a six-pack of beer and has to make several trips to the bathroom. This increase in urination reflects an increase in what process occurring in the kidneys? (a) diffusion, (b) osmosis, (c) solute pumping, (d) filtration.

5. The term used to describe the type of solution in which cells will lose water to their environment is (a) isotonic, (b) hypertonic, (c) hypotonic, (d) catatonic.

6. Osmosis always involves (a) a differentially permeable membrane, (b) a difference in solvent concentration, (c) diffusion, (d) active transport, (e) a, b, and c.

7. A physiologist observes that the concentration of sodium inside a cell is decidedly lower than that outside the cell. Yet sodium does diffuse easily across the plasma membrane of such cells when they are dead, but *not* when they are alive. Which of the following terms applies best to this cellular function that is lacking in dead cells? (a) osmosis, (b) diffusion, (c) active transport (solute pumping), (d) dialysis.

8. The solute-pumping variety of active transport is accomplished by (a) exocytosis, (b) phagocytosis, (c) electrical forces in the cell membrane, (d) conformational-positional changes in carrier molecules in the plasma membrane.

9. The endocytotic process in which a sampling of particulate matter is engulfed and brought into the cell is called (a) phagocytosis, (b) bulk-phase endocytosis, (c) exocytosis.

10. The nuclear substance composed of histone proteins and DNA is (a) chromatin, (b) the nucleolus, (c) nuclear sap, or nucleoplasm, (d) nuclear pores.

11. The information sequence that determines the nature of a protein is the (a) nucleotide, (b) gene, (c) triplet, (d) codon.

12. Mutations may be caused by (a) X rays, (b) certain chemicals, (c) radiation from ionizing radioisotopes, (d) all of these.

13. The phase of mitosis during which centrioles reach the poles and chromosomes attach to the spindle is (a) anaphase, (b) metaphase, (c) prophase, (d) telophase.

14. Final preparations for cell division are made during the life cycle subphase called (a) G_1, (b) G_2, (c) M, (d) S.

15. The RNA synthesized on one of the DNA strands is (a) mRNA, (b) tRNA, (c) rRNA, (d) all of these.

16. The RNA species that carries the coded message, specifying the sequence of amino acids in the protein to be made, from the nucleus to the cytoplasm is (a) mRNA, (b) tRNA, (c) rRNA, (d) all of these.

17. If DNA has a sequence of AAA, then a segment of mRNA synthesized on it will have a sequence of (a) TTT, (b) UUU, (c) GGG, (d) CCC.

18. A nerve cell and a lymphocyte are presumed to differ in their (a) specialized structure, (b) suppressed genes and embryonic history, (c) genetic information, (d) a and b, (e) a and c.

Short Answer Essay Questions

19. (a) Name the organelle that is the major site of ATP synthesis. (b) Name three organelles involved in protein synthesis or modification or both. (c) Name two organelles that contain enzymes, and describe their relative functions.

20. Explain why mitosis can be thought of as cellular immortality.

21. If a cell loses or ejects its nucleus, what is its fate and why?

22. The external faces of some of the proteins in the plasma membrane have carbohydrate groups attached to them. What role do such "sugar-coated" proteins play in the life of the cell?

23. Cells are living units. However, three classes of nonliving substances are found outside the cells. What are these classes, and what functional roles do they play?

24. Comment on the role of the sodium-potassium pump in maintaining a cell's resting membrane potential.

25. Differentiate clearly between primary and secondary active transport processes.

4

TISSUE: THE LIVING FABRIC

Epithelial Tissue (pp. 106–114)

1. List several structural and functional characteristics of epithelial tissue.
2. Name, classify, and describe the various types of epithelia; also indicate their chief function(s) and location(s).
3. Define gland. Differentiate between exocrine and endocrine glands and multicellular and unicellular glands.
4. Describe how multicellular exocrine glands are classified structurally and functionally.

Connective Tissue (pp. 114–126)

5. Indicate common characteristics of connective tissue, and list and describe its structural elements.
6. Describe the types of connective tissue found in the body, and indicate their characteristic functions.

Covering and Lining Membranes (pp. 126–128)

7. Describe the structure and function of cutaneous, mucous, and serous membranes.

Nervous Tissue (p. 128)

8. Indicate the general characteristics of nervous tissue.

Muscle Tissue (pp. 128–130)

9. Compare and contrast the structures and body locations of the three types of muscle tissue.

Tissue Repair (pp. 130–132)

10. Outline the process of tissue repair involved in normal healing of a superficial wound.

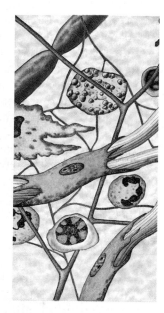

Amoebas and other unicellular (one-cell) organisms are rugged individualists. Each cell alone obtains and digests its food, ejects its wastes, and carries out all the other activities necessary to keep itself alive and "buzzin' around on all cylinders." But in the multicellular human body, cells do not operate independently. Instead, they form tight cell communities that live and work together.

Individual body cells are specialized, with each type performing specific functions that help maintain homeostasis and benefit the body as a whole. Cell specialization is obvious: How muscle cells look and act differs greatly from skin cells, which in turn are easy to distinguish from brain cells. Cell specialization allows the body to function in sophisticated ways, but division of labor has certain hazards. When a particular group of cells is indispensable, its loss or injury can severely disable or even destroy the body.

Groups of cells that are similar in structure and perform a common or related function are called **tissues** (*tissu* = woven). Four primary tissue types interweave to form the "fabric" of the body. These basic tissues are epithelial (ep"i-the'le-ul), connective, muscle, and nervous tissue, and each has numerous subclasses or varieties. If we had to assign a single term to each primary tissue type that would best describe its general role, the terms would most likely be *covering* (epithelial), *support* (connective), *movement* (muscle), and *control* (nervous). However, these terms reveal only a fraction of the functions that each tissue performs.

As explained in Chapter 1, tissues are organized into organs such as the kidneys and the heart. Most organs contain all four tissue types, and their arrangement determines the organ's structure and capabilities. The study of tissues, or **histology,** complements the study of gross anatomy. Together they provide the structural basis for understanding organ physiology.

Epithelial Tissue

Epithelial tissue, or an **epithelium** (plural: epithelia), is a sheet of cells that covers a body surface or lines a body cavity (*epithe* = laid on, covering). It occurs in the body as (1) *covering and lining epithelium* and (2) *glandular epithelium.* Covering and lining epithelium forms the outer layer of the skin, dips into and lines the open cavities of the cardiovascular, digestive, and respiratory systems, and covers the walls and organs of the closed ventral body cavity. Glandular epithelium fashions the glands of the body.

Epithelia form boundaries between different environments. For example, the epidermis of the skin lies between the inside and the outside of the body, and epithelium lining the urinary bladder separates underlying cells of the bladder wall from urine. Furthermore, nearly all substances received or given off by the body must pass through an epithelium.

In its role as an interface tissue, epithelium accomplishes many functions, including (1) protection, (2) absorption, (3) filtration, (4) excretion, (5) secretion, and (6) sensory reception. Each of these functions is described in detail later, but to illustrate these functions briefly: The epithelium of the skin protects underlying tissues from mechanical and chemical injury and bacterial invasion and contains nerve endings that respond to various stimuli acting at the skin surface (pressure, heat, etc.). The epithelium lining the digestive tract is specialized to absorb substances; and that found in the kidneys performs nearly the whole functional "menu"—excretion, absorption, secretion, and filtration. Secretion is the specialty of glands.

Special Characteristics of Epithelium

Epithelial tissues have many characteristics that distinguish them from other tissue types.

1. Cellularity. Epithelial tissue is composed almost entirely of close-packed cells. Only a tiny amount of extracellular material lies in the narrow spaces between them.

2. Specialized contacts. Epithelial cells fit close together to form continuous sheets. Adjacent cells are bound together at many points by lateral contacts, including *tight junctions* and *desmosomes* (see Chapter 3).

3. Polarity. All epithelia have an **apical surface,** an upper free surface exposed to the body exterior or the cavity of an internal organ, and a lower attached **basal surface.** Hence, all epithelia exhibit *polarity,* meaning that cell regions near the apical surface differ from those near the basal surface in both structure and function.

Although some apical surfaces are smooth and slick, most have **microvilli,** fingerlike extensions of the plasma membrane. Microvilli tremendously increase the exposed surface area, and in epithelia that absorb or secrete substances (those lining the intestine or kidney tubules, for instance), the microvilli are often so dense that the cell apices have a fuzzy appearance called a *brush border.* Some epithelia, such as that lining the trachea, have motile **cilia** that propel substances along their free surface.

Lying adjacent to the basal surface of an epithelium is a thin supporting sheet called the **basal lamina** (lam'ĭ-nah; "sheet"). This noncellular, adhesive sheet consists largely of glycoproteins secreted by the epithelial cells. The basal lamina acts as a selective filter that determines which molecules diffusing from the underlying connective tissue will be allowed to enter the epithelium. The basal lamina also acts as a scaffolding along which epithelial cells can migrate to repair a wound.

4. Supported by connective tissue. All epithelial sheets rest upon and are supported by connective tissue. Just deep to the basal lamina is the **reticular lamina,** a layer of extracellular material containing a fine network of collagen protein fibers that "belongs to" the underlying connective tissue. Together the two laminae form the **basement membrane.** The basement membrane reinforces the epithelial sheet, helping it to resist stretching and tearing forces, and defines the epithelial boundary.

Ⓗ HOMEOSTATIC IMBALANCE

An important characteristic of cancerous epithelial cells is their failure to respect this boundary, which they penetrate to invade the tissues beneath. ●

5. Avascular but innervated. Although epithelium is *innervated* (supplied by nerve fibers), it is *avascular* (contains no blood vessels). Epithelial cells are nourished by substances diffusing from blood vessels in the underlying connective tissue.

6. Regeneration. Epithelium has a high regenerative capacity. Some epithelia are exposed to friction and their surface cells rub off. Others are damaged by hostile substances in the external environment (bacteria, acids, smoke). As long as epithelial cells receive adequate nutrition, they can replace lost cells rapidly by cell division.

Classification of Epithelia

Each epithelium is given two names. The first name indicates the number of cell layers present; the second describes the shape of its cells (Figure 4.1). Based on the number of cell layers, there are simple and stratified epithelia. **Simple epithelia** are composed of a single cell layer. They are typically found where absorption and filtration occur and a thin epithelial barrier is desirable. **Stratified epithelia,** consisting of two or more cell layers stacked one on top of the other, are common in high-abrasion areas where protection is important, such as the skin surface and the lining of the mouth.

In cross section, all epithelial cells have six (somewhat irregular) sides, and an apical surface view of an epithelial sheet looks like a honeycomb. This polyhedral shape allows the cells to be closely packed. However, epithelial cells vary in height, and on that basis, there are three common shapes of epithelial cells. **Squamous** (skwa'mus) **cells** are flattened and scalelike (*squam* = scale); **cuboidal** (ku-boi'dahl) **cells** are boxlike, approximately as tall as

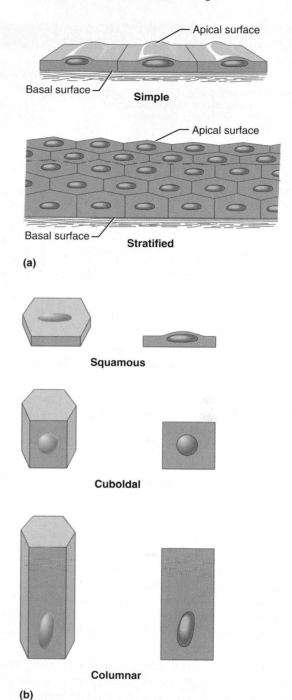

(a)

(b)

FIGURE 4.1 **Classification of epithelia. (a)** Classification on the basis of number of layers. **(b)** Classification on the basis of cell shape. For each category, a whole cell is shown on the left and a longitudinal section is shown on the right.

they are wide; and **columnar** (kŏ-lum'nar) **cells** are tall and column shaped. In each case, the shape of the nucleus conforms to that of the cell. The nucleus of a squamous cell is a flattened disc; that of a cuboidal cell is spherical; and a columnar cell nucleus is elongated from top to bottom and usually located close to the cell base. Keep nuclear shape in mind when you attempt to identify epithelial types.

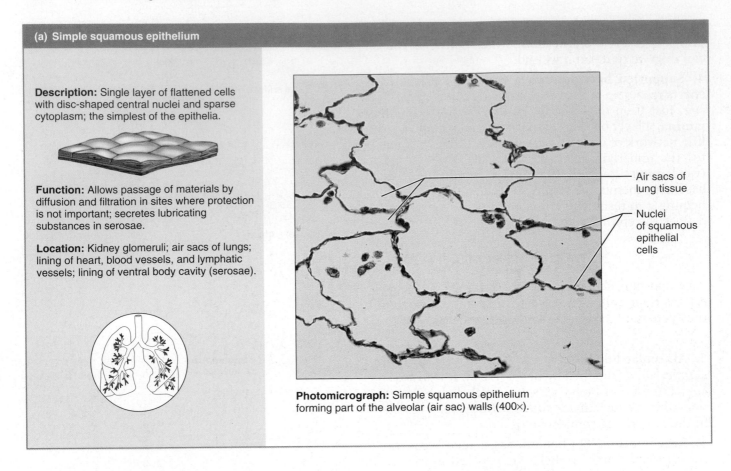

(a) Simple squamous epithelium

Description: Single layer of flattened cells with disc-shaped central nuclei and sparse cytoplasm; the simplest of the epithelia.

Function: Allows passage of materials by diffusion and filtration in sites where protection is not important; secretes lubricating substances in serosae.

Location: Kidney glomeruli; air sacs of lungs; lining of heart, blood vessels, and lymphatic vessels; lining of ventral body cavity (serosae).

Air sacs of lung tissue

Nuclei of squamous epithelial cells

Photomicrograph: Simple squamous epithelium forming part of the alveolar (air sac) walls (400×).

FIGURE 4.2   **Epithelial tissues.** Simple epithelia **(a** and **b).**

Simple epithelia are easy to classify by cell shape because all cells in the layer usually have the same shape. In stratified epithelia, however, the cell shapes usually differ among the different cell layers. To avoid ambiguity, stratified epithelia are named according to the shape of the cells in the *apical* layer. This naming system will become clearer as we explore the specific epithelial types.

As you read about the epithelial classes, study Figure 4.2. Using the photomicrographs, try to pick out the individual cells within each epithelium. This is not always easy, because the boundaries between epithelial cells often are indistinct. Furthermore, the nucleus of a particular cell may or may not be visible, depending on the precise plane of the cut made to prepare the tissue slides.

Simple Epithelia

The simple epithelia are most concerned with absorption, secretion, and filtration. Because they consist of a single cell layer and are usually very thin, protection is not one of their specialties.

Simple Squamous Epithelium   The cells of a **simple squamous epithelium** are flattened laterally, and their cytoplasm is sparse (Figure 4.2a). In a surface view, the close-fitting cells resemble a tiled floor. When the cells are cut perpendicular to their free surface, they resemble fried eggs seen from the side, with their cytoplasm wisping out from the slightly bulging nucleus. Thin and often permeable, this epithelium is found where filtration or the exchange of substances by rapid diffusion is a priority. In the kidneys, simple squamous epithelium forms part of the filtration membrane; in the lungs, it forms the walls of the air sacs across which gas exchange occurs.

Two simple squamous epithelia in the body have special names that reflect their location. **Endothelium** (en"do-the'le-um; "inner covering") provides a slick, friction-reducing lining in lymphatic vessels and in all hollow organs of the cardiovascular system—blood vessels and the heart. Capillaries consist exclusively of endothelium, where its exceptional thinness encourages the efficient exchange of nutrients and wastes between the bloodstream and surrounding tissue cells. **Mesothelium** (mez"o-the'le-um; "middle covering") is the epithelium found in serous membranes lining the ventral body cavity and covering its organs.

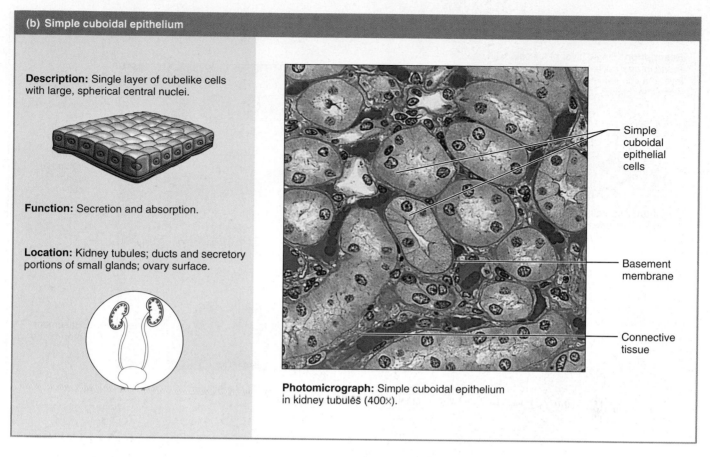

(b) Simple cuboidal epithelium

Description: Single layer of cubelike cells with large, spherical central nuclei.

Function: Secretion and absorption.

Location: Kidney tubules; ducts and secretory portions of small glands; ovary surface.

Simple cuboidal epithelial cells

Basement membrane

Connective tissue

Photomicrograph: Simple cuboidal epithelium in kidney tubules (400×).

FIGURE 4.2 *(continued)* Simple epithelium **(b)**.

Simple Cuboidal Epithelium   Simple cuboidal epithelium consists of a single layer of cells as tall as they are wide (Figure 4.2b). The spherical nuclei stain darkly, causing the cell layer to look like a string of beads when viewed microscopically. Important functions of simple cuboidal epithelium are secretion and absorption. This epithelium forms the walls of the smallest ducts of glands and of many kidney tubules.

Simple Columnar Epithelium   **Simple columnar epithelium** is seen as a single layer of tall, closely packed cells, aligned like soldiers in a row (Figure 4.2c). It lines the digestive tract from the stomach through the rectum. Columnar cells are mostly associated with absorption and secretion, and the digestive tract lining has two distinct modifications that make it ideal for that dual function: (1) dense microvilli on the apical surface of absorptive cells and (2) **goblet cells** that secrete a protective lubricating mucus. The goblet cells are named for their goblet-shaped "cups" of mucus that occupy most of the apical cell volume (see Figure 4.2d). Some simple columnar epithelia display cilia on their free surfaces, which help to move substances or cells through an internal passageway.

Pseudostratified Columnar Epithelium   The cells of **pseudostratified columnar epithelium** (soo"do-stră′tĭ-fīd) vary in height (Figure 4.2d). All of its cells rest on the basement membrane, but only the tallest reach the free surface of the epithelium. Because the cell nuclei lie at different levels above the basement membrane, it gives the false (pseudo) impression that several cell layers are present, hence "pseudostratified." The short cells are relatively unspecialized and give rise to the taller cells. This epithelium, like the simple columnar variety, secretes or absorbs substances. A ciliated version containing goblet cells lines most of the respiratory tract. Here the motile cilia propel sheets of dust-trapping mucus superiorly away from the lungs.

Stratified Epithelia

Stratified epithelia contain two or more cell layers. They regenerate from below; that is, the basal cells divide and push apically to replace the older surface cells. Considerably more durable than the simple epithelia, the major (but not the only) role of stratified epithelia is protection.

(c) Simple columnar epithelium

Description: Single layer of tall cells with *round* to *oval* nuclei; some cells bear cilia; layer may contain mucus-secreting unicellular glands (goblet cells).

Function: Absorption; secretion of mucus, enzymes, and other substances; ciliated type propels mucus (or reproductive cells) by ciliary action.

Location: Nonciliated type lines most of the digestive tract (stomach to anal canal), gallbladder, and excretory ducts of some glands; ciliated variety lines small bronchi, uterine tubes, and some regions of the uterus.

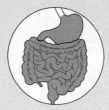

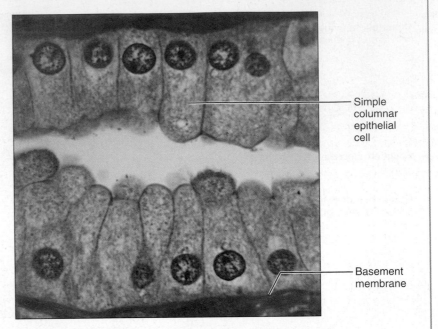

Simple columnar epithelial cell

Basement membrane

Photomicrograph: Simple columnar epithelium of the stomach mucosa (1300×).

(d) Pseudostratified columnar epithelium

Description: Single layer of cells of differing heights, some not reaching the free surface; nuclei seen at different levels; may contain goblet cells and bear cilia.

Function: Secretion, particularly of mucus; propulsion of mucus by ciliary action.

Location: Nonciliated type in male's sperm-carrying ducts and ducts of large glands; ciliated variety lines the trachea, most of the upper respiratory tract.

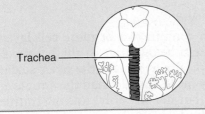

Trachea

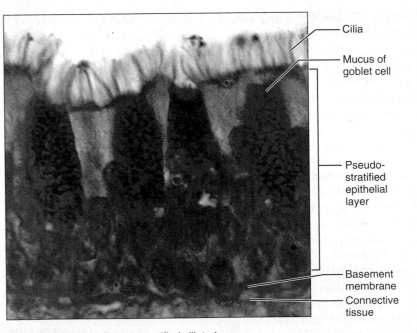

Cilia

Mucus of goblet cell

Pseudo-stratified epithelial layer

Basement membrane

Connective tissue

Photomicrograph: Pseudostratified ciliated columnar epithelium lining the human trachea (400×).

FIGURE 4.2 *(continued)* Simple epithelia (**c** and **d**).

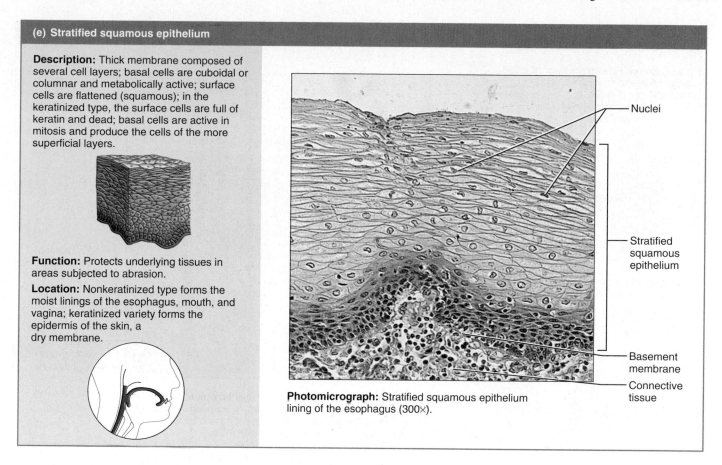

(e) Stratified squamous epithelium

Description: Thick membrane composed of several cell layers; basal cells are cuboidal or columnar and metabolically active; surface cells are flattened (squamous); in the keratinized type, the surface cells are full of keratin and dead; basal cells are active in mitosis and produce the cells of the more superficial layers.

Function: Protects underlying tissues in areas subjected to abrasion.

Location: Nonkeratinized type forms the moist linings of the esophagus, mouth, and vagina; keratinized variety forms the epidermis of the skin, a dry membrane.

Nuclei

Stratified squamous epithelium

Basement membrane

Connective tissue

Photomicrograph: Stratified squamous epithelium lining of the esophagus (300×).

FIGURE 4.2 *(continued)* Stratified epithelium **(e)**.

Stratified Squamous Epithelium Stratified **squamous epithelium** is the most widespread of the stratified epithelia (Figure 4.2e). Composed of several layers, it is thick and well suited for its protective role in the body. Its free surface cells are squamous; cells of the deeper layers are cuboidal or columnar. This epithelium is found in areas subjected to wear and tear, and its surface cells are constantly being rubbed away and replaced by division of its basal cells. Since epithelium depends on diffusion of nutrients from a deeper connective tissue layer, the epithelial cells farther from the basement membrane are less viable and those at the apical surface are often flattened and atrophied.

To avoid memorizing all its locations simply remember that this epithelium forms the external part of the skin and extends a short distance into every body opening that is directly continuous with the skin. The outer layer, or *epidermis*, of the skin is *keratinized* (ker′ah-tin″īzd), meaning its surface cells contain *keratin*, a tough protective protein. (The epidermis is discussed in Chapter 5.) The other stratified squamous epithelia of the body are *nonkeratinized*.

Stratified Cuboidal and Columnar Epithelia **Stratified cuboidal epithelium** is quite rare in the body, mostly found in the ducts of some of the larger glands (sweat glands, mammary glands). Where it is found, there are typically two layers of cuboidal cells.

Stratified columnar epithelium also has a limited distribution in the body. Small amounts are found in the pharynx, the male urethra, and lining some glandular ducts. This epithelium also occurs at transition areas or junctions between two other types of epithelia. Only its apical layer of cells is columnar. Because of their relative scarcity in the body, these two stratified epithelia are not illustrated in Figure 4.2.

Transitional Epithelium **Transitional epithelium** forms the lining of hollow urinary organs, which stretch as they fill with urine (Figure 4.2f). Cells of its basal layer are cuboidal or columnar. The apical cells vary in appearance, depending on the degree of distension of the organ. When the organ is distended with urine, the transitional epithelium thins from about six cell layers to three, and its domelike apical cells flatten and become squamous-like. The ability of transitional cells to change their shape (undergo "transitions") allows a greater

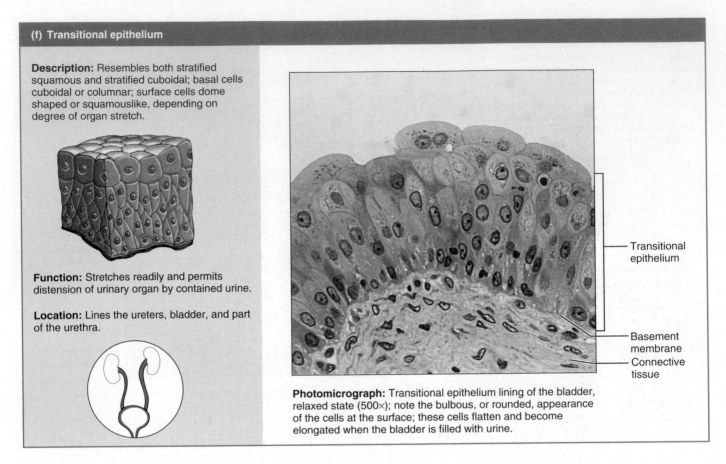

(f) Transitional epithelium

Description: Resembles both stratified squamous and stratified cuboidal; basal cells cuboidal or columnar; surface cells dome shaped or squamouslike, depending on degree of organ stretch.

Function: Stretches readily and permits distension of urinary organ by contained urine.

Location: Lines the ureters, bladder, and part of the urethra.

Transitional epithelium

Basement membrane

Connective tissue

Photomicrograph: Transitional epithelium lining of the bladder, relaxed state (500×); note the bulbous, or rounded, appearance of the cells at the surface; these cells flatten and become elongated when the bladder is filled with urine.

FIGURE 4.2 *(continued)*   Stratified epithelia **(f)**.

volume of urine to flow through a tubelike organ. In the bladder, it allows more urine to be stored.

Glandular Epithelia

A **gland** consists of one or more cells that make and secrete (export) a particular product. This product, called a **secretion,** is an aqueous (water-based) fluid that usually contains proteins, but there is variation—for example, some glands release a lipid- or steroid-rich secretion. Secretion is an active process; glandular cells obtain needed substances from the blood and transform them chemically into a product that is then discharged from the cell. Notice that the term *secretion* can refer to both the gland's *product* and the *process* of making and releasing that product.

Glands are classified as *endocrine* ("internally secreting") or *exocrine* ("externally secreting") depending on where they release their product, and as *unicellular* ("one-celled") or *multicellular* ("many-celled") based on the relative cell number making up the gland. Unicellular glands are scattered within epithelial sheets. By contrast, most multicellular epithelial glands form by invagination (inward growth) or evagination (outward growth) from an epithelial

sheet and, at least initially, most have **ducts,** tubelike connections to the epithelial sheets.

Endocrine Glands

Because **endocrine glands** eventually lose their ducts, they are often called **ductless glands.** They produce **hormones,** regulatory chemicals that they secrete by exocytosis directly into the extracellular space. From there the hormones enter the blood or lymphatic fluid and travel to specific target organs. Each hormone prompts its target organ(s) to respond in some characteristic way. For example, hormones produced by certain intestinal cells cause the pancreas to release enzymes that help digest food in the digestive tract.

Endocrine glands are structurally diverse, so one description does not fit all. Although most endocrine glands are compact multicellular organs, some individual hormone-producing cells are scattered in the digestive tract mucosa and in the brain, giving rise to their collective description as the *diffuse endocrine system.* Their secretions are also varied, ranging from modified amino acids to peptides, glycoproteins, and steroids. Since not all endocrine glands are epithelial derivatives, we defer consideration of their structure and function to Chapter 15.

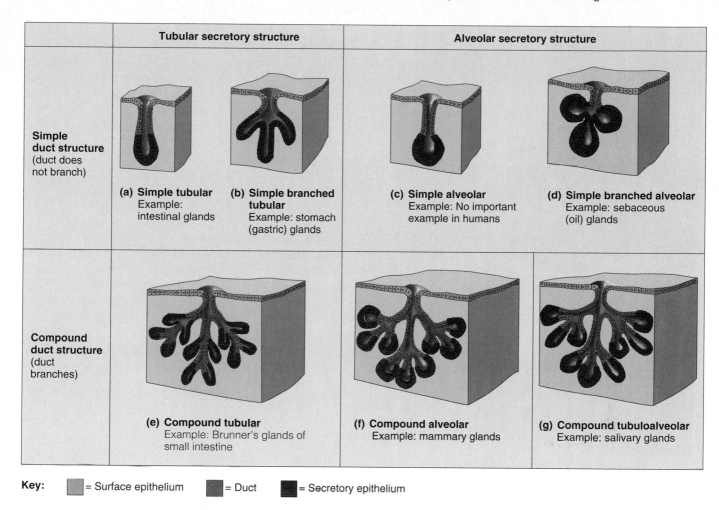

Key: ■ = Surface epithelium ■ = Duct ■ = Secretory epithelium

FIGURE 4.3 **Types of multicellular exocrine glands.** Multicellular glands are classified according to duct type (simple or compound) and the structure of their secretory units (tubular, alveolar, or tubuloalveolar).

Exocrine Glands

Exocrine glands are far more numerous than endocrine glands, and many of their products are familiar. All exocrine glands secrete their products onto body surfaces (skin) or into body cavities—the unicellular glands directly (by exocytosis) and the multicellular glands via an epithelial-walled duct that transports the secretion to the epithelial surface. Exocrine glands are a diverse lot. They include mucous, sweat, oil, and salivary glands, the liver (which secretes bile), the pancreas (which synthesizes digestive enzymes), and many others.

Unicellular Exocrine Glands   The only important example of a **unicellular** (or one-celled) **gland** is the *goblet cell.* True to its name, a goblet cell is shaped like a goblet (a drinking glass with a stem). Goblet cells are sprinkled in the epithelial linings of the intestinal and respiratory tracts amid columnar cells with other functions (see Figure 4.2d). In humans, all such glands produce **mucin** (mu'sin), a complex glycoprotein that dissolves in water when secreted. Once dissolved, mucin forms **mucus,** a slimy coating that both protects and lubricates surfaces.

Multicellular Exocrine Glands   Compared to the unicellular glands, **multicellular exocrine glands** are structurally more complex. They have two basic parts: an epithelium-derived *duct* and a *secretory unit* consisting of secretory cells (acini). In all but the simplest glands, *supportive connective tissue* surrounds the secretory unit and supplies it with blood vessels and nerve fibers, and forms a *fibrous capsule* that extends into the gland proper and divides the gland into *lobes.*

▪ **Structural classification.** On the basis of their duct structures, multicellular exocrine glands are either simple or compound (Figure 4.3). **Simple glands** have an unbranched duct, whereas **compound glands** have a branched duct. The glands are further categorized by their secretory units as (1) **tubular** if the secretory cells form tubes; (2) **alveolar**

Which of the gland types illustrated would have the highest rate of cellular mitosis? Why?

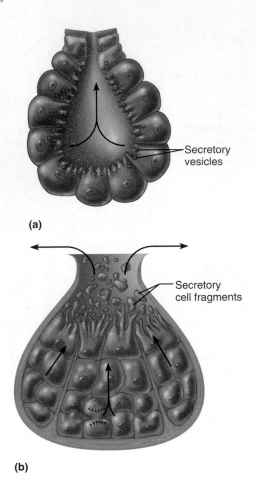

(a)

(b)

FIGURE 4.4   **Chief modes of secretion in human exocrine glands. (a)** Merocrine glands secrete their products by exocytosis. **(b)** In holocrine glands, the entire secretory cell ruptures, releasing secretions and dead cell fragments.

(al-ve'o-lar) if the secretory cells form small, flask-like sacs (*alveolus* = "small hollow cavity"); and (3) **tubuloalveolar** if they have both types of secretory units. Note that the term **acinar** (as'ĭ-nar; "berrylike") is used interchangeably with alveolar.

■ **Modes of secretion.** Multicellular exocrine glands secrete their products in different ways, so they can also be described functionally. Most are **merocrine glands** (mer'o-krin), which secrete their products by exocytosis as they are produced. The secretory cells are not altered in any way. The pancreas, most sweat glands, and salivary glands belong to this class (Figure 4.4a).

The holocrine gland, because cells that are fragmented and secreted have to be replaced. ■

Secretory cells of **holocrine glands** (hol'o-krin) accumulate their products within them until they rupture. (They are replaced by the division of underlying cells.) Because holocrine gland secretions include the synthesized product plus dead cell fragments (*holos* = all), you could say that their cells "die for their cause." Sebaceous (oil) glands of the skin are the only true example of holocrine glands (Figure 4.4b).

Although *apocrine* (ap'o-krin) *glands* are definitely present in other animals, there is some controversy over whether humans have this third gland type. Like holocrine glands, apocrine glands accumulate their products, but in this case only just beneath the free surface. Eventually, the apex of the cell pinches off (*apo* = from, off), releasing the secretory granules and a small amount of cytoplasm. The cell repairs its damage and the process repeats again and again. The best possibility in humans is the release of lipid droplets by the mammary glands, but most histologists classify mammary glands as merocrine glands because this is the means by which milk proteins are secreted.

Connective Tissue

Connective tissue is found everywhere in the body. It is the most abundant and widely distributed of the primary tissues, but its amount in particular organs varies. For example, skin consists primarily of connective tissue, while the brain contains very little. There are four main classes of connective tissue and several subclasses. The main classes are (1) *connective tissue proper* (which includes fat and the fibrous tissue of ligaments), (2) *cartilage*, (3) *bone tissue*, and (4) *blood*.

Connective tissue does much more than just *connect* body parts; it has many forms and functions. Its major functions include (1) *binding and support*, (2) *protection*, (3) *insulation*, and as blood, (4) *transportation* of substances within the body. For example, bone and cartilage support and protect body organs by providing the hard underpinnings of the skeleton; fat cushions insulate and protect body organs and provide reserve energy fuel.

Common Characteristics of Connective Tissue

Despite their many and diverse functions in the body, connective tissues have some common properties that set them apart from other primary tissues:

1. Common origin. All connective tissues arise from **mesenchyme** (an embryonic tissue) and hence have a common kinship (Figure 4.5).

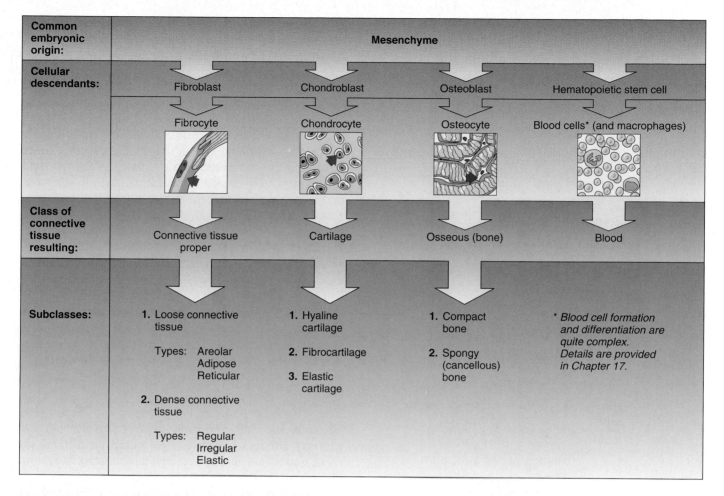

Common embryonic origin:	Mesenchyme			
Cellular descendants:	Fibroblast	Chondroblast	Osteoblast	Hematopoietic stem cell
	Fibrocyte	Chondrocyte	Osteocyte	Blood cells* (and macrophages)
Class of connective tissue resulting:	Connective tissue proper	Cartilage	Osseous (bone)	Blood
Subclasses:	**1.** Loose connective tissue Types: Areolar Adipose Reticular **2.** Dense connective tissue Types: Regular Irregular Elastic	**1.** Hyaline cartilage **2.** Fibrocartilage **3.** Elastic cartilage	**1.** Compact bone **2.** Spongy (cancellous) bone	*Blood cell formation and differentiation are quite complex. Details are provided in Chapter 17.*

FIGURE 4.5 **Major classes of connective tissue.** All these classes arise from the same embryonic tissue type (mesenchyme).

2. Degrees of vascularity. Connective tissues run the entire gamut of vascularity. Cartilage is avascular; dense connective tissue is poorly vascularized; and the other types of connective tissue have a rich supply of blood vessels.

3. Extracellular matrix. Whereas all other primary tissues are composed mainly of cells, connective tissues are largely nonliving **extracellular matrix** (ma′triks; "womb"), which separates, often widely, the living cells of the tissue. Because of its matrix, connective tissue is able to bear weight, withstand great tension, and endure abuses, such as physical trauma and abrasion, that no other tissue would be able to tolerate.

Structural Elements of Connective Tissue

Connective tissues have three main elements: *ground substance, fibers,* and *cells.* Ground substance and fibers make up the extracellular matrix. (Note that some authors use the term *matrix* to indicate the ground substance only.)

The properties of the cells and the composition and arrangement of extracellular matrix elements vary tremendously. The result is an amazing diversity of connective tissues, each adapted to perform its specific function in the body. For example, the matrix can be delicate and fragile to form a soft "packing" around an organ, or it can form "ropes" (tendons and ligaments) of incredible strength. Nonetheless, connective tissues have a common structural plan, and we use *areolar* (ah-re′o-lar) *connective tissue* as our *prototype,* or model, for this group of tissues (see Figure 4.7). All other subclasses are simply variants of this common tissue type.

Ground Substance

Ground substance is the unstructured material that fills the space between the cells and contains the fibers. It is composed of *interstitial (tissue) fluid, cell adhesion proteins,* and *proteoglycans* (pro″te-o-gli′kanz). Cell adhesion proteins *(fibronectin, laminin,* and others) serve mainly as a connective tissue glue that allows connective tissue cells to

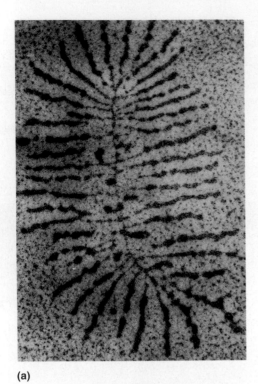

(a)

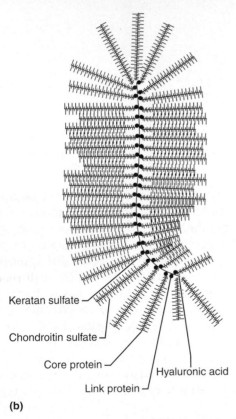

Keratan sulfate

Chondroitin sulfate

Core protein

Link protein

Hyaluronic acid

(b)

FIGURE 4.6 **Proteoglycan structure in bovine cartilage. (a)** An electron micrograph of a proteoglycan aggregate. **(b)** A schematic drawing of the same structure. Keratan sulfate and chondroitin sulfate are linked to core protein molecules, which attach to a long hyaluronic acid molecule with the aid of a link protein.

attach themselves to matrix elements. The proteoglycans consist of a protein core to which *glycosaminoglycans* (gli"kos-ah-me"no-gli'kanz) (GAGs) are attached. The strandlike GAGs are large, negatively charged polysaccharides that stick out from the core protein like the fibers of a bottle brush (Figure 4.6). Important examples of GAGs in connective tissues include *chondroitin* and *keratan sulfates* and *hyaluronic* (hi"ah-lu-ron'ik) *acid.* The proteoglycans tend to form huge aggregates (often attached to a hyaluronic acid molecule). The GAGs intertwine and trap water, forming a substance that varies from a fluid to a viscous gel. In general, the higher its GAG content, the more viscous the ground substance is.

The ground substance holds large amounts of fluid and functions as a molecular sieve, or medium, through which nutrients and other dissolved substances can diffuse between the blood capillaries and the cells. The fibers embedded in the ground substance make it less pliable and impede diffusion somewhat.

Fibers

The *fibers* of connective tissue provide support. Three types of fibers are found in connective tissue matrix: collagen, elastic, and reticular fibers. Of these, collagen fibers are by far the strongest and most abundant.

Collagen fibers are constructed primarily of the fibrous protein *collagen.* Collagen molecules are secreted into the extracellular space, where they assemble spontaneously into cross-linked fibers, which in turn are bundled together into the thick collagen fibers seen with a microscope. Collagen fibers are extremely tough and provide high tensile strength (that is, the ability to resist longitudinal stress) to the matrix. Indeed, stress tests show that collagen fibers are stronger than steel fibers of the same size! When fresh, they have a glistening white appearance; they are therefore also called *white fibers.*

Elastic fibers are long, thin fibers that form branching networks in the extracellular matrix. These fibers contain a rubberlike protein, *elastin,* that allows them to stretch and recoil like rubber bands. Connective tissue can stretch only so much before its thick, ropelike collagen fibers become taut. Then, when the tension lets up, elastic fibers snap the connective tissue back to its normal length and shape. Elastic fibers are found where greater elasticity is needed, for example, in the skin, lungs, and blood vessel walls. Because fresh elastic fibers appear yellow, they are sometimes called *yellow fibers.*

Reticular fibers are fine collagenous fibers (with a slightly different chemistry and form) and are continuous with collagen fibers. They branch

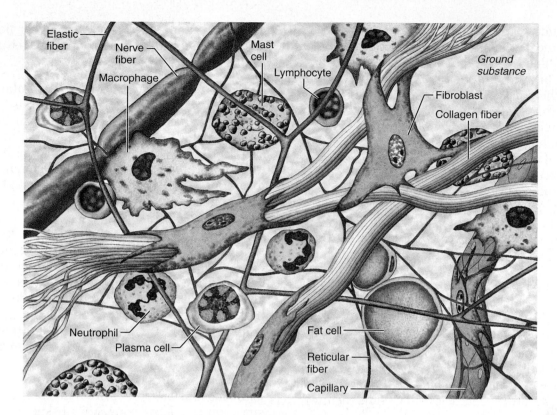

FIGURE 4.7 **Areolar connective tissue: A prototype (model) connective tissue.** This tissue underlies epithelia and surrounds capillaries. Note the various cell types and the three classes of fibers (collagen, reticular, elastic) embedded in the ground substance. (See Figure 4.8b for a less idealized version.)

extensively, forming delicate networks (*reticul* = network) that surround small blood vessels and support the soft tissue of organs. They are particularly abundant where connective tissue abuts other tissue types, for example, in the basement membrane of epithelial tissues, and around capillaries, where they form fuzzy "nets" that allow more "give" than the larger collagen fibers.

Cells

Each major class of connective tissue has a fundamental cell type that exists in immature and mature forms (see Figure 4.5). The undifferentiated cells, indicated by the suffix *blast* (literally, "bud," or "sprout," but meaning "forming"), are actively mitotic cells that secrete the ground substance and the fibers characteristic of their particular matrix. The primary blast cell types by connective tissue class are (1) connective tissue proper: **fibroblast;** (2) cartilage: **chondroblast** (kon'dro-blast"); (3) bone: **osteoblast** (os'te-o-blast"); and (4) blood: **hematopoietic stem cell.**

Once they synthesize the matrix, the blast cells assume their less active, mature mode, indicated by the suffix *cyte* (see Figure 4.5). The mature cells maintain the health of the matrix. However, if the

matrix is injured, they can easily revert to their more active state to repair and regenerate the matrix. (The blood-forming hematopoietic stem cell found in bone marrow is always actively mitotic.)

Additionally, connective tissue is home to an assortment of other cell types, such as nutrient-storing *fat cells* and mobile cells that migrate into the connective tissue matrix from the bloodstream. The latter include defensive **white blood cells** (neutrophils, eosinophils, lymphocytes) and other cell types concerned with tissue response to injury, such as *mast cells, macrophages* (mak'ro-fāj-es), and antibody-producing **plasma cells.** This wide variety of cells is particularly obvious in our prototype, areolar connective tissue (Figure 4.7).

All of these accessory cell types are described in later chapters, but mast cells and macrophages are so important to overall body defense that they deserve a brief mention here. The oval **mast cells** typically cluster along blood vessels. These cells act as sensitive sentinels to detect foreign substances (e.g., bacteria, fungi) and initiate local inflammatory responses against them. In the mast cell cytoplasm are conspicuous secretory granules (mast = "stuffed full of granules") containing several chemicals that mediate inflammation, especially in severe allergies.

These include (1) *heparin* (hep'ah-rin), an anticoagulant chemical that prevents blood clotting when free in the bloodstream (but in human mast cells it appears to bind to and regulate the action of other mast cell chemicals); (2) *histamine* (his'tah-mēn), a substance that makes capillaries leaky; and (3) *proteases* (protein-degrading enzymes).

Macrophages (*macro* = large; *phago* = eat) are large, irregularly shaped cells that avidly phagocytize a broad variety of foreign materials, ranging from foreign molecules to entire bacteria to dust particles. These "big eaters" also dispose of dead tissue cells, and they are central actors in the immune system. In connective tissues, they may be attached to connective tissue fibers (fixed) or may migrate freely through the matrix.

Macrophages are peppered throughout loose connective tissue, bone marrow, and lymphatic tissue. Those in certain sites are given specific names; for example, those in loose connective tissue are called *histiocytes* (his'te-o-sītz"). Some macrophages have selective appetites; for example, those of the spleen primarily dispose of aging red blood cells, but they will not turn down other "delicacies" that come their way.

Types of Connective Tissue

As noted, all classes of connective tissue consist of living cells surrounded by a matrix. Their major differences reflect cell type, relative type, and amount of fibers. The connective tissues described in this section are illustrated in Figure 4.8. Mature connective tissues arise from a common embryonic tissue, which we describe here as well.

Embryonic Connective Tissue: Mesenchyme

Mesenchyme (mes'en-kīm) is the first definitive tissue formed from the mesoderm germ layer. Mesenchyme is composed of star-shaped mesenchymal cells and a fluid ground substance containing fine fibrils (Figure 4.8a). It arises during the early weeks of embryonic development and eventually differentiates (specializes) into all other connective tissues. However, some mesenchymal cells remain and provide a source of new cells in mature connective tissues.

Mucous connective tissue is a temporary tissue derived from mesenchyme and similar to it. *Wharton's jelly,* which supports the umbilical cord of the fetus, is the best example of this scant embryonic tissue.

Connective Tissue Proper

Connective tissue proper has two subclasses: the **loose connective tissues** (areolar, adipose, and reticular) and **dense connective tissues** (dense regular, dense irregular, and elastic). Except for bone, cartilage, and blood, all mature connective tissues belong to this class.

Areolar Connective Tissue Areolar connective tissue, as you may have guessed, is special. Its functions, shared by some but not all connective tissues, include (1) supporting and binding other tissues (the job of the fibers); (2) holding body fluids (the ground substance's role); (3) defending against infection (via the activity of white blood cells and macrophages); (4) storing nutrients as fat (in fat cells) (Figure 4.8b).

Fibroblasts, flat, branching cells that appear spindle shaped in profile, predominate, but numerous macrophages are also seen and present a formidable barrier to invading microorganisms. Fat cells appear singly or in small clusters, and occasional mast cells are identified easily by the large, darkly stained cytoplasmic granules that often obscure their nuclei. Other cell types are scattered throughout.

The most obvious structural feature of this tissue is the loose *arrangement* of its fibers; hence, its classification as a *loose* connective tissue is fitting. The rest of the matrix, occupied by ground substance, appears to be empty space when viewed through the microscope; in fact, the Latin term *areola* means "a small open space." Because of its loose nature, areolar connective tissue provides a reservoir of water and salts for surrounding body tissues, always holding approximately as much fluid as there is in the entire bloodstream. Essentially all body cells obtain their nutrients from and release their wastes into this "tissue fluid." However, its high content of hyaluronic acid makes the ground substance quite viscous, like molasses, which may hinder the movement of cells through it. Some white blood cells, which protect the body from disease-causing microorganisms, secrete the enzyme hyaluronidase, to liquefy the ground substance and ease their passage. (Unhappily, some potentially harmful bacteria have the same ability.) When a body region is inflamed, the areolar tissue in the area soaks up excess fluids like a sponge, and the affected area swells and becomes puffy, a condition called **edema** (ĕ-de'mah).

Areolar connective tissue is the most widely distributed connective tissue in the body and it serves as a kind of universal packing material between other tissues. It binds body parts together while allowing them to move freely over one another; wraps small blood vessels and nerves; surrounds glands; and forms the subcutaneous tissue, which cushions and attaches the skin to underlying structures. It is present in all mucous membranes as the *lamina propria.* (Mucous membranes line body cavities open to the exterior.)

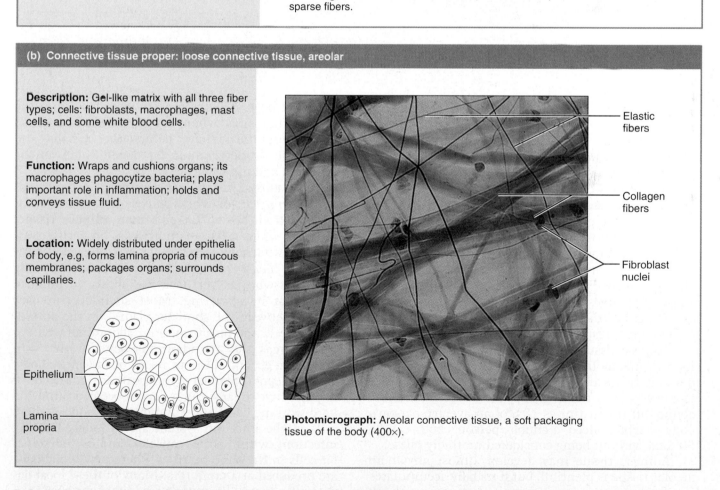

(a) Embryonic connective tissue: mesenchyme

Description: Embryonic connective tissue; gel-like ground substance containing fibers; star-shaped mesenchymal cells.

Function: Gives rise to all other connective tissue types.

Location: Primarily in embryo.

Mesenchymal cell

Ground substance

Fibers

Photomicrograph: Mesenchymal tissue, an embryonic connective tissue (400×); the clear-appearing background is the fluid ground substance of the matrix; notice the fine, sparse fibers.

(b) Connective tissue proper: loose connective tissue, areolar

Description: Gel-like matrix with all three fiber types; cells: fibroblasts, macrophages, mast cells, and some white blood cells.

Function: Wraps and cushions organs; its macrophages phagocytize bacteria; plays important role in inflammation; holds and conveys tissue fluid.

Location: Widely distributed under epithelia of body, e.g, forms lamina propria of mucous membranes; packages organs; surrounds capillaries.

Epithelium

Lamina propria

Elastic fibers

Collagen fibers

Fibroblast nuclei

Photomicrograph: Areolar connective tissue, a soft packaging tissue of the body (400×).

FIGURE 4.8 Connective tissues. Embryonic connective tissue **(a)** and connective tissue proper **(b)**.

(c) Connective tissue proper: loose connective tissue, adipose

Description: Matrix as in areolar, but very sparse; closely packed adipocytes, or fat cells, have nucleus pushed to the side by large fat droplet.

Function: Provides reserve food fuel; insulates against heat loss; supports and protects organs.

Location: Under skin; around kidneys and eyeballs; within abdomen; in breasts.

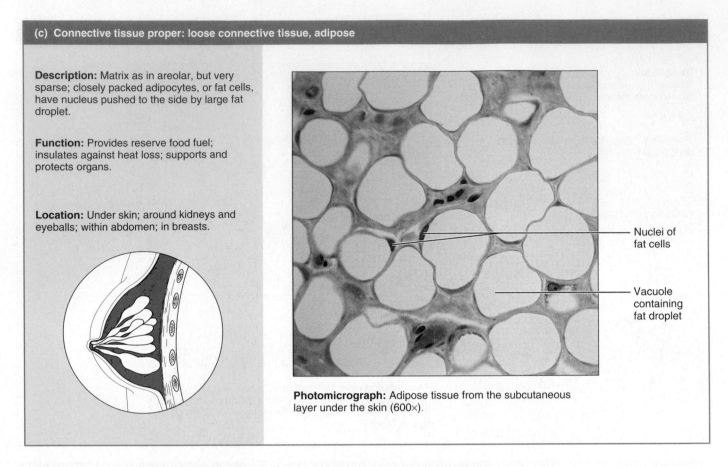

Nuclei of
fat cells

Vacuole
containing
fat droplet

Photomicrograph: Adipose tissue from the subcutaneous layer under the skin (600×).

FIGURE 4.8 *(continued)* **Connective tissues.** Connective tissue proper **(c)**.

Adipose (Fat) Tissue **Adipose tissue** (ad′ĭ-pōs) is similar to areolar tissue in structure and function but its nutrient-storing ability is much greater. Consequently, **adipocytes** (ad′ĭ-po-sītz), commonly called *adipose* or *fat cells,* predominate and account for 90% of this tissue's mass. The matrix is scanty and the cells are packed closely together, giving a chicken wire appearance to the tissue. A glistening oil droplet (almost pure neutral fat) occupies most of a fat cell's volume and displaces the nucleus to one side so that only a thin rim of surrounding cytoplasm is seen (Figure 4.8c). Mature adipocytes are among the largest cells in the body. As they take up or release fat, they become plumper or more wrinkled looking, respectively.

Adipose tissue is richly vascularized, indicating its high metabolic activity. Without the fat stores in our adipose tissue, we could not live for more than a few days without eating. Adipose tissue is certainly abundant: It constitutes 18% of an average person's body weight. Indeed, a chubby person's body can be 50% fat without being considered morbidly obese.

Adipose tissue may develop almost anywhere areolar tissue is plentiful, but it usually accumulates in subcutaneous tissue, where it also acts as a shock absorber and as insulation. Because fat is a poor conductor of heat, it helps prevent heat loss from the body. Other sites where fat accumulates include surrounding the kidneys, behind the eyeballs, and at genetically determined fat depots such as the abdomen and hips.

The adipose tissue just described is sometimes called *white fat,* or *white adipose tissue,* to distinguish it from **brown fat** or **brown adipose tissue.** Whereas white fat stores nutrients, brown fat consumes its nutrient stores to generate heat to warm the body. The richly vascular brown fat occurs only in babies who (as yet) lack the ability to produce body heat by shivering. Most such deposits are located between the shoulder blades, on the anterolateral neck, and on the anterior abdominal wall.

Whereas the abundant fat beneath the skin serves the general nutrient needs of the entire body, smaller depots of fat serve the local nutrient needs of highly active organs. Such depots occur around the hard-working heart and around lymph nodes (where cells of the immune system are furiously fighting infection), within some muscles, and as individual fat cells in the bone marrow, where new blood cells are produced at a rapid rate. Many of these local depots offer special lipids that are highly enriched.

(d) Connective tissue proper: loose connective tissue, reticular

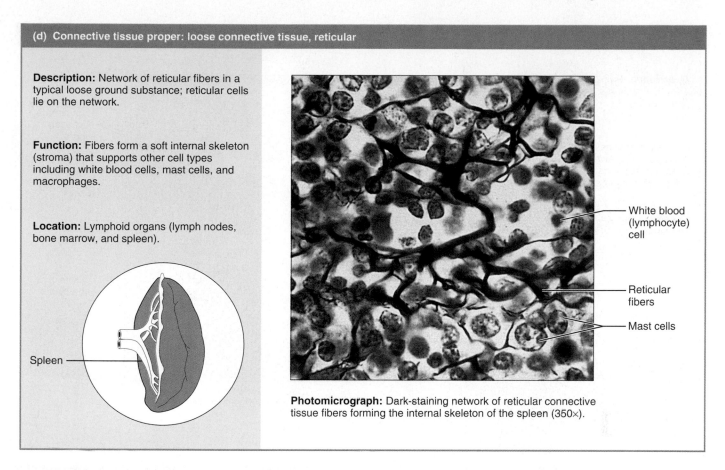

Description: Network of reticular fibers in a typical loose ground substance; reticular cells lie on the network.

Function: Fibers form a soft internal skeleton (stroma) that supports other cell types including white blood cells, mast cells, and macrophages.

Location: Lymphoid organs (lymph nodes, bone marrow, and spleen).

Spleen

White blood (lymphocyte) cell

Reticular fibers

Mast cells

Photomicrograph: Dark-staining network of reticular connective tissue fibers forming the internal skeleton of the spleen (350×).

FIGURE 4.8 *(continued)*   Connective tissue proper **(d)**.

Reticular Connective Tissue

Reticular connective tissue resembles areolar connective tissue, but the only fibers in its matrix are reticular fibers, which form a delicate network along which fibroblasts called **reticular cells** lie scattered (Figure 4.8d). Although reticular fibers are widely distributed in the body, reticular tissue is limited to certain sites. It forms a labyrinth-like **stroma** (literally, "bed" or "mattress"), or internal framework, that can support many free blood cells (largely lymphocytes) in lymph nodes, the spleen, and bone marrow.

Dense Regular Connective Tissue

Dense regular connective tissue (Figure 4.8e) is one variety of the dense connective tissues, all of which have fibers as their predominant element. For this reason, the dense connective tissues are often referred to as **fibrous connective tissues.**

Dense regular connective tissue contains closely packed bundles of collagen fibers running in the same direction, parallel to the direction of pull. This results in white, flexible structures with great resistance to tension (pulling forces) where the tension is exerted in a single direction. Crowded between the collagen fibers are rows of fibroblasts that continuously manufacture the fibers and scant ground substance. As seen in Figure 4.8e, collagen fibers are slightly wavy.

This allows the tissue to stretch a little, but once the fibers are straightened out by a pulling force, there is no further "give" to this tissue. Unlike our model (areolar) connective tissue, this tissue has few cells other than fibroblasts and it is poorly vascularized.

With its enormous tensile strength, dense regular connective tissue forms the *tendons,* cords that attach muscles to bones, and flat, sheetlike tendons called *aponeuroses* (ap"o-nu-ro'sēz) that attach muscles to other muscles or to bones. It also forms the *ligaments* that bind bones together at joints. Ligaments contain more elastic fibers than tendons and are slightly more stretchy. A few ligaments, such as the *ligamenta nuchae* and *flava* connecting adjacent vertebrae, are very elastic. Indeed, their content of elastic fibers is so high that the connective tissue in those structures is referred to as **elastic connective tissue.**

Dense Irregular Connective Tissue

Dense irregular connective tissue has the same structural elements as the regular variety. However, the bundles of collagen fibers are much thicker and they are arranged irregularly; that is, they run in more than one plane (Figure 4.8f). This type of tissue forms sheets in body areas where tension is exerted from many different directions. It is found in the skin as

(e) Connective tissue proper: dense connective tissue, dense regular

Description: Primarily parallel collagen fibers; a few elastin fibers; major cell type is the fibroblast.

Function: Attaches muscles to bones or to muscles; attaches bones to bones; withstands great tensile stress when pulling force is applied in one direction.

Location: Tendons, most ligaments, aponeuroses.

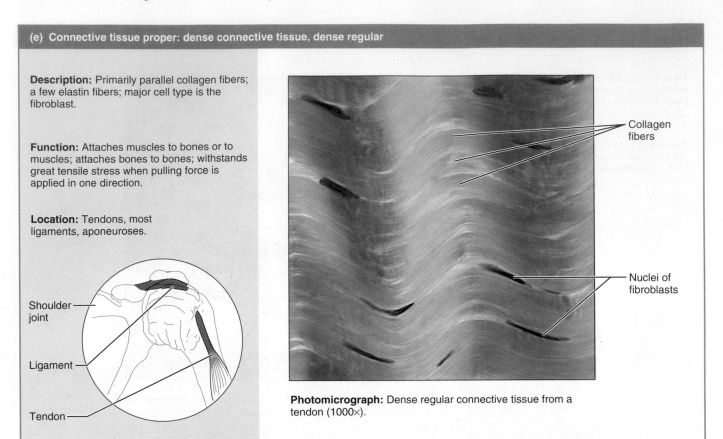

Shoulder joint

Ligament

Tendon

Collagen fibers

Nuclei of fibroblasts

Photomicrograph: Dense regular connective tissue from a tendon (1000×).

(f) Connective tissue proper: dense connective tissue, dense irregular

Description: Primarily irregularly arranged collagen fibers; some elastic fibers; major cell type is the fibroblast.

Function: Able to withstand tension exerted in many directions; provides structural strength.

Location: Dermis of the skin; submucosa of digestive tract; fibrous capsules of organs and of joints.

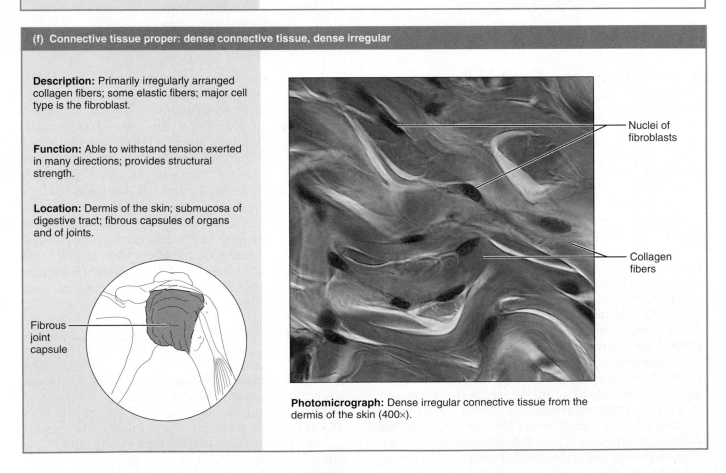

Fibrous joint capsule

Nuclei of fibroblasts

Collagen fibers

Photomicrograph: Dense irregular connective tissue from the dermis of the skin (400×).

FIGURE 4.8 *(continued)*   Connective tissue proper (**e** and **f**).

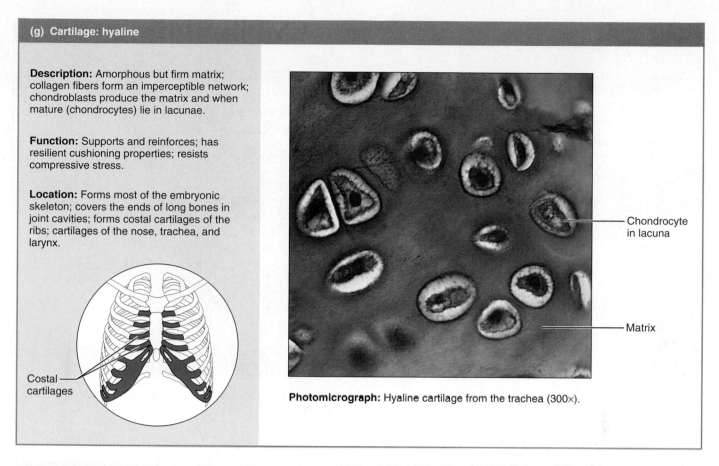

(g) Cartilage: hyaline

Description: Amorphous but firm matrix; collagen fibers form an imperceptible network; chondroblasts produce the matrix and when mature (chondrocytes) lie in lacunae.

Function: Supports and reinforces; has resilient cushioning properties; resists compressive stress.

Location: Forms most of the embryonic skeleton; covers the ends of long bones in joint cavities; forms costal cartilages of the ribs; cartilages of the nose, trachea, and larynx.

Costal cartilages

Chondrocyte in lacuna

Matrix

Photomicrograph: Hyaline cartilage from the trachea (300×).

FIGURE 4.8 *(continued)* Cartilage **(g)**.

the leathery *dermis,* and it forms fibrous joint capsules and the fibrous coverings that surround some organs (testes, kidneys, bones, cartilages, muscles, and nerves).

Cartilage

Cartilage (kar′tĭ-lij), which stands up to both tension *and* compression, has qualities intermediate between dense connective tissue and bone. It is tough but flexible, providing a resilient rigidity to the structures it supports. Cartilage lacks nerve fibers and is avascular. It receives its nutrients by diffusion from blood vessels located in the connective tissue membrane (perichondrion) surrounding it. Its ground substance contains large amounts of the GAGs chrondroitin sulfate and hyaluronic acid. The ground substance contains firmly bound collagen fibers and in some cases elastic fibers, and is usually quite firm. Cartilage matrix also contains an exceptional amount of tissue fluid; in fact, cartilage is up to 80% water! The movement of tissue fluid in its matrix enables cartilage to rebound after being compressed and also helps to nourish the cartilage cells.

Chondroblasts, the predominant cell type in growing cartilage, produce new matrix until the skeleton stops growing at the end of adolescence. The firm cartilage matrix prevents the cells from becom-

ing widely separated, so **chondrocytes,** or mature cartilage cells, are typically found in small groups within cavities called **lacunae** (lah-ku′ne; "pits").

HOMEOSTATIC IMBALANCE

Because cartilage is avascular and aging cartilage cells lose their ability to divide, cartilages heal slowly when injured. This phenomenon is excruciatingly familiar to those who have experienced sports injuries. During later life, cartilages tend to calcify or even ossify (become bony). In such cases, the chondrocytes are poorly nourished and die. ●

There are three varieties of cartilage: *hyaline cartilage, elastic cartilage,* and *fibrocartilage,* each dominated by a particular fiber type.

Hyaline Cartilage Hyaline cartilage (hi′ah-lin), or *gristle,* is the most abundant cartilage type in the body. Although it contains large numbers of collagen fibers, they are not apparent and the matrix appears amorphous and glassy (*hyalin* = glass) blue-white when viewed by the unaided eye (Figure 4.8g). Chondrocytes account for only 1–10% of the cartilage volume.

(h) Cartilage: elastic

Description: Similar to hyaline cartilage, but more elastic fibers in matrix.

Function: Maintains the shape of a structure while allowing great flexibility.

Location: Supports the external ear (pinna); epiglottis.

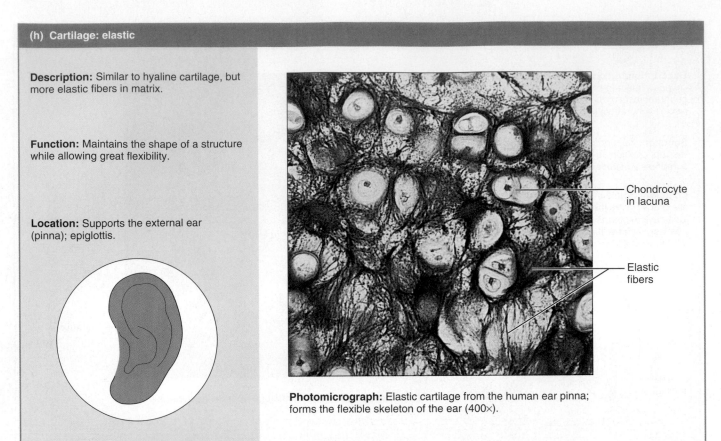

Chondrocyte in lacuna

Elastic fibers

Photomicrograph: Elastic cartilage from the human ear pinna; forms the flexible skeleton of the ear (400×).

(i) Cartilage: fibrocartilage

Description: Matrix similar to but less firm than that in hyaline cartilage; thick collagen fibers predominate.

Function: Tensile strength with the ability to absorb compressive shock.

Location: Intervertebral discs; pubic symphysis; discs of knee joint.

Intervertebral discs

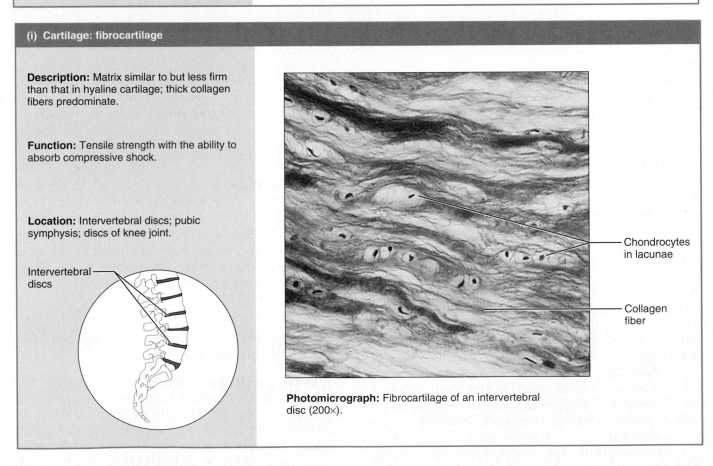

Chondrocytes in lacunae

Collagen fiber

Photomicrograph: Fibrocartilage of an intervertebral disc (200×).

FIGURE 4.8 *(continued)* Cartilage (**h** and **i**).

(j) Others: bone (osseous tissue)

Description: Hard, calcified matrix containing many collagen fibers; osteocytes lie in lacunae. Very well vascularized.

Function: Bone supports and protects (by enclosing); provides levers for the muscles to act on; stores calcium and other minerals and fat; marrow inside bones is the site for blood cell formation (hematopoiesis).

Location: Bones

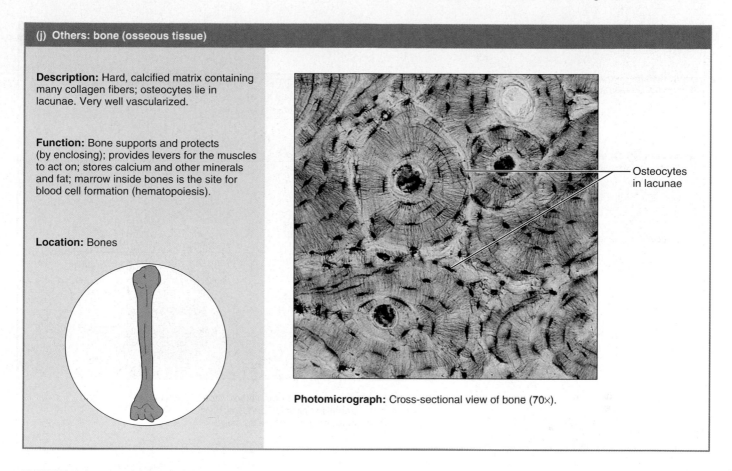

Osteocytes in lacunae

Photomicrograph: Cross-sectional view of bone (70×).

FIGURE 4.8 *(continued)*   Bone **(j)**.

Hyaline cartilage provides firm support with some pliability. It covers the ends of long bones as *articular cartilage,* providing springy pads that absorb compression at joints. Hyaline cartilage also supports the tip of the nose, connects the ribs to the sternum, and supports most of the respiratory system passages. Most of the embryonic skeleton is formed of hyaline cartilage before bone is formed. Hyaline cartilage persists during childhood as the *epiphyseal* (e″pĭ-fis′e-ul) *plates,* actively growing regions near the end of long bones that provide for continued growth in length.

Elastic Cartilage   Histologically, **elastic cartilage** (Figure 4.8h) is nearly identical to hyaline cartilage. However, there are many more elastin fibers in elastic cartilage. Found where strength and exceptional stretchability are needed, elastic cartilage forms the "skeletons" of the external ear and the epiglottis. (The epiglottis is the flap that covers the opening to the respiratory passageway when we swallow, preventing food or fluids from entering the lungs.)

Fibrocartilage   **Fibrocartilage** is often found where hyaline cartilage meets a true ligament or a tendon. Fibrocartilage is a perfect structural inter-

mediate between hyaline cartilage and dense regular connective tissues. Its rows of chondrocytes (a cartilage feature) alternate with rows of thick collagen fibers (a feature of dense regular connective tissue) (Figure 4.8i). Because it is compressible and resists tension well, fibrocartilage is found where strong support and the ability to withstand heavy pressure are required. For example, the intervertebral discs (resilient cushions between the bony vertebrae) and the spongy cartilages of the knee are fibrocartilage structures (see Figure 6.1, p. 154).

Bone (Osseous Tissue)

Because of its rocklike hardness, **bone,** or **osseous tissue** (os′e-us), has an exceptional ability to support and protect body structures. Bones of the skeleton also provide cavities for fat storage and synthesis of blood cells. Bone matrix is similar to that of cartilage but is harder and more rigid because, in addition to its more abundant collagen fibers, bone has an added matrix element—inorganic calcium salts (bone salts).

Osteoblasts produce the organic portion of the matrix; then bone salts are deposited on and between the fibers. Mature bone cells, or **osteocytes,** reside in the lacunae within the matrix they have made (Figure 4.8j). Unlike cartilage, the next firmest

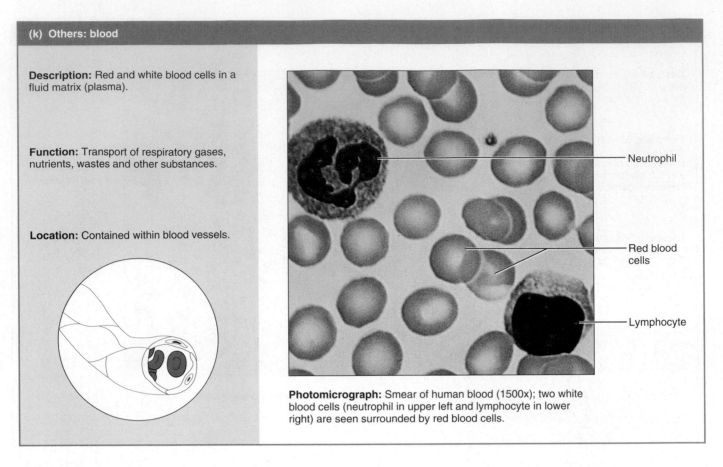

(k) Others: blood

Description: Red and white blood cells in a fluid matrix (plasma).

Function: Transport of respiratory gases, nutrients, wastes and other substances.

Location: Contained within blood vessels.

Neutrophil

Red blood cells

Lymphocyte

Photomicrograph: Smear of human blood (1500x); two white blood cells (neutrophil in upper left and lymphocyte in lower right) are seen surrounded by red blood cells.

FIGURE 4.8 (continued)   Blood **(k)**.

connective tissue, bone is very well supplied by invading blood vessels.

Blood

Blood, the fluid within blood vessels, is the most atypical connective tissue. It does *not* connect things or give mechanical support. It is classified as a connective tissue because it develops from mesenchyme and consists of blood *cells,* surrounded by a nonliving fluid matrix called *blood plasma* (Figure 4.8k). The "fibers" of blood are soluble protein molecules that become visible only during blood clotting. Blood functions as the transport vehicle for the cardiovascular system, carrying nutrients, wastes, respiratory gases, and many other substances throughout the body.

Covering and Lining Membranes

Now that we have described both connective and epithelial tissues, we can consider the membranes that incorporate both types of tissue. The covering and lining membranes are of three types: *cutaneous, mucous,* or *serous.* Essentially they all are continuous multicellular sheets composed of at least two primary tissue types: an epithelium bound to an underlying layer of connective tissue proper. Hence, these membranes are simple organs. The *synovial membranes,* which line joint cavities and consist of connective tissue only, are described in Chapter 8.

Cutaneous Membrane

The **cutaneous membrane** (ku-ta′ne-us; *cutis* = skin) is your skin (Figure 4.9a). It is an organ system consisting of a keratinized stratified squamous epithelium (epidermis) firmly attached to a thick layer of dense irregular connective tissue (dermis). Unlike other epithelial membranes, the cutaneous membrane is exposed to the air and is a dry membrane. Chapter 5 is devoted to this unique organ system.

Mucous Membranes

Mucous membranes, or **mucosae** (mu-ko′se), line body cavities that open to the exterior, such as those of the hollow organs of the digestive, respiratory, and urogenital tracts (Figure 4.9b). In all cases, they are "wet," or moist, membranes bathed by secretions or, in the case of the urinary mucosa, urine. Notice that the term *mucosa* refers to the location of the membrane, *not* its cell composition, which varies. However, most mucosae contain either stratified

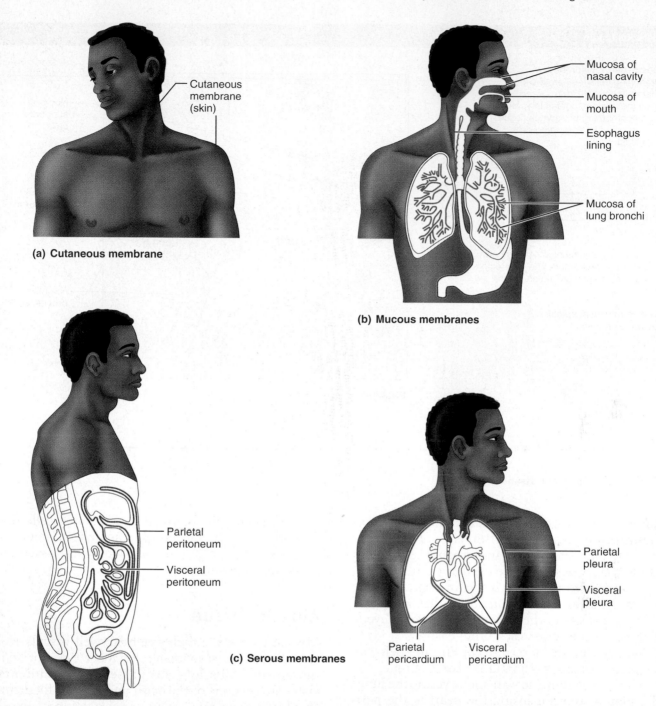

FIGURE 4.9 **Classes of membranes. (a)** Cutaneous membrane, or skin. **(b)** Mucous membranes line body cavities that are open to the exterior. **(c)** Serous membranes line ventral body cavities that are closed to the exterior.

squamous or simple columnar epithelia. The epithelial sheet is directly underlain by a layer of loose connective tissue called the **lamina propria** (lam'ĭ-nah pro'pre-ah; "one's own layer"). In some mucosae, the lamina propria rests on a third (deeper) layer of smooth muscle cells.

Mucous membranes are often adapted for absorption and secretion. Although many mucosae secrete mucus, this is not a requirement. The mu-

cosae of both the digestive and respiratory tracts secrete copious amounts of lubricating mucus; that of the urinary tract does not.

Serous Membranes

Serous membranes, or **serosae** (se-ro'se), introduced in Chapter 1, are the moist membranes found in closed ventral body cavities (Figure 4.9c). A serous

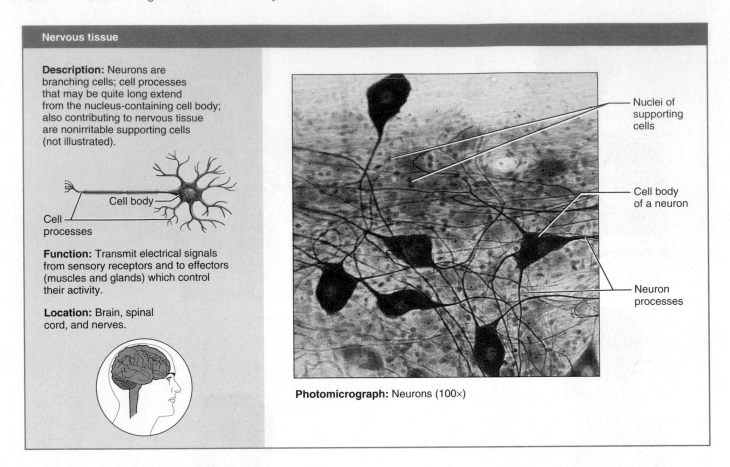

Nervous tissue

Description: Neurons are branching cells; cell processes that may be quite long extend from the nucleus-containing cell body; also contributing to nervous tissue are nonirritable supporting cells (not illustrated).

Cell body

Cell processes

Function: Transmit electrical signals from sensory receptors and to effectors (muscles and glands) which control their activity.

Location: Brain, spinal cord, and nerves.

Nuclei of supporting cells

Cell body of a neuron

Neuron processes

Photomicrograph: Neurons (100×)

FIGURE 4.10   **Nervous tissue.**

membrane consists of simple squamous epithelium (a mesothelium) resting on a thin layer of loose connective (areolar) tissue. The mesothelial cells enrich the fluid that filters from the capillaries in the associated connective tissue with hyaluronic acid. The result is the thin, clear *serous fluid* that lubricates the facing surfaces of the parietal and visceral layers, so that they slide across each other easily.

The serosae are named according to their site and specific organ associations. For example, the serosa lining the thoracic wall and covering the lungs is the **pleura;** that enclosing the heart is the **pericardium;** and those of the abdominopelvic cavity and viscera are the **peritoneums.**

Nervous Tissue

Nervous tissue is the main component of the nervous system—the brain, spinal cord, and nerves—which regulates and controls body functions. It contains two major cell types. **Neurons** are highly specialized nerve cells that generate and conduct nerve impulses (Figure 4.10). Typically, they are branching cells. Their cytoplasmic extensions, or processes, allow them to transmit electrical impulses over substantial distances within the body. The balance of nervous tissue consists of various

types of **supporting cells,** nonconducting cells that support, insulate, and protect the delicate neurons. A more complete discussion of nervous tissue appears in Chapter 11.

Muscle Tissue

Muscle tissues are highly cellular, well-vascularized tissues that are responsible for most types of body movement. Muscle cells possess **myofilaments,** elaborate versions of the *actin* and *myosin* filaments that bring about movement or contraction in all cell types. There are three kinds of muscle tissue: skeletal, cardiac, and smooth.

Skeletal muscle is packaged by connective tissue sheets into organs called *skeletal muscles* that are attached to the bones of the skeleton. These muscles form the flesh of the body, and as they contract, they pull on bones or skin, causing body movements. Skeletal muscle cells, also called **muscle fibers,** are long, cylindrical cells that contain many nuclei. Their obvious banded, or *striated*, appearance reflects the precise alignment of their myofilaments (Figure 4.11a).

Cardiac muscle is found only in the wall of the heart. Its contractions help propel blood through the blood vessels to all parts of the body. Like skeletal

(a) Skeletal muscle

Description: Long, cylindrical, multinucleate cells; obvious striations.

Function: Voluntary movement; locomotion; manipulation of the environment; facial expression; voluntary control.

Location: In skeletal muscles attached to bones or occasionally to skin.

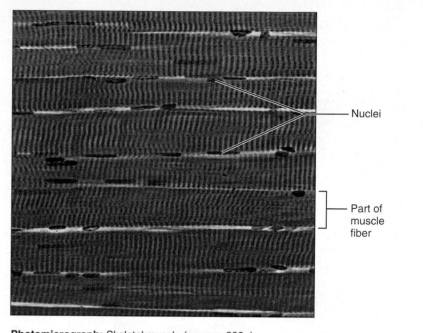

Nuclei

Part of muscle fiber

Photomicrograph: Skeletal muscle (approx. 300×). Notice the obvious banding pattern and the fact that these large cells are multinucleate.

(b) Cardiac muscle

Description: Branching, striated, generally uninucleate cells that interdigitate at specialized junctions (intercalated discs).

Function: As it contracts, it propels blood into the circulation; involuntary control.

Location: The walls of the heart.

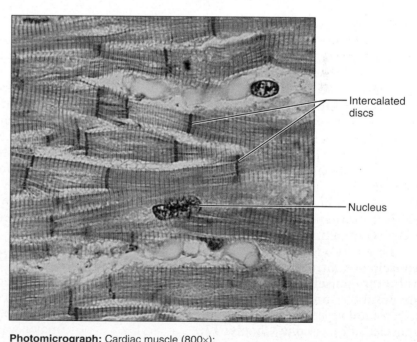

Intercalated discs

Nucleus

Photomicrograph: Cardiac muscle (800×); notice the striations, branching of cells, and the intercalated discs.

FIGURE 4.11   **Muscle tissues (a** and **b).**

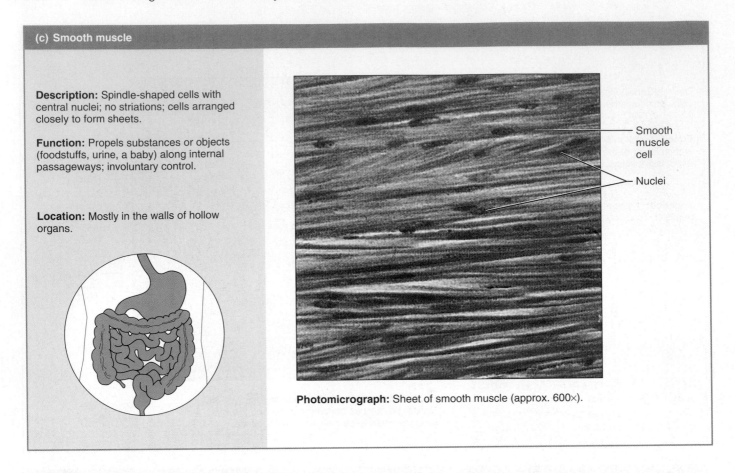

(c) Smooth muscle

Description: Spindle-shaped cells with central nuclei; no striations; cells arranged closely to form sheets.

Function: Propels substances or objects (foodstuffs, urine, a baby) along internal passageways; involuntary control.

Location: Mostly in the walls of hollow organs.

Smooth muscle cell

Nuclei

Photomicrograph: Sheet of smooth muscle (approx. 600×).

FIGURE 4.11 *(continued)*   **Muscle tissues (c)**.

muscle cells, cardiac muscle cells are striated. However, they differ structurally in that cardiac cells (1) are uninucleate and (2) are branching cells that fit together tightly at unique junctions called **intercalated discs** (in-ter'kah-la"ted) (Figure 4.11b).

   Smooth muscle is so named because its cells have no visible striations. Individual smooth muscle cells are spindle shaped and contain one centrally located nucleus (Figure 4.11c). Smooth muscle is found mainly in the walls of hollow organs other than the heart (digestive and urinary tract organs, uterus, and blood vessels). It generally acts to squeeze substances through these organs by alternately contracting and relaxing.

   Because skeletal muscle contraction is under our conscious control, skeletal muscle is often called **voluntary muscle,** and the other two types are called **involuntary muscle.** Skeletal muscle and smooth muscle are described in detail in Chapter 9; cardiac muscle is discussed in Chapter 17.

Tissue Repair

The body has many techniques for protecting itself from uninvited "guests" or injury. Intact mechanical barriers such as the skin and mucosae, the cilia of epithelial cells lining the respiratory tract, and the strong acid (chemical barrier) produced by stomach glands represent three defenses exerted at the body's external boundaries. When tissue injury occurs these barriers are penetrated; this stimulates the body's inflammatory and immune responses, which wage their battles largely in the connective tissues of the body. The inflammatory response is a relatively nonspecific reaction that develops quickly wherever tissues are injured, while the immune response is extremely specific, but takes longer to swing into action. The inflammatory and immune responses are considered in detail in Chapter 20.

Steps of Tissue Repair

Tissue repair requires that cells divide and migrate, activities that are initiated by growth factors (wound hormones) released by injured cells. It occurs in two major ways: by regeneration and by fibrosis. Which of these occurs depends on (1) the type of tissue damaged and (2) the severity of the injury. **Regeneration** is replacement of destroyed tissue with the same kind of tissue, whereas **fibrosis** involves proliferation of fibrous connective tissue called **scar tissue.** In skin, the tissue we will use as our example, repair involves both activities.

(1) What is the makeup of granulation tissue formed at the injured site? **(2)** Where do the new capillaries that invade the injured area come from? **(3)** How would this pictorial series look different if repair by regeneration was not possible?

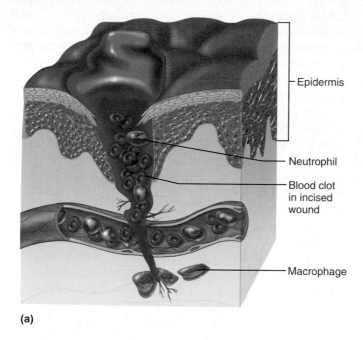

Epidermis

Neutrophil

Blood clot in incised wound

Macrophage

(a)

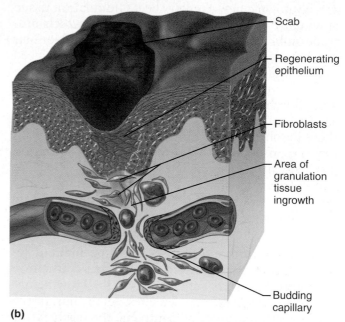

Scab

Regenerating epithelium

Fibroblasts

Area of granulation tissue ingrowth

Budding capillary

(b)

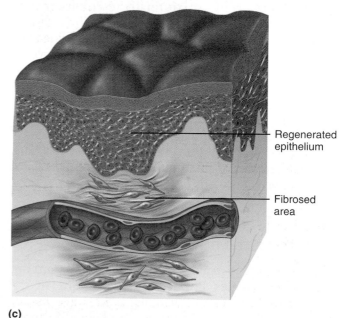

Regenerated epithelium

Fibrosed area

(c)

FIGURE 4.12 Tissue repair of a nonextensive skin wound: regeneration and fibrosis.
(a) Severed blood vessels bleed. Inflammatory chemicals are released, and local blood vessels dilate and become more permeable, allowing white blood cells, fluid, clotting proteins, and other plasma proteins to invade the injured site. Clotting proteins initiate clotting; surface dries and forms a scab. **(b)** Granulation tissue is formed. Capillary buds invade the clot, restoring a vascular supply. Fibroblasts secrete soluble collagen, which constructs collagen fibers that bridge the gap. Macrophages phagocytize dead and dying cell debris. Surface epithelial cells proliferate and migrate over the granulation tissue. **(c)** Approximately one week later, the fibrosed area (scar) has contracted and regeneration of the epithelium is in progress.

1. Inflammation sets the stage. Let us briefly examine the inflammatory events that tissue injury sets into motion. First, the tissue trauma causes injured tissue cells, macrophages, mast cells, and others to release inflammatory chemicals, which cause the capillaries to dilate and become very permeable. This allows white blood cells (neutrophils, monocytes) and plasma fluid rich in clotting proteins, antibodies, and other substances to seep into the injured area. The leaked clotting proteins construct a clot, which stops the loss of blood, holds the edges of the wound together, and effectively walls in, or isolates, the injured area, preventing bacteria, toxins, or

other harmful substances from spreading to surrounding tissues (Figure 4.12a). The part of the clot exposed to air quickly dries and hardens, forming a scab. The inflammatory events leave excess fluid, bits of destroyed cells, and other debris in the area, which are eventually removed via lymphatic vessels or phagocytized by macrophages.

2. Organization restores the blood supply. Even while the inflammatory process is going on, **organization**, the first phase of tissue repair, begins. During organization the blood clot is replaced by granulation tissue (Figure 4.12b). **Granulation tissue** is a delicate pink tissue composed of several elements. It

(1) Budding capillaries and loose connective tissue. **(2)** They bud from uninjured blood vessels in the area. **(3)** The epidermis would not be continuous. The tissue in the injured region would be totally replaced by scar tissue. ■

contains capillaries that grow in from nearby areas and lay down a new capillary bed. Granulation tissue is actually named for these capillaries, which protrude nublike from its surface, giving it a granular appearance. These capillaries are fragile and bleed freely, as demonstrated when someone picks at a scab. Also present in granulation tissue are proliferating fibroblasts that produce growth factors as well as new collagen fibers to bridge the gap. Some of these fibroblasts have contractile properties that pull the margins of the wound together. As organization proceeds, macrophages digest the original blood clot and collagen fiber deposit continues. The granulation tissue, destined to become scar tissue (a permanent fibrous patch), is highly resistant to infection because it produces bacteria-inhibiting substances.

3. Regeneration and fibrosis effect permanent repair. During organization, the surface epithelium begins to *regenerate* (see Figure 4.12b), growing under the scab, which soon detaches. As the fibrous tissue beneath matures and contracts, the regenerating epithelium thickens until it finally resembles that of the adjacent skin (Figure 4.12c). The end result is a fully regenerated epithelium, and an underlying area of scar tissue. The scar may be invisible, or visible as a thin white line, depending on the severity of the wound.

The repair process described above follows healing of a wound (cut, scrape, puncture) that breaches an epithelial barrier. In pure *infections* (a pimple or sore throat), healing is solely by regeneration. There is usually no clot formation or scarring. Only severe (destructive) infections lead to scarring.

Regenerative Capacity of Different Tissues

The capacity for regeneration varies widely among the different tissues. Epithelial tissues regenerate extremely well, as do bone, areolar connective tissue, dense irregular connective tissue, and blood-forming

tissue. Smooth muscle and dense regular connective tissue have a moderate capacity for regeneration, but skeletal muscle and cartilage have a weak regenerative capacity. Cardiac muscle and the nervous tissue in the brain and spinal cord have virtually no *functional* regenerative capacity; hence they are routinely replaced by scar tissue. However, recent studies have shown that some unexpected (and highly selective) cellular division occurs in both these tissues after damage, and efforts are under way to coax them to regenerate better.

In nonregenerating tissues and in exceptionally severe wounds, fibrosis totally replaces the lost tissue. Over a period of months, the fibrous mass shrinks and becomes more and more compact. The resulting scar appears as a pale, often shiny area composed mostly of collagen fibers. Scar tissue is strong, but it lacks the flexibility and elasticity of most normal tissues. Also, it cannot perform the normal functions of the tissue it has replaced.

🅗 *HOMEOSTATIC IMBALANCE*

Scar tissue that forms in the wall of the urinary bladder, heart, or other muscular organ may severely hamper the function of that organ. The normal shrinking of the scar reduces the internal volume of an organ and may hinder or even block movement of substances through a hollow organ. Scar tissue hampers muscle's ability to contract and may interfere with its normal excitation by the nervous system. In the heart, these problems may lead to progressive heart failure. In irritated visceral organs, particularly following abdominal surgery, *adhesions* may form as the newly forming scar tissue connects adjacent organs together. Such adhesions can prevent the normal shifting about (churning) of loops of the intestine, dangerously obstructing the flow of foodstuffs through it. Adhesions can also restrict heart movements and immobilize joints. ●

Review Questions

Multiple Choice/Matching

(Some questions have more than one correct answer. Select the best answer or answers from the choices given.)

1. Use the key to classify each of the following described tissue types into one of the four major tissue categories.

Key: **(a)** connective tissue   **(c)** muscle
       (b) epithelium   **(d)** nervous tissue

____ **(1)** Tissue type composed largely of nonliving extracellular matrix; important in protection and support

____ **(2)** The tissue immediately responsible for body movement

____ **(3)** The tissue that enables us to be aware of the external environment and to react to it

____ **(4)** The tissue that lines body cavities and covers surfaces

2. An epithelium that has several layers, with an apical layer of flattened cells, is called (choose all that apply): (a) ciliated, (b) columnar, (c) stratified, (d) simple, (e) squamous.

3. Match the epithelial types named in column B with the appropriate description(s) in column A.

Column A	Column B
____ **(1)** Lines most of the digestive tract	**(a)** pseudostratified ciliated columnar
____ **(2)** Lines the esophagus	**(b)** simple columnar
____ **(3)** Lines much of the respiratory tract	**(c)** simple cuboidal
____ **(4)** Forms the walls of the air sacs of the lungs	**(d)** simple squamous
____ **(5)** Found in urinary tract organs	**(e)** stratified columnar
____ **(6)** Endothelium and mesothelium	**(f)** stratified squamous
	(g) transitional

4. The gland type that secretes products such as milk, saliva, bile, or sweat through a duct is (a) an endocrine gland, (b) an exocrine gland.

5. The membrane lining body cavities that open to the exterior is a(n) (a) endothelium, (b) cutaneous membrane, (c) mucous membrane, (d) serous membrane.

6. Scar tissue is a variety of (a) epithelium, (b) connective tissue, (c) muscle tissue, (d) nerve tissue, (e) all of these.

Short Answer Essay Questions

7. Define tissue.

8. Name four important functions of epithelial tissue and provide at least one example of a tissue that exemplifies each function.

9. Describe the criteria used to classify covering and lining epithelia.

10. Explain the functional classification of multicellular exocrine glands and supply an example for each class.

11. Name four important functions of connective tissue and provide examples from the body that illustrate each function.

12. Name the primary cell type in connective tissue proper; in cartilage; in bone.

13. Name the two major components of matrix and, if applicable, subclasses of each component.

14. Matrix is extracellular. How does the matrix get to its characteristic position?

15. Name the specific connective tissue type found in the following body locations: (a) forming the soft packing around organs, (b) supporting the ear pinna, (c) forming "stretchy" ligaments, (d) first connective tissue in the embryo, (e) forming the intervertebral discs, (f) covering the ends of bones at joint surfaces, (g) main component of subcutaneous tissue.

16. What is the function of macrophages?

17. Differentiate clearly between the roles of neurons and the supporting cells of nervous tissue.

18. Compare and contrast skeletal, cardiac, and smooth muscle tissue relative to structure, body location, and specific function.

19. Describe the process of tissue repair, making sure you indicate factors that influence this process.

5

The Integumentary System

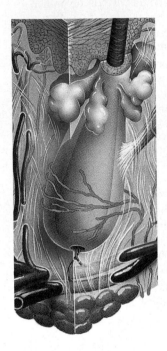

The Skin (pp. 135–140)

1. Name the tissue types composing the epidermis and dermis. List the major layers of each and describe the functions of each layer.

2. Describe the factors that normally contribute to skin color. Briefly describe how changes in skin color may be used as clinical signs of certain disease states.

Appendages of the Skin (pp. 140–145)

3. Compare the structure and locations of sweat and oil glands. Also compare the composition and functions of their secretions.

4. Compare and contrast eccrine and apocrine glands.

5. List the parts of a hair follicle and explain the function of each part. Also describe the functional relationship of arrector pili muscles to the hair follicle.

6. Name the regions of a hair and explain the basis of hair color. Describe the distribution, growth, replacement, and changing nature of hair during the life span.

7. Describe the structure of nails.

Functions of the Integumentary System (pp. 145–147)

8. Describe how the skin accomplishes at least five different functions.

Homeostatic Imbalances of Skin (pp. 147–150)

9. Summarize the characteristics of the three major types of skin cancers.

10. Explain why serious burns are life threatening. Describe how to determine the extent of a burn and differentiate first-, second-, and third-degree burns.

ould you be enticed by an advertisement for a coat that is waterproof, stretchable, washable, and permanent-press, that automatically repairs small cuts, rips, and burns, and that is guaranteed to last a lifetime with reasonable care? Sounds too good to be true, but you already have such a coat—your skin. The skin and its derivatives (sweat and oil glands, hairs, and nails) make up a complex set of organs that serves several functions, mostly protective. Together, these organs form the **integumentary** (in-teg"u-men'tar-e) **system.**

The Skin

The skin ordinarily receives very little respect from its inhabitants, but architecturally it is a marvel. It covers the entire body, has a surface area of 1.2 to 2.2 square meters, weighs 4 to 5 kilograms (4–5 kg = 9–11 pounds), and accounts for about 7% of total body weight in the average adult. Also called the integument, which simply means "covering," the skin's functions go well beyond serving as a large, opaque bag for the body contents. It is pliable yet tough, allowing it to take constant punishment from external agents. Without our skin, we would quickly fall prey to bacteria and perish from water and heat loss.

The skin, which varies in thickness from 1.5 to 4.0 millimeters (mm) or more in different parts of the body, is composed of two distinct regions, the *epidermis* (ep"ĭ-der'mis) and the *dermis* (Figure 5.1). The epidermis (*epi* = upon), composed of epithelial cells, is the outermost protective shield of the body. The underlying dermis, making up the bulk of the skin, is a tough leathery layer composed of fibrous connective tissue. Only the dermis is vascularized. Nutrients reach the epidermis by diffusing through the tissue fluid from blood vessels in the dermis.

The subcutaneous tissue just deep to the skin is known as the **hypodermis** (Figure 5.1). Strictly speaking, the hypodermis is not really part of the skin, but it shares some of the skin's protective functions. The hypodermis, also called **superficial fascia** because it is superficial to the tough connective tissue wrapping (fascia) of the skeletal muscles, consists of mostly adipose tissue. Besides storing fat, the hypodermis anchors the skin to the underlying structures (mostly to muscles), but loosely enough that the skin can slide relatively freely over those structures. Sliding skin protects us by ensuring that many blows just glance off our bodies. Because of its fatty composition, the hypodermis also acts as a shock absorber and an insulator that prevents heat loss from the body. The hypodermis thickens markedly when one gains weight. In females, this "extra" subcutaneous fat accumulates first in the thighs and breasts, but in males it first collects in the anterior abdomen (as a "beer belly").

Epidermis

Structurally, the **epidermis** is a keratinized stratified squamous epithelium consisting of four distinct cell types and four or five distinct layers.

Cells of the Epidermis

The cells populating the epidermis include *keratinocytes, melanocytes, Merkel cells,* and *Langerhans' cells* (Figure 5.2). Most epidermal cells are keratinocytes, so we will consider them first. The chief role of **keratinocytes** (kĕ-rat'ĭ-no-sītz"; "keratin cells") is to produce **keratin,** the fibrous protein that helps give the epidermis its protective properties (*kera* = horn in Greek). Tightly connected to one another by desmosomes, the keratinocytes arise in the deepest part of the epidermis from a layer of cells (stratum basale) that undergo almost continuous mitosis. As these cells are pushed upward by the production of new cells beneath them, they make the keratin that eventually dominates their cell contents. By the time the keratinocytes reach the free surface of the skin, they are dead, scalelike structures that are little more than keratin-filled plasma membranes. Millions of these dead cells rub off every day, giving us a totally new epidermis every 25 to 45 days. In body areas regularly subjected to friction, such as the hands and feet, both cell production and keratin formation are accelerated. Persistent friction (from a poorly fitting shoe, for example) causes a thickening of the epidermis called a *callus.*

Melanocytes (mel'ah-no-sītz), the spider-shaped epithelial cells that synthesize the pigment **melanin** (mel'ah-nin; *melan* = black), are found in the deepest layer of the epidermis (Figure 5.2). As melanin is made and accumulated in membrane-bound granules called *melanosomes,* it is moved along actin filaments by motor proteins to the ends of the melanocyte's processes (the "spider arms") from where they are taken up by nearby keratinocytes. The melanin granules accumulate on the superficial, or "sunny," side of the keratinocyte nucleus, forming a pigment shield that protects the nucleus from the damaging effects of ultraviolet (UV) radiation in sunlight.

The star-shaped **Langerhans'** (lahng'er-hanz) **cells** arise from bone marrow and migrate to the epidermis. Also called **epidermal dendritic cells,** these cells are macrophages that help activate our immune system as described later in this chapter (p. 146). Their slender processes extend among the surrounding keratinocytes, forming a more or less continuous network (Figure 5.2).

Occasional **Merkel cells** are present at the epidermal-dermal junction. Shaped like a spiky hemisphere (Figure 5.2), each Merkel cell is intimately associated with a disclike sensory nerve ending. The combination, called a *Merkel disc,* functions as a sensory receptor for touch.

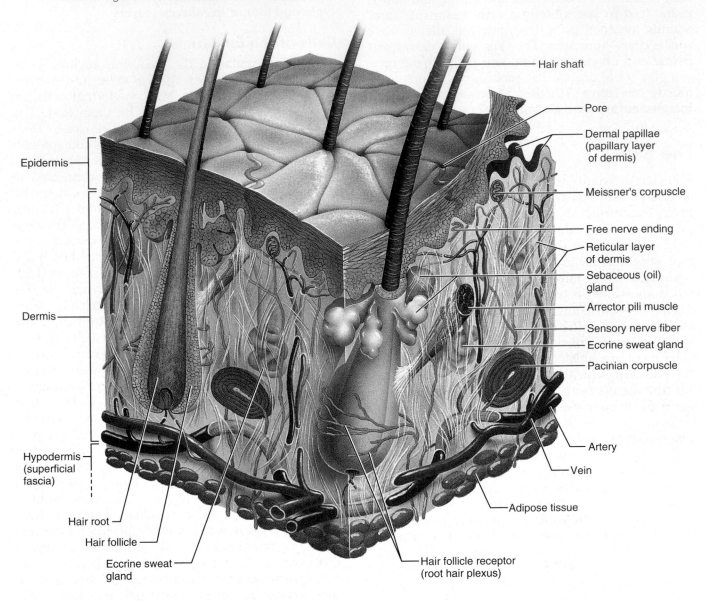

- Hair shaft
- Pore
- Dermal papillae (papillary layer of dermis)
- Meissner's corpuscle
- Free nerve ending
- Reticular layer of dermis
- Sebaceous (oil) gland
- Arrector pili muscle
- Sensory nerve fiber
- Eccrine sweat gland
- Pacinian corpuscle
- Artery
- Vein
- Adipose tissue

Epidermis

Dermis

Hypodermis (superficial fascia)

Hair root

Hair follicle

Eccrine sweat gland

Hair follicle receptor (root hair plexus)

FIGURE 5.1 Skin structure. Three-dimensional view of the skin and underlying subcutaneous tissue. The epidermal and dermal layers have been pulled apart at the right corner to reveal the dermal papillae.

Layers of the Epidermis

Variation in epidermal thickness determines if skin is *thick* or *thin*. In **thick skin,** which covers the palms, fingertips, and soles of the feet, the epidermis consists of five layers, or *strata* (stra′tah; "bed sheets"). From deep to superficial, these layers are

stratum basale, stratum spinosum, stratum granulosum, stratum lucidum, and stratum corneum. In **thin skin,** which covers the rest of the body, the stratum lucidum appears to be absent and the other strata are thinner (Figure 5.2).

Stratum Basale (Basal Layer)   The **stratum basale** (stra′tum ba-să′le), the deepest epidermal layer, is attached to the underlying dermis along a wavy borderline that reminds one of corrugated cardboard. For the most part, it consists of a single

(1) The stratum basale or basal layer of the epidermis. (2) Adipose cells. ■

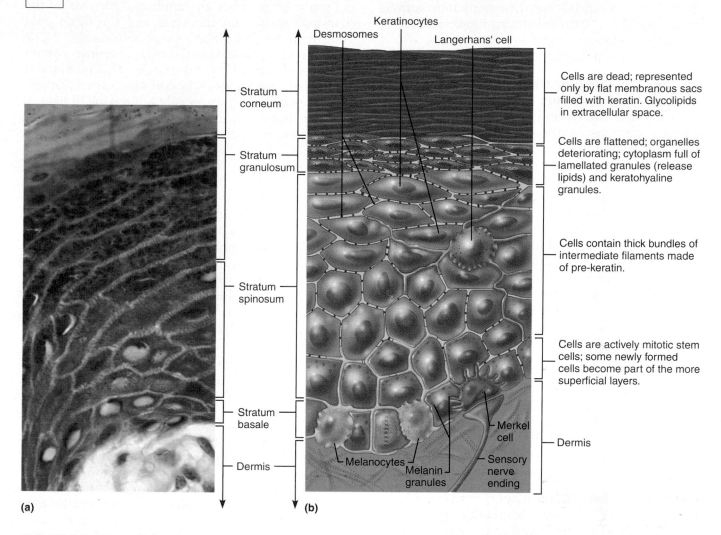

FIGURE 5.2 **The main structural features in epidermis of the skin.**
(a) Photomicrograph depicting the four major epidermal layers. (b) Diagram showing the layers and relative distribution of the different cell types. The keratinocytes (tan) form the bulk of the epidermis. Less numerous are the

melanocytes (gray), which produce the pigment melanin; Langerhans' cells (blue), which function as macrophages; and Merkel cells (purple). A sensory nerve ending (yellow), extending to the Merkel cell from the dermis (pink), is shown associated with the Merkel cell forming a Merkel disc (touch receptor).

Notice that the keratinocytes, but not the other cell types, are joined by numerous desmosomes (indicated by connections between membranes of adjacent cells). The stratum lucidum, present in thick skin, is not illustrated here. Only a portion of the stratum corneum is illustrated in each case.

row of cells representing the youngest keratinocytes. The many mitotic nuclei seen in this layer reflect the rapid division of these cells and account for its alternate name, **stratum germinativum** (jer′mĭ-nă″tiv-um; "germinating layer").

Some 10 to 25% of the cells in the stratum basale are melanocytes, and their branching processes extend among the surrounding cells,

reaching well into the more superficial stratum spinosum layer. Occasional Merkel cells are also seen in this stratum.

Stratum Spinosum (Prickly Layer) The **stratum spinosum** (spi′no-sum; "prickly") is several cell layers thick. These cells contain a weblike system of intermediate filaments, which span their cytosol to attach to desmosomes. These intermediate filaments consist mainly of tension-resisting bundles of pre-keratin filaments. The keratinocytes in this layer appear irregular (spiny) in shape, causing them to be

The desmosomes connecting the keratinocytes are important to maintain continuity of the epidermis.

called *prickle cells*. However, the spines do not exist in the living cells; they arise during tissue preparation when these cells shrink but their numerous desmosomes hold tight. Scattered among the keratinocytes are melanin granules and Langerhans' cells, which are most abundant in this epidermal layer.

Stratum Granulosum (Granular Layer)   The thin **stratum granulosum** (gran"u-lo'sum) consists of three to five cell layers in which keratinocyte appearance changes drastically. These cells flatten, their nuclei and organelles begin to disintegrate, and they accumulate *keratohyaline granules* and *lamellated granules.* The keratohyaline (ker"ah-to-hi'ah-lin) granules help to form keratin in the upper layers, as we will see. The lamellated (lam'ĭ-la-ted; "plated") granules contain a waterproofing glycolipid that is spewed into the extracellular space and is a major factor in slowing water loss across the epidermis. The plasma membranes of these cells thicken as cytosol proteins bind to the inner membrane face and lipids released by the lamellated granules coat their external surfaces. This makes them more resistant to destruction, so you might say that the keratinocytes are "toughening up" to make the outer strata the strongest skin region.

Like all epithelia, the epidermis relies on capillaries in the underlying connective tissue (the dermis in this case) for its nutrients. Above the stratum granulosum, the epidermal cells are too far from the dermal capillaries, so they die. This is a completely normal sequence of events.

Stratum Lucidum (Clear Layer)   Through the light microscope, the **stratum lucidum** (loo'sid-um; "light") appears as a thin translucent band just above the stratum granulosum. It consists of a few rows of clear, flat, dead keratinocytes with indistinct boundaries. Here, or in the stratum corneum above, the gummy substance of the keratohyaline granules clings to the keratin filaments in the cells, causing them to aggregate in parallel arrays. As mentioned before, the stratum lucidum is visible only in thick skin.

Stratum Corneum (Horny Layer)   The outermost **stratum corneum** (kor'ne-um) is a broad zone 20 to 30 cell layers thick that accounts for up to three-quarters of the epidermal thickness. Keratin and the thickened plasma membranes of cells in this stratum protect the skin against abrasion and penetration, and the glycolipid between its cells waterproofs this layer. Hence, the stratum corneum provides a durable "overcoat" for the body, protecting deeper cells from the hostile external environment (air) and from water loss, and rendering the body relatively insensitive to biological, chemical, and physical assaults. It is amazing that a layer of dead cells can still play so many roles.

The shingle-like cell remnants of the stratum corneum are referred to as *cornified,* or *horny, cells* (*cornu* = horn). They are familiar to everyone as the dandruff shed from the scalp and the flakes that slough off dry skin. [The average person sheds 18 kg (40 pounds) of these skin flakes in a lifetime, providing a lot of fodder for the dust mites that inhabit our homes and bed linens.] The common saying "Beauty is only skin deep" is especially interesting in light of the fact that nearly everything we see when we look at someone is dead!

Dermis

The **dermis** (*derm* = skin), the second major skin region, is strong, flexible connective tissue. Its cells are typical of those found in any connective tissue proper: fibroblasts, macrophages, and occasional mast cells and white blood cells, and its semifluid matrix is heavily embedded with collagen, elastin, and reticular fibers. The dermis binds the entire body together like a body stocking. It is your "hide" and corresponds exactly to animal hides used to make leather products.

The dermis is richly supplied with nerve fibers, blood vessels, and lymphatic vessels. The major portions of hair follicles, as well as oil and sweat glands, are derived from epidermal tissue but reside in the dermis.

The dermis has two layers: the papillary and reticular. The thin superficial **papillary** (pap'il-er-e) **layer** is areolar connective tissue in which the collagen and elastin fibers form a loosely woven mat that is heavily invested with blood vessels. Its superior surface is thrown into peglike projections called **dermal papillae** (pah-pil'e; *papill* = nipple) that indent the overlying epidermis (see Figure 5.1). Many dermal papillae contain capillary loops; others house free nerve endings (pain receptors) and touch receptors called *Meissner's corpuscles* (mīs'nerz kor'pus-lz). On the palms of the hands and soles of the feet, these papillae lie atop larger mounds called *dermal ridges,* which in turn cause the overlying epidermis to form *epidermal ridges* that increase friction and enhance the gripping ability of the fingers and feet. Epidermal ridge patterns are genetically determined and unique to each of us. Because sweat pores open along their crests, our fingertips leave identifying films of sweat called *fingerprints* on almost anything they touch.

The deeper **reticular layer,** accounting for about 80% of the thickness of the dermis, is dense irregular connective tissue. Its extracellular matrix contains thick bundles of interlacing collagen fibers that run in various planes; however, most run parallel to the skin surface. Separations, or less dense regions, between these bundles form *cleavage,* or *tension, lines,* in the skin. These externally invisible lines tend to run longitudinally in the skin of the head and limbs and in circular patterns around the neck and trunk. Cleavage lines are important to both surgeons and their patients. When an incision is made *parallel* to these lines, the skin gapes less and heals more readily than when the incision is made *across* cleavage lines.

The collagen fibers of the dermis give skin strength and resiliency that prevent most jabs and scrapes from penetrating the dermis. In addition, collagen binds water, helping to keep skin hydrated. Elastin fibers provide the stretch-recoil properties of skin.

In addition to the epidermal ridges and cleavage lines, a third type of skin marking, *flexure lines*, reflects dermal modifications. Flexure lines are dermal folds that occur at or near joints, where the dermis is tightly secured to deeper structures (notice the deep creases on your palms). Since the skin cannot slide easily to accommodate joint movement in such regions, the dermis folds and deep skin creases form. Flexure lines are also visible on the wrists, fingers, soles, and toes.

HOMEOSTATIC IMBALANCE

Extreme stretching of the skin, such as occurs during pregnancy, can tear the dermis. Dermal tearing is indicated by silvery white scars called *striae* (stri'e; "streaks"), commonly called "stretch marks." Short-term but acute trauma (as from a burn or wielding a hoe) can cause a *blister,* the separation of the epidermal and dermal layers by a fluid-filled pocket. ●

Skin Color

Three pigments contribute to skin color: melanin, carotene, and hemoglobin. Of these, only melanin is made in the skin. **Melanin,** a polymer made of tyrosine amino acids, ranges in color from yellow to reddish-brown to black. Its synthesis depends on an enzyme in melanocytes called tyrosinase (ti-ro'sĭ-nās) and, as noted earlier, it passes from melanocytes to the basal keratinocytes. Since all humans have the same relative number of melanocytes, individual and racial differences in skin coloring reflect the relative kind and amount of melanin made and retained. Melanocytes of black- and brown-skinned people produce many more and darker melanosomes than those of fair-skinned individuals, and their keratinocytes retain it longer. *Freckles* and *pigmented moles* are local accumulations of melanin.

Melanocytes are stimulated to greater activity when we expose our skin to sunlight. Prolonged sun exposure causes a substantial melanin buildup, which helps protect the DNA of viable skin cells

from UV radiation by absorbing the light and dissipating the energy as heat. Indeed, the initial signal for speeding up melanin synthesis seems to be a faster rate of repair of photodamaged DNA. In all but the darkest people, this response causes visible darkening of the skin (a tan).

HOMEOSTATIC IMBALANCE

Despite melanin's protective effects, excessive sun exposure eventually damages the skin. It causes clumping of elastin fibers, which results in leathery skin; temporarily depresses the immune system; and can alter the DNA of skin cells and in this way lead to skin cancer. The fact that dark-skinned people get skin cancer less often than fair-skinned people and get it in areas with less pigment—the soles of the feet and nail beds—attests to melanin's effectiveness as a natural sunscreen.

Ultraviolet radiation has other consequences as well. Many chemicals induce photosensitivity; that is, they heighten the skin's sensitivity to UV radiation, setting sun worshippers up for a skin rash the likes of which they would rather not see. Such substances include some antibiotic and antihistamine drugs, many chemicals in perfumes and detergents, and a chemical found in limes and celery. Small, itchy, blisterlike lesions erupt all over the body; then the peeling begins, in sheets! ●

Carotene (kar'o-tēn) is a yellow to orange pigment found in certain plant products such as carrots. It tends to accumulate in the stratum corneum and in fatty tissue of the hypodermis. Its color is most obvious in the palms and soles, where the stratum corneum is thickest (for example the skin of the heels), and most intense when large amounts of carotene-rich foods are eaten. However, the yellowish tinge of the skin of some Asian peoples is due to variations in melanin, as well as to carotene.

The pinkish hue of fair skin reflects the crimson color of oxygenated **hemoglobin** (he'mo-glo"bin) in the red blood cells circulating through the dermal capillaries. Because Caucasian skin contains only small amounts of melanin, the epidermis is nearly transparent and allows hemoglobin's color to show through.

HOMEOSTATIC IMBALANCE

When hemoglobin is poorly oxygenated, both the blood and the skin of Caucasians appear blue, a condition called *cyanosis* (si"ah-no'sis; *cyan* = dark blue). Skin often becomes cyanotic during heart failure and severe respiratory disorders. In dark-skinned individuals, the skin does not appear cyanotic because of the masking effects of melanin, but

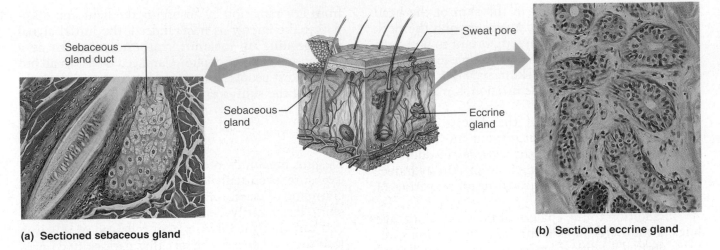

(a) Sectioned sebaceous gland

(b) Sectioned eccrine gland

FIGURE 5.3   **Cutaneous glands. (a)** Photomicrograph of a sebaceous gland (120×). **(b)** Photomicrograph of eccrine sweat glands (200×).

cyanosis is apparent in their mucous membranes and nail beds (the same sites where the red cast of normally oxygenated blood is visible).

The skin color of some people is also influenced by emotional stimuli, and many alterations in skin color signal certain disease states:

- *Redness*, or *erythema* (er"ĭ-the'mah): Reddened skin may indicate embarrassment (blushing), fever, hypertension, inflammation, or allergy.

- *Pallor*, or *blanching:* During fear, anger, and certain other types of emotional stress, some people become pale. Pale skin may also signify anemia or low blood pressure.

- *Jaundice* (jawn'dis), or *yellow cast:* An abnormal yellow skin tone usually signifies a liver disorder, in which yellow bile pigments accumulate in the blood and are deposited in body tissues. [Normally, the liver cells secrete the bile pigments (bilirubin) as a component of bile.]

- *Bronzing:* A bronze, almost metallic appearance of the skin is a sign of Addison's disease, hypofunction of the adrenal cortex.

- *Black-and-blue marks*, or *bruises:* Black-and-blue marks reveal where blood escaped from the circulation and clotted beneath the skin. Such clotted blood masses are called *hematomas* (he"mah-to'mah; "blood swelling"). ●

Appendages of the Skin

Along with the skin itself, the integumentary system includes several derivatives of the epidermis. These **skin appendages** include the nails, sweat glands, sebaceous (oil) glands, and hair follicles and hair. Each of these plays a unique role in maintaining body homeostasis.

Sweat (Sudoriferous) Glands

Sweat glands, also called **sudoriferous** (su"do-rif'er-us; *sudor* = sweat) **glands,** are distributed over the entire skin surface except the nipples and parts of the external genitalia. Their number is staggering—more than 2.5 million per person. There are two types of sweat glands: eccrine and apocrine.

Eccrine (ek'rin; "secreting") **sweat glands,** also called **merocrine sweat glands,** are far more numerous and are particularly abundant on the palms, soles of the feet, and forehead. Each is a simple, coiled, tubular gland. The secretory part lies coiled in the dermis; the duct extends to open in a funnel-shaped *pore* (*por* = channel) at the skin surface (see Figure 5.3b). (These sweat pores are different from the so-called pores of one's complexion, which are actually the external outlets of hair follicles.)

Eccrine gland secretion, commonly called **sweat,** is a hypotonic filtrate of the blood that passes through the secretory cells of the sweat glands and is released by exocytosis. It is 99% water, with some salts (mostly sodium chloride), vitamin C, antibodies, a microbe-killing peptide dubbed *dermicidin*, traces of metabolic wastes (urea, uric acid, ammonia), and lactic acid (the chemical that attracts mosquitoes). The exact composition depends on heredity and diet. Small amounts of ingested drugs may also be excreted by this route. Normally, sweat is acidic with a pH between 4 and 6.

Sweating is regulated by the sympathetic division of the autonomic nervous system, over which we have little control. Its major role is to prevent overheating of the body. Heat-induced sweating begins on the forehead and then spreads inferiorly over the remainder of the body. Emotionally induced sweating—the so-called "cold sweat" brought on by fright, embarrassment, or nervousness—begins on the palms, soles, and axillae (armpits) and then spreads to other body areas.

Apocrine (ap'o-krin) **sweat glands*** are largely confined to the axillary and anogenital areas. They are larger than eccrine glands, and their ducts empty into hair follicles. Apocrine secretion contains the same basic components as true sweat, plus fatty substances and proteins. Consequently, it is quite viscous and sometimes has a milky or yellowish color. The secretion is odorless, but when its organic molecules are decomposed by bacteria on the skin, it takes on a musky and generally unpleasant odor, the basis of body odor.

Apocrine glands begin functioning at puberty under the influence of androgens and have little role to play in thermoregulation. Their precise function is not yet known, but they are activated by sympathetic nerve fibers during pain and stress. Because their activity is increased by sexual foreplay, and they enlarge and recede with the phases of a woman's menstrual cycle, they may be analogous to the sexual scent glands of other animals.

Ceruminous (sĕ-roo'mĭ-nus; *cera* = wax) **glands** are modified apocrine glands found in the lining of the external ear canal. They secrete a sticky, bitter substance called *cerumen*, or earwax, that is thought to deter insects and block entry of foreign material.

Mammary glands, another variety of specialized sweat glands, secrete milk. Although they are properly part of the integumentary system, we consider the mammary glands in Chapter 26, along with female reproductive organs.

Sebaceous (Oil) Glands

The **sebaceous** (se-ba'shus; "greasy") **glands,** or **oil glands** (see Figure 5.3a), are simple alveolar glands that are found all over the body except on the palms and soles. They are small on the body trunk and limbs, but quite large on the face, neck, and upper chest. These glands secrete an oily secretion called **sebum** (se'bum). The central cells of the alveoli accumulate oily lipids until they become so engorged that they burst, so functionally these glands are *holocrine glands* (see p. 114). The accumulated lipids and cell fragments constitute sebum. Sebum is secreted into a hair follicle, or occasionally to a pore on the skin surface. Sebum softens and lubricates the hair and skin, prevents hair from becoming brittle, and slows water loss from the skin when the exter-

nal humidity is low. Perhaps even more important is its *bactericidal* (bacterium-killing) action.

The secretion of sebum is stimulated by hormones, especially androgens. These glands are relatively inactive during childhood but are activated in both sexes during puberty, when androgen production begins to rise.

HOMEOSTATIC IMBALANCE

If a sebaceous gland duct is blocked by accumulated sebum, a *whitehead* appears on the skin surface. If the material oxidizes and dries, it darkens to form a *blackhead. Acne* is an active inflammation of the sebaceous glands accompanied by "pimples" (pustules or cysts) on the skin. It is usually caused by bacterial infection, particularly by staphylococcus, and can be mild or extremely severe, leading to permanent scarring. **Seborrhea** (seb"o-re'ah; "fast-flowing sebum"), known as "cradle cap" in infants, is caused by overactive sebaceous glands. It begins on the scalp as pink, raised lesions that gradually become yellow to brown and begin to slough off oily scales. Careful washing to remove the excessive oil often helps. ●

Nails

A **nail** is a scalelike modification of the epidermis that forms a clear protective covering on the dorsal surface of the distal part of a finger or toe (Figure 5.4). Nails, which correspond to the hooves or claws of other animals, are particularly useful as "tools" to help pick up small objects and to scratch an itch. In contrast to *soft keratin* of the epidermis, nails contain *hard keratin*. Each nail has a *free edge*, a *body* (visible attached portion), and a proximal *root* (embedded in the skin). The deeper layers of the epidermis extend beneath the nail as the *nail bed*; the nail itself corresponds to the superficial keratinized layers. The thickened proximal portion of the nail bed, called the **nail matrix,** is responsible for nail growth. As the nail cells produced by the matrix become heavily keratinized, the nail body slides distally over the nail bed.

Nails normally appear pink because of the rich bed of capillaries in the underlying dermis. However, the region that lies over the thick nail matrix appears as a white crescent called the *lunula* (lu'nu-lah; "little moon"). The proximal and lateral borders of the nail are overlapped by skin folds, called **nail folds.** The proximal nail fold projects onto the nail body as the **cuticle** or **eponychium** (ep"o-nik'e-um; "on the nail"). The region beneath the free edge of the nail where dirt and debris tend to accumulate is the **hyponychium** ("below nail"), informally called the quick.

---

* The term *apocrine sweat glands* is a misnomer. These glands were originally thought to release their product by the apocrine mode (in which the apical surface pinches off) and were named accordingly. Subsequent studies proved them to be merocrine glands, which release their product by exocytosis like the eccrine sweat glands.

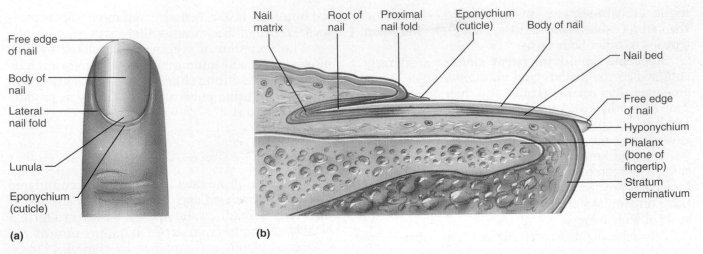

Free edge of nail

Body of nail

Lateral nail fold

Lunula

Eponychium (cuticle)

Nail matrix

Root of nail

Proximal nail fold

Eponychium (cuticle)

Body of nail

Nail bed

Free edge of nail

Hyponychium

Phalanx (bone of fingertip)

Stratum germinativum

(a)

(b)

FIGURE 5.4   Structure of a nail. (a) Surface view of the distal part of a finger showing nail parts. (b) Sagittal section of the fingertip. The nail matrix that forms the nail lies beneath the lunula; the epidermis of the nail bed underlies the nail.

Hairs and Hair Follicles

Hair is an important part of our body image—consider, for example, the spiky hair style of punk rockers and the flowing, glossy manes of some high fashion models. Millions of hairs are distributed over our entire skin surface except our palms, soles, lips, nipples, and parts of the external genitalia (the head of the penis, for instance). Although hair helps to keep other mammals warm, our sparse body hair is far less luxuriant and useful. Its main function in humans is to sense insects on the skin before they sting us. Hair on the scalp guards the head against physical trauma, heat loss, and sunlight. Eyelashes shield the eyes, and nose hairs filter large particles like lint and insects from the air we inhale.

Structure of a Hair

Hairs, or **pili** (pi'li), are flexible strands produced by hair follicles that consist largely of dead, keratinized cells. The *hard keratin* that dominates hairs and nails has two advantages over the *soft keratin* found in typical epidermal cells: (1) It is tougher and more durable, and (2) its individual cells do not flake off.

The chief regions of a hair are the *shaft*, which projects from the skin (Figure 5.5), and the *root*, the part embedded in the skin (Figure 5.6). If the shaft is flat and ribbonlike in cross section, the hair is kinky; if it is oval, the hair is silky and wavy; if it is perfectly round, the hair is straight and tends to be coarse.

A hair has three concentric layers of keratinized cells (Figure 5.6a and b). Its central core, the *medulla* (mĕ-dul'ah; "middle"), consists of large cells and air spaces. The medulla is absent in fine hairs. The *cortex*, a bulky layer surrounding the medulla,

FIGURE 5.5   Hair shaft emerging from a follicle at the epidermal surface. Notice how the scalelike cells of the cuticle overlap one another (scanning electron micrograph, 1300×).

consists of several layers of flattened cells. The outermost **cuticle** is formed from a single layer of cells that overlap one another from below like shingles on a roof (Figure 5.5). This arrangement helps to keep neighboring hairs apart so that the hair does not mat. (Hair conditioners smooth out the rough surface of the cuticle and make our hair look shiny.) The most heavily keratinized part of the hair, the cuticle, provides strength and helps keep the inner layers tightly compacted. Because it is subjected to the most abrasion, the cuticle tends to wear away at the tip of the hair shaft, allowing the keratin fibrils in the cortex and medulla to frizz out, creating "split ends."

Hair pigment is made by melanocytes at the base of the hair follicle and transferred to the cortical cells. Various proportions of melanins of different colors (yellow, rust, brown, and black) combine to produce hair color from blond to pitch black. Additionally, red hair is colored by the iron-containing pigment called *trichosiderin*. Gray or white hair results from decreased melanin production (mediated by delayed-action genes) and from the replacement of melanin by air bubbles in the hair shaft.

Structure of a Hair Follicle

Hair follicles (*folli* = bag) fold down from the epidermal surface into the dermis. In the scalp, they may even extend into the hypodermis. The deep end of the follicle, located about 1/6 in. below the skin surface, is expanded, forming a **hair bulb** (Figure 5.6c and d). A knot of sensory nerve endings called a **hair follicle receptor,** or **root hair plexus,** wraps around each hair bulb (see Figure 5.1), and bending the hair stimulates these endings. Consequently, our hairs act as sensitive touch receptors.

▪ Feel the tickle as you run your hand over the hairs on your forearm.

A *hair papilla*, a nipplelike bit of dermal tissue, protrudes into the hair bulb. This papilla contains a knot of capillaries that supplies nutrients to the growing hair and signals it to grow. Except for its specific location, this papilla is similar to the dermal papillae underlying other epidermal regions.

The wall of a hair follicle is composed of an outer **connective tissue root sheath,** derived from the dermis, a thickened basement membrane called the *glassy membrane,* and an inner **epithelial root sheath,** derived mainly from an invagination of the epidermis (Figure 5.6c and d). The epithelial root sheath, which has external and internal parts, thins as it approaches the hair bulb, so that only a single layer of epithelial cells covers the papilla. However, the cells that compose the **hair matrix,** or actively dividing area of the hair bulb that produces the hair, originate in a region called the *hair bulge* located a fraction of a millimeter above the hair bulb. When chemical signals diffusing from the papilla reach the hair bulge, some of its cells migrate toward the papilla, where they divide to produce the hair cells. As new hair cells are produced by the matrix, the older part of the hair is pushed upward, and its fused cells become increasingly keratinized and die.

Associated with each hair follicle is a bundle of smooth muscle cells called an **arrector pili** (ah-rek′tor pi′li; "raiser of hair") muscle. As you can see in Figure 5.1, most hair follicles approach the skin surface at a slight angle. The arrector pili muscles are attached in such a way that their contraction pulls the hair follicle into an upright position and dimples the skin surface to produce goose bumps in response to cold external temperatures or fear. Although this "hair-raising" response is not very useful to humans, with our short sparse hairs, it is an important heat retention and protection mechanism in other animals. Furry animals can stay warmer by trapping a layer of insulating air in their fur; and a scared animal with its hair on end looks larger and more formidable to its enemy.

Types and Growth of Hair

Hairs come in various sizes and shapes, but as a rule, they can be classified as vellus or terminal. The body hair of children and adult females is of the pale, fine **vellus** (vel′us; *vell* = wool, fleece) **hair** variety. The coarser, longer hair of the eyebrows and scalp is **terminal hair;** terminal hair may also be darker. At puberty, terminal hairs appear in the axillary and pubic regions of both sexes and on the face and chest (and typically the arms and legs) of males. These terminal hairs grow in response to the stimulating effects of male sex hormones called *androgens* (of which *testosterone* is the most important).

Hair growth and density are influenced by many factors, but most importantly by nutrition and hormones. Poor nutrition means poor hair growth, whereas conditions that increase local dermal blood flow (such as chronic physical irritation or inflammation) may enhance local hair growth. Many old-time bricklayers who carried their hod on one shoulder all the time developed one hairy shoulder. As noted, testosterone also encourages hair growth, and when male hormones are present in large amounts, terminal hair growth is luxuriant. Undesirable hair growth (such as on a woman's upper lip) may be arrested by *electrolysis* or laser treatments, which use electricity or light energy, respectively, to destroy the hair roots.

HOMEOSTATIC IMBALANCE

In women, small amounts of androgens are normally produced by both the ovaries and the adrenal

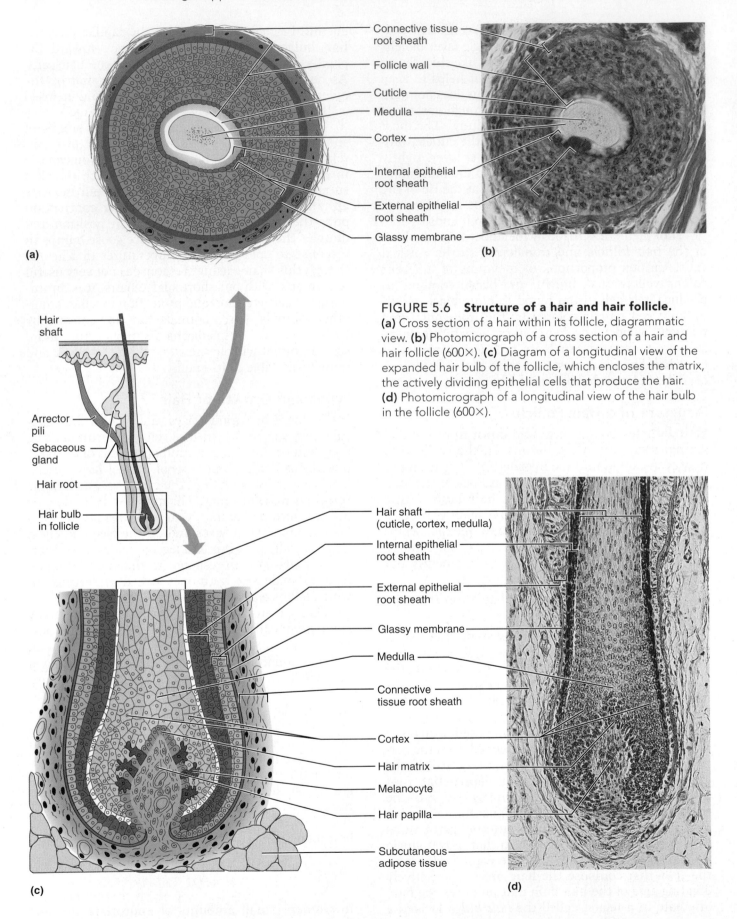

(a)

(b)

FIGURE 5.6   Structure of a hair and hair follicle.
(a) Cross section of a hair within its follicle, diagrammatic
view. **(b)** Photomicrograph of a cross section of a hair and
hair follicle (600×). **(c)** Diagram of a longitudinal view of the
expanded hair bulb of the follicle, which encloses the matrix,
the actively dividing epithelial cells that produce the hair.
(d) Photomicrograph of a longitudinal view of the hair bulb
in the follicle (600×).

(c)

(d)

glands. Excessive hairiness, or *hirsutism* (her'soot-izm; *hirsut* = hairy), as well as other signs of masculinization, may result from an adrenal gland or ovarian tumor that secretes abnormally large amounts of androgens. Since few women want a beard or hairy chest, such tumors are surgically removed as soon as possible. ●

The rate of hair growth varies from one body region to another and with sex and age, but it averages 2.5 mm per week. Each follicle goes through *growth cycles*. In each cycle, an active growth phase *(anagen)*, ranging from weeks to years, is followed by a regressive phase *(catagen)*, when the hair matrix cells die and the follicle base and hair bulb shrivel somewhat. The follicle then enters a resting phase *(telogen)* for one to three months. After the resting phase, the matrix proliferates again and forms a new hair to replace the old one that has fallen out or will be pushed out by the new hair.

The life span of hairs varies and appears to be under control of a slew of proteins. The follicles of the scalp remain active for six to ten years before becoming inactive for a few months. Because only a small percentage of the hair follicles are shed at any one time, we lose an average of 90 scalp hairs daily. The follicles of the eyebrow hairs remain active for only three to four months, which explains why your eyebrows are never as long as the hairs on your head.

Hair Thinning and Baldness

A follicle has only a limited number of cycles in it. Given ideal conditions, hair grows fastest from the teen years to the 40s; then its growth slows. The fact that hairs are not replaced as fast as they are shed leads to hair thinning and some degree of baldness, or **alopecia** (al"o-pe'she-ah), in both sexes. Much less dramatic in women, the process usually begins at the anterior hairline and progresses posteriorly. Coarse terminal hairs are replaced by vellus hairs, and the hair becomes increasingly wispy.

True, or *frank, baldness* is a different story entirely. The most common type, **male pattern baldness,** is a genetically determined, sex-influenced condition. It is thought to be caused by a delayed-action gene that "switches on" in adulthood and changes the response of the hair follicles to DHT (dihydrotestosterone), a metabolite of testosterone. As a result, the follicular growth cycles become so short that many hairs never even emerge from their follicles before shedding, and those that do are fine vellus hairs that look like peach fuzz in the "bald" area. Until recently, the only cure for male pattern baldness was drugs that inhibit testosterone production, but they also cause loss of sex drive—a trade-off few men would choose. Quite by accident, it was discovered that minoxidil, a drug used to reduce high blood pressure, has an interesting side effect in some bald men; it stimulates hair regrowth. Although its results are variable, minoxidil is available over the counter in dropper bottles or spray form for application to the scalp. Finasteride, according to some the most promising cure ever developed for male pattern baldness, hit pharmacy shelves in early 1998 and has had moderate success. Available only by prescription in once-a-day pill form, it must be taken for the rest of one's life. Once the patient stops taking it, all of the new growth falls out, and so does what remains of the "old" hair.

Ⓗ HOMEOSTATIC IMBALANCE

Hair thinning can be induced by a number of factors that upset the normal balance between hair loss and replacement. Outstanding examples are acutely high fever, surgery, or severe emotional trauma, and certain drugs (excessive vitamin A, some antidepressants and blood thinners, anabolic steroids, and most chemotherapy drugs). Protein-deficient diets and lactation lead to hair thinning because new hair growth stops when protein needed for keratin synthesis is not available or is being used for milk production. In all of these cases, hair regrows if the cause of thinning is removed or corrected. In the rare condition called *alopecia areata* the follicles are attacked by the immune system and the hair falls out in patches. But again, the follicles survive. However, hair loss due to severe burns, excessive radiation, or other factors that eliminate the follicles is permanent. ●

Functions of the Integumentary System

The skin and its derivatives perform a variety of functions that affect body metabolism, and prevent external factors from upsetting body homeostasis. Given its superficial location it is our most vulnerable organ system, exposed to bacteria, abrasion, temperature extremes, and harmful chemicals.

Protection

The skin constitutes at least three types of barriers: chemical, physical, and biological.

Chemical Barriers

The chemical barriers include skin secretions and melanin. Although the skin's surface teems with bacteria, the low pH of skin secretions, or the so-called **acid mantle,** retards their multiplication. In addition, many bacteria are killed outright by bactericidal substances in sebum. Skin cells also secrete a natural antibiotic called *human defensin* that literally punches holes in bacteria, making them look like sieves. Wounded skin releases large quantities of protective peptides called *cathelicidins* that are particularly effective in preventing infection by group A streptococcus bacteria. As discussed earlier, melanin provides a chemical pigment shield to prevent UV damage to the viable skin cells.

Physical/Mechanical Barriers

Physical, or mechanical, barriers are provided by the continuity of skin and the hardness of its keratinized cells. As a physical barrier, the skin is a remarkable compromise. A thicker epidermis would be more impenetrable, but we would pay the price in loss of suppleness and agility. Epidermal continuity works hand in hand with the acid mantle to ward off bacterial invasion. The waterproofing glycolipids of the epidermis block the diffusion of water and water-soluble substances between cells, preventing both their loss from and entry into the body through the skin. Substances that *do* penetrate the skin in limited amounts include (1) *lipid-soluble substances*, such as oxygen, carbon dioxide, fat-soluble vitamins (A, D, E, and K), and steroids; (2) *oleoresins* (o"le-o-rez'inz) of certain plants, such as poison ivy and poison oak; (3) *organic solvents*, such as acetone, dry-cleaning fluid, and paint thinner, which dissolve the cell lipids; (4) *salts of heavy metals*, such as lead, mercury, and nickel; and (5) drug agents called *penetration enhancers* (dialkylamino acetates) that help ferry other drugs into the body.

⬤H HOMEOSTATIC IMBALANCE

Organic solvents and heavy metals are devastating to the body and can be lethal. Passage of organic solvents through the skin into the blood can cause the kidneys to shut down and can also cause brain damage. Absorption of lead results in anemia and neurological defects. These substances should never be handled with bare hands. ●

Biological Barriers

Biological barriers include the Langerhans' cells of the epidermis, macrophages in the dermis, and DNA itself. Langerhans' cells are active elements of the immune system. For the immune response to be activated, the foreign substances, or *antigens*, must be presented to specialized white blood cells called lymphocytes. In the epidermis, it is the Langerhans' cells that play this role. Dermal macrophages constitute a second line of defense to dispose of viruses and bacteria that have managed to penetrate the epidermis. They, too, act as antigen "presenters." Although melanin provides a fairly good chemical sun-

screen, DNA itself is a remarkably effective biologically based sunscreen. Electrons in DNA molecules absorb UV radiation and transfer it to the atomic nuclei, which heat up and vibrate vigorously. However, since the heat dissipates to surrounding water molecules instantaneously, the DNA converts potentially destructive radiation into harmless heat.

Body Temperature Regulation

The body works best when its temperature remains within homeostatic limits. Like car engines, we need to get rid of the heat generated by our internal reactions. As long as the external temperature is lower than body temperature, the skin surface loses heat to the air and to cooler objects in its environment, just as a car radiator loses heat to the air and other nearby engine parts.

Under normal resting conditions, and as long as the environmental temperature is below 31–32°C (88–90°F), sweat glands continuously secrete unnoticeable amounts of sweat [about 500 ml (0.5 L) of sweat per day]. When body temperature rises, dermal blood vessels dilate and the sweat glands are stimulated into vigorous secretory activity. Sweat becomes noticeable and can account for the loss of up to 12 L of body water in one day. Evaporation of sweat from the skin surface dissipates body heat and efficiently cools the body, thus preventing overheating.

When the external environment is cold, dermal blood vessels constrict. This causes the warm blood to bypass the skin temporarily and allows skin temperature to drop to that of the external environment. Once this has happened, passive heat loss from the body is slowed, thus conserving body heat. Body temperature regulation is discussed in greater detail in Chapter 23.

Cutaneous Sensation

The skin is richly supplied with **cutaneous sensory receptors,** which are actually part of the nervous system. The cutaneous receptors are classified as *exteroceptors* (ek"ster-o-sep'torz) because they respond to stimuli arising outside the body. For example, Meissner's corpuscles (in the dermal papillae) and Merkel discs allow us to become aware of a caress or the feel of our clothing against our skin, whereas Pacinian receptors (in the deeper dermis or hypodermis) alert us to bumps or contacts involving deep pressure. Hair follicle receptors report on wind blowing through our hair and a playful tug on a pigtail. Painful stimuli (irritating chemicals, extreme heat or cold, and others) are sensed by bare nerve endings that meander throughout the skin. Detailed discussion of these cutaneous receptors is deferred to Chapter 13, but those mentioned above are illustrated in Figure 5.1. One, Merkel's disc, is also shown in Figure 5.2a.

Metabolic Functions

When sunlight bombards the skin, modified cholesterol molecules circulating through dermal blood vessels are converted to a vitamin D precursor, and transported via the blood to other body areas to play various roles in calcium metabolism. For example, calcium cannot be absorbed from the digestive tract without vitamin D.

Besides synthesizing vitamin D, the epidermis has a host of other metabolic functions. It makes chemical conversions that supplement those of the liver—for example, keratinocyte enzymes can (1) "disarm" many cancer-causing chemicals that penetrate the epidermis; (2) convert some harmless chemicals into carcinogens; and (3) activate some steroid hormones, for example, they can transform cortisone applied to irritated skin into hydrocortisone, a potent anti-inflammatory drug. Skin cells also make several biologically important proteins, including collagenase, an enzyme that aids the natural turnover of collagen (and deters wrinkles).

Blood Reservoir

The dermal vascular supply is extensive and can hold large volumes of blood (about 5% of the body's entire blood volume). When other body organs, such as vigorously working muscles, need a greater blood supply, the nervous system constricts the dermal blood vessels. This shunts more blood into the general circulation, making it available to the muscles and other body organs.

Excretion

Limited amounts of nitrogen-containing wastes (ammonia, urea, and uric acid) are eliminated from the body in sweat, although most such wastes are excreted in urine. Profuse sweating is an important avenue for water and salt (sodium chloride) loss.

Homeostatic Imbalances of Skin

When skin rebels, it is quite a visible revolution. Loss of homeostasis in body cells and organs reveals itself on the skin in ways that are sometimes almost too incredible to believe. The skin can develop more than 1000 different conditions and ailments. The most common skin disorders are bacterial, viral, or yeast infections. Less common, but far more damaging to body well-being, are skin cancer and burns, considered next.

Skin Cancer

Most tumors that arise in the skin are benign and do not spread (metastasize) to other body areas. (A wart, a neoplasm caused by a virus, is one example.) However, some skin tumors are malignant, or cancerous, and invade other body areas. A crucial risk factor for the nonmelanoma skin cancers is overexposure to the UV radiation in sunlight, which appears to disable a tumor suppressor gene [*p53* or the patched (*ptc*) gene]. In limited numbers of cases, however, frequent irritation of the skin by infections, chemicals, or physical trauma seems to be a predisposing factor.

Interestingly, sunburned skin accelerates its production of Fas, a protein that causes genetically damaged skin cells to commit suicide, thus decreasing the risk of mutations that will cause sun-linked skin cancer. It is the death of these gene-damaged cells that causes the skin to peel after a sunburn. Good news for sun worshipers is the newly developed skin lotions that can fix damaged DNA before the involved cells can develop into cancer cells. These lotions contain tiny oily vesicles (liposomes) filled with enzymes that initiate repair of the DNA mutations most commonly caused by sunlight. The liposomes penetrate the epidermis and enter the keratinocytes, ultimately making they way into their nuclei to bind to specific sites where two DNA bases have fused. There, by selective cutting of the DNA strands, they begin a DNA repair process that is completed by cellular enzymes.

Basal Cell Carcinoma

Basal cell carcinoma (kar″sĭ-no′mah) is the least malignant and most common skin cancer; over 30% of all white people get it in their lifetime. Stratum basale cells proliferate, invading the dermis and hypodermis. The cancer lesions occur most often on sun-exposed areas of the face and appear as shiny, dome-shaped nodules that later develop a central ulcer with a pearly, beaded edge (Figure 5.7a). Basal cell carcinoma is relatively slow-growing, and metastasis seldom occurs before it is noticed. Full cure by surgical excision is the rule in 99% of cases.

Squamous Cell Carcinoma

Squamous cell carcinoma arises from the keratinocytes of the stratum spinosum. The lesion appears as a scaly reddened papule (small, rounded elevation) that arises most often on the head (scalp, ears, and lower lip), and hands (Figure 5.7b). It tends to grow rapidly and metastasize if not removed. If it is caught early and removed surgically or by radiation therapy, chance of complete cure is good.

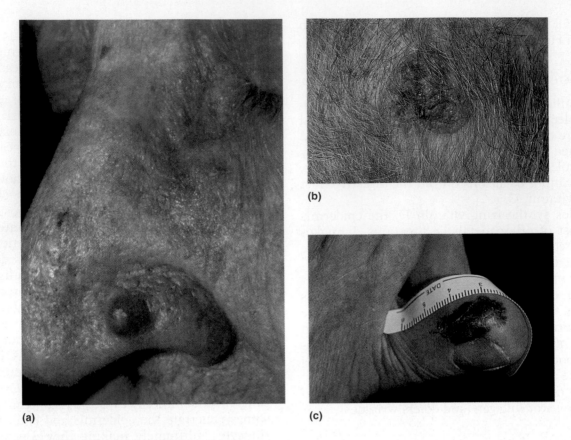

FIGURE 5.7   Photographs of skin cancers. (a) Basal cell carcinoma.
(b) Squamous cell carcinoma. **(c)** Melanoma.

Melanoma

Melanoma (mel"ah-no'mah), cancer of melanocytes, is the most dangerous skin cancer because it is highly metastatic and resistant to chemotherapy. It accounts for only about 5% of skin cancers, but its incidence is increasing rapidly (by 3 to 8% per year in the United States). Melanoma can begin wherever there is pigment. Most such cancers appear spontaneously; about one-third develop from preexisting moles. It usually appears as a spreading brown to black patch (Figure 5.7c) that metastasizes rapidly to surrounding lymph and blood vessels. The key to surviving melanoma is early detection. The chance of survival is poor if the lesion is over 4 mm thick. The usual therapy for melanoma is wide surgical excision accompanied by immunotherapy (immunizing the body against its cancer cells).

The American Cancer Society suggests that sun worshipers regularly examine their skin for new moles or pigmented spots and apply the **ABCD rule** for recognizing melanoma. **A. Asymmetry:** The two sides of the pigmented spot or mole do not match. **B. Border irregularity:** The borders of the lesion exhibit indentations. **C. Color:** The pigmented spot contains several colors (blacks, browns, tans, and sometimes blues and reds). **D. Diameter:** The spot is larger than 6 mm in diameter (the size of a pencil eraser). Some experts have found that adding an **E,** for elevation above the skin surface, improves diagnosis, so they use the ABCD(E) rule.

Burns

Burns are a devastating threat to the body primarily because of their effects on the skin. A **burn** is tissue damage inflicted by intense heat, electricity, radiation, or certain chemicals, all of which denature cell proteins and cause cell death in the affected areas.

The immediate threat to life resulting from severe burns is a catastrophic loss of body fluids containing proteins and electrolytes, resulting in dehydration and electrolyte imbalance. These, in turn, lead to renal shutdown and circulatory shock (inadequate blood circulation due to reduced blood volume). To save the patient, the lost fluids must be replaced immediately. In adults, the volume of fluid lost can be estimated by computing the percentage of body surface burned (extent of the burns) using the **rule of nines.** This method divides the body into 11 areas, each accounting for 9% of total body area, plus an additional area surrounding the genitals accounting for 1% of body surface area (Figure 5.8). This method

?   *Although the anterior head and face represent a very small proportion of the body surface, burns to this region are often much more serious than burns to the trunk. Why?*

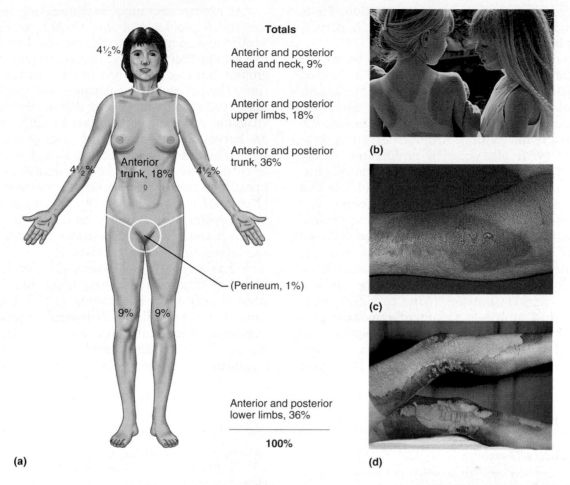

Totals

Anterior and posterior head and neck, 9%

Anterior and posterior upper limbs, 18%

Anterior and posterior trunk, 36%

(Perineum, 1%)

Anterior and posterior lower limbs, 36%

_____

100%

4½%

Anterior trunk, 18%

4½% 4½%

9% 9%

(a)

(b)

(c)

(d)

FIGURE 5.8 Estimating the extent and severity of burns. (a) The extent of burns is estimated by using the rule of nines. Surface area values for the anterior body surface are indicated on the human figure. Total surface area (anterior and posterior body surfaces) for each body region is indicated to the right of the figure. Severity of burns depends on the depth of tissue damage. **(b)** In first-degree burns only the epidermis is destroyed. **(c)** In second-degree burns, the epidermis and part of the dermis are involved and blistering occurs. **(d)** In third-degree burns the epidermis and dermis (and often part of the hypodermis) are destroyed.

is only approximate, so special tables are used when greater accuracy is desired.

    Burn patients also need thousands of extra food calories daily to replace lost proteins and allow tissue repair. No one can eat enough food to provide these calories, so burn patients are given supplementary nutrients through gastric tubes and intravenous (IV) lines. After the initial crisis has passed, infection becomes the main threat and is the leading cause of death in burn victims. Burned skin is sterile for about 24 hours. Thereafter, bacteria, fungi, and other pathogens easily invade areas where the skin barrier is destroyed, and they multiply rapidly in the nutrient-rich environment of dead tissues. Adding to this problem is the fact that the immune system becomes deficient within one to two days after severe burn injury.

    Burns are classified according to their severity (depth) as first-, second-, or third-degree burns (Figure 5.8b–d). In **first-degree burns,** only the epidermis is damaged. Symptoms include localized redness, swelling, and pain. First-degree burns tend to heal in two to three days without special attention. Sunburn is usually a first-degree burn. **Second-degree**

Burn injury to the facial area may burn the respiratory passages, putting the patient at risk for suffocation as the burned tissues swell. ■

burns injure the epidermis and the upper region of the dermis. Symptoms mimic those of first-degree burns, but blisters also appear. Skin regeneration occurs with little or no scarring within three to four weeks if care is taken to prevent infection. First- and second-degree burns are referred to as *partial-thickness* **burns.**

Third-degree burns are *full-thickness* **burns,** involving the entire thickness of the skin. The burned area appears gray-white, cherry red, or blackened, and initially there is little or no edema. Since the nerve endings in the area have been destroyed, the burned area is not painful. Although skin regeneration might eventually occur by proliferation of epithelial cells at the edges of the burn, it is usually impossible to wait this long because of fluid loss and infection. Thus, skin grafting is usually necessary.

To prepare the burned area for grafting, the *eschar* (es'kar), or burned skin, must first be debrided (removed). To prevent infection and fluid loss, the area is then flooded with antibiotics and covered temporarily with a synthetic membrane, animal (pig) skin, cadaver skin, or "living bandage" made from the thin amniotic sac membrane that surrounds a fetus. Then healthy skin is transplanted to the burned site. Unless the graft is taken from the patient (an autograft), however, there is a good chance that it will be rejected by the patient's immune system (see p. 702 in Chapter 20). Even if the graft "takes," extensive scar tissue often forms in the burned areas.

An exciting technique is eliminating many of the traditional problems of skin grafting and rejection. Synthetic skin made of a silicone "epidermis" bound to a spongy "dermal" layer composed of collagen and ground cartilage is applied to the debrided area. In time, the patient's own dermal tissue replaces and reabsorbs the artificial one. While dermal reconstruction is going on, tiny bits of epidermal tissue are harvested from unburned parts of the patient's body. The cells are isolated and coaxed to proliferate in culture dishes. Once the new dermis is complete (usually in two to three months), the silicone sheet is stripped off, and sheets of smooth, pink, bottle-grown epidermis are grafted on the dermal surface.

In general, burns are considered critical if any of the following conditions exists: (1) over 25% of the body has second-degree burns, (2) over 10% of the body has third-degree burns, or (3) there are third-degree burns of the face, hands, or feet. Facial burns introduce the possibility of burned respiratory passageways, which can swell and cause suffocation. Burns at joints are also troublesome because scar tissue formation can severely limit joint mobility.

Review Questions

Multiple Choice/Matching

(Some questions have more than one correct answer. Select the best answer or answers from the choices given.)

1. Which epidermal cell type is most numerous? (a) keratinocyte, (b) melanocyte, (c) Langerhans' cell, (d) Merkel cell.

2. Which is a macrophage? (a) a keratinocyte, (b) a melanocyte, (c) a Langerhans' cell, (d) a Merkel cell.

3. The epidermis provides a physical barrier due largely to the presence of (a) melanin, (b) carotene, (c) collagen, (d) keratin.

4. Skin color is determined by (a) the amount of blood, (b) pigments, (c) oxygenation level of the blood, (d) all of these.

5. The sensations of touch and pressure are picked up by receptors located in (a) the stratum germinativum, (b) the dermis, (c) the hypodermis, (d) the stratum corneum.

6. Which is not a true statement about the papillary layer of the dermis? (a) It produces the pattern for fingerprints, (b) it is most responsible for the toughness of the skin, (c) it contains nerve endings that respond to stimuli, (d) it is highly vascular.

7. Skin surface markings that reflect points of tight dermal attachment to underlying tissues are called (a) tension lines, (b) papillary ridges, (c) flexure lines, (d) epidermal papillae.

8. Which of the following is not an epidermal derivative? (a) hair, (b) sweat gland, (c) sensory receptor, (d) sebaceous gland.

9. You can cut hair without feeling pain because (a) there are no nerves associated with hair, (b) the shaft of the hair consists of dead cells, (c) hair follicles develop from epidermal cells and the epidermis lacks a nerve supply, (d) hair follicles have no source of nourishment and therefore cannot respond.

10. An arrector pili muscle (a) is associated with each sweat gland, (b) can cause the hair to stand up straight, (c) enables each hair to be stretched when wet, (d) provides new cells for continued growth of its associated hair.

11. The product of this type of sweat gland includes protein and lipid substances that become odoriferous as a result of bacterial action: (a) apocrine gland, (b) eccrine gland, (c) sebaceous gland, (d) pancreatic gland.

12. Sebum (a) lubricates the surface of the skin and hair, (b) consists of dead cells and fatty substances, (c) in excess may cause seborrhea, (d) all of these.

13. The rule of nines is helpful clinically in (a) diagnosing skin cancer, (b) estimating the extent of a burn, (c) estimating how serious a cancer is, (d) preventing acne.

Short Answer Essay Questions

14. Which epidermal cells are also called prickle cells? Which contain keratohyaline and lamellated granules?

15. Is a bald man really hairless? Explain.

16. You go to the beach to swim on an extremely hot, sunshiny summer afternoon. Describe two ways in which your integumentary system acts to preserve homeostasis during your outing.

17. Distinguish clearly between first-, second-, and third-degree burns.

18. Describe the process of hair formation, and list several factors that may influence (a) growth cycles and (b) hair texture.

19. What is cyanosis and what does it reflect?

20. Why does skin wrinkle and what factors accelerate the wrinkling process?

21. Explain each of these familiar phenomena in terms of what you learned in this chapter: (a) pimples, (b) dandruff, (c) greasy hair and "shiny nose," (d) stretch marks from gaining weight, (e) freckles, (f) leaving fingerprints.

22. Why are there no skin cancers that originate from stratum corneum cells?

23. A man got his finger caught in a machine at the factory. The damage was less serious than expected, but the entire nail was torn off his right index finger. The parts lost were the body, root, bed, matrix, and eponychium of the nail. First, define each of these parts. Then, tell if this nail is likely to grow back.

24. On a diagram of the human body, mark off various regions according to the rule of nines. What percentage of the total body surface is affected if the skin over the following body parts is burned? (a) the entire posterior trunk and buttocks, (b) an entire lower limb, (c) the entire front of the left upper limb.

6

BONES AND SKELETAL TISSUES

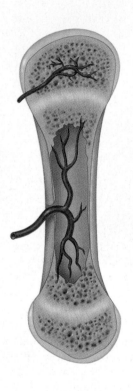

Skeletal Cartilages (p. 153)

1. Describe the functional properties of the three types of cartilage tissue.
2. Locate the major cartilages of the adult skeleton.
3. Explain how cartilage grows.

Classification of Bones (pp. 153–155)

4. Name the major regions of the skeleton and describe their relative functions.
5. Compare and contrast the structure of the four bone classes and provide examples of each class.

Functions of Bones (pp. 155–156)

6. List and describe five important functions of bones.

Bone Structure (pp. 156–160)

7. Describe the gross anatomy of a typical long bone and flat bone. Indicate the locations and functions of red and yellow marrow, articular cartilage, periosteum, and endosteum.
8. Indicate the functional importance of bone markings.
9. Describe the histology of compact and spongy bone.
10. Discuss the chemical composition of bone and the relative advantages conferred by the organic and inorganic components.

Bone Development (pp. 160–164)

11. Compare and contrast intramembranous ossification and endochondral ossification.
12. Describe the process of long bone growth that occurs at the epiphyseal plates.

Bone Homeostasis: Remodeling and Repair (pp. 164–170)

13. Compare the locations and remodeling functions of the osteoblasts, osteocytes, and osteoclasts.
14. Explain how hormones and physical stress regulate bone remodeling.
15. Describe the steps of fracture repair.

Homeostatic Imbalances of Bone (pp. 170–171)

16. Contrast the disorders of bone remodeling seen in osteoporosis, osteomalacia, and Paget's disease.

All of us have heard the expressions "bone tired" and "bag of bones"—rather unflattering and inaccurate images of one of our most phenomenal tissues and our main skeletal elements. Our brains, not our bones, convey feelings of fatigue. As for "bag of bones," they are indeed more prominent in some of us, but without bones to form our internal skeleton we would all creep along the ground like slugs, lacking any definite shape or form. Along with its bones, the skeleton contains resilient cartilages, which we briefly discuss in this chapter. However, our major focus is the structure and function of bone tissue and the dynamics of its formation and remodeling throughout life.

Skeletal Cartilages

Although the human skeleton is initially made up of cartilages and fibrous membranes, most of these early supports are soon replaced by bone. The few cartilages that remain in adults are found mainly in regions where flexible skeletal tissue is needed.

Basic Structure, Types, and Locations

A **skeletal cartilage** is made of some variety of *cartilage tissue*, which consists primarily of water. The high water content of cartilage accounts for its resilience, that is, its ability to spring back to its original shape after being compressed. The cartilage, which contains no nerves or blood vessels, is surrounded by a layer of dense irregular connective tissue, the *perichondrium* (per"ĭ-kon'drĭ-um; "around the cartilage"). The perichondrium acts like a girdle to resist outward expansion when the cartilage is compressed. Additionally, the perichondrium contains the blood vessels from which nutrients diffuse through the matrix to reach the cartilage cells. This mode of nutrient delivery limits cartilage thickness.

As described in Chapter 4, there are three types of cartilage tissue in the body: hyaline, elastic, and fibrocartilage. All three types have the same basic components—cells called *chondrocytes*, encased in small cavities (lacunae) within an *extracellular matrix* containing a jellylike ground substance and fibers. The skeletal cartilages contain representatives from all three types (Figure 6.1).

Hyaline cartilages, which look like frosted glass when freshly exposed, provide support with flexibility and resilience. They are the most abundant skeletal cartilages. When viewed under the microscope, their chondrocytes appear spherical (see Figure 4.8g). The only fiber type in their matrix is fine collagen fibers (which, however, are not detectable microscopically). Skeletal hyaline cartilages include (1) *articular cartilages*, which cover the ends of most bones at movable joints; (2) *costal cartilages*, which connect the ribs to the sternum (breastbone); (3) *respiratory cartilages*, which form the skeleton of the larynx (voicebox), and reinforce other respiratory passageways; and (4) *nasal cartilages*, which support the external nose.

Elastic cartilages look very much like hyaline cartilages (see Figure 4.8h), but they contain more stretchy elastic fibers and so are better able to stand up to repeated bending. They are found in only two skeletal locations (Figure 6.1)—the external ear and the epiglottis (the flap that bends to cover the opening of the larynx each time we swallow).

Fibrocartilages are highly compressible and have great tensile strength. The perfect intermediate between hyaline and elastic cartilages, fibrocartilages consist of roughly parallel rows of chondrocytes alternating with thick collagen fibers (see Figure 4.8i). Fibrocartilages occur in sites that are subjected to both heavy pressure and stretch, such as the padlike cartilages (menisci) of the knee and the discs between vertebrae (Figure 6.1).

Growth of Cartilage

Cartilage grows in two ways. In **appositional** (ap"o-zish'un-al) **growth** ("growth from outside"), cartilage-forming cells in the surrounding perichondrium secrete new matrix against the external face of the existing cartilage tissue. In **interstitial** (in"ter-stish'al) **growth** ("growth from inside"), the lacunae-bound chondrocytes divide and secrete new matrix, expanding the cartilage from within. Typically, cartilage growth ends during adolescence when the skeleton stops growing.

Under certain conditions—during normal bone growth in youth and during old age—calcium salts may be deposited in the matrix and cause it to harden, a process called calcification. Note, however, that calcified cartilage is *not* bone; cartilage and bone are always distinct tissues.

Classification of Bones

The 206 named bones of the human skeleton are divided into two groups: axial and appendicular. The **axial skeleton** forms the long axis of the body and includes the bones of the skull, vertebral column, and rib cage. Generally speaking these bones are most involved in protecting, supporting, or carrying other body parts. The **appendicular** (ap"en-dik'u-lar) **skeleton** consists of the bones of the upper and lower limbs and the girdles (shoulder bones and hip bones) that attach the limbs to the axial skeleton. Bones of the limbs help to get us from place to place (locomotion) and to manipulate our environment. An overview of the human skeleton appears in Figure 6.1.

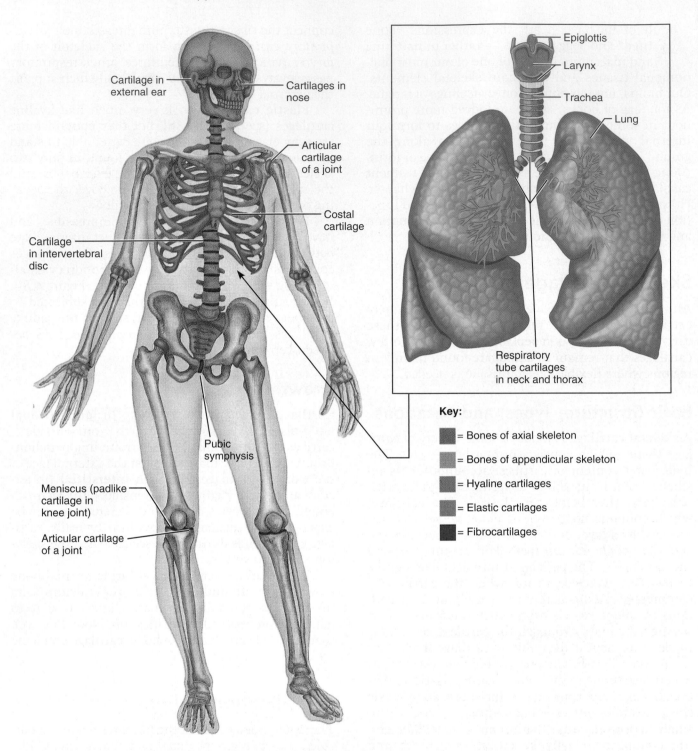

FIGURE 6.1 **The bones and cartilages of the human skeleton.**

Bones come in many sizes and shapes. For example, the pisiform bone of the wrist is the size and shape of a pea, whereas the femur (thigh bone) is nearly 2 feet long in some people and has a large ball-shaped head. The unique shape of each bone fulfills a particular need. The femur, for example, withstands great weight and pressure, and its hollow-cylinder design provides maximum strength with minimum weight.

For the most part, bones are classified by their shape as long, short, flat, and irregular (Figure 6.2).

1. Long bones, as their name suggests, are considerably longer than they are wide. A long bone has a shaft plus two ends. All limb bones except the

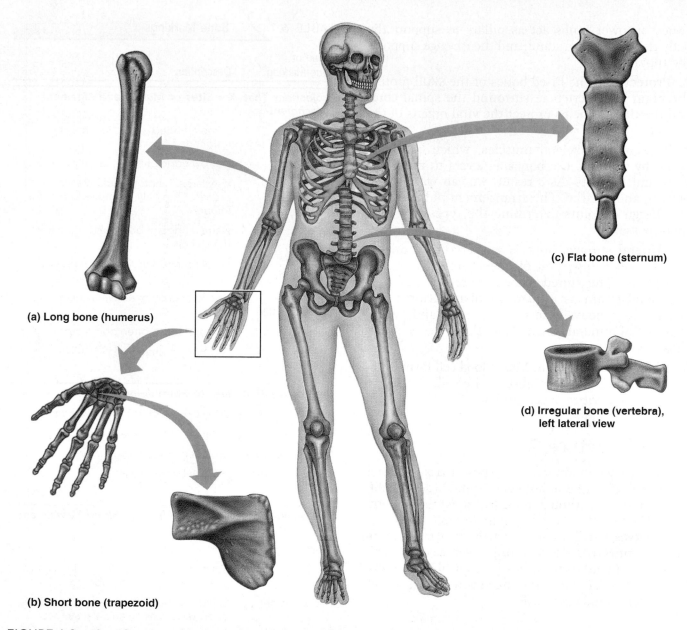

(a) Long bone (humerus)

(b) Short bone (trapezoid)

(c) Flat bone (sternum)

(d) Irregular bone (vertebra),
left lateral view

FIGURE 6.2 **Classification of bones on the basis of shape.**

patella (kneecap) and the wrist and ankle bones are long bones. Notice that these bones are named for their elongated shape, *not* their overall size. The three bones in each of your fingers are long bones, even though they are very small.

2. Short bones are roughly cube shaped. The bones of the wrist and ankle are examples.

Sesamoid (ses'ah-moid; "shaped like a sesame seed") **bones** are a special type of short bone that form in a tendon (for example, the patella). They vary in size and number in different individuals. Some sesamoid bones clearly act to alter the direction of pull of a tendon; the function of others is not known.

3. Flat bones are thin, flattened, and usually a bit curved. The sternum (breastbone), scapulae (shoulder blades), ribs, and most skull bones are flat bones.

4. Irregular bones. Irregular bones have complicated shapes that fit none of the preceding classes. Examples include the vertebrae and the hip bones.

Functions of Bones

Besides contributing to body shape and form, our bones perform several important functions:

1. Support. Bones provide a framework that supports the body and cradles its soft organs. For example, bones of lower limbs act as pillars to support the body trunk when we stand, and the rib cage supports the thoracic wall.

2. **Protection.** The fused bones of the skull protect the brain. The vertebrae surround the spinal cord, and the rib cage helps protect the vital organs of the thorax.

3. **Movement.** Skeletal muscles, which attach to bones by tendons, use bones as levers to move the body and its parts. As a result, we can walk, grasp objects, and breathe. The arrangement of bones and the design of joints determine the types of movement possible.

4. **Mineral storage.** Bone is a reservoir for minerals, the most important of which are calcium and phosphate. The stored minerals are released into the bloodstream as needed for distribution to all parts of the body. Indeed, "deposits" and "withdrawals" of minerals to and from the bones go on almost continuously.

5. **Blood cell formation.** Most blood cell formation, or *hematopoiesis* (hem"ah-to-poi-e'sis), occurs in the marrow cavities of certain bones.

Bone Structure

Because they contain various types of tissue, bones are *organs*. (Recall that an organ contains several different tissues.) Although bone (osseous) tissue dominates bones, they also contain nervous tissue in their nerves, cartilage in their articular cartilages, fibrous connective tissue lining their cavities, and muscle and epithelial tissues in their blood vessels. We will consider bone structure at three levels: gross, microscopic, and chemical.

Gross Anatomy

Bone Markings

The external surfaces of bones are rarely smooth and featureless. Instead, they display projections, depressions, and openings that serve as sites of muscle, ligament, and tendon attachment, as joint surfaces, or as conduits for blood vessels and nerves. These **bone markings** are named in different ways. Projections (bulges) that grow outward from the bone surface include heads, trochanters, spines, and others, each having distinguishing features and functions. Depressions and openings include fossae, sinuses, foramina, and grooves. The most important types of bone markings are described in Table 6.1. You should familiarize yourself with these terms because you will meet them again as identifying marks of the individual bones studied in the lab.

TABLE 6.1 ❭ Bone Markings

Name of Bone Marking	Description
Projections That Are Sites of Muscle and Ligament Attachment	
Tuberosity (too"bĕ-ros'ĭ-te)	Large rounded projection; may be roughened
Crest	Narrow ridge of bone; usually prominent
Trochanter (tro-kan'ter)	Very large, blunt, irregularly shaped process (The only examples are on the femur.)
Line	Narrow ridge of bone; less prominent than a crest
Tubercle (too'ber-kl)	Small rounded projection or process
Epicondyle (ep"ĭ-kon'dĭl)	Raised area on or above a condyle
Spine	Sharp, slender, often pointed projection
Process	Any bony prominence
Projections That Help to Form Joints	
Head	Bony expansion carried on a narrow neck
Facet	Smooth, nearly flat articular surface
Condyle (kon'dĭl)	Rounded articular projection
Ramus (ra'mus)	Armlike bar of bone
Depressions and Openings Allowing Blood Vessels and Nerves to Pass	
Meatus (me-a'tus)	Canal-like passageway
Sinus	Cavity within a bone, filled with air and lined with mucous membrane
Fossa (fos'ah)	Shallow, basinlike depression in a bone, often serving as an articular surface
Groove	Furrow
Fissure	Narrow, slitlike opening
Foramen (fo-ra'men)	Round or oval opening through a bone

Bone Textures: Compact and Spongy Bone

Every bone has a dense outer layer that looks smooth and solid to the naked eye. This external layer is **compact bone** (Figures 6.3 and 6.4). Internal to this is **spongy bone** (also called *cancellous bone*), a honeycomb of small needle-like or flat pieces called *trabeculae* (trah-bek'u-le; "little beams"). In living bones the open spaces between trabeculae are filled with red or yellow bone marrow.

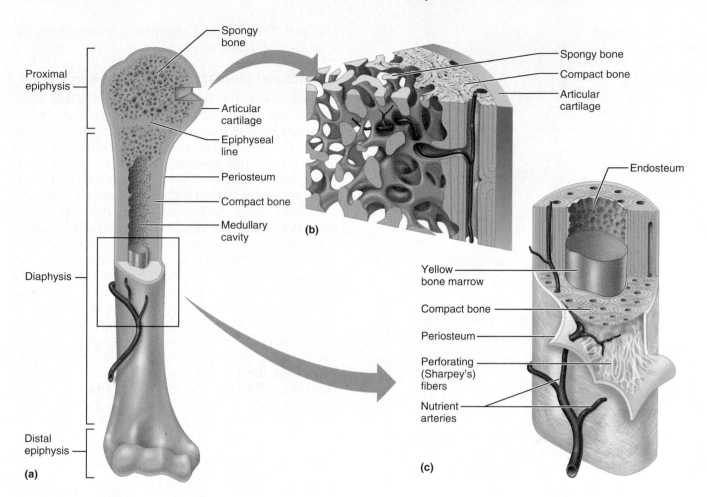

FIGURE 6.3   The structure of a long bone (humerus of arm). (a) Anterior view with bone sectioned frontally to show the interior at the proximal end. (b) Enlarged view of spongy bone and compact bone of the epiphysis of (a). (c) Enlarged cross-sectional view of the shaft (diaphysis) of (a). Note that the external surface of the diaphysis is covered by periosteum, but the articular surface of the epiphysis is covered with hyaline cartilage.

Structure of a Typical Long Bone

With few exceptions, all long bones have the same general structure (Figure 6.3).

Diaphysis   A tubular **diaphysis** (di-af′ ĭ-sis; *dia* = through, *physis* = growth), or shaft, forms the long axis of the bone. It is constructed of a relatively thick *collar* of compact bone that surrounds a central **medullary cavity** (med′u-lar-e; "middle") or *marrow cavity*. In adults, the medullary cavity contains fat (yellow marrow) and is called the **yellow bone marrow cavity.**

Epiphyses   The **epiphyses** (e-pif′ ĭ-sēz; singular = epiphysis) are the bone ends (*epi* = upon). In many cases, they are more expanded than the diaphysis. Compact bone forms the exterior of epiphyses; their interior contains spongy bone. The joint surface of each epiphysis is covered with a thin layer of articular (hyaline) cartilage, which cushions the op-posing bone ends during joint movement and absorbs stress. Between the diaphysis and each epiphysis of an adult long bone is an **epiphyseal line**, a remnant of the **epiphyseal plate**, a disc of hyaline cartilage that grows during childhood to lengthen the bone. The region where the diaphysis and epiphysis meet, whether it is the epiphyseal plate or line, is sometimes called the *metaphysis.*

Membranes   A third structural feature of long bones is membranes. The external surface of the entire bone except the joint surfaces is covered by a glistening white, double-layered membrane called the **periosteum** (per″e-os′te-um; *peri* = around, *osteo* = bone). The outer *fibrous layer* is dense irregular connective tissue. The inner *osteogenic layer,* abutting the bone surface, consists primarily of bone-forming cells, **osteoblasts** (os′te-o-blasts; "bone germinators"), and bone-destroying cells, **osteoclasts** ("bone breakers").

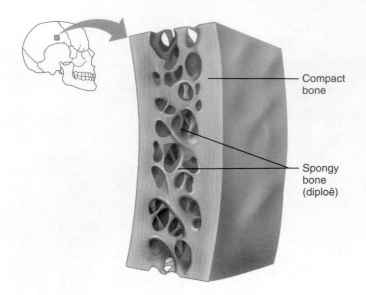

FIGURE 6.4   Structure of a flat bone. Flat bones consist of a layer of spongy bone (the diploë) sandwiched between two thin layers of compact bone.

The periosteum is richly supplied with nerve fibers, lymphatic vessels, and blood vessels, which enter the diaphysis via a **nutrient foramen** (fo-ra'min; "opening").

The periosteum is secured to the underlying bone by **perforating (Sharpey's) fibers** (Figure 6.3), tufts of collagen fibers that extend from its fibrous layer into the bone matrix. The periosteum also provides anchoring points for tendons and ligaments. At these points the perforating fibers are exceptionally dense.

Internal bone surfaces are covered with a delicate connective tissue membrane called the **endosteum** (en-dos'te-um; "within the bone"). The endosteum (Figure 6.3) covers the trabeculae of spongy bone and lines the canals that pass through the compact bone. Like the periosteum, the endosteum contains both osteoblasts and osteoclasts.

Structure of Short, Irregular, and Flat Bones

Short, irregular, and flat bones share a simple design: They all consist of thin plates of periosteum-covered compact bone on the outside and endosteum-covered spongy bone within. However, these bones are not cylindrical and so they have no shaft or epiphyses. They contain bone marrow (between their trabeculae), but no marrow cavity is present.

Figure 6.4 shows a typical flat bone of the skull. In flat bones, the spongy bone is called the **diploë** (dip'lo-e; "folded") and the whole arrangement resembles a stiffened sandwich.

Location of Hematopoietic Tissue in Bones

In long bones, the hematopoietic tissue, **red marrow,** is typically found within the trabecular cavities of spongy bone of long bones and in the diploë of flat bones. For this reason, both these cavities are often referred to as **red marrow cavities.** In newborn infants, the medullary cavity of the diaphysis and all areas of spongy bone contain red bone marrow. In most adult long bones, the fat-containing medullary cavity extends well into the epiphysis, and little red marrow is present in the spongy bone cavities. Hence, blood cell production in adult long bones routinely occurs only in the head of the femur and humerus (the long bone of the arm). The red marrow found in the diploë of flat bones (such as the sternum) and in some irregular bones (such as the hip bone) is much more active in hematopoiesis, and these are the sites routinely used for obtaining red marrow samples when problems with the blood-forming tissue are suspected. However, yellow marrow in the medullary cavity can revert to red marrow if a person becomes very anemic and needs enhanced red blood cell production.

Microscopic Anatomy of Bone

Compact Bone

Although compact bone looks dense and solid, a microscope reveals that it is riddled with passageways that serve as conduits for nerves, blood vessels, and lymphatic vessels (see Figure 6.6). The structural unit of compact bone is called either the **osteon** (os'te-on) or the **Haversian** (ha-ver'shan) **system.** Each osteon is an elongated cylinder oriented parallel to the long axis of the bone. Functionally, osteons are tiny weight-bearing pillars. As shown in the "exploded" view in Figure 6.5, an osteon is a group of hollow tubes of bone matrix, one placed outside the next like the growth rings of a tree trunk. Each matrix tube is a **lamella** (lah-mel'ah; "little plate"), and for this reason compact bone is often called **lamellar bone.** Although all of the collagen fibers in a particular lamella run in a single direction, the collagen fibers in adjacent lamellae always run in opposite directions. This alternating pattern is beautifully designed to withstand torsion stresses—the adjacent lamellae reinforce one another to resist twisting. You can think of the osteon's design as a "twister resister." Collagen fibers are not the only part of bone lamellae that are beautifully ordered. The tiny crystals of bone salts align with the collagen fibers and thus also alternate their direction in adjacent lamellae.

Running through the core of each osteon is the **central,** or **Haversian, canal** containing small blood vessels and nerve fibers that serve the needs of the osteon's cells. Canals of a second type called **perforating,** or **Volkmann's** (folk'mahnz), **canals,** lie at right angles to the long axis of the bone and connect the blood and nerve supply of the periosteum to those in the central canals and the medullary cavity (see Figure 6.6a). Like all other internal bone cavities, these canals are lined with endosteum.

Spider-shaped **osteocytes** (mature bone cells) occupy **lacunae** at the junctions of the lamellae. Hairlike canals called **canaliculi** (kan"ah-lik'u-li) connect the lacunae to each other and to the central canal. The manner in which canaliculi are formed is interesting. When bone is being formed, osteoblasts secreting bone matrix surround blood vessels and maintain contact with one another by tentacle-like projections containing gap junctions. Then, as the matrix hardens and the maturing cells become trapped within it, a system of tiny canals—the canaliculi, filled with tissue fluid and containing the osteocyte extensions—is formed. The canaliculi tie all the osteocytes in an osteon together, permitting nutrients and wastes to be relayed from one osteocyte to the next throughout the osteon. Although bone matrix is hard and impermeable to nutrients, its canaliculi and cell-to-cell relays (via gap junctions) allow bone cells to be well nourished. The function of osteocytes is to maintain the bone matrix. If they die, the surrounding matrix is resorbed.

Not all the lamellae in compact bone are part of osteons. Lying between intact osteons are incomplete lamellae called **interstitial** (in"ter-stish'al) **lamellae** (Figure 6.6c). These either fill the gaps between forming osteons or are remnants of osteons that have been cut through by bone remodeling (discussed later). Additionally, there are **circumferential lamellae,** located just deep to the periosteum and just superficial to the endosteum, that extend around the entire circumference of the diaphysis and effectively resist twisting of the long bone.

Spongy Bone

In contrast to compact bone, spongy bone looks like a poorly organized, even haphazard, tissue (Figure 6.4 and Figure 6.3b). However, the trabeculae in spongy bone align precisely along lines of stress and help the bone resist stress as much as possible. Thus, these tiny bone struts are as carefully positioned as the flying buttresses of a Gothic cathedral.

Only a few cells thick, trabeculae contain irregularly arranged lamellae and osteocytes interconnected by canaliculi. No osteons are present. Nutrients reach the osteocytes of spongy bone by diffusing through the canaliculi from capillaries in the endosteum surrounding the trabeculae.

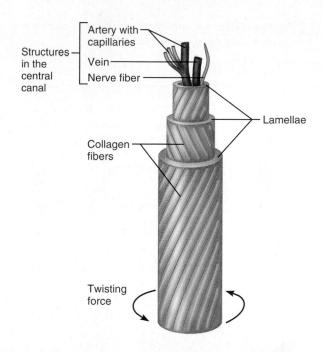

FIGURE 6.5 **A single osteon drawn as if pulled out like a telescope to illustrate the individual lamellae.** The slanted lines in each lamella indicate the direction of its collagen fibers.

Chemical Composition of Bone

Bone has both organic and inorganic components. Its *organic components* include the cells (osteoblasts, osteocytes, and osteoclasts) and **osteoid** (os'te-oid), the organic part of the matrix. Osteoid, which makes up approximately one-third of the matrix, includes ground substance (composed of proteoglycans and glycoproteins) and collagen fibers, both of which are made and secreted by osteoblasts. These organic substances, particularly collagen, contribute not only to a bone's structure but also to the flexibility and great tensile strength that allow the bone to resist stretch and twisting. Bone's exceptional toughness and tensile strength has been the subject of intense research. It now appears that this resilience comes from the presence of *sacrificial bonds* in or between collagen molecules that break easily on impact dissipating energy to prevent the force from rising to a fracture value. In the absence of continued or additional trauma, most of the sacrificial bonds reform.

The balance of bone tissue (65% by mass) consists of inorganic **hydroxyapatites** (hi-drok"se-ap'ah-tītz), or *mineral salts,* largely calcium phosphates present in the form of tiny crystals surrounding the collagen fibers in the extracellular matrix. The crystals are tightly packed and account for the most notable characteristic of bone—its exceptional hardness, which allows it to resist compression.

? *What membrane lines the internal canals and covers the trabeculae of spongy bone?*

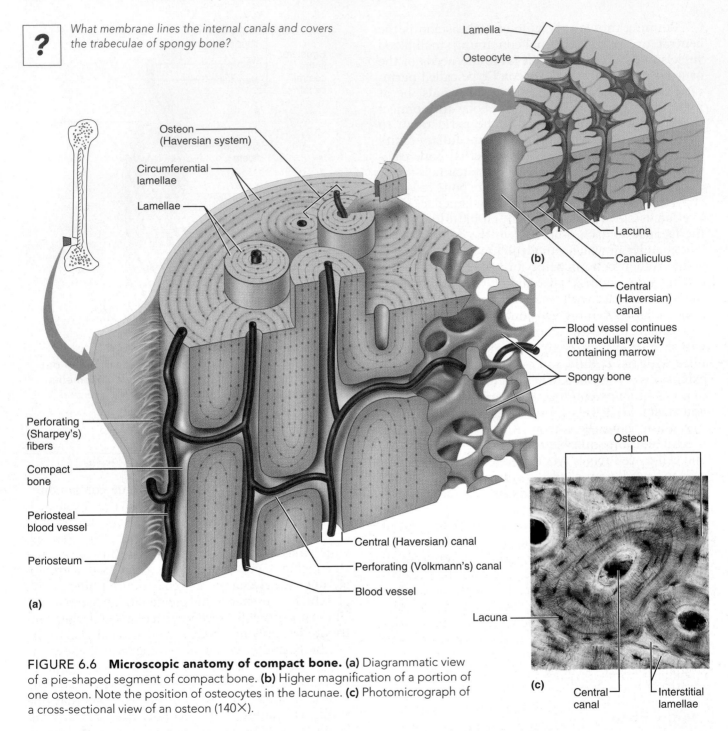

FIGURE 6.6 Microscopic anatomy of compact bone. (a) Diagrammatic view of a pie-shaped segment of compact bone. **(b)** Higher magnification of a portion of one osteon. Note the position of osteocytes in the lacunae. **(c)** Photomicrograph of a cross-sectional view of an osteon (140✕).

The proper combination of organic and inorganic matrix elements allows bones to be exceedingly durable and strong without being brittle. Healthy bone is half as strong as steel in resisting compression and fully as strong as steel in resisting tension.

Because of the salts they contain, bones last long after death and provide an enduring "monument." In fact, skeletal remains many centuries old have revealed the shapes and sizes of ancient peoples, the kinds of work they did, and many of the ailments they suffered from (arthritis, for example).

Bone Development

Ossification and **osteogenesis** (os"te-o-jen'ĕ-sis) are synonyms meaning the process of bone formation (*os* = bone, *genesis* = beginning). In embryos this process leads to the *formation of the bony skeleton*. Later another form of ossification known as *bone*

■ Endosteum.

growth goes on until early adulthood as the body continues to increase in size. Bones are capable of growing in thickness throughout life. However, ossification in adults serves mainly for bone *remodeling* and repair.

Formation of the Bony Skeleton

Before week 8, the skeleton of a human embryo is constructed entirely from fibrous membranes and hyaline cartilage. Bone tissue begins to develop at about this time and eventually replaces most of the existing fibrous or cartilage structures. When a bone develops from a fibrous membrane, the process is *intramembranous ossification,* and the bone is called a **membrane bone.** Bone development by replacing hyaline cartilage is called *endochondral ossification* (*endo* = within; *chondro* = cartilage), and the resulting bone is called a **cartilage,** or **endochondral, bone.**

Intramembranous Ossification

Intramembranous ossification (Figure 6.7) results in the formation of cranial bones of the skull (frontal, parietal, occipital, and temporal bones) and the clavicles. All bones formed by this process are flat bones. Fibrous connective tissue membranes formed by *mesenchymal cells* are the supporting structures on which ossification begins at about week 8 of development. Essentially, the process involves four major steps, depicted in Figure 6.7.

Endochondral Ossification

Except for the clavicles, essentially all bones of the skeleton below the base of the skull form by **endochondral** (en"do-kon'dral) **ossification.** This process, which begins in the second month of development, uses hyaline cartilage "bones" formed earlier as models, or patterns, for bone construction. It is more complex than intramembranous ossification because the hyaline cartilage must be broken down as ossification proceeds. We will use a forming long bone as our example.

The formation of a long bone typically begins in the center of the hyaline cartilage shaft at a region called the **primary ossification center.** First, the perichondrium covering the hyaline cartilage "bone" is infiltrated with blood vessels, converting it to a vascularized periosteum. As a result of this change in nutrition, the underlying mesenchymal cells specialize into osteoblasts. The stage is now set for ossification to begin, as illustrated in Figure 6.8.

FIGURE 6.7 Intramembranous ossification. Diagrams ③ and ④ represent much lower magnification than diagrams ① and ②.

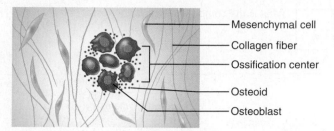

① **An ossification center appears in the fibrous connective tissue membrane.**

- Selected centrally located mesenchymal cells cluster and differentiate into osteoblasts, forming an ossification center.

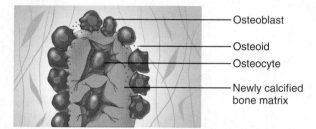

② **Bone matrix (osteoid) is secreted within the fibrous membrane.**

- Osteoblasts begin to secrete osteoid, which is mineralized within a few days.
- Trapped osteoblasts become osteocytes.

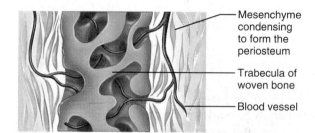

③ **Woven bone and periosteum form.**

- Accumulating osteoid is laid down between embryonic blood vessels, which form a random network. The result is a network (instead of lamellae) of trabeculae.
- Vascularized mesenchyme condenses on the external face of the woven bone and becomes the periosteum.

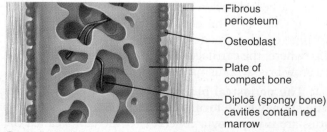

④ **Bone collar of compact bone forms and red marrow appears.**

- Trabeculae just deep to the periosteum thicken, forming a woven bone collar that is later replaced with mature lamellar bone.
- Spongy bone (diploë), consisting of distinct trabeculae, persists internally and its vascular tissue becomes red marrow.

?   *What is the composition of the periosteal bud?*

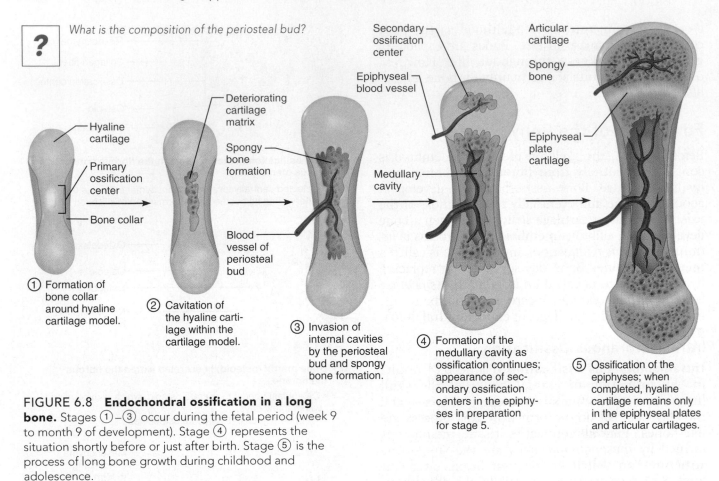

FIGURE 6.8 Endochondral ossification in a long bone. Stages ①–③ occur during the fetal period (week 9 to month 9 of development). Stage ④ represents the situation shortly before or just after birth. Stage ⑤ is the process of long bone growth during childhood and adolescence.

① **A bone collar forms around the diaphysis of the hyaline cartilage model.** Osteoblasts of the newly converted periosteum secrete osteoid against the hyaline cartilage diaphysis, encasing it in a bone collar.

② **Cartilage in the center of the diaphysis calcifies and then cavitates.** As the bone collar forms, chondrocytes within the shaft hypertrophy (enlarge) and signal the surrounding cartilage matrix to calcify. Because calcified cartilage matrix is impermeable to diffusing nutrients, the chondrocytes die and the matrix begins to deteriorate. Although this deterioration opens up cavities, the hyaline cartilage model is stabilized by the bone collar. Elsewhere, the cartilage remains healthy and continues to grow briskly, causing the cartilage model to elongate.

③ **The periosteal bud invades the internal cavities and spongy bone forms.** In month 3, the forming cavities are invaded by a collection of elements called the **periosteal bud,** which contains a nutrient artery and vein, lymphatics, nerve fibers, red marrow elements, osteoblasts, and osteoclasts. The entering osteoclasts partially erode the calcified cartilage matrix, and the os-

teoblasts secrete osteoid around the remaining fragments of hyaline cartilage, forming bone-covered cartilage trabeculae. In this way, the earliest version of spongy bone in a developing long bone forms.

④ **The diaphysis elongates and a medullary cavity forms.** As the primary ossification center enlarges, osteoclasts break down the newly formed spongy bone and open up a medullary cavity in the center of the diaphysis. Throughout the fetal period, the rapidly growing epiphyses consist only of cartilage, and the hyaline cartilage models continue to elongate by division of viable cartilage cells at the epiphyses. Because cartilage is calcifying, being eroded, and then being replaced by bony spicules on the epiphyseal surfaces facing the medullary cavity, ossification "chases" cartilage formation along the length of the shaft.

⑤ **The epiphyses ossify.** When we are born, most of our long bones have a bony diaphysis surrounding remnants of spongy bone, a widening medullary cavity, and two cartilaginous epiphyses. Shortly before or after birth, **secondary ossification centers** appear in one or both epiphyses, and the epiphyses gain bony tissue. (Typically, the large long bones form secondary centers in both epiphyses, whereas the small long bones form only one secondary ossification center.) The cartilage

■ *Blood vessels, nerve fibers, lymph vessels, osteoblasts, osteo- clasts, and red marrow elements.*

in the center of the epiphysis calcifies and deteriorates, opening up cavities that allow a periosteal bud to enter. Then bone trabeculae appear, just as they did earlier in the primary ossification center. (In short bones, only the primary ossification center is formed. Most irregular bones develop from several distinct ossification centers.)

Secondary ossification reproduces almost exactly the events of primary ossification, except that the spongy bone in the interior is retained and no medullary cavity forms in the epiphyses. When secondary ossification is complete, hyaline cartilage remains only at two places: (1) on the epiphyseal surfaces, as the *articular cartilages*, and (2) at the junction of the diaphysis and epiphysis, where it forms the *epiphyseal plates*.

Postnatal Bone Growth

During infancy and youth, long bones lengthen entirely by interstitial growth of the epiphyseal plates, and all bones grow in thickness by appositional growth. Most bones stop growing during adolescence; however, some facial bones, such as those of the nose and lower jaw, continue to grow almost imperceptibly throughout life.

Growth in Length of Long Bones

Longitudinal bone growth mimics many of the events of endochondral ossification. On the side of the epiphyseal plate facing the epiphysis, the cartilage is relatively quiescent and inactive. But the epiphyseal plate cartilage abutting the diaphysis organizes into a pattern that allows fast, efficient growth. The cartilage cells here form tall columns, like coins in a stack. The cells at the "top" (epiphysis-facing side) of the stack (the *growth zone* in Figure 6.9) divide quickly, pushing the epiphysis away from the diaphysis, causing the entire long bone to lengthen (see left side of Figure 6.10).

Meanwhile, the older chondrocytes in the stack, which are closer to the diaphysis (*transformation zone* in Figure 6.9), hypertrophy, and their lacunae erode and enlarge. Subsequently, the surrounding cartilage matrix calcifies and these chondrocytes die and deteriorate. This leaves long spicules of calcified cartilage at the epiphysis-diaphysis junction, which look like stalactites hanging from the roof of a cave. These become part of the *osteogenic zone*, and are invaded by marrow elements from the medullary cavity. The cartilage spicules are partly eroded by osteoclasts and then quickly covered with bone matrix by osteoblasts, forming spongy bone. The spicule tips are eventually digested by osteoclasts; thus, the medullary cavity also grows longer as the long bone lengthens. During growth, the epiphyseal plate

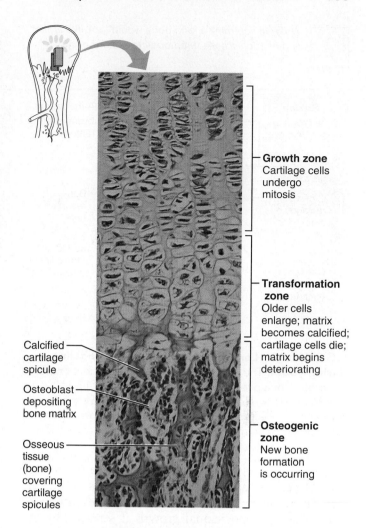

Growth zone
Cartilage cells undergo mitosis

Transformation zone
Older cells enlarge; matrix becomes calcified; cartilage cells die; matrix begins deteriorating

Calcified cartilage spicule

Osteoblast depositing bone matrix

Osteogenic zone
New bone formation is occurring

Osseous tissue (bone) covering cartilage spicules

FIGURE 6.9   Growth in length of a long bone. The side of the epiphyseal plate facing the epiphysis (distal face) contains resting cartilage cells. The cells of the epiphyseal plate proximal to the resting cartilage area are arranged in three zones—growth, transformation, and osteogenic (80✕).

maintains a constant thickness because the rate of cartilage growth on its epiphysis-facing side is balanced by its replacement with bony tissue on its diaphysis-facing side.

Longitudinal growth is accompanied by almost continuous remodeling of the epiphyseal ends to maintain the proper proportions between the diaphysis and epiphyses (Figure 6.10). Bone remodeling, involving both new bone formation and bone resorption (destruction), is described in more detail below in conjunction with the changes that occur in adult bones.

As adolescence draws to an end, the chondroblasts of the epiphyseal plates divide less often and the plates become thinner and thinner until they are entirely replaced by bone tissue. Longitudinal bone growth ends when the bone of the epiphysis and diaphysis fuses. This process, called *epiphyseal plate*

What do you think a long bone would look like at the end of adolescence if remodeling did not occur?

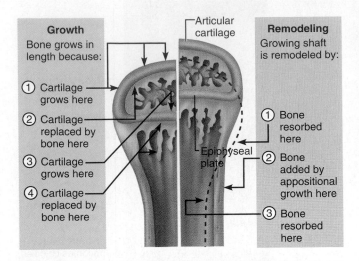

Growth
Bone grows in length because:

① Cartilage grows here

② Cartilage replaced by bone here

③ Cartilage grows here

④ Cartilage replaced by bone here

Articular cartilage

Epiphyseal plate

Remodeling
Growing shaft is remodeled by:

① Bone resorbed here

② Bone added by appositional growth here

③ Bone resorbed here

FIGURE 6.10   Long bone growth and remodeling during youth. The events at the left depict endochondral ossification that occurs at the articular cartilages and epiphyseal plates as the bone grows in length. Events at the right show bone remodeling during long bone growth to maintain proper bone proportions.

closure, happens at about 18 years of age in females and 21 years of age in males. However, as noted earlier, an adult bone can still increase in diameter or thickness by appositional growth if stressed by excessive muscle activity or body weight.

Growth in Width (Thickness)

Growing bones widen as they lengthen. As with cartilages, bones increase in thickness or, in the case of long bones, diameter, by appositional growth (Figure 6.11). Osteoblasts beneath the periosteum secrete bone matrix on the external bone surface as osteoclasts on the endosteal surface of the diaphysis remove bone (see Figure 6.10). However, there is normally slightly less breaking down than building up. This unequal process produces a thicker, stronger bone but prevents it from becoming too heavy.

Hormonal Regulation of Bone Growth

The growth of bones that occurs until young adulthood is exquisitely controlled by a symphony of hormones. During infancy and childhood, the single most important stimulus of epiphyseal plate activity is *growth hormone* released by the anterior pituitary gland. Thyroid hormones modulate the activity of

growth hormone, ensuring that the skeleton has proper proportions as it grows. At puberty, male and female sex hormones (testosterone and estrogens, respectively) are released in increasing amounts. Initially these sex hormones promote the growth spurt typical of adolescence, as well as the masculinization or feminization of specific parts of the skeleton. Later the hormones induce epiphyseal plate closure, ending longitudinal bone growth.

Excesses or deficits of any of these hormones can result in obviously abnormal skeletal growth. For example, hypersecretion of growth hormone in children results in excessive height (gigantism), and deficits of growth hormone or thyroid hormone produce characteristic types of dwarfism.

Bone Homeostasis: Remodeling and Repair

Bones appear to be the most lifeless of body organs, and may even summon images of a graveyard. But as you have just learned, appearances can be deceiving. Bone is a dynamic and active tissue, and small-scale changes in bone architecture occur continually. Every week we recycle 5 to 7% of our bone mass, and as much as half a gram of calcium may enter or leave the adult skeleton each day! Spongy bone is replaced every 3 to 4 years; compact bone, every 10 years or so. This is fortunate because when bone remains in place for long periods the calcium crystallizes and becomes more brittle—ripe conditions for fracture. And when we break bones—the most common disorder of bone homeostasis—they undergo a remarkable process of self-repair.

Bone Remodeling

In the adult skeleton, bone deposit and bone resorption (removal) occur both at the surface of the periosteum and the surface of the endosteum. Together, the two processes constitute **bone remodeling,** and they are coupled and coordinated by "packets" of adjacent osteoblasts and osteoclasts called *remodeling units.* In healthy young adults, total bone mass remains constant, an indication that the rates of bone deposit and resorption are essentially equal. Remodeling does not occur uniformly, however. For example, the distal part of the femur, or thigh bone, is fully replaced every five to six months, whereas its shaft is altered much more slowly.

Bone deposit occurs wherever bone is injured or added bone strength is required. For optimal bone deposit, a healthy diet rich in proteins, vitamin C, vitamin D, vitamin A, and several minerals (calcium, phosphorus, magnesium, and manganese, to name a few) is essential.

Its shaft would still be its original length, and it would have two very elongated and expanded ends (epiphyses). ■

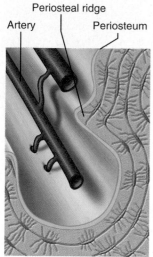

Periosteal ridge
Artery   Periosteum   Penetrating canal

① Osteoblasts beneath the periosteum secrete bone matrix, forming ridges that follow the course of periosteal blood vessels.

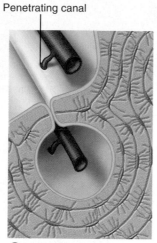

② As the bony ridges enlarge and meet, the groove containing the blood vessel becomes a tunnel.

③ The periosteum lining the tunnel is transformed into an endosteum and the osteoblasts just deep to the tunnel endosteum secrete bone matrix, narrowing the canal.

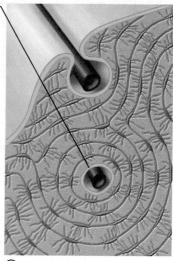

Central canal of osteon

④ As the osteoblasts beneath the endosteum form new lamellae, a new osteon is created. Meanwhile new circumferential lamellae are elaborated beneath the periosteum and the process is repeated, continuing to enlarge bone diameter.

FIGURE 6.11 **Appositional growth of bone.**

New matrix deposits by osteocytes are marked by the presence of an **osteoid seam,** an unmineralized band of gauzy-looking bone matrix 10 to 12 μm wide. Between the osteoid seam and the older mineralized bone, there is an abrupt transition called the **calcification front.** Because the osteoid seam is always of constant width and the change from unmineralized to mineralized matrix is sudden, it seems that the osteoid must mature for about a week before it can calcify. The precise trigger for calcification is still controversial. However, one critical factor is the product of the local concentrations of calcium and phosphate (P_i) ions (the $Ca^{2+} \cdot P_i$ product). When this product reaches a certain level, tiny crystals of hydroxyapatite form spontaneously and then catalyze further crystallization of calcium salts in the area. Other factors involved are matrix proteins that bind and concentrate calcium, and the enzyme **alkaline phosphatase** (shed by the osteoblasts), which is essential for mineralization. Once proper conditions are present, calcium salts are deposited all at once and with great precision throughout the "matured" matrix.

Bone resorption is accomplished by osteoclasts, giant multinucleate cells that arise from the same *hematopoietic stem cells* that differentiate into macrophages. Osteoclasts move along a bone surface, digging grooves called *resorption bays* as they break down the bone matrix. The part of the osteoclast that touches the bone is highly folded to form a ruffled membrane that clings tightly to the bone, sealing off the area of bone destruction. The ruffled border secretes (1) *lysosomal enzymes* that digest the organic matrix and (2) *hydrochloric acid* that converts the calcium salts into soluble forms that pass easily into solution. Osteoclasts may also phagocytize the demineralized matrix and dead osteocytes. The digested matrix end products and dissolved minerals are then endocytosed, transported across the osteoclast (by transcytosis), and released at the opposite side where they enter first the interstitial fluid and then the blood. Although there is much to learn about osteoclast activation, proteins secreted by T cells of the immune system appear to be important.

Control of Remodeling

The remodeling that goes on continuously in the skeleton is regulated by two control loops that serve different "masters." One is a negative feedback hormonal mechanism that maintains Ca^{2+} homeostasis in the blood. The other involves responses to mechanical and gravitational forces acting on the skeleton.

The hormonal mechanism becomes much more meaningful when you understand calcium's importance in the body. Ionic calcium is necessary for an amazing number of physiological processes, including transmission of nerve impulses, muscle contraction, blood coagulation, secretion by glands and nerve

? *What is the result of the negative feedback of calcitonin?*

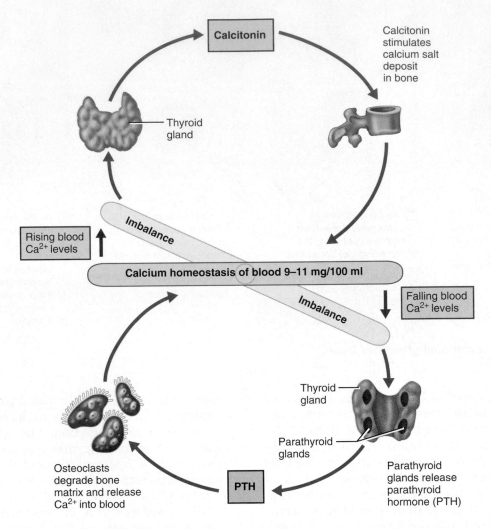

FIGURE 6.12 **Hormonal controls of blood calcium levels.** PTH and calcitonin operate in negative feedback control systems that influence each other.

cells, and cell division. The human body contains 1200–1400 g of calcium, more than 99% present as bone minerals. Most of the remainder is in body cells. Less than 1.5 g is present in blood, and the hormonal control loop normally maintains blood Ca²⁺ within the very narrow range of 9–11 mg per 100 ml of blood. Calcium is absorbed from the intestine under the control of vitamin D metabolites. The daily calcium requirement is 400–800 mg from birth until the age of 10, and 1200–1500 mg from ages 11 to 24.

Calcium salts are deposited in bone, and blood Ca²⁺ levels fall.

Hormonal Mechanism   The hormonal mechanism involves **parathyroid hormone (PTH)**, produced by the parathyroid glands, and **calcitonin** (kal″sĭ-to′nin), produced by parafollicular cells (C cells) of the thyroid gland (Figure 6.12). PTH is released when blood levels of ionic calcium decline. The increased PTH level stimulates osteoclasts to resorb bone, releasing calcium to the blood. Osteoclasts are no respecters of matrix age. When activated, they break down both old and new matrix. Only osteoid, which lacks calcium salts, escapes digestion. As blood concentrations of calcium rise, the stimulus for PTH release ends.

Calcitonin, secreted when blood calcium levels rise, inhibits bone resorption and encourages calcium salt deposit in bone matrix, effectively reducing blood calcium levels. As blood calcium levels fall, calcitonin release wanes.

These hormonal controls act not to preserve the skeleton's strength or well-being but rather to maintain blood calcium homeostasis. In fact, if blood calcium levels are low for an extended time, the bones become so demineralized that they develop large, punched-out-looking holes. Thus, the bones serve as a storehouse from which ionic calcium is drawn as needed.

🄷 HOMEOSTATIC IMBALANCE

Minute changes from the homeostatic range for blood calcium can lead to severe neuromuscular problems ranging from hyperexcitability (when blood Ca^{2+} levels are too low) to nonresponsiveness and inability to function (with high blood Ca^{2+} levels). In addition, sustained high blood levels of Ca^{2+}, a condition known as *hypercalcemia* (hi"per-kal-se'me-ah), can lead to undesirable deposits of calcium salts in the blood vessels, kidneys, and other soft organs, which may hamper the functioning of these organs. ●

Response to Mechanical Stress The second set of controls regulating bone remodeling, bone's response to mechanical stress (muscle pull) and gravity, serves the needs of the skeleton by keeping the bones strong where stressors are acting. *Wolff's law* holds that a bone grows or remodels in response to the demands placed on it. The first thing to understand is that a bone's anatomy reflects the common stresses it encounters. For example, a bone is loaded (stressed) whenever weight bears down on it or muscles pull on it. This loading is usually off center, however, and tends to *bend* the bone. Bending compresses the bone on one side and subjects it to tension (stretching) on the other (Figure 6.13). But both forces are minimal toward the center of the bone (they cancel each other out), so a bone can "hollow out" for lightness (using spongy bone instead of compact bone) without jeopardy.

Other observations explained by Wolff's law include these: (1) long bones are thickest midway along the diaphysis, exactly where bending stresses are greatest (bend a stick and it will split near the middle); (2) curved bones are thickest where they are most likely to buckle; (3) the trabeculae of spongy bone form trusses, or struts, along lines of compression; and (4) large, bony projections occur where heavy, active muscles attach. (The bones of weight lifters have enormous thickenings at the attachment sites of the most used muscles.) Wolff's law also explains the featureless bones of the fetus and the atrophied bones of bedridden people—situations in which bones are not stressed.

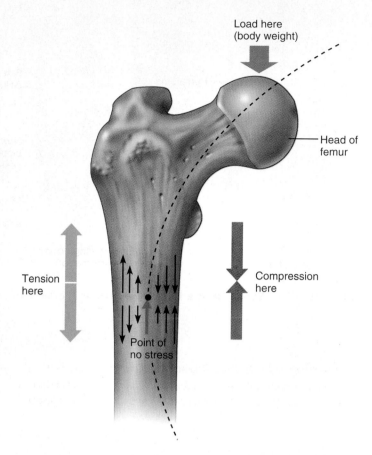

FIGURE 6.13 Bone anatomy and stress. When loaded, bones are subjected to bending stress. In this example using the femur of the thigh, body weight transmitted to the head of the femur threatens to bend the bone along the indicated arc. This bending compresses the bone on one side (converging arrows), and stretches it on the other side (diverging arrows). Because these two forces cancel each other internally, much less bone material is needed internally than superficially.

How do mechanical forces communicate with the cells responsible for remodeling? Although the mechanisms by which bone responds to mechanical stimuli are still uncertain, we do know that deforming a bone produces an electrical current. Because compressed and stretched regions are oppositely charged, it has been suggested that electrical signals direct remodeling.

The skeleton is continuously subjected to both hormonal influences and mechanical forces. At the risk of constructing too large a building on too small a foundation, we can speculate that the hormonal loop determines *whether* and *when* remodeling occurs in response to changing blood calcium levels, and mechanical stress determines *where* it occurs. For example, when bone must be broken down to increase blood calcium levels, PTH is released and targets the osteoclasts. However, mechanical forces

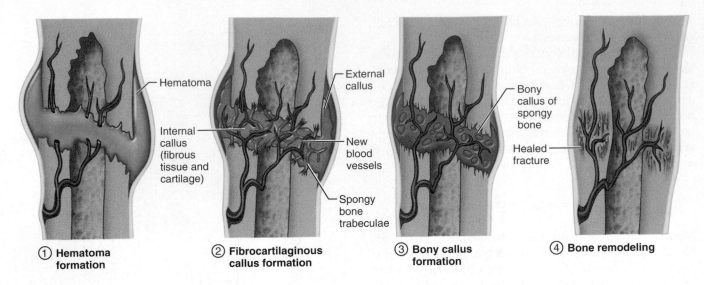

① **Hematoma formation**

Hematoma

Internal callus (fibrous tissue and cartilage)

② **Fibrocartilaginous callus formation**

External callus

New blood vessels

Spongy bone trabeculae

③ **Bony callus formation**

Bony callus of spongy bone

Healed fracture

④ **Bone remodeling**

FIGURE 6.14   **Stages in the healing of a bone fracture.**

determine *which* osteoclasts are most sensitive to PTH stimulation, so that bone in the *least* stressed areas (which is temporarily dispensable) is broken down.

Bone Repair

Despite their remarkable strength, bones are susceptible to **fractures,** or breaks. During youth, most fractures result from exceptional trauma that twists or smashes the bones (sports injuries, automobile accidents, and falls, for example). In old age, most fractures occur as bones thin and weaken.

Fractures may be classified by:

1. Position of the bone ends after fracture. In **nondisplaced fractures** the bone ends retain their normal position; in **displaced fractures** the bone ends are out of normal alignment.

2. Completeness of the break. If the bone is broken through, the fracture is a **complete fracture;** if not, it is an **incomplete fracture.**

3. Orientation of the break relative to the long axis of the bone. If the break parallels the long axis, the fracture is **linear;** if the break is perpendicular to the bone's long axis, it is **transverse.**

4. Whether the bone ends penetrate the skin. If so, the fracture is an **open** *(compound)* **fracture;** if not it is a **closed** *(simple)* **fracture.**

In addition to these four either-or classifications, all fractures can be described in terms of the location of the fracture, the external appearance of the fracture, and/or the nature of the break. Table 6.2 summarizes the various descriptions.

A fracture is treated by *reduction,* the realignment of the broken bone ends. In **closed reduction,** the bone ends are coaxed into position by the physician's hands. In **open reduction,** the bone ends are secured together surgically with pins or wires. After the broken bone is reduced, it is immobilized either by a cast or traction to allow the healing process to begin. For a simple fracture the healing time is six to eight weeks for small or medium-sized bones in young adults, but it is much longer for large, weight-bearing bones and for bones of elderly people (because of their poorer circulation).

Repair in a simple fracture involves four major stages (Figure 6.14).

① **Hematoma formation.** When a bone breaks, blood vessels in the bone and periosteum, and perhaps in surrounding tissues, are torn and hemorrhage. As a result, a **hematoma** (he"mah-to′mah), a mass of clotted blood, forms at the fracture site. Soon, bone cells deprived of nutrition die, and the tissue at the site becomes swollen, painful, and inflamed.

② **Fibrocartilaginous callus formation.** Within a few days, several events lead to the formation of soft *granulation tissue,* also called the *soft callus* (kal′us; "hard skin"). Capillaries grow into the hematoma and phagocytic cells invade the area and begin cleaning up the debris. Meanwhile, fibroblasts and osteoblasts invade the fracture site from the nearby periosteum and endosteum and begin reconstructing the bone. The fibroblasts produce collagen fibers that span the break and connect the broken bone ends, and some differentiate into chondroblasts that secrete cartilage matrix. Within this mass of repair tissue, osteoblasts begin forming

TABLE 6.2 Common Types of Fractures

Fracture Type	Description and Comments	Fracture Type	Description and Comments
Comminuted	Bone fragments into three or more pieces Particularly common in the aged, whose bones are more brittle	Compression	Bone is crushed Common in porous bones (i.e., osteoporotic bones) subjected to extreme trauma, as in a fall

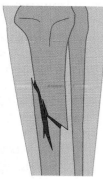 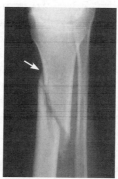

—Crushed vertebra

Fracture Type	Description and Comments	Fracture Type	Description and Comments
Spiral	Ragged break occurs when excessive twisting forces are applied to a bone Common sports fracture	Epiphyseal	Epiphysis separates from the diaphysis along the epiphyseal plate Tends to occur where cartilage cells are dying and calcification of the matrix is occurring

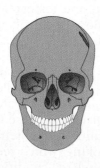 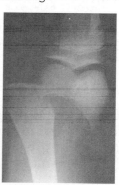

Fracture Type	Description and Comments	Fracture Type	Description and Comments
Depressed	Broken bone portion is pressed inward Typical of skull fracture	Greenstick	Bone breaks incompletely, much in the way a green twig breaks. Only one side of the shaft breaks; the other side bends Common in children, whose bones have relatively more organic matrix and are more flexible than those of adults

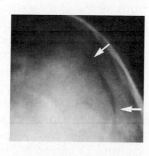  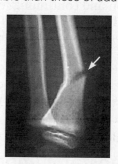

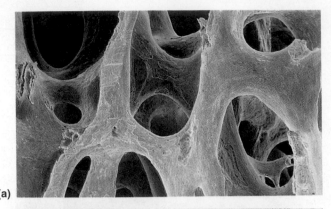

(a)

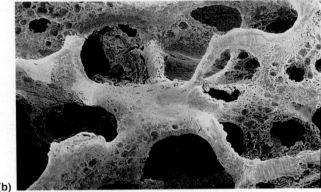

(b)

FIGURE 6.15   Osteoporosis: the contrasting architecture of (a) normal bone and (b) osteoporotic bone (both 300X).

spongy bone, but those farthest from the capillary supply secrete an externally bulging cartilaginous matrix that later calcifies. This entire mass of repair tissue, now called the **fibrocartilaginous callus**, splints the broken bone.

③ **Bony callus formation.** Within a week, new bone trabeculae begin to appear in the fibrocartilaginous callus and gradually convert it to a **bony (hard) callus** of spongy bone. Bony callus formation continues until a firm union is formed about two months later.

④ **Bone remodeling.** Beginning during bony callus formation and continuing for several months after, the bony callus is remodeled. The excess material on the diaphysis exterior and within the medullary cavity is removed, and compact bone is laid down to reconstruct the shaft walls. The final structure of the remodeled area resembles that of the original unbroken bony region because it responds to the same set of mechanical stressors.

Homeostatic Imbalances of Bone

Imbalances between bone deposit and bone resorption underlie nearly every disease that affects the adult skeleton.

Osteomalacia and Rickets

Osteomalacia (os"te-o-mah-la'she-ah; "soft bones") includes a number of disorders in which the bones are inadequately mineralized. Osteoid is produced, but calcium salts are not deposited, so bones soften and weaken. The main symptom is pain when weight is put on the affected bones.

Rickets is the analogous disease in children. Because young bones are still growing rapidly, rickets is much more severe than adult osteomalacia. Bowed legs and deformities of the pelvis, skull, and rib cage are common. Because the epiphyseal plates cannot be calcified, they continue to widen, and the ends of long bones become visibly enlarged and abnormally long.

Osteomalacia and rickets are caused by insufficient calcium in the diet or by a vitamin D deficiency. Hence, drinking vitamin D–fortified milk and exposing the skin to sunlight (which spurs the body to form vitamin D) usually cure these disorders.

Osteoporosis

For most of us, the phrase "bone problems of the elderly" brings to mind the stereotype of a victim of osteoporosis—a hunched-over old woman shuffling behind her walker. **Osteoporosis** (os"te-o-po-ro'sis) refers to a group of diseases in which bone resorption outpaces bone deposit. The bones become so fragile that something as simple as a hearty sneeze or stepping off a curb can cause them to break. The composition of the matrix remains normal but bone mass is reduced, and the bones become more porous and lighter (Figure 6.15). Even though osteoporosis affects the entire skeleton, the spongy bone of the spine is most vulnerable, and compression fractures of the vertebrae are common. The femur, particularly its neck, is also very susceptible to fracture (called a *broken hip*) in people with osteoporosis.

Osteoporosis occurs most often in the aged, particularly in postmenopausal women, but cessation of sex hormone synthesis can promote the disorder in both sexes. Both estrogen and testosterone help to maintain the health and normal density of the skeleton by restraining osteoclast activity and by promoting deposit of new bone. After menopause, however, estrogen secretion wanes and estrogen deficiency is strongly implicated in osteoporosis in older women. Other factors that contribute to osteoporosis include a petite body form, insufficient exercise to stress the bones, a diet poor in calcium and protein, abnormal vitamin D receptors, smoking (which reduces estrogen levels), and hormone-related conditions (such as hyperthyroidism and diabetes mellitus). In addition, osteoporosis can occur at any age as a result of immobility.

Osteoporosis has traditionally been treated with calcium and vitamin D supplements, weight-bearing exercise, and *hormone (estrogen) replacement therapy (HRT)*. Frustratingly, HRT only slows the loss of bone; it does not reverse it. For those who can't or won't take estrogen, other drugs are available. These include alendronate, a drug that suppresses osteoclast activity and shows promise in reversing osteoporosis in the spine; and raloxifene, dubbed "estrogen light" because it mimics estrogen's beneficial bone-sparing properties without targeting the uterus or breast. Additionally, natural progesterone cream prompts the body to make new bone, and *statins*, drugs used by tens of thousands of people to lower cholesterol levels, have been shown to have an unexpected side effect of increasing bone mineral density up to 8% over four years. Although not a substitute for HRT, estrogenic compounds in soy protein (principally the isoflavones daidzein and genistein) offer a good addition or adjunct for some patients.

Osteoporosis can be prevented (or at least delayed). The first requirement is to get enough calcium while your bones are still increasing in density (long bones reach their peak density between the ages of 35 and 40; bones with relatively more spongy bone are densest between the ages of 25 and 30). Second, drinking fluoridated water hardens bones (as well as teeth). Finally, getting plenty of weight-bearing exercise (walking, jogging, tennis, etc.) throughout life will increase bone mass above normal values and provide a greater buffer against age-related bone loss.

Paget's Disease

Often discovered by accident when X rays are taken for some other reason, **Paget's** (paj'ets) **disease** is characterized by excessive bone deposit and resorption. The newly formed bone, called *Pagetic bone*, is hastily made and has an abnormally high ratio of spongy bone to compact bone. This, along with reduced mineralization, causes a spotty weakening of the bones. Late in the disease, osteoclast activity wanes, but osteoblasts continue to work, often forming irregular bone thickenings or filling the marrow cavity with Pagetic bone.

Paget's disease may affect any part of the skeleton, but it is usually a localized condition. The spine, pelvis, femur, and skull are most often involved and become increasingly deformed and painful. It rarely occurs before the age of 40, and it affects about 3% of North American elderly people. Its cause is unknown, but it may be initiated by a virus. Drug therapies include etidronate, used for three to six months at a time; calcitonin (now administered by a nasal inhaler); and the newer alendronate, which is successful in preventing bone breakdown.

Review Questions

Multiple Choice/Matching

(Some questions have more than one correct answer. Select the best answer or answers from the choices given.)

 1. Which is a function of the skeletal system? (a) support, (b) hematopoietic site, (c) storage, (d) providing levers for muscle activity, (e) all of these.

 2. A bone with approximately the same width, length, and height is most likely (a) a long bone, (b) a short bone, (c) a flat bone, (d) an irregular bone.

 3. The shaft of a long bone is properly called the (a) epiphysis, (b) periosteum, (c) diaphysis, (d) compact bone.

 4. Sites of hematopoiesis include all but (a) red marrow cavities of spongy bone, (b) the diploë of flat bones, (c) medullary cavities in bones of infants, (d) medullary cavities in bones of a healthy adult.

 5. An osteon has (a) a central canal carrying blood vessels, (b) concentric lamellae, (c) osteocytes in lacunae, (d) canaliculi that connect lacunae to the central canal, (e) all of these.

 6. The organic portion of matrix is important in providing *all but* (a) tensile strength, (b) hardness, (c) ability to resist stretch, (d) flexibility.

 7. The flat bones of the skull develop from (a) areolar tissue, (b) hyaline cartilage, (c) fibrous connective tissue, (d) compact bone.

 8. The remodeling of bone is a function of which cells? (a) chondrocytes and osteocytes, (b) osteoblasts and osteoclasts, (c) chondroblasts and osteoclasts, (d) osteoblasts and osteocytes.

 9. Bone growth during childhood and in adults is regulated and directed by (a) growth hormone, (b) thyroxine, (c) sex hormones, (d) mechanical stress, (e) all of these.

10. Where within the epiphyseal plate are the *dividing* cartilage cells located? (a) nearest the shaft, (b) in the marrow cavity, (c) farthest from the shaft, (d) in the primary ossification center.

11. Wolff's law is concerned with (a) calcium homeostasis of the blood, (b) the thickness and shape of a bone being determined by mechanical and gravitational stresses placed on it, (c) the electrical charge on bone surfaces.

12. Formation of the bony callus in fracture repair is followed by (a) hematoma formation, (b) fibrocartilaginous callus formation, (c) bone remodeling to convert woven bone to compact bone, (d) formation of granulation tissue.

13. The fracture type in which the bone ends are incompletely separated is (a) greenstick, (b) compound, (c) simple, (d) comminuted, (e) compression.

14. The disorder in which bones are porous and thin but bone composition is normal is (a) osteomalacia, (b) osteoporosis, (c) Paget's disease.

Short Answer Essay Questions

15. Compare bone to cartilage tissue relative to its resilience, speed of regeneration, and access to nutrients.

16. Describe in proper order the events of endochondral ossification.

17. Compare compact and spongy bone in macroscopic appearance, microscopic structure, and relative location.

18. As we grow, our long bones increase in diameter, but the thickness of the bony collar of the shaft remains relatively constant. Explain this phenomenon.

19. Describe the process of new bone formation in an adult bone. Use the terms *osteoid seam* and *calcification front* in your discussion.

20. Compare and contrast controls of bone remodeling exerted by hormones and by mechanical and gravitational forces relative to the actual purpose of each control system and changes in bone architecture that might occur.

21. (a) During what period of life does skeletal mass increase dramatically? Begin to decline? (b) Why are fractures most common in elderly individuals? (c) Why are greenstick fractures most common in children?

7

THE SKELETON

1. Name the major parts of the axial and appendicular skeletons and describe their relative functions.

PART 1: THE AXIAL SKELETON

The Skull (pp. 174–189)

2. Name, describe, and identify the skull bones. Identify their important markings.

3. Compare and contrast the major functions of the cranium and the facial skeleton.

4. Define the bony boundaries of the orbits, nasal cavity, and paranasal sinuses.

The Vertebral Column (pp. 190–197)

5. Describe the structure of the vertebral column, list its components, and describe its curvatures.

6. Indicate a common function of the spinal curvatures and the intervertebral discs.

7. Discuss the structure of a typical vertebra and describe regional features of cervical, thoracic, and lumbar vertebrae.

The Bony Thorax (pp. 197–199)

8. Name and describe the bones of the bony thorax.

9. Differentiate true from false ribs.

PART 2: THE APPENDICULAR SKELETON

The Pectoral (Shoulder) Girdle (pp. 199–203)

10. Identify bones forming the pectoral girdle and relate their structure and arrangement to the function of this girdle.

11. Identify important bone markings on the pectoral girdle.

The Upper Limb (pp. 203–208)

12. Identify or name the bones of the upper limb and their important markings.

The Pelvic (Hip) Girdle (pp. 208–211)

13. Name the bones contributing to the os coxa and relate the pelvic girdle's strength to its function.

14. Describe differences in the male and female pelves and relate these to functional differences.

The Lower Limb (pp. 212–218)

15. Identify the lower limb bones and their important markings.

16. Name the arches of the foot and explain their importance.

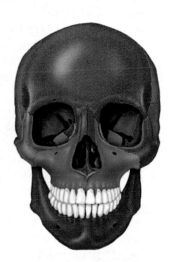

The word *skeleton* comes from the Greek word meaning "dried-up body" or "mummy," a rather unflattering description. Nonetheless, the human skeleton is a triumph of design and engineering that puts most skyscrapers to shame. It is strong, yet light, and almost perfectly adapted for the protective, locomotor, and manipulative functions it performs.

The **skeleton,** composed of bones, cartilages, joints, and ligaments, accounts for about 20% of body mass (about 30 pounds in a 160-pound person). Bones make up most of the skeleton. Cartilages occur only in isolated areas, such as the nose, parts of the ribs, and the joints. Ligaments connect bones and reinforce joints, allowing required movements while restricting motions in other directions. Joints, the junctions between bones, provide for the remarkable mobility of the skeleton. Joints and ligaments are discussed separately in Chapter 8.

PART 1: THE AXIAL SKELETON

As described in Chapter 6, the skeleton is divided into *axial* and *appendicular* portions (see Figure 6.1) for descriptive purposes. The **axial skeleton** is structured from 80 bones segregated into three major regions: the *skull, vertebral column,* and *bony thorax* (Figure 7.1). This part of the skeleton supports the head, neck, and trunk, and it protects the brain, spinal cord, and the organs in the thorax.

The Skull

The **skull** is the body's most complex bony structure. It is formed by *cranial* and *facial bones,* 22 in all. The cranial bones, or **cranium** (kra'ne-um), enclose and protect the fragile brain and furnish a site for attachment of head and neck muscles. The facial bones (1) form the framework of the face, (2) contain cavities for the special sense organs of sight, taste, and smell, (3) provide openings for air and food passage, (4) secure the teeth, and (5) anchor the facial muscles of expression, which we use to show our feelings. As you will see, the individual skull bones are well suited to their assignments.

Most skull bones are flat bones. Except for the mandible, which is connected to the rest of the skull by a freely movable joint, all bones of the adult skull are firmly united by interlocking joints called **sutures** (soo'cherz). The suture lines have a saw-toothed or serrated appearance. The major skull sutures, the *coronal, sagittal, squamous,* and *lambdoid sutures,* connect cranial bones (Figures 7.2b and 7.3a). Most other skull sutures connect facial bones and are named according to the specific bones they connect.

Overview of Skull Geography

It is worth surveying basic skull "geography" before describing the individual bones. With the lower jaw removed, the skull resembles a lopsided, hollow, bony sphere. The facial bones form its anterior aspect, and the cranium forms the rest of the skull (see Figure 7.3a). The cranium can be divided into a vault and a base. The *cranial vault,* also called the *calvaria* (kal-va're-ah), forms the superior, lateral, and posterior aspects of the skull, as well as the forehead. The *cranial base,* or *floor,* forms the skull's inferior aspect. Internally, prominent bony ridges divide the base into three distinct "steps" or fossae—the *anterior, middle,* and *posterior cranial fossae* (see Figure 7.4c on p. 179). The brain sits snugly in these cranial fossae, completely enclosed by the cranial vault. Overall, the brain is said to occupy the *cranial cavity.*

In addition to the large cranial cavity, the skull has many smaller cavities. These include the middle and inner ear cavities (carved into the lateral side of its base) and, anteriorly, the nasal cavity and the orbits (see Figure 1.13, p. 20). The *orbits* house the eyeballs. Several bones of the skull contain air-filled sinuses, which lighten the skull.

The skull also has about 85 named openings (foramina, canals, fissures, etc.). The most important of these provide passageways for the spinal cord, the major blood vessels serving the brain, and the 12 pairs of cranial nerves (numbered I through XII), which transmit impulses to and from the brain.

As you read about the bones of the skull, locate each bone on the different skull views in Figures 7.2, 7.3, and 7.4. The skull bones and their important markings are also summarized in Table 7.1 (pp. 188–189). The color-coded box enclosing a bone's name in the table corresponds to the color of that bone in the figures.

Cranium

The eight cranial bones are the paired parietal and temporal bones and the unpaired frontal, occipital, sphenoid, and ethmoid bones. Together, these construct the brain's protective bony "helmet." Because its superior aspect is curved, the cranium is self-bracing. This allows the bones to be thin, and, like an eggshell, the cranium is remarkably strong for its weight.

Frontal Bone

The shell-shaped **frontal bone** (Figures 7.2a, 7.3, and 7.4b) forms the anterior cranium. It articulates posteriorly with the paired parietal bones via the prominent *coronal suture.*

The most anterior part of the frontal bone is the vertical *frontal squama,* commonly called the

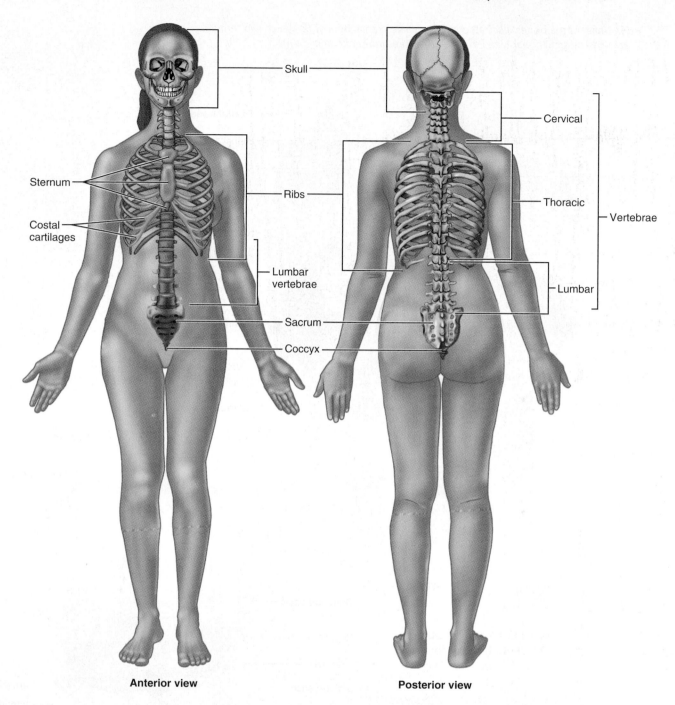

Anterior view

Posterior view

FIGURE 7.1   **Bones of the axial skeleton.** The three major subdivisions are the skull, bony thorax (sternum and ribs), and vertebral column (vertebrae, sacrum, and coccyx).

forehead. The frontal squama ends inferiorly at the **supraorbital margins,** the thickened superior margins of the orbits that lie under the eyebrows. From here, the frontal bone extends posteriorly, forming the superior wall of the *orbits* and most of the **anterior cranial fossa** (Figure 7.4b and c). This fossa supports the frontal lobes of the brain. Each supraorbital margin is pierced by a **supraorbital fora-**

men (notch), which allows the supraorbital artery and nerve to pass to the forehead.

The smooth portion of the frontal bone between the orbits is the **glabella** (glah-bel'ah). Just inferior to this the frontal bone meets the nasal bones at the *frontonasal suture* (Figure 7.2a). The areas lateral to the glabella are riddled internally with sinuses, called the **frontal sinuses** (Figures 7.3b and 7.11).

(1) Which of the bones illustrated in view (a) are cranial bones? *(2)* Which two terms are used interchangeably to indicate the entire group of cranial bones?

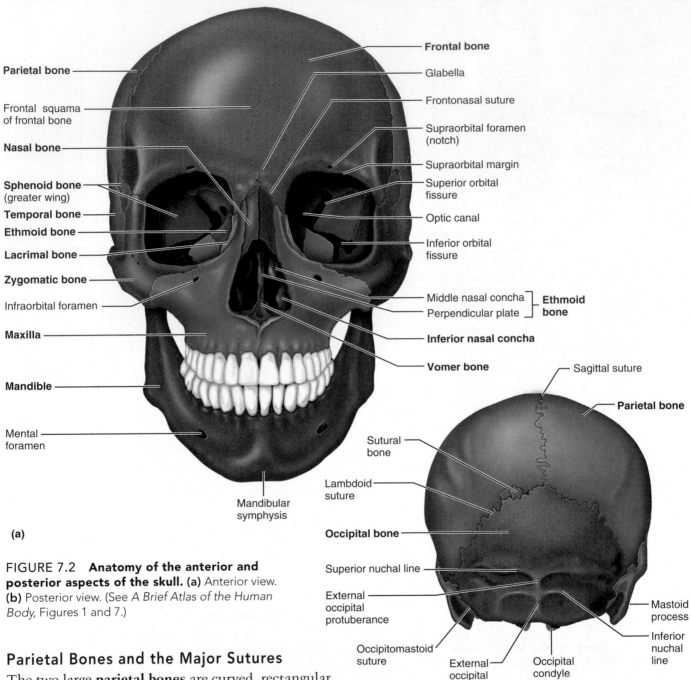

FIGURE 7.2 **Anatomy of the anterior and posterior aspects of the skull.** **(a)** Anterior view. **(b)** Posterior view. (See *A Brief Atlas of the Human Body*, Figures 1 and 7.)

Parietal Bones and the Major Sutures

The two large **parietal bones** are curved, rectangular bones that form most of the superior and lateral aspects of the skull; hence they form the bulk of the cranial vault. The four largest sutures occur where the parietal bones articulate (form a joint) with other cranial bones:

1. The **coronal** (kŏ-ro′nul) **suture,** where the parietal bones meet the frontal bone anteriorly (Figure 7.3)

2. The **sagittal suture,** where the parietal bones meet superiorly at the cranial midline (Figure 7.2b)

3. The **lambdoid** (lam′doid) **suture,** where the parietal bones meet the occipital bone posteriorly (see Figures 7.2b and 7.3)

4. The **squamous** (or **squamosal**) **suture** (one on each side), where a parietal and temporal bone meet on the lateral aspect of the skull (see Figure 7.3)

(1) The temporals, parietals, frontal, and sphenoid are all cranial bones. (2) The cranial vault or calvaria. ■

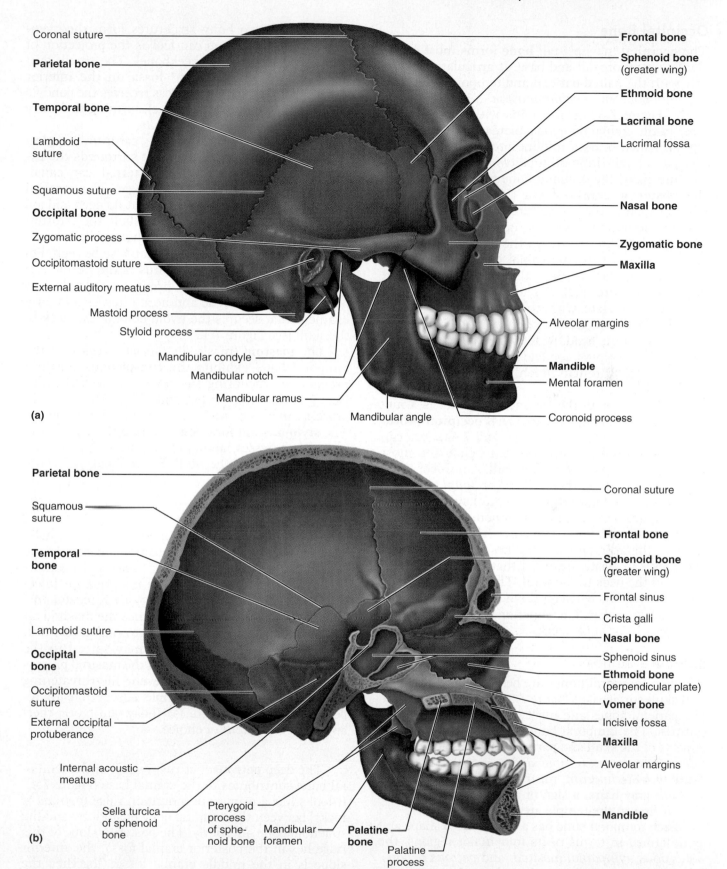

FIGURE 7.3 **Anatomy of the lateral aspects of the skull. (a)** External anatomy of the right lateral aspect of the skull. **(b)** Midsagittal view of the skull showing the internal anatomy of the left side of the skull. (See *A Brief Atlas of the Human Body,* Figures 2 and 3.)

Occipital Bone

The **occipital** (ok-sip'ĭ-tal) **bone** forms most of the skull's posterior wall and base. It articulates anteriorly with the paired parietal and temporal bones via the *lambdoid* and *occipitomastoid sutures*, respectively (Figure 7.3). It also joins with the sphenoid bone in the cranial floor via a plate called the basioccipital, which bears a midline projection called the *pharyngeal* (fah-rin'je-ul) *tubercle* (Figure 7.4a).

Internally, the occipital bone forms the walls of the **posterior cranial fossa** (Figure 7.4b and c), which supports the cerebellum of the brain. In the base of the occipital bone is the **foramen magnum** ("large hole") through which the inferior part of the brain connects with the spinal cord. The foramen magnum is flanked laterally by two **occipital condyles** (Figure 7.4a). The rockerlike occipital condyles articulate with the first vertebra of the spinal column in a way that permits a nodding movement of the head. Hidden medially and superiorly to each occipital condyle is a **hypoglossal canal** (Figure 7.4b), through which a nerve of the same name passes.

Just superior to the foramen magnum is a median protrusion called the **external occipital protuberance** (see Figures 7.2b, 7.3b, and 7.4a). You can feel this knoblike projection just below the most bulging part of your posterior skull. A number of inconspicuous ridges, the *external occipital crest* and the *superior* and *inferior nuchal* (nu'kal) *lines*, mark the occipital bone near the foramen magnum. The external occipital crest secures the *ligamentum nuchae* (lig"ah-men'tum noo'ke; *nucha* = back of the neck), an elastic ligament that connects the vertebrae of the neck to the skull. The nuchal lines, and the bony regions between them, anchor many neck and back muscles. The superior nuchal line marks the upper limit of the neck.

Temporal Bones

The two **temporal bones** are best viewed on the lateral skull surface (see Figure 7.3). They lie inferior to the parietal bones and meet them at the squamous sutures. The temporal bones form the inferolateral aspects of the skull and parts of the cranial floor. The use of the terms *temple* and *temporal*, from the Latin word *temporum*, meaning "time," came about because gray hairs, a sign of time's passing, usually appear first in the temple areas.

Each temporal bone has a complicated shape and is described in terms of its four major regions, the *squamous*, *tympanic*, *mastoid*, and *petrous regions*. The flaring **squamous region** abuts the squamous suture. It has a barlike **zygomatic process** that meets the zygomatic bone of the face anteriorly. To-gether, these two bony structures form the **zygomatic arch**, which you can feel as the projection of your cheek (*zygoma* = cheekbone). The small, oval **mandibular** (man-dib'u-lar) **fossa** on the inferior surface of the zygomatic process receives the condyle of the mandible (lower jawbone), forming the freely movable *temporomandibular joint*.

The **tympanic** (tim-pan'ik; "eardrum") **region** (Figure 7.5) of the temporal bone surrounds the **external acoustic meatus**, or external ear canal, through which sound enters the ear. The external acoustic meatus and the eardrum at its deep end are part of the *external ear*. In a dried skull, the eardrum has been removed; thus, part of the middle ear cavity deep to the external meatus can also be seen. Below the external acoustic meatus is the needle-like **styloid** (sti'loid; "stakelike") **process**, an attachment point for several tongue and neck muscles and for a ligament that secures the hyoid bone of the neck to the skull (see Figure 7.12).

The **mastoid** (mas'toid; "breast") **region** of the temporal bone exhibits the conspicuous **mastoid process**, an anchoring site for some neck muscles (see Figures 7.3a, 7.4a, and 7.5). The mastoid process can be felt as a lump just posterior to the ear. The **stylomastoid foramen**, between the styloid and mastoid processes, allows cranial nerve VII (the facial nerve) to leave the skull (Figure 7.4a).

Ⓗ HOMEOSTATIC IMBALANCE

The mastoid process is full of air cavities, the **mastoid sinuses**, or **air cells**. Its position adjacent to the middle ear cavity (a high-risk area for infections spreading from the throat) puts it at risk for infection itself. A mastoid sinus infection, or *mastoiditis*, is notoriously difficult to treat. Since the mastoid air cells are separated from the brain by only a very thin bony plate, mastoid infections may spread to the brain as well. Surgical removal of the mastoid process was once the best way to prevent life-threatening brain inflammations in people susceptible to repeated bouts of mastoiditis. Today, antibiotic therapy is the treatment of choice. ●

The deep **petrous** (pet'rus) **region** of the temporal bone contributes to the cranial base (Figure 7.4). It looks like a miniature mountain ridge (*petrous* = rocky) between the occipital bone posteriorly and the sphenoid bone anteriorly. The posterior slope of this ridge lies in the posterior cranial fossa; the anterior slope is in the middle cranial fossa. Together, the sphenoid bone and the petrous portions of the temporal bones construct the **middle cranial fossa** (see Figure 7.4b and c), which supports the temporal

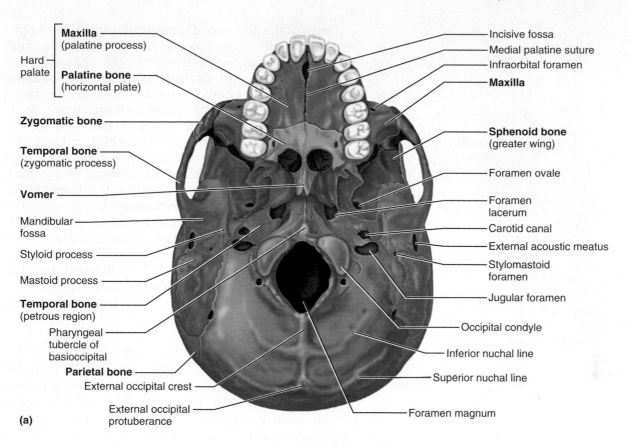

(a)

Maxilla (palatine process)
Hard palate
Palatine bone (horizontal plate)
Zygomatic bone
Temporal bone (zygomatic process)
Vomer
Mandibular fossa
Styloid process
Mastoid process
Temporal bone (petrous region)
Pharyngeal tubercle of basioccipital
Parietal bone
External occipital crest
External occipital protuberance

Incisive fossa
Medial palatine suture
Infraorbital foramen
Maxilla
Sphenoid bone (greater wing)
Foramen ovale
Foramen lacerum
Carotid canal
External acoustic meatus
Stylomastoid foramen
Jugular foramen
Occipital condyle
Inferior nuchal line
Superior nuchal line
Foramen magnum

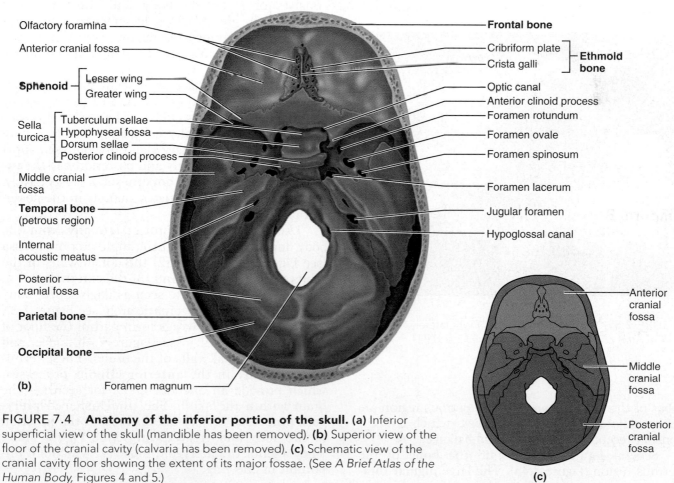

(b)

Olfactory foramina
Anterior cranial fossa
Sphenoid — Lesser wing / Greater wing
Sella turcica — Tuberculum sellae / Hypophyseal fossa / Dorsum sellae / Posterior clinoid process
Middle cranial fossa
Temporal bone (petrous region)
Internal acoustic meatus
Posterior cranial fossa
Parietal bone
Occipital bone
Foramen magnum

Frontal bone
Cribriform plate — Ethmoid bone
Crista galli
Optic canal
Anterior clinoid process
Foramen rotundum
Foramen ovale
Foramen spinosum
Foramen lacerum
Jugular foramen
Hypoglossal canal

(c)

Anterior cranial fossa
Middle cranial fossa
Posterior cranial fossa

FIGURE 7.4 **Anatomy of the inferior portion of the skull. (a)** Inferior superficial view of the skull (mandible has been removed). **(b)** Superior view of the floor of the cranial cavity (calvaria has been removed). **(c)** Schematic view of the cranial cavity floor showing the extent of its major fossae. (See *A Brief Atlas of the Human Body,* Figures 4 and 5.)

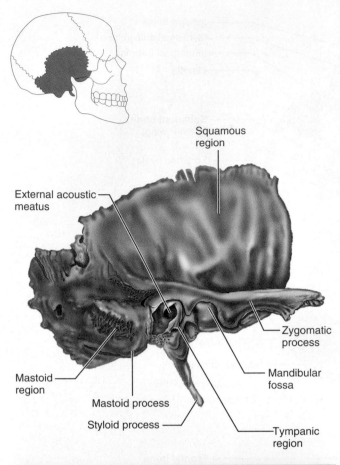

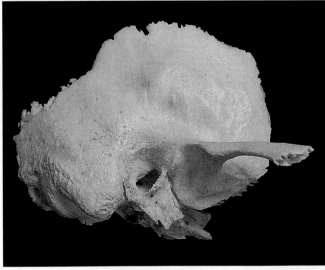

FIGURE 7.5 **The temporal bone.** Right lateral view. (See *A Brief Atlas of the Human Body,* Figures 3 and 8.)

lobes of the brain. Housed in the petrous region are the *middle* and *inner ear cavities*, which contain sensory receptors for hearing and balance.

Several foramina penetrate the bone of the petrous region (Figure 7.4a). The large **jugular fora-men** at the junction of the occipital and petrous temporal bones allows passage of the internal jugular vein and three cranial nerves. The **carotid** (kar-rot'id) **canal,** just anterior to the jugular foramen, transmits the internal carotid artery into the cranial cavity. The two internal carotid arteries supply blood to over 80% of the cerebral hemispheres of the brain; their closeness to the inner ear cavities explains why, during excitement or exertion, we sometimes hear our rapid pulse as a thundering sound in the head. The **foramen lacerum** (la'ser-um) is a jagged open-ing (*lacerum* = torn or lacerated) between the petrous temporal bone and the sphenoid bone. It is almost completely closed by cartilage in a living per-son, but it is conspicuous in a dried skull, and stu-dents usually ask its name. The **internal acoustic meatus,** positioned superolateral to the jugular fora-men (Figures 7.3b and 7.4b), transmits cranial nerves VII and VIII.

Sphenoid Bone

The butterfly-shaped **sphenoid** (sfe'noid; *sphen* = wedge) **bone** spans the width of the middle cranial fossa (Figure 7.4b). The sphenoid is considered the keystone of the cranium because it forms a central wedge that articulates with all other cranial bones. It consists of a central body and three pairs of processes: the greater wings, lesser wings, and ptery-goid (ter'ĭ-goid) processes (Figure 7.6). Within the **body** of the sphenoid are the paired **sphenoid si-nuses** (see Figures 7.3b and 7.11).

The superior surface of the body bears a saddle-shaped prominence, the **sella turcica** (sel'ah ter'sĭ-kah), meaning "Turk's saddle." The seat of this sad-dle, called the **hypophyseal fossa,** forms a snug enclosure for the pituitary gland (hypophysis). This fossa is abutted anteriorly and posteriorly by the **tu-berculum sellae** and the **dorsum sellae,** respectively. The dorsum sellae terminates laterally in the **poste-rior clinoid processes.**

The **greater wings** project laterally from the body, forming parts of (1) the middle cranial fossa (see Figure 7.4b and c), (2) the dorsal walls of the orbits (see Figure 7.2), and (3) the external wall of the skull, where they are seen as flag-shaped, bony areas medial to the zygomatic arch (see Figure 7.3). The hornlike **lesser wings** form part of the floor of the anterior cranial fossa (Figure 7.4b and c) and part of the medial walls of the orbits. These termi-nate medially in the **anterior clinoid processes,** which provide an anchoring site for securing the brain within the skull. The trough-shaped **ptery-goid processes** project inferiorly from the junction of the body and greater wings (see Figure 7.6b). They anchor the pterygoid muscles, which are im-portant in chewing.

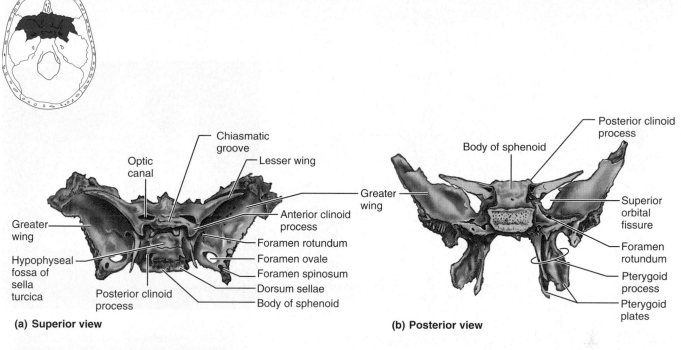

(a) Superior view

Chiasmatic groove
Optic canal
Lesser wing
Greater wing
Anterior clinoid process
Greater wing
Foramen rotundum
Hypophyseal fossa of sella turcica
Foramen ovale
Foramen spinosum
Posterior clinoid process
Dorsum sellae
Body of sphenoid

(b) Posterior view

Posterior clinoid process
Body of sphenoid
Greater wing
Superior orbital fissure
Foramen rotundum
Pterygoid process
Pterygoid plates

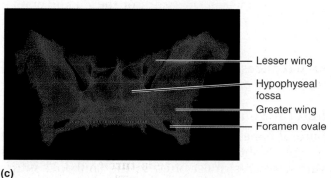

Lesser wing
Hypophyseal fossa
Greater wing
Foramen ovale

(c)

FIGURE 7.6　The sphenoid bone. (a) Superior view. **(b)** Posterior view. **(c)** High-resolution microradiograph of the sphenoid bone. Note in particular its body, contained sinuses, the greater and lesser wings, and the pterygoid processes. (See *A Brief Atlas of the Human Body*, Figures 3 and 9.)

A number of openings in the sphenoid bone are visible in Figures 7.4b and 7.6. The **optic canals,** connected by the *chiasmatic groove,* lie anterior to the sella turcica; they allow the optic nerves to pass to the eyes. On each side of the sphenoid body is a crescent-shaped row of four openings. The anteriormost of these, the **superior orbital fissure,** is a long slit between the greater and lesser wings. It allows cranial nerves that control eye movements (III, IV, VI) to enter the orbit. This fissure is most obvious in an anterior view of the skull (see Figure 7.2). The foramen rotundum and foramen ovale provide passageways for branches of cranial nerve V (namely the maxillary and mandibular nerves) to reach the face. The **foramen rotundum** is in the medial part of the greater wing and is usually oval, despite its name

meaning "round opening." The **foramen ovale** (o-vă′le), a large, oval foramen posterior to the foramen rotundum, is visible in an inferior view of the skull (see Figure 7.4a). Posterolateral to the foramen ovale is the small **foramen spinosum** (Figure 7.4b); it transmits the *middle meningeal artery,* which serves the internal faces of some of the cranial bones.

Ethmoid Bone

Like the temporal and sphenoid bones, the delicate **ethmoid bone** has a complex shape (Figure 7.7). Lying between the sphenoid and the nasal bones of the face, it is the most deeply situated bone of the skull. It forms most of the bony area between the nasal cavity and the orbits.

?   *Of what importance is the crista galli?*

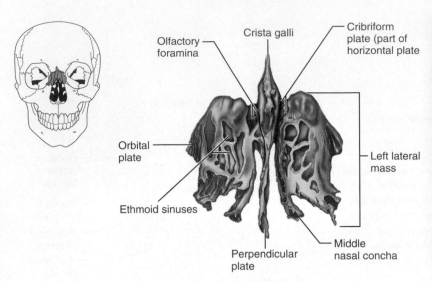

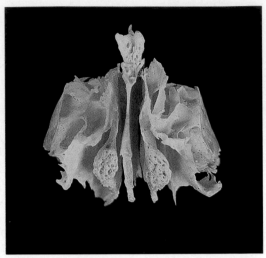

FIGURE 7.7   The ethmoid bone. Anterior view. (See *A Brief Atlas of the Human Body*, Figures 3 and 10.)

The superior surface of the ethmoid, called the **cribriform** (krib′rĭ-form) **plate** (see also Figure 7.4b), helps form the roof of the nasal cavities and the floor of the anterior cranial fossa. The cribriform plate is punctured by tiny holes (*cribr* = sieve) called *olfactory foramina* that allow the olfactory nerves to pass from the smell receptors in the nasal cavities to the brain. Projecting superiorly from the midline of the cribriform plate is a triangular process called the **crista galli** (kris′tah gah′le; "rooster's comb"). The outermost covering of the brain (the dura mater) attaches to the crista galli and helps secure the brain in the cranial cavity.

The **perpendicular plate** of the ethmoid bone projects inferiorly from the cribriform plate and forms the superior part of the nasal septum, which divides the nasal cavity into right and left halves (see Figure 7.3b). Flanking the perpendicular plate on each side is a **lateral mass** riddled with the **ethmoid sinuses** (Figures 7.7 and 7.11) for which the bone itself is named (*ethmos* = sieve). Extending medially from the lateral masses are the delicately coiled **superior** and **middle nasal conchae** (kong′ke; *concha* = shell), or **turbinates,** which protrude into the nasal cavity (Figures 7.7 and 7.10a). The lateral surfaces of the ethmoid's lateral masses are called

orbital plates because they contribute to the medial walls of the orbits.

Sutural Bones

Sutural bones, also called **Wormian bones,** are tiny irregularly shaped bones or bone clusters that appear within sutures, most often in the lambdoid suture (see Figure 7.2b). Structurally unimportant, their number varies, and not all skulls exhibit them. Sutural bones represent additional ossification centers that appeared when the skull was expanding very rapidly during fetal development.

Facial Bones

The facial skeleton is made up of 14 bones (see Figures 7.2a and 7.3a), of which only the mandible and the vomer are unpaired. The maxillae, zygomatics, nasals, lacrimals, palatines, and inferior conchae are paired bones. As a rule, the facial skeleton of men is more elongated than that of women; therefore, women's faces tend to be rounder and less angular.

Mandible

The U-shaped **mandible** (man′dĭ-bl), or lower jawbone (Figures 7.2, 7.3, and 7.8a), is the largest, strongest bone of the face. It has a *body*, which forms the chin, and two upright *rami* (*rami* = branches). Each ramus meets the body posteriorly at a **mandibular angle.** At the superior margin of each

■ *It is the site to which the membranes surrounding the brain attach to secure the brain in the cranial cavity.*

? Which of the bones illustrated is the keystone of the facial skeleton? Which contributes to the hard palate?

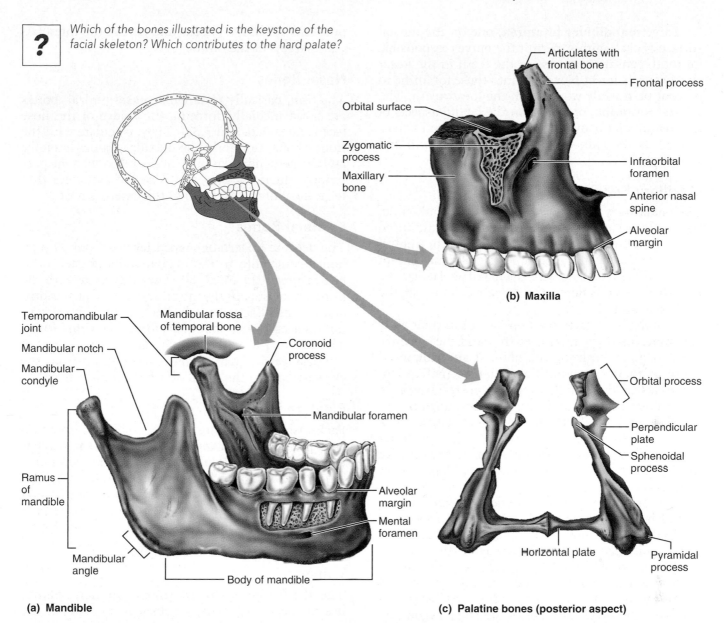

(a) **Mandible**

Temporomandibular joint
Mandibular notch
Mandibular condyle
Ramus of mandible
Mandibular angle
Mandibular fossa of temporal bone
Coronoid process
Mandibular foramen
Alveolar margin
Mental foramen
Body of mandible

(b) **Maxilla**

Articulates with frontal bone
Frontal process
Orbital surface
Zygomatic process
Maxillary bone
Infraorbital foramen
Anterior nasal spine
Alveolar margin

(c) **Palatine bones (posterior aspect)**

Orbital process
Perpendicular plate
Sphenoidal process
Horizontal plate
Pyramidal process

FIGURE 7.8 **Detailed anatomy of some isolated facial bones. (a)** The mandible (and its articulation with the temporal bone). **(b)** The maxilla (and its sites of articulation with the frontal and zygomatic bones). **(c)** Posterior view of the palatine bones. Note that the bones in (c) are enlarged more than those in (a) and (b). (See *A Brief Atlas of the Human Body*, Figures 11, 12, and 13.)

ramus are two processes separated by the **mandibular notch.** The anterior **coronoid** (kor'o-noid; "crown-shaped") **process** is an insertion point for the large temporalis muscle that elevates the lower jaw during chewing. The posterior **mandibular condyle** articulates with the mandibular fossa of the temporal bone, forming the *temporomandibular joint* on the same side.

The mandibular **body** anchors the lower teeth. Its superior border, called the **alveolar** (al-ve'o-lar) **margin,** contains the sockets (*alveoli*) in which the teeth are embedded. In the midline of the mandibular body is a slight depression, the **mandibular symphysis** (sim'fih-sis), indicating where the two mandibular bones fused during infancy (see Figure 7.2a).

The maxilla is the keystone of the facial skeleton and forms the anterior portion of the hard palate. The palatine bones form the posterior hard palate. ■

Large **mandibular foramina,** one on the medial surface of each ramus, permit the nerves responsible for tooth sensation to pass to the teeth in the lower jaw. Dentists inject Novocain into these foramina to prevent pain while working on the lower teeth. The **mental foramina,** openings on the lateral aspects of the mandibular body, allow blood vessels and nerves to pass to the skin of the chin (*ment* = chin) and lower lip.

Maxillary Bones

The **maxillary bones,** or **maxillae** (mak-sil′le; "jaws") (see Figures 7.2 to 7.4 and 7.8b), are fused medially. They form the upper jaw and the central portion of the facial skeleton. All facial bones except the mandible articulate with the maxillae. Hence, the maxillae are considered the keystone bones of the facial skeleton.

The maxillae carry the upper teeth in their **alveolar margins.** Just inferior to the nose the maxillae meet medially, forming the pointed **anterior nasal spine** at their junction. The **palatine** (pă′lah-tīn) **processes** of the maxillae project posteriorly from the alveolar margins and fuse medially, forming the anterior two-thirds of the hard palate, or bony roof of the mouth (see Figures 7.3b and 7.4a). Just posterior to the teeth is a midline foramen, called the **incisive fossa,** which serves as a passageway for blood vessels and nerves.

The **frontal processes** extend superiorly to the frontal bone, forming part of the lateral aspects of the bridge of the nose (see Figures 7.2a and 7.8b). The regions that flank the nasal cavity laterally contain the **maxillary sinuses** (Figure 7.11), the largest of the paranasal sinuses. They extend from the orbits to the upper teeth. Laterally, the maxillae articulate with the zygomatic bones via their **zygomatic processes.**

The **inferior orbital fissure** is located deep within the orbit (see Figure 7.2a) at the junction of the maxilla with the greater wing of the sphenoid. It permits the zygomatic nerve, the maxillary nerve (a branch of cranial nerve V), and blood vessels to pass to the face. Just below the eye socket on each side is an **infraorbital foramen** that allows the infraorbital nerve and artery to reach the face.

Zygomatic Bones

The irregularly shaped **zygomatic bones** (see Figures 7.2a, 7.3a, and 7.4a) are commonly called the cheekbones (*zygoma* = cheekbone). They articulate with the zygomatic processes of the temporal bones posteriorly, the zygomatic process of the frontal bone superiorly, and with the zygomatic processes of the maxillae anteriorly. The zygomatic bones form the prominences of the cheeks and part of the inferolateral margins of the orbits.

Nasal Bones

The thin, basically rectangular **nasal** (na′zal) **bones** are fused medially, forming the bridge of the nose (see Figures 7.2a and 7.3a). They articulate with the frontal bone superiorly, the maxillary bones laterally, and the perpendicular plate of the ethmoid bone posteriorly. Inferiorly they attach to the cartilages that form most of the skeleton of the external nose.

Lacrimal Bones

The delicate fingernail-shaped **lacrimal** (lak′rĭ-mal) **bones** contribute to the medial walls of each orbit (see Figures 7.2a and 7.3a). They articulate with the frontal bone superiorly, the ethmoid bone posteriorly, and the maxillae anteriorly. Each lacrimal bone contains a deep groove that helps form a **lacrimal fossa.** The lacrimal fossa houses the *lacrimal sac,* part of the passageway that allows tears to drain from the eye surface into the nasal cavity (*lacrima* = tears).

Palatine Bones

Each L-shaped **palatine bone** is fashioned from two bony plates, the *horizontal* and *perpendicular,* and has three important articular processes, the *pyramidal, sphenoidal,* and *orbital* (Figures 7.10a, 7.4a, and 7.8c). The **horizontal plates** complete the posterior portion of the hard palate. The superiorly projecting **perpendicular (vertical) plates** form part of the posterolateral walls of the nasal cavity and a small part of the orbits.

Vomer

The slender, plow-shaped **vomer** (vo′mer; "plow") lies in the nasal cavity, where it forms part of the nasal septum (see Figures 7.2a and 7.10b). It is discussed below in connection with the nasal cavity.

Inferior Nasal Conchae

The paired **inferior nasal conchae** are thin, curved bones in the nasal cavity. They project medially from the lateral walls of the nasal cavity, just inferior to the middle nasal conchae of the ethmoid bone (see Figures 7.2a and 7.10a). They are the largest of the three pairs of conchae and, like the others, they form part of the lateral walls of the nasal cavity.

Special Characteristics of the Orbits and Nasal Cavity

Two restricted skull regions, the orbits and the nasal cavity, are formed from an amazing number of bones. Thus, even though the individual bones

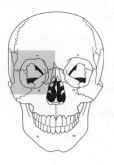

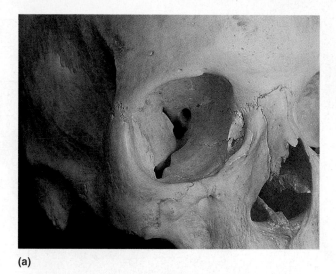

(a)

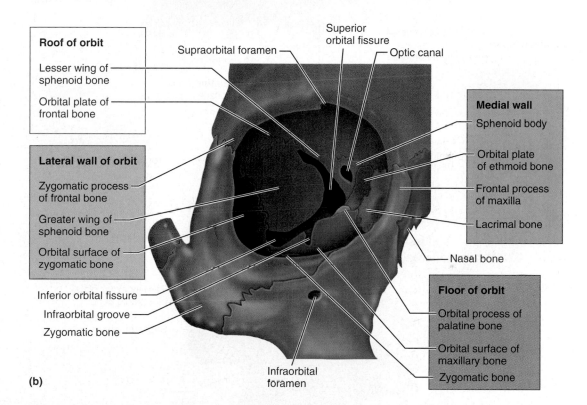

(b)

FIGURE 7.9 **Special anatomical characteristics of the orbits.** The contribution of each of the seven bones forming the right orbit is illustrated. **(a)** Photograph. **(b)** Diagrammatic view. (See *A Brief Atlas of the Human Body*, Figure 14.)

forming these structures have been described, a brief summary is provided here to pull the parts together.

The Orbits

The **orbits** are bony cavities in which the eyes are firmly encased and cushioned by fatty tissue. The muscles that move the eyes and the tear-producing lacrimal glands are also housed in the orbits. The walls of each orbit are formed by parts of seven bones—the frontal, sphenoid, zygomatic, maxilla, palatine, lacrimal, and ethmoid bones. Their relationships are shown in Figure 7.9. Also seen in the orbits are the superior and inferior orbital fissures and the optic canals, described earlier.

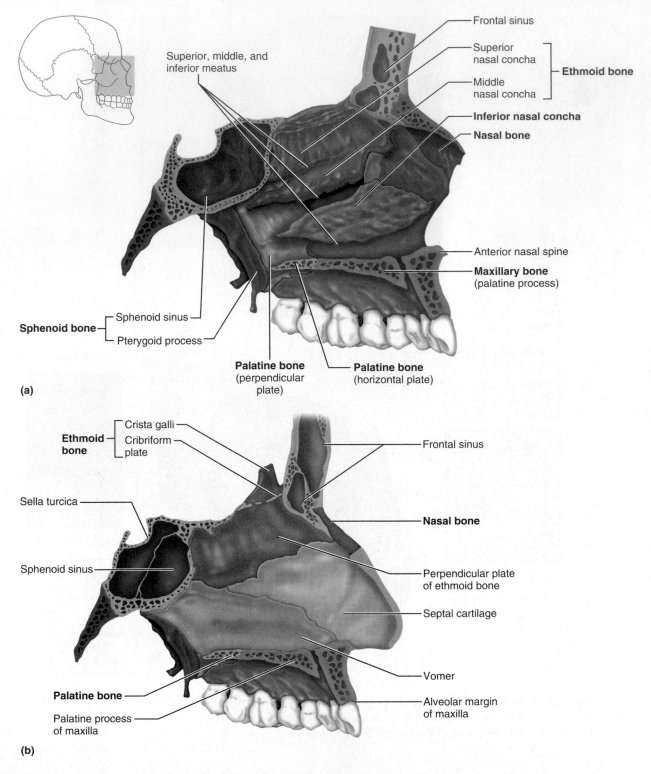

FIGURE 7.10   Special anatomical characteristics of the nasal cavity.
(a) Bones forming the left lateral wall of the nasal cavity. **(b)** The contribution of the ethmoid, vomer, and cartilage to the nasal septum. (See *A Brief Atlas of the Human Body*, Figure 15.)

The Nasal Cavity

The **nasal cavity** is constructed of bone and hyaline cartilage (Figure 7.10a). The *roof* of the nasal cavity is formed by the cribriform plate of the ethmoid. The *lateral walls* are largely shaped by the superior and middle conchae of the ethmoid bone, the perpendicular plates of the palatine bones, and the inferior nasal conchae. The depressions under cover of the conchae on the lateral walls are called *meatuses* (*meatus* = passage), so there are superior, middle,

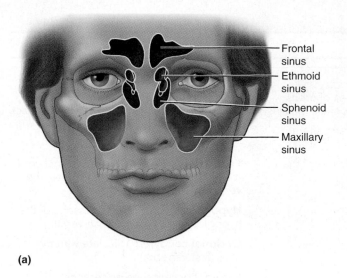

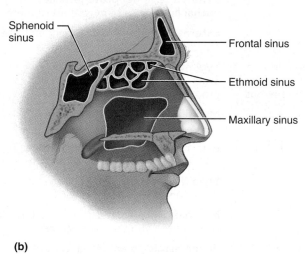

(a)

(b)

FIGURE 7.11 **Paranasal sinuses.** **(a)** Anterior aspect. **(b)** Medial aspect.

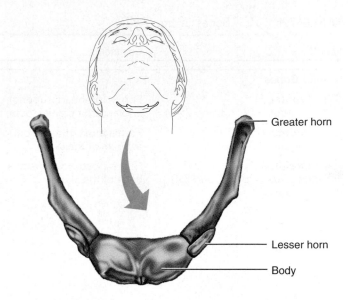

FIGURE 7.12 **Anatomical location and structure of the hyoid bone.** The hyoid bone is suspended in the mid-anterior neck by ligaments attached to the lesser horns (cornua) and the styloid processes of the temporal bones.

Paranasal Sinuses

Five skull bones—the frontal, sphenoid, ethmoid, and paired maxillary bones—contain mucosa-lined, air-filled sinuses that cause them to look rather moth-eaten in an X-ray image. These particular sinuses are called **paranasal sinuses** because they cluster around the nasal cavity (Figure 7.11). Small openings connect the sinuses to the nasal cavity and act as "two-way streets"; air enters the sinuses from the nasal cavity, and mucus formed by the sinus mucosae drains into the nasal cavity. The mucosae of the sinuses also help to warm and humidify inspired air. The paranasal sinuses lighten the skull and enhance the resonance of the voice.

The Hyoid Bone

Though not really part of the skull, the **hyoid** (hi'oid; "U-shaped") **bone** (Figure 7.12) lies just inferior to the mandible in the anterior neck. The hyoid bone is unique in that it is the only bone of the body that does not articulate directly with any other bone. Instead, it is anchored by the narrow *stylohyoid ligaments* to the styloid processes of the temporal bones. Horseshoe-shaped, with a *body* and two pairs of *horns*, or *cornua*, the hyoid bone acts as a movable base for the tongue. Its body and greater horns are attachment points for neck muscles that raise and lower the larynx during swallowing and speech.

★ ★ ★

Table 7.1 summarizes the bones of the skull.

and inferior meatuses. The *floor* of the nasal cavity is formed by the palatine processes of the maxillae and the palatine bones. The nasal cavity is divided into right and left parts by the *nasal septum.* The bony portion of the septum is formed by the vomer inferiorly and the perpendicular plate of the ethmoid bone superiorly (Figure 7.10b). A sheet of cartilage called the *septal cartilage* completes the septum anteriorly.

The nasal septum and conchae are covered with a mucus-secreting mucosa that moistens and warms the entering air and helps cleanse it of debris. The scroll-shaped conchae increase the turbulence of air flowing through the nasal cavity. This swirling forces more of the inhaled air into contact with the warm, damp mucosa and encourages trapping of airborne particles (dust, pollen, bacteria) in the sticky mucus.

TABLE 7.1   Bones of the Skull

Bone*	Comments	Important Markings
Cranial Bones		
Frontal (1) (Figures 7.2a, 7.3, and 7.4b)	Forms forehead; superior part of orbits and anterior cranial fossa; contains sinuses	**Supraorbital foramina (notches):** allow the supraorbital arteries and nerves to pass
Parietal (2) (Figures 7.2 and 7.3)	Forms most of the superior and lateral aspects of the skull	
Occipital (1) (Figures 7.2b, 7.3, and 7.4)	Forms posterior aspect and most of the base of the skull	**Foramen magnum:** allows the spinal cord to issue from the brain stem to enter the vertebral canal
		Hypoglossal canal: allows passage of the hypoglossal nerve (cranial nerve XII)
		Occipital condyles: articulate with the atlas (first vertebra)
		External occipital protuberance and **nuchal lines:** sites of muscle attachment
		External occipital crest: attachment site of ligamentum nuchae
Temporal (2) (Figures 7.3, 7.4, and 7.5)	Forms inferolateral aspects of the skull and contributes to the middle cranial fossa; has squamous, mastoid, tympanic, and petrous regions	**Zygomatic process:** helps to form the zygomatic arch, which forms the prominence of the cheek
		Mandibular fossa: articular point of the mandibular condyle
		External acoustic meatus: canal leading from the external ear to the eardrum
		Styloid process: attachment site for hyoid bone and several neck muscles
		Mastoid process: attachment site for several neck and tongue muscles
		Stylomastoid foramen: allows cranial nerve VII (facial nerve) to pass
		Jugular foramen: allows passage of the internal jugular vein and cranial nerves IX, X, and XI
		Internal acoustic meatus: allows passage of cranial nerves VII and VIII
		Carotid canal: allows passage of the internal carotid artery
Sphenoid (1) (Figures 7.2a, 7.3, 7.4, and 7.6)	Keystone of the cranium; contributes to the middle cranial fossa and orbits; main parts are the body, greater wings, lesser wings, and pterygoid processes	**Sella turcica:** hypophyseal fossa portion is the seat of the pituitary gland
		Optic canals: allow passage of cranial nerve II and the ophthalmic arteries
		Superior orbital fissures: allow passage of cranial nerves III, IV, VI, part of V (ophthalmic division), and ophthalmic vein
		Foramen rotundum: allows passage of the maxillary division of cranial nerve V
		Foraman ovale: allows passage of the mandibular division of cranial nerve V
		Foramen spinosum: allows passage of the middle meningeal artery

*The color code beside each bone name corresponds to the bone's color in Figures 7.2 to 7.10. The number in parentheses ( ) following the bone name denotes the total number of such bones in the body.

TABLE 7.1 *(continued)*

Bone*	Comments	Important Markings
Cranial Bones		
Ethmoid (1) (Figures 7.2a, 7.3, 7.4b, 7.7, and 7.10)	Helps to form the anterior cranial fossa; forms part of the nasal septum and the lateral walls and roof of the nasal cavity; contributes to the medial wall of the orbit	**Crista galli:** attachment point for the falx cerebri, a dural membrane **Cribriform plate:** allows passage of nerve filaments of the olfactory nerves (cranial nerve I) **Superior** and **middle nasal conchae:** form part of lateral walls of nasal cavity; increase turbulence of air flow
Ear ossicles (malleus, incus, and stapes) (2 each)	Found in middle ear cavity; involved in sound transmission; see Chapter 15	
Facial Bones		
Mandible (1) (Figures 7.2a, 7.3, and 7.8a)	The lower jaw	**Coronoid processes:** insertion points for the temporalis muscles **Mandibular condyles:** articulate with the temporal bones in freely movable joints (temporomandibular joints) of the jaw **Mandibular symphysis:** medial fusion point of the mandibular bones **Alveoli:** sockets for the teeth **Mandibular foramina:** permit the inferior alveolar nerves to pass **Mental foramina:** allow blood vessels and nerves to pass to the chin and lower lip
Maxilla (2) (Figures 7.2a, 7.3, 7.4a, and 7.8b)	Keystone bones of the face; form the upper jaw and parts of the hard palate, orbits, and nasal cavity walls	**Alveoli:** sockets for the teeth **Zygomatic processes:** help form the zygomatic arches **Palatine processes:** form the anterior hard palate; meet in medial palatine suture **Frontal processes:** form part of lateral aspect of bridge of nose **Incisive fossa:** permits blood vessels and nerves to pass through hard palate (fused palatine processes) **Inferior orbital fissures:** permit maxillary branch of cranial nerve V, the zygomatic nerve, and blood vessels to pass **Infraorbital foramen:** allows passage of infraorbital nerve to skin of face
Zygomatic (2) (Figures 7.2a, 7.3a, and 7.4a)	Form the cheek and part of the orbit	
Nasal (2) (Figures 7.2a and 7.3)	Construct the bridge of the nose	
Lacrimal (2) (Figures 7.2a and 7.3a)	Form part of the medial orbit wall	**Lacrimal fossa:** houses the lacrimal sac, which helps to drain tears into the nasal cavity
Palatine (2) (Figures 7.3b, 7.4a, and 7.8c)	Form posterior part of the hard palate and a small part of nasal cavity and orbit walls	
Vomer (1) (Figures 7.2a, 7.3b, and 7.10)	Part of the nasal septum	
Inferior nasal concha (2) (Figures 7.2a and 7.10a)	Form part of the lateral walls of the nasal cavity	

Why would it be anatomically "stupid" to have the sturdiest vertebrae in the cervical instead of the lumbar region of the spine?

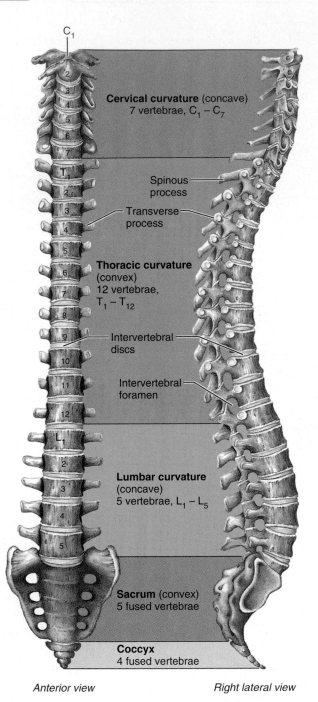

Anterior view   *Right lateral view*

FIGURE 7.13   The vertebral column. Notice the curvatures in the lateral view. (The terms *convex* and *concave* refer to the curvature of the posterior aspect of the vertebral column.) (See *A Brief Atlas of the Human Body,* Figure 17.)

■ *Because the lumbar region bears the greatest weight, its vertebrae are the sturdiest.* ■

The Vertebral Column
General Characteristics

Some people think of the **vertebral column** as a rigid supporting rod, but this is inaccurate. Also called the **spine,** the vertebral column is formed from 26 irregular bones connected in such a way that a flexible, curved structure results (Figure 7.13). Serving as the axial support of the trunk, the spine extends from the skull to the pelvis, where it transmits the weight of the trunk to the lower limbs. It also surrounds and protects the delicate spinal cord and provides attachment points for the ribs and for the muscles of the back and neck. In the fetus and infant, the vertebral column consists of 33 separate bones, or **vertebrae** (ver'tě-bre). Inferiorly, nine of these eventually fuse to form two composite bones, the sacrum and the tiny coccyx. The remaining 24 bones persist as individual vertebrae separated by intervertebral discs.

Divisions and Curvatures

The vertebral column is about 70 cm (28 inches) long in an average adult and has five major divisions (Figure 7.13). The seven vertebrae of the neck are the **cervical** (ser'vi-kal) **vertebrae,** the next 12 are the **thoracic** (tho-ras'ik) **vertebrae,** and the five supporting the lower back are the **lumbar** (lum'bar) **vertebrae.** Remembering common meal times—7 AM, 12 noon, and 5 PM—will help you recall the number of bones in these three regions of the spine. The vertebrae become progressively larger from the cervical to the lumbar region, as they must support greater and greater weight.

Inferior to the lumbar vertebrae is the **sacrum** (sa'krum), which articulates with the hip bones of the pelvis. The terminus of the vertebral column is the tiny **coccyx** (kok'siks). All of us have the same number of cervical vertebrae. Variations in numbers of vertebrae in other regions occur in about 5% of people.

When you view the vertebral column from the side, you can see the four curvatures that give it its S, or sinusoid, shape. The **cervical** and **lumbar curvatures** are concave posteriorly; the **thoracic** and **sacral curvatures** are convex posteriorly. These curvatures increase the resilience and flexibility of the spine, allowing it to function like a spring rather than as a rigid rod.

H   HOMEOSTATIC IMBALANCE

There are several types of abnormal spinal curvatures. Some are congenital (present at birth); others result from disease, poor posture, or unequal muscle pull on the spine. *Scoliosis* (sko"le-o'sis), literally,

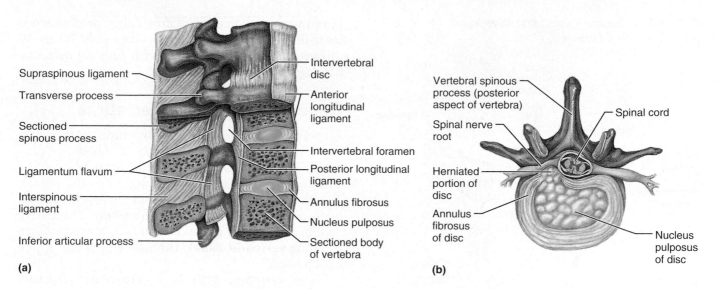

FIGURE 7.14   Ligaments and fibrocartilage discs uniting the vertebrae.
(a) Median section of three vertebrae, illustrating the composition of the discs and the ligaments. **(b)** Superior view of a sectioned herniated disc (slipped disc).

"twisted disease," is an abnormal *lateral* curvature that occurs most often in the thoracic region. It is quite common during late childhood, particularly in girls, for some unknown reason. Other, more severe cases result from abnormal vertebral structure, lower limbs of unequal length, or muscle paralysis. If muscles on one side of the body are nonfunctional, those of the opposite side exert an unopposed pull on the spine and force it out of alignment. Scoliosis is treated (with body braces or surgically) before growth ends to prevent permanent deformity and breathing difficulties due to a compressed lung.

Kyphosis (ki-fo'sis), or hunchback, is a *dorsally* exaggerated *thoracic* curvature. It is particularly common in aged individuals because of osteoporosis, but may also reflect tuberculosis of the spine, rickets, or osteomalacia. *Lordosis,* or swayback, is an accentuated *lumbar* curvature. It, too, can result from spinal tuberculosis or osteomalacia. Temporary lordosis is common in those carrying a "large load up front," such as men with "potbellies" and pregnant women. In an attempt to preserve their center of gravity, these individuals automatically throw back their shoulders, accentuating their lumbar curvature. ●

Ligaments

Like a tall, tremulous TV transmitting tower, the vertebral column cannot possibly stand upright by itself. It must be held in place by an elaborate system of cablelike supports. In the case of the vertebral column, straplike ligaments and the trunk muscles assume this role. The major supporting ligaments are the **anterior** and **posterior longitudinal ligaments** (Figure 7.14). These run as continuous bands down the front and back surfaces of the spine from the neck to the sacrum. The broad anterior ligament is strongly attached to both the bony vertebrae and the discs. Along with its supporting role, it prevents hyperextension of the spine (bending too far backward). The posterior ligament, which resists hyperflexion of the spine (bending too sharply forward), is narrow and relatively weak. It attaches only to the discs. However, the *ligamentum flavum,* which contains elastic connective tissue, is especially strong. It stretches on bending forward and recoils when we resume an erect posture. Short ligaments connect each vertebra to those immediately above and below.

Intervertebral Discs

Each **intervertebral disc** is a cushionlike pad composed of two parts. The inner gelatinous **nucleus pulposus** (pul-po'sus; "pulp") acts like a rubber ball, giving the disc its elasticity and compressibility. Surrounding the nucleus pulposus is a strong collar composed of collagen fibers superficially and fibrocartilage (internally), the **annulus fibrosus** (an'u-lus fi-bro'sus; "ring of fibers") (see Figure 7.14a). The annulus fibrosus limits the expansion of the nucleus pulposus when the spine is compressed. It also holds together successive vertebrae and resists tension in the spine.

The intervertebral discs act as shock absorbers during walking, jumping, and running, and they allow the spine to flex and extend, and to a lesser extent to

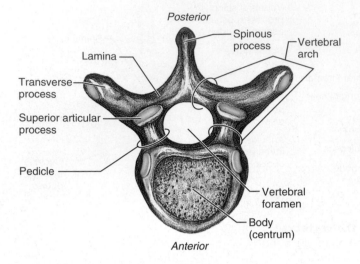

? *What structural parts of the vertebra are sites of muscle attachment?*

Posterior

Spinous process

Vertebral arch

Lamina

Transverse process

Superior articular process

Pedicle

Vertebral foramen

Body (centrum)

Anterior

FIGURE 7.15 Structure of a typical vertebra. Superior view (inferior articulating surfaces not shown).

bend laterally. At points of compression, the discs flatten and bulge out a bit between the vertebrae. The discs are thickest in the lumbar and cervical regions, which enhances the flexibility of these regions. Collectively the discs account for about 25% of the height of the vertebral column. They flatten somewhat during the course of the day, so we are always a few centimeters shorter at night than when we awake in the morning.

H   HOMEOSTATIC IMBALANCE

Severe or sudden physical trauma to the spine—for example, from bending forward while lifting a heavy object—may result in herniation of one or more discs. A **herniated (prolapsed) disc** (commonly called a *slipped disc*) usually involves rupture of the annulus fibrosus followed by protrusion of the spongy nucleus pulposus through the annulus (see Figure 7.14b). If the protrusion presses on the spinal cord or on spinal nerves exiting from the cord, numbness or excruciating pain may result. Herniated discs are generally treated with moderate exercise, massage, heat therapy, and painkillers. If this fails, the protruding disc may have to be removed surgically and a bone graft done to fuse the adjoining vertebrae. For those preferring to avoid general anesthesia, the disc can be partially vaporized with a

laser in an outpatient procedure called percutaneous laser disc decompression that takes only 30 to 40 minutes. The patient leaves with only an adhesive bandage to mark the spot. ●

General Structure of Vertebrae

All vertebrae have a common structural pattern (Figure 7.15). Each vertebra consists of a **body**, or **centrum**, anteriorly and a **vertebral arch** posteriorly. The disc-shaped body is the weight-bearing region. Together, the body and vertebral arch enclose an opening called the **vertebral foramen**. Successive vertebral foramina of the articulated vertebrae form the long **vertebral canal**, through which the spinal cord passes.

The vertebral arch is a composite structure formed by two pedicles and two laminae. The **pedicles** (ped'ĭ-kel; "little foot"), short bony pillars projecting posteriorly from the vertebral body, form the sides of the arch. The **laminae** (lam'ĭ-ne), flattened plates that fuse in the median plane, complete the arch posteriorly. Seven processes project from the vertebral arch. The **spinous process** is a median posterior projection arising at the junction of the two laminae. A **transverse process** extends laterally from each side of the vertebral arch. The spinous and transverse processes are attachment sites for muscles that move the vertebral column and for ligaments that stabilize it. The paired **superior** and **inferior articular processes** protrude superiorly and inferiorly, respectively, from the pedicle-lamina junctions. The smooth joint surfaces of the articular processes, called *facets* ("little faces"), are covered with hyaline cartilage. The inferior articular processes of each vertebra form movable joints with the superior articular processes of the vertebra immediately below. Thus, successive vertebrae join both at their bodies and at their articular processes.

The pedicles have notches on their superior and inferior borders, providing lateral openings between adjacent vertebrae called **intervertebral foramina** (see Figure 7.13). The spinal nerves issuing from the spinal cord pass through these foramina.

Regional Vertebral Characteristics

Beyond their common structural features, vertebrae exhibit variations that allow different regions of the spine to perform slightly different functions and movements. In general, movements that can occur between vertebrae are (1) flexion and extension (anterior bending and posterior straightening of the spine), (2) lateral flexion (bending the *upper body* to the right or left), and (3) rotation (in which vertebrae rotate on one another in the longitudinal axis of the spine). The regional vertebral characteristics

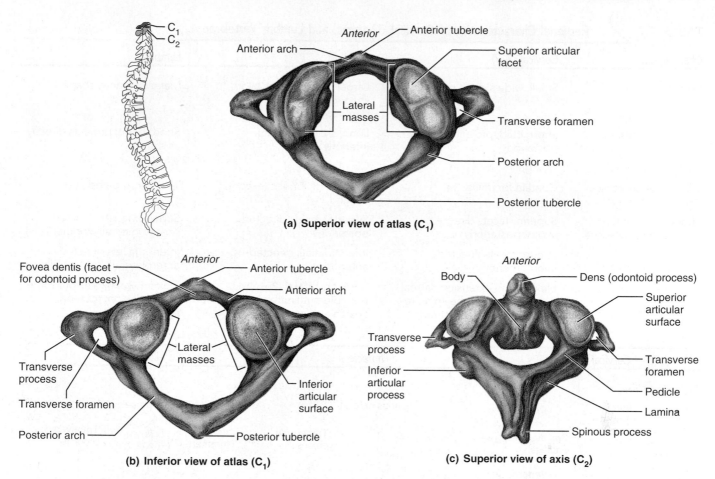

(a) Superior view of atlas (C₁)

(b) Inferior view of atlas (C₁)

(c) Superior view of axis (C₂)

FIGURE 7.16 **The first and second cervical vertebrae.** (See *A Brief Atlas of the Human Body*, Figure 18.)

described in this section are illustrated and summarized in Table 7.2.

Cervical Vertebrae

The seven **cervical vertebrae,** identified as $C_1–C_7$, are the smallest, lightest vertebrae. The first two (C_1 and C_2) are unusual and we will skip them for the moment. The "typical" cervical vertebrae ($C_3–C_7$) have the following distinguishing features:

1. The body is oval—wider from side to side than in the anteroposterior dimension.

2. Except in C_7, the spinous process is short, projects directly back, and is *bifid* (bi′fid), or split at its tip.

3. The vertebral foramen is large and generally triangular.

4. Each transverse process contains a **transverse foramen** through which the vertebral arteries pass to service the brain.

The spinous process of C_7 is not bifid and is much larger than those of the other cervical verte-

brae (see Figure 7.17a). Because its spinous process is visible through the skin, C_7 can be used as a landmark for counting the vertebrae and is called the **vertebra prominens** ("prominent vertebra").

The first two cervical vertebrae, the atlas and the axis, have no intervertebral disc between them, and they are highly modified, reflecting their special functions. The **atlas** (C_1) has no body and no spinous process (Figure 7.16a and b). Essentially, it is a ring of bone consisting of *anterior* and *posterior arches* and a *lateral mass* on each side. Each lateral mass has articular facets on both its superior and inferior surfaces. The superior articular facets receive the occipital condyles of the skull; thus, they "carry" the skull, just as Atlas supported the heavens in Greek mythology. These joints allow you to nod "yes." The inferior articular surfaces form joints with the axis (C_2) below.

The **axis,** which has a body, spine, and the other typical vertebral processes, is not as specialized as the atlas. In fact, its only unusual feature is the knoblike **dens** (denz; "tooth"), or **odontoid** (o-don′-toid) **process,** projecting superiorly from its body.

TABLE 7.2   Regional Characteristics of Cervical, Thoracic, and Lumbar Vertebrae

Characteristic	Cervical (3–7)	Thoracic	Lumbar
Body	Small, wide side to side	Larger than cervical; heart shaped; bears two costal demifacets	Massive; kidney shaped
Spinous process	Short; bifid; projects directly posteriorly	Long; sharp; projects inferiorly	Short; blunt; projects directly posteriorly
Vertebral foramen	Triangular	Circular	Triangular
Transverse processes	Contain foramina	Bear facets for ribs (except T_{11} and T_{12})	Thin and tapered
Superior and inferior articulating processes	Superior facets directed superoposteriorly	Superior facets directed posteriorly	Superior facets directed posteromedially (or medially)
	Inferior facets directed inferoanteriorly	Inferior facets directed anteriorly	Inferior facets directed anterolaterally (or laterally)
Movements allowed	Flexion and extension; lateral flexion; rotation; the spine region with the greatest range of movement	Rotation; lateral flexion possible but limited by ribs; flexion and extension prevented	Flexion and extension; some lateral flexion; rotation prevented

Cervical (3–7)	Thoracic	Lumbar

Superior View

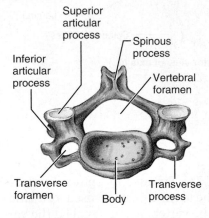

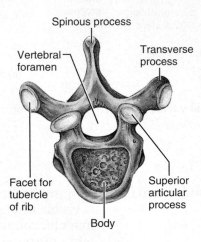

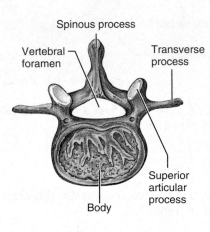

Right Lateral View

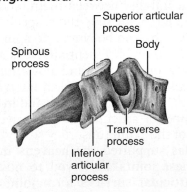

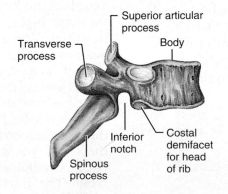

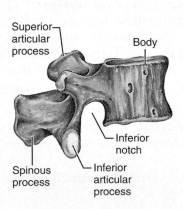

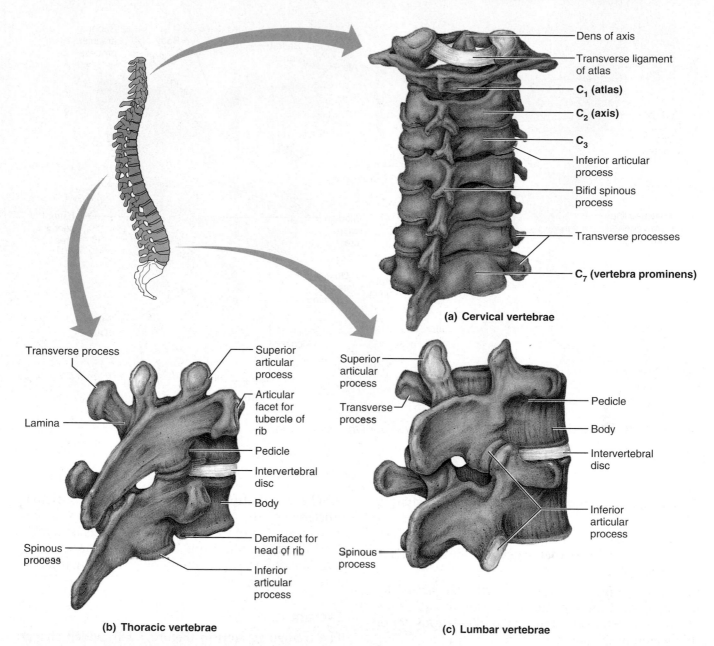

- Dens of axis
- Transverse ligament of atlas
- **C₁ (atlas)**
- **C₂ (axis)**
- **C₃**
- Inferior articular process
- Bifid spinous process
- Transverse processes
- **C₇ (vertebra prominens)**

(a) Cervical vertebrae

- Transverse process
- Lamina
- Spinous process
- Superior articular process
- Articular facet for tubercle of rib
- Pedicle
- Intervertebral disc
- Body
- Demifacet for head of rib
- Inferior articular process

(b) Thoracic vertebrae

- Superior articular process
- Transverse process
- Spinous process
- Pedicle
- Body
- Intervertebral disc
- Inferior articular process

(c) Lumbar vertebrae

FIGURE 7.17 **Posterolateral views of articulated vertebrae.** Notice the bulbous tip on the spinous process of C₇, the vertebra prominens. (See *A Brief Atlas of the Human Body*, Figures 19, 20, and 21.)

The dens is actually the "missing" body of the atlas, which fuses with the axis during embryonic development. Cradled in the anterior arch of the atlas by the transverse ligaments (see Figure 7.17a), the dens acts as a pivot for the rotation of the atlas. Hence, this joint allows you to rotate your head from side to side to indicate "no."

Thoracic Vertebrae

The 12 **thoracic vertebrae** (T₁–T₁₂) all articulate with the ribs (see Table 7.2 and Figure 7.17b). The first looks much like C₇, and the last four show a progression toward lumbar vertebral structure. The thoracic vertebrae increase in size from the first to the last. Unique characteristics of these vertebrae include the following:

1. The body is roughly heart shaped. It typically bears two *facets*, or more typically *demifacets* (half-facets), on each side, one at the superior edge and the other at the inferior edge. The demifacets receive the heads of the ribs. (The bodies of T₁₀–T₁₂ vary from this pattern by having only a single facet to receive their respective ribs.)

2. The vertebral foramen is circular.

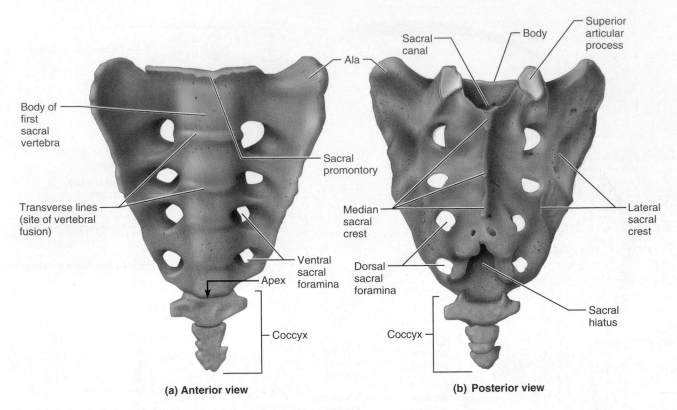

(a) Anterior view

(b) Posterior view

FIGURE 7.18   The sacrum and coccyx. (See *A Brief Atlas of the Human Body*, Figure 22.)

3. The spinous process is long and points sharply downward.

4. With the exception of T_{11} and T_{12}, the transverse processes have facets that articulate with the *tubercles* of the ribs.

5. The superior and inferior articular facets lie mainly in the frontal plane, a situation that prevents flexion and extension, but which allows rotation of this region of the spine.

Lumbar Vertebrae

The lumbar region of the vertebral column, commonly referred to as the small of the back, receives the most stress. The enhanced weight-bearing function of the five **lumbar vertebrae** ($L_1 - L_5$) is reflected in their sturdier structure. Their bodies are massive and kidney shaped (see Table 7.2 and Figure 7.17c). Other characteristics typical of these vertebrae:

1. The pedicles and laminae are shorter and thicker than those of other vertebrae.

2. The spinous processes are short, flat, and hatchet shaped and are easily seen when a person bends forward. These processes are robust and project directly backward, adaptations for the attachment of the large back muscles.

3. The vertebral foramen is triangular.

4. The orientation of the facets of the articular processes of the lumbar vertebrae differs substantially from that of the other vertebra types (see Table 7.2). These modifications lock the lumbar vertebrae together and provide stability by preventing rotation of the lumbar spine.

Sacrum

The triangular **sacrum** (Figure 7.18), which shapes the posterior wall of the pelvis, is formed by five fused vertebrae ($S_1 - S_5$) in adults. It articulates superiorly (via its **superior articular processes**) with L_5 and inferiorly with the coccyx. Laterally, the sacrum articulates, via its **auricular surfaces,** with the two hip bones to form the **sacroiliac** (sa"kro-il'e-ak) **joints** of the pelvis.

The **sacral promontory** (prom'on-tor"e), the anterosuperior margin of the first sacral vertebra, bulges anteriorly into the pelvic cavity. The body's center of gravity lies about 1 cm posterior to this landmark. Four ridges, the **transverse lines,** cross its concave anterior aspect, marking the lines of fusion of the sacral vertebrae. The **ventral sacral foramina** penetrate the sacrum at the lateral ends of these ridges and transmit blood vessels and nerves. The regions lateral to these foramina expand superiorly as the winglike **alae.**

In its dorsal midline the sacral surface is roughened by the **median sacral crest** (the fused spinous processes of the sacral vertebrae). This is flanked laterally by the **dorsal sacral foramina,** and then the **lateral sacral crests** (remnants of the transverse processes of $S_1 - S_5$).

The vertebral canal continues inside the sacrum as the **sacral canal.** Since the laminae of the fifth (and sometimes the fourth) sacral vertebrae fail to fuse medially, an enlarged external opening called the **sacral hiatus** (hi-a'tus; "gap") is obvious at the inferior end of the sacral canal.

Coccyx

The **coccyx,** our tailbone, is a small triangular bone (Figure 7.18). It consists of four (or in some cases three or five) vertebrae fused together. The coccyx articulates superiorly with the sacrum. (The name *coccyx* is from the Greek word meaning "cuckoo" and was so named because of its fancied resemblance to a bird's beak.) Except for the slight support the coccyx affords the pelvic organs, it is a nearly useless bone. Occasionally, a baby is born with an unusually long coccyx. In most such cases, this bony "tail" is discreetly snipped off by the physician.

* * *

The characteristics of the regional vertebrae are summarized in Table 7.2.

The Bony Thorax

Anatomically, the thorax is the chest, and its bony underpinnings are called the **bony thorax** or **thoracic cage.** Elements of the bony thorax include the thoracic vertebrae dorsally, the ribs laterally, and the sternum and costal cartilages anteriorly. The costal cartilages secure the ribs to the sternum (Figure 7.19a). Roughly cone shaped with its broad dimension inferiorly, the bony thorax forms a protective cage around the vital organs of the thoracic cavity (heart, lungs, and great blood vessels), supports the shoulder girdles and upper limbs, and provides attachment points for many muscles of the neck, back, chest, and shoulders. The *intercostal spaces* between the ribs are occupied by the intercostal muscles, which lift and depress the thorax during breathing.

Sternum

The **sternum** (breastbone) lies in the anterior midline of the thorax. Vaguely resembling a dagger, it is a flat bone approximately 15 cm (6 inches) long, resulting from the fusion of three bones: the manubrium, the body, and the xiphoid process. The **manubrium** (mah-nu'bre-um), the superior portion shaped like the knot in a necktie, articulates via its **clavicular** (klah-vik'u-lar) **notches** with the clavicles (collarbones) laterally. It also articulates with the first two pairs of ribs. The **body,** or midportion, forms the bulk of the sternum. The sides of the body are notched where it articulates with the cartilages of the second to seventh ribs. The **xiphoid** (zif'oid; "swordlike") **process** forms the inferior end of the sternum. This small, variably shaped process is a plate of hyaline cartilage in youth, but it is usually ossified in adults. The xiphoid process articulates only with the sternal body and serves as an attachment point for some abdominal muscles.

🅗 HOMEOSTATIC IMBALANCE

In some people, the xiphoid process projects dorsally. This may present a problem because blows to the chest can push such a xiphoid into the underlying heart or liver, causing massive hemorrhage. ●

The sternum has three important anatomical landmarks: the jugular notch, the sternal angle, and the xiphisternal joint (see Figure 7.19). The easily palpated **jugular (suprasternal) notch** is the central indentation in the superior border of the manubrium. It is generally in line with the disc between the second and third thoracic vertebrae and the point where the left common carotid artery issues from the aorta (see Figure 7.19b). The **sternal angle** is felt as a horizontal ridge across the front of the sternum, where the manubrium joins the sternal body. This cartilaginous joint acts like a hinge, allowing the sternal body to swing forward when we inhale. The sternal angle is in line with the disc between the fourth and fifth thoracic vertebrae and at the level of the second pair of ribs. It is a handy reference point for finding the second rib and thus for counting the ribs during a physical examination and for listening to sounds made by specific heart valves. The **xiphisternal** (zif"ĭ-ster'nul) **joint** is the point where the sternal body and xiphoid process fuse. It lies opposite the ninth thoracic vertebra.

Ribs

Twelve pairs of **ribs** form the flaring sides of the thoracic cage (Figure 7.19a). All ribs attach posteriorly to the thoracic vertebrae (bodies and transverse processes) and curve inferiorly toward the anterior body surface. The superior seven rib pairs attach directly to the sternum by individual costal cartilages (bars of hyaline cartilage). These are **true** or **vertebrosternal** (ver"tĕ-bro-ster'nal) **ribs.** (Notice that the

?   *Which ribs are also known as vertebrosternal ribs?*

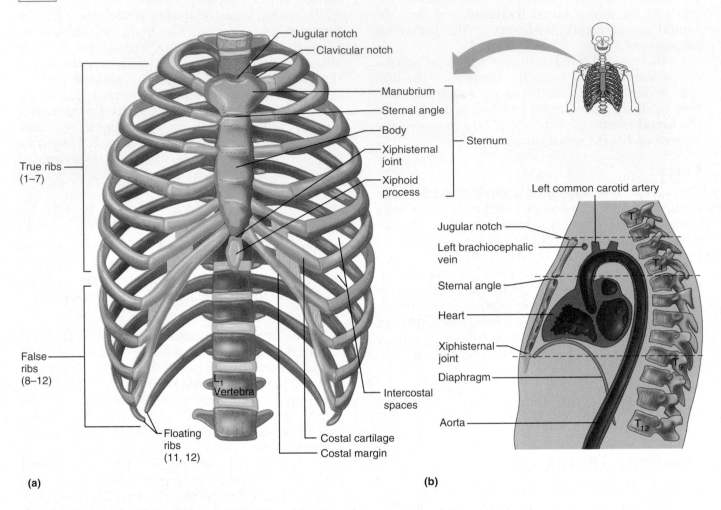

FIGURE 7.19   The bony thorax. (a) Skeleton of the bony thorax, anterior view (costal cartilages are shown in blue). **(b)** Left lateral view of the thorax, illustrating the relationship of the surface anatomical landmarks of the thorax to the vertebral column (thoracic portion). (See *A Brief Atlas of the Human Body*, Figure 23.)

anatomical name indicates the two attachment points of a rib—the posterior attachment given first.)

The remaining five pairs of ribs are called **false ribs;** they either attach indirectly to the sternum or entirely lack a sternal attachment. Rib pairs 8–10 attach to the sternum indirectly; each joins the costal cartilage immediately above it. These ribs are also called **vertebrochondral** (ver″tĕ-bro-kon′dral) **ribs.** The inferior margin of the rib cage, or **costal margin,** is formed by the costal cartilages of ribs 7–10. Rib pairs 11 and 12 are called **vertebral ribs** or **floating ribs** because they have no anterior attachments. Instead, their costal cartilages lie embedded in the

muscles of the lateral body wall. The ribs increase in length from pair 1 to pair 7, then decrease in length from pair 8 to pair 12.

The first pair of ribs is quite atypical. They are flattened superiorly to inferiorly and are quite broad, forming a horizontal table that supports the subclavian blood vessels that serve the upper limbs. There are also other exceptions to the typical rib pattern. Ribs 1 and 10–12 articulate with only one vertebral body, and ribs 11 and 12 do not articulate with a vertebral transverse process. Except for the first rib, which lies deep to the clavicle, the ribs are easily felt in people of normal weight.

■ *Rib pairs 1–7, the true ribs.*

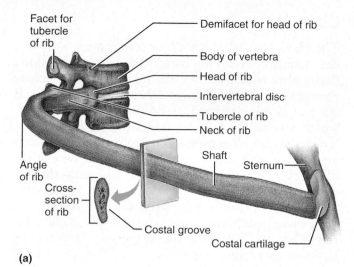

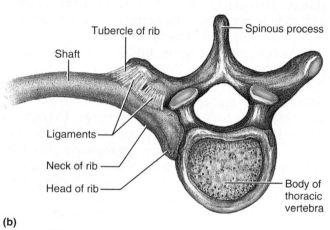

FIGURE 7.20   **Structure of a "typical" true rib and its articulations. (a)** Vertebral and sternal articulations of a typical true rib. **(b)** Superior view of the articulation between a rib and a thoracic vertebra. (See *A Brief Atlas of the Human Body,* Figure 23e and f.)

A typical rib is a bowed flat bone (Figure 7.20). The bulk of a rib is simply called the *shaft.* Its superior border is smooth, but its inferior border is sharp and thin and has a *costal groove* on its inner face that lodges the intercostal nerves and blood vessels. In addition to the shaft, each rib has a head, neck, and tubercle. The wedge-shaped *head,* the posterior-most end, articulates with the vertebral bodies by two facets: One joins the body of the same-numbered thoracic vertebra, the other articulates with the body of the vertebra immediately superior. The *neck* is the constricted portion of the rib just beyond the head. Lateral to this, the knoblike *tubercle* articulates with the transverse process of the same-numbered thoracic vertebra. Beyond the tubercle, the shaft angles sharply forward (at the angle of the rib) and then extends to attach to its costal cartilage anteriorly. The costal cartilages provide secure but flexible rib attachments to the sternum.

PART 2: THE APPENDICULAR SKELETON

Bones of the limbs and their girdles are collectively called the **appendicular skeleton** because they are *appended to* the axial skeleton that forms the longitudinal axis of the body (see Figure 7.21). The yoke-like *pectoral* (pek'tor-al; "chest") *girdles* attach the upper limbs to the body trunk. The more sturdy *pelvic girdle* secures the lower limbs. Although the bones of the upper and lower limbs differ in their functions and mobility, they have the same fundamental plan: Each limb is composed of three major segments connected by movable joints.

The appendicular skeleton enables us to carry out the movements typical of our freewheeling and manipulative lifestyle. Each time we take a step, throw a ball, or pop a caramel into our mouth, we are making good use of our appendicular skeleton.

The Pectoral (Shoulder) Girdle

The **pectoral,** or **shoulder, girdle** consists of the *clavicle* (klav'ĭ-kl) anteriorly and the *scapula* (skap'u-lah) posteriorly (Figure 7.22 and Table 7.3 on p. 202). The paired pectoral girdles and their associated muscles form your shoulders. Although the term *girdle* usually signifies a beltlike structure encircling the body, a single pectoral girdle, or even the pair, does not quite satisfy this description. Anteriorly, the medial end of each clavicle joins the sternum; the distal ends of the clavicles meet the scapulae laterally. However, the scapulae fail to complete the ring posteriorly, because their medial borders do not join each other or the axial skeleton. Instead, the scapulae are attached to the thorax and vertebral column only by the muscles that clothe their surfaces.

The pectoral girdles attach the upper limbs to the axial skeleton and provide attachment points for many of the muscles that move the upper limbs. These girdles are very light and allow the upper limbs a degree of mobility not seen anywhere else in the body. This mobility is due to the following factors:

1. Because only the clavicle attaches to the axial skeleton, the scapula can move quite freely across the thorax, allowing the arm to move with it.

2. The socket of the shoulder joint (the scapula's glenoid cavity) is shallow and poorly reinforced, so it does not restrict the movement of the humerus (arm bone). Although this arrangement is good for flexibility, it is bad for stability: Shoulder dislocations are fairly common.

? *The skeleton shown here is incomplete. What major bony regions are missing?*

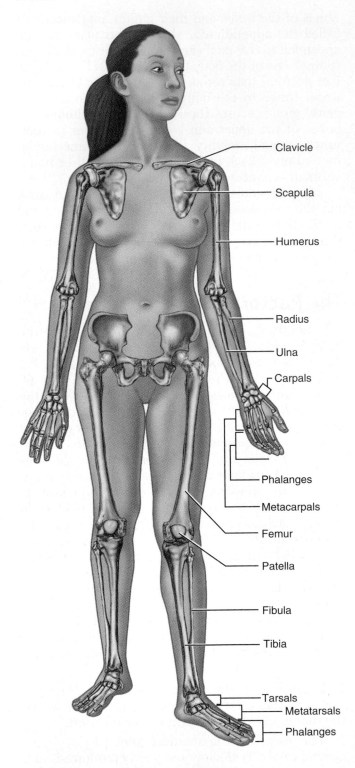

Clavicle

Scapula

Humerus

Radius

Ulna

Carpals

Phalanges

Metacarpals

Femur

Patella

Fibula

Tibia

Tarsals

Metatarsals

Phalanges

FIGURE 7.21   **The appendicular skeleton.**

The skull, the bony thorax, and the vertebral column, all parts of the axial skeleton. ■

Clavicles

The **clavicles** ("little keys"), or collarbones, are slender, doubly curved long bones that can be felt along their entire course as they extend horizontally across the superior thorax (Figures 7.22a–c and 7.25). Each clavicle is cone shaped at its medial **sternal end,** which attaches to the sternal manubrium, and flattened at its lateral **acromial** (ah-kro'me-al) **end,** which articulates with the scapula. The medial two-thirds of the clavicle is convex anteriorly; its lateral third is concave anteriorly. Its superior surface is smooth, but the inferior surface is ridged and grooved.

Besides anchoring many muscles, the clavicles act as braces: They hold the scapulae and arms out laterally, away from the narrower superior part of the thorax. This bracing function becomes obvious when a clavicle is fractured: The entire shoulder region collapses medially. The clavicles also transmit compression forces from the upper limbs to the axial skeleton, as when someone pushes a car to a gas station.

The clavicles are not very strong and are likely to fracture, for example, when a person uses outstretched arms to break a fall. The curves in the clavicle ensure that it usually fractures anteriorly (outward). If it were to collapse posteriorly (inward), bone splinters would damage the subclavian artery, which passes just deep to the clavicle to serve the upper limb. The clavicles are exceptionally sensitive to muscle pull and become noticeably larger and stronger in those who perform manual labor or athletics involving the shoulder and arm muscles.

Scapulae

The **scapulae,** or *shoulder blades,* are thin, triangular flat bones (Figures 7.22d–f and 7.25). Interestingly, their name derives from a word meaning "spade" or "shovel," for ancient cultures made spades from the shoulder blades of animals. The scapulae lie on the dorsal surface of the rib cage, between ribs 2 and 7. Each scapula has three borders. The **superior border** is the shortest, sharpest border. The **medial,** or *vertebral,* **border** parallels the vertebral column. The thick **lateral,** or *axillary,* **border** abuts the armpit and ends superiorly in a small, shallow fossa, the **glenoid** (gle'noid) **cavity.** This cavity articulates with the humerus of the arm, forming the shoulder joint.

Like all triangles, the scapula has three corners or *angles.* The superior scapular border meets the medial border at the **superior angle** and the lateral border at the **lateral angle.** The medial and lateral borders join at the **inferior angle.** The inferior angle moves extensively as the arm is raised and lowered, and is an important landmark for studying scapular movements.

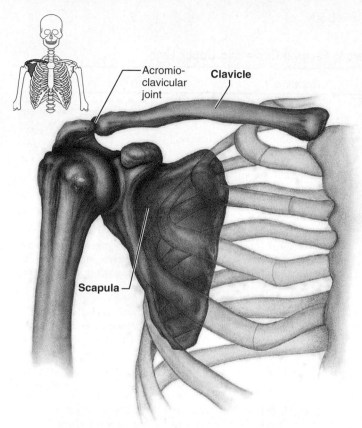

Acromio-clavicular joint

Clavicle

Scapula

(a) Articulated pectoral girdle

Posterior

Sternal (medial) end

Acromial (lateral) end

Anterior

(b) Right clavicle, superior view

Acromial end

Anterior

Sternal end

Posterior

(c) Right clavicle, inferior view

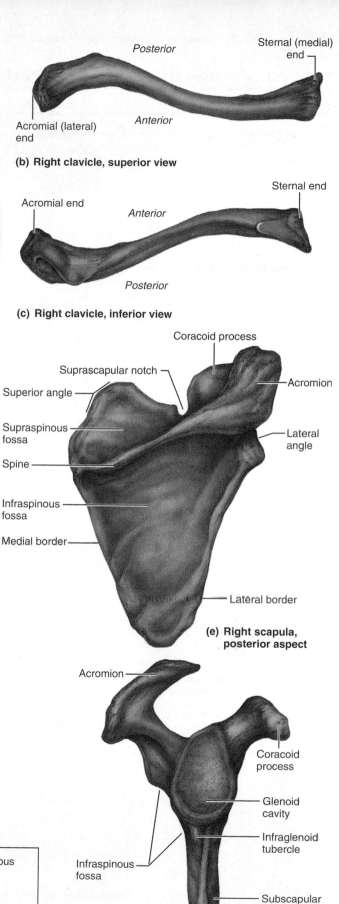

Coracoid process

Suprascapular notch

Superior angle

Supraspinous fossa

Spine

Infraspinous fossa

Medial border

Acromion

Lateral angle

Lateral border

(e) Right scapula, posterior aspect

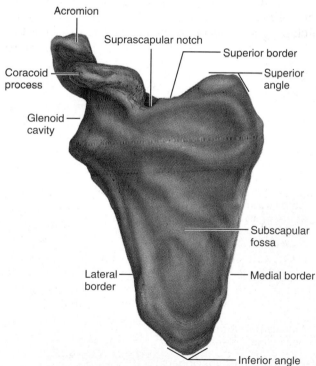

Acromion

Suprascapular notch

Coracoid process

Glenoid cavity

Superior border

Superior angle

Subscapular fossa

Lateral border

Medial border

Inferior angle

(d) Right scapula, anterior aspect

FIGURE 7.22 **Bones of the pectoral girdle.** In view **(a)**, the scapula is positioned posterior to the ribs. View **(f)** is accompanied by a schematic representation of its orientation. (See *A Brief Atlas of the Human Body*, Figure 24.)

Supraspinous fossa

Infraspinous fossa

Subscapular fossa

Posterior   *Anterior*

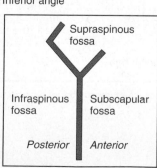

Acromion

Coracoid process

Glenoid cavity

Infraglenoid tubercle

Infraspinous fossa

Subscapular fossa

Inferior angle

(f) Right scapula, lateral aspect

TABLE 7.3   **Bones of the Appendicular Skeleton Part 1: Pectoral Girdle and Upper Limb**

Body Region	Bones*	Illustration	Location	Markings
Pectoral girdle (Figures 7.22, 7.25)	Clavicle (2)		Clavicle is in superoanterior thorax; articulates medially with sternum and laterally with scapula	Acromial end; sternal end
	Scapula (2)		Scapula is in posterior thorax; forms part of the shoulder; articulates with humerus and clavicle	Glenoid cavity; spine; acromion; coracoid process; infraspinous, supraspinous, and subscapular fossae
Free upper limb Arm (Figures 7.23, 7.25)	Humerus (2)		Humerus is the sole bone of arm; between scapula and elbow	Head; greater and lesser tubercles; intertubercular groove; deltoid tuberosity; trochlea; capitulum; coronoid and olecranon fossae; radial groove; epicondyles
Forearm (Figures 7.24, 7.25)	Ulna (2)		Ulna is the medial bone of forearm between elbow and wrist; forms elbow joint	Coronoid process; olecranon process; radial notch; trochlear notch; styloid process; head
	Radius (2)		Radius is lateral bone of forearm; carries wrist	Radial tuberosity; styloid process; head; ulnar notch
Hand (Figure 7.26)	8 Carpals (16) • scaphoid • lunate • triquetral • pisiform • trapezium • trapezoid • capitate • hamate		Carpals form a bony crescent at the wrist; arranged in two rows of four bones each	
	5 Metacarpals (10)		Metacarpals form the palm; one in line with each digit	
	14 Phalanges (28) • distal • middle • proximal		Phalanges form the fingers; three in digits 2–5; two in digit 1 (the thumb)	

Anterior view of pectoral girdle and upper limb

*The number in parentheses ( ) following the bone name denotes the total number of such bones in the body.

The anterior, or costal, surface of the scapula is concave and relatively featureless. Its posterior surface bears a prominent **spine** that is easily felt through the skin. The spine ends laterally in an enlarged, roughened triangular projection called the **acromion** (ah-kro′me-on), literally, the "point of the shoulder." The acromion articulates with the acromial end of the clavicle, forming the **acromioclavicular joint.**

Projecting anteriorly from the superior scapular border is the **coracoid** (kor′ah-coid) **process;** *corac* means "beaklike," but this process looks more like a bent little finger. The coracoid process helps anchor the biceps muscle of the arm. It is bounded by the **suprascapular notch** (a nerve passage) medially and by the glenoid cavity laterally.

Several large fossae appear on both sides of the scapula and are named according to location. The **infraspinous** and **supraspinous fossae** are respectively inferior and superior to the spine. The **subscapular fossa** is the shallow concavity formed by the entire anterior scapular surface.

The Upper Limb

Thirty separate bones form the bony framework of each upper limb (see Figures 7.23 to 7.26, and Table 7.3). Each of these bones may be described regionally as a bone of the arm, forearm, or hand. (Keep in mind that anatomically the "arm" is only that part of the upper limb between the shoulder and elbow.)

Arm

The **humerus** (hu′mer-us), the sole bone of the arm, is a typical long bone (Figures 7.23 and 7.25). The largest, longest bone of the upper limb, it articulates with the scapula at the shoulder and with the radius and ulna (forearm bones) at the elbow.

At the proximal end of the humerus is its smooth, hemispherical **head,** which fits into the glenoid cavity of the scapula in a manner that allows the arm to hang freely at one's side. Immediately inferior to the head is a slight constriction, the **anatomical neck.** Just inferior to this are the lateral **greater tubercle** and the more medial **lesser tubercle,** separated by the **intertubercular,** or *bicipital,* **groove** (bi-sip′ĭ-tal). The tubercles are sites where muscles attach. The intertubercular groove guides a tendon of the biceps muscle of the arm to its attachment point at the rim of the glenoid cavity. Just distal to the tubercles is the **surgical neck,** so named because it is the most frequently fractured part of the humerus. About midway down the shaft on its lateral side is the **deltoid tuberosity,** the roughened attachment site for the deltoid muscle of the shoulder. Nearby, the **radial groove** runs obliquely down the posterior aspect of the shaft, marking the course of the radial nerve, an important nerve of the upper limb.

At the distal end of the humerus are two condyles, a medial **trochlea** (trok′le-ah; "pulley"), which looks like an hourglass tipped on its side, and the lateral ball-like **capitulum** (kah-pit′u-lum). These condyles articulate with the ulna and the radius, respectively. The condyle pair is flanked by the **medial** and **lateral epicondyles** (muscle attachment sites). Directly above these epicondyles are the supracondylar ridges. The ulnar nerve runs behind the medial epicondyle and is responsible for the painful, tingling sensation you experience when you hit your "funny bone."

Superior to the trochlea on the anterior surface is the **coronoid fossa;** on the posterior surface is the deeper **olecranon fossa** (o-lek′rah-non). These two depressions allow the corresponding processes of the ulna to move freely when the elbow is flexed and extended. A small **radial fossa,** lateral to the coronoid fossa, receives the head of the radius when the elbow is flexed.

Forearm

Two parallel long bones, the radius and the ulna, form the skeleton of the forearm, or *antebrachium* (an″te-bra′ke-um) (Figures 7.24 and 7.25). Unless a person's forearm muscles are very bulky, these bones are easily palpated along their entire length. Their proximal ends articulate with the humerus; their distal ends form joints with bones of the wrist. The radius and ulna articulate with each other both proximally and distally at small **radioulnar** (ra″de-o-ul′nar) **joints,** and they are connected along their entire length by a flexible **interosseous** (in″ter-os′e-us) **membrane.** In the anatomical position, the radius lies laterally (on the thumb side) and the ulna medially. However, when you rotate your forearm so that the palm faces posteriorly (a movement called pronation), the distal end of the radius crosses over the ulna and the two bones form an X.

Ulna

The **ulna** (ul′nah; "elbow") is slightly longer than the radius. It has the main responsibility for forming the elbow joint with the humerus. Its proximal end looks like the adjustable end of a monkey wrench; it bears two prominent processes, the **olecranon** (elbow) and **coronoid processes,** separated by a deep concavity, the **trochlear notch** (see Figures 7.24 and

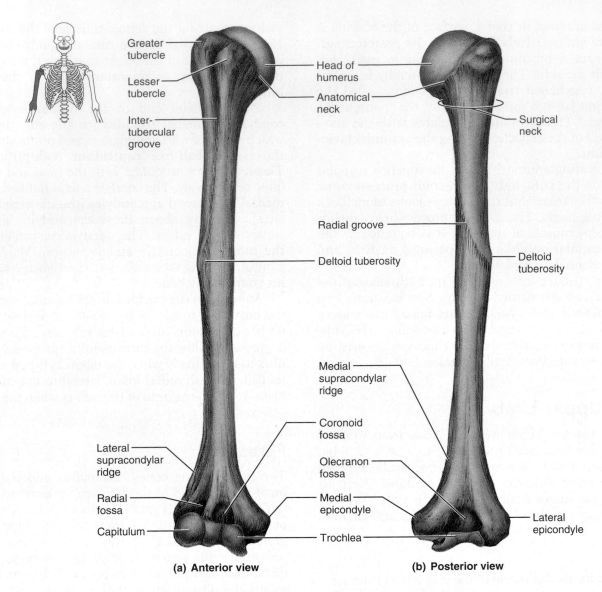

Greater tubercle

Lesser tubercle

Intertubercular groove

Head of humerus

Anatomical neck

Surgical neck

Radial groove

Deltoid tuberosity

Deltoid tuberosity

Medial supracondylar ridge

Coronoid fossa

Olecranon fossa

Lateral supracondylar ridge

Radial fossa

Medial epicondyle

Capitulum

Trochlea

Lateral epicondyle

(a) Anterior view

(b) Posterior view

FIGURE 7.23   The humerus of the arm. (a) Anterior view of the right humerus. **(b)** Posterior view of the right humerus. (See *A Brief Atlas of the Human Body,* Figure 25.)

7.25). Together, these two processes grip the trochlea of the humerus, forming a hinge joint that allows the forearm to be bent upon the arm (flexed), then straightened again (extended). When the forearm is fully extended, the olecranon process "locks" into the olecranon fossa, keeping the forearm from hyperextending (moving posteriorly beyond the elbow joint). The posterior olecranon process forms the angle of the elbow when the forearm is flexed and is the bony part that rests on the table when you lean on your elbows. On the lateral side of the coronoid process is a small depression, the **radial notch,** where the ulna articulates with the head of the radius.

Distally the ulnar shaft narrows and ends in a knoblike **head.** Medial to the head is a **styloid process** ("stake-shaped"), from which a ligament runs to the wrist. The ulnar head is separated from the bones of the wrist by a disc of fibrocartilage and plays little or no role in hand movements.

Radius

The **radius** ("rod") is thin at its proximal end and wide distally—the opposite of the ulna. The **head** of the radius is shaped somewhat like the head of a nail (see Figures 7.24 and 7.25). The superior surface of this head is concave, and it articulates with the capitulum of the humerus. Medially, the head articu-

(1) Which of these bones "carries" the hand? (2) Which plays the major role in forming the elbow joint with the humerus?

?

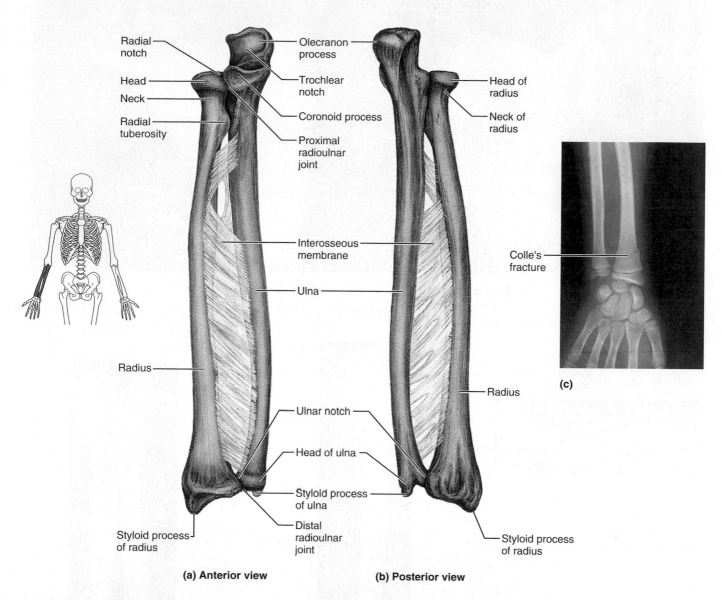

Radial notch
Head
Neck
Radial tuberosity
Olecranon process
Trochlear notch
Coronoid process
Proximal radioulnar joint
Head of radius
Neck of radius
Interosseous membrane
Ulna
Radius
Colle's fracture
Radius
(c)
Ulnar notch
Head of ulna
Styloid process of ulna
Distal radioulnar joint
Styloid process of radius
Styloid process of radius

(a) Anterior view   **(b) Posterior view**

FIGURE 7.24   Bones of the forearm. (a) Anterior view of the radius and ulna of the right forearm, anatomical position. Interosseous membrane also shown. **(b)** Posterior view of the radius and ulna of the right forearm. **(c)** X ray of Colle's fracture of the radius. (See *A Brief Atlas of the Human Body,* Figure 26.)

lates with the radial notch of the ulna. Just inferior to the head is the rough **radial tuberosity,** which anchors the biceps muscle of the arm. Distally, where the radius is expanded, it has a medial **ulnar notch,** which articulates with the ulna, and a lateral **styloid process** (an anchoring site for ligaments that run to

the wrist). Between these two markings, the radius is concave where it articulates with carpal bones of the wrist. While the ulna contributes more heavily to the elbow joint, the radius is the major forearm bone contributing to the wrist joint. When the radius moves, the hand moves with it.

(1) The radius carries the hand. (2) The ulna forms the elbow joint with the humerus.

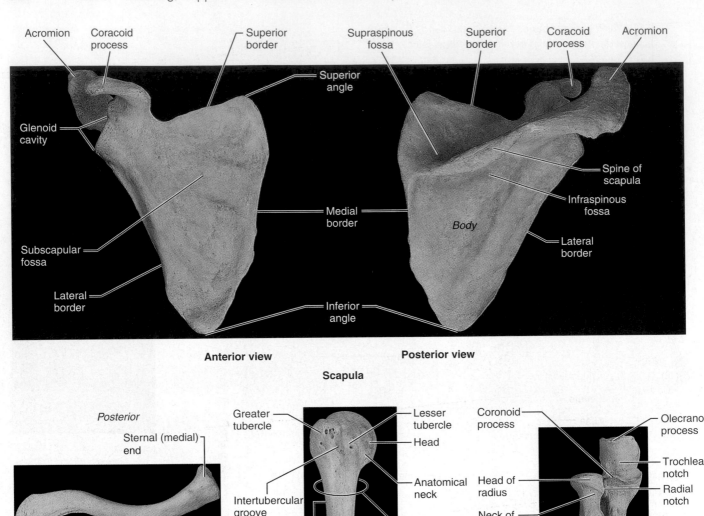

Acromion   Coracoid process   Superior border   Superior angle

Suprspinous fossa   Superior border   Coracoid process   Acromion

Glenoid cavity

Subscapular fossa

Lateral border

Medial border

Inferior angle

Spine of scapula

Infraspinous fossa

Body

Lateral border

Anterior view   **Posterior view**

Scapula

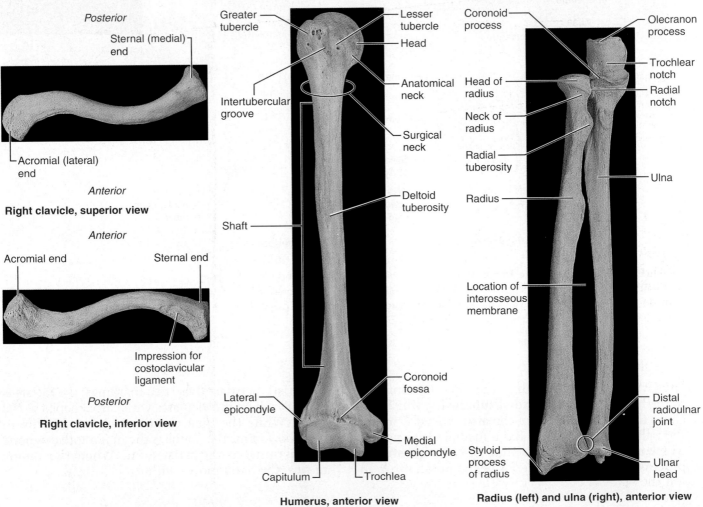

Posterior

Sternal (medial) end

Acromial (lateral) end

Anterior

Right clavicle, superior view

Anterior

Acromial end   Sternal end

Impression for costoclavicular ligament

Posterior

Right clavicle, inferior view

Greater tubercle   Lesser tubercle   Head

Intertubercular groove   Anatomical neck

Surgical neck

Shaft   Deltoid tuberosity

Lateral epicondyle   Coronoid fossa

Capitulum   Trochlea   Medial epicondyle

Humerus, anterior view

Coronoid process   Olecranon process

Head of radius   Trochlear notch   Radial notch

Neck of radius

Radial tuberosity   Ulna

Radius

Location of interosseous membrane

Distal radioulnar joint

Styloid process of radius   Ulnar head

Radius (left) and ulna (right), anterior view

FIGURE 7.25 **Photographs of selected bones of the pectoral girdle and right upper limb.** (See *A Brief Atlas of the Human Body*, Figures 24, 25, and 26.)

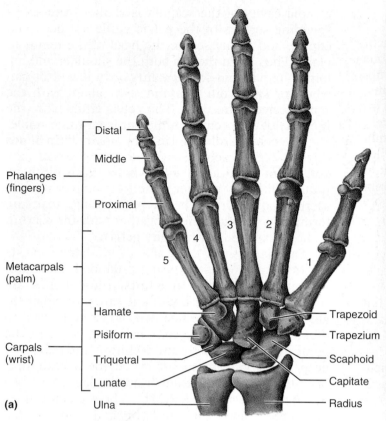

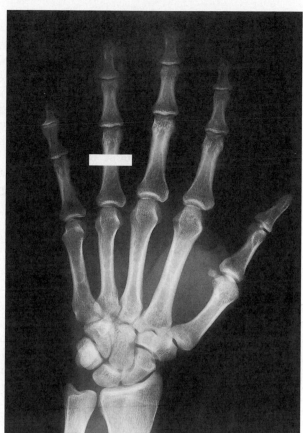

FIGURE 7.26 Bones of the hand. (a) Ventral view of the right hand, illustrating the anatomical relationships of the carpals, metacarpals, and phalanges. **(b)** X ray of the right hand. Notice the white bar on the proximal phalanx of finger 4, showing the position at which a ring would be worn. (See *A Brief Atlas of the Human Body,* Figure 27.)

ⓗ HOMEOSTATIC IMBALANCE

Colle's fracture is a break in the distal end of the radius. It is a common fracture when a falling person attempts to break his or her fall with outstretched hands (see Figure 7.24c). ●

Hand

The skeleton of the hand (Figure 7.26) includes the bones of the *carpus* (wrist); the bones of the *metacarpus* (palm); and the *phalanges* (bones of the fingers).

Carpus (Wrist)

A "wrist" watch is actually worn on the distal forearm (over the lower ends of the radius and ulna), not on the wrist at all. The true wrist, or **carpus,** is the proximal part of the structure we generally call our "hand." The carpus consists of a group of eight mar-ble-size short bones, or **carpals** (kar'palz), closely united by ligaments. Because gliding movements occur between these bones, the carpus as a whole is quite flexible. The carpals are arranged in two irregular rows of four bones each (Figure 7.26). In the proximal row (lateral to medial) are the **scaphoid** (skaf'oid), **lunate** (lu'nāt), **triquetral** (tri-kwe'trul), and **pisiform** (pi'sĭ-form). Only the scaphoid and lunate articulate with the radius to form the wrist joint. The carpals of the distal row (lateral to medial) are the **trapezium** (trah-pe'ze-um), **trapezoid** (tra'peh-zoid), **capitate,** and **hamate** (ham'āt). There are numerous memory-jogging phrases to help you recall the carpals in the order given above. If you don't have one, try: **S**ally **l**eft **t**he **p**arty **t**o **t**ake **C**athy **h**ome. As with all such memory jogs, the first letter of each word is the first letter of the term you need to remember.

H HOMEOSTATIC IMBALANCE

The carpus is concave anteriorly and a ligament roofs over this concavity, forming the notorious *carpal tunnel.* Besides the median nerve (which supplies the lateral side of the hand), several long muscle tendons crowd into this tunnel. Overuse and inflammation of the tendons cause them to swell, compressing the median nerve, which causes numbness of the areas served. Those who repeatedly flex their wrists and fingers, such as those who work at computer keyboards all day, are particularly susceptible to this nerve impairment, called *carpal tunnel syndrome.* ●

Metacarpus (Palm)

Five **metacarpal bones** radiate from the wrist like spokes to form the palm of the hand (*meta* = beyond). These small long bones are not named, but instead are numbered 1 to 5 from thumb to little finger. The **bases** of the metacarpals articulate with the carpals proximally and each other medially and laterally (see Figure 7.26). Their bulbous **heads** articulate with the proximal phalanges of the fingers. When you clench your fist, the heads of the metacarpals become prominent as your *knuckles.*

Metacarpal 1, associated with the thumb, is the shortest and most mobile. It occupies a more anterior position than the other metacarpals. Consequently, the joint between metacarpal 1 and the trapezium is a unique saddle joint that allows *opposition,* the action of touching your thumb to the tips of your other fingers.

Phalanges (Fingers)

The **fingers,** or **digits** of the upper limb, are numbered 1 to 5 beginning with the thumb, or **pollex** (pol'eks). In most people, the third finger is the longest. Each hand contains 14 miniature long bones called **phalanges** (fah-lan'jēz). Except for the thumb, each finger has three phalanges: *distal, middle,* and *proximal.* The thumb has no middle phalanx. [Phalanx (fa'langks) is the singular term for phalanges.]

The Pelvic (Hip) Girdle

The **pelvic girdle,** or **hip girdle,** attaches the lower limbs to the axial skeleton, transmits the weight of the upper body to the lower limbs, and supports the visceral organs of the pelvis (Figures 7.27 and 7.30 and Table 7.4, p. 211). Whereas the pectoral girdle is sparingly attached to the thoracic cage, the pelvic girdle is secured to the axial skeleton by some of the strongest ligaments in the body. And whereas the glenoid cavity of the scapula is shallow, the corresponding sockets of the pelvic girdle are deep and cuplike and firmly secure the head of the femur in place. Thus, even though both the shoulder and hip joints are ball-and-socket joints, very few of us can wheel or swing our legs and arms about with the same degree of freedom. The pelvic girdle lacks the mobility of the pectoral girdle but is far more stable.

The pelvic girdle is formed by a pair of **hip bones** (Figure 7.27), each also called an **os coxae** (ahs kok'se), or **coxal** (*coxa* = hip) **bone.** Each hip bone unites with its partner anteriorly and with the sacrum posteriorly. The deep, basinlike structure formed by the hip bones, together with the sacrum and coccyx, is called the **bony pelvis.**

Each large, irregularly shaped hip bone consists of three separate bones during childhood: the ilium, ischium, and pubis. In adults, these bones are firmly fused and their boundaries are indistinguishable. Their names are retained, however, to refer to different regions of the composite hip bone. At the point of fusion of the ilium, ischium, and pubis is a deep hemispherical socket called the **acetabulum** (as"ĕ-tab'u-lum; "wine cup") on the lateral surface of the pelvis (see Figure 7.27b). The acetabulum receives the head of the femur, or thigh bone, at this *hip joint.*

Ilium

The **ilium** (il'e-um; "flank") is a large flaring bone that forms the superior region of a coxal bone. It consists of a **body** and a superior winglike portion called the **ala** (a'lah). When you rest your hands on your hips, you are resting them on the thickened superior margins of the alae, the **iliac crests.** Thickest at the **tubercle of the iliac crest,** each iliac crest ends anteriorly in the blunt **anterior superior iliac spine** and posteriorly in the sharp **posterior superior iliac spine.** Located below these are the less prominent **anterior** and **posterior inferior iliac spines.** All of these spines are attachment points for the muscles of the trunk, hip, and thigh. The anterior superior iliac spine is an especially important anatomical landmark. It is easily felt through the skin and is visible in thin people. The posterior superior iliac spine is difficult to palpate, but its position is revealed by a skin dimple in the sacral region.

Just inferior to the posterior inferior iliac spine, the ilium indents deeply to form the **greater sciatic** (si-at'ik) **notch,** through which the thick cordlike sciatic nerve passes to enter the thigh. The broad posterolateral surface of the ilium, the **gluteal** (gloo'te-al) **surface,** is crossed by three ridges, the **posterior, anterior,** and **inferior gluteal lines,** to which the gluteal (buttock) muscles attach.

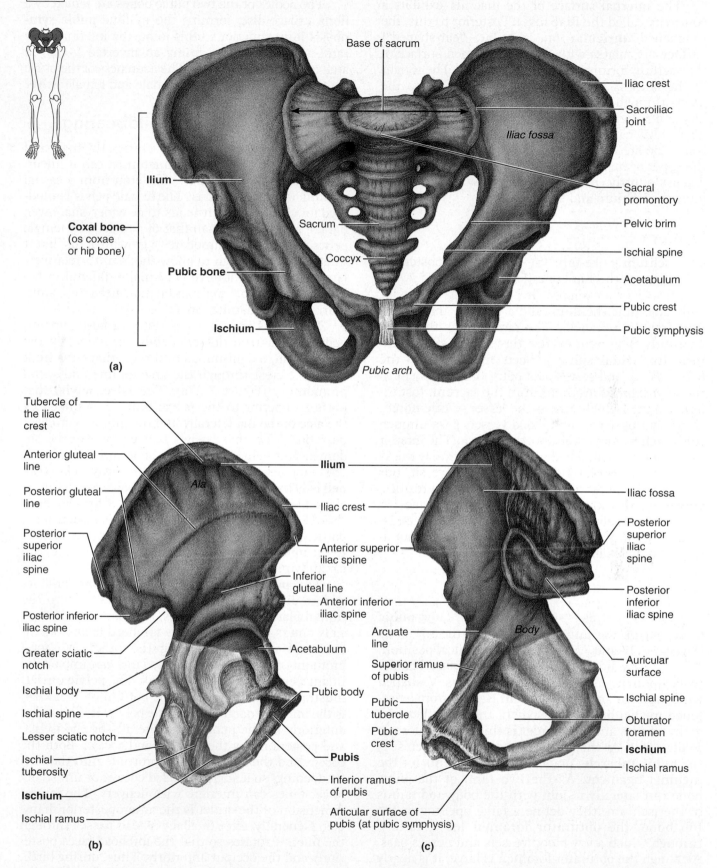

(a)

- Base of sacrum
- Iliac crest
- Sacroiliac joint
- *Iliac fossa*
- Sacral promontory
- Pelvic brim
- Ischial spine
- Acetabulum
- Pubic crest
- Pubic symphysis
- Ilium
- Coxal bone (os coxae or hip bone)
- Pubic bone
- Sacrum
- Coccyx
- Ischium
- *Pubic arch*

(b)

- Tubercle of the iliac crest
- Anterior gluteal line
- *Ala*
- Posterior gluteal line
- Posterior superior iliac spine
- Posterior inferior iliac spine
- Greater sciatic notch
- Ischial body
- Ischial spine
- Lesser sciatic notch
- Ischial tuberosity
- Ischium
- Ischial ramus
- Ilium
- Iliac crest
- Anterior superior iliac spine
- Inferior gluteal line
- Anterior inferior iliac spine
- Acetabulum
- Pubic body
- Pubis
- Inferior ramus of pubis

(c)

- Iliac fossa
- Posterior superior iliac spine
- Posterior inferior iliac spine
- Auricular surface
- Ischial spine
- Obturator foramen
- Ischium
- Ischial ramus
- Arcuate line
- *Body*
- Superior ramus of pubis
- Pubic tubercle
- Pubic crest
- Articular surface of pubis (at pubic symphysis)

FIGURE 7.27 Bones of the bony pelvis. (a) Articulated pelvis showing the two hip (coxal) bones (which together form the pelvic girdle) and the sacrum. **(b)** Lateral view of right hip bone showing the point of fusion of the ilium (gold), ischium (blue), and pubic (pink) bones at the acetabulum. **(c)** Right hip bone, medial view. (See *A Brief Atlas of the Human Body,* Figure 28.)

The internal surface of the iliac ala exhibits a concavity called the **iliac fossa.** Posterior to this, the roughened **auricular** (aw-rik'u-lar; "ear-shaped") **surface** articulates with the same-named surface of the sacrum, forming the *sacroiliac joint.* The weight of the body is transmitted from the spine to the pelvis through the sacroiliac joints. Running inferiorly and anteriorly from the auricular surface is a robust ridge called the **arcuate** (ar'ku-at; "bowed") **line.** The arcuate line helps define the **pelvic brim,** the superior margin of the *true pelvis,* which we will discuss shortly. Anteriorly, the body of the ilium joins the ischium and pubis.

Ischium

The **ischium** (is'ke-um; "hip") forms the posteroinferior part of the hip bone (Figures 7.27 and 7.30). Roughly L- or arc-shaped, it has a thicker, superior **body** adjoining the ilium and a thinner, inferior **ramus** (*ramus* = branch). The ramus joins the pubis anteriorly. The ischium has three important markings. Its **ischial spine** projects medially into the pelvic cavity and serves as a point of attachment of the *sacrospinous ligament* from the sacrum. Just inferior to the ischial spine is the **lesser sciatic notch.** A number of nerves and blood vessels pass through this notch to supply the anogenital area. The inferior surface of the ischial body is rough and grossly thickened as the **ischial tuberosity.** When we sit, our weight is borne entirely by the ischial tuberosities, which are the strongest parts of the hip bones. A massive ligament runs from the sacrum to each ischial tuberosity. This *sacrotuberous ligament* (not illustrated) helps hold the pelvis together.

Pubis

The **pubis** (pu'bis; "sexually mature"), or **pubic bone,** forms the anterior portion of the hip bone (Figures 7.27 and 7.30). In the anatomical position, it lies nearly horizontally and the urinary bladder rests upon it. Essentially, the pubis is V shaped with **superior** and **inferior rami** issuing from a flattened medial **body.** The body of the pubis lies medially and its anterior border is thickened to form a **pubic crest.** At the lateral end of the pubic crest is the **pubic tubercle,** one of the attachments for the inguinal ligament. As the two rami of the pubic bone run laterally to join with the body and ramus of the ischium, they define a large opening in the hip bone, the **obturator foramen** (ob"tu-ra'tor), through which a few blood vessels and nerves pass. Although the obturator foramen is large, it is nearly closed by a fibrous membrane in life (*obturator* = closed up).

The bodies of the two pubic bones are joined by a fibrocartilage disc, forming the midline **pubic symphysis joint.** Inferior to this joint, the inferior pubic rami angle laterally, forming an inverted V-shaped arch called the **pubic arch.** The acuteness of the pubic arch helps to differentiate the male and female pelve.

Pelvic Structure and Childbearing

So striking are the differences between the male and female pelves that a trained anatomist can immediately determine the sex of a skeleton from a casual examination of the pelvis. The female pelvis is modified for childbearing: It tends to be wider, shallower, lighter, and rounder than that of a male. The female pelvis not only accommodates a growing fetus, but it must be large enough to allow the infant's relatively large head to exit at birth. The major differences between the "typical" male and female pelves are summarized and illustrated in Table 7.4.

The pelvis is said to consist of a false (greater) pelvis and a true (lesser) pelvis separated by the **pelvic brim,** a continuous oval ridge that runs from the pubic crest through the arcuate line and sacral promontory (Figure 7.27a). The **false pelvis,** that portion superior to the pelvic brim, is bounded by the alae of the ilia laterally and the lumbar vertebrae posteriorly. The false pelvis is really part of the abdomen and helps support the abdominal viscera. It does not restrict childbirth in any way. The **true pelvis** is the region inferior to the pelvic brim that is almost entirely surrounded by bone. It forms a deep "bowl" containing the pelvic organs. Its dimensions, particularly those of its *inlet* and *outlet,* are critical to the uncomplicated delivery of a baby, and they are carefully measured by an obstetrician.

The **pelvic inlet** *is* the **pelvic brim,** and its widest dimension is from right to left along the frontal plane. As labor begins, an infant's head typically enters the inlet with its forehead facing one ilium and its occiput facing the other. A sacral promontory that is particularly large can impair the infant's entry into the true pelvis. The **pelvic outlet,** illustrated in the photos at the bottom of Table 7.4, is the inferior margin of the true pelvis. It is bounded anteriorly by the pubic arch, laterally by the ischia, and posteriorly by the sacrum and coccyx. Both the coccyx and the ischial spines protrude into the outlet opening, so a sharply angled coccyx or unusually large spines can interfere with delivery. The largest dimension of the outlet is the anteroposterior diameter. Generally, after the baby's head passes through the inlet, it rotates so that the forehead faces posteriorly and the occiput anteriorly. Thus, during birth, the infant's head makes a quarter turn to follow the widest dimensions of the true pelvis.

TABLE 7.4   Comparison of the Male and Female Pelves

Characteristic	Female	Male
General structure and functional modifications	Tilted forward; adapted for childbearing; true pelvis defines the birth canal; cavity of the true pelvis is broad, shallow, and has a greater capacity	Tilted less far forward; adapted for support of a male's heavier build and stronger muscles; cavity of the true pelvis is narrow and deep
Bone thickness	Less; bones lighter, thinner, and smoother	Greater; bones heavier and thicker, and markings are more prominent
Acetabula	Smaller; farther apart	Larger; closer
Pubic arch/angle	Broader (80–90°); more rounded	More acute (50–60°)
Anterior view		

Pelvic brim — Pubic arch

Sacrum	Wider; shorter; sacral curvature is accentuated	Narrow; longer; sacral promontory more ventral
Coccyx	More movable; straighter	Less movable; curves ventrally
Left lateral view		
Pelvic inlet (brim)	Wider; oval from side to side	Narrow; basically heart shaped
Pelvic outlet	Wider; ischial tuberosities shorter, farther apart and everted	Narrower; ischial tuberosities longer, sharper, and point more medially
Posteroinferior view		

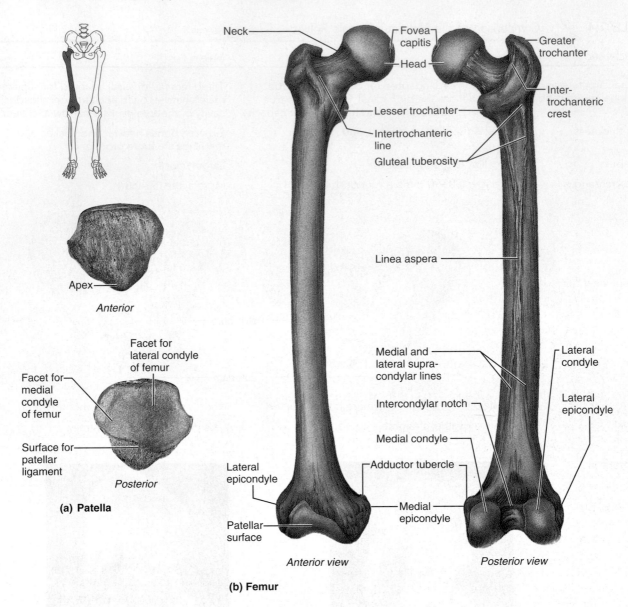

FIGURE 7.28   **Bones of the right thigh and knee. (a)** The patella (kneecap).
(b) The femur (thigh bone). (See *A Brief Atlas of the Human Body*, Figure 29.)

The Lower Limb

The lower limbs carry the entire weight of the erect body and are subjected to exceptional forces when we jump or run. Thus, it is not surprising that the bones of the lower limbs are thicker and stronger than comparable bones of the upper limbs. The three segments of each lower limb are the thigh, the leg, and the foot (see Table 7.5 on p. 217).

Thigh

The **femur** (fe′mur; "thigh"), the single bone of the thigh (Figures 7.28 and 7.30), is the largest, longest, strongest bone in the body. Its durable structure reflects the fact that the stress on the femur during vigorous jumping can reach 280 kg/cm² (about 2 tons per square inch)! The femur is clothed by bulky muscles that prevent us from palpating its course down the length of the thigh. Its length is roughly one-quarter of a person's height.

Proximally, the femur articulates with the hip bone and then courses medially as it descends toward the knee. This arrangement allows the knee joints to be closer to the body's center of gravity and provides for better balance. The medial course of the two femurs is more pronounced in women because of their wider pelvis.

The ball-like **head** of the femur has a small central pit called the **fovea capitis** (fo′ve-ah kă′pĭ-tis; "pit of the head"). The short *ligament of the head of the femur*, also called the *ligamentum teres* (tĕ′rēz), runs from this pit to the acetabulum, where it helps secure the femur. The head is carried on a *neck* that angles *laterally* to join the shaft. This arrangement reflects the fact that the femur articulates with the lateral aspect (rather than the inferior region) of the pelvis. The neck is the weakest part of the femur and is often fractured, an injury commonly called a broken hip. At the junction of the shaft and neck are the lateral **greater trochanter** (tro-kan'ter) and posteromedial **lesser trochanter.** These projections serve as sites of attachment for thigh and buttock muscles. The two trochanters are connected by the **intertrochanteric line** anteriorly and by the prominent **intertrochanteric crest** posteriorly.

Inferior to the intertrochanteric crest on the posterior shaft is the **gluteal tuberosity,** which blends into a long vertical ridge, the **linea aspera** (lin'e-ah as'per-ah; "rough line"), inferiorly. Distally, the linea aspera diverges, forming the medial and lateral supracondylar lines. All of these markings are sites of muscle attachment. Except for the linea aspera, the femur shaft is smooth and rounded.

Distally, the femur broadens and ends in the wheel-like **lateral** and **medial condyles,** which articulate with the tibia of the leg. The **medial** and **lateral epicondyles** (sites of muscle attachment) flank the condyles superiorly. On the superior part of the medial epicondyle is a bump, the **adductor tubercle.** The smooth **patellar surface,** between the condyles on the anterior femoral surface, articulates with the patella (pah-tel'ah), or kneecap (see Figures 7.28a, 7.30, and 7.21).

Between the condyles on the posterior aspect of the femur is the deep, U-shaped **intercondylar notch,** and superior to that on the shaft is the smooth popliteal surface.

The **patella** (pah-tel'ah; "small pan") is a triangular sesamoid bone enclosed in the (quadriceps) tendon that secures the anterior thigh muscles to the tibia. It protects the knee joint anteriorly and improves the leverage of the thigh muscles acting across the knee.

Leg

Two parallel bones, the tibia and fibula, form the skeleton of the leg, the region of the lower limb between the knee and the ankle (Figures 7.29 and 7.30). These two bones are connected by an *interosseous membrane* and articulate with each other both proximally and distally. However, unlike the joints between the radius and ulna of the forearm, the *tibiofibular joints* (tib"e-o-fib'u-lar) of the leg allow essentially no movement. The bones of the leg thus form a less flexible but stronger and more stable limb than those of the forearm. The medial tibia articulates proximally with the femur to form the modified hinge joint of the knee and distally with the talus bone of the foot at the ankle. The fibula, by contrast, does not contribute to the knee joint and merely helps stabilize the ankle joint.

Tibia

The **tibia** (tib'e-ah; "shinbone") receives the weight of the body from the femur and transmits it to the foot. It is second only to the femur in size and strength. At its broad proximal end are the concave **medial** and **lateral condyles,** which look like two huge checkers lying side by side. These are separated by an irregular projection, the **intercondylar eminence.** The tibial condyles articulate with the corresponding condyles of the femur. The inferior region of the lateral tibial condyle bears a facet that indicates the site of the *proximal tibiofibular joint.* Just inferior to the condyles, the tibia's anterior surface displays the rough **tibial tuberosity,** to which the patellar ligament attaches.

The tibial shaft is triangular in cross section. Its anterior border is the sharp **anterior crest.** Neither this crest nor the tibia's medial surface is covered by muscles, so they can be felt just deep to the skin along their entire length. The anguish of a "bumped" shin is an experience familiar to nearly everyone. Distally the tibia is flat where it articulates with the talus bone of the foot. Medial to that joint surface is an inferior projection, the **medial malleolus** (mahle'o-lus; "little hammer"), which forms the medial bulge of the ankle. The **fibular notch,** on the lateral surface of the tibia, participates in the *distal tibiofibular joint.*

Fibula

The **fibula** (fib'u-lah; "pin") is a sticklike bone with slightly expanded ends. It articulates proximally and distally with the lateral aspects of the tibia. Its proximal end is its **head;** its distal end is the **lateral malleolus.** The lateral malleolus forms the conspicuous lateral ankle bulge and articulates with the talus. The fibular shaft is heavily ridged and appears to have been twisted a quarter turn. The fibula does not bear weight, but several muscles originate from it.

 HOMEOSTATIC IMBALANCE

A *Pott's fracture* occurs at the distal end of the fibula, the tibia, or both. It is a common sports injury. (See Figure 7.29b.) ●

Foot

The skeleton of the foot includes the bones of the *tarsus,* the bones of the *metatarsus,* and the *phalanges,* or toe bones (Figure 7.31). The foot has two important functions: It supports our body weight,

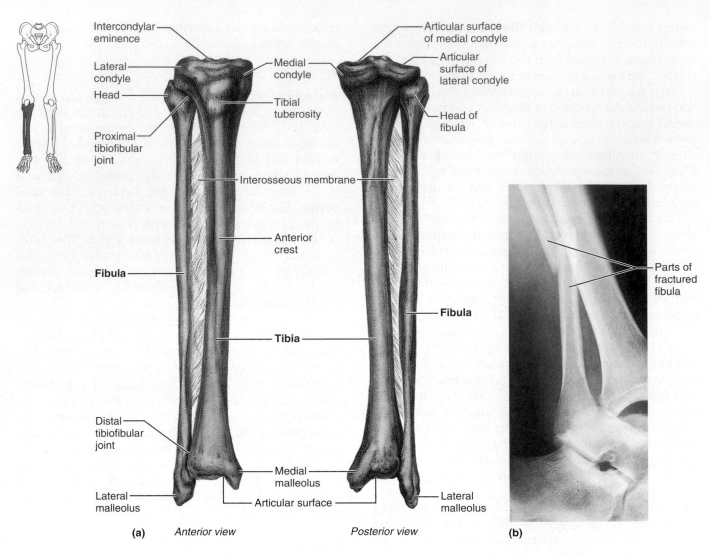

FIGURE 7.29   The tibia and fibula of the right leg. (a) Anterior (left) and posterior (right) views. **(b)** X ray of Pott's fracture of the fibula. (See *A Brief Atlas of the Human Body*, Figure 30.)

and it acts as a lever to propel the body forward when we walk and run. A single bone could serve both purposes, but it would adapt poorly to uneven ground. Segmentation makes the foot pliable, avoiding this problem.

Tarsus

The **tarsus** is made up of seven **tarsal** (tar′sal) **bones** that form the posterior half of the foot. It corresponds to the carpus of the hand. Body weight is carried primarily by the two largest, most posterior tarsals: the **talus** (ta′lus; "ankle"), which articulates with the tibia and fibula superiorly, and the strong **calcaneus** (kal-ka′ne-us; "heel bone"), which forms the heel of the foot and carries the talus on its superior surface. The thick *calcaneal*, or *Achilles, tendon* of the calf muscles attaches to the posterior surface of the calcaneus. The part of the calcaneus that touches the ground is the *tuber calcanei* (kal-ka′ne-

i), the calcaneal tuberosity, and its shelflike projection that supports part of the talus is the *sustentaculum tali* (sus″ten-tak′u-lum ta′le; "supporter of the talus"). The remaining tarsals are the lateral **cuboid**, the medial **navicular** (nah-vik′u-lar), and the anterior **medial, intermediate,** and **lateral cuneiform** (ku-ne′ĭ-form; "wedge-shaped") **bones.** The cuboid and cuneiform bones articulate with the metatarsal bones anteriorly.

Metatarsus

The **metatarsus** consists of five, small, long bones called **metatarsal bones.** These are numbered 1 to 5 beginning on the medial (great toe) side of the foot. The first metatarsal is short and thick and its plantar surface rests on paired sesamoid bones (not shown), which play an important role in supporting body weight. The arrangement of the metatarsals is more parallel than that of the metacarpals of the hands.

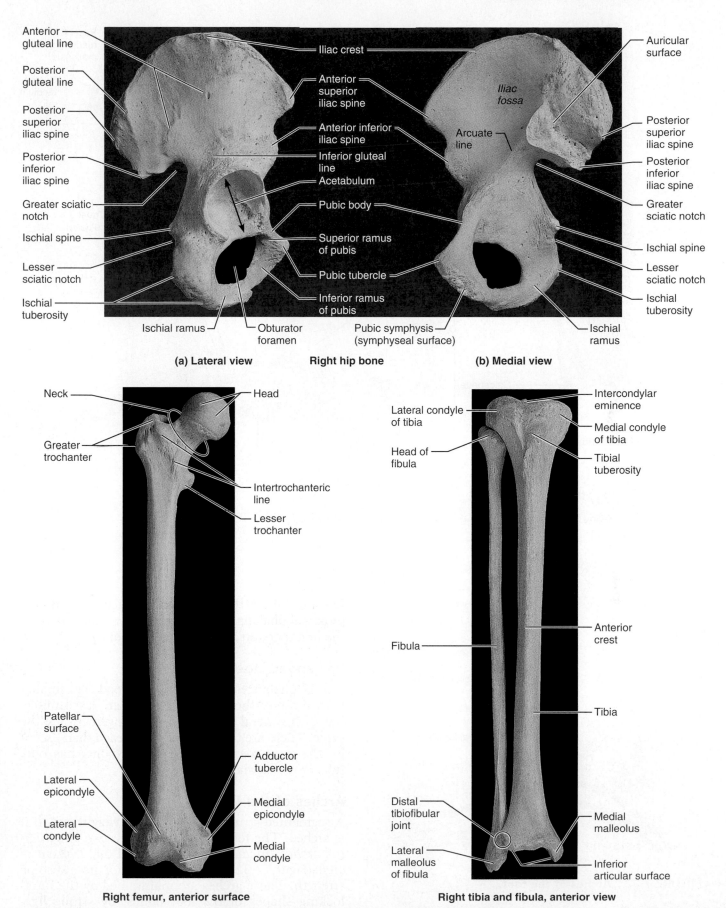

Anterior gluteal line

Posterior gluteal line

Posterior superior iliac spine

Posterior inferior iliac spine

Greater sciatic notch

Ischial spine

Lesser sciatic notch

Ischial tuberosity

Ischial ramus

Obturator foramen

Iliac crest

Anterior superior iliac spine

Anterior inferior iliac spine

Inferior gluteal line

Acetabulum

Pubic body

Superior ramus of pubis

Pubic tubercle

Inferior ramus of pubis

Pubic symphysis (symphyseal surface)

(a) Lateral view

Right hip bone

Auricular surface

Iliac fossa

Arcuate line

Posterior superior iliac spine

Posterior inferior iliac spine

Greater sciatic notch

Ischial spine

Lesser sciatic notch

Ischial tuberosity

Ischial ramus

(b) Medial view

Neck

Greater trochanter

Patellar surface

Lateral epicondyle

Lateral condyle

Head

Intertrochanteric line

Lesser trochanter

Adductor tubercle

Medial epicondyle

Medial condyle

Right femur, anterior surface

Lateral condyle of tibia

Head of fibula

Fibula

Distal tibiofibular joint

Lateral malleolus of fibula

Intercondylar eminence

Medial condyle of tibia

Tibial tuberosity

Anterior crest

Tibia

Medial malleolus

Inferior articular surface

Right tibia and fibula, anterior view

FIGURE 7.30 **Photographs of selected bones of the pelvic girdle and lower limb.** (See *A Brief Atlas of the Human Body,* Figures 28, 29, and 30.)

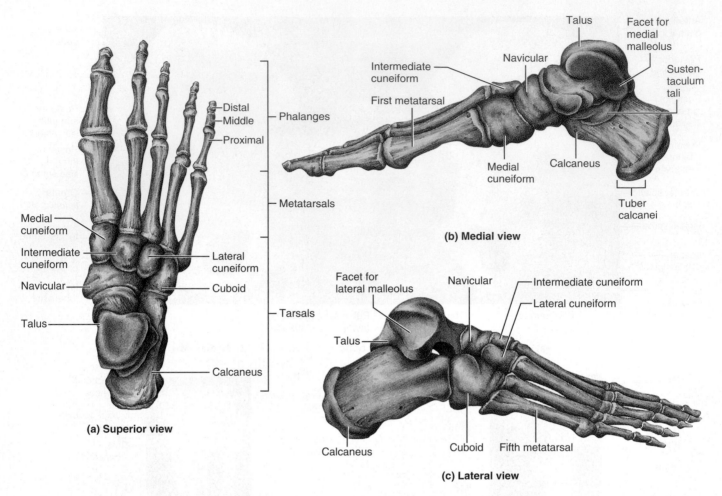

FIGURE 7.31 **Bones of the right foot.** (See *A Brief Atlas of the Human Body*, Figure 31a, c, and d.)

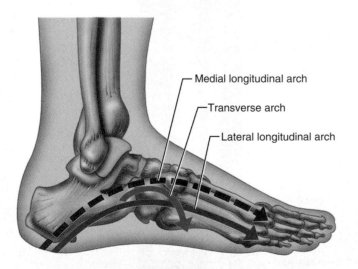

FIGURE 7.32 **Arches of the foot.**

Distally, where the metatarsals articulate with the proximal phalanges of the toes, the enlarged head of the first metatarsal forms the "ball" of the foot.

Phalanges (Toes)

The 14 phalanges of the toes are a good deal smaller than those of the fingers and thus are less nimble. But their general structure and arrangement are the same. There are three phalanges in each digit except for the great toe, the **hallux.** The hallux has only two, proximal and distal.

Arches of the Foot

A segmented structure can hold up weight only if it is arched. The foot has three arches: two *longitudinal arches* (*medial* and *lateral*) and one *transverse arch* (Figure 7.32), which account for its awesome strength. These arches are maintained by the interlocking shapes of the foot bones, by strong ligaments, and by the pull of some tendons during muscle activity. The ligaments and muscle tendons

TABLE 7.5	Bones of the Appendicular Skeleton Part 2: Pelvic Girdle and Lower Limb			
Body Region	*Bones**	*Illustration*	*Location*	*Markings*
Pelvic girdle (Figures 7.27, 7.30)	Coxal (2) (hip)		Each hip bone is formed by the fusion of an ilium, an ischium, and a pubic bone; the hip bones fuse anteriorly at the pubic symphysis and form sacroiliac joints with the sacrum posteriorly; girdle consisting of both hip bones is basinlike	Iliac crest; anterior and posterior iliac spines; auricular surface; greater and lesser sciatic notches; obturator foramen; ischial tuberosity and spine; acetabulum; pubic arch; pubic crest; pubic tubercle
Free lower limb Thigh (Figures 7.28b, 7.30)	Femur (2)		Femur is the only bone of thigh; between hip joint and knee; largest bone of the body	Head; greater and lesser trochanters; neck; lateral and medial condyles and epicondyles; adductor tubercle; gluteal tuberosity; linea aspera
Kneecap (Figure 7.28a)	Patella (2)		Patella is a sesamoid bone lodged in the tendon of the quadriceps (anterior thigh) muscles	
Leg (Figures 7.29, 7.30)	Tibia (2)		Tibia is the larger and more medial bone of leg; between knee and foot	Medial and lateral condyles; tibial tuberosity; anterior crest; medial malleolus
	Fibula (2)		Fibula is the lateral bone of leg; sticklike	Head; lateral malleolus
Foot (Figure 7.31)	7 Tarsals (14) • talus • calcaneus • navicular • cuboid • lateral cuneiform • intermediate cuneiform • medial cuneiform		Seven tarsal bones form the proximal part of the foot; the talus articulates with the leg bones at the ankle joint; the calcaneus, the largest tarsal, forms the heel	
	5 Metatarsals (10)		Metatarsals are five bones numbered 1–5 from the great toe	
	14 Phalanges (28) • distal • middle • proximal		Phalanges form the toes; three in digits 2–5, two in digit 1 (the great toe)	

Anterior view of pelvic girdle and left lower limb

*The number in parentheses ( ) following the bone name denotes the total number of such bones in the body.

provide a certain amount of springiness. In general, the arches "give" or stretch slightly when weight is applied to the foot and spring back when the weight is removed, which makes walking and running more economical in terms of energy use than would otherwise be the case.

If you examine your wet footprints, you will see that the medial margin from the heel to the head of the first metatarsal leaves no print. This is because the **medial longitudinal arch** curves well above the ground. The talus is the keystone of this arch, which originates at the calcaneus, rises to the talus, and then descends to the three medial metatarsals. The **lateral longitudinal arch** is very low. It elevates the lateral part of the foot just enough to redistribute some of the weight to the calcaneus and the head of the fifth metatarsal (to the ends of the arch). The cuboid is the keystone bone of this arch. The two longitudinal arches serve as pillars for the **transverse arch,** which runs obliquely from one side of the foot to the other, following the line of the joints between the tarsals and metatarsals. Together, the arches of the foot form a half-dome that distributes about half our standing and walking weight to the heel bones and half to the heads of the metatarsals.

H HOMEOSTATIC IMBALANCE

Standing immobile for extended periods places excessive strain on the tendons and ligaments of the feet (because the muscles are inactive) and can result in fallen arches, or "flat feet," particularly if one is overweight. Running on hard surfaces can also cause arches to fall unless one wears shoes that give proper arch support. ●

★ ★ ★

The bones of the thigh, leg, and foot are summarized in Table 7.5.

Review Questions

Multiple Choice/Matching

(Some questions have more than one correct answer. Select the best answer or answers from the choices given.)

1. Using the letters from column B, match the bone descriptions in column A. (Note that some require more than a single choice.)

Column A	Column B
____ (1) connected by the coronal suture	(a) ethmoid
____ (2) keystone bone of cranium	(b) frontal
____ (3) keystone bone of the face	(c) mandible
____ (4) form the hard palate	(d) maxillary
____ (5) allows the spinal cord to pass	(e) occipital
____ (6) forms the chin	(f) palatine
____ (7) contain paranasal sinuses	(g) parietal
____ (8) contains mastoid sinuses	(h) sphenoid
	(i) temporal

2. Match the key terms with the bone descriptions that follow.

Key: (a) clavicle   (b) ilium   (c) ischium   (d) pubis
      (e) sacrum   (f) scapula   (g) sternum

____ (1) bone of the axial skeleton to which the pectoral girdle attaches
____ (2) markings include glenoid cavity and acromion
____ (3) features include the ala, crest, and greater sciatic notch
____ (4) doubly curved; acts as a shoulder strut
____ (5) pelvic girdle bone that articulates with the axial skeleton
____ (6) the "sit-down" bone
____ (7) anteriormost bone of the pelvic girdle
____ (8) part of the vertebral column

3. Use key choices to identify the bone descriptions that follow.

Key: (a) carpals   (b) femur   (c) fibula   (d) humerus
      (e) radius   (f) tarsals   (g) tibia   (h) ulna

____ (1) articulates with the acetabulum and the tibia
____ (2) forms the lateral aspect of the ankle
____ (3) bone that "carries" the hand
____ (4) the wrist bones
____ (5) end shaped like a monkey wrench
____ (6) articulates with the capitulum of the humerus
____ (7) largest bone of this "group" is the calcaneus

Short Answer Essay Questions

4. Name the cranial and facial bones and compare and contrast the functions of the cranial and facial skeletons.

5. Name and diagram the normal vertebral curvatures. Which are primary and which are secondary curvatures?

6. List at least two specific anatomical characteristics each for typical cervical, thoracic, and lumbar vertebrae that would allow *anyone* to identify each type correctly.

7. (a) What is the function of the intervertebral discs? (b) Distinguish between the annulus fibrosis and nucleus pulposus regions of a disc. (c) Which provides durability and strength? (d) Which provides resilience? (e) Which part herniates in a "slipped" disc?

8. Name the major components of the bony thorax.

9. (a) What is a true rib? A false rib? (b) Is a floating rib a true rib or a false rib? (c) Why are floating ribs easily broken?

10. The major function of the shoulder girdle is flexibility. What is the major function of the pelvic girdle? Relate these functional differences to anatomical differences seen in these girdles.

11. List three important differences between the male and female pelvis.

12. Describe the function of the arches of the foot.

13. Briefly describe the anatomical characteristics and impairment of function seen in cleft palate and hip dysplasia.

14. Peter Howell, a teaching assistant in the anatomy class, picked up a hip bone and pretended it was a telephone. He held the big hole in this bone right up to his ear and said, "Hello, obturator, obturator (operator, operator)." Name the structure he was helping the student to learn.

8

JOINTS

1. Define joint or articulation.

Classification of Joints (p. 220)

2. Classify joints structurally and functionally.

Fibrous Joints (pp. 220–221)

3. Describe the general structure of fibrous joints. Name and give an example of each of the three common types of fibrous joints.

Cartilaginous Joints (pp. 221–222)

4. Describe the general structure of cartilaginous joints. Name and give an example of each of the two common types of cartilaginous joints.

Synovial Joints (pp. 222–238)

5. Describe the structural characteristics of synovial joints.

6. List three natural factors that stabilize synovial joints.

7. Compare the structures and functions of bursae and tendon sheaths.

8. Name and describe (or perform) the common body movements.

9. Name and provide examples of the six types of synovial joints based on the movement(s) allowed.

10. Describe the elbow, knee, hip, and shoulder joints relative to articulating bones, anatomical characteristics of the joint, movements allowed, and joint stability.

Homeostatic Imbalances of Joints (pp. 239–241)

11. Name the most common joint injuries and discuss the symptoms and problems associated with each.

12. Compare and contrast the common types of arthritis.

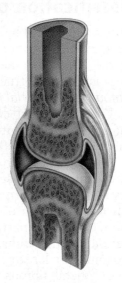

The graceful movements of ballet dancers and the rough-and-tumble grapplings of football players demonstrate the great variety of motion allowed by **joints,** or **articulations**—the sites where two or more bones meet. Our joints have two fundamental functions: They give our skeleton mobility, and they hold it together, sometimes playing a protective role in the process.

Joints are the weakest parts of the skeleton. Nonetheless, their structure resists various forces, such as crushing or tearing, that threaten to force them out of alignment.

Classification of Joints

Joints are classified by structure and by function. The *structural classification* focuses on the material binding the bones together and whether or not a joint cavity is present. Structurally, there are *fibrous, cartilaginous,* and *synovial joints* (Table 8.1 on p. 223).

The *functional classification* is based on the amount of movement allowed at the joint. On this basis, there are **synarthroses** (sin"ar-thro'sēz), which are immovable joints (*syn* = together; *arthro* = joint); **amphiarthroses** (am"fe-ar-thro'sēz), slightly movable joints (*amphi* = on both sides); and **diarthroses** (di"ar-thro'sēz), or freely movable joints (*dia* = through, apart). Freely movable joints predominate in the limbs; immovable and slightly movable joints are largely restricted to the axial skeleton.

In general, fibrous joints are immovable, and synovial joints are freely movable. However, cartilaginous joints have both rigid and slightly movable examples. Since the structural categories are more clear-cut, we will use the structural classification in this discussion, indicating functional properties where appropriate.

Fibrous Joints

In **fibrous joints,** the bones are joined by fibrous tissue; no joint cavity is present. The amount of movement allowed depends on the length of the connective tissue fibers uniting the bones. Although a few are slightly movable, most fibrous joints are immovable. The three types of fibrous joints are *sutures, syndesmoses,* and *gomphoses.*

Sutures

Sutures, literally "seams," occur only between bones of the skull (Figure 8.1a). The wavy articulating bone edges interlock, and the junction is completely filled by a minimal amount of very short connective tissue fibers that are continuous with the periosteum. The result is nearly rigid splices that bind the bones tightly together, yet allow the bones to grow at their

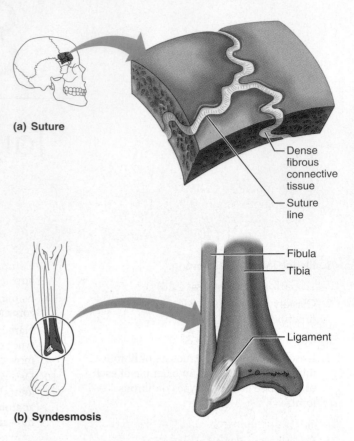

(a) Suture

— Dense fibrous connective tissue
— Suture line

(b) Syndesmosis

— Fibula
— Tibia
— Ligament

FIGURE 8.1 Examples of fibrous joints. (a) In sutures of the skull, interconnecting connective tissue fibers are very short, and the bone edges interlock so that the joint is immovable (synarthrotic). **(b)** In the syndesmosis at the distal tibiofibular joint, the fibrous tissue (ligament) connecting the bones is longer than that in sutures; "give," but no true movement, is permitted.

edges during youth. During middle age, the fibrous tissue ossifies and the skull bones fuse into a single unit. At this stage, the sutures are more precisely called **synostoses** (sin"os-to'sēz), literally, "bony junctions." Because movement of the cranial bones would damage the brain, the immovable nature of sutures is a protective adaptation.

Syndesmoses

In **syndesmoses** (sin"des-mo'sēz), the bones are connected by a *ligament* (*syndesmos* = ligament), a cord or band of fibrous tissue. Although the connecting fibers are always longer than those in sutures, they vary quite a bit in length. The amount of movement allowed depends on the length of the connecting fibers, and slight to considerable movement is possible. For example, the ligament connecting the distal ends of the tibia and fibula is short (Figure 8.1b), and this joint allows only slightly more movement than a suture, a characteristic best described as "give." True movement is still prevented, so the joint is

? *Which one of the joint types shown here has the greatest flexibility and why?*

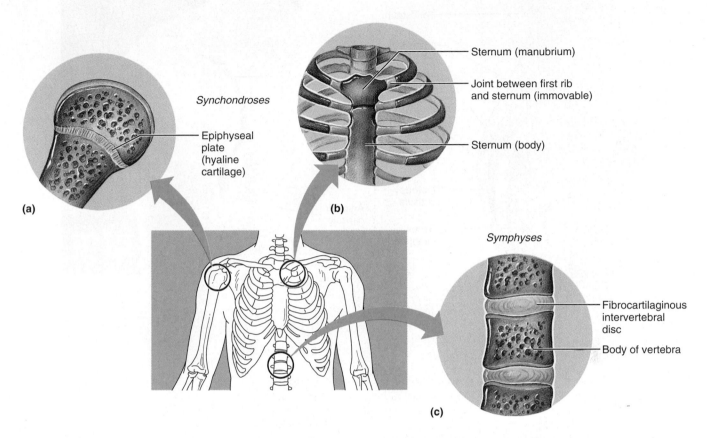

FIGURE 8.2 **Cartilaginous joints. (a)** The epiphyseal plate seen in a growing long bone is a temporary synchondrosis; the shaft and epiphysis are connected by hyaline cartilage that is later completely ossified. **(b)** The sternocostal joint between rib 1 and the manubrium of the sternum is an immovable hyaline cartilage joint, or synchondrosis. **(c)** The intervertebral joints, in which the vertebrae are connected by fibrocartilaginous discs, are symphyses.

classed functionally as an immovable joint, or synarthrosis. (Note, however, that some authorities classify this joint as an amphiarthrosis.) On the other hand, the fibers of the ligament-like interosseous membrane connecting the radius and ulna along their length (Figure 7.24, p. 205) are long enough to permit rotation of the radius around the ulna.

Gomphoses

A **gomphosis** (gom-fo'sis) is a peg-in-socket fibrous joint. The only example is the articulation of a tooth with its bony alveolar socket. The term *gomphosis* comes from the Greek *gompho,* meaning "nail" or "bolt," and refers to the way teeth are embedded in

their sockets (as if hammered in). The fibrous connection in this case is the short **periodontal ligament** (see Figure 22.11 on p. 764).

Cartilaginous Joints

In **cartilaginous joints** (kar"ti-laj'ĭ-nus) the articulating bones are united by cartilage. Like fibrous joints, they lack a joint cavity. The two types of cartilaginous joints are *synchondroses* and *symphyses.*

Synchondroses

A bar or plate of *hyaline cartilage* unites the bones at a **synchondrosis** (sin"kon-dro'sis; "junction of cartilage"). Virtually all synchondroses are synarthrotic.

The most common examples of synchondroses are the epiphyseal plates connecting the diaphysis and epiphysis regions in long bones of children (Figure 8.2a). Epiphyseal plates are temporary joints and

The symphyses, represented by the intervertebral joints in (c), are most flexible because the uniting cartilage in this joint type is the resilient fibrocartilage, versus the hyaline cartilage seen in (a) and (b). ■

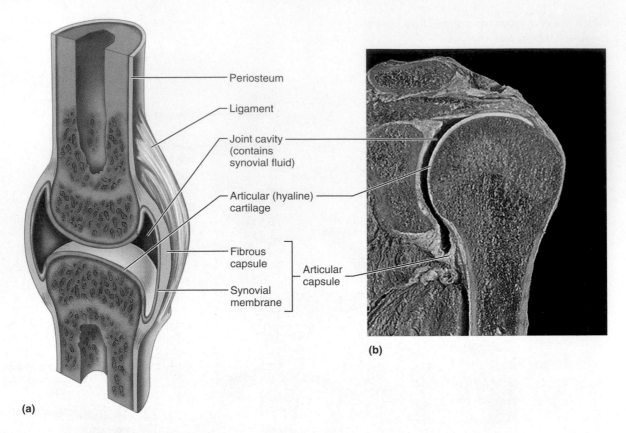

Periosteum

Ligament

Joint cavity
(contains
synovial fluid)

Articular (hyaline)
cartilage

Fibrous
capsule

Synovial
membrane

Articular
capsule

(a)

(b)

FIGURE 8.3   **General structure of a synovial joint.** **(a)** The articulating bone ends are covered with articular cartilage and enclosed within an articular capsule. The exterior (fibrous) portion of the articular capsule is typically reinforced by ligaments. Internally, the fibrous capsule is lined with a smooth synovial membrane that secretes synovial fluid. **(b)** Photo of frontally sectioned shoulder joint.

eventually become synostoses. Another example of a synchondrosis is the immovable joint between the costal cartilage of the first rib and the manubrium of the sternum (Figure 8.2b).

Symphyses

In **symphyses** (sim′fih-sēz; "growing together") the articular surfaces of the bones are covered with articular (hyaline) cartilage, which in turn is fused to an intervening pad, or plate, of *fibrocartilage*. Since fibrocartilage is compressible and resilient, it acts as a shock absorber and permits a limited amount of movement at the joint. Symphyses are amphiarthrotic joints designed for strength with flexibility. Examples include the intervertebral joints (Figure 8.2c) and the pubic symphysis of the pelvis (see Table 8.2 on p. 226).

Synovial Joints

Synovial joints are those in which the articulating bones are separated by a fluid-containing joint cav-

ity. This arrangement permits substantial freedom of movement, and all synovial joints are freely movable diarthroses. All joints of the limbs—indeed, most joints of the body—fall into this class.

General Structure

Synovial joints have five distinguishing features (Figure 8.3a):

1. Articular cartilage. Glassy-smooth articular (hyaline) cartilage covers the opposing bone surfaces. These thin (1 mm or less) but spongy cushions absorb compression placed on the joint and thereby keep the bone ends from being crushed.

2. Joint (synovial) cavity. A feature unique to synovial joints, the joint cavity is really just a potential space that contains a small amount of synovial fluid.

3. Articular capsule. The joint cavity is enclosed by a two-layered articular (joint) capsule. The external layer is a tough **fibrous capsule,** composed of dense irregular connective tissue, that is continuous with the periostea of the articulating bones. It strengthens the joint so that the bones are not pulled apart. The

TABLE 8.1 Summary of Joint Classes

Structural Class	Structural Characteristics	Types		Mobility
Fibrous	Bone ends/parts united by collagenic fibers	(1) Suture (short fibers)		Immobile (synarthrosis)
		(2) Syndesmosis (longer fibers)		Slightly mobile (amphiarthrosis) and immobile
		(3) Gomphosis (periodontal ligament)		Immobile
Cartilaginous	Bone ends/parts united by cartilage	(1) Synchondrosis (hyaline cartilage)		Immobile
		(2) Symphysis (fibrocartilage)		Slightly movable
Synovial	Bone ends/parts covered with articular cartilage and enclosed within an articular capsule lined with synovial membrane	(1) Plane (2) Hinge (3) Pivot	(4) Condyloid (5) Saddle (6) Ball and socket	Freely movable (diarthrosis; movements depend on design of joint)

inner layer of the joint capsule is a **synovial membrane** composed of loose connective tissue. Besides lining the fibrous capsule internally, it covers all internal joint surfaces that are not hyaline cartilage.

4. Synovial fluid. A small amount of slippery synovial fluid occupies all free spaces within the joint capsule. This fluid is derived largely by filtration from blood flowing through the capillaries in the synovial membrane. Synovial (*synovi* = joint egg) fluid has a viscous, egg-white consistency due to its content of hyaluronic acid secreted by cells in the synovial membrane, but it thins, becoming less viscous, as it warms during joint activity. Synovial fluid, which is also found *within* the articular cartilages, provides a slippery weight-bearing film that reduces friction between the cartilages. The synovial fluid is forced from the cartilages when a joint is compressed; then as pressure on the joint is relieved, synovial fluid seeps back into the articular cartilages like water into a sponge, ready to be squeezed out again the next time the joint is loaded (put under pressure). This mechanism, called **weeping lubrication,** lubricates the free surfaces of the cartilages and nourishes their cells. Synovial fluid also contains phagocytic cells that rid the joint cavity of microbes and cellular debris.

5. Reinforcing ligaments. Synovial joints are reinforced and strengthened by a number of bandlike **ligaments.** Most often, these are **capsular,** or **intrinsic ligaments;** that is, they are thickened parts of the fibrous capsule. In other cases, they remain distinct and are found outside the capsule (**extracapsular ligaments)** or deep to it (**intracapsular ligaments).** Since intracapsular ligaments are covered with synovial membrane, they do not actually lie *within* the joint cavity.

People said to be double-jointed amaze the rest of us by placing both heels behind their neck. However, they have the normal number of joints; it's just that their joint capsules and ligaments are more stretchy and looser than average.

The articular capsule and ligaments are richly supplied with sensory nerve endings that monitor joint position and help to maintain muscle tone. Stretching these structures sends nerve impulses to the central nervous system, resulting in reflexive contraction of muscles surrounding the joint. Synovial joints are also richly supplied with blood vessels, most of which supply the synovial membrane.

Besides the basic components described above, certain synovial joints have other structural features. Some, such as the hip and knee joints, have cushioning **fatty pads** between the fibrous capsule and the synovial membrane or bone. Others have discs or wedges of fibrocartilage separating the articular surfaces. Where present, these so-called **articular discs,** or **menisci** (mě-nis'ki; "crescents"), extend inward from the articular capsule and partially or completely divide the synovial cavity in two (see the menisci in Figure 8.8). Articular discs improve the fit between articulating bone ends, making the joint more stable and minimizing wear and tear on the joint surfaces. Articular discs occur in the knee, jaw, and a few other joints (see Table 8.2).

Bursae and Tendon Sheaths

Bursae and tendon sheaths are not strictly part of synovial joints, but they are often found closely associated with them (Figure 8.4). Essentially bags of lubricant, they act as "ball bearings" to reduce friction between adjacent structures during joint activity. **Bursae** (ber'se; "purse,"), are flattened fibrous

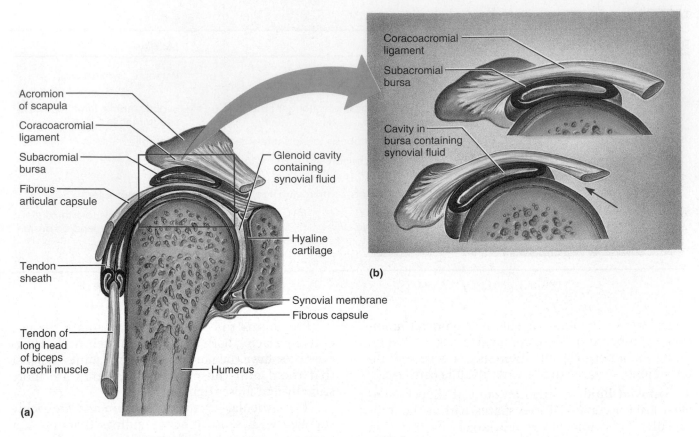

FIGURE 8.4 Friction-reducing structures: Bursae and tendon sheaths.
(a) Frontal section through the right shoulder joint showing a bursa and a tendon sheath around a muscle tendon. **(b)** Enlargement of part (a), showing the manner in which a bursa eliminates friction where a tendon (or other structure) is liable to rub against a bone. The synovial fluid within the bursa acts as a lubricant that allows its walls to slide easily across each other.

sacs lined with synovial membrane and containing a thin film of synovial fluid. They are common where ligaments, muscles, skin, tendons, or bones rub together. Many people have never heard of a bursa, but most have heard of bunions (bun'yunz). A *bunion* is an enlarged bursa at the base of the big toe, swollen from rubbing of a tight or poorly fitting shoe.

A **tendon sheath** is essentially an elongated bursa that wraps completely around a tendon subjected to friction, like a bun around a hot dog.

Factors Influencing the Stability of Synovial Joints

Because joints are constantly stretched and compressed, they must be stabilized so that they do not dislocate (come out of alignment). The stability of a synovial joint depends chiefly on three factors: the shapes of the articular surfaces; the number and positioning of ligaments; and muscle tone.

Articular Surfaces

The shapes of articular surfaces determine what movements are possible at a joint, but surprisingly, articular surfaces play only a minor role in joint stability. Many joints have shallow sockets or noncomplementary articulating surfaces ("misfits") that actually hinder joint stability. But when articular surfaces are large and fit snugly together, or when the socket is deep, stability is vastly improved. The ball and deep socket of the hip joint provide the best example of a joint made extremely stable by the shape of its articular surfaces.

Ligaments

The capsules and ligaments of synovial joints unite the bones and prevent excessive or undesirable motion. As a rule, the more ligaments a joint has, the stronger it is. However, when other stabilizing factors are inadequate, undue tension is placed on the ligaments and they stretch. Stretched ligaments stay stretched, like taffy, and a ligament can stretch only

TABLE 8.2 Structural and Functional Characteristics of Body Joints

Illustration	Joint	Articulating Bones	Structural Type*	Functional Type; Movements Allowed
	Skull	Cranial and facial bones	Fibrous; suture	Synarthrotic; no movement
	Temporo-mandibular	Temporal bone of skull and mandible	Synovial; modified hinge† (contains articular disc)	Diarthrotic; gliding and uniaxial rotation; slight lateral movement, elevation, depression, protraction, and retraction of mandible
	Atlanto-occipital	Occipital bone of skull and atlas	Synovial; condyloid	Diarthrotic; biaxial; flexion, extension, lateral flexion, circumduction of head on neck
	Atlantoaxial	Atlas (C_1) and axis (C_2)	Synovial; pivot	Diarthrotic; uniaxial; rotation of the head
	Intervertebral	Between adjacent vertebral bodies	Cartilaginous; symphysis	Amphiarthrotic; slight movement
	Intervertebral	Between articular processes	Synovial; plane	Diarthrotic; gliding
	Vertebrocostal	Vertebrae (transverse processes or bodies) and ribs	Synovial; plane	Diarthrotic; gliding of ribs
	Sternoclavicular	Sternum and clavicle	Synovial; shallow saddle (contains articular disc)	Diarthrotic; multiaxial (allows clavicle to move in all axes)
	Sternocostal	Sternum and rib 1	Cartilaginous; synchondrosis	Synarthrotic; no movement
	Sternocostal	Sternum and ribs 2–7	Synovial; double plane	Diarthrotic; gliding
	Acromio-clavicular	Acromion of scapula and clavicle	Synovial; plane (contains articular disc)	Diarthrotic; gliding and rotation of scapula on clavicle
	Shoulder (glenohumeral)	Scapula and humerus	Synovial; ball and socket	Diarthrotic; multiaxial; flexion, extension, abduction, adduction, circumduction, rotation of humerus
	Elbow	Ulna (and radius) with humerus	Synovial; hinge	Diarthrotic; uniaxial; flexion, extension of forearm
	Radioulnar (proximal)	Radius and ulna	Synovial; pivot	Diarthrotic; uniaxial; rotation of radius around long axis of forearm to allow pronation and supination
	Radioulnar (distal)	Radius and ulna	Synovial; pivot (contains articular disc)	Diarthrotic; uniaxial; rotation (convex head of ulna rotates in ulnar notch of radius)
	Wrist (radiocarpal)	Radius and proximal carpals	Synovial; condyloid	Diarthrotic; biaxial; flexion, extension, abduction, adduction, circumduction of hand
	Intercarpal	Adjacent carpals	Synovial; plane	Diarthrotic; gliding
	Carpometacarpal of digit 1 (thumb)	Carpal (trapezium) and metacarpal 1	Synovial; saddle	Diarthrotic; biaxial; flexion, extension, abduction, adduction, circumduction, opposition of metacarpal 1
	Carpometacarpal of digits 2–5	Carpal(s) and metacarpal(s)	Synovial; plane	Diarthrotic; gliding of metacarpals
	Knuckle (metacarpo-phalangeal)	Metacarpal and proximal phalanx	Synovial; condyloid	Diarthrotic; biaxial; flexion, extension, abduction, adduction, circumduction of fingers
	Finger (interphalangeal)	Adjacent phalanges	Synovial; hinge	Diarthrotic; uniaxial; flexion, extension of fingers

TABLE 8.2   **Structural and Functional Characteristics of Body Joints** (continued)

Illustration	Joint	Articulating Bones	Structural Type*	Functional Type; Movements Allowed
	Sacroiliac	Sacrum and coxal bone	Synovial; plane	Diarthrotic; little movement, slight gliding possible (more during pregnancy)
	Pubic symphysis	Pubic bones	Cartilaginous; symphysis	Amphiarthrotic; slight movement (enhanced during pregnancy)
	Hip (coxal)	Hip bone and femur	Synovial; ball and socket	Diarthrotic; multiaxial; flexion, extension, abduction, adduction, rotation, circumduction of thigh
	Knee (tibiofemoral)	Femur and tibia	Synovial; modified hinge† (contains articular discs)	Diarthrotic; biaxial; flexion, extension of leg, some rotation allowed
	Knee (femoropatellar)	Femur and patella	Synovial; plane	Diarthrotic; gliding of patella
	Tibiofibular	Tibia and fibula (proximally)	Synovial; plane	Diarthrotic; gliding of fibula
	Tibiofibular	Tibia and fibula (distally)	Fibrous; syndesmosis	Synarthrotic; slight "give" during dorsiflexion
	Ankle	Tibia and fibula with talus	Synovial; hinge	Diarthrotic; uniaxial; dorsiflexion, and plantar flexion of foot
	Intertarsal	Adjacent tarsals	Synovial; plane	Diarthrotic; gliding; inversion and eversion of foot
	Tarsometatarsal	Tarsal(s) and metatarsal(s)	Synovial; plane	Diarthrotic; gliding of metatarsals
	Metatarso-phalangeal	Metatarsal and proximal phalanx	Synovial; condyloid	Diarthrotic; biaxial; flexion, extension, abduction, adduction, circumduction of great toe
	Toe (interpha-langeal)	Adjacent phalanges	Synovial; hinge	Diarthrotic; uniaxial; flexion; extension of toes

*Fibrous joints indicated by orange circles; **cartilaginous joints** by blue circles; **synovial joints** by purple circles.
†These modified hinge joints are structurally bicondylar.

about 6% of its length before it snaps. Thus, when ligaments are the major means of bracing a joint, the joint is not very stable.

Muscle Tone

For most joints, the muscle tendons that cross the joint are the most important stabilizing factor. These tendons are kept taut at all times by the tone of their muscles. (Muscle tone is defined as low levels of contractile activity in relaxed muscles that keep the muscles healthy and ready to react to stimulation.) Muscle tone is extremely important in reinforcing the shoulder and knee joints and the arches of the foot.

Movements Allowed by Synovial Joints

Every skeletal muscle of the body is attached to bone or other connective tissue structures at no fewer than two points. The muscle's **origin** is attached to the immovable (or less movable) bone. Its other end, the **insertion**, is attached to the movable bone. Body movement occurs when muscles contract across joints and their insertion moves toward their origin. The movements can be described in directional terms relative to the lines, or *axes*, around which the body part moves and the planes of space along which the movement occurs, that is, along the transverse, frontal, or sagittal plane. (These planes were described in Chapter 1.)

Range of motion allowed by synovial joints varies from **nonaxial movement** (slipping movements only, since there is no axis around which movement can occur) to **uniaxial movement** (movement in one plane) to **biaxial movement** (movement in two planes) to **multiaxial movement** (movement in or around all three planes of space and axes). Range of motion varies greatly in different people. In some, such as trained gymnasts or acrobats, range of joint movement may be extraordinary. The ranges of motion at the major joints are given in Table 8.2.

There are three general types of movements: *gliding, angular movements,* and *rotation.* The most common body movements allowed by synovial joints are described next and illustrated in Figure 8.5.

Gliding Movements

Gliding movements (Figure 8.5a), also known as *translation,* are the simplest joint movements. One flat, or nearly flat, bone surface glides or slips over another (back-and-forth and side-to-side) without appreciable angulation or rotation. Gliding movements occur at the intercarpal and intertarsal joints, and between the flat articular processes of the vertebrae, as well as in combination with other movements (see Table 8.2).

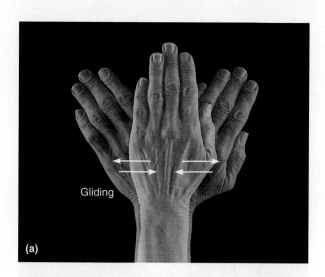

FIGURE 8.5   Movement allowed by synovial joints. (a) Gliding movements.

Angular Movements

Angular movements (Figure 8.5b–f) increase or decrease the angle between two bones. These movements may occur in any plane of the body and include flexion, extension, hyperextension, abduction, adduction, and circumduction.

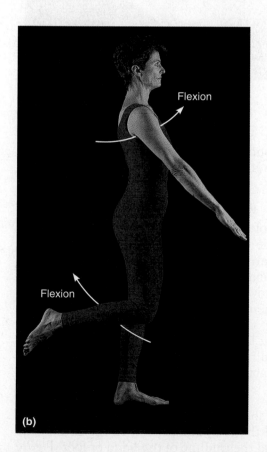
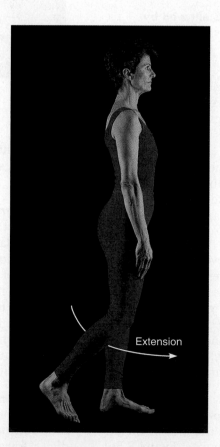
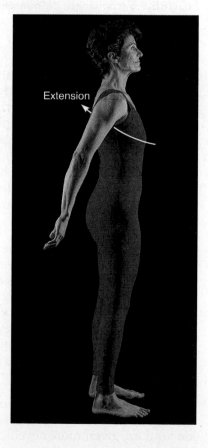

FIGURE 8.5 *(continued)*   **Movement allowed by synovial joints. (b)** Angular movements.

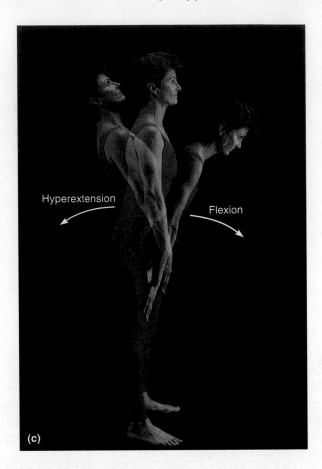

(c)

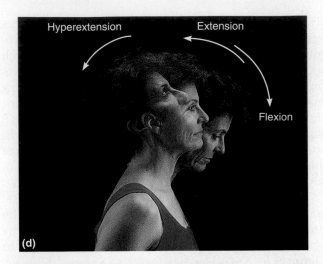

(d)

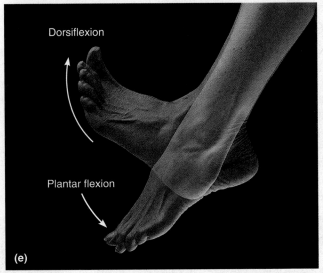

(e)

FIGURE 8.5 *(continued)* **Movements allowed by synovial joints. (c–e)** Angular movements.

Flexion   Flexion (flek'shun) is a bending movement, usually along the sagittal plane, that *decreases the angle* of the joint and brings the articulating bones closer together. Examples include bending the head forward on the chest (Figure 8.5d) and bending the body trunk or the knee from a straight to an angled position (Figure 8.5b and c). As a less obvious example, the arm is flexed at the shoulder when the arm is lifted in an anterior direction (Figure 8.5b).

Extension   Extension is the reverse of flexion and occurs at the same joints. It involves movement along the sagittal plane that *increases the angle* between the articulating bones, such as straightening a flexed neck, body trunk, elbow, or knee (Figure 8.5b–d). Bending the head backward beyond its straight (upright) position is called **hyperextension**. At the shoulder, extension carries the arm to a point posterior to the shoulder joint.

Dorsiflexion and Plantar Flexion of the Foot
The up-and-down movements of the foot at the ankle joint are given more specific names (Figure 8.5e). Lifting the foot so that its superior surface ap-

proaches the shin is **dorsiflexion** (corresponds to wrist extension), whereas depressing the foot (pointing the toes) is **plantar flexion** (corresponds to wrist flexion).

Abduction   Abduction ("moving away") is movement of a limb *away* from the midline or median plane of the body, along the frontal plane. Raising the arm (Figure 8.5f) or thigh laterally is an example of abduction. When the term is used to indicate the movement of the fingers or toes, it means spreading them apart. In this case "midline" is the longest digit: the third finger or second toe. Notice, however, that lateral bending of the trunk away from the body midline in the frontal plane is called *lateral flexion*, not abduction.

Adduction   Adduction ("moving toward") is the opposite of abduction, so it is the movement of a limb *toward* the body midline or, in the case of the digits, toward the midline of the hand or foot (Figure 8.5f).

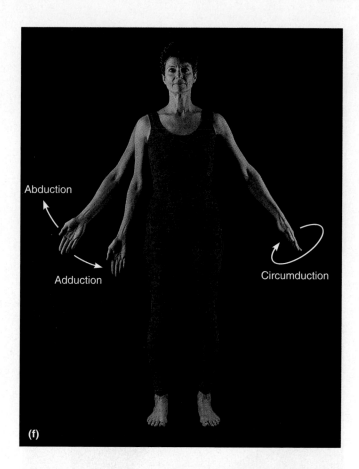

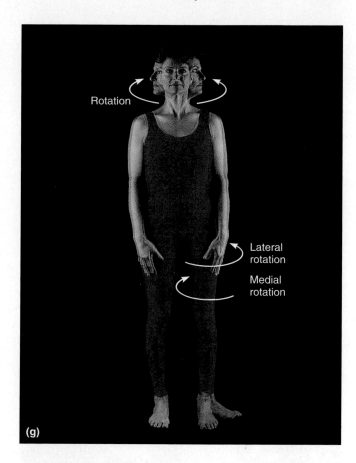

FIGURE 8.5 *(continued)* **Movements allowed by synovial joints. (f)** Angular movements, continued. **(g)** Rotation.

Circumduction   Circumduction (Figure 8.5f) is moving a limb so that it describes a cone in space (*circum* = around; *duco* = to draw). The distal end of the limb moves in a circle, while the point of the cone (the shoulder or hip joint) is more or less stationary. A pitcher winding up to throw a ball is actually circumducting his or her pitching arm. Because circumduction consists of flexion, abduction, extension, and adduction performed in succession, it is the quickest way to exercise the many muscles that move the hip and shoulder ball-and-socket joints.

Rotation

Rotation is the turning of a bone around its own long axis. It is the only movement allowed between the first two cervical vertebrae and is common at the hip (Figure 8.5g) and shoulder joints. Rotation may be directed toward the midline or away from it. For example, in *medial rotation* of the thigh, the femur's anterior surface moves toward the median plane of the body; *lateral rotation* is the opposite movement.

Special Movements

Certain movements do not fit into any of the above categories and occur at only a few joints. Some of these special movements are illustrated in Figure 8.6.

Supination and Pronation   The terms **supination** (soo″pĭ-na′shun; "turning backward") and **pronation** (pro-na′shun; "turning forward") refer to the movements of the radius around the ulna (Figure 8.6a). Rotating the forearm laterally so that the palm faces anteriorly or superiorly is supination. In the anatomical position, the hand is supinated and the radius and ulna are parallel. In pronation, the forearm rotates medially and the palm faces posteriorly or inferiorly. Pronation moves the distal end of the radius across the ulna so that the two bones form an X. This is the forearm's position when we are standing in a relaxed manner. Pronation is a much weaker movement than supination.

Tricks to help you keep these terms straight: If you lifted a cup of soup up to your mouth *on your palm,* you would be supinating ("soup"-inating), and a *pro* basketball player pronates his or her forearm to dribble the ball.

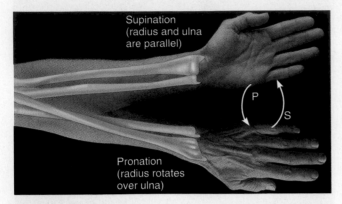

(a) Supination (S) and pronation (P)

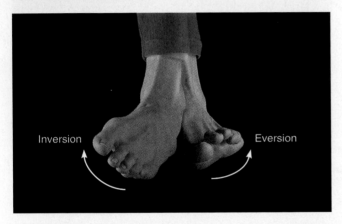

(b) Inversion and eversion

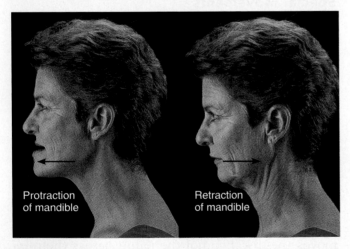

(c) Protraction and retraction

FIGURE 8.6   **Special body movements. (a)** Supination and pronation. **(b)** Inversion and eversion. **(c)** Protraction and retraction. **(d)** Elevation and depression. **(e)** Opposition.

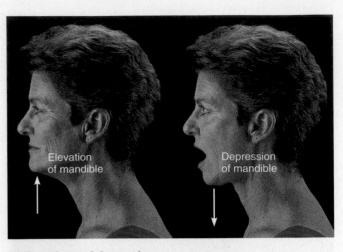

(d) Elevation and depression

(e) Opposition

Inversion and Eversion   Inversion and eversion are special movements of the foot (Figure 8.6b). In inversion, the sole of the foot turns medially. In eversion, the sole faces laterally.

Protraction and Retraction   Nonangular anterior and posterior movements in a transverse plane are called **protraction** and **retraction,** respectively (Figure 8.6c). The mandible is protracted when you jut out your jaw and retracted when you move it back to its original position.

Elevation and Depression   **Elevation** means lifting a body part superiorly (Figure 8.6d). For example, the scapulae are elevated when you shrug your shoulders. Moving the elevated part inferiorly is **depression.** During chewing, the mandible is alternately elevated and depressed.

Opposition   The saddle joint between metacarpal 1 and the carpals allows a movement called **opposition** of the thumb (Figure 8.6e). This movement is the action taken when you touch your thumb to the

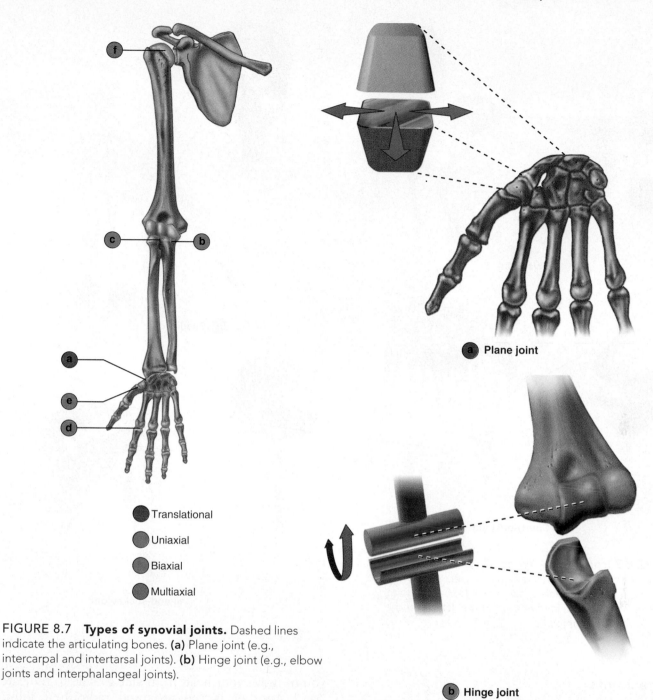

- ● Translational
- ● Uniaxial
- ● Biaxial
- ● Multiaxial

a Plane joint

b Hinge joint

FIGURE 8.7   Types of synovial joints. Dashed lines indicate the articulating bones. **(a)** Plane joint (e.g., intercarpal and intertarsal joints). **(b)** Hinge joint (e.g., elbow joints and interphalangeal joints).

tips of the other fingers on the same hand. It is opposition that makes the human hand such a fine tool for grasping and manipulating objects.

Types of Synovial Joints

Although all synovial joints have structural features in common, they do not have a common structural plan. Based on the shape of their articular surfaces, which in turn determine the movements allowed, synovial joints can be classified further into six major categories—plane, hinge, pivot, condyloid, saddle, and ball-and-socket joints.

Plane Joints

In **plane joints** (Figure 8.7a) the articular surfaces are essentially flat, and they allow only short gliding or translational movements. Examples are the gliding joints introduced earlier—the intercarpal and intertarsal joints, and the joints between vertebral articular processes. Gliding does not involve rotation around any axis, and gliding joints are the only examples of nonaxial joints.

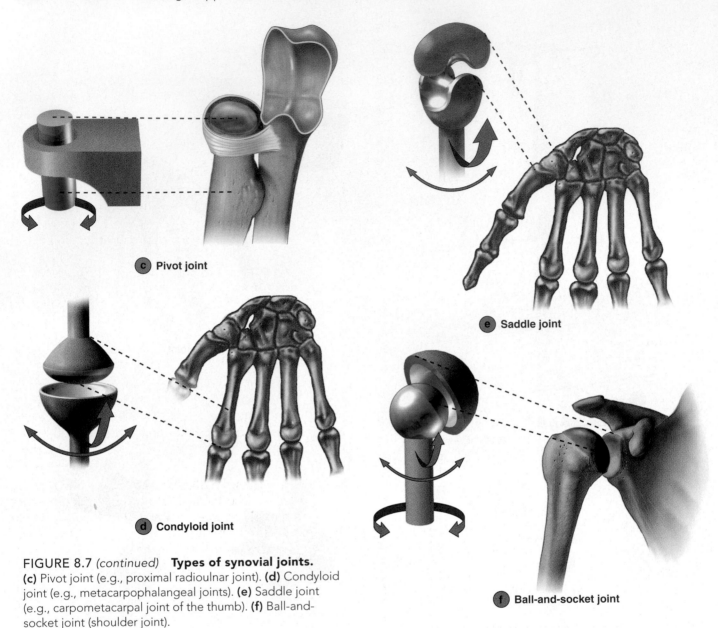

FIGURE 8.7 *(continued)* **Types of synovial joints.**
(c) Pivot joint (e.g., proximal radioulnar joint). **(d)** Condyloid
joint (e.g., metacarpophalangeal joints). **(e)** Saddle joint
(e.g., carpometacarpal joint of the thumb). **(f)** Ball-and-
socket joint (shoulder joint).

Hinge Joints

In **hinge joints** (Figure 8.7b), a cylindrical projection
of one bone fits into a trough-shaped surface on another. Motion is along a single plane and resembles
that of a mechanical hinge. The uniaxial hinge
joints permit flexion and extension only, typified by
bending and straightening the elbow and interphalangeal joints.

Pivot Joints

In a **pivot joint** (Figure 8.7c), the rounded end of one
bone protrudes into a "sleeve" or ring composed of
bone (and possibly ligaments) of another. The only
movement allowed is uniaxial rotation of one bone
around its own long axis. An example is the joint between the atlas and dens of the axis, which allows

you to move your head from side to side to indicate
"no." Another is the proximal radioulnar joint,
where the head of the radius rotates within a ringlike
ligament secured to the ulna.

Condyloid Joints

In **condyloid joints** (kon′dĭ-loid; "knuckle-like"), or
ellipsoidal joints, the oval articular surface of one
bone fits into a complementary depression in
another (Figure 8.7d). The important characteristic
is that both articulating surfaces are oval. The biaxial condyloid joints permit all *angular* motions, that
is, flexion and extension, abduction and adduction,
and circumduction. The radiocarpal (wrist) joints
and the metacarpophalangeal (knuckle) joints are
typical condyloid joints.

Saddle Joints

Saddle joints (Figure 8.7e) resemble condyloid joints, but they allow greater freedom of movement. Each articular surface has *both* concave and convex areas; that is, it is shaped like a saddle. The articular surfaces then fit together, concave to convex surfaces. The most clear-cut examples of saddle joints in the body are the carpometacarpal joints of the thumbs, and the movements allowed by these joints are clearly demonstrated by twiddling your thumbs.

Ball-and-Socket Joints

In **ball-and-socket joints** (Figure 8.7f), the spherical or hemispherical head of one bone articulates with the cuplike socket of another. These joints are multiaxial and the most freely moving synovial joints. Universal movement is allowed (that is, in all axes and planes, including rotation). The shoulder and hip joints are examples.

Selected Synovial Joints

In this section, we examine four joints (knee, shoulder, hip, and elbow) in detail. All of these joints have the five distinguishing characteristics of synovial joints, and we will not discuss these common features again. Instead, we will emphasize the unique structural features, functional abilities, and, in certain cases, functional weaknesses of each of these joints.

Knee Joint

The knee joint is the largest and most complex joint in the body (Figure 8.8). It allows extension, flexion, and some rotation. Despite its single joint cavity, the knee consists of three joints in one: an intermediate one between the patella and the lower end of the femur (the **femoropatellar joint**), and lateral and medial joints (collectively known as the **tibiofemoral joint**) between the femoral condyles above and the C-shaped **menisci**, or *semilunar cartilages*, of the tibia below. Besides deepening the shallow tibial articular surfaces, the menisci help prevent side-to-side rocking of the femur on the tibia and absorb shock transmitted to the knee joint. However, the menisci are attached only at their outer margins and are frequently torn free.

The tibiofemoral joint acts primarily as a hinge, permitting flexion and extension. However, structurally it is a bicondylar joint. Some rotation is possible when the knee is partly flexed, but when it is extended, side-to-side movements and rotation are strongly resisted by ligaments and the menisci. The femoropatellar joint is a plane joint, and the patella glides across the distal end of the femur during knee movements.

The knee joint is unique in that its joint cavity is only partially enclosed by a capsule. The relatively thin articular capsule is present only on the sides and posterior aspects of the knee, where it covers the bulk of the femoral and tibial condyles. Anteriorly, where the capsule is absent, three broad ligaments run from the patella to the tibia below. These are the **patellar ligament** flanked by the **medial** and **lateral patellar retinacula** (ret"ĭ-nak'u-lah; "retainers"), which merge imperceptibly into the articular capsule on each side (Figure 8.8c). The patellar ligament and retinacula are actually continuations of the tendon of the bulky quadriceps muscle of the anterior thigh. Physicians tap the patellar ligament to test the knee-jerk reflex.

The synovial cavity of the knee joint has a complicated shape, with several extensions that lead into "blind alleys." At least a dozen bursae are associated with this joint, some of which are shown in Figure 8.8a. For example, notice the *subcutaneous prepatellar bursa*, which is often injured when the knee is bumped.

All three types of joint ligaments stabilize and strengthen the capsule of the knee joint. The capsular and extracapsular ligaments all act to prevent hyperextension of the knee and are stretched taut when the knee is extended. These include the following:

1. The extracapsular **fibular** and **tibial collateral ligaments** are also critical in preventing lateral or medial rotation when the knee is extended. The broad flat tibial collateral ligament runs from the medial epicondyle of the femur to the medial condyle of the tibial shaft below and is fused to the medial meniscus.

2. The **oblique popliteal ligament** (pop"lĭ-te'al) is actually part of the tendon of the semimembranosus muscle that fuses with the capsule and strengthens the posterior aspect of the knee joint (Figure 8.8e).

3. The **arcuate popliteal ligament** arcs superiorly from the head of the fibula and reinforces the joint capsule posteriorly.

The knee's *intracapsular ligaments* are called *cruciate ligaments* (kroo'she-āt) because they cross each other, forming an X (*cruci* = cross) in the notch between the femoral condyles. They help prevent anterior-posterior displacement of the articular surfaces and secure the articulating bones when we stand (see Figure 8.8b). Although these ligaments are in the joint capsule, they are *outside* the synovial cavity, and synovial membrane nearly covers their surfaces. Note that the two cruciate ligaments both run up to the femur and are named for their *tibial* attachment site. The **anterior cruciate ligament** attaches to the *anterior* intercondylar area of the tibia. From there it passes posteriorly, laterally, and upward to attach to the femur on the medial side of its

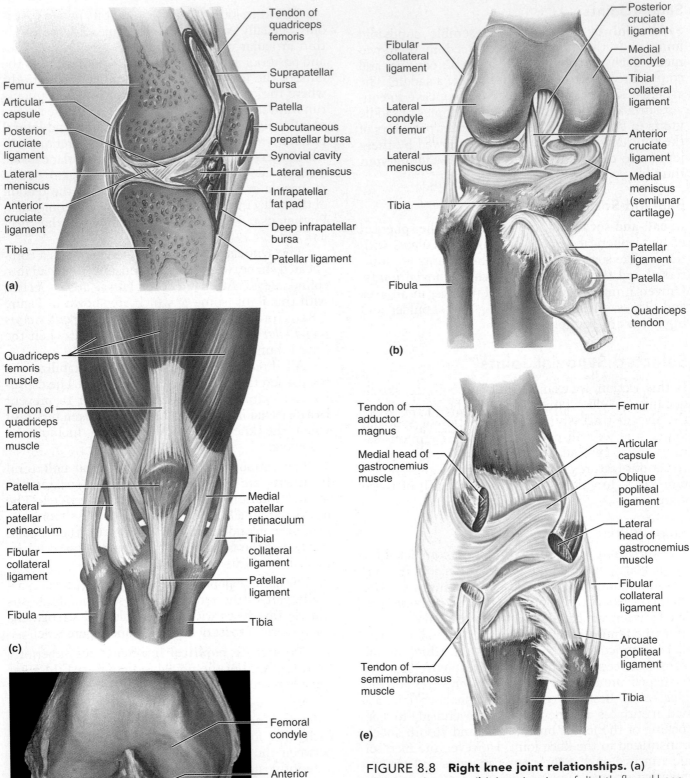

(a)

Tendon of quadriceps femoris
Suprapatellar bursa
Femur
Articular capsule
Patella
Posterior cruciate ligament
Subcutaneous prepatellar bursa
Synovial cavity
Lateral meniscus
Lateral meniscus
Anterior cruciate ligament
Infrapatellar fat pad
Tibia
Deep infrapatellar bursa
Patellar ligament

(b)

Fibular collateral ligament
Posterior cruciate ligament
Medial condyle
Tibial collateral ligament
Lateral condyle of femur
Lateral meniscus
Anterior cruciate ligament
Medial meniscus (semilunar cartilage)
Tibia
Patellar ligament
Fibula
Patella
Quadriceps tendon

(c)

Quadriceps femoris muscle
Tendon of quadriceps femoris muscle
Patella
Medial patellar retinaculum
Lateral patellar retinaculum
Tibial collateral ligament
Fibular collateral ligament
Patellar ligament
Fibula
Tibia

(d)

Femoral condyle
Anterior cruciate ligament
Medial tibial condyle
Medial meniscus

(e)

Tendon of adductor magnus
Femur
Medial head of gastrocnemius muscle
Articular capsule
Oblique popliteal ligament
Lateral head of gastrocnemius muscle
Fibular collateral ligament
Arcuate popliteal ligament
Tendon of semimembranosus muscle
Tibia

FIGURE 8.8 Right knee joint relationships. (a)
Midsagittal section. **(b)** Anterior view of slightly flexed knee joint showing the cruciate ligaments. Articular capsule has been removed; the quadriceps tendon is cut and reflected distally. **(c)** Anterior view. **(d)** Photograph of an opened knee joint corresponds to view in (b). **(e)** Posterior superficial view of the ligaments clothing the knee joint.

lateral condyle. This ligament prevents forward sliding of the tibia on the femur and checks hyperextension of the knee. It is somewhat lax when the knee is flexed and taut when the knee is extended. The stronger **posterior cruciate ligament** is attached to the *posterior* intercondylar area of the tibia and passes anteriorly, medially, and upward to attach to the lateral side of the medial femoral condyle. This ligament prevents backward displacement of the tibia or forward sliding of the femur.

The knee capsule is heavily reinforced by muscle tendons. Most important are the strong tendons of the quadriceps muscles of the anterior thigh and the tendon of the semimembranosus muscle posteriorly. Since the muscles associated with the joint are the main knee stabilizers, the greater their strength and tone, the less the chance of knee injury.

The knees have a built-in locking device that provides steady support for the body in the standing position. As we begin to stand up, the wheel-shaped femoral condyles roll like ball bearings across the flat condyles of the tibia and the flexed leg begins to extend at the knee. Because the lateral femoral condyle stops rolling before the medial condyle stops, the femur *spins* (rotates) medially on the tibia, until all major ligaments of the knee are twisted and taut and the menisci are compressed. The tension in the ligaments effectively locks the joint into a rigid structure that cannot be flexed again until it is unlocked. This unlocking is accomplished by the popliteus muscle (see Table 10.15, p. 332), which rotates the femur laterally on the tibia, causing the ligaments to become untwisted and slack.

HOMEOSTATIC IMBALANCE

Of all body joints, the knees are most susceptible to sports injuries because of their high reliance on nonarticular factors for stability and the fact that they carry the body's weight. The knee can absorb a vertical force equal to nearly seven times body weight. However, it is very vulnerable to *horizontal* blows, such as those that occur during blocking and tackling in football. When thinking of common knee injuries, remember the 3 C's: collateral ligaments, cruciate ligaments, and cartilages (menisci). Most dangerous are *lateral* blows to the extended knee. These forces tear the tibial collateral ligament and the medial meniscus attached to it, as well as the anterior cruciate ligament (Figure 8.9). It is estimated that 50% of all professional football players have serious knee injuries during their careers.

Although less devastating than the injury just described, injuries that affect only the anterior cruciate ligament (ACL) are becoming more common, particularly as women's sports become more vigor-

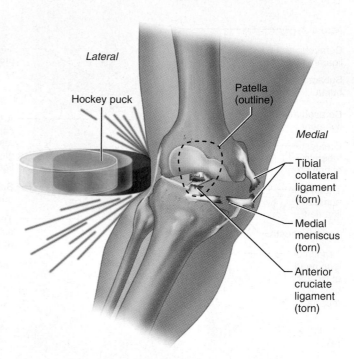

FIGURE 8.9 **A common knee injury.** Anterior view of a knee being hit by a hockey puck. By separating the femur from the tibia medially, such blows to the lateral side tear both the tibial collateral ligament and the medial meniscus because the two are attached. The anterior cruciate ligament also tears.

ous and competitive. Most ACL injuries occur when a runner changes direction quickly, twisting a hyperextended knee. A torn ACL heals poorly, so repair usually requires a ligament graft using connective tissue taken from one of the larger ligaments (e.g., patellar, Achilles, or semitendinosus). ●

Shoulder (Glenohumeral) Joint

In the shoulder joint, stability has been sacrificed to provide the most freely moving joint of the body. The shoulder joint is a ball-and-socket joint. The large hemispherical head of the humerus fits in the small, shallow glenoid cavity of the scapula (Figure 8.10), like a golf ball sitting on a tee. Although the glenoid cavity is slightly deepened by a rim of fibrocartilage, the **glenoid labrum** (*labrum* = lip), it is only about one-third the size of the humeral head and contributes little to joint stability.

The articular capsule enclosing the joint cavity (from the margin of the glenoid cavity to the anatomical neck of the humerus) is remarkably thin and loose, qualities that contribute to this joint's freedom of movement. The few ligaments reinforcing the shoulder joint are located primarily on its anterior aspect. The superiorly located **coracohumeral ligament** (kor'ah-ko-hu'mer-ul) provides the only strong thickening of the capsule and helps support the weight of the upper limb. Three **glenohumeral**

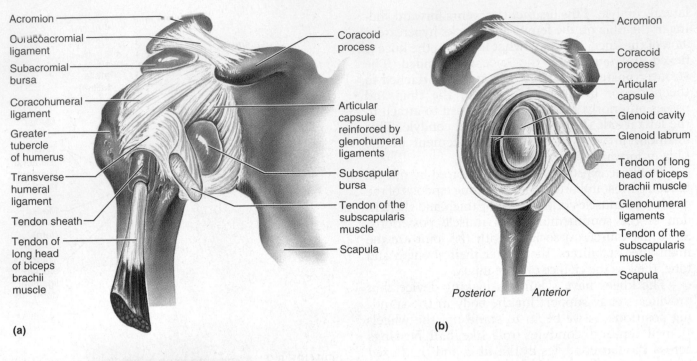

Acromion

Coracoacromial ligament

Subacromial bursa

Coracohumeral ligament

Greater tubercle of humerus

Transverse humeral ligament

Tendon sheath

Tendon of long head of biceps brachii muscle

Coracoid process

Articular capsule reinforced by glenohumeral ligaments

Subscapular bursa

Tendon of the subscapularis muscle

Scapula

(a)

Acromion

Coracoid process

Articular capsule

Glenoid cavity

Glenoid labrum

Tendon of long head of biceps brachii muscle

Glenohumeral ligaments

Tendon of the subscapularis muscle

Scapula

Posterior   Anterior

(b)

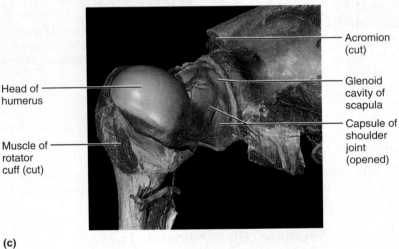

Acromion (cut)

Glenoid cavity of scapula

Capsule of shoulder joint (opened)

Head of humerus

Muscle of rotator cuff (cut)

(c)

FIGURE 8.10 Right shoulder joint relationships. (a) Anterior superficial view illustrating some of the reinforcing ligaments, associated muscles, and bursae. **(b)** The shoulder joint, cut open and viewed from the lateral aspect; the humerus has been removed. **(c)** Photograph of the interior of the shoulder joint, anterior view.

ligaments (glĕ"no-hu'mer-ul) strengthen the front of the capsule somewhat but are weak and may even be absent.

Muscle tendons that cross the shoulder joint contribute most to this joint's stability. The "super-stabilizer" is the tendon of the long head of the biceps brachii muscle of the arm (Figure 8.10a). This tendon attaches to the superior margin of the glenoid labrum, travels through the joint cavity, and then runs within the intertubercular groove of the humerus. It secures the head of the humerus against the glenoid cavity. Four other tendons (and the associated muscles) make up the **rotator cuff.** This cuff encircles the shoulder joint and blends with the articular capsule. The muscles include the subscapularis, supraspinatus, infraspinatus, and teres minor. (The rotator cuff muscles are illus-

trated in Figure 10.14, p. 312.) The rotator cuff can be severely stretched when the arm is vigorously circumducted; this is a common injury of baseball pitchers. As noted in Chapter 7, shoulder dislocations are fairly common. Because the shoulder's reinforcements are weakest anteriorly and inferiorly, the humerus tends to dislocate in the forward and downward direction.

Hip (Coxal) Joint

The hip joint, like the shoulder joint, is a ball-and-socket joint. It has a good range of motion, but not nearly as wide as the shoulder's range. Movements occur in all possible planes but are limited by the joint's strong ligaments and its deep socket. The hip joint is formed by the articulation of the spherical

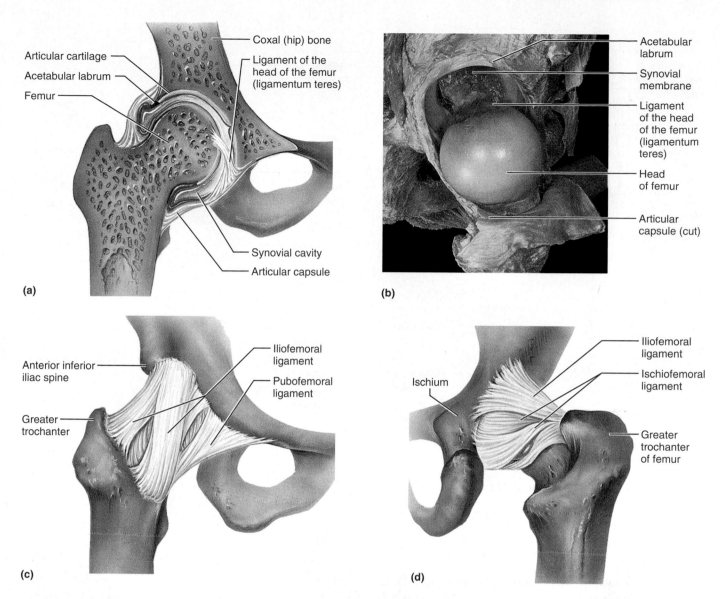

FIGURE 8.11   Right hip joint relationships. (a) Frontal section through the hip joint. **(b)** Photograph of the joint interior lateral view. **(c)** Anterior superficial view. **(d)** Posterior superficial view.

head of the femur with the deeply cupped acetabulum of the hip bone (Figure 8.11). The depth of the acetabulum is enhanced by a circular rim of fibrocartilage called the **acetabular labrum** (as″ĕ-tab′u-lar) (Figure 8.11a and b). The labrum's diameter is less than that of the head of the femur, and these articular surfaces fit snugly together, so hip joint dislocations are rare.

The thick articular capsule extends from the rim of the acetabulum to the neck of the femur and completely encloses the joint. Several strong ligaments reinforce the capsule of the hip joint. These include the **iliofemoral ligament** (il″e-o-fem′o-ral), a strong V-shaped ligament anteriorly; the **pubofemoral ligament** (pu″bo-fem′o-ral), a triangular thickening of the inferior part of the capsule; and the **ischiofemoral ligament** (is″ke-o-fem′o-ral), a spiraling posteriorly located ligament. These ligaments are arranged in such a way that they "screw" the femur head into the acetabulum when a person stands up straight, thereby providing more stability.

The **ligament of the head of the femur,** also called the **ligamentum teres,** is a flat intracapsular band that runs from the femur head to the lower lip of the acetabulum. This ligament is slack during most hip movements, so it is not important in stabilizing the joint. In fact, its mechanical function (if any) is unclear, but it does contain an artery that helps supply the head of the femur. Damage to this artery may lead to severe arthritis of the hip joint.

Muscle tendons that cross the joint and the bulky hip and thigh muscles that surround it contribute to

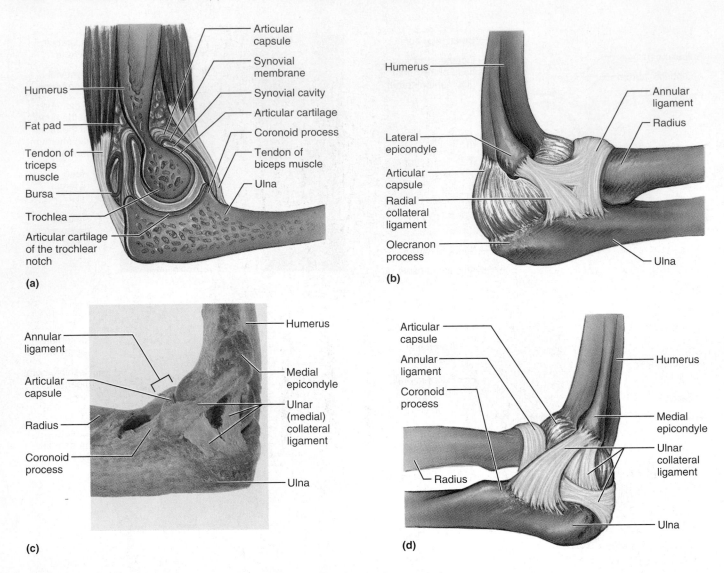

(a)

(b)

(c)

(d)

FIGURE 8.12 **Right elbow joint relationships. (a)** Midsagittal section of elbow joint. **(b)** Lateral view. **(c)** Photograph of elbow joint showing the important reinforcing ligaments, medial view. **(d)** Medial view.

its stability and strength. However, the deep socket that securely encloses the femoral head and the strong ligaments contribute most to the stability of the hip joint.

Elbow Joint

Our upper limbs are flexible extensions that permit us to reach out and manipulate things in our environment. Besides the shoulder joint, the most prominent of the upper limb joints is the elbow. The elbow joint provides a stable and smoothly operating hinge that allows flexion and extension only (Figure 8.12). Within the joint, both the radius and ulna articulate with the condyles of the humerus, but it is the close gripping of the trochlea by the ulna's trochlear notch that forms the "hinge" and stabilizes this joint. A relatively lax articular capsule

extends inferiorly from the humerus to the ulna and to the **annular ligament** (an'u-lar) surrounding the head of the radius.

Anteriorly and posteriorly, the articular capsule is thin and allows substantial freedom for elbow flexion and extension. However, side-to-side movements are restricted by two strong capsular ligaments: the **ulnar collateral ligament** medially, and the **radial collateral ligament,** a triangular ligament on the lateral side. Additionally, tendons of several arm muscles, such as the biceps and triceps, cross the elbow joint and provide security.

The radius is a passive "onlooker" in the angular elbow movements. However, its head rotates within the annular ligament during supination and pronation of the forearm.

Homeostatic Imbalances of Joints

Few of us pay attention to our joints unless something goes wrong with them. Joint pain and malfunction can be caused by a number of factors, but most joint problems result from injuries and inflammatory or degenerative conditions.

Common Joint Injuries

For most of us, sprains and dislocations are the most common trauma-induced joint injuries, but cartilage injuries are equally threatening to athletes.

Sprains

In a **sprain,** the ligaments reinforcing a joint are stretched or torn. The lumbar region of the spine, the ankle, and the knee are common sprain sites. Partially torn ligaments will repair themselves, but they heal slowly because ligaments are so poorly vascularized. Sprains tend to be painful and immobilizing. Completely ruptured ligaments require prompt surgical repair because inflammation in the joint will break down the neighboring tissues and turn the injured ligament to "mush." Surgical repair can be difficult: A ligament consists of hundreds of fibrous strands, and sewing one back together has been compared to trying to sew two hairbrushes together.

When important ligaments are too severely damaged to be repaired, they must be removed and replaced with grafts or substitute ligaments. For example, a piece of tendon from a muscle, or woven collagen bands, can be stapled to the articulating bones. Alternatively, carbon fibers are implanted in the torn ligament to form a supporting mesh, which is then invaded by fibroblasts that reconstruct the ligament.

Cartilage Injuries

Many aerobics devotees, encouraged to "feel the burn" during their workout, may feel the snap and pop of their overstressed cartilage instead. Although most cartilage injuries involve tearing of the knee menisci, overuse damage to the articular cartilages of other joints is becoming increasingly common in competitive young athletes.

Cartilage is avascular and it rarely can obtain sufficient nourishment to repair itself; thus, it usually stays torn. Because cartilage fragments (called *loose bodies*) can interfere with joint function by causing the joint to lock or bind, most sports physicians recommend that the central (nonvascular) part of a damaged cartilage be removed. Today, this can be done by **arthroscopic surgery** (ar-thro-skop'ik; "looking into joints"), a procedure that enables patients to be out of the hospital the same

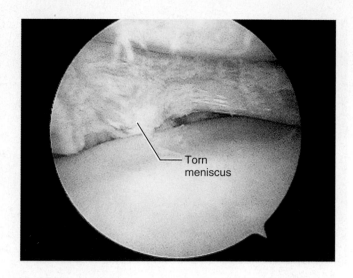

FIGURE 8.13 Arthroscopic photograph of a torn medial meniscus (Courtesy of the author's tennis game).

day. The arthroscope, a small instrument bearing a tiny lens and fiber-optic light source, enables the surgeon to view the joint interior (Figure 8.13), repair a ligament, or remove cartilage fragments through one or more tiny slits, minimizing tissue damage and scarring. Removal of part of a meniscus does not severely impair knee joint mobility, but the joint is definitely less stable.

Dislocations

A **dislocation (luxation)** occurs when bones are forced out of alignment. It is usually accompanied by sprains, inflammation, and joint immobilization. Dislocations may result from serious falls and are common contact sports injuries. Joints of the jaw, shoulders, fingers, and thumbs are most commonly dislocated. Like fractures, dislocations must be *reduced;* that is, the bone ends must be returned to their proper positions by a physician. **Subluxation** is a partial dislocation of a joint.

Repeat dislocations of the same joint are common because the initial dislocation stretches the joint capsule and ligaments. The resulting loose capsule provides poor reinforcement for the joint.

Inflammatory and Degenerative Conditions

Inflammatory conditions that affect joints include bursitis, tendonitis, and various forms of arthritis.

Bursitis and Tendonitis

Bursitis is inflammation of a bursa and is usually caused by a blow or friction. Falling on one's knee

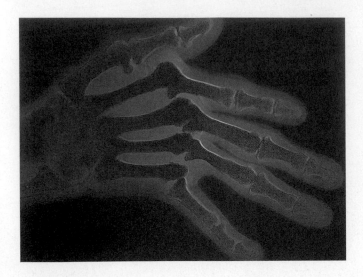

FIGURE 8.14 X ray of a hand deformed by rheumatoid arthritis.

may result in a painful bursitis of the prepatellar bursa, known as *housemaid's knee* or *water on the knee*. Prolonged leaning on one's elbows may damage the bursa close to the olecranon process, producing *student's elbow*, or *olecranon bursitis*. Severe cases are treated by injecting anti-inflammatory drugs into the bursa. If excessive fluid accumulates, removing some fluid by needle aspiration may relieve the pressure.

Tendonitis is inflammation of tendon sheaths, typically caused by overuse. Its symptoms (pain and swelling) and treatment (rest, ice, and anti-inflammatory drugs) mirror those of bursitis.

Arthritis

The term **arthritis** describes over 100 different types of inflammatory or degenerative diseases that damage the joints. In all its forms, arthritis is the most widespread crippling disease in the United States. One out of seven Americans suffers its ravages. To a greater or lesser degree, all forms of arthritis have the same initial symptoms: pain, stiffness, and swelling of the joint.

Acute forms of arthritis usually result from bacterial invasion and are treated with antibiotics. The synovial membrane thickens and fluid production decreases, causing increased friction and pain. Chronic forms of arthritis include osteoarthritis, rheumatoid arthritis, and gouty arthritis.

Osteoarthritis   **Osteoarthritis (OA)** is the most common chronic arthritis. A chronic (long-term) degenerative condition, OA is often called "wear-and-tear arthritis." OA is most prevalent in the aged and is probably related to the normal aging process (although it is seen occasionally in younger people

and some forms have a genetic basis). More women than men are affected, but 85% of all Americans develop this condition.

Current theory holds that normal joint use prompts the release of (metalloproteinase) enzymes that break down articular cartilage. In healthy individuals, this damaged cartilage is eventually replaced, but in people with OA, more cartilage is destroyed than replaced. Although its specific cause is unknown, OA may reflect the cumulative effects of years of compression and abrasion acting at joint surfaces, causing excessive amounts of the cartilage-destroying enzymes to be released. The result is softened, roughened, pitted, and eroded articular cartilages.

As the disease progresses, the exposed bone tissue thickens and forms bony spurs (osteophytes) that enlarge the bone ends and may restrict joint movement. Patients complain of stiffness on arising that lessens somewhat with activity. The affected joints may make a crunching noise as they move. This sound, called *crepitus* (krep'ĭ-tus), results as the roughened articular surfaces rub together. The joints most often affected are those of the cervical and lumbar spine and the fingers, knuckles, knees, and hips.

The course of osteoarthritis is usually slow and irreversible. In many cases, its symptoms are controllable with a mild pain reliever like aspirin or acetaminophen, along with moderate activity to keep the joints mobile. Magnetic therapy (assumed to stimulate the growth and repair of articular cartilage) is reported to provide relief to about 70% of patients treated. Glucosamine sulfate, a nutritional supplement, appears to decrease pain and inflammation and may help to preserve the articular cartilage. Osteoarthritis is rarely crippling, but it can be, particularly when the hip or knee joints are involved.

Rheumatoid Arthritis   **Rheumatoid arthritis (RA;** roo'mah-toid) is a chronic inflammatory disorder with an insidious onset. Though it usually arises between the ages of 40 and 50, it may occur at any age. It affects three times as many women as men. While not as common as osteoarthritis, rheumatoid arthritis causes disability in millions (Figure 8.14). It occurs in more than 1% of Americans.

In the early stages of RA, joint tenderness and stiffness are common. Many joints, particularly the small joints of the fingers, wrists, ankles, and feet, are afflicted at the same time and bilaterally. For example, if the right elbow is affected, most likely the left elbow is also affected. The course of RA is variable and marked by flare-ups (exacerbations) and remissions (*rheumat* = susceptible to change). Other manifestations include anemia, osteoporosis, muscle atrophy, and cardiovascular problems.

RA is an *autoimmune disease*—a disorder in which the body's immune system attacks its own tissues. The initial trigger for this reaction is unknown, but the streptococcus bacterium and viruses have been suspect. Perhaps these microorganisms bear molecules similar to some naturally present in the joints, and the immune system, once activated, attempts to destroy both.

RA begins with inflammation of the synovial membrane (*synovitis*) of the affected joints. Inflammatory cells (lymphocytes, neutrophils, and others) migrate into the joint cavity from the blood and unleash a deluge of inflammatory chemicals that destroy body tissues when released inappropriately in large amounts as in RA. Synovial fluid accumulates, causing joint swelling and in time, the inflamed synovial membrane thickens into a **pannus** ("rag"), an abnormal tissue that clings to the articular cartilages. The pannus erodes the cartilage (and sometimes the underlying bone) and eventually scar tissue forms and connects the bone ends. Later this scar tissue ossifies and the bone ends fuse together, immobilizing the joint. This end condition, called *ankylosis* (ang'kĭ-lo'sis; "stiff condition"), often produces bent, deformed fingers (see Figure 8.14). Not all cases of RA progress to the severely crippling ankylosis stage, but all cases do involve restriction of joint movement and extreme pain.

A wonder drug for RA sufferers is still undiscovered. Currently the pendulum is swinging from conservative RA therapy utilizing aspirin, long-term antibiotic therapy, and physical therapy to a more progressive treatment course using anti-inflammatory drugs or immunosuppressants. Particularly promising are etanercept (Enbrel) and infliximab (Remicade), the first in a class of drugs called *biologic response modifiers* that neutralize some of the harmful properties of the inflammatory chemicals. Joint prostheses, if available, are the last resort for severely crippled RA patients.

Gouty Arthritis   Uric acid, a normal waste product of nucleic acid metabolism, is ordinarily excreted in urine without any problems. However, when blood levels of uric acid rise excessively (due to its excessive production or slow excretion), it may be deposited as needle-shaped urate crystals in the soft tissues of joints. An inflammatory response follows, leading into an agonizingly painful attack of **gouty arthritis** (gow'te), or gout. The initial attack typically affects one joint, often at the base of the great toe.

Gout is far more common in males than in females because males naturally have higher blood levels of uric acid. Because gout seems to run in families, genetic factors are definitely implicated.

Untreated gout can be very destructive; the articulating bone ends fuse and immobilize the joint. Fortunately, several drugs (colchicine, nonsteroidal anti-inflammatory drugs, glucocorticoids, and others) that terminate or prevent gout attacks are available. Patients are advised to avoid alcohol excess (which promotes uric acid overproduction), and foods high in purine-containing nucleic acids, such as liver, kidneys, and sardines.

Review Questions

Multiple Choice/Matching

(Some questions have more than one correct answer. Select the best answer or answers from the choices given.)

1. Match the key terms to the appropriate descriptions.

Key: **(a)** fibrous joints   **(b)** cartilaginous joints   **(c)** synovial joints

_____ **(1)** exhibit a joint cavity
_____ **(2)** types are sutures and syndesmoses
_____ **(3)** bones connected by collagen fibers
_____ **(4)** types include synchondroses and symphyses
_____ **(5)** all are diarthrotic
_____ **(6)** many are amphiarthrotic
_____ **(7)** bones connected by a disc of hyaline cartilage or fibrocartilage
_____ **(8)** nearly all are synarthrotic
_____ **(9)** shoulder, hip, jaw, and elbow joints

2. Freely movable joints are (a) synarthroses, (b) diarthroses, (c) amphiarthroses.

3. Anatomical characteristics of a synovial joint include (a) articular cartilage, (b) a joint cavity, (c) an articular capsule, (d) all of these.

4. Factors that influence the stability of a synovial joint include (a) shape of articular surfaces, (b) presence of strong reinforcing ligaments, (c) tone of surrounding muscles, (d) all of these.

5. The following description—"Articular surfaces deep and secure; capsule heavily reinforced by ligaments and muscle tendons; extremely stable joint"—best describes (a) the elbow joint, (b) the hip joint, (c) the knee joint, (d) the shoulder joint.

6. Ankylosis means (a) twisting of the ankle, (b) tearing of ligaments, (c) displacement of a bone, (d) immobility of a joint due to fusion of its articular surfaces.

7. An autoimmune disorder in which joints are affected bilaterally and which involves pannus formation and gradual joint immobilization is (a) bursitis, (b) gout, (c) osteoarthritis, (d) rheumatoid arthritis.

Short Answer Essay Questions

8. Define joint.

9. Discuss the relative value (to body homeostasis) of immovable, slightly movable, and freely movable joints.

10. Compare the structure, function, and common body locations of bursae and tendon sheaths.

11. Joint movements may be nonaxial, uniaxial, biaxial, or multiaxial. Define what each of these terms means.

12. Compare and contrast the paired movements of flexion and extension with adduction and abduction.

13. How does rotation differ from circumduction?

14. Name two types of uniaxial, biaxial, and multiaxial joints.

15. What is the specific role of the menisci of the knee? of the anterior and posterior cruciate ligaments?

16. The knee has been called "a beauty and a beast." Provide several reasons that might explain the negative (beast) part of this description.

17. Why are sprains and cartilage injuries a particular problem?

18. List the functions of the following elements of a synovial joint: fibrous part of the capsule, synovial fluid, articular disc.

9

MUSCLES AND MUSCLE TISSUE

Overview of Muscle Tissues (pp. 244–245)

1. Compare and contrast the basic types of muscle tissue.
2. List four important functions of muscle tissue.

Skeletal Muscle (pp. 245–272)

3. Describe the gross structure of a skeletal muscle.
4. Describe the microscopic structure and functional roles of the myofibrils, sarcoplasmic reticulum, and T tubules of muscle fibers (cells).
5. Explain the sliding filament mechanism.
6. Define motor unit and explain how muscle fibers are stimulated to contract.
7. Define muscle twitch and describe the events occurring during its three phases.
8. Explain how smooth, graded contractions of a skeletal muscle are produced.
9. Differentiate between isometric and isotonic contractions.
10. Describe three ways in which ATP is regenerated during skeletal muscle contraction.

11. Define oxygen debt and muscle fatigue. List possible causes of muscle fatigue.
12. Describe factors that influence the force, velocity, and duration of skeletal muscle contraction.
13. Describe three types of skeletal muscle fibers and explain the relative value of each type.
14. Compare and contrast the effects of aerobic and resistance exercise on skeletal muscles and on other body systems.

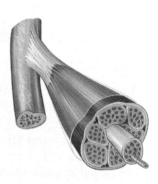

Smooth Muscle (pp. 272–278)

15. Compare the gross and microscopic anatomy of smooth muscle cells to that of skeletal muscle cells.
16. Compare and contrast the contractile mechanisms and the means of activation of skeletal and smooth muscles.
17. Distinguish between single-unit and multiunit smooth muscle structurally and functionally.

Because flexing muscles look like mice scurrying beneath the skin, some scientist long ago dubbed them *muscles,* from the Latin *mus* meaning "little mouse." Indeed, the rippling muscles of professional boxers or weight lifters are often the first thing that comes to mind when one hears the word *muscle.* But muscle is also the dominant tissue in the heart and in the walls of other hollow organs. In all its forms, muscle tissue makes up nearly half the body's mass. The most distinguishing functional characteristic of muscles is their ability to transform chemical energy (ATP) into directed mechanical energy. In so doing, they become capable of exerting force.

Overview of Muscle Tissues

Types of Muscle Tissue

The three types of muscle tissue are *skeletal, cardiac,* and *smooth.* Before exploring their characteristics, however, let us introduce some terminology. First, skeletal and smooth muscle cells (but not cardiac muscle cells) are elongated and, for this reason, are called **muscle fibers.** Second, muscle contraction depends on two kinds of **myofilaments,** which are the muscle equivalents of the actin- or myosin-containing microfilaments described in Chapter 3. As you will recall, these two proteins play a role in motility and shape changes in virtually every cell in the body, but this property reaches its highest development in the contractile muscle fibers. Third, whenever you see the prefixes **myo** or **mys** (both are word roots meaning "muscle") and **sarco** (flesh), the reference is to muscle. For example, the plasma membrane of muscle fibers is called the *sarcolemma* (sar"ko-lem'ah), literally "muscle" (sarco) "husk" (lemma), and muscle fiber cytoplasm is called *sarcoplasm.* Now we are ready to describe the three types of muscle tissue.

Skeletal muscle tissue is packaged into the *skeletal muscles,* organs that attach to and cover the bony skeleton. Skeletal muscle fibers are the longest muscle cells; they have obvious stripes called **striations** and can be controlled voluntarily. Although it is often activated by reflexes, skeletal muscle is called **voluntary muscle** because it is the *only type subject to* conscious control. Therefore, when you think of skeletal muscle tissue, the key words to keep in mind are *skeletal, striated,* and *voluntary.* Skeletal muscle is responsible for overall body mobility. It can contract rapidly, but it tires easily and must rest after short periods of activity. Nevertheless, it can exert tremendous power, a fact revealed by reports of people lifting cars to save their loved ones. Skeletal muscle is also remarkably adaptable. For example, your hand muscles can exert a force of a fraction of an ounce to pick up a dropped paper clip and the same muscles can exert a force of about 70 pounds to pick up this book!

Cardiac muscle tissue occurs only in the heart (the body's blood pump), where it constitutes the bulk of the heart walls. Like skeletal muscle cells, cardiac muscle cells are striated, but cardiac muscle is not voluntary. Most of us have no conscious control over how fast our heart beats. Key words to remember for this muscle type are *cardiac, striated,* and *involuntary.* Cardiac muscle usually contracts at a fairly steady rate set by the heart's pacemaker, but neural controls allow the heart to "shift into high gear" for brief periods, as when you race across the tennis court to make that overhead smash.

Smooth muscle tissue is found in the walls of hollow visceral organs, such as the stomach, urinary bladder, and respiratory passages. Its role is to force fluids and other substances through internal body channels. It has no striations, and like cardiac muscle, it is not subject to voluntary control. We can describe smooth muscle tissue most precisely as *visceral, nonstriated,* and *involuntary.* Contractions of smooth muscle fibers are slow and sustained.

Skeletal and smooth muscles are discussed in this chapter. Cardiac muscle is discussed in Chapter 17 (The Heart). Table 9.3 on pp. 277–278 summarizes the most important characteristics of each type of muscle tissue.

Functional Characteristics of Muscle Tissue

Muscle tissue is endowed with some special functional properties that enable it to perform its duties.

Excitability, or **irritability,** is the ability to receive and respond to a stimulus, that is, any change in the environment whether inside or outside the body. In the case of muscle, the stimulus is usually a chemical—for example, a neurotransmitter released by a nerve cell, or a local change in pH. The response is generation of an electrical impulse that passes along the sarcolemma (plasma membrane) of the muscle cell and causes the cell to contract.

Contractility is the ability to shorten forcibly when adequately stimulated. This property sets muscle apart from all other tissue types.

Extensibility is the ability to be stretched or extended. Muscle fibers shorten when contracting, but they can be stretched, even beyond their resting length, when relaxed.

Elasticity is the ability of a muscle fiber to recoil and resume its resting length after being stretched.

Muscle Functions

Muscle performs four important functions for the body: It produces movement, maintains posture, stabilizes joints, and generates heat.

Producing Movement

Just about all movements of the human body and its parts are a result of muscle contraction. Skeletal muscles are responsible for all locomotion and manipulation. They enable you to respond quickly to changes in the external environment, for example, by jumping out of the way of a runaway car, directing your eyeballs, and smiling or frowning.

The coursing of blood through your body is evidence of the work of the rhythmically beating cardiac muscle of your heart and the smooth muscle in the walls of your blood vessels, which helps maintain blood pressure. Smooth muscle in organs of the digestive, urinary, and reproductive tracts propels, or squeezes, substances (foodstuffs, urine, a baby) through the organs and along the tract.

Maintaining Posture

We are rarely aware of the workings of the skeletal muscles that maintain body posture. Yet these muscles function almost continuously, making one tiny adjustment after another that enables us to maintain an erect or seated posture despite the neverending downward pull of gravity.

Stabilizing Joints

Even as muscles pull on bones to cause movements, they stabilize and strengthen the joints of the skeleton (Chapter 8).

Generating Heat

Finally, muscles generate heat as they contract. This heat is vitally important in maintaining normal body temperature. Because skeletal muscle accounts for at least 40% of body mass, it is the muscle type most responsible for generating heat.

* * *

In the following section, we examine the structure and functioning of skeletal muscle in detail. Then we consider smooth muscle more briefly, largely by comparing it with skeletal muscle. Because cardiac muscle is described in Chapter 17, its treatment in this chapter is limited to the summary of its characteristics provided in Table 9.3.

Skeletal Muscle

The levels of skeletal muscle organization, gross to microscopic, are summarized in Table 9.1.

Gross Anatomy of a Skeletal Muscle

Each **skeletal muscle** is a discrete organ, made up of several kinds of tissues. Although skeletal muscle fibers predominate, blood vessels, nerve fibers, and substantial amounts of connective tissue are also

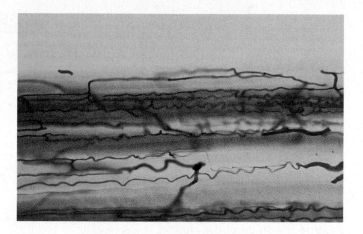

FIGURE 9.1 **Photomicrograph of the capillary network surrounding skeletal muscle fibers.** The arterial supply was injected with dark red gelatin to demonstrate the capillary bed. The long muscle fibers are stained orange (30×).

present. A skeletal muscle's shape and its attachments in the body can be examined easily without the help of a microscope.

Nerve and Blood Supply

In general, each muscle is served by one nerve, an artery, and by one or more veins, all of which enter or exit near the central part of the muscle and branch profusely through its connective tissue sheaths (described below). Unlike cells of cardiac and smooth muscle tissues, which can contract in the absence of nerve stimulation, each skeletal muscle fiber is supplied with a nerve ending that controls its activity.

Contracting muscle fibers use huge amounts of energy, a situation that requires more or less continuous delivery of oxygen and nutrients via the arteries. Muscle cells also give off large amounts of metabolic wastes that must be removed through veins if contraction is to remain efficient. Muscle capillaries, the smallest of the body's blood vessels, are long and winding and have numerous cross-links, features that accommodate changes in muscle length (Figure 9.1). They straighten when the muscle is stretched and contort when the muscle contracts.

Connective Tissue Sheaths

In an intact muscle, the individual muscle fibers are wrapped and held together by several different connective tissue sheaths. Together these connective tissue sheaths support each cell and reinforce the muscle as a whole. We will consider these from internal to external (Figure 9.2).

1. Endomysium. Each individual muscle fiber is surrounded by a fine sheath of connective tissue consisting mostly of reticular fibers. This is the endomysium (en"do-mis'e-um; "within the muscle").

TABLE 9.1   Structure and Organizational Levels of Skeletal Muscle

Structure and Organizational Level	Description	Connective Tissue Wrappings
Muscle (organ) 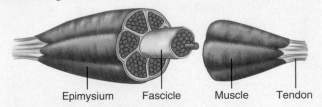 Epimysium Fascicle Muscle Tendon	Consists of hundreds to thousands of muscle cells, plus connective tissue wrappings, blood vessels, and nerve fibers	Covered externally by the epimysium
Fascicle (a portion of the muscle) 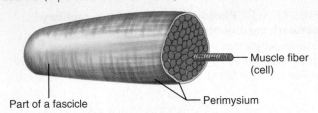 Part of a fascicle Muscle fiber (cell) Perimysium	Discrete bundle of muscle cells, segregated from the rest of the muscle by a connective tissue sheath	Surrounded by a perimysium
Muscle fiber (cell) 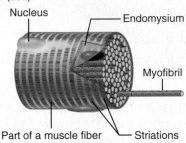 Nucleus Endomysium Myofibril Part of a muscle fiber Striations	Elongated multinucleate cell; has a banded (striated) appearance	Surrounded by the endomysium
Myofibril or fibril (complex organelle composed of bundles of myofilaments) 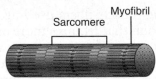 Sarcomere Myofibril	Rodlike contractile element; myofibrils occupy most of the muscle cell volume; appear banded, and bands of adjacent myofibrils are aligned; composed of sarcomeres arranged end to end	
Sarcomere (a segment of a myofibril) 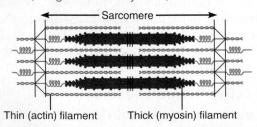 Sarcomere Thin (actin) filament Thick (myosin) filament	The contractile unit, composed of myofilaments made up of contractile proteins	
Myofilament or filament (extended macromolecular structure) 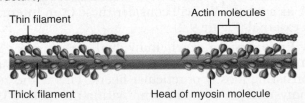 Thin filament Actin molecules Thick filament Head of myosin molecule	Contractile myofilaments are of two types—thick and thin: the thick filaments contain bundled myosin molecules; the thin filaments contain actin molecules (plus other proteins); the sliding of the thin filaments past the thick filaments produces muscle shortening. Elastic filaments (not shown here) maintain the organization of the A band and provide for elastic recoil when muscle contraction ends	

Consider the term epimysium. What is the meaning of epi? Of mys? How do these word stems relate to the role and position of this connective tissue sheath?

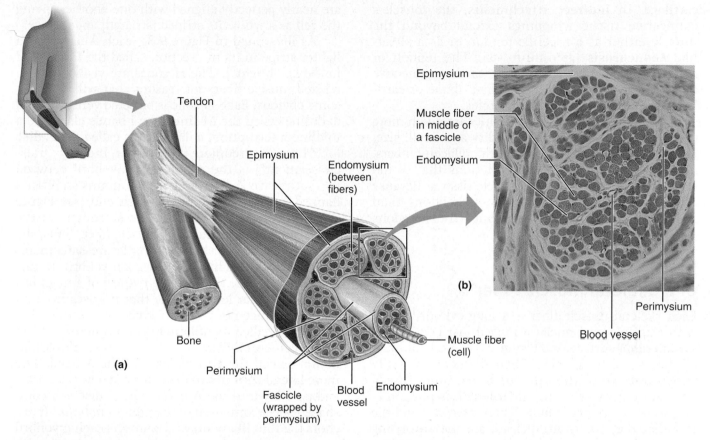

FIGURE 9.2 **Connective tissue sheaths of skeletal muscle.** **(a)** Each muscle fiber is wrapped with a delicate connective tissue sheath, the endomysium. Bundles, or fascicles, of muscle fibers are bound by a collagenic sheath called a perimysium. The entire muscle is strengthened and wrapped by a coarse epimysium sheath. **(b)** Photomicrograph of a cross section of part of a skeletal muscle (90×).

2. Perimysium and fascicles. Within each skeletal muscle, the endomysium-wrapped muscle fibers are grouped into fascicles (fas'i-klz; "bundles") that resemble bundles of sticks. Surrounding each fascicle is a layer of fibrous connective tissue called perimysium [per"i-mis'e-um; "around the muscle (fascicles)"].

3. Epimysium. An "overcoat" of dense irregular connective tissue surrounds the whole muscle. This coat is the epimysium (ep"i-mis'e-um), a name that means "outside the muscle." Sometimes the epimysium blends with the deep fascia that lies between neighboring muscles or the superficial fascia deep to the skin.

As shown in Figure 9.2, all of these connective tissue sheaths are continuous with one another as well as with the tendons that join muscles to bones. Therefore, when muscle fibers contract, they pull on these sheaths, which in turn transmit the pulling force to the bone to be moved. They also contribute to the natural elasticity of muscle tissue, and for this reason these elements are sometimes referred to collectively as the *series elastic components.* The wrappings also provide entry and exit routes for the blood vessels and nerve fibers that serve the muscle.

Attachments

Recall from Chapter 8 that (1) most skeletal muscles span joints and are attached to bones (or other structures) in at least two places and (2) when a muscle contracts, the movable bone, the muscle's **insertion,** moves toward the immovable or less movable bone, the muscle's **origin.** In the muscles of the limbs, the origin typically lies proximal to the insertion.

epi = outer; mys = muscle. The epimysium covers the external surface of the muscle. ■

Muscle attachments, whether origin or insertion, may be direct or indirect. In **direct,** or **fleshy, attachments,** the epimysium of the muscle is fused to the periosteum of a bone or perichondrium of a cartilage. In **indirect attachments,** the muscle's connective tissue wrappings extend beyond the muscle either as a ropelike tendon or as a sheet-like **aponeurosis** (ap"o-nu-ro'sis). The tendon or aponeurosis anchors the muscle to the connective tissue covering of a skeletal element (bone or cartilage) or to the fascia of other muscles.

Of the two, indirect attachments are much more common because of their durability and small size. Because tendons are mostly tough collagenic fibers, they can cross rough bony projections that would tear apart the more delicate muscle tissues. Because of their relatively small size, more tendons than fleshy muscles can pass over a joint—thus, tendons also conserve space.

Microscopic Anatomy of a Skeletal Muscle Fiber

Each skeletal muscle fiber is a long cylindrical cell with multiple oval nuclei arranged just beneath its **sarcolemma** surface (see Figure 9.3a). Skeletal muscle fibers are huge cells. Their diameter typically ranges from 10 to 100 µm—up to ten times that of an average body cell—and their length is phenomenal, some up to 30 cm long. Their large size and the fact that they are multinucleate are not surprising once you learn that each fiber is actually a *syncytium* (sin-sish'e-um; "fused cells") produced by the fusion of hundreds of embryonic cells.

The **sarcoplasm** of a muscle fiber is similar to the cytoplasm of other cells, but it contains unusually large amounts of *glycosomes* (granules of stored glycogen) and substantial amounts of an oxygen-binding protein called *myoglobin.* **Myoglobin,** a red pigment that stores oxygen, is similar to hemoglobin, the pigment that transports oxygen in blood. The usual organelles are present, along with some that are highly modified in muscle fibers: myofibrils and the sarcoplasmic reticulum. T tubules are unique modifications of the sarcolemma. Let's look at these special structures more closely.

Myofibrils

Each muscle fiber contains a large number of rodlike **myofibrils** that run parallel to its length (see Figure 9.3b). With a diameter of 1–2 µm each, the myofibrils are so densely packed in the fiber that mitochondria and other organelles appear to be squeezed between them. Hundreds to thousands of myofibrils are in a single muscle fiber, depending on its size, and they account for about 80% of cellular volume. The myofibrils contain the contractile elements of skeletal muscle cells.

Striations, Sarcomeres, and Myofilaments

Striations, a repeating series of dark **A bands** and light **I bands,** are evident along the length of each myofibril. In an intact muscle fiber, the A and I bands are nearly perfectly aligned with one another, giving the cell as a whole its striped (striated) appearance.

As illustrated in Figure 9.3c, each A band has a lighter stripe in its midsection called the **H zone** (*H* for *helle;* "bright"). The H zones are visible only in relaxed muscle fibers for reasons that will soon become obvious. Each H zone is bisected vertically by a dark line called the **M line.** The I bands also have a midline interruption, a darker area called the **Z disc** (or Z line). A **sarcomere** (sar'ko-měr; literally, "muscle segment") is the region of a myofibril between two successive Z discs, that is, it contains an A band flanked by half an I band at each end (see Figure 9.3c). Averaging 2 µm long, the sarcomere is the smallest contractile unit of a muscle fiber. Thus, the *functional units* of skeletal muscle are sarcomeres aligned end-to-end like boxcars in a myofibril "train."

If we examine the *banding pattern* of a myofibril at the molecular level, we see that it arises from an orderly arrangement of two types of even smaller structures, called **myofilaments** or **filaments,** within the sarcomeres (Figure 9.3d). The central **thick filaments** extend the entire length of the A band. The more lateral **thin filaments** extend across the I band and partway into the A band. The Z disc is a coin-shaped sheet composed of the protein *nebulin.* It anchors the thin filaments and connects each myofibril to the next throughout the width of the muscle cell. The H zone of the A band appears less dense because the thin filaments do not extend into this region. The M line in the center of the H zone is slightly darker because of the presence of fine protein strands that hold adjacent thick filaments together in that area. The third type of myofilament illustrated in Figure 9.3d, the *elastic filament,* is described in the next section.

A longitudinal view of the myofilaments, such as that in Figure 9.3d, is a bit misleading because it gives the impression that each thick filament interdigitates with only four thin filaments. In areas where thick and thin filaments overlap, each thick filament is actually surrounded by a hexagonal arrangement of six thin filaments.

Ultrastructure and Molecular Composition of the Myofilaments

Thick filaments (about 16 nm in diameter) are composed primarily of the protein **myosin** (Figure 9.4a). Each myosin molecule has a rodlike *tail* terminating in two globular *heads.* The tail consists of two interwoven *heavy* polypeptide chains. Its globular heads are the ends of the heavy chains, and the "business end" of myosin. The heads link the thick and thin filaments together (form **cross bridges**) during contraction. As

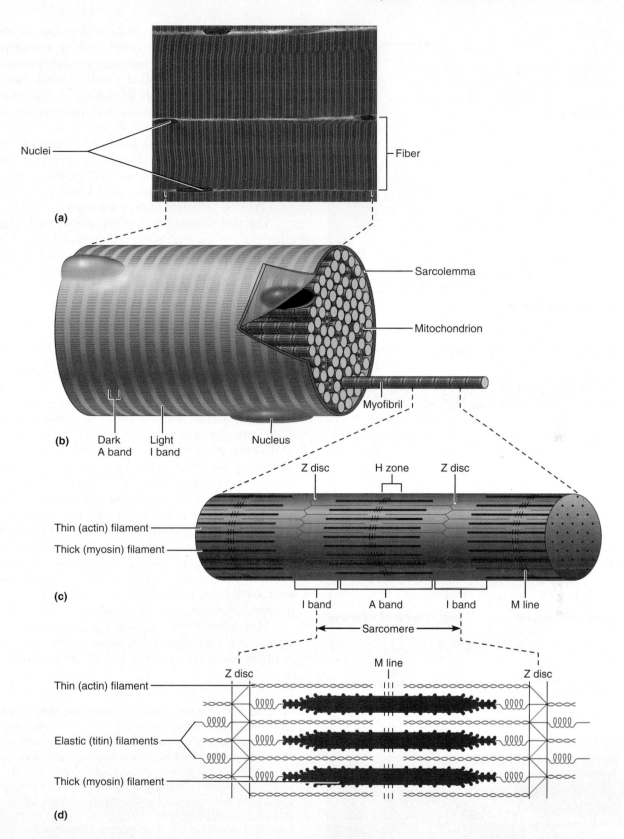

FIGURE 9.3 **Microscopic anatomy of a skeletal muscle fiber. (a)** Photomicrograph of portions of two isolated muscle fibers (700×). Notice the obvious striations (alternating light and dark bands). **(b)** Diagram of part of a muscle fiber showing the myofibrils. One myofibril extends from the cut end of the fiber. **(c)** A small portion of one myofibril enlarged to show the myofilaments responsible for the banding pattern. Each sarcomere extends from one Z disc to the next. **(d)** Enlargement of one sarcomere (sectioned lengthwise). Notice the myosin heads on the thick filaments.

Tail   Heads

(a) **Myosin molecule**

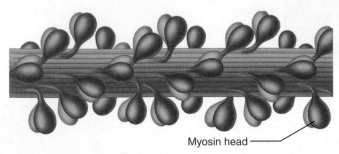

Myosin head

(b) **Portion of a thick filament**

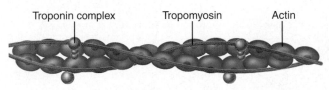

Troponin complex   Tropomyosin   Actin

(c) **Portion of a thin filament**

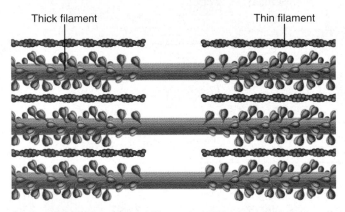

Thick filament   Thin filament

(d) **Longitudinal section of filaments within one sarcomere
   of a myofibril**

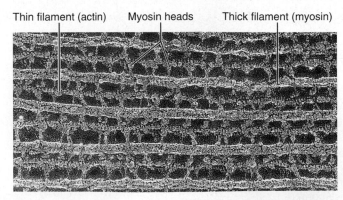

Thin filament (actin)   Myosin heads   Thick filament (myosin)

(e) **Transmission electron micrograph of part of a sarcomere**

explained shortly, these cross bridges act as motors to generate the tension developed by a contracting muscle cell. Each thick filament contains about 200 myosin molecules bundled together with their tails forming the central part of the thick filament and their heads facing outward and in opposite directions at each end (Figure 9.4b and d). As a result, the central portion of a thick filament is smooth, but its ends are studded with a staggered array of myosin heads. Besides bearing actin binding sites, the heads contain ATP binding sites and ATPase enzymes that split ATP to generate energy for muscle contraction.

The thin filaments (7–8 nm thick) are composed chiefly of the protein **actin** (Figure 9.4c). The kidney-shaped polypeptide subunits of actin, called *globular actin* or *G actin*, bear the active sites to which the myosin heads attach during contraction. G actin monomers are polymerized into long actin filaments called *fibrous,* or *F, actin*. The backbone of each thin filament appears to be formed by an actin filament that coils back on itself, forming a helical structure that looks like a twisted double strand of pearls.

Several regulatory proteins are also present in the thin filament. Two strands of **tropomyosin** (tro″po-mi′o-sin), a rod-shaped protein, spiral about the actin core and help stiffen it. Successive tropomyosin molecules are arranged end-to-end along the actin filaments, and in a relaxed muscle fiber, they block actin's active sites so that the myosin heads cannot bind to the thin filaments. The other major protein in the thin filament, **troponin** (tro′po-nin), is a three-polypeptide complex. One of these polypeptides (TnI) is an inhibitory subunit that binds to actin; another (TnT) binds to tropomyosin and helps position it on actin. The third (TnC) binds calcium ions. Both troponin and tropomyosin help control the myosin-actin interactions involved in contraction.

The discoveries of additional types of muscle filaments during the past decade or so demand that the established view of striated muscle as a two-filament

FIGURE 9.4   Composition of thick and thin filaments.
(a) An individual myosin molecule has a rodlike tail, from which two "heads" protrude. **(b)** Each thick filament consists of many myosin molecules, whose heads protrude at opposite ends of the filament, as shown in (d). **(c)** A thin filament contains a strand of actin twisted into a helix. Each strand is made up of actin subunits. Tropomyosin molecules coil around the actin filaments, helping to reinforce them. A troponin complex is attached to each tropomyosin molecule. **(d)** Arrangement of the filaments in a sarcomere (longitudinal view). In the center of the sarcomere, the thick filaments are devoid of myosin heads, but at points of thick and thin filament overlap, the heads extend toward the actin, with which they interact during contraction. **(e)** Transmission electron micrograph of part of a sarcomere clearly showing the myosin heads that generate the contractile force.

Which of the following structures shown in this diagram would contain the highest concentration of Ca^{2+} in a resting muscle cell: T tubule, mitochondrion, or SR?

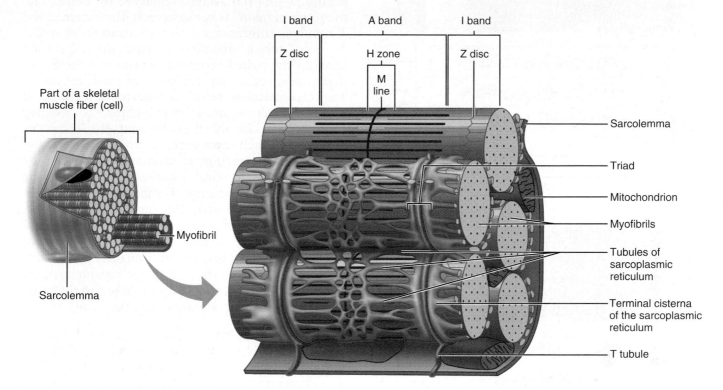

FIGURE 9.5 Relationship of the sarcoplasmic reticulum and T tubules to myofibrils of skeletal muscle. The tubules of the SR encircle each myofibril like a "holey" sleeve. The tubules fuse to form communicating channels at the level of the H zone and abutting the A-I junctions, where the saclike elements called terminal cisternae are formed. The T tubules, inward invaginations of the sarcolemma, run deep into the cell between the terminal cisternae. Sites of close contact of these three elements (terminal cisterna, T tubule, and terminal cisterna) are called triads.

system be rewritten. One of these newly discovered filament types, the **elastic filament** referred to earlier, is composed of the giant protein **titin,** which extends from the Z disc to the thick filament, and then runs within the latter to attach to the M line. It has two basic functions: (1) holding the thick filaments in place, thus maintaining the organization of the A band, and (2) assisting the muscle cell to spring back into shape after being stretched. It can perform the second task because the part of the titin that spans the I bands is extensible, unfolding when the muscle is stretched and recoiling when the tension is released. Titin does not resist stretching in the ordinary range of extension, but it stiffens as it uncoils, helping the muscle to resist excessive stretching, which might pull the sarcomeres apart.

■ *The SR, which is a calcium storage depot.*

Sarcoplasmic Reticulum and T Tubules

Skeletal muscle fibers (cells) contain two sets of intracellular tubules that participate in regulation of muscle contraction: (1) the sarcoplasmic reticulum and (2) the T tubules (Figure 9.5).

Sarcoplasmic Reticulum The **sarcoplasmic reticulum (SR)** is an elaborate smooth endoplasmic reticulum (see pp. 81–82). Its interconnecting tubules surround each myofibril the way the sleeve of a loosely crocheted sweater surrounds your arm. Most of these tubules run longitudinally along the myofibril. Others form larger, perpendicular cross channels at the A band–I band junctions. These channels are called **terminal cisternae** ("end sacs") and they always occur in pairs.

The major role of the SR is to regulate intracellular levels of ionic calcium: It stores calcium and releases it on demand when the muscle fiber is stimulated to contract. As you will see, calcium provides the final "go" signal for contraction.

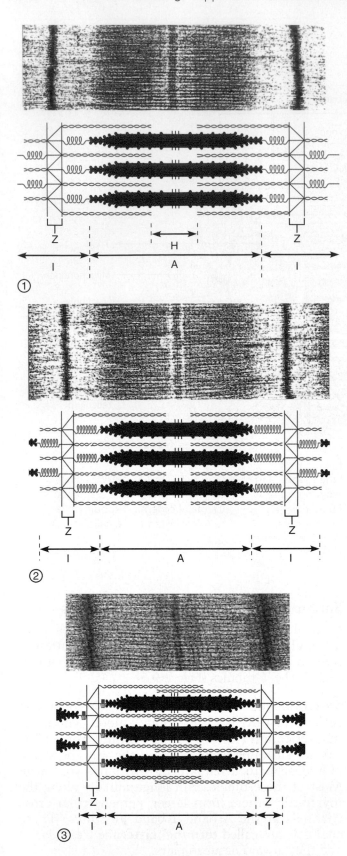

FIGURE 9.6   Sliding filament model of contraction.
The numbers indicate the sequence of events, with
① being relaxed and ③ fully contracted. At full contraction,
the Z discs abut the thick filaments and the thin filaments
overlap each other. The photomicrographs (top view in each
case) show enlargements of 25,000×.

T Tubules   At each A band–I band junction, the sarcolemma of the muscle cell penetrates into the cell interior to form an elongated tube called the **T tubule** (T for "transverse"). Believed by some to be the result of fusing tubelike caveoli, the lumen of the T tubule is continuous with the extracellular space. As each T tubule protrudes deep into the cell, it runs between the paired terminal cisternae of the SR so that **triads,** successive groupings of the three membranous structures (terminal cisterna, T tubule, and terminal cisterna), are formed (Figure 9.5). As they pass from one myofibril to the next, the T tubules also encircle each sarcomere.

Muscle contraction is ultimately controlled by nerve-initiated electrical impulses that travel along the sarcolemma. Because T tubules are continuations of the sarcolemma, they can conduct impulses to the deepest regions of the muscle cell and to every sarcomere. These impulses signal for the release of calcium from the adjacent terminal cisternae. Thus, the T tubules can be thought of as a rapid telegraphy system that ensures that every myofibril in the muscle fiber contracts at virtually the same time.

Triad Relationships   The roles of the T tubules and SR in providing signals for contraction are tightly linked. At the triads, where these organelles come into closest contact, something that resembles a *double zipper* of integral proteins protrudes into the intermembrane spaces. The protruding integral proteins of the T tubule act as voltage sensors; and those of the SR, called *foot proteins,* are receptors that regulate the release of Ca^{2+} from the SR cisternae. We will return to consider their interaction shortly.

Sliding Filament Model of Contraction

Although we almost always think "shortening" when we hear the word **contraction,** to physiologists the term refers only to the activation of myosin's cross bridges, which are the force-generating sites. Shortening occurs when the tension generated by the cross bridges on the thin filaments exceeds the forces opposing shortening. Contraction ends when the cross bridges become inactive and the tension generated declines, inducing **relaxation** of the muscle fiber. Proposed in 1954 by Hugh Huxley, the **sliding filament theory of contraction** states that during contraction the thin filaments slide past the thick ones so that the actin and myosin filaments overlap to a greater degree (Figure 9.6). In a relaxed muscle fiber, the thick and thin filaments overlap only slightly. But when muscle fibers are stimulated by the nervous system, the cross bridges latch on to myosin binding sites on actin in the thin filaments, and the sliding begins. Each cross bridge attaches and detaches several times during a contraction, acting like a tiny ratchet to generate tension and propel the thin filaments toward the center of the sarcomere. As this event occurs simultaneously in sarcomeres throughout the cell, the muscle cell shortens. Notice that as the thin filaments slide centrally, the Z discs to which they are attached are pulled *toward* the thick filaments. Overall, the distance between successive Z discs is reduced, the I bands shorten, the H zones disappear, and the contiguous A bands move closer together but do not change in length.

Physiology of a Skeletal Muscle Fiber

For a skeletal muscle fiber to contract, it must be stimulated by a nerve ending and must propagate an electrical current, or **action potential,** along its sarcolemma. This electrical event causes the short-lived rise in intracellular calcium ion levels that is the final trigger for contraction. The series of events linking the electrical signal to contraction is called *excitation-contraction coupling.* Let's consider these events in order.

The Neuromuscular Junction and the Nerve Stimulus

Skeletal muscle cells are stimulated by *motor neurons** of the somatic nervous system. Although these motor neurons "reside" in the brain or spinal cord, their long threadlike extensions called *axons* travel, bundled within nerves, to the muscle cells they serve. The axon of each motor neuron divides profusely as it enters the muscle, and each axonal ending forms a branching **neuromuscular junction** with a single muscle fiber (Figure 9.7). As a rule, each muscle fiber has only one neuromuscular junction,

---

*Somatic motor neurons are nerve cells that activate skeletal muscle cells.

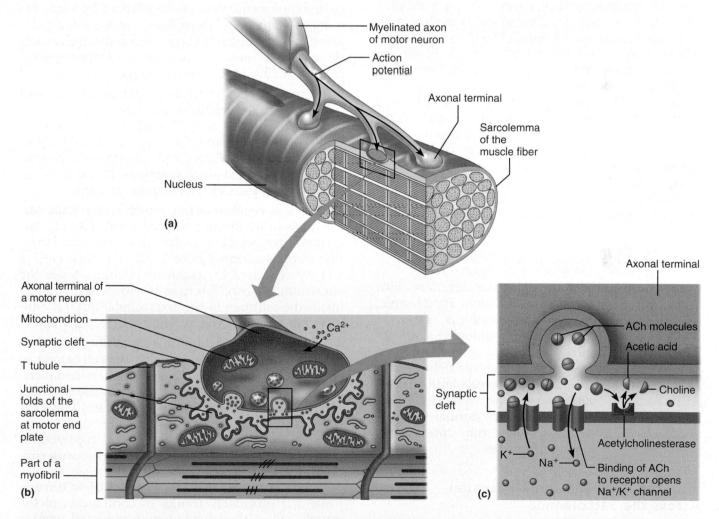

FIGURE 9.7 The neuromuscular junction. (a) Axonal ending of a motor neuron forming a neuromuscular junction with a muscle fiber. **(b)** Each axonal terminal contains vesicles filled with the neurotransmitter acetylcholine (ACh), which is released when the action potential reaches the axonal terminal. The sarcolemma is highly folded adjacent to the synaptic cleft, and ACh receptors are present in these junctional folds. **(c)** ACh diffuses across the synaptic cleft and attaches to receptors on the sarcolemma, opening ion channels and initiating depolarization of the sarcolemma.

located approximately midway along the fiber's length. Although the axonal ending and the muscle fiber are exceedingly close (1–2 nm apart), they remain separated by a space, the **synaptic cleft,** filled with a gel-like extracellular substance rich in glycoproteins. Within the flattened moundlike axonal ending are **synaptic vesicles,** small membranous sacs containing the neurotransmitter **acetylcholine** (as"ĕ-til-ko'lēn), or **ACh.** The **motor end plate,** the troughlike part of the muscle fiber's sarcolemma that helps form the neuromuscular junction, is highly folded. These *junctional folds* provide a large surface area for the millions of **ACh receptors** located there.

When a nerve impulse reaches the end of an axon, voltage-gated calcium channels in its membrane open, allowing Ca^{2+} to flow in from the extracellular fluid. The presence of calcium inside the axon terminal causes some of the synaptic vesicles to fuse with the axonal membrane and release ACh into the synaptic cleft by *exocytosis.* ACh diffuses across the cleft and attaches to the flowerlike ACh receptors on the sarcolemma. The electrical events triggered in a sarcolemma when ACh binds are similar to those that take place in excited nerve cell membranes (see Chapter 11).

After ACh binds to the ACh receptors, it is swiftly broken down to its building blocks, acetic acid and choline, by **acetylcholinesterase** (as"ĕ-til-ko"lin-es'ter-ās), an enzyme located on the sarcolemma at the neuromuscular junction and in the synaptic cleft (Figure 9.7c). This destruction of ACh prevents continued muscle fiber contraction in the absence of additional nervous system stimulation.

Ⓗ HOMEOSTATIC IMBALANCE

Many toxins, drugs, and diseases interfere with events at the neuromuscular junction. For example, *myasthenia gravis* (*asthen* = weakness; *gravi* = heavy), a disease characterized by drooping of the upper eyelids, difficulty swallowing and talking, and generalized muscle weakness, involves a shortage of ACh receptors. Serum analysis reveals antibodies to ACh receptors, suggesting that myasthenia gravis is an autoimmune disease. Although normal numbers of receptors are initially present, they appear to be destroyed as the disease progresses. ●

Generation of an Action Potential Across the Sarcolemma

Like the plasma membranes of all cells, a resting sarcolemma is *polarized.* That is, a voltmeter would show there is a potential difference (voltage) across the membrane and the inside is negative relative to the outer membrane face. (The resting membrane potential is described on pp. 76–77.) Binding of ACh

molecules to ACh receptors at the motor end plate opens chemically (ligand) gated ion channels housed in the ACh receptors that allow both Na^+ and K^+ to pass (Figure 9.7). Because more Na^+ diffuses in than K^+ diffuses out, a transient change in membrane potential occurs such that the interior of the sarcolemma becomes slightly less negative, an event called **depolarization.**

Initially, depolarization is a local electrical event called an *end plate potential,* but it ignites the action potential that spreads in all directions from the neuromuscular junction across the sarcolemma, just as ripples move away from pebbles dropped into a stream.

The action potential (AP) is the result of a predictable sequence of electrical changes that once initiated occurs along the entire length of the sarcolemma (Figure 9.8). Essentially three steps are involved.

1. First the membrane areas adjacent to the depolarized motor end plate are depolarized by local currents that spread to them from the neuromuscular junction. This opens voltage-gated sodium channels there, so Na^+ enters, following its electrochemical gradient, and initiates the action potential.

2. During step 2, the action potential is propagated (moves along the length of the sarcolemma) as the local depolarization wave spreads to adjacent areas of the sarcolemma and opens voltage-gated sodium channels there (see Figure 9.8c). Again, sodium ions, normally restricted from entering, diffuse into the cell following their electrochemical gradient.

3. Step 3 is **repolarization,** which restores the sarcolemma to its initial polarized state. The repolarization wave, which quickly follows the depolarization wave, is a consequence of Na^+ channels closing and voltage-gated K^+ channels opening. Since the potassium ion concentration is substantially higher inside the cell than in the extracellular fluid, K^+ diffuses rapidly out of the muscle fiber (Figure 9.8d).

During repolarization, a muscle fiber is said to be in a **refractory period,** because the cell cannot be stimulated again until repolarization is complete. Notice that repolarization restores only the *electrical conditions* of the resting (polarized) state. The ATP-dependent Na^+-K^+ pump restores the ionic conditions of the resting state, but several contractions can occur before ionic imbalances interfere with contractile activity.

Once initiated, the action potential is unstoppable and ultimately results in contraction of the muscle cell. Although the action potential itself is very brief [1–2 milliseconds (ms)], the contraction phase of a muscle fiber may persist for 100 ms or more and far outlasts the electrical event that triggers it because active transport of Ca^{2+} back into the SR takes substantially longer than its release.

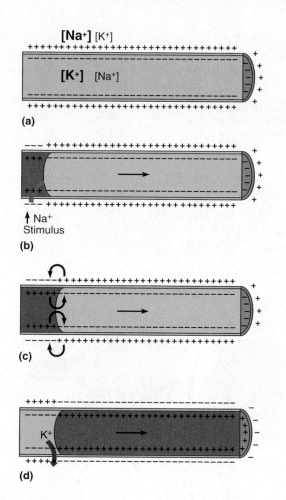

(a) **Electrical conditions of a resting (polarized) sarcolemma.** The outside face is positive, while the inside face is negative. The predominant extracellular ion is sodium (Na⁺); the predominant intracellular ion is potassium (K⁺). The sarcolemma is relatively impermeable to both ions.

(b) **Step 1: Depolarization and generation of the action potential.** Production of an end plate potential at the motor end plate causes adjacent areas of the sarcolemma to become permeable to sodium (voltage-gated sodium channels open). As sodium ions diffuse rapidly into the cell, the resting potential is decreased (i.e., depolarization occurs). If the stimulus is strong enough, an action potential is initiated.

(c) **Step 2: Propagation of the action potential.** The positive charge inside the initial patch of sarcolemma changes the permeability of an adjacent patch, opening voltage-gated Na⁺ channels there. Consequently the membrane potential in that region decreases and depolarization occurs there as well. Thus, the action potential travels rapidly over the entire sarcolemma.

(d) **Step 3: Repolarization.** Immediately after the depolarization wave passes, the sarcolemma's permeability changes once again: Na⁺ channels close and K⁺ channels open, allowing K⁺ to diffuse from the cell. This restores the electrical conditions of the resting (polarized) state. Repolarization occurs in the same direction as depolarization, and must occur before the muscle fiber can be stimulated again. The ionic concentrations of the resting state are restored later by the sodium-potassium pump.

FIGURE 9.8 **Summary of events in the generation and propagation of an action potential in a skeletal muscle fiber.**

Excitation-Contraction Coupling

Excitation-contraction (E-C) coupling is the sequence of events by which transmission of an action potential along the sarcolemma leads to the sliding of myofilaments. The action potential is brief and ends well before any signs of contraction are obvious. During the *latent period* (*laten* = hidden), between action potential initiation and the beginning of mechanical activity (shortening), the events of excitation-contraction coupling occur. As you will see, the electrical signal does not act directly on the myofilaments; rather, it causes the rise in intracellular calcium ion concentration that allows the filaments to slide (Figure 9.9).

Excitation-contraction coupling consists of the following steps:

① The action potential propagates along the sarcolemma and down the T tubules.

② Transmission of the action potential past the triads causes the terminal cisternae of the SR to release Ca^{2+} into the sarcoplasm, where it becomes avail-

able to the myofilaments. The "double zippers" (p. 252) at the T-SR junctions of the triads are involved in this action as follows: The T tubule proteins are sensitive to voltage and change shape in response to the arrival of the action potential. This voltage-regulated change is communicated to the SR foot proteins, which in turn undergo shape changes that open their calcium channels (and perhaps other mechanically linked Ca^{2+} channels nearby). Because these events occur at every triad in the cell, within 1 ms massive amounts of Ca^{2+} flood into the sarcoplasm from the SR cisternae.

③ Some of this calcium binds to troponin, which changes shape and removes the blocking action of tropomyosin.

④ When the intracellular calcium is about $10^{-5}\,M$, the myosin heads attach and pull the thin filaments toward the center of the sarcomere.

⑤ The short-lived Ca^{2+} signal ends, usually within 30 ms after the action potential is over. The fall in Ca^{2+} levels reflects the operation of a continuously

(1) Why is Ca²⁺ called the final trigger for contraction? (2) What constitutes the initial trigger?

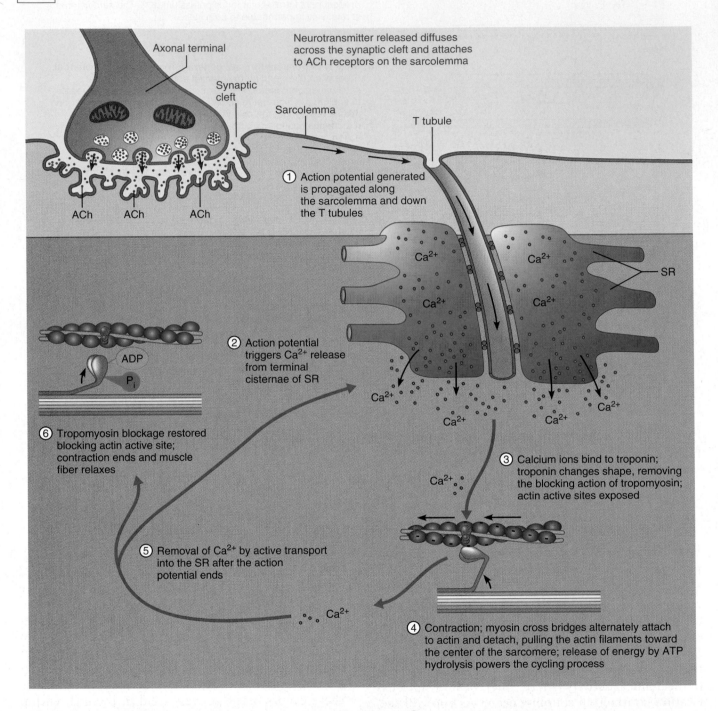

FIGURE 9.9 Excitation-contraction coupling. Events ① through ⑤ indicate the sequence of events in the coupling process. Contraction continues until the calcium signal ends ⑥ .

active, ATP-dependent calcium pump that moves Ca²⁺ back into the SR to be stored once again.

⑥ When intracellular Ca²⁺ levels drop too low to allow contraction, the tropomyosin blockade is reestablished and myosin ATPases are inhibited. Cross bridge activity ends and relaxation occurs.

(1) Because calcium binding to troponin frees the actin active sites to bind with myosin cross bridges. (2) The initial trigger is neurotransmitter binding, which initiates generation of an action potential along the sarcolemma. ■

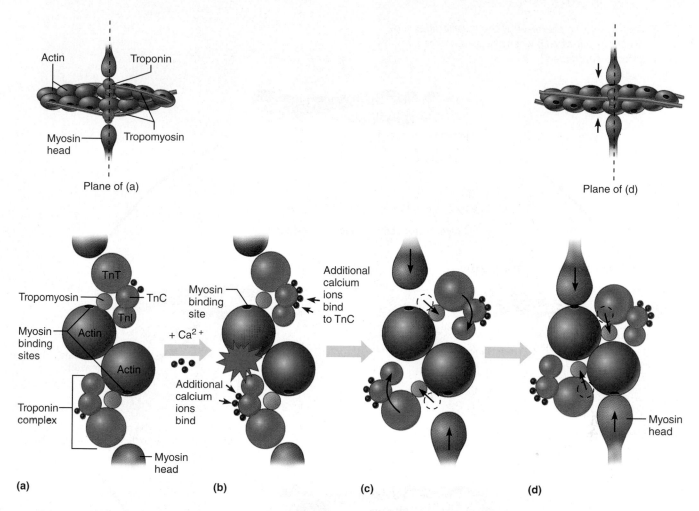

FIGURE 9.10   Role of ionic calcium in the contraction mechanism. Views (a–d) are cross-sectional views of the thin (actin) filament. **(a)** At low intracellular Ca^{2+} concentration, tropomyosin blocks the binding sites on actin, preventing attachment of myosin cross bridges and enforcing the relaxed muscle state. **(b)** At higher intracellular Ca^{2+} concentrations, additional calcium binds to (TnC) of troponin. **(c)** Calcium-activated troponin undergoes a conformational change that moves the tropomyosin away from actin's binding sites. **(d)** This displacement allows the myosin heads to bind and cycle, and contraction (sliding of the thin filaments by the myosin cross bridges) begins.

This sequence of events is repeated when another nerve impulse arrives at the neuromuscular junction. When the impulses are delivered rapidly, intracellular Ca^{2+} levels increase greatly due to successive "puffs" or rounds of Ca^{2+} release from the SR. In such cases, the muscle cells do not completely relax between successive stimuli and contraction is stronger and more sustained (within limits) until nervous stimulation ceases.

Summary of Roles of Ionic Calcium in Muscle Contraction

Except for the brief period following muscle cell excitation, calcium ion concentrations in the sarcoplasm are kept almost undetectably low. There is a reason for this: ATP provides the cell's energy source and its hydrolysis yields inorganic phosphates (P_i). If the intracellular level of Ca^{2+} were always high, calcium and phosphate ions would combine to form hydroxyapatite crystals, the stony-hard salts found in bone matrix. Such calcified cells would die. Also, because calcium's physiological roles are so vital (i.e., besides triggering neurotransmitter secretion, release of Ca^{2+} from the SR, and sliding of the myofilaments, it promotes breakdown of glycogen and ATP synthesis), its cytoplasmic concentration is exquisitely regulated by intracellular proteins. These include **calsequestrin** (found within the SR cisternae) and **calmodulin**, which can alternately bind and release calcium to provide a metabolic signal.

Muscle Fiber Contraction

As noted, cross bridge attachment to actin requires Ca^{2+}. When intracellular calcium levels are low, the muscle cell is relaxed, and the active (myosin binding) sites on actin are physically blocked by tropomyosin molecules (Figure 9.10a). As Ca^{2+} levels rise, the ions bind to regulatory sites on troponin

What would be the result if the muscle fiber suddenly ran out of ATP when the sarcomeres had only partially contracted?

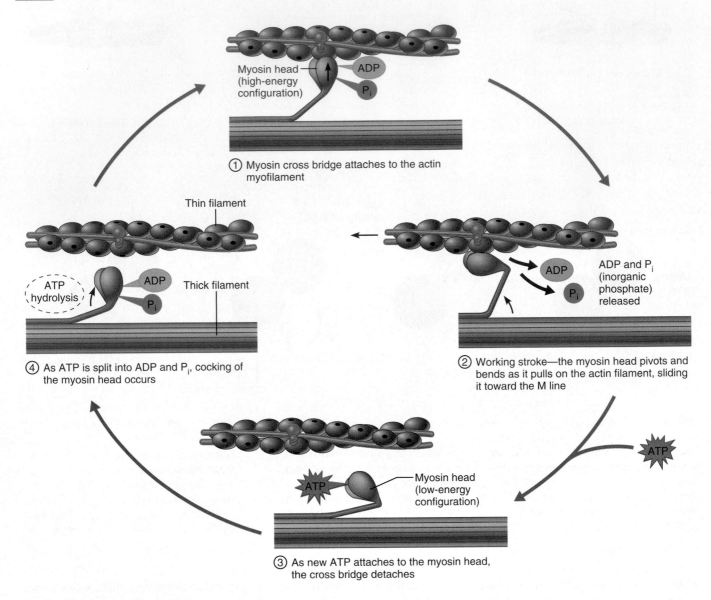

① Myosin cross bridge attaches to the actin myofilament

Myosin head (high-energy configuration)

ADP

P_i

Thin filament

ATP hydrolysis

ADP

P_i

Thick filament

④ As ATP is split into ADP and P_i, cocking of the myosin head occurs

ADP and P_i (inorganic phosphate) released

ADP

P_i

② Working stroke—the myosin head pivots and bends as it pulls on the actin filament, sliding it toward the M line

ATP

Myosin head (low-energy configuration)

ATP

③ As new ATP attaches to the myosin head, the cross bridge detaches

FIGURE 9.11 Sequence of events involved in the sliding of the thin filaments during contraction. A small section of adjacent thick and thin filaments is used to illustrate the interactions that occur between the two types of myofilaments. These events occur only in the presence of ionic calcium (Ca^{2+}), which releases tropomyosin's blockade of actin's active sites.

TnC (Figure 9.10b), causing it to change shape. This event moves tropomyosin deeper into the groove of the actin helix and away from the myosin binding sites (Figure 9.10c and d). Thus, the tropomyosin "blockade" is removed when sufficient calcium is present.

Once binding sites on actin are exposed, the following events occur in rapid succession (Figure 9.11).

① **Cross bridge formation.** The activated myosin heads are strongly attracted to the exposed binding sites on actin and cross bridges form.

② **The working (power) stroke.** As a myosin head binds, it pivots (moves through an angle of about 70°) changing from its high-energy configuration to its bent, low-energy shape, which pulls on the thin

■ *The sarcomeres would remain in their partially contracted state.*

filament, sliding it toward the center of the sarcomere. At the same time, inorganic phosphate (P_i) and ADP generated during the *prior* contraction cycle are released sequentially from the myosin head. Each working stroke of myosin produces a movement or step of 5–15 nm.

③ **Cross bridge detachment.** As a new ATP molecule binds to the myosin head, myosin's hold on actin loosens and the cross bridge detaches from actin.

④ **"Cocking" of the myosin head.** The ATPase in the myosin head hydrolyzes ATP to ADP and P_i which provides the energy needed to return the myosin head to its prestroke high-energy, or "cocked," position. This provides the potential energy needed for its *next* sequence of attachment and working stroke. (The ADP and P_i remain attached to the myosin head during this phase.)

At this point, the cycle is back where it started. The myosin head is in its upright high-energy configuration, ready to take another "step" and attach to an actin site farther along the thin filament. This "walking" of the myosin heads along the adjacent thin filaments during muscle shortening is much like a centipede's gait. Because some myosin heads ("legs") are always in contact with actin (the "ground"), the thin filaments cannot slide backward as the cycle is repeated again and again.

A single working stroke of all the cross bridges in a muscle results in a shortening of only about 1%. Because contracting muscles routinely shorten 30 to 35% of their total resting length, each myosin cross bridge must attach and detach many times during a single contraction. It is likely that only half of the myosin heads of a thick filament are exerting a pulling force at the same instant; the balance are randomly seeking their next binding site. Sliding of thin filaments continues as long as the calcium signal and adequate ATP are present. As the Ca^{2+} pumps of the SR reclaim calcium ions from the sarcoplasm and troponin again changes shape, the actin active sites are covered by tropomyosin, the contraction ends, and the muscle fiber relaxes.

HOMEOSTATIC IMBALANCE

Rigor mortis (death rigor) illustrates the fact that cross bridge detachment is ATP driven. Most muscles begin to stiffen 3 to 4 hours after death. Peak rigidity occurs at 12 hours and then gradually dissipates over the next 48 to 60 hours. Dying cells are unable to exclude calcium (which is in higher concentration in the extracellular fluid), and the calcium influx into muscle cells promotes formation of myosin cross bridges. Shortly after breathing stops, however, ATP synthesis ceases, and cross bridge detachment is impossible. Actin and myosin become irreversibly cross-linked, producing the stiffness of rigor mortis, which then disappears as muscle proteins break down several hours after death. ●

Contraction of a Skeletal Muscle

In its relaxed state, a muscle is soft and unimpressive; not at all what you would expect of a prime mover of the body. However, within a few milliseconds, it can contract to become a hard elastic structure with dynamic characteristics that intrigue not only biologists but engineers and physicists as well.

Before we consider muscle contraction on the organ level, let's review a few principles of muscle mechanics.

1. The principles governing contraction of a muscle fiber (cell) and of a skeletal muscle consisting of a huge number of cells are pretty much the same.

2. The force exerted by a contracting muscle on an object is called **muscle tension,** and the opposing force exerted on the muscle by the weight of the object to be moved is called the **load.**

3. A contracting muscle does not always shorten and move the load. If muscle tension develops but the load is not moved, the contraction is called *isometric* ("same measure"). If the muscle tension developed overcomes the load and muscle shortening occurs, the contraction is *isotonic.* These major types of contraction will be described in detail shortly, but for now the important thing to remember when reading the accompanying graphs is that *increasing muscle tension* is measured in isometric contractions, whereas the *amount of shortening* (distance in millimeters) is measured in isotonic contractions.

4. A skeletal muscle contracts with varying force and for different periods of time in response to stimuli of varying frequencies and intensities. To understand how this occurs, we must first look at the nerve-muscle functional unit called a *motor unit.* This is our next topic.

The Motor Unit

Each muscle is served by at least one motor nerve, which contains axons (fibrous extensions) of hundreds of motor neurons. As an axon enters a muscle, it branches into a number of terminals, each of which forms a neuromuscular junction with a single muscle fiber. A motor neuron and all the muscle fibers it supplies is called a **motor unit** (Figure 9.12). When a motor neuron fires (transmits an electrical impulse), all the muscle fibers it innervates contract. The number of muscle fibers per motor unit may be as high as several hundred or as few as four. Muscles that exert very fine control (such as those controlling

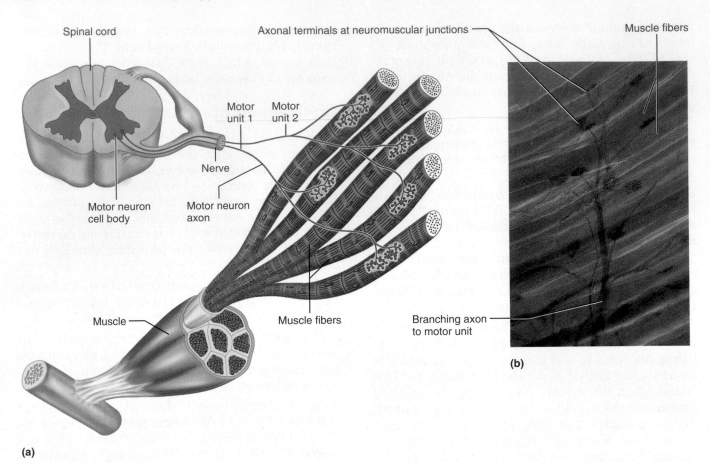

FIGURE 9.12 A motor unit consists of a motor neuron and all the muscle fibers it innervates. **(a)** Schematic view of portions of two motor units. The cell bodies of the motor neurons reside in the spinal cord, and their axons extend to the muscle. In the muscle, each axon divides into a number of axonal terminals that are distributed to muscle fibers scattered throughout the muscle. **(b)** Photomicrograph of a portion of a motor unit (80×). Notice the diverging axonal terminals and the neuromuscular junctions formed by the terminals and muscle fibers.

the fingers and eyes) have small motor units. By contrast, large, weight-bearing muscles, whose movements are less precise (such as the hip muscles), have large motor units. The muscle fibers in a single motor unit are not clustered together but are spread throughout the muscle. As a result, stimulation of a single motor unit causes a weak contraction of the *entire* muscle.

The Muscle Twitch

Muscle contraction is easily investigated in the laboratory using an isolated muscle. The muscle is attached to an apparatus that produces a **myogram,** a graphic recording of contractile activity. (The line recording the activity is called a *tracing.*)

The response of a motor unit to a single action potential of its motor neuron is called a **muscle twitch.** The muscle fibers contract quickly and then relax. Every twitch myogram has three distinct phases (Figure 9.13a).

1. **Latent period.** The latent period is the first few milliseconds following stimulation when excitation-contraction coupling is occurring. During this period, muscle tension is beginning to increase but no response is seen on the myogram.

2. **Period of contraction.** The period of contraction is when cross bridges are active, from the onset to the peak of tension development, and the myogram tracing rises to a peak. This period lasts 10–100 ms. If the tension (pull) becomes great enough to overcome the resistance of a load, the muscle shortens.

3. **Period of relaxation.** The period of contraction is followed by the period of relaxation. This final phase, lasting 10–100 ms, is initiated by reentry of Ca^{2+} into the SR. Because contractile force is no longer being generated, muscle tension decreases to zero and the tracing returns to the baseline. If the muscle shortened during contraction, it now returns to its initial length.

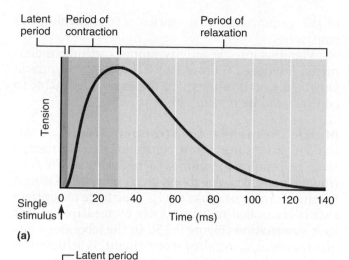

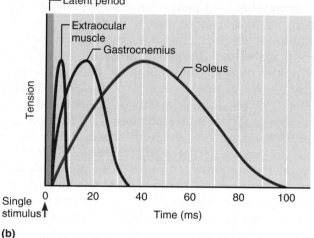

FIGURE 9.13 The muscle twitch. (a) Myogram of an isometric twitch contraction, showing its three phases: the latent period, the period of contraction, and the period of relaxation. **(b)** Comparison of the twitch responses of extraocular, gastrocnemius, and soleus muscles.

As you can see in Figure 9.13b, twitch contractions of some muscles are rapid and brief, as with the muscles controlling eye movements. In contrast, the fibers of fleshy calf muscles (gastrocnemius and soleus) contract more slowly and remain contracted for much longer periods. These differences between muscles reflect metabolic properties of the myofibrils and enzyme variations.

Graded Muscle Responses

Although *muscle twitches*—like those single, jerky contractions provoked in a laboratory—may occur as a result of certain neuromuscular problems, this is *not* the way our muscles normally operate. Instead, healthy muscle contractions are relatively smooth and vary in strength as different demands are placed on them. These variations (an obvious requirement for proper control of skeletal movement) are referred to as **graded muscle responses.** In general, muscle contraction can be graded in two ways: (1) by changing the frequency of stimulation and (2) by changing the strength of the stimulus.

Muscle Response to Changes in Stimulation Frequency   If two identical stimuli (electrical shocks or nerve impulses) are delivered to a muscle in rapid succession, the second twitch will be stronger than the first. On a myogram the second twitch will appear to ride on the shoulders of the first (Figure 9.14). This phenomenon, called **wave summation,** occurs because the second contraction occurs before the muscle has completely relaxed. Because the muscle is already partially contracted when the next stimulus arrives and more calcium is being released to replace that being reclaimed by the SR, muscle tension produced during the second contraction causes more shortening than the first. In other words, the contractions are summed. (However, the

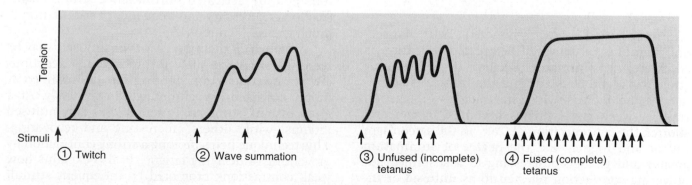

FIGURE 9.14 Wave summation and tetanus in a whole muscle. In ① a single stimulus is delivered, and the muscle contracts and relaxes (twitch contraction). In ② stimuli are delivered more frequently, so that the muscle does not have adequate time to relax completely, and contraction force increases (wave summation). In ③ more complete twitch fusion (unfused or incomplete tetanus) occurs as stimuli are delivered more rapidly. In ④, fused or complete tetanus, a smooth, continuous contraction without any evidence of relaxation occurs.

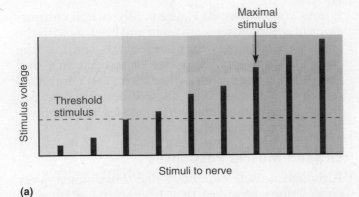

(a)

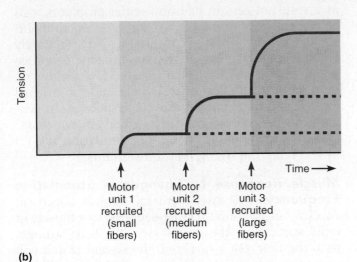

(b)

FIGURE 9.15 Relationship between stimulus intensity and muscle tension. (a) Below threshold voltage, no muscle response is seen on the tracing. Once threshold is reached, increases in voltage excite (recruit) more and more motor units until the maximal stimulus is reached. **(b)** At lower voltages, small motor neurons controlling motor units containing small-diameter muscle fibers are recruited. As the voltage is increased, recruitment proceeds to larger and larger motor neurons, which in turn stimulate increasingly larger diameter muscle fibers. Consequently, the contractions get stronger and stronger.

refractory period is *always* honored. Thus, if a second stimulus is delivered before repolarization is complete, no summation occurs.) If the stimulus strength is held constant and the muscle is stimulated at an increasingly faster rate, the relaxation time between the twitches becomes shorter and shorter, the concentration of Ca^{2+} in the sarcoplasm higher and higher, and the degree of summation greater and greater, progressing to a sustained but quivering contraction referred to as **unfused** or **incomplete tetanus.** Finally, all evidence of muscle relaxation disappears and the contractions fuse into a smooth, sustained contraction called **fused** or **complete tetanus** (tet'ah-nus; *tetan* = rigid, tense). (Tetanus is often confused with the bacterial disease

of the same name that causes severe involuntary contractions.)

Vigorous muscle activity cannot continue indefinitely. Prolonged tetanus inevitably leads to *muscle fatigue,* a situation in which the muscle is unable to contract and its tension drops to zero.

Muscle Response to Stronger Stimuli   Although wave summation contributes to contractile force, its primary function is to produce smooth, continuous muscle contractions by rapidly stimulating a specific number of muscle cells. The force of contraction is controlled more precisely by **multiple motor unit summation** (Figure 9.15). In the laboratory this phenomenon, also called **recruitment,** is achieved by delivering shocks of increasing voltage to the muscle, calling more and more muscle fibers into play. The stimulus at which the first observable contraction occurs is called the **threshold stimulus.** Beyond this point, the muscle contracts more and more vigorously as the stimulus strength is increased (Figure 9.15a). The **maximal stimulus** is the strongest stimulus that produces increased contractile force. It represents the point at which all the muscle's motor units are recruited. Increasing the stimulus intensity beyond the maximal stimulus does not produce a stronger contraction. In the body, the same phenomenon is caused by neural activation of an increasingly large number of motor units serving the muscle.

In weak and precise muscle contractions, relatively small motor units are stimulated. Conversely, when larger forces are needed, the larger motor units are activated. This explains how the same hand that pats your cheek can deliver a stinging slap. In any muscle, the smallest motor units (those with the fewest and smallest muscle fibers) are controlled by small, highly excitable motor neurons and these motor units tend to be activated first. The larger motor units, containing large coarse muscle fibers, are controlled by larger, less excitable neurons and are activated only when a stronger contraction is necessary and frequency of stimulation is much greater (Figure 9.15b).

Although *all* the motor units of a muscle may be recruited simultaneously to produce an exceptionally strong contraction, motor units are more commonly activated asynchronously in the body. At a given instant, some are in tetanus (usually unfused tetanus) while others are resting and recovering. This technique helps prolong a strong contraction by preventing or delaying fatigue. It also explains how weak contractions promoted by infrequent stimuli can remain smooth.

Treppe: The Staircase Effect   When a muscle begins to contract, its contractions may be only half as strong as those that occur later in response to stimuli of the same strength. During these periods, a tracing shows a staircase pattern called **treppe** (trep'ĕ) (Figure 9.16). Treppe probably reflects the increasing availability of Ca^{2+} in the sarcoplasm; more Ca^{2+} ions expose more active sites on the thin filaments for cross bridge attachment. Additionally, as the muscle begins to work and liberates heat, its enzymes become more efficient and the muscle becomes more pliable. Together, these factors produce a slightly stronger contraction with each successive stimulus during the initial phase of muscle activity. This is the basis of the warm-up period required of athletes.

Muscle Tone

Skeletal muscles are described as "being voluntary," but even relaxed muscles are almost always slightly contracted, a phenomenon called **muscle tone.** This is due to spinal reflexes that activate first one group of motor units and then another in response to activation of stretch receptors in the muscles and tendons. Muscle tone does not produce active movements, but it keeps the muscles firm, healthy, and ready to respond to stimulation. Skeletal muscle tone also helps stabilize joints and maintain posture.

Isotonic and Isometric Contractions

As noted earlier, there are two main categories of contractions—*isotonic* and *isometric*. In **isotonic contractions** (*iso* = same; *ton* = tension), muscle length changes (decreasing the angle at the joint) and moves the load. Once sufficient tension has developed to move the load, the tension remains rela-

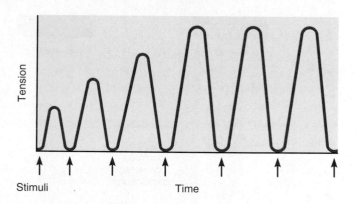

FIGURE 9.16 **Myogram of treppe.** Notice that although the stimuli are of the same intensity and the muscle is not being stimulated rapidly, the first few contractile responses get stronger and stronger.

tively constant through the rest of the contractile period (Figure 9.17a).

Isotonic contractions come in two "flavors"—*concentric* and *eccentric*. **Concentric contractions** in which the muscle *shortens* and does work—picking up a book or kicking a ball, for instance—are probably more familiar. However, **eccentric contractions,** in which the muscle contracts as it *lengthens*, are equally important for coordination and purposeful movements. Eccentric contractions occur in your calf muscle, for example, as you walk up a steep hill. Eccentric contractions are about 50% more forceful than concentric ones at the same load and more often cause delayed-onset muscle soreness. (Consider

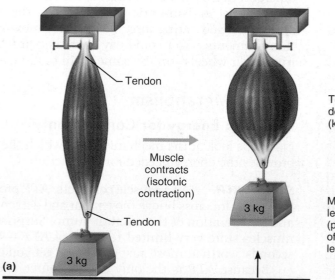

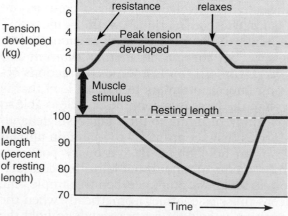

FIGURE 9.17 **Isotonic (concentric) and isometric contractions. (a)** On stimulation, this muscle develops enough tension (force) to lift the load (weight). Once the resistance is overcome, tension remains constant for the rest of the contraction and the muscle shortens. This is a concentric isotonic contraction.

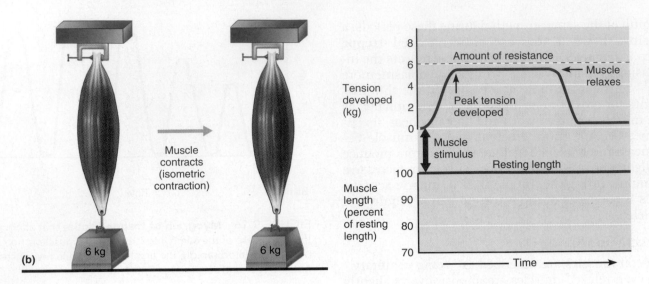

FIGURE 9.17 *(continued)* **Isotonic (concentric) and isometric contractions. (b)** This muscle is attached to a weight that exceeds the muscle's peak tension-developing capabilities. When stimulated, the tension increases to the muscle's peak tension-developing capability, but the muscle does not shorten. This is an isometric contraction.

how your calf muscles *feel* the day after hiking up that hill.) Just why this is so is unclear, but it may be that the muscle stretching that occurs during such contractions causes microtears in the muscles.

Squats, or deep knee bends, provide a simple example of how concentric and eccentric contractions work together in our everyday activities. As the knees flex, the powerful quadriceps muscles of the anterior thigh lengthen (are stretched), but at the same time they also contract (eccentrically) to counteract the force of gravity and control the descent of the torso ("muscle braking") and prevent joint injury. Raising the body back to its starting position requires that the same muscles contract concentrically as they shorten to extend the knees again. As you can see, eccentric contractions put the body in position to contract concentrically. All jumping and throwing activities involve both types of contraction.

In **isometric contractions** (*metric* = measure), tension builds to the muscle's peak tension-producing capacity, but the muscle *neither shortens nor lengthens* (Figure 9.17b). Isometric contractions occur when a muscle attempts to move a load that is greater than the force (tension) the muscle is able to develop—think of trying to lift a piano singlehandedly. Muscles that act primarily to maintain upright posture or to hold joints in stationary positions while movements occur at other joints are contracting isometrically. In our knee bend example, the quadriceps muscles contract isometrically when the squat position is held for a few seconds to hold the knee in the flexed position. They also contract isometrically when we begin to rise to the upright position until their tension exceeds the load (weight of the upper body). At that point muscle shortening

(concentric contraction) begins. So the quadriceps contractile sequence for a deep knee bend from start to finish is (1) knee flex (eccentric), (2) hold squat position (isometric), (3) knee extend (isometric, then concentric). Of course, this does not even begin to consider the isometric contractions of the posterior thigh muscles or of the trunk muscles that maintain a relatively erect trunk posture during the movement.

Electrochemical and mechanical events occurring within a muscle are identical in both isotonic and isometric contractions. However, the result is different. In isotonic contractions, the thin filaments are sliding; in isometric contractions, the cross bridges are generating force but are not moving the thin filaments. (You could say that they are "spinning their wheels" on the same actin binding site.)

Muscle Metabolism

Providing Energy for Contraction

Now let's look at the mechanisms by which the body provides the energy needed for contraction.

Stored ATP   As a muscle contracts, ATP provides the energy for cross bridge movement and detachment and for operation of the calcium pump. Surprisingly, muscles store very limited reserves of ATP—4 to 6 seconds' worth at most, just enough to get you going.

Because ATP is the *only* energy source used directly for contractile activities, it must be regenerated as fast as it is broken down if contraction is to continue. Fortunately, after ATP is hydrolyzed to ADP and inorganic phosphate, it is regenerated within a fraction of a second by three pathways

> **?** *Which of these energy-producing pathways would pre-dominate in the leg muscles of a long-distance cyclist?*

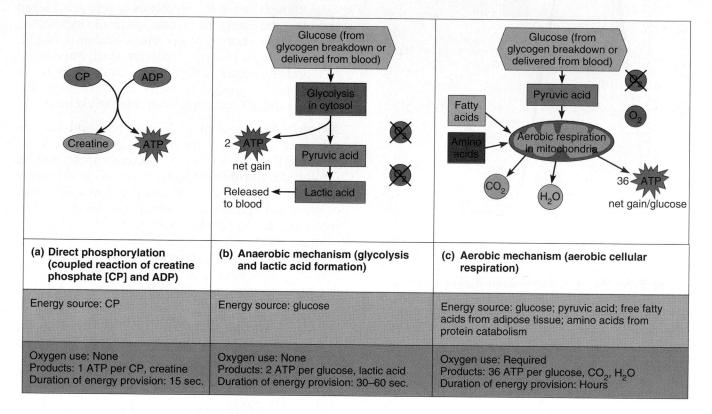

(a) Direct phosphorylation (coupled reaction of creatine phosphate [CP] and ADP)

(b) Anaerobic mechanism (glycolysis and lactic acid formation)

(c) Aerobic mechanism (aerobic cellular respiration)

Energy source: CP	Energy source: glucose	Energy source: glucose; pyruvic acid; free fatty acids from adipose tissue; amino acids from protein catabolism
Oxygen use: None Products: 1 ATP per CP, creatine Duration of energy provision: 15 sec.	Oxygen use: None Products: 2 ATP per glucose, lactic acid Duration of energy provision: 30–60 sec.	Oxygen use: Required Products: 36 ATP per glucose, CO_2, H_2O Duration of energy provision: Hours

FIGURE 9.18 Methods of regenerating ATP during muscle activity. The fastest mechanism is direct phosphorylation **(a)**; the slowest is the aerobic mechanism **(c)**.

(Figure 9.18): (1) by interaction of ADP with creatine phosphate, (2) from stored glycogen via the anaerobic pathway called glycolysis, and (3) by aerobic respiration. The second and third of these metabolic pathways, which are used to produce ATP in all body cells, are just touched on here, as they are described in detail in Chapter 23.

Direct Phosphorylation of ADP by Creatine Phosphate

As we begin to exercise vigorously, ATP stored in working muscles is consumed within a few twitches. Then **creatine phosphate (CP)** (kre'ah-tin), a unique high-energy molecule stored in muscles, is tapped to regenerate ATP while the metabolic pathways are adjusting to the suddenly higher demands for ATP. The result of coupling CP with ADP is almost instant transfer of energy and a phosphate group from CP to ADP to form ATP:

Creatine phosphate + ADP → creatine + ATP

Muscle cells store much more CP than ATP, and the CP-ADP reaction, catalyzed by the enzyme **creatine kinase**, is so efficient that the amount of ATP in muscle cells changes very little during the initial period of contraction.

Together, stored ATP and CP provide for maximum muscle power for 10 to 15 seconds—long enough to energize a 100-meter dash. The coupled reaction is readily reversible, and to keep CP "on tap," CP reserves are replenished during periods of inactivity.

Anaerobic Mechanism: Glycolysis and Lactic Acid Formation

As stored ATP and CP are used, more ATP is generated by catabolism of glucose obtained from the blood or by breakdown of glycogen stored in the muscle. The initial phase of glucose respiration is **glycolysis** (gli-kol'ĭ-sis; "sugar splitting"). This pathway occurs in both the presence *and* the absence of oxygen, but because it does not *use* oxygen, it is an *anaerobic* (an-a'er-ob-ik; "without oxygen") pathway (see Figure 9.18b). During glycolysis, glucose is broken down to two *pyruvic acid* molecules, releasing enough energy to form small amounts of ATP (2 ATPs per glucose).

Ordinarily, pyruvic acid produced in glycolysis then enters the mitochondria and reacts with oxygen to produce still more ATP in the oxygen-using

■ (c) *The aerobic pathway.* ■

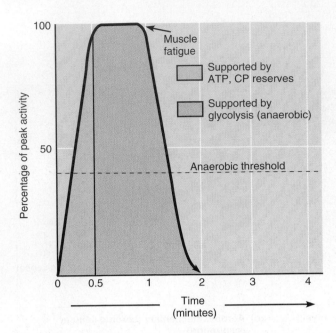

FIGURE 9.19 Energy system used during sports at peak activity levels. Skeletal muscles rely largely on glycolysis (anaerobic respiration) for ATP synthesis. The initial activity burst is energized by ATP and CP reserves. Muscles operating at peak levels fatigue rapidly as lactic acid accumulates.

pathway called aerobic respiration, described shortly. But when muscles contract vigorously and contractile activity reaches about 70% of the maximum possible (e.g., running 600 meters with maximal effort), the bulging muscles compress the blood vessels within them, impairing blood flow and hence oxygen delivery. Under these anaerobic conditions, most of the pyruvic acid produced during glycolysis is converted into **lactic acid,** and the overall process is referred to as **anaerobic glycolysis.** Thus, during oxygen deficit, lactic acid is the end product of cellular metabolism of glucose. Most of the lactic acid diffuses out of the muscles into the bloodstream and is completely gone from the muscle tissue within 30 minutes after exercise stops. Subsequently, the lactic acid is picked up by liver, heart, or kidney cells, which can use it as an energy source. Additionally, liver cells can reconvert it to pyruvic acid or glucose and release it back into the bloodstream for muscle use, or convert it to glycogen for storage.

The anaerobic pathway harvests only about 5% as much ATP from each glucose molecule as the aerobic pathway; however, it produces ATP about 2½ times faster. Thus, when large amounts of ATP are needed for moderate periods (30–40 seconds) of strenuous muscle activity, glycolysis can provide most of the ATP needed as long as the required fuels

and enzymes are available. Together, stored ATP and CP and the glycolysis–lactic acid system can support strenuous muscle activity for nearly a minute.

Although anaerobic glycolysis is very effective in providing the energy to sustain spurts of vigorous exercise, it has shortcomings. Huge amounts of glucose are used to produce relatively small harvests of ATP, and the accumulating lactic acid contributes to muscle fatigue and is partially responsible for the muscle soreness resulting from intense exercise.

Aerobic Respiration   During rest and light to moderate exercise, even if prolonged, 95% of the ATP used for muscle activity comes from aerobic respiration. **Aerobic respiration** occurs in the mitochondria, requires oxygen, and involves a sequence of chemical reactions in which the bonds of fuel molecules are broken and the energy released is used to make ATP.

During aerobic respiration, which includes glycolysis and the reactions that take place in the mitochondria, glucose is broken down entirely, yielding water, carbon dioxide, and large amounts of ATP as the final products (Figure 9.18c).

Glucose + oxygen → carbon dioxide + water + ATP

The carbon dioxide released diffuses out of the muscle tissue into the blood and is removed from the body by the lungs.

As exercise begins, muscle glycogen provides most of the fuel. After that, bloodborne glucose, pyruvic acid from glycolysis, free fatty acids, and in some cases even amino acids are the major sources of fuels for oxidation. Aerobic respiration provides a high yield of ATP (about 36 ATPs per glucose), but it is relatively sluggish because of its many steps and it requires continuous delivery of oxygen and nutrient fuels to keep it going.

Energy Systems Used During Sports Activities
As long as it has enough oxygen, a muscle cell will form ATP by aerobic reactions. When ATP demands are within the capacity of the aerobic pathway, light to moderate muscular activity can continue for several hours in well-conditioned individuals. However, when exercise demands begin to exceed the ability of the muscle cells to carry out the necessary reactions quickly enough, glycolysis begins to contribute more and more of the total ATP generated (Figure 9.19). The length of time a muscle can continue to contract using aerobic pathways is called **aerobic endurance,** and the point at which muscle metabolism converts to anaerobic glycolysis is called **anaerobic threshold.**

Exercise physiologists have been able to estimate the relative importance of each energy-producing system to athletic performance. Activities that require a surge of power but last only a few seconds, such as weight lifting, diving, and sprinting, rely entirely on ATP and CP stores. The more on-and-off or burstlike activities of tennis, soccer, and a 100-meter swim appear to be fueled almost entirely by anaerobic glycolysis. Prolonged activities such as marathon runs and jogging, where endurance rather than power is the goal, depend mainly on aerobic respiration.

Muscle Fatigue

The relatively large amounts of glycogen stored in muscle cells provide some independence from blood-delivered glucose, but with continued exertion even those reserves are exhausted. When oxygen is limited and ATP production fails to keep pace with ATP use, muscles contract less and less effectively and ultimately muscle fatigue sets in. **Muscle fatigue** is a state of *physiological inability to contract* even though the muscle still may be receiving stimuli. This is quite different from psychological fatigue, in which the flesh is still able to perform but we feel tired. It is the will to win in the face of psychological fatigue that sets athletes apart from the rest of us. Note that muscle fatigue results from a *relative deficit* of ATP, not its total absence. When no ATP is available, **contractures,** which are states of continuous contraction, result because the cross bridges are unable to detach (not unlike what happens in rigor mortis). Writer's cramp is a familiar example of temporary contractures.

Excessive intracellular accumulation of lactic acid (which causes the muscles to ache and raises H^+) and other ionic imbalances also contribute to muscle fatigue. The drop in muscle pH limits the usefulness of anaerobic ATP production. As action potentials are transmitted, potassium is lost from the muscle cells, and excess sodium enters. So long as ATP is available to energize the Na^+-K^+ pump, these slight ionic imbalances are corrected. However, fatigued muscle cells seem to lose more K^+ than can be explained by normal action potential generation, and the Na^+-K^+ pumps are insufficient to reverse the ionic imbalances. In general, intense exercise of short duration produces fatigue rapidly via ionic disturbances that alter E-C coupling, but recovery is also rapid. By contrast the slow-developing fatigue of prolonged low-intensity exercise may require several hours for complete recovery. It appears that this type of exercise damages the SR, interfering with Ca^{2+} regulation and release, and therefore with muscle activation.

Oxygen Debt

Whether or not fatigue occurs, vigorous exercise causes a muscle's chemistry to change dramatically.

For a muscle to return to its resting state, its oxygen reserves must be replenished, the accumulated lactic acid must be reconverted to pyruvic acid, glycogen stores must be replaced, and ATP and creatine phosphate reserves must be resynthesized. Additionally, the liver must convert any lactic acid persisting in blood to glucose or glycogen. During anaerobic muscle contraction, all of these oxygen-requiring activities occur more slowly and are (at least partially) deferred until oxygen is again available. Thus, we say an *oxygen debt* is incurred, which must be repaid. **Oxygen debt** is defined as the extra amount of oxygen that the body must take in for these restorative processes. It represents the difference between the amount of oxygen needed for totally aerobic muscle activity and the amount actually used. All nonaerobic sources of ATP used during muscle activity contribute to this debt.

A simple example will help illustrate oxygen debt. If you ran the 100-yard dash in 12 seconds, your body would require about 6 L of oxygen for totally aerobic respiration. However, the VO_2 max (the amount of oxygen that can be delivered to and used by your muscles) during that 12-second interval is only about 1.2 L, far short of the required amount. Thus, you would have incurred an oxygen debt of 4.8 L, which you would repay by breathing rapidly and deeply once you stopped running. This heavy breathing is triggered primarily by high levels of H^+ in the blood, which indirectly stimulates the respiratory center of the brain. The amount of oxygen used during exertion depends on several factors, including age, size, athletic training, and health. In general, the more exercise to which a person is accustomed, the higher his or her oxygen delivery (and anaerobic threshold) during exercise and the lower the oxygen debt incurred. For example, the VO_2 max of most athletes is at least 10% greater than that of a sedentary person, and that of a trained marathon runner may be as much as 50% greater.

Heat Production During Muscle Activity

Only about 40% of the energy released during muscle contraction is converted to useful work. The rest is given off as heat, which has to be dealt with if body homeostasis is to be maintained. When you exercise vigorously, you start to feel hot as your blood is warmed by the liberated heat. Ordinarily, heat buildup is prevented from reaching dangerous levels by several homeostatic mechanisms, including sweating and radiation of heat from the skin surface. Shivering represents the opposite end of homeostatic balance, in which muscle contractions are used to produce more heat.

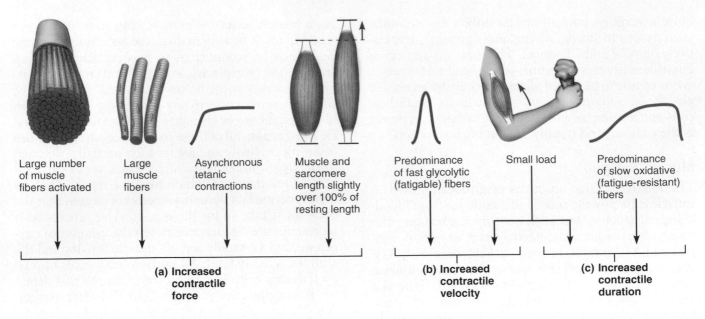

| | | | | | | |
| Large number of muscle fibers activated | Large muscle fibers | Asynchronous tetanic contractions | Muscle and sarcomere length slightly over 100% of resting length | Predominance of fast glycolytic (fatigable) fibers | Small load | Predominance of slow oxidative (fatigue-resistant) fibers |

(a) Increased contractile force

(b) Increased contractile velocity

(c) Increased contractile duration

FIGURE 9.20 **Factors influencing force, velocity, and duration of skeletal muscle contraction.**

Force of Muscle Contraction

The force of muscle contraction is affected by (1) the number of muscle fibers stimulated, (2) the relative size of the fibers, (3) frequency of stimulation, and (4) the degree of muscle stretch (Figure 9.20a). Let's briefly examine the role of each of these factors.

Number of Muscle Fibers Stimulated

As already discussed, the more motor units that are recruited, the greater the muscle force will be.

Size of the Muscle Fibers Stimulated

The bulkier the muscle (the greater its cross-sectional area), the more tension it can develop and the greater its strength, but there is more to it than this. As noted earlier, the large fibers of large motor units are very effective in producing the most powerful movements. Regular exercise increases muscle force by causing muscle cells to *hypertrophy* (hi-per'tro-fe) or increase in size.

Frequency of Stimulation

As a muscle begins to contract, the force generated by the cross bridges (myofibrils)—the **internal tension**—stretches the series elastic (noncontractile) components. These in turn become taut and transfer their tension, called the **external tension,** to the load (muscle insertion), and when the contraction ends, their recoil helps to return the muscle to its resting length.

The point is that time is required to take up slack and stretch the series elastic components, and while this is happening, the internal tension is already declining. So, in brief twitch contractions, the external tension is always less than the internal tension (Figure 9.21a). However, when a muscle is stimulated rapidly, contractions are summed up, becoming stronger and more vigorous and ultimately producing tetanus. During tetanic contractions more time is available to stretch the series elastic components, and external tension approaches the internal tension (Figure 9.21b). So, the more rapidly a muscle is stimulated, the greater the force it exerts.

Degree of Muscle Stretch

The optimal resting length for muscle fibers is the length at which they can generate maximum force (Figures 9.20a and 9.22). Within a sarcomere, the ideal **length-tension relationship** occurs when a muscle is slightly stretched and the thin and thick filaments barely overlap, because this permits sliding along nearly the entire length of the thin filaments. If a muscle fiber is stretched to the extent that the filaments do not overlap (Figure 9.22c), the cross bridges have nothing to attach to and cannot generate tension. Alternatively, if the sarcomeres are so compressed and cramped that the Z discs abut the thick myofilaments and the thin filaments touch and interfere with one another (Figure 9.22a), little or no further shortening can occur.

Identical relationships exist in a whole muscle. A severely stretched muscle (say one at 180% of its optimal length) cannot develop tension. Likewise, once a muscle contracts to 75% of its resting length, further shortening is limited. Muscles, like muscle

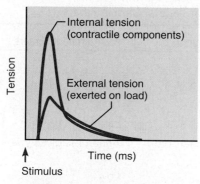

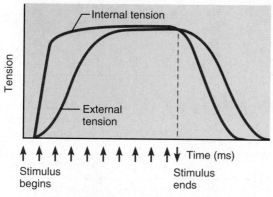

(a) Twitch

(b) Tetanic contraction

FIGURE 9.21 **Relationship between stimulus frequency and external tension. (a)** During a single-twitch contraction, the internal tension developed by the cross bridges peaks and begins to drop well before the series elastic components are stretched to equal tension. As a result, the external tension exerted on the load is always less than the internal tension. **(b)** When the muscle is stimulated tetanically, the internal tension lasts long enough for the series elastic components to be stretched to similar tension, so that external tension approaches and finally equals the internal tension.

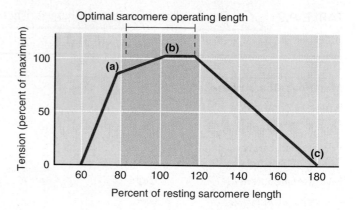

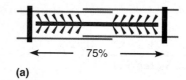

(a)

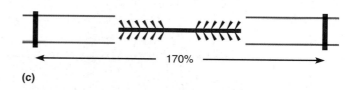

(b)

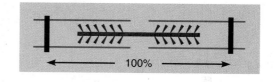

(c)

FIGURE 9.22 **Length-tension relationships in skeletal muscles.** Maximum force is generated when the muscle is between 80 and 120% of its resting length. Increases and decreases beyond this optimal range result in decreased force and inability to generate tension. Depicted here is the relative sarcomere length in a muscle that is **(a)** strongly contracted, **(b)** at normal resting length, and **(c)** excessively stretched.

fibers, are at optimal operational length from about 80% to about 120% of their normal resting length. In the body, skeletal muscles are maintained near that optimum by the way they are attached to bones; that is, the joints normally prevent bone movements that would stretch attached muscles beyond their optimal range.

Velocity and Duration of Contraction

Muscles vary in how fast they can contract and in how long they can continue to contract before they fatigue. These characteristics are influenced by muscle fiber type, load, and recruitment.

Muscle Fiber Type

There are several ways of classifying muscle fibers, but learning about these classes will be easier if you initially pay attention to just two major functional characteristics:

- **Speed of contraction.** On the basis of speed of shortening or contraction, there are **slow fibers** and **fast fibers.** The difference in speed of these fibers reflects how fast their myosin ATPases split ATP.

- **The major pathways for forming ATP.** The cells that rely mostly on the oxygen-using aerobic pathways for ATP generation are **oxidative fibers;** those that rely more on anaerobic glycolysis are **glycolytic fibers.**

On the basis of these two criteria, we can classify skeletal muscle cells as being **slow oxidative fibers,**

TABLE 9.2   Structural and Functional Characteristics of the Three Types of Skeletal Muscle Fibers

	Slow Oxidative Fibers	Fast Oxidative Fibers	Fast Glycolytic Fibers
Metabolic Characteristics			
Speed of contraction	Slow	Fast	Fast
Myosin ATPase activity	Slow	Fast	Fast
Primary pathway for ATP synthesis	Aerobic	Aerobic (some anaerobic glycolysis)	Anaerobic glycolysis
Myoglobin content	High	High	Low
Glycogen stores	Low	Intermediate	High
Recruitment order	First	Second	Third
Rate of fatigue	Slow (fatigue-resistant)	Intermediate (moderately fatigue-resistant)	Fast (fatigable)
Activities Best Suited For			
	Endurance-type activities— e.g., running a marathon; maintaining posture (antigravity muscles)	Sprinting, walking	Short-term intense or powerful movements, e.g., hitting a baseball
Structural Characteristics			
Color	Red	Red to pink	White (pale)
Fiber diameter	Small	Intermediate	Large
Mitochondria	Many	Many	Few
Capillaries	Many	Many	Few

fast oxidative fibers, or **fast glycolytic fibers.** Details about each group are given in Table 9.2, but a word to the wise: Do not approach this information by rote memorization—you'll just get frustrated. Instead, start with what you know for any category and see how the characteristics listed support that. For example, a *slow oxidative fiber*

- Contracts relatively slowly because its myosin ATPases are slow (a criterion)

- Depends on oxygen delivery and aerobic mechanisms (high oxidative capacity—a criterion)

- Is fatigue resistant and has high endurance (typical of fibers that depend on aerobic metabolism)

- Is thin (a large amount of cytoplasm impedes diffusion of O_2 and nutrients from the blood)

- Has relatively little power (a thin cell can contain only a limited number of myofibrils)

- Has many mitochondria (actual sites of oxygen use)

- Has a rich capillary supply (the better to deliver bloodborne O_2)

- Is red (its color stems from an abundant supply of myoglobin, muscle's oxygen-binding pigment that stores O_2 reserves in the cell and aids diffusion of O_2 through the cell)

Conversely, a *fast glycolytic fiber* contracts rapidly (Figure 9.20b), depends on plentiful glycogen reserves for fuel rather than on a blood-delivered supply, and does not use oxygen (Table 9.2). Consequently, compared with an oxidative cell, it has few mitochondria, little myoglobin and low capillary density (and so is white), and tends to be a much larger cell (because it doesn't depend on continuous oxygen and nutrient diffusion from the blood). Because glycogen reserves are short-lived and lactic acid accumulates quickly in these cells, they tire quickly and hence are fatigable fibers. However, their large diameter reflects their plentiful contractile myofilaments that allow them to contract powerfully before they "poop out." Thus, the fast glycolytic fibers are best suited for short-term, rapid, intense movements (moving furniture across the room, for example). Details about the intermediate muscle fiber type, called the fast oxidative fiber, are listed in Table 9.2.

Although some muscles have a predominance of one fiber type, most contain a mixture of fiber types, which gives them a range of contractile speeds and fatigue resistance. For example, a calf muscle can propel us in a sprint (using mostly its white fast glycolytic fibers), or a long-distance race (making good use of its intermediate fast oxidative fibers), or may simply help maintain our standing posture (using slow oxidative fibers). As might be expected, all muscle fibers in a particular motor unit are of the same type.

Although everyone's muscles contain mixtures of the three fiber types, some people have relatively more of one variety. These differences are genetically controlled and no doubt determine athletic capabilities to a large extent. For example, muscles of marathon runners have a high percentage of slow oxidative fibers (about 80%), while those of sprinters contain a higher percentage (about 60%) of fast oxidative fibers. Weight lifters appear to have approximately equal amounts of both.

Load   Because muscles are attached to bones, they are always pitted against some resistance, or load, when they contract. As you might expect, they contract fastest when there is no added load on them. The greater the load, the longer the latent period, the slower the contraction, and the shorter the duration of contraction (Figure 9.23). If the load exceeds the muscle's maximum tension, the speed of shortening is zero and the contraction is isometric.

Recruitment   Just as many hands on a project can get a job done more quickly and also can keep working longer, the more motor units that are contracting, the faster and more prolonged the contraction.

Effect of Exercise on Muscles

The amount of work a muscle does is reflected in changes in the muscle itself. When used actively or strenuously, muscles may increase in size or strength or become more efficient and fatigue resistant. Muscle inactivity, on the other hand, *always* leads to muscle weakness and wasting.

Adaptations to Exercise

Aerobic, or **endurance, exercise** such as swimming, jogging, fast walking, and biking results in several recognizable changes in skeletal muscles. There is an increase in the number of capillaries surrounding the muscle fibers, and in the number of mitochondria within them, and the fibers synthesize more myoglobin. These changes occur in all fiber types, but are most dramatic in slow oxidative fibers, which depend primarily on aerobic pathways. The changes result in more efficient muscle metabolism and in greater endurance, strength, and resistance to fatigue.

The moderately weak but sustained muscle activity required for endurance exercise does not pro-

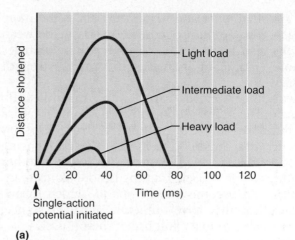

(a)

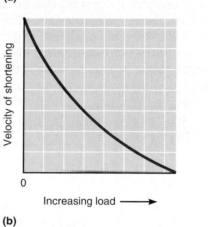

(b)

FIGURE 9.23 **Influence of load on contraction velocity and duration.** **(a)** Relationship of load to degree and duration of contraction (shortening). **(b)** Relationship of load to velocity of shortening. As the load increases, the speed of contraction decreases.

mote significant skeletal muscle hypertrophy, even though the exercise may go on for hours. Muscle hypertrophy, illustrated by the bulging biceps and chest muscles of a professional weight lifter, results mainly from high-intensity **resistance exercise** (typically under anaerobic conditions) such as weight lifting or isometric exercise, in which the muscles are pitted against high-resistance or immovable forces. Here strength, not stamina, is important; a few minutes every other day is sufficient to allow even a proverbial weakling to put on 50% more muscle within a year. The increased muscle bulk largely reflects increases in the size of individual muscle fibers (particularly the fast glycolytic variety) rather than an increased number of muscle fibers. [However, some of the increased muscle size may result either from longitudinal splitting or tearing of the fibers and subsequent growth of these "split" cells, or from the proliferation and fusion of satellite cells. The controversy is still raging.] Vigorously stressed muscle fibers contain more mitochondria, form more myofilaments and myofibrils, and store more glycogen. The amount of connective tissue between the cells also increases. Collectively these changes promote significant increases in muscle strength and size.

Resistance training can produce magnificently bulging muscles, but if done unwisely, some muscles may develop more than others. Because muscles work in antagonistic pairs (or groups), opposing muscles must be equally strong to work together smoothly. When muscle training is not balanced, individuals can become *muscle-bound*, which means they lack flexibility, have a generally awkward stance, and are unable to make full use of their muscles.

Endurance and resistance exercises produce different patterns of muscular response, so it is important to know what your exercise goals are. Lifting weights will not improve your endurance for a triathlon. By the same token, jogging will do little to improve your muscle definition, nor will it enhance your strength for moving furniture. A program that alternates aerobic activities with anaerobic ones provides the best program for optimal health.

Ⓗ HOMEOSTATIC IMBALANCE

To remain healthy, muscles must be active. Complete immobilization due to enforced bed rest or loss of neural stimulation results in *disuse atrophy* (degeneration and loss of mass), which begins almost as soon as the muscles are immobilized. Under such conditions, muscle strength can decrease at the rate of 5% per day!

Even at rest, muscles receive weak intermittent stimuli from the nervous system. When totally deprived of neural stimulation, a paralyzed muscle may atrophy to one-quarter of its initial size. Lost muscle tissue is replaced by fibrous connective tissue, making muscle rehabilitation impossible. Atrophy of a denervated muscle may be delayed by electrically stimulating the muscle periodically while waiting to see whether the damaged nerve fibers regenerate. ●

Training Smart to Prevent Overuse Injuries

Don't expect to get in shape by playing a sport; you've *got to get in shape to play* the sport. Regardless of your choice—running, lifting weights, or tennis, for example—exercise stresses muscles. Muscle fibers tear, tendons stretch, and accumulation of lactic acid in the muscle causes pain.

Effective training walks a fine line between working hard enough to improve and preventing overuse injuries. Whatever the activity, exercise gains adhere to the *overload principle*. Forcing a muscle to work hard promotes increased muscle strength and endurance, and as muscles adapt to the increased demands, they must be overloaded even more to produce further gains. To become faster, you must train at an increasingly fast pace. A heavy-workout day should be followed by one of rest or an easy workout to allow the muscles to recover and repair themselves. Doing too much too soon, or ignoring the warning signs of muscle or joint pain, increases the risk of **overuse injuries** that may prevent future sports activities, or even lead to lifetime disability. Changing or restricting activity, using ice packs to prevent or reduce inflammation, and taking nonsteroidal anti-inflammatory drugs (aspirin, ibuprofen, acetaminophen) to reduce pain are the main treatment modes for virtually all overuse injuries.

Smooth Muscle

Except for the heart, which is made of cardiac muscle, the muscle in the walls of all the body's hollow organs is almost entirely smooth muscle. Although the chemical and mechanical events of contraction are essentially the same in all muscle tissues, smooth muscle is distinctive in several ways (see Table 9.3).

Microscopic Structure of Smooth Muscle Fibers

Smooth muscle fibers are spindle-shaped cells, each with one centrally located nucleus (Figure 9.24). Typically, they have a diameter of 2–5 μm and are 100 to 400 μm in length. Skeletal muscle fibers are some 20 times wider and thousands of times longer.

Smooth muscle lacks the coarse connective tissue sheaths seen in skeletal muscle. However, a small amount of fine connective tissue (endomysium), secreted by the smooth muscles themselves and containing blood vessels and nerves, is found between smooth muscle fibers. Most smooth muscle is organized into sheets of closely apposed fibers. These sheets occur in the walls of all but the smallest blood vessels and in the walls of hollow organs of the respiratory, digestive, urinary, and reproductive tracts. In most cases, two sheets of smooth muscle are present, with their fibers oriented at right angles to each other (Figure 9.24). In the *longitudinal layer*, the muscle fibers run parallel to the long axis of the organ. Consequently, when the muscle contracts, the organ dilates and shortens. In the *circular layer*, the fibers run around the circumference of the organ. Contraction of this layer constricts the lumen (cavity) of the organ and causes the organ to elongate.

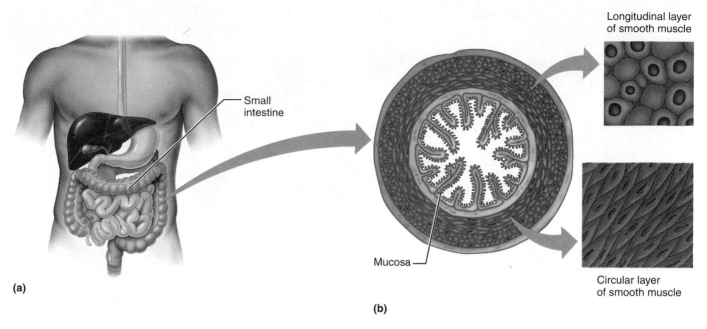

(a)

Small intestine

Mucosa

Longitudinal layer of smooth muscle

Circular layer of smooth muscle

(b)

FIGURE 9.24 **Arrangement of smooth muscle in the walls of hollow organs.** Simplified cross-sectional view of the intestine showing two smooth muscle layers (one circular and the other longitudinal) running at right angles to each other.

The alternating contraction and relaxation of these opposing layers mixes substances in the lumen and squeezes them through the organ's internal pathway. This phenomenon is called **peristalsis** (per″i-stal′sis; "around contraction"). Contraction of smooth muscle in the rectum, urinary bladder, and uterus helps those organs to expel their contents. Smooth muscle contraction also accounts for the constricted breathing of asthma and for stomach cramps.

Smooth muscle lacks the highly structured neuromuscular junctions of skeletal muscle. Instead, the innervating nerve fibers, which are part of the autonomic nervous system, have numerous bulbous swellings, called **varicosities** (Figure 9.25). The varicosities release neurotransmitter into a wide synaptic cleft in the general area of the smooth muscle cells. Such junctions are called **diffuse junctions.**

The sarcoplasmic reticulum of smooth muscle fibers is less developed than that of skeletal muscle and lacks a specific pattern relative to the myofilaments. Some SR tubules of smooth muscle touch the sarcolemma at several sites, forming what resembles half-triads that may couple the action potential to calcium release from the SR. T tubules are notably absent, but the sarcolemma of the smooth muscle fiber has multiple caveoli, pouchlike infoldings that enclose bits of extracellular fluid and allow a high concentration of Ca^{2+} to be sequestered close to the membrane. Consequently, when calcium

channels open, Ca^{2+} influx occurs rapidly. Although the SR *does* release some of the calcium ions that trigger contraction, most gain entry via calcium channels directly from the extracellular space. Contraction ends when calcium is actively transported into the SR and out of the cell.

There are no striations, as the name *smooth muscle* indicates, and no sarcomeres. Smooth muscle fibers *do* contain interdigitating thick and thin filaments, but the thick filaments are much longer than those in skeletal muscle. The proportion and organization of the myofilaments are also different:

1. The ratio of thick to thin filaments is much lower in smooth muscle than in skeletal muscle (1:13 compared to 1:2). However, thick filaments of smooth muscle contain actin-gripping heads along their *entire length*, a feature that allows smooth muscle to be as powerful as a skeletal muscle of the same size.

2. As with skeletal muscle, tropomyosin is associated with the thin filament of smooth muscle, but no troponin complex is present.

3. Smooth muscle thick and thin filaments are arranged diagonally within the cell so that they spiral down the long axis of the cell like the stripes on a barber pole. Because of this arrangement, the smooth muscle cells contract in a twisting way so that they look like tiny corkscrews (Figure 9.26b).

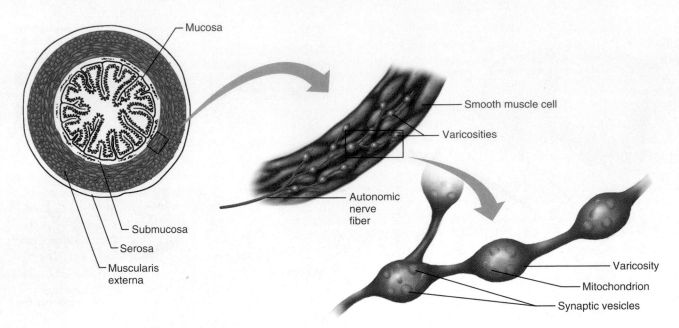

FIGURE 9.25   **Innervation of smooth muscle.** Most smooth muscle cells are innervated by autonomic nervous system fibers that release their neurotransmitters from varicosities into a wide synaptic cleft (a diffuse junction).

4. Smooth muscle fibers contain longitudinal bundles of noncontractile *intermediate filaments* that resist tension. These attach at regular intervals to structures called **dense bodies.** The dense bodies, which are also tethered to the sarcolemma, act as anchoring points for thin filaments and therefore correspond to Z discs of skeletal muscle. The intermediate filament–dense body network forms a strong, cable-like intracellular cytoskeleton that harnesses the pull generated by the sliding of the thick and thin filaments (Figure 9.26). During contraction, areas of the sarcolemma between the dense bodies bulge outward, giving the cell a puffy appearance (see Figure 9.26b). Dense bodies at the sarcolemma surface also bind the muscle cell to the connective tissue fibers outside the cell (endomysium) and to adjacent cells, an arrangement that transmits the pulling force to the surrounding connective tissue and that partly accounts for the synchronous contraction of most smooth muscle.

Contraction of Smooth Muscle

Mechanism and Characteristics of Contraction

In most cases, adjacent smooth muscle fibers exhibit slow, synchronized contractions, the whole sheet responding to a stimulus in unison. This phenomenon reflects electrical coupling of smooth muscle cells by *gap junctions*, specialized cell connections described in Chapter 3. Skeletal muscle fibers are electrically isolated from one another, each stimulated to contract by its own neuromuscular junction. By contrast, gap junctions allow smooth muscles to transmit action potentials from fiber to fiber. Some smooth muscle fibers in the stomach and small intestine are *pacemaker cells* and, once excited, act as "drummers" to set the contractile pace for the entire

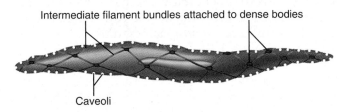

(a) Relaxed smooth muscle cell

(b) Contracted smooth muscle cell

FIGURE 9.26   **Intermediate filaments and dense bodies of smooth muscle fibers harness the pull generated by the activity of myosin cross bridges.** Intermediate filaments attach to dense bodies scattered throughout the sarcoplasm and occasionally anchor to the sarcolemma. **(a)** A relaxed smooth muscle cell. **(b)** A contracted smooth muscle cell. [Caveoli are not illustrated in (b).]

muscle sheet. Additionally, some of these pacemakers are self-excitatory, that is, they depolarize spontaneously in the absence of external stimuli. However, both the rate and the intensity of smooth muscle contraction may be modified by neural and chemical stimuli.

The contraction mechanism in smooth muscle is like that in skeletal muscle in the following ways: (1) actin and myosin interact by the sliding filament mechanism; (2) the final trigger for contraction is a rise in the intracellular calcium ion level; and (3) the sliding process is energized by ATP.

During excitation-contraction coupling, Ca^{2+} is released by the tubules of the SR, but, as mentioned above, it also moves into the cell from the extracellular space. Calcium ions have the same triggering role in all muscle types, but in smooth muscle, they activate myosin by interacting with two regulatory molecules: **calmodulin,** a cytoplasmic calcium-binding protein, and a kinase enzyme called **myosin light chain kinase.** The thin filaments lack troponin and so are always ready for contraction. Apparently the sequence of events is:

1. Ca^{2+} binds to calmodulin, activating it.
2. Activated calmodulin activates the kinase enzyme.
3. The activated kinase catalyzes transfer of phosphate from ATP to myosin cross bridges.
4. Phosphorylated cross bridges interact with actin of the thin filaments, shortening the fiber. As with the other muscle types, ATP powers the cross bridge cycle.
5. The muscle relaxes when intracellular Ca^{2+} levels drop.

Smooth muscle takes 30 times longer to contract and relax than does skeletal muscle and can maintain the same contractile tension for prolonged periods at less than 1% of the energy cost. If skeletal muscle is like a speedy windup car that quickly runs down, then smooth muscle is like a steady, heavy-duty engine that lumbers along tirelessly. Part of the striking energy economy of smooth muscle is the sluggishness of its ATPases (the myosin light chain kinases) compared to those in skeletal muscle. Moreover, smooth muscle myofilaments may latch together during prolonged contractions, saving energy in that way as well.

The ATP-efficient contraction of smooth muscle is extremely important to overall body homeostasis. The smooth muscle in small arterioles and other visceral organs routinely maintains a moderate degree of contraction, called *smooth muscle tone*, day in and day out without fatiguing. Smooth muscle has low energy requirements. Typically, enough ATP is made via aerobic pathways to keep up with the demand.

Regulation of Contraction

Neural Regulation   In some cases, the activation of smooth muscle by a neural stimulus is identical to that in skeletal muscle: An action potential is generated by neurotransmitter binding, and is coupled to a rise in calcium ions in the sarcoplasm. However, some types of smooth muscle respond to neural stimulation with graded potentials (local electrical signals) only.

All somatic nerve endings, that is, nerve endings that serve skeletal muscle, release acetylcholine, which excites the skeletal muscle. However, different autonomic nerves serving the smooth muscle of visceral organs release different neurotransmitters, each of which may excite or inhibit a particular group of smooth muscle cells. The effect of a specific neurotransmitter on a given smooth muscle cell depends on the type of receptor molecules on its sarcolemma. For example, when acetylcholine binds to ACh receptors on smooth muscle in the bronchioles (small air passageways of the lungs), the muscle contracts strongly, narrowing the bronchioles. When norepinephrine, released by a different type of autonomic nerve fiber, binds to norepinephrine receptors on the *same* smooth muscle cells, the effect is inhibitory; the muscle relaxes, dilating the air passageways. However, when norepinephrine binds to smooth muscle in the walls of most blood vessels, it stimulates the smooth muscle cells to contract and constrict the vessel.

Hormones and Local Factors   Not all smooth muscle activation results from neural signals. Some smooth muscle layers have no nerve supply at all; instead they depolarize spontaneously or in response to chemical stimuli that bind to G protein–linked receptors. Others respond to both neural and chemical stimuli. Chemical factors that cause smooth muscle contraction or relaxation without an action potential (by enhancing or inhibiting Ca^{2+} entry into the sarcoplasm) include certain hormones, lack of oxygen, histamine, excess carbon dioxide, and low pH. The direct response to these chemical stimuli alters smooth muscle activity according to local tissue needs and probably is most responsible for smooth muscle tone. For example, the hormone gastrin stimulates contraction of stomach smooth muscle so that it can churn foodstuffs more efficiently. We will consider how the smooth muscle of specific organs is activated as we discuss each organ in subsequent chapters.

Special Features of Smooth Muscle Contraction

Smooth muscle is intimately involved in the functioning of most hollow organs and has a number of unique characteristics. Some of these—smooth muscle tone; slow, prolonged contractile activity; and low energy requirements—have already been considered. But smooth muscle also responds differently to stretch and can shorten more than other muscle types. Let's take a look.

Response to Stretch   When cardiac muscle is stretched, it responds with more vigorous contractions. So does skeletal muscle up to a point (about 120% of resting length). Stretching of smooth muscle also provokes contraction, which automatically moves substances along an internal tract. However, the increased tension persists only briefly; soon the muscle adapts to its new length and relaxes, while still retaining the ability to contract on demand. This **stress-relaxation response** allows a hollow organ to fill or expand slowly (within certain limits) to accommodate a greater volume without promoting strong contractions that would expel their contents. This is an important attribute, because organs such as the stomach and intestines must be able to store their contents temporarily to provide sufficient time for digestion and absorption of the nutrients. Likewise, your urinary bladder must be able to store the continuously made urine until it is convenient to empty your bladder or you would spend all your time in the bathroom.

Length and Tension Changes   Smooth muscle stretches much more than skeletal muscle and generates more tension than skeletal muscles stretched to a comparable extent. As Figure 9.22c shows, precise, highly organized sarcomeres limit how far a skeletal muscle can be stretched before it is unable to generate force. In contrast, the lack of sarcomeres and the irregular, overlapping arrangement of smooth muscle filaments allow them to generate considerable force, even when they are substantially stretched. The total length change that skeletal muscles can undergo and still function efficiently is about 60% (from 30% shorter to 30% longer than resting length), but smooth muscle can contract from twice to half its resting length—a total length change of 150%. This allows hollow organs to tolerate tremendous changes in volume without becoming flabby when they empty.

Hyperplasia   Besides the ability to hypertrophy (increase in cell size), which is common to all muscle cells, certain smooth muscle fibers can divide to increase their numbers, that is, they undergo *hyperplasia* (hi"per-pla'ze-ah). One example is the response of the uterus to estrogen. At puberty, girls'

plasma estrogen levels rise. As estrogen binds to uterine smooth muscle receptors, it stimulates the synthesis of more uterine smooth muscle, causing the uterus to grow to adult size. During pregnancy, high blood levels of estrogen stimulate uterine hyperplasia to accommodate the growing fetus.

Types of Smooth Muscle

The smooth muscle in different body organs varies substantially in its (1) fiber arrangement and organization, (2) responsiveness to various stimuli, and (3) innervation. For simplicity, however, smooth muscle is usually categorized into two major types: *single-unit* and *multiunit*.

Single-Unit Smooth Muscle

Single-unit smooth muscle, commonly called **visceral muscle,** is far more common. Its cells (1) contract rhythmically and as a unit, (2) are electrically coupled to one another by *gap junctions*, and (3) often exhibit spontaneous action potentials. All the smooth muscle characteristics described so far pertain to single-unit smooth muscle. Thus, the cells of single-unit smooth muscle are arranged in opposing sheets, exhibit the stress-relaxation response, and so on.

Multiunit Smooth Muscle

The smooth muscles in the large airways to the lungs and in large arteries, the arrector pili muscles attached to hair follicles, and the internal eye muscles that adjust pupil size and allow the eye to focus visually are all examples of **multiunit smooth muscle.**

   In contrast to what we see in single-unit muscle, gap junctions are rare, and spontaneous synchronous depolarizations are infrequent. Like skeletal muscle, multiunit smooth muscle:

1. Consists of muscle fibers that are structurally independent of one another

2. Is richly supplied with nerve endings, each of which forms a motor unit with a number of muscle fibers

3. Responds to neural stimulation with graded contractions

However, while skeletal muscle is served by the somatic (voluntary) division of the nervous system, multiunit smooth muscle (like single-unit smooth muscle) is innervated by the autonomic (involuntary) division and is also responsive to hormonal controls.

TABLE 9.3	Comparison of Skeletal, Cardiac, and Smooth Muscle		
Characteristic	Skeletal	Cardiac	Smooth
Body location	Attached to bones or (some facial muscles) to skin	Walls of the heart	Single-unit muscle in walls of hollow visceral organs (other than the heart); multiunit muscle in intrinsic eye muscles
Cell shape and appearance	Single, very long, cylindrical, multinucleate cells with very obvious striations	Branching chains of cells; uni- or binucleate; striations	Single, fusiform, uninucleate; no striations
Connective tissue components	Epimysium, perimysium, and endomysium	Endomysium attached to fibrous skeleton of heart	Endomysium
Presence of myofibrils composed of sarcomeres	Yes	Yes, but myofibrils are of irregular thickness	No, but actin and myosin filaments are present throughout; dense bodies anchor actin filaments
Presence of T tubules and site of invagination	Yes; two in each sarcomere at A-I junctions	Yes; one in each sarcomere at Z disc; larger diameter than those of skeletal muscle	No; only caveoli
Elaborate sarcoplasmic reticulum	Yes	Less than skeletal muscle (1–8% of cell volume); scant terminal cisternae	Equivalent to cardiac muscle (1–8% of cell volume); some SR contacts the sarcolemma

TABLE 9.3   **Comparison of Skeletal, Cardiac, and Smooth Muscle** (continued)

Characteristic	Skeletal	Cardiac	Smooth
Presence of gap junctions	No	Yes; at intercalated discs	Yes; in single-unit muscle
Cells exhibit individual neuromuscular junctions	Yes	No	Not in single-unit muscle; yes in multiunit muscle
Regulation of contraction	Voluntary via axonal endings of the somatic nervous system	Involuntary; intrinsic system regulation; also autonomic nervous system controls; hormones; stretch	Involuntary; autonomic nerves, hormones, local chemicals; stretch
Source of Ca^{2+} for calcium pulse	Sarcoplasmic reticulum (SR)	SR and from extracellular fluid	SR and from extracellular fluid
Site of calcium regulation	Troponin on actin-containing thin filaments	Troponin on actin-containing thin filaments	Calmodulin in the sarcoplasm
Presence of pacemaker(s)	No	Yes	Yes (in single-unit muscle only)
Effect of nervous system stimulation	Excitation	Excitation or inhibition	Excitation or inhibition
Speed of contraction	Slow to fast	Slow	Very slow
Rhythmic contraction	No	Yes	Yes in single-unit muscle
Response to stretch	Contractile strength increases with degree of stretch (to a point)	Contractile strength increases with degree of stretch	Stress-relaxation response
Respiration	Aerobic and anaerobic	Aerobic	Mainly aerobic

Review Questions

Multiple Choice/Matching

(Some questions have more than one correct answer. Select the best answer or answers from the choices given.)

1. The connective tissue covering that encloses the sarcolemma of an individual muscle fiber is called the (a) epimysium, (b) perimysium, (c) endomysium, (d) periosteum.

2. A fascicle is a (a) muscle, (b) bundle of muscle fibers enclosed by a connective tissue sheath, (c) bundle of myofibrils, (d) group of myofilaments.

3. Thick and thin myofilaments have different compositions. For each descriptive phrase, indicate whether the filament is (a) thick or (b) thin.

_____ **(1)** contains actin
_____ **(2)** contains ATPases
_____ **(3)** attaches to the Z disc
_____ **(4)** contains myosin
_____ **(5)** contains troponin
_____ **(6)** does not lie in the I band

4. The function of the T tubules in muscle contraction is to (a) make and store glycogen, (b) release Ca^{2+} into the cell interior and then pick it up again, (c) transmit the action potential deep into the muscle cells, (d) form proteins.

5. The sites where the motor nerve impulse is transmitted from the nerve endings to the skeletal muscle cell membranes are the (a) neuromuscular junctions, (b) sarcomeres, (c) myofilaments, (d) Z discs.

6. Contraction elicited by a single brief stimulus is called (a) a twitch, (b) wave summation, (c) multiple motor unit summation, (d) tetanus.

7. A smooth, sustained contraction resulting from very rapid stimulation of the muscle, in which no evidence of relaxation is seen, is called (a) a twitch, (b) wave summation, (c) multiple motor unit summation, (d) fused tetanus.

8. Characteristics of isometric contractions include all but (a) shortening, (b) increased muscle tension throughout, (c) absence of shortening, (d) use in resistance training.

9. During muscle contraction, ATP is provided by (a) a coupled reaction of creatine phosphate with ADP, (b) aerobic respiration of glucose, and (c) anaerobic glycolysis.

_____ **(1)** Which provides ATP fastest?
_____ **(2)** Which does (do) not require that oxygen be available?
_____ **(3)** Which (aerobic or anaerobic pathway) provides the highest yield of ATP per glucose molecule?
_____ **(4)** Which results in the formation of lactic acid?
_____ **(5)** Which has carbon dioxide and water products?
_____ **(6)** Which is most important in endurance sports?

10. The neurotransmitter released by somatic motor neurons is (a) acetylcholine, (b) acetylcholinesterase, (c) norepinephrine.

11. The ions that enter the muscle cell during action potential generation are (a) calcium ions, (b) chloride ions, (c) sodium ions, (d) potassium ions.

12. Myoglobin has a special function in muscle tissue. It (a) breaks down glycogen, (b) is a contractile protein, (c) holds a reserve supply of oxygen in the muscle.

13. Aerobic exercise results in all of the following except (a) increased cardiovascular system efficiency, (b) more mitochondria in the muscle cells, (c) increased size and strength of existing muscle cells, (d) increased neuromuscular system coordination.

14. The smooth muscle type found in the walls of digestive and urinary system organs and that exhibits gap junctions and pacemaker cells is (a) multiunit, (b) single-unit.

Short Answer Essay Questions

15. Name and describe the four special functional characteristics of muscle that are the basis for muscle response.

16. Distinguish between (a) direct and indirect muscle attachments and (b) a tendon and an aponeurosis.

17. (a) Describe the structure of a sarcomere and indicate the relationship of the sarcomere to the myofilament. (b) Explain the sliding filament theory of contraction using appropriately labeled diagrams of a relaxed and a contracted sarcomere.

18. What is the importance of acetylcholinesterase in muscle cell contraction?

19. Explain how a slight (but smooth) contraction differs from a vigorous contraction of the same muscle using the understandings of multiple motor unit summation.

20. Explain what is meant by the term excitation-contraction coupling.

21. Define motor unit.

22. Describe the three distinct types of skeletal muscle fibers.

23. True or false: Most muscles contain a predominance of one skeletal muscle fiber type. Explain the reasoning behind your choice.

24. Describe the cause(s) of muscle fatigue and define this term clearly.

25. Define oxygen debt.

26. Name four factors that influence contractile force and two that influence velocity and duration of contraction.

27. Smooth muscle has some unique properties, such as low energy usage, ability to maintain contraction over long periods, and the stress-relaxation response. Tie these properties to the function of smooth muscle in the body.

10

THE MUSCULAR SYSTEM

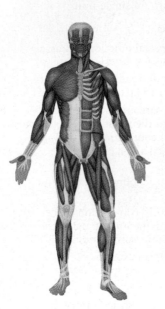

Interactions of Skeletal Muscles in the Body (p. 281)

1. Describe the function of prime movers, antagonists, synergists, and fixators.

Naming Skeletal Muscles (pp. 281–282)

2. List the criteria used in naming muscles. Provide an example to illustrate the use of each criterion.

Muscle Mechanics: Importance of Fascicle Arrangement and Leverage (pp. 282–286)

3. Define lever, and explain how a lever operating at a mechanical advantage differs from one operating at a mechanical disadvantage.

4. Name the three types of lever systems and indicate the arrangement of effort, fulcrum, and load in each. Also note the advantages of each type of lever system.

5. Name the common patterns of muscle fascicle arrangement and relate these to power generation.

Major Skeletal Muscles of the Body (pp. 286–339)

6. Name and identify the muscles described in Tables 10.1 to 10.17. State the origin, insertion, and action of each.

The human body enjoys an incredibly wide range of movements. The gentle blinking of your eye, standing on tiptoe, and wielding a sledgehammer are just a small sample of different activities promoted by the muscular system. Although muscle tissue includes all contractile tissues (skeletal, cardiac, and smooth muscle), when we study the muscular system, **skeletal muscles** take center stage. These muscular "machines" that enable us to perform so many different activities are the focus of this chapter. Before describing the individual muscles in detail, we will describe the manner in which muscles "play" with or against each other to bring about movements, consider the criteria used for naming muscles, and explain the principles of leverage.

Interactions of Skeletal Muscles in the Body

The arrangement of body muscles permits them to work either together or in opposition to achieve a wide variety of movements. As you eat, for example, you alternately raise your fork to your lips and lower it to your plate, and both sets of actions are accomplished by your arm and hand muscles. But muscles can only *pull;* they never *push.* Generally as a muscle shortens, its *insertion* (attachment on the movable bone) moves toward its *origin* (its fixed or immovable point of attachment). Thus, whatever one muscle or muscle group can do, there is another muscle or group of muscles that "undoes" the action.

Muscles can be classified into four *functional* groups: prime movers, antagonists, synergists, and fixators. A muscle that provides the major force for producing a specific movement is a **prime mover,** or **agonist** (ag'o-nist; "leader"), of that movement. The biceps brachii muscle, which fleshes out the anterior arm (and inserts on the radius), is a prime mover of elbow flexion.

Muscles that oppose, or reverse, a particular movement are called **antagonists** (an-tag'o-nists). When a prime mover is active, the antagonist muscles are often stretched and may be relaxed. Antagonists can also help to regulate the action of a prime mover by contracting to provide some resistance, thus helping to prevent overshoot or to slow or stop the movement. As you might expect, a prime mover and its antagonist are located on opposite sides of the joint across which they act. Antagonists can also be prime movers in their own right. For example, flexion of the forearm by the biceps brachii muscle of the arm is antagonized by the triceps brachii, the prime mover for extending the forearm.

In addition to agonists and antagonists, most movements involve the action of one or more **synergists** (sin'er-jists; *syn* = together, *erg* = work). Synergists help prime movers by (1) adding a little extra force to the same movement or (2) reducing undesirable or unnecessary movements that might occur as the prime mover contracts. This latter function deserves more explanation. When a muscle crosses two or more joints, its contraction causes movement at all of the spanned joints unless other muscles act as joint stabilizers. For example, the finger flexor muscles cross both the wrist and the phalangeal joints, but you can make a fist without bending your wrist because synergistic muscles stabilize the wrist. Additionally, as some flexors act, undesirable rotatory movements occur; synergists can prevent this, allowing all of the prime mover's force to be exerted in the desired direction.

When synergists immobilize a bone, or a muscle's origin, they are more specifically called **fixators** (fik'sa-terz). Recall from Chapter 7 that the scapula is held to the axial skeleton only by muscles and is quite freely movable. The fixator muscles that run from the axial skeleton to the scapula can immobilize the scapula so that only the desired movements occur at the mobile shoulder joint. Additionally, muscles that help to maintain upright posture are fixators.

In summary, although prime movers seem to get all the credit for causing certain movements, antagonistic and synergistic muscles are also important in producing smooth, coordinated, and precise movements. Furthermore, a muscle may act as a prime mover in one movement, an antagonist for another movement, a synergist for a third movement, and so on.

Naming Skeletal Muscles

Skeletal muscles are named according to a number of criteria, each of which describes the muscle in some way. Paying attention to these cues can simplify the task of learning muscle names and actions.

1. Location of the muscle. Some muscle names indicate the bone or body region with which the muscle is associated. For example, the temporalis (tem"por-ă'lis) muscle overlies the temporal bone, and intercostal (*costal* = rib) muscles run between the ribs.

2. Shape of the muscle. Some muscles are named for their distinctive shapes. For example, the deltoid (del'toid) muscle is roughly triangular (*deltoid* = triangle), and together the right and left trapezius (trah-pe'ze-us) muscles form a trapezoid.

3. Relative size of the muscle. Terms such as *maximus* (largest), *minimus* (smallest), *longus* (long), and *brevis* (short) are often used in muscle names—as in gluteus maximus and gluteus minimus (the large and small gluteus muscles, respectively).

4. Direction of muscle fibers. The names of some muscles reveal the direction in which their fibers (and fascicles) run in reference to some imaginary line, usually the midline of the body or the longitudinal axis of a limb bone. In muscles with the term *rectus* (straight) in their names, the fibers run parallel to that imaginary line (axis), whereas the terms *transversus* and *oblique* indicate that the muscle fibers run respectively at right angles and obliquely to that line. Specific examples include the rectus femoris (straight muscle of the thigh, or femur) and transversus abdominis (transverse muscle of the abdomen).

5. Number of origins. When *biceps*, *triceps*, or *quadriceps* forms part of a muscle's name, you can assume that the muscle has two, three, or four origins, respectively. For example, the biceps brachii (bra′ke-i) muscle of the arm has two origins, or *heads*.

6. Location of the attachments. Some muscles are named according to their points of origin and insertion. The origin is always named first. For instance, the sternocleidomastoid (ster″no-kli″do-mas′toid) muscle of the neck has a dual origin on the sternum *(sterno)* and clavicle *(cleido)*, and it inserts on the *mastoid* process of the temporal bone.

7. Action. When muscles are named for their action, action words such as *flexor, extensor,* or *adductor* appear in the muscle's name. For example, the adductor longus, located on the medial thigh, brings about thigh adduction, and the supinator (soo′pĭ-na″tor) muscle supinates the forearm.

Often, several criteria are combined in the naming of a muscle. For instance, the name *extensor carpi radialis longus* tells us the muscle's action (extensor), what joint it acts on (*carpi* = wrist), and that it lies close to the radius of the forearm (radialis); it also hints at its size (longus) relative to other wrist extensor muscles. Unfortunately, not all muscle names are this descriptive.

Muscle Mechanics: Importance of Fascicle Arrangement and Leverage

Most factors contributing to muscle force and speed (load, fiber type, etc.) have already been covered in Chapter 9 with two important exceptions—the patterns of fascicle arrangement in muscles and lever systems. These are attended to next.

Arrangement of Fascicles

All skeletal muscles consist of fascicles, but fascicle arrangements vary, resulting in muscles with different shapes and functional capabilities. The most common patterns of fascicle arrangement are parallel, pennate, convergent, and circular (Figure 10.1).

The fascicular pattern is **circular** when the fascicles are arranged in concentric rings (Figure 10.1a). Muscles with this arrangement surround external body openings, which they close by contracting. A general term for such muscles is *sphincters* ("squeezers"). Examples are the orbicularis muscles surrounding the eyes and the mouth.

A **convergent** muscle has a broad origin, and its fascicles *converge* toward a single tendon of insertion. Such a muscle is triangular or fan shaped like the pectoralis major muscle of the anterior thorax (Figure 10.1b).

In a **parallel** arrangement, the long axes of the fascicles run parallel to the long axis of the muscle. Such muscles are either *straplike* (Figure 10.1c), or spindle shaped with an expanded belly (midsection), like the biceps brachii muscle of the arm (Figure 10.1f). However, some authorities classify the spindle-shaped muscles into a separate class as **fusiform muscles.** This is the approach used here.

In a **pennate** (pen′āt) pattern, the fascicles are short and they attach obliquely (*penna* = feather) to a central tendon that runs the length of the muscle. If, as seen in the extensor digitorum muscle of the leg, the fascicles insert into only one side of the tendon, the muscle is *unipennate* (Figure 10.1d). If the fascicles insert into the tendon from opposite sides, so that the muscle's "grain" resembles a feather, the arrangement is *bipennate* (Figure 10.1g). The rectus femoris of the thigh is bipennate. A *multipennate* arrangement looks like many feathers situated side by side, with all their quills inserted into one large tendon. The deltoid muscle, which forms the roundness of the shoulder, is multipennate (Figure 10.1e).

The arrangement of a muscle's fascicles determines its range of motion and power. Because skeletal muscle fibers shorten to about 70% of their resting length when they contract, the longer and the more nearly parallel the muscle fibers are to a muscle's long axis, the more the muscle can shorten. Muscles with parallel fascicle arrangement shorten the most, but they are not usually very powerful. Muscle power depends more on the total number of muscle cells in the muscle; the greater the number, the greater the power. The stocky bipennate and multipennate muscles, which "pack in" the most fibers, shorten very little but are very powerful.

Lever Systems: Bone-Muscle Relationships

The operation of most skeletal muscles involves the use of leverage and **lever systems** (partnerships between the *muscular* and skeletal systems). A **lever** is a rigid bar that moves on a fixed point, or **fulcrum,**

? | *Of the muscles illustrated, which could shorten most? Which two would probably be most powerful? Why?*

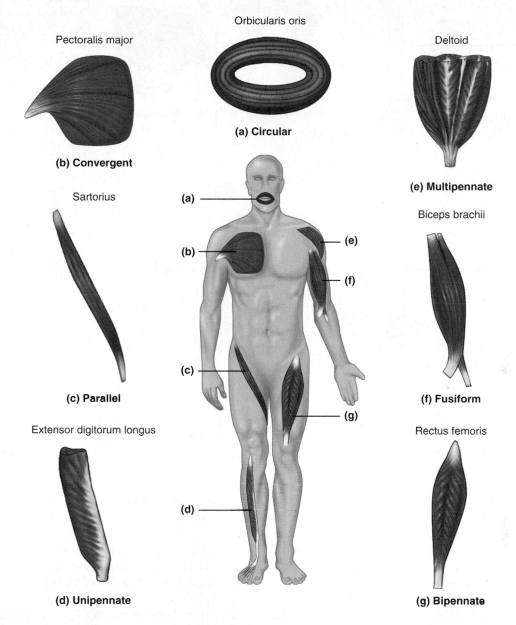

FIGURE 10.1   **Relationship of fascicle arrangement to muscle structure.**

when a force is applied to it. The applied force, or **effort,** is used to move a resistance, or **load.** In your body, your joints are the fulcrums, and your bones act as levers. Muscle contraction provides the effort which is applied at the muscle's insertion point on a

bone. The load is the bone itself, along with overlying tissues and anything else you are trying to move with that lever.

A lever allows a given effort to move a heavier load, or to move a load farther or faster, than it otherwise could. If, as shown in Figure 10.2a, the load is close to the fulcrum and the effort is applied far from the fulcrum, a small effort exerted over a relatively large distance can be used to move a large load over a small distance. Such a lever is said to operate at a **mechanical advantage** and is commonly called a

■ *The biceps brachii and sartorius could shorten most because they have the longest fibers and fibers that run generally parallel to the long axis of the bone with which it is closely associated. The pectoralis major, deltoid, and rectus femoris would be the most powerful because they are the most fleshy.*

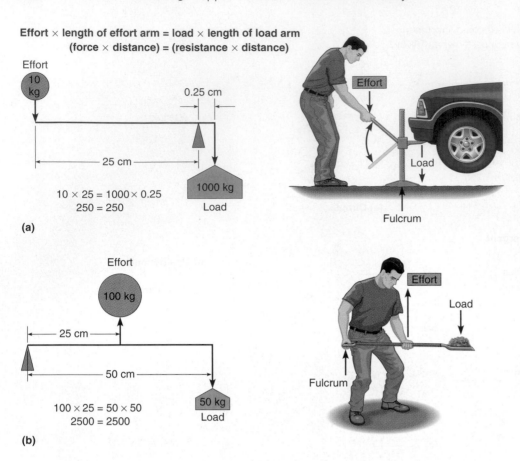

Effort × length of effort arm = load × length of load arm
(force × distance) = (resistance × distance)

$10 \times 25 = 1000 \times 0.25$
$250 = 250$

(a)

$100 \times 25 = 50 \times 50$
$2500 = 2500$

(b)

FIGURE 10.2 **Lever systems operating at a mechanical advantage and a mechanical disadvantage.** The equation at the top expresses the relationships among the forces and distances in any lever system. **(a)** 10 kg of force (the effort) is used to lift a 1000-kg car (the load). This lever system, which employs a jack, operates at a mechanical advantage: The load lifted is greater than the applied muscular effort. **(b)** Using a shovel to lift dirt is an example of a lever system that operates at a mechanical disadvantage. A muscular force (effort) of 100 kg is used to lift 50 kg of dirt (the load). Levers operating at a mechanical disadvantage are common in the body.

power lever. For example, as shown to the right in Figure 10.2a, a person can lift a car with such a lever, in this case, a jack. The car moves up only a small distance with each downward "push" of the jack handle, but relatively little muscle effort is needed. If, on the other hand, the load is far from the fulcrum and the effort is applied near the fulcrum, the force exerted by the muscle must be greater than the load moved or supported (Figure 10.2b). This lever system operates at a **mechanical disadvantage** and is a *speed lever.* These levers are useful because they provide rapid contractions with a wide range of motion. Wielding a shovel is an example. As you can see, small differences in the site of a muscle's insertion (relative to the fulcrum or joint) can translate into large differences in the amount of force a muscle must generate to move a given load or resistance. Regardless of type, all levers follow the same basic principle: effort farther than load from fulcrum = mechanical advantage; effort nearer than load to fulcrum = mechanical disadvantage.

Depending on the relative position of the three elements—effort, fulcrum, and load—a lever belongs to one of three classes. In **first-class levers,** the effort is applied at one end of the lever and the load is at the other, with the fulcrum somewhere between (Figure 10.3a). Seesaws and scissors are first-class levers. First-class leverage also occurs when you lift your head off your chest. Some first-class levers in the body operate at a mechanical advantage, but others, such as the action of the triceps muscle in extending the forearm against resistance, operate at a mechanical disadvantage.

In a **second-class lever,** the effort is applied at one end of the lever and the fulcrum is located at the other, with the load between them (Figure 10.3b). A wheelbarrow demonstrates this type of lever system. Second-class levers are uncommon in the body; the best example is the act of standing on your toes. All second-class levers in the body work at a mechanical advantage because the muscle insertion is always farther from the fulcrum than is the load to be moved. Second-class levers are levers of strength, but speed and range of motion are sacrificed for that strength.

In **third-class levers,** the effort is applied between the load and the fulcrum (Figure 10.3c). These levers operate with great speed and *always* at a mechanical disadvantage. Tweezers or forceps provide this type of leverage. Most skeletal muscles of the body act in third-class lever systems; an example is the activity of the biceps muscle of the arm, lifting the distal forearm and anything carried in the hand. Third-class lever systems permit a muscle to be inserted very close to the joint across which movement occurs, which allows rapid, extensive movements with relatively little shortening of the muscle.

Which of these lever systems as demonstrated at far left in the diagrams would be the fastest lever?

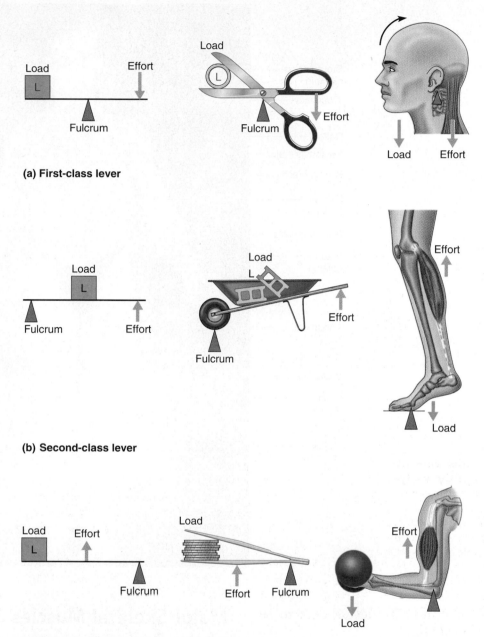

(a) First-class lever

(b) Second-class lever

(c) Third-class lever

FIGURE 10.3   Lever systems. (a) In a first-class lever, the arrangement of the elements is load-fulcrum-effort. Scissors are in this lever class. In the body, a first-class lever system is activated when you raise your head off your chest. The posterior neck muscles provide the effort, the atlanto-occipital joint is the fulcrum, and the weight to be lifted is the facial skeleton. **(b)** In second-class levers, the arrangement is fulcrum-load-effort, as exemplified by a wheelbarrow. In the body, second-class leverage is exerted when you stand on tiptoe. The joints of the ball of the foot are the fulcrum, the weight of the body is the load, and the effort is exerted by the calf muscles pulling upward on the heel (calcaneus) of the foot. **(c)** In a third-class lever, the arrangement is load-effort-fulcrum. A pair of tweezers or forceps uses this type of leverage. The operation of the biceps brachii muscle in flexing the forearm exemplifies third-class leverage. The load is the hand and distal end of the forearm, the effort is exerted on the proximal radius of the forearm, and the fulcrum is the elbow joint.

■ *(c) The third-class lever*

Which two muscles identified in (a) are used to do biceps curls (flex the elbow)? Which would be active during abdominal crunches (sit-ups)?

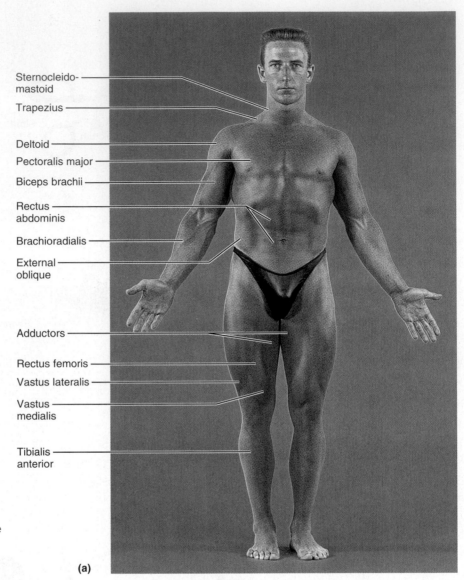

Sternocleido-mastoid

Trapezius

Deltoid

Pectoralis major

Biceps brachii

Rectus abdominis

Brachioradialis

External oblique

Adductors

Rectus femoris

Vastus lateralis

Vastus medialis

Tibialis anterior

(a)

FIGURE 10.4   Anterior view of superficial muscles of the body.
(a) Photograph of surface anatomy.
(b) Diagrammatic view. The abdominal surface has been partially dissected on the right side to show somewhat deeper muscles. (See *A Brief Atlas of the Human Body*, Figure 62.)

Muscles involved in third-class levers tend to be thicker and more powerful.

In conclusion, differences in the positioning of the three elements modify muscle activity with respect to (1) speed of contraction, (2) range of movement, and (3) the weight of the load that can be lifted. In lever systems that operate at a mechanical disadvantage (speed levers), force is lost but speed and range of movement are gained, and this can be a distinct benefit. Systems that operate at a mechanical advantage (power levers) are slower, more stable, and used where strength is a priority.

Brachioradialis and biceps brachii would be used to do curls. The external obliques and rectus abdominis muscle would be working during crunches. ■

Major Skeletal Muscles of the Body

The grand plan of the muscular system is all the more impressive because of the sheer number of skeletal muscles in the body—there are over 600 of them! Obviously, trying to remember all the names, locations, and actions of these muscles is a monumental task. Take heart; only the principal muscles (approximately 125 pairs of them) are considered here. Although this number is far fewer than 600, the job of learning about all these muscles will still require a concerted effort on your part. Memorization will be easier if you can apply what you have learned in a practical, or clinical, way; that is, with a *functional anatomy focus*. Once you are satisfied that you

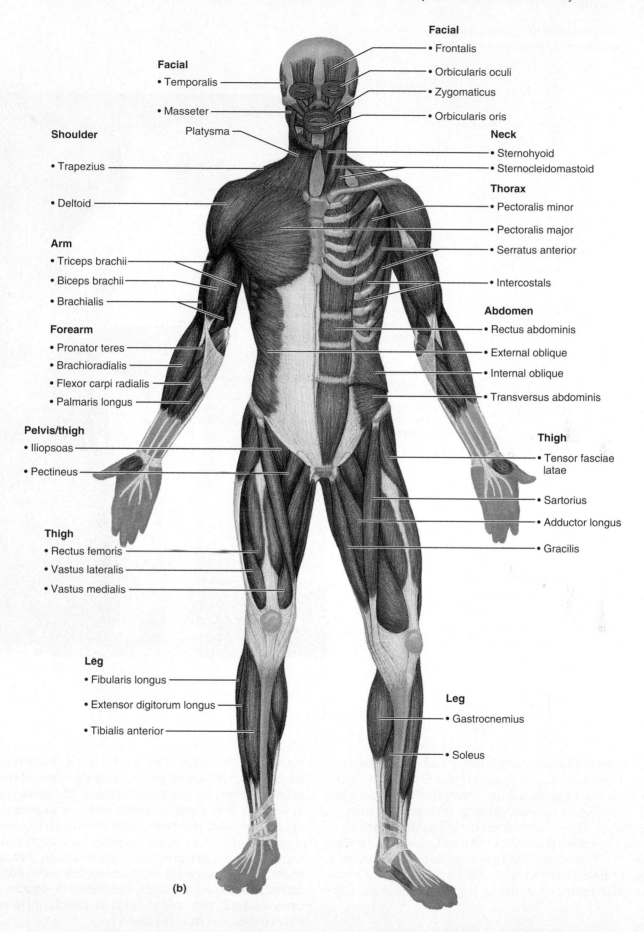

Facial
- Temporalis
- Masseter

Platysma

Facial
- Frontalis
- Orbicularis oculi
- Zygomaticus
- Orbicularis oris

Shoulder
- Trapezius
- Deltoid

Arm
- Triceps brachii
- Biceps brachii
- Brachialis

Forearm
- Pronator teres
- Brachioradialis
- Flexor carpi radialis
- Palmaris longus

Pelvis/thigh
- Iliopsoas
- Pectineus

Thigh
- Rectus femoris
- Vastus lateralis
- Vastus medialis

Leg
- Fibularis longus
- Extensor digitorum longus
- Tibialis anterior

Neck
- Sternohyoid
- Sternocleidomastoid

Thorax
- Pectoralis minor
- Pectoralis major
- Serratus anterior
- Intercostals

Abdomen
- Rectus abdominis
- External oblique
- Internal oblique
- Transversus abdominis

Thigh
- Tensor fasciae latae
- Sartorius
- Adductor longus
- Gracilis

Leg
- Gastrocnemius
- Soleus

(b)

Which muscles would you contract to shrug your shoulders? To retract your scapulae?

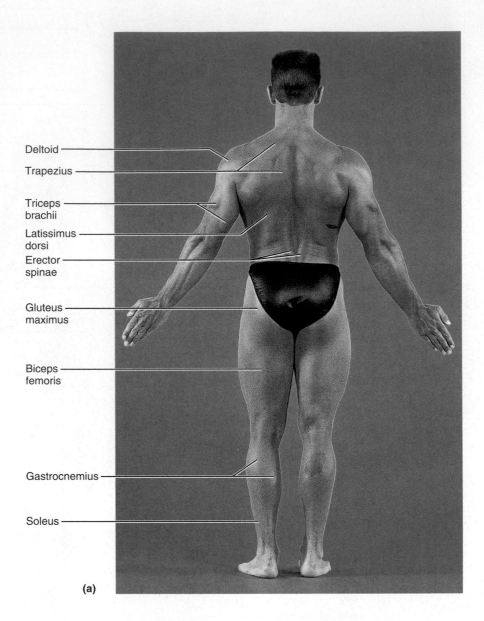

Deltoid

Trapezius

Triceps brachii

Latissimus dorsi

Erector spinae

Gluteus maximus

Biceps femoris

Gastrocnemius

Soleus

FIGURE 10.5   **Posterior view of superficial muscles of the body.**
(a) Photograph of surface anatomy.
(b) Diagrammatic view. (See *A Brief Atlas of the Human Body*, Figures 61, 63, and 64.)

(a)

have learned the name of a muscle and can identify it on a cadaver, model, or diagram, you must then flesh out your learning by asking yourself, "What does it do?" (It might be a good idea to review body movements (pp. 226–230) before you get to this point.)

In the tables that follow, the muscles of the body have been grouped by function and by location, roughly from head to foot. Each table is keyed to a particular figure or group of figures illustrating the

muscles it describes. The legend at the beginning of each table provides an overview of the types of movements effected by the muscles listed and gives pointers on the way those muscles interact with one another. The table itself describes each muscle's shape, location relative to other muscles, origin and insertion, primary actions, and innervation. (Because some instructors want students to defer learning the muscle innervations until the nervous system has been studied, you might want to check on what is expected of you in this regard.)

■ *The trapezius. The trapezius and rhomboids.*

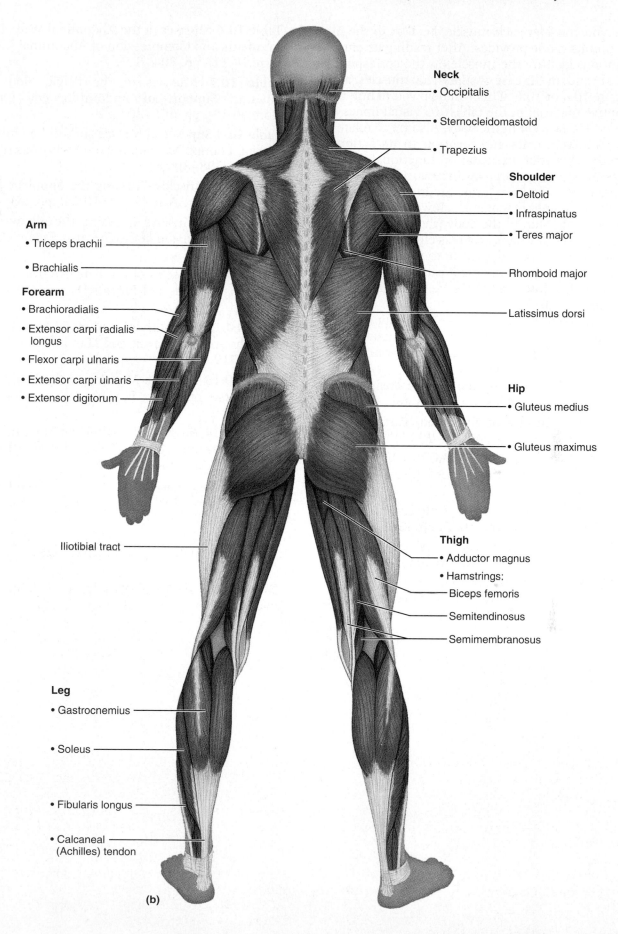

Neck
- Occipitalis
- Sternocleidomastoid
- Trapezius

Shoulder
- Deltoid
- Infraspinatus
- Teres major

Rhomboid major

Latissimus dorsi

Arm
- Triceps brachii
- Brachialis

Forearm
- Brachioradialis
- Extensor carpi radialis longus
- Flexor carpi ulnaris
- Extensor carpi uinaris
- Extensor digitorum

Hip
- Gluteus medius
- Gluteus maximus

Thigh
- Adductor magnus
- Hamstrings:
  - Biceps femoris
  - Semitendinosus
  - Semimembranosus

Iliotibial tract

Leg
- Gastrocnemius
- Soleus
- Fibularis longus
- Calcaneal (Achilles) tendon

(b)

As you consider each muscle, be alert to the information its name provides. After reading its entire description, identify the muscle on the corresponding figure and, in the case of superficial muscles, also on Figure 10.4 or 10.5. This will help you to link the descriptive material in the table to a visual image of the muscle's location in the body. Also try to relate a muscle's attachments and location to its actions. This will focus your attention on functional details that often escape student awareness. For example, both the elbow and knee joints are hinge joints that allow flexion and extension. However, the knee flexes to the dorsum of the body (the calf moves toward the posterior thigh), whereas elbow flexion carries the forearm toward the anterior aspect of the arm. Therefore, leg flexors are located on the posterior thigh, while forearm flexors are found on the anterior aspect of the humerus.

Finally, keep in mind that the *best* way to learn muscle actions is to act out their movements yourself while feeling for the muscles contracting (bulging) beneath your skin.

The following list summarizes the organization and sequence of the tables in this chapter:

Muscle Gallery
TABLE 10.1   Muscles of the Head, Part I: Facial Expression (Figure 10.6)

The muscles that promote facial expression lie in the scalp and face just deep to the skin. They are thin and variable in shape and strength, and adjacent muscles tend to be fused. They are unusual muscles in that they insert into skin (or other muscles), not bones. In the scalp, the main muscle is the **epicranius,** which has distinct anterior and posterior parts; the lateral scalp muscles are vestigial in humans. Muscles clothing the facial bones lift the eyebrows, flare the nostrils, open and close the eyes and mouth, and provide one of the best tools for influencing others—the smile. The tremendous importance of facial muscles in nonverbal communication becomes especially clear when they are paralyzed, as in some stroke victims. All muscles listed in this table are innervated by *cranial nerve VII* (the *facial nerve*). The external muscles of the eyes, which act to direct the eyeballs, and the levator palpebrae superioris muscles that raise the eyelids are described in Chapter 13.

Muscle	Description	Origin (O) and Insertion (I)	Action	Nerve Supply
Muscles of the Scalp				
Epicranius (occipitofrontalis) (ep"ĭ-kra'ne-us; ok-sip"ĭ-to-fron-ta'lis) (*epi* = over; *cran* = skull)	Bipartite muscle consisting of the frontalis and occipitalis muscles connected by a cranial aponeurosis, the galea aponeurotica; the alternate actions of these two muscles pull scalp forward and backward			
▪ **Frontalis** (fron-ta'lis) (*front* = forehead)	Covers forehead and dome of skull; no bony attachments	O—galea aponeurotica I—skin of eyebrows and root of nose	With aponeurosis fixed, raises the eyebrows (as in surprise); wrinkles forehead skin horizontally	Facial nerve (cranial VII)
▪ **Occipitalis** (ok-sip"ĭ-tal'is) (*occipito* = base of skull)	Overlies posterior occiput; by pulling on the galea, fixes origin of frontalis	O—occipital and temporal bones I—galea aponeurotica	Fixes aponeurosis and pulls scalp posteriorly	Facial nerve
Muscles of the Face				
Corrugator supercilii (kor'ah-ga-ter soo"per-sĭ'le-i) (*corrugo* = wrinkle; *supercilium* = eyebrow)	Small muscle; activity associated with that of orbicularis oculi	O—arch of frontal bone above nasal bone I—skin of eyebrow	Draws eyebrows together and inferiorly; wrinkles skin of forehead vertically (as in frowning)	Facial nerve
Orbicularis oculi (or-bik'u-lar-is ok'u-li) (*orb* = circular; *ocul* = eye)	Thin, tripartite sphincter muscle of eyelid; surrounds rim of the orbit	O—frontal and maxillary bones and ligaments around orbit I—tissue of eyelid	Protects eyes from intense light and injury; various parts can be activated individually; produces blinking, squinting, and draws eyebrows inferiorly	Facial nerve
Zygomaticus—major and minor (zi-go-mat'ĭ-kus), (*zygomatic* = cheekbone)	Muscle pair extending diagonally from cheekbone to corner of mouth	O—zygomatic bone I—skin and muscle at corner of mouth	Raises lateral corners of mouth upward (smiling muscle)	Facial nerve
Risorius (ri-zor'e-us) (*risor* = laughter)	Slender muscle inferior and lateral to zygomaticus	O—lateral fascia associated with masseter muscle I—skin at angle of mouth	Draws corner of lip laterally; tenses lips; synergist of zygomaticus	Facial nerve

Muscle Gallery
TABLE 10.1   **Muscles of the Head, Part I: Facial Expression (Figure 10.6)** *(continued)*

Muscle	Description	Origin (O) and Insertion (I)	Action	Nerve Supply
Levator labii superioris (lĕ-va'tor la'be-i soo-per"e-or'is) (*leva* = raise; *labi* = lip; *superior* = above, over)	Thin muscle between orbicularis oris and inferior eye margin	O—zygomatic bone and infraorbital margin of maxilla I—skin and muscle of upper lip	Opens lips; raises and furrows the upper lip	Facial nerve
Depressor labii inferioris (de-pres'or la'be-i in-fer"e-or'is) (*depressor* = depresses; *infer* = below)	Small muscle running from mandible to lower lip	O—body of mandible lateral to its midline I—skin and muscle of lower lip	Draws lower lip inferiorly (as in a pout)	Facial nerve
Depressor anguli oris (ang'gu-li or-is) (*angul* = angle, corner; *or* = mouth)	Small muscle lateral to depressor labii inferioris	O—body of mandible below incisors I—skin and muscle at angle of mouth below insertion of zygomaticus	Zygomaticus antagonist; draws corners of mouth downward and laterally (as in a "tragedy mask" grimace)	Facial nerve
Orbicularis oris	Complicated, multilayered muscle of the lips with fibers that run in many different directions; most run circularly	O—arises indirectly from maxilla and mandible; fibers blended with fibers of other facial muscles associated with the lips I—encircles mouth; inserts into muscle and skin at angles of mouth	Closes lips; purses and protrudes lips; kissing and whistling muscle	Facial nerve
Mentalis (men-ta'lis) (*ment* = chin)	One of the muscle pair forming a V-shaped muscle mass on chin	O—mandible below incisors I—skin of chin	Protrudes lower lip; wrinkles chin	Facial nerve
Buccinator (bu'sĭ-na"ter) (*bucc* = cheek or "trumpeter")	Thin, horizontal cheek muscle; principal muscle of cheek; deep to masseter (see also Figure 10.7)	O—molar region of maxilla and mandible I—orbicularis oris	Draws corner of mouth laterally; compresses cheek (as in whistling and sucking); holds food between teeth during chewing; well developed in nursing infants	Facial nerve
Platysma (plah-tiz'mah) (*platy* = broad, flat)	Unpaired, thin, sheetlike superficial neck muscle; not strictly a head muscle, but plays a role in facial expression	O—fascia of chest (over pectoral muscles and deltoid) I—lower margin of mandible, and skin and muscle at corner of mouth	Helps depress mandible; pulls lower lip back and down, i.e., produces downward sag of mouth; tenses skin of neck (e.g., during shaving)	Facial nerve

Muscle Gallery
TABLE 10.1   *(continued)*

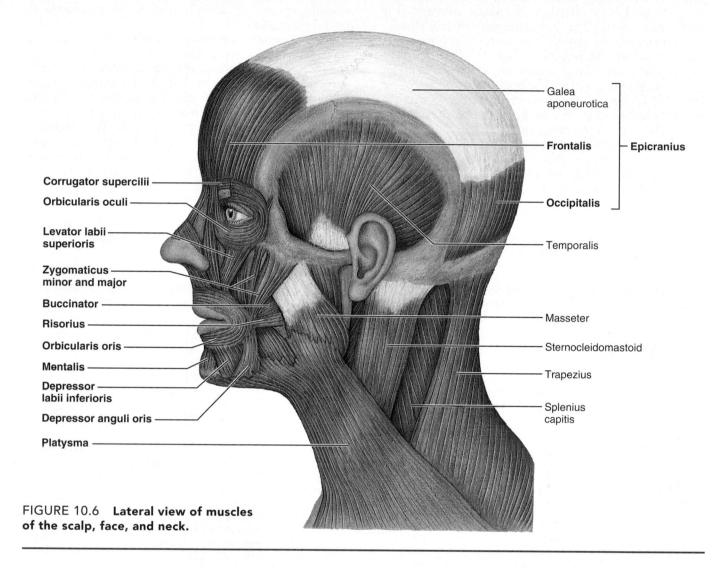

Galea
aponeurotica ⎤
                ⎥
Frontalis   ⎥ ⎦ Epicranius
                ⎥
Occipitalis ⎦

Temporalis

Corrugator supercilii

Orbicularis oculi

Levator labii
superioris

Zygomaticus
minor and major

Buccinator

Risorius

Orbicularis oris

Mentalis

Depressor
labii inferioris

Depressor anguli oris

Platysma

Masseter

Sternocleidomastoid

Trapezius

Splenius
capitis

FIGURE 10.6 **Lateral view of muscles
of the scalp, face, and neck.**

Muscle Gallery
TABLE 10.2

Muscles of the Head, Part II: Mastication and Tongue Movement (Figure 10.7)

Four pairs of muscles are involved in mastication (chewing and biting) activities, and all are innervated by the mandibular division of *cranial nerve V* (the *trigeminal nerve*). The prime movers of jaw closure (and biting) are the powerful **masseter** and **temporalis** muscles, which can be palpated easily when the teeth are clenched. Grinding movements are brought about by the **pterygoid** muscles. The **buccinator** muscles (see Table 10.1) also play a role in chewing. Normally, gravity is sufficient to depress the mandible, but if there is resistance to jaw opening, neck muscles such as the digastric and mylohyoid muscles (see Table 10.3) are activated. In 1996, researchers claimed to have discovered a tiny new muscle (not illustrated) extending from the lateral sphenoid to the medial aspect of the

mandibular ramus and coronoid process. It is still uncertain whether this *sphenomandibular muscle* is a chewing muscle or actually part of the temporalis or lateral pterygoid.

The tongue is composed of muscle fibers that curl, squeeze, and fold the tongue during speaking and chewing. These **intrinsic tongue muscles**, arranged in several planes, change the shape of the tongue and contribute to its exceptional nimbleness, but they do not really move the tongue. They are considered in Chapter 22 with the digestive system. Only the **extrinsic tongue muscles**, which anchor and move the tongue, are considered in this table. The extrinsic tongue muscles are all innervated by *cranial nerve XII* (the *hypoglossal nerve*).

Muscle	Description	Origin (O) and Insertion (I)	Action	Nerve Supply
Muscles of Mastication				
Masseter (mah-se′ter) (*maseter* = chewer)	Powerful muscle that covers lateral aspect of mandibular ramus	O—zygomatic arch and maxilla I—angle and ramus of mandible	Prime mover of jaw closure; elevates mandible	Trigeminal nerve (cranial V)
Temporalis (tem″por-ă′lis) (*tempora* = time; pertaining to the temporal bone)	Fan-shaped muscle that covers parts of the temporal, frontal, and parietal bones	O—temporal fossa I—coronoid process of mandible via a tendon that passes deep to zygomatic arch	Closes jaw; elevates and retracts mandible; synergist of pterygoids (see below) in side-to-side movements; maintains position of the mandible at rest	Trigeminal nerve
Medial pterygoid (me′de-ul ter′ĭ-goid) (*medial* = toward median plane; *pterygoid* = winglike)	Deep two-headed muscle that runs along internal surface of mandible and is largely concealed by that bone	O—medial surface of lateral pterygoid plate of sphenoid bone, maxilla, and palatine bone I—medial surface of mandible near its angle	Synergist of temporalis and masseter muscles in elevation of the mandible; acts with the lateral pterygoid muscle to protrude mandible and to promote side-to-side (grinding) movements	Trigeminal nerve
Lateral pterygoid (*lateral* = away from median plane)	Deep two-headed muscle; lies superior to medial pterygoid muscle	O—greater wing and lateral pterygoid plate of sphenoid bone I—condyle of mandible and capsule of temporomandibular joint	Protrudes mandible (pulls it anteriorly); provides forward sliding and side-to-side grinding movements of the lower teeth	Trigeminal nerve
Buccinator	See Table 10.1	See Table 10.1	Trampoline-like action of buccinator muscles helps keep food between grinding surfaces of teeth during chewing	Facial nerve (cranial VII)

Muscle Gallery
TABLE 10.2 (continued)

Muscle	Description	Origin (O) and Insertion (I)	Action	Nerve Supply
Muscles Promoting Tongue Movements (Extrinsic Muscles)				
Genioglossus (je"ne-o-glah'sus) (geni = chin; glossus = tongue)	Fan-shaped muscle; forms bulk of inferior part of tongue; its attachment to mandible prevents tongue from falling backward and obstructing respiration	O—internal surface of mandible near symphysis I—inferior aspect of the tongue and body of hyoid bone	Primarily protrudes tongue, but can depress or act in concert with other extrinsic muscles to retract tongue	Hypoglossal nerve (cranial XII)
Hyoglossus (hi'o-glos"us) (hyo = pertaining to hyoid bone)	Flat, quadrilateral muscle	O—body and greater horn of hyoid bone I—inferolateral tongue	Depresses tongue and draws its sides downward	Hypoglossal nerve
Styloglossus (sti-lo-glah'sus) (stylo = pertaining to styloid process)	Slender muscle running superiorly to and at right angles to hyoglossus	O—styloid process of temporal bone I—inferolateral tongue	Retracts (and elevates) tongue	Hypoglossal nerve

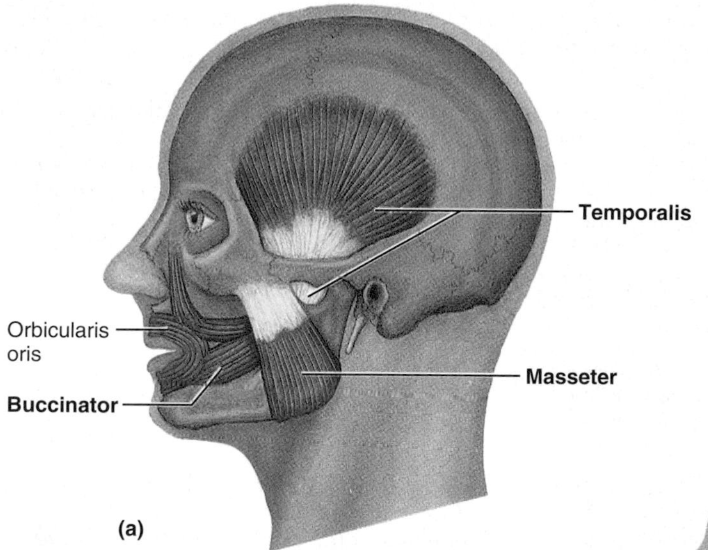

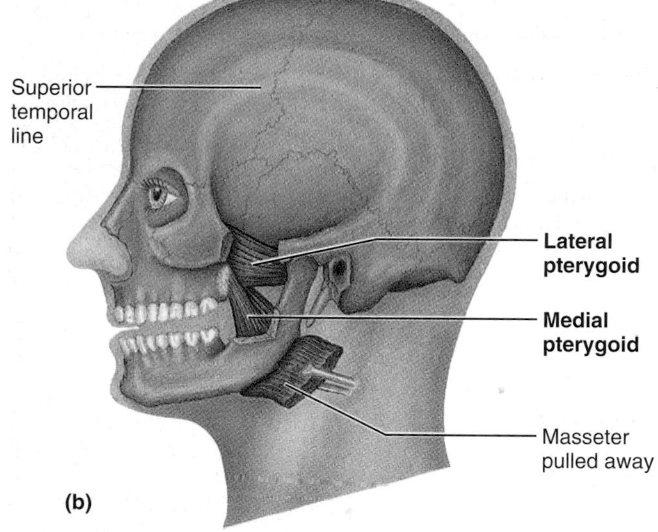

FIGURE 10.7 **Muscles promoting mastication and tongue movements.**
(a) Lateral view of the temporalis, masseter, and buccinator muscles.
(b) Lateral view of the deep chewing muscles, the medial and lateral pterygoid muscles.
(c) Extrinsic muscles of the tongue. Some suprahyoid muscles of the throat are also illustrated.

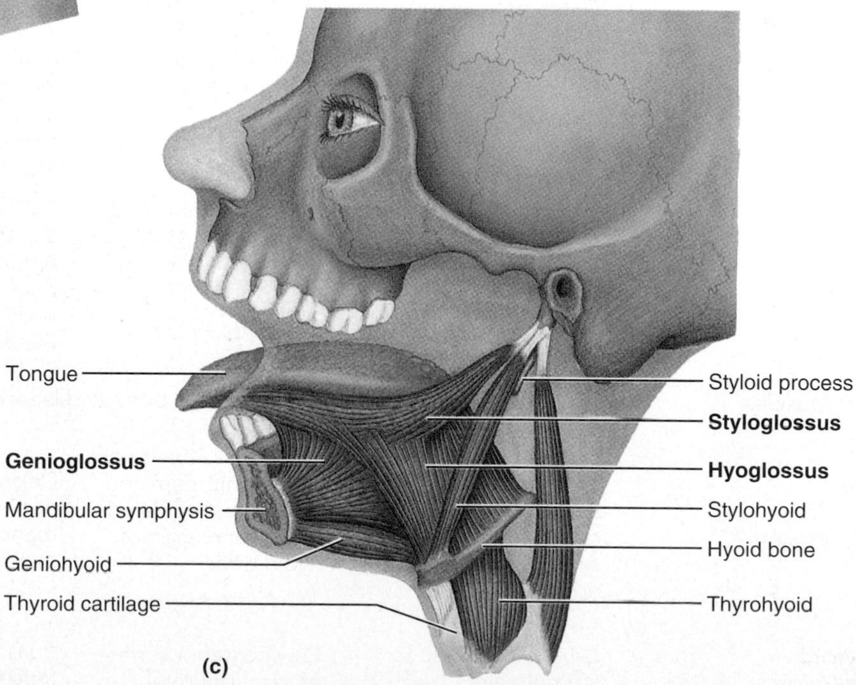

The neck is divided into two triangles (anterior and posterior) by the sternocleidomastoid muscle (Figure 10.8a). This table considers the muscles of the *anterior* triangle, which are divided into **suprahyoid** and **infrahyoid muscles** (above and below the hyoid bone respectively). Most of these muscles are deep (throat) muscles involved in swallowing.

Swallowing begins when the tongue and buccinator muscles of the cheeks squeeze the food back along the roof of the mouth toward the pharynx. Then a rapid series of muscular movements in the posterior mouth and pharynx complete the process. Events of swallowing include: (1) Widening the pharynx to receive the food, and closing the respiratory passageway (larynx) anteriorly to prevent aspiration of food. This is accom-

plished by the *suprahyoid muscles,* which pull the hyoid bone upward and forward toward the mandible. Since the hyoid bone is attached by the thyrohyoid membrane to the larynx, the larynx is also pulled upward and forward, a maneuver that widens the pharynx and closes the respiratory passageway. (2) Closing off the nasal passages to prevent food from entering the superior nasal cavity, accomplished by small muscles that elevate the soft palate. (These muscles, the *tensor* and *levator veli palatini,* are not described in the table but are illustrated in Figure 10.8b.) (3) Food is propelled through the pharynx into the esophagus inferiorly by the **pharyngeal constrictor muscles.** (4) Returning the hyoid bone and larynx to their more inferior positions after swallowing, promoted by the *infrahyoid muscles.*

Muscle	Description	Origin (O) and Insertion (I)	Action	Nerve Supply
Suprahyoid Muscles (soo"prah-hi'oid)	Muscles that help form floor of oral cavity, anchor tongue, elevate hyoid, and move larynx superiorly during swallowing; lie superior to hyoid bone			
Digastric (di-gas'trik) (*di* = two; *gaster* = belly)	Consists of two bellies united by an intermediate tendon, forming a V shape under the chin	O—lower margin of mandible (anterior belly) and mastoid process of the temporal bone (posterior belly) I—by a connective tissue loop to hyoid bone	Acting in concert, the digastric muscles elevate hyoid bone and steady it during swallowing and speech; acting from behind, they open mouth and depress mandible	Mandibular branch of trigeminal nerve (cranial V) for anterior belly; facial nerve (cranial VII) for posterior belly
Stylohyoid (sti"lo-hi'oid) (also see Figure 10.7)	Slender muscle below angle of jaw; parallels posterior belly of diagastric muscle	O—styloid process of temporal bone I—hyoid bone	Elevates and retracts hyoid, thereby elongating floor of mouth during swallowing	Facial nerve
Mylohyoid (mi"lo-hi'oid) (*myle* = molar)	Flat, triangular muscle just deep to digastric muscle; this muscle pair forms a sling that forms the floor of the anterior mouth	O—medial surface of mandible I—hyoid bone and median raphe	Elevates hyoid bone and floor of mouth, enabling tongue to exert backward and upward pressure that forces food bolus into pharynx	Mandibular branch of trigeminal nerve
Geniohyoid (je'ne-o-hy"oid) (also see Figure 10.7) (*geni* = chin)	Narrow muscle in contact with its partner medially; runs from chin to hyoid bone	O—inner surface of mandibular symphysis I—hyoid bone	Pulls hyoid bone superiorly and anteriorly, shortening floor of mouth and widening pharynx for receiving food	First cervical spinal nerve via hypoglossal nerve (cranial XII)
Infrahyoid Muscles	Straplike muscles that depress the hyoid bone and larynx during swallowing and speaking (see also Figure 10.9c)			
Sternohyoid (ster"no-hi'oid) (*sterno* = sternum)	Most medial muscle of the neck: thin; superficial except inferiorly, where covered by sternocleidomastoid	O—manubrium and medial end of clavicle I—lower margin of hyoid bone	Depresses larynx and hyoid bone if mandible is fixed; may also flex skull	Cervical spinal nerves 1–3 (C_1–C_3) through ansa cervicalis (slender nerve root in cervical plexus)
Sternothyroid (ster"no-thi'roid) (*thyro* = thyroid cartilage)	Lateral and deep to sternohyoid	O—posterior surface of manubrium of sternum I—thyroid cartilage	Pulls thyroid cartilage (plus larynx and hyoid bone) inferiorly	As for sternohyoid

Muscle	Description	Origin (O) and Insertion (I)	Action	Nerve Supply
Omohyoid (o"mo-hi'oid) (*omo* = shoulder)	Straplike muscle with two bellies united by an intermediate tendon; lateral to sternohyoid	O—superior surface of scapula I—hyoid bone, lower border	Depresses and retracts hyoid bone	As for sternohyoid
Thyrohyoid (thi"ro-hi'oid) (also see Figure 10.7)	Appears as a superior continuation of sternothyroid muscle	O—thyroid cartilage I—hyoid bone	Depresses hyoid bone and elevates larynx if hyoid is fixed	First cervical nerve via hypoglossal
Pharyngeal constrictor muscles—superior, middle, and inferior (far-rin'je-al)	Composite of three paired muscles whose fibers run circularly in pharynx wall; superior muscle is innermost and inferior one is outermost; substantial overlap	O—attached anteriorly to mandible and medial pterygoid plate (superior), hyoid bone (middle), and laryngeal cartilages (inferior) I—posterior median raphe of pharynx	Working as a group and in sequence, all constrict pharynx during swallowing, which propels a food bolus to esophagus (via a massagelike action called peristalsis)	Pharyngeal plexus [branches of vagus (X)]

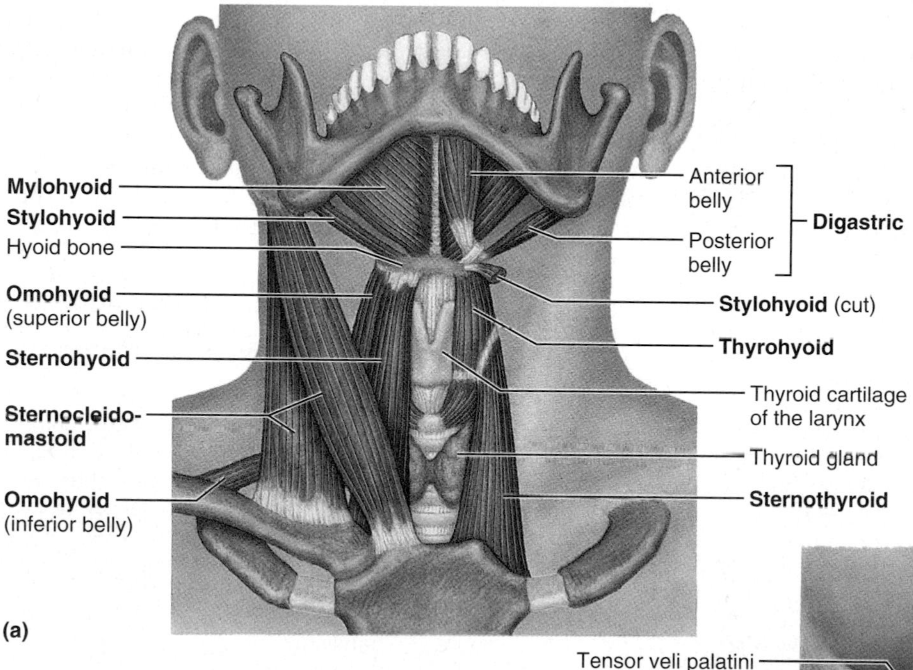

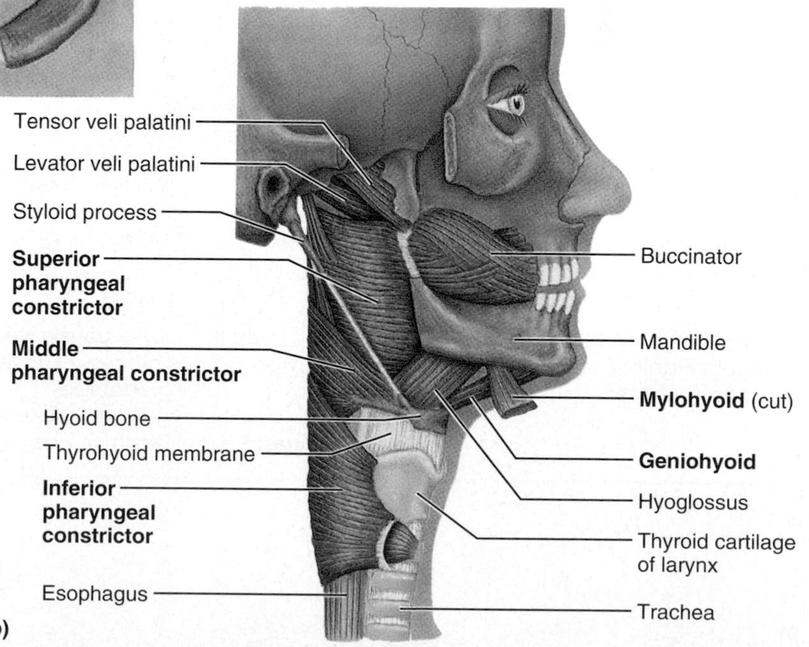

(a)

(b)

FIGURE 10.8 Muscles of the anterior neck and throat that promote swallowing.
(a) Anterior view of the suprahyoid and infrahyoid muscles. The sternocleidomastoid muscle (not involved in swallowing) is shown on the left side of the illustration to provide an anatomical landmark. Deeper neck muscles are illustrated on the right side. **(b)** Lateral view of the constrictor muscles of the pharynx shown in their proper anatomical relationship to the buccinator (a chewing muscle) and the hyoglossus muscle (which promotes tongue movements).

Muscle Gallery
TABLE 10.4   Muscles of the Neck and Vertebral Column: Head and Trunk Movements (Figure 10.9)

Head movements. The head is moved by muscles originating from the axial skeleton. The major head flexors are the **sternocleidomastoid muscles,** but the suprahyoid and infrahyoid muscles described in Table 10.3 act as synergists in this action. Lateral head movements are effected by the sternocleidomastoids and a number of deeper neck muscles, including the **scalenes,** and by several straplike muscles of the vertebral column at the back of the neck. Head extension is aided by the superficial trapezius muscles of the back, but the **splenius** muscles deep to the trapezius muscles bear most of the responsibility for head extension.

Trunk movements. Trunk extension is effected by the *deep* or *intrinsic back muscles* associated with the bony vertebral column. As opposed to these deep back muscles, superficial back muscles are concerned primarily with movements of the shoulder girdle and upper limbs (see Tables 10.8 and 10.9).

The deep muscles of the back form a broad, thick column extending from the sacrum to the skull. Many muscles of varying length contribute to this mass. It helps to regard each of these individual muscles as a string that when pulled causes one or several vertebrae to extend or to rotate on the vertebrae below. The largest of the deep back muscle groups is the **erector spinae** ("spine erectors"). Since the origins and insertions of the different muscle groups overlap extensively, entire regions of the vertebral column can be moved simultaneously and smoothly. Acting in concert, the deep back muscles extend (or hyperextend) the spine, but contraction of the muscles on only one side causes lateral bending of the back, neck, or head. Lateral flexion is automatically accompanied by some degree of rotation of the vertebral column. During vertebral movements, the articular facets of the vertebrae glide on each other.

In addition to the long back muscles, there are a number of short muscles that extend from one vertebra to the next (Figure 10.9e). These small muscles (the rotatores, multifidus, interspinales, and intertransversarii muscles) act primarily as synergists in extension and rotation of the spine and as spine stabilizers. They are not described in the table but you can deduce their actions by examining their origins and insertions in the figure.

Trunk muscles also maintain the normal curvatures of the spine, acting as postural muscles. It is these deep trunk *extensors* that are considered in this table. The more superficial muscles, which have other functions, are considered in subsequent tables. For example, the anterior muscles of the abdominal wall that cause trunk *flexion* are described in Table 10.6.

Muscle	Description	Origin (O) and Insertion (I)	Action	Nerve Supply
Anterolateral Neck Muscles (Figure 10.9a and c)				
Sternocleidomastoid (ster"no-kli"do-mas'toid) (*sterno* = breastbone; *cleido* = clavicle; *mastoid* = mastoid process)	Two-headed muscle located deep to platysma on anterolateral surface of neck; fleshy parts on either side of neck delineate limits of anterior and posterior triangles; key muscular landmark in neck; spasms of one of these muscles may cause torticollis (wryneck)	O—manubrium of sternum and medial portion of clavicle I—mastoid process of temporal bone and superior nuchal line of occipital bone	Prime mover of active head flexion; simultaneous contraction of both muscles causes neck flexion, generally against resistance as when one raises head when lying on back; acting alone, each muscle rotates head toward shoulder on opposite side and tilts or laterally flexes head to its own side	Accessory nerve (cranial nerve XI) and branches of cervical spinal nerves 2–4
Scalenes (ska'lēnz)— anterior, middle, and posterior (*scalene* = uneven)	Located more laterally than anteriorly on neck; deep to platysma and sternocleidomastoid	O—transverse processes of cervical vertebrae I—anterolaterally on first two ribs	Elevate first two ribs (aid in inspiration); flex and rotate neck	Cervical spinal nerves

Muscle Gallery
TABLE 10.4 (continued)

Muscle	Description	Origin (O) and Insertion (I)	Action	Nerve Supply
Intrinsic Muscles of the Back (Figure 10.9b, d, e)				
Splenius (sple'ne-us)—capitis and cervicis portions (kah-pit'us; ser-vis'us) (*splenion* = bandage; *caput* = head; *cervi* = neck) (Figures 10.9b and 10.6)	Broad bipartite superficial muscle (capitis and cervicis parts) extending from upper thoracic vertebrae to skull; capitis portion known as "bandage muscle" because it covers and holds down deeper neck muscles	O—ligamentum nuchae,* spinous processes of vertebrae C_7–T_6 I—mastoid process of temporal bone and occipital bone (capitis); transverse processes of C_2–C_4 vertebrae (cervicis)	Act as a group to extend or hyperextend head; when splenius muscles on one side are activated, head is rotated and bent laterally toward same side	Cervical spinal nerves (dorsal rami)

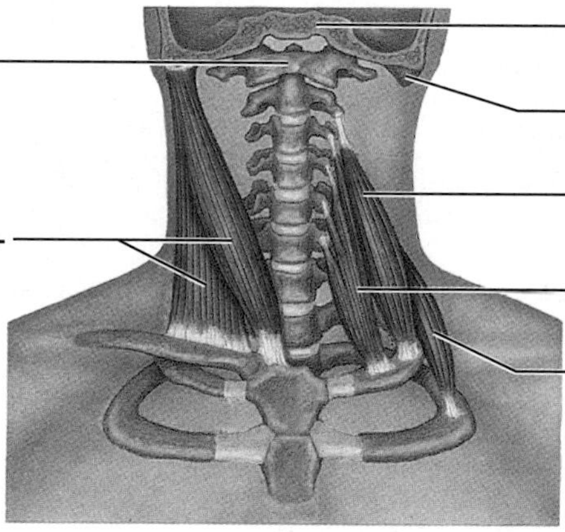

(a) Anterior

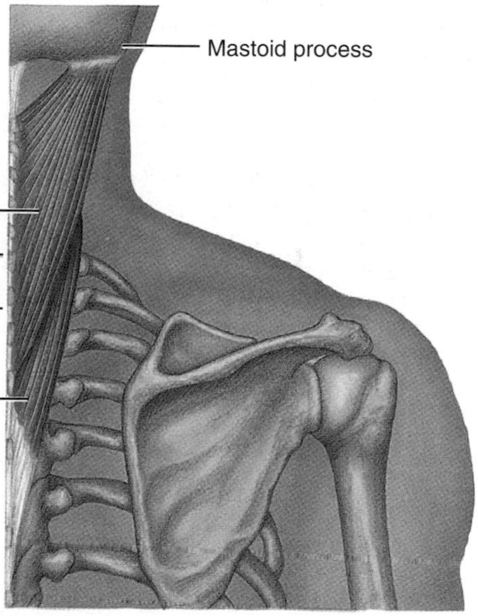

1st cervical vertebra
Sternocleido-mastoid
Base of occipital bone
Mastoid process
Middle scalene
Anterior scalene
Posterior scalene

(b) Posterior

Mastoid process
Splenius capitis
Spinous processes of the vertebrae
Splenius cervicis

FIGURE 10.9 **Muscles of the neck and vertebral column causing movements of the head and trunk. (a)** Muscles of the anterolateral neck. The superficial platysma muscle and the deeper neck muscles have been removed to show the origins and insertions of the sternocleidomastoid and scalene muscles. **(b)** Deep muscles of the posterior neck; superficial muscles have been removed. **(c)** Photograph of the anterior and lateral regions of the neck. (Figure continues on p. 301)

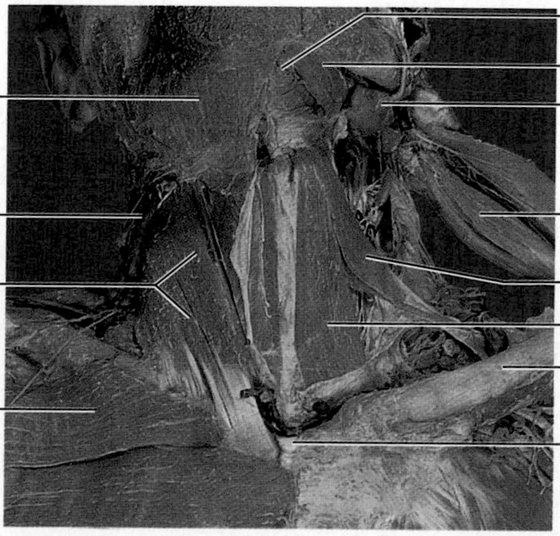

Platysma
External jugular vein
Sternocleidomastoid
Pectoralis major
Mylohyoid
Digastric
Submandibular gland
Sternocleidomastoid (reflected)
Omohyoid
Sternohyoid
Left clavicle
Sternal manubrium

(c)

*The ligamentum nuchae (lig"ah-men'tum noo'ke) is a strong, elastic ligament extending from the occipital bone of the skull along the tips of the spinous processes of the cervical vertebrae. It binds the cervical vertebrae together and inhibits excessive head and neck flexion, thus preventing damage to the spinal cord in the vertebral canal.

Muscle Gallery
TABLE 10.4   **Muscles of the Neck and Vertebral Column: Head and Trunk Movements (Figure 10.9)** *(continued)*

Muscle	Description	Origin (O) and Insertion (I)	Action	Nerve Supply
Erector spinae (e-rek'tor spi'ne) Also called **sacrospinalis** (Figure 10.9d, left side)	Prime mover of back extension; erector spinae muscles, each consisting of three columns—the iliocostalis, longissimus, and spinalis muscles—form intermediate layer of intrinsic back muscles; erector spinae provide resistance that helps control action of bending forward at the waist and act as powerful extensors to promote return to erect position; during full flexion (i.e., when touching fingertips to floor), erector spinae are relaxed and strain is borne entirely by ligaments of back; on reversal of the movement, these muscles are initially inactive, and extension is initiated by hamstring muscles of thighs and gluteus maximus muscles of buttocks. As a result of this peculiarity, lifting a load or moving suddenly from a bent-over position is potentially dangerous (in terms of possible injury) to muscles and ligaments of back and intervertebral discs; erector spinae muscles readily go into painful spasms following injury to back structures			
▪ **Iliocostalis** (il"e-o-kos-tă'lis)—lumborum, thoracis, and cervicis portions (lum'bor-um; tho-ra'sis) (*ilio* = ilium; *cost* = rib)	Most lateral muscle group of erector spinae muscles; extend from pelvis to neck	O—iliac crests (lumborum); inferior 6 ribs (thoracis); ribs 3 to 6 (cervicis) I—angles of ribs (lumborum and thoracis); transverse processes of cervical vertebrae C_6–C_4 (cervicis)	Extend vertebral column, maintain erect posture; acting on one side, bend vertebral column to same side	Spinal nerves (dorsal rami)
▪ **Longissimus** (lon-jis'ĭ-mus)—thoracis, cervicis, and capitis parts (*longissimus* = longest)	Intermediate tripartite muscle group of erector spinae; extend by many muscle slips from lumbar region to skull; mainly pass between transverse processes of the vertebrae	O—transverse processes of lumbar through cervical vertebrae I—transverse processes of thoracic or cervical vertebrae and to ribs superior to origin as indicated by name; capitis inserts into mastoid process of temporal bone	Thoracis and cervicis act together to extend vertebral column and acting on one side, bend it laterally; capitis extends head and turns the face toward same side	Spinal nerves (dorsal rami)
▪ **Spinalis** (spi-nă'lis)—thoracis and cervicis parts (*spin* = vertebral column, spine)	Most medial muscle column of erector spinae; cervicis usually rudimentary and poorly defined	O—spines of upper lumbar and lower thoracic vertebrae I—spines of upper thoracic and cervical vertebrae	Extends vertebral column	Spinal nerves (dorsal rami)
Semispinalis (sem'e-spĭ-nă'lis)—thoracis, cervicis, and capitis regions (*semi* = half; *thorac* = thorax) (Figure 10.9d, right side)	Composite muscle forming part of deep layer of intrinsic back muscles; extends from thoracic region to head	O—transverse processes of C_7–T_{12} I—occipital bone (capitis) and spinous processes of cervical (cervicis) and thoracic vertebrae T_1–T_4 (thoracis)	Extends vertebral column and head and rotates them to opposite side; acts synergistically with sternocleidomastoid muscles of opposite side	Spinal nerves (dorsal rami)
Quadratus lumborum (kwod-ra'tus lum-bor'um) (*quad* = four-sided; *lumb* = lumbar region) (See also Figure 10.19a)	Fleshy muscle forming part of posterior abdominal wall	O—iliac crest and lumbar fascia I—transverse processes of upper lumbar vertebrae and lower margin of 12th rib	Flexes vertebral column laterally when acting separately; when pair acts jointly, lumbar spine is extended and 12th rib is fixed; maintains upright posture; assists in forced inspiration	T_{12} and upper lumbar spinal nerves (ventral rami)

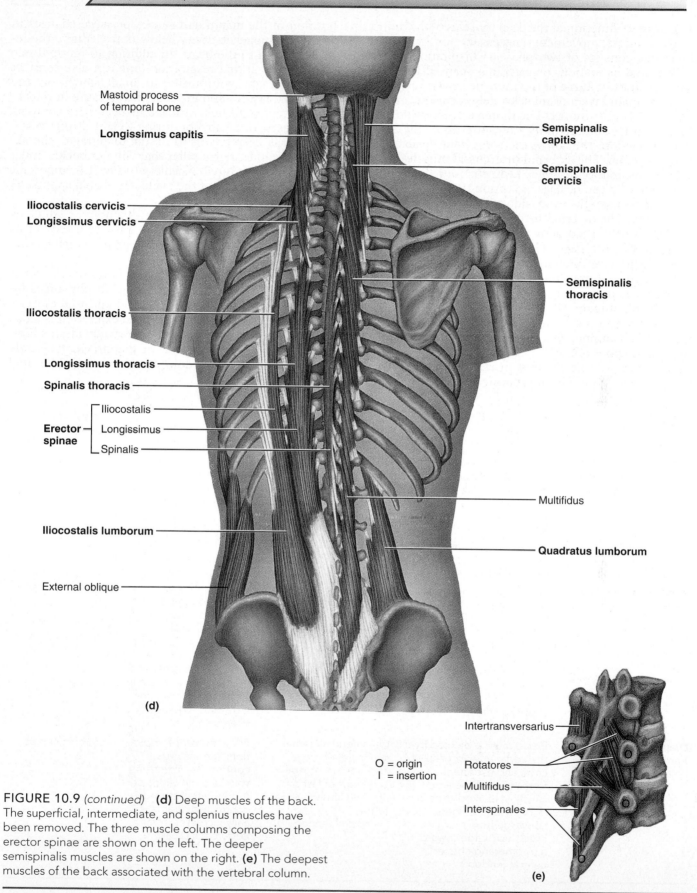

Mastoid process of temporal bone

Longissimus capitis

Iliocostalis cervicis

Longissimus cervicis

Iliocostalis thoracis

Longissimus thoracis

Spinalis thoracis

Erector spinae
— Iliocostalis
— Longissimus
— Spinalis

Iliocostalis lumborum

External oblique

(d)

Semispinalis capitis

Semispinalis cervicis

Semispinalis thoracis

Multifidus

Quadratus lumborum

O = origin
I = insertion

Intertransversarius

Rotatores

Multifidus

Interspinales

(e)

FIGURE 10.9 *(continued)* **(d)** Deep muscles of the back. The superficial, intermediate, and splenius muscles have been removed. The three muscle columns composing the erector spinae are shown on the left. The deeper semispinalis muscles are shown on the right. **(e)** The deepest muscles of the back associated with the vertebral column.

Muscle Gallery
TABLE 10.5   Muscles of the Thorax: Breathing (Figure 10.10)

The primary function of the deep muscles of the thorax is to promote movements necessary for breathing. Breathing consists of two phases—inspiration, or inhaling, and expiration, or exhaling—caused by cyclic changes in the volume of the thoracic cavity.

Two main layers of muscles help form the anterolateral wall of the thorax. The thoracic muscles are very short, most extending only from one rib to the next. On contraction, they draw the somewhat flexible ribs closer together. The **external intercostal muscles,** considered inspiratory muscles, form the more superficial layer. They lift the rib cage, an action that increases the anterior to posterior and side-to-side dimensions of the thorax. The **internal intercostal muscles** form the deeper layer and aid active (forced) expiration by depressing the rib cage. (However, quiet expiration is largely a passive phenomenon, resulting from relaxation of the external intercostals and diaphragm and elastic recoil of the lungs.)

The **diaphragm,** the most important muscle of inspiration, forms a muscular partition between the thoracic and abdominopelvic cavities. In the relaxed state, the diaphragm is dome shaped, but when it contracts it moves inferiorly and flattens, increasing the volume of the thoracic cavity. The alternating contraction and relaxation of the diaphragm causes pressure changes in the abdominopelvic cavity below that facilitate the return of blood to the heart. In addition to its rhythmic contractions during respiration, one can also contract the diaphragm voluntarily to push down on the abdominal viscera and increase the pressure in the abdominal cavity to help evacuate pelvic organ contents (urine, feces, or a baby) or during weight lifting. When one takes a deep breath to fix the diaphragm, the abdomen becomes a firm pillar that will not buckle under the weight being lifted. Needless to say, it is important to have good control of the bladder and anal sphincters during such maneuvers.

With the exception of the diaphragm, which is served by the *phrenic nerves,* the muscles listed in this table are served by the *intercostal nerves,* which run between the ribs.

Forced breathing calls into play a number of other muscles that insert into the ribs; e.g., during forced inspiration the scalene and sternocleidomastoid muscles of the neck help lift the ribs. Forced expiration is aided by muscles that pull the ribs inferiorly (quadratus lumborum) and those that push the diaphragm superiorly by compressing the abdominal contents (abdominal wall muscles).

Muscle	Description	Origin (O) and Insertion (I)	Action	Nerve Supply
External intercostals (in"ter-kos'talz) (*external* = toward the outside; *inter* = between; *cost* = rib)	11 pairs lie between ribs; fibers run obliquely (down and forward) from each rib to rib below; in lower intercostal spaces, fibers are continuous with external oblique muscle, forming part of abdominal wall	O—inferior border of rib above I—superior border of rib below	With first ribs fixed by scalene muscles, pull ribs toward one another to elevate rib cage; aids in inspiration; synergists of diaphragm	Intercostal nerves
Internal intercostals (*internal* = toward the inside, deep)	11 pairs lie between ribs; fibers run deep to and at right angles to those of external intercostals (i.e., run downward and posteriorly); lower internal intercostal muscles are continuous with fibers of internal oblique muscle of abdominal wall	O—superior border of rib below I—inferior border (costal groove) of rib above	With 12th ribs fixed by quadratus lumborum, muscles of posterior abdominal wall, and oblique muscles of the abdominal wall, they draw ribs together and depress rib cage; aid in forced expiration; antagonistic to external intercostals	Intercostal nerves
Diaphragm (di'ah-fram) (*dia* = across; *phragm* = partition)	Broad muscle pierced by the aorta, inferior vena cava, and esophagus, forms floor of thoracic cavity; in relaxed state is dome shaped; fibers converge from margins of thoracic cage toward a boomerang-shaped central tendon	O—inferior, internal surface of rib cage and sternum, costal cartilages of last six ribs and lumbar vertebrae I—central tendon	Prime mover of inspiration; flattens on contraction, increasing vertical dimensions of thorax; when strongly contracted, dramatically increases intra-abdominal pressure	Phrenic nerves

Muscle Gallery
TABLE 10.5 *(continued)*

FIGURE 10.10 **Muscles of respiration. (a)** Deep muscles of the thorax. The external intercostals (inspiratory muscles) are shown on the left and the internal intercostals (expiratory muscles) are shown on the right. These two muscle layers run obliquely and at right angles to each other. **(b)** Inferior view of the diaphragm, the prime mover of inspiration. Notice that its muscle fibers converge toward a central tendon, an arrangement that causes the diaphragm to flatten and move inferiorly as it contracts. **(c)** Photograph of the diaphragm, superior view. (See *A Brief Atlas of the Human Body,* Figure 45.)

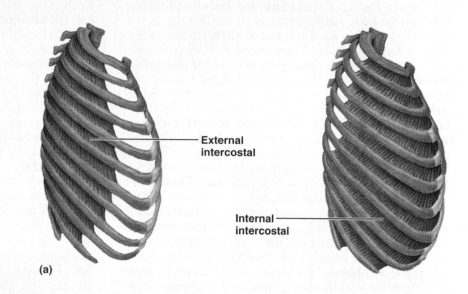

External intercostal

Internal intercostal

(a)

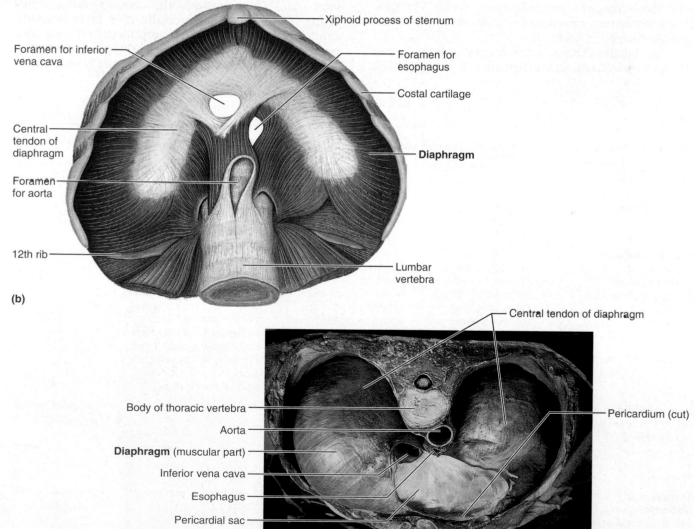

Xiphoid process of sternum

Foramen for inferior vena cava

Foramen for esophagus

Costal cartilage

Central tendon of diaphragm

Diaphragm

Foramen for aorta

12th rib

Lumbar vertebra

(b)

Central tendon of diaphragm

Body of thoracic vertebra

Aorta

Diaphragm (muscular part)

Inferior vena cava

Esophagus

Pericardial sac

Pericardium (cut)

(c)

Muscle Gallery
TABLE 10.6 ⟩ Muscles of the Abdominal Wall: Trunk Movements and Compression of Abdominal Viscera (Figure 10.11)

Unlike the thorax, the anterior and lateral abdominal wall has no bony reinforcements (ribs). Instead, it is a composite of four paired muscles, their investing fasciae, and aponeuroses. Three broad flat muscle pairs, layered one atop the next, form the lateral abdominal walls: The fibers of the **external oblique muscle** run inferomedially and at right angles to those of the **internal oblique**, which immediately underlies it. The fibers of the deep **transversus abdominis muscle** run horizontally across the abdomen at an angle to both. This alternation of fascicle directions is similar to the construction of plywood (which is made of sheets with different grains) and provides great strength. These three muscles blend into broad insertion aponeuroses anteriorly. The aponeuroses, in turn, enclose a fourth muscle pair medially, the straplike **rectus abdominis muscles,** and then fuse, forming the **linea alba** ("white line"), a tendinous raphe (seam) that runs from the sternum to the pubic symphysis. Enclosure of the rectus abdominis muscles within the aponeuroses prevents them from "bowstringing" (protruding anteriorly). The quadratus lumborum muscles of the *posterior* abdominal wall are covered in Table 10.4.

The abdominal muscles protect and support the viscera most effectively when they are well toned. When weak or severely stretched (as during pregnancy), they allow the abdomen to become pendulous (i.e., to form a potbelly). Additional functions include lateral flexion and rotation of the trunk and anterior flexion of the trunk against resistance (as in sit-ups). During quiet expiration, the abdominal muscles relax, allowing the abdominal viscera to be pushed inferiorly by the descending diaphragm. When all these abdominal muscles contract in unison, several different activities may be effected. For example, when all the abdominal muscles are contracted, the ribs are pulled inferiorly and the abdominal contents compressed. This pushes the visceral organs upward on the diaphragm, aiding forced expiration. When the abdominal muscles contract with the diaphragm and the glottis is closed (an action called the Valsalva maneuver), the increased intra-abdominal pressure helps to promote urination, defecation, childbirth, vomiting, coughing, screaming, sneezing, burping, and nose blowing. (Next time you perform one of these activities, feel your abdominal muscles contract under your skin.) These muscles also contract during heavy lifting—sometimes so forcefully that hernias result. Contraction of the abdominal muscles along with contraction of the deep back muscles helps prevent hyperextension of the spine and splints the entire body trunk.

Muscle	Description	Origin (O) and Insertion (I)	Action	Nerve Supply
Muscles of the Anterior and Lateral Abdominal Wall				
Rectus abdominis (rek′tus ab-dom′ĭ-nis) (*rectus* = straight; *abdom* = abdomen)	Medial superficial muscle pair; extend from pubis to rib cage; ensheathed by aponeuroses of lateral muscles; segmented by 3 tendinous intersections	O—pubic crest and symphysis I—xiphoid process and costal cartilages of ribs 5–7	Flex and rotate lumbar region of vertebral column; fix and depress ribs, stabilize pelvis during walking, increase intra-abdominal pressure	Intercostal nerves (T_6 or T_7–T_{12})
External oblique (o-blēk′) (*external* = toward outside; *oblique* = running at an angle)	Largest and most superficial of the three lateral muscles; fibers run downward and medially (same direction outstretched fingers take when hands put into pants pockets); aponeurosis turns under inferiorly, forming inguinal ligament	O—by fleshy strips from outer surfaces of lower eight ribs I—most fibers insert anteriorly via a broad aponeurosis into linea alba; some into pubic crest and tubercle and iliac crest	When pair contract simultaneously, aid rectus abdominis muscles in flexing vertebral column and in compressing abdominal wall and increasing intra-abdominal pressure; acting individually, aid muscles of back in trunk rotation and lateral flexion	Intercostal nerves (T_7–T_{12})
Internal oblique (*internal* = toward the inside; deep)	Most fibers run upward and medially; however, the muscle fans so its inferior fibers run downward and medially	O—lumbar fascia, iliac crest, and inguinal ligament I—linea alba, pubic crest, last three or four ribs, and costal margin	As for external oblique	Intercostal nerves (T_7–T_{12}) and L_1
Transversus abdominis (transver′sus) (*transverse* = running straight across)	Deepest (innermost) muscle of abdominal wall; fibers run horizontally	O—inguinal ligament, lumbar fascia, cartilages of last six ribs; iliac crest I—linea alba, pubic crest	Compresses abdominal contents	Intercostal nerves (T_7–T_{12}) and L_1

Muscle Gallery
TABLE 10.6   (continued)

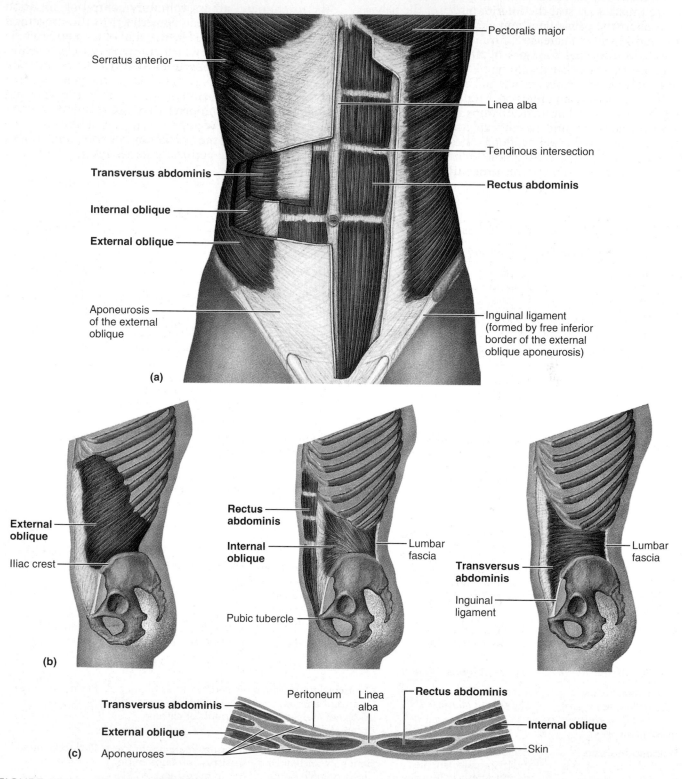

FIGURE 10.11 **Muscles of the abdominal wall. (a)** Anterior view of the muscles forming the anterolateral abdominal wall. The superficial muscles have been partially cut away to reveal the deeper muscles. **(b)** Lateral view of the trunk, illustrating the fiber direction and attachments of the external and internal obliques, and transversus abdominis muscles. Although shown in the same view, the rectus abdominis is deep to the fascia of the internal oblique muscle. **(c)** Transverse section through the anterolateral abdominal wall (midregion), showing how the aponeuroses of the lateral abdominal muscles contribute to the rectus abdominis sheath.

Two paired muscles, the **levator ani** and **coccygeus,** form the funnel-shaped pelvic floor, or **pelvic diaphragm.** These muscles (1) seal the inferior outlet of the pelvis, (2) support the pelvic floor and elevate it to release feces, and (3) resist increased intra-abdominal pressure (which would expel contents of the bladder, rectum, and uterus). The pelvic diaphragm is pierced by the rectum and urethra (urinary tube), and by the vagina in females. The body region inferior to the pelvic diaphragm is the *perineum.* The relationships of the perineum are a bit complex and worth explaining. Inferior to the muscles of the pelvic floor, and stretching between the two sides of the pubic arch in the anterior half of the perineum, is the **urogenital diaphragm.**

This thin triangular sheet of muscle contains the **sphincter urethrae,** a sphincter muscle that surrounds the urethra and allows voluntary control of urination when it is inconvenient. Superficial to the urogenital diaphragm, and covered by the skin of the perineum, is the *superficial space,* which contains muscles (**ischiocavernosus** and **bulbospongiosus**) that help maintain erection of the penis and clitoris. In the posterior half of the perineum encircling the anus is the **external anal sphincter,** a sphincter that allows voluntary control of defecation. Just anterior to this sphincter is the **central tendon of the perineum,** a strong tendon into which many of the perineal muscles insert.

Muscle	Description	Origin (O) and Insertion (I)	Action	Nerve Supply
Muscles of the Pelvic Diaphragm (Figure 10.12a)				
Levator ani (lĕ-va′tor a′ne) (*levator* = raises; *ani* = anus)	Broad, thin, tripartite muscle (pubococcygeus, puborectalis, and iliococcygeus parts); its fibers extend inferomedially, forming a muscular "sling" around male prostate (or female vagina), urethra, and anorectal junction before meeting in the median plane	O—extensive linear origin inside pelvis from pubis to ischial spine I—inner surface of coccyx, levator ani of opposite side, and (in part) into the structures that penetrate it	Supports and maintains position of pelvic viscera; resists downward thrusts that accompany rises in intrapelvic pressure during coughing, vomiting, and expulsive efforts of abdominal muscles; forms sphincters at anorectal junction and vagina; lifts anal canal during defecation	S_4 and inferior rectal nerve (branch of pudendal nerve)
Coccygeus (kok-sij′e-us) (*coccy* = coccyx)	Small triangular muscle lying posterior to levator ani; forms posterior part of pelvic diaphragm	O—spine of ischium I—sacrum and coccyx	Supports pelvic viscera; supports coccyx and pulls it forward after it has been reflected posteriorly by defecation and childbirth	S_4 and S_5
Muscles of the Urogenital Diaphragm (Figure 10.12b)				
Deep transverse perineus (per″ĭ-ne′us) (*deep* = far from surface; *transverse* = across; *perine* = near anus)	Together the pair spans distance between ischial rami; in females, lies posterior to vagina	O—ischial rami I—midline central tendon of perineum; some fibers into vaginal wall in females	Supports pelvic organs; steadies central tendon	Pudendal nerve
Sphincter urethrae (*sphin* = squeeze)	Muscle encircling urethra and vagina (female)	O—ischiopubic rami I—midline raphe	Constricts urethra; helps support pelvic organs	Pudendal nerve
Muscles of the Superficial Space (Figure 10.12c)				
Ischiocavernosus (is′ke-o-kav′ern-o′sus) (*ischi* = hip; *caverna* = hollow chamber)	Runs from pelvis to base of penis or clitoris	O—ischial tuberosities I—crus of corpus cavernosa of male penis or female clitoris	Retards venous drainage and maintains erection of penis or clitoris	Pudendal nerve
Bulbospongiosus (bul″bo-spun″je-o′sus) (*bulbon* = bulb; *spongio* = sponge)	Encloses base of penis (bulb) in males and lies deep to labia in females	O—central tendon of perineum and midline raphe of male penis I—anteriorly into corpus cavernosa of penis or clitoris	Empties male urethra; assists in erection of penis in males and of clitoris in females	Pudendal nerve
Superficial transverse perineus (*superficial* = closer to surface)	Paired muscle bands posterior to urethral (and in females, vaginal) opening; variable; sometimes absent	O—ischial tuberosity I—central tendon of perineum	Stabilizes and strengthens midline tendon of perineum	Pudendal nerve

Muscle Gallery
TABLE 10.7   (continued)

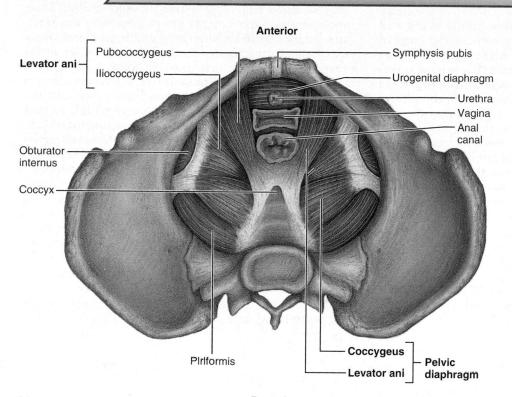

Anterior

Levator ani — [Pubococcygeus
               Iliococcygeus

Symphysis pubis

Urogenital diaphragm

Urethra

Vagina

Anal
canal

Obturator
internus

Coccyx

Plrlformis

Coccygeus —] **Pelvic
Levator ani —] diaphragm**

(a)

Posterior

FIGURE 10.12 **Muscles of the pelvic floor and perineum.** (a) Superior view of muscles of the pelvic diaphragm (levator ani and coccygeus) in a female pelvis. (b) Inferior view of muscles of the urogenital diaphragm of the perineum (sphincter urethrae and deep transverse perineus). (c) Inferior view of muscles of the superficial space of the perineum (ischiocavernosus, bulbospongiosus, and superficial transverse perineus), which lie just deep to the skin of the perineum. Note that a raphe is a seam of fibrous tissue.

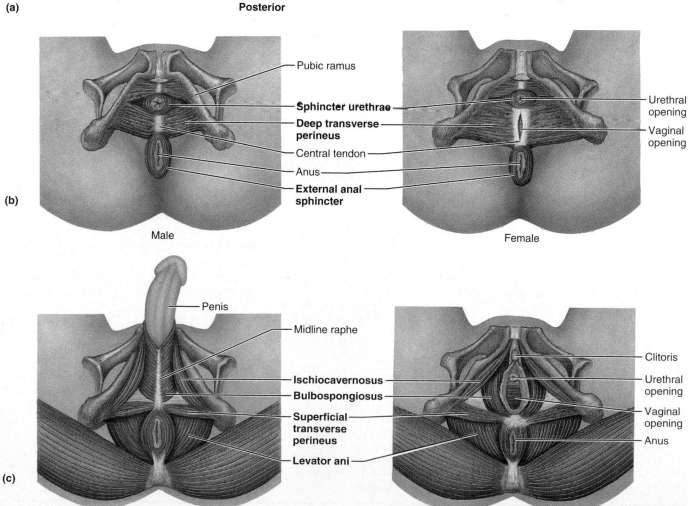

Pubic ramus

Sphincter urethrae

Deep transverse perineus

Central tendon

Anus

External anal sphincter

Urethral opening

Vaginal opening

(b)

Male

Female

Penis

Midline raphe

Ischiocavernosus

Bulbospongiosus

Superficial transverse perineus

Levator ani

Clitoris

Urethral opening

Vaginal opening

Anus

(c)

Superficial Muscles of the Anterior and Posterior Thorax: Movements of the Scapula (Figure 10.13)

Most superficial thorax muscles are *extrinsic shoulder muscles*, which run from the ribs and vertebral column to the shoulder girdle. They fix the scapula to the wall of the thorax; they also move it to increase the range of arm movements. The muscles of the anterior thorax include the **pectoralis major, pectoralis minor, serratus anterior,** and **subclavius.** Except for the pectoralis major, which inserts into the humerus, all muscles of the anterior group insert into the pectoral girdle. Extrinsic muscles of the posterior thorax include the **latissimus dorsi** and **trapezius muscles** superficially and the underlying **levator scapulae** and **rhomboids.** The latissimus dorsi, like the pectoralis major muscles anteriorly, insert into the humerus and are more concerned with movements of the arm than of the scapula. We will defer consideration of these two muscle pairs to Table 10.9 (arm-moving muscles).

The important movements of the pectoral girdle involve displacements of the scapula, i.e., its elevation and depression, rotation, lateral (forward) movements, and medial (backward) movements. The clavicles rotate around their own axes to provide both stability and precision to scapular movements.

Except for the serratus anterior, the anterior muscles stabilize and depress the shoulder girdle. Thus, most scapular movements are promoted by the serratus anterior muscles anteriorly and by posterior muscles. The arrangement of muscle attachments to the scapula is such that one muscle cannot bring about a simple (linear) movement on its own. To effect scapular movements, several muscles must act in combination.

The prime movers of shoulder elevation are the trapezius and levator scapulae. When acting together to shrug the shoulder, their opposite rotational effects counterbalance each other. The scapula is depressed largely by gravity (weight of the arm), but when it is depressed against resistance, the trapezius and serratus anterior (along with the latissimus dorsi, Table 10.9) are active. Forward movements (abduction) of the scapula on the thorax wall, as in pushing or punching movements, mainly reflect serratus anterior activity. Retraction (adduction) of the scapula is effected by the trapezius and rhomboids. Although the serratus anterior and trapezius muscles are antagonists in forward/backward movements of the scapulae, they act together to coordinate *rotational* scapular movements.

Muscle	Description	Origin (O) and Insertion (I)	Action	Nerve Supply
Muscles of the Anterior Thorax (Figure 10.13a)				
Pectoralis minor (pek"to-ra'lis mi'nor) (*pectus* = chest, breast; *minor* = lesser)	Flat, thin muscle directly beneath and obscured by pectoralis major	O—anterior surfaces of ribs 3–5 (or 2–4) I—coracoid process of scapula	With ribs fixed, draws scapula forward and downward; with scapula fixed, draws rib cage superiorly	Both pectoral nerves (C_6–C_8)
Serratus anterior (ser-a'tus) (*serratus* = saw)	Lies deep to scapula, beneath and inferior to pectoral muscles on lateral rib cage; forms medial wall of axilla; origins have serrated, or sawtooth, appearance; paralysis results in "winging" of vertebral border of scapula away from chest wall, making arm elevation impossible	O—by a series of muscle slips from ribs 1–8 (or 9) I—entire anterior surface of vertebral border of scapula	Prime mover to protract and hold scapula against chest wall; rotates scapula so that its inferior angle moves laterally and upward; raises point of shoulder; important role in abduction and raising of arm and in horizontal arm movements (pushing, punching); called "boxer's muscle"	Long thoracic nerve (C_5–C_7)
Subclavius (sub-kla've-us) (*sub* = under, beneath; *clav* = clavicle)	Small cylindrical muscle extending from rib 1 to clavicle	O—costal cartilage of rib 1 I—groove on inferior surface of clavicle	Helps stabilize and depress pectoral girdle	Nerve to subclavius (C_5 and C_6)
Muscles of the Posterior Thorax (Figure 10.13b)				
Trapezius (trah-pe'ze-us) (*trapezion* = irregular four-sided figure)	Most superficial muscle of posterior thorax; flat, and triangular in shape; upper fibers run inferiorly to scapula; middle fibers run horizontally to scapula; lower fibers run superiorly to scapula	O—occipital bone, ligamentum nuchae, and spines of C_7 and all thoracic vertebrae I—a continuous insertion along acromion and spine of scapula and lateral third of clavicle	Stabilizes, raises, retracts, and rotates scapula; middle fibers retract (adduct) scapula; superior fibers elevate scapula or can help extend head with scapula fixed; inferior fibers depress scapula (and shoulder)	Accessory nerve (cranial nerve XI); C_3 and C_4

Muscle	Description	Origin (O) and Insertion (I)	Action	Nerve Supply
Levator scapulae (skap'u-le) (*levator* = raises)	Located at back and side of neck, deep to trapezius; thick, straplike muscle	O—transverse processes of C_1–C_4 I—medial border of the scapula, superior to the spine	Elevates/adducts scapula in concert with superior fibers of trapezius; tilts glenoid cavity downward when scapula is fixed, flexes neck to same side	Cervical spinal nerves and dorsal scapular nerve (C_3–C_5)
Rhomboids (rom'boidz)—major and minor (*rhomboid* = diamond shaped)	Two rectangular muscles lying deep to trapezius and inferior to levator scapulae; rhomboid minor is the more superior muscle	O—spinous processes of C_7 and T_1 (minor) and spinous processes of T_2–T_5 (major) I—medial border of scapula	Act together (and with middle trapezius fibers) to retract scapula, thus "squaring shoulders"; rotate glenoid cavity of scapula downward (as when arm is lowered against resistance; e.g., paddling a canoe); stabilize scapula	Dorsal scapular nerve (C_4 and C_5)

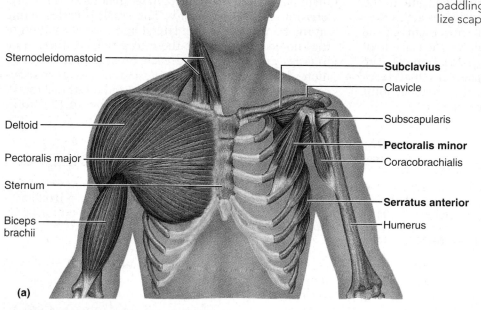

(a)

FIGURE 10.13 Superficial muscles of the thorax and shoulder acting on the scapula and arm. **(a)** Anterior view. The superficial muscles, which effect arm movements, are shown on the left side of the illustration. These muscles are removed on the right to show the muscles that stabilize or move the pectoral girdle. **(b)** Posterior view. The superficial muscles are shown for the left side of the illustration, with a corresponding photograph. The superficial muscles are removed on the right side to reveal the deeper muscles acting on the scapula, and the rotator cuff muscles that help stabilize the shoulder joint.

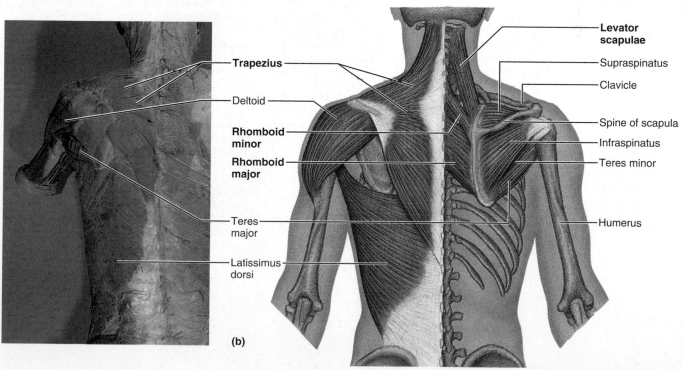

(b)

Muscle Gallery
TABLE 10.9   Muscles Crossing the Shoulder Joint: Movements of the Arm (Figure 10.14)

Recall that the ball-and-socket shoulder joint is the most flexible joint in the body, but pays the price of instability. A total of nine muscles cross each shoulder joint to insert on the humerus. All muscles acting on the humerus originate from the pectoral girdle; however, the latissimus dorsi and pectoralis major primarily originate on the axial skeleton.

Of these nine muscles, only the superficial **pectoralis major, latissimus dorsi,** and **deltoid muscles** are prime movers of arm movements. The remaining six are synergists and fixators. Four of these, the **supraspinatus, infraspinatus, teres minor,** and **subscapularis,** are known as *rotator cuff muscles.* They originate on the scapula, and their tendons blend with the fibrous capsule of the shoulder joint en route to the humerus. Although the rotator cuff muscles act as synergists in the angular and rotational movements of the arm, their main function is to reinforce the capsule of the shoulder joint to prevent dislocation of the humerus. The remaining two muscles, the small **teres major** and **coracobrachialis,** cross the shoulder joint but do not contribute to its reinforcement.

Generally speaking, muscles that originate *anterior* to the shoulder joint (pectoralis major, coracobrachialis, and anterior fibers of the deltoid) *flex* the arm, i.e., lift it anteriorly. The prime mover of arm flexion is the pectoralis major. The biceps brachii of the arm (see Table 10.10) also assists in this action. Muscles originating *posterior* to the shoulder joint extend the arm. These include the latissimus dorsi and posterior fibers of the deltoid muscles (both prime movers of arm extension) and the teres major. Thus, the latissimus dorsi and pectoralis muscles are *antagonists* of one another in the flexion-extension movements of the arm.

The middle region of the fleshy deltoid muscle of the shoulder is the prime mover of arm abduction. The main adductors are the pectoralis major anteriorly and latissimus dorsi posteriorly. The small muscles acting on the humerus promote lateral and medial rotation of the shoulder joint. Since the interactions of these nine muscles are complex and each muscle contributes to more than a single movement, a summary of muscles contributing to the various angular and rotational movements of the humerus is provided in Table 10.12 (Part I).

Muscle	Description	Origin (O) and Insertion (I)	Action	Nerve Supply
Pectoralis major (pek″to-ra′lis ma′jer) (*pectus* = breast, chest; *major* = larger)	Large, fan-shaped muscle covering upper portion of chest; forms anterior axillary fold; divided into clavicular and sternal parts	O—sternal end of clavicle, sternum, cartilage of ribs 1–6 (or 7), and aponeurosis of external oblique muscle I—fibers converge to insert by a short tendon into intertubercular groove of humerus	Prime mover of arm flexion; rotates arm medially; adducts arm against resistance; with scapula (and arm) fixed, pulls rib cage upward, thus can help in climbing, throwing, pushing, and in forced inspiration	Lateral and medial pectoral nerves (C_5–C_8, and T_1)
Latissimus dorsi (lah-tis′ĭ-mus dor′si) (*latissimus* = widest; *dorsi* = back)	Broad, flat, triangular muscle of lower back (lumbar region); extensive superficial origins; covered by trapezius superiorly; contributes to the posterior wall of axilla	O—indirect attachment via lumbodorsal fascia into spines of lower six thoracic vertebrae, lumbar vertebrae, lower 3 to 4 ribs, and iliac crest I—spirals around teres major to insert in floor of intertubercular groove of humerus	Prime mover of arm extension; powerful arm adductor; medially rotates arm at shoulder; depresses scapula; because of its power in these movements, it plays an important role in bringing the arm down in a power stroke, as in striking a blow, hammering, swimming, and rowing; with arms fixed overhead, it pulls the rest of the body upward and forward	Thoracodorsal nerve (C_6–C_8)

Muscle Gallery
TABLE 10.9   *(continued)*

Muscle	Description	Origin (O) and Insertion (I)	Action	Nerve Supply
Deltoid (del'toid)) (*delta* = triangular)	Thick, multipennate muscle forming rounded shoulder muscle mass; responsible for roundness of shoulder; a site commonly used for intramuscular injection, particularly in males, where it tends to be quite fleshy	O—embraces insertion of the trapezius; lateral third of clavicle; acromion and spine of scapula I—deltoid tuberosity of humerus	Prime mover of arm abduction when all its fibers contract simultaneously; antagonist of pectoralis major and latissimus dorsi, which adduct the arm; if only anterior fibers are active, can act powerfully in flexion and medial rotation of humerus, therefore synergist of pectoralis major; if only posterior fibers are active, effects extension and lateral rotation of arm; active during rhythmic arm swinging movements during walking	Axillary nerve (C5 and C6)
Subscapularis (sub-scap"u-lar'is) (*sub* = under; *scapular* = scapula)	Forms part of posterior wall of axilla; tendon of insertion passes in front of shoulder joint; a rotator cuff muscle	O—subscapular fossa of scapula I—lesser tubercle of humerus	Chief medial rotator of humerus; assisted by pectoralis major; helps to hold head of humerus in glenoid cavity, thereby stabilizing shoulder joint	Subscapular nerves (C5–C7)
Supraspinatus (soo"prah-spi-nah'tus) (*supra* = above, over; *spin* = spine)	Named for its location on posterior aspect of scapula; deep to trapezius; a rotator cuff muscle	O—supraspinous fossa of scapula I—superior part of greater tubercle of humerus	Stabilizes shoulder joint; helps to prevent downward dislocation of humerus, as when carrying a heavy suitcase; assists in abduction	Suprascapular nerve
Infraspinatus (in"frah-spi-nah'tus) (*infra* = below)	Partially covered by deltoid and trapezius; named for its scapular location; a rotator cuff muscle	O—infraspinous fossa of scapula I—greater tubercle of humerus posterior to insertion of supraspinatus	Helps to hold head of humerus in glenoid cavity, stabilizing the shoulder joint; rotates humerus laterally	Suprascapular nerve
Teres minor (te'rēz) (*teres* = round; *minor* = lesser)	Small, elongated muscle; lies inferior to infraspinatus and may be Inseparable from that muscle; a rotator cuff muscle	O—lateral border of dorsal scapular surface I—greater tubercle of humerus inferior to infraspinatus insertion	Same action(s) as infraspinatus muscle; adducts arm	Axillary nerve
Teres major	Thick, rounded muscle; located inferior to teres minor; helps to form posterior wall of axilla (along with latissimus dorsi and subscapularis)	O—posterior surface of scapula at inferior angle I—intertubercular groove of the humerus; insertion tendon fused with that of latissimus dorsi	Posteromedially extends, medially rotates, and adducts humerus; synergist of latissimus dorsi	Lower subscapular nerve (C6 and C7)
Coracobrachialis (kor"ah-ko-bra"ke-al'is) (*coraco* = coracoid; *brachi* = arm)	Small, cylindrical muscle	O—coracoid process of scapula I—medial surface of humerus shaft	Flexion and adduction of the humerus; synergist of pectoralis major	Musculocutaneous nerve (C5–C7)

Muscle Gallery
TABLE 10.9

Muscles Crossing the Shoulder Joint: Movements of the Arm (Figure 10.14) *(continued)*

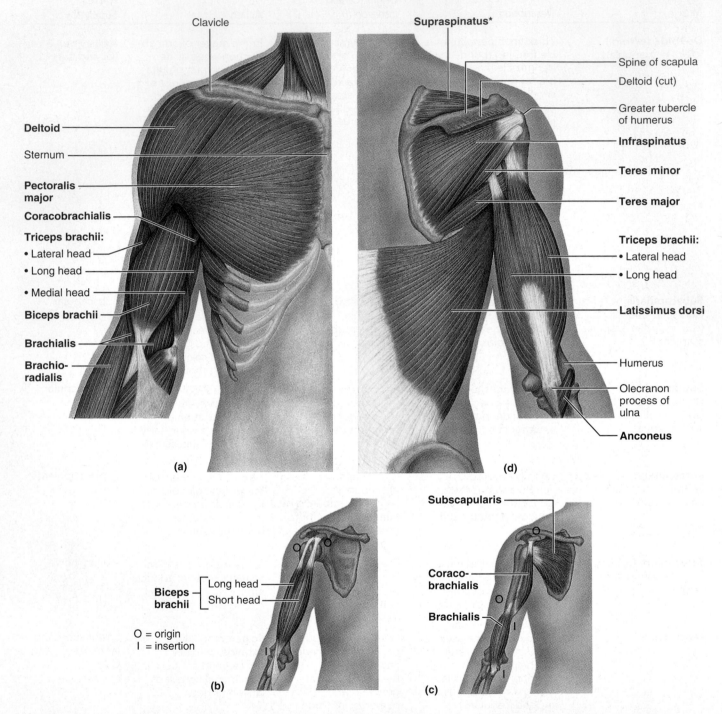

(a)

Clavicle

Deltoid

Sternum

Pectoralis major

Coracobrachialis

Triceps brachii:
• Lateral head
• Long head
• Medial head

Biceps brachii

Brachialis

Brachio-radialis

(d)

Supraspinatus*

Spine of scapula

Deltoid (cut)

Greater tubercle of humerus

Infraspinatus

Teres minor

Teres major

Triceps brachii:
• Lateral head
• Long head

Latissimus dorsi

Humerus

Olecranon process of ulna

Anconeus

(b)

Biceps brachii
Long head
Short head

O = origin
I = insertion

(c)

Subscapularis

Coraco-brachialis

Brachialis

FIGURE 10.14   Muscles crossing the shoulder and elbow joint, respectively causing movements of the arm and forearm. (a) Superficial muscles of the anterior thorax, shoulder, and arm, anterior view. (See *A Brief Atlas of the Human Body,* Figures 63 and 64.) **(b)** The biceps brachii muscle of the anterior arm, shown in isolation. **(c)** The brachialis muscle and the coraco-brachialis and subscapularis muscles shown in isolation. **(d)** The extent of the triceps brachii muscle of the posterior arm, shown in relation to the deep scapular muscles. The deltoid muscle of the shoulder has been removed.

*Rotator cuff muscles

Muscle Gallery
TABLE 10.10) **Muscles Crossing the Elbow Joint: Flexion and Extension of the Forearm (Figure 10.14)**

Muscles fleshing out the arm cross the elbow joint to insert on the forearm bones. Since the elbow is a hinge joint, movements promoted by these muscles are limited almost entirely to flexion and extension of the forearm. Walls of fascia divide the arm into two muscle compartments—the *posterior extensors* and *anterior flexors*. The prime mover of forearm extension is the bulky **triceps brachii** muscle, which forms nearly the entire musculature of the posterior compartment. It is assisted (minimally) by the tiny **anconeus** muscle, which just about spans the elbow joint posteriorly.

All anterior arm muscles cause elbow flexion. In order of decreasing strength, these are the **brachialis, biceps brachii,** and **brachioradialis.** The brachialis and biceps insert (respectively) into the ulna and radius and contract simultaneously during flexion; they are the chief forearm flexors. The biceps brachii, a muscle that bulges when the forearm is flexed, is familiar to almost everyone; the brachialis, which lies deep to the biceps, is less known, but is as important in flexing the elbow. The biceps muscle also supinates the forearm and is ineffective in flexing the elbow when the forearm *must* stay pronated. (This is why doing chin-ups with palms facing anteriorly is harder than with palms facing posteriorly.) Since the brachioradialis arises from the distal humerus and inserts on the distal forearm, it resides mainly in the forearm (rather than in the arm). Because its force is exerted far from the fulcrum, the brachioradialis is a weak forearm flexor. It becomes active only when the elbow has been partially flexed and is semipronated.

The actions of the muscles described here are summarized in Table 10.12 (Part II).

Muscle	Description	Origin (O) and Insertion (I)	Action	Nerve Supply
Posterior Muscles				
Triceps brachii (tri'seps bra'ke-i) (*triceps* = three heads; *brachi* = arm)	Large fleshy muscle; the only muscle of posterior compartment of arm; three-headed origin; long and lateral heads lie superficial to medial head	O—long head: infraglenoid tubercle of scapula; lateral head: posterior shaft of humerus; medial head: posterior humeral shaft distal to radial groove I—by common tendon into olecranon process of ulna	Powerful forearm extensor (prime mover, particularly medial head); antagonist of forearm flexors; long and lateral heads mainly active in extension against resistance; long head tendon may help stabilize shoulder joint and assist in arm adduction	Radial nerve (C_6–C_8)
Anconeus (an-ko'ne-us) (*ancon* = elbow) (see Figure 10.16)	Short triangular muscle; closely associated (blended) with distal end of triceps on posterior humerus	O—lateral epicondyle of humerus I—lateral aspect of olecranon process of ulna	Abducts ulna during forearm pronation; synergist of triceps brachii in elbow extension	Radial nerve
Anterior Muscles				
Biceps brachii (bi'seps) ss (*biceps* = two heads)	Two-headed fusiform muscle; bellies unite as insertion point is approached; tendon of long head helps stabilize shoulder joint	O—short head: coracoid process; long head: tubercle above and lip of glenoid cavity; tendon of long head runs within capsule and into intertubercular groove of humerus I—by common tendon into radial tuberosity	Flexes elbow joint and supinates forearm; these actions usually occur at same time (e.g., when you open a bottle of wine, it turns the corkscrew and pulls the cork); weak flexor of arm at shoulder	Musculocutaneous nerve (C_5 and C_6)
Brachialis (bra'ke-al-is)	Strong muscle that is immediately deep to biceps brachii on distal humerus	O—front of distal humerus; embraces insertion of deltoid muscle I—coronoid process of ulna	A major forearm flexor (lifts ulna as biceps lifts the radius)	Musculocutaneous nerve
Brachioradialis (bra"ke-o-ra"de-a'lis) (*radi* = radius, ray) (also see Figure 10.15)	Superficial muscle of lateral forearm; forms lateral boundary of antecubital fossa; extends from distal humerus to distal forearm	O—lateral supracondylar ridge at distal end of humerus I—base of styloid process of radius	Synergist in forearm flexion; acts to best advantage when forearm is partially flexed and semipronated; during rapid flexion *and* extension, stabilizes the elbow	Radial nerve (an important exception: the radial nerve typically serves extensor muscles)

Muscle Gallery
TABLE 10.11)   Muscles of the Forearm: Movements of the Wrist, Hand, and Fingers (Figures 10.15 and 10.16)

The forearm muscles are functionally divided into two approximately equal groups: those causing wrist movements and those moving the fingers and thumb. In most cases, their fleshy portions contribute to the roundness of the proximal forearm and then they taper to long insertion tendons. Their insertions are securely anchored by strong ligaments called **flexor** and **extensor retinacula** ("retainers"). These "wrist bands" keep the tendons from jumping outward when tensed. Crowded together in the wrist and palm, the muscle tendons are surrounded by slippery tendon sheaths that minimize friction as they slide against one another.

Although many forearm muscles arise from the humerus (and thus cross both the elbow and wrist joints), their actions on the elbow are slight. Flexion and extension are the movements typically effected at both the wrist and finger joints. In addition, the wrist can be abducted and adducted.

The forearm muscles are subdivided by fascia into two main compartments (the *anterior flexors* and *posterior extensors*), each with superficial and deep muscle layers. Most flexors in the anterior compartment arise from a common tendon on the humerus and are innervated largely by the median nerve. Two

anterior compartment muscles are not flexors but pronators, the **pronator teres** and **pronator quadratus.** Pronation is one of the most important forearm movements.

Muscles of the posterior compartment extend the wrist and fingers. An exception is the **supinator** muscle, which assists the biceps brachii muscle of the arm in supinating the forearm. (Also residing in the posterior compartment is the brachioradialis muscle, the weak elbow flexor considered in Table 10.10.) Like those of the anterior compartment, most muscles of the posterior compartment arise from a common origin tendon on the humerus. All posterior forearm muscles are supplied by the radial nerve.

As described above, most muscles that move the hand are located in the forearm and "operate" the fingers via their long tendons, like operating a puppet by strings. This design makes the hand less bulky and enables it to perform finer movements. The hand movements promoted by the forearm muscles are assisted and made more precise by the small *intrinsic* muscles of the hand (see Table 10.13). A summary of the actions of the forearm muscles is provided in Table 10.12 (Parts II and III).

Muscle	Description	Origin (O) and Insertion (I)	Action	Nerve Supply
Part I: Anterior Muscles (Figure 10.15)	These eight muscles of the anterior fascial compartment are listed from the lateral to the medial aspect. Most arise from a common flexor tendon attached to the medial epicondyle of the humerus and have additional origins as well. Most of the tendons of insertion of these flexors are held in place at the wrist by a thickening of deep fascia called the *flexor retinaculum*.			
Superficial Muscles				
Pronator teres (pro'na'tor te'rēz) (*pronation* = turning palm posteriorly, or down; *teres* = round)	Two-headed muscle; seen in superficial view between proximal margins of brachioradialis and flexor carpi radialis; forms medial boundary of antecubital fossa	O—medial epicondyle of humerus; coronoid process of ulna I—by common tendon into lateral radius, midshaft	Pronates forearm; weak flexor of elbow	Median nerve
Flexor carpi radialis (flek'sor kar'pe ra"de-a'lis) (*flex* = decrease angle between two bones; *carpi* = wrist; *radi* = radius)	Runs diagonally across forearm; midway, its fleshy belly is replaced by a flat tendon that becomes cordlike at wrist	O—medial epicondyle of humerus I—base of second and third metacarpals; insertion tendon easily seen and provides guide to position of radial artery (used for pulse taking) at wrist	Powerful flexor of wrist; abducts hand; weak synergist of elbow flexion	Median nerve
Palmaris longus (pahl-ma'ris lon'gus) (*palma* = palm; *longus* = long)	Small fleshy muscle with a long insertion tendon; often absent; may be used as guide to find median nerve that lies lateral to it at wrist	O—medial epicondyle of humerus I—palmar aponeurosis; skin and fascia of palm	Weak wrist flexor; tenses skin and fascia of palm during hand movements; weak synergist for elbow flexion	Median nerve
Flexor carpi ulnaris (ul-na'ris) (*ulnar* = ulna)	Most medial muscle of this group; two-headed; ulnar nerve lies lateral to its tendon	O—medial epicondyle of humerus; olecranon process and posterior surface of ulna I—pisiform and hamate bones and base of fifth metacarpal	Powerful flexor of wrist; also adducts hand in concert with extensor carpi ulnaris (posterior muscle); stabilizes wrist during finger extension	Ulnar nerve (C_7 and C_8)

Muscle Gallery
TABLE 10.11 *(continued)*

Muscle	Description	Origin (O) and Insertion (I)	Action	Nerve Supply
Flexor digitorum superficialis (di″ji-tor′um soo″per-fish″e-a′lis) (*digit* = finger, toe; *superficial* = close to surface)	Two-headed muscle; more deeply placed (therefore, actually forms an intermediate layer); overlain by muscles above but visible at distal end of forearm	O—medial epicondyle of humerus, coronoid process of ulna; shaft of radius I—by four tendons into middle phalanges of fingers 2–5	Flexes wrist and middle phalanges of fingers 2–5; the important finger flexor when speed and flexion against resistance are required	Median nerve (C_7, C_8, and T_1)

Deep Muscles

Muscle	Description	Origin (O) and Insertion (I)	Action	Nerve Supply
Flexor pollicis longus (pah′li-kis) (*pollix* = thumb)	Partly covered by flexor digitorum superficialis; parallels flexor digitorum profundus laterally	O—anterior surface of radius and interosseous membrane I—distal phalanx of thumb	Flexes distal phalanx of thumb	Branch of median nerve (C_8, T_1)
Flexor digitorum profundus (pro-fun′dus) (*profund* = deep)	Extensive origin; overlain entirely by flexor digitorum superficialis	O—coronoid process, anteromedial surface of ulna, and interosseous membrane I—by four tendons into distal phalanges of fingers 2–5	Slow-acting flexor of any or all fingers; assists in flexing wrist; the only muscle that can flex distal interphalangeal joints	Medial half by ulnar nerve; lateral half by median nerve

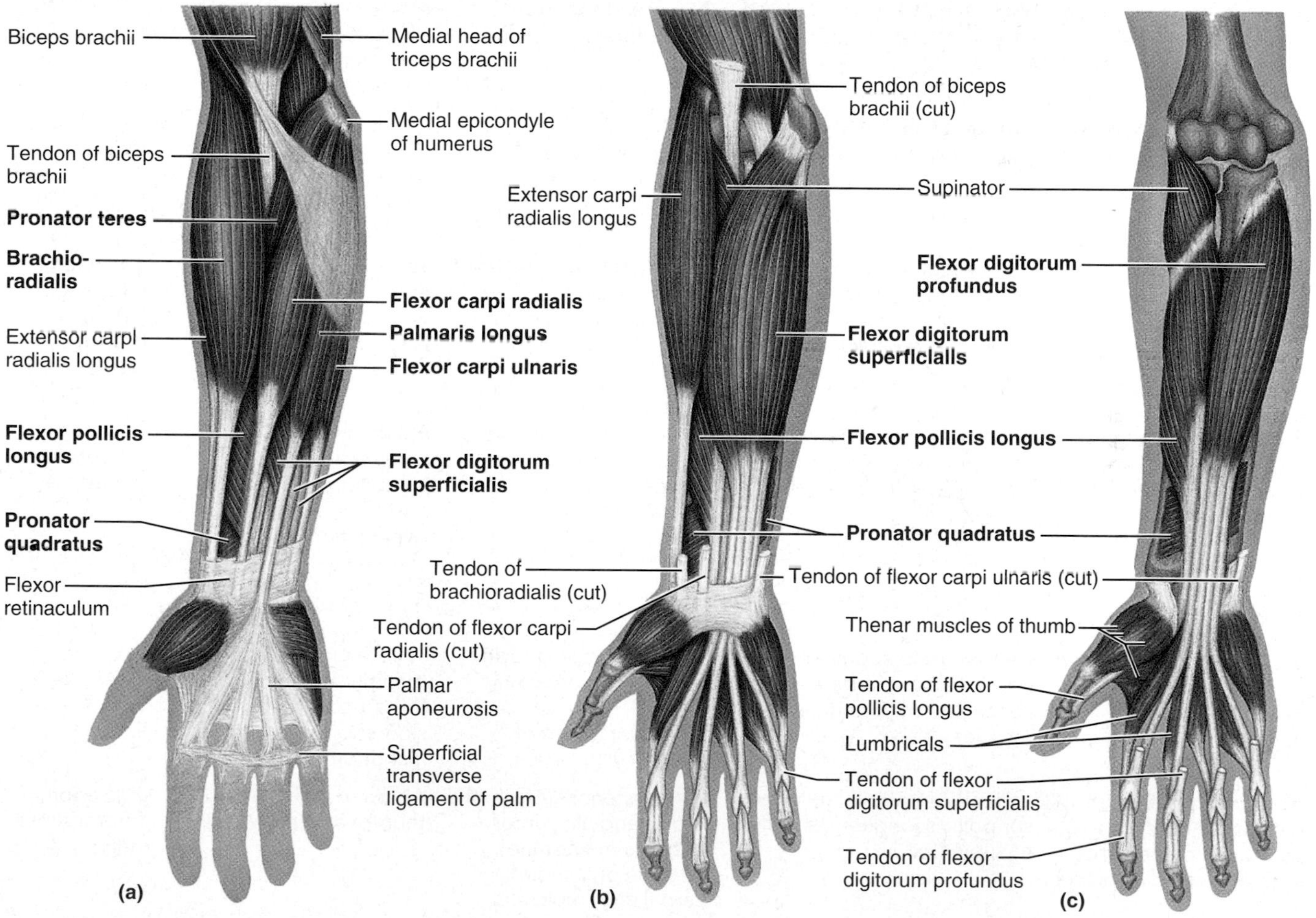

FIGURE 10.15 **Muscles of the anterior compartment of the forearm acting on the right wrist and fingers.** (a) Superficial view. (b) The brachioradialis, flexors carpi radialis and ulnaris, and palmaris longus muscles have been removed to reveal the position of the flexor digitorum superficialis. (c) Deep muscles of the anterior compartment. The lumbricals and thenar muscles (intrinsic hand muscles) are also illustrated. (See *A Brief Atlas of the Human Body*, Figure 65.)

Muscle Gallery
TABLE 10.11

**Muscles of the Forearm: Movements of the Wrist, Hand, and Fingers
(Figures 10.15 and 10.16)** *(continued)*

Muscle	Description	Origin (O) and Insertion (I)	Action	Nerve Supply
Pronator quadratus (kwod-ra'tus) (*quad* = square, four-sided)	Deepest muscle of distal forearm; passes downward and laterally; only muscle that arises solely from ulna and inserts solely into radius	O—distal portion of anterior ulnar shaft I—distal surface of anterior radius	Prime mover of forearm pronation; acts with pronator teres; also helps hold ulna and radius together	Median nerve (C_8 and T_1)
Part II: Posterior Muscles (Figure 10.16)	These muscles of the posterior fascial compartment are listed from the lateral to the medial aspect. They are all innervated by the radial nerve or its branches. More than half of the posterior compartment muscles arise from a common extensor origin tendon attached to the posterior surface of the lateral epicondyle of the humerus and adjacent fascia. The extensor tendons are held in place at the posterior aspect of the wrist by the *extensor retinaculum,* which prevents "bowstringing" of these tendons when the wrist is hyperextended. The extensor muscles of the fingers end in a broad hood over the dorsal side of the digits, the extensor expansion.			

Superficial Muscles

Brachioradialis (see Table 10.10)

Muscle	Description	Origin (O) and Insertion (I)	Action	Nerve Supply
Extensor carpi radialis longus (ek-sten'sor) (*extend* = increase angle between two bones)	Parallels brachioradialis on lateral forearm, and may blend with it	O—lateral supracondylar ridge of humerus I—base of second metacarpal	Extends wrist in conjunction with the extensor carpi ulnaris and abducts wrist in conjunction with the flexor carpi radialis	Radial nerve (C_6 and C_7)
Extensor carpi radialis brevis (brē'vis) (*brevis* = short)	Somewhat shorter than extensor carpi radialis longus and lies deep to it	O—lateral epicondyle of humerus I—base of third metacarpal	Extends and abducts wrist; acts synergistically with extensor carpi radialis longus to steady wrist during finger flexion	Deep branch of radial nerve
Extensor digitorum	Lies medial to extensor carpi radialis brevis; a detached portion of this muscle, called *extensor digiti minimi,* extends little finger	O—lateral epicondyle of humerus I—by four tendons into extensor expansions and distal phalanges of fingers 2–5	Prime mover of finger extension; extends wrist; can abduct (flare) fingers	Posterior interosseous nerve, a branch of radial nerve (C_5 and C_6)
Extensor carpi ulnaris	Most medial of superficial posterior muscles; long, slender muscle	O—lateral epicondyle of humerus and posterior border of ulna I—base of fifth metacarpal	Extends wrist in conjunction with the extensor carpi radialis and adducts wrist in conjunction with flexor carpi ulnaris	Posterior interosseous nerve

Deep Muscles

Muscle	Description	Origin (O) and Insertion (I)	Action	Nerve Supply
Supinator (soo"pĭ-na'tor) (*supination* = turning palm anteriorly or upward)	Deep muscle at posterior aspect of elbow; largely concealed by superficial muscles	O—lateral epicondyle of humerus; proximal ulna I—proximal end of radius	Assists biceps brachii to forcibly supinate forearm; works alone in slow supination; antagonist of pronator muscles	Posterior interosseous nerve
Abductor pollicis longus (ab-duk'tor) (*abduct* = movement away from median plane)	Lateral and parallel to extensor pollicis longus; just distal to supinator	O—posterior surface of radius and ulna; interosseous membrane I—base of first metacarpal and trapezium	Abducts and extends thumb; abducts wrist	Posterior interosseous nerve
Extensor pollicis brevis and **longus**	Deep muscle pair with a common origin and action; overlain by extensor carpi ulnaris	O—dorsal shaft of radius and ulna; interosseous membrane I—base of proximal (brevis) and distal (longus) phalanx of thumb	Extends thumb	Posterior interosseous nerve

Muscle	Description	Origin (O) and Insertion (I)	Action	Nerve Supply
Extensor indicis (in'dĭ-kis) (*indicis* = index finger)	Tiny muscle arising close to wrist	O—posterior surface of distal ulna; interosseous membrane I—extensor expansion of index finger; joins tendon of extensor digitorum	Extends index finger and assists in wrist extension	Posterior interosseous nerve

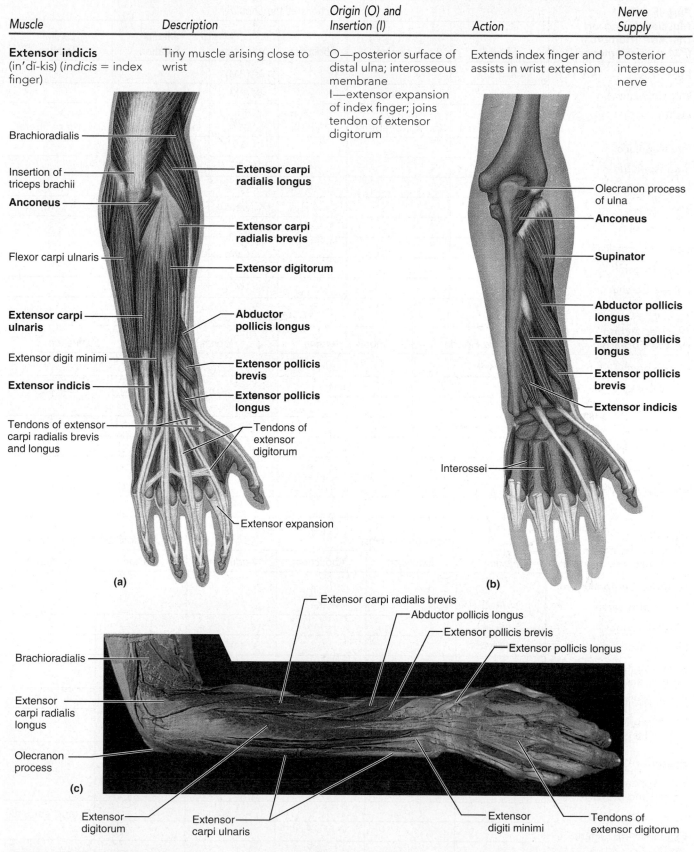

Brachioradialis

Insertion of triceps brachii

Anconeus

Flexor carpi ulnaris

Extensor carpi ulnaris

Extensor digit minimi

Extensor indicis

Tendons of extensor carpi radialis brevis and longus

Extensor carpi radialis longus

Extensor carpi radialis brevis

Extensor digitorum

Abductor pollicis longus

Extensor pollicis brevis

Extensor pollicis longus

Tendons of extensor digitorum

Extensor expansion

(a)

Olecranon process of ulna

Anconeus

Supinator

Abductor pollicis longus

Extensor pollicis longus

Extensor pollicis brevis

Extensor indicis

Interossei

(b)

Extensor carpi radialis brevis

Abductor pollicis longus

Extensor pollicis brevis

Extensor pollicis longus

Brachioradialis

Extensor carpi radialis longus

Olecranon process

(c)

Extensor digitorum

Extensor carpi ulnaris

Extensor digiti minimi

Tendons of extensor digitorum

FIGURE 10.16 Muscles of the posterior compartment of the right forearm acting on the wrist and fingers. (a) Superficial muscles, posterior view. (See *A Brief Atlas of the Human Body*, Figure 65b.) **(b)** Deep posterior muscles, superficial muscles removed. The interossei, the deepest layer of intrinsic hand muscles, are also illustrated. **(c)** Photograph of the deep posterior muscles.

Summary of Actions of Muscles Acting on the Arm, Forearm, and Hand (Figure 10.17)

Part I: Muscles Acting on the Arm (Humerus) (PM = prime mover)

	Actions at the Shoulder					
	Flexion	Extension	Abduction	Adduction	Medial Rotation	Lateral Rotation
Pectoralis	X (PM)			X (PM)	X	
Latissimus dorsi		X (PM)		X (PM)	X	
Deltoid	X (PM) (anterior fibers)	X (PM) (posterior fibers)	X (PM)		X (anterior fibers)	X (posterior fibers)
Subscapularis					X (PM)	
Supraspinatus			X			
Infraspinatus					X	X (PM)
Teres minor				X (weak)		X (PM)
Teres major		X		X	X	
Coracobrachialis	X			X		
Biceps brachii	X					
Triceps brachii				X		

Part II: Muscles Acting on the Forearm

	Actions			
	Elbow Flexion	Elbow Extension	Pronation	Supination
Biceps brachii	X (PM)			X
Triceps brachii		X (PM)		
Anconeus		X		
Brachialis	X (PM)			
Brachioradialis	X			
Pronator teres	X (weak)		X	
Pronator quadratus			X (PM)	
Supinator				X

Part III: Muscles Acting on the Wrist and Fingers

	Actions on the Wrist				Actions on the Fingers	
	Flexion	Extension	Abduction	Adduction	Flexion	Extension
Anterior Compartment						
Flexor carpi radialis	X (PM)		X			
Palmaris longus	X (weak)					
Flexor carpi ulnaris	X (PM)			X		
Flexor digitorum superficialis	X (PM)				X	
Flexor pollicis longus					X (thumb)	
Flexor digitorum profundus	X				X	
Posterior Compartment						
Extensor carpi radialis longus and brevis		X	X			
Extensor digitorum		X (PM)				X (and abducts)
Extensor carpi ulnaris		X		X		
Abductor pollicis longus			X		(abducts thumb)	
Extensor pollicis longus and brevis						X (thumb)
Extensor indicis						X (index finger)

Key:

■ = Extensors
■ = Flexors
■ = Others

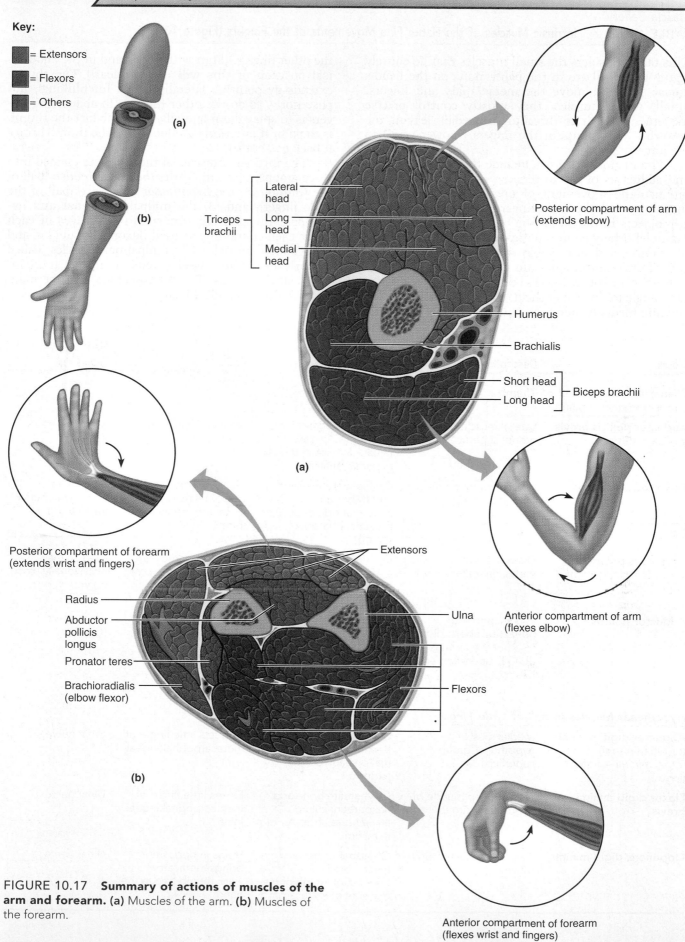

(a)

(b)

Triceps brachii
Lateral head
Long head
Medial head

Posterior compartment of arm (extends elbow)

Humerus
Brachialis
Short head
Long head — Biceps brachii

(a)

Anterior compartment of arm (flexes elbow)

Posterior compartment of forearm (extends wrist and fingers)

Radius
Abductor pollicis longus
Pronator teres
Brachioradialis (elbow flexor)

Extensors
Ulna
Flexors

(b)

Anterior compartment of forearm (flexes wrist and fingers)

FIGURE 10.17   Summary of actions of muscles of the arm and forearm. (a) Muscles of the arm. **(b)** Muscles of the forearm.

Muscle Gallery
TABLE 10.13   Intrinsic Muscles of the Hand: Fine Movements of the Fingers (Figure 10.18)

This table considers the small muscles that lie entirely in the hand. All are in the palm, none on the hand's dorsal side. All move the metacarpals and fingers. Small, weak muscles, they mostly control precise movements (such as threading a needle), leaving the powerful movements of the fingers ("power grip") to the forearm muscles.

The intrinsic muscles include the main abductors and adductors of the fingers, as well as muscles that produce the movement of opposition—moving the thumb toward the little finger—that enables one to grip objects in the palm (the handle of a hammer, for example). Many palm muscles are specialized to move the thumb, and surprisingly many move the little finger. Thumb movements are defined differently from movements of other fingers because the thumb lies at a right angle to the rest of the hand. The thumb flexes by bending medially along the palm, not anteriorly, as do the other fingers. (Start with your hand in the anatomical position or this will not be clear!) The thumb extends by pointing laterally (as in hitchhiking), not posteriorly, as do the other fingers. To abduct the fingers is to splay them laterally, but to abduct the thumb is to point it anteriorly. Adduction of the thumb brings it back posteriorly.

The intrinsic muscles of the palm are divided into three groups, those in (1) the *thenar eminence* (ball of the thumb); (2) the *hypothenar eminence* (ball of the little finger); and (3) the midpalm. Thenar and hypothenar muscles are almost mirror images of each other, each containing a small flexor, an abductor, and an opponens muscle. The midpalmar muscles, called **lumbricals** and **interossei**, extend our fingers at the interphalangeal joints. The interossei are also the main finger abductors and adductors.

Muscle	Description	Origin (O) and Insertion (I)	Action	Nerve Supply
Thenar Muscles in Ball of Thumb (the'nar) (*thenar* = palm)				
Abductor pollicis brevis (*pollex* = thumb)	Lateral muscle of thenar group; superficial	O—flexor retinaculum and nearby carpals I—lateral base of thumb's proximal phalanx	Abducts thumb (at carpometacarpal joint)	Median nerve (C_8, T_1)
Flexor pollicis brevis	Medial and deep muscle of thenar group	O—flexor retinaculum and trapezium I—lateral side of base of proximal phalanx of thumb	Flexes thumb (at carpometacarpal and metacarpophalangeal joints)	Median (or occasionally ulnar) nerve (C_8, T_1)
Opponens pollicis (o-pōn'enz) (*opponens* = opposition)	Deep to abductor pollicis brevis, on metacarpal 1	O—flexor retinaculum and trapezium I—whole anterior side of metacarpal 1	Opposition: moves thumb to touch tip of little finger	Median (or occasionally ulnar) nerve
Adductor pollicis	Fan-shaped with horizontal fibers; distal to other thenar muscles; oblique and transverse heads	O—capitate bone and bases of metacarpals 2–4; front of metacarpal 3 I—medial side of base of proximal phalanx of thumb	Adducts and helps to oppose thumb	Ulnar nerve (C_8, T_1)
Hypothenar Muscles in Ball of Little Finger				
Abductor digiti minimi (dǐ'jǐ-ti min'ǐ-mi) (*digiti minimi* = little finger)	Medial muscle of hypothenar group; superficial	O—pisiform bone I—medial side of proximal phalanx of little finger	Abducts little finger at metacarpophalangeal joint	Ulnar nerve
Flexor digiti minimi brevis	Lateral deep muscle of hypothenar group	O—hamate bone and flexor retinaculum I—same as abductor digiti minimi	Flexes little finger at metacarpophalangeal joint	Ulnar nerve
Opponens digiti minimi	Deep to abductor digiti minimi	O—same as flexor digiti minimi brevis I—most of length of medial side of metacarpal 5	Helps in opposition: brings metacarpal 5 toward thumb to cup the hand	Ulnar nerve

Muscle Gallery
TABLE 10.13   *(continued)*

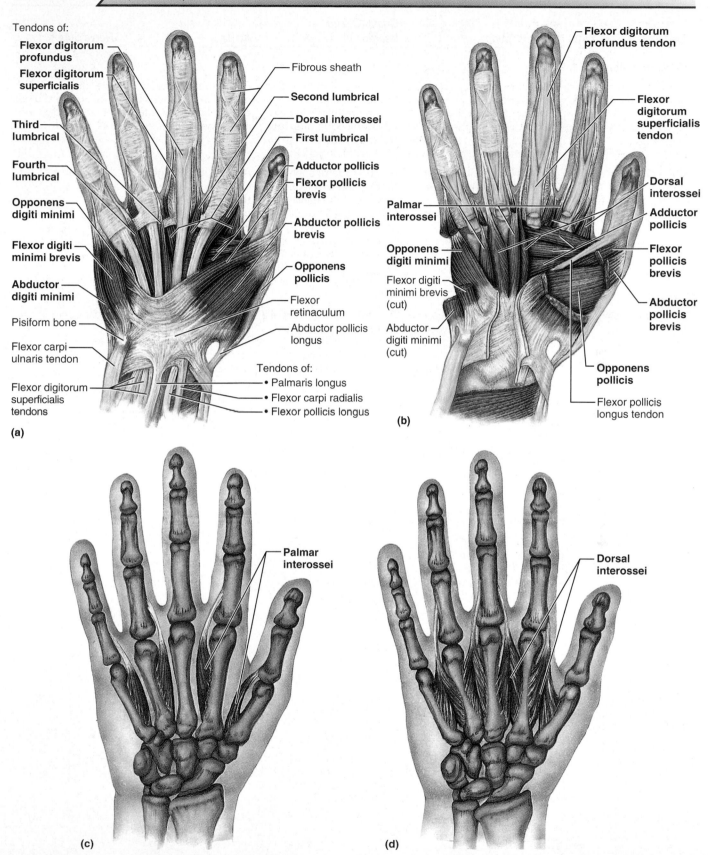

FIGURE 10.18   Hand muscles, ventral view of right hand. (a) First superficial layer. **(b)** Second layer. **(c)** Palmar interossei (isolated). **(d)** Dorsal interossei (isolated).

Muscle Gallery
TABLE 10.13   /   **Intrinsic Muscles of the Hand: Fine Movements of the Fingers (Figure 10.18)** *(continued)*

Muscle	Description	Origin (O) and Insertion (I)	Action	Nerve Supply
Midpalmar Muscles				
Lumbricals (lum'brĭ-klz) (*lumbric* = earthworm)	Four worm-shaped muscles in palm, one to each finger (except thumb); odd because they originate from the tendons of another muscle	O—lateral side of each tendon of flexor digitorum profundus in palm I—lateral edge of extensor expansion on first phalanx of fingers 2–5	Flex fingers at metacarpophalangeal joints but extend fingers at interphalangeal joints	Median nerve (lateral two) and ulnar nerve (medial two)
Palmar interossei (in"ter-os'e-i) (*interossei* = between bones)	Four long, cone-shaped muscles in the spaces between the metacarpals; lie ventral to the dorsal interossei	O—the side of each metacarpal that faces the midaxis of the hand (metacarpal 3, where it's absent) I—extensor expansion on first phalanx of each finger (except finger 3), on side facing midaxis of hand	Adductors of fingers: pull fingers in toward third digit; act with lumbricals to extend fingers at interphalangeal joints and flex them at metacarpophalangeal joints	Ulnar nerve
Dorsal interossei	Four bipennate muscles filling spaces between the metacarpals; deepest palm muscles, also visible on dorsal side of hand (Figure 10.16b)	O—sides of metacarpals I—extensor expansion over first phalanx of fingers 2–4 on side opposite midaxis of hand (finger 3), but on *both* sides of finger 3	Abduct (diverge) fingers; extend fingers at interphalangeal joints and flex them at metacarpophalangeal joint	Ulnar nerve

Muscle Gallery
TABLE 10.14 ⟩ Muscles Crossing the Hip and Knee Joints: Movements of the Thigh and Leg
(Figures 10.19 and 10.20)

The muscles fleshing out the thigh are difficult to segregate into groups on the basis of action. Some thigh muscles act only at the hip joint, others only at the knee, while still others act at both joints. Attempts to classify these muscles solely on the basis of location are equally frustrating because different muscles in a particular location often have very different actions. However, the *most anterior* muscles of the hip and thigh flex the femur at the hip and extend the leg at the knee—producing the foreswing phase of walking. The *posterior* muscles of the hip and thigh, by contrast, mostly extend the thigh and flex the leg—the backswing phase of walking. The third group of muscles in this region, the *medial*, or *adductor*, muscles, all adduct the thigh; they have no effect on the leg. In the thigh, the anterior, posterior, and adductor muscles are separated by walls of fascia into *anterior, posterior,* and *medial compartments* (see Figure 10.24a). The deep fascia of the thigh, the *fascia lata,* surrounds and encloses all three groups of muscles like a support stocking.

Movements of the thigh (occurring at the hip joint) are accomplished largely by muscles anchored to the pelvic girdle. Like the shoulder joint, the hip joint is a ball-and-socket joint permitting flexion, extension, abduction, adduction, circumduction, and rotation. Muscles effecting these movements of the thigh are among

the most powerful muscles of the body. For the most part, the thigh *flexors* pass in front of the hip joint. The most important of these are the **iliopsoas, tensor fasciae latae,** and **rectus femoris.** They are assisted in this action by the **adductor muscles** of the medial thigh and the straplike **sartorius.** The prime mover of thigh flexion is the iliopsoas.

Thigh *extension* is effected primarily by the massive **hamstring muscles** of the posterior thigh. However, during forceful extension, the **gluteus maximus** of the buttock is activated. Buttock muscles that lie lateral to the hip joint (**gluteus medius** and **minimus**) *abduct* the thigh. Thigh adduction is the role of the adductor muscles of the medial thigh. Abduction and adduction of the thighs are extremely important during walking to keep the body's weight balanced over the limb that is on the ground. Many different muscles bring about medial and lateral rotation of the thigh.

At the knee joint, flexion and extension are the main movements. The sole knee *extensor* is the **quadriceps femoris** muscle of the anterior thigh, the most powerful muscle in the body. The quadriceps is antagonized by the hamstrings of the posterior compartment, which are prime movers of knee flexion.

The actions of the muscles presented here are further summarized in Table 10.16 (Part I).

Muscle	Description	Origin (O) and Insertion (I)	Action	Nerve Supply
Part I: Anterior and Medial Muscles (Figure 10.19)				
Origin on the Pelvis				
Iliopsoas (il″e-o-so′us)	Iliopsoas is a composite of two closely related muscles (iliacus and psoas major) whose fibers pass under the inguinal ligament [see Table 10.6 (external oblique muscle) and Figure 10.11] to insert via a common tendon on the femur.			
■ **Iliacus** (il-e-ak′us) (*iliac* = ilium)	Large, fan-shaped, more lateral muscle	O—iliac fossa and crest, lateral sacrum I—femur on and immediately below lesser trochanter of femur via iliopsoas tendon	Illiopsoas is the prime mover for flexing thigh or for flexing trunk on thigh during a bow	Femoral nerve (L$_2$ and L$_3$)
■ **Psoas major** (so′us) (*psoa* = loin muscle; *major* = larger)	Longer, thicker, more medial muscle of the pair (Butchers refer to this muscle as the tenderloin)	O—by fleshy slips from transverse processes, bodies, and discs of lumbar vertebrae and T$_{12}$ I—lesser trochanter of femur via iliopsoas tendon	As above; also effects lateral flexion of vertebral column; important postural muscle	Ventral rami L$_1$–L$_3$
Sartorius (sar-tor′e-us) (*sartor* = tailor)	Straplike superficial muscle running obliquely across anterior surface of thigh to knee; longest muscle in body; crosses both hip and knee joints	O—anterior superior iliac spine I—winds around medial aspect of knee and inserts into medial aspect of proximal tibia	Flexes, abducts, and laterally rotates thigh; flexes knee (weak); known as "tailor's muscle" because it helps effect cross-legged position in which tailors are often depicted	Femoral nerve

▶

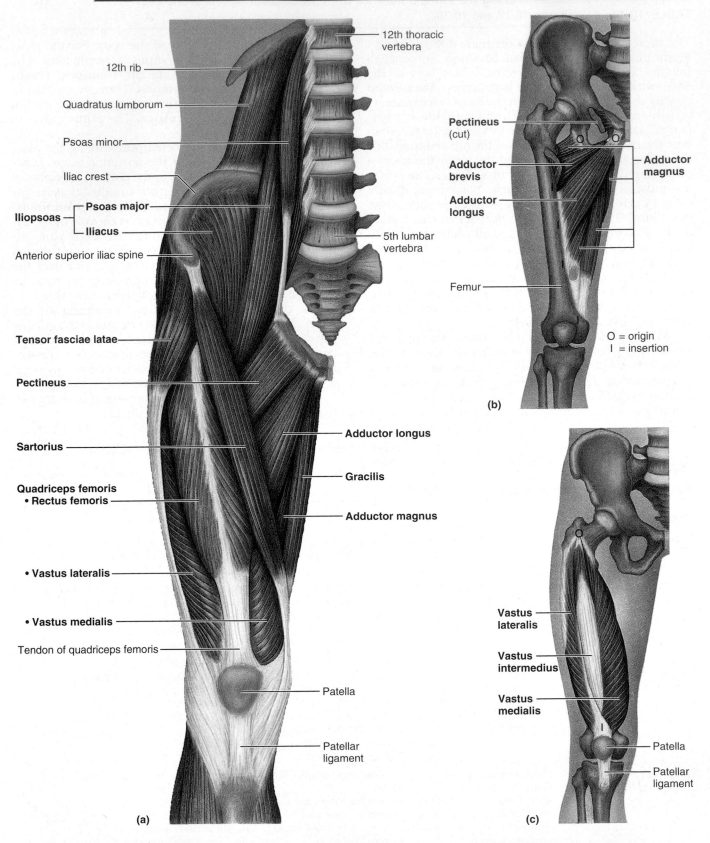

12th thoracic vertebra

12th rib

Quadratus lumborum

Psoas minor

Iliac crest

Pectineus (cut)

Adductor brevis

Adductor longus

Adductor magnus

Psoas major

Iliopsoas

Iliacus

Anterior superior iliac spine

5th lumbar vertebra

Femur

O = origin
I = insertion

Tensor fasciae latae

Pectineus

(b)

Sartorius

Adductor longus

Gracilis

Quadriceps femoris
• **Rectus femoris**

Adductor magnus

Vastus lateralis

Vastus lateralis

Vastus intermedius

• **Vastus medialis**

Vastus medialis

Tendon of quadriceps femoris

Patella

Patella

Patellar ligament

Patellar ligament

(a)

(c)

FIGURE 10.19 **Anterior and medial muscles promoting movements of the thigh and leg.** (a) Anterior view of the deep muscles of the pelvis and superficial muscles of the right thigh. (See *A Brief Atlas of the Human Body*, Figure 68.) **(b)** Adductor muscles of the medial compartment of the thigh isolated. **(c)** The vastus muscles of the quadriceps group. The rectus femoris muscle of the quadriceps group and surrounding muscles have been removed.

Muscle Gallery
TABLE 10.14 *(continued)*

Muscle	Description	Origin (O) and Insertion (I)	Action	Nerve Supply

Muscles of the Medial Compartment of the Thigh

Adductors (ah-duk'torz): Large muscle mass consisting of three muscles (magnus, longus, and brevis) forming medial aspect of thigh; arise from inferior part of pelvis and insert at various levels on femur; all used in movements that press thighs together, as when astride a horse; important in pelvic tilting movements that occur during walking and in fixing the hip when the knee is flexed and the foot is off the ground; entire group innervated by obturator nerve. Strain or stretching of this muscle group is called a "pulled groin."

Muscle	Description	Origin (O) and Insertion (I)	Action	Nerve Supply
▪ **Adductor magnus** (mag'nus) (*adduct* = move toward midline; *magnus* = large)	A triangular muscle with a broad insertion; is a composite muscle that is part adductor and part hamstring in action	O—ischial and pubic rami and ischial tuberosity I—linea aspera and adductor tubercle of femur	Anterior part adducts and medially rotates and flexes thigh; posterior part is a synergist of hamstrings in thigh extension	Obturator nerve and sciatic nerve (L_2–L_4)
▪ **Adductor longus** (*longus* = long)	Overlies middle aspect of adductor magnus; most anterior of adductor muscles	O—Pubis near pubic symphysis I—linea aspera	Adducts, flexes, and medially rotates thigh	Anterior division of obturator nerve (L_2–L_4)
▪ **Adductor brevis** (*brevis* = short)	In contact with obturator externus muscle; largely concealed by adductor longus and pectineus	O—body and inferior ramus of pubis I—linea aspera above adductor longus	Adducts and medially rotates thigh	Obturator nerve
Pectineus (pek-tin'e-us) (*pecten* = comb)	Short, flat muscle; overlies adductor brevis on proximal thigh; abuts adductor longus medially	O—pectineal line of pubis (and superior ramus) I—inferior from lesser trochanter to the linea aspera of posterior aspect of femur	Adducts, flexes, and medially rotates thigh	Femoral and sometimes obturator nerve
Gracilis (grah-si'lis) (*gracilis* = slender)	Long, thin, superficial muscle of medial thigh	O—inferior ramus and body of pubis and adjacent ischial ramus I—medial surface of tibia just inferior to its medial condyle	Adducts thigh, flexes and medially rotates thigh, especially during walking; flexes knee	Obturator nerve

Muscles of the Anterior Compartment of the Thigh

Quadriceps femoris (kwod'ri-seps fem'o-ris): Quadriceps femoris arises from four separate heads (*quadriceps* = four heads) that form the flesh of front and sides of thigh; these heads (rectus femoris, and lateral, medial, and intermediate vastus muscles) have a common insertion tendon, the quadriceps tendon, which inserts into the patella and then via the patellar ligament into tibial tuberosity. The quadriceps is a powerful knee extensor used in climbing, jumping, running, and rising from seated position; group is innervated by femoral nerve; the tone of quadriceps plays important role in strengthening knee joint.

Muscle	Description	Origin (O) and Insertion (I)	Action	Nerve Supply
▪ **Rectus femoris** (rek'tus) (*rectus* = straight; *femoris* = femur)	Superficial muscle of anterior thigh; runs straight down thigh; longest head and only muscle of group to cross hip joint	O—anterior inferior iliac spine and superior margin of acetabulum I—patella and tibial tuberosity via patellar ligament	Extends knee and flexes thigh at hip	Femoral nerve (L_2–L_4)
▪ **Vastus lateralis** (vas'tus lat"er-a'lis) (*vastus* = large; *lateralis* = lateral)	Largest head of the group, forms lateral aspect of thigh; a common intramuscular injection site, particularly in infants (who have poorly developed buttock and arm muscles)	O—greater trochanter, intertrochanteric line, linea aspera I—as for rectus femoris	Extends and stabilizes knee	Femoral nerve
▪ **Vastus medialis** (me"de-a'lis) (*medialis* = medial)	Forms inferomedial aspect of thigh	O—linea aspera, intertrochanteric line I—as for rectus femoris	Extends knee; inferior fibers stabilize patella	Femoral nerve

▶

Muscle Gallery
TABLE 10.14) Muscles Crossing the Hip and Knee Joints: Movements of the Thigh and Leg (Figures 10.19 and 10.20) *(continued)*

Muscle	Description	Origin (O) and Insertion (I)	Action	Nerve Supply
■ **Vastus intermedius** (in"ter-me'de-us) (*intermedius* = intermediate)	Obscured by rectus femoris; lies between vastus lateralis and vastus medialis on anterior thigh	O—anterior and lateral surfaces of proximal femur shaft I—as for rectus femoris	Extends knee	Femoral nerve
Tensor fasciae latae (ten'sor fä'she-e la'te) (*tensor* = to make tense; *fascia* = band; *lata* = wide)	Enclosed between fascia layers of anterolateral aspect of thigh; functionally associated with medial rotators and flexors of thigh	O—anterior aspect of iliac crest and anterior superior iliac spine I—iliotibial tract*	Flexes and abducts thigh (thus a synergist of the iliopsoas and gluteus medius and minimus muscles); rotates thigh medially; steadies the knee and trunk on thigh by making iliotibial tract taut	Superior gluteal nerve (L_4 and L_5)

Part II: Posterior Muscles (Figure 10.20)

Gluteal Muscles—Origin on Pelvis

Muscle	Description	Origin (O) and Insertion (I)	Action	Nerve Supply
Gluteus maximus (gloo'te-us mak'sĭ-mus) (*glutos* = buttock; *maximus* = largest)	Largest and most superficial of gluteus muscles; forms bulk of buttock mass; fibers are thick and coarse; important site of intramuscular injection (dorsal gluteal site); overlies large sciatic nerve; covers ischial tuberosity only when standing; when sitting, moves superiorly, leaving ischial tuberosity exposed in the subcutaneous position	O—dorsal ilium, sacrum, and coccyx I—gluteal tuberosity of femur; iliotibial tract	Major extensor of thigh; complex, powerful, and most effective when thigh is flexed and force is necessary, as in rising from a forward flexed position and in thrusting the thigh posteriorly in climbing stairs and running; generally inactive during standing and walking; laterally rotates and abducts thigh	Inferior gluteal nerve (L_5, S_1, and S_2)
Gluteus medius (me'de-us) (*medius* = middle)	Thick muscle largely covered by gluteus maximus; important site for intramuscular injections (ventral gluteal site); considered safer than dorsal gluteal site because there is less chance of injuring sciatic nerve	O—between anterior and posterior gluteal lines on lateral surface of ilium I—by short tendon into lateral aspect of greater trochanter of femur	Abducts and medially rotates thigh; steadies pelvis; its action is extremely important in walking; e.g., muscle of limb planted on ground tilts or holds pelvis in abduction so that pelvis on side of swinging limb does not sag; the foot of swinging limb can thus clear the ground	Superior gluteal nerve (L_5, S_1)
Gluteus minimus (mĭ'nĭ-mus) (*minimus* = smallest)	Smallest and deepest of gluteal muscles	O—between anterior and inferior gluteal lines on external surface of ilium I—anterior border of greater trochanter of femur	As for gluteus medius	Superior gluteal nerve (L_5, S_1)

Lateral Rotators

Muscle	Description	Origin (O) and Insertion (I)	Action	Nerve Supply
Piriformis (pir'ĭ-form-is) (*piri* = pear; *forma* = shape)	Pyramidal muscle located on posterior aspect of hip joint; inferior to gluteus minimus; issues from pelvis via greater sciatic notch	O—anterolateral surface of sacrum (opposite greater sciatic notch) I—superior border of greater trochanter of femur	Rotates extended thigh laterally; since inserted above head of femur, can also assist in abduction of thigh when hip is flexed; stabilizes hip joint	S_1 and S_2, L_5

*The iliotibial tract is a thickened lateral portion of the *fascia lata* (the fascia that ensheathes all the muscles of the thigh). It extends as a tendinous band from the iliac crest to the knee (see Figure 10.20a).

Muscle Gallery
TABLE 10.14   *(continued)*

Muscle	Description	Origin (O) and Insertion (I)	Action	Nerve Supply
Obturator externus (ob"tu-ra'tor ek-ster'-nus) (*obturator* = obturator foramen; *externus* = outside)	Flat, triangular muscle deep in upper medial aspect of thigh	O—outer surfaces of obturator membrane, pubis, and ischium, margins of obturator foramen I—by a tendon into trochanteric fossa of posterior femur	As for piriformis	Obturator nerve

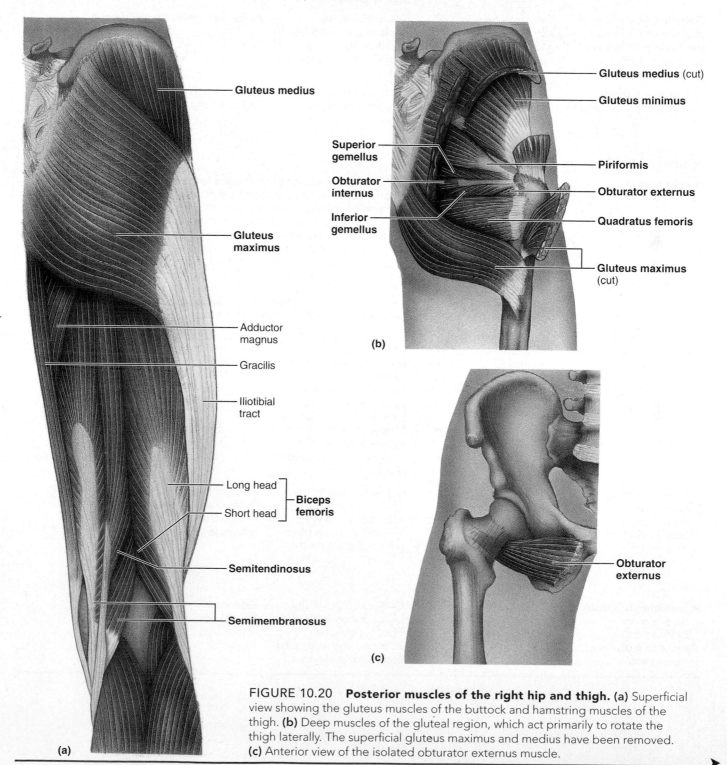

(a)

Gluteus medius

Gluteus maximus

Adductor magnus

Gracilis

Iliotibial tract

Long head
Short head — **Biceps femoris**

Semitendinosus

Semimembranosus

(b)

Gluteus medius (cut)
Gluteus minimus

Superior gemellus

Obturator internus

Inferior gemellus

Piriformis

Obturator externus

Quadratus femoris

Gluteus maximus (cut)

(c)

Obturator externus

FIGURE 10.20   **Posterior muscles of the right hip and thigh. (a)** Superficial view showing the gluteus muscles of the buttock and hamstring muscles of the thigh. **(b)** Deep muscles of the gluteal region, which act primarily to rotate the thigh laterally. The superficial gluteus maximus and medius have been removed. **(c)** Anterior view of the isolated obturator externus muscle.

Muscle Gallery
TABLE 10.14

Muscles Crossing the Hip and Knee Joints: Movements of the Thigh and Leg
(Figures 10.19 and 10.20) *(continued)*

Muscle	Description	Origin (O) and Insertion (I)	Action	Nerve Supply
Obturator internus (in-ter'nus) (*internus* = inside)	Surrounds obturator foramen within pelvis; leaves pelvis via lesser sciatic notch and turns acutely forward to insert in femur	O—inner surface of obturator membrane, greater sciatic notch, and margins of obturator foramen I—greater trochanter in front of piriformis	As for piriformis	L_5 and S_1
Gemellus (jĕ-mĕ'lis)— superior and inferior (*gemin* = twin, double; *superior* = above; *inferior* = below)	Two small muscles with common insertions and actions; considered extrapelvic portions of obturator internus	O—ischial spine (superior); ischial tuberosity (inferior) I—greater trochanter of femur	As for piriformis	L_5 and S_1
Quadratus femoris (*quad* = four-sided square)	Short, thick muscle; most inferior of lateral rotator muscles; extends laterally from pelvis	O—ischial tuberosity I—trochanteric crest of femur	Rotates thigh laterally and stabilizes hip joint	L_5 and S_1

Muscles of the Posterior Compartment of the Thigh

Hamstrings				
	The hamstrings are fleshy muscles of the posterior thigh (biceps femoris, semitendinosus, and semimembranosus); they cross both the hip and knee joints and are prime movers of thigh extension and knee flexion; group has a common origin site and is innervated by sciatic nerve; ability of hamstrings to act on one of the two joints spanned depends on which joint is fixed; i.e., if knee is fixed (extended), they promote hip extension; if hip is extended, they promote knee flexion; however, when hamstrings are stretched, they tend to restrict full accomplishment of antagonistic movements; e.g., if knees are fully extended, it is difficult to flex the hip fully (and touch your toes), and when the thigh is fully flexed as in kicking a football, it is almost impossible to extend the knee fully at the same time (without considerable practice); name of this muscle group comes from old butchers' practice of using their tendons to hang hams for smoking; "pulled hamstrings" are common sports injuries in those who run very hard, e.g., football halfbacks.			
■ **Biceps femoris** (*biceps* = two heads)	Most lateral muscle of the group; arises from two heads	O—ischial tuberosity (long head); linea aspera and distal femur (short head) I—common tendon passes downward and laterally (forming lateral border of popliteal fossa) to insert into head of fibula and lateral condyle of tibia	Extends thigh and flexes knee; laterally rotates leg, especially when knee is flexed	Sciatic nerve (L_5–S_2)
■ **Semitendinosus** (sem"e-ten"dĭ-no'sus) (*semi* = half; *tendinosus* = tendon)	Lies medial to biceps femoris; although its name suggests that this muscle is largely tendinous, it is quite fleshy; its long slender tendon begins about two-thirds of the way down thigh	O—ischial tuberosity in common with long head of biceps femoris I—medial aspect of upper tibial shaft	Extends thigh at hip; flexes knee; with semimembranosus, medially rotates leg	Sciatic nerve
■ **Semimembranosus** (sem"e-mem"-brah-no'sus) (*membranosus* = membrane)	Deep to semitendinosus	O—ischial tuberosity I—medial condyle of tibia; via oblique popliteal ligament to lateral condyle of femur	Extends thigh and flexes knee; medially rotates leg	Sciatic nerve

The deep fascia of the leg is continuous with the fascia lata that ensheathes the thigh. Like a snug knee sock beneath the skin, the leg fascia binds the leg muscles tightly, helping to prevent excessive swelling of muscles during exercise and aiding venous return. Its inward extensions segregate the leg muscles into *anterior, lateral,* and *posterior compartments* (see Figure 10.24b), each with its own nerve and blood supply. Distally the leg fascia thickens to form the **flexor, extensor,** and **fibular** (or **peroneal**) **retinaculae,** "ankle brackets" that hold the tendons in place where they run to the foot.

The various muscles of the leg promote movements at the ankle joint (dorsiflexion and plantar flexion), the intertarsal joints (inversion and eversion of the foot), and/or the toes (flexion and extension). Muscles in the *anterior extensor compartment* of the leg—**tibialis anterior, extensor digitorum longus, extensor hallucis longus,** and **fibularis tertius**—are primarily toe extensors and ankle dorsiflexors. Although dorsiflexion is not

a powerful movement, it is important in preventing the toes from dragging during walking. Lateral compartment muscles are the **fibular,** formerly **peroneal** (*peron* = fibula), **muscles** that plantar flex and evert the foot. Muscles of the *posterior flexor compartment*—**gastrocnemius, soleus, tibialis posterior, flexor digitorum longus,** and **flexor hallucis longus**—primarily plantar flex the foot and flex the toes. Plantar flexion is the most powerful movement of the ankle (and foot) because it lifts the entire weight of our body. It is essential for standing on tiptoe and provides the forward thrust when walking and running. The **popliteus muscle,** which crosses the knee, "unlocks" the extended knee in preparation for flexion.

The tiny intrinsic muscles of the sole of the foot (lumbricals, interossei, and others) are considered separately in Table 10.17.

The actions of the muscles in this table are summarized in Table 10.16 (Part II).

Muscle	Description	Origin (O) and Insertion (I)	Action	Nerve Supply
Part I: Muscles of the Anterior Compartment (Figures 10.21 and 10.22)				
	All muscles of the anterior compartment are dorsiflexors of the ankle and have a common innervation, the deep fibular nerve. Paralysis of the anterior muscle group causes *foot drop,* which requires that the leg be lifted unusually high during walking to prevent tripping over one's toes. "Shin splints" is a painful inflammatory condition of the muscles of the anterior compartment.			
Tibialis anterior (tib"e-a'lis) (*tibial* = tibia; *anterior* = toward the front)	Superficial muscle of anterior leg; laterally parallels sharp anterior margin of tibia	O—lateral condyle and upper ⅔ of tibial shaft; interosseous membrane I—by tendon into inferior surface of medial cuneiform and first metatarsal bone	Prime mover of dorsiflexion; inverts foot; assists in supporting medial longitudinal arch of foot	Deep fibular nerve (L_4 and L_5)
Extensor digitorum longus (*extensor* = increases angle at a joint; *digit* = finger or toe; *longus* = long)	Unipennate muscle on anterolateral surface of leg; lateral to tibialis anterior muscle	O—lateral condyle of tibia; proximal ¾ of fibula; interosseous membrane I—middle and distal phalanges of toes 2–5 via extensor expansion	Prime mover of toe extension (acts mainly at metatarsophalangeal joints); dorsiflexes foot (in conjunction with the tibialis anterior and extensor hallucis longus)	Deep fibular nerve (L_5 and S_1)
Fibularis (peroneus) tertius (fib-u-lar'ris ter'shus) (*perone* = fibula; *tertius* = third)	Small muscle; usually continuous and fused with distal part of extensor digitorum longus; not always present	O—distal anterior surface of fibula and interosseous membrane I—tendon inserts on dorsum of fifth metatarsal	Dorsiflexes and everts foot	Deep fibular nerve (L_5 and S_1)
Extensor hallucis longus (hal'yu-kis) (*hallux* = great toe)	Deep to extensor digitorum longus and tibialis anterior; narrow origin	O—anteromedial fibula shaft and interosseous membrane I—tendon inserts on distal phalanx of great toe	Extends great toe; dorsiflexes foot	Deep fibular nerve (L_5 and S_1)

►

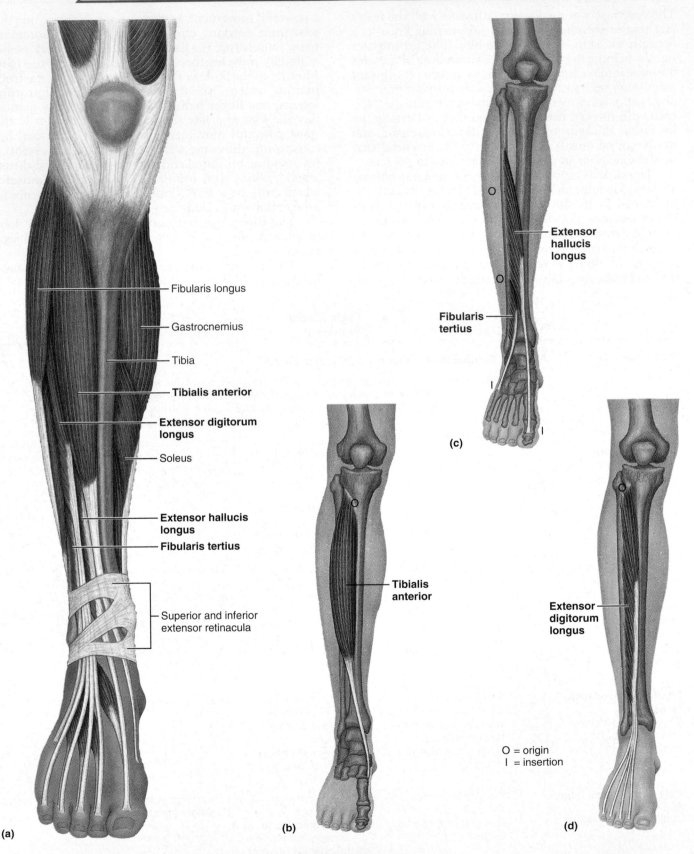

(a)

Fibularis longus

Gastrocnemius

Tibia

Tibialis anterior

Extensor digitorum longus

Soleus

Extensor hallucis longus

Fibularis tertius

Superior and inferior extensor retinacula

(b)

Tibialis anterior

(c)

Extensor hallucis longus

Fibularis tertius

I

I

(d)

Extensor digitorum longus

O = origin
I = insertion

FIGURE 10.21   **Muscles of the anterior compartment of the right leg.**
(a) Superficial view of anterior leg muscles. **(b–d)** Some of the same muscles shown in isolation to allow visualization of their origins and insertions.

Muscle Gallery
TABLE 10.15   *(continued)*

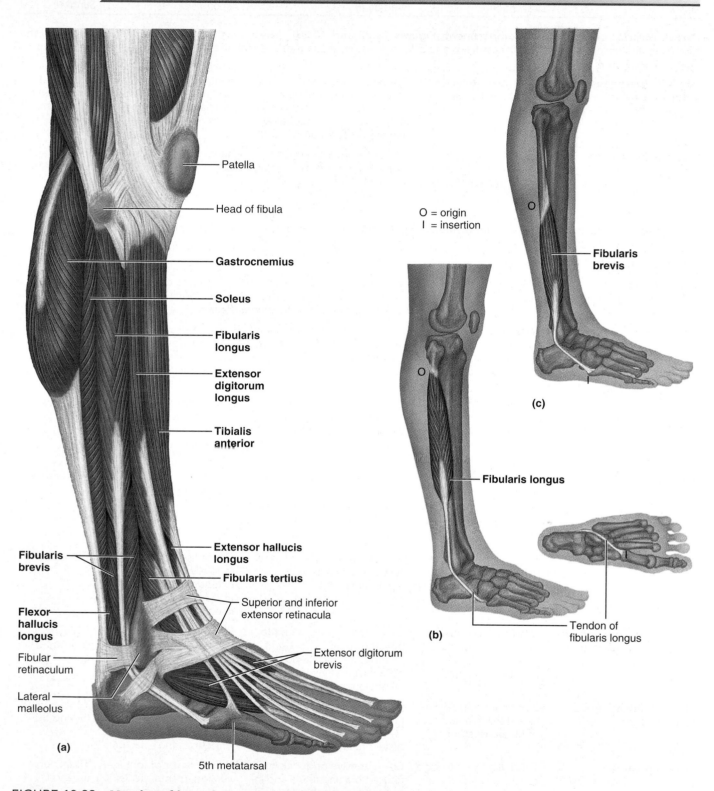

Patella

Head of fibula

Gastrocnemius

Soleus

Fibularis longus

Extensor digitorum longus

Tibialis anterior

Fibularis brevis

Flexor hallucis longus

Fibular retinaculum

Lateral malleolus

Extensor hallucis longus

Fibularis tertius

Superior and inferior extensor retinacula

Extensor digitorum brevis

5th metatarsal

(a)

O = origin
 I = insertion

Fibularis brevis

O

(c)

O

Fibularis longus

Tendon of fibularis longus

(b)

I

I

FIGURE 10.22   Muscles of lateral compartment of the right leg.
(a) Superficial view of lateral aspect of the leg, illustrating positions of lateral compartment muscles (fibularis longus and brevis) relative to anterior and posterior leg muscles. **(b)** Isolated view of fibularis longus; inset illustrates the insertion of the fibularis longus on the plantar surface of the foot. **(c)** Isolated view of the fibularis brevis muscle. (See *A Brief Atlas of the Human Body,* Figure 70.)

Muscle	Description	Origin (O) and Insertion (I)	Action	Nerve Supply
Part II: Muscles of the Lateral Compartment (Figures 10.22 and 10.23) These muscles have a common innervation, the superficial fibular nerve. Besides plantar flexion and foot eversion, these muscles stabilize the lateral ankle and lateral longitudinal arch of the foot.				
Fibularis (peroneus) longus (See also Figure 10.21)	Superficial lateral muscle; overlies fibula	O—head and upper portion of lateral aspect of fibula I—by long tendon that curves under foot to first metatarsal and medial cuneiform	Plantar flexes and everts foot; may help keep foot flat on ground	Superficial fibular nerve (L_5–S_2)
Fibularis (peroneus) brevis (*brevis* = short)	Smaller muscle; deep to fibularis longus; enclosed in a common sheath	O—distal fibula shaft I—by tendon running behind lateral malleolus to insert on proximal end of fifth metatarsal	Plantar flexes and everts foot	Superficial fibular nerve (L_5–S_2)
Part III: Muscles of the Posterior Compartment (Figure 10.23) The muscles of the posterior compartment have a common innervation, the tibial nerve. They act in concert to plantar flex the ankle.				
Superficial Muscles				
Triceps surae (tri"seps sur'e) (See also Figure 10.22)	Refers to muscle pair (gastrocnemius and soleus) that shapes the posterior calf and inserts via a common tendon into the calcaneus of the heel; this calcaneal or Achilles tendon is the largest tendon in the body; prime movers of ankle plantar flexion			
■ **Gastrocnemius** (gas"truk-ne'me-us) (*gaster* = belly; *kneme* = leg)	Superficial muscle of pair; two prominent bellies that form proximal curve of calf	O—by two heads from medial and lateral condyles of femur I—posterior calcaneus via calcaneal tendon	Plantar flexes foot when knee is extended; since it also crosses knee joint, it can flex knee when foot is dorsiflexed	Tibial nerve (S_1, S_2)
■ **Soleus** (so'le-us) (*soleus* = fish)	Broad, flat muscle, deep to gastrocnemius on posterior surface of calf	O—extensive cone-shaped origin from superior tibia, fibula, and interosseous membrane I—as for gastrocnemius	Plantar flexes foot; important locomotor and postural muscle during walking, running, and dancing	Tibial nerve
Plantaris (plan-tar'is) (*planta* = sole of foot)	Generally a small feeble muscle, but varies in size and extent; may be absent	O—posterior femur above lateral condyle I—via a long, thin tendon into calcaneus or calcaneal tendon	Assists in knee flexion and plantar flexion of foot	Tibial nerve
Deep Muscles				
Popliteus (pop-lit'e-us) (*poplit* = back of knee)	Thin, triangular muscle at posterior knee; passes inferomedially to tibial surface	O—lateral condyle of femur and lateral meniscus I—proximal tibia	Flexes and rotates leg medially to unlock extended knee; with tibia fixed, rotates thigh laterally	Tibial nerve (L_4–S_1)
Flexor digitorum longus (*flexor* = decreases angle at a joint)	Long, narrow muscle; runs medial to and partially overlies tibialis posterior	O—extensive origin on the posterior tibia I—tendon runs behind medial malleolus and inserts into distal phalanges of toes 2–5	Plantar flexes and inverts foot; flexes toes; helps foot "grip" ground	Tibial nerve (S_2 and S_3)
Flexor hallucis longus (See also Figure 10.22)	Bipennate muscle; lies lateral to inferior aspect of tibialis posterior	O—midshaft of fibula; interosseous membrane I—tendon runs under foot to distal phalanx of great toe	Plantar flexes and inverts foot; flexes great toe at all joints; "push off" muscle during walking	Tibial nerve (S_2 and S_3)
Tibialis posterior (*posterior* = toward the back)	Thick, flat muscle deep to soleus; placed between posterior flexors	O—extensive origin from superior tibia and fibula and interosseous membrane I—tendon passes behind medial malleolus and under arch of foot; inserts into several tarsals and metatarsals 2–4	Prime mover of foot inversion; plantar flexes foot; stabilizes medial longitudinal arch of foot (as during ice skating)	Tibial nerve (L_4 and L_5)

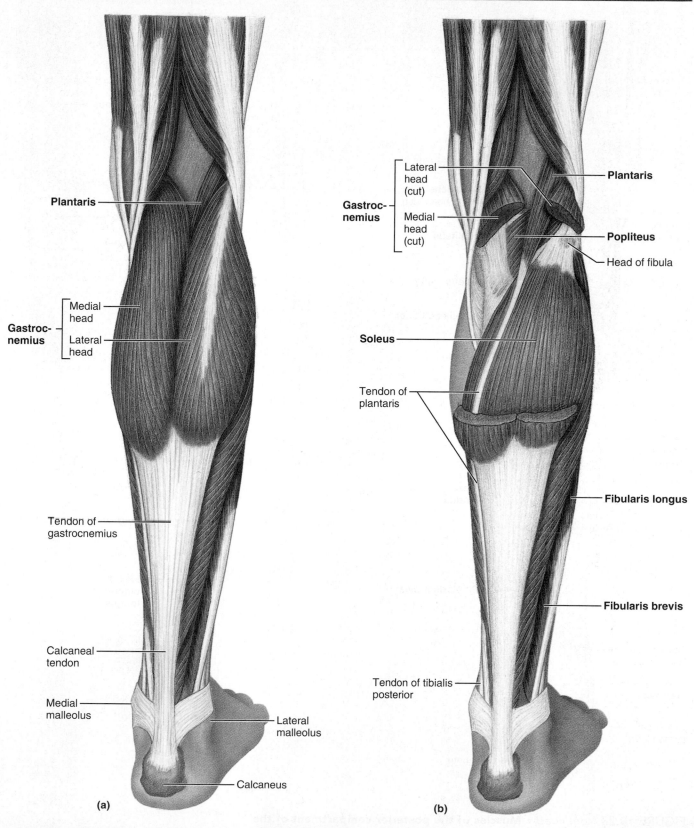

Plantaris

Gastroc-
nemius
  { Medial
    head
    Lateral
    head }

Tendon of
gastrocnemius

Calcaneal
tendon

Medial
malleolus

Lateral
malleolus

Calcaneus

(a)

Gastroc-
nemius
  { Lateral
    head
    (cut)
    Medial
    head
    (cut) }

Plantaris

Popliteus

Head of fibula

Soleus

Tendon of
plantaris

Fibularis longus

Fibularis brevis

Tendon of tibialis
posterior

(b)

FIGURE 10.23 **Muscles of the posterior compartment of the right leg. (a)**
Superficial view of the posterior leg. **(b)** The gastrocnemius has been removed to
show the soleus immediately deep to it.

Muscle Gallery
TABLE 10.15   Muscles of the Leg: Movements of the Ankle and Toes (Figures 10.21 to 10.23) (continued)

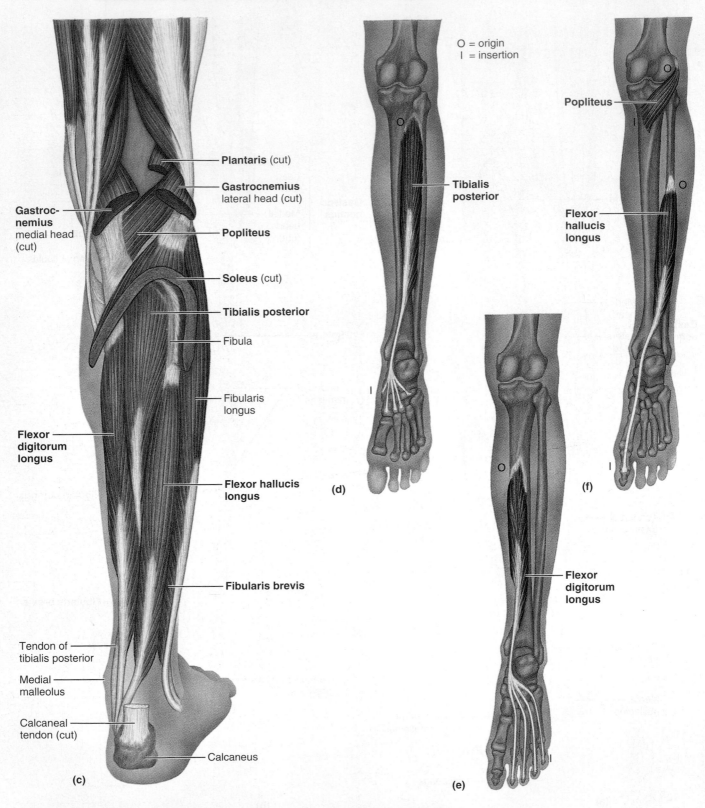

O = origin
I = insertion

Plantaris (cut)

Gastrocnemius
lateral head (cut)

Gastroc-
nemius
medial head
(cut)

Popliteus

Soleus (cut)

Tibialis posterior

Fibula

Fibularis
longus

Flexor
digitorum
longus

Flexor hallucis
longus

Fibularis brevis

Tendon of
tibialis posterior

Medial
malleolus

Calcaneal
tendon (cut)

Calcaneus

(c)

Tibialis
posterior

(d)

Popliteus

Flexor
hallucis
longus

(f)

Flexor
digitorum
longus

(e)

FIGURE 10.23 (continued)   **Muscles of the posterior compartment of the
right leg. (c)** The triceps surae has been removed to show the deep muscles of the
posterior compartment. **(d–f)** Individual deep muscles are shown in isolation.

Muscle Gallery
TABLE 10.16 — Summary of Actions of Muscles Acting on the Thigh, Leg, and Foot (Figure 10.24)

Part I: Muscles Acting on the Thigh and Leg (PM = prime mover)

	Actions at the Hip joint						Actions at the Knee	
	Flexion	Extension	Abduction	Adduction	Medial Rotation	Lateral Rotation	Flexion	Extension
Anterior and Medial Muscles								
Iliopsoas	X (PM)							
Sartorius	X		X			X	X	
Adductor magnus	X	X		X	X			
Adductor longus	X			X	X			
Adductor brevis	X			X	X			
Pectineus	X			X	X			
Gracilis				X	X		X	
Rectus femoris	X							X (PM)
Vastus muscles								X (PM)
Tensor fasciae latae	X		X		X			
Posterior Muscles								
Gluteus maximus		X (PM)	X			X		
Gluteus medius			X (PM)		X			
Gluteus minimus			X		X			
Piriformis			X			X		
Obturator internus						X		
Obturator externus						X		
Gemelli						X		
Quadratus femoris						X		
Biceps femoris		X (PM)				X	X (PM)	
Semitendinosus		X					X (PM) and rotates leg medially	
Semimembranosus		X					X (PM)	
Gastrocnemius							X	
Plantaris							X	
Popliteus						X	X (and rotates leg medially)	

Part II: Muscles Acting on the Ankle and Toes

	Actions at the Ankle Joint				Actions at the Toes	
	Plantar Flexion	Dorsiflexion	Inversion	Eversion	Flexion	Extension
Anterior Compartment						
Tibialis anterior		X (PM)	X			
Extensor digitorum longus		X				X (PM)
Fibularis tertius		X		X		
Extensor hallucis longus		X	X (weak)			X (great toe)
Lateral Compartment						
Fibularis longus and brevis	X			X		
Posterior Compartment						
Gastrocnemius	X (PM)					
Soleus	X (PM)					
Plantaris	X					
Flexor digitorum longus	X		X		X (PM)	
Flexor hallucis longus	X		X		X (great toe)	
Tibialis posterior	X		X (PM)			

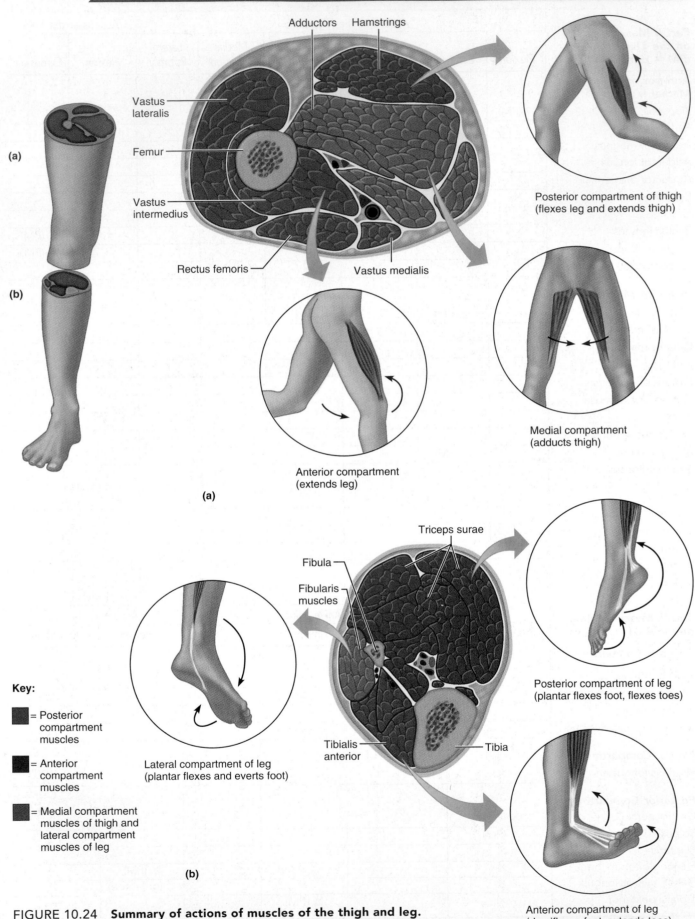

(a)

(b)

Adductors

Hamstrings

Vastus lateralis

Femur

Vastus intermedius

Rectus femoris

Vastus medialis

Posterior compartment of thigh
(flexes leg and extends thigh)

Anterior compartment
(extends leg)

Medial compartment
(adducts thigh)

(a)

Triceps surae

Fibula

Fibularis muscles

Tibialis anterior

Tibia

Posterior compartment of leg
(plantar flexes foot, flexes toes)

Lateral compartment of leg
(plantar flexes and everts foot)

Anterior compartment of leg
(dorsiflexes foot, extends toes)

Key:

■ = Posterior compartment muscles

■ = Anterior compartment muscles

■ = Medial compartment muscles of thigh and lateral compartment muscles of leg

(b)

FIGURE 10.24 **Summary of actions of muscles of the thigh and leg.**
(a) Muscles of the thigh. **(b)** Muscles of the leg.

Muscle Gallery
TABLE 10.17   Intrinsic Muscles of the Foot: Toe Movement and Arch Support (Figure 10.25)

The intrinsic muscles of the foot help to flex, extend, abduct, and adduct the toes. Collectively, along with the tendons of some leg muscles that enter the sole, the foot muscles help support the arches of the foot. There is a single muscle on the foot's dorsum (supe-rior aspect), and several muscles on the plantar aspect (the sole). The plantar muscles occur in four layers, from superficial to deep. Overall, the foot muscles are remarkably similar to those in the palm of the hand.

Muscle	Description	Origin (O) and Insertion (I)	Action	Nerve Supply
Muscle on Dorsum of Foot				
Extensor digitorum brevis (Figure 10.22a)	Small, four-part muscle on dorsum of foot; deep to the tendons of extensor digitorum longus; corresponds to the extensor indicis and extensor pollicis muscles of forearm	O—anterior part of calcaneus bone; extensor retinaculum I—base of proximal phalanx of great toe; extensor expansions on toes 2–5	Helps extend toes at metatarsophalangeal joints	Deep fibular nerve (S_1 and S_2)
Muscles on Sole of Foot—First Layer (Most Superficial) (Figure 10.25)				
Flexor digitorum brevis	Bandlike muscle in middle of sole; corresponds to flexor digitorum superficialis of forearm and inserts into digits in the same way	O—tuber calcanei I—middle phalanx of toes 2–4	Helps flex toes	Medial plantar nerve (S_2 and S_3)
Abductor hallucis (hal'yu-kis) (*hallux* = great toe)	Lies medial to flexor digitorum brevis (recall the similar thumb muscle, abductor pollicis brevis)	O—tuber calcanei and flexor retinaculum I—proximal phalanx of great toe, medial side, in the tendon of flexor hallucis brevis (see below)	Abducts great toe	Medial plantar nerve
Abductor digiti minimi	Most lateral of the three superficial sole muscles; (recall similar abductor muscle in palm)	O—tuber calcanei I—lateral side of base of little toe's proximal phalanx	Abducts and flexes little toe	Lateral plantar nerve (S_2 and S_3)
Muscles on Sole of Foot—Second Layer				
Flexor accessorius (quadratus plantae)	Rectangular muscle just deep to flexor digitorum brevis in posterior half of sole; two heads (see also Figure 10.25c)	O—medial and lateral sides of calcaneus I—tendon of flexor digitorum longus in midsole	Straightens out the oblique pull of flexor digitorum longus	Lateral plantar nerve
Lumbricals	Four little "worms" (like lumbricals in hand)	O—from each tendon of flexor digitorum longus I—extensor expansion on proximal phalanx of toes 2–5, medial side	By pulling on extensor expansion, flex toes at metatarsophalangeal joints and extend toes at interphalangeal joints	Medial plantar nerve (first lumbrical) and lateral plantar nerve (second to fourth lumbrical)

▶

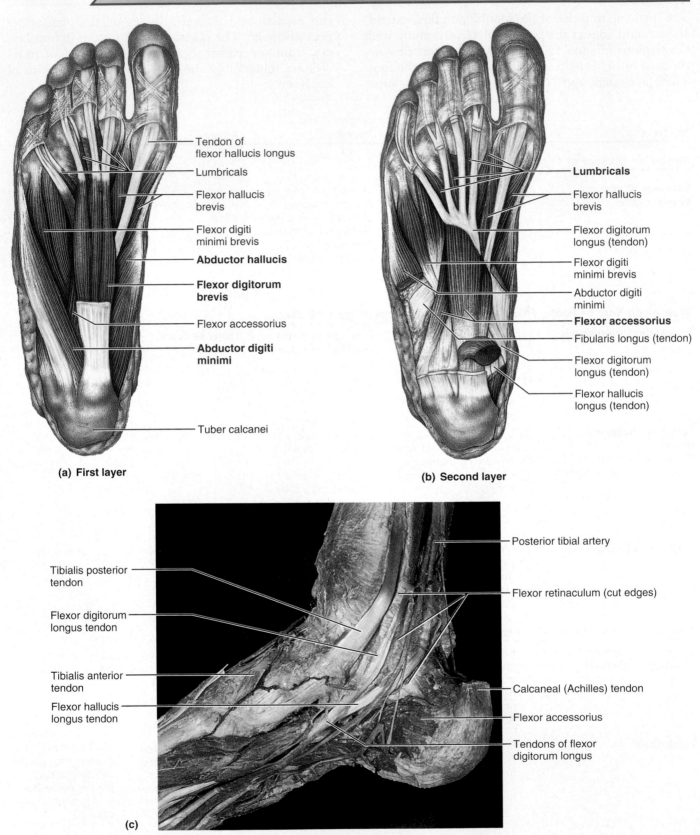

Tendon of
flexor hallucis longus

Lumbricals

Flexor hallucis
brevis

Flexor digiti
minimi brevis

Abductor hallucis

**Flexor digitorum
brevis**

Flexor accessorius

**Abductor digiti
minimi**

Tuber calcanei

(a) First layer

Lumbricals

Flexor hallucis
brevis

Flexor digitorum
longus (tendon)

Flexor digiti
minimi brevis

Abductor digiti
minimi

Flexor accessorius

Fibularis longus (tendon)

Flexor digitorum
longus (tendon)

Flexor hallucis
longus (tendon)

(b) Second layer

Tibialis posterior
tendon

Flexor digitorum
longus tendon

Tibialis anterior
tendon

Flexor hallucis
longus tendon

Posterior tibial artery

Flexor retinaculum (cut edges)

Calcaneal (Achilles) tendon

Flexor accessorius

Tendons of flexor
digitorum longus

(c)

FIGURE 10.25   **Muscles of the right foot, plantar and medial aspects.**

Muscles on Sole of Foot—Third Layer

Flexor hallucis brevis	Covers metatarsal 1; splits into two bellies—recall flexor pollicis brevis of thumb	O—lateral cuneiform and cuboid bones I—via two tendons onto base of the proximal phalanx of great toe	Flexes great toe's metatarsophalangeal joint	Medial plantar nerve
Adductor hallucis	Oblique and transverse heads; deep to lumbricals (recall adductor pollicis in thumb)	O—from bases of metatarsals 2–4 and from fibularis longus tendon sheath (oblique head); from a ligament across metatarsophalangeal joints (transverse head) I—base of proximal phalanx of great toe, lateral side	Helps maintain the transverse arch of foot; weak adductor of great toe	Lateral plantar nerve (S$_2$ and S$_3$)
Flexor digiti minimi brevis	Covers metatarsal 5 (recall same muscle in hand)	O—base of metatarsal 5 and tendon sheath of fibularis longus I—base of proximal phalanx of toe 5	Flexes little toe at metatarsophalangeal joint	Lateral plantar nerve

Muscles on Sole of Foot—Fourth Layer (Deepest)

Plantar (3) and dorsal interossei (4)	Similar to the palmar and dorsal interossei of hand in locations, attachments, and actions; however, the long axis of foot around which these muscles orient is the second digit, not the third	See palmar and dorsal interossei (Table 10.13)	See palmar and dorsal interossei (Table 10.13)	Lateral plantar nerve

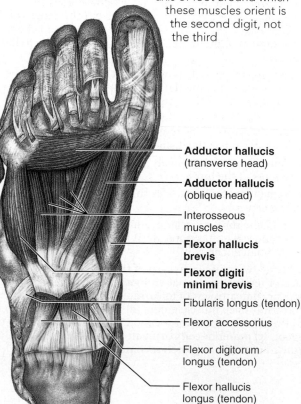

Adductor hallucis (transverse head)

Adductor hallucis (oblique head)

Interosseous muscles

Flexor hallucis brevis

Flexor digiti minimi brevis

Fibularis longus (tendon)

Flexor accessorius

Flexor digitorum longus (tendon)

Flexor hallucis longus (tendon)

(d) Third layer

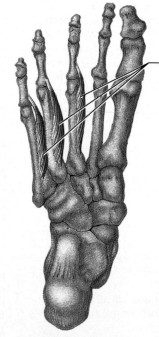

Plantar interossei

(e) Fourth layer: plantar interossei

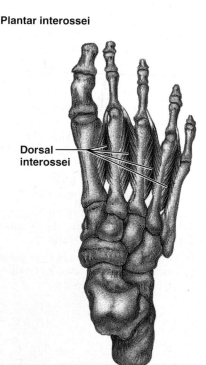

Dorsal interossei

(f) Fourth layer: dorsal interossei

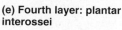

FIGURE 10.25 *(continued)* **Muscles of the right foot, plantar aspect.**

Review Questions

Multiple Choice/Matching

(Some questions have more than one correct answer. Select the best answer or answers from the choices given.)

1. A muscle that assists an agonist by causing a like movement or by stabilizing a joint over which an agonist acts is (a) an antagonist, (b) a prime mover, (c) a synergist, (d) an agonist.

2. A muscle in which the fibers are arranged at an angle to a central longitudinal tendon has a _____ arrangement. (a) circular, (b) longitudinal, (c) pennate, (d) parallel.

3. Match the muscle names in column B to the facial muscles described in column A.

Column A	Column B
____ (1) squints the eyes	(a) corrugator supercilii
____ (2) raises the eyebrows	(b) depressor anguli oris
____ (3) smiling muscle	(c) frontalis
____ (4) puckers the lips	(d) occipitalis
____ (5) pulls the scalp posteriorly	(e) orbicularis oculi
	(f) orbicularis oris
	(g) zygomaticus

4. The prime mover of inspiration is the (a) diaphragm, (b) internal intercostals, (c) external intercostals, (d) abdominal wall muscles.

5. The arm muscle that both flexes the elbow and supinates the forearm is the (a) brachialis, (b) brachioradialis, (c) biceps brachii, (d) triceps brachii.

6. The chewing muscles that protrude the mandible and produce side-to-side grinding movements are the (a) buccinators, (b) masseters, (c) temporalis, (d) pterygoids.

7. Muscles that depress the hyoid bone and larynx include all but the (a) sternohyoid, (b) omohyoid, (c) geniohyoid, (d) sternothyroid.

8. Intrinsic muscles of the back that promote extension of the spine (or head) include all but (a) splenius muscles, (b) semispinalis muscles, (c) scalene muscles, (d) erector spinae.

9. Several muscles act to move and/or stabilize the scapula. Which of the following are small rectangular muscles that square the shoulders as they act together to retract the scapula? (a) levator scapulae, (b) rhomboids, (c) serratus anterior, (d) trapezius.

10. The quadriceps include all but (a) vastus lateralis, (b) vastus intermedius, (c) vastus medialis, (d) biceps femoris, (e) rectus femoris.

11. A prime mover of hip flexion is the (a) rectus femoris, (b) iliopsoas, (c) vasti muscles, (d) gluteus maximus.

12. The prime mover of hip extension *against* resistance is the (a) gluteus maximus, (b) gluteus medius, (c) biceps femoris, (d) semimembranosus.

13. Muscles that cause plantar flexion include all but the (a) gastrocnemius, (b) soleus, (c) tibialis anterior, (d) tibialis posterior, (e) fibularis muscles.

14. In walking, which two lower limb muscles keep the forward-swinging foot from dragging on the ground? (a) pronator teres and popliteus, (b) flexor digitorum longus and popliteus, (c) adductor longus and abductor digiti minimi in foot, (d) gluteus medius and tibialis anterior.

15. List four criteria used in naming muscles, and name a specific muscle that illustrates each criterion. (Do not use the examples given in the text.)

16. Which of the following is a large, deep muscle that protracts the scapula during punching? (a) serratus anterior, (b) rhomboids, (c) levator scapulae, (d) subscapularis.

Short Answer Essay Questions

17. Name four criteria used in naming muscles, and provide an example (other than those used in the text) that illustrates each criterion.

18. Differentiate between the arrangement of elements (load, fulcrum, and effort) in first-, second-, and third-class levers.

19. What does it mean when we say that a lever operates at a mechanical disadvantage, and what benefits does such a lever system provide?

20. What muscles act to propel a food bolus down the length of the pharynx to the esophagus?

21. Name and describe the action of muscles used to shake your head no. To nod yes.

22. (a) Name the four muscle pairs that act in unison to compress the abdominal contents. (b) How does their arrangement (fiber direction) contribute to the strength of the abdominal wall? (c) Which of these muscles can effect lateral rotation of the spine? (d) Which can act alone to flex the spine?

23. List all (six) possible movements that can occur at the shoulder joint and name the prime mover(s) of each movement. Then name their antagonists.

24. (a) Name two forearm muscles that are powerful extensors and abductors of the wrist. (b) Name the sole forearm muscle that can flex the distal interphalangeal joints.

25. Name the muscles usually grouped together as the lateral rotators of the hip.

26. Name three thigh muscles that help you keep your seat astride a horse.

27. (a) Name three muscles or muscle groups used as sites for intramuscular injections. (b) Which of these is used most often in infants, and why?

28. Name two muscles in each of the following compartments or regions: (a) thenar eminence (ball of thumb), (b) posterior compartment of forearm, (c) anterior compartment of forearm—deep muscle group, (d) anterior muscle group in the arm, (e) muscles of mastication, (f) third muscle layer of the foot, (g) posterior compartment of leg, (h) medial compartment of thigh, (i) posterior compartment of thigh.

11

FUNDAMENTALS OF THE NERVOUS SYSTEM AND NERVOUS TISSUE

1. List the basic functions of the nervous system.

Organization of the Nervous System (pp. 342–343)

2. Explain the structural and functional divisions of the nervous system.

Histology of Nervous Tissue (pp. 343–351)

3. List the types of neuroglia and cite their functions.

4. Define neuron, describe its important structural components, and relate each to a functional role.

5. Differentiate between a nerve and a tract, and between a nucleus and a ganglion.

6. Explain the importance of the myelin sheath and describe how it is formed in the central and peripheral nervous systems.

7. Classify neurons structurally and functionally.

Neurophysiology (pp. 351–375)

8. Define resting membrane potential and describe its electrochemical basis.

9. Compare and contrast graded and action potentials.

10. Explain how action potentials are generated and propagated along neurons.

11. Define absolute and relative refractory periods.

12. Define saltatory conduction and contrast it to conduction along unmyelinated fibers.

13. Define synapse. Distinguish between electrical and chemical synapses structurally and in their mechanisms of information transmission.

14. Distinguish between excitatory and inhibitory postsynaptic potentials.

15. Describe how synaptic events are integrated and modified.

16. Define neurotransmitter and name several classes of neurotransmitters.

Basic Concepts of Neural Integration (pp. 375–377)

17. Describe common patterns of neuronal organization and processing.

18. Distinguish between serial and parallel processing.

You are driving down the freeway, and a horn blares to your right. You immediately swerve to your left. Charlie leaves a note on the kitchen table: "See you later. Have the stuff ready at 6." You know the "stuff" is chili with taco chips. You are dozing but you awaken instantly as your infant son makes a soft cry. What do these three events have in common? They are all everyday examples of the functioning of your nervous system, which has your body cells humming with activity nearly all the time.

The **nervous system** is the master controlling and communicating system of the body. Every thought, action, and emotion reflects its activity. Its cells communicate by electrical signals, which are rapid and specific, and usually cause almost immediate responses.

The nervous system has three overlapping functions (Figure 11.1): (1) It uses its millions of sensory receptors to monitor changes occurring both inside and outside the body. The gathered information is called **sensory input.** (2) It processes and interprets sensory input and decides what should be done at each moment—a process called **integration.** (3) It causes a *response*, called **motor output**, by activating *effector organs*. For example, when you are driving and see a red light ahead (sensory input), your nervous system integrates this information (red light means "stop"), and your foot goes for the brake (motor output).

This chapter begins with a brief overview of the organization of the nervous system. It then focuses on the functional anatomy of nervous tissue, especially that of nerve cells, or *neurons*, which are the key to neural communication.

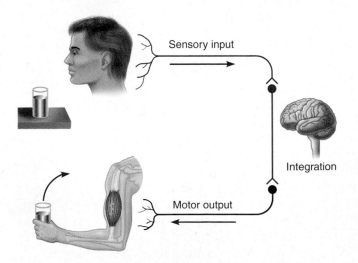

FIGURE 11.1 **The nervous system's functions.**

Organization of the Nervous System

We have only one highly integrated nervous system. However, for convenience, it can be divided into two principal parts (Figure 11.2). The **central nervous system (CNS)** consists of the *brain* and *spinal cord*, which occupy the dorsal body cavity. The CNS is the integrating and command center of the nervous system. It interprets sensory input and dictates motor responses based on past experience, reflexes, and current conditions. The **peripheral nervous system (PNS)**, the part of the nervous system *outside* the CNS, consists mainly of the nerves (bundles of axons) that extend from the brain and spinal cord. *Spinal nerves* carry impulses to and from the spinal cord; *cranial nerves* carry impulses to and from the brain. These peripheral nerves serve as the communication lines that link all parts of the body to the CNS.

The PNS has two functional subdivisions (see Figure 11.2). The **sensory,** or **afferent, division** (af′er-ent; "carrying toward") consists of nerve fibers that convey impulses *to* the central nervous system from sensory receptors located throughout the body. Sensory fibers conveying impulses from the skin, skeletal muscles, and joints are called *somatic afferent fibers* (*soma* = body), and those transmitting impulses from the visceral organs (organs within the ventral body cavity) are called *visceral afferent fibers*. The sensory division keeps the CNS constantly informed of events going on both inside and outside the body.

The **motor,** or **efferent, division** (ef′er-ent; "carrying away") of the PNS transmits impulses *from* the CNS to effector organs, which are the muscles and glands. These impulses activate muscles to contract and glands to secrete; that is, they *effect* (bring about) a motor response.

The motor division also has two main parts:

1. The **somatic nervous system** is composed of somatic motor nerve fibers (axons) that conduct impulses from the CNS to skeletal muscles. It is often referred to as the **voluntary nervous system** because it allows us to consciously control our skeletal muscles.

2. The **autonomic nervous system (ANS)** consists of visceral motor nerve fibers that regulate the activity of smooth muscles, cardiac muscles, and glands. *Autonomic* means "a law unto itself," and because we generally cannot control such activities as the pumping of our heart or the movement of food through our digestive tract, the ANS is also referred to as the **involuntary nervous system.** As indicated in Figure 11.2 and described in Chapter 14, the ANS has two functional subdivisions, the **sympathetic**

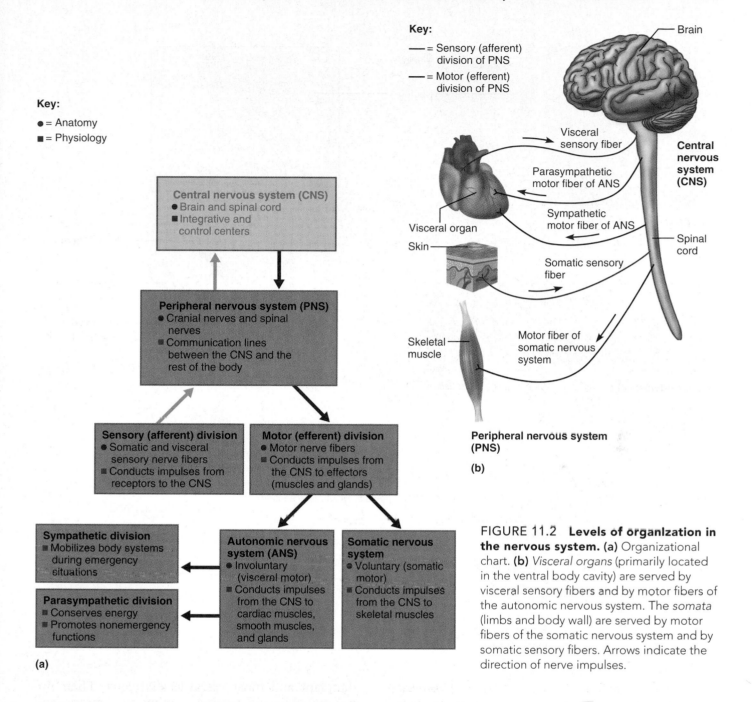

Key:
- ● = Anatomy
- ■ = Physiology

Key:
- ——= Sensory (afferent) division of PNS
- ——= Motor (efferent) division of PNS

Central nervous system (CNS)
- ● Brain and spinal cord
- ■ Integrative and control centers

Peripheral nervous system (PNS)
- ● Cranial nerves and spinal nerves
- ■ Communication lines between the CNS and the rest of the body

Sensory (afferent) division
- ● Somatic and visceral sensory nerve fibers
- ■ Conducts impulses from receptors to the CNS

Motor (efferent) division
- ● Motor nerve fibers
- ■ Conducts impulses from the CNS to effectors (muscles and glands)

Sympathetic division
- ■ Mobilizes body systems during emergency situations

Parasympathetic division
- ■ Conserves energy
- ■ Promotes nonemergency functions

Autonomic nervous system (ANS)
- ● Involuntary (visceral motor)
- ■ Conducts impulses from the CNS to cardiac muscles, smooth muscles, and glands

Somatic nervous system
- ● Voluntary (somatic motor)
- ■ Conducts impulses from the CNS to skeletal muscles

(a)

Brain

Visceral sensory fiber

Parasympathetic motor fiber of ANS

Central nervous system (CNS)

Sympathetic motor fiber of ANS

Visceral organ

Skin

Spinal cord

Somatic sensory fiber

Skeletal muscle

Motor fiber of somatic nervous system

Peripheral nervous system (PNS)

(b)

FIGURE 11.2 **Levels of organization in the nervous system. (a)** Organizational chart. **(b)** *Visceral organs* (primarily located in the ventral body cavity) are served by visceral sensory fibers and by motor fibers of the autonomic nervous system. The *somata* (limbs and body wall) are served by motor fibers of the somatic nervous system and by somatic sensory fibers. Arrows indicate the direction of nerve impulses.

and the **parasympathetic,** which typically work in opposition to each other—what one subdivision stimulates, the other inhibits.

Histology of Nervous Tissue

The nervous system consists mostly of nervous tissue, which is highly cellular. For example, less than 20% of the CNS is extracellular space, which means that the cells are densely packed and tightly intertwined. Although it is very complex, nervous tissue is made up of just two principal types of cells: (1) *supporting cells,* smaller cells that surround and

wrap the more delicate neurons, and (2) *neurons,* the excitable nerve cells that transmit electrical signals.

Neuroglia

Neurons associate closely with much smaller cells called **neuroglia** (nu-rog′le-ah; "nerve glue") or simply **glial cells** (gle′al). There are six types of neuroglia—four in the CNS and two in the PNS (Figure 11.3). Each type has a unique function, but in general, these cells provide a supportive scaffolding for neurons. Some segregate and insulate neurons so that electrical activities of adjacent neurons don't

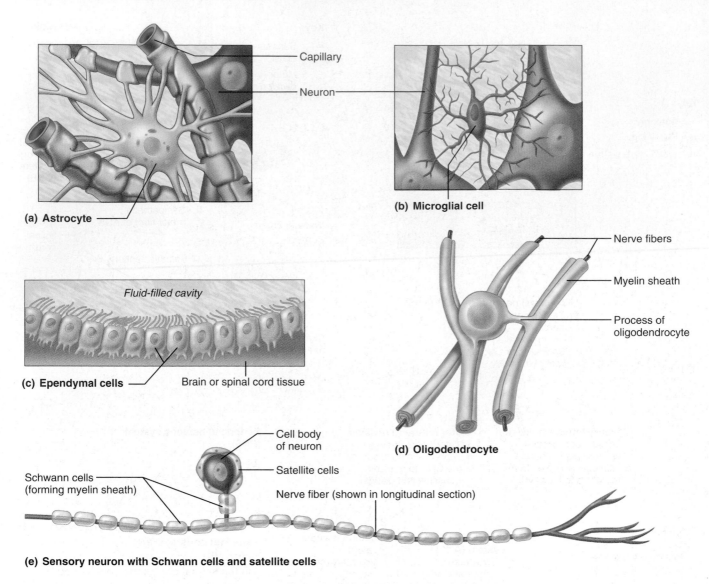

(a) **Astrocyte**

Capillary

Neuron

(b) **Microglial cell**

Fluid-filled cavity

(c) **Ependymal cells**   Brain or spinal cord tissue

Nerve fibers

Myelin sheath

Process of oligodendrocyte

(d) **Oligodendrocyte**

Cell body of neuron

Satellite cells

Schwann cells (forming myelin sheath)

Nerve fiber (shown in longitudinal section)

(e) **Sensory neuron with Schwann cells and satellite cells**

FIGURE 11.3   **Neuroglia. (a–d)** Types of neuroglia found in the CNS. Notice in (d) that it is the processes of the oligodendrocytes that form the myelin sheaths around CNS nerve fibers. **(e)** The relationships of Schwann cells (myelinating cells) and satellite cells to a sensory neuron (nerve cell) in the peripheral nervous system.

interfere with each other. Others produce chemicals that guide young neurons to the proper connections, and promote neuron health and growth.

Neuroglia in the CNS

Neuroglia in the CNS include *astrocytes, microglia, ependymal cells,* and *oligodendrocytes.* Like neurons, most glial cells have branching processes (extensions) and a central cell body (Figure 11.3a–d). Neuroglia can be distinguished, however, by their much smaller size and by their darker-staining nuclei. They outnumber neurons in the CNS by about 10 to 1, and make up about half the mass of the brain.

Shaped like delicate branching sea anemones, **astrocytes** (as′tro-sītz; "star cells") are the most

abundant and most versatile glial cells. Their numerous radiating processes cling to neurons and their synaptic endings, and cover nearby capillaries, supporting and bracing the neurons and anchoring them to their nutrient supply lines, the blood capillaries (Figure 11.3a). Astrocytes have a role in making exchanges between capillaries and neurons (they take up glucose from the bloodstream and deliver it to neurons in the form of lactic acid), in guiding the migration of young neurons, in synapse formation, and in helping to determine capillary permeability. They also control the chemical environment around neurons, where their most important job is "mopping up" leaked potassium ions and recapturing (and recycling) released neurotransmitters. Furthermore, astrocytes (connected together by gap junctions) have been shown to signal one another (and perhaps neurons) via slow-paced intracellular calcium pulses. According to recent research, these calcium pulses elicit surges of Ca^{2+} in adjacent neurons and influence the neurons' electrical signaling.

Microglia (mi-kro′gle-ah) are small ovoid cells with relatively long "thorny" processes (Figure 11.3b). Their processes touch nearby neurons, monitoring their health, and when they sense that certain neurons are injured or in other trouble, the microglia migrate toward them. Where invading microorganisms or dead neurons are present, the microglia transform into a special type of macrophage that phagocytizes the microorganisms or neuronal debris. This protective role of the microglia is important because cells of the immune system are denied access to the CNS.

Ependymal cells (ĕ-pen′dĭ-mul; "wrapping garment") range in shape from squamous to columnar, and many are ciliated. They line the central cavities of the brain and the spinal cord, where they form a fairly permeable barrier between the cerebrospinal fluid that fills those cavities and the tissue fluid bathing the cells of the CNS. The beating of their cilia helps to circulate the cerebrospinal fluid that cushions the brain and spinal cord (Figure 11.3c).

Though they also branch, the **oligodendrocytes** (ol″ĭ-go-den′dro-sīts) have fewer processes (*oligo* = few; *dendr* = branch) than astrocytes. Oligodendrocytes line up along the thicker neuron fibers in the CNS and wrap their processes tightly around the fibers, producing insulating coverings called *myelin sheaths* (Figure 11.3d).

Neuroglia in the PNS

The two kinds of PNS neuroglia—*satellite cells* and *Schwann cells*—differ mainly in location. **Satellite cells** surround neuron cell bodies within ganglia (Figure 11.3e), but their function is still largely unknown. Their name comes from a fancied resemblance to the moons (satellites) around a planet.

Schwann cells (also called *neurolemmocytes*) surround and form myelin sheaths around the larger nerve fibers in the peripheral nervous system (Figures 11.3e and 11.4b). Hence, they are functionally similar to oligodendrocytes. (The formation of myelin sheaths is described later in this chapter.) Schwann cells are vital to regeneration of peripheral nerve fibers.

Neurons

The billions of **neurons,** also called **nerve cells,** are the structural units of the nervous system. They are highly specialized cells that conduct messages in the form of nerve impulses from one part of the body to another. Besides their ability to conduct nerve impulses, neurons have some other special characteristics:

1. They have *extreme longevity*. Given good nutrition, neurons can function optimally for a lifetime (over 100 years).

2. They are largely *amitotic*. As neurons assume their roles as communicating links of the nervous system, they lose their ability to divide. We pay a high price for this neuron feature because they cannot replace themselves if destroyed. There *are* exceptions to this rule. For example, olfactory epithelium and some hippocampal regions contain stem cells that can produce new neurons throughout life. (The hippocampus is a brain region involved in memory.)

3. They have an exceptionally *high metabolic rate* and require continuous and abundant supplies of oxygen and glucose. Neurons cannot survive for more than a few minutes without oxygen.

Neurons are typically large, complex cells. Although they vary in structure, they all have a *cell body* from which one or more slender *processes* project (Figure 11.4). The plasma membrane of neurons is the site of electrical signaling, and it plays a crucial role in cell-to-cell interactions that occur during development.

Cell Body

The **neuron cell body** consists of a transparent, spherical nucleus with a conspicuous nucleolus surrounded by cytoplasm. Also called the **perikaryon** (*peri* = around, *kary* = nucleus) or **soma,** the cell body ranges in diameter from 5 to 140 μm. The cell body is the major *biosynthetic center* of a neuron. Except for centrioles, it contains the usual organelles. (Centrioles play an important role in forming the mitotic spindle. Their apparent absence reflects the amitotic nature of most neurons.)

The neuron cell body's protein- and membrane-making machinery, consisting of clustered free ribosomes and rough endoplasmic reticulum (ER), is probably the most active and best developed in the body. This rough ER, referred to as **Nissl bodies** (nis′l) or **chromatophilic substance** (chromatophilic = "color loving"), stains darkly with basic dyes. The Golgi apparatus is also well developed and forms an arc or a complete circle around the nucleus. Mitochondria are scattered among the other organelles. Microtubules and **neurofibrils,** bundles of intermediate filaments (*neurofilaments*), which are important in maintaining cell shape and integrity, are seen throughout the cell body. The cell body of some neurons also contains pigment inclusions. For example, some contain a black melanin, a red iron-containing pigment, or a golden-brown pigment

Neurons with the longest axons tend to have large cell bodies, while those with short axons usually have small cell bodies. Can you think of an explanation for this relationship?

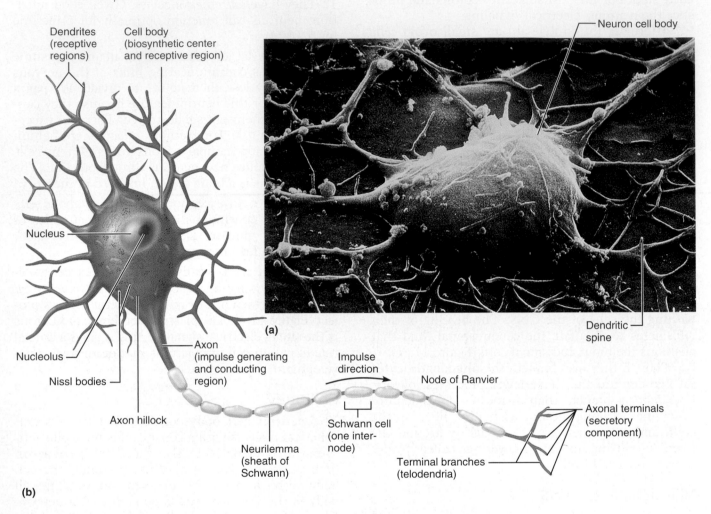

FIGURE 11.4   Structure of a motor neuron. (a) Scanning electron micrograph showing the cell body and dendrites with obvious dendritic spines (6000×).
(b) Diagrammatic view.

called *lipofuscin* (lip″o-fu′sin). Lipofuscin, a harmless by-product of lysosomal activity, is sometimes called the "aging pigment" because it accumulates in neurons of elderly individuals.

The cell body is the focal point for the outgrowth of neuron processes during embryonic development. In most neurons, the plasma membrane of the cell body also acts as *part of the receptive surface* (Table 11.1, pp. 350–351) that receives information from other neurons.

Most neuron cell bodies are located in the CNS, where they are protected by the bones of the skull and vertebral column. Clusters of cell bodies in the CNS are called **nuclei,** whereas those that lie along the nerves in the PNS are called **ganglia** (gang′gle-ah; *ganglion* = knot on a string, swelling).

Processes

Armlike **processes** extend from the cell body of all neurons. The brain and spinal cord (CNS) contain both neuron cell bodies and their processes. The PNS, for the most part, consists chiefly of neuron processes. Bundles of neuron processes are called **tracts** in the CNS and **nerves** in the PNS.

The two types of neuron processes, *dendrites* and *axons* (ak′sonz), differ from each other in the structure and function of their plasma membranes. The convention is to describe these processes using

The longer the axon, the larger the cell body needed to service it. ■

a motor neuron as an example of a typical neuron. We shall follow this practice, but keep in mind that many sensory neurons and some tiny CNS neurons differ from the "typical" pattern we present here.

Dendrites   **Dendrites** of motor neurons are short, tapering, diffusely branching extensions. Typically, motor neurons have hundreds of lacy dendrites clustering close to the cell body. Virtually all organelles present in the cell body also occur in dendrites. Dendrites are the main **receptive** or **input regions** (see Table 11.1). They provide an enormous surface area for receiving signals from other neurons. In many brain areas, the finer dendrites are highly specialized for information collection. They bristle with thorny appendages having bulbous or spiky ends called *dendritic spines* (Figure 11.4a), which represent points of close contact (synapses) with other neurons. Dendrites convey incoming messages *toward* the cell body. These electrical signals are *not* nerve impulses (action potentials) but are short-distance signals called *graded potentials*, as described shortly.

The Axon   Each neuron has a single **axon** (*axo* = axis, axle). The initial region of the axon arises from a cone-shaped area of the cell body called the **axon hillock** ("little hill") and then narrows to form a slender process that is uniform in diameter for the rest of its length (Figure 11.4). In some neurons, the axon is very short or absent; in others it is long and accounts for nearly the entire length of the neuron. For example, axons of the motor neurons controlling the skeletal muscles of your great toe extend from the lumbar region of your spine to your foot, a distance of a meter or more (3–4 feet), making them the longest cells in the body. Any long axon is called a **nerve fiber.**

Each neuron has only one axon, but axons may have occasional branches along their length. These rare branches, called **axon collaterals,** extend from the axon at more or less right angles. Whether an axon is undivided or has collaterals, it usually branches profusely at its end (terminus): 10,000 or more **terminal branches,** or **telodendria,** per neuron is not unusual. The knoblike distal endings of the terminal branches are variously called **axonal terminals, synaptic knobs,** or **boutons** (boo-tonz; "buttons"). Take your pick!

Functionally, the axon is the **conducting component** of the neuron (Table 11.1). It *generates nerve impulses* and *transmits them,* typically away from the cell body. In motor neurons, the nerve impulse is generated at the junction of the axon hillock and axon (which is therefore called the *trigger zone*) and conducted along the axon to the axonal terminals, which are the **secretory component** of the neuron.

When the impulse reaches the axonal terminals, it causes *neurotransmitters,* signaling chemicals stored in vesicles there, to be released into the extracellular space. The neurotransmitters either excite or inhibit neurons (or effector cells) with which the axon is in close contact. Because each neuron both receives signals from and sends signals to scores of other neurons, it carries on "conversations" with many different neurons at the same time.

An axon contains the same organelles found in the dendrites and cell body with two important exceptions—it lacks Nissl bodies and a Golgi apparatus, the structures involved with protein synthesis and packaging. Consequently, an axon depends (1) on its cell body to renew the necessary proteins and membrane components, and (2) on efficient transport mechanisms to distribute them. Axons quickly decay if cut or severely damaged.

Because axons are often very long, moving molecules along their length might appear to be a problem. However, through the cooperative effort of several types of cytoskeletal elements (microtubules, actin filaments, and so on), substances travel continuously along the axon both away from and toward the cell body. Movement toward the axonal terminals is *anterograde movement,* and that in the opposite direction is *retrograde movement.*

Substances moved in the anterograde direction include mitochondria, cytoskeletal elements, membrane components used to renew the axon plasma membrane, or **axolemma** (ak"so-lem'ah), and enzymes needed for synthesis of certain neurotransmitters. (Some neurotransmitters are synthesized in the cell body and then transported to the axonal terminals.)

Substances transported through the axon in the retrograde direction are mostly organelles being returned to the cell body for degradation or recycling. Retrograde transport is also an important means of intracellular communication for "advising" the cell body of conditions at the axonal terminals, and for delivering to the cell body vesicles containing signal molecules (like nerve growth factor, which activates certain nuclear genes promoting growth).

There are two or three transport mechanisms operating in axons, the fastest of which is ATP dependent, bidirectional, and uses "motor" proteins—ATPases such as kinesin, dynein, and perhaps others. These proteins propel membranous particles along the microtubules like trains along tracks.

HOMEOSTATIC IMBALANCE

Certain viruses and bacterial toxins that damage neural tissues use retrograde axonal transport to reach the cell body. This transport mechanism has

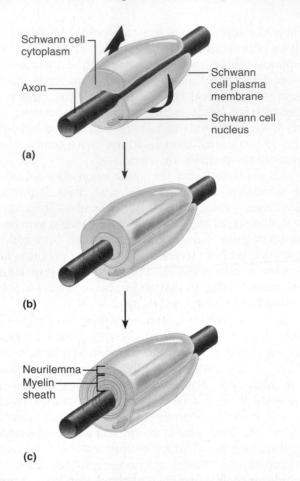

(a)

(b)

(c)

Schwann cell cytoplasm

Axon

Schwann cell plasma membrane

Schwann cell nucleus

Neurilemma

Myelin sheath

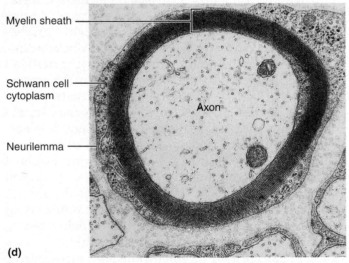

Myelin sheath

Schwann cell cytoplasm

Axon

Neurilemma

(d)

FIGURE 11.5   **Relationship of Schwann cells to axons in the PNS. (a–c)** Myelination of a nerve fiber (axon). As illustrated, a Schwann cell envelops an axon in a trough. It then begins to rotate around the axon, enveloping the axon loosely in successive layers of its plasma membrane. Eventually, the Schwann cell cytoplasm is forced from between the membranes and comes to lie peripherally just beneath the exposed portion of the Schwann cell plasma membrane. The tight membrane wrappings surrounding the axon form the myelin sheath; the area of Schwann cell cytoplasm and its exposed membrane is referred to as the neurilemma. **(d)** Electron micrograph of a myelinated axon, cross-sectional view (20,000×).

been demonstrated for polio, rabies, and herpes simplex viruses and for tetanus toxin. Its use as a tool to treat genetic diseases by introducing viruses containing "corrected" genes is under investigation. ●

Myelin Sheath and Neurilemma   Many nerve fibers, particularly those that are long or large in diameter, are covered with a whitish, fatty (protein-lipoid), segmented **myelin sheath** (mi'ĕ-lin). Myelin protects and electrically insulates fibers from one another, and it increases the speed of transmission of nerve impulses. **Myelinated fibers** (axons bearing a myelin sheath) conduct nerve impulses rapidly, whereas **unmyelinated fibers** conduct impulses quite slowly. Note that myelin sheaths are associated only with axons. Dendrites are *always* unmyelinated.

Myelin sheaths in the PNS are formed by Schwann cells, which indent to receive an axon and then wrap themselves around it in a jelly roll fashion (Figure 11.5). Initially the wrapping is loose, but the Schwann cell cytoplasm is gradually squeezed from between the membrane layers. When the wrapping process is complete, many concentric layers of Schwann cell plasma membrane enclose the axon, much like gauze wrapped around an injured finger. This tight coil of wrapped membranes *is* the myelin sheath, and its thickness depends on the number of spirals. Plasma membranes of myelinating cells contain much less protein than the plasma membranes of most body cells. Channel and carrier proteins are notably absent, a characteristic that makes myelin sheaths exceptionally good electrical insulators. Another unique characteristic of these membranes is the presence of specific protein molecules that interlock to form a sort of molecular Velcro between adjacent myelin membranes.

The nucleus and most of the cytoplasm of the Schwann cell end up as a bulge just external to the myelin sheath. This portion of the Schwann cell, which includes the exposed part of its plasma membrane, is called the **neurilemma** ("neuron husk"). Adjacent Schwann cells along an axon do not touch one another, so there are gaps in the sheath. These gaps, called **nodes of Ranvier** (ran'vēr) or **neurofibril nodes,** occur at regular intervals (about 1 mm apart) along the myelinated axon. It is at these nodes that axon collaterals can emerge from the axon.

Sometimes Schwann cells surround peripheral nerve fibers but the coiling process does not occur. In such instances, a single Schwann cell can partially enclose 15 or more axons, each of which occupies a separate recess in the Schwann cell surface. Nerve fibers associated with Schwann cells in this manner are said to be *unmyelinated* and are typically thin fibers.

Both myelinated and unmyelinated axons are also found in the central nervous system. However, it is oligodendrocytes that form CNS myelin sheaths (Figure 11.3d). In contrast to Schwann cells, each of which forms only one segment (internode) of a myelin sheath, oligodendrocytes have multiple flat processes that can coil around as many as 60 axons at the same time. Nodes of Ranvier are present, but are more widely spaced than in the PNS. CNS myelin sheaths lack a neurilemma because cell extensions are doing the coiling and the squeezed-out cytoplasm is forced not peripherally but back toward the centrally located nucleus. As in the PNS, the smallest diameter axons are unmyelinated. These unmyelinated axons are covered by the long extensions of adjacent glial cells.

Regions of the brain and spinal cord containing dense collections of myelinated fibers are referred to as **white matter** and are primarily fiber tracts. **Gray matter** contains mostly nerve cell bodies and unmyelinated fibers.

Classification of Neurons

Neurons are classified both structurally and functionally. We describe both classifications here but use the functional classification in most discussions.

Structural Classification   Neurons are grouped structurally according to the number of processes extending from their cell body. Three major neuron groups make up this classification: multipolar (*polar* = end, pole), bipolar, and unipolar neurons (Table 11.1).

Multipolar neurons have three or more processes. They are the most common neuron type in humans (more than 99% of neurons belong to this class) and the major neuron type in the CNS.

Bipolar neurons have two processes—an axon and a dendrite—that extend from opposite sides of the cell body. These rare neurons are found only in some of the special sense organs, where they act as receptor cells. Examples include some neurons in the retina of the eye and in the olfactory mucosa.

Unipolar neurons have a single short process that emerges from the cell body and divides T-like into proximal and distal branches. The more distal process, which is often associated with a sensory receptor, is the **peripheral process,** whereas that entering the CNS is the **central process** (Table 11.1). Unipolar neurons are more accurately called **pseudounipolar neurons** (*pseudo* = false) because they originate as bipolar neurons. Then, during early embryonic development, the two processes converge and partially fuse to form the short single process that issues from the cell body. Unipolar neurons are found chiefly in ganglia in the PNS, where they function as sensory neurons.

The fact that the fused peripheral and central processes of unipolar neurons are continuous and function as a single fiber might make you wonder whether they are axons or dendrites. The central process is definitely an axon because it conducts impulses away from the cell body (one definition of axon). However, the peripheral process is perplexing. Facts that favor classifying it as an axon are: (1) It generates and conducts an impulse (functional definition of axon); (2) when large, it is heavily myelinated; and (3) it has a uniform diameter and is indistinguishable microscopically from an axon. However, the older definition of a dendrite as a process that transmits impulses *toward* the cell body interferes with that conclusion. So which is it? In this book, we have chosen to emphasize the newer definition of an axon as generating and transmitting an impulse. Hence, *for unipolar neurons*, we will refer to the combined length of the peripheral and central process as an axon. In place of "dendrites," unipolar neurons have *receptive endings* (sensory terminals) at the end of the peripheral process.

Functional Classification   This scheme groups neurons according to the direction in which the nerve impulse travels relative to the central nervous system. Based on this criterion, there are sensory neurons, motor neurons, and interneurons (Table 11.1).

Sensory, or **afferent, neurons** transmit impulses from sensory receptors in the skin or internal organs *toward* or *into* the central nervous system. Except for the bipolar neurons found in some special sense organs, virtually all sensory neurons are unipolar, and their cell bodies are located in sensory ganglia *outside* the CNS. Only the most distal parts of these unipolar neurons act as impulse receptor sites and the peripheral processes are often very long. For example, fibers carrying sensory impulses from the skin of your great toe travel for more than a meter before they reach their cell bodies in a ganglion close to the spinal cord.

Although the receptive endings of some sensory neurons are naked, in which case those terminals themselves function as sensory receptors, many bear receptors that include other cell types. The various types of general sensory receptor end organs, such as those of the skin, are described in Chapter 13. The special sensory receptors (of the ear, eye, etc.) are the topic of Chapter 13.

Motor, or **efferent, neurons** carry impulses *away from* the CNS to the effector organs (muscles and glands) of the body periphery. Motor neurons are multipolar, and except for some neurons of the autonomic nervous system, their cell bodies are located in the CNS.

Interneurons, or **association neurons,** lie between motor and sensory neurons in neural

TABLE 11.1   **Comparison of Structural Classes of Neurons**

Neuron Type		
Multipolar	Bipolar	Unipolar (Pseudounipolar)

Structural Class: Neuron Type According to the Number of Processes Extending from the Cell Body

Many processes extend from the cell body; all dendrites except for a single axon.	Two processes extend from the cell body: one is a fused dendrite, the other is an axon.	One process extends from the cell body and forms central and peripheral processes, which together comprise an axon.

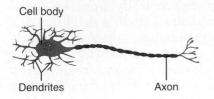

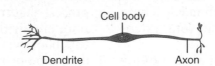

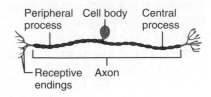

Relationship of Anatomy to the Three Functional Regions

▪ = Receptive region (receives stimulus). Plasma membrane exhibits chemically gated ion channels.	▪ = Conducting region (generates/ transmits action potential). Plasma membrane exhibits voltage-gated Na⁺ and K⁺ channels.	▪ = Secretory region (axonal terminals release neurotransmitters). Plasma membrane exhibits voltage-gated Ca⁺ channels.

▪ = Receptive region (receives stimulus). Plasma membrane exhibits chemically gated ion channels.

▪ = Conducting region (generates/transmits action potential). Plasma membrane exhibits voltage-gated Na^+ and K^+ channels.

▪ = Secretory region (axonal terminals release neurotransmitters). Plasma membrane exhibits voltage-gated Ca^+ channels.

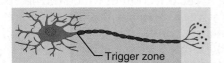

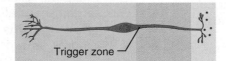

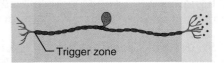

(Many bipolar neurons do not generate action potentials and, in those that do, the location of the trigger zone is not universal.)

Relative Abundance and Location in Human Body

Most abundant in body. Major neuron type in the CNS.	Rare. Are found in some special sensory organs (olfactory mucosa, eye).	Found mainly in the PNS. Common only in dorsal root ganglia of the spinal cord and sensory ganglia of cranial nerves.

Structural Variations

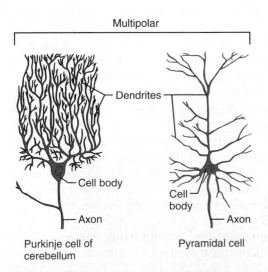

Purkinje cell of cerebellum   Pyramidal cell

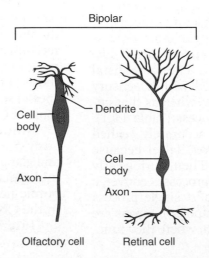

Olfactory cell   Retinal cell

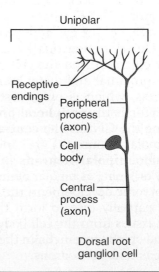

Dorsal root ganglion cell

TABLE 11.1 *(continued)*

Neuron Type		
Multipolar	*Bipolar*	*Unipolar (Pseudounipolar)*

Functional Class: Neuron Type According to Direction of Impulse Conduction

1. Some multipolar neurons are **motor neurons** that conduct impulses along the efferent pathways from the CNS to an effector (muscle/gland). **2.** Some multipolar neurons are higher level sensory neurons that convey sensory input from the first-order sensory neurons to higher CNS levels. **3.** Most multipolar neurons are **interneurons (association neurons)** that conduct impulses within the CNS; may be one of a chain of CNS neurons, or single neuron connecting sensory and motor neurons.	Essentially all bipolar neurons are **sensory neurons** that are located in some special sense organs. For example, bipolar cells of the retina are involved with the transmission of visual inputs from the eye to the brain (via an intermediate chain of neurons).	Most unipolar neurons are **sensory neurons** that conduct impulses along afferent pathways to the CNS for interpretation. (These sensory neurons are primary or first-order sensory neurons.)

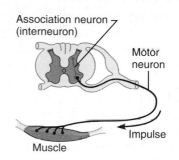

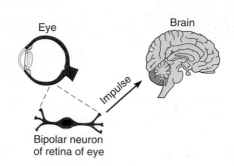

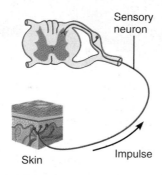

pathways and shuttle signals through CNS pathways where integration occurs. Most interneurons are confined within the CNS. They make up over 99% of the neurons of the body, including most of those in the CNS. Almost all interneurons are multipolar, but there is considerable diversity in both size and fiber-branching patterns. The Purkinje and pyramidal cells illustrated as structural variations in Table 11.1 are just two examples of their variety.

Neurophysiology

Neurons are highly *irritable* (responsive to stimuli). When a neuron is adequately stimulated, an electrical impulse is generated and conducted along the length of its axon. This response, called the *action potential (nerve impulse)*, is always the same, regardless of the source or type of stimulus, and it underlies virtually all functional activities of the nervous system.

In this section, we will consider how neurons become excited or inhibited and how they communicate with other cells. First, however, we need to ex-

plore some basic principles of electricity and revisit the resting membrane potential.

Basic Principles of Electricity

The human body is electrically neutral; it has the same number of positive and negative charges. However, there are areas where one type of charge predominates, making such regions positively or negatively charged. Because opposite charges attract each other, energy must be used (work must be done) to separate them. On the other hand, the coming together of opposite charges liberates energy that can be used to do work. Thus, situations in which there are separated electrical charges of opposite sign have potential energy.

Some Definitions: Voltage, Resistance, Current

The measure of potential energy generated by separated charge is called **voltage** and is measured in either *volts* or *millivolts* (1 mV = 0.001 V). Voltage is always measured between two points and is called

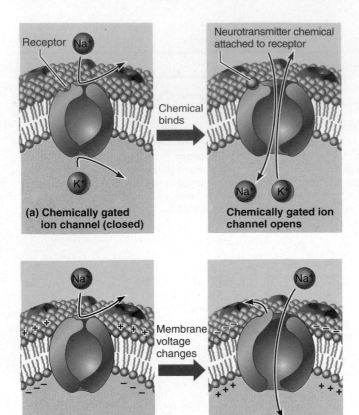

(a) Chemically gated ion channel (closed)

Receptor Na⁺

K⁺

Chemical binds

Neurotransmitter chemical attached to receptor

Chemically gated ion channel opens

Na⁺ K⁺

(b) Voltage-gated ion channel (closed)

Na⁺

Membrane voltage changes

Voltage-gated ion channel opens

Na⁺

FIGURE 11.6 Operation of gated channels. (a) Chemically (ligand) gated channels (a Na⁺–K⁺ channel in this example) open when the appropriate neurotransmitter binds to the receptor, allowing simultaneous movement of Na⁺ and K⁺ through the channel. **(b)** Voltage-gated channels open or close in response to changes in membrane voltage. In this example, a Na⁺ channel opens as the membrane interior changes from negative to positive.

the **potential difference** or simply the **potential** between the points. The greater the difference in charge between two points, the higher the voltage.

The flow of electrical charge from one point to another is called a **current,** and it can be used to do work—for example, to power a flashlight. The amount of charge that moves between the two points depends on two factors: voltage and resistance. **Resistance** is the hindrance to charge flow provided by substances through which the current must pass. Substances with high electrical resistance are called *insulators;* those with low resistance are called *conductors.*

The relationship between voltage, current, and resistance is given by **Ohm's law:**

$$\text{Current } (I) = \frac{\text{voltage } (V)}{\text{resistance } (R)}$$

which tells us that current (I) is directly proportional to voltage: The greater the voltage (potential difference), the greater the current. This relationship also tells us there is no net current flow between points that have the same potential, as you can see by inserting a value of 0 V into the equation. A third thing Ohm's law tells us is that current is inversely related to resistance: The greater the resistance, the smaller the current.

In the body, electrical currents reflect the flow of ions (rather than free electrons) across cellular membranes. (There are no free electrons "running around" in a living system.) As described in Chapter 3, there is a slight difference in the numbers of positive and negative ions on the two sides of cellular plasma membranes (there is a charge separation), so there is a potential across those membranes. The resistance to current flow is provided by the plasma membranes.

Role of Membrane Ion Channels

Plasma membranes are peppered with a variety of *ion channels* made up of membrane proteins. Some of these channels are *passive,* or *leakage, channels* that are always open. Others are *active,* or *gated, channels* (Figure 11.6). Gated channels have a molecular "gate," usually one or more protein molecules, that can change shape to open or close the channel in response to various signals. **Chemically gated,** or **ligand-gated, channels** open when the appropriate chemical (in this case a neurotransmitter) binds. **Voltage-gated channels** open and close in response to changes in the membrane potential. **Mechanically gated** channels open in response to physical deformation of the receptor (seen in sensory receptors for touch and pressure)—more on this later. Each type of channel is selective as to the type of ion (or ions) it allows to pass. For example, a potassium ion channel allows only potassium ions to pass.

When gated ion channels are open, ions diffuse quickly across the membrane following their electrochemical gradients, creating electrical currents and voltage changes across the membrane according to the rearranged Ohm's law equation:

$$\text{Voltage } (V) = \text{current } (I) \times \text{resistance } (R)$$

Ions move along *chemical gradients* when they diffuse passively from an area of their higher concentration to an area of lower concentration, and along *electrical gradients* when they move toward an area of opposite electrical charge. Together, electrical and chemical gradients constitute the **electrochemical gradient.** It is ion flows along electrochemical gradients that underlie all electrical phenomena in neurons.

The Resting Membrane Potential

The potential difference between two points is measured with a voltmeter (Figure 11.7). When one microelectrode of the voltmeter is inserted into the neuron and the other rests on the neuron's outside surface, a voltage across the membrane of approximately −70 mV is recorded. The minus sign indicates that the cytoplasmic side (inside) of the membrane is negatively charged relative to the outside. This potential difference in a resting neuron (V_r) is called the **resting membrane potential**, and the membrane is said to be **polarized**. The value of the resting membrane potential varies (from −40 mV to −90 mV) in different types of neurons.

The resting potential exists only across the membrane; that is, the bulk solutions inside and outside the cell are electrically neutral. As shown in Figure 11.8, the resting membrane potential is generated by differences in the ionic makeup of the intracellular and extracellular fluids and the differential permeability of the plasma membrane to those ions. The cell cytosol contains a lower concentration of Na^+ and a higher concentration of K^+ than the extracellular fluid. In the extracellular fluid, the positive charges of sodium and other cations are balanced chiefly by chloride ions (Cl^-). Negatively charged (anionic) proteins (A^-) help to balance the positive charges of intracellular cations (primarily K^+). Although there are many other solutes (glucose, urea, and other ions) in both fluids,

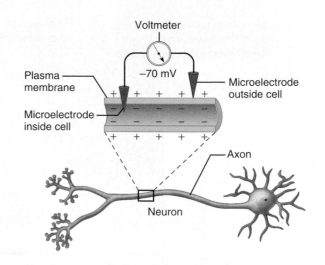

FIGURE 11.7 Measuring potential difference between two points in neurons. When one electrode of a voltmeter is placed on the external membrane and the other electrode is inserted just inside the membrane, a voltage (membrane potential) of approximately −70 mV (inside negative) is recorded.

potassium plays the most important role in generating the membrane potential.

At rest the membrane is impermeable to the large anionic cytoplasmic proteins, very slightly permeable to sodium, approximately 75 times more permeable to potassium than to sodium, and quite freely permeable to chloride ions. These resting

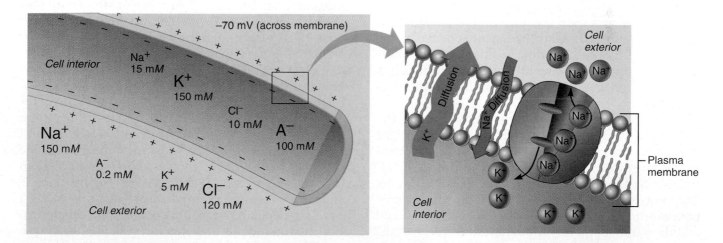

FIGURE 11.8 The basis of the resting membrane potential. Approximate ion concentrations of mammalian cells are indicated in millimoles per liter (mM). Steady diffusion of K^+ to the cell exterior through leakage channels is strongly promoted by its concentration gradient. Na^+ is strongly attracted to the cell interior by its concentration gradient, but it is far less able to cross the cell membrane because of the near lack of Na^+ leakage channels. The relative membrane permeability to Na^+ and K^+ is indicated by diffusion arrows of different thickness. The resulting net outward diffusion of positive charge (K^+) leads to a state of relative negativity on the inner membrane face. This membrane potential (−70 mV, inside negative) is maintained by the sodium-potassium pump, which transports three Na^+ out of the cell for each two K^+ transported back into the cell. The electrical gradient thus maintained resists K^+ efflux but enhances the drive for Na^+ entry.

? *Hyper means greater than or more than. When we say the membrane becomes hyperpolarized, what value has increased or become greater?*

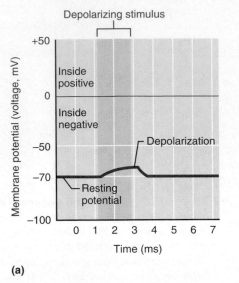

(a)

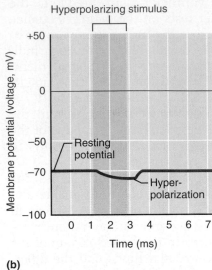

(b)

FIGURE 11.9 Depolarization and hyperpolarization of the membrane. The resting membrane potential is approximately −70 mV (inside negative) in neurons. Changes in this potential result in depolarization or hyperpolarization of the membrane. **(a)** In depolarization, the membrane potential moves toward 0 mV, the inside becoming less negative (more positive). **(b)** In hyperpolarization, the membrane potential increases, the inside becoming more negative.

permeabilities reflect the properties of the passive (leakage) ion channels in the membrane. Potassium ions diffuse out of the cell along their *concentration gradient* much more easily and quickly than sodium ions can enter the cell along theirs, and K^+ loss from the cell continues until the force of its concentration gradient is balanced exactly by the pull created by the negativity of the cell interior. At this point K^+ diffusion across the membrane into and out of the cell is equalized and the resting membrane potential is established.

Because some K^+ is always leaking out of the cell and some Na^+ is always leaking in, you might think that the concentration gradients would eventually "run down," resulting in equal concentrations of Na^+ and K^+ inside and outside the cell. This does not happen, because the ATP-driven sodium-potassium pump first ejects three Na^+ from the cell and then transports two K^+ back into the cell. Thus, the sodium-potassium pump stabilizes the resting membrane potential by maintaining the concentration gradients for sodium and potassium.

Membrane Potentials That Act as Signals

Neurons use changes in their membrane potential as communication signals for receiving, integrating,

and sending information. A change in membrane potential can be produced by (1) anything that changes membrane permeability to any ion or (2) anything that alters ion concentrations on the two sides of the membrane. Two types of signals are produced by a change in membrane potential: *graded potentials*, which signal over short distances, and *action potentials*, which are long-distance signals.

The terms *depolarization* and *hyperpolarization* are used in the following sections to describe membrane potential changes *relative to resting membrane potential*, so it is important to understand these terms clearly. **Depolarization** is a reduction in membrane potential: The inside of the membrane becomes *less negative* (moves closer to zero) than the resting potential (Figure 11.9a). For instance, a change in resting potential from −70 mV to −65 mV is a depolarization. By convention, depolarization also includes events in which the membrane potential reverses and moves above zero to become positive.

Hyperpolarization occurs when the membrane potential increases, becoming *more negative* than the resting potential. For example, a change from

■ *The negativity of the internal membrane surface.*

−70 mV to −75 mV is hyperpolarization (Figure 11.9b). As described shortly, depolarization increases the probability of producing nerve impulses, whereas hyperpolarization reduces this probability.

Graded Potentials

Graded potentials are short-lived, local changes in membrane potential that can be either depolarizations or hyperpolarizations. These changes cause current flows that decrease in magnitude with distance. Graded potentials are called "graded" because their magnitude varies directly with stimulus strength. The stronger the stimulus, the more the voltage changes and the farther the current flows.

Graded potentials are triggered by some change (a stimulus) in the neuron's environment that causes gated ion channels to open. Graded potentials are given different names, depending on where they occur and the functions they perform. When the receptor of a sensory neuron is excited by some form of energy (heat, light, or others), the resulting graded potential is called a *generator potential*. Generator potentials are considered in Chapter 13. When the stimulus is a neurotransmitter released by another neuron, the graded potential is called a *postsynaptic potential*, because the neurotransmitter is released into a fluid-filled gap called a synapse and influences the neuron beyond (post) the synapse.

Fluids inside and outside cells are fairly good conductors, and current, carried by ions, flows through these fluids whenever voltage changes occur. Let us assume that a small area of a neuron's plasma membrane has been depolarized by a stimulus (Figure 11.10a). Current will flow on both sides of the membrane between the depolarized (active) membrane area and the adjacent polarized (resting) areas. Positive ions migrate toward more negative areas (the direction of cation movement is the direction of current flow), and negative ions simultaneously move toward more positive areas (Figure 11.10b). Thus, inside the cell, positive ions (mostly K$^+$) move away from the depolarized area and accumulate on the neighboring membrane areas, where they displace negative ions. Meanwhile, positive ions on the outer membrane face are moving toward the region of reversed membrane polarity (the depolarized region), which is momentarily less positive. As these positive ions move, their "places" on the membrane become occupied by negative ions (such as Cl$^-$ and HCO^{3-}), sort of like ionic musical chairs. Thus, at regions abutting the depolarized region, the inside becomes less negative and the outside becomes less positive; that is, the neighboring membrane is, in turn, depolarized.

To simplify the explanation, virtually all diagrams that illustrate local currents give the

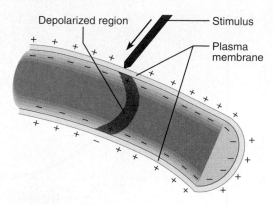

(a) Depolarization

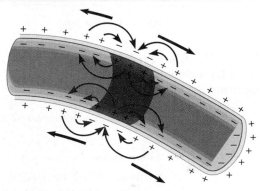

(b) Spread of depolarization

FIGURE 11.10   **The mechanism of a graded potential.**
(a) A small patch of the membrane has become depolarized, resulting in a change in polarity at that point. **(b)** As positive ions flow toward negative areas (and negative ions flow toward adjacent more positive areas), local currents are created that depolarize adjacent membrane areas and allow the wave of depolarization to spread.

impression that the circuit is completed by ions passing into and out of the cell through the membrane. This is *not* the case. Only the inward current across the membrane is caused by the flow of ions through gated channels. The outward current, the so-called *capacitance current*, reflects changes in the charge distribution as the ions migrate *along* the two membrane faces as described above. The capacitance current reflects the fact that the fatty membrane interior is a poor conductor of current. That is, it is a *capacitor* that *temporarily* stores the charge, forcing the ions of opposite charge to accumulate opposite one another on either side of the membrane.

As just explained, the flow of current to adjacent membrane areas changes the membrane potential there as well. However, the plasma membrane is permeable like a leaky water hose, and most of the charge is quickly lost through the membrane. Consequently, the current is *decremental*; that is, it dies

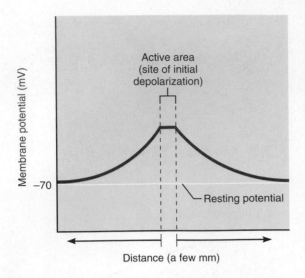

Active area
(site of initial
depolarization)

Membrane potential (mV)

−70

Resting potential

Distance (a few mm)

FIGURE 11.11 Changes in membrane potential produced by a depolarizing graded potential. Such voltage changes are decremental because the current is quickly dissipated by loss of ions through the "leaky" plasma membrane. Consequently, graded potentials are short-distance signals.

out within a few millimeters of its origin (Figure 11.11). Because the current dissipates quickly and dies out with increasing distance from the site of initial depolarization, graded potentials can act as signals only over very short distances. Nonetheless, they are essential in initiating action potentials, the long-distance signals.

Action Potentials

The principal way neurons communicate is by generating and propagating **action potentials (APs)**, and for the most part, only cells with *excitable membranes*—neurons and muscle cells—can generate action potentials. As illustrated in the graph in Figure 11.12, an action potential is a brief reversal of membrane potential with a total amplitude (change in voltage) of about 100 mV (from −70 mV to +30 mV). A depolarization phase is followed by a repolarization phase and often a short period of hyperpolarization. The whole event is over in a few milliseconds. Unlike graded potentials, action potentials do not decrease in strength with distance.

The events of action potential generation and transmission are identical in skeletal muscle cells and neurons. In a neuron, an action potential is also called a **nerve impulse,** and *only axons can generate one.* A neuron transmits a nerve impulse only when it is adequately stimulated. The stimulus changes the permeability of the neuron's membrane by opening specific voltage-gated channels on the axon. These channels open and close in response to changes in the membrane potential and are activated

by local currents (graded potentials) that spread toward the axon along the dendritic and cell body membranes. In many neurons, the transition from local graded potential to action potential takes place at the axon hillock. In sensory neurons, the action potential is generated by the peripheral (axonal) process just proximal to the receptor region. However, for simplicity, we will just use the term *axon* in our discussion.

Generation of an Action Potential   Generating an action potential involves three consecutive but overlapping changes in membrane permeability resulting from the opening and closing of active ion gates, all induced by depolarization of the axonal membrane (Figure 11.12). Sequentially, these permeability changes are a transient increase in Na^+ permeability, followed by restoration of Na^+ impermeability, and then a short-lived increase in K^+ permeability. The first two permeability changes occur during the *depolarization phase* of action potential generation, indicated by the upward-rising part of the AP curve or spike of Figure 11.12. The third permeability change is responsible for both the *repolarization* (the downward part of the AP spike) and *hyperpolarization phases* shown in the Figure 11.12 graph. Let's examine each of these phases more carefully—we will start with a neuron in the resting (polarized) state.

① **Resting state: Voltage-gated channels closed.** Virtually all the voltage-gated Na^+ and K^+ channels are closed, but, small amounts of K^+ leave the cell via leakage channels and even smaller amounts of Na^+ diffuse in.

Each Na^+ channel has two voltage-sensitive gates: an *activation gate* that is closed at rest and responds to depolarization by opening rapidly, and an *inactivation gate* that is open at rest and responds to depolarization by closing slowly. Thus, *depolarization opens and then closes sodium channels.* Both gates must be open in order for Na^+ to enter, but the closing of *either* gate effectively closes the channel. By contrast, each active potassium channel has a single voltage-sensitive gate; it is closed in the resting state and opens slowly in response to depolarization.

② **Depolarizing phase: Increase in Na^+ permeability and reversal of membrane potential.** As the axonal membrane is depolarized by local currents, the sodium channel activation gates open quickly and Na^+ rushes into the cell. This influx of positive charge depolarizes that local "patch" of membrane further, opening more activation gates so that the cell interior becomes progressively less negative. When depolarization at the stimulation site reaches a certain critical level called **threshold** (often between −55 and −50 mV), depolarization becomes

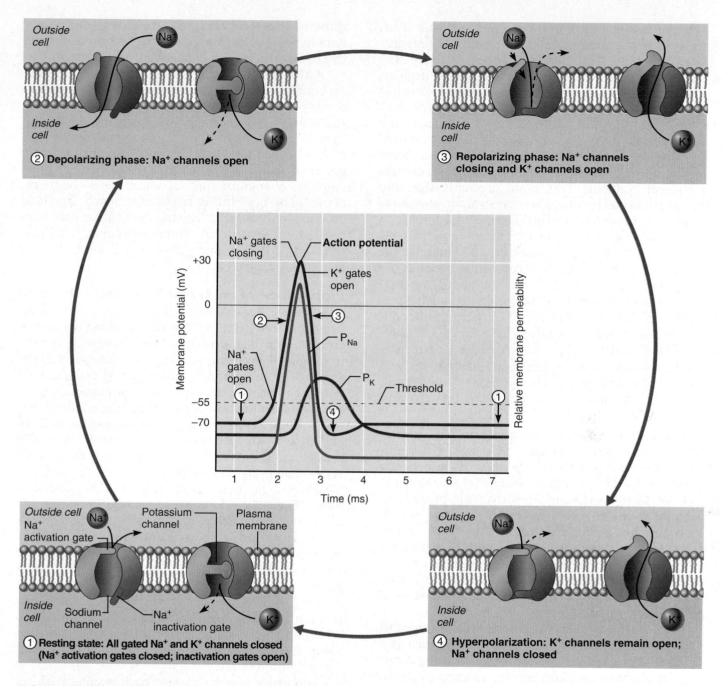

② Depolarizing phase: Na⁺ channels open

③ Repolarizing phase: Na⁺ channels closing and K⁺ channels open

Na⁺ gates closing

Action potential

K⁺ gates open

P_{Na}

Na⁺ gates open

P_K

Threshold

+30

0

−55

−70

Membrane potential (mV)

Relative membrane permeability

1 2 3 4 5 6 7

Time (ms)

Outside cell
Na⁺ activation gate
Potassium channel
Plasma membrane
Inside cell
Sodium channel
Na⁺ inactivation gate

① Resting state: All gated Na⁺ and K⁺ channels closed (Na⁺ activation gates closed; inactivation gates open)

④ Hyperpolarization: K⁺ channels remain open; Na⁺ channels closed

FIGURE 11.12 Phases of the action potential and the role of gated ion channels. The action potential scan in the center of the diagram can be divided into four phases, during which the voltage-sensitive gates controlling Na⁺ and K⁺ channels are in different states and membrane permeability to sodium (P_{Na}) and potassium (P_K) is changing.

① Resting state: neither channel is open. (Sodium inactivation gates are open, but activation gates are closed.)
② Depolarizing phase of the action potential: sodium gates are open, but potassium channels remain closed.
③ Repolarizing phase of the action potential: sodium inactivation gates close the sodium channels, and potassium channels open.

④ Hyperpolarization: both sodium channel gates have closed, but potassium channels remain open temporarily because the relatively slow gates of those channels have not had time to respond to repolarization. Within another millisecond or two, the resting state ① is restored, and the system is ready to respond to the next stimulus.

self-generating, urged on by positive feedback. That is, after being initiated by the stimulus, depolarization is driven by the ionic currents created by Na^+ influx. As more Na^+ enters, the membrane depolarizes further and opens still more activation gates until all Na^+ channels are open. At this point, Na^+ permeability is about 1000 times greater than in a resting neuron. As a result, the membrane potential becomes less and less negative and then overshoots to about +30 mV as Na^+ rushes in along its electrochemical gradient. This rapid depolarization and polarity reversal produce the sharply upward *spike* of the action potential (see the Figure 11.12 graph).

Earlier, we stated that membrane potential depends on membrane permeability, but here we are saying that membrane permeability depends on membrane potential. Can both statements be true? Yes, because these two relationships establish a *positive feedback* cycle (increased Na^+ permeability due to increased channel openings leads to greater depolarization, which leads to increased Na^+ permeability, and so on). It is this explosive positive feedback cycle that is responsible for the rising (depolarizing) phase of action potentials and is the property that puts the "action" in the action potential.

③ **Repolarizing phase: Decrease in Na^+ permeability.** The explosively rising phase of the action potential persists for only about 1 ms and is self-limiting. As the membrane potential passes 0 mV and becomes increasingly positive, the positive intracellular charge resists further Na^+ entry. In addition, the slow inactivation gates of the Na^+ channels begin to close after a few milliseconds of depolarization. As a result, the membrane permeability to Na^+ declines to resting levels, and the net influx of Na^+ stops completely. Consequently, the AP spike stops rising and reverses direction.

Repolarizing phase: Increase in K^+ permeability. As Na^+ entry declines, the slow voltage-sensitive K^+ gates open and K^+ rushes out of the cell, following its electrochemical gradient. Consequently, internal negativity of the resting neuron is restored, an event called **repolarization** (see the Figure 11.12 graph). Both the abrupt decline in Na^+ permeability and the increased permeability to K^+ contribute to repolarization.

④ **Hyperpolarization: K^+ permeability continues.** Because potassium gates are sluggish gates that are slow to respond to the depolarization signal, the period of increased K^+ permeability typically lasts longer than needed to restore the resting state. As a result of the excessive K^+ efflux, an **after-hyperpolarization**, also called the *undershoot*, is seen on the AP curve as a slight dip following the spike (and before the potassium gates close). Notice that both the activation and inactivation

gates of the Na^+ channels are closed during the after-hyperpolarization; hence the neuron is insensitive to a stimulus and depolarization at this time.

Although repolarization restores resting electrical conditions, it does *not* restore resting ionic conditions. The ion redistribution is accomplished by the **sodium-potassium pump** following repolarization. While it might appear that tremendous numbers of Na^+ and K^+ ions change places during action potential generation, this is not the case. Only small amounts of sodium and potassium cross the membrane. (The Na^+ influx required to reach threshold produces only a 0.012% change in cellular Na^+ concentration.) Because an axonal membrane has thousands of Na^+-K^+ pumps, these small ionic changes are quickly corrected.

Propagation of an Action Potential If it is to serve as the neuron's signaling device, an AP must be **propagated** (sent or transmitted) along the axon's entire length (Figure 11.13). As we have seen, the AP is generated by the influx of Na^+ through a given area of the membrane. This establishes local currents that depolarize adjacent membrane areas in the forward direction (away from the origin of the nerve impulse), which opens voltage-gated channels and triggers an action potential there. Because the area where the AP originated has just generated an action potential, the sodium gates in that area are closed and no new action potential is generated there. Thus, the AP propagates away from its point of origin. (If an *isolated* axon is stimulated by an electrode, or an axon is stimulated at a node of Ranvier, the nerve impulse will move away from the point of stimulus in *all* directions along the membrane.) In the body, action potentials are initiated at one end of the axon and conducted away from that point toward the axon's terminals. Once initiated, an action potential is *self-propagating* and continues along the axon at a constant velocity—something like a domino effect.

Following depolarization, each segment of axonal membrane repolarizes, which restores the resting membrane potential in that region. Because these electrical changes also set up local currents, the repolarization wave chases the depolarization wave down the length of the axon. The propagation process just described occurs on unmyelinated axons. Propagation that occurs along myelinated axons, called *saltatory conduction*, is described shortly.

Although the phrase *conduction of a nerve impulse* is commonly used, nerve impulses are not really conducted in the same way that an insulated wire conducts current. In fact, neurons are fairly poor conductors, and as noted earlier, local current flows decline with distance because the charges leak through the membrane. The expression *propagation*

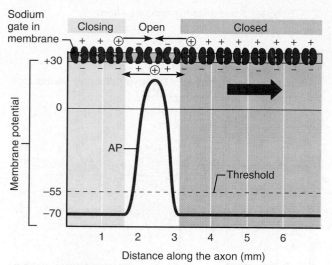

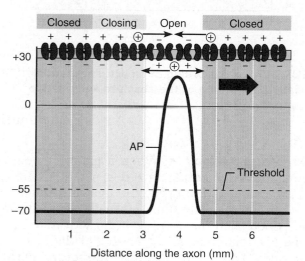

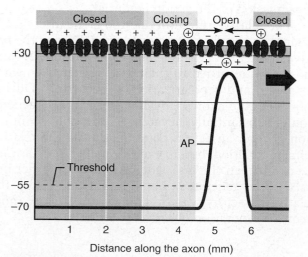

FIGURE 11.13 **Propagation of an action potential (AP).** An action potential propagating along an axon at 0 ms, 1 ms, and 2 ms. The sodium gates are labeled closed, open, or closing. Small arrows indicate local currents generated by the movement of positive ions. The large arrows indicate the direction of action potential propagation. Not depicted, the current flows created by the opening of potassium channels (and subsequent repolarization) occur where sodium gates are shown as closing.

of a nerve impulse is more accurate, because the action potential is *regenerated anew* at each membrane patch, and every subsequent action potential is identical to that generated initially.

Threshold and the All-or-None Phenomenon

Not all local depolarization events produce action potentials. The depolarization must reach threshold values if an axon is to "fire." What determines the *threshold point*? One explanation is that threshold is the membrane potential at which the outward current created by K^+ movement is exactly equal to the inward current created by Na^+ movement. Threshold is typically reached when the membrane has been depolarized by 15 to 20 mV from the resting value. This depolarization status seems to represent an unstable equilibrium state at which one of two things can happen. If one more Na^+ enters, further depolarization occurs, opening more Na^+ channels and allowing more sodium ions entry. If, on the other hand, one more K^+ leaves, the membrane potential is driven away from threshold, Na^+ channels close, and K^+ continues to diffuse outward until the potential returns to its resting value.

Recall that local depolarizations are graded potentials and that their magnitude increases with increasing stimulus intensity. Brief weak stimuli *(subthreshold stimuli)* produce subthreshold depolarizations that are not translated into nerve impulses. On the other hand, stronger *threshold stimuli* produce depolarizing currents that push the membrane potential toward and beyond the threshold voltage. As a result, Na^+ permeability is increased to such an extent that entering sodium ions "swamp" (exceed) the outward movement of K^+, allowing the positive feedback cycle to become established and generating an action potential. The critical factor here is the total amount of current that flows through the membrane during a stimulus (electrical charge × time). Strong stimuli depolarize the membrane to threshold quickly. Weaker stimuli must be applied for longer periods to provide the crucial amount of current flow. Very weak stimuli do not trigger an action potential because the local current flows they produce are so slight that they dissipate long before threshold is reached.

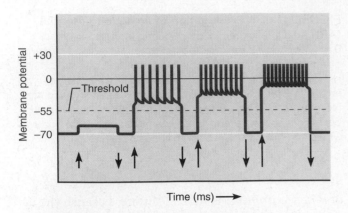

FIGURE 11.14 **Relationship between stimulus strength, local potential, and action potential frequency.** Action potentials are shown as vertical lines. Upward arrows (↑) indicate points of stimulus application; downward arrows (↓) indicate stimulus cessation; arrow length indicates strength of stimulus. Notice that a subthreshold stimulus does not generate an action potential, but once threshold voltage is reached, the stronger the stimulus, the more frequently action potentials are generated.

The action potential is an **all-or-none phenomenon;** it either happens completely or doesn't happen at all. Generation of an action potential can be compared to lighting a match under a small dry twig. The changes occurring where the twig is being heated are analogous to the change in membrane permeability that initially allows more Na⁺ to enter the cell. When that part of the twig becomes hot enough (when enough Na⁺ has entered the cell), the flash point (threshold) is reached and the flame consumes the entire twig, even if you blow out the match (the action potential is generated and propagated whether or not the stimulus continues). But if the match is extinguished just before the twig has reached the critical temperature, ignition will not take place. Likewise, if the number of Na⁺ entering the cell is too low to achieve threshold, no action potential will occur.

Coding for Stimulus Intensity

Once generated, all action potentials are independent of stimulus strength, and all action potentials are alike. So how can the CNS determine whether a particular stimulus is intense or weak—information it needs to initiate an appropriate response? The answer is really quite simple: Strong stimuli cause nerve impulses to be generated more *often* in a given time interval than do weak stimuli (Figure 11.14). Thus, stimulus intensity is coded for by the number of impulses generated per second—that is, by the *frequency of impulse transmission*—rather than by increases in the strength (amplitude) of the individual action potentials.

? *What causes after-hyperpolarization?*

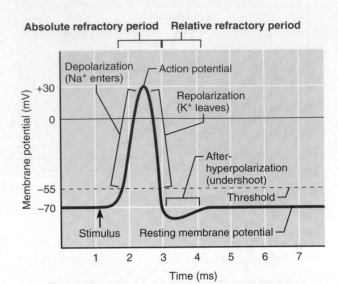

FIGURE 11.15 **Recording of an action potential indicating the timing of the absolute and relative refractory periods.**

Refractory Periods

When a patch of neuron membrane is generating an action potential and its sodium channels are open, the neuron cannot respond to another stimulus, no matter how strong. This period from the opening of the activation gates of the Na⁺ channels to the closing of the inactivation gates, called the **absolute refractory period** (Figure 11.15), ensures that each action potential is a separate, *all-or-none event* and enforces one-way transmission of the action potential.

The **relative refractory period** is the interval following the absolute refractory period. During the relative refractory period, Na⁺ gates are closed and most have returned to their resting state, K⁺ gates are open, and repolarization is occurring. During this time, the axon's threshold for impulse generation is substantially elevated. A threshold stimulus won't trigger an action potential during the relative refractory period, but an exceptionally strong stimulus can reopen the Na⁺ gates and allow another impulse to be generated. Thus, by intruding into the relative refractory period, strong stimuli cause more frequent generation of action potentials.

■ *After-hyperpolarization occurs because potassium channel gates are slow to respond to the membrane depolarization. Hence more K⁺ leaves the cell than is absolutely necessary to restore the neuron to its resting membrane potential.*

? *How does the location of voltage-gated Na⁺ channels differ in a myelinated axon (as depicted here) and in an unmyelinated axon?*

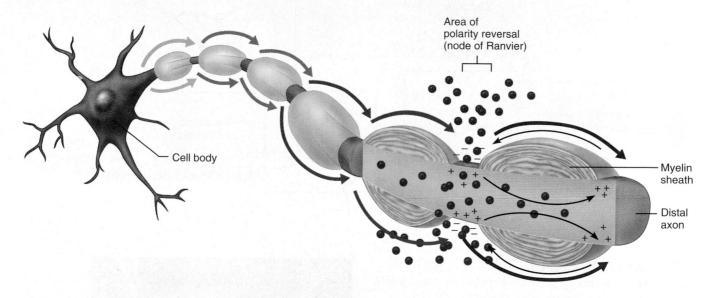

FIGURE 11.16 Saltatory conduction in a myelinated axon. In myelinated nerve fibers, local currents (thin black arrows) give rise to a propagated action potential (pink and red arrows) that appears to jump from node to node. Notice that while action potentials are generated only at the nodes, the current flows along the axon from node to node.

Conduction Velocities Conduction velocities of neurons vary widely. Nerve fibers that transmit impulses most rapidly (100 m/s or more) are found in neural pathways where speed is essential, such as those that mediate some postural reflexes. Axons that conduct impulses more slowly typically serve internal organs (the gut, glands, blood vessels), where slower responses are not a handicap. The rate of impulse propagation depends largely on two factors:

1. **Axon diameter.** Axons vary considerably in diameter and, as a rule, the larger the axon's diameter, the faster it conducts impulses. This is because larger axons offer less resistance to the flow of local currents, and so adjacent areas of the membrane can more quickly be brought to threshold.

2. **Degree of myelination.** On unmyelinated axons, action potentials are generated at sites immediately adjacent to each other and conduction is relatively slow, a type of AP propagation called **continuous conduction**. The presence of a myelin sheath dramatically increases the rate of impulse propagation because myelin acts as an insulator to prevent almost all leakage of charge from the axon. Current can pass through the membrane of a myelinated axon *only* at the nodes of Ranvier, where the myelin sheath is interrupted and the axon is bare, and essentially all the voltage-gated Na⁺ channels are concentrated at the nodes. Thus, when an action potential is generated in a myelinated fiber, the local depolarizing current does not dissipate through the adjacent (nonexcitable) membrane regions but instead is maintained and moves to the next node, a distance of approximately 1 mm, where it triggers another action potential. Consequently, action potentials are triggered only at the nodes, a type of conduction called **saltatory conduction** (*saltare* = to leap) because the electrical signal jumps from node to node along the axon (Figure 11.16). Saltatory conduction is much faster than continuous conduction.

⒣ HOMEOSTATIC IMBALANCE

The importance of myelin to nerve transmission is painfully clear to those with demyelinating diseases such as **multiple sclerosis (MS).** This autoimmune disease affects mostly young adults. Common symptoms are visual disturbances (including blindness), problems controlling muscles (weakness, clumsiness, and ultimately paralysis), speech disturbances,

■ *In a myelinated axon, voltage-gated sodium channels are located only at the nodes of Ranvier, as opposed to being located along the entire length of an unmyelinated axon.*

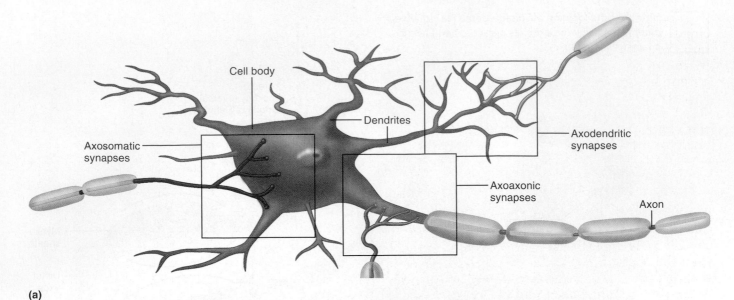

(a)

FIGURE 11.17 **Synapses. (a)** Axodendritic, axosomatic, and axoaxonic synapses. **(b)** Scanning electron micrograph of incoming fibers at axosomatic synapses (4000×).

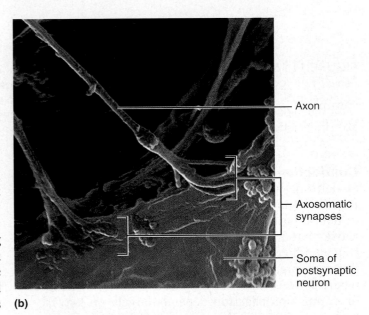

(b)

and urinary incontinence. In this disease, myelin sheaths in the CNS are gradually destroyed, reduced to nonfunctional hardened lesions called *scleroses.* The loss of myelin (a result of the immune system's attack on myelin proteins) causes such substantial shunting and short-circuiting of the current that successive nodes are excited more and more slowly, and eventually impulse conduction ceases. However, the axons themselves are not damaged and growing numbers of Na^+ channels appear spontaneously in the demyelinated fibers. This may account for the remarkably variable cycles of relapse (disability) and remission (symptom-free periods) typical of this disease.

Until a few years ago, little could be done to help MS victims. This situation is changing with the advent of the so-called disease-modifying drugs including interferon beta-1a and -1b, Avonex, Betaseran, and Copaxone. These drugs seem to hold the symptoms at bay, reducing complications and the disability that often occurs with MS. ●

Nerve fibers may be classified based on diameter, degree of myelination, and conduction speed. **Group A fibers** are mostly somatic sensory and motor fibers serving the skin, skeletal muscles, and joints. They have the largest diameter and thick myelin sheaths, and conduct impulses at speeds ranging up to 150 m/s (over 300 mph). Autonomic nervous system motor fibers serving the visceral organs; visceral sensory fibers; and the smaller somatic sensory fibers transmitting afferent impulses from the skin (such

as pain and small touch fibers) belong in the B and C fiber groups. **Group B fibers,** lightly myelinated fibers of intermediate diameter, transmit impulses at an average rate of 15 m/s (about 40 mph). **Group C fibers** have the smallest diameter and are unmyelinated. Hence, they are incapable of saltatory conduction and conduct impulses at a leisurely pace—1 m/s (2 mph) or less.

Ⓗ *HOMEOSTATIC IMBALANCE*

A number of chemical and physical factors impair impulse propagation. Although their mechanisms of action differ, alcohol, sedatives, and injected anesthetics all block nerve impulses by reducing membrane permeability to ions, mainly Na^+. As we have seen, no Na^+ entry—no action potential.

Cold and continuous pressure interrupt blood circulation (and hence the delivery of oxygen and nutrients) to neuron processes, impairing their ability to conduct impulses. For example, your fingers get numb when you hold an ice cube for more than a few seconds, and your foot "goes to sleep" when you sit on it. When you remove the cold object or pressure, impulses are transmitted again, leading to an unpleasant prickly feeling. ●

The Synapse

The operation of the nervous system depends on the flow of information through chains of neurons functionally connected by synapses. A **synapse** (sin'aps), from the Greek *syn*, "to clasp or join," is a junction that mediates information transfer from one neuron to the next or from a neuron to an effector cell—it's where the action is.

Synapses between the axonal endings of one neuron and the dendrites of other neurons are **axodendritic synapses** (Figure 11.17). Those between axonal endings of one neuron and cell bodies of other neurons are **axosomatic synapses.** Less common (and far less understood) are synapses between axons *(axoaxonic)*, between dendrites *(dendrodendritic)*, or between dendrites and cell bodies *(dendrosomatic)*.

The neuron conducting impulses toward the synapse is the **presynaptic neuron** and the neuron transmitting the electrical signal away from the synapse is the **postsynaptic neuron.** At a *given* synapse, the presynaptic neuron is the information sender, and the postsynaptic neuron is the information receiver. As you might anticipate, most neurons function both as presynaptic and postsynaptic neurons. Neurons have anywhere from 1000 to 10,000 axonal terminals making synapses and are stimulated by an equal number of other neurons. In the body periphery, the postsynaptic cell may be either another neuron or an effector cell (a muscle cell or gland cell).

There are two varieties of synapses: *electrical* and *chemical*. These are described next.

Electrical Synapses

Electrical synapses, the less common variety, correspond to the gap junctions found between certain other body cells (Figure 11.18). They contain protein channels, made of connexin subunits, that connect the cytoplasm of adjacent neurons and allow ions to flow directly from one neuron to the next. Neurons joined in this way are said to be *electrically coupled*, and transmission across these synapses is very rapid. Depending on the nature of the synapse, communication may be unidirectional or bidirectional.

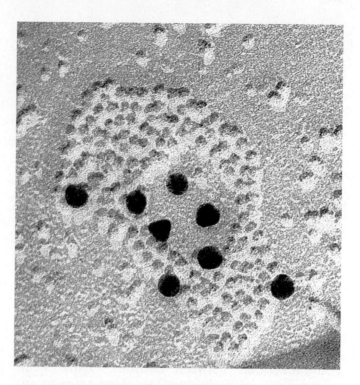

FIGURE 11.18 Freeze-fracture image of an electrical synapse in an adult rat hippocampus. The dark spheres are immunogold labels for the protein connexin36, which is the first connexin established as present in CNS electrical synapses (gap junctions) (200,000×). Photograph courtesy of John Rash, Ph.D.; Colorado State University.

A key feature of electrical synapses between neurons is that they provide a simple means of synchronizing the activity of all interconnected neurons. They appear to be important in CNS arousal from sleep and in mental attention and conscious perception. In adults, electrical synapses are found in regions of the brain responsible for certain stereotyped movements, such as the normal jerky movements of the eyes, and in axoaxonic synapses in the hippocampus, a region intimately involved in emotions and memory. They are far more abundant in embryonic nervous tissue, where they permit exchange of guiding clues during early neuronal development so that neurons can connect properly with one another. As the nervous system develops, some electrical synapses are replaced by chemical synapses. Electrical synapses also exist between glial cells of the CNS, where they play a role in ion and water homeostasis.

Chemical Synapses

In contrast to electrical synapses, which are specialized to allow the flow of ions between neurons,

chemical synapses are specialized for release and reception of chemical neurotransmitters. A typical chemical synapse is made up of two parts: (1) a knoblike *axonal terminal* of the presynaptic neuron, which contains many tiny, membrane-bounded sacs called **synaptic vesicles,** each containing thousands of neurotransmitter molecules; and (2) a neurotransmitter *receptor region* on the membrane of a dendrite or the cell body of the postsynaptic neuron.

Although close to each other, presynaptic and postsynaptic membranes are always separated by the **synaptic cleft,** a fluid-filled space approximately 30 to 50 nm (about one-millionth of an inch) wide. Because the current from the presynaptic membrane dissipates in the fluid-filled cleft, chemical synapses effectively prevent a nerve impulse from being *directly* transmitted from one neuron to another. Instead, transmission of signals across these synapses is a *chemical event* that depends on the release, diffusion, and receptor binding of neurotransmitter molecules and results in *unidirectional communication* between neurons. Thus, while transmission of nerve impulses along an axon and across electrical synapses is a purely electrical event, chemical synapses convert the electrical signals to chemical signals (neurotransmitters) that travel across the synapse to the postsynaptic cells, where they are converted back into electrical signals.

Information Transfer Across Chemical Synapses

When a nerve impulse reaches the axonal terminal, it sets into motion a chain of events that triggers neurotransmitter release. The neurotransmitter crosses the synaptic cleft and, on binding to receptors on the postsynaptic membrane, causes changes in the postsynaptic membrane permeability. The steps are shown in Figure 11.19:

① **Calcium channels open in the presynaptic axonal terminal.** When the nerve impulse reaches the axonal terminal, membrane depolarization opens not only Na^+ channels but voltage-gated Ca^{2+} channels as well. During the brief time the Ca^{2+} gates are open, Ca^{2+} floods into the terminal from the extracellular fluid.

② **Neurotransmitter is released.** The surge of Ca^{2+} into the axonal terminal acts as an intracellular messenger, directing docked synaptic vesicles to fuse with the axonal membrane and empty their contents by exocytosis into the synaptic cleft. The Ca^{2+} is then quickly removed from the terminal, either taken up into the mitochondria or ejected from the neuron by an active Ca^{2+} pump. The precise Ca^{2+} sensor that initiates neurotransmitter exocytosis is still a question, but a Ca^{2+}-binding protein called *synaptotagmin* found in the synaptic vesicles seems a likely candidate.

③ **Neurotransmitter binds to postsynaptic receptors.** The neurotransmitter diffuses across the synaptic cleft and binds reversibly to specific protein receptors clustered on the postsynaptic membrane.

④ **Ion channels open in the postsynaptic membrane.** As the receptor proteins bind neurotransmitter molecules, the three-dimensional shape of the proteins changes. This causes ion channels to open, and the resulting current flows produce local changes in the membrane potential. Depending on the receptor protein to which the neurotransmitter binds and the type of channel the receptor controls, the postsynaptic neuron may be either excited or inhibited.

For each nerve impulse reaching the presynaptic terminal, many vesicles (perhaps 300) are emptied into the synaptic cleft. The higher the impulse frequency (that is, the more intense the stimulus), the greater the number of synaptic vesicles that fuse and spill their contents, and the greater the effect on the postsynaptic cell.

Termination of Neurotransmitter Effects As long as it is bound to a postsynaptic receptor, a neurotransmitter continues to affect membrane permeability and to block reception of additional "messages" from presynaptic neurons. Thus, some means of "wiping the postsynaptic slate clean" is necessary. The effects of neurotransmitters last a few milliseconds before being terminated by one of three mechanisms. Depending on the particular neurotransmitter, the terminating mechanism may be

1. Degradation by enzymes associated with the postsynaptic membrane or present in the synapse. This is the case for acetylcholine.

2. Reuptake by astrocytes or the presynaptic terminal, where the neurotransmitter is stored or destroyed by enzymes, as with norepinephrine.

3. Diffusion away from the synapse.

Synaptic Delay Although some neurons can transmit impulses at 150 m/s (300 mph), neural transmission across a chemical synapse is comparatively slow and reflects the time required for neurotransmitter release, diffusion across the synapse, and binding to receptors. Typically, this **synaptic delay,** which lasts 0.3–5.0 ms, is the *rate-limiting* (slowest) step of neural transmission. Synaptic delay helps explain why transmission along short neural pathways involving only two or three neurons occurs rapidly, but transmission along multisynaptic pathways typical of higher mental functioning occurs much more slowly. However, in practical terms these differences are not noticeable.

Why may such axonal terminals be referred to as "biological transducers"?

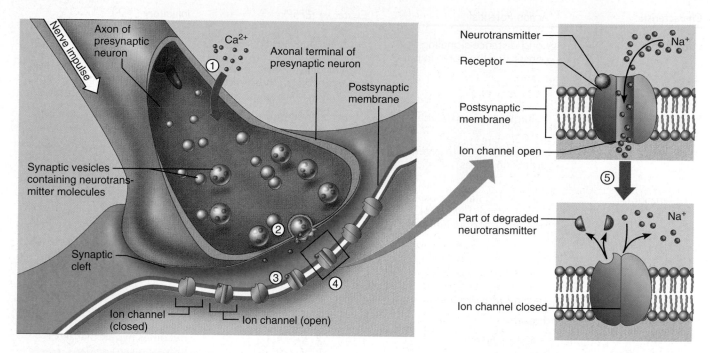

FIGURE 11.19 Events at a chemical synapse in response to depolarization. ① Arrival of the depolarization wave (nerve impulse) opens calcium channels and allows Ca^{2+} influx into the axonal terminal. ② Synaptic vesicles fuse with the presynaptic membrane and neurotransmitter is released into the synapse. ③ The neurotransmitter diffuses across the synaptic cleft and attaches to receptors on the postsynaptic membrane. ④ Binding of neurotransmitter opens ion channels in the postsynaptic membrane, resulting in voltage changes in that membrane. ⑤ Neurotransmitter is quickly destroyed by enzymes present at the synapse or taken back into the presynaptic terminal; depletion of neurotransmitter closes the ion channels and terminates the synaptic response.

Postsynaptic Potentials and Synaptic Integration

Many receptors present on postsynaptic membranes at chemical synapses are specialized to open ion channels, thereby converting chemical signals to electrical signals. Unlike the voltage-gated ion channels responsible for action potentials, however, these chemically gated channels are relatively insensitive to changes in membrane potential. Consequently, channel opening at postsynaptic membranes cannot possibly become self-amplifying or self-generating. Instead, neurotransmitter receptors mediate local changes in membrane potential that are *graded* according to the amount of neurotransmitter released and the time it remains in the area. Action potentials are compared with postsynaptic potentials in Table 11.2.

Chemical synapses are either excitatory or inhibitory, depending on how they affect the membrane potential of the postsynaptic neuron.

Excitatory Synapses and EPSPs

At excitatory synapses, neurotransmitter binding causes depolarization of the postsynaptic membrane. However, in contrast to what happens on axonal membranes, only a single type of channel opens on postsynaptic membranes (those of dendrites and neuronal cell bodies). This channel allows Na^+ and K^+ to diffuse *simultaneously* through the membrane in opposite directions. Although this two-way cation flow may appear to be self-defeating when depolarization is the goal, remember that the electrochemical gradient for sodium is much steeper than that for potassium. Hence, Na^+ influx is greater than K^+ efflux, and *net* depolarization occurs.

If enough neurotransmitter binds, depolarization of the postsynaptic membrane can successfully

■ *Because they change the signal from an electrical current to a chemical signal (neurotransmitter), which in turn initiates an electrical current in the postsynaptic neuron.* ■

TABLE 11.2 Comparison of Action Potentials with Postsynaptic Potentials

Characteristic	Action Potential	Postsynaptic Potential	
		Excitatory (EPSP)	Inhibitory (IPSP)
Function	Long-distance signaling; constitutes the nerve impulse	Short-distance signaling; depolarization that spreads to axon hillock; moves membrane potential *toward* threshold for generation of action potential	Short-distance signaling; hyperpolarization that spreads to axon hillock; moves potential *away from* threshold for generation of action potential
Stimulus for opening of ionic gates	Voltage (depolarization)	Chemical (neurotransmitter)	Chemical (neurotransmitter)
Initial effect of stimulus	First opens sodium gates, then potassium gates	Opens channels that allow simultaneous sodium and potassium fluxes	Opens potassium or chloride channels or both
Repolarization	Voltage regulated; closing of sodium gates followed by opening of potassium gates	Dissipation of membrane changes with time and distance	
Conduction distance	Not conducted by local current flows; continually regenerated (propagated) along entire axon; intensity does not decline with distance	1–2 mm; local electrical events; intensity declines with distance	
Positive feedback cycle	Present	Absent	Absent
Peak membrane potential	+40 to +50 mV	0 mV	Becomes hyperpolarized; moves toward −90 mV
Summation	None; an all-or-none phenomenon	Present; produces graded depolarization	Present; produces graded hyperpolarization
Refractory period	Present	Absent	Absent

reach 0 mV, which is well above an axon's threshold (about −50 mV) for "firing off" an action potential. However, *postsynaptic membranes do not generate action potentials; only axons* (with their voltage-gated channels) *have this capability.* The dramatic polarity reversal seen in axons never occurs in membranes containing *only* chemically gated channels because the opposite movements of K^+ and Na^+ prevent accumulation of excessive positive charge inside the cell. Hence, instead of action potentials, local graded depolarization events called **excitatory postsynaptic potentials (EPSPs)** occur at excitatory postsynaptic membranes (see Figure 11.20a). Each EPSP lasts a few milliseconds and then the membrane returns to its resting potential. The only function of EPSPs is to help trigger an action potential distally at the axon hillock of the postsynaptic neuron. Although currents created by individual EPSPs decline with distance, they can and often *do* spread all the way to the axon hillock. If currents reaching the hillock are strong enough to depolarize the axon to threshold, axonal voltage-gated channels open and an action potential is generated.

Inhibitory Synapses and IPSPs

Binding of neurotransmitters at inhibitory synapses *reduces* a postsynaptic neuron's ability to generate an action potential. Most inhibitory neurotransmitters induce hyperpolarization of the postsynaptic membrane by making the membrane more permeable to K^+ and/or Cl^-. Sodium ion permeability is not affected. If K^+ channels are opened, K^+ moves out of the cell; if Cl^- channels are opened, Cl^- moves in. In either case, the charge on the inner face of the membrane becomes more negative. As the membrane potential increases and is driven farther from the axon's threshold, the postsynaptic neuron becomes less and less likely to "fire" and larger depolarizing currents are required to induce an action potential. Such changes in potential are called **inhibitory postsynaptic potentials (IPSPs)** (see Figure 11.20b).

Integration and Modification of Synaptic Events

Summation by the Postsynaptic Neuron A single EPSP cannot induce an action potential in the postsynaptic neuron. But if thousands of excitatory axonal terminals are firing on the same postsynaptic membrane, or if a smaller number of terminals are delivering impulses rapidly, the probability of reaching threshold depolarization increases greatly. Thus, EPSPs can add together, or **summate,** to influence the activity of a postsynaptic neuron (Figure 11.21). Nerve impulses would never be initiated if this were not so.

Two types of summation occur. **Temporal summation** occurs when one or more presynaptic neurons transmit impulses in rapid-fire order and bursts of neurotransmitter are released in quick succession. The first impulse produces a slight EPSP, and before it dissipates, successive impulses trigger more EPSPs. These summate, producing a much greater depolarization of the postsynaptic membrane than would result from a single EPSP.

Spatial summation occurs when the postsynaptic neuron is stimulated at the same time by a large number of terminals from the same or, more commonly, different neurons. Huge numbers of its receptors bind neurotransmitter and simultaneously initiate EPSPs, which summate and dramatically enhance depolarization.

Although we have focused on EPSPs here, IPSPs also summate, both temporally and spatially. In this case, the postsynaptic neuron is inhibited to a greater degree.

Most neurons receive both stimulatory and inhibitory inputs from thousands of other neurons. Additionally, the same fiber may form different types of synapses (in terms of biochemical and electrical characteristics) with different types of target neurons. How is all this conflicting information sorted out? Each neuron's axon hillock appears to keep a running account of all the signals it receives (Figure 11.21d). Not only do EPSPs summate and IPSPs summate, but also EPSPs summate with IPSPs. If the stimulatory effects of EPSPs dominate the membrane potential enough to reach threshold, the neuron will fire. If summation yields only subthreshold depolarization or hyperpolarization, the neuron fails to generate an action potential. However, partially depolarized neurons are **facilitated**—that is, more easily excited by successive depolarization events—because they are already near threshold. Thus, axon hillock membranes function as *neural integrators,* and their potential at any time reflects the sum of all incoming neural information.

Because EPSPs and IPSPs are graded potentials that diminish in strength the farther they spread, we might expect that the most effective synapses are those closest to the axon hillock. What we find, however, is that distal synapses do count as much as more proximal ones (those on the cell body), because synapses get stronger with increasing distance from

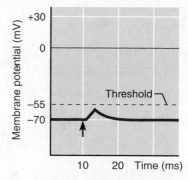

(a) Excitatory postsynaptic potential (EPSP)

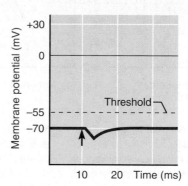

(b) Inhibitory postsynaptic potential (IPSP)

FIGURE 11.20   Postsynaptic potentials. (a) An excitatory postsynaptic potential (EPSP) is a local depolarization of the postsynaptic membrane that brings the neuron closer to threshold for AP generation. It is mediated by neurotransmitter binding that opens channels allowing the simultaneous passage of Na^+ and K^+ through the postsynaptic membrane. **(b)** An inhibitory postsynaptic potential results in hyperpolarization of the postsynaptic neuron and drives the neuron away from the threshold for firing. It is mediated by neurotransmitter binding that opens K^+ or Cl^- gates or both. The red vertical arrows represent stimulation.

the cell body. Consequently, amplitude of synaptic responses at the cell body is homogeneous regardless of the site of input.

Synaptic Potentiation   Repeated or continuous use of a synapse (even for short periods) enhances the presynaptic neuron's ability to excite the postsynaptic neuron, producing larger than expected postsynaptic potentials. This phenomenon is called **synaptic potentiation.** The presynaptic terminals at such synapses contain relatively high Ca^{2+} concentrations, a condition that (presumably) triggers the release of more neurotransmitter, which in turn produces larger EPSPs. Furthermore, synaptic potentiation increases Ca^{2+} influx via dendritic spines in the postsynaptic neuron as well. Brief high-frequency

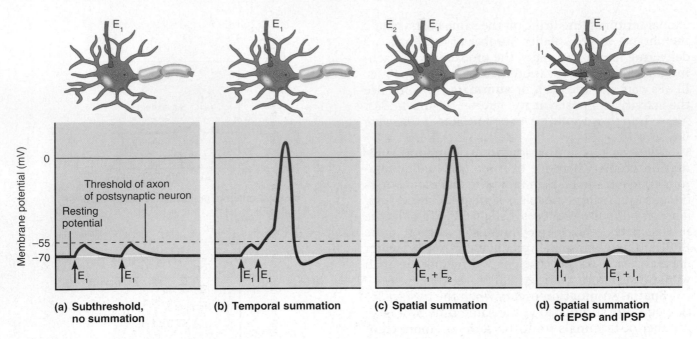

(a) Subthreshold, no summation

(b) Temporal summation

(c) Spatial summation

(d) Spatial summation of EPSP and IPSP

FIGURE 11.21   **Neural integration of EPSPs and IPSPs.** Synapses E_1 and E_2 are excitatory; synapse I_1 is inhibitory. Each individual impulse is subthreshold. **(a)** Synapse E_1 is stimulated and then stimulated again shortly thereafter. The two EPSPs do not overlap in time, so no summation occurs; threshold is not reached in the axon of the postsynaptic neuron. **(b)** Synapse E_1 is stimulated a second time before the initial EPSP has died away; temporal summation occurs, and the axon's threshold is reached, causing an action potential to be generated. **(c)** Synapses E_1 and E_2 are stimulated simultaneously (spatial summation), resulting in a threshold depolarization. **(d)** Synapse I_1 is stimulated, resulting in a short-lived IPSP (hyperpolarization). When E_1 and I_1 are simultaneously stimulated, the changes in potential cancel each other.

stimulation specifically activates voltage-regulated receptors called *NMDA (N-methyl-D-aspartate) receptors*, which are located on the postsynaptic membrane and couples the depolarization to increased Ca^{2+} entry. Theoretically, as Ca^{2+} floods into the cell, it activates certain kinase enzymes that promote changes that result in more efficient responses to subsequent stimuli.

Synaptic potentiation, also called *post-tetanic potentiation*, can be viewed as a learning process that increases the efficiency of neurotransmission along a particular pathway. Indeed, the hippocampus of the brain, which plays a special role in memory and learning, exhibits especially long post-tetanic potentiations.

Another facet of synaptic potentiation appears to occur under certain conditions and at certain brain synapses. Dendrites of pyramidal neurons of the cerebral cortex and hippocampal neurons have fast voltage-gated sodium channels and high-threshold gates for Ca^{2+} currents that act in synaptic integration and influence the strength and plasticity of their synapses. Apparently as a neuron shoots off an AP, a back-propagating AP travels from the cell body to the dendrites. This current flow alters the efficiency of the synapses, promoting synaptic potentiation.

Presynaptic Inhibition and Neuromodulation

Postsynaptic activity can also be influenced by events occurring at the presynaptic membrane.

Presynaptic inhibition occurs when the release of excitatory neurotransmitter by one neuron is inhibited by the activity of another neuron via an axoaxonic synapse. More than one mechanism is involved, but the end result is that less neurotransmitter is released and bound, and smaller EPSPs are formed. Notice that this is the opposite of what we see with synaptic potentiation. In contrast to postsynaptic inhibition by IPSPs, which decreases the excitability of the postsynaptic neuron, presynaptic inhibition is more like a functional synaptic "pruning." Events at the presynaptic neuron reduce excitatory stimulation of the postsynaptic neuron.

Neuromodulation, another presynaptic event that affects postsynaptic activity, occurs when a neurotransmitter acts via slow changes in target cell metabolism or when chemicals other than neurotransmitters modify neuronal activity. Some *neuromodulators* influence the synthesis, release, degradation, or reuptake of neurotransmitter by a presynaptic neuron. Others alter the sensitivity of the postsynaptic membrane to the neurotransmitter. Some neuromodulators are hormones that act at sites far from their release site.

Neurotransmitters and Their Receptors

Neurotransmitters, along with electrical signals, are the "language" of the nervous system—the means by which each neuron communicates with others to process and send messages to the rest of the body. Sleep, thought, rage, hunger, memory, movement, and even your smile reflect the "doings" of these versatile molecules. Most factors that affect synaptic transmission do so by enhancing or inhibiting neurotransmitter release or destruction, or by blocking their binding to receptors. Just as speech defects may hinder interpersonal communication, interferences with neurotransmitter activity may short-circuit the brain's "conversations" or internal talk.

At present, more than 50 neurotransmitters or neurotransmitter candidates have been identified. Although some neurons produce and release only one kind of neurotransmitter, most make two or more and may release any one or all of them. It appears that in most cases, different neurotransmitters are released at different stimulation frequencies, a restriction that avoids producing a jumble of nonsense messages. However, co-release of two neurotransmitters from the same vesicles has been documented. The coexistence of more than one neurotransmitter in a single neuron makes it possible for that cell to exert several influences rather than one discrete effect.

Neurotransmitters are classified chemically and functionally. Table 11.3 provides a fairly detailed overview of neurotransmitters, some of which are described here. No one expects you to memorize this table at this point, but it will be a handy reference for you to look back to when neurotransmitters are mentioned in subsequent chapters.

Classification of Neurotransmitters by Chemical Structure

Neurotransmitters fall into several chemical classes based on molecular structure.

Acetylcholine (ACh) Acetylcholine (as″ĕ-til-ko′lēn) was the first neurotransmitter identified. It is still the best understood because it is released at neuromuscular junctions, which are much easier to study than synapses buried in the CNS. ACh is synthesized and enclosed in synaptic vesicles in axonal terminals in a reaction catalyzed by the enzyme *choline acetyltransferase*. Acetic acid is bound to coenzyme A (CoA) to form acetyl-CoA, which then combines with choline. Then coenzyme A is released.

$$\text{Acetyl-CoA} + \text{choline} \xrightarrow{\text{choline acetyltransferase}} \text{ACh} + \text{CoA}$$

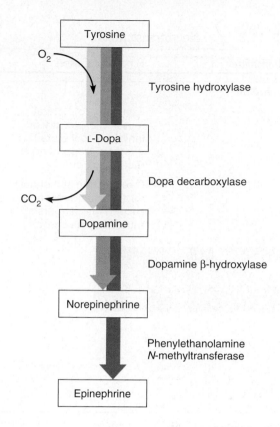

FIGURE 11.22 **Common pathway for synthesis of dopamine, norepinephrine, and epinephrine.** How far synthesis proceeds along the pathway depends on the enzymes present in the cell.

Once released by the presynaptic terminal, ACh binds to the postsynaptic receptors briefly. Then it is released and degraded to acetic acid and choline by the enzyme **acetylcholinesterase (AChE)**, located in the synaptic cleft and on postsynaptic membranes. The released choline is recaptured by the presynaptic terminals and reused to synthesize more ACh.

ACh is released by all neurons that stimulate skeletal muscles and by some neurons of the autonomic nervous system. ACh-releasing neurons are also prevalent in the CNS.

Biogenic Amines The **biogenic amines** (bi″o-jen′ik) include the **catecholamines** (kat″ĕ-kol′ah-mēnz), such as dopamine, norepinephrine (NE), and epinephrine, and the **indolamines**, which include serotonin and histamine. As illustrated in Figure 11.22, *dopamine* and *NE* are synthesized from the amino acid tyrosine in a common pathway consisting of several steps. Apparently, neurons contain only the enzymes needed to produce their own neurotransmitter(s). Thus, the sequence stops at dopamine in dopamine-releasing neurons but continues on to NE in NE-releasing neurons. The same pathway is used by the epinephrine-releasing cells of

TABLE 11.3 / Neurotransmitters

Neurotransmitter	Functional Classes	Sites Where Secreted	Comments
Acetylcholine	Excitatory to skeletal muscles; excitatory or inhibitory to visceral effectors, depending on receptors bound Direct action at nicotinic receptors; indirect action via second messengers at muscarinic receptors	CNS: basal nuclei and some neurons of motor cortex of brain PNS: all neuromuscular junctions with skeletal muscle; some autonomic motor endings (all preganglionic and parasympathetic postganglionic fibers)	Effects prolonged (leading to tetanic muscle spasms and neural "frying") by nerve gas and organophosphate insecticides (malathion); release inhibited by botulinus toxin and barbiturates; binding to receptors inhibited by curare (a muscle paralytic agent) and some snake venoms; decreased ACh levels in certain brain areas in Alzheimer's disease; ACh receptors destroyed in myasthenia gravis; binding of nicotine to nicotinic ACh receptors in the brain enhances excitatory neurotransmitter (glutamate, ACh) release by enhancing presynaptic Ca^{2+} levels; may account for behavioral effects of nicotine in smokers; atropine competes with ACh for binding sites

$$H_3C - \overset{\overset{\displaystyle O}{\|}}{C} - O - CH_2 - CH_2 - \overset{+}{N} - (CH_3)_3$$

Biogenic Amines			
Norepinephrine	Excitatory or inhibitory, depending on receptor type bound Indirect action via second messengers	CNS: brain stem, particularly in the locus coeruleus of the midbrain; limbic system; some areas of cerebral cortex PNS: main neurotransmitter of ganglion cells in the sympathetic nervous system	A "feeling good" neurotransmitter; release enhanced by amphetamines; removal from synapse blocked by tricyclic antidepressants [amitriptyline (Elavil) and others] and cocaine; brain levels reduced by reserpine (an antihypertensive drug), leading to depression
Dopamine	Excitatory or inhibitory depending on the receptor type bound Indirect action via second messengers	CNS: substantia nigra of midbrain; hypothalamus; is the principal neurotransmitter of extrapyramidal system PNS: some sympathetic ganglia	A "feeling good" neurotransmitter; release enhanced by L-dopa and amphetamines; reuptake blocked by cocaine; deficient in Parkinson's disease; may be involved in pathogenesis of schizophrenia
Serotonin (5-HT)	Mainly inhibitory Indirect action via second messengers; direct action at 5-HT₃ receptors	CNS: brain stem, especially midbrain; hypothalamus; limbic system; cerebellum; pineal gland; spinal cord	Activity blocked by LSD; may play a role in sleep, appetite, nausea, migraine headaches, and regulation of mood; drugs that block its uptake (Prozac) relieve anxiety and depression
Histamine	Indirect action via second messengers	CNS: hypothalamus	Also released by mast cells during inflammation and acts as powerful vasodilator

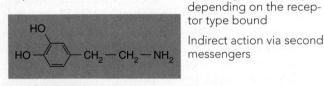

TABLE 11.3 *(continued)*

Neurotransmitter	Functional Classes	Sites Where Secreted	Comments
ATP			
	Excitatory or inhibitory depending on receptor type bound Direct and indirect actions via second messengers	CNS: basal nuclei, induces Ca^{2+} wave propagation in astrocytes PNS: dorsal root ganglion neurons	ATP released by sensory neurons (as well as that released by injured cells) provokes pain sensation

Amino Acids			
GABA (γ-aminobutyric acid)	Generally inhibitory Direct action	CNS: hypothalamus; Purkinje cells of cerebellum; spinal cord; granule cells of olfactory bulb; retina	Principal inhibitory neurotransmitter in the brain; important in presynaptic inhibition at axoaxonal synapses; inhibitory effects augmented by alcohol (resulting in impaired motor coordination) and antianxiety drugs of the benzodiazepine class (e.g., Valium); substances that block its synthesis, release, or action induce convulsions

$$H_2N - CH - CH_2 - CH_2 - COOH$$

Glutamate	Generally excitatory Direct action	CNS: spinal cord; widespread in brain where it represents the major excitatory neurotransmitter	Important in learning and memory; the "stroke neurotransmitter" —excessive release produces excitotoxicity: neurons literally stimulated to death; most commonly caused by ischemia (oxygen deprivation, usually due to a blocked blood vessel); when released by gliomas aids tumor advance

$$H_2N - CH - CH_2 - CH_2 - COOH$$
$$\qquad\quad | $$
$$\qquad\; COOH$$

Glycine	Generally inhibitory Direct action	CNS: spinal cord and brain stem; retina	Strychnine blocks glycine receptors, resulting in uncontrolled convulsions and respiratory arrest

$$H_2N - CH_2 - COOH$$

Peptides			
Endorphins, dynorphin, enkephalins (illustrated) Tyr Gly Gly Phe Met	Generally inhibitory Indirect action via second messengers	CNS: widely distributed in brain; hypothalamus; limbic system; pituitary; spinal cord	Natural opiates; inhibit pain by inhibiting substance P; effects mimicked by morphine, heroin, and methadone
Tachykinins: Substance P (illustrated), neurokinin A (NKA) Arg Pro Lys Pro Gln Gln Phe Phe Gly Leu Met	Excitatory Indirect action via second messengers	CNS: basal nuclei; midbrain; hypothalamus; cerebral cortex PNS: certain sensory neurons of dorsal root ganglia of spinal cord (pain afferents)	Substance P mediates pain transmission in the PNS; in the CNS tachykinins are involved in respiratory and cardiovascular controls and in mood

TABLE 11.3   Neurotransmitters *(continued)*

Neurotransmitter	Functional Classes	Sites Where Secreted	Comments
Somatostatin	Generally inhibitory Indirect action via second messengers	CNS: hypothalamus; retina and other parts of brain Pancreas	Inhibits release of growth hormone; a gut-brain peptide

Ala–Gly–Cys–Lys–Asn–Phe–Phe–Trp–Lys–Thr–Phe–Thr–Ser–Cys

| Cholecystokinin (CCK) | Possible neurotransmitter | Cerebral cortex

Small intestine | May be related to feeding behaviors; a gut-brain peptide |

Asp–Tyr–Met–Gly–Trp–Met–Asp–Phe
|
SO₄ → SO_4

Dissolved Gases

| Nitric oxide (NO) | Excitatory

Indirect action via second messengers | CNS: brain spinal cord

PNS: adrenal gland; nerves to penis | Its release potentiates stroke damage; some types of male impotence treated by stimulating NO release |
| Carbon monoxide (CO) | Excitatory

Indirect action via second messengers | Brain and some neuromuscular and neuroglandular synapses | — |

the brain and the adrenal medulla. *Serotonin* is synthesized from the amino acid tryptophan. *Histamine* is synthesized from the amino acid histidine.

Biogenic amine neurotransmitters are broadly distributed in the brain, where they play a role in emotional behavior and help regulate the biological clock. Additionally, catecholamines (particularly NE) are released by some motor neurons of the autonomic nervous system. Imbalances of these neurotransmitters are associated with mental illness; for example, overproduction of dopamine occurs in schizophrenia. Additionally, certain psychoactive drugs (LSD and mescaline) can bind to biogenic amine receptors and induce hallucinations.

Amino Acids   It is difficult to prove a neurotransmitter role when the suspect is an *amino acid,* because amino acids occur in all cells of the body and are important in many biochemical reactions. The amino acids for which a neurotransmitter role is certain include **gamma (γ)-aminobutyric acid (GABA),** **glycine, aspartate,** and **glutamate,** but there are others. Amino acid neurotransmitters have so far been found only in the CNS.

Peptides   The **neuropeptides,** essentially strings of amino acids, include a broad spectrum of molecules with diverse effects. For example, a neuropep-

tide called **substance P** is an important mediator of pain signals. By contrast, **endorphins,** which include **beta endorphin, dynorphin,** and **enkephalins** (en-kef'ah-linz), act as natural opiates, reducing our perception of pain under certain stressful conditions. Enkephalin activity increases dramatically in pregnant women in labor. Endorphin release is enhanced when an athlete gets a so-called second wind and is probably responsible for the "runner's high." Additionally, some researchers claim that the placebo effect is due to endorphin release. These pain-killing neurotransmitters remained undiscovered until investigators began to ask why morphine and other opiates reduce anxiety and pain, and found that these drugs attach to the same receptors that bind natural opiates, producing similar but stronger effects.

Some neuropeptides, such as somatostatin and cholecystokinin, are also produced by nonneural body tissues and are widespread in the gastrointestinal tract. Such peptides are commonly referred to as **gut-brain peptides.**

Novel Messengers   Just a few years ago, it would have been scientific suicide to suggest that ATP, nitric oxide, and carbon monoxide—all ubiquitous molecules—might be neurotransmitters. Nonetheless, the discovery of these unlikely messengers has opened up a whole new chapter in the story of neurotransmission.

Although it was well known that **adenosine triphosphate (ATP)**, the universal form of cellular energy, was stored in synaptic vesicles, its presence there was thought to promote synthesis or reuptake of other neurotransmitters. ATP is now recognized as a major neurotransmitter (perhaps the most primitive one) in both the CNS and PNS. Like glutamate and acetylcholine it produces a fast excitatory response at certain receptors. Depending on the ATP receptor type it binds to, ATP binding can mediate fast excitatory responses or trigger slow second-messenger responses.

Even more preposterous was the idea that **nitric oxide (NO)**, a short-lived toxic gas, could be a neurotransmitter. It defies all the official descriptions of neurotransmitters in that rather than being stored in vesicles and released by exocytosis, it is synthesized on demand and diffuses out of the cells making it. Instead of attaching to surface receptors, it zooms through the plasma membrane of nearby cells to bind with a peculiar intracellular receptor—iron in *guanylyl cyclase,* the enzyme that makes the second messenger *cyclic GMP.* The precise role of NO in the brain has been hard to pin down, but when the neurotransmitter glutamate binds to NMDA receptors, Ca^{2+} channels open to allow an influx of Ca^{2+} into the cell. The Ca^{2+} triggers reactions that activate *nitric oxide synthase (NOS),* the enzyme needed for NO synthesis. Many believe NO is the retrograde messenger involved in long-term potentiation (learning and memory)—that is, it may travel back from the postsynaptic neuron to the presynaptic neuron where it activates guanylyl cyclase. Excessive release of NO is responsible for much of the brain damage seen in stroke patients (see pp. 467–468). In the myenteric plexus of the intestine, NO causes intestinal smooth muscle to relax.

It may be that NO is just the first of a soon-to-be-discovered class of signaling gases that pass swiftly into cells, bind briefly to metal-containing enzymes, and then vanish. **Carbon monoxide (CO)**, another airy messenger, also stimulates synthesis of cyclic GMP, and some researchers speculate that CO is the main regulator of cyclic GMP levels in the brain. Though NO and CO are found in different brain regions and appear to act in different pathways, their mode of action is very similar. The proposed role of CO in neurotransmission leads to an interesting speculation: Heavy smokers who quit often bemoan the fact that they can't concentrate for weeks after. Because CO displaces oxygen in the blood, it would seem that they should think better after quitting. Perhaps the higher blood CO levels they were accustomed to enhanced neurotransmission in certain circuits involved in logic. Time will tell.

Classification of Neurotransmitters by Function

This text cannot begin to describe the incredible diversity of functions that neurotransmitters mediate. Therefore, we limit our discussion to two broad ways of classifying neurotransmitters according to function; more details will be added as appropriate in subsequent chapters.

Effects: Excitatory Versus Inhibitory   We can summarize this classification scheme by saying that some neurotransmitters are excitatory (cause depolarization), some are inhibitory (cause hyperpolarization), and others exert both effects, depending on the specific receptor types with which they interact. For example, the amino acids GABA and glycine are usually inhibitory, while glutamate is typically excitatory (see Table 11.3). On the other hand, ACh and NE each bind to at least two receptor types that cause opposite effects. For example, acetylcholine is excitatory at neuromuscular junctions with skeletal muscle and inhibitory in cardiac muscle.

Mechanism of Action: Direct Versus Indirect
Neurotransmitters that open ion channels are said to act *directly.* These neurotransmitters provoke rapid responses in postsynaptic cells by promoting changes in membrane potential. ACh and the amino acid neurotransmitters are direct-acting neurotransmitters.

Neurotransmitters that act *indirectly* promote broader, longer-lasting effects by acting through intracellular *second-messenger* molecules [typically via G protein mechanisms as described in Chapter 3 (p. 78)]. In this way their mechanism of action is similar to that of many hormones. The biogenic amines, neuropeptides, and the dissolved gases are indirect neurotransmitters.

Neurotransmitter Receptors

In Chapter 3, we introduced the various receptors involved in cell signaling. Now we are ready to pick up that thread again as we examine the action of receptors that bind neurotransmitters. For the most part, neurotransmitter receptors are either channel-linked receptors, which mediate fast synaptic transmission, or G protein–linked receptors, which oversee slow synaptic responses.

Mechanism of Action of Channel-Linked Receptors   Equivalent to ligand-gated ion channels, **channel-linked receptors** mediate direct transmitter action. Also called *ionotropic receptors,* these receptors are composed of several protein subunits arranged in a "rosette" around a central pore. As the ligand binds to one (or more) receptor subunits, the proteins change shape. This event opens the central

Why is cyclic AMP called a second messenger?

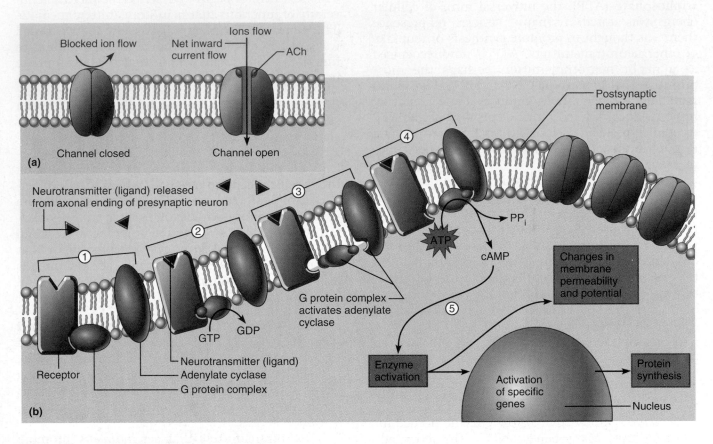

FIGURE 11.23 Neurotransmitter receptor mechanisms. (a) Channel-linked receptors open in response to ligand (ACh in this case) binding. When no ligand is bound, the channel is closed and no current passes through it. As soon as ligand binds, the channel opens and ions flow through it. The precise ion current is determined by the structure and charge of the channel proteins. (In ACh channels, Na^+, K^+, and Ca^{2+} pass, resulting in a depolarizing current.) **(b)** A G protein–linked receptor. ① Unbound receptor. ② Binding of the neurotransmitter (ligand) to the receptor results in receptor–G protein interaction and activation of the G protein. Once GTP replaces GDP in the G protein complex, ③ the G protein can interact with and activate adenylate cyclase. ④ Activated adenylate cyclase catalyzes the formation of cyclic AMP (cAMP) from ATP. ⑤ cAMP, acting as an intracellular second messenger, mediates events which activate enzymes that bring about the postsynaptic neuron's response (changes in membrane potential, protein synthesis, etc.).

channel and allows ions to pass (Figure 11.23a), altering the membrane potential of the target cell. Channel-linked receptors are always located precisely opposite sites of neurotransmitter release, and their ion channels open instantly upon ligand binding and remain open 1 ms or less while the ligand is bound. At excitatory receptor sites (nicotinic ACh channels and receptors for glutamate, aspartate, and ATP), the channel-linked receptors are *cation* channels that allow small cations (Na^+, K^+, Ca^{2+}) to pass, but the greatest drive is for Na^+ entry, which contributes to membrane depolarization. Channel-linked receptors that respond to GABA and glycine, and allow K^+ or Cl^- to pass, mediate fast inhibition (hyperpolarization).

Mechanism of Action of G Protein-Linked Receptors

Unlike responses to neurotransmitter binding at channel-linked receptors, which are immediate, simple, brief, and highly localized at a single postsynaptic cell, the activity mediated by

The term first messenger designates the original chemical messenger (the stimulus), which in this case is the neurotransmitter. The binding of the first messenger promotes events that generate the second messenger, which acts within the cell to bring about the desired response.

G protein–linked receptors is indirect, slow (hundreds of ms or more), complex, prolonged, and often diffuse—ideal as a basis for some types of learning. Receptors that fall into this class are transmembrane protein complexes and include muscarinic ACh receptors and those that bind the biogenic amines and neuropeptides. When a neurotransmitter binds to a G protein–linked receptor, the G protein is activated (Figure 11.23b). (You might like to refer back to the simpler G protein explanation in Figure 3.16 on p. 78 to orient yourself.) Activated G proteins typically work by controlling the production of second messengers such as **cyclic AMP, cyclic GMP, diacylglycerol,** or **Ca^{2+}**. These, in turn, act as go-betweens to regulate (open *or* close) ion channels or activate kinase enzymes that initiate a cascade of enzymatic reactions in the target cells. Some of these second messengers modify (activate or inactivate) other proteins, including channel proteins, by attaching phosphate groups to them. Others interact with nuclear proteins that activate genes and induce synthesis of new proteins in the target cell. Because the effects produced tend to bring about widespread metabolic changes, G protein–linked receptors are commonly called *metabotropic receptors.*

Basic Concepts of Neural Integration

Until now, we have been concentrating on the activities of individual neurons; however, neurons function in groups, and each group contributes to still broader neural functions. Thus, the organization of the nervous system is hierarchical, or ladderlike.

Any time you have a large number of *anything*—people included—there must be *integration;* that is, the parts must be fused into a smoothly operating whole. In this section, we begin to ascend the ladder of **neural integration,** dealing only with the first rung, *neuronal pools* and their patterns of communicating with other parts of the nervous system. Chapter 13 picks up the thread of neural integration once again to examine how sensory inputs interface with motor activity. The highest levels of neural integration—how we think and remember—are discussed in Chapter 12.

Organization of Neurons: Neuronal Pools

The millions of neurons in the CNS are organized into **neuronal pools,** functional groups of neurons that integrate incoming information received from receptors or different neuronal pools and then forward the processed information to other destinations.

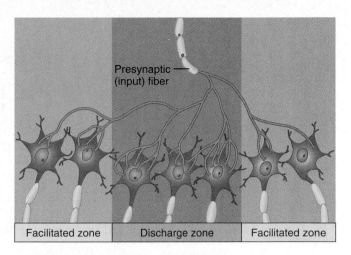

Presynaptic (input) fiber

Facilitated zone | Discharge zone | Facilitated zone

FIGURE 11.24   Simple neuronal pool. This simplified representation shows the relative position of postsynaptic neurons in the discharge and facilitated zones. Notice that the presynaptic fiber makes more synapses per neuron with neurons in the discharge zone.

In a simple type of neuronal pool, shown in Figure 11.24, one incoming presynaptic fiber branches profusely as it enters the pool and then synapses with several different neurons in the pool. When the incoming fiber is excited, it will excite some postsynaptic neurons and facilitate others. Neurons most likely to generate impulses are those closely associated with the incoming fiber, because they receive the bulk of the synaptic contacts. Those neurons are said to be in the *discharge zone* of the pool. Neurons farther from the center are not usually excited to threshold by EPSPs induced by the incoming fiber, but they are facilitated and can easily be brought to threshold by stimuli from another source. Thus, the periphery of the pool is the *facilitated zone.* Keep in mind, however, that our figure is a gross oversimplification. Most neuronal pools consist of thousands of neurons and include inhibitory as well as excitatory neurons.

Types of Circuits

Individual neurons in a neuronal pool both send and receive information, and synaptic contacts may cause either excitation or inhibition. The patterns of synaptic connections in neuronal pools are called **circuits,** and they determine the pool's functional capabilities. Four basic circuit patterns are shown in simplified form in Figure 11.25.

In **diverging circuits,** one incoming fiber triggers responses in ever-increasing numbers of neurons farther and farther along in the circuit. Thus, diverging circuits are often *amplifying* circuits. Divergence can occur along a single pathway or along several (Figure 11.25a and b). These circuits are common in

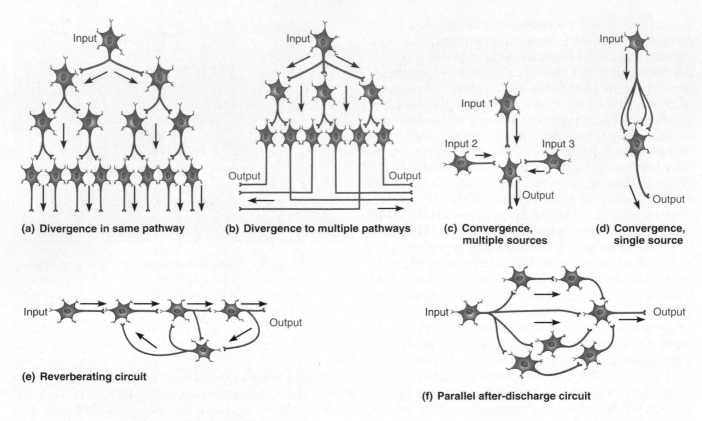

(a) **Divergence in same pathway**

(b) **Divergence to multiple pathways**

(c) **Convergence, multiple sources**

(d) **Convergence, single source**

(e) **Reverberating circuit**

(f) **Parallel after-discharge circuit**

FIGURE 11.25   **Types of circuits in neuronal pools.**

both sensory and motor systems. For example, impulses traveling from a single neuron of the brain can activate a hundred or more motor neurons in the spinal cord and, consequently, thousands of skeletal muscle fibers.

The pattern of **converging circuits** is opposite that of diverging circuits, but they too are common in both sensory and motor pathways. In a converging circuit, the pool receives inputs from several presynaptic neurons, and the circuit has a funneling, or *concentrating*, effect. Incoming stimuli may converge from one area or from many different areas (Figure 11.25c and d), which results in strong stimulation or inhibition. The former condition helps explain how different types of sensory stimuli can have the same ultimate effect. For instance, seeing the smiling face of their infant, smelling the baby's freshly powdered skin, or hearing the baby gurgle can all trigger a flood of loving feelings in parents.

In **reverberating, or oscillating, circuits** (Figure 11.25e), the incoming signal travels through a chain of neurons, each of which makes collateral synapses with neurons in a previous part of the pathway. As a result of the positive feedback, the impulses *reverberate* (are sent through the circuit again and again), giving a continuous output signal until one neuron in the circuit fails to fire. Reverberating circuits are involved in control of rhythmic activities, such as the sleep-wake cycle, breathing, and certain motor

activities (such as arm swinging when walking). Some researchers believe that such circuits underlie short-term memory. Depending on the specific circuit, reverberating circuits may continue to oscillate for seconds, hours, or (in the case of the circuit controlling the rhythm of breathing) a lifetime.

In **parallel after-discharge circuits,** the incoming fiber stimulates several neurons arranged in parallel arrays that eventually stimulate a common output cell (Figure 11.25f). Impulses reach the output cell at different times, creating a burst of impulses called an *after discharge* that lasts 15 ms or more after the initial input has ended. This type of circuit has no positive feedback, and once all the neurons have fired, circuit activity ends. Parallel after-discharge circuits may be involved in complex, exacting types of mental processing, such as working on mathematical problems.

Patterns of Neural Processing

Input processing is both *serial* and *parallel*. In serial processing, the input travels along one pathway to a specific destination. In parallel processing, the input travels along several different pathways to be integrated in different CNS regions. Each mode has unique advantages in the overall scheme of neural functioning, but as an information processor, the brain derives its power from its ability to process in parallel.

Serial Processing

In **serial processing,** the whole system works in a predictable all-or-nothing manner. One neuron stimulates the next, which stimulates the next, and so on, eventually causing a specific, anticipated response. The most clear-cut examples of serial processing are spinal reflexes, but straight-through sensory pathways from receptors to the brain are also examples. Because reflexes are the functional units of the nervous system, it is important that you understand them early on.

Reflexes are rapid, automatic responses to stimuli, in which a particular stimulus always causes the *same* motor response. Reflex activity, which produces the simplest of behaviors, is stereotyped and dependable. For example, jerking away your hand after touching a hot object is the norm, and an object approaching the eye triggers a blink. Reflexes occur over neural pathways called **reflex arcs** that have five essential components—receptor, sensory neuron, CNS integration center, motor neuron, and effector (Figure 11.26).

Parallel Processing

In **parallel processing,** inputs are segregated into many pathways, and information delivered by each pathway is dealt with simultaneously by different parts of the neural circuitry. For example, smelling a pickle (the input) may cause you to remember picking cucumbers on a farm; or it may remind you that you don't like pickles or that you must buy some at the market; or perhaps it will call to mind *all* these thoughts. For each person, parallel processing triggers some pathways that are unique. The same stimulus—pickle smell, in our example—promotes many responses beyond simple awareness of the smell. Parallel processing is not repetitious because

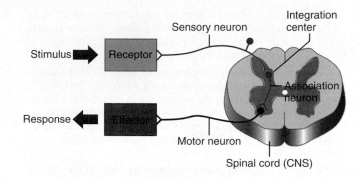

FIGURE 11.26 **A simple reflex arc.**

the circuits do different things with the information, and each "channel" is decoded in relation to all the others to produce a total picture.

Think, for example, about what happens when you step on something sharp while walking barefoot. The serially processed withdrawal reflex causes instantaneous removal of your injured foot from the sharp object (painful stimulus). At the same time, pain and pressure impulses are speeding up to the brain along parallel pathways that allow you to decide whether to simply rub the hurt spot to soothe it or to seek first aid.

Parallel processing is extremely important for higher level mental functioning—for putting the parts together to understand the whole. Because of parallel processing, you can recognize a dollar bill in a split second, a task that takes a serial-based computer a fairly long time. This is because parallel processing, which allows a single neuron to send information along several pathways instead of just one, allows a large amount of information to be packed into a small volume. Hence, logic systems can work much faster.

Review Questions

Multiple Choice/Matching

(Some questions have more than one correct answer. Select the best answer or answers from the choices given.)

1. Which of the following structures is not part of the central nervous system? (a) the brain, (b) a nerve, (c) the spinal cord, (d) a tract.

2. Match the names of the supporting cells found in column B with the appropriate descriptions in column A.

Column A	Column B
____ **(1)** myelinates nerve fibers in the CNS	**(a)** astrocyte
____ **(2)** lines brain cavities	**(b)** ependymal cell
____ **(3)** myelinates nerve fibers in the PNS	**(c)** microglia
	(d) oligodendrocyte
	(e) satellite cell
____ **(4)** CNS phagocytes	**(f)** Schwann cell
____ **(5)** may help regulate the ionic composition of the extracellular fluid	

3. What type of current flows through the axolemma during the steep phase of repolarization? (a) chiefly a sodium current, (b) chiefly a potassium current, (c) sodium and potassium currents of approximately the same magnitude.

4. Assume that an EPSP is being generated on the dendritic membrane. Which will occur? (a) specific Na^+ channels will open, (b) specific K^+ channels will open, (c) a single type of channel will open, permitting simultaneous flow of Na^+ and K^+, (d) Na^+ channels will open first and then close as K^+ channels open.

5. The velocity of nerve impulse conduction is greatest in (a) heavily myelinated, large-diameter fibers, (b) myelinated, small-diameter fibers, (c) unmyelinated, small-diameter fibers, (d) unmyelinated, large-diameter fibers.

6. Chemical synapses are characterized by all of the following except (a) the release of neurotransmitter by the presynaptic membranes, (b) postsynaptic membranes bearing receptors that bind neurotransmitter, (c) ions flowing through protein channels from the presynaptic to the postsynaptic neuron, (d) a fluid-filled gap separating the neurons.

7. Biogenic amine neurotransmitters include all but (a) norepinephrine, (b) acetylcholine, (c) dopamine, (d) serotonin.

8. The neuropeptides that act as natural opiates are (a) substance P, (b) somatostatin, (c) cholecystokinin, (d) enkephalins.

9. Inhibition of acetylcholinesterase by poisoning blocks neurotransmission at the neuromuscular junction because (a) ACh is no longer released by the presynaptic terminal, (b) ACh synthesis in the presynaptic terminal is blocked, (c) ACh is not degraded, hence prolonged depolarization is enforced on the postsynaptic neuron, (d) ACh is blocked from attaching to the postsynaptic ACh receptors.

10. The anatomical region of a multipolar neuron that has the lowest threshold for generating an action potential is the (a) soma, (b) dendrites, (c) axon hillock, (d) distal axon.

11. An IPSP is inhibitory because (a) it hyperpolarizes the postsynaptic membrane, (b) it reduces the amount of neurotransmitter released by the presynaptic terminal, (c) it prevents calcium ion entry into the presynaptic terminal, (d) it changes the threshold of the neuron.

12. Identify the neuronal circuits described by choosing the correct response from the key.

Key: (a) converging, **(b)** diverging, **(c)** parallel after-discharge, **(d)** reverberating

_____ **(1)** Impulses continue around and around the circuit until one neuron stops firing.
_____ **(2)** One or a few inputs ultimately influence large numbers of neurons.
_____ **(3)** Many neurons influence a few neurons.
_____ **(4)** May be involved in exacting types of mental activity.

Short Answer Essay Questions

13. Explain both the anatomical and functional divisions of the nervous system. Include the subdivisions of each.

14. (a) Describe the composition and function of the cell body. (b) How are axons and dendrites alike? In what ways (structurally and functionally) do they differ?

15. (a) What is myelin? (b) How does the myelination process differ in the CNS and PNS?

16. (a) Contrast unipolar, bipolar, and multipolar neurons structurally. (b) Indicate where each is most likely to be found.

17. What is the polarized membrane state? How is it maintained? (Note the relative roles of both passive and active mechanisms.)

18. Describe the events that must occur to generate an action potential. Indicate how the ionic gates are controlled, and explain why the action potential is an all-or-none phenomenon.

19. Since all action potentials generated by a given nerve fiber have the same magnitude, how does the CNS "know" whether a stimulus is strong or weak?

20. (a) Explain the difference between an EPSP and an IPSP. (b) What specifically determines whether an EPSP or IPSP will be generated at the postsynaptic membrane?

21. Since at any moment a neuron is likely to have thousands of neurons releasing neurotransmitters at its surface, how is neuronal activity (to fire or not to fire) determined?

22. The effects of neurotransmitter binding are very brief. Explain.

23. During a neurobiology lecture, a professor repeatedly refers to type A and type B fibers, absolute refractory period, and nodes of Ranvier. Define these terms.

24. Distinguish between serial and parallel processing.

12

THE CENTRAL NERVOUS SYSTEM

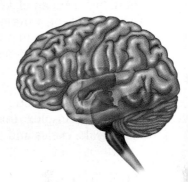

The Brain (pp. 380–405)

1. Describe the process of brain development.
2. Name the major regions of the adult brain.
3. Name and locate the ventricles of the brain.
4. List the major lobes, fissures, and functional areas of the cerebral cortex.
5. Explain lateralization of hemisphere function.
6. Differentiate between commissures, association fibers, and projection fibers.
7. Describe the general function of the basal nuclei (basal ganglia).
8. Describe the location of the diencephalon, and name its subdivisions.
9. Identify the three major regions of the brain stem, and note the functions of each area.
10. Describe the structure and function of the cerebellum.
11. Locate the limbic system and the reticular formation, and explain the role of each functional system.

Higher Mental Functions (pp. 405–412)

12. Define EEG and distinguish between alpha, beta, theta, and delta brain waves.
13. Compare and contrast the events and importance of slow-wave and REM sleep, and indicate how their patterns change through life.
14. Describe consciousness clinically.
15. Compare and contrast the stages and categories of memory.
16. Describe the relative roles of the major brain structures believed to be involved in fact and skill memories.

Protection of the Brain (pp. 412–417)

17. Describe how meninges, cerebrospinal fluid, and the blood-brain barrier protect the CNS.
18. Describe the formation of cerebrospinal fluid, and follow its circulatory pathway.
19. Describe the cause (if known) and major signs and symptoms of cerebrovascular accidents, Alzheimer's disease, Huntington's disease, and Parkinson's disease.

The Spinal Cord (pp. 417–428)

20. Describe the gross and microscopic structure of the spinal cord.
21. List the major spinal cord tracts, and classify each as a motor or sensory tract.
22. Distinguish between flaccid and spastic paralysis and between paralysis and paresthesia.

Diagnostic Procedures for Assessing CNS Dysfunction (p. 428)

23. List and explain several techniques used to diagnose brain disorders.

Historically, the **central nervous system (CNS)**—brain and spinal cord—has been compared to the central switchboard of a telephone system that interconnects and directs a dizzying number of incoming and outgoing calls. Nowadays, many people compare it to a supercomputer. These analogies may explain some workings of the spinal cord, but neither does justice to the fantastic complexity of the human brain. Whether we view the brain as an evolved biological organ, an impressive computer, or simply a miracle, it is one of the most amazing things known.

During the course of animal evolution, **cephalization** (se″fah-lĭ-za′shun) has occurred. That is, there has been an elaboration of the rostral ("toward the snout"), or anterior, portion of the CNS, along with an increase in the number of neurons in the head. This phenomenon reaches its highest level in the human brain.

In this chapter, we examine the structure of the CNS and the functions associated with its various regions. Complex integrative functions, such as sleep-wake cycles and memory, are also touched on.

The Brain

The unimpressive appearance of the human **brain** gives few hints of its remarkable abilities. It is about two good fistfuls of quivering pinkish gray tissue, wrinkled like a walnut, and somewhat the consistency of cold oatmeal. The average adult man's brain has a mass of about 1600 g (3.5 pounds); that of a woman averages 1450 g. In terms of brain mass per body mass, however, males and females have equivalent brain sizes.

Embryonic Development

We will begin our study of the brain with brain embryology, as the terminology used for the structural divisions of the adult brain is easier to follow when you understand brain development.

The earliest phase of brain development is shown in Figure 12.1. Starting in the three-week-old embryo, the ectoderm (cell layer at the dorsal surface) thickens along the dorsal midline axis of the embryo to form the **neural plate.** The neural plate then invaginates, forming a groove flanked by **neural folds.** As this **neural groove** deepens, the superior edges of the neural folds fuse, forming the **neural tube,** which soon detaches from the surface ectoderm and sinks to a deeper position. The neural tube, formed by the fourth week of pregnancy, differentiates rapidly into the CNS. The brain forms anteriorly (rostrally) and the spinal cord develops from the caudal ("toward the tail") or posterior portion of the neural

Anterior (rostral) end

(a) 19 days

(b) 20 days

(c) 22 days

(d) 26 days

Level of section

Surface ectoderm

Neural plate

Neural folds

Neural groove

Neural crest

Surface ectoderm

Neural tube

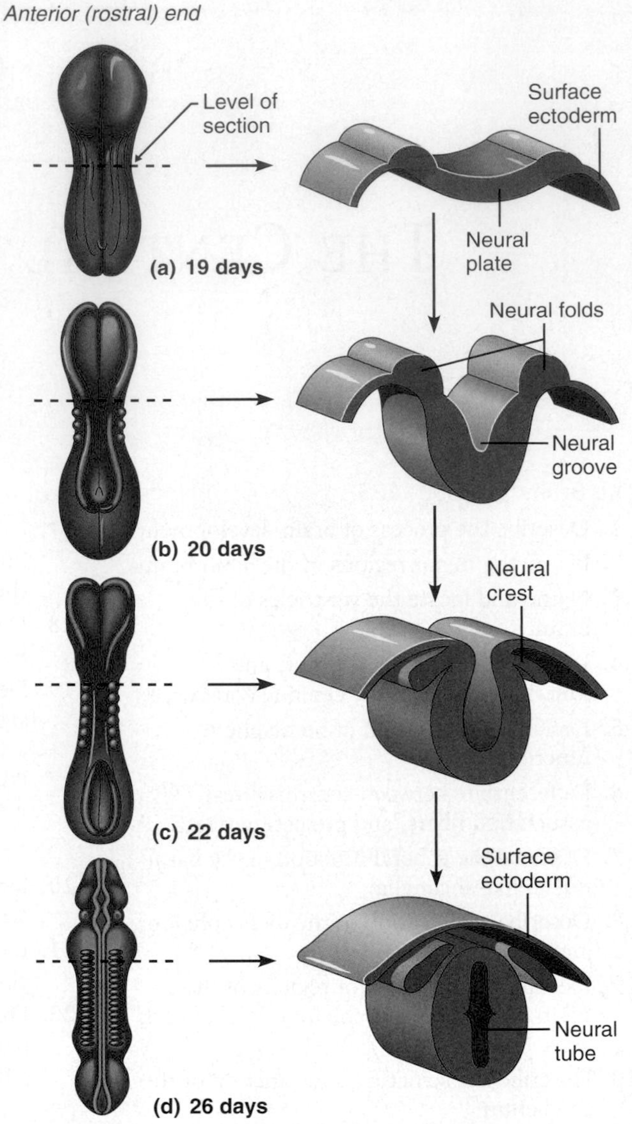

FIGURE 12.1   Development of the neural tube from embryonic ectoderm. Left: dorsal surface views of the embryo; right: transverse sections. **(a)** Formation of the neural plate from surface ectoderm. **(b–d)** Development of neural plate into neural groove (flanked by neural folds) and then neural tube. The neural tube forms CNS structures; neural crest cells form some PNS structures.

tube. Small groups of neural fold cells migrate laterally from between the surface ectoderm and the neural tube, forming the **neural crest** (Figure 12.1c), which gives rise to some neurons destined to reside in ganglia.

As soon as the neural tube forms, its anterior end immediately begins to expand and constrictions appear that mark off the three **primary brain vesicles** (Figure 12.2): the **prosencephalon** (pros″en-sef′ah-lon), or **forebrain;** the **mesencephalon** (mes″en-sef′ah-lon), or **midbrain;** and the **rhombencephalon** (romb″en-sef′ah-lon), or **hindbrain.** (Note that

(a) Neural tube	(b) Primary brain vesicles	(c) Secondary brain vesicles		(d) Adult brain structures	(e) Adult neural canal regions
Anterior (rostral)	Prosencephalon (forebrain)	Telencephalon		Cerebrum: Cerebral hemispheres (cortex, white matter, basal nuclei)	Lateral ventricles
		Diencephalon		Diencephalon (thalamus, hypothalamus, epithalamus)	Third ventricle
	Mesencephalon (midbrain)	Mesencephalon		Brain stem: midbrain	Cerebral aqueduct
	Rhombencephalon (hindbrain)	Metencephalon		Brain stem: pons	Fourth ventricle
				Cerebellum	
		Myelencephalon		Brain stem: medulla oblongata	
Posterior (caudal)				Spinal cord	Central canal

FIGURE 12.2 **Embryonic development of the human brain.** (a) Formed by week 4, the neural tube quickly subdivides into (b) the primary brain vesicles, which subsequently form (c) the secondary brain vesicles by week 5. These five vesicles differentiate into (d) the adult brain structures. (e) The adult structures derived from the neural canal.

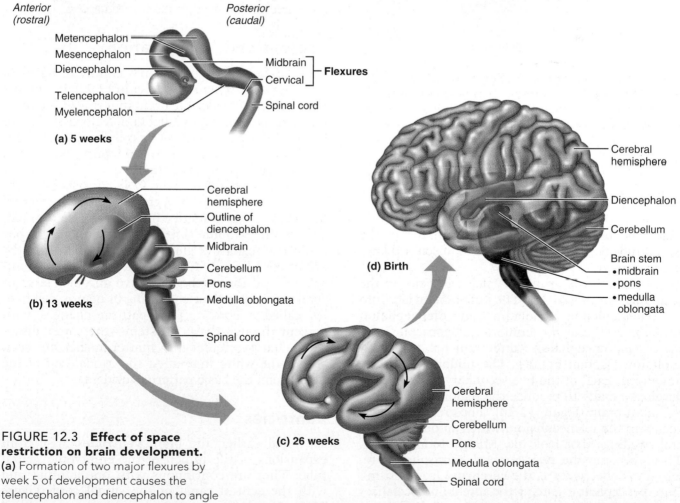

FIGURE 12.3 **Effect of space restriction on brain development.**
(a) Formation of two major flexures by week 5 of development causes the telencephalon and diencephalon to angle toward the brain stem. Development of the cerebral hemispheres at (b) 13 weeks, (c) 26 weeks, and (d) birth. Initially, the cerebral surface is smooth; the folding begins in month 6, and convolutions become more obvious as development continues. The posterolateral growth of the cerebral hemispheres ultimately encloses the diencephalon and superior aspect of the brain stem (seen through the cerebral hemispheres in this see-through view).

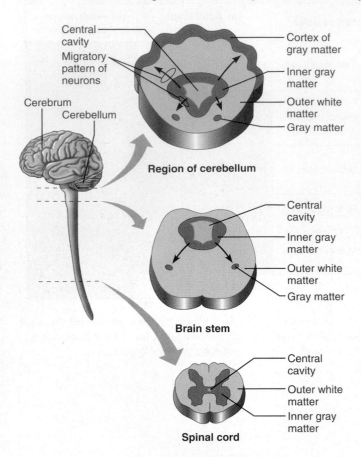

FIGURE 12.4 Arrangement of gray and white matter in the CNS (highly simplified). Cross sections at three CNS levels. In each section, the dorsal aspect is at the top. In general, white matter lies external to gray matter; however, collections of gray matter migrate externally into the white matter in the developing brain (see black arrows). The cerebrum resembles the cerebellum in its external cortex of gray matter.

encephalo means "brain.") The remainder of the neural tube becomes the spinal cord (discussed later in the chapter).

In week 5, the primary vesicles give rise to the **secondary brain vesicles.** The forebrain divides into the **telencephalon** ("endbrain") and **diencephalon** ("interbrain"), and the hindbrain constricts, forming the **metencephalon** ("afterbrain") and **myelencephalon** ("spinal brain"). The midbrain remains undivided. Each of the five secondary vesicles then develops rapidly to produce the major structures of the adult brain (Figure 12.2d). The greatest change occurs in the telencephalon, which sprouts two lateral swellings that look like Mickey Mouse's ears. These become the two *cerebral hemispheres*, referred to collectively as the **cerebrum** (ser'ĕ-brum). The diencephalon part of the forebrain specializes to form the *hypothalamus* (hi"po-thal'ah-mus), *thalamus,* and *epithalamus.* Less dramatic changes occur in the mesencephalon, metencephalon, and

myelencephalon as these regions are transformed into the *midbrain,* the *pons and cerebellum,* and the *medulla oblongata,* respectively. All these midbrain and hindbrain structures, except the cerebellum, form the **brain stem.** The central cavity of the neural tube remains continuous and enlarges in four areas to form the fluid-filled *ventricles (ventr = little belly)* of the brain. The ventricles are described shortly.

Because the brain grows more rapidly than the membranous skull that contains it, two major flexures develop—the *midbrain* and *cervical flexures*—which bend the forebrain toward the brain stem (Figure 12.3a). A second consequence of restricted space is that the cerebral hemispheres are forced to take a horseshoe-shaped course and grow posteriorly and laterally (indicated by black arrows in Figure 12.3b and c). As a result, they grow back over and almost completely envelop the diencephalon and midbrain. By week 26, the continued growth of the cerebral hemispheres causes their surfaces to crease and fold (Figure 12.3c and d), producing *convolutions* and increasing their surface area, which allows more neurons to occupy the limited space.

Regions and Organization

Some textbooks discuss brain anatomy in terms of the *embryonic scheme* (see Figure 12.2c), but in this text, we will consider the brain in terms of the medical scheme and the adult brain regions shown in Figure 12.3d: (1) cerebral hemispheres, (2) diencephalon, (3) brain stem (midbrain, pons, and medulla), and (4) cerebellum.

The basic pattern of the CNS consists of a central cavity surrounded by a gray matter core, external to which is white matter (myelinated fiber tracts). The brain exhibits this basic design but it has additional regions of gray matter not present in the spinal cord (Figure 12.4). Both the cerebral hemispheres and the cerebellum have an outer layer or "bark" of gray matter consisting of neuron cell bodies called a *cortex.* This pattern changes with descent through the brain stem—the cortex disappears, but scattered gray matter nuclei are seen within the white matter. At the caudal end of the brain stem, the basic pattern is evident.

Ventricles

As noted earlier, the brain **ventricles** arise from expansions of the lumen of the embryonic neural tube. They are continuous with one another and with the central canal of the spinal cord (Figure 12.5). The hollow ventricular chambers are filled with cerebrospinal fluid and lined by *ependymal cells,* a type of neuroglia (see Figure 11.3c on p. 344).

? *Why are the lateral ventricles horn-shaped rather than vertically erect like the third and fourth ventricles?*

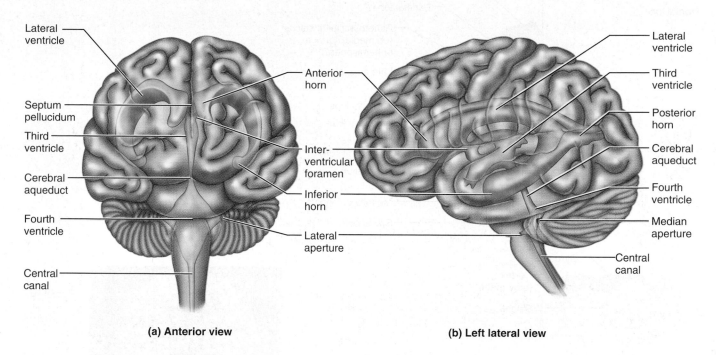

(a) Anterior view   **(b) Left lateral view**

FIGURE 12.5 **Ventricles of the brain.** Note that different regions of the large lateral ventricles are indicated by the terms anterior horn, posterior horn, and inferior horn.

The paired **lateral ventricles,** one deep within each cerebral hemisphere, are large C-shaped chambers that reflect the pattern of cerebral growth. Anteriorly, the lateral ventricles lie close together, separated only by a thin median membrane called the **septum pellucidum** (pĕ-lu'sid-um; "transparent wall"). (See Figure 12.12, p. 394.) Each lateral ventricle communicates with the narrow **third ventricle** in the diencephalon via a channel called an **interventricular foramen** (*foramen of Monro*). The third ventricle is continuous with the **fourth ventricle** via the canal-like **cerebral aqueduct** that runs through the midbrain. The fourth ventricle lies in the hindbrain dorsal to the pons and superior medulla. It is continuous with the central canal of the spinal cord inferiorly. Three openings mark the walls of the fourth ventricle: the paired **lateral apertures** in its side walls and the **median aperture** in its roof. These apertures connect the ventricles to the *subarachnoid* (sub"ah-rak'noid) *space,* a fluid-filled space surrounding the brain.

■ *Because their growth is restricted by the forming embryonic skull, the cerebral hemispheres are forced to grow posteriorly and inferiorly. Hence their cavities (ventricles) become horn-shaped.*

Cerebral Hemispheres

The **cerebral hemispheres** form the superior part of the brain (Figure 12.6). Together they account for about 83% of total brain mass and are the most conspicuous parts of an intact brain. Picture how a mushroom cap covers the top of its stalk, and you have a fairly good idea of how the paired cerebral hemispheres cover and obscure the diencephalon and the top of the brain stem (see Figure 12.3d).

Nearly the entire surface of the cerebral hemispheres is marked by elevated ridges of tissue called **gyri** (ji'ri; "twisters"), separated by shallow grooves called **sulci** (sul'ki; "furrows"). The singular forms of these terms are *gyrus* and *sulcus.* Deeper grooves, called **fissures,** separate large regions of the brain. The more prominent gyri and sulci are similar in all people and are important anatomical landmarks. The median **longitudinal fissure** separates the cerebral hemispheres (Figure 12.6b). Another large fissure, the **transverse cerebral fissure,** separates the cerebral hemispheres from the cerebellum below (see Figure 12.6a).

Some sulci divide each hemisphere into five lobes—frontal, parietal, temporal, occipital, and insula—all but the last named for the cranial bones that overlie them (Figure 12.6a). The **central sulcus,**

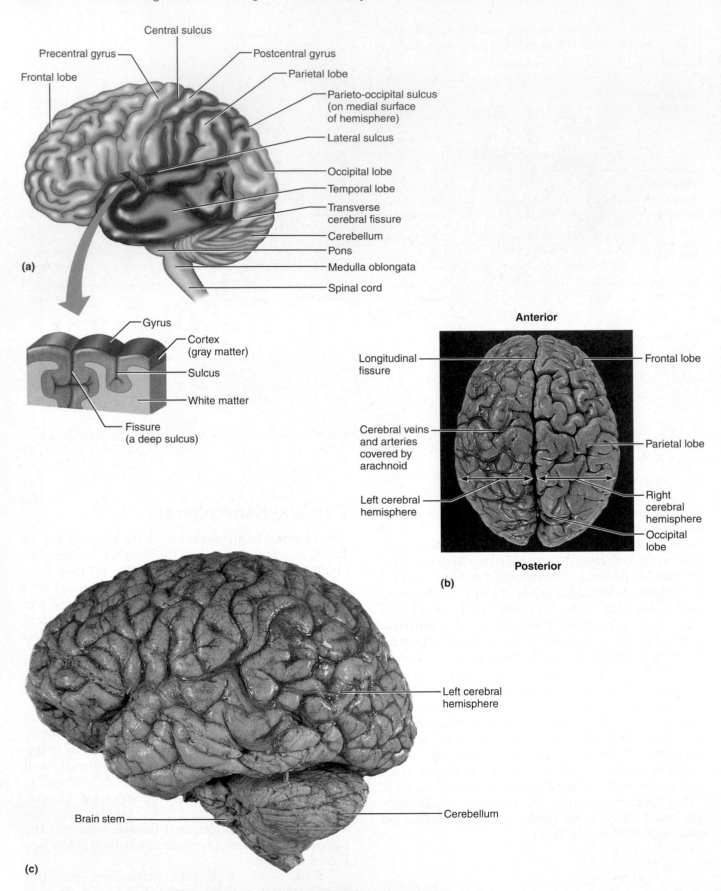

FIGURE 12.6 Lobes and fissures of the cerebral hemispheres. (a) Diagram of the lobes and major sulci and fissures of the brain. **(b)** Superior surface of the cerebral hemispheres. **(c)** Photograph of the left lateral view of the brain.

which lies in the frontal plane, separates the **frontal lobe** from the **parietal lobe.** Bordering the central sulcus are the **precentral gyrus** anteriorly and the **postcentral gyrus** posteriorly. More posteriorly, the **occipital lobe** is separated from the parietal lobe by the **parieto-occipital sulcus** (pah-ri″ĕ-to-ok-sip′ĭ-tal), located on the medial surface of the hemisphere.

The deep **lateral sulcus** outlines the flaplike **temporal lobe** and separates it from the overlying parietal and frontal lobes. A fifth lobe of the cerebral hemisphere, the **insula** (in′su-lah; "island"), is buried deep within the lateral sulcus and forms part of its floor. The insula is covered by portions of the temporal, parietal, and frontal lobes.

The cerebral hemispheres fit snugly in the skull. Rostrally, the frontal lobes lie in the anterior cranial fossa. The anterior parts of the temporal lobes fill the middle cranial fossa. The posterior cranial fossa, however, houses the brain stem and cerebellum; the occipital lobes are located well superior to that cranial fossa.

Each cerebral hemisphere has three basic regions: a superficial *cortex* of gray matter, which looks gray in fresh brain tissue; an internal *white matter*; and the *basal nuclei*, islands of gray matter situated deep within the white matter. We consider these regions next.

Cerebral Cortex

The **cerebral cortex** is the "executive suite" of the nervous system, where our *conscious mind* is found. It enables us to be aware of ourselves and our sensations, to communicate, remember, and understand, and to initiate voluntary movements. Because it is composed of gray matter, the cerebral cortex consists of neuron cell bodies, dendrites, and unmyelinated axons (plus associated glia and blood vessels), but no fiber tracts. It contains billions of neurons arranged in six layers, and accounts for roughly 40% of total brain mass. Although it is only 2–4 mm (about 1/8 inch) thick, its many convolutions effectively triple its surface area.

In the late 1800s, anatomists mapped subtle variations in the thickness and structure of the cerebral cortex. Most successful in these efforts was K. Brodmann, who in 1906 produced an elaborate numbered mosaic of 52 cortical areas, now called **Brodmann areas.** With a structural map emerging, early neurologists were eager to localize *functional* regions of the cortex as well. Modern imaging techniques—PET scans to show maximal metabolic activity in the brain, or functional MRI scans to reveal blood flow (Figure 12.7)—have shown that specific motor and sensory functions are localized in discrete cortical areas called *domains*. However, many higher mental functions, such as memory and language, appear to have overlapping domains

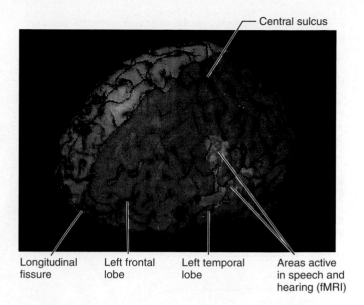

FIGURE 12.7 Functional neuroimaging of the cerebral cortex. The rostral direction is to the left. Functional magnetic resonance image (fMRI) of the cerebral cortex of a person speaking and hearing reveals activity (blood flow) in the posterior frontal and superior temporal lobes, respectively.

and are spread over large areas of the cortex. Some of the most important Brodmann areas are numbered in Figure 12.8.

Before we examine the functional regions of the cerebral cortex, let's consider some generalizations about this region of the brain:

1. The cerebral cortex contains three kinds of functional areas: *motor areas*, *sensory areas*, and *association areas*. As you read about these areas, do not confuse the sensory and motor areas of the cortex with sensory and motor neurons. All neurons in the cortex are interneurons.

2. Each hemisphere is chiefly concerned with the sensory and motor functions of the opposite (contralateral) side of the body.

3. Although largely symmetrical in structure, the two hemispheres are not entirely equal in function. Instead, there is a lateralization (specialization) of cortical functions.

4. The final, and perhaps most important, generalization to keep in mind is that our approach is a gross oversimplification; *no* functional area of the cortex acts alone, and conscious behavior involves the entire cortex in one way or another.

Motor Areas   As shown in Figure 12.8a, the following **motor areas** of the cortex, which control voluntary movement, lie in the posterior part of the frontal lobes: primary motor cortex, premotor cortex, Broca's area, and the frontal eye field.

? *What anatomical landmark separates motor areas of the cerebral cortex from sensory areas?*

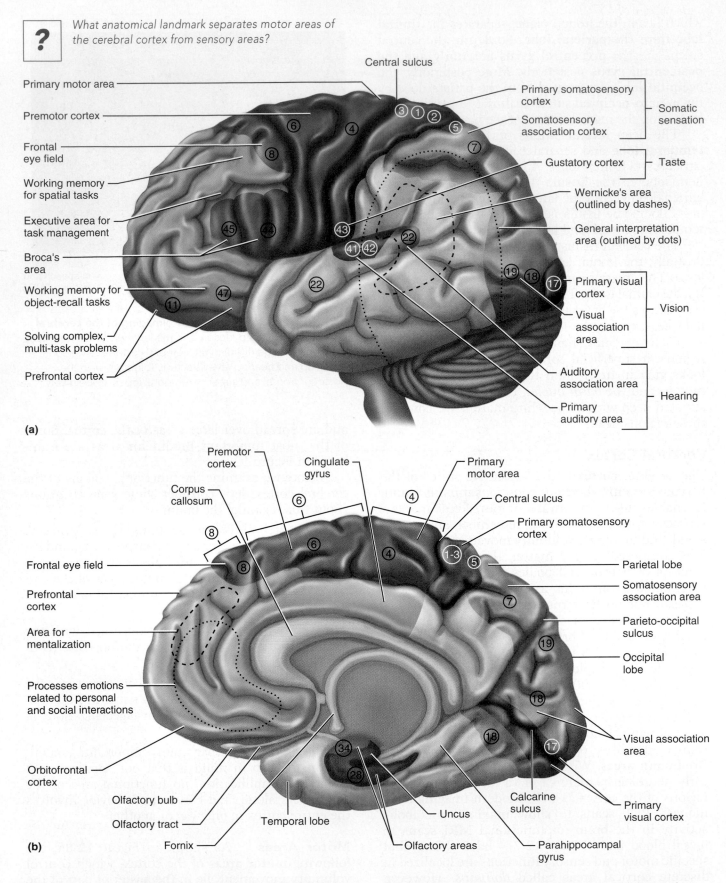

Primary motor area

Premotor cortex

Frontal eye field

Working memory for spatial tasks

Executive area for task management

Broca's area

Working memory for object-recall tasks

Solving complex, multi-task problems

Prefrontal cortex

(a)

Central sulcus

Primary somatosensory cortex

Somatosensory association cortex

Somatic sensation

Gustatory cortex — Taste

Wernicke's area (outlined by dashes)

General interpretation area (outlined by dots)

Primary visual cortex

Visual association area

Vision

Auditory association area

Primary auditory area

Hearing

Premotor cortex

Cingulate gyrus

Primary motor area

Corpus callosum

Central sulcus

Primary somatosensory cortex

Frontal eye field

Prefrontal cortex

Area for mentalization

Processes emotions related to personal and social interactions

Orbitofrontal cortex

Olfactory bulb

Olfactory tract

Fornix

Temporal lobe

Uncus

Olfactory areas

Parietal lobe

Somatosensory association area

Parieto-occipital sulcus

Occipital lobe

Visual association area

Primary visual cortex

Calcarine sulcus

Parahippocampal gyrus

(b)

FIGURE 12.8   Functional and structural areas of the cerebral cortex. (a) Lateral view, left cerebral hemisphere. Different colors define functional regions of the cortex. The olfactory area, which is deep to the temporal lobe on the medial hemispheric surface, is not identified. Numbers indicate Brodmann structural areas, whereas colors and outlines represent functional regions of the cortex. **(b)** Parasagittal view, right hemisphere.

■ *Central sulcus.*

? *Why are the motor and sensory homunculi anatomically "disformed"?*

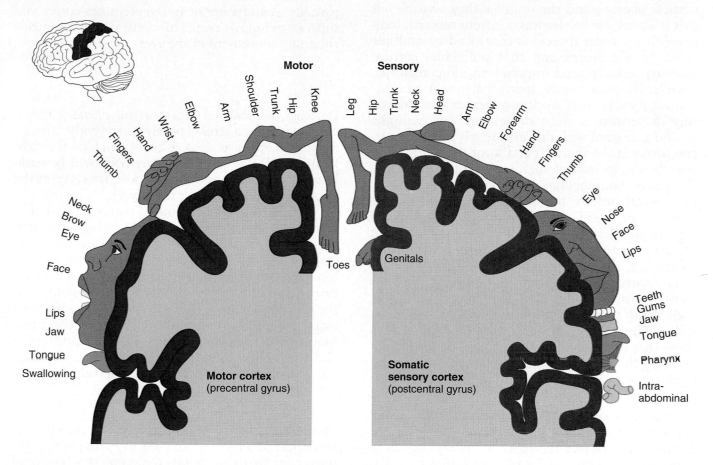

FIGURE 12.9 **Motor and sensory areas of the cerebral cortex.** The relative amount of cortical tissue devoted to each function is indicated by the amount of the gyrus occupied by the body diagrams. The primary motor cortex in the precentral gyrus is represented on the left, and the somatic sensory cortex in the postcentral gyrus is represented on the right.

1. Primary (somatic) motor cortex. The primary motor cortex is located in the precentral gyrus of the frontal lobe of each hemisphere (Brodmann area 4). Large neurons, called **pyramidal cells,** in these gyri allow us to consciously control the precise or skilled voluntary movements of our skeletal muscles. Their long axons, which project to the spinal cord, form the massive voluntary motor tracts called the **pyramidal tracts,** or **corticospinal tracts** (kor"tĭ-ko-spi'nal). All other descending motor tracts issue from brain stem nuclei and consist of chains of two or more neurons.

The entire body is represented spatially in the primary motor cortex of each hemisphere. In other words, the pyramidal cells that control foot movements are in one place and those that control hand movements are in another. Such a mapping of the body in CNS structures is called **somatotopy** (so"mah-to-to'pe). As illustrated in Figure 12.9, the body is typically represented upside down—with the head at the inferolateral part of the precentral gyrus, and the toes at the superomedial end. Most of the neurons in these gyri control muscles in body areas having the most precise motor control—that is, the face, tongue, and hands. Consequently, these regions of the caricature-like **motor homunculis** (ho-mung'ku-lis; singular = homunculus; "little man") drawn in Figure 12.9 are shown disproportionately large. The motor innervation of the body is contralateral; that is, the left primary motor gyrus controls muscles on the right side of the body, and vice versa.

■ *Because the relative size of each body area in the drawing represents the relative amount of the cerebral cortex dedicated to that body area.*

The *motor homunculus* view of the primary motor cortex, shown at the left in Figure 12.9, implies a one-to-one correspondence between certain cortical neurons and the muscles they control, but this is somewhat misleading. Current research indicates that a given muscle is controlled by multiple spots on the cortex and that individual cortical neurons actually send impulses to more than one muscle. In other words, individual motor neurons control muscles that work together in a synergistic way to perform a given movement. For example, reaching forward with one arm involves some muscles acting at the shoulder and some acting at the elbow. Thus, instead of the discrete map offered by the motor homunculus, the primary motor cortex map is an orderly but fuzzy map with neurons arranged in useful ways to control and coordinate sets of muscles. Neurons controlling the arm, for instance, are intermingled and overlap with those controlling the hand and shoulder. However, neurons controlling unrelated movements, such as those controlling the arm and those controlling body trunk muscles, do not cooperate in motor activity. Thus, the motor homunculus is useful to show that broad areas of the primary cortex are devoted to the leg, arm, torso, and head, but neuron organization within those broad areas is much more diffuse than initially imagined.

2. Premotor cortex. Just anterior to the precentral gyrus in the frontal lobe (Brodmann area 6) is the premotor cortex (see Figure 12.8). This region controls learned motor skills of a repetitious or patterned nature, such as playing a musical instrument and typing. The premotor cortex coordinates the movement of several muscle groups either simultaneously or sequentially, mainly by sending activating impulses to the primary motor cortex. However, the premotor cortex also influences motor activity more directly by supplying about 15% of pyramidal tract fibers. Think of this region as the memory bank for skilled motor activities.

This region also appears to be involved in planning movements. Using highly processed sensory information received from other cortical areas, it can control voluntary actions that depend on sensory feedback, such as moving an arm through a maze to grasp a hidden object.

3. Broca's area (bro'kahz). Broca's area lies anterior to the inferior region of the premotor area and overlaps Brodmann areas 44 and 45. It has long been considered to be (1) present in one hemisphere only (usually the left) and (2) a special *motor speech area* that directs the muscles involved in speech production. However, recent studies using PET scans to "light up" active areas of the brain indicate that Broca's area also becomes active as we prepare to speak and even as we think about (plan) many voluntary motor activities other than speech.

4. Frontal eye field. The frontal eye field is located partially in and anterior to the premotor cortex and superior to Broca's area. This cortical region controls voluntary movement of the eyes.

H HOMEOSTATIC IMBALANCE

Damage to localized areas of the *primary motor cortex* (as from a stroke) paralyzes the body muscles controlled by those areas. If the lesion is in the right hemisphere, the left side of the body will be paralyzed. Only *voluntary* control is lost, however, as the muscles can still contract reflexively.

Destruction of the *premotor cortex*, or part of it, results in a loss of the motor skill(s) programmed in that region, but muscle strength and the ability to perform the discrete individual movements are not hindered. For example, if the premotor area controlling the flight of your fingers over a computer keyboard were damaged, you couldn't type with your usual speed, but you could still make the same movements with your fingers. Reprogramming the skill into another set of premotor neurons would require practice, just as the initial learning process did. ●

Sensory Areas   Areas concerned with conscious awareness of sensation, the **sensory areas** of the cortex, occur in the parietal, temporal, and occipital lobes (see Figure 12.8).

1. Primary somatosensory cortex. This cortex resides in the postcentral gyrus of the parietal lobe, just posterior to the primary motor cortex (Brodmann areas 1–3). Neurons in this gyrus receive information from the general (somatic) sensory receptors in the skin and from proprioceptors in skeletal muscles. The neurons then identify the body region being stimulated, an ability called **spatial discrimination.** As with the primary motor cortex, the body is represented spatially and upside-down according to the site of stimulus input, and the right hemisphere receives input from the left side of the body. The amount of sensory cortex devoted to a particular body region is related to that region's sensitivity (that is, to how many receptors it has), not to the size of the body region. In humans, the face (especially the lips) and fingertips are the most sensitive body areas; hence these are the largest parts of the **somatosensory homunculus** shown in the right half of Figure 12.9.

2. Somatosensory association cortex. The somatosensory association cortéx lies just posterior to the primary somatosensory cortex and has many connections with it. The major function of this area is to integrate sensory inputs (temperature, pressure, and so forth) relayed to it via the primary somatosensory cortex to produce an understanding of an object being felt: its size, texture, and the relationship of its parts. For example, when you reach into your pocket, your somatosensory association cortex draws upon stored memories of past sensory experiences to perceive the objects you feel as coins or keys. Someone with damage to this area could not recognize these objects without looking at them.

3. Visual areas. The **primary visual (striate) cortex** is seen on the extreme posterior tip of the occipital lobe, but most of it is buried deep in the *calcarine sulcus* (Figure 12.8b) in the medial aspect of the occipital lobe. The largest of all cortical sensory areas, the primary visual cortex receives visual information that originates on the retina of the eye. There is a contralateral map of visual space on the primary visual cortex, analogous to the body map on the somatosensory cortex.

The **visual association area** surrounds the primary visual cortex and covers much of the occipital lobe. Communicating with the primary visual cortex, the visual association area uses past visual experiences to interpret visual stimuli (color, form, and movement), enabling us to recognize a flower or a person's face and to appreciate what we are seeing. We do our "seeing" with these cortical neurons. However, recent experiments on monkeys indicate that complex visual processing involves the entire posterior half of the cerebral hemispheres. Particularly important are two visual "streams"—one running along the top of the brain and handling object identity, the other taking the lower road and focusing on object locations.

4. Auditory areas. Each **primary auditory cortex** is located in the superior margin of the temporal lobe abutting the lateral sulcus. Sound energy exciting the inner ear hearing receptors causes impulses to be transmitted to the primary auditory cortex, where they are related to pitch, rhythm, and loudness.

The more posterior **auditory association area** then permits the perception of the sound stimulus, which we "hear" as speech, a scream, music, thunder, noise, and so on. Memories of sounds heard in the past appear to be stored here for reference. Wernicke's area, described on the next page, is also part of the auditory cortex.

5. Olfactory (smell) cortex. Each olfactory cortex is found in a small area of the frontal lobe just above the orbit, and in the medial aspect of the temporal lobe in a small region called the *piriform lobe* which is dominated by the hooklike *uncus* (Figure 12.8b). Afferent fibers from smell receptors in the superior nasal cavities send impulses along the olfactory tracts that are ultimately relayed to the olfactory cortices. The outcome is conscious awareness of different odors.

The olfactory cortex is part of the primitive **rhinencephalon** (ri"nen-sef'ah-lon; "nose brain"), which includes all parts of the cerebrum that receive olfactory signals—the orbitofrontal cortex, the uncus and associated regions located on or in the medial aspects of the temporal lobes, and the protruding olfactory tracts and bulbs that extend to the nose. During the course of evolution, most of the "old" rhinencephalon has taken on new functions concerned chiefly with emotions and memory. This "newer" emotional brain, called the *limbic system*, is considered later in this chapter. The only portions of the human rhinencephalon still devoted to smell are the olfactory bulbs and tracts (described in Chapter 13) and the greatly reduced olfactory cortices.

6. Gustatory (taste) cortex. The gustatory cortex (gus'tah-tor-e) (see Figure 12.8a), a region involved in the perception of taste stimuli, is located in the parietal lobe just deep to the temporal lobe. Logically, it occurs at the tip of the tongue of the somatosensory homunculus.

7. Vestibular (equilibrium) cortex. It has been difficult to pin down the part of the cortex responsible for conscious awareness of balance, that is, of the position of the head in space. However, medical imaging studies now locate this region in the posterior part of the insula, deep to the temporal lobe.

HOMEOSTATIC IMBALANCE

Damage to the *primary visual cortex* results in functional blindness. By contrast, those with damage to the visual association area can see, but they do not comprehend what they are looking at. ●

Association Areas   An **association area** is any cortical area that doesn't have the word *primary* in its name. As already described, the primary somatosensory cortex and each of the special sensory cortices have nearby association areas with which they communicate. These association areas, in turn, communicate ("associate") with the motor cortex and with other sensory association areas to analyze and act on sensory inputs in the light of past experience. Each association area has multiple inputs and outputs quite independent of the primary sensory and motor areas, indicating that their function is complex indeed. The remaining association areas, not connected with any of the sensory cortices, are described next.

1. Prefrontal cortex. The prefrontal cortex, in the anterior portion of the frontal lobe (Figure 12.8b), is the most complicated cortical region of all. It is involved with intellect, complex learning abilities (called cognition), recall, and personality. It is necessary for the production of abstract ideas, judgment, reasoning, persistence, long-term planning, concern for others, and conscience. That these qualities develop slowly in children implies that the prefrontal cortex matures slowly and is heavily dependent on positive and negative feedback from one's social environment. This cortex is closely linked to the emotional part of the brain (limbic system) and plays a role in intuitive judgments and mood. It is the tremendous elaboration of this region that sets human beings apart from other animals.

Tumors or other lesions of the *prefrontal cortex* may cause mental and personality disorders including mood swings and loss of judgment, attentiveness, and inhibitions. The affected individual may be oblivious to social restraints, perhaps becoming careless about personal appearance, or rashly attacking a 7-foot opponent rather than running.

2. Language areas. A large continuous area for language comprehension and articulation surrounds the lateral sulcus in the left hemisphere, which is the language-dominant hemisphere. The best-known parts of this area are: (1) **Wernicke's area** (Figure 12.8a), formerly believed to be responsible for understanding written and spoken language, but now thought to be primarily involved in sounding out unfamiliar words; (2) Broca's area for speech production; (3) the *lateral prefrontal cortex*, involved in language comprehension and word analysis; and (4) most of the lateral and ventral parts of the temporal lobe, which coordinate the auditory and visual aspects of language as when naming objects or reading. (This is only part of the story, but you've probably heard enough.)

The corresponding areas in the right or non-language-dominant hemisphere are involved in "body language"—the nonverbal emotional (affective) components of language. These areas allow the lilt or tone of our voice and our gestures to express our emotions when we speak, and permit us to comprehend the emotional content of what we hear. For example, a soft melodious response to your question conveys quite a different meaning than a sharp reply.

3. General (common) interpretation area. The general interpretation area is an ill-defined region encompassing parts of the temporal, parietal, and occipital lobes. It is found in one hemisphere only, usually the left. This region receives input from all sensory association areas and integrates incoming signals into a single thought or understanding of the situation. Suppose, for example, you drop a bottle of acid in the chem lab and it splashes on you. You see the bottle shatter; hear the crash; feel your skin burning; and smell the acid fumes. However, these individual perceptions do not dominate your consciousness. What *does* is the overall message "danger"—by which time your leg muscles are propelling you to the safety shower.

So the traditional story goes—but some modern sources are abandoning this idea of a multisensory interpretation area or at least confining it to a much smaller region near the top and back of Wernicke's area. This change in theory reflects newer studies indicating that most of the relevant area is involved in the processing of spatial relationships.

4. Visceral association area. The cortex of the insula may be involved in conscious perception of visceral sensations (upset stomach, full bladder, and the like). However, a small part of the insular cortex functions in language.

Lateralization of Cortical Functioning We use both cerebral hemispheres for almost every activity, and the hemispheres appear nearly identical. Nonetheless, there *is* a division of labor, and each hemisphere has unique abilities not shared by its partner. This phenomenon is called **lateralization.** Although one cerebral hemisphere or the other "dominates" each task, the term **cerebral dominance** designates the hemisphere that is *dominant for language*. In most people (about 90%), the left hemisphere has greater control over language abilities, math, and logic. This so-called dominant hemisphere is working when we compose a sentence, balance a checkbook, and memorize a list. The other hemisphere (usually the right) is more free-spirited, involved in visual-spatial skills, intuition, emotion, and artistic and musical skills. It is the poetic, creative, and the "Ah-ha!" (insightful) side of our nature, and it is far better at recognizing faces. Most individuals with left cerebral dominance are right-handed.

In the remaining 10% of people, the roles of the hemispheres are reversed or the hemispheres share their functions equally. Typically, right-cerebral-dominant people are left-handed and male. Some "lefties" who have a cerebral cortex that functions bilaterally are ambidextrous. In other cases, however, this phenomenon results in cerebral confusion ("Is it your turn, or mine?") and learning disabilities, such as the reading disorder *dyslexia*, in which otherwise intelligent people may reverse the order of letters in words (and the order of words in sentences).

The two cerebral hemispheres have perfect and almost instantaneous communication with one another via connecting fiber tracts, as well as complete functional integration. Furthermore, although

How can you explain the observation that the two cerebral hemispheres have instantaneous communication with each other when it is obvious from diagram (b) that the projection fibers are part of crossed pathways and that information flows from each side of the body to only one (the opposite) hemisphere?

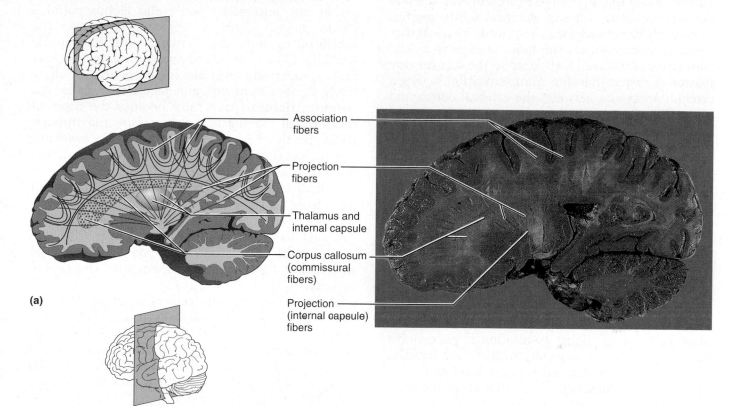

Association fibers

Projection fibers

Thalamus and internal capsule

Corpus callosum (commissural fibers)

Projection (internal capsule) fibers

(a)

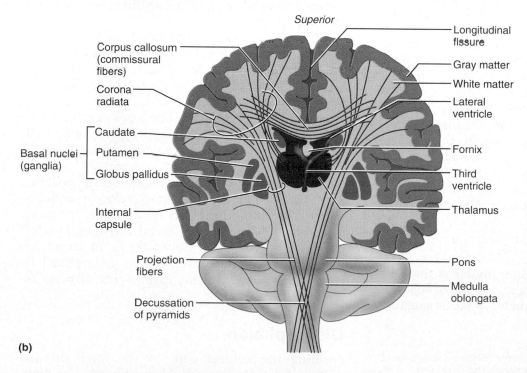

Superior

Corpus callosum (commissural fibers)

Corona radiata

Basal nuclei (ganglia)
- Caudate
- Putamen
- Globus pallidus

Internal capsule

Projection fibers

Decussation of pyramids

Longitudinal fissure

Gray matter

White matter

Lateral ventricle

Fornix

Third ventricle

Thalamus

Pons

Medulla oblongata

(b)

FIGURE 12.10 Types of fiber tracts in white matter. (a) Midsagittal views of the right cerebral hemisphere showing the association tract fibers (tracts connecting different parts of the same hemisphere), and the corpus callosum, a commissure that connects the hemispheres. (b) Frontal section showing commissural fibers and projection fibers running between the cerebrum and lower CNS centers. Notice the tight band of projection fibers called the internal capsule that passes between the thalamus and the basal nuclei, and then fans out as the corona radiata.

Commissures allow the cerebral hemispheres to "talk to each other." ■

lateralization means that each hemisphere is better than the other at certain functions, neither side is better at everything.

Cerebral White Matter

The second of the three basic regions of each cerebral hemisphere is the internal **cerebral white matter.** From what has already been described, it is clear that communication within the brain is extensive. The white matter (Figure 12.10) deep to the cortical gray matter is responsible for communication between cerebral areas and between the cerebral cortex and lower CNS centers. White matter consists largely of myelinated fibers bundled into large tracts. These fibers and tracts are classified according to the direction in which they run as *commissural, association,* or *projection.*

Commissures (kom'ĭ-shūrz), composed of **commissural fibers,** connect corresponding gray areas of the two hemispheres, enabling them to function as a coordinated whole. The largest commissure is the **corpus callosum** (kah-lo'sum; "thickened body"), which lies superior to the lateral ventricles, deep within the longitudinal fissure. Less important examples are the **anterior** and **posterior commissures.**

Association fibers connect different parts of the same hemisphere. Short association fibers connect adjacent gyri. Long association fibers are bundled into tracts and connect different cortical lobes.

Projection fibers are those that enter the cerebral hemispheres from lower brain or cord centers, and those that leave the cortex to travel to lower areas. They tie the cortex to the rest of the nervous system and to the body's receptors and effectors. In contrast to commissural and association fibers, which run horizontally, projection fibers run vertically as Figure 12.10b shows.

At the top of the brain stem, the projection fibers on each side form a compact band, the **internal capsule,** that passes between the thalamus and some of the basal nuclei. Beyond that point, the fibers radiate fanlike through the cerebral white matter to the cortex. This distinctive arrangement of projection tract fibers is known as the **corona radiata** ("radiating crown").

Basal Nuclei

Deep within the cerebral white matter is the third basic region of each hemisphere, a group of subcortical nuclei called the **basal nuclei** or **basal ganglia.** *

---

* Because a nucleus is a collection of nerve cell bodies within the CNS, the term *basal nuclei* is technically correct. The more frequently used but misleading historical term *basal ganglia* is a misnomer and should be abandoned, because ganglia are PNS structures.

Although the definition of the precise structures forming the basal nuclei is controversial, most anatomists agree that the **caudate nucleus** (kaw'dāt), **putamen** (pu-ta'men), and **globus pallidus** (glo'bis pal'ĭ-dus) constitute most of the mass of each group of basal nuclei (Figure 12.11). Together, the putamen ("pod") and globus pallidus ("pale globe") form a lens-shaped mass, the **lentiform nucleus,** that flanks the internal capsule laterally. The comma-shaped caudate nucleus arches superiorly over the diencephalon. Collectively, the lentiform and caudate nuclei are called the **corpus striatum** (stri-a'tum) because the fibers of the internal capsule that course past and through them give them a striped appearance. The basal nuclei are functionally associated with the *subthalamic nuclei* (located in the lateral "floor" of the diencephalon) and the *substantia nigra* of the midbrain (see Figure 12.16a).

The almond-shaped **amygdala** (ah-mig'dah-lah; "almond") sits on the tail of the caudate nucleus. Traditionally, it has been grouped with the basal nuclei, but functionally it belongs to the limbic system.

The corpus striatum receives input from the entire cerebral cortex, as well as from other subcortical nuclei and each other. Via relays through the thalamus, the output nuclei of the basal nuclei (globus pallidus and the substantia nigra) project to the premotor and prefrontal cortices and so influence muscle movements directed by the primary motor cortex. The basal nuclei have no direct access to motor pathways.

The precise role of the basal nuclei has been elusive because of their inaccessible location and because their functions overlap with those of the cerebellum. Their role in motor control is very complex and there is evidence they play a part in regulating attention and in cognition. The basal nuclei are particularly important in starting, stopping, and monitoring the intensity of movements executed by the cortex, especially those that are relatively slow or stereotyped, such as arm-swinging during walking. Additionally, they inhibit antagonistic or unnecessary movements; thus their input seems necessary to our ability to perform several activities at once. Disorders of the basal nuclei result in either too much or too little movement as exemplified by Huntington's chorea and Parkinson's disease, respectively (see p. 417).

Diencephalon

Forming the central core of the forebrain and surrounded by the cerebral hemispheres, the **diencephalon** consists largely of three paired structures—

? *Why are the lentiform nucleus and caudate nucleus collectively referred to as the corpus striatum?*

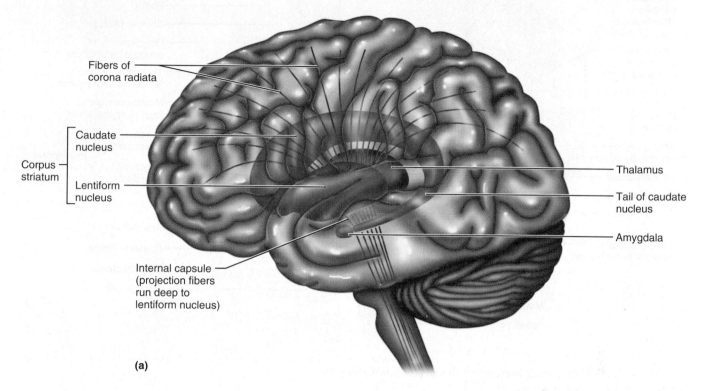

Fibers of corona radiata

Corpus striatum
- Caudate nucleus
- Lentiform nucleus

Internal capsule (projection fibers run deep to lentiform nucleus)

Thalamus

Tail of caudate nucleus

Amygdala

(a)

FIGURE 12.11 **Basal nuclei. (a)** Three-dimensional view of the basal nuclei (basal ganglia), showing their position in the cerebrum. **(b)** Transverse section of cerebrum and diencephalon showing the relationship of the basal nuclei to the thalamus and the lateral and third ventricles.

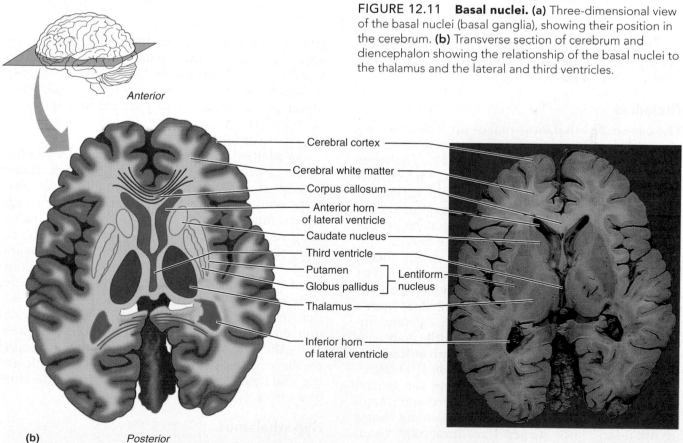

Anterior

Cerebral cortex

Cerebral white matter

Corpus callosum

Anterior horn of lateral ventricle

Caudate nucleus

Third ventricle

Putamen

Globus pallidus

Thalamus

Inferior horn of lateral ventricle

Lentiform nucleus

(b)   Posterior

■ *Because the fibers of the corona radiata running through them make them appear striped.*

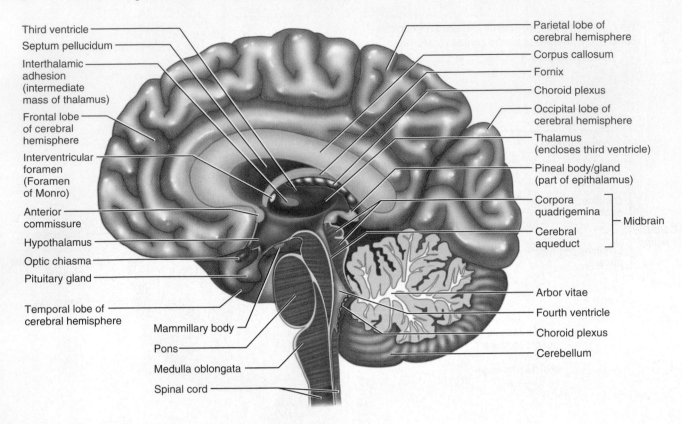

FIGURE 12.12 Midsagittal section of the brain illustrating the diencephalon and brain stem.

the thalamus, hypothalamus, and epithalamus. These gray matter areas collectively enclose the third ventricle (Figure 12.12).

Thalamus

The egg-shaped **thalamus** makes up 80% of the diencephalon and forms the superolateral walls of the third ventricle (Figures 12.10 and 12.12). *Thalamus* is a Greek word meaning "inner room," which well describes this deep, well-hidden brain region. In some people, its bilateral masses of gray matter are held together by a midline connection called the **interthalamic adhesion (intermediate mass).**

The thalamus contains about a dozen nuclei, most named according to their relative location (Figure 12.13a). Each nucleus has a functional specialty, and each projects fibers to and receives fibers from a specific region of the cerebral cortex. Afferent impulses from all senses and all parts of the body converge on the thalamus and synapse with at least one of its nuclei. For example, the *ventral posterior lateral nucleus* receives impulses from the general somatic sensory receptors (touch, pressure, pain, etc.), and the *lateral* and *medial geniculate bodies* (jĕ-nik'u-lāt; "knee shaped") are important visual and auditory relay centers, respectively. Within the thalamus, information is sorted out and "edited." Impulses having to do with similar functions are

relayed as a group via the internal capsule to the appropriate area of the sensory cortex as well as to specific cortical association areas. As the afferent impulses reach the thalamus, we have a crude recognition of the sensation as pleasant or unpleasant. However, specific stimulus localization and discrimination occur in the cerebral cortex.

In addition to sensory inputs, virtually *all* other inputs ascending to the cerebral cortex funnel through thalamic nuclei. These include impulses participating in the regulation of emotion and visceral function from the hypothalamus (via the anterior nuclei), and impulses that help direct the activity of the motor cortices from the cerebellum and basal nuclei (via the ventral lateral and ventral anterior nuclei, respectively). Several thalamic nuclei (pulvinar, lateral dorsal, and lateral posterior nuclei) are involved in integration of sensory information and project to specific association cortices. Thus the thalamus plays a key role in mediating sensation, motor activities, cortical arousal, learning, and memory. It is truly the gateway to the cerebral cortex.

Hypothalamus

Named for its position below *(hypo)* the thalamus, the **hypothalamus** caps the brain stem and forms the inferolateral walls of the third ventricle (Figure 12.12).

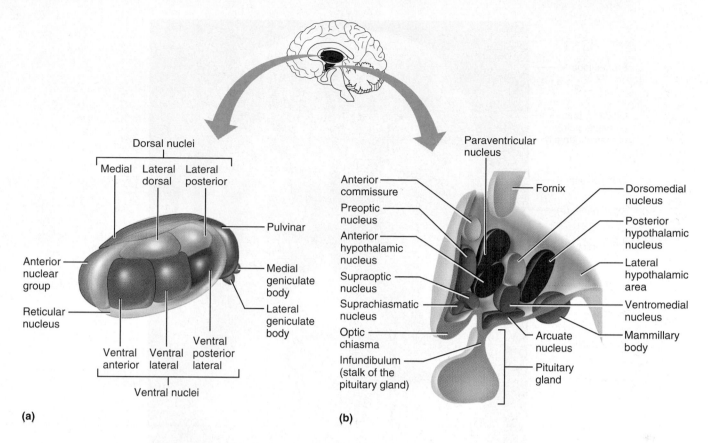

FIGURE 12.13 Selected structures of the diencephalon. (a) Thalamus showing the main thalamic nuclei. (The reticular nuclei that "cap" the thalamic nuclei laterally are depicted as curving translucent structures.) **(b)** Main hypothalamic nuclei.

Merging into the midbrain inferiorly, the hypothalamus extends from the optic chiasma (crossover point of the optic nerves) to the posterior margin of the mammillary bodies. The **mammillary bodies** (mam′mil-er-e; "little breast"), paired pealike nuclei that bulge anteriorly from the hypothalamus, are relay stations in the olfactory pathways. Between the optic chiasma and mammillary bodies is the **infundibulum** (in″fun-dib′u-lum), a stalk of hypothalamic tissue that connects the **pituitary gland** to the base of the hypothalamus. Like the thalamus, the hypothalamus contains many functionally important nuclei (Figure 12.13b).

Despite its small size, the hypothalamus is the main visceral control center of the body and is vitally important to overall body homeostasis. Few tissues in the body escape its influence. Its chief homeostatic roles are summarized next.

1. Autonomic control center. As you will remember, the autonomic nervous system (ANS) is a system of peripheral nerves that regulates cardiac and smooth muscle and secretion by the glands. The hypothalamus regulates ANS activity by controlling the activity of centers in the brain stem and spinal cord. In this role, the hypothalamus influences blood pressure, rate and force of heartbeat, digestive tract motility, respiratory rate and depth, eye pupil size, and many other visceral activities.

2. Center for emotional response. The hypothalamus lies at the "heart" of the limbic system (the emotional part of the brain). Nuclei involved in the perception of pleasure, fear, and rage, as well as those involved in biological rhythms and drives (such as the sex drive), are found in the hypothalamus.

The hypothalamus acts through ANS pathways to initiate most physical expressions of emotion. For example, a fearful person has a pounding heart, high blood pressure, pallor, sweating, and a dry mouth.

3. Body temperature regulation. The body's thermostat is in the hypothalamus. Thermoreceptors located in other parts of the brain as well as in the body periphery, and certain hypothalamic neurons monitor blood temperature. Accordingly, the hypothalamus initiates cooling (sweating) or heat-retention mechanisms (shivering) as needed to maintain a relatively constant body temperature.

4. Regulation of food intake. In response to changing blood levels of certain nutrients (glucose and perhaps amino acids) or hormones (insulin and

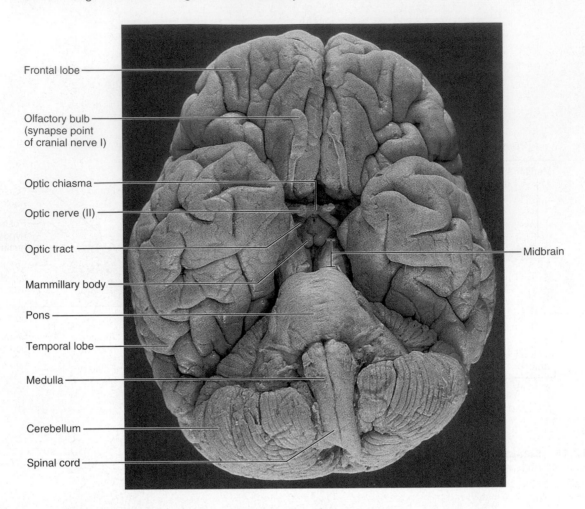

Frontal lobe

Olfactory bulb
(synapse point
of cranial nerve I)

Optic chiasma

Optic nerve (II)

Optic tract

Mammillary body

Pons

Temporal lobe

Medulla

Cerebellum

Spinal cord

Midbrain

FIGURE 12.14   Ventral aspect of the human brain, showing the three regions of the brain stem. Only a small portion of the midbrain can be seen; the rest is surrounded by other brain regions.

others), the hypothalamus regulates feelings of hunger and satiety.

5. Regulation of water balance and thirst. When body fluids become too concentrated, hypothalamic neurons called *osmoreceptors* are activated. These neurons excite hypothalamic nuclei that trigger the release of antidiuretic hormone (ADH) from the posterior pituitary. ADH causes the kidneys to retain water. The same conditions also stimulate hypothalamic neurons in the *thirst center*, causing us to feel thirsty and, therefore, to drink more fluids.

6. Regulation of sleep-wake cycles. Acting with other brain regions, the hypothalamus helps regulate sleep. Through the operation of its *suprachiasmatic nucleus* (our biological clock), it sets the timing of the sleep cycle in response to daylight-darkness cues received from the visual pathways.

7. Control of endocrine system functioning. The hypothalamus acts as the helmsman of the endocrine system in two important ways. First, its *releasing hormones* control the secretion of hormones by the anterior pituitary gland. Second, its *supraoptic* and

paraventricular nuclei produce the hormones ADH and oxytocin.

 HOMEOSTATIC IMBALANCE

Hypothalamic disturbances cause a number of disorders including severe body wasting, obesity, sleep disturbances, dehydration, and a broad range of emotional imbalances. For example, infants deprived of a warm, nurturing relationship may develop sleep disorders and fail to thrive. ●

Epithalamus

The **epithalamus** is the most dorsal portion of the diencephalon and forms the roof of the third ventricle. Extending from its posterior border and visible externally is the **pineal gland** or **body** (pin′e-al; "pine cone shaped") (see Figures 12.12 and 12.15). The pineal gland secretes the hormone *melatonin* (the sleep-inducing signal) and, along with hypothalamic nuclei, helps regulate the sleep-wake cycle and some

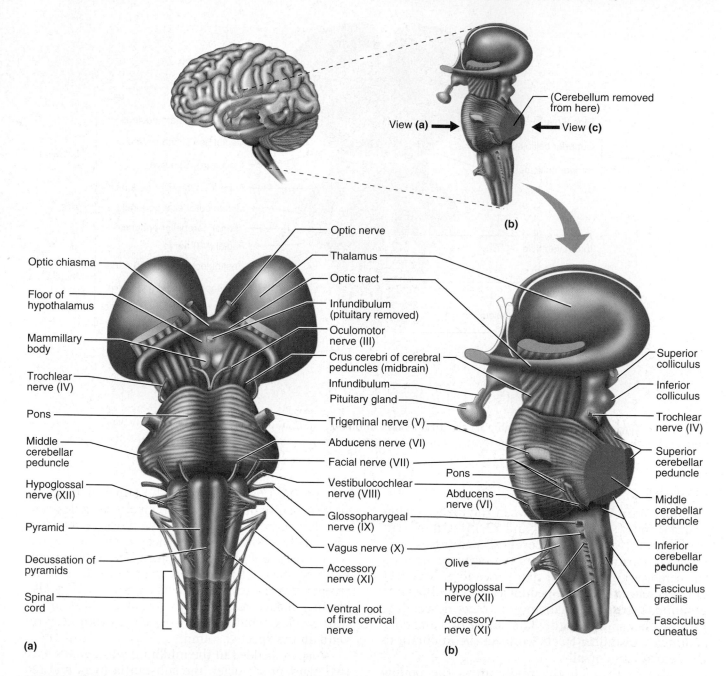

FIGURE 12.15 Relationship of the brain stem and the diencephalon.
(a) Ventral view. **(b)** Lateral view.

aspects of mood. A cerebrospinal fluid–forming structure called a **choroid plexus** (ko′roid plek′sus) is also part of the epithalamus (see Figure 12.12).

Brain Stem

From superior to inferior, the brain stem regions are midbrain, pons, and medulla oblongata (Figures 12.12, 12.14, and 12.15), each roughly an inch long. Histologically, the organization of the brain stem is similar (but not identical) to that of the spinal cord—deep gray matter surrounded by white matter

fiber tracts (see Figure 12.4). However, in the brain stem, there are nuclei of gray matter embedded in the white matter, a feature not found in the spinal cord.

Brain stem centers produce the rigidly programmed, automatic behaviors necessary for survival. Positioned between the cerebrum and the spinal cord, the brain stem also provides a pathway for fiber tracts running between higher and lower neural centers. Additionally, brain stem nuclei are associated with 10 of the 12 pairs of cranial nerves (described in Chapter 13), so it is heavily involved with innervation of the head.

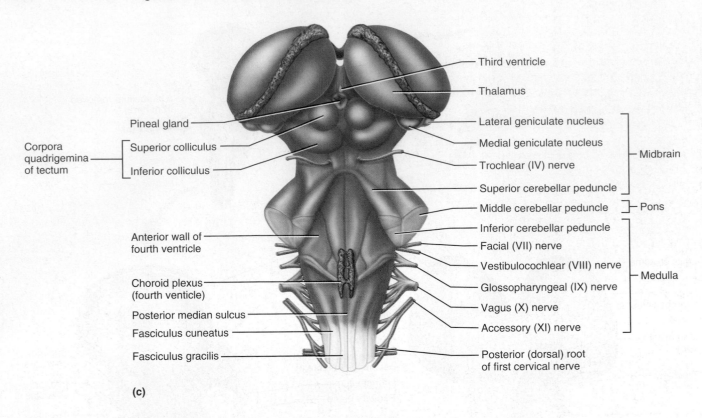

Labels on figure:
- Pineal gland
- Corpora quadrigemina of tectum
  - Superior colliculus
  - Inferior colliculus
- Anterior wall of fourth ventricle
- Choroid plexus (fourth venticle)
- Posterior median sulcus
- Fasciculus cuneatus
- Fasciculus gracilis
- Third ventricle
- Thalamus
- Lateral geniculate nucleus
- Medial geniculate nucleus
- Trochlear (IV) nerve
- Superior cerebellar peduncle
- Middle cerebellar peduncle — Pons
- Inferior cerebellar peduncle
- Facial (VII) nerve
- Vestibulocochlear (VIII) nerve
- Glossopharyngeal (IX) nerve — Medulla
- Vagus (X) nerve
- Accessory (XI) nerve
- Posterior (dorsal) root of first cervical nerve
- Midbrain

(c)

FIGURE 12.15 *(continued)* **Relationship of the brain stem and the diencephalon. (c)** Dorsal view. Cerebral hemispheres have been removed.

Midbrain

The **midbrain** is located between the diencephalon and the pons (Figures 12.14 and 12.15). On its ventral aspect two bulging **cerebral peduncles** (pĕ-dung'klz) form vertical pillars that seem to hold up the cerebrum, hence their name meaning "little feet of the cerebrum." These peduncles contain the large pyramidal (corticospinal) motor tracts descending toward the spinal cord. The *superior cerebellar peduncles,* also fiber tracts, connect the midbrain to the cerebellum dorsally.

Running through the midbrain is the hollow **cerebral aqueduct** (Figures 12.12 and 12.16a), which connects the third and fourth ventricles, and delineates the cerebral peduncles ventrally from the *tectum,* the midbrain's roof. Surrounding the aqueduct is the *periaqueductal gray matter,* which is involved in pain suppression and serves as the link between the fear-perceiving amygdala and ANS pathways that control the "fight-or-flight" response. The periaqueductal gray matter also includes nuclei that control two cranial nerves, the *oculomotor* and the *trochlear nuclei* (trok'le-ar).

Nuclei are also scattered in the surrounding white matter of the midbrain. The largest of these are the **corpora quadrigemina** (kor'por-ah kwod"ri-jem'i-nah; "quadruplets"), which raise four dome-like protrusions on the dorsal midbrain surface

(Figures 12.12 and 12.15). The superior pair, the **superior colliculi** (kŏ-lik'u-li), are visual reflex centers that coordinate head and eye movements when we visually follow a moving object, even if we are not consciously looking at the object. The **inferior colliculi** are part of the auditory relay from the hearing receptors of the ear to the sensory cortex. They also act in reflexive responses to sound, such as in the *startle reflex,* which causes you to turn your head toward an unexpected sound.

Also embedded in the midbrain white matter are two pigmented nuclei, the substantia nigra and red nucleus (Figure 12.16a). The bandlike **substantia nigra** (sub-stan'she-ah ni'grah) is located deep to the cerebral peduncle. Its dark (*nigr* = black) color reflects a high content of melanin pigment, a precursor of the neurotransmitter (dopamine) released by these neurons. The substantia nigra is functionally linked to the basal nuclei (its axons project to the globus pallidus), and is considered part of the basal nuclear complex by many authorities. Degeneration of the dopamine-releasing neurons of the substantia nigra is the ultimate cause of Parkinson's disease.

The oval **red nucleus** lies deep to the substantia nigra. Its reddish hue is due to its rich blood supply and to the presence of iron pigment in its neurons. The red nuclei are relay nuclei in some descending motor pathways that effect limb flexion, and they are

Just what are the pyramids of the medulla?

?

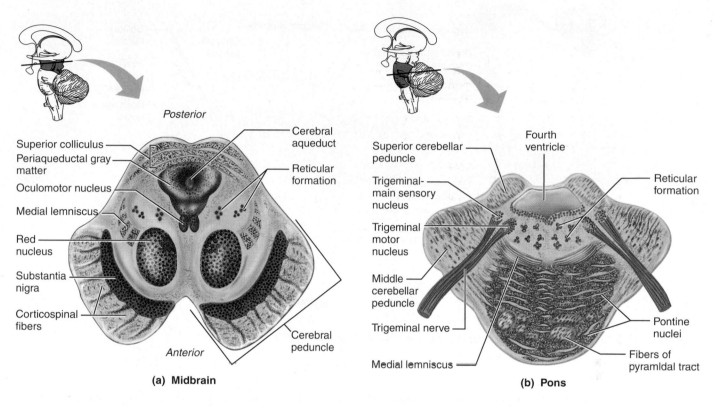

FIGURE 12.16 Important brain stem nuclei. Horizontal sections through **(a)** midbrain at the level of superior colliculi, **(b)** pons at the level of its cranial nerve nuclei.

the largest nuclei of the *reticular formation,* a system of small nuclei scattered through the core of the brain stem (see p. 404).

Pons

The **pons** is the bulging brain stem region wedged between the midbrain and the medulla oblongata (see Figures 12.12, 12.14, and 12.15). Dorsally, it forms part of the anterior wall of the fourth ventricle.

As its name suggests (*pons* = bridge), the pons is chiefly composed of conduction tracts which course in two directions. The deep projection fibers run longitudinally and complete the pathway between higher brain centers and the spinal cord. The more superficial ventral fibers issue from numerous *pontine nuclei,* which act as relays for "conversations" between the motor cortex and cerebellum. These fibers are oriented transversely and dorsally as the *middle cerebellar peduncles* (Figure 12.15) and connect the pons bilaterally with the two sides of the cerebellum posteriorly.

The corticospinal (pyramidal) tracts. The large voluntary motor tracts descending from the motor cortex. ■

Several cranial nerve pairs issue from pons nuclei (Figure 12.16b), including the *trigeminal* (tri-jem'ĭ-nal), the *abducens* (ab-du'senz), and *facial nerves.* The cranial nerves and their functions are discussed in Chapter 13. Other important pons nuclei are part of the reticular formation. The *pneumotaxic center,* for example, is a respiratory center. Together with medullary respiratory centers, it helps maintain the normal rhythm of breathing.

Medulla Oblongata

The conical **medulla oblongata** (me-dul'ah ob"long-gah'tah), or simply **medulla,** is the most inferior part of the brain stem. It blends imperceptibly into the spinal cord at the level of the foramen magnum of the skull (Figures 12.12 and 12.14). The central canal of the spinal cord continues upward into the medulla, where it broadens out to form the cavity of the fourth ventricle. Thus, both the medulla and the pons help form the ventral wall of the fourth ventricle. [The dorsal ventricular wall is formed by a thin capillary-rich membrane called a choroid plexus which abuts the cerebellum dorsally (Figure 12.12).]

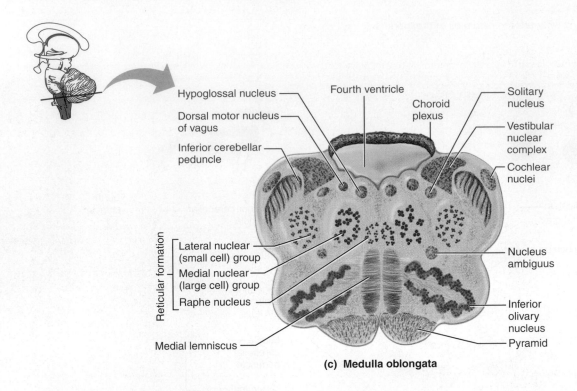

Hypoglossal nucleus

Dorsal motor nucleus of vagus

Inferior cerebellar peduncle

Reticular formation

Lateral nuclear (small cell) group

Medial nuclear (large cell) group

Raphe nucleus

Medial lemniscus

Fourth ventricle

Choroid plexus

Solitary nucleus

Vestibular nuclear complex

Cochlear nuclei

Nucleus ambiguus

Inferior olivary nucleus

Pyramid

(c) Medulla oblongata

FIGURE 12.16 *(continued)* **Important brain stem nuclei.** Horizontal section through **(c)** superior medulla oblongata.

Flanking the midline on the medulla's ventral aspect are two longitudinal ridges called **pyramids,** formed by the large pyramidal (corticospinal) tracts descending from the motor cortex (Figure 12.16c). Just above the medulla–spinal cord junction, most of these fibers cross over to the opposite side before continuing into the spinal cord. This crossover point is called the **decussation of the pyramids** (de"kus-sa'shun; "a crossing"). As mentioned earlier, the consequence of this crossover is that each cerebral hemisphere chiefly controls the voluntary movements of muscles on the opposite side of the body.

Also obvious externally are several other structures. The *inferior cerebellar peduncles* are fiber tracts that connect the medulla to the cerebellum dorsally. Situated lateral to the pyramids, the **olives** are oval swellings (which *do* resemble olives) produced mainly by the underlying **inferior olivary nuclei,** wavy folds of gray matter (Figure 12.16c). These nuclei relay sensory information on the state of stretch of muscles and joints to the cerebellum. The rootlets of the *hypoglossal nerves* emerge from the groove between the pyramid and olive on each side of the brain stem. Other cranial nerves associated with the medulla are the *glossopharyngeal nerves* and *vagus nerves,* and portions of the *accessory nerves.* Additionally, the fibers of the *vestibulocochlear nerves* (ves-tib"u-lo-kok'le-ar) synapse with the **cochlear nuclei** (auditory relays), and with numerous vestibular nuclei in both the pons and

medulla. Collectively, the vestibular nuclei, called the **vestibular nuclear complex,** mediate responses that maintain equilibrium.

Also housed in the medulla are several nuclei associated with ascending sensory tracts. The most prominent are the dorsally located **nucleus gracilis** (grah-sĭ'lis) and **nucleus cuneatus** (ku'ne-āt-us), associated with the **medial lemniscal tract.** These serve as relay nuclei in a pathway by which general somatic sensory information ascends from the spinal cord to the somatosensory cortex.

The small size of the medulla belies its crucial role as an autonomic reflex center involved in maintaining body homeostasis. Important visceral motor nuclei found in the medulla include the following:

1. The **cardiovascular center,** which includes cardiac and vasomotor centers. The *cardiac center* adjusts the force and rate of heart contraction to meet the body's needs. The *vasomotor center* changes blood vessel diameter to regulate blood pressure.

2. The **respiratory centers,** which control the rate and depth of breathing and (in a negative feedback interaction with pons centers) maintain respiratory rhythm.

3. Various other centers. Additional centers regulate such activities as vomiting, hiccuping, swallowing, coughing, and sneezing.

Notice that many functions listed above were also attributed to the hypothalamus (pp. 394–396).

TABLE 12.1 — Functions of Major Brain Regions

Region	Function

Cerebral Hemispheres (pp. 383–392)

Cortical gray matter localizes and interprets sensory inputs, controls voluntary and skilled skeletal muscle activity, and functions in intellectual and emotional processing; basal nuclei (ganglia) are subcortical motor centers important in initiation of skeletal muscle movements

Diencephalon (pp. 392–397)

Thalamic nuclei are relay stations in conduction of (1) sensory impulses to cerebral cortex for interpretation, and (2) impulses to and from cerebral motor cortex and lower (subcortical) motor centers, including cerebellum; thalamus is also involved in memory processing

Hypothalamus is chief integration center of autonomic (involuntary) nervous system; it functions in regulation of body temperature, food intake, water balance, thirst, and biological rhythms and drives; regulates hormonal output of anterior pituitary gland and is an endocrine organ in its own right (produces ADH and oxytocin); part of limbic system

Limbic system (pp. 403–404) — A functional system involving cerebral and diencephalon structures that mediates emotional response; also involved in memory processing

Brain Stem (pp. 397–401)

Midbrain: Conduction pathway between higher and lower brain centers (e.g., cerebral peduncles contain the fibers of the pyramidal tracts); its superior and inferior colliculi are visual and auditory reflex centers; substantia nigra and red nuclei are subcortical motor centers; contains nuclei for cranial nerves III and IV

Pons: Conduction pathway between higher and lower brain centers; pontine nuclei relay information from the cerebrum to the cerebellum; its respiratory nuclei cooperate with the medullary respiratory centers to control respiratory rate and depth; houses nuclei of cranial nerves V–VII

Medulla oblongata: Conduction pathway between higher brain centers and spinal cord; site of decussation of the pyramidal tracts; houses nuclei of cranial nerves VIII–XII; contains nuclei cuneatus and gracilis (synapse points of ascending sensory pathways transmitting sensory impulses from skin and proprioceptors), and visceral nuclei controlling heart rate, blood vessel diameter, respiratory rate, vomiting, coughing, etc.; its inferior olivary nuclei provide the sensory relay to the cerebellum

Reticular formation (pp. 404–405) — A functional brain stem system that maintains cerebral cortical alertness (reticular activating system) and filters out repetitive stimuli; its motor nuclei help regulate skeletal and visceral muscle activity

Cerebellum (pp. 401–403)

Processes information from cerebral motor cortex and from proprioceptors and visual and equilibrium pathways, and provides "instructions" to cerebral motor cortex and subcortical motor centers that result in proper balance and posture and smooth, coordinated skeletal muscle movements

---

The overlap is easily explained. The hypothalamus exerts its control over many visceral functions by relaying its instructions through medullary reticular centers, which carry them out.

Cerebellum

The cauliflower-like **cerebellum** (ser"ĕ-bel'um; "small brain"), exceeded in size only by the cerebrum, accounts for about 11% of total brain mass.

The cerebellum is located dorsal to the pons and medulla (and to the intervening fourth ventricle). It protrudes under the occipital lobes of the cerebral hemispheres, from which it is separated by the transverse cerebral fissure (see Figure 12.6a).

By processing inputs received from the cerebral motor cortex, various brain stem nuclei, and sensory receptors, the cerebellum provides the precise timing and appropriate patterns of skeletal muscle contraction, for smooth, coordinated movements and agility

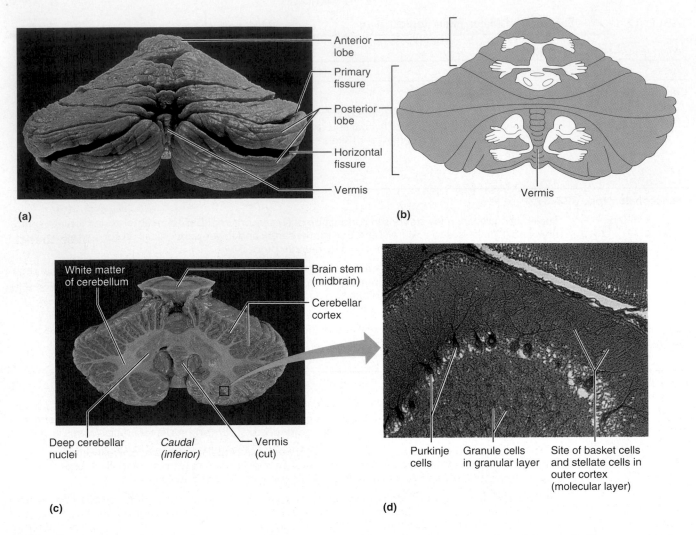

FIGURE 12.17 Cerebellum.
(a) Photograph of the posterior aspect of the cerebellum. The flocculonodular lobe, which lies deep to the vermis, is not visible in this view. **(b)** Positioning of overlapping motor and sensory maps of the body (in the form of three homunculi). **(c)** Posterior view sectioned frontally to reveal the three cerebellar layers. **(d)** Photomicrograph of a small portion of the cerebellar cortex showing the relative positions of granule cells, Purkinje cells, and stellate and basket cells (200×).

needed for our daily living—driving, typing, and for some of us, playing the tuba. Cerebellar activity occurs subconsciously; that is, we have no awareness of its functioning.

Anatomy

The cerebellum is bilaterally symmetrical; its two apple-sized **cerebellar hemispheres** are connected medially by the wormlike **vermis** (Figure 12.17). Its surface is heavily convoluted, with fine, transversely oriented pleatlike gyri known as **folia** ("leaves"). Deep fissures subdivide each hemisphere into **anterior, posterior,** and **flocculonodular lobes** (flok"u-lo-nod'u-lar). The small propeller-shaped flocculonodular lobes, situated deep to the vermis and posterior lobe, cannot be seen in a surface view.

Like the cerebrum, the cerebellum has a thin outer cortex of gray matter, internal white matter, and small, deeply situated, paired masses of gray matter, the most familiar of which are the *dentate nuclei.* Several types of neurons populate the cerebellar cortex, including stellate, basket, granule, and Purkinje cells (Figure 12.17d). Of these, the large **Purkinje cells,** with their extensively branched dendrites, are the only cortical neurons that send axons through the white matter to synapse with the central nuclei of the cerebellum. The distinctive pattern of white matter in the cerebellum resembles a branching tree, a pattern fancifully called the **arbor vitae** (ar'bor vi'te; "tree of life").

The anterior and posterior lobes of the cerebellum, which coordinate body movements, have completely overlapping sensory and motor maps of the entire body as indicated by the three homunculi in Figure 12.17b. The medial portions influence the motor activities of the trunk and girdle muscles. The intermediate parts of each hemisphere are more concerned with the distal parts of the limbs and skilled movements. The lateralmost parts of each hemisphere integrate information from the association areas of the cerebral cortex and appear to play a role in planning rather than executing movements. The flocculonodular lobes receive inputs from the equilibrium apparatus of the inner ears, and adjust posture to maintain balance.

Cerebellar Peduncles

As noted earlier, three paired fiber tracts—the cerebellar peduncles—connect the cerebellum to the brain stem (see Figure 12.15). Unlike the contralateral fiber distribution to and from the cerebral cortex, virtually all fibers entering and leaving the cerebellum are **ipsilateral** (*ipsi* = same)—from and to the *same* side of the body. The **superior cerebellar peduncles** connecting cerebellum and midbrain carry instructions from neurons in the deep cerebellar nuclei to the cerebral motor cortex via thalamic relays. Like the basal nuclei, the cerebellum has no *direct* connections with the cerebral cortex.

The **middle cerebellar peduncles** carry one-way communication from the pons to the cerebellum, advising the cerebellum of voluntary motor activities initiated by the motor cortex (via relays in the pontine nuclei). The **inferior cerebellar peduncles** connect medulla and cerebellum. These peduncles convey sensory information to the cerebellum from (1) muscle proprioceptors throughout the body and (2) the vestibular nuclei of the brain stem, which are concerned with equilibrium and balance.

Cerebellar Processing

The functional scheme of cerebellar processing for motor activity seems to be as follows:

1. The frontal motor association area of the cerebral cortex, via collateral fibers of the pyramidal tracts, notifies the cerebellum of its intent to initiate voluntary muscle contractions.

2. At the same time, the cerebellum receives information from proprioceptors throughout the body (regarding tension in the muscles and tendons, and joint position) and from visual and equilibrium pathways. This information enables the cerebellum to evaluate body position and momentum, that is, where the body is and where it is going.

3. The cerebellar cortex calculates the best way to coordinate the force, direction, and extent of muscle contraction to prevent overshoot, maintain posture, and ensure smooth, coordinated movements.

4. Then, via the superior peduncles, the cerebellum dispatches to the cerebral motor cortex its "blueprint" for coordinating movement. Cerebellar fibers also send information to brain stem nuclei,

which in turn influence motor neurons of the spinal cord.

Just as its automatic pilot compares a plane's instrument settings with the planned course, the cerebellum continually compares the higher brain's intention with the body's performance and sends out messages to initiate the appropriate corrective measures. Cerebellar injury results in loss of muscle tone and clumsy, unsure movements.

Cognitive Function of the Cerebellum

Functional imaging studies indicate that the cerebellum plays a role in cognition and in language and problem solving. Researchers surmise that the cerebellum's overall cognitive function may be to recognize and predict sequences of events so that it may adjust for the multiple forces exerted on a limb during complex movements involving several joints.

Functional Brain Systems

Functional brain systems are networks of neurons that work together but span relatively large distances in the brain, so they cannot be localized to specific brain regions. The *limbic system* and the *reticular formation* are excellent examples (Table 12.1).

The Limbic System

The **limbic system** is a group of structures located on the medial aspect of each cerebral hemisphere and diencephalon. Its cerebral structures encircle (*limbus* = ring) the upper part of the brain stem (Figure 12.18), and include parts of the rhinencephalon (*septal nuclei, cingulate gyrus, parahippocampal gyrus, dentate gyrus,* and C-shaped *hippocampus*), and part of the *amygdala.* In the diencephalon, the main limbic structures are the *hypothalamus* and the *anterior nucleus of the thalamus.* The **fornix** ("arch") and other fiber tracts link these limbic system regions together.

The limbic system is our *emotional,* or *affective* (feelings), *brain.* Two parts seem especially important in emotions—the amygdala and the anterior part of the **cingulate gyrus.** The former recognizes angry or fearful facial expressions, assesses danger, and elicits the fear response. The latter plays a role in expressing our emotions through gestures and in resolving mental conflicts when we are frustrated.

Odors often trigger emotional reactions and memories, which reflect the origin of much of the limbic system in the primitive "smell brain" (rhinencephalon). Our reactions to odors are rarely neutral (a skunk smells *bad* and repulses us), and odors often recall memories of emotionally laden events.

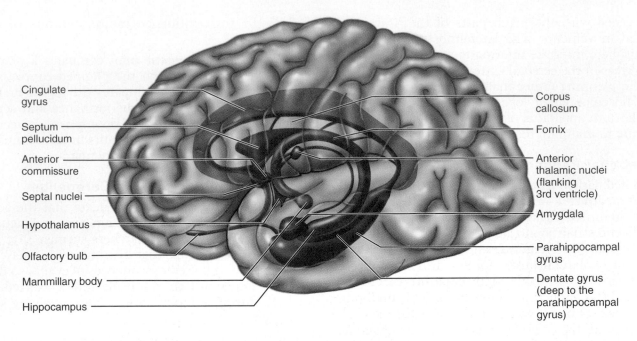

Cingulate gyrus

Septum pellucidum

Anterior commissure

Septal nuclei

Hypothalamus

Olfactory bulb

Mammillary body

Hippocampus

Corpus callosum

Fornix

Anterior thalamic nuclei (flanking 3rd ventricle)

Amygdala

Parahippocampal gyrus

Dentate gyrus (deep to the parahippocampal gyrus)

FIGURE 12.18   **The limbic system.** Lateral view of the brain, showing some of the structures of the limbic system, the emotional-visceral brain. The brain stem is not illustrated.

Extensive connections between the limbic system and lower and higher brain regions allow the system to integrate and respond to a variety of environmental stimuli. Most limbic system output is relayed through the hypothalamus. Because the hypothalamus is the neural clearinghouse for both autonomic (visceral) function and emotional response, it is not surprising that some people under acute or unrelenting emotional stress fall prey to visceral illnesses, such as high blood pressure and heartburn. Such emotion-induced illnesses are called **psychosomatic illnesses.**

Because the limbic system interacts with the prefrontal lobes, there is an intimate relationship between our feelings (mediated by the emotional brain) and our thoughts (mediated by the cognitive brain). As a result, we (1) react emotionally to things we consciously understand to be happening, and (2) are consciously aware of the emotional richness of our lives. Communication between the cerebral cortex and limbic system explains why emotions sometimes override logic and, conversely, why reason can stop us from expressing our emotions in inappropriate situations. Particular limbic system structures—the **hippocampal structures** and amygdala—also play a role in memory.

Reticular Formation

The **reticular formation** extends through the central core of the medulla oblongata, pons, and midbrain (Figure 12.19). It is composed of loosely clustered neurons in what is otherwise white matter.

These neurons form three broad columns along the length of the brain stem (Figure 12.16c): (1) the midline **raphe nuclei** (ra′fe; *raphe* = seam or crease), which are flanked laterally by (2) the **medial (large cell) group** and then (3) the **lateral (small cell) group of nuclei.** You will be hearing more about some of these nuclei when we discuss sleep later in the chapter.

The outstanding feature of the reticular neurons is their far-flung axonal connections. Individual reticular neurons project to the hypothalamus, thalamus, cerebellum, and spinal cord, making reticular neurons ideal for governing the arousal of the brain as a whole. For example, certain reticular neurons, unless inhibited by other brain areas, send a continuous stream of impulses (via thalamic relays) to the cerebral cortex, which keeps the cortex alert and conscious and enhances its excitability. This arm of the reticular formation is known as the **reticular activating system (RAS).** Impulses from all the great ascending sensory tracts synapse with RAS neurons, keeping them active and enhancing their arousing effect on the cerebrum. (This may explain why many students, stimulated by a bustling environment, like to study in a crowded cafeteria.) The RAS also acts like a filter for this flood of sensory inputs. Repetitive, familiar, or weak signals are filtered out, but unusual, significant, or strong impulses do reach consciousness. For example, you are probably unaware of your watch encircling your wrist, but would immediately notice it if the clasp broke. Between them, the RAS and the cerebral cortex disregard

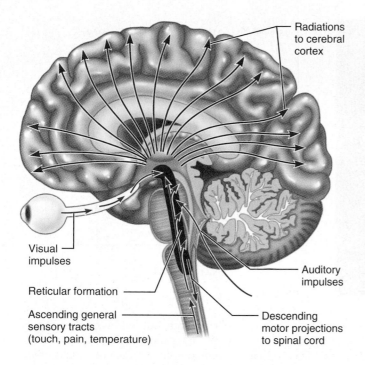

Radiations to cerebral cortex

Visual impulses

Reticular formation

Ascending general sensory tracts (touch, pain, temperature)

Auditory impulses

Descending motor projections to spinal cord

FIGURE 12.19 The reticular formation. The reticular formation extends the length of the brain stem. A portion of this formation, the reticular activating system, maintains alert wakefulness of the cerebral cortex. Ascending arrows indicate input of sensory systems to the RAS and then reticular output via thalamic relays to the cerebral cortex. Other reticular nuclei are involved in muscle coordination. Their output is indicated by the red arrow descending the brain stem.

perhaps 99% of all sensory stimuli as unimportant. If this did not occur, the sensory overload would drive us crazy. The drug LSD interferes with these sensory dampers, promoting an often overwhelming sensory overload.

■ Take a moment to become aware of all the stimuli in your environment. Notice all the colors, shapes, odors, sounds, and so on. How many of these sensory stimuli are you usually aware of?

The RAS is inhibited by sleep centers located in the hypothalamus and other neural regions, and is depressed by alcohol, sleep-inducing drugs, and tranquilizers. Severe injury to this system, as might follow a knockout punch that twists the brain stem, results in permanent unconsciousness (irreversible *coma*).

The reticular formation also has a *motor* arm. Some of its motor nuclei project to motor neurons in the spinal cord via the *reticulospinal tracts,* and help control skeletal muscles during coarse limb movements. Other reticular motor nuclei, such as the vasomotor, cardiac, and respiratory centers of the medulla, are autonomic centers that regulate visceral motor functions.

Higher Mental Functions

During the last four decades, an exciting exploration of our "inner space," or what we commonly call *the mind,* has been going on. But researchers in the field of cognition are still struggling to understand how the mind's presently incomprehensible qualities might spring from living tissue and electrical impulses. Souls and synapses are hard to reconcile!

Because brain waves reflect the electrical activity on which higher mental functions are based, we will consider them first, along with the related topics of consciousness and sleep. We will then (hesitantly) examine memory, an area in which we are still forced to hypothesize or speculate about what must (or should) happen.

Brain Wave Patterns and the EEG

Normal brain function involves continuous electrical activity of neurons. An **electroencephalogram** (e-lek″tro-en-sef′ah-lo-gram), or **EEG,** records some aspects of this activity. An EEG is made by placing electrodes on the scalp and then connecting the electrodes to an apparatus that measures electrical potential differences between various cortical areas (Figure 12.20a). The patterns of neuronal electrical activity recorded are called **brain waves.**

Each of us has a brain wave pattern that is as unique as our fingerprints. For simplicity, however, we can group brain waves into the four frequency classes shown in Figure 12.20b. Each wave is a continuous train of peaks and troughs, and the wave frequency, expressed in hertz, is the number of peaks that pass a given point in one second. A frequency of 1 Hz means that one peak passes the reference point each second.

■ **Alpha waves** (8–13 Hz) are relatively regular and rhythmic, low-amplitude, synchronous waves. In most cases, they indicate a brain that is "idling"—a calm, relaxed state of wakefulness.

■ **Beta waves** (14–25 Hz) are also rhythmic, but they are not as regular as alpha waves and have a higher frequency. Beta waves occur when we are mentally alert, as when concentrating on some problem or visual stimulus.

■ **Theta waves** (4–7 Hz) are still more irregular. Though common in children, theta waves are abnormal in awake adults.

■ **Delta waves** (4 Hz or less) are high-amplitude waves seen during deep sleep and when the reticular activating system is damped, such as during anesthesia. In awake adults, they indicate brain damage.

The amplitude or intensity of any wave is represented by how high the wave peaks rise and how low the troughs dip. The amplitude of brain waves

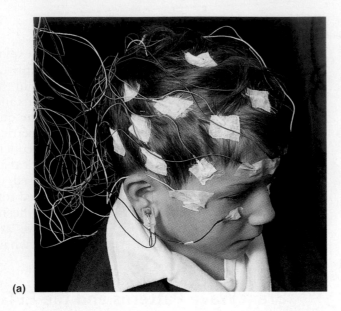

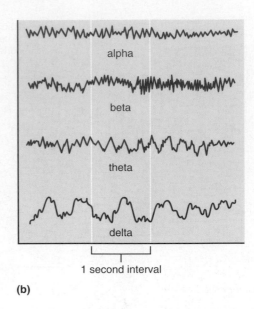

(a)   (b)

FIGURE 12.20   Electroencephalography and brain waves. (a) To obtain a recording of brain wave activity (an EEG), electrodes are positioned on the patient's scalp and attached to a recording device called an electro-encephalograph. **(b)** Typical EEGs. Alpha waves are typical of the awake but relaxed state; beta waves occur in the awake, alert state; theta waves are common in children but not adults; delta waves occur during deep sleep.

reflects the number of neurons firing together in synchrony, not the degree of electrical activity of individual neurons. When the brain is active, brain waves are complex and low amplitude. But when the brain is inactive, as during sleep, neurons tend to fire synchronously, producing similar, high-amplitude brain waves.

Brain waves change with age, sensory stimuli, brain disease, and the chemical state of the body. EEGs are used for diagnosing and localizing many types of brain lesions, such as tumors, infarcts, and epileptic lesions. Interference with cerebral cortical functions is suggested when brain waves are too fast or too slow, and unconsciousness occurs at both extremes. Because spontaneous brain waves are always present, even during unconsciousness and coma, their absence—called a "flat EEG"—is clinical evidence of brain death.

🇭 *HOMEOSTATIC IMBALANCE*

Almost without warning, a victim of epilepsy may lose consciousness and fall stiffly to the ground, body wracked by uncontrollable jerking. These **epileptic seizures** reflect a torrent of electrical discharges of groups of brain neurons, and while their uncontrolled activity is occurring, no other messages can get through. Epilepsy, manifested by one out of 25 of us, is not associated with, nor does it cause, intellectual impairment. Some cases of epilepsy are induced by genetic factors, but it can also result from brain injuries caused by blows to the head, stroke, infections, high fever, or tumors.

Epileptic seizures can vary tremendously in their expression and severity. *Absence seizures,* formerly known as *petit mal,* are mild forms in which the facial muscles twitch and the expression goes blank for a few seconds. These are typically seen in young children and usually disappear by the age of ten. In the most severe, convulsive form of epileptic seizures, *grand mal,* the person loses consciousness. Bones are often broken during the intense convulsions, showing the incredible strength of the muscle contractions that occur. Loss of bowel and bladder control and severe biting of the tongue are common. The seizure lasts for a few minutes, then the muscles relax and the person awakens but remains disoriented for several minutes. Many seizure sufferers experience a sensory hallucination, such as a taste, smell, or flashes of light, just before the seizure begins. This phenomenon, called an **aura,** is helpful because it gives the person time to lie down and avoid falling to the floor.

Epilepsy can usually be controlled by anticonvulsive drugs. Newer on the scene is the *vagus nerve stimulator,* which is implanted under the skin of the chest and delivers pulses via the vagus nerve to the brain at predetermined intervals to keep the electrical activity of the brain from becoming chaotic. A current line of research seeks to use implanted electrodes programmed to detect coming seizures in time to deliver customized drug or shock therapy. ●

Consciousness

Consciousness encompasses conscious perception of sensations, voluntary initiation and control of movement, and capabilities associated with higher mental processing (memory, logic, judgment, perseverance, and so on). Clinically, consciousness is defined on a continuum that grades behavior in response to stimuli as (1) *alertness*, (2) *drowsiness* or *lethargy* (which proceeds to sleep), (3) *stupor*, and (4) *coma*. Alertness is the highest state of consciousness and cortical activity, and coma the most depressed. Consciousness is difficult to define. A sleeping person obviously lacks something that he or she has when awake, and we call this "something" consciousness. The current suppositions about consciousness are as follows:

1. **It involves simultaneous activity of large areas of the cerebral cortex.**

2. **It is superimposed on other types of neural activity.** At any time, specific neurons and neuronal pools are involved both in localized activities (such as motor control) and in cognition.

3. **It is holistic and totally interconnected.** Information for "thought" can be claimed from many locations in the cerebrum simultaneously. For example, retrieval of a specific memory can be triggered by several routes—a smell, a place, a particular person, and so on.

HOMEOSTATIC IMBALANCE

Except when a person is sleeping, unconsciousness is always a signal that brain function is impaired. A brief loss of consciousness is called **fainting** or **syncope** (sing'ko-pe; "cut short"). Most often, it indicates inadequate cerebral blood flow due to low blood pressure, as might follow hemorrhage or sudden emotional stress.

Total unresponsiveness to sensory stimuli for an extended period is called **coma**. Coma is *not* deep sleep. During sleep, the brain is active and oxygen consumption resembles that of the waking state. In coma patients, in contrast, oxygen use is always below resting levels.

Blows to the head may induce coma by causing widespread cerebral or brain stem trauma. Coma may also be produced by tumors or infections that invade the brain stem. Metabolic disturbances such as hypoglycemia (abnormally low blood sugar levels), drug overdose, or liver or kidney failure interfere with overall brain function and can result in coma. Cerebral infarctions, unless massive and accompanied by extreme swelling of the brain, rarely cause coma.

When the brain has suffered irreparable damage, irreversible coma occurs, even though life-support measures may have restored vitality to other body organs. The result is **brain death,** a dead brain in an otherwise living body. Because life support can be removed only after death, physicians must determine whether a patient in an irreversible coma is legally alive or dead. ●

Sleep and Sleep-Awake Cycles

Sleep is defined as a state of partial unconsciousness from which a person can be aroused by stimulation. This distinguishes sleep from *coma*, a state of unconsciousness from which a person *cannot* be aroused by even the most vigorous stimuli. Cortical activity is depressed during sleep, but brain stem functions, such as control of respiration, heart rate, and blood pressure, continue. Even environmental monitoring continues to some extent, as illustrated by the fact that strong stimuli ("things that go bump in the night") immediately arouse us. In fact, people who sleepwalk can avoid objects and navigate stairs while truly asleep.

Types of Sleep

The two major types of sleep, which alternate through most of the sleep cycle, are **non–rapid eye movement (NREM) sleep** and **rapid eye movement (REM) sleep,** defined in terms of their EEG patterns. During the first 30 to 45 minutes of the sleep cycle, we pass through the four stages of NREM sleep (Table 12.2), culminating in **slow-wave sleep.** As we pass through these stages and slip into deeper and deeper sleep, the frequency of the EEG waves declines, but their amplitude increases.

About 90 minutes after sleep begins, once NREM stage 4 has been achieved, the EEG pattern changes abruptly. It becomes very irregular and appears to backtrack quickly through the stages until alpha waves (more typical of the awake state) appear, indicating the onset of REM sleep. This brain wave change is coupled with increases in body temperature, heart rate, respiratory rate, and blood pressure and a decrease in gastrointestinal motility. Oxygen use by the brain is tremendous during REM—greater than during the awake state.

Although the eyes move rapidly under the lids during REM, most of the body's skeletal muscles are actively inhibited and go limp. This temporary paralysis prevents us from acting out our dreams. Most dreaming occurs during REM sleep, and some suggest that the flitting eye movements are following the visual imagery of our dreams. (Note, however, that most nightmares and night terrors occur during NREM stages 3 and 4.) In adolescent and adult males, REM episodes are frequently associated with erection of the penis.

TABLE 12.2 / Types and Stages of Sleep

Non–Rapid Eye Movement (NREM) Sleep

Stage 1 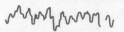 Eyes are closed and relaxation begins. Thoughts flit in and out and a drifting sensation occurs; vital signs (body temperature, respiration, pulse, and blood pressure) are normal; the EEG pattern shows alpha waves; stimulation causes immediate arousal.

Stage 2 The EEG pattern becomes more irregular; *sleep spindles*—sudden, short high-voltage wave bursts occurring at 12–14 Hz—appear, and arousal is more difficult.

Stage 3 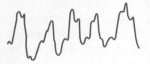 Sleep deepens, and theta and delta waves appear; vital signs begin to decline, and the skeletal muscles are very relaxed; dreaming is common; usually reached about 20 minutes after the onset of stage 1.

Stage 4  Called slow-wave sleep because the EEG pattern is dominated by delta waves (1–4 Hz); vital signs reach their lowest normal levels, and digestive system motility increases; skeletal muscles are relaxed, but normal sleepers turn approximately every 20 minutes; arousal is difficult; bed-wetting and sleepwalking occur during this phase.

Rapid Eye Movement (REM) Sleep

EEG pattern reverts through the NREM stages to the stage 1 pattern. Vital signs increase and digestive system activity declines; skeletal muscles (except for ocular muscles) are inhibited; the stage of most dreaming.

Sleep Patterns

The alternating cycles of sleep and wakefulness reflect a natural *circadian*, or 24-hour, *rhythm*. In the awake state, alertness of the cerebral cortex is mediated by the RAS (see Figure 12.19). When RAS activity declines, cerebral cortical activity declines as well; thus, lesions of the RAS nuclei result in unconsciousness. However, sleep is much more than simply turning off the arousal (RAS) mechanism. RAS centers not only help maintain the awake state but also mediate some sleep stages, especially dreaming sleep. The hypothalamus is responsible for the timing of the sleep cycle: Its *suprachiasmatic nucleus* (a biological clock) regulates its *preoptic nucleus* (a sleep-inducing center).

In young and middle-aged adults, a typical night's sleep alternates between REM and NREM sleep. Following each REM episode, the sleeper descends to stage 4 again. REM recurs about every 90 minutes, with each REM period getting longer. The first REM of the night lasts 5–10 minutes and the final one 20 to 50 minutes. Hence, our longest dreams occur just before we awake. When the neurons of the dorsal raphe nuclei of the midbrain reticular formation fire at maximal rates, we awaken for the day.

One cause of sleepiness appears to be adenosine increases in the brain following periods of high ATP usage by the brain. Unless inhibited by adenosine, the activity of ACh-secreting cells of the giant cell nucleus, part of the pons reticular formation, is believed to cause the arousal pattern of REM sleep. Norepinephrine released by the locus coeruleus of the pons is thought to induce the transient paralysis of REM sleep. NREM sleep areas include the nucleus of the solitary tract, reticular nucleus of the thalamus, hypothalamus, and basal forebrain.

Importance of Sleep

Slow-wave (NREM stage 4) and REM sleep seem to be important in different ways. Slow-wave sleep is presumed to be restorative—the time when most neural mechanisms can wind down to basal levels. When deprived of sleep, we spend more time than usual in slow-wave sleep during the next sleep episode.

A person persistently deprived of REM sleep becomes moody and depressed, and exhibits various personality disorders. REM sleep may give the brain an opportunity to analyze the day's events and to work through emotional problems in dream imagery. Another idea is that REM sleep is reverse learning. According to this hypothesis, accidental, repetitive, and meaningless communications continually occur, and they must be eliminated from the neural networks by dreaming if the cortex is to remain a well-behaved and efficient thinking system. In other words, we dream to forget.

Alcohol and most sleep medications (barbiturates and others) suppress REM sleep but not slow-wave sleep. On the other hand, certain tranquilizers, such as diazepam (Valium) reduce slow-wave sleep much more than REM sleep.

Whatever its importance, the daily sleep requirement declines steadily from 16 hours or so in infants to approximately 7½ to 8½ hours in early adulthood. It then levels off before declining once again in old age. Sleep patterns also change throughout life. REM sleep occupies about half the total sleeping time in infants and then declines until the age of ten years, when it stabilizes at about 25%. In contrast, stage 4 sleep declines steadily from birth and often disappears completely in those over 60.

Ⓗ HOMEOSTATIC IMBALANCE

People with **narcolepsy** lapse abruptly into sleep from the awake state. These sleep episodes last about 15 minutes, can occur without warning at any time, and are often triggered by a pleasurable event—a good joke, a game of poker. This condition can be extremely hazardous when the person must drive, operate machinery, or take a bath. In all daytime episodes, narcoleptics show an EEG typical of REM sleep. During the nighttime sleep cycle, however, they spend a below-normal amount of time in REM sleep.

Insomnia is a chronic inability to obtain the *amount* or *quality* of sleep needed to function adequately during the day. Sleep requirements vary from four to nine hours a day in healthy people, so there is no way to determine the "right" amount. Self-professed insomniacs tend to exaggerate the extent of their sleeplessness, and many have a notorious tendency toward self-medication and barbiturate abuse.

True insomnia often reflects normal age-related changes, but perhaps the most common cause is psychological disturbance. We have difficulty falling asleep when we are anxious or upset, and depression is often accompanied by early awakening.

Sleep apnea, a temporary cessation of breathing during sleep, is scary. The victim awakes abruptly due to hypoxia (lack of oxygen)—a condition that may occur repeatedly throughout the night. Its cause varies and is uncertain. However, it tends to be more common in the elderly (the stimulating effects of carbon dioxide buildup on breathing decline with age), and in those with some sort of blockage of the upper airway (tonsillitis, for example). Sleep apnea is enhanced by alcohol, high blood pressure, obesity, and conditions that affect CNS respiratory centers. Treatment is focused on the cause (if known). ●

Memory

Memory is the storage and retrieval of information. Memories are essential for learning and incorporating our experiences into behavior and are part and parcel of our consciousness. Stored somewhere in your 3 pounds of wrinkled brain are zip codes, the face of your grandfather, and the taste of yesterday's pizza. Your memories reflect your lifetime.

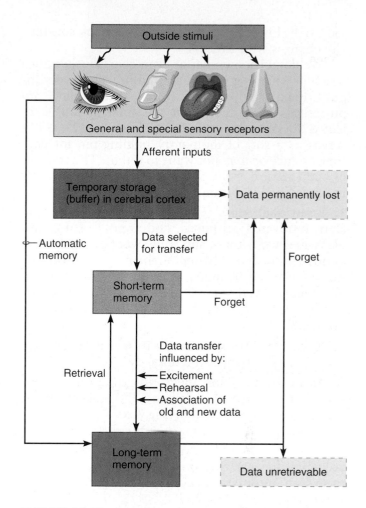

FIGURE 12.21 **Memory processing.** Sensory inputs are processed by the cerebral cortex (indicated as the temporary storage area), which selects what is to be sent to short-term memory. Rehearsal and other factors help the transfer of information from short-term memory to long-term memory. For long-term memory to become permanent, consolidation must occur. Some information not attended to consciously goes directly into long-term memory (automatic memory).

Stages of Memory

Memory storage involves two distinct stages: short-term memory and long-term memory (Figure 12.21). **Short-term memory (STM),** also called *working memory,* is the preliminary step, as well as the power that lets you look up a telephone number, dial it, and then never think of it again. The capacity of STM is limited to seven or eight chunks of information, such as the digits of a telephone number or the sequence of words in an elaborate sentence.

In contrast to STM, **long-term memory (LTM)** seems to have a limitless capacity. Although our STM devices cannot recall numbers much longer than a telephone number, we can remember scores of telephone numbers by committing them to LTM. However, our ability to store and to retrieve information declines with aging. Thus, long-term memories too can be forgotten, and our memory bank continually changes with time.

We do not remember or even consciously notice much of what is going on around us. As sensory inputs flood into our cerebral cortex, they are processed, and some 5% of this information is selected for transfer to STM (Figure 12.21). STM serves as a sort of temporary holding bin for data that we may or may not want to retain. The transfer of information from STM to LTM is affected by many factors, including:

1. Emotional state. We learn best when we are alert, motivated, surprised, and aroused. For example, when we witness shocking events, transferal is almost immediate. Norepinephrine, a neurotransmitter involved in memory processing of emotionally charged events, is released when we are excited or "stressed out," which helps to explain this phenomenon.

2. Rehearsal. An opportunity to repeat or rehearse the material enhances memory.

3. Association. Tying "new" information to "old" information already stored in LTM appears to be very important in remembering facts.

4. Automatic memory. Not all impressions that become part of LTM are consciously formed. A student concentrating on a lecturer's speech may record an automatic memory of the pattern of the lecturer's tie.

Memories transferred to LTM take time to become permanent. The process of **memory consolidation** apparently involves fitting new facts into the various categories of knowledge already stored in the cerebral cortex.

Categories of Memory

The brain distinguishes between factual knowledge and skills, and we process and store these different kinds of information in different ways. **Fact (declarative) memory** entails learning explicit information, such as names, faces, words, and dates. It is related to our conscious thoughts and our ability to manipulate symbols and language. When fact memories are committed to LTM, they are usually filed along with the context in which they were learned. For instance, when you think of your new friend Joe, you probably picture him at the basketball game where you met him.

Skill (procedural) memory is less conscious learning and usually involves motor activities. It is acquired through practice, as when we learn to ride a bike. Skill memories do not preserve the circumstances of learning; in fact, they are best remembered in the doing. You do not have to think through how to tie your shoes. Once learned, skill memories are hard to unlearn.

Brain Structures Involved in Memory

Much of what scientists know about learning and memory comes from experiments with macaque monkeys and studies of amnesia in humans. Such studies have revealed that different brain structures are involved in the two categories of memory.

It appears that specific pieces of each memory are stored near regions of the brain that need them so that new inputs can be quickly associated with the old. Accordingly, visual memories would be stored in the occipital cortex, memories of music in the temporal cortex, and so on.

But how are the memory connections made? The proposed scheme of information flow for fact memory is shown in Figure 12.22a. When a sensory perception is formed in the cerebral sensory cortex, the cortical neurons dispatch impulses along two parallel circuits to the hippocampus and the amygdala, both of which play a major role in memory consolidation and memory access and have connections with the diencephalon, basal forebrain, and prefrontal cortex. The basal forebrain then closes the memory loop by sending impulses back to the sensory cortical areas initially forming the perception. This repeated feedback presumably transforms the new perception into a more durable memory. Later recall of the new memory occurs when the same cortical neurons are stimulated.

The prefrontal cortex seems to be necessary for retrieving fact learning from LTM storage depots elsewhere in the brain for use in present tasks. The hippocampus oversees the circuitry for learning and remembering spatial relationships. The amygdala, with its widespread connections to all cortical sensory areas and the emotional response centers of the hypothalamus, seems to associate memories formed through different senses and link them to emotional states generated in the hypothalamus.

H HOMEOSTATIC IMBALANCE

Damage to either the hippocampus or the amygdala results in only slight memory loss, but bilateral destruction of both structures causes widespread amnesia. Consolidated memories are not lost, but new sensory inputs cannot be associated with old, and the person lives in the here and now from that point on. This phenomenon is called *anterograde* (an'ter-o-grād") *amnesia*, in contrast to *retrograde amnesia*, which is the loss of memories formed in the distant past. You could carry on an animated conversation with a person with anterograde amnesia, excuse yourself, return five minutes later, and that person would not remember you. ●

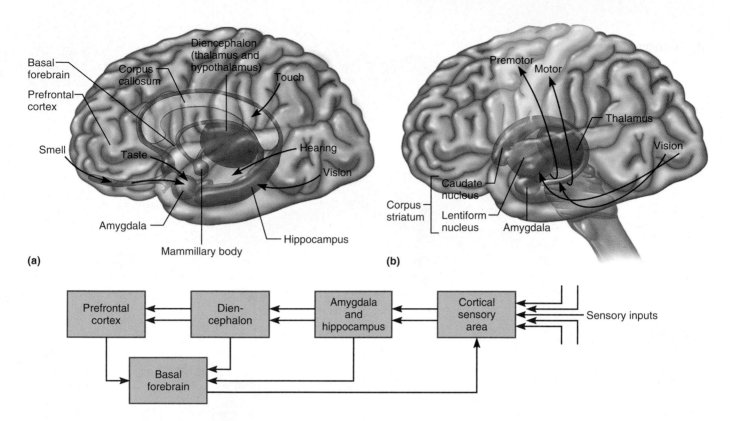

(a)

(b)

FIGURE 12.22 Proposed memory circuits. (a) Essential structures of fact memory circuits. The flowchart below indicates how these structures may interact in memory formation. Sensory inputs from the cortex flow through parallel circuits, one involving the hippocampus and the other the amygdala. Both circuits encompass parts of the diencephalon, basal forebrain, and prefrontal cortex. The basal forebrain feeds back to the sensory cortex, closing the memory loop. **(b)** Major structures involved in skills memory. The corpus striatum mediates the automatic connections between a stimulus (visual here) and a motor response.

Individuals suffering from anterograde amnesia can still learn skills. Consequently, a second learning circuit, independent of the pathways used in fact memory, has been suggested (Figure 12.22b). When the cerebral cortex is activated by sensory inputs, it signals the corpus striatum of its intent to mobilize a skill memory. The corpus striatum then communicates with one or more brain stem nuclei as well as with the cerebral cortex to promote the desired movement. Thus, the corpus striatum is the link between a perceived stimulus and a motor response. The cerebellum also plays a role in skill memory.

Mechanisms of Memory

Human memory is notoriously difficult to study, but experimental studies reveal that during learning, (1) neuronal RNA content is altered and newly synthesized mRNAs are delivered to axons and dendrites, (2) dendritic spines change shape, (3) unique extracellular proteins are deposited at synapses involved in LTM, (4) the number and size of presynaptic terminals may increase, (5) more neurotransmitter is released by the presynaptic

neurons, and (6) new neurons appear in the hippocampus that seem to have a role in the timing of learned responses.

Long-term potentiation (LTP) is involved in some cases. **NMDA receptors,** named after the chemical N-methyl-D-aspartate used to detect them, act as calcium channels and mediate synaptic LTP. Recall that LTP refers to the persistent changes in synaptic strength first identified in hippocampal memory pathways that use the amino acid glutamate as their transmitter.

The NMDA receptors appear to be activated by both voltage (a depolarizing current) and a chemical (binding of glutamate). These signals, which apparently must occur sequentially ("click-click"), lead to Ca^{2+} influx, which provides the signal necessary for LTP induction. Details of synaptic events are murky, but binding of a local neurotropin called brain-derived neurotropic factor (BDNF) postsynaptically causes depolarization (presumably by Na^+ entry) followed by Ca^{2+} channel opening. Depolarization enhances NMDA channel opening to which glutamate binds, thereby removing the Mg^{2+} blockage of the receptor channel. Ca^{2+} appears to promote activation

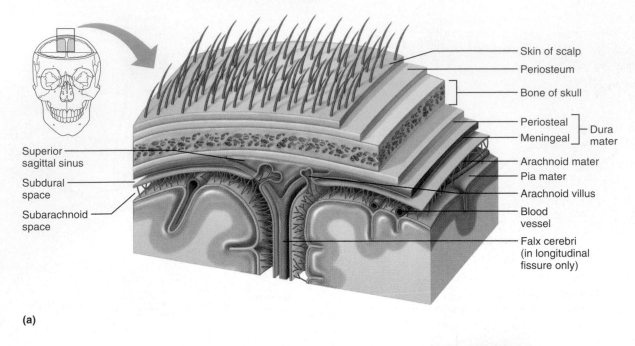

(a)

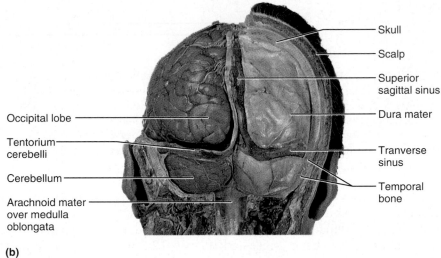

(b)

FIGURE 12.23 Meninges. (a) Frontal section showing the relationship of the dura mater, arachnoid, and pia mater. The meningeal dura forms the falx cerebri fold. A dural sinus, the superior sagittal sinus, is enclosed by the dural membranes superiorly. Arachnoid villi, which return cerebrospinal fluid to the dural sinus, are also shown. **(b)** Posterior view of the brain in place, surrounded by the dura mater and sinuses.

of several enzymes in the postsynaptic cells. It's a good bet that at least some of these "reshaped proteins" bring about changes in the NMDA receptors that increase their sensitivity to glutamate.

Protection of the Brain

Nervous tissue is soft and delicate, and neurons are injured by even slight pressure. However, the brain is protected by bone (the skull), membranes (the meninges), and a watery cushion (cerebrospinal fluid). Furthermore, the brain is protected from harmful substances in the blood by the blood-brain barrier. We described the cranium, the brain's bony encasement, in Chapter 7. Here we will attend to the other protective devices.

Meninges

The **meninges** (mě-nin'jēz; *mening* = membrane) are three connective tissue membranes that lie just external to the CNS organs. They (1) cover and protect the CNS, (2) protect blood vessels and enclose venous sinuses, (3) contain cerebrospinal fluid, and (4) form partitions in the skull. From external to internal, the meninges (singular, **meninx**) are the dura mater, arachnoid mater, and pia mater (Figure 12.23).

Dura Mater

The leathery **dura mater** (du'rah ma'ter), meaning "tough mother," is the strongest meninx. Where it surrounds the brain, it is a two-layered sheet of fibrous connective tissue. The more superficial *periosteal layer* is attached to the inner surface of the skull (the periosteum). (There is no dural periosteal layer surrounding the spinal cord.) The deeper *meningeal layer* forms the true external covering of the brain and continues caudally in the vertebral canal as the dural sheath of the spinal cord. The brain's two dural layers are fused together except in certain areas, where they separate to enclose **dural sinuses** that collect venous blood from the brain and direct it into the internal jugular veins of the neck.

In several places, the meningeal dura mater extends inward to form flat partitions that subdivide the cranial cavity. These **dural septa,** which limit excessive movement of the brain within the cranium, include the following (Figure 12.24):

- **Falx cerebri** (falks ser'ĕ-bri). A large sickle-shaped (*falks* = sickle) fold that dips into the longitudinal fissure between the cerebral hemispheres. Anteriorly, it attaches to the crista galli of the ethmoid bone.

- **Falx cerebelli** (ser"ĕ-bel'i). Continuing inferiorly from the posterior falx cerebri, this small midline partition runs along the vermis of the cerebellum.

- **Tentorium cerebelli** (ten-to're-um; "tent"). Resembling a tent over the cerebellum, this nearly horizontal dural fold extends into the transverse fissure between the cerebral hemispheres (which it helps to support) and the cerebellum.

Arachnoid Mater

The middle meninx, the **arachnoid mater,** or simply the **arachnoid** (ah-rak'noid), forms a loose brain covering, never dipping into the sulci at the cerebral surface. It is separated from the dura mater by a narrow serous cavity, the **subdural space,** which contains a film of fluid. Beneath the arachnoid membrane is the wide **subarachnoid space.** Weblike extensions span this space and secure the arachnoid mater to the underlying pia mater. (*Arachnida* means "spider," and this membrane was named for its weblike extensions.) The subarachnoid space is filled with cerebrospinal fluid and also contains the largest blood vessels serving the brain. Because the arachnoid is fine and elastic, these blood vessels are poorly protected.

Knoblike projections of the arachnoid mater called **arachnoid villi** (vil'i) protrude superiorly through the dura mater and into the superior sagittal sinus (see Figure 12.23a). Cerebrospinal fluid is absorbed into the venous blood of the sinus by these valvelike villi.

Pia Mater

The **pia mater** (pi'ah), meaning "gentle mother," is composed of delicate connective tissue and is richly invested with tiny blood vessels. It is the only meninx that clings tightly to the brain, following its

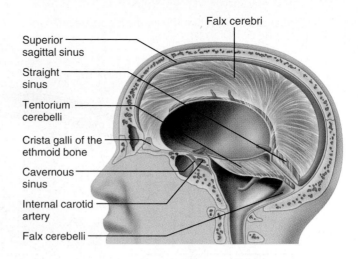

FIGURE 12.24 **Partitioning folds of dura mater in the cranial cavity.** Some dural sinuses are also shown.

every convolution. Small arteries entering the brain tissue carry ragged sheaths of pia mater inward with them for short distances.

 HOMEOSTATIC IMBALANCE

Meningitis, inflammation of the meninges, is a serious threat to the brain because a bacterial or viral meningitis may spread to the CNS. This condition of brain inflammation is called *encephalitis* (en'sef-ah-li'tis). Meningitis is usually diagnosed by taking a sample of cerebrospinal fluid from the subarachnoid space and examining it for microbes. ●

Cerebrospinal Fluid

Cerebrospinal fluid (CSF), found in and around the brain and spinal cord, forms a liquid cushion that gives buoyancy to the CNS organs. By floating the jellylike brain, the CSF effectively reduces brain weight by 97% and prevents the brain from crushing under its own weight. CSF also protects the brain and spinal cord from blows and other trauma. Additionally, although the brain has a rich blood supply, CSF helps nourish the brain, and there is some evidence that it carries chemical signals (such as hormones and sleep- and appetite-inducing molecules) from one part of the brain to another.

CSF is a watery "broth" similar in composition to blood plasma, from which it is formed. However, it contains less protein than plasma and its ion concentration is different. For example, CSF contains more Na^+, Cl^-, and H^+, than blood plasma, and less Ca^{2+} and K^+. CSF composition, particularly its pH, is important in the control of cerebral blood flow and breathing.

The **choroid plexuses** that hang from the roof of each ventricle form CSF (Figure 12.25a). These

What structures return CSF to the bloodstream?

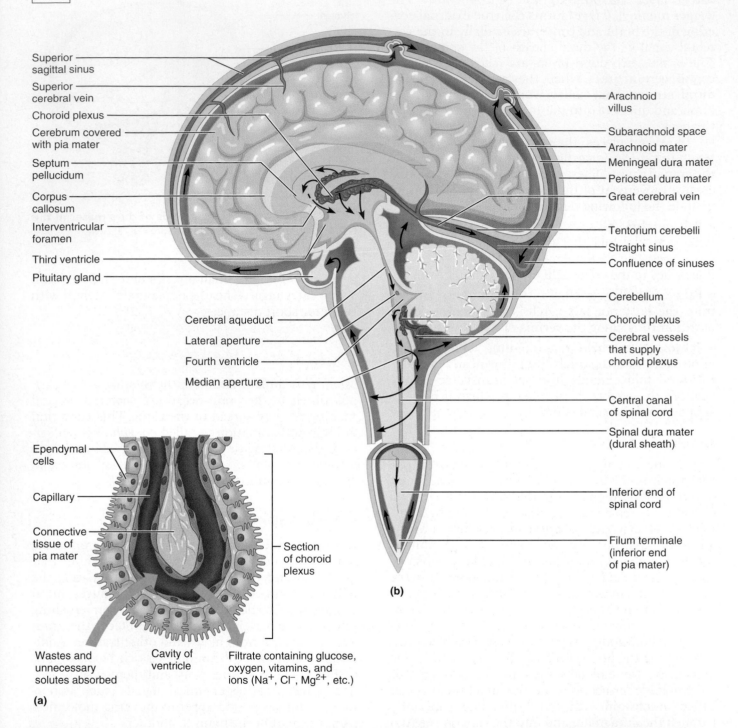

Superior sagittal sinus

Superior cerebral vein

Choroid plexus

Cerebrum covered with pia mater

Septum pellucidum

Corpus callosum

Interventricular foramen

Third ventricle

Pituitary gland

Arachnoid villus

Subarachnoid space

Arachnoid mater

Meningeal dura mater

Periosteal dura mater

Great cerebral vein

Tentorium cerebelli

Straight sinus

Confluence of sinuses

Cerebral aqueduct

Lateral aperture

Fourth ventricle

Median aperture

Cerebellum

Choroid plexus

Cerebral vessels that supply choroid plexus

Central canal of spinal cord

Spinal dura mater (dural sheath)

Inferior end of spinal cord

Filum terminale (inferior end of pia mater)

(b)

Ependymal cells

Capillary

Connective tissue of pia mater

Section of choroid plexus

Wastes and unnecessary solutes absorbed

Cavity of ventricle

Filtrate containing glucose, oxygen, vitamins, and ions (Na$^+$, Cl$^-$, Mg^{2+}, etc.)

(a)

FIGURE 12.25 Formation, location, and circulation of CSF. (a) Choroid plexus forms CSF. Each choroid plexus consists of a knot of porous capillaries surrounded by a single layer of ependymal cells joined by tight junctions and bearing long cilia with blunted ends. Although the filtrate moves easily from the capillaries, it must move through and be processed by the ependymal cells before it is allowed to enter the ventricles as cerebrospinal fluid. **(b)** Location and circulatory pattern of CSF. Direction of circulation is indicated by the arrows. (The relative position of the right lateral ventricle is indicated by the pale blue area deep to the corpus callosum and septum pellucidum.)

■ *Arachnoid villi.*

plexuses are frond-shaped clusters of broad, thin-walled capillaries (*plex* = interwoven) enclosed first by pia mater and then by a layer of ependymal cells lining the ventricles. These capillaries are fairly permeable, and tissue fluid filters continuously from the bloodstream. However, the choroid plexus ependymal cells are joined by tight junctions and have ion pumps that allow them to modify this filtrate by actively transporting only certain ions across their membranes into the CSF pool. This sets up ionic gradients that cause water to diffuse into the ventricles as well. In adults, the total CSF volume of about 150 ml (about half a cup) is replaced every 8 hours or so; hence about 500 ml of CSF is formed daily. The choroid plexuses also help cleanse the CSF by removing waste products and unnecessary solutes.

Once produced, CSF moves freely through the ventricles. Some CSF circulates into the central canal of the spinal cord, but most enters the subarachnoid space via the lateral and medial apertures in the walls of the fourth ventricle (Figure 12.25b). The constant motion of the CSF is aided by movement of the long cilia of the ependymal cells lining the ventricles. In the subarachnoid space, CSF bathes the outer surfaces of the brain and spinal cord and then returns to the blood in the dural sinuses via the arachnoid villi.

H HOMEOSTATIC IMBALANCE

Ordinarily, CSF is produced and drained at a constant rate. However, if something (such as a tumor) obstructs its circulation or drainage, it accumulates and exerts pressure on the brain. This condition is called *hydrocephalus* ("water on the brain"). Hydrocephalus in a newborn baby causes its head to enlarge; this is possible because the skull bones have not yet fused. In adults, however, hydrocephalus is likely to result in brain damage because the skull is rigid and hard, and accumulating fluid compresses blood vessels serving the brain and crushes the soft nervous tissue. Hydrocephalus is treated by inserting a shunt into the ventricles to drain the excess fluid into a vein in the neck. ●

Blood-Brain Barrier

The **blood-brain barrier** is a protective mechanism that helps maintain a stable environment for the brain. No other body tissue is so absolutely dependent on a constant internal milieu as is the brain. In other body regions, the extracellular concentrations of hormones, amino acids, and ions are in constant flux, particularly after eating or exercise. If the brain were exposed to such chemical variations, the neurons would fire uncontrollably, because some hor-

mones and amino acids serve as neurotransmitters and certain ions (particularly K⁺) modify the threshold for neuronal firing.

Bloodborne substances in the brain's capillaries are separated from the extracellular space and neurons by (1) the continuous endothelium of the capillary wall, (2) a relatively thick basal lamina surrounding the external face of each capillary, (3) the bulbous "feet" of the astrocytes clinging to the capillaries. The capillary endothelial cells are almost seamlessly joined all around by *tight junctions* (Chapter 18), making them the least permeable capillaries in the body, a characteristic that constitutes most (if not all) of the blood-brain barrier. Although the astrocytes' feet contribute to the blood-brain barrier, it now appears that their major role is to provide the signals that stimulate the capillary endothelial cells to *form* the tight junctions characteristic of the blood-brain barrier.

The blood-brain barrier is selective, rather than absolute. Nutrients such as glucose, essential amino acids, and some electrolytes move passively by facilitated diffusion through the endothelial cell membranes. Bloodborne metabolic wastes, proteins, certain toxins, and most drugs are denied entry to brain tissue. Small nonessential amino acids and potassium ions not only are prevented from entering the brain, but also are actively pumped from the brain across the capillary endothelium.

The barrier is ineffective against fats, fatty acids, oxygen, carbon dioxide, and other fat-soluble molecules that diffuse easily through all plasma membranes. This explains why bloodborne alcohol, nicotine, and anesthetics can affect the brain.

The structure of the blood-brain barrier is not completely uniform. The capillaries of the choroid plexuses are very porous, but the ependymal cells surrounding them have tight junctions. In some brain areas surrounding the third and fourth ventricles, the blood-brain barrier is entirely absent and the capillary endothelium is quite permeable, allowing bloodborne molecules easy access to the neural tissue. One such region is the vomiting center of the brain stem, which monitors the blood for poisonous substances. Another is in the hypothalamus, which regulates water balance, body temperature, and many other metabolic activities; lack of a blood-brain barrier here is essential to allow the hypothalamus to sample the chemical composition of the blood. The barrier is incomplete in newborn and premature infants, and potentially toxic substances can enter the CNS and cause problems not seen in adults.

Injury to the brain, whatever the cause, may result in a localized breakdown of the blood-brain barrier. Most likely, this reflects some change in the capillary endothelial cells or their tight junctions.

Homeostatic Imbalances of the Brain

Brain dysfunctions are unbelievably varied and extensive. We have mentioned some of them already; here, we will focus on traumatic brain injuries, cerebrovascular accidents, and degenerative brain disorders.

Traumatic Brain Injury

Head injuries are the leading cause of accidental death in the United States. Consider, for example, what happens if you forget to fasten your seat belt and then rear-end another car. Your head is moving and then is suddenly stopped as it hits the windshield. Brain damage is caused not only by localized injury at the site of the blow (the *coup* injury), but also by the ricocheting effect as the brain hits the opposite end of the skull (the *contrecoup* injury).

A **concussion** occurs when brain injury is slight and the symptoms are mild and short-lived. The victim may be dizzy, or lose consciousness briefly, but there is no permanent neurological damage.

Marked brain tissue destruction yields a **contusion.** In cortical contusions, the individual may remain conscious, but severe brain stem contusions always cause coma, lasting from hours to a lifetime because of injury to the reticular activating system.

Following a head blow, death may result from **subdural** or **subarachnoid hemorrhage** (bleeding from ruptured vessels into those spaces). Individuals who are initially lucid and then begin to deteriorate neurologically are, in all probability, hemorrhaging intracranially. Blood accumulating in the skull increases intracranial pressure and compresses brain tissue. If the pressure forces the brain stem inferiorly through the foramen magnum, control of blood pressure, heart rate, and respiration is lost. Intracranial hemorrhages are treated by surgical removal of the hematoma (localized blood mass) and repair of the ruptured vessels.

Another consequence of traumatic head injury is **cerebral edema,** swelling of the brain. Anti-inflammatory drugs are routinely administered to head injury patients in an attempt to prevent cerebral edema from aggravating brain injury.

Cerebrovascular Accidents

The single most common nervous system disorder and the third leading cause of death in the United States are **cerebrovascular accidents (CVAs)** (ser"ĕ-bro-vas'ku-lar), also called *strokes* or *brain attacks.* CVAs occur when blood circulation to a brain area is blocked and brain tissue dies. [Deprivation of blood supply to *any* tissue is called **ischemia** (is-ke'me-ah; "to hold back blood") and results in deficient oxygen and nutrient delivery to cells.] The most common cause of CVA is blockage of a cerebral artery by a blood clot. Other causes include compression of brain tissue by hemorrhage or edema and narrowing of brain vessels by atherosclerosis.

Those who survive a CVA are typically paralyzed on one side of the body. Many exhibit sensory deficits or have difficulty in understanding or vocalizing speech. Whatever the cause, fewer than 35% of patients who survive a CVA are alive three years later. Even so, the picture is not hopeless. Some patients recover at least part of their lost faculties, because undamaged neurons sprout new branches that spread into the injured area and take over some lost functions. Physical therapy is usually started as soon as possible to prevent muscle contractures (abnormal shortening of the paralyzed muscles due to lack of exercise).

Not all strokes are "completed." Temporary episodes of reversible cerebral ischemia, called **transient ischemic attacks (TIAs),** are common. TIAs last from 5 to 50 minutes and are characterized by temporary numbness, paralysis, and impaired speech. While these deficits are not permanent, TIAs do constitute "red flags" that warn of impending, more serious CVAs.

CVAs are like undersea earthquakes. It's not the initial temblor that does most of the damage, it's the tsunami wave that floods the coast later. Similarly, the initial vascular blockage during a stroke is not usually disastrous because there are many blood vessels in the brain that can pick up the slack; it's the events that lead to killing of neurons not in the initial ischemic zone that wreak the most havoc.

Some recent evidence indicates that the main culprit is *glutamate,* an excitatory neurotransmitter in learning and memory. Normally, glutamate binding to NMDA receptors opens gates that allow Ca^{2+} to enter the stimulated neuron. After brain injury, neurons totally deprived of oxygen begin to disintegrate, unleashing the cellular equivalent of "buckets" of glutamate that acts as an *excitotoxin* to mediate changes in ion transport by activating more NMDA receptors. The initial change, which occurs within minutes, consists of neuronal uptake of Na^+, Cl^-, and water. The delayed changes, which take place over the next few hours, involve unregulated Ca^{2+} influx and generation of free radicals including NO that kill or injure thousands of surrounding, still-healthy neurons. This in turn activates the microglia, which phagocytize dead and injured neurons, and initiates inflammation, which compounds the damage. These new insights have launched an intense search for neuroprotective drugs, such as NMDA and NO antagonists that can prevent this neurotoxic domino effect. At present, the only approved treatment for stroke is tissue plasminogen activator (TPA), which dissolves blood clots in the brain. A still experimental procedure involves injecting LBS-Neurons (immature neurons grown from human cancer cells) into stroke-damaged brain regions in hope that they will take on properties of nearby mature neurons.

Degenerative Brain Disorders

Alzheimer's Disease   **Alzheimer's disease (AD)** (altz'hi-merz) is a progressive degenerative disease of the brain that ultimately results in dementia (mental deterioration). Alzheimer's patients represent nearly half of the people living in nursing homes. Between 5 and 15% of people over 65 develop this condition, and up to half of those over 85 die of it.

Its victims exhibit memory loss (particularly for recent events), shortened attention span, disorientation, and eventual language loss. Over a period of several years, formerly good-natured people may become irritable, moody, and confused. Ultimately, hallucinations occur.

AD is associated with an ACh shortage and with structural changes in the brain, particularly in areas involved with cognition and memory. The gyri shrink and many neurons are lost. Some cases of AD run in families, and several genetic mutations appear to be involved.

Examinations of brain tissue reveal senile plaques (aggregations of cells and degenerated fibers around a *beta amyloid peptide core*) littering the brain like shrapnel between the neurons, plus neurofibrillar tangles, twisted fibrils within neuron cell bodies. It has been frustratingly difficult for researchers to uncover how beta amyloid peptide acts as a neurotoxin, particularly seeing that it is also present in healthy brain cells (but in lower amounts). Just what tips the biochemical pathways off balance to favor production of more beta amyloid is not understood, but it is known that this tiny peptide does its damage by enhancing Ca^{2+} influx (mediated by glutamate and NMDA receptors) into neurons.

Another line of research has implicated a protein called *tau*, which appears to function like railroad ties to bind microtubule "tracks" together. In the brains of AD victims, tau abandons its microtubule-stabilizing role and grabs onto other tau molecules, forming spaghetti-like neurofibrillary tangles. Hopefully the lines of investigation will eventually merge and point to a treatment, but presently only drugs that ease symptoms by inhibiting ACh breakdown are useful.

Parkinson's Disease   Typically striking people in their 50s and 60s, **Parkinson's disease** results from a degeneration of the dopamine-releasing neurons of the substantia nigra. As those neurons deteriorate, the dopamine-deprived basal nuclei they target become overactive, causing the well-known symptoms of the disease. Afflicted individuals have a persistent tremor at rest (exhibited by head nodding and a "pill-rolling" movement of the fingers), a forward-bent walking posture and shuffling gait, and a stiff facial expression, and they are slow in initiating and executing movement.

The cause of Parkinson's disease is still unknown. The drug L-*dopa* often helps to alleviate some symptoms; however, it is not curative, and as more and more neurons die off, L-dopa becomes ineffective. It also has undesirable side effects: severe nausea, dizziness, and in some, liver damage. Newer to the drug therapy scene is deprenyl. When given early on in the disease, deprenyl slows the neurological deterioration to some extent and delays the need to adminster L-dopa for up to 18 months.

Deep brain stimulation via implanted electrodes has proved helpful in alleviating tremors but little else. More promising are intrabrain transplants of embryonic substantia nigra tissue, genetically engineered adult nigral cells, or fetal pig dopamine-producing cells, all of which have produced some regression of disease symptoms. However, the use of embryonic or fetal tissue is controversial and riddled with ethical and legal roadblocks.

Huntington's Disease   **Huntington's disease** is a fatal hereditary disorder that strikes during middle age. Tangles of mutant *huntingtin* protein accumulate in brain cells and the tissue dies, leaving gaping holes that lead to massive degeneration of the basal nuclei and later of the cerebral cortex. Its initial symptoms in many are wild, jerky, almost continuous "flapping" movements called *chorea* (Greek for "dance"). Although the movements appear to be voluntary they are not. Late in the disease, marked mental deterioration occurs. Huntington's disease is progressive and usually fatal within 15 years of onset of symptoms.

The hyperkinetic manifestations of Huntington's disease are essentially the opposite of those of Parkinson's disease (overstimulation rather than inhibition of the motor drive), and Huntington's is usually treated with drugs that block, rather than enhance, dopamine's effects. As with Parkinson's disease, fetal tissue implants may provide promise for its treatment in the future.

The Spinal Cord

Gross Anatomy and Protection

The spinal cord, enclosed in the vertebral column, extends from the foramen magnum of the skull to the level of the first or second lumbar vertebra, just inferior to the ribs (Figure 12.26). About 42 cm (17 inches) long and 1.8 cm (3/4 of an inch) thick, the glistening-white **spinal cord** provides a two-way

? *What is the explanation for the cervical and lumbar enlargements of the cord?*

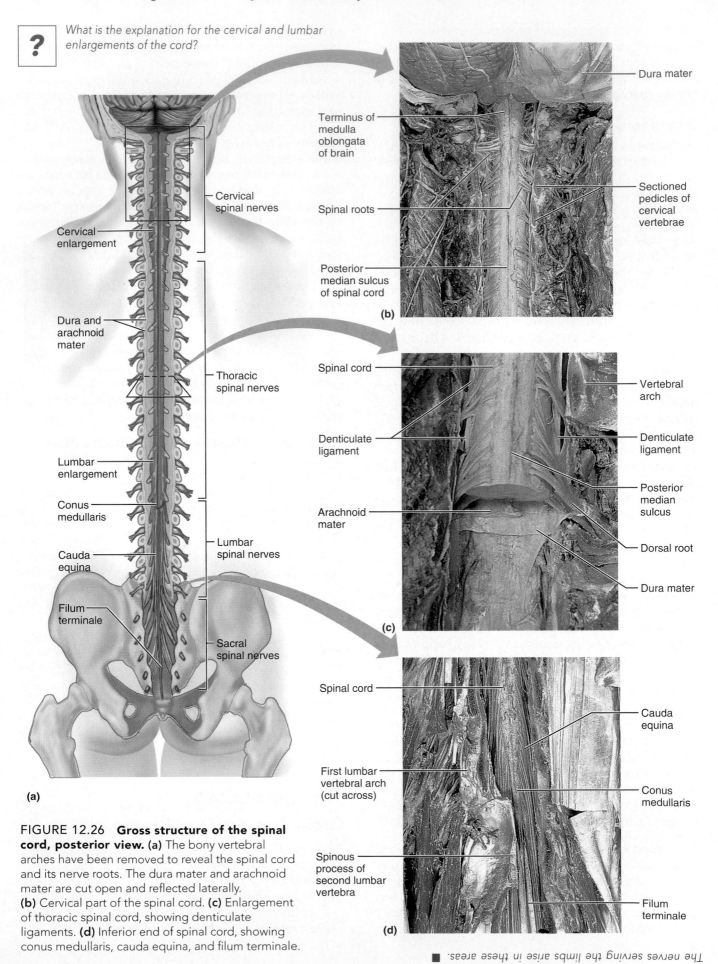

(a)

Cervical spinal nerves

Cervical enlargement

Dura and arachnoid mater

Thoracic spinal nerves

Lumbar enlargement

Conus medullaris

Cauda equina

Filum terminale

Lumbar spinal nerves

Sacral spinal nerves

Dura mater

Terminus of medulla oblongata of brain

Spinal roots

Sectioned pedicles of cervical vertebrae

Posterior median sulcus of spinal cord

(b)

Spinal cord

Vertebral arch

Denticulate ligament

Denticulate ligament

Posterior median sulcus

Arachnoid mater

Dorsal root

Dura mater

(c)

Spinal cord

Cauda equina

First lumbar vertebral arch (cut across)

Conus medullaris

Spinous process of second lumbar vertebra

Filum terminale

(d)

FIGURE 12.26 Gross structure of the spinal cord, posterior view. (a) The bony vertebral arches have been removed to reveal the spinal cord and its nerve roots. The dura mater and arachnoid mater are cut open and reflected laterally. **(b)** Cervical part of the spinal cord. **(c)** Enlargement of thoracic spinal cord, showing denticulate ligaments. **(d)** Inferior end of spinal cord, showing conus medullaris, cauda equina, and filum terminale.

■ *The nerves serving the limbs arise in these areas.*

FIGURE 12.27 **Diagrammatic view of a lumbar tap.**

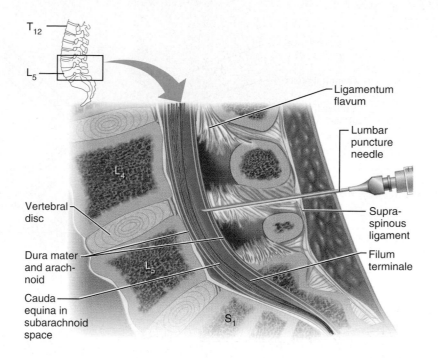

the coccyx, where it anchors the spinal cord in place so it is not jostled by body movements (Figure 12.26d). Furthermore, the spinal cord is secured to the bony walls of the vertebral canal throughout its length by saw-toothed shelves of pia mater (Figure 12.26c) called **denticulate ligaments** (den-tik'u-lāt; "toothed").

In humans, 31 pairs of *spinal nerves* attach to the cord by paired roots and exit from the vertebral column via the intervertebral foramina to travel to the body regions they serve. The spinal cord, like the vertebral column, is segmented. Each cord segment is defined by a pair of spinal nerves that lie superior to their corresponding vertebra.

The spinal cord is about the width of a thumb for most of its length, but it has obvious enlargements in the cervical and lumbosacral regions, where the nerves serving the upper and lower limbs arise. These enlargements are the **cervical** and **lumbar enlargements,** respectively (Figure 12.26a). Because the cord does not reach the end of the vertebral column, the lumbar and sacral spinal nerve roots angle sharply downward and travel inferiorly through the vertebral canal for some distance before reaching their intervertebral foramina. The collection of nerve roots at the inferior end of the vertebral canal is named the **cauda equina** (kaw'da e-kwi'nuh) because of its resemblance to a horse's tail. This strange arrangement reflects the fact that during fetal development, the vertebral column grows faster than the spinal cord, forcing the lower spinal nerve roots to "chase" their exit points inferiorly through the vertebral canal.

conduction pathway to and from the brain. It is a major reflex center: Spinal reflexes are initiated and completed at the spinal cord level. We discuss reflex functions and motor activity of the cord in subsequent chapters. In this section we focus on the anatomy of the cord and on the location and naming of its ascending and descending tracts.

Like the brain, the spinal cord is protected by bone, meninges, and cerebrospinal fluid. The single-layered dura mater of the spinal cord, called the **spinal dural sheath** (Figure 12.25), is not attached to the bony walls of the vertebral column. Between the bony vertebrae and the dural sheath is an **epidural space** filled with a soft padding of fat and a network of veins. Cerebrospinal fluid fills the subarachnoid space between the *arachnoid* and *pia mater* meninges. Inferiorly, the dural and arachnoid membranes extend to the level of S_2, well beyond the end of the spinal cord. The spinal cord typically ends at L_1 (Figure 12.26a). Thus, the subarachnoid space within the meningeal sac inferior to that point provides a nearly ideal spot for removing cerebrospinal fluid for testing, a procedure called a **lumbar puncture** or **tap** (Figure 12.27). Because the spinal cord is absent there and the delicate nerve roots drift away from the point of needle insertion, there is little or no danger of damaging the cord (or spinal roots) beyond L_3.

Inferiorly, the spinal cord terminates in a tapering cone-shaped structure called the **conus medullaris** (ko'nus me"dul-ar'is). A fibrous extension of the pia mater, the **filum terminale** (fi'lum ter"mĭ-nah'le; "terminal filament") extends inferiorly from the conus medullaris to the posterior surface of

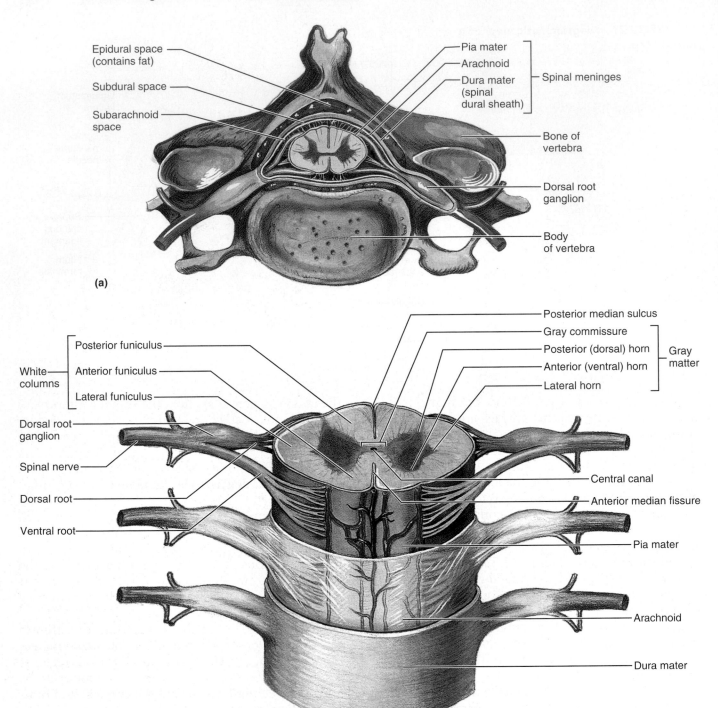

FIGURE 12.28   Anatomy of the spinal cord. (a) Cross section through the spinal cord illustrating its relationship to surrounding vertebral column. **(b)** Three-dimensional view of the adult spinal cord and its meningeal coverings.

Cross-Sectional Anatomy

The spinal cord is somewhat flattened from front to back and two grooves mark its surface: the **anterior median fissure** and the shallower **posterior median sulcus** (see Figure 12.28b). These grooves run the length of the cord and partially divide it into right and left halves. The gray matter of the cord is located in its core, the white matter outside.

Gray Matter and Spinal Roots

In cross section the gray matter of the cord looks like the letter H or like a butterfly (Figure 12.28b). It

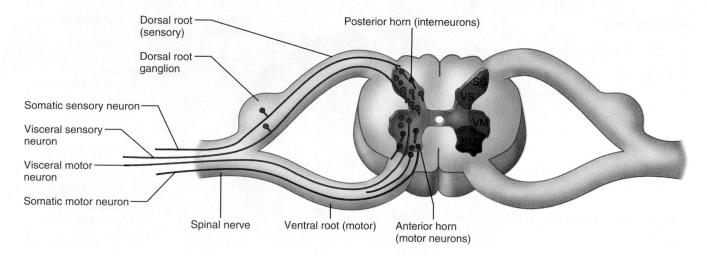

Dorsal root (sensory)

Dorsal root ganglion

Posterior horn (interneurons)

Somatic sensory neuron

Visceral sensory neuron

Visceral motor neuron

Somatic motor neuron

Spinal nerve

Ventral root (motor)

Anterior horn (motor neurons)

SS

VS

VM

SM

FIGURE 12.29 Organization of the gray matter of the spinal cord. The gray matter of the spinal cord is divided into a sensory half dorsally and a motor half ventrally. SS: interneurons receiving input from somatic sensory neurons. VS: interneurons receiving input from visceral sensory neurons. VM: visceral motor (autonomic) neurons. SM: somatic motor neurons. Note that the dorsal and ventral roots are part of the PNS, not of the spinal cord.

consists of mirror-image lateral gray masses connected by a cross-bar of gray matter, the **gray commissure,** that encloses the central canal. The two posterior projections of the gray matter are the **posterior (dorsal) horns;** the anterior pair are the **anterior (ventral) horns.** An additional pair of gray matter columns, the small **lateral horns,** is present in the thoracic and superior lumbar segments of the cord.

All neurons whose cell bodies are in the spinal cord gray matter are multipolar. The posterior horns consist entirely of interneurons. The anterior horns have some interneurons but mainly house cell bodies of somatic motor neurons. These send their axons out via the **ventral roots** of the spinal cord (Figure 12.28b) to the skeletal muscles (their effector organs). The amount of ventral gray matter present at a given level of the spinal cord reflects the amount of skeletal muscle innervated at that level. Thus, the anterior horns are largest in the limb-innervating cervical and lumbar regions of the cord and are responsible for the cord enlargements seen in those regions.

The lateral horn neurons are autonomic (sympathetic division) motor neurons that serve visceral organs. Their axons leave the cord via the ventral root along with those of the somatic motor neurons. Because the ventral roots contain both somatic and autonomic efferents, they serve both motor divisions of the peripheral nervous system (Figure 12.29).

Afferent fibers carrying impulses from peripheral sensory receptors form the **dorsal roots** of the spinal cord (Figure 12.28). The cell bodies of the associated sensory neurons are found in an enlarged region of the dorsal root called the **dorsal root ganglion** or **spinal ganglion.** After entering the cord, the axons of these neurons may take a number of routes. Some enter the posterior white matter of the cord directly and travel to synapse at higher cord or brain levels. Others synapse with interneurons in the posterior horns of the spinal cord gray matter at their entry level.

The dorsal and ventral roots are very short and fuse laterally to form the **spinal nerves.** The spinal nerves, which are part of the peripheral nervous system, are considered further in Chapter 13.

The spinal gray matter can be divided further according to its neurons' relative involvement in the innervation of the somatic and visceral regions of the body. The following four zones are evident within this gray matter (Figure 12.29): **somatic sensory (SS), visceral** (autonomic) **sensory (VS), visceral motor (VM), somatic motor (SM).**

White Matter

The white matter of the spinal cord is composed of myelinated and unmyelinated nerve fibers that allow communication between different parts of the spinal cord and between the cord and brain. These fibers run in three directions: (1) *ascending*—up to higher centers (sensory inputs), (2) *descending*—down to the cord from the brain or within the cord to lower levels (motor outputs), and (3) *transversely*—across from one side of the cord to the other (commissural fibers). Ascending and descending tracts make up most of the white matter.

The white matter on each side of the cord is divided into three **white columns,** or **funiculi** (fu-nik'u-li; "long ropes"), named according to their position as **posterior, lateral,** and **anterior funiculi** (Figure 12.28b). Each funiculus contains several

Ascending tracts

Descending tracts

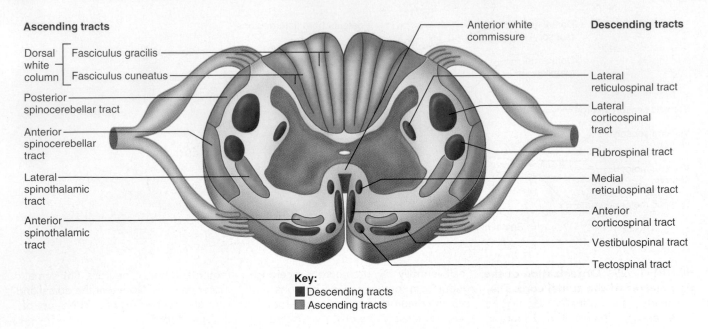

Dorsal white column { Fasciculus gracilis — Fasciculus cuneatus —

Posterior spinocerebellar tract —

Anterior spinocerebellar tract —

Lateral spinothalamic tract —

Anterior spinothalamic tract —

Anterior white commissure

Lateral reticulospinal tract

Lateral corticospinal tract

Rubrospinal tract

Medial reticulospinal tract

Anterior corticospinal tract

Vestibulospinal tract

Tectospinal tract

Key:
■ Descending tracts
■ Ascending tracts

FIGURE 12.30 Major ascending (sensory) and descending (motor) tracts of the spinal cord, cross-sectional view.

fiber tracts, and each tract is made up of axons with similar destinations and functions. With a few exceptions, the names of the spinal tracts reveal both their origin and destination. The principal ascending and descending tracts of the spinal cord are illustrated schematically in the cross-sectional view of the spinal cord in Figure 12.30.

All major spinal tracts are *part of multineuron pathways* that connect the brain to the body periphery. These great ascending and descending pathways contain not only spinal cord neurons but also parts of peripheral neurons and neurons in the brain. Before we get specific about the individual tracts, we will make some generalizations about them and the pathways to which they contribute:

1. Most pathways cross from one side of the CNS to the other (decussate) at some point along their journey.

2. Most consist of a chain of two or three neurons that contribute to successive tracts of the pathway.

3. Most exhibit *somatotopy,* a precise spatial relationship among the tract fibers that reflects the orderly mapping of the body. For example, in an ascending sensory tract, somatotopy refers to the fact that fibers transmitting inputs from sensory receptors in the superior parts of the body lie lateral to those conveying sensory information from inferior body regions.

4. All pathways and tracts are paired (right and left) with a member of the pair present on each side of the spinal cord or brain.

Ascending Pathways to the Brain

Neuronal Composition   The ascending pathways conduct sensory impulses upward, typically through chains of three successive neurons (first-, second-, and third-order neurons) to various areas of the brain (Figure 12.31). (Note that both second- and third-order neurons are interneurons.)

▪ **First-order neurons,** whose cell bodies reside in a ganglion (dorsal root or cranial), conduct impulses from the cutaneous receptors of the skin and from proprioceptors to the spinal cord or brain stem, where they synapse with second-order neurons. Impulses from the facial area are transmitted by cranial nerves; spinal nerves conduct somatic sensory impulses from the rest of the body to the CNS.

▪ **Second-order neurons,** whose cell bodies reside in the dorsal horn of the spinal cord or in medullary nuclei, transmit impulses to the thalamus or to the cerebellum where they synapse.

▪ **Third-order neurons** are located in the thalamus and conduct impulses to the somatosensory cortex of the cerebrum. (There are no third-order neurons in the cerebellum.)

The Main Ascending Pathways   In general, somatosensory information is conveyed along three main pathways on each side of the spinal cord. Two of these pathways (the *nonspecific* and *specific ascending pathways*) transmit impulses to the sensory cortex. Collectively the inputs of these sister tracts provide *discriminative touch* and *conscious proprioception.* Both pathways decussate—the first in the spinal cord and the second in the medulla.

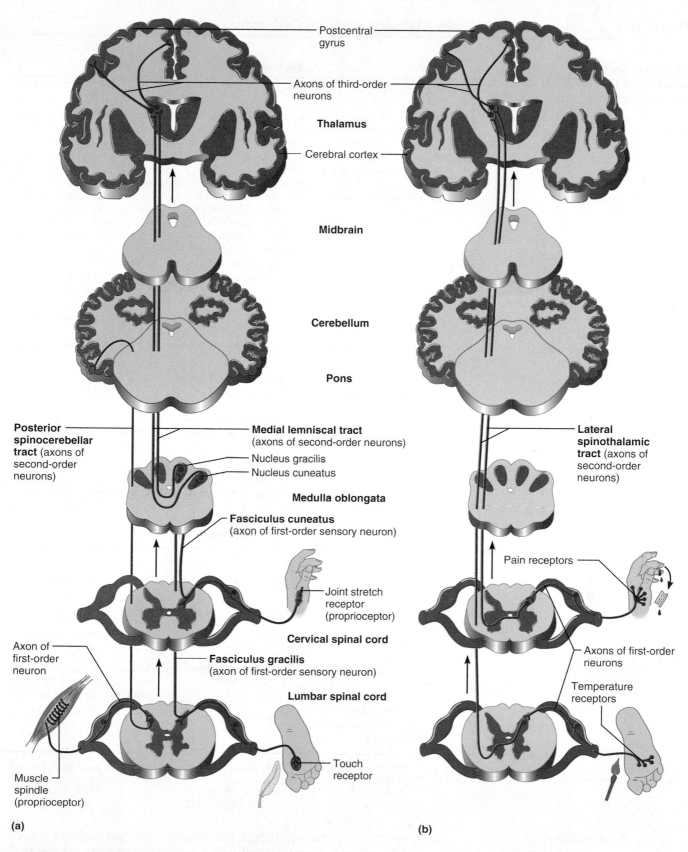

FIGURE 12.31 **Pathways of selected ascending spinal cord tracts.** (a) Right: Specific pathways for discriminative touch and conscious proprioception in the fasciculus gracilis and fasciculus cuneatus tracts and their continuation, the medial lemniscal tracts. Left: Posterior spinocerebellar tract (extends only to cerebellum). (b) Nonspecific pathways for pain, temperature, and crude touch in the lateral spinothalamic tract of the anterolateral pathways. The entire ascending pathway is shown in each case.

TABLE 12.3   Major Ascending (Sensory) Pathways and Spinal Cord Tracts

Spinal Cord Tract	Location (Funiculus)	Origin	Termination	Function
Specific Ascending (Lemniscal) Pathways				
Fasciculus cuneatus and fasciculus gracilis (dorsal white column)	Posterior	Central axons of sensory (first-order) neurons enter dorsal root of the spinal cord and branch; branches enter posterior white column on same side without synapsing	By synapse with second-order neurons in nucleus cuneatus and nucleus gracilis in medulla; fibers of medullary neurons cross over and ascend in medial lemniscal tracts to thalamus, where they synapse with third-order neurons; thalamic neurons then transmit impulses to somatosensory cortex	Both tracts transmit sensory impulses from general sensory receptors of skin and proprioceptors, which are interpreted as discriminative touch, pressure, and "body sense" (limb and joint position) in opposite somatosensory cortex; cuneatus transmits afferent impulses from upper limbs, upper trunk, and neck; it is not present in spinal cord below level of T_6; gracilis carries impulses from lower limbs and inferior body trunk
Nonspecific Ascending (Anterolateral) Pathways				
Lateral spinothalamic	Lateral	Association (second-order) neurons of posterior horn; fibers cross to opposite side before ascending	By synapse with third-order neurons in thalamus; impulses then conveyed to somatosensory cortex by thalamic neurons	Transmits impulses concerned with pain and temperature to opposite side of brain; eventually interpreted in somatosensory cortex
Anterior spinothalamic	Anterior	Association neurons in posterior horns; fibers cross to opposite side before ascending	By synapse with third-order neurons in thalamus; impulses eventually conveyed to somatosensory cortex by thalamic neurons	Transmits impulses concerned with crude touch and pressure to opposite side of brain for interpretation by somatosensory cortex
Spinocerebellar Pathways				
Posterior spinocerebellar*	Lateral (posterior part)	Association (second-order) neurons in posterior horn on same side of cord; fibers ascend without crossing	By synapse in cerebellum	Transmits impulses from trunk and lower limb proprioceptors on one side of body to same side of cerebellum; subconscious proprioception
Anterior spinocerebellar*	Lateral (anterior part)	Association (second-order) neurons of posterior horn; contains crossed fibers that cross back to the opposite side in the pons	By synapse in cerebellum	Transmits impulses from the trunk and lower limb on the same side of body to cerebellum; subconscious proprioception

*The posterior and anterior spinocerebellar tracts carry information from the lower limbs and trunk only. The corresponding tracts for the upper limb and neck (rostral spinocerebellar and others) are beyond the scope of this book.

The third pathway, consisting of the *spinocerebellar tracts*, tracks to the cerebellum, and hence does not contribute to sensory perception. Let's examine these pathways more closely.

1. Nonspecific ascending pathways. The evolutionarily older **nonspecific ascending pathways** receive inputs from many different types of sensory receptors and make multiple synapses in the brain stem. These pathways, also called the **anterolateral pathways** (because they are located in the anterior and lateral white columns of the spinal cord), are largely formed by the **lateral** and **anterior spinotha-**

lamic tracts (see Figure 12.31b and Table 12.3). Crossover of fibers occurs in the spinal cord.

Most of the fibers in these pathways transmit pain, temperature, and coarse touch impulses, sensations that we are aware of but have difficulty localizing precisely on the body surface.

2. **Specific ascending pathways.** Also called the **medial lemniscal system** (lem-nis′kul; *lemnisc* = a ribbon), the **specific ascending pathways** mediate precise, straight-through transmission of inputs from a single type (or a few related types) of sensory receptor that *can* be localized precisely on the body surface, such as discriminative touch and vibrations. These pathways are formed by the paired tracts of the **dorsal white column—fasciculus cuneatus** and **fasciculus gracilis**—of the spinal cord and the **medial lemniscal tracts.** The latter arise in the medulla and terminate in the *ventral posterior nuclei* of the thalamus (see Figure 12.31a and Table 12.3). From the thalamus, impulses are forwarded to specific areas of the somatosensory cortex.

3. **Spinocerebellar tracts.** The last pair of ascending pathways, the **anterior** and **posterior spinocerebellar tracts,** convey information about muscle or tendon stretch to the cerebellum, which uses this information to coordinate skeletal muscle activity. As noted earlier, these pathways do *not* contribute to conscious sensation. The fibers of the spinocerebellar pathways either do not decussate or else cross over twice (thus "undoing" the decussation).

Descending Pathways and Tracts

The descending tracts that deliver efferent impulses from the brain to the spinal cord are divided into two groups: (1) the *direct pathways* equivalent to the pyramidal tracts and (2) the *indirect pathways,* essentially all others. Motor pathways involve two neurons, referred to as the upper and lower motor neurons. The pyramidal cells of the motor cortex, as well as the neurons in subcortical motor nuclei that give rise to other descending motor pathways, are called *upper motor neurons.* The anterior horn motor neurons, which actually innervate the skeletal muscles (their effectors), are called *lower motor neurons.* These tracts are briefly overviewed here; more information is provided in Table 12.4.

The Direct (Pyramidal) System The direct pathways originate mainly with the pyramidal neurons located in the precentral gyri. These neurons send impulses through the brain stem via the large **pyramidal (corticospinal) tracts** (Figure 12.32a). The direct pathways are so called because their axons descend without synapsing from the pyramidal neurons to the spinal cord. There they synapse either with interneurons or with anterior horn motor neurons. Stimulation of the anterior horn neurons activates the skeletal muscles with which they are associated. The direct pathway primarily regulates fast and fine (or skilled) movements such as doing needlework and writing.

The Indirect (Extrapyramidal) System The indirect system includes brain stem motor nuclei and *all motor pathways except* the pyramidal pathways. These tracts were formerly lumped together as the **extrapyramidal system** because their nuclei of origin were presumed to be independent of ("extra to") the pyramidal tracts. This term is still widely used clinically; however, pyramidal tract neurons are now known to project to and influence the activity of most "extrapyramidal" nuclei, so modern anatomists prefer to use the term **indirect,** or **multineuronal, pathways,** or even just the names of the individual motor pathways.

These motor pathways are complex and multisynaptic. They are most involved in regulating (1) the axial muscles that maintain balance and posture, (2) the muscles controlling coarse limb movements, and (3) head, neck, and eye movements that follow objects in the visual field. Many of the activities controlled by subcortical motor nuclei are heavily dependent on reflex activity. One of these tracts, the rubrospinal tract, is illustrated in Figure 12.32b.

Overall, the **reticulospinal tracts** (see Table 12.4) maintain balance by varying the tone of postural muscles. The **rubrospinal tracts** control flexor muscles, whereas the **superior colliculi** and the **tectospinal tracts** mediate head movements in response to visual stimuli.

Spinal Cord Trauma and Disorders

Spinal Cord Trauma

The spinal cord is elastic, stretching with every turn of the head or bend of the trunk, but it is exquisitely sensitive to direct pressure. Any localized damage to the spinal cord or its roots leads to some functional loss, either **paralysis** (loss of motor function) or **paresthesias** (par″es-the′ze-ahz) (sensory loss). Severe damage to ventral root or anterior horn cells results in a **flaccid paralysis** (flak′sid) of the skeletal muscles served. Nerve impulses do not reach these muscles, which consequently cannot move either voluntarily or involuntarily. Without stimulation, the muscles atrophy. When only the upper motor neurons of the primary motor cortex are damaged, **spastic paralysis** occurs. In this case, the spinal motor neurons remain intact and the muscles continue to be stimulated irregularly by spinal reflex activity. Thus, the muscles remain healthy longer, but their

TABLE 12.4 / Major Descending (Motor) Pathways and Spinal Cord Tracts

Spinal Cord Tract	Location (Funiculus)	Origin	Termination	Function
Direct (Pyramidal)				
Lateral corticospinal	Lateral	Pyramidal neurons of motor cortex of the cerebrum; decussate in pyramids of medulla	Anterior horn interneurons that influence motor neurons and occasionally directly with anterior horn motor neurons	Transmits motor impulses from cerebrum to spinal cord motor neurons (which activate skeletal muscles on opposite side of body); voluntary motor tract
Anterior corticospinal	Anterior	Pyramidal neurons of motor cortex; fibers cross over at the spinal cord level	Anterior horn (as above)	Same as lateral corticospinal tract
Indirect (Extrapyramidal) Pathways				
Tectospinal	Anterior	Superior colliculus of midbrain of brain stem (fibers cross to opposite side of cord)	Anterior horn (as above)	Transmits motor impulses from midbrain that are important for coordinated movement of head and eyes toward visual targets
Vestibulospinal	Anterior	Vestibular nuclei in medulla of brain stem (fibers descend without crossing)	Anterior horn (as above)	Transmits motor impulses that maintain muscle tone and activate ipsilateral limb and trunk extensor muscles and muscles that move head; in this way helps maintain balance during standing and moving
Rubrospinal	Lateral	Red nucleus of midbrain of brain stem (fibers cross to opposite side just inferior to the red nucleus)	Anterior horn (as above)	In experimental animals, transmits motor impulses concerned with muscle tone of distal limb muscles (mostly flexors) on opposite side of body; function in humans is controversial: may be vestigial with function largely assumed by corticospinal tracts
Reticulospinal (anterior, medial, and lateral)	Anterior and lateral	Reticular formation of brain stem (medial nuclear group of pons and medulla); both crossed and uncrossed fibers	Anterior horn (as above)	Transmits impulses concerned with muscle tone and many visceral motor functions; may control most unskilled movements

movements are no longer subject to voluntary control. In many such cases, the muscles become permanently shortened and fibrotic.

Transection (cross sectioning) of the spinal cord at any level results in total motor and sensory loss in body regions inferior to the site of damage. If the transection occurs between T_1 and L_1, both lower limbs are affected, resulting in **paraplegia** (par"ah-ple'je-ah; *para* = beside, *plegia* = a blow). If the injury occurs in the cervical region, all four limbs are affected and the result is **quadriplegia.** *Hemiplegia,*

paralysis of one side of the body, usually reflects brain injury rather than spinal cord injury.

Anyone with traumatic spinal cord injury must be watched for symptoms of **spinal shock,** a transient period of functional loss that follows the injury. Spinal shock results in immediate depression of all reflex activity caudal to the lesion site. Bowel and bladder reflexes stop, blood pressure falls, and all muscles (somatic and visceral alike) below the injury are paralyzed and insensitive. Neural function usually returns within a few hours following injury. If

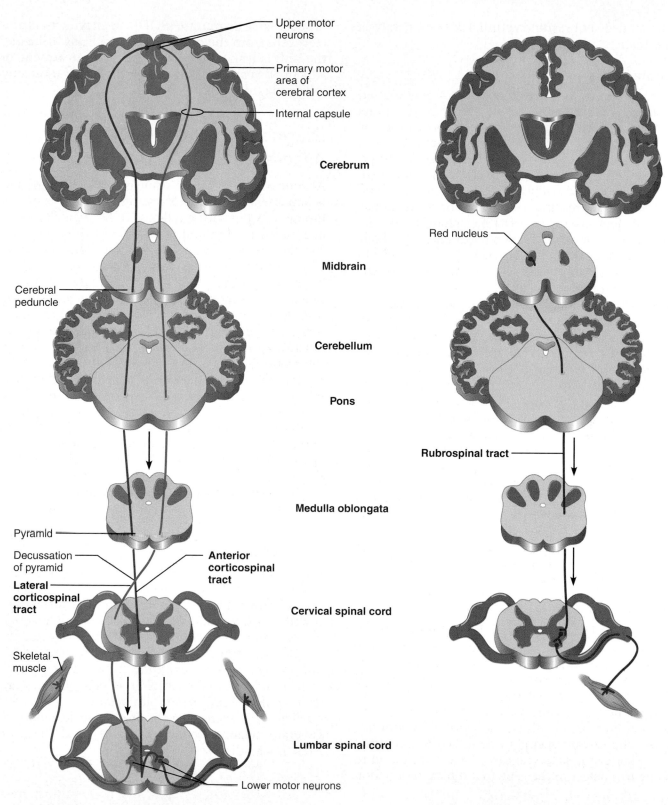

(a) Pyramidal (lateral and anterior corticospinal) tracts

(b) Rubrospinal tract (an extrapyramidal tract)

FIGURE 12.32   **Pathways of selected descending spinal cord tracts.**
(a) Direct pathways, that is, the pyramidal tracts (lateral and anterior corticospinal tracts), carrying motor impulses to skeletal muscles. **(b)** The rubrospinal tract, one of the indirect or extrapyramidal tracts, helps regulate tone of muscles on opposite side of body.

function does not resume within 48 hours, paralysis is permanent in most cases.

Poliomyelitis

Poliomyelitis (po"le-o-mi"ĕ-li'tis; *polio* = gray matter; *myelitis* = inflammation of the spinal cord) results from destruction of anterior horn motor neurons by the poliovirus. Early symptoms include fever, headache, muscle pain and weakness, and loss of certain somatic reflexes. Later, paralysis develops and the muscles served atrophy. The victim may die from paralysis of the respiratory muscles or from cardiac arrest if neurons in the medulla oblongata are destroyed. In most cases, the poliovirus enters the body in feces-contaminated water (such as might occur in a public swimming pool), and the incidence of the disease has traditionally been highest in children and during the summer months. Fortunately, vaccines have nearly eliminated this disease and a global effort is ongoing to eradicate it completely from the world.

However, many of the "recovered" survivors of the great polio epidemic of the late 1940s and 1950s have begun to experience extreme lethargy, sharp burning pains in their muscles, and progressive muscle weakness and atrophy. Initially dismissed as the flu or psychosomatic illness, these disturbing symptoms are now referred to as **postpolio syndrome.** The cause of postpolio syndrome is not known, but a likely explanation is that its victims, like the rest of us, continue to lose neurons throughout life. While a healthy nervous system can recruit nearby neurons to compensate for the losses, polio survivors have already drawn on that "pool" and have few neurons left to take over. Ironically, those who worked hardest to overcome their disease are its newest victims.

Amyotrophic Lateral Sclerosis

Amyotrophic lateral sclerosis (ALS) (ah-mi"o-trof'ik), also called Lou Gehrig's disease, is a devastating neuromuscular condition that involves progressive destruction of anterior horn motor neurons and fibers of the pyramidal tract. As the disease progresses, the sufferer loses the ability to speak, swallow, and breathe. Death typically occurs within five years. The disease seems to be linked to abnormal genes. For example, some cases have been found to be due to a defect in the gene that directs production of superoxide dismutase (SOD), an enzyme that protects cells from the ravages of free radicals. Riluzole, the only advance in pharmacological treatment of ALS in 50 years, appears to prolong life by indirectly inhibiting free radical formation.

Diagnostic Procedures for Assessing CNS Dysfunction

Anyone who has had a routine physical examination is familiar with the reflex tests done to assess neural function. A tap with a reflex hammer stretches your quadriceps tendon and your anterior thigh muscles contract, which results in the knee-jerk response. This response shows that the spinal cord and upper brain centers are functioning normally. Abnormal reflex test responses may indicate such serious disorders as intracranial hemorrhage, multiple sclerosis, or hydrocephalus and suggest that more sophisticated neurological tests are needed to identify the problem.

Pneumoencephalography (nu"mo-en-sef"ah-log'-rah-fe) provides a fairly clear X-ray picture of the brain ventricles and has been the procedure of choice for diagnosing hydrocephalus. Cerebrospinal fluid is withdrawn, and air injected into the subarachnoid space floats upward into the ventricles, allowing them to be visualized.

In individuals who have suffered a stroke or TIA, a *cerebral angiogram* (an'je-o-gram") is commonly ordered to assess the condition of the cerebral arteries serving the brain (or the carotid arteries of the neck, which feed most of those vessels). A dye is injected into an artery and allowed to disperse in the brain. Then an X ray is taken of the arteries of interest. The dye allows visualization of arteries narrowed by arteriosclerosis or blocked by blood clots.

New imaging techniques have revolutionized the diagnosis of brain lesions. Together the various *CT* and *MRI scanning techniques* allow most tumors, intracranial lesions, multiple sclerosis plaques, and areas of dead brain tissue (infarcts) to be identified quickly. PET scans can localize brain lesions that generate seizures (epileptic tissue) and diagnose Alzheimer's disease.

Review Questions

Multiple Choice/Matching

(Some questions have more than one correct answer. Select the best answer or answers from the choices given.)

1. The primary motor cortex, Broca's area, and the premotor area are located in which lobe? (a) frontal, (b) parietal, (c) temporal, (d) occipital.

2. The innermost layer of the meninges, delicate and closely apposed to the brain tissue, is the (a) dura mater, (b) corpus callosum, (c) arachnoid, (d) pia mater.

3. Cerebrospinal fluid is formed by (a) arachnoid villi, (b) the dura mater, (c) choroid plexuses, (d) all of these.

4. A patient has suffered a cerebral hemorrhage that has caused dysfunction of the precentral gyrus of his right cerebral cortex. As a result, (a) he cannot voluntarily move his left arm or leg, (b) he feels no sensation on the left side of his body, (c) he feels no sensation on his right side.

5. Choose the proper term from the key to respond to the statements describing various brain areas.

Key:
(a) cerebellum   (d) corpus striatum   (g) midbrain
(b) corpora quadrigemina (e) hypothalamus   (h) pons
(c) corpus callosum   (f) medulla   (i) thalamus

_____ **(1)** basal nuclei involved in fine control of motor activities

_____ **(2)** region where there is a gross crossover of fibers of descending pyramidal tracts

_____ **(3)** control of temperature, autonomic nervous system reflexes, hunger, and water balance

_____ **(4)** houses the substantia nigra and cerebral aqueduct

_____ **(5)** relay stations for visual and auditory stimuli input; found in midbrain

_____ **(6)** houses vital centers for control of the heart, respiration, and blood pressure

_____ **(7)** brain area through which all the sensory input must travel to get to the cerebral cortex

_____ **(8)** brain area most concerned with equilibrium, body posture, and coordination of motor activity

6. Ascending pathways in the spinal cord convey (a) motor impulses, (b) sensory impulses, (c) commissural impulses, (d) all of these.

7. Destruction of the anterior horn cells of the spinal cord results in loss of (a) integrating impulses, (b) sensory impulses, (c) voluntary motor impulses, (d) all of these.

8. Fiber tracts that allow neurons within the same cerebral hemisphere to communicate are (a) association tracts, (b) commissures, (c) projection tracts.

9. A number of brain structures are listed below. If an area is primarily gray matter, write **a** in the answer blank; if mostly white matter, respond with **b**.

_____ **(1)** cerebral cortex
_____ **(2)** corpus callosum and corona radiata
_____ **(3)** red nucleus
_____ **(4)** medial and lateral nuclear groups
_____ **(5)** medial lemniscal tract
_____ **(6)** cranial nerve nuclei
_____ **(7)** spinothalamic tract
_____ **(8)** fornix
_____ **(9)** cingulate and precentral gyri

10. A professor unexpectedly blew a loud horn in his anatomy and physiology class. The students looked up, startled. The reflexive movements of their eyes were mediated by the (a) cerebral cortex, (b) inferior olives, (c) raphe nuclei, (d) superior colliculi, (e) nucleus gracilis.

11. Identify the stage of sleep described by using choices from the key. (Note that responses a–d refer to NREM sleep.)

Key: **(a)** stage 1, **(b)** stage 2, **(c)** stage 3, **(d)** stage 4, **(e)** REM

_____ **(1)** the stage when vital signs (blood pressure, heart rate, and body temperature) reach their lowest levels

_____ **(2)** indicated by movement of the eyes under the lids; dreaming occurs

_____ **(3)** when sleepwalking is likely to occur

_____ **(4)** when the sleeper is very easily awakened; EEG shows alpha waves

Short Answer Essay Questions

12. Make a diagram showing the three primary (embryonic) brain vesicles. Name each and then use clinical terminology to name the resulting adult brain regions.

13. (a) What is the advantage of having a cerebrum that is highly convoluted? (b) What term is used to indicate its grooves? Its outward folds? (c) What groove divides the cerebrum into two hemispheres? (d) What divides the parietal from the frontal lobe? The parietal from the temporal lobe?

14. (a) Make a rough drawing of the lateral aspect of the left cerebral hemisphere. (b) You may be thinking, "But I just can't draw!" So, name the hemisphere involved with most people's ability to draw. (c) On your drawing, locate the following areas and provide the major function of each: primary motor cortex, premotor cortex, somatosensory association area, primary sensory area, visual and auditory areas, prefrontal cortex, Wernicke's and Broca's areas.

15. (a) What does lateralization of cortical functioning mean? (b) Why is the term *cerebral dominance* a misnomer?

16. (a) What is the function of the basal nuclei? (b) Which basal nuclei form the lentiform nucleus? (c) Which arches over the diencephalon?

17. (a) Explain how the cerebellum is physically connected to the brain stem. (b) List ways in which the cerebellum is very similar to the cerebrum.

18. Describe the role of the cerebellum in maintaining smooth, coordinated skeletal muscle activity.

19. (a) Where is the limbic system located? (b) What structures make up this system? (c) How is the limbic system important in behavior?

20. (a) Localize the reticular formation in the brain. (b) What does RAS mean, and what is its function?

21. What is an aura?

22. How do sleep patterns, extent of sleeping time, and amount of time spent in REM and NREM sleep change through life?

23. Compare and contrast short-term memory (STM) and long-term memory (LTM) relative to storage capacity and duration of the memory.

24. (a) Name several factors that can enhance transfer of information from STM to LTM. (b) Define memory consolidation.

25. Compare and contrast fact and skill memory relative to the types of things remembered and the importance of conscious retrieval.

26. List four ways in which the CNS is protected.

27. (a) How is cerebrospinal fluid formed and drained? Follow its pathway within and around the brain. (b) What happens if CSF is not drained properly? Why is this consequence more harmful in adults?

28. What constitutes the blood-brain barrier?

29. A brain surgeon is about to make an incision. Name all the tissue layers that he cuts through from the skin to the brain.

30. (a) Differentiate between a concussion and a contusion. (b) Why do severe brain stem contusions result in unconsciousness?

31. Describe the spinal cord, depicting its extent, its composition of gray and white matter, and its spinal roots.

32. All of the following descriptions refer to specific ascending pathways except one: (a) they include the fasciculus gracilis and fasciculus cuneatus, which terminate in the thalamus; (b) they include a chain of three neurons; (c) their connections are diffuse and polymodal; (d) they are concerned with precise transmission of one or a few related sensory modalities.

33. Describe the functional problems that would be experienced by a person in which these fiber tracts have been cut: (a) lateral spinothalamic, (b) anterior and posterior spinocerebellar, (c) tectospinal.

34. Differentiate between spastic and flaccid paralysis.

35. How do the conditions paraplegia, hemiplegia, and quadriplegia differ?

36. (a) Define cerebrovascular accident or CVA. (b) Describe its possible causes and consequences.

37. (a) What factors account for brain growth after birth? (b) List some structural brain changes observed with aging.

13

The Peripheral Nervous System and Reflex Activity

1. Define peripheral nervous system and list its components.

PART 1: SENSORY RECEPTORS AND SENSATION (pp. 432–471)

Overview: From Sensation to Perception (pp. 432–434)

2. Outline the events that lead to sensation and perception.

3. Describe receptor and generator potentials and sensory adaptation.

4. Describe the main aspects of sensory perception.

Sensory Receptors (pp. 434–471)

5. Classify general sensory receptors by stimulus type, body location, and structural complexity.

6. Describe the location, structure, and afferent pathways of taste and smell receptors, and explain how these receptors are activated.

7. Describe the structure and function of accessory eye structures, eye tunics, lens, and humors of the eye.

8. Trace the pathway of light through the eye to the retina, and explain how light is focused for distant and close vision.

9. Describe the events involved in the stimulation of photoreceptors by light, and compare and contrast the roles of rods and cones in vision.

10. Note the cause and consequences of astigmatism, cataract, glaucoma, hyperopia, myopia, and color blindness.

11. Compare and contrast light and dark adaptation.

12. Trace the visual pathway to the visual cortex, and briefly describe the steps in visual processing.

13. Describe the structure and general function of the outer, middle, and inner ears.

14. Describe the sound conduction pathway to the fluids of the inner ear, and follow the auditory pathway from the organ of Corti to the temporal cortex.

15. Explain how one is able to differentiate pitch and loudness and localize the source of sounds.

16. Explain how the balance organs of the semicircular canals and the vestibule help maintain dynamic and static equilibrium.

17. List possible causes and symptoms of otitis media, deafness, Ménière's syndrome, and motion sickness.

PART 2: TRANSMISSION LINES: NERVES AND THEIR STRUCTURE AND REPAIR (pp. 471–491)

Nerves and Associated Ganglia (pp. 471–473)

18. Define ganglion and indicate the general body location of ganglia.

19. Describe the general structure of a nerve, and follow the process of nerve regeneration.

Cranial Nerves (pp. 473–475)

20. Name the 12 pairs of cranial nerves; indicate the body region and structures innervated by each.

Spinal Nerves (pp. 475–491)

21. Describe the formation of a spinal nerve and the general distribution of its rami.

22. Define plexus. Name the major plexuses and describe the distribution and function of the peripheral nerves arising from each plexus.

PART 3: MOTOR ENDINGS AND MOTOR ACTIVITY (pp. 492–493)

Peripheral Motor Endings (p. 492)

23. Compare and contrast the motor endings of somatic and autonomic nerve fibers.

Overview of Motor Integration: From Intention to Effect (pp. 492–493)

24. Outline the three levels of the motor hierarchy.

25. Compare the roles of the cerebellum and basal nuclei in controlling motor activity.

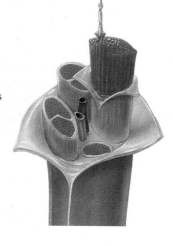

PART 4: REFLEX ACTIVITY (pp. 493–499)

The Reflex Arc (pp. 493–494)

26. Name the components of a reflex arc and distinguish between autonomic and somatic reflexes.

Spinal Reflexes (pp. 494–499)

27. Compare and contrast stretch, flexor, and crossed extensor reflexes.

The human brain, for all its sophistication, would be useless without its links to the outside world. In one experiment that illustrates this point, blindfolded volunteers suspended in warm water in a sensory deprivation tank (a situation that limited sensory inputs) hallucinated. One saw charging pink and purple elephants. Another heard a chorus, still others had taste hallucinations. Our very sanity depends on a continuous flow of information from the outside.

No less important to our well-being are the orders sent from the CNS to voluntary muscles and other effectors of the body, which allows us to move and to take care of our own needs. The **peripheral nervous system (PNS)** provides these links from and to the world outside our bodies. Ghostly white nerves thread through virtually every part of the body, enabling the CNS to receive information and to carry out its decisions.

The PNS includes all neural structures outside the brain and spinal cord, that is, the *sensory receptors*, peripheral *nerves* and their associated *ganglia*, and efferent *motor endings*. Its basic components are diagrammed in Figure 13.1. The first portion of this chapter deals with the functional anatomy of each PNS element. Then we consider the components of reflex arcs and some important somatic reflexes, played out almost entirely in PNS structures, that help to maintain homeostasis.

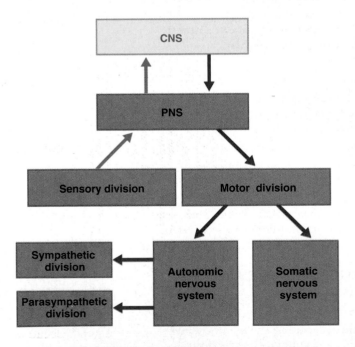

FIGURE 13.1   **Place of the PNS in the structural organization of the nervous system.**

PART 1: SENSORY RECEPTORS AND SENSATION

Overview: From Sensation to Perception

Our survival depends not only on **sensation** (awareness of changes in the internal and external environments) but also on **perception** (conscious interpretation of those stimuli). For example, a pebble kicked up into my shoe causes the *sensation* of localized deep pressure, but my *perception* of it is an awareness of discomfort. Perception in turn determines how we will respond to stimuli. In the case of the pebble-in-my-shoe example, I'm taking off my shoe to get rid of the pesky pebble in a hurry.

General Organization of the Somatosensory System

The **somatosensory system,** or that part of the sensory system serving the body wall and limbs, receives inputs from exteroceptors, proprioceptors, and interoceptors; consequently, it transmits information about several different sensory modalities.

There are three main levels of neural integration in the somatosensory (or any sensory) system:

① **Receptor level:** sensory receptors.

② **Circuit level:** ascending pathways.

③ **Perceptual level:** neuronal circuits in the cerebral cortex. Although sensory input is generally relayed toward the head, it is also processed along the way.

Let's examine the events that must occur at each level along the pathway.

Processing at the Receptor Level

For sensation to occur, a stimulus must excite a receptor. For this to happen:

■ The receptor must have *specificity* for the stimulus energy. For example, a given touch receptor may be sensitive to mechanical pressure, stretch, and vibration, but not to light energy (which is the province of receptors in the eye). The more complex the sensory receptor, the greater its specificity.

■ The stimulus must be applied within a sensory receptor's *receptive field*, the particular area monitored by the receptor. Typically, the smaller the receptive field, the greater the ability of the brain to accurately localize the stimulus site.

■ The stimulus energy must be converted into the energy of a *graded potential* called a **receptor potential,** a process called **transduction.** This may be a depolarizing or hyperpolarizing graded potential similar to the EPSPs or IPSPs generated at postsynaptic membranes in response to neurotransmitter binding. Membrane depolarizations that summate and directly lead to generation of action potentials in an afferent fiber are called **generator potentials.** When the receptor region is part of a sensory neuron (as with free dendrites or the encapsulated receptors of most general sense receptors), the terms receptor potential and generator potential are synonomous. When the receptor is a separate (non-neuron) cell, the receptor and generator potentials are completely separate events. With an adequate stimulus, such a depolarizing receptor cell produces a receptor potential, which (if strong enough) causes it to release a neurotransmitter which, in turn, excites the associated afferent neuron.

▪ A generator potential in the associated sensory neuron (a first-order neuron) must reach threshold so that voltage-gated sodium channels on the axon (these are usually just proximal to the receptor membrane, often at the first node of Ranvier) are opened and nerve impulses are generated and propagated to the CNS.

Information about the stimulus—its strength, duration, and pattern—is encoded in the frequency of arriving nerve impulses (the faster, the stronger the stimulus). Some receptors are **tonic receptors;** that is, they generate nerve impulses at a constant rate unless inhibited or activated to a greater degree. This is true of equilibrium receptors in the inner ears. Other receptors, called **phasic receptors,** are normally "off" unless activated by some change in the environment. These act mainly to report changes in the internal or external environment.

Receptors for hearing and smell, and many but not all general sensory receptors, exhibit **adaptation,** a reduction in sensitivity (and nerve impulse generation) in the presence of a constant stimulus. Generally, phasic receptors are *fast-adapting* and tonic receptors are *slowly adapting receptors.* Because of the protective importance of pain receptors and proprioceptors, they do not exhibit adaptation.

Processing at the Circuit Level

The task at the circuit level is to deliver impulses to the appropriate region of the cerebral cortex for stimulus localization and perception. For general senses, taste, and hearing, this occurs via the ascending sensory pathways. After the central processes of first-order neurons (located in dorsal root or cranial ganglia) enter the spinal cord or medulla, they branch diffusely. Some take part in local spinal cord reflexes; others synapse with second-order neurons in the dorsal horn of the cord or continue upward to synapse in medullary nuclei. These second-order neurons transmit impulses to the thalamus or to the cerebellum where they synapse. Third-order thalamic neurons conduct impulses to the somatosensory cortex of the cerebrum. (There are no third-order neurons in the cerebellum.)

Impulses ascending in the specific and nonspecific ascending pathways (see p. 423) reach conscious awareness in the sensory cortex. In general, fibers in the nonspecific ascending pathways transmit pain, temperature, and coarse touch impulses. They give off numerous branches to the reticular formation and thalamus on the way up; and information sent to "headquarters" is fairly general, nondiscriminatory, and heavily involved in emotional aspects of perception (pleasure, pain, etc.).

The specific ascending pathways are more involved in the discriminative aspects of touch (tactile discrimination), vibration, pressure, and conscious proprioception (limb and joint position).

Proprioceptive impulses conducted via the *spinocerebellar tracts* end at the cerebellum, which uses this information to coordinate skeletal muscle activity. These tracts do not contribute to conscious sensation.

Processing at the Perceptual Level

Interpretation of sensory input occurs in the cerebral cortex. The ability to identity and appreciate the kinds of sensations transmitted depends on the specific location of the target neurons in the sensory cortex, not on the nature of the message, which is *always* action potentials. Each sensory fiber is analogous to a "labeled line" that tells the brain "who" is calling and from "where," whether it is a taste bud or a pressure receptor.

The brain also interprets the activity of a specific sensory receptor as a specific modality (touch, taste, and so on), regardless of how the receptor is activated. For example, whether a pressure receptor in the index finger is pressed or jolted with electricity, or the area of the cortex that recognizes it is stimulated via electrode, our perception is deep touch or pressure coming from the index finger. This phenomenon, by which the brain refers sensations to their usual point of stimulation, is called **projection.**

The main aspects of sensory perception include the following:

▪ **Perceptual detection** is the ability to detect that a stimulus has occurred and is the simplest level of perception. As a general rule, several receptor impulses must be summated for perceptual detection to occur.

▪ **Magnitude estimation** is the ability to detect how *much* of the stimulus is acting on the body. Because of frequency coding, perception increases as stimulus intensity increases.

▪ **Spatial discrimination** allows us to identify the site or pattern of stimulation. A common tool for studying this quality in the laboratory is the **two-point** discrimination test. The test determines how close together two points on the skin can be and still be perceived as two points rather than as one. This test provides a crude map of the density of tactile receptors in the various regions of the skin. The distance between perceived points varies from less than 5 mm on highly sensitive body areas (tip of the tongue) to more than 50 mm on less sensitive areas (back).

- **Feature abstraction** is the mechanism by which a neuron or circuit is tuned to one feature in preference to others. Sensation usually involves an interplay of several stimulus properties or features. For example, one touch tells us that velvet is warm, compressible, and smooth but not completely continuous, each a feature that contributes to our perception of the velvet fabric. Feature abstraction enables us to identify a substance or object that has specific texture or shape.

- **Quality discrimination** is the ability to differentiate the submodalities of a particular sensation. Each sensory modality has several **qualities,** or submodalities. For example, the submodalities of taste include the qualities sweet, salt, bitter, and sour.

- **Pattern recognition** is the ability to take in the scene around us and recognize a familiar pattern, an unfamiliar one, or one that has special significance for us. For example, a figure made of dots may be recognized as a familiar face, and when we listen to music, we hear the melody, not just a string of notes.

Overview of Sensory Receptors

Sensory receptors are specialized to respond to changes in their environment; such environmental changes are called **stimuli.** Typically, activation of a sensory receptor by an adequate stimulus results in graded potentials that in turn trigger nerve impulses along the afferent PNS fibers coursing to the CNS. *Sensation* (awareness of the stimulus) and *perception* (interpretation of the meaning of the stimulus) occur in the brain. But we are getting ahead of ourselves here. For now, let us just examine how sensory receptors are classified. Basically, there are three ways to classify sensory receptors: (1) by the type of stimulus they detect; (2) by their body location; and (3) by their structural complexity.

Classification by Stimulus Type

The receptor classes named according to the activating stimulus are easy to remember because the class name usually indicates the stimulus.

1. **Mechanoreceptors** generate nerve impulses when they, or adjacent tissues, are deformed by a mechanical force such as touch, pressure (including blood pressure), vibration, stretch, and itch.

2. **Thermoreceptors** are sensitive to temperature changes.

3. **Photoreceptors,** such as those of the retina of the eye, respond to light energy.

4. **Chemoreceptors** respond to chemicals in solution (molecules smelled or tasted, or changes in blood chemistry).

5. **Nociceptors** (no"se-sep'torz; *noci* = harm) respond to potentially damaging stimuli that result in pain.

Because overstimulation of any receptor is painful, essentially all receptors function as nociceptors at one time or another. For example, searing heat, extreme cold, excessive pressure, and inflammatory chemicals are all interpreted as painful.

Classification by Location

Three receptor classes are recognized according to either their location or the location of the activating stimulus.

1. **Exteroceptors** (ek"ster-o-sep'torz) are sensitive to stimuli arising outside the body (*extero* = outside), so most exteroceptors are near or at the body surface. They include touch, pressure, pain, and temperature receptors in the skin and most receptors of the special senses (vision, hearing, equilibrium, taste, smell).

2. **Interoceptors** (in"ter-o-sep'torz), also called **visceroceptors,** respond to stimuli within the body (*intero* = inside), such as from the internal viscera and blood vessels. They monitor a variety of stimuli, including chemical changes, tissue stretch, and temperature. Sometimes their activity causes us to feel pain, discomfort, hunger, or thirst. However, we are usually unaware of their workings.

3. **Proprioceptors** (pro"pre-o-sep'torz), like interoceptors, respond to internal stimuli; however, their location is much more restricted. Proprioceptors occur in skeletal muscles, tendons, joints, and ligaments and in connective tissue coverings of bones and muscles. (Some authorities include the equilibrium receptors of the inner ear in this class.) Proprioceptors constantly advise the brain of our body movements (*propria* = one's own) by monitoring how much the organs containing these receptors are stretched.

Classification by Structural Complexity

On the basis of overall receptor structure, there are **simple** and **complex receptors,** but the overwhelming majority are simple. The simple receptors are modified dendritic endings of sensory neurons. They are found througout the body and monitor most types of general sensory information. Complex receptors are actually **sense organs,** localized collections of cells (usually of many types) associated with the **special senses** (vision, hearing, equilibrium, smell, and taste). For example, the sense organ we know as the eye is composed not only of sensory neurons but also of nonneural cells that form its supporting wall, lens, and other associated structures. Though the complex sense organs are most familiar to us, the simple sensory receptors associated with the **general senses** are no less important, and we will concentrate on their structure and function in this chapter.

Simple Receptors of the General Senses

The widely distributed **general sensory receptors** are involved in tactile sensation (a mix of touch, pressure, stretch, and vibration), temperature monitoring and pain, as well as the "muscle sense" provided by proprioceptors. Anatomically, these receptors are either *free nerve endings* or *encapsulated nerve endings.* The general sensory receptors are illustrated and classified in Table 13.1.

Free Nerve Endings   **Free,** or **naked, nerve endings** of sensory neurons are present nearly everywhere in the body, but they are particularly abundant in epithelia and connective tissues. Most of these sensory fibers are unmyelinated, small-diameter C fibers, and their distal endings (the sensory terminals) usually have small knoblike swellings. They respond chiefly to pain and temperature, but some respond to tissue movements caused by pressure as well. Thus, although the bare nerve endings are considered "the" pain receptors (nociceptors) that warn of tissue injury, they also function as mechanoreceptors and thermoreceptors.

Certain free nerve endings associate with enlarged, disc-shaped epidermal cells *(Merkel cells)* to form **Merkel (tactile) discs,** which lie in the deeper layers of the epidermis and function as light touch receptors. **Hair follicle receptors,** free nerve endings that wrap basketlike around hair follicles, are light touch receptors that detect bending of hairs. The tickle of a mosquito landing on your skin is mediated by hair follicle receptors. Another receptor consisting of free dendritic endings is the *itch receptor.* Located in the dermis, it escaped detection until 1997 because of its thin diameter. Its stimulus appears to be a chemical—such as *histamine* or *bradykinin*—present at inflamed sites.

Encapsulated Dendritic Endings   All **encapsulated dendritic endings** consist of one or more fiber terminals of sensory neurons enclosed in a connective tissue capsule. Virtually all the encapsulated

receptors are mechanoreceptors; however, they vary greatly in shape, size, and distribution in the body.

Meissner's corpuscles (mīs'nerz kor'pus″lz) are small receptors in which a few spiraling sensory terminals are surrounded by Schwann cells and then by a thin egg-shaped connective tissue capsule. Meissner's corpuscles are found just beneath the epidermis in the dermal papillae and are especially numerous in sensitive and hairless skin areas such as the nipples, fingertips, and soles of the feet. Also called **tactile corpuscles,** they are receptors for discriminative touch, and apparently play the same role in light touch reception in hairless skin as do the hair follicle receptors in hairy skin.

Pacinian corpuscles, also called **lamellated corpuscles,** are scattered deep in the dermis, and in subcutaneous tissue underlying the skin. Although they are mechanoreceptors stimulated by deep pressure, they respond only when the pressure is first applied, and thus are best suited to monitoring vibration (an "on/off" pressure stimulus). They are the largest corpuscular receptors. Some are over 3 mm long and half as wide and are visible to the naked eye as white, egg-shaped bodies. In section, a Pacinian corpuscle resembles a cut onion. Its single dendrite is surrounded by a capsule containing up to 60 layers of collagen fibers and flattened supporting cells.

Ruffini's corpuscles, which lie in the dermis, subcutaneous tissue, and joint capsules, contain a spray of receptor endings enclosed by a flattened capsule. They bear a striking resemblance to Golgi tendon organs (which monitor tendon stretch) and probably play a similar role in other dense connective tissues where they respond to deep and *continuous* pressure.

Muscle spindles are fusiform (spindle-shaped) proprioceptors found throughout the perimysia of skeletal muscles. Each muscle spindle consists of a bundle of modified skeletal muscle fibers, called **intrafusal fibers** (in″trah-fu′zal; *intra* = within, *fusal* = the spindle), enclosed in a connective tissue capsule (Table 13.1). They detect when a muscle is stretched and initiate a reflex that resists the stretch. Details of muscle spindle innervation are considered later when the stretch reflex is described (see pp. 494–497).

Golgi tendon organs are proprioceptors located in tendons, close to the skeletal muscle insertion. They consist of small bundles of tendon (collagen) fibers enclosed in a layered capsule, with sensory terminals coiling between and around the fibers. Golgi tendon organs are stimulated when the associated muscle contracts and stretches the tendon. When Golgi tendon organs are activated, the contracting muscle is inhibited, which causes it to relax.

Joint kinesthetic receptors (kin″es-thet′ik) are proprioceptors that monitor stretch in the articular capsules that enclose synovial joints. At least four

TABLE 13.1   **General Sensory Receptors Classified by Structure and Function**

Structural Class	Illustration	Functional Classes According to Location (L) and Stimulus Type (S)	Body Location
Unencapsulated			
Free nerve endings of sensory neurons		L: Exteroceptors, interoceptors, and proprioceptors S: Nociceptors (pain), thermoreceptors (heat and cold), possible mechanoreceptors (pressure)	Most body tissues; most dense in connective tissues (ligaments, tendons, dermis, joint capsules, periostea) and epithelia (epidermis, cornea, mucosae, and glands)
Modified free nerve endings: Merkel discs		L: Exteroceptors S: Mechanoreceptors (light pressure); slowly adapting	Basal layer of epidermis of skin
Hair follicle receptors		L: Exteroceptors S: Mechanoreceptors (hair deflection); rapidly adapting	In and surrounding hair follicles
Encapsulated			
Meissner's corpuscles		L: Exteroceptors S: Mechanoreceptors (light pressure, discriminative touch, vibration of low frequency); rapidly adapting	Dermal papillae of hairless skin, particularly nipples, external genitalia, fingertips, soles of feet, eyelids
Pacinian corpuscles		L: Exteroceptors, interoceptors, and some proprioceptors S: Mechanoreceptors (deep pressure, stretch, vibration of high frequency); rapidly adapting	Subcutaneous tissue of the skin; periostea, mesentery, tendons, ligaments, joint capsules, most abundant on fingers, soles of feet, external genitalia, nipples
Ruffini's corpuscles		L: Exteroceptors and proprioceptors S: Mechanoreceptors (deep pressure and stretch); slowly or nonadapting	Deep in dermis, hypodermis, and joint capsules
Muscle spindles	Intrafusal fibers	L: Proprioceptors S: Mechanoreceptors (muscle stretch)	Skeletal muscles, particularly those of the extremities
Golgi tendon organs		L: Proprioceptors S: Mechanoreceptors (tendon stretch)	Tendons
Joint kinesthetic receptors		L: Proprioceptors S: Mechanoreceptors and nociceptors	Joint capsules of synovial joints

receptor types (Pacinian and Ruffini's corpuscles, free nerve endings, and receptors resembling Golgi tendon organs) contribute to this receptor category. Together these receptors provide information on joint position and motion (*kines* = movement), a sensation of which we are highly conscious.

■ Close your eyes and flex and extend your fingers—you can *feel* exactly which joints are moving.

The Chemical Senses: Taste and Smell

Taste and smell are gritty, primitive senses that alert us to whether that "stuff" nearby (or in our mouth) is to be savored or avoided. The receptors for taste (gustation) and smell (olfaction) are **chemoreceptors** (respond to chemicals in an aqueous solution); they complement each other and respond to many of the same stimuli. Taste receptors are excited by food chemicals dissolved in saliva; smell receptors, by airborne chemicals that dissolve in fluids coating nasal membranes.

Taste Buds and the Sense of Taste

The word *taste* comes from the Latin *taxare,* meaning "to touch, estimate, or judge." When we taste things, we are in fact intimately testing or judging our environment, and the sense of taste is considered by many to be the most pleasurable of the special senses.

Localization and Structure of Taste Buds

The **taste buds,** the sensory receptor organs for taste, are located primarily in the oral cavity. Of our 10,000 or so taste buds, most are on the tongue. A few taste buds are scattered on the soft palate, inner surface of the cheeks, pharynx, and epiglottis of the larynx, but most are found in **papillae** (pah-pil′e), peglike projections of the tongue mucosa that give the tongue surface a slightly abrasive feel. Taste buds are located mainly on the tops of the mushroom-shaped **fungiform papillae** (fun′jĭ-form) (which are scattered over the entire tongue surface) and in the epithelium of the side walls of the large round **circumvallate** (ser″kum-val′āt), or simply **vallate, papillae.** The vallate papillae are the largest and least numerous papillae; 7 to 12 of these form an inverted V at the back of the tongue (see Figure 13.2a).

Each gourd-shaped taste bud consists of 50 to 100 *epithelial cells* of three major types: supporting cells, receptor cells, and basal cells (Figure 13.2c). **Supporting cells** form the bulk of the taste bud. They insulate the receptor cells called **gustatory,** or **taste, cells** from each other and from the surrounding tongue epithelium. Long microvilli called **gustatory hairs** project from the tips of both cell types and extend through a **taste pore** to the surface of the epithelium, where they are bathed by saliva. Apparently, the gustatory hairs are the sensitive portions *(receptor membranes)* of the gustatory cells. Coiling intimately around the gustatory cells are sensory dendrites that represent the initial part of the gustatory pathway to the brain. Each afferent fiber receives signals from several receptor cells within the taste bud. Because of their location, taste bud cells are subjected to huge amounts of friction and are routinely burned by hot foods. Luckily, they are among the most dynamic cells in the body, and they are replaced every seven to ten days. The **basal cells** act as stem cells, dividing and differentiating into supporting cells, which in turn give rise to new gustatory cells.

Basic Taste Sensations

Normally, our taste sensations are complicated mixtures of qualities. However, when taste is tested with pure chemical compounds, taste sensations can all be grouped into one of five basic qualities: *sweet, sour, salty, bitter,* and *umami.* The sweet taste is elicited by many organic substances including sugars, saccharin, alcohols, some amino acids, and some lead salts (such as those found in lead paint). Sour taste is produced by acids, specifically their hydrogen ions (H^+) in solution. Salty taste is produced by metal ions (inorganic salts); table salt (sodium chloride) tastes the "saltiest." The bitter taste is elicited by alkaloids (such as quinine, nicotine, caffeine, morphine, and strychnine) as well as a number of nonalkaloid substances, such as aspirin. Umami (u-mam′e; "delicious"), a new taste discovered by the Japanese, is elicited by the amino acid glutamate, which appears to be responsible for the "beef taste" of steak, the characteristic tang of aging cheese, and the flavor of the food additive monosodium glutamate. However, these differences are not absolute: Most taste buds respond to two or more taste qualities, and many substances produce a mixture of the basic taste sensations.

Taste maps that spatially assign sweet receptors to the tip of the tongue, salty and sour receptors to the sides, bitter receptors to the back, and umami receptors to the pharynx are common in the literature. However, researchers have known for years that these mapped areas are misrepresentations, or are dubious at the very least. In reality, there are only slight differences in the localization of specific taste receptors in different regions of the tongue, and all modalities of taste can be elicited from all areas that contain taste buds.

Think back to your studies of the cranial nerves as you view Figure 13.2a. Sensitivity from which part of the tongue would be most impaired by injury to the glossopharyngeal nerve?

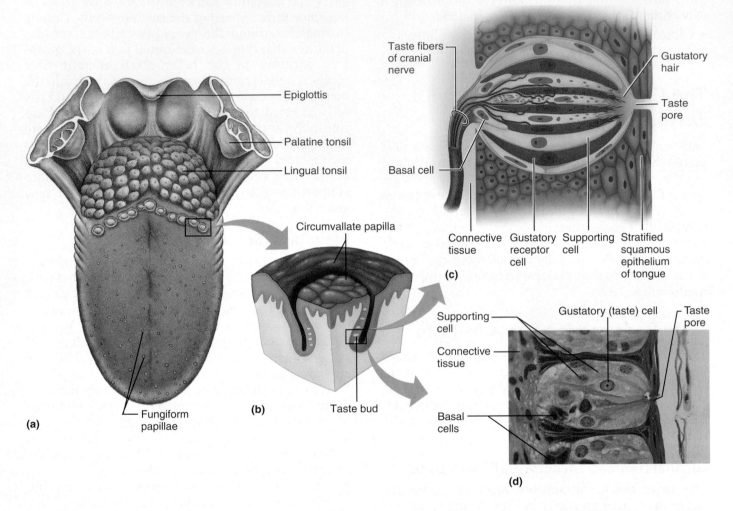

FIGURE 13.2 Location and structure of taste buds on the tongue. (a) Taste buds on the tongue are associated with fungiform and vallate (circumvallate) papillae, peglike projections of the tongue mucosa. (See *A Brief Atlas of the Human Body*, Figure 36.) **(b)** A sectioned circumvallate papilla shows the positioning of the taste buds in its lateral walls. **(c)** An enlarged view of one taste bud, illustrating gustatory (taste), supporting, and basal cells. **(d)** Photomicrograph of a taste bud.

■ *Posterior tongue.*

Physiology of Taste

Activation of Taste Receptors   For a chemical to be tasted it must dissolve in saliva, diffuse into the taste pore, and contact the gustatory hairs. The gustatory cells contain synaptic vesicles, and binding of the food chemical, or *tastant,* to the gustatory cell membrane induces a depolarizing potential that causes the release of the synaptic vesicle contents, neurotransmitter. Binding of the neurotransmitter to the associated sensory dendrites triggers generator potentials that elicit action potentials in these fibers.

The different gustatory cells have different thresholds for activation. The bitter receptors (in line with their protective nature) detect substances present in very minute amounts; the other receptors are less sensitive. Taste receptors adapt rapidly, with partial adaptation in 3–5 seconds and complete adaptation in 1–5 minutes.

Salty taste is due to Na^+ influx through sodium channels followed by Ca^{2+} influx; sour is mediated by H^+, which appears to act on a taste cell in one of three ways—by directly entering the cell, by opening channels that allow other cations to enter, or by blockade of K^+ channels. Bitter, sweet, and umami responses are mediated by G protein–dependent mechanisms that act via second messengers to promote depolarization by increasing intracellular levels of Ca^{2+} (bitter) or closing K^+ channels (sweet).

The Gustatory Pathway

Afferent fibers carrying taste information from the tongue are found primarily in two cranial nerve pairs. A branch of the **facial nerve** (VII), the *chorda tympani*, transmits impulses from taste receptors in the anterior two-thirds of the tongue, whereas the lingual branch of the **glossopharyngeal nerve** (IX) services the posterior third and the pharynx just behind. Taste impulses from the few taste buds in the epiglottis and the lower pharynx are conducted primarily by the **vagus nerve** (X). These afferent fibers synapse in the **solitary nucleus** of the medulla, and from there impulses stream to the thalamus and ultimately to the gustatory cortex in the parietal lobes. Fibers also project to the hypothalamus and limbic system structures, regions that determine our appreciation of what we are tasting.

The Olfactory Epithelium and the Sense of Smell

Although our olfactory sense (*olfact* = to smell) is far less acute than that of many other animals, the human nose is still no slouch in picking up small differences in odors. Some people capitalize on this ability by becoming wine tasters.

Localization and Structure of Olfactory Receptors

Like taste, olfaction detects odor chemicals (*odorants*) in solution. The organ of smell is a yellow-tinged patch (about 5 cm^2) of pseudostratified epithelium, called the **olfactory epithelium,** located in the roof of the nasal cavity (Figure 13.3). Air entering the nasal cavity must make a hairpin turn to stimulate olfactory receptors before entering the respiratory passageway below, so the human olfactory epithelium is in a poor position for doing its job. (This is why sniffing, which draws more air superiorly across the olfactory epithelium, intensifies the sense of smell.) The olfactory epithelium covers the superior nasal concha on each side of the nasal septum, and contains millions of bowling pin–shaped **olfactory receptor cells.** These are surrounded and cushioned by columnar **supporting cells,** which make up the bulk of the penny-thin epithelial membrane (Figure 13.3). The supporting cells contain a yellow-brown pigment similar to lipofuscin, which gives the olfactory epithelium its yellow hue. At the base of the epithelium lie the short **basal cells.**

The olfactory receptor cells are unusual bipolar neurons. Each has a thin apical dendrite that termi-

nates in a knob from which several long cilia radiate. These **olfactory cilia,** which substantially increase the receptive surface area, typically lie flat on the nasal epithelium and are covered by a coat of thin mucus produced by the supporting cells and by olfactory glands in the underlying connective tissue. This mucus is a solvent that "captures" and dissolves airborne odorants. Unlike other cilia in the body, which beat rapidly and in a coordinated manner, olfactory cilia are largely nonmotile. The slender, unmyelinated axons of the olfactory receptor cells are gathered into small fascicles that collectively form the **filaments of the olfactory nerve** (cranial nerve I). They project superiorly through the openings in the cribriform plate of the ethmoid bone, where they synapse in the overlying olfactory bulbs.

Olfactory receptor cells are unique in that they are one of the few types of *neurons* that undergo noticeable turnover throughout adult life. Their superficial location puts them at risk for damage and their typical life span is 60 days, after which they are replaced by differentiation of the basal cells in the olfactory epithelium.

Specificity of the Olfactory Receptors

Taste has been neatly packaged into four or five taste qualities, but science has yet to discover any similar means for classifying smell. Most promising in throwing light on this matter are the studies done in the early 1990s suggesting that there are at least 1000 "smell genes" that are active only in the nose. Each such gene encodes an *odorant binding protein,* a unique receptor protein, and it appears that each of the receptor proteins responds to several different odors and each odor binds to several different receptor types. However, each receptor cell has only one type of receptor protein. Olfactory neurons are exquisitely sensitive—in some cases, just a few molecules activate them.

Physiology of Smell

Activation of the Olfactory Receptors   For us to smell a particular odorant, it must be *volatile*—that is, it must be in the gaseous state as it enters the nasal cavity. Additionally, it must dissolve in the fluid coating the olfactory epithelium. The dissolved odorants stimulate the olfactory receptors by binding to protein receptors in the olfactory cilium membranes and opening specific ion (Na^+) channels. This leads to a receptor potential and ultimately (assuming threshold stimulation) to an action potential that is conducted to the first relay station in the olfactory bulb.

The Olfactory Pathway

As noted above, axons of the olfactory receptor cells constitute the olfactory nerves that synapse in the

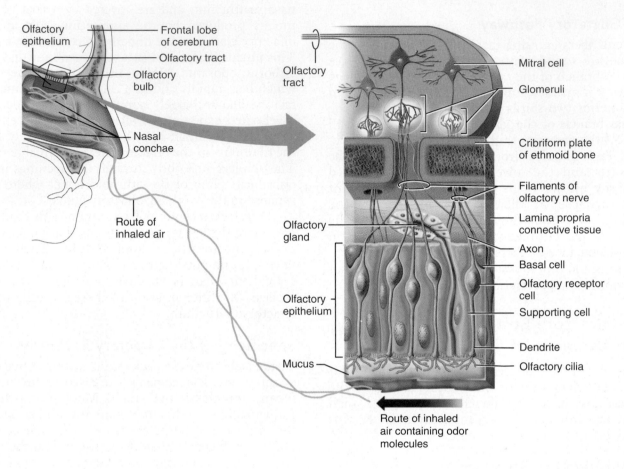

FIGURE 13.3 Olfactory receptors. Site of olfactory epithelium in the superior nasal cavity is shown on the left. On the right is an enlarged view of the olfactory epithelium and the course of the fibers (filaments) of the olfactory nerve (I) through the ethmoid bone to synapse in the overlying olfactory bulb. The glomeruli and mitral cells (output cells) within the olfactory bulb are also shown.

overlying **olfactory bulbs,** the distal ends of the olfactory tracts (see Table 13.2). There, the filaments of the olfactory nerves synapse with **mitral cells** (mi′tral), which are second-order neurons, in complex structures called **glomeruli** (glo-mer′u-li; "little balls") (see Figure 13.3). Axons from neurons bearing the same kind of receptor converge on a given type of glomerulus. Hence, each glomerulus represents a "file" that receives only one type of odor signal, and different odors (which would excite a variety of receptor types) would be expected to activate different subsets of glomeruli. The mitral cells refine the signal, amplify it, and then relay it. The olfactory bulbs also house *granule cells,* GABA-releasing cells that inhibit mitral cells (via dendrodendritic synapses), so that only highly excitatory olfactory impulses are transmitted. This inhibition contributes to the mechanism of *olfactory adaptation* and helps explain how a person working in a paper mill or sewage treatment plant can still enjoy lunch!

When the mitral cells are activated, impulses flow from the olfactory bulbs via the **olfactory tracts** (composed mainly of mitral cell axons) to two main destinations. The first pathway travels via the thalamus to the piriform lobe of the olfactory cortex and the part of the frontal lobe just above the orbit, where smells are consciously interpreted and identified. Each olfactory cortical neuron receives input from up to 50 receptors to analyze. The second pathway flows via the subcortical route to the hypothalamus, amygdala, and other regions of the limbic system, which elicits emotional responses to odors. Smells associated with danger—smoke, cooking gas,

FIGURE 13.4 **The eye and associated accessory structures.** (a) Surface anatomy. The eyelids are open more widely than normal to show the entire iris. (b) Sagittal section.

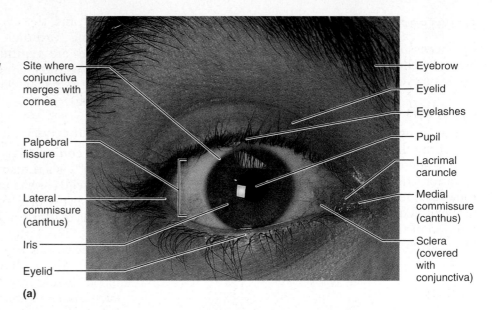

(a)

or skunk scent—trigger the sympathetic fight-or-flight response. As with tastes, appetizing odors cause increased salivation and stimulation of the digestive tract, and unpleasant odors can trigger protective reflexes such as sneezing and choking.

Homeostatic Imbalances of the Chemical Senses

Most olfactory disorders, or *anosmias* (an-oz'me-ahz; "without smells"), result from head injuries that tear the olfactory nerves, the aftereffects of nasal cavity inflammation, and aging. The culprit in one-third of all cases of chemical sense loss is zinc deficiency, and the cure is rapid once the right form of zinc supplement is prescribed. Zinc is a known growth factor for the receptors of the chemical senses.

Brain disorders can distort the sense of smell. Some people have *uncinate fits* (uns'ih-nāt), olfactory hallucinations during which they experience a particular (usually unpleasant) odor, such as rotting meat. Some such cases are undeniably psychological, but many result from irritation of the olfactory pathway by brain surgery or head trauma. *Olfactory auras* experienced by some epileptics just before they have a seizure are transient uncinate fits.

The Eye and Vision

Vision is our dominant sense: Some 70% of all the sensory receptors in the body are in the eyes, and nearly half of the cerebral cortex is involved in some aspect of visual processing.

The adult **eye** is a sphere with a diameter of about 2.5 cm (1 inch). Only the anterior one-sixth of the eye's surface is visible (Figure 13.4); the rest is

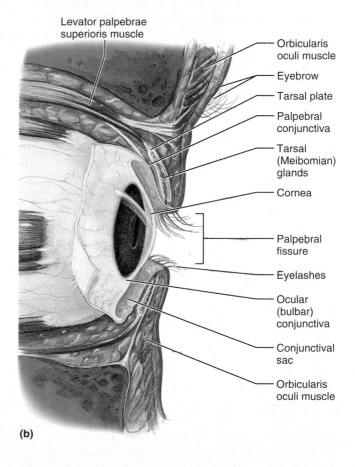

(b)

enclosed and protected by a cushion of fat and the walls of the bony orbit. The fat pad occupies nearly all of the orbit not occupied by the eye itself. The eye is a complex structure and only a small portion of its tissues are actually involved in photoreception. Before turning our attention to the eye itself, let us consider the accessory structures that protect it or aid its functioning.

Accessory Structures of the Eye

The **accessory structures** of the eye include the eyebrows, eyelids, conjunctiva, lacrimal apparatus, and extrinsic eye muscles.

Eyebrows

The **eyebrows** are short, coarse hairs that overlie the supraorbital margins of the skull (Figure 13.4). They help shade the eyes from sunlight and prevent perspiration trickling down the forehead from reaching the eyes. Deep to the skin of the eyebrows are parts of the orbicularis oculi and corrugator muscles. Contraction of the orbicularis muscle depresses the eyebrow, whereas the corrugator moves the eyebrow medially.

Eyelids

Anteriorly, the eyes are protected by the mobile **eyelids** or **palpebrae** (pal'pĕ-bre). The eyelids are separated by the **palpebral fissure** ("eyelid slit") and meet at the medial and lateral angles of the eye—the **medial** and **lateral commissures (canthi),** respectively (Figure 13.4a). The medial canthus sports a fleshy elevation called the **lacrimal caruncle** (kar'ung-kl; "a bit of flesh"). The caruncle contains sebaceous and sweat glands and produces the whitish, oily secretion (fancifully called the Sandman's eye-sand) that sometimes collects at the medial canthus, especially during sleep.

The eyelids are thin, skin-covered folds supported internally by connective tissue sheets called **tarsal plates** (Figure 13.4b). The tarsal plates also anchor the orbicularis oculi and levator palpebrae superioris muscles that run within the eyelid. The orbicularis muscle encircles the eye; when it contracts, the eye closes. Of the two eyelids, the larger upper one is much more mobile, mainly because of the **levator palpebrae superioris** muscle, which raises that eyelid to open the eye. The eyelid muscles are activated reflexively to cause blinking every 3–7 seconds and to protect the eye when it is threatened by foreign objects. Reflex blinking helps prevent drying of the eyes because each time we blink, accessory structure secretions (oil, mucus, and saline solution) are spread across the eyeball surface.

Projecting from the free margin of each eyelid are the **eyelashes.** The follicles of the eyelash hairs are richly innervated by nerve endings (hair follicle receptors), and anything that touches the eyelashes (even a puff of air) triggers reflex blinking.

Several types of glands are associated with the eyelids. The **tarsal,** or **Meibomian** (mi-bo'me-an), **glands** are embedded in the tarsal plates (see Figure 13.4b), and their ducts open at the eyelid edge just posterior to the eyelashes. These modified sebaceous glands produce an oily secretion that lubricates the eyelid and the eye and prevents the eyelids from sticking together. Associated with the eyelash follicles are a number of smaller, more typical sebaceous glands, and modified sweat glands called *ciliary glands* lie between the hair follicles (*cilium* = eyelash).

Conjunctiva

The **conjunctiva** (kon"junk-ti'vah; "joined together") is a transparent mucous membrane. It lines the eyelids as the **palpebral conjunctiva** and reflects (folds back) over the anterior surface of the eyeball as the **ocular,** or **bulbar, conjunctiva** (see Figure 13.4b). The latter covers only the white of the eye, not the cornea (the clear "window" over the iris and pupil). The ocular conjunctiva is very thin, and blood vessels are clearly visible beneath it. (They are even more visible in irritated "bloodshot" eyes.) When the eye is closed, a slitlike space occurs between the conjunctiva-covered eyeball and eyelids. This so-called **conjunctival sac** is where a contact lens lies, and eye medications are often administered into its inferior recess. The major function of the conjunctiva is to produce a lubricating mucus that prevents the eyes from drying out.

🅗   HOMEOSTATIC IMBALANCE

Inflammation of the conjunctiva, called *conjunctivitis*, results in reddened, irritated eyes. *Pinkeye*, a conjunctival infection caused by bacteria or viruses, is highly contagious. ●

Lacrimal Apparatus

The **lacrimal apparatus** (lak'rĭ-mal; "tear") consists of the lacrimal gland and the ducts that drain excess lacrimal secretions into the nasal cavity (Figure 13.5). The **lacrimal gland** lies in the orbit above the lateral end of the eye and is visible through the conjunctiva when the lid is everted. It continually releases a dilute saline solution called **lacrimal secretion**—or, more commonly, **tears**—into the superior part of the conjunctival sac through several small excretory ducts. Blinking spreads the tears downward and across the eyeball to the medial commissure, where they enter the paired **lacrimal canaliculi (canals)** via two tiny openings called **lacrimal puncta** (literally, "prick points"), visible as tiny red dots on the medial margin of each eyelid. From the lacrimal canals, the tears drain into the **lacrimal sac** and then into the **nasolacrimal duct,** which empties into the nasal cavity at the inferior nasal meatus.

Lacrimal fluid contains mucus, antibodies, and **lysozyme,** an enzyme that destroys bacteria. Thus, it cleanses and protects the eye surface as it moistens and lubricates it. When lacrimal secretion increases substantially, tears spill over the eyelids and fill the nasal cavities, causing congestion and the "sniffles."

⊞ HOMEOSTATIC IMBALANCE

Because the nasal cavity mucosa is continuous with that of the lacrimal duct system, a cold or nasal inflammation often causes the lacrimal mucosa to become inflamed and swell. This constricts the ducts and prevents tears from draining from the eye surface, causing "watery" eyes. ●

Extrinsic Eye Muscles

The movement of each eyeball is controlled by six straplike **extrinsic eye muscles,** which originate from the bony orbit and insert into the outer surface of the eyeball (Figure 13.6). These muscles allow the eyes to follow a moving object, and provide external "guy-wires" that help to maintain the shape of the eyeball and hold it in the orbit.

The four *rectus muscles* originate from a common tendinous ring, the **annular ring,** at the back of the orbit and run straight to their insertion on the eyeball. Their locations and the movements that they promote are clearly indicated by their names: **superior, inferior, lateral,** and **medial rectus muscles.** The actions of the two *oblique muscles* are less easy to deduce because they take rather strange paths through the orbit. They move the eye in the vertical plane when the eyeball is already turned medially by the rectus muscle. The **superior oblique muscle** originates in common with the rectus muscles, runs along the medial wall of the orbit, and then makes a right-angle turn and passes through a fibrocartilaginous loop suspended from the frontal bone called the **trochlea** (trok′le-ah; "pulley") before inserting on the superolateral aspect of the eyeball. It rotates the eye downward and somewhat laterally. The **inferior oblique muscle** originates from the medial orbit surface and runs laterally and obliquely to insert on the inferolateral eye surface. It rotates the eye up and laterally.

The four rectus muscles would seem to provide all the eye movements we require—medial, lateral, superior, and inferior—so why the two oblique muscles? The simplest way to answer this question is to point out that the superior and inferior recti cannot elevate or depress the eye *without also turning it medially* because they approach the eye from a posteromedial direction. For an eye to be *directly* elevated or depressed, the lateral pull of the oblique muscles is necessary to cancel the medial pull of the superior and inferior recti.

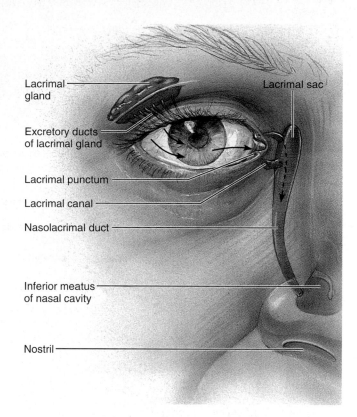

Lacrimal gland

Lacrimal sac

Excretory ducts of lacrimal gland

Lacrimal punctum

Lacrimal canal

Nasolacrimal duct

Inferior meatus of nasal cavity

Nostril

FIGURE 13.5 The lacrimal apparatus. Arrows indicate the direction of flow of lacrimal fluid after its release by the lacrimal gland.

Except for the lateral rectus and superior oblique muscles, which are innervated respectively by the *abducens* and *trochlear nerves,* all extrinsic eye muscles are served by the *oculomotor nerves.* The actions and nerve supply of these muscles are summarized in Figure 13.6c. The courses of the associated cranial nerves are illustrated in Table 13.2.

The extrinsic eye muscles are among the most precisely and rapidly controlled skeletal muscles in the entire body. This reflects their high axon-to-muscle-fiber ratio: The motor units of these muscles contain only 8 to 12 muscle cells and in some cases as few as two or three.

⊞ HOMEOSTATIC IMBALANCE

When movements of the external muscles of the two eyes are not perfectly coordinated, a person cannot properly focus the images of the same area of the visual field from each eye and so sees two images instead of one. This condition is called **diplopia** (dĭ-plo′pe-ah), or *double vision.* It can result from paralysis or weakness of certain extrinsic muscles, or it may be a temporary consequence of acute alcohol intoxication.

Which three extrinsic eye muscles turn the eyeball laterally?

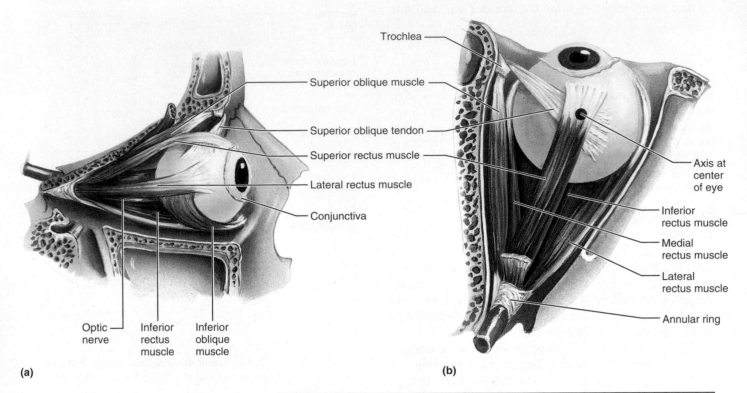

(a)

(b)

FIGURE 13.6 Extrinsic eye muscles.
(a) Lateral view of the right eye. **(b)** Superior view of the right eye. **(c)** Summary of innervating cranial nerves and muscle actions.

Name	Action	Controlling cranial nerve
Lateral rectus	Moves eye laterally	VI (abducens)
Medial rectus	Moves eye medially	III (oculomotor)
Superior rectus	Elevates eye	III (oculomotor)
Inferior rectus	Depresses eye	III (oculomotor)
Inferior oblique	Elevates eye and turns it laterally	III (oculomotor)
Superior oblique	Depresses eye and turns it laterally	IV (trochlear)

(c)

Congenital weakness of the external eye muscles may cause **strabismus** (strah-biz′mus; "cross-eyed"), a condition in which the affected eye rotates medially or laterally. To compensate, the eyes may alternate in focusing on objects. In other cases, only the controllable eye is used, and the brain begins to disregard inputs from the deviant eye, which then becomes functionally blind. Strabismus is treated either with eye exercises to strengthen the weak muscles or by temporarily placing a patch on the stronger eye, which forces the child to use the weaker eye. Surgery is needed for unyielding conditions. ●

■ *The lateral rectus and the superior and inferior obliques.*

Structure of the Eyeball

The eye itself, commonly called the **eyeball,** is a slightly irregular hollow sphere (Figure 13.7). Because the eyeball is shaped roughly like the globe of the earth, it is said to have poles. Its most anterior point is the **anterior pole;** its most posterior point, the **posterior pole.** Its wall is composed of three coats, or tunics, the fibrous, vascular, and sensory tunics. Its internal cavity is filled with fluids called *humors* that help to maintain its shape. The lens, the adjustable focusing apparatus of the eye, is supported vertically within the internal cavity, dividing it into *anterior* and *posterior segments,* or *cavities.*

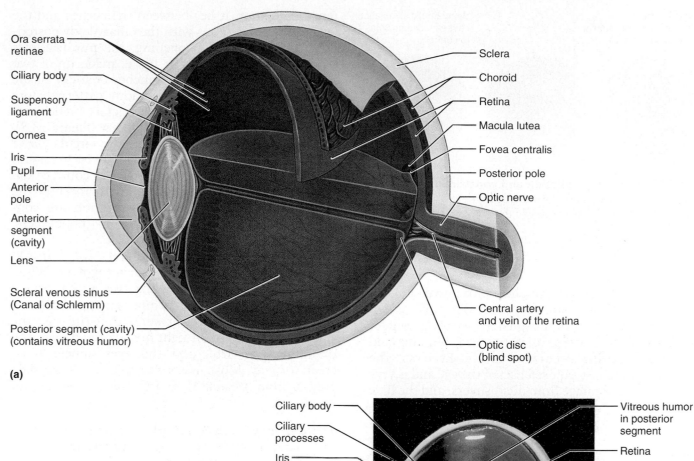

Ora serrata retinae
Ciliary body
Suspensory ligament
Cornea
Iris
Pupil
Anterior pole
Anterior segment (cavity)
Lens
Scleral venous sinus (Canal of Schlemm)
Posterior segment (cavity) (contains vitreous humor)

Sclera
Choroid
Retina
Macula lutea
Fovea centralis
Posterior pole
Optic nerve
Central artery and vein of the retina
Optic disc (blind spot)

(a)

Ciliary body
Ciliary processes
Iris
Margin of pupil
Anterior segment
Lens
Cornea
Suspensory ligament of lens

Vitreous humor in posterior segment
Retina
Choroid
Sclera
Fovea centralis
Optic disc
Optic nerve

(b)

FIGURE 13.7 Internal structure of the eye (sagittal section). (a) Diagrammatic view. The vitreous humor is illustrated only in the bottom half of the eyeball. **(b)** Photograph of the human eye.

Tunics Forming the Wall of the Eyeball

The Fibrous Tunic   The outermost coat of the eye, the **fibrous tunic,** is composed of dense avascular connective tissue. It has two obviously different regions: the sclera and the cornea. The **sclera** (skle′rah), forming the posterior portion and the bulk of the fibrous tunic, is glistening white and opaque. Seen anteriorly as the "white of the eye," the tough, tendonlike sclera (*sclera* = hard) protects and shapes the eyeball and provides a sturdy anchoring site for the extrinsic eye muscles. Posteriorly, where the sclera is pierced by the optic nerve, it is continuous with the dura mater of the brain. The anterior sixth of the fibrous tunic is modified to form the transparent **cornea,** which bulges anteri-orly from its junction with the sclera. The crystal-clear cornea forms a window that lets light enter the eye, and is also part of the light-bending apparatus of the eye.

The cornea is covered by epithelial sheets on both faces. The external sheet, a stratified squamous epithelium that helps protect the cornea from abrasion, merges with the ocular conjunctiva at the sclera-cornea junction. Epithelial cells that continually renew the cornea are located here. The deep *corneal endothelium,* composed of simple squamous epithelium, lines the inner face of the cornea. Its cells have active sodium pumps that maintain the clarity of the cornea by keeping the water content of the cornea low.

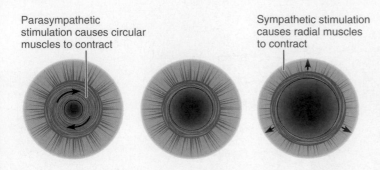

Parasympathetic stimulation causes circular muscles to contract

Sympathetic stimulation causes radial muscles to contract

FIGURE 13.8   Pupil dilation and constriction, anterior view. The circular and radial muscles of the iris respond to stimulation from different parts of the autonomic nervous system.

The cornea is well supplied with nerve endings, most of which are pain receptors. (For this reason, some people can never adjust to wearing contact lenses.) When the cornea is touched, blinking and increased tearing occur reflexively. Even so, the cornea is the most exposed part of the eye and is very vulnerable to damage from dust, slivers, and the like. Luckily, its capacity for regeneration and repair is extraordinary. Furthermore, the cornea is the only tissue in the body that can be transplanted from one person to another with little or no possibility of rejection.

The Vascular Tunic (Uvea)   The **vascular tunic,** or middle coat of the eyeball, is also called the **uvea** (u've-ah; "grape"). The pigmented uvea has three regions: choroid, ciliary body, and iris (see Figure 13.7).

The **choroid** is a highly vascular, dark brown membrane (*choroid* = membranelike) that forms the posterior five-sixths of the uvea. Its blood vessels provide nutrition to all eye tunics. Its brown pigment, produced by melanocytes, helps absorb light, preventing it from scattering and reflecting within the eye (which would cause visual confusion). The choroid is incomplete posteriorly where the optic nerve leaves the eye. Anteriorly, it becomes the **ciliary body,** a thickened ring of tissue that encircles the lens. The ciliary body consists chiefly of interlacing smooth muscle bundles called **ciliary muscles,** which are important in controlling lens shape. Near the lens, its posterior surface is thrown into radiating folds called **ciliary processes,** which contain the capillaries that secrete the fluid that fills the cavity of the anterior segment of the eyeball. The **suspensory ligament,** or **zonule,** extends from the ciliary processes to the lens; this halo of fine fibers encircles and helps hold the lens in its upright position in the eye.

The **iris,** the visible colored part of the eye, is the most anterior portion of the uvea. Shaped like a flat-tened doughnut, it lies between the cornea and the lens and is continuous with the ciliary body posteriorly. Its round central opening, the **pupil,** allows light to enter the eye. The iris is made up of two smooth muscle layers with bunches of sticky elastic fibers that congeal into a random pattern before birth. Its muscle fibers allow it to act as a reflexively activated diaphragm to vary pupil size (Figure 13.8). In close vision and bright light, the circular layer muscles contract and the pupil constricts. In distant vision and dim light, the radial muscles contract and the pupil dilates, allowing more light to enter the eye. Pupillary dilation and constriction are controlled by sympathetic and parasympathetic fibers, respectively.

Although irises come in different colors (*iris* = rainbow), they contain only brown pigment. When they have a lot of pigment, the eyes appear brown or black. If the amount of pigment is small and restricted to the posterior surface of the iris, the shorter wavelengths of light are scattered from the unpigmented parts, and the eyes appear blue, green, or gray. Most newborn babies' eyes are slate gray or blue because their iris pigment is not yet developed.

The Sensory Tunic (Retina)   The innermost tunic is the delicate, two-layered **retina** (ret'ĭ-nah). The outer **pigmented layer,** a single-cell-thick lining, abuts the choroid, and extends anteriorly to cover the ciliary body and the posterior face of the iris. Its pigmented epithelial cells, like those of the choroid, absorb light and prevent it from scattering in the eye. They also act as phagocytes and store vitamin A needed by the photoreceptor cells. The transparent inner **neural (nervous) layer** extends anteriorly to the posterior margin of the ciliary body. This junction is called the **ora serrata retinae,** literally the saw-toothed margin of the retina (see Figure 13.7). Actually an outpocketing of the brain, the retina contains millions of photoreceptors that transduce light energy as well as other neurons involved in the processing of light stimuli and glia. Although these two layers are very close together, they are not fused. Notice that while the retina is commonly called the **sensory tunic,** only its neural layer plays a direct role in vision.

From posterior to anterior, the neural layer is composed of three main types of neurons: **photoreceptors, bipolar cells,** and **ganglion cells** (Figure 13.9). Local currents are produced in response to light and spread from the photoreceptors (abutting the pigmented layer) to the bipolar neurons and then to the innermost ganglion cells, where action potentials are generated. The ganglion cell axons make a right-angle turn at the inner face of the retina, then leave the posterior aspect of the eye as

? Which retinal cells are the output cells?

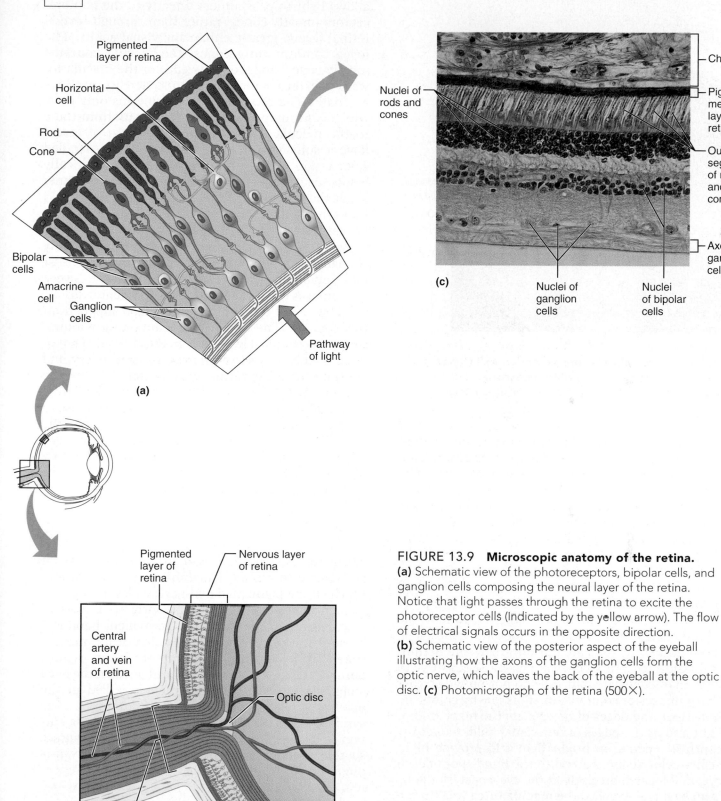

(a)

(b)

(c)

FIGURE 13.9 **Microscopic anatomy of the retina.**
(a) Schematic view of the photoreceptors, bipolar cells, and ganglion cells composing the neural layer of the retina. Notice that light passes through the retina to excite the photoreceptor cells (Indicated by the yellow arrow). The flow of electrical signals occurs in the opposite direction.
(b) Schematic view of the posterior aspect of the eyeball illustrating how the axons of the ganglion cells form the optic nerve, which leaves the back of the eyeball at the optic disc. **(c)** Photomicrograph of the retina (500✕).

■ Ganglion cells.

Which of the areas shown here would have the highest cone density?

Central artery and vein
emerging from the optic disc

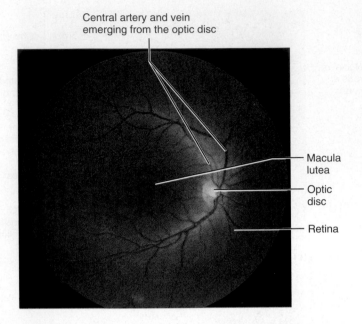

Macula
lutea

Optic
disc

Retina

FIGURE 13.10   Part of the posterior wall (fundus) of the eye as seen with an ophthalmoscope. Note the optic disc from which the retinal blood vessels radiate.

the thick optic nerve. The retina also contains other types of neurons—horizontal cells and amacrine cells—which play a role in visual processing. The **optic disc,** where the optic nerve exits the eye, is a weak spot in the **fundus** (posterior wall) of the eye because it is not reinforced by the sclera. The optic disc is also called the **blind spot** because it lacks photoreceptors and light focused on it cannot be seen. Nonetheless, we do not usually notice these gaps in our vision because the brain uses a sophisticated process called *filling in* to deal with absence of input.

The quarter-billion photoreceptors found in the neural retinas are of two types: rods and cones. The more numerous **rods** are our dim-light and peripheral vision receptors. They are far more sensitive to light than cones are, but they do not provide either sharp images or color vision. This is why colors are indistinct and edges of objects appear fuzzy in dim light and at the edges of our visual field. **Cones,** by contrast, operate in bright light and provide high-acuity color vision. Lateral to the blind spot of each eye, and located precisely at the eye's posterior pole, is an oval region called the **macula lutea** (mak'u-lah lu'te-ah; "yellow spot") with a minute (0.4 mm) pit in its center called the **fovea centralis** (see Figure

13.7). In this region, the retinal structures abutting the vitreous humor are displaced to the sides. This allows light to pass almost directly to the photoreceptors (mostly cones) rather than through several retinal layers, greatly enhancing visual acuity. The fovea contains only cones; the macula contains mostly cones; and from the edge of the macula toward the retina periphery, cone density declines gradually. The retina periphery contains only rods, which continuously decrease in density from there to the macula. Only the foveae (plural of fovea) have a sufficient cone density to provide detailed color vision, so anything we wish to view critically is focused on the foveae. Because each fovea is only about the size of the head of a pin, not more than a thousandth of the entire visual field is in *hard focus* (foveal focus) at a given moment. Consequently, for us to visually comprehend a scene that is rapidly changing (as when we drive in traffic), our eyes must flick rapidly back and forth to provide the foveae with images of different parts of the visual field.

The neural retina receives its blood supply from two sources. The outer third (containing photoreceptors) is supplied by vessels in the choroid. The inner two-thirds is served by the **central artery** and **central vein of the retina,** which enter and leave the eye through the center of the optic nerve. Radiating outward from the optic disc, these vessels give rise to a rich vascular network that is clearly seen when the eyeball interior is examined with an ophthalmoscope (Figure 13.10). The fundus of the eye is the only place in the body where small blood vessels can be observed directly in a living person.

🅗 HOMEOSTATIC IMBALANCE

The pattern of vascularization of the retina makes it susceptible to *retinal detachment.* This condition, in which the pigmented and nervous layers separate (detach) and allow the jellylike vitreous humor to seep between them, can cause permanent blindness because it deprives the neural retina of its nutrient source. It usually happens when the retina is torn during a traumatic blow to the head or when the head stops moving suddenly and then is jerked in the opposite direction (as in bungee jumping). The symptom that victims most often describe is "a curtain being drawn across the eye," but some people see sootlike spots or light flashes. If diagnosed early, it is often possible to reattach the retina with a laser before photoreceptor damage becomes permanent. ●

Internal Chambers and Fluids

As noted earlier, the lens and its halolike suspensory ligament divide the eye into two segments, the anterior segment in front of the lens and the larger

■ *The macula lutea.* ■

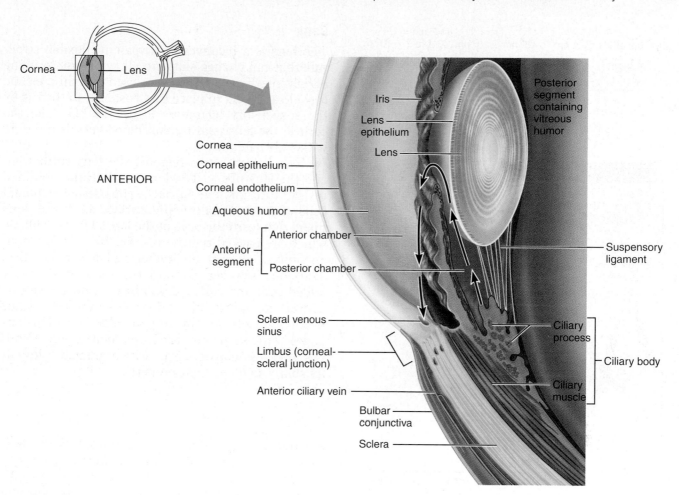

FIGURE 13.11 Circulation of aqueous humor. The anterior segment in front of the lens is incompletely divided into anterior and posterior chambers, which are continuous through the pupil. Aqueous humor fills the chambers and diffuses freely through the vitreous humor in the posterior segment. It is formed by filtration from capillaries in the ciliary processes and is reabsorbed into the venous blood by the scleral venous sinus. The arrows indicate the circulation pathway.

posterior segment behind it (Figure 13.7a). The **posterior segment** (cavity) is filled with a clear gel called **vitreous humor** (*vitre* = glassy) that binds tremendous amounts of water. Vitreous humor (1) transmits light, (2) supports the posterior surface of the lens and holds the neural retina firmly against the pigmented layer, and (3) contributes to intraocular pressure, helping to counteract the pulling force of the extrinsic eye muscles.

The **anterior segment** (cavity) (Figure 13.11) is partially subdivided by the iris into the **anterior chamber** (between the cornea and the iris) and the **posterior chamber** (between the iris and the lens). The *entire* anterior segment is filled with **aqueous humor,** a clear fluid similar in composition to blood plasma. Unlike the vitreous humor, which forms in the embryo and lasts for a lifetime, aqueous humor forms and drains continually and is in constant motion. It filters from the capillaries of the ciliary processes into the posterior chamber and freely dif-

fuses through the vitreous humor in the posterior segment before returning to the anterior segment. After flowing through the pupil into the anterior chamber, it drains into the venous blood via the **scleral venous sinus (canal of Schlemm),** an unusual venous channel that encircles the eye in the angle at the sclera-cornea junction. Normally, aqueous humor is produced and drained at the same rate, maintaining a constant intraocular pressure of about 16 mm Hg, which helps to support the eyeball internally. Aqueous humor supplies nutrients and oxygen to the lens and cornea and to some cells of the retina, and it carries away their metabolic wastes.

HOMEOSTATIC IMBALANCE

If the drainage of aqueous humor is blocked, fluid backs up like in a clogged sink. Pressure within the eye may increase to dangerous levels and compress

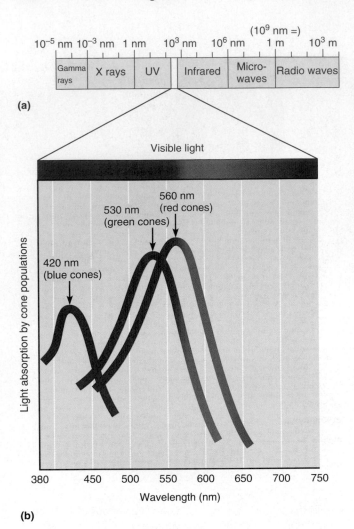

(a)

(b)

FIGURE 13.12 **The electromagnetic spectrum and cone cell sensitivities.** (a) The electromagnetic spectrum extends from the very short gamma waves to the long radio waves. The wavelengths of the electromagnetic spectrum are indicated in nanometers. The spectrum of visible light constitutes only a small portion of this energy range. (b) Sensitivities of the three cone types to the different wavelengths of the visible spectrum.

the retina and optic nerve—a condition called **glaucoma** (glaw-ko′mah). The eventual result is blindness (*glaucoma* = vision growing gray) unless the condition is detected early. Unfortunately, many forms of glaucoma steal sight so slowly and painlessly that people do not realize they have a problem until the damage is done. Late signs include seeing halos around lights and blurred vision. The glaucoma examination is simple. The intraocular pressure is determined by directing a puff of air at the cornea and measuring the amount of corneal deformation it causes. This exam should be done yearly after the age of 40. Treatment begins with eye drops (miotics) that increase the rate of aqueous humor drainage and progresses to surgery. ●

Lens

The **lens** is a biconvex, transparent, flexible structure that can change shape to allow precise focusing of light on the retina. It is enclosed in a thin, elastic capsule and held in place just posterior to the iris by the suspensory ligament (Figure 13.11). Like the cornea, the lens is avascular; blood vessels interfere with transparency.

The lens has two regions: the lens epithelium and the lens fibers. The **lens epithelium,** confined to the anterior lens surface, consists of cuboidal cells that eventually differentiate into the **lens fibers** that form the bulk of the lens. The lens fibers, which are packed tightly together like the layers in an onion, contain no nuclei and few organelles. They do, however, contain transparent, precisely folded proteins called **crystallins.** The crystallins convert sugar into energy for use by the lens. Since new lens fibers are continually added, the lens enlarges throughout life, becoming denser, more convex, and less elastic, all of which gradually impair its ability to focus light properly.

🄷   *HOMEOSTATIC IMBALANCE*

A **cataract** ("waterfall") is a clouding of the lens that causes the world to appear distorted, as if seen through frosted glass. Some cataracts are congenital, but most result from age-related hardening and thickening of the lens or are a secondary consequence of diabetes mellitus. Heavy smoking and frequent exposure to intense sunlight increase the risk for cataracts, whereas long-term dietary supplementation with vitamin C appears to significantly decrease the risk. Whatever the promoting factors, the *direct* cause of cataracts seems to be inadequate delivery of nutrients to the deeper lens fibers. The metabolic changes that result promote clumping of the crystallin proteins. Fortunately, the offending lens can be surgically removed and an artificial lens implanted to save the patient's sight. ●

Physiology of Vision

Overview: Light and Optics

To really comprehend the function of the eye as a photoreceptor organ, we need to have some understanding of the properties of light.

Wavelength and Color   Our eyes respond to the part of the spectrum called **visible light,** which has a wavelength range of approximately 400–700 nm (Figure 13.12a). (1 nm = 10^{-9} m, or one-billionth of a meter.) Visible light travels in the form of waves, and its wavelengths can be measured very accurately.

However, light is actually composed of small particles or packets of energy called **photons** or **quanta.** Attempts to reconcile these two findings have led to the present concept of light as packets of energy (photons) traveling in a wavelike fashion at very high speeds (186,000 miles per second or about 300,000 km/s). Thus, light can be thought of as a vibration of pure energy ("a bright wiggle") rather than a material substance.

When visible light passes through a prism, each of its component waves bends to a different degree, so that the beam of light is dispersed and a **visible spectrum,** or band of colors, is seen (Figure 13.12b). (In the same manner, the rainbow seen during a summer shower represents the collective prismatic effects of all the tiny water droplets suspended in air.) Red wavelengths are the longest and have the lowest energy, whereas the violet wavelengths are the shortest and most energetic.

Refraction and Lenses   Light travels in straight lines and is easily blocked by any nontransparent object. Like sound, light can reflect, or bounce, off a surface. **Reflection** of light by objects in our environment accounts for most of the light reaching our eyes.

When light is traveling in a given medium, its speed is constant. But when it passes from one transparent medium into another with a different density, its speed changes. Light speeds up as it passes into a less dense medium and slows as it passes into a denser medium. Because of these changes in speed, bending or **refraction** of a light ray occurs when it meets the surface of a different medium at an oblique angle rather than at a right angle (perpendicular). The greater the incident angle, the greater the amount of bending. To demonstrate the consequence of light refraction at home place a spoon in a half-full glass of water. The spoon appears to break at the air-water interface.

A lens is a transparent object curved on one or both surfaces. Since light hits the curve at an angle, it is refracted. If the lens surface is convex, that is, thickest in the center like a camera lens, the light rays are bent so that they converge (come together) or intersect at a single point called the **focal point** (Figure 13.13). In general, the thicker (more convex) the lens, the more the light is bent and the shorter the focal distance (distance between the lens and focal point). The image formed by a convex lens, called a **real image,** is upside down and reversed from left to right. Concave lenses, which are thicker at the edges than at the center, like those of magnifying glasses, diverge the light (bend it outward) so that the light rays move away from each other. Consequently, concave lenses prevent light from focusing and extend the focal distance.

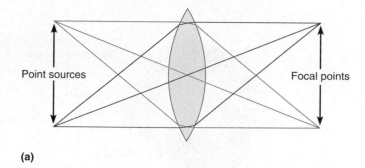

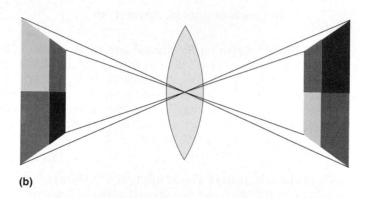

FIGURE 13.13   Bending of light points by a convex lens. (a) How two light points are focused on the opposite side of the lens. **(b)** Formation of an image by a convex lens. Notice that the image is upside down and reversed from right to left.

Focusing of Light on the Retina

As light passes from air into the eye, it moves sequentially through the cornea, aqueous humor, lens, and vitreous humor, and then passes *through the entire thickness of the neural layer of the retina* to excite the photoreceptors that abut the pigmented layer (see Figures 13.7 and 13.9). During its passage, light is bent three times: as it enters the cornea and on entering and leaving the lens. The refractory power of the humors and cornea is constant. On the other hand, the lens is highly elastic, and its curvature and light-bending power can be actively changed to allow fine focusing of the image.

Focusing for Distant Vision   Our eyes are best adapted ("preset to focus") for distant vision. To look at distant objects, we need only aim our eyeballs so that they are both fixated on the same spot. The **far point of vision** is that distance beyond which no change in lens shape (accommodation) is needed for focusing. For the normal or **emmetropic** (em″ĕ-tro′pik) eye, the far point is 6 m (20 feet).

Any object being viewed can be said to consist of many small points, with light radiating outward in all directions from each point. However, because distant objects appear smaller, light from an object at or

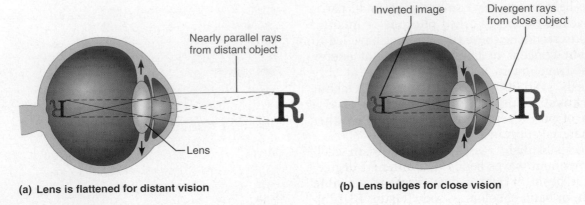

(a) Lens is flattened for distant vision

(b) Lens bulges for close vision

FIGURE 13.14   Focusing for distant and close vision. (a) Light from distant objects (over 6 m away) approaches as parallel rays and, in the normal eye, need not be adjusted for proper focusing on the retina. **(b)** Light from close objects (less than 6 m away) tends to diverge and lens convexity must be increased (accommodation) for proper focusing. Notice in both cases that the image formed on the retina is inverted and reversed from left to right (i.e., it is a real image).

beyond the far point of vision approaches the eyes as nearly parallel rays and is focused precisely on the retina by the fixed refractory apparatus (cornea and humors) and the at-rest lens (Figure 13.14a). During distant vision, the ciliary muscles are completely relaxed, and the lens (stretched flat by tension in the suspensory ligaments) is as thin as it gets. Consequently, it is at its lowest refractory power.

Focusing for Close Vision   Light from objects less than 6 m away diverges as it approaches the eyes and it comes to a focal point farther from the lens. Thus, close vision demands that the eye make active adjustments. To restore focus, three processes—accommodation of the lenses, constriction of the pupils, and convergence of the eyeballs—must occur simultaneously. The signal that induces this trio of reflex responses appears to be a blurring of the retinal image.

1. Accommodation of the lenses. Accommodation is the process that increases the refractory power of the lens. As the ciliary muscles contract, the ciliary body is pulled anteriorly and inward toward the pupil and releases tension in the suspensory ligament. No longer stretched, the elastic lens recoils and bulges, providing the shorter focal length needed to focus the image of a close object on the retina (Figure 13.14b). Contraction of the ciliary muscles is controlled mainly by the parasympathetic fibers of the oculomotor nerves.

The closest point on which we can focus clearly is called the **near point of vision,** and it represents the maximum bulge the lens can achieve. In young adults with emmetropic vision, the near point is 10 cm (4 inches) from the eye. However, it is closer in children and gradually recedes with age, explaining why children can hold their books very close to

their faces while many elderly people must hold the newspaper at arm's length. The gradual loss of accommodation with age reflects the lens's decreasing elasticity. In many people over the age of 50, the lens is nonaccommodating, a condition known as *presbyopia* (pres"be-o'pe-ah), literally "old person's vision."

2. Constriction of the pupils. The circular (constrictor) muscles of the iris enhance the effect of accommodation by reducing the size of the pupil (see Figure 13.8) toward 2 mm. This **accommodation pupillary reflex,** mediated by parasympathetic fibers of the oculomotor nerves, prevents the most divergent light rays from entering the eye. Such rays would pass through the extreme edge of the lens and would not be focused properly, and so would cause blurred vision.

3. Convergence of the eyeballs. The visual goal is always to keep the object being viewed focused on the retinal foveae. When we look at distant objects, both eyes are directed either straight ahead or to one side to the same degree, but when we fixate on a close object, our eyes converge. **Convergence,** controlled by somatic motor fibers of the oculomotor nerves, is medial rotation of the eyeballs by the medial rectus muscles so that each is directed toward the object being viewed. The closer that object, the greater the degree of convergence required; when you focus on the tip of your nose, you "go cross-eyed."

Reading or other close work requires almost continuous accommodation, pupillary constriction, and convergence. This is why prolonged periods of reading tire the eye muscles and can result in *eyestrain.*

Homeostatic Imbalances of Refraction Visual problems related to refraction can result from a hyperrefractive (overconverging) or hyporefractive (underconverging) lens or from structural abnormalities of the eyeball.

Myopia (mi-o'pe-ah; "short vision") occurs when distant objects are focused not on, but in front of, the retina. Myopic people see close objects without problems because of the active ability of their lens to accommodate to the extent necessary, but distant objects are blurred. The common name for myopia is *nearsightedness.* (Notice that this terminology names the aspect of vision that is *unimpaired.*) Myopia typically results from an eyeball that is too long. Correction has traditionally involved use of concave lenses that diverge the light before it enters the eye, but procedures to flatten the cornea slightly—a painless 10-minute surgery called *radial keratotomy,* or PRK and LASIK procedures using a laser—have offered other treatment options.

Hyperopia (hy'per-o"pe-ah; "far vision"), or *farsightedness,* occurs when the parallel light rays from distant objects are focused *behind* the retina. Hyperopic individuals can see distant objects perfectly well because their ciliary muscles contract almost continuously to increase the light-bending power of the lens, which moves the focal point forward onto the retina. However, diverging light rays from *nearby* objects are focused so far behind the retina that the lens cannot bring the focal point onto the retina even at its full refractory power. Thus, close objects appear blurry, and convex corrective lenses are needed to converge the light more strongly for close vision. Hyperopia usually results from an eyeball that is too short or a lazy lens (one with poor refractory power).

Unequal curvatures in different parts of the lens (or cornea) lead to blurry images. This refractory problem is **astigmatism** (*astigma* = not a point). Special cylindrically ground lenses and laser procedures are used to correct this problem.

Functional Anatomy of the Photoreceptors

Once light is focused on the retina, the photoreceptors come into play. **Photoreception** is the process by which the eye detects light energy. Although photoreceptors are modified neurons, structurally they resemble tall epithelial cells turned upside down with their "tips" immersed in the pigmented layer of the retina (Figure 13.15a). Moving from the pigmented layer into the neural layer, both the rods and cones have an **outer segment** (the receptor region) joined to an **inner segment** by a stalk containing a cilium. The inner segment connects to the *cell body,* which is continuous with an *inner fiber* bearing synaptic endings. The outer segment of the rods is slender and rod shaped (hence their name) and their inner segment connects to the cell body by the *outer fiber.* By contrast, the fatter cones have a short conical outer segment and the inner segment directly joins the cell body.

The outer segments contain an elaborate array of **visual pigments (photopigments)** that change shape as they absorb light. These pigments are embedded in areas of the plasma membrane that forms discs. This coupling of photoreceptor pigments to membranes magnifies the surface area available for trapping light (Figure 13.15b). In rods, the discs are discontinuous—stacked like a row of pennies in a coin wrapper. In cones the disc membranes are continuous with the plasma membrane; thus, the interiors of the cone discs are continuous with the extracellular space.

In rods, new discs are assembled at the proximal end of the outer segment from materials synthesized in the cell body at the end of each night. As new discs are made, they push the others peripherally. The discs at the tip of the outer segment continually fragment off and are phagocytized by cells of the pigmented layer. The tips of the outer segments of cones are also removed and renewed (but at the end of each day); however, the precise way the membrane elements migrate into the outer segment is still unclear.

Because the rods and each of the three cone types contain unique visual pigments, they absorb different wavelengths of light and have different thresholds for activation. Rods, for example, (1) are very sensitive (respond to very dim light), making them best suited for night vision and peripheral vision, and (2) absorb all wavelengths of visible light, but their inputs are perceived only in gray tones. Cones (1) need bright light for activation (have low sensitivity), but (2) have pigments that furnish a vividly colored view of the world.

Rods and cones are also "wired" differently to other retinal neurons. Rods participate in converging pathways, and as many as 100 rods may ultimately feed into each ganglion cell. As a result, rod effects are summated and considered collectively, resulting in vision that is fuzzy and indistinct. (The visual cortex has no way of knowing exactly *which* rods of the large number influencing a particular ganglion cell are actually activated.) In contrast, each cone in the fovea (or at most a few) has a straight-through pathway via its "own personal bipolar cell" to a ganglion cell (see Figure 15.10). Hence, each cone essentially has its own "labeled line" to the higher visual centers.

The Chemistry of Visual Pigments

How does light affect the photoreceptors so that it is ultimately translated into electrical signals? The key is a light-absorbing molecule called **retinal** that combines with proteins called **opsins** to form four types of visual pigments. Depending on the type of opsin to which it is bound, retinal preferentially absorbs

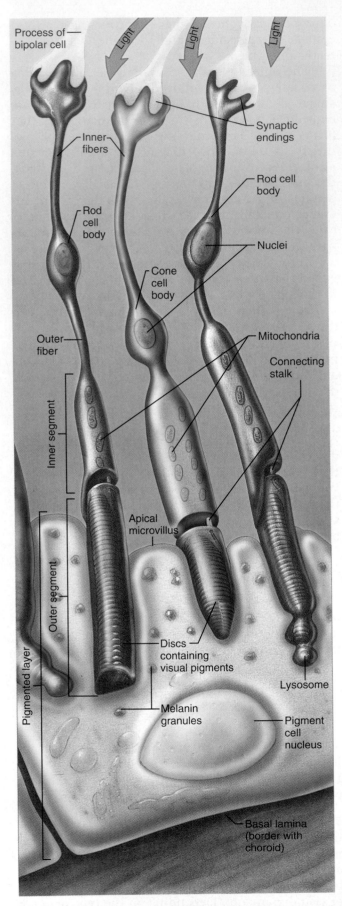

(a)

What is the advantage of having the photoreceptor pigment form part of the disc membranes?

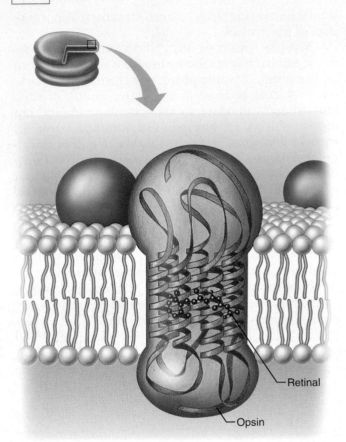

(b)

FIGURE 13.15   Photoreceptors of the retina.
(a) Diagram of the photoreceptors (rods and cones). The relationship of the outer segments of the photoreceptors to the pigmented layer of the retina is also illustrated.
(b) Diagram of a small section of the membrane of a visual pigment—containing disc of the outer segment of a rod cell. The visual pigments consist of a light-absorbing molecule called retinal bound to an opsin protein. Each type of photoreceptor has a characteristic kind of opsin protein, which affects the absorption spectrum of the retinal. In rods, the pigment-opsin complex is called rhodopsin. Notice that the light-absorbing retinal occupies the core of the rhodopsin molecule.

different wavelengths of the visible spectrum. Retinal is chemically related to vitamin A and is made from it.

Retinal can assume a variety of distinct three-dimensional forms, each form called an isomer. When bound to opsin, retinal has a bent, or kinked, shape

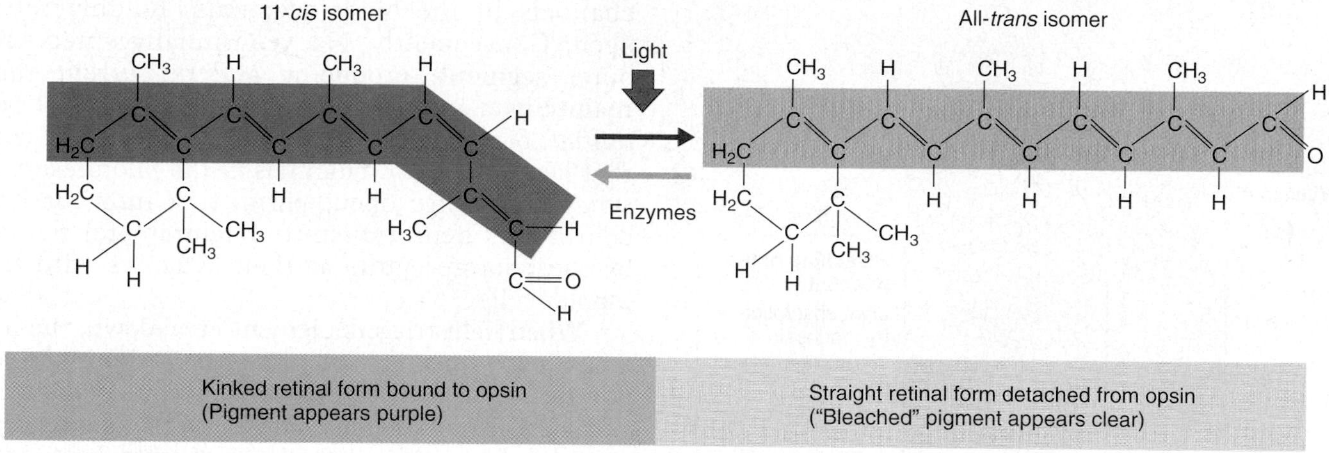

FIGURE 13.16   **Structure of retinal isomers involved in photoreception.**
When the visual pigment absorbs light, 11-*cis* retinal is transformed to the all-*trans*
isomer and the pigment is bleached.

called the **11-*cis* isomer** (Figure 13.16). However, when the pigment is struck by light and absorbs photons, retinal twists and snaps into a new configuration, its **all-*trans* isomer,** which causes retinal to detach from opsin. This is the *only* light-dependent stage, and this simple photochemical event initiates a whole chain of chemical and electrical reactions in rods and cones that ultimately causes electrical impulses to be transmitted along the optic nerve.

Stimulation of the Photoreceptors

1. Excitation of rods. The visual pigment of rods is a deep purple pigment called **rhodopsin** (ro-dop'sin; *rhodo* = rose, *opsis* = vision). (Do you suppose the person who coined the expression "looking at the world through rose-colored glasses" knew the meaning of "rhodopsin"?) Rhodopsin molecules are arranged in a single layer in the membranes of each of the thousands of discs in the rods' outer segments (see Figure 13.15b). Although rhodopsin absorbs light throughout the entire visible spectrum, it maximally absorbs green light.

Rhodopsin forms and accumulates in the dark in the sequence of reactions shown on the left side of Figure 13.17. As illustrated, vitamin A is oxidized to the 11-*cis* retinal form and then combined with opsin to form rhodopsin. When rhodopsin absorbs light, retinal changes shape to its all-*trans* isomer, and the retinal-opsin combination breaks down, allowing retinal and opsin to separate. This breakdown is known as the **bleaching of the pigment.** Actually, the story is a good deal more complex than this, and the breakdown of rhodopsin that triggers the transduction process involves a rapid cascade of intermediate steps (right side of Figure 13.17) that takes just a few milliseconds. Once the light-struck all-*trans* retinal detaches from opsin, it is recon-

verted by enzymes within the pigmented epithelium to its 11-*cis* isomer in an ATP-requiring process. Then, retinal heads "homeward" again to the photoreceptor cells' outer segments. Rhodopsin is regenerated when 11-*cis* retinal is rejoined to opsin.

2. Excitation of cones. The chemistry of cone function is not completely worked out, but breakdown and regeneration of their visual pigments is essentially the same as for rhodopsin. However, the threshold for cone activation is much higher than that for rods because cones respond only to high-intensity (bright) light.

Visual pigments of the three types of cones, like those of rods, are a combination of retinal and opsins. However, the cone opsins differ both from the opsin of the rods and from one another. The naming of cones reflects the colors (that is, wavelengths) of light that each cone variety absorbs best. The blue cones respond maximally to wavelengths around 420 nm, the green cones to wavelengths of 530 nm, and the red cones to wavelengths at or close to 560 nm (see Figure 13.12b). However, their absorption spectra overlap, and perception of intermediate hues, such as orange, yellow, and purple, results from differential activation of more than one type of cone at the same time. For example, yellow light stimulates both red and green cone receptors, but if the red cones are stimulated more strongly than the green cones, we see orange instead of yellow. When all cones are stimulated equally, we see white.

HOMEOSTATIC IMBALANCE

Color blindness is due to a congenital lack of one or more of the cone types. Inherited as a sex-linked condition, it is far more common in males than in

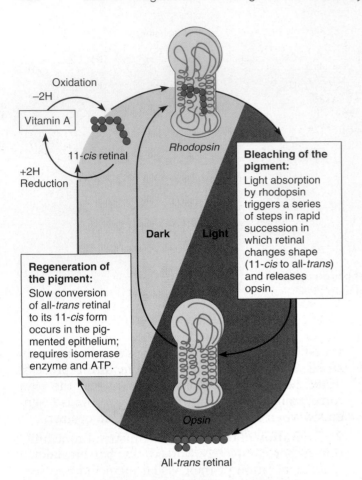

Oxidation
−2H

Vitamin A

11-*cis* retinal

+2H
Reduction

Rhodopsin

Dark   **Light**

Bleaching of the pigment:
Light absorption by rhodopsin triggers a series of steps in rapid succession in which retinal changes shape (11-*cis* to all-*trans*) and releases opsin.

Regeneration of the pigment:
Slow conversion of all-*trans* retinal to its 11-*cis* form occurs in the pigmented epithelium; requires isomerase enzyme and ATP.

Opsin

All-*trans* retinal

FIGURE 13.17   Sequence of chemical events in the formation and breakdown of rhodopsin. When struck by light, rhodopsin breaks down and is converted to its precursors via a number of intermediate steps. Absorption of light converts 11-*cis* retinal to all-*trans* retinal, which separates from opsin, an event that is responsible for transduction of the light stimulus. The 11-*cis* retinal is regenerated in energy (ATP)-requiring reactions by enzymatic action from all-*trans* retinal or from vitamin A and recombined with opsin to form rhodopsin.

females. As many as 8–10% of males have some form of color blindness. The most common type is red-green color blindness, resulting from a deficit or absolute absence of either red or green cones. Red and green are seen as the same color—either red or green, depending on the cone type present. Many color-blind people are unaware of their condition because they have learned to rely on other cues—such as differences in intensities of the same color—to distinguish something green from something red, such as traffic signals. ●

Light Transduction in Photoreceptors   How is the light stimulus transduced into an electrical event, that is, a change in membrane potential? In the dark, **cyclic GMP** (cGMP) binds to sodium

channels in the outer segments, holding them open. Consequently, Na$^+$ continually enters the outer segment, producing a *dark current* that maintains a transmembrane potential (a *dark potential*) of about −40 mV. This depolarizing current keeps the Ca^{2+} channels *at the photoreceptor synaptic endings* open, permitting more or less continuous neurotransmitter (glutamate) release by the photoreceptors at their synapses with the bipolar cells.

When light triggers pigment breakdown, the net effect is to "turn off" sodium entry. Sodium permeability now decreases dramatically but potassium permeability remains unchanged, so the photoreceptors develop a *hyperpolarizing receptor potential* (−70 mV) that inhibits their release of neurotransmitter. This is bewildering, to say the least. Here we have receptors built to detect light that depolarize in the dark and hyperpolarize when exposed to light! Nevertheless, turning off neurotransmitter release by hyperpolarization conveys information as effectively as depolarization.

The photoreceptors do not generate action potentials. In fact, none of the retinal neurons generate action potentials except the ganglion cells, which is consistent with their function as the output neurons of the retina. In all other cases, only local currents (graded potentials) are produced.

Light and Dark Adaptation   Rhodopsin is amazingly sensitive; even starlight causes some of the molecules to become bleached. As long as the light is low intensity, relatively little rhodopsin is bleached and the retina continues to respond to light stimuli. However, in high-intensity light there is wholesale bleaching of the pigment, plus rhodopsin breaks down nearly as rapidly as it is made. At this point, the rods become nonfunctional and cones begin to respond. Hence, retinal sensitivity automatically adjusts to the amount of light present.

Light adaptation occurs when we move from darkness into bright light, as when leaving a movie matinee. We are momentarily dazzled—all we see is white light—because the sensitivity of the retina is still "set" for dim light. Both rods and cones are strongly stimulated, and large amounts of the photopigments are broken down almost instantaneously, producing a flood of signals that accounts for the glare. Under such conditions, compensations occur. (1) Retinal sensitivity decreases dramatically, and (2) retinal neurons adapt rapidly, switching from the rod to the cone system. Within about 60 seconds, the cones are sufficiently excited by the bright light to take over. Visual acuity and color vision continue to improve over the next 5–10 minutes. Thus, during light adaptation, retinal sensitivity (rod function) is lost, but visual acuity is gained.

Dark adaptation, essentially the reverse of light adaptation, occurs when we go from a well-lit area into a dark one. Initially, we see nothing but velvety blackness because (1) our cones stop functioning in low-intensity light, and (2) our rod pigments have been bleached out by the bright light, and the rods are still initially inhibited. But once we are in the dark, rhodopsin accumulates and retinal sensitivity increases. Dark adaptation is much slower than light adaptation and can go on for hours. However, there is usually enough rhodopsin within 20–30 minutes to allow adequate dim-light vision.

During these adaptations, reflexive changes also occur in pupil size. Bright light shining in one or both eyes causes both pupils to constrict (elicits the *consensual* and *pupillary light reflexes*). These pupillary reflexes are mediated by the pretectal nucleus of the midbrain and by parasympathetic fibers. In dim light, the pupils dilate, allowing more light to enter the eye interior.

HOMEOSTATIC IMBALANCE

Night blindness, or *nyctalopia* (nic"tă-lo'pe-uh), is a condition in which rod function is seriously hampered, impairing one's ability to drive safely at night. The most common cause of night blindness is prolonged vitamin A deficiency, which leads to rod degeneration. Vitamin A supplements restore function if they are administered before degenerative changes occur. ●

The Visual Pathway to the Brain

As described earlier, the axons of the retinal ganglion cells issue from the back of the eyeballs in the **optic nerves** (Figure 13.18). At the X-shaped **optic chiasma** (*chiasm* = cross), fibers from the medial aspect of each eye cross over to the opposite side and then continue on via the **optic tracts.** Thus, each optic tract (1) contains fibers from the lateral (temporal) aspect of the eye on the same side and fibers from the medial (nasal) aspect of the opposite eye, and (2) carries all the information from the same half of the visual field. Also notice that, because the lens system of each eye reverses all images, the medial half of each retina receives light rays from the *temporal* (lateral-most) part of the visual field (that is, from the far left or far right rather than from straight ahead), and the lateral half of each retina receives an image of the *nasal* (central) part of the visual field. Consequently, the left optic tract carries (and sends on) a complete representation of the right half of the visual field, and the opposite is true for the right optic tract.

The paired optic tracts sweep posteriorly around the hypothalamus and send most of their axons to synapse with neurons in the **lateral geniculate body** of the thalamus, which maintains the fiber separation established at the chiasma, but balances and combines the retinal input for delivery to the visual cortex. Axons of the thalamic neurons project through the internal capsule to form the **optic radiation** of fibers in the cerebral white matter. These fibers project to the **primary visual cortex** in the occipital lobes, where conscious perception of visual images (seeing) occurs.

Some nerve fibers in the optic tracts send branches to the midbrain. These fibers end in the **superior colliculi,** visual reflex centers controlling the extrinsic muscles of the eyes. A small subset of neurons donating fibers to the visual pathway contain the visual pigment *melanopsin,* dubbed the circadian pigment, and respond directly to light stimuli independent of vision. These project to the **pretectal nuclei,** which mediate pupillary light reflexes, and to the **suprachiasmatic nucleus** of the hypothalamus, which functions as the "timer" to set our daily biorhythms.

Notice that although both eyes are set anteriorly and look in approximately the same direction, their visual fields, each about 170 degrees, overlap to a considerable extent, and each eye sees a slightly different view (see Figure 13.18). Cortical "fusion" of the slightly different images delivered by the two eyes provides us with **depth perception,** an accurate means of locating objects in space. This faculty is also called **three-dimensional vision.**

HOMEOSTATIC IMBALANCE

The relationships described above explain patterns of blindness that follow damage to different visual structures. Loss of an eye or destruction of one optic nerve eliminates true depth perception, and peripheral vision is lost on the side of damage. For example, if the "left eye" in Figure 13.18 was lost in a hunting accident, nothing would be seen in the visual field area colored gold in that figure. If neural destruction occurs beyond the optic chiasma—in an optic tract, the thalamus, or visual cortex—then part or all of the opposite half of the visual field is lost. For example, a stroke affecting the left visual cortex leads to blindness in the right half of the visual field, but since the right (undamaged) visual cortex still receives inputs from both eyes, depth perception in the remaining half of the visual field is retained. ●

Visual Processing

How does information received by the rods and cones become vision? In the past few years, a tremendous amount of research has been devoted to this question. Some of the more basic information that has come out of these studies is presented here.

Where on the retina of the right eye would light from an object in the nasal portion of its visual field be projected?

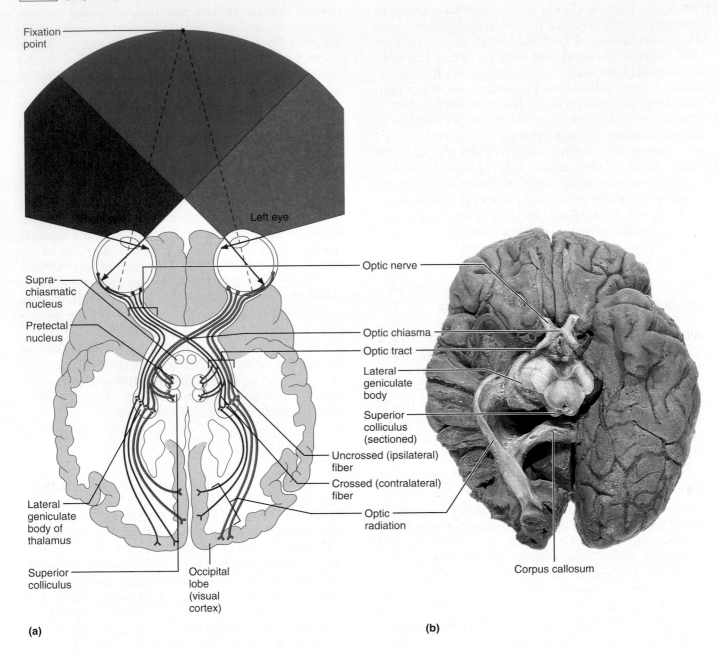

FIGURE 13.18 Visual fields of the eyes and visual pathway to the brain, inferior view. (a) Diagram. **(b)** Photograph. In (a) notice that the visual fields of the two eyes overlap considerably. Also notice the retinal sites at which a real image would be focused when both eyes are fixated on a close, pointlike object. The full lateral extension of the visual fields is not illustrated here.

To the temporal portion of the retina.

Retinal Processing   The retinal ganglion cells generate action potentials at a fairly steady rate (20–30 per second), even in the dark. Surprisingly, illuminating the entire retina evenly has no effect on that basal rate. However, the activity of individual ganglion cells changes dramatically when a tiny spot of light falls just on certain portions of their receptive field. The *receptive field* of a ganglion cell consists of the area of the retina that, when stimulated, influences the activity of that ganglion cell (the rods or cones that funnel their impulses to it).

When researchers mapped the receptive fields of ganglion cells receiving inputs only from rods, they found two types of doughnut-shaped (circle-within-a-circle) receptive fields. These are called **on-center** and **off-center fields,** based on what happens to the ganglion cell when the center of its receptive field is illuminated with a spot of light. Ganglion cells with on-center fields are stimulated (depolarized) by light hitting the field center (the "doughnut hole") and are inhibited by light hitting rods in the periphery (the doughnut itself). The opposite situation occurs in ganglion cells with off-center fields. Equal illumination of "on" and "off" regions causes little change in the basal rate of ganglion cell firing.

The mechanisms of retinal processing as currently understood can be summarized briefly as follows:

1. The action of light on photoreceptors hyperpolarizes them.

2. Bipolar neurons in the "on" regions are excited (depolarize) and excite the associated ganglion cell when the rods feeding into them are illuminated (and hyperpolarized). Bipolar neurons in the "off" regions are inhibited (hyperpolarize) and inhibit the ganglion cell when the rods feeding into them are stimulated. The opposite responses of the bipolar neurons in the "on" versus the "off" regions reflect the fact that they have different receptor types for glutamate.

3. Bipolar neurons receiving signals from cones feed directly into excitatory synapses on ganglion cells. Hence, cone inputs are perceived as sharp and clear (and in color).

4. Bipolar neurons receiving inputs from rods excite amacrine cells via gap junctions. These local integrator neurons modify rod inputs and ultimately direct excitatory inputs to appropriate ganglion cells. Thus, rod inputs are not only summated but are subject to "detours" before reaching the output (ganglion) cells. Together these factors provide a more smeary picture than cone inputs.

5. Rod inputs are also modified and subjected to lateral inhibition by synaptic (gap junction) contacts with horizontal cells. Inputs from these local integrator cells allow the retina to convert inputs that are points of light into perceptually more meaningful contour information by accentuating bright/dark contrasts, or edges.

6. Two varieties of ganglion cells are identified. M cells, which receive rod input, respond best to large, moving objects at the edge of the image. P cells, the target of cone input, relay information concerning small nonmoving objects in the center of the visual field, that is, their color and details.

Thalamic Processing   The lateral geniculate nuclei (LGN) of the thalamus relay information on movement, segregate the retinal axons in preparation for depth perception, emphasize visual inputs from the cones, and sharpen the contrast information received from the retina. The separation of signals from the two eyes is relayed accurately to the visual cortex.

Cortical Processing   Two types of areas for processing retinal inputs are found in the visual cortex. The *primary visual cortex,* also called the **striate cortex,** is thick with fibers coming in from the LGN. This area contains an accurate topographical map of the retina; the left visual cortex receives input from the right visual field and vice versa. Visual processing here occurs at a relatively basic level, with the processing neurons responding to dark and bright edges (contrast information) and object orientation.

The striate cortex also provides form, color, and motion inputs, with temporary storage areas called "blobs," which forward information on to *visual association areas* collectively called the prestriate cortices. The more anterior **prestriate cortices** are occipital lobe centers that continue the processing of visual information concerned with form, color, and movement. Functional neuroimaging of humans has revealed that complex visual processing extends well forward into the temporal, parietal, and frontal lobes.

The Ear: Hearing and Balance

At first glance, the machinery for hearing and balance appears very crude. Fluids must be stirred to stimulate the mechanoreceptors of the inner ear. Nevertheless, our hearing apparatus allows us to hear an extraordinary range of sound, and our equilibrium receptors keep the nervous system continually informed of head movements and position. Although the organs serving these two senses are structurally interconnected within the ear, their receptors respond to different stimuli and are activated independently of one another.

Structure of the Ear

The ear is divided into three major areas: outer ear, middle ear, and inner ear (Figure 13.19). The outer and middle ear structures are involved with hearing only and are rather simply engineered. The inner ear functions in both equilibrium and hearing and is extremely complex.

The Outer (External) Ear

The **outer (external) ear** consists of the auricle and the external auditory canal. The **auricle,** or **pinna,** is what most people call the ear—the shell-shaped projection surrounding the opening of the external auditory canal. The auricle is composed of elastic

Besides the bony boundaries, what structure separates the outer from the middle ear? The middle from the inner ear?

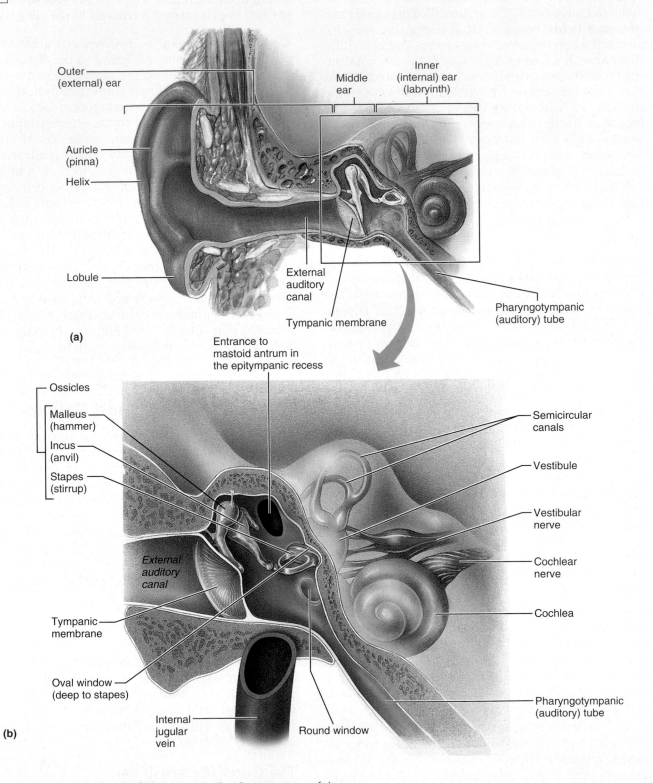

FIGURE 13.19 Structure of the ear. (a) The three regions of the ear.
(b) Enlarged view of the middle and inner ear. Notice that the semicircular canals, vestibule, and cochlea of the inner ear form the bony labyrinth.

■ *The tympanic membrane. Secondary tympanic membrane.*

cartilage covered with thin skin and an occasional hair. Its rim, the **helix,** is somewhat thicker, and its fleshy, dangling **lobule** ("earlobe") lacks supporting cartilage. The function of the auricle is to direct sound waves into the external auditory canal.

The **external auditory canal (meatus)** is a short, curved tube (about 2.5 cm long by 0.6 cm wide) that extends from the auricle to the eardrum. Near the auricle, its framework is elastic cartilage; the remainder of the canal is carved into the temporal bone. The entire canal is lined with skin bearing hairs, sebaceous glands, and modified apocrine sweat glands called **ceruminous glands** (se-roo'mĭ-nus). These glands secrete yellow-brown waxy **cerumen,** or earwax (*cere* = wax), which provides a sticky trap for foreign bodies and repels insects. Occasionally, cerumen builds up excessively and becomes compacted, a condition that can impair hearing.

Sound waves entering the external auditory canal eventually hit the **tympanic membrane,** or **eardrum** (*tympanum* = drum), the boundary between the outer and middle ears. The eardrum is a thin, translucent, connective tissue membrane, covered by skin on its external face and by a mucosa internally. It is shaped like a flattened cone, with its apex protruding medially into the middle ear. Sound waves make the eardrum vibrate. The eardrum, in turn, transfers the sound energy to the tiny bones of the middle ear and sets them into vibration.

The Middle Ear

The **middle ear,** or **tympanic cavity,** is a small, air-filled, mucosa-lined cavity in the petrous portion of the temporal bone. It is flanked laterally by the eardrum and medially by a bony wall with two openings, the superior **oval (vestibular) window** and the inferior **round (cochlear) window.** The round window is closed by the *secondary tympanic membrane.* Superiorly the tympanic cavity arches upward as the **epitympanic recess,** the "roof" of the middle ear cavity. The **mastoid antrum,** a canal in the posterior wall of the tympanic cavity, allows it to communicate with *mastoid air cells* housed in the mastoid process. The anterior wall of the middle ear abuts the internal carotid artery (the main artery supplying the brain) and contains the opening of the **pharyngotympanic (auditory) tube.** This tube, formerly called the eustachian tube, runs obliquely downward to link the middle ear cavity with the nasopharynx (the superiormost part of the throat), and the mucosa of the middle ear is continuous with that lining the pharynx (throat).

Normally, the pharyngotympanic tube is flattened and closed, but swallowing or yawning opens it briefly to equalize pressure in the middle ear cavity with external air pressure. This is important because the eardrum vibrates freely only if the pressure on both of its surfaces is the same; otherwise sounds are distorted. The ear-popping sensation of the pressures equalizing is familiar to anyone who has flown in an airplane.

🅗 HOMEOSTATIC IMBALANCE

Otitis media (me'de-ah), or middle ear inflammation, is a fairly common result of a sore throat, especially in children, whose auditory tubes are shorter and run more horizontally. Otitis media is the most frequent cause of hearing loss in children. In acute infectious forms, the eardrum bulges and becomes inflamed and red. Most cases of otitis media are treated with antibiotics. When large amounts of fluid or pus accumulate in the cavity, an emergency *myringotomy* (lancing of the eardrum) may be required to relieve the pressure, and a tiny tube implanted in the eardrum permits pus to drain into the external ear. The tube falls out by itself within the year. Noninfectious cases of otitis media, in which clear fluid accumulates in the tympanic cavity, are often due to food allergies (typically milk or wheat). In such cases, the cure is restricting the offending food. ●

The tympanic cavity is spanned by the three smallest bones in the body: the **ossicles** (see Figure 13.19). These bones, named for their shape, are the **malleus** (mal'e-us), or **hammer;** the **incus** (ing'kus), or **anvil;** and the **stapes** (sta'pēz), or **stirrup.** The "handle" of the malleus is secured to the eardrum, and the base of the stapes fits into the oval window.

Tiny ligaments suspend the ossicles, and mini-synovial joints link them together into a chain that spans the middle ear cavity. The incus articulates with the malleus laterally and the stapes medially. The ossicles transmit the vibratory motion of the eardrum to the oval window, which in turn sets the fluids of the inner ear into motion, eventually exciting the hearing receptors.

Two tiny skeletal muscles are associated with the ossicles. The **tensor tympani** (ten'sor tim'pah-ni) arises from the wall of the auditory tube and inserts on the malleus. The **stapedius** (stah-pe'de-us) runs from the posterior wall of the middle ear cavity to the stapes. When the ears are assaulted by very loud sounds, these muscles contract reflexively to prevent damage to the hearing receptors. Specifically, the tensor tympani tenses the eardrum by pulling it medially; the stapedius checks vibration of the whole ossicle chain and limits the movement of the stapes in the oval window.

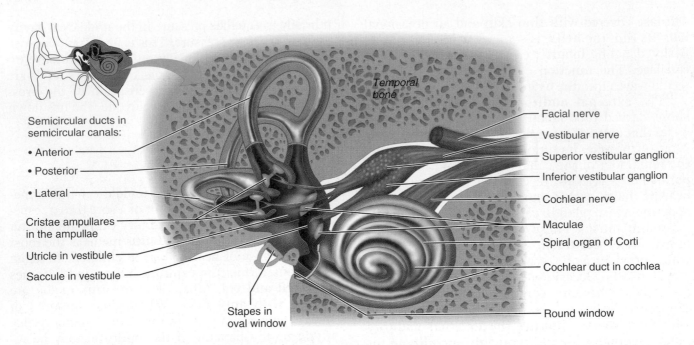

Semicircular ducts in
semicircular canals:

• Anterior

• Posterior

• Lateral

Cristae ampullares
in the ampullae

Utricle in vestibule

Saccule in vestibule

Stapes in
oval window

Temporal
bone

Facial nerve

Vestibular nerve

Superior vestibular ganglion

Inferior vestibular ganglion

Cochlear nerve

Maculae

Spiral organ of Corti

Cochlear duct in cochlea

Round window

FIGURE 13.20   Membranous labyrinth of the inner ear, shown in relation to the chambers of the bony labyrinth. The locations of the specialized receptors for hearing (organ of Corti) and equilibrium (maculae and cristae ampullares) are also indicated.

The Inner (Internal) Ear

The **inner (internal) ear** is also called the **labyrinth** ("maze") because of its complicated shape (see Figure 13.19). It lies deep in the temporal bone behind the eye socket and provides a secure site for all of the delicate receptor machinery housed there. The inner ear has two major divisions: the bony labyrinth and the membranous labyrinth. The **bony, or osseous, labyrinth** is a system of tortuous channels worming through the bone. Its three regions are the *vestibule,* the *cochlea* (kok'le-ah), and the *semicircular canals.* The views of the bony labyrinth typically seen in textbooks, including this one, are somewhat misleading because we are talking about a *cavity* here. The representation seen in Figure 13.19 can be compared to a plaster of paris cast of the cavity or hollow space inside the bony labyrinth. The **membranous labyrinth** is a continuous series of membranous sacs and ducts contained within the bony labyrinth and (more or less) following its contours (Figure 13.20).

The bony labyrinth is filled with **perilymph,** a fluid similar to cerebrospinal fluid and continuous with it. The membranous labyrinth is surrounded by and floats in the perilymph; its interior contains **endolymph,** which is chemically similar to K^+-rich intracellular fluid. These two fluids conduct the sound vibrations involved in hearing and respond to the mechanical forces occurring during changes in body position and acceleration.

The Vestibule   The **vestibule** is the central egg-shaped cavity of the bony labyrinth. It lies posterior to the cochlea, anterior to the semicircular canals, and flanks the middle ear medially. In its lateral wall is the oval window. Suspended in its perilymph and united by a small duct are two membranous labyrinth sacs, the **saccule** and **utricle** (u'tri-kl) (Figure 13.20). The smaller saccule is continuous with the membranous labyrinth extending anteriorly into the cochlea (the *cochlear duct*), whereas the utricle is continuous with the semicircular ducts extending into the semicircular canals posteriorly. The saccule and utricle house equilibrium receptor regions called *maculae* that respond to the pull of gravity and report on changes of head position.

The Semicircular Canals   The **semicircular canals** lie posterior and lateral to the vestibule and each of these canals defines about 2/3 of a circle. The cavities of the bony semicircular canals project from the posterior aspect of the vestibule, each oriented in one of the three planes of space. Accordingly, there is an *anterior, posterior,* and *lateral* semicircular canal in each inner ear. The anterior and posterior canals are oriented at right angles to each other in the vertical plane, whereas the lateral canal lies horizontally (see Figure 13.20). Snaking through each semicircular canal is a corresponding membranous **semicircular duct,** which communicates with the utricle anteriorly. Each of these ducts has an enlarged swelling at one end called an **ampulla,** which houses an equilibrium receptor region called a *crista ampullaris* (literally, crest of the ampulla). These receptors respond to angular (rotational) movements of the head.

The Cochlea   The **cochlea,** from the Latin "snail," is a spiral, conical, bony chamber about half the size of a split pea. It extends from the anterior part of the vestibule and coils for about 2½ turns around a bony pillar called the **modiolus** (mo-di'o-lus) (Figure 13.21a). Running through its center like a wedge-shaped worm is the membranous **cochlear duct,** which ends blindly at the cochlear apex. The cochlear duct houses the **spiral organ of Corti,** the receptor organ for hearing (Figures 13.20 and 13.21b). The cochlear duct and the **spiral lamina,** a thin shelflike extension of bone that spirals up the modiolus like the thread on a screw, together divide the cavity of the bony cochlea into three separate chambers or **scalas** (*scala* = ladder). The **scala vestibuli** (ska'lah věs-tǐ'bu-li), which lies superior to the cochlear duct, is continuous with the vestibule and abuts the oval window. The middle **scala media** is the cochlear duct itself. The **scala tympani,** which terminates at the round window, is inferior to the cochlear duct. Since the scala media is part of the membranous labyrinth, it is filled with endolymph. The scala vestibuli and the scala tympani, both part of the bony labyrinth, contain perilymph. The perilymph-containing chambers are continuous with each other at the cochlear apex, a region called the **helicotrema** (hel"ĭ-ko-tre'mah; "the hole in the spiral").

The "roof" of the cochlear duct, separating the scala media from the scala vestibuli, is the **vestibular membrane** (see Figure 13.21b). Its external wall, the **stria vascularis,** is composed of an unusual richly vascularized mucosa that secretes endolymph. The "floor" of the cochlear duct is composed of the bony spiral lamina and the flexible, fibrous **basilar membrane,** which supports the organ of Corti. (We will describe the organ of Corti when we discuss the mechanism of hearing.) The basilar membrane, which plays a critical role in sound reception, is narrow and thick near the oval window and gradually widens and thins as it approaches the cochlear apex. The *cochlear nerve,* a division of the *vestibulocochlear* (eighth cranial) *nerve,* runs from the organ of Corti through the modiolus on its way to the brain.

Physiology of Hearing

Overview: Properties of Sound

Light can be transmitted through a vacuum (for instance, outer space), but sound depends on an *elastic* medium for its transmission. **Sound** is a pressure disturbance—alternating areas of high and low pressure—produced by a vibrating object and propagated by the molecules of the medium. Consider a sound arising from a vibrating tuning fork. If the tuning fork is struck on the left, its prongs will move first to the right, creating an area of high pressure by compressing the air molecules there. Then, as the prongs rebound to the left, the air on the left will be compressed, and the region on the right will be a rarefied, or low-pressure, area (since most of its air molecules have been pushed farther to the right). As the fork vibrates alternately from right to left, it produces a series of compressions and rarefactions, collectively called a *sound wave,* which moves outward in all directions. However, the individual air molecules just vibrate back and forth for short distances as they bump other molecules and rebound. Because the outward-moving molecules give up kinetic energy to the molecules they bump, energy is always transferred in the direction the sound wave is traveling. Thus, with time and distance, the energy of the wave declines, and the sound dies a natural death.

We can illustrate a sound wave as an S-shaped curve, or *sine wave,* in which the compressed areas are crests and the rarefied areas are troughs (Figure 13.22b). Sound can be described in terms of two physical properties inferred from this sine wave graph: frequency and amplitude.

Frequency is defined as the number of waves that pass a given point in a given time. The sine wave of a pure tone is *periodic;* that is, its crests and troughs repeat at definite distances. The distance between two consecutive crests (or troughs) is called the **wavelength** of the sound and is constant for a particular tone. The shorter the wavelength, the higher the frequency of the sound (Figure 13.22a).

The frequency range of human hearing is from 20 to 20,000 waves per second, or *hertz (Hz).* Our ears are most sensitive to frequencies between 1500 and 4000 Hz and, in that range, we can distinguish frequencies differing by only 2–3 Hz. We perceive different sound frequencies as differences in **pitch:** The higher the frequency, the higher the pitch.

The **amplitude,** or height, of the sine wave crests reveals a sound's intensity, which is related to its energy, or the pressure differences between its compressed and rarefied areas (Figure 13.22b).

Loudness refers to our subjective interpretation of sound intensity. Because we can hear such an enormous range of intensities, from the proverbial pin drop to a steam whistle 100 trillion times as loud, sound intensity (and loudness) is measured in logarithmic units called **decibels (dB)** (des'ĭ-belz). On a clinical audiometer, the decibel scale is arbitrarily set to begin at 0 dB, which is the threshold of hearing (barely audible sound) for normal ears. Each 10-dB increase represents a tenfold increase in sound intensity. A sound of 10-dB has ten times more energy than one of 0 dB, and a 20 dB sound has

Which scala contains endolymph?

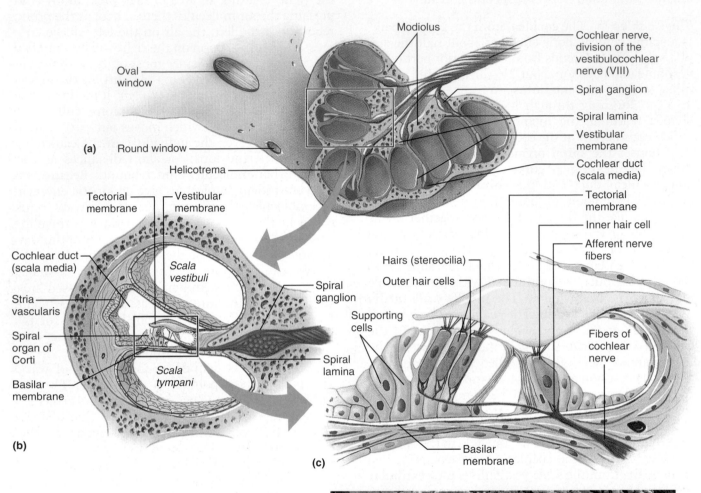

Modiolus

Oval window

Round window

Helicotrema

(a)

Cochlear nerve, division of the vestibulocochlear nerve (VIII)

Spiral ganglion

Spiral lamina

Vestibular membrane

Cochlear duct (scala media)

Tectorial membrane

Vestibular membrane

Cochlear duct (scala media)

Stria vascularis

Spiral organ of Corti

Basilar membrane

Scala vestibuli

Scala tympani

Spiral ganglion

Spiral lamina

(b)

Hairs (stereocilia)

Outer hair cells

Supporting cells

Tectorial membrane

Inner hair cell

Afferent nerve fibers

Fibers of cochlear nerve

Basilar membrane

(c)

(d)

FIGURE 13.21   Anatomy of the cochlea. (a) A section through the cochlea. **(b)** Magnified cross-sectional view of one turn of the cochlea, showing the relationship of the three scalae. The scalae vestibuli and tympani contain perilymph; the cochlear duct (scala media) contains endolymph. **(c)** Detailed structure of the spiral organ of Corti. **(d)** Electron micrograph of cochlear hair cells (2000×).

■ *The scala media.*

100 times (10 × 10) more energy than one of 0 dB. However, the same 10-dB increase represents only a twofold increase in loudness. In other words, most people would report that a 20-dB sound seems about twice as loud as a 10-dB sound. The normal range of hearing (from barely audible to the loudest sound we can process without excruciating pain) extends over a range of 120 dB. (The threshold of pain is 130 dB.)

Severe hearing loss occurs with frequent or prolonged exposure to sounds with intensities greater than 90 dB, and in the U.S., employees exposed to occupational noise over that range must wear ear (hearing) protection. This figure becomes more meaningful when you realize that a normal conversation is in the 50-dB range, a noisy restaurant has 70-dB levels, and amplified rock music is 120 dB or more, far above the 90-dB danger zone.

Transmission of Sound to the Inner Ear

Hearing occurs when the auditory area of the temporal lobe cortex is stimulated. However, before this can happen, sound waves must be propagated through air, membranes, bones, and fluids to reach and stimulate receptor cells in the organ of Corti.

Airborne sound entering the external auditory canal strikes the tympanic membrane and sets it vibrating at the same frequency. The greater the intensity, the farther the membrane is displaced in its vibratory motion. The motion of the tympanic membrane is amplified and transferred to the oval window by the ossicle lever system, which acts much like a hydraulic press or piston to transfer the same total force hitting the eardrum to the oval window. Because the tympanic membrane is 17 to 20 times larger than the oval window, the pressure (force per unit area) actually exerted on the oval window is about 20 times that on the tympanic membrane.

As the stapes rocks back and forth against the oval window, it sets the perilymph in the scala vestibuli into a similar back-and-forth motion, and a pressure wave travels through the perilymph from the basal end toward the helicotrema, much as a piece of rope held horizontally can be set into wave motion by movements initiated at one end. Sounds of very low frequency (below 20 Hz) create pressure waves that take the complete route through the cochlea—up the scala vestibuli, around the helicotrema, and back toward the round window through the scala tympani (Figure 13.23a). Such sounds do not activate the organ of Corti (are below the threshold of hearing). But sounds of higher frequency (and shorter wavelengths) create pressure waves that take a "shortcut" and are transmitted through the cochlear duct into the perilymph of the scala tympani. Fluids are incompressible. Think of what happens when you sit on a water bed—you sit

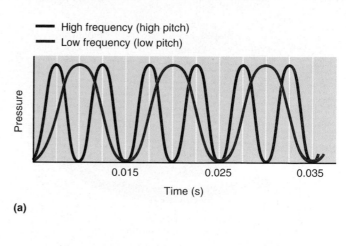

(a)

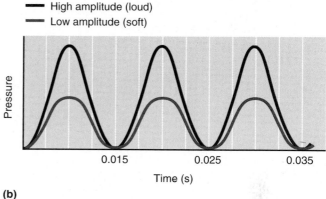

(b)

FIGURE 13.22 **Frequency and amplitude of sound waves. (a)** The wave shown in red has a shorter wavelength, and thus a greater frequency, than that shown in blue. The frequency of sound is perceived as pitch. **(b)** The wave shown in purple has a greater amplitude (intensity) than that shown in green and is perceived as being louder.

"here" and it bulges over "there." Thus, each time the fluid adjacent to the oval window is forced medially by the stapes, the membrane of the round window bulges laterally into the middle ear cavity and acts as a pressure valve.

As a pressure wave descends through the flexible cochlear duct, it sets the entire basilar membrane into vibrations. Maximal displacement of the membrane occurs where the fibers of the basilar membranes are "tuned" to a particular sound frequency (Figure 13.23b). (This characteristic of many natural substances is called *resonance*.) The fibers of the basilar membrane span its width like the strings of a harp. The fibers near the oval window (cochlear base) are short and stiff, and they resonate in response to high-frequency pressure waves (Figure 13.23b). The longer, more floppy basilar membrane fibers near the cochlear apex resonate in time with lower-frequency pressure waves. Thus, sound signals are mechanically processed by the resonance of the basilar membrane, before ever reaching the receptors.

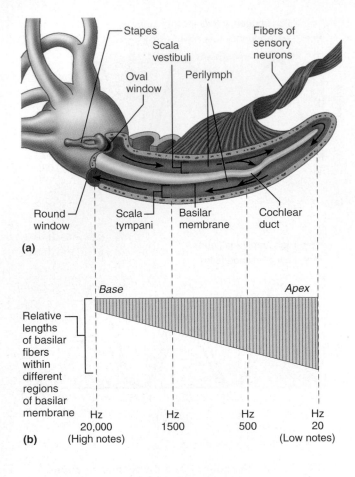

(a)

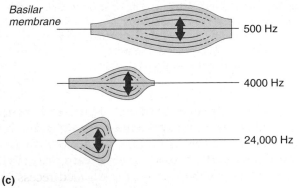

Relative lengths of basilar fibers within different regions of basilar membrane

| Hz 20,000 (High notes) | Hz 1500 | Hz 500 | Hz 20 (Low notes) |

(b)

Basilar membrane

500 Hz

4000 Hz

24,000 Hz

(c)

FIGURE 13.23   Resonance of the basilar membrane and activation of the cochlear hair cells. (a) The cochlea is depicted as if uncoiled to make the events of sound transmission occurring there easier to follow. Sound waves of low frequency that are below the level of hearing travel around the helicotrema without exciting hair cells. But sounds of higher frequency result in pressure waves that penetrate through the cochlear duct and set up vibrations in the basilar membrane, causing it to vibrate maximally in certain areas in response to certain frequencies of sound. **(b)** Fibers span the width of the basilar membrane. These fibers vary in length, being shorter near the base of the membrane and longer near its apex. The length of the fibers "tunes" specific regions of the basilar membrane to vibrate at specific frequencies. **(c)** Different frequencies of pressure waves in the cochlea cause certain places along the basilar membrane to vibrate, stimulating particular hair cells and sensory neurons. The differential stimulation of hair cells is perceived in the brain as sound of a certain pitch.

Excitation of Hair Cells in the Organ of Corti

The organ of Corti, which rests atop the basilar membrane, is composed of supporting cells and hearing receptor cells called *cochlear hair cells.* The hair cells are arranged functionally—specifically, one row of **inner hair cells** and three rows of **outer hair cells**—sandwiched between the tectorial and basilar membranes. Afferent fibers of the **cochlear nerve** [a division of the vestibulocochlear nerve (VIII)] are coiled about the bases of the hair cells. The "hairs" (stereocilia) of the hair cells are stiffened by actin filaments and linked together by fine fibers called *tip-links.* They protrude into the K^+-rich endolymph, and the longest of them are enmeshed in the overlying, gel-like **tectorial membrane** (see Figure 13.21c).

Transduction of sound stimuli occurs after the trapped stereocilia of the hair cells are "tweaked" or deflected by localized movements of the basilar membrane. Bending the cilia toward the tallest cilium puts tension on the tip-links which in turn opens cation channels in the adjacent shorter stereocilia. This results in an inward K^+ (and Ca^{2+}) current and a graded depolarization (receptor potential). Bending the cilia away from the tallest cilium relaxes the tip-links, closes the mechanically gated ion channels, and allows repolarization and even a graded hyperpolarization. Depolarization and the rise in intracellular Ca^{2+} increases the hair cells' release of neurotransmitter (probably glutamate), which causes the afferent cochlear fibers to transmit a faster stream of impulses to the brain for auditory interpretation. Hyperpolarization produces the exact opposite effect. As you might guess, activation of the hair cells occurs at points of vigorous basilar membrane vibration.

Although the outer hair cells are much more numerous, they instigate relatively few action potentials. In fact, 90–95% of the sensory fibers of the spiral ganglion service the inner hair cells, which shoulder nearly the entire responsibility for sending auditory messages to the brain. By contrast, most fibers coiling around the outer hair cells are *efferent* fibers that convey messages from brain to ear. What is the brain "telling" the outer hair cells? Sound activates a feedback loop from the outer hair cells to the brain stem and back that causes these hair cells to stretch and contract (on hyperpolarization and depolarization respectively)—in a type of cellular boogie called *fast motility.* Their strange behavior changes the stiffness of the basilar membrane, which alters its motion and amplifies the responsiveness of the inner hair cells—kind of a cochlear tuning. Outer hair cell mobility is also responsible for producing ear sounds *(otoacoustic emissions)* that are so audible in some people that they can actually be heard by others.

The Auditory Pathway to the Brain

Impulses generated in the cochlea pass through the **spiral ganglion,** where the bipolar sensory neurons for the sense of hearing reside, and along the afferent fibers of the cochlear nerve to the **cochlear nuclei** of the medulla. From there, impulses are sent on to the **superior olivary nucleus** and then via the **lateral lemniscal tract** to the **inferior colliculus** (auditory reflex center in the midbrain) and the **auditory cortex** (which provides for conscious awareness of sound) in the temporal lobe. Relays also communicate with the **medial geniculate body** of the thalamus and the **superior colliculus** of the midbrain. The colliculi act together to initiate auditory reflexes to sound, such as the startle reflex and head turning. The auditory pathway is unusual in that not all of the fibers from each ear decussate (cross over); therefore each auditory cortex receives impulses from both ears.

Cortical processing of sound stimuli is complex. For example, certain cortical cells depolarize at the beginning of a particular tone, others depolarize when the tone ends. Some cortical cells depolarize continuously, others appear to have high thresholds (low sensitivity), and so on. Here we will concentrate on the more straightforward aspects of cortical perception of pitch, loudness, and sound location.

Perception of Pitch   As explained, hair cells in different parts of the organ of Corti are activated by sound waves of different frequencies and impulses from specific hair cells are interpreted as specific pitches. When the sound is composed of tones of many frequencies, several populations of cochlear hair cells and cortical cells are activated simultaneously, resulting in the perception of multiple tones.

Detection of Loudness   Our perception of loudness suggests that certain cochlear cells have higher thresholds than others for responding to a tone of the same frequency. As the intensity of the sound increases, the basilar membrane vibrates more vigorously. As a result, more hair cells would begin to respond and more impulses would reach the auditory cortex and be recognized as a louder sound of the same pitch.

Localization of Sound   Several brain stem nuclei (most importantly the superior olivary nuclei) process signals that help us localize a sound's source in space by means of two cues: the *relative intensity* and the *relative timing* of sound waves reaching the two ears. If the sound source is directly in front, in back, or over the midline of the head, the intensity and timing cues are the same for both ears. However, when sound comes from one side, the receptors of the nearer ear are activated slightly earlier and more vigorously (because of the greater intensity of the sound waves entering that ear).

Homeostatic Imbalances of Hearing

Deafness

Any hearing loss, no matter how slight, is **deafness** of some sort. Deafness is classified as conduction or sensorineural deafness, according to its cause.

Conduction deafness occurs when something hampers sound conduction to the fluids of the inner ear, for example, impacted earwax blocking the auditory canal or a *perforated (ruptured)* eardrum, which prevents sound conduction from the eardrum to the ossicles. But the most common causes of conduction deafness are middle ear inflammations (otitis media) and **otosclerosis** (o"to-sklĕ-ro'sis) of the ossicles. Otosclerosis ("hardening of the ear"), a common age-related problem, occurs when overgrowth of bony tissue fuses the stapes foot plate to the oval window or welds the ossicles to one another. In such cases, sound is conducted to the receptors of that ear through vibrations of the skull bones, which is far less satisfactory. Otosclerosis is treated surgically.

Sensorineural deafness results from damage to neural structures at any point from the cochlear hair cells to and including the auditory cortical cells. This type of deafness typically results from the gradual loss of the hearing receptor cells throughout life. These cells can also be destroyed at an earlier age by a single explosively loud noise or prolonged exposure to high-intensity sounds, such as rock bands or airport noise, which causes them to stiffen or tears their cilia. Degeneration of the cochlear nerve, cerebral infarcts, and tumors in the auditory cortex are other causes. For age- or noise-related cochlear damage, cochlear implants (devices that convert sound energy into electrical signals) can be inserted into a drilled recess in the temporal bone.

Tinnitus

Tinnitus (tĭ-ni'tus) is a ringing or clicking sound in the ears in the absence of auditory stimuli. It is more a symptom of pathology than a disease. For example, tinnitus is one of the first symptoms of cochlear nerve degeneration. It may also result from inflammation of the middle or inner ears and is a side effect of some medications, such as aspirin.

Ménière's Syndrome

Classic **Ménière's syndrome** (men"ĕ-ār'z) is a labyrinth disorder that affects both the semicircular canals and the cochlea. The afflicted person has repeated attacks of vertigo, nausea, and vomiting.

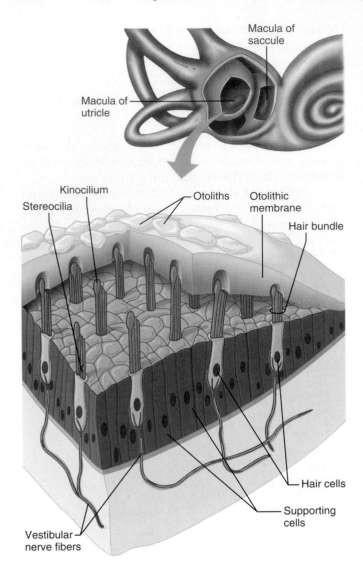

FIGURE 13.24 Structure of a macula. The "hairs" of the receptor cells of a macula project into the gelatinous otolithic membrane. Vestibular nerve fibers surround the base of the hair cells.

Balance is so disturbed that standing erect is nearly impossible. A "howling" tinnitus is common, so hearing is impaired (and ultimately lost) as well. The cause is uncertain, but it may result from distortion of the membranous labyrinth by excessive endolymph accumulation, or rupture of membranes which allows the perilymph and endolymph to mix, an event that generates ototoxins (chemicals poisonous to the ear). Mild cases can usually be managed by antimotion drugs or a low-salt diet and diuretics to decrease endolymph fluid volume. In severe cases, draining the excess endolymph from the inner ear may help. A last resort is removal of the entire malfunctioning labyrinth, which is usually deferred until hearing loss is complete.

Mechanisms of Equilibrium and Orientation

The equilibrium sense is not easy to describe because it does not "see," "hear," or "feel," but responds (frequently without our awareness) to various head movements. Under normal conditions the equilibrium receptors in the semicircular canals and vestibule, collectively called the **vestibular apparatus,** send signals to the brain that initiate reflexes needed to make the simplest changes in position as well as to serve a tennis ball precisely to the right spot (your opponent's backhand). The equilibrium receptors of the inner ear can be divided into two functional arms, with the receptors in the vestibule and semicircular canals monitoring **static** and **dynamic equilibrium,** respectively.

The Maculae and Static Equilibrium

The sensory receptors for static equilibrium are the **maculae** (mak'u-le; "spots"), one in each saccule wall and one in each utricle wall. These receptors monitor the position of the head in space, and in so doing, they play a key role in control of posture. They respond to *linear* acceleration forces, that is, straight-line changes in speed and direction, but not to rotation.

Anatomy of the Maculae   Each macula is a flat epithelial patch containing **supporting cells** and scattered receptor cells called **hair cells** (Figure 13.24). The hair cells have numerous *stereocilia* (actually long microvilli) and a single *kinocilium* (a true cilium) protruding from their apices. These "hairs" are embedded in the overlying **otolithic membrane** (o'to-lith-ik), a jellylike mass studded with tiny stones (calcium carbonate crystals) called **otoliths** ("ear stones"). The otoliths, though small, are dense and they increase the membrane's weight and its inertia (resistance to change in motion). In the utricle, the macula is horizontal (Figure 13.24), and the hairs are vertically oriented when the head is upright. In the saccule, the macula is nearly vertical, and the hairs protrude horizontally into the otolithic membrane. The saccular maculae respond best to vertical movements, such as the sudden acceleration of an elevator. The receptor cells release neurotransmitter continually, but movement of their hairs modifies the amount they release, causing an increase or decrease in the rate of impulse generation by the **vestibular nerve** endings coiling around their bases. [Like the cochlear nerve, the vestibular nerve is a subdivision of the vestibulocochlear nerve (VIII)]. The cell bodies of the sensory neurons are located in the nearby **superior** and **inferior vestibular ganglia.**

Activating Maculae Receptors Let's look more closely at the happenings in the maculae that lead to sensory transduction. When your head starts or stops moving in a linear direction, inertia causes the otolithic membrane to slide backward or forward like a greased plate over the hair cells, bending the hairs. For example, when you run, the otolithic membranes of the utricle maculae lag behind, bending the hairs backward. When you suddenly stop, the otolithic membrane slides abruptly forward (just as you slide forward in your car when you brake), bending the hair cells forward. Likewise, when you nod your head or fall, the otoliths roll inferiorly, bending the hairs of the maculae in the saccules. Fibers of the receptor cells release neurotransmitter continuously but movement of their hairs modifies the amount they release. When the hairs are bent *toward the kinocilium*, the hair cells depolarize, stepping up their pace of neurotransmitter release, and a faster stream of impulses is sent to the brain. When the hairs are bent in the opposite direction, the receptors hyperpolarize, and neurotransmitter release and impulse generation decline. In either case, the brain is informed of the changing position of the head in space.

It is important to understand that the maculae respond *only to changes* in acceleration or velocity of head movement. Because the hair cells adapt quickly (resuming their basic level of neurotransmitter release), they do not report on unchanging head positions. Thus, the maculae help us to maintain normal head position with respect to gravity.

The Crista Ampullaris and Dynamic Equilibrium

The receptor for dynamic equilibrium, called the **crista ampullaris,** or simply *crista*, is a minute elevation in the ampulla of each semicircular canal (see Figure 13.25). Like the maculae, the cristae are excited by head movement (acceleration and deceleration), but in this case the major stimuli are rotatory (angular) movements. When you twirl on the dance floor or suffer through a rough boat ride, these gyroscope-like receptors are working overtime.

Anatomy of the Crista Ampullaris Each crista is composed of supporting cells and hair cells. These hair cells, like those of the maculae, have stereocilia plus one kinocilium that project into a gel-like mass. In this case, the gelled mass is a **cupula** (ku′pu-lah), which resembles a pointed cap (Figure 13.25b and c). The cupula is a delicate, loosely organized network of gelatinous strands that radiate outward to contact the "hairs" of each hair cell. Dendrites of vestibular nerve fibers encircle the base of the hair cells.

Activating Crista Ampullaris Receptors The cristae respond to *changes* in the velocity of rotatory movements of the head. Because of its inertia, the endolymph in the semicircular ducts moves briefly in the direction *opposite* the body's rotation, deforming the crista in the duct. As the hairs are bent, the hair cells depolarize and impulses reach the brain at a faster rate. Bending the cilia in the opposite direction causes hyperpolarization and reduces impulse generation. Because the axes of the hair cells in the complementary semicircular ducts are opposite, rotation in a given direction causes depolarization of the receptors in one ampulla of the pair, and hyperpolarization of the receptors in the other (Figure 13.25d).

If the body continues to rotate at a constant rate, the endolymph eventually comes to rest—it moves along at the same speed as the body and stimulation of the hair cells ends. Consequently, if we are blindfolded, we cannot tell whether we are moving at a constant speed or not moving at all after the first few seconds of rotation. However, when we suddenly stop moving, the endolymph keeps on going, in effect reversing its direction within the canal. This sudden reversal in the direction of hair bending results in membrane voltage changes in the receptor cells and modifies the rate of impulse transmission, which tells the brain that we have slowed or stopped.

The key point to remember when considering both types of equilibrium receptors is that the rigid bony labyrinth moves with the body, while the fluids (and gels) within the membranous labyrinth are free to move at various rates, depending on the forces (gravity, acceleration, and so on) acting on them.

Impulses transmitted from the semicircular canals are particularly important to reflex movements of the eyes. **Vestibular nystagmus** is a complex of rather strange eye movements that occurs during and immediately after rotation. As you rotate, your eyes slowly drift in the opposite direction, as though fixed on some object in the environment. This reaction relates to the backflow of endolymph in the semicircular canals. Then, because of CNS compensating mechanisms, the eyes jump rapidly toward the direction of rotation to establish a new fixation point. These alternating eye movements continue until the endolymph comes to rest. When you stop rotating, at first your eyes continue to move in the direction of the previous spin, and then they jerk rapidly in the opposite direction. This sudden change is caused by the change in the direction in which the cristae are bent after you stop. Nystagmus is often accompanied by vertigo.

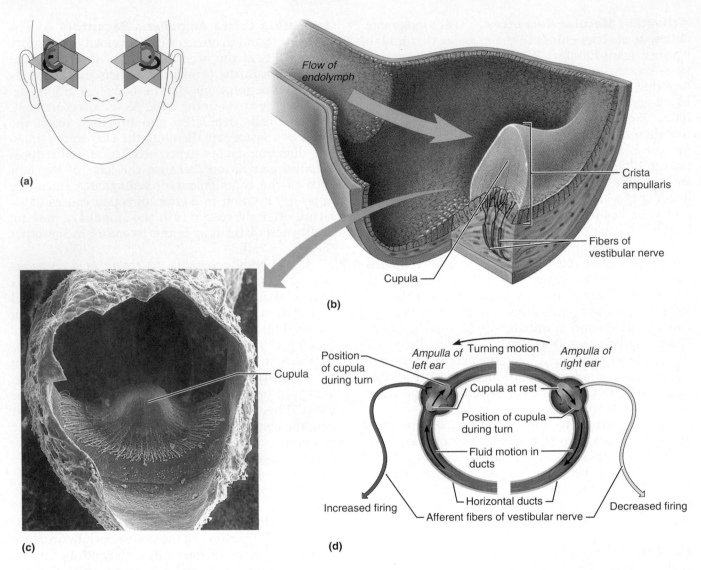

FIGURE 13.25 Location and structure of a crista ampullaris. (a) Location of the semicircular canals within the temporal bones of the skull. (b) The ampulla of a semicircular duct has been sectioned to show the position of a crista ampullaris receptor. (c) Scanning EM of a crista ampullaris (1000×). (d) View of the horizontal ducts from above shows how paired semicircular canals work together to provide bilateral information on rotatory head movement.

The Equilibrium Pathway to the Brain

Our responses to body imbalance, such as during stumbling, must be fast and reflexive. By the time we "thought about" correcting our fall, we would already be on the ground. Hence, information from the balance receptors goes directly to reflex centers in the brain stem, rather than to the cerebral cortex as with the other special senses. Impulses generated along the vestibular nerve (a subdivision of the vestibulocochlear nerve [VIII]), travel initially to one of two destinations: the **vestibular nuclear complex** in the brain stem or the **cerebellum.** The vestibular nuclei, the major integrative center for balance, also receive inputs from the visual and somatic receptors, particularly from proprioceptors in neck muscles that report on the position of the head. They integrate this information and then send commands to brain stem motor centers that control the extrinsic eye muscles (cranial nerve nuclei III, IV, and VI) and reflex movements of the neck, limb, and trunk muscles (via the vestibulospinal tracts). The ensuing reflex movements of the eyes and body allow us to remain focused on the visual field and to quickly adjust our body position to maintain or regain balance.

The cerebellum also integrates inputs from the eyes and somatic receptors (as well as from the cerebrum). It coordinates skeletal muscle activity and regulates muscle tone so that head position, posture, and balance are maintained, often in the face of rapidly changing inputs.

Notice that the vestibular apparatus does *not automatically compensate* for forces acting on the body. Its job is to send warning signals to the central nervous system, which initiates the appropriate compensations (righting) to keep your body balanced, your weight evenly distributed, and your eyes focused on what you were looking at when the disturbance occurred.

🇭 *HOMEOSTATIC IMBALANCE*

Responses to equilibrium signals are totally reflexive, and usually we are aware of vestibular apparatus activity only when its functioning is impaired. Equilibrium problems are usually obvious and unpleasant. Nausea, dizziness, and loss of balance are common and there may be nystagmus in the absence of rotational stimuli.

Motion sickness has been difficult to explain, but it appears to be due to sensory input mismatch. For example, if you are inside a ship during a storm, visual inputs indicate that your body is fixed with reference to a stationary environment (your cabin). But as the ship is tossed about by the rough seas, your vestibular apparatus detects movement and sends impulses that disagree with the visual information. The brain thus receives conflicting information, and its "confusion" somehow leads to motion sickness. Warning signals, which precede nausea and vomiting, include excessive salivation, pallor, rapid deep breathing, and profuse sweating. Removal of the stimulus usually ends the symptoms. Over-the-counter antimotion drugs, such as meclizine HCl (Bonine), depress vestibular inputs and help alleviate the symptoms. ●

PART 2: TRANSMISSION LINES: NERVES AND THEIR STRUCTURE AND REPAIR

Nerves and Associated Ganglia

Structure and Classification

A **nerve** is a cordlike organ that is part of the peripheral nervous system. Nerves vary in size, but every nerve consists of parallel bundles of peripheral axons (some myelinated and some not) enclosed by successive wrappings of connective tissue (Figure 13.26).

Within a nerve, each axon is surrounded by **endoneurium** (en"do-nu're-um), a delicate layer of loose connective tissue that also encloses the fiber's associated myelin sheath or neurilemma.

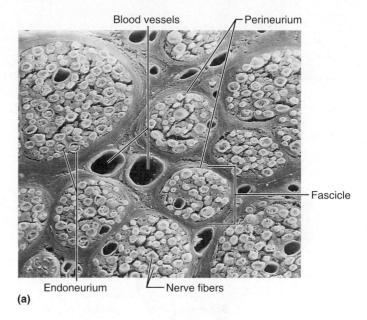

(a)

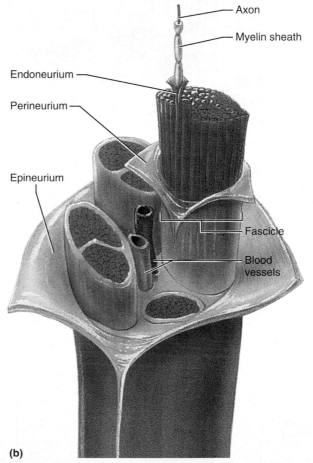

(b)

FIGURE 13.26   **Structure of a nerve. (a)** Scanning electron micrograph of a cross section of a portion of a nerve (500×). (© R. G. Kessel and R. H. Kardon, *Tissues and Organs: A Text-Atlas of Scanning Electron Microscopy,* W. H. Freeman and Company, 1979, all rights reserved.) **(b)** Three-dimensional view of a portion of a nerve, showing connective tissue wrappings.

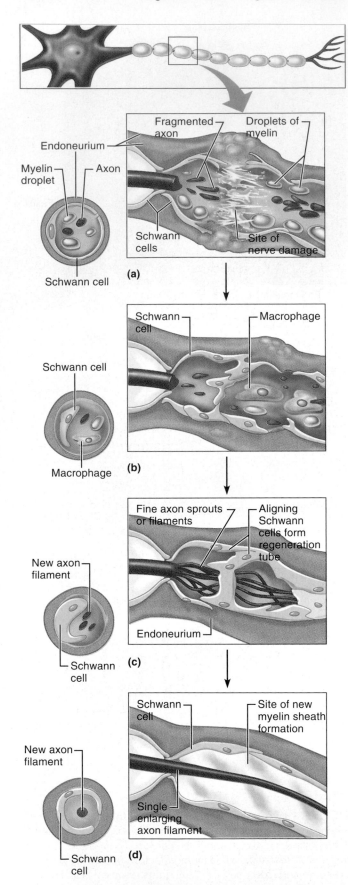

Groups of fibers are bound into bundles or **fascicles** by a coarser connective tissue wrapping, the **perineurium.** Finally, all the fascicles are enclosed by a tough fibrous sheath, the **epineurium,** to form the nerve. Neuron processes constitute only a small fraction of a nerve's bulk. The balance consists chiefly of myelin, the protective connective tissue wrappings, blood vessels, and lymphatic vessels.

Recall that the PNS is divided into *sensory* (afferent) and *motor* (efferent) divisions. Hence, nerves are classified according to the direction in which they transmit impulses. Nerves containing both sensory and motor fibers and transmitting impulses both to and from the central nervous system are called **mixed nerves.** Those that carry impulses only toward the CNS are **sensory (afferent) nerves;** and those carrying impulses only away from the CNS are **motor (efferent) nerves.** Most nerves are mixed. Purely sensory or motor nerves are rare.

Because mixed nerves often carry both somatic and autonomic (visceral) nervous system fibers, the fibers in them may be classified according to the region they innervate as *somatic afferent, somatic efferent, visceral afferent,* and *visceral efferent.*

For convenience, the peripheral nerves are classified as *cranial* or *spinal* depending on whether they arise from the brain or the spinal cord. Although autonomic efferents of cranial nerves are mentioned, this chapter focuses on somatic functions. The discussion of the autonomic nervous system and the visceral functions it serves is deferred to Chapter 14.

Ganglia are collections of neuron cell bodies associated with nerves in the PNS. Ganglia associated with *afferent* nerve fibers contain cell bodies of sensory neurons. (These are the *dorsal root ganglia* studied in Chapter 12.) Ganglia associated with *efferent* nerve fibers mostly contain cell bodies of autonomic motor neurons. These more complicated ganglia are described in Chapter 14.

Regeneration of Nerve Fibers

Damage to nervous tissue is serious because, as a rule, mature neurons do not divide. If the damage is severe or close to the cell body, the entire neuron may die, and other neurons that are normally stimulated by its axon may die as well. However, if the cell body remains intact, cut or compressed axons on peripheral nerves can regenerate successfully.

Almost immediately after a peripheral axon has been severed or crushed, the separated ends seal themselves off and then swell as substances being transported along the axon begin to accumulate in the sealed ends (Figure 13.27a). Within a few hours,

FIGURE 13.27   **Events leading to regeneration of nerve fibers in peripheral nerves.**

the axon and its myelin sheath distal to the injury site begin to disintegrate because they cannot receive nutrients from the cell body. This process, **Wallerian degeneration,** spreads distally from the injury site, completely fragmenting the axon (Figure 13.27b). Generally, the entire axon distal to the injury is degraded by phagocytes within a week, but the neurilemma remains intact within the endoneurium. After the debris has been disposed of, surviving Schwann cells proliferate in response to mitosis-stimulating chemicals released by the macrophages, and migrate into the injury site. Once there, they release growth factors and begin to express cell surface adhesion molecules (CAMs) that encourage axonal growth. Additionally, they form a *regeneration tube,* a system of cellular cords that guide the regenerating axon "sprouts" across the gap and to their original contacts (Figure 13.27c and d). The same Schwann cells protect, support, and remyelinate the regenerating axons.

Changes also occur in the neuronal cell body after axonal destruction. Within two days, its chromatophilic substance breaks apart, and then the cell body swells as protein synthesis revs up to support regeneration of its axon.

Axons regenerate at the approximate rate of 1.5 mm a day. The greater the distance between the severed endings, the less the chance of recovery because adjacent tissues block growth by protruding into the gaps, and axonal sprouts escape into surrounding areas. Neurosurgeons align cut nerve endings surgically to enhance the chance of successful regeneration, and scaffolding devices have been successful in guiding axon growth. Whatever the measures taken, post-trauma axon regrowth is never exactly the same as what existed before the injury, and much of the recovery protocol involves retraining the nervous system to respond appropriately so that stimulus and response are coordinated.

Unlike peripheral nerve fibers, most CNS fibers never regenerate. Consequently, damage to the brain or spinal cord has been viewed as irreversible. This difference in regenerative capacity seems to have less to do with the neurons themselves than with the "company they keep"—their supporting cells. After CNS injury, "cleanup" is much slower. Moreover, the oligodendrocytes surrounding the injured fibers die and fail to guide fiber regrowth. To make matters worse, the CNS is soaked through with growth-inhibiting proteins (Nogo and others) that cause growth cone collapse and repulsion. However, neutralization of the myelin-bound growth inhibitors by antibodies can induce substantial fiber regeneration, and dousing CNS neurons with growth factors can prompt mature neurons to divide and form new nervous tissue.

Cranial Nerves

Twelve pairs of **cranial nerves** are associated with the brain (Figure 13.28). The first two pairs attach to the forebrain; the rest originate from the brain stem. Other than the vagus nerves, which extend into the abdomen, cranial nerves serve only head and neck structures.

In most cases, the names of the cranial nerves reveal either the structures they serve or their functions. The nerves are also numbered (using Roman numerals) from the most rostral to the most caudal. A mini-introduction to the cranial nerves follows:

I. Olfactory. These are the tiny sensory nerves (filaments) of smell, which run from the nasal mucosa to synapse with the olfactory bulbs (Figure 13.28a).

II. Optic. Because this sensory nerve of vision develops as an outgrowth of the brain, it is really a brain tract.

III. Oculomotor. This nerve has a name meaning "eye mover" because it supplies four of the six extrinsic muscles that move the eyeball in the orbit.

IV. Trochlear. This nerve's name means "pulley" and it innervates an extrinsic eye muscle that loops through a pulley-shaped ligament in the orbit.

V. Trigeminal. Three *(tri)* branches spring from this, the largest of the cranial nerves. It supplies sensory fibers to the face and motor fibers to the chewing muscles.

VI. Abducens. This nerve controls the extrinsic eye muscle that *abducts* the eyeball (turns it laterally).

VII. Facial. A large nerve that innervates muscles of *facial* expression (among other things).

VIII. Vestibulocochlear. This sensory nerve for hearing and balance was formerly called the *auditory nerve.*

IX. Glossopharyngeal. Its name, meaning "tongue and pharynx," reveals the structures it helps to innervate.

X. Vagus. This nerve's name means "wanderer" or "vagabond," and it is the only cranial nerve to extend beyond the head and neck to the thorax and abdomen.

XI. Accessory. Considered an *accessory* part of the vagus nerve, this nerve was formerly called the *spinal accessory nerve.*

XII. Hypoglossal. This nerve runs inferior to the tongue and innervates some tongue-moving muscles, as reflected by its name meaning "under the tongue."

I suggest you make up your own saying to remember the first letters of the cranial nerves in order, or use the following memory jog sent to me by a student: "**O**n **o**ccasion, **o**ur **t**rusty **t**ruck **a**cts **f**unny—**v**ery **g**ood **v**ehicle **a**ny**h**ow."

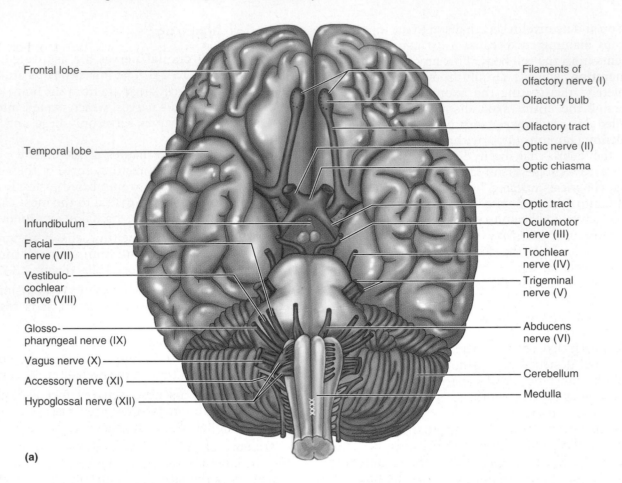

(a)

(b) *PS = parasympathetic

Cranial nerves I – VI	Sensory function	Motor function	PS* fibers
I Olfactory	Yes (smell)	No	No
II Optic	Yes (vision)	No	No
III Oculomotor	No	Yes	Yes
IV Trochlear	No	Yes	No
V Trigeminal	Yes (general sensation)	Yes	No
VI Abducens	No	Yes	No

Cranial nerves VII – XII	Sensory function	Motor function	PS* fibers
VII Facial	Yes (taste)	Yes	Yes
VIII Vestibulocochlear	Yes (hearing and balance)	No	No
IX Glossopharyngeal	Yes (taste)	Yes	Yes
X Vagus	Yes (taste)	Yes	Yes
XI Accessory	No	Yes	No
XII Hypoglossal	No	Yes	No

FIGURE 13.28 Location and function of cranial nerves. (a) Ventral view of the human brain, showing the cranial nerves. **(b)** Summary of cranial nerves by function. Three cranial nerves (I, II, and VIII) have sensory function only; no motor function. Four nerves (III, VII, IX, and X) carry parasympathetic fibers that serve visceral muscles and glands. All cranial nerves that have a motor function also carry afferent fibers from proprioceptors in the muscles served; only sensory functions other than proprioception are indicated.

In the last chapter, we described how spinal nerves are formed by the fusion of ventral (motor) and dorsal (sensory) roots. Cranial nerves, on the other hand, vary markedly in their composition. Most cranial nerves are mixed nerves as shown in Figure 13.28b. However, three nerve pairs (the olfactory, optic, and vestibulocochlear) are associated with special sense organs and are generally considered purely sensory. The cell bodies of the sensory neurons of the olfactory and optic nerves are located *within* their respective special sense organs. In all other cases of sensory neurons contributing to cranial nerves (V, VII, IX, and X) the cell bodies are located in **cranial sensory ganglia** just outside the brain. Some cranial nerves have a single sensory ganglion, others have several, and still others have none.

Several of the mixed cranial nerves contain both somatic and autonomic motor fibers and hence serve both skeletal muscles and visceral organs. Except for some autonomic motor neurons located in ganglia, the cell bodies of motor neurons contributing to the cranial nerves are located in the ventral gray matter regions (nuclei) of the brain stem.

Having read this overview, you are now ready to tackle Table 13.2, which provides a more detailed description of the origin, course, and function of the cranial nerves. Notice that the pathways of the purely sensory nerves (I, II, and VIII) are described from the receptors to the brain, while those of the other nerves are described in the opposite direction (from the brain distally). Remembering the functions of the cranial nerves (as sensory, motor, or both) can be a problem; this sentence might help: "**S**ome **s**ay **m**arry **m**oney, **b**ut **m**y **b**rother **s**ays (it's) **b**ad **b**usiness (to) **m**arry **m**oney."

Spinal Nerves

General Features

Thirty-one pairs of **spinal nerves,** each containing thousands of nerve fibers, arise from the spinal cord and supply all parts of the body except the head and some areas of the neck. All are mixed nerves. As

Text continues on page 481.

TABLE 13.2 **Cranial Nerves**

I The Olfactory Nerves (ol-fak'to-re)

Origin and course: Olfactory nerve fibers arise from olfactory receptor cells located in olfactory epithelium of nasal cavity and pass through cribriform plate of ethmoid bone to synapse in olfactory bulb. Fibers of olfactory bulb neurons extend posteriorly as olfactory tract, which runs beneath frontal lobe to enter cerebral hemispheres and terminates in primary olfactory cortex. See also Figure 13.3.
Function: Purely sensory; carry afferent impulses for sense of smell
Clinical testing: Person is asked to sniff aromatic substances, such as oil of cloves and vanilla, and to identify each.

🜨 Homeostatic imbalance: Fracture of ethmoid bone or lesions of olfactory fibers may result in partial or total loss of smell, a condition known as *anosmia* (an-oz'me-ah). ●

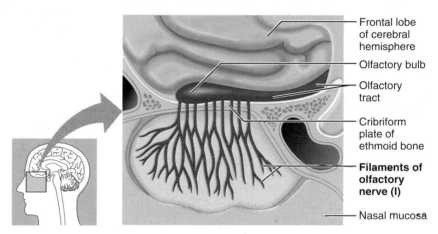

- Frontal lobe of cerebral hemisphere
- Olfactory bulb
- Olfactory tract
- Cribriform plate of ethmoid bone
- **Filaments of olfactory nerve (I)**
- Nasal mucosa

II The Optic Nerves

Origin and course: Fibers arise from retina of eye to form optic nerve, which passes through optic foramen of orbit. The optic nerves converge to form the optic chiasma (ki-az'mah) where fibers partially cross over, continue on as optic tracts, enter thalamus, and synapse there. Thalamic fibers run (as the optic radiation) to occipital (visual) cortex, where visual interpretation occurs. See also Figure 13.18.
Function: Purely sensory; carry afferent impulses for vision
Clinical testing: Vision and visual field are determined with eye chart and by testing the point at which the person first sees an object (finger) moving into the visual field. Fundus of eye viewed with ophthalmoscope to detect papilledema (swelling of optic disc, the site where the optic nerve leaves the eyeball), as well as for routine examination of the optic disc and retinal blood vessels.

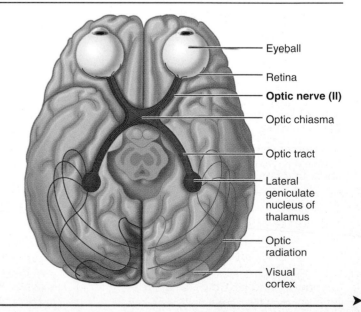

- Eyeball
- Retina
- **Optic nerve (II)**
- Optic chiasma
- Optic tract
- Lateral geniculate nucleus of thalamus
- Optic radiation
- Visual cortex

▶

TABLE 13.2 **Cranial Nerves** *(continued)*

II The Optic Nerves *(continued)*

Homeostatic imbalance: Damage to optic nerve results in blindness in eye served by nerve; damage to visual pathway beyond the optic chiasma results in partial visual losses; visual defects are called *anopsias* (ahnop'se-ahz). ●

III The Oculomotor Nerves *(ok"u-lo-mo'tor)*

Origin and course: Fibers extend from ventral midbrain (near its junction with pons) and pass through bony orbit, via superior orbital fissure, to eye.

Function: Chiefly motor nerves (*oculomotor* = motor to the eye); contain a few proprioceptive afferents; each nerve includes:

■ Somatic motor fibers to four of the six extrinsic eye muscles (inferior oblique and superior, inferior, and medial rectus muscles) that help direct eyeball, and to levator palpebrae superioris muscle, which raises upper eyelid

■ Parasympathetic (autonomic) motor fibers to constrictor muscles of iris, which cause pupil to constrict, and to ciliary muscle, controlling lens shape for visual focusing; some parasympathetic cell bodies are in the ciliary ganglia

■ Sensory (proprioceptor) afferents, which run from same four extrinsic eye muscles to midbrain

Clinical testing: Pupils are examined for size, shape, and equality. Pupillary reflex is tested with penlight (pupils should constrict when illuminated). Convergence for near vision is tested, as is subject's ability to follow objects with the eyes.

Homeostatic imbalance: In oculomotor nerve paralysis, eye cannot be moved up, down, or inward, and at rest, eye rotates laterally [*external strabismus* (strahbiz'mus)] because the actions of the two extrinsic eye muscles not served by cranial nerve III are unopposed; upper eyelid droops (*ptosis*), and the person has double vision and trouble focusing on close objects. ●

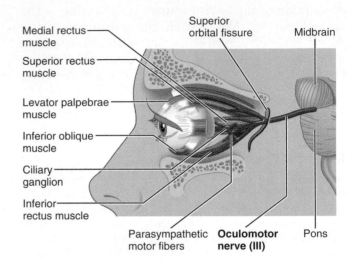

IV The Trochlear Nerves *(trok'le-ar)*

Origin and course: Fibers emerge from dorsal midbrain and course ventrally around midbrain to enter orbits through superior orbital fissures along with oculomotor nerves.

Function: Primarily motor nerves; supply somatic motor fibers to (and carry proprioceptor fibers from) one of the extrinsic eye muscles, the superior oblique muscle

Clinical testing: Tested in common with cranial nerve III.

Homeostatic imbalance: Trauma to, or paralysis of, a trochlear nerve results in double vision and reduced ability to rotate eye inferolaterally. ●

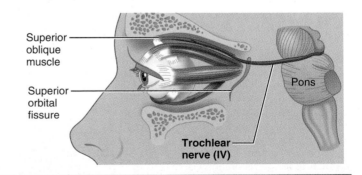

TABLE 13.2 *(continued)*

V The Trigeminal Nerves

Largest of cranial nerves; fibers extend from pons to face, and form three divisions (*trigemina* = threefold): ophthalmic, maxillary, and mandibular divisions. As major general sensory nerves of face, transmit afferent impulses from touch, temperature, and pain receptors. Cell bodies of sensory neurons of all three divisions are located in large *trigeminal* (also called *semilunar* or *gasserian*) *ganglion*. The mandibular division also contains some motor fibers that innervate chewing muscles.

Dentists desensitize upper and lower jaws by injecting local anesthetic (such as Novocain) into alveolar branches of maxillary and mandibular divisions, respectively; since this blocks pain-transmitting fibers of teeth, the surrounding tissues become numb.

	Ophthalmic division (V_1)	Maxillary division (V_2)	Mandibular division (V_3)
Origin and course	Fibers run from face to pons via superior orbital fissure	Fibers run from face to pons via foramen rotundum	Fibers pass through skull via foramen ovale
Function	Conveys sensory impulses from skin of anterior scalp, upper eyelid, and nose, and from nasal cavity mucosa, cornea, and lacrimal gland	Conveys sensory impulses from nasal cavity mucosa, palate, upper teeth, skin of cheek, upper lip, lower eyelid	Conveys sensory impulses from anterior tongue (except taste buds), lower teeth, skin of chin, temporal region of scalp; supplies motor fibers to, and carries proprioceptor fibers from, muscles of mastication
Clinical testing	Corneal reflex tested: touching cornea with wisp of cotton should elicit blinking	Sensations of pain, touch, and temperature are tested with safety pin and hot and cold objects	Motor branch assessed by asking person to clench his teeth, open mouth against resistance, and move jaw side to side

🔴 **Homeostatic imbalance:** *Tic douloureux* (tik doo"looroo'), or *trigeminal neuralgia* (nu-ral'je-ah), caused by inflammation of trigeminal nerve, is widely considered to produce most excruciating pain known; the stabbing pain lasts for a few seconds to a minute, but it can be relentless, occurring a hundred times a day (the term *tic* refers to victim's wincing during pain). Usually provoked by some sensory stimulus, such as brushing teeth or even a passing breeze hitting the face. It seems to be caused by pressure on the trigeminal nerve root. Analgesics and carbamazepine (an anticonvulsant) are only partially effective. In severe cases, nerve is cut proximal to trigeminal ganglion; this relieves the agony, but also results in loss of sensation on that side of face. ●

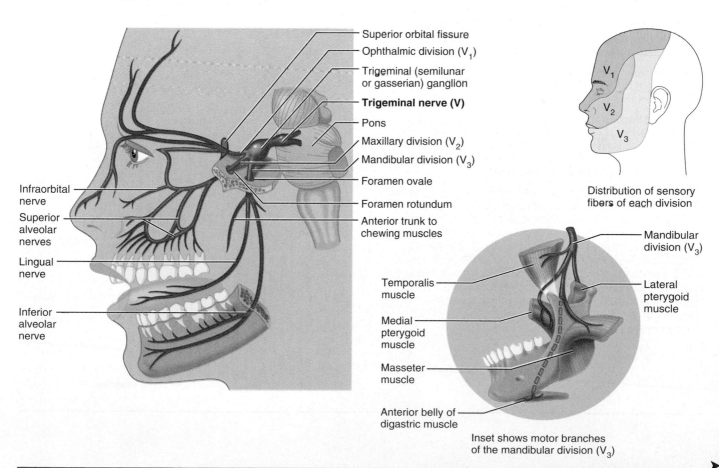

Superior orbital fissure
Ophthalmic division (V_1)
Trigeminal (semilunar or gasserian) ganglion
Trigeminal nerve (V)
Pons
Maxillary division (V_2)
Mandibular division (V_3)
Foramen ovale
Foramen rotundum
Anterior trunk to chewing muscles

Infraorbital nerve
Superior alveolar nerves
Lingual nerve
Inferior alveolar nerve

Distribution of sensory fibers of each division

Mandibular division (V_3)
Lateral pterygoid muscle
Temporalis muscle
Medial pterygoid muscle
Masseter muscle
Anterior belly of digastric muscle

Inset shows motor branches of the mandibular division (V_3)

▶

TABLE 13.2 **Cranial Nerves** (continued)

VI The Abducens Nerves (ab-du'senz)

Origin and course: Fibers leave inferior pons and enter orbit via superior orbital fissure to run to eye.

Function: Mixed nerve, primarily motor. Supplies somatic motor fibers to lateral rectus muscle, an extrinsic muscle of the eye, and conveys proprioceptor impulses from same muscle to brain.

Clinical testing: Tested in common with cranial nerve III.

⊞ Homeostatic imbalance: In abducens nerve paralysis, eye cannot be moved laterally; at rest, affected eyeball rotates medially (internal strabismus). ●

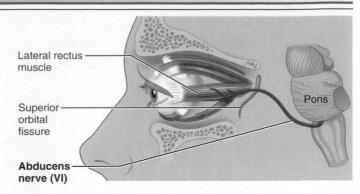

VII The Facial Nerves

Origin and course: Fibers issue from pons, just lateral to abducens nerves (see Figure 13.28), enter temporal bone via internal acoustic meatus, and run within bone (and through inner ear cavity) before emerging through stylomastoid foramen; nerve then courses to lateral aspect of face.

Function: Mixed nerves that are the chief motor nerves of face; have five major branches: temporal, zygomatic, buccal, mandibular, and cervical (see **c**)

- Convey motor impulses to skeletal muscles of face (muscles of facial expression), except for chewing muscles served by trigeminal nerves, and transmit proprioceptor impulses from same muscles to pons (see **b**)

- Transmit parasympathetic (autonomic) motor impulses to lacrimal (tear) glands, nasal and palatine glands, and submandibular and sublingual salivary glands. Some of the cell bodies of these parasympathetic motor neurons are in pterygopalatine (ter"eh-go-pal'ah-tīn) and submandibular ganglia on the trigeminal nerve (see **a**)

- Convey sensory impulses from taste buds of anterior two-thirds of tongue; cell bodies of these sensory neurons are in geniculate ganglion (see **a**)

Clinical testing: Anterior two-thirds of tongue is tested for ability to taste sweet (sugar), salty, sour (vinegar), and bitter (quinine) substances. Symmetry of face is checked. Subject is asked to close eyes, smile, whistle, and so on. Tearing is assessed with ammonia fumes.

⊞ Homeostatic imbalance: Bell's palsy, characterized by paralysis of facial muscles on affected side and partial loss of taste sensation, may develop rapidly (often overnight). Caused by herpes simplex I viral infection, which causes swelling and inflammation of facial nerve. Lower eyelid droops, corner of mouth sags (making it difficult to eat or speak normally), tears drip continuously from eye and eye cannot be completely closed (conversely, dry-eye syndrome may occur). Condition may disappear spontaneously without treatment. ●

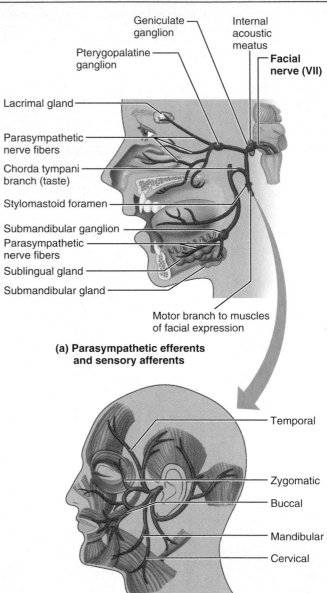

(a) Parasympathetic efferents and sensory afferents

(b) Motor branches to muscles of facial expression and scalp muscles

TABLE 13.2 *(continued)*

VII *The Facial Nerves (continued)*

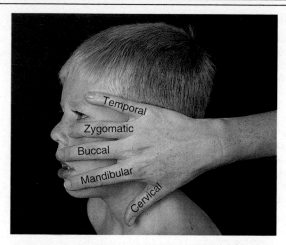

(c) A simple method of remembering the courses of the five major motor nerves of the face

VIII *The Vestibulocochlear Nerves* (ves-tib"u-lo-kok'le-ar)

Origin and course: Fibers arise from hearing and equilibrium apparatus located within inner ear of temporal bone and pass through internal acoustic meatus to enter brain stem at pons-medulla border. Afferent fibers from hearing receptor in cochlea form the *cochlear division;* those from equilibrium receptors in semicircular canals and vestibule form the *vestibular division* (vestibular nerve); the two divisions merge to form vestibulocochlear nerve.

Function: Purely sensory. Vestibular branch transmits afferent impulses for sense of equilibrium, and sensory nerve cell bodies are located in *vestibular ganglia.* Cochlear branch transmits afferent impulses for sense of hearing, and sensory nerve cell bodies are located in *spiral ganglion* within cochlea.

Clinical testing: Hearing is checked by air and bone conduction using tuning fork.

⚕ Homeostatic imbalance: Lesions of cochlear nerve or cochlear receptors result in *central* or *nerve deafness,* whereas damage to vestibular division produces dizziness, rapid involuntary eye movements, loss of balance, nausea, and vomiting. ●

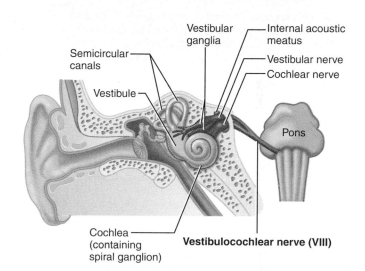

IX *The Glossopharyngeal Nerves* (glos"o-fah-rin'je-al)

Origin and course: Fibers emerge from medulla and leave skull via *jugular foramen* to run to throat.

Function: Mixed nerves that innervate part of tongue and pharynx. Provide motor fibers to, and carry proprioceptor fibers from, a superior pharyngeal muscle called the *stylopharyngeus,* which elevates the pharynx in swallowing. Provide parasympathetic motor fibers to parotid salivary gland (some of the nerve cell bodies of these parasympathetic motor neurons are located in *otic ganglion*).

TABLE 13.2   Cranial Nerves (continued)

IX The Glossopharyngeal Nerves (continued)

Sensory fibers conduct taste and general sensory (touch, pressure, pain) impulses from pharynx and posterior tongue, from chemoreceptors in the carotid body (which monitor O_2 and CO_2 levels in the blood and help regulate respiratory rate and depth), and from pressure receptors of carotid sinus (which help to regulate blood pressure by providing feedback information). Sensory neuron cell bodies are located in *superior* and *inferior ganglia.*

Clinical testing: Position of the uvula is checked. Gag and swallowing reflexes are checked. Subject is asked to speak and cough. Posterior third of tongue may be tested for taste.

Homeostatic imbalance: Injury or inflammation of glossopharyngeal nerves impairs swallowing and taste, particularly for sour and bitter substances. ●

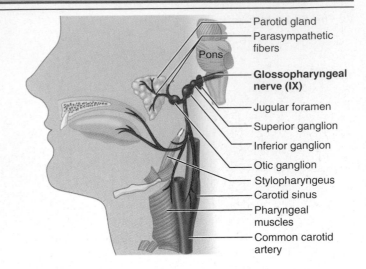

- Parotid gland
- Parasympathetic fibers
- Pons
- **Glossopharyngeal nerve (IX)**
- Jugular foramen
- Superior ganglion
- Inferior ganglion
- Otic ganglion
- Stylopharyngeus
- Carotid sinus
- Pharyngeal muscles
- Common carotid artery

X The Vagus Nerves (va′gus)

Origin and course: The only cranial nerves to extend beyond head and neck region. Fibers emerge from medulla, pass through skull via jugular foramen, and descend through neck region into thorax and abdomen. See also Figure 14.4.

Function: Mixed nerves; nearly all motor fibers are parasympathetic efferents, except those serving skeletal muscles of pharynx and larynx (involved in swallowing). Parasympathetic motor fibers supply heart, lungs, and abdominal viscera and are involved in regulation of heart rate, breathing, and digestive system activity. Transmit sensory impulses from thoracic and abdominal viscera, from the carotid sinus (pressoreceptor for blood pressure) and the carotid and aortic bodies (chemoreceptors for respiration), and taste buds of posterior tongue and pharynx. Carry proprioceptor fibers from muscles of larynx and pharynx.

Clinical testing: As for cranial nerve IX (IX and X are tested in common, since they both innervate muscles of throat and mouth).

Homeostatic imbalance: Since nearly all muscles of the larynx ("voice box") are innervated by laryngeal branches of the vagus, vagal nerve paralysis can lead to hoarseness or loss of voice; other symptoms are difficulty swallowing and impaired digestive system mobility. Total destruction of both vagus nerves is incompatible with life, because these parasympathetic nerves are crucial in maintaining normal state of visceral organ activity; without their influence, the activity of the sympathetic nerves, which mobilize and accelerate vital body processes (and shut down digestion), would be unopposed. ●

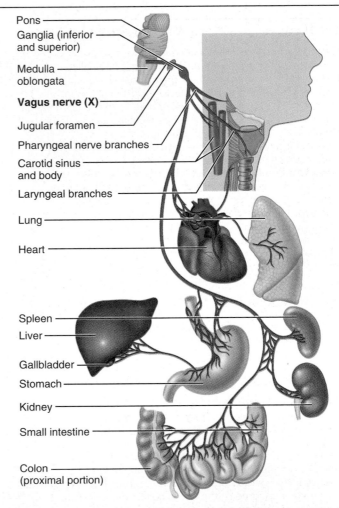

- Pons
- Ganglia (inferior and superior)
- Medulla oblongata
- **Vagus nerve (X)**
- Jugular foramen
- Pharyngeal nerve branches
- Carotid sinus and body
- Laryngeal branches
- Lung
- Heart
- Spleen
- Liver
- Gallbladder
- Stomach
- Kidney
- Small intestine
- Colon (proximal portion)

TABLE 13.2   *(continued)*

XI The Accessory Nerves

Origin and course: Unique in that they are formed by union of a *cranial root* and a *spinal root*. Cranial root emerges from lateral aspect of medulla of brain stem; spinal root arises from superior region (C_1–C_5) of spinal cord. Spinal portion passes upward along spinal cord, enters skull via foramen magnum, and temporarily joins cranial root; the resulting accessory nerve exits from skull through *jugular foramen*. Cranial and spinal fibers then diverge; cranial root fibers join vagus nerve, and spinal root runs to the largest neck muscles.

Function: Mixed nerves, but primarily motor in function. Cranial division joins with fibers of vagus nerve (X) to supply motor fibers to larynx, pharynx, and soft palate. Spinal root supplies motor fibers to trapezius and sternocleidomastoid muscles, which together move head and neck, and conveys proprioceptor impulses from same muscles.

Clinical testing: Sternocleidomastoid and trapezius muscles are checked for strength by asking person to rotate head and shrug shoulders against resistance.

Homeostatic imbalance: Injury to the spinal root of one accessory nerve causes head to turn toward injury side as result of sternocleidomastoid muscle paralysis; shrugging of that shoulder (role of trapezius muscle) becomes difficult. ●

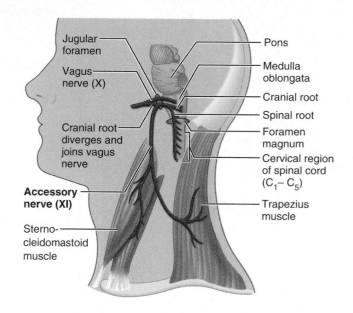

XII The Hypoglossal Nerves *(hi"po-glos'al)*

Origin and course: As their name implies (*hypo* = beneath; *glossal* = tongue), hypoglossal nerves mainly serve the tongue. Fibers arise by a series of roots from medulla and exit from skull via *hypoglossal canal* to travel to tongue. See also Figure 13.28.

Function: Mixed nerves, but primarily motor in function. Carry somatic motor fibers to intrinsic and extrinsic muscles of tongue, and proprioceptor fibers from same muscles to brain stem. Hypoglossal nerve control allows not only food mixing and manipulation by tongue during chewing, but also tongue movements that contribute to swallowing and speech.

Clinical testing: Person is asked to protrude and retract tongue. Any deviations in position are noted.

Homeostatic imbalance: Damage to hypoglossal nerves causes difficulties in speech and swallowing; if both nerves are impaired, the person cannot protrude tongue; if only one side is affected, tongue deviates (leans) toward affected side; eventually paralyzed side begins to atrophy. ●

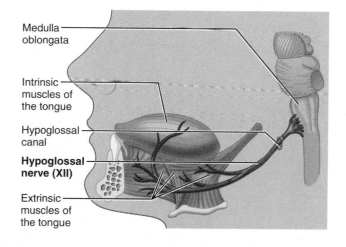

illustrated in Figure 13.29, these nerves are named according to their point of issue from the spinal cord. There are 8 pairs of cervical spinal nerves (C_1–C_8), 12 pairs of thoracic nerves (T_1–T_{12}), 5 pairs of lumbar nerves (L_1–L_5), 5 pairs of sacral nerves (S_1–S_5), and 1 pair of tiny coccygeal nerves (C_0).

Notice that there are eight pairs of cervical nerves but only seven cervical vertebrae. This "discrepancy" is easily explained. The first seven pairs exit the vertebral canal *superior to* the vertebrae for which they are named, but C_8 emerges *inferior to* the seventh cervical vertebra (between C_7 and T_1). Below the cervical level, each spinal nerve leaves the vertebral column *inferior to* the same-numbered vertebra.

As mentioned in Chapter 12, each spinal nerve connects to the spinal cord by a dorsal root and a ventral root. Each root forms from a series of

? *Why are the lumbar and cervical regions of the spinal cord enlarged?*

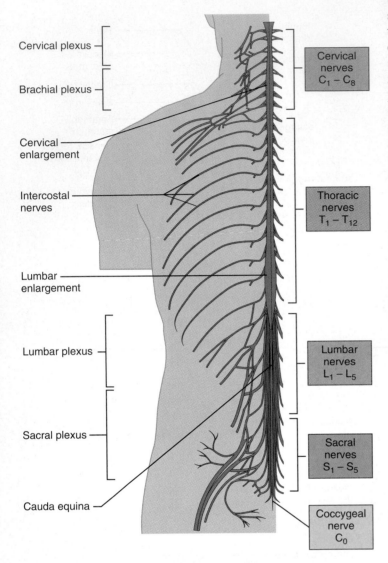

Cervical plexus

Brachial plexus

Cervical enlargement

Intercostal nerves

Lumbar enlargement

Lumbar plexus

Sacral plexus

Cauda equina

Cervical nerves C₁ – C₈

Thoracic nerves T₁ – T₁₂

Lumbar nerves L₁ – L₅

Sacral nerves S₁ – S₅

Coccygeal nerve C₀

FIGURE 13.29 Distribution of spinal nerves. Distribution of spinal nerves, posterior view. Note that the spinal nerves are named according to their points of issue.

rootlets that attach along the length of the corresponding spinal cord segment (Figure 13.30). The **ventral roots** contain *motor* (efferent) fibers that arise from anterior horn motor neurons and extend to and innervate the skeletal muscles. (Autonomic nervous system efferents also contained in the ventral roots are described in Chapter 14.) **Dorsal roots** contain *sensory* (afferent) fibers that arise from sensory neurons in the dorsal root ganglia and conduct

impulses from peripheral receptors to the spinal cord.

The spinal roots pass laterally from the cord and unite just distal to the dorsal root ganglion to form a spinal nerve before emerging from the vertebral column via their respective intervertebral foramina. Because motor and sensory fibers mingle together in a spinal nerve, it contains both efferent and afferent fibers. The length of the spinal roots increases progressively from the superior to the inferior aspect of the cord. In the cervical region, the roots are short and run horizontally, but the roots of the lumbar and sacral nerves extend inferiorly for some distance through the lower vertebral canal as the *cauda equina* before exiting the vertebral column (Figure 13.29).

A spinal nerve is quite short (only 1–2 cm) because almost immediately after emerging from its foramen, it divides into a small **dorsal ramus,** a larger **ventral ramus** (ra′mus; "branch"), and a tiny **meningeal branch** (mĕ-nin′je-al) that reenters the vertebral canal to innervate the meninges and blood vessels within. Each ramus, like the spinal nerve itself, is mixed. Finally, joined to the base of the ventral rami of the thoracic spinal nerves are special rami called **rami communicantes,** which contain autonomic (visceral) nerve fibers.

Innervation of Specific Body Regions

This section explores how the rami and their branches innervate the following regions of the body: the back, the thorax and abdominal wall, the neck, the limbs, the joints, and the skin.

Before getting into the specifics, however, some important points about the ventral rami of the spinal nerves need to be understood.

Except for T₂–T₁₂, all ventral rami branch and join one another lateral to the vertebral column, forming complicated **nerve plexuses** (Figure 13.29). Such interlacing nerve networks occur in the cervical, brachial, lumbar, and sacral regions and primarily serve the limbs. Notice that *only ventral rami form plexuses.* Within a plexus, fibers from the various ventral rami crisscross one another and become redistributed so that (1) each resulting branch of the plexus contains fibers from several spinal nerves and (2) fibers from each ventral ramus travel to the body periphery via several routes. Thus, each muscle in a limb receives its nerve supply from more than one spinal nerve. An advantage of this fiber regrouping is that damage to one spinal segment or root cannot completely paralyze any limb muscle.

Throughout this section, we mention major groups of skeletal muscles served. For more specific information on muscle innervations, see Tables 10.1–10.17.

The peripheral nerves serving the limbs arise in the lumbar and cervical regions. ■

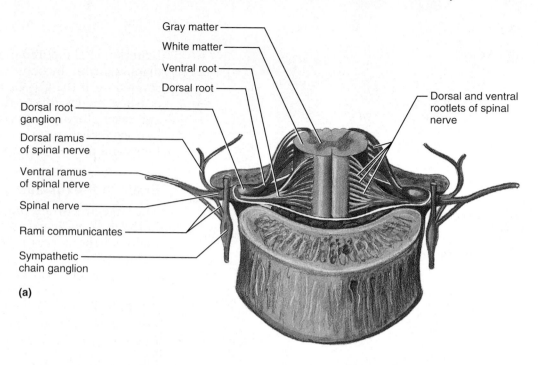

Gray matter
White matter
Ventral root
Dorsal root

Dorsal and ventral rootlets of spinal nerve

Dorsal root ganglion
Dorsal ramus of spinal nerve
Ventral ramus of spinal nerve
Spinal nerve
Rami communicantes
Sympathetic chain ganglion

(a)

FIGURE 13.30 Formation of spinal nerves and rami distribution. (a) A diagrammatic view of a spinal cord segment, illustrating the formation of one pair of spinal nerves by the union of the ventral and dorsal roots of the spinal cord. **(b)** Cross-sectional view of the left side of the body at the level of the thorax, showing the distribution of the dorsal and ventral rami of the spinal nerve. Notice also the rami communicantes branches of the spinal nerve. (The small meningeal branch is not illustrated.)

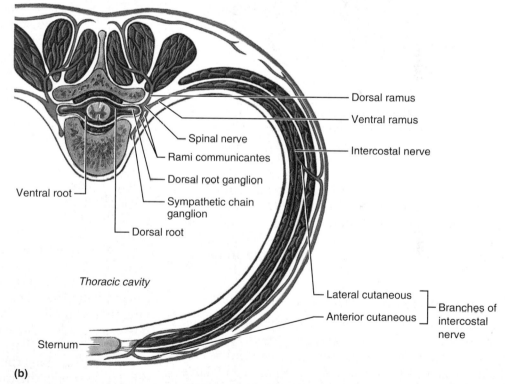

Dorsal ramus
Ventral ramus
Intercostal nerve

Spinal nerve
Rami communicantes
Dorsal root ganglion
Sympathetic chain ganglion
Ventral root
Dorsal root

Thoracic cavity

Lateral cutaneous
Anterior cutaneous
Branches of intercostal nerve

Sternum

(b)

Back

The innervation of the posterior body trunk by the dorsal rami follows a neat, segmented plan. Via its several branches (see Figure 13.30), each dorsal ramus innervates the narrow strip of muscle (and skin) in line with its emergence point from the spinal column.

Anterolateral Thorax and Abdominal Wall

Only in the thorax are the ventral rami arranged in a simple segmental pattern corresponding to that of the dorsal rami. The ventral rami of T_1–T_{12} mostly course anteriorly, deep to each rib, as the **intercostal nerves.** Along their course, these nerves give off *cutaneous branches* to the skin (Figures 13.29 and 13.30b). Two thoracic nerves are unusual: the tiny T_1

Key:

■ = Ventral
rami

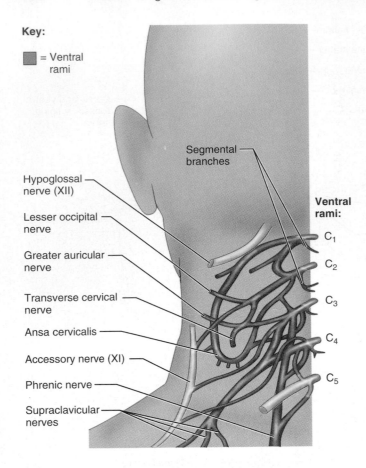

Hypoglossal
nerve (XII)

Lesser occipital
nerve

Greater auricular
nerve

Transverse cervical
nerve

Ansa cervicalis

Accessory nerve (XI)

Phrenic nerve

Supraclavicular
nerves

Segmental
branches

Ventral
rami:

C_1

C_2

C_3

C_4

C_5

FIGURE 13.31 The cervical plexus. The nerves colored gray connect to the plexus but do not belong to it. (See Table 13.3.)

(most fibers enter the brachial plexus) and T_{12}, which lies inferior to the twelfth rib, making it a **subcostal nerve.** The intercostal nerves and their branches supply the intercostal muscles lying between the ribs, the muscle and skin of the anterolateral thorax, and most of the abdominal wall.

Cervical Plexus and the Neck

Buried deep in the neck under the sternocleidomastoid muscle, the looping **cervical plexus** is formed by the ventral rami of the first four cervical nerves (Figure 13.31). Its branches are summarized in Table 13.3. Most branches are **cutaneous nerves** (nerves that supply only the skin) that transmit sensory impulses from the skin of the neck, the ear area, the back of the head, and the shoulder. Other branches innervate muscles of the anterior neck.

The single most important nerve from this plexus is the **phrenic nerve** (fren'ik) (which receives its major input from C_3 and C_4). The phrenic nerve runs inferiorly through the thorax and supplies both motor and sensory fibers to the diaphragm (*phren* = diaphragm), which is the chief muscle causing breathing movements.

 HOMEOSTATIC IMBALANCE

Irritation of the phrenic nerve causes spasms of the diaphragm, or hiccups. If both phrenic nerves are severed, or if the C_3–C_5 region of the spinal cord is crushed or destroyed, the diaphragm is paralyzed and respiratory arrest occurs. Victims are kept alive by mechanical respirators that force air into their lungs and do their breathing for them. ●

Brachial Plexus and Upper Limb

The large, important **brachial plexus,** situated partly in the neck and partly in the axilla, gives rise to virtually all the nerves that innervate the upper limb (Table 13.4). It can be palpated (felt) in a living person just superior to the clavicle at the lateral border of the sternocleidomastoid muscle.

This plexus is formed by intermixing of ventral rami of C_5–C_8 and most of the T_1 ramus. Additionally, it often receives fibers from C_4 or T_2 or both.

The brachial plexus is very complex (some consider it to be the anatomy student's nightmare). Perhaps the simplest approach is to master the terms used for its four major groups of branches (Figure 13.32). From medial to lateral, these are (1) the ventral rami, misleadingly called *roots,* which form (2) *trunks,* which form (3) *divisions,* which form (4) *cords.* You might want to use the saying "**R**eally **t**ired? **D**rink **c**offee" to help you remember this branching sequence.

The five **roots** (ventral rami C_5–T_1) of the brachial plexus lie deep to the sternocleidomastoid muscle. At the lateral border of that muscle, these roots unite to form **upper, middle,** and **lower trunks,** each of which divides almost immediately into an **anterior** and a **posterior division.** These divisions, which generally indicate which fibers serve the front or back of the limb, pass deep to the clavicle and enter the axilla, where they give rise to three large fiber bundles called the **lateral, medial,** and **posterior cords.** (The cords are named for their relationship to the axillary artery, which runs through the axilla.) All along the plexus, small nerves branch off. These supply the muscles and skin of the shoulder and superior thorax.

 HOMEOSTATIC IMBALANCE

Injuries to the brachial plexus are common; when severe, they cause weakness or paralysis of the entire upper limb. Such injuries may occur when the upper limb is pulled hard and the plexus is stretched (as when a football tackler yanks the arm of the halfback), and by blows to the top of the shoulder that force the humerus inferiorly (as when a cyclist is

TABLE 13.3	Branches of the Cervical Plexus (See Figure 13.31)		
Nerves		Spinal Roots (Ventral Rami)	Structures Served
Cutaneous Branches (Superficial)			
Lesser occipital		C_2 (C_3)	Skin on posterolateral aspect of neck
Greater auricular		C_2, C_3	Skin of ear, skin over parotid gland
Transverse cervical		C_2, C_3	Skin on anterior and lateral aspect of neck
Supraclavicular (anterior, middle, and posterior)		C_3, C_4	Skin of shoulder and anterior aspect of chest
Motor Branches (Deep)			
Ansa cervicalis (superior and inferior roots)		C_1–C_3	Infrahyoid muscles of neck (omohyoid, sternohyoid, and sternothyroid)
Segmental and other muscular branches		C_1–C_5	Deep muscles of neck (geniohyoid and thyrohyoid) and portions of scalenes, levator scapulae, trapezius, and sternocleidomastoid muscles
Phrenic		C_3–C_5	Diaphragm (sole motor nerve supply)

pitched headfirst off a motorcycle and his shoulder grinds into the pavement). ●

The brachial plexus ends in the axilla, where its three cords wind along the axillary artery and then give rise to the main nerves of the upper limb (Figure 13.32c). Five of these nerves are especially important: the axillary, musculocutaneous, median, ulnar, and radial nerves. Their distribution and targets are described briefly here, and in more detail in Table 13.4.

The **axillary nerve** branches off the posterior cord and runs posterior to the surgical neck of the humerus. It innervates the deltoid and teres minor muscles and the skin and joint capsule of the shoulder.

The **musculocutaneous nerve,** the major end branch of the lateral cord, courses inferiorly in the anterior arm, supplying motor fibers to the biceps brachii and brachialis muscles. Distal to the elbow, it provides for cutaneous sensation of the lateral forearm.

The **median nerve** descends through the arm to the anterior forearm, where it gives off branches to the skin and to most flexor muscles. On reaching the hand, it innervates five intrinsic muscles of the lateral palm. The median nerve activates muscles that pronate the forearm, flex the wrist and fingers, and oppose the thumb.

🅗 HOMEOSTATIC IMBALANCE

Median nerve injury makes it difficult to use the pincer grasp (opposed thumb and index finger) and thus to pick up small objects. Because this nerve runs down the midline of the forearm and wrist, it is a frequent casualty of wrist-slashing suicide attempts. ●

The **ulnar nerve** branches off the medial cord of the plexus. It descends along the medial aspect of the arm toward the elbow, swings behind the medial epicondyle, and then follows the ulna along the medial forearm. There it supplies the flexor carpi ulnaris and the medial part of the flexor digitorum profundus (the flexors not supplied by the median nerve). It continues into the hand, where it innervates most intrinsic hand muscles and the skin of the medial aspect of the hand. It produces wrist and finger flexion and (with the median nerve) adduction and abduction of the medial fingers.

🅗 HOMEOSTATIC IMBALANCE

Where it takes a superficial course, the ulnar nerve is very vulnerable to injury. Striking the "funny bone"—the spot where this nerve rests against the medial epicondyle—causes tingling of the little finger. Severe or chronic damage can lead to sensory loss, paralysis, and muscle atrophy. Affected individuals have trouble making a fist and gripping objects. As the little and ring fingers become hyperextended at the knuckles and flexed at the distal interphalangeal joints, the hand contorts into a *clawhand.* ●

The **radial nerve,** the largest branch of the brachial plexus, is a continuation of the posterior

TABLE 13.4   Branches of the Brachial Plexus (See Figure 13.32)

Nerves	Cord and Spinal Roots (Ventral Rami)	Structures Served
Axillary	Posterior cord (C_5, C_6)	Muscular branches: deltoid and teres minor muscles Cutaneous branches: some skin of shoulder region
Musculocutaneous	Lateral cord (C_5–C_7)	Muscular branches: flexor muscles in anterior arm (biceps brachii, brachialis, coracobrachialis) Cutaneous branches: skin on anterolateral forearm (extremely variable)
Median	By two branches, one from medial cord (C_8, T_1) and one from the lateral cord (C_5–C_7)	Muscular branches to flexor group of anterior forearm (palmaris longus, flexor carpi radialis, flexor digitorum superficialis, flexor pollicis longus, lateral half of flexor digitorum profundus, and pronator muscles); intrinsic muscles of lateral palm and digital branches to the fingers Cutaneous branches: skin of lateral two-thirds of hand, palm side and dorsum of fingers 2 and 3
Ulnar	Medial cord (C_8, T_1)	Muscular branches: flexor muscles in anterior forearm (flexor carpi ulnaris and medial half of flexor digitorum profundus); most intrinsic muscles of hand Cutaneous branches: skin of medial third of hand, both anterior and posterior aspects
Radial	Posterior cord (C_5–C_8, T_1)	Muscular branches: posterior muscles of arm, forearm, and hand (triceps brachii, anconeus, supinator, brachioradialis, extensors carpi radialis longus and brevis, extensor carpi ulnaris, and several muscles that extend the fingers) Cutaneous branches: skin of posterolateral surface of entire limb (except dorsum of fingers 2 and 3)
Dorsal scapular	Branches of C_5 rami	Rhomboid muscles and levator scapulae
Long thoracic	Branches of C_5–C_7 rami	Serratus anterior muscle
Subscapular	Posterior cord; branches of C_5 and C_6 rami	Teres major and subscapular muscles
Suprascapular	Upper trunk (C_5, C_6)	Shoulder joint; supraspinatus and infraspinatus muscles
Pectoral (lateral and medial)	Branches of lateral and medial cords (C_5–T_1)	Pectoralis major and minor muscles

cord. This nerve wraps around the humerus (in the radial groove), and then runs anteriorly around the lateral epicondyle at the elbow. There it divides into a superficial branch that follows the lateral edge of the radius to the hand and a deep branch (not illustrated) that runs posteriorly. It supplies the posterior skin of the limb along its entire course. Its motor branches innervate essentially all the extensor muscles of the upper limb. The radial nerve produces elbow extension, forearm supination, wrist and finger extension, and thumb abduction.

ⓗ HOMEOSTATIC IMBALANCE

Trauma to the radial nerve results in *wrist drop,* an inability to extend the hand at the wrist. Improper use of a crutch or "Saturday night paralysis," in which an intoxicated person falls asleep with an arm draped over the back of a chair or sofa edge, cause

radial nerve compression and ischemia (deprivation of blood supply). ●

Lumbosacral Plexus and Lower Limb

The sacral and lumbar plexuses overlap substantially, and because many fibers of the lumbar plexus contribute to the sacral plexus via the **lumbosacral trunk,** the two plexuses are often referred to as the **lumbosacral plexus.** Although the lumbosacral plexus serves mainly the lower limb, it also sends some branches to the abdomen, pelvis, and buttock.

Lumbar Plexus   The **lumbar plexus** arises from the spinal nerves L_1–L_4 (Figure 13.33) and lies within the psoas major muscle. Its proximal branches innervate parts of the abdominal wall muscles and the psoas muscle, but its major branches descend to innervate the anterior and

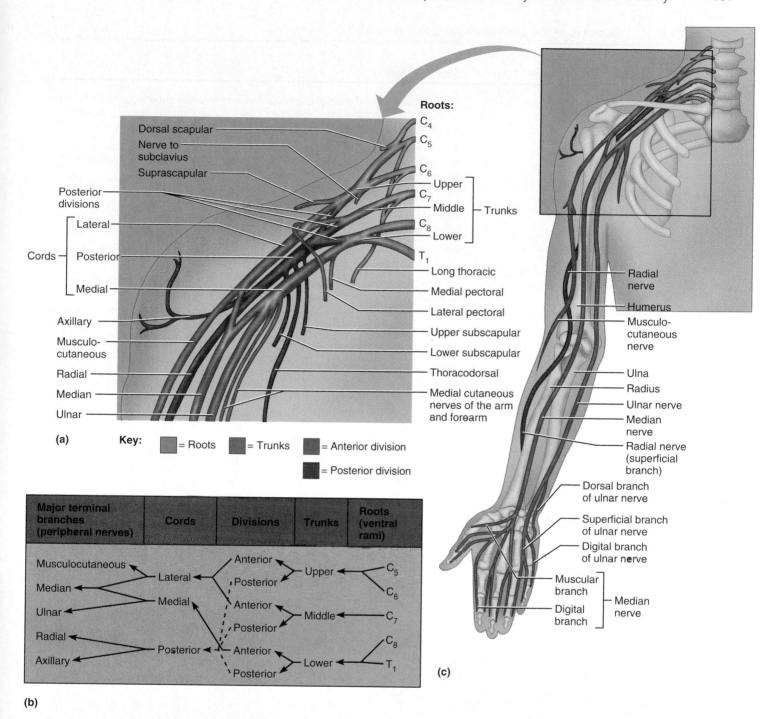

FIGURE 13.32 The brachial plexus. (a) Roots, trunks, divisions, and cords. **(b)** Flowchart of consecutive branches formed in the brachial plexus from the spinal roots (ventral rami) to the major nerves formed from the cords. **(c)** Distribution of the major peripheral nerves of the upper limb. (See Table 13.4.)

medial thigh. The **femoral nerve,** the largest terminal nerve of this plexus, runs deep to the inguinal ligament to enter the thigh and then divides into several large branches. The motor branches innervate anterior thigh muscles (quadriceps), which are the principal thigh flexors and knee extensors. The cutaneous branches serve the skin of the anterior

thigh and the medial surface of the leg from knee to foot. The **obturator nerve** (ob"tu-ra'tor) enters the medial thigh via the obturator foramen and innervates the adductor muscles. These and other smaller branches of the lumbar plexus are summarized in Table 13.5.

TABLE 13.5   Branches of the Lumbar Plexus (See Figure 13.33)

Nerves	Spinal Roots (Ventral Rami)	Structures Served
Femoral	L₂–L₄	Skin of anterior and medial thigh via *anterior femoral cutaneous* branch; skin of medial leg and foot, hip and knee joints via *saphenous* branch; motor to anterior muscles (quadriceps and sartorius) of thigh; pectineus, iliacus
Obturator	L₂–L₄	Motor to adductor magnus (part), longus, and brevis muscles, gracilis muscle of medial thigh, obturator externus; sensory for skin of medial thigh and for hip and knee joints
Lateral femoral cutaneous	L₂, L₃	Skin of lateral thigh; some sensory branches to peritoneum
Iliohypogastric	L₁	Skin of lower abdomen, lower back, and hip; muscles of anterolateral abdominal wall (obliques and transversus) and pubic region
Ilioinguinal	L₁	Skin of external genitalia and proximal medial aspect of the thigh; inferior abdominal muscles
Genitofemoral	L₁, L₂	Skin of scrotum in males, of labia majora in females, and of anterior thigh inferior to middle portion of inguinal region; cremaster muscle in males

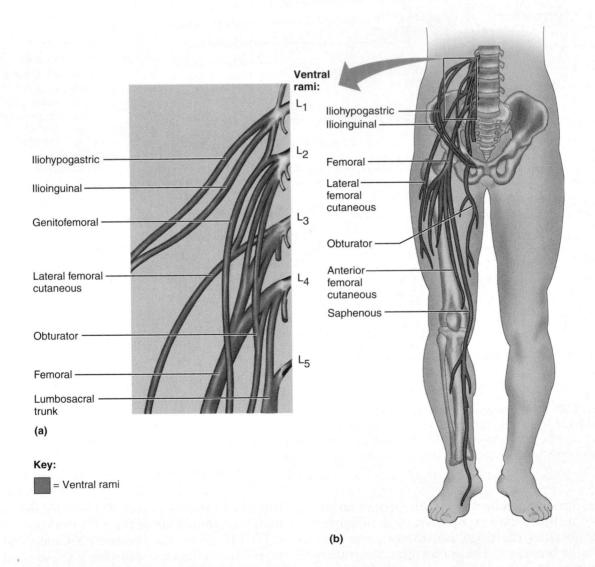

Key:

■ = Ventral rami

FIGURE 13.33   The lumbar plexus. (a) Spinal roots (anterior rami) and major branches. **(b)** Distribution of the major peripheral nerves in the lower limb (anterior view). (See Table 13.5.)

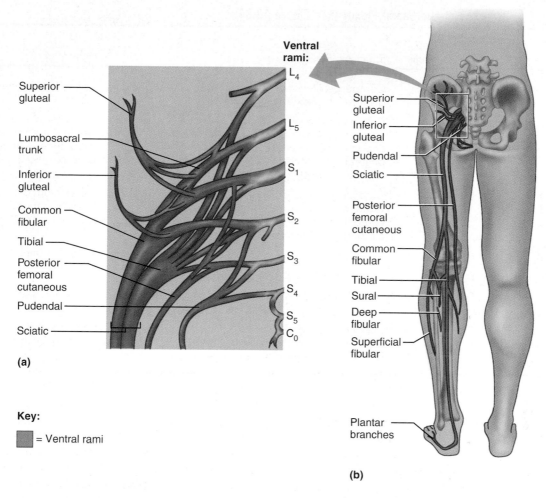

FIGURE 13.34   The sacral plexus. (a) The spinal roots (ventral rami) and major branches. **(b)** Distribution of the major peripheral nerves in the lower limb (posterior view). (See Table 13.6.)

(H) HOMEOSTATIC IMBALANCE

Compression of the spinal roots of the lumbar plexus, as by a herniated disc, results in gait problems because the femoral nerve serves the prime movers of both hip flexion and knee extension. Anesthesia of the anterior thigh and pain in the medial thigh occur if the obturator nerve is impaired. ●

Sacral Plexus   The **sacral plexus** arises from spinal nerves L_4–S_4 and lies immediately caudal to the lumbar plexus (Figure 13.34). Some fibers of the lumbar plexus contribute to the sacral plexus via the *lumbosacral trunk*, as mentioned earlier. The sacral plexus has about a dozen named branches. About half of these serve the buttock and lower limb; the others innervate pelvic structures and the perineum. The most important branches are described here. Table 13.6 summarizes all but the smallest ones.

The largest branch of the sacral plexus is the **sciatic nerve** (si-at'ik), the thickest and longest nerve in the body. It supplies the entire lower limb, except the anteromedial thigh.

Actually two nerves (the tibial and common fibular) wrapped in a common sheath, the sciatic nerve leaves the pelvis via the greater sciatic notch. It courses deep to the gluteus maximus muscle and enters the posterior thigh just medial to the hip joint (*sciatic* = of the hip). There it gives off motor branches to the hamstring muscles (all thigh extensors and knee flexors) and to the adductor magnus. Immediately above the knee, the two divisions of the sciatic nerve diverge.

The **tibial nerve** courses through the popliteal fossa (the region just posterior to the knee joint) and supplies the posterior compartment muscles of the leg and the skin of the posterior calf and sole of the foot. In the vicinity of the knee, the tibial nerve gives off the **sural nerve**, which serves the skin of the posterolateral leg, and at the ankle the tibial nerve divides into the **medial** and **lateral plantar nerves**, which serve most of the foot. The **common fibular (peroneal) nerve**

TABLE 13.6   Branches of the Sacral Plexus (See Figure 13.34)

Nerves	Spinal Roots (Ventral Rami)	Structures Served
Sciatic nerve	L_4, L_5, S_1–S_3	Composed of two nerves (tibial and common fibular) in a common sheath that diverge just proximal to the knee
■ Tibial (including sural branch and medial and lateral plantar branches)	L_4–S_3	Cutaneous branches: to skin of posterior surface of leg and sole of foot Motor branches: to muscles of back of thigh, leg, and foot [hamstrings (except short head of biceps femoris), posterior part of adductor magnus, triceps surae, tibialis posterior, popliteus, flexor digitorum longus, flexor hallucis longus, and intrinsic muscles of foot]
■ Common fibular (superficial and deep branches)	L_4–S_2	Cutaneous branches: to skin of anterior surface of leg and dorsum of foot Motor branches: to short head of biceps femoris of thigh, fibular muscles of lateral compartment of leg, tibialis anterior, and extensor muscles of toes (extensor hallucis longus, extensors digitorum longus and brevis)
Superior gluteal	L_4, L_5, S_1	Motor branches: to gluteus medius and minimus and tensor fasciae latae
Inferior gluteal	L_5–S_2	Motor branches: to gluteus maximus
Posterior femoral cutaneous	S_1–S_3	Skin of buttock, posterior thigh, and popliteal region; length variable; may also innervate part of skin of calf and heel
Pudendal	S_2–S_4	Supplies most of skin and muscles of perineum (region encompassing external genitalia and anus and including clitoris, labia, and vaginal mucosa in females, and scrotum and penis in males); external anal sphincter

descends from its point of origin, wraps around the head of the fibula (fibula = *perone*), and then divides into superficial and deep branches. These branches innervate the knee joint, skin of the lateral calf and dorsum of the foot, and muscles of the anterolateral leg (the extensors that dorsiflex the foot).

The next largest sacral plexus branches are the **superior** and **inferior gluteal nerves.** Together, they innervate the buttock (gluteal) and tensor fasciae latae muscles. The **pudendal nerve** (pu-den'dal; "shameful") innervates the muscles and skin of the perineum, mediates the act of erection, and is involved in voluntary control of urination (see Table 10.7). Other branches of the sacral plexus supply the thigh rotators and muscles of the pelvic floor.

🄷 HOMEOSTATIC IMBALANCE

Injury to the proximal part of the sciatic nerve, as might follow a fall, disc herniation, or improper administration of an injection into the buttock, results in a number of lower limb impairments, depending on the precise nerve roots injured. *Sciatica* (si-at'ĭ-kah), characterized by stabbing pain radiating over the course of the sciatic nerve, is common. When the nerve is transected, the leg is nearly useless. The leg cannot be flexed (because the hamstrings are paralyzed), and all foot and ankle movement is lost. The foot drops into plantar flexion (it dangles), a condition called *footdrop.* Recovery from sciatic nerve injury is usually slow and incomplete.

If the lesion occurs below the knee, thigh muscles are spared. When the tibial nerve is injured, the paralyzed calf muscles cannot plantar flex the foot and a shuffling gait develops. The common fibular nerve is susceptible to injury largely because of its superficial location at the head and neck of the fibula. Even a tight leg cast, or remaining too long in a side-lying position on a firm mattress, can compress this nerve and cause footdrop. ●

Innervation of Skin: Dermatomes

The area of skin innervated by the cutaneous branches of a single spinal nerve is called a **dermatome** (der'mah-tōm; "skin segment"). Every spinal nerve except C_1 innervates dermatomes. Adjacent dermatomes on the body trunk are fairly uniform in width, almost horizontal, and in direct line with their spinal nerves (Figure 13.35). The dermatome arrangement in the limbs is less obvious. (It is also less well worked out and different clinicians have mapped a variety of areas for the same dermatomes.) The skin of the upper limbs is supplied by ventral rami of C_5–T_1 (or T_2). The lumbar nerves supply most of the anterior surfaces of the thighs and legs, and the sacral nerves serve most of the posterior surfaces of the lower limbs. (This distribution basically reflects the areas supplied by the lumbar and sacral plexuses, respectively.)

Dermatome regions are not as cleanly separated as a typical dermatome map indicates. Because

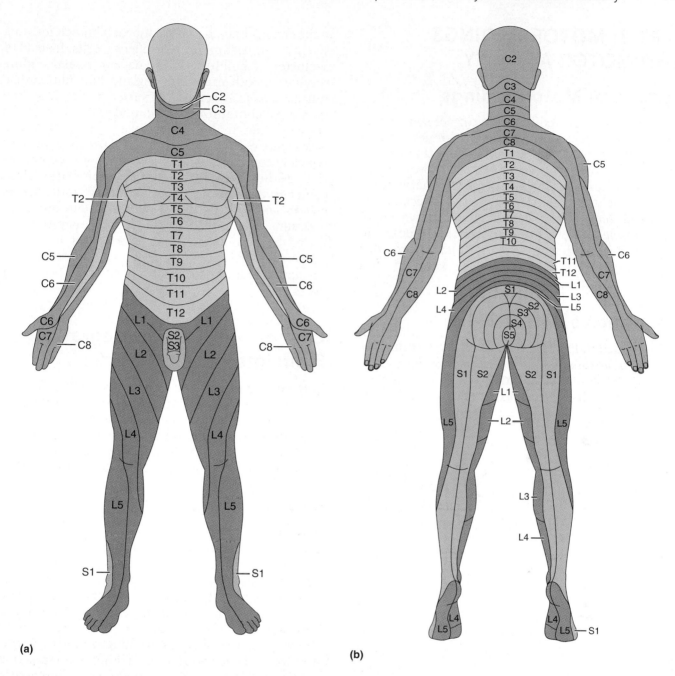

FIGURE 13.35 Dermatomes (skin segments) of the body are related to the sensory innervation regions of the spinal nerves. All spinal nerves but C_1 participate in the innervation of the dermatomes. **(a)** Anterior view. **(b)** Posterior view.

those of the trunk overlap considerably (about 50%), destruction of a single spinal nerve will not result in complete numbness anywhere. In the limbs, the overlap is less complete and some skin regions are innervated by just one spinal nerve.

Innervation of Joints

The easiest way to remember which nerves serve which synovial joint is to use **Hilton's law** which says: *Any nerve serving a muscle that produces movement at a joint also innervates the joint and the skin over the joint.* Hence, once you learn which nerves serve the various major muscles and muscle groups, no *new* learning is necessary. For example, the knee is crossed by the quadriceps, gracilis, and hamstring muscles. The nerves to these muscles are the femoral nerve anteriorly and branches of the sciatic and obturator nerves posteriorly. Consequently, these nerves innervate the knee joint as well.

PART 3: MOTOR ENDINGS AND MOTOR ACTIVITY

Peripheral Motor Endings

So far we have covered the structure of sensory receptors that detect stimuli, and that of nerves containing the afferent and efferent fibers that deliver impulses to and from the CNS. We now turn to **motor endings**, the PNS elements that activate effectors by releasing neurotransmitters. Because we discussed that topic earlier with the innervation of body muscles, all we need to do here is recap. To balance the overview of sensory function provided earlier in the chapter, this will be followed by a brief overview of motor integration.

Innervation of Skeletal Muscle

As illustrated in Figure 9.7 (p. 253), the terminals of somatic motor fibers that innervate voluntary muscles form elaborate **neuromuscular junctions** with their effector cells. As each axonal ending reaches its target, a single muscle fiber, the ending splits into a cluster of *axonal terminals (boutons)* that branch treelike over the motor end plate of the muscle fiber. The axonal terminals contain mitochondria and synaptic vesicles filled with the neurotransmitter acetylcholine (ACh). When a nerve impulse reaches an axonal terminal, ACh is released by exocytosis, diffuses across the fluid-filled synaptic cleft (about 50 nm wide), and attaches to ACh receptors on the highly infolded sarcolemma at the junction. ACh binding results in the opening of ligand-gated channels that allow both Na^+ and K^+ to pass. Because more Na^+ enters the cell than K^+ leaves, the muscle cell interior at that point depolarizes, producing an end plate potential. The end plate potential spreads to adjacent areas of the membrane where it triggers the opening of voltage-gated sodium channels. This event leads to propagation of an action potential across the sarcolemma that stimulates the muscle fiber to contract. The synaptic cleft at somatic neuromuscular junctions is filled with a glycoprotein-rich basal lamina (a structure not seen at other synapses). The basal lamina contains *acetylcholinesterase*, the enzyme that breaks down ACh almost immediately after it binds.

Innervation of Visceral Muscle and Glands

The junctions between autonomic motor endings and the visceral effectors (which are smooth and cardiac muscle and the visceral glands) are much simpler than the junctions formed between somatic fibers and skeletal muscle cells. The autonomic motor axons branch repeatedly, each branch forming *synapses en passant* with its effector cells. Instead of a cluster of bulblike terminals, an axonal ending serving smooth muscle or a gland (but not cardiac muscle) has a series of **varicosities**, knoblike swellings containing mitochondria and synaptic vesicles, that make it look like a string of beads (see Figure 9.25). The autonomic synaptic vesicles typically contain either acetylcholine or norepinephrine. Although some varicosities come into close contact with the effector cells, the synaptic cleft is always wider than that at somatic neuromuscular junctions. Consequently, the neurotransmitters released by autonomic fibers take longer to diffuse across the cleft, and visceral motor responses tend to be slower than those induced by somatic motor fibers.

Overview of Motor Integration: From Intention to Effect

In the motor system, we have motor endings serving effectors (muscle fibers) instead of sensory receptors, descending efferent circuits instead of ascending afferent circuits, and motor behavior instead of perception. However, like sensory systems, the basic mechanisms of motor systems operate at three levels.

Levels of Motor Control

The cerebral cortex is at the highest level of the conscious motor pathways, but it is *not* the ultimate planner and coordinator of complex motor activities. The cerebellum and basal nuclei (ganglia) play this role and are therefore put at the top of the motor control hierarchy. Motor control exerted by lower levels is mediated by *reflex arcs* in some cases, but complex motor behavior, such as walking and swimming, appears to depend on more complex fixed-action patterns. **Fixed-action patterns** are stereotyped sequential motor actions triggered internally or by appropriate environmental stimuli. Once triggered, the entire sequence is released in an all-or-none fashion. Currently, we define three levels of motor control: the *segmental level*, the *projection level*, and the *precommand level*.

The Segmental Level

The lowest level of the motor hierarchy, the **segmental level**, consists of the spinal cord circuits. A segmental circuit activates a network of anterior horn neurons of a single cord segment, causing them to stimulate a specific group of muscle fibers. Those circuits that control locomotion and other specific and oft-repeated motor activities are called **central pattern generators (CPGs)**.

The Projection Level

The spinal cord is under the direct control of the **projection level** of motor control. The projection level consists of *upper motor neurons* of the motor cortex which initiate the *direct (pyramidal) system* and of brain stem motor areas that oversee the *indirect (multineuronal) system.* Axons of these neurons project to the spinal cord where they help control reflex and *fixed-action* motor actions and produce discrete voluntary movements of the skeletal muscles. The brain stem areas house the command neurons that modify and control the activity of the segmental apparatus.

Axons of the pyramidal neurons forming the *pyramidal tracts* (see Table 12.4, p. 426) descend without synapsing to the associated anterior horn neurons *(lower motor neurons).* Although their axons act on all anterior horn cells, they influence mainly the most lateral motor neurons that control skeletal muscles of the distal extremities. Hence the direct pathway primarily regulates fast and fine (or skilled) movements such as those used for threading a needle and writing. Part of the direct pathway called the *corticobulbar tracts* innervates the cranial nerve nuclei of the brain stem.

The *indirect* or *extrapyramidal tracts* arise from several brain stem nuclei including the reticular, vestibular, red, and superior colliculi. These transmit impulses via the reticulospinal, vestibulospinal, rubrospinal, and tectospinal tracts respectively (Table 12.4). The roles of these nuclei are highly interrelated and complex. They act to maintain muscle tone and support the body against gravity, mediate head movements used in vision, and presumably control the CPGs of the spinal cord during locomotion and other rhythmic activities, such as arm swinging and scratching.

Projection motor pathways convey information to lower motor neurons, and send a copy of that information *(internal feedback)* back to higher command levels, continually informing them of what is happening. The direct and indirect systems provide separate and parallel pathways for controlling the spinal cord, but these systems are interrelated at all levels.

The Precommand Level

Two other systems of brain neurons, located in the cerebellum and basal nuclei, regulate motor activity. They precisely start or stop movements, coordinate movements with posture, block unwanted movements, and monitor muscle tone. Collectively called **precommand areas,** these systems *control the outputs* of the cortex and brain stem motor centers and stand at the highest level of the motor hierarchy.

The key center for "online" sensorimotor integration and control is the **cerebellum.** Remember the cerebellum is the ultimate target of ascending proprioceptor, tactile, equilibrium, and visual inputs—feedback that it needs for rapid correction of "errors" in motor activity. It also receives information via branches from descending pyramidal tracts, and from various brain stem nuclei. The cerebellum lacks direct connections to the spinal cord; it acts on motor pathways through the projection areas of the brain stem and on the motor cortex via the thalamus to fine-tune motor activity.

The **basal nuclei** receive inputs from *all* cortical areas and send their output back mainly to premotor and prefrontal cortical areas. Compared to the cerebellum, the basal nuclei appear to be involved in more complex aspects of motor control. Under resting conditions, the basal nuclei inhibit various motor centers of the brain, but when the motor centers are released from inhibition, coordinated motions can begin. Cells in both the basal nuclei and the cerebellum are involved in this unconscious planning and discharge *in advance* of willed movements. When you actually move your fingers, both the precommand areas and the primary motor cortex are active. At the risk of oversimplifying, it appears that the cortex says, "I want to do this," and then lets the precommand areas take over to provide the proper timing and patterns to execute the movements desired. The precommand areas control the motor cortex and provide its readiness to initiate a voluntary act. The conscious cortex then chooses to act or not act, but the groundwork has already been laid.

PART 4: REFLEX ACTIVITY

The Reflex Arc

Many of the body's control systems belong to a general category known as reflexes, which can be either inborn or learned. In the most restricted sense, an *inborn,* or *intrinsic, reflex* is a rapid, predictable motor response to a stimulus. It is unlearned, unpremeditated, and involuntary, and may be considered as built into our neural anatomy.

One example of an inborn reflex is what happens when you splash a pot of boiling water on your arm; you are likely to drop the pot instantly and involuntarily even before feeling any pain. This response is triggered by a spinal reflex without any help from the brain. In many cases we are aware of the final response of a basic reflex activity (you know you've dropped the pot of boiling water). In other cases, reflex activities go on without any awareness on our part. This is typical of many visceral activities, which are regulated by the subconscious lower regions of the CNS, specifically the brain stem and spinal cord.

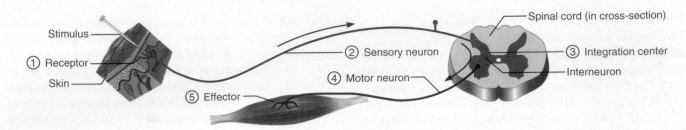

FIGURE 13.36 **The basic components of all human reflex arcs.** The reflex arc illustrated is polysynaptic.

In addition to these basic, inborn types of reflexes, there are *learned,* or *acquired, reflexes* that result from practice or repetition. Take, for instance, the complex sequence of reactions that occurs when an experienced driver drives a car. The process is largely automatic, but only because substantial time and effort were expended to acquire the driving skill. In reality, the distinction between inborn and learned reflexes is not clear-cut and most inborn reflex actions are subject to modification by learning and conscious effort. For instance, if a three-year-old child was standing by your side when you scalded your arm, you most likely would set the pot down (rather than just letting go) because of your conscious recognition of the danger to the child.

Recall the discussion in Chapter 11 about serial and parallel processing of sensory input. What happens when you scald your arm is a good example of how these two processing mechanisms work together. You drop the pot before feeling any pain, but the pain signals picked up by the interneurons of the spinal cord are quickly transmitted to the brain, so that within the next few seconds you *do* become aware of pain, and you also know what happened to cause it. The withdrawal reflex is serial processing mediated by the spinal cord, and pain awareness reflects simultaneous parallel processing of the sensory input.

Components of a Reflex Arc

As you learned in Chapter 11, reflexes occur over highly specific neural paths called **reflex arcs,** all of which have five essential components (Figure 13.36):

① **Receptor:** Site of the stimulus action.

② **Sensory neuron:** Transmits afferent impulses to the CNS.

③ **Integration center:** In simple reflex arcs, may be a single synapse between a sensory neuron and a motor neuron **(monosynaptic reflex).** More complex reflex arcs involve multiple synapses with chains of interneurons **(polysynaptic reflex).** The integration center is always within the CNS.

④ **Motor neuron:** Conducts efferent impulses from the integration center to an effector organ.

⑤ **Effector:** Muscle fiber or gland cell that responds to the efferent impulses (by contracting or secreting).

Reflexes are classified functionally as **somatic reflexes** if they activate skeletal muscle, and as **autonomic (visceral) reflexes** if they activate visceral effectors (smooth or cardiac muscle or glands). Here we describe some common somatic reflexes mediated by the spinal cord. Autonomic reflexes are considered in later chapters with the visceral processes they help to regulate.

Spinal Reflexes

Somatic reflexes mediated by the spinal cord are called **spinal reflexes.** Many spinal reflexes occur without the involvement of higher brain centers. These reflexes work equally well in decerebrate animals (those in which the brain has been destroyed) as long as the spinal cord is functional. Other spinal reflexes require brain activity for their successful completion. Additionally, the brain is "advised" of most spinal reflex activity and therefore can facilitate or inhibit it. Moreover, continuous facilitating signals from the brain are required for normal spinal reflex activity because, as noted in Chapter 12, if the spinal cord is transected, *spinal shock* occurs (all functions controlled by the cord are immediately depressed).

Testing of somatic reflexes is important clinically to assess the condition of the nervous system. Exaggerated, distorted, or absent reflexes indicate degeneration or pathology of specific nervous system regions, often before other signs are apparent.

Stretch and Deep Tendon Reflexes

There are two requirements for skeletal muscles to perform normally: (1) The brain must be continuously informed of the current state of the muscles, and (2) the muscles must have healthy *tone* (resistance to active or passive stretch at rest). The first requirement depends on transmission of information from *muscle spindles* and *Golgi tendon organs* (proprioceptors located in the skeletal muscles and their associated tendons) to the cerebellum and cerebral cortex. The second requirement depends on **stretch reflexes** initiated by the muscle spindles, which monitor changes in muscle length.

Functional Anatomy of Muscle Spindles

Each muscle spindle consists of 3 to 10 **intrafusal muscle fibers** (*intra* = within; *fusal* = the spindle) enclosed in a connective tissue capsule (Figure 13.37). These modified skeletal muscle fibers are less than one-quarter the size of **extrafusal muscle fibers** (the effector fibers of the muscle). The central regions of the intrafusal fibers, which lack myofilaments and are noncontractile, are the receptive surfaces of the spindle. These regions are wrapped by two types of afferent endings that send sensory inputs to the CNS. The **primary sensory endings** of large **type Ia fibers**, which innervate the spindle center, are stimulated by both the rate and degree of stretch. The **secondary sensory endings** of small **type II fibers** supply the spindle ends and are stimulated only by degree of stretch. The contractile regions of the intrafusal muscle fibers are limited to their ends because these are the only areas that contain actin and myosin myofilaments. These regions are innervated by **gamma (γ) efferent fibers** that arise from small motor neurons in the ventral horn of the spinal cord. These γ motor fibers, which maintain spindle sensitivity (as described shortly), are distinct from the **alpha (α) efferent fibers** of the large **alpha (α) motor neurons** that stimulate the extrafusal muscle fibers to contract.

The Stretch Reflex

The muscle spindle is stretched (and excited) in one of two ways: (1) by applying an external force that lengthens the entire muscle, such as occurs when we carry a heavy weight or when antagonistic muscles contract *(external stretch)*; (2) by activating the γ motor neurons that stimulate the distal ends of the intrafusal fibers to contract, thus stretching the middle of the spindle *(internal stretch)*. Whichever the stimulus, once spindles are activated the associated sensory neurons transmit impulses at a higher frequency to the spinal cord (Figure 13.38a). There the sensory neurons synapse directly with α motor neurons, which rapidly excite the extrafusal muscle fibers of the stretched muscle (Figure 13.39a). The reflexive muscle contraction that follows (an example of serial processing) resists further muscle stretching. Branches of the afferent fibers also synapse with interneurons that inhibit motor neurons controlling antagonistic muscles (parallel processing), and the resulting inhibition is called **reciprocal inhibition.**

There is a major difference in the functions of the intrafusal and extrafusal muscle fibers. What is it?

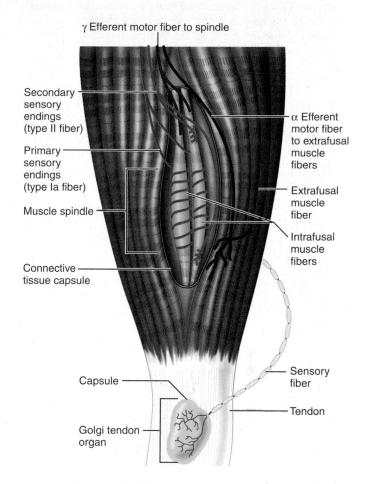

FIGURE 13.37 Anatomy of the muscle spindle and Golgi tendon organ. Notice the afferent fibers from and efferent fibers to the muscle spindle.

Consequently, the stretch stimulus causes the antagonists to relax so that they cannot resist the shortening of the "stretched" muscle caused by the main reflex arc.

While this spinal reflex is occurring, information on muscle length and the speed of muscle shortening is being relayed (mainly via the dorsal white columns) to higher brain centers (more parallel processing). This sequence of events is how muscle tone is maintained and adjusted reflexively to the requirements of posture and movement. If either afferent or efferent fibers are cut, the muscle immediately loses its tone and becomes flaccid. The stretch reflex is most important in the large

■ *The extrafusal muscle fibers are effectors; when adequately stimulated, they contract, shortening the muscle. The intrafusal fibers are part of a receptor apparatus that provides input that helps regulate the activity of the muscle.*

FIGURE 13.38   **Operation of the muscle spindle.**
(a) Stretching the muscle activates the muscle spindle and increases the rate of action potential generation in the associated sensory (Ia) fiber. **(b)** Muscle contraction reduces tension on the spindle and lowers the rate of action potential generation. Arrows indicate the direction of force exerted on the muscle spindles.

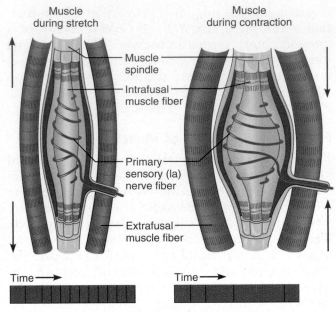

Muscle during stretch

Muscle during contraction

- Muscle spindle
- Intrafusal muscle fiber
- Primary sensory (Ia) nerve fiber
- Extrafusal muscle fiber

Time⟶

Time⟶

(a) Action potential frequency increases during stretch

(b) Action potential frequency declines during contraction

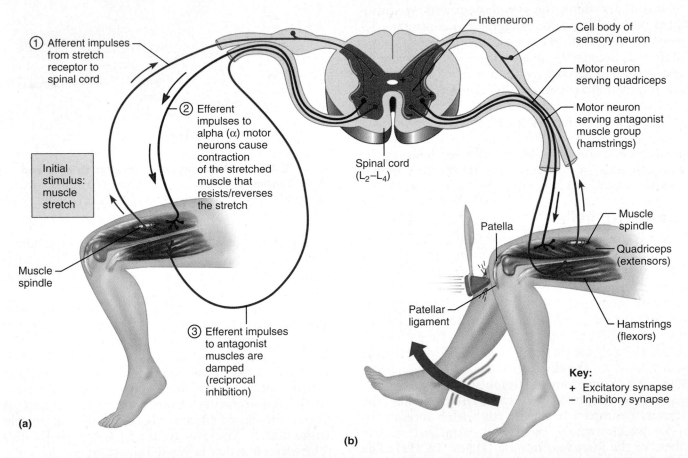

① Afferent impulses from stretch receptor to spinal cord

Interneuron

Cell body of sensory neuron

Motor neuron serving quadriceps

Motor neuron serving antagonist muscle group (hamstrings)

② Efferent impulses to alpha (α) motor neurons cause contraction of the stretched muscle that resists/reverses the stretch

Spinal cord (L₂–L₄)

Initial stimulus: muscle stretch

Patella

Muscle spindle

Muscle spindle

Quadriceps (extensors)

Patellar ligament

③ Efferent impulses to antagonist muscles are damped (reciprocal inhibition)

Hamstrings (flexors)

Key:
+ Excitatory synapse
– Inhibitory synapse

(a)

(b)

FIGURE 13.39   **The stretch reflex.**
(a) Events of the stretch reflex by which muscle stretch is damped. The events are shown in circular fashion. **(b)** The patellar reflex. Tapping the patellar tendon excites muscle spindles in the quadriceps muscle. Afferent impulses travel to the spinal cord, where synapses occur with motor neurons and interneurons. The motor neurons send activating impulses to the quadriceps causing it to contract, which causes extension of the knee and forward movement of the foot, counteracting the initial stretch. The interneurons make inhibitory synapses with anterior horn neurons that serve the antagonistic muscle group (hamstrings), thus preventing the hamstrings from resisting the contraction.

extensor muscles that sustain upright posture and in postural muscles of the trunk. For example, contractions of the postural muscles of the spine are almost continuously regulated by stretch reflexes initiated first on one side of the spine and then on the other.

The stretch reflex is essential for normal muscle tone and activity, but essentially it never acts alone. It is always accompanied by the **gamma motor neuron reflex arc.** There is a good reason for this: As a muscle shortens, the spindle's rate of firing declines, reducing impulse generation by the alpha motor neurons (Figure 13.38b). Thus, the stretch reflex alone would result in crude, jerky muscle contractions. Our muscles do not operate in this spastic fashion, however, because contractions promoted by stretch reflexes are refined and smoothed out by gamma motor neuron reflex arcs, which regulate how the spindle intrafusal muscle fibers respond. Descending fibers of motor pathways synapse with both α and γ motor neurons, and motor impulses are simultaneously sent to the large extrafusal fibers and to muscle spindle intrafusal fibers. Stimulating the intrafusal fibers maintains the spindle's tension (and sensitivity) during muscle contraction, so that the brain continues to be notified of conditions in the muscle. Without such a system, information on changes in muscle length would cease to flow from contracting muscles.

The motor supply to the muscle spindle also allows us to voluntarily control the stretch reflex response and the firing rate of α motor neurons. When the γ neurons are vigorously stimulated by impulses from the brain, the spindle is stretched and highly sensitive, and muscle contraction force is maintained or increased. When the γ motor neurons are inhibited, the spindle resembles a loose rubber band and is nonresponsive, and the extrafusal muscles relax.

The ability to modify the stretch reflex is important in many situations. For example, if you want to wind up to pitch a baseball, it is essential to suppress the stretch reflex so that your muscles can produce a large degree of motion (i.e., circumduct your pitching arm). On the other hand, when maximum force is desired, the muscle should be stretched as much and as quickly as possible just before the movement. This advantage is demonstrated by the crouch that athletes assume just before jumping or running.

The most familiar clinical example of the stretch reflex is the **patellar** (pah-tel'ar), or **knee-jerk, reflex,** elicited by striking the patellar tendon (Figure 13.39b). The sudden jolt to the tendon stretches the quadriceps, stimulates the muscle spindles, and results in contraction of the quadriceps and inhibition of its antagonist, the hamstrings. Stretch reflexes can be elicited in any skeletal muscle by suddenly striking its tendon or the muscle itself. All stretch reflexes are **monosynaptic** and **ipsilateral;** that is, they involve a single synapse and motor activity on the same side of the body. Although stretch reflexes *themselves* are monosynaptic, the reflex arcs that inhibit the motor neurons serving the antagonistic muscles are polysynaptic.

A positive knee jerk (or a positive result for any other stretch reflex test) provides two important pieces of information. First, it proves that the sensory and motor connections between that muscle and the spinal cord are intact. Second, the vigor of the response indicates the degree of excitability of the spinal cord. When the spinal motor neurons are highly facilitated by impulses descending from higher centers, just touching the muscle tendon produces a vigorous reflex response. On the other hand, when the lower motor neurons are bombarded by inhibitory signals, even pounding on the tendon may fail to cause the reflex response.

H   HOMEOSTATIC IMBALANCE

Stretch reflexes tend to be either hypoactive or absent in cases of peripheral nerve damage or ventral horn injury involving the tested area. These reflexes are absent in those with chronic diabetes mellitus and neurosyphilis and during coma. However, they are hyperactive when lesions of the corticospinal tract reduce the inhibitory effect of the brain on the spinal cord (as in polio and stroke patients). ●

The Deep Tendon Reflex   Whereas stretch reflexes cause muscle contraction in response to increased muscle length (stretch), the polysynaptic **deep tendon reflexes** produce exactly the opposite effect: muscle relaxation and lengthening in response to contraction. When muscle tension increases substantially during contraction or passive stretching, high-threshold Golgi tendon organs in the tendon may be activated, and afferent impulses are transmitted to the spinal cord, and then to the cerebellum, where the information is used to adjust muscle tension. Simultaneously, motor neurons in spinal cord circuits supplying the contracting muscle are inhibited and antagonist muscles are activated, a phenomenon called **reciprocal activation.** As a result, the contracting muscle relaxes as its antagonist is activated.

Golgi tendon organs protect muscles and tendons subjected to possibly damaging stretching force from tearing and help ensure smooth onset and termination of muscle contraction. Golgi tendon organs are also stimulated during a clinically induced stretch reflex, such as the flexor reflex, but the brevity of the stimulus and the fact that the stretched muscle is already relaxed prevent it from being inhibited by the deep tendon reflex.

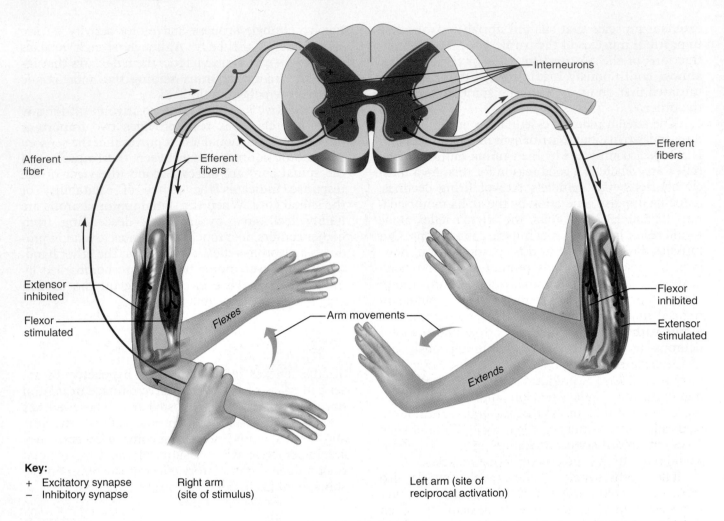

Interneurons

Afferent
fiber

Efferent
fibers

Efferent
fibers

Extensor
inhibited

Flexor
stimulated

Flexes

Arm movements

Flexor
inhibited

Extensor
stimulated

Extends

Key:
+   Excitatory synapse
−   Inhibitory synapse

**Right arm
(site of stimulus)**

**Left arm (site of
reciprocal activation)**

FIGURE 13.40 The crossed extensor reflex. The crossed extensor reflex causes reflex withdrawal of the body part on the stimulated side (flexor reflex) as shown on the left side of the diagram, and extension of the muscles on the opposite side of the body (extensor reflex) as shown on the right side of the diagram. The example illustrated might occur if a stranger suddenly grabbed your arm. The flexor reflex would cause immediate withdrawal of your clutched arm, while the extensor reflex would simultaneously push the intruder away with the other arm by activating extensor muscles on the opposite side of the body.

The Flexor Reflex

The **flexor,** or **withdrawal, reflex** is initiated by a painful stimulus (actual or perceived) and causes automatic withdrawal of the threatened body part from the stimulus (Figure 13.40). The response that occurs when you prick your finger is a good example. So, too, is the trunk flexion that occurs when someone pretends to "throw a punch" at your abdomen. Flexor reflexes are ipsilateral and polysynaptic, the latter a necessity when several muscles must be recruited to withdraw the injured body part. Because flexor reflexes are protective reflexes important to our survival, they override the spinal pathways and prevent any other reflexes from using them at the same time.

The Crossed Extensor Reflex

The **crossed extensor reflex** is a complex spinal reflex consisting of an ipsilateral withdrawal reflex and a contralateral extensor reflex. Incoming afferent fibers synapse with interneurons that control the flexor withdrawal response on the same side of the body and with other interneurons that control the extensor muscles on the opposite side. This reflex is obvious when someone unexpectedly grabs for your arm (Figure 13.40). A more common example occurs when you step barefoot on broken glass. The ipsilateral response causes rapid lifting of the cut foot, while the contralateral response activates the extensor muscles of the opposite leg to support the weight suddenly shifted to it. Crossed extensor reflexes are particularly important in maintaining balance.

Superficial Reflexes

Superficial reflexes are elicited by gentle cutaneous stimulation, such as that produced by stroking the skin with a tongue depressor. These clinically important reflexes depend both on functional upper motor pathways and on cord-level reflex arcs. The best known of these are the plantar and abdominal reflexes.

The **plantar reflex** tests the integrity of the spinal cord from L_4 to S_2 and indirectly determines if the corticospinal tracts are functioning properly. It is elicited by drawing a blunt object downward along the lateral aspect of the plantar surface (sole) of the foot. The normal response is a downward flexion (curling) of the toes. However, if the primary motor cortex or corticospinal tract is damaged, the plantar reflex is replaced by an abnormal reflex called **Babinski's sign,** in which the great toe dorsiflexes and the smaller toes fan laterally. Infants exhibit Babinski's sign until they are about a year old because their nervous systems are incompletely myelinated. Despite its clinical significance, the physiological mechanism of Babinski's sign is not understood.

Stroking the skin of the lateral abdomen above, to the side, or below the umbilicus may induce a reflex contraction of the abdominal muscles in which the umbilicus moves toward the stimulated site. These reflexes, called **abdominal reflexes,** check the integrity of the spinal cord and ventral rami from T_8 to T_{12}. The abdominal reflexes vary in intensity from one person to another. They are absent when corticospinal tract lesions are present.

Review Questions

Multiple Choice/Matching

(Some questions have more than one correct answer. Select the best answer or answers from the choices given.)

1. The large onion-shaped receptors that are found deep in the dermis and in subcutaneous tissue and that respond to deep pressure are (a) Merkel discs, (b) Pacinian receptors, (c) free nerve endings, (d) muscle spindles.

2. Proprioceptors include all of the following except (a) muscle spindles, (b) Golgi tendon organs, (c) Merkel discs, (d) joint kinesthetic receptors.

3. The connective tissue sheath that surrounds a fascicle of nerve fibers is the (a) epineurium, (b) endoneurium, (c) perineurium, (d) neurilemma.

4. Match the receptor type from column B to the correct description in column A.

Column A	Column B
____ (1) pain, itch, and temperature receptors	(a) Ruffini's corpuscle
____ (2) contains intrafusal fibers and type Ia and II sensory endings	(b) Golgi tendon organ
	(c) muscle spindle
____ (3) discriminative touch receptor in hairless skin (fingertips)	(d) free dendritic endings
____ (4) contains receptor endings wrapped around thick collagen bundles	(e) Pacinian corpuscle
	(f) Meissner's corpuscle
____ (5) rapidly adapting deep-pressure receptor	
____ (6) slowly adapting deep-pressure receptor	

5. Match the names of the cranial nerves in column B to the appropriate description in column A.

Column A	Column B
____ (1) causes pupillary constriction	(a) abducens
	(b) accessory
____ (2) is the major sensory nerve of the face	(c) facial
	(d) glossopharyngeal
____ (3) serves the sternocleido-mastoid and trapezius muscles	(e) hypoglossal
	(f) oculomotor
____ (4) are purely sensory (three nerves)	(g) olfactory
	(h) optic
____ (5) serves the tongue muscles	(i) trigeminal
	(j) trochlear
____ (6) allows you to chew your food	(k) vagus
	(l) vestibulocochlear
____ (7) is impaired in tic douloureux	
____ (8) helps to regulate heart activity	
____ (9) helps you to hear and to maintain your balance	
____ (10) contain parasympathetic motor fibers (four nerves)	

6. For each of the following muscles or body regions, identify the plexus and the peripheral nerve(s) (or branch of one) involved. Use choices from keys A and B.

		Key A: Plexuses
____; ____ (1) the diaphragm		(a) brachial
____; ____ (2) muscles of the posterior thigh and leg		(b) cervical
____; ____ (3) anterior thigh muscles		(c) lumbar
____; ____ (4) medial thigh muscles		(d) sacral
____; ____ (5) anterior arm muscles that flex the forearm		**Key B: Nerves**
____; ____ (6) muscles that flex the wrist and digits (two nerves)		(1) common fibular
		(2) femoral
____; ____ (7) muscles that extend the wrist and digits		(3) median
		(4) musculocutaneous
____; ____ (8) skin and extensor muscles of the posterior arm		(5) obturator
		(6) phrenic
____; ____ (9) fibularis muscles, tibialis anterior, and toe extensors		(7) radial
		(8) tibial
____; ____, ____, ____, ____ (10) elbow joint		(9) ulnar

7. Characterize each receptor activity described below by choosing the appropriate letter and number from keys A and B.

	Key A:
____, ____ (1) You are enjoying an ice cream cone.	(a) exteroceptor
____, ____ (2) You have just scalded yourself with hot coffee.	(b) interoceptor
	(c) proprioceptor
____, ____ (3) The retinas of your eyes are stimulated.	**Key B:**
	(1) chemoreceptor
____, ____ (4) You bump (lightly) into someone.	(2) mechanoreceptor
	(3) nociceptor
____, ____ (5) You have been lifting weights and experience a sensation in your upper limbs.	(4) photoreceptor
	(5) thermoreceptor

8. Olfactory tract damage would probably affect your ability to (a) see, (b) hear, (c) feel pain, (d) smell.

9. Sensory impulses transmitted over the facial, glossopharyngeal, and vagus nerves are involved in the sensation of (a) taste, (b) touch, (c) equilibrium, (d) smell.

10. Taste buds are found on the (a) anterior part of the tongue, (b) posterior part of the tongue, (c) palate, (d) all of these.

11. Gustatory cells are stimulated by (a) movement of otoliths, (b) stretch, (c) substances in solution, (d) photons of light.

12. Olfactory nerve filaments are found (a) in the optic bulbs, (b) passing through the cribriform plate of the ethmoid bone, (c) in the optic tracts, (d) in the olfactory cortex.

13. The accessory glands that produce an oily secretion are the (a) conjunctiva, (b) lacrimal glands, (c) tarsal glands.

14. The portion of the fibrous tunic that is white and opaque is the (a) choroid, (b) cornea, (c) retina, (d) sclera.

15. Which sequence best describes a normal route for the flow of tears from the eyes into the nasal cavity? (a) lacrimal canals, nasolacrimal ducts, nasal cavity; (b) lacrimal ducts, lacrimal canals, nasolacrimal ducts; (c) nasolacrimal ducts, lacrimal canals, lacrimal sacs.

16. Four refractory media of the eye, listed in the sequence in which they refract light, are (a) vitreous humor, lens, aqueous humor, cornea; (b) cornea, aqueous humor, lens, vitreous humor; (c) cornea, vitreous humor, lens, aqueous humor; (d) lens, aqueous humor, cornea, vitreous humor.

17. Damage to the medial recti muscles would probably affect (a) accommodation, (b) refraction, (c) convergence, (d) pupil constriction.

18. The phenomenon of light adaptation is best explained by the fact that (a) rhodopsin does not function in dim light, (b) rhodopsin breakdown occurs slowly, (c) rods exposed to intense light need time to generate rhodopsin, (d) cones are stimulated to function by bright light.

19. Blockage of the scleral venous sinus might result in (a) a sty, (b) glaucoma, (c) conjunctivitis, (d) a cataract.

20. Nearsightedness is more properly called (a) myopia, (b) hyperopia, (c) presbyopia, (d) emmetropia.

21. Of the neurons in the retina, the axons of which of these form the optic nerve? (a) bipolar neurons, (b) ganglion cells, (c) cone cells, (d) horizontal cells.

22. Which sequence of reactions occurs when a person looks at a distant object? (a) pupils constrict, suspensory ligament relaxes, lenses become less convex, (b) pupils dilate, suspensory ligament becomes taut, lenses become less convex, (c) pupils dilate, suspensory ligament becomes taut, lenses become more convex, (d) pupils constrict, suspensory ligament relaxes, lenses become more convex.

23. The blind spot of the eye is (a) where more rods than cones are found, (b) where the macula lutea is located, (c) where only cones occur, (d) where the optic nerve leaves the eye.

24. Conduction of sound from the middle ear to the inner ear occurs via vibration of the (a) malleus against the tympanic membrane, (b) stapes in the oval window, (c) incus in the round window, (d) stapes against the tympanic membrane.

25. Which of the following statements does not correctly describe the organ of Corti? (a) sounds of high frequency stimulate hair cells at the basal end, (b) the "hairs" of the receptor cells are embedded in the tectorial membrane, (c) the basilar membrane acts as a resonator, (d) sounds of high frequency stimulate cells at the apex of the basilar membrane.

26. Pitch is to frequency of sound as loudness is to (a) quality, (b) intensity, (c) overtones, (d) all of these.

27. The structure that allows pressure in the middle ear to be equalized with atmospheric pressure is the (a) pinna, (b) pharyngotympanic tube, (c) tympanic membrane, (d) secondary tympanic membrane.

28. Which of the following is important in maintaining the balance of the body? (a) visual cues, (b) semicircular canals, (c) the saccule, (d) proprioceptors, (e) all of these.

29. Static equilibrium receptors that report the position of the head in space relative to the pull of gravity are (a) organ of Corti, (b) maculae, (c) cristae ampullares.

30. Which of the following is *not* a possible cause of conduction deafness? (a) impacted cerumen, (b) middle ear infection, (c) cochlear nerve degeneration, (d) otosclerosis.

31. Otoliths (ear stones) are (a) a cause of deafness, (b) a type of hearing aid, (c) important in equilibrium, (d) the rock-hard petrous temporal bones.

32. An ipsilateral reflex that causes rapid withdrawal of a body part from a painful stimulus is the (a) crossed extensor, (b) flexor, (c) Golgi tendon, (d) muscle stretch.

Short Answer Essay Questions

33. List the structural components of the peripheral nervous system, and describe the function of each component.

34. Describe the functional problems that would be experienced by a person in which these fiber tracts have been cut: (a) lateral spinothalamic, (b) anterior and posterior spinocerebellar, (c) tectospinal.

35. (a) Define plexus. (b) Indicate the spinal roots of origin of the four major nerve plexuses, and name the general body regions served by each.

36. What is the homeostatic value of flexor reflexes?

37. Compare and contrast flexor and crossed extensor reflexes.

38. What clinical information can be gained by conducting somatic reflex tests?

39. Where are the olfactory receptors, and why is that site poorly suited for their job?

40. How do rods and cones differ functionally?

41. Where is the fovea centralis, and why is it important?

42. Describe the response of rhodopsin to light stimuli. What is the outcome of this cascade of events?

43. Since there are only three types of cones, how can you explain the fact that we see many more colors?

44. Central pattern generators (CPGs) are found at the segmental level of motor control. (a) What is the job of the CPGs? (b) What controls them, and where is this control localized?

45. How do the types of motor activity controlled by the direct (pyramidal) and indirect systems differ?

46. Why are the cerebellum and basal nuclei called *precommand* areas?

14

THE AUTONOMIC NERVOUS SYSTEM

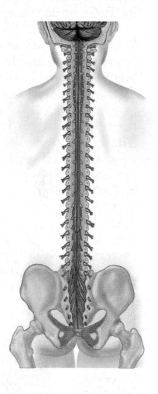

Introduction (pp. 502–504)

1. Define autonomic nervous system and explain its relationship to the peripheral nervous system.

2. Compare the somatic and autonomic nervous systems relative to effectors, efferent pathways, and neurotransmitters released.

3. Compare and contrast the functions of the parasympathetic and sympathetic divisions.

ANS Anatomy (pp. 504–511)

4. For the parasympathetic and sympathetic divisions, describe the site of CNS origin, locations of ganglia, and general fiber pathways.

ANS Physiology (pp. 511–517)

5. Define cholinergic and adrenergic fibers, and list the different types of their receptors.

6. Describe the clinical importance of drugs that mimic or inhibit adrenergic or cholinergic effects.

7. State the effects of the parasympathetic and sympathetic divisions on the following organs: heart, blood vessels, gastrointestinal tract, lungs, adrenal medulla, and external genitalia.

8. Describe autonomic nervous system controls.

Homeostatic Imbalances of the ANS (pp. 517–518)

9. Explain the relationship of some types of hypertension, Raynaud's disease, and the mass reflex reaction to disorders of autonomic functioning.

In its never-ending attempt to maintain balance, the body is far more sensitive to changing events than any human invention or cultural institution. Although all body systems contribute, the stability of our internal environment depends largely on the **autonomic nervous system (ANS),** the system of motor neurons that innervates smooth and cardiac muscle and glands (Figure 14.1). At every moment, signals flood from visceral organs into the CNS, and autonomic nerves make adjustments as necessary to ensure optimal support for body activities. In response to changing conditions, the ANS shunts blood to "needy" areas, speeds or slows heart rate, adjusts blood pressure and body temperature, and increases or decreases stomach secretions.

Most of this fine-tuning occurs without our awareness or attention. Can you tell when your arteries are constricting or when your pupils are dilating? Probably not, but if you've ever been stuck in a check-out line, and your full bladder was contracting as if it had a mind of its own, you've been very aware of your body's fine-tuning activity. These functions, both those we're aware of and those that occur without our awareness or attention, are controlled by the ANS. Indeed, as the term *autonomic* (*auto* = self; *nom* = govern) implies, this motor subdivision of the peripheral nervous system has a certain amount of functional independence. The ANS is also called the **involuntary nervous system,** which reflects its subconscious control, and the **general visceral motor system,** which indicates the location of most of its effectors.

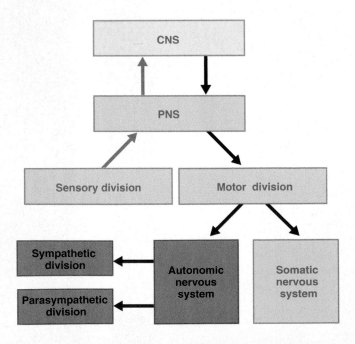

FIGURE 14.1 **Place of the ANS in the structural organization of the nervous system.**

Introduction

Comparison of the Somatic and Autonomic Nervous Systems

Our previous discussions of motor nerves have focused largely on the activity of the somatic nervous system. So, before describing autonomic nervous system anatomy, we will point out the major differences between the somatic and autonomic systems as well as some areas of functional overlap. Although both systems have motor fibers, the somatic and autonomic nervous systems differ (1) in their effectors, (2) in their efferent pathways, and (3) to some degree in target organ responses to their neurotransmitters. The differences are summarized in Figure 14.2.

Effectors

The somatic nervous system stimulates skeletal muscles, whereas the ANS innervates cardiac and smooth muscle and glands. Differences in the physiology of the effector organs account for most of the remaining differences between somatic and autonomic effects on their target organs.

Efferent Pathways and Ganglia

In the somatic nervous system, the motor neuron cell bodies are in the CNS, and their axons extend in spinal nerves all the way to the skeletal muscles they activate. Somatic motor fibers are typically thick, heavily myelinated type A fibers that conduct nerve impulses rapidly.

In contrast, the motor unit of the ANS is a *two-neuron chain.* The cell body of the first neuron, the **preganglionic neuron,** resides in the brain or spinal cord. Its axon, the **preganglionic axon,** synapses with the second motor neuron, the **ganglionic neuron,** in an **autonomic ganglion** outside the CNS. The axon of the ganglionic neuron, called the **postganglionic axon,** extends to the effector organ. If you think about the meanings of all these terms while referring to Figure 14.2, understanding the rest of the chapter will be much easier. Preganglionic axons are lightly myelinated, thin fibers; postganglionic axons are even thinner and are unmyelinated. Consequently, conduction through the autonomic efferent chain is slower than conduction in the somatic motor system. For most of their course, many pre- and postganglionic fibers are incorporated into spinal or cranial nerves.

Keep in mind that autonomic ganglia are *motor* ganglia, containing the cell bodies of motor neurons. Technically, they are sites of synapse and information transmission from preganglionic to ganglionic neurons. Also, remember that the somatic

? *Why are the axons of the first and second neurons of the two-neuron motor autonomic chain named the preganglionic and postganglionic axons respectively?*

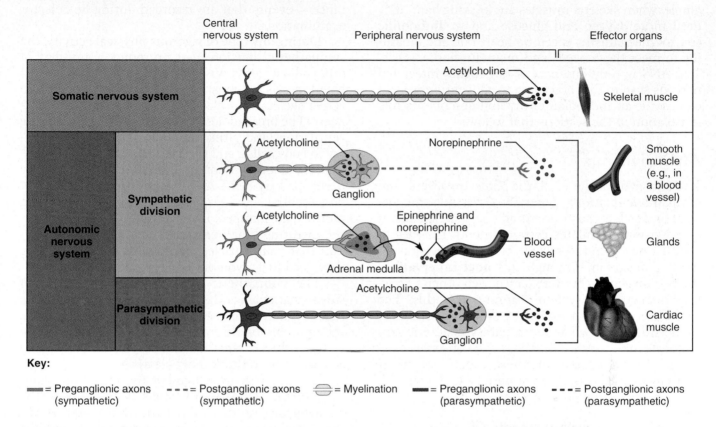

Key:

━━ = Preganglionic axons (sympathetic)   - - - = Postganglionic axons (sympathetic)   ⬭ = Myelination   ━━ = Preganglionic axons (parasympathetic)   - - - = Postganglionic axons (parasympathetic)

FIGURE 14.2 Comparison of somatic and autonomic nervous systems. *Somatic Division:* Axons of somatic motor neurons extend from the CNS to their effectors (skeletal muscle cells). These axons are typically heavily myelinated. Somatic motor neurons release acetylcholine, and the effect is always stimulatory.
Autonomic Division: Axons of most preganglionic neurons run from the CNS to synapse in a peripheral autonomic ganglion with a ganglionic neuron. A few sympathetic preganglionic axons synapse with cells of the adrenal medulla. Postganglionic axons run from the ganglion to effectors (cardiac and smooth muscle fibers and glands). Preganglionic axons are lightly myelinated; postganglionic axons are unmyelinated. All preganglionic fibers release acetylcholine; all parasympathetic postganglionic fibers release acetylcholine; most sympathetic postganglionic fibers release norepinephrine. Stimulated adrenal medullary cells release norepinephrine and epinephrine into the blood. Autonomic effects are stimulatory or inhibitory, depending on the postganglionic neurotransmitter released and the receptor types on the effector organs.

motor division *lacks* ganglia entirely. The dorsal root ganglia are part of the sensory, not the motor, division of the PNS.

Neurotransmitter Effects

All somatic motor neurons release **acetylcholine (ACh)** at their synapses with skeletal muscle fibers. The effect is always *excitatory,* and if stimulation reaches threshold, the muscle fibers contract.

Neurotransmitters released onto visceral effector organs by postganglionic autonomic fibers include **norepinephrine (NE)** secreted by most sympathetic fibers, and ACh released by parasympathetic fibers. Depending on the type of receptors present on the target organ (Figure 14.2 and Table 14.3 on p. 513), the organ's response may be either excitation or inhibition.

Overlap of Somatic and Autonomic Function

Higher brain centers regulate and coordinate both somatic and autonomic motor activities, and nearly all spinal nerves (and many cranial nerves) contain both somatic and autonomic fibers. Moreover, most of the

The axon of the preganglionic neuron spans the distance between the CNS and the autonomic ganglion, whereas the ganglionic neuron's axon lies distal to ("post") the ganglion. ∎

body's adaptations to changing internal and external conditions involve both skeletal muscle activity and enhanced responses of certain visceral organs. For example, when skeletal muscles are working hard, they need more oxygen and glucose and so autonomic control mechanisms speed up heart rate and breathing to meet these needs and maintain homeostasis. The ANS is only one part of our highly integrated nervous system, but according to convention we will consider it an individual entity and describe its role in isolation in the sections that follow.

ANS Divisions

The two arms of the ANS, the *parasympathetic* and *sympathetic divisions,* generally serve the same visceral organs but cause essentially opposite effects. If one division stimulates certain smooth muscles to contract or a gland to secrete, the other division inhibits that action. Through this **dual innervation,** the two divisions counterbalance each other's activities to keep body systems running smoothly. The sympathetic division mobilizes the body during extreme situations, whereas the parasympathetic arm performs maintenance activities and conserves body energy. Let's elaborate on these functional differences by focusing briefly on situations in which each division is exerting primary control.

Role of the Parasympathetic Division

The **parasympathetic division,** sometimes called the "resting and digesting" system, keeps body energy use as low as possible, even as it directs vital "housekeeping" activities like digestion and elimination of feces and urine. (This explains why it is a good idea to relax after a heavy meal: so that digestion is not interfered with by sympathetic activity.) Parasympathetic activity is best illustrated in a person who relaxes after a meal and reads the newspaper. Blood pressure, heart rate, and respiratory rate are regulated at low normal levels, the gastrointestinal tract is actively digesting food, and the skin is warm (indicating that there is no need to divert blood to skeletal muscles or vital organs). In the eyes, the pupils are constricted to protect the retinas from excessive light, and the lenses are accommodated for close vision.

Role of the Sympathetic Division

The **sympathetic division** is often referred to as the "fight-or-flight" system. Its activity is evident when we are excited or find ourselves in emergency or threatening situations, such as being frightened by street toughs late at night. A pounding heart; rapid, deep breathing; cold, sweaty skin; and dilated eye pupils are sure signs of sympathetic nervous system

mobilization. Not as obvious, but equally characteristic, are changes in brain wave patterns and in the electrical resistance of the skin (galvanic skin resistance)—events that are recorded during lie detector examinations.

During any type of vigorous physical activity, the sympathetic division also promotes a number of other adjustments. Visceral (and perhaps cutaneous) blood vessels are constricted, and blood is shunted to active skeletal muscles and the vigorously working heart. The bronchioles in the lungs dilate, increasing ventilation (and ultimately increasing oxygen delivery to body cells), and the liver releases more glucose into the blood to accommodate the increased energy needs of body cells. At the same time, temporarily nonessential activities, such as gastrointestinal tract motility, are damped. If you are running from a mugger, digesting lunch can wait! It is far more important that your muscles be provided with everything they need to get you out of danger.

The sympathetic division generates a head of steam that enables the body to cope with situations that threaten homeostasis. Its function is to provide the optimal conditions for an appropriate response to some threat, whether that response is to run, to see better, or to think more clearly.

An easy way to remember the most important roles of the two ANS divisions is to think of the parasympathetic division as the **D** division [digestion, defecation, and diuresis (urination)], and the sympathetic division as the **E** division (exercise, excitement, emergency, embarrassment). Remember, however, while it is easiest to think of the two ANS divisions as working in an all-or-none fashion, this is rarely the case. A dynamic antagonism exists between the divisions, and fine adjustments are made continuously by both.

ANS Anatomy

The sympathetic and parasympathetic divisions are distinguished by

1. Their unique origin sites: Parasympathetic fibers emerge from the brain and sacral spinal cord (are craniosacral); sympathetic fibers originate in the thoracolumbar region of the spinal cord.

2. The relative lengths of their fibers: The parasympathetic division has long preganglionic and short postganglionic fibers; the sympathetic division has the opposite condition.

3. The location of their ganglia: Most parasympathetic ganglia are located in the visceral effector organs; sympathetic ganglia lie close to the spinal cord.

These and other differences are summarized in Table 14.1 and illustrated in Figure 14.3.

TABLE 14.1) **Anatomical and Physiological Differences Between the Parasympathetic and Sympathetic Divisions**

Characteristic	Parasympathetic	Sympathetic
Origin	Craniosacral outflow: brain stem nuclei of cranial nerves III, VII, IX, and X; spinal cord segments S_2–S_4	Thoracolumbar outflow: lateral horn of gray matter of spinal cord segments T_1–L_2
Location of ganglia	Ganglia in (intramural) or close to visceral organ served	Ganglia within a few centimeters of CNS: alongside vertebral column (chain or paravertebral ganglia) and anterior to vertebral column (collateral or prevertebral ganglia)
Relative length of pre- and postganglionic fibers	Long preganglionic; short postganglionic	Short preganglionic; long postganglionic
Rami communicantes	None	Gray and white rami communicantes. White rami contain myelinated preganglionic fibers; gray contain unmyelinated postganglionic fibers
Degree of branching of preganglionic fibers	Minimal	Extensive
Functional role	Maintenance functions; conserves and stores energy	Prepares body to cope with emergencies and intense muscular activity
Neurotransmitters	All fibers release ACh (cholinergic fibers)	All preganglionic fibers release ACh; most postganglionic fibers release norepinephrine (adrenergic fibers); postganglionic fibers serving sweat glands and some blood vessels of skeletal muscles release ACh; neurotransmitter activity augmented by release of adrenal medullary hormones (norepinephrine and epinephrine)

Parasympathetic (Craniosacral) Division

We begin our exploration of the ANS with the anatomically simpler parasympathetic division. Because its preganglionic fibers spring from opposite ends of the CNS—the brain stem and the sacral region of the spinal cord—the parasympathetic division is also called the **craniosacral division** (Figure 14.4). The preganglionic axons extend from the CNS nearly all the way to the structures to be innervated. There the axons synapse with ganglionic neurons located in **terminal ganglia** that lie very close to or within the target organs. Very short postganglionic axons issue from the terminal ganglia and synapse with effector cells in their immediate area.

Cranial Outflow

Preganglionic fibers run in the oculomotor, facial, glossopharyngeal, and vagus cranial nerves. Their cell bodies lie in associated motor cranial-nerve

FIGURE 14.3 Overview of the subdivisions of the ANS. Sites of origin of their nerves and major organs served are indicated. Synapse sites indicate the relative locations of the sympathetic and parasympathetic ganglia.

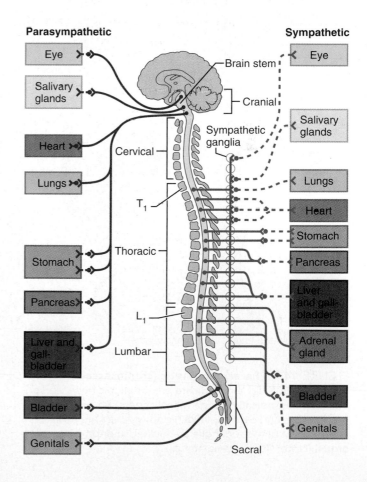

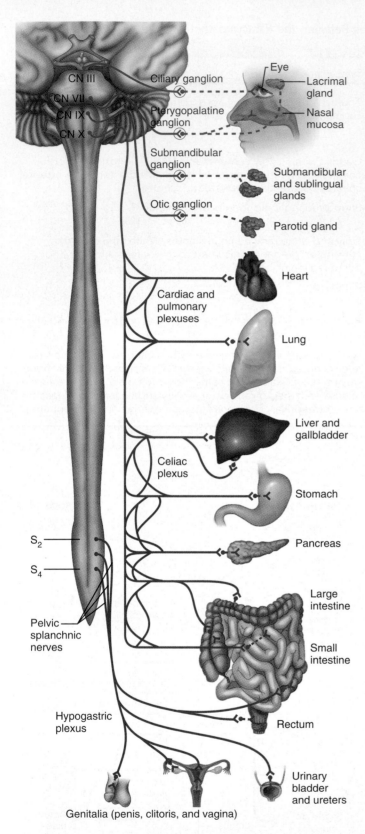

FIGURE 14.4 Parasympathetic (craniosacral) division of the ANS. Solid lines indicate preganglionic nerve fibers. Dashed lines indicate postganglionic fibers. Terminal ganglia of the vagus and pelvic splanchnic nerve fibers are not shown; most of these ganglia are located in or on the target organ. (Note: CN = cranial nerve.)

nuclei in the brain stem (see Figures 12.15 and 12.16). The precise locations of the neurons of the cranial parasympathetics are described next.

1. Oculomotor nerves (III). The parasympathetic fibers of the oculomotor nerves innervate smooth muscles in the eyes that cause the pupils to constrict and the lenses to bulge—actions needed to focus on close objects. The preganglionic axons found in the oculomotor nerves issue from the *accessory oculomotor (Edinger-Westphal) nuclei* in the midbrain. The cell bodies of the ganglionic neurons are in the **ciliary ganglia** within the eye orbits (see Table 13.2, p. 476).

2. Facial nerves (VII). The parasympathetic fibers of the facial nerves stimulate many large glands in the head. Fibers that activate the nasal glands and the lacrimal glands of the eyes originate in the *lacrimal nuclei* of the pons. The preganglionic fibers synapse with ganglionic neurons in the **pterygopalatine ganglia** (ter″eh-go-pal′ah-tīn) just posterior to the maxillae. The preganglionic neurons that stimulate the submandibular and sublingual salivary glands originate in the *superior salivatory nuclei* of the pons and synapse with ganglionic neurons in the **submandibular ganglia,** deep to the mandibular angles (see Table 13.2, p. 478).

3. Glossopharyngeal nerves (IX). The parasympathetics in the glossopharyngeal nerves originate in the *inferior salivatory nuclei* of the medulla and synapse in the **otic ganglia,** located just inferior to the foramen ovale of the skull. The postganglionic fibers course to and activate the parotid salivary glands anterior to the ears (see Table 13.2 pp. 479–480).

Cranial nerves III, VII, and IX supply the entire parasympathetic innervation of the head, however only the *preganglionic fibers* lie within these three pairs of cranial nerves—postganglionic fibers do not. The distal ends of the preganglionic fibers "jump over" to branches of the *trigeminal nerves (V)* to synapse; then the postganglionic fibers travel in the trigeminal nerves to reach the face. This "hitchhiking" takes advantage of the fact that the trigeminal nerves have the broadest facial distribution of all the cranial nerves and are well suited for this "delivery" role to the widely separated glands and smooth muscles in the head.

4. Vagus nerves (X). The remaining and major portion of the parasympathetic cranial outflow is via the vagus (X) nerves. Between them, the two vagus nerves account for about 90% of all preganglionic parasympathetic fibers in the body. They provide fibers to the neck and to nerve plexuses (interweaving networks of nerves) that serve virtually every organ in the thoracic and abdominal cavities. The vagal nerve fibers (preganglionic axons) arise mostly from the *dorsal motor nuclei* of the medulla and synapse in terminal ganglia usually located in the walls of the target organ. Most terminal ganglia are not individually named; instead they are collectively called *intramural ganglia*, literally "ganglia within the walls." As the vagus nerves pass into the thorax, they send branches to the **cardiac plexuses** supplying fibers to the heart that slow heart rate, the **pulmonary plexuses** serving the lungs and bronchi, and the **esophageal plexuses** (ĕ-sof"ah-je'al) supplying the esophagus.

When the main trunks of the vagus nerves reach the esophagus, their fibers intermingle, forming the **anterior** and **posterior vagal trunks,** each containing fibers from both vagus nerves. These vagal trunks then "ride" the esophagus down to the abdominal cavity. There they send fibers *through* the large **aortic plexus** [formed by a number of smaller plexuses (e.g., *celiac, superior mesenteric,* and *hypogastric*) that run along the aorta] before giving off branches to the abdominal viscera. The vagus nerves innervate the liver, gallbladder, stomach, small intestine, kidneys, pancreas, and the proximal half of the large intestine.

Sacral Outflow

The rest of the large intestine and the pelvic organs are served by the sacral outflow, which arises from neurons located in the lateral gray matter of spinal cord segments S_2–S_4. Axons of these neurons run in the ventral roots of the spinal nerves to the ventral rami and then branch off to form the **pelvic (splanchnic) nerves** (Figure 14.4), which pass through the **inferior hypogastric (pelvic) plexus** in the pelvic floor. Some preganglionic fibers synapse with ganglia in this plexus, but most synapse in intramural ganglia in the walls of the following organs: distal half of the large intestine, urinary bladder, ureters, and reproductive organs.

Sympathetic (Thoracolumbar) Division

The sympathetic division is more complex than the parasympathetic division, partly because it innervates more organs. It supplies not only the visceral organs in the internal body cavities but also all visceral structures in the superficial (somatic) part of the body. This sounds impossible, but there is an explanation—some glands and smooth muscle structures in the soma (sweat glands and the hair-raising arrector pili muscles of the skin) require autonomic innervation and are served only by sympathetic fibers. Moreover, all arteries and veins (be they deep or superficial) have smooth muscle in their walls that is innervated by sympathetic fibers. But these matters will be explained later—let us get on with the anatomy of the sympathetic division.

All preganglionic fibers of the sympathetic division arise from cell bodies of preganglionic neurons

in spinal cord segments T_1 through L_2 (Figure 14.5). For this reason, the sympathetic division is also referred to as the **thoracolumbar division** (tho"rah-ko-lum'bar). The presence of numerous preganglionic sympathetic neurons in the gray matter of the spinal cord produces the **lateral horns**—the so-called **visceral motor zones** (see Figures 12.28b, p. 420, and 12.29, p. 421). The lateral horns are just posterolateral to the ventral horns that house somatic motor neurons. (Parasympathetic preganglionic neurons in the sacral cord are far less abundant than the comparable sympathetic neurons in the thoracolumbar regions, and *lateral horns are absent* in the sacral region of the spinal cord. This is a major anatomical difference between the two divisions.)

After leaving the cord via the ventral root, preganglionic sympathetic fibers pass through a **white ramus communicans** [plural, **rami communicantes** (kom-mu'nĭ-kan"tēz)] to enter an adjoining **chain (paravertebral) ganglion** forming part of the **sympathetic trunk** or **chain** (Figure 14.6). Looking like strands of glistening white beads, the sympathetic trunks flank each side of the vertebral column. Thus, these ganglia are named for their location.

Although the sympathetic *trunks* extend from neck to pelvis, sympathetic *fibers* arise only from the thoracic and lumbar cord segments, as shown in Figure 14.5. The ganglia vary in size, position, and number, but typically there are 23 in each sympathetic chain—3 cervical, 11 thoracic, 4 lumbar, 4 sacral, and 1 coccygeal.

Once a preganglionic axon reaches a chain ganglion, one of three things can happen to the axon:

1. It can synapse with a ganglionic neuron in the same chain ganglion (pathway ① in Figure 14.6).

2. It can ascend or descend the sympathetic chain to synapse in another chain ganglion (pathway ② in Figure 14.6). (It is these fibers running from one ganglion to another that connect the ganglia into the sympathetic trunk.)

3. It can pass through the chain ganglion and emerge from the sympathetic chain without synapsing (pathway ③ in Figure 14.6).

Preganglionic fibers following pathway ③ help form the **splanchnic nerves** (splank'nik) that synapse with **prevertebral,** or **collateral, ganglia** located anterior to the vertebral column. Unlike chain (paravertebral) ganglia, the prevertebral ganglia are neither paired nor segmentally arranged and occur only in the abdomen and pelvis.

Regardless of where the synapse occurs, all sympathetic ganglia are close to the spinal cord, and their postganglionic fibers are typically much longer than their preganglionic fibers. Recall that the opposite condition exists in the parasympathetic division, an anatomical distinction that is functionally important.

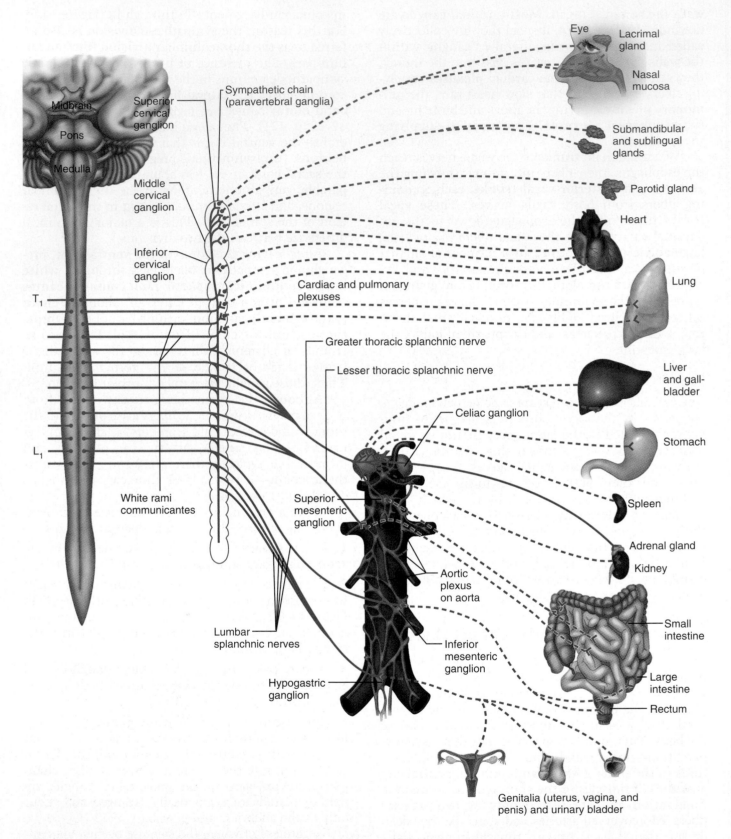

FIGURE 14.5 Sympathetic (thoracolumbar) division of the ANS. Solid lines indicate preganglionic fibers; dashed lines indicate postganglionic fibers. The least thoracic and sacral splanchnic nerves are not shown.

? On what basis are the rami communicantes distinguished as gray or white?

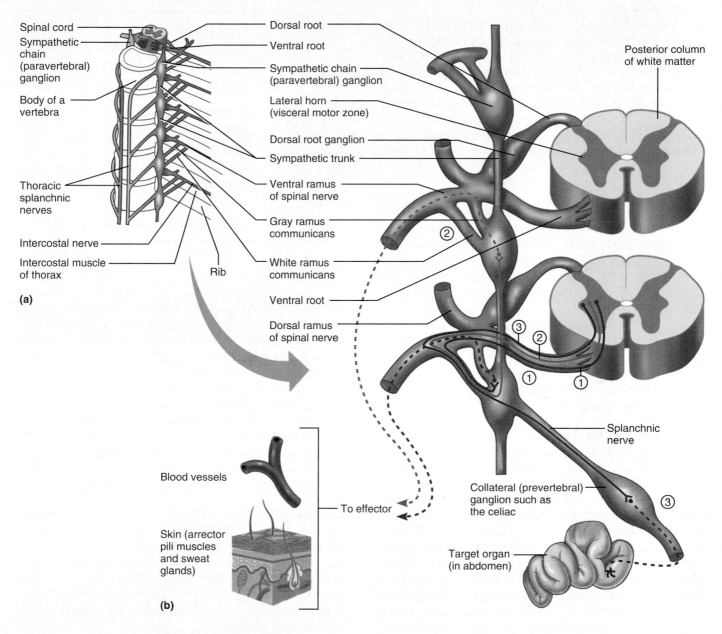

FIGURE 14.6 Sympathetic trunks and pathways. (a) The organs of the anterior thorax have been removed to allow visualization of the chain ganglia of the sympathetic trunks. **(b)** Three sympathetic pathways. ① Synapse in a chain (paravertebral) ganglion at same level. ② Synapse in a chain ganglion at a different level. ③ Synapse in a collateral (prevertebral) ganglion anterior to the vertebral column.

Pathways with Synapses in Chain Ganglia

When synapses are made in chain ganglia, the postganglionic axons enter the ventral (or dorsal) ramus of the adjoining spinal nerves by way of communicating branches called **gray rami communicantes**

The gray rami are unmyelinated. The white rami are myelinated, which gives them a distinguishing white color. ■

(Figure 14.6). From there they travel via branches of the rami to sweat glands and arrector pili muscles of the skin. Anywhere along their path, the postganglionic axons may transfer over to nearby blood vessels and innervate the vascular smooth muscle all the way to their final branches.

Notice that the naming of the rami communicantes as *white* or *gray* reflects their appearance, revealing whether or not their fibers are myelinated (and has *no* relationship to the white and gray matter of the CNS). Preganglionic fibers composing the white rami are myelinated. Postganglionic axons forming the gray rami are not.

The white rami, which carry preganglionic axons to the sympathetic trunks, are found only in the T_1–T_2 cord segments, regions of sympathetic outflow. However, gray rami carrying postganglionic fibers headed for the periphery issue from *every* chain ganglion from the cervical to the sacral region, allowing sympathetic output to reach all parts of the body. Because parasympathetic fibers never run in spinal nerves, *rami communicantes are associated only with the sympathetic division.*

Pathways to the Head

Sympathetic preganglionic fibers serving the head emerge from spinal cord segments T_1–T_4 and ascend the sympathetic chain to synapse with ganglionic neurons in the **superior cervical ganglion.** This ganglion contributes sympathetic fibers that run in several cranial nerves and with the upper three or four cervical spinal nerves. Besides serving the skin and blood vessels of the head, its fibers stimulate the dilator muscles of the irises of the eyes, inhibit the nasal and salivary glands (the reason your mouth goes dry when you are scared), and innervate the smooth (tarsal) muscle that lifts the upper eyelid. The superior cervical ganglion also sends direct branches to the heart.

Pathways to the Thorax

Sympathetic preganglionic fibers innervating the thoracic organs originate at T_1–T_6. From there the preganglionic fibers run to synapse in the cervical chain ganglia. Postganglionic fibers emerging from the **middle and inferior cervical ganglia** enter cervical nerves C_4–C_8. Some of these fibers innervate the heart via the cardiac plexus, and some innervate the thyroid gland, but most serve the skin. Additionally, some T_1–T_6 preganglionic fibers synapse in the nearest chain ganglion, and the postganglionic fibers pass directly to the organ served. Fibers to the heart, aorta, lungs, and esophagus take this direct route. Along the way, they run into the plexuses associated with those organs. Sympathetic innervation by spinal cord segment is summarized in Table 14.2.

Pathways with Synapses in Collateral Ganglia

The preganglionic fibers from T_5 down synapse in collateral ganglia; thus these fibers enter and leave the sympathetic chains without synapsing. They form several nerves called splanchnic nerves, including the **thoracic splanchnic nerves** (greater, lesser, and least) and the **lumbar** and **sacral splanchnic nerves.** The splanchnic nerves contribute to a number of interweaving nerve plexuses

TABLE 14.2   Segmental Sympathetic Supplies

Spinal Cord Segment	Organs Served
T_1–T_5	Head and neck, heart
T_2–T_4	Bronchi and lungs
T_2–T_5	Upper limb
T_5–T_6	Esophagus
T_6–T_{10}	Stomach, spleen, pancreas
T_7–T_9	Liver
T_9–T_{10}	Small intestine
T_{10}–L_1	Kidney, reproductive organs (uterus, testis, ovary, etc.)
T_{10}–L_2	Lower limb
T_{11}–L_2	Large intestine, ureter, urinary bladder

known collectively as the **abdominal aortic plexus** (a-or′tik), which clings to the surface of the abdominal aorta (Figure 14.5). This complex plexus contains several ganglia that together serve the abdominopelvic viscera (*splanchni* = viscera). From superior to inferior, the most important of these ganglia (and related subplexuses) are the **celiac, superior mesenteric, inferior mesenteric,** and **hypogastric,** named for the arteries with which they most closely associate. Postganglionic fibers issuing from these ganglia generally travel to their target organs in the company of the arteries serving these organs.

Pathways to the Abdomen

Sympathetic innervation of the abdomen is via preganglionic fibers from T_5 to L_2, which travel in the *thoracic splanchnic nerves* to synapse mainly at the celiac and superior mesenteric ganglia. Postganglionic fibers issuing from these ganglia serve the stomach, intestines (except the distal half of the large intestine), liver, spleen, and kidneys.

Pathways to the Pelvis

Preganglionic fibers innervating the pelvis originate from T_{10} to L_2 and then descend in the sympathetic trunk to the lumbar and sacral chain ganglia. Some fibers synapse there, but most go directly out from the cord via the *lumbar* and *sacral splanchnic nerves,* which send the bulk of their fibers to the inferior mesenteric and hypogastric ganglia. From these ganglia issue postganglionic fibers serving the distal half of the large intestine, the urinary bladder, and the reproductive organs. For the most part, sympathetic fibers *inhibit* the activity of the muscles and glands in these visceral organs.

Pathways with Synapses in the Adrenal Medulla

Some fibers traveling in the thoracic splanchnic nerves pass through the celiac ganglion without synapsing and terminate by synapsing with the hormone-producing medullary cells of the adrenal gland (Figure 14.5). When stimulated by preganglionic fibers, the medullary cells secrete NE and *epinephrine* (also called *noradrenaline* and *adrenaline*, respectively) into the blood, producing the excitatory effects we have all felt as a "surge of adrenaline." Embryologically, sympathetic ganglia and the adrenal medulla arise from the same tissue. For this reason, the adrenal medulla is sometimes viewed as a "misplaced" sympathetic ganglion, and its hormone-releasing cells, although lacking nerve processes, are considered equivalent to ganglionic sympathetic neurons.

Visceral Reflexes

Because most anatomists consider the ANS to be a visceral motor system, the presence of sensory fibers (mostly visceral pain afferents) is often overlooked. However, **visceral sensory neurons**, which send information concerning chemical changes, stretch, and irritation of the viscera, are the first link in autonomic reflexes. **Visceral reflex arcs** have essentially the same components as somatic reflex arcs—receptor, sensory neuron, integration center, motor neuron, effector—except that a visceral reflex arc has a *two-neuron* motor chain (Figure 14.7).

Nearly all the sympathetic and parasympathetic fibers described thus far are accompanied by afferent fibers conducting sensory impulses from glands or muscles. Thus, peripheral processes of visceral sensory neurons are found in cranial nerves VII, IX, and X, splanchnic nerves, and the sympathetic trunk, as well as in spinal nerves. Like sensory neurons serving somatic structures (skeletal muscles and skin), the cell bodies of visceral sensory neurons are located either in the sensory ganglia of associated cranial nerves or in dorsal root ganglia of the spinal cord. Visceral sensory neurons are also found in sympathetic ganglia where preganglionic neurons synapse. Furthermore, complete three-neuron reflex arcs (with sensory, motor, and intrinsic neurons) exist entirely within the walls of the gastrointestinal tract. Neurons composing these reflex arcs make up the *enteric nervous system*, which plays an important role in controlling gastrointestinal tract activity.

The fact that visceral pain afferents travel along the same pathways as somatic pain fibers helps explain the phenomenon of **referred pain**, in which pain stimuli arising in the viscera are perceived as somatic in origin. For example, a heart attack may

 According to the figure legend, the afferent neuron shown in this diagram reaches the CNS via a spinal nerve. How would you know that if it was not mentioned?

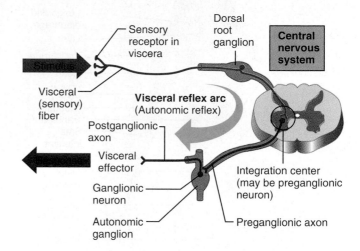

FIGURE 14.7 Visceral reflexes. Visceral reflexes have the same elements as somatic reflexes but occur over polysynaptic pathways because of the two-motor-neuron efferent pathway. The integration center may involve a dorsal horn interneuron as shown, or it may just be a synapse with a preganglionic neuron. The visceral afferent fibers are found both in spinal nerves (as depicted here) and in autonomic nerves.

produce a sensation of pain that radiates to the superior thoracic wall and along the medial aspect of the left arm. Because the same spinal segments (T_1–T_5) innervate both the heart and the regions to which pain signals from heart tissue are referred, the brain interprets most such inputs as coming from the more common somatic pathway. Cutaneous areas to which visceral pain is commonly referred are shown in Figure 14.8.

ANS Physiology

Neurotransmitters and Receptors

Acetylcholine and *norepinephrine* are the major neurotransmitters released by ANS neurons. ACh, the same neurotransmitter secreted by somatic motor neurons, is released by (1) all ANS preganglionic axons and (2) all parasympathetic postganglionic axons at synapses with their effectors. ACh-releasing fibers are called **cholinergic fibers** (ko"lin-er'jik).

The visceral afferent fiber enters the spinal cord via a dorsal root ganglion, found only on the dorsal root of spinal nerves. ∎

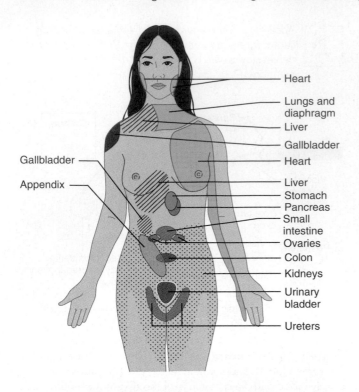

Gallbladder

Appendix

Heart

Lungs and
diaphragm

Liver

Gallbladder

Heart

Liver

Stomach

Pancreas

Small
intestine

Ovaries

Colon

Kidneys

Urinary
bladder

Ureters

FIGURE 14.8 Referred pain. Anterior cutaneous areas to which pain from certain visceral organs is referred.

In contrast, most sympathetic postganglionic axons release NE and are classified as **adrenergic fibers** (ad″ren-er′jik). The only exceptions are sympathetic postganglionic fibers innervating sweat glands and some blood vessels in skeletal muscles and the external genitalia; these fibers secrete ACh.

Unfortunately for memorization purposes, the effects of ACh and NE on their effectors are not consistently either excitation or inhibition. The response of visceral effectors to these neurotransmitters depends not only on the neurotransmitters but also on the receptors to which they attach. The two or more kinds of receptors for each autonomic neurotransmitter allow them to exert different effects (activation or inhibition) at different body targets (Table 14.3).

Cholinergic Receptors

The two types of receptors that bind ACh are named for drugs that bind to them and mimic acetylcholine's effects. The first of these receptors identified were the **nicotinic receptors** (nik″o-tin′ik). A mushroom poison, *muscarine* (mus′kah-rin), activates a different set of ACh receptors, named **muscarinic receptors.** All ACh receptors are either nicotinic or muscarinic.

Nicotinic receptors are found on (1) motor end plates of skeletal muscle cells (which, however, are somatic and not autonomic targets), (2) *all* ganglionic neurons, both sympathetic and parasympathetic, and (3) the hormone-producing cells of the adrenal medulla. The effect of ACh binding to nicotinic receptors is *always* stimulatory.

Muscarinic receptors occur on all effector cells stimulated by postganglionic cholinergic fibers—that is, all parasympathetic target organs and a few sympathetic targets, such as eccrine sweat glands and some blood vessels of skeletal muscles. The effect of ACh binding to muscarinic receptors can be either inhibitory or stimulatory, depending on the target organ. For example, binding of ACh to cardiac muscle receptors slows heart activity, whereas ACh binding to receptors on smooth muscle of the gastrointestinal tract increases its motility.

Adrenergic Receptors

There are also two major classes of adrenergic (NE-binding) receptors: **alpha** (α) and **beta** (β). Organs that respond to NE (or epinephrine) have one or both types of receptor. In general, NE or epinephrine binding to α receptors is stimulatory, and their binding to β receptors is inhibitory. But there are exceptions. For example, binding of NE to the β receptors of cardiac muscle prods the heart into more vigorous activity. These differences reflect the fact that both α and β receptors have two or three receptor subclasses (α_1 and α_2; β_1, β_2, and β_3), and each receptor type predominates in certain target organs. An overview of the locations and effects of these receptor subclasses is provided in Table 14.3.

The Effects of Drugs

Knowing the locations of the cholinergic and adrenergic receptor subtypes allows specific drugs to be prescribed to obtain the desired inhibitory or stimulatory effects on selected target organs. For example, atropine, an anticholinergic drug that blocks parasympathetic effects, is routinely administered before surgery to prevent salivation and to dry up respiratory system secretions. It is also used by ophthalmologists to dilate the pupils for eye examination. The anticholinesterase drug *neostigmine* inhibits acetylcholinesterase, thus preventing enzymatic breakdown of ACh and allowing it to accumulate in the synapses. This drug is used to treat myasthenia gravis, a condition in which skeletal muscle activity is impaired for lack of ACh stimulation.

As described in Chapter 11, NE is one of our "feeling good" neurotransmitters, and drugs that prolong the activity of NE on the postsynaptic membrane help to relieve depression. Hundreds of over-the-counter drugs used to treat colds, coughs, allergies, and nasal congestion contain sympathomimetics (ephedrine, phenylephrine, and others),

TABLE 14.3 Cholinergic and Adrenergic Receptors

Neurotransmitter	Receptor Type	Major Locations	Effect of Binding
Acetylcholine	**Cholinergic**		
	Nicotinic	All ganglionic neurons; adrenal medullary cells (also neuromuscular junctions of skeletal muscle)	Excitation
	Muscarinic	All parasympathetic target organs	Excitation in most cases; inhibition of cardiac muscle
		Selected sympathetic targets:	
		▪ Eccrine sweat glands	Activation
		▪ Blood vessels in skeletal muscles	Inhibition (causes vasodilation)
Norepinephrine (and epinephrine released by adrenal medulla)	**Adrenergic**		
	β_1	Heart and coronary blood vessels predominantly, but also kidneys, liver, and adipose tissue	Increases heart rate and strength; dilates coronary arterioles; stimulates renin release by kidneys
	β_2	Lungs and most other sympathetic target organs; abundant on blood vessels serving the heart	Stimulates secretion of insulin by pancreas; other effects mostly inhibitory: dilation of blood vessels and bronchioles; relaxes smooth muscle walls of digestive and urinary visceral organs; relaxes pregnant uterus
	β_3	Adipose tissue	Stimulates lipolysis by fat cells
	α_1	Most importantly blood vessels serving the skin, mucosae, abdominal viscera, kidneys, and salivary glands; but virtually all sympathetic target organs except heart	Constricts blood vessels and visceral organ sphincters; dilates pupils of the eyes
	α_2	Membrane of adrenergic axon terminals; blood platelets	Mediates inhibition of NE release from adrenergic terminals; promotes blood clotting

sympathetic-mimicking drugs that stimulate α-adrenergic receptors. Much pharmaceutical research is directed toward finding drugs that affect only one subclass of receptor without upsetting the whole adrenergic or cholinergic system. An important breakthrough was finding adrenergic blockers that attach mainly to cardiac β_1 receptors. Also called *beta-blockers*, these drugs and others of their class are used when the goal is to reduce heart rate and prevent arrhythmias (irregular heartbeats) without interfering with other sympathetic effects. Selected drug classes that influence ANS activity are summarized in Table 14.4.

Interactions of the Autonomic Divisions

As mentioned earlier, most visceral organs receive *dual innervation*. Normally, both ANS divisions are partially active, producing a dynamic antagonism that allows visceral activity to be precisely controlled. However, one division or the other usually exerts the predominant effects in given circumstances, and in a few cases, the two divisions actually cooperate with each other. The major effects of the two divisions are summarized in Table 14.5.

Antagonistic Interactions

Antagonistic effects, described earlier, are most clearly seen on the activity of the heart, respiratory system, and gastrointestinal organs. In a fight-or-flight situation, the sympathetic division increases respiratory and heart rates while inhibiting digestion and elimination. When the emergency is over, the parasympathetic division restores heart and breathing rates to resting levels and then attends to processes that refuel body cells and discard wastes.

Sympathetic and Parasympathetic Tone

Although we have described the parasympathetic division as the "resting and digesting" division, the sympathetic division is the major actor in controlling blood pressure, even at rest. With few exceptions, the vascular system is entirely innervated by

TABLE 14.4 ⟩ **Selected Drug Classes That Influence the Activity of the Autonomic Nervous System**

Drug Class	Receptor Bound	Effects	Example	Clinical Use
Nicotinic agents (little therapeutic value, but important because of presence of nicotine in tobacco)	Nicotinic ACh receptors on all ganglionic neurons	Typically stimulation of sympathetic effects; heart rhythm becomes less regular; blood pressure increases	Nicotine	Smoking cessation products
Ganglionic blocking agents	Nicotinic ACh receptors on all ganglionic neurons	ACh blocking agents; typically block sympathetic effects, overshadowing blocking effects on the PNS	Trimethaphan camsylate	To treat high blood pressure by producing vasodilation
Parasympathomimetic agents (muscarinic agents)	Muscarinic ACh receptors	Mimics effects of ACh, enhances PNS effects	Pilocarpine	To constrict pupils after they've been dilated for an eye exam; to treat wide-angle glaucoma (opens aqueous humor drainage pores)
Acetylcholinesterase inhibitors	None; binds to the enzyme (AChE) that degrades ACh	Indirect effect at all ACh receptors; prolongs the effect of ACh	Neostigmine	To treat myasthenia gravis, increasing muscle strength by increasing availability of ACh
Sympathomimetic agents	Adrenergic receptors	Enhances sympathetic activity by increasing NE release or binding to adrenergic receptors	Phenylephrine	Dilate bronchioles by binding to α_1 receptors
Sympatholytic agents	Adrenergic receptors	Decreases sympathetic activity by blocking adrenergic receptors or inhibiting NE release	Propranolol	To treat hypertension; a β_1 blocker that decreases heart rate and blood pressure

sympathetic fibers that keep the blood vessels in a continual state of partial constriction called **sympathetic** or **vasomotor tone.** When faster blood delivery is needed, these fibers deliver impulses more rapidly, causing blood vessels to constrict and blood pressure to rise. When blood pressure is to be decreased, the vessels are prompted to dilate. *Alphablocker drugs* that interfere with the activity of these **vasomotor fibers** are used to treat hypertension. During circulatory shock (inadequate blood delivery to body tissues), or when more blood is needed to meet the increased needs of working skeletal muscles, blood vessels serving the skin and abdominal viscera are strongly constricted. This blood "shunting" helps maintain circulation to vital organs or enhance blood delivery to skeletal muscles.

On the other hand, parasympathetic effects normally dominate the heart and the smooth muscle of digestive and urinary tract organs. Thus, these organs exhibit **parasympathetic tone.** The parasympathetic division slows the heart and dictates the normal activity levels of the digestive and urinary tracts. However, the sympathetic division can override these parasympathetic effects during times of stress. Drugs that block parasympathetic responses increase heart rate and cause fecal and urinary

retention. Except for the adrenal glands and sweat glands of the skin, most glands are activated by parasympathetic fibers.

Cooperative Effects

The best example of cooperative ANS effects is seen in controls of the external genitalia. Parasympathetic stimulation causes vasodilation of blood vessels in the external genitalia and is responsible for erection of the male penis or female clitoris during sexual excitement. (This may explain why sexual performance is sometimes impaired when people are anxious or upset and the sympathetic division is in charge.) Sympathetic stimulation then causes ejaculation of semen by the penis or reflex peristalsis of a female's vagina.

Unique Roles of the Sympathetic Division

The adrenal medulla, sweat glands and arrector pili muscles of the skin, the kidneys, and most blood vessels receive only sympathetic fibers. It is easy to remember that the sympathetic system innervates these structures because most of us sweat under stress, our scalp "prickles" during fear, and our blood pressure skyrockets (from widespread vasoconstriction) when

TABLE 14.5 Effects of the Parasympathetic and Sympathetic Divisions on Various Organs

Target Organ or System	Parasympathetic Effects	Sympathetic Effects
Eye (iris)	Stimulates constrictor muscles; constricts eye pupils	Stimulates dilator muscles; dilates eye pupils
Eye (ciliary muscle)	Stimulates muscles, which results in bulging of the lens for accommodation and close vision	No effect (no innervation)
Glands (nasal, lacrimal, salivary, gastric, pancreas)	Stimulates secretory activity	Inhibits secretory activity; causes vasoconstriction of blood vessels supplying the glands
Sweat glands	No effect (no innervation)	Stimulates copious sweating (cholinergic fibers)
Adrenal medulla	No effect (no innervation)	Stimulates medulla cells to secrete epinephrine and norepinephrine
Arrector pili muscles attached to hair follicles	No effect (no innervation)	Stimulates to contract (erects hairs and produces "goosebumps")
Heart muscle	Decreases rate; slows and steadies heart	Increases rate and force of heartbeat
Heart: coronary blood vessels	Constricts coronary vessels	Causes vasodilation
Bladder/urethra	Causes contraction of smooth muscle of bladder wall; relaxes urethral sphincter; promotes voiding	Causes relaxation of smooth muscle of bladder wall; constricts urethral sphincter; inhibits voiding
Lungs	Constricts bronchioles	Dilates bronchioles and mildly constricts blood vessels
Digestive tract organs	Increases motility (peristalsis) and amount of secretion by digestive organs; relaxes sphincters to allow movement of foodstuffs along tract	Decreases activity of glands and muscles of digestive system and constricts sphincters (e.g., anal sphincter)
Liver	No effect (no innervation)	Epinephrine stimulates liver to release glucose to blood
Gallbladder	Excites (gallbladder contracts to expel bile)	Inhibits (gallbladder is relaxed)
Kidney	No effect (no innervation)	Causes vasoconstriction; decreases urine output; promotes renin formation
Penis	Causes erection (vasodilation)	Causes ejaculation
Vagina/clitoris	Causes erection (vasodilation) of clitoris	Causes reverse peristalsis (contraction) of vagina; increases mucus secretion
Blood vessels	Little or no effect	Constricts most vessels and increases blood pressure; constricts vessels of abdominal viscera and skin to divert blood to muscles, brain, and heart when necessary; dilates vessels of the skeletal muscles (via cholinergic fibers and epinephrine) during exercise
Blood coagulation	No effect (no innervation)	Increases coagulation
Cellular metabolism	No effect (no innervation)	Increases metabolic rate
Adipose tissue	No effect (no innervation)	Stimulates lipolysis (fat breakdown)
Mental activity	No effect (no innervation)	Increases alertness

we get excited. We have already described how sympathetic control of blood vessels regulates blood pressure and shunting of blood in the vascular system. We will now mention several other uniquely sympathetic functions.

Thermoregulatory Responses to Heat   The sympathetic division mediates reflexes that regulate body temperature. For example, applying heat to the skin causes reflexive dilation of blood vessels in that area. When systemic body temperature is elevated, sympathetic nerves cause the skin's blood vessels to dilate, allowing the skin to become flushed with warm blood, and activate the sweat glands to help cool the body. When body temperature falls, skin blood vessels are constricted, and, as a result, blood is restricted to deeper, more vital organs.

Release of Renin from the Kidneys   Sympathetic impulses stimulate the kidneys to release renin, an enzyme that promotes an increase in blood pressure. This renin-angiotensin mechanism is described in Chapter 24.

Metabolic Effects   Through both direct neural stimulation and release of adrenal medullary hormones, the sympathetic division promotes a number of metabolic effects not reversed by parasympathetic activity. It (1) increases the metabolic rate of body cells; (2) raises blood glucose levels; (3) mobilizes fats for use as fuels; and (4) increases mental alertness by stimulating the reticular activating system of the brain stem. These medullary hormones also cause skeletal muscle to contract more strongly and quickly. As a side effect, muscle spindles are stimulated more often and, consequently, nerve impulses traveling to the muscles occur more synchronously. These neutral bursts, which put muscle contractions on a "hair trigger," are great if you have to make a quick jump or run, but they can be embarrassing or even disabling to the nervous musician or surgeon.

Localized Versus Diffuse Effects

In the parasympathetic division, one preganglionic neuron synapses with one (or at most a few) ganglionic neurons. Additionally, all parasympathetic fibers release ACh, which is quickly destroyed (hydrolyzed) by acetylcholinesterase. Consequently, the parasympathetic division exerts short-lived, highly localized control over its effectors. In contrast, preganglionic sympathetic axons branch profusely as they enter the sympathetic trunk, and they synapse with ganglionic neurons at several levels. Thus, when the sympathetic division is activated, it responds in a diffuse and highly interconnected way. Indeed, the literal translation of sympathetic (*sym* = together; *pathos* = feeling) relates to the body-wide mobilization this division provokes.

Effects produced by sympathetic activation are much longer-lasting than parasympathetic effects. This reflects three phenomena: (1) NE is inactivated more slowly than ACh because it must be taken back up into the presynaptic ending before being hydrolyzed (or stored); (2) NE acts indirectly through a second-messenger system, so it exerts its effects more slowly than ACh, which acts directly; and (3) when the sympathetic division is mobilized, NE and epinephrine are secreted into the blood by adrenal medullary cells. Although epinephrine is more potent at increasing heart rate and raising blood sugar levels and metabolic rate, these hormones have essentially the same effects as NE released by sympathetic neurons. In fact, circulating adrenal medullary hormones produce 25 to 50% of all the sympathetic effects acting on the body at a given time. These effects continue for several minutes until the hormones are destroyed by the liver. Thus, although sympathetic nerve impulses act only briefly, the hormonal effects they provoke linger. The widespread and prolonged effect of sympathetic activation helps explain why we need time to "come down" after an extremely stressful experience.

Control of Autonomic Functioning

Although the ANS is not usually considered to be under voluntary control, its activity *is* regulated by CNS controls in the spinal cord, brain stem, hypothalamus, and cerebral cortex (Figure 14.9). In general, the hypothalamus is like a "head ganglion" at the top of the ANS control hierarchy. From there, orders flow to lower and lower CNS centers for execution. Although the cerebral cortex may modify the workings of the ANS, it does so at the subconscious level and by acting through limbic system structures on hypothalamic centers.

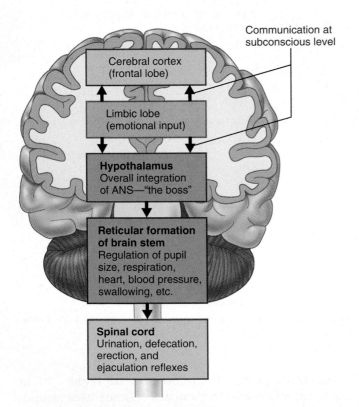

FIGURE 14.9   Levels of ANS control. The hypothalamus stands at the top of the control hierarchy as the integrator of ANS activity. However, subconscious cerebral inputs via limbic lobe connections influence hypothalamic functioning.

Brain Stem and Spinal Cord Controls

The hypothalamus is the "boss," but the brain stem reticular formation appears to exert the most *direct* influence over autonomic functions. For example, certain motor centers in the ventrolateral medulla (*cardiac* and *vasomotor centers*) reflexively regulate heart rate and blood vessel diameter and respiration. Other medullary regions oversee gastrointestinal activities. Most sensory impulses involved in eliciting these autonomic reflexes reach the brain stem via vagus nerve afferents. The pons contains *respiratory centers* that interact with those of the medulla, and midbrain centers (*oculomotor nuclei*) are concerned with controlling the size of the eye pupils. Defecation and micturition reflexes that promote emptying of the rectum and bladder are integrated at the spinal cord level but are subject to conscious inhibition. All of these autonomic reflexes are described in later chapters in relation to the organ systems they serve.

Hypothalamic Controls

As noted, the main integration center of the autonomic nervous system is the hypothalamus. Medial and anterior hypothalamic regions direct parasympathetic functions, whereas lateral and posterior areas direct sympathetic functions. These centers exert their effects both directly and via relays through the *reticular formation,* which in turn influences the preganglionic motor neurons in the brain stem and spinal cord (Figure 14.9). The hypothalamus contains centers that coordinate heart activity, blood pressure, body temperature, water balance, and endocrine activity. It also contains centers that mediate various emotional states (rage, pleasure) and biological drives (thirst, hunger, sex).

Via its associations with the periaqueductal gray matter and the amygdaloid nucleus, the hypothalamus also mediates our reactions to fear. Emotional responses of the limbic lobe of the cerebrum to danger and stress signal the hypothalamus to activate the sympathetic system to fight-or-flight status. Thus, the hypothalamus serves as the keystone of the emotional and visceral brain, and through its centers emotions influence autonomic nervous system functioning and behavior.

Cortical Controls

It was originally believed that the ANS is not subject to voluntary controls. However, we have all had occasions when remembering a frightening event made our heart race (sympathetic response) or just the thought of a favorite food, pecan pie for example, made our mouth water (a parasympathetic response). These inputs converge on the hypothalamus through its connections to the limbic lobe.

Additionally, studies have shown that voluntary cortical control of visceral activities is possible—a capability untapped by most people.

Influence of Biofeedback on Autonomic Function

During **biofeedback training,** subjects are connected to monitoring devices that provide an awareness of what is happening in their body. This awareness is called **biofeedback.** The devices detect and amplify changes in physiological processes such as heart rate and blood pressure, and these data are "fed back" in the form of flashing lights or audible tones. Subjects are asked to try to alter or control some "involuntary" function by concentrating on calming, pleasant thoughts. The monitor allows them to identify changes in the desired direction, so they can recognize the feelings associated with these changes and learn to produce the changes at will.

Biofeedback techniques have been successful in helping individuals plagued by migraine headaches. They are also used by cardiac patients to manage stress and reduce their risk of heart attack. However, biofeedback training is time-consuming and often frustrating, and the training equipment is expensive and difficult to use.

Homeostatic Imbalances of the ANS

Because the ANS is involved in nearly every important process that goes on in the body, it is not surprising that abnormalities of autonomic functioning can have far-reaching effects and threaten life itself. Most autonomic disorders reflect exaggerated or deficient controls of smooth muscle activity. Of these, the most devastating involve blood vessels and include conditions such as hypertension, Raynaud's disease, and the mass reflex reaction.

Hypertension, or high blood pressure, may result from an overactive sympathetic vasoconstrictor response promoted by continuous high levels of stress. Hypertension is always serious because it increases the workload on the heart, which may precipitate heart disease, and increases the wear and tear on artery walls. Stress-induced hypertension is treated with adrenergic receptor–blocking drugs.

Raynaud's disease is characterized by intermittent attacks causing the skin of the fingers and toes to become pale, then cyanotic and painful. Commonly provoked by exposure to cold or emotional stress, it is an exaggerated vasoconstriction response. The severity of this disease ranges from merely uncomfortable to such severe blood vessel constriction that ischemia and gangrene (tissue death) results. To treat severe cases, preganglionic sympathetic fibers serving the affected regions are severed (a procedure called *sympathectomy*). The involved vessels then dilate, reestablishing adequate blood delivery to the region.

The *mass reflex reaction* is a life-threatening condition involving uncontrolled activation of both autonomic and somatic motor neurons. It occurs in a majority of individuals with quadriplegia and in others with spinal cord injuries above the T_6 level. The initial cord injury is followed by *spinal shock* (see p. 426). When reflex activity returns, it is usually exaggerated because of lack of inhibitory input from higher centers. Then episodes of mass reflex, surges of nervous output from large regions of the spinal cord, begin. The usual trigger is a painful stimulus to the skin or overfilling of a visceral organ, such as the bladder. The body goes into flexor spasms, the colon and bladder empty, and profuse sweating begins. Arterial blood pressure rises to life-threatening levels, which may rupture a blood vessel in the brain, precipitating stroke. The precise mechanism of the mass reflex is unknown, but it is envisioned as a type of epilepsy of the spinal cord.

Review Questions

Multiple Choice/Matching

(Some questions have more than one correct answer. Select the best answer or answers from the choices given.)

1. All of the following characterize the ANS except (a) two-neuron efferent chain, (b) presence of nerve cell bodies in the CNS, (c) presence of nerve cell bodies in the ganglia, (c) innervation of skeletal muscles.

2. Relate each of the following terms or phrases to either the sympathetic (S) or parasympathetic (P) division of the autonomic nervous system:

_____ **(1)** short preganglionic, long postganglionic fibers
_____ **(2)** intramural ganglia
_____ **(3)** craniosacral outflow
_____ **(4)** adrenergic fibers
_____ **(5)** cervical ganglia
_____ **(6)** otic and ciliary ganglia
_____ **(7)** more specific control
_____ **(8)** increases heart rate, respiratory rate, and blood pressure
_____ **(9)** increases gastric motility and secretion of lacrimal, salivary, and digestive juices
_____ **(10)** innervates blood vessels
_____ **(11)** most active when you are swinging in a hammock
_____ **(12)** active when you are running in the Boston Marathon

3. Preganglionic neurons develop from (a) neural crest cells, (b) neural tube cells, (c) alar plate cells, (d) endoderm.

4. The white rami communicantes contain what kind of fibers? (a) preganglionic parasympathetic, (b) postganglionic parasympathetic, (c) preganglionic sympathetic, (d) postganglionic sympathetic.

5. Prevertebral sympathetic ganglia are involved with the innervation of the (a) abdominal organs, (b) thoracic organs, (c) head, (d) arrector pili, (e) all of these.

Short Answer Essay Questions

6. Briefly explain why the following terms are sometimes used to refer to the autonomic nervous system: involuntary nervous system and emotional-visceral system.

7. Describe the anatomical relationship of the white and gray rami to the spinal nerve, and indicate the kind of fibers found in each ramus type.

8. Indicate the results of sympathetic activation of the following structures: sweat glands, eye pupils, adrenal medulla, heart, lungs, liver, blood vessels of vigorously working skeletal muscles, blood vessels of digestive viscera, salivary glands.

9. Which of the effects listed in response to question 8 would be reversed by parasympathetic activity?

10. Which ANS fibers release acetylcholine? Which release norepinephrine?

11. Describe the meaning and importance of sympathetic tone and parasympathetic tone.

12. List the receptor subtypes for ACh and NE, and indicate the major sites where each type is found.

13. What area of the brain is most directly involved in mediating autonomic reflexes?

14. Describe the importance of the hypothalamus in controlling the autonomic nervous system.

15. Describe the basis and uses of biofeedback training.

16. Ganglionic neurons were initially called postganglionic neurons. Why is this a misnomer?

15

THE ENDOCRINE SYSTEM

1. Indicate important differences between hormonal and neural controls of body functioning.

The Endocrine System: An Overview (pp. 520–521)

2. List the major endocrine organs, and describe their body locations.

3. Distinguish between circulating hormones and local hormones.

Hormones (pp. 521–527)

4. Describe how hormones are classified chemically.

5. Describe the two major mechanisms by which hormones bring about their effects on their target tissues.

6. List three kinds of interaction that different hormones acting on the same target cell can have.

7. Explain how hormone release is regulated.

Major Endocrine Organs (pp. 527–550)

8. Describe structural and functional relationships between the hypothalamus and the pituitary gland.

9. List and describe the chief effects of adenohypophyseal hormones.

10. Discuss the structure of the neurohypophysis, and describe the effects of the two hormones it releases.

11. Describe important effects of the two groups of hormones produced by the thyroid gland. Follow the process of thyroxine formation and release.

12. Indicate general functions of parathyroid hormone.

13. List hormones produced by the adrenal gland, and cite their physiological effects.

14. Compare and contrast the effects of the two major pancreatic hormones.

15. Describe the functional roles of hormones of the testes and ovaries.

16. Briefly describe the importance of thymic hormones in immunity.

Other Hormone-Producing Structures (pp. 550–551)

17. Name a hormone produced by the heart, and localize enteroendocrine cells.

18. Briefly explain the hormonal functions of the placenta, kidney, skin, and adipose tissue.

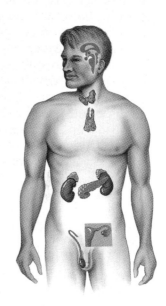

You don't have to watch *ER* to experience action-packed drama. Molecules and cells inside your body have dynamic adventures on microscopic levels all the time. For instance, when insulin molecules, carried passively along in the blood, attach to protein receptors of nearby cells, the response is dramatic: Glucose molecules begin to disappear from the blood into the cells, and cellular activity accelerates. Such is the power of the second great control system of the body, the **endocrine system,** which interacts with the nervous system to coordinate and integrate the activity of body cells.

The means of control and speed of the endocrine system are very different from those of the nervous system, however. The nervous system regulates the activity of muscles and glands via electrochemical impulses delivered by neurons, and those organs respond within milliseconds. The endocrine system influences metabolic activity by means of **hormones** (*hormon* = to excite), which are chemical messengers released into the blood to be transported throughout the body, and responses to hormones typically occur after a lag period of seconds or even days. But, once initiated, those responses tend to be much more prolonged than those induced by the nervous system.

Hormonal targets ultimately include most cells of the body, and hormones have widespread and diverse effects. The major processes controlled and integrated by these "mighty molecules" are reproduction; growth and development; mobilization of body defenses; maintenance of electrolyte, water, and nutrient balance of the blood; and regulation of cellular metabolism and energy balance. As you can see, the endocrine system orchestrates processes that go on for relatively long periods, in some instances continuously. The scientific study of hormones and the endocrine organs is called **endocrinology.**

The Endocrine System: An Overview

Compared with other organs of the body, those of the endocrine system are small and unimpressive. Indeed, to collect 1 kg (2.2 lbs) of hormone-producing tissue, you would need to collect *all* the endocrine tissue from eight or nine adults! In addition, the anatomical continuity typical of most organ systems does not exist in the endocrine system. Instead, endocrine organs are widely scattered about the body.

As explained in Chapter 4, there are two kinds of glands. *Exocrine glands* produce nonhormonal substances, such as sweat and saliva, and have ducts through which these substances are routed to a membrane surface. **Endocrine glands,** also called *ductless glands,* produce hormones and lack ducts. They release their hormones into the surrounding

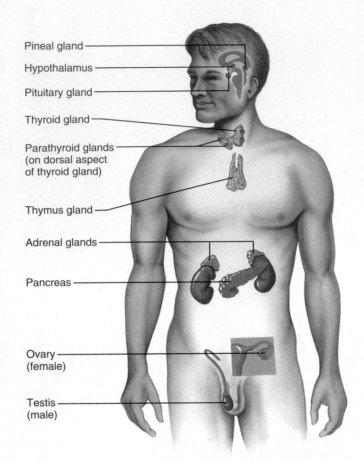

Pineal gland
Hypothalamus
Pituitary gland
Thyroid gland
Parathyroid glands (on dorsal aspect of thyroid gland)
Thymus gland
Adrenal glands
Pancreas
Ovary (female)
Testis (male)

FIGURE 15.1 Location of the major endocrine organs of the body.

tissue fluid (*endo* = within; *crine* = to secrete), and they typically have a rich vascular and lymphatic drainage that receives their hormones. Most of the hormone-producing cells in endocrine glands are arranged in cords and branching networks—a situation that maximizes contact between them and the capillaries surrounding them.

The endocrine glands include the pituitary, thyroid, parathyroid, adrenal, pineal, and thymus glands (Figure 15.1). In addition, several organs contain discrete areas of endocrine tissue and produce hormones as well as exocrine products. Such organs, which include the pancreas and gonads (ovaries and testes), are also major endocrine glands. The hypothalamus also falls into this latter category. Along with its neural functions, it produces and releases hormones, so we can consider the hypothalamus a **neuroendocrine organ.**

Besides the major endocrine organs, various other tissues and organs produce hormones. For example, adipose cells release leptin, and pockets of hormone-producing cells are found in the walls of the small intestine, stomach, kidneys, and heart—organs whose chief functions have little to do with hormone production. These other hormone-producing structures are described on p. 550.

Although some biologists include local hormones—autocrines and paracrines—as part of the realm of the endocrine system, that is not the consensus. Hormones are long-distance chemical signals that travel in blood or lymph throughout the body. *Autocrines* are chemicals that exert their effects on the same cells that secrete them, for example, certain prostaglandins released by smooth muscle cells cause the smooth muscle cells to contract. *Paracrines* also act locally but affect cell types other than those releasing the paracrine chemicals. Somatostatin released by one population of pancreatic cells inhibits the release of insulin by a different population of pancreatic cells.

 HOMEOSTATIC IMBALANCE

Certain tumor cells, such as those of some cancers of the lung or pancreas, synthesize hormones identical to those made in normal endocrine glands. However, they do so in an excessive and uncontrolled fashion. ●

Hormones

The Chemistry of Hormones

Hormones are chemical substances, secreted by cells into the extracellular fluids, that regulate the metabolic function of other cells in the body. Although a large variety of hormones are produced, nearly all of them can be classified chemically as either amino acid based or steroids.

Most hormones are **amino acid based.** Molecular size varies widely in this group—from simple amino acid derivatives (which include the amines and thyroxine, both constructed from the amino acid tyrosine), to peptides (short chains of amino acids), to proteins (long polymers of amino acids). The **steroids** are synthesized from cholesterol. Of the hormones produced by the major endocrine organs, only gonadal and adrenocortical hormones are steroids.

If we also consider the **eicosanoids** (i-ko′să-noyds), which include *leukotrienes* and *prostaglandins*, we must add a third chemical class. These local hormones are biologically active lipids (made from arachidonic acid) released by nearly all cell membranes. Leukotrienes are signaling chemicals that mediate inflammation and some allergic reactions. Prostaglandins have multiple targets and effects, ranging from raising blood pressure and increasing the expulsive uterine contractions of birth to enhancing blood clotting, pain, and inflammation. Because the effects of eicosanoids are typically highly localized, affecting only nearby cells, these substances do not fit the definition of the true *circulating hormones*, which influence distant targets.

Hence, this class of hormonelike chemicals will not be considered here, but its members are considered in later chapters as appropriate.

Mechanisms of Hormone Action

Even though all major hormones circulate to virtually all tissues, a given hormone influences the activity of only certain tissue cells, referred to as its **target cells.** Hormones bring about their characteristic effects on target cells by *altering* cell activity; that is, they increase or decrease the rates of normal cellular processes. The precise response depends on the target cell type. For example, when the hormone epinephrine binds to smooth muscle cells in blood vessel walls, it stimulates them to contract. Epinephrine binding to cells other than muscle cells may have a different effect, but it does *not* cause those cells to contract.

A hormonal stimulus typically produces one or more of the following changes:

1. Alters plasma membrane permeability or membrane potential, or both, by opening or closing ion channels

2. Stimulates synthesis of proteins or regulatory molecules such as enzymes within the cell

3. Activates or deactivates enzymes

4. Induces secretory activity

5. Stimulates mitosis

Two main mechanisms account for how a hormone communicates with its target cell, that is, how hormone receptor binding is harnessed to the intracellular machinery needed for hormone action. The mechanism used by amino acid–based hormones involves regulatory molecules called G proteins and one or more intracellular second messengers, which mediate the target cell's response to the hormone. The steroid hormone mechanism typically involves direct gene activation by the hormone.

Amino Acid–Based Hormones and Second-Messenger Systems

Because proteins and peptides cannot penetrate the plasma membranes of tissue cells, virtually all amino acid–based hormones exert their signaling effects through intracellular **second messengers** generated when a hormone binds to a receptor on the plasma membrane. Of the second messengers, **cyclic AMP** is by far the best understood, so it will receive most of our attention.

The Cyclic AMP Signaling Mechanism   In this mechanism, three plasma membrane components interact to determine intracellular levels of cyclic AMP (cAMP): a hormone receptor, a signal transducer (a G protein), and an effector enzyme (adenylate cyclase).

Which of the membrane proteins acts as the signal transducer in these schemes? Why are reactions initiated by second-messenger mechanisms called cascades?

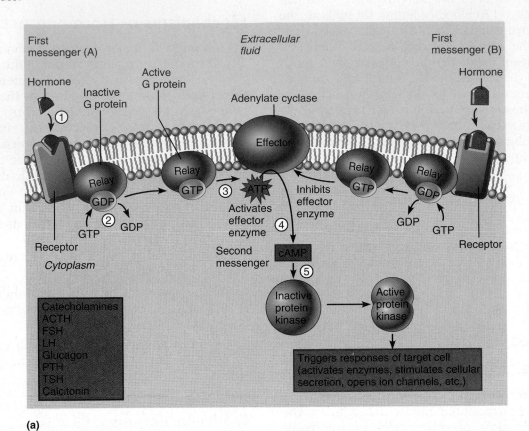

(a)

FIGURE 15.2 **Second-messenger mechanisms of amino acid–based hormones. (a)** Mechanisms that generate cyclic AMP by activation of adenylate cyclase are mediated by G proteins, which are activated when a hormone (first messenger, A in this case) binds to plasma membrane receptors. Cyclic AMP (the second messenger) acts intracellularly to activate protein kinase enzymes that mediate the cell's responses to the hormone. Notice that not all hormones provoke the activation of the effector enzyme. As depicted on the right, binding of hormone B to a membrane receptor is inhibitory.

① The hormone, acting as the **first messenger,** binds to its receptor (Figure 15.2a). This causes the receptor to change shape, allowing it to bind a nearby inactive **G protein.**

② The G protein is activated as the guanosine diphosphate (GDP) bound to it is displaced by the high-energy compound *guanosine triphosphate (GTP)*. The G protein behaves like a light switch; it is "off" when GDP is bound to it, and "on" when GTP is bound.

③ The activated G protein (moving along the membrane) binds to and activates the effector enzyme **adenylate cyclase.** At this point the GTP bound to the G protein is hydrolyzed to GDP and the G protein becomes inactive once again. (The G protein possesses GTPase activity and cleaves the terminal phosphate group off GTP in much the same way that ATPase enzymes hydrolyze ATP.)

④ The activated adenylate cyclase generates the second-messenger cAMP from ATP.

⑤ cAMP, which is free to diffuse throughout the cell, triggers a cascade of chemical reactions in which one or more enzymes, called **protein kinases,** are activated. The protein kinases *phosphorylate* (add a phosphate group to) various proteins, many of which are other enzymes. Because phosphorylation activates some of these proteins and inhibits others, a variety of reactions may occur in the same target cell at the same time.

This type of intracellular enzymatic cascade has a huge amplification effect. Each activated adenylate

■ *The G protein(s). Because each step in the pathway has a huge amplification effect and the number of product molecules increases dramatically at each step.*

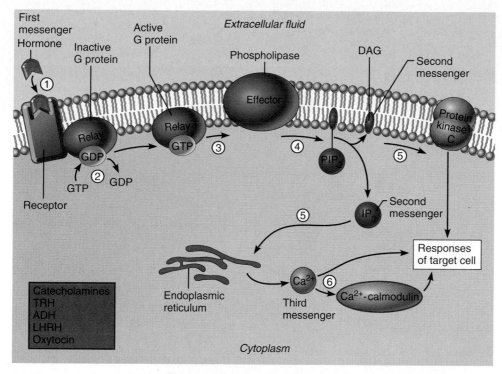

(b)

FIGURE 15.2 *(continued)*
Second-messenger mechanisms of amino acid–based hormones.
(b) Some mechanisms use Ca^{2+} as an intracellular messenger. The activated G protein stimulates the effector molecule,

phospholipase, to cleave PIP_2 into two fragments—inositol triphosphate (IP_3) and diacylglycerol (DAG), which act as intracellular second messengers to activate protein kinases and increase the cytoplasmic concentration of Ca^{2+}. Ionic

calcium also acts as a second messenger to modify the activity of cellular proteins. A list of hormones that act through these mechanisms is provided.

cyclase generates large numbers of cAMP molecules, and a single kinase enzyme can catalyze hundreds of reactions. Hence, as the reaction cascades through one enzyme intermediate after another, the number of product molecules increases dramatically at each step. Theoretically, receptor binding of a single hormone molecule could generate millions of final product molecules!

The sequence of reactions set into motion by cAMP depends on the type of target cell, the specific protein kinases it contains, and the hormone acting as first messenger. For example, in thyroid cells, binding of thyroid-stimulating hormone promotes synthesis of the thyroid hormone thyroxine; in bone and muscle cells, binding of growth hormone activates anabolic reactions in which amino acids are built into tissue proteins. Notice on the right of Figure 15.2a that some G proteins inhibit rather than activate adenylate cyclase, thus reducing the cytoplasmic concentration of cAMP. Such opposing effects permit even slight changes in levels of antagonistic hormones to influence a target cell's activity.

Because cAMP is rapidly degraded by the intracellular enzyme **phosphodiesterase,** its action persists only briefly. While at first glance this may

appear to be a problem, it is quite the opposite. Because of the amplification effect, most hormones need to be present only briefly to cause the desired results. Continued production of hormones then prompts continued cellular activity; no extracellular controls are necessary to stop the activity.

The PIP-Calcium Signal Mechanism Although cyclic AMP is the activating second messenger in some tissues for at least ten amino acid–based hormones, some of the same hormones (e.g., epinephrine) act through a different second-messenger system in other tissues. In one such mechanism, called the PIP-calcium signal mechanism, intracellular calcium ions act as the final mediator. Let's follow this process (Figure 15.2b).

① Hormone docking on the receptor causes it to bind the nearby inactive G protein.

② The G protein is activated as GTP binds, displacing GDP.

③ The activated G protein then binds to and activates membrane-bound **phospholipase** (the effector enzyme). The G protein then becomes inactive.

What is the crucial difference between the signaling mechanism depicted here and those shown in Figure 15.2?

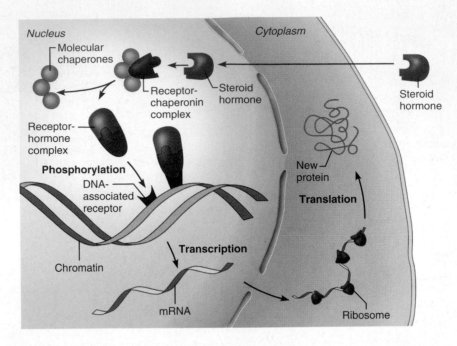

FIGURE 15.3 Direct gene activation mechanism of steroid hormones. A lipid-soluble steroid diffuses through the plasma membrane of the target cell and binds to a receptor-chaperonin complex in the nucleus. Once hormone is bound, the chaperonin dissociates from the receptor and the hormone-receptor complex is activated by phosphorylation and binds to a specific receptor protein on the chromatin, initiating transcription of a certain gene. The messenger RNA formed migrates into the cytoplasm, where it directs synthesis of a specific protein.

④ Phospholipase splits a plasma-membrane phospholipid called **PIP$_2$** (phosphatidyl inositol biphosphate) into **diacylglycerol (DAG)** and **inositol triphosphate (IP$_3$)**, and both these molecules act as second messengers.

⑤ DAG activates specific protein kinases, and IP$_3$ triggers the release of Ca^{2+} from the endoplasmic reticulum and other intracellular storage sites.

⑥ The liberated Ca^{2+} takes on a second-messenger role, either by directly altering the activity of specific enzymes and plasma membrane Ca^{2+} channels or by binding to the intracellular regulatory protein **calmodulin.** Once Ca^{2+} binds to calmodulin, enzymes are activated that amplify the cellular response.

■ *Here, the hormone can enter the cell (because it is soluble in the plasma membrane lipids) and directly stimulates the desired effect by acting on DNA to cause transcription. In the Figure 15.2 mechanisms, the hormone cannot enter the cell because it is insoluble in membrane lipids.*

Hormones known to act on their target cells via G protein receptors which act via cAMP or the PIP mechanism are listed in Figure 15.2. Other amino acid–based hormones act on their target cells through different (and in some cases, unknown) mechanisms. For example, cyclic guanosine monophosphate (cGMP) is a second messenger for selected hormones. Insulin (and other growth factors) appear to work without second messengers. The insulin receptor is a *tyrosine kinase* enzyme that is activated by autophosphorylation (addition of phosphate to several of its own tyrosines) when insulin binds. The activated receptor provides docking sites for intracellular *relay proteins* that, in turn, initiate a series of protein phosphorylations that trigger specific cell responses. In certain instances, any of the second messengers mentioned—and the hormone receptor itself—can cause changes in intracellular Ca^{2+} levels.

Steroid Hormones and Direct Gene Activation

Being lipid soluble, steroid hormones (and, strangely, thyroid hormone, a small iodinated amine) can diffuse into their target cells. Once inside, they bind to an intracellular receptor that is activated by the coupling (Figure 15.3). The activated hormone-receptor complex then makes its way to the nuclear chromatin, where the hormone binds to a DNA-associated *receptor protein* specific for it. (The exception to these generalizations is that thyroid hormone receptors are always bound to DNA even in the absence of thyroid hormone.) This interaction "turns on" a gene, that is, prompts transcription of DNA to produce a messenger RNA (mRNA). The mRNA is then translated on the cytoplasmic ribosomes, producing specific protein molecules. These proteins include enzymes that promote the metabolic activities induced by that particular hormone and, in some cases, promote synthesis of either structural proteins or proteins to be exported from the target cell.

In the absence of hormone, the nuclear receptors are bound up in receptor-chaperonin complexes, associations that seem to keep the receptors from binding to DNA and perhaps protect them from proteolysis. (Check back to Chapter 2, p. 50, if your recall of molecular chaperones needs a jog.) When the hormone *is* present, the complex dissociates and the receptor is phosphorylated, which allows it to bind to DNA and influence transcription.

Target Cell Specificity

In order for a target cell to respond to a hormone, the cell must have *specific* protein receptors on its plasma membrane or in its interior to which that hormone can bind. For example, receptors for adrenocorticotropic hormone (ACTH) are normally found only on certain cells of the adrenal cortex. By contrast, thyroxine is the principal hormone stimulating cellular metabolism, and nearly all body cells have thyroxine receptors.

A hormone receptor responds to hormone binding by prompting the cell to perform, or turn on, some gene-determined "preprogrammed" function. Hence, hormones are molecular triggers rather than informational molecules. Although binding of a hormone to a receptor is the crucial first step, target cell activation by hormone-receptor interaction depends equally on three factors: (1) blood levels of the hormone, (2) relative numbers of receptors for that hormone on or in the target cells, and (3) *affinity* (strength) of the bond between the hormone and the receptor. All three factors change rapidly in response to various stimuli and changes within the body. As a rule, for a given level of hormone in the blood, a large number of high-affinity receptors produces a pronounced hormonal effect, and a smaller number of low-affinity receptors

results in reduced target cell response or outright endocrine dysfunction.

Receptors are dynamic structures. In some instances, target cells form more receptors in response to rising blood levels of the specific hormones to which they respond, a phenomenon called **up-regulation.** In other cases, prolonged exposure to high hormone concentrations desensitizes the target cells, so that they respond less vigorously to hormonal stimulation. This **down-regulation** involves loss of receptors and prevents the target cells from overreacting to persistently high hormone levels. Hormones influence the number and affinity not only of their own receptors but also of receptors that respond to other hormones. For example, progesterone induces a loss of estrogen receptors in the uterus, thus antagonizing estrogen's actions. On the other hand, estrogen causes the same cells to produce more progesterone receptors, enhancing their ability to respond to progesterone.

Half-Life, Onset, and Duration of Hormone Activity

Hormones are potent chemicals, and they exert profound effects on their target organs at very low concentrations. Hormones circulate in the blood in two forms—free or bound to a protein carrier. In general, lipid-soluble hormones (steroids and thyroid hormone) travel in the bloodstream attached to plasma proteins. All others circulate unencumbered by carriers. The concentration of a circulating hormone in blood at any time reflects (1) its rate of release, and (2) the speed at which it is inactivated and removed from the body. Some hormones are rapidly degraded by enzymes in their target cells, but most are removed from the blood by the kidneys or liver, and their breakdown products are excreted from the body in urine or, to a lesser extent, in feces. As a result, the length of time a hormone remains in the blood, referred to as its **half-life,** is usually brief—from a fraction of a minute to 30 minutes, with the water-soluble hormones exhibiting the shortest half-lives.

The time required for hormone effects to appear varies greatly. Some hormones provoke target organ responses almost immediately, while others, particularly the steroid hormones, require hours to days before their effects are seen. Additionally, some hormones are secreted in a relatively inactive form and must be activated in the target cells.

The duration of hormone action is limited, ranging from 10 seconds to several hours, depending on the hormone. Effects may disappear rapidly as blood levels drop, or they may persist for hours after very low hormone levels have been reached. Because of these many variations, hormonal blood levels must be precisely and individually controlled to meet the continuously changing needs of the body.

Interaction of Hormones at Target Cells

Understanding hormonal effects is a bit more complicated than you might expect because multiple hormones may act on the same target cells at the same time and in many cases the result of such an interaction is not predictable even when you know the effects of the individual hormones. Here we will look at three types of hormone interaction—permissiveness, synergism, and antagonism.

Permissiveness is the situation when one hormone cannot exert its full effects without another hormone being present. For example, the development of the reproductive system is largely regulated by reproductive system hormones, as we might expect. However, thyroid hormone is necessary (has a permissive effect) for normal *timely* development of reproductive structures; without thyroid hormone, reproductive system development is delayed.

Synergism of hormones occurs in situations where more than one hormone produces the same effects at the target cell and their combined effects are amplified. For example, both glucagon (produced by the pancreas) and epinephrine cause the liver to release glucose to the blood; when they act together, the amount of glucose released is about 150% of what is released when each hormone acts alone.

When one hormone opposes the action of another hormone, the interaction is called **antagonism.** For example, insulin, which lowers blood sugar levels, is antagonized by the action of glucagon, which acts to raise blood sugar levels. Antagonists may compete for the same receptors, act through different metabolic pathways, or even, as noted in the progesterone–estrogen interaction at the uterus, cause down-regulation of the receptors for the antagonistic hormone.

Control of Hormone Release

The synthesis and release of most hormones are regulated by some type of **negative feedback system** (see Chapter 1). In such a system, hormone secretion is triggered by some internal or external stimulus. As hormone levels rise, they cause target organ effects and inhibit further hormone release. As a result, blood levels of many hormones vary only within a narrow range.

Endocrine Gland Stimuli

Endocrine glands are stimulated to manufacture and release their hormones by three major types of stimuli: *humoral* (hu'mer-ul), *neural,* and *hormonal.*

Humoral Stimuli   Some endocrine glands secrete their hormones in direct response to changing blood levels of certain ions and nutrients. These stimuli are called *humoral stimuli* to distinguish them from hormonal stimuli, which are also blood-borne chemicals. The term *humoral* harks back to the ancient use of the term *humor* to refer to various body fluids (blood, bile, and others). This is the simplest of the endocrine control systems. For example, cells of the parathyroid glands monitor blood Ca^{2+} levels, and when they detect a decline from normal values, they secrete parathyroid hormone (PTH). Because PTH acts by several routes to reverse that decline, blood Ca^{2+} levels soon rise, ending the initiative for PTH release (Figure 15.4a). Other hormones released in response to humoral stimuli include insulin, produced by the pancreas, and aldosterone, one of the adrenal cortex hormones.

Neural Stimuli   In a few cases, nerve fibers stimulate hormone release. The classic example of neural stimuli is sympathetic nervous system stimulation of the adrenal medulla to release catecholamines (norepinephrine and epinephrine) during periods of stress (Figure 15.4b).

Hormonal Stimuli   Finally, many endocrine glands release their hormones in response to hormones produced by other endocrine organs, and the stimuli in these cases are called hormonal stimuli. For example, release of most anterior pituitary hormones is regulated by releasing and inhibiting hormones produced by the hypothalamus, and many anterior pituitary hormones in turn stimulate other endocrine organs to release their hormones (Figure 15.4c). As blood levels of the hormones produced by the final target glands increase, they inhibit the release of anterior pituitary hormones and thus their own release. This hypothalamic–pituitary–target endocrine organ feedback loop lies at the very core of endocrinology, and it will come up many times in this chapter. Hormonal stimuli promote rhythmic hormone release, with hormone blood levels rising and falling in a specific pattern.

Although these three mechanisms typify most systems that control hormone release, they are by no means all-inclusive or mutually exclusive, and some endocrine organs respond to multiple stimuli.

Nervous System Modulation

Both "turn on" factors (hormonal, humoral, and neural stimuli) and "turn off" factors (feedback inhibition and others) may be modified by the nervous system. Without this added safeguard,

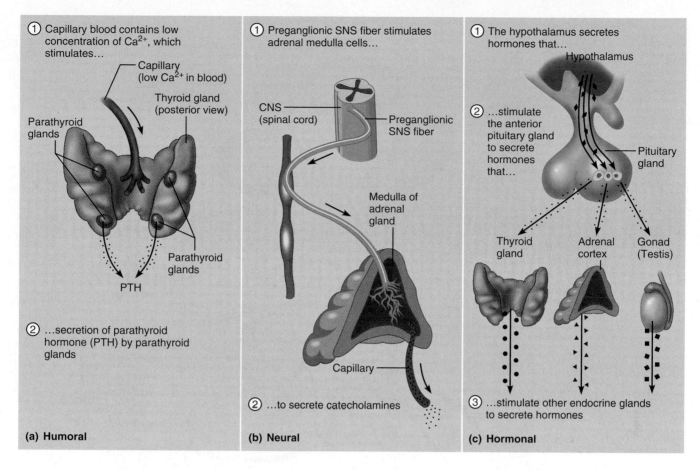

① Capillary blood contains low concentration of Ca²⁺, which stimulates...

Capillary (low Ca²⁺ in blood)

Thyroid gland (posterior view)

Parathyroid glands

Parathyroid glands

PTH

② ...secretion of parathyroid hormone (PTH) by parathyroid glands

(a) Humoral

① Preganglionic SNS fiber stimulates adrenal medulla cells...

CNS (spinal cord)

Preganglionic SNS fiber

Medulla of adrenal gland

Capillary

② ...to secrete catecholamines

(b) Neural

① The hypothalamus secretes hormones that...

Hypothalamus

② ...stimulate the anterior pituitary gland to secrete hormones that...

Pituitary gland

Thyroid gland

Adrenal cortex

Gonad (Testis)

③ ...stimulate other endocrine glands to secrete hormones

(c) Hormonal

FIGURE 15.4 **Three types of endocrine gland stimuli. (a)** Humoral stimulus: Low blood calcium levels trigger parathyroid hormone release from the parathyroid glands, which causes blood Ca²⁺ levels to rise by stimulating release of Ca²⁺ from bone. **(b)** Neural stimulus: The stimulation of adrenal medullary cells by sympathetic nervous system (SNS) fibers triggers the release of catecholamines to the blood. **(c)** Hormonal stimulus: Hormones released by the hypothalamus stimulate the anterior pituitary to release hormones that stimulate other endocrine organs to secrete hormones.

endocrine system activity would be strictly mechanical, much like a household thermostat. A thermostat can maintain the temperature at or around its set value, but it cannot sense that your grandmother visiting from Florida feels cold at that temperature and reset itself accordingly. *You* must make that adjustment. This is not usually the case in your body, however, where the nervous system can, in certain cases, override normal endocrine controls as needed to maintain homeostasis. For example, the action of insulin and several other hormones normally keeps blood sugar levels in the range of 90–110 mg glucose per 100 ml of blood. However, when the body is under severe stress, blood sugar levels rise because the hypothalamus and sympathetic nervous system centers are strongly activated. This ensures that body cells have sufficient fuel for the more vigorous activity required during such periods.

Major Endocrine Organs
The Pituitary Gland (Hypophysis)

Securely seated in the sella turcica of the sphenoid bone, the tiny **pituitary gland,** or **hypophysis** (hi-pof′ĭ-sis; "to grow under"), secretes at least nine hormones. Usually said to be the size and shape of a pea, this gland is more accurately described as a pea on a stalk. Its stalk, the funnel-shaped **infundibulum,** connects the gland to the hypothalamus superiorly (Figure 15.1). In humans, the pituitary gland has two major lobes; one is neural tissue and the other glandular. The **posterior pituitary** (lobe) (Figure 15.5) is composed largely of pituicytes (glia-like supporting cells) and nerve fibers. It releases **neurohormones** (hormones secreted by neurons) received ready-made from the hypothalamus. Thus, this lobe is a hormone-storage area and not a true endocrine gland in the precise sense. The posterior lobe plus

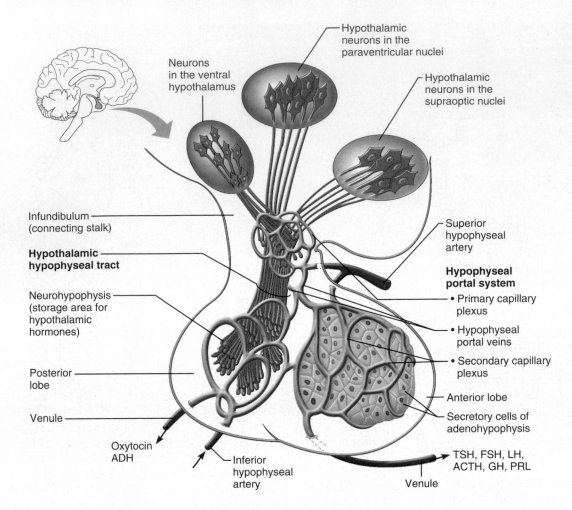

FIGURE 15.5 Relationships of the pituitary gland and hypothalamus. Hypothalamic neurons in the supraoptic and paraventricular nuclei synthesize ADH and oxytocin, which are transported along the hypothalamic- hypophyseal tract to the neurohypophysis for storage. Neurons in the ventral hypothalamus have short axons that discharge releasing and inhibiting hormones to the capillaries of the hypophyseal portal system, which runs to the adenohypophysis. These factors influence the adenohypophyseal secretory cells to release (or not release) their hormones.

the infundibulum make up the region called the **neurohypophysis** (nu"ro-hi-pof'ĭ-sis), a term commonly used (incorrectly) to indicate the posterior lobe alone. The **anterior pituitary (lobe),** or **adenohypophysis** (ad"ĕ-no-hi-pof'ĭ-sis), is composed of glandular tissue (adeno = gland), and it manufactures and releases a number of hormones (Table 15.1 on pp. 532–533).

Arterial blood is delivered to the pituitary via hypophyseal branches of the internal carotid arteries. The veins leaving the pituitary drain into the dural sinuses.

Pituitary-Hypothalamic Relationships

The contrasting histology of the two pituitary lobes reflects the dual origin of this tiny gland. The posterior lobe is actually part of the brain. It derives from a downgrowth of hypothalamic tissue and maintains its neural connection with the hypothalamus via a nerve bundle called the **hypothalamic-hypophyseal tract,** which runs through the infundibulum (Figure 15.5). This tract arises from neurons in the **supraoptic** and **paraventricular nuclei** of the hypothalamus. These neurosecretory cells synthesize two neurohormones and transport them along their axons to the posterior pituitary. Oxytocin (ok"sĭ-to'sin) is made by the paraventricular neurons, and antidiuretic hormone by the supraoptic neurons. When these hypothalamic neurons fire, they release the stored hormones into a capillary bed in the posterior pituitary for distribution throughout the body.

The glandular anterior lobe originates from a superior outpocketing of the oral mucosa *(Rathke's pouch)* and is formed from epithelial tissue. After touching the posterior lobe, the anterior lobe loses its connection with the oral mucosa and adheres to the neurohypophysis. There is no direct neural connection between the anterior lobe and hypothalamus, but there is a vascular connection. Specifically, the **primary capillary plexus** in the infundibulum communicates inferiorly via the small **hypophyseal portal veins** with a **secondary capillary plexus** in the anterior lobe. The primary and secondary capillary plexuses and the intervening hypophyseal portal veins make up the **hypophyseal portal system*** (Figure 15.5). Via this portal system, **releasing** and **inhibiting hormones** secreted by neurons in the ventral hypothalamus circulate to the adenohypophysis, where they regulate secretion of its hormones. All these hypothalamic regulatory hormones are amino acid based, but they vary in size from small peptides to proteins.

Adenohypophyseal Hormones

Although the adenohypophysis has traditionally been called the "master endocrine gland" because of its numerous hormonal products, many of which regulate the activity of other endocrine glands, it has in recent years been dethroned by the hypothalamus, which is now known to control anterior pituitary activity. Researchers have identified six distinct adenohypophyseal hormones, all of them proteins (Table 15.1). In addition, a large molecule with the tongue-twisting name **pro-opiomelanocortin (POMC)** (pro"o"pe-o-mah-lan'o-kor"tin) has been isolated from the anterior pituitary. POMC is a *prohormone*, that is, a large precursor molecule that can be split enzymatically into one or more active hormones. POMC is the source of adrenocorticotropic hormone, two natural opiates (an enkephalin and a beta endorphin, described in Chapter 11), and *melanocyte-stimulating hormone (MSH)*. In amphibians, reptiles, and other animals MSH stimulates melanocytes to increase synthesis of melanin pigment, but in humans MSH plasma levels are insignificant and this hormone is probably more important as a CNS neurotransmitter than as a hormone. Additionally, MSH release is tonically suppressed by dopamine-releasing hypothalamic neurons.

When the adenohypophysis receives an appropriate chemical stimulus from the hypothalamus, one or more of its hormones are released by certain of its cells. Although many different hormones pass from the hypothalamus to the anterior lobe, each target cell in the anterior lobe distinguishes the messages directed to it and responds in kind—secreting the proper hormone in response to specific

releasing hormones, and shutting off hormone release in response to specific inhibiting hormones. The hypothalamic releasing hormones are far more important as regulatory factors because only very little hormone is stored by secretory cells of the anterior lobe.

Four of the six anterior pituitary hormones—thyroid-stimulating hormone, adrenocorticotropic hormone, follicle-stimulating hormone, and luteinizing hormone—are **tropins** or **tropic hormones** (*tropi* = turn on, change), which are hormones that regulate the secretory action of other endocrine glands. All six hormones affect their target cells via a cyclic AMP second-messenger system.

Growth Hormone   Growth hormone (GH) is produced by the **somatotropic cells** of the anterior lobe (Figure 15.6). Although GH stimulates most body cells to increase in size and divide, its major targets are the bones and skeletal muscles. Stimulation of the epiphyseal plate leads to long bone growth; stimulation of skeletal muscles promotes increased muscle mass.

Essentially an anabolic (tissue building) hormone, GH promotes protein synthesis, and it encourages the use of fats for fuel, thus conserving glucose. Most growth-promoting effects of GH are mediated indirectly by **insulin-like growth factors (IGFs)**, also known as **somatomedins** (so"mah-to-me'dinz), a family of growth-promoting proteins produced by the liver, skeletal muscle, bone, and other tissues. Specifically, IGFs (1) stimulate uptake of amino acids from the blood and their incorporation into cellular proteins throughout the body; and (2) stimulate uptake of sulfur (needed for the synthesis of chondroitin sulfate) into cartilage matrix. Acting directly, GH mobilizes fats from fat depots for transport to cells, increasing blood levels of fatty acids. It also decreases the rate of glucose uptake and metabolism. In the liver, it encourages glycogen breakdown and release of glucose to the blood. The elevation in blood sugar levels that occurs as a result of this *glucose sparing* is called the **diabetogenic effect** of GH, because it mimics the high blood sugar levels typical of diabetes mellitus.

Secretion of GH is regulated chiefly by two hypothalamic hormones with antagonistic effects. **Growth hormone–releasing hormone (GHRH)** stimulates GH release, while **growth hormone–inhibiting hormone (GHIH)**, also called **somatostatin** (so"mah-to-stat'in), inhibits it. GHIH release is (presumably) triggered by the feedback of GH and IGFs. Rising levels of GH also feed back to inhibit its own release. As indicated in Table 15.1, a number of secondary triggers also influence GH release. Typically, GH secretion has a daily cycle, with the highest levels occurring during evening sleep, but

---

* A *portal system* is an unusual arrangement of blood vessels in which a capillary bed feeds into veins, which in turn feed into another capillary bed.

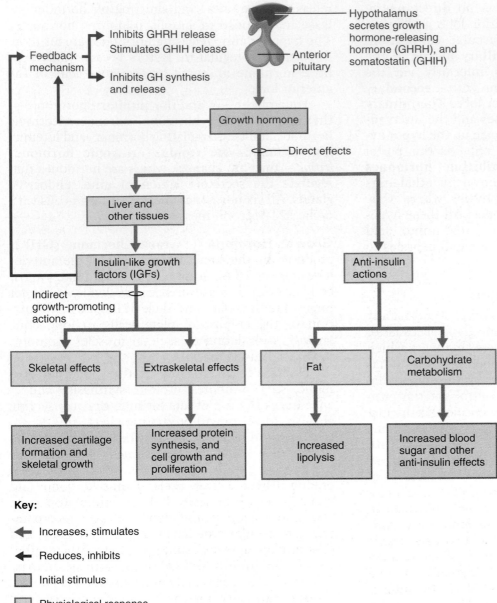

Hypothalamus secretes growth hormone-releasing hormone (GHRH), and somatostatin (GHIH)

Anterior pituitary

Feedback mechanism
- Inhibits GHRH release Stimulates GHIH release
- Inhibits GH synthesis and release

Growth hormone

Direct effects

Liver and other tissues

Insulin-like growth factors (IGFs)

Indirect growth-promoting actions

Anti-insulin actions

Skeletal effects

Extraskeletal effects

Fat

Carbohydrate metabolism

Increased cartilage formation and skeletal growth

Increased protein synthesis, and cell growth and proliferation

Increased lipolysis

Increased blood sugar and other anti-insulin effects

Key:

◄ Increases, stimulates

◄ Reduces, inhibits

☐ Initial stimulus

☐ Physiological response

☐ Result

FIGURE 15.6 Metabolic actions of growth hormone (GH). The direct actions of GH on target cells stimulate liver, skeletal muscle, bone, and cartilage cells to produce and release insulin-like growth factors, which mediate the indirect, largely anabolic effects of GH. GH acts directly on other tissue cells to promote fat breakdown (lipolysis) and release from adipose tissues and to hinder glucose uptake from the blood by tissue cells. (Because these actions antagonize those of the pancreatic hormone insulin, they are referred to as anti-insulin actions.) Elevated levels of GH and IGF feed back to promote GHIH release (and depress GHRH release) by the hypothalamus and to inhibit GH release by the anterior pituitary gland.

the total amount secreted daily peaks during adolescence and then declines with age.

Besides inhibiting growth hormone secretion, GHIH blocks the release of several other adenohypophyseal hormones. Additionally, it inhibits the release of virtually all gastrointestinal and pancreatic secretions—both endocrine and exocrine.

🇭 HOMEOSTATIC IMBALANCE

Both hypersecretion and hyposecretion of GH may result in structural abnormalities. Hypersecretion in children results in **gigantism** as the still-active epiphyseal (growth) plates are targeted by GH. The person becomes abnormally tall, often reaching a

height of 2.4 m (8 feet), but has relatively normal body proportions. If excessive amounts of GH are secreted after the epiphyseal plates have closed, **acromegaly** (ak"ro-meg'ah-le) results. Literally translated as "enlarged extremities," this condition is characterized by overgrowth of bony areas still responsive to GH, namely bones of the hands, feet, and face. Hypersecretion usually results from an adenohypophyseal tumor that churns out excessive amounts of GH and the usual treatment is surgical removal of the tumor; however, anatomical changes that have already occurred are not reversible.

Hyposecretion of GH in adults usually causes no problems. GH deficiency in children results in slowed long bone growth, a condition called **pituitary dwarfism.** Such individuals attain a maximum height of 1.2 m (4 feet), but usually have fairly normal body proportions. Lack of GH is typically accompanied by deficiencies of other adenohypophyseal hormones, and if thyroid-stimulating hormone and gonadotropins are lacking, the individual will be malproportioned and will fail to mature sexually as well. Fortunately, human GH is produced commercially by genetic engineering techniques and when pituitary dwarfism is diagnosed before puberty, growth hormone replacement therapy can promote nearly normal somatic growth. ●

Thyroid-Stimulating Hormone   Thyroid-stimulating hormone (TSH), or **thyrotropin**, is a tropic hormone that stimulates normal development and secretory activity of the thyroid gland. TSH release from the **thyrotrope cells** of the anterior pituitary is triggered by the hypothalamic peptide **thyrotropin-releasing hormone (TRH).** Rising blood levels of thyroid hormones act on both the pituitary and the hypothalamus to inhibit TSH secretion. The hypothalamus, in response, releases GHIH, which reinforces the blockade of TSH release.

Adrenocorticotropic Hormone   Adrenocorticotropic hormone (ACTH) (ah-dre"no-kor"tĭ-ko-trōp'ik) or **corticotropin**, is secreted by the **corticotrope cells** of the adenohypophysis. ACTH stimulates the adrenal cortex to release corticosteroid hormones, most importantly glucocorticoids that help the body to resist stressors. ACTH release, elicited by hypothalamic **corticotropin-releasing hormone (CRH),** has a daily rhythm, with levels peaking in the morning, shortly after one arises. Rising levels of glucocorticoids feed back and block secretion of CRH and consequently ACTH release. Internal and external factors that alter the normal ACTH rhythm by triggering CRH release include fever, hypoglycemia, and stressors of all types.

Gonadotropins   Follicle-stimulating hormone **(FSH)** and **luteinizing hormone (LH)** lu'te-in-īz"ing), referred to collectively as **gonadotropins,** regulate the function of the gonads (ovaries and testes). In both sexes, FSH stimulates gamete (sperm or egg) production and LH promotes production of gonadal hormones. In females, LH works with FSH to cause maturation of an egg-containing ovarian follicle. LH then independently triggers ovulation and promotes synthesis and release of ovarian hormones. In males, LH stimulates the interstitial cells of the testes to produce the male hormone testosterone. For this reason, LH is sometimes called **interstitial cell–stimulating hormone (ICSH)** (in"ter-stish'al) in males.

Gonadotropins are virtually absent from the blood of prepubertal boys and girls. During puberty, the **gonadotrope cells** of the adenohypophysis are activated and gonadotropin levels begin to rise, causing the gonads to mature. In both sexes, gonadotropin release by the adenohypophysis is prompted by **gonadotropin-releasing hormone (GnRH)** produced by the hypothalamus. Gonadal hormones, produced in response to the gonadotropins, feed back to suppress FSH and LH release.

Prolactin   Prolactin (PRL) is a protein hormone structurally similar to GH. Produced by the **lactotropes**, PRL stimulates the gonads of some animals (other than humans) and is considered a gonadotropin by some researchers. Its only well-documented effect in humans is to stimulate milk production by the breasts (pro = for; lact = milk). However, there is some evidence that PRL enhances testosterone production in males.

As with GH, PRL release is controlled by hypothalamic releasing and inhibiting hormones. The exact nature of **prolactin-releasing hormone (PRH)** is unknown, but it is thought to be serotonin. PRH causes prolactin synthesis and release, whereas **prolactin-inhibiting hormone (PIH),** now known to be the neurotransmitter dopamine (DA), prevents prolactin secretion. In males, the influence of PIH predominates, but in females, prolactin levels rise and fall in rhythm with estrogen blood levels. Low estrogen levels stimulate PIH release, whereas high estrogen levels promote release of PRH and, thus, prolactin. A brief rise in prolactin levels just before the menstrual period partially accounts for the breast swelling and tenderness some women experience at that time, but because this PRL stimulation is so brief, the breasts do not produce milk. In pregnant women, PRL blood levels rise dramatically toward the end of pregnancy, and milk production becomes possible. After birth, the infant's suckling stimulates PRH release in the mother, encouraging continued milk production and availability.

Ⓗ HOMEOSTATIC IMBALANCE

Hypersecretion of prolactin is more common than hyposecretion (which is not a problem in anyone except women who choose to nurse). In fact, hyperprolactinemia is the most frequent abnormality of adenohypophyseal tumors. Clinical signs include inappropriate lactation, lack of menses, infertility in females, and breast enlargement and impotence in males. ●

TABLE 15.1   Pituitary Hormones: Summary of Regulation and Effects

Hormone (Chemical Structure and Cell Type)	Regulation of Release	Target Organ and Effects	Effects of Hyposecretion ↓ and Hypersecretion ↑
Anterior Pituitary Hormones			
Growth hormone (GH) (Protein, somatotrope)	**Stimulated** by GHRH* release, which is triggered by low blood levels of GH as well as by a number of secondary triggers including estrogens, hypoglycemia, increases in blood levels of amino acids, low levels of fatty acids, exercise and other types of stressors **Inhibited** by feedback inhibition exerted by GH and IGFs, and by hyperglycemia, hyperlipidemia, obesity, and emotional deprivation, all of which elicit GHIH* (somatostatin) release	 Liver, muscle, bone, cartilage, and other tissues: anabolic hormone; stimulates somatic growth; mobilizes fats; spares glucose Most effects mediated indirectly by IGFs	↓ Pituitary dwarfism in children ↑ Gigantism in children; acromegaly in adults
Thyroid-stimulating hormone (TSH) (Glycoprotein, thyrotrope)	**Stimulated** by TRH* and indirectly by pregnancy and cold temperature **Inhibited** by feedback inhibition exerted by thyroid hormones on anterior pituitary and hypothalamus and by GHIH*	 Thyroid gland: stimulates thyroid gland to release thyroid hormone	↓ Cretinism in children; myxedema in adults ↑ Graves' disease; exophthalmos
Adrenocorticotropic hormone (ACTH) (Polypeptide, 39 amino acids, corticotrope)	**Stimulated** by CRH*; stimuli that increase CRH release include fever, hypoglycemia, and other stressors **Inhibited** by feedback inhibition exerted by glucocorticoids	 Adrenal cortex: promotes release of glucocorticoids and androgens (mineralocorticoids to a lesser extent)	↓ Rare ↑ Cushing's disease
Follicle-stimulating hormone (FSH) (Glycoprotein, gonadotrope)	**Stimulated** by GnRH* **Inhibited** by feedback inhibition exerted by estrogen in females and testosterone and inhibin in males	 Ovaries and testes: in females, stimulates ovarian follicle maturation and estrogen production; in males, stimulates sperm production	↓ Failure of sexual maturation ↑ No important effects

The Posterior Pituitary and Hypothalamic Hormones

The posterior pituitary, made largely of the axons of hypothalamic neurons, stores antidiuretic hormone (ADH) and oxytocin that have been synthesized and forwarded by hypothalamic neurons of the supraoptic and paraventricular nuclei. These hormones are later released "on demand" in response to nerve impulses from the same hypothalamic neurons.

ADH and oxytocin, each composed of nine amino acids, are almost identical. They differ in only two amino acids, and yet they have dramatically different physiological effects. ADH influences body water balance, and oxytocin stimulates contraction of smooth muscle, particularly that of the uterus and breasts. Both hormones use the PIP-calcium second-messenger mechanism.

TABLE 15.1 *(continued)*

Hormone (Chemical Structure and Cell Type)	Regulation of Release	Target Organ and Effects	Effects of Hyposecretion ↓ and Hypersecretion ↑
Luteinizing hormone (LH) (Glycoprotein, gonadotrope)	**Stimulated** by GnRH* **Inhibited** by feedback inhibition exerted by estrogen and progesterone in females and testosterone in males	Ovaries and testes: in females, triggers ovulation and stimulates ovarian production of estrogen and progesterone; in males, promotes testosterone production	As for FSH
Prolactin (PRL) (Protein, lactotrope)	**Stimulated** by PRH*; PRH release enhanced by estrogens, birth control pills, opiates, and breastfeeding **Inhibited** by PIH* (dopamine)	Breast secretory tissue: promotes lactation	↓ Poor milk production in nursing women ↑ Inappropriate milk production (galactorrhea); cessation of menses in females; impotence and breast enlargement (gynecomastia) in males

 Posterior Pituitary Hormones (made by hypothalamic neurons and stored in posterior pituitary)

Oxytocin (Peptide, neurons in paraventricular nucleus of hypothalamus)	**Stimulated** by impulses from hypothalamic neurons in response to cervical/uterine stretching and suckling of infant at breast **Inhibited** by lack of appropriate neural stimuli	Uterus: stimulates uterine contractions; initiates labor; breast: initiates milk ejection	Unknown
Antidiuretic hormone (ADH) or **vasopressin** (Peptide, neurons in supraoptic nucleus of hypothalamus)	**Stimulated** by impulses from hypothalamic neurons in response to increased osmolarity of blood or decreased blood volume; also stimulated by pain, some drugs, low blood pressure **Inhibited** by adequate hydration of the body and by alcohol	Kidneys: stimulates kidney tubule cells to reabsorb water	↓ Diabetes insipidus ↑ Syndrome of inappropriate ADH secretion (SIADH)

*Indicates hypothalamic releasing and inhibiting hormones: GHRH = growth hormone–releasing hormone; GHIH = growth hormone–inhibiting hormone; TRH = thyrotropin-releasing hormone; CRH = corticotropin-releasing hormone; GnRH = gonadotropin-releasing hormone; PRH = prolactin-releasing hormone; PIH = prolactin-inhibiting hormone.

Oxytocin A strong stimulant of uterine contraction, **oxytocin** is released in significantly higher amounts during childbirth (*ocytocia* = childbirth) and in nursing women. The number of oxytocin receptors in the uterus peaks near the end of pregnancy, and uterine smooth muscle becomes more and more sensitive to the hormone's stimulatory effects. Stretching of the uterus and cervix as birth nears dispatches afferent impulses to the hypothalamus, which responds by synthesizing oxytocin and triggering its release from the neurohypophysis. As blood levels of oxytocin rise, the expulsive contractions of labor gain momentum and finally end in birth.

Oxytocin also acts as the hormonal trigger for milk ejection (the "letdown" reflex) in women whose breasts are producing milk in response to prolactin. Suckling causes a reflex-initiated release of oxytocin, which targets specialized myoepithelial cells surrounding the milk-producing glands. As these cells contract, milk is forced from the breast into the infant's mouth. Both of these mechanisms are *positive feedback mechanisms*.

Both natural and synthetic oxytocic drugs are used to induce labor or to hasten normal labor that is progressing slowly. Less frequently, oxytocics are used to stop postpartum bleeding (by compressing ruptured blood vessels at the placental site) and to stimulate the milk ejection reflex.

Until recently, oxytocin's role in males and nonpregnant, nonlactating females was unknown, but new studies reveal that this potent peptide plays a role in sexual arousal and orgasm when the body is already primed for reproduction by sex hormones. Then, it is responsible for the feeling of sexual satisfaction that results from that interaction. In nonsexual relationships, it is thought to promote nurturing and affectionate behavior, that is, it acts as a "cuddle hormone."

Antidiuretic Hormone   *Diuresis* (di"u-re'sis) is urine production. Thus, an *antidiuretic* is a substance that inhibits or prevents urine formation. **Antidiuretic hormone (ADH)** prevents wide swings in water balance, helping the body avoid dehydration and water overload. Hypothalamic neurons, called *osmoreceptors*, continually monitor the solute concentration (and thus the water concentration) of the blood. When solutes threaten to become too concentrated (as might follow excessive perspiration or inadequate fluid intake), the osmoreceptors transmit excitatory impulses to the hypothalamic neurons in the supraoptic nucleus, which synthesize and release ADH. Liberated into the blood by the neurohypophysis, ADH targets the kidney tubules. The tubule cells respond by reabsorbing more water from the forming urine and returning it to the bloodstream. As a result, less urine is produced and blood volume increases. As the solute concentration of the blood declines, the osmoreceptors stop depolarizing, effectively ending ADH release. Other stimuli triggering ADH release include pain, low blood pressure, and such drugs as nicotine, morphine, and barbiturates.

Drinking alcoholic beverages inhibits ADH secretion and causes copious urine output. The dry mouth and intense thirst of the "morning after" reflect this dehydrating effect of alcohol. As might be expected, ADH release is also inhibited by drinking excessive amounts of water. *Diuretic drugs* antagonize the effects of ADH and cause water to be flushed from the body. These drugs are used to man-age some cases of hypertension and the edema (water retention in tissues) typical of congestive heart failure.

At high blood concentrations, ADH causes vasoconstriction, primarily of the visceral blood vessels, a response that indicates targeting of different receptors found on vascular smooth muscle. Under certain conditions, such as severe blood loss, exceptionally large amounts of ADH are released, causing a rise in blood pressure. The alternative name for this hormone (**vasopressin**) reflects this particular effect.

🅗 *HOMEOSTATIC IMBALANCE*

One result of ADH deficiency is **diabetes insipidus,** a syndrome marked by the output of huge amounts of urine and intense thirst. The name of this condition (*diabetes* = overflow; *insipidus* = tasteless) distinguishes it from diabetes mellitus (*mel* = honey), in which insulin deficiency causes large amounts of blood sugar to be lost in the urine. At one time, urine was tasted to determine which type of diabetes the patient was suffering from.

Diabetes insipidus can be caused by a blow to the head that damages the hypothalamus or the posterior pituitary. In either case, ADH release is deficient. Though inconvenient, the condition is not serious when the thirst center is operating properly and the person drinks enough water to prevent dehydration. However, it can be life threatening in unconscious or comatose patients, so accident victims with head trauma must be carefully monitored.

Hypersecretion of ADH occurs in children with meningitis, may follow neurosurgery or hypothalamic injury, or results from ectopic ADH secretion by cancer cells (particularly pulmonary cancers). It also may occur after general anesthesia or administration of certain drugs. The resulting condition, *syndrome of inappropriate ADH secretion (SIADH),* is marked by retention of fluid, headache and disorientation due to brain edema, weight gain, and hypo-osmolarity of the blood. SIADH management requires fluid restriction and careful monitoring of blood sodium levels. ●

The Thyroid Gland

Location and Structure

The butterfly-shaped **thyroid gland** is located in the anterior neck, on the trachea just inferior to the larynx (Figures 15.1 and 15.7a). Its two lateral *lobes* are connected by a median tissue mass called the *isthmus*. The thyroid gland is the largest pure endocrine gland in the body. Its prodigious blood supply (from the *superior* and *inferior thyroid arteries*) makes thyroid surgery a painstaking (and bloody) endeavor.

? *Which of the cell populations shown in view (b) produces calcitonin?*

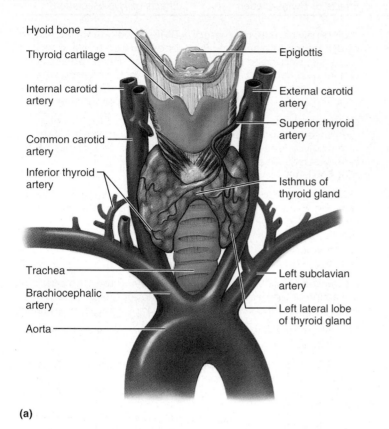

Hyoid bone

Thyroid cartilage

Internal carotid artery

Common carotid artery

Inferior thyroid artery

Trachea

Brachiocephalic artery

Aorta

Epiglottis

External carotid artery

Superior thyroid artery

Isthmus of thyroid gland

Left subclavian artery

Left lateral lobe of thyroid gland

(a)

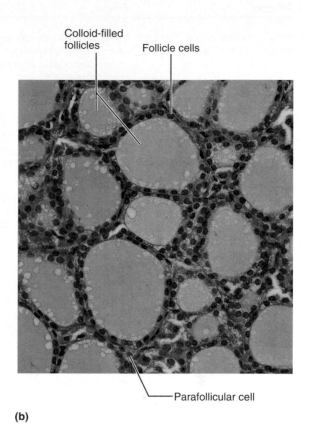

Colloid-filled follicles

Follicle cells

Parafollicular cell

(b)

FIGURE 15.7   **Gross and microscopic anatomy of the thyroid gland.**
(a) Location and arterial blood supply of the thyroid gland, anterior view.
(b) Photomicrograph of the thyroid gland follicles (250×).

Internally, the gland is composed of hollow, spherical **follicles** (Figure 15.7b). The walls of each follicle are formed largely by cuboidal or squamous epithelial cells called *follicle cells*, which produce the glycoprotein **thyroglobulin** (thi"ro-glob'u-lin). The central cavity, or lumen, of the follicle stores **colloid**, an amber-colored, sticky material consisting of thyroglobulin molecules with attached iodine atoms. *Thyroid hormone* is derived from this iodinated thyroglobulin. The *parafollicular cells*, another population of endocrine cells in the thyroid gland, produce *calcitonin*, an entirely different hormone. The parafollicular cells lie in the follicular epithelium but protrude into the soft connective tissue that separates and surrounds the thyroid follicles.

Thyroid Hormone

Often referred to as the body's major metabolic hormone, **thyroid hormone (TH)** is actually two iodine-containing amine hormones, **thyroxine** (thi-rok'sin), or **T$_4$,** and **triiodothyronine** (tri"i-o"do-thi'ro-nēn), or **T$_3$**. T$_4$ is the major hormone secreted by the thyroid follicles; most T$_3$ is formed at the target tissues by conversion of T$_4$ to T$_3$. Very much like one another, the hormones are constructed from two linked tyrosine amino acids. The principal difference is that T$_4$ has four bound iodine atoms, and T$_3$ has three (thus, T$_4$ and T$_3$).

Except for the adult brain, spleen, testes, uterus, and the thyroid gland itself, TH affects virtually every cell in the body (Table 15.2). By stimulating enzymes concerned with glucose oxidation, it increases basal metabolic rate and body heat production; this is known as the hormone's **calorigenic effect** (*calorigenic* = heat producing). Because TH provokes an increase in the number of adrenergic receptors in blood vessels, it plays an important role in maintaining blood pressure. Additionally, it is an important regulator of tissue growth and development. It is especially critical for normal skeletal and nervous system development and maturation and for reproductive capabilities.

■ *The parafollicular cells.*

TABLE 15.2 Major Effects of Thyroid Hormone (T$_4$ and T$_3$) in the Body

Process or System Affected	Normal Physiological Effects	Effects of Hyposecretion	Effects of Hypersecretion
Basal metabolic rate (BMR)/ temperature regulation	Promotes normal oxygen use and BMR; calorigenesis; enhances effects of sympathetic nervous system	BMR below normal; decreased body temperature and cold intolerance; decreased appetite; weight gain; reduced sensitivity to catecholamines	BMR above normal; increased body temperature and heat intolerance; increased appetite; weight loss; increased sensitivity to catecholamines may lead to high blood pressure
Carbohydrate/lipid/protein metabolism	Promotes glucose catabolism; mobilizes fats; essential for protein synthesis; enhances liver's synthesis of cholesterol	Decreased glucose metabolism; elevated cholesterol/triglyceride levels in blood; decreased protein synthesis; edema	Enhanced catabolism of glucose, proteins, and fats; weight loss; loss of muscle mass
Nervous system	Promotes normal development of nervous system in fetus and infant; promotes normal adult nervous system function	In infant, slowed/deficient brain development, retardation; in adult, mental dulling, depression, paresthesias, memory impairment, hypoactive reflexes	Irritability, restlessness, insomnia, exophthalmos, personality changes
Cardiovascular system	Promotes normal functioning of the heart	Decreased efficiency of pumping action of the heart; low heart rate and blood pressure	Rapid heart rate and possible palpitations; high blood pressure; if prolonged, heart failure
Muscular system	Promotes normal muscular development, and function	Sluggish muscle action; muscle cramps; myalgia	Muscle atrophy and weakness
Skeletal system	Promotes normal growth and maturation of the skeleton	In child, growth retardation, skeletal stunting and retention of child's body proportions; in adult, joint pain	In child, excessive skeletal growth initially, followed by early epiphyseal closure and short stature; in adult, demineralization of skeleton
Gastrointestinal system	Promotes normal GI motility and tone; increases secretion of digestive juices	Depressed GI motility, tone, and secretory activity; constipation	Excessive GI motility; diarrhea; loss of appetite
Reproductive system	Promotes normal female reproductive ability and lactation	Depressed ovarian function; sterility; depressed lactation	In females, depressed ovarian function; in males, impotence
Integumentary system	Promotes normal hydration and secretory activity of skin	Skin pale, thick, and dry; facial edema; hair coarse and thick	Skin flushed, thin, and moist; hair fine and soft; nails soft and thin

Synthesis Thyroid hormone synthesis involves six interrelated processes that begin when TSH secreted by the anterior pituitary binds to follicle cell receptors. The following numbers correspond to steps 1–6 in Figure 15.8:

① **Formation and storage of thyroglobulin.** After being synthesized on the ribosomes, thyroglobulin is transported to the Golgi apparatus, where sugar residues are attached and the molecules are packed into vesicles. These transport vesicles move to the apex of the follicle cell, where their contents are discharged into the follicle lumen and become part of the stored colloid.

② **Iodide trapping and oxidation to iodine.** To produce the functional iodinated hormones, the follicle cells must accumulate iodides (anions of iodine, I$^-$) from the blood. Because the intracellular concentration of I$^-$ is over 30 times higher than that in blood, iodide trapping depends on active transport. Once they enter the follicle cell, iodides are oxidized (by removal of electrons) and converted to iodine (I$_2$).

③ **Iodination.** Once formed, iodine is attached to tyrosine amino acids that form part of the thyroglobulin colloid. This iodination reaction occurs at the apical follicle cell–colloid junction and is mediated by peroxidase enzymes.

④ **Coupling of T$_2$ and T$_1$.** Attachment of one iodine to a tyrosine produces **monoiodotyrosine** (**MIT** or **T$_1$**); attachment of two iodines produces

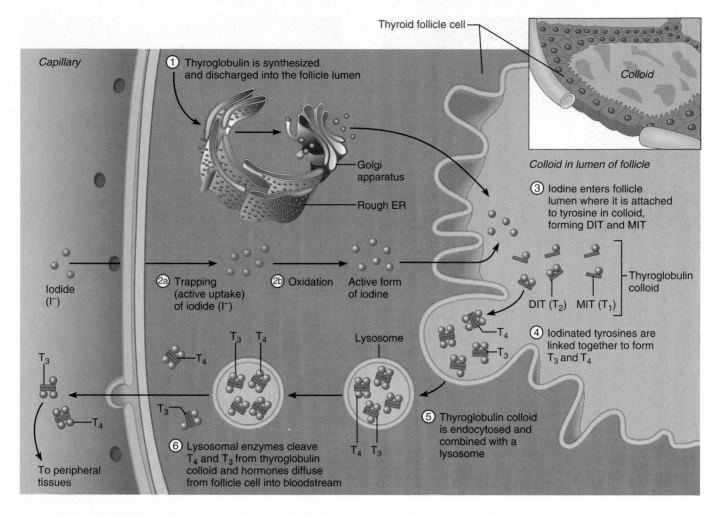

Thyroid follicle cell

① Thyroglobulin is synthesized and discharged into the follicle lumen

Capillary

Colloid

Golgi apparatus

Rough ER

Colloid in lumen of follicle

③ Iodine enters follicle lumen where it is attached to tyrosine in colloid, forming DIT and MIT

Iodide (I⁻)

②a Trapping (active uptake) of iodide (I⁻)

②b Oxidation

Active form of iodine

Thyroglobulin colloid

DIT (T₂)   MIT (T₁)

④ Iodinated tyrosines are linked together to form T₃ and T₄

T₃   T₄

Lysosome

T₄

T₃

T₃

T₄

T₃

T₄

⑤ Thyroglobulin colloid is endocytosed and combined with a lysosome

T₄   T₃

T₃

T₄

To peripheral tissues

⑥ Lysosomal enzymes cleave T₄ and T₃ from thyroglobulin colloid and hormones diffuse from follicle cell into bloodstream

FIGURE 15.8 Synthesis of thyroid hormones. Binding of TSH to follicle cell receptors (not illustrated) stimulates synthesis of the thyroglobulin colloid, the six steps illustrated here, and the release of thyroxine (T_4) and triiodothyronine (T_3) from the cell. (Only iodinated tyrosine amino acids of the thyroglobulin colloid are illustrated. The rest of the colloid is represented by the unstructured yellow substance.)

diiodotyrosine (**DIT** or **T₂**). Then, enzymes in the colloid link T_1 and T_2 together. Two linked DITs result in T_4; coupling of MIT and DIT produces T_3. At this point, the hormones are still part of the thyroglobulin colloid.

⑤ **Colloid endocytosis.** Hormone secretion requires that the follicle cells reclaim iodinated thyroglobulin by endocytosis and combine the vesicles with lysosomes.

⑥ **Cleavage of the hormones for release.** In the lysosomes, the hormones are cleaved out of the colloid by lysosomal enzymes. The hormones then diffuse from the follicle cells into the bloodstream. The main hormonal product secreted is T_4. Some T_4 is converted to T_3 before secretion, but most T_3 is generated in the peripheral tissues.

Although we have followed TH synthesis from beginning to end (secretion), the *initial response* to TSH binding is secretion of thyroid hormone. Then more colloid is synthesized to "restock" the follicle lumen. As a general rule, TSH levels are lower during the day, peak just before sleep, and remain high during the night.

The thyroid gland is unique among the endocrine glands in its ability to store its hormone extracellularly and in large quantities. In the normal thyroid gland, the amount of stored colloid remains relatively constant and is sufficient to provide normal levels of hormone release for two to three months.

Transport and Regulation   Most released T_4 and T_3 immediately bind to transport proteins, most importantly *thyroxine-binding globulins (TBGs)* produced by the liver. Both T_4 and T_3 bind to target tissue receptors, but T_3 binds much more avidly and is about ten times more active. Most peripheral tissues have the enzymes needed to convert T_4 to T_3, a process that entails enzymatic removal of one iodine group.

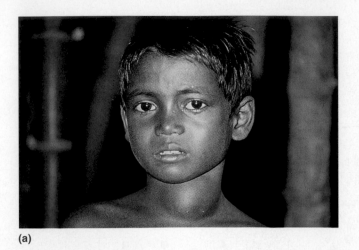

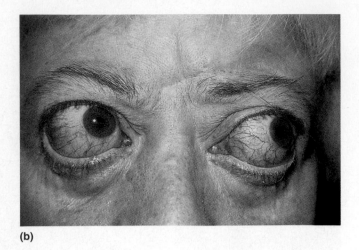

(a)   **(b)**

FIGURE 15.9   **Thyroid disorders. (a)** An enlarged thyroid (goiter) of a
Bangladeshi boy. **(b)** Exophthalmos of Graves' disease.

Several mechanisms of TH activity probably
exist. However, what we know for sure is that, like
steroids, T_3 enters a target cell and binds to intracellular receptors within the cell's nucleus and initiates
transcription of mRNA.

Falling T_4 blood levels trigger release of *thyroid-
stimulating hormone (TSH)*, and ultimately of
more T_4. Rising T_4 levels feed back to inhibit the
hypothalamic-adenohypophyseal axis, temporarily
shutting off the stimulus for TSH release. Conditions that increase body energy requirements, such
as pregnancy and prolonged cold, stimulate the hypothalamus to secrete *thyrotropin-releasing hormone
(TRH)*, which triggers TSH release, and in such instances, TRH overcomes the negative feedback controls. As the thyroid gland releases larger amounts of
thyroid hormones, body metabolism and heat production are enhanced. Factors that inhibit TSH
release include somatostatin, rising levels of glucocorticoids and sex hormones (estrogens or testosterone),
and excessively high blood iodide concentration.

⒣ *HOMEOSTATIC IMBALANCE*

Both overactivity and underactivity of the thyroid
gland can cause severe metabolic disturbances. Hypothyroid disorders may result from some thyroid
gland defect or secondarily from inadequate TSH or
TRH release. They also occur when the thyroid
gland is removed surgically and when dietary iodine
is inadequate.

In adults, the full-blown hypothyroid syndrome
is called **myxedema** (mik″sĕ-de′mah; "mucous
swelling"). Symptoms include a low metabolic rate;
feeling chilled; constipation; thick, dry skin and
puffy eyes; edema; lethargy; and mental sluggishness
(but not mental retardation). If myxedema results
from lack of iodine, the thyroid gland enlarges and

protrudes, a condition called **endemic** (en-dem′ik),
or *colloidal*, **goiter** (Figure 15.9a). The follicle cells
produce colloid but cannot iodinate it or make functional hormones. The pituitary gland secretes
increasing amounts of TSH in a futile attempt to
stimulate the thyroid to produce TH, but the only
result is that the follicles accumulate more and more
unusable colloid. Untreated, the thyroid cells eventually "burn out" from frantic activity, and the gland
atrophies. Before the marketing of iodized salt, parts
of the midwestern United States were called the
"goiter belt." Because these areas had iodine-poor
soil and no access to iodine-rich shellfish, goiters
were very common there. Depending on the cause,
myxedema can be reversed by iodine supplements or
hormone replacement therapy.

Severe hypothyroidism in infants is called
cretinism (kre′tĭ-nizm). The child is mentally re-
tarded and has a short, disproportionately sized body
and a thick tongue and neck. Cretinism may reflect
a genetic deficiency of the fetal thyroid gland or maternal factors, such as lack of dietary iodine. It is
preventable by thyroid hormone replacement therapy if diagnosed early enough, but once developmental abnormalities and mental retardation appear,
they are not reversible.

The most common (and puzzling) hyperthyroid
pathology is **Graves' disease.** Because the serum of
patients with this condition often contains abnormal antibodies that mimic TSH and continuously
stimulate TH release, Graves' disease is believed to
be an autoimmune disease. Typical symptoms
include an elevated metabolic rate; sweating; rapid,
irregular heartbeat; nervousness; and weight loss despite adequate food intake. *Exophthalmos*, protrusion of the eyeballs, may occur if the tissue behind
the eyes becomes edematous and then fibrous
(Figure 15.9b). Treatments include surgical removal

of the thyroid gland or ingestion of radioactive iodine (^{131}I), which selectively destroys the most active thyroid cells. ●

Calcitonin

Calcitonin is a polypeptide hormone produced by the **parafollicular,** or **C, cells** of the thyroid gland. Because its most important effect is to lower blood Ca^{2+} levels, calcitonin is a direct antagonist of parathyroid hormone, produced by the parathyroid glands. Calcitonin targets the skeleton, where it (1) inhibits osteoclast activity and hence bone resorption and release of Ca^{2+} from the bony matrix and (2) stimulates Ca^{2+} uptake and incorporation into bone matrix. Thus, calcitonin has a bone-sparing effect.

Excessive blood levels of Ca^{2+} (approximately 20% above normal) act as a humoral stimulus for calcitonin release, whereas declining blood Ca^{2+} levels inhibit C cell secretory activity. Calcitonin regulation of blood Ca^{2+} levels is short-lived but extremely rapid.

Calcitonin appears to be important only in childhood, when the skeleton grows quickly and the bones are changing dramatically in mass, size, and shape. In adults, it is at best a weak hypocalcemic agent.

The Parathyroid Glands

The tiny, yellow-brown **parathyroid glands** are nearly hidden from view in the posterior aspect of the thyroid gland (Figure 15.10a). There are usually four of these glands, but the precise number varies from one individual to another. As many as eight have been reported, and some may be located in other regions of the neck or even in the thorax. The parathyroid's glandular cells are arranged in thick, branching cords containing scattered *oxyphil cells* and large numbers of smaller **chief cells** (Figure 15.10b). The chief cells secrete PTH. The function of the oxyphil cells is unclear.

Discovery of the parathyroid glands was accidental. Years ago, surgeons were baffled by the observation that while most patients recovered uneventfully after partial (or even total) thyroid gland removal, others suffered uncontrolled muscle spasms and severe pain, and subsequently died. It was only after several such tragic deaths that the parathyroid glands were discovered and their hormonal function, quite different from that of the thyroid gland hormones, became known.

Parathyroid hormone (PTH), or **parathormone,** the protein hormone of these glands, is the single most important hormone controlling the calcium balance of the blood. PTH release is triggered by

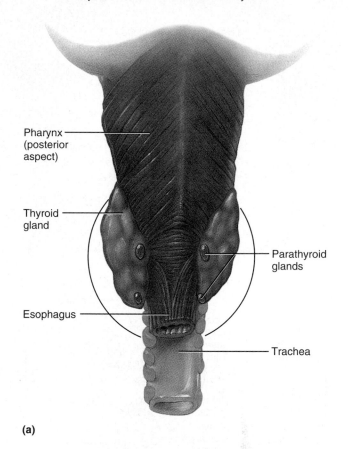

(a)

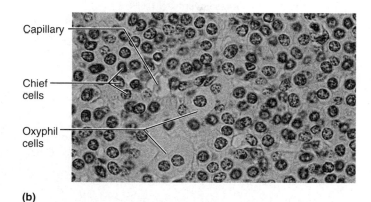

(b)

FIGURE 15.10 The parathyroid glands. (a) Location of the parathyroid glands on the posterior aspect of the thyroid gland. These tiny glands may be more inconspicuous than depicted. **(b)** Micrograph of a section of the parathyroid gland (1500×).

falling blood Ca^{2+} levels and inhibited by hypercalcemia. PTH increases Ca^{2+} levels in blood by stimulating three target organs: the skeleton (which contains considerable amounts of calcium salts in its matrix), the kidneys, and the intestine (Figure 15.11).

PTH release (1) stimulates osteoclasts (bone-resorbing cells) to digest some of the bony matrix and release ionic calcium and phosphates to the blood;

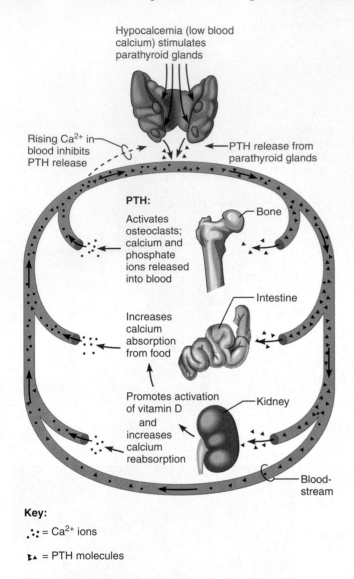

Key:

∴ = Ca²⁺ ions

▸▲ = PTH molecules

FIGURE 15.11 Effect of parathyroid hormone on bone, the intestine, and the kidneys. The effect on the intestine is indirect via activated vitamin D.

(2) enhances reabsorption of Ca^{2+} (and excretion of PO_4^{3-}) by the kidneys; and (3) increases absorption of Ca^{2+} by the intestinal mucosal cells. Calcium absorption by the intestine is enhanced indirectly by PTH's effect on vitamin D activation. Vitamin D is required for absorption of Ca^{2+} from ingested food, but the form in which the vitamin is ingested or produced by the skin is relatively inactive. For vitamin D to exert its physiological effects, it must first be converted by the kidneys to its active vitamin D_3 form, **calcitriol** (1,25-dihydroxycholecalciferol), a transformation stimulated by PTH.

Because plasma calcium ion homeostasis is essential for so many functions, including transmission of nerve impulses, muscle contraction, and blood clotting, precise control of Ca^{2+} levels is critical.

ⓗ HOMEOSTATIC IMBALANCE

Hyperparathyroidism is rare and usually the result of a parathyroid gland tumor. When it occurs, calcium is leached from the bones, and the bones soften and deform as their mineral salts are replaced by fibrous connective tissue. In *osteitis cystica fibrosa*, a severe example of this disorder, the bones have a moth-eaten appearance on X rays and tend to fracture spontaneously. The resulting abnormally elevated blood Ca^{2+} level (hypercalcemia) has many outcomes, but the two most notable are (1) depression of the nervous system, which leads to abnormal reflexes and weakness of the skeletal muscles, and (2) formation of kidney stones as excess calcium salts precipitate in the kidney tubules. Calcium deposits may also form in soft tissues throughout the body and severely impair vital organ functioning, a condition called *metastatic calcification*.

Hypoparathyroidism, or PTH deficiency, most often follows parathyroid gland trauma or removal during thyroid surgery. However, an extended deficiency of dietary magnesium (required for PTH secretion) can cause functional hypoparathyroidism. The resulting hypocalcemia increases the excitability of neurons and accounts for the classical symptoms of *tetany* such as loss of sensation, muscle twitches, and convulsions. Untreated, the symptoms progress to respiratory paralysis and death. ●

The Adrenal (Suprarenal) Glands

The paired **adrenal glands** are pyramid-shaped organs perched atop the kidneys (*ad* = near; *renal* = kidney), where they are enclosed in a fibrous capsule and a cushion of fat (see Figures 15.1 and 15.12). They are often referred to as the **suprarenal glands** (*supra* = above) as well.

Each adrenal gland is structurally and functionally two endocrine glands. The inner **adrenal medulla**, more like a knot of nervous tissue than a gland, is part of the sympathetic nervous system. The outer **adrenal cortex**, encapsulating the medullary region and forming the bulk of the gland, is glandular tissue derived from embryonic mesoderm. Each region produces its own set of hormones (Table 15.3), but all adrenal hormones help us cope with stressful situations.

The Adrenal Cortex

Well over two dozen *steroid* hormones, collectively called **corticosteroids,** are synthesized from cholesterol by the adrenal cortex. The pathway is multistep and involves varying intermediates depending on the hormone being formed. Unlike the amino acid based hormones, steroid hormones are not stored in cells. Therefore, their rate of release in response to stimulation depends on their rate of synthesis. The

Suppose your lab professor springs a "pop" quiz and has adrenal cortical tissue under the microscope for identification. What information about the cell arrangements would help you in this quiz?

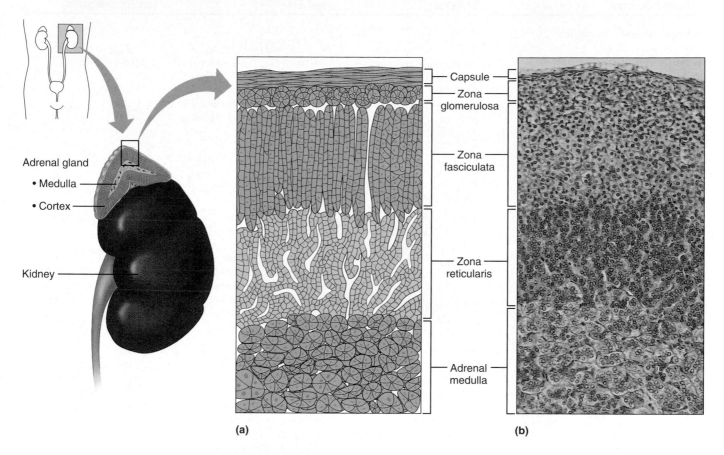

FIGURE 15.12 Microscopic structure of the adrenal gland. (a) Diagram and **(b)** photomicrograph of the zona glomerulosa, zona fasciculata, and zona reticularis of the adrenal cortex (200×). (Part of the adrenal medulla is also shown.)

structures of several corticosteroids are illustrated in Table 15.3.

The large, lipid-laden cortical cells are arranged in three layers or zones (Figure 15.12). The cell clusters forming the superficial **zona glomerulosa** (zo'nah glo-mer"u-lo'sah) mainly produce mineralocorticoids, hormones that help control the balance of minerals and water in the blood. The principal products of the middle **zona fasciculata** (fah-sik"u-la'tah) cells, arranged in more or less linear cords, are the metabolic hormones called glucocorticoids. The cells of the innermost **zona reticularis** (rĕ-tik'u-lar-is), abutting the adrenal medulla, have a netlike arrangement. These cells mainly produce small amounts of adrenal sex hormones, or gonadocorticoids. Al-

though, as just described, there is a division of labor in corticosteroid production, the entire spectrum of corticosteroids is produced by all three cortical layers.

Mineralocorticoids   The essential function of **mineralocorticoids** is regulation of the electrolyte (mineral salt) concentrations in extracellular fluids, particularly of Na^+ and K^+. The single most abundant cation in extracellular fluid is Na^+, and although this ion is vital to homeostasis, excessive sodium intake and retention may promote high blood pressure (hypertension) in susceptible individuals. Although there are several mineralocorticoids, **aldosterone** (al-dos'ter-ōn) is the most potent and accounts for more than 95% of the mineralocorticoids produced. As you might guess, maintaining sodium ion balance is aldosterone's primary job, which it performs by stimulating transcription of the Na^+, K^+-ATPase, the sodium pump that exchanges K^+ for Na^+.

■ *Zona glomerulosa cells are in clusters, zona fasciculata cells are in parallel cords, and zona reticularis cells have a netlike arrangement.*

TABLE 15.3   Adrenal Gland Hormones: Summary of Regulation and Effects

Hormone, Structure	Regulation of Release	Target Organ and Effects	Effects of Hyposecretion ↓ and Hypersecretion ↑
Adrenocortical Hormones			
Mineralocorticoids (chiefly aldosterone) Aldosterone	Stimulated by renin-angiotensin mechanism (activated by decreasing blood volume or blood pressure), elevated K^+ or low Na^+ blood levels, and ACTH (minor influence); inhibited by increased blood volume and pressure, increased Na^+ and decreased K^+ blood levels	Kidneys: increased blood levels of Na^+ and decreased blood levels of K^+; since water reabsorption accompanies sodium retention, blood volume and blood pressure rise	↑ Aldosteronism ↓ Addison's disease
Glucocorticoids (chiefly cortisol) Cortisol	Stimulated by ACTH; inhibited by feedback inhibition exerted by cortisol	Body cells: promote gluconeogenesis and hyperglycemia; mobilize fats for energy metabolism; stimulate protein catabolism; assist body to resist stressors; depress inflammatory and immune responses	↑ Cushing's disease ↓ Addison's disease
Gonadocorticoids (chiefly androgens such as testosterone) Testosterone	Stimulated by ACTH; mechanism of inhibition incompletely understood, but feedback inhibition not seen	Insignificant effects in adults; the adrenally produced hormones may be responsible for female libido and source of estrogen after menopause	↑ Virilization of females (androgenital syndrome) ↓ No effects known
Adrenal Medullary Hormones			
Catecholamines (epinephrine and norepinephrine) Epinephrine	Stimulated by preganglionic fibers of the sympathetic nervous system	Sympathetic nervous system target organs: effects mimic sympathetic nervous system activation; increase heart rate and metabolic rate; increase blood pressure by promoting vasoconstriction	↑ Prolonged fight-or-flight response; hypertension ↓ Unimportant

Aldosterone reduces excretion of Na^+ from the body. Its primary target is the distal parts of the kidney tubules, where it stimulates Na^+ reabsorption from the forming urine and its return to the bloodstream. Aldosterone also enhances Na^+ reabsorption from perspiration, saliva, and gastric juice. The mechanism of aldosterone activity appears to involve the synthesis of an enzyme required for Na^+ transport.

Regulation of a number of other ions, including K^+, H^+, HCO_3^- (bicarbonate), and Cl^- (chloride), is coupled to that of Na^+; and where Na^+ goes, water follows—an event that leads to changes in blood volume and blood pressure. Hence, Na^+ regulation is crucial to overall body homeostasis. Briefly, aldosterone's effects on the renal tubules cause sodium and water retention accompanied by elimination of K^+ and, in some instances, alterations in the acid-base balance of the blood (by H^+ excretion). Because aldosterone's regulatory effects are brief (lasting approximately 20 minutes), plasma electrolyte balance can be precisely controlled and modified continuously.

Although this discussion represents the major roles of bloodborne aldosterone produced by the adrenal cortex, aldosterone is also secreted by cardiovascular organs, where it is a paracrine and plays a completely different role in cardiac regulation.

Aldosterone secretion is stimulated by rising blood levels of K^+, low blood levels of Na^+, and decreasing blood volume and blood pressure. The reverse conditions inhibit aldosterone secretion. Four mechanisms regulate aldosterone secretion (Figure 15.13); these mechanisms are described next.

1. The renin-angiotensin mechanism (re′nin an″je-o-ten′sin). The renin-angiotensin mechanism, the major regulator of aldosterone release, influences both the electrolyte-water balance of the blood and blood pressure. Specialized cells of the *juxtaglomerular apparatus* in the kidneys become excited when blood pressure (or blood volume) declines or plasma osmolarity (solute concentration) drops. These cells respond by releasing **renin** into the blood. Renin cleaves off part of the plasma protein **angiotensinogen** (an″je-o-ten′sin-o-gen), triggering an enzymatic cascade leading to the formation of **angiotensin II**, a potent stimulator of aldosterone release by the glomerulosa cells. However, the renin-angiotensin mechanism does much more than trigger aldosterone release, and all of its effects are ultimately involved in raising the blood pressure. These additional effects are described in detail in Chapters 24 and 25.

2. Plasma concentration of sodium and potassium ions. Fluctuating blood levels of Na^+ and K^+ directly influence the zona glomerulosa cells. Increased K^+ and decreased Na^+ are stimulatory; the opposite conditions are inhibitory.

3. ACTH. Under normal circumstances, ACTH released by the anterior pituitary has little or no effect on aldosterone release. However, when a person is severely stressed, the hypothalamus secretes more **corticotropin-releasing hormone (CRH)**, and the rise in ACTH blood levels that follows steps up the rate of aldosterone secretion to a small extent. The increase in blood volume and blood pressure that results helps ensure adequate delivery of nutrients and respiratory gases during the stressful period.

4. Atrial natriuretic peptide (ANP). Atrial natriuretic peptide, a hormone secreted by the heart when blood pressure rises, fine-tunes blood pressure and sodium-water balance of the body. Its major effect is to inhibit the renin-angiotensin mechanism: It blocks renin and aldosterone secretion and inhibits other angiotensin-induced mechanisms that enhance water and Na^+ reabsorption. Consequently, ANP's overall influence is to decrease blood pressure by allowing Na^+ (and water) to flow out of the body in urine (*natriuretic* = producing salty urine).

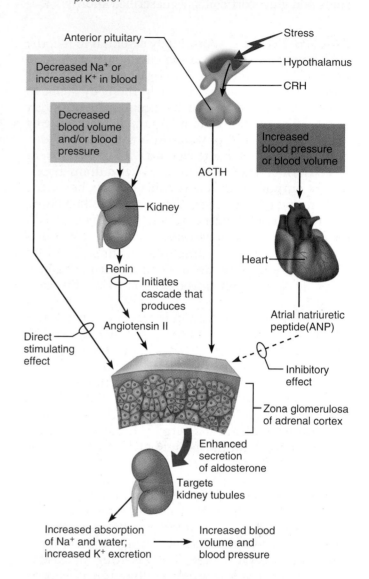

? *Generally speaking, how does the response time for this hormonal mechanism compare to that triggered when sympathetic nerves act to control blood pressure?*

FIGURE 15.13 Major mechanisms controlling aldosterone release from the adrenal cortex.

Ⓗ *HOMEOSTATIC IMBALANCE*

Hypersecretion of aldosterone, a condition called *aldosteronism*, typically results from adrenal neoplasms. Two major sets of problems result: (1) hypertension and edema due to excessive Na^+ and water retention and (2) accelerated excretion of potassium ions. If K^+ loss is extreme, neurons become nonresponsive and muscle weakness (and

This response is much slower because the "messengers" are bloodborne hormones instead of nerve impulses and neurotransmitters. ■

eventually paralysis) occurs. *Addison's disease,* a hyposecretory disease of the adrenal cortex, generally involves a deficient output of both mineralocorticoids and glucocorticoids, as described shortly. ●

Glucocorticoids Absolutely essential to life, the **glucocorticoids** influence the energy metabolism of most body cells and help us to resist stressors. Under normal circumstances, they help the body adapt to intermittent food intake by keeping blood sugar levels fairly constant, and maintain blood volume by preventing the shift of water into tissue cells. However, severe stress due to hemorrhage, infections, or physical or emotional trauma evokes a dramatically higher output of glucocorticoids, which helps the body negotiate the crisis. Glucocorticoid hormones include **cortisol (hydrocortisone)** (see Table 15.3), **cortisone,** and **corticosterone,** but only cortisol is secreted in significant amounts in humans. As for all steroid hormones, the basic mechanism of glucocorticoid activity on target cells is to modify gene activity.

Glucocorticoid secretion is regulated by negative feedback. Cortisol release is promoted by ACTH, triggered in turn by the hypothalamic releasing hormone CRH. Rising cortisol levels feed back to act on both the hypothalamus and the anterior pituitary, preventing CRH release and shutting off ACTH and cortisol secretion. Cortisol secretory bursts, driven by patterns of eating and activity, occur in a definite pattern throughout the day and night. Cortisol blood levels peak shortly after we arise in the morning. The lowest levels occur in the evening just before and shortly after we fall asleep. The normal cortisol rhythm is interrupted by acute stress of any variety as the sympathetic nervous system overrides the (usually) inhibitory effects of elevated cortisol levels and triggers CRH release. The resulting increase in ACTH blood levels causes an outpouring of cortisol from the adrenal cortex.

Stress results in a dramatic rise in blood levels of glucose, fatty acids, and amino acids, all provoked by cortisol. Cortisol's prime metabolic effect is to provoke *gluconeogenesis,* that is, the formation of glucose from fats and proteins. In order to "save" glucose for the brain, cortisol mobilizes fatty acids from adipose tissue and encourages their increased use for energy. Under cortisol's influence, stored proteins are broken down to provide building blocks for repair or for making enzymes to be used in metabolic processes. Cortisol enhances epinephrine's vasoconstrictive effects, and the rise in blood pressure and circulatory efficiency that results helps ensure that these nutrients are quickly distributed to the cells.

Although *ideal amounts of glucocorticoids promote normal function,* significant anti-inflammatory and anti-immune effects are associated with cortisol excess. Excessive levels of glucocorticoids (1) depress cartilage and bone formation, (2) inhibit inflammation by stabilizing lysosomal membranes and preventing vasodilation, (3) depress the immune system, and (4) promote changes in cardiovascular, neural, and gastrointestinal function. Recognition of the effects of glucocorticoid hypersecretion has led to widespread use of glucocorticoid drugs to control symptoms of many chronic inflammatory disorders, such as rheumatoid arthritis or allergic responses. However, these potent drugs are a double-edged sword. Although they relieve some of the symptoms of these disorders, they also cause the undesirable effects of excessive levels of these hormones.

🅗 HOMEOSTATIC IMBALANCE

The pathology of glucocorticoid excess, **Cushing's disease,** or **syndrome,** may be caused by an ACTH-releasing pituitary tumor; by an ACTH-releasing malignancy of the lungs, pancreas, or kidneys; or by a tumor of the adrenal cortex. However, it most often results from the clinical administration of pharmacological doses (doses higher than those found in the body) of glucocorticoid drugs. The syndrome is characterized by persistent hyperglycemia *(steroid diabetes),* dramatic losses in muscle and bone protein, and water and salt retention, leading to hypertension and edema. The so-called *cushingoid signs* (Figure 15.14) include a swollen "moon" face, redistribution of fat to the abdomen and the posterior neck (causing a "buffalo hump"), a tendency to bruise, and poor wound healing. Because of enhanced anti-inflammatory effects, infections may become overwhelmingly severe before producing recognizable symptoms. Eventually, muscles weaken and spontaneous fractures force the person to become bedridden. The only treatment is removal of the cause—be it surgical removal of the offending tumor or discontinuation of the drug.

Addison's disease, the major hyposecretory disorder of the adrenal cortex, usually involves deficits in both glucocorticoids and mineralocorticoids. Its victims tend to lose weight; their plasma glucose and sodium levels drop and potassium levels rise. Severe dehydration and hypotension are common. Corticosteroid replacement therapy at physiological doses (doses typical of those normally found in the body) is the usual treatment. ●

Gonadocorticoids (Sex Hormones) The bulk of the **gonadocorticoids** secreted by the adrenal cortex are weak **androgens,** or male sex hormones, such as *androstenedione* and *dehydroepiandrosterone (DEHA).* These are converted to the more potent

(a)

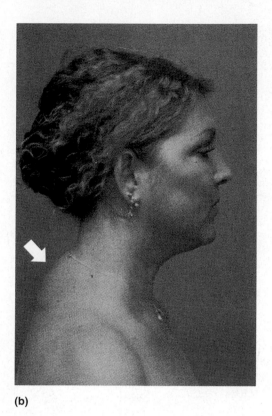

(b)

FIGURE 15.14   **A patient (a) before and (b) during Cushing's disease.** In (b), the characteristic "buffalo hump" of fat on the upper back is indicated by a white arrow.

male hormone, *testosterone,* in the tissue cells or to estrogens (female sex homones) in females (see Table 15.3). The adrenal cortex also makes small amounts of female hormones (estradiol and other estrogens). The amount of gonadocorticoids produced by the adrenal cortex is insignificant compared with the amounts made by the gonads during late puberty and adulthood. The exact role of the adrenal sex hormones is still in question, but because adrenal androgen levels rise continuously between the ages of 7 and 13 in both boys and girls, it is assumed that these hormones contribute to the onset of puberty and the appearance of axillary and pubic hair during that time. In adult women adrenal androgens are thought to be responsible for the sex drive, and they may account for the estrogens produced after menopause when ovarian estrogens are no longer produced. Control of gonadocorticoid secretion is not completely understood. Release seems to be stimulated by ACTH, but the gonadocorticoids do not appear to exert feedback inhibition on ACTH release.

HOMEOSTATIC IMBALANCE

Since androgens predominate, hypersecretion of gonadocorticoids causes *androgenital syndrome* (mas-

culinization). In adult males, these effects of elevated gonadocorticoid levels may be obscured, since testicular testosterone has already produced virilization; but in prepubertal males and in females, the results can be dramatic. In the young man, maturation of the reproductive organs and appearance of the secondary sex characteristics occur rapidly, and the sex drive emerges with a vengeance. Females develop a beard and a masculine pattern of body hair distribution, and the clitoris grows to resemble a small penis.   ●

The Adrenal Medulla

Because it is discussed in Chapter 14 as part of the autonomic nervous system, the **adrenal medulla** is covered only briefly here. Its spherical **chromaffin cells** (kro-maf'in), which crowd around blood-filled capillaries and sinusoids, are modified ganglionic sympathetic neurons that synthesize the *catecholamines* **epinephrine** and **norepinephrine** via a molecular sequence from tyrosine to dopamine to NE to epinephrine.

When the body is activated to fight-or-flight status by some short-term stressor, the sympathetic nervous system is mobilized: Blood sugar levels rise, blood vessels constrict and the heart beats faster (together raising the blood pressure), blood is diverted

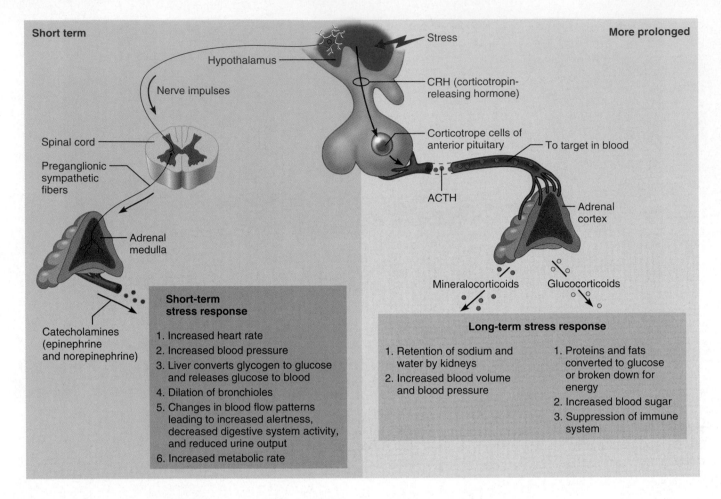

FIGURE 15.15 Stress and the adrenal gland. Stressful stimuli cause the hypothalamus to activate the adrenal medulla via sympathetic nerve impulses and the adrenal cortex via hormonal signals. The medulla mediates short-term responses to stress by secreting epinephrine and norepinephrine. The cortex controls more prolonged responses by secreting steroid hormones.

from temporarily nonessential organs to the brain, heart, and skeletal muscles, and preganglionic sympathetic nerve endings weaving through the adrenal medulla signal for release of catecholamines, which reinforce and prolong the fight-or-flight response.

Unequal amounts of the two hormones are stored and released; approximately 80% is epinephrine and 20% norepinephrine. With a few exceptions, the two hormones exert the same effects (Table 14.3, p. 513). Epinephrine is the more potent stimulator of the heart and metabolic activities, but norepinephrine has the greater influence on peripheral vasoconstriction and blood pressure. Epinephrine is used clinically as a heart stimulant and to dilate the bronchioles during acute asthmatic attacks.

Unlike the adrenocortical hormones, which promote long-lasting body responses to stressors, catecholamines cause fairly brief responses. The interrelationships of the adrenal hormones and the hypothalamus, the "director" of the stress response, are depicted in Figure 15.15.

H   HOMEOSTATIC IMBALANCE

Because hormones of the adrenal medulla merely intensify activities set into motion by the sympathetic nervous system neurons, a deficiency of these hormones is not a problem. Unlike glucocorticoids, adrenal catecholamines are not essential for life. However, hypersecretion of catecholamines, sometimes arising from a chromaffin cell tumor called a *pheochromocytoma* (fe-o-kro"mo-si-to'mah), produces symptoms of uncontrolled sympathetic nervous system activity—hyperglycemia, increased metabolic rate, rapid heartbeat and palpitations, hypertension, intense nervousness, and sweating. ●

The Pancreas

Located partially behind the stomach in the abdomen, the soft, triangular **pancreas** is a mixed gland composed of both endocrine and exocrine gland cells (Figure 15.1). Along with the thyroid and parathyroids, it develops as an outpocketing of the epithelial lining of the gastrointestinal tract. *Acinar cells*, forming the bulk of the gland, produce an enzyme-rich juice that is ducted into the small intestine during food digestion.

Scattered among the acinar cells are approximately a million **pancreatic islets** (also called **islets of Langerhans**), tiny cell clusters that produce pancreatic hormones (Figure 15.16). The islets contain two major populations of hormone-producing cells, the glucagon-synthesizing **alpha (α) cells** and the more numerous insulin-producing **beta (β) cells**. These cells act as tiny fuel sensors, secreting glucagon and insulin appropriately during the fasting and fed states. Thus insulin and glucagon are intimately but independently involved in the regulation of blood glucose levels. Their effects are antagonistic: Insulin is a *hypoglycemic* hormone, whereas glucagon is a *hyperglycemic* hormone (Figure 15.17). Some islet cells also synthesize other peptides in small amounts. These include *somatostatin, pancreatic polypeptide (PP)*, and others. However, these will not be dealt with here.

Glucagon

Glucagon (gloo'kah-gon), a 29-amino-acid polypeptide, is an extremely potent hyperglycemic agent. One molecule of this hormone can cause the release of 100 million molecules of glucose into the blood! The major target of glucagon is the liver, where it promotes the following actions (Figure 15.17):

1. Breakdown of glycogen to glucose *(glycogenolysis)*

2. Synthesis of glucose from lactic acid and from noncarbohydrate molecules *(gluconeogenesis)*

3. Release of glucose to the blood by liver cells, which causes blood sugar levels to rise

A secondary effect is a fall in the amino acid concentration in the blood as the liver cells sequester these molecules to make new glucose molecules.

Secretion of glucagon by the alpha cells is prompted by humoral stimuli, mainly falling blood sugar levels. However, sympathetic nervous system stimulation and rising amino acid levels (as might follow a protein-rich meal) are also stimulatory. Glucagon release is suppressed by rising blood sugar levels and somatostatin. Because glucagon is such an important hyperglycemic agent, it has been speculated that people with persistently low blood sugar levels (hypoglycemics) are deficient in glucagon.

Insulin

Insulin is a small (51-amino-acid) protein consisting of two amino acid chains linked by disulfide (—S—S—) bonds. It is synthesized as part of a

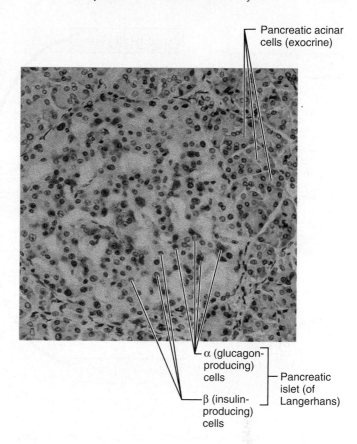

FIGURE 15.16 **Photomicrograph of pancreatic tissue, differentially stained.** A pancreatic islet is surrounded by blue-gray acinar cells, which produce the exocrine product (enzyme-rich pancreatic juice). In this preparation, the β cells of the islets that produce insulin are stained pale pink, and the α cells that produce glucagon are stained bright pink (150×).

larger polypeptide chain called **proinsulin.** The middle portion of this chain is then excised by enzymes, releasing functional insulin. This "clipping" process occurs in the secretory vesicles just before insulin is released from the beta cell.

Insulin's effects are most obvious when we have just eaten. Its main effect is to lower blood sugar levels (Figure 15.17), but it also influences protein and fat metabolism. Circulating insulin lowers blood sugar levels by enhancing membrane transport of glucose (and other simple sugars) into body cells, especially muscle and fat cells. (It does *not* accelerate glucose entry into liver, kidney, and brain* tissue, all of which have easy access to blood glucose regardless of insulin levels.) Insulin inhibits the breakdown of

---

* However, the cerebral cortex and hippocampus are well supplied with insulin receptors, hinting that insulin may play a different role in the brain, perhaps in cognition, memory, or neuronal growth.

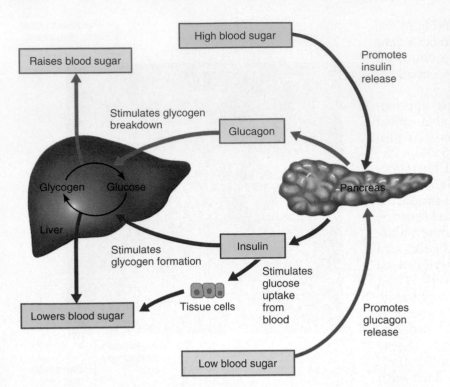

Raises blood sugar

Stimulates glycogen breakdown

Glucagon

Glycogen   Glucose

Liver

Stimulates glycogen formation

Insulin

Stimulates glucose uptake from blood

Tissue cells

Lowers blood sugar

High blood sugar

Promotes insulin release

Pancreas

Promotes glucagon release

Low blood sugar

FIGURE 15.17 Regulation of blood sugar levels by insulin and glucagon. When blood sugar levels are high, the pancreas releases the hypoglycemic hormone insulin, which stimulates sugar uptake by cells and glycogen formation in the liver, so that blood sugar levels are lowered. When blood sugar levels are low, the pancreas releases the hyperglycemic hormone glucagon, which stimulates glycogen breakdown and thereby increases the amount of blood sugar.

glycogen to glucose and the conversion of amino acids or fats to glucose; thus, it counters any metabolic activity that would increase plasma levels of glucose. Insulin's mechanism of action is still being investigated, but as noted earlier, its receptor is a tyrosine kinase enzyme.

After glucose enters a target cell, insulin binding triggers enzymatic activities that

1. Catalyze the oxidation of glucose for ATP production

2. Join glucose molecules together to form glycogen

3. Convert glucose to fat (particularly in adipose tissue)

As a rule, energy needs are met first, followed by glycogen formation. Finally, if excess glucose is still available, it is converted to fat. Insulin also stimulates amino acid uptake and protein synthesis in muscle tissue. In summary, insulin sweeps glucose out of the blood, causing it to be used for energy or converted to other forms (glycogen or fats), and it promotes protein synthesis and fat storage.

Pancreatic beta cells are stimulated to secrete insulin chiefly by elevated blood sugar levels, but rising plasma levels of amino acids and fatty acids, and release of acetylcholine by parasympathetic nerve fibers, also trigger insulin release. As body cells take up sugar and other nutrients, and plasma levels of these substances drop, insulin secretion is suppressed. Other hormones also influence insulin release. For example, any hyperglycemic hormone (such as glucagon, epinephrine, growth hormone,

thyroxine, or glucocorticoids) called into action as blood sugar levels drop indirectly stimulates insulin release by promoting glucose entry into the bloodstream. Somatostatin depresses insulin release. Thus, blood sugar levels represent a balance of humoral and hormonal influences. Insulin and (indirectly) somatostatin are the hypoglycemic factors that counterbalance the many hyperglycemic hormones.

⒣   HOMEOSTATIC IMBALANCE

Diabetes mellitus (DM) results from either hyposecretion or hypoactivity of insulin. When insulin is absent or deficient, blood sugar levels remain high after a meal because glucose is unable to enter most tissue cells. Ordinarily, when blood sugar levels rise, hyperglycemic hormones are not released, but when hyperglycemia becomes excessive, the person begins to feel nauseated, which precipitates the fight-or-flight response. This results, inappropriately, in all the reactions that normally occur in the hypoglycemic (fasting) state to make glucose available—that is, glycogenolysis, lipolysis (breakdown of fat), and gluconeogenesis. Thus, the already high blood sugar levels soar even higher, and excesses of glucose begin to be lost from the body in the urine *(glycosuria)*.

When sugars cannot be used as cellular fuel, more fats are mobilized, resulting in high fatty acid levels in the blood, a condition called lipidemia or lipemia. In severe cases of diabetes mellitus, blood

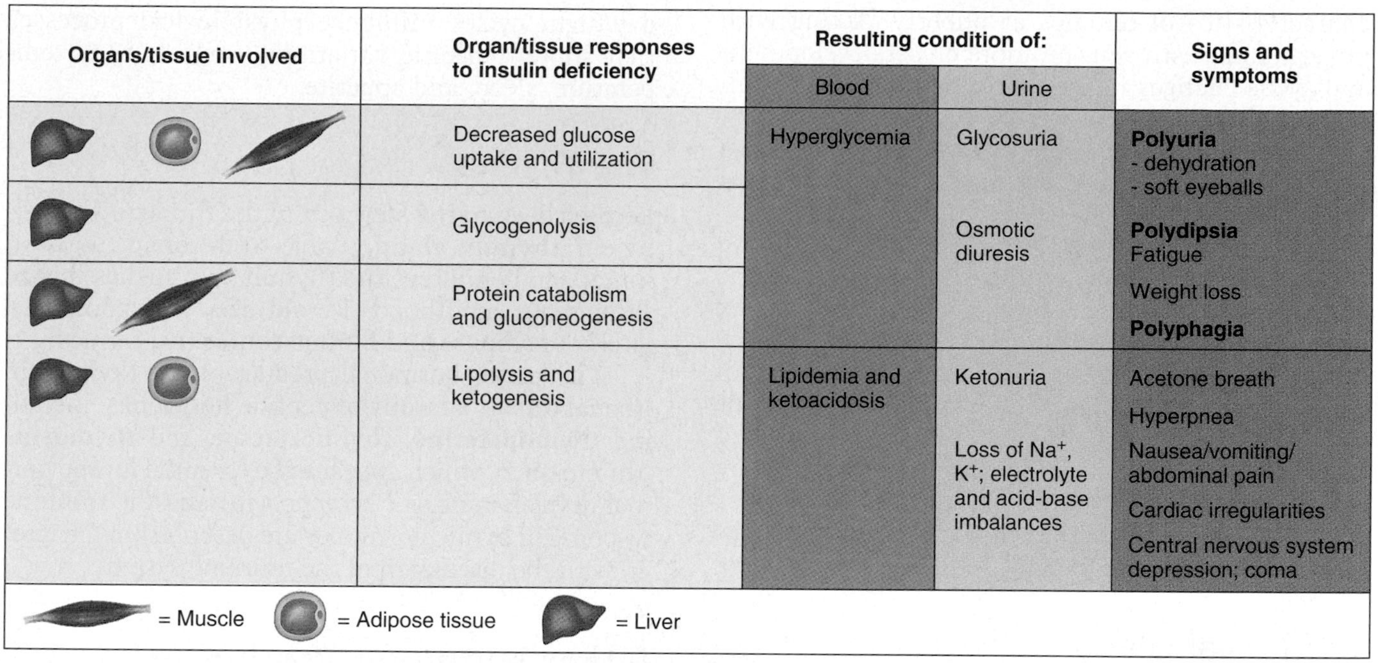

Organs/tissue involved	Organ/tissue responses to insulin deficiency	Resulting condition of:		Signs and symptoms
		Blood	Urine	
	Decreased glucose uptake and utilization	Hyperglycemia	Glycosuria	**Polyuria** - dehydration - soft eyeballs
	Glycogenolysis		Osmotic diuresis	**Polydipsia** Fatigue
	Protein catabolism and gluconeogenesis			Weight loss **Polyphagia**
	Lipolysis and ketogenesis	Lipidemia and ketoacidosis	Ketonuria	Acetone breath Hyperpnea
			Loss of Na⁺, K⁺; electrolyte and acid-base imbalances	Nausea/vomiting/ abdominal pain Cardiac irregularities Central nervous system depression; coma

= Muscle   = Adipose tissue   = Liver

FIGURE 15.18 **Symptomatic results of insulin deficit (diabetes mellitus).**

levels of fatty acids and their metabolites (acetoacetic acid, acetone, and others) rise dramatically. The fatty acid metabolites, collectively called **ketones** (ke′tōnz) or **ketone bodies,** are organic acids. When they accumulate in the blood, the blood pH drops, resulting in **ketoacidosis,** and ketone bodies begin to spill into the urine *(ketonuria).* Severe ketoacidosis is life threatening. The nervous system responds by initiating rapid deep breathing to blow off carbon dioxide from the blood and increase blood pH. (The physiological basis of this mechanism is explained in Chapter 21.) If untreated, ketoacidosis disrupts heart activity and oxygen transport, and severe depression of the nervous system leads to coma and death.

The three cardinal signs of diabetes mellitus are polyuria, polydipsia, and polyphagia. The excessive glucose in the kidney filtrate acts as an osmotic diuretic, that is, it inhibits water reabsorption by the kidney tubules, resulting in **polyuria,** a huge urine output that leads to decreased blood volume and dehydration. Serious electrolyte losses also occur as the body rids itself of excess ketone bodies. Ketone bodies are negatively charged and carry positive ions out with them; as a result, sodium and potassium ions are also lost from the body. Because of the electrolyte imbalance, the person gets abdominal pains and may vomit, and the stress reaction spirals even higher. Dehydration stimulates hypothalamic thirst centers, causing **polydipsia,** or excessive thirst. The final sign, **polyphagia,** refers to excessive hunger and

food consumption, a sign that the person is "starving in the land of plenty." Although plenty of glucose is available, it cannot be used, and the body starts to utilize its fat and protein stores for energy metabolism. Figure 15.18 summarizes the consequences of insulin deficiency.

Hyperinsulinism, or excessive insulin secretion, results in low blood sugar levels, or **hypoglycemia.** This condition triggers the release of hyperglycemic hormones, which cause anxiety, nervousness, tremors, and weakness. Insufficient glucose delivery to the brain causes disorientation, progressing to convulsions, unconsciousness, and even death. In rare cases, hyperinsulinism results from an islet cell tumor. More commonly, it is caused by an overdose of insulin and is easily treated by ingesting some sugar. ●

The Gonads

The male and female **gonads** (see Figure 15.1) produce steroid sex hormones, identical to those produced by adrenal cortical cells. The major distinction is the source and relative amounts produced. The paired *ovaries* are small, oval organs located in the female's abdominopelvic cavity. Besides producing ova, or eggs, the ovaries produce several hormones, most importantly **estrogens** and **progesterone** (pro-jes′tĕ-rōn). Alone, the estrogens are responsible for maturation of the reproductive organs and the appearance of the secondary sex characteristics of females at puberty. Acting with progesterone, estrogens promote breast development and cyclic changes in the uterine mucosa (the menstrual cycle).

The male *testes,* located in an extra-abdominal skin pouch called the scrotum, produce sperm and male sex hormones, primarily **testosterone** (tes-tos′tĕ-rōn). During puberty, testosterone initiates the maturation of the male reproductive organs and the appearance of secondary sex characteristics and sex drive. In addition, testosterone is necessary for normal sperm production and maintains the reproductive organs in their mature functional state in adult males.

The release of gonadal hormones is regulated by gonadotropins, as described earlier. We will discuss the roles of the gonadal hormones in detail in Chapter 26, where we consider the reproductive system.

The Pineal Gland

The tiny, pine cone–shaped **pineal gland** hangs from the roof of the third ventricle in the diencephalon (see Figure 15.1). Its secretory cells, called **pinealocytes,** are arranged in compact cords and clusters. Lying between pinealocytes in adults are dense particles containing calcium salts ("brain-sand" or "pineal sand"). These salts are radiopaque; hence, the pineal gland is a handy landmark for determining brain orientation in X rays.

The endocrine function of the pineal gland is still somewhat of a mystery. Although many peptides and amines have been isolated from this minute gland, its only major secretory product is **melatonin** (mel″ah-to′nin), an amine hormone derived from serotonin. Melatonin concentrations in the blood rise and fall in a diurnal cycle. Peak levels occur during the night and make us drowsy; lowest levels occur around noon.

The pineal gland indirectly receives input from the visual pathways (retina → suprachiasmatic nucleus of hypothalamus → superior cervical ganglion → pineal gland) concerning the intensity and duration of daylight. In some animals, mating behavior and gonadal size vary with changes in the relative lengths of light and dark periods, and melatonin mediates these effects. In children, melatonin may have an antigonadotropic effect, that is, it may inhibit precocious (too early) sexual maturation and thus affect the timing of puberty.

The *suprachiasmatic nucleus* of the hypothalamus, an area referred to as our "biological clock," is richly supplied with melatonin receptors, and exposure to bright light (known to suppress melatonin secretion) can reset the clock timing. Hence, changing melatonin levels may also be a means by which the day/night cycles influence physiological processes that show rhythmic variations, such as body temperature, sleep, and appetite.

The Thymus

Located deep to the sternum in the thorax is the lobulated **thymus gland.** Large and conspicuous in infants and children, the thymus diminishes in size throughout adulthood. By old age, it is composed largely of adipose and fibrous connective tissues.

The major hormonal product of the thymic epithelial cells is a family of peptide hormones, including **thymopoietins, thymic factor,** and **thymosins** (thi′mo-sinz), which appear to be essential for the normal development of *T lymphocytes* and the immune response. Thymic hormones are described in Chapter 20 with the discussion of the immune system.

Other Hormone-Producing Structures

Other hormone-producing cells occur in various organs of the body, including the following (see Table 15.4):

1. Heart. The atria contain some specialized cardiac muscle cells that secrete **atrial natriuretic peptide.** By signaling the kidneys to increase their production of salty urine and by inhibiting aldosterone release by the adrenal cortex, ANP reduces blood volume, blood pressure, and blood sodium concentration (see Figure 15.13).

2. Gastrointestinal tract. *Enteroendocrine cells* are hormone-secreting cells sprinkled in the mucosa of the gastrointestinal (GI) tract. These scattered cells release several amine and peptide hormones that help regulate a wide variety of digestive functions, some of which are summarized in Table 15.4. Enteroendocrine cells are sometimes referred to as *paraneurons* because they are similar in certain ways to neurons. Many of their hormones are chemically identical to neurotransmitters, and in some cases, the hormones simply diffuse to and influence nearby target cells without first entering the bloodstream, that is, they act as local hormones or *paracrines.*

3. Placenta. Besides sustaining the fetus during pregnancy, the placenta secretes several steroid and protein hormones that influence the course of pregnancy. Placental hormones include estrogens and progesterone (hormones more often associated with the ovary), and human chorionic gonadotropin (hCG).

4. Kidneys. As yet unidentified cells in the kidneys secrete **erythropoietin** (ĕ-rith″ro-poi′ĕ-tin; "red-

TABLE 15.4 / **Selected Examples of Hormones Produced by Organs Other Than the Major Endocrine Organs**

Source	Hormone	Chemical Composition	Trigger	Target Organ and Effects
GI tract mucosa				
▪ Stomach	Gastrin	Peptide	Secreted in response to food	Stomach: stimulates glands to release hydrochloric acid (HCl)
▪ Stomach	Serotonin	Amine	Secreted in response to food	Stomach: causes contraction of stomach muscle
▪ Duodenum of small intestine	Intestinal gastrin	Peptide	Secreted in response to food, especially fats	Stomach: inhibits HCl secretion and gastrointestinal tract mobility
▪ Duodenum	Secretin	Peptide	Secreted in response to food	Pancreas and liver: stimulates release of bicarbonate-rich juice; stomach: inhibits secretory activity
▪ Duodenum	Cholecystokinin (CCK)	Peptide	Secreted in response to food	Pancreas: stimulates release of enzyme-rich juice; gallbladder: stimulates expulsion of stored bile; sphincter of Oddi: causes sphincter to relax, allowing bile and pancreatic juice to enter duodenum
Kidney	Erythropoietin (EPO)	Glycoprotein	Erythropoietin secreted in response to hypoxia	Bone marrow: stimulates production of red blood cells
Skin (epidermal cells)	Cholecalciferol (provitamin D_3)	Steroid	Cholecalciferol activated by the kidneys to active vitamin D_3 [1,25(OH) D_3] and released in response to parathyroid hormone	Intestine: stimulates active transport of dietary calcium across intestinal cell membranes
Heart (atria)	Atrial natriuretic peptide	Peptide	Secreted in response to stretching of atria (by rising blood pressure)	Kidney: inhibits sodium ion reabsorption and renin release; adrenal cortex: inhibits secretion of aldosterone; decreases blood pressure
Adipose tissue	Leptin	Peptide	Secreted in response to fatty foods	Brain: suppresses appetite; increases energy expenditure
Adipose tissue	Resistin	Peptide	Same as leptin	Fat, muscle, liver: antagonizes insulin's action on fat, muscle, and liver cells

maker"), a protein hormone that signals the bone marrow to increase production of red blood cells.

5. Skin. The skin produces **cholecalciferol**, an inactive form of vitamin D_3, when modified cholesterol molecules in epidermal cells are exposed to ultraviolet radiation. This compound then enters the blood via the dermal capillaries, is modified in the liver, and becomes fully activated in the kidney. The active form of vitamin D_3, **calcitriol**, is an essential part of the carrier system that intestinal cells

use to absorb Ca^{2+} from ingested food. Without this vitamin, the bones become weak and soft.

6. Adipose tissue. Adipose cells release **leptin** following their uptake of glucose and lipids, which they store as fat. Leptin binds to CNS neurons concerned with appetite control, producing a sensation of satiety, and appears to stimulate increased energy expenditure. *Resistin*, also secreted by adipose cells, is an insulin antagonist.

Review Questions

Multiple Choice/Matching

(Some questions have more than one correct answer. Select the best answer or answers from the choices given.)

1. The major stimulus for release of parathyroid hormone is (a) hormonal, (b) humoral, (c) neural.

2. The anterior pituitary secretes all but (a) antidiuretic hormone, (b) growth hormone, (c) gonadotropins, (d) TSH.

3. A hormone *not* involved in sugar metabolism is (a) glucagon, (b) cortisone, (c) aldosterone, (d) insulin.

4. Parathyroid hormone (a) increases bone formation and lowers blood calcium levels, (b) increases calcium excretion from the body, (c) decreases calcium absorption from the gut, (d) demineralizes bone and raises blood calcium levels.

5. Choose from the following key to identify the hormones described.

Key: (a) aldosterone   (e) oxytocin
          (b) antidiuretic hormone   (f) prolactin
          (c) growth hormone   (g) T_4 and T_3
          (d) luteinizing hormone   (h) TSH

____ (1) important anabolic hormone; many of its effects mediated by IGFs

____ (2) involved in water balance; causes the kidneys to conserve water

____ (3) stimulates milk production

____ (4) tropic hormone that stimulates the gonads to secrete sex hormones

____ (5) increases uterine contractions during birth

____ (6) major metabolic hormone(s) of the body

____ (7) causes reabsorption of sodium ions by the kidneys

____ (8) tropic hormone that stimulates the thyroid gland to secrete thyroid hormone

____ (9) a hormone secreted by the neurohypophysis (two possible choices)

____ (10) the only steroid hormone in the list

6. A hypodermic injection of epinephrine would (a) increase heart rate, increase blood pressure, dilate the bronchi of the lungs, and increase peristalsis, (b) decrease heart rate, decrease blood pressure, constrict the bronchi, and increase peristalsis, (c) decrease heart rate, increase blood pressure, constrict the bronchi, and decrease peristalsis, (d) increase heart rate, increase blood pressure, dilate the bronchi, and decrease peristalsis.

7. Testosterone is to the male as what hormone is to the female? (a) luteinizing hormone, (b) progesterone, (c) estrogen, (d) prolactin.

8. If anterior pituitary secretion is deficient in a growing child, the child will (a) develop acromegaly, (b) become a dwarf but have fairly normal body proportions, (c) mature sexually at an earlier than normal age, (d) be in constant danger of becoming dehydrated.

9. If there is adequate carbohydrate intake, secretion of insulin results in (a) lower blood sugar levels, (b) increased cell utilization of glucose, (c) storage of glycogen, (d) all of these.

10. Hormones (a) are produced by exocrine glands, (b) are carried to all parts of the body in blood, (c) remain at constant concentration in the blood, (d) affect only non-hormone producing organs.

11. Some hormones act by (a) increasing the synthesis of enzymes, (b) converting an inactive enzyme into an active enzyme, (c) affecting only specific target organs, (d) all of these.

12. Absence of thyroxine would result in (a) increased heart rate and increased force of heart contraction, (b) depression of the CNS and lethargy, (c) exophthalmos, (d) high metabolic rate.

13. Chromaffin cells are found in the (a) parathyroid gland, (b) anterior pituitary gland, (c) adrenal gland, (d) pineal gland.

14. Atrial natriuretic hormone secreted by the heart has exactly the opposite function of this hormone secreted by the zona glomerulosa: (a) antidiuretic hormone, (b) epinephrine, (c) calcitonin, (d) aldosterone, (e) androgens.

Short Answer Essay Questions

15. Define hormone.

16. Which type of hormone receptor—plasma membrane bound or intracellular—would be expected to provide the most long-lived response to hormone binding and why?

17. (a) Describe the body location of each of the following endocrine organs: anterior pituitary, pineal gland, pancreas, ovaries, testes, and adrenal glands. (b) List the hormones produced by each organ.

18. Name two endocrine glands (or regions) that are important in the stress response, and explain why they are important.

19. The anterior pituitary is often referred to as the master endocrine organ, but it, too, has a "master." What controls the release of anterior pituitary hormones?

20. The posterior pituitary is not really an endocrine gland. Why not? What is it?

21. A colloidal, or endemic, goiter is not really the result of malfunction of the thyroid gland. What does cause it?

22. Name a hormone secreted by a muscle cell and two hormones secreted by neurons.

23. How are the hyperglycemia and lipidemia of insulin deficiency linked?

16

BLOOD

Overview: Blood Composition and Functions (pp. 554–555)

1. Describe the composition and physical characteristics of whole blood. Explain why it is classified as a connective tissue.

2. List six functions of blood.

Blood Plasma (pp. 555–556)

3. Discuss the composition and functions of plasma.

Formed Elements (pp. 556–570)

4. Describe the structure, function, and production of erythrocytes.

5. Describe the chemical makeup of hemoglobin.

6. List the classes, structural characteristics, and functions of leukocytes. Also describe leukocyte genesis.

7. Describe the structure and function of platelets.

8. Give examples of disorders caused by abnormalities of each of the formed elements. Explain the mechanism of each disorder.

Hemostasis (pp. 570–575)

9. Describe the process of hemostasis. List factors that limit clot formation and prevent undesirable clotting.

10. Give examples of hemostatic disorders. Indicate the cause of each condition.

Transfusion and Blood Replacement (pp. 575–578)

11. Describe the ABO and Rh blood groups. Explain the basis of transfusion reactions.

12. Describe the function of blood expanders and the circumstances for their use.

Diagnostic Blood Tests (p. 578)

13. Explain the diagnostic importance of blood testing.

Blood is the river of life that surges within us, transporting nearly everything that must be carried from one place to another. Long before modern medicine, blood was viewed as magical—an elixir that held the mystical force of life—because when it drained from the body, life departed as well. Today, centuries later, blood still has enormous importance in the practice of medicine. Clinicians examine it more often than any other tissue when trying to determine the cause of disease in their patients.

In this chapter, we describe the composition and functions of this life-sustaining fluid that serves as a transport "vehicle" for the organs of the cardiovascular system. To get started, we need a brief overview of blood circulation, which is initiated by the pumping action of the heart. Blood exits the *heart* via *arteries*, which branch repeatedly until they become tiny *capillaries*. By diffusing across the capillary walls, oxygen and nutrients leave the blood and enter the body tissues, and carbon dioxide and wastes move from the tissues to the bloodstream. As oxygen-deficient blood leaves the capillary beds, it flows into *veins*, which return it to the heart. The returning blood then flows from the heart to the lungs, where it picks up oxygen and then returns to the heart to be pumped throughout the body once again. Now let us look more closely at the nature of blood.

Overview: Blood Composition and Functions

Components

Blood is the only fluid tissue in the body. Although blood appears to be a thick, homogeneous liquid, the microscope reveals that it has both cellular and liquid components. Blood is a specialized type of connective tissue in which living blood cells, the *formed elements*, are suspended in a nonliving fluid matrix called *plasma* (plaz'mah). The collagen and elastic fibers typical of other connective tissues are absent from blood, but dissolved fibrous proteins become visible as fibrin strands during blood clotting.

If a sample of blood is spun in a centrifuge, the heavier formed elements are packed down by centrifugal force and the less dense plasma remains at the top (Figure 16.1). Most of the reddish mass at the bottom of the tube is *erythrocytes* (ĕ-rith'ro-sīts; *erythro* = red), the red blood cells that transport oxygen. A thin, whitish layer called the **buffy coat** is present at the erythrocyte-plasma junction. This layer contains *leukocytes* (*leuko* = white), the white blood cells that act in various ways to protect the body, and *platelets*, cell fragments that help stop bleeding. Erythrocytes normally constitute about 45% of the total volume of a blood sample, a percentage known as the **hematocrit** (he-mat'o-krit;

?   *Which of the percentages given in this figure represents the measurement called the hematocrit?*

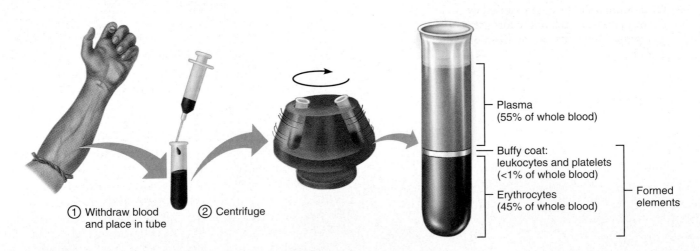

① Withdraw blood and place in tube   ② Centrifuge

Plasma (55% of whole blood)

Buffy coat: leukocytes and platelets (<1% of whole blood)

Erythrocytes (45% of whole blood)

Formed elements

FIGURE 16.1   The major components of whole blood. When a blood sample is centrifuged, it separates into layers, the heaviest particles (erythrocytes) moving to the bottom of the test tube. The least dense component is plasma, which is the top layer.

■ *45%, the percentage of the erythrocytes.*

"blood fraction"). Normal hematocrit values vary. In healthy males the norm is 47% ± 5%; in females it is 42% ± 5%. Leukocytes and platelets contribute less than 1% of blood volume. Plasma makes up most of the remaining 55% of whole blood.

Physical Characteristics and Volume

Blood is a sticky, opaque fluid with a characteristic metallic taste. As children, we discover its saltiness the first time we stick a cut finger into our mouth. Depending on the amount of oxygen it is carrying, the color of blood varies from scarlet (oxygen-rich) to dark red (oxygen-poor). Blood is more dense than water and about five times more viscous, largely because of its formed elements. Blood is slightly alkaline, with a pH between 7.35 and 7.45, and its temperature (38°C or 100.4°F) is always slightly higher than body temperature.

Blood accounts for approximately 8% of body weight. Its average volume in healthy adult males is 5–6 L (about 1.5 gallons), somewhat greater than in healthy adult females (4–5 L).

Functions

Blood performs a number of functions, all concerned in one way or another with substance distribution, regulating blood levels of particular substances, or body protection.

Distribution

Distribution functions of blood include:

- Delivering oxygen from the lungs and nutrients from the digestive tract to all body cells.

- Transporting metabolic waste products from cells to elimination sites (to the lungs for elimination of carbon dioxide, and to the kidneys for disposal of nitrogenous wastes in urine).

- Transporting hormones from the endocrine organs to their target organs.

Regulation

Regulatory functions of blood include:

- Maintaining appropriate body temperature by absorbing and distributing heat throughout the body and to the skin surface to encourage heat loss.

- Maintaining normal pH in body tissues. Many blood proteins and other bloodborne solutes act as buffers to prevent excessive or abrupt changes in blood pH that could jeopardize normal cell activities. Additionally, blood acts as the reservoir for the body's "alkaline reserve" of bicarbonate atoms.

- Maintaining adequate fluid volume in the circulatory system. Salts (sodium chloride and others) and

blood proteins act to prevent excessive fluid loss from the bloodstream into the tissue spaces. As a result, the fluid volume in the blood vessels remains ample to support efficient blood circulation to all parts of the body.

Protection

Protective functions of blood include:

- Preventing blood loss. When a blood vessel is damaged, platelets and plasma proteins initiate clot formation, halting blood loss.

- Preventing infection. Drifting along in blood are antibodies, complement proteins, and white blood cells, all of which help defend the body against foreign invaders such as bacteria and viruses.

Blood Plasma

Blood **plasma** is a straw-colored, sticky fluid (see Figure 16.1). Although it is mostly water (about 90%), plasma contains over 100 different dissolved solutes, including nutrients, gases, hormones, wastes and products of cell activity, ions, and proteins. Table 16.1 summarizes the major plasma components.

Plasma proteins, accounting for about 8% by weight of plasma volume, are the most abundant plasma solutes. Except for hormones and gamma globulins, most plasma proteins are produced by the liver. Plasma proteins serve a variety of functions, but they are *not* taken up by cells to be used as fuels or metabolic nutrients as are most other plasma solutes, such as glucose, fatty acids, and oxygen. **Albumin** (al-bu'min) accounts for some 60% of plasma protein. It acts as a carrier to shuttle certain molecules through the circulation, is an important blood buffer, and is the major blood protein contributing to the plasma osmotic pressure (the pressure that helps to keep water in the bloodstream). (Sodium ions are the other major solute contributing to blood osmotic pressure.)

The makeup of plasma varies continuously as cells remove or add substances to the blood. However, assuming a healthy diet, plasma composition is kept relatively constant by various homeostatic mechanisms. For example, when blood protein levels drop undesirably, the liver makes more proteins; and when the blood starts to become too acidic (acidosis), both the respiratory system and the kidneys are called into action to restore plasma's normal, slightly alkaline pH. Body organs make dozens of adjustments, day in and day out, to maintain the many plasma solutes at life-sustaining levels. In addition to transporting various solutes around the body, plasma distributes heat (a by-product of cellular metabolism) throughout the body.

TABLE 16.1   Composition of Plasma

Constituent	Description and Importance
Water	90% of plasma volume; dissolving and suspending medium for solutes of blood; absorbs heat
Solutes	
Proteins	8% (by weight) of plasma volume
▪ Albumin	60% of plasma proteins; produced by liver; exerts osmotic pressure to maintain water balance between blood and tissues
▪ Globulins	36% of plasma proteins
alpha, beta	Produced by liver; transport proteins that bind to lipids, metal ions, and fat-soluble vitamins
gamma	Antibodies released primarily by plasma cells during immune response
▪ Clotting proteins	4% of plasma proteins; include fibrinogen and prothrombin produced by liver; act in blood clotting
▪ Others	Metabolic enzymes, antibacterial proteins (such as complement), hormones
Nonprotein nitrogenous substances	By-products of cellular metabolism, such as urea, uric acid, creatinine, and ammonium salts
Nutrients (organic)	Materials absorbed from digestive tract and transported for use throughout body; include glucose and other simple carbohydrates, amino acids (digestion products of proteins), fatty acids, glycerol and triglycerides (fat products), cholesterol, and vitamins
Electrolytes	Cations include sodium, potassium, calcium, magnesium; anions include chloride, phosphate, sulfate, and bicarbonate; help to maintain plasma osmotic pressure and normal blood pH
Respiratory gases	Oxygen and carbon dioxide; some dissolved oxygen (most bound to hemoglobin inside RBCs); carbon dioxide transported bound to hemoglobin in RBCs and as bicarbonate ion dissolved in plasma

Formed Elements

The **formed elements** of blood, *erythrocytes, leukocytes,* and *platelets,* have some unusual features. (1) Two of the three are not even true cells: Erythrocytes have no nuclei or organelles, and platelets are cell fragments. Only leukocytes are complete cells. (2) Most of the formed elements survive in the bloodstream for only a few days. (3) Most blood cells do not divide. Instead, they are continuously renewed by division of cells in bone marrow, where they originate.

If you examine a stained smear of human blood under the light microscope, you will see disc-shaped

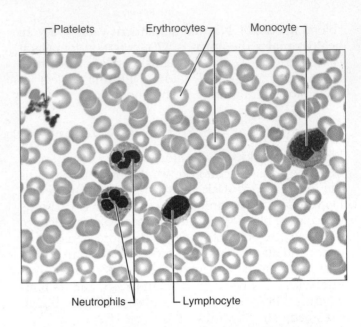

FIGURE 16.2   **Photomicrograph of a human blood smear stained with Wright's stain.**

red blood cells, a variety of gaudily stained spherical white blood cells, and some scattered platelets that look like debris (Figure 16.2). Erythrocytes vastly outnumber the other types of formed elements. Table 16.2 on p. 565 summarizes the important characteristics of the formed elements.

Erythrocytes

Structural Characteristics

Erythrocytes or **red blood cells (RBCs)** are small cells, about 7.5 μm in diameter. Shaped like biconcave discs—flattened discs with depressed centers (Figure 16.3)—their thin centers appear lighter in color than their edges. Consequently, erythrocytes look like miniature doughnuts when viewed with a microscope. Mature erythrocytes are bound by a plasma membrane, but lack a nucleus (are *anucleate*) and have essentially no organelles. In fact, they are little more than "bags" of *hemoglobin (Hb)*, the RBC protein that functions in gas transport. Other proteins are present such as antioxidant enzymes that rid the body of harmful oxygen radicals, but most function mainly to maintain the plasma membrane or promote changes in RBC shape. For example, the biconcave shape of an erythrocyte is maintained by a network of proteins, especially one called *spectrin*, attached to the cytoplasmic face of its plasma membrane. Because the spectrin net is deformable, it gives erythrocytes flexibility to change shape as necessary—to twist, turn, and become cup-shaped as they are carried passively through capillaries with diameters smaller than themselves—and then to resume their biconcave shape.

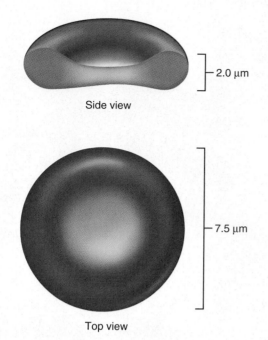

Side view

— 2.0 μm

— 7.5 μm

Top view

FIGURE 16.3 **Structure of erythrocytes.** Cut side view above; a superficial surface view below. Notice the distinctive biconcave shape.

The erythrocyte is a superb example of complementarity of structure and function. It picks up oxygen in the capillary beds of the lungs and releases it to tissue cells across other capillaries throughout the body. It also transports some 20% of the carbon dioxide released by tissue cells back to the lungs. Each erythrocyte structural characteristic contributes to its gas transport functions: (1) Its small size and biconcave shape provide a huge surface area relative to volume (about 30% more surface area than comparable spherical cells). Because no point within its cytoplasm is far from the surface, the biconcave disc shape is ideally suited for gas exchange. (2) Discounting water content, an erythrocyte is over 97% hemoglobin, the molecule that binds to and transports respiratory gases. (3) Because erythrocytes lack mitochondria and generate ATP by anaerobic mechanisms, they do not consume any of the oxygen they are transporting, making them very efficient oxygen transporters indeed.

Erythrocytes are the major factor contributing to blood viscosity. Women typically have a lower red blood cell count than men (4.3–5.2 million cells per cubic millimeter of blood versus 5.1–5.8 million cells per cubic millimeter respectively). When the number of red blood cells increases beyond the normal range, blood viscosity rises and blood flows more slowly. Similarly, as the number of red blood cells drops below the lower end of the range, the blood thins and flows more rapidly.

Function

Erythrocytes are completely dedicated to their job of respiratory gas (oxygen and carbon dioxide) transport. **Hemoglobin,** the protein that makes red blood cells red, binds easily and reversibly with oxygen, and most oxygen carried in blood is bound to hemoglobin. Normal values for hemoglobin are 14–20 grams per 100 milliliters of blood (g/100 ml) in infants, 13–18 g/100 ml in adult males, and 12–16 g/100 ml in adult females.

Hemoglobin is made up of the protein **globin** bound to the red **heme** pigment. Globin consists of four polypeptide chains—two alpha (α) and two beta (β)—each bound to a ringlike heme group. Each heme group bears an atom of iron set like a jewel in its center (Figure 16.4). Since each iron atom can combine reversibly with one molecule of oxygen, a hemoglobin molecule can transport four molecules of oxygen. A single red blood cell contains about 250 million hemoglobin molecules, so each of these tiny cells can scoop up about 1 billion molecules of oxygen!

The fact that hemoglobin is contained in erythrocytes, rather than existing free in plasma, prevents it (1) from breaking into fragments that would leak out of the bloodstream (through the rather porous capillary membranes) and (2) from contributing to blood viscosity and osmotic pressure.

Oxygen loading occurs in the lungs and the direction of transport is from lungs to tissue cells. As oxygen-deficient blood moves through the lungs, oxygen diffuses from the air sacs of the lungs into the blood and then into the erythrocytes, where it binds to hemoglobin. When oxygen binds to iron, the hemoglobin, now called **oxyhemoglobin,** assumes a new three-dimensional shape and becomes ruby red. In the tissues, the process is reversed. Oxygen detaches from iron, hemoglobin resumes its former shape, and the resulting **deoxyhemoglobin,** or *reduced hemoglobin,* becomes dark red. The released oxygen diffuses from the blood into the tissue fluid and then into the tissue cells.

About 20% of the carbon dioxide transported in the blood combines with hemoglobin, but it binds to globin's amino acids rather than with the heme group. This formation of **carbaminohemoglobin** (kar-bam″ĭ-no-he″mo-glo′bin) occurs more readily when hemoglobin is in the reduced state (dissociated from oxygen). Carbon dioxide loading occurs in the tissues and the direction of transport is from tissues to lungs, where carbon dioxide is eliminated from the body. The mechanisms of loading and unloading these respiratory gases are described in Chapter 21.

How many molecules of oxygen can a hemoglobin molecule transport?

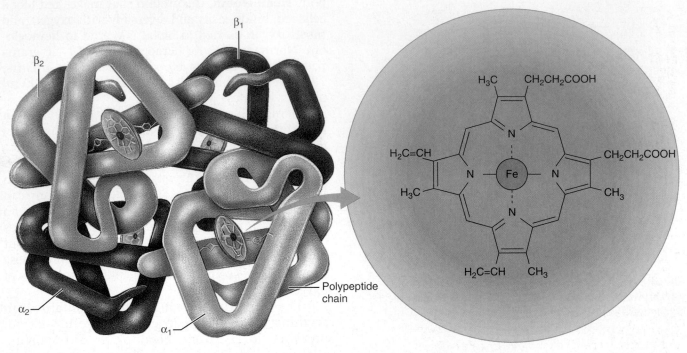

(a) Hemoglobin

(b) Iron-containing heme group

FIGURE 16.4   Structure of hemoglobin. (a) Hemoglobin is composed of the protein globin bound to the iron-containing heme pigments. Each globin molecule has four polypeptide chains: two alpha (α) chains and two beta (β) chains. Each chain is complexed with a heme group, shown as a circular green structure with iron at its center. **(b)** Structure of a heme group.

Production of Erythrocytes

Blood cell formation is referred to as **hematopoiesis** (hem"ah-to-poi-e'sis), or **hemopoiesis** (*hemo, hemato* = blood; *poiesis* = to make). This process occurs in the **red bone marrow,** which is composed largely of a soft network of reticular connective tissue bordering on wide blood capillaries called *blood sinusoids.* Within this network are immature blood cells, macrophages, fat cells, and *reticular cells* (which secrete the fibers). In adults, red marrow is found chiefly in the bones of the axial skeleton and girdles, and in the proximal epiphyses of the humerus and femur. Each type of blood cell is produced in different numbers in response to changing body needs and different regulatory factors. As they mature, they migrate through the thin walls of the sinusoids to enter the bloodstream. On average, the marrow turns out an ounce of new blood containing some 100 billion new cells each and every day.

Although the various formed elements have different functions, there are similarities in their life histories. All arise from the same type of *stem cell,* the pleuripotential **hematopoietic stem cell,** or **hemocytoblast** (*cyte* = cell, *blast* = bud), which resides in the red bone marrow. However, their maturation pathways differ; and once a cell is *committed* to a specific blood cell pathway, it cannot change. This commitment is signaled by the appearance of membrane surface receptors that respond to specific hormones or growth factors, which in turn "push" the cell toward further specialization.

Erythrocyte production, or **erythropoiesis** (ĕ-rith"ro-poi-e'sis) begins when a hemocytoblast descendant called a **myeloid stem cell** is transformed into a **proerythroblast** (Figure 16.5). Proerythroblasts, in turn, give rise to the **early (basophilic) erythroblasts** that produce huge numbers of ribosomes. During these first two phases, the cells divide many times. Hemoglobin synthesis and iron accumulation occur as the early erythroblast is

■ *Four O_2 per hemoglobin molecule.*

? *What does the term "committed cell" refer to?*

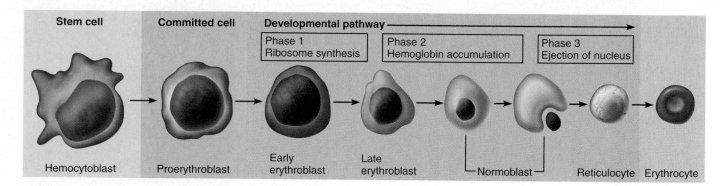

FIGURE 16.5 Erythropoiesis: genesis of red blood cells.
Erythropoiesis is a sequence involving proliferation and differentiation of committed red marrow cells through the erythroblast and normoblast stages to the reticulocytes that are released into the bloodstream, and finally become erythrocytes. (The myeloid stem cell, the phase intermediate between the hemocytoblast and the proerythroblast, is not illustrated.)

transformed into a **late erythroblast** and then a **normoblast.** The "color" of the cell cytoplasm changes as the blue-staining ribosomes become masked by the pink color of hemoglobin. When a normoblast has accumulated a hemoglobin concentration of about 34%, it ejects most of its organelles. Additionally, its nuclear functions end and its nucleus degenerates and is pinched off, causing the cell to collapse inward and assume the biconcave shape. The result is the *reticulocyte.* **Reticulocytes** (essentially young erythrocytes) are so named because they still contain a scant network of clumped ribosomes and rough endoplasmic reticulum. The entire process from hemocytoblast to reticulocyte takes three to five days. The reticulocytes, filled almost to bursting with hemoglobin, enter the bloodstream to begin their task of oxygen transport. Usually they become fully mature erythrocytes within two days of release as their ribosomes are degraded by intracellular enzymes. Reticulocytes account for 1–2% of all erythrocytes in the blood of healthy people. **Reticulocyte counts** provide a rough index of the *rate* of RBC formation—reticulocyte counts below or above this percentage range indicate abnormal rates of erythrocyte formation.

Regulation and Requirements for Erythropoiesis

The number of circulating erythrocytes in a given individual is remarkably constant and reflects a balance between red blood cell production and destruction. This balance is important because too few erythrocytes leads to tissue hypoxia (oxygen deprivation), whereas too many makes the blood undesirably viscous. To ensure that the number of erythrocytes in blood remains within the homeostatic range, new cells are produced at the incredibly rapid rate of more than 2 million per second in healthy people. This process is controlled hormonally and depends on adequate supplies of iron, amino acids, and certain B vitamins.

Hormonal Controls The direct stimulus for erythrocyte formation is provided by **erythropoietin (EPO),** a glycoprotein hormone. Normally, a small amount of EPO circulates in the blood at all times and sustains red blood cell production at a basal rate (Figure 16.6). Although the liver produces some, the kidneys play the major role in EPO production. When certain kidney cells become hypoxic (i.e., have inadequate oxygen), they accelerate their release of erythropoietin. The drop in normal blood oxygen levels that triggers EPO formation can result from:

1. Reduced numbers of red blood cells due to hemorrhage or excess RBC destruction

2. Reduced availability of oxygen, as might occur at high altitudes or during pneumonia

3. Increased tissue demands for oxygen (common in those who engage in aerobic exercise)

Conversely, too many erythrocytes or excessive oxygen in the bloodstream depresses erythropoietin production. The easily missed point to keep in mind is that it is *not* the number of erythrocytes in blood

■ *A committed cell is a blood cell precursor whose path of specialization is determined. For example, the proerythroblast can only become an erythrocyte, not one of the leukocytes.*

How does blood doping, practiced by some athletes (see p. 563), affect the negative feedback cycle outlined here?

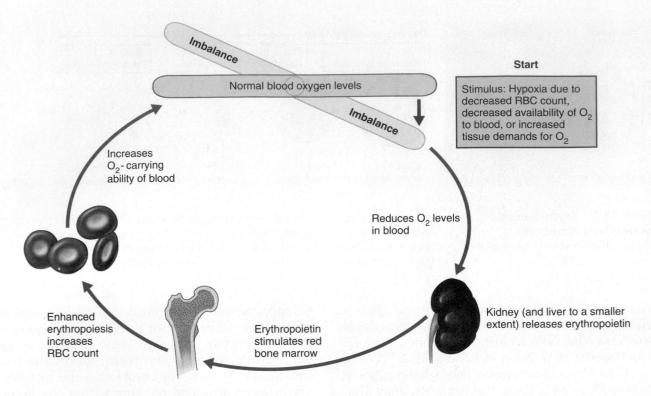

FIGURE 16.6   Erythropoietin mechanism for regulating erythropoiesis. Notice that increased erythropoietin release occurs when oxygen levels in the blood become inadequate to support normal cellular activity, whatever the cause.

that controls the rate of erythropoiesis. Control is based on their ability to transport enough oxygen to meet tissue demands.

Bloodborne erythropoietin stimulates red marrow cells that *are already committed* to becoming erythrocytes, causing them to mature more rapidly. One to two days after erythropoietin levels rise in the blood, a marked increase in the rate of reticulocyte release (and hence reticulocyte count) occurs. Notice that hypoxia (oxygen deficit) does not activate the bone marrow directly. Instead it stimulates the kidneys, which in turn provide the hormonal stimulus that activates the bone marrow.

🄷 HOMEOSTATIC IMBALANCE

Renal dialysis patients whose kidneys have failed produce too little EPO to support normal erythro-

poiesis. Consequently, they routinely have red blood cell counts less than half that of healthy individuals. Genetically engineered (recombinant) EPO has helped such patients immeasurably and has also become a substance of abuse in athletes—particularly in professional bike racers and marathon runners seeking increased stamina and performance. However, the consequences can be deadly. By injecting EPO, healthy athletes increase their normal RBC volume from 45% to as much as 65%. Then, with the dehydration that occurs in a long race, the blood concentrates even further, becoming a thick, sticky "sludge" that can cause clotting, stroke, and even heart failure. ●

The male sex hormone *testosterone* also enhances EPO production by the kidneys. Because female sex hormones do not have similar stimulatory effects, testosterone may be at least partially responsible for the higher RBC counts and hemoglobin levels seen in males. Also, a wide variety of chemicals released by leukocytes, platelets, and even reticular cells stimulates bursts of RBC production.

Blood doping, which increases the number of RBCs in the circulation and thus increases the amount of oxygen being transported, would inhibit erythropoietin release by ending the stimulus for its formation. ■

Dietary Requirements   The raw materials required for erythropoiesis include the usual nutrients and structural materials—proteins, lipids, and carbohydrates. Iron is essential for hemoglobin synthesis (Figure 16.7). Iron is available from the diet, and its absorption into the bloodstream is precisely controlled by intestinal cells in response to changing body stores of iron.

Approximately 65% of the body's iron supply (about 4000 mg) is in hemoglobin. Most of the remainder is stored in the liver, spleen, and (to a much lesser extent) bone marrow. Because free iron ions (Fe^{2+}, Fe^{3+}) are toxic, iron is stored inside cells as protein-iron complexes such as **ferritin** (fer′ĭ-tin) and **hemosiderin** (he″mo-sid′er-in). In blood, iron is transported loosely bound to a transport protein called **transferrin,** and developing erythrocytes take up iron as needed to form hemoglobin. Small amounts of iron are lost each day in feces, urine, and perspiration. The average daily loss of iron is 1.7 mg in women and 0.9 mg in men. In women, the menstrual flow accounts for the additional losses.

Two B-complex vitamins—vitamin B_{12} and folic acid—are necessary for normal DNA synthesis. Thus, even slight deficits jeopardize rapidly dividing cell populations, such as developing erythrocytes.

Fate and Destruction of Erythrocytes

The anucleate condition of erythrocytes carries with it some important limitations. Red blood cells are unable to synthesize new proteins, to grow, or to divide. Erythrocytes become "old" as they lose their flexibility and become increasingly rigid and fragile, and their contained hemoglobin begins to degenerate. Red blood cells have a useful life span of 100 to 120 days, after which they become trapped and fragment in smaller circulatory channels, particularly in those of the spleen. For this reason, the spleen is sometimes called the "red blood cell graveyard."

Dying erythrocytes are engulfed and destroyed by macrophages. The heme of their hemoglobin is split off from globin. Its core of iron is salvaged, bound to protein (as ferritin or hemosiderin), and stored for reuse. The balance of the heme group is degraded to **bilirubin** (bil″i-roo′bin), a yellow pigment that is released to the blood and binds to albumin for transport (Figure 16.7). Bilirubin is picked up by liver cells, which in turn secrete it (in bile) into the intestine, where it is metabolized to *urobilinogen.* Most of this degraded pigment leaves the body in feces, as a brown pigment called *stercobilin.* The

FIGURE 16.7   Life cycle of red blood cells.

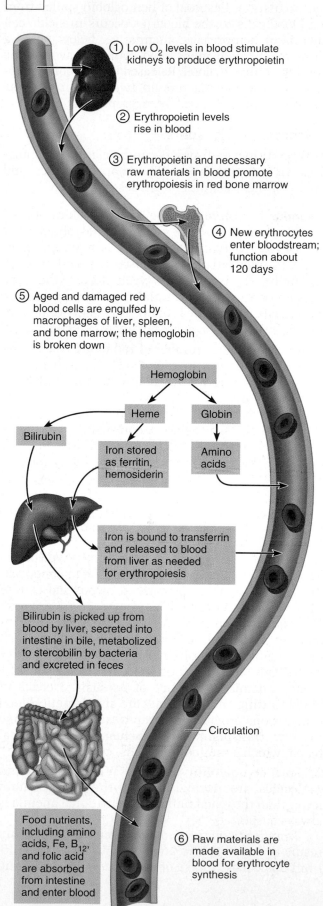

protein (globin) part of hemoglobin is metabolized or broken down to amino acids, which are released to the circulation. Disposal of hemoglobin spilled from red blood cells to the blood (as occurs in sickle-cell anemia or hemorrhagic anemia; see below) takes a similar but much more rapid course to avoid toxic buildup of iron in blood. Released hemoglobin is captured by the plasma protein *haptoglobin* and the complex is phagocytized by macrophages.

Erythrocyte Disorders

Most erythrocyte disorders can be classified as anemias or polycythemias. The many varieties and causes of these conditions are described next.

Anemias   Anemia (ah-ne′me-ah; "lacking blood") is a condition in which the blood has abnormally low oxygen-carrying capacity. It is a *symptom* of some disorder rather than a disease in and of itself. Its hallmark is blood oxygen levels that are inadequate to support normal metabolism. Anemic individuals are fatigued, often pale, short of breath, and chilly. Common causes of anemia include:

1. An insufficient number of red blood cells. Conditions that reduce the red blood cell count include blood loss, excessive RBC destruction, and bone marrow failure.

Hemorrhagic anemias (hem″o-raj′ik) result from blood loss. In acute hemorrhagic anemia, blood loss is rapid (as might follow a severe stab wound); it is treated by blood replacement. Slight but persistent blood loss (due to hemorrhoids or an undiagnosed bleeding ulcer, for example) causes chronic hemorrhagic anemia. Once the primary problem is resolved, normal erythropoietic mechanisms replace the deficient blood cells.

In *hemolytic anemias* (he″mo-lit′ik), erythrocytes rupture, or lyse, prematurely. Hemoglobin abnormalities, transfusion of mismatched blood, and certain bacterial and parasitic infections are possible causes.

Aplastic anemia results from destruction or inhibition of the red marrow by certain bacterial toxins, drugs, and ionizing radiation. Because marrow destruction impairs formation of *all* formed elements, anemia is just one of its signs. Defects in blood clotting and immunity are also present. Blood transfusions provide a stopgap treatment until bone marrow transplants or transfusion of umbilical blood, which contains stem cells, can be done.

2. Low hemoglobin content. When hemoglobin molecules are normal, but erythrocytes contain fewer than the usual number, a nutritional anemia is always suspected.

Iron-deficiency anemia is generally a secondary result of hemorrhagic anemias, but it also results from inadequate intake of iron-containing foods and impaired iron absorption. The erythrocytes produced, called **microcytes,** are small and pale. The obvious treatment is iron supplements, but if chronic hemorrhage is the cause, blood transfusions may also be needed.

When athletes exercise vigorously, their blood volume expands and can increase by as much as 15%. Since this in effect dilutes the blood components, a test for the iron content of the blood at such times would indicate iron-deficiency anemia. However, this iron deficiency illusion, called **athlete's anemia,** is quickly reversed as blood components return to physiological levels within a day or so after resuming a normal level of activity.

Pernicious anemia is due to a deficiency of vitamin B_{12}. Because meats, poultry, and fish provide ample amounts of the vitamin, diet is rarely the problem except for strict vegetarians. A substance called **intrinsic factor,** produced by the stomach mucosa, must be present for vitamin B_{12} to be absorbed by intestinal cells. In most cases of pernicious anemia, intrinsic factor is deficient. Consequently, the developing erythrocytes grow but do not divide, and large, pale cells called **macrocytes** result. Because the stomach mucosa atrophies with age, the elderly are particularly at risk for this type of anemia. Treatment involves regular intramuscular injections of vitamin B_{12} or application of a B_{12}-containing gel (Nascobal) to the nasal lining once a week.

3. Abnormal hemoglobin. Production of abnormal hemoglobin usually has a genetic basis. Two such examples, thalassemia and sickle-cell anemia, can be serious, incurable, and sometimes fatal diseases. In both diseases the globin part of hemoglobin is abnormal and the erythrocytes produced are fragile and rupture prematurely.

Thalassemias (thal″ah-se′me-ahs; "sea blood") are typically seen in people of Mediterranean ancestry, such as Greeks and Italians. One of the globin chains is absent or faulty, and the erythrocytes are thin, delicate, and deficient in hemoglobin. The red blood cell count is generally less than 2 million cells per cubic millimeter. In most cases the reduced RBC count is not a major problem and no treatment is required. Severe cases require monthly blood transfusions.

In **sickle-cell anemia,** the havoc caused by the abnormal hemoglobin formed, *hemoglobin S (HbS),* result from a change in just *one* of the 287 amino acids in a beta chain of the globin molecule! This alteration causes the beta chains to link together under low-oxygen conditions, forming stiff rods so that hemoglobin S becomes spiky and sharp. This, in turn, causes the red blood cells to become crescent shaped (Figure 16.8) when they unload oxygen molecules or when the oxygen content of the blood is

FIGURE 16.8 **Comparison of (a) a normal erythrocyte to (b) a sickled erythrocyte** (6700×).

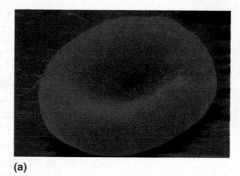

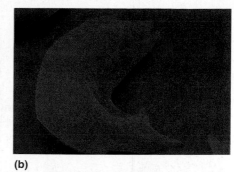

(a) (b)

lower than normal, as during vigorous exercise and other activities that increase metabolic rate. The stiff, deformed erythrocytes rupture easily and tend to dam up in small blood vessels. These events interfere with oxygen delivery, leaving the victims gasping for air and in extreme pain. Bone and chest pain are particularly severe; infection and stroke are common sequels. The standard treatment for an acute sickle-cell crisis is blood transfusion.

Sickle-cell anemia occurs chiefly in black people who live in the malaria belt of Africa and among their descendants. It strikes nearly 1 of every 400 black newborns in the United States. Globally, 300–500 million people are infected and more than a million die each year. Apparently, the gene that causes sickling of red blood cells also causes erythrocytes infected by the malaria-causing parasite to stick to capillary walls. Then, during oxygen deficit, these abnormal erythrocytes lose potassium, an essential ingredient for survival of the parasite. Thus, individuals with the sickle-cell gene have a better chance of surviving in regions where malaria is prevalent. Only those carrying two copies of the defective gene have sickle-cell anemia. Those carrying only one sickling gene have sickle-cell trait; they generally do not display the symptoms, but can transmit the gene to their offspring.

Since fetal hemoglobin (HbF) does not "sickle," even in those destined to have sickle-cell anemia, investigators have been looking for ways to switch the fetal hemoglobin gene back on. *Hydroxyurea*, a drug used to treat chronic leukemia, appears to do just that. This drug dramatically reduces the excruciating pain and overall severity and complications of sickle-cell anemia (by 50%). Additionally, a number of products are now being clinically tested for their effectiveness against this scourge, including several vaccines; G25, an injectable drug that prevents the parasite's replication; and a small portion of the HIV virus used as a vector to deliver therapeutic β^A-globin genes for synthesizing beta chains that resist linking together or polymerization.

Polycythemia Polycythemia (pol″e-si-the′me-ah; "many blood cells") is an abnormal excess of erythrocytes that increases blood viscosity, causing it

to sludge, or flow sluggishly. *Polycythemia vera*, most often a result of bone marrow cancer, is characterized by dizziness and an exceptionally high RBC count (8–11 million cells per cubic millimeter). The hematocrit may be as high as 80% and blood volume may double, causing the vascular system to become engorged with blood and severely impairing circulation.

Secondary polycythemias result when less oxygen is available or EPO production increases. The secondary polycythemia that appears in individuals living at high altitudes is a normal physiological response to the reduced atmospheric pressure and lower oxygen content of the air in such areas. RBC counts of 6–8 million per cubic millimeter are common in such people. Severe polycythemia is treated by blood dilution, that is, removing some blood and replacing it with saline.

Blood doping, practiced by some athletes competing in aerobic events, is artificially induced polycythemia. Some of the athlete's red blood cells are drawn off and then reinjected a few days before the event. Because the erythropoietin mechanism is triggered shortly after blood removal, the erythrocytes are quickly replaced. Then, when the stored blood is reinfused, a temporary polycythemia results. Since red blood cells carry oxygen, the additional infusion should translate into increased oxygen-carrying capacity due to a higher hematocrit, and hence greater endurance and speed. Other than problems that might derive from increased blood viscosity, such as temporary high blood pressure or reduced blood delivery to body tissues, blood doping seems to work. However, the practice is considered unethical and has been banned from the Olympic Games.

Leukocytes

General Structural and Functional Characteristics

Leukocytes (*leuko* = white), or **white blood cells (WBCs),** are the only formed elements that are complete cells, with nuclei and the usual organelles. Accounting for less than 1% of total blood volume, leukocytes are far less numerous than red blood

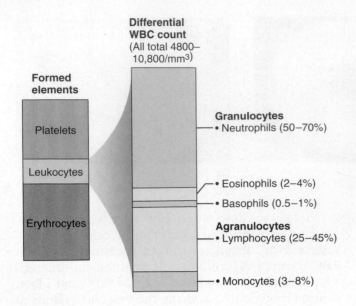

Differential WBC count (All total 4800–10,800/mm³)

Formed elements

Platelets

Leukocytes

Erythrocytes

Granulocytes
• Neutrophils (50–70%)

• Eosinophils (2–4%)

• Basophils (0.5–1%)

Agranulocytes
• Lymphocytes (25–45%)

• Monocytes (3–8%)

FIGURE 16.9   Types and relative percentages of leukocytes in normal blood. (The relationship of the formed elements in the left column is not shown to scale because erythrocytes comprise nearly 98% of the formed elements whereas leukocytes and platelets together account for the remaining 2+%.)

cells. On average, there are 4800–10,800 WBCs per cubic millimeter (mm³) of blood.

Leukocytes are crucial to our defense against disease. They form a mobile army that helps protect the body from damage by bacteria, viruses, parasites, toxins, and tumor cells. As such, they have some very special functional characteristics. Red blood cells are confined to the bloodstream, and they carry out their functions in the blood. But white blood cells are able to slip out of the capillary blood vessels—a process called **diapedesis** (di"ah-pě-de'sis; "leaping across")—and the circulatory system is simply their means of transport to areas of the body (mostly loose connective tissues or lymphoid tissues) where they are needed to mount inflammatory or immune responses. As explained in more detail in Chapter 20, the signals that prompt WBCs to leave the bloodstream at specific locations are adhesion molecules (selectins) displayed by endothelial cells forming the capillary walls at sites of inflammation. Once out of the bloodstream, leukocytes move through the tissue spaces by **amoeboid motion** (they form flowing cytoplasmic extensions that move them along). By following the chemical trail of molecules released by damaged cells or other leukocytes, a phenomenon called **positive chemotaxis,** they can pinpoint and gather in large numbers at areas of tissue damage and infection to destroy foreign substances or dead cells.

Whenever white blood cells are mobilized for action, the body speeds up their production and

twice the normal number may appear in the blood within a few hours. A *white blood cell count* of over 11,000 cells per cubic millimeter is **leukocytosis.** This condition is a normal homeostatic response to an infection in the body.

Leukocytes are grouped into two major categories on the basis of structural and chemical characteristics: *Granulocytes* contain obvious membrane-bound cytoplasmic granules; *agranulocytes* lack obvious granules. General information about the various leukocytes is provided next. More details appear in Figure 16.9 and Table 16.2.

Students are often asked to list the leukocytes in order from most abundant to least abundant. The following phrase may help you with this task: **N**ever **l**et **m**onkeys **e**at **b**ananas (neutrophils, lymphocytes, monocytes, eosinophils, basophils).

Granulocytes

Granulocytes (gran'u-lo-sīts), which include neutrophils, basophils, and eosinophils, are all roughly spherical in shape. They are larger and much shorter lived (in most cases) than erythrocytes. They characteristically have lobed nuclei (rounded nuclear masses connected by thinner strands of nuclear material), and their membrane-bound cytoplasmic granules stain quite specifically with Wright's stain. Functionally, all granulocytes are phagocytes.

Neutrophils   Neutrophils (nu'tro-filz), the most numerous of the white blood cells, account for 50 to 70% of the WBC population. Neutrophils are about twice as large as erythrocytes.

The neutrophil cytoplasm stains pale lilac and contains very fine granules (of two varieties) that are difficult to see (see Table 16.2 and Figure 16.10a). They are called neutrophils (literally, "neutral-loving") because their granules take up both *basic* (blue) and *acidic* (red) *dyes*. Together, the two types of granules give the cytoplasm a lilac color. Some of these granules contain hydrolytic enzymes, and are regarded as lysosomes. Others, especially the smaller granules, contain a potent "brew" of antibiotic-like proteins, called **defensins.** Neutrophil nuclei consist of three to six lobes. Because of this nuclear variability, they are often called **polymorphonuclear leukocytes (PMNs)** or simply **polys** (polymorphonuclear = many shapes of the nucleus). (However, some authorities use this term to refer to all granulocytes.)

Neutrophils are chemically attracted to sites of inflammation and are active phagocytes. They are especially partial to bacteria and some fungi, and bacterial killing is promoted by a process called a **respiratory burst.** In the respiratory burst, oxygen is actively metabolized to produce potent germ-killer oxidizing substances such as bleach and hydrogen

TABLE 16.2 Summary of Formed Elements of the Blood

Cell Type	Illustration	Description*	Cells/mm³ (μl) of Blood	Duration of Development (D) and Life Span (LS)	Function
Erythrocytes (red blood cells, RBCs)		Biconcave, anucleate disc; salmon-colored; diameter 7–8 μm	4–6 million	D: 5–7 days LS: 100–120 days	Transport oxygen and carbon dioxide
Leukocytes (white blood cells, WBCs)		Spherical, nucleated cells	4800–10,800		
Granulocytes					
• Neutrophil		Nucleus multilobed; inconspicuous cytoplasmic granules; diameter 10–12 μm	3000–7000	D: 6–9 days LS: 6 hours to a few days	Phagocytize bacteria
• Eosinophil		Nucleus bilobed; red cytoplasmic granules; diameter 10–14 μm	100–400	D: 6–9 days LS: 8–12 days	Kill parasitic worms; destroy antigen-antibody complexes; inactivate some inflammatory chemicals of allergy
• Basophil		Nucleus lobed; large blue-purple cytoplasmic granules; diameter 8–10 μm	20–50	D: 3–7 days LS: ? (a few hours to a few days)	Release histamine and other mediators of inflammation; contain heparin, an anticoagulant
Agranulocytes					
• Lymphocyte		Nucleus spherical or indented; pale blue cytoplasm; diameter 5–17 μm	1500–3000	D: days to weeks LS: hours to years	Mount immune response by direct cell attack or via antibodies
• Monocyte		Nucleus U or kidney shaped; gray-blue cytoplasm; diameter 14–24 μm	100–700	D: 2–3 days LS: months	Phagocytosis; develop into macrophages in tissues
Platelets		Discoid cytoplasmic fragments containing granules; stain deep purple; diameter 2–4 μm	150,000–400,000	D: 4–5 days LS: 5–10 days	Seal small tears in blood vessels; instrumental in blood clotting

*Appearance when stained with Wright's stain.

peroxide, and defensin-mediated lysis occurs. It appears that when the granules containing defensins are merged with a microbe-containing phagosome, the defensins form peptide "spears" that pierce holes in the membrane of the ingested "foe." Neutrophils are our body's bacteria slayers, and their numbers increase explosively during acute bacterial infections such as meningitis and appendicitis.

Eosinophils Eosinophils (e"o-sin'o-filz) account for 2 to 4% of all leukocytes and are approximately the size of neutrophils. Their deep red nucleus usually resembles an old-fashioned telephone receiver; that is, it has two lobes connected by a broad band of nuclear material (see Table 16.2 and Figure 16.10b). Large, coarse granules that stain from brick red to crimson with acid (eosin) dyes pack the cytoplasm. These granules are lysosome-like and filled with a unique variety of digestive enzymes. However, unlike typical lysosomes, they lack enzymes that specifically digest bacteria.

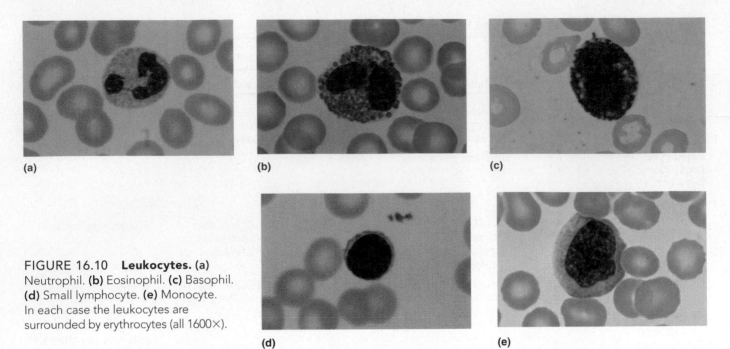

FIGURE 16.10 **Leukocytes.** **(a)** Neutrophil. **(b)** Eosinophil. **(c)** Basophil. **(d)** Small lymphocyte. **(e)** Monocyte. In each case the leukocytes are surrounded by erythrocytes (all 1600×).

The most important role of eosinophils is to lead the counterattack against parasitic worms, such as flatworms (tapeworms and flukes) and roundworms (pinworms and hookworms) that are too large to be phagocytized. These worms are ingested in food (especially raw fish) or invade the body via the skin and then typically burrow into the intestinal or respiratory mucosae. Eosinophils reside in the loose connective tissues at the same body sites, and when a parasitic worm "prey" is encountered, they gather around and release the enzymes from their cytoplasmic granules onto the parasite's surface, digesting it away. Eosinophils also lessen the severity of allergies by phagocytizing immune (antigen-antibody) complexes involved in allergy attacks and inactivating certain inflammatory chemicals released during allergic reactions.

Basophils   **Basophils** are the rarest white blood cells, averaging only 0.5 to 1% of the leukocyte population. Basophils are slightly smaller than neutrophils. Their cytoplasm contains large, coarse histamine-containing granules that have an affinity for the basic dyes (*basophil* = "base loving") and stain purplish-black. *Histamine* is an inflammatory chemical that acts as a vasodilator (makes blood vessels dilate) and attracts other white blood cells to the inflamed site; drugs called antihistamines counter this effect. The deep purple nucleus is generally U or S shaped with two or three conspicuous constrictions.

Granulated cells similar to basophils, called *mast cells,* are found in connective tissues. Although mast cell nuclei tend to be more oval than lobed, the cells are similar microscopically, and both cell types

bind to a particular antibody (immunoglobulin E) that causes the cells to release histamine. However, they arise from different cell lines.

Agranulocytes

The **agranulocytes** include lymphocytes and monocytes, WBCs that lack *visible* cytoplasmic granules. Although they are similar structurally, they are functionally distinct and unrelated cell types. Their nuclei are typically spherical or kidney shaped.

Lymphocytes   **Lymphocytes,** accounting for 25% or more of the WBC population, are the second most numerous leukocytes in the blood. When stained, a typical lymphocyte has a large, dark-purple nucleus that occupies most of the cell volume. The nucleus is usually spherical but may be slightly indented, and it is surrounded by a thin rim of pale-blue cytoplasm (see Table 16.2 and Figure 16.10d). Lymphocyte diameter ranges from 5 to 17 μm, but they are often classified according to size as small (5–8 μm), medium (10–12 μm), and large (14–17μm).

Although large numbers of lymphocytes exist in the body, only a small proportion of them (mostly the small lymphocytes) is found in the bloodstream. In fact, lymphocytes are so called because most are firmly enmeshed in lymphoid tissues (lymph nodes, spleen, etc.), where they play a crucial role in immunity. **T lymphocytes (T cells)** function in the immune response by acting directly against virus-infected cells and tumor cells. **B lymphocytes (B cells)** give rise to *plasma cells,* which produce **antibodies** (immunoglobulins) that are released to the blood. (B and T lymphocyte functions are described in Chapter 20.)

Monocytes   **Monocytes,** which account for 3–8% of WBCs, have an average diameter of 18 μm and are the largest leukocytes. They have abundant pale-blue cytoplasm and a darkly staining purple nucleus, which is distinctively U or kidney shaped (see Table 16.2 and Figure 16.10e). When circulating monocytes leave the bloodstream and enter the tissues, they differentiate into highly mobile **macrophages** with prodigious appetites. Macrophages are actively phagocytic, and they are crucial in the body's defense against viruses, certain intracellular bacterial parasites, and *chronic* infections such as tuberculosis. As explained in Chapter 20, macrophages are also important in activating lymphocytes to mount the immune response.

Production and Life Span of Leukocytes

Like erythropoiesis, **leukopoiesis,** or the production of white blood cells, is hormonally stimulated. These hormones, released mainly by macrophages and T lymphocytes, are glycoproteins that fall into two families of hematopoietic factors, **interleukins** and **colony-stimulating factors,** or **CSFs.** While the interleukins are numbered (e.g., IL-3, IL-5), most CSFs are named for the leukocyte population they stimulate—thus *granulocyte-CSF (G-CSF)* stimulates production of granulocytes. The hematopoietic factors not only prompt the white blood cell precursors to divide and mature, but also enhance the protective potency of mature leukocytes.

Many of the hematopoietic hormones (EPO and several of the CSFs) are used clinically to stimulate the bone marrow of cancer patients who are receiving chemotherapy (which suppresses the marrow) and of those who have received marrow transplants, and to beef up the protective responses of AIDS patients.

Figure 16.11 shows the pathways of leukocyte differentiation. An early branching of the pathway divides the **lymphoid stem cells,** which produce lymphocytes, from the **myeloid stem cells,** which give rise to all other formed elements. In each granulocyte line, the committed cells, called **myeloblasts** (mi′ĕ-lo-blasts″), accumulate lysosomes, becoming **promyelocytes.** The distinctive granules of each granulocyte type appear next in the **myelocyte** stage and then cell division stops. In the subsequent stage, the nuclei become arc-like, producing the **band cell** stage. Just before granulocytes leave the marrow and enter the circulation, their nuclei constrict, beginning the process of nuclear segmentation. The bone marrow stores mature granulocytes and usually contains 10 to 20 times more granulocytes than are found in the blood. The normal ratio of granulocytes to erythrocytes produced is about 3:1, which reflects the much shorter life span (0.5 to 9.0 days) of the granulocytes, most of which die combating invading microorganisms.

Despite their similar appearances, the two types of agranulocytes have very different lineages. Like granulocytes, monocytes diverge from common pleuripotent myeloid stem cells. They then progress through **monoblast** and **promonocyte** stages (see Figure 16.11). Lymphocytes derive from the lymphoid stem cell and progress through the **lymphoblast** and **prolymphocyte** stages. The promonocytes and prolymphocytes leave the bone marrow and travel to the lymphoid tissues, where their further differentiation occurs (as described in Chapter 20). Monocytes may live for several months, whereas the life span of lymphocytes varies from a few days to decades.

Leukocyte Disorders

Overproduction of abnormal leukocytes occurs in leukemia and infectious mononucleosis. At the opposite pole, **leukopenia** (loo″ko-pe′ne-ah) is an abnormally low white blood cell count (*penia* = poverty), commonly induced by drugs, particularly glucocorticoids and anticancer agents.

Leukemias   The term *leukemia,* literally "white blood," refers to a group of cancerous conditions involving white blood cells. As a rule, the renegade leukocytes are members of a single *clone* (descendants of a single cell) that remain unspecialized and proliferate out of control, impairing normal bone marrow function. The leukemias are named according to the abnormal cell type primarily involved. For example, *myelocytic leukemia* involves myeloblast descendants, whereas *lymphocytic leukemia* involves the lymphocytes. Leukemia is *acute* (quickly advancing) if it derives from blast-type cells like lymphoblasts, and *chronic* (slowly advancing) if it involves proliferation of later cell stages like myelocytes. The more serious acute forms primarily affect children. Chronic leukemia is seen more often in elderly people. Without therapy, all leukemias are fatal; only the time course differs.

In all leukemias, the bone marrow becomes almost totally occupied by cancerous leukocytes and immature WBCs flood into the bloodstream. Because the other blood cell lines are crowded out, severe anemia and bleeding problems also result. Other symptoms include fever, weight loss, and bone pain. Although tremendous numbers of leukocytes are produced, they are nonfunctional and cannot defend the body in the usual way. The most common causes of death are internal hemorrhage and overwhelming infections.

Irradiation and administration of antileukemic drugs to destroy the rapidly dividing cells have successfully induced remissions (symptom-free periods) lasting from months to years. Bone marrow or umbilical cord blood transplants are used in selected patients when compatible donors are available.

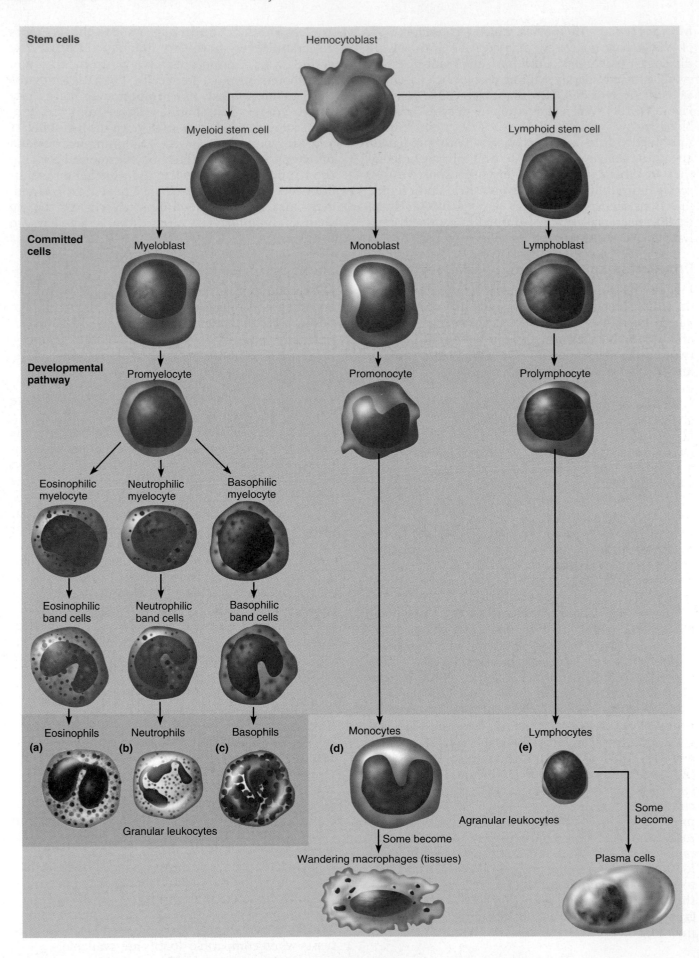

Stem cells

Hemocytoblast

Myeloid stem cell

Lymphoid stem cell

Committed cells

Myeloblast

Monoblast

Lymphoblast

Developmental pathway

Promyelocyte

Promonocyte

Prolymphocyte

Eosinophilic myelocyte

Neutrophilic myelocyte

Basophilic myelocyte

Eosinophilic band cells

Neutrophilic band cells

Basophilic band cells

Eosinophils

Neutrophils

Basophils

Monocytes

Lymphocytes

(a)

(b)

(c)

(d)

(e)

Granular leukocytes

Agranular leukocytes

Some become

Wandering macrophages (tissues)

Some become

Plasma cells

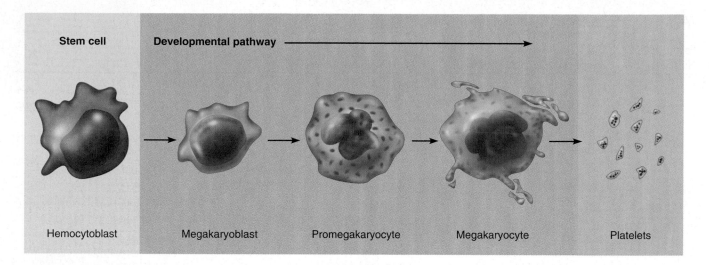

Stem cell   **Developmental pathway**

Hemocytoblast   Megakaryoblast   Promegakaryocyte   Megakaryocyte   Platelets

FIGURE 16.12   Genesis of platelets. The hemocytoblast gives rise to cells that undergo several mitotic divisions unaccompanied by cytoplasmic division to produce megakaryocytes. The cytoplasm of the megakaryocyte becomes compartmentalized by membranes, and the plasma membrane then fragments, liberating the platelets. (Intermediate stages between the hemocytoblast and megakaryoblast are not illustrated.)

Infectious Mononucleosis   Once called the kissing disease, *infectious mononucleosis* is a highly contagious viral disease most often seen in children and young adults. Caused by the Epstein-Barr virus, its hallmark is excessive numbers of agranulocytes, many of which are atypical. The affected individual complains of being tired and achy, and has a chronic sore throat and a low-grade fever. There is no cure, but with rest the condition typically runs its course to recovery in a few weeks.

Platelets

Platelets are not cells in the strict sense. About 1/4 the diameter of a lymphocyte, they are cytoplasmic fragments of extraordinarily large cells (up to 60 μm in diameter) called **megakaryocytes** (meg"ah-kar′e-o-sītz). In blood smears, each platelet exhibits a blue-staining outer region and an inner area containing granules that stain purple. The granules contain an impressive array of chemicals that act in the clotting process, including serotonin, Ca^{2+}, a

FIGURE 16.11   Leukocyte formation. Leukocytes arise from ancestral stem cells called hemocytoblasts. **(a–c)** Granular leukocytes develop via a sequence involving myeloblasts. The developmental pathway is common until the granules typical of each granulocyte begin to form. **(d–e)** Monocytes, like granular leukocytes, are progeny of the myeloid stem cell. Only lymphocytes arise via the lymphoid stem cell line.

variety of enzymes, ADP, and platelet-derived growth factor (PDGF).

Platelets are essential for the clotting process that occurs in plasma when blood vessels are ruptured or their lining is injured. By sticking to the damaged site, platelets form a temporary plug that helps seal the break. (This mechanism is explained shortly.) Because platelets are anucleate, they age quickly and degenerate in about ten days if they are not involved in clotting. In the meantime, they circulate freely, kept mobile but inactive by molecules (nitric oxide, prostaglandin I_2) secreted by endothelial cells lining the blood vessels.

Platelet formation is regulated by a hormone called **thrombopoietin.** Their immediate ancestral cells, the megakaryocytes, are progeny of the hemocytoblast and the myeloid stem cell, but their formation is quite unusual (Figure 16.12). In this line, repeated mitoses of the **megakaryoblast** occur, but cytokinesis does not. The final result is the megakaryocyte (literally "big nucleus cell"), a bizarre cell with a huge, multilobed nucleus and a large cytoplasmic mass. When formed, the megakaryocyte presses up against a sinusoid (the specialized type of capillary in the marrow) and sends cytoplasmic extensions through the sinusoid wall into the bloodstream. These extensions rupture, releasing the platelet fragments like stamps being torn from a sheet of postage stamps and seeding the blood with platelets. The plasma membranes associated with each fragment quickly seal around the cytoplasm to form the grainy, roughly disc-shaped platelets (see Table 16.2), each with a diameter of 2–4 μm. Each

TABLE 16.3   **Blood Clotting Factors (Procoagulants)**

Factor Number	Factor Name	Nature/Origin	Function or Pathway
I	Fibrinogen	Plasma protein; synthesized by liver	Common pathway; converted to fibrin, weblike substance of clot
II	Prothrombin	Plasma protein; synthesized by liver; formation requires vitamin K	Common pathway; converted to thrombin, which enzymatically converts fibrinogen to fibrin
III	Tissue factor (TF) or tissue thromboplastin	Lipoprotein complex; released from damaged tissues	Activates extrinsic pathway
IV	Calcium ions (Ca^{2+})	Inorganic ion present in plasma; acquired from diet or released from bone	Needed for essentially all stages of coagulation process
V	Proaccelerin, labile factor, or platelet accelerator	Plasma protein; synthesized in liver; also released by platelets	Both extrinsic and intrinsic mechanisms
VI	Number no longer used; substance now believed to be same as factor V		
VII	Proconvertin or serum prothrombin conversion accelerator (SPCA)	Plasma protein; synthesized in liver in process that requires vitamin K	Both extrinsic and intrinsic mechanisms
VIII	Antihemophilic factor (AHF)	Globulin synthesized in liver; deficiency causes hemophilia A	Intrinsic mechanism
IX	Plasma thromboplastin component (PTC) or Christmas factor	Plasma protein; synthesized in liver; deficiency results in hemophilia B; synthesis requires vitamin K	Intrinsic mechanism
X	Stuart factor, Stuart-Prower factor, or thrombokinase	Plasma protein; synthesized in liver; synthesis requires vitamin K	Both extrinsic and intrinsic pathways
XI	Plasma thromboplastin antecedent (PTA)	Plasma protein; synthesized in liver; deficiency results in hemophilia C	Intrinsic mechanism
XII	Hageman factor, glass factor	Plasma protein; proteolytic enzyme; synthesized in the liver	Intrinsic mechanism; activates plasmin; known to be activated by contact with glass and may initiate clotting in vitro
XIII	Fibrin stabilizing factor (FSF)	Plasma protein; synthesized in liver and present in platelets	Cross-links fibrin and renders it insoluble

cubic millimeter of blood contains between 150,000 and 400,000 of the tiny platelets.

Hemostasis

Normally, blood flows smoothly past the intact blood vessel lining (endothelium). But if a blood vessel wall breaks, a whole series of reactions is set in motion to accomplish **hemostasis** (he"mo-sta'sis), or stoppage of bleeding (*stasis* = halting). Without this plug-the-hole defensive reaction, we would quickly bleed out our entire blood volume from even the smallest cuts.

The hemostasis response, which is fast, localized, and carefully controlled, involves many blood coagulation factors (Table 16.3) normally present in plasma as well as some substances that are released by platelets and injured tissue cells. During hemostasis, three phases occur in rapid sequence: (1) vascular spasms, (2) platelet plug formation, and (3) coagulation, or blood clotting. Blood loss at the site is permanently prevented when fibrous tissue grows into the clot and seals the hole in the blood vessel.

Vascular Spasms

The immediate response to blood vessel injury is constriction of the damaged blood vessel (vasoconstriction). Factors that trigger this **vascular spasm** include direct injury to vascular smooth muscle, chemicals released by endothelial cells and platelets, and reflexes initiated by local pain receptors. The spasm mechanism becomes more and more efficient as the amount of tissue damage increases, and is most effective in the smaller blood vessels. The value of the spasm response is obvious: A strongly constricted artery can significantly reduce blood loss for 20–30 minutes, allowing time for platelet plug formation and blood clotting to occur.

Platelet Plug Formation

Platelets play a key role in hemostasis by forming a plug that temporarily seals the break in the vessel wall. They also help to orchestrate subsequent events that lead to blood clot formation. These events are shown in simplified form in Figure 16.13a.

As a rule, platelets do not stick to each other or to the smooth endothelial linings of blood vessels. However, when the endothelium is damaged and underlying collagen fibers are exposed, platelets, with the help of a large plasma protein called von Willebrand factor (VWF) synthesized by endothelial cells, adhere tenaciously to the collagen fibers and undergo some remarkable changes. They swell, form spiked processes, and become sticky.

Once attached, the platelets are activated by the enzyme thrombin and their granules begin to break down and release several chemicals. Some, like **serotonin,** enhance the vascular spasm. Others, like **adenosine diphosphate (ADP),** are potent aggregating agents that attract more platelets to the area and cause them to release their contents. **Thromboxane A₂** (throm-boks'ān), a short-lived prostaglandin derivative that is generated and released, stimulates both events.

Thus, a positive feedback cycle that activates and attracts greater and greater numbers of platelets to the area begins and, within one minute, a platelet plug is built up, which further reduces blood loss. The platelet plug is limited to the immediate area where it is needed by **PGI₂,** a prostaglandin produced by the endothelial cells. Also called **prostacyclin,** PGI₂ is a strong inhibitor of platelet aggregation. Platelet plugs are loosely knit, but when reinforced by fibrin threads to act as a "molecular glue" for the aggregated platelets, they are quite effective in sealing the small tears in a blood vessel that occur with normal activity. Once the platelet plug is formed, the next stage, coagulation, comes into play.

Coagulation

Coagulation or **blood clotting** (Figure 16.13a), during which blood is transformed from a liquid to a gel, is a multistep process that leads to its critically important last three phases. The final reactions are:

1. A complex substance called *prothrombin activator* is formed.

2. Prothrombin activator converts a plasma protein called *prothrombin* into *thrombin,* an enzyme.

3. Thrombin catalyzes the joining of *fibrinogen* molecules present in plasma to a *fibrin mesh,* which traps blood cells and effectively seals the hole until the blood vessel can be permanently repaired.

The complete coagulation process is much more complicated, however. Over 30 different substances are involved. Factors that enhance clot formation are called **clotting factors** or **procoagulants.** Although vitamin K is not directly involved in coagulation, this fat-soluble vitamin is required for the synthesis of four of the procoagulants made by the liver (see Table 16.3). Factors that inhibit clotting are called **anticoagulants.** Whether or not blood clots depends on a delicate balance between these two groups of factors. Normally, anticoagulants dominate and clotting is prevented; but when a vessel is ruptured, procoagulant activity in that area increases dramatically and clot formation begins. The procoagulants are numbered I to XIII (Table 16.3) according to the order of their discovery; hence the numerical order does *not* reflect the reaction sequence. Tissue factor (III) and Ca²⁺ (IV) are usually indicated by their names, rather than by numerals. Most of these factors are plasma proteins made by the liver that circulate in an inactive form in blood until mobilized.

Phase 1: Two Pathways to Prothrombin Activator

Clotting may be initiated by either the **intrinsic** or the **extrinsic pathway** (Figure 16.13b), and in the body both pathways are usually triggered by the same tissue-damaging events. Clotting of blood outside the body (such as in a test tube) is initiated *only* by the intrinsic mechanism. Let's examine the "why" of these differences.

A pivotal molecule in both mechanisms is **PF₃,** a phospholipid associated with the external surfaces of aggregated platelets. Apparently, many intermediates of both pathways can be activated only in the presence of PF₃. In the slower intrinsic pathway, all factors needed for clotting are present in (intrinsic to) the blood. By contrast, when blood is exposed to an additional factor released by injured cells called **tissue factor (TF), factor III,** or **tissue thromboplastin,** the "shortcut" extrinsic mechanism, which bypasses several steps of the intrinsic pathway, is triggered.

Each pathway requires ionic calcium and involves the activation of a *series* of procoagulants, each functioning as an enzyme to activate the next procoagulant in the sequence. The intermediate steps of each pathway *cascade* toward a common intermediate, factor X (Figure 16.13b). Once factor X

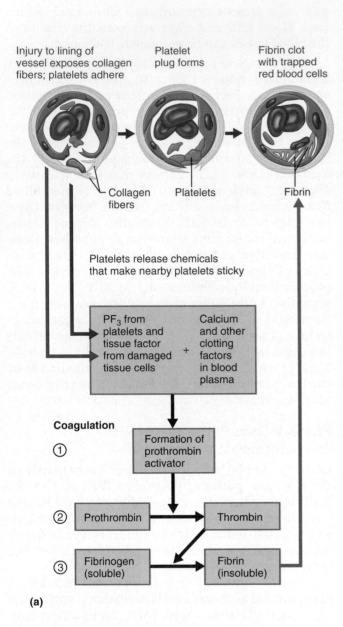

Injury to lining of vessel exposes collagen fibers; platelets adhere

Platelet plug forms

Fibrin clot with trapped red blood cells

Collagen fibers

Platelets

Fibrin

Platelets release chemicals that make nearby platelets sticky

| PF₃ from platelets and tissue factor from damaged tissue cells | + | Calcium and other clotting factors in blood plasma |

Coagulation

① Formation of prothrombin activator

② Prothrombin → Thrombin

③ Fibrinogen (soluble) → Fibrin (insoluble)

(a)

FIGURE 16.13   Events of platelet plug formation and blood clotting. (a) Simplified schematic of events. Steps 1–3 are the major events of coagulation. The color of the arrows indicates their source or destination: red from tissue, purple from platelets, and yellow to fibrin. **(b)** Detailed flowchart indicating the intermediates and events involved in platelet plug formation and the intrinsic and extrinsic mechanisms of blood clotting (coagulation). (The subscript "a" indicates the activated procoagulant.)

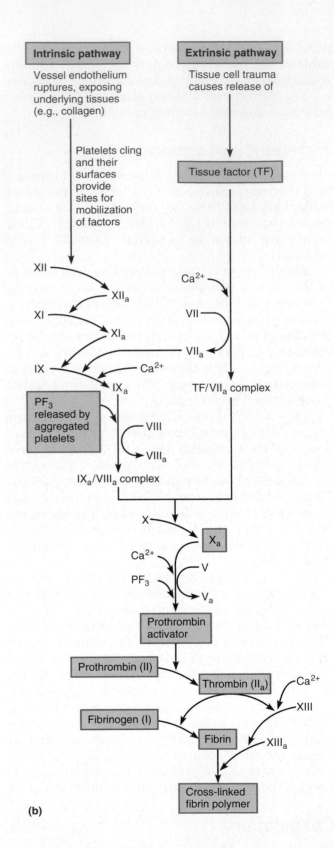

Intrinsic pathway
Vessel endothelium ruptures, exposing underlying tissues (e.g., collagen)

Platelets cling and their surfaces provide sites for mobilization of factors

XII → XIIₐ

XI → XIₐ

IX → IXₐ

PF₃ released by aggregated platelets

VIII → VIIIₐ

IXₐ/VIIIₐ complex

Extrinsic pathway
Tissue cell trauma causes release of

Tissue factor (TF)

Ca²⁺

VII → VIIₐ

Ca²⁺

TF/VIIₐ complex

X → Xₐ

Ca²⁺
PF₃

V → Vₐ

Prothrombin activator

Prothrombin (II)

Thrombin (IIₐ)   Ca²⁺

Fibrinogen (I)

Fibrin

XIII

XIIIₐ

Cross-linked fibrin polymer

(b)

has been activated, it complexes with calcium ions, PF₃, and factor V to form **prothrombin activator.** This step is usually the slowest step of the blood clotting process, but once prothrombin activator is present, the clot forms in 10 to 15 seconds.

Phase 2: Common Pathway to Thrombin

Prothrombin activator catalyzes the transformation of the plasma protein **prothrombin** to the active enzyme **thrombin.**

Phase 3: Common Pathway to the Fibrin Mesh

Thrombin catalyzes the polymerization of **fibrinogen** (another plasma protein made by the liver). As the fibrinogen molecules are aligned into long, hairlike, insoluble **fibrin** strands, they glue the platelets together and make a web that forms the structural basis of the clot. In the presence of fibrin, plasma becomes gel-like and traps formed elements that try to pass through it (Figure 16.14).

In the presence of calcium ions, thrombin also activates **factor XIII (fibrin stabilizing factor),** a cross-linking enzyme that binds the fibrin strands tightly together and strengthens and stabilizes the clot. Clot formation is normally complete within 3 to 6 minutes after blood vessel damage. Because the extrinsic pathway involves fewer steps it is more rapid than the intrinsic pathway; in cases of severe tissue trauma it can promote clot formation within 15 seconds.

Clot Retraction and Repair

Within 30 to 60 minutes, the clot is stabilized further by a platelet-induced process called **clot retraction.** Platelets contain contractile proteins (actin and myosin), and they contract in much the same manner as muscle cells. As the platelets contract, they pull on the surrounding fibrin strands, squeezing **serum** (plasma minus the clotting proteins) from the mass, compacting the clot and drawing the ruptured edges of the blood vessel more closely together. Even as clot retraction is occurring, vessel healing is taking place. **Platelet-derived growth factor (PDGF)** released by platelet degranulation stimulates smooth muscle and fibroblasts to divide and rebuild the wall. As fibroblasts form a connective tissue patch in the injured area, endothelial cells, stimulated by vascular endothelial growth factor (VEGF), multiply and restore the endothelial lining.

Fibrinolysis

A clot is not a permanent solution to blood vessel injury, and a process called **fibrinolysis** removes unneeded clots when healing has occurred. Because small clots are formed continually in vessels throughout the body, this cleanup detail is crucial. Without fibrinolysis, blood vessels would gradually become completely blocked.

The critical natural "clot buster" is a fibrin-digesting enzyme called **plasmin,** which is produced when the blood protein **plasminogen** is activated. Large amounts of plasminogen are incorporated into a forming clot, where it remains inactive until appropriate signals reach it. The presence of a clot in

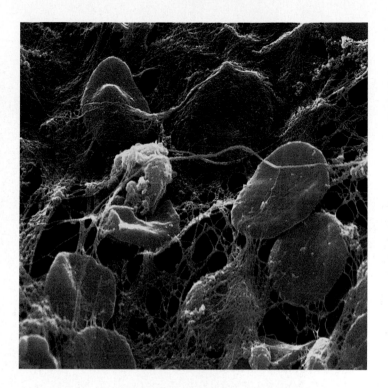

FIGURE 16.14 **Scanning electron micrograph of erythrocytes trapped in a fibrin mesh.** (3000×)

and around the blood vessel causes the endothelial cells to secrete **tissue plasminogen activator (TPA).** Activated factor XII and thrombin released during clotting also serve as plasminogen activators. As a result, most plasmin activity is confined to the clot, and any plasmin that strays into the plasma is quickly destroyed by circulating enzymes. Fibrinolysis begins within two days and continues slowly over several days until the clot is finally dissolved.

Factors Limiting Clot Growth or Formation

Factors Limiting Normal Clot Growth

Once the clotting cascade has begun, it continues until a clot is formed. Normally, two homeostatic mechanisms prevent clots from becoming unnecessarily large: (1) swift removal of clotting factors, and (2) inhibition of activated clotting factors. For clotting to occur in the first place, the concentration of activated procoagulants must reach certain critical levels. Clot formation in rapidly moving blood is usually curbed because the activated clotting factors are diluted and washed away. For the same reasons, further growth of a forming clot is hindered when it contacts normally flowing blood.

Other mechanisms block the final step in which fibrinogen is polymerized into fibrin by restricting thrombin to the clot or by inactivating it if it escapes into the general circulation. As a clot forms, almost all of the thrombin produced is bound onto the fibrin threads. This is an important safeguard because thrombin also exerts positive feedback effects on the coagulation process prior to the common pathway. Not only does it speed up the production of prothrombin activator by acting indirectly through factor V, but it also accelerates the earliest steps of the intrinsic pathway by activating platelets. Thus, fibrin effectively acts as an anticoagulant to prevent enlargement of the clot and prevents thrombin from acting elsewhere. Thrombin not bound to fibrin is quickly inactivated by **antithrombin III,** a protein present in plasma. Antithrombin III and **protein C,** another protein produced in the liver, also inhibit the activity of other intrinsic pathway procoagulants.

Heparin, the natural anticoagulant contained in basophil and mast cell granules and also produced by endothelial cells, is ordinarily secreted in small amounts into the plasma. It inhibits thrombin by enhancing the activity of antithrombin III. Like most other clotting inhibitors, heparin also inhibits the intrinsic pathway.

Factors Preventing Undesirable Clotting

As long as the endothelium is smooth and intact, platelets are prevented from clinging and piling up. Also, antithrombic substances—heparin and PGI_2—secreted by the endothelial cells normally prevent platelet adhesion. Additionally, it has been found that vitamin E quinone, a molecule formed in the body when vitamin E reacts with oxygen, is a potent anticoagulant.

Disorders of Hemostasis

Blood clotting is one of nature's most elegant creations, but it sometimes goes awry. The two major disorders of hemostasis are at opposite poles. **Thromboembolytic disorders** result from conditions that cause undesirable clot formation. **Bleeding disorders** arise from abnormalities that prevent normal clot formation.

Thromboembolytic Conditions

Despite the body's many safeguards, undesirable intravascular clotting, called "hemostasis in the wrong place" by some, sometimes occurs. A clot that develops and persists in an *unbroken* blood vessel is called a **thrombus.** If the thrombus is large enough, it may block circulation to the cells beyond the occlusion and lead to death of those tissues. For example, if the blockage occurs in the coronary circulation of the heart (coronary thrombosis), the consequences may be death of heart muscle and a fatal heart attack. If the thrombus breaks away from the vessel wall and floats freely in the bloodstream, it becomes an **embolus** (plural, *emboli*). An embolus ("wedge") is usually no problem until it encounters a blood vessel too narrow for it to pass through; then it becomes an **embolism,** obstructing the vessel. For example, emboli that become trapped in the lungs (pulmonary embolisms) dangerously impair the ability of the body to obtain oxygen. A cerebral embolism may cause a stroke.

Conditions that roughen the vessel endothelium, such as arteriosclerosis, severe burns, or inflammation, cause thromboembolytic disease by allowing platelets to gain a foothold. Slowly flowing blood or blood stasis is another risk factor, particularly in bedridden patients and those taking a long flight in economy-class seats. In this case, clotting factors are *not* washed away as usual and accumulate so that clot formation finally becomes possible.

A number of drugs, most importantly aspirin, heparin, and dicumarol, are used clinically to prevent undesirable clotting in patients at risk for heart attack or stroke. **Aspirin** is an antiprostaglandin drug that inhibits thromboxane A_2 formation (hence, it blocks platelet aggregation and platelet plug formation). Clinical studies of men taking low-dose aspirin (one aspirin every two days) over several years demonstrated a 50% reduction in (anticipated) incidence of heart attack. Heparin (see above) is also prescribed as an anticoagulant drug, as is warfarin, an ingredient in rat poison. Administered in injectable form, heparin is the anticoagulant most used clinically (i.e., for preoperative and postoperative cardiac patients and for those receiving blood transfusions). Taken orally, **warfarin** (Coumadin) is a mainstay in the treatment of those prone to atrial fibrillation, a condition in which blood pools in the heart, to reduce the risk of stroke. It works via a different mechanism than heparin; it interferes with the action of vitamin K in the production of some procoagulants (see Impaired Liver Function below).

HOMEOSTATIC IMBALANCE

Disseminated Intravascular Coagulation

Disseminated intravascular coagulation (DIC) is a situation in which widespread clotting occurs in intact blood vessels and the residual blood becomes unable to clot. Blockage of blood flow accompanied by severe bleeding follows. DIC is most commonly encountered as a complication of pregnancy or a result of septicemia or incompatible blood transfusions. ●

Bleeding Disorders

Anything that interferes with the clotting mechanism can result in abnormal bleeding. The most common causes are platelet deficiency (thrombocytopenia) and deficits of some procoagulants, which can result from impaired liver function or certain genetic conditions.

Thrombocytopenia A condition in which the number of circulating platelets is deficient, **thrombocytopenia** (throm″bo-si″to-pe′ne-ah) causes spontaneous bleeding from small blood vessels all over the body. Even normal movement leads to widespread hemorrhage, evidenced by many small purplish blotches, called *petechiae* (pe-te′ke-e), on the skin. Thrombocytopenia can arise from any condition that suppresses or destroys the bone marrow, such as bone marrow malignancy, exposure to ionizing radiation, or certain drugs. A platelet count of under 50,000 platelets per cubic millimeter of blood is usually diagnostic for this condition. Whole blood transfusions provide temporary relief from bleeding.

Impaired Liver Function When the liver is unable to synthesize its usual supply of procoagulants, abnormal, and often severe, bleeding occurs. The causes can range from an easily resolved vitamin K deficiency (common in newborns and after taking systemic antibiotics) to nearly total impairment of liver function (as in hepatitis or cirrhosis). Vitamin K is required by the liver cells for production of the clotting factors, and because vitamin K is produced by bacteria that reside in the intestines, dietary deficiencies are rarely a problem. However, vitamin K deficiency can occur if fat absorption is impaired, because vitamin K is a fat-soluble vitamin that is absorbed into the blood along with fats. In liver disease, the nonfunctional liver cells fail to produce not only the procoagulants but also bile, which is required for fat and vitamin K absorption.

Hemophilias The term **hemophilia** refers to several different hereditary bleeding disorders that have similar signs and symptoms. *Hemophilia A*, or *classical hemophilia*, results from a deficiency of **factor VIII (antihemophilic factor)**. It accounts for 83% of cases. *Hemophilia B* results from a deficiency of factor IX. Both types are sex-linked conditions occurring primarily in males. *Hemophilia C*, a less severe form of hemophilia seen in both sexes, is due to a lack of factor XI. The relative mildness of this form, as compared to the A and B forms, reflects the fact that the procoagulant (factor IX) that factor XI activates may also be activated by factor VII (see Figure 16.13b).

Symptoms of hemophilia begin early in life; even minor tissue trauma causes prolonged bleeding into tissues that can be life threatening. Commonly, the person's joints become seriously disabled and painful because of repeated bleeding into the joint cavities after exercise or trauma. Hemophilias are managed clinically by transfusions of fresh plasma or injections of the appropriate purified clotting factor. These therapies provide relief for several days but are expensive and inconvenient. Because hemophiliacs are absolutely dependent on blood transfusions or factor injections, many have become infected by the hepatitis virus and, since the early 1980s, by HIV, a blood-transmitted virus that depresses the immune system and causes AIDS. (See Chapter 20.) This infection problem has been resolved because of new testing methods for HIV and availability of genetically engineered factor VIII.

Transfusion and Blood Replacement

The human cardiovascular system is designed to minimize the effects of blood loss by (1) reducing the volume of the affected blood vessels, and (2) stepping up the production of red blood cells. However, the body can compensate for only so much blood loss. Losses of 15–30% cause pallor and weakness. Loss of more than 30% of blood volume results in severe shock, which can be fatal.

Transfusion of Whole Blood

Whole blood transfusions are routine when blood loss is substantial and when treating thrombocytopenia. Infusions of **packed red cells** (whole blood from which most of the plasma has been removed) are preferred to treat anemia. The usual blood bank procedure involves collecting blood from a donor and then mixing it with an anticoagulant, such as certain citrate or oxalate salts, which prevents clotting by binding with calcium ions. The shelf life of the collected blood at 4°C is about 35 days. When freshly collected blood is transfused, heparin is the anticoagulant used.

Human Blood Groups

People have different blood types and transfusion of incompatible blood can be fatal. RBC plasma membranes, like those of all body cells, bear highly specific glycoproteins (antigens) at their external surfaces, which identify each of us as unique from all others. One person's RBC proteins may be recognized as foreign if transfused into someone with a different red blood cell type, and the transfused cells may be agglutinated (clumped together) and destroyed. Since these RBC antigens promote agglutination, they are more specifically called **agglutinogens** (ag″loo-tin′o-jenz).

TABLE 16.4 ABO Blood Groups

Blood Group	Frequency (% U.S. Population)				RBC Antigens (Agglutinogens)	Illustration	Plasma Antibodies (Agglutinins)	Blood That Can Be Received
	White	Black	Asian	Native American				
AB	4	4	5	<1	A B		None	A, B, AB, O Universal recipient
B	11	20	27	4	B		Anti-A (a)	B, O
A	40	27	28	16	A		Anti-B (b)	A, O
O	45	49	40	79	None		Anti-A (a) Anti-B (b)	O Universal donor

At least 30 varieties of naturally occurring RBC antigens are common in humans. Besides these, perhaps 100 others occur in individual families ("private antigens") rather than in the general population. The presence or absence of each antigen allows each person's blood cells to be classified into several different blood groups. Antigens determining the ABO and Rh blood groups cause vigorous transfusion reactions (in which the foreign erythrocytes are destroyed) when they are improperly transfused. Thus, blood typing for these antigens is *always* done before blood is transfused. Other antigens (such as the M, N, Duffy, Kell, and Lewis factors) are mainly of legal or academic importance. Because these factors cause weak or no transfusion reactions, blood is not specifically typed for them unless the person is expected to need several transfusions, in which case the many weak transfusion reactions could hve cumulative effects. Only the ABO and Rh blood groups are described here.

ABO Blood Groups

As shown in Table 16.4, the **ABO blood groups** are based on the presence or absence of two agglutinogens, type A and type B. Depending on which of these a person inherits, his or her ABO blood group will be one of the following: A, B, AB, or O. The O blood group, which has neither agglutinogen, is the most common ABO blood group in white, black, Asian, and Native Americans; AB, with both antigens, is least prevalent. The presence of either the A or the B agglutinogen results in group A or B, respectively.

Unique to the ABO blood groups is the presence in the plasma of *preformed antibodies* called **agglutinins.** The agglutinins act against RBCs carrying ABO antigens that are *not* present on a person's own red blood cells. A newborn lacks these antibodies, but they begin to appear in the plasma within two months. They reach peak levels between eight and ten years of age and then slowly decline throughout the rest of life. As indicated in Table 16.4, a person with neither the A nor the B antigen (group O) possesses both anti-A and anti-B antibodies, also called *a* and *b* agglutinins respectively. Those with group A blood have anti-B antibodies, while those with group B have anti-A antibodies. Neither antibody is produced by AB individuals.

Rh Blood Groups

There are at least eight different types of Rh agglutinogens, each of which is called an **Rh factor.** Only three of these, the C, D, and E antigens, are fairly common. The Rh blood typing system is so named because one Rh antigen (agglutinogen D) was originally identified in *rhesus* monkeys. Later, the same antigen was discovered in humans. Most Americans (about 85%) are Rh^+ (Rh positive), meaning that their RBCs carry the Rh antigen. As a rule, a person's ABO and Rh blood groups are reported together, for example, O^+, A^-, and so on.

Unlike the ABO system, anti-Rh antibodies are not spontaneously formed in the blood of Rh⁻ (Rh negative) individuals. However, if an Rh⁻ person receives Rh⁺ blood, the immune system becomes sensitized and begins producing anti-Rh antibodies against the foreign antigen soon after the transfusion. Hemolysis does not occur after the first such transfusion because it takes time for the body to react and start making antibodies. But the second time, and every time thereafter, a typical transfusion reaction occurs in which the recipient's antibodies attack and rupture the donor RBCs.

H   HOMEOSTATIC IMBALANCE

An important problem related to the Rh factor occurs in pregnant Rh⁻ women who are carrying Rh⁺ babies. The first such pregnancy usually results in the delivery of a healthy baby. But, when bleeding occurs as the placenta detaches from the uterus, the mother may be sensitized by her baby's Rh⁺ antigens that pass into her bloodstream. If so, she will form anti-Rh antibodies unless treated with RhoGAM before or shortly after she has given birth. (The same precautions are taken in women who have miscarried or aborted the fetus.) RhoGAM is a serum containing anti-Rh agglutinins. Because it agglutinates the Rh factor, it blocks the mother's immune response and prevents her sensitization. If the mother is not treated and becomes pregnant again with an Rh⁺ baby, her antibodies will cross through the placenta and destroy the baby's RBCs, producing a condition known as **hemolytic disease of the newborn,** or **erythroblastosis fetalis.** The baby becomes anemic and hypoxic. Brain damage and even death may result unless transfusions are done *before* birth to provide the fetus with more erythrocytes for oxygen transport. Additionally, one or two *exchange transfusions* are done after birth. The baby's Rh⁺ blood is removed, and Rh⁻ blood is infused. Within six weeks, the transfused Rh⁻ erythrocytes have been broken down and replaced with the baby's own Rh⁺ cells. ●

Transfusion Reactions: Agglutination and Hemolysis

When mismatched blood is infused, a **transfusion reaction** occurs in which the donor's red blood cells are attacked by the recipient's plasma agglutinins. (Note that the donor's plasma antibodies may also be agglutinating the host's RBCs, but they are so diluted in the recipient's circulation that this does not usually present a serious problem.)

The initial event, agglutination of the foreign red blood cells, clogs small blood vessels throughout the body. During the next few hours, the clumped red blood cells begin to rupture or are destroyed by phagocytes, and their hemoglobin is released into the bloodstream. (When the transfusion reaction is exceptionally severe, the RBCs are lysed almost immediately.) These events lead to two easily recognized problems: (1) The oxygen-carrying capability of the transfused blood cells is disrupted, and (2) the clumping of red blood cells in small vessels hinders blood flow to tissues beyond those points. Less apparent, but more devastating, is the consequence of hemoglobin escaping into the bloodstream. Circulating hemoglobin passes freely into the kidney tubules, and in high concentration it precipitates, blocking the kidney tubules and causing renal shutdown. If shutdown is complete (acute renal failure), the person may die.

Transfusion reactions can also cause fever, chills, low blood pressure, rapid heartbeat, nausea, vomiting, and general toxicity; but in the absence of renal shutdown, these reactions are rarely lethal. Treatment of transfusion reactions is directed toward preventing kidney damage by infusing alkaline fluids to dilute and dissolve the hemoglobin and wash it out of the body. Diuretics, which increase urine output, are also given.

As indicated in Table 16.4, group O red blood cells bear neither the A nor the B antigen; thus, theoretically this blood group is the **universal donor.** Indeed, some medical centers are enzymatically converting type B blood to type O by clipping off the extra (B-specific) sugar residue. Since group AB plasma is devoid of antibodies to both A and B antigens, group AB people are theoretically **universal recipients** and can receive blood transfusions from any of the ABO groups. However, these classifications are misleading, because they do not take into account the other agglutinogens in blood that can trigger transfusion reactions.

Because pooled blood transfusions carry the risk of transfusion reactions and transmission of life-threatening infections (particularly with HIV), public interest in **autologous transfusions** has risen. Autologous (*auto* = self) transfusions are particularly sought by those considering elective surgery when they are in no immediate danger. The patient *predonates* his or her own blood, and it is stored and immediately available if needed during or after the operation. Iron supplements are given, and as long as the patient's preoperative hematocrit is at least 30%, one unit (400–500 ml) of blood can be collected every 4 days, with the last unit taken 72 hours prior to surgery.

Blood Typing

The importance of determining the blood group of both the donor and the recipient *before* blood is transfused is glaringly obvious. The general procedure

Is an agglutinin a plasma membrane glycoprotein or an antibody in the plasma?

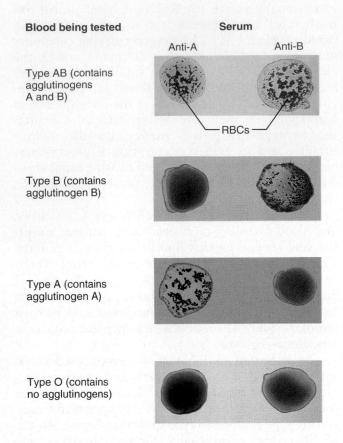

Blood being tested	Serum	
	Anti-A	Anti-B

Type AB (contains agglutinogens A and B)

RBCs

Type B (contains agglutinogen B)

Type A (contains agglutinogen A)

Type O (contains no agglutinogens)

FIGURE 16.15 Blood typing of ABO blood types. When serum containing anti-A or anti-B agglutinins is added to a blood sample diluted with saline, agglutination will occur between the agglutinin and the corresponding agglutinogen (A or B). As illustrated, agglutination occurs with both sera in blood group AB, with anti-B serum in blood group B, with anti-A serum in blood group A, and with neither serum in blood group O.

for determining ABO blood type is briefly outlined in Figure 16.15. Because it is critical that blood groups be compatible, cross matching is also done. *Cross matching* tests for agglutination of donor RBCs by the recipient's serum, and of the recipient's RBCs by the donor serum. Typing for the Rh factors is done in the same manner as ABO blood typing.

Plasma and Blood Volume Expanders

When a patient's blood volume is so low that death from shock is imminent, there may not be time to type blood, or appropriate whole blood may be unavailable. Such emergencies demand that blood *volume* be replaced immediately to restore adequate circulation. Although there is not yet a satisfactory whole blood substitute, research is ongoing.

Plasma can be administered to anyone without concern about a transfusion reaction because the antibodies it contains become harmlessly diluted in the recipient's blood. Except for red blood cells, plasma provides a complete and natural blood replacement. When plasma is not available, various colloidal solutions, or **plasma expanders,** such as *purified human serum albumin, plasminate,* and *dextran,* can be infused. All of these have osmotic properties that directly increase the fluid volume of blood. Still another option is to infuse isotonic salt solutions. *Normal saline* or a *multiple electrolyte solution* that mimics the electrolyte composition of plasma (for example, *Ringer's solution*) is a common choice.

Diagnostic Blood Tests

A laboratory examination of blood yields information that can be used to evaluate a person's current state of health. For example, in some anemias, the blood is pale and yields a low hematocrit. A high fat content *(lipidemia)* gives blood plasma a yellowish hue and forecasts possible problems in those with heart disease. How well a diabetic is controlling diet and blood sugar levels is routinely determined by blood glucose tests. Infections are signaled by leukocytosis and, if severe, larger than normal buffy coats in the hematocrit.

Microscopic studies of blood can reveal variations in the size and shape of erythrocytes that predict iron deficiency or pernicious anemia. Furthermore, a determination of the relative proportion of individual leukocyte types, a **differential white blood cell count,** is a valuable diagnostic tool. For example, a high eosinophil count may indicate a parasitic infection or an allergic response somewhere in the body.

A number of tests provide information on the status of the hemostasis system. The amount of prothrombin present in blood is assessed by determining the **prothrombin time,** and a **platelet count** is done when thrombocytopenia is suspected.

Two batteries of tests—a SMAC, SMA12–60, or similar series, and a **complete blood count (CBC)**—are routinely ordered during physical examinations and before hospital admissions. SMAC is a blood *chemistry* profile. The CBC includes counts of the different types of formed elements, a hematocrit, and tests for clotting factors. Together these tests provide a comprehensive picture of one's general health status in relation to normal blood values.

Review Questions

Multiple Choice/Matching

(Some questions have more than one correct answer. Select the best answer or answers from the choices given.)

1. The blood volume in an adult averages approximately (a) 1 L, (b) 3 L, (c) 5 L, (d) 7 L.

2. The hormonal stimulus that prompts red blood cell formation is (a) serotonin, (b) heparin, (c) erythropoietin, (d) thrombopoietin.

3. All of the following are true of RBCs except (a) biconcave disc shape, (b) life span of approximately 120 days, (c) contain hemoglobin, (d) contain nuclei.

4. The most numerous WBC is the (a) eosinophil, (b) neutrophil, (c) monocyte, (d) lymphocyte.

5. Blood proteins play an important part in (a) blood clotting, (b) immunity, (c) maintenance of blood volume, (d) all of the above.

6. The white blood cell that releases histamine and other inflammatory chemicals is the (a) basophil, (b) neutrophil, (c) monocyte, (d) eosinophil.

7. The blood cell that is said to be immunologically competent is the (a) lymphocyte, (b) megakaryocyte, (c) neutrophil, (d) basophil.

8. The normal erythrocyte count (per cubic millimeter) for adults is (a) 3 to 4 million, (b) 4.5 to 5 million, (c) 8 million, (d) 500,000.

9. The normal pH of the blood is about (a) 8.4, (b) 7.8, (c) 7.4, (d) 4.7.

10. Suppose your blood was found to be AB positive. This means that (a) agglutinogens A and B are present on your red blood cells, (b) there are no anti-A or anti-B agglutinins in your plasma, (c) your blood is Rh⁺, (d) all of the above.

Short Answer Essay Questions

11. (a) Define formed elements and list their three major categories. (b) Which is least numerous? (c) Which comprise(s) the buffy coat in a hematocrit tube?

12. Discuss hemoglobin relative to its chemical structure, its function, and the color changes it undergoes during loading and unloading of oxygen.

13. If you had a high hematocrit, would you expect your hemoglobin determination to be low or high? Why?

14. What nutrients are needed for erythropoiesis?

15. (a) Describe the process of erythropoiesis. (b) What name is given to the immature cell type released to the circulation? (c) How does it differ from a mature erythrocyte?

16. Besides the ability to move by amoeboid action, what other physiological attributes contribute to the function of white blood cells in the body?

17. (a) If you had a severe infection, would you expect your WBC count to be closest to 5000, 10,000, or 15,000 per cubic millimeter? (b) What is this condition called?

18. (a) Describe the appearance of platelets and state their major function. (b) Why should platelets not be called "cells"?

19. (a) Define hemostasis. (b) List the three major steps of coagulation. Explain what initiates each step and what the step accomplishes. (c) In what general way do the intrinsic and extrinsic mechanisms of clotting differ? (d) What ion is essential to virtually all stages of coagulation?

20. (a) Define fibrinolysis. (b) What is the importance of this process?

21. (a) How is clot overgrowth usually prevented? (b) List two conditions that may lead to unnecessary (and undesirable) clot formation.

22. How can liver dysfunction cause bleeding disorders?

23. (a) What is a transfusion reaction and why does it happen? (b) What are its possible consequences?

24. How can poor nutrition lead to anemia?

17

THE CARDIOVASCULAR SYSTEM: THE HEART

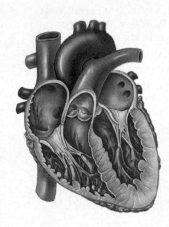

Heart Anatomy (pp. 581–592)

1. Describe the size, shape, location, and orientation of the heart in the thorax.
2. Name the coverings of the heart.
3. Describe the structure and function of each of the three layers of the heart wall.
4. Describe the structure and functions of the four heart chambers. Name each chamber and provide the name and general route of its associated great vessel(s).
5. Trace the pathway of blood through the heart.
6. Name the major branches and describe the distribution of the coronary arteries.
7. Name the heart valves and describe their location, function, and mechanism of operation.

Properties of Cardiac Muscle Fibers (pp. 592–595)

8. Describe the structural and functional properties of cardiac muscle, and explain how it differs from skeletal muscle.
9. Briefly describe the events of cardiac muscle cell contraction.

Heart Physiology (pp. 595–608)

10. Name the components of the conduction system of the heart, and trace the conduction pathway.
11. Draw a diagram of a normal electrocardiogram tracing; name the individual waves and intervals, and indicate what each represents. Name some abnormalities that can be detected on an ECG tracing.
12. Describe the timing and events of the cardiac cycle.
13. Describe normal heart sounds, and explain how heart murmurs differ.
14. Name and explain the effects of various factors regulating stroke volume and heart rate.
15. Explain the role of the autonomic nervous system in regulating cardiac output.

The ceaselessly beating heart in the chest has intrigued people for centuries. The ancient Greeks believed the heart was the seat of intelligence. Others thought it was the source of emotions. While these theories have proved false, we do know that emotions affect heart rate. When your heart pounds or occasionally skips a beat, you become acutely aware of how much you depend on this dynamic organ for your very life.

Despite its vital importance, the heart does not work alone. Indeed, it is part of the cardiovascular system, which also includes the miles of blood vessels that run through the body. Day and night, tissue cells take in nutrients and oxygen and excrete wastes. Because cells can make such exchanges only with their immediate environment, some means of changing and renewing that environment is necessary to ensure a continual supply of nutrients and to prevent a buildup of wastes. The cardiovascular system provides the transport system "hardware" that keeps blood continuously circulating to fulfill this critical homeostatic need.

Stripped of its romantic cloak, the **heart** is no more than the transport system pump; the hollow blood vessels are the delivery routes. Using blood as the transport medium, the heart continuously propels oxygen, nutrients, wastes, and many other substances into the interconnecting blood vessels that service body cells. Blood and blood vessels are considered separately in Chapters 16 and 18, respectively. This chapter focuses on the structure and function of the heart.

Heart Anatomy

Size, Location, and Orientation

The modest size and weight of the heart belie its incredible strength and endurance. About the size of a fist, the hollow, cone-shaped heart (Figure 17.1) has a mass of between 250 and 350 grams—less than a pound.

Snugly enclosed within the **mediastinum** (me"de-ah-sti'num), the medial cavity of the thorax, the heart extends obliquely for 12 to 14 cm (about 5 inches) from the second rib to the fifth intercostal space (Figure 17.1). As it rests on the superior surface of the diaphragm, the heart lies anterior to the vertebral column and posterior to the sternum. The lungs flank the heart laterally and partially obscure it. Approximately two-thirds of its mass lies to the left of the midsternal line; the balance projects to the right. Its broad, flat **base,** or posterior surface, is about 9 cm (3.5 in) wide and directed toward the right shoulder. Its **apex** points inferiorly toward the left hip. If you press your fingers between the fifth and sixth ribs just below the left nipple, you can easily feel your heart beating where the apex contacts the chest wall. Hence, this site is referred to as the **point of maximal intensity (PMI).**

Coverings of the Heart

The heart is enclosed in a double-walled sac called the **pericardium** (per"ĭ-kar'de-um; *peri* = around, *cardi* = heart) (Figure 17.2). The loosely fitting superficial part of this sac is the **fibrous pericardium.** This tough, dense connective tissue layer (1) protects the heart, (2) anchors it to surrounding structures, and (3) prevents overfilling of the heart with blood.

Deep to the fibrous pericardium is the **serous pericardium,** a thin, slippery, two-layer serous membrane. Its **parietal layer** lines the internal surface of the fibrous pericardium. At the superior margin of the heart, the parietal layer attaches to the large arteries exiting the heart, and then turns inferiorly and continues over the external heart surface as the **visceral layer,** also called the **epicardium** ("upon the heart"), which is an integral part of the heart wall.

Between the parietal and visceral layers is the slitlike **pericardial cavity,** which contains a film of serous fluid. The serous membranes, lubricated by the fluid, glide smoothly past one another during heart activity, allowing the mobile heart to work in a relatively friction-free environment.

🅗 *HOMEOSTATIC IMBALANCE*

Pericarditis, inflammation of the pericardium, hinders production of serous fluid and roughens the serous membrane surfaces. Consequently, as the beating heart rubs against its pericardial sac, it creates a creaking sound *(pericardial friction rub)* that can be heard with a stethoscope. Pericarditis is characterized by pain deep to the sternum. Over time, it may lead to adhesions in which the pericardia stick together and impede heart activity. In severe cases, the situation changes and *large* amounts of inflammatory fluid seep into the pericardial cavity. This excess fluid compresses the heart, limiting its ability to pump blood. This condition in which the heart is compressed by fluid is called *cardiac tamponade* (tam"pŏ-nād'), literally, "heart plug." Physicians treat it by inserting a syringe into the pericardial cavity and draining off the excess fluid. ●

Layers of the Heart Wall

The heart wall is composed of three layers, all richly supplied with blood vessels (Figure 17.2).

The title for this figure points out that the heart is in the mediastinum. Just what is the mediastinum?

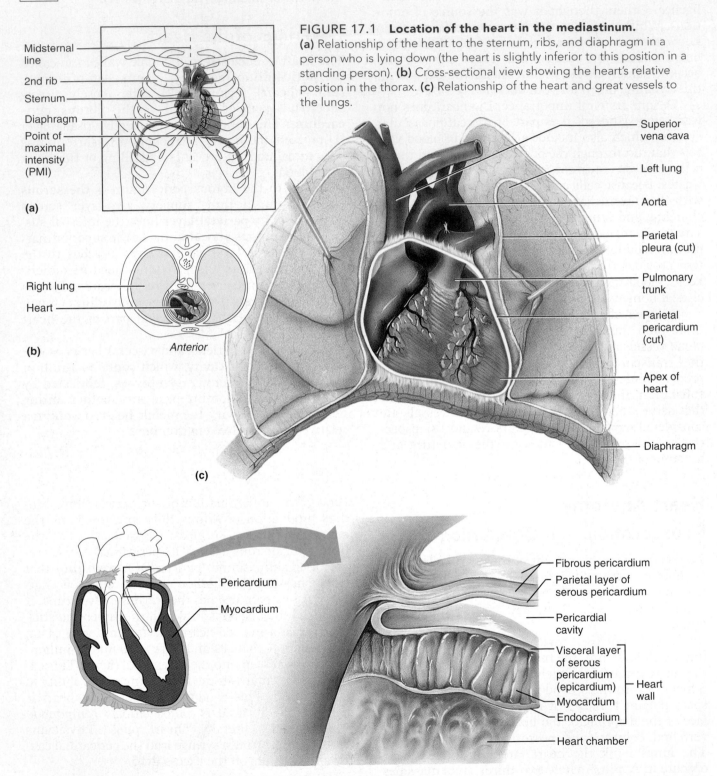

Midsternal line

2nd rib

Sternum

Diaphragm

Point of maximal intensity (PMI)

(a)

FIGURE 17.1   **Location of the heart in the mediastinum.**
(a) Relationship of the heart to the sternum, ribs, and diaphragm in a person who is lying down (the heart is slightly inferior to this position in a standing person). **(b)** Cross-sectional view showing the heart's relative position in the thorax. **(c)** Relationship of the heart and great vessels to the lungs.

Superior vena cava

Left lung

Aorta

Parietal pleura (cut)

Pulmonary trunk

Parietal pericardium (cut)

Apex of heart

Diaphragm

Right lung

Heart

Anterior

(b)

(c)

Pericardium

Myocardium

Fibrous pericardium

Parietal layer of serous pericardium

Pericardial cavity

Visceral layer of serous pericardium (epicardium)

Myocardium

Endocardium

Heart wall

Heart chamber

FIGURE 17.2   **The pericardial layers and layers of the heart wall.**

The medial cavity of the thorax within which the heart, great vessels, and trachea are found. ■

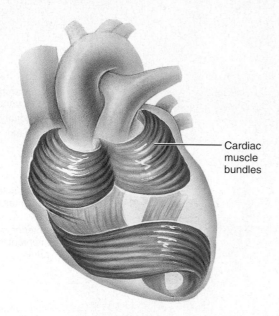

FIGURE 17.3 Packaging of cardiac muscle in the heart. Longitudinal view showing the spiral and circular arrangement of cardiac muscle bundles.

Cardiac muscle bundles

The superficial *epicardium* is the visceral layer of the serous pericardium. It is often infiltrated with fat, especially in older people.

The middle layer, the **myocardium** ("muscle heart"), is composed mainly of cardiac muscle and forms the bulk of the heart. It is the layer that contracts. In this layer, the branching cardiac muscle cells are tethered to one another by crisscrossing connective tissue fibers and arranged in spiral or circular *bundles* (Figure 17.3). These interlacing bundles effectively link all parts of the heart together. The connective tissue fibers form a dense network, the **fibrous skeleton of the heart,** that reinforces the myocardium internally and anchors the cardiac muscle fibers. This network of collagen and elastin fibers is thicker in some areas than others. For example, it constructs ropelike rings that provide additional support where the great vessels issue from the heart and around the heart valves (see Figure 17.8a, p. 590). Without this support, the vessels and valves might eventually become stretched because of the continuous stress of blood pulsing through them. Additionally, because connective tissue is not electrically excitable, the fibrous skeleton limits the direct spread of action potentials across the heart to specific pathways.

The third layer, the **endocardium** ("inside the heart"), is a glistening white sheet of endothelium (squamous epithelium) resting on a thin connective tissue layer. Located on the inner myocardial surface, it lines the heart chambers and covers the fibrous skeleton of the valves. The endocardium is continuous with the endothelial linings of the blood vessels leaving and entering the heart.

Chambers and Associated Great Vessels

The heart has four chambers (Figure 17.4e)—two superior **atria** (a'tre-ah) and two inferior **ventricles** (ven'trĭ-klz). The internal partition that divides the heart longitudinally is called the **interatrial septum** where it separates the atria, and the **interventricular septum** where it separates the ventricles. The right ventricle forms most of the anterior surface of the heart. The left ventricle dominates the infero-posterior aspect of the heart and forms the heart apex. Two grooves visible on the heart surface indicate the boundaries of its four chambers and carry the blood vessels supplying the myocardium. The **atrioventricular groove** (Figure 17.4d), or **coronary sulcus,** encircles the junction of the atria and ventricles like a crown (*corona* = crown). The **anterior interventricular sulcus,** cradling the anterior interventricular artery, marks the anterior position of the septum separating the right and left ventricles. It continues as the **posterior interventricular sulcus,** which provides a similar landmark on the heart's posteroinferior surface.

Atria: The Receiving Chambers

Except for small, wrinkled, protruding appendages called **auricles** (or'ĭ-klz; *auricle* = little ear), which increase the atrial volume somewhat, the right and left atria are remarkably free of distinguishing surface features. Internally, each atrium has two basic parts (Figure 17.4c): a smooth-walled posterior part and an anterior portion in which the walls are ridged by bundles of muscle tissue. Because the walls look as if they were raked by a comb, these muscle bundles are called **pectinate muscles** (*pectin* = comb). The posterior and anterior regions of each atrium are separated by a C-shaped ridge called the *crista terminalis* ("terminal crest"). The interatrial septum bears a shallow depression, the **fossa ovalis** (o-vă'lis), that marks the spot where an opening, the *foramen ovale*, existed in the fetal heart (see Figures 17.4c and 17.4e).

Functionally, the atria are receiving chambers for blood returning to the heart from the circulation (*atrium* = entryway). Because they need contract only minimally to push blood "next door" into the ventricles, the atria are relatively small, thin-walled chambers. As a rule, they contribute little to the propulsive pumping activity of the heart.

Blood enters the *right atrium* via three veins: (1) The **superior vena cava** returns blood from body regions superior to the diaphragm; (2) the **inferior vena cava** returns blood from body areas below the diaphragm; and (3) the **coronary sinus** collects blood draining from the myocardium. Four **pulmonary**

Which heart chamber has the thickest walls? What is the significance of this structural difference?

FIGURE 17.4 Gross anatomy of the heart. In diagrammatic views, vessels transporting oxygen-rich blood are red; those transporting oxygen-poor blood are blue. **(a)** Photograph of anterior aspect (pericardium removed). **(b)** Anterior view. (See *A Brief Atlas of the Human Body*, Figure 47.)

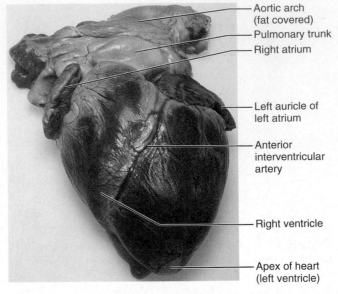

- Aortic arch (fat covered)
- Pulmonary trunk
- Right atrium
- Left auricle of left atrium
- Anterior interventricular artery
- Right ventricle
- Apex of heart (left ventricle)

(a)

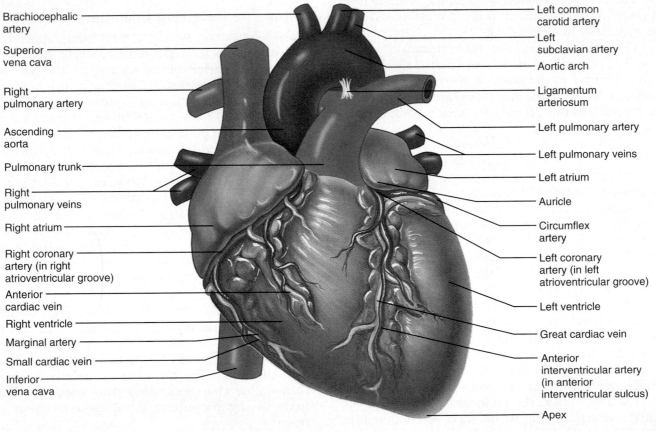

- Brachiocephalic artery
- Superior vena cava
- Right pulmonary artery
- Ascending aorta
- Pulmonary trunk
- Right pulmonary veins
- Right atrium
- Right coronary artery (in right atrioventricular groove)
- Anterior cardiac vein
- Right ventricle
- Marginal artery
- Small cardiac vein
- Inferior vena cava

- Left common carotid artery
- Left subclavian artery
- Aortic arch
- Ligamentum arteriosum
- Left pulmonary artery
- Left pulmonary veins
- Left atrium
- Auricle
- Circumflex artery
- Left coronary artery (in left atrioventricular groove)
- Left ventricle
- Great cardiac vein
- Anterior interventricular artery (in anterior interventricular sulcus)
- Apex

(b)

Left ventricle. It's the systemic pump that has to pump blood through the entire systemic circulation against high resistance. ∎

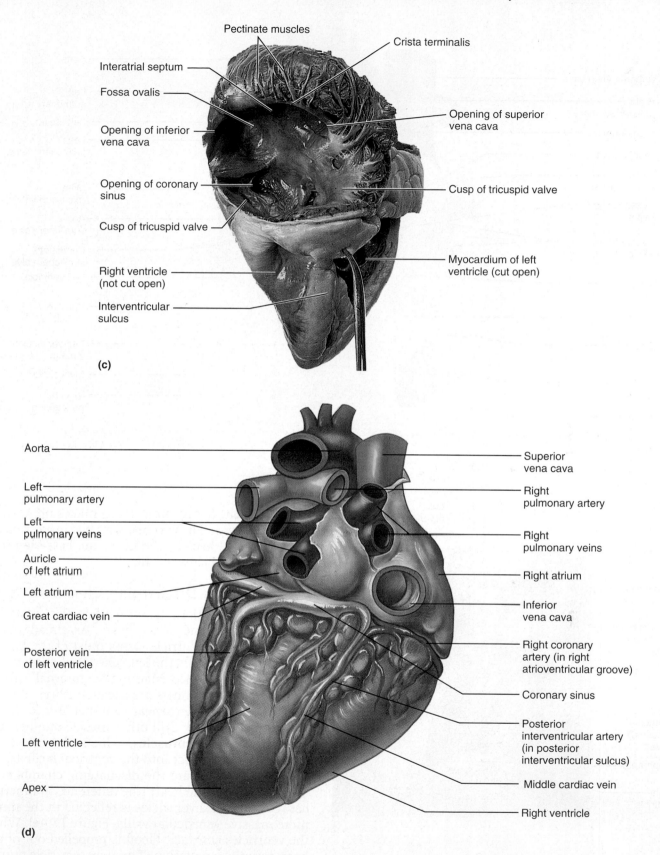

Pectinate muscles

Crista terminalis

Interatrial septum

Fossa ovalis

Opening of inferior vena cava

Opening of superior vena cava

Opening of coronary sinus

Cusp of tricuspid valve

Cusp of tricuspid valve

Right ventricle (not cut open)

Myocardium of left ventricle (cut open)

Interventricular sulcus

(c)

Aorta

Superior vena cava

Left pulmonary artery

Right pulmonary artery

Left pulmonary veins

Auricle of left atrium

Right pulmonary veins

Left atrium

Right atrium

Great cardiac vein

Inferior vena cava

Posterior vein of left ventricle

Right coronary artery (in right atrioventricular groove)

Coronary sinus

Posterior interventricular artery (in posterior interventricular sulcus)

Left ventricle

Apex

Middle cardiac vein

Right ventricle

(d)

FIGURE 17.4 (continued) **Gross anatomy of the heart. (c)** Right anterior view of the internal aspect of the right atrium. The anterior wall is incised near its left margin and folded back to the right. **(d)** Posterior surface view. (See *A Brief Atlas of the Human Body,* Figure 49.)

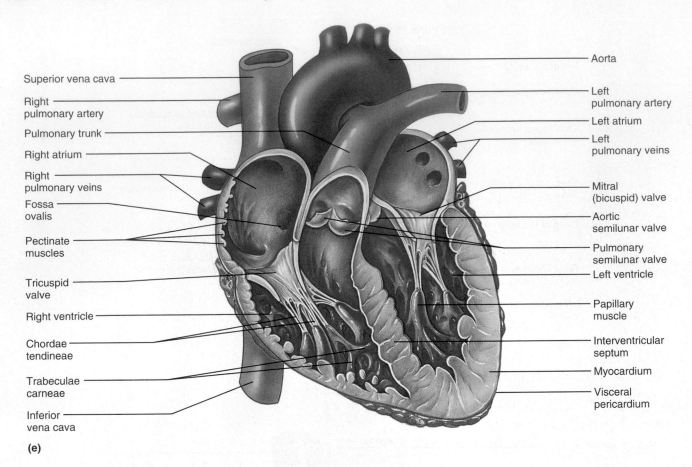

Superior vena cava

Right pulmonary artery

Pulmonary trunk

Right atrium

Right pulmonary veins

Fossa ovalis

Pectinate muscles

Tricuspid valve

Right ventricle

Chordae tendineae

Trabeculae carneae

Inferior vena cava

Aorta

Left pulmonary artery

Left atrium

Left pulmonary veins

Mitral (bicuspid) valve

Aortic semilunar valve

Pulmonary semilunar valve

Left ventricle

Papillary muscle

Interventricular septum

Myocardium

Visceral pericardium

(e)

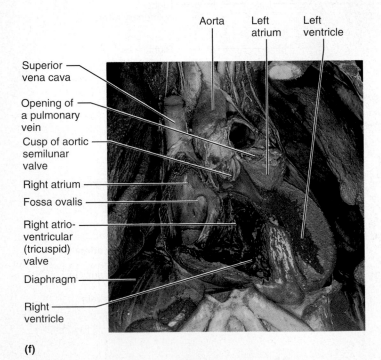

Aorta   Left atrium   Left ventricle

Superior vena cava

Opening of a pulmonary vein

Cusp of aortic semilunar valve

Right atrium

Fossa ovalis

Right atrio-ventricular (tricuspid) valve

Diaphragm

Right ventricle

(f)

FIGURE 17.4 *(continued)*   **Gross anatomy of the heart. (e)** Frontal section showing interior chambers and valves. **(f)** Photograph of a cadaver heart comparable to diagrammatic view of Figure 17.4e.

veins enter the *left atrium*, which makes up most of the heart's base. These veins, which transport blood from the lungs back to the heart, are best seen in a posterior view (Figure 17.4d).

Ventricles: The Discharging Chambers

Together the ventricles (*ventr* = underside) make up most of the volume of the heart. As already mentioned, the right ventricle forms most of the heart's anterior surface while the left ventricle dominates its posteroinferior surface. Marking the internal walls of the ventricular chambers are irregular ridges of muscle called **trabeculae carneae** (trah-bek'u-le kar'ne-e; "crossbars of flesh"). Still other muscle bundles, the conelike **papillary muscles**, which play a role in valve function, project into the ventricular cavity.

The ventricles are the discharging chambers or actual pumps of the heart (the difference in function between atria and ventricles is reflected in the much more massive ventricular walls; Figure 17.4e). When the ventricles contract, blood is propelled out of the heart into the circulation. The right ventricle pumps blood into the **pulmonary trunk**, which routes the blood to the lungs where gas exchange occurs. The left ventricle ejects blood into the **aorta** (a-or'tah), the largest artery in the body.

Pathway of Blood Through the Heart

Until the sixteenth century, it was believed that blood moved from one side of the heart to the other by seeping through pores in the septum. We now know that the heart passages open not from one side to the other but vertically and that the heart is actually two side-by-side pumps, each serving a separate blood circuit (Figure 17.5). The blood vessels that carry blood to and from the lungs form the **pulmonary circuit** (*pulmonos* = lung), which serves gas exchange. The blood vessels that carry the functional blood supply to and from all body tissues constitute the **systemic circuit.**

The right side of the heart is the *pulmonary circuit pump.* Blood returning from the body is relatively oxygen-poor and carbon dioxide–rich. It enters the right atrium and passes into the right ventricle, which pumps it to the lungs via the pulmonary trunk (Figure 17.4a, b). In the lungs, the blood unloads carbon dioxide and picks up oxygen. The freshly oxygenated blood is carried by the pulmonary veins back to the left side of the heart. Notice how unique this circulation is. Typically, we think of veins as vessels that carry blood that is relatively oxygen-poor to the heart and arteries as transporters of oxygen-rich blood from the heart to the rest of the body. Exactly the opposite condition exists in the pulmonary circuit.

The left side of the heart is the *systemic circuit pump.* Freshly oxygenated blood leaving the lungs is returned to the left atrium and passes into the left ventricle, which pumps it into the aorta. From there the blood is transported via smaller systemic arteries to the body tissues, where gases and nutrients are exchanged across the capillary walls. Then the blood, once again loaded with carbon dioxide and depleted of oxygen, returns through the systemic veins to the right side of the heart, where it enters the right atrium through the superior and inferior venae cavae. This cycle repeats itself continuously.

Although equal volumes of blood are pumped to the pulmonary and systemic circuits at any moment, the two ventricles have very unequal workloads. The pulmonary circuit, served by the right ventricle, is a short, low-pressure circulation, whereas the systemic circuit, associated with the left ventricle, takes a long pathway through the entire body and encounters about five times as much friction, or resistance to blood flow. This functional difference is revealed in the anatomy of the two ventricles (Figures 17.4e and 17.6). The walls of the left ventricle are three times as thick as those of the right ventricle, and its cavity is nearly circular. The right ventricular cavity is flattened into a crescent shape that partially encloses the left ventricle, much the way a hand might loosely grasp a clenched fist.

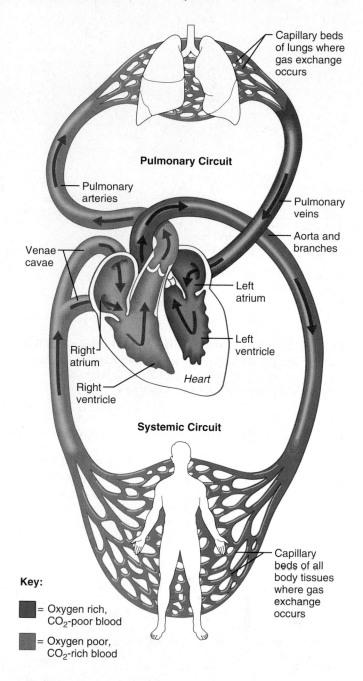

Key:

■ = Oxygen rich, CO_2-poor blood

■ = Oxygen poor, CO_2-rich blood

FIGURE 17.5 The systemic and pulmonary circuits. The right side of the heart pumps blood through the pulmonary circuit* (to the lungs and back to the left side of the heart). The left side of the heart pumps blood via the systemic circuit to all body tissues and back to the right side of the heart. Blood flowing through the pulmonary circuit gains oxygen and loses carbon dioxide, indicated by the color change from blue to red. Blood flowing through the systemic circuit loses oxygen and picks up carbon dioxide (red to blue color change).

*For simplicity, the actual number of two pulmonary arteries and four pulmonary veins has been reduced to one each.

Consequently, the left ventricle can generate much more pressure than the right and is a far more powerful pump.

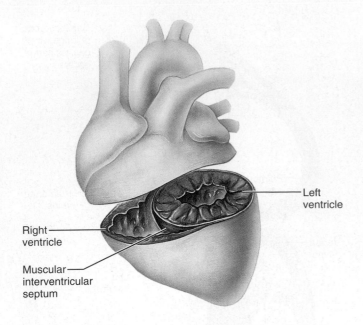

Right ventricle

Left ventricle

Muscular interventricular septum

FIGURE 17.6   Anatomical differences in right and left ventricles. The left ventricle has a thicker wall and its cavity is basically circular; the right ventricle cavity is crescent shaped and wraps around the left ventricle.

Coronary Circulation

Although the heart is more or less continuously filled with blood, this blood provides little nourishment to heart tissue. (The myocardium is too thick to make diffusion a practical means of nutrient delivery.) The **coronary circulation**, the functional blood supply of the heart, is the shortest circulation in the body. The arterial supply of the coronary circulation is provided by the *right* and *left coronary arteries*, both arising from the base of the aorta and encircling the heart in the atrioventricular groove (Figure 17.7a). The **left coronary artery** runs toward the left side of the heart and then divides into its major branches: the **anterior interventricular artery**, which follows the anterior interventricular sulcus and supplies blood to the interventricular septum and anterior walls of both ventricles; and the **circumflex artery**, which supplies the left atrium and the posterior walls of the left ventricle.

The **right coronary artery** courses to the right side of the heart, where it also divides into two branches: the **marginal artery**, which serves the myocardium of the lateral right side of the heart, and the **posterior interventricular artery** which runs to the heart apex and supplies the posterior ventricular walls. Near the apex of the heart, this artery merges (anastomoses) with the anterior interventricular artery. Together the branches of the right coronary artery supply the right atrium and nearly all the right ventricle.

The arterial supply of the heart varies considerably. For example, in 15% of people, the left coronary artery gives rise to *both* the anterior and posterior

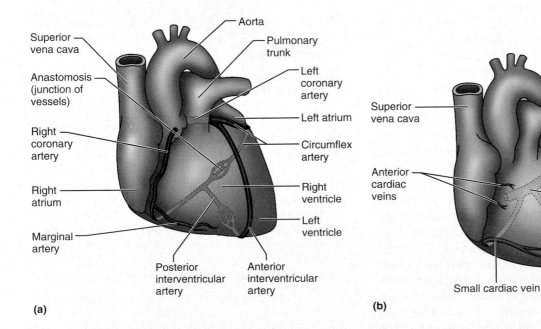

Superior vena cava

Anastomosis (junction of vessels)

Right coronary artery

Right atrium

Marginal artery

Aorta

Pulmonary trunk

Left coronary artery

Left atrium

Circumflex artery

Right ventricle

Left ventricle

Posterior interventricular artery

Anterior interventricular artery

(a)

Superior vena cava

Anterior cardiac veins

Great cardiac vein

Coronary sinus

Small cardiac vein

Middle cardiac vein

(b)

FIGURE 17.7   Coronary circulation. (a) Arterial supply. **(b)** Venous supply. (In both drawings, lighter-tinted vessels are more posterior in the heart.)

interventricular arteries; in about 4% of people, a single coronary artery supplies the whole heart. Additionally, there may be both right and left marginal arteries. There are many anastomoses among the coronary arterial branches. These fusing networks provide additional *(collateral)* routes for blood delivery to the heart muscle, which explains how the heart can receive adequate nutrition even when one of its coronary arteries is almost entirely occluded. Even so, complete blockage of a coronary artery leads to tissue death and heart attack.

The coronary arteries provide an intermittent, pulsating blood flow to the myocardium. These vessels and their main branches lie in the epicardium and send branches inward to nourish the myocardium. They deliver blood when the heart is relaxed, but are fairly ineffective when the ventricles are contracting because (1) they are compressed by the contracting myocardium, and (2) their entrances are partly blocked by the flaps of the open aortic semilunar valve. Although the heart represents only about $\frac{1}{200}$ of the body's weight, it requires about $\frac{1}{20}$ of the body's blood supply. As might be expected, the left ventricle receives the most plentiful blood supply.

After passing through the capillary beds of the myocardium, the venous blood is collected by the **cardiac veins,** whose paths roughly follow those of the coronary arteries. These veins join together to form an enlarged vessel called the **coronary sinus,** which empties the blood into the right atrium. The coronary sinus is obvious on the posterior aspect of the heart (Figure 17.7b). The sinus has three large tributaries: the **great cardiac vein,** in the anterior interventricular sulcus; the **middle cardiac vein** in the posterior interventricular sulcus; and the **small cardiac vein,** running along the heart's right inferior margin. Additionally, several **anterior cardiac veins** empty directly into the right atrium anteriorly.

✋ HOMEOSTATIC IMBALANCE

Blockage of the coronary arterial circulation can be serious and sometimes fatal. **Angina pectoris** (an-ji'nah pek'tor-is; "choked chest") is thoracic pain caused by a fleeting deficiency in blood delivery to the myocardium. It may result from stress-induced spasms of the coronary arteries or from increased physical demands on the heart. The myocardial cells are weakened by the temporary lack of oxygen but do not die. Far more serious is prolonged coronary blockage, which can lead to a **myocardial infarction (MI),** commonly called a **heart attack** or **coronary.** Because adult cardiac muscle is essentially amitotic, most areas of cell

death are repaired with noncontractile scar tissue. Whether or not a person survives a myocardial infarction depends on the extent and location of the damage. Damage to the left ventricle, which is the systemic pump, is most serious. ●

Heart Valves

Blood flows through the heart in one direction: from atria to ventricles and out the great arteries leaving the superior aspect of the heart. This one-way traffic is enforced by four valves (Figures 17.4e and 17.8) that open and close in response to differences in blood pressure on their two sides.

Atrioventricular Valves

The two **atrioventricular (AV) valves,** one located at each atrial-ventricular junction, prevent backflow into the atria when the ventricles are contracting. The right AV valve, the **tricuspid valve** (tri-kus'pid), has three flexible cusps (flaps of endocardium reinforced by connective tissue cores). The left AV valve, with two flaps, is the **bicuspid valve,** but is more commonly called the **mitral valve** (mi'tral) because of its resemblance to the two-sided bishop's miter or hat. Attached to each AV valve flap are tiny white collagen cords called **chordae tendineae** (kor'de ten"di'ne-e; "tendonous cords"), "heart strings" which anchor the cusps to the papillary muscles protruding from the ventricular walls.

When the heart is completely relaxed, the AV valve flaps hang limply into the ventricular chambers below and blood flows into the atria and then through the open AV valves into the ventricles. When the ventricles contract, compressing the blood in their chambers, the intraventricular pressure rises, forcing the blood superiorly against the valve flaps. As a result, the flap edges meet, closing the valve (Figure 17.9). The chordae tendineae and the papillary muscles serve as guy-wires to anchor the valve flaps in their *closed* position. If the cusps were not anchored in this manner, they would be blown upward into the atria, in the same way an umbrella is blown inside out by a gusty wind. The papillary muscles contract *before* the other ventricular musculature so that they take up the slack on the chordae tendineae before the full force of ventricular contraction hurls the blood against the AV valve flaps.

Semilunar Valves

The **aortic** and **pulmonary semilunar (SL) valves** guard the bases of the large arteries issuing from the ventricles (aorta and pulmonary trunk, respectively) and prevent backflow into the associated ventricles. Each SL valve is fashioned from three pocketlike

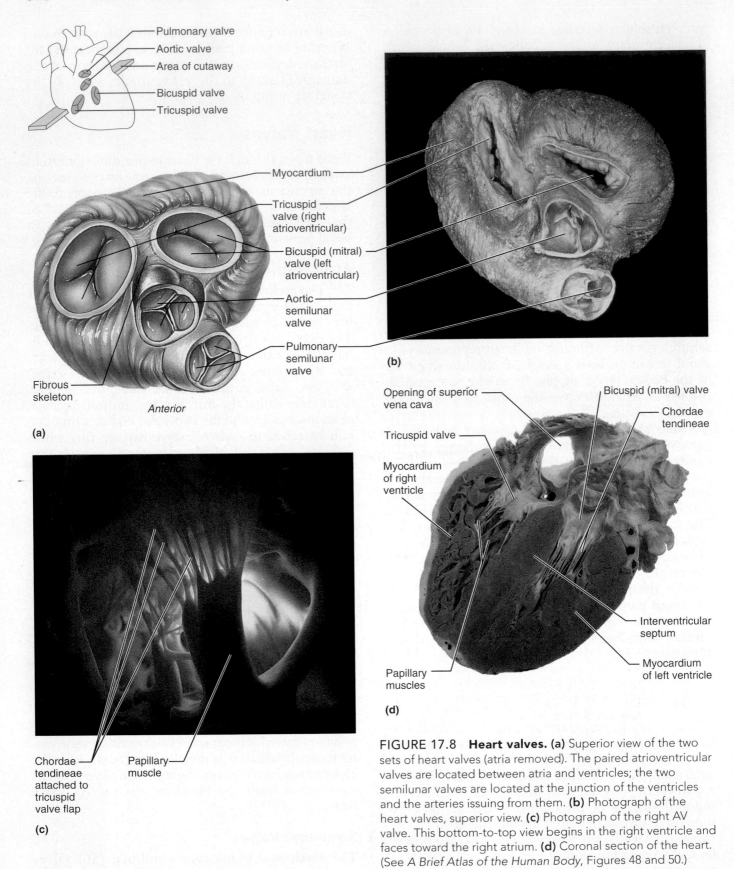

(a)

Pulmonary valve
Aortic valve
Area of cutaway
Bicuspid valve
Tricuspid valve

Myocardium
Tricuspid valve (right atrioventricular)
Bicuspid (mitral) valve (left atrioventricular)
Aortic semilunar valve
Pulmonary semilunar valve
Fibrous skeleton
Anterior

(b)

(c)
Chordae tendineae attached to tricuspid valve flap
Papillary muscle

(d)
Opening of superior vena cava
Tricuspid valve
Myocardium of right ventricle
Bicuspid (mitral) valve
Chordae tendineae
Interventricular septum
Myocardium of left ventricle
Papillary muscles

FIGURE 17.8   Heart valves. (a) Superior view of the two sets of heart valves (atria removed). The paired atrioventricular valves are located between atria and ventricles; the two semilunar valves are located at the junction of the ventricles and the arteries issuing from them. **(b)** Photograph of the heart valves, superior view. **(c)** Photograph of the right AV valve. This bottom-to-top view begins in the right ventricle and faces toward the right atrium. **(d)** Coronal section of the heart. (See *A Brief Atlas of the Human Body*, Figures 48 and 50.)

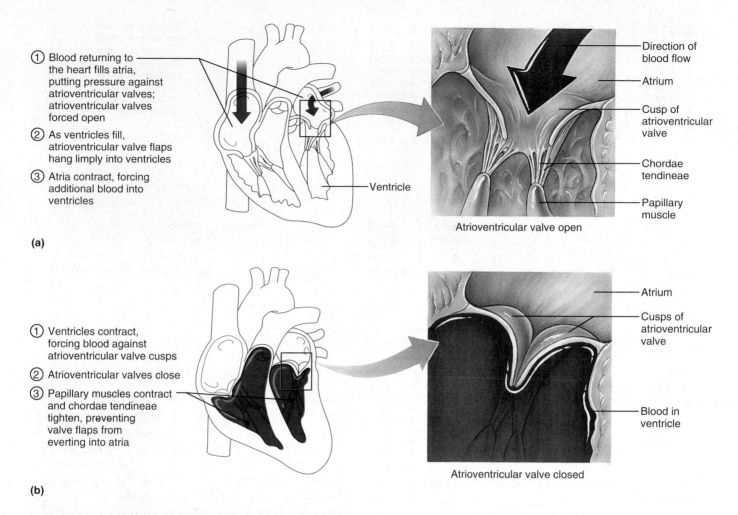

① Blood returning to the heart fills atria, putting pressure against atrioventricular valves; atrioventricular valves forced open

② As ventricles fill, atrioventricular valve flaps hang limply into ventricles

③ Atria contract, forcing additional blood into ventricles

Ventricle

Direction of blood flow

Atrium

Cusp of atrioventricular valve

Chordae tendineae

Papillary muscle

Atrioventricular valve open

(a)

① Ventricles contract, forcing blood against atrioventricular valve cusps

② Atrioventricular valves close

③ Papillary muscles contract and chordae tendineae tighten, preventing valve flaps from everting into atria

Atrium

Cusps of atrioventricular valve

Blood in ventricle

Atrioventricular valve closed

(b)

FIGURE 17.9 **The atrioventricular valves. (a)** The valves open when the blood pressure exerted on their atrial side is greater than that exerted on their ventricular side. **(b)** The valves are forced closed when the ventricles contract and intraventricular pressure rises, moving the contained blood superiorly. The action of the papillary muscles and chordae tendineae keeps the valve flaps closed.

cusps, each shaped roughly like a crescent moon (*semilunar* = half-moon). Their mechanism of action differs from that of the AV valves. In the SL case, when the ventricles are contracting and intraventricular pressure *rises above* the pressure in the aorta and pulmonary trunk, the SL valves are forced open and their cusps flatten against the arterial walls as the blood rushes past them (Figure 17.10). When the ventricles relax, and the blood (no longer propelled forward by the pressure of ventricular contraction) flows backward toward the heart, it fills the cusps and closes the valves.

We complete the valve story by mentioning what seems to be an important omission—there are no valves guarding the entrances of the venae cavae and pulmonary veins into the right and left atria, respectively. Small amounts of blood *do* spurt back into these vessels during atrial contraction, but backflow is minimal because as it contracts, the atrial myo-

cardium compresses (and collapses) these venous entry points.

 HOMEOSTATIC IMBALANCE

Heart valves are simple devices, and the heart—like any mechanical pump—can function with "leaky" valves as long as the impairment is not too great. However, severe valve deformities can seriously hamper cardiac function. An *incompetent valve* forces the heart to repump the same blood over and over because the valve does not close properly and blood backflows. In valvular *stenosis* ("narrowing"), the valve flaps become stiff (typically because of scar tissue formation following endocarditis or calcium salt deposit) and constrict the opening. This stiffness compels the heart to contract more forcibly than normal. In both instances, the heart's work load increases and, ultimately, the heart may be

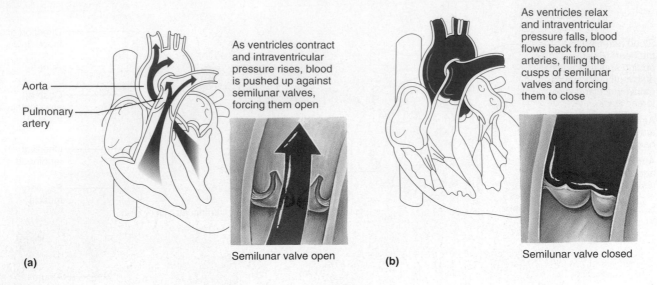

Aorta

Pulmonary artery

As ventricles contract and intraventricular pressure rises, blood is pushed up against semilunar valves, forcing them open

(a)

Semilunar valve open

As ventricles relax and intraventricular pressure falls, blood flows back from arteries, filling the cusps of semilunar valves and forcing them to close

(b)

Semilunar valve closed

FIGURE 17.10 **The semilunar valves. (a)** During ventricular contraction, the valves are open and their flaps flattened against the artery walls. **(b)** When the ventricles relax, the backflowing blood fills the cusps and closes the valves.

severely weakened. Under such conditions, the faulty valve (most often the mitral valve) is replaced with a synthetic valve, a pig heart valve chemically treated to prevent rejection, or cryopreserved valves from human cadavers. Tissue-engineered heart valves are being developed using biodegradable polymer scaffolds. ●

Properties of Cardiac Muscle Fibers

Although similar to skeletal muscle, cardiac muscle displays some special anatomical features that reflect its unique blood-pumping role.

Microscopic Anatomy

Cardiac muscle, like skeletal muscle, is striated, and it contracts by the sliding filament mechanism. However, in contrast to the long, cylindrical, multinucleate skeletal muscle fibers, cardiac cells are short, fat, branched, and interconnected. Each fiber contains one or at most two large, pale, *centrally* located nuclei (Figure 17.11a). The intercellular spaces are filled with a loose connective tissue matrix (the *endomysium*) containing numerous capillaries. This delicate matrix is connected to the fibrous skeleton, which acts both as a tendon and as an insertion, giving the cardiac cells something to pull or exert their force against.

Skeletal muscle fibers are independent of one another both structurally and functionally. By contrast, the plasma membranes of adjacent cardiac cells in-

terlock like the ribs of two sheets of corrugated cardboard at dark-staining junctions called **intercalated discs** (in-ter′kah-la″ted; *intercala* = insert) (Figure 17.11b). These discs contain anchoring *desmosomes* and *gap junctions* (cell junctions discussed in Chapter 3). The desmosomes prevent adjacent cells from separating during contraction, and the gap junctions allow ions to pass from cell to cell, transmitting a depolarizing current across the entire heart. Because cardiac cells are electrically coupled by the gap junctions, the entire myocardium *behaves* as a single coordinated unit, or **functional syncytium.**

Large mitochondria account for about 25% of the volume of cardiac cells (compared with only 2% in skeletal muscle) and give cardiac cells a high resistance to fatigue. Most of the remaining volume is occupied by myofibrils composed of fairly typical sarcomeres. The sarcomeres have Z discs, A bands, and I bands that reflect the arrangement of the thick (myosin) and thin (actin) filaments composing them. However, in contrast to what is seen in skeletal muscle, the myofibrils of cardiac muscle cells vary greatly in diameter and branch extensively, accommodating the abundant mitochondria that lie between them. This produces a banding pattern less dramatic than that seen in skeletal muscle.

The system for delivering Ca^{2+} is less elaborate in cardiac muscle cells. The T tubules are wider and fewer than in skeletal muscle and they enter the cells once per sarcomere at the Z discs. (Remember that T tubules are invaginations of the sarcolemma. In skeletal muscle, the T tubules invaginate twice per sarcomere and at the A-I junctions.) The cardiac sar-

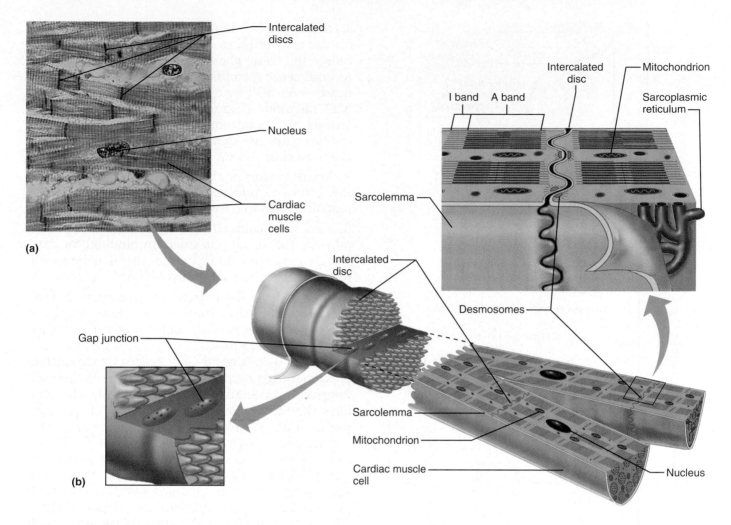

FIGURE 17.11 Microscopic anatomy of cardiac muscle. (a) Photomicrograph of cardiac muscle (700×). Notice that the cardiac muscle cells are short, branched, and striated. The dark-staining areas are intercalated discs, or junctions, between adjacent cells. **(b)** Cardiac cell relationships at the intercalated discs.

coplasmic reticulum is simpler and lacks the large terminal cisternae seen in skeletal muscle. Hence, no *triads* are seen in cardiac muscle fibers.

Mechanism and Events of Contraction

Although heart muscle and skeletal muscle are contractile tissues, they have some fundamental differences:

1. Means of stimulation. Each skeletal muscle fiber must be stimulated to contract by a nerve ending, but some cardiac muscle cells are self-excitable and can initiate their own depolarization, and that of the rest of the heart, in a spontaneous and rhythmic way. This property, called **automaticity**, or **autorhythmicity**, is described in the next section.

2. Organ versus motor unit contraction. In skeletal muscle, all cells of a given motor unit (but not necessarily all motor units of the muscle) are stimulated and contract at the same time. Impulses do not spread from cell to cell. In cardiac muscle, the heart either contracts as a unit or doesn't contract at all. This is ensured by the transmission of the depolarization wave across the heart from cell to cell via ion passage through gap junctions, which tie all cardiac muscle cells together into a single contractile unit.

3. Length of absolute refractory period. The absolute refractory period (the inexcitable period when Na^+ channels are still open or are closed or inactivated) lasts approximately 250 ms in cardiac muscle cells, nearly as long as the contraction (Figure 17.12a). Contrast this to the 1- to 2-ms refractory periods of skeletal muscle fibers, in which contractions last 20 to 100 ms. The long cardiac refractory period normally prevents tetanic contractions, which would stop the heart's pumping action.

Having probed their major differences, let's now look at cardiac and skeletal muscle similarities. As

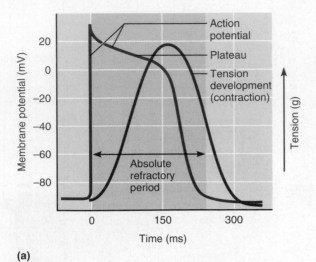

(a)

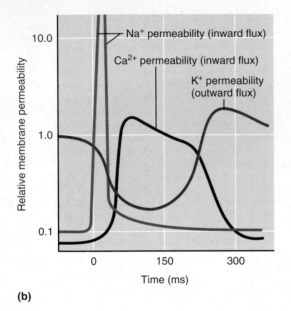

(b)

FIGURE 17.12 **Changes in membrane potential and permeability during action potentials of contractile cardiac muscle cells.** **(a)** Relationship between the action potential, period of contraction, and absolute refractory period in a single ventricular cell. (The tracing for an atrial contractile cell is similar but the plateau phase is shorter.) **(b)** Membrane permeability changes during the action potential. (The Na^+ permeability rises to a point off the scale during the action potential spike.)

with skeletal muscle, cardiac muscle contraction is triggered by action potentials that sweep across cell membranes. Although about 1% of cardiac fibers are *autorhythmic* ("self-rhythm") and have the special ability to depolarize spontaneously and thus pace the heart, the bulk of heart muscle is composed of *contractile muscle fibers* responsible for the heart's pumping activity. In these contractile cells, the sequence of electrical events is similar to that in skeletal muscle fibers:

1. Influx of Na^+ from the extracellular fluid into cardiac cells initiates a positive feedback cycle that causes the rising phase of the action potential (and reversal of the membrane potential from -90 mV to nearly $+30$ mV) by opening **voltage-regulated fast Na^+ channels** (Figure 17.12). The period of increased Na^+ permeability is very brief, because the sodium gates are quickly inactivated and the Na^+ channels close.

2. Transmission of the depolarization wave down the T tubules (ultimately) causes the sarcoplasmic reticulum (SR) to release Ca^{2+} into the sarcoplasm.

3. Excitation-contraction coupling occurs as Ca^{2+} provides the signal (via troponin binding) for cross bridge activation and couples the depolarization wave to the sliding of the myofilaments.

Although these three steps are common to both skeletal and cardiac muscle cells, how the SR is stimulated to release Ca^{2+} differs in the two muscle types. Let's take a look.

Some 10–20% of the Ca^{2+} needed for the calcium pulse that triggers contraction enters the cardiac cells from the extracellular space. Once inside, it stimulates the SR to release the other 80% of the Ca^{2+} needed. Ca^{2+} is barred from entering nonstimulated cardiac fibers, but when Na^+-dependent membrane depolarization occurs, the voltage change *also* opens channels that allow Ca^{2+} entry from the extracellular space. These channels are called **slow Ca^{2+} channels** because their opening is delayed a bit (Figure 17.12b). The local influxes of Ca^{2+} through these channels trigger opening of nearby Ca^{2+}-sensitive (ryanodine) channels in the SR tubules, which liberate bursts of Ca^{2+} ("calcium sparks") that dramatically increase the intracellular Ca^{2+} concentration.

Although Na^+ permeability has plummeted to its resting levels and repolarization has begun by this point, the calcium surge across the membrane prolongs the depolarization potential briefly, producing a **plateau** in the action potential tracing (Figure 17.12a). At the same time, K^+ permeability decreases, which also prolongs the plateau and prevents rapid repolarization. As long as Ca^{2+} is entering, the cells continue to contract. Notice in Figure 17.12a that muscle tension develops during the plateau, and peaks just after the plateau ends. Notice also that the duration of the action potential and contractile phase is much greater in cardiac muscle than in skeletal muscle. In skeletal muscle, the action potential typically lasts 1–5 ms and the contraction (for a single stimulus) 15–100 ms. In cardiac muscle, the action potential lasts 200 ms or more (because of the plateau), and tension development persists for 200 ms or more, providing the sustained contraction needed to eject blood from the heart.

After about 200 ms, the slope of the action potential tracing falls rapidly. This results from closure of Ca^{2+} channels, Ca^{2+} transport from the cytosol into the extracellular space or SR (or mitochondria), and opening of voltage-regulated K^+ channels, which allows a rapid loss of potassium from the cell that restores the resting membrane potential. During repolarization, Ca^{2+} is pumped back into the SR and the extracellular space.

Energy Requirements

Cardiac muscle has more mitochondria than skeletal muscle does, reflecting its greater dependence on oxygen for its energy metabolism. Unlike skeletal muscle, which can contract for prolonged periods, even during oxygen deficits, by carrying out anaerobic respiration, the heart relies almost exclusively on aerobic respiration. Hence, cardiac muscle cannot incur much of an oxygen debt and still operate effectively.

Both types of muscle tissue use multiple fuel molecules, including glucose and fatty acids. But cardiac muscle is much more adaptable and readily switches metabolic pathways to use whatever nutrient supply is available, including lactic acid generated by skeletal muscle activity. Thus, the real danger of an inadequate blood supply to the myocardium is lack of oxygen, not of nutrient fuels.

H HOMEOSTATIC IMBALANCE

When a region of heart muscle is deprived of blood (is ischemic), the oxygen-starved cells begin to metabolize anaerobically, producing lactic acid. The rising H^+ level that results hinders the cardiac cells' ability to produce the ATP they need to pump Ca^{2+} into the extracellular fluid. The resulting increase in intracellular H^+ and Ca^{2+} levels causes the gap junctions (which are usually open) to close, electrically isolating the damaged cells and forcing generated action potentials to find alternate routes to the cardiac cells beyond them. If the ischemic area is large, the pumping activity of the heart as a whole may be severely impaired, leading to a heart attack. ●

Heart Physiology
Electrical Events

The ability of cardiac muscle to depolarize and contract is intrinsic; that is, it is a property of heart muscle and does not depend on the nervous system. Even if all nerve connections to the heart are severed, the heart continues to beat rhythmically, as demonstrated by transplanted hearts. Nevertheless,

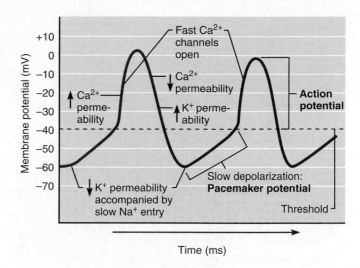

FIGURE 17.13 Pacemaker and action potentials of autorhythmic cells of the heart.

the healthy heart is amply supplied with nerve fibers that can alter the basic rhythm of heart activity set by intrinsic factors.

Setting the Basic Rhythm: The Intrinsic Conduction System

The independent, but coordinated, activity of the heart is a function of (1) the presence of gap junctions, and (2) the activity of the heart's "in-house" conduction system. The latter, called the **intrinsic cardiac conduction system,** consists of noncontractile cardiac cells specialized to initiate and distribute impulses throughout the heart, so that it depolarizes and contracts in an orderly, sequential manner. Thus, the heart beats as a coordinated unit. Let's look at how this system works.

Action Potential Initiation by Autorhythmic Cells Unlike unstimulated contractile cells of the heart (and neurons and skeletal muscle fibers), the **autorhythmic cells** making up the intrinsic conduction system do *not* maintain a stable resting membrane potential. Instead, they have an **unstable resting potential** that continuously depolarizes, drifting slowly toward threshold. These spontaneously changing membrane potentials, called **pacemaker potentials** or **prepotentials,** initiate the action potentials that spread throughout the heart to trigger its rhythmic contractions (Figure 17.13). The precise mechanism of the pacemaker potential is still in doubt, but it is believed to result from gradually reduced membrane permeability to K^+. Then, because Na^+ permeability is unchanged and Na^+ continues to diffuse into the cell at a slow rate, the balance between K^+ loss and Na^+ entry is upset and the membrane interior becomes less and less

What ionic event produces the plateau in the AP of the contractile muscle cells?

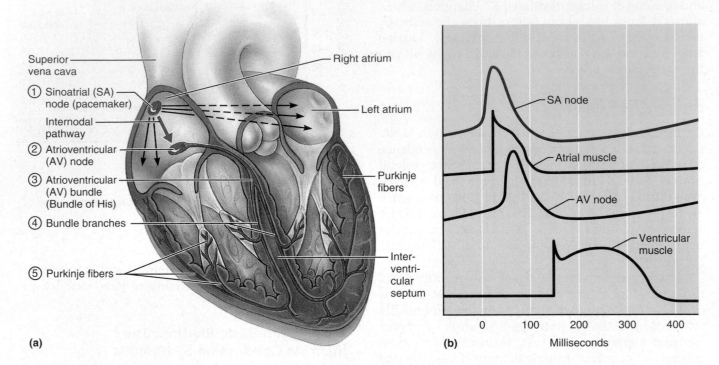

(a)

Superior vena cava

① Sinoatrial (SA) node (pacemaker)

Internodal pathway

② Atrioventricular (AV) node

③ Atrioventricular (AV) bundle (Bundle of His)

④ Bundle branches

⑤ Purkinje fibers

Right atrium

Left atrium

Purkinje fibers

Inter-ventri-cular septum

(b)

SA node

Atrial muscle

AV node

Ventricular muscle

0 100 200 300 400

Milliseconds

FIGURE 17.14 Cardiac intrinsic conduction system and action potential succession during one heartbeat. (a) The depolarization wave initiated by the SA node passes successively through the atrial myocardium to the AV node, the AV bundle, the right and left bundle branches, and the Purkinje fibers in the ventricular myocardium. **(b)** Comparison of the tracings of potentials generated across the heart is shown from top to bottom beginning with the pacemaker potential generated by the SA node and ending with an action potential (with a broad plateau) typical of the contractile cardiac cells of the ventricles.

negative (more positive). Ultimately, at threshold (approximately −40 mV), **fast Ca^{2+} channels** open, allowing explosive entry of Ca^{2+} (as well as some Na$^+$) from the extracellular space. Thus, in autorhythmic cells, it is the influx of Ca^{2+} (rather than Na$^+$) that produces the rising phase of the action potential and reverses the membrane potential.

As in other excitable cells, the falling phase of the action potential and repolarization reflect increased K$^+$ permeability and efflux from the cell. Once repolarization is complete, K$^+$ channels are inactivated, K$^+$ permeability declines, and the slow depolarization to threshold begins again.

Sequence of Excitation   Autorhythmic cardiac cells are found in the following areas (Figure 17.14): sinoatrial (si"no-a'tre-al) node, atrioventricular node, atrioventricular bundle, right and left bundle branches, and ventricular walls (as Purkinje fibers). Impulses pass across the heart in the same order.

■ *Calcium entry.*

① **Sinoatrial node.** The crescent-shaped sinoatrial (SA) node is located in the right atrial wall, just inferior to the entrance of the superior vena cava. A minute cell mass with a mammoth job, the SA node typically generates impulses about 75 times every minute. (However, its inherent rate in the absence of extrinsic neural and hormonal factors is closer to 100 times per minute.) Because no other region of the conduction system or the myocardium has a faster depolarization rate, the SA sets the pace for the heart as a whole. Hence, it is the heart's **pacemaker,** and its characteristic rhythm, called **sinus rhythm,** determines heart rate.

② **Atrioventricular node.** From the SA node, the depolarization wave spreads via gap junctions throughout the atria and via the *internodal pathway* to the atrioventricular (AV) node, located in the inferior portion of the interatrial septum immediately above the tricuspid valve. At the AV node, the impulse is delayed for about 0.1 s, allowing the atria to respond and complete their contraction before the ventricles contract. This delay reflects the smaller diameter of the fibers here and the fact that they have fewer gap junctions for current flow. Thus, the AV node conducts impulses more slowly than other parts of the system, just as traffic slows when cars are forced to merge into two lanes from four. Once through the AV node, the signaling impulse passes rapidly through the rest of the system.

③ **Atrioventricular bundle.** From the AV node, the impulse sweeps to the atrioventricular (AV) bundle (also called the **bundle of His**) in the superior part of the interventricular septum. Although the atria and ventricles abut each other, they are *not* connected by gap junctions. The AV bundle is the *only* electrical connection between them. The balance of the AV junction is insulated by the nonconducting fibrous skeleton of the heart.

④ **Right and left bundle branches.** The AV bundle persists only briefly before splitting into two pathways—the right and left bundle branches which course along the interventricular septum toward the heart apex.

⑤ **Purkinje fibers.** Essentially long strands of barrel-shaped cells with few myofibrils, the Purkinje (pur-kin′je) fibers complete the pathway through the interventricular septum, penetrate into the heart apex, and then turn superiorly into the ventricular walls. The bundle branches excite the septal cells, but the bulk of ventricular depolarization depends on the large Purkinje fibers and, ultimately, on cell-to-cell transmission of the impulse via gap junctions between the ventricular muscle cells. Because the left ventricle is much larger than the right, the Purkinje network is more elaborate in that side of the heart.

Interestingly, the Purkinje fibers directly supply the papillary muscles, which are excited to contract before the rest of the ventricular muscles. As mentioned before, the papillary muscles are already tightening the chordae tendineae before the full force of the ventricular contraction pushes blood against the AV valve flaps.

The total time between initiation of an impulse by the SA node and depolarization of the last of the ventricular muscle cells is approximately 0.22 s (220 ms) in a healthy human heart.

Ventricular contraction almost immediately follows the ventricular depolarization wave. A wringing contraction begins at the heart apex and moves toward the atria, following the direction of the excitation wave through the ventricle walls. This ejects some of the contained blood *superiorly* into the large arteries leaving the ventricles.

The various autorhythmic cardiac cells have different rates of spontaneous depolarization. Although the SA node normally drives the heart at a rate of 75 beats per minute, the AV node depolarizes only about 50 times per minute, the AV bundle and the Purkinje fibers (though they conduct very rapidly) depolarize at the rate of only about 30 times per minute. However, these slower "pacemakers" cannot dominate the heart unless everything faster than them becomes nonfunctional.

The cardiac conduction system not only coordinates and synchronizes heart activity, it also forces the heart to beat faster. Without it, the impulse would travel much more slowly through the myocardium—at the rate of 0.3 to 0.5 m/s as opposed to several meters per second in most parts of the conduction system. This slower rate would allow some muscle fibers to contract well before others, resulting in reduced pumping effectiveness.

H *HOMEOSTATIC IMBALANCE*

Defects in the intrinsic conduction system can cause irregular heart rhythms, or **arrhythmias** (ah-rith′me-ahz), uncoordinated atrial and ventricular contractions, and even **fibrillation,** a condition of rapid and irregular or out-of-phase contractions in which control of heart rhythm is taken away from the SA node by rapid activity in other heart regions. The heart in fibrillation has been compared with a squirming bag of worms. Fibrillating ventricles are useless as pumps; and unless the heart is defibrillated quickly, circulation stops and brain death occurs. Defibrillation is accomplished by electrically shocking the heart, which interrupts its chaotic twitching by depolarizing the entire myocardium. The hope is that "with the slate wiped clean" the SA node will begin to function normally and sinus rhythm will be reestablished. Implantable cardioverter defibrillators (ICDs) are now available that continually monitor heart rhythms and can slow an abnormally fast heart rate or emit an electrical shock if the heart begins to fibrillate.

A defective SA node may have several consequences. An **ectopic focus** (ek-top′ik), which is an abnormal pacemaker, may appear and take over the pacing of heart rate, or the AV node may become the pacemaker. The pace set by the AV node (**junctional rhythm**) is 40 to 60 beats per minute, slower than sinus rhythm but still adequate to maintain circulation. Occasionally, ectopic pacemakers appear even when the conduction system is operating normally. A small region of the heart becomes hyperexcitable, sometimes as a result of too much caffeine (several cups of coffee) or nicotine (excessive smoking), and generates impulses more quickly than the SA node. This leads to a *premature contraction* or **extrasystole** (ek″strah-sis′to-le) before the SA node initiates the next contraction. Then, because the heart has a longer time for filling, the next (normal) contraction is felt as a thud. As you

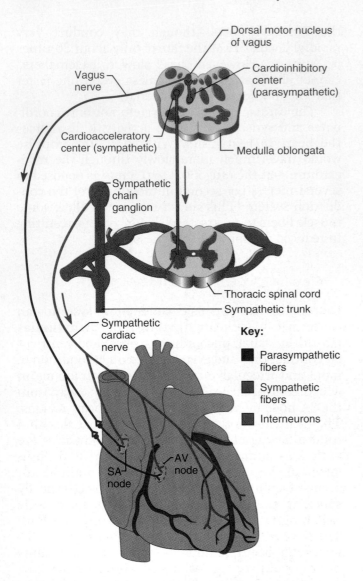

FIGURE 17.15   **Autonomic innervation of the heart.**

Key:

■ Parasympathetic fibers

■ Sympathetic fibers

■ Interneurons

Modifying the Basic Rhythm: Extrinsic Innervation of the Heart

Although the basic heart rate is set by the intrinsic conduction system, fibers of the autonomic nervous system modify the marchlike beat and introduce a subtle variability from one beat to the next. The sympathetic nervous system (the "accelerator") increases both the rate and the force of heartbeat; parasympathetic activation (the "brakes") slows the heart. These neural controls are explained later. Here we discuss the anatomy of the nerve supply to the heart.

The cardiac centers are located in the medulla oblongata. The sympathetic **cardioacceleratory center** projects to motor neurons in the T_1–T_5 level of the spinal cord. These neurons, in turn, synapse with ganglionic neurons in the cervical and upper thoracic sympathetic chain ganglia (Figure 17.15). From there, postganglionic fibers run through the cardiac plexus to the heart where they innervate the SA and AV nodes, heart muscle, and the coronary arteries. The parasympathetic or **cardioinhibitory center** sends impulses to the dorsal vagus nucleus in the medulla, which in turn sends inhibitory impulses to the heart via branches of the vagus nerves. Most parasympathetic ganglionic motor neurons lie in ganglia in the heart wall and their fibers project most heavily to the SA and AV nodes.

Electrocardiography

The electrical currents generated in and transmitted through the heart spread throughout the body and can be amplified with an **electrocardiograph.** The graphic recording of heart activity obtained is called an **electrocardiogram** (**ECG** or **EKG**; *kardio* is German for "heart"). An ECG is a composite of all of the APs generated by nodal and contractile cells at a given time and *not,* as sometimes assumed, a tracing of a single AP. Recording electrodes (leads) are positioned at various sites on the body surface; typically in a clinical setting, 12 leads are used. Three are bipolar leads that measure the voltage difference either between the arms or between an arm and a leg, and nine are unipolar leads. Together the 12 leads provide a fairly comprehensive picture of the heart's electrical activity.

A typical ECG has three distinguishable waves called **deflection waves** (Figure 17.16). The first, the small **P wave,** lasts about 0.08 s and results from movement of the depolarization wave from the SA node through the atria. Approximately 0.1 s after the P wave begins, the atria contract.

The large **QRS complex** results from ventricular depolarization and precedes ventricular contraction. It has a complicated shape because the paths of the depolarization waves through the ventricular walls change continuously, producing corresponding

might guess, premature *ventricular* contractions (PVCs) are most problematic.

Because the only route for impulse transmission from atria to ventricles is through the AV node, any damage to the AV node, referred to as a **heart block,** interferes with the ability of the ventricles to receive pacing impulses. In total heart block no impulses get through and the ventricles beat at their intrinsic rate, which is too slow to maintain adequate circulation. In such cases, a *fixed-rate* artificial pacemaker, set to deliver impulses at a constant rate, is usually implanted. Those suffering from partial heart block, in which some of the atrial impulses reach the ventricles, commonly receive *demand-type* pacemakers, which deliver impulses only when the heart is not transmitting on its own. ●

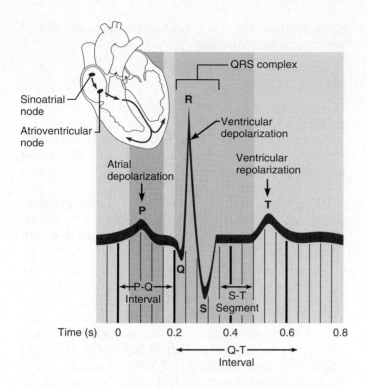

FIGURE 17.16 **An electrocardiogram tracing (lead I). The three normally recognizable deflection waves and the important intervals are labeled.**

changes in current direction. Additionally, the time required for each ventricle to depolarize depends on its size relative to the other ventricle. Average duration of the QRS complex is 0.08 s.

The **T wave** is caused by ventricular repolarization and typically lasts about 0.16 s. Repolarization is slower than depolarization, so the T wave is more spread out and has a lower amplitude (height) than the QRS wave. Because atrial repolarization takes place during the period of ventricular excitation, the wave representing atrial repolarization is normally obscured by the large QRS complex being recorded at the same time.

The **P-Q interval** is the time (about 0.16 s) from the beginning of atrial excitation to the beginning of ventricular excitation. Sometimes called the **P-R interval** because the Q wave tends to be very small, it includes atrial depolarization (and contraction) as well as the passage of the depolarization wave through the rest of the conduction system. During the **S-T** segment of the ECG, when the action potential is in its plateau phase, the entire ventricular myocardium is depolarized. The **Q-T interval**, lasting about 0.38 s, is the period from the beginning of ventricular depolarization through ventricular repolarization. The relationship of the parts of an ECG to movement of the action potential over the heart is illustrated in Figure 17.17.

? *Knowing that ventricular contraction follows the pattern of ventricular excitation, describe the path you expect the contraction wave to follow across the ventricles.*

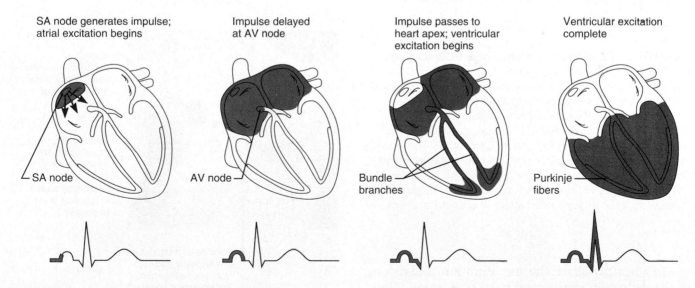

FIGURE 17.17 **The sequence of excitation of the heart related to the deflection waves of an ECG tracing.**

From the apex of the heart superiorly toward the atria. ■

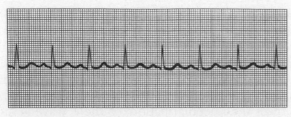

(a)

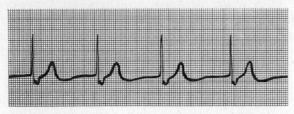

(b)

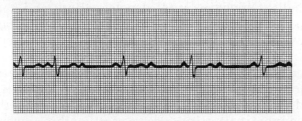

(c)

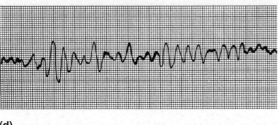

(d)

FIGURE 17.18 Normal and abnormal ECG tracings.
(a) Normal sinus rhythm. **(b)** Junctional rhythm. The SA node is nonfunctional, P waves are absent, and heart rate is paced by the AV node at 40–60 beats/min. **(c)** Second-degree heart block. Some P waves are not conducted through the AV node; hence more P than QRS waves are seen. Where P waves are conducted normally, the P:QRS ratio is 1:1. In total heart block, there is no whole number ratio between P and QRS waves, and the ventricles are no longer paced by the SA node. **(d)** Ventricular fibrillation. Chaotic, grossly irregular ECG deflections. Seen in acute heart attack and electrical shock.

In a healthy heart, the size, duration, and timing of the deflection waves tend to be consistent. Thus, changes in the pattern or timing of the ECG may reveal a diseased or damaged heart or problems with the heart's conduction system (Figure 17.18). For example, an enlarged R wave hints of enlarged ventricles, a flattened T wave indicates cardiac ischemia, and a prolonged Q-T interval reveals a repolarization abnormality that increases the risk of ventricular arrhythmias.

Heart Sounds

During each heartbeat, two sounds can be distinguished when the thorax is auscultated (listened to) with a stethoscope. These **heart sounds,** often described as lub-dup, are associated with closing of heart valves. (The timing of heart sounds in the cardiac cycle is shown in Figure 17.20a.)

The basic rhythm of the heart sounds is lub-dup, pause, lub-dup, pause, and so on, with the pause indicating the quiescent period. The first sound, which occurs as the AV valves close, signifies the beginning of systole when ventricular pressure rises above atrial pressure. The first sound tends to be louder, longer, and more resonant than the second, which is a short, sharp sound heard as the SL valves snap shut at the beginning of ventricular diastole. Because the mitral valve closes slightly before the tricuspid valve does, and the aortic SL valve generally snaps shut just before the pulmonary valve, it is possible to distinguish the individual valve sounds by auscultating four specific regions of the thorax (Figure 17.19). Notice that these four points, while not directly superficial to the valves (because the sounds take oblique paths to reach the chest wall), do handily

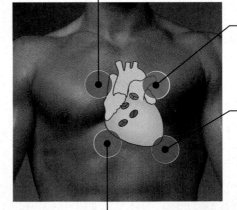

Sounds of aortic semilunar valve are heard in 2nd intercostal space at right sternal margin

Sounds of pulmonary semilunar valve are heard in 2nd intercostal space at left sternal margin

Sounds of mitral valve are heard over heart apex, in 5th intercostal space in line with middle of clavicle

Sounds of tricuspid valve are typically heard in right sternal margin of 5th intercostal space; variations include over sternum or over left sternal margin in 5th intercostal space

FIGURE 17.19 Areas of the thoracic surface where the heart sounds can be best detected.

define the four corners of the normal heart. Knowing normal heart size and location is essential for recognizing an enlarged (and often diseased) heart.

H *HOMEOSTATIC IMBALANCE*

Abnormal heart sounds are called **heart murmurs.** Blood flows silently as long as the flow is smooth and uninterrupted. If it strikes obstructions, however, its flow becomes turbulent and generates heart murmurs that can be heard with a stethoscope. Heart murmurs are fairly common in young children (and some elderly people) with perfectly healthy hearts, probably because their heart walls are relatively thin and vibrate with rushing blood. Most often, however, murmurs indicate valve problems. If a valve is *incompetent,* a swishing sound is heard as the blood backflows or regurgitates through the partially open valve, *after* the valve has (supposedly) closed. A stenotic valve, in which the valvular opening is narrowed, restricts blood flow *through* the valve. In a stenotic aortic valve, for instance, a high-pitched sound or click can be detected when the valve should be wide open during systole, but is not. ●

Mechanical Events: The Cardiac Cycle

The heart undergoes some dramatic writhing movements as it alternately contracts, forcing blood out of its chambers, and then relaxes, allowing its chambers to refill with blood. The terms **systole** (sis'to-le) and **diastole** (di-as'to-le) refer respectively to these *contraction* and *relaxation* periods. The **cardiac cycle** includes *all* events associated with the blood flow through the heart during one complete heartbeat, that is, atrial systole and diastole followed by ventricular systole and diastole.

The cardiac cycle is marked by a succession of pressure and blood volume changes in the heart. Because blood circulates endlessly, we must choose an arbitrary starting point for one turn of the cardiac cycle. As shown in Figure 17.20, which outlines what happens in the left side of the heart, we begin our explanation with the heart in total relaxation: Atria and ventricles are quiet, and it is mid-to-late diastole.

① **Ventricular filling: mid-to-late diastole.** Pressure in the heart is low, blood returning from the circulation is flowing passively through the atria and the open AV valves into the ventricles, and the aortic and pulmonary semilunar valves are closed. Approximately 70% of ventricular filling occurs during this period, and the AV valve flaps begin to drift toward the closed position. (The remaining 30% is delivered to the ventricles when the atria contract toward the end of this phase.) Now the stage is set for atrial systole. Following depolarization (P wave of ECG), the atria contract, com-

pressing the blood in their chambers. This causes a sudden slight rise in atrial pressure, which propels residual blood out of the atria into the ventricles. At this point the ventricles are in the last part of their diastole and have the maximum volume of blood they will contain in the cycle, a volume called the *end diastolic volume (EDV).* Then the atria relax and the ventricles depolarize (QRS complex). Atrial diastole persists through the rest of the cycle.

② **Ventricular systole.** As the atria relax, the ventricles begin contracting. Their walls close in on the blood in their chambers, and ventricular pressure rises rapidly and sharply, closing the AV valves. Because, for a split second, the ventricles are completely closed chambers and blood volume in the chambers remains constant, this is called the **isovolumetric contraction phase** (i"so-vol"u-met'rik). Ventricular pressure continues to rise and when it finally exceeds the pressure in the large arteries issuing from the ventricles, the isovolumetric stage ends as the SL valves are forced open and blood is expelled from the ventricles into the aorta and pulmonary trunk. During this **ventricular ejection phase,** the pressure in the aorta normally reaches about 120 mm Hg.

③ **Isovolumetric relaxation: early diastole.** During this brief phase following the T wave, the ventricles relax. Because the blood remaining in their chambers, referred to as the *end systolic volume (ESV),* is no longer compressed, ventricular pressure drops rapidly and blood in the aorta and pulmonary trunk backflows toward the heart, closing the SL valves. Closure of the aortic SL valve causes a brief rise in aortic pressure as backflowing blood rebounds off the closed valve cusps, an event called the **dicrotic notch.** Once again the ventricles are totally closed chambers.

All during ventricular systole, the atria have been in diastole; they have been filling with blood and the intra-atrial pressure has been rising. When blood pressure on the atrial side of the AV valves exceeds that in the ventricles, the AV valves are forced open and ventricular filling, phase 1, begins again. Atrial pressure drops to its lowest point and ventricular pressure begins to rise, completing the cycle.

Assuming the average heart beats 75 times each minute, the length of the cardiac cycle is about 0.8 s, with atrial systole accounting for 0.1 s and ventricular systole 0.3 s. The remaining 0.4 s is a period of total heart relaxation, the **quiescent period.**

Notice two important points: (1) blood flow through the heart is controlled entirely by pressure changes and (2) blood flows down a pressure gradient through any available opening. The pressure changes, in turn, reflect the alternating contraction and relaxation of the myocardium and cause the heart valves to open, which keeps blood flowing in the forward direction.

Are the ventricular cardiac cells in isotonic or isometric contraction during phase 2a of part (b)?

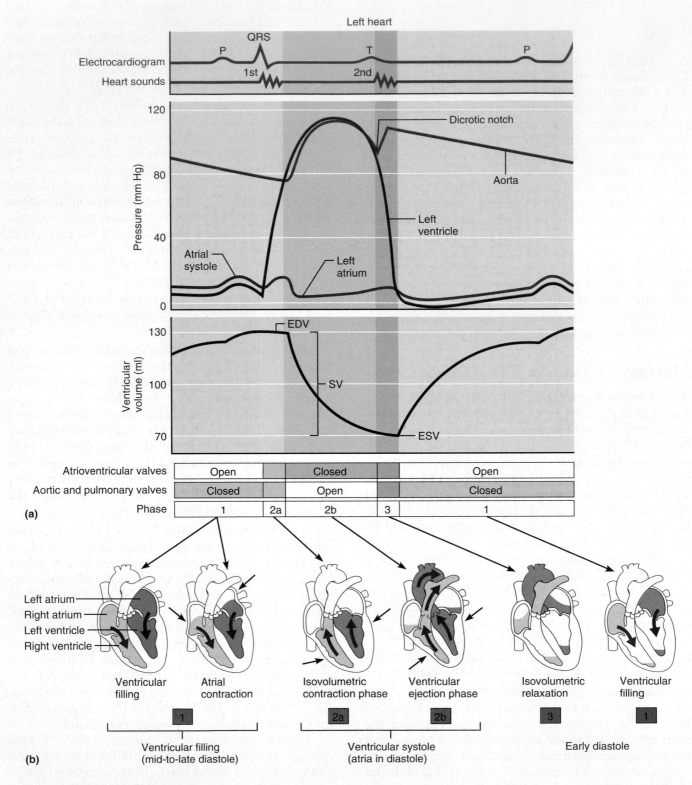

FIGURE 17.20   Summary of events during the cardiac cycle in the left side of the heart. (a) An ECG tracing correlated with graphs of pressure and volume changes. Time occurrence of heart sounds is also indicated. **(b)** Events of phases 1 through 3 of the cardiac cycle.

■ *They are contracting isometrically until ventricular pressure rises above the pressure in the aorta.*

The situation in the right side of the heart is essentially the same as in the left side *except* for pressure. The pulmonary circulation is a low-pressure circulation as evidenced by the much thinner myocardium of its right ventricle. Hence, typical systolic and diastolic pressures for the pulmonary artery are 24 and 8 mm Hg as compared to systemic aortic pressures of 120 and 80 mm Hg, respectively. However, the two sides of the heart eject the same blood volume with each heartbeat.

Cardiac Output

Cardiac output (CO) is the amount of blood pumped out by *each* ventricle in 1 minute. It is the product of heart rate (HR) and stroke volume (SV). **Stroke volume** is defined as the volume of blood pumped out by one ventricle with each beat. In general, stroke volume is correlated with the force of ventricular contraction.

Using normal resting values for heart rate (75 beats/min) and stroke volume (70 ml/beat), the average adult cardiac output can be computed:

$$CO = HR \times SV = \frac{75 \text{ beats}}{\text{min}} \times \frac{70 \text{ ml}}{\text{beat}}$$

$$= \frac{5250 \text{ ml}}{\text{min}} \left(\frac{5.25 \text{ L}}{\text{min}} \right)$$

The normal adult blood volume is about 5 L (a little more than 1 gallon). Thus, the entire blood supply passes through each side of the heart once each minute. Notice that cardiac output varies directly with SV and HR. Thus CO increases when the stroke volume increases or the heart beats faster or both, and decreases when either or both of these factors decrease.

Cardiac output is highly variable and increases markedly in response to special demands, such as running to catch a bus. The difference between resting and maximal CO is referred to as **cardiac reserve**. In nonathletic people, cardiac reserve is typically four to five times resting CO (20–25 L/min), but in trained athletes, CO during competition may reach 35 L/min (a seven-times increase). To understand how the heart accomplishes such tremendous increases in output, let's look at how stroke volume and heart rate are regulated.

Regulation of Stroke Volume

Essentially, SV represents the difference between **end diastolic volume (EDV)**, the amount of blood that collects in a ventricle during diastole, and **end systolic volume (ESV)**, the volume of blood remaining in a ventricle *after* it has contracted. The EDV, determined by how long ventricular diastole lasts and by venous pressure (an increase in either factor *raises* EDV), is normally about 120 ml. The ESV,

What problem might ensue relative to cardiac output if the heart beats far too rapidly for an extended period, that is, if tachycardia occurs? Why?

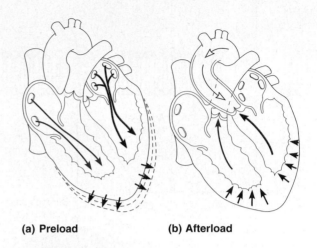

(a) Preload   **(b) Afterload**

FIGURE 17.21 Preload and afterload influence stroke volume. (a) Preload is related to the amount of blood stretching the ventricular fibers just before systole. **(b)** Afterload is the pressure that the ventricles must overcome to force open the aortic and pulmonary valves. Black arrows indicate the direction the ventricular walls are moving.

determined by arterial blood pressure (the higher the pressure the higher the ESV) and the force of ventricular contraction, is approximately 50 ml. Thus, to figure normal stroke volume, these values are simply plugged into this equation:

$$SV = EDV - ESV = \frac{120 \text{ ml}}{\text{beat}} - \frac{50 \text{ ml}}{\text{beat}} = \frac{70 \text{ ml}}{\text{beat}}$$

Hence, with each beat, each ventricle pumps out about 70 ml of blood, which is about 60% of the blood in its chambers.

So what is important here—how do we make sense out of this alphabet soup (SV, ESV, EDV)? Although many factors affect SV by causing changes in EDV or ESV, the three most important are *preload*, *contractility*, and *afterload*. As described in detail next, preload affects EDV, whereas contractility and afterload affect the ESV.

Preload: Degree of Stretch of Heart Muscle

According to the **Frank-Starling law of the heart**, the critical factor controlling SV is the **preload**, which is the degree to which cardiac muscle cells are stretched just before they contract (Figure 17.21a). Recall that stretching muscle fibers (and sarcomeres) (1) increases the number of active cross bridge attachments possible between actin and myosin, and (2) increases the force of contraction within

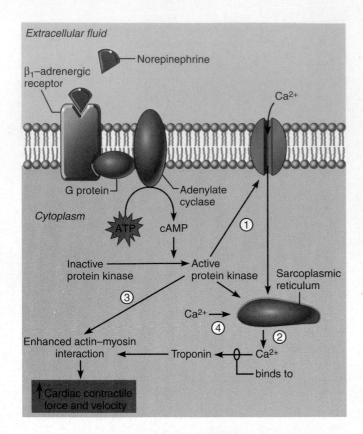

FIGURE 17.22 Mechanism by which norepinephrine influences heart contractility. Cyclic AMP activation of protein kinases phosphorylates ① slow calcium channels, promoting entry of more Ca^{2+} from the extracellular space; ② an SR protein that causes the SR to release more Ca^{2+}; and ③ myosin, which increases the rate of myosin cross bridge cycling. ④ In addition, phosphorylation of the SR calcium uptake pump removes Ca^{2+} more rapidly from the sarcoplasm, thus speeding relaxation.

physiological limits. Cardiac muscle, like skeletal muscle, exhibits a *length-tension relationship.* Resting skeletal muscle fibers are kept at or near optimal length for developing maximal tension while resting cardiac cells are normally *shorter* than optimal length. Therefore, stretching cardiac cells can produce dramatic increases in contractile force. The most important factor stretching cardiac muscle is the amount of blood returning to the heart and distending its ventricles, called **venous return.**

Anything that increases the volume or speed of venous return, such as a slow heart rate or exercise, increases EDV and, consequently, SV and contraction force. A slow heartbeat allows more time for ventricular filling. Exercise speeds venous return because of an increased heart rate and the squeezing action of the skeletal muscles on the veins returning blood to the heart. During vigorous exercise, SV may double as a result of these factors. Conversely, low venous return, such as might result from severe

blood loss or an extremely rapid heart rate, decreases EDV, causing the heart to beat less forcefully and the SV to drop (see Figure 17.23 on p. 606).

Because the systemic and pulmonary circulations are in series, the intrinsic mechanism just described ensures equal outputs of the two ventricles and proper distribution of blood volume between the two circuits. If one side of the heart suddenly begins to pump more blood than the other, the increased venous return to the opposite ventricle forces that ventricle to pump out an equal volume, thus preventing backup or accumulation of blood in the circulation.

Contractility   Although EDV is the major *intrinsic factor* influencing SV, the latter can also be enhanced by *extrinsic factors* that increase heart muscle contractility. **Contractility** is defined as an increase in contractile strength that is independent of muscle stretch and EDV. The more vigorous contractions are a direct consequence of a greater Ca^{2+} influx into the cytoplasm from the extracellular fluid and the SR. Enhanced contractility results in ejection of more blood from the heart, hence a lower ESV and greater SV. This is the result of increased sympathetic stimulation of the heart. As noted on page 598, sympathetic fibers serve not only the intrinsic conduction system but the entire heart, and one effect of norepinephrine or epinephrine binding is to initiate a cyclic AMP second-messenger system that increases Ca^{2+} entry, which in turn promotes more cross bridge binding and enhances ventricular contractility (Figure 17.22).

A battery of other chemicals also influence contractility. For example, contractility is enhanced by Ca^{2+}, by the hormones glucagon, thyroxine, and epinephrine, and by the drug digitalis. Factors that increase contractility are called *positive inotropic agents* (ino = muscle, fiber). Those that impair or decrease contractility are called *negative inotropic agents*, and include acidosis (excess H^+), rising extracellular K^+ levels, and drugs called calcium channel blockers.

Afterload: Back Pressure Exerted by Arterial Blood   The pressure that must be overcome for the ventricles to eject blood is called **afterload.** It is essentially the back pressure exerted on the aortic and pulmonary valves by arterial blood, about 80 mm Hg in the aorta and 8 mm Hg in the pulmonary trunk (Figure 17.21b). In healthy individuals, afterload is not a major determinant of stroke volume because it is relatively constant. However, in those with hypertension (high blood pressure), it is important indeed because it reduces the ability of the ventricles to eject blood. Consequently, more blood remains in the heart after systole, resulting in increased ESV and reduced stroke volume.

Regulation of Heart Rate

Given a healthy cardiovascular system, SV tends to be relatively constant. However, when blood volume drops sharply or when the heart is seriously weakened, SV declines and CO is maintained by increasing HR and contractility. Temporary stressors can also influence HR—and consequently CO—by acting through homeostatic mechanisms induced neurally, chemically, and physically. Factors that increase HR are called *positive chronotropic* (*chrono* = time) factors; those decreasing HR are *negative chronotropic* factors.

Autonomic Nervous System Regulation
The most important extrinsic controls affecting heart rate are exerted by the autonomic nervous system. When the sympathetic nervous system is activated by emotional or physical stressors, such as fright, anxiety, or exercise (Figure 17.23), sympathetic nerve fibers release norepinephrine at their cardiac synapses. Norepinephrine binds to β_1 adrenergic receptors in the heart, causing threshold to be reached more quickly and accelerating relaxation. As a result, the pacemaker fires more rapidly and the heart responds by beating faster. Sympathetic stimulation also enhances contractility by enhancing Ca^{2+} entry into the contractile cells as described above and in Figure 17.22. Because ESV falls as a result of this increased contractility, SV does not decline, as it would if only heart *rate* were increased. (Remember, when the heart beats faster, there is less time for ventricular filling and so a lower EDV.)

The parasympathetic division opposes sympathetic effects and effectively reduces heart rate when a stressful situation has passed. However, the parasympathetic division may be persistently activated in certain emotional conditions, such as grief and severe depression. Parasympathetic-initiated cardiac responses are mediated by acetylcholine, which hyperpolarizes the membranes of its effector cells by *opening* K^+ channels. Because vagal innervation of the ventricles is sparse, parasympathetic activity has little or no effect on cardiac contractility.

Under resting conditions, both autonomic divisions continuously send impulses to the SA node of the heart, but the *dominant* influence is inhibitory. Thus, the heart is said to exhibit **vagal tone,** and heart rate is generally slower than it would be if the vagal nerves were not innervating it. Cutting the vagal nerves results in an almost immediate increase in heart rate of about 25 beats/min, reflecting the inherent rate (100 beats/min) of the pacemaking SA node.

When either division of the autonomic nervous system is stimulated more strongly by sensory inputs relayed from various parts of the cardiovascular system, the other division is temporarily inhibited. Most such sensory input is generated by *baroreceptors* which respond to changes in systemic blood pressure. For example, the **atrial (Bainbridge) reflex** is a sympathetic reflex initiated by increased venous return and blood congestion in the atria. Stretching the atrial walls increases heart rate and force by stimulating both the SA node and the atrial baroreceptors, which trigger reflexive adjustments that result in increased sympathetic stimulation of the heart.

Because increased CO results in increased systemic blood pressure and vice versa, blood pressure regulation often involves reflexive controls of heart rate. Neural mechanisms that regulate blood pressure are described in more detail in Chapter 18.

Chemical Regulation
Chemicals normally present in the blood and other body fluids may influence heart rate (Figure 17.23), particularly if they become excessive or deficient.

1. Hormones. *Epinephrine,* liberated by the adrenal medulla during sympathetic nervous system activation, produces the same cardiac effects as does norepinephrine released by the sympathetic nerves; that is, it enhances heart rate and contractility.

Thyroxine is a thyroid gland hormone that increases metabolic rate and body heat production. When released in large quantities, it causes a slower but more sustained increase in heart rate than that caused by epinephrine. Because thyroxine also *enhances* the effect of epinephrine and norepinephrine on the heart, chronically hyperthyroid individuals may develop a weakened heart.

2. Ions. Physiological relationships between intracellular and extracellular ions must be maintained for normal heart function. Plasma electrolyte imbalances pose real dangers to the heart.

HOMEOSTATIC IMBALANCE

Reduced Ca^{2+} blood levels *(hypocalcemia)* depress the heart. Conversely, above-normal levels *(hypercalcemia)* tightly couple the excitation-contraction mechanism and prolong the plateau phase of the action potential. These events dramatically increase heart irritability, which can lead to spastic heart contractions that permit the heart little rest. Many drugs used to treat heart conditions focus on calcium transport into cardiac cells.

Excesses of Na^+ and K^+ are equally dangerous. Too much Na^+ *(hypernatremia)* inhibits transport of Ca^{2+} into the cardiac cells, thus blocking heart contraction. Excessive K^+ *(hyperkalemia)* interferes with depolarization by lowering the resting potential, and may lead to heart block and cardiac arrest.

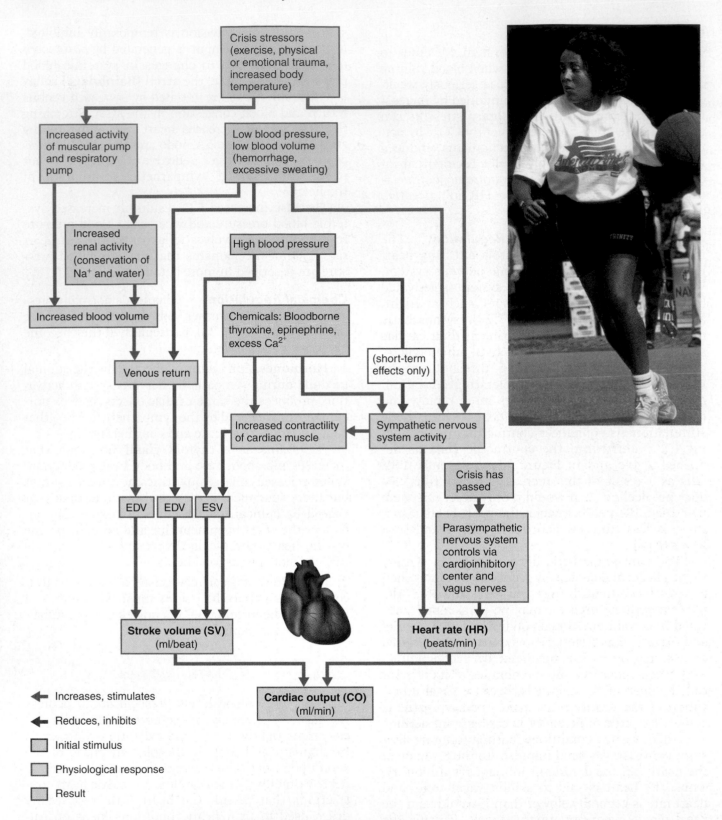

FIGURE 17.23   Factors involved in regulation of cardiac output.

Hypokalemia is also life threatening, in that the heart beats feebly and arrhythmically. ●

Other Factors Although less important than neural factors, age, gender, exercise, and body temperature also influence HR. Resting heart rate is fastest in the fetus (between 140 and 160 beats/min) and gradually declines throughout life. Average heart rate is faster in females (72 to 80 beats/min) than in males (64 to 72 beats/min).

Exercise raises HR by acting through the sympathetic nervous system (Figure 17.23). Exercise also increases systemic blood pressure and routes more blood to the working muscles. However, resting HR in the physically fit tends to be substantially lower than in those who are out of condition, and in trained athletes, may be as slow as 40 beats/min. This apparent paradox is explained below.

Heat increases HR by enhancing the metabolic rate of cardiac cells. This explains the rapid, pounding heartbeat you feel when you have a high fever and also accounts, in part, for the effect of exercise on HR (remember, working muscles generate heat). Cold directly decreases heart rate.

⊞ HOMEOSTATIC IMBALANCE

Although HR varies with changes in activity, marked and persistent rate changes usually signal cardiovascular disease. **Tachycardia** (tak″e-kar′de-ah; "heart hurry") is an abnormally fast heart rate (more than 100 beats per minute) that may result from elevated body temperature, stress, certain drugs, or heart disease. Because tachycardia occasionally promotes fibrillation, persistent tachycardia is considered pathological.

Bradycardia (brad″e-kar′de-ah; *brady* = slow) is a heart rate slower than 60 beats/min. It may result from low body temperature, certain drugs, or parasympathetic nervous activation. It is a known, and desirable, consequence of endurance training. As physical and cardiovascular conditioning increases, the heart hypertrophies and SV increases; thus, resting heart rate can be lower and still provide the same cardiac output. However, persistent bradycardia in poorly conditioned people may result in grossly inadequate blood circulation to body tissues, and bradycardia is a frequent warning of brain edema after head trauma. ●

Homeostatic Imbalance of Cardiac Output

The heart's pumping action ordinarily maintains a balance between cardiac output and venous return. Were this not so, a dangerous damming up of blood (blood congestion) would occur in the veins returning blood to the heart.

When the pumping efficiency (CO) of the heart is so low that blood circulation is inadequate to meet tissue needs, the heart is said to be in **congestive heart failure (CHF)**. This progressively worsening condition reflects weakening of the myocardium by various conditions which damage it in different ways. Let's take a look.

1. Coronary atherosclerosis. This condition, essentially a clogging of the coronary vessels with fatty buildup, impairs blood and oxygen delivery to the cardiac cells. The heart becomes increasingly hypoxic and begins to contract ineffectively.

2. Persistent high blood pressure. Normally, pressure in the aorta during diastole is 80 mm Hg, and the left ventricle exerts only slightly over that amount of force to eject blood from its chamber. When aortic diastolic blood pressure rises to 90 mm Hg or more, the myocardium must exert more force to open the aortic valve and pump out the same amount of blood. If this situation of enhanced afterload becomes chronic, the ESV rises and the myocardium hypertrophies. Eventually, the stress takes its toll and the myocardium becomes progressively weaker.

3. Multiple myocardial infarcts. A succession of MIs depress pumping efficiency because the dead heart cells are replaced by noncontractile fibrous (scar) tissue.

4. Dilated cardiomyopathy (DCM) (kar″de-o-my-ah′path-e). The cause of this condition, in which the ventricles stretch and become flabby and the myocardium deteriorates, is often unknown. Drug toxicity (alcohol, cocaine, excess catecholamines, chemotherapeutic agents), hyperthyroidism, and inflammation of the heart are implicated in some cases, as is congestive heart failure. The heart's attempts to work harder result in increasing levels of Ca^{2+} in the cardiac cells. The Ca^{2+} activates calcineurin, a calcium-sensitive enzyme that initiates a cascade which switches on genes that cause heart enlargement. Because ventricular contractility is impaired, CO is poor and the condition progressively worsens.

Because the heart is a double pump, each side can initially fail independently of the other. If the left side fails, **pulmonary congestion** occurs. The right side continues to propel blood to the lungs, but the left side does not adequately eject the returning blood into the systemic circulation. Blood vessels in the lungs become engorged with blood, the pressure in them increases, and fluid leaks from the circulation into the lung tissue, causing pulmonary edema. If the congestion is untreated, the person suffocates.

If the right side of the heart fails, **peripheral congestion** occurs. Blood stagnates in body organs, and pooled fluids in the tissue spaces impair the ability of body cells to obtain adequate amounts of nutrients and oxygen and to rid themselves of wastes. The resulting edema is most noticeable in the extremities (feet, ankles, and fingers).

Failure of one side of the heart puts a greater strain on the other side, and ultimately the whole heart fails. A seriously weakened, or *decompensated*, heart is irreparable. Treatment is directed primarily toward (1) conserving heart energy with digitalis derivatives (which reduce HR), (2) removing the excess leaked fluid with *diuretics* (drugs that increase excretion of Na^+ and water by the kidneys), and (3) reducing afterload with drugs that drive down blood pressure. However, heart transplants and other surgical or mechanical remedies to replace damaged heart muscle are providing additional hope for some cardiac patients.

Review Questions

Multiple Choice/Matching

(Some questions have more than one correct answer. Select the best answer or answers from the choices given.)

1. When the semilunar valves are open, which of the following are occurring? (a) 2, 3, 5, 6, (b) 1, 2, 3, 7, (c) 1, 3, 5, 6, (d) 2, 4, 5, 7.

(1) coronary arteries fill
(2) AV valves are closed
(3) ventricles are in systole
(4) ventricles are in diastole
(5) blood enters aorta
(6) blood enters pulmonary arteries
(7) atria contract

2. The portion of the intrinsic conduction system located in the interventricular septum is the (a) AV node, (b) SA node, (c) AV bundle, (d) Purkinje fibers.

3. An ECG provides information about (a) cardiac output, (b) movement of the excitation wave across the heart, (c) coronary circulation, (d) valve impairment.

4. The sequence of contraction of the heart chambers is (a) random, (b) left chambers followed by right chambers, (c) both atria followed by both ventricles, (d) right atrium, right ventricle, left atrium, left ventricle.

5. The fact that the left ventricular wall is thicker than the right reveals that it (a) pumps a greater volume of blood, (b) pumps blood against greater resistance, (c) expands the thoracic cage, (d) pumps blood through a smaller valve.

6. The chordae tendineae (a) close the atrioventricular valves, (b) prevent the AV valve flaps from everting, (c) contract the papillary muscles, (d) open the semilunar valves.

7. In the heart, which of the following apply? (1) Action potentials are conducted from cell to cell across the myocardium via gap junctions, (2) the SA node sets the pace for the heart as a whole, (3) spontaneous depolarization of cardiac cells can occur in the absence of nerve stimulation, (4) cardiac muscle can continue to contract for long periods in the absence of oxygen. (a) all of the above, (b) 1, 3, 4, (c) 1, 2, 3, (d) 2, 3.

8. The activity of the heart depends on intrinsic properties of cardiac muscle and on neural factors. Thus, (a) vagus nerve stimulation of the heart reduces heart rate, (b) sympathetic nerve stimulation of the heart decreases time available for ventricular filling, (c) sympathetic stimulation of the heart increases its force of contraction, (d) all of the above.

9. Freshly oxygenated blood is first received by the (a) right atrium, (b) left atrium, (c) right ventricle, (d) left ventricle.

Short Answer Essay Questions

10. Describe the location and position of the heart in the body.

11. Describe the pericardium and distinguish between the fibrous and the serous pericardia relative to histological structure and location.

12. Trace one drop of blood from the time it enters the right atrium until it enters the left atrium. What is this circuit called?

13. (a) Describe how heart contraction and relaxation influence coronary blood flow. (b) Name the major branches of the coronary arteries, and note the heart regions served by each.

14. The refractory period of cardiac muscle is much longer than that of skeletal muscle. Why is this a desirable functional property?

15. (a) Name the elements of the intrinsic conduction system of the heart in order, beginning with the pacemaker. (b) What is the important function of this conduction system?

16. Draw a normal ECG pattern. Label and explain the significance of its deflection waves.

17. Define cardiac cycle, and follow the events of one cycle.

18. What is cardiac output, and how is it calculated?

19. Discuss how the Frank-Starling law of the heart helps to explain the influence of venous return on stroke volume.

18

THE CARDIOVASCULAR SYSTEM: BLOOD VESSELS

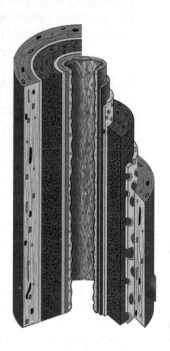

PART 1: OVERVIEW OF BLOOD VESSEL STRUCTURE AND FUNCTION
(pp. 610–617)

1. Describe the three layers that typically form the wall of a blood vessel, and state the function of each.

2. Define vasoconstriction and vasodilation.

3. Compare and contrast the structure and function of the three types of arteries.

4. Describe the structure and function of veins, and explain how veins differ from arteries.

5. Describe the structure and function of a capillary bed.

PART 2: PHYSIOLOGY OF CIRCULATION
(pp. 617–635)

6. Define blood flow, blood pressure, and resistance, and explain the relationships between these factors.

7. List and explain the factors that influence blood pressure, and describe how blood pressure is regulated.

8. Define hypertension. Describe its symptoms and consequences.

9. Explain how blood flow is regulated in the body in general and in its specific organs.

10. Outline factors involved in capillary dynamics, and explain the significance of each.

11. Define circulatory shock. List several possible causes.

PART 3: CIRCULATORY PATHWAYS: BLOOD VESSELS OF THE BODY
(pp. 635–659)

12. Trace the pathway of blood through the pulmonary circuit, and state the importance of this special circulation.

13. Describe the general functions of the systemic circuit. Name and give the location of the major arteries and veins in the systemic circulation.

14. Describe the structure and special function of the hepatic portal system.

Although blood vessels are sometimes compared to a system of pipes within which the blood circulates, this analogy is only a starting point. Unlike rigid pipes, blood vessels are dynamic structures that pulsate, constrict, relax, and even proliferate. This chapter examines the structure and function of these important circulatory passageways.

The **blood vessels** of the body form a closed delivery system that begins and ends at the heart. The idea that blood circulates in the body dates back to the 1620s with the inspired experiments of William Harvey, an English physician. Prior to that time, it was thought, as proposed by the ancient Greek physician Galen, that blood moved through the body like an ocean tide, first moving out from the heart and then ebbing back in the same vessels.

PART 1: OVERVIEW OF BLOOD VESSEL STRUCTURE AND FUNCTION

The three major types of blood vessels are *arteries*, *capillaries*, and *veins*. As the heart contracts, it forces blood into the large arteries leaving the ventricles. The blood then moves into successively smaller arteries, finally reaching their smallest branches, the *arterioles* (ar-te're-ōlz; "little arteries"), which feed into the capillary beds of body organs and tissues. Blood drains from the capillaries into *venules* (ven-ulz), the smallest veins, and then on into larger and larger veins that merge to form the large veins that ultimately empty into the heart. Altogether, the blood vessels in the adult human stretch for about 100,000 km (60,000 miles) through the internal body landscape!

Because **arteries** carry blood *away from* the heart, they are said to "branch," "diverge," or "fork" as they form smaller and smaller divisions. **Veins,** by contrast, carry blood *toward* the heart and so are said to "join," "merge," and "converge" into the successively larger vessels approaching the heart. In the systemic circulation, arteries always carry oxygenated blood and veins always carry oxygen-poor blood. The opposite is true in the pulmonary circulation, where the arteries, still defined as the vessels leading away from the heart, carry oxygen-poor blood to the lungs, and the veins carry oxygen-rich blood from the lungs to the heart. The special umbilical vessels of a fetus also differ in the roles of veins and arteries.

Of all the blood vessels, only the capillaries have intimate contact with tissue cells and directly serve cellular needs. Exchanges between the blood and tissue cells occur primarily through the gossamer-thin capillary walls.

Structure of Blood Vessel Walls

The walls of all blood vessels, except the very smallest, have three distinct layers, or *tunics* ("cloaks"), that surround a central blood-containing space, the vessel **lumen** (Figure 18.1).

The innermost tunic is the **tunica interna,** or **tunica intima** (in'tǐ-mah). The second name is easy to remember once you know that this tunic is in *intimate* contact with the blood in the lumen. This tunic contains the *endothelium*, the simple squamous epithelium that lines the lumen of all vessels. The endothelium is a continuation of the endocardial lining of the heart, and its flat cells fit closely together, forming a slick surface that minimizes friction as blood moves through the lumen. In vessels larger than 1 mm in diameter, a *subendothelial layer* of loose connective tissue (a basement membrane) supports the endothelium.

The middle tunic, the **tunica media** (me'de-ah), is mostly circularly arranged smooth muscle cells and sheets of elastin. The activity of the smooth muscle is regulated by sympathetic *vasomotor nerve fibers* of the autonomic nervous system and a whole battery of chemicals. Depending on the body's needs at any given moment, either **vasoconstriction** (reduction in lumen diameter as the smooth muscle contracts) or **vasodilation** (increase in lumen diameter as the smooth muscle relaxes) can be effected. Because small changes in vessel diameter greatly influence blood flow and blood pressure, the activities of the tunica media are critical in regulating circulatory dynamics. Generally, the tunica media is the bulkiest layer in arteries, which bear the chief responsibility for maintaining blood pressure and continuous blood circulation.

The outermost layer of a blood vessel wall, the **tunica externa,** or **tunica adventitia** (ad"ven-tish'e-ah; "coming from outside"), is composed largely of loosely woven collagen fibers that protect and reinforce the vessel, and anchor it to surrounding structures. The tunica externa is infiltrated with nerve fibers, lymphatic vessels, and, in larger veins, a network of elastin fibers. In larger vessels, the tunica externa contains a system of tiny blood vessels, the **vasa vasorum** (va'sah va-sor'um)—literally, "vessels of the vessels" that nourish the more external tissues of the blood vessel wall. The innermost or luminal portion of the vessel obtains its nutrients directly from blood in the lumen.

The three vessel types vary in length, diameter, relative wall thickness, and tissue makeup (Table 18.1). The relationship of the various blood vascular channels to one another and to vessels of the lymphatic system (which recovers fluids that leak from the circulation) is summarized in Figure 18.2.

What is the functional value of having capillaries that consist of a single layer of squamous cells underlain by sparse connective tissue?

FIGURE 18.1 **Generalized structure of arteries, veins, and capillaries. (a)** Light photomicrograph of a muscular artery and the corresponding vein in cross section (30×). **(b)** The walls of arteries and veins are composed of the tunica interna (endothelium underlain by a layer of loose connective tissue), the tunica media (smooth muscle cells and elastic fibers), and the tunica externa (largely collagen fibers). Capillaries are composed of only the endothelium and a sparse basal lamina. The tunica media is thick in arteries and thin in veins, while the tunica externa is thin in arteries and relatively thicker in veins.

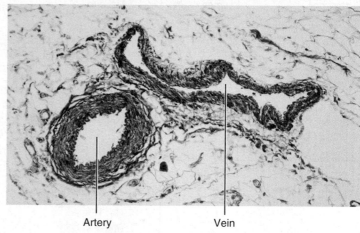

Artery   Vein

(a)

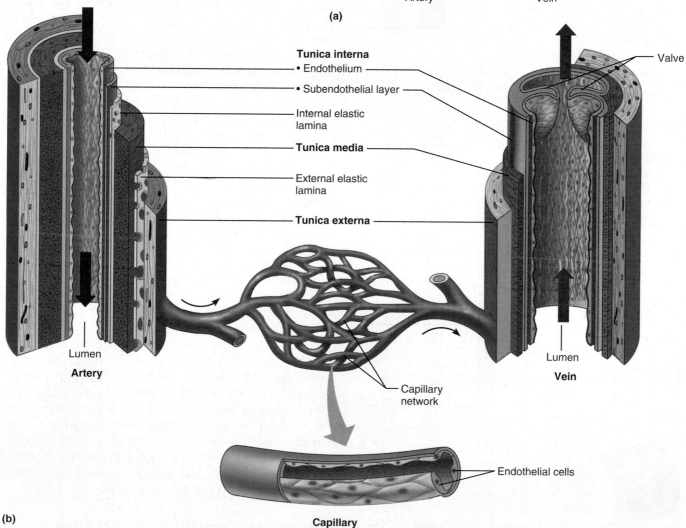

Tunica interna
• Endothelium
• Subendothelial layer

Internal elastic lamina

Tunica media

External elastic lamina

Tunica externa

Lumen

Artery

Capillary network

Lumen

Vein

Valve

Endothelial cells

Capillary

(b)

TABLE 18.1	Summary of Blood Vessel Anatomy	

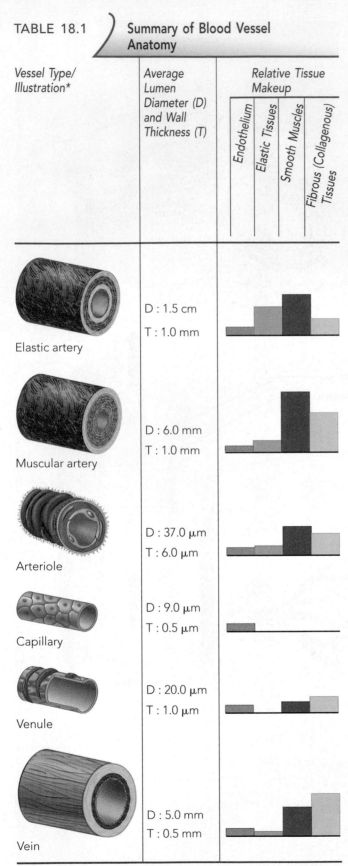

Vessel Type/ Illustration*	Average Lumen Diameter (D) and Wall Thickness (T)	Relative Tissue Makeup (Endothelium, Elastic Tissues, Smooth Muscles, Fibrous (Collagenous) Tissues)
Elastic artery	D : 1.5 cm T : 1.0 mm	
Muscular artery	D : 6.0 mm T : 1.0 mm	
Arteriole	D : 37.0 μm T : 6.0 μm	
Capillary	D : 9.0 μm T : 0.5 μm	
Venule	D : 20.0 μm T : 1.0 μm	
Vein	D : 5.0 mm T : 0.5 mm	

(*Size relationships are not proportional. Smaller vessels are drawn relatively larger so detail can be seen. See column 2 for actual dimensions.)

Arterial System

In terms of relative size and function, arteries can be divided into three groups—elastic arteries, muscular arteries, and arterioles.

Elastic (Conducting) Arteries

Elastic arteries are the thick-walled arteries near the heart—the aorta and its major branches. These arteries are the largest in diameter, ranging from 2.5 cm to 1 cm, and the most elastic (Table 18.1). Because their large lumen makes them low-resistance pathways that conduct blood from the heart to medium-sized arteries, elastic arteries are sometimes referred to as **conducting arteries** (Figure 18.2). They contain more elastin than any other vessel type. It is present in all three tunics, but the tunica media contains the most. There the elastin constructs concentric "holey" laminae (sheets) of elastic connective tissue that look like slices of Swiss cheese interspersed between the layers of smooth muscle cells. The abundant elastin enables these arteries to withstand and smooth out large pressure fluctuations by expanding when the heart forces blood into them, and then recoiling to propel blood onward into the circulation when the heart relaxes. Elastic arteries also contain substantial amounts of smooth muscle, but they are relatively inactive in vasoconstriction. Thus, in terms of function, they can be visualized as simple elastic tubes.

The elastic arteries expand and recoil passively to accommodate changes in blood volume. Consequently, blood flows fairly continuously rather than starting and stopping with the pulsating rhythm of the heartbeat. If the blood vessels become hard and unyielding, as in arteriosclerosis, blood flows more intermittently, similar to the way water flows through a hard rubber garden hose attached to a faucet. When the faucet is on, the high pressure makes the water gush out of the hose. But when the faucet is shut off, the water flow abruptly becomes a trickle and then stops, because the hose walls cannot recoil to keep the water under pressure. Also, without the pressure-smoothing effect of the elastic arteries, the walls of arteries throughout the body experience higher pressures. Battered by high pressures, the arteries eventually weaken and may balloon out or even burst.

Muscular (Distributing) Arteries

Distally the elastic arteries give way to the **muscular,** or **distributing, arteries,** which deliver blood to specific body organs and account for most of the named arteries studied in the anatomy laboratory. Their internal diameter ranges from that of a little finger

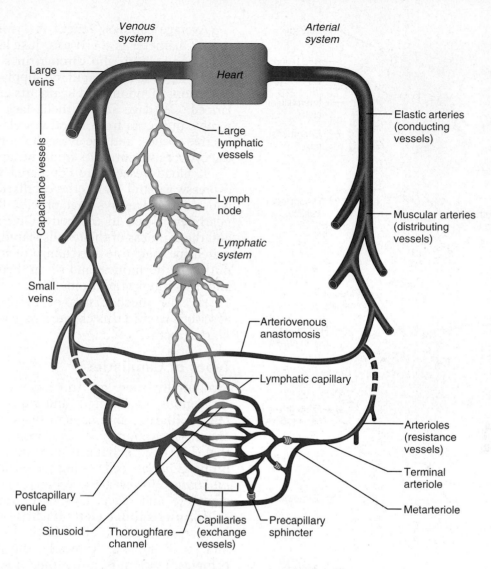

FIGURE 18.2 **Relationship of vessels of the blood vascular system to one another and to lymphatic vessels.**

(1 cm) to that of a pencil lead (about 0.3 mm). Proportionately, they have the thickest media of all vessels. Their tunica media contains relatively more smooth muscle and less elastic tissue than do elastic arteries (Table 18.1); therefore, they are more active in vasoconstriction and less distensible. In muscular arteries, however, there *is* an *elastic lamina* on each face of the tunica media.

Arterioles

The smallest of the arteries, **arterioles** have a lumen diameter ranging from 0.3 mm down to 10 μm. Larger arterioles have all three tunics, but their tunica media is chiefly smooth muscle with a few scattered elastic fibers. Smaller arterioles, which lead into the capillary beds, are little more than a single layer of smooth muscle cells spiraling around the endothelial lining.

As described shortly, minute-to-minute blood flow into the capillary beds is determined by arteriole diameter, which varies in response to changing neural stimuli and local chemical influences. When arterioles constrict, the tissues served are largely bypassed. When arterioles dilate, blood flow into the local capillaries increases dramatically.

Capillaries

The microscopic **capillaries** are the smallest blood vessels. Their exceedingly thin walls consist of just a thin tunica interna (see Figure 18.1b). In some cases, one endothelial cell forms the entire circumference of the capillary wall. Here and there along the outer surface of the capillaries are spider-shaped *pericytes*, smooth muscle-like cells that stabilize the capillary wall.

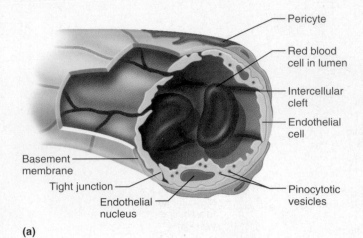

(a)

Pericyte
Red blood cell in lumen
Intercellular cleft
Endothelial cell
Pinocytotic vesicles
Basement membrane
Tight junction
Endothelial nucleus

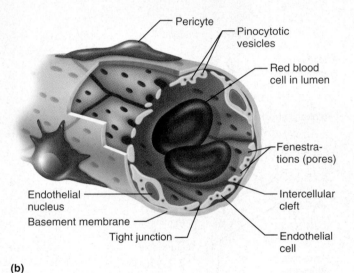

(b)

Pericyte
Pinocytotic vesicles
Red blood cell in lumen
Fenestrations (pores)
Intercellular cleft
Endothelial nucleus
Basement membrane
Tight junction
Endothelial cell

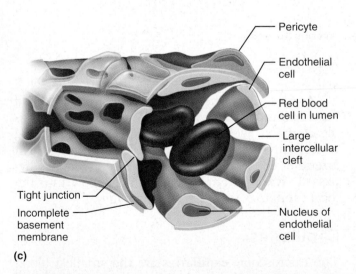

(c)

Pericyte
Endothelial cell
Red blood cell in lumen
Large intercellular cleft
Nucleus of endothelial cell
Tight junction
Incomplete basement membrane

FIGURE 18.3   Capillary structure. Three-dimensional views of **(a)** continuous capillary, **(b)** fenestrated capillary, **(c)** discontinuous sinusoidal capillary. [Note: The basement membrane is incomplete only in (c).]

Average capillary length is 1 mm and average lumen diameter is 8–10 μm, just large enough for red blood cells to slip through in single file. Most tissues have a rich capillary supply, but there are exceptions. Tendons and ligaments are poorly vascularized. Cartilage and epithelia lack capillaries, but receive nutrients from blood vessels in nearby connective tissues, and the avascular cornea and lens of the eye receive nutrients from the aqueous humor.

If blood vessels are compared to a system of expressways and roads, the capillaries are the back alleys and driveways that provide direct access to nearly every cell in the body. Given their location and the thinness of their walls, capillaries are ideally suited for their role—exchange of materials (gases, nutrients, hormones, and so on) between the blood and the interstitial fluid (Figure 18.2 and Table 18.1). The mechanisms of these exchanges are described later in this chapter; here we focus on capillary structure.

Types of Capillaries

Structurally, there are three types of capillaries—*continuous, fenestrated,* and *sinusoidal.* **Continuous capillaries,** abundant in the skin and muscles, are most common. They are continuous in the sense that their endothelial cells provide an uninterrupted lining, adjacent cells being joined laterally by *tight junctions.* However, these junctions are usually incomplete and leave gaps of unjoined membrane called **intercellular clefts** (Figure 18.3a), which are just large enough to allow limited passage of fluids and small solutes. Typically, the endothelial cell cytoplasm contains numerous pinocytotic vesicles believed to ferry fluids across the capillary wall. However, brain capillaries are unique. There the tight junctions of the continuous capillaries are complete and extend around the entire perimeter of the endothelial cells, constituting the structural basis of the *blood-brain barrier,* described in Chapter 12.

Fenestrated capillaries (fen′es-tra-tid) are similar to the continuous variety except that some of the endothelial cells in fenestrated capillaries are riddled with oval *pores,* or *fenestrations* (*fenestra* = window) (Figure 18.3b). The fenestrations are usually covered by a delicate membrane, or diaphragm (probably condensed basal lamina material), but even so, fenestrated capillaries are much more permeable to fluids and small solutes than continuous capillaries are. Fenestrated capillaries are found wherever active capillary absorption or filtrate formation occurs. For example, fenestrated capillaries in the small intestine receive nutrients from digested food, and those in endocrine organs allow hormones rapid entry into the blood. Fenestrated capillaries with perpetually open pores occur in the kidneys, where rapid filtration of blood plasma is essential.

Sinusoids (si'nŭ-soyds), or **sinusoidal capillaries,** are highly modified, leaky capillaries found only in the liver, bone marrow, lymphoid tissues, and some endocrine organs. Sinusoids have large, irregularly shaped lumens and are usually fenestrated. Their endothelial lining has fewer tight junctions and larger intercellular clefts than ordinary capillaries (Figure 18.3c). These structural adaptations allow large molecules and even blood cells to pass between the blood and surrounding tissues. In the liver, the endothelium of the sinusoids is *discontinuous* and large macrophages called **Kupffer cells,** which remove and destroy any contained bacteria, form part of the lining. In other organs, such as the spleen, phagocytes located just outside the sinusoids stick cytoplasmic extensions through the intercellular clefts into the sinusoidal lumen to get at their "prey." Blood flows sluggishly through the tortuous sinusoid channels, allowing time for it to be modified in various ways.

Capillary Beds

Capillaries do not function independently. Instead they tend to form interweaving networks called **capillary beds.** The flow of blood from an arteriole to a venule—that is, through a capillary bed—is called the **microcirculation.** In most body regions, a capillary bed consists of two types of vessels: (1) a *vascular shunt* (metarteriole–thoroughfare channel), a short vessel that directly connects the arteriole and venule at opposite ends of the bed, and (2) *true capillaries*, the actual *exchange vessels* (Figure 18.4). The **terminal arteriole** feeding the bed leads into a **metarteriole** (a vessel structurally intermediate between an arteriole and a capillary), which is continuous with the **thoroughfare channel** (intermediate between a capillary and a venule). The thoroughfare channel, in turn, joins the **postcapillary venule** that drains the bed.

The **true capillaries** number 10 to 100 per capillary bed, depending on the organ or tissues served. They usually branch off the metarteriole (proximal end of the shunt) and return to the thoroughfare channel (the distal end), but occasionally they spring from the terminal arteriole and empty directly into the venule. A cuff of smooth muscle fibers, called a **precapillary sphincter,** surrounds the root of each true capillary at the metarteriole and acts as a valve to regulate blood flow into the capillary. Blood flowing through a terminal arteriole may go either through the true capillaries or through the shunt. When the precapillary sphincters are relaxed (open), blood flows through the true capillaries and takes part in exchanges with tissue cells. When the sphincters are contracted (closed), blood flows through the shunts and bypasses the tissue cells.

Assume the capillary bed depicted is in your calf muscle. Which condition would the bed be in—(a) or (b)—if you were doing calf raises at the gym?

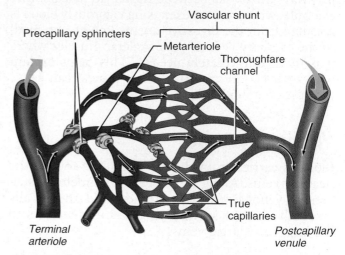

Precapillary sphincters

Vascular shunt

Metarteriole

Thoroughfare channel

True capillaries

Terminal arteriole

Postcapillary venule

(a) Sphincters open

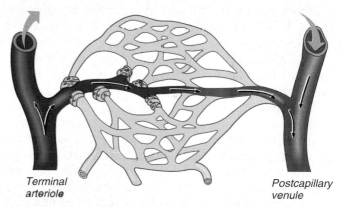

Terminal arteriole

Postcapillary venule

(b) Sphincters closed

FIGURE 18.4   Anatomy of a capillary bed. When precapillary sphincters are constricted, the metarteriole–thoroughfare channel acts as a shunt to bypass the true capillaries.

The relative amount of blood entering a capillary bed is regulated by vasomotor nerve fibers and local chemical conditions. A bed may be flooded with blood or almost completely bypassed, depending on conditions in the body or in that specific organ. For example, suppose you have just eaten and are sitting relaxed, listening to your favorite musical group. Food digestion is going on, and blood is circulating

■ *Condition (a): The true capillaries would be flushed with blood to ensure that the working calf muscles could receive the needed nutrients and dispose of their metabolic wastes.*

freely through the true capillaries of your gastrointestinal organs to receive the breakdown products of digestion. However, most of these same capillary pathways are closed between meals. To take another example, when you are exercising vigorously blood is rerouted from the digestive organs (food or no food) to the capillary beds of your skeletal muscles where it is more immediately needed. This helps explain why vigorous exercise right after a meal can cause indigestion or abdominal cramps.

Venous System

Blood is carried from the capillary beds toward the heart by veins. Along the route, the diameter of successive venous vessels increases, and their walls gradually thicken as they progress from venules to the larger and larger veins.

Venules

Ranging from 8 to 100 μm in diameter, **venules** are formed when capillaries unite. The smallest venules, the **postcapillary venules,** consist entirely of endothelium around which a few pericytes congregate. They are extremely porous (more like capillaries than veins in this way), and fluid and white blood cells move easily from the bloodstream through their walls. Indeed, a well-recognized sign of an inflamed area is adhesion of white blood cells to the postcapillary venule endothelium, followed by their migration through the wall into the inflamed tissue. The larger venules have one or two layers of smooth muscle cells (a scanty tunica media) and thin externa as well.

Veins

Venules join to form **veins.** Veins usually have three distinct tunics, but their walls are always thinner and their lumens larger than those of corresponding arteries (Figure 18.1 and Table 18.1). Consequently, in the view seen in routine histological preparations, veins are usually collapsed and their lumens appear slitlike. There is relatively little smooth muscle or elastin in the tunica media, which is poorly developed and tends to be thin even in the largest veins. The tunica externa is the heaviest wall layer. Consisting of thick longitudinal bundles of collagen fibers and elastic networks, it is often several times thicker than the tunica media. In the largest veins—the venae cavae, which return blood directly to the heart—the tunica externa is further thickened by longitudinal bands of smooth muscle.

With their large lumens and thin walls, veins can accommodate a fairly large blood volume. Because up to 65% of the body's blood supply is found in the veins at any time, veins are called **capacitance vessels** and **blood reservoirs.** Even so, they are normally only partially filled with blood.

Because the blood pressure in veins is low, the walls of these vessels can be much thinner than arterial walls without danger of bursting. However, the low-pressure condition demands some adaptations to ensure that blood is returned to the heart at the same rate it was pumped into the circulation. The large-diameter lumens of veins (which offer relatively little resistance to blood flow) are one structural adaptation. Another is valves that prevent blood from flowing backward. **Venous valves** are formed from folds of the tunica interna (Figure 18.1), and resemble the semilunar valves of the heart in both structure and function. Venous valves are most abundant in the veins of the limbs, where the upward flow of blood is opposed by gravity. They are absent in veins of the ventral body cavity.

A simple experiment demonstrates the effectiveness of venous valves. Hang one hand by your side until the blood vessels on its dorsal aspect distend with blood. Next place two fingertips against one of the distended veins, and pressing firmly, move the superior finger proximally along the vein and then release that finger. The vein will remain flattened and collapsed, despite the pull of gravity. Finally, remove the distal fingertip and watch the vein refill with blood.

🅗 *HOMEOSTATIC IMBALANCE*

Varicose veins are veins that have become tortuous and dilated because of incompetent valves. More than 15% of adults suffer from varicose veins, usually in the lower limbs. Several factors contribute, including heredity and conditions that hinder venous return, such as prolonged standing in one position, obesity, or pregnancy. Both the "potbelly" of an overweight person and the enlarged uterus of a pregnant woman exert downward pressure on vessels of the groin, and restrict return of blood to the heart. Consequently, blood pools in the lower limbs, and with time, the valves weaken and the venous walls stretch and become floppy. Superficial veins, which receive little support from surrounding tissues, are especially susceptible. Another cause of varicose veins is elevated venous pressure. For example, straining to deliver a baby or have a bowel movement raises intra-abdominal pressure, preventing blood from draining from the veins of the anal canal. The resulting varicosities in the anal veins are called *hemorrhoids* (hem′ŏ-roidz). ●

Venous sinuses, such as the *coronary sinus* of the heart and the *dural sinuses* of the brain, are highly specialized, flattened veins with extremely thin walls composed only of endothelium. They are supported by the tissues that surround them, rather than by any additional tunics. The dural sinuses, which receive cerebrospinal fluid and blood draining from the brain, are reinforced by the tough dura mater that covers the brain surface.

Vascular Anastomoses

Where vascular channels unite, they form **vascular anastomoses** (ah-nas"to-mo'sēz; "coming together"). Most organs receive blood from more than one arterial branch, and arteries supplying the same territory often merge, forming **arterial anastomoses.** These anastomoses provide alternate pathways, called **collateral channels,** for blood to reach a given body region. If one branch is cut or blocked by a clot, the collateral channel can often provide the area with an adequate blood supply. Arterial anastomoses occur around joints, where active movement may hinder blood flow through one channel. They are also common in abdominal organs, the brain, and the heart. Arteries that do not anastomose or that have a poorly developed collateral circulation supply the retina, kidneys, and spleen. If their blood flow is interrupted, cells supplied by such vessels die.

The metarteriole–thoroughfare channel shunts of capillary beds that connect arterioles and venules are examples of **arteriovenous anastomoses.** Veins interconnect much more freely than arteries, and **venous anastomoses** are common. (You may be able to see venous anastomoses through the skin on the dorsum of your hand.) Because venous anastomoses are abundant, occlusion of a vein rarely blocks blood flow or leads to tissue death.

PART 2: PHYSIOLOGY OF CIRCULATION

Have you ever climbed a mountain? Well, get ready to climb a hypothetical mountain as you learn about circulatory dynamics. Like scaling a mountain, tackling blood pressure regulation and other topics of cardiovascular physiology is challenging while you're doing it, and exhilarating when you succeed. Let's begin the climb.

To sustain life, blood must be kept circulating. By now, you are aware that the heart is the pump, the aorta is the pressure reservoir, the arteries are conduits, the arterioles are resistance vessels, the capillaries are exchange sites, and the veins are conduits and blood reservoirs. Now for the dynamics of this system.

Introduction to Blood Flow, Blood Pressure, and Resistance

First we need to define three physiologically important terms—blood flow, blood pressure, and resistance—and examine how these factors relate to the physiology of blood circulation.

Definition of Terms

1. Blood flow is the volume of blood flowing through a vessel, an organ, or the entire circulation in a given period (ml/min). If we consider the entire vascular system, blood flow is equivalent to cardiac output (CO), and under resting conditions, it is relatively constant. At any given moment, however, blood flow through *individual* body organs may vary widely and is intimately related to their immediate needs.

2. Blood pressure (BP), the force per unit area exerted on a vessel wall by the contained blood, is expressed in millimeters of mercury (mm Hg). For example, a blood pressure of 120 mm Hg is equal to the pressure exerted by a column of mercury 120 mm high. Unless stated otherwise, the term *blood pressure* means systemic arterial blood pressure in the largest arteries near the heart. It is the pressure gradient—the *differences* in blood pressure within the vascular system—that provides the driving force that keeps blood moving—always from an area of higher pressure to an area of lower pressure—through the body.

3. Resistance is opposition to flow and is a measure of the amount of friction blood encounters as it passes through the vessels. Because most friction is encountered in the peripheral (systemic) circulation, well away from the heart, we generally use the term **peripheral resistance.**

There are three important sources of resistance: blood viscosity, vessel length, and vessel diameter.

▪ **Blood viscosity.** *Viscosity* (vis-kos'ĭ-te) is the internal resistance to flow that exists in all fluids and is related to the thickness or "stickiness" of a fluid. The greater the viscosity, the less easily molecules slide past one another and the more difficult it is to get and keep the fluid moving. Because it contains formed elements and plasma proteins, blood is much more viscous than water; hence it flows more slowly under the same conditions. Blood viscosity is fairly constant, but infrequent conditions such as polycythemia (excessive numbers of red blood cells) can increase blood viscosity and, hence, resistance. On the other hand, if the red blood cell count is low, as in some anemias, blood is less viscous and peripheral resistance declines.

■ **Total blood vessel length.** The relationship between total blood vessel length and resistance is straightforward: the longer the vessel, the greater the resistance. An extra pound or two of fat requires that miles of small vessels be added to service it. This significantly increases the peripheral resistance.

■ **Blood vessel diameter.** Because blood viscosity and vessel length are normally unchanging, the influence of these factors can be considered constant in healthy people. Changes in blood vessel diameter are frequent, however, and significantly alter peripheral resistance. How so? The answer lies in principles of fluid flow. Fluid close to the wall of a tube or channel is slowed by friction as it passes along the wall, whereas fluid in the center of the tube flows more freely and faster. You can verify this by watching the flow of water in a river. Water close to the bank hardly seems to move, while that in the middle of the river flows quite rapidly.

In a tube of a given size, the relative speed and position of fluid in the different regions of the tube's cross section remain constant, a phenomenon called *laminar flow* or *streamlining*. The smaller the tube, the greater the friction, because relatively more of the fluid contacts the tube wall where its movement is impeded. Resistance varies *inversely* with the *fourth power* of the vessel radius (one-half the diameter). This means, for example, that if the radius of a vessel is doubled, the resistance drops to one-sixteenth of its original value ($r^4 = 2 \times 2 \times 2 \times 2 = 16$ and $1/r^4 = 1/16$). Thus, the large arteries close to the heart, which do not change dramatically in diameter, contribute little to peripheral resistance, and the small-diameter arterioles, which can enlarge or constrict in response to neural and chemical controls, are the major determinants of peripheral resistance.

When blood encounters either an abrupt change in the tube size or rough or protruding areas of the tube wall (such as the fatty plaques of atherosclerosis), the smooth laminar blood flow is replaced by *turbulent flow*, that is, irregular fluid motion where blood from the different laminae mixes. *Turbulence* dramatically increases resistance.

Relationship Between Flow, Pressure, and Resistance

Now that we have defined these terms, let us summarize briefly the relationships between them. Blood flow (*F*) is *directly* proportional to the difference in blood pressure (ΔP) between two points in the circulation, that is, the blood pressure, or hydrostatic pressure, gradient. Thus, when ΔP increases, blood flow speeds up, and when ΔP decreases, blood flow declines. Blood flow is *inversely* proportional to the peripheral resistance (*R*) in the systemic circulation; if *R* increases, blood flow decreases. These relationships are expressed by the formula

$$F = \frac{\Delta P}{R}$$

Of these two factors influencing blood flow, *R* is far more important than ΔP in influencing local blood flow. This is because when the arterioles serving a particular tissue dilate (thus decreasing the resistance), blood flow to that tissue increases, even though the systemic pressure is unchanged or may actually be falling.

Systemic Blood Pressure

Any fluid driven by a pump through a circuit of closed channels operates under pressure, and the nearer the fluid is to the pump, the greater the pressure exerted on the fluid. The dynamics of blood flow in blood vessels is no exception, and blood flows through the blood vessels along a pressure gradient, always moving from higher- to lower-pressure areas. Fundamentally, *the pumping action of the heart generates blood flow. Pressure results when flow is opposed by resistance.*

As illustrated in Figure 18.5, systemic blood pressure is highest in the aorta and declines throughout the pathway to finally reach 0 mm Hg in the right atrium. The steepest drop in blood pressure occurs in the arterioles, which offer the greatest resistance to blood flow. However, as long as a pressure gradient exists, no matter how small, blood continues to flow until it completes the circuit back to the heart.

Arterial Blood Pressure

Arterial blood pressure reflects two factors: (1) how much the elastic arteries close to the heart can be stretched (their *compliance* or *distensibility*) and (2) the volume of blood forced into them at any time. If the amounts of blood entering and leaving the elastic arteries in a given period were equal, arterial pressure would be constant. Instead, as Figure 18.5 reveals, blood pressure rises and falls in a regular fashion in the elastic arteries near the heart, that is, it is obviously *pulsatile*.

As the left ventricle contracts and expels blood into the aorta, it imparts kinetic energy to the blood, which stretches the elastic aorta as aortic pressure reaches its peak. Indeed, if the aorta were opened during this period, blood would spurt upward 5 or 6 feet! This pressure peak, called the **systolic pressure** (sis-tah'lik), averages 120 mm Hg in healthy adults. Blood moves forward into the arterial bed because the pressure in the aorta is higher than the pressure in the more distal vessels. During diastole, the aortic semilunar valve closes, preventing blood from flowing back into the heart, and the walls of the aorta (and other elastic arteries) recoil, maintaining adequate pressure on the reducing blood volume to keep the blood flowing forward into the smaller vessels. During this time, aortic pressure drops to its lowest level (approximately 70 to 80 mm Hg in healthy adults), called the **diastolic pressure** (di-as-tah'lik). Thus, you can picture the elastic arteries as pressure reservoirs that operate as auxiliary pumps to keep blood circulating throughout the period of diastole, when the heart is relaxing. Essentially, the volume and energy of blood stored in the elastic arteries during systole are given back during diastole.

The difference between the systolic and diastolic pressures is called the **pulse pressure.** It is felt as a throbbing pulsation in an artery (a *pulse*) during systole, as the elastic arteries are expanded by the blood being forced into them by ventricular contraction. Increased stroke volume and faster blood ejection from the heart (a result of increased contractility) cause *temporary* increases in the pulse pressure. Pulse pressure is chronically increased by arteriosclerosis because the elastic arteries become less stretchy. Because aortic pressure fluctuates up and down with each heartbeat, the important pressure figure to consider is the **mean arterial pressure (MAP)**; it is this pressure that propels the blood to the tissues. Because diastole usually lasts longer than systole, the MAP is not simply the value halfway between systolic and diastolic pressures. Instead, it is roughly equal to the diastolic pressure plus one-third of the pulse pressure.

$$MAP = \frac{diastolic}{pressure} + \frac{pulse\ pressure}{3}$$

Thus, for a person with a systolic blood pressure of 120 mm Hg and a diastolic pressure of 80 mm Hg:

$$MAP = 80\ mm\ Hg + \frac{40\ mm\ Hg}{3} = 93\ mm\ Hg$$

MAP and pulse pressure both decline with increasing distance from the heart. The MAP loses ground to the never-ending friction between the blood and the vessel walls, and the pulse pressure is gradually phased out in the less elastic muscular arteries, where elastic rebound of the vessels ceases to occur. At the end of the arterial tree, blood flow is steady and the pulse pressure has disappeared.

Capillary Blood Pressure

As Figure 18.5 shows, by the time blood reaches the capillaries, blood pressure has dropped to approximately 40 mm Hg and by the end of the capillary

What is it about the anatomy of the largest arteries that leads to the pulsatile pressure changes shown on the left side of the graph?

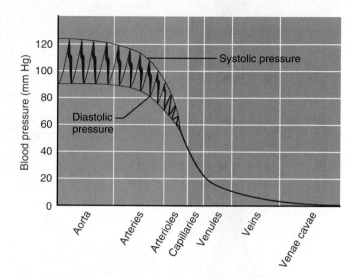

FIGURE 18.5   **Blood pressure in various blood vessels of the systemic circulation.**

beds is only 20 mm Hg or less. Such low capillary pressures are desirable because (1) capillaries are fragile and high pressures would rupture them, and (2) most capillaries are extremely permeable and thus even the low capillary pressure forces solute-containing fluids (filtrate) out of the bloodstream into the interstitial space. As described later in this chapter, these fluid flows help to distribute nutrients, gases, and hormones between blood and tissue cells and continuously refresh the interstitial fluid.

Venous Blood Pressure

Unlike arterial pressure, which pulsates with each contraction of the left ventricle, venous blood pressure is steady and changes very little during the cardiac cycle. The pressure gradient in the veins, from venules to the termini of the venae cavae, is only about 20 mm Hg (that from the aorta to the ends of the arterioles is about 60 mm Hg). This pressure difference between arteries and veins becomes very clear when the vessels are cut. If a vein is cut, the blood flows evenly from the wound; a lacerated artery produces rapid spurts of blood. The very low pressure in the venous system reflects the cumulative effects of peripheral resistance, which dissipates most of the energy of blood pressure (as heat) during each circuit.

■ *The largest arteries stretch and recoil as blood is pumped into them and then runs distally into the circulation.*

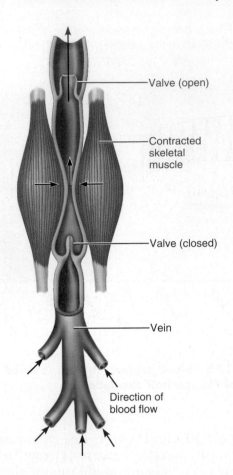

Valve (open)

Contracted skeletal muscle

Valve (closed)

Vein

Direction of blood flow

FIGURE 18.6 The muscular pump. When contracting skeletal muscles press against a vein, the valves proximal to the area of contraction are forced open and blood is propelled toward the heart. The valves distal to the area of contraction are closed by the backflowing blood.

Despite the structural modifications of veins (large lumens and valves), venous pressure is normally too low to promote adequate venous return. Hence, two functional adaptations are critically important to venous return. The first is the **respiratory "pump."** Pressure changes occurring in the ventral body cavity during breathing create the respiratory pump that moves blood up toward the heart. As we inhale, abdominal pressure increases, squeezing the local veins and forcing blood toward the heart. At the same time, the pressure in the chest decreases, allowing thoracic veins to expand and speeding blood entry into the right atrium. The second adaptation is the **muscular "pump."** Skeletal muscle activity, or the so-called muscular pump, is the more important pumping mechanism. As the skeletal muscles surrounding the deep veins contract and relax, they "milk" blood toward the heart, and once blood passes each successive valve, it cannot flow back (Figure 18.6). People who earn their living in "stand-

ing professions," such as hairdressers and dentists, often have swollen ankles because blood pools in the feet and legs during prolonged periods of skeletal muscle inactivity.

Maintaining Blood Pressure

Maintaining a steady flow of blood from the heart to the toes is vital for proper organ function. But making sure a person jumping out of bed in the morning does not keel over from inadequate blood flow to the brain requires the finely tuned cooperation of the heart, blood vessels, and kidneys—all supervised by the brain. Central among the homeostatic mechanisms that regulate cardiovascular dynamics are those that maintain blood pressure, principally *cardiac output, peripheral resistance,* and *blood volume.* If we rearrange the formula pertaining to blood flow presented on p. 618, we can see how cardiac output (blood flow of the entire circulation) and peripheral resistance relate to blood pressure.

$$F = \Delta P/R \quad \text{or} \quad CO = \Delta P/R \quad \text{or} \quad \Delta P = CO \times R$$

Because CO depends on blood volume (the heart can't pump out what doesn't enter its chambers), it is clear that blood pressure varies *directly* with CO, R, and blood volume. Thus, in theory, a change (increase or decrease) in any of these variables causes a corresponding change in blood pressure. However, what *really* happens in the body is that changes in one variable that threaten blood pressure homeostasis are quickly compensated for by changes in the other variables.

As described in Chapter 17, CO is equal to *stroke volume* (ml/beat) times *heart rate* (beats/min), and normal CO is 5.0 to 5.5 L/min. Figure 18.7 shows the main factors determining cardiac output—venous return and the neural and hormonal controls. Remember that the parasympathetic cardioinhibitory center in the medulla is "in charge" of heart rate most of the time and, via the vagus nerves, maintains the *resting heart rate.* During "resting" periods, stroke volume is controlled mainly by venous return (end diastolic volume). During stress, the sympathetic cardioacceleratory center takes over and increases both heart rate (by acting on the SA node) and stroke volume (by enhancing cardiac muscle contractility, which decreases end systolic volume). The enhanced CO, in turn, results in an increase in MAP.

The following discussion focuses on factors that regulate blood pressure by altering peripheral resistance and blood volume, but a flowchart showing the influence of nearly all other important factors is provided in Figure 18.10 (p. 626).

Short-Term Mechanisms: Neural Controls

The *short-term controls* of blood pressure, mediated by the nervous system and bloodborne chemicals, counteract moment-to-moment fluctuations in blood pressure by altering peripheral resistance.

Neural controls of peripheral resistance are directed at two main goals: (1) Altering blood distribution to respond to specific demands of various organs. For example, during exercise blood is shunted temporarily from the digestive organs to the skeletal muscles. (2) Maintaining adequate MAP by altering blood vessel diameter. (Remember, very small changes in blood vessel diameter cause substantial changes in peripheral resistance, hence in systemic blood pressure.) Under conditions of low blood volume, all vessels except those supplying the heart and brain are constricted to allow as much blood as possible to flow to those vital organs.

Most neural controls operate via reflex arcs involving *baroreceptors* and associated afferent fibers, the vasomotor center of the medulla, vasomotor fibers, and vascular smooth muscle. Occasionally, inputs from *chemoreceptors* and higher brain centers also influence the neural control mechanism.

Role of the Vasomotor Center

The neural center that oversees changes in the diameter of blood vessels is the **vasomotor center,** a cluster of sympathetic neurons in the medulla. This center plus the cardiac centers described earlier make up the **cardiovascular center** that integrates blood pressure control by altering cardiac output and blood vessel diameter. The vasomotor center transmits impulses at a fairly steady rate along sympathetic efferents called **vasomotor fibers,** which exit from the T_1 through L_2 levels of the spinal cord and run to innervate the smooth muscle of blood vessels, mainly arterioles. As a result, the arterioles are almost always in a state of moderate constriction, called **vasomotor tone.**

The degree of tonic vasoconstriction varies from organ to organ. Generally, arterioles of the skin and digestive viscera receive vasomotor impulses more frequently and tend to be more strongly constricted than those of skeletal muscles. Any increase in sympathetic activity produces generalized vasoconstriction and a rise in blood pressure. Decreased sympathetic activity allows the vascular muscle to relax somewhat and causes blood pressure to decline to basal levels. Most vasomotor fibers release norepinephrine, which is a potent vasoconstrictor. In skeletal muscle, however, some vasomotor fibers release acetylcholine, causing va-

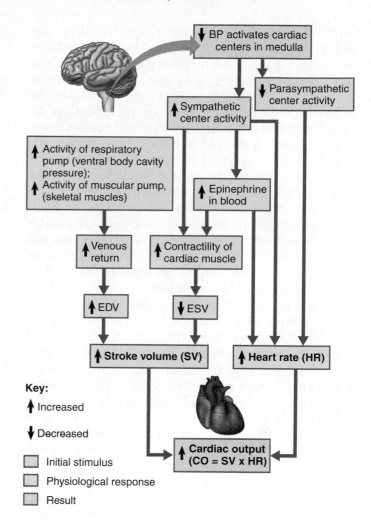

Key:

↑ Increased

↓ Decreased

☐ Initial stimulus

☐ Physiological response

☐ Result

FIGURE 18.7 Major factors enhancing cardiac output. (Efferent impulses from cardiac centers travel along autonomic nerves. EDV = end diastolic volume, ESV = end systolic volume.)

sodilation. Though important to local blood flow, such vasodilator fibers are *not* important to overall regulation of systemic blood pressure.

Vasomotor activity is modified by inputs from (1) baroreceptors (pressure-sensitive mechanoreceptors that respond to changes in arterial pressure and stretch), (2) chemoreceptors (receptors that respond to changes in blood levels of oxygen, carbon dioxide, and H^+), and (3) higher brain centers, as well as by certain hormones and other bloodborne chemicals. Let's take a look.

Baroreceptor-Initiated Reflexes

When arterial blood pressure rises, it stretches **baroreceptors,** neural receptors located in the *carotid sinuses* (dilations in the internal carotid arteries, which provide the major blood supply to the brain), in the *aortic arch,* and in the walls of nearly every large artery of the neck and thorax. When stretched, baroreceptors send a rapid stream of impulses to the vasomotor center. This input inhibits the vasomotor center, resulting in vasodilation of not only arterioles but also veins, and a decline in blood pressure (Figure 18.8). While dilation of the arterioles substantially reduces peripheral resistance, venodilation shifts blood to the venous reservoirs, causing a decline in both venous return and cardiac output. Afferent impulses from the baroreceptors also reach the cardiac centers, where the impulses stimulate parasympathetic activity and inhibit the cardioacceleratory center, reducing heart rate and contractile force. A decline in MAP initiates reflex vasoconstriction and increases cardiac output, causing blood pressure to rise. Thus, peripheral resistance and cardiac output are regulated in tandem so that changes in blood pressure are minimized.

The function of rapidly responding baroreceptors is to protect the circulation against short-term (acute) changes in blood pressure, such as those occurring when you change your posture. For example, blood pressure falls (particularly in the head) when one stands up after reclining. Baroreceptors taking part in the **carotid sinus reflex** protect the blood supply to the brain, whereas those activated in the **aortic reflex** help maintain adequate blood pressure in the systemic circuit as a whole. Baroreceptors are relatively *ineffective* in protecting us against sustained pressure changes, as evidenced by the fact that some people do have chronic hypertension. In such cases, the baroreceptors are apparently "reprogrammed" to monitor pressure changes at a higher set point.

Chemoreceptor-Initiated Reflexes

When the oxygen content or pH of the blood drops sharply or the carbon dioxide levels rise, **chemoreceptors** in the aortic arch and large arteries of the neck transmit impulses to the cardioacceleratory center, which then increases cardiac output, and to the vasomotor center, which causes reflex vasoconstriction. The rise in blood pressure that follows speeds the return of blood to the heart and lungs.

The most prominent chemoreceptors are the *carotid* and *aortic bodies* located close by the baroreceptors in the carotid sinus and aortic arch. Because chemoreceptors are more important in regulating respiratory rate than blood pressure, their function is considered in Chapter 21.

Influence of Higher Brain Centers

Reflexes that regulate blood pressure are integrated in the medulla of the brain stem. Although the cerebral cortex and hypothalamus are not involved in routine controls of blood pressure, these higher brain centers can modify arterial pressure via relays to the medullary centers. For example, the fight-or-flight response mediated by the hypothalamus has profound effects on blood pressure. (Even the simple act of speaking can make your blood pressure jump if the person you are talking to makes you anxious.) The hypothalamus also mediates the redistribution of blood flow and other cardiovascular responses that occur during exercise and changes in body temperature.

Short-Term Mechanisms: Chemical Controls

As we have seen, changing levels of oxygen and carbon dioxide help regulate blood pressure via the chemoreceptor reflexes. In addition, numerous other bloodborne chemicals influence blood pressure by acting directly on vascular smooth muscle or on the vasomotor center. The most important such agents are hormones (Table 18.2).

- **Adrenal medulla hormones.** During periods of stress, the adrenal gland releases norepinephrine (NE) and epinephrine to the blood, and both hormones enhance the sympathetic fight-or-flight response. As noted earlier, NE has a vasoconstrictive action. Epinephrine increases cardiac output and promotes generalized vasoconstriction (except in skeletal and cardiac muscle, where it generally causes vasodilation). It is interesting to note that *nicotine*, an important chemical in tobacco and one of the strongest toxins known, mimics the effects of NE and epinephrine. It causes intense vasoconstriction not only by directly stimulating ganglionic sympathetic neurons but also by prompting release of large amounts of epinephrine and NE.

- **Atrial natriuretic peptide (ANP).** The atria of the heart produce the hormone atrial natriuretic peptide, which causes blood volume and blood pressure to decline. As noted in Chapter 15, ANP antagonizes aldosterone and prods the kidneys to excrete more sodium and water from the body, causing blood volume to drop. It also causes a generalized vasodilation and reduces cerebrospinal fluid formation in the brain.

- **Antidiuretic hormone (ADH).** This hormone is produced by the hypothalamus and stimulates the kidneys to conserve water. It is not usually important in short-term blood pressure regulation, but, when blood pressure falls to dangerously low levels (as during severe hemorrhage), much more ADH is released and helps restore arterial pressure by causing intense vasoconstriction.

- **Angiotensin II.** When renal perfusion is inadequate, the kidneys release renin, an enzyme. Release of renin causes the generation of angiotensin II (an"je-o-ten'sin), which stimulates intense vasocon-

? Which part of this feedback loop occurs when you rise from a seated position?

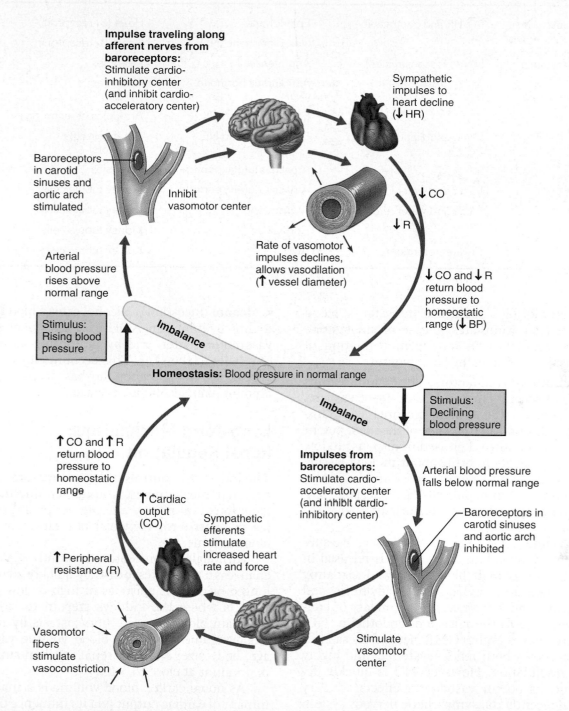

FIGURE 18.8 **Baroreceptor reflexes that help maintain blood pressure homeostasis.** (CO = cardiac output; R = peripheral resistance; HR = heart rate; BP = blood pressure.)

Arterial blood pressure initially drops, resulting in stimulation of the vasomotor and cardioacceleratory centers. The result is increased heart rate and vasoconstriction, which increases arterial pressure, restoring homeostasis. ■

TABLE 18.2 Influence of Selected Hormones on Variables Affecting Blood Pressure

Variable Affecting BP	Effect on Variable	Hormone(s)	Site of Action	Result
Cardiac output	↑ (↑ HR and contractility)	Epinephrine	Heart (β₁ receptors)	↑ BP
		Norepinephrine (NE)	Heart (β₁ receptors)	↑ BP
Peripheral resistance	↑ (via vasoconstriction)	Angiotensin II	Arterioles	↑ BP
		Antidiuretic hormone (ADH)	Arterioles	↑ BP
		Epinephrine and NE	Arterioles (α receptors)	↑ BP
	↓ (via vasodilation)	Epinephrine and NE	Large veins (β₂ receptors)	↓ BP
		Atrial natriuretic peptide	Arterioles	↓ BP
Blood volume	↓ (salt and water loss)	Atrial natriuretic peptide	Kidney tubule cells	↓ BP
	↑ (salt and water retention)	Aldosterone	Kidney tubule cells	↑ BP
		Cortisol	Kidney tubule cells	↑ BP
	↑ (water retention)	ADH	Kidney tubule cells	↑ BP

striction, promoting a rapid rise in systemic blood pressure. It also stimulates release of aldosterone and ADH, which act in long-term regulation of blood pressure by enhancing blood volume.

- **Endothelium-derived factors.** The endothelium is the source of several chemicals that affect vascular smooth muscle (as well as blood clotting). For example, the peptide **endothelin** is one of the most potent vasoconstrictors known. Released in response to low blood flow, it appears to bring about its long-lasting effects by enhancing calcium entry into vascular smooth muscle. Endothelial cells also release **PDGF** (prostaglandin-derived growth factor), another vasoconstrictor chemical.

   Nitric oxide (NO) is yet another vasoactive substance secreted by endothelial cells. It is released in direct response to high blood flow and signaling molecules such as acetylcholine, bradykinin, and nitroglycerine, and its release is a secondary or consequent response to the release of endothelin. NO, which acts via a cyclic GMP second-messenger system, promotes both reflex (systemic) and highly localized vasodilation. However, NO is quickly destroyed and its potent vasodilator effects are very brief. Until recently the sympathetic nervous system was thought to rule blood vessel diameter, but we now know that its primary vascular role is to produce vasoconstriction, and that NO plays the major role in causing vasodilation.

- **Inflammatory chemicals.** Histamine, prostacyclin, kinins, and certain other chemicals released during the inflammatory response and certain allergic responses are potent vasodilators. They also promote fluid loss from the bloodstream by increasing capillary permeability.

- **Alcohol.** Ingestion of alcohol causes blood pressure to drop by inhibiting ADH release, by depressing the vasomotor center, and by promoting vasodilation, especially in the skin. This accounts for the flushed appearance of someone who has drunk a generous amount of an alcoholic beverage.

Long-Term Mechanisms: Renal Regulation

The long-term controls of blood pressure, mediated by renal mechanisms, counteract fluctuations in blood pressure not by altering peripheral resistance (as in short-term controls) but rather by altering blood volume.

Although baroreceptors respond to short-term changes in blood pressure, they quickly adapt to prolonged or chronic episodes of high or low pressure. This is where the kidneys step in to restore and maintain blood pressure homeostasis by regulating blood volume. Although blood volume varies with age, body size, and sex, renal mechanisms usually maintain it at close to 5 L.

As noted earlier, blood volume is a major determinant of cardiac output (via its influence on venous return, EDV, and stroke volume). An increase in blood volume is followed by a rise in blood pressure, and anything that increases blood volume—such as excessive salt intake, which promotes water retention—raises MAP because of the greater fluid load in the vascular tree. By the same token, decreased blood volume translates to a fall in blood pressure. Blood loss and the garden-variety dehydration that occurs during vigorous exercise are common causes of reduced blood volume. A sudden drop in blood pressure often signals internal bleeding and blood volume too low to support normal circulation. However, these assertions—increased blood volume increases BP and decreased blood volume decreases BP—do not tell the whole story because we are dealing with a dynamic system. Increases in blood volume that cause a rise in blood pressure also stimulate the kidneys to eliminate water, which reduces blood volume and consequently blood pressure. Likewise, falling blood volume triggers renal mechanisms that increase blood volume and blood pressure. As you can see, blood pressure can be stabilized or maintained within normal limits only when blood volume is stable.

The kidneys act both directly and indirectly to regulate arterial pressure and provide the major long-term mechanism of blood pressure control. The *direct renal mechanism* alters blood volume. When either blood volume or blood pressure rises, the rate at which fluid filters from the bloodstream into the kidney tubules is speeded up. In such situations, the kidneys cannot process the filtrate rapidly enough, and more of it leaves the body in urine. As a result, blood volume and blood pressure fall. When blood pressure or blood volume is low, water is conserved and returned to the bloodstream, and blood pressure rises (Figures 18.9 and 18.10). As blood volume goes, so goes the arterial blood pressure.

The *indirect renal mechanism* is the **renin-angiotensin mechanism.** When arterial blood pressure declines, the kidneys release the enzyme *renin* into the blood. Renin triggers a series of reactions that produce angiotensin II, as mentioned previously. **Angiotensin II** is a potent vasoconstrictor that, by increasing systemic blood pressure, increases the rate of blood delivery to the kidneys and the rate of renal perfusion. It also stimulates the adrenal cortex to secrete **aldosterone,** a hormone that enhances renal reabsorption of sodium, and prods the posterior pituitary to release ADH, which promotes more water reabsorption (Figure 18.9). As sodium moves into the bloodstream, water follows; thus, both blood volume and blood pressure rise.

Monitoring Circulatory Efficiency

The efficiency of a person's circulation can be assessed by taking pulse and blood pressure measurements. These measurements, along with measurements of respiratory rate and body temperature, are referred to collectively as the body's **vital signs.** Let's examine how vital signs are determined or measured.

Taking a Pulse

The alternating expansion and recoil of elastic arteries during each cardiac cycle create a pressure wave—a **pulse**—that is transmitted through the ar-

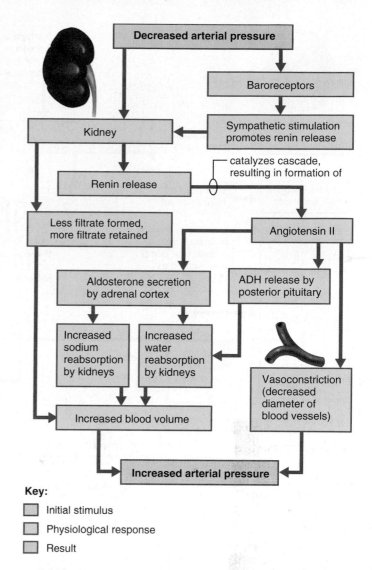

Key:
- Initial stimulus
- Physiological response
- Result

FIGURE 18.9 Hormonal mechanisms for renal control of blood pressure. Sympathetic nervous system stimulation and action, both part of the more rapidly acting short-term control pathway, also participate in this system by triggering renin release.

terial tree. You can feel a pulse in any artery that lies close to the body surface by compressing the artery against firm tissue, and this provides an easy way to count heart rate. Because it is so accessible, the point where the radial artery surfaces at the wrist, the *radial pulse*, is routinely used to take a pulse measurement, but there are several other clinically important arterial pulse points (Figure 18.11). Because these same points are compressed to stop blood flow into distal tissues during hemorrhage, they are also called **pressure points.** For example, if you seriously lacerate your hand, you can slow or stop the bleeding by compressing the radial or the brachial artery.

Monitoring pulse rate is an easy way to assess the effects of activity, postural changes, and emotions on heart rate. For example, the pulse of a

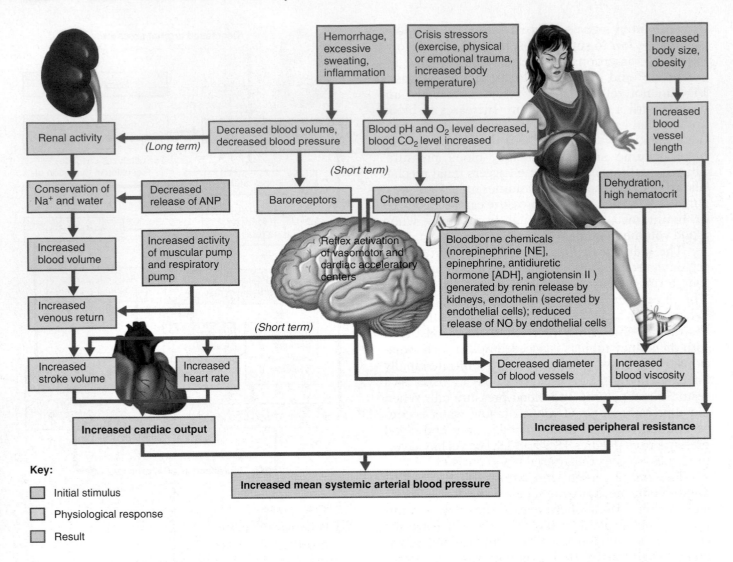

FIGURE 18.10 **Factors causing an increase in MAP.**

healthy man may be around 66 beats per minute when he is lying down, rise to 70 when he sits up, and rise to 80 when he suddenly stands. During vigorous exercise or emotional upset, pulse rates between 140 and 180 are not unusual because of sympathetic nervous system effects on the heart.

Measuring Blood Pressure

Most often, systemic arterial blood pressure is measured indirectly in the brachial artery of the arm by the **auscultatory method** (aw-skul′tah-to″re). The *blood pressure cuff,* or *sphygmomanometer* (sfig″mo-mah-nom′ĕ-ter; *sphygmo* = pulse), is wrapped snugly around the arm just superior to the elbow and inflated until the cuff pressure exceeds systolic pressure. At this point, blood flow into the arm is stopped and a brachial pulse cannot be felt or heard. As the cuff pressure is gradually reduced, the examiner listens (auscultates) with a stethoscope for sounds in the brachial artery. The pressure read as

the first soft tapping sounds are heard (the first point at which a small amount of blood is spurting through the constricted artery) is systolic pressure. As the cuff pressure is reduced further, these sounds, called the *sounds of Korotkoff,* become louder and more distinct, but when the artery is no longer constricted and blood flows freely, the sounds can no longer be heard. The pressure at which the sounds disappear is the diastolic pressure.

In normal adults at rest, systolic pressure varies between 110 and 140 mm Hg, and diastolic pressure between 75 and 80 mm Hg. Blood pressure also cycles over a 24-hour period, peaking in the morning in response to waxing and waning levels of retinoic acid in blood. [It appears that blood vessels, like certain brain cells, contain "clock" proteins that regulate circadian rhythms in response to retinoic acid (a vitamin A derivative) binding to the vessels' retinoic acid receptors.]

Extrinsic factors also play a role: Blood pressure varies with age, sex, weight, race, mood, physical activity, posture, and socioeconomic status. What is "normal" for you may not be normal for your grandfather or your neighbor. Nearly all of these variations can be explained in terms of the factors affecting blood pressure discussed earlier.

Alterations in Blood Pressure

Hypotension

Hypotension, or low blood pressure, is a systolic pressure below 100 mm Hg. In many cases, hypotension simply reflects individual variations and is no cause for concern. In fact, low blood pressure is often associated with long life and an old age free of illness.

🄗 HOMEOSTATIC IMBALANCE

Elderly people are prone to *orthostatic hypotension*—temporary low blood pressure and dizziness when they rise suddenly from a reclining or sitting position. Because the aging sympathetic nervous system does not respond as quickly as it once did to postural changes, blood pools briefly in the lower limbs, reducing blood pressure and consequently blood delivery to the brain. Making postural changes slowly to give the nervous system time to adjust usually prevents this problem.

Chronic hypotension may hint at poor nutrition because poorly nourished people are often anemic and have inadequate levels of blood proteins. Because blood viscosity is low, their blood pressure is also lower than normal. Chronic hypotension also warns of Addison's disease (inadequate adrenal cortex function), hypothyroidism, or severe tissue wasting. *Acute hypotension* is one of the most important signs of circulatory shock (p. 635) and a threat to patients undergoing surgery and those in intensive care units. •

Hypertension

Hypertension, high blood pressure, may be transient or persistent. Transient elevations in systolic pressure occur as normal adaptations during fever, physical exertion, and emotional upset. Persistent hypertension is common in obese people because the total length of their blood vessels is relatively greater than that in thinner individuals.

🄗 HOMEOSTATIC IMBALANCE

Chronic hypertension is a common and dangerous disease that warns of increased peripheral resistance. An estimated 30% of people over the age of 50 years are hypertensive. Although this "silent killer" is usually asymptomatic for the first 10 to 20 years, it slowly but surely strains the heart and damages the arteries. Prolonged hypertension is the major cause of

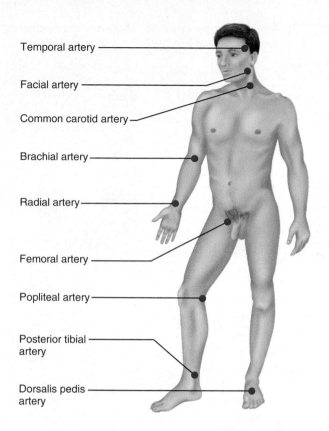

Temporal artery

Facial artery

Common carotid artery

Brachial artery

Radial artery

Femoral artery

Popliteal artery

Posterior tibial artery

Dorsalis pedis artery

FIGURE 18.11 Body sites where the pulse is most easily palpated. (The specific arteries indicated are discussed on pp. 638–649.)

heart failure, vascular disease, renal failure, and stroke. Because the heart is forced to pump against greater resistance, it must work harder, and in time the myocardium enlarges. When finally strained beyond its capacity to respond, the heart weakens and its walls become flabby. Hypertension also ravages the blood vessels, causing small tears in the endothelium that accelerate the progress of atherosclerosis. As the vessels become increasingly blocked, blood flow to the tissues becomes inadequate and vascular complications appear in the brain, heart, kidneys, and retinas of the eyes.

Although hypertension and atherosclerosis are often linked, it is often difficult to blame hypertension on any distinct anatomical pathology. Hypertension is defined physiologically as a condition of sustained elevated arterial pressure of 140/90 or higher, and the higher the pressure, the greater the risk for serious cardiovascular problems. As a rule, elevated diastolic pressures are more significant medically, because they always indicate progressive occlusion and/or hardening of the arterial tree.

About 90% of hypertensive people have **primary, or essential, hypertension,** in which no underlying cause has been identified. However, the following factors are believed to be involved:

? *Why is blood delivery to the skin vasculature increased during strenuous exercise?*

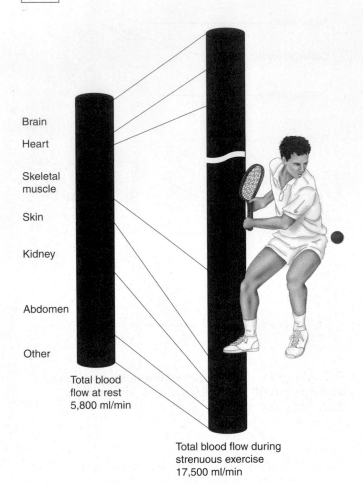

Total blood flow at rest 5,800 ml/min

Total blood flow during strenuous exercise 17,500 ml/min

FIGURE 18.12 Distribution of blood flow at rest and during strenuous exercise.

1. Diet. Dietary factors that contribute to hypertension include high sodium, saturated fat, and cholesterol intake and deficiencies in certain metal ions (K$^+$, Ca^{2+}, and Mg^{2+}).

2. Obesity.

3. Age. Clinical signs of the disease usually appear after age 40.

4. Race. More blacks than whites are hypertensive, and the course of the disease varies in different population groups.

5. Heredity. Hypertension runs in families. Children of hypertensive parents are twice as likely to develop hypertension as are children of normotensive parents.

Muscle activity generates heat, which must be dissipated if homeostasis is to be maintained. The skin is the body's heat exchange site. ■

6. Stress. Particularly at risk are "hot reactors," people whose blood pressure zooms upward during every stressful event.

7. Smoking. Nicotine enhances the sympathetic nervous system's vasoconstrictor effects.

Primary hypertension cannot be cured, but most cases can be controlled by restricting salt, fat, and cholesterol intake, losing weight, stopping smoking, managing stress, and taking antihypertensive drugs. Drugs commonly used are diuretics, beta-blockers, calcium channel blockers, and angiotensin-converting enzyme (ACE) inhibitors. Inhibiting ACE suppresses the renin-angiotensin mechanism.

Secondary hypertension, which accounts for 10% of cases, is due to identifiable disorders, such as excessive renin secretion by the kidneys, arteriosclerosis, and endocrine disorders such as hyperthyroidism and Cushing's disease. Treatment for secondary hypertension is directed toward correcting the causative problem. ●

Blood Flow Through Body Tissues: Tissue Perfusion

Blood flow through body tissues, or **tissue perfusion,** is involved in (1) delivery of oxygen and nutrients to, and removal of wastes from, tissue cells, (2) gas exchange in the lungs, (3) absorption of nutrients from the digestive tract, and (4) urine formation in the kidneys. The rate of blood flow to each tissue and organ is almost exactly the right amount to provide for proper function—no more, no less. When the body is at rest, the brain receives about 13% of total blood flow, the heart 4%, kidneys 20%, and abdominal organs 24%. Skeletal muscles, which make up almost half of body mass, normally receive about 20% of total blood flow. During exercise, however, nearly all of the increased cardiac output flushes into the skeletal muscles and blood flow to the kidneys and digestive organs is reduced (Figure 18.12).

Velocity of Blood Flow

As shown in Figure 18.13, the speed or velocity of blood flow changes as blood travels through the systemic circulation. All other factors being constant, it is fastest in the aorta and other large arteries, slowest in the capillaries, and then picks up speed again in the veins.

Velocity is *inversely* related to the *cross-sectional area* of the blood vessels to be filled. Blood flows fastest where the total cross-sectional area is least. As the arterial system branches, the total cross-sectional area of the vascular bed increases, and the velocity of blood flow declines proportionately. Even though the individual branches have smaller lumens, their combined cross-sectional areas and thus the volume of blood they can hold, are much greater than that of the aorta. For example, the cross-sectional area of the aorta is 2.5 cm² but that of the capillaries is 4500 cm². This difference in cross-sectional area results in a fast blood flow in the aorta (40–50 cm/s) and a slow blood flow in the capillaries (about 0.03 cm/s). Slow capillary flow is beneficial because it allows adequate time for exchanges between the blood and tissue cells.

As capillaries combine to form first venules and then veins, total cross-sectional area declines and velocity increases. The cross-sectional area of the venae cavae is 8 cm², and the velocity of blood flow varies from 10 to 30 cm/s in those vessels, depending on the activity of the skeletal muscle pump.

Autoregulation: Local Regulation of Blood Flow

Autoregulation is the automatic adjustment of blood flow to each tissue in proportion to the tissue's requirements at any instant. This process is regulated by local conditions and is largely independent of systemic factors. The MAP is the same everywhere in the body and homeostatic mechanisms adjust cardiac output as needed to maintain that constant pressure. Changes in blood flow through individual organs are controlled *intrinsically* by modifying the diameter of local arterioles feeding the capillaries. You can compare blood flow autoregulation to water use in your home. To get water in a sink, or a garden hose, you have to turn on a faucet. Whether you have several taps open or none, the pressure in the main water pipe in the street remains relatively constant, as it does in the even larger water lines closer to the pumping station. Similarly, what is happening in the arterioles feeding the capillary beds of a given organ has little effect on pressure in the muscular artery feeding the organ, or in the large elastic arteries. The pumping station is, of course, the heart. The beauty of this system is that as long as the water company (circulatory feedback mechanisms) maintains a relatively constant water pressure (MAP), local demand regulates the amount of fluid (blood) delivered to various areas.

Thus, organs regulate their own blood flows by varying the resistance of their arterioles. As described next, these intrinsic control mechanisms may be classed as *metabolic* or *myogenic*.

Metabolic Controls

In most tissues, declining levels of nutrients, particularly oxygen, are the strongest stimuli for autoregu-

Does the speed of blood flow vary directly or indirectly with the cross-sectional area of the vascular bed?

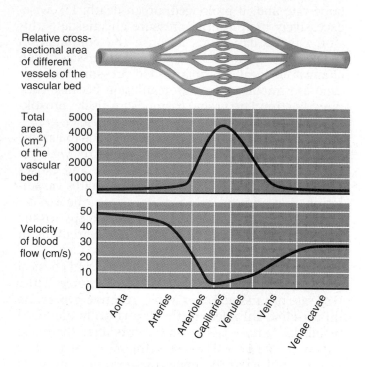

FIGURE 18.13 **Relationship between blood flow velocity and total cross-sectional area in various blood vessels of the systemic circulation.**

lation. An interesting recent finding is that nitric oxide produced by endothelial cells induces vasodilation at the capillaries to help get oxygen to the tissue cells where it is most needed. As oxygen is being loaded onto hemoglobin in the lungs, NO latches onto hemoglobin as well and is delivered to the tissue capillaries, where it is shed and converted to a biologically active *S*-nitrosothiol molecule, which causes vasodilation as oxygen is unloaded. Certain other substances released by metabolically active tissues (such as K⁺, H⁺, adenosine, and lactic acid), prostaglandins, and inflammatory chemicals (histamine and kinins) also serve as autoregulation stimuli.

Whatever the stimulus, the net result of metabolically controlled autoregulation is immediate vasodilation of the arterioles serving the capillary beds of the "needy" tissues, and therefore a temporary increase in blood flow to the area. This is accompanied by relaxation of the precapillary sphincters, which allows blood to surge through the true capillaries and become available to the tissue cells.

Indirectly: As cross-sectional area of the vascular bed increases, flow velocity decreases. ∎

Myogenic Controls

Inadequate blood perfusion through an organ is quickly followed by a decline in the organ's metabolic rate and, if prolonged, organ death. Likewise, excessively high arterial pressure and tissue perfusion can be dangerous because the combination may rupture the more fragile blood vessels. Such changes in local arteriolar blood pressure and volume are important in autoregulation because they directly stimulate vascular smooth muscle, provoking **myogenic responses** (*myo* = muscle; *gen* = origin). Vascular smooth muscle responds to passive stretch (increased intravascular pressure) with increased tone, which resists the stretch and causes vasoconstriction. Reduced stretch promotes vasodilation and increases blood flow into the tissue. Hence, the myogenic mechanism keeps tissue perfusion fairly constant despite most variations in systemic pressure.

Generally, both chemical (metabolic) and physical (myogenic) factors determine the final autoregulatory response of a tissue. For example, **reactive hyperemia** (hi″per-e′me-ah) refers to the dramatically increased blood flow into a tissue that occurs after the blood supply to the area has been temporarily blocked. It results both from the myogenic response and from an accumulation of metabolic wastes in the area during occlusion.

Long-Term Autoregulation

If the nutrient requirements of a tissue are greater than the short-term autoregulatory mechanism can easily supply, a long-term autoregulation mechanism may evolve over a period of weeks or months to enrich the local blood flow still more. The number of blood vessels in the region increases, and existing vessels enlarge. This phenomenon, called *angiogenesis*, is particularly common in the heart when a coronary vessel is partially occluded. It occurs throughout the body in people who live in high-altitude areas, where the air contains less oxygen.

Blood Flow in Special Areas

Each organ has special requirements and functions that are revealed in its pattern of autoregulation. Autoregulation in the brain, heart, and kidneys is extraordinarily efficient. In those organs, adequate perfusion is maintained even when MAP is fluctuating.

Skeletal Muscles

Blood flow in skeletal muscle varies with muscle activity and fiber type. Resting skeletal muscles receive about 1 L of blood per minute, and only about 25% of their capillaries are open. Generally speaking, capillary density and blood flow is greater in red (slow oxidative) fibers than in white (fast glycolytic) fibers. During such periods, myogenic and general neural mechanisms predominate. When muscles become active, blood flow increases (*hyperemia*) in direct proportion to their greater *metabolic* activity, a phenomenon called **active** or **exercise hyperemia.**

The arterioles in skeletal muscle have cholinergic receptors and both alpha and beta (α, β) adrenergic receptors, which bind epinephrine. When epinephrine levels are low, epinephrine mediates vasodilation by binding chiefly to the β receptors. Likewise, "occupied" cholinergic receptors are believed to promote vasodilation. Consequently, blood flow can increase tenfold or more during physical activity (see Figure 18.12), and virtually all capillaries in the active muscles open to accommodate the increased flow. By contrast, the high levels of epinephrine typical of massive sympathetic nervous system activation (and extremely vigorous exercise involving large numbers of skeletal muscles) cause intense vasoconstriction mediated by the alpha-adrenergic receptors. This protective response, believed to be initiated by muscle chemoreflexes when muscle oxygen delivery falls below some critical level, ensures that muscle demands for blood do not exceed cardiac pumping ability and that vital organs continue to receive an adequate blood supply. Without question, strenuous exercise is one of the most demanding conditions the cardiovascular system faces.

Muscular autoregulation occurs almost entirely in response to the decreased oxygen concentrations that result from the "revved-up" metabolism of working muscles. However, systemic adjustments mediated by the vasomotor center must also occur to ensure that blood delivery to the muscles is both faster and more abundant. Strong vasoconstriction of the vessels of blood reservoirs such as those of the digestive viscera and skin diverts blood away from these regions temporarily, ensuring that more blood reaches the muscles. Ultimately, the major factor determining how long muscles can continue to contract vigorously is the ability of the cardiovascular system to deliver adequate oxygen and nutrients.

The Brain

Blood flow to the brain averages 750 ml/min and is maintained at a relatively constant level. The necessity for constant cerebral blood flow becomes crystal clear when one takes into account that neurons are totally intolerant of ischemia.

Although the brain is the most metabolically active organ in the body, it is the least able to store essential nutrients. Cerebral blood flow is regulated by one of the body's most precise autoregulatory systems and is tailored to local neuronal need. Thus, when you make a fist with your right hand, the neurons in the left cerebral motor cortex controlling that movement receive a more abundant blood supply than the adjoining neurons. Brain tissue is exceptionally sensitive to declining pH, and increased blood carbon dioxide levels (resulting in acidic conditions in brain tissue) cause marked vasodilation. Oxygen deficit is a much less potent stimulus for autoregulation. However, very high carbon dioxide levels abolish autoregulatory mechanisms and severely depress brain activity.

Besides metabolic controls, the brain also has a myogenic mechanism that protects it from possibly damaging changes in blood pressure. When MAP declines, cerebral vessels dilate to ensure adequate brain perfusion. When MAP rises, cerebral vessels constrict, protecting the small, more fragile vessels farther along the pathway from rupture due to excessive pressure. Under certain circumstances, such as brain ischemia caused by rising intracranial pressure (as with a brain tumor), the brain (via the medullary cardiovascular centers) regulates its own blood flow by triggering a rise in systemic blood pressure. However, when systemic pressure changes are extreme, the brain becomes vulnerable. Fainting, or *syncope* (sin'cuh-pe; "cutting short"), occurs when MAP falls below 60 mm Hg. Cerebral edema is the usual result of pressures over 160 mm Hg, which dramatically increase brain capillary permeability.

The Skin

Blood flow through the skin (1) supplies nutrients to the cells, (2) aids in body temperature regulation, and (3) provides a blood reservoir. The first function is served by autoregulation in response to the need for oxygen; the second and third require neural intervention. The primary function of the cutaneous circulation is to help maintain body temperature, so we will concentrate on the skin's temperature regulation function here.

Below the skin surface are extensive venous plexuses, in which the blood flow velocity can change from 50 ml/min to as much as 2500 ml/min, depending on body temperature. This capability reflects neural adjustments of blood flow through arterioles and through unique coiled arteriovenous anastomoses. These tiny A-V shunts are located mainly in the fingertips, palms of the hands, toes, soles of the feet, ears, nose, and lips. They are richly supplied with sympathetic nerve endings (a characteristic that sets them apart from the shunts of most other capillary beds), and are controlled by reflexes initiated by temperature receptors or signals from higher CNS centers. The arterioles, in addition, are responsive to local and metabolic autoregulatory stimuli.

When the skin surface is exposed to heat, or body temperature rises for other reasons (such as vigorous exercise), the hypothalamic "thermostat" signals for reduced vasomotor stimulation of the skin vessels. As a result, warm blood flushes into the capillary beds and heat radiates from the skin surface. Vasodilation of the arterioles is enhanced even more when we sweat, because an enzyme in perspiration acts on a protein present in tissue fluid to produce *bradykinin*, which stimulates the vessel's endothelial cells to release the potent vasodilator NO.

When the ambient temperature is cold and body temperature drops, superficial skin vessels are strongly constricted. Hence, blood almost entirely bypasses the capillaries associated with the arteriovenous anastomoses, diverting the warm blood to the deeper, more vital organs. Paradoxically, the skin may stay quite rosy because some blood gets "trapped" in the superficial capillary loops as the shunts swing into operation; also, the chilled skin cells take up less O_2.

The Lungs

Blood flow through the pulmonary circuit to and from the lungs is unusual in many ways. The pathway is relatively short, and pulmonary arteries and arterioles are structurally like veins and venules. That is, they have thin walls and large lumens. Because resistance to blood flow is low in the pulmonary arterial system, less pressure is needed to propel blood through those vessels. Consequently, arterial pressure in the pulmonary circulation is much lower than in the systemic circulation (24/8 versus 120/80).

Another unusual feature of the pulmonary circulation is that the autoregulatory mechanism is the *opposite* of what is seen in most tissues: Low pulmonary oxygen levels cause vasoconstriction, and high levels promote vasodilation. While this may seem odd, it is perfectly consistent with the gas exchange role of this circulation. When the air sacs of the lungs are flooded with oxygen-rich air, the pulmonary capillaries become flushed with blood and ready to receive the oxygen load. If the air sacs are collapsed or blocked with mucus, the oxygen content in those areas is low, and blood largely bypasses those nonfunctional areas.

The Heart

Movement of blood through the smaller vessels of the coronary circulation is influenced by aortic pressure and by the pumping activity of the ventricles. When the ventricles contract and compress the coronary vessels, blood flow through the myocardium stops. As the heart relaxes, the high aortic pressure forces blood through the coronary circulation. Under normal circumstances, the myoglobin in cardiac cells stores sufficient oxygen to satisfy the cells' oxygen needs during systole. However, an abnormally

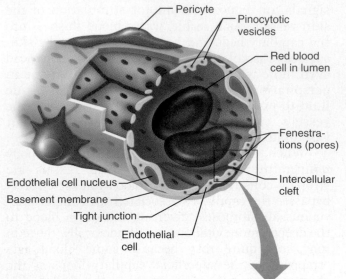

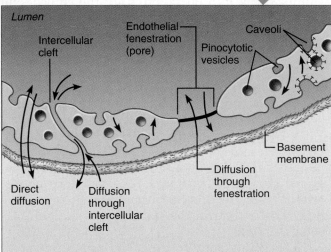

FIGURE 18.14 Capillary transport mechanisms. The four possible pathways or routes of transport across the endothelial cell wall of a fenestrated capillary. (The endothelial cell is drawn as if cut in cross section.)

rapid heartbeat seriously reduces the ability of the myocardium to receive adequate oxygen and nutrients during diastole.

Under resting conditions, blood flow through the heart is about 250 ml/min and is probably controlled by a myogenic mechanism. During strenuous exercise, the coronary vessels dilate in response to local accumulation of carbon dioxide (leading to acidosis), and blood flow may increase three to four times (see Figure 18.12). Additionally, any event that decreases the oxygen content of the blood causes release of a vasodilator that adjusts the O_2 supply to the O_2 demand. Consequently, blood flow remains fairly constant despite wide variations (50 to 140 mm Hg) in coronary perfusion pressure. This enhanced blood flow during increased heart activity is important because under resting conditions, cardiac cells use as much as 65% of the oxygen carried to them in blood.

(Most other tissue cells use about 25% of the delivered oxygen.) Thus, increasing the blood flow is the only way to make sufficient additional oxygen available to a more vigorously working heart.

Blood Flow Through Capillaries and Capillary Dynamics

Blood flow through capillary networks is slow and intermittent. This phenomenon, called **vasomotion,** reflects the on/off opening and closing of precapillary sphincters in response to local autoregulatory controls.

Capillary Exchange of Respiratory Gases and Nutrients

Oxygen, carbon dioxide, most nutrients, and metabolic wastes pass between the blood and interstitial fluid by diffusion. Recall that in **diffusion,** movement always occurs along a concentration gradient—each substance moving from an area of its higher concentration to an area of its lower concentration. Hence, oxygen and nutrients pass from the blood, where their concentration is fairly high, through the interstitial fluid to the tissue cells. Carbon dioxide and metabolic wastes leave the cells, where their content is higher, and diffuse into the capillary blood. In general, small water-soluble solutes, such as amino acids and sugars, pass through fluid-filled intercellular capillary clefts (and sometimes through fenestrations), while lipid-soluble molecules, such as respiratory gases, diffuse directly through the lipid bilayer of the endothelial cell plasma membranes (Figure 18.14). Newly forming caveoli translocate some larger molecules, such as small proteins, and pinocytotic vesicles imbibe solute-containing fluid. As mentioned earlier, capillaries differ in their "leakiness," or permeability. Liver capillaries, for instance, have large fenestrations that allow even proteins to pass freely, whereas brain capillaries are impermeable to most substances.

Fluid Movements

All the while nutrient and gas exchanges are occurring across the capillary walls by diffusion, bulk fluid flows are also going on. Fluid is forced out of the capillaries through the clefts at the arterial end of the bed, but most of it returns to the bloodstream at the venous end. These fluid flows are relatively unimportant to capillary exchange; instead they help determine the relative fluid volumes in the bloodstream and the extracellular space. As described next, the *direction and amount* of fluid that flows across the capillary walls reflect the balance between two dynamic and opposing forces—hydrostatic and colloid osmotic pressures (Figure 18.15).

How would fluid flows change if the OP of the interstitial fluid rose dramatically—say because of a severe bacterial infection in the surrounding tissues?

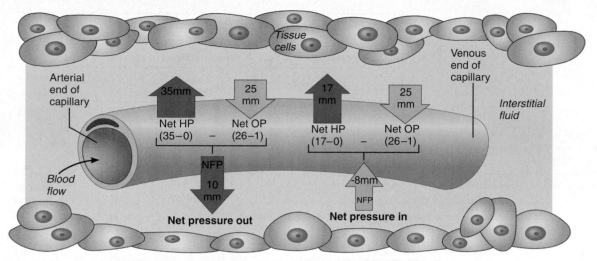

Key to pressure values:
HP_c at arterial end = 35 mm Hg HP_{if} = 0 mm Hg OP_{if} = 1 mm Hg
HP_c at venous end = 17 mm Hg OP_c = 26 mm Hg

FIGURE 18.15 Fluid flows at capillaries. The direction of fluid movement depends on the difference between net hydrostatic pressure (HP), the force that tends to push fluid out of the capillary, and net colloid osmotic pressure (OP), the force that draws fluid back into the capillary. Because HP_{if} is taken to be zero, net HP ($HP_c - HP_{if}$) is equal to the hydrostatic pressure of the blood (HP_c), which varies along the length of the capillary. Net OP ($OP_c - OP_{if}$) is constant, reflecting the fact that blood has a much higher content of nondiffusible solutes (proteins) than does interstitial fluid. At the arterial end of the capillary bed, net HP forcing fluid outward exceeds OP drawing water inward, resulting in a NFP (net filtration pressure) that promotes a net fluid loss from the capillary: NFP = 10 mm Hg. At the venule end of the bed, HP is overpowered by OP (NFP = −8 mm Hg), and fluid returns to the bloodstream.

Hydrostatic Pressures **Hydrostatic pressure** is the force exerted by a fluid pressing against a wall. In capillaries, hydrostatic pressure is the same as *capillary blood pressure*—the pressure exerted by blood on the capillary walls. **Capillary hydrostatic pressure (HP_c)** tends to force fluids through the capillary walls. Because blood pressure drops as blood flows along the length of a capillary bed, HP_c is higher at the arterial end of the bed (35 mm Hg) than at the venous end (17 mm Hg).

In theory, blood pressure—which forces fluid out of the capillaries—is opposed by the **interstitial fluid hydrostatic pressure (HP_{if})** acting outside the capillaries and pushing fluid in. Thus, the *net* hydrostatic pressure acting on the capillaries at any point is the difference between HP_c and HP_{if}. However, there is usually very little fluid in the interstitial space, because any fluid there is constantly withdrawn by the lymphatic vessels. Although HP_{if} may vary from slightly negative to slightly positive, traditionally it is assumed to be zero. For simplicity, that is the value we use here. The *net effective hydrostatic pressures* at the arterial and venous ends of the capillary bed are essentially equal to HP_c (in other words, to blood pressure) at those locations. This information is summarized in Figure 18.15.

Colloid Osmotic Pressures **Colloid osmotic pressure,** the force opposing hydrostatic pressure, is created by the presence in a fluid of large nondiffusible molecules, such as plasma proteins, that are prevented from moving through the capillary membrane. Such molecules draw water toward themselves; that is, they encourage osmosis whenever the water concentration in their vicinity is lower than it is on the opposite side of the capillary membrane. The abundant plasma proteins in capillary blood (primarily albumin molecules) develop a **capillary colloid osmotic pressure (OP_c),** also called *oncotic pressure,* of approximately 26 mm Hg. Because interstitial fluid contains few proteins, its colloid

■ *Fluid that would ordinarily reenter the circulation at the venous end of the capillary bed would remain in the tissue spaces, held by the elevated OP of the interstitial fluid.*

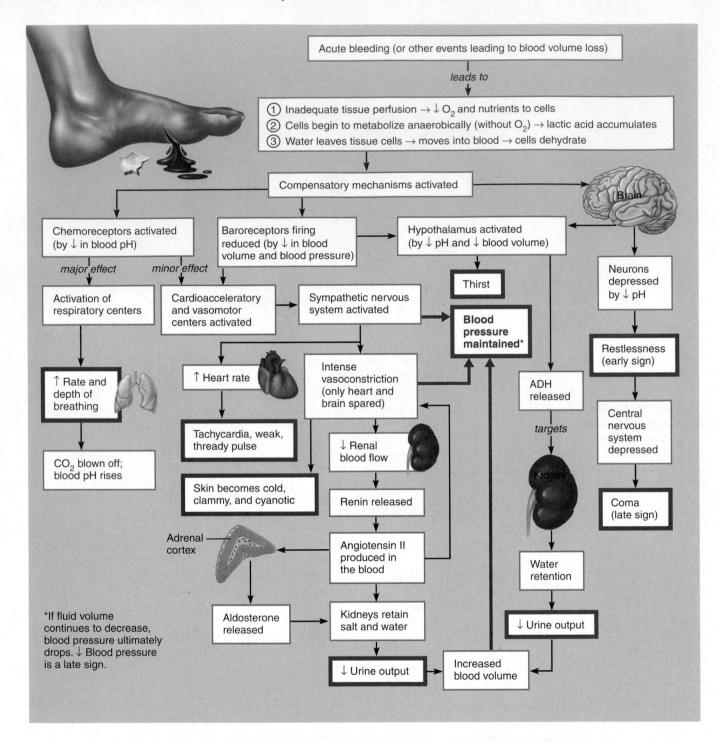

FIGURE 18.16 **Events and signs of compensated (nonprogressive) hypovolemic shock.** (Recognizable clinical signs are shown in *red-bordered boxes*.)

osmotic pressure (**OP$_{if}$**) is substantially lower—from 0.1 to 5 mm Hg. We use a value of 1 mm Hg for the OP$_{if}$ in Figure 18.15. Unlike HP, OP does not vary significantly from one end of the capillary bed to the other. Thus, in our example, the *net osmotic pressure* that pulls fluid back into the capillary blood is OP$_c$ − OP$_{if}$ = 26 mm Hg − 1 mm Hg = 25 mm Hg.

Hydrostatic-Osmotic Pressure Interactions

To determine whether there is a net gain or net loss of fluid from the blood, we have to calculate the **net filtration pressure (NFP),** which considers all the forces acting at the capillary bed. At any point along a capillary, fluids will leave the capillary if net HP is greater than net OP, and fluids will enter the capillary if net OP exceeds net HP. As shown in Figure 18.15, hydrostatic forces dominate at the arterial end (all values are in millimeters of Hg):

$$NFP = (HP_c - HP_{if}) - (OP_c - OP_{if})$$
$$= (35 - 0) - (26 - 1)$$
$$= (35 - 25) = 10 \text{ mm Hg}$$

Thus, in this example, a pressure of 10 mm Hg (net excess of HP) is forcing fluid out of the capillary. At the other, venous end, osmotic forces dominate:

$$NFP = (17 - 0) - (26 - 1)$$
$$= 17 - 25 = -8 \text{ mm Hg}$$

The negative pressure value we see here indicates that the NFP (due to net excess of OP) is driving fluid *into* the capillary bed. Thus, net fluid flow is *out* of the circulation at the arterial ends of capillary beds and *into* the circulation at the venous ends. However, more fluid enters the tissue spaces than is returned to the blood, resulting in a net loss of fluid from the circulation of about 1.5 ml/min. This fluid and any leaked proteins are picked up by the lymphatic vessels and returned to the vascular system, which accounts for the relatively low levels of both fluid and proteins in the interstitial space. Were this not so, this "insignificant" fluid loss would empty your blood vessels of plasma in about 24 hours!

Circulatory Shock

Circulatory shock is any condition in which blood vessels are inadequately filled and blood cannot circulate normally. This results in inadequate blood flow to meet tissue needs. If the condition persists, cells die and organ damage follows.

The most common form of shock is **hypovolemic shock** (hi"po-vo-le'mik; *hypo* = low, deficient; *volemia* = blood volume), which results from large-scale loss of blood, as might follow acute hemorrhage, severe vomiting or diarrhea, or extensive burns. If blood volume drops rapidly, heart rate increases in an attempt to correct the problem. Thus, a weak, "thready" pulse is often the first sign of hypovolemic shock. Intense vasoconstriction also occurs, which shifts blood from the various blood reservoirs into the major circulatory channels and enhances venous return. Blood pressure is stable at first, but eventually drops if blood loss continues. A sharp drop in blood pressure is a serious, and late, sign of hypovolemic shock. The key to managing hypovolemic shock is to replace fluid volume as quickly as possible.

Although many of the body's responses to hypovolemic shock have yet to be explored, acute bleeding is such a threat to life that it seems important to have a comprehensive flowchart of its recognizable signs and symptoms and an accounting of the body's attempt to restore homeostasis. Figure 18.16 provides such a resource. Study it in part now, and then in more detail later once you have studied the remaining body systems.

In **vascular shock,** blood volume is normal and constant, but there is poor circulation as a result of an abnormal expansion of the vascular bed caused by extreme vasodilation. The huge drop in peripheral resistance that follows is revealed by rapidly falling blood pressure. The most common causes of vascular shock are loss of vasomotor tone due to anaphylaxis (anaphylactic shock), a systemic allergic reaction in which bodywide vasodilation is triggered by the massive release of histamine; failure of autonomic nervous system regulation (also referred to as *neurogenic shock*); and septicemia *(septic shock)*, a severe systemic bacterial infection (bacterial toxins are notorious vasodilators).

Transient vascular shock may occur when you sunbathe for a prolonged time. The heat of the sun on your skin causes cutaneous blood vessels to dilate. Then, if you stand up abruptly, blood pools briefly (because of gravity) in the dilated vessels of your lower limbs rather than returning promptly to the heart. Consequently, your blood pressure falls. The dizziness you feel at this point is a signal that your brain is not receiving enough oxygen.

Cardiogenic shock, or pump failure, occurs when the heart is so inefficient that it cannot sustain adequate circulation. Its usual cause is myocardial damage, as might follow numerous myocardial infarcts.

PART 3: CIRCULATORY PATHWAYS: BLOOD VESSELS OF THE BODY

When referring to the body's complex network of blood vessels, the term **vascular system** is often used. However, the heart is actually a double pump that serves two distinct circulations, each with its own set of arteries, capillaries, and veins. The *pulmonary circulation* is the short loop that runs from the heart to the lungs and back to the heart. The *systemic circulation* routes blood through a long loop to all parts of the body before returning it to the heart. Both circuits are shown schematically in Table 18.3 on p. 636.

The principal arteries and veins of the systemic circulation are described in Tables 18.4 through 18.13. Although there are many similarities between

Text continues on page 659.

TABLE 18.3 Pulmonary and Systemic Circulations

Pulmonary Circulation

The pulmonary circulation (Figure 18.17a) functions only to bring blood into close contact with the alveoli (air sacs) of the lungs so that gases can be exchanged. It does not directly serve the metabolic needs of body tissues.

Oxygen-poor, dark red blood enters the pulmonary circulation as it is pumped from the right ventricle into the large **pulmonary trunk** (Figure 18.17b), which runs diagonally upward for about 8 cm and then divides abruptly to form the **right** and **left pulmonary arteries.** In the lungs, the pulmonary arteries subdivide into the **lobar arteries** (lo'bar) (three in the right lung and two in the left lung), each of which serves one lung lobe. The lobar arteries accompany the main bronchi into the lungs and then branch profusely, forming first arterioles and then the dense networks of **pulmonary capillaries** that surround and cling to the delicate air sacs. It is here that oxygen moves from the alveolar air to the blood and carbon dioxide moves from the blood to the alveolar air. As gases are exchanged and the oxygen content of the blood rises, the blood becomes bright red. The pulmonary capillary beds drain into venules, which join to form the two **pulmonary veins** exiting from each lung. The four pulmonary veins complete the circuit by unloading their precious cargo into the left atrium of the heart. Note that any vessel with the term *pulmonary* or *lobar* in its name is part of the pulmonary circulation. All others are part of the systemic circulation.

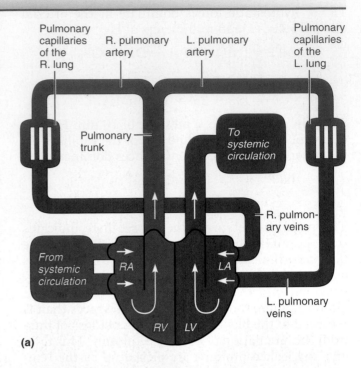

(a)

FIGURE 18.17 Pulmonary circulation. (a) Schematic flowchart. **(b)** Illustration. The arterial system is shown in blue to indicate that the blood carried is oxygen poor; the venous drainage is shown in red to indicate that the blood transported is oxygen rich.

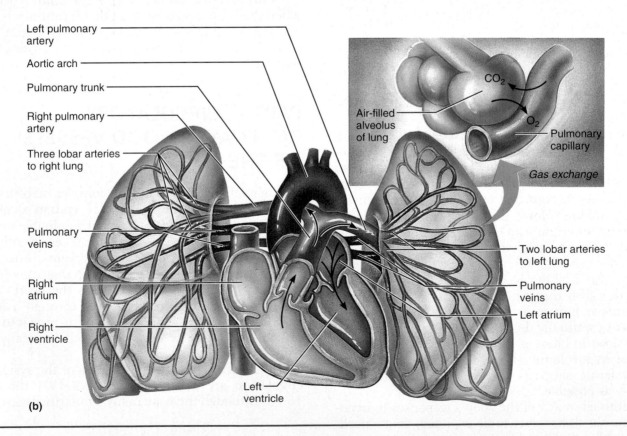

(b)

TABLE 18.3 (continued)

Pulmonary arteries carry oxygen-poor, carbon dioxide-rich blood, and pulmonary veins carry oxygen-rich blood.* This is opposite the situation in the systemic circulation, where arteries carry oxygen-rich blood and veins carry carbon dioxide–rich, relatively oxygen-poor blood.

Systemic Circulation

The systemic circulation provides the *functional blood supply* to all body tissues; that is, it delivers oxygen, nutrients, and other needed substances while carrying away carbon dioxide and other metabolic wastes. Freshly oxygenated blood* returning from the pulmonary circuit is pumped out of the left ventricle into the aorta (Figure 18.18). From the aorta, blood can take various routes, because essentially all systemic arteries branch from this single great vessel. The aorta arches upward from the heart and then curves and runs downward along the body midline to its terminus in the pelvis, where it splits to form the two large arteries serving the lower extremities. The branches of the aorta continue to subdivide to produce the arterioles and, finally, the capillaries that ramify through the organs. Venous blood draining from organs inferior to the diaphragm ultimately enters the inferior vena cava.† Except for some thoracic venous drainage (which enters the azygos system of veins), body regions above the diaphragm are drained by the superior vena cava. The venae cavae empty the carbon dioxide–laden blood into the right atrium of the heart.

Two points concerning the two major circulations must be emphasized: (1) Blood passes from systemic veins to systemic arteries only after first moving through the pulmonary circuit (Figure 18.17a), and (2) although the entire cardiac output of the right ventricle passes through the pulmonary circulation, only a small fraction of the output of the left ventricle flows through any single organ (Figure 18.18). The systemic circulation can be viewed as multiple circulatory channels functioning in parallel to distribute blood to all body organs.

As you examine the tables that follow and locate the various systemic arteries and veins in the illustrations, be aware of cues that make your memorization task easier. In many cases, the name of a vessel reflects the body region traversed (axillary, brachial, femoral, etc.), the organ served (renal, hepatic, gonadal), or the bone followed (vertebral, radial, tibial). Also, notice that arteries and veins tend to run together side by side and, in many places, they also run with nerves. Finally, be alert to the fact that the systemic vessels do not always match on the right and left sides of the body. Thus, while almost all vessels in the head and limbs are bilaterally symmetrical, some of the large, deep vessels of the trunk region are asymmetrical or unpaired.

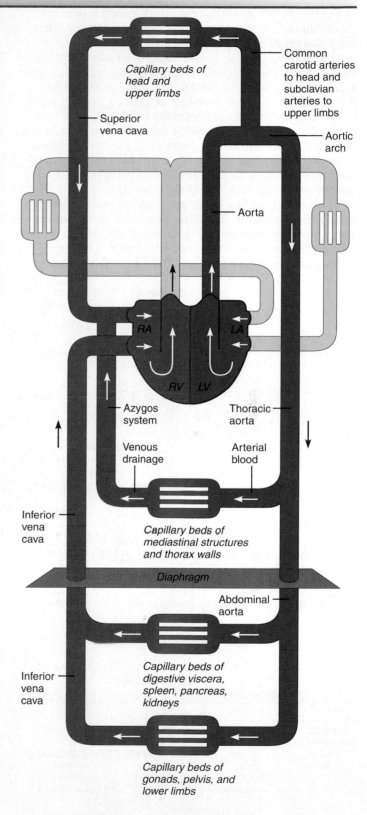

FIGURE 18.18 Schematic flowchart showing an overview of the systemic circulation. The pulmonary circulation is shown in gray for comparison.

---

*By convention, oxygen-rich blood is shown *red* and oxygen-poor blood is shown *blue*.
†Venous blood from the digestive viscera passes through the hepatic portal circulation (liver and associated veins) before entering the inferior vena cava.

TABLE 18.4   The Aorta and Major Arteries of the Systemic Circulation

The distribution of the aorta and major arteries of the systemic circulation is diagrammed in flowchart form in Figure 18.19a and illustrated in Figure 18.19b. Fine points about the various vessels arising from the aorta are provided in Tables 18.5 through 18.8.

The **aorta** is the largest artery in the body. In adults, the aorta (a-or'tah) is approximately the size of a garden hose where it issues from the left ventricle of the heart. Its internal diameter is 2.5 cm, and its wall is about 2 mm thick. It decreases in size slightly as it runs to its terminus. The aortic semilunar valve guards the base of the aorta and prevents backflow of blood during diastole. Opposite each semilunar valve cusp is an *aortic sinus*, which contains baroreceptors important in reflex regulation of blood pressure.

Different portions of the aorta are named according to shape or location. The first portion, the **ascending aorta**, runs posteriorly and to the right of the pulmonary trunk. It persists for only about 5 cm before curving to the left as the aortic arch. The only branches

of the ascending aorta are the **right** and **left coronary arteries**, which supply the myocardium. The **aortic arch**, deep to the sternum, begins and ends at the sternal angle (T_4 level). Its three major branches (R to L) are: (1) the **brachiocephalic trunk** (bra'ke-o-sě-fal''ik; "armhead"), which passes superiorly under the right clavicle and branches into the **right common carotid artery** (kah-rot'id) and the **right subclavian artery**, (2) the **left common carotid artery**, and (3) the **left subclavian artery**. These three vessels provide the arterial supply of the head, neck, upper limbs, and part of the thorax wall. The **thoracic**, or **descending, aorta** runs along the anterior spine from T_5 to T_{12}, sending off numerous small arteries to the thorax wall and viscera before piercing the diaphragm. As it enters the abdominal cavity, it becomes the **abdominal aorta**. This portion supplies the abdominal walls and viscera and ends at the L_4 level, where it splits into the **right** and **left common iliac arteries**, which supply the pelvis and lower limbs.

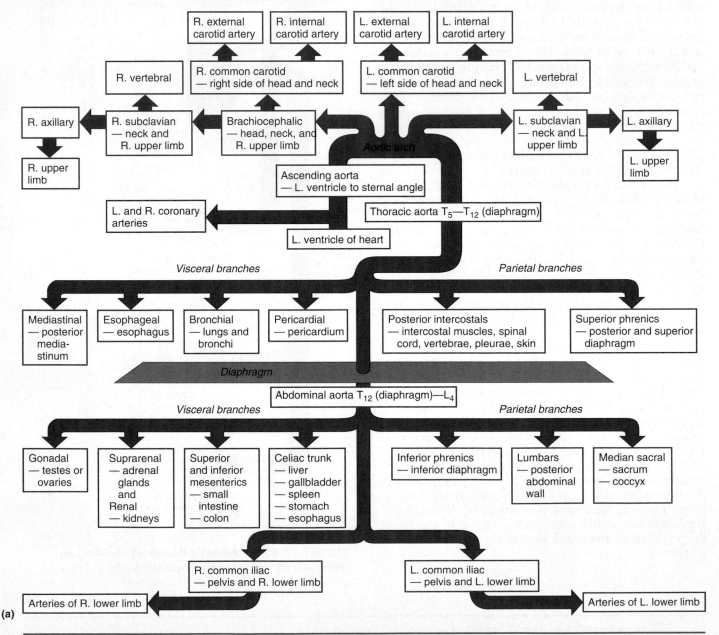

(a)

TABLE 18.4   *(continued)*

FIGURE 18.19   Major arteries of the systemic circulation. (a) Schematic flowchart. **(b)** Illustration, anterior view.

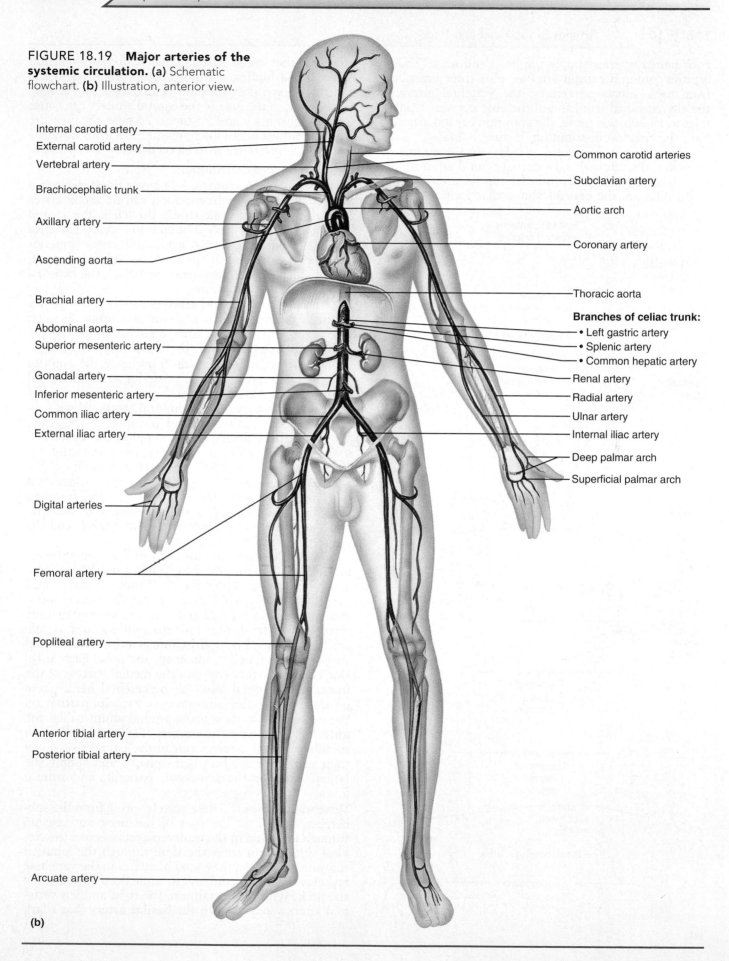

Internal carotid artery

External carotid artery

Vertebral artery

Brachiocephalic trunk

Axillary artery

Ascending aorta

Brachial artery

Abdominal aorta

Superior mesenteric artery

Gonadal artery

Inferior mesenteric artery

Common iliac artery

External iliac artery

Digital arteries

Femoral artery

Popliteal artery

Anterior tibial artery

Posterior tibial artery

Arcuate artery

Common carotid arteries

Subclavian artery

Aortic arch

Coronary artery

Thoracic aorta

Branches of celiac trunk:
• Left gastric artery
• Splenic artery
• Common hepatic artery

Renal artery

Radial artery

Ulnar artery

Internal iliac artery

Deep palmar arch

Superficial palmar arch

(b)

TABLE 18.5   Arteries of the Head and Neck

Four paired arteries supply the head and neck. These are the common carotid arteries, plus three branches from each subclavian artery: the vertebral arteries, the thyrocervical trunks, and the costocervical trunks (Figure 18.20b). Of these, the common carotid arteries have the broadest distribution (Figure 18.20a).

Each common carotid divides into two major branches (the internal and external carotid arteries). At the division point, each internal carotid artery has a slight dilation, the **carotid sinus**, that contains baro-

receptors that assist in reflex blood pressure control. The **carotid bodies**, chemoreceptors involved in the control of respiratory rate, are located close by. Pressing on the neck in the area of the carotid sinuses can cause unconsciousness (*carot* = stupor) because the pressure created mimics high blood pressure, eliciting vasodilation, which interferes with blood delivery to the brain.

Description and Distribution

Common carotid arteries.   The origins of these two arteries differ: The right common carotid artery arises from the brachiocephalic trunk; the left is the second branch of the aortic arch. The common carotid arteries ascend through the lateral neck, and at the superior border of the larynx (the level of the "Adam's apple"), each divides into its two major branches, the *external* and *internal carotid arteries.*

The **external carotid arteries** supply most tissues of the head except for the brain and orbit. As each artery runs superiorly, it sends branches to the thyroid gland and larynx (**superior thyroid artery**), the tongue (**lingual artery**), the skin and muscles of the anterior face (**facial artery**), and the posterior scalp (**occipital artery**). Each external carotid artery terminates by splitting into a **superficial temporal artery**, which supplies the parotid salivary gland and most of the scalp, and a **maxillary artery**, which supplies the upper and lower jaws and chewing muscles, the teeth, and the nasal cavity. A clinically important branch of the maxillary artery is the *middle meningeal artery* (not illustrated). It enters the skull through the foramen spinosum and supplies the inner surface of the parietal bone, squamous region of the temporal bone, and the underlying dura mater.

The larger **internal carotid arteries** supply the orbits and more than 80% of the cerebrum. They assume a deep course and enter the skull through the carotid canals of the temporal bones. Once inside the cranium, each artery gives off one main branch, the ophthalmic artery, and then divides into the anterior and middle cerebral arteries. The **ophthalmic arteries** (of-thal'mik) supply the eyes, orbits, forehead, and nose. Each **anterior cerebral artery** supplies the medial surface of the frontal and parietal lobes of the cerebral hemisphere on its side and also anastomoses with its partner on the opposite side via a short arterial shunt called the **anterior communicating artery** (Figure 18.20d). The **middle cerebral arteries** run in the lateral fissures of their respective cerebral hemispheres and supply the lateral parts of the temporal, parietal, and frontal lobes.

Vertebral arteries.   These vessels spring from the subclavian arteries at the root of the neck and ascend through foramina in the transverse processes of the cervical vertebrae to enter the skull through the foramen magnum. En route, they send branches to the vertebrae and cervical spinal cord and to some deep structures of the neck. Within the cranium, the right and left vertebral arteries join to form the **basilar artery** (bas'ĭ-lar),

(a)

TABLE 18.5 (continued)

which ascends along the anterior aspect of the brain stem, giving off branches to the cerebellum, pons, and inner ear (Figure 18.20b and d). At the pons-midbrain border, the basilar artery divides into a pair of **posterior cerebral arteries,** which supply the occipital lobes and the inferior parts of the temporal lobes.

Arterial shunts called **posterior communicating arteries** connect the posterior cerebral arteries to the middle cerebral arteries anteriorly. The two posterior and single anterior communicating arteries complete the formation of an arterial anastomosis called the **circle of Willis.** This structure encircles the pituitary gland and optic chiasma and unites the brain's anterior and posterior blood supplies. It also equalizes blood pressure in the two brain areas and provides alternate routes for blood to reach the brain tissue if a carotid or vertebral artery becomes occluded.

Thyrocervical and costocervical trunks. These short vessels arise from the subclavian artery just lateral to the vertebral arteries on each side (Figures 18.20b and Figure 18.21). The thyrocervical trunk mainly supplies the thyroid gland, portions of the cervical vertebrae and spinal cord, and some scapular muscles. The costocervical trunk serves deep neck and superior intercostal muscles.

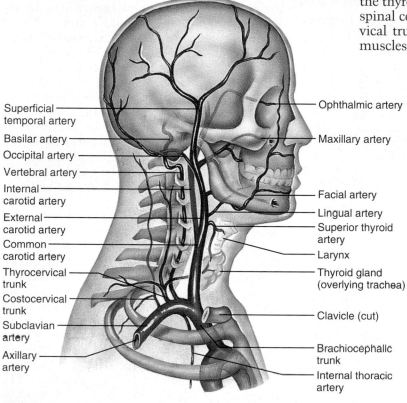

(b)

FIGURE 18.20 **Arteries of the head, neck, and brain. (a)** Schematic flowchart. **(b)** Arteries of the head and neck, right aspect. **(c)** Arteriograph of the arterial supply of the brain. **(d)** Major arteries serving the brain and circle of Willis. In this inferior view of the brain, the right side of the cerebellum and part of the right temporal lobe have been removed to show the distribution of the middle and posterior cerebral arteries.

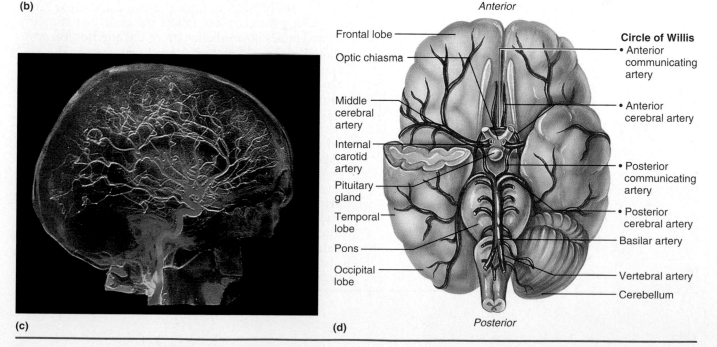

(c)

(d)

TABLE 18.6   Arteries of the Upper Limbs and Thorax

The upper limbs are supplied entirely by arteries arising from the **subclavian arteries** (Figure 18.21a). After giving off branches to the neck, each subclavian artery courses laterally between the clavicle and first rib to enter the axilla, where its name changes to axillary artery. The thorax wall is supplied by an array of vessels that arise either directly from the thoracic aorta or from branches of the subclavian arteries. Most visceral organs of the thorax receive their functional blood supply from small branches issuing from the thoracic aorta. Because these vessels are so small and tend to vary in number (except for the bronchial arteries), they are not illustrated in Figures 18.21a and b, but several of them are listed at the end of this table.

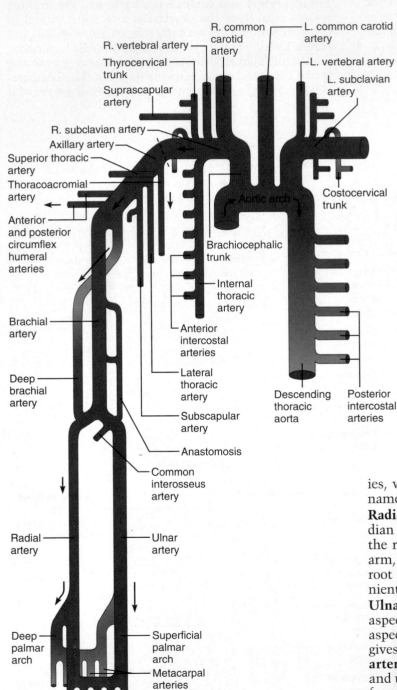

(a)

Description and Distribution
Arteries of the Upper Limb

Axillary artery. As it runs through the axilla accompanied by cords of the brachial plexus, each axillary artery gives off branches to the axilla, chest wall, and shoulder girdle. These branches include the **thoracoacromial artery** (tho"rah-ko-ah-kro'me-al), which supplies the deltoid muscle and pectoral region; the **lateral thoracic artery**, which serves the lateral chest wall and breast; the **subscapular artery** to the scapula, dorsal thorax wall, and part of the latissimus dorsi muscle; and the **anterior and posterior circumflex humeral arteries,** which wrap around the humeral neck and help supply the shoulder joint and the deltoid muscle. As the axillary artery emerges from the axilla, it becomes the brachial artery.

Brachial artery. The brachial artery runs down the medial aspect of the humerus and supplies the anterior flexor muscles of the arm. One major branch, the **deep brachial artery,** serves the posterior triceps brachii muscle. As it nears the elbow, the brachial artery gives off several small branches that contribute to an anastomosis serving the elbow joint and connecting it to the arteries of the forearm. As the brachial artery crosses the anterior midline aspect of the elbow, it provides an easily palpated pulse point (brachial pulse) (see Figure 18.11). Immediately beyond the elbow, the brachial artery splits to form the radial and ulnar arteries, which more or less follow the course of similarly named bones down the length of the anterior forearm.

Radial artery. The radial artery runs from the median line of the cubital fossa to the styloid process of the radius. It supplies the lateral muscles of the forearm, the wrist, and the thumb and index finger. At the root of the thumb, the radial artery provides a convenient site for taking the radial pulse.

Ulnar artery. The ulnar artery supplies the medial aspect of the forearm, fingers 3–5, and the medial aspect of the index finger. Proximally, the ulnar artery gives off a short branch, the **common interosseous artery** (in"ter-os'e-us), which runs between the radius and ulna to serve the deep flexors and extensors of the forearm.

Palmar arches. In the palm, branches of the radial and ulnar arteries anastomose to form the **superficial** and **deep palmar arches.** The **metacarpal arteries** and the

TABLE 18.6 (continued)

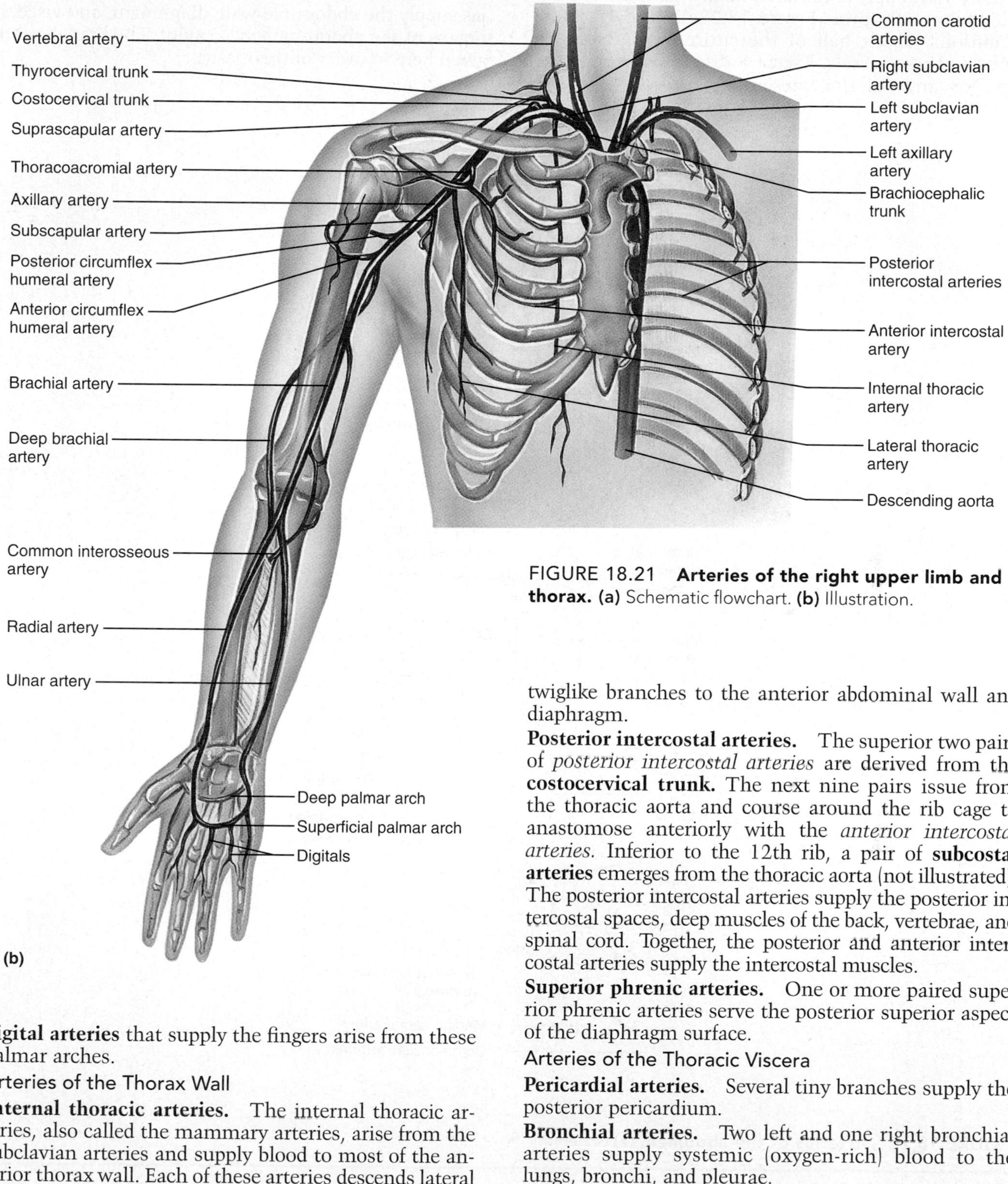

Vertebral artery

Thyrocervical trunk

Costocervical trunk

Suprascapular artery

Thoracoacromial artery

Axillary artery

Subscapular artery

Posterior circumflex humeral artery

Anterior circumflex humeral artery

Brachial artery

Deep brachial artery

Common interosseous artery

Radial artery

Ulnar artery

Deep palmar arch

Superficial palmar arch

Digitals

Common carotid arteries

Right subclavian artery

Left subclavian artery

Left axillary artery

Brachiocephalic trunk

Posterior intercostal arteries

Anterior intercostal artery

Internal thoracic artery

Lateral thoracic artery

Descending aorta

(b)

FIGURE 18.21 **Arteries of the right upper limb and thorax.** **(a)** Schematic flowchart. **(b)** Illustration.

digital arteries that supply the fingers arise from these palmar arches.

Arteries of the Thorax Wall

Internal thoracic arteries. The internal thoracic arteries, also called the mammary arteries, arise from the subclavian arteries and supply blood to most of the anterior thorax wall. Each of these arteries descends lateral to the sternum and gives off **anterior intercostal arteries,** which supply the intercostal spaces anteriorly. The internal thoracic artery also sends superficial branches to the skin and mammary glands and terminates in

twiglike branches to the anterior abdominal wall and diaphragm.

Posterior intercostal arteries. The superior two pairs of *posterior intercostal arteries* are derived from the **costocervical trunk.** The next nine pairs issue from the thoracic aorta and course around the rib cage to anastomose anteriorly with the *anterior intercostal arteries.* Inferior to the 12th rib, a pair of **subcostal arteries** emerges from the thoracic aorta (not illustrated). The posterior intercostal arteries supply the posterior intercostal spaces, deep muscles of the back, vertebrae, and spinal cord. Together, the posterior and anterior intercostal arteries supply the intercostal muscles.

Superior phrenic arteries. One or more paired superior phrenic arteries serve the posterior superior aspect of the diaphragm surface.

Arteries of the Thoracic Viscera

Pericardial arteries. Several tiny branches supply the posterior pericardium.

Bronchial arteries. Two left and one right bronchial arteries supply systemic (oxygen-rich) blood to the lungs, bronchi, and pleurae.

Esophageal arteries. Four to five esophageal arteries supply the esophagus.

Mediastinal arteries. Many small mediastinal arteries serve the posterior mediastinum.

TABLE 18.7   Arteries of the Abdomen

The arterial supply to the abdominal organs arises from the abdominal aorta (Figure 18.22a). Under resting conditions, about half of the entire arterial flow is found in these vessels. Except for the celiac trunk, the superior and inferior mesenteric arteries, and the median sacral artery, all are paired vessels. These arteries supply the abdominal wall, diaphragm, and visceral organs of the abdominopelvic cavity. The branches are given here in order of their issue.

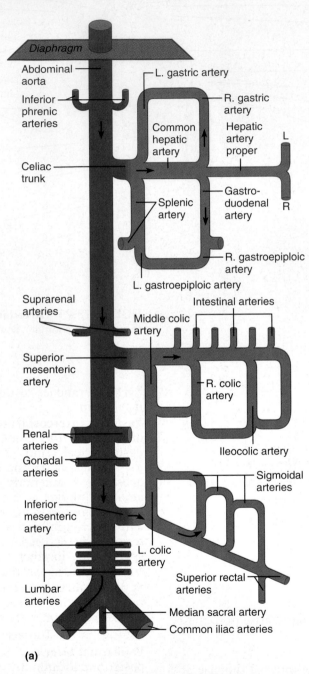

(a)

FIGURE 18.22   Arteries of the abdomen. (a) Schematic flowchart.

TABLE 18.7   *(continued)*

Description and Distribution

Inferior phrenic arteries. The inferior phrenics emerge from the aorta at T$_{12}$, just inferior to the diaphragm. They serve the inferior diaphragm surface.

Celiac trunk. This very large unpaired branch of the abdominal aorta divides almost immediately into three branches: the common hepatic, splenic, and left gastric arteries (Figure 18.22b). The **common hepatic artery** (hĕ-pat'ik) gives off branches to the stomach, duodenum, and pancreas. Where the **gastroduodenal artery** branches off, it becomes the **hepatic artery proper,** which splits into right and left branches that serve the liver. As the **splenic artery** (splen'ik) passes deep to the stomach, it sends branches to the pancreas and stomach and terminates in branches to the spleen. The **left gastric artery** (*gaster* = stomach) supplies part of the stomach and the inferior esophagus. The **right** and **left gastroepiploic arteries** (gas"tro-ep"ĭ-plo'ik), branches of the gastroduodenal and splenic arteries, respectively, serve the left (greater) curvature of the stomach. A **right gastric artery,** which supplies the stomach's right (lesser) curvature, may arise from the common hepatic artery or from the hepatic artery proper.

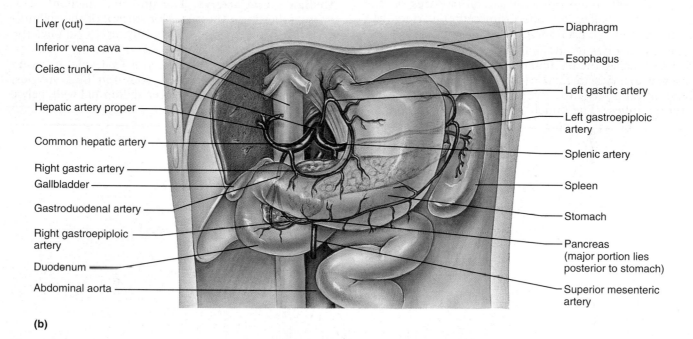

Liver (cut)
Inferior vena cava
Celiac trunk
Hepatic artery proper
Common hepatic artery
Right gastric artery
Gallbladder
Gastroduodenal artery
Right gastroepiploic artery
Duodenum
Abdominal aorta

Diaphragm
Esophagus
Left gastric artery
Left gastroepiploic artery
Splenic artery
Spleen
Stomach
Pancreas (major portion lies posterior to stomach)
Superior mesenteric artery

(b)

FIGURE 18.22 *(continued)* **Arteries of the abdomen. (b)** The celiac trunk and its major branches.

TABLE 18.7 ╱ **Arteries of the Abdomen** (continued)

Superior mesenteric artery (mes-en-ter'ik).   This large, unpaired artery arises from the abdominal aorta at the L_1 level immediately below the celiac trunk (Figure 18.22d). It runs deep to the pancreas and then enters the mesentery, where its numerous anastomosing branches serve virtually all of the small intestine via the **intestinal arteries,** and most of the large intestine—the appendix, cecum, ascending colon (via the **ileocolic artery**), and part of the transverse colon (via the **right** and **middle colic arteries**).

Suprarenal arteries (soo"prah-re'nal).   The suprarenal arteries flank the origin of the superior mesenteric artery as they emerge from the abdominal aorta (Figure 18.22c). They supply blood to the adrenal (suprarenal) glands overlying the kidneys.

Renal arteries.   The short but wide renal arteries, right and left, issue from the lateral surfaces of the aorta slightly below the superior mesenteric artery (between L_1 and L_2). Each serves the kidney on its side.

Gonadal arteries (go-nă'dul).   The paired gonadal arteries are called the **testicular arteries** in males and the **ovarian arteries** in females. The ovarian arteries extend into the pelvis to serve the ovaries and part of the uterine tubes. The much longer testicular arteries descend through the pelvis and inguinal canal to enter the scrotal sac, where they serve the testes.

Inferior mesenteric artery.   This final major branch of the abdominal aorta is unpaired and arises from the anterior aortic surface at the L_3 level. It serves the distal part of the large intestine—from the midpart of the transverse colon to the midrectum—via its **left colic, sigmoidal,** and **superior rectal branches** (Figure 18.22d). Looping anastomoses between the superior and inferior mesenteric arteries help ensure that blood will continue to reach the digestive viscera in cases of trauma to one of these abdominal arteries.

Lumbar arteries.   Four pairs of lumbar arteries arise from the posterolateral surface of the aorta in the lumbar region. These segmental arteries supply the posterior abdominal wall.

Median sacral artery.   The unpaired median sacral artery issues from the posterior surface of the abdominal aorta at its terminus. This tiny artery supplies the sacrum and coccyx.

Common iliac arteries.   At the L_4 level, the aorta splits into the right and left common iliac arteries, which supply blood to the lower abdominal wall, pelvic organs, and lower limbs (Figure 18.22c).

TABLE 18.7   *(continued)*

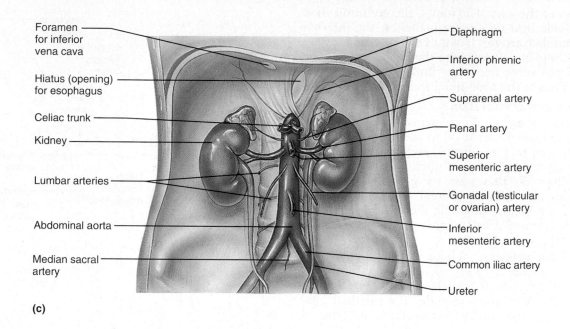

(c)

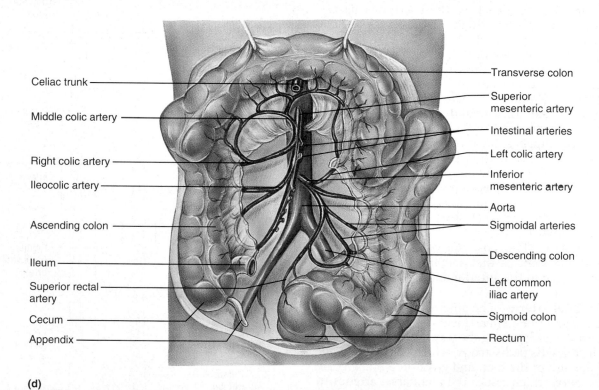

(d)

FIGURE 18.22 *(continued)*   **Arteries of the abdomen. (c)** Major branches of the abdominal aorta. **(d)** Distribution of the superior and inferior mesenteric arteries. (The transverse colon has been reflected superiorly to provide a better view of these arteries.)

TABLE 18.8 Arteries of the Pelvis and Lower Limbs

At the level of the sacroiliac joints, the **common iliac arteries** divide into two major branches, the internal and external iliac arteries (Figure 18.23a). The internal iliacs distribute blood mainly to the pelvic region. The external iliacs serve the lower limbs; they also send some branches to the abdominal wall.

Description and Distribution

Internal iliac arteries. These paired arteries run into the pelvis and distribute blood to the pelvic walls and viscera (bladder, rectum, uterus, and vagina in the female and prostate gland and ductus deferens in the male). Additionally they serve the gluteal muscles via the **superior** and **inferior gluteal arteries,** adductor muscles of the medial thigh via the **obturator artery,** and external genitalia and perineum via the **internal pudendal artery** (not illustrated).

External iliac arteries. These arteries supply the lower limbs (Figure 18.23b). As they course through the pelvis, they give off branches to the anterior abdominal wall. After passing under the inguinal ligaments to enter the thigh, they become the femoral arteries.

Femoral arteries. As each of these arteries passes down the anteromedial thigh, it gives off several branches to the thigh muscles. The largest of the deep branches is the **deep femoral artery** (also more simply called **deep artery of the thigh**) that is the main supply to the thigh muscles (hamstrings, quadriceps, and adductors). Proximal branches of the deep femoral artery, the **lateral** and **medial circumflex femoral arteries,** encircle the neck of the femur. The medial circumflex artery supplies the head and neck of the femur. A long descending branch of the posterior circumflex artery supplies the vastus lateralis muscle. Near the knee the femoral artery passes posteriorly and through a gap in the adductor magnus muscle, the *adductor hiatus,* to enter the popliteal fossa, where its name changes to popliteal artery.

Popliteal artery. This posterior vessel contributes to an arterial anastomosis that supplies the knee region and then splits into the anterior and posterior tibial arteries of the leg.

Anterior tibial artery. The anterior tibial artery runs through the anterior compartment of the leg, supplying the extensor muscles along the way. At the ankle, it becomes the **dorsalis pedis artery,** which supplies the ankle and dorsum of the foot, and gives off a branch, the **arcuate artery,** which issues the metatarsal arteries to the metatarsus of the foot. The superficial dorsalis pedis ends by penetrating into the sole where it forms the medial part of the **plantar arch.** The dorsalis pedis artery provides a clinically important pulse point, the pedal pulse. If the pedal pulse is easily felt, it is fairly certain that the blood supply to the leg is good.

Posterior tibial artery. This large artery courses through the posteromedial part of the leg and supplies the flexor muscles. Proximally, it gives off a large branch, the **fibular (peroneal) artery,** which supplies the lateral

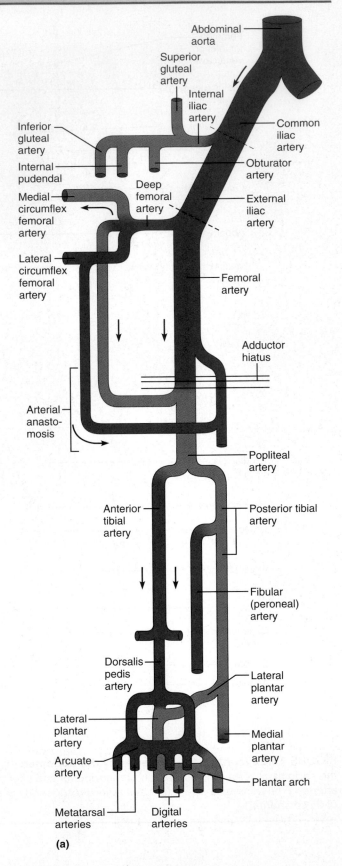

(a)

FIGURE 18.23 Arteries of the right pelvis and lower limb. (a) Schematic flowchart.

TABLE 18.8 *(continued)*

fibularis (peroneal) muscles of the leg. At the ankle, the posterior tibial artery divides into **lateral** and **medial plantar arteries** that serve the plantar surface of the foot. The lateral plantar artery forms the lateral end of the plantar arch. The **digital arteries** serving the toes arise from the plantar arch formed by the lateral plantar artery.

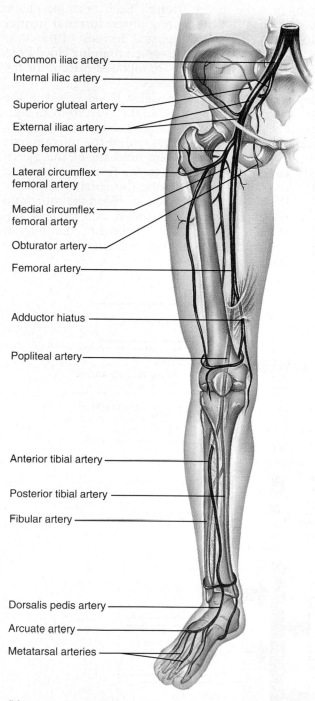

Common iliac artery
Internal iliac artery
Superior gluteal artery
External iliac artery
Deep femoral artery
Lateral circumflex femoral artery
Medial circumflex femoral artery
Obturator artery
Femoral artery
Adductor hiatus
Popliteal artery
Anterior tibial artery
Posterior tibial artery
Fibular artery
Dorsalis pedis artery
Arcuate artery
Metatarsal arteries

(b)

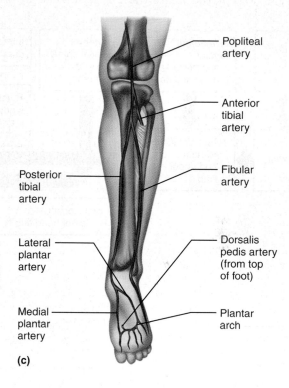

Popliteal artery
Anterior tibial artery
Posterior tibial artery
Fibular artery
Lateral plantar artery
Dorsalis pedis artery (from top of foot)
Medial plantar artery
Plantar arch

(c)

FIGURE 18.23 *(continued)* **Arteries of the right pelvis and lower limb.**
(b) Illustration, anterior view. **(c)** Posterior view of the leg and foot.

In our survey of the systemic veins, the major tributaries (branches) of the venae cavae are noted first in Figure 18.24, followed by a description in Tables 18.10 through 18.13 of the venous pattern of the various body regions. Because veins run toward the heart, the most distal veins are named first and those closest to the heart last. Because deep veins generally drain the same areas served by their companion arteries, they are not described in detail.

Description and Areas Drained

Superior vena cava. This great vein receives systemic blood draining from all areas superior to the diaphragm, except the heart wall. It is formed by the union of the **right** and **left brachiocephalic veins** and empties into the right atrium (Figure 18.24b). Notice that there are two brachiocephalic veins, but only one brachiocephalic artery (trunk). Each brachiocephalic vein is formed by the joining of the **internal jugular** and **subclavian veins** on its side. In most of the flowcharts that follow, only the vessels draining blood from the right side of the body are followed (except for the azygos circulation of the thorax).

Inferior vena cava. The widest blood vessel in the body, this vein returns blood to the heart from all body regions below the diaphragm. The abdominal aorta lies directly to its left. The distal end of the inferior vena cava is formed by the junction of the paired **common iliac veins** at L_5. From this point, it courses superiorly along the anterior aspect of the spine, receiving venous blood draining from the abdominal walls, gonads, and kidneys. Immediately above the diaphragm, the inferior vena cava ends as it enters the inferior aspect of the right atrium.

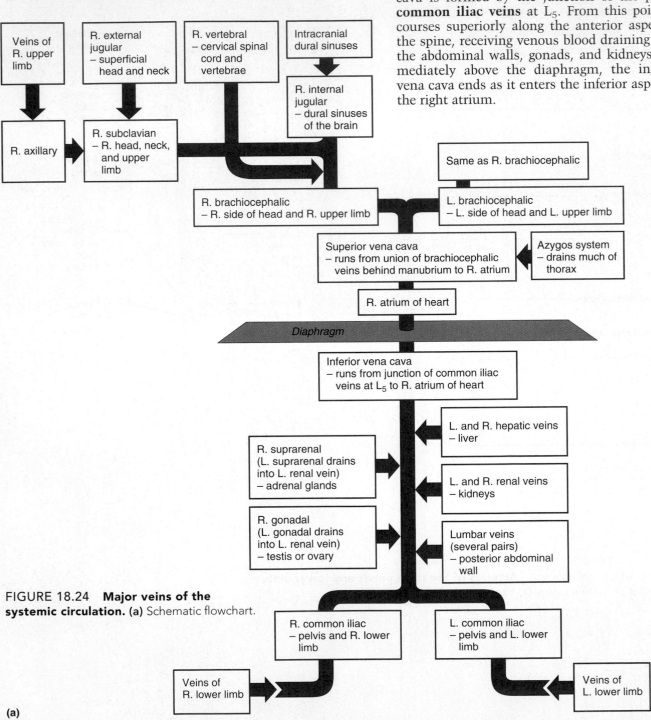

FIGURE 18.24 **Major veins of the systemic circulation. (a)** Schematic flowchart.

(a)

TABLE 18.9   *(continued)*

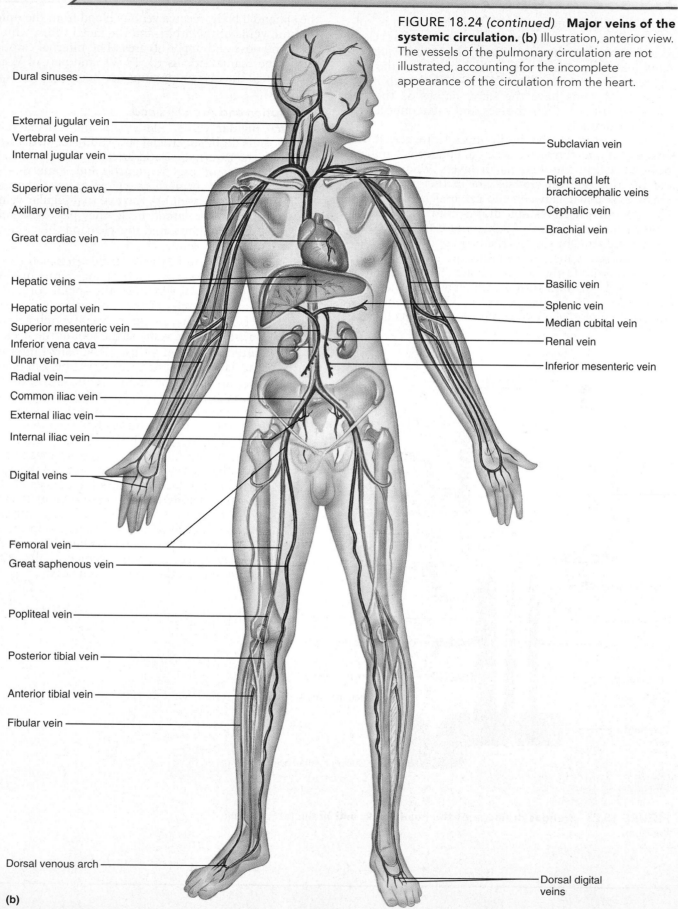

FIGURE 18.24 *(continued)* **Major veins of the systemic circulation. (b)** Illustration, anterior view. The vessels of the pulmonary circulation are not illustrated, accounting for the incomplete appearance of the circulation from the heart.

Dural sinuses

External jugular vein
Vertebral vein
Internal jugular vein

Superior vena cava

Axillary vein

Great cardiac vein

Hepatic veins

Hepatic portal vein
Superior mesenteric vein
Inferior vena cava
Ulnar vein
Radial vein
Common iliac vein
External iliac vein
Internal iliac vein

Digital veins

Femoral vein
Great saphenous vein

Popliteal vein

Posterior tibial vein

Anterior tibial vein

Fibular vein

Dorsal venous arch

Subclavian vein

Right and left
brachiocephalic veins
Cephalic vein
Brachial vein

Basilic vein
Splenic vein
Median cubital vein
Renal vein

Inferior mesenteric vein

Dorsal digital
veins

(b)

TABLE 18.10   Veins of the Head and Neck

Most blood draining from the head and neck is collected by three pairs of veins: the external jugular veins, which empty into the subclavians, the internal jugular veins, and the vertebral veins, which drain into the brachiocephalic vein (see Figure 18.25a). Although most extracranial veins have the same names as the extracranial arteries, their courses and interconnections differ substantially.

Most veins of the brain drain into the **dural sinuses**, an interconnected series of enlarged chambers located between the dura mater layers. The **superior** and **inferior sagittal sinuses** are in the falx cerebri, which dips down between the cerebral hemispheres. The inferior sagittal sinus drains into the **straight sinus** posteriorly (Figures 18.25a and c). The superior sagittal and straight sinuses then empty into the **transverse sinuses**, which run in shallow grooves on the internal surface of the occipital bone. These drain into the S-shaped **sigmoid sinuses**, which become the *internal jugular veins* as they leave the skull through the jugular foramen. The **cavernous sinuses**, which flank the sphenoid body, receive venous blood from the **ophthalmic veins** of the orbits and the facial veins, which drain the nose and upper lip area. The internal carotid artery and cranial nerves III, IV, VI, and part of V, all run *through* the cavernous sinus on their way to the orbit and face.

Description and Area Drained

External jugular veins.   The right and left external jugular veins drain superficial scalp and face structures served by the external carotid arteries. However, their tributaries anastomose frequently, and some of the superficial drainage from these regions enters the internal jugular veins as well. As the external jugular veins descend through the lateral neck, they pass obliquely over the sternocleidomastoid muscles and then empty into the subclavian veins.

Vertebral veins.   Unlike the vertebral arteries, the vertebral veins do not serve much of the brain. Instead they drain the cervical vertebrae, the spinal cord, and some small neck muscles. They run inferiorly through the transverse foramina of the cervical vertebrae and join the brachiocephalic veins at the root of the neck.

Internal jugular veins.   The paired internal jugular veins, which receive the bulk of blood draining from the brain, are the largest of the paired veins draining the head and neck. They arise from the dural venous sinuses, exit the skull via the *jugular foramina*, and then descend through the neck alongside the internal carotid arteries. As they move inferiorly, they receive blood from some of the deep veins of the face and neck—branches of the **facial** and **superficial temporal veins** (Figure 18.25b). At the base of the neck, each internal jugular vein joins the subclavian vein on its own side to form a brachiocephalic vein. As already noted, the two brachiocephalic veins unite to form the superior vena cava.

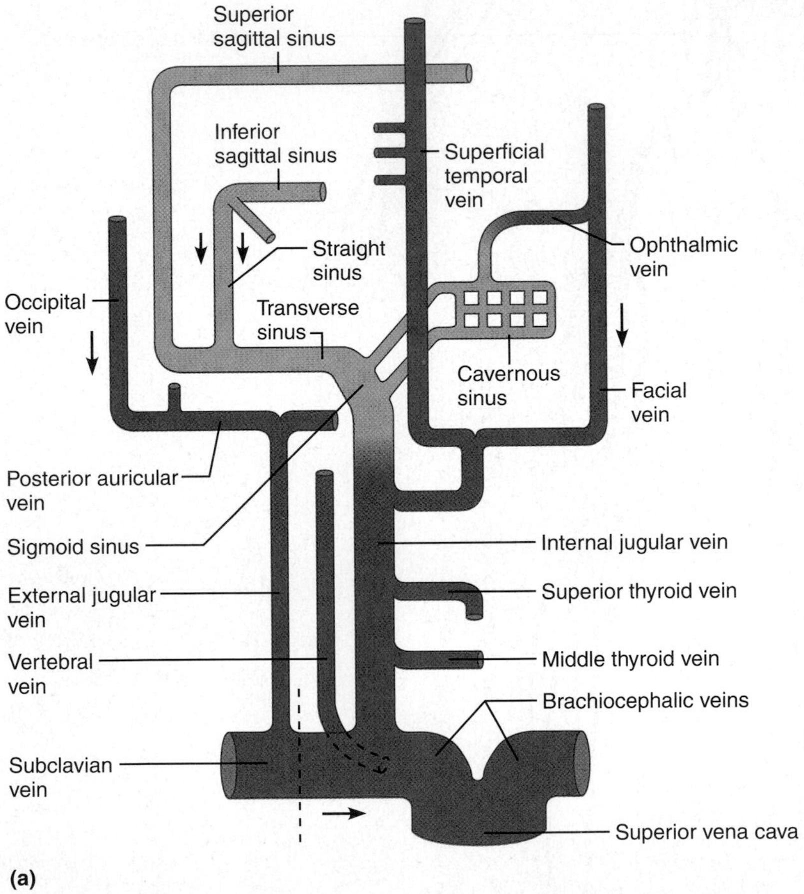

(a)

FIGURE 18.25   **Venous drainage of the head, neck, and brain. (a)** Schematic flowchart.

TABLE 18.10   (continued)

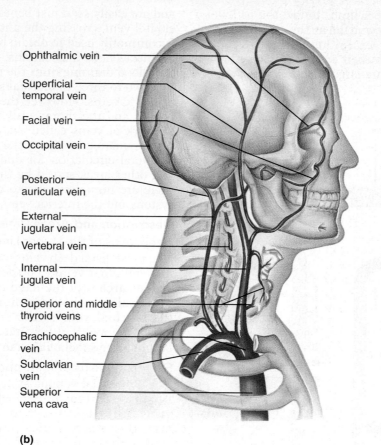

Ophthalmic vein

Superficial
temporal vein

Facial vein

Occipital vein

Posterior
auricular vein

External
jugular vein

Vertebral vein

Internal
jugular vein

Superior and middle
thyroid veins

Brachiocephalic
vein

Subclavian
vein

Superior
vena cava

(b)

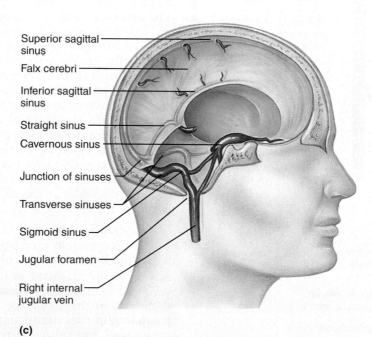

Superior sagittal
sinus

Falx cerebri

Inferior sagittal
sinus

Straight sinus

Cavernous sinus

Junction of sinuses

Transverse sinuses

Sigmoid sinus

Jugular foramen

Right internal
jugular vein

(c)

FIGURE 18.25 *(continued)* **Venous drainage of the head, neck, and brain.**
(b) Veins of the head and neck, right superficial aspect. **(c)** Dural sinuses of the
brain, right aspect.

TABLE 18.11   Veins of the Upper Limbs and Thorax

The deep veins of the upper limbs follow the paths of their companion arteries and have the same names (Figure 18.26a). However, except for the largest, most are paired veins that flank their artery. The superficial veins of the upper limbs are larger than the deep veins and are easily seen just beneath the skin. The median cubital vein, crossing the anterior aspect of the elbow, is commonly used to obtain blood samples or administer intravenous medications.

Blood draining from the mammary glands and the first two to three intercostal spaces enters the **brachiocephalic veins.** However, the vast majority of thoracic tissues and the thorax wall are drained by a complex network of veins called the **azygos system** (az'ĭ-gos). The branching nature of the azygos system provides a collateral circulation for draining the abdominal wall and other areas served by the inferior vena cava, and there are numerous anastomoses between the azygos system and the inferior vena cava.

Description and Areas Drained
Deep Veins of the Upper Limbs
The most distal deep veins of the upper limb are the radial and ulnar veins. The deep and superficial **palmar venous arches** of the hand empty into the **radial** and **ulnar veins** of the forearm, which then unite to form the **brachial vein** of the arm. As the brachial vein enters the axilla, it becomes the **axillary vein,** which becomes the **subclavian vein** at the level of the first rib.

Superficial Veins of the Upper Limbs
The superficial venous system begins with the **dorsal venous arch** (not illustrated), a plexus of superficial veins in the dorsum of the hand. In the distal forearm, this plexus drains into three major superficial veins—the cephalic and basilic veins and the median antebrachial vein of the forearm—which anastomose frequently as they course upward (see Figure 18.26b). The **cephalic vein** coils around the radius as it travels superiorly and then continues up the lateral superficial aspect of the arm to the shoulder, where it runs in the groove between the deltoid and pectoralis muscles to join the axillary vein. The **basilic vein** courses along the posteromedial aspect of the forearm, crosses the elbow, and then takes a deep course. In the axilla, it joins the brachial vein, forming the axillary vein. At the anterior aspect of the elbow, the **median cubital vein** connects the basilic and cephalic veins. The **median antebrachial vein of the forearm** lies between the radial and ulnar veins in the forearm and terminates (variably) at the elbow by entering either the basilic or the cephalic vein.

The Azygos System
The azygos system consists of the following vessels which flank the vertebral column laterally:

Azygos vein. Located against the right side of the vertebral column, the **azygos vein** (*azygos* = unpaired) originates in the abdomen, from the **right ascending lumbar vein** that drains most of the right abdominal cavity wall and from the **right posterior intercostal veins** (except the first) that drain the chest muscles. At the T_4 level, it arches over the great vessels that run to the right lung and empties into the superior vena cava.

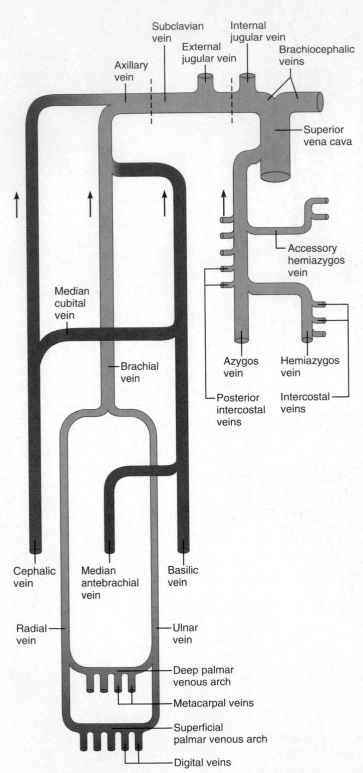

(a)

TABLE 18.11 *(continued)*

Hemiazygos vein (hĕ'me-a-zi'gus; "half the azygos"). This vessel ascends on the left side of the vertebral column. Its origin, from the **left ascending lumbar vein** and the lower (9th–11th) **posterior intercostal veins**, mirrors that of the inferior portion of the azygos vein on the right. About midthorax, the hemiazygos vein passes in front of the vertebral column and joins the azygos vein.

Accessory hemiazygos vein. The accessory hemiazygos completes the venous drainage of the left (middle) thorax and can be thought of as a superior continuation of the hemiazygos vein. It receives blood from the 4th–8th posterior intercostal veins and then crosses to the right to empty into the azygos vein. Like the azygos, it receives venous blood from the lungs *(bronchial veins)*.

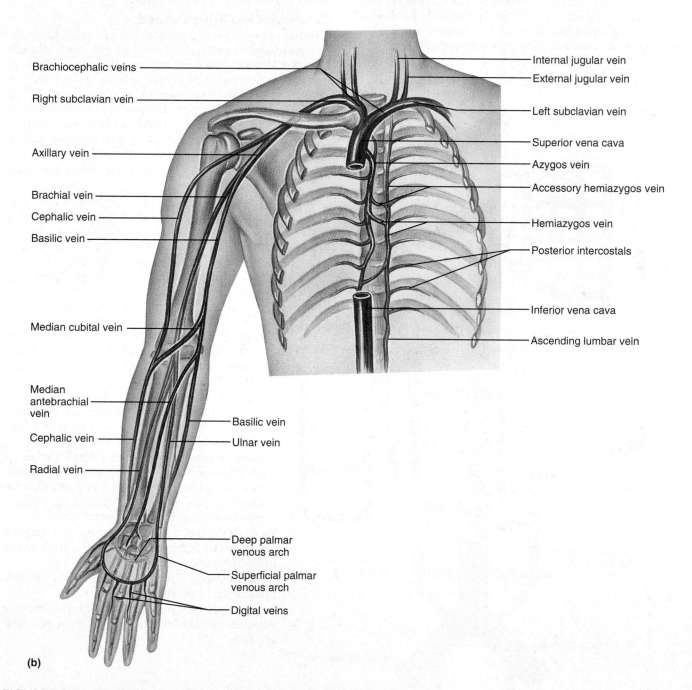

(b)

FIGURE 18.26 **Veins of the right upper limb and shoulder.** **(a)** Schematic flowchart. **(b)** Illustration. For clarity, the abundant branching and anastomoses of these vessels are not shown.

TABLE 18.12   Veins of the Abdomen

Blood draining from the abdominopelvic viscera and abdominal walls is returned to the heart by the **inferior vena cava** (Figure 18.27a). Most of its venous tributaries have names that correspond to the arteries serving the abdominal organs.

Veins draining the digestive viscera empty into a common vessel, the *hepatic portal vein,* which transports this venous blood into the liver before it is allowed to enter the major systemic circulation via the hepatic veins (Figure 18.27b). Such a venous system— veins to capillaries (or sinusoids) to veins—is called a *portal system* and always serves very specific regional

tissue needs. The hepatic portal system carries nutrient-rich blood from the digestive organs to the liver. As the blood percolates slowly through the liver sinusoids, hepatic parenchymal cells remove the nutrients they need for their various metabolic functions, and phagocytic cells lining the sinusoids rid the blood of bacteria and other foreign matter that has penetrated the digestive mucosa. The veins of the abdomen are listed in inferior to superior order.

Description and Areas Drained

Lumbar veins.   Several pairs of lumbar veins drain the posterior abdominal wall. They empty both directly into the inferior vena cava and into the ascending lumbar veins of the azygos system of the thorax.

Gonadal (testicular or ovarian) veins.   The right gonadal vein drains the ovary or testis on the right side of the body and empties into the inferior vena cava. The left member drains into the left renal vein superiorly.

Renal veins.   The right and left renal veins drain the kidneys.

Suprarenal veins.   The right suprarenal vein drains the adrenal gland on the right and empties into the inferior vena cava. The left suprarenal vein drains into the left renal vein.

Hepatic portal system.   The short **hepatic portal vein** begins at the L_2 level. Numerous tributaries from the stomach and pancreas contribute to the hepatic portal system (Figure 18.27c), but the major vessels are as follows:

- **Superior mesenteric vein:** Drains the entire small intestine, part of the large intestine (ascending and transverse regions), and stomach.
- **Splenic vein:** Collects blood from the spleen, parts of the stomach and pancreas, and then joins the superior mesenteric vein to form the hepatic portal vein.
- **Inferior mesenteric vein:** Drains the distal portions of the large intestine and rectum and joins the splenic vein just before that vessel unites with the superior mesenteric vein to form the hepatic portal vein.

Hepatic veins.   The right and left hepatic veins carry venous blood from the liver to the inferior vena cava.

Cystic veins.   The cystic veins drain the gallbladder and join the hepatic veins.

Inferior phrenic veins.   The inferior phrenic veins drain the inferior surface of the diaphragm.

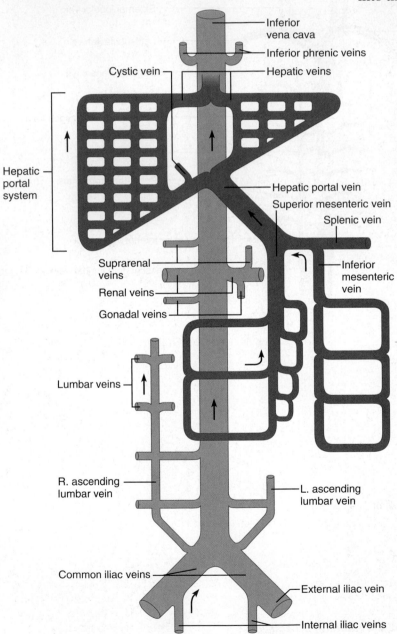

(a)

TABLE 18.12 *(continued)*

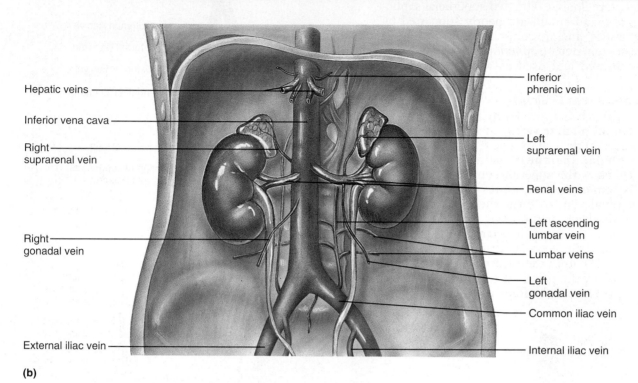

Hepatic veins

Inferior vena cava

Right suprarenal vein

Right gonadal vein

External iliac vein

Inferior phrenic vein

Left suprarenal vein

Renal veins

Left ascending lumbar vein

Lumbar veins

Left gonadal vein

Common iliac vein

Internal iliac vein

(b)

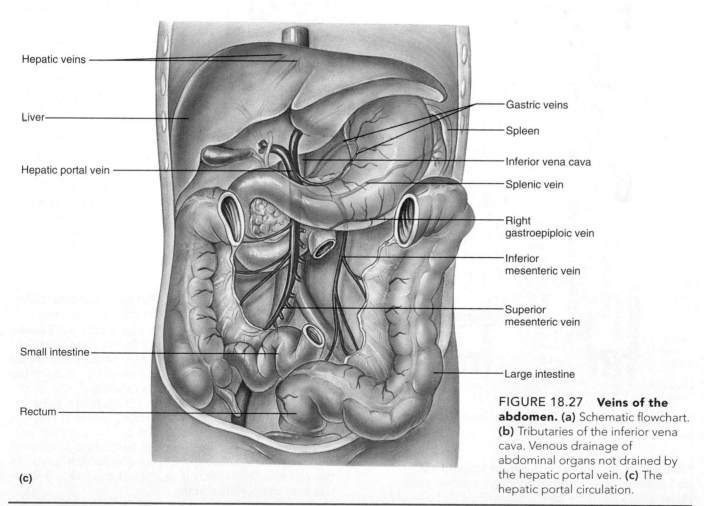

Hepatic veins

Liver

Hepatic portal vein

Small intestine

Rectum

Gastric veins

Spleen

Inferior vena cava

Splenic vein

Right gastroepiploic vein

Inferior mesenteric vein

Superior mesenteric vein

Large intestine

(c)

FIGURE 18.27   Veins of the abdomen. (a) Schematic flowchart. **(b)** Tributaries of the inferior vena cava. Venous drainage of abdominal organs not drained by the hepatic portal vein. **(c)** The hepatic portal circulation.

TABLE 18.13 Veins of the Pelvis and Lower Limbs

As in the upper limbs, most deep veins of the lower limbs have the same names as the arteries they accompany and many are double. The two superficial saphenous veins (great and small) are poorly supported by surrounding tissues, and are common sites of varicosities. The great saphenous (*saphenous* = obvious) vein is frequently excised and used as a coronary bypass vessel.

Description and Areas Drained

Deep veins. After being formed by the union of the **medial** and **lateral plantar veins**, the **posterior tibial vein** ascends deep in the calf muscle (Figure 18.28) and receives the **fibular (peroneal) vein**. The **anterior tibial vein**, which is the superior continuation of the **dorsalis pedis vein** of the foot, unites at the knee with the posterior tibial vein to form the **popliteal vein**, which crosses the back of the knee. As the popliteal vein emerges from the knee, it becomes the **femoral vein**, which drains the deep structures of the thigh. The femoral vein becomes the **external iliac vein** as it enters the pelvis. In the pelvis, the external iliac vein unites with the **internal iliac vein** to form the

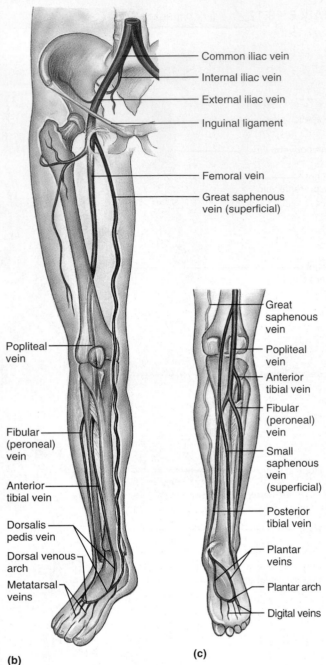

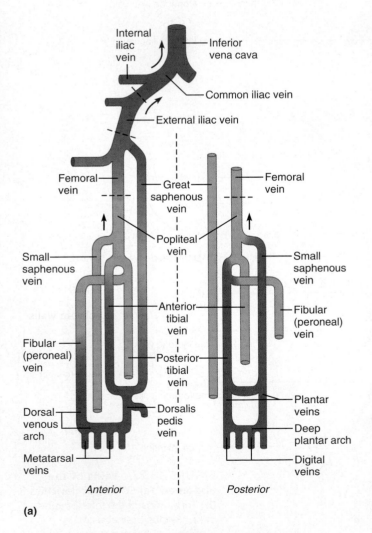

FIGURE 18.28 **Veins of the right lower limb.**
(a) Schematic flowchart of anterior and posterior vessels.
(b) Anterior view of the lower limb. **(c)** Posterior view of the leg and foot.

common iliac vein. The distribution of the internal iliac veins parallels that of the internal iliac arteries.

Superficial veins. The **great** and **small saphenous veins** (sah-fe'nus) issue from the **dorsal venous arch** of the foot (Figure 18.28b and c). These veins anastomose frequently with each other and with the deep veins along their course. The great saphenous vein is the longest vein in the body. It travels superiorly along the medial aspect of the leg to the thigh, where it empties into the femoral vein just distal to the inguinal ligament. The small saphenous vein runs along the lateral aspect of the foot and then through the deep fascia of the calf muscles, which it drains. At the knee, it empties into the popliteal vein.

the systemic arteries and veins, there are also important differences:

1. Whereas the heart pumps all of its blood into a single systemic artery—the aorta—blood returning to the heart is delivered largely by two terminal systemic veins, the superior and inferior venae cavae. The single exception to this is the blood draining from the myocardium of the heart, which is collected by the cardiac veins and reenters the right atrium via the coronary sinus.

2. All arteries run deep and are well protected by body tissues along most of their course, but both deep and superficial veins exist. Deep veins parallel the course of the systemic arteries, and with a few exceptions, the naming of these veins is identical to that of their companion arteries. Superficial veins run just beneath the skin and are readily seen, especially in the limbs, face, and neck. Because there are no superficial arteries, the names of the superficial veins do not correspond to the names of any of the arteries.

3. Unlike the fairly clear arterial pathways, venous pathways tend to have numerous interconnections, and many veins are represented by not one but two similarly named vessels. As a result, venous pathways are more difficult to follow.

4. In most body regions, there is a similar and predictable arterial supply and venous drainage. However, the venous drainage pattern in at least two important body areas is unique. First, venous blood draining from the brain enters large *dural sinuses* rather than typical veins. Second, blood draining from the digestive organs enters a special subcirculation, the *hepatic portal circulation*, and perfuses through the liver before it reenters the general systemic circulation.

Notice that by convention, oxygen-rich blood is shown red, while blood that is relatively oxygen-poor is depicted blue, regardless of vessel types. A unique convention used in the schematic flowcharts (pipe diagrams) that accompany each table is that the vessels that would be closer to the viewer are shown in brighter, more intense colors than those deeper or farther from the viewer—for example, for veins, the darker blue vessels would be closer to the viewer in the body region shown.

Review Questions

Multiple Choice/Matching

(Some questions have more than one correct answer. Select the best answer or answers from the choices given.)

1. Which statement does not accurately describe veins? (a) They have less elastic tissue and smooth muscle than arteries, (b) they contain more fibrous tissue than arteries, (c) most veins in the extremities have valves, (d) they always carry deoxygenated blood.

2. Which of the following tissues is mainly responsible for vasoconstriction? (a) elastic tissue, (b) smooth muscle, (c) collagenic tissue, (d) adipose tissue.

3. Peripheral resistance (a) is inversely related to the diameter of the arterioles, (b) tends to increase if blood viscosity increases, (c) is directly proportional to the length of the vascular bed, (d) all of these.

4. Which of the following can lead to decreased venous return of blood to the heart? (a) an increase in blood volume, (b) an increase in venous pressure, (c) damage to the venous valves, (d) increased muscular activity.

5. Arterial blood pressure increases in response to: (a) increasing stroke volume, (b) increasing heart rate, (c) arteriosclerosis, (d) rising blood volume, (e) all of these.

6. Which of the following would *not* result in the dilation of the feeder arterioles and opening of the precapillary sphincters in any capillary bed? (a) a decrease in O_2 content of the blood, (b) an increase in CO_2 content of the blood, (c) a local increase in histamine, (d) a local increase in pH.

7. The structure of a capillary wall differs from that of a vein or an artery because (a) it has two tunics instead of three, (b) there is less smooth muscle, (c) it has a single tunic—only the tunica interna, (d) none of these.

8. The baroreceptors in the carotid and aortic bodies are sensitive to (a) a decrease in carbon dioxide, (b) changes in arterial pressure, (c) a decrease in oxygen, (d) all of these.

9. The myocardium receives its blood supply directly from (a) the aorta, (b) the coronary arteries, (c) the coronary sinus, (d) the pulmonary arteries.

10. Blood flow in the capillaries is steady despite the rhythmic pumping of the heart because of the (a) elasticity of the large arteries, (b) small diameter of capillaries, (c) thin walls of the veins, (d) venous valves.

11. Tracing the blood from the heart to the right hand, we find that blood leaves the heart and passes through the aorta, the right subclavian artery, the axillary and brachial arteries, and through either the radial or ulnar artery to arrive at the hand. Which artery is missing from this sequence? (a) coronary, (b) brachiocephalic, (c) cephalic, (d) right common carotid.

12. Which of the following do not drain directly into the inferior vena cava? (a) lumbar veins, (b) hepatic veins, (c) inferior mesenteric vein, (d) renal veins.

13. In atherosclerosis, which layer of the vessel wall thickens most? (a) tunica media, (b) tunica interna, (c) tunica adventitia, (d) tunica externa.

Short Answer Essay Questions

14. How is the anatomy of capillaries and capillary beds well suited to their function?

15. Distinguish between elastic arteries, muscular arteries, and arterioles relative to location, histology, and functional adaptations.

16. Write an equation showing the relationship between peripheral resistance, blood flow, and blood pressure.

17. (a) Define blood pressure. Differentiate between systolic and diastolic blood pressure. (b) What is the normal blood pressure value for a young adult?

18. Describe the neural mechanisms responsible for controlling blood pressure.

19. Explain the reasons for the observed changes in blood flow velocity in the different regions of the circulation.

20. How does the control of blood flow to the skin for the purpose of regulating body temperature differ from the control of nutrient blood flow to skin cells?

21. Describe neural and chemical (both systemic and local) effects exerted on the blood vessels when one is fleeing from a mugger. (Be careful, this is more involved than it appears at first glance.)

22. How are nutrients, wastes, and respiratory gases transported to and from the blood and tissue spaces?

23. (a) What blood vessels contribute to the formation of the hepatic portal circulation? (b) What is the function of this circulation? (c) Why is a portal circulation a "strange" circulation?

24. Physiologists often consider capillaries and postcapillary venules together. (a) What functions do these vessels share? (b) Structurally, how do they differ?

19

THE LYMPHATIC SYSTEM

Lymphatic Vessels (pp. 662–664)

1. Describe the structure and distribution of lymphatic vessels, and note their important functions.
2. Describe the source of lymph and mechanism(s) of lymph transport.

Lymphoid Cells and Tissues (p. 665)

3. Describe the basic structure and cellular population of lymphoid tissue, and name the major lymphoid organs.

Lymph Nodes (pp. 666–667)

4. Describe the general location, histological structure, and functions of lymph nodes.

Other Lymphoid Organs (pp. 667–670)

5. Name and describe the other lymphoid organs of the body. Compare and contrast them with lymph nodes, structurally and functionally.

They can't all be superstars! When we mentally tick off the names of the body's organ systems, the lymphatic (lim-fat'ik) system is probably not the first to come to mind. Yet without this quietly working system, our cardiovascular system would stop working and our immune system would be hopelessly impaired. The **lymphatic system** actually consists of two semi-independent parts: (1) a meandering network of *lymphatic vessels* and (2) various *lymphoid tissues* and *organs* scattered throughout the body. The lymphatic vessels transport back to the blood any fluids that have escaped from the blood vascular system. The lymphoid organs house phagocytic cells and lymphocytes, which play essential roles in the body is defense mechanisms and its resistance to disease.

Lymphatic Vessels

As blood circulates through the body, nutrients, wastes, and gases are exchanged between the blood and the interstitial fluid. As explained in Chapter 18, the hydrostatic and colloid osmotic pressures operating at capillary beds force fluid out of the blood at the arterial ends of the beds ("upstream") and cause most of it to be reabsorbed at the venous ends ("downstream"). The fluid that remains behind in the tissue spaces, as much as 3 L daily, becomes part of the interstitial fluid. This leaked fluid, plus any plasma proteins that escape from the bloodstream, must be carried back to the blood to ensure that the cardiovascular system has sufficient blood volume to operate properly. This problem of circulatory dynamics is resolved by the **lymphatic vessels,** or **lymphatics,** an elaborate system of drainage vessels that collect the excess protein-containing interstitial fluid and return it to the bloodstream. Once interstitial fluid enters the lymphatics, it is called **lymph** (*lymph* = clear water).

Distribution and Structure of Lymphatic Vessels

The lymphatic vessels form a one-way system in which lymph flows only toward the heart. This transport system begins in microscopic blind-ended **lymphatic capillaries** (Figure 19.1a), which weave between the tissue cells and blood capillaries in the loose connective tissues of the body. Lymph capillaries are widespread; however, they are absent from bones and teeth, bone marrow, and the entire central nervous system (where the excess tissue fluid drains into the cerebrospinal fluid).

Although similar to blood capillaries, lymphatic capillaries are so remarkably permeable that they were once thought to be open at one end like a straw.

We now know that they owe their permeability to two unique structural modifications:

1. The endothelial cells forming the walls of lymphatic capillaries are not tightly joined; instead, the edges of adjacent cells overlap each other loosely, forming easily opened, flaplike *minivalves* (Figure 19.1b).

2. Collagen filaments anchor the endothelial cells to surrounding structures so that any increase in interstitial fluid volume opens the minivalves, rather than causing the lymphatic capillaries to collapse.

So, what we have is a system analogous to one-way swinging doors in the lymphatic capillary wall. When fluid pressure in the interstitial space is greater than the pressure in the lymphatic capillary, the minivalve flaps gape open, allowing fluid to enter the lymphatic capillary. However, when the pressure is greater *inside* the lymphatic capillary, the endothelial minivalve flaps are forced closed, preventing lymph from leaking back out as the pressure moves it along the vessel.

Proteins in the interstitial space are unable to enter blood capillaries, but they enter lymphatic capillaries easily. In addition, when tissues are inflamed, lymphatic capillaries develop openings that permit uptake of even larger particles such as cell debris, pathogens (disease-causing microorganisms such as bacteria and viruses), and cancer cells. The pathogenic agents and cancer cells can then use the lymphatics to travel throughout the body. This threat to the body is partly resolved by the fact that the lymph takes "detours" through the lymph nodes, where it is cleansed of debris and "examined" by cells of the immune system.

Highly specialized lymphatic capillaries called **lacteals** (lak'te-alz) are present in the fingerlike villi of the intestinal mucosa. The lymph draining from the digestive viscera is milky white (*lacte* = milk) rather than clear because the lacteals play a major role in absorbing digested fats from the intestine. This fatty lymph, called **chyle** ("juice"), is also delivered to the blood via the lymphatic stream.

From the lymphatic capillaries, lymph flows through successively larger and thicker-walled channels—first collecting vessels, then trunks, and finally the largest of all, the ducts (Figure 19.2). The **lymphatic collecting vessels** have the same three tunics as veins, but the collecting vessels are thinner-walled, have more internal valves, and anastomose more. In general these vessels in the skin travel along with superficial *veins*, and the deep lymphatic vessels of the trunk and digestive viscera travel with the deep *arteries*.

The **lymphatic trunks** are formed by the union of the largest collecting vessels, and drain fairly large areas of the body. The major trunks, named mostly for the regions from which they collect lymph, are the paired **lumbar, bronchomediastinal, subclavian,** and **jugular trunks,** and the single **intestinal trunk** (Figure 19.2b).

Lymph is eventually delivered to one of two large *ducts* in the thoracic region. The **right lymphatic duct** drains lymph from the right upper arm and the right side of the head and thorax (Figure 19.2a). The much larger **thoracic duct** receives lymph from the rest of the body. It arises anterior to the first two lumbar vertebrae as an enlarged sac, the **cisterna chyli** (sis-ter'nah ki'li), that collects lymph from the two large lumbar trunks that drain the lower limbs and from the intestinal trunk that drains the digestive organs. As the thoracic duct runs superiorly, it receives lymphatic drainage from the left side of the thorax, left upper limb, and the head region. Each terminal duct empties its lymph into the venous circulation at the junction of the internal jugular vein and subclavian vein on its own side of the body (Figure 19.2b).

 HOMEOSTATIC IMBALANCE

Like the larger blood vessels, the larger lymphatics receive their nutrient blood supply from a branching vasa vasorum. When lymphatic vessels are severely inflamed, the related vessels of the vasa vasorum become congested with blood. As a result, the pathway of the associated superficial lymphatics becomes visible through the skin as red lines that are tender to the touch. This unpleasant condition is called *lymphangitis* (lim"fan-ji'tis; *angi* = vessel). ●

Lymph Transport

The lymphatic system lacks an organ that acts as a pump. Under normal conditions, lymphatic vessels are low-pressure conduits, and the same mechanisms that promote venous return in blood vessels act here as well—the milking action of active skeletal muscles, pressure changes in the thorax during breathing, and valves to prevent backflow. Lymphatics are usually bundled together in connective tissue sheaths along with blood vessels, and pulsations of nearby arteries also promote lymph flow. In addition to these mechanisms, smooth muscle in the walls of the lymphatic trunks and thoracic duct contracts

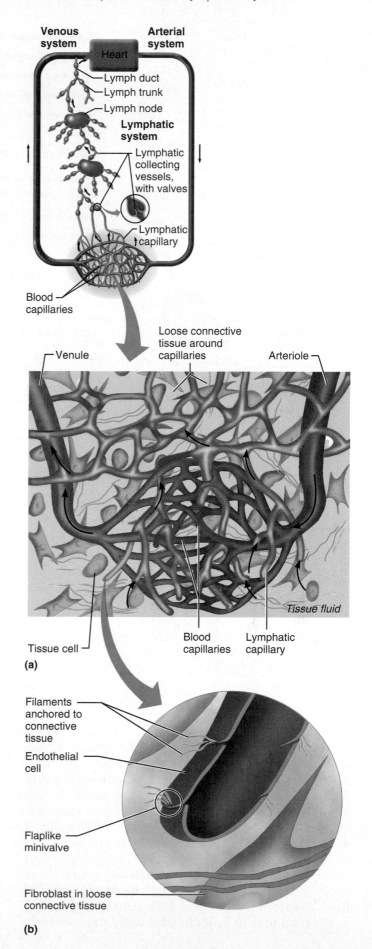

FIGURE 19.1 Distribution and special structural features of lymphatic capillaries. (a) Structural relationship between a capillary bed of the blood vascular system and lymphatic capillaries. Arrows indicate direction of fluid movement. **(b)** Lymphatic capillaries begin as blind-ended tubes. Adjacent endothelial cells in a lymphatic capillary overlap each other, forming flaplike minivalves.

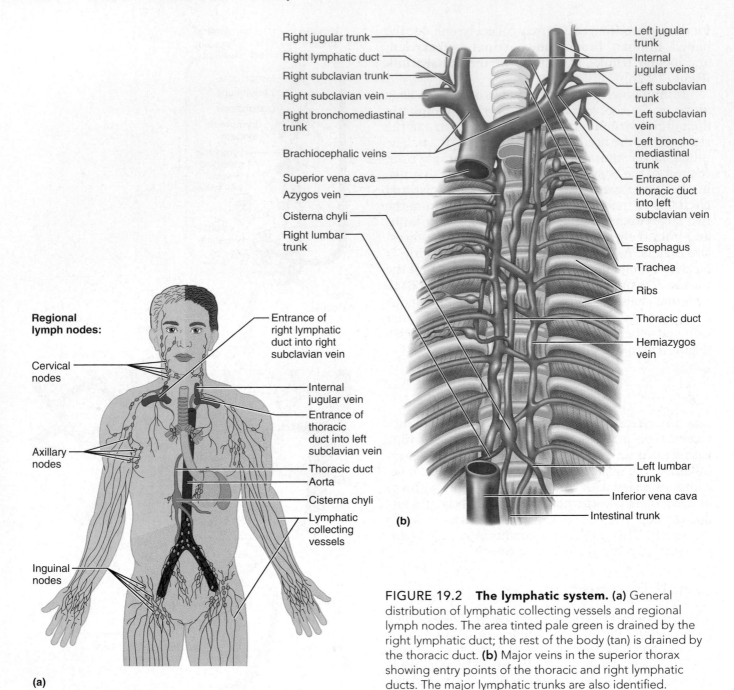

Regional lymph nodes:

Right jugular trunk
Right lymphatic duct
Right subclavian trunk
Right subclavian vein
Right bronchomediastinal trunk
Brachiocephalic veins
Superior vena cava
Azygos vein
Cisterna chyli
Right lumbar trunk

Left jugular trunk
Internal jugular veins
Left subclavian trunk
Left subclavian vein
Left broncho-mediastinal trunk
Entrance of thoracic duct into left subclavian vein
Esophagus
Trachea
Ribs
Thoracic duct
Hemiazygos vein

Cervical nodes

Entrance of right lymphatic duct into right subclavian vein

Internal jugular vein
Entrance of thoracic duct into left subclavian vein
Thoracic duct
Aorta
Cisterna chyli
Lymphatic collecting vessels

Axillary nodes

Inguinal nodes

Left lumbar trunk
Inferior vena cava
Intestinal trunk

(b)

(a)

FIGURE 19.2   **The lymphatic system. (a)** General distribution of lymphatic collecting vessels and regional lymph nodes. The area tinted pale green is drained by the right lymphatic duct; the rest of the body (tan) is drained by the thoracic duct. **(b)** Major veins in the superior thorax showing entry points of the thoracic and right lymphatic ducts. The major lymphatic trunks are also identified.

rhythmically, helping to pump the lymph along. Even so, lymph transport is sporadic and slow. About 3 L of lymph enters the bloodstream every 24 hours, a volume almost exactly equal to the amount of fluid lost to the tissue spaces from the bloodstream in the same time period. Movement of adjacent tissues is extremely important in propelling lymph through the lymphatics. When physical activity or passive movements increase, lymph flows much more rapidly (balancing the greater rate of fluid loss from the blood in such situations). Hence, it is a good idea to immobilize a badly infected body

part to hinder flow of inflammatory material from that region.

HOMEOSTATIC IMBALANCE

Anything that prevents the normal return of lymph to the blood, such as blockage of the lymphatics by tumors or removal of lymphatics during cancer surgery, results in short-term but severe localized edema *(lymphedema)*. However, lymphatic drainage is eventually reestablished by regrowth from the vessels remaining in the area.   ●

Lymphoid Cells and Tissues

In order to understand the basic aspects of the lymphatic system's role in the body, we investigate the components of lymphoid organs—lymphoid cells and lymphoid tissues—before considering the organs themselves.

Lymphoid Cells

Infectious microorganisms that manage to penetrate the body's epithelial barriers quickly proliferate in the underlying loose connective tissues. These invaders are fought off by the inflammatory response, by phagocytes (macrophages), and by lymphocytes.

Lymphocytes, the main warriors of the immune system, arise in red bone marrow (along with other formed elements). They then mature into one of the two main varieties of immunocompetent cells—**T cells (T lymphocytes)** or **B cells (B lymphocytes)**—that protect the body against antigens. (*Antigens* are anything the body perceives as foreign, such as bacteria and their toxins, viruses, mismatched RBCs, or cancer cells.) Activated T cells manage the immune response and some of them directly attack and destroy foreign cells. B cells protect the body by producing **plasma cells,** daughter cells that secrete antibodies into the blood (or other body fluids). Antibodies immobilize antigens until they can be destroyed by phagocytes or other means. The precise roles of the lymphocytes in immunity are explored in Chapter 20.

Lymphoid **macrophages** play a crucial role in body protection and in the immune response by phagocytizing foreign substances and by helping to activate T cells. So, too, do the spiny-looking **dendritic cells** found in lymphoid tissue. Last but not least are the **reticular cells,** fibroblastlike cells that produce the reticular fiber **stroma** (stro′mah), which is the network that supports the other cell types in the lymphoid organs (Figure 19.3).

Lymphoid Tissue

Lymphoid (lymphatic) tissue is an important component of the immune system, mainly because it (1) houses and provides a proliferation site for lymphocytes and (2) furnishes an ideal surveillance vantage point for lymphocytes and macrophages. Lymphoid tissue, largely composed of a type of loose connective tissue called **reticular connective tissue,** dominates all the lymphoid organs except the thymus. Macrophages live on the fibers of the reticular network, and in the spaces of the network are huge numbers of lymphocytes that have squeezed through the walls of postcapillary venules coursing through this tissue. The lymphocytes reside temporarily

How is reticular connective tissue classified?

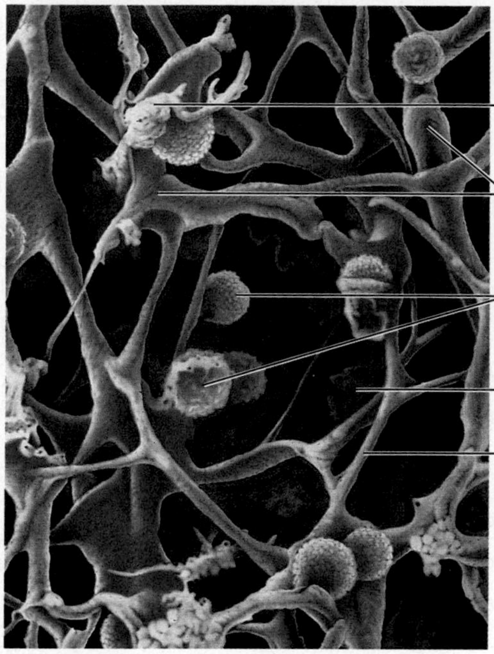

FIGURE 19.3   Reticular tissue in a human lymph node. Scanning electron micrograph (1100×).

Labels on figure: Macrophage · Reticular cells on reticular fibers · Lymphocytes · Medullary sinus · Reticular fiber

in the lymphoid tissue (Figure 19.3), and then leave to patrol the body again. The cycling of lymphocytes between the circulatory vessels, lymphoid tissues, and loose connective tissues of the body ensures that lymphocytes reach infected or damaged sites quickly.

Lymphoid tissue comes in various "packages." **Diffuse lymphatic tissue,** consisting of a few scattered reticular tissue elements, is found in virtually every body organ, but larger collections appear in the lamina propria of mucous membranes and in lymphoid organs. **Lymphoid follicles (nodules)** represent another way lymphoid tissue is organized. Like diffuse lymphatic tissue, they lack a capsule, but follicles are solid, spherical bodies consisting of tightly packed reticular elements and cells. Follicles often have lighter-staining centers, called **germinal centers.** Follicular dendritic cells and B cells predominate in germinal centers, and these centers enlarge dramatically when the B cells are dividing rapidly and producing plasma cells. In many cases, the follicles are found forming part of larger lymphoid organs, such as lymph nodes. However, isolated aggregations of lymphatic follicles occur in the intestinal wall as Peyer's patches and in the appendix.

■ Loose connective tissue proper.

What is the benefit of having fewer efferent than afferent lymphatics in lymph nodes?

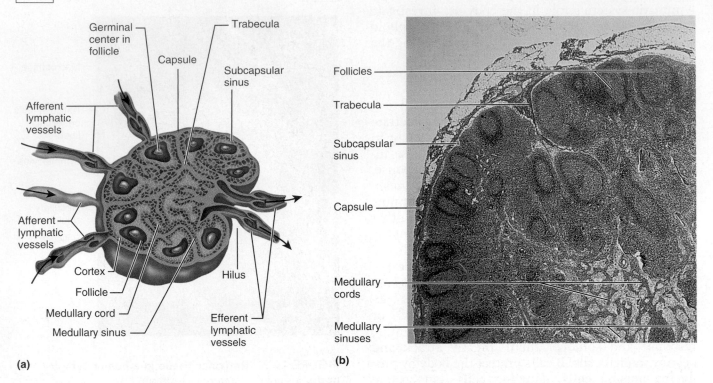

FIGURE 19.4   Lymph node.
(a) Longitudinal view of the internal structure of a lymph node and associated lymphatics. Notice that several afferent lymphatics converge on its convex side, whereas fewer efferent lymphatics exit at its hilus. Arrows indicate the direction of lymph flow into and out of the node. **(b)** Photomicrograph of part of a lymph node (60×).

Lymph Nodes

The principal lymphoid organs in the body are the **lymph nodes,** which cluster along the lymphatic vessels of the body. As lymph is transported back to the bloodstream, it is filtered through the lymph nodes. There are hundreds of these small organs, but because they are usually embedded in connective tissue, they are not ordinarily seen. Large clusters of lymph nodes occur near the body surface in the inguinal, axillary, and cervical regions, places where the lymphatic collecting vessels converge to form trunks (see Figure 19.2a).

Lymph nodes have two basic functions, both concerned with body protection. (1) They act as lymph "filters." Macrophages in the nodes remove and destroy microorganisms and other debris that enter the lymph from the loose connective tissues,

effectively preventing them from being delivered to the blood and spreading to other parts of the body. (2) They help activate the immune system. Lymphocytes, also strategically located in the lymph nodes, monitor the lymphatic stream for the presence of *antigens* and mount an attack against them. Let's look at how the structure of a lymph node supports these defensive functions.

Structure of a Lymph Node

Lymph nodes vary in shape and size, but most are bean shaped and less than 2.5 cm (1 inch) in length. Each node is surrounded by a dense fibrous **capsule** from which connective tissue strands called **trabeculae** extend inward to divide the node into a number of compartments (Figure 19.4). The node's internal framework or stroma of reticular fibers physically supports the ever-changing population of lymphocytes.

A lymph node has two histologically distinct regions, the **cortex** and the **medulla.** The superficial

■ *Having fewer efferents causes lymph to accumulate in lymph nodes, allowing more time for its cleansing.*

part of the cortex contains densely packed follicles, many with germinal centers heavy with dividing B cells. Dendritic cells nearly encapsulate the follicles and abut the deeper part of the cortex, which primarily houses T cells in transit. The T cells circulate continuously between the blood, lymph nodes, and lymph, performing their surveillance role.

Medullary cords, which are thin inward extensions from the cortical lymphoid tissue, contain both types of lymphocytes plus plasma cells and they define the medulla. Throughout the node are **lymph sinuses,** large lymph capillaries spanned by crisscrossing reticular fibers. Numerous macrophages reside on these reticular fibers and phagocytize foreign matter in the lymph as it flows by in the sinuses. Additionally, some of the lymph-borne antigens in the percolating lymph leak into the surrounding lymphoid tissue, where they activate lymphocytes to mount an immune attack against them.

Circulation in the Lymph Nodes

Lymph enters the convex side of a lymph node through a number of **afferent lymphatic vessels.** It then moves through a large, baglike sinus, the **subcapsular sinus,** into a number of smaller sinuses that cut through the cortex and enter the medulla. The lymph meanders through these sinuses and finally exits the node at its **hilus** (hi′lus), the indented region on the concave side, via **efferent lymphatic vessels.** Because there are fewer efferent vessels draining the node than afferent vessels feeding it, the flow of lymph through the node stagnates somewhat, allowing time for the lymphocytes and macrophages to carry out their protective functions. Lymph passes through several nodes before it is completely cleansed.

 HOMEOSTATIC IMBALANCE

Sometimes lymph nodes are overwhelmed by the agents they are trying to destroy. For example, when large numbers of bacteria are trapped in the nodes, the nodes become inflamed, swollen, and tender to the touch, a condition often referred to (erroneously) as swollen glands. Such infected lymph nodes are called *buboes* (bu′boz). (Buboes are the most obvious symptom of bubonic plague, the "Black Death" that killed much of Europe's population in the late Middle Ages.) Lymph nodes can also become secondary cancer sites, particularly in metastasizing cancers that enter lymphatic vessels and become trapped there. The fact that cancer-infiltrated lymph nodes are swollen but not painful helps distinguish cancerous lymph nodes from those infected by microorganisms. ●

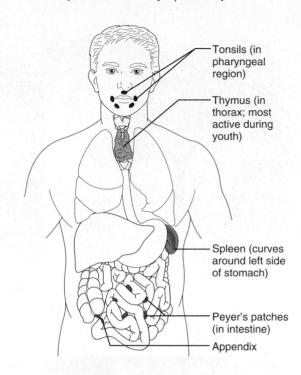

FIGURE 19.5   Lymphoid organs. Tonsils, spleen, thymus, and Peyer's patches.

Other Lymphoid Organs

Lymph nodes are just one example of the many types of **lymphoid organs** or aggregates of lymphatic tissue in the body. Others are the spleen, thymus gland, tonsils, and Peyer's patches of the intestine (Figure 19.5), as well as bits of lymphatic tissue scattered in the connective tissues. The common feature of all these organs is their tissue makeup: All are composed of *reticular connective tissue.* Although all lymphoid organs help protect the body, only the lymph nodes filter lymph. The other lymphoid organs and tissues typically have efferent lymphatics draining them, but lack afferent lymphatics.

Spleen

The soft, blood-rich **spleen** is about the size of a fist and is the largest lymphoid organ. Located in the left side of the abdominal cavity just beneath the diaphragm, it curls around the anterior aspect of the stomach (Figures 19.5 and 19.6). It is served by the large *splenic artery* and *vein,* which enter and exit the *hilus* on its slightly concave anterior surface.

The spleen provides a site for lymphocyte proliferation and immune surveillance and response. But perhaps even more important are its blood-cleansing functions. Besides extracting aged and defective

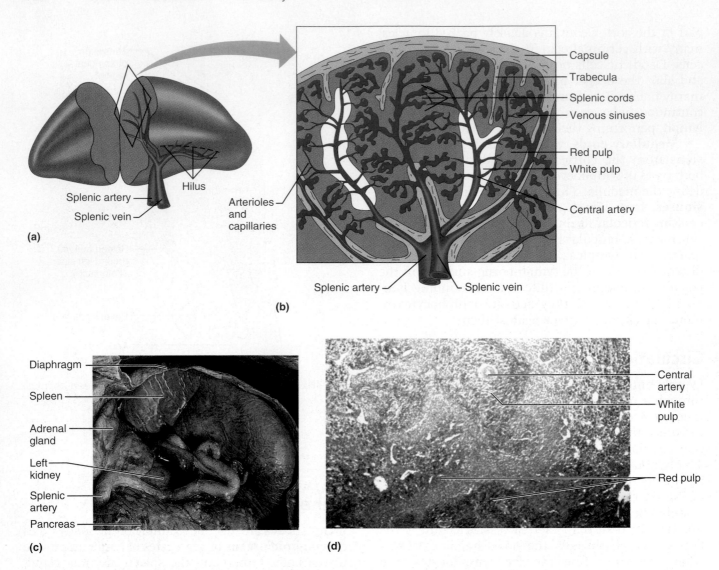

FIGURE 19.6   **The spleen.** **(a)** Gross structure. **(b)** Diagram of the histological structure. **(c)** Photograph of the spleen in its normal position in the abdominal cavity, anterior view. **(d)** Photomicrograph of spleen tissue showing white and red pulp regions (30×).

blood cells and platelets from the blood, its macrophages remove debris and foreign matter from blood flowing through its sinuses. The spleen also performs three additional, and related, functions.

1. It stores some of the breakdown products of red blood cells for later reuse (for example, it salvages iron for making hemoglobin) and releases others to the blood for processing by the liver.

2. It is a site of erythrocyte production in the fetus (a capability that normally ceases after birth).

3. It stores blood platelets.

Like lymph nodes, the spleen is surrounded by a fibrous capsule, has trabeculae that extend inward, and contains both lymphocytes and macrophages. Consistent with its blood-processing functions, it also contains huge numbers of erythrocytes. Areas composed mostly of lymphocytes suspended on reticular fibers are called **white pulp.** The white pulp clusters or forms "cuffs" around the *central arteries* (small branches of the splenic artery) in the organ and forms what appear to be islands in a sea of red pulp. **Red pulp** is essentially all remaining splenic tissue, that is, the venous sinuses (blood sinusoids) and the **splenic cords,** regions of reticular connective tissue exceptionally rich in macrophages. Red pulp is most concerned with disposing of worn-out red blood cells and bloodborne pathogens, whereas white pulp is involved with the immune functions of the spleen. The naming of the pulp regions reflects their appearance in fresh spleen tissue rather than their staining properties. Indeed, as can

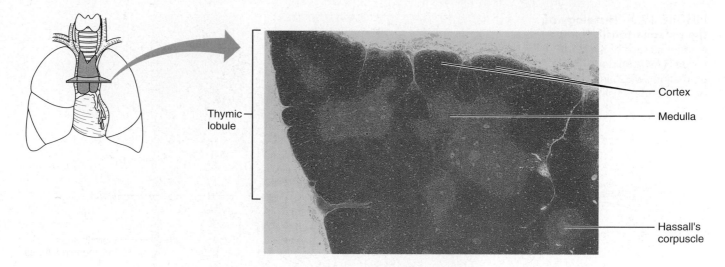

Thymic lobule

Cortex

Medulla

Hassall's corpuscle

FIGURE 19.7   **The thymus.** The photomicrograph of a portion of the thymus shows its lobules with cortical and medullary regions (20×).

be seen in the photomicrograph in Figure 19.6d, the white pulp takes on a purplish hue and sometimes appears darker than the red pulp.

ⓗ HOMEOSTATIC IMBALANCE

Because the spleen's capsule is relatively thin, a direct blow or severe infection may cause it to rupture, spilling blood into the peritoneal cavity. Under such conditions the spleen must be removed quickly (a procedure called a *splenectomy*) and the splenic artery tied off to prevent life-threatening hemorrhage and shock. Surgical removal of the spleen seems to create few problems because the liver and bone marrow take over most of its functions. In children younger than 12, the spleen will regenerate if a small part of it is left in the body. ●

Thymus

The bilobed **thymus** (thi′mus) has important functions primarily during the early years of life. It is found in the inferior neck and extends into the superior thorax, where it partially overlies the heart deep to the sternum (see Figures 19.5 and 19.7). By secreting the hormones thymosin and thymopoietin, the thymus causes T lymphocytes to become immunocompetent; that is, it enables them to function against specific pathogens in the immune response. Prominent in newborns, the thymus continues to increase in size during childhood, when it is most active. During adolescence its growth stops, and it starts to atrophy gradually. By old age it has been replaced almost entirely by fibrous and fatty tissue and is difficult to distinguish from surrounding connective tissue.

To understand thymic histology, it helps to compare the thymus to a cauliflower head—the flowerets represent *thymic lobules*, each containing an outer cortex and an inner medulla (Figure 19.7). Most thymic cells are lymphocytes. In the cortical regions the rapidly dividing lymphocytes are densely packed, but a few macrophages are scattered among them. The lighter-staining medullary areas contain fewer lymphocytes plus some bizarre structures called **Hassall's** or **thymic corpuscles.** Hassall's corpuscles appear to be areas of degenerating cells, but their significance is unknown. Because the thymus lacks B cells, it has no follicles.

The thymus differs from other lymphoid organs in two other important ways. First, it functions strictly in T lymphocyte maturation and thus is the only lymphoid organ that does not *directly* fight antigens. In fact, the so-called *blood-thymus barrier* keeps blood-borne antigens from leaking into the cortical regions to prevent premature activation of the immature lymphocytes. Second, the stroma of the thymus consists of epithelial cells rather than reticular fibers. These **thymocytes** secrete the hormones that stimulate the lymphocytes to become immunocompetent.

Tonsils

The **tonsils** are the simplest lymphoid organs. They form a ring of lymphatic tissue around the entrance to the pharynx (throat), where they appear as swellings of the mucosa (Figure 19.5). The tonsils are named according to location. The paired **palatine tonsils** are located on either side at the posterior end of the oral cavity. These are the largest of the tonsils and the ones most often infected. The **lingual tonsils,** paired lumpy collections of lymphoid follicles, lie at the base

FIGURE 19.8 **Histology of the palatine tonsil.** The exterior surface of the tonsil is covered by squamous epithelium, which invaginates deep into the tonsil (60×).

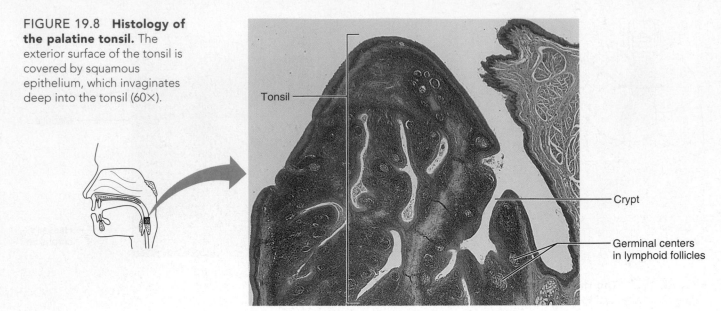

Tonsil

Crypt

Germinal centers in lymphoid follicles

of the tongue. The **pharyngeal tonsil** (referred to as the *adenoids* if enlarged) is in the posterior wall of the nasopharynx. The tiny **tubal tonsils** surround the openings of the auditory tubes into the pharynx. The tonsils gather and remove many of the pathogens entering the pharynx in food or in inhaled air.

The lymphoid tissue of the tonsils contains follicles with obvious germinal centers surrounded by diffusely scattered lymphocytes. The tonsils are not fully encapsulated, and the epithelium overlying them invaginates deep into their interior, forming

blind-ended **crypts** (Figure 19.8). The crypts trap bacteria and particulate matter, and the bacteria work their way through the mucosal epithelium into the lymphoid tissue, where most are destroyed. It seems a bit dangerous to "invite" infection this way, but this strategy produces a wide variety of immune cells that have a "memory" for the trapped pathogens. Thus, the body takes a calculated risk early on (during childhood) for the benefits of heightened immunity and better health later.

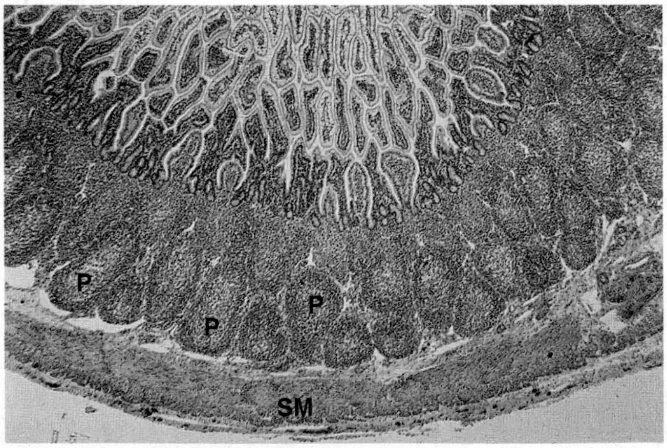

Aggregates of Lymphoid Follicles

Peyer's patches (pi'erz) are large isolated clusters of lymphoid follicles, structurally similar to the tonsils, that are located in the wall of the distal portion of the small intestine (Figures 19.5 and 19.9). Lymphoid follicles are also heavily concentrated in the wall of the **appendix,** a tubular offshoot of the first part of the large intestine. Peyer's patches and the appendix are in an ideal position (1) to destroy bacteria (which are present in large numbers in the intestine), thereby preventing these pathogens from breaching the intestinal wall, and (2) to generate many "memory" lymphocytes for long-term immunity. Peyer's patches, the appendix, and the tonsils—all located in the digestive tract—and lymphoid follicles in the walls of the bronchi (organs of the respiratory tract), are part of the collection of small lymphoid tissues referred to as **mucosa-associated lymphatic tissue (MALT).** Collectively, MALT protects the digestive and respiratory tracts from the never-ending onslaughts of foreign matter entering those cavities.

FIGURE 19.9 **Peyer's patches.** Histological structure of aggregated lymphoid follicles—Peyer's patches (P)—in the wall of the ileum of the small intestine (SM = submucosa) (20×).

P

P

P

SM

Review Questions

Multiple Choice/Matching

(Some questions have more than one correct answer. Select the best answer or answers from the choices given.)

1. Lymphatic vessels (a) serve as sites for immune surveillance, (b) filter lymph, (c) transport leaked plasma proteins and fluids to the cardiovascular system, (d) are represented by vessels that resemble arteries, capillaries, and veins.

2. The saclike initial portion of the thoracic duct is the (a) lacteal, (b) right lymphatic duct, (c) cisterna chyli, (d) lymph sac.

3. Entry of lymph into the lymphatic capillaries is promoted by which of the following? (a) one-way minivalves formed by overlapping endothelial cells, (b) the respiratory pump, (c) the skeletal muscle pump, (d) greater fluid pressure in the interstitial space.

4. The structural framework of lymphoid organs is (a) areolar connective tissue, (b) hematopoietic tissue, (c) reticular tissue, (d) adipose tissue.

5. Lymph nodes are densely clustered in all of the following body areas *except* (a) the brain, (b) the axillae, (c) the groin, (d) the cervical region.

6. The germinal centers in lymph nodes are largely sites of (a) macrophages, (b) proliferating B lymphocytes, (c) T lymphocytes, (d) all of these.

7. The red pulp areas of the spleen are sites of (a) venous sinuses, macrophages, and red blood cells, (b) clustered lymphocytes, (c) connective tissue septa.

8. The lymphoid organ that functions primarily during youth and then begins to atrophy is the (a) spleen, (b) thymus, (c) palatine tonsils, (d) bone marrow.

9. Collections of lymphoid tissue (MALT) that guard mucosal surfaces include all of the following *except* (a) appendix nodules, (b) the tonsils, (c) Peyer's patches, (d) the thymus.

Short Answer Essay Questions

10. Compare and contrast blood, interstitial fluid, and lymph.

11. Compare the structure and functions of a lymph node to those of the spleen.

12. (a) What anatomical characteristic ensures that the flow of lymph through a lymph node is slow? (b) Why is this desirable?

13. Why doesn't the absence of lymphatic *arteries* represent a problem?

20

THE IMMUNE SYSTEM: INNATE AND ADAPTIVE BODY DEFENSES

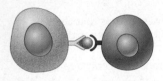

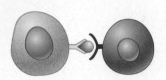

PART 1: INNATE DEFENSES (pp. 673–682)

Surface Barriers: Skin and Mucosae (pp. 673–674)

1. Describe surface membrane barriers and their protective functions.

Internal Defenses: Cells and Chemicals (pp. 674–682)

2. Explain the importance of phagocytosis and natural killer cells in nonspecific body defense.

3. Describe the inflammatory process. Identify several inflammatory chemicals and indicate their specific roles.

4. Name the body's antimicrobial substances and describe their function.

5. Explain how fever helps protect the body.

PART 2: ADAPTIVE DEFENSES (pp. 682–708)

Antigens (pp. 683–684)

6. Define antigen and describe how antigens affect the immune system.

7. Define complete antigen, hapten, and antigenic determinant.

8. Follow antigen processing in the body.

Cells of the Adaptive Immune System: An Overview (pp. 684–687)

9. Compare and contrast the origin, maturation process, and general function of B and T lymphocytes.

10. Describe the role of macrophages and other phagocytes.

11. Define immunocompetence and self-tolerance.

Humoral Immune Response (pp. 687–693)

12. Define humoral immunity.

13. Describe the process of clonal selection of a B cell.

14. Recount the roles of plasma cells and memory cells in humoral immunity.

15. Compare and contrast active and passive humoral immunity.

16. Describe the structure of an antibody monomer, and name the five classes of antibodies.

17. Explain the function(s) of antibodies and describe clinical uses of monoclonal antibodies.

Cell-Mediated Immune Response (pp. 693–704)

18. Define cell-mediated immunity and describe the process of activation and clonal selection of T cells.

19. Describe T cell functions in the body.

20. Indicate the tests ordered before an organ transplant is done, and methods used to prevent transplant rejection.

Homeostatic Imbalances of Immunity (pp. 704–708)

21. Give examples of immune deficiency diseases and of hypersensitivity states.

22. Cite factors involved in autoimmune disease.

Every second of every day, an army of hostile bacteria, fungi, and viruses swarms on our skin and yet we stay amazingly healthy most of the time. The body seems to have evolved a single-minded approach to such foes—if you're not with us, you're against us! To implement that stance, it relies heavily on two intrinsic defense systems that act both independently and cooperatively to provide resistance to disease, or **immunity** (*immun* = free).

1. The **innate (nonspecific) system,** like a lowly foot soldier, is always prepared, responding within minutes to protect the body from all foreign substances. This system has two "barricades." The *first line of defense* is the external body membranes—intact skin and mucosae. The *second line of defense* called into action whenever the first line has been penetrated, uses antimicrobial proteins, phagocytes, and other cells to inhibit the invaders' spread throughout the body. The hallmark of the second line of defense is inflammation.

2. The **adaptive** (or **specific**) **defense system*** is more like an elite fighting force equipped with high-tech weapons that attacks *particular* foreign substances and provides the body's *third line of defense*. This defensive response takes considerably more time to mount than the innate response. Although we consider them separately, the adaptive and innate systems always work hand in hand.

Although certain organs of the body (notably lymphoid organs) are intimately involved in the immune response, the **immune system** is a *functional system* rather than an organ system in an anatomical sense. Its "structures" are a diverse array of molecules plus trillions of immune cells (especially lymphocytes), that inhabit lymphoid tissues and circulate in body fluids.

When the immune system is operating effectively, it protects the body from most infectious microorganisms, cancer cells, and transplanted organs or grafts. It does this both directly, by cell attack, and indirectly, by releasing mobilizing chemicals and protective antibody molecules.

---

* Sometimes the term *immune system* is equated with the adaptive system only. However, recent research has shown that (1) many defensive molecules are released and recognized by both the innate and adaptive arms; (2) the innate responses are not as nonspecific as once thought and may have specific pathways to target certain foreign substances; and (3) proteins released during innate responses alert cells of the adaptive system to the presence of specific foreign molecules in the body.

PART 1: INNATE DEFENSES

Because they are part and parcel of our anatomy, you could say we come fully equipped with innate or nonspecific defenses. The mechanical barriers that cover body surfaces and the cells and chemicals that act on the initial internal battlefronts are in place at birth, ready to ward off invading **pathogens** (harmful or disease-causing microorganisms) and infection. Many times, our innate defenses alone are able to destroy pathogens and ward off infection. In others, the adaptive immune system is called into action to reinforce and enhance the nonspecific mechanisms. In either case, the innate defenses reduce the workload of the adaptive system by preventing entry and spread of microorganisms in the body.

Surface Barriers: Skin and Mucosae

The body's first line of defense—the *skin* and the *mucous membranes,* along with the secretions these membranes produce—is highly effective. As long as the epidermis is unbroken, this heavily keratinized epithelial membrane presents a formidable physical barrier to most microorganisms that swarm on the skin. Keratin is also resistant to most weak acids and bases and to bacterial enzymes and toxins. Intact mucosae provide similar mechanical barriers within the body. Recall that mucous membranes line all body cavities that open to the exterior: the digestive, respiratory, urinary, and reproductive tracts. Besides serving as physical barriers, these epithelial membranes produce a variety of protective chemicals:

1. The acidity of skin secretions (pH 3 to 5) inhibits bacterial growth and sebum contains chemicals that are toxic to bacteria. Vaginal secretions of adult females are also very acidic.

2. The stomach mucosa secretes a concentrated hydrochloric acid solution and protein-digesting enzymes. Both kill microorganisms.

3. Saliva, which cleanses the oral cavity and teeth, and lacrimal fluid of the eye contain **lysozyme,** an enzyme that destroys bacteria.

4. Sticky mucus traps many microorganisms that enter the digestive and respiratory passageways.

The respiratory tract mucosae also have structural modifications that counteract potential invaders. Tiny mucus-coated hairs inside the nose trap inhaled particles, and cilia on the mucosa of the upper respiratory tract sweep dust- and bacteria-laden mucus toward the mouth, preventing it from entering the lower respiratory passages, where the warm, moist environment provides an ideal site for bacterial growth.

Although the surface barriers are quite effective, they are breached occasionally by small nicks and cuts resulting, for example, from brushing your teeth or shaving. When this happens and microorganisms invade deeper tissues, the *internal* innate defenses come into play.

Internal Defenses: Cells and Chemicals

The body uses an enormous number of nonspecific cellular and chemical devices to protect itself, including phagocytes, natural killer cells, antimicrobial proteins, and fever. The inflammatory response enlists macrophages, mast cells, all types of white blood cells, and dozens of chemicals that kill pathogens and help repair tissue. All these protective ploys identify potentially harmful substances by recognizing surface carbohydrates unique to infectious organisms (bacteria, viruses, and fungi). Fever is also an innate protective response.

Phagocytes

Pathogens that get through the skin and mucosae into the underlying connective tissue are confronted by *phagocytes* (*phago* = eat). The chief phagocytes are **macrophages** ("big eaters"), which derive from white blood cells called **monocytes** that leave the bloodstream, enter the tissues, and develop into macrophages.

Free macrophages, like the alveolar macrophages of the lungs and dendritic cells of the epidermis, wander throughout the tissue spaces in search of cellular debris or "foreign invaders." *Fixed macrophages* like *Kupffer cells* in the liver and microglia of the brain are permanent residents of particular organs. Whatever their mobility, all macrophages are similar structurally and functionally.

Neutrophils, the most abundant type of white blood cell, become phagocytic on encountering infectious material in the tissues. **Eosinophils,** another type of white blood cell, are only weakly phagocytic, but they are important in defending the body against parasitic worms. When they encounter such parasites, eosinophils position themselves against the worm and discharge the destructive contents of their large cytoplasmic granules all over their prey. Recently **mast cells,** more associated with their role in allergies, have been shown to have a striking ability to bind with, ingest, and kill a wide range of bacteria. Although mast cells are not normally included in listings of professional phagocytes, they share their capabilities.

Mechanism of Phagocytosis

A phagocyte engulfs particulate matter much the way an amoeba ingests a food particle. Flowing cytoplasmic extensions bind to the particle and then pull it inside, enclosed within a membrane-lined vacuole. The **phagosome** thus formed is then fused with a *lysosome* to form a **phagolysosome** (steps ①–③ in Figure 20.1b).

Phagocytic attempts are not always successful. In order for a phagocyte to accomplish ingestion, **adherence** must occur. The phagocyte must first *adhere* or cling to the pathogen, a feat made possible by recognizing the pathogen's carbohydrate "signature." Recognition is particularly difficult with microorganisms such as pneumococcus, which have an external capsule made of complex sugars. These pathogens can sometimes elude capture because phagocytes cannot bind to their capsules. Adherence is both more probable and more efficient when complement proteins and antibodies coat foreign particles, a process called **opsonization** ("to make tasty"), because the coating provides "handles" to which phagocyte receptors can bind.

Sometimes the way neutrophils and macrophages kill ingested prey is more than simple digestion by lysosomal enzymes. For example, pathogens such as the tuberculosis bacillus and certain parasites are resistant to lysosomal enzymes and can even multiply within the phagolysosome. However, when the macrophage is stimulated by chemicals released by immune cells, additional enzymes are activated that produce the **respiratory burst,** an event that liberates a deluge of free radicals (including nitric oxide), that have potent cell-killing ability. More widespread cell killing is caused by oxidizing chemicals (H_2O_2 and a substance identical to household bleach) released into the extracellular space. These, in turn, promote K^+ entry into the phagolysosome, causing its pH to rise and creating hyperosmolar conditions which activate protein-digesting enzymes that digest the invader. Neutrophils also produce antibiotic-like chemicals, called *defensins* (see p. 564), that pierce the pathogen's membrane. Unhappily, the neutrophils also destroy themselves in the process, whereas macrophages, which rely only on intracellular killing, can go on to kill another day.

Natural Killer Cells

Natural killer (NK) cells, which "police" the body in blood and lymph, are a unique group of defensive cells that can lyse and kill cancer cells and virus-infected body cells before the adaptive immune system is activated. Sometimes called the "pit bulls" of the defense system, NK cells are part of a small group of *large granular lymphocytes.* Unlike lymphocytes of

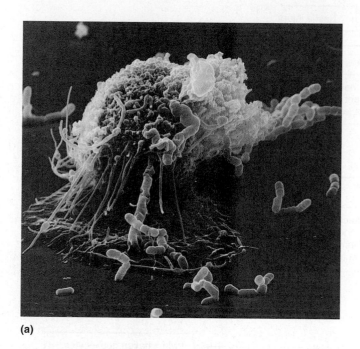

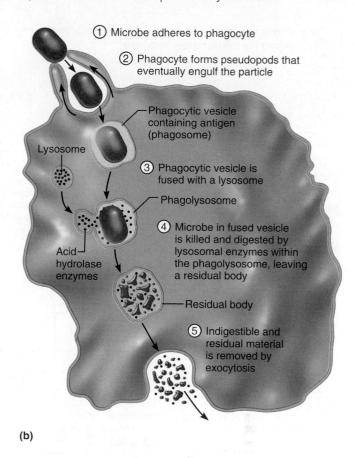

① Microbe adheres to phagocyte

② Phagocyte forms pseudopods that eventually engulf the particle

Phagocytic vesicle containing antigen (phagosome)

Lysosome

③ Phagocytic vesicle is fused with a lysosome

Phagolysosome

④ Microbe in fused vesicle is killed and digested by lysosomal enzymes within the phagolysosome, leaving a residual body

Acid hydrolase enzymes

Residual body

⑤ Indigestible and residual material is removed by exocytosis

(a)

(b)

FIGURE 20.1   Phagocytosis. (a) In this scanning electron micrograph (2600×), a macrophage uses its long cytoplasmic extensions to pull sausage-shaped *E. coli* bacteria toward it. **(b)** Events of phagocytosis.

the adaptive immune system, which recognize and react only against *specific* virus-infected or tumor cells, NK cells are far less picky. They can eliminate a variety of infected or cancerous cells, apparently by detecting the lack of "self" cell surface receptors and by recognizing certain surface sugars on the target cell. The name "natural" killer cells reflects this non-specificity of NK cells.

NK cells are not phagocytic. Their mode of killing involves an attack on the target cell's membrane and release of cytolytic chemicals called *perforins.* Shortly after perforin release, channels appear in the target cell's membrane and its nucleus disintegrates. NK cells also secrete potent chemicals that enhance the inflammatory response.

Inflammation: Tissue Response to Injury

The **inflammatory response** is triggered whenever body tissues are injured by physical trauma (a blow), intense heat, irritating chemicals, or infection by viruses, fungi, or bacteria. The inflammatory response has several beneficial effects:

1. Prevents the spread of damaging agents to nearby tissues

2. Disposes of cell debris and pathogens

3. Sets the stage for repair

The four *cardinal signs* of short-term, or acute, inflammation are *redness, heat* (*inflam* = set on fire), *swelling,* and *pain* (Figure 20.2). If the inflamed area is a joint, joint movement may be hampered temporarily. This forces the injured part to rest, which aids healing. Some authorities consider *impairment of function* to be the fifth cardinal sign of acute inflammation.

Vasodilation and Increased Vascular Permeability

The inflammatory process begins with a chemical "alarm" as a flood of inflammatory chemicals are released into the extracellular fluid. Macrophages (and cells of certain boundary tissues such as epithelial cells lining the gastrointestinal and respiratory tracts) bear surface membrane receptors, called **Toll-like receptors (TLRs),** that play a central role in triggering immune responses. So far ten types of

Why is it important that the capillaries become leaky during the inflammatory response?

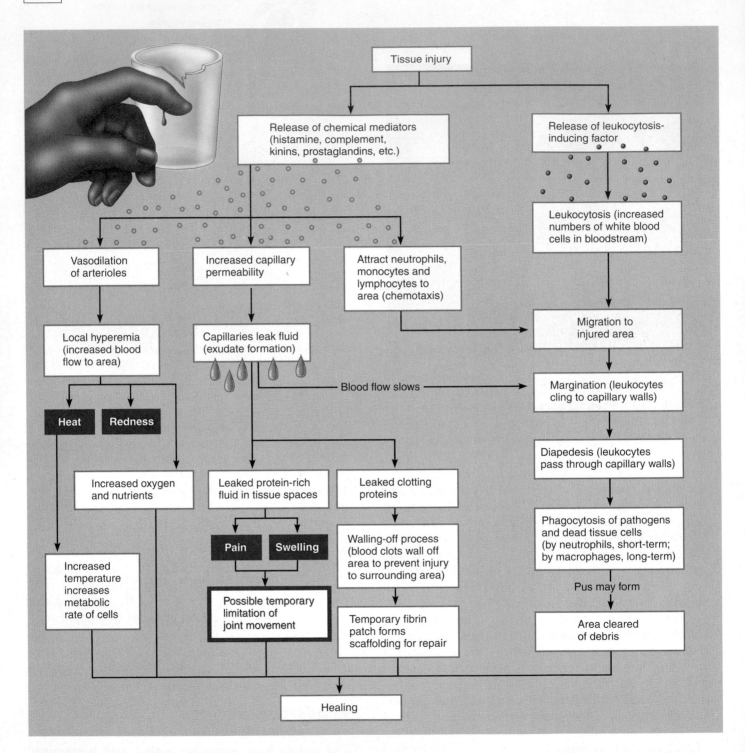

FIGURE 20.2 Flowchart of events in inflammation. The four cardinal signs of acute inflammation are shown in red boxes, as is limitation of joint movement, which in some cases constitutes a fifth cardinal sign (impairment of function).

This allows more filtrate containing oxygen, nutrients, blood-clotting proteins, and antibodies to enter the inflamed area. ■

TABLE 20.1   Inflammatory Chemicals

Chemical	Source	Physiological Effects
Histamine	Granules of basophils and mast cells; released in response to mechanical injury, presence of certain microorganisms, and chemicals released by neutrophils	Promotes vasodilation of local arterioles; increases permeability of local capillaries, promoting exudate formation
Kinins (bradykinin and others)	A plasma protein, kininogen, is cleaved by the enzyme kallikrein found in plasma, urine, saliva, and in lysosomes of neutrophils and other types of cells; cleavage releases active kinin peptides	Same as for histamine; also induce chemotaxis of leukocytes and prompt neutrophils to release lysosomal enzymes, thereby enhancing generation of more kinins; induce pain
Prostaglandins (PGs)	Fatty acid molecules produced from arachidonic acid; found in all cell membranes; generated by lysosomal enzymes of neutrophils and other cell types	Sensitize blood vessels to effects of other inflammatory mediators; one of the intermediate steps of prostaglandin generation produces free radicals, which themselves can cause inflammation; induce pain
Platelet-derived growth factor	Secreted by platelets and endothelial cells	Stimulates fibroblast activity and repair of damaged tissues
Complement	See Table 20.2 (p. 679)	
Cytokines	See Table 20.4 (pp. 699–700)	

human TLRs have been identified, each recognizing a specific class of attacking microbe. For example, one type responds to a glycolipid in cell walls of the tuberculosis bacterium and another to a component of gram-negative bacteria such as salmonella. Once activated, a TLR triggers the release of chemicals called **cytokines** that promote inflammation and attract WBCs to the scene. But macrophages are not the only sort of recognition "tool" in the innate system. Injured and stressed tissue cells, phagocytes, lymphocytes, mast cells, and blood proteins are all sources of inflammatory mediators, the most important of which are **histamine** (his'tah-mēn), **kinins** (ki'ninz), **prostaglandins (PGs)** (pros"tah-glan'dinz), and **complement**, as well as cytokines. Though some of these mediators have individual inflammatory roles as well (see Table 20.1), they all cause small blood vessels in the injured area to dilate. As more blood flows into the area, local **hyperemia** (congestion with blood) occurs, accounting for the *redness* and *heat* of an inflamed region.

The liberated chemicals also increase the permeability of local capillaries. Consequently, **exudate**—fluid containing clotting factors and antibodies—seeps from the blood into the tissue spaces. This exudate causes the local edema *(swelling)* that presses on adjacent nerve endings, contributing to a sensation of *pain*. Pain also results from the release of bacterial toxins, lack of nutrition to cells in the area, and the sensitizing effects of released prostaglandins and kinins. Aspirin and some other anti-inflammatory drugs produce their analgesic (pain-reducing) effects by inhibiting prostaglandin synthesis.

Although edema may seem detrimental, it isn't. The surge of protein-rich fluids into the tissue spaces (1) helps to dilute harmful substances that may be present, (2) brings in the large quantities of oxygen and nutrients needed for repair, and (3) allows the entry of clotting proteins (Figure 20.2). The clotting proteins in the tissue space form a gel-like fibrin mesh that forms a scaffolding for permanent repair. This isolates the injured area and prevents the spread of bacteria and other harmful agents into surrounding tissues.

At inflammatory sites where an epithelial barrier has been breached, an additional chemical enters the battle—β-defensins. These broad-spectrum antibiotic-like chemicals are continuously present in epithelial mucosal cells in small amounts and help maintain the sterile environment of the body's internal passageways (urinary tract, respiratory bronchi, etc.). However, when the mucosal surface is abraded or penetrated and the underlying connective tissue becomes inflamed, β-defensin output increases dramatically, helping to control bacterial and fungal colonization in the exposed area.

Phagocyte Mobilization

Soon after inflammation begins, the damaged area is invaded by more phagocytes—neutrophils lead, followed by macrophages. If the inflammation was provoked by pathogens, a group of plasma proteins known as complement (discussed shortly) is activated and elements of adaptive immunity (lymphocytes and antibodies) also invade the injured site. The process by which phagocytes are mobilized to infiltrate the injured site is illustrated in Figure 20.3.

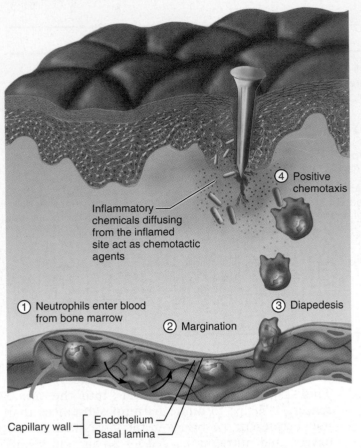

Inflammatory chemicals diffusing from the inflamed site act as chemotactic agents

④ Positive chemotaxis

① Neutrophils enter blood from bone marrow

② Margination

③ Diapedesis

Capillary wall — Endothelium
Basal lamina

FIGURE 20.3   Phagocyte mobilization. When leukocytosis-inducing factors are released by injured cells at an inflamed site, ① neutrophils are released from the bone marrow to the blood. Because blood at the inflammatory site loses fluid, blood flow slows locally and the neutrophils roll along the vascular endothelium. ② Margination begins as neutrophil cell adhesion molecules (CAMs) cling to those of the capillary endothelial cells, and then ③ diapedesis occurs as the neutrophils squeeze through the capillary walls. ④ Positive chemotaxis is the continued migration toward the site where inflammatory chemicals are released by neutrophils. Once at the inflamed site, the neutrophils help rid the area of pathogens and cellular debris.

① **Leukocytosis.** Chemicals called **leukocytosis-inducing factors** released by injured cells promote rapid release of neutrophils from red bone marrow, and within a few hours the number of neutrophils in blood increases four to five fold. This increase in WBCs, called **leukocytosis,** is a characteristic of inflammation.

② **Margination.** Because of the tremendous outpouring of fluid from the blood into the injured area, blood flow in the region slows and the neutrophils begin to roll along the capillary endothelium as if "tasting" the local environment. In inflamed areas, the endothelial cells sprout cell adhesion molecules (CAMs) called *selectins.* These provide footholds (signal that "this is the place") for CAMs (integrins) on the surfaces of the neutrophils. When the complementary CAMs bind together, the neutrophils cling to the inner walls of the capillaries and postcapillary venules. This phenomenon is known as **margination** or **pavementing.**

③ **Diapedesis.** The continued chemical signaling prompts the neutrophils to squeeze through the capillary walls, a process called **diapedesis** or **emigration.**

④ **Chemotaxis.** Neutrophils usually migrate randomly, but inflammatory chemicals act as homing devices, or more precisely **chemotactic agents,** that attract the neutrophils and other white blood cells to the site of the injury. Within an hour after the inflammatory response has begun, neutrophils have collected at the site and are devouring any foreign material present.

As the counterattack continues, monocytes follow neutrophils into the injured area. Monocytes are fairly poor phagocytes, but within 12 hours of leaving the blood and entering the tissues, they swell and develop large numbers of lysosomes, becoming macrophages with insatiable appetites. These late-arriving macrophages replace the neutrophils on the battlefield. Macrophages are the central actors in the final disposal of cell debris as an acute inflammation subsides, and they predominate at sites of prolonged, or *chronic,* inflammation. The ultimate goal of an inflammatory response is to clear the injured area of pathogens, dead tissue cells, and any other debris so that tissue can be repaired. Once this is accomplished, healing usually occurs quickly.

🜂 HOMEOSTATIC IMBALANCE

In severely infected areas, the battle takes a considerable toll on both sides, and creamy, yellow ***pus,*** a mixture of dead or dying neutrophils, broken-down

TABLE 20.2 Summary of Nonspecific Body Defenses

Category/Associated Elements	Protective Mechanism
First Line of Defense: Surface Membrane Barriers	
Intact skin epidermis	Forms mechanical barrier that prevents entry of pathogens and other harmful substances into body
▪ Acid mantle	Skin secretions (perspiration and sebum) make epidermal surface acidic, which inhibits bacterial growth; sebum also contains bactericidal chemicals
▪ Keratin	Provides resistance against acids, alkalis, and bacterial enzymes
Intact mucous membranes	Form mechanical barrier that prevents entry of pathogens
▪ Mucus	Traps microorganisms in respiratory and digestive tracts
▪ Nasal hairs	Filter and trap microorganisms in nasal passages
▪ Cilia	Propel debris-laden mucus away from lower respiratory passages
▪ Gastric juice	Contains concentrated hydrochloric acid and protein-digesting enzymes that destroy pathogens in stomach
▪ Acid mantle of vagina	Inhibits growth of most bacteria and fungi in female reproductive tract
▪ Lacrimal secretion (tears); saliva	Continuously lubricate and cleanse eyes (tears) and oral cavity (saliva); contain lysozyme, an enzyme that destroys microorganisms
▪ Urine	Normally acid pH inhibits bacterial growth; cleanses the lower urinary tract as it flushes from the body
Second Line of Defense: Nonspecific Cellular and Chemical Defenses	
Phagocytes	Engulf and destroy pathogens that breach surface membrane barriers; macrophages also contribute to immune response
Natural killer cells	Promote cell lysis by direct cell attack against virus-infected or cancerous body cells; do not depend on specific antigen recognition; do not exhibit a memory response
Inflammatory response	Prevents spread of injurious agents to adjacent tissues, disposes of pathogens and dead tissue cells, and promotes tissue repair; chemical mediators released attract phagocytes (and immunocompetent cells) to the area
Antimicrobial proteins	
▪ Interferons (α, β, γ)	Proteins released by virus-infected cells that protect uninfected tissue cells from viral takeover; mobilize immune system
▪ Complement	Lyses microorganisms, enhances phagocytosis by opsonization, and intensifies inflammatory and immune responses
Fever	Systemic response initiated by pyrogens; high body temperature inhibits microbial multiplication and enhances body repair processes

tissue cells, and living and dead pathogens, may accumulate in the wound. If the inflammatory mechanism fails to clear the area of debris, the sac of pus may be walled off by collagen fibers, forming an *abscess*. Surgical drainage of abscesses is often necessary before healing can occur.

Bacteria, such as tuberculosis bacilli, that are resistant to digestion by macrophages escape the effects of prescription antibiotics by remaining snugly enclosed within their macrophage hosts. In such cases, *infectious granulomas* form. These tumorlike growths contain a central region of infected macrophages surrounded by uninfected macrophages and an outer fibrous capsule. A person may harbor pathogens walled off in granulomas for years without displaying any symptoms. However, if the person's resistance to infection is ever compromised, the bacteria may be activated and break out, leading to clinical disease symptoms. ●

Antimicrobial Proteins

A variety of **antimicrobial proteins** enhance the innate defenses by attacking microorganisms directly or by hindering their ability to reproduce. The most important of these are interferon and complement proteins (Table 20.2).

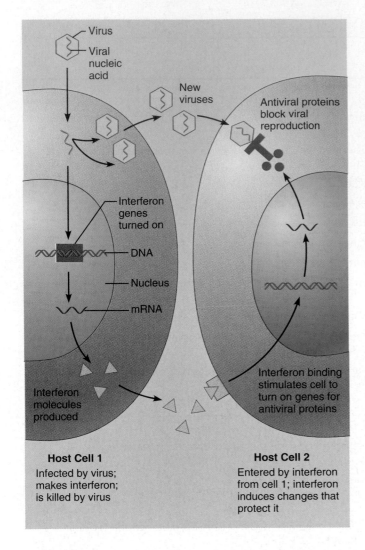

FIGURE 20.4   **The interferon mechanism against viruses.**

The IFNs are a family of related proteins, produced by a variety of body cells, each having a slightly different physiological effect. Lymphocytes secrete gamma (γ) or immune, interferon, but most other leukocytes secrete alpha (α) interferon. Fibroblasts secrete beta (β) interferon which is particularly active in reducing inflammation. Besides their antiviral effects, interferons activate macrophages and mobilize NK cells. Because both macrophages and NK cells can act directly against malignant cells, the interferons play some anticancer role.

The IFNs have found a niche as antiviral agents. *Alpha IFN* is used to treat genital warts (caused by the herpes virus) and it is the first drug to have some success in combating hepatitis C (spread via blood and sexual intercourse), the most common and most dreaded form of hepatitis. The IFNs are also used against devastating viral infections in organ transplant patients.

Complement

The term **complement system,** or simply **complement,** refers to a group of at least 20 plasma proteins that normally circulate in the blood in an inactive state. These proteins include C1 through C9, factors B, D, and P, plus several regulatory proteins. Complement provides a major mechanism for destroying foreign substances in the body. Its activation unleashes chemical mediators that amplify virtually all aspects of the inflammatory process. Another effect of complement activation is that bacteria and certain other cell types are killed by cell lysis. (Luckily our own cells are equipped with proteins that inactivate complement.) Although complement is a nonspecific defensive mechanism, it "complements" (enhances) the effectiveness of *both* innate and adaptive defenses.

Complement can be activated by the two pathways outlined in Figure 20.5. The **classical pathway** involves *antibodies,* water-soluble protein molecules that the adaptive immune system produces to fight off foreign invaders. The classical pathway depends on the binding of antibodies to the invading organisms and the subsequent binding of C1 to the microorganism-antibody complexes, a step called **complement fixation** (described on p. 692). The **alternative pathway** is triggered when factors B, D, and P interact with polysaccharide molecules present on the surface of certain microorganisms.

Each pathway involves a cascade in which complement proteins are activated in an orderly sequence—each step causing catalysis of the next step. The two pathways converge on C3, cleaving it into C3a and C3b. This event initiates a common terminal pathway that causes cell lysis, promotes phagocytosis, and enhances inflammation.

Interferon

Viruses—essentially nucleic acids surrounded by a protein coat—lack the cellular machinery to generate ATP or synthesize proteins. They do their "dirty work" or damage in the body by invading tissue cells and taking over the cellular metabolic machinery needed to reproduce themselves. Although the infected cells can do little to save themselves, some can secrete small proteins called **interferons (IFNs;** in"ter-fēr'onz) to help protect cells that have not yet been infected. The IFNs diffuse to nearby cells, where they stimulate synthesis of a protein known as *PKR,* which then "interferes" with viral replication in the still-healthy cells by blocking protein synthesis at the ribosomes (Figure 20.4). Because IFN protection is not *virus-specific,* IFNs produced against a particular virus protect against a variety of other viruses.

Which pathway can occur without the need for an adaptive immune response?

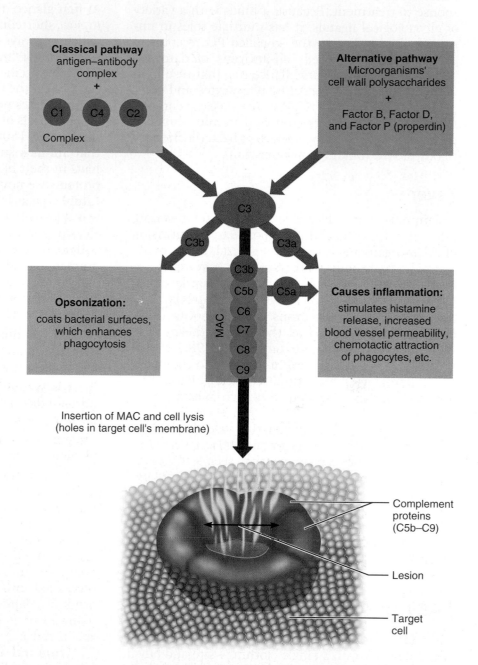

FIGURE 20.5 **Complement activation.** The classical pathway, mediated by 11 complement proteins designated C1–C9 (C1 incorporates three proteins) requires antigen-antibody interactions. The alternative pathway occurs when plasma proteins called factors B, D, and P interact with polysaccharides on the cell walls of certain bacteria and fungi. The two pathways converge to activate C3 (cleaving it into C3a and C3b), a step that initiates a common terminal sequence. Once bound to the target cell's surface, C3b initiates the remaining steps of complement activation. These steps result in the incorporation of MAC, the membrane attack complex (components C5b, and C6 to C9), into the target cell's membrane, creating a funnel-shaped lesion that causes cell lysis. Bound C3b also enhances phagocytosis. Released C3a, C5a, and other products of complement activation promote the events of inflammation (histamine release, increased vascular permeability, and so on).

The lytic events begin when C3b binds to the target cell's surface, triggering the insertion of a group of complement proteins called **MAC (membrane attack complex)** into the cell's membrane. MAC forms and stabilizes a hole in the membrane that ensures lysis of the target cell by inducing a massive influx of Ca^{2+}.

The C3b molecules that coat the microorganism provide "handles" that receptors on macrophages and neutrophils can adhere to, allowing them to engulf the particle more rapidly. As noted earlier, this process is called *opsonization*. C3a and other cleavage products formed during complement fixation amplify the inflammatory response by stimulating mast cells and basophils to release histamine and by attracting neutrophils and other inflammatory cells to the area.

C-reactive protein (CRP), produced by the liver in response to inflammatory molecules, is used as a

■ *The alternative pathway.*

clinical marker to assess for the presence of an acute infection or an inflammatory condition and its response to treatment. Because it binds with a variety of physiological ligands, it has multiple roles in immunity. By binding to the so-called PC receptor of pathogens and exposed self-antigens of damaged body cells, it plays a surveillance role that results in their targeting for disposal by phagocytes and complement. CRP binding to C1 of the classical complement pathway activates complement and results in the deposit of C3 on the surface of bacteria after antibody or CRP binding (opsonization).

Fever

Inflammation is a localized response to infection, but sometimes the body's response to the invasion of microorganisms is more widespread. **Fever,** or abnormally high body temperature, is a systemic response to invading microorganisms. As described further in Chapter 23, body temperature is regulated by a cluster of neurons in the hypothalamus, commonly referred to as the body's thermostat. Normally set at approximately 36.2°C (98.2°F), the thermostat is reset upward in response to chemicals called **pyrogens** (*pyro* = fire), secreted by leukocytes and macrophages exposed to foreign substances in the body.

High fevers are dangerous because excess heat denatures enzymes. Mild or moderate fever, however, is an adaptive response that seems to benefit the body. In order to multiply, bacteria require large amounts of iron and zinc, but during a fever the liver and spleen sequester these nutrients, making them less available. Fever also increases the metabolic rate of tissue cells in general, speeding up repair processes.

PART 2: ADAPTIVE DEFENSES

Most of us would find it wonderfully convenient if we could walk into a single clothing store and buy a complete wardrobe—hat to shoes—that fit perfectly regardless of any special figure problems. We *know* that such a service would be next to impossible to find. And yet, we take for granted our **adaptive immune system,** the body's built-in *specific defensive system* that stalks and eliminates with nearly equal precision almost any type of pathogen that intrudes into the body.

When it operates effectively, the adaptive immune system protects us from a wide variety of infectious agents, as well as from abnormal body cells. When it fails, or is disabled, the result is such devastating diseases as cancer, rheumatoid arthritis, and AIDS. The activity of the adaptive immune system

tremendously amplifies the inflammatory response and is responsible for most complement activation. At first glance, the adaptive system seems to have a major shortcoming. Unlike the innate system, which is always ready and able to react, the adaptive system must "meet" or be primed by an initial exposure to a specific foreign substance (antigen) before it can protect the body against that substance, and this priming takes precious time.

The basis of this specific immunity was revealed in the late 1800s, when researchers demonstrated that animals surviving a serious bacterial infection have in their blood protective factors (which are the proteins we now call *antibodies*) that defend against future attacks by the same pathogen. Furthermore, it was shown that if antibody-containing serum from the surviving animals was injected into animals that had *not* been exposed to the pathogen, those injected animals would also be protected. These landmark experiments were exciting because they revealed three important aspects of the adaptive immune response:

1. It is specific: It recognizes and is directed against *particular* pathogens or foreign substances that initiate the immune response.

2. It is systemic: Immunity is not restricted to the initial infection site.

3. It has "memory": After an initial exposure, it recognizes and mounts even stronger attacks on the previously encountered pathogens.

At first antibodies were thought to be the sole artillery of the adaptive immune system, but in the mid-1900s it was discovered that injection of antibody-containing serum did *not* always protect the recipient from diseases the serum donor had survived. In such cases, however, injection of the donor's lymphocytes *did* provide immunity. As the pieces fell into place, two separate but overlapping arms of adaptive immunity were recognized, each using a variety of attack mechanisms that vary with the intruder.

Humoral immunity (hu'mor-ul), also called **antibody-mediated immunity,** is provided by antibodies present in the body's "humors," or fluids (blood, lymph, etc.). Though they are produced by lymphocytes (or their offspring), antibodies circulate freely in the blood and lymph, where they bind primarily to bacteria, to bacterial toxins, and to free viruses, inactivating them temporarily and marking them for destruction by phagocytes or complement.

When lymphocytes themselves rather than antibodies defend the body, the immunity is called **cellular** or **cell-mediated immunity** because the protective factor is living cells. Cellular immunity also has cellular targets—virus-infected or parasite-infected tissue cells, cancer cells, and cells of foreign grafts. The lymphocytes act against such targets either *directly,* by lysing the foreign cells, or *indirectly,* by releasing chemical mediators that enhance the inflammatory response or activate other lymphocytes or macrophages. Thus, as you can see, although both arms of the adaptive immune system respond to virtually the same foreign substances, they do it in very different ways.

Before describing the humoral and cell-mediated responses, we will consider the *antigens* that trigger the activity of the remarkable cells involved in these immune responses.

Antigens

Antigens (an'tĭ-jenz) are substances that can mobilize the immune system and provoke an immune response. As such, they are the ultimate targets of all immune responses. Most antigens are large, complex molecules (both natural and synthetic) that are not normally present in the body. Consequently, as far as our immune system is concerned, they are intruders, or **nonself.**

Complete Antigens and Haptens

Antigens can be *complete* or *incomplete.* **Complete antigens** have two important functional properties:

1. Immunogenicity, which is the ability to stimulate proliferation of specific lymphocytes and antibodies. (*Antigen* is a contraction of "*anti*body *gen*erating," which refers to this particular antigenic property.)

2. Reactivity, which is the ability to react with the activated lymphocytes and the antibodies released by immunogenic reactions.

An almost limitless variety of foreign molecules can act as complete antigens, including virtually all foreign proteins, nucleic acids, some lipids, and many large polysaccharides. Of these, proteins are the strongest antigens. Pollen grains and microorganisms—such as bacteria, fungi, and virus particles—are all immunogenic because their surfaces bear many different foreign macromolecules.

As a rule, small molecules—such as peptides, nucleotides, and many hormones—are not immunogenic. But, if they link up with the body's own proteins, the adaptive immune system may recognize the combination as foreign and mount an attack that is harmful rather than protective. (These reactions, called *allergies,* are described later in the chapter.) In such cases, the troublesome small molecule is called a **hapten** (hap'ten; *haptein* = grasp) or **incomplete antigen.** Unless attached to protein carriers, haptens have reactivity but not immunogenicity. Besides cer-

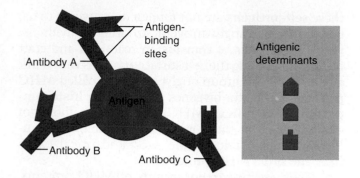

FIGURE 20.6   Antigenic determinants. Antibodies bind to antigenic determinants on the antigen surface. Most antigens have several different antigenic determinants, which means that different antibodies can bind to a given antigen. In this example, three types of antibodies react with different antigenic determinants on the same antigen molecule. (On a real antigen molecule, each antigenic determinant would be represented many times.)

tain drugs (particularly penicillin), chemicals that act as haptens are found in poison ivy, animal dander, detergents, cosmetics, and a number of common household and industrial products.

Antigenic Determinants

The ability of a molecule to act as an antigen depends on both its size and its complexity. Only certain parts of an entire antigen, called **antigenic determinants,** are immunogenic. Free antibodies or activated lymphocytes bind to these sites in much the same manner an enzyme binds to a substrate.

Most naturally occurring antigens have a variety of antigenic determinants (Figure 20.6) on their surfaces, some more potent than others in provoking an immune response. Because different antigenic determinants are "recognized" by different lymphocytes, a single antigen may mobilize several lymphocyte populations and may stimulate formation of many kinds of antibodies. Large proteins have hundreds of chemically different antigenic determinants, which accounts for their high immunogenicity and reactivity. However, large simple molecules such as plastics, which have many identical, regularly repeating units (and thus are not chemically complex), have little or no immunogenicity. Such substances are used to make artificial implants because the substances are not seen as foreign and rejected by the body.

Self-Antigens: MHC Proteins

The external surfaces of all our cells are dotted with a huge variety of protein molecules. Assuming our immune system has been properly "programmed," these **self-antigens** are not foreign or antigenic to us, but they are strongly antigenic to other individuals. (This is the basis of transfusion reactions and graft rejection.) Among the cell surface proteins that mark a cell as *self* is a group of glycoproteins called **MHC proteins,** coded for by genes of the **major histocompatibility complex (MHC).** Because millions of combinations of these genes are possible, it is unlikely that any two people except identical twins have the same MHC proteins.

There are two major groups of MHC proteins, distinguished by location. Class I MHC proteins are found on virtually *all* body cells, but class II MHC proteins are found *only on certain cells* that act in the immune response.

Each MHC molecule has a deep groove that typically displays a peptide. In healthy cells, all the displayed peptides come from the breakdown of cellular proteins during normal protein recycling and tend to be quite diverse. However, in infected cells, MHC proteins also bind to fragments of (foreign) antigens and, as described shortly, this event plays a crucial role in mobilizing the immune system.

Cells of the Adaptive Immune System: An Overview

The three crucial cell types of the adaptive immune system are two distinct populations of lymphocytes, and **antigen-presenting cells (APCs). B lymphocytes** or **B cells** oversee humoral immunity. **T lymphocytes** or **T cells** are non-antibody-producing lymphocytes that constitute the cell-mediated arm of adaptive immunity. Unlike the lymphocytes, APCs do not respond to specific antigens but instead play essential auxiliary roles.

Lymphocytes

Like all other blood cells, lymphocytes originate in red bone marrow from hematopoietic stem cells. When released from bone marrow, the immature lymphocytes are essentially identical. Whether a given lymphocyte matures into a B cell or a T cell depends on where in the body it becomes **immunocompetent,** that is, able to recognize a specific antigen by binding to it. T cells undergo this two- to three-day process in the thymus under the direction of thymic hormones. In the thymus, the immature lymphocytes divide rapidly and their numbers increase enormously, but only the maturing T cells with the sharpest ability to identify *foreign* antigens survive.

This "education" of fetal T cells, which involves both positive and negative selection processes, is expensive indeed—only about 2 percent of T cells survive it.

Positive selection, which occurs in the thymic cortex is essentially an **MHC restriction process.** It identifies T cells whose receptors are capable of recognizing self MHC molecules, thus producing an army of self-MHC restricted T cells (Figure 20.7). Those thymocytes unable to recognize self MHC molecules are selected against and undergo apoptosis, dying within 3–4 days.

T cells that make it through positive selection are then tested by a **negative selection** process that occurs throughout the thymus. T cells that bind too strongly to self MHC or to self-MHC bound to self peptides are eliminated. This ensures that the T cells surviving this second screening process exhibit **self tolerance** (relative unresponsiveness to self antigens).

B cells become immunocompetent and self-tolerant in bone marrow, but little is known about the factors that control B cell maturation in humans. However, some self-reactive B cells are inactivated (a phenomenon called *anergy*), while others are killed outright or physically eliminated through apoptosis *(clonal deletion).* Recently, *receptor editing*, in which there is a secondary rearrangement of the antigen-binding part of the receptor, has been proposed as another way autoreactive B cells are eliminated.

The lymphoid organs where lymphocytes become immunocompetent (thymus and bone marrow) are called **primary lymphoid organs.** All other lymphoid organs are referred to as **secondary lymphoid organs.**

When B or T cells become immunocompetent, they display a unique type of receptor on their surface. These receptors (some 10^4 to 10^5 per cell) enable the lymphocyte to recognize and bind to a specific antigen. Once these receptors appear, the lymphocyte is committed to react to one distinct antigenic determinant, and one only, because *all* of its antigen receptors are the same. Although the antigen receptors of T cells and B cells differ in overall structure, both types of receptor proteins are from the same immunoglobulin supergene family, and both cell types are capable of responding to the same antigens.

While many of the details of lymphocyte processing are still unknown, we do know that lymphocytes become immunocompetent *before* meeting the antigens they may later attack. Thus, *it is our genes, not antigens, that determine what specific foreign substances our immune system will be able to recognize and resist.* Another way of saying the same thing is that it is the immune cell receptors that constitute our genetically acquired knowledge of the microbes that are likely to be in our environment. An antigen determines only which of the existing T or B

? *What cell types act as APCs?*

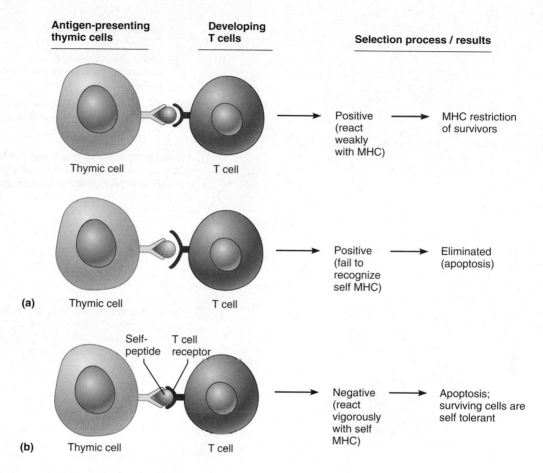

Antigen-presenting   Developing
thymic cells   T cells   Selection process / results

Thymic cell   T cell   Positive   →   MHC restriction
                                     (react   of survivors
                                     weakly
                                     with MHC)

(a) Thymic cell   T cell   Positive   →   Eliminated
                                     (fail to   (apoptosis)
                                     recognize
                                     self MHC)

Self-   T cell
peptide   receptor

(b) Thymic cell   T cell   Negative   →   Apoptosis;
                                     (react   surviving cells are
                                     vigorously   self tolerant
                                     with self
                                     MHC)

FIGURE 20.7   T cell selection in the thymus. (a) During positive selection, only those T cells that are able to recognize self MHC (with or without bound peptide) are allowed to continue the maturation process. Those that fail this test undergo apoptosis. **(b)** Negative selection identifies T cells that are self-tolerant. Those that react too vigorously with self MHC are selected against and eliminated.

cells will proliferate and mount the attack against it. Only some of the antigens our lymphocytes are programmed to resist will ever invade our bodies. Consequently, only some members of our army of immunocompetent cells are mobilized in our lifetime. The others are forever idle.

After becoming immunocompetent, the naive (still immature) T cells and B cells are exported to the lymph nodes, spleen, and other secondary lymphoid organs, where the encounters with antigens occur (Figure 20.8). Then, when the lymphocytes bind with recognized antigens, the lymphocytes complete their differentiation into fully functional—mature, antigen-activated—T cells and B cells.

Antigen-Presenting Cells

The major role of **antigen-presenting cells** in immunity is to engulf antigens and then present fragments of these antigens, like signal flags, on their own surfaces where they can be recognized by T cells—that is, they *present antigens* to the cells that will destroy the antigens. The major types of cells acting as APCs are *dendritic cells* present in connective tissues, *Langerhans' cells* of the skin epidermis, *macrophages*, and *activated B lymphocytes*.

Notice that all these cell types are in sites that make it very easy to encounter and process antigens. Dendritic cells are at the body's frontiers, best situated to act as mobile sentinels. Macrophages are widely distributed throughout the lymphoid organs and connective tissues. Besides their antigen-presenting role, dendritic cells and macrophages also

■ *Macrophages, dendritic cells, and B cells to name a few.*

What is the advantage of having lymphocytes that are mobile rather than fixed in the lymphoid organs?

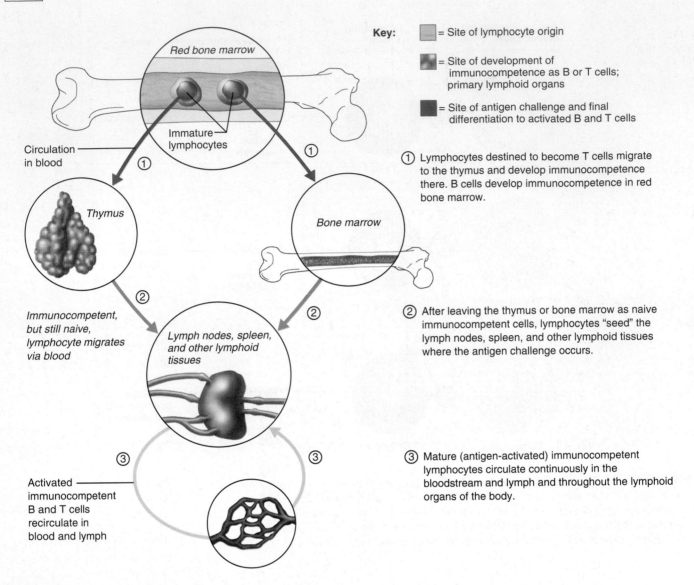

Key:
= Site of lymphocyte origin

= Site of development of immunocompetence as B or T cells; primary lymphoid organs

= Site of antigen challenge and final differentiation to activated B and T cells

① Lymphocytes destined to become T cells migrate to the thymus and develop immunocompetence there. B cells develop immunocompetence in red bone marrow.

② After leaving the thymus or bone marrow as naive immunocompetent cells, lymphocytes "seed" the lymph nodes, spleen, and other lymphoid tissues where the antigen challenge occurs.

③ Mature (antigen-activated) immunocompetent lymphocytes circulate continuously in the bloodstream and lymph and throughout the lymphoid organs of the body.

FIGURE 20.8 Lymphocyte traffic. Immature lymphocytes arise in red bone marrow. (Note that red marrow is not found in the medullary cavity of the diaphysis of long bones in adults.)

secrete soluble proteins that activate T cells. Activated T cells, in turn, release chemicals that rev up the mobilization and maturation of dendritic cells and prod macrophages to become *activated macrophages*, true "killers" that are insatiable phagocytes and secrete bactericidal chemicals. As you will see, interactions between various lymphocytes, and between lymphocytes and APCs, underlie virtually all phases of the immune response.

It allows them to come into contact with a broader spectrum of antigens and other body cells. ■

Although APCs and lymphocytes are found throughout each lymphoid organ, specific cells tend to be more numerous in certain areas. For example, the paracortical areas of lymph nodes house mostly T cells, and dendritic cells and B cells tend to populate the *germinal centers* (see Figure 19.4, p. 666), whereas relatively more macrophages are clustered around the medullary sinuses. Macrophages tend to remain fixed in the lymphoid organs, as if waiting for antigens to come to them. But lymphocytes, especially the T cells (which account for 65–85% of bloodborne lymphocytes), circulate continuously throughout the body. This circulation greatly increases a lymphocyte's chance of coming into contact with antigens located in different parts of the body, as well as with huge numbers of macrophages and other lymphocytes. Although lymphocyte recirculation appears to be random, the lymphocyte emigration to the tissues where their protective services are needed is highly specific, regulated by homing signals (CAMs) displayed on vascular endothelial cells.

Because lymph capillaries pick up proteins and pathogens from nearly all body tissues, immune cells in lymph nodes are in a strategic position to encounter a large variety of antigens. Lymphocytes and APCs in the tonsils act primarily against microorganisms that invade the oral and nasal cavities, whereas the spleen acts as a filter to trap bloodborne antigens.

In addition to T cell recirculation and passive delivery of antigens to lymphoid organs by lymphatics, a third delivery mechanism—migration of dendritic cells to secondary lymphoid organs—may be an even more important way of ensuring that the immune cells encounter invading antigens. With their long veil-like extensions, dendritic cells are very efficient antigen catchers, and once they have internalized antigens by phagocytosis, they enter nearby lymphatics to get to the lymphoid organ where they will present the antigens to T and B cells. Indeed, recent research suggests that the dendritic cells are the true initiators of adaptive immunity.

In summary, the two-fisted adaptive immune system uses lymphocytes, APCs, and specific molecules to identify and destroy all substances—both living and nonliving—that are in the body but not recognized as self. The system's response to such threats depends on the ability of its cells (1) to recognize antigens in the body by binding to them and (2) to communicate with one another so that the whole system mounts a response specific to those antigens.

Humoral Immune Response

The **antigen challenge,** the first encounter between an immunocompetent but naive lymphocyte and an invading antigen, usually takes place in the spleen or in a lymph node, but it may happen in any lymphoid tissue. If the lymphocyte is a B cell, the challenging antigen provokes the *humoral immune response,* in which antibodies are produced against the challenger.

Clonal Selection and Differentiation of B Cells

An immunocompetent but naive B lymphocyte is *activated* (stimulated to complete its differentiation) when antigens bind to its surface receptors and cross-link adjacent receptors together. Antigen binding is quickly followed by receptor-mediated endocytosis of the cross-linked antigen-receptor complexes. These events (plus some interactions with T cells described shortly) trigger **clonal selection** (klo'nul). That is, the events stimulate the B cell to grow and then multiply rapidly to form an army of cells all exactly like itself and bearing the same antigen-specific receptors (Figure 20.9). The resulting family of identical cells, all descended from the *same* ancestor cell, is called a **clone.** It is the antigen that does the selecting in clonal selection by "choosing" a lymphocyte with complementary receptors.

Most cells of the clone become **plasma cells,** the antibody-secreting effector cells of the humoral response. Although B cells secrete limited amounts of antibodies, plasma cells develop the elaborate internal machinery (largely rough endoplasmic reticulum) needed to secrete antibodies at the unbelievable rate of about 2000 molecules per second. Each plasma cell functions at this breakneck pace for 4 to 5 days and then dies. The secreted antibodies, each with the same antigen-binding properties as the receptor molecules on the surface of the parent B cell, circulate in the blood or lymph, where they bind to free antigens and mark them for destruction by other specific or nonspecific mechanisms. Clone cells that do not differentiate into plasma cells become long-lived **memory cells** that can mount an almost immediate humoral response if they encounter the same antigen again at some future time (see Figure 20.9).

Immunological Memory

The cellular proliferation and differentiation just described constitute the **primary immune response,** which occurs on first exposure to a particular antigen. The primary response typically has a lag period of 3 to 6 days after the antigen challenge. This lag period mirrors the time required for the few B cells specific for that antigen to proliferate and for their offspring to differentiate into plasma cells. After the mobilization period, plasma antibody levels rise, reach peak levels in about 10 days, and then decline (Figure 20.10).

If (and when) someone is reexposed to the same antigen, whether it's the second or the twenty-second time, a **secondary immune response** occurs. Secondary immune responses are faster, more prolonged, and more effective, because the immune system has already been primed to the antigen, and sensitized memory cells are already in place "on alert." These memory cells provide what is commonly called **immunological memory.**

Within hours after recognition of the "old enemy" antigen, a new army of plasma cells is being generated. Within 2 to 3 days the antibody

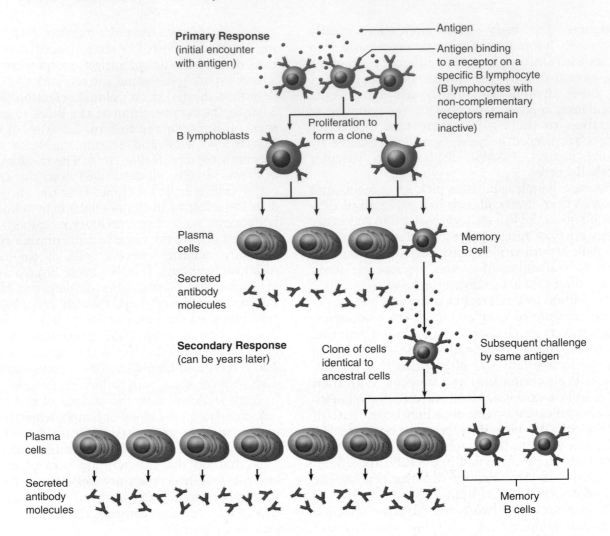

Primary Response
(initial encounter
with antigen)

Antigen

Antigen binding
to a receptor on a
specific B lymphocyte
(B lymphocytes with
non-complementary
receptors remain
inactive)

B lymphoblasts

Proliferation to
form a clone

Plasma
cells

Memory
B cell

Secreted
antibody
molecules

Secondary Response
(can be years later)

Clone of cells
identical to
ancestral cells

Subsequent challenge
by same antigen

Plasma
cells

Secreted
antibody
molecules

Memory
B cells

FIGURE 20.9 **Clonal selection of a B cell.** The initial meeting between a B cell and antigen stimulates the primary response in which the B cell proliferates rapidly, forming many identical B cells (a clone), most of which differentiate into antibody-producing plasma cells. (Though not indicated in the figure, the production of mature plasma cells takes about 5 days and eight cell generations.) Cells that do not differentiate into plasma cells become memory cells primed to respond to subsequent exposures to the same antigen. Should such a meeting occur, the memory cells quickly produce more memory cells and larger numbers of effector plasma cells with the same antigen specificity. Responses generated by memory cells are called secondary responses.

titer (concentration) in the blood rises steeply to reach much higher levels than were achieved in the primary response. Secondary response antibodies not only bind with greater affinity (more tightly), but their blood levels remain high for weeks to months. (Hence, when the appropriate chemical signals are present, plasma cells can keep functioning for much longer than the 4 to 5 days seen in primary responses.) Memory cells persist for long periods in humans and many retain their capacity to produce powerful secondary humoral responses for life.

The same general phenomena occur in the cellular immune response: A primary response sets up a pool of activated lymphocytes (in this case, T cells) and generates memory cells that can then mount secondary responses.

Active and Passive Humoral Immunity

When your B cells encounter antigens and produce antibodies against them, you are exhibiting **active humoral immunity** (Figure 20.11). Active immunity is (1) *naturally acquired* when you get a bacterial or viral infection, during which time you may develop symptoms of the disease and suffer a little (or a lot), and (2) *artificially acquired* when you receive **vaccines.** Indeed, once it was realized that secondary responses are so much more vigorous than primaries, the race was on to develop vaccines to "prime" the immune response by providing a first meeting with the antigen.

Most vaccines contain dead or *attenuated* (living, but extremely weakened) pathogens, or their components. Vaccines provide two benefits:

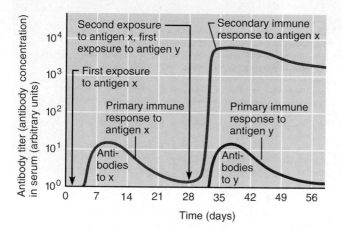

FIGURE 20.10 **Primary and secondary humoral responses.** The blue graph line shows a primary and a secondary response to antigen X. In the primary response, there is a 4-day time lag followed by a rapid rise and fairly rapid decline in the level of antibodies in the blood. A second exposure to antigen X on day 28 elicits a secondary response that has a very short lag time and is both more rapid and more intense. Additionally, antibody levels remain high for a much longer time. If a second antigen, antigen Y, were also encountered at day 28, the reaction to that antigen would be a primary response, not a secondary one, and its graph line (red) would show a pattern similar to that seen during the primary response to antigen X. (Time shown on the horizontal axis is given for example only. The time for response to different antigens varies greatly.)

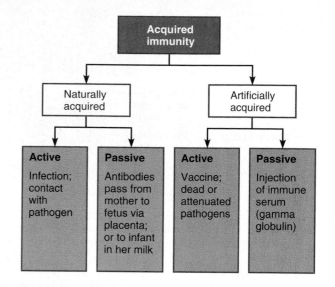

FIGURE 20.11 **Types of acquired immunity.** Immunological memory is established via active immunity, never via passive immunity.

1. They spare us most of the symptoms and discomfort of the disease that would otherwise occur during the primary response.

2. Their weakened antigens provide functional antigenic determinants that are both immunogenic and reactive.

Vaccine *booster shots*, which may intensify the immune response at later meetings with the same antigen, are also available.

Vaccines have wiped out smallpox and have substantially lessened the illness caused by such former childhood killers as whooping cough, polio, and measles. Although vaccines have dramatically reduced hepatitis B, tetanus, and pneumonia in adults, immunization of adults in the U.S. has a much lower priority than that of children and as a result more than 50,000 Americans die each year from vaccine-preventable infections (pneumonia, influenza, and hepatitis).

Conventional vaccines have shortcomings. Although it was originally believed that the immune response was about the same regardless of how an antigen got into the body (under its own power or via a vaccine), that has proved not to be the case. Apparently vaccines mainly target the type of helper T cell (more details shortly) that revs up B cell defenses and antibody formation (the T_H2 cell) as opposed to the type (T_H1) that generates strong cell-mediated responses. As a result, lots of antibodies are formed that provide immediate protection, but cellular immunological memory is only poorly established. (The immune system is deprived of the learning experience that comes with clearing an infection via a T_H1-mediated response.) In rare cases, vaccines cause the very disease they are trying to prevent because the attenuated antigen isn't weakened enough. In other cases, vaccines may trigger an allergic response. The new "naked DNA" antiviral vaccines, blasted into the skin with a gene gun, and *edible* vaccines taken orally, appear to circumvent these problems.

Passive humoral immunity differs from active immunity, both in the antibody source and in the degree of protection it provides (see Figure 20.11).

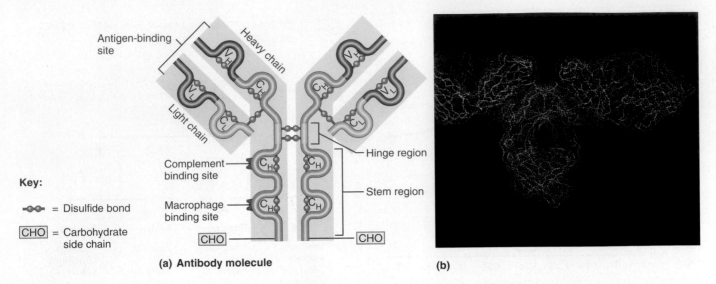

FIGURE 20.12 Antibody structure.
(a) Basic antibody structure consists of four polypeptide chains joined together by disulfide bonds (S—S). Two of the chains are short *light chains*; the other two are long *heavy chains*. Each chain has a V (variable) region (different in different antibodies) and a C (constant) region (essentially identical in different antibodies of the same class). Because the variable regions are the antigen-binding sites, each antibody monomer has two antigen-binding sites.
(b) Computer-generated image of antibody structure. Each tiny colored dot represents an amino acid of the polypeptide chains.

Instead of being made by your plasma cells, the antibodies are harvested from the serum of an immune human or animal donor. As a result, your B cells are *not* challenged by antigens, immunological memory does *not* occur, and the protection provided by the "borrowed" antibodies ends when they naturally degrade in the body.

Passive immunity is conferred *naturally* on a fetus when the mother's antibodies cross the placenta and enter the fetal circulation. For several months after birth, the baby is protected from all the antigens to which the mother has been exposed. Passive immunity is *artificially* conferred via a serum such as *gamma globulin*, which is administered after exposure to hepatitis. Other immune sera are used to treat poisonous snake bites (antivenom), botulism, rabies, and tetanus (antitoxin) because these rapidly fatal diseases would kill a person before active immunity could be established. The donated antibodies provide immediate protection, but their effect is short-lived (two to three weeks).

Antibodies

Antibodies, also called **immunoglobulins** (Igs; im″u-no-glob′u-linz), constitute the **gamma globulin** part of blood proteins. As mentioned earlier, antibodies are proteins secreted by activated B cells or plasma cells in response to an antigen, and they are capable of binding specifically with that antigen. They are formed in response to an incredible number of different antigens. Despite their variety, all antibodies can be grouped into one of five Ig classes, each slightly different in structure and function. Before seeing how these Ig classes differ from one another, let's take a look at how all antibodies are alike.

Basic Antibody Structure

Regardless of its class, each antibody has a basic structure consisting of four looping polypeptide chains linked together by disulfide (sulfur-to-sulfur) bonds (Figure 20.12). Two of these chains, the **heavy (H) chains,** are identical to each other and contain approximately 400 amino acids each. The other two chains, the **light (L) chains,** are also identical to each other, but only about half as long as each H chain. The heavy chains have a flexible *hinge* region at their approximate "middles." The "loops" on each chain are created by disulfide bonds between amino acids that are in the same chain but about 110 amino acids apart, which cause the intervening parts of the polypeptide chains to loop out. The four chains combined form a molecule, called an **antibody monomer** (mon′o-mer), with two identical halves, each consisting of a heavy and a light chain. The molecule as a whole is T or Y shaped.

Each chain forming an antibody has a **variable (V) region** at one end and a much larger **constant (C) region** at the other end. Antibodies responding to different antigens have very different V regions, but their C regions are the same (or nearly so) in all antibodies of a given class. The V regions of the heavy and light chains in each arm of the monomer combine to form an **antigen-binding site** shaped to "fit" a specific antigenic determinant. Hence, each antibody monomer has two such antigen-binding regions.

The C regions that form the *stem* of the antibody monomer determine the antibody class and serve common functions in all antibodies: These are the *effector regions* of the antibody that dictate (1) the cells and chemicals the antibody can bind to, and (2) how the antibody class functions in antigen elimination. For example, some antibodies can fix complement, some circulate in blood and others are found primarily in body secretions, some cross the placental barrier, and so on.

Antibody Classes

The five major immunoglobulin classes are designated IgM, IgA, IgD, IgG, and IgE, on the basis of the C regions in their heavy chains. (Remember the name MADGE to recall the five Ig types.) As illustrated in Table 20.3, IgD, IgG, and IgE have the same basic Y-shaped structure and thus are *monomers*. IgA occurs in both monomer and *dimer* (two linked monomers) forms. (Only the dimer is shown in the table.) Compared to the other antibodies, IgM is a huge antibody, a *pentamer* (penta = five) constructed from five linked monomers.

The antibodies of each class have different biological roles and locations in the body. IgM is the first antibody class *released* to the blood by plasma cells. IgG is the most abundant antibody in plasma and the only Ig class that crosses the placental barrier. Hence, the passive immunity that a mother transfers to her fetus is "with the compliments" of her IgG antibodies. *Only* IgM and IgG can fix complement. The IgA dimer, also called **secretory IgA,** is found primarily in mucus and other secretions that bathe body surfaces. It plays a major role in preventing pathogens from gaining entry into the body. IgD is always bound to a B cell surface; hence it acts as a B cell receptor. IgE antibodies, almost never found in blood, are the "troublemaker" antibodies involved in some allergies. These and other characteristics unique to each immunoglobulin class are summarized in Table 20.3.

Mechanisms of Antibody Diversity

It is estimated that our various plasma cells can make over a billion different types of antibodies. Because antibodies, like all other proteins, are specified by genes, it would seem that an individual must have billions of genes. Not so; each body cell contains perhaps 100,000 genes that code for all the proteins the cell must make.

How can a limited number of genes generate a seemingly limitless number of different antibodies? Recombinant DNA studies have shown that

TABLE 20.3   **Immunoglobulin Classes**

IgD
(monomer)

IgD is virtually always attached to the external surface of a B cell, where it functions as the antigen receptor of the B cell; important in B cell activation.

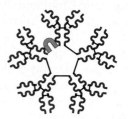

IgM
(pentamer)

IgM exists in monomer and pentamer (five united monomers) forms. The monomer, which is attached to the B cell surface, serves as an antigen receptor. The pentamer (illustrated) circulates in blood plasma and is the first Ig class released by plasma cells during the primary response. (This fact is diagnostically useful because presence of IgM in plasma usually indicates current infection by the pathogen eliciting IgM's formation.) Because of its numerous antigen binding sites, IgM is a potent agglutinating agent and readily fixes and activates complement.

IgG
(monomer)

IgG is the most abundant and diverse antibody in plasma, accounting for 75–85% of circulating antibodies. It protects against bacteria, viruses, and toxins circulating in blood and lymph, readily fixes complement, and is the main antibody of both primary and secondary responses. It crosses the placenta and confers passive immunity from the mother to the fetus.

IgA
(dimer)

IgA monomer exists in limited amounts in plasma. The dimer (illustrated), referred to as secretory IgA, is found in body secretions such as saliva, sweat, intestinal juice, and milk, and helps prevent attachment of pathogens to epithelial cell surfaces (including mucous membranes and the epidermis).

IgE
(monomer)

IgE is slightly larger than the IgG antibody. It is secreted by plasma cells in skin, mucosae of the gastrointestinal and respiratory tracts, and tonsils. Its stem region becomes bound to mast cells and basophils, and when its receptor ends are triggered by an antigen, it causes the cells to release histamine and other chemicals that mediate inflammation and an allergic reaction. Typically only traces of IgE are found in plasma, but levels rise during severe allergic attacks or chronic parasitic infections of the gastrointestinal tract.

the genes that dictate the structure of each antibody are not present as such in embryonic cells. Instead of a complete set of "antibody genes," embryonic cells contain a few hundred genetic bits and pieces that can be thought of as an erector set or Legos for antibody genes. These gene segments are shuffled and combined in different ways (a process called **somatic recombination**) by each B cell as it becomes immunocompetent. The information of the newly assembled genes is then expressed in the surface receptors of B cells and in the antibodies later released by their plasma cell "offspring."

In an immature B cell, the gene segments (exons) coding for the L and H chains are physically separated by noncoding DNA segments (introns), but are located on the same chromosome. (See Chapter 3 if you need a refresher on this terminology.) The L and H chains are manufactured separately and then joined to form the antibody monomer. The events forming the chains are similar, but the process for the heavy chains is more complex because of the larger variety of gene segments needed for the C regions of the H chains (remember, the C regions of heavy chains determine antibody class).

The random mixing of gene segments to make unique antibody genes specifying the H and L chains accounts for only part of the huge variability seen in antibodies. Certain areas of the V gene segments, the so-called *hypervariable regions*, are "hot spots" for somatic mutations that enormously increase antibody variation.

A single plasma cell can switch from making one kind of H chain to making another kind, thereby producing two or more different antibody classes having the same antigen specificity. For example, the first antibody released in the primary response is IgM; then the plasma cell begins to secrete IgG. During secondary responses, almost all of the Ig protein is IgG.

Antibody Targets and Functions

Though antibodies cannot themselves destroy antigens, they can inactivate them and tag them for destruction (Figure 20.13). The common event in all antibody-antigen interactions is formation of **antigen-antibody** (or **immune**) **complexes.** Defensive mechanisms used by antibodies include neutralization, agglutination, precipitation, and complement fixation, with the first two most important.

Complement fixation and activation is the chief antibody ammunition used against cellular antigens, such as bacteria or mismatched red blood

cells. When antibodies bind to cells, the antibodies change shape to expose complement-binding sites on their C regions (see Figure 20.12a). This triggers complement fixation into the antigenic cell's surface, followed by cell lysis. Additionally, as described earlier, molecules released during complement activation tremendously amplify the inflammatory response and promote phagocytosis via opsonization. Hence, a positive feedback cycle that enlists more and more defensive elements is set into motion.

Neutralization, the simplest effector mechanism, occurs when antibodies block specific sites on viruses or bacterial exotoxins (toxic chemicals secreted by bacteria). As a result, the virus or exotoxin loses its toxic effect because it cannot bind to receptors on tissue cells to cause injury. The antigen-antibody complexes are eventually destroyed by phagocytes.

Because antibodies have more than one antigen-binding site, they can bind to the same determinant on more than one antigen at a time; consequently, antigen-antibody complexes can be cross-linked into large lattices. When cell-bound antigens are cross-linked the process causes clumping, or **agglutination,** of the foreign cells. IgM, with ten antigen-binding sites (see Table 20.3), is an especially potent agglutinating agent. Recall from Chapter 16 that this type of reaction occurs when mismatched blood is transfused (the foreign red blood cells are clumped) and is the basis of tests used for blood typing.

In **precipitation,** soluble molecules (instead of cells) are cross-linked into large complexes that settle out of solution. Like agglutinated bacteria, precipitated antigen molecules are much more easily captured and engulfed by phagocytes than are freely moving antigens.

A quick and dirty way to remember how antibodies work is to remember they have a PLAN of action—**p**recipitation, **l**ysis (by complement), **a**gglutination, and **n**eutralization.

Monoclonal Antibodies

In addition to their role in providing passive immunity, commercially prepared antibodies are used in research, clinical testing, and treating certain cancers. **Monoclonal antibodies,** used for such purposes, are produced by descendants of a single cell, and are pure antibody preparations specific for a single antigenic determinant.

A current technology for making monoclonal antibodies involves fusion of tumor cells and B lymphocytes. The resulting cell hybrids, called **hybridomas** (hi"brĭ-do'mahz), have desirable traits of both parent cells. Like tumor cells, hybridomas proliferate indefinitely in culture; like B cells, they produce a single type of antibody.

? *Both complement fixation and agglutination aid phagocytosis, but they do so in different ways. What is this difference?*

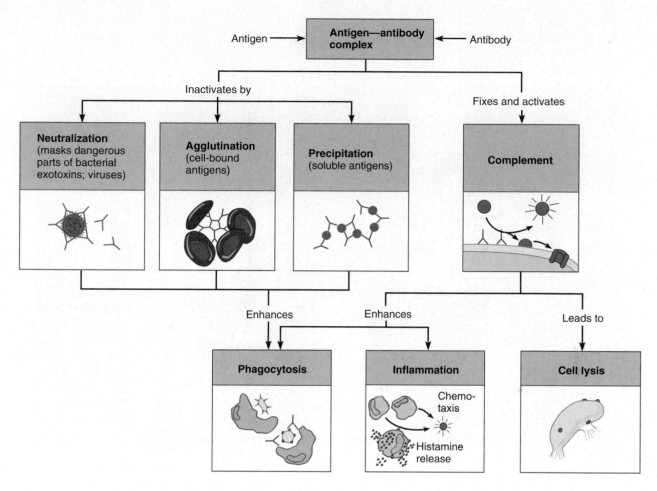

FIGURE 20.13 **Mechanisms of antibody action.** Antibodies act against free viruses, red blood cell antigens, bacterial toxins, and intact bacteria.

Monoclonal antibodies are used to diagnose pregnancy, certain sexually transmitted diseases, some types of cancer, hepatitis, and rabies. These monoclonal antibody tests are more specific, more sensitive, and faster than their conventional counterparts. Monoclonal antibodies are also being used to treat leukemia and lymphomas, both cancers that are present in the circulation and thus easily accessible to injected antibodies, and as "guided missiles" to deliver anticancer drugs only to cancerous tissue.

■ *With complement fixation, C3b molecules coat cellular antigens, providing rough spots on the antigen's surface, to which phagocytes can bind. Agglutination forms large clumps of antigen which precipitate and can't go anywhere, making the clumps fair game for the phagocytes.* ■

Cell-Mediated Immune Response

Despite their immense versatility, antibodies provide only partial immunity. Their prey is the *obvious* pathogen. They are fairly useless against infectious microorganisms like the tuberculosis bacillus which quickly slips inside body cells to multiply there. In these cases, the cell-mediated arm of adaptive immunity comes into play.

The T cells that mediate cellular immunity are a diverse lot, much more complex than B cells in both classification and function. There are two major populations of effector T cells based on which of a pair of structurally related *cell differentiation glycoproteins*—CD4 or CD8—is displayed by a mature T cell. These glycoprotein surface receptors, which are distinct from the T cell antigen receptors, play a role in interactions between a tissue cell and other cells or foreign antigens. **CD4 cells,** also called

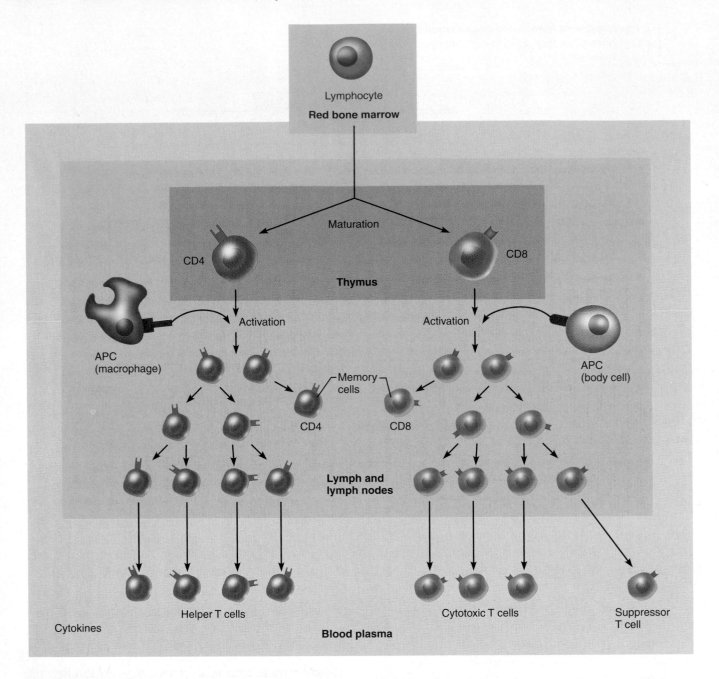

FIGURE 20.14 **Major types of T cells based on cell surface cell differentiation glycoproteins (CD4, CD8) displayed.** (Origin of suppressor T cells from the CD8 line is now being questioned. Many investigators believe they arise from the CD4 line.)

T4 cells, are primarily *helper T cells* (T_H), whereas most **CD8 cells** (T8 cells) are *cytotoxic T cells* (T_C), whose role is to destroy any cells in the body that harbor anything foreign. In addition to these two major groups of T cells, there are suppressor T cells (T_S), memory T cells, and some fairly rare subgroups (Figure 20.14). Details of T cells' roles are provided shortly.

Before going into the details of the cell-mediated immune response, we will recap and compare the relative importance of the humoral and cellular responses in adaptive immunity. Antibodies, produced

by plasma cells, are in many ways the simplest ammunition of the immune response. They are specialized to latch onto intact bacteria and soluble foreign molecules *in extracellular environments*—that is, free in body secretions and tissue fluid and circulating in blood and lymph. Antibodies never invade solid tissues unless a lesion is present. At the most basic level, antibody production and pathogen multiplication are in a race against each other and the fate of antibody targets is a case of "the quick or the dead." Remember, however, that forming antibody-antigen complexes does *not* destroy the antigens;

instead, it prepares them for destruction by innate defenses and activated T cells.

In contrast to B cells and antibodies, T cells cannot "see" either free antigens or antigens that are in their natural state. T cells can recognize and respond only to *processed* fragments of protein antigens displayed on body cell surfaces (APCs and others) and then only under specific circumstances. Consequently, T cells are best suited for cell-to-cell interactions, and most of their direct attacks on antigens (mediated by the cytotoxic T cells) target body cells infected by viruses or bacteria; abnormal or cancerous body cells; and cells of infused or transplanted foreign tissues.

Clonal Selection and Differentiation of T Cells

The stimulus for clonal selection and differentiation is the same in B cells and T cells—binding of antigen. However, the mechanism by which T cells recognize "their" antigen is very different than that seen in B cells and has some unique restrictions.

Antigen Recognition and MHC Restriction

Like B cells, immunocompetent T cells are activated when the variable regions of their surface receptors bind to a "recognized" antigen. However, T cells must accomplish *double recognition:* They must simultaneously recognize nonself (the antigen) and self (a MHC protein of a body cell).

Two types of MHC proteins are important to T cell activation. **Class I MHC proteins** are displayed by virtually all body cells except red blood cells and are always recognized by CD8 T cells. After being synthesized at the endoplasmic reticulum (Figure 20.15a), a class I MHC protein binds with a peptide 8 or 9 amino acids long ferried into the ER from the cytosol by transport proteins aptly called TAPs (**t**ransporter **a**ssociated with **a**ntigen **p**rocessing). The "loaded" class I MHC protein then migrates to the plasma membrane to display its attached protein fragment. This attached fragment is always part of some protein synthesized *in the cell*—either a bit of a cellular (self) protein or a peptide derived from an endogenous antigen that is broken down to peptides within a proteosome. **Endogenous antigens** are foreign proteins synthesized in a body cell. Two examples are the viral proteins produced by virus-infected cells and alien (mutated) proteins made by a body cell that has become cancerous.

Unlike the widely distributed class I proteins, **class II MHC proteins** are typically found only on the surfaces of mature B cells, some T cells, and antigen-presenting cells, where they enable the cells of the immune system to recognize one another.

Like their class I MHC counterparts, class II MHC proteins are synthesized at the ER and bind to peptide fragments (Figure 20.15b). However, the peptide fragments they bind are longer (14–17 amino acids) and come from **exogenous antigens** (foreign antigens) that have been phagocytosed and broken down in the phagolysosome vesicle. While still in the ER, a class II MHC molecule binds to what is called an *invariant protein,* a coupling that prevents the MHC molecule from binding to a peptide while in the ER. The class II MHC protein then moves from the ER through the Golgi apparatus and into a phagolysosome. It is there that the invariant chain loosens its hold on the MHC molecule, allowing the "jaws" of MHC to close around a degraded protein fragment. Vesicles are then recycled to the cell surface, where the class II MHC proteins display their booty for recognition by CD4 cells.

The role of MHC proteins in the immune response is strikingly important—they provide the means for signaling to immune system cells that infectious microorganisms are hiding in body cells. Without such a mechanism, viruses and certain bacteria that thrive in cells could multiply unbothered and unnoticed. When MHC proteins are complexed with fragments of our own (self) proteins, T cells passing by get the signal "Leave this cell alone, it's ours!" and ignore them. When MHC proteins are complexed to antigenic peptides (either exogenous or endogenous), they betray the invaders and "sound a molecular alarm" that signals invasion by (1) acting as antigen holders and (2) forming the self part of the self-nonself complexes that T cells must recognize in order to be activated.

T Cell Activation

T cell activation is actually a two-step process involving antigen binding and co-stimulation.

Antigen Binding This first step mostly entails what has already been described: T cell antigen receptors (**TCRs**) bind to an antigen-MHC complex on the surface of a body cell. Like B cell receptors, a TCR has variable and constant regions, but it consists of two rather than four polypeptide chains.

Helper and cytotoxic T cells have different preferences for the class of MHC protein that helps deliver the activation signal. This constraint, acquired during the "education" process in the thymus, is called *MHC restriction.* Helper T cells (CD4) can bind only to antigens linked to class II MHC proteins, which are typically displayed on APC surfaces. Dendritic cells are the most potent APCs for T$_H$ cells. As described earlier, APCs attach very small parts of the antigens to the class II MHC proteins for T cell recognition. In addition, mobile APCs like Langerhans' cells (dendritic cells of the epidermis)

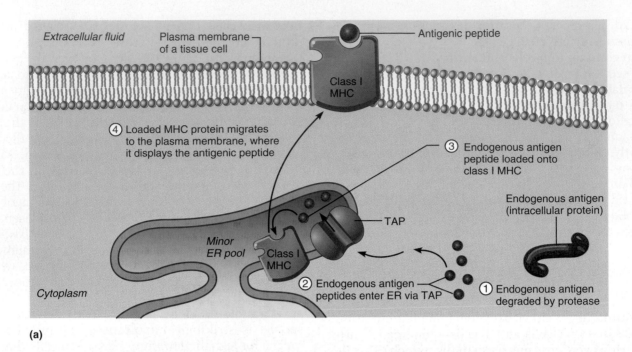

(a)

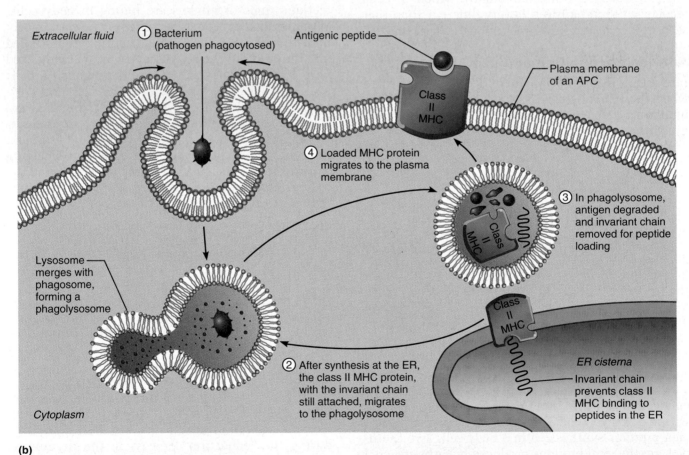

(b)

FIGURE 20.15 MHC proteins, and antigen processing and display.
(a) Class I MHC proteins pick up peptide fragments in the ER. These peptides may be derived from self-proteins or from foreign (viral or cancer) proteins made within the cell. The latter are called endogenous antigens. (b) Class II MHC proteins pick up peptide fragments from endocytosed vesicles where phagocytized foreign debris is degraded. Both MHC classes display the peptides at the cell surface.

actually migrate to the lymph nodes and other lymphoid tissues to present their antigens. As a result of this early alert, the body is spared a good deal of tissue damage that might otherwise occur before the antigens entered the blood and were carried to the lymph nodes for recognition.

Cytotoxic T cells (CD8) are activated by antigen fragments complexed with class I MHC proteins (Figure 20.16). Because virtually all body cells exhibit class I MHC proteins, T_C cells do not require *specialized* antigen-presenting cells—any body cell displaying a recognizable self-nonself complex will do. Nonetheless, antigen-presenting cells produce co-stimulatory molecules that *are* required for T_C cell activation. Hence, whether a T_H or a T_C cell is involved, the process of antigen presentation is essentially the same—the only things that differ are the presenting cell type and the MHC class recognized. However, slight variations in the antigenic peptides displayed can ultimately result in very different degrees of T cell activation. The process in which the T cells adhere to and crawl over the surface of other cells examining them for antigens they might recognize is sometimes called **immunologic surveillance.** (Note, however, that this term is commonly used to indicate *only* the process by which NK and dendritic cells constantly monitor body cells for foreign antigens.)

The TCR that recognizes the nonself-self complex is linked to multiple intracellular signaling pathways. Besides the TCR, other T cell surface proteins are involved in this first step. For example, the CD4 and CD8 proteins used to identify the two major T cell groups (T_H and T_C) are adhesion molecules that help maintain the coupling during antigen recognition. Additionally, the CD4 and CD8 proteins are associated with kinase enzymes located inside the T cells that phosphorylate cell proteins, activating some and inactivating others when antigen binding occurs. Once antigen binding has occurred, the T cell is stimulated but is still "idling," like a car that has been started but not put into gear. Such T cells are referred to as *naive T cells* at this stage.

Co-stimulation The story isn't over yet because step 2 comes next. Before a T cell can proliferate and form a clone, it must recognize one or more **co-stimulatory signals.** Sometimes the added requirement is T cell binding to still other surface receptors on an APC. For example, macrophages begin to sprout *B7 proteins* on their surfaces when the nonspecific defenses are being mobilized. B7 binding to the CD_{28} receptor on a T cell is a crucial co-stimulatory signal, particularly to T_H cells. Other co-stimulators include the cytokine chemicals released by macrophages or T cells; and interleukins 1 and 2. As you might guess, there are several types

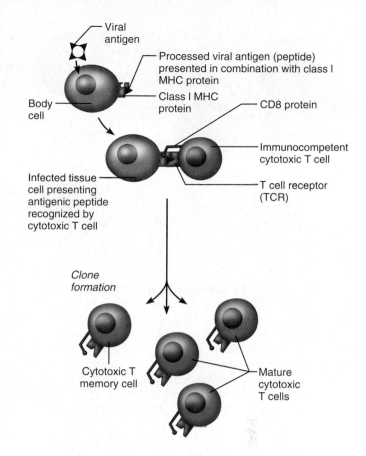

FIGURE 20.16 Clonal selection of T_C and T_H cells involves simultaneous recognition of self and nonself. Both T_C and T_H cells are stimulated to proliferate (clone) and complete their differentiation when they bind to parts of foreign antigens complexed to MHC proteins. Immunocompetent T_C cells are activated when they bind to endogenous antigens (nonself)—part of a virus in this example—complexed to a class I MHC protein. Activation of T_H cells is similar, except that the processed antigen is complexed with a class II MHC protein.

of co-stimulators, each type promoting a different response in the activated T cells. The important thing to understand is that, depending on the receptor to which a co-stimulator binds, it nudges the T cell either to complete activation or to abort it entirely. Or, to go back to our idling car analogy, the co-stimulator tells the T cell to (1) put the car in gear and step on the gas, or (2) turn off the ignition.

A T cell that binds to antigen without receiving the co-stimulatory signal becomes tolerant to that antigen and is unable to divide or to secrete cytokines. This state of unresponsiveness to antigen is called **anergy.** Just why this happens is uncertain, but the current hypothesis is that the two-signal sequence is a safeguard to prevent the immune system from destroying healthy cells.

Once activated, a T cell enlarges and proliferates to form a clone of cells that differentiate and perform

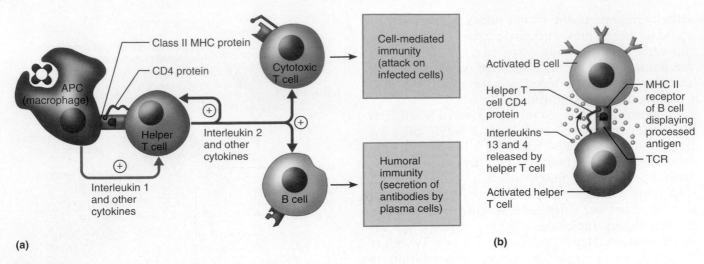

(a)   **(b)**

FIGURE 20.17   The central role of T_H cells. Helper T cells mobilize both arms (cellular and humoral) of the immune response. Once an immuno-competent T_H cell binds to an antigen-presenting cell (APC) and recognizes the self-nonself complex, the T_H cell produces a clone of T_H cells (not illustrated), all with identical receptors keyed to class II MHC receptors complexed with the same antigen. **(a)** In addition, the macrophage releases interleukin 1, a co-stimulator that enhances T cell activation. Activated T_H cells release interleukin 2, which stimulates proliferation and activity of other T_H cells (specific for the same antigenic determinant). Interleukin 2 also helps activate T_C cells and B cells. **(b)** T_H and B cells must sometimes interact directly for full B cell activation to occur. In such cases, T_H cells bind with the self-nonself complexes on activated B cell surfaces, and then release interleukins as co-stimulatory signals.

functions according to their T cell class. This primary response peaks within a week of exposure to the triggering antigen. A period of apoptosis then occurs between days 7 and 30, during which time the activated T cells die off and effector activity wanes as the amount of antigen declines. This wholesale disposal of T cells has a critical protective role because activated T cells are potential hazards. They produce huge amounts of inflammatory cytokines, which contribute to infection-driven hyperplasia of lymph nodes (an important forerunner of lymphoid malignancies). Additionally, once they've done their job, the effector T cells are unnecessary and thus disposable. Thousands of clone members become memory T cells, persisting perhaps for life, and providing a reservoir of T cells that can later mediate secondary responses to the same antigen.

Cytokines

As in the inflammatory response, chemical mediators enhance the immune response. **Cytokines,** the mediators involved in cellular immunity, include hormonelike glycoproteins released by activated T cells and macrophages. As mentioned previously, some cytokines act as co-stimulators of T cells and T cell proliferation (Figure 20.17). For example, **interleukin 1 (IL-1),** released by macrophages, co-stimulates bound T cells to liberate **interleukin 2 (IL-2)** and to synthesize more IL-2 receptors. IL-2 is a key growth factor. Acting as a local hormone, it sets up a positive feedback cycle that encourages activated T cells to divide even more rapidly. (Therapeutically IL-2 is used to treat melanoma and kidney cancers.)

Additionally, all activated T cells secrete one or more other cytokines that help amplify and regulate a variety of adaptive and innate immune responses. Some (such as tumor necrosis factor) are cell toxins; others (e.g., interferon) enhance the killing power of macrophages; and still others are inflammatory factors. Cytokines and their effects on target cells are summarized in Table 20.4.

Specific T Cell Roles

Helper T Cells

Helper T cells are regulatory cells that play a central role in adaptive immunity. Once T_H cells have been primed by APC presentation of antigen, their major function is to stimulate proliferation of other T cells and of B cells that have already become bound to antigen. In fact, without the "director" T_H cells, there is *no* immune response. Their cytokines furnish the chemical help needed to recruit other immune cells to fight off intruders (Figure 20.17a).

Helper T cells interact directly with B cells displaying antigen fragments bound to class II MHC receptors (Figure 20.17b), prodding the B cells into more rapid division and then, like the boss of an

TABLE 20.4	Cells and Molecules of the Adaptive Immune Response

Element	Function in Immune Response
Cells	
B cell	Lymphocyte that resides in lymph nodes, spleen, or other lymphoid tissues, where it is induced to replicate by antigen binding and helper T cell interactions; its progeny (clone members) form memory cells and plasma cells
Plasma cell	Antibody-producing "machine"; produces huge numbers of antibodies (immunoglobulins) with the same antigen specificity; represents further specialization of B cell clone descendants
Helper T cell (T_H)	A regulatory T cell that binds with a specific antigen presented by an APC; upon circulating into spleen and lymph nodes, it stimulates production of other cells (cytotoxic T cells and B cells) to help fight invader; acts both directly and indirectly by releasing cytokines
Cytotoxic T cell (T_C)	Also called a cytolytic (CTL) or killer T cell; activated by antigen presented by any body cell; recruited and activity enhanced by helper T cells; its specialty is killing virus-invaded body cells and cancer cells; it is involved in rejection of foreign tissue grafts
Suppressor T cell (T_S)	Slows or stops activity of B and T cells once infection (or onslaught by foreign cells) has been conquered
Memory cell	Descendant of activated B cell or any class of T cell; generated during initial immune response (primary response); may exist in body for years after, enabling it to respond quickly and efficiently to subsequent infections or meetings with same antigen
Antigen-presenting cell (APC)	One of several cell types (dendritic cell, macrophage, activated B cell) that engulfs and digests antigens that it encounters and presents parts of them on its plasma membrane (bound to a MHC protein) for recognition by T cells bearing receptors for same antigen; this function, antigen presentation, is essential for normal cell-mediated responses; macrophages also release chemicals (cytokines) that activate T cells and prevent viral multiplication
Molecules	
Antibody (Immunoglobulin)	Protein produced by B cell or by plasma cell; antibodies produced by plasma cells are released into body fluids (blood, lymph, saliva, mucus, etc.), where they attach to antigens, causing complement fixation, neutralization, precipitation, or agglutination, which "mark" the antigens for destruction by complement or phagocytes
Cytokines	
Interferons (IFNs)	
▪ Alpha (α)	Secreted by most leukocytes; antiviral effects; activates macrophages and NK cells
▪ Beta (β)	Secreted by fibroblasts; effects as above, but also acts to reduce inflammation
▪ Gamma (γ)	Secreted by lymphocytes; effects as for alpha IFN; stimulates synthesis and expression of more class I and II MHC proteins; enhances activity of B cells and T_C cells

assembly line, signaling for antibody formation to begin. Whenever a T_H cell binds to a B cell, the T cell releases interleukins 4 and 13 (and other cytokines). Although B cells may be activated solely by binding to certain antigens called **T cell–independent antigens,** most antigens require T cell co-stimulation to activate the B cells to which they bind. This more common variety of antigens is called **T cell–dependent antigens.** In general, T cell–independent antigen responses are weak and short-lived. B cell division continues as long as it is stimulated by the T_H cell. Thus, helper T cells help unleash the protective potential of B cells.

Cytokines released by T_H cells not only mobilize lymphocytes and macrophages but also attract other types of white blood cells into the area and tremendously amplify nonspecific defenses. As the released chemicals summon more and more cells into the battle, the immune response gains momentum, and the antigens are overwhelmed by the sheer numbers of immune elements acting against them.

It is interesting, but not surprising, that exposure to different cytokines during T_H cell differentiation induces different subsets of helper T cells. For example, IL-12 induces T_H1 differentiation, while IL-4 drives T_H2 differentiation. Generally speaking, the T_H1 cells stimulate inflammation, activate macrophages, and promote differentiation of cytotoxic T cells, that is, they mediate classical cell-mediated immunity. By contrast, T_H2 cells

TABLE 20.4   **Cells and Molecules of the Adaptive Immune Response** (continued)

Element	Function in Immune Response
Interleukins (ILS)	
■ IL-1	Secreted by activated macrophages; co-stimulates T and B cells to proliferate; promotes inflammation; causes fever (believed to be the pyrogen that resets the thermostat of the hypothalamus)
■ IL-2	Secreted by T_H cells; stimulates proliferation of T and B cells; activates NK cells; also called T cell growth factor
■ IL-3	Stimulates production of leukocytes and mast cells
■ IL-4	Secreted by T_H cells; co-stimulates activated B cells; causes plasma cells to secrete more IgE antibodies
■ IL-5	Secreted by some T_H cells and mast cells; co-stimulates B cells; causes plasma cells to secrete IgA antibodies; attracts eosinophils
■ IL-6	Induces differentiation of B cells into plasma cells; enhances proliferation and activity of T cells; stimulates liver to secrete mannose-binding protein, which triggers complement binding to bacteria with mannose sugar in their capsules
■ IL-7	B cell stimulator; influences macrophage activity
■ IL-8	Stimulates angiogenesis
■ IL-9	Stimulates myeloid tissue to produce blood cells
■ IL-10	Inhibits or turns down the immune response
■ IL-12	Secreted by granulocytes, dendritic cells, NK cells, and macrophages; stimulates T_C activity and NK cell activity; promotes T_H1 differentiation
■ IL-13	Secreted by T_H cells on binding to antigen-activated B cells
Lymphotoxin (LT)	Released by T_C cells; causes DNA fragmentation; promotes inflammation
Macrophage migration inhibitory factor (MIF)	Inhibits macrophage migration and keeps them in the area of antigen deposition
Perforin	Released by T_C cells; causes cell lysis by creating large pores in the target cell's membrane
Suppressor factors	Released by T_S cells; suppresses the immune response
Transforming growth factor beta (TGF-β)	Inhibits activation and proliferation of macrophages, T cells, and B cells; a suppressor of the immune response
Tumor necrosis factors (TNFs)	Produced by lymphocytes and in large amounts by macrophages; enhance nonspecific killing; slow tumor growth; cause selective damage to blood vessels; enhance granulocyte chemotaxis; help activate T cells, phagocytes, and eosinophils
Complement	Group of bloodborne proteins activated after binding to antibody-covered antigens; causes lysis of microorganism and enhances inflammatory response
Antigen	Substance capable of provoking an immune response; typically are large complex molecules (e.g., proteins and modified proteins) that are not normally present in the body

mainly promote the migration of eosinophils and basophils to the battlefield and activate immune response that depend on B cells and antibody formation.

Cytotoxic T Cells

Cytotoxic T cells, also called **killer T cells,** are the only T cells that can directly attack and kill other cells. Activated T_C cells roam the body, circulating in and out of the blood and lymph and through lymphoid organs in search of body cells displaying antigens to which the T_C cells have been sensitized.

Their main targets are virus-infected cells, but they also attack tissue cells infected by certain intracellular bacteria or parasites, cancer cells, and foreign cells introduced into the body by blood transfusions or organ transplants.

Before the onslaught can begin, the cytotoxic T cell must "dock" on the target cell by binding to a self-nonself complex. Remember, all body cells display class I MHC antigens, so all infected or abnormal body cells can be destroyed by these T cells as long as the appropriate antigen and co-stimulatory signal (typically IL-2 released by T_H cells) are also

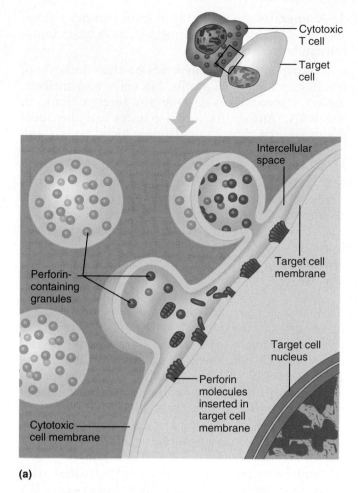

(a)

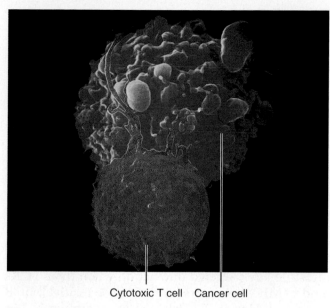

Cytotoxic T cell   Cancer cell

(b)

FIGURE 20.18 Cytotoxic T cells attack infected and cancerous cells. (a) A proposed mechanism of target cell lysis by T_C cells. The initial event is the tight binding of the cytotoxic cell to the target cell. During this binding period, granules in the T_C cell fuse with its plasma membrane and release perforin molecules which insert into the target cell membrane and later polymerize, forming cylindrical holes that allow free exchange of ions and water. The pores allow granzymes (also contained in the T_C granules, but not illustrated) to enter the target cell. Target cell apoptosis and lysis occurs after the T_C cell has detached. **(b)** Scanning electron micrograph of a cancer cell lysed by a T_C cell. (8300×)

present. The attack on foreign human cells, such as those of a graft, is more difficult to explain because here *all* of the antigens are nonself. However, apparently the T_C cells sometimes "see" the foreign class I MHC antigens as a combination of class I MHC protein bound to antigen, and NK cells are always on surveillance. In contrast to the T_C cells that recognize antigen-bearing class I MHC molecules, NK cells activate their killing machinery when they bind to a MICA receptor, which is a MHC-related cell surface protein that is switched on in cancer cells and in cells under stress, such as virus-infected cells and cells of transplanted organs and tissues. Thus, NK cells stalk abnormal or foreign cells in the body that T_C cells can't "see." They bypass cells when their Ly49 receptor (an inhibitory receptor) can bind to those cells' class I MHC receptors.

The mechanism of the cytotoxic cell's **lethal hit** which leads to cell lysis is unclear, but the following events occur in at least some cases:

1. The T_C cell binds tightly to the target cell and, during this period, the chemicals **perforin** and **granzymes** are released from the T cell's granules. The perforins

insert into the plasma membrane of the target cell (Figure 20.18).

2. In the presence of Ca^{2+}, the perforin molecules in the target cell membrane polymerize and create transmembrane pores very much like those produced by complement activation.

3. Granzymes enter the target cell via the pores and once inside these proteases degrade cellular contents, stimulating apoptosis.

4. The T_C cell then detaches and searches for another prey.

Other T_C cells apparently use different or additional signals to induce target cell death. For example, some secrete **lymphotoxin,** which causes fragmentation of the target cell's DNA. Still others secrete **gamma interferon,** which stimulates macrophages to killer status and indirectly enhances the killing process. Because of their ability to induce target cell lysis, T_C cells are often called **cytolytic T cells.**

Other T Cells

Like T_H cells, mature **suppressor T (T_S) cells** are regulatory cells. Because they release cytokines that suppress the activity of both B cells and other types of T cells, T_S cells are thought to be vital for winding down and finally stopping the immune response after an antigen has been inactivated or destroyed. This helps prevent runaway or unnecessary immune system activity. Because of their inhibitory role, suppressors are presumed to be important in preventing autoimmune reactions. However, because these T cells are very difficult to study, their activation process is obscure and controversial. Most of what is written about them is still hypothetical.

Gamma delta T cells (T_{gd}) represent about 10% of all T cells. These oddball T cells, which live in the intestine, are triggered into action when their TCRs bind to MICA receptors (or their close relative MICB receptors). Hence, T_{gd} cells are more similar to NK cells than to other T cells.

★   ★   ★

In summary, each type of T cell has unique roles to play in the immune response, yet is heavily enmeshed in interactions with other immune cells and elements as summarized in the overview of the entire primary response in Figure 20.19. The lesson to take away with you is that without helper T cells, *there is no adaptive immune response* because the helper cells direct or help complete the activation of *all* other immune cells. Their crucial role in immunity is painfully evident when they are destroyed, as in AIDS (see Immunodeficiencies on p. 704).

Organ Transplants and Prevention of Rejection

Organ transplants, a viable treatment option for many patients with end-stage cardiac or renal disease, have been done with mixed success for over 30 years. Immune rejection presents a particular problem when the goal is to provide such patients with functional organs from a living or recently deceased donor.

Essentially, there are four major varieties of grafts:

1. Autografts are tissue grafts transplanted from one body site to another in the *same person*.

2. Isografts are grafts donated to a patient by a *genetically identical individual*, the only example being identical twins.

3. Allografts are grafts transplanted from individuals that are *not genetically identical* but belong to the *same species*.

4. Xenografts are grafts taken *from another animal species*, such as transplanting a baboon heart into a human being.

Transplant success depends on the similarity of the tissues because T_C cells, NK cells, and antibodies act vigorously to destroy any foreign tissue in the body. Autografts and isografts are the ideal donor tissues. Given an adequate blood supply and no infection, they are always successful because the MHC-encoded proteins are identical. Xenografts are as yet only temporarily successful.* Consequently, the problematic graft type, and the type that is also most frequently used, is the allograft, with the organ usually obtained (harvested) from a human donor who has just died (in the case of heart or lung), or from living donors (kidney, liver, bone marrow).

Before an allograft is attempted, the ABO and other blood group antigens of donor and recipient must be determined, because these antigens are also present on most body cells. Next, recipient and donor tissues are typed to determine their MHC antigen match. MHC variation among human tissues is tremendous, so good tissue matches between unrelated individuals are next to impossible to obtain. However, the donor and recipient tissues are matched as closely as possible (at least a six-antigen match is essential).

Following surgery the patient is treated with *immunosuppressive therapy* involving drugs of the following categories: (1) corticosteroid drugs to suppress inflammation; (2) antiproliferative drugs, and (3) immunosuppressant drugs. Many of these drugs kill rapidly dividing cells (such as activated lymphocytes), and all of them have severe side effects.

FIGURE 20.19   The primary immune response. Activities of the humoral and cellular arms of the immune system are indicated by different colored backgrounds. Some events known to be co-stimulated by cytokines are noted. (Other co-stimulatory signals are also involved in many of these events.) Although complement, NK cells, and phagocytes are nonspecific defenses, they are enlisted in the fight by cytokines released by immune cells. (Immune cell receptors are not illustrated here for simplicity.)

---

* Though still true at publication of this book, this situation is about to change. By injecting human genes into fertilized pig eggs, researchers have been able to neutralize the attack of complement and inhibit the hyperacute rejection response which can destroy a transplanted organ within minutes by cutting off its blood supply. Genetically altered pigs are now being bred and human trials using these animals' organs are ongoing.

? *What are the possible consequences of co-stimulation with different co-stimulatory factors?*

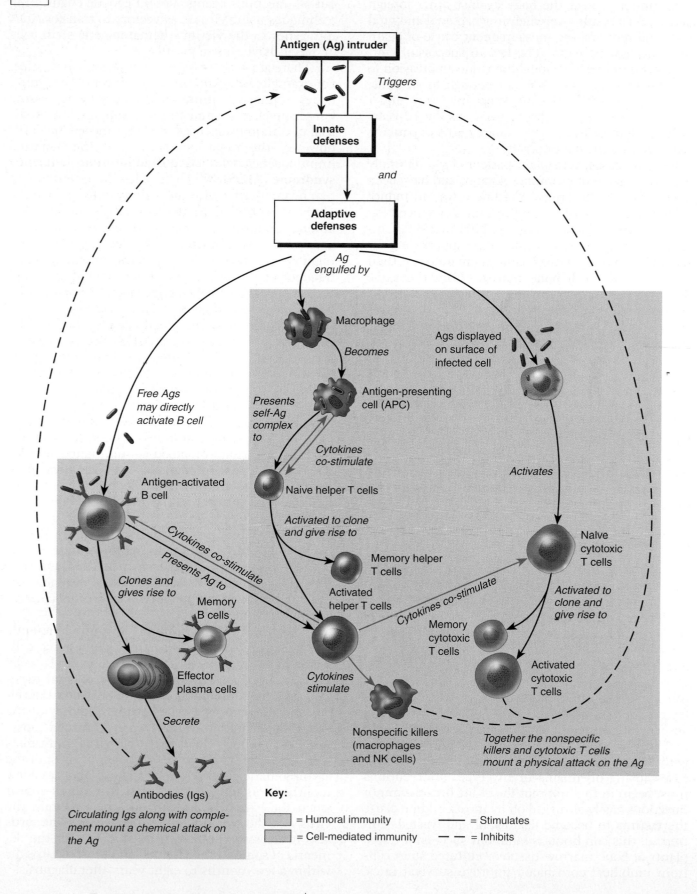

Antigen (Ag) intruder

Triggers

Innate defenses

and

Adaptive defenses

Ag engulfed by

Macrophage

Becomes

Ags displayed on surface of infected cell

Presents self-Ag complex to

Antigen-presenting cell (APC)

Cytokines co-stimulate

Activates

Naive helper T cells

Free Ags may directly activate B cell

Antigen-activated B cell

Cytokines co-stimulate

Presents Ag to

Activated to clone and give rise to

Memory helper T cells

Nalve cytotoxic T cells

Clones and gives rise to

Memory B cells

Activated helper T cells

Cytokines co-stimulate

Memory cytotoxic T cells

Activated to clone and give rise to

Effector plasma cells

Cytokines stimulate

Activated cytotoxic T cells

Secrete

Nonspecific killers (macrophages and NK cells)

Together the nonspecific killers and cytotoxic T cells mount a physical attack on the Ag

Antibodies (Igs)

Circulating Igs along with complement mount a chemical attack on the Ag

Key:

= Humoral immunity   —— = Stimulates

= Cell-mediated immunity   - - - = Inhibits

Completion or abortion of T cell activation.

The major problem with immunosuppressive therapy is that the patient's suppressed immune system cannot protect the body against other foreign agents. As a result, overwhelming bacterial and viral infection remains the most frequent cause of death in transplant patients. The key to successful graft survival is to provide enough immunosuppression to prevent graft rejection but not enough to be toxic, and to use antibiotics to keep infection under control. Even with the best conditions, by 10 years after receiving a transplant, roughly 50% of patients have rejected the donor organ.

In rare cases, transplant patients have naturally achieved a state of immune tolerance and have been able to "ditch the drugs." Finding a way to induce tolerance is the goal of the Immune Tolerance Network, a project launched by the NIH in 1999. Currently their surgeons are temporarily suppressing the recipient's bone marrow and then dousing their bone marrow with bone marrow from the same donors as their new organ in the hope that the transplanted marrow cells will establish themselves and create a **chimeric immune system** that treats the transplanted organ as self.

Homeostatic Imbalances of Immunity

Certain circumstances can cause the immune system to become depressed, to fail, or to act in a way that damages the body. Most such problems can be classified as immunodeficiencies, autoimmune diseases, or hypersensitivities.

Immunodeficiencies

Any congenital or acquired condition that causes immune cells, phagocytes, or complement to behave abnormally is called an **immunodeficiency.**

The most devastating *congenital condition* is **severe combined immunodeficiency (SCID) syndromes,** which result from various genetic defects that produce a marked deficit of B and T cells. One SCID defect causes abnormalities in the receptors for several interleukins. Another results in a defective *adenosine deaminase (ADA)* enzyme. In the absence of normal ADA, metabolites that are lethal to T cells accumulate in the body. Children afflicted with SCID have little or no protection against disease-causing organisms of any type. Interventions must begin in the first months of life because minor infections easily shrugged off by most children cause the victims to become deathly ill and wasted. Untreated, this condition is fatal, but successful transplants of bone marrow tissue or cultured stem cells from umbilical cord blood improve survival rates.

Lacking such a treatment, the only hope for survival has been living behind protective barriers that keep out all infectious agents. Recent genetic engineering techniques using viruses as vectors to transfer ADA minigenes to the victim's hematopoietic stem cells or T cells have shown promise.

There are various *acquired immunodeficiencies.* For example, *Hodgkin's disease,* cancer of the lymph nodes, can lead to immunodeficiency by depressing lymph node cells, and immunosuppression is the goal of certain drugs used to treat cancer. Currently, however, the most devastating of the acquired immunodeficiencies is **acquired immune deficiency syndrome (AIDS),** which cripples the immune system by interfering with the activity of helper T cells.

First identified in the United States in 1981 among homosexual men and intravenous drug users, AIDS is characterized by severe weight loss, night sweats, swollen lymph nodes, and increasingly frequent opportunistic infections, including a rare type of pneumonia called *pneumocystis pneumonia,* and the bizarre malignancy *Kaposi's sarcoma,* a cancerlike condition of the blood vessels evidenced by purple skin lesions. Some AIDS victims develop severe dementia. The course of AIDS is often grim, finally ending in complete debilitation and death from cancer or overwhelming infection.

AIDS is caused by a virus transmitted in body secretions—especially blood, semen, and vaginal secretions. The virus commonly enters the body via blood transfusions or blood-contaminated needles and during sexual intercourse. It is also present in saliva and tears, and there are documented cases of transmission by oral sex. Early in the epidemic, hemophiliacs were also at particular risk because the blood factors they need were isolated from pooled blood donations. Although manufacturers began taking measures to kill the virus in 1984 and new genetically engineered factors became available, an estimated 60% of the hemophiliacs in this country were already infected.

The virus, **HIV (human immunodeficiency virus),** destroys T_H cells, thus depressing cell-mediated immunity. Although B cells and T_C cells initially mount a vigorous response to viral exposure, in time a profound deficit of antibodies develops and the cytotoxic cells become the prey of the virus. The whole immune system is turned topsy-turvy. It is now clear that the virus multiplies steadily in the lymph nodes throughout most of the asymptomatic period, which can be as short as a few months or as long as ten years. Symptoms appear when the lymph nodes can no longer contain the virus and the immune system collapses. The virus also invades the brain (which accounts for the dementia of some AIDS patients) and most victims die within a few months to eight years after diagnosis.

The infectious specificity of HIV reflects the fact that CD4 proteins provide the avenue of attack. Researchers have identified an HIV coat glycoprotein (gp120) that fits into the CD4 receptor like a plug fits into a socket. However, HIV also needs a nearby protein called gp41 to gain entry to a body cell. Once gp120 has attached, gp41 emerges and fuses the virus to the target cell. Once inside, HIV "sets up housekeeping," using the enzyme *reverse transcriptase* to produce DNA from the information encoded in its (viral) RNA. This DNA copy, now called a *provirus*, then inserts itself into the target cell's DNA and directs the cell to crank out new copies of viral RNA and proteins so that the virus can multiply and infect other cells. Although T_H cells are the main HIV targets, other body cells displaying CD4 proteins (macrophages, monocytes, and dendritic cells) are also at risk. The HIV reverse transcriptase enzyme is not very accurate and produces errors rather frequently, causing HIV's relatively high mutation rate and its changing resistance to drugs.

The years since 1981 have witnessed a global AIDS epidemic. As of 2001, more than 40 million people were infected worldwide, and almost 90% of them live in the developing countries of Asia and sub-Saharan Africa. Half of the victims are women and nearly a quarter are children, a case distribution indicating that in the most heavily hit countries, most HIV transmission occurs via heterosexual contacts. The virus can also be transmitted from an infected mother to her fetus.

By 2001, recorded cases of Americans infected with HIV topped 700,000. Anti-HIV antibodies may appear in the blood as early as two weeks or as long as six months after infection. So for every diagnosed case, there are probably many more asymptomatic carriers of the virus. Not only has the number of identified cases in the United States jumped exponentially in the at-risk populations, but the "face of AIDS" is changing too. Though homosexual men still account for the bulk of cases transmitted by sexual contact, more and more heterosexuals are contracting this disease. Particularly disturbing is the near-epidemic increase in diagnosed cases among teenagers and young adults, with AIDS now the fifth leading killer of all Americans ages 25 to 44.

No cure for AIDS has yet been found. More than 100 drugs are now in the Food and Drug Administration pipeline, and over 20 prototype vaccines, including one that binds to gp120, and T-20, a fusion inhibitor that targets gp41, are undergoing clinical trials, but it is unlikely that an approved vaccine will be available soon. Several antiviral drugs are now clinically available. *Reverse transcriptase inhibitors*, such as AZT and ddCs, were early on the scene. In late 1995 and early 1996, *protease inhibitors* (saquinavir, ritonavir, and others) were approved. Initially, it appeared that combination therapy using drugs from each class delivers a one-two punch to the virus; that is, it postpones drug resistance and causes the *viral load* (amount of HIV virus per millimeter of blood) to plummet while boosting the number of T_H cells. Several patients in the combination drug regimen seemed to revive from near death. Sadly, the treatments are beginning to fail in about half of those treated. New hope comes from integrase, a new drug now being developed which blocks integration of the HIV provirus into the target cell's DNA. However, drug research is slow and laborious and the clock is ticking.

Autoimmune Diseases

Occasionally the immune system loses its ability to distinguish friend (self) from foe (foreign antigens). When this happens, the artillery of the immune system, like friendly fire, turns against itself. The body produces antibodies *(autoantibodies)* and sensitized T_C cells that destroy its own tissues. This puzzling phenomenon is called **autoimmunity.** If a disease state results, it is referred to as **autoimmune disease.**

Some 5% of adults in North America—two-thirds of them women—are afflicted with autoimmune disease. Most common are:

- *Multiple sclerosis*, which destroys the white matter of the brain and spinal cord (see pp. 361–362)
- *Myasthenia gravis*, which impairs communication between nerves and skeletal muscles (see p. 254)
- *Graves' disease*, which prompts the thyroid gland to produce excessive amounts of thyroxine
- *Type I (juvenile) diabetes mellitus*, which destroys pancreatic beta cells, resulting in a deficit of insulin and inability to use carbohydrates (see pp. 548–549)
- *Systemic lupus erythematosus (SLE)*, a systemic disease that particularly affects the kidneys, heart, lungs, and skin
- *Glomerulonephritis*, a severe impairment of renal function
- *Rheumatoid arthritis*, which systematically destroys joints (see pp. 240–241)

One line of therapy depresses certain aspects of the immune response. For example, injections of genetically engineered antibodies to the CD4 receptors on T_H cells seem to stabilize some multiple sclerosis victims. More recent is the discovery that *thalidomide* is a potential boon to those with autoimmune disease. This drug, used in the 1950s to alleviate morning sickness in pregnant women until it was found to cause tragic birth defects, inhibits the immune system's production of alpha-TNF.

How does the normal state of self-tolerance break down? It appears that one or more of the following events may be triggers:

1. **Lymphocyte programming is ineffective.** Self-reactive T cells or B cells, which should be silenced or eliminated during their programming phase in the thymus and bone marrow, instead escape to the rest of the body. This is believed to occur in multiple sclerosis.

2. **New self-antigens appear.** Self-proteins not previously exposed to the immune system may appear in the circulation. They may be generated (1) by gene mutations that cause new proteins to appear at the external cell surface and (2) by structural changes in self-antigens caused by hapten attachment or infectious damage. These newly generated proteins then become immune system targets.

3. **Foreign antigens resemble self-antigens.** If the determinants on a self-antigen resemble those on a foreign antigen, antibodies made against the foreign antigen can cross-react with the self-antigen. For instance, antibodies produced during a streptococcal infection react with heart antigens, causing lasting damage to the heart muscle and valves, as well as to joints and kidneys. This age-old disease is called *rheumatic fever.*

Hypersensitivities

At first, the immune response was thought to be purely protective. However, it was not long before its dangerous potentials were discovered. **Hypersensitivities** or **allergies** (*allo* = altered; *erg* = reaction) are the result when the immune system causes tissue damage as it fights off a perceived threat (such as pollen or animal dander) that would otherwise be harmless to the body. The term **allergen** is used to distinguish this type of antigen from those producing essentially normal responses. People rarely die of allergies; they are just miserable with them.

The different types of hypersensitivity reactions are distinguished by (1) their time course, and (2) whether antibodies or T cells are involved. Allergies mediated by antibodies are the *immediate* and *subacute hypersensitivities.* The most important allergic state mediated by T cells is *delayed hypersensitivity.*

Immediate Hypersensitivities

The **immediate hypersensitivities,** also called **acute** or **type I hypersensitivities,** begin within seconds after contact with the allergen and last about half an hour.

Anaphylaxis   The most common type of immediate hypersensitivity is **anaphylaxis** (an"ah-fi-lak'sis; "against protection"). The initial meeting with an allergen produces no symptoms but it sensitizes the person. APCs digest the allergen and present its fragments to T cells as usual. The steps that follow are obscure, but appear to be mediated by IL-4 secreted by T cells. IL-4 stimulates B cells to mature into IgE-secreting plasma cells, which spew out huge amounts of that antibody. When the IgE molecules attach to **mast cells** and **basophils,** sensitization is complete. Anaphylaxis is triggered at later encounters with the same allergen, which promptly binds and cross-links the IgE antibodies on the surfaces of the mast cells and basophils. This event induces an enzymatic cascade that causes the mast cells and basophils to degranulate, releasing a flood of **histamine** and other inflammatory chemicals that together induce the inflammatory response typical of anaphylaxis (Figure 20.20).

Anaphylactic reactions may be local or systemic. Mast cells are abundant in connective tissues of the skin and beneath the mucosa of respiratory passages and the gastrointestinal tract, and these areas are common sites of local allergic reactions. Histamine causes blood vessels to become dilated and leaky, and is largely to blame for the best recognized symptoms of anaphylaxis: runny nose, itching reddened skin (hives), and watery eyes. When the allergen is inhaled, symptoms of *asthma* appear because smooth muscle in the walls of the bronchioles contracts, constricting those small passages and restricting air flow. When the allergen is ingested in food or via drugs, gastrointestinal discomfort (cramping, vomiting, or diarrhea) occurs. Over-the-counter antiallergy drugs contain antihistamines that counteract these effects.

The bodywide or systemic response known as **anaphylactic shock** is fairly rare. It typically occurs when the allergen directly enters the blood and circulates rapidly through the body, as might happen with certain bee stings or spider bites. It may also follow injection of a foreign substance (such as penicillin or other drugs which act as haptens). The mechanism of anaphylactic shock is essentially the same as that of local responses; but when mast cells and basophils are enlisted throughout the entire body, the outcome is life threatening. The bronchioles constrict (and the tongue may swell), making it difficult to breathe, and the sudden vasodilation and fluid loss from the bloodstream may cause circulatory collapse (hypotensive shock) and death within minutes. Epinephrine is the drug of choice to reverse these histamine-mediated effects.

Atopy Though the term *anaphylaxis* covers most immediate hypersensitivities, it does not include the fairly common type called **atopy** (literally, "out of place"). About 10% of people in the United States have inherited a tendency to spontaneously (without prior sensitization) develop immediate-type allergies to certain environmental antigens (such as ragweed or house dust mites). Consequently, when they encounter even very small amounts of the appropriate allergen, they promptly develop hives, hay fever, or asthmatic symptoms.

Subacute Hypersensitivities

Like the immediate types, **subacute hypersensitivities** are caused by antibodies (IgG and IgM rather than IgE) and can be transferred via blood plasma or serum. However, their onset is slower (1–3 hours after antigen exposure) and the duration of the reaction is longer (10–15 hours).

Cytotoxic (type II) reactions occur when antibodies bind to antigens on specific body cells and subsequently stimulate phagocytosis and complement-mediated lysis of the cellular antigens. Type II hypersensitivity may occur after a patient has received a transfusion of mismatched blood and the foreign red blood cells are lysed by complement.

Immune complex (type III) hypersensitivity results when antigens are widely distributed through the body or blood and the insoluble antigen-antibody complexes formed cannot be cleared from a particular area. (This may reflect a persistent infection or a situation in which huge amounts of antigen-antibody complexes are formed.) An intense inflammatory reaction occurs, complete with complement-mediated cell lysis and cell killing by neutrophils that severely damages local tissues. One example of type III allergy is *farmer's lung* (induced by inhaling moldy hay). Additionally, many type III allergic responses are involved in autoimmune disorders, such as glomerulonephritis, systemic lupus erythematosus, and rheumatoid arthritis.

Delayed Hypersensitivities

Delayed hypersensitivity (type IV) reactions are slower to appear (1–3 days) than antibody-mediated hypersensitivity reactions. The mechanism of the former is basically that of a cell-mediated immune response. The *usual* reaction to most cellular pathogens depends on T_C cells directed against specific antigens. However, delayed hypersensitivity reactions involve both cytotoxic cells and T_H1 cells, and depend to a greater extent on cytokine-activated macrophages and nonspecific killing. Delayed hypersensitivity to antigens can be passively transferred from one person to another by transfusions of whole blood containing the T cells that initiate the response.

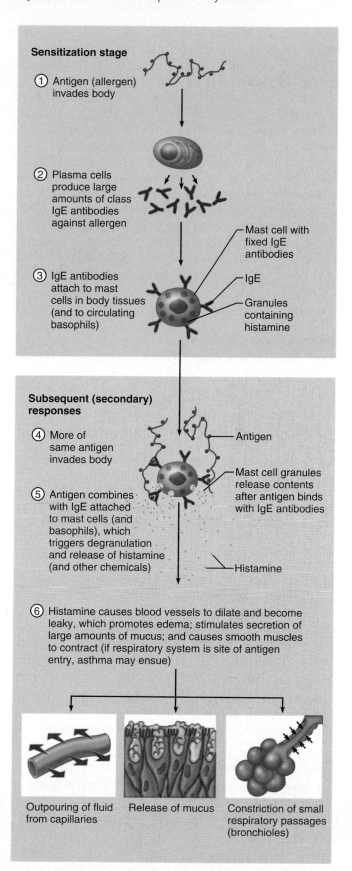

Sensitization stage

① Antigen (allergen) invades body

② Plasma cells produce large amounts of class IgE antibodies against allergen

③ IgE antibodies attach to mast cells in body tissues (and to circulating basophils)

— Mast cell with fixed IgE antibodies

— IgE

— Granules containing histamine

Subsequent (secondary) responses

④ More of same antigen invades body

— Antigen

⑤ Antigen combines with IgE attached to mast cells (and basophils), which triggers degranulation and release of histamine (and other chemicals)

— Mast cell granules release contents after antigen binds with IgE antibodies

— Histamine

⑥ Histamine causes blood vessels to dilate and become leaky, which promotes edema; stimulates secretion of large amounts of mucus; and causes smooth muscles to contract (if respiratory system is site of antigen entry, asthma may ensue)

Outpouring of fluid from capillaries   Release of mucus   Constriction of small respiratory passages (bronchioles)

FIGURE 20.20 **Mechanism of an acute allergic (immediate hypersensitivity) response.**

Cytokines released by activated T cells are the major mediators of these inflammatory responses; thus antihistamine drugs are not helpful. Corticosteroid drugs are used to provide relief.

The most familiar examples of delayed hypersensitivity reactions are those classified as **allergic contact dermatitis** which follow skin contact with poison ivy, some heavy metals (lead, mercury, and others), and certain cosmetic and deodorant chemicals. These agents act as haptens, and after diffusing through the skin and attaching to self-proteins, they are perceived as foreign by the immune system. The *Mantoux* and *tine tests,* two skin tests for tuberculosis, depend on delayed hypersensitivity reactions. When the tubercle antigens are introduced just under the skin, a small hard lesion forms that persists for days if the person has been sensitized to the antigen.

Delayed hypersensitivity reactions also involve a large number of protective reactions including (1) protection against viruses, bacteria, fungi, and protozoa; (2) resistance against cancer; and (3) rejection of foreign grafts or transplanted organs. They are particularly effective against **facultative intracellular pathogens (FIPs)**, which include salmonella bacteria and some yeasts. FIPs are readily phagocytized by macrophages but are not killed and may multiply in a host unless the macrophages are activated to killer status by gamma interferon and certain other cytokines. Thus, the enhanced release of cytokines during delayed hypersensitivity reactions plays an important protective role.

Review Questions

Multiple Choice/Matching

(Some questions have more than one correct answer. Select the best answer or answers from the choices given.)

1. All of the following are considered innate or nonspecific body defenses *except* (a) complement, (b) phagocytosis, (c) antibodies, (d) lysozyme, (e) inflammation.

2. The process by which neutrophils squeeze through capillary walls in response to inflammatory signals is called (a) diapedesis, (b) chemotaxis, (c) margination, (d) opsonization.

3. Antibodies released by plasma cells are involved in (a) humoral immunity, (b) immediate hypersensitivity reactions, (c) autoimmune disorders, (d) all of the above.

4. Which of the following antibodies can fix complement? (a) IgA, (b) IgD, (c) IgE, (d) IgG, (e) IgM.

5. Which antibody class is abundant in body secretions? (Use the choices from question 4.)

6. Small molecules that must combine with large proteins to become immunogenic are called (a) complete antigens, (b) reagins, (c) idiotypes, (d) haptens.

7. Lymphocytes that develop immunocompetence in the thymus are (a) B lymphocytes, (b) T lymphocytes, (c) NK cells.

8. Cells that can directly attack target cells include all of the following *except* (a) macrophages, (b) cytotoxic T cells, (c) helper T cells, (d) natural killer cells.

9. Which of the following is involved in the activation of a B cell? (a) antigen, (b) helper T cell, (c) cytokine, (d) all of the above.

Short Answer Essay Questions

10. Besides acting as mechanical barriers, the skin epidermis and mucosae of the body have other attributes that contribute to their protective roles. Cite the common body locations and the importance of mucus, lysozyme, keratin, acid pH, and cilia.

11. Explain why attempts at phagocytosis are not always successful; cite factors that increase the likelihood of success.

12. What is complement? How does it cause bacterial lysis? What are some of the other roles of complement?

13. Interferons are referred to as antiviral proteins. What stimulates their production, and how do they protect uninfected cells? What cells of the body secrete interferons?

14. Differentiate between humoral and cell-mediated adaptive immunity.

15. Although the adaptive immune system has two arms, it has been said, "no T cells, no immunity." Explain.

16. Define immunocompetence. What event (or observation) signals that a B or a T cell has developed immunocompetence?

17. Describe the process of activation of a helper T cell.

18. Differentiate between a primary and a secondary immune response. Which is more rapid and why?

19. Define antibody. Using an appropriately labeled diagram, describe the structure of an antibody monomer. Indicate and label variable and constant regions, heavy and light chains.

20. What is the role of the variable regions of an antibody? Of the constant regions?

21. Name the five antibody classes and describe where each is most likely to be found in the body.

22. How do antibodies help defend the body?

23. Do vaccines produce active or passive humoral immunity? Explain your answer. Why is passive immunity less satisfactory?

24. Describe the specific roles of helper, suppressor, and cytotoxic T cells in normal cell-mediated immunity.

25. Name several cytokines and describe their role in the immune response.

26. Define hypersensitivity. List three types of hypersensitivity reactions. For each, note whether antibodies or T cells are involved and provide two examples.

27. What events can result in autoimmune disease?

21

THE RESPIRATORY SYSTEM

Functional Anatomy of the Respiratory System (pp. 710–725)

1. Identify the organs forming the respiratory passageway(s) in descending order until the alveoli are reached. Distinguish between conducting and respiratory zone structures.

2. List and describe several protective mechanisms of the respiratory system.

3. Describe the makeup of the respiratory membrane, and relate structure to function.

4. Describe the gross structure of the lungs and pleurae.

Mechanics of Breathing (pp. 725–733)

5. Relate Boyle's law to events of inspiration and expiration.

6. Explain the relative roles of the respiratory muscles and lung elasticity in producing the volume changes that cause air to flow into and out of the lungs.

7. Explain the functional importance of the partial vacuum that exists in the intrapleural space.

8. List several physical factors that influence pulmonary ventilation.

9. Explain and compare the various lung volumes and capacities. Indicate types of information that can be gained from pulmonary function tests.

10. Define dead space.

Basic Properties of Gases (pp. 733–734)

11. State Dalton's law of partial pressures and Henry's law.

Composition of Alveolar Gas (p. 734)

12. Describe how atmospheric and alveolar air differ in composition, and explain these differences.

Gas Exchanges Between the Blood, Lungs, and Tissues (pp. 734–737)

13. Relate Dalton's and Henry's laws to events of external and internal respiration.

Transport of Respiratory Gases by Blood (pp. 737–742)

14. Describe how oxygen is transported in the blood, and explain how oxygen loading and unloading is affected by temperature, pH, BPG, and P_{CO_2}.

15. Describe carbon dioxide transport in the blood.

Control of Respiration (pp. 742–747)

16. Describe the neural controls of respiration.

17. Compare and contrast the influences of lung reflexes, volition, emotions, arterial pH, and arterial partial pressures of oxygen and carbon dioxide on respiratory rate and depth.

Respiratory Adjustments (pp. 747–748)

18. Compare and contrast the hyperpnea of exercise with involuntary hyperventilation.

19. Describe the process and effects of acclimatization to high altitude.

Homeostatic Imbalances of the Respiratory System (pp. 748–751)

20. Compare the causes and consequences of chronic bronchitis, emphysema, asthma, and lung cancer.

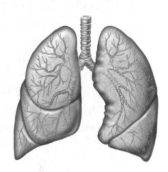

ar from self-sustaining, the body depends on the external environment, both as a source of substances the body needs to survive and as a catch basin for its wastes. The trillions of cells making up the body require a continuous supply of oxygen to carry out their vital functions. We cannot do without oxygen for even a little while, as we can without food or water. As cells use oxygen, they give off carbon dioxide, a waste product the body must get rid of. They also generate dangerous free radicals, the inescapable by-products of living in a world full of oxygen. But let's get to the topic of this chapter, the respiratory system.

The major function of the **respiratory system** is to supply the body with oxygen and dispose of carbon dioxide. To accomplish this function, at least four processes, collectively called **respiration,** must happen:

1. **Pulmonary ventilation:** movement of air into and out of the lungs so that the gases there are continuously changed and refreshed (commonly called breathing).

2. **External respiration:** movement of oxygen from the lungs to the blood and of carbon dioxide from the blood to the lungs.

3. **Transport of respiratory gases:** transport of oxygen from the lungs to the tissue cells of the body, and of carbon dioxide from the tissue cells to the lungs. This is accomplished by the cardiovascular system using blood as the transporting fluid.

4. **Internal respiration:** movement of oxygen from blood to the tissue cells and of carbon dioxide from tissue cells to blood.*

Only the first two processes are the special responsibility of the respiratory system, but it cannot accomplish its primary goal of obtaining oxygen and eliminating carbon dioxide unless the third and fourth processes also occur. Thus, the respiratory and circulatory systems are closely coupled, and if either system fails, the body's cells begin to die from oxygen starvation.

Because it moves air, the respiratory system is also involved with the sense of smell and with speech.

Functional Anatomy of the Respiratory System

The respiratory system includes the *nose* and *nasal cavity, pharynx, larynx, trachea, bronchi* and their smaller branches, and the *lungs,* which contain the

---

* The actual use of oxygen and production of carbon dioxide by tissue cells, that is, *cellular respiration,* is the cornerstone of all energy-producing chemical reactions in the body. Cellular respiration, which occurs in all body cells, is discussed more appropriately in the metabolism section of Chapter 23.

terminal air sacs, or *alveoli* (Figure 21.1). Functionally, the system consists of two zones. The **respiratory zone,** the actual site of gas exchange, is composed of the respiratory bronchioles, alveolar ducts, and alveoli, all microscopic structures. The **conducting zone** includes all other respiratory passageways, which provide fairly rigid conduits for air to reach the gas exchange sites. The conducting zone organs also cleanse, humidify, and warm incoming air. Thus, air reaching the lungs has fewer irritants (dust, bacteria, etc.) than when it entered the system, and it is warm and damp, like the air of the tropics. The functions of the major organs of the respiratory system are summarized in Table 21.1 on p. 717.

In addition to these organs, some authorities also include the respiratory muscles (diaphragm, etc.) as part of this system. Although we will consider how these skeletal muscles bring about the volume changes that promote ventilation, we continue to classify them as part of the *muscular system.*

The Nose and Paranasal Sinuses

The nose is the only externally visible part of the respiratory system. Unlike the eyes and lips, facial features often referred to poetically, the nose is usually an irreverent target. We are urged to keep our nose to the grindstone and to keep it out of other people's business. However, considering its important functions, it deserves more esteem. The nose (1) provides an airway for respiration, (2) moistens and warms entering air, (3) filters and cleans inspired air, (4) serves as a resonating chamber for speech, and (5) houses the olfactory (smell) receptors.

The structures of the nose are divided into the *external nose* and the internal *nasal cavity* for ease of consideration. The surface features of the external nose (Figure 21.2a) include the *root* (area between the eyebrows), *bridge,* and *dorsum nasi* (anterior margin), the latter terminating in the *apex* (tip of the nose). Just inferior to the apex is a shallow vertical groove called the *philtrum* (fil'trum). The external openings of the nose, the *nostrils* or *external nares* (na'rez), are bounded laterally by the flared *alae*. The skeletal framework of the external nose is fashioned by the nasal and frontal bones superiorly (forming the bridge and root, respectively), the maxillary bones laterally, and flexible plates of hyaline cartilage (the lateral, septal, and alar cartilages) inferiorly (Figure 21.2b). Noses vary a great deal in size and shape, largely because of differences in the nasal cartilages. The skin covering the nose's dorsal and lateral aspects is thin and contains many sebaceous glands.

The internal **nasal cavity** lies in and posterior to the external nose. During breathing, air enters the cavity by passing through the **nostrils,** or **external nares** (Figure 21.2a and Figure 21.3b). The nasal

? *Which of the organs shown contains both conducting zone and respiratory zone structures?*

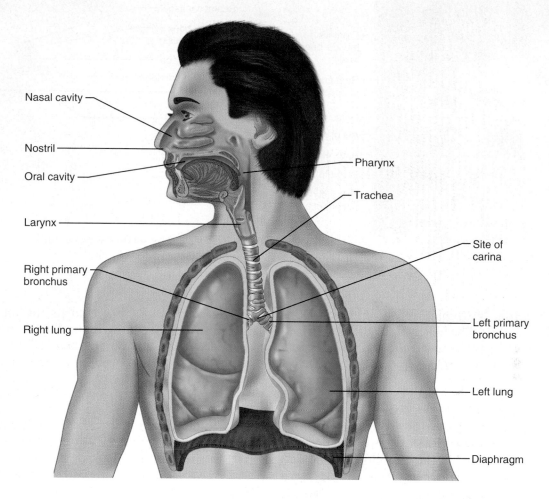

FIGURE 21.1 **The major respiratory organs in relation to surrounding structures.**

cavity is divided by a midline **nasal septum,** formed anteriorly by the septal cartilage and posteriorly by the vomer bone and perpendicular plate of the ethmoid bone (see Figure 7.10b, p. 186). The nasal cavity is continuous posteriorly with the nasal portion of the pharynx through the **posterior nasal apertures,** also called the *internal nares,* or *choanae* (ko-a'ne; "funnels").

The roof of the nasal cavity is formed by the ethmoid and sphenoid bones of the skull. The floor is formed by the *palate,* which separates the nasal cavity from the oral cavity below. Anteriorly, where the palate is supported by the maxillary processes and palatine bones, it is called the **hard palate.** The unsupported posterior portion is the muscular **soft palate.**

The part of the nasal cavity just superior to the nostrils, called the **vestibule,** is lined with skin containing sebaceous and sweat glands and numerous hair follicles. The hairs, or **vibrissae** (vi-bris'e; *vibro* = to quiver), filter coarse particles (dust, pollen) from inspired air. The rest of the nasal cavity is lined with two types of mucous membrane. The **olfactory mucosa,** lining the slitlike superior region of the nasal cavity, contains smell receptors. The balance of the nasal cavity mucosa, the **respiratory mucosa,** is a pseudostratified ciliated columnar epithelium, containing scattered *goblet cells,* that rests on a lamina propria richly supplied with *mucous* and *serous glands.* (Mucous cells secrete mucus, and serous cells secrete a watery fluid containing enzymes.) Each day, these glands secrete about a quart of mucus containing *lysozyme,* an antibacterial enzyme. The sticky mucus traps inspired dust, bacteria, and

The lungs. ■

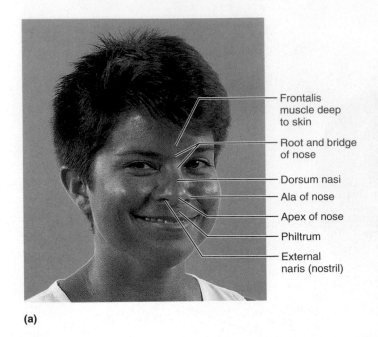

(a)

- Frontalis muscle deep to skin
- Root and bridge of nose
- Dorsum nasi
- Ala of nose
- Apex of nose
- Philtrum
- External naris (nostril)

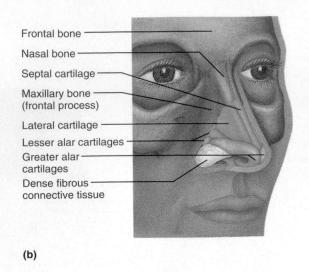

- Frontal bone
- Nasal bone
- Septal cartilage
- Maxillary bone (frontal process)
- Lateral cartilage
- Lesser alar cartilages
- Greater alar cartilages
- Dense fibrous connective tissue

(b)

FIGURE 21.2 **The external nose. (a)** Surface anatomy. **(b)** External skeletal framework.

other debris, while lysozyme attacks and destroys bacteria chemically. The epithelial cells of the respiratory mucosa also secrete *defensins,* natural antibiotics that help get rid of invading microbes. Additionally, the high water content of the mucus film acts to humidify the inhaled air.

The ciliated cells of the respiratory mucosa create a gentle current that moves the sheet of contaminated mucus posteriorly toward the throat, where it is swallowed and digested by stomach juices. We are usually unaware of this important action of our nasal

cilia, but when exposed to cold air, they become sluggish, allowing mucus to accumulate in the nasal cavity and then dribble out the nostrils. This along with the fact that water vapor in expired air tends to condense at lower temperatures helps explain why you might have a "runny" nose on a crisp, wintry day.

The nasal mucosa is richly supplied with sensory nerve endings, and contact with irritating particles (dust, pollen, and the like) triggers a sneeze reflex. The sneeze forces air outward in a violent burst—a somewhat crude way of expelling irritants from the nose.

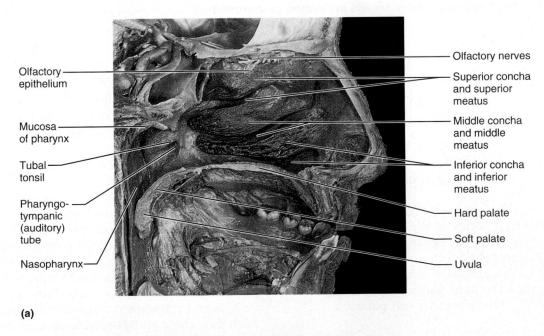

- Olfactory epithelium
- Mucosa of pharynx
- Tubal tonsil
- Pharyngo-tympanic (auditory) tube
- Nasopharynx

- Olfactory nerves
- Superior concha and superior meatus
- Middle concha and middle meatus
- Inferior concha and inferior meatus
- Hard palate
- Soft palate
- Uvula

(a)

FIGURE 21.3 **The upper respiratory tract.** Midsagittal section of the head and neck: **(a)** Photograph. (See *A Brief Atlas of the Human Body,* Figure 35.)

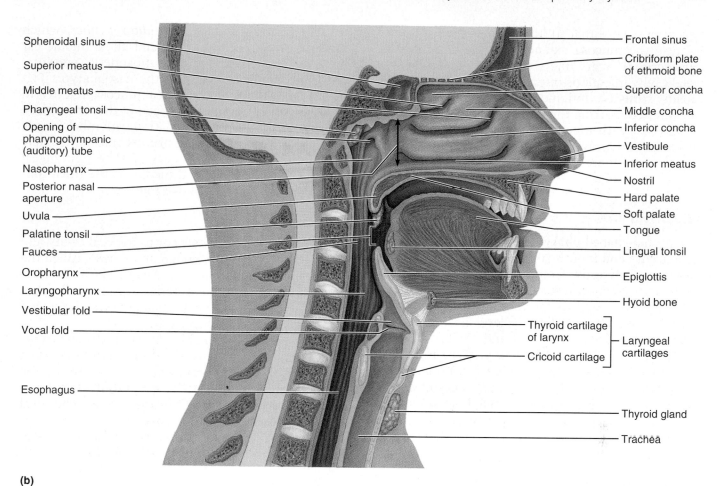

Sphenoidal sinus

Superior meatus

Middle meatus

Pharyngeal tonsil

Opening of pharyngotympanic (auditory) tube

Nasopharynx

Posterior nasal aperture

Uvula

Palatine tonsil

Fauces

Oropharynx

Laryngopharynx

Vestibular fold

Vocal fold

Esophagus

Frontal sinus

Cribriform plate of ethmoid bone

Superior concha

Middle concha

Inferior concha

Vestibule

Inferior meatus

Nostril

Hard palate

Soft palate

Tongue

Lingual tonsil

Epiglottis

Hyoid bone

Thyroid cartilage of larynx

Cricoid cartilage

Laryngeal cartilages

Thyroid gland

Trachea

(b)

FIGURE 21.3 *(continued)*   **(b)** Illustration.

Rich plexuses of capillaries and thin-walled veins underlie the nasal epithelium and warm incoming air as it flows across the mucosal surface. When the inspired air is cold, the vascular plexus becomes engorged with blood, thereby intensifying the air-heating process. Because of the abundance and superficial location of these blood vessels, nosebleeds are common and often profuse.

Protruding medially from each lateral wall of the nasal cavity are three scroll-like mucosa-covered projections, the *superior, middle,* and *inferior conchae* (kong′ke) (Figure 21.3). The groove inferior to each concha is a *meatus* (me-a′tus). The curved conchae greatly increase the mucosal surface area exposed to the air and enhance air turbulence in the cavity. The gases in inhaled air swirl through the twists and turns, but heavier, nongaseous particles tend to be deflected onto the mucus-coated surfaces, where they become trapped. As a result, few particles larger than 4 μm make it past the nasal cavity.

The conchae and nasal mucosa not only function during inhalation to filter, heat, and moisten the air, but also act during exhalation to reclaim this heat and moisture. That is, the inhaled air cools the

conchae, then during exhalation these cooled conchae precipitate moisture and extract heat from the humid air flowing over them. This reclamation mechanism minimizes the amount of moisture and heat lost from the body through breathing, helping us to survive in dry and cold climates.

The nasal cavity is surrounded by a ring of **paranasal sinuses** (Figure 21.3b) located in the frontal, sphenoid, ethmoid, and maxillary bones (see also Figure 7.11, p. 187). The sinuses lighten the skull, and together with the nasal cavity they warm and moisten the air. The mucus they produce ultimately flows into the nasal cavity and the suctioning effect created by nose blowing helps drain the sinuses.

HOMEOSTATIC IMBALANCE

Cold viruses, streptococcal bacteria, and various allergens can cause *rhinitis* (ri-ni′tis), inflammation of the nasal mucosa accompanied by excessive mucus production, nasal congestion, and postnasal drip. The nasal mucosa is continuous with that of the rest of the respiratory tract, explaining the typical nose to throat to lungs progression of colds. Because the mucosa extends tentacle-like into the nasolacrimal (tear) ducts and paranasal sinuses, nasal cavity infections often spread to those regions, causing **sinusitis** (inflamed sinuses). When the passageways connecting the sinuses to the nasal cavity are blocked with mucus or infectious material, the air in the sinus cavities is absorbed. The result is a partial vacuum and a *sinus headache* localized over the inflamed areas. ●

The Pharynx

The funnel-shaped **pharynx** (far'ingks) connects the nasal cavity and mouth superiorly to the larynx and esophagus inferiorly. Commonly called the *throat*, the pharynx vaguely resembles a short length of garden hose as it extends for about 13 cm (5 inches) from the base of the skull to the level of the sixth cervical vertebra (see Figures 21.1 and 21.3b).

From superior to inferior, the pharynx is divided into three regions—the *nasopharynx, oropharynx,* and *laryngopharynx.* The muscular pharynx wall is composed of skeletal muscle throughout its length (see Table 10.3, pp. 296–297), but the cellular composition of its mucosa varies from one pharyngeal region to another.

The Nasopharynx

The **nasopharynx** is posterior to the nasal cavity, inferior to the sphenoid bone, and superior to the level of the soft palate. Because it lies above the point where food enters the body, it serves *only* as an air passageway. During swallowing, the soft palate and its pendulous *uvula* (u'vu-lah; "little grape") move superiorly, an action that closes off the nasopharynx and prevents food from entering the nasal cavity. (When we giggle, this sealing action fails and fluids being swallowed can end up spraying out the nose.)

The nasopharynx is continuous with the nasal cavity through the posterior nasal apertures (see Figure 21.3b), and its ciliated pseudostratified epithelium takes over the job of propelling mucus where the nasal mucosa leaves off. High on its posterior wall is the **pharyngeal tonsil** (far-rin'je-al) (or **adenoids**) which traps and destroys pathogens entering the nasopharynx in air.

ⓗ HOMEOSTATIC IMBALANCE

Infected and swollen adenoids block air passage in the nasopharynx, making it necessary to breathe through the mouth. As a result, the air is not properly moistened, warmed, or filtered before reaching the lungs. When the adenoids are chronically enlarged, both speech and sleep may be disturbed. ●

The *pharyngotympanic (auditory) tubes,* which drain the middle ear cavities and allow middle ear pressure to equalize with atmospheric pressure, open into the lateral walls of the nasopharynx (Figure 21.3a). A ridge of pharyngeal mucosa, referred to as a *tubal tonsil,* arches over each of these openings. Because of their strategic location, the tubal tonsils help protect the middle ear against infections likely to spread from the nasopharynx. The pharyngeal tonsil, superoposterior and medial to the tubal tonsils, also plays this protective role.

The Oropharynx

The **oropharynx** lies posterior to the oral cavity and is continuous with it through an archway called the **fauces** (faw'sēz; "throat") (see Figure 21.3b). Because the oropharynx extends inferiorly from the level of the soft palate to the epiglottis, both swallowed food and inhaled air pass through it.

As the nasopharynx blends into the oropharynx, the epithelium changes from pseudostratified columnar to a more protective stratified squamous epithelium. This structural adaptation accommodates the increased friction and greater chemical trauma accompanying food passage.

Two kinds of tonsils lie embedded in the oropharyngeal mucosa. The paired **palatine tonsils** lie in the lateral walls of the fauces. The **lingual tonsil** covers the base of the tongue.

The Laryngopharynx

Like the oropharynx above it, the **laryngopharynx** (lah-ring"go-far'ingks) serves as a passageway for food and air and is lined with a stratified squamous epithelium. It lies directly posterior to the upright epiglottis and extends to the larynx, where the respiratory and digestive pathways diverge. At that point the laryngopharynx is continuous with the esophagus posteriorly. The esophagus conducts food and fluids to the stomach; air enters the larynx anteriorly. During swallowing, food has the "right of way," and air passage temporarily stops.

The Larynx

Basic Anatomy

The **larynx** (lar'ingks), or voice box, extends for about 5 cm (2 inches) from the level of the fourth to the sixth cervical vertebra. Superiorly it attaches to the hyoid bone and opens into the laryngopharynx. Inferiorly it is continuous with the trachea (Figure 21.3b).

The larynx has three functions. Its two main tasks are to provide a *patent* (open) airway and to act as a switching mechanism to route air and food into the proper channels. Because it houses the vocal

? Which structure seals the larynx when we swallow?

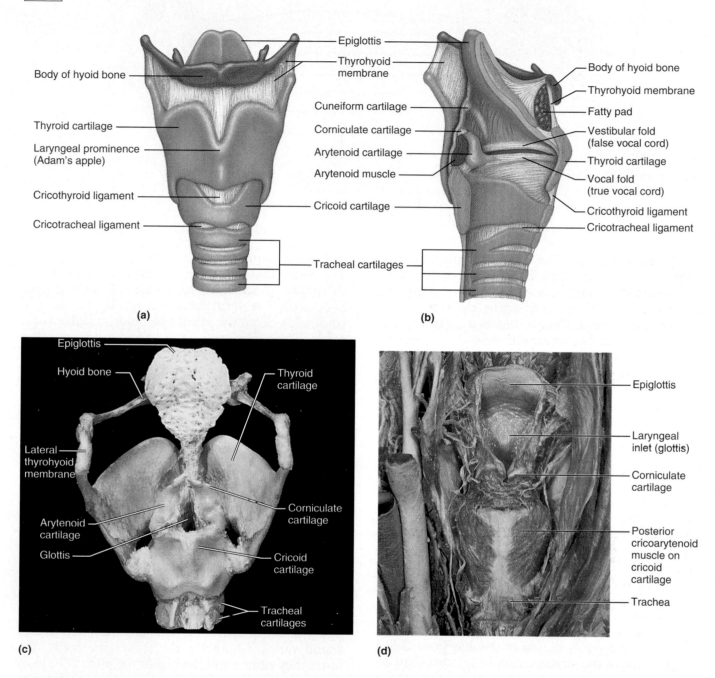

(a)

Epiglottis
Thyrohyoid membrane
Body of hyoid bone
Thyroid cartilage
Laryngeal prominence (Adam's apple)
Cricothyroid ligament
Cricotracheal ligament
Cuneiform cartilage
Corniculate cartilage
Arytenoid cartilage
Arytenoid muscle
Cricoid cartilage
Tracheal cartilages

(b)

Epiglottis
Thyrohyoid membrane
Body of hyoid bone
Thyrohyoid membrane
Fatty pad
Vestibular fold (false vocal cord)
Thyroid cartilage
Vocal fold (true vocal cord)
Cricothyroid ligament
Cricotracheal ligament
Tracheal cartilages

(c)

Epiglottis
Hyoid bone
Lateral thyrohyoid membrane
Arytenoid cartilage
Glottis
Thyroid cartilage
Corniculate cartilage
Cricoid cartilage
Tracheal cartilages

(d)

Epiglottis
Laryngeal inlet (glottis)
Corniculate cartilage
Posterior cricoarytenoid muscle on cricoid cartilage
Trachea

FIGURE 21.4   **The larynx. (a)** Anterior superficial view. **(b)** Sagittal view; anterior surface to the right. **(c)** Photograph of the cartilaginous framework of the larynx, posterior view. **(d)** Photograph of posterior aspect.

cords, the third function of the larynx is voice production.

The framework of the larynx is an intricate arrangement of nine cartilages connected by membranes and ligaments (Figure 21.4). Except for the epiglottis, all laryngeal cartilages are hyaline cartilages. The large, shield-shaped **thyroid cartilage** is formed by the fusion of two cartilage plates. The midline **laryngeal prominence** (lah-rin'je-al), which marks the fusion point, is obvious externally

■ *The epiglottis.*

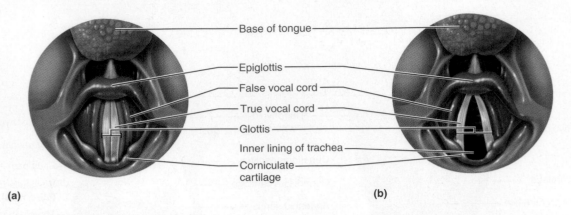

FIGURE 21.5   Movements of the vocal cords. Superior views of the larynx and vocal cords, mucous membranes in place. **(a)** Vocal cords in closed position. **(b)** Vocal cords in open position.

as the *Adam's apple.* The thyroid cartilage is typically larger in males than in females because male sex hormones stimulate its growth during puberty. Inferior to the thyroid cartilage is the ring-shaped **cricoid cartilage** (kri'koid), perched atop and anchored to the trachea inferiorly.

Three pairs of small cartilages, **arytenoid** (ar"ĭ-te'noid), **cuneiform** (ku-ne'ĭ-form), and **corniculate cartilages,** form part of the lateral and posterior walls of the larynx. The most important of these are the pyramid-shaped arytenoid cartilages, which anchor the vocal cords.

The ninth cartilage, the flexible, spoon-shaped **epiglottis** (ep"ĭ-glot'is; "above the glottis"), is composed of elastic cartilage and is almost entirely covered by mucosa-containing taste buds. The epiglottis extends from the posterior aspect of the tongue to its anchoring point on the anterior rim of the thyroid cartilage (see Figure 21.4b and c). When only air is flowing into the larynx, the inlet to the larynx is open wide and the free edge of the epiglottis projects upward. During swallowing, the larynx is pulled superiorly and the epiglottis tips to cover the laryngeal inlet. Because this action keeps food out of the lower respiratory passages, the epiglottis has been called the guardian of the airways. Anything other than air entering the larynx initiates the cough reflex, which acts to expel the substance. Because this protective reflex does not work when we are unconscious, it is never a good idea to administer liquids when attempting to revive an unconscious person.

Lying under the laryngeal mucosa on each side are the **vocal ligaments,** which attach the arytenoid cartilages to the thyroid cartilage. These ligaments, composed largely of elastic fibers, form the core of mucosal folds called the **vocal folds,** or **true vocal cords,** which appear pearly white because they lack blood vessels (Figure 21.5). The vocal folds vibrate, producing sounds as air rushes up from the lungs. The vocal folds and the medial opening between them through which air passes are called the **glottis.** Superior to the vocal folds is a similar pair of mucosal folds called the **vestibular folds,** or **false vocal cords.** These play no part in sound production but help to close the glottis when we swallow.

The superior portion of the larynx, an area subject to food contact, is lined by stratified squamous epithelium. Below the vocal folds the epithelium is a pseudostratified ciliated columnar type that acts as a dust filter. The power stroke of its cilia is directed upward toward the pharynx so that mucus is continually moved *away* from the lungs. We help to move mucus up and out of the larynx when we "clear our throat."

Voice Production

Speech involves the intermittent release of expired air and the opening and closing of the glottis. The length of the true vocal cords and the size of the glottis change with the action of the intrinsic laryngeal muscles that clothe the cartilages. Most of these muscles move the arytenoid cartilages. As the length and tension of the cords change, the pitch of the sound varies. Generally, the tenser the cords, the faster they vibrate and the higher the pitch. The glottis is wide (Figure 21.5b) when we produce deep tones and narrows to a slit for high-pitched sounds. As a boy's larynx enlarges during puberty, his true vocal cords become longer and thicker. Because this causes them to vibrate more slowly, his voice becomes deeper. Until the young man learns to control his newly enlarged true vocal cords, his voice "cracks."

Loudness of the voice depends on the force with which the airstream rushes across the vocal cords. The greater the force, the stronger the vibration and the louder the sound. The vocal cords do not move at all when we whisper, but they vibrate vigorously when

TABLE 21.1   Principal Organs of the Respiratory System

Structure	Description, General and Distinctive Features	Function
Nose	Jutting external portion supported by bone and cartilage; internal nasal cavity divided by midline nasal septum and lined with mucosa	Produces mucus; filters, warms, and moistens incoming air; resonance chamber for speech
	Roof of nasal cavity contains olfactory epithelium	Receptors for sense of smell
	Paranasal sinuses around nasal cavity	Same as for nasal cavity; also lighten skull
Pharynx	Passageway connecting nasal cavity to larynx and oral cavity to esophagus; three subdivisions: nasopharynx, oropharynx, and laryngopharynx	Passageway for air and food
	Houses tonsils (lymphoid tissue masses involved in body protection against pathogens)	Facilitates exposure of immune system to inhaled antigens
Larynx	Connects pharynx to trachea; framework of cartilage and dense connective tissue; opening (glottis) can be closed by epiglottis or vocal folds	Air passageway; prevents food from entering lower respiratory tract
	Houses true vocal cords	Voice production
Trachea	Flexible tube running from larynx and dividing inferiorly into two primary bronchi; walls contain C-shaped cartilages that are incomplete posteriorly where connected by trachealis muscle	Air passageway; cleans, warms, and moistens incoming air
Bronchial tree	Consists of right and left primary bronchi, which subdivide within the lungs to form secondary and tertiary bronchi and bronchioles; bronchiolar walls contain complete layer of smooth muscle; constriction of this muscle impedes expiration	Air passageways connecting trachea with alveoli; cleans, warms, and moistens incoming air
Alveoli	Microscopic chambers at termini of bronchial tree; walls of simple squamous epithelium underlain by thin basement membrane; external surfaces intimately associated with pulmonary capillaries	Main sites of gas exchange
	Special alveolar cells produce surfactant	Reduces surface tension; helps prevent lung collapse
Lungs	Paired composite organs located within pleural cavities of thorax; composed primarily of alveoli and respiratory passageways; stroma is fibrous elastic connective tissue, allowing lungs to recoil passively during expiration	House respiratory passages smaller than the primary bronchi
Pleurae	Serous membranes; parietal pleura lines thoracic cavity; visceral pleura covers external lung surfaces	Produce lubricating fluid and compartmentalize lungs

we yell. The power source for creating the airstream is the muscles of the chest, abdomen, and back.

The vocal folds actually produce buzzing sounds. The perceived quality of the voice depends on the co-ordinated activity of many structures above the glottis. For example, the entire length of the pharynx acts as a resonating chamber, to amplify and enhance the sound quality. The oral, nasal, and sinus cavities also contribute to vocal resonance. In addition, good enunciation depends on the "shaping" of sound into recognizable consonants and vowels by muscles in the pharynx, tongue, soft palate, and lips.

H   HOMEOSTATIC IMBALANCE

Inflammation of the vocal folds, or **laryngitis,** causes the vocal folds to swell, interfering with their vibra-tion. This produces a change in the voice tone, hoarseness, or in severe cases inability to speak above a whisper. Laryngitis is also caused by overuse of the voice, very dry air, bacterial infections, tumors on the vocal folds, and inhalation of irritating chemicals. ●

Sphincter Functions of the Larynx

Under certain conditions, the vocal folds act as a sphincter that prevents air passage. During abdominal straining associated with defecation, the glottis closes to prevent exhalation and the abdomi-nal muscles contract, causing the intra-abdominal pressure to rise. These events, collectively known as **Valsalva's maneuver,** help empty the rectum and can also splint (stabilize) the body trunk when one lifts a heavy load.

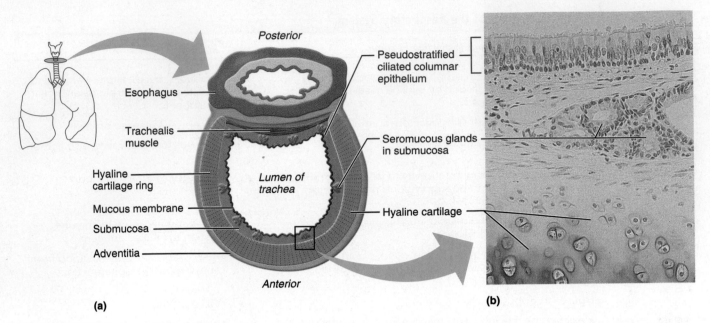

Posterior

Esophagus

Trachealis muscle

Hyaline cartilage ring

Mucous membrane

Submucosa

Adventitia

Lumen of trachea

Anterior

(a)

Pseudostratified ciliated columnar epithelium

Seromucous glands in submucosa

Hyaline cartilage

(b)

FIGURE 21.6   Tissue composition of the tracheal wall.
(a) Cross-sectional view of the trachea, illustrating its relationship to the esophagus, the position of the supporting hyaline cartilage rings, and the trachealis muscle connecting the free ends of the cartilage rings. **(b)** Photomicrograph of a portion of the tracheal wall (cross-sectional view; 225×). **(c)** Scanning electron micrograph of cilia in the trachea (13,500×). The cilia appear as yellow, grasslike projections. Mucus-secreting goblet cells (orange) with short microvilli are interspersed between the ciliated cells.

(c)

The Trachea

The **trachea** (tra′ke-ah), or *windpipe,* descends from the larynx through the neck and into the mediastinum. It ends by dividing into the two primary bronchi at midthorax (see Figure 21.1). In humans, it is 10–12 cm (about 4 inches) long and 2.5 cm (1 inch) in diameter, and very flexible and mobile. Interestingly, early anatomists mistook the trachea for a rough-walled artery (*trachea* = rough).

The tracheal wall consists of several layers that are common to many tubular body organs—the *mucosa, submucosa,* and *adventitia* (Figure 21.6). The **mucosa** has the same goblet cell–containing pseudostratified epithelium that occurs throughout most of the respiratory tract. Its cilia continually propel debris-laden mucus toward the pharynx. This epithelium rests on a fairly thick lamina propria that has a rich supply of elastic fibers.

H   HOMEOSTATIC IMBALANCE

Smoking inhibits and ultimately destroys cilia, after which coughing is the only means of preventing mucus from accumulating in the lungs. For this reason, smokers with respiratory congestion should avoid medications that inhibit the cough reflex. ●

The **submucosa,** a connective tissue layer deep to the mucosa, contains seromucous glands that help produce the mucus "sheets" within the trachea. The outermost **adventitia** layer is a connective tissue layer reinforced internally by 16 to 20 C-shaped rings of hyaline cartilage (Figure 21.6). Because of its elastic elements, the trachea is flexible enough to stretch and move inferiorly during inspiration and recoil during expiration, but the cartilage rings prevent it from collapsing and keep the airway patent despite the pressure changes that occur during breathing. The open posterior parts of the cartilage rings, which abut the esophagus (see Figure 21.6a), are connected by smooth muscle

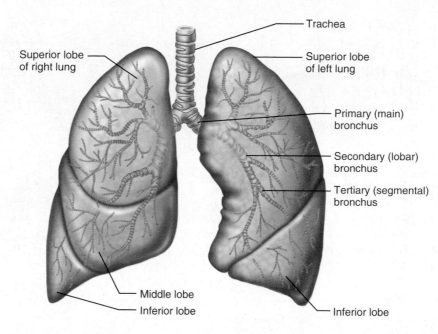

FIGURE 21.7 Conducting zone passages. The air pathway inferior to the larynx consists of the trachea and the primary, secondary, and tertiary bronchi, which branch into the smaller bronchi and bronchioles until the terminal bronchioles of the lungs are reached.

fibers of the **trachealis muscle** and by soft connective tissue. Because this portion of the tracheal wall is not rigid, the esophagus can expand anteriorly as swallowed food passes through it. Contraction of the trachealis muscle decreases the trachea's diameter, causing expired air to rush upward from the lungs with greater force. This action helps to expel mucus from the trachea when we cough by accelerating the exhaled air to speeds of 100 mph! The last tracheal cartilage is expanded, and a spar of cartilage, called the **carina** (kar-ri′nah; "keel"), projects posteriorly from its inner face, marking the point where the trachea branches into the two *primary bronchi*. The mucosa of the carina is highly sensitive and violent coughing is triggered when a foreign object makes contact with it.

HOMEOSTATIC IMBALANCE

Tracheal obstruction is life threatening. Many people have suffocated after choking on a piece of food that suddenly closed off their trachea. The **Heimlich maneuver,** a procedure in which air in the victim's lungs is used to "pop out," or expel, an obstructing piece of food, has saved many people from becoming victims of "café coronaries." The maneuver is simple to learn and easy to do. However, it is best learned by demonstration because cracked ribs are a distinct possibility when it is done incorrectly. ●

The Bronchi and Subdivisions: The Bronchial Tree

The bronchial tree (Figure 21.7) is the site where conducting zone structures give way to respiratory zone structures (Figure 21.8).

Conducting Zone Structures

The **right** and **left primary (main) bronchi** (brong′ki) are formed by the division of the trachea approximately at the level of T_7 in an erect (standing) person. Each bronchus runs obliquely in the mediastinum before plunging into the medial depression (hilus) of the lung on its own side (Figure 21.7). The right primary bronchus is wider, shorter, and more vertical than the left and is the more common site for an inhaled foreign object to become lodged. By the time incoming air reaches the bronchi, it is warm, cleansed of most impurities, and saturated with water vapor.

Once inside the lungs, each primary bronchus subdivides into **secondary (lobar) bronchi**—three on the right and two on the left—each supplying one lung lobe. The secondary bronchi branch into third-order **tertiary (segmental) bronchi**, which divide repeatedly into smaller and smaller bronchi (fourth-order, fifth-order, etc.). Overall, there are about 23 orders of branching air passageways in the lungs. Passages smaller than 1 mm in diameter are called

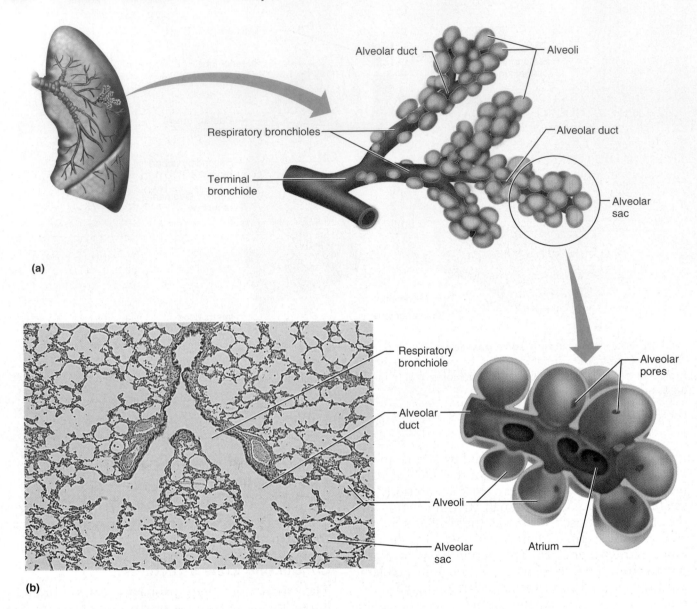

(a)

(b)

FIGURE 21.8   **Respiratory zone structures. (a)** Diagrammatic view of respiratory bronchioles, alveolar ducts, alveolar sacs, and alveoli.
(b) Photomicrograph of a section of human lung, showing the respiratory structures that form the final divisions of the bronchial tree (40×). Notice the thinness of the alveolar walls.

bronchioles ("little bronchi"), and the tiniest of these, the **terminal bronchioles**, are less than 0.5 mm in diameter. Because of this branching pattern, the conducting network within the lungs is often called the **bronchial** or **respiratory tree.**

The tissue composition of the walls of the primary bronchi mimics that of the trachea, but as the conducting tubes become smaller, the following structural changes occur:

1. Support structures change. The cartilage rings are replaced by irregular *plates* of cartilage, and by the time the bronchioles are reached, supportive cartilage is no longer present in the tube walls. However, elastic fibers are found in the tube walls throughout the bronchial tree.

2. Epithelium type changes. The mucosal epithelium thins as it changes from pseudostratified columnar to columnar and then to cuboidal in the terminal bronchioles. Cilia are sparse, and mucus-producing cells are absent in the bronchioles. Thus, most airborne debris found at or below the level of the bronchioles is removed by macrophages in the alveoli.

3. Amount of smooth muscle increases. The relative amount of smooth muscle in the tube walls increases as the passageways become smaller. A complete layer of circular smooth muscle in the bronchioles and the lack of supporting cartilage (which would hinder constriction) allows the bronchioles to provide substantial resistance to air passage under certain conditions (described later).

Respiratory Zone Structures

Defined by the presence of thin-walled air sacs called **alveoli** (al-ve′o-li; *alveol* = small cavity), the respiratory zone begins as the terminal bronchioles feed into **respiratory bronchioles** within the lung (Figure 21.8). Protruding from these smallest bronchioles are scattered alveoli. The respiratory bronchioles lead into winding **alveolar ducts**, whose walls consist of diffusely arranged rings of smooth muscle cells, connective tissue fibers, and outpocketing alveoli. The alveolar ducts lead into terminal clusters of alveoli called **alveolar sacs.** Many people mistakenly equate alveoli, the site of gas exchange, with alveolar sacs, but they are not the same thing. The alveolar sac is analogous to a bunch of grapes, and the alveoli are the individual grapes. The 300 million or so gas-filled alveoli in the lungs account for most of the lung volume and provide a tremendous surface area for gas exchange.

The Respiratory Membrane The walls of the alveoli are composed primarily of a single layer of squamous epithelial cells, called **type I cells,** surrounded by a flimsy basal lamina. The thinness of their walls is hard to imagine, but a sheet of tissue paper is much thicker. The external surfaces of the alveoli are densely covered with a "cobweb" of pulmonary capillaries (Figure 21.9). Together, the alveolar and capillary walls and their fused basal laminae form the **respiratory membrane,** an **air-blood barrier** that has gas on one side and blood flowing past on the other (Figure 21.9d). Gas exchanges occur readily by simple diffusion across the respiratory membrane—O_2 passes from the alveolus into the blood, and CO_2 leaves the blood to enter the gas-filled alveolus. The type I cells also are the primary source of *angiotensin converting enzyme,* which plays a role in blood pressure regulation.

Scattered amid the type I squamous cells that form the major part of the alveolar walls are cuboidal **type II cells** (Figure 21.9c). The type II cells secrete a fluid containing surfactant that coats the gas-exposed alveolar surfaces. (Surfactant's role in reducing the surface tension of the alveolar fluid is described later in this chapter.)

The alveoli have three other significant features: (1) They are surrounded by fine elastic fibers of the same type that surround the entire bronchial tree. (2) Open **alveolar pores** connecting adjacent alveoli allow air pressure throughout the lung to be equalized and provide alternate air routes to any alveoli whose bronchi have collapsed due to disease. (3) Remarkably efficient **alveolar macrophages** crawl freely along the internal alveolar surfaces. Although huge numbers of infectious microorganisms are continuously carried into the alveoli, alveolar surfaces are usually sterile. Because the alveoli are "dead ends," aged and dead macrophages must be prevented from accumulating in them. Most macrophages simply get swept up by the ciliary current of superior regions and carried passively to the pharynx. In this manner, we clear and swallow over 2 million alveolar macrophages per hour!

The Lungs and Pleurae
Gross Anatomy of the Lungs

The paired **lungs** occupy all of the thoracic cavity except the mediastinum, which houses the heart, great blood vessels, bronchi, esophagus, and other organs (Figure 21.10). Each cone-shaped lung is suspended in its own pleural cavity and connected to the mediastinum by vascular and bronchial attachments, collectively called the lung **root.** The anterior, lateral, and posterior lung surfaces lie in close contact with the ribs and form the continuously curving **costal surface.** Just deep to the clavicle is the **apex,** the narrow superior tip of the lung. The concave, inferior surface that rests on the diaphragm is the **base.** On the mediastinal surface of each lung is an indentation, the **hilus,** through which pulmonary and systemic blood vessels enter and leave the lungs. Each primary bronchus also plunges into the hilus on its own side and begins to branch almost immediately. All conducting and respiratory passageways distal to the primary bronchi are found in the lungs.

Because the apex of the heart is slightly to the left of the median plane, the two lungs differ slightly in shape and size. The left lung is smaller than the right, and the **cardiac notch**—a concavity in its medial aspect—is molded to and accommodates the heart (see Figure 21.10a). The left lung is subdivided into upper and lower **lobes** by the *oblique fissure,* whereas the right lung is partitioned into upper, middle, and lower lobes by the *oblique* and *horizontal fissures.* Each lobe contains a number of pyramid-shaped **bronchopulmonary segments** separated from one another by connective tissue septa. Each segment is served by its own

What is the role of surfactant, produced by the type II cells?

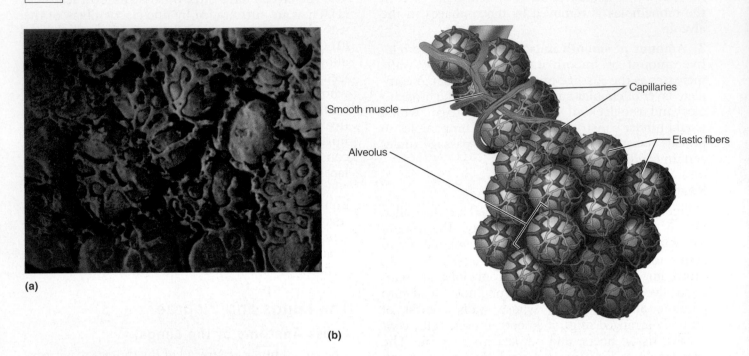

Smooth muscle

Alveolus

Capillaries

Elastic fibers

(b)

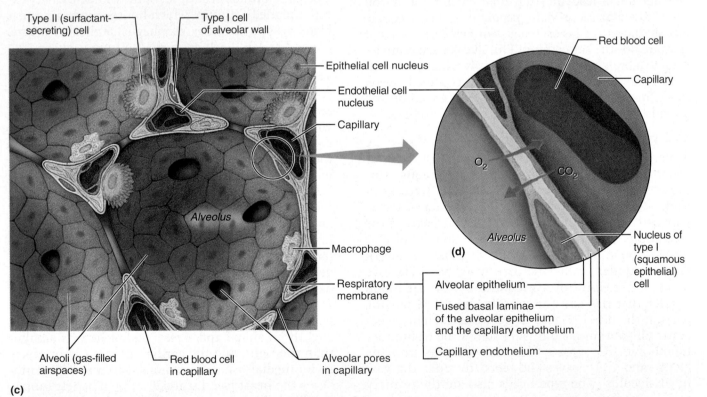

Type II (surfactant-secreting) cell

Type I cell of alveolar wall

Epithelial cell nucleus

Endothelial cell nucleus

Capillary

Alveolus

Macrophage

Respiratory membrane

Alveoli (gas-filled airspaces)

Red blood cell in capillary

Alveolar pores in capillary

(c)

Red blood cell

Capillary

O_2

CO_2

Alveolus

(d)

Nucleus of type I (squamous epithelial) cell

Alveolar epithelium

Fused basal laminae of the alveolar epithelium and the capillary endothelium

Capillary endothelium

FIGURE 21.9   The respiratory membrane. (a) Scanning electron micrograph of casts of alveoli and associated pulmonary capillaries (1100×). From *Tissues and Organs* by R. G. Kessel and R. H. Kardon. © 1979 W. H. Freeman. **(b)** Diagrammatic view of the pulmonary capillary-alveoli relationship. **(c, d)** Detailed anatomy of the respiratory membrane composed of the alveolar squamous epithelial cells (type I cells), the capillary endothelium, and the scant basement membranes (fused basal laminae) intervening between. Type II (surfactant-secreting) alveolar cells are also shown.

■ *Reduces the surface tension of the watery film in the alveoli.*

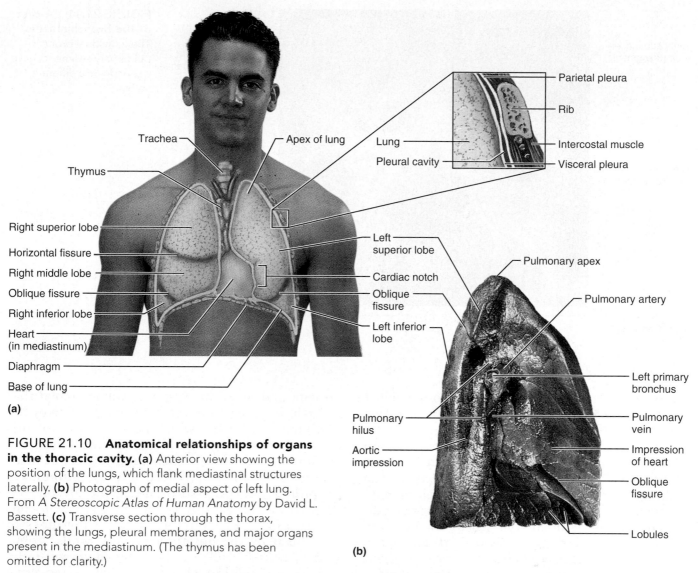

(a)

Trachea

Thymus

Right superior lobe

Horizontal fissure

Right middle lobe

Oblique fissure

Right inferior lobe

Heart
(in mediastinum)

Diaphragm

Base of lung

Apex of lung

Parietal pleura

Rib

Lung

Pleural cavity

Intercostal muscle

Visceral pleura

Left
superior lobe

Cardiac notch

Oblique
fissure

Left inferior
lobe

Pulmonary apex

Pulmonary artery

Left primary
bronchus

Pulmonary
vein

Impression
of heart

Oblique
fissure

Lobules

Pulmonary
hilus

Aortic
impression

(b)

FIGURE 21.10 Anatomical relationships of organs in the thoracic cavity. (a) Anterior view showing the position of the lungs, which flank mediastinal structures laterally. **(b)** Photograph of medial aspect of left lung. From *A Stereoscopic Atlas of Human Anatomy* by David L. Bassett. **(c)** Transverse section through the thorax, showing the lungs, pleural membranes, and major organs present in the mediastinum. (The thymus has been omitted for clarity.)

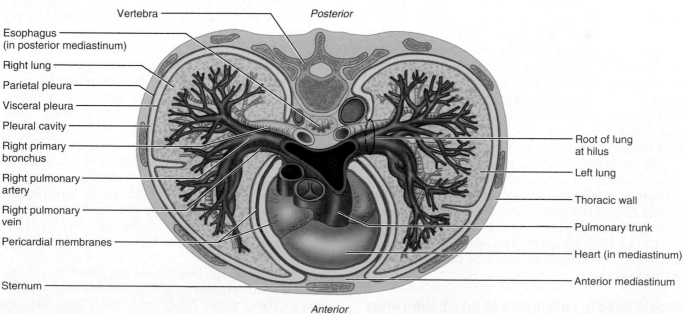

Vertebra

Posterior

Esophagus
(in posterior mediastinum)

Right lung

Parietal pleura

Visceral pleura

Pleural cavity

Right primary
bronchus

Right pulmonary
artery

Right pulmonary
vein

Pericardial membranes

Sternum

Root of lung
at hilus

Left lung

Thoracic wall

Pulmonary trunk

Heart (in mediastinum)

Anterior mediastinum

Anterior

(c)

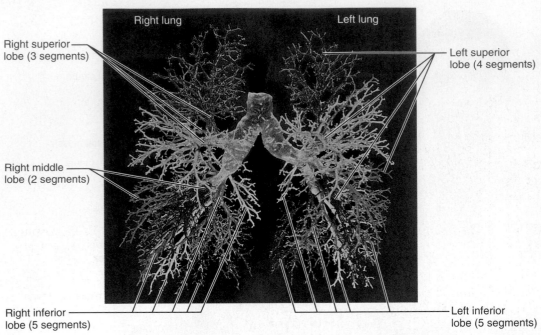

Right lung

Left lung

Right superior lobe (3 segments)

Left superior lobe (4 segments)

Right middle lobe (2 segments)

Right inferior lobe (5 segments)

Left inferior lobe (5 segments)

FIGURE 21.11 A cast of the bronchial tree. The individual broncho-pulmonary segments have been painted different colors.

artery and vein and receives air from an individual segmental (tertiary) bronchus. Initially each lung contains ten bronchopulmonary segments arranged in similar (but not identical) patterns (Figure 21.11), but subsequent fusion of adjacent segmental arteries reduces the number in the left lung to eight or nine segments. The bronchopulmonary segments are clinically important because pulmonary disease is often confined to one or a few segments. Their connective tissue partitions allow diseased segments to be surgically removed without damaging neighboring healthy segments or impairing their blood supply.

The smallest subdivisions of the lung visible with the naked eye are the **lobules,** which appear at the lung surface as hexagons ranging from the size of a pencil eraser to the size of a penny (Figure 21.10b). Each lobule is served by a large bronchiole and its branches. In most city dwellers and in smokers, the connective tissue that separates the individual lobules is blackened with carbon.

As mentioned earlier, the lungs consist largely of air spaces. The balance of lung tissue, or its **stroma** (literally "mattress" or "bed"), is mostly elastic connective tissue. As a result, the lungs are soft, spongy, elastic organs that together weigh just over 1 kg (2.5 pounds). The elasticity of healthy lungs helps to reduce the work of breathing, as described shortly.

Blood Supply and Innervation of the Lungs

The lungs are perfused by two circulations, the pulmonary and the bronchial, which differ in size,

origin, and function. Systemic venous blood that is to be oxygenated in the lungs is delivered by the **pulmonary arteries,** which lie anterior to the primary bronchi (Figure 21.10c). In the lungs, the pulmonary arteries branch profusely along with the bronchi and finally feed into the **pulmonary capillary networks** surrounding the alveoli (see Figure 21.9a). Freshly oxygenated blood is conveyed from the respiratory zones of the lungs to the heart by the **pulmonary veins,** whose tributaries course back to the hilus both with the corresponding bronchi and in the connective tissue septa separating the bronchopulmonary segments.

The large-volume, low-pressure venous blood entering the lungs via the pulmonary arteries contrasts with the small-volume, high-pressure input via the bronchial arteries. The **bronchial arteries,** which provide systemic blood to lung tissue, arise from the aorta, enter the lungs at the hilus, and then run along the branching bronchi, supplying all lung tissues except the alveoli (these are supplied by the pulmonary circulation). Although some systemic venous blood is drained from the lungs by the tiny bronchial veins, there are multiple anastomoses between the two circulations, and most venous blood returns to the heart via the pulmonary veins.

The lungs are innervated by parasympathetic and sympathetic motor fibers, and visceral sensory fibers. These nerve fibers enter each lung through the **pulmonary plexus** on the lung root and run along the bronchial tubes and blood vessels in the lungs. Parasympathetic fibers constrict the air tubes, and sympathetic fibers dilate them.

The Pleurae

The **pleurae** (ploo're; "sides") form a thin, double-layered serosa (see Figure 21.10). The layer called the **parietal pleura** covers the thoracic wall and superior face of the diaphragm. It continues around the heart and between the lungs, forming the lateral walls of the mediastinal enclosure and snugly enclosing the lung root. From here, the pleura extends as the layer called the **visceral pleura** to cover the external lung surface, dipping into and lining its fissures.

The pleurae produce **pleural fluid,** which fills the slitlike **pleural cavity** between them. This lubricating secretion allows the lungs to glide easily over the thorax wall during our breathing movements. Although the pleurae slide easily across each other, their separation is strongly resisted by the surface tension of the pleural fluid. Consequently, the lungs cling tightly to the thorax wall and are forced to expand and recoil passively as the volume of the thoracic cavity alternately increases and decreases during breathing.

The pleurae also help divide the thoracic cavity into three chambers—the central mediastinum and the two lateral pleural compartments, each containing a lung. This compartmentalization helps prevent one mobile organ (for example, the lung or heart) from interfering with another. It also limits the spread of local infections.

ⓗ HOMEOSTATIC IMBALANCE

Pleurisy (ploo'rĭ-se), inflammation of the pleurae, often results from pneumonia. Inflamed pleurae produce less pleural fluid, and the pleural surfaces become dry and rough, resulting in friction and stabbing pain with each breath. Conversely, the pleurae may produce an excessive amount of fluid, which exerts pressure on the lungs. This type of pleurisy hinders breathing movements, but it is much less painful than the dry rubbing type.

Other fluids that may accumulate in the pleural cavity include blood (leakage from damaged blood vessels) and blood filtrate (the watery fluid that oozes from the lung capillaries when right-sided heart failure occurs). The general term for this type of fluid accumulation in the pleural space is *pleural effusion.* ●

Mechanics of Breathing

Breathing, or **pulmonary ventilation,** consists of two phases: **inspiration,** the period when air flows into the lungs, and **expiration,** the period when gases exit the lungs.

Pressure Relationships in the Thoracic Cavity

Before we can begin to describe the breathing process, it is important to understand that *respiratory pressures are always described relative to* **atmospheric pressure (P_{atm}),** which is the pressure exerted by the air (gases) surrounding the body. At sea level, atmospheric pressure is 760 mm Hg (the pressure exerted by a column of mercury 760 mm high). This pressure can also be expressed in atmosphere units: atmospheric pressure = 760 mm Hg = 1 atm. A negative respiratory pressure in any respiratory area, such as −4 mm Hg, indicates that the pressure in that area is lower than atmospheric pressure by 4 mm Hg (760 − 4 = 756 mm Hg). A positive respiratory pressure is higher than atmospheric pressure and zero respiratory pressure is equal to atmospheric pressure. Now, we are ready to examine the pressure relationships that normally exist in the thoracic cavity.

Intrapulmonary Pressure

The **intrapulmonary** (intra-alveolar) **pressure (P_{pul})** is the pressure in the alveoli. Intrapulmonary pressure rises and falls with the phases of breathing, but it *always* eventually equalizes with the atmospheric pressure (Figure 21.12).

Intrapleural Pressure

The pressure in the pleural cavity, the **intrapleural pressure (P_{ip}),** also fluctuates with breathing phases. However, it is always about 4 mm Hg less than P_{pul}. Hence, P_{ip} is negative relative to *both* the intrapulmonary and atmospheric pressures.

The question often asked is "How is this negative intrapleural pressure established?" or "What causes it?" Let's examine the forces that exist in the thorax to see if we can answer these questions. First of all, we know there are opposing forces acting. Two forces act to pull the lungs (visceral pleura) away from the thorax wall (parietal pleura) and cause lung collapse:

1. The lungs' natural tendency to recoil. Because of their elasticity, lungs always assume the smallest size possible.

2. The surface tension of the alveolar fluid. The surface tension of the alveolar fluid constantly acts to draw the alveoli to their smallest possible dimension.

However, these lung-collapsing forces are opposed by the natural elasticity of the chest wall, a force that

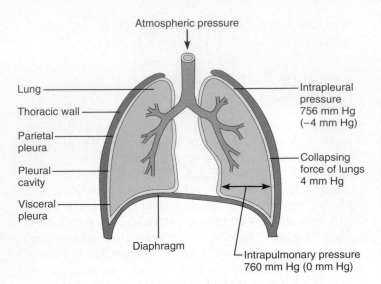

FIGURE 21.12 **Intrapulmonary and intrapleural pressure relationships.** Differences in pressure relative to atmospheric pressure (760 mm Hg) are given in parentheses.

tends to pull the thorax outward and to enlarge the lungs.

So which force wins? The answer is neither in a healthy person, because of the strong adhesive force between the parietal and visceral pleura. Pleural fluid secures the pleurae together in the same way a drop of water holds two glass slides together. The pleurae slide from side to side easily, but they remain closely apposed, and separating them requires extreme force. The net result of the dynamic interplay between these forces is a negative P_{ip}.

The amount of pleural fluid in the pleural cavity must remain minimal in order for the negative P_{ip} to be maintained. Active pumping of the pleural fluid out of the pleural cavity into the lymphatics occurs almost continuously. If it didn't, fluid would accumulate in the intrapleural space (remember, fluids move from high to low pressure), producing a positive pressure in the pleural cavity.

The importance of negative pressure in the intrapleural space and the tight coupling of the lungs to the thorax wall cannot be overemphasized. Any condition that equalizes P_{ip} with the intrapulmonary (or atmospheric) pressure causes *immediate lung collapse*. It is the **transpulmonary pressure**—the difference between the intrapulmonary and intrapleural pressures ($P_{pul} - P_{ip}$)—that keeps the air spaces of the lungs open or, phrased another way, keeps the lungs from collapsing.

⊞ HOMEOSTATIC IMBALANCE

Atelectasis (at"ĕ-lik'tah-sis), or lung collapse, commonly occurs when air enters the pleural cavity through a chest wound, but it may result from a rupture of the visceral pleura, which allows air to enter the pleural cavity from the respiratory tract. It is a common sequel to pneumonia.

The presence of air in the intrapleural space is referred to as a **pneumothorax** (nu"mo-tho'raks; "air thorax"). The condition is reversed by closing the "hole" and drawing air out of the intrapleural space with chest tubes, which allows the lung to reinflate and resume its normal function. Note that because the lungs are in separate cavities, one lung can collapse without interfering with the function of the other. ●

Pulmonary Ventilation: Inspiration and Expiration

Pulmonary ventilation is a mechanical process that depends on volume changes in the thoracic cavity. A rule to keep in mind throughout the following discussion is that *volume changes* lead to *pressure changes*, and pressure changes lead to the *flow of gases* to equalize the pressure.

The relationship between the pressure and volume of a gas is given by **Boyle's law:** At constant temperature, the pressure of a gas varies inversely with its volume. That is:

$$P_1V_1 = P_2V_2$$

where P is the pressure of the gas in millimeters of mercury, V is its volume in cubic millimeters, and subscripts 1 and 2 represent the initial and resulting conditions respectively.

Gases always *fill* their container. Therefore, in a large container, the molecules in a given amount of gas will be far apart and the pressure will be low. But if the volume of the container is reduced, the gas molecules will be forced closer together and the pressure will rise. A good example is an inflated automobile tire. The tire is hard and strong enough to bear the weight of the car because air is compressed to about one-third of its atmospheric volume in the tire, providing the high pressure. Now let us see how this relates to inspiration and expiration.

Inspiration

Visualize the thoracic cavity as a gas-filled box with a single entrance at the top, the tubelike trachea. The volume of this box is changeable and can be increased by enlarging all of its dimensions, thereby decreasing the gas pressure inside it. This drop in pressure causes air to rush into the box from the atmosphere, because gases always flow down their pressure gradients.

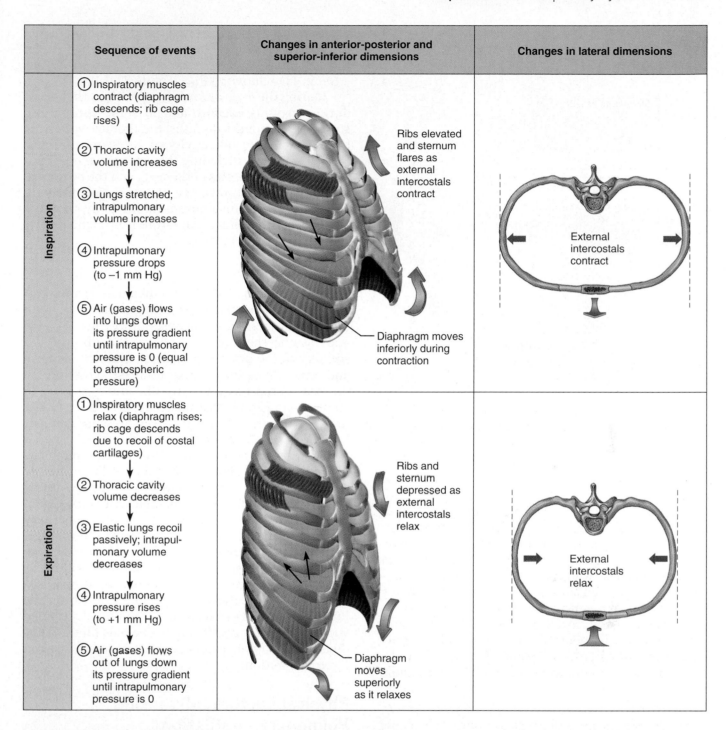

Sequence of events	Changes in anterior-posterior and superior-inferior dimensions	Changes in lateral dimensions
Inspiration ① Inspiratory muscles contract (diaphragm descends; rib cage rises) ↓ ② Thoracic cavity volume increases ↓ ③ Lungs stretched; intrapulmonary volume increases ↓ ④ Intrapulmonary pressure drops (to –1 mm Hg) ↓ ⑤ Air (gases) flows into lungs down its pressure gradient until intrapulmonary pressure is 0 (equal to atmospheric pressure)	Ribs elevated and sternum flares as external intercostals contract Diaphragm moves inferiorly during contraction	External intercostals contract
Expiration ① Inspiratory muscles relax (diaphragm rises; rib cage descends due to recoil of costal cartilages) ↓ ② Thoracic cavity volume decreases ↓ ③ Elastic lungs recoil passively; intrapulmonary volume decreases ↓ ④ Intrapulmonary pressure rises (to +1 mm Hg) ↓ ⑤ Air (gases) flows out of lungs down its pressure gradient until intrapulmonary pressure is 0	Ribs and sternum depressed as external intercostals relax Diaphragm moves superiorly as it relaxes	External intercostals relax

FIGURE 21.13 Changes in thoracic volume during inspiration (top) and expiration (bottom). The sequence of events in the left column includes flowcharts indicating volume changes during inspiration and expiration. The lateral views in the middle column show changes in the superior-inferior dimension (as the diaphragm alternately contracts and relaxes) and in the anterior-posterior dimension (as the external intercostal muscles alternately contract and relax). The superior views of transverse thoracic sections in the right column show lateral dimension changes resulting from alternate contraction and relaxation of the external intercostal muscles.

The same thing happens during normal quiet **inspiration,** when the **inspiratory muscles**—the diaphragm and external intercostal muscles—are activated. Here's how quiet inspiration works:

1. Action of the diaphragm. When the dome-shaped diaphragm contracts, it moves inferiorly and flattens out (Figure 21.13, top). As a result, the superior-inferior dimension (height) of the thoracic cavity increases.

As thoracic volume increases, what happens to the intrapulmonary pressure? To the intrapleural pressure?

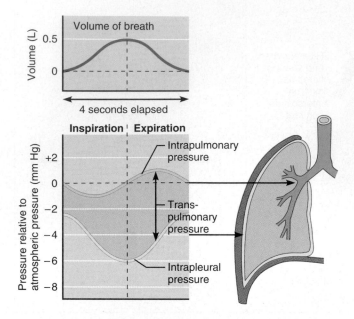

FIGURE 21.14 Changes in intrapulmonary and intrapleural pressures during inspiration and expiration. Notice that normal atmospheric pressure (760 mm Hg) is given a value of 0 on the scale.

2. Action of the intercostal muscles. Contraction of the external intercostal muscles lifts the rib cage and pulls the sternum superiorly (Figure 21.13, top). Because the ribs curve downward as well as forward around the chest wall, the broadest lateral and anteroposterior dimensions of the rib cage are normally directed obliquely downward. But when the ribs are raised and drawn together, they swing outward, expanding the diameter of the thorax both laterally and in the anteroposterior plane. This is much like the action that occurs when a curved bucket handle is raised.

Although these actions expand the thoracic dimensions by only a few millimeters along each plane, this is enough to increase thoracic volume by almost 500 ml—the usual volume of air that enters the lungs during a normal quiet inspiration. Of the two types of inspiratory muscles, the diaphragm is far more important in producing the volume changes that lead to normal quiet inspiration.

As the thoracic dimensions increase during inspiration, the lungs are stretched and the intrapulmonary volume increases. As a result, P_{pul} drops about 1 mm Hg relative to P_{atm}. Anytime the intrapulmonary pressure is less than the atmospheric

pressure ($P_{pul} < P_{atm}$), air rushes into the lungs along the pressure gradient. Inspiration ends when $P_{pul} = P_{atm}$. During the same period, P_{ip} declines to about −6 mm Hg relative to P_{atm} (Figure 21.14).

During the *deep* or *forced inspirations* that occur during vigorous exercise and in some chronic obstructive pulmonary diseases, the thoracic volume is further increased by activity of accessory muscles. Several muscles, including the scalenes and sternocleidomastoid muscles of the neck and the pectoralis minor of the chest, raise the ribs even more than occurs during quiet inspiration. Additionally, the back extends as the thoracic curvature is straightened by the erector spinae muscles.

Expiration

Quiet **expiration** in healthy individuals is a passive process that depends more on lung elasticity than on muscle contraction. As the inspiratory muscles relax and resume their resting length, the rib cage descends and the lungs recoil. Thus, both the thoracic and intrapulmonary volumes decrease. This volume decrease compresses the alveoli, and P_{pul} rises to about 1 mm Hg above atmospheric pressure (see Figure 21.14). When $P_{pul} > P_{atm}$, the pressure gradient forces gases to flow out of the lungs.

Forced expiration is an active process produced by contraction of abdominal wall muscles, primarily the oblique and transversus muscles. These contractions (1) increase the intra-abdominal pressure, which forces the abdominal organs superiorly against the diaphragm, and (2) depress the rib cage. The internal intercostal muscles may also help to depress the rib cage and decrease thoracic volume.

Control of accessory muscles of expiration is important when precise regulation of air flow from the lungs is desired. For instance, the ability of a trained vocalist to hold a musical note depends on the coordinated activity of several muscles normally used in forced expiration.

Physical Factors Influencing Pulmonary Ventilation

As we have seen, the lungs are stretched during inspiration and recoil passively during expiration. The inspiratory muscles consume energy to enlarge the thorax. Energy is also used to overcome various factors that hinder air passage and pulmonary ventilation. These factors are examined next.

Airway Resistance

The major *nonelastic* source of resistance to gas flow is friction, or drag, encountered in the respiratory passageways. The relationship between gas flow (*F*), pressure *(P)*, and resistance *(R)* is given by the following equation:

$$F = \frac{\Delta P}{R}$$

Notice that the factors determining gas flow in the respiratory passages and blood flow in the cardiovascular system are equivalent. The amount of gas flowing into and out of the alveoli is directly proportional to ΔP, the *difference* in pressure, or the pressure gradient, between the external atmosphere and the alveoli. Normally, very small differences in pressure produce large changes in the volume of gas flow. The average pressure gradient during normal quiet breathing is 2 mm Hg or less, and yet it is sufficient to move 500 ml of air in and out of the lungs with each breath.

But, as the equation also indicates, gas flow changes *inversely* with resistance. That is, gas flow decreases as resistance due to friction increases. As with the cardiovascular system, resistance in the respiratory tree is determined mostly by the diameters of the conducting tubes. However, as a rule, airway resistance is insignificant for two reasons:

1. Airway diameters in the first part of the conducting zone are huge, relatively speaking.

2. Gas flow stops at the terminal bronchioles (where small airway diameters might start to be a problem) and diffusion takes over as the main force driving gas movement.

Hence, as shown in Figure 21.15, the greatest resistance to gas flow occurs in the medium-sized bronchi.

(H) HOMEOSTATIC IMBALANCE

Smooth muscle of the bronchiolar walls is exquisitely sensitive to neural controls and certain chemicals. For example, inhaled irritants and inflammatory chemicals such as histamine activate a reflex of the parasympathetic division of the nervous system that causes vigorous constriction of the bronchioles and dramatically reduces air passage. Indeed, the strong bronchoconstriction occurring during an *acute asthma attack* can stop pulmonary ventilation almost completely, regardless of the pressure gradient. Conversely, epinephrine released during sympathetic nervous system activation or administered as a drug dilates bronchioles and reduces airway resistance. Local accumulations of mucus, infectious material, or solid tumors in the passageways are important sources of airway resistance in those with respiratory disease.

Whenever airway resistance rises, breathing movements become more strenuous, but such com-

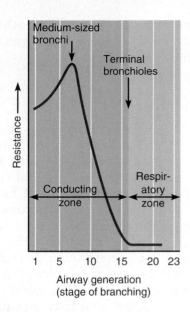

FIGURE 21.15 Resistance in respiratory passageways. Airway resistance peaks in the medium-sized bronchi and then declines sharply as the total cross-sectional area of the airway increases rapidly.

pensation has its limits. When the bronchioles are severely constricted or obstructed, even the most magnificent respiratory efforts cannot restore ventilation to life-sustaining levels. ●

Alveolar Surface Tension Forces

At any gas-liquid boundary, the molecules of the liquid are more strongly attracted to each other than to the gas molecules. This unequal attraction produces a state of tension at the liquid surface, called **surface tension**, that (1) draws the liquid molecules closer together and reduces their contact with the dissimilar gas molecules, and (2) resists any force that tends to increase the surface area of the liquid.

Water is composed of highly polar molecules and has a very high surface tension. Because water is the major component of the liquid film that coats the alveolar walls, it is always acting to reduce the alveoli to their smallest possible size. If the film was pure water, the alveoli would collapse between breaths. But the alveolar film contains **surfactant** (ser-fak'tant), a detergent-like complex of lipids and proteins produced by the type II alveolar cells. Surfactant decreases the cohesiveness of water molecules, much the way a laundry detergent reduces the attraction of water for water, allowing water to interact with and pass through fabric. As a result, the surface tension of alveolar fluid is reduced, and less energy is needed to overcome those forces to expand the lungs and discourage alveolar collapse.

According to some authorities, breaths that are deeper than normal stimulate type II cells to secrete more surfactant.

 HOMEOSTATIC IMBALANCE

When too little surfactant is present, surface tension forces can collapse the alveoli. Once this happens, the alveoli must be completely reinflated during each inspiration, an effort that uses tremendous amounts of energy. This is the problem faced by newborns with **infant respiratory distress syndrome (IRDS),** a condition peculiar to premature babies. Since inadequate pulmonary surfactant is produced until the last two months of fetal development, babies born prematurely often are unable to keep their alveoli inflated between breaths. IRDS is treated with positive-pressure respirators that force air into the alveoli, keeping them open between breaths. Spraying natural or synthetic surfactant into the newborn's respiratory passageways also helps.

Many IRDS survivors suffer from *bronchopulmonary dysplasia,* a chronic lung disease, during childhood and beyond. This condition is believed to result from inflammatory injury to respiratory zone structures caused by use of the respirator on the newborn's delicate lungs. ●

Lung Compliance

Healthy lungs are unbelievably stretchy, and this distensibility is referred to as **lung compliance.** Specifically, lung compliance (C_L) is a measure of the change in lung volume (ΔV_L) that occurs with a given change in the transpulmonary pressure [$\Delta(P_{pul} - P_{ip})$]. This is stated as

$$C_L = \frac{\Delta V_L}{\Delta(P_{pul} - P_{ip})}$$

The more a lung expands for a given rise in transpulmonary pressure, the greater its compliance. Said another way, the higher the lung compliance, the easier it is to expand the lungs at any given transpulmonary pressure.

Lung compliance is determined largely by two factors: (1) distensibility of the lung tissue and of the surrounding thoracic cage, and (2) alveolar surface tension. Because lung (and thoracic) distensibility is generally high and alveolar surface tension is kept low by surfactant, the lungs of healthy people tend to have high compliance, which favors efficient ventilation.

Compliance is diminished by factors with any of the following effects:

1. Reduce the natural resilience of the lungs, such as fibrosis (e.g., development of nonelastic scar tissue in tuberculosis)

2. Block the smaller respiratory passages (e.g., with fluid or thick mucus, as in pneumonia or chronic bronchitis, respectively)

3. Reduce the production of surfactant

4. Decrease the flexibility of the thoracic cage or its ability to expand

The lower the lung compliance, the more energy is needed just to breathe.

 HOMEOSTATIC IMBALANCE

Deformities of the thorax, ossification of the costal cartilages (common during old age), and paralysis of the intercostal muscles all reduce lung compliance by hindering thoracic expansion. ●

Respiratory Volumes and Pulmonary Function Tests

The amount of air flushed in and out of the lungs depends on the conditions of inspiration and expiration. Consequently, several respiratory volumes can be described. Specific combinations of these respiratory volumes, called *respiratory capacities,* are measured to gain information about a person's respiratory status.

Respiratory Volumes

The four **respiratory volumes** of interest are tidal, inspiratory reserve, expiratory reserve, and residual. The values recorded in Figure 21.16a represent normal values for a healthy 20-year-old male weighing about 70 kg (155 pounds). Figure 21.16b provides average values for males and females.

During normal quiet breathing, about 500 ml of air moves into and then out of the lungs with each breath. This respiratory volume is the **tidal volume (TV).** The amount of air that can be inspired forcibly beyond the tidal volume (2100 to 3200 ml) is called the **inspiratory reserve volume (IRV).**

The **expiratory reserve volume (ERV)** is the amount of air—normally 1000 to 1200 ml—that can be evacuated from the lungs after a tidal expiration. Even after the most strenuous expiration, about 1200 ml of air remains in the lungs; this is the **residual volume (RV),** which helps to keep the alveoli patent (open) and to prevent lung collapse.

Respiratory Capacities

The **respiratory capacities** include inspiratory capacity, functional residual capacity, vital capacity, and total lung capacity (Figure 21.16). As noted, the respiratory capacities always consist of two or more lung volumes.

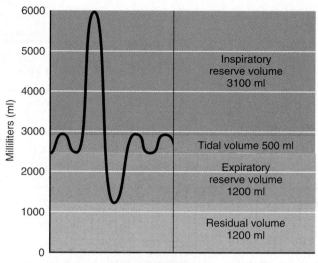
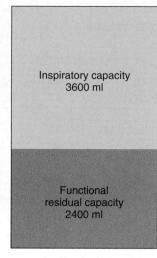
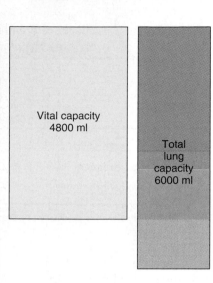

(a) Spirographic record for a male

	Measurement	Adult male average value	Adult female average value	Description
Respiratory volumes	Tidal volume (TV)	500 ml	500 ml	Amount of air inhaled or exhaled with each breath under resting conditions
	Inspiratory reserve volume (IRV)	3100 ml	1900 ml	Amount of air that can be forcefully inhaled after a normal tidal volume inhalation
	Expiratory reserve volume (ERV)	1200 ml	700 ml	Amount of air that can be forcefully exhaled after a normal tidal volume exhalation
	Residual volume (RV)	1200 ml	1100 ml	Amount of air remaining in the lungs after a forced exhalation
Respiratory capacities	Total lung capacity (TLC)	6000 ml	4200 ml	Maximum amount of air contained in lungs after a maximum inspiratory effort: TLC = TV + IRV + ERV + RV
	Vital capacity (VC)	4800 ml	3100 ml	Maximum amount of air that can be expired after a maximum inspiratory effort: VC = TV + IRV + ERV (should be 80% TLC)
	Inspiratory capacity (IC)	3600 ml	2400 ml	Maximum amount of air that can be inspired after a normal expiration: IC = TV + IRV
	Functional residual capacity (FRC)	2400 ml	1800 ml	Volume of air remaining in the lungs after a normal tidal volume expiration: FRC = ERV + RV

(b) Summary of respiratory volumes and capacities for males and females

FIGURE 21.16 **Respiratory volumes and capacities. (a)** Idealized spirographic record of respiratory volumes, in a healthy young 70-kg adult male. **(b)** Respiratory volumes and capacities for males and females.

The **inspiratory capacity (IC)** is the total amount of air that can be inspired after a tidal expiration; thus, it is the sum of TV and IRV. The **functional residual capacity (FRC)** is the combined RV and ERV and represents the amount of air remaining in the lungs after a tidal expiration.

Vital capacity (VC) is the total amount of exchangeable air. It is the sum of TV, IRV, and ERV. In healthy young males, VC is approximately 4800 ml. The **total lung capacity (TLC)** is the sum of all lung volumes and is normally around 6000 ml in males. As indicated in Figure 21.16b, lung volumes and capacities (with the possible exception of TV) tend to be smaller in women than in men because of women's smaller size.

Dead Space

Some of the inspired air fills the conducting respiratory passageways and never contributes to gas exchange in the alveoli. The volume of these conducting zone conduits, which make up the **anatomical dead space,** typically amounts to about 150 ml. (The rule of thumb is that the anatomical dead space volume in a healthy young adult is equal to 1 ml per pound of body weight.) This means that if TV is 500 ml, only 350 ml of it is involved in alveolar

TABLE 21.2 **Effects of Breathing Rate and Depth on Alveolar Ventilation of Three Hypothetical Patients**

Breathing Pattern of Hypothetical Patient	Dead Space Volume (DSV)	Tidal Volume (TV)	Respiratory Rate*	Minute Ventilation	Alveolar Ventilation	% of TV = Dead Space Volume
I—Normal rate and depth	150 ml	500 ml	20/min	10,000 ml/min	7000 ml/min	30%
II—Slow, deep breathing	150 ml	1000 ml	10/min	10,000 ml/min	8500 ml/min	15%
III—Rapid, shallow breathing	150 ml	250 ml	40/min	10,000 ml/min	4000 ml/min	60%

*Respiratory rate values are artificially adjusted to provide equivalent minute respiratory volumes as a baseline for comparison of alveolar ventilation.

ventilation. The remaining 150 ml of the tidal breath is in the anatomical dead space.

If some alveoli cease to act in gas exchange (due to alveolar collapse or obstruction by mucus, for example), the **alveolar dead space** is added to the anatomical dead space, and the sum of the nonuseful volumes is referred to as **total dead space.**

Pulmonary Function Tests

Because the various lung volumes and capacities are often abnormal in people with pulmonary disorders, they are routinely measured in such patients. The measuring device, a **spirometer** (spi-rom'ĕ-ter), is a simple instrument utilizing a hollow bell inverted over water. The bell moves as the patient breathes into a connecting mouthpiece, and a graphic recording is made on a rotating drum. Spirometry is most useful for evaluating losses in respiratory function and for following the course of certain respiratory diseases. Although it cannot provide a specific diagnosis, it can distinguish between *obstructive pulmonary disease* involving increased airway resistance (such as chronic bronchitis) and *restrictive disorders* involving a reduction in total lung capacity resulting from structural or functional changes in the lungs (due to diseases such as tuberculosis, or to fibrosis due to exposure to certain environmental agents such as asbestos). Increases in TLC, FRC, and RV may occur as a result of hyperinflation of the lungs in obstructive disease, whereas VC, TLC, FRC, and RV are reduced in restrictive diseases, which limit lung expansion.

More information can be obtained about a patient's ventilation status by assessing the rate at which gas moves into and out of the lungs. The **minute ventilation** is the total amount of gas that flows into or out of the respiratory tract in 1 minute. During normal quiet breathing, the minute ventilation in healthy people is about 6 L/min (500 ml per breath multiplied by 12 breaths per minute). During vigorous exercise, the minute ventilation may reach 200 L/min.

Two other useful tests are FVC and FEV. **FVC,** or **forced vital capacity,** measures the amount of gas expelled when a subject takes a deep breath and then forcefully exhales maximally and as rapidly as possible. **FEV,** or **forced expiratory volume,** determines the amount of air expelled during specific time intervals of the FVC test. For example, the volume exhaled during the first second is FEV_1. Those with healthy lungs can exhale about 80% of the FVC within 1 second. Those with obstructive pulmonary disease have a low FEV_1, and restrictive disease produces a low FVC.

Alveolar Ventilation

Minute ventilation values provide a rough yardstick for assessing respiratory efficiency, but the **alveolar ventilation rate (AVR)** is a better index of effective ventilation. The AVR takes into account the volume of air wasted in the dead space and measures the flow of fresh gases in and out of the alveoli during a particular time interval. AVR is computed with this equation:

$$\underset{\text{(ml/min)}}{\text{AVR}} = \underset{\text{(breaths/min)}}{\text{frequency}} \times \underset{\text{(ml/breath)}}{\text{(TV} - \text{dead space)}}$$

In healthy people, AVR is usually about 12 breaths per minute times the difference of 500 − 150 ml per breath, or 4200 ml/min.

Because anatomical dead space is constant in a particular individual, increasing the volume of each inspiration (breathing depth) enhances AVR and gas exchange more than raising the respiratory rate. AVR drops dramatically during rapid shallow breathing because most of the inspired air never reaches the exchange sites. Furthermore, as tidal volume approaches the dead space value, effective ventilation approaches zero, regardless of how fast a person is breathing. The effects of breathing rate and breathing depth on alveolar ventilation are summarized for three hypothetical patients in Table 21.2.

TABLE 21.3	Nonrespiratory Air (Gas) Movements
Movement	Mechanism and Result
Cough	Taking a deep breath, closing glottis, and forcing air superiorly from lungs against glottis; glottis opens suddenly and a blast of air rushes upward; can dislodge foreign particles or mucus from lower respiratory tract and propel such substances superiorly
Sneeze	Similar to a cough, except that expelled air is directed through nasal cavities as well as through oral cavity; depressed uvula closes oral cavity off from pharynx and routes air upward through nasal cavities; sneezes clear upper respiratory passages
Crying	Inspiration followed by release of air in a number of short expirations; primarily an emotionally induced mechanism
Laughing	Essentially same as crying in terms of air movements produced; also an emotionally induced response
Hiccups	Sudden inspirations resulting from spasms of diaphragm; believed to be initiated by irritation of diaphragm or phrenic nerves, which serve diaphragm; sound occurs when inspired air hits vocal folds of closed glottis
Yawn	Very deep inspiration, taken with jaws wide open; formerly believed to be triggered by need to increase amount of oxygen in blood but this theory is now being questioned; ventilates all alveoli (not the case in normal quiet breathing)

Nonrespiratory Air Movements

Many processes other than breathing move air into or out of the lungs, and these processes may modify the normal respiratory rhythm. Most of these **nonrespiratory air movements** result from reflex activity, but some are produced voluntarily. The most common of these movements are described in Table 21.3.

Basic Properties of Gases

Gas exchange in the body occurs by bulk flow of gases (and solutions of gases) and by diffusion of gases through tissues. To understand these processes, we must examine some of the physical properties of gases and their behavior in liquids. Two more gas laws—*Dalton's law of partial pressures* and *Henry's law*—provide most of the information we need.

Dalton's Law of Partial Pressures

Dalton's law of partial pressures states that the total pressure exerted by a mixture of gases is the sum of the pressures exerted independently by each gas in the mixture. Further, the pressure exerted by each gas—its **partial pressure**—is directly proportional to the percentage of that gas in the gas mixture.

As indicated in Table 21.4, nitrogen makes up about 79% of air, and the partial pressure of nitrogen (P_{N_2}) is 78.6% × 760 mm Hg, or 597 mm Hg. Oxygen gas, which accounts for nearly 21% of the atmosphere, has a partial pressure (P_{O_2}) of 159 mm Hg (20.9% × 760 mm Hg). Thus, nitrogen and oxygen together contribute about 99% of the total atmospheric pressure. Air also contains 0.04% carbon dioxide, up to 0.5% water vapor, and insignificant amounts of inert gases (such as argon and helium).

At high altitudes, where the atmosphere is less influenced by gravity, partial pressures decline in direct proportion to the decrease in atmospheric pressure. For example, at 10,000 feet above sea level where the atmospheric pressure is 563 mm Hg, P_{O_2} is 110 mm Hg. Moving in the opposite direction, atmospheric pressure increases by 1 atm (760 mm Hg) for each 33 feet of descent (in water) below sea level. Thus, at 99 feet below sea level, the total pressure exerted on the body is equivalent to 4 atm, or 3040 mm Hg, and the partial pressure exerted by each component gas is also quadrupled.

Henry's Law

According to **Henry's law,** when a mixture of gases is in contact with a liquid, each gas will dissolve in the liquid in proportion to its partial pressure. Thus the greater the concentration of a particular gas in the gas phase, the more and the faster that gas will go into solution in the liquid. At equilibrium, the gas partial pressures in the two phases are the same. If, however, the partial pressure of one of the gases later becomes greater in the liquid than in the adjacent gas phase, some of the dissolved gas molecules will reenter the gaseous phase. So the direction and amount of movement of each gas is determined by its partial pressure in

TABLE 21.4 Comparison of Gas Partial Pressures and Approximate Percentages in the Atmosphere and in the Alveoli

Gas	Atmosphere (Sea Level)		Alveoli	
	Approximate Percentage	Partial Pressure (mm Hg)	Approximate Percentage	Partial Pressure (mm Hg)
N_2	78.6	597	74.9	569
O_2	20.9	159	13.7	104
CO_2	0.04	0.3	5.2	40
H_2O	0.46	3.7	6.2	47
	100.0%	760	100.0%	760

the two phases. This flexible situation is exactly what occurs when gases are exchanged in the lungs and tissues.

How much of a gas will dissolve in a liquid at any given partial pressure also depends on the solubility of the gas in the liquid and on the temperature of the liquid. The gases in air have very different solubilities in water (and in plasma). Carbon dioxide is most soluble. Oxygen is only 1/20 as soluble as CO_2, and N_2 is nearly insoluble. Thus, at a given partial pressure, much more CO_2 than O_2 dissolves in water, and practically no N_2 goes into solution. The effect of increasing the liquid's temperature is to decrease gas solubility. Think of club soda, which is produced by forcing CO_2 gas to dissolve in water under high pressure. If you take the cap off a refrigerated bottle of club soda and allow it to stand at room temperature, in just a few minutes you will have plain water—all the CO_2 gas will have escaped from solution.

Hyperbaric oxygen chambers provide clinical applications of Henry's law. These chambers contain O_2 gas at pressures higher than 1 atm and are used to force greater-than-normal amounts of O_2 into the blood of patients suffering from carbon monoxide poisoning, circulatory shock, or asphyxiation. Hyperbaric therapy is also used to treat individuals with gas gangrene or tetanus poisoning because the anaerobic bacteria causing these infections cannot live in the presence of high O_2 levels. Scuba diving provides another illustration of Henry's law.

H HOMEOSTATIC IMBALANCE

Although breathing O_2 gas at 2 atm is not a problem for short periods, **oxygen toxicity** develops rapidly when P_{O_2} is greater than 2.5–3 atm. Excessively high O_2 concentrations generate huge amounts of harmful free radicals, resulting in profound CNS disturbances, coma, and death. ●

Composition of Alveolar Gas

As shown in Table 21.4, the gaseous makeup of the atmosphere is quite different from that in the alveoli. The atmosphere is almost entirely O_2 and N_2; the alveoli contain more CO_2 and water vapor and much less O_2. These differences reflect the effects of: (1) gas exchanges occurring in the lungs (O_2 diffuses from the alveoli into the pulmonary blood and CO_2 diffuses in the opposite direction), (2) humidification of air by conducting passages, and (3) the mixing of alveolar gas that occurs with each breath. Because only 500 ml of air is inspired with each tidal inspiration, gas in the alveoli is actually a mixture of newly inspired gases and gases remaining in the respiratory passageways between breaths.

The partial pressures of O_2 and CO_2 are easily changed by increasing breathing depth and rate. A high AVR brings more O_2 into the alveoli, increasing alveolar P_{O_2}, and rapidly eliminates CO_2 from the lungs.

Gas Exchanges Between the Blood, Lungs, and Tissues

As described, during *external respiration* oxygen enters and carbon dioxide leaves the blood in the lungs. At the body tissues, where the process is called *internal respiration*, the same gases move in opposite directions by the same mechanism (diffusion).

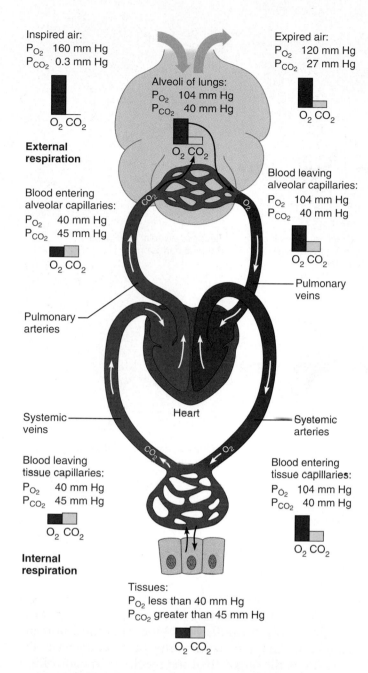

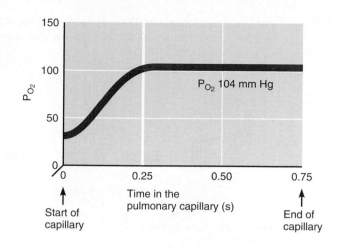

FIGURE 21.18 Oxygenation of blood in the pulmonary capillaries. Note that the time from blood entering the pulmonary capillaries (indicated by 0) until the P_{O_2} is 104 mm Hg is approximately 0.25 second.

FIGURE 21.17 Partial pressure gradients promoting gas movements in the body. Top: Gradients promoting O_2 and CO_2 exchange across the respiratory membrane in the lungs. Bottom: Gradients promoting gas movements across systemic capillary membranes in body tissues. (Note that the composition of alveolar air and expired air are different. This is because some dead space air mixes with the air being expired.)

External Respiration: Pulmonary Gas Exchange

During external respiration, dark red blood flowing through the pulmonary circuit is transformed into the scarlet river that is returned to the heart for distribution by systemic arteries to all body tissues. Although this color change is due to O_2 uptake and binding to hemoglobin in red blood cells (RBCs), CO_2 exchange (unloading) is occurring equally fast.

Three factors influencing the movement of oxygen and carbon dioxide across the respiratory membrane are:

1. Partial pressure gradients and gas solubilities

2. Matching of alveolar ventilation and pulmonary blood perfusion

3. Structural characteristics of the respiratory membrane

Partial Pressure Gradients and Gas Solubilities

Because the P_{O_2} of venous blood in the pulmonary arteries is only 40 mm Hg, as opposed to a P_{O_2} of approximately 104 mm Hg in the alveoli, a steep oxygen partial pressure gradient exists, and O_2 diffuses rapidly from the alveoli into the pulmonary capillary blood (Figure 21.17). Equilibrium—that is, a P_{O_2} of 104 mm Hg on both sides of the respiratory membrane—usually occurs in 0.25 second, which is about one-third the time a red blood cell is in a pulmonary capillary (Figure 21.18). The lesson here is that the blood can flow through the pulmonary capillaries three times as quickly and still be adequately oxygenated. Carbon dioxide moves in the opposite direction along a much gentler partial pressure gradient of about 5 mm Hg (45 mm Hg to 40 mm Hg) until equilibrium occurs at 40 mm Hg. Carbon dioxide is then expelled gradually from the alveoli during expiration. Even though the O_2 pressure gradient for

Suppose a patient is receiving oxygen by mask. What is the condition of the arterioles leading into O_2-enriched alveoli? What is the advantage of this response?

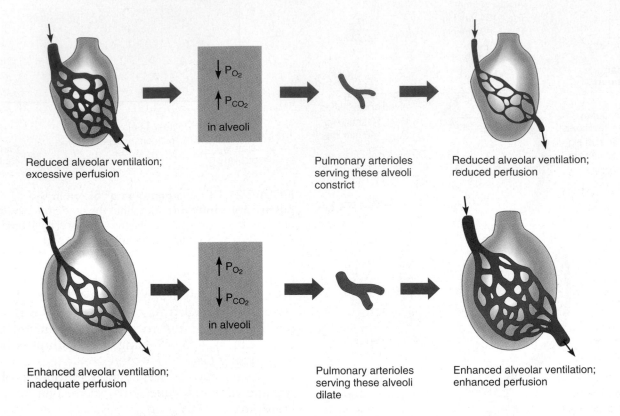

Reduced alveolar ventilation; excessive perfusion

Pulmonary arterioles serving these alveoli constrict

Reduced alveolar ventilation; reduced perfusion

Enhanced alveolar ventilation; inadequate perfusion

Pulmonary arterioles serving these alveoli dilate

Enhanced alveolar ventilation; enhanced perfusion

FIGURE 21.19 Ventilation-perfusion coupling. Autoregulatory events that result in local matching of blood flow (perfusion) through the pulmonary capillaries and conditions of alveolar ventilation.

oxygen diffusion is much steeper than the CO_2 gradient, equal amounts of these gases are exchanged because CO_2 is 20 times more soluble in plasma and alveolar fluid than O_2.

Ventilation-Perfusion Coupling

For gas exchange to be efficient, there must be a close match, or coupling, between *ventilation* (the amount of gas reaching the alveoli) and *perfusion* (the blood flow in pulmonary capillaries). As explained in Chapter 18 and illustrated in Figure 21.19, local autoregulatory mechanisms continuously respond to alveolar conditions.

In alveoli where ventilation is inadequate, P_{O_2} is low. As a result, the terminal arterioles constrict, and blood is redirected to respiratory areas where P_{O_2} is high and oxygen pickup may be more efficient. In

alveoli where ventilation is maximal, pulmonary arterioles dilate, increasing blood flow into the associated pulmonary capillaries. Notice that the autoregulatory mechanism controlling pulmonary vascular muscle is the opposite of the mechanism controlling most arterioles in the systemic circulation.

While changes in alveolar P_{O_2} affect the diameter of pulmonary blood vessels (arterioles), changes in alveolar P_{CO_2} cause changes in the diameters of the *bronchioles*. Passageways servicing areas where alveolar CO_2 levels are high dilate, allowing CO_2 to be eliminated from the body more rapidly, while those serving areas where P_{CO_2} is low constrict.

As a result of modifications made by these two systems, alveolar ventilation and pulmonary perfusion are synchronized. Poor alveolar ventilation results in low oxygen and high carbon dioxide levels in the alveoli; consequently, the pulmonary arterioles constrict and the airways dilate, bringing air flow and blood flow into closer physiological match. High P_{O_2} and low P_{CO_2} in the alveoli cause respiratory passageways serving the alveoli to constrict, and

They would be dilated. This response allows matching of blood flow to availability of oxygen. ∎

promote flushing of blood into the pulmonary capillaries. Although these homeostatic mechanisms provide appropriate conditions for efficient gas exchange, they never completely balance ventilation and perfusion in every alveolus because of (1) the shunting of blood from the bronchial and coronary veins, (2) the effects of gravity, and (3) the occasional alveolar duct plugged with mucus. Consequently, blood in the pulmonary veins actually has a slightly lower P_{O_2} (100 mm Hg) than alveolar air (104 mm Hg) rather than the equality indicated in Figure 21.17.

Thickness and Surface Area of the Respiratory Membrane

In healthy lungs, the respiratory membrane is only 0.5 to 1 μm thick, and gas exchange is usually very efficient.

 ### HOMEOSTATIC IMBALANCE

If the lungs become waterlogged and edematous (as in pneumonia), the exchange membrane thickens dramatically. Under such conditions, even the total time (0.75 s) that red blood cells are in transit through the pulmonary capillaries may not be enough for adequate gas exchange, and body tissues begin to suffer from oxygen deprivation. ●

The greater the surface area of the respiratory membrane, the more gas can diffuse across it in a given time period. The alveolar surface area is enormous in healthy lungs. Spread flat, the total gas exchange surface of these tiny sacs in an adult male's lungs is about 60 m^2, approximately 40 times greater than the surface area of his skin!

 ### HOMEOSTATIC IMBALANCE

In certain pulmonary diseases, the alveolar surface area actually functioning in gas exchange is drastically reduced. This occurs in emphysema, when the walls of adjacent alveoli break through and the alveolar chambers become larger. It also occurs when tumors, mucus, or inflammatory material blocks gas flow into the alveoli. ●

Internal Respiration: Capillary Gas Exchange in the Body Tissues

In internal respiration, the partial pressure and diffusion gradients are reversed from the situation described for external respiration and pulmonary gas exchange. However, the factors promoting gas exchanges between the systemic capillaries and the tissue cells are essentially identical to those acting in the lungs (Figure 21.17). Tissue cells continuously use O_2 for their metabolic activities and produce CO_2. Because P_{O_2} in the tissues is always lower than that in the systemic arterial blood (40 mm Hg versus 104 mm Hg), O_2 moves rapidly from the blood into the tissues until equilibrium is reached, and CO_2 moves quickly along its pressure gradient into the blood. As a result, venous blood draining the tissue capillary beds and returning to the heart has a P_{O_2} of 40 mm Hg and a P_{CO_2} of 45 mm Hg.

In summary, the gas exchanges that occur between the blood and the alveoli and between the blood and the tissue cells take place by simple diffusion driven by the partial pressure gradients of O_2 and CO_2 that exist on the opposite sides of the exchange membranes.

Transport of Respiratory Gases by Blood

Oxygen Transport

External and internal respiration have been considered consecutively to emphasize their similarities, but keep in mind that it is the blood that transports O_2 and CO_2 between these two exchange sites.

Molecular oxygen is carried in blood in two ways, bound to hemoglobin within red blood cells and dissolved in plasma (see Figure 21.22, p. 741). Oxygen is poorly soluble in water, so only about 1.5% of the oxygen transported is carried in the dissolved form. Indeed, if this were the *only* means of oxygen transport, a P_{O_2} of 3 atm or a cardiac output of 15 times normal would be required to provide the oxygen levels needed by body tissues! This problem, of course, has been solved by hemoglobin, and 98.5% of the oxygen ferried from the lungs to the tissues is carried in a loose chemical combination with hemoglobin.

Association of Oxygen and Hemoglobin

As described in Chapter 16, hemoglobin (Hb) is composed of four polypeptide chains, each bound to an iron-containing heme group (see Figure 16.4). Because the iron atoms bind oxygen, each hemoglobin molecule can combine with four molecules of O_2, and oxygen loading is rapid and reversible.

The hemoglobin-oxygen combination, called **oxyhemoglobin** (ok"sĭ-he"mo-glo'bin), is written **HbO$_2$**. Hemoglobin that has released oxygen is called **reduced hemoglobin,** or **deoxyhemoglobin,** and is written **HHb.** Loading and unloading of O_2

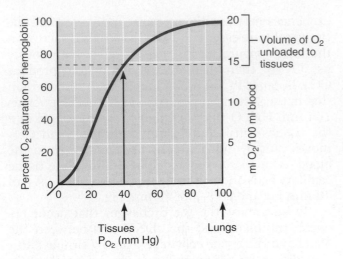

FIGURE 21.20 Oxygen-hemoglobin dissociation curve. Changes in Hb saturation and blood O_2 content as P_{O_2} changes. Notice that hemoglobin is almost completely saturated at a P_{O_2} of 70 mm Hg. Rapid loading and unloading of O_2 to and from hemoglobin occur at P_{O_2} values in the steep portion of the curve. During the systemic circuit, approximately 25% of Hb-bound O_2 is unloaded to the tissues (that is, approximately 5 ml of the 20 ml of O_2 per 100 ml of blood is released). Thus, hemoglobin of venous blood is still about 75% saturated with oxygen after one round through the body.

can be indicated by a single reversible equation:

$$HHb + O_2 \underset{\text{Tissues}}{\overset{\text{Lungs}}{\rightleftharpoons}} HbO_2 + H^+$$

After the first O_2 molecule binds to iron, the Hb molecule changes shape. As a result, it more readily takes up two more O_2 molecules, and uptake of the fourth is even more facilitated. When all four heme groups are bound to O_2, a hemoglobin molecule is said to be *fully saturated*. When one, two, or three oxygen molecules are bound, hemoglobin is *partially saturated*. By the same token, unloading of one oxygen molecule enhances the unloading of the next, and so on. Thus, the *affinity* of hemoglobin for oxygen changes with the extent of oxygen saturation, and both loading and unloading of oxygen are very efficient.

The rate at which Hb reversibly binds or releases O_2 is regulated by P_{O_2}, temperature, blood pH, P_{CO_2}, and blood concentration of an organic chemical called BPG. These factors interact to ensure adequate deliveries of O_2 to tissue cells.

Influence of P_{O_2} on Hemoglobin Saturation

Because of the cooperation just described, the relationship between the degree of hemoglobin saturation and the P_{O_2} of blood is not linear, as the **oxygen-hemoglobin dissociation curve** of Figure 21.20

shows. This S-shaped curve has a steep slope between 10 and 50 mm Hg P_{O_2} and then flattens out (plateaus) between 70 and 100 mm Hg.

Under normal resting conditions (P_{O_2} = 104 mm Hg), arterial blood hemoglobin is 98% saturated, and 100 ml of systemic arterial blood contains about 20 ml of O_2. This *oxygen content* of arterial blood is written as 20 vol % (volume percent). As arterial blood flows through the systemic capillaries, about 5 ml O_2 per 100 ml of blood is released, yielding an Hb saturation of 75% (and an O_2 content of 15 vol %) in venous blood.

The nearly complete saturation of Hb in arterial blood explains why breathing deeply (which increases both the alveolar and arterial blood P_{O_2} above 104 mm Hg) causes very little increase in the O_2 saturation of hemoglobin. Remember, P_{O_2} measurements indicate only the amount of O_2 dissolved in plasma, not the amount bound to hemoglobin. However, P_{O_2} values are a good index of lung function, and when arterial P_{O_2} is significantly less than alveolar P_{O_2}, some degree of respiratory impairment exists.

A hemoglobin saturation curve (Figure 21.20) reveals two other important facts. First, Hb is almost completely saturated at a P_{O_2} of 70 mm Hg, and further increases in P_{O_2} produce only small increases in O_2 binding. The adaptive value of this is that O_2 loading and delivery to the tissues can still be adequate when the P_{O_2} of inspired air is well below its usual levels, a situation common at higher altitudes and in those with cardiopulmonary disease. Moreover, because most O_2 *unloading* occurs on the steep portion of the curve, where the partial pressure changes very little, only 20–25% of bound oxygen is unloaded during one systemic circuit, and substantial amounts of O_2 are still available in venous blood (the *venous reserve*). Thus, if O_2 drops to very low levels in the tissues, as might occur during vigorous exercise, much more O_2 can dissociate from hemoglobin to be used by the tissue cells without any increase in respiratory rate or cardiac output.

Influence of Other Factors on Hemoglobin Saturation

Temperature, blood pH, P_{CO_2}, and the amount of BPG in the blood all influence hemoglobin saturation at a given P_{O_2}. BPG (2,3-bisphosphoglycerate), which binds reversibly with hemoglobin, is produced by red blood cells (RBCs) as they break down glucose by the anaerobic process called glycolysis.

All of these factors influence Hb saturation by modifying hemoglobin's three-dimensional structure, and thereby its affinity for O_2. Generally speaking, an *increase* in temperature, P_{CO_2}, H^+ content of the blood (Figure 21.21), or BPG levels in blood decreases Hb's affinity for O_2 and causes the oxygen-hemoglobin dissociation curve to shift to the right. This enhances oxygen unloading from the blood. Conversely, a *decrease* in any of these factors increases hemoglobin's affinity for oxygen and shifts the dissociation curve to the left.

If you give a little thought to how these factors might be related, you'll realize that they all tend to be at their highest levels in the systemic capillaries where oxygen unloading is the goal. As cells metabolize glucose and use O_2, they release CO_2, which increases the P_{CO_2} and H^+ levels in the capillary blood. Because declining blood pH (acidosis) weakens the Hb–O_2 bond, a phenomenon called the **Bohr effect,** oxygen unloading is accelerated where it is most needed. Additionally, heat is a by-product of metabolic activity, and active tissues are usually warmer than less active ones. A rise in temperature affects hemoglobin's affinity for O_2 both directly and indirectly (via its influence on RBC metabolism and BPG synthesis). Collectively, these factors see to it that Hb unloads much more O_2 in the vicinity of hard-working tissue cells.

The Hemoglobin–Nitric Oxide Partnership in Gas Exchange

Nitric oxide (NO), secreted by lung and vascular endothelial cells, is a well-known vasodilator that plays an important role in blood pressure regulation. Hemoglobin, on the other hand, has a formidable reputation as a vasoconstrictor because it is a NO scavenger—its iron-containing heme group destroys NO. Yet—and here's the paradox—local vessels dilate where gases are being unloaded and loaded.

A second (nonrespiratory) Hb cycle, one that involves NO and aids respiratory gas loading and unloading, seems to speak to this mystery. In the lungs, as O_2 binds to Hb's heme groups, Hb changes in a way that enables NO to latch onto a dangling cysteine amino acid. The result of this binding is a thiol group that protects the NO from being degraded by the iron in hemoglobin. The enriched hemoglobin circulates and as it unloads O_2 it also unloads NO, which dilates the local vessels and aids oxygen delivery. Then as deoxygenated hemoglobin picks up CO_2 (a process described shortly), it also picks up any circulating NO in the area and carries these gases to the lungs where they are unloaded. What we seem to have here is Hb carrying its own vasodilator!

H **HOMEOSTATIC IMBALANCE**

Whatever the cause, inadequate oxygen delivery to body tissues is called **hypoxia** (hi-pok'se-ah). This

Do increasing temperature and P_{CO_2} affect O_2 unloading in the same or opposite directions?

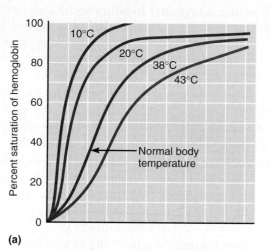

(a)

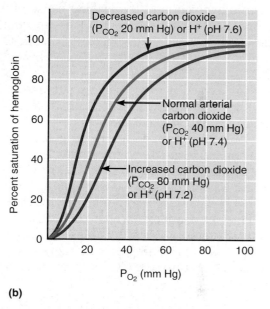

(b)

FIGURE 21.21 **Effect of temperature, P_{CO_2}, and blood pH on the oxygen-hemoglobin dissociation curve.** Oxygen unloading is accelerated at conditions of **(a)** increased temperature, **(b)** increased P_{CO_2}, and/or hydrogen ion concentration (decreased pH), causing the dissociation curve to shift to the right. This response to decreased pH is called the Bohr effect.

condition is easy to recognize in fair-skinned people when Hb saturation falls below 75%, because their skin and mucosae take on a bluish cast (become cyanotic). In dark-skinned individuals, this color change can be observed only in the mucosae and nailbeds.

Classification of hypoxia based on cause is as follows:

The same direction. ■

1. Anemic hypoxia reflects poor O_2 delivery resulting from too few RBCs or from RBCs that contain abnormal or too little Hb.

2. Ischemic (stagnant) hypoxia results when blood circulation is impaired or blocked. Congestive heart failure may cause body-wide ischemic hypoxia, whereas emboli or thrombi block oxygen delivery only to tissues distal to the obstruction.

3. Histotoxic hypoxia occurs when body cells are unable to use O_2 even though adequate amounts are delivered. This variety of hypoxia is the consequence of metabolic poisons, such as cyanide.

4. Hypoxemic hypoxia is indicated by reduced arterial P_{O_2}. Possible causes include imbalances in the ventilation-perfusion coupling mechanism, pulmonary diseases that impair ventilation, and breathing air containing scant amounts of O_2.

Carbon monoxide poisoning is a unique type of hypoxemic hypoxia, and the leading cause of death from fire. Carbon monoxide (CO) is an odorless, colorless gas that competes vigorously with O_2 for heme binding sites. Moreover, because Hb's affinity for CO is more than 200 times greater than its affinity for oxygen, CO is a highly successful competitor. Even at minuscule partial pressures, carbon monoxide can displace oxygen.

CO poisoning is particularly dangerous because it does not produce the characteristic signs of hypoxia—cyanosis and respiratory distress. Instead the victim is confused and has a throbbing headache. In rare cases, fair skin becomes cherry red (the color of the Hb–CO complex), which the eye of the beholder interprets as a healthy "blush." Those with CO poisoning are given hyperbaric therapy (if available) or 100% O_2 until the CO has been cleared from the body. ●

Carbon Dioxide Transport

Normally active body cells produce about 200 ml of CO_2 each minute—exactly the amount excreted by the lungs. Blood transports CO_2 from the tissue cells to the lungs in three forms (Figure 21.22):

1. Dissolved in plasma (7–10%). The smallest amount of CO_2 is transported simply dissolved in plasma.

2. Chemically bound to hemoglobin (just over 20%). In this form, CO_2 is carried in RBCs as **carbaminohemoglobin** (kar-bam″ĭ-no-he″mo-glo′bin):

$$CO_2 + Hb \rightleftharpoons HbCO_2$$
$$\text{(carbaminohemoglobin)}$$

This reaction is rapid and does not require a catalyst. Because carbon dioxide binds directly to the amino acids of *globin* (and not to the heme), carbon dioxide transport in RBCs does not compete with the oxyhemoglobin (or NO) transport mechanism.

CO_2 loading and unloading to and from Hb are directly influenced by the P_{CO_2} and the degree of Hb oxygenation. Carbon dioxide rapidly dissociates from hemoglobin in the lungs, where the P_{CO_2} of alveolar air is lower than that in blood. Carbon dioxide is loaded in the tissues, where the P_{CO_2} is higher than that in the blood. Deoxygenated hemoglobin combines more readily with carbon dioxide than does oxygenated hemoglobin (see the discussion of the Haldane effect on p. 742).

3. As bicarbonate ion in plasma (about 70%). Most carbon dioxide molecules entering the plasma quickly enter the RBCs, where most of the reactions that prepare carbon dioxide for transport as **bicarbonate ions (HCO_3^-)** in plasma occur. As illustrated in Figure 21.22a, when CO_2 diffuses into the RBCs, it combines with water, forming carbonic acid (H_2CO_3). H_2CO_3 is unstable and quickly dissociates into hydrogen ions and bicarbonate ions:

$$CO_2 + H_2O \rightleftharpoons H_2CO_3 \rightleftharpoons H^+ + HCO_3^-$$

Carbon dioxide   Water   Carbonic acid   Hydrogen ion   Bicarbonate ion

Although this reaction also occurs in plasma, it is thousands of times faster in RBCs because they (and not plasma) contain **carbonic anhydrase** (kar-bon′ik an-hi′drās), an enzyme that reversibly catalyzes the conversion of carbon dioxide and water to carbonic acid. Hydrogen ions released during the reaction bind to Hb, triggering the Bohr effect; thus, O_2 release is enhanced by CO_2 loading (as HCO_3^-). Because of the buffering effect of Hb, the liberated H^+ causes little change in pH under resting conditions. Hence, blood becomes only slightly more acidic (the pH declines from 7.4 to 7.34) as it passes through the tissues.

Once generated, HCO_3^- diffuses quickly from the RBCs into the plasma, where it is carried to the lungs. To counterbalance the rapid outrush of these anions from the RBCs, chloride ions (Cl^-) move from the plasma into the RBCs. This ion exchange process is called the **chloride shift.**

In the lungs, the process is reversed (Figure 21.22b). As blood moves through the pulmonary capillaries, its P_{CO_2} declines from 45 mm Hg to 40 mm Hg. For this to occur, CO_2 must first be freed from its "bicarbonate housing." HCO_3^- reenters the RBCs (and Cl^- moves into the plasma) and binds with H^+ to form carbonic acid, which is then split by carbonic anhydrase to release CO_2 and water. This CO_2, along with that released from hemoglobin and from solution in plasma, then diffuses along

In diagram (a), as O_2 is being unloaded from hemoglobin, the hemoglobin also sheds some nitric oxide. What possible role does the NO play in this gas exchange process?

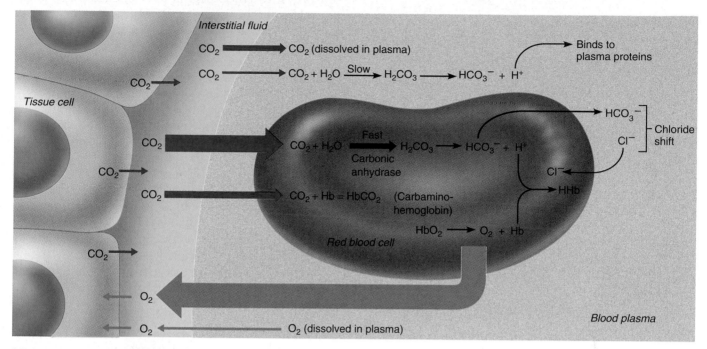

(a) Oxygen release and carbon dioxide pickup at the tissues

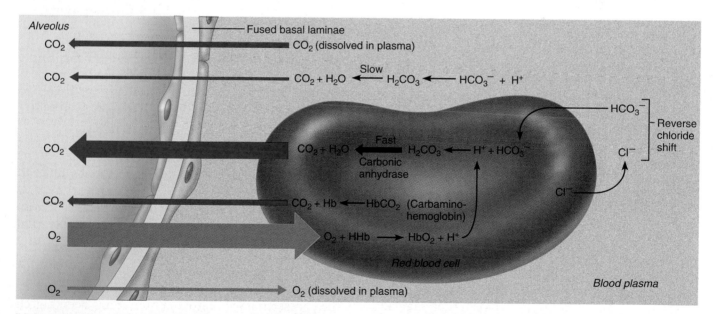

(b) Oxygen pickup and carbon dioxide release in the lungs

FIGURE 21.22   Transport and exchange of CO_2 and O_2. Gas exchanges occurring **(a)** at the tissues and **(b)** in the lungs. CO_2 is transported primarily as bicarbonate ion (HCO_3^-) in plasma (70%); smaller amounts are transported bound to hemoglobin (indicated as $HbCO_2$) in the red blood cells (22%) or in solution in plasma (7%). Nearly all O_2 transported in blood is bound to Hb as oxyhemoglobin (HbO_2) in RBCs. Very small amounts (about 1.5%) are carried dissolved in plasma. (Reduced hemoglobin is HHb.) Relative sizes of the transport arrows indicate the proportionate amounts of O_2 and CO_2 moved by each method.

■ *Causes vasodilation of the local blood vessels, aiding oxygen delivery to the tissue cells and CO_2 loading into the blood.*

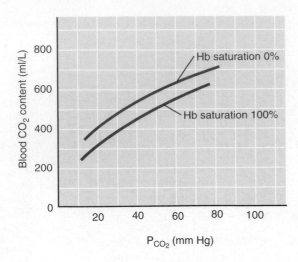

FIGURE 21.23 The Haldane effect. Equilibrium curves for blood CO_2 content relative to fully oxygenated hemoglobin (Hb) ($P_{O_2} = 100\%$) and deoxygenated Hb. The lower the degree of Hb saturation, the greater the amount of CO_2 that can be transported by the blood.

its partial pressure gradient from the blood into the alveoli.

The Haldane Effect

The amount of carbon dioxide transported in blood is markedly affected by the degree of oxygenation of the blood. The lower the P_{O_2} and the lower the extent of Hb saturation with oxygen, the more CO_2 that can be carried in the blood (Figure 21.23). This phenomenon, called the **Haldane effect,** reflects the greater ability of reduced hemoglobin to form carbaminohemoglobin and to buffer H^+ by combining with it. As CO_2 enters the systemic bloodstream, it causes more oxygen to dissociate from Hb (Bohr effect), which allows more CO_2 to combine with Hb and more HCO_3^- to be formed (Haldane effect).

In the pulmonary circulation (Figure 21.22b), the situation is reversed—uptake of O_2 facilitates release of CO_2. As Hb becomes saturated with O_2, the H^+ released combines with HCO_3^-, helping to unload CO_2 from the pulmonary blood. The Haldane effect encourages CO_2 exchange in both the tissues and lungs.

Influence of CO_2 on Blood pH

Typically, the H^+ released during carbonic acid dissociation is buffered by Hb or other proteins within the RBCs or in plasma. The HCO_3^- generated in the red blood cells diffuses into the plasma, where it acts as the *alkaline reserve* part of the blood's carbonic acid–bicarbonate buffer system. The **carbonic acid–bicarbonate buffer system** is very

important in resisting shifts in blood pH (see the equation in point 3 on p. 740 concerning CO_2 transport). For example, if the hydrogen ion concentration in blood begins to rise, excess H^+ is removed by combining with HCO_3^- to form carbonic acid (a weak acid that dissociates very little at either physiological or acidic pH). If H^+ concentration drops below desirable levels in blood, carbonic acid dissociates, releasing hydrogen ions and lowering the pH again.

Changes in respiratory rate or depth can produce dramatic changes in blood pH by altering the amount of carbonic acid in the blood. Slow, shallow breathing allows CO_2 to accumulate in the blood. As a result, carbonic acid levels increase and blood pH drops. Conversely, rapid, deep breathing quickly flushes CO_2 out of the blood, reducing carbonic acid levels and increasing blood pH. Thus, respiratory ventilation can provide a fast-acting system to adjust blood pH (and P_{CO_2}) when it is disturbed by metabolic factors. Respiratory adjustments play a major role in the acid-base balance of the blood, as discussed in Chapter 25.

Control of Respiration

Neural Mechanisms and Generation of Breathing Rhythm

Although our tidelike breathing seems so beautifully simple, its control is fairly complex, involving neurons in the reticular formation of the medulla and pons. Because the medulla sets the respiratory rhythm, we will begin there.

Medullary Respiratory Centers

Clustered neurons in two areas of the medulla oblongata appear to be critically important in respiration. These are (1) the **dorsal respiratory group (DRG),** located dorsally near the root of cranial nerve IX, and (2) the **ventral respiratory group (VRG),** a network of neurons that extends in the ventral brain stem from the spinal cord to the pons-medulla junction. Because the DRG appears to be the pacesetting respiratory center, it has been tagged the **inspiratory center.** When its neurons fire, a burst of impulses travels along the **phrenic** and **intercostal nerves** to excite the diaphragm and external intercostal muscles, respectively (Figure 21.24). As a result, the thorax expands and air rushes into the lungs. The DRG then becomes dormant, and expiration occurs passively as the inspiratory muscles relax and the lungs recoil. This cyclic on/off activity of the inspiratory neurons repeats continuously and produces a respiratory rate of 12–15 breaths per minute, with inspiratory phases

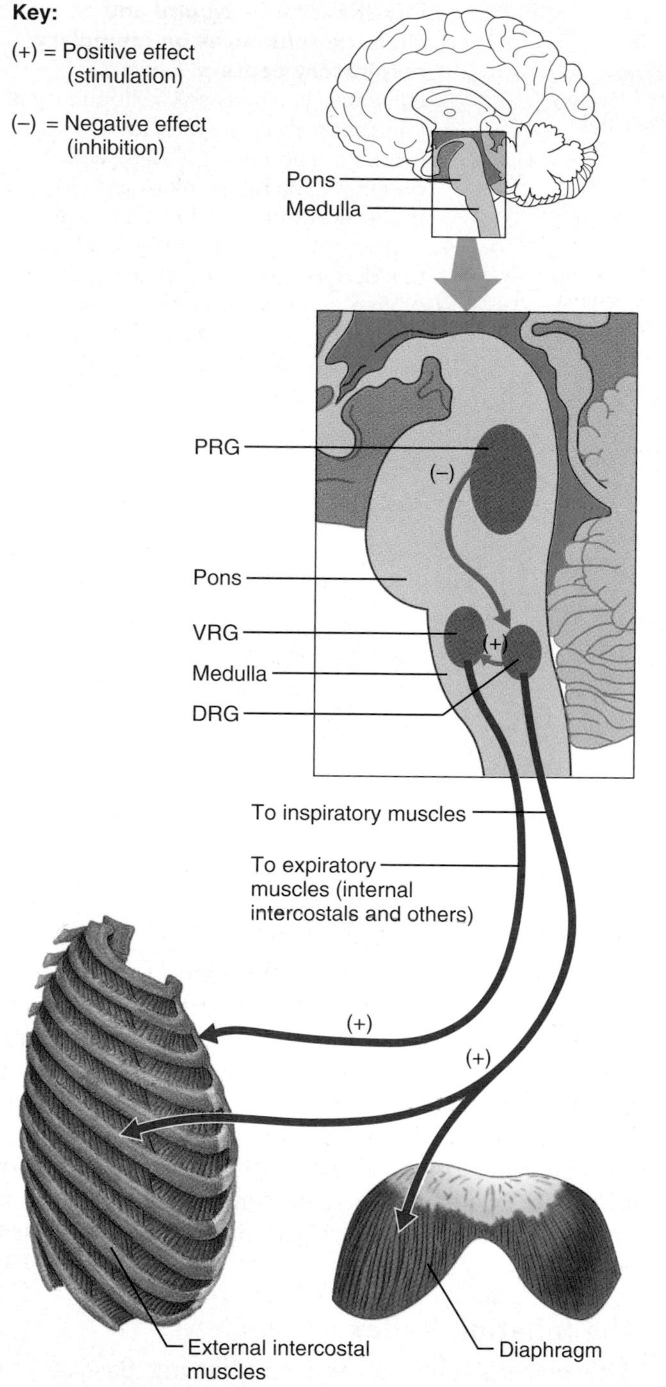

Key:

(+) = Positive effect (stimulation)

(−) = Negative effect (inhibition)

Pons

Medulla

PRG

(−)

Pons

VRG

(+)

Medulla

DRG

To inspiratory muscles

To expiratory muscles (internal intercostals and others)

(+)

(+)

External intercostal muscles

Diaphragm

FIGURE 21.24 Postulated neural pathways involved in respiratory rhythm. During normal quiet breathing, neurons of the dorsal respiratory group (DRG) set the pace by alternately (1) depolarizing and sending impulses to the inspiratory muscles and then (2) becoming quiescent, allowing passive expiration to occur. Neurons of the ventral respiratory group (VRG) are largely inactive during normal quiet breathing. When forced breathing is required, the DRG enlists neurons in the VRG, which help activate the muscles of forced expiration. The pontine respiratory group interacts with the medullary centers in such a way that the breathing pattern is smooth and normally limits the inspiratory phase. (The efferent pathway is incomplete. Medullary neurons communicate with lower motor neurons in the spinal cord. For simplicity, the lower motor neurons, which innervate the muscles of respiration, are not illustrated.)

lasting for about 2 seconds followed by expiratory phases lasting about 3 seconds. This normal respiratory rate and rhythm is referred to as **eupnea** (ūp-ne'ah; *eu* = good, *pne* = breath). During severe hypoxia, DRG neurons generate gasping (perhaps in a last-ditch effort to restore O$_2$ to the brain). When the inspiratory center is completely suppressed, as by an overdose of sleeping pills, morphine, or alcohol, respiration stops completely.

Unlike the DRG, which contains neurons mostly involved in inspiration, the VRG contains a more even mix of neurons involved in inspiration and expiration. Both neuronal networks appear to direct the activity of the respiratory muscles, but the contribution of the VRG is murky. It is called into play mainly during forced breathing (especially forced expiration) when more strenuous breathing movements are needed.

Pons Respiratory Centers

Although the inspiratory center generates the basic respiratory rhythm, the pontine respiratory centers influence and modify the activity of medullary neurons. For example, pons centers appear to smooth out the transitions from inspiration to expiration, and vice versa; and when lesions are made in its superior region, inspirations become very prolonged, a phenomenon called *apneustic breathing*.

The **pontine respiratory group (PRG)**, formerly called the **pneumotaxic center** (noo"mo-tak'sik), continuously transmits inhibitory impulses to the inspiratory center of the medulla (Figure 21.24). When its signals are particularly strong, the period of inspiration is shortened. The most important role of the PRG is to fine-tune the breathing rhythm and prevent lung overinflation.

Genesis of the Respiratory Rhythm

Although there is little question that breathing is rhythmic, we still cannot fully explain the origin of its rhythm. One theory is that the inspiratory neurons themselves or other reticular system neurons are *pacemaker neurons*, which have intrinsic automaticity and rhythmicity. Though pacemaker activity has been demonstrated in rostral VRG neurons in newborns, no such activity has been found in adults.

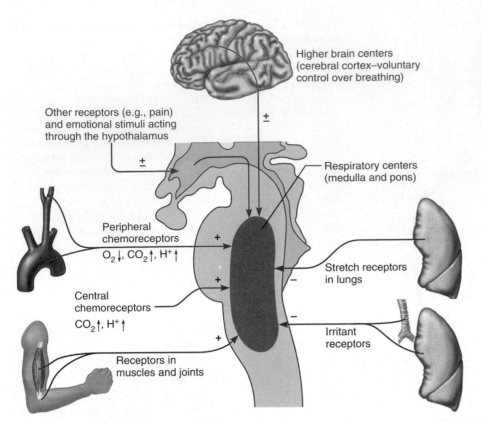

FIGURE 21.25 **Neural and chemical influences on medullary respiratory centers.** Excitatory influences (+) increase the frequency of impulses sent to the muscles of respiration and result in deeper, faster breathing. Inhibitory influences (−) have the reverse effect. In some cases, the impulses may be excitatory or inhibitory (±), depending on which receptors or brain regions are activated.

Another hypothesis is that inputs from stretch receptors in the lungs help establish respiratory rhythm. However, these influences seem relatively unimportant under resting conditions. The most popular theory is that normal respiratory rhythm is a result of reciprocal inhibition of interconnected neuronal networks in the medulla. Whatever the influences acting on the medulla, one thing is sure—the medullary centers are themselves capable of maintaining the normal rhythm of breathing.

Factors Influencing Breathing Rate and Depth

Inspiratory depth is determined by how actively the respiratory center stimulates the motor neurons serving the respiratory muscles. The greater the frequency of stimulation, the greater the number of motor units excited and the greater the force of respiratory muscle contractions. Respiratory rate is determined by how long the inspiratory center is active or how quickly it is switched off.

Depth and rate of breathing can be modified in response to changing body demands. The respiratory centers in the medulla and pons are sensitive to both excitatory and inhibitory stimuli. These stimuli are described next and summarized in Figure 21.25.

Pulmonary Irritant Reflexes

The lungs contain receptors that respond to an enormous variety of irritants. When activated, these receptors communicate with the respiratory centers via vagal nerve afferents. Accumulated mucus, inhaled debris such as dust, or noxious fumes stimulate receptors in the bronchioles that promote reflex constriction of those air passages. The same irritants stimulate a cough when present in the trachea or bronchi, and a sneeze when present in the nasal cavity.

The Inflation Reflex

The visceral pleurae and conducting passages in the lungs contain numerous stretch receptors (baroreceptors) that are vigorously stimulated when the lungs are inflated. These receptors signal the medullary respiratory centers via afferent fibers of the vagus nerves, sending inhibitory impulses that end inspiration and allow expiration to occur. As the lungs recoil, the stretch receptors become quiet, and inspiration is initiated once again. This reflex, called the **inflation reflex,** or **Hering-Breuer reflex** (her′ing broy′er), is thought to be more a protective response (to prevent excessive stretching of the lungs) than a normal regulatory mechanism.

Influence of Higher Brain Centers

Hypothalamic Controls   Acting through the limbic system, strong emotions and pain activate sympathetic centers in the hypothalamus that can modify respiratory rate and depth by sending signals to the respiratory centers. For example, have you ever touched something cold and clammy and gasped? That response was mediated through the hypothalamus. So too is the breath holding that occurs when we are angry and the increased respiratory rate that occurs when we are excited. A rise in body temperature acts to increase the respiratory rate, while a drop in body temperature produces the opposite effect; and sudden chilling of the body (a dip in the north Atlantic Ocean in late October) can cause cessation of breathing (apnea)—or at the very least, gasping.

Cortical Controls   Although breathing is normally regulated involuntarily by the brain stem respiratory centers, we can also exert conscious (volitional) control over the rate and depth of our breathing—choosing to hold our breath or to take an extra deep breath, for example. During voluntary control, the cerebral motor cortex sends signals to the motor neurons that stimulate the respiratory muscles, bypassing the medullary centers. Our ability to voluntarily hold our breath is limited, however, because the brain stem respiratory centers automatically reinitiate breathing when the blood concentration of CO_2 reaches critical levels. That explains why drowning victims typically have water in their lungs.

Chemical Factors

Among the factors that influence breathing rate and depth, the most important are changing levels of CO_2, O_2, and H^+ in arterial blood. Sensors responding to such chemical fluctuations, called **chemoreceptors,** are found in two major body locations. The **central chemoreceptors** are located bilaterally in the ventrolateral medulla. The **peripheral chemoreceptors** are found in the great vessels of the neck.

Influence of P_{CO_2}   Of all the chemicals influencing respiration, CO_2 is the most potent and the most closely controlled. Normally, arterial P_{CO_2} is 40 mm Hg and is maintained within ±3 mm Hg of this level by an exquisitely sensitive homeostatic mechanism that is mediated mainly by the effect that rising CO_2 levels have on the central chemoreceptors of the brain stem (Figure 21.26). CO_2 diffuses easily from the blood into the cerebrospinal fluid, where it is hydrated and forms carbonic acid. As the acid dissociates, H^+ is liberated. This is the same reaction that occurs when CO_2 enters RBCs (see p. 740). Unlike RBCs or plasma, however, cerebrospinal fluid contains virtually no proteins that can buffer the

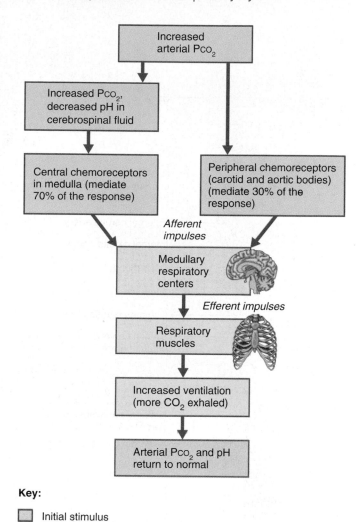

Key:

▢ Initial stimulus

▢ Physiological response

▢ Result

FIGURE 21.26   Negative feedback mechanism by which changes in P_{CO_2} and blood pH regulate ventilation.

added H^+. Thus, as P_{CO_2} levels rise, a condition referred to as **hypercapnia** (hi″per-kap′ne-ah), the cerebrospinal fluid pH drops, exciting the central chemoreceptors, which make abundant synapses with the respiratory regulatory centers. As a result, the depth and perhaps the rate of breathing are increased. This breathing pattern, called **hyperventilation,** enhances alveolar ventilation and quickly flushes CO_2 out of the blood, increasing blood pH. An elevation of only 5 mm Hg in arterial P_{CO_2} results in a doubling of alveolar ventilation, even when arterial O_2 levels and pH are unchanged. When P_{O_2} and pH are below normal, the response to elevated P_{CO_2} is even greater. Hyperventilation is normally self-limiting, ending when homeostatic blood P_{CO_2} levels are restored.

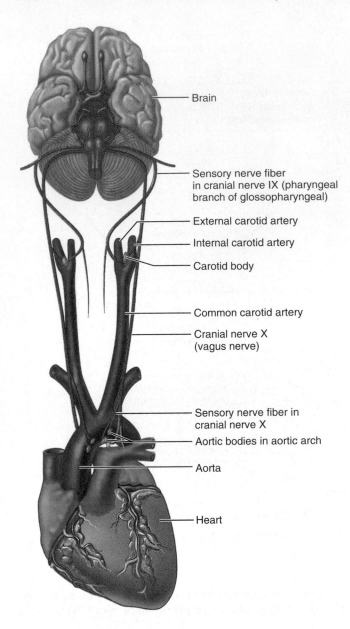

Brain

Sensory nerve fiber
in cranial nerve IX (pharyngeal
branch of glossopharyngeal)

External carotid artery

Internal carotid artery

Carotid body

Common carotid artery

Cranial nerve X
(vagus nerve)

Sensory nerve fiber in
cranial nerve X

Aortic bodies in aortic arch

Aorta

Heart

FIGURE 21.27 Location of the peripheral chemoreceptors in the carotid and aortic bodies. Also shown are the sensory pathways from these receptors through cranial nerves IX and X to the respiratory center in the medulla.

Notice that while rising CO_2 levels act as the initial stimulus, it is rising H^+ levels that prod the *central* chemoreceptors into activity. In the final analysis, control of breathing during rest is aimed primarily at *regulating the H^+ concentration in the brain.*

HOMEOSTATIC IMBALANCE

A person experiencing an anxiety attack may hyperventilate involuntarily to the point where he or she becomes dizzy or faints. This happens because low CO_2 levels in the blood (**hypocapnia**) cause cerebral blood vessels to constrict, reducing brain perfusion and producing cerebral ischemia. Such attacks may be averted by breathing into a paper bag because then the air being inspired is expired air, rich in carbon dioxide, which causes carbon dioxide to be retained in the blood. ●

When P_{CO_2} is abnormally low, respiration is inhibited and becomes slow and shallow; that is, **hypoventilation** occurs. In fact, periods of **apnea** (breathing cessation) may occur until arterial P_{CO_2} rises and again stimulates respiration.

Sometimes swimmers voluntarily hyperventilate so that they can hold their breath longer during swim meets. This is incredibly dangerous for the following reasons. Blood O_2 content rarely drops much below 60% of normal during regular breath-holding, because as P_{O_2} drops, P_{CO_2} rises enough to make breathing unavoidable. However, strenuous hyperventilation can lower P_{CO_2} so much that a lag period occurs before it rebounds enough to stimulate respiration again. This lag may allow oxygen levels to fall well below 50 mm Hg, causing the swimmer to black out (and perhaps drown) before he or she has the urge to breathe.

Influence of P_{O_2}   Cells sensitive to arterial O_2 levels are found in the peripheral chemoreceptors, that is, in the **aortic bodies** of the aortic arch and in the **carotid bodies** at the bifurcation of the common carotid arteries (Figure 21.27). Those in the carotid bodies are the main oxygen sensors.

Under normal conditions, the effect of declining P_{O_2} on ventilation is slight and mostly limited to enhancing the sensitivity of central receptors to increased P_{CO_2}. Arterial P_{O_2} must drop *substantially*, to at least 60 mm Hg, before O_2 levels become a major stimulus for increased ventilation. This is not as strange as it may appear. Remember, there is a huge reservoir of O_2 bound to Hb, and Hb remains almost entirely saturated unless or until the P_{O_2} of alveolar gas and arterial blood falls below 60 mm Hg. As central chemoreceptors then begin to suffer from O_2 starvation, the oxygen-poor blood produces NO-derived compounds called *S*-nitrosothiols, or SNOs, which act on the respiratory center to induce rapid breathing. At the same time, the peripheral chemoreceptors become excited and stimulate the respiratory centers to increase ventilation, even if P_{CO_2} is normal. Thus, the peripheral chemoreceptor system can maintain ventilation when alveolar O_2 levels are low even though brain stem centers are depressed by hypoxia.

H *HOMEOSTATIC IMBALANCE*

In people who retain CO_2 because of pulmonary disease (e.g., emphysema and chronic bronchitis), arterial P_{CO_2} is chronically elevated and, as a result, chemoreceptors become unresponsive to this chemical stimulus. In such cases, a declining P_{O_2} acting on the oxygen-sensitive peripheral chemoreceptors provides the principal respiratory stimulus, the **hypoxic drive.** Thus, gas mixtures administered to such patients during respiratory distress are only slightly enriched with O_2 because inspiration of pure oxygen would stop their breathing by removing their respiratory stimulus (low P_{O_2} levels). ●

Influence of Arterial pH   Changes in arterial pH can modify respiratory rate and rhythm even when CO_2 and O_2 levels are normal. Because little H^+ diffuses from the blood into the cerebrospinal fluid, the direct effect of arterial H^+ concentration on central chemoreceptors is insignificant compared to the effect of H^+ generated by elevations in P_{CO_2}. The increased ventilation that occurs in response to falling arterial pH is mediated through the peripheral chemoreceptors.

Although changes in P_{CO_2} and H^+ concentration are interrelated, they are distinct stimuli. A drop in blood pH may reflect CO_2 retention, but it may also result from metabolic causes, such as accumulation of lactic acid during exercise or of fatty (or other organic) acids in patients with poorly controlled diabetes mellitus. Regardless of cause, as arterial pH declines, respiratory system controls attempt to compensate and raise the pH by eliminating CO_2 (and carbonic acid) from the blood; thus, respiratory rate and depth increase.

Summary of Interactions of P_{CO_2}, P_{O_2}, and Arterial pH   Although every cell in the body must have O_2 to live, the body's need to rid itself of CO_2 is the most important stimulus for breathing in a healthy person. However, CO_2 does not act in isolation, and various chemical factors enforce or inhibit one another's effects. These interactions are summarized here:

1. Rising CO_2 levels are the most powerful respiratory stimulant. As CO_2 is hydrated in cerebrospinal fluid, liberated H^+ acts directly on the central chemoreceptors, causing a reflexive increase in breathing rate and depth. Low P_{CO_2} levels depress respiration.

2. Under normal conditions, blood P_{O_2} affects breathing only indirectly by influencing chemoreceptor sensitivity to changes in P_{CO_2}. Low P_{O_2} augments P_{CO_2} effects; high P_{O_2} levels diminish the effectiveness of CO_2 stimulation.

3. When arterial P_{O_2} falls below 60 mm Hg, it becomes the major stimulus for respiration, and ventilation is increased via reflexes initiated by the peripheral chemoreceptors. This may increase O_2 loading into the blood, but it also causes hypocapnia (low P_{CO_2} blood levels) and an increase in blood pH, both of which inhibit respiration.

4. Changes in arterial pH resulting from CO_2 retention or metabolic factors act indirectly through the peripheral chemoreceptors to promote changes in ventilation, which in turn modify arterial P_{CO_2} and pH. Arterial pH does not influence the central chemoreceptors directly.

Respiratory Adjustments
Adjustments During Exercise

Respiratory adjustments during exercise are geared both to intensity and duration of the exercise. Working muscles consume tremendous amounts of O_2 and produce large amounts of CO_2; thus, ventilation can increase 10 to 20 fold during vigorous exercise. Breathing becomes deeper and more vigorous, but the respiratory rate may not change significantly. This breathing pattern is called **hyperpnea** (hi"perp-ne'ah) to distinguish it from the deep and often rapid pattern of hyperventilation. Also, the respiratory changes seen in hyperpnea match metabolic demands and so do not lead to significant changes in blood O_2 and CO_2 levels. By contrast, hyperventilation may provoke excessive ventilation, resulting in low P_{CO_2} and alkalosis.

Exercise-enhanced ventilation does *not* appear to be prompted by rising P_{CO_2} and declining P_{O_2} and pH in the blood for two reasons. First, ventilation increases abruptly as exercise begins, followed by a gradual increase, and then a steady state of ventilation. When exercise stops, there is an initial small but abrupt decline in ventilation rate, followed by a gradual decrease to the pre-exercise value. Second, although venous levels change, arterial P_{CO_2} and P_{O_2} levels remain surprisingly constant during exercise. In fact, P_{CO_2} may decline to below normal and P_{O_2} may rise slightly because of the efficiency of the respiratory adjustments. Our present understanding of the mechanisms that produce these observations is sketchy, but the most accepted explanation is as follows.

The abrupt increase in ventilation that occurs as exercise begins reflects interaction of three neural factors:

1. Psychic stimuli (our conscious anticipation of exercise)

2. Simultaneous cortical motor activation of skeletal muscles and respiratory centers

3. Excitatory impulses reaching respiratory centers from proprioceptors in moving muscles, tendons, and joints.

The subsequent gradual increase and then plateauing of respiration probably reflect the rate of CO_2 delivery to the lungs (the "CO_2 flow").

The small but abrupt decrease in ventilation that occurs as exercise ends reflects the shutting off of the neural control mechanisms. The subsequent gradual decline to baseline ventilation likely reflects a decline in the CO_2 flow that occurs as the oxygen debt is being repaid. The rise in lactic acid levels that contributes to O_2 debt is *not* a result of inadequate respiratory function, because alveolar ventilation and pulmonary perfusion are as well matched during exercise as during rest. Rather, it reflects cardiac output limitations or inability of the skeletal muscles to further increase their oxygen consumption. In light of this fact, the practice of inhaling pure O_2 by mask, used by some football players to replenish their "oxygen-starved" bodies as quickly as possible, is useless. The panting athlete *does* have an O_2 deficit, but inspiring extra oxygen will not help, because the shortage is in the muscles—not the lungs.

Adjustments at High Altitude

Most Americans live between sea level and an altitude of approximately 8000 feet. In this range, differences in atmospheric pressure are not great enough to cause healthy people any problems when they spend brief periods in the higher-altitude areas. However, when you travel quickly from sea level to elevations above 8000 ft, where air density and P_{O_2} are lower, your body responds with symptoms of *acute mountain sickness (AMS)*—headaches, shortness of breath, nausea, and dizziness. AMS is common in travelers to ski resorts such as Vail, Colorado (8120 ft), and Brian Head, Utah (a heart-pounding 9600 ft). In severe cases of AMS, lethal pulmonary and cerebral edema may occur.

When you move on a *long-term* basis from sea level to the mountains, your body begins to make respiratory and hematopoietic adjustments via an adaptive response called **acclimatization.** As already explained, decreases in arterial P_{O_2} cause the central chemoreceptors to become more responsive to increases in P_{CO_2}, and a substantial decline in P_{O_2} directly stimulates the peripheral chemoreceptors. As a result, ventilation increases as the brain attempts to restore gas exchange to previous levels. Within a few days, the minute respiratory volume stabilizes at a level 2–3 L/min higher than the sea level rate. Because increased ventilation also reduces arterial CO_2 levels, the P_{CO_2} of individuals living at high altitudes is typically below 40 mm Hg (its value at sea level).

Because less O_2 is available to be loaded, high-altitude conditions always result in lower-than-normal hemoglobin saturation levels. For example, at about 19,000 ft above sea level, O_2 saturation of arterial blood is only 67% (compared to nearly 98% at sea level). But Hb unloads only 20–25% of its oxygen at sea level, which means that even at the reduced saturations at high altitudes, the O_2 needs of the tissues are still met adequately under resting conditions. Additionally, at high altitudes Hb releases more NO to the capillary blood in the tissues and hemoglobin's affinity for O_2 is reduced because of increases in BPG concentration, with the result that more O_2 is released to the tissues during each circulatory round.

Although the tissues in a person at high altitude receive adequate oxygen under normal conditions, problems arise when all-out efforts are demanded of the cardiovascular and respiratory systems (as discovered by U.S. athletes competing in the 1968 summer Olympics on the high mesa of Mexico City, at 7370 ft). Unless one is fully acclimatized, such conditions almost guarantee that body tissues will become severely hypoxic.

When blood O_2 levels decline, the kidneys accelerate production of erythropoietin, which stimulates bone marrow production of RBCs (see Chapter 16). This phase of acclimatization, which occurs slowly, provides long-term compensation for living at high altitudes.

Homeostatic Imbalances of the Respiratory System

The respiratory system is particularly vulnerable to infectious diseases because it is wide open to airborne pathogens. Many of these inflammatory conditions, such as rhinitis and laryngitis, were considered earlier in the chapter. Here we turn our attention to the most disabling disorders: *chronic obstructive pulmonary disease (COPD), asthma, tuberculosis,* and *lung cancer.* COPD and lung cancer are living proof of the devastating effects of cigarette smoking on the body. Long known to promote cardiovascular disease, cigarettes are perhaps even more effective at destroying the lungs.

Chronic Obstructive Pulmonary Disease

The **chronic obstructive pulmonary diseases (COPD),** exemplified best by chronic bronchitis and obstructive emphysema, are a major cause of death and disability in the United States and are becoming increasingly prevalent. These diseases have certain features in common (Figure 21.28):

1. Patients almost invariably have a history of smoking.

2. **Dyspnea** (disp-ne'ah), difficult or labored breathing often referred to as "air hunger," occurs and gets progressively more severe.

3. Coughing and frequent pulmonary infections are common.

4. Most COPD victims develop respiratory failure (accompanied by hypoxemia, CO_2 retention, and respiratory acidosis).

Obstructive emphysema is distinguished by permanent enlargement of the alveoli, accompanied by deterioration of the alveolar walls. Chronic inflammation leads to lung fibrosis, and invariably the lungs lose their elasticity. Arterial O_2 and CO_2 levels remain essentially normal until late in the disease. As the lungs become less elastic, the airways collapse during expiration and obstruct the outflow of air. This has two important consequences: (1) Accessory muscles must be enlisted to breathe, and victims are perpetually exhausted because breathing requires 15–20% of their total body energy supply (as opposed to 5% in healthy individuals). (2) For complex reasons, the bronchioles open during inspiration but collapse during expiration, trapping huge volumes of air in the alveoli. This hyperinflation leads to development of a permanently expanded "barrel chest" and flattens the diaphragm, thus reducing ventilation efficiency.

Damage to the pulmonary capillaries increases resistance in the pulmonary circuit, forcing the right ventricle to overwork and consequently become enlarged. Emphysema victims are sometimes called "pink puffers" because their breathing is labored, but they do not become cyanotic (blue) because gas exchange remains surprisingly adequate until late in the disease. In addition to cigarette smoking, hereditary factors (e.g., alpha-1 antitrypsin deficiency) may help to cause emphysema in some patients.

In **chronic bronchitis**, inhaled irritants lead to chronic excessive mucus production by the mucosa of the lower respiratory passageways and to inflammation and fibrosis of that mucosa. These responses obstruct the airways and severely impair lung ventilation and gas exchange. Pulmonary infections are frequent because bacteria thrive in the stagnant pools of mucus. Patients are sometimes called "blue bloaters" because hypoxia and CO_2 retention occur early in the disease and cyanosis is common. However, the degree of dyspnea is usually moderate compared to that of emphysema sufferers. In addition to the major factor of cigarette smoking, environmental pollution may contribute to the development of chronic bronchitis.

COPD is routinely treated with bronchodilators and corticosteroids in aerosol form (inhalers). Severe

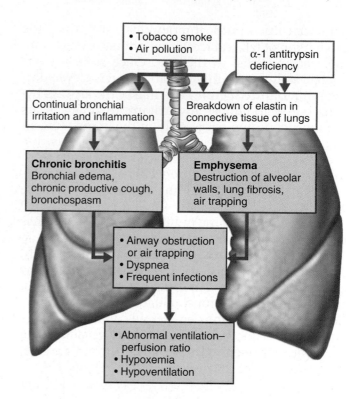

FIGURE 21.28 The pathogenesis of COPD.

dyspnea and hypoxia mandate oxygen use. A highly controversial surgical treatment introduced in 1994, called *lung volume reduction surgery*, has been performed on some emphysema patients. Part of the grossly enlarged lungs is removed to give the remaining lung tissue room to expand. At least in the short term, enough breathing capacity is restored to allow those patients to live something close to a normal life. (The average increase in lung function is 55%.)

Asthma

Asthma is characterized by episodes of coughing, dyspnea, wheezing, and chest tightness—alone or in combination. Most acute attacks are accompanied by a sense of panic. Although sometimes classed with COPD because it is an obstructive disorder, asthma is marked by acute exacerbations followed by symptom-free periods.

The cause of asthma has been hard to pin down. Initially it was viewed as a consequence of bronchospasm triggered by various factors such as cold air, exercise, or allergens. However, when it was discovered that bronchoconstriction has relatively little effect on air flow through the lungs, researchers probed more deeply and found that in asthma, active inflammation of the airways comes first. The airway inflammation is an immune response under the control of T_H2 cells, a subset of T lymphocytes that, by secreting certain interleukins, stimulate the production of IgE and recruit inflammatory cells (notably eosinophils) to the site. Once someone has allergic asthma, the inflammation persists even during symptom-free periods and makes the airways hypersensitive to almost any irritant. (It now appears the most common triggers are in the home—the allergens from dust mites, cockroaches, cats, dogs, and fungi.) Once the airway walls are thickened with inflammatory exudate, the effect of bronchospasm is vastly magnified and can dramatically reduce air flow.

Since the 1980s the number of asthma cases has been increasing, and in the decade from 1979 to 1989, the number of asthma deaths doubled. Sadly, this increase in morbidity and mortality has come about despite changes in drug therapy—a change in focus from inhaled bronchodilators, which bring relief in minutes, to inhaled steroids, which act more slowly but reduce the inflammatory response and the frequency of episodes—that came from a greater understanding of the disease.

Tuberculosis

Tuberculosis (TB), the infectious disease caused by the bacterium *Mycobacterium tuberculosis,* is spread by coughing and primarily enters the body in inhaled air. TB mostly affects the lungs but can spread through the lymphatics to affect other organs. One-third of the world's population is infected, but most people never develop active TB because a massive inflammatory and immune response usually contains the primary infection in fibrous, or calcified, nodules (tubercles) in the lungs. However, the bacteria survive in the nodules and when the person's immunity is low, they may break out and cause symptomatic TB, involving fever, night sweats, weight loss, a racking cough, and spitting up blood.

Until the 1930s, TB was responsible for one-third of all deaths among 20- to 45-year-old U.S. adults. With the advent of antibiotics in the 1940s, this killer was put into retreat and its prevalence declined dramatically. However, since 1985 there has been an alarming increase in TB, and it has become a leading cause of death from infectious disease. Most of this increase is associated with HIV-infected individuals, especially those who abuse intravenous drugs and reside in shelters for the homeless. The TB bacterium grows very slowly and drug therapy entails a 12-month course of antibiotics. Most individuals infected with HIV and TB test negative on TB screening tests because of their weakened immune system. (The TB test depends on detecting anti-TB antibodies in the patient.) Then when they become symptomatic for TB, many fail to complete the drug course and pass on their microbes to others.

Even more alarming is that deadly drug-resistant TB strains are emerging in patients who stop taking the drugs (about 20% of the total treated). The threat of TB epidemics is so real that health centers in some cities are detaining such patients in sanitariums against their will for as long as it takes to complete a cure.

Lung Cancer

Lung cancer accounts for fully one-third of cancer deaths in the United States. Its incidence, strongly associated with cigarette smoking (more than 90% of lung cancer patients were smokers), is increasing daily. In both sexes, lung cancer is the most prevalent type of malignancy. Its cure rate is notoriously low. Most victims die within one year of diagnosis; overall five-year survival of those with lung cancer is about 7%. Because lung cancer is aggressive and metastasizes rapidly and widely, most cases are not diagnosed until they are well advanced.

Ordinarily, nasal hairs, sticky mucus, and cilia do a fine job of protecting the lungs from chemical and biological irritants, but when one smokes, these devices are overwhelmed and eventually stop functioning. Continuous irritation prompts the production of more mucus, but smoking paralyzes the cilia that clear this mucus. Mucus pooling leads to frequent pulmonary infections, including pneumonia, and COPD. However, it is the irritant effects of the "cocktail" of free radicals and carcinogens in tobacco smoke that eventually translate into lung cancer.

The three most common types of lung cancer are (1) **squamous cell carcinoma** (20–40% of cases), which arises in the epithelium of the bronchi or their larger subdivisions and tends to form masses that cavitate (hollow out) and bleed, (2) **adenocarcinoma** (25–35%), which originates in peripheral lung areas as solitary nodules that develop from bronchial glands and alveolar cells, and (3) **small cell carcinoma** (20–25%), which contains lymphocyte-like cells that originate in the primary bronchi and grow aggressively in small grapelike clusters within the mediastinum. Subsequent metastasis from the mediastinum is especially rapid. Some small cell carcinomas cause problems beyond their effects on the lungs because they become ectopic sites of hormone production. For example, some secrete ACTH (leading to Cushing's disease) or calcitonin (which results in hypocalcemia).

Complete removal of the diseased lung has the greatest potential for prolonging life and providing a cure. However, this choice is open to few lung cancer patients because the cancer has often metastasized before it is discovered. In most cases, radiation therapy and chemotherapy are the only options, but only small cell carcinoma is responsive to chemotherapy.

However, this picture may change soon. Most lung cancers other than small cell carcinoma result from absence or mutation of the tumor suppressor gene *p53* or through action of the *K-ras* oncogene. By infusing tumor cells with retroviruses carrying working *p53* genes or *K-ras* gene inhibitors, researchers have achieved an 80% cure rate in mice and are optimistic that similar results might be obtained in humans.

Review Questions

Multiple Choice/Matching

(Some questions have more than one correct answer. Select the best answer or answers from the choices given.)

1. Cutting the phrenic nerves will result in (a) air entering the pleural cavity, (b) paralysis of the diaphragm, (c) stimulation of the diaphragmatic reflex, (d) paralysis of the epiglottis.

2. Following the removal of his larynx, an individual would (a) be unable to speak, (b) be unable to cough, (c) have difficulty swallowing, (d) be in respiratory difficulty or arrest.

3. Under ordinary circumstances, the inflation reflex is initiated by (a) the inspiratory center, (b) ventral respiratory group, (c) overinflation of the alveoli and bronchioles, (d) the pontine respiratory group.

4. The detergent-like substance that keeps the alveoli from collapsing between breaths because it reduces the surface tension of the water film in the alveoli is called (a) lecithin, (b) bile, (c) surfactant, (d) reluctant.

5. Which of the following determines the *direction* of gas movement? (a) solubility in water, (b) partial pressure gradient, (c) temperature, (d) molecular weight and size of the gas molecule.

6. When the inspiratory muscles contract, (a) the size of the thoracic cavity is increased in diameter, (b) the size of the thoracic cavity is increased in length, (c) the volume of the thoracic cavity is decreased, (d) the size of the thoracic cavity is increased in both length and diameter.

7. The nutrient blood supply of the lungs is provided by (a) the pulmonary arteries, (b) the aorta, (c) the pulmonary veins, (d) the bronchial arteries.

8. Oxygen and carbon dioxide are exchanged in the lungs and through all cell membranes by (a) active transport, (b) diffusion, (c) filtration, (d) osmosis.

9. Which of the following would not normally be treated by 100% oxygen therapy? (Choose all that apply.) (a) anoxia, (b) carbon monoxide poisoning, (c) respiratory crisis in an emphysema patient, (d) eupnea.

10. Most oxygen carried in the blood is (a) in solution in the plasma, (b) combined with plasma proteins, (c) chemically combined with the heme in red blood cells, (d) in solution in the red blood cells.

11. Which of the following has the greatest stimulating effect on the respiratory center in the brain? (a) oxygen, (b) carbon dioxide, (c) calcium, (d) willpower.

12. In mouth-to-mouth artificial respiration, the rescuer blows air from his or her own respiratory system into that of the victim. Which of the following statements are correct?

   (1) Expansion of the victim's lungs is brought about by blowing air in at higher than atmospheric pressure (positive-pressure breathing).

   (2) During inflation of the lungs, the intrapleural pressure increases.

   (3) This technique will not work if the victim has a hole in the chest wall, even if the lungs are intact.

   (4) Expiration during this procedure depends on the elasticity of the alveolar and thoracic walls.

(a) all of these, (b) 1, 2, 4, (c) 1, 2, 3, (d) 1, 4.

13. A baby holding its breath will (a) have brain cells damaged because of low blood oxygen levels, (b) automatically start to breathe again when the carbon dioxide levels in the blood reach a high enough value, (c) suffer heart damage because of increased pressure in the carotid sinus and aortic arch areas, (d) be called a "blue baby."

14. Under ordinary circumstances, which of the following blood components is of no physiological significance? (a) bicarbonate ions, (b) carbaminohemoglobin, (c) nitrogen, (d) chloride.

15. Damage to which of the following would result in cessation of breathing? (a) the pontine respiratory group, (b) the medulla, (c) the stretch receptors in the lungs, (d) the apneustic center.

16. The bulk of carbon dioxide is carried (a) chemically combined with the amino acids of hemoglobin as carbaminohemoglobin in the red blood cells, (b) as the ion HCO_3^- in the plasma after first entering the red blood cell, (c) as carbonic acid in the plasma, (d) chemically combined with the heme portion of Hb.

Short Answer Essay Questions

17. Trace the route of air from the external nares to an alveolus. Name subdivisions of organs where applicable, and differentiate between conducting and respiratory zone structures.

18. (a) Why is it important that the trachea is reinforced with cartilage rings? (b) Of what advantage is it that the rings are incomplete posteriorly?

19. Briefly explain the anatomical "reason" why most men have deeper voices than boys or women.

20. The lungs are mostly passageways and elastic tissue. (a) What is the role of the elastic tissue? (b) Of the passageways?

21. Describe the functional relationships between volume changes and gas flow into and out of the lungs.

22. What is it about the structure of the respiratory membrane that makes the alveoli ideal sites for gas exchange?

23. Discuss how airway resistance, lung compliance, and alveolar surface tension influence pulmonary ventilation.

24. (a) Differentiate clearly between minute respiratory volume and alveolar ventilation rate. (b) Which provides a more accurate measure of ventilatory efficiency, and why?

25. State Dalton's law of partial pressures and Henry's law.

26. (a) Define hyperventilation. (b) If you hyperventilate, do you retain or expel more carbon dioxide? (c) What effect does hyperventilation have on blood pH?

22

THE DIGESTIVE SYSTEM

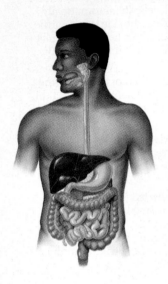

PART 1: OVERVIEW OF THE DIGESTIVE SYSTEM (pp. 753–759)

1. Describe the function of the digestive system, and differentiate between organs of the alimentary canal and accessory digestive organs.

2. List and define the major processes occurring during digestive system activity.

3. Describe the location and function of the peritoneum. Define retroperitoneal and name the retroperitoneal organs.

4. Describe the tissue composition and the general function of each of the four layers of the alimentary canal.

PART 2: FUNCTIONAL ANATOMY OF THE DIGESTIVE SYSTEM (pp. 759–796)

5. Describe the anatomy and basic function of each organ and accessory organ of the alimentary canal.

6. Explain the dental formula and differentiate clearly between deciduous and permanent teeth.

7. Describe the composition and functions of saliva, and explain how salivation is regulated.

8. Describe the mechanisms of chewing and swallowing.

9. Identify structural modifications of the wall of the stomach and small intestine that enhance the digestive process in these regions.

10. Describe the composition of gastric juice, name the cell types responsible for secreting its components, and indicate the importance of each component in stomach activity.

11. Explain regulation of gastric secretion and stomach motility.

12. Describe the function of local intestinal hormones.

13. State the roles of bile and of pancreatic juice in digestion.

14. Describe how entry of pancreatic juice and bile into the small intestine is regulated.

15. List the major functions of the large intestine, and describe the regulation of defecation.

PART 3: PHYSIOLOGY OF CHEMICAL DIGESTION AND ABSORPTION (pp. 796–802)

16. List the enzymes involved in chemical digestion; name the foodstuffs on which they act and the end products of protein, fat, carbohydrate, and nucleic acid digestion.

17. Describe the process of absorption of digested foodstuffs that occurs in the small intestine.

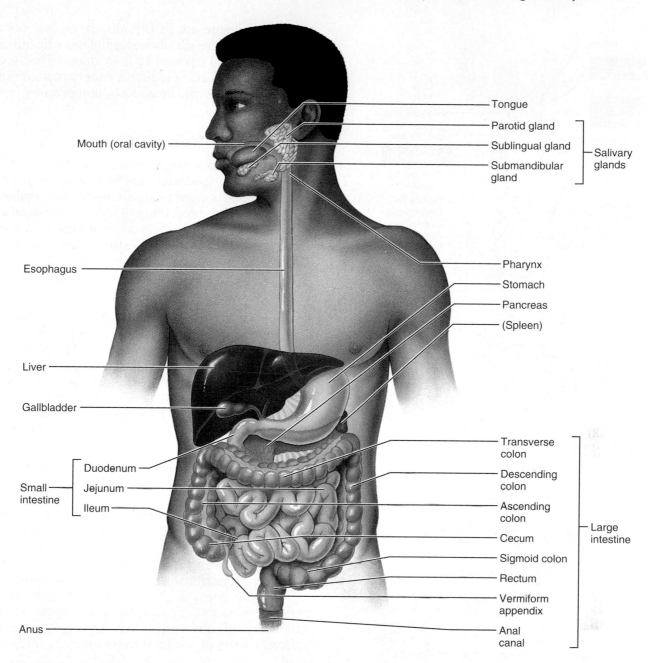

FIGURE 22.1 **Alimentary canal and related accessory digestive organs.** (See *A Brief Atlas of the Human Body*, Figure 56a.)

C hildren are fascinated by the workings of the digestive system. They relish crunching a potato chip, delight in making "mustaches" with milk, and giggle when their stomach "growls." As adults, we know that a healthy digestive system is essential to maintaining life, because it converts foods into the raw materials that build and fuel our body's cells. Specifically, the **digestive system** takes in food, breaks it down into nutrient molecules, absorbs these molecules into the bloodstream, and then rids the body of the indigestible remains.

PART 1: OVERVIEW OF THE DIGESTIVE SYSTEM

The organs of the digestive system (Figure 22.1) fall into two main groups: (1) those of the *alimentary canal* (al"ĭ-men'tar-e; *aliment* = nourish) and (2) *accessory digestive organs*.

The **alimentary canal,** also called the **gastrointestinal (GI) tract,** is the continuous, muscular digestive tube that winds through the body. It **digests** food—breaks it down into smaller fragments (*digest* = dissolved)—and **absorbs** the digested fragments

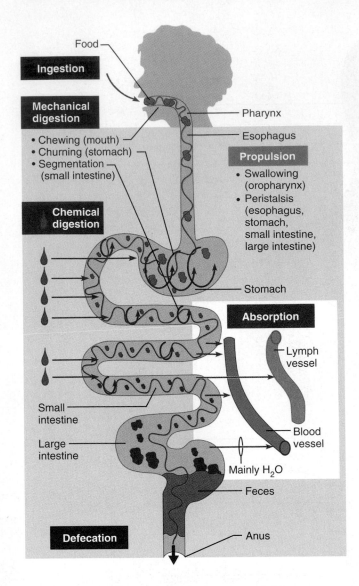

Food

Ingestion

Mechanical digestion

- Chewing (mouth)
- Churning (stomach)
- Segmentation (small intestine)

Chemical digestion

Defecation

Pharynx

Esophagus

Propulsion

- Swallowing (oropharynx)
- Peristalsis (esophagus, stomach, small intestine, large intestine)

Stomach

Absorption

Lymph vessel

Blood vessel

Mainly H_2O

Feces

Small intestine

Large intestine

Anus

FIGURE 22.2   Gastrointestinal tract activities.
Gastrointestinal tract activities include ingestion, mechanical digestion, chemical (enzymatic) digestion, propulsion, absorption, and defecation. Sites of chemical digestion are also sites that produce enzymes or that receive enzymes or other secretions made by accessory organs outside the alimentary canal. The mucosa of the GI tract secretes mucus, which protects and lubricates.

through its lining into the blood. The organs of the alimentary canal are the *mouth, pharynx, esophagus, stomach, small intestine,* and *large intestine*. The large intestine leads to the terminal opening, or *anus*. In a cadaver, the alimentary canal is approximately 9 m (about 30 feet) long, but in a living person, it is considerably shorter because of its muscle tone. Food material in this tube is technically outside the body because the canal is open to the external environment at both ends.

The **accessory digestive organs** are the *teeth, tongue, gallbladder,* and a number of large digestive glands—the *salivary glands, liver,* and *pancreas*. The teeth and tongue are in the mouth, or oral cavity, while the digestive glands and gallbladder lie outside the GI tract and connect to it by ducts. The accessory digestive glands produce a variety of secretions that contribute to the breakdown of foodstuffs.

Digestive Processes

The digestive tract can be viewed as a "disassembly line" in which food becomes less complex at each step of processing and its nutrients become available to the body. The processing of food by the digestive system involves six essential activities: ingestion, propulsion, mechanical digestion, chemical digestion, absorption, and defecation (Figure 22.2).

1. **Ingestion** is simply taking food into the digestive tract, usually via the mouth.

2. **Propulsion,** which moves food through the alimentary canal, includes *swallowing*, which is initiated voluntarily, and *peristalsis* (per"ĭ-stal'sis), an involuntary process. **Peristalsis** (*peri* = around; *stalsis* = constriction), the major means of propulsion, involves alternate waves of contraction and relaxation of muscles in the organ walls (Figure 22.3a). Its main effect is to squeeze food along the tract, but some mixing occurs as well. In fact, peristaltic waves are so powerful that, once swallowed, food and fluids will reach your stomach even if you stand on your head.

3. **Mechanical digestion** physically prepares food for chemical digestion by enzymes. Mechanical processes include chewing, mixing of food with saliva by the tongue, churning food in the stomach, and **segmentation,** or rhythmic local constrictions of the intestine (Figure 22.3b). Segmentation mixes food with digestive juices and increases the efficiency of absorption by repeatedly moving different parts of the food mass over the intestinal wall.

4. **Chemical digestion** is a series of catabolic steps in which complex food molecules are broken down to their chemical building blocks by enzymes secreted into the lumen of the alimentary canal. Chemical digestion of foodstuffs begins in the mouth and is essentially complete in the small intestine.

5. **Absorption** is the passage of digested end products (plus vitamins, minerals, and water) from the lumen of the GI tract through the mucosal cells by active or passive transport into the blood or lymph. The small intestine is the major absorptive site.

6. **Defecation** eliminates indigestible substances from the body via the anus in the form of feces.

Some of these processes are the job of a single organ. For example, only the mouth ingests and only

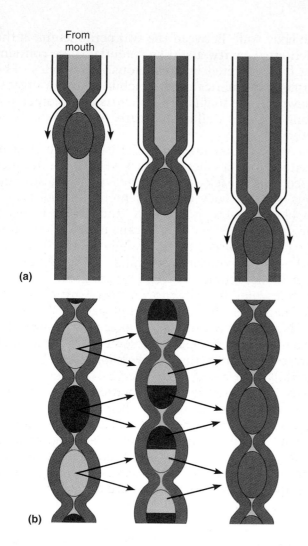

(a)

(b)

creates an optimal environment for its functioning in the lumen (cavity) of the GI tract, an area that is actually *outside* the body, and essentially all digestive tract regulatory mechanisms act to control luminal conditions so that digestion and absorption can occur there as effectively as possible.

Two facts apply to these regulatory mechanisms:

1. Digestive activity is provoked by a range of mechanical and chemical stimuli. Sensors (mechanoreceptors and chemoreceptors) involved in controls of GI tract activity are located in the walls of the tract organs (Figure 22.4). These sensors respond to several stimuli; the most important are stretching of the organ by food in the lumen, osmolarity (solute concentration) and pH of the contents, and the presence of substrates and end products of digestion. When stimulated, these receptors initiate reflexes that (1) activate or inhibit glands that secrete digestive juices into the lumen or hormones into the blood or (2) mix lumen contents and move them along the tract by stimulating smooth muscle of the GI tract walls.

2. Controls of digestive activity are both extrinsic and intrinsic. Many of the controlling systems of the digestive tract are intrinsic—a product of "in-house" nerve plexuses or local hormone-producing cells. The wall of the alimentary canal contains nerve plexuses which extend the entire length of the GI tract and influence each other both in the same and in different digestive organs. As a

FIGURE 22.3 Peristalsis and segmentation. (a) In peristalsis, adjacent segments of the intestine (or other alimentary tract organs) alternately contract and relax, which moves food along the tract distally. **(b)** In segmentation, nonadjacent segments of the intestine alternately contract and relax, moving the food now forward and then backward. This results in food mixing rather than food propulsion.

the large intestine defecates. But most digestive system activities require the cooperation of several organs and occur bit by bit as food moves along the tract. Later, when we discuss the function of each GI tract organ, we will consider which of these specific processes it performs and the neural or hormonal factors that regulate these processes.

Basic Functional Concepts

A theme stressed in this book is the body's efforts to maintain the constancy of its internal environment. Most organ systems respond to changes in that environment either by attempting to restore some plasma variable to its former levels or by changing their own function. The digestive system, however,

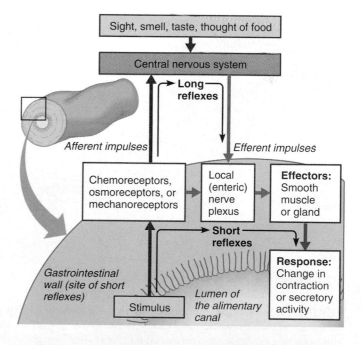

FIGURE 22.4 Neural reflex pathways initiated by stimuli inside or outside the gastrointestinal tract.

result, two kinds of reflex activity occur, short and long. *Short reflexes* are mediated entirely by the local *(enteric)* plexuses (the so-called gut brain) in response to GI tract stimuli. *Long reflexes* are initiated by stimuli arising inside or outside the GI tract and involve CNS centers and extrinsic autonomic nerves (Figure 22.4).

The stomach and small intestine also contain hormone-producing cells that, when appropriately stimulated, release their products to the extracellular space. These hormones are distributed via blood and interstitial fluid to their target cells in the same or different digestive tract organs, which they prod to secrete or contract.

Digestive System Organs: Relationship and Structural Plan

Relationship of the Digestive Organs to the Peritoneum

Most digestive system organs reside in the abdominopelvic cavity. Recall from Chapter 1 that all ventral body cavities contain slippery *serous membranes*. The **peritoneum** of the abdominopelvic cavity is the most extensive of these membranes (Figure 22.5a). The **visceral peritoneum** covers the external surfaces of most digestive organs and is continuous with the **parietal peritoneum** that lines the body wall. Between the two peritoneums is the **peritoneal cavity,** a slitlike potential space containing fluid secreted by the serous membranes. The serous fluid lubricates the mobile digestive organs, allowing them to glide easily across one another and along the body wall as they carry out their digestive activities.

A **mesentery** (mes'en-ter"e) is a double layer of peritoneum—a sheet of two serous membranes fused back to back—that extends to the digestive organs from the body wall. Mesenteries provide routes for blood vessels, lymphatics, and nerves to reach the digestive viscera; hold organs in place; and store fat. In most places the mesentery is *dorsal* and attaches to the posterior abdominal wall, but there are *ventral* mesenteries as well (Figure 22.5a). Some of the mesenteries, or peritoneal folds, are given specific names (such as the *omenta*), as described later.

Not all alimentary canal organs are suspended by a mesentery. For example, some parts of the small intestine adhere to the dorsal abdominal wall (Figure 22.5b). In so doing, they lose their mesentery and come to lie posterior to the peritoneum. These organs, which include most of the pancreas and parts of the large intestine, are called **retroperitoneal organs** (*retro* = behind). By contrast, digestive organs (like the stomach) that keep their mesentery and remain in the peritoneal cavity are called **intraperitoneal** or **peritoneal organs.**

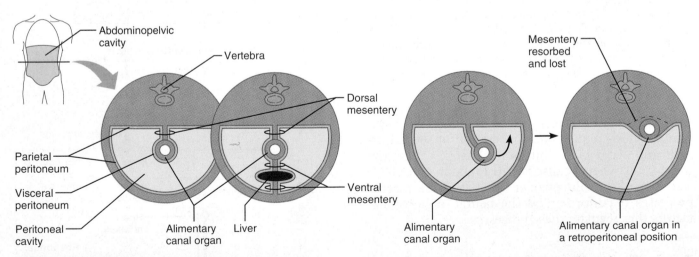

(a) **Transverse section of abdominal cavity**

Abdominopelvic cavity
Vertebra
Dorsal mesentery
Parietal peritoneum
Visceral peritoneum
Peritoneal cavity
Alimentary canal organ
Liver
Ventral mesentery

(b) **Some organs become retroperitoneal**

Mesentery resorbed and lost
Alimentary canal organ
Alimentary canal organ in a retroperitoneal position

FIGURE 22.5 **The peritoneum and the peritoneal cavity. (a)** Simplified cross sections through the abdominal cavity showing the relative locations of the visceral and parietal peritoneums, and dorsal (left) and ventral (right) mesenteries. Note that the peritoneal cavity is much smaller than depicted here, as it is nearly filled by the organs within it. **(b)** Some alimentary canal organs lose their mesentery during development and become retroperitoneal.

Peritonitis, or inflammation of the peritoneum, can arise from a piercing abdominal wound or from a perforating ulcer that leaks stomach juices into the peritoneal cavity, but most commonly it results from a burst appendix (that sprays bacteria-containing feces all over the peritoneum). In peritonitis, the peritoneal coverings tend to stick together around the infection site. This localizes the infection, providing time for macrophages to attack to prevent the inflammation from spreading. If peritonitis becomes widespread within the peritoneal cavity, it is dangerous and often lethal. Treatment includes removing as much infectious debris as possible from the peritoneal cavity and administering megadoses of antibiotics. ●

Blood Supply: The Splanchnic Circulation

The **splanchnic circulation** includes those arteries that branch off the abdominal aorta to serve the digestive organs and the *hepatic portal circulation*. The arterial supply—the hepatic, splenic, and left gastric branches of the celiac trunk that serve the spleen, liver, and stomach, and the mesenteric arteries (superior and inferior) that serve the small and large intestines (see pp. 644 and 647)—normally receives one-quarter of the cardiac output. This percentage (blood volume) increases after a meal has been eaten. The hepatic portal circulation (described on pp. 656–657) collects nutrient-rich venous blood draining from the digestive viscera and delivers it to the liver. The liver collects the absorbed nutrients for metabolic processing or for storage before releasing them back to the bloodstream for general cellular use.

Histology of the Alimentary Canal

Each digestive organ has only a share of the work of digestion. Consequently, it helps to consider structural characteristics that promote similar functions in all parts of the alimentary canal before we consider the functional anatomy of the digestive system.

From the esophagus to the anal canal, the walls of the alimentary canal have the same four basic layers, or *tunics* (Figure 22.6)—*mucosa, submucosa, muscularis externa,* and *serosa*—each containing a predominant tissue type that plays a specific role in food breakdown.

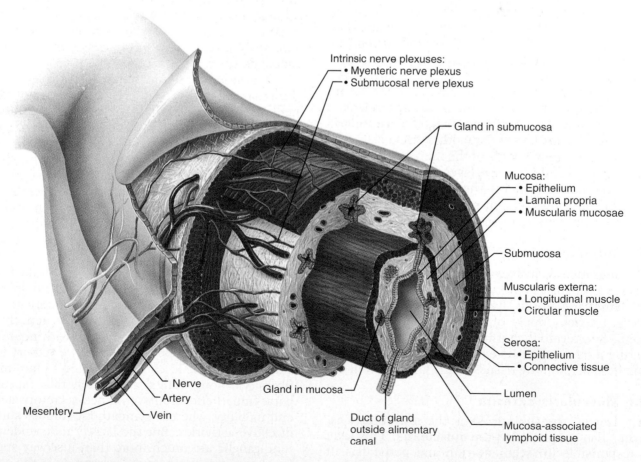

Intrinsic nerve plexuses:
• Myenteric nerve plexus
• Submucosal nerve plexus

Gland in submucosa

Mucosa:
• Epithelium
• Lamina propria
• Muscularis mucosae

Submucosa

Muscularis externa:
• Longitudinal muscle
• Circular muscle

Serosa:
• Epithelium
• Connective tissue

Lumen

Mucosa-associated lymphoid tissue

Duct of gland outside alimentary canal

Gland in mucosa

Mesentery

Nerve

Artery

Vein

FIGURE 22.6 Basic structure of the alimentary canal. Its four basic layers are the mucosa, submucosa, muscularis externa, and serosa.

The Mucosa

The **mucosa**, or **mucous membrane**—the innermost layer—is a moist epithelial membrane that lines the alimentary canal lumen from mouth to anus. Its major functions are (1) *secretion* of mucus, digestive enzymes, and hormones, (2) *absorption* of the end products of digestion into the blood, and (3) *protection* against infectious disease. The mucosa in a particular region of the GI tract may express one or all three of these capabilities.

More complex than most other mucosae in the body, the typical digestive mucosa consists of three sublayers: (1) a lining epithelium, (2) a lamina propria, and (3) a muscularis mucosae. Typically, the **epithelium** of the mucosa is a *simple columnar epithelium* rich in mucus-secreting *goblet cells*. The slippery mucus it produces protects certain digestive organs from being digested themselves by enzymes working within their cavities and eases food passage along the tract. In the stomach and small intestine, the mucosa also contains both enzyme-secreting and hormone-secreting cells. Thus, in such sites, the mucosa is a diffuse kind of endocrine organ as well as part of the digestive organ.

The **lamina propria** (*proprius* = one's own), which underlies the epithelium, is loose areolar connective tissue. Its capillaries nourish the epithelium and absorb digested nutrients. Its isolated lymph nodules, part of **MALT**, the mucosa-associated lymphatic tissue described on p. 670, help defend us against bacteria and other pathogens, which have rather free access to our digestive tract. Particularly large collections of lymphoid follicles occur within the pharynx (as the tonsils) and in the appendix.

External to the lamina propria is the **muscularis mucosae**, a scant layer of smooth muscle cells that produces local movements of the mucosa. For example, twitching of this muscle layer dislodges food particles that have adhered to the mucosa. In the small intestine, it throws the mucosa into a series of small folds that immensely increase its surface area.

The Submucosa

The **submucosa**, just external to the mucosa, is a moderately dense connective tissue containing blood and lymphatic vessels, lymphoid follicles, and nerve fibers. Its rich supply of elastic fibers enables the stomach to regain its normal shape after temporarily storing a large meal. Its extensive vascular network supplies surrounding tissues of the GI tract wall.

The Muscularis Externa

Just deep to the submucosa is the **muscularis externa**, also simply called the **muscularis**. This layer is responsible for segmentation and peristalsis. It typically has an inner *circular layer* and an outer *longitudinal layer* of smooth muscle cells (Figure 22.6).

In several places along the tract, the circular layer thickens, forming *sphincters* that act as valves to prevent backflow and control food passage from one organ to the next.

The Serosa

The **serosa**, the protective outermost layer of the intraperitoneal organs, is the *visceral peritoneum*. It is formed of areolar connective tissue covered with *mesothelium*, a single layer of squamous epithelial cells.

In the esophagus, which is located in the thoracic instead of the abdominopelvic cavity, the serosa is replaced by an **adventitia** (ad"ven-tish'e-ah). The adventitia is ordinary fibrous connective tissue that binds the esophagus to surrounding structures. Retroperitoneal organs have *both* a serosa (on the side facing the peritoneal cavity) and an adventitia (on the side abutting the dorsal body wall).

Enteric Nervous System of the Alimentary Canal

The alimentary canal has its own in-house nerve supply, staffed by the so-called **enteric neurons** (*enter* = gut), which communicate widely with one another to regulate digestive system activity. These enteric neurons constitute the bulk of the two major *intrinsic nerve plexuses* found in the walls of the alimentary canal: the submucosal and myenteric nerve plexuses (Figure 22.6).

The **submucosal nerve plexus** occupies the submucosa and chiefly regulates the activity of glands and smooth muscle in the mucosa. The large **myenteric nerve plexus** (mi-en'ter-ik; "intestinal muscle") lies between the circular and longitudinal layers of smooth muscle of the muscularis externa. Enteric neurons of this plexus provide the major nerve supply to the GI tract wall and control GI tract mobility. Control of the patterns of segmentation and peristalsis is largely automatic, involving local reflex arcs between enteric neurons in the same or different plexuses or (even) organs.

The enteric nervous system is also linked to the central nervous system by afferent visceral fibers and by sympathetic and parasympathetic branches (motor fibers) of the autonomic nervous system that enter the intestinal wall and synapse with neurons in the intrinsic plexuses. Thus, digestive activity is also subject to extrinsic controls exerted by autonomic fibers via long reflex arcs. Generally speaking, parasympathetic inputs enhance secretory activity and mobility, whereas sympathetic impulses inhibit digestive activities. But the largely independent enteric ganglia are much more than just way stations for the autonomic nervous system as is the case in other organ systems. Indeed, the enteric nervous system contains some 100 million neurons, as many as the entire spinal cord.

PART 2: FUNCTIONAL ANATOMY OF THE DIGESTIVE SYSTEM

Now that we have summarized some points that unify the digestive system organs, we are ready to consider the special structural and functional capabilities of each organ of this system. Most of the digestive organs are shown in their normal body positions in Figure 22.1, so you may find it helpful to refer back to that illustration from time to time as you read the following sections.

The Mouth, Pharynx, and Esophagus

The mouth is the only part of the alimentary canal involved in ingestion. However, most digestive functions associated with the mouth reflect the activity of the related accessory organs, such as the teeth, salivary glands, and tongue, because in the mouth food is chewed and mixed with saliva containing enzymes that begin the process of chemical digestion. The mouth also begins the propulsive process of swallowing, which carries food through the pharynx and esophagus to the stomach.

The Mouth and Associated Organs

The Mouth

The **mouth,** a mucosa-lined cavity, is also called the **oral cavity,** or **buccal cavity** (buk'al). Its boundaries are the lips anteriorly, cheeks laterally, palate superiorly, and tongue inferiorly (Figure 22.7). Its anterior opening is the **oral orifice.** Posteriorly, the oral cavity is continuous with the *oropharynx*. The walls of the mouth are lined with stratified squamous epithelium which can withstand considerable friction. The epithelium on the gums, hard palate, and dorsum of the tongue is slightly keratinized for extra protection against abrasion during eating. Like all moist surface linings, the oral mucosa responds to

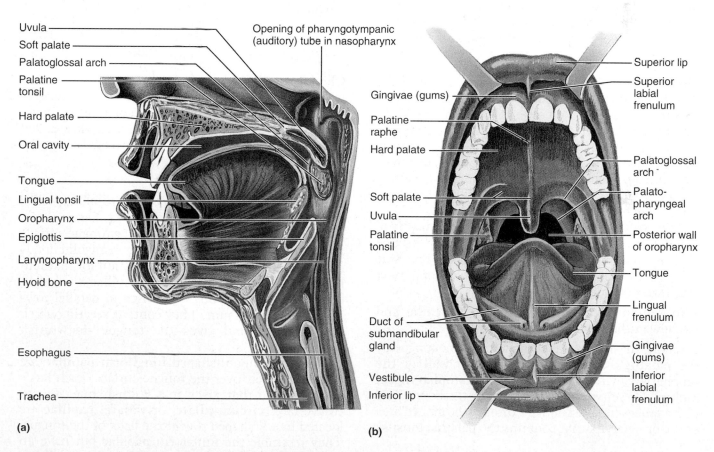

(a)

(b)

FIGURE 22.7 **Anatomy of the oral cavity (mouth). (a)** Sagittal section of the oral cavity and pharynx. **(b)** Anterior view.

injury by producing antimicrobial peptides called *defensins*, which helps to explain how the mouth, a site teeming with disease-causing microbes, remains so remarkably healthy.

The Lips and Cheeks   The **lips (labia)** and the **cheeks** have a core of skeletal muscle covered externally by skin. The *orbicularis oris muscle* forms the fleshy lips; the cheeks are formed largely by the *buccinators*. The lips and cheeks help keep food between the teeth when we chew and play a small role in speech. The recess bounded externally by the lips and cheeks and internally by the gums and teeth is called the **vestibule** ("porch"). The area that lies within the teeth and gums is the **oral cavity proper.**

The lips are much larger than most people think; anatomically they extend from the inferior margin of the nose to the superior boundary of the chin. The reddened area where one applies lipstick or lands a kiss is called the **red margin.** This transitional zone, where keratinized skin meets the oral mucosa, is poorly keratinized and translucent, allowing the red color of blood in the underlying capillaries to show through. The lack of sweat or sebaceous glands in this region means it must be moistened with saliva periodically to prevent it from becoming dry and cracked (chapped lips). The **labial frenulum** (fren'u-lum) is a median fold that joins the internal aspect of each lip to the gum (Figure 22.7b).

The Palate   The **palate,** forming the roof of the mouth, has two distinct parts: the hard palate anteriorly and the soft palate posteriorly (see Figure 22.7). The **hard palate** is underlain by the palatine bones and the palatine processes of the maxillae, and it forms a rigid surface against which the tongue forces food during chewing. The mucosa on either side of its *raphe* (ra'fe), a midline ridge, is slightly corrugated, which helps to create friction.

The **soft palate** is a mobile fold formed mostly of skeletal muscle. Projecting downward from its free edge is the fingerlike **uvula** (u'vu-lah). The soft palate rises reflexively to close off the nasopharynx when we swallow.

▪ To demonstrate this action, try to breathe and swallow at the same time.

Laterally, the soft palate is anchored to the tongue by the **palatoglossal arches** and to the wall of the oropharynx by the more posterior **palatopharyngeal arches.** These two paired folds form the boundaries of the **fauces** (faw'sēz; *fauc* = throat), the arched area of the oropharynx that contains the palatine tonsils.

The Tongue

The **tongue** occupies the floor of the mouth and fills most of the oral cavity when the mouth is closed (see Figure 22.7). The tongue is composed of interlacing bundles of skeletal muscle fibers, and during chewing, it grips the food and constantly repositions it between the teeth. The tongue also mixes food with saliva and forms it into a compact mass called a **bolus** (bo'lus; "a lump"), and then initiates swallowing by pushing the bolus posteriorly into the pharynx. The versatile tongue also helps us form consonants (k, d, t, and so on) when we speak.

The tongue has both intrinsic and extrinsic skeletal muscle fibers. The **intrinsic muscles** are confined in the tongue and are not attached to bone. Their muscle fibers, which run in several different planes, allow the tongue to change its shape (but not its position), becoming thicker, thinner, longer, or shorter as needed for speech and swallowing. The **extrinsic muscles** extend to the tongue from their points of origin on bones of the skull or the soft palate, as described in Chapter 10 (see Table 10.2 and Figure 10.7). The extrinsic muscles alter the tongue's position; they protrude it, retract it, and move it from side to side. The tongue has a median septum of connective tissue, and each half contains identical muscle groups. A fold of mucosa, called the **lingual frenulum,** secures the tongue to the floor of the mouth and limits posterior movements of the tongue.

H   *HOMEOSTATIC IMBALANCE*

Children born with an extremely short lingual frenulum are often referred to as "tongue-tied" because of speech distortions that result when tongue movement is restricted. This congenital condition, called *ankyloglossia* ("fused tongue"), is corrected surgically by snipping the frenulum. ●

The superior tongue surface bears papillae, peglike projections of the underlying mucosa (Figure 22.8). The conical **filiform papillae** give the tongue surface a roughness that aids in licking semisolid foods (such as ice cream) and provide friction for manipulating foods in the mouth. These papillae, the smallest and most numerous type, align in parallel rows on the tongue dorsum. They contain keratin, which stiffens them and gives the tongue its whitish appearance.

The mushroom-shaped **fungiform papillae** are scattered widely over the tongue surface. Each has a vascular core that gives it a reddish hue. Ten to twelve large **circumvallate,** or **vallate, papillae** are located in a V-shaped row at the back of the tongue. They resemble the fungiform papillae but have an additional surrounding furrow. Both the fungiform and circumvallate papillae house taste buds.

Immediately posterior to the circumvallate papillae is the **sulcus terminalis,** a groove that

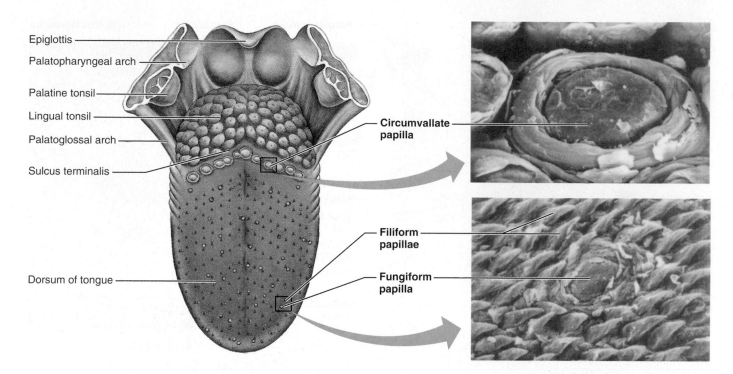

Epiglottis
Palatopharyngeal arch
Palatine tonsil
Lingual tonsil
Palatoglossal arch
Sulcus terminalis
Dorsum of tongue

Circumvallate papilla

Filiform papillae

Fungiform papilla

FIGURE 22.8   Dorsal surface of the tongue. Also shown are the tonsils, the locations and detailed structures of the circumvallate, fungiform, and filiform papillae. (Top right, 300×; bottom right, 100×.) (See *A Brief Atlas of the Human Body*, Figure 36.)

distinguishes the anterior two-thirds of the tongue that lies in the oral cavity from its posterior third residing in the oropharynx. The mucosa covering the root of the tongue lacks papillae, but it is still bumpy because of the nodular *lingual tonsil*, which lies just deep to its mucosa (see Figure 22.8).

The Salivary Glands

A number of glands associated with the oral cavity secrete **saliva.** Saliva (1) cleanses the mouth, (2) dissolves food chemicals so that they can be tasted, (3) moistens food and aids in compacting it into a bolus, and (4) contains enzymes that begin the chemical breakdown of starchy foods.

Most saliva is produced by **extrinsic salivary glands** that lie outside the oral cavity and empty their secretions into it. Their output is augmented slightly by small **intrinsic salivary glands,** also called **buccal glands,** scattered throughout the oral cavity mucosa.

The extrinsic salivary glands are paired compound tubuloalveolar glands that develop from the oral mucosa and remain connected to it by ducts (Figure 22.9). The large **parotid gland** (pah-rot'id; *par* = near, *otid* = the ear) lies anterior to the ear between the masseter muscle and the skin. The prominent **parotid duct** parallels the zygomatic arch, pierces the buccinator muscle, and opens into the vestibule next to the second upper molar. Branches of the facial nerve run through the parotid gland on their way to the muscles of facial expression. For this reason, surgery on this gland can result in facial paralysis.

HOMEOSTATIC IMBALANCE

Mumps, a common children's disease, is an inflammation of the parotid glands caused by the mumps virus *(myxovirus),* which spreads from person to person in saliva. If you check the location of the parotid glands in Figure 22.9a, you can understand why people with mumps complain that it hurts to open their mouth or chew. Besides that discomfort, other signs and symptoms of mumps include moderate fever and pain when swallowing acid foods (sour pickles, grapefruit juice, etc.). Mumps viral infections in adult males carry a 25% risk that the testes may become infected as well, leading to sterility. ●

About the size of a walnut, the **submandibular gland** lies along the medial aspect of the mandibular body. Its duct runs beneath the mucosa of the oral cavity floor and opens at the base of the lingual frenulum (see Figure 22.7). The small **sublingual gland** lies anterior to the submandibular gland under

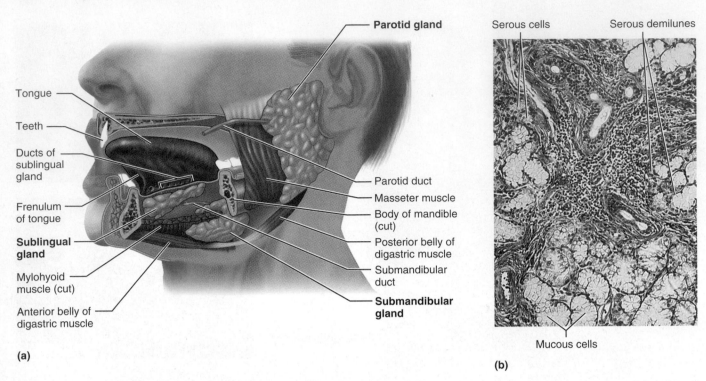

FIGURE 22.9 **The salivary glands. (a)** The parotid, submandibular, and sublingual salivary glands associated with the left aspect of the oral cavity. **(b)** Photomicrograph of the sublingual salivary gland (40×), which is a mixed salivary gland. Mucus-producing cells stain light blue and serous-secreting units stain purple. The serous cells sometimes form demilunes (caps) around the bases of the mucous cells. Copyright © Science Photo Library/Photo Researchers, Inc.

the tongue and opens via 10–12 ducts into the floor of the mouth.

To a greater or lesser degree, the salivary glands are composed of two types of secretory cells: mucous and serous. **Serous cells** produce a watery secretion containing enzymes, ions, and a tiny bit of mucin, whereas the **mucous cells** produce **mucus,** a stringy, viscous solution. The parotid glands contain only serous cells. The submandibular and buccal glands have approximately equal numbers of serous and mucous cells. The sublingual glands contain mostly mucous cells.

Composition of Saliva Saliva is largely water—97 to 99.5%—and therefore is hypo-osmotic. Its osmolarity depends on the precise glands that are active and the nature of the stimulus for salivation. As a rule, saliva is slightly acidic (pH 6.75 to 7.00), but its pH may vary. Its solutes include electrolytes (Na^+, K^+, Cl^-, PO_4^-, and HCO_3^-); the digestive enzyme salivary amylase; the proteins mucin (mu′sin), lysozyme, and IgA; and metabolic wastes (urea and uric acid). When dissolved in water, the glycoprotein *mucin* forms thick mucus that lubricates the oral cavity and hydrates foodstuffs. Protection against microorganisms is provided by (1) *IgA antibodies;* (2) *lysozyme,* a bacteriostatic enzyme

that inhibits bacterial growth in the mouth and may help to prevent tooth decay; (3) a cyanide compound; and (4) *defensins* (see p. 564.) Besides acting as a local antibiotic, defensins function as cytokines to call defensive cells (lymphocytes, neutrophils, etc.) into the mouth for battle.

In addition to these four protectors, the friendly bacteria that live on the back of the tongue convert food-derived nitrates in saliva into nitrites which, in turn, are converted into *nitric oxide* in an acid environment. This transformation occurs around the gums, where acid-producing bacteria tend to cluster, and in the hydrochloric acid–rich secretions of the stomach. The highly toxic nitric oxide is believed to act as a bactericidal agent in these locations.

Control of Salivation The intrinsic salivary glands secrete saliva continuously in amounts just sufficient to keep the mouth moist. But when food enters the mouth, the extrinsic glands are activated and copious amounts of saliva pour out. The average output of saliva is 1000–1500 ml per day.

Salivation is controlled primarily by the parasympathetic division of the autonomic nervous system. When we ingest food, chemoreceptors and pressoreceptors in the mouth send signals to the **salivatory nuclei** in the brain stem (pons and medulla). As a result, parasympathetic nervous system activity increases and impulses sent via motor fibers in the *facial (VII)* and *glossopharyngeal (IX) nerves* trigger a dramatically increased output of watery (serous), enzyme-rich saliva. The chemoreceptors are activated most strongly by acidic substances such as vinegar and citrus juice. The pressoreceptors are activated by virtually any mechanical stimulus in the mouth—even rubber bands.

Sometimes just the sight or smell of food is enough to get the juices flowing. The mere thought of hot fudge sauce on peppermint stick ice cream will make many a mouth water! Irritation of the lower regions of the GI tract by bacterial toxins, spicy foods, or hyperacidity—particularly when accompanied by a feeling of nausea—also increases salivation. This response may help wash away or neutralize the irritants.

In contrast to parasympathetic controls, the sympathetic division causes release of a thick mucin-rich saliva. Extremely strong activation of the sympathetic division constricts blood vessels serving the salivary glands and almost completely inhibits saliva release, causing a dry mouth. Dehydration also inhibits salivation because low blood volume results in reduced filtration pressure at capillary beds.

HOMEOSTATIC IMBALANCE

Any disease process that inhibits saliva secretion causes difficulty in talking, swallowing, and eating. Because decomposing food particles are allowed to accumulate and bacteria flourish, *halitosis* (hal"i-to'sis; "bad breath") can result.

The Teeth

The **teeth** lie in sockets (alveoli) in the gum-covered margins of the mandible and maxilla. The role of the teeth in food processing needs little introduction. We *masticate,* or chew, by opening and closing our jaws and moving them from side to side while continually using our tongue to move the food between our teeth. In the process, the teeth tear and grind the food, breaking it down into smaller fragments.

Dentition and the Dental Formula Ordinarily by age 21, two sets of teeth, the **primary** and **permanent dentitions,** have formed (Figure 22.10). The primary dentition consists of the **deciduous teeth** (de-sid'u-us; *decid* = falling off), also called **milk** or **baby teeth.** The first teeth to appear, at about age six months, are the lower central incisors. Additional pairs of teeth erupt at one- to two-month intervals until about 24 months, when all 20 milk teeth have emerged.

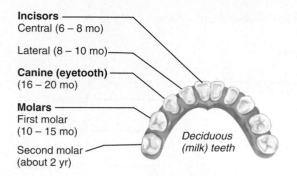

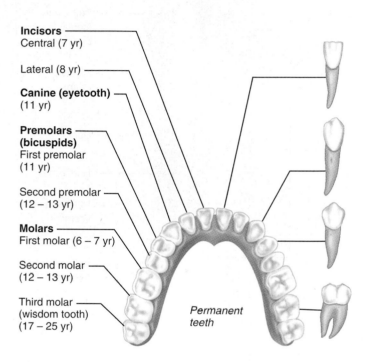

FIGURE 22.10 Human deciduous and permanent teeth of the lower jaw. Approximate time of tooth eruption is shown in parentheses. The shapes of individual teeth are shown on the right.

As the deep-lying **permanent teeth** enlarge and develop, the roots of the milk teeth are resorbed from below, causing them to loosen and fall out between the ages of 6 and 12 years. Generally, all the teeth of the permanent dentition but the third molars have erupted by the end of adolescence. The third molars, also called the *wisdom teeth,* emerge between the ages of 17 and 25 years. There are usually 32 permanent teeth in a full set, but sometimes the wisdom teeth never erupt or are completely absent.

HOMEOSTATIC IMBALANCE

When a tooth remains embedded in the jawbone, it is said to be *impacted.* Impacted teeth can cause a good deal of pressure and pain and must be removed surgically. Wisdom teeth are most commonly impacted.

? *Which material forms the bulk of the tooth?*

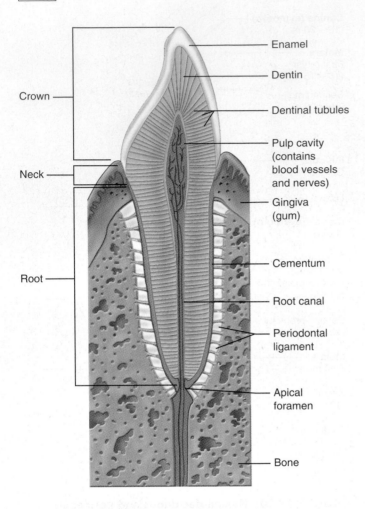

Crown

Neck

Root

— Enamel

— Dentin

— Dentinal tubules

— Pulp cavity (contains blood vessels and nerves)

— Gingiva (gum)

— Cementum

— Root canal

— Periodontal ligament

— Apical foramen

— Bone

FIGURE 22.11   Longitudinal section of a canine tooth within its bony alveolus.

Teeth are classified according to their shape and function as incisors, canines, premolars, and molars (Figure 22.10). The chisel-shaped **incisors** are adapted for cutting or nipping off pieces of food. The conical or fanglike **canines** (cuspids or eyeteeth) tear and pierce. The **premolars** (bicuspids) and **molars** have broad crowns with rounded cusps (tips) and are best suited for grinding or crushing. The molars (literally, "millstones"), with four or five cusps, are the best grinders. During chewing, the upper and lower molars repeatedly lock together, an action that generates tremendous crushing forces.

The **dental formula** is a shorthand way of indicating the numbers and relative positions of the different types of teeth in the mouth. This formula is written as a ratio, uppers over lowers, for *one-half* of the mouth. Since the other side is a mirror image,

the total dentition is obtained by multiplying the dental formula by 2. The primary dentition consists of two incisors (I), one canine (C), and two molars (M) on each side of each jaw, and its dental formula is written as

$$\frac{2I, \ 1C, \ 2M \ (\text{upper jaw})}{2I, \ 1C, \ 2M \ (\text{lower jaw})} \times 2 \ (20 \ \text{teeth})$$

Similarly, the permanent dentition [two incisors, one canine, two premolars (PM), and three molars] is

$$\frac{2I, \ 1C, \ 2PM, \ 3M}{2I, \ 1C, \ 2PM, \ 3M} \times 2 \ (32 \ \text{teeth})$$

Tooth Structure   Each tooth has two major regions: the crown and the root (Figure 22.11). The enamel-covered **crown** is the exposed part of the tooth above the **gingiva** (jin′jĭ-vah), or **gum,** which surrounds the tooth like a tight collar. **Enamel,** an acellular, brittle material that directly bears the force of chewing, is the hardest substance in the body. It is heavily mineralized with calcium salts, and its densely packed hydroxyapatite (mineral) crystals are oriented in force-resisting columns perpendicular to the tooth's surface. The cells that produce enamel degenerate when the tooth erupts; consequently, any decayed or cracked areas of the enamel will not heal and must be artificially filled.

The portion of the tooth embedded in the jawbone is the **root.** Canine teeth, incisors, and premolars have one root, although the first upper premolars commonly have two. As for molars, the first two upper molars have three roots, while the corresponding lower molars have two. The root pattern of the third molar varies, but a fused single root is most common.

The crown and root are connected by a constricted tooth region called the **neck.** The outer surface of the root is covered by **cementum,** a calcified connective tissue, which attaches the tooth to the thin **periodontal ligament** (per″e-o-don′tal; "around the tooth"). This ligament anchors the tooth in the bony alveolus of the jaw, forming a fibrous joint called a *gomphosis*. Where the gingiva borders on a tooth, it dips downward to form a shallow groove called the *gingival sulcus*. In youth, the gingiva adheres tenaciously to the enamel covering the crown. But as the gums begin to recede with age, the gingiva adheres to the more sensitive cementum covering the superior region of the root. As a result, the teeth *appear* to get longer in old age—hence the expression "long in the tooth" sometimes applied to elderly people.

Dentin, a bonelike material, underlies the enamel cap and forms the bulk of a tooth. It surrounds a central **pulp cavity** containing a number of soft tissue structures (connective tissue, blood vessels, and nerve fibers) collectively called **pulp.** Pulp supplies nutrients to the tooth tissues and provides for tooth sensation. Where the pulp cavity extends into the root, it becomes the **root canal.** At the proximal end of each root canal is an **apical foramen** that provides a route for blood vessels, nerves, and other structures to enter the pulp cavity of the tooth. The teeth are served by the superior and inferior alveolar nerves, branches of the trigeminal nerve (see Table 13.2, p. 477) and by the superior and inferior alveolar arteries, branches of the maxillary artery (see Figure 18.20, p. 641).

Dentin contains unique radial striations called *dentinal tubules* (Figure 22.11). Each tubule contains an elongated process of an **odontoblast** (o-don'to-blast; "tooth former"), the cell type that secretes and maintains the dentin. The cell bodies of odontoblasts line the pulp cavity just deep to the dentin. Dentin is formed throughout adult life and gradually encroaches on the pulp cavity. New dentin can also be laid down fairly rapidly to compensate for tooth damage or decay.

HOMEOSTATIC IMBALANCE

Death of a tooth's nerve and consequent darkening of the tooth is commonly caused by a blow to the jaw. Swelling in the local area pinches off the blood supply to the tooth and the nerve dies. Typically the pulp becomes infected by bacteria some time later and must be removed by *root canal therapy*. After the cavity is sterilized and filled with an inert material, the tooth is capped. ●

Although enamel, dentin, and cementum are all calcified and resemble bone, they differ from bone in that they are avascular. Enamel also differs from cementum and dentin because it lacks collagen as its main organic component.

Tooth and Gum Disease   Dental caries (kar'ēz; "rottenness"), or **cavities**, result from gradual demineralization of enamel and underlying dentin by bacterial action. Decay begins when **dental plaque** (a film of sugar, bacteria, and other mouth debris) adheres to the teeth. Bacterial metabolism of the trapped sugars produces acids, which can dissolve the calcium salts of the teeth. Once the salts are leached out, the remaining organic matrix of the tooth is readily digested by protein-digesting enzymes released by the bacteria. Frequent brushing and flossing daily help prevent damage by removing forming plaque.

More serious than tooth decay is the effect of unremoved plaque on the gums. As dental plaque accumulates, it calcifies, forming **calculus** (kal'ku-lus; "stone") or tartar, which disrupts the seals between the gingivae and the teeth, putting the gums at risk for infection. In the early stages of such an infection,

called *gingivitis* (jin"ji-vi'tis), the gums are red, sore, swollen, and may bleed. Gingivitis is reversible if the calculus is removed, but if it is neglected the bacteria eventually invade the bone around the teeth, forming pockets of infection. The immune system attacks not only the intruders but also the body tissues, carving deep pockets around the teeth and dissolving the bone away. This more serious condition, called **periodontal disease,** or **periodontitis,** affects up to 95% of all people over the age of 35 and accounts for 80–90% of tooth loss in adults.

Still, tooth loss is not inevitable. Even advanced cases of periodontitis can be treated by scraping the teeth, cleaning the infected pockets, then cutting the gums to shrink the pockets, and following up with antibiotic therapy. Together, these treatments alleviate the bacterial infestations and encourage reattachment of the surrounding tissues to the teeth and bone. Much less painful are (1) a new laser approach for destroying the diseased tissues and (2) a nonsurgical therapy now in clinical trials in which an antibiotic-impregnated film is temporarily glued to the exposed root surface. Clinical treatment is followed up by a home regimen of plaque removal by consistent frequent brushing and flossing and hydrogen peroxide rinses.

Although periodontal disease has traditionally been viewed as a self-limiting low-grade infection, there may be more at risk than teeth. Some contend that it increases the risk of heart disease in at least two ways: (1) the chronic inflammation promotes atherosclerotic plaque formation, and (2) bacteria entering the blood from infected gums stimulate clot formation that helps to clog coronary arteries.

The Pharynx

From the mouth, food passes posteriorly into the **oropharynx** and then the **laryngopharynx** (see Figure 22.8), both common passageways for food, fluids, and air. (The nasopharynx has no digestive role.)

The histology of the pharyngeal wall resembles that of the oral cavity. The mucosa contains a friction-resistant stratified squamous epithelium well supplied with mucus-producing glands. The external muscle layer consists of two *skeletal muscle* layers. The cells of the inner layer run longitudinally. Those of the outer layer, the *pharyngeal constrictor muscles*, encircle the wall like three stacked fists (see Figure 10.8). Contractions of these muscles propel food into the esophagus below.

The Esophagus

The **esophagus** (ĕ-sof'ah-gus; "carry food"), a muscular tube about 25 cm (10 inches) long, is collapsed when not involved in food propulsion (Figure 22.12).

? *What is the functional significance of the epithelial change seen in part (b)?*

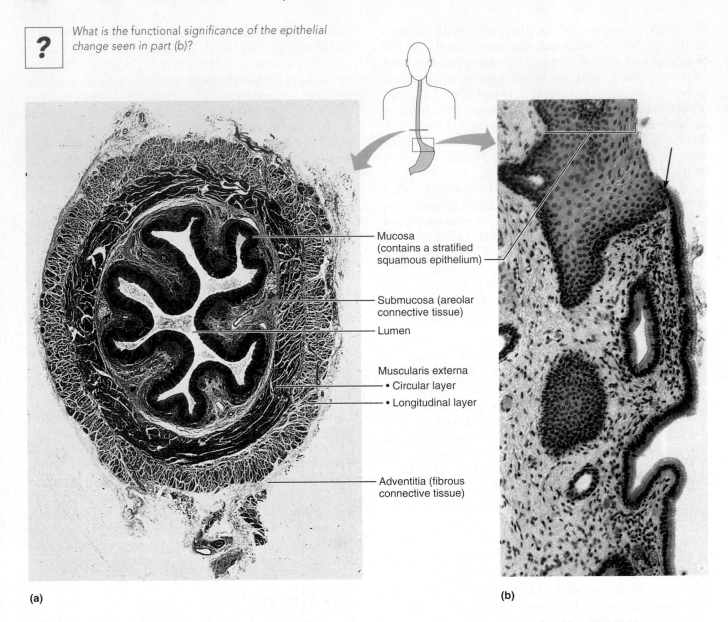

Mucosa (contains a stratified squamous epithelium)

Submucosa (areolar connective tissue)

Lumen

Muscularis externa
• Circular layer
• Longitudinal layer

Adventitia (fibrous connective tissue)

(a)

(b)

FIGURE 22.12 Microscopic structure of the esophagus.
(a) Cross-sectional view of the esophagus showing its four tunics (5×). The section shown is taken from the region close to the stomach junction, so the muscularis is composed of smooth muscle. (b) Longitudinal section through the esophagus-stomach junction (80×). Arrow shows the point of abrupt transition from the stratified squamous epithelium of the esophagus (top) to the simple columnar epithelium of the stomach (bottom).

After food moves through the laryngopharynx, it is routed into the esophagus posteriorly as the epiglottis closes off the larynx to food entry.

As shown in Figure 22.1, the esophagus takes a fairly straight course through the mediastinum of the thorax and pierces the diaphragm at the **esophageal hiatus** (hi-a'tus; "gap") to enter the abdomen. It joins the stomach at the **cardiac orifice.** The cardiac orifice is surrounded by the **cardiac** or **gastroesophageal sphincter** (gas"tro-ĕsof"ah-je'al) (see Figure 22.13), which is a *physiological* sphincter. That is, it acts as a valve, but the only structural evidence of this sphincter is a slight thickening of the circular smooth muscle at that point. The muscular diaphragm, which surrounds this sphincter, helps keep it closed when food is not being swallowed.

The presence of the stratified squamous epithelium indicates the esophagus is mainly a chute and thus must accommodate high friction, whereas the stomach has a secretory simple columnar epithelium, reflecting its secretory role in chemical digestion. ■

Heartburn, the first symptom of *gastroesophageal reflux disease (GERD),* is the burning, radiating substernal pain that occurs when the acidic gastric juice regurgitates into the esophagus. Symptoms are so similar to those of a heart attack that many first-time sufferers of heartburn are rushed to the hospital emergency room. Heartburn is most likely to happen when one has eaten or drunk to excess, and in conditions that force abdominal contents superiorly, such as extreme obesity, pregnancy, and running, which causes stomach contents to splash upward with each step (runner's reflux). It is also common in those with a **hiatal hernia,** a structural abnormality in which the superior part of the stomach protrudes slightly above the diaphragm. Since the diaphragm no longer reinforces the cardiac sphincter, gastric juice may flow into the esophagus, particularly when lying down. If the episodes are frequent and prolonged, *esophagitis* (inflammation of the esophagus) and *esophageal ulcers* may result. An even more threatening sequel is esophageal cancer. However, these consequences can usually be prevented or managed by avoiding late-night snacks and by using antacid preparations. ●

Unlike the mouth and pharynx, the esophagus wall has all four of the basic alimentary canal layers described earlier. Some features of interest:

1. The esophageal mucosa contains a nonkeratinized stratified squamous epithelium. At the esophagus-stomach junction, that abrasion-resistant epithelium changes abruptly to the simple columnar epithelium of the stomach, which is specialized for secretion (Figure 22.12b).

2. When the esophagus is empty, its mucosa and submucosa are thrown into longitudinal folds (Figure 22.12a). When food is in transit in the esophagus, these folds flatten out.

3. The submucosa contains mucus-secreting *esophageal glands.* As a bolus moves through the esophagus, it compresses these glands, causing them to secrete mucus that "greases" the esophageal walls and aids food passage.

4. The muscularis externa is skeletal muscle in its superior third, a mixture of skeletal and smooth muscle in its middle third, and entirely smooth muscle in its inferior third.

5. Instead of a serosa, the esophagus has a fibrous adventitia composed entirely of connective tissue, which blends with surrounding structures along its route.

Digestive Processes Occurring in the Mouth, Pharynx, and Esophagus

The mouth and its accessory digestive organs are involved in most digestive processes. The mouth (1) ingests, (2) begins mechanical digestion by chewing, and (3) initiates propulsion by swallowing. Salivary amylase, the enzyme in saliva, starts the chemical breakdown of polysaccharides (starch and glycogen) into smaller fragments of linked glucose molecules. (If you chew a piece of bread for a few minutes, it will begin to taste sweet as sugars are released.) Except for a few drugs that are absorbed through the oral mucosa (for example, nitroglycerine), essentially no absorption occurs in the mouth.

In contrast to the multifunctional mouth, the pharynx and esophagus merely serve as conduits to pass food from the mouth to the stomach. Their single digestive function is food propulsion, accomplished by the role they play in swallowing.

Since chemical digestion is covered in a special physiology section later in the chapter, only the mechanical processes of chewing and swallowing are discussed here.

Mastication (Chewing)

As food enters the mouth, its mechanical breakdown begins with **mastication,** or chewing. The cheeks and closed lips hold food between the teeth, the tongue mixes food with saliva to soften it, and the teeth cut and grind solid foods into smaller morsels. Mastication is partly voluntary and partly reflexive. We voluntarily put food into our mouths and contract the muscles that close our jaws. Continued jaw movements are controlled mainly by stretch reflexes and in response to pressure inputs from receptors in the cheeks, gums, and tongue, but they can also be voluntary as desired.

Deglutition (Swallowing)

To send food on its way from the mouth, it is first compacted by the tongue into a bolus and then swallowed. **Deglutition** (deg"loo-tish'un), or swallowing, is a complicated process that involves coordinated activity of over 22 separate muscle groups. It has two major phases, the buccal and the pharyngeal-esophageal.

The **buccal phase** occurs in the mouth and is voluntary. In the buccal phase, we place the tip of the tongue against the hard palate, and then contract the tongue to force the bolus into the oropharynx (Figure 22.13a). As food enters the pharynx and stimulates tactile receptors there, it passes out of our control and into the realm of involuntary reflex activity.

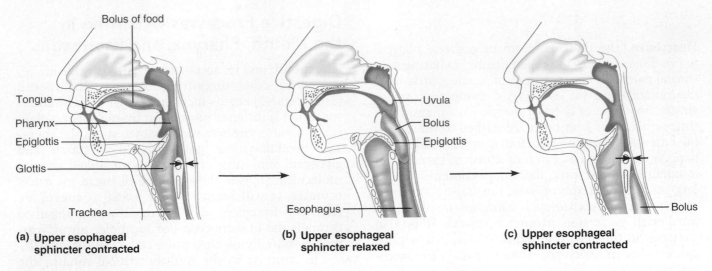

(a) Upper esophageal sphincter contracted

(b) Upper esophageal sphincter relaxed

(c) Upper esophageal sphincter contracted

FIGURE 22.13 **Deglutition (swallowing).** The process of swallowing consists of a voluntary (buccal) phase (a) and involuntary (pharyngeal-esophageal) phases (b–e). **(a)** During the buccal phase, the tongue rises and presses against the hard palate, forcing the food bolus into the oropharynx where the involuntary phase of swallowing begins. **(b)** The uvula and larynx rise to prevent food from entering respiratory passageways. Relaxation of the upper esophageal sphincter allows food to enter the esophagus. **(c)** The constrictor muscles of the pharynx contract, forcing food into the esophagus inferiorly, and the upper esophageal sphincter contracts after entry. **(d)** Food is moved through the esophagus to the stomach by peristalsis. **(e)** The gastroesophageal sphincter opens, and food enters the stomach.

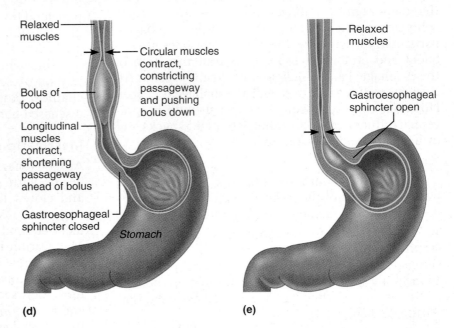

The involuntary **pharyngeal-esophageal phase** of swallowing is controlled by the swallowing center located in the medulla and lower pons. Motor impulses from that center are transmitted via various cranial nerves, most importantly the vagus nerves, to the muscles of the pharynx and esophagus. As illustrated in Figure 22.13b, once food enters the pharynx, all routes except the desired one—into the digestive tract—are blocked off:

- The tongue blocks off the mouth.
- The soft palate rises to close off the nasopharynx.
- The larynx rises so that the epiglottis covers its opening into the respiratory passageways, and the upper esophageal sphincter relaxes.

Food is squeezed through the pharynx and into the esophagus by wavelike peristaltic contractions (Figure 22.13c–e). Solid foods pass from the oropharynx to the stomach in 4 to 8 seconds; fluids pass in

1 to 2 seconds. Just before the peristaltic wave (and food) reaches the end of the esophagus, the gastroesophageal sphincter relaxes reflexively to allow food to enter the stomach.

HOMEOSTATIC IMBALANCE

If we try to talk or inhale while swallowing, the various protective mechanisms may be short-circuited and food may enter the respiratory passageways instead. This event typically triggers the cough reflex in an attempt to expel the food. ●

The Stomach

Below the esophagus, the GI tract expands to form the **stomach** (see Figure 22.1), a temporary "storage tank" where chemical breakdown of proteins begins

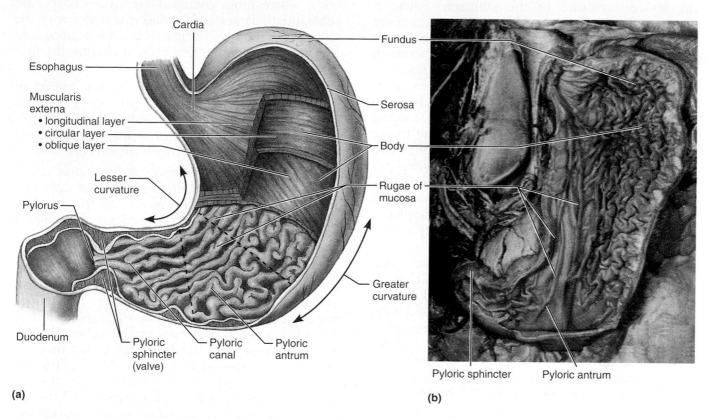

FIGURE 22.14 Anatomy of the stomach. (a) Gross internal anatomy (frontal section). **(b)** Photograph of internal aspect of stomach. (See *A Brief Atlas of the Human Body,* Figure 53a.)

and food is converted to a creamy paste called **chyme** (kīm; "juice"). The stomach lies in the upper left quadrant of the peritoneal cavity, nearly hidden by the liver and diaphragm. Specifically, it lies in the left hypochondriac, epigastric, and umbilical regions of the abdomen. Though relatively fixed at both ends, the stomach is quite movable in between. It tends to lie high and run horizontally in short, stout people (a steer-horn stomach) and is often elongated vertically in tall, thin people (a J-shaped stomach).

Gross Anatomy

The adult stomach varies from 15 to 25 cm (6 to 10 inches) long, but its diameter and volume depend on how much food it contains. An empty stomach has a volume of about 50 ml and a cross-sectional diameter only slightly larger than the large intestine, but when it is really distended it can hold about 4 L

(1 gallon) of food and may extend nearly all the way to the pelvis! When empty, the stomach collapses inward, throwing its mucosa (and submucosa) into large, longitudinal folds called **rugae** (roo′ge; *ruga* = wrinkle, fold).

The major regions of the stomach are shown in Figure 22.14a. The small **cardiac region,** or **cardia** ("near the heart"), surrounds the cardiac orifice through which food enters the stomach from the esophagus. The **fundus** is its dome-shaped part, tucked beneath the diaphragm, that bulges superolaterally to the cardia. The **body,** the midportion of the stomach, is continuous inferiorly with the funnel-shaped **pyloric region.** The wider and more superior part of the pyloric region, the **pyloric antrum** (*antrum* = cave) narrows to form the **pyloric canal,** which terminates at the **pylorus.** The pylorus is continuous with the duodenum (the first part of the small intestine) through the **pyloric sphincter,** which controls stomach emptying (*pylorus* = gatekeeper).

The convex lateral surface of the stomach is its **greater curvature,** and its concave medial surface is the **lesser curvature.** Extending from these curvatures are two mesenteries, called *omenta*

It has a third layer of smooth muscle in its muscularis in which the fibers run obliquely, allowing it to exhibit churning movements that physically break down the food. ■

(o-men′tah), that help tether the stomach to other digestive organs and the body wall (see Figure 22.30, p. 794). The **lesser omentum** runs from the liver to the lesser curvature of the stomach, where it becomes continuous with the visceral peritoneum covering the stomach. The **greater omentum** drapes inferiorly from the greater curvature of the stomach to cover the coils of the small intestine. It then runs dorsally and superiorly (enclosing the spleen on its way) to wrap the transverse portion of the large intestine before blending with the *mesocolon*, a dorsal mesentery that secures the large intestine to the parietal peritoneum of the posterior abdominal wall. The greater omentum is riddled with fat deposits (*oment* = fatty skin) that give it the appearance of a lacy apron. It also contains large collections of lymph nodes. The immune cells and macrophages in these nodes "police" the peritoneal cavity and intraperitoneal organs.

The stomach is served by the autonomic nervous system. Sympathetic fibers from thoracic splanchnic nerves are relayed through the celiac plexus. Parasympathetic fibers are supplied by the vagus nerve. The arterial supply of the stomach is provided by branches (gastric and splenic) of the celiac trunk (see Figure 18.22). The corresponding veins are part of the hepatic portal system (see Figure 18.27c) and ultimately drain into the hepatic portal vein.

Microscopic Anatomy

The stomach wall contains the four tunics typical of most of the alimentary canal, but its muscularis and mucosa are modified for the special roles of the stomach. Besides the usual circular and longitudinal layers of smooth muscle, the muscularis externa has an innermost smooth muscle layer that runs *obliquely* (Figure 22.14a and Figure 22.15). This arrangement allows the stomach not only to move food along the tract, but also to churn, mix, and pummel the food, physically breaking it down into smaller fragments.

The lining epithelium of the stomach mucosa is a simple columnar epithelium composed entirely of goblet cells, which produce a protective two-layer coat of alkaline mucus in which the surface layer consists of viscous mucus that traps a layer of bicarbonate-rich fluid beneath it. This otherwise smooth lining is dotted with millions of deep **gastric pits** (Figure 22.15), which lead into the **gastric glands** that produce the stomach secretion called **gastric juice.** The cells forming the walls of the gastric pits are primarily goblet cells, but those composing the gastric glands vary in different stomach regions. For example, the cells in the glands of the cardia and pylorus are primarily mucus secreting, whereas cells of

the pyloric antrum produce mucus and several hormones including most of the stimulatory hormone called gastrin. Glands of the stomach fundus and body, where most chemical digestion occurs, are substantially larger and produce the majority of the stomach secretions. The glands in these regions contain a variety of secretory cells, including the four types described here:

1. **Mucous neck cells,** found in the upper, or "neck," regions of the glands, produce a different type of mucus from that secreted by the goblet cells of the surface epithelium. It is not yet understood what special function this *acidic* mucus performs.

2. **Parietal cells,** found mainly in the middle region of the glands scattered among the chief cells (described next), secrete *hydrochloric acid (HCl)* and *intrinsic factor.* Although the parietal cells appear spherical when viewed with a light microscope, they actually have three prongs that exhibit dense microvilli (they look like fuzzy pitchforks!). This structure provides a huge surface area for secreting H^+ and Cl^- into the stomach lumen. HCl makes the stomach contents extremely acidic (pH 1.5–3.5), a condition necessary for activation and optimal activity of pepsin, and harsh enough to kill many of the bacteria ingested with foods. The acidity also helps in food digestion by denaturing proteins and breaking down cell walls of plant foods. Intrinsic factor is a glycoprotein required for vitamin B_{12} absorption in the small intestine.

3. **Chief cells** produce *pepsinogen* (pep-sin′o-jen), the inactive form of the protein-digesting enzyme **pepsin.** The chief cells occur mainly in the basal regions of the gastric glands. When chief cells are stimulated, the first pepsinogen molecules they release are activated by HCl encountered in the apical region of the gland (Figure 22.15c). But once pepsin is present, it also catalyzes the conversion of pepsinogen to pepsin. This positive feedback process is limited only by the amount of pepsinogen present. The activation process involves removal of a small peptide fragment from the pepsinogen molecule, causing it to change shape and expose its active site. Chief cells also secrete insignificant amounts of lipases (fat-digesting enzymes).

4. **Enteroendocrine cells** (en″ter-o-en′do-krin; "gut endocrine") release a variety of hormones or hormonelike products directly into the lamina propria. These products, including **gastrin, histamine, endorphins** (natural opiates), **serotonin, cholecystokinin,** and **somatostatin,** diffuse into the blood capillaries, and ultimately influence several digestive system target organs (see Table 22.1, p. 776). Gastrin, in particular, plays essential roles in regulating stomach secretion and mobility, as described shortly.

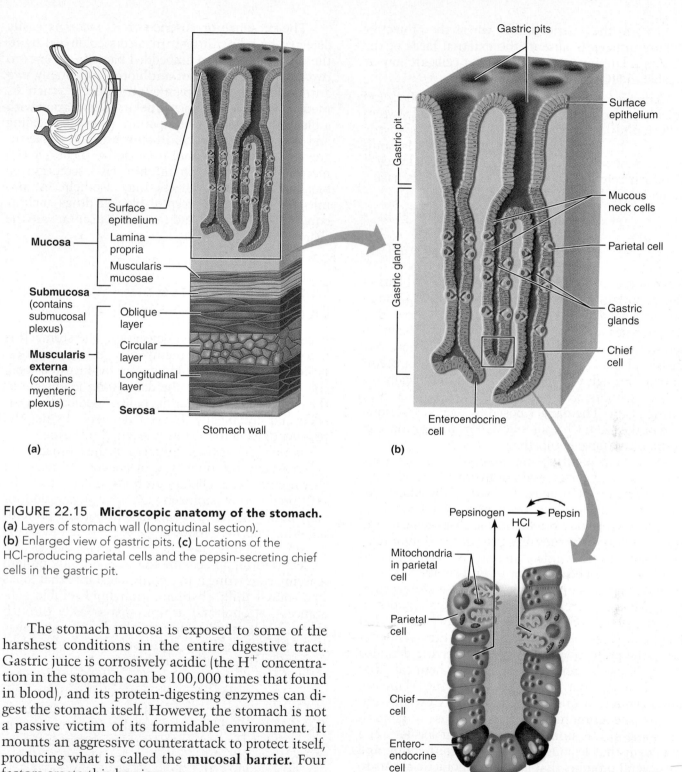

FIGURE 22.15 **Microscopic anatomy of the stomach.**
(a) Layers of stomach wall (longitudinal section).
(b) Enlarged view of gastric pits. **(c)** Locations of the
HCl-producing parietal cells and the pepsin-secreting chief
cells in the gastric pit.

The stomach mucosa is exposed to some of the harshest conditions in the entire digestive tract. Gastric juice is corrosively acidic (the H^+ concentration in the stomach can be 100,000 times that found in blood), and its protein-digesting enzymes can digest the stomach itself. However, the stomach is not a passive victim of its formidable environment. It mounts an aggressive counterattack to protect itself, producing what is called the **mucosal barrier**. Four factors create this barrier:

1. A thick coating of bicarbonate-rich mucus is built up on the stomach wall.

2. The epithelial cells of the mucosa are joined together by tight junctions that prevent gastric juice from leaking into the underlying tissue layers.

3. Deep in the gastric glands, where the protective alkaline mucus is absent, the external faces of the plasma membranes of the glandular cells are impermeable to HCl.

4. Damaged epithelial mucosal cells are shed and quickly replaced by division of *undifferentiated stem cells* that reside where the gastric pits join the gastric glands. The stomach surface epithelium is completely renewed every three to six days. (However, glandular cells deep within the gastric glands have a much longer life span.)

H HOMEOSTATIC IMBALANCE

Anything that breaches the gel-like mucosal barrier causes inflammation of the stomach wall, a condition called *gastritis*. Persistent damage to the underlying tissues can promote **gastric ulcers,** erosions of the stomach wall. The most distressing symptom of gastric ulcers is gnawing epigastric pain that seems to bore through to your back. The pain typically occurs 1–3 hours after eating and is often relieved by eating again. The danger posed by ulcers is perforation of the stomach wall followed by peritonitis and, perhaps, massive hemorrhage.

Common predisposing factors for ulcer formation include hypersecretion of hydrochloric acid and hyposecretion of mucus. For years, the blame for causing ulcers was put on factors that favor high HCl or low mucus production such as aspirin and nonsteroidal anti-inflammatory drugs (ibuprofen), smoking, alcohol, coffee, and stress. Although acid conditions *are* necessary for ulcer formation, acidity in and of itself is not sufficient to cause ulcer formation. Most recurrent ulcers (90%) are the work of acid-resistant, corkscrew-shaped *Helicobacter pylori* bacteria, which burrow beneath the mucus and destroy the protective mucosal layer, leaving denuded areas. These bacteria release several chemicals that help them do their "dirty work" including (1) *urease,* an enzyme that breaks down urea to CO_2 and ammonia (the ammonia then acts as a base to neutralize some of the stomach acid in their locale), (2) a *cytotoxin* that lesions the stomach epithelium, and (3) several proteins that act as chemotactic agents to attract macrophages and other defensive cells into the area. This bacterial-causal theory has been difficult to prove because the bacterium is found not only in some 70–90% of ulcer and gastritis sufferers but in more than 33% of healthy people as well. Even more troubling are studies that link this bacterium to some stomach cancers.

The presence or absence of *H. pylori* is easily detected by a breath test. In ulcers colonized by it, the goal is to kill the embedded bacteria. A one- to two-week-long course of antibiotics (preferably two antibiotics with complementary effects, such as metronidazole and tetracycline) in combination with a bismuth-containing compound, promotes healing and prevents recurrence. (Bismuth is the active ingredient in Pepto-Bismol.) For active ulcers, a H_2-receptor blocker, which inhibits HCl secretion by blocking histamine's effects, may also help. In noninfectious cases, H_2-receptor blocker drugs such as cimetidine (Tagamet) and ranitidine (Zantac) are the therapy of choice. ●

Digestive Processes Occurring in the Stomach

Except for ingestion and defecation, the stomach is involved in the whole "menu" of digestive activities. Besides serving as a holding area for ingested food, the stomach continues the demolition job begun in the oral cavity by further degrading food both physically and chemically. It then delivers chyme, the product of its activity, into the small intestine.

Protein digestion is initiated in the stomach and is essentially the only type of enzymatic digestion that occurs there. Dietary proteins are denatured by HCl produced by stomach glands in preparation for enzymatic digestion. The most important protein-digesting enzyme produced by the gastric mucosa is pepsin. In infants, however, the stomach glands also secrete **rennin,** an enzyme that acts on milk protein (casein), converting it to a curdy substance that looks like soured milk. Two common lipid-soluble substances—alcohol and aspirin—pass easily through the stomach mucosa into the blood, and may cause gastric bleeding; thus, these substances should be avoided by those with gastric ulcers.

Despite the obvious benefits of preparing food to enter the intestine, the only stomach function essential to life is secretion of intrinsic factor. **Intrinsic factor** is required for intestinal absorption of vitamin B_{12}, needed to produce mature erythrocytes; in its absence, *pernicious anemia* results. However, if vitamin B_{12} is administered by injection, individuals can survive with minimal digestive problems even after total gastrectomy (stomach removal). (The stomach's activities are summarized in Table 22.2, p. 782).

Since chemical digestion and absorption are described later, here we will focus on events that (1) control secretory activity of the gastric glands and (2) regulate stomach motility and emptying.

? Distension of the stomach and duodenal walls have differing effects on stomach secretory activity. What are these effects?

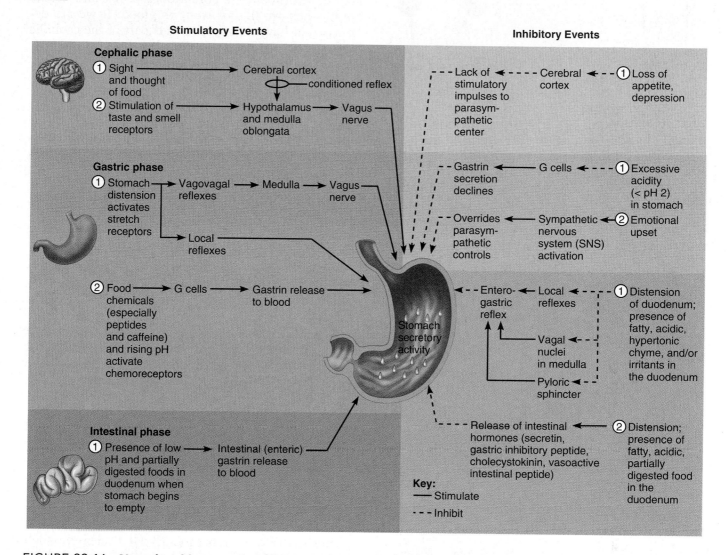

Stimulatory Events   **Inhibitory Events**

Cephalic phase
① Sight and thought of food ⟶ Cerebral cortex
                              ⊖ conditioned reflex
② Stimulation of taste and smell receptors ⟶ Hypothalamus and medulla oblongata ⟶ Vagus nerve

Lack of stimulatory impulses to parasympathetic center ⟵ Cerebral cortex ⟵ ① Loss of appetite, depression

Gastric phase
① Stomach distension activates stretch receptors ⟶ Vagovagal reflexes ⟶ Medulla ⟶ Vagus nerve
                                                 ⟶ Local reflexes
② Food chemicals (especially peptides and caffeine) and rising pH activate chemoreceptors ⟶ G cells ⟶ Gastrin release to blood

Gastrin secretion declines ⟵ G cells ⟵ ① Excessive acidity (< pH 2) in stomach

Overrides parasympathetic controls ⟵ Sympathetic nervous system (SNS) activation ⟵ ② Emotional upset

Enterogastric reflex ⟵ Local reflexes ⟵ ① Distension of duodenum; presence of fatty, acidic, hypertonic chyme, and/or irritants in the duodenum
                     ⟵ Vagal nuclei in medulla
                     ⟵ Pyloric sphincter

Stomach secretory activity

Intestinal phase
① Presence of low pH and partially digested foods in duodenum when stomach begins to empty ⟶ Intestinal (enteric) gastrin release to blood

Release of intestinal hormones (secretin, gastric inhibitory peptide, cholecystokinin, vasoactive intestinal peptide) ⟵ ② Distension; presence of fatty, acidic, partially digested food in the duodenum

Key:
— Stimulate
--- Inhibit

FIGURE 22.16 Neural and hormonal mechanisms that regulate release of gastric juice. Stimulatory factors are shown on the left; inhibitory factors are shown on the right.

Regulation of Gastric Secretion

Gastric secretion is controlled by both neural and hormonal mechanisms. Under normal conditions the gastric mucosa pours out as much as 3 L of gastric juice—an acid brew so potent it can dissolve nails—every day. Nervous control is provided by long (vagus nerve–mediated) and short (local enteric) nerve reflexes. When the vagus nerves stimulate the stomach, secretory activity of virtually all of its glands increases. (By contrast, activation of sympathetic nerves depresses secretory activity.) Hormonal control of gastric secretion is largely the province of gastrin, which stimulates secretion of enzymes and HCl, and of hormones produced by the small intestine, which are mostly gastrin antagonists.

Stimuli acting at three distinct sites—the head, stomach, and small intestine—provoke or inhibit gastric secretory activity and, accordingly, the three phases of gastric secretion are called the *cephalic, gastric,* and *intestinal phases* (Figure 22.16). However, the effector site is the stomach in all cases and, once initiated, one or all three phases may be occurring at the same time.

■ Distension of the stomach promotes increased secretory activity whereas stretching the wall of the duodenum causes hormones to be released that inhibit stomach secretory activity.

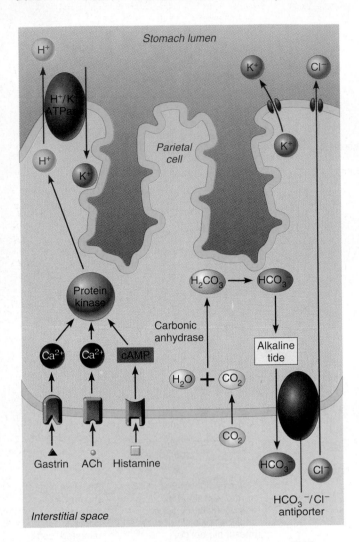

FIGURE 22.17 Regulation and mechanism of HCl secretion. Binding of histamine, gastrin, and acetylcholine (ACh) to parietal cell membrane receptors initiates intracellular events (mediated by second-messenger systems) that lead to HCl secretion into the stomach lumen. H^+ and HCO_3^- (bicarbonate ions) are generated from the dissociation of carbonic acid (H_2CO_3). As H^+/K^+ ATPase pumps H^+ into the lumen, K^+ enters the cell, HCO_3^- is pumped into the interstitial space in exchange for chloride ions (Cl^-). Cl^- and K^+ then diffuse into the lumen through membrane channels.

Phase 1: Cephalic (Reflex)

The **cephalic,** or **reflex, phase** of gastric secretion occurs *before* food enters the stomach. Only a few minutes long, this phase is triggered by the aroma, taste, sight, or thought of food, and it gets the stomach ready for its upcoming digestive chore. Inputs from activated olfactory receptors and taste buds are relayed to the hypothalamus, which in turn stimulates the vagal nuclei of the medulla oblongata, causing motor impulses to be transmitted via the vagus nerves to parasympathetic enteric ganglia. Enteric ganglionic neurons in turn stimulate the stomach glands. The enhanced secretory activity that results when we see or think of food is a *conditioned reflex* and occurs only when we like or want the food. If we are depressed or have no appetite, this part of the cephalic reflex is suppressed.

Phase 2: Gastric

Once food reaches the stomach, local neural and hormonal mechanisms initiate the **gastric phase.** This phase is three to four hours long and provides about two-thirds of the gastric juice released. The most important stimuli are distension, peptides, and low acidity. Stomach distension activates stretch receptors and initiates both local (myenteric) reflexes and the long vago-vagal reflexes. In the latter type of reflex, impulses travel to the medulla and then back to the stomach via vagal fibers. Both types of reflexes lead to acetylcholine (ACh) release, which in turn stimulates the output of more gastric juice by the secretory cells.

Though neural influences initiated by stomach distension are important, the hormone gastrin probably plays a greater role in stimulating stomach gland secretion during the gastric phase. Chemical stimuli provided by partially digested proteins, caffeine, and rising pH directly activate gastrin-secreting enteroendocrine cells called **G cells.** Although gastrin also stimulates the release of enzymes, its main target is the HCl-secreting parietal cells, which it prods to spew out even more HCl (Figure 22.17). Highly acidic (pH below 2) gastric contents *inhibit* gastrin secretion.

When protein foods are in the stomach, the pH of the gastric contents generally rises because proteins act as buffers to tie up H^+. The rise in pH stimulates gastrin and subsequently HCl release, which in turn provides the acidic conditions needed for protein digestion. The more protein in the meal, the greater the amount of gastrin and HCl released. As proteins are digested, the gastric contents gradually become more acidic, which again inhibits the gastrin-secreting cells. This negative feedback mechanism helps maintain optimal pH and working conditions for the gastric enzymes.

G cells are also activated by the neural reflexes already described. Emotional upsets, fear, anxiety, or anything that triggers the fight-or-flight response inhibits gastric secretion because (during such times) the sympathetic division overrides parasympathetic controls of digestion (Figure 22.16).

The control of the HCl-secreting parietal cells is multifaceted. HCl secretion is stimulated by three chemicals, all of which work through second-messenger systems (Figure 22.17): *ACh* released by parasympathetic nerve fibers and *gastrin* secreted by G cells bring about their effects by increasing intracellular Ca^{2+} levels. *Histamine,* released by mast cells in the lamina propria, acts through cyclic AMP (cAMP). When only one of the three chemicals binds to the parietal cells, HCl secretion is scanty, but when all three bind, HCl pours forth as if shot out by a high-pressure hose. (As noted earlier, antihistamines, such as cimetidine, which bind to and block the H_2 receptors of parietal cells, are used to treat gastric ulcers due to hyperacidity.)

The process of HCl formation within the parietal cells is complicated. The consensus is that H^+ is actively pumped into the stomach lumen against a tremendous concentration gradient by H^+/K^+ ATPases as K^+ ions are moved into the cell from the interstitial fluid. Chloride ions (Cl^-) are also pumped into the lumen to maintain an electrical balance in the stomach. The Cl^- is obtained from blood plasma, while H^+ comes from the breakdown of carbonic acid (formed by the combination of carbon dioxide and water) within the parietal cell (Figure 22.17):

$$CO_2 + H_2O \rightarrow H_2CO_3 \rightarrow H^+ + HCO_3^-$$

As H^+ is pumped from the cell and HCO_3^- (bicarbonate ion) accumulates within the cell, HCO_3^- is ejected through the basal cell membrane into the capillary blood. As a result, blood draining from the stomach is more alkaline than the blood serving it. This phenomenon is called the **alkaline tide.** Notice that HCO_3^- and Cl^- are moved by a Cl^-/HCO_3^- antiporter in the basolateral membrane, and this is the means of entry of the Cl^- that moves into the lumen as the chloride part of the HCl product. K^+ and Cl^- move into the lumen by diffusing through membrane channels.

Phase 3: Intestinal

The **intestinal phase** of gastric secretion has two components—one excitatory and the other inhibitory (Figure 22.16). The *excitatory* aspect is set into motion as partially digested food fills the initial part (duodenum) of the small intestine. This stimulates intestinal mucosal cells to release a hormone that encourages the gastric glands to continue their secretory activity. The effects of this hormone mirror those of gastrin, so it has been named **intestinal (enteric) gastrin.** However, this stimulatory effect is brief because as the intestine distends with chyme containing large amounts of H^+, fats, partially digested proteins, and various irritating substances, the *inhibitory* component is triggered in the form of the **enterogastric reflex.**

The enterogastric reflex is actually a trio of reflexes that (1) inhibit the vagal nuclei in the medulla, (2) inhibit local reflexes, and (3) activate sympathetic fibers that cause the pyloric sphincter to tighten and prevent further food entry into the small intestine. As a result, gastric secretory activity declines. These "brakes" on gastric activity protect the small intestine from excessive acidity and match the small intestine's processing abilities to the amount of chyme entering it at a given time.

In addition, the factors just named trigger the release of several intestinal hormones, collectively called **enterogastrones,** which include **secretin** (se-kre'tin), **cholecystokinin (CCK)** (ko"le-sis"to-ki'nin), **vasoactive intestinal peptide (VIP),** and **gastric inhibitory peptide (GIP).** All of these hormones inhibit gastric secretion when the stomach is very active and also play other roles, which are summarized in Table 22.1.

Gastric Motility and Emptying

Stomach contractions not only cause its emptying but also compress, knead, twist, and continually mix the food with gastric juice to produce chyme. Because the mixing movements are accomplished by a unique type of peristalsis (for example, in the pylorus it is bidirectional rather than unidirectional), the processes of mechanical digestion and propulsion are inseparable in the stomach.

Response of the Stomach to Filling The stomach stretches to accommodate incoming food, but internal stomach pressure remains constant until about 1 L of food has been ingested. Thereafter, the pressure rises. The relatively unchanging pressure in a filling stomach is due to (1) the reflex-mediated relaxation of the stomach muscle and (2) the plasticity of visceral smooth muscle.

Reflexive relaxation of stomach smooth muscle in the fundus and body occurs both in anticipation of and in response to food entry into the stomach. As food moves through the esophagus, the stomach muscles relax. This **receptive relaxation** is coordinated by the swallowing center of the brain stem and mediated by the vagus nerves. The stomach also actively dilates in response to gastric filling, which activates stretch receptors in the wall. This phenomenon, called **adaptive relaxation,** appears to depend on local reflexes involving *nitric oxide (NO)–releasing neurons.*

Plasticity is the intrinsic ability of visceral smooth muscle to exhibit the stress-relaxation response, that is, to be stretched without greatly increasing its tension and contracting expulsively. As described in Chapter 9, this capability is very important in hollow organs, like the stomach, that must serve as temporary reservoirs.

TABLE 22.1   Hormones and Hormonelike Products That Act in Digestion*

Hormone	Site of Production	Stimulus for Production	Target Organ	Activity
Gastrin	Stomach mucosa	Food (particularly partially digested proteins) in stomach (chemical stimulation); acetylcholine released by nerve fibers	Stomach	▪ Causes gastric glands to increase secretory activity; most pronounced effect is on HCl secretion ▪ Stimulates gastric emptying
			Small intestine	▪ Stimulates contraction of intestinal muscle
			Ileocecal valve	▪ Relaxes ileocecal valve
			Large intestine	▪ Stimulates mass movements
Serotonin	Stomach mucosa	Food in stomach	Stomach	▪ Causes contraction of stomach muscle
Histamine	Stomach mucosa	Food in stomach	Stomach	▪ Activates parietal cells to release HCl
Somatostatin	Stomach mucosa; duodenal mucosa	Food in stomach; stimulation by sympathetic never fibers	Stomach	▪ Inhibits gastric secretion of all products; inhibits gastric motility and emptying
			Pancreas	▪ Inhibits secretion
			Small intestine	▪ Inhibits GI blood flow; thus inhibits intestinal absorption
			Gallbladder	▪ Inhibits contraction and bile release
Intestinal gastrin	Duodenal mucosa	Acidic and partially digested foods in duodenum	Stomach	▪ Stimulates gastric glands and motility
Secretin	Duodenal mucosa	Acidic chyme (also partially digested proteins, fats, hypertonic or hypotonic fluids, or irritants in chyme)	Stomach	▪ Inhibits gastric gland secretion and gastric motility during gastric phase of secretion
			Pancreas	▪ Increases output of pancreatic juice rich in bicarbonate ions; potentiates CCK's action
			Liver	▪ Increases bile output
Cholecystokinin (CCK)	Duodenal mucosa	Fatty chyme, in particular, but also partially digested proteins	Liver/pancreas	▪ Potentiates secretin's actions on these organs
			Pancreas	▪ Increases output of enzyme-rich pancreatic juice
			Gallbladder	▪ Stimulates organ to contract and expel stored bile
			Hepatopancreatic sphincter (of Oddi)	▪ Relaxes to allow entry of bile and pancreatic juice into duodenum
Gastric inhibitory peptide (GIP)[†]	Duodenal mucosa	Fatty and/or glucose-containing chyme	Stomach	▪ Inhibits gastric gland secretion and gastric motility during gastric phase
Vasoactive intestinal peptide (VIP)	Duodenal mucosa	Chyme containing partially digested foods	Duodenum	▪ Stimulates buffer secretion; dilates intestinal capillaries
			Stomach	▪ Inhibits HCl production
			Small intestine	▪ Relaxes intestinal smooth muscle

*Except for somatostatin, all of these polypeptides also stimulate the growth (particularly of the mucosa) of the organs they affect.
[†]Also called gastric insulinotropic peptide because it stimulates the release of insulin from the pancreatic islets.

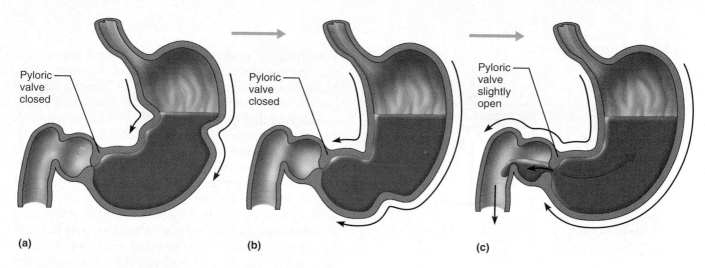

Pyloric valve closed

Pyloric valve closed

Pyloric valve slightly open

(a)   (b)   (c)

FIGURE 22.18 **Peristaltic waves in the stomach** (a) Peristaltic waves move toward the pylorus. (b) The most vigorous peristalsis and mixing action occurs close to the pylorus. (c) The pyloric end of the stomach acts as a pump that delivers small amounts of chyme into the duodenum, simultaneously forcing most of its contained material backward into the stomach, where it undergoes further mixing.

Gastric Contractile Activity   After a meal, peristalsis begins near the cardiac sphincter, where it produces gentle rippling movements of the stomach wall. But as the contractions approach the pylorus, where the stomach musculature is thicker, they become much more powerful. Consequently, the contents of the fundus remain relatively undisturbed, while foodstuffs close to the pylorus receive a truly lively pummeling and mixing.

The pyloric region of the stomach, which holds about 30 ml of chyme, acts as a "dynamic filter" that allows only liquids and small particles to pass through the barely open pyloric valve during the digestive period. Normally, each peristaltic wave reaching the pyloric muscle "spits" or squirts 3 ml or less of chyme into the small intestine. Because the contraction also *closes* the valve, which is normally partially relaxed, the rest (about 27 ml) is propelled backward into the stomach, where it is mixed further (Figure 22.18). This back-and-forth pumping action effectively breaks up solids in the gastric contents.

Although the intensity of the stomach's peristaltic waves can be modified, their rate is constant—always around three per minute. This contractile rhythm is set by the spontaneous activity of *pacemaker cells* located in the longitudinal smooth muscle layer. The pacemaker cells, musclelike noncontractile cells called *interstitial cells of Cajal* (ka-hal'), depolarize and repolarize spontaneously three times each minute, establishing the so-called *cyclic slow waves* of the stomach, or its **basic electrical rhythm (BER)**. Since the pacemakers are electrically coupled to the rest of the smooth muscle sheet by gap junctions, their "beat" is transmitted efficiently and quickly to the entire muscularis. The pacemakers set the maximum rate of contraction, but they do not initiate the contractions or regulate their force. Instead, they generate subthreshold depolarization waves, which are then "ignited" (enhanced by further depolarization and brought to threshold) by neural and hormonal factors.

Factors that increase the strength of stomach contractions are the same factors that enhance gastric secretory activity. Distension of the stomach wall by food activates stretch receptors and gastrin-secreting cells, which both ultimately stimulate gastric smooth muscle and so increase gastric motility. Thus, the more food there is in the stomach, the more vigorous the stomach mixing and emptying movements will be—within certain limits—as described next.

Regulation of Gastric Emptying   The stomach usually empties completely within four hours after a meal. However, the larger the meal (the greater the stomach distension) and the more liquid its contents, the faster the stomach empties. Fluids pass quickly through the stomach. Solids linger, remaining until they are well mixed with gastric juice and converted to the liquid state.

The rate of gastric emptying depends as much— and perhaps more—on the contents of the duodenum as on what is happening in the stomach. The

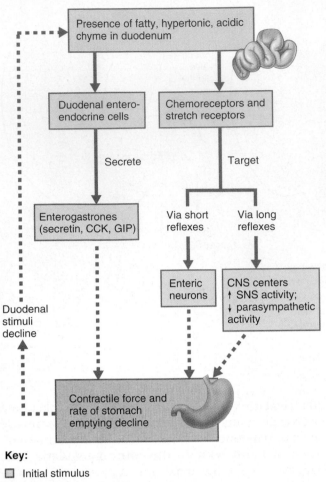

Key:
☐ Initial stimulus
☐ Physiological response
☐ Result
◄— Stimulate
◄· · Inhibit

FIGURE 22.19   Neural and hormonal factors inhibiting gastric emptying. These controls ensure that the food will be well liquefied in the stomach and prevent the small intestine from being overwhelmed.

stomach and duodenum act in tandem like a "coupled meter" that functions at less than full capacity. As chyme enters the duodenum, receptors in its wall respond to chemical signals and to stretch, initiating the enterogastric reflex and the hormonal (enterogastrone) mechanisms described earlier. These mechanisms inhibit gastric secretory activity and prevent further duodenal filling by reducing the force of pyloric contractions (Figure 22.19).

A carbohydrate-rich meal moves through the duodenum rapidly, but fats form an oily layer at the top of the chyme and are digested more slowly by enzymes acting in the intestine. Thus, when chyme entering the duodenum is fatty, food may remain in the stomach six hours or more.

HOMEOSTATIC IMBALANCE

Vomiting, or **emesis,** is an unpleasant experience that causes stomach emptying by a different route. Many factors signal the stomach to "launch lunch," but the most common are extreme stretching of the stomach or intestine or the presence of irritants such as bacterial toxins, excessive alcohol, spicy foods, and certain drugs in those organs. Sensory impulses stream from the irritated sites to the **emetic center** (e-met′ik) of the medulla and initiate a number of motor responses. The diaphragm and abdominal wall muscles contract, increasing intra-abdominal pressure, the cardiac sphincter relaxes, and the soft palate rises to close off the nasal passages. As a result, the stomach (and perhaps duodenal) contents are forced upward through the esophagus and pharynx and out the mouth. Before vomiting, an individual typically is pale, feels nauseated, and salivates. Excessive vomiting can cause dehydration and may lead to severe disturbances in the electrolyte and acid-base balance of the body. Since large amounts of HCl are lost in vomitus, the blood becomes alkaline as the stomach attempts to replace its lost acid. ●

The Small Intestine and Associated Structures

In the small intestine, usable food is finally prepared for its journey into the cells of the body. However, this vital function cannot be accomplished without the aid of secretions from the liver (bile) and pancreas (digestive enzymes). Thus, these necessary accessory organs will also be considered in this section.

The Small Intestine

The **small intestine** is the body's major digestive organ. Within its twisted passageways, digestion is completed and virtually all absorption occurs.

Gross Anatomy

The small intestine is a convoluted tube extending from the pyloric sphincter in the epigastric region to the **ileocecal valve** (il″e-o-se′kal) in the right iliac region where it joins the large intestine (see Figure 22.1). It is the longest part of the alimentary tube, but is only about half the diameter of the large intestine, ranging from 2.5 to 4 cm (1–1.6 in.). Although it is 6–7 m long (approximately 20 feet, or as tall as a two-story building) in a cadaver, the small intestine is only about 2–4 m (8–13 feet) long during life because of muscle tone.

The small intestine has three subdivisions: the duodenum, which is mostly retroperitoneal, and the jejunum and ileum, both intraperitoneal organs.

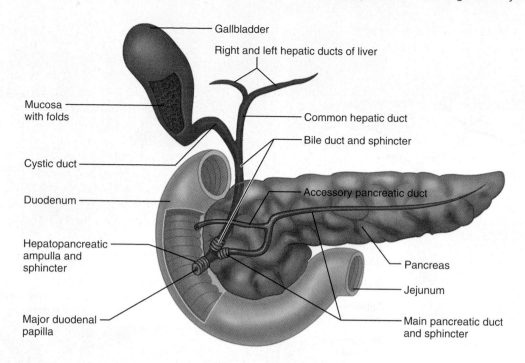

FIGURE 22.20 The duodenum of the small intestine, and related organs.
Emptying into the duodenum are ducts from the pancreas, gallbladder, and liver.

The relatively immovable **duodenum** (du″o-de′num; "twelve finger widths long"), which curves around the head of the pancreas (see Figure 22.20), is about 25 cm (10 inches) long. Although it is the shortest intestinal subdivision, the duodenum has the most features of interest. The bile duct, delivering bile from the liver, and the main pancreatic duct, carrying pancreatic juice from the pancreas, unite in the wall of the duodenum in a bulblike point called the **hepatopancreatic ampulla** (hep″ah-to-pan″kre-at′ik am-pul′ah; *ampulla* = flask). The ampulla opens into the duodenum via the volcano-shaped **major duodenal papilla.** The entry of bile and pancreatic juice is controlled by a muscular valve called the **hepatopancreatic sphincter, or sphincter of Oddi.**

The **jejunum** (jĕ-joo′num; "empty"), about 2.5 m (8 feet) long, extends from the duodenum to the ileum. The **ileum** (il′e-um; "twisted"), approximately 3.6 m (12 feet) in length, joins the large intestine at the ileocecal valve. The jejunum and ileum hang in sausagelike coils in the central and lower part of the abdominal cavity, suspended from the posterior abdominal wall by the fan-shaped **mesentery** (see Figure 22.30). These more distal parts of the small intestine are encircled and framed by the large intestine.

Nerve fibers serving the small intestine include parasympathetics from the vagus and sympathetics from the thoracic splanchnic nerves, both relayed through the superior mesenteric (and celiac) plexus. The arterial supply is primarily from the superior mesenteric artery (p. 646). The veins parallel the arteries and typically drain into the superior mesenteric vein. From there, the nutrient-rich venous blood from the small intestine drains into the hepatic portal vein, which carries it to the liver.

Microscopic Anatomy

Modifications for Absorption

The small intestine is highly adapted for nutrient absorption. Its length alone provides a huge surface area, and its wall has three structural modifications—plicae circulares, villi, and microvilli—that amplify its absorptive surface enormously. One estimate is that the intestinal surface area is about 200 m² (equal to the floor space of an average two-story house)! Most absorption occurs in the proximal part of the small intestine, so these specializations decrease in number toward its distal end.

The **circular folds,** or **plicae circulares** (pli′ke ser″ku-lar′ēs) are deep, permanent folds of the mucosa and submucosa (Figure 22.21a). Nearly 1 cm tall, these folds force chyme to spiral through the lumen, slowing its movement and allowing time for full nutrient absorption.

Villi (vil′i; "tufts of hair") are fingerlike projections of the mucosa, over 1 mm high, that give it a velvety texture, much like the soft nap of a towel (Figures 22.21b and 22.22a). The epithelial cells of the villi are chiefly absorptive columnar cells. In the core of each villus is a dense capillary bed and a wide lymph capillary called a **lacteal** (lak′te-al). Digested

?   *What is a lacteal and what is its function?*

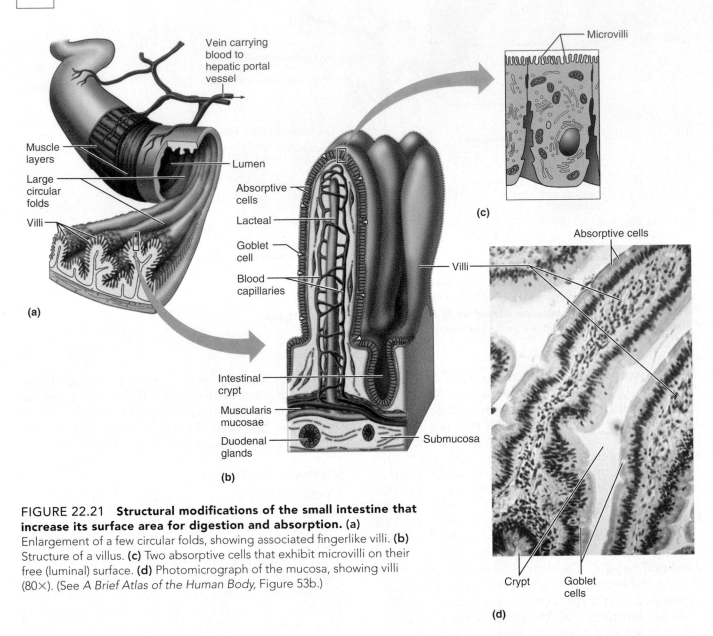

FIGURE 22.21   Structural modifications of the small intestine that increase its surface area for digestion and absorption. (a) Enlargement of a few circular folds, showing associated fingerlike villi. **(b)** Structure of a villus. **(c)** Two absorptive cells that exhibit microvilli on their free (luminal) surface. **(d)** Photomicrograph of the mucosa, showing villi (80×). (See *A Brief Atlas of the Human Body,* Figure 53b.)

foodstuffs are absorbed through the epithelial cells into both the capillary blood and the lacteal. The villi are large and leaflike in the duodenum (the intestinal site of most active absorption) and gradually narrow and shorten along the length of the small intestine. A "slip" of smooth muscle in the villus core allows it to alternately shorten and lengthen, pulsations that (1) increase the contact between the villus and the contents of the intestinal lumen, making absorption more efficient, and (2) "milk" lymph along through the lacteals.

Microvilli, tiny projections of the plasma membrane of the absorptive cells of the mucosa, give the mucosal surface a fuzzy appearance called the **brush border** (Figures 22.21c and 22.22b). The plasma membranes of the microvilli bear enzymes referred to as **brush border enzymes,** which complete the digestion of carbohydrates and proteins in the small intestine.

Histology of the Wall   Externally the subdivisions of the small intestine appear to be nearly identical, but their internal and microscopic anatomies

A lymph capillary that receives some of the fatty materials absorbed from the intestine. ∎

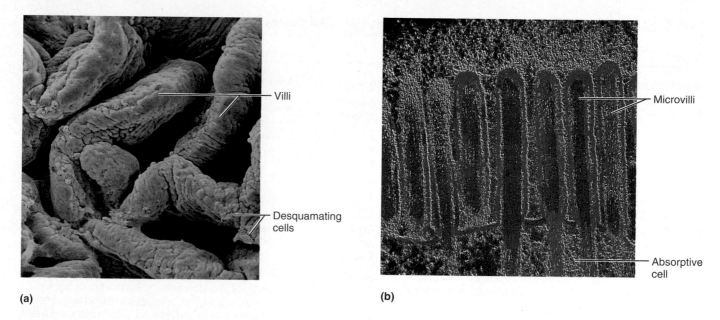

FIGURE 22.22 **Villi and microvilli of the small intestine. False-color scanning electron micrographs. (a)** Villi (700×). Dead desquamating cells seen on the ridges of the villi are colored yellow and orange. **(b)** Microvilli of absorptive cells appear as red projections from the surface of the absorptive cell. Yellow granules are mucus granules (40,000×).

reveal some important differences. The four tunics typical of the GI tract are also seen here, but the mucosa and submucosa are modified to reflect the intestine's functions in the digestive pathway.

The epithelium of the mucosa is largely simple columnar *absorptive cells* bound by tight junctions and richly endowed with microvilli. It also has many mucus-secreting *goblet cells;* scattered *enteroendocrine cells* (the source of the enterogastrones—secretin and cholecystokinin to name two), and interspersed T cells called *intraepithelial lymphocytes* (IELs), which represent an important immunological compartment. Unlike thymus-derived T cells, IELs are generated locally and do *not* need "priming." Upon encountering antigens, they immediately release cytokines that cause killing of infected target cells.

Between the villi, the mucosa is studded with *pits* that lead into tubular intestinal glands called **intestinal crypts,** or **crypts of Lieberkühn** (le'ber-kun). The epithelial cells that line these crypts secrete *intestinal juice,* a watery mixture containing mucus that serves as a carrier fluid for absorbing nutrients from chyme. Deep in the crypts are specialized secretory cells called *Paneth cells,* which fortify the small intestine's defenses by releasing *lysozyme,* an antibacterial enzyme. The crypts decrease in number along the length of the small intestine, but the goblet cells become more abundant.

The various epithelial cells arise from rapidly dividing stem cells at the base of the crypts. The daughter cells gradually migrate up the villi, where they are shed from the villus tips, renewing the villus epithelium every three to six days. The rapid replacement of intestinal (and gastric) epithelial cells has clinical as well as physiological significance. Treatments for cancer, such as radiation therapy and chemotherapy, preferentially target rapidly dividing cells. This kills cancer cells, but it also nearly obliterates the GI tract epithelium, causing nausea, vomiting, and diarrhea after each treatment.

The submucosa is typical areolar connective tissue, and it contains both individual and *aggregated lymphoid follicles,* the latter called **Peyer's patches** (pi'erz). Peyer's patches increase in abundance toward the end of the small intestine, reflecting the fact that this region of the small intestine contains huge numbers of bacteria that must be prevented from entering the bloodstream. Elaborate mucus-secreting **duodenal *(Brunner's)* glands** are found in the submucosa of the duodenum only. These glands (see Figure 22.21b) produce an alkaline (bicarbonate-rich) mucus that helps neutralize the acidic chyme moving in from the stomach. When this protective mucus barrier is inadequate, the intestinal wall erodes and *duodenal ulcers* result.

The muscularis is typical and bilayered. Except for the bulk of the duodenum, which is retroperitoneal and has an adventitia, the external intestinal surface is covered by visceral peritoneum (serosa).

TABLE 22.2   Overview of the Functions of the Gastrointestinal Organs

Organ	Major Functions*	Comments/Additional Functions
Mouth and associated accessory organs	■ Ingestion: food is voluntarily placed into oral cavity ■ Propulsion: voluntary (buccal) phase of deglutition (swallowing) initiated by tongue; propels food into pharynx ■ Mechanical digestion: mastication (chewing) by teeth and mixing movements by tongue ■ Chemical digestion: chemical breakdown of starch is begun by salivary amylase present in saliva produced by salivary glands	Mouth serves as a receptacle; most functions performed by associated accessory organs; mucus present in saliva helps dissolve foods so they can be tasted and moistens food so that tongue can compact it into a bolus that can be swallowed; oral cavity and teeth cleansed and lubricated by saliva
Pharynx and esophagus	■ Propulsion: peristaltic waves move food bolus to stomach, thus accomplishing involuntary (pharyngeal-esophageal) phase of deglutition	Primarily food chutes; mucus produced helps to lubricate food passageways
Stomach	■ Mechanical digestion and propulsion: peristaltic waves mix food with gastric juice and propel it into the duodenum ■ Chemical digestion: digestion of proteins begun by pepsin ■ Absorption: absorbs a few fat-soluble substances (aspirin, alcohol, some drugs)	Also serves as storage site for food until it can be moved into the duodenum; hydrochloric acid produced is a bacteriostatic agent and activates protein-digesting enzymes; mucus produced helps lubricate and protect stomach from self-digestion; intrinsic factor produced is required for intestinal absorption of vitamin B_{12}
Small intestine and associated accessory organs (liver, gallbladder, pancreas)	■ Mechanical digestion and propulsion: segmentation by smooth muscle of the small intestine continually mixes contents with digestive juices and moves food along tract and through ileocecal valve at a slow rate, allowing sufficient time for digestion and absorption ■ Chemical digestion: digestive enzymes conveyed in from pancreas and brush border enzymes attached to villi membranes complete digestion of all classes of foods ■ Absorption: breakdown products of carbohydrate, protein, fat, and nucleic acid digestion; vitamins; electrolytes; and water are absorbed by active and passive mechanisms	Small intestine is highly modified for digestion and absorption (circular folds, villi, and microvilli); alkaline mucus produced by intestinal glands and bicarbonate-rich juice ducted in from pancreas help neutralize acidic chyme and provide proper environment for enzymatic activity; bile produced by liver emulsifies fats and enhances (1) fat digestion and (2) absorption of fatty acids, monoglycerides, cholesterol, phospholipids, and fat-soluble vitamins; gallbladder stores and concentrates bile; bile is released to small intestine in response to hormonal signals
Large intestine	■ Chemical digestion: some remaining food residues are digested by enteric bacteria (which also produce vitamin K and some B vitamins) ■ Absorption: absorbs most remaining water, electrolytes (largely NaCl), and vitamins produced by bacteria ■ Propulsion: propels feces toward rectum by peristalsis, haustral churning, and mass movements ■ Defecation: reflex triggered by rectal distension; eliminates feces from body	Temporarily stores and concentrates residues until defecation can occur; copious mucus produced by goblet cells eases passage of feces through colon

*The colored boxes beside the functions correspond to the color coding of digestive functions (gastrointestinal tract activities) illustrated in Figure 22.2.

Intestinal Juice: Composition and Control

The intestinal glands normally secrete 1 to 2 L of **intestinal juice** daily. The major stimulus for its production is distension or irritation of the intestinal mucosa by hypertonic or acidic chyme. Normally, intestinal juice is slightly alkaline (7.4–7.8), and isotonic with blood plasma. Intestinal juice is largely water but it also contains some mucus, which is secreted both by the duodenal glands and by goblet cells of the mucosa. Intestinal juice is relatively enzyme-poor because intestinal enzymes are largely limited to the bound enzymes of the brush border.

The Liver and Gallbladder

The *liver* and *gallbladder* are accessory organs associated with the small intestine. The liver, one of the body's most important organs, has many metabolic and regulatory roles. However, its *digestive* function is to produce bile for export to the duodenum. Bile is a fat emulsifier; that is, it breaks up fats into tiny particles so that they are more accessible to digestive enzymes. We will describe bile and the emulsification process when we discuss the digestion and absorption of fats later in the chapter. Although the liver also processes nutrient-laden venous blood delivered to it from the digestive organs, this is a metabolic rather than a digestive role. (The metabolic functions of the liver are described in Chapter 23.) The gallbladder is chiefly a storage organ for bile.

Gross Anatomy of the Liver

The ruddy, blood-rich **liver** is the largest gland in the body, weighing about 1.4 kg (3 pounds) in the average adult. Shaped like a wedge, it occupies most of the right hypochondriac and epigastric regions, extending farther to the right of the body midline than to the left. Located under the diaphragm, the liver lies almost entirely within the rib cage, which provides some protection (see Figures 22.1 and 22.23c).

Typically, the liver is said to have four primary lobes. The largest of these, the *right lobe*, is visible on all liver surfaces and separated from the smaller *left lobe* by a deep fissure (Figure 22.23a). The posteriormost *caudate lobe* and the *quadrate lobe*, which lies inferior to the left lobe, are visible in an inferior view of the liver (Figure 22.23b). A mesentery, the **falciform ligament**, separates the right and left lobes anteriorly and suspends the liver from the diaphragm and anterior abdominal wall (see also Figure 22.30a and d, p. 794). Running along the inferior edge of the falciform ligament is the **round ligament**, or **ligamentum teres** (te'rēz; "round"), a fibrous remnant of the fetal umbilical vein. Except for the superiormost liver area (the *bare area*), which touches the diaphragm, the entire liver is enclosed by the visceral peritoneum. As mentioned earlier, a dorsal mesentery, the lesser omentum (see Figure 22.30b), anchors the liver to the lesser curvature of the stomach. The **hepatic artery** and the **hepatic portal vein**, which enter the liver at the **porta hepatis** ("gateway to the liver"), and the common hepatic duct, which runs inferiorly from the liver, all travel through the lesser omentum to reach their destinations. The gallbladder rests in a recess on the inferior surface of the right liver lobe.

The traditional scheme of defining liver lobes (outlined above) has been criticized because it is based on superficial features of the liver. Some anatomists emphasize that the primary lobes of the liver should be defined as the territories served by the right and left hepatic ducts. These two territories are delineated by a plane drawn from the indentation (sulcus) of the inferior vena cava to the gallbladder recess (see Figure 22.23b). Those to the right of the plane are the *right lobe* and those to its left constitute the *left lobe*. According to this scheme, the small quadrate and caudate lobes are part of the left lobe.

Bile leaves the liver through several bile ducts that ultimately fuse to form the large **common hepatic duct**, which travels downward toward the duodenum. Along its course, that duct fuses with the **cystic duct** draining the gallbladder to form the **bile duct** (Figure 22.20).

Microscopic Anatomy of the Liver

The liver is composed of sesame seed–sized structural and functional units called **liver lobules** (Figure 22.24). Each lobule is a roughly hexagonal (six-sided) structure consisting of plates of *liver cells*, or **hepatocytes** (hep'ah-to-sīts), organized like bricks in a garden wall. The hepatocyte plates radiate outward from a **central vein** running in the longitudinal axis of the lobule. To make a rough "model" of a liver lobule, open a thick paperback book until its two covers meet: The pages represent the plates of hepatocytes and the hollow cylinder formed by the rolled spine represents the central vein.

If you keep in mind that the liver's main function is to filter and process the nutrient-rich blood delivered to it, the description of its anatomy that follows will make a lot of sense. At each of the six corners of a lobule is a **portal triad** (*portal tract* region), so named because three basic structures are always present there: a branch of the *hepatic artery* (supplying oxygen-rich arterial blood to the liver), a branch of the *hepatic portal vein* (carrying venous blood laden with nutrients from the digestive viscera), and a *bile duct*. Between the hepatocyte plates are enlarged, leaky capillaries, the **liver sinusoids.** Blood from both the hepatic portal vein and the hepatic artery percolates from the triad regions through these sinusoids and empties into the central vein.

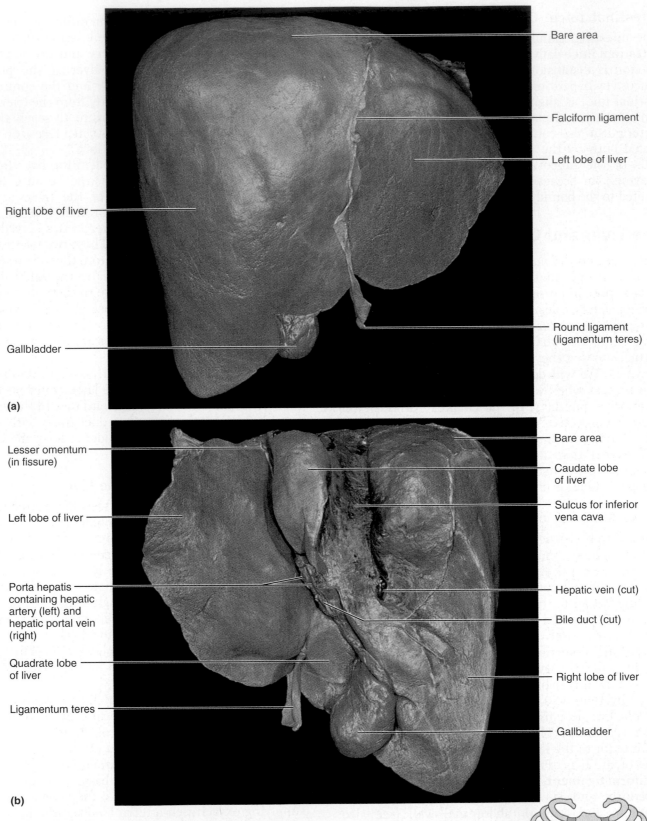

Bare area

Falciform ligament

Left lobe of liver

Right lobe of liver

Round ligament
(ligamentum teres)

Gallbladder

(a)

Lesser omentum
(in fissure)

Bare area

Caudate lobe
of liver

Sulcus for inferior
vena cava

Left lobe of liver

Porta hepatis
containing hepatic
artery (left) and
hepatic portal vein
(right)

Hepatic vein (cut)

Bile duct (cut)

Quadrate lobe
of liver

Right lobe of liver

Ligamentum teres

Gallbladder

(b)

FIGURE 22.23 Gross anatomy of the human liver.
(a) Anterior view of the liver. **(b)** Posteroinferior aspect (visceral
surface) of the liver. The four liver lobes are separated by a
group of fissures in this view. The porta hepatis is a deep
fissure that contains the hepatic portal vein, hepatic artery,
hepatic duct, and lymphatics. (Note: Another scheme classifies
the caudate and quadrate lobes as part of the left liver lobe.)
(c) Diagrammatic view of the liver's location in the body. (See *A
Brief Atlas of the Human Body,* Figures 54 and 56.)

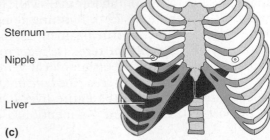

Sternum

Nipple

Liver

(c)

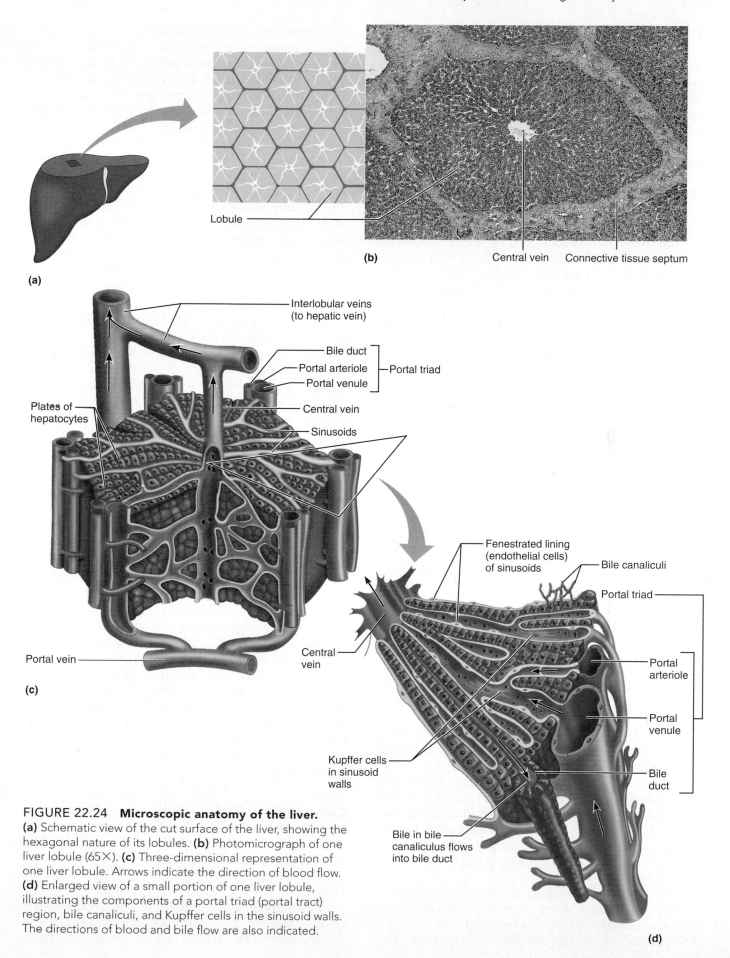

(a)

(b)

Lobule

Central vein   Connective tissue septum

Interlobular veins
(to hepatic vein)

Bile duct
Portal arteriole — Portal triad
Portal venule

Central vein

Sinusoids

Plates of
hepatocytes

Portal vein

(c)

Fenestrated lining
(endothelial cells)
of sinusoids

Bile canaliculi

Portal triad

Central
vein

Portal
arteriole

Portal
venule

Kupffer cells
in sinusoid
walls

Bile
duct

Bile in bile
canaliculus flows
into bile duct

(d)

FIGURE 22.24 Microscopic anatomy of the liver.
(a) Schematic view of the cut surface of the liver, showing the
hexagonal nature of its lobules. **(b)** Photomicrograph of one
liver lobule (65×). **(c)** Three-dimensional representation of
one liver lobule. Arrows indicate the direction of blood flow.
(d) Enlarged view of a small portion of one liver lobule,
illustrating the components of a portal triad (portal tract)
region, bile canaliculi, and Kupffer cells in the sinusoid walls.
The directions of blood and bile flow are also indicated.

From the central veins blood eventually enters the hepatic veins, which drain the liver, and empty into the inferior vena cava. Forming part of the sinusoid walls are star-shaped **hepatic macrophages,** also called **Kupffer cells** (koop'fer) (Figure 22.24d), which remove debris such as bacteria and worn-out blood cells from the blood as it flows past.

The versatile hepatocytes have large amounts of both rough and smooth ER, Golgi apparatuses, peroxisomes, and mitochondria. Thus equipped, the hepatocytes not only can produce bile but can also (1) process the bloodborne nutrients in various ways (e.g., they store glucose as glycogen and use amino acids to make plasma proteins); (2) store fat-soluble vitamins; and (3) play important roles in detoxification, such as ridding the blood of ammonia by converting it to urea (Chapter 23). Thanks to these various jobs, the blood leaving the liver contains fewer nutrients and waste materials than the blood that entered it.

Secreted bile flows through tiny canals, called **bile canaliculi** (kan"ah-lik'u-li; "little canals"), that run between adjacent hepatocytes toward the bile duct branches in the portal triads (see Figure 22.24d). Notice that blood and bile flow in opposite directions in the liver lobule. Bile entering the bile ducts eventually leaves the liver via the common hepatic duct to travel toward the duodenum.

Ⓗ HOMEOSTATIC IMBALANCE

Hepatitis (hep"ah-ti'tis), or inflammation of the liver, is most often due to viral infection. Thus far the catalogue of hepatitis-causing viruses numbers six—from the hepatitis A virus (HVA) to the hepatitis F virus (HVF). Of these, two (HVA and HVE) are transmitted enterically and the infections they cause tend to be self-limiting. Those transmitted via blood—most importantly HVB and HVC—are linked to chronic hepatitis and liver cirrhosis (see discussion below). HVD is a mutated virus that needs HVB to be infectious. Thus far, little is known about HVF.

In the United States, over 40% of hepatitis cases are due to HVB, which is transmitted via blood transfusions, contaminated needles, or sexual contact. A serious problem in its own right, hepatitis B carries with it a greater menace—an elevated risk of liver cancer. However, childhood immunization using vaccine produced in bacteria is sweeping the feet out from under HVB, and the incidence of acute hepatitis from this strain has fallen dramatically since its peak in 1985.

Hepatitis A, accounting for about 32% of cases, is a more benign form frequently observed in day-care centers. It is transmitted via sewage-contaminated food, raw shellfish, water, and by the feces-mouth route, which explains why it is important for restaurant employees to scrub their hands after using the washroom. Hepatitis E is transmitted in a similar way. It causes waterborne epidemics, largely in developing countries, and is a major cause of death in pregnant women. It is relatively insignificant in the United States.

Hepatitis C has emerged as the most important liver disease in the United States because it produces persistent or chronic liver infections (as opposed to acute infections). More than 10,000 Americans die annually due to sequels of HVC infection. However, the life-threatening C form of hepatitis is now being successfully treated by combination drug therapy (either the immune-suppressing steroid prednisone and genetically engineered alpha interferon or ribavirin, an antiviral drug, and interferon, depending on the hepatitis strain).

Nonviral causes of acute hepatitis include drug toxicity and wild mushroom poisoning.

Cirrhosis (sĭ-ro'sis; "orange colored") is a diffuse and progressive chronic inflammation of the liver that typically results from chronic alcoholism or severe chronic hepatitis. The alcohol-poisoned or damaged hepatocytes regenerate, but the liver's connective (scar) tissue regenerates faster. As a result, the liver becomes fatty and fibrous and its activity is depressed.

As the scar tissue shrinks, it obstructs blood flow throughout the hepatic portal system, causing **portal hypertension.** Fortunately, some veins of the portal system anastomose with veins that drain into the venae cavae *(portal-caval anastomoses)*. The main ones are (1) veins in the inferior esophagus, (2) hemorrhoidal veins in the anal canal, and (3) superficial veins around the umbilicus. However, these connecting veins are small and tend to burst when forced to carry large volumes of blood. Signs of their failure include vomiting blood, and a snakelike network of distended veins surrounding the navel. This network is called *caput medusae* (kap'ut mĕ-du'se; "medusa head") after a monster in Greek mythology whose hair was made of writhing snakes. Other complications due to portal hypertension include swollen veins in the esophagus (esophageal varices) and *ascites*, an accumulation of fluid in the peritoneal cavity. ●

Composition of Bile Bile is a yellow-green, alkaline solution containing bile salts, bile pigments, cholesterol, neutral fats, phospholipids (lecithin and others), and a variety of electrolytes. Of these *only* bile salts and phospholipids aid the digestive process.

Bile salts, primarily cholic acid and chenodeoxycholic acids, are cholesterol derivatives. Their role is to *emulsify* fats—that is, to distribute them

FIGURE 22.25 **Mechanisms promoting secretion of bile.** When digestion is not occurring, bile is stored and concentrated in the gallbladder. When acidic chyme enters the small intestine, several mechanisms are initiated that accelerate the liver's output of bile and cause the gallbladder to contract and the hepatopancreatic sphincter to relax. This allows bile (and pancreatic juice) to enter the small intestine. The single most important stimulus of bile secretion is an increased level of bile salts in the enterohepatic circulation.

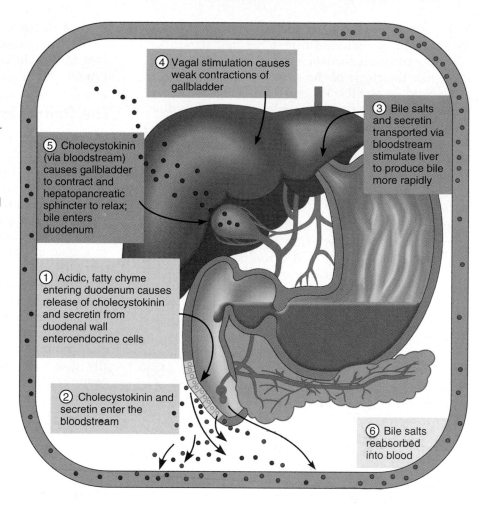

④ Vagal stimulation causes weak contractions of gallbladder

③ Bile salts and secretin transported via bloodstream stimulate liver to produce bile more rapidly

⑤ Cholecystokinin (via bloodstream) causes gallbladder to contract and hepatopancreatic sphincter to relax; bile enters duodenum

① Acidic, fatty chyme entering duodenum causes release of cholecystokinin and secretin from duodenal wall enteroendocrine cells

② Cholecystokinin and secretin enter the bloodstream

⑥ Bile salts reabsorbed into blood

throughout the watery intestinal contents. (Another example of emulsification is homogenization, which distributes cream throughout the watery phase of milk.) As a result, large fat globules entering the small intestine are physically separated into millions of small fatty droplets that provide large surface areas for the fat-digesting enzymes to work on. Bile salts also facilitate fat and cholesterol absorption (discussed later) and help solubilize cholesterol, both that contained in bile and that entering the small intestine in food. Although many substances secreted in bile leave the body in feces, bile salts are not among them. Instead, bile salts are conserved by means of a recycling mechanism called the **enterohepatic circulation.** In this process, bile salts are (1) reabsorbed into the blood by the ileum, (2) returned to the liver via the hepatic portal blood, and then (3) resecreted in newly formed bile.

The chief bile pigment is **bilirubin** (bil″ĭ-roo'bin), a waste product of the heme of hemoglobin formed during the breakdown of worn-out erythrocytes. The globin and iron parts of hemoglobin are saved and recycled, but bilirubin is absorbed from the blood by the liver cells, excreted into bile, and metabolized in the small intestine by resident

bacteria. One of its breakdown products, *urobilinogen* (u″ro-bi-lin'o-jen), gives feces a brown color. In the absence of bile, feces are gray-white in color and have fatty streaks (because essentially no fats are digested or absorbed).

The liver produces 500 to 1000 ml of bile daily, and production is stepped up when the GI tract contains fatty chyme. Bile salts themselves are the major stimulus for enhanced bile secretion (Figure 22.25), and when the enterohepatic circulation is returning large amounts of bile salts to the liver, its output of bile rises dramatically. *Secretin*, released by intestinal cells exposed to fatty chyme, stimulates liver cells to secrete bile.

The Gallbladder

The **gallbladder** is a thin-walled green muscular sac about 10 cm (4 inches) long, roughly the size of a kiwi fruit, that snuggles in a shallow fossa on the ventral surface of the liver (see Figures 22.1 and 22.23). Its rounded fundus protrudes from the inferior margin of the liver. The gallbladder stores bile that is not immediately needed for digestion and concentrates it by absorbing some of its water and ions. (In some cases, bile released from the gallbladder is ten times as concentrated as that entering it.) When empty, or when storing only small amounts of bile, its mucosa is thrown into honeycomb-like folds that, like the rugae of the stomach, allow the organ to expand as it fills. When its muscular wall contracts, bile is expelled into its duct, the **cystic duct,** and then flows into the bile duct. The gallbladder, like most of the liver, is covered by visceral peritoneum.

Regulation of Bile Release into the Small Intestine

When no digestion is occurring, the hepatopancreatic sphincter (guarding the entry of bile and pancreatic juice into the duodenum) is closed and the released bile backs up the cystic duct into the gallbladder, where it is stored until needed. Although the liver makes bile continuously, bile does not usually enter the small intestine until the gallbladder contracts. The major stimulus for gallbladder contraction is *cholecystokinin (CCK),* an intestinal hormone (Figure 22.25). CCK is released to the blood when acidic, fatty chyme enters the duodenum. Besides causing the gallbladder to contract, CCK has two other important effects: (1) it stimulates secretion of pancreatic juice, and (2) it relaxes the hepatopancreatic sphincter so that bile and pancreatic juice can enter the duodenum. CCK and the other digestive hormones are summarized in Table 22.1. Parasympathetic impulses delivered by the vagus nerves are a minor stimulus for gallbladder contraction.

ⓗ   *HOMEOSTATIC IMBALANCE*

Bile is the major vehicle for cholesterol excretion from the body, and bile salts keep the cholesterol dissolved within bile. Too much cholesterol or too few bile salts leads to cholesterol crystallization, forming **gallstones,** or *biliary calculi* (bil′e-a″re kal′ku-li), which obstruct the flow of bile from the gallbladder. Then, when the gallbladder or its duct contracts, the sharp crystals cause agonizing pain that radiates to the right thoracic region. Treatments for gallstones include dissolving the crystals with drugs, pulverizing them with ultrasound vibrations (lithotripsy), vaporizing them with lasers, and the classical treatment, surgically removing the gallbladder. When the gallbladder is removed, the bile duct enlarges to assume the bile-storing role. Gallstones are easy to diagnose because they show up well with ultrasound imaging.

Bile duct blockage prevents both bile salts and bile pigments from entering the intestine. As a result, yellow bile pigments accumulate in blood and eventually are deposited in the skin, causing it to become yellow, or *jaundiced.* Jaundice caused by blocked ducts is called *obstructive jaundice,* but jaundice may also reflect liver disease (in which the liver is unable to carry out its normal metabolic duties).   ●

The Pancreas

The **pancreas** (pan′kre-as; *pan* = all, *creas* = flesh, meat) is a soft, tadpole-shaped gland that extends across the abdomen from its *tail* (abutting the spleen) to its *head,* which is encircled by the C-shaped duodenum (see Figures 22.1 and 22.20). Most of the pancreas is retroperitoneal and lies deep to the greater curvature of the stomach.

An accessory digestive organ, the pancreas is important to the digestive process because it produces enzymes that break down all categories of foodstuffs, which the pancreas then delivers to the duodenum. This exocrine product, called **pancreatic juice,** drains from the pancreas via the centrally located **main pancreatic duct.** The pancreatic duct generally fuses with the bile duct just as it enters the duodenum (at the hepatopancreatic ampulla). A smaller *accessory pancreatic duct* empties directly into the duodenum just proximal to the main duct.

Within the pancreas are the **acini** (as′ĭ-ni; singular = acinus), clusters of secretory cells surrounding ducts (Figure 22.26). These cells are full of rough endoplasmic reticulum and exhibit deeply staining **zymogen granules** (zi′mo-jen; "fermenting") containing the digestive enzymes they manufacture.

Scattered amidst the acini are the more lightly staining *pancreatic islets (islets of Langerhans).* These mini-endocrine glands release insulin and glucagon, hormones that play an important role in carbohydrate metabolism, as well as several other hormones (see Chapter 15).

Composition of Pancreatic Juice

Approximately 1200 to 1500 ml of clear pancreatic juice is produced daily. It consists mainly of water, and contains enzymes and electrolytes (primarily bicarbonate ions). The acinar cells produce the enzyme-rich component of pancreatic juice. The epithelial cells lining the smallest pancreatic ducts release the bicarbonate ions that make it alkaline (about pH 8). The high pH of pancreatic fluid enables it to neutralize acid chyme entering the duodenum and provides the optimal environment for activity of intestinal and pancreatic enzymes. Like pepsin of the stomach, pancreatic proteases (protein-digesting enzymes) are produced and released in inactive forms, which are activated in the duodenum, where they do their work. This prevents the pancreas from self-digestion. For example, within the duodenum, *trypsinogen* is activated to **trypsin** by **enterokinase,** an intestinal brush border enzyme.

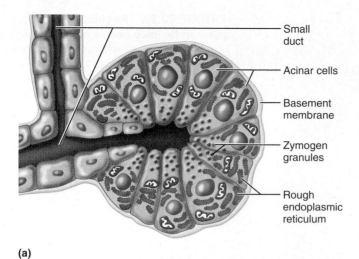

- Small duct
- Acinar cells
- Basement membrane
- Zymogen granules
- Rough endoplasmic reticulum

(a)

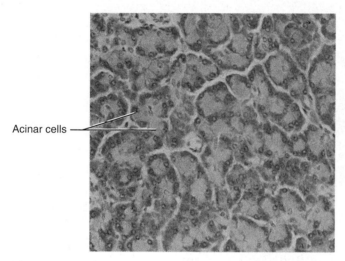

Acinar cells

(b)

FIGURE 22.26 Structure of the enzyme-producing tissue of the pancreas. (a) One acinus (a secretory unit). The cell apices contain abundant zymogen (enzyme-containing) granules. **(b)** Photomicrograph of pancreatic acinar tissue (200×).

Trypsin, in turn, activates two other pancreatic proteases (*procarboxypeptidase* and *chymotrypsinogen*) to their active forms, **carboxypeptidase** (kar-bok″se-pep′tĭ-dās) and **chymotrypsin** (ky″mo-trip′sin), respectively (Figure 22.27). Other pancreatic enzymes—**amylase, lipases,** and **nucleases**—are secreted in active form, but require that ions or bile be present in the intestinal lumen for optimal activity.

Regulation of Pancreatic Secretion

Secretion of pancreatic juice is regulated both by local hormones and by the parasympathetic nervous system (Figure 22.28). The more important hormonal mechanism is exerted by two intestinal

What is the significance of the fact that pancreatic enzymes are activated in the small intestine?

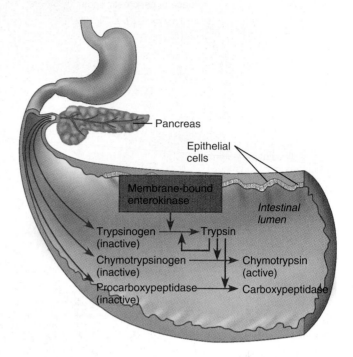

- Pancreas
- Epithelial cells
- Membrane-bound enterokinase
- *Intestinal lumen*
- Trypsinogen (inactive) → Trypsin
- Chymotrypsinogen (inactive) → Chymotrypsin (active)
- Procarboxypeptidase (inactive) → Carboxypeptidase

FIGURE 22.27 Activation of pancreatic proteases in the small intestine. Pancreatic proteases are secreted in an inactive form and are activated in the duodenum.

hormones. Both hormones act on the pancreas, but secretin released in response to the presence of HCl in the intestine targets the duct cells, prompting their release of watery *bicarbonate-rich* pancreatic juice, whereas CCK, released in response to the entry of proteins and fats, stimulates the acini to release *enzyme-rich* pancreatic juice. Vagal stimulation causes release of pancreatic juice primarily during the cephalic and gastric phases of gastric secretion.

Normally, the amount of HCl produced in the stomach is exactly balanced by the amount of bicarbonate (HCO_3^-) secreted by the pancreas, and as HCO_3^- is secreted into the pancreatic juice, H^+ enters the blood. Consequently, the pH of venous blood returning to the heart remains relatively unchanged because alkaline blood draining from the stomach is neutralized by the acidic blood draining the pancreas.

They are activated in the small intestine where they work and where the cells exposed to them are protected by mucus. No such protection exists in the pancreas. ■

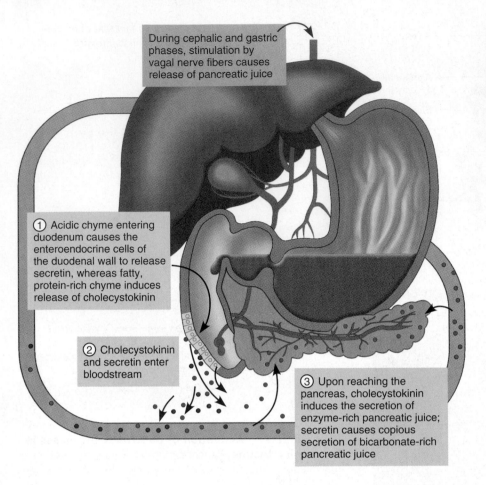

During cephalic and gastric phases, stimulation by vagal nerve fibers causes release of pancreatic juice

① Acidic chyme entering duodenum causes the enteroendocrine cells of the duodenal wall to release secretin, whereas fatty, protein-rich chyme induces release of cholecystokinin

② Cholecystokinin and secretin enter bloodstream

③ Upon reaching the pancreas, cholecystokinin induces the secretion of enzyme-rich pancreatic juice; secretin causes copious secretion of bicarbonate-rich pancreatic juice

FIGURE 22.28   **Regulation of pancreatic juice secretion.** Hormonal controls, exerted by secretin and cholecystokinin (steps 1–3), are the main regulatory factors. Neural control is mediated by parasympathetic fibers of the vagus nerves.

Digestive Processes Occurring in the Small Intestine

Food reaching the small intestine is unrecognizable, but it is far from being digested chemically. Carbohydrates and proteins are partially degraded, but virtually no fat digestion has occurred to this point. The process of food digestion is accelerated during the chyme's tortuous journey of 3 to 6 hours through the small intestine, and it is here that virtually all nutrient absorption occurs. Like the stomach, the small intestine plays no part in ingestion or defecation.

Requirements for Optimal Intestinal Digestive Activity

Although the primary functions of the small intestine are digestion and absorption, intestinal juice provides little of what is needed to perform these functions. Most substances required for chemical digestion—bile, digestive enzymes (except for the brush border enzymes), and bicarbonate ions (to provide the proper pH for enzymatic catalysis)—are *imported* from the liver and pancreas. Hence, anything that impairs liver or pancreatic function or delivery of their juices to the

small intestine severely hinders our ability to digest food and absorb nutrients.

Optimal digestive activity in the small intestine also depends on a slow, measured delivery of chyme from the stomach. Why is this so? Entering chyme is usually hypertonic. Thus, if large amounts of chyme were rushed into the small intestine, the osmotic water loss from the blood into the intestinal lumen would result in dangerously low blood volume. Additionally, the low pH of entering chyme must be adjusted upward and the chyme must be well mixed with bile and pancreatic juice for digestion to continue. These modifications take time. Thus, food movement into the small intestine is carefully controlled by the pumping action of the stomach pylorus (see pp. 777–778) to prevent the duodenum from being overwhelmed. Since the actual chemistry of digestion and absorption is covered in detail later, next we will examine how the small intestine mixes and propels food along its length, and how its motility is regulated.

Motility of the Small Intestine

Intestinal smooth muscle mixes chyme thoroughly with bile and pancreatic and intestinal juices, and moves food residues through the ileocecal valve into the large intestine. In contrast to the peristaltic waves of the stomach, which both mix and propel food, *segmentation* is the most common motion of the small intestine.

If X-ray fluoroscopy is used to examine the small intestine after it is "loaded" with a meal, it looks like the intestinal contents are being massaged—that is, the chyme is simply moved backward and forward in the lumen a few centimeters at a time by alternating contraction and relaxation of rings of smooth muscle. These segmenting movements of the intestine (see Figure 22.3b), like the peristalsis of the stomach, are initiated by intrinsic pacemaker cells in the longitudinal smooth muscle layer. However, unlike the stomach pacemakers, which have only one rhythm, the pacemakers in the duodenum depolarize more frequently (12–14 contractions per minute) than those in the ileum (8 or 9 contractions per minute). As a result, segmentation moves intestinal contents slowly and steadily toward the ileocecal valve at a rate that allows ample time to complete digestion and absorption. The intensity of segmentation is altered by long and short reflexes (which parasympathetic activity enhances and sympathetic activity decreases) and hormones. The more intense the contractions, the greater the mixing effect; however, the basic contractile rhythms of the various intestinal regions remain unchanged.

True peristalsis occurs only after most nutrients have been absorbed. At this point, segmenting movements wane, and peristaltic waves initiated in the duodenum begin to sweep slowly along the intestine, moving 10–70 cm before dying out. Each successive wave is initiated a bit more distally, and this pattern of peristaltic activity is called the **migrating mobility complex.** A complete "trip " from duodenum to ileum takes about two hours. The process then repeats itself, sweeping the last remnants of the meal plus bacteria, sloughed-off mucosal cells, and other debris into the large intestine. This "housekeeping" function is critical for preventing the overgrowth of bacteria that migrate from the large intestine into the small intestine. As food again enters the stomach with the next meal, peristalsis is replaced by segmentation.

The local enteric neurons of the GI tract wall coordinate these intestinal mobility patterns. The physiological diversity of the enteric neurons allows a variety of effects to occur depending on which neurons are activated or inhibited. For example, a given ACh-releasing (cholinergic) sensory neuron in the small intestine, once activated, may simultaneously send messages to several different interneurons in the myenteric plexus that regulate peristalsis:

- Impulses sent proximally by cholinergic neurons cause contraction and shortening of the circular muscle layer.

- Impulses sent distally to certain interneurons cause shortening of the longitudinal muscle layer and distension of the intestine, in response to ACh-releasing neurons.

- Other impulses sent distally by activated VIP or by NO-releasing enteric neurons cause relaxation of the circular muscle.

As a result, as the proximal area constricts and forces chyme along the tract, the lumen of the distal part of the intestine enlarges to receive it.

Most of the time, the ileocecal sphincter is constricted and closed. However, two mechanisms—one neural and the other hormonal—cause it to relax and allow food residues to enter the cecum when ileal mobility increases. Enhanced activity of the stomach initiates the **gastroileal reflex** (gas"tro-il'e-ul), a long reflex that enhances the force of segmentation in the ileum. In addition, gastrin released by the stomach increases the motility of the ileum and relaxes the ileocecal sphincter. Once the chyme has passed through, it exerts backward pressure that closes the valve's flaps, preventing regurgitation into the ileum.

The Large Intestine

The **large intestine** (see Figure 22.1) frames the small intestine on three sides and extends from the ileocecal valve to the anus. Its diameter, at about 7 cm, is greater than that of the small intestine (hence, *large* intestine), but it is less than half as long (1.5 m versus 6 m). In terms of nutrient absorption, its major function is to absorb water from indigestible food residues (delivered to it in a fluid state) and eliminate them from the body as semisolid **feces** (fe'sēz).

Gross Anatomy

The large intestine exhibits three features not seen elsewhere—teniae coli, haustra, and epiploic appendages. Except for its terminal end, the longitudinal muscle layer of its muscularis is reduced to three bands of smooth muscle called **teniae coli** (ten'ne-e ko'li; "ribbons of the colon"). Their tone causes the wall of the large intestine to pucker into pocketlike sacs called **haustra** (haw'strah; "to draw up"). Another obvious feature of the large intestine is its **epiploic appendages** (ep"ĭ-plo'ik), small fat-filled pouches of visceral peritoneum that hang from its surface (Figure 22.29a). Their significance is not known.

Subdivisions

The large intestine has the following subdivisions: cecum, appendix, colon, rectum, and anal canal.

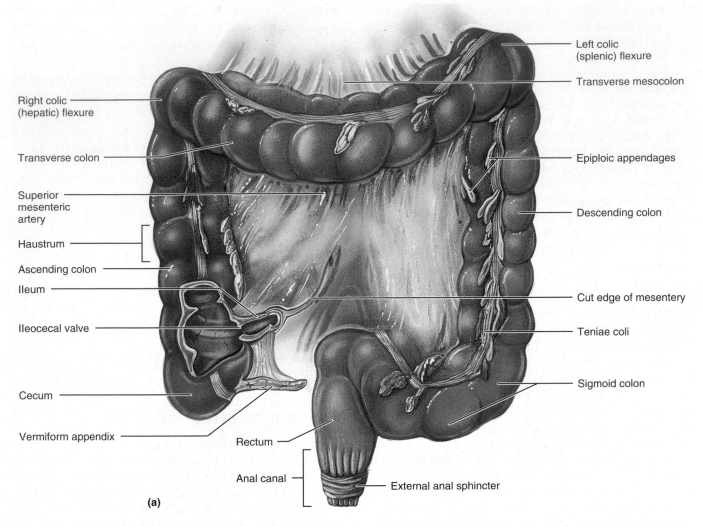

Left colic
(splenic) flexure

Transverse mesocolon

Right colic
(hepatic) flexure

Transverse colon

Superior
mesenteric
artery

Haustrum

Ascending colon

Ileum

Ileocecal valve

Cecum

Vermiform appendix

Epiploic appendages

Descending colon

Cut edge of mesentery

Teniae coli

Sigmoid colon

Rectum

Anal canal

External anal sphincter

(a)

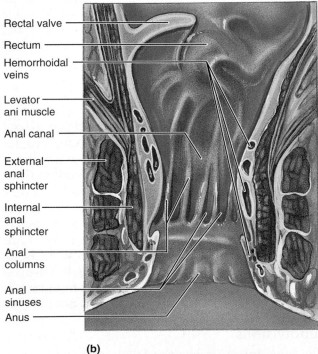

Rectal valve

Rectum

Hemorrhoidal
veins

Levator
ani muscle

Anal canal

External
anal
sphincter

Internal
anal
sphincter

Anal
columns

Anal
sinuses

Anus

(b)

FIGURE 22.29   Gross anatomy of the large intestine.
(a) Diagrammatic view. **(b)** Structure of the anal canal.

The saclike **cecum** (se'kum; "blind pouch"), which lies below the ileocecal valve in the right iliac fossa, is the first part of the large intestine (Figure 22.29a). Attached to its posteromedial surface is the blind, wormlike **vermiform appendix.** The appendix contains masses of lymphoid tissue, and as part of MALT (see p. 670), it plays an important role in body immunity. However, it has an important structural shortcoming—its twisted structure provides an ideal location for enteric bacteria to accumulate and multiply.

HOMEOSTATIC IMBALANCE

Acute inflammation of the appendix, or **appendicitis,** results from a blockage (often by feces) that traps infectious bacteria in its lumen. Unable to empty its contents, the appendix swells, squeezing off venous drainage, which may lead to ischemia and necrosis (death and decay) of the appendix. If the appendix ruptures, feces containing bacteria spray over the abdominal contents, causing *peritonitis*. Although the symptoms of appendicitis are variable, the first is usually pain in the umbilical region. Loss of appetite, nausea and vomiting, and relocalization of pain to the lower right abdominal quadrant follow. Immediate surgical removal of the appendix (appendectomy) is the accepted treatment for suspected appendicitis. Appendicitis is most common during adolescence, when the entrance to the appendix is at its widest. ●

The **colon** has several distinct regions. As the **ascending colon,** it travels up the right side of the abdominal cavity to the level of the right kidney. Here it makes a right-angle turn—the **right colic,** or **hepatic, flexure**—and travels across the abdominal cavity as the **transverse colon.** Directly anterior to the spleen, it bends acutely at the **left colic (splenic) flexure** and descends down the left side of the posterior abdominal wall as the **descending colon.** Inferiorly, it enters the pelvis, where it becomes the S-shaped **sigmoid colon.** Except for the transverse and sigmoid parts of the colon, which are anchored to the posterior abdominal wall by mesentery sheets called **mesocolons** (see Figure 22.30c and d), the colon is retroperitoneal.

In the pelvis, at the level of the third sacral vertebra, the sigmoid colon joins the **rectum,** which runs posteroinferiorly just in front of the sacrum. The position of the rectum allows a number of pelvic organs (e.g., the prostate gland of males) to be examined digitally (with a finger) through the anterior rectal wall. This is called a **rectal exam.** Despite its name (*rectum* = straight), the rectum has three lateral curves or bends, represented internally as three transverse folds called **rectal valves** (Figure 22.29b). These valves separate feces from flatus, that is, they stop feces from being passed along with gas.

The **anal canal,** the last segment of the large intestine, lies entirely external to the abdominopelvic cavity. About 3 cm long, the anal canal begins where the rectum penetrates the levator ani muscle of the pelvic floor and it opens to the body exterior at the **anus.** The anal canal has two sphincters, an involuntary **internal anal sphincter** composed of smooth muscle (part of the muscularis), and a voluntary **external anal sphincter** composed of skeletal muscle. The sphincters, which act rather like purse strings to open and close the anus, are ordinarily closed except during defecation.

Microscopic Anatomy

The wall of the large intestine (Figure 22.31) differs in several ways from that of the small intestine. The colon *mucosa* is simple columnar epithelium except in the anal canal. Because most food is absorbed before reaching the large intestine, there are no circular folds, no villi, and virtually no cells that secrete digestive enzymes. However, its mucosa is thicker, its abundant crypts are deeper, and there are tremendous numbers of goblet cells in the crypts (Figure 22.31). Mucus produced by goblet cells eases the passage of feces and protects the intestinal wall from irritating acids and gases released by resident bacteria in the colon.

The mucosa of the anal canal, a stratified squamous epithelium, merges with the true skin surrounding the anus and is quite different from that of the rest of the colon, reflecting the greater abrasion that this region receives. It hangs in long ridges or folds called **anal columns. Anal sinuses** (Figure 22.29b), recesses between the anal columns, exude mucus when compressed by feces, which aids in emptying the anal canal. The horizontal line that parallels the inferior margins of the anal sinuses is called the *pectinate line.* Superior to this line, the mucosa is innervated by visceral sensory fibers and is relatively insensitive to pain. The area inferior to the pectinate line is very sensitive to pain, a reflection of the somatic sensory fibers serving it. Two superficial venous plexuses are associated with the anal canal, one with the anal columns and the other with the anus itself. If these (hemorrhoidal) veins are inflamed, itchy varicosities called *hemorrhoids* result.

In contrast to the more proximal regions of the large intestine, teniae coli and haustra are absent in the rectum and anal canal. Consistent with its need to generate strong contractions to perform its expulsive role in defecation, the rectum's muscularis muscle layers are complete and well developed.

Bacterial Flora

Although most bacteria entering the cecum from the small intestine are dead (having been killed by the action of lysozyme, defensins, HCl, and protein-digesting enzymes), some are still "alive and kicking." Together with bacteria that enter the GI tract via the anus, these constitute the **bacterial flora** of the large intestine. These bacteria colonize the colon and ferment some of the indigestible carbohydrates (cellulose and others), releasing irritating acids and a mixture of gases (including dimethyl sulfide, H_2, N_2, CH_4, and CO_2). Some of these gases (such as dimethyl sulfide) are quite odorous. About 500 ml of gas (flatus) is produced each day, much more when certain carbohydrate-rich foods (such as beans) are eaten. The bacterial flora also synthesize B complex vitamins and most of the vitamin K the liver requires to synthesize some of the clotting proteins.

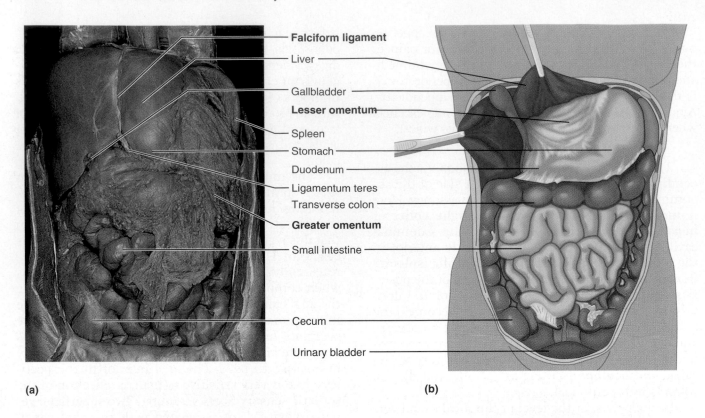

(a)

Falciform ligament
Liver
Gallbladder
Lesser omentum
Spleen
Stomach
Duodenum
Ligamentum teres
Transverse colon
Greater omentum
Small intestine

Cecum

Urinary bladder

(b)

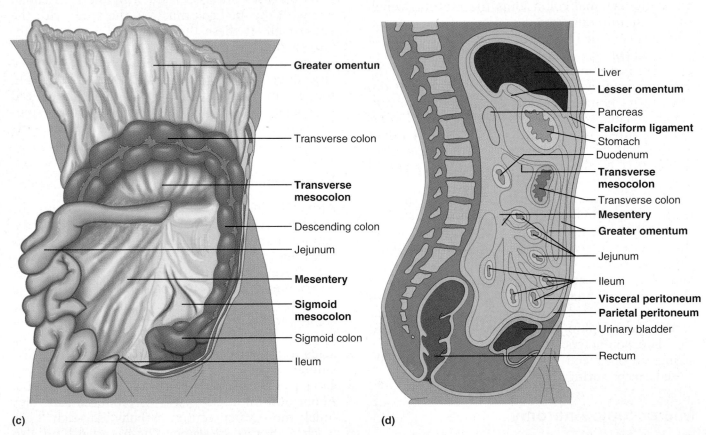

(c)

Greater omentun

Transverse colon

Transverse mesocolon

Descending colon

Jejunum

Mesentery

Sigmoid mesocolon

Sigmoid colon

Ileum

(d)

Liver
Lesser omentum
Pancreas
Falciform ligament
Stomach
Duodenum
Transverse mesocolon
Transverse colon
Mesentery
Greater omentum
Jejunum
Ileum
Visceral peritoneum
Parietal peritoneum
Urinary bladder
Rectum

FIGURE 22.30 Mesenteries of the abdominal digestive organs. (a) The greater omentum, a dorsal mesentery, is shown in its normal position covering the abdominal viscera. **(b)** The lesser omentum, a ventral mesentery attaching the liver to the lesser curvature of the stomach (the liver and gallbladder have been reflected superiorly). **(c)** The greater omentum has been reflected superiorly to reveal the mesentery attachments of the small and large intestine. **(d)** Sagittal section of the abdominopelvic cavity of a male.

Digestive Processes Occurring in the Large Intestine

What is finally delivered to the large intestine contains few nutrients, but it still has 12 to 24 hours more to spend there. Except for a small amount of digestion of that residue by the enteric bacteria, no further food breakdown occurs in the large intestine.

Although the large intestine harvests vitamins made by the bacterial flora and reclaims most of the remaining water and some of the electrolytes (particularly sodium and chloride), absorption is not the *major* function of this organ. As considered in the next section, the primary concerns of the large intestine are propulsive activities that force the fecal material toward the anus and then eliminate it from the body (defecation).

While the large intestine is important for our comfort, it is not essential for life. If the colon is removed, as may be necessitated by colon cancer, the terminal ileum is brought out to the abdominal wall in a procedure called an *ileostomy* (il"e-os'to-me), and food residues are eliminated from there into a sac attached to the abdominal wall. Another surgical technique, *ileoanal juncture,* links the ileum directly to the anal canal.

Motility of the Large Intestine

The large intestine musculature is inactive much of the time, and when it is mobile, its contractions are sluggish or short lived. The movements most seen in the colon are **haustral contractions,** slow segmenting movements that occur every 30 minutes or so. These contractions reflect local controls of smooth muscle within the walls of the individual haustra. As a haustrum fills with food residue, the distension stimulates its muscle to contract, which propels the luminal contents into the next haustrum. These movements also mix the residue, which aids in water absorption.

Mass movements (mass peristalsis) are long, slow-moving, but powerful contractile waves that move over large areas of the colon three or four times daily and force the contents toward the rectum. Typically, they occur during or just after eating, which indicates that the presence of food in the stomach activates the gastroileal reflex in the small intestine and the propulsive **gastrocolic reflex** in the colon. Bulk, or fiber, in the diet increases the strength of colon contractions and softens the stool, allowing the colon to act like a well-oiled machine.

🅗 *HOMEOSTATIC IMBALANCE*

When the diet lacks bulk and the volume of residues in the colon is small, the colon narrows and its con-

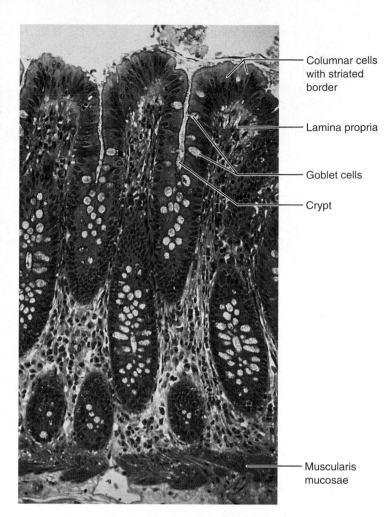

FIGURE 22.31   The mucosa of the large intestine.
Light photomicrograph (500×).

Labels on figure:
- Columnar cells with striated border
- Lamina propria
- Goblet cells
- Crypt
- Muscularis mucosae

tractions become more powerful, increasing the pressure on its walls. This promotes formation of *diverticula* (di"ver-tik'u-lah), small herniations of the mucosa through the colon walls. This condition, called **diverticulosis,** most commonly occurs in the sigmoid colon, and affects over half of people over the age of 70. **Diverticulitis,** in which the diverticula become inflamed, can be life threatening if the diverticula rupture. ●

The semisolid product delivered to the rectum, feces, or the stool, contains undigested food residues, mucus, sloughed-off epithelial cells, millions of bacteria, and just enough water to allow its smooth passage. Of the 500 ml or so of food residue entering the cecum daily, approximately 150 ml becomes feces.

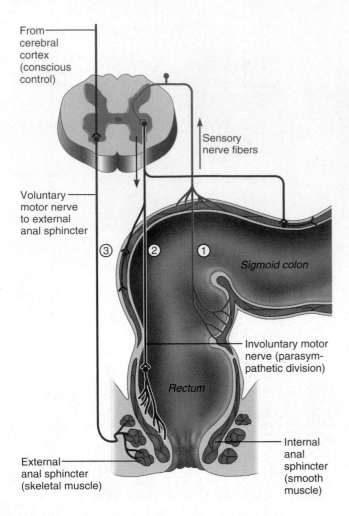

FIGURE 22.32 Defecation reflex. ① Distension, or stretch, of the rectal walls caused by movement of feces into the rectum triggers the depolarization of sensory (afferent) fibers, which synapse with spinal cord neurons. ② Parasympathetic motor (efferent) fibers, in turn, stimulate contraction of the rectal walls and relaxation of the internal anal sphincter. ③ If it is convenient to defecate, voluntary signals stimulate relaxation of the external anal sphincter. Defecation may be delayed temporarily by conscious (cortical) controls, which allow voluntary constriction of the external sphincter.

Defecation

The rectum is usually empty, but when feces are forced into it by mass movements, stretching of the rectal wall initiates the **defecation reflex** (Figure 22.32). This is a spinal cord–mediated parasympathetic reflex that causes the sigmoid colon and the rectum to contract and the anal sphincters to relax. As feces are forced into the anal canal, messages reach the brain allowing us to decide whether the external (voluntary) anal sphincter should remain open or be constricted to stop feces passage temporarily. If defecation is delayed, the reflex contractions end within a few seconds, and

the rectal walls relax. With the next mass movement, the defecation reflex is initiated again—and so on, until one chooses to defecate or the urge to defecate becomes unavoidable.

During defecation, the muscles of the rectum contract to expel the feces. We aid this process voluntarily by closing the glottis and contracting our diaphragm and abdominal wall muscles to increase the intra-abdominal pressure (a procedure called *Valsalva's maneuver*). We also contract the levator ani muscle (pp. 306–307), which lifts the anal canal superiorly. This lifting action leaves the feces below the anus—and outside the body. Involuntary or automatic defecation (incontinence of feces) occurs in infants because they have not yet gained control of their external anal sphincter. It also occurs in those with spinal cord transections.

⟨H⟩ *HOMEOSTATIC IMBALANCE*

Watery stools, or **diarrhea,** result from any condition that rushes food residue through the large intestine before that organ has had sufficient time to absorb the remaining water (as in irritation of the colon by bacteria). Prolonged diarrhea may result in dehydration and electrolyte imbalance. Conversely, when food remains in the colon for extended periods, too much water is absorbed and the stool becomes hard and difficult to pass. This condition, called **constipation,** may result from lack of fiber in the diet, improper bowel habits (failing to heed the "call"), lack of exercise, emotional upset, or laxative abuse. ●

PART 3: PHYSIOLOGY OF CHEMICAL DIGESTION AND ABSORPTION

So far in this chapter, we have examined the structure and overall function of the organs that make up the digestive system. Now we will investigate the entire chemical processing (enzymatic breakdown and absorption) of each class of foodstuffs as it moves through the GI tract. These matters are summarized in Figure 22.33.

Chemical Digestion

After foodstuffs have spent even a short time in the stomach, they are unrecognizable. Nonetheless, they are still mostly the starchy foods, meat proteins, butter, and so on that we ingested. Only their appearance has been changed by mechanical digestive activities. By contrast, the products of chemical

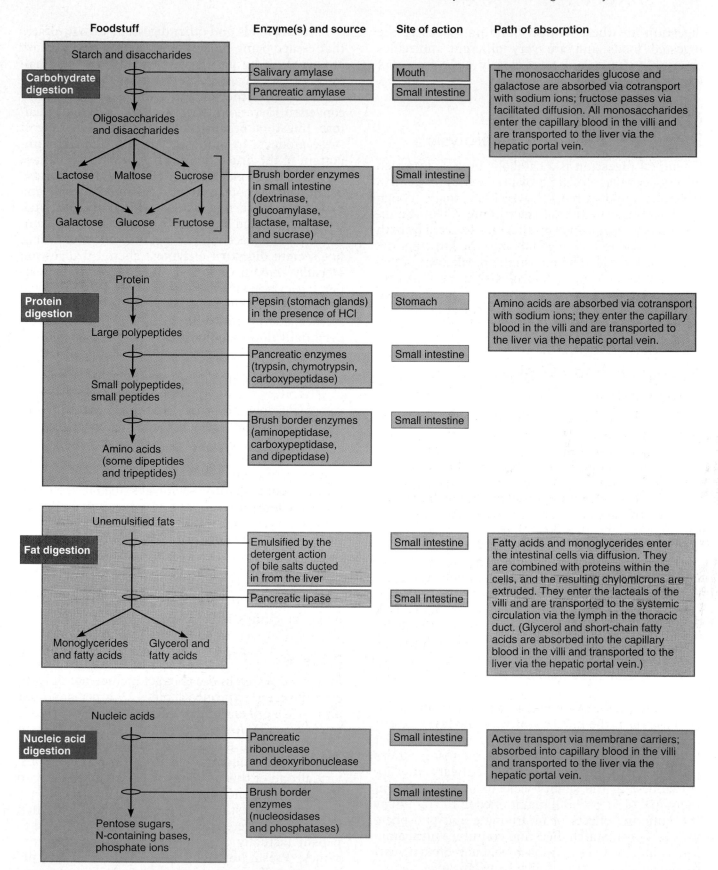

FIGURE 22.33 **Flowchart of chemical digestion and absorption of foodstuffs.**

digestion are the chemical building blocks of the ingested foods and are very different molecules chemically. Let us look more closely at the process of chemical digestion.

Mechanism of Chemical Digestion: Enzymatic Hydrolysis

Chemical digestion is a catabolic process in which large food molecules are broken down to *monomers* (chemical building blocks), which are small enough to be absorbed by the GI tract lining. Chemical digestion is accomplished by enzymes secreted by both intrinsic and accessory glands into the lumen of the alimentary canal. The enzymatic breakdown of any type of food molecule is **hydrolysis** (hi-drol'ĭ-sis) because it involves addition of a water molecule to each molecular bond to be broken (lysed).

Chemical Digestion of Specific Food Groups

Carbohydrates

Most of us ingest between 200 and 600 grams of carbohydrate foods each day. **Monosaccharides** (simple sugars), the monomers of carbohydrates, are absorbed immediately without further ado. Only three of these are common in our diet: *glucose, fructose,* and *galactose.* The more complex carbohydrates that our digestive system is able to break down to monosaccharides are the disaccharides *sucrose* (table sugar), *lactose* (milk sugar), and *maltose* (grain sugar) and the polysaccharides *glycogen* and *starch.* In the average diet, most digestible carbohydrates are in the form of starch, with smaller amounts of disaccharides and monosaccharides. Humans lack enzymes capable of breaking down most other polysaccharides, such as cellulose. Thus, indigestible polysaccharides do not nourish us but they do help move the food along the GI tract by providing bulk, or fiber.

Chemical digestion of starch (and perhaps glycogen) begins in the mouth. **Salivary amylase,** present in saliva, splits starch into *oligosaccharides,* smaller fragments of two to eight linked monosaccharides (in this case glucose molecules). Salivary amylase works best in the slightly acid to neutral environment (pH of 6.75–7.00) maintained in the mouth by the buffering effects of bicarbonate and phosphate ions in saliva. Starch digestion continues until amylase is inactivated by stomach acid and broken apart by the stomach's protein-digesting enzymes. Generally speaking, the larger the meal, the longer amylase continues to work in the stomach because foodstuffs in its relatively immobile fundus are poorly mixed with gastric juices.

Starchy foods and other digestible carbohydrates that escape being broken down by salivary amylase are acted on by **pancreatic amylase** in the small intestine (see Figure 22.33). Within about ten minutes of entering the small intestine, starch is entirely converted to various oligosaccharides, mostly maltose. Intestinal brush border enzymes further digest these products to monosaccharides. The most important of the brush border enzymes are **dextrinase** and **glucoamylase,** which act on oligosaccharides composed of more than three simple sugars, and **maltase, sucrase,** and **lactase,** which hydrolyze maltose, sucrose, and lactose respectively into their constituent monosaccharides. Because the colon does not secrete digestive enzymes, chemical digestion *officially ends* in the small intestine. As noted earlier, however, resident colon bacteria do break down and metabolize the residual complex carbohydrates further, adding much to their own nutrition but essentially nothing to ours.

🅗 HOMEOSTATIC IMBALANCE

In some people, intestinal lactase is present at birth but then becomes deficient, probably due to genetic factors. In such cases, the person becomes intolerant of milk products (the source of lactose), and undigested lactose creates osmotic gradients that not only prevent water from being absorbed in the small and large intestines but also pull water from the interstitial space into the intestines. The result is diarrhea. Bacterial metabolism of the undigested solutes produces large amounts of gas that result in bouts of bloating, flatulence, and cramping pain. The solution to this problem is simple—add lactase enzyme "drops" to your milk or take a lactose tablet before meals containing milk products. ●

Proteins

Proteins digested in the GI tract include not only dietary proteins (typically about 125 g per day), but also 15–25 g of enzyme proteins secreted into the GI tract lumen by its various glands and (probably) an equal amount of protein derived from sloughed and disintegrating mucosal cells. In healthy individuals, virtually all of this protein is digested all the way to its **amino acid** monomers.

Protein digestion begins in the stomach when pepsinogen secreted by the chief cells is activated to **pepsin** (actually a group of protein-digesting enzymes). Pepsin functions optimally in the acidic pH range found in the stomach: 1.5–2.5. It preferentially cleaves bonds involving the amino acids tyrosine and phenylalanine so that proteins are broken into polypeptides and small numbers of free amino acids (see Figure 22.33). Pepsin, which hydrolyzes 10–15% of ingested protein, is inactivated by the high pH in the duodenum, so its proteolytic activity is restricted to the stomach. **Rennin** (the enzyme that coagulates milk protein) is not produced in adults.

Protein fragments entering the small intestine from the stomach are greeted by a host of proteolytic enzymes (Figure 22.33). **Trypsin** and **chymotrypsin** secreted by the pancreas cleave the proteins into smaller peptides, which in turn become the grist for other enzymes. The pancreatic and brush border enzyme **carboxypeptidase** splits off one amino acid at a time from the end of the polypeptide chain that bears the carboxyl group. Other brush border enzymes such as **aminopeptidase** and **dipeptidase** liberate the final amino acid products (Figure 22.34). Aminopeptidase digests a protein, one amino acid at a time, by working from the amine end. Both carboxypeptidase and aminopeptidase can independently dismantle a protein, but the teamwork between these enzymes and between trypsin and chymotrypsin, which attack the more internal parts of the protein, speeds up the process tremendously.

Lipids

Although the American Heart Association recommends a low-fat diet, the amount of lipids (fats) ingested daily varies tremendously among American adults, ranging from 30 g to 150 g or more. The small intestine is essentially the sole site of lipid digestion because the pancreas is the only significant source of fat-digesting enzymes, or **lipases** (Figure 22.33). Neutral fats (triglycerides or triacylglycerols) are the most abundant fats in the diet.

Because triglycerides and their breakdown products are insoluble in water, fats need special "pretreatment" with bile salts to be digested and absorbed in the watery environment of the small intestine. In aqueous solutions, triglycerides aggregate to form large fat globules, and only the triglyceride molecules at the surfaces of such fatty masses are accessible to the water-soluble lipase enzymes. However, this problem is quickly resolved because as the fat globules enter the duodenum, they are coated with detergent-like bile salts (Figure 22.35). Bile salts have both nonpolar and polar regions. Their nonpolar (hydrophobic) parts cling to the fat molecules, and their polar (ionized hydrophilic) parts allow them to repel each other and to interact with water. As a result, fatty droplets are pulled off the large fat globules, and a stable *emulsion*—an aqueous suspension of fatty droplets, each about 1 μm in diameter—is formed. The process of emulsification does *not* break chemical bonds. It just reduces the attraction between fat molecules so that they can be

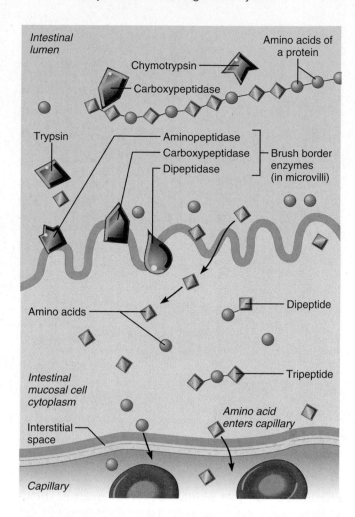

FIGURE 22.34 Protein digestion and absorption in the small intestine. Proteins and protein fragments are digested to amino acids by the action of pancreatic proteases (trypsin, chymotrypsin, and carboxypeptidase), and by brush border enzymes of the intestinal mucosal cells. The amino acids are then absorbed by active transport into the capillary blood of the villi.

more widely dispersed. This process vastly increases the number of triglyceride molecules exposed to the pancreatic lipases. Without bile, lipids would be incompletely digested in the time food is in the small intestine.

The pancreatic lipases catalyze the breakdown of fats by cleaving off two of the fatty acid chains, thus yielding free **fatty acids** and **monoglycerides** (glycerol with one fatty acid chain attached). Fat-soluble vitamins that ride with fats require no digestion.

Nucleic Acids

Both DNA and RNA, present in small amounts in foods, are hydrolyzed to their **nucleotide** monomers by **pancreatic nucleases** present in pancreatic juice. The nucleotides are then broken apart by intestinal

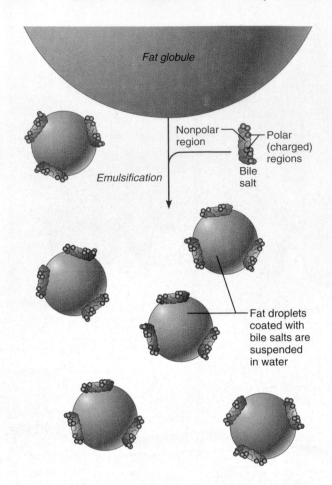

FIGURE 22.35 Role of bile salts in fat emulsification. As large aggregates of fats enter the small intestine, bile salts cling via their nonpolar parts to the fat molecules (triglycerides). Their polar parts, facing the aqueous phase, interact with water and repel each other, causing the fatty globule to be physically broken up into smaller fat droplets.

brush border enzymes (**nucleosidases** and **phosphatases**), which release their free bases, pentose sugars, and phosphate ions (see Figure 22.33).

Absorption

Up to 10 L of food, drink, and GI secretions enter the alimentary canal daily, but only 1 L or less reaches the large intestine. Virtually all of the food-stuffs, 80% of the electrolytes, and most of the water are absorbed in the small intestine. Although absorption occurs all along the length of the small intestine, most of it is completed by the time chyme reaches the ileum. Hence, the major absorptive role of the ileum is to reclaim bile salts to be recycled back to the liver for re-secretion. At the end of the ileum, all that remains is some water, indigestible food materials (largely plant fibers such as cellulose), and millions of bacteria. This debris is passed on to the large intestine.

Most nutrients are absorbed through the mucosa of the intestinal villi by *active transport* processes driven directly or indirectly (secondarily) by metabolic energy (ATP). They then enter the capillary blood in the villus to be transported in the hepatic portal vein to the liver. The exception is some of the lipid digestion products, which are absorbed passively by diffusion and then enter the lacteal in the villus to be carried to the blood via lymphatic fluid. Because the epithelial cells of the intestinal mucosa are joined at their luminal surfaces by tight junctions, substances cannot move *between* the cells. Consequently, materials must pass *through* the epithelial cells and into the interstitial fluid abutting their basal membranes (via *transepithelial transport*) if they are to enter the blood capillaries. Absorption of each nutrient class is described next.

Absorption of Specific Nutrients

Carbohydrates

The monosaccharides glucose and galactose, liberated by the breakdown of starch and disaccharides, are transported into the epithelial cells by common protein carriers and then move by facilitated diffusion into the capillary blood. The carriers, which are located very close to the disaccharidase enzymes on the microvilli, combine with the monosaccharides as soon as the disaccharides are broken down. These sugars are conveyed by secondary active transport coupled to sodium ion transport (cotransport). By contrast, fructose moves *entirely* by facilitated diffusion.

Proteins

Several types of carriers transport the different amino acids resulting from protein digestion. These carriers, like those for glucose and galactose, are coupled to the active transport of sodium. Short chains of two or three amino acids (dipeptides and tripeptides, respectively) are also actively absorbed, but are digested to their amino acids within the epithelial cells before entering the capillary blood by diffusion.

HOMEOSTATIC IMBALANCE

Whole proteins are not usually absorbed, but in rare cases intact proteins are taken up by endocytosis and released on the opposite side of the epithelial cell by exocytosis. This process, most common in newborn infants, reflects the immaturity of their intestinal mucosa and accounts for many early food allergies. The immune system "sees" the intact proteins as antigenic and mounts an attack. These food allergies usually disappear as the mucosa matures. This mechanism may also provide a route for IgA antibodies present in breast milk to reach an infant's bloodstream. These antibodies confer some passive immunity on the infant (temporary protection against antigens to which the mother has been sensitized). ●

Lipids

Just as bile salts accelerate lipid digestion, they are also essential for the absorption of its end products. As the water-insoluble products of fat digestion—the monoglycerides and free fatty acids—are liberated by lipase activity, they quickly become associated with bile salts and *lecithin* (a phospholipid found in bile) to form micelles. **Micelles** (mi-selz′) are collections of fatty elements clustered together with bile salts in such a way that the polar (hydrophilic) ends of the molecules face the water and the nonpolar portions form the core. Also nestled in the hydrophobic core are cholesterol molecules and fat-soluble vitamins. Although micelles are similar to emulsion droplets, they are much smaller "vehicles" and easily diffuse between microvilli to come into close contact with the mucosal cell surface (Figure 22.36). The various lipid substances then leave the micelles and move through the lipid phase of the plasma membrane by simple diffusion. Without the micelles, the lipids simply float on the surface of the chyme (like oil on water), inaccessible to the absorptive surfaces of the epithelial cells. Generally, fat absorption is completed in the ileum, but in the absence of bile (as might occur when a gallstone blocks the cystic duct) it happens so slowly that most of the fat passes into the large intestine and is lost in feces.

Once inside the epithelial cells, the free fatty acids and monoglycerides are resynthesized into triglycerides. The triglycerides are then combined with phospholipids and cholesterol, and coated with a "skin" of proteins to form water-soluble lipoprotein droplets called **chylomicrons** (ki″lo-mi′kronz). These are processed by the Golgi apparatus for extrusion from the cell. This series of events is quite different from the absorption of amino acids and simple sugars, which pass through the epithelial cells unchanged.

Although a few free fatty acids enter the capillary blood, the milky-white chylomicrons are too large to pass through the basement membranes of the blood capillaries and instead enter the more permeable lacteals. Thus, most fat enters the lymphatic stream and is eventually emptied into the venous blood in the neck region via the thoracic duct, which drains the digestive viscera. While in the bloodstream, the triglycerides of the chylomicrons are hydrolyzed to free fatty acids and glycerol by **lipoprotein lipase,** an

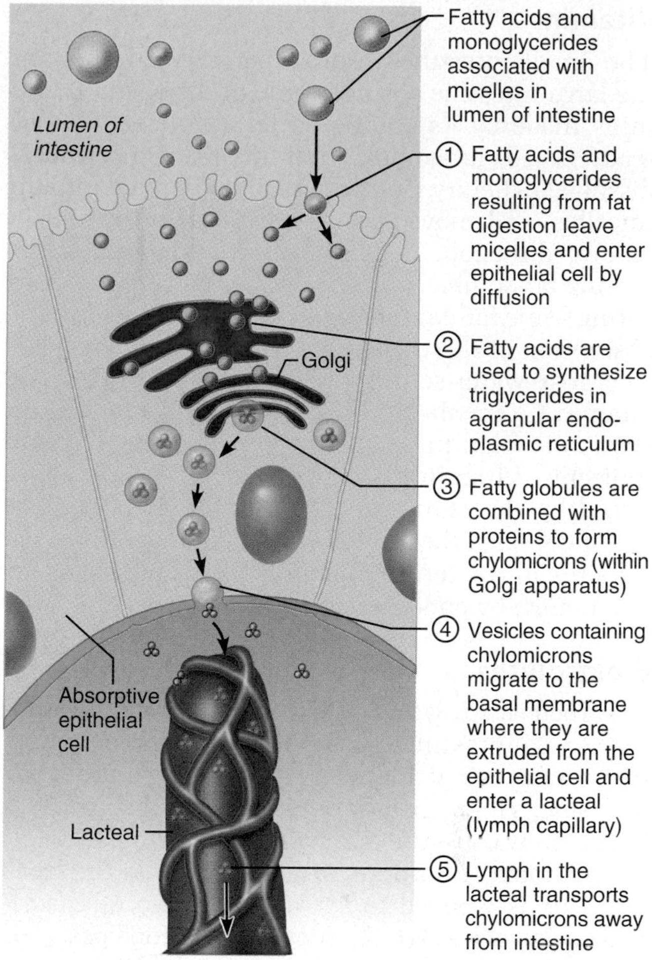

FIGURE 22.36 **Fatty acid absorption.** Digestion products of fat breakdown, lecithin, and cholesterol associate with bile salts to form micelles, which serve to "ferry" them to the intestinal mucosa. They then dissociate and enter the mucosal cells by diffusion. In the mucosal epithelial cells, they are recombined to lipids and packaged with other lipoid substances and protein to form chylomicrons. The chylomicrons are extruded from the epithelial cells by exocytosis and enter the lacteal for distribution in the lymph. Free fatty acids and monoglycerides enter the capillary bed.

enzyme associated with the capillary endothelium. The fatty acids and glycerol can then pass through the capillary walls to be used by tissue cells for energy or stored as fats in adipose tissue. The residual chylomicron material is combined with proteins by the liver cells, and these "new" lipoproteins are used to transport cholesterol in the blood.

Nucleic Acids

The pentose sugars, nitrogenous bases, and phosphate ions resulting from nucleic acid digestion are transported actively across the epithelium by special carriers in the villus epithelium. They then enter the blood.

Vitamins

The small intestine absorbs dietary vitamins, and the large intestine absorbs some of the K and B vitamins made by its enteric bacterial "guests." As already noted, fat-soluble vitamins (A, D, E, and K) dissolve in dietary fats, become incorporated into the micelles, and move across the villus epithelium by passive diffusion. It follows that gulping pills containing fat-soluble vitamins without simultaneously eating some fat-containing food results in little or no absorption of these vitamins.

Most water-soluble vitamins (B vitamins and vitamin C) are absorbed easily by diffusion. The exception is vitamin B_{12}, which is a very large, charged molecule. Vitamin B_{12} binds to *intrinsic factor*, produced by the stomach. The vitamin B_{12}–intrinsic factor complex then binds to specific mucosal receptor sites in the terminal ileum, which trigger its active uptake by endocytosis.

Electrolytes

Absorbed electrolytes come from both ingested foods and gastrointestinal secretions. Most ions are actively absorbed along the entire length of the small intestine. However, iron and calcium absorption is largely limited to the duodenum.

As noted, absorption of sodium ions in the small intestine is coupled to active absorption of glucose and amino acids. For the most part, anions passively follow the electrical potential established by sodium transport. That is, Na^+ is actively pumped out of the epithelial cells by a Na^+-K^+ pump after entering those cells. Chloride ions are also transported actively, and in the terminus of the small intestine HCO_3^- is actively secreted into the lumen in exchange for Cl^-.

Potassium ions move across the intestinal mucosa by simple diffusion in response to osmotic gradients. As water is absorbed from the lumen, the resulting rise in potassium levels in chyme creates a concentration gradient for its absorption. Thus, anything that interferes with water absorption (resulting in diarrhea) not only reduces potassium absorption but also "pulls" K^+ from the interstitial space into the intestinal lumen.

For most nutrients, the amount *reaching* the intestine is the amount absorbed, regardless of the nutritional state of the body. In contrast, absorption of iron and calcium is intimately related to the body's need for them at the time.

Ionic iron, essential for hemoglobin production, is actively transported into the mucosal cells, where it binds to the protein **ferritin** (fer'ĭ-tin). This phenomenon is called the *mucosal iron barrier*. The intracellular iron-ferritin complexes then serve as local storehouses for iron. When body reserves of iron are adequate, little is allowed to pass into the portal blood, and most of the stored iron is lost as the epithelial cells later slough off. However, when iron reserves are depleted (as during acute or chronic hemorrhage), iron uptake from the intestine and its release to the blood are accelerated. Menstrual bleeding is a major route of iron loss in females, and the intestinal epithelial cells of women have about four times as many iron transport proteins as do those of men. In the blood, iron binds to **transferrin,** a plasma protein that transports it in the circulation.

Calcium absorption is closely related to blood levels of ionic calcium. It is locally regulated by the active form of **vitamin D,** which acts as a cofactor to facilitate active calcium absorption. Decreased blood levels of ionic calcium prompt *parathyroid hormone (PTH)* release from the parathyroid glands. Besides facilitating the release of calcium ions from bone matrix and enhancing the reabsorption of calcium by the kidneys, PTH stimulates activation of vitamin D by the kidneys, which in turn accelerates calcium ion absorption in the small intestine.

Water

Approximately 9 L of water, mostly derived from GI tract secretions, enter the small intestine daily. Water is the most abundant substance in chyme, and 95% of it is absorbed in the small intestine by osmosis. The normal rate of water absorption is 300 to 400 ml per hour. Water moves freely in both directions across the intestinal mucosa, but *net osmosis* occurs whenever a concentration gradient is established by the active transport of solutes (particularly Na^+) into the mucosal cells. Thus, water uptake is effectively coupled to solute uptake and, in turn, affects the rate of absorption of substances that normally pass by diffusion. As water moves into the mucosal cells, these substances follow along their concentration gradients.

Malabsorption of Nutrients

Malabsorption, or impaired nutrient absorption, has many and varied causes. It can result from anything that interferes with the delivery of bile or pancreatic juice to the small intestine, as well as factors that damage the intestinal mucosa (severe bacterial infections and antibiotic therapy with neomycin) or reduce its absorptive surface area. A common but poorly understood malabsorption syndrome is *gluten enteropathy*, also called *adult celiac disease.* In this condition, gluten, a protein plentiful in some grains (wheat, rye, barley, oats), damages the intestinal villi and reduces the length of the microvilli of the brush border. The resulting diarrhea, pain, and malnutrition are usually controlled by eliminating gluten-containing grains (all grains but rice and corn) from the diet.

Review Questions

Multiple Choice/Matching

(Some questions have more than one correct answer. Select the best answer or answers from the choices given.)

1. The peritoneal cavity (a) is the same thing as the abdominopelvic cavity, (b) is filled with air, (c) like the pleural and pericardial cavities is a potential space containing serous fluid, (d) contains the pancreas and all of the duodenum.

2. Obstruction of the hepatopancreatic sphincter impairs digestion by reducing the availability of (a) bile and HCl, (b) HCl and intestinal juice, (c) pancreatic juice and intestinal juice, (d) pancreatic juice and bile.

3. The action of an enzyme is influenced by (a) its chemical surroundings, (b) its specific substrate, (c) the presence of needed cofactors or coenzymes, (d) all of these.

4. Carbohydrates are acted on by (a) peptidases, trypsin, and chymotrypsin, (b) amylase, maltase, and sucrase, (c) lipases, (d) peptidases, lipases, and galactase.

5. The parasympathetic nervous system influences digestion by (a) relaxing smooth muscle, (b) stimulating peristalsis and secretory activity, (c) constricting sphincters, (d) none of these.

6. The digestive juice product containing enzymes capable of digesting all four major foodstuff categories is (a) pancreatic, (b) gastric, (c) salivary, (d) biliary.

7. The vitamin associated with calcium absorption is (a) A, (b) K, (c) C, (d) D.

8. Someone has eaten a meal of buttered toast, cream, and eggs. Which of the following would you expect to happen? (a) Compared to the period shortly after the meal, gastric motility and secretion of HCl decrease when the food reaches the duodenum; (b) gastric motility increases even as the person is chewing the food (before swallowing); (c) fat will be emulsified in the duodenum by the action of bile; (d) all of these.

9. The site of production of GIP and cholecystokinin is (a) the stomach, (b) the small intestine, (c) the pancreas, (d) the large intestine.

10. Which of the following is not characteristic of the large intestine? (a) It is divided into ascending, transverse, and descending portions; (b) it contains abundant bacteria, some of which synthesize certain vitamins; (c) it is the main absorptive site; (d) it absorbs much of the water and salts remaining in the wastes.

11. The gallbladder (a) produces bile, (b) is attached to the pancreas, (c) stores and concentrates bile, (d) produces secretin.

12. The sphincter between the stomach and duodenum is (a) the pyloric sphincter, (b) the cardiac sphincter, (c) the hepatopancreatic sphincter, (d) the ileocecal sphincter.

In items 13–17, trace the path of a single protein molecule that has been ingested.

13. The protein molecule will be digested by enzymes secreted by (a) the mouth, stomach, and colon, (b) the stomach, liver, and small intestine, (c) the small intestine, mouth, and liver, (d) the pancreas, small intestine, and stomach.

14. The protein molecule must be digested before it can be transported to and utilized by the cells because (a) pro-

tein is only useful directly, (b) protein has a low pH, (c) proteins in the circulating blood produce an adverse osmotic pressure, (d) the protein is too large to be readily absorbed.

15. The products of protein digestion enter the bloodstream largely through cells lining (a) the stomach, (b) the small intestine, (c) the large intestine, (d) the bile duct.

16. Before the blood carrying the products of protein digestion reaches the heart, it first passes through capillary networks in (a) the spleen, (b) the lungs, (c) the liver, (d) the brain.

17. Having passed through the regulatory organ selected above, the products of protein digestion are circulated throughout the body. They will enter individual body cells as a result of (a) active transport, (b) diffusion, (c) osmosis, (d) phagocytosis.

Short Answer Essay Questions

18. Make a simple line drawing of the organs of the alimentary tube and label each organ. Then add three labels to your drawing—salivary glands, liver, and pancreas—and use arrows to show where each of these organs empties its secretion into the alimentary tube.

19. Sara was on a diet but she could not eat less and kept claiming her stomach had a mind of its own. She was joking, but indeed, there is a "gut brain" called the enteric nervous system. Is it part of the parasympathetic and sympathetic nervous system? Explain.

20. Name the layers of the alimentary tube wall. Note the tissue composition and major function of each layer.

21. What is a mesentery? Mesocolon? Greater omentum?

22. Name the six functional activities of the digestive system.

23. (a) Describe the boundaries of the oral cavity. (b) Why do you suppose its mucosa is stratified squamous epithelium rather than the more typical simple columnar epithelium?

24. (a) What is the normal number of permanent teeth? Of deciduous teeth? (b) What substance covers the tooth crown? Its root? (c) What substance makes up the bulk of a tooth? (d) What and where is pulp?

25. Describe the two phases of swallowing, noting the organs involved and the activities that occur.

26. Describe the role of these cells found in gastric glands: parietal, chief, mucous neck, and enteroendocrine.

27. Describe the regulation of the cephalic, gastric, and intestinal phases of gastric secretion.

28. (a) What is the relationship between the cystic, common hepatic, bile, and pancreatic ducts? (b) What is the point of fusion of the bile and pancreatic ducts called?

29. Explain why fatty stools result from the absence of bile or pancreatic juice.

30. Indicate the function of the Kupffer cells and the hepatocytes of the liver.

31. What are (a) brush border enzymes? (b) chylomicrons?

32. Name one inflammatory condition particularly common to adolescents, two common in middle age, and one common in old age.

23

NUTRITION, METABOLISM, AND BODY TEMPERATURE REGULATION

Nutrition (pp. 805–817)

1. Define nutrient, essential nutrient, and calorie.
2. List the six major nutrient categories. Note important sources and main cellular uses.
3. Distinguish between nutritionally complete and incomplete proteins.
4. Define nitrogen balance and indicate possible causes of positive and negative nitrogen balance.
5. Distinguish between fat- and water-soluble vitamins, and list the vitamins in each group.
6. For each vitamin, list important sources, body functions, and consequences of its deficit or excess.
7. List minerals essential for health; note important dietary sources and describe how each is used.

Metabolism (pp. 818–843)

8. Define metabolism. Explain how catabolism and anabolism differ.
9. Define oxidation and reduction and note the importance of these reactions in metabolism. Indicate the role of coenzymes used in cellular oxidation reactions.
10. Explain the difference between substrate-level phosphorylation and oxidative phosphorylation.
11. Follow the oxidation of glucose in body cells. Summarize important events and products of glycolysis, the Krebs cycle, and electron transport.
12. Define glycogenesis, glycogenolysis, and gluconeogenesis.

13. Describe the process by which fatty acids are oxidized for energy.
14. Define ketone bodies, and indicate the stimulus for their formation.
15. Describe how amino acids are metabolized for energy.
16. Describe the need for protein synthesis in body cells.
17. Explain the concept of amino acid or carbohydrate-fat pools and describe pathways by which substances in these pools can be interconverted.
18. List important events of absorptive and postabsorptive states, and explain how these events are regulated.
19. Describe several metabolic functions of the liver.
20. Differentiate between LDLs and HDLs relative to their structures and major roles in the body.

Energy Balance (pp. 843–850)

21. Explain what is meant by body energy balance.
22. Describe some current theories of food intake regulation.
23. Define basal metabolic rate and total metabolic rate. Name factors that influence each.
24. Describe how body temperature is regulated, and indicate the common mechanisms regulating heat production/ retention and heat loss from the body.

Are you a food lover? People can be divided into two camps according to their reactions to food—those who live to eat and those who eat to live. The saying "you are what you eat" is true in that part of the food we eat is converted to our living flesh. In other words, some nutrients are used to build cell structures, replace worn-out parts, and synthesize functional molecules. However, most nutrients we ingest are used as metabolic fuel. That is, they are oxidized and transformed to **ATP**, the chemical energy form used by cells. The energy value of foods is measured in **kilocalories** (kcal) or "large calories (C)." One kilocalorie is the amount of heat energy needed to raise the temperature of 1 kilogram of water 1°C (1.8°F) and is the unit conscientiously counted by dieters.

In Chapter 22, we considered how foods are digested and absorbed, but what happens to these foods once they enter the blood? Why do we need bread, meat, and fresh vegetables? Why does everything we eat seem to turn to fat? This chapter will try to answer these questions as it explains both the nature of nutrients and their metabolic roles.

Nutrition

A **nutrient** is a substance in food that is used by the body to promote normal growth, maintenance, and repair. The nutrients needed for health divide neatly into six categories. Three of these—carbohydrates, lipids, and proteins—are collectively called the **major nutrients** and make up the bulk of what we eat. The fourth and fifth categories, vitamins and minerals, though equally crucial for health, are required in minute amounts. In a strict sense, water, which accounts for about 60% by volume of the food

we eat, is also a major nutrient. However, because its importance in the body is described in Chapter 2, only the five nutrient categories listed above will be considered in detail here.

Most foods offer a combination of nutrients. For example, cream of mushroom soup contains all the major nutrients plus some vitamins and minerals. A diet consisting of foods from each of the five food groups—grains, fruits, vegetables, meats and fish, and milk products—normally guarantees adequate amounts of all the needed nutrients (Figure 23.1).

The ability of cells, especially those of the liver, to convert one type of molecule to another is truly remarkable. These interconversions allow the body to use the wide range of chemicals found in foods and to adjust to varying food intakes. But there are limits to this ability to conjure up new molecules from old. At least 45 and possibly 50 molecules, called **essential nutrients,** cannot be made by the body and so must be provided by the diet. As long as all the essential nutrients are ingested, the body can synthesize the hundreds of additional molecules required for life and good health. The use of the word "essential" to describe the chemicals that must be obtained from outside sources is unfortunate and misleading, to say the least, because both the essential and the nonessential nutrients are equally vital for normal functioning.

Carbohydrates

Dietary Sources

Except for milk sugar (lactose) and small amounts of glycogen in meats, all the carbohydrates we ingest are derived from plants. Sugars (monosaccharides and disaccharides) come from fruits, sugar cane,

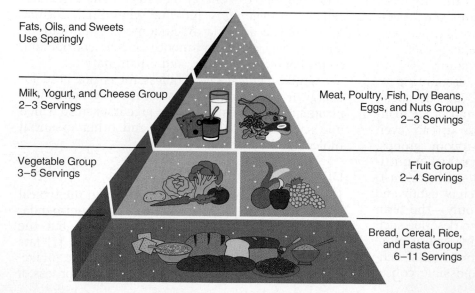

Fats, Oils, and Sweets
Use Sparingly

Milk, Yogurt, and Cheese Group
2–3 Servings

Meat, Poultry, Fish, Dry Beans,
Eggs, and Nuts Group
2–3 Servings

Vegetable Group
3–5 Servings

Fruit Group
2–4 Servings

Bread, Cereal, Rice,
and Pasta Group
6–11 Servings

FIGURE 23.1 Official food guide pyramid. Issued by the U.S. Department of Agriculture in 1992, this food pyramid suggests eating a variety of foods (with the emphasis on fruits, vegetables, and starches) for good health. Caloric intake ranges from 1500 calories to 2800 calories daily.

sugar beets, honey, and milk. The polysaccharide starch is found in grains, legumes, and root vegetables. Cellulose, a polysaccharide plentiful in most vegetables, is not digested by humans but provides roughage, or fiber, which increases the bulk of the stool and facilitates defecation.

Uses in the Body

The monosaccharide **glucose** is the carbohydrate molecule ultimately used by body cells. Carbohydrate digestion also yields fructose and galactose, but these monosaccharides are converted to glucose by the liver before they enter the general circulation. Glucose is a major body fuel and is readily used to make ATP. Although many body cells use fats as energy sources, neurons and red blood cells rely almost entirely on glucose for their energy needs. Because even a temporary shortage of blood glucose can severely depress brain function and lead to death of neurons, the body carefully monitors and regulates blood glucose levels. Any glucose in excess of what the body needs for ATP synthesis is converted to glycogen or fat and stored for later use.

Other uses of monosaccharides are meager. Small amounts of pentose sugars are used to synthesize nucleic acids, and a variety of sugars are attached to externally facing plasma membrane proteins and lipids.

Dietary Requirements

The low-carbohydrate diet of the Inuit (Eskimos) and the high-carbohydrate diet of peoples in the Far East indicate that humans can be healthy even with wide variations in carbohydrate intake. The minimum requirement for carbohydrates is not known, but 100 grams per day is presumed to be the smallest amount needed to maintain adequate blood glucose levels. The current recommendation is 125 to 175 grams of carbohydrate daily with the emphasis on *complex* carbohydrates (whole grains and vegetables). When less than 50 grams per day is consumed, tissue proteins and fats are used for energy fuel.

American adults typically consume 200 to 300 grams of carbohydrate each day, accounting for about 46% of dietary food energy. Because starchy foods are less expensive than meat and other high-protein foods, carbohydrates make up an even greater percentage of the diet in low-income groups. Starchy foods and milk have many valuable nutrients, such as vitamins and minerals. By contrast, highly refined carbohydrate foods such as candy and soft drinks provide energy sources only—the term "empty calories" is commonly used to describe such foods. Eating refined, sugary foods instead of more complex carbohydrates may cause nutritional deficiencies as well as obesity. Other possible conse-

quences of excessive intake of simple carbohydrates are listed in Table 23.1.

Lipids
Dietary Sources

The most abundant dietary lipids are neutral fats—triglycerides or triacylglycerols (Chapter 2). We eat saturated fats in animal products such as meat and dairy foods and in a few plant products such as coconut. Unsaturated fats are present in seeds, nuts, and most vegetable oils. Fats are digested to fatty acids and monoglycerides and then reconverted to triglycerides for transport in the lymph. Major sources of cholesterol are egg yolk, meats, and milk products.

Although the liver is adept at converting one fatty acid to another, it cannot synthesize *linoleic acid* (lin"o-le'ik), a fatty acid component of *lecithin* (les'ĭ-thin). Thus, linoleic acid is an *essential fatty acid* that must be ingested. Recent research indicates that linolenic acid may also be essential. Fortunately, these two fatty acids are found in most vegetable oils.

Uses in the Body

Fats have fallen into disfavor, particularly among those for whom the "battle of the bulge" is constant. But fats make foods tender, flaky, or creamy, and make us feel full and satisfied. Furthermore, dietary fats *are* essential for several reasons. They help the body absorb fat-soluble vitamins; triglycerides are the major energy fuel of hepatocytes and skeletal muscle; and phospholipids are an integral component of myelin sheaths and cellular membranes. Fatty deposits in adipose tissue provide (1) a protective cushion around body organs, (2) an insulating layer beneath the skin, and (3) an easy-to-store concentrated source of energy fuel. Regulatory molecules called *prostaglandins* (pros"tah-glan'dinz), formed from linoleic acid via arachidonic acid (ah"rah-kĭ-don'ik), play a role in smooth muscle contraction, control of blood pressure, and inflammation.

Unlike neutral fats, cholesterol is not used for energy. It is important as a stabilizing component of plasma membranes and is the precursor from which bile salts, steroid hormones, and other essential molecules are formed.

Dietary Requirements

Fats represent over 40% of the calories in the typical American diet. There are no precise recommendations on amount or type of dietary fats, but the American Heart Association suggests that (1) fats should represent 30% or less of total caloric intake, (2) saturated fats should be limited to 10% or less of

TABLE 23.1 Summary of Carbohydrate, Lipid, and Protein Nutrients

Food Sources	Recommended Daily Amounts (RDA) for Adults	Problems	
		Excesses	Deficits
Carbohydrates			
▪ *Complex carbohydrates (starches):* bread, cereal, crackers, flour, pasta, nuts, rice, potatoes	125–175 g 55–60% of total caloric intake	Obesity; nutritional deficits; dental caries; gastrointestinal irritation; elevated triglycerides in plasma	Tissue wasting (in extreme deprivation); metabolic acidosis resulting from accelerated fat use for energy
▪ *Simple carbohydrates (sugars):* carbonated drinks, candy, fruit, ice cream, pudding, young (immature) vegetables			
▪ *Both complex and simple carbohydrates:* pastries (pies, cookies, cakes)			
Lipids			
▪ *Animal sources:* lard, meat, poultry, eggs, milk, milk products	80–100 g 30% or less of total caloric intake	Obesity and increased risk of cardiovascular disease (particularly if excesses of saturated fat)	Weight loss; fat stores and tissue proteins catabolized to provide metabolic energy; problems controlling heat loss (due to depletion of subcutaneous fat)
▪ *Plant sources:* chocolate; corn, soy, cottonseed, olive oils; coconut; corn; peanuts			
▪ *Essential fatty acids:* corn, cottonseed, soy oils; wheat germ; vegetable shortenings	6000 mg	Not known	Poor growth; skin lesions (eczema-like)
▪ *Cholesterol:* organ meats (liver, kidneys, brains), egg yolks, fish roe; smaller concentrations in milk products and meat	250 mg or less	Increased levels of plasma cholesterol and LDL lipoproteins; correlated with increased risk of cardiovascular disease	Increased risk of stroke (CVA) in susceptible individuals
Proteins			
▪ *Complete proteins:* eggs, milk, milk products, meat (fish, poultry, pork, beef, lamb)	0.8 g/kg body weight	Obesity; possibly aggravation of chronic disease states	Profound weight loss and tissue wasting; growth retardation in children; anemia; edema (due to deficits of plasma proteins) During pregnancy: miscarriage or premature birth
▪ *Incomplete proteins:* legumes (soybeans, lima beans, kidney beans, lentils); nuts and seeds; grains and cereals; vegetables			

total fat intake, and (3) daily cholesterol intake should be no more than 200 mg (the amount in one egg yolk). Because a diet high in saturated fats and cholesterol may contribute to cardiovascular disease, these are wise guidelines. Sources of the various lipid classes and consequences of their deficient or excessive intake are summarized in Table 23.1.

Fat Substitutes

In an attempt to reduce fat intake without losing fat's appetizing aspects, many people have turned to fat substitutes or foods prepared with them. Perhaps the oldest fat substitute is air (beaten into a product to make it fluffy). Other fat substitutes are modified starches and gums, and more recently whey protein. Such products are metabolized and provide calories, but the newest substitutes, sucrose polyesters, are not metabolized because they are not absorbed at all.

Most fat substitutes have two drawbacks: (1) they don't stand up to the intense heat needed to fry foods, and (2) although manufacturers claim otherwise, they don't taste nearly as good as the "real thing." The ones that are not absorbed tend to cause flatus (gas) and may interfere with absorption of fat-soluble drugs and vitamins.

John eats nothing but baked bean sandwiches. Is he getting all the essential amino acids in this rather restricted diet?

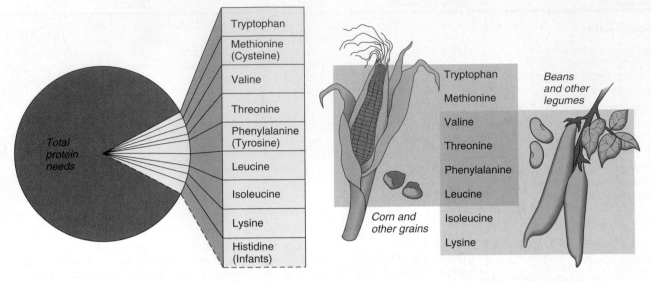

(a) Essential amino acids

(b) Vegetarian diets providing the eight essential amino acids for humans

FIGURE 23.2 Essential amino acids. In order for protein synthesis to occur, eight amino acids must be available simultaneously and in the correct relative amounts. (The ninth, histidine, is essential in infants but not adults.) **(a)** Relative amounts of the essential amino acids and total proteins needed by adults. Notice that the essential amino acids represent only a small percentage of the total recommended protein intake. Histidine is graphed with dashed lines because its requirement in adults has not been established. The amino acids shown in parentheses are *not* essential but can substitute in part for methionine and phenylalanine. **(b)** Vegetarian diets must be carefully constructed to provide all essential amino acids. As shown, corn has little isoleucine and lysine. Beans have ample isoleucine and lysine but little tryptophan and methionine. All essential amino acids can be obtained by consuming a meal of corn and beans.

Proteins

Dietary Sources

Animal products contain the highest-quality proteins, that is, those with the greatest amount and best ratios of essential amino acids (Figure 23.2). Proteins in eggs, milk, and most meats (Table 23.1) are **complete proteins** that meet all the body's amino acid requirements for tissue maintenance and growth. Legumes (beans and peas), nuts, and cereals are protein-rich but their proteins are nutritionally incomplete because they are low in one or more of the essential amino acids. Leafy green vegetables are well balanced in all essential amino acids except methionine, but contain only small amounts of protein. Strict vegetarians must carefully plan their diets to obtain all the essential amino acids and prevent protein malnutrition. When ingested together, cereal grains and legumes provide all the essential amino acids (Figure 23.2b), and some combination of these foods is found in the diets of all cultures (most obviously in the rice and beans seen on nearly every plate in a Mexican restaurant). For nonvegetarians, grains and legumes are useful as partial substitutes for more expensive animal proteins.

Uses in the Body

Proteins are important structural materials of the body, including, for example, keratin in skin, collagen and elastin in connective tissues, and muscle proteins. In addition, functional proteins such as enzymes and some hormones regulate an incredible variety of body functions. Whether amino acids are used to synthesize new proteins or are burned for energy depends on a number of factors:

1. The all-or-none rule. All amino acids needed to make a particular protein must be present in a cell at the same time and in sufficient amounts. If one is

Yes. He has both grain and beans, which together contain all the essential amino acids.

missing, the protein cannot be made. Because essential amino acids cannot be stored, those not used immediately to build proteins are oxidized for energy or converted to carbohydrates or fats.

2. **Adequacy of caloric intake.** For optimal protein synthesis, the diet must supply sufficient carbohydrate or fat calories for ATP production. When it doesn't, dietary and tissue proteins are used for energy.

3. **Nitrogen balance.** In healthy adults the rate of protein synthesis equals the rate of protein breakdown and loss. This homeostatic state is reflected in the body's **nitrogen balance.** The body is in nitrogen balance when the amount of nitrogen ingested in proteins equals the amount excreted in urine and feces.

The body is in *positive nitrogen balance* when the amount of protein being incorporated into tissue is greater than the amount being broken down and used for energy—the normal situation in growing children and pregnant women. A positive balance also occurs when tissues are being repaired following illness or injury.

In *negative nitrogen balance*, protein breakdown for energy exceeds the amount of protein being incorporated into tissues. This occurs during physical and emotional stress (for example, infection, injury, or burns), when the quality of dietary protein is poor, and during starvation.

4. **Hormonal controls.** Certain hormones, called **anabolic hormones,** accelerate protein synthesis and growth. The effects of these hormones vary continually throughout life. For example, pituitary *growth hormone* stimulates tissue growth during childhood and conserves protein in adults, and the *sex hormones* trigger the growth spurt of adolescence. Other hormones, such as the adrenal *glucocorticoids* released during stress, enhance protein breakdown and conversion of amino acids to glucose.

Dietary Requirements

Besides supplying essential amino acids, dietary proteins furnish the raw materials for making nonessential amino acids and various nonprotein nitrogen-containing substances. The amount of protein a person needs reflects his or her age, size, metabolic rate, and current state of nitrogen balance. As a rule of thumb, nutritionists recommend a daily intake of 0.8 g per kilogram of body weight. About 30 g of protein is supplied by a small serving of fish and a glass of milk. Most Americans eat far more protein than they need.

Vitamins

Vitamins (*vita* = life) are potent organic compounds needed in minute amounts for growth and good health. Unlike other organic nutrients, vitamins are not used for energy and do not serve as building blocks, but they are crucial in helping the body use those nutrients that do. Without vitamins, all the carbohydrates, proteins, and fats we eat would be useless.

Most vitamins function as **coenzymes** (or parts of coenzymes), which act with an enzyme to accomplish a particular chemical task. For example, the B vitamins riboflavin and niacin act as coenzymes in the oxidation of glucose for energy. We will describe the roles of some vitamins in the metabolism discussion.

Because most vitamins are not made in the body, they must be taken in via foods or vitamin supplements. The exceptions are vitamin D made in the skin, and vitamin K synthesized by intestinal bacteria. In addition, the body can convert *beta-carotene* (kar'o-tēn), the orange pigment in carrots and other foods, to vitamin A. (Thus, beta-carotene and substances like it are called *provitamins*.) Vitamins are found in all major food groups, but no one food contains all the required vitamins. Thus, a balanced diet is the best way to ensure a full vitamin complement.

Initially vitamins were given a convenient letter designation that indicated the order of their discovery. Although more chemically descriptive names have been assigned to the vitamins, this earlier terminology is still commonly used.

Vitamins are either **fat soluble** or **water soluble.** The water-soluble vitamins, which include B-complex vitamins and vitamin C, are absorbed along with water from the gastrointestinal tract. (The exception is vitamin B_{12}, which must bind to gastric intrinsic factor to be absorbed.) Because only insignificant amounts of water-soluble vitamins are stored in the body, any ingested amounts not used within an hour or so are excreted in urine. Consequently, few conditions resulting from excessive levels of these vitamins (*hypervitaminosis*) are known.

Fat-soluble vitamins (A, D, E, and K) bind to ingested lipids and are absorbed along with their digestion products. Anything that interferes with fat absorption also interferes with the uptake of fat-soluble vitamins. Except for vitamin K, fat-soluble vitamins are stored in the body, and pathologies due to fat-soluble vitamin toxicity, particularly vitamin A hypervitaminosis, are well documented clinically.

As described in the next section, metabolism uses oxygen, and during these reactions some potentially harmful free radicals are generated. Vitamins A, C, and E are *antioxidants* that disarm tissue-damaging free radicals and thereby appear to have anticancer effects. Broccoli, cabbage, cauliflower, and brussels sprouts are all good sources of vitamins A and C. The whole story of how antioxidants interact

TABLE 23.2   Vitamins

Vitamin	Description, Comments	Sources and Recommended Daily Amounts (RDA) for Adults	Importance in Body	Problems Excesses	Deficits

Fat-Soluble Vitamins

Vitamin	Description, Comments	Sources and Recommended Daily Amounts (RDA) for Adults	Importance in Body	Excesses	Deficits
A (retinol)	Group of compounds including retinol and retinal; 90% is stored in liver, which can supply body needs for a year; stable to heat, acids, alkalis; easily oxidized; rapidly destroyed by exposure to light	Formed from provitamin beta-carotene in intestine, liver, kidneys; carotene found in deep-yellow and deep-green leafy vegetables; vitamin A found in fish liver oils, egg yolk, liver, fortified foods (milk, margarine) RDA: males 5000 IU (1.5 mg), females 4000 IU	Required for synthesis of photoreceptor pigments, integrity of skin and mucosae, normal tooth and bone development, normal reproductive capabilities; important antioxidant (hence, anticancer and antiatherosclerosis effects)	Toxic when ingested in excess of 50,000 IU daily for months; symptoms: nausea, vomiting, anorexia, headache, hair loss, bone and joint pain, bone fragility, enlargement of liver and spleen; may increase smokers' risk of lung cancer	Night blindness; epithelial changes; dry skin and hair, skin sores; increased respiratory, digestive, urogenital infections; drying of conjunctiva; clouding of cornea; in pregnant women, developmental defects of the embryo
D (antirachitic factor)	Group of chemically distinct sterols; concentrated in liver and to lesser extent in skin, kidneys, spleen, other tissues; stable to heat, light, acids, alkalis, oxidation	Vitamin D_3 is chief form in body cells; produced in skin by irradiation of 7-dehydrocholesterol by UV light; active form (1,25-dihydroxyvitamin D_3) produced by chemical modification of vitamin D_3 in liver, then kidneys; major food sources: fish liver oils, egg yolk, fortified milk RDA: 400 IU	Functionally a hormone; increases calcium blood levels by enhancing absorption of calcium; with PTH mobilizes calcium from bones; both mechanisms serve calcium homeostasis of blood (essential for normal neuromuscular function, blood clotting, bone and tooth formation)	1800 IU/day may be toxic to children, massive doses induce toxicity in adults; symptoms: vomiting, diarrhea, fatigue, weight loss, hypercalcemia and calcification of soft tissues, irreversible cardiac and renal damage	Faulty mineralization of bones and teeth; rickets in children, osteomalacia in adults; poor muscle tone, restlessness, irritability
E (antisterility factor)	Group of related compounds called tocopherols, chemically related to sex hormones; stored primarily in muscle and adipose tissues; resistant to heat, light, acids; unstable in presence of oxygen	Wheat germ and vegetable oils, nuts, whole grains, dark-green leafy vegetables RDA: males 15 IU; females 12 IU (actual RDA depends on diet; high-fat diet requires a higher RDA)	An antioxidant that disarms free radicals; helps prevent oxidation of unsaturated fatty acids and cholesterol, thus helps prevent oxidative damage to cell membranes and atherosclerosis	Few side effects even at massive doses; slow wound healing; decreased platelet adhesion and increased clotting time	Extremely rare, precise effects uncertain: possible decreased life span and hemolysis of RBCs, fragile capillaries; spinocerebellar degeneration
K (coagulation vitamin)	Number of related compounds known as quinones; small amounts stored in liver; heat resistant; destroyed by acids, alkalis, light, oxidizing agents; antagonized by certain anticoagulants and antibiotics that interfere with synthetic activity of enteric bacteria	Most synthesized by bacteria in large intestine; food sources: leafy green vegetables, broccoli, cabbage, cauliflower, pork liver RDA: males 70 μg, females 55 μg; obtaining adequate amount is not a problem	Essential for formation of clotting proteins and some other proteins made by liver; as intermediate in electron transport chain, participates in oxidative phosphorylation in all body cells	None known, not stored in appreciable amounts	Easy bruising and bleeding (prolonged clotting time)

TABLE 23.2   *(continued)*

Vitamin	Description, Comments	Sources and Recommended Daily Amounts (RDA) for Adults	Importance in Body	Problems Excesses	Deficits

Water-Soluble Vitamins

Vitamin	Description, Comments	Sources and Recommended Daily Amounts (RDA) for Adults	Importance in Body	Excesses	Deficits
C (ascorbic acid)	Simple 6-carbon crystalline compound derived from glucose; rapidly destroyed by heat, light, alkalis; about 1500 mg is stored in body, particularly in adrenal gland, retina, intestine, pituitary; when tissues are saturated, excess excreted by kidneys	Fruits and vegetables, particularly citrus, cantaloupe, strawberries, tomatoes, fresh potatoes, leafy green vegetables RDA: 75 mg for females, 90 mg for males, and an extra 35 mg for those who smoke	Antioxidant; acts in hydroxylation reactions in formation of nearly all connective tissues; in conversion of tryptophan to serotonin (vasoconstrictor) and of cholesterol to bile salts; enhances iron absorption and use; required for activation of folacin (a B vitamin)	Result of megadoses (10 or more times RDA); diarrhea; enhanced mobilization of bone minerals and blood coagulation; high uric acid levels and exacerbation of gout, kidney stone formation	Defective formation of intercellular cement; joint pains, poor tooth and bone growth; poor wound healing, increased susceptibility to infection; extreme deficit causes scurvy (bleeding gums, anemia, degeneration of muscle and cartilage, weight loss)
B_1 (thiamine)	Rapidly destroyed by heat; very limited amount stored in body; excess eliminated in urine	Lean meats, liver, fish, eggs, whole grains, leafy green vegetables, legumes RDA: 1.5 mg	Part of coenzyme cocarboxylase, which acts in carbohydrate metabolism; required for transformation of pyruvic acid to acetyl CoA, for synthesis of pentose sugars and acetylcholine; for oxidation of alcohol	None known	Beriberi; decreased appetite, vision disturbances; unsteadiness when standing or walking, confusion, loss of memory; profound fatigue; heart enlargement, tachycardia
B_2 (riboflavin)	Named for its similarity to ribose sugar: has green-yellow fluorescence; quickly decomposed by UV, visible light, alkalis; body stores are carefully guarded; excess eliminated in urine	Widely varying sources such as liver, yeast, egg white, whole grains, meat, poultry, fish, legumes; major source is milk RDA: 1.7 mg	Present in body as coenzymes FAD and FMN (flavin mononucleotide), both of which act as hydrogen acceptors in body; also is component of amino acid oxidases	None known	Dermatitis; cracking of lips at corners (cheilosis); lips and tongue become purple-red and shiny; ocular problems: light sensitivity, blurred vision; one of the most common vitamin deficiencies
Niacin (nicotinamide)	Simple organic compounds stable to acids, alkalis, heat, light, oxidation (even boiling does not decrease potency); very limited amount stored in body, day-to-day supply is desirable; excess secreted in urine	Diets that provide adequate protein usually provide adequate niacin because amino acid tryptophan is easily converted to niacin; preformed niacin provided by poultry, meat, fish; less important sources: liver, yeast, peanuts, potatoes, leafy green vegetables RDA: 20 mg	Constituent of NAD^+ (nicotinamide adenine dinucleotide) coenzyme involved in glycolysis, oxidative phosphorylation, fat breakdown; inhibits cholesterol synthesis; peripheral vascular dilator	Result of megadoses; hyperglycemia; vasodilation leading to flushing of skin, tingling sensations; possible liver damage; gout	Pellagra after months of deprivation (rare in U.S.); early signs are vague: listlessness, headache, weight loss, loss of appetite; progresses to sore, red tongue and lips; nausea, vomiting, diarrhea; photosensitivity: skin becomes rough, cracked, may ulcerate; neurological symptoms also occur: characterized by the 4 Ds: dermatitis, diarrhea, dementia, and death (in final stage)

TABLE 23.2   **Vitamins** (continued)

Water-Soluble Vitamins

Vitamin	Description, Comments	Sources and Recommended Daily Amounts (RDA) for Adults	Importance in Body	Problems Excesses	Deficits
B_6 (pyridoxine)	Group of three pyridines occurring in both free and phosphorylated forms in body; stable to heat, acids; destroyed by alkalis, light; body stores very limited	Meat, poultry, fish, whole grains, bananas; less important sources: potatoes, sweet potatoes, tomatoes, spinach RDA: 2–3 mg	Active form is coenzyme pyridoxal phosphate, which functions with several enzymes involved in amino acid metabolism; required for conversion of tryptophan to niacin, for glycogenolysis, for formation of antibodies and hemoglobin, and for breakdown of homocysteine	Depressed deep tendon reflexes, numbness and loss of sensation in extremities; difficulty walking; nerve damage	Infants: nervous irritability, convulsions, anemia, vomiting, weakness, abdominal pain; adults: seborrhea lesions around eyes and mouth; increased risk of heart disease
B_5 (pantothenic acid)	Quite stable, little loss of activity with cooking except in acidic or alkaline solutions; liver, kidneys, brain, adrenal, heart tissues contain large amounts	Name derived from Greek *panthos* meaning "everywhere"; widely distributed in animal foods, whole grains, legumes; liver, yeast, egg yolk, meat especially good sources; some produced by enteric bacteria RDA: 10 mg	Functions in form of coenzyme A in reactions that remove or transfer acetyl group, e.g., formation of acetyl CoA from pyruvic acid, oxidation and synthesis of fatty acids; also involved in synthesis of steroids and heme of hemoglobin	None known	Symptoms vague: loss of appetite, abdominal pain, mental depression, pains in arms and legs, muscle spasms, neuromuscular degeneration (neuropathy of alcoholics is thought to be related to deficits)
Biotin	Urea derivative containing sulfur; crystalline in its free form; stable to heat, light acids; in tissues, is usually combined with protein; stored in minute amounts, particularly in liver, kidneys, brain, adrenal glands	Liver, egg yolk, legumes, nuts; some synthesized by bacteria in gastrointestinal tract RDA: not established, probably about 0.3 mg (assumed that formation of the vitamin by enteric bacteria provides far more than is needed)	Functions as coenzyme for a number of enzymes that catalyze carboxylation, decarboxylation, deamination reactions; essential for reactions of Krebs cycle, for formation of purines and nonessential amino acids, for use of amino acids for energy	None known	Scaly skin, muscle pains, pallor, anorexia, nausea, fatigue; elevated blood cholesterol levels

TABLE 23.2   (continued)

Vitamin	Description, Comments	Sources and Recommended Daily Amounts (RDA) for Adults	Importance in Body	Problems	
				Excesses	Deficits
B$_{12}$ (cyano-cobalamin)	Most complex vitamin; contains cobalt; stable to heat; inactivated by light and strongly acidic or basic solutions; intrinsic factor required for transport across intestinal membrane; stored principally in liver; liver stores of 2000–3000 μg sufficient to provide body needs for 3–5 years	Liver, meat, poultry, fish, dairy foods except butter, eggs; not found in plant foods RDA: 3–6 μg	Functions as coenzyme in all cells, particularly in gastrointestinal tract, nervous system, and bone marrow; in bone marrow, acts in synthesis of DNA; when absent, erythrocytes do not divide but continue to enlarge; essential for synthesis of methionine and choline	None known	Pernicious anemia, signified by pallor, anorexia, dyspnea, weight loss, neurological disturbances; most cases reflect impaired absorption rather than actual deficit
Folic acid (folacin)	Pure vitamin is bright-yellow crystalline compound; stable to heat; easily oxidized in acidic solutions and light; stored mainly in liver	Liver, orange juice, deep-green vegetables, yeast, lean beef, eggs, veal, whole grains; synthesized by enteric bacteria RDA: 0.4 μg	Basis of coenzymes that act in synthesis of methionine and certain other amino acids, choline, DNA; essential for formation of red blood cells and for normal neural tube development in the embryo; helps breakdown of homocysteine	None known	Macrocytic or megaloblastic anemia; gastrointestinal disturbances; diarrhea; spina bifida risk in newborn; low birth weight and disabling neurological deficits; increased risk of heart attack and stroke

Notes

1. Each vitamin has specific functions; one vitamin cannot substitute for another. Many reactions in body require several vitamins, and lack of one can interfere with activity of others.

2. Diet that includes recommended amounts of the five food groups will furnish adequate amounts of all vitamins except vitamin D. Each food group makes a special vitamin contribution. For example, fruits and vegetables are principal sources of vitamin C; dark-green leafy vegetables and deep-yellow vegetables and fruits are primary sources of vitamin A (carotene); milk is principal source of riboflavin; meat, poultry, and fish are outstanding sources of niacin, thiamine; and vitamins B$_6$ and B$_{12}$. Whole grains are also important sources of niacin and thiamine.

3. Because vitamin D is present in natural foods in only very small amounts, infants, pregnant and lactating women, and people who have little exposure to sunlight should use vitamin D supplements (or supplemented foods, e.g., fortified milk).

4. Vitamins A and D are toxic in excessive amounts and should be used in supplementary form only when prescribed by a physician. Because most water-soluble vitamins are excreted in urine when ingested in excess, the effectiveness of taking supplements is questionable.

5. Many vitamin deficiencies are secondary to disease (including anorexia, vomiting, diarrhea, or malabsorption diseases) or reflect increased metabolic requirements due to fever or stress factors. Specific vitamin deficiencies require therapy with the vitamins that are lacking.

in the body is still murky, but chemists propose that they act much like a bucket brigade to pass the dangerous chemical property from one molecule to the next. First, vitamin E returns a free radical to its less harmful state. This reaction converts vitamin E to a free radical that is inactivated by carotenoids (bases of vitamin A). The carotenoid radicals thus formed are inactivated by vitamin C, and the water-soluble radicals formed during this reaction flush from the body in urine.

Controversy abounds concerning the ability of vitamins to work wonders, such as the idea that huge doses of vitamin C will prevent colds. The notion that megadoses of vitamin supplements are the road to eternal youth and glowing health is useless at best and, at worst, may cause serious health problems, particularly in the case of fat-soluble vitamins. Table 23.2 contains an overview of the roles of vitamins in the body.

Minerals

The body requires moderate amounts of seven **minerals** (calcium, phosphorus, potassium, sulfur, sodium, chlorine, magnesium) and trace amounts of about a dozen others (Table 23.3). Minerals make up about 4% of the body by weight, with calcium and phosphorus (as bone salts) accounting for about three-quarters of this amount.

Minerals, like vitamins, are not used for fuel but work with other nutrients to ensure a smoothly functioning body. Incorporation of minerals into structures gives added strength. For example, calcium, phosphorus, and magnesium salts harden the teeth and strengthen the skeleton. Most minerals are ionized in body fluids or bound to organic compounds to form phospholipids, hormones, enzymes, and other functional proteins. For example, iron is essential to the oxygen-binding heme of hemoglobin, and sodium and chloride ions are the major electrolytes in blood. The amount of a particular mineral in the body gives very few clues to its importance in body function. For example, just a few milligrams of iodine can make a critical difference to health.

A fine balance between uptake and excretion is crucial for retaining needed amounts of minerals while preventing toxic overload. For example, sodium, present in virtually all natural foods (and added in large amounts to processed foods), may contribute to high blood pressure if ingested in excess.

Fats and sugars are practically devoid of minerals, and highly refined cereals and grains are poor sources. The most mineral-rich foods are vegetables, legumes, milk, and some meats.

TABLE 23.3 Minerals

Mineral	Distribution in Body, Comments	Sources and Recommended Daily Amounts (RDA) for Adults	Importance in Body	Problems	
				Excesses	Deficits
Calcium (Ca)	Most stored in salt form in bones; the most abundant cation in body; absorbed from intestine in presence of vitamin D; excess excreted in feces; blood levels regulated by PTH and calcitonin	Milk, milk products, leafy green vegetables, egg yolk, shellfish RDA: 1200–1500 mg, dropping to 800–1000 mg after age 25	In salt form, required for hardness of bones, teeth; ionic calcium in blood and cells essential for normal membrane permeability, transmission of nerve impulses, muscle contraction, normal heart rhythm, blood clotting; activates certain enzymes; helps prevent hypertension	Depressed neural function; lethargy and confusion; muscle pain and weakness; calcium salt deposit in soft tissues; kidney stones	Muscle tetany; osteomalacia, osteoporosis; retarded growth and rickets in children
Chlorine (Cl)	Exists in body almost entirely as chloride ion; principal anion of extracellular fluid; highest concentrations in cerebrospinal fluid and gastric juice; excreted in urine	Table salts (as for sodium); usually ingested in excess RDA: not established; normal diet contains 3–9 g	With sodium, helps maintain osmotic pressure and pH of extracellular fluid; required for HCl formation by stomach glands; activates salivary amylase; aids in transport of CO_2 by blood (chloride shift)	Vomiting	Severe vomiting or diarrhea leads to Cl loss and alkalosis; muscle cramps; apathy

TABLE 23.3 *(continued)*

Mineral	Distribution in Body, Comments	Sources and Recommended Daily Amounts (RDA) for Adults	Importance in Body	Problems	
				Excesses	Deficits
Sulfur (S)	Widely distributed: particularly abundant in hair, skin, nails; excreted in urine	Meat, milk, eggs, legumes (all rich in sulfur-containing amino acids) RDA: not established; diet adequate in proteins meets body's needs	Essential constituent of many proteins (insulin), some vitamins (thiamine and biotin); found in mucopolysaccharides present in cartilage, tendons, bone	Not known	Not known
Potassium (K)	Principal intracellular cation, 97% within cells; fixed proportion of K is bound to proteins, and measurements of body K are used to determine lean body mass; K^+ leaves cells during protein catabolism, dehydration, glycogenolysis; most excreted in urine	Widely distributed in foods; normal diet provides 2–6 g/day; avocados, dried apricots, meat, fish, fowl, cereals RDA: not established; diet adequate in calories provides ample amount, i.e., 2500 mg	Helps maintain intracellular osmotic pressure; needed for normal nerve impulse conduction, muscle contraction, glycogenesis, protein synthesis	Usually a complication of renal failure or severe dehydration, but may result from severe alcoholism; paresthesias, muscular weakness, cardiac abnormalities	Rare but may result from severe diarrhea or vomiting; muscular weakness, paralysis, nausea, vomiting, tachycardia, heart failure
Sodium (Na)	Widely distributed: 50% found in extracellular fluid, 40% in bone salts, 10% within cells; absorption is rapid and almost complete; excretion, chiefly in urine, controlled by aldosterone	Table salt (1 tsp = 2000 mg); cured meats (ham, etc.), sauerkraut, cheese; diet supplies substantial excess RDA: not established; probably about 2500 mg	Most abundant cation in extracellular fluid: principal electrolyte maintaining osmotic pressure of extracellular fluids and water balance; as part of bicarbonate buffer system, aids in acid-base balance of blood; needed for normal neuromuscular function; part of pump for transport of glucose and other nutrients	Hypertension; edema	Rare but can occur with excessive vomiting, diarrhea, sweating, or poor dietary intake; nausea, abdominal and muscle cramping, convulsions
Magnesium (Mg)	In all cells, particularly abundant in bones; absorption parallels that of calcium; excreted chiefly in urine	Milk, dairy products, whole-grain cereals, nuts, legumes, leafy green vegetables RDA: 350 mg	Constituent of many coenzymes that play a role in conversion of ATP to ADP; required for normal muscle and nerve irritability	Diarrhea	Neuromuscular problems, tremors, muscle weakness, irregular heartbeat; vasospasm and hypertension; sudden cardiac death; seen in alcoholism, severe renal disease, prolonged diarrhea, and with diuretic drugs

TABLE 23.3   **Minerals** *(continued)*

Mineral	Distribution in Body, Comments	Sources and Recommended Daily Amounts (RDA) for Adults	Importance in Body	Problems	
				Excesses	Deficits
Phosphorus (P)	About 80% found in inorganic salts of bones, teeth; remainder in muscle, nervous tissue, blood; absorption aided by vitamin D; about 1/3 dietary intake excreted in feces; metabolic by-products excreted in urine	Diets rich in proteins are usually rich in phosphorus; plentiful in milk, eggs, meat, fish, poultry, legumes, nuts, whole grains RDA: 800 mg	Component of bones and teeth, nucleic acids, proteins, phospholipids, ATP, phosphates (buffers) of body fluids; thus, important for energy storage and transfer, muscle and nerve activity, cell permeability	Not known, but excess in diet may depress absorption of iron and manganese	Rickets, poor growth
Trace Minerals*					
Fluorine (F)	Component of bones, teeth, other body tissues; excreted in urine	Fluoridated water RDA: 1.5–4 mg	Important for tooth structure; may help prevent dental caries (particularly in children) and osteoporosis in adults	Mottling of teeth	Not known
Cobalt (Co)	Found in all cells; larger amounts in bone marrow	Liver, lean meat, poultry, fish, milk RDA: not established	A constituent of vitamin B_{12}, which is needed for normal maturation of red blood cells	Goiter, polycythemia, heart disease	See deficit of vitamin B_{12} (Table 23.2)
Chromium (Cr)	Widely distributed	Liver, meat, cheese, whole grains, brewer's yeast, wine RDA: 0.05–2 mg	Required for synthesis of glucose tolerance factor (GTF) needed for proper glucose metabolism; enhances effectiveness of insulin on carbohydrate metabolism; may increase blood levels of HDLs while decreasing levels of LDLs	Not known	Impairs ability of insulin to work, hence increases insulin secretion and risk of adult-onset diabetes mellitus
Copper (Cu)	Concentrated in liver, heart, brain, spleen; excreted in feces	Liver, shellfish, whole grains, legumes, meat; typical diet provides 2–5 mg daily RDA: 2–3 mg	Required for synthesis of hemoglobin; essential for manufacture of melanin, myelin, some intermediates of electron transport chain	Rare; abnormal storage results in Wilson's disease	Rare

*Trace minerals together account for less than 0.005% of body weight.

TABLE 23.3 (continued)

Mineral	Distribution in Body, Comments	Sources and Recommended Daily Amounts (RDA) for Adults	Importance in Body	Problems Excesses	Problems Deficits
Iodine (I)	Found in all tissues, but in high concentrations only in thyroid gland; absorption controlled by blood levels of protein-bound iodine; excreted in urine	Cod liver oil, iodized salt, shellfish, vegetables grown in iodine-rich soil RDA: 0.15 mg	Required to form thyroid hormones (T_3 and T_4), which are important in regulating cellular metabolic rate	Depressed synthesis of thyroid hormones	Hypothyroidism: cretinism in infants, myxedema in adults (if less severe, simple goiter); impaired learning and motivation in children
Iron (Fe)	60–70% in hemoglobin; remainder in skeletal muscle, liver, spleen, bone marrow bound to ferritin; only 2–10% of dietary iron is absorbed because of mucosal barrier; lost from body in urine, perspiration, menstrual flow, shed hair; sloughed skin and mucosal cells	Best sources: meat, liver, also in shellfish, egg yolk, dried fruit, nuts, legumes, molasses RDA: males 10 mg, females 15 mg	Part of heme of hemoglobin, which binds bulk of oxygen transported in blood; component of cytochromes, which function in oxidative phosphorylation	Hemochromatosis (inherited condition of iron excess); damage to liver (cirrhosis and liver cancer), heart, pancreas (causing diabetes)	Iron-deficiency anemia; pallor, lethargy, flatulence, anorexia, paresthesias, impaired cognitive performance in children; inability to maintain body temperature; depressed killing by phagocytes (due to inability to produce the respiratory burst)
Manganese (Mn)	Most concentrated in liver, kidneys, spleen; excreted largely in feces	Nuts, legumes, whole grains, leafy green vegetables, fruit RDA: 2.5–5 mg	Acts with enzymes catalyzing synthesis of fatty acids, cholesterol, urea, hemoglobin; needed for normal neural function, lactation, oxidation of carbohydrates, protein hydrolysis	Appears to contribute to obsessive behavior, hallucinations, and violent behavior	Not known
Selenium (Se)	Stored in liver, kidneys	Meat, seafood, cereals RDA: 0.05–2 mg	Antioxidant; constituent of certain enzymes; spares vitamin E	Nausea, vomiting, irritability, fatigue, weight loss, loss of hair	Oxidative damage
Zinc (Zn)	Concentrated in liver, kidneys, brain; excreted in feces	Seafood, meat, cereals, legumes, nuts, wheat germ, yeast RDA: 15 mg	Constituent of several enzymes (e.g., carbonic anhydrase); plays a structural role in a variety of proteins (e.g., transcription factors and tumor suppressor protein p53); required for normal growth, wound healing, taste, smell, sperm production	Difficulty in walking; slurred speech; tremors; interferes with body's ability to absorb copper, which can lead to weakened immunity	Loss of taste and smell; growth retardation; learning impairment; depressed immunity

Metabolism

Overview

Once inside body cells, nutrients become involved in an incredible variety of biochemical reactions known collectively as **metabolism** (*metabol* = change). During metabolism, substances are constantly being built up and torn down. Cells use energy to extract more energy from foods, and then use some of this extracted energy to drive their activities. Even at rest, the body uses energy on a grand scale.

Anabolism and Catabolism

Metabolic processes are either *anabolic* (synthetic, building up) or *catabolic* (degradative, tearing down). **Anabolism** (ah-nab′o-lizm) is the general term for all reactions in which larger molecules or structures are built from smaller ones; for example, the bonding together of amino acids to make proteins. **Catabolism** (kah-tab′o-lizm) refers to all processes that break down complex structures to simpler ones, one example being hydrolysis of foods in the digestive tract. In the group of catabolic reactions collectively called **cellular respiration,** food fuels (particularly glucose) are broken down in cells and some of the energy released is captured to form ATP, the cells′ energy currency. ATP then serves as the "chemical drive shaft" that links energy-releasing catabolic reactions to cellular work.

Recall from Chapter 2 that reactions driven by ATP are coupled. ATP is never hydrolyzed directly. Instead enzymes shift its high-energy phosphate groups to other molecules, which are then said to be **phosphorylated** (fos″for-ĭ-la′ted). Phosphorylation primes the molecule to change in a way that increases its activity, produces motion, or does work. For example, many regulatory enzymes that catalyze key steps in metabolic pathways are activated by phosphorylation.

Three major stages are involved in the processing of energy-containing nutrients in the body (Figure 23.3 on p. 819):

- Stage 1 is digestion in the gastrointestinal tract as described in Chapter 22. The absorbed nutrients are then transported in blood to the tissue cells.

- In stage 2, which occurs in the cell cytoplasm, newly delivered nutrients are either built into lipids, proteins, and glycogen by anabolic pathways or broken down by catabolic pathways to *pyruvic acid* (pi-roo′vik) and *acetyl CoA* (as′ĕ-til ko-a′).

- Stage 3, which is almost entirely catabolic, occurs in the mitochondria, requires oxygen, and completes the breakdown of foods, producing carbon dioxide and water and harvesting large amounts of ATP.

The primary function of *cellular respiration,* which consists of glycolysis of stage 2 and all events of stage 3, is to generate ATP, which traps some of the chemical energy of the original food molecules in its own high-energy bonds. Thus, food fuels, such as glycogen and fats, *store* energy in the body, and these stores are later *mobilized to produce ATP for cellular use.*

You do not need to memorize Figure 23.3, but as you will soon see, it provides a cohesive summary of nutrient processing and metabolism in the body.

Oxidation-Reduction Reactions and the Role of Coenzymes

Many of the reactions that take place within cells are **oxidation reactions.** *Oxidation* was originally defined as the combination of oxygen with other elements. Examples are the rusting of iron (the slow formation of iron oxide) and the burning of wood and other fuels. In burning, oxygen combines rapidly with carbon, producing carbon dioxide, water, and an enormous amount of energy, which is liberated as heat and light. Later it was discovered that oxidation *also* occurs when hydrogen atoms are *removed* from compounds and so the definition was expanded to its current form: *Oxidation is the gain of oxygen or the loss of hydrogen.* As explained in Chapter 2, whichever way oxidation occurs, the oxidized substance always *loses* (or nearly loses) electrons as they move to (or toward) a substance that more strongly attracts them.

This loss of electrons can be explained by reviewing the consequences of different electron-attracting abilities of atoms (see p. 31). Consider a molecule made up of a hydrogen atom plus some other kinds of atoms. Because hydrogen is very electropositive, its lone electron usually spends more time orbiting the other atoms of the molecule. But when a hydrogen *atom* is removed, its electron goes with it, and the molecule as a whole loses that electron. Conversely, oxygen is very electron-hungry (electronegative), and so when oxygen binds with other atoms the shared electrons spend more time in oxygen's vicinity. Again, the molecule as a whole loses electrons. As you will soon see, essentially all oxidation of food fuels involves the step-by-step removal of pairs of hydrogen atoms (and thus pairs of electrons) from the substrate molecules, eventually leaving only carbon dioxide (CO_2). Molecular oxygen (O_2) is the *final* electron acceptor. It combines with the removed hydrogen atoms at the very end of the process, to form water (H_2O).

Whenever one substance loses electrons (is oxidized), another substance gains them (is reduced). Thus, oxidation and reduction are coupled reactions and we speak of **oxidation-reduction reactions,** or more commonly, **redox reactions.** The key understanding about redox reactions is that "oxidized"

FIGURE 23.3 **Three stages of metabolism of energy-containing nutrients.** In stage 1, foods are degraded to absorbable forms by digestive enzymes in the GI tract.

In stage 2, the absorbed nutrients are transported in blood to the body cells, where they may be incorporated into molecules (anabolism) or broken down via glycolysis and other reactions to pyruvic acid and/or acetyl CoA, and then funneled into the pathways of stage 3.

Stage 3 consists of the catabolic pathways of the Krebs cycle and oxidative phosphorylation, both of which take place in the mitochondria. During the Krebs cycle, acetyl CoA is broken apart: Its carbon atoms are liberated as carbon dioxide (CO_2), and the hydrogen atoms removed are delivered to a chain of receptors (the electron transport chain), which ultimately delivers them (as protons and free electrons) to molecular oxygen so that water is formed. Some of the energy released during the electron transport chain reactions is used to form ATP. Glycolysis (stage 2) and the Krebs cycle and oxidative phosphorylation (stage 3) collectively constitute cellular respiration.

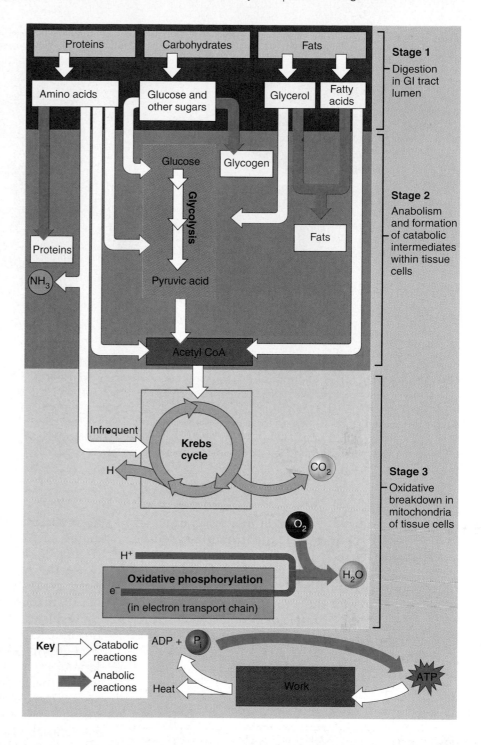

substances *lose* energy and "reduced" substances *gain* energy as energy-rich electrons are transferred from the former to the latter. Consequently, as food fuels are oxidized, their energy is transferred to a "bucket brigade" of other molecules and ultimately to ADP to form energy-rich ATP.

Like all other chemical reactions in the body, redox reactions are catalyzed by enzymes. Those that catalyze redox reactions in which hydrogen atoms are removed are called **dehydrogenases** (de-hi'dro-jen-ās"ez), while those catalyzing the transfer of oxygen are **oxidases.** Most of these enzymes require the help of a specific coenzyme, typically derived from one of the B vitamins. Although the enzymes catalyze the removal of hydrogen atoms to oxidize a substance, they cannot *accept* the hydrogen (hold on or bond to it). Their **coenzymes,** however, act as hydrogen (or electron) acceptors, becoming reduced each time a

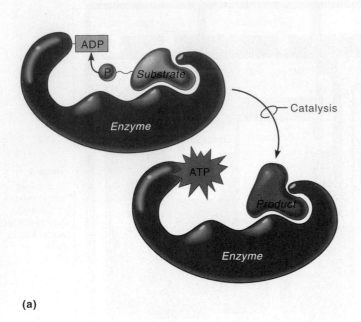

(a)

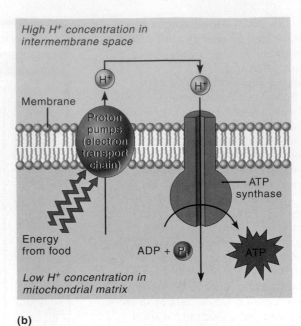

(b)

FIGURE 23.4 **Mechanisms of phosphorylation. (a)** Substrate-level phosphorylation occurs when a high-energy phosphate group is transferred directly from a substrate to ADP to form ATP. This reaction occurs both in the cytoplasm and in the mitochondrial matrix. **(b)** Oxidative phosphorylation, which occurs in mitochondria, reflects the activity of electron transport proteins that act as proton "pumps" to create a proton gradient across the cristae membranes. The source of energy for this pumping is energy released during oxidation of food molecules. As the protons flow passively back into the mitochondrial fluid matrix, the energy of the diffusion gradient is used to bind phosphate groups to ADP.

substrate is oxidized. Two very important coenzymes of the oxidative pathways are **nicotinamide adenine dinucleotide (NAD$^+$)** (nik″o-tin′ah-mīd), based on *niacin*, and **flavin adenine dinucleotide (FAD)**, derived from *riboflavin*. The oxidation of succinic acid to fumaric acid and the simultaneous reduction of FAD to FADH$_2$, an example of a coupled redox reaction, is:

FAD (oxidized)   FAD(H$_2$) (reduced)

Succinic acid → Oxidation [−2H] → Fumaric acid

Mechanisms of ATP Synthesis

How do our cells capture some of the energy liberated during cellular respiration to make ATP molecules? There appear to be two mechanisms—substrate-level phosphorylation and oxidative phosphorylation.

Substrate-level phosphorylation occurs when high-energy phosphate groups are transferred directly from phosphorylated substrates (metabolic intermediates such as glyceraldehyde phosphate) to ADP (Figure 23.4a). Essentially, this process occurs because the high-energy bonds attaching the phosphate groups to the substrates are even more unstable than those in ATP. ATP is synthesized by this route during one of the steps in glycolysis, and once during each turn of the Krebs cycle (Figure 23.5). The enzymes catalyzing substrate-level phosphorylations are located in both the cytoplasm and in the watery matrix inside the mitochondria.

Oxidative phosphorylation is much more complicated, but it also releases most of the energy that is eventually captured in ATP bonds during cellular respiration. This process, which is carried out by electron transport proteins forming part of the mitochondria cristae, is an example of a **chemiosmotic process**. Chemiosmotic processes couple the movement of substances across membranes to chemical reactions. In this case, some of the energy released during the oxidation of food fuels (the "chemi" part of the term) is used to pump (*osmo* = push) protons (H$^+$) across the cristae membrane into the intermembrane space (Figure 23.4b). This creates a steep diffusion gradient for protons across the membrane. Then, when H$^+$ does flow back

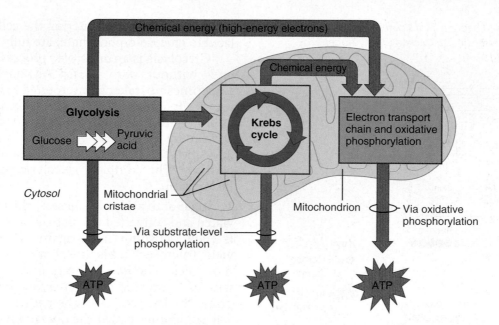

FIGURE 23.5 Sites of ATP formation during cellular respiration. Glycolysis occurs in the cytosol. The Krebs cycle and the electron transport chain reactions occur in the mitochondria. During glycolysis, each glucose molecule is broken down to two molecules of pyruvic acid. The pyruvic acid enters the mitochondrial matrix, where the Krebs cycle decomposes it to CO_2. During glycolysis and the Krebs cycle, small amounts of ATP are formed by substrate-level phosphorylation. Chemical energy from glycolysis and the Krebs cycle, in the form of energy-rich electrons picked up by coenzymes, is then transferred to the electron transport chain, which is built into the membrane of the cristae. The electron transport chain carries out oxidative phosphorylation, which accounts for most of the ATP generated by cellular respiration.

across the membrane (through a membrane channel protein called *ATP synthase*), some of this gradient energy is captured and used to attach phosphate groups to ADP.

Carbohydrate Metabolism

Because all food carbohydrates are eventually transformed to glucose, the story of carbohydrate metabolism is really a tale of glucose metabolism. Glucose enters the tissue cells by facilitated diffusion, a process that is greatly enhanced by insulin. Immediately upon entry into the cell, glucose is phosphorylated to *glucose-6-phosphate* by transfer of a phosphate group to its sixth carbon during a coupled reaction with ATP:

$$\text{Glucose} + \text{ATP} \rightarrow \text{glucose-6-PO}_4 + \text{ADP}$$

Most body cells lack the enzymes needed to reverse this reaction, so it effectively traps glucose inside the cells. Because glucose-6-phosphate is a *different* molecule from simple glucose, the reaction also keeps intracellular glucose levels low, maintaining a diffusion gradient for glucose entry. Only intestinal mucosa cells, kidney tubule cells, and liver cells have the enzymes needed to reverse this phosphorylation reaction, which reflects their central roles in glucose uptake *and* release. The catabolic and anabolic pathways for carbohydrates all begin with glucose-6-phosphate.

Oxidation of Glucose

Glucose is the pivotal fuel molecule in the oxidative (ATP-producing) pathways. Glucose is catabolized via the reaction

$$\text{C}_6\text{H}_{12}\text{O}_6 + 6\text{O}_2 \rightarrow 6\text{H}_2\text{O} + 6\text{CO}_2 + 36 \text{ ATP} + \text{heat}$$
glucose   oxygen   water   carbon
                               dioxide

This equation gives few hints that glucose breakdown is complex and involves three of the pathways included in Figures 23.3 and 23.5:

1. Glycolysis (color-coded light orange through the chapter)

2. The Krebs cycle (color-coded pale-green)

3. The electron transport chain and oxidative phosphorylation (color-coded violet)

These metabolic pathways occur in a definite order, and we will consider them sequentially.

Glycolysis   A series of ten chemical steps by which glucose is converted to two *pyruvic acid* molecules, **glycolysis** (gli-kol′ĭ-sis; "sugar splitting") occurs in the cytosol of cells. All steps except the first,

What would happen in this pathway if NADH + H⁺ could not transfer its "picked up" hydrogens to pyruvic acid?

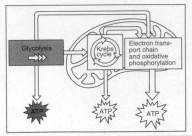

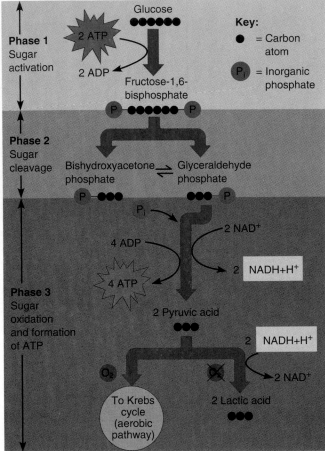

FIGURE 23.6   The three major phases of glycolysis. During phase 1, glucose is activated by phosphorylation and converted to fructose-1,6-bisphosphate. In phase 2, fructose-1,6-bisphosphate is cleaved into two 3-carbon fragments (reversible isomers). In phase 3, the 3-carbon fragments are oxidized (by removal of hydrogen) and 4 ATP molecules are formed. The fate of pyruvic acid depends on whether or not molecular O_2 is available.

■ *Glycolysis would come to a halt because the oxidase enzyme cannot hold onto the H⁺ removed during substrate oxidation. If all the coenzymes are already reduced and can't release hydrogen, the oxidase enzyme couldn't continue to function.*

during which glucose entering the cell is phosphorylated to glucose-6-phosphate, are fully *reversible*.

Glycolysis is an *anaerobic process* (an-a′er-ōb-ik; *an* = without, *aero* = air). Although this term is sometimes mistakenly interpreted to mean the pathway occurs only in the absence of oxygen, the correct interpretation is that glycolysis *does not use oxygen and occurs whether or not oxygen is present.* Figure 23.6 shows the three major phases of the glycolytic pathway. The complete glycolytic pathway appears in Appendix A.

1. **Sugar activation.** In phase 1, glucose is phosphorylated and converted to fructose-6-phosphate, which is then phosphorylated again. These three steps yield fructose-1,6-bisphosphate and use two ATP molecules. The two separate reactions of the sugar with ATP provide the *activation energy* needed to prime the later stages of the pathway—hence phase 1 is sometimes called the *energy investment phase.* (The importance of activation energy is described in Chapter 2.)

2. **Sugar cleavage.** During phase 2, fructose-1,6-bisphosphate is split into two 3-carbon fragments that exist (reversibly) as one of two isomers: glyceraldehyde (glis″er-al′dĕ-hīd) 3-phosphate or bishydroxyacetone (bīs″hi-drok″se-as′ĕ-tōn) phosphate.

3. **Oxidation and ATP formation.** In phase 3, consisting of six steps, two major events happen. First, the two 3-carbon fragments are oxidized by the removal of hydrogen, which is picked up by NAD⁺. Hence, some of glucose's energy is transferred to NAD⁺. Second, inorganic phosphate groups (P_i) are attached to each oxidized fragment by high-energy bonds. Later, as these terminal phosphates are cleaved off, enough energy is captured to form four ATP molecules. As noted earlier, formation of ATP this way is called *substrate-level phosphorylation.*

The final products of glycolysis are two molecules of **pyruvic acid** and two molecules of reduced NAD⁺ (NADH + H⁺),* with a net gain of two ATP molecules per glucose molecule. Four ATPs are produced, but remember that two are consumed in phase 1 to "prime the pump." Each pyruvic acid molecule has the formula $C_3H_4O_3$, and glucose is $C_6H_{12}O_6$. Thus, between them the two pyruvic acid molecules have lost 4 hydrogen atoms, which are now bound to two molecules of NAD⁺. Although a small amount of ATP has been harvested, the other two products of glucose oxidation (H_2O and CO_2) have yet to appear.

The fate of pyruvic acid, which still contains most of glucose's chemical energy, depends on the

---

*NAD carries a positive charge (NAD⁺); thus, when it accepts the hydrogen pair, NADH + H⁺ is the resulting reduced product.

availability of oxygen at the time the pyruvic acid is produced. Because the supply of NAD^+ is limited, glycolysis can continue only if the reduced coenzymes ($NADH + H^+$) formed during glycolysis are relieved of their extra hydrogen. Only then can they continue to act as hydrogen acceptors. When oxygen is readily available, this is no problem. $NADH + H^+$ delivers its burden of hydrogen atoms to the enzymes of the electron transport chain in the mitochondria, which deliver them to O_2, forming water. However, when oxygen is not present in sufficient amounts, as might occur during strenuous exercise, $NADH + H^+$ unloads its hydrogen atoms *back onto pyruvic acid*, thus reducing it. This addition of two hydrogen atoms to pyruvic acid yields **lactic acid** (see bottom right of Figure 23.6), some of which diffuses out of the cells and is transported to the liver. When oxygen is again available, lactic acid is oxidized back to pyruvic acid and enters the **aerobic pathways** (the oxygen-requiring Krebs cycle and electron transport chain within the mitochondria), and is completely oxidized to water and carbon dioxide. The liver may also convert lactic acid all the way back to glucose-6-phosphate (reverse glycolysis) and then store it as glycogen or free it of its phosphate and release it to the blood if blood sugar levels are low.

Although glycolysis generates ATP rapidly, only 2 ATP molecules are produced per glucose molecule, as compared to the 36 ATP per glucose harvested when glucose is completely oxidized. Except for red blood cells (which typically carry out *only* glycolysis), prolonged anaerobic metabolism ultimately results in acid-base problems. Consequently, *totally* anaerobic conditions resulting in lactic acid formation provide only a temporary route for rapid ATP production. It can go on without tissue damage for the longest periods in skeletal muscle, for much shorter periods in cardiac muscle, and almost not at all in the brain.

Krebs Cycle   Named after its discoverer Hans Krebs, the **Krebs cycle** is the next stage of glucose oxidation. The Krebs cycle occurs in the mitochondrial matrix and is fueled largely by pyruvic acid produced during glycolysis and by fatty acids resulting from fat breakdown.

After pyruvic acid enters the mitochondria, the first order of business is to convert it to *acetyl CoA* via a three-step process (Figure 23.7):

1. Decarboxylation, in which one of pyruvic acid's carbons is removed and released as carbon dioxide gas. CO_2 diffuses out of the cells into the blood to be expelled by the lungs.

2. Oxidation by the removal of hydrogen atoms, which are picked up by NAD^+.

3. Combination of the resulting acetic acid with *coenzyme A* to produce the final product, **acetyl**

coenzyme A (acetyl CoA). Coenzyme A (CoA-SH) is a sulfur-containing coenzyme derived from pantothenic acid, a B vitamin.

Acetyl CoA is now ready to enter the Krebs cycle and be broken down completely by mitochondrial enzymes. Coenzyme A shuttles the 2-carbon acetic acid to an enzyme that condenses it with a 4-carbon acid called **oxaloacetic acid** (ok″sah-lo″ah-sēt′ik) to produce the 6-carbon **citric acid.** Because citric acid is the first substrate of the cycle, biochemists prefer to call the Krebs cycle the **citric acid cycle.**

As the cycle moves through its eight successive steps, the atoms of citric acid are rearranged to produce different intermediate molecules, most called **keto acids.** The acetic acid that enters the cycle is broken apart carbon by carbon (decarboxylated) and oxidized, simultaneously generating $NADH + H^+$ and $FADH_2$. At the end of the cycle, acetic acid has been totally disposed of and oxaloacetic acid, the *pickup molecule*, is regenerated. Because two *decarboxylations* and four *oxidations* occur, the products of the Krebs cycle are two CO_2 molecules and four molecules of reduced coenzymes (3 $NADH + H^+$ and 1 $FADH_2$). The addition of water at certain steps accounts for some of the released hydrogen. One molecule of ATP is formed (via substrate-level phosphorylation) during each turn of the cycle. The detailed events of each of the eight steps of the Krebs cycle are described in Appendix A.

Now let's account for the pyruvic acid molecules entering the mitochondria. Each pyruvic acid yields three CO_2 molecules and five molecules of reduced coenzymes—1 $FADH_2$ and 4 $NADH + H^+$ (equal to the removal of 10 hydrogen atoms). The products of glucose oxidation in the Krebs cycle are twice that (remember 1 glucose = 2 pyruvic acids): six CO_2, ten molecules of reduced coenzymes, and two ATP molecules. Notice that it is these Krebs cycle reactions that produce the CO_2 evolved during glucose oxidation. The reduced coenzymes, which carry their extra electrons in high-energy linkages, must now be oxidized if the Krebs cycle and glycolysis are to continue.

Although the glycolytic pathway is exclusive to carbohydrate oxidation, breakdown products of carbohydrates, fats, and proteins can feed into the Krebs cycle to be oxidized for energy. On the other hand, some Krebs cycle intermediates can be siphoned off to make fatty acids and nonessential amino acids. Thus the Krebs cycle, besides serving as the final common pathway for the oxidation of food fuels, is a source of building materials for anabolic reactions.

Electron Transport Chain and Oxidative Phosphorylation   Like glycolysis, none of the reactions of the Krebs cycle use oxygen directly. This is the exclusive function of the **electron transport**

What two major kinds of chemical reactions occur in this cycle, and how are these reactions indicated symbolically?

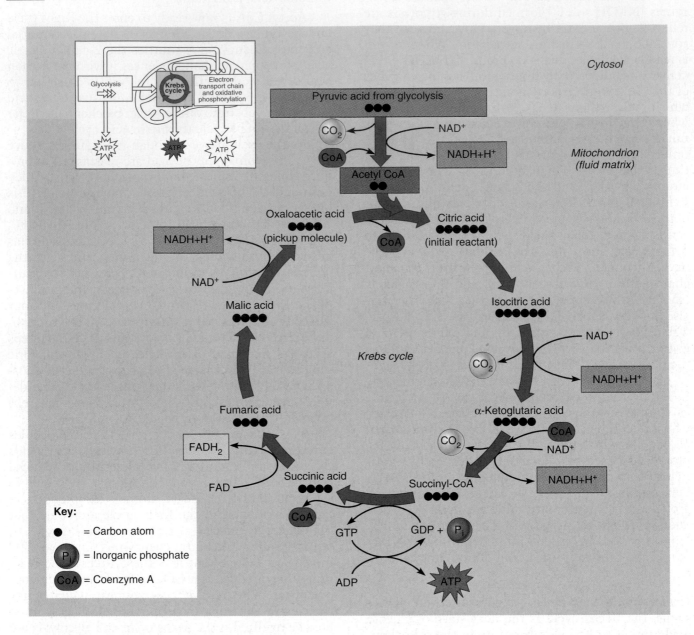

FIGURE 23.7 **Simplified version of the Krebs cycle.** During each turn of the cycle, two carbon atoms are removed from the substrates as CO_2 (decarboxylation reactions); four oxidations by removal of hydrogen atoms occur, producing four molecules of reduced coenzymes (3 NADH + H^+ and 1 $FADH_2$); and one ATP is synthesized by substrate-level phosphorylation. An additional decarboxylation and an oxidation reaction occur to convert pyruvic acid, the product of glycolysis, to acetyl CoA, the molecule that enters the Krebs cycle pathway.

Decarboxylation, shown as removal of CO_2, and oxidation, shown as reduction of coenzymes, e.g., $NAD^+ \rightarrow NADH + H^+$. ∎

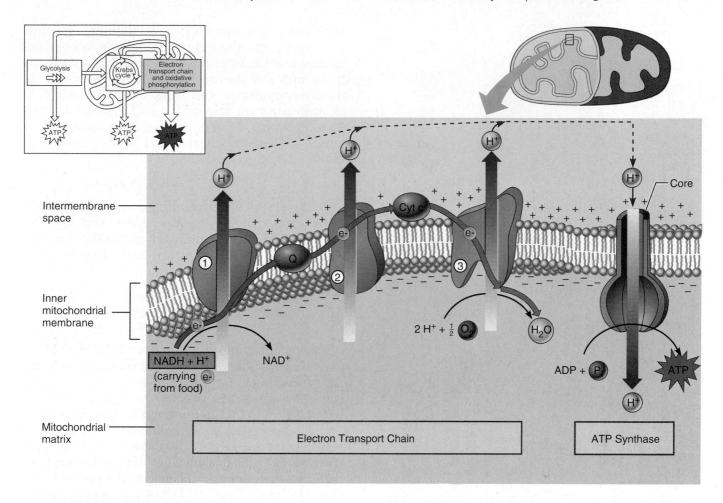

FIGURE 23.8 Mechanism of oxidative phosphorylation.
Schematic diagram showing the flow of electrons through the three major respiratory enzyme complexes—① NADH dehydrogenase (FMN, Fe-S), ② cytochrome b-c_1, and ③ cytochrome oxidase (a-a_3)—of the electron transport chain during the transfer of two electrons from reduced NAD^+ to oxygen. Coenzyme Q and cytochrome c are mobile and act as carriers between the complexes.

The electron transport chain is an energy converter, transforming chemical energy to the energy of a H^+ gradient. As electrons flow along their energy gradient, some of the energy is used by each complex to pump H^+ from the mitochondrial matrix into the intermembrane space. These H^+ ions create an electrochemical proton gradient that drives them back across the inner membrane through the ATP synthase complex. ATP synthase uses the energy of H^+ flow (electrical energy) to synthesize ATP from ADP and P_i. Oxidation of each $NADH + H^+$ to NAD^+ yields 3 ATP. Because $FADH_2$ unloads its H atoms beyond the first respiratory complex, less energy (2 ATP) is captured as a result of its oxidation.

chain, which oversees the final catabolic reactions that occur on the mitochondrial cristae. However, because the reduced coenzymes produced in the Krebs cycle are the substrates for the electron transport chain, these pathways are coupled, and both phases are considered to be oxygen requiring (aerobic).

In the electron transport chain, the hydrogens removed during the oxidation of food fuels are combined with O_2, and the energy released during those reactions is harnessed to attach P_i groups to ADP. As noted earlier, this type of phosphorylation process is called *oxidative phosphorylation*. Let us peek under the hood of a cell's power plant and look more closely at this rather complicated process.

Most components of the electron transport chain are proteins that are bound to metal atoms *(cofactors)*. These proteins, which form part of the mitochondrial cristae, vary in composition (Figure 23.8). For example, some of the proteins, the **flavins,** contain flavin mononucleotide (FMN) derived from the vitamin riboflavin, and others contain both sulfur (S) and iron (Fe). Most, however, are brightly colored iron-containing pigments called **cytochromes** (si'to-krōmz; *cyto* = cell, *chrom* = color). Neighboring carriers are clustered together to form three major **respiratory enzyme complexes** that are alternately reduced and oxidized by picking up electrons and passing them on to the next complex in the

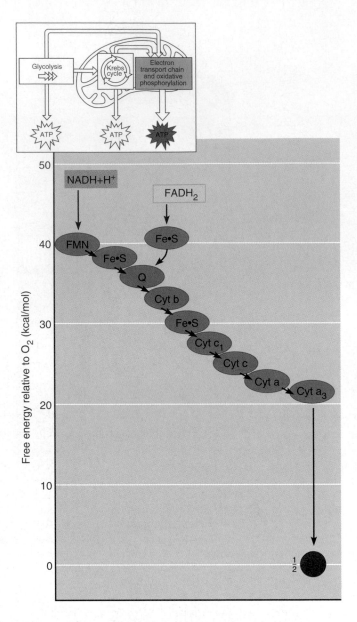

FIGURE 23.9   Electronic energy gradient in the electron transport chain. Each member of the chain oscillates between a reduced state and an oxidized state. A member becomes reduced by accepting electrons from its "uphill" neighbor and then reverts to its oxidized form as it passes electrons to its "downhill" neighbor (which has a greater affinity for the electrons). At the "bottom" of the chain is oxygen, which is *very* electronegative. The overall energy drop for electrons traveling from NADH to oxygen is 53 kcal/mol, but this fall is broken up into smaller steps by the electron transport chain.

membrane from one acceptor to the next. The protons escape into the watery matrix only to be picked up and deposited ("pumped") across the crista membrane into the intermembrane space by one of the three respiratory enzyme complexes. Ultimately the electron pairs are delivered to half a molecule of O_2 (in other words, to an oxygen atom), creating oxygen ions (O^-) that strongly attract H^+ and form water as indicated by the reaction

$$2H^+ + 2e^- + \tfrac{1}{2}O_2 \rightarrow H_2O$$

Virtually all the water resulting from glucose oxidation is formed during oxidative phosphorylation. Because NADH + H^+ and $FADH_2$ are oxidized as they release their burden of picked-up hydrogen atoms, the net reaction for the electron transport chain is

$$\text{Coenzyme-2H} + \tfrac{1}{2}O_2 \rightarrow \text{coenzyme} + H_2O$$
reduced   oxidized
coenzyme   coenzyme

The transfer of electrons from NADH + H^+ to oxygen releases large amounts of energy. If hydrogen combined directly with molecular oxygen, the energy would be released in one big burst and most of it would be lost to the environment as heat. Instead energy is released in many small steps as the electrons stream from one electron acceptor to the next. Each successive carrier has a greater affinity for electrons than those preceding it. Therefore, the electrons cascade "downhill" from NADH + H^+ to progressively lower energy levels until they are finally delivered to oxygen, which has the greatest affinity of all for electrons (Figure 23.9). You could say that oxygen "pulls" the electrons down the chain.

By using the stepwise release of electronic energy to pump protons from the gel-like matrix into the intermembrane space, the electron transport chain functions as an energy converter. Because the crista membrane is nearly impermeable to H^+, this chemiosmotic process creates an **electrochemical proton (H^+) gradient** across that membrane that temporarily has potential energy. This energy source is referred to as *proton motive force* to emphasize the capacity of the gradient to do work.

The proton gradient (1) creates a pH gradient, with the H^+ concentration in the matrix much lower than that in the intermembrane space; and (2) generates a voltage across the membrane that is negative on the matrix side and positive between the mitochondrial membranes. Both conditions strongly attract H^+ back into the matrix. However, the only areas of the membrane freely permeable to H^+ are at-large enzyme-protein complexes called **ATP synthases.** As the protons take this "route" they create an electrical current, and ATP synthase harnesses this electrical energy to catalyze attachment of a phosphate group to ADP to form ATP (Figure 23.8).

sequence. The first such complex accepts hydrogen atoms from NADH + H^+, oxidizing it to NAD^+. $FADH_2$ transfers its hydrogen atoms slightly farther along the chain. The hydrogen atoms delivered to the electron transport chain by the reduced coenzymes are quickly split into protons (H^+) plus electrons. The electrons are shuttled along the crista

The enzyme's subunits appear to work together like gears. As the ATP synthase core rotates, ADP and inorganic phosphate are pulled in and ATP is churned out, thus completing the process of oxidative phosphorylation.

Studies of the molecular structure of ATP synthase are providing an understanding of how it works (Figure 23.10). The enzyme complex consists of three major parts, each with several protein subunits: (1) a *rotor* embedded in the crista membrane, (2) a *knob* extending into the mitochondrial matrix, and (3) a *rod* connecting the two. The current created by the downhill flow of H^+ causes the rotor and rod to rotate, just as flowing water turns a water wheel. This rotation activates catalytic sites in the knob where ADP and P_i are combined to make ATP.

Notice something here. The ATP synthase works like an ion pump running in reverse. Recall from Chapter 3 that ion pumps use ATP as their energy source to transport ions against an electrochemical gradient. Here we have ATP synthases using the energy of a proton gradient to power ATP synthesis. The proton gradient also supplies energy to pump needed metabolites (ADP, pyruvic acid, inorganic phosphate) and calcium ions across the relatively impermeable crista membrane. The outer membrane is quite freely permeable to these substances, so no "help" is needed there. However, the supply of energy from oxidation is not limitless; so when more of the proton motive force or gradient energy is used to drive these transport processes, less is available to make ATP.

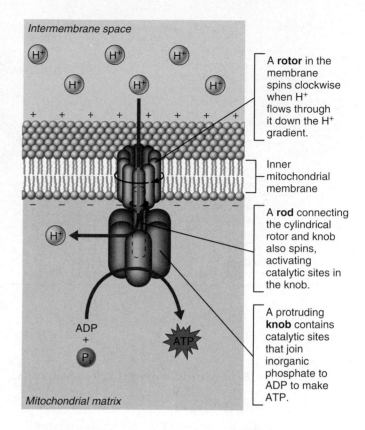

FIGURE 23.10 **Structure of ATP synthase.**

HOMEOSTATIC IMBALANCE

Studies of metabolic poisons support the chemiosmotic model of oxidative phosphorylation. For example, cyanide (the gas used in gas chambers) disrupts the process by binding to cytochrome oxidase and blocking electron flow from cytochrome a_3 to oxygen. Poisons commonly called "uncouplers" abolish the proton gradient by making the crista membrane permeable to H^+. Consequently, although the electron transport chain continues to deliver electrons to oxygen at a furious pace, and oxygen consumption rises, no ATP is made. ●

The stimulus for ATP production is entry of *ADP* into the mitochondrial fluid matrix. As ADP is transported in, ATP is moved out in a coupled transport process.

Summary of ATP Production When O_2 is present, cellular respiration is remarkably efficient. Of the 686 kilocalories (kcal) of energy present in 1 mole of glucose, as much as 262 kcal can be cap-

tured in ATP bonds. (The rest is liberated as heat.) This corresponds to an energy capture of about 38%, making cells far more efficient than any human-made machines, which use only 10–30% of the energy available to them.

During cellular respiration, most energy flows in this sequence: Glucose → NADH + H^+ → electron transport chain → proton motive force → ATP. Let's do a little bookkeeping to summarize the net energy gain from one glucose molecule. Once we have tallied the net gain of 4 ATP produced directly by substrate-level phosphorylations (2 during glycolysis and 2 during the Krebs cycle), all that remains to be calculated is the number of ATP molecules produced by oxidative phosphorylation (Figure 23.11).

Each NADH + H^+ that transfers a pair of high-energy electrons to the electron transport chain contributes enough energy to the proton gradient to generate between 2 and 3 ATP molecules. (We will round this yield off to 3 to simplify our calculations.) The oxidation of $FADH_2$ is less efficient because it doesn't donate electrons to the "top" of the electron transport chain as does NADH + H^+, but to a lower energy level. So, for each 2 H delivered by $FADH_2$, just 2 ATPs are produced (instead of 2+ or 3). Thus, the 8 NADH + H^+ and the 2 $FADH_2$ produced during the Krebs cycle are "worth" 24 and 4 ATPs respectively. The 2 NADH + H^+ generated during

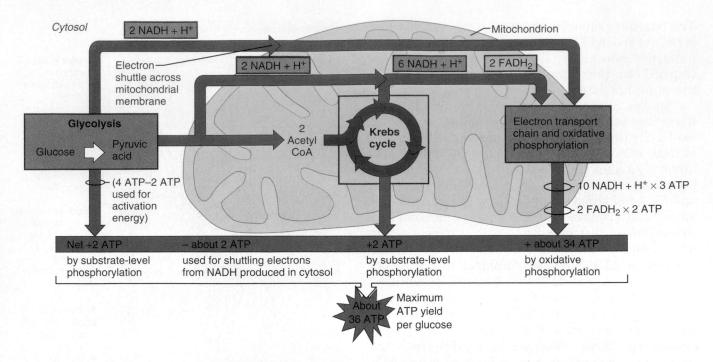

FIGURE 23.11   Energy yield during cellular respiration.

glycolysis yields 4 (or 6) ATP molecules. Overall, complete oxidation of 1 glucose molecule to CO_2 and H_2O yields 38 or 36 molecules of ATP (Figure 23.11).

The alternative figures represent the present uncertainty about the energy yield of reduced NAD^+ generated *outside* the mitochondria by glycolysis. The crista membrane is *not* permeable to reduced NAD^+ generated in the cytosol, so $NADH + H^+$ formed during glycolysis uses a *shuttle molecule* to deliver its extra electron pair to the electron transport chain. As we know, shuttles cost money, and the money used for this shuttle is ATP. At present, the consensus is that the net energy yield for reoxidation of this reduced NAD^+ is probably the same as for $FADH_2$, that is, 2 ATP per electron pair. Thus, we have deducted 2 ATP to cover the "fare" of the shuttle (Figure 23.11), and our bookkeeping comes up with a grand total of 36 ATP per glucose as the maximum possible energy yield. (Actually our figures are probably still too high because, as mentioned earlier, the proton motive force is also used to do other work.)

Glycogenesis and Glycogenolysis

Although most glucose is used to generate ATP molecules, unlimited amounts of glucose do *not* result in unlimited ATP synthesis, because cells cannot store large amounts of ATP. When more glucose is available than can be immediately oxidized, rising intracellular ATP concentrations eventually inhibit glucose catabolism and initiate processes that store glucose as either glycogen or fat. Because the body

can store much more fat than glycogen, fats account for 80–85% of stored energy.

When glycolysis is "turned off" by high ATP levels, glucose molecules are combined in long chains to form glycogen, the animal carbohydrate storage product. This process, called **glycogenesis** (*glyco* = sugar; *genesis* = origin), begins when glucose entering cells is phosphorylated to glucose-6-phosphate and then converted to its isomer, *glucose-1-phosphate*. The terminal phosphate group is cleaved off as the enzyme *glycogen synthase* catalyzes the attachment of glucose to the growing glycogen chain (Figure 23.12). Liver and skeletal muscle cells are most active in glycogen synthesis and storage.

When blood glucose levels drop, glycogen lysis (splitting) occurs. This process is known as **glycogenolysis** (gli″ko-jĕ-nol′ĭ-sis). The enzyme *glycogen phosphorylase* oversees phosphorylation and cleavage of glycogen to release glucose-1-phosphate, which is then converted to glucose-6-phosphate, a form that can enter the glycolytic pathway to be oxidized for energy.

In muscle cells and most other cells, the glucose-6-phosphate resulting from glycogenolysis is trapped because it cannot cross the cell membrane. However, hepatocytes (and some kidney and intestinal cells) contain *glucose-6-phosphatase*, an enzyme that removes the terminal phosphate, producing free glucose. Because glucose readily diffuses from the cell into the blood, the liver can use its glycogen stores to provide blood sugar for the benefit of other organs when blood glucose levels drop. Liver glycogen is also an important energy source for skeletal muscles that have depleted their own glycogen reserves.

A common misconception is that athletes need to eat large amounts of protein to improve their performance and maintain their muscle mass. Actually, a diet rich in complex carbohydrates, which stores more muscle glycogen, is much more effective in sustaining intense muscle activity than are high-protein meals. Notice that the emphasis is on *complex* carbohydrates. Eating a candy bar before an athletic event to provide "quick" energy does more harm than good because it stimulates insulin secretion, which favors glucose use and retards fat use at a time when fat use should be maximal. Building muscle protein or avoiding its loss requires not only extra protein, but also extra (protein-sparing) calories to meet the greater energy needs of the increasingly massive muscles.

Long-distance runners in particular are well aware of the practice of glycogen loading, popularly called "carbo loading," for endurance events. Glycogen loading "tricks" the muscles into storing more glycogen than they normally would. It involves (1) eating meals high in protein and fat while exercising heavily over a period of several days (this depletes muscle glycogen stores) and then (2) two to three days before the event decreasing exercise intensity (to about 40% of the norm) and switching abruptly to a high-carbohydrate diet. This causes a rebound in muscle glycogen stores of two to four times the normal amount. However, glycogen retains water, and the resulting weight increase and fluid retention may hamper the ability of muscle cells to obtain adequate oxygen. Some athletes using the procedure have complained of cardiac and skeletal muscle pain. Consequently, its use is cautioned against.

Gluconeogenesis

When too little glucose is available to stoke the "metabolic furnace," glycerol and amino acids are converted to glucose. **Gluconeogenesis,** the process of forming new *(neo)* glucose from *noncarbohydrate* molecules, occurs in the liver. It takes place when dietary sources and glucose reserves have been depleted and blood glucose levels are beginning to drop. Gluconeogenesis protects the body, the nervous system in particular, from the damaging effects of low blood sugar *(hypoglycemia)* by ensuring that ATP synthesis can continue.

Lipid Metabolism

Fats are the body's most concentrated source of energy. They contain very little water, and the energy yield from fat catabolism is approximately twice that from either glucose or protein catabolism—9 kcal per gram of fat versus 4 kcal per gram of carbohydrate or protein. Most products of fat digestion are

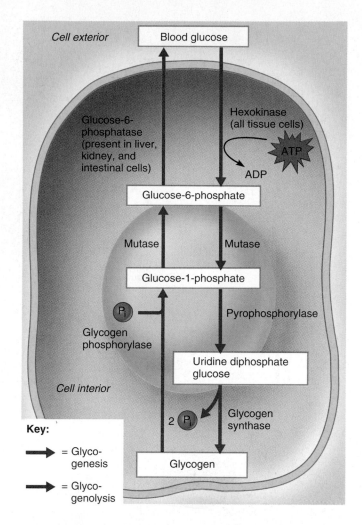

FIGURE 23.12 Glycogenesis and glycogenolysis. When glucose supplies exceed demands, glycogenesis, the conversion of glucose to glycogen for storage, occurs. Glycogenolysis, the breakdown of glycogen to release glucose, is stimulated by falling blood glucose levels. Notice that glycogen is synthesized and degraded by different enzymatic pathways.

transported in lymph in the form of fatty-protein droplets called *chylomicrons* (see Chapter 22). Eventually, the lipids in the chylomicrons are hydrolyzed by plasma enzymes, and the resulting fatty acids and glycerol are taken up by body cells and processed in various ways.

Oxidation of Glycerol and Fatty Acids

Of the various lipids, only neutral fats are routinely oxidized for energy. Their catabolism involves the separate oxidation of their two different building blocks: glycerol and fatty acid chains. Most body cells easily convert glycerol to glyceraldehyde phosphate, a glycolysis intermediate that enters the Krebs cycle. Glyceraldehyde is equal to half a glucose molecule, and ATP energy harvest from its complete

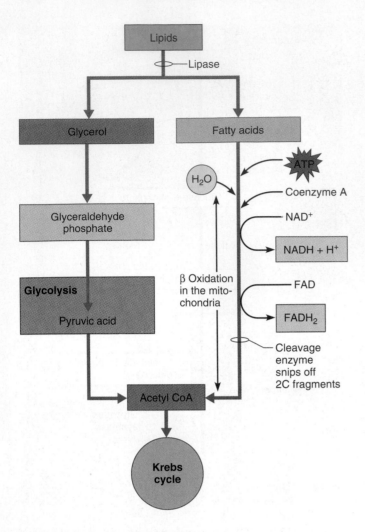

FIGURE 23.13   Initial phase of lipid oxidation. The glycerol portion is converted to glyceraldehyde phosphate, a glycolysis intermediate, and completes the glycolytic pathway through pyruvic acid to acetyl CoA. The fatty acids undergo beta oxidation. In this process the fatty acids are first activated by a coupled reaction with ATP and combined with coenzyme A. After being oxidized twice (reducing NAD$^+$ and FAD), the acetyl CoA created in β oxidation is cleaved off and the process begins again.

oxidation is approximately half that of glucose (18 ATP/glycerol).

Beta oxidation, the initial phase of fatty acid oxidation, occurs in the mitochondria. Although oxidation, dehydration, and other reactions are involved, the net result is that the fatty acid chains are broken apart into two-carbon *acetic acid* fragments, and coenzymes are reduced (Figure 23.13). Each acetic acid molecule is fused to coenzyme A, forming acetyl CoA. The term "beta oxidation" reflects the fact that the carbon in the beta (third) position is oxidized during the process and cleavage of the fatty acid in each case occurs between the alpha and beta carbons. Acetyl CoA is then picked up by oxaloacetic

acid and enters the aerobic pathways to be oxidized to CO_2 and H_2O.

Notice that unlike glycerol, which enters the glycolytic pathway, acetyl CoA resulting from fatty acid breakdown *cannot* be used for gluconeogenesis because the metabolic pathway is irreversible past pyruvic acid.

Lipogenesis and Lipolysis

There is a continuous turnover of neutral fats in adipose tissue. New fats are "put in the larder" for later use, while stored fats are broken down and released to the blood. That bulge of fatty tissue you see today does *not* contain the same fat molecules it did a month ago.

Glycerol and fatty acids from dietary fats not immediately needed for energy are recombined into triglycerides and stored. About 50% ends up in subcutaneous tissue; the balance is stockpiled in other fat depots of the body. Triglyceride synthesis, or **lipogenesis** (Figure 23.14), occurs when cellular ATP and glucose levels are high. Excess ATP also leads to an accumulation of acetyl CoA and glyceraldehyde-PO_4, two intermediates of glucose metabolism that would otherwise feed into the Krebs cycle. But when these two metabolites are present in excess, they are channeled into triglyceride synthesis pathways. Acetyl CoA molecules are condensed together, forming fatty acid chains that grow two carbons at a time. (This accounts for the fact that almost all fatty acids in the body contain an even number of carbon atoms.) Because acetyl CoA, an intermediate in glucose catabolism, is also the *starting point* for fatty acid synthesis glucose is easily converted to fat. Glyceraldehyde-PO_4 is converted to glycerol, which is condensed with fatty acids to form triglycerides. Thus, even if the diet is fat-poor, carbohydrate intake can provide *all the raw materials* needed to form neutral fats. When blood sugar is high, lipogenesis is the major activity in adipose tissues and is also an important liver function.

Lipolysis (lĭ-pol'ĭ-sis; "fat splitting"), the breakdown of stored fats into glycerol and fatty acids, is essentially lipogenesis in reverse. The fatty acids and glycerol are released to the blood, helping to ensure that body organs have continuous access to fat fuels for aerobic respiration. (The liver, cardiac muscle, and resting skeletal muscles actually prefer fatty acids as an energy fuel.) The meaning of the adage "fats burn in the flame of carbohydrates" becomes clear when carbohydrate intake is inadequate. Under such conditions, lipolysis is accelerated as the body attempts to fill the fuel gap with fats. However, the ability of acetyl CoA to enter the Krebs cycle depends on the availability of oxaloacetic acid to act as the pickup molecule. When carbohydrates are deficient, oxaloacetic acid is converted to glucose (to fuel the

? What is the central molecule in lipid metabolism?

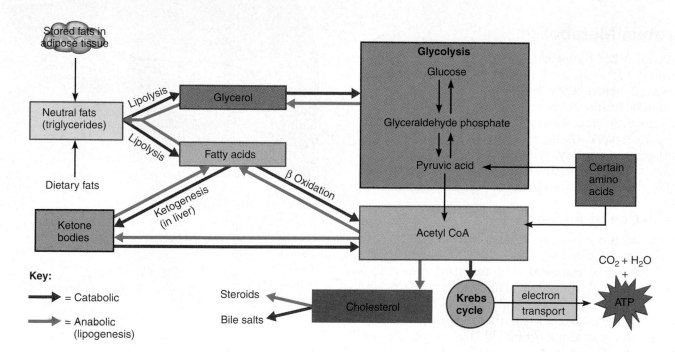

FIGURE 23.14 Metabolism of triglycerides. When needed for energy, dietary or stored fats enter the catabolic pathways. Glycerol enters the glycolytic pathway (as glyceraldehyde phosphate) and the fatty acids are broken down by beta oxidation to acetyl CoA, which enters the Krebs cycle.

When fats are to be synthesized (lipogenesis) and stored in fat depots, the intermediates are drawn from glycolysis and the Krebs cycle in a reversal of the processes noted above. Likewise, excess dietary fats are stored in adipose tissues. When triglycerides are in excess or are the primary energy source, the liver releases their breakdown products (fatty acids → acetyl CoA) in the form of ketone bodies. Excessive amounts of carbohydrates and amino acids are also converted to triglycerides (lipogenesis).

brain). Without oxaloacetic acid, fat oxidation is incomplete, acetyl CoA accumulates, and via a process called **ketogenesis,** the liver converts acetyl CoA molecules to **ketones,** or **ketone bodies,** which are released into the blood. Ketone bodies include acetoacetic acid, β-hydroxybutyric acid, and acetone, all formed from acetic acid. (The *keto acids* cycling through the Krebs cycle and the *ketone bodies* resulting from fat metabolism are quite different and should not be confused.)

🄷 HOMEOSTATIC IMBALANCE

When ketone bodies accumulate in the blood, *ketosis* results and large amounts of ketone bodies are excreted in the urine. Ketosis is a common consequence of starvation, unwise dieting (in which inadequate amounts of carbohydrates are eaten), and diabetes mellitus. Because most ketone bodies are organic acids, the outcome of ketosis is *metabolic acidosis*. The body's buffer systems cannot tie up the acids (ketones) fast enough, and blood pH drops to dangerously low levels. The person's breath smells fruity as acetone vaporizes from the lungs, and breathing becomes more rapid as the respiratory system tries to reduce blood carbonic acid by blowing off CO_2 to force the blood pH up. In severe untreated cases, the person may become comatose or even die as the acid pH depresses the nervous system. ●

Synthesis of Structural Materials

All body cells use phospholipids and cholesterol to build their membranes. Phospholipids are also important components of myelin sheaths of neurons. In addition, the liver (1) synthesizes lipoproteins for transport of cholesterol, fats, and other substances in the blood; (2) makes the clotting factor called tissue factor; (3) synthesizes cholesterol from acetyl

■ *Acetyl CoA.*

CoA; and (4) uses cholesterol to form bile salts. The ovaries, testes, and adrenal cortex use cholesterol to synthesize their steroid hormones.

Protein Metabolism

Like all other biological molecules, proteins have a limited life span and must be broken down and replaced before they begin to deteriorate. Newly ingested amino acids transported in the blood are taken up by cells by active transport processes and used to *replace* tissue proteins at the rate of about 100 grams each day. When more protein is ingested than is needed for these anabolic purposes, amino acids are oxidized for energy or converted to fat.

Oxidation of Amino Acids

Before amino acids can be oxidized for energy, they must be *deaminated*, that is, their amine group (NH_2) must be removed. The resulting molecule is then converted to pyruvic acid or to one of the keto acid intermediates in the Krebs cycle. The key molecule in these conversions is the nonessential amino acid *glutamic acid* (gloo-tam'ik). The events that occur (Figure 23.15) are:

1. **Transamination** (trans"am-ĭ-na'shun). A number of amino acids can transfer their amine group to α-ketoglutaric acid (a Krebs cycle keto acid), thereby transforming α-ketoglutaric acid to glutamic acid. In the process, the original amino acid becomes a keto acid (that is, it has an oxygen atom where the amine group formerly was). This reaction is fully reversible.

2. **Oxidative deamination.** In the liver, the amine group of glutamic acid is removed as **ammonia** (NH_3), and α-ketoglutaric acid is regenerated. The liberated NH_3 molecules are combined with CO_2, yielding **urea** and water. The urea is released to the blood and removed from the body in urine. Because ammonia is toxic to body cells, the ease with which glutamic acid funnels amine groups into the **urea cycle** is extremely important. This mechanism rids the body not only of NH_3 produced during oxidative deamination, but also of bloodborne NH_3 produced by intestinal bacteria.

3. **Keto acid modification.** The goal of amino acid degradation is to produce molecules that can be either oxidized in the Krebs cycle or converted to glucose. Hence, keto acids resulting from transamination are altered as necessary to produce metabolites that can enter the Krebs cycle. The most important of these metabolites are pyruvic acid, acetyl CoA, α-ketoglutaric acid, and oxaloacetic acid (Figure 23.7). Because the reactions of glycolysis are reversible, deaminated amino acids that are converted to pyruvic acid can be reconverted to glucose and contribute to gluconeogenesis.

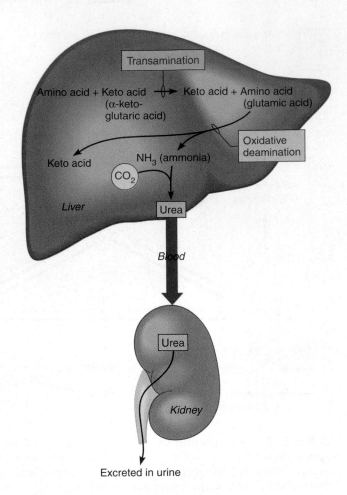

FIGURE 23.15 Transamination and oxidative processes that occur when amino acids are oxidized for energy. Transamination involves the switching of an amine group from an amino acid to a keto acid, usually α-ketoglutaric acid. As a result the original amino acid becomes a keto acid, and the keto acid becomes an amino acid, typically glutamic acid. In oxidative deamination the amine group of glutamic acid is released as ammonia and then combined with CO_2 by the liver cells to form urea.

Protein Synthesis

Amino acids are the most important anabolic nutrients. Not only do they form all protein structures, but they form the bulk of the body's functional molecules as well. As described in Chapter 3, protein synthesis occurs on ribosomes, where cytoplasmic enzymes oversee the formation of peptide bonds linking the amino acids together into protein polymers. The amount and type of protein synthesized are precisely controlled by hormones (growth hormone, thyroxine, sex hormones, and others), and so protein anabolism reflects the hormonal balance at each stage of life.

During your lifetime, your cells will have synthesized 225–450 kg (about 500–1000 pounds) of proteins, depending on your size. However, you do not need to consume anywhere near that amount of

TABLE 23.4	Thumbnail Summary of Metabolic Reactions

Carbohydrates

Cellular respiration	Reactions that together complete the oxidation of glucose, yielding CO_2, H_2O, and ATP
Glycolysis	Conversion of glucose to pyruvic acid
Glycogenesis	Polymerization of glucose to form glycogen
Glycogenolysis	Hydrolysis of glycogen to glucose monomers
Gluconeogenesis	Formation of glucose from noncarbohydrate precursors
Krebs cycle	Complete breakdown of pyruvic acid to CO_2, yielding small amounts of ATP and reduced coenzymes
Electron transport chain	Energy-yielding reactions that split H removed during oxidations to H^+ and e^- and create a proton gradient used to bond ADP to P_i, forming ATP

Lipids

Beta oxidation	Conversion of fatty acids to acetyl CoA
Lipolysis	Breakdown of lipids to fatty acids and glycerol
Lipogenesis	Formation of lipids from acetyl CoA and glyceraldehyde phosphate

Proteins

Transamination	Transfer of an amine group from an amino acid to α-ketoglutaric acid, thereby transforming α-ketoglutaric acid to glutamic acid
Oxidative deamination	Removal of an amine group from glutamic acid as ammonia and regenerating α-ketoglutaric acid (NH_3 is converted to urea by the liver)

protein because nonessential amino acids are easily formed by siphoning keto acids from the Krebs cycle and transferring amine groups to them. Most of these transformations occur in the liver, which provides nearly all the nonessential amino acids needed to produce the relatively small amount of protein that the body synthesizes each day. However, a complete set of amino acids must be present for protein synthesis to take place, so all essential amino acids must be provided by the diet. If some are not, the rest are oxidized for energy even though they may be needed for anabolism. In such cases, negative nitrogen balance results because body protein is broken down to supply the essential amino acids needed.

The various metabolic reactions described thus far are summarized briefly in Table 23.4.

Catabolic-Anabolic Steady State of the Body

The body exists in a *dynamic catabolic-anabolic state* as organic molecules are continuously broken down and rebuilt—frequently at a head-spinning rate.

The blood is the transport pool for all body cells, and it contains many kinds of energy sources—glucose, ketone bodies, fatty acids, glycerol, and lactic acid. Some organs routinely use blood energy sources other than glucose, thus saving glucose for tissues with stricter glucose requirements (see Table 23.5 on p. 838).

The body **nutrient pools**—amino acid, carbohydrate, and fat stores—can be drawn on to meet its varying needs (Figure 23.16). These pools are interconvertible because their pathways are linked by key intermediates (see Figure 23.17). The liver, adipose tissue, and skeletal muscles are the primary effector organs determining the amounts and direction of the conversions shown in the figure.

The **amino acid pool** is the body's total supply of free amino acids. Small amounts of animo acids and proteins are lost daily in urine and in sloughed hairs and skin cells. Typically, these lost molecules are replaced via the diet; otherwise, amino acids arising from tissue breakdown return to the pool. This pool is the source of amino acids used for protein synthesis and in the formation of amino acid derivatives. In addition, as described above, deaminated amino acids can participate in gluconeogenesis. Not all events of amino acid metabolism occur in all cells. For example, *only* the liver forms urea. Nonetheless, the concept of a common amino acid pool is valid because all cells are connected by the blood.

Because carbohydrates are easily and frequently converted to fats, the **carbohydrate** and **fat pools** are usually considered together (Figure 23.17). There are two major differences between this pool and the amino acid pool: (1) Fats and carbohydrates are

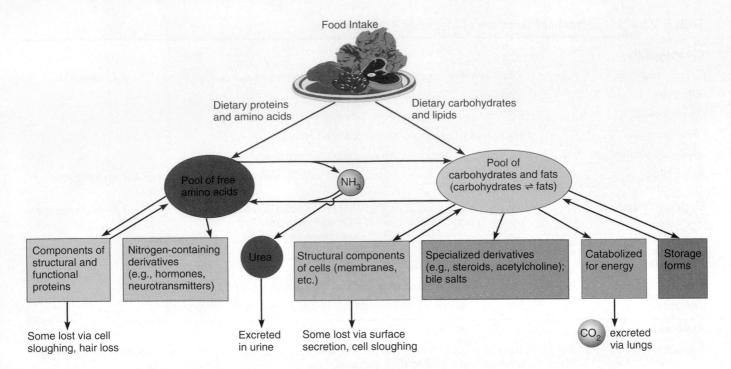

FIGURE 23.16 **Carbohydrate/fat and amino acid pools.**

oxidized directly to produce cellular energy, whereas amino acids can be used to supply energy *only after being converted to a carbohydrate intermediate* (a keto acid). (2) Excess carbohydrate and fat can be stored as such, whereas excess amino acids are *not* stored as protein. Instead, they are oxidized for energy or converted to fat or glycogen for storage.

Absorptive and Postabsorptive States

Metabolic controls act to equalize blood concentrations of energy sources between two nutritional states. Sometimes referred to as the *fed state,* the **absorptive state** is the time during and shortly after eating, when nutrients are flushing into the blood

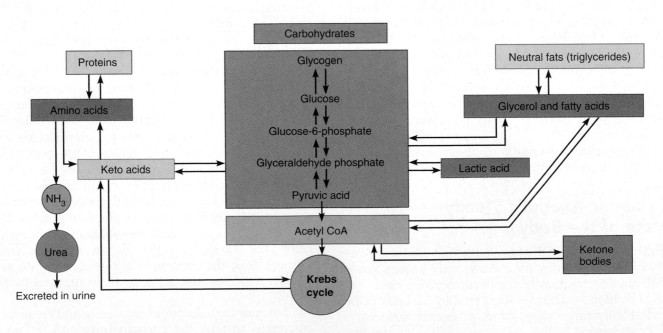

FIGURE 23.17 **Interconversion of carbohydrates, fats, and proteins.** The liver, adipose tissue, and skeletal muscles are the primary effector organs determining the amounts and direction of the conversions shown.

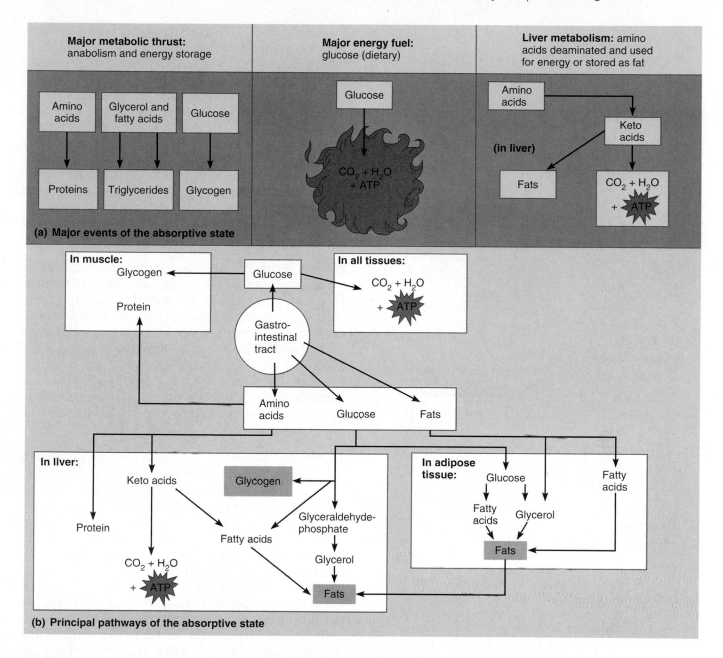

FIGURE 23.18 **Major events and principal metabolic pathways of the absorptive state.** Although not indicated in (b), amino acids are also taken up by tissue cells and used for protein synthesis, and fats are the primary energy fuel of muscle, liver cells, and adipose tissue.

from the gastrointestinal tract. The **postabsorptive** *(fasting)* **state** is the period when the GI tract is empty and energy sources are supplied by the breakdown of body reserves. Those who eat "three squares" a day are in the absorptive state for the four hours during and after each meal and in the postabsorptive state in the late morning, late afternoon, and all night. However, postabsorptive mechanisms can sustain the body for much longer intervals if necessary—even to accommodate weeks of fasting—as long as water is taken in.

Absorptive State

During the absorptive state (Figure 23.18) anabolism exceeds catabolism. Glucose is the major energy fuel. Dietary amino acids and fats are used to remake degraded body protein or fat, and small amounts are oxidized to provide ATP. Excess metabolites, regardless of source, are transformed to fat if not used for anabolism. We will consider the fate and hormonal control of each nutrient group during this phase.

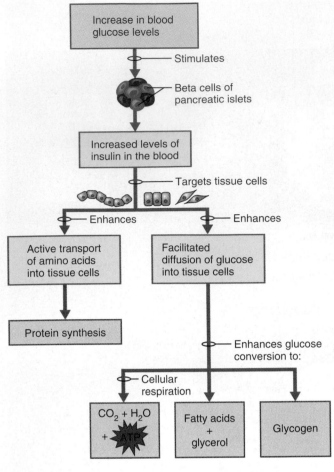

Key:

- [ ] Initial stimulus
- [ ] Physiological response
- [ ] Result

FIGURE 23.19 **Effects of insulin on metabolism.**
When insulin concentrations are low (during the postabsorptive period), these normal effects of insulin are inhibited and glycogenolysis and gluconeogenesis occur, that is, the insulin-mediated intracellular reactions are reversed. (Note: Not all effects shown occur in all cells.)

Carbohydrates   Absorbed monosaccharides are delivered directly to the liver, where fructose and galactose are converted to glucose. Glucose, in turn, is released to the blood or converted to glycogen and fat. Glycogen formed in the liver is stored there, but most fat synthesized there is released to the blood and picked up for storage by adipose tissues. Blood-borne glucose not sequestered by the liver enters body cells to be metabolized for energy; any excess is stored in skeletal muscle cells as glycogen or in adipose cells as fat.

Triglycerides   Nearly all products of fat digestion enter the lymph in the form of chylomicrons, which are hydrolyzed to fatty acids and glycerol before they can pass through the capillary walls. *Lipoprotein lipase,* the enzyme that catalyzes fat hydrolysis, is particularly active in the capillaries of muscle and fat tissues. Adipose cells, skeletal muscle cells, and liver cells use triglycerides as their primary energy source, and when dietary carbohydrates are limited, other cells begin to oxidize more fat for energy. Although some fatty acids and glycerol are used for anabolic purposes by tissue cells, most enter adipose tissue to be reconverted to triglycerides and stored.

Amino Acids   Absorbed amino acids are delivered to the liver, which deaminates some of them to keto acids. The keto acids may flow into the Krebs cycle to be used for ATP synthesis, or they may be converted to liver fat stores. The liver also uses some of the amino acids to synthesize plasma proteins, including albumin, clotting proteins, and transport proteins. However, most amino acids flushing through the liver sinusoids remain in the blood for uptake by other body cells, where they are used for protein synthesis.

Hormonal Control   **Insulin** directs essentially all events of the absorptive state (Figure 23.19). Rising blood glucose levels after a carbohydrate-containing meal act as a humoral stimulus that prods the beta cells of the pancreatic islets to secrete more insulin. (This glucose-induced stimulation of insulin release is enhanced by several GI tract hormones including gastrin, CCK, and secretin.) A second important stimulus for insulin release is elevated amino acid levels in the blood. As insulin binds to membrane receptors of its target cells, it activates carrier-mediated facilitated diffusion of glucose into the cells. Within minutes, the rate of glucose entry into tissue cells increases about 20-fold. (The exception is brain cells, which take up glucose whether or not insulin is present.) Once glucose enters tissue cells, insulin enhances glucose oxidation for energy and stimulates its conversion to glycogen and, in adipose tissue, to triglycerides. Insulin also "revs up" the active transport of amino acids into cells, promotes protein synthesis, and inhibits virtually all liver enzymes that promote gluconeogenesis.

As you can see, insulin is a **hypoglycemic hormone** (hi"po-gli-se'mik). It sweeps glucose out of the blood into the tissue cells, thereby lowering blood glucose levels. Additionally, it enhances glucose oxidation or storage while simultaneously inhibiting any process that might increase blood glucose levels.

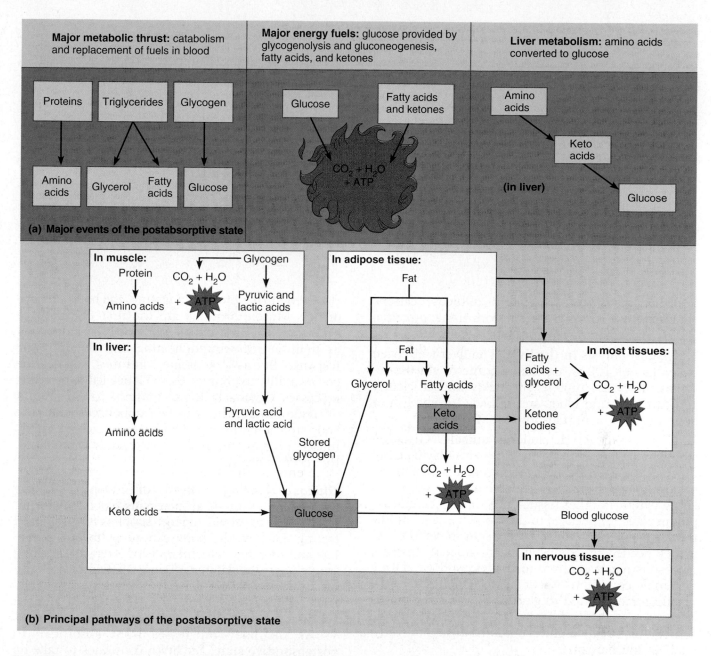

FIGURE 23.20 **Major events and principal metabolic pathways of the postabsorptive state.**

 HOMEOSTATIC IMBALANCE

Diabetes mellitus is a consequence of inadequate insulin production or abnormal insulin receptors. Without insulin or receptors that "recognize" it, glucose becomes unavailable to most body cells. Thus, blood glucose levels remain high, and large amounts of glucose are excreted in urine. Metabolic acidosis, protein wasting, and weight loss occur as large amounts of fats and tissue proteins are used for energy. (Diabetes mellitus is described in more detail in Chapter 15.) ●

Postabsorptive State

The primary goal during the postabsorptive state, that is, between meals when blood glucose levels are dropping, is to maintain blood glucose levels within the homeostatic range (80–100 mg glucose/100 ml). The importance of constant blood glucose has already been explained, the most important factor being that the brain almost always uses glucose as its energy source. Most events of the postabsorptive state either make glucose available to the blood or save glucose for the organs that need it most (Figure 23.20).

TABLE 23.5 ⟋ Profiles of the Major Body Organs in Fuel Metabolism

Tissue	Fuel Stores	Preferred Fuel	Fuel Sources Exported
Brain	None	Glucose (ketone bodies during starvation)	None
Skeletal muscle (resting)	Glycogen	Fatty acids	None
Skeletal muscle (during exertion)	None	Glucose	Lactate
Heart muscle	None	Fatty acids	None
Adipose tissue	Triglycerides	Fatty acids	Fatty acids, glycerol
Liver	Glycogen, triglycerides	Amino acids, glucose, fatty acids	Fatty acids, glucose, ketone bodies

Adapted from Mathews, van Holde, and Ahern, 1999, *Biochemistry* 3/e, San Francisco: Addison Wesley Longman, p. 832.

Sources of Blood Glucose   Glucose can be obtained from stored glycogen, from tissue proteins, and in limited amounts from fats (see Table 23.5).

1. Glycogenolysis in the liver. The liver's glycogen stores (about 100 g) are the first line of glucose reserves. They are mobilized quickly and efficiently and can maintain blood sugar levels for about four hours during the postabsorptive state.

2. Glycogenolysis in skeletal muscle. Glycogen stores in skeletal muscle are approximately equal to those of the liver. Before liver glycogen is exhausted, glycogenolysis begins in skeletal muscle (and to a lesser extent in other tissues). However, the glucose produced is not released to the blood because, unlike the liver, skeletal muscle does not have the enzymes needed to dephosphorylate glucose. Instead, glucose is partly oxidized to pyruvic acid (or, during anaerobic conditions, lactic acid), which enters the blood, is reconverted to glucose by the liver, and is released to the blood again. Thus, skeletal muscle contributes to blood glucose homeostasis indirectly, via liver mechanisms.

3. Lipolysis in adipose tissues and the liver. Adipose and liver cells produce glycerol by lipolysis, and the liver converts the glycerol to glucose (gluconeogenesis), which is released to the blood. Because acetyl CoA, a product of the beta oxidation of fatty acids, is produced beyond the *reversible* steps of glycolysis, fatty acids *cannot* be used to bolster blood glucose levels.

4. Catabolism of cellular protein. Tissue proteins become the major source of blood glucose when fasting is prolonged and glycogen and fat stores are nearly exhausted. Cellular amino acids (mostly from muscle) are deaminated and converted to glucose in the liver. During fasts lasting several weeks, the kidneys also carry out gluconeogenesis and contribute fully as much glucose to the blood as does the liver.

Even during prolonged fasting, the body sets priorities. Muscle proteins are the first to go (to be catabolized). Movement is not nearly as important as maintaining wound healing and the immune response. But as long as life continues, so does the body's ability to produce the ATP needed to drive life processes. Obviously, there are limits to the amount of tissue protein that can be catabolized before the body stops functioning. The heart is almost entirely muscle protein, and when it is severely catabolized, the result is death.

Glucose Sparing   Even collectively, all of the manipulations aimed at increasing blood glucose are not enough to provide energy supplies for prolonged fasting. Luckily, the body can adapt to burn more fats and proteins, which enter the Krebs cycle along with glucose breakdown products. The increased use of noncarbohydrate fuel molecules (especially triglycerides) to conserve glucose is called **glucose sparing.**

As the body progresses from absorptive to postabsorptive state, the brain continues to take its share of blood glucose; but virtually every other organ switches to fatty acids as its major energy source, thus sparing glucose for the brain. During this transition phase, lipolysis begins in adipose tissues and released fatty acids are picked up by tissue cells and oxidized for energy. In addition, the liver oxidizes fats to ketone bodies and releases them to the blood for use by tissue cells. If fasting continues for longer than four or five days, the brain too begins to use large quantities of ketone bodies as well as glucose as its energy fuel (Table 23.5). The survival value of the brain's ability to use an alternative fuel source is obvious—much less tissue protein has to be ravaged to form glucose.

Hormonal and Neural Controls   The sympathetic nervous system and several hormones interact to control events of the postabsorptive state.

Consequently, regulation of this state is much more complex than that of the absorptive state when a single hormone, insulin, holds sway.

An important trigger for initiating postabsorptive events is damping of insulin release, which occurs as blood glucose levels drop. As insulin levels decline, all insulin-induced cellular responses are inhibited as well.

Declining glucose levels also stimulate the alpha cells of the pancreatic islets to release the insulin antagonist **glucagon.** Like other hormones acting during the postabsorptive state, glucagon is a *hyperglycemic hormone*, that is, it promotes a rise in blood glucose levels. Glucagon targets the liver and adipose tissue (Figure 23.21). The hepatocytes respond by accelerating glycogenolysis and gluconeogenesis. Adipose cells mobilize their fatty stores (lipolysis) and release fatty acids and glycerol to the blood. Thus, glucagon "refurbishes" blood energy sources by enhancing both glucose and fatty acid levels. Glucagon release is inhibited after the next meal or whenever blood glucose levels rise and insulin secretion begins again.

Thus far, the picture is pretty straightforward. Increasing blood glucose levels trigger insulin release, which "pushes" glucose out of the blood and into the cells. This drop in blood glucose stimulates secretion of glucagon, which "pulls" glucose from the cells into the blood. However, there is more than a push-pull mechanism here because *both* insulin and glucagon release are strongly stimulated by rising amino acid levels in the blood. This effect is insignificant when we eat a balanced meal, but it has an important adaptive role when we eat a high-protein, low-carbohydrate meal. In this instance, the stimulus for insulin release is strong, and if it were not counterbalanced, the brain might be damaged by the abrupt onset of hypoglycemia as glucose is rushed out of the blood. Simultaneous release of glucagon modulates the effects of insulin and helps stabilize blood glucose levels.

The sympathetic nervous system plays a crucial role in supplying fuel quickly when blood glucose levels drop suddenly. Adipose tissue is well supplied with sympathetic fibers, and epinephrine released by the adrenal medulla in response to sympathetic activation acts on the liver, skeletal muscle, and adipose tissues. Together, these stimuli mobilize fat and promote glycogenolysis—essentially the same effects prompted by glucagon. Injury, anxiety, or any other stressor that mobilizes the fight-or-flight response will trigger this control pathway.

In addition to glucagon and epinephrine, a number of other hormones—including growth hormone, thyroxine, sex hormones, and corticosteroids—influence metabolism and nutrient flow. Growth hormone secretion is enhanced by prolonged fasting

? What hormone is glucagon's main antagonist?

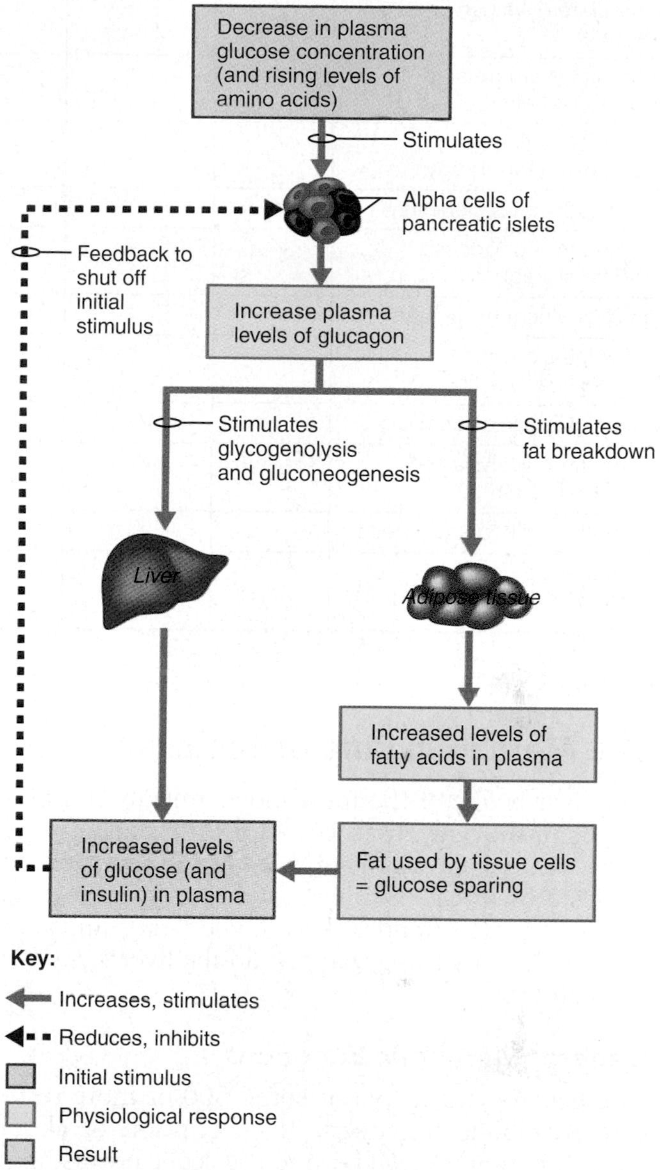

FIGURE 23.21 **Influence of glucagon on blood glucose levels.** Negative feedback control exerted by rising plasma glucose levels on glucagon secretion is indicated by the dashed arrow.

or rapid declines in blood glucose levels, and GH exerts important anti-insulin effects. However, the release and activity of most of these hormones are not specifically related to absorptive or postabsorptive metabolic events. Typical metabolic effects of various hormones are described in Table 23.6.

■ *Insulin.*

TABLE 23.6   Summary of Normal Hormonal Influences on Metabolism

Hormone's Effects	Insulin	Glucagon	Epinephrine	Growth Hormone	Thyroxine	Cortisol	Testosterone
Stimulates glucose uptake by cells	✓				✓		
Stimulates amino acid uptake by cells	✓			✓			
Stimulates glucose catabolism for energy	✓				✓		
Stimulates glycogenesis	✓						
Stimulates lipogenesis and fat storage	✓						
Inhibits gluconeogenesis	✓						
Stimulates protein synthesis (anabolic)	✓			✓	✓		✓
Stimulates glycogenolysis		✓	✓				
Stimulates lipolysis and fat mobilization		✓	✓	✓	✓	✓	
Stimulates gluconeogenesis		✓	✓	✓		✓	
Stimulates protein breakdown (catabolic)						✓	

The Metabolic Role of the Liver

The liver is one of the most biochemically complex organs in the body. It processes nearly every class of nutrients and plays a major role in regulating plasma cholesterol levels. While mechanical contraptions can, in a pinch, stand in for a failed heart, lungs, or kidney, the only thing that can do the liver's work is a hepatocyte.

General Metabolic Functions

The hepatocytes carry out some 500 or more intricate metabolic functions. A description of all of these functions is well beyond the scope of this text, but a brief summary is provided in Table 23.7.

Cholesterol Metabolism and Regulation of Blood Cholesterol Levels

Cholesterol, though an important dietary lipid, has received little attention in this discussion so far, primarily because it is not used as an energy source. It serves instead as the structural basis of bile salts, steroid hormones, and vitamin D and as a major component of plasma membranes. Additionally, cholesterol is part of a key signaling molecule (the *hedgehog protein*) that helps direct embryonic development. About 15% of blood cholesterol comes from the diet. The other 85% is made from acetyl CoA by the liver, and to a lesser extent other body cells, particularly intestinal cells. Cholesterol is lost from the body when it is catabolized and secreted in bile salts, which are eventually excreted in feces.

Cholesterol Transport   Because triglycerides and cholesterol are insoluble in water, they do not circulate free in the blood. Instead, they are transported to and from tissue cells bound to small lipid-protein complexes called **lipoproteins.** These complexes solubilize the hydrophobic lipids, and the protein part of the complexes contains signals that regulate lipid entry and exit at specific target cells.

Lipoproteins vary considerably in their relative fat-protein composition, but they all contain triglycerides, phospholipids, and cholesterol in addition to protein (Figure 23.22). In general, the higher the percentage of lipid in the lipoprotein, the lower its density; and the greater the proportion of protein, the higher its density. On this basis, there are **high-density lipoproteins (HDLs), low-density lipoproteins (LDLs),** and **very low density lipoproteins (VLDLs).** *Chylomicrons,* which transport absorbed lipids from the GI tract, are a separate class and have the lowest density of all.

The liver is the primary source of VLDLs, which transport triglycerides from the liver to the peripheral tissues, but mostly to *adipose tissues.* Once the triglycerides are unloaded, the VLDL residues are converted to LDLs, which are cholesterol-rich. The role of the LDLs is to transport cholesterol to *peripheral tissues*, making it available to the tissue

TABLE 23.7   Summary of Metabolic Functions of the Liver

Metabolic Processes Targeted	Functions
Carbohydrate Metabolism Particularly important in maintaining blood glucose homeostasis	▪ Converts galactose and fructose to glucose ▪ Glucose buffer function: stores glucose as glycogen when blood glucose levels are high; in response to hormonal controls, performs glycogenolysis and releases glucose to blood ▪ Gluconeogenesis; converts amino acids and glycerol to glucose when glycogen stores are exhausted and blood glucose levels are falling ▪ Converts glucose to fats for storage
Fat metabolism Although most cells are capable of some fat metabolism, liver bears the major responsibility	▪ Primary body site of beta oxidation (breakdown of fatty acids to acetyl CoA) ▪ Converts excess acetyl CoA to ketone bodies for release to tissue cells ▪ Stores fats ▪ Forms lipoproteins for transport of fatty acids, fats, and cholesterol to and from tissues ▪ Synthesizes cholesterol from acetyl CoA; catabolizes cholesterol to bile salts, which are secreted in bile
Protein metabolism Other metabolic functions of the liver could be dispensed with and body could still survive; however, without liver metabolism of proteins, severe survival problems ensue: many essential clotting proteins would not be made, and ammonia would not be disposed of, for example	▪ Deaminates amino acids (required for their conversion to glucose or use for ATP synthesis); amount of deamination that occurs outside the liver is unimportant ▪ Forms urea for removal of ammonia from body; inability to perform this function (e.g., in cirrhosis or hepatitis) results in accumulation of ammonia in blood ▪ Forms most plasma proteins (exceptions are gamma globulins and some hormones and enzymes); plasma protein depletion causes rapid mitosis of the hepatocytes and actual growth of liver, which is coupled with increase in synthesis of plasma proteins until blood values are again normal ▪ Transamination: intraconversion of nonessential amino acids; amount that occurs outside liver is inconsequential
Vitamin/mineral storage	▪ Stores vitamin A (1–2 years' supply) ▪ Stores sizable amounts of vitamins D and B_{12} (1–4 months' supply) ▪ Stores iron; other than iron bound to hemoglobin, most of body's supply is stored in liver as ferritin until needed; releases iron to blood as blood levels drop
Biotransformation functions	▪ Relative to alcohol and drug metabolism, performs synthetic reactions which yield inactive products that can be secreted by the kidneys and nonsynthetic reactions which may result in products that are more active, changed in activity, or less active ▪ Processes bilirubin resulting from RBC breakdown and excretes bile pigments in bile ▪ Metabolizes bloodborne hormones to forms that can be excreted in urine

cells for membrane or hormone synthesis and for storage for later use. The LDLs also regulate cholesterol synthesis in the tissue cells. Docking of LDL to the LDL receptor triggers receptor-mediated endocytosis of the entire particle.

The major function of HDLs, which are particularly rich in phospholipids and cholesterol, is to transport excess cholesterol *from peripheral tissues to the liver*, where it is broken down and becomes part of bile. The liver makes the protein envelopes of the HDL particles and then ejects them into the bloodstream in collapsed form, rather like deflated beach balls. Once in the blood, these still-incomplete HDL particles fill with cholesterol picked up from the tissue cells and "pulled" from the artery walls.

HDL also services the needs of steroid-producing organs, like the ovaries and adrenal glands, for their raw material (cholesterol). These organs have the ability to selectively remove cholesterol from the HDL particles without engulfing them.

High blood cholesterol levels (above 200 mg cholesterol/100 ml blood) have been linked to risk of atherosclerosis, which clogs the arteries and causes strokes and heart attacks. However, it is not enough to simply measure total cholesterol. The form in which cholesterol is transported in the blood is more important clinically. As a rule, high levels of HDLs are considered *good* because the transported cholesterol is destined for degradation. An HDL level between 35 and 60 is considered okay; levels above 60

Are the lipoprotein particles with the most healthy effects those with high protein and low cholesterol or those with high cholesterol and low protein?

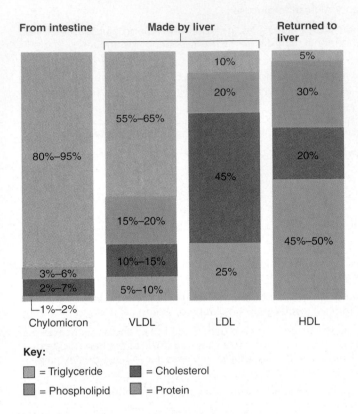

FIGURE 23.22 Approximate composition of lipoproteins that transport lipids in body fluids. VLDL = very low density lipoprotein; LDL = low-density lipoprotein; HDL = high-density lipoprotein.

are thought to protect against heart disease. High LDL levels (160 or above) are considered *bad* because when LDLs are excessive, potentially lethal cholesterol deposits are laid down in the artery walls.

If LDLs are "bad" cholesterol, then one variety of LDL—lipoprotein (a)—is "really nasty." This lipid appears to promote plaque formation that thickens and stiffens the blood vessel walls. High levels of lipoprotein (a) in the blood can double a man's risk of a heart attack before the age of 55. It is estimated that one in five males has elevated levels of this lipoprotein in the blood.

Factors Regulating Plasma Cholesterol Levels

A negative feedback loop partially adjusts the amount of cholesterol produced by the liver according to cholesterol in the diet. A high cholesterol in-

take inhibits its synthesis by the liver, but it is not a one-to-one relationship because the liver produces a certain basal amount of cholesterol even when dietary intake is excessive. For this reason, severe restriction of dietary cholesterol, although helpful, does not lead to a steep reduction in plasma cholesterol levels.

The relative amounts of saturated and unsaturated fatty acids in the diet have an important effect on blood cholesterol levels. Saturated fatty acids *stimulate liver synthesis* of cholesterol and *inhibit its excretion* from the body. Thus, moderate decreases in intake of saturated fats (found chiefly in animal fats and coconut oil) can reduce cholesterol levels as much as 20%. In contrast, unsaturated fatty acids (found in most vegetable oils) *enhance excretion* of cholesterol and its catabolism to bile salts, thereby reducing total cholesterol levels. The unhappy exception to this good news about unsaturated fats concerns "healthy" oils that have been hardened by hydrogenation to make them more solid. Hydrogenation changes the fatty acids in the oils to *trans fatty acids*, which cause serum changes worse than those caused by saturated fats—they spark a greater increase in LDLs and a greater reduction in HDLs, thus producing the unhealthiest ratio of total cholesterol to HDL.

The unsaturated omega-3 fatty acids found in especially large amounts in some cold-water fish lower the proportions of both saturated fats and cholesterol. The omega-3 fatty acids have a powerful antiarrhythmic effect on the heart and also make blood platelets less sticky, thus helping prevent spontaneous clotting that can block blood vessels. They also appear to lower blood pressure (even in nonhypertensive people). In those with moderate to high cholesterol levels, replacing half of the lipid- and cholesterol-rich animal proteins in the diet with soy protein lowers cholesterol significantly.

Factors other than diet also influence blood cholesterol levels. For example, cigarette smoking, coffee drinking, and stress appear to increase LDL levels, whereas regular aerobic exercise lowers LDL levels and increases HDL levels. Interestingly, body shape, that is, the distribution of body fat, provides clues to risky blood levels of cholesterol and fats. "Apples" (people with upper body and abdominal fat distribution, seen more often in men) tend to have higher levels of cholesterol in general and LDLs in particular than "pears" (fat localized in the hips and thighs, a pattern more common in women).

Most cells other than liver and intestinal cells obtain the bulk of the cholesterol they need for membrane synthesis from the blood. When a cell needs cholesterol, it makes receptor proteins for LDL and inserts them in its plasma membrane. LDL binds to the receptors and is engulfed in coated pits

■ *High protein and low cholesterol.*

by endocytosis. Within 15 minutes the endocytotic vesicles fuse with lysosomes, where the cholesterol is freed for use. When excessive cholesterol accumulates in a cell, it inhibits both the cell's own cholesterol synthesis and its synthesis of LDL receptors.

⒣ HOMEOSTATIC IMBALANCE

High cholesterol levels undoubtedly put us at risk for cardiovascular disease and heart attack. Cholesterol-lowering drugs such as the *statins* (e.g., lovastatin and pravastatin) are routinely prescribed for cardiac patients with high cholesterol, and tests of a vaccine that lowers LDL while raising HDL are under way. Now, however, studies indicate that cholesterol levels below 190 for males and 178 for females may be equally devastating because they may enhance the risk of "bleeding" strokes and death from cerebral hemorrhage. Additionally, almost half of those who get heart disease have normal cholesterol levels, while others with poor lipid profiles remain free of heart disease. Although most Americans would probably benefit from reducing their intake of saturated fats and cholesterol-rich foods, at least for now, the moderate approach to cholesterol control may be the wisest. ●

Energy Balance

When any fuel is burned, it consumes oxygen and liberates heat. The "burning" of food fuels by our cells is no exception. As described in Chapter 2, energy can be neither created nor destroyed—only converted from one form to another. If we apply this principle (actually the *first law of thermodynamics*) to cell metabolism, it means that bond energy released as foods are catabolized must be precisely balanced by the total energy output of the body. Thus, a dynamic balance exists between the body's energy intake and its energy output:

Energy intake = energy output
(heat + work + energy storage)

Energy intake is the energy liberated during food oxidation. (Undigested foods are not part of the equation because they contribute no energy.) **Energy output** includes energy (1) immediately lost as heat (about 60% of the total), (2) used to do work (driven by ATP), and (3) stored as fat or glycogen. (Because losses of organic molecules in urine, feces, and perspiration are very small in healthy people, they are usually ignored in calculating energy output.) A close look at this situation reveals that *nearly all the energy derived from foodstuffs is eventually converted to heat.* Heat is lost during every cellular activity—when ATP bonds are formed and when they are

cleaved to do work, as muscles contract, and through friction as blood flows through blood vessels. Though cells cannot use this energy to do work, the heat warms the tissues and blood and helps maintain the homeostatic body temperature that allows metabolic reactions to occur efficiently. Energy storage is an important part of the equation only during periods of growth and net fat deposit.

Regulation of Food Intake

When energy intake and energy output are balanced, body weight remains stable. When they are not, weight is either gained or lost. Body weight in most people is surprisingly stable. This fact indicates the existence of physiological mechanisms that control food intake (and hence, the amount of food oxidized) or heat production or both.

Control of food intake poses difficult questions to researchers. For example, what type of receptor could sense the body's total calorie content and alert one to start eating or to put down that fork? Despite heroic research efforts, no such receptor type has been found.

It has been known for some time that the hypothalamus releases several peptides that influence feeding behavior. For example, a pair of peptides called *orexins* are powerful appetite enhancers, *neuropeptide Y* causes us to crave carbohydrates, and *galanin* produces a yen for fats, while *GLP-1* (glucagon-like peptide) and *serotonin* make us feel full and satisfied. Current theories of how feeding behavior and hunger are regulated focus on one or more of five factors: neural signals from the digestive tract, bloodborne signals related to body energy stores, hormones, body temperature, and psychological factors. All these factors appear to operate through feedback signals to the feeding centers of the brain. Brain receptors include thermoreceptors, chemoreceptors (for glucose, insulin, and others), and receptors that respond to a number of peptides (leptin, discussed shortly; neuropeptide Y; and others). Although the hypothalamic nuclei play an essential role in regulating hunger and satiety, brain stem areas are also involved. Sensors in peripheral locations have also been suggested, with the liver and gut itself (alimentary canal) as the prime candidates.

Neural Signals from the Digestive Tract

One way the brain evaluates the contents of the gut depends on vagal nerve fibers that carry on a two-way conversation between gut and brain. Vagal afferents transmit distinctively different patterns of neural impulses in response to the presence of carbohydrates and proteins in the gut. For example, clinical tests show that two calories' worth of protein

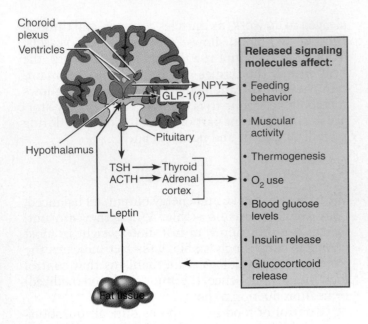

FIGURE 23.23 Hypothetical scheme of controls of feeding behavior and satiety.

Hormones

Insulin, released during food absorption, depresses hunger and is presumed to be an important satiety signal, as is leptin, released by fat cells. In contrast, glucagon levels rise during fasting and stimulate hunger. Other hormonal controls include epinephrine (released during fasting), which triggers hunger, and cholecystokinin, an intestinal hormone secreted during food digestion, which depresses hunger. Interestingly, elderly people have more CCK in their bloodstream than younger people do, which may explain in part the greater susceptibility of elders to anorexia.

Body Temperature

Increased body temperature may inhibit eating. Such a thermal signal would help explain why people in cold climates normally eat more than those in warm climates.

Psychological Factors

Although the mechanisms described thus far are reflexive, their final outcome may be enhanced or inhibited by psychological factors that have little to do with caloric balance. Psychological factors are thought to be very important in obese people. However, even when psychological factors are the underlying cause of obesity, individuals do *not* continue to gain weight endlessly. Their feeding controls still operate, but they have a higher "set point," which maintains total energy content at higher-than-normal levels.

produces a 30–40% larger and longer response in vagal afferents than does two calories' worth of glucose. Using these signals, together with others it receives, the brain can decode what is eaten and how much.

Nutrient Signals Related to Energy Stores

At any time, blood levels of glucose, amino acids, and fatty acids provide information to the brain that may help adjust energy intake to energy output. For example:

1. When we eat, blood glucose levels rise and glucose catabolism increases. Subsequent activation of glucose receptors in the brain ultimately depresses eating. During fasting, this signal is absent, resulting in hunger and a "turning on" of food-seeking behavior.

2. Elevated blood levels of amino acids depress eating, whereas low amino acid levels in blood stimulate eating, but the precise mechanism mediating these effects is unknown.

3. Blood concentrations of fatty acids and of *leptin* ("thin"), a peptide released by adipose cells that circulates at levels related to fat reserves, serve as indicators of the body's total energy stores (in adipose tissue) and provide a mechanism for controlling hunger. According to this theory, the greater the fat reserves, the larger the baseline amount of fatty acids and leptin released to the blood and the more eating behavior is inhibited.

A Hypothetical Model: Interrelationship of Factors in Control of Food Intake

Whew! How does all this information fit together? The best current hypothesis based on animal studies is shown in Figure 23.23.

The overall satiety signal appears to be **leptin,** which over a period of hours is secreted exclusively by fat tissue in response to an increase in body fat mass. Blood levels of glucocorticoids and insulin play a role in regulating leptin release (inhibit it), but the precise mechanism is yet to be worked out.

Leptin binds to receptors in the choroid plexuses of the ventricles, where it gains entry to the brain. There, it acts on the hypothalamus to regulate the amount of body fat via controls of appetite and energy output. Its main target is the ventromedial hypothalamus, where it suppresses the secretion of neuropeptide Y (NPY), the most potent appetite stimulant known. Hence, leptin decreases food intake and cranks up activity and heat production. Whether the NPY receptors are the final targets of leptin is not known. The hypothalamus also has receptors for GLP-1, a powerful inhibitor of feeding behavior, and some researchers believe leptin may

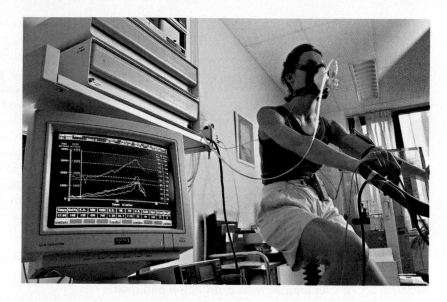

FIGURE 23.24 Indirect measurement of metabolic rate. The method is based on the fact that O_2 use and heat liberation during food oxidation are directly proportional. The person breathes into a respirometer, and the total amount of O_2 consumed during testing (in this case, while pedaling on a stationary bicycle) is measured. The average amount of O_2 consumed per hour is multiplied by 4.83 (the amount of energy liberated when 1 L of O_2 is consumed in the oxidation of protein, carbohydrate, or fat) to calculate the metabolic rate. If, for example, O_2 use is 16 L/h, the metabolic rate is 16 L/h $\times$ 4.83 kcal/L = 77.28 kcal/h.

also act through GLP-1. So far we only have bits and pieces of this story.

Metabolic Rate and Heat Production

The body's rate of energy output is called the **metabolic rate.** It is the total heat produced by all the chemical reactions and mechanical work of the body. It can be measured directly or indirectly. In the *direct method*, a person enters a chamber called a **calorimeter** and heat liberated by the body is absorbed by water circulating around the chamber. The rise in water temperature is directly related to the heat produced by the person's body. The *indirect method* uses a **respirometer** (Figure 23.24) to measure oxygen consumption, which is directly proportional to heat production. For each liter of oxygen used, the body produces about 4.8 kcal of heat.

Because many factors influence metabolic rate, it is usually measured under standardized conditions. The person is in a postabsorptive state (has not eaten for at least 12 hours), is reclining, and is mentally and physically relaxed. The temperature of the room is a comfortable 20–25°C. The measurement obtained under these circumstances is called the **basal metabolic rate (BMR)** and reflects the energy the body needs to perform only its most essential activities, such as breathing and maintaining resting levels of organ function. Although named the *basal* metabolic rate, this measurement is *not* the lowest metabolic state of the body. That situation occurs during sleep, when the skeletal muscles are completely relaxed. The BMR, often referred to as the "energy cost of living," is reported in kilocalories per square meter of body surface per hour ($kcal/m^2/h$).

A 70-kg adult has a BMR of approximately 66 kcal/h. You can approximate your BMR by multiplying weight in kilograms (2.2 pounds = 1 kg) by 1 if you are male and by 0.9 if you are female.

Factors influencing BMR include body surface area, age, gender, stress, and hormones. Although BMR is related to overall weight and size, the critical factor is body surface area. This reflects the fact that as the ratio of body surface area to body volume increases, heat loss to the environment increases and the metabolic rate must be higher to replace the lost heat. Hence, if two people weigh the same, the taller or thinner person will have the higher BMR.

In general, the younger a person, the higher the BMR. Children and adolescents require large amounts of energy for growth. In old age, BMR declines dramatically as skeletal muscles begin to atrophy. (This helps explain why those elderly who fail to reduce their caloric intake gain weight.) Gender also plays a role. Metabolic rate is disproportionately higher in males than in females. Males typically have more muscle, which is very active metabolically even during rest. Fatty tissue, present in greater relative amounts in females, is metabolically more sluggish than muscle. Surprisingly, physical training has little effect on BMR. Although it would appear that athletes, especially those with greatly enhanced muscle mass, should have much higher BMRs than nonathletes, there is little difference between those of the same sex and surface area.

Body temperature and BMR tend to rise and fall together. Thus, fever (hyperthermia) results in a marked increase in metabolic rate. Stress, whether physical or emotional, increases BMR by mobilizing the sympathetic nervous system. Bloodborne norepinephrine and epinephrine (released by adrenal

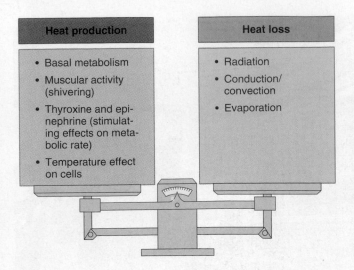

Heat production	Heat loss
• Basal metabolism • Muscular activity (shivering) • Thyroxine and epinephrine (stimulating effects on metabolic rate) • Temperature effect on cells	• Radiation • Conduction/ convection • Evaporation

FIGURE 23.25 As long as heat production and heat loss are balanced, body temperature remains constant.

medullary cells) provoke a rise in BMR primarily by stimulating fat catabolism.

The amount of **thyroxine** produced by the thyroid gland is probably the most important hormonal factor in determining BMR; hence, thyroxine has been dubbed the "metabolic hormone." Its direct effect on most body cells (except brain cells) is to increase O_2 consumption, presumably by accelerating the use of ATP to operate the sodium-potassium pump. As ATP reserves decline, cellular respiration accelerates. Thus, the more thyroxine produced, the higher the BMR.

Ⓗ *HOMEOSTATIC IMBALANCE*

Hyperthyroidism causes a host of problems resulting from the excessive BMR it produces. The body catabolizes stored fats and tissue proteins, and, despite increased hunger and food intake, the person often loses weight. Bones weaken and muscles, including the heart, begin to atrophy. In contrast, *hypothyroidism* results in slowed metabolism, obesity, and diminished thought processes. ●

The **total metabolic rate (TMR)** is the rate of kilocalorie consumption needed to fuel *all* ongoing activities—involuntary and voluntary. BMR accounts for a surprisingly large part of TMR. For example, a woman whose total energy needs per day are about 2000 kcal may spend 1400 kcal or so supporting vital body activities. Skeletal muscle activity causes the most dramatic short-term changes in TMR, reflecting the fact that skeletal muscles make up nearly half of

body mass. Even slight increases in muscular work can cause remarkable leaps in TMR and heat production. When a well-trained athlete exercises vigorously for several minutes, TMR may increase 20-fold, and it remains elevated for several hours after exercise stops. Food ingestion also induces a rapid increase in TMR. This effect, called **dietary,** or **food-induced, thermogenesis,** is greatest when proteins are eaten. The heightened metabolic activity of the liver during such periods probably accounts for the bulk of additional energy use. In contrast, fasting or very low caloric intake depresses TMR and results in a slower breakdown of body reserves.

Regulation of Body Temperature

As shown in Figure 23.25, body temperature represents the balance between heat production and heat loss. Although all body tissues produce heat, those most active metabolically produce the greatest amounts. When the body is at rest, most heat is generated by the liver, heart, brain, and endocrine organs, with the inactive skeletal muscles accounting for only 20–30%. This situation changes dramatically with even slight alterations in muscle tone, however; and during vigorous exercise, heat production by the skeletal muscles can be 30 to 40 times that of the rest of the body. Clearly, a change in muscle activity is one of the most important means of modifying body temperature.

Body temperature averages 36.2°C (98.2°F) and is usually maintained within the range 35.6–37.8°C (96–100°F), despite considerable change in external (air) temperature. A healthy individual's body temperature fluctuates approximately 1°C (1.8°F) in 24 hours, with the low occurring in early morning and the high in late afternoon or early evening.

The adaptive value of this precise temperature homeostasis becomes obvious when we consider the effect of temperature on the rate of enzymatic activity. At normal body temperature, conditions are optimal for enzymatic activity. As body temperature rises, enzymatic catalysis is accelerated: With each rise of 1°C, the rate of chemical reactions increases about 10%. As the temperature spikes above the homeostatic range, neurons are depressed and proteins begin to denature. Most adults go into convulsions when body temperature reaches 41°C (106°F), and 43°C (about 110°F) appears to be the absolute limit for life. In contrast, most body tissues can withstand marked reductions in temperature if other conditions are carefully controlled. This fact underlies the use of body cooling during open heart surgery when the heart must be stopped. Low body temperature reduces metabolic rate (and consequently nutrient requirements of body tissues and the heart), allowing more time for surgery without incurring tissue damage.

Core and Shell Temperatures

Different body regions have different resting temperatures. The body's **core** (organs within the skull and the thoracic and abdominal cavities) has the highest temperature and its **shell** (essentially the skin) has the lowest temperature in most circumstances. Of the two body sites used routinely to obtain body temperature clinically, the rectum typically has a temperature about 0.4°C (0.7°F) higher than the mouth and is a better indicator of core temperature.

It is core temperature that is precisely regulated. Blood serves as the major *agent of heat exchange* between the core and shell. Whenever the shell is warmer than the external environment, heat is lost from the body as warm blood is allowed to flush into the skin capillaries. On the other hand, when heat must be conserved, blood largely bypasses the skin. This reduces heat loss and allows the shell temperature to fall toward that of the environment. Thus, while the core temperature stays relatively constant, the temperature of the shell may fluctuate substantially [between 20°C (68°F) and 40°C (104°F) for instance] as its "thickness" is adapted to changes in body activity and external temperature. (You really *can* have cold hands and a warm heart.)

Mechanisms of Heat Exchange

The mechanisms that govern heat exchange between our skin and the external environment (Figure 23.26) are the same as those governing heat exchange between inanimate objects. It helps to think of an object's temperature—whether that object is the skin or a radiator—as a guide to its heat content (*think* "heat concentration"). Then just remember that heat always flows down its concentration gradient from a warmer region to a cooler region. The body uses four mechanisms of heat transfer—radiation, conduction, convection, and evaporation.

Radiation is the loss of heat in the form of infrared waves (thermal energy). Any object that is warmer than objects in its environment—for example, a radiator and (usually) the body—will transfer heat to those objects. Under normal conditions, close to half of body heat loss occurs by radiation.

Because the direction of radiant energy flow is always from warmer to cooler, radiation explains why an initially cold room warms up shortly after it is filled with people (the "body heat furnace," so to speak). The body also gains heat by radiation, as demonstrated by the warming of the skin during sunbathing.

Conduction is the transfer of heat from a warmer object to a cooler one when the two are in direct contact with each other. For example, when we step into a hot tub, some of the heat of the water

FIGURE 23.26 **Mechanisms of heat exchange.** Passengers enjoying a hot tub on a cruise to Alaska illustrate several mechanisms of heat exchange between the body and its environment: **(1)** conduction—heat transfers from the hot water to the skin; **(2)** radiation—heat transfers from the exposed part of the body to the cooler air; **(3)** convection—warmed air moves away from the body via breezes; and perhaps **(4)** evaporation of perspiration—excess body heat is carried away.

is transferred to our skin, and warm buttocks transfer heat to the seat of a chair by conduction.

The process called **convection** occurs because warm air expands and rises and cool air (being denser) falls. Thus, the warmed air enveloping the body is continually being replaced by cooler air molecules. Convection substantially enhances heat transfer from the body surface to the air because the cooler air absorbs heat by conduction more rapidly than the already-warmed air. Together, conduction and convection account for 15–20% of heat loss to the environment. These processes are enhanced by anything that moves air more rapidly across the body surface such as wind or a fan, that is, by *forced convection*.

The fourth mechanism by which the body loses heat is **evaporation.** Water evaporates because its molecules absorb heat from the environment and become energetic enough (that is, vibrate fast enough) to escape as gas (water vapor). The heat absorbed by water during evaporation is called **heat of vaporization.** Because water absorbs a great deal of heat before vaporizing, its evaporation from body surfaces removes large amounts of body heat. For every gram of water evaporated, about 0.58 kcal of heat is removed from the body.

There is a basal level of body heat loss due to the continuous evaporation of water from the lungs, from the oral mucosa, and through the skin. The unnoticeable water loss occurring via these routes is called **insensible water loss** and the heat loss that accompanies it is called **insensible heat loss.** Insensible heat loss dissipates about 10% of the basal heat production of the body and is a constant not subject to body temperature controls. When necessary, however, the control mechanisms do initiate heat-promoting activities to counterbalance this insensible heat loss.

Evaporative heat loss becomes an active *(sensible)* process when body temperature rises and sweating produces increased amounts of water for vaporization. Extreme emotional states activate the sympathetic nervous system, causing body temperature to rise one degree or so, and vigorous exercise can thrust body temperature upward as much as 2–3°C (5–6°F). During vigorous muscular activity, 1–2 L/h of perspiration can be produced and evaporated, removing 600–1200 kcal of heat from the body each hour. This is more than 30 times the amount of heat lost via insensible heat loss!

 HOMEOSTATIC IMBALANCE

When sweating is heavy and prolonged, especially in untrained individuals, losses of water and NaCl may cause painful muscle spasms called *heat cramps.* This situation is easily rectified by ingesting fluids. ●

Role of the Hypothalamus

Although other brain regions contribute, the hypothalamus, particularly its *preoptic* region, is the main integrating center for thermoregulation. Together the **heat-loss center** (located more anteriorly) and the **heat-promoting center** make up the brain's **thermoregulatory centers.**

The hypothalamus receives afferent input from (1) **peripheral thermoreceptors** located in the shell (the skin), and (2) **central thermoreceptors** sensitive to blood temperature and located in the body core including the anterior portion of the hypothalamus. Much like a thermostat, the hypothalamus responds to this input by reflexively initiating appropriate heat-promoting or heat-loss activities via autonomic effector pathways. Although the central thermoreceptors are more critically located than the peripheral ones, varying inputs from the shell probably alert the hypothalamus of its need to act to pre-

vent temperature changes in the core; that is, they allow the hypothalamus to anticipate possible changes to be made.

Heat-Promoting Mechanisms

When the external temperature is low or blood temperature falls for any reason, the heat-promoting center is activated and triggers one or more of the following mechanisms to maintain or increase core body temperature (Figure 23.27).

1. **Constriction of cutaneous blood vessels.** Activation of the sympathetic vasoconstrictor fibers serving the blood vessels of the skin causes strong vasoconstriction. As a result, blood is restricted to deep body areas and largely bypasses the skin. Because the skin is separated from deeper organs by a layer of insulating subcutaneous (fatty) tissue, heat loss from the shell is dramatically reduced and shell temperature drops toward that of the external environment.

 HOMEOSTATIC IMBALANCE

Restricting blood flow to the skin is not a problem for a brief period, but if it is extended (as during prolonged exposure to very cold weather), skin cells deprived of oxygen and nutrients begin to die. This extremely serious condition is called *frostbite.* ●

2. **Increase in metabolic rate.** Cold stimulates the release of norepinephrine by sympathetic nerve fibers, which elevates the metabolic rate, which enhances heat production. This mechanism is called **chemical,** or **nonshivering, thermogenesis.**

3. **Shivering.** If the mechanisms described above are not enough to handle the situation, shivering is triggered. Brain centers controlling muscle tone are activated and when muscle tone reaches sufficient levels to alternately stimulate stretch receptors in antagonistic muscles, the muscles begin involuntary shuddering contractions. Shivering is very effective in increasing body temperature because muscle activity produces large amounts of heat.

4. **Enhanced thyroxine release.** When environmental temperature decreases gradually, as in the transition from summer to winter, the hypothalamus releases *thyrotropin-releasing hormone.* This activates the anterior pituitary to release *thyroid-stimulating hormone,* which induces the thyroid to liberate larger amounts of thyroid hormone to the blood. Because thyroid hormone increases metabolic rate, body heat production increases, allowing us to maintain a constant body temperature in cold environments.

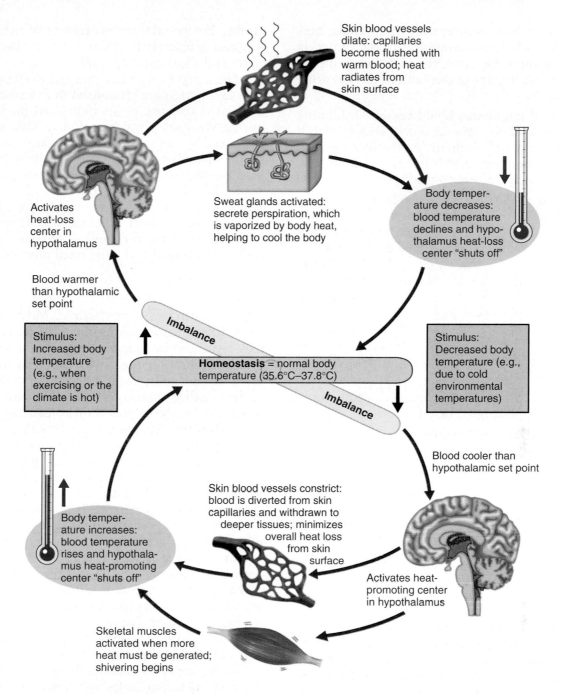

Skin blood vessels dilate: capillaries become flushed with warm blood; heat radiates from skin surface

Sweat glands activated: secrete perspiration, which is vaporized by body heat, helping to cool the body

Body temperature decreases: blood temperature declines and hypothalamus heat-loss center "shuts off"

Activates heat-loss center in hypothalamus

Blood warmer than hypothalamic set point

Stimulus: Increased body temperature (e.g., when exercising or the climate is hot)

Imbalance

Homeostasis = normal body temperature (35.6°C–37.8°C)

Imbalance

Stimulus: Decreased body temperature (e.g., due to cold environmental temperatures)

Blood cooler than hypothalamic set point

Body temperature increases: blood temperature rises and hypothalamus heat-promoting center "shuts off"

Skin blood vessels constrict: blood is diverted from skin capillaries and withdrawn to deeper tissues; minimizes overall heat loss from skin surface

Activates heat-promoting center in hypothalamus

Skeletal muscles activated when more heat must be generated; shivering begins

FIGURE 23.27   Mechanisms of body temperature regulation.

Beside these involuntary adjustments, we humans make a number of *behavioral modifications* to prevent overcooling of our body core:

■ Putting on more or warmer clothing to restrict heat loss (a hat, gloves, and "insulated" outer garments)

■ Drinking hot fluids

■ Changing posture to reduce exposed body surface area (hunching over or clasping the arms across the chest)

■ Increasing physical activity to generate more heat (jumping up and down, clapping the hands)

Heat-Loss Mechanisms

Heat-loss mechanisms protect the body from excessively high temperatures. Most heat loss occurs through the skin via radiation, conduction, convection, and evaporation. How do the heat-exchange mechanisms fit into the heat-loss temperature regulation scheme? The answer is quite simple.

Whenever core body temperature rises above normal, the hypothalamic heat-promoting center is inhibited. At the same time, the heat-loss center is activated and so triggers one or both of the following (Figure 23.27):

1. Dilation of cutaneous blood vessels. Inhibiting the vasomotor fibers serving blood vessels of the skin allows the vessels to dilate. As the blood vessels swell with warm blood, heat is lost from the shell by radiation, conduction, and convection.

2. Enhanced sweating. If the body is extremely overheated or if the environment is so hot [over 61°C (about 92°F)] that heat cannot be lost by other means, evaporation becomes necessary. The sweat glands are strongly activated by sympathetic fibers and spew out large amounts of perspiration. Evaporation of perspiration is an efficient means of ridding the body of surplus heat as long as the air is dry. However, when the relative humidity is high, evaporation occurs much more slowly. In such cases, the heat-liberating mechanisms cannot work well, and we feel miserable and irritable.

Behavioral or voluntary measures commonly taken to reduce body heat include:

- Reducing activity ("laying low")

- Seeking a cooler environment (a shady spot) or using a device to increase convection (a fan) or cooling (an air conditioner)

- Wearing light-colored, loose clothing that reflects radiant energy. (This is actually cooler than being nude because bare skin absorbs most of the radiant energy striking it)

🅗 HOMEOSTATIC IMBALANCE

When normal heat loss processes become ineffective, the **hyperthermia** (elevated body temperature) that ensues depresses the hypothalamus. As a result, heat-control mechanisms are suspended, creating a vicious positive feedback cycle. Increasing temperatures increase the metabolic rate, which increases heat production. The skin becomes hot and dry and, as the temperature continues to spiral upward, multiple organ damage becomes a distinct possibility, including brain damage. This condition, called **heat stroke,** can be fatal unless corrective measures are initiated immediately (immersing the body in cool water and administering fluids).

The terms *heat exhaustion* and *exertion-induced heat exhaustion* are often used to describe the heat-associated collapse of an individual during or following vigorous physical activity. This condition, evidenced by elevated body temperature and mental confusion and/or fainting, is due to dehydration and consequent low blood pressure. In contrast to heat stroke, heat-loss mechanisms are still functional in heat exhaustion; indeed, the symptoms are a direct consequence of those mechanisms. However, heat exhaustion can rapidly progress to heat stroke if the body is not cooled and rehydrated promptly. ●

Fever

Fever is *controlled hyperthermia*. Most often, it results from infection somewhere in the body, but it may be caused by cancer, allergic reactions, or CNS injuries. Whatever the cause, white blood cells, injured tissue cells, and macrophages release cytokines called *pyrogens* (literally, fire starters), which act on the hypothalamus, causing release of prostaglandins. The prostaglandins reset the hypothalamic thermostat to a higher-than-normal temperature, causing heat-promoting mechanisms to kick in. As a result of vasoconstriction, heat loss from the body surface declines, the skin cools, and shivering begins to generate heat. These "chills" are a sure sign body temperature is rising. The temperature rises until it reaches the new setting, and then is maintained at that setting until natural body defenses or antibiotics reverse the disease process. The thermostat is then reset to a lower (or normal) level, and heat-loss mechanisms swing into action. Sweating begins and the skin becomes flushed and warm. Physicians have long recognized these signs as signals that body temperature is falling.

As explained in Chapter 20, fever, by increasing the metabolic rate, helps speed healing, and it also appears to inhibit bacterial growth. The danger of fever is that if the body thermostat is set too high, proteins may be denatured and permanent brain damage may occur.

Review Questions

Multiple Choice/Matching

(Some questions have more than one correct answer. Select the best answer or answers from the choices given.)

1. Which of the following reactions would liberate the most energy? (a) complete oxidation of a molecule of sucrose to CO_2 and water, (b) conversion of a molecule of ADP to ATP, (c) respiration of a molecule of glucose to lactic acid, (d) conversion of a molecule of glucose to carbon dioxide and water.

2. The formation of glucose from glycogen is (a) gluconeogenesis, (b) glycogenesis, (c) glycogenolysis, (d) glycolysis.

3. The net gain of ATP from the complete metabolism (aerobic) of glucose is closest to (a) 2, (b) 30, (c) 36, (d) 4.

4. Which of the following *best* defines cellular respiration? (a) intake of carbon dioxide and output of oxygen by cells, (b) excretion of waste products, (c) inhalation of oxygen and exhalation of carbon dioxide, (d) oxidation of substances by which energy is released in usable form to the cells.

5. During aerobic respiration, electrons are passed down the electron transport chain and _____ is formed. (a) oxygen, (b) water, (c) glucose, (d) NADH + H^+.

6. Metabolic rate is relatively low in (a) youth, (b) physical exercise, (c) old age, (d) fever.

7. In a temperate climate under ordinary conditions, the greatest loss of body heat occurs through (a) radiation, (b) conduction, (c) evaporation, (d) none of the above.

8. Which of the following is *not* a function of the liver? (a) glycogenolysis and gluconeogenesis, (b) synthesis of cholesterol, (c) detoxification of alcohol and drugs, (d) synthesis of glucagon, (e) deamination of amino acids.

9. Amino acids are essential (and important) to the body for all the following *except* (a) production of some hormones, (b) production of antibodies, (c) formation of most structural materials, (d) as a source of quick energy.

10. A person has been on a hunger strike for seven days. Compared to normal, he has (a) increased release of fatty acids from adipose tissue, ketosis, and ketonuria, (b) elevated glucose concentration in the blood, (c) increased plasma insulin concentration, (d) increased glycogen synthetase (enzyme) activity in the liver.

11. Transamination is a chemical process by which (a) protein is synthesized, (b) an amine group is transferred from an amino acid to a keto acid, (c) an amine group is cleaved from the amino acid, (d) amino acids are broken down for energy.

12. Three days after removal of the pancreas from an animal, the researcher finds a persistent increase in (a) acetoacetic acid concentration in the blood, (b) urine volume, (c) blood glucose, (d) all of the above.

13. Hunger, appetite, obesity, and physical activity are interrelated. Thus, (a) hunger sensations arise *primarily* from the stimulation of receptors in the stomach and intestinal tract in response to the absence of food in these organs; (b) obesity, in most cases, is a result of the abnormally high enzymatic activity of the fat-synthesizing enzymes in adipose tissue; (c) in all cases of obesity, the energy content of the ingested food has exceeded the energy expenditure of the body; (d) in a normal individual, increasing blood glucose concentration increases hunger sensations.

14. Body temperature regulation is (a) influenced by temperature receptors in the skin, (b) influenced by the temperature of the blood perfusing the heat regulation centers of the brain, (c) subject to both neural and hormonal control, (d) all of the above.

15. Which of the following yields the greatest caloric value per gram? (a) fats, (b) proteins, (c) carbohydrates, (d) all are equal in caloric value.

Short Answer Essay Questions

16. What is cellular respiration? What is the common role of FAD and NAD^+ in cellular respiration?

17. Describe the site, major events, and outcomes of glycolysis.

18. Pyruvic acid is a product of glycolysis, but it is not the substance that enters the Krebs cycle. What is the substance, and what must occur if pyruvic acid is to be transformed into this molecule?

19. Define glycogenesis, glycogenolysis, gluconeogenesis, and lipogenesis. Which is (are) likely to be occurring (a) shortly after a carbohydrate-rich meal, (b) just before waking up in the morning?

20. What is the harmful result when excessive amounts of fats are burned for energy? Name two conditions that might lead to this result.

21. Make a flowchart that indicates the pivotal intermediates through which glucose can be converted to fat.

22. Distinguish between the role of HDLs and that of LDLs.

23. List some factors that influence plasma cholesterol levels. Also list the sources and fates of cholesterol in the body.

24. What is meant by "body energy balance," and what happens if the balance is not precise?

25. Explain the effect of the following on metabolic rate: thyroxine levels, eating, body surface area, muscular exercise, emotional stress, starvation.

26. Explain the terms "core" and "shell" relative to body temperature balance. What serves as the heat-transfer agent from one to the other?

27. Compare and contrast mechanisms of heat loss with mechanisms of heat promotion, and explain how these mechanisms determine body temperature.

24

THE URINARY SYSTEM

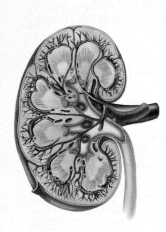

Kidney Anatomy (pp. 853–861)

1. Describe the gross anatomy of the kidney and its coverings.
2. Trace the blood supply through the kidney.
3. Describe the anatomy of a nephron.

Kidney Physiology: Mechanisms of Urine Formation (pp. 862–876)

4. List several kidney functions that help maintain body homeostasis.
5. Identify the nephron parts responsible for filtration, reabsorption, and secretion, and describe mechanisms underlying each of these functional processes.
6. Explain the role of aldosterone and of atrial natriuretic peptide in sodium and water balance.
7. Describe the mechanism responsible for the medullary osmotic gradient.
8. Explain formation of dilute versus concentrated urine.

Urine (p. 876)

9. Describe the normal physical and chemical properties of urine.
10. List several abnormal urine components, and name the condition when each is present in detectable amounts.

Ureters (pp. 876–877)

11. Describe the general location, structure, and function of the ureters.

Urinary Bladder (pp. 878–879)

12. Describe the general location, structure, and function of the urinary bladder.

Urethra (p. 879)

13. Describe the general location, structure, and function of the urethra.
14. Compare the course, length, and functions of the male urethra with those of the female.

Micturition (pp. 880–881)

15. Define micturition and describe its neural control.

Every day the kidneys filter nearly 200 liters of fluid from the bloodstream, allowing toxins, metabolic wastes, and excess ions to leave the body in urine while returning needed substances to the blood. Much like a water purification plant that keeps a city's water drinkable and disposes of its wastes, the kidneys are usually unappreciated until they malfunction and body fluids become contaminated. Although the lungs and skin also participate in excretion, the kidneys are the major excretory organs.

As the kidneys perform these excretory functions, they simultaneously regulate the volume and chemical makeup of the blood, maintaining the proper balance between water and salts and between acids and bases. Frankly, this would be tricky work for a chemical engineer, but the kidneys do it efficiently most of the time.

Other renal functions include:

- Gluconeogenesis (see p. 829) during prolonged fasting

- Producing the enzyme *renin* (re'nin; *ren* = kidney), which helps regulate blood pressure and kidney function, and the hormone *erythropoietin* (ĕ-rith"ro-poi'ĕ-tin), which stimulates red blood cell production (see Chapter 16)

- Metabolizing vitamin D to its active form (see Chapter 15)

Besides the urine-forming kidneys, the **urinary system** includes the *urinary bladder*, a temporary storage reservoir for urine, plus three tubelike organs—the paired *ureters* (u-re'terz) and the *urethra* (u-re'thrah), all three of which furnish transportation channels for urine (Figure 24.1).

Kidney Anatomy

Location and External Anatomy

The bean-shaped kidneys lie in a retroperitoneal position (between the dorsal body wall and the parietal peritoneum) in the *superior* lumbar region (Figure 24.2). Extending approximately from T_{12} to L_3, the kidneys receive some protection from the lower part of the rib cage (Figure 24.2b). The right kidney is crowded by the liver and lies slightly lower than the left. An adult's kidney has a mass of about 150 g (5 ounces) and its average dimensions are 12 cm long, 6 cm wide, and 3 cm thick—about the size of a large bar of soap. The lateral surface is convex. The medial surface is concave and has a vertical cleft called the **renal hilus** that leads into an internal space within the kidney called the *renal sinus*. The ureters, renal blood vessels, lymphatics, and nerves all join the kidney at the hilus and occupy the sinus. Atop each kidney is an *adrenal* (or *suprarenal*) *gland*, an endocrine gland that is functionally unrelated to the kidney.

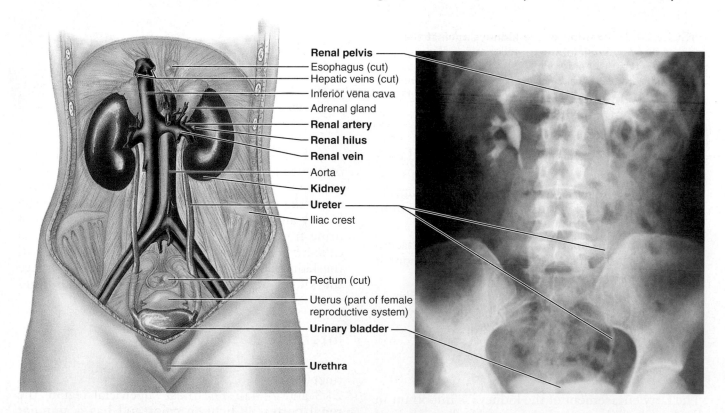

Renal pelvis
Esophagus (cut)
Hepatic veins (cut)
Inferior vena cava
Adrenal gland
Renal artery
Renal hilus
Renal vein
Aorta
Kidney
Ureter
Iliac crest

Rectum (cut)
Uterus (part of female reproductive system)
Urinary bladder

Urethra

FIGURE 24.1 **The urinary system.** Anterior view of the female urinary organs. (Most unrelated abdominal organs have been omitted.) **(a)** Diagrammatic view. **(b)** X ray of urinary organs after injection of contrast medium. (See *A Brief Atlas of the Human Body*, Figure 55.)

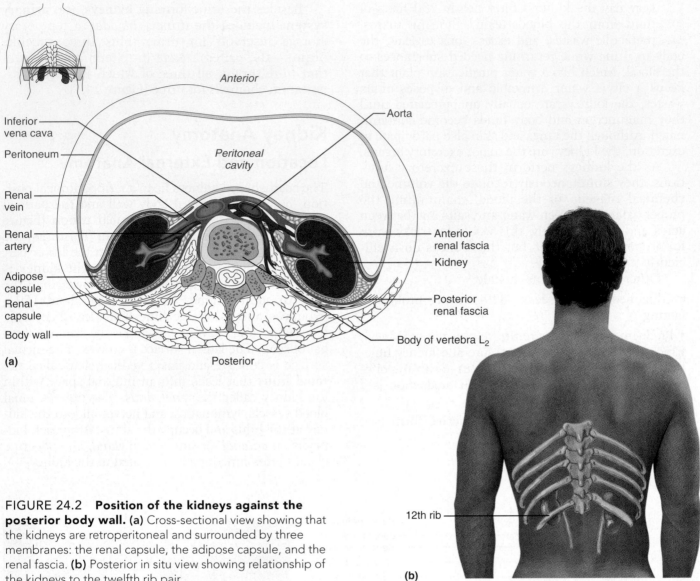

Anterior

Inferior vena cava

Peritoneum

Renal vein

Renal artery

Adipose capsule

Renal capsule

Body wall

(a)

Peritoneal cavity

Aorta

Anterior renal fascia

Kidney

Posterior renal fascia

Body of vertebra L₂

Posterior

12th rib

(b)

FIGURE 24.2   Position of the kidneys against the posterior body wall. (a) Cross-sectional view showing that the kidneys are retroperitoneal and surrounded by three membranes: the renal capsule, the adipose capsule, and the renal fascia. **(b)** Posterior in situ view showing relationship of the kidneys to the twelfth rib pair.

Three layers of supportive tissue surround each kidney:

1. The **renal capsule,** a fibrous transparent capsule that prevents infections in surrounding regions from spreading to the kidneys

2. The **adipose capsule,** a fatty mass that attaches the kidney to the posterior body wall and cushions it against blows

3. The **renal fascia,** an outer layer of dense fibrous connective tissue that anchors the kidney and the adrenal gland to surrounding structures

Ⓗ HOMEOSTATIC IMBALANCE

The fatty encasement of the kidneys is important in holding the kidneys in their normal body position. If the amount of fatty tissue dwindles (as with extreme emaciation or rapid weight loss), one or both kidneys may drop to a lower position, an event called *renal ptosis* (to'sis; "a fall"). Renal ptosis may cause a ureter to become kinked, which creates problems because the urine, unable to drain, backs up into the kidney and exerts pressure on its tissue. Backup of urine from ureteral obstruction or other causes is called *hydronephrosis* (hi"dro-nĕ-fro'sis; "water in the kidney"). Hydronephrosis can severely damage the kidney, leading to necrosis (tissue death) and renal failure. ●

Internal Anatomy

A frontal section through a kidney reveals three distinct regions: *cortex, medulla,* and *pelvis* (Figures 24.3 and 24.4a). The most superficial region, the **renal cortex,** is light in color and has a granular appearance. Deep to the cortex is the darker, reddish-brown **renal medulla,** which exhibits cone-shaped

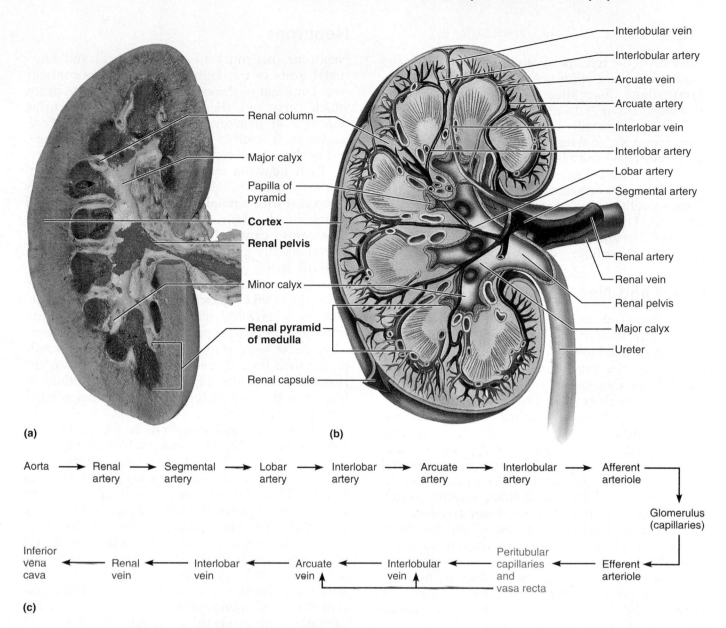

FIGURE 24.3 **Internal anatomy of the kidney. (a)** Photograph of a frontal section of a triple-injected kidney. (Arteries and veins are injected with red and blue latex respectively; pelvis and ureter are injected with yellow latex.) **(b)** Diagrammatic view of a coronally sectioned kidney, illustrating major blood vessels. **(c)** Summary of renal vasculature pathway.

tissue masses called **medullary** or **renal pyramids.** The broad *base* of each pyramid faces toward the cortex, and its apex, or *papilla* ("nipple"), points internally. The pyramids appear striped because they are formed almost entirely of parallel bundles of microscopic urine-collecting tubules. The **renal columns,** inward extensions of cortical tissue, separate the pyramids. Each pyramid and its surrounding cortical tissue constitutes one of approximately eight **lobes** of a kidney.

The **renal pelvis,** a flat, funnel-shaped tube, is continuous with the ureter leaving the hilus.

Branching extensions of the pelvis form two or three **major calyces** (ka'lih-sēz; singular, *calyx*), each of which subdivides to form several **minor calyces,** cup-shaped areas that enclose the papillae. The calyces collect urine, which drains continuously from the papillae, and empty it into the renal pelvis. The urine then flows through the renal pelvis and into the ureter, which transports it to the bladder to be stored. The walls of the calyces, pelvis, and ureter contain smooth muscle that contracts rhythmically to propel urine along its course by peristalsis.

H   HOMEOSTATIC IMBALANCE

Infection of the renal pelvis and calyces produces the condition called *pyelitis* (pi″ĕ-li′tis). Infections or inflammations that affect the entire kidney are *pyelonephritis* (pi″ĕ-lo-nĕ-fri′tis). Kidney infections in females are usually caused by fecal bacteria that spread from the anal region to the urinary tract. Less often they result from bloodborne bacteria (traveling from other infected sites) that lodge and multiply in the kidneys. In severe cases of pyelonephritis, the kidney swells, abscesses form, and the pelvis fills with pus. Untreated, the kidneys may be severely damaged, but antibiotic therapy can usually treat the infection successfully. ●

Blood and Nerve Supply

The kidneys continuously cleanse the blood and adjust its composition, and so it is not surprising that they have a rich blood supply. Under normal resting conditions, the large **renal arteries** deliver one-fourth of the total cardiac output (about 1200 ml) to the kidneys each minute. The renal arteries issue at right angles from the abdominal aorta; because the aorta lies to the left of the midline, the right renal artery is longer than the left. As it approaches a kidney, each renal artery divides into five **segmental arteries.** Within the renal sinus, each segmental artery branches into several **lobar arteries,** which then divide to form several **interlobar arteries.**

At the medulla-cortex junction, the interlobar arteries branch into the **arcuate arteries** (ar′ku-āt) that arch over the bases of the medullary pyramids. Small **interlobular arteries** radiate outward from the arcuate arteries to supply the cortical tissue (see Figure 24.3a). More than 90% of the blood entering the kidney perfuses the renal cortex.

Veins pretty much trace the pathway of the arterial supply in reverse. Blood leaving the renal cortex drains sequentially into the **interlobular, arcuate, interlobar,** and finally **renal veins.** (There are no lobar or segmental veins.) The renal veins issue from the kidneys and empty into the inferior vena cava. Because the inferior vena cava lies to the right of the vertebral column, the left renal vein is about twice as long as the right.

The **renal plexus,** a variable network of autonomic nerve fibers and ganglia, provides the nerve supply of the kidney and its ureter. An offshoot of the celiac plexus, the renal plexus is largely supplied by sympathetic fibers from the least thoracic and first lumbar splanchnic nerves, which course along with the renal artery to reach the kidney. These sympathetic vasomotor fibers regulate renal blood flow by adjusting the diameter of renal arterioles and also influence the urine-forming role of the nephrons.

Nephrons

Nephrons (nef′ronz) are the structural and functional units of the kidneys. Each kidney contains over 1 million of these tiny blood-processing units, which carry out the processes that form urine (Figure 24.4). In addition, there are thousands of *collecting ducts,* each of which collects urine from several nephrons and conveys it to the renal pelvis.

Each **nephron** consists of a **glomerulus** (glomer′u-lus; *glom* = ball of yarn), a tuft of capillaries associated with a **renal tubule.** The cup-shaped end of the renal tubule, the **glomerular capsule** (or **Bowman's capsule**) is blind and completely surrounds the glomerulus (much as a well-worn baseball glove encloses a ball). Collectively, the glomerular capsule and the enclosed glomerulus are called the **renal corpuscle.** The glomerular endothelium is *fenestrated* (penetrated by many pores), which makes these capillaries exceptionally porous. They allow large amounts of solute-rich, virtually protein-free fluid to pass from the blood into the glomerular capsule. This plasma-derived fluid or **filtrate** is the raw material that the renal tubules process to form urine.

The external *parietal layer* of the glomerular capsule (Figures 24.4b, 24.6, and 24.7) is simple squamous epithelium. This layer simply contributes to the capsule structure and plays no part in forming filtrate. The *visceral layer,* which clings to the glomerulus, consists of highly modified, branching epithelial cells called **podocytes** (pod′o-sīts; "foot cells"). The octopus-like podocytes terminate in **foot processes,** which intertwine as they cling to the basement membrane of the glomerulus. The clefts or openings between the foot processes, called **filtration slits** or **slit pores,** allow the filtrate to enter the **capsular space** inside the glomerular capsule.

The remainder of the renal tubule is about 3 cm (1.25 inches) long and has three named parts. It leaves the glomerular capsule as the elaborately coiled **proximal convoluted tubule (PCT),** makes a hairpin loop called the **loop of Henle,** and then winds and twists again as the **distal convoluted tubule (DCT)** before emptying into a collecting duct. The terms *proximal* and *distal* indicate relationship to the loop—the PCT is before and the DCT after. The meandering nature of the renal tubule increases its length and enhances its filtrate processing capabilities.

The **collecting ducts,** each of which receives filtrate from many nephrons, run through the medullary pyramids and give them their striped appearance. As the collecting ducts approach the renal pelvis, they fuse to form the large **papillary ducts,** which deliver urine into the minor calyces via papillae of the pyramids.

What path would a creatinine molecule in the glomerular blood take to reach the renal pelvis?

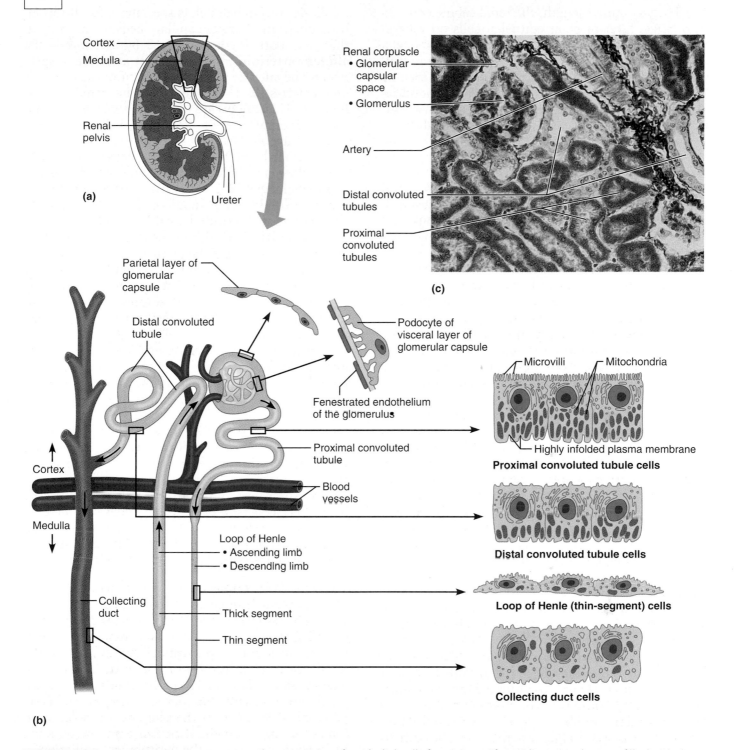

(a)

Cortex
Medulla
Renal pelvis
Ureter

Renal corpuscle
• Glomerular capsular space
• Glomerulus

Artery

Distal convoluted tubules

Proximal convoluted tubules

(c)

Parietal layer of glomerular capsule

Distal convoluted tubule

Podocyte of visceral layer of glomerular capsule

Fenestrated endothelium of the glomerulus

Cortex

Medulla

Proximal convoluted tubule

Blood vessels

Loop of Henle
• Ascending limb
• Descending limb

Collecting duct

Thick segment

Thin segment

Microvilli   Mitochondria

Highly infolded plasma membrane

Proximal convoluted tubule cells

Distal convoluted tubule cells

Loop of Henle (thin-segment) cells

Collecting duct cells

(b)

FIGURE 24.4 Location and structure of nephrons. (a) Diagram showing the orientation of nephrons; such a segment contains more than 100,000 nephrons. **(b)** Schematic view of a nephron depicting the structural characteristics of epithelial cells forming its various regions. **(c)** Photomicrograph of renal cortical tissue. Notice the renal corpuscles and the sections through the various tubules. In the proximal convoluted tubules, the lumens appear "fuzzy" because they are filled with the long microvilli of the epithelial cells that form the tubule walls. In the distal tubules, by contrast, the lumens are clear (300×).

It would go from the glomerular blood through the filtration membrane into Bowman's capsule, then through the renal tubule (PCT → loop of Henle → DCT), then into the collecting duct which carries the urine through the cortex and medulla, and into the minor calyx, into the major calyx, and into the pelvis. ■

Throughout its length, the renal tubule consists of a single layer of polar epithelial cells on a basement membrane, but each of its regions has a unique cellular anatomy that reflects its role in processing filtrate. The walls of the PCT are formed by cuboidal epithelial cells with large mitochondria; their luminal (exposed) surfaces bear dense microvilli (Figure 24.4b and c). This so-called *brush border* dramatically increases their surface area and capacity for reabsorbing water and solutes from the filtrate and secreting substances into it.

The U-shaped *loop of Henle* has *descending* and *ascending limbs*. The proximal part of the descending limb is continuous with the proximal tubule and its cells are similar. The rest of the descending limb, called the **thin segment,** is a simple squamous epithelium freely permeable to water. The epithelium becomes cuboidal or even low columnar in the ascending part of the loop of Henle, which therefore becomes the **thick segment.** In some nephrons, the thin segment is found only in the descending limb. In others, it extends into the ascending limb as well.

The epithelial cells of the DCT, like those of the PCT, are cuboidal and confined to the cortex, but they are thinner and almost entirely lack microvilli. These structural features reveal that the DCT tubules may play a larger role in secreting solutes into the filtrate than in reabsorbing substances from it. Beginning late in the DCT, the tubule cells become heterogeneous, more like those in the collecting ducts. (For this reason, this part of the DCT is sometimes called the *connecting tubule.*) The two most important cell types seen here and in the collecting ducts are *intercalated cells,* cuboidal cells with abundant microvilli, and the more numerous *principal cells,* which lack microvilli. The two varieties (alpha and beta) of intercalated cells play a major role in maintaining the acid-base balance of the blood. The principal cells help maintain the body's water and Na^+ balance.

Cortical nephrons represent 85% of the nephrons in the kidneys. Except for small parts of their loops of Henle that dip into the outer medulla, they are located entirely in the cortex. The remaining **juxtamedullary nephrons** (juks"tah-mĕ'dul-ah-re) are located close to (*juxta* = near to) the cortex-medulla junction, and they play an important role in the kidney's ability to produce concentrated urine. Their loops of Henle deeply invade the medulla, and their thin segments are much more extensive than those of cortical nephrons (Figure 24.5).

Nephron Capillary Beds

Every nephron is closely associated with two capillary beds: the *glomerulus* and the *peritubular capillaries* (Figure 24.5). The glomerulus, in which the capillaries run in parallel, is specialized for filtration. It differs from all other capillary beds in the body in that it is both fed and drained by arterioles—the **afferent arteriole** and the **efferent arteriole,** respectively. The afferent arterioles arise from the *interlobular arteries* that run through the renal cortex. Because (1) arterioles are high-resistance vessels and (2) the afferent arteriole has a larger diameter than the efferent, the blood pressure in the glomerulus is extraordinarily high for a capillary bed and easily forces fluid and solutes out of the blood into the glomerular capsule. Most of this filtrate (99%) is reabsorbed by the renal tubule cells and returned to the blood in the peritubular capillary beds.

The **peritubular capillaries** arise from the efferent arterioles draining the glomeruli. These capillaries cling closely to adjacent renal tubules and empty into nearby venules. They are low-pressure, porous capillaries that readily absorb solutes and water from the tubule cells as these substances are reclaimed from the filtrate.

Notice that the efferent arterioles serving the juxtamedullary nephrons tend *not* to break up into peritubular capillaries. Instead they form bundles of long straight vessels called **vasa recta** (va'sah rek'tah; "straight vessels") that extend deep into the medulla paralleling the longest loops of Henle (Figure 24.5b). The thin-walled vasa recta play an important role in forming concentrated urine, as described shortly.

In summary, the microvasculature of the nephrons consists of two capillary beds separated by intervening efferent arterioles. The first capillary bed (glomerulus) produces the filtrate. The second (peritubular capillaries) reclaims most of that filtrate.

Vascular Resistance in the Microcirculation

Blood flowing through the renal circulation encounters high resistance, first in the afferent and then in the efferent arterioles. As a result, renal blood pressure declines from approximately 95 mm Hg in the renal arteries to 8 mm Hg or less in the renal veins. The resistance of the afferent arterioles protects the glomeruli from large fluctuations in systemic blood pressure. Resistance in the efferent arterioles reinforces the high glomerular pressure and reduces the hydrostatic pressure in the peritubular capillaries.

Juxtaglomerular Apparatus

Each nephron has a region called a **juxtaglomerular apparatus (JGA)** (juks"tah-glo-mer'u-lar), where the initial portion of its coiling DCT lies against the afferent arteriole feeding the glomerulus (and sometimes the efferent arteriole) (Figure 24.6). Both structures are modified at the point of contact.

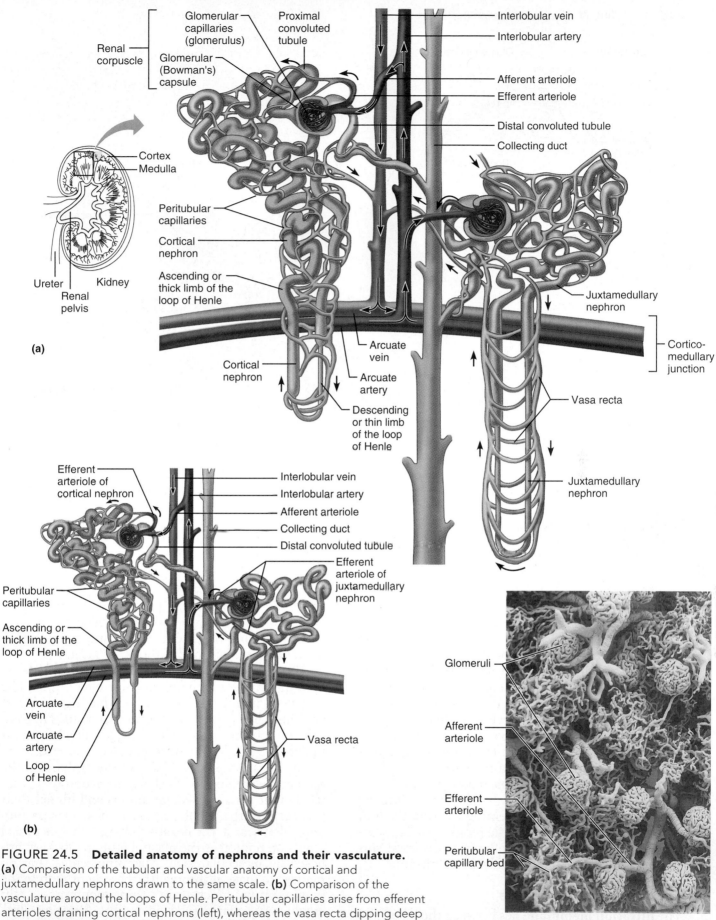

(a)

Renal corpuscle
- Glomerular capillaries (glomerulus)
- Glomerular (Bowman's) capsule

Proximal convoluted tubule

Cortex
Medulla

Ureter
Renal pelvis
Kidney

Peritubular capillaries

Cortical nephron

Ascending or thick limb of the loop of Henle

Cortical nephron

Arcuate vein

Arcuate artery

Descending or thin limb of the loop of Henle

Interlobular vein
Interlobular artery

Afferent arteriole
Efferent arteriole

Distal convoluted tubule

Collecting duct

Juxtamedullary nephron

Cortico-medullary junction

Vasa recta

Juxtamedullary nephron

(b)

Efferent arteriole of cortical nephron

Interlobular vein
Interlobular artery
Afferent arteriole
Collecting duct
Distal convoluted tubule

Efferent arteriole of juxtamedullary nephron

Peritubular capillaries

Ascending or thick limb of the loop of Henle

Arcuate vein

Arcuate artery

Loop of Henle

Vasa recta

(c)

Glomeruli

Afferent arteriole

Efferent arteriole

Peritubular capillary bed

FIGURE 24.5 Detailed anatomy of nephrons and their vasculature. **(a)** Comparison of the tubular and vascular anatomy of cortical and juxtamedullary nephrons drawn to the same scale. **(b)** Comparison of the vasculature around the loops of Henle. Peritubular capillaries arise from efferent arterioles draining cortical nephrons (left), whereas the vasa recta dipping deep into the medulla arise from efferent arterioles of juxtamedullary nephrons (right). Arrows in (a) and (b) indicate direction of blood flow. **(c)** Scanning electron micrograph of a cast of blood vessels associated with nephrons (90×). (© R. G. Kessel and R. H. Kardon, *Tissues and Organs: A Text-Atlas of Scanning Electron Microscopy,* W. H. Freeman and Company, 1979, all rights reserved.)

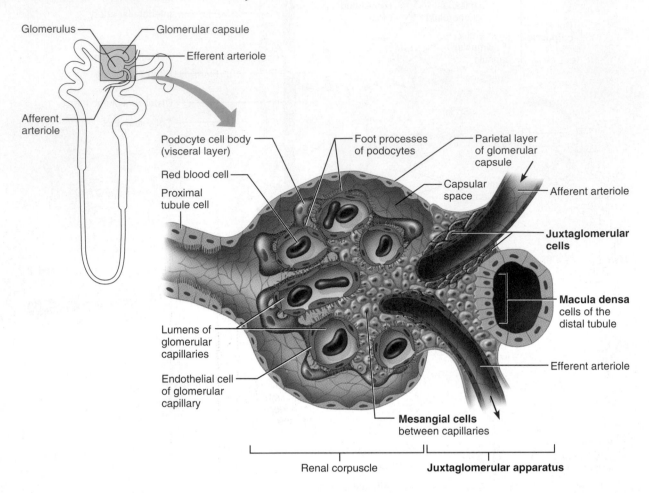

Glomerulus — — Glomerular capsule

— Efferent arteriole

Afferent arteriole

Podocyte cell body (visceral layer)

Red blood cell

Proximal tubule cell

Foot processes of podocytes

Parietal layer of glomerular capsule

Capsular space

Afferent arteriole

Juxtaglomerular cells

Lumens of glomerular capillaries

Endothelial cell of glomerular capillary

Macula densa cells of the distal tubule

Efferent arteriole

Mesangial cells between capillaries

Renal corpuscle   **Juxtaglomerular apparatus**

FIGURE 24.6   Juxtaglomerular apparatus of a nephron. The JGA is made up of the juxtaglomerular cells of the arterioles, the macula densa cells of the distal tubule, and the mesangial cells which surround the glomerular capillaries.

In the arteriole walls are the granular **juxtaglomerular (JG) cells**—enlarged, smooth muscle cells with prominent secretory granules containing renin. JG cells act as mechanoreceptors that sense the blood pressure in the afferent arteriole. The **macula densa** (mak′u-lah den′sah; "dense spot") is a group of tall, closely packed DCT cells that lies adjacent to the JG cells. The macula densa cells are chemoreceptors (or osmoreceptors) that respond to changes in the solute content of the filtrate. These two cell populations play important roles in regulating the rate of filtrate formation and systemic blood pressure as described shortly. The *mesangial cells* surrounding the glomerular capillaries and seemingly part of the JGA have phagocytic and contractile properties. The contractile state of these cells influences the total surface area of the capillaries available for filtration.

The Filtration Membrane

The **filtration membrane** lies between the blood and the interior of the glomerular capsule (Figure 24.7). It is a porous membrane that allows free passage of water and solutes smaller than plasma proteins. Its three layers are: (1) the fenestrated endothelium of the glomerular capillaries, (2) the visceral membrane of the glomerular capsule made of podocytes, and (3) the intervening basement membrane composed of the fused basal laminae of the other layers.

The capillary pores (fenestrations) allow passage of all plasma components but not blood cells. The basement membrane restricts all but the smallest proteins while permitting most other solutes to pass. The structural makeup of the gel-like basement membrane also seems to confer electrical selectivity on the filtration process. Most of the proteins in the membrane are negatively charged glycoproteins that repel other macromolecular anions and hinder their passage into the tubule. Because most plasma proteins also bear a net negative charge, this electrical repulsion reinforces the plasma protein blockage imposed by molecular size. Macromolecules that do manage to make it through the basement membrane may still be blocked by thin membranes (*slit diaphragms*) that extend across the filtration slits. Although it is unclear what happens to macromolecules that get "hung up" in the filtration membrane, the assumption is that they are engulfed by podocytes and degraded.

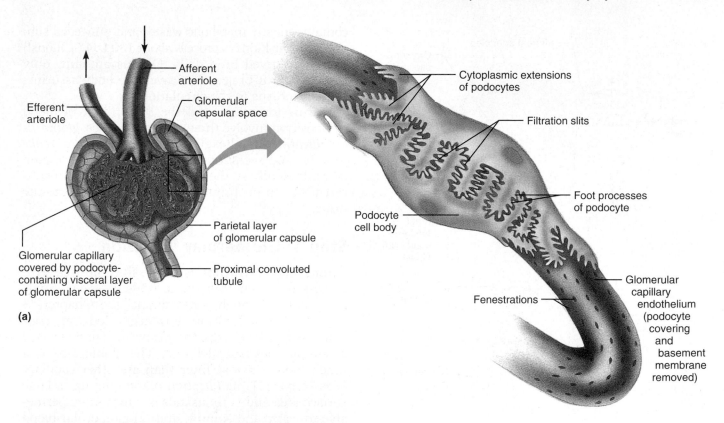

(a)

- Efferent arteriole
- Afferent arteriole
- Glomerular capsular space
- Glomerular capillary covered by podocyte-containing visceral layer of glomerular capsule
- Parietal layer of glomerular capsule
- Proximal convoluted tubule

- Cytoplasmic extensions of podocytes
- Filtration slits
- Podocyte cell body
- Foot processes of podocyte
- Fenestrations
- Glomerular capillary endothelium (podocyte covering and basement membrane removed)

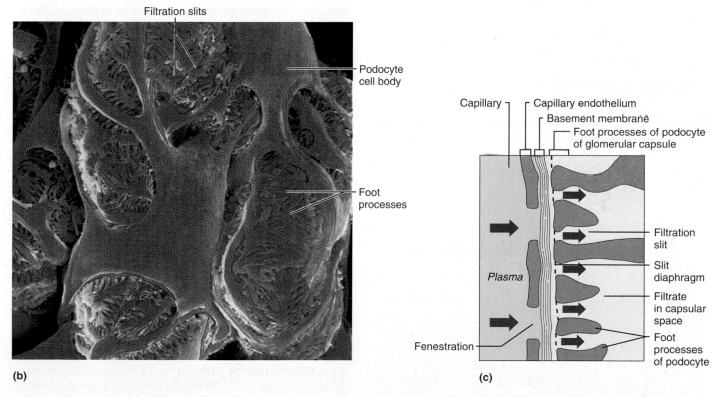

(b)

- Filtration slits
- Podocyte cell body
- Foot processes

(c)

- Capillary
- Capillary endothelium
- Basement membrane
- Foot processes of podocyte of glomerular capsule
- Plasma
- Filtration slit
- Slit diaphragm
- Filtrate in capsular space
- Fenestration
- Foot processes of podocyte

FIGURE 24.7 **The filtration membrane.** The filtration membrane is composed of three layers: the fenestrated endothelium of the glomerular capillaries, the podocyte-containing visceral layer of the glomerular capsule, and the intervening basement membrane. **(a)** Diagrammatic view of the relationship of the visceral layer of the glomerular capsule to the glomerular capillaries. The visceral epithelium is drawn incompletely to show the fenestrations in the underlying capillary wall. **(b)** Scanning electron micrograph of the visceral layer. Filtration slits between the podocyte foot processes are evident (3000×). **(c)** Diagrammatic view of a section through the filtration membrane showing all three structural elements.

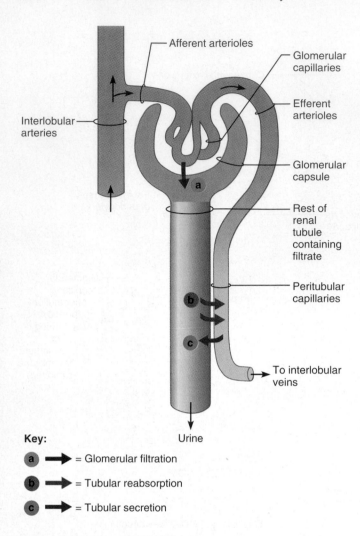

Key:

a ⟶ = Glomerular filtration

b ⟶ = Tubular reabsorption

c ⟶ = Tubular secretion

FIGURE 24.8   The kidney depicted schematically as a single, uncoiled nephron. A kidney actually has millions of nephrons acting in parallel. The three major mechanisms by which the kidneys adjust plasma composition are **(a)** glomerular filtration, **(b)** tubular reabsorption, and **(c)** tubular secretion.

Kidney Physiology: Mechanisms of Urine Formation

Of the approximately 1200 ml of blood that passes through the glomeruli each minute, some 650 ml is plasma, and about one-fifth of this (120–125 ml) is forced into the renal tubules. This is equivalent to filtering out your entire plasma volume more than 60 times each day! Considering the magnitude of their task, it is not surprising that the kidneys (which account for only 1% of body weight) consume 20–25% of all oxygen used by the body at rest.

Filtrate and urine are quite different. Filtrate contains everything found in blood plasma except proteins, but by the time filtrate has percolated into the collecting ducts, it has lost most of its water, nutrients, and ions. What remains, now called **urine,** contains mostly metabolic wastes and unneeded substances. The kidneys process about 180 L (47 gallons!) of blood-derived fluid daily. Of this amount, only about 1% (1.8 L) actually leaves the body as urine; the rest returns to the circulation.

Urine formation and the adjustment of blood composition involve three major processes: *glomerular filtration* by the glomeruli, and *tubular reabsorption* and *secretion* in the renal tubules (Figure 24.8). In addition, the collecting ducts work in concert with the nephrons to concentrate or dilute the urine.

Step 1: Glomerular Filtration

Glomerular filtration is a passive, nonselective process in which hydrostatic pressure forces fluids and solutes through a membrane (see Chapters 3 and 18). Because filtrate formation does not consume metabolic energy, the glomeruli can be viewed as simple mechanical filters. The glomerulus is a much more efficient filter than are other capillary beds because (1) its *filtration membrane* has a large surface area and is thousands of times more permeable to water and solutes, and (2) glomerular blood pressure is much higher than that in other capillary beds (approximately 55 mm Hg as opposed to 18 mm Hg or less), resulting in a much higher *net filtration pressure.* As a result of these differences, the kidneys produce about 180 L of filtrate daily, in contrast to the 3 to 4 L formed daily by all other capillary beds of the body combined.

Molecules smaller than 3 nm in diameter—such as water, glucose, amino acids, and nitrogenous wastes—pass freely from the blood into the renal tubule. Hence, these substances usually show similar concentrations in the blood and the glomerular filtrate. Larger molecules pass with greater difficulty, and those larger than 7–9 nm are generally barred from entering the tubule. Keeping the plasma proteins *in* the capillaries maintains the colloid osmotic (oncotic) pressure of the glomerular blood, preventing the loss of all its water to the renal tubules. The presence of proteins or blood cells in the urine usually indicates a problem with the filtration membrane.

Net Filtration Pressure

The **net filtration pressure (NFP),** responsible for filtrate formation, involves forces acting at the glomerular bed (Figure 24.9). **Glomerular hydrostatic pressure (HP$_g$)** (essentially glomerular blood pressure) is the chief force pushing water and solutes out of the blood and across the filtration membrane. Although theoretically the colloid osmotic pressure in the intracapsular space of the glomerular capsule "pulls" the filtrate into the tubule, this pressure is essentially zero because virtually no proteins enter the capsule. The HP_g is opposed by two forces that drive fluids back into glomerular capillaries. These filtration-opposing forces are (1) **colloid osmotic (oncotic) pressure of glomerular blood (OP_g)** and (2) **capsular hydrostatic pressure (HP_c)** exerted by fluids in the glomerular capsule. Thus, using the values shown in Figure 24.9, the NFP responsible for forming renal filtrate from plasma is 10 mm Hg:

$$NFP = HP_g - (OP_g + HP_c)$$
$$= 55 \text{ mm Hg} - (30 \text{ mm Hg} + 15 \text{ mm Hg})$$
$$= 10 \text{ mm Hg}$$

Glomerular Filtration Rate

The **glomerular filtration rate** or **GFR** is the volume of filtrate formed each minute by the combined activity of all 2 million glomeruli of the kidneys. Factors governing filtration rate at the capillary beds are (1) total surface area available for filtration, (2) filtration membrane permeability, and (3) NFP. In adults the normal GFR in both kidneys is 120–125 ml/min. Because glomerular capillaries are exceptionally permeable and have a huge surface area (collectively equal to the surface area of the skin), huge amounts of filtrate can be produced even with the usual modest NFP of 10 mm Hg. The opposite side of this "coin" is that a drop in glomerular pressure of only 15% stops filtration altogether.

Because the GFR is *directly proportional* to the NFP, any change in any of the pressures acting at the filtration membrane (Figure 24.9) changes both the NFP and the GFR. An increase in arterial (and glomerular) blood pressure in the kidneys increases the GFR, whereas dehydration (which causes an increase in glomerular osmotic pressure) inhibits filtrate formation.

Regulation of Glomerular Filtration

Maintaining a fairly constant GFR is important because reabsorption of water and other substances from the filtrate depends partly on the *rate* at which it flows through the renal tubules. If massive amounts of filtrate form the flow is too rapid for needed substances to be reabsorbed fast enough and some are lost in urine. When filtrate is scanty and flows slowly, nearly all of it is reabsorbed, including most of the wastes that are normally disposed of. In either case, a diseased state exists.

In humans, the GFR is held relatively constant by both intrinsic *(renal autoregulation)* and extrinsic *(neural and hormonal)* controls which regulate renal blood flow. These plus some less well-defined factors are examined next.

Intrinsic Controls: Renal Autoregulation   By adjusting its own resistance to blood flow, a process called **renal autoregulation**, the kidney can

How would liver disease affect the process shown here?

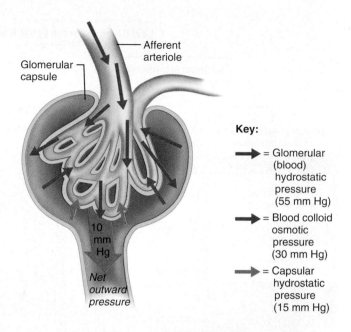

Key:
- → = Glomerular (blood) hydrostatic pressure (55 mm Hg)
- → = Blood colloid osmotic pressure (30 mm Hg)
- → = Capsular hydrostatic pressure (15 mm Hg)

FIGURE 24.9 Forces determining glomerular filtration and filtration pressure. The glomerular hydrostatic (blood) pressure is the major factor forcing fluids and solutes out of the blood. This pressure is opposed by the colloid osmotic pressure of the blood and the hydrostatic pressure in the glomerular capsule. The pressure values cited in the diagram are approximate.

maintain a nearly constant GFR despite fluctuations in systemic arterial blood pressure. Renal autoregulation entails two types of controls: (1) a *myogenic mechanism* and (2) a *tubuloglomerular feedback mechanism* (Figure 24.10).

The **myogenic mechanism** (mi"o-jen'ik) reflects the tendency of vascular smooth muscle to contract when stretched. Increasing systemic blood pressure causes the afferent arterioles to constrict, which restricts blood flow into the glomerulus and prevents glomerular blood pressure from rising to damaging levels. Declining systemic blood pressure causes dilation of afferent arterioles and raises glomerular hydrostatic pressure. Both responses help maintain a normal GFR.

Autoregulation by the flow-dependent **tubuloglomerular feedback mechanism** is "directed" by the *macula densa cells* of the *juxtaglomerular apparatus* (see Figure 24.6). These cells, located in the walls of the distal tubules, respond to filtrate flow

With liver disease you probably wouldn't have a normal supply of plasma proteins. Hence, more filtrate would be formed due to lower osmotic pressure *in the blood.* ■

? *Why are the myogenic and tubuloglomerular mechanisms called autoregulatory mechanisms?*

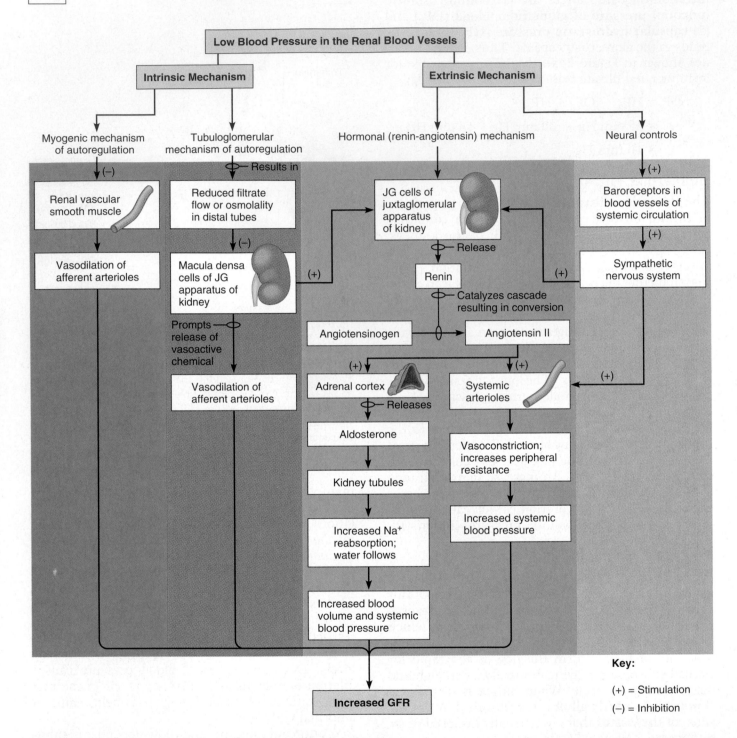

FIGURE 24.10 **Physiological mechanisms regulating the glomerular filtration rate (GFR) in the kidneys.** (Note that while the focus of this flowchart is regulation of the GFR, remember that the main thrust of the extrinsic mechanisms is to regulate systemic blood volume and blood pressure. For them, regulation of the GFR is just a tool.)

GFR. ■

Because they are initiated by and within the kidney and allow the kidney to control its own (auto) rate of blood flow and

rate and osmotic signals. When the macula densa cells are exposed to slowly flowing filtrate or filtrate with low osmolality (Figure 24.10), their signals promote vasodilation of the afferent arterioles. This allows more blood to flow into the glomerulus, thus increasing the NFP and GFR. On the other hand, when the filtrate is flowing rapidly and/or it has a high sodium and chloride content (or high osmolality in general), the macula densa cells prompt generation of a vasoconstrictor chemical that causes intense renal vasconstriction. This hinders blood flow into the glomerulus, which decreases the GFR and allows more time for filtrate processing. The precise stimulus that is sensed by the macula densa cells and the effector chemical that alters resistance of the afferent arteriole are still unidentified. The macula densa cells also send a signal to the JG cells of the juxtaglomerular apparatus that sets the renin-angiotensin mechanism into motion.

Autoregulatory mechanisms maintain a relatively constant blood flow through the kidneys over an arterial pressure range from about 80 to 180 mm Hg. Consequently, large changes in water and solute excretion are prevented. However, the intrinsic controls cannot handle extremely low systemic blood pressure, such as might result from serious hemorrhage *(hypovolemic shock)*. Once mean systemic blood pressure drops below 90 mm Hg, autoregulation ceases.

Extrinsic Controls: Neural and Hormonal Mechanisms

The purpose of the extrinsic controls regulating the GFR is to maintain systemic blood pressure—sometimes to the detriment of the kidneys.

- **Sympathetic nervous system controls.** Neural renal controls serve the needs of the body as a whole. When the volume of the extracellular fluid is normal and the sympathetic nervous system is at rest, the renal blood vessels are maximally dilated and renal autoregulation mechanisms prevail. However, during extreme stress or emergency when it is necessary to shunt blood to vital organs, neural controls may overcome renal autoregulatory mechanisms. Norepinephrine, released by sympathetic nerve fibers (and epinephrine released by the adrenal medulla) acts on alpha-adrenergic receptors on vascular smooth muscle, strongly constricting afferent arterioles, thereby inhibiting filtrate formation. This, in turn, indirectly trips the renin-angiotensin mechanism by stimulating the macula densa cells. The sympathetic nervous system also directly stimulates the JG cells to release renin, as discussed next.

- **Renin-angiotensin mechanism.** The renin-angiotensin mechanism is triggered when various stimuli cause the JG cells to release renin. The enzyme **renin** acts on **angiotensinogen,** a plasma globulin made both by the liver and locally by PCT cells, to release **angiotensin I,** which is converted to **angiotensin II** by **angiotensin converting enzyme (ACE)** associated with the capillary endothelium in various body tissues, particularly the lungs. Angiotensin II, a potent vasoconstrictor, activates smooth muscle of arterioles throughout the body, raising mean arterial blood pressure. By binding to receptors on the luminal membranes of the kidney tubule cells, angiotensin II activates a Na^+/H^+ exchanger, thus increasing Na^+ reabsorption in the PCT cells. It also stimulates the adrenal cortex to release aldosterone, which causes the renal tubules to reclaim more sodium ions from the filtrate. Because water follows sodium osmotically, blood volume and blood pressure rise (Figure 24.10). Because the afferent arterioles sport fewer angiotensin receptors than do the efferent arterioles, angiotensin II causes the efferent arterioles to constrict to a greater extent, thereby increasing the glomerular hydrostatic (blood) pressure. This defensive mechanism partially restores the GFR to normal levels. Angiotensin II also targets the mesangial cells associated with the glomerulus, causing them to contract and reduce the GFR by decreasing the total surface area of glomerular capillaries available for filtration.

Several factors acting independently or collectively, can trigger renin release.

1. Reduced stretch of the granular JG cells. A drop in mean systemic blood pressure below 80 mm Hg (as might be due to hemorrhage, dehydration, etc.) reduces the stretch of the JG cells and stimulates them to release more renin.

2. Stimulation of the JG cells by input from activated macula densa cells (Figure 24.10). Under conditions in which the macula densa cells prompt a *reduced* release of the vasoconstrictor chemical (promoting vasodilation of the afferent arteriole), they also stimulate the JG cells to release renin.

3. Direct stimulation of JG cells via β_1-adrenergic receptors by renal sympathetic nerves.

The main thrust of the renin-angiotensin mechanism is to stabilize systemic blood pressure and extracellular fluid volume. To that end, angiotensin II also stimulates the hypothalamus to release antidiuretic hormone and activates the hypothalamic thirst center (see Chapter 25).

- **Other factors.** Renal cells produce a battery of chemicals, many of which act locally as signaling molecules (paracrines):

1. Prostaglandins: PGE_2 and PGI_2, both vasodilators produced in response to sympathetic stimulation and angiotensin II, are believed to counteract

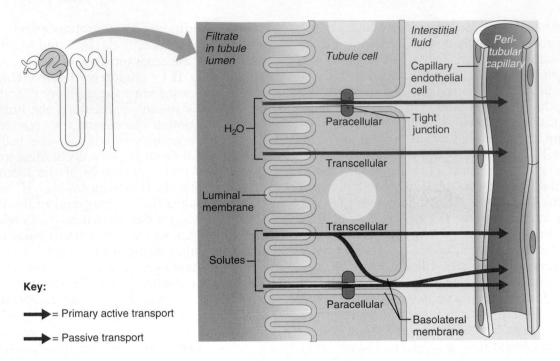

FIGURE 24.11 Routes of water and solute reabsorption across a renal tubule cell. Water is reabsorbed by both the paracellular route (through tight junctions) and the transcellular route. Solutes move via the transcellular route, and often partially through the lateral intercellular space (because ATPase pumps and other transporters are located on the basolateral membrane). NaCl may also move by the paracellular route. (Transporters and aquaporins are not depicted.)

the effect of norepinephrine and angiotensin II on the kidney. The adaptive value of these opposing actions is to prevent renal damage while responding to body demands to increase peripheral resistance.

2. Nitric oxide: A potent vasodilator produced by the vascular endothelium.

3. Adenosine: Although it functions as a vasodilator systemically, adenosine constricts the renal vasculature. It is the most likely candidate for the vasoconstrictor of the tubuloglomerular mechanism.

4. Endothelin: Secreted by the vascular endothelium and selected tubule cells, endothelin is a powerful vasoconstrictor.

HOMEOSTATIC IMBALANCE

Abnormally low urinary output (less than 50 ml/day), called *anuria* (ah-nu're-ah), may indicate that glomerular blood pressure is too low to cause filtration. However, renal failure and anuria usually result from situations in which the nephrons cease to function for a variety of reasons, including acute nephritis, transfusion reactions, and crush injuries. ●

Step 2: Tubular Reabsorption

Our total blood volume filters into the renal tubules about every 45 minutes, so all our plasma would be drained away as urine within an hour were it not for the fact that most of the tubule contents are quickly reclaimed and returned to the blood. This reclamation process, called **tubular reabsorption,** is a *transepithelial process* that begins as soon as the filtrate enters the proximal tubules. To reach the blood, transported substances move through *three* membrane barriers—the *luminal* and *basolateral membranes* of the tubule cells and the *endothelium* of the peritubular capillaries. Because the tubule cells are connected by tight junctions, movement of substances between the cells is limited, although some important ions (Ca^{2+}, Mg^{2+}, K^+, and some Na^+) may use the *paracellular pathway* (Figure 24.11).

Given healthy kidneys, virtually all organic nutrients such as glucose and amino acids are completely reabsorbed to maintain or restore normal plasma concentrations. On the other hand, the reabsorption of water and many ions is continuously regulated and adjusted in response to hormonal signals. Depending on the substances transported, the

? How are primary and secondary active transport processes (both shown here) different?

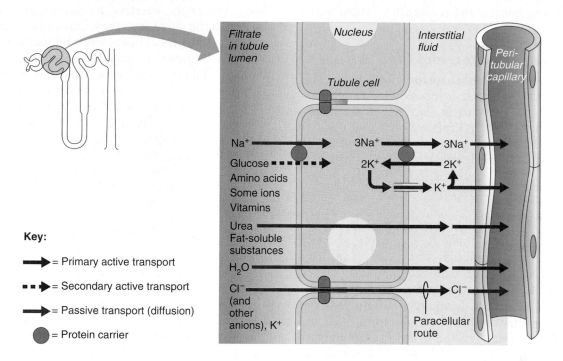

Key:

⟶ = Primary active transport

■ ■ ▶ = Secondary active transport

⟶ = Passive transport (diffusion)

● = Protein carrier

FIGURE 24.12 Reabsorption by PCT cells. Most often Na$^+$ entry at the luminal surface is coupled to the transport of another solute (a cotransport process referred to as secondary active transport). After Na$^+$ enters, it diffuses to the basolateral membrane, where it is pumped into the interstitial space by a sodium pump (a Na$^+$–K$^+$ ATPase). From there it diffuses into the peritubular capillary. Active pumping of Na$^+$ at the basolateral membrane creates concentration and osmotic gradients that drive reabsorption of water by osmosis, of anions and fat-soluble substances by diffusion, and of organic nutrients and selected cations by secondary active transport (symport with Na$^+$ at the luminal membrane). Though not illustrated here, most organic nutrients move through the basolateral membrane by facilitated diffusion to be reabsorbed in the PCT. (Apical microvilli not illustrated.)

reabsorption process may be *passive* (no ATP required) or *active* (at least one of its steps is driven by ATP directly or indirectly).

Sodium Reabsorption

Sodium ions are the single most abundant cation in the filtrate, and about 80% of the energy used for active transport is devoted to their reabsorption. Sodium reabsorption is almost always active and via the transcellular (transepithelial) route. In general, two basic processes that promote active Na$^+$ reabsorption occur in each tubule segment: Na$^+$ (1) enters the tubule cells from the filtrate at the luminal

membrane and then (2) is actively transported out of the tubule cell by a Na$^+$-K$^+$ ATPase pump present in the basolateral membrane (Figure 24.12). From there, Na$^+$ diffuses into adjacent peritubular capillaries. Movement of Na$^+$ and other absorbed substances into the peritubular capillaries is rapid because of that blood's low hydrostatic pressure and high osmotic pressure (remember, most proteins are not filtered out of the blood into the tubule).

Active pumping of Na$^+$ from the tubule cells results in a strong electrochemical gradient that favors its *passive* entry at the luminal face via cotransport or antitransport carriers or via facilitated diffusion through channels. This occurs because (1) the pump maintains the intracellular Na$^+$ concentration at low levels, and (2) the K$^+$ pumped into the tubule cells almost immediately diffuses out into the interstitial fluid via leakage channels, leaving the interior of the tubule cell with a net negative charge.

Because each tubule segment plays a slightly different role in reabsorption, the precise mechanism

In primary active transport, the energy for the process is provided directly by the cleavage of ATP. In secondary active transport, the energy for the process is provided by the Na$^+$ concentration gradient established by the active pumping of Na$^+$ occurring elsewhere in the cell; it drives the movement of another substance (e.g., glucose) against its concentration gradient.

by which Na^+ is reabsorbed at the luminal membrane varies. However, the net effect is that Na^+ reabsorption by primary active transport provides the energy and the means for reabsorbing most other solutes.

Reabsorption of Water, Ions, and Nutrients

In **passive tubular reabsorption,** which encompasses diffusion, facilitated diffusion, and osmosis, substances move along their electrochemical gradient without the use of ATP. As they move through the tubule cells into the peritubular capillary blood, Na^+ ions establish an electrical gradient that favors passive reabsorption of anions (Cl^- and HCO_3^- for example) to restore electrical neutrality in the filtrate and plasma.

As we have seen, Na^+ movement also establishes a strong osmotic gradient, and water moves by osmosis into the peritubular capillaries, a process aided by water-filled pores called *aquaporins*. In continuously water-permeable regions of the renal tubules, such as the PCT, aquaporins are constant components of the tubule cell membranes. Aquaporins are virtually absent in the distal tubule and collecting duct membranes unless ADH levels are increased. Because water is "obliged" to follow salt, this sodium-linked water flow is referred to as **obligatory water reabsorption.**

As water leaves the tubules, the concentration of solutes in the filtrate increases and, if able, they too begin to follow their concentration gradients into the peritubular capillaries (Figure 24.12). This phenomenon of solutes following solvent explains the passive reabsorption of a number of solutes present in the filtrate, such as cations, fatty acids, and some urea. It also explains in part why lipid-soluble drugs and environmental toxins are difficult to excrete: Although the gradient provides the driving force, solute size and lipid solubility may limit which molecules can be reabsorbed.

Substances reabsorbed by **secondary active transport** (the "push" comes from the gradient created by Na^+-K^+ pumping at the basolateral membrane) include glucose, amino acids, lactate, vitamins, and most cations. In nearly all these cases, a common carrier moves Na^+ down its concentration gradient as it cotransports (symports) another solute (Figure 24.12). Cotransported solutes diffuse through the basolateral membrane and into the peritubular capillaries. Although there is some overlap of carriers, the transport systems for the various solutes are quite specific *and limited.*

Excluding Na^+, there is a **transport maximum (T_m)** for nearly every substance that is actively reabsorbed. The T_m (reported in mg/min) reflects the number of carriers in the renal tubules available to ferry each particular substance. In general, there are plenty of carriers and therefore high T_m values for substances such as glucose that need to be retained, and few or no carriers for substances of no use to the body. When the carriers are saturated—that is, all bound to the substance they transport—the excess is excreted in urine. This is what happens in individuals who become hyperglycemic because of uncontrolled diabetes mellitus. As levels of glucose approach 400 mg/100 ml of plasma, the glucose T_m of 375 mg/min is exceeded and large amounts of glucose are lost in the urine even though the renal tubules are still functioning normally.

Any plasma proteins that squeeze through the filtration membrane are removed from the filtrate in the proximal tubule by endocytosis, digested to their amino acids, and moved into the peritubular blood.

Nonreabsorbed Substances

Some substances are either not reabsorbed or are reabsorbed incompletely because they (1) lack carriers, (2) are not lipid soluble, or (3) are too large to pass through plasma membrane pores of the tubular cells. The most important of these are the nitrogenous end products of protein and nucleic acid metabolism: **urea, creatinine,** and **uric acid.** Urea is small enough to diffuse through the membrane pores, and 50–60% of the urea present in filtrate is reclaimed. Creatinine, a large, lipid-insoluble molecule, is not reabsorbed at all. Its plasma concentration remains stable as long as muscle mass is stable, which makes it useful for measuring the GFR and glomerular function.

Absorptive Capabilities of the Renal Tubules and Collecting Ducts

Proximal Convoluted Tubule Table 24.1 compares the reabsorptive abilities of various regions of the renal tubules and collecting ducts. Although the entire renal tubule is involved in reabsorption to some degree, the PCT cells are by far the most active "reabsorbers" and the events just described occur mainly in this tubular segment. Normally, the PCT reabsorbs all of the glucose, lactate, and amino acids in the filtrate and 65% of the Na^+ and water. Additionally, 90% of the filtered bicarbonate (HCO_3^-), 50% of the Cl^-, and about 55% of the K^+ are reclaimed in the PCT. The bulk of the *selective,* or active transport–dependent, reabsorption of electrolytes is accomplished by the time the filtrate reaches the loop of Henle. For the most part, the reabsorption of electrolytes such as Ca^{2+} and PO_4^{3-} is hormonally controlled and closely related to the need to regulate their plasma levels. (These controls are described in Chapter 25.) Nearly all of the uric acid is reabsorbed in the proximal tubule, but it is later secreted back into the filtrate.

TABLE 24.1 | Reabsorption Capabilities of Different Segments of the Renal Tubules and Collecting Ducts

Tubule Segment	Substance Reabsorbed	Mechanism
Proximal convoluted tubule	Sodium ions (Na^+)	Primary active transport via ATP-dependent Na^+-K^+ carrier; sets up electrochemical gradient for passive solute diffusion, osmosis, and secondary active transport (cotransport) with Na^+
	Virtually all nutrients (glucose, amino acids, vitamins)	Active transport; cotransport with Na^+
	Cations (K^+, Mg^{2+}, Ca^{2+}, and others)	Passive transport driven by electrochemical gradient for most; K^+ mainly by the paracellular route
	Anions (Cl^-, HCO_3^-)	Passive transport; paracellular diffusion driven by electrochemical gradient for Cl^-; active transport (cotransport with Na^+ for HCO_3^-)
	Water	Osmosis; driven by solute reabsorption (obligatory)
	Urea and lipid-soluble solutes	Passive diffusion driven by the electrochemical gradient created by osmotic movement of water
	Small proteins	Endocytosed by tubule cells and digested to amino acids within tubule cells
Loop of Henle		
Descending limb	Water	Osmosis
Ascending limb	Na^+, Cl^-, K^+	Active transport of Cl^-, Na^+, and K^+ via a Na^+-K^+-$2Cl^-$ cotransporter in thick portion; also paracellular transport; Na^+-H^+ antitransport
	Ca^{2+}, Mg^{2+}	Passive transport driven by electrochemical gradient; paracellular route
Distal convoluted tubule	Na^+	Primary active transport; requires aldosterone
	Ca^{2+}	PTH-mediated primary active transport via ATP-dependent Ca^{2+} carrier; Na^+/Ca^{2+} exchanger in basal membrane
	Cl^-	Diffusion; follow electrochemical gradient created by active reabsorption of Na^+; also cotransport with Na^+
	Water	Osmosis; facultative water reabsorption; depends on ADH to increase porosity of tubule epithelium in the most distal portion
Collecting duct	Na^+, H^+, K^+, HCO_3^-, and Cl^-	Aldosterone-mediated primary active transport of Na^+ and the medullary gradient create the conditions for passive transport of some HCO_3^- and Cl^- and cotransport of H^+, K^+, Cl^-, and HCO_3^-
	Water	Osmosis; facultative water reabsorption; depends on ADH to increase porosity of tubule epithelium
	Urea	Facilitated diffusion in response to concentration gradient in the deep medulla region; most remains in medullary interstitial space

Loop of Henle   Beyond the PCT, the permeability of the tubule epithelium changes dramatically. Here, for the first time, water reabsorption is not coupled to solute reabsorption. Water can leave the descending limb of the loop of Henle but *not* the ascending limb, where aquaporins are scarce or absent in the tubule membrane. For reasons that will be explained shortly, these permeability differences play a vital role in the kidneys' ability to form dilute and concentrated urine. More electrolytes are reabsorbed in the loop of Henle (see Table 24.1); however, K^+ recycles—it is reabsorbed in the ascending limb and secreted from the descending limb.

A Na^+-K^+-$2Cl^-$ symporter is the main means of Na^+ entry at the luminal surface in the thick portion of the ascending limb. However, this segment also has Na^+-H^+ antiporters like the PCT does, and some 50% of Na^+ passes via the paracellular route in this region. An Na^+/K^- ATPase operates at the basolateral membrane to create the ionic gradient that drives the symporter. In the thin portion of the ascending limb Na^+ moves passively, but the mechanism is unclear.

Distal Convoluted Tubule and Collecting Duct By the time the DCT is reached, only about 10% of the originally filtered NaCl and 25% of the water remain in the tubule. Na^+-Cl^- symporters absorb Na^+ and Cl^-, but most reabsorption from this point on depends largely on the body's needs at the time and is regulated by hormones. If necessary, nearly all of the water and Na^+ reaching these regions can be reclaimed. In the absence of regulatory hormones, the DCT and collecting duct are relatively impermeable to water. Reabsorption of more water depends on the presence of antidiuretic hormone (ADH), which makes the collecting ducts more permeable to water by inserting aquaporin subunits in the collecting duct apical membranes.

Aldosterone "fine-tunes" reabsorption of the remaining Na^+. Decreased blood volume or blood pressure, low extracellular Na^+ concentration (hyponatremia) or high extracellular K^+ concentration (hyperkalemia) can cause the adrenal cortex to release aldosterone to the blood. Except for hyperkalemia (which *directly* stimulates the adrenal cortex to secrete aldosterone), these conditions promote the renin-angiotensin mechanism, which in turn prompts the release of aldosterone (see Figure 24.10). Aldosterone targets the principal cells of the collecting ducts (prodding them to open or synthesize more luminal Na^+ channels, and more basolateral Na^+-K^+ transporters and K^+ channels). As a result, little or no Na^+ leaves the body in urine. In the absence of aldosterone, essentially no Na^+ is reabsorbed by these segments, resulting in Na^+ losses of about 2% of Na^+ filtered daily, an amount incompatible with life.

A second effect of aldosterone is to facilitate water absorption because as Na^+ is reabsorbed, water follows it back into the blood (if it can). Aldosterone also reduces blood K^+ concentrations because aldosterone-induced reabsorption of Na^+ is coupled to K^+ secretion in the principal cells, that is, as Na^+ enters, K^+ moves into the lumen.

In contrast to aldosterone, which acts to conserve Na^+, atrial natriuretic peptide (ANP) reduces blood Na^+, thereby decreasing blood volume and blood pressure. Released by cardiac atrial cells when blood volume or blood pressure is elevated, ANP exerts several effects that lower blood Na^+ levels. For example,

it (1) acts directly on medullary collecting ducts to inhibit Na^+ reabsorption, (2) indirectly inhibits Na^+ reabsorption by counteracting the stimulatory effect of angiotensin II on aldosterone secretion by the adrenal cortex, and (3) indirectly stimulates an increased GFR by targeting the renal arterioles, thereby reducing water reabsorption and blood volume.

Step 3: Tubular Secretion

The failure of tubule cells to reabsorb some solutes is an important means of clearing plasma of unwanted substances. A second such mechanism is **tubular secretion**—essentially, reabsorption in reverse. Substances such as H^+, K^+, NH_4^+, creatinine, and certain organic acids move either from blood of the peritubular capillaries through the tubule cells or directly from the tubule cells into the filtrate. Thus, the urine eventually excreted contains *both filtered and secreted substances*. With one major exception (K^+), the PCT is the main site of secretion, but the cortical parts of the collecting ducts and the late regions of the distal tubules are also active (see Figure 24.16, p. 875).

Tubular secretion is important for:

1. Disposing of substances not already in the filtrate, such as certain drugs

2. Eliminating undesirable substances or end products that have been reabsorbed by passive processes, (urea and uric acid)

3. Ridding the body of excess K^+

4. Controlling blood pH (see Chapter 25)

Because virtually all K^+ present in the filtrate is reabsorbed in the PCT and ascending loop of Henle, nearly all K^+ in urine is from aldosterone-driven active tubular secretion into the collecting ducts. When blood pH drops toward the acidic end of its homeostatic range, the renal tubule cells actively secrete more H^+ into the filtrate and retain more HCO_3^- and K^+. As a result, the blood pH rises and the urine drains off the excess H^+. Conversely, when blood pH approaches the alkaline end of its range, more Cl^- rather than HCO_3^- is reabsorbed and bicarbonate is allowed to leave the body in urine.

Regulation of Urine Concentration and Volume

A solution's **osmolality** (oz″mo-lal′ĭ-te) is the number of solute particles dissolved in one liter (1000 g) of water and reflects the solution's ability to cause osmosis. For any solution interfacing with a semipermeable membrane, this ability, called *osmotic activity*, is determined only by the number of nonpenetrating solute particles (solute particles unable to pass through the membrane) and is independent of their type. Ten sodium ions have the same osmotic activity as ten glucose molecules or ten amino acids in the same volume of water. Because the *osmol* (equivalent to 1 mole of a nonionizing substance in 1 L of water) is a fairly large unit, the **milliosmol (mOsm)** (mil"e-oz'mōl), equal to 0.001 osmol, is used to describe the solute concentration of body fluids.

One crucial renal function is to keep the solute load of body fluids constant at about 300 mOsm, the osmotic concentration of blood plasma, by regulating urine concentration and volume. The kidneys accomplish this feat by a **countercurrent mechanism.** The term *countercurrent* means that something flows in opposite directions through adjacent channels. In the kidneys the countercurrent mechanism involves the interaction between the flow of filtrate through the long loops of Henle of juxtamedullary nephrons (the *countercurrent multiplier*), and the flow of blood through the adjacent vasa recta blood vessels (the *countercurrent exchanger*). The countercurrent mechanism establishes an osmotic gradient extending from the cortex through the depths of the medulla that allows the kidneys to vary urine concentration dramatically.

The osmolality of the filtrate entering the PCT is identical to that of plasma, about 300 mOsm. As described earlier, because of PCT reabsorption of water *and* solutes, the filtrate is still isosmotic with plasma by the time the descending limb of the loop of Henle is reached. However, its osmolality increases from 300 to about 1200 mOsm in the deepest part of the medulla (Figure 24.13). How does this increase in concentration occur? The answer lies in the unique workings of the long loop of Henle of the juxtamedullary nephrons, and the vasa recta. Notice in Figure 24.13 that in each case the fluids involved—filtrate in the loop of Henle and blood in the vasa recta—first descend and then ascend through parallel limbs.

The Countercurrent Multiplier

First, we will follow filtrate processing through the loop of Henle, as portrayed in Figure 24.14a, to see how the loop functions as a **countercurrent multiplier** to establish the osmotic gradient. The countercurrent multiplier functions because of three factors:

1. **The descending limb of the loop of Henle is relatively impermeable to solutes and freely permeable to water.** Because the osmolality of the medullary interstitial fluid increases all along the descending limb (the mechanism of this increase is explained shortly), water passes osmotically out of the filtrate all along this course. Thus, the filtrate osmolality reaches its highest point (1200 mOsm) at the "elbow" of the loop.

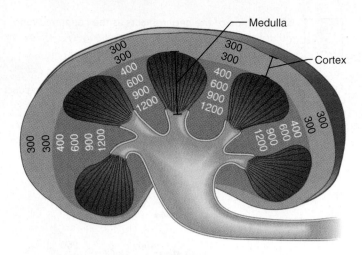

FIGURE 24.13   Osmotic gradient in the renal medulla. The osmolality of the interstitial fluid in the renal cortex is isotonic at 300 mOsm/L, but the osmolality of the interstitial fluid in the renal medulla increases progressively from 300 mOsm/L at the boundary with the cortico-medullary junction to 1200 mOsm/L at the medullary-pelvis junction.

2. **The ascending limb is permeable to solutes, but not to water.** As the filtrate rounds the corner into the ascending limb, the tubule permeability changes, becoming impermeable to water and selectively permeable to salt. The Na^+ and Cl^- concentration in the filtrate entering the ascending limb is very high (and interstitial fluid concentrations of these two ions are lower). Although the thin segment of the ascending limb of the loop of Henle also contributes to Na^+ and Cl^- reabsorption, most of the NaCl reabsorption takes place in the thick segment via its Na^+-K^+-$2Cl^-$ cotransporter. As Na^+ and Cl^- are extruded from the filtrate into the medullary interstitial fluid, they contribute to the high osmolality there. Because it loses salt but not water, the filtrate in the ascending limb becomes increasingly dilute until, at 100 mOsm at the DCT, it is hypoosmotic, or hypotonic, to blood plasma and cortical fluids.

Notice in Figure 24.14 that there is a constant difference in filtrate concentration between the two limbs of the loop of Henle. Because the ascending limb is relatively more permeable to Na^+ and Cl^-, some salt diffuses back into the filtrate in that portion of the tubule, even as NaCl is being actively reabsorbed. Consequently, the solute content in the ascending limb filtrate is always about 200 mOsm lower than the filtrate concentration in the descending limb and in the surrounding medullary interstitial fluid. However, because of countercurrent flow, the loop of Henle is able to "multiply" these small changes in solute concentration into a gradient change along the vertical length of the loop (both

What portion of the nephron acts as the countercurrent multiplier? As the countercurrent exchanger?

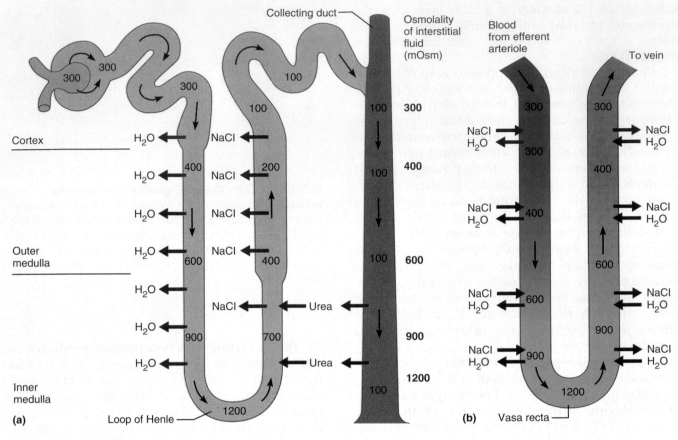

Key:

➤ = Active transport

➤ = Passive transport

FIGURE 24.14 Countercurrent mechanism for establishing and maintaining the medullary osmotic gradient. (a) Filtrate entering the descending limb of the loop of Henle is isosmotic to both blood plasma and cortical interstitial fluid. As the filtrate flows from the cortex to the medulla in the descending limb, water leaves the tubule by osmosis and the filtrate osmolality increases from 300 to 1200 mOsm. As the filtrate flows into the ascending limb, the permeability of the tubule epithelium changes from being permeable to water and impermeable to salt to the opposite condition: water impermeable, salt permeable. Consequently, NaCl leaves the ascending limb, diluting the filtrate as it approaches the cortex. Thus the descending limb produces increasingly salty filtrate, while the ascending limb uses this high salt concentration to establish the high osmolality of the interstitial fluid in the medulla and the medullary osmotic gradient. Diffusion of urea from the lower portion of the collecting duct also contributes to the high osmolality in the medulla. **(b)** Dissipation of the medullary osmotic gradient is prevented because the blood in the vasa recta continuously becomes isosmotic to the interstitial fluid—becoming more concentrated as it follows the descending limb of the loop of Henle and less concentrated as it approaches the cortical region. This occurs because of the high porosity and sluggish blood flow in the vasa recta.

■ *vasa recta.*

The countercurrent multiplier is the loop of Henle of juxtamedullary nephrons. The countercurrent exchanger is the

inside and outside) that is closer to 900 mOsm (1200 mOsm − 300 mOsm).

Although the two limbs of the loop of Henle are not in direct contact with each other, they are close enough to influence each other's exchanges with the interstitial fluid they share. Hence, water diffusing out of the descending limb produces the increasingly "salty" filtrate that the ascending limb uses to raise the osmolality of the medullary interstitial fluid. Furthermore, the more NaCl the ascending limb extrudes, the more water diffuses out of the descending limb and the saltier the filtrate in the descending limb becomes. This establishes a positive feedback mechanism that produces the high osmolality of the fluids in the descending limb and the interstitial fluid.

3. **The collecting ducts in the deep medullary regions are permeable to urea.** The amount of urea in the filtrate remains high because most nephron segments beyond the PCT are impermeable to it. However, as urine passes through the collecting duct in the deep medullary region where the duct is highly permeable to urea, urea diffuses out of the duct into the medullary interstitial fluid, where it contributes to the high osmolality in that region. Urea continues to move out of the duct passively until its concentration inside and outside the duct is equal. Even though the ascending limb of the loop of Henle is poorly permeable to urea, when urea concentration in the medullary interstitial space is high, some urea does enter the limb. However, it is recycled back to the collecting duct, where it diffuses out again. Unlike the active reabsorption of NaCl in the thick part of the ascending limb, which is a crucial element of the countercurrent multiplier system, urea's cycling simply equalizes its concentration inside and outside the renal tubules.

The Countercurrent Exchanger

As shown in Figure 24.14b, the vasa recta function as a **countercurrent exchanger**, maintaining the osmotic gradient established by the cycling of salt while delivering blood to cells in the area. Because these vessels receive only about 10% of the renal blood supply, blood flow through the vasa recta is sluggish. Moreover, they are freely permeable to water and NaCl, allowing blood to make passive exchanges with the surrounding interstitial fluid and achieve equilibrium. Consequently, as the blood flows into the medullary depths, it loses water and gains salt (becomes hypertonic). Then, as it emerges from the medulla into the cortex, the process is reversed: It picks up water and loses salt. Because blood leaving and reentering the cortex via the vasa recta has the same solute concentration,

the vessels of the vasa recta act as a countercurrent exchanger. This system does not create the medullary gradient, but it protects it by preventing rapid removal of salt from the medullary interstitial space.

Formation of Dilute Urine

Because tubular filtrate is diluted as it travels through the ascending limb of the loop of Henle, all the kidney needs to do to secrete dilute (hypoosmotic) urine is allow the filtrate to continue on its way into the renal pelvis (Figure 24.15a). When antidiuretic hormone is not being released by the posterior pituitary, that is exactly what happens. The collecting ducts remain essentially impermeable to water due to the absence of aquaporins at the luminal cell membranes, and no further water reabsorption occurs. Moreover, as noted, Na$^+$ and selected other ions can be removed from the filtrate by DCT and collecting duct cells so that urine becomes even more dilute before entering the renal pelvis. The osmolality of urine can plunge as low as 50 mOsm, about one-sixth the concentration of glomerular filtrate or blood plasma.

Formation of Concentrated Urine

As its name reveals, **antidiuretic hormone** inhibits *diuresis* (di"u-re'sis), or urine output. It accomplishes this via a second-messenger system using cyclic AMP that increases the number of water-filled channels in the principal cells of the collecting ducts by stimulating shuttling of the channels' protein subunits (aquaporin 2) to the luminal surfaces, where they are inserted. Consequently water passes easily into and through the cells into the interstitial space, and the osmolality of the filtrate becomes equal to that of the interstitial fluid.

In the distal tubules, which are in the cortex, the filtrate osmolality is approximately 100 mOsm, but as the filtrate flows through the collecting ducts and is subjected to the hyperosmolar conditions in the medulla, water, followed by urea, rapidly leaves the filtrate (Figures 24.15b and 24.16e). Depending on the amount of ADH released (which is keyed to the level of body hydration), urine concentration may rise as high as 1200 mOsm, the concentration of interstitial fluid in the deepest part of the medulla. With maximal ADH secretion, up to 99% of the water in the filtrate is reabsorbed and returned to the blood, and less than 1 L/day (about 1 ml/min) of highly concentrated urine is excreted. The ability of our kidneys to produce such concentrated urine is critically tied to our ability to survive without water. Water reabsorption that depends on the presence of ADH is called **facultative water reabsorption.**

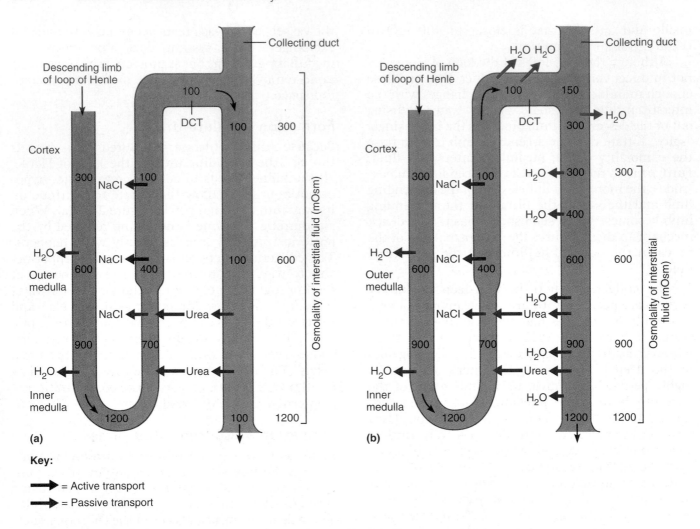

Key:

⟶ = Active transport

➡ = Passive transport

FIGURE 24.15 Mechanisms for forming dilute and concentrated urine.
(a) Because the filtrate is diluted by the workings of the countercurrent mechanism, all that is needed to secrete dilute urine is to allow the filtrate reaching the DCT to be excreted. **(b)** Concentrated urine is excreted in the presence of ADH. ADH causes insertion of aquaporins in luminal membranes of principal cells of the late DCT and collecting duct. Consequently water rapidly leaves the filtrate in the collecting duct.

ADH is released more or less continuously unless the blood solute concentration drops too low. Release of ADH is enhanced by any event that raises plasma osmolality above 300 mOsm/L, such as sweating or diarrhea, or reduced blood volume or blood pressure (see Chapter 25). Although release of ADH is the "signal" to produce concentrated urine that opens the door (pores) for water reabsorption, the kidneys' ability to respond to this signal depends on the high medullary osmotic gradient.

Diuretics

There are several types of **diuretics,** chemicals that enhance urinary output. An *osmotic diuretic* is a substance that is not reabsorbed and that carries water out with it (for example, the high blood glucose levels of a diabetes mellitus patient). Alcohol, essentially a sedative, encourages diuresis by inhibiting release of ADH. Other diuretics increase urine flow by inhibiting Na^+ reabsorption and the obligatory water reabsorption that normally follows. Examples include caffeine (found in coffee, tea, and colas) and many drugs prescribed for hypertension or the edema of congestive heart failure. Common diuretics, like Lasix and Diuril, inhibit Na^+-associated symporters in the ascending loop of Henle or the DCT.

Renal Clearance

Renal clearance refers to the volume of plasma that is cleared of a particular substance in a given time, usually 1 minute. Renal clearance tests are done to determine the GFR, which provides information about the amount of functioning renal tissue. Renal clearance tests are also done to detect glomerular damage and to follow the progress of renal disease.

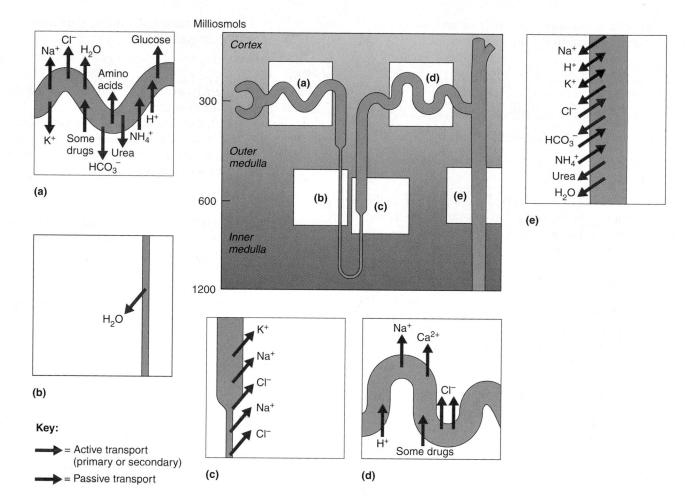

FIGURE 24.16 Summary of nephron functions. The glomerulus provides the filtrate processed by the renal tubule. The various regions of the renal tubule carry out reabsorption and secretion and maintain a gradient of osmolality within the medullary interstitial fluid. Varying osmolality at different points in the tubule and interstitial fluid is symbolized in the central figure by gradients of color. The inserts that accompany the central figure describe the main transport functions of the four regions of the nephron tubule and of the collecting duct. **(a) Proximal tubule.** Filtrate that enters the PCT from the glomerular capsule has about the same osmolality as blood plasma. The main activity of the PCT transport epithelium is reabsorption of certain solutes from the filtrate back into the blood. Nearly all nutrients and about 65% of Na^+ are actively transported out of the PCT and enter the peritubular capillaries; Cl^- and water follow passively. PCT cells also secrete the ammonium ion and other nitrogenous wastes into the filtrate and help maintain a constant pH in blood and interstitial fluid by the controlled secretion of H^+ and reabsorption of HCO_3^-. By the end of the proximal

tubule, the filtrate volume has been reduced by 65%. **(b) Descending limb.** The descending limb of the loop of Henle is freely permeable to water but not to NaCl. As filtrate in this limb descends into the medulla, the filtrate loses water by osmosis to the interstitial fluid, which is increasingly hypertonic in that direction. Consequently, salt and other solutes become more concentrated in the filtrate. **(c) Ascending limb.** The filtrate enters the ascending limb of the loop of Henle, which is impermeable to water but permeable to Na^+ and Cl^-. These ions, which became concentrated in the descending limb, move passively out of the thin portion of the ascending limb, are actively pumped out of the thick portion of the ascending limb, and contribute to the high osmolality of interstitial fluid in the inner medulla. K^+ is cotransported with Na^+ and Cl^-. The tubule epithelium here is not permeable to water, so the filtrate becomes more and more dilute as the exodus of salt from the filtrate continues. **(d) Distal tubule.** The DCT, like the PCT, is specialized for selective secretion and reabsorption. Na^+ and Cl^- are cotransported, H^+ may be secreted, and

in the presence of aldosterone, more Na^+ is reabsorbed. The water permeability of the DCT is extremely low and almost no further H_2O absorption occurs there. **(e) Collecting duct.** The urine, normally quite dilute at this point, begins its journey via the collecting duct back into the medulla, with its increasing osmolality gradient. In the cortical collecting duct, K^+, H^+, and/or HCO_3^- ions may be reabsorbed or secreted depending on what is required to maintain blood pH. The wall of the medullary region of the collecting duct is permeable to urea and is made more so by the presence of ADH. Some urea diffuses out of the collecting duct and contributes to the high osmolality of the inner medulla. In the absence of ADH, the collecting duct is nearly impermeable to water, and the dilute urine passes out of the kidney. In the presence of ADH, more aquaporins are inserted into the collecting duct, and the filtrate loses water by osmosis as it passes through medullary regions of increasing osmolality. Consequently, water is conserved, and concentrated urine is excreted.

The renal clearance rate (RC) of any substance, in ml/min, is calculated from the equation

$$RC = UV/P$$

where:

U = concentration of the substance in urine (mg/ml)

V = flow rate of urine formation (ml/min)

P = concentration of the substance in plasma (mg/ml)

Because it is not reabsorbed, stored, or secreted by the kidneys, *inulin* (in′u-lin) is often used as the standard to determine the GFR. A polysaccharide with a molecular weight of approximately 5000, inulin's renal clearance value is equal to the GFR. When inulin is infused such that its plasma concentration is 1 mg/ml ($P = 1$ mg/ml), then generally $U = 125$ mg/ml, and $V = 1$ ml/min. Therefore, its renal clearance is RC = $(125 \times 1)/1 = 125$ ml/min, meaning that in 1 minute the kidneys have removed (cleared) all the inulin present in 125 ml of plasma.

A clearance value less than that of inulin means that a substance is partially reabsorbed. An example is urea with a clearance value of 70 ml/min, meaning that of the 125 ml of glomerular filtrate formed each minute, approximately 70 ml is completely cleared of urea, while the urea in the remaining 55 ml is recovered and returned to the plasma. If the renal clearance value is zero (such as for glucose in healthy individuals), reabsorption is complete or the substance is not filtered. If clearance is greater than that of inulin, the tubule cells are secreting the substance into the filtrate. This is the case with creatinine, which has an RC of 140 ml/min, and with most drug metabolites. Knowing a drug's renal clearance value is essential because if it is high, the drug dosage must also be high and administered frequently to maintain a therapeutic level.

Urine

Physical Characteristics

Color and Transparency

Freshly voided urine is clear and pale to deep yellow. Its yellow color is due to **urochrome** (u′ro-krōm), a pigment that results from the body's destruction of hemoglobin (via bilirubin or bile pigments). The more concentrated the urine, the deeper the yellow color. An abnormal color such as pink or brown, or a smoky tinge, may result from eating certain foods (beets, rhubarb) or may be due to the presence in the urine of bile pigments or blood. Additionally, some commonly prescribed drugs and vitamin supplements alter the color of urine. Cloudy urine may indicate a urinary tract infection.

Odor

Fresh urine is slightly aromatic, but if allowed to stand, it develops an ammonia odor as bacteria metabolize its urea solutes. Some drugs and vegetables alter the usual odor of urine, as do some diseases. For example, in diabetes mellitus the urine smells fruity because of its acetone content.

pH

Urine is usually slightly acidic (around pH 6), but changes in body metabolism or diet may cause the pH to vary from about 4.5 to 8.0. A predominantly *acidic* diet that contains large amounts of protein and whole wheat products produces acidic urine. A vegetarian *(alkaline)* diet, prolonged vomiting, and bacterial infection of the urinary tract all cause the urine to become alkaline.

Specific Gravity

Because urine is water plus solutes, a given volume has a greater mass than the same volume of distilled water. The term used to compare the mass of a substance to the mass of an equal volume of distilled water is **specific gravity.** The specific gravity of distilled water is 1.0 and that of urine ranges from 1.001 to 1.035, depending on its solute concentration.

Chemical Composition

Water accounts for about 95% of urine volume; the remaining 5% consists of solutes. The largest component of urine by weight, apart from water, is *urea*, which is derived from the normal breakdown of amino acids. Other **nitrogenous wastes** in urine include *uric acid* (an end product of nucleic acid metabolism) and *creatinine* (a metabolite of creatine phosphate, which stores energy for the regeneration of ATP and is found in large amounts in skeletal muscle tissue). Normal solute constituents of urine, in order of decreasing concentration, are urea, Na^+, K^+, PO_4^{3-}, SO_4^{2-}, creatinine, and uric acid. Much smaller but highly variable amounts of Ca^{2+}, Mg^{2+}, and HCO_3^- are also present in urine. Unusually high concentrations of any solute, or the presence of abnormal substances such as blood proteins, WBCs (pus), or bile pigments, may indicate pathology (Table 24.2).

Ureters

The **ureters** are slender tubes that convey urine from the kidneys to the bladder (see Figure 24.1). Each ureter begins at the level of L_2 as a continuation of the renal pelvis. From there, it descends behind the

TABLE 24.2   Abnormal Urinary Constituents

Substance	Name of Condition	Possible Causes
Glucose	Glycosuria	Nonpathological: excessive intake of sugary foods Pathological: diabetes mellitus
Proteins	Proteinuria, or albuminuria	Nonpathological: excessive physical exertion; pregnancy; high-protein diet Pathological (over 250 mg/day): heart failure, severe hypertension; glomerulonephritis; often initial sign of asymptomatic renal disease
Ketone bodies	Ketonuria	Excessive formation and accumulation of ketone bodies, as in starvation and untreated diabetes mellitus
Hemoglobin	Hemoglobinuria	Various: transfusion reaction, hemolytic anemia, severe burns, etc.
Bile pigments	Bilirubinuria	Liver disease (hepatitis, cirrhosis) or obstruction of bile ducts from liver or gallbladder
Erythrocytes	Hematuria	Bleeding urinary tract (due to trauma, kidney stones, infection, or neoplasm)
Leukocytes (pus)	Pyuria	Urinary tract infection

peritoneum and runs obliquely through the posterior bladder wall. This arrangement prevents backflow of urine during bladder filling because any increase in bladder pressure compresses and closes the distal ends of the ureters.

Histologically, the ureter wall is trilayered. The transitional epithelium of its lining *mucosa* is continuous with that of the kidney pelvis superiorly and the bladder medially. Its middle *muscularis* is composed chiefly of two smooth muscle sheets: the internal longitudinal layer and the external circular layer. An additional smooth muscle layer, the external longitudinal layer, appears in the lower third of the ureter. The *adventitia* covering the ureter's external surface is typical fibrous connective tissue (Figure 24.17).

The ureter plays an active role in transporting urine. Incoming urine distends the ureter and stimulates its muscularis to contract, propelling urine into the bladder. (Urine does *not* reach the bladder through gravity alone.) The strength and frequency of the peristaltic waves are adjusted to the rate of urine formation. Although each ureter is innervated by both sympathetic and parasympathetic fibers, neural control of peristalsis appears to be insignificant compared to the way ureteral smooth muscle responds to stretch.

H   HOMEOSTATIC IMBALANCE

On occasion, calcium, magnesium, or uric acid salts in urine may crystallize and precipitate in the renal pelvis, forming **renal calculi** (kal'ku-li; *calculus* = little stone), or kidney stones. Most calculi are under 5 mm in diameter and pass through the urinary tract without causing problems. However, larger calculi can obstruct a ureter and block urine drainage. Increasing pressure in the kidney causes excruciating

pain, which radiates from the flank to the anterior abdominal wall on the same side. Pain also occurs when the contracting ureter wall closes in on the sharp calculi as they are being eased through a ureter by peristalsis.

Predisposing conditions are frequent bacterial infections of the urinary tract, urine retention, high blood levels of calcium, and alkaline urine. Surgical removal of calculi has been almost entirely replaced by *shock wave lithotripsy*, a noninvasive procedure that uses ultrasonic shock waves to shatter the calculi. The pulverized, sandlike remnants of the calculi are then painlessly eliminated in the urine. People with a history of kidney stones are encouraged to acidify their urine by drinking cranberry juice and to ingest large quantities of water to keep the urine dilute. ●

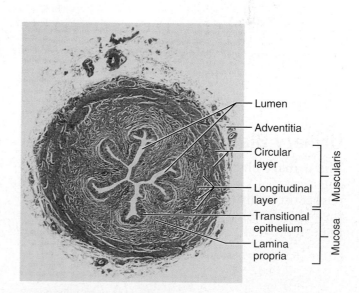

FIGURE 24.17   **Cross-sectional view of the ureter wall, (15×).**

How do the internal and external urethral sphincters differ structurally and functionally?

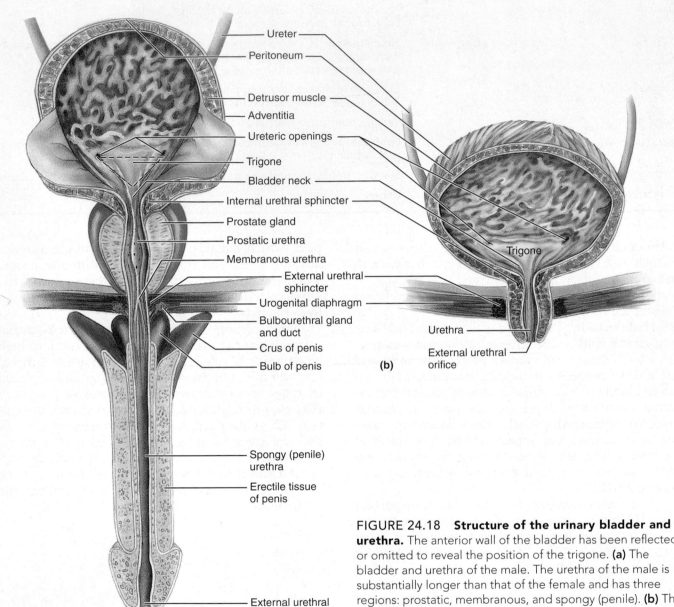

(a)

Ureter
Peritoneum
Detrusor muscle
Adventitia
Ureteric openings
Trigone
Bladder neck
Internal urethral sphincter
Prostate gland
Prostatic urethra
Membranous urethra
External urethral sphincter
Urogenital diaphragm
Bulbourethral gland and duct
Crus of penis
Bulb of penis
Spongy (penile) urethra
Erectile tissue of penis
External urethral orifice

Trigone
Urethra
External urethral orifice

(b)

FIGURE 24.18   Structure of the urinary bladder and urethra. The anterior wall of the bladder has been reflected or omitted to reveal the position of the trigone. **(a)** The bladder and urethra of the male. The urethra of the male is substantially longer than that of the female and has three regions: prostatic, membranous, and spongy (penile). **(b)** The bladder and urethra of the female.

Urinary Bladder

The **urinary bladder** is a smooth, collapsible, muscular sac that stores urine temporarily. It is located retroperitoneally on the pelvic floor just posterior to the pubic symphysis. The prostate gland (part of the male reproductive system) surrounds the bladder

neck inferiorly where it empties into the urethra. In females, the bladder is anterior to the vagina and uterus.

The interior of the bladder has openings for both ureters and the urethra (Figure 24.18). The smooth, triangular region of the bladder base outlined by these three openings is the **trigone** (tri′gōn; *trigon* = triangle), important clinically because infections tend to persist in this region.

The bladder wall has three layers: a mucosa containing transitional epithelium, a thick muscular layer, and a fibrous adventitia (except on its superior

The internal sphincter, composed of smooth muscle, is controlled involuntarily; the external sphincter, of skeletal muscle, is controlled voluntarily. ■

surface, where it is covered by the peritoneum). The muscular layer, called the **detrusor muscle** (de-tru'sor; "to thrust out"), consists of intermingled smooth muscle fibers arranged in inner and outer longitudinal layers and a middle circular layer.

The bladder is very distensible and uniquely suited for its function of urine storage. When empty, the bladder collapses into its basic pyramidal shape and its walls are thick and thrown into folds *(rugae)*. As urine accumulates, the bladder expands, becomes pear-shaped, and rises superiorly in the abdominal cavity (Figure 24.19). The muscular wall stretches and thins, and rugae disappear. These changes allow the bladder to store more urine without a significant rise in internal pressure. A moderately full bladder is about 12 cm (5 inches) long and holds approximately 500 ml (1 pint) of urine, but it can hold more than double that if necessary. When tense with urine, it can be palpated well above the pubic symphysis. The maximum capacity of the bladder is 800–1000 ml and when it is overdistended, it may burst. Although urine is formed continuously by the kidneys, it is usually stored in the bladder until its release is convenient.

Urethra

The **urethra** is a thin-walled muscular tube that drains urine from the bladder and conveys it out of the body. The epithelium of its mucosal lining is mostly pseudostratified columnar epithelium. However, near the bladder it becomes transitional epithelium, and near the external opening it changes to a protective stratified squamous epithelium.

At the bladder-urethra junction a thickening of the detrusor smooth muscle forms the **internal urethral sphincter** (Figure 24.18). This involuntary sphincter keeps the urethra closed when urine is not being passed and prevents leaking between voidings. The **external urethral sphincter (sphincter urethrae)** surrounds the urethra as it passes through the *urogenital diaphragm*. This sphincter is formed of skeletal muscle and is voluntarily controlled. The *levator ani* muscle of the pelvic floor also serves as a voluntary constrictor of the urethra (see Table 10.7, p. 306).

The length and functions of the urethra differ in the two sexes. In females the urethra is only 3–4 cm (1.5 inches) long and tightly bound to the anterior vaginal wall by fibrous connective tissue. Its external opening, the **external urethral orifice (meatus),** lies anterior to the vaginal opening and posterior to the clitoris.

In males the urethra is approximately 20 cm (8 inches) long and has three regions. The **prostatic urethra,** about 2.5 cm (1 inch) long, runs within the prostate gland. The **membranous urethra,**

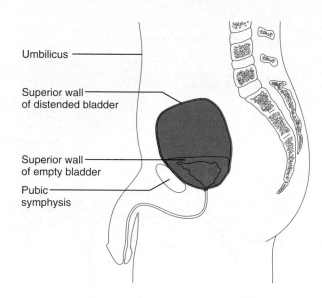

Umbilicus

Superior wall of distended bladder

Superior wall of empty bladder

Pubic symphysis

FIGURE 24.19   Position and shape of a distended and an empty urinary bladder in an adult male.

which runs through the urogenital diaphragm, extends about 2 cm from the prostate gland to the beginning of the penis. The **spongy,** or **penile, urethra,** about 15 cm long, passes through the penis and opens at its tip via the **external urethral orifice.** The male urethra has a double function: It carries semen as well as urine out of the body. The reproductive function of the male urethra is discussed in Chapter 26.

HOMEOSTATIC IMBALANCE

Because the female's urethra is very short and its external orifice is close to the anal opening, improper toilet habits (wiping back to front after defecation) can easily carry fecal bacteria into the urethra. Actually, most urinary tract infections occur in sexually active women, because intercourse drives bacteria from the vagina and external genital region toward the bladder. The use of spermicides magnifies this problem, because the spermicide kills helpful bacteria, allowing infectious fecal bacteria to colonize the vagina. Overall, 40% of all women get urinary tract infections.

The urethral mucosa is continuous with that of the rest of the urinary tract, and an inflammation of the urethra *(urethritis)* can ascend the tract to cause bladder inflammation *(cystitis)* or even renal inflammations *(pyelitis* or *pyelonephritis)*. Symptoms of urinary tract infection include *dysuria* (painful urination), urinary *urgency* and *frequency*, fever, and sometimes cloudy or blood-tinged urine. When the kidneys are involved, back pain and a severe headache often occur. Most urinary tract infections are easily cured by antibiotics. ●

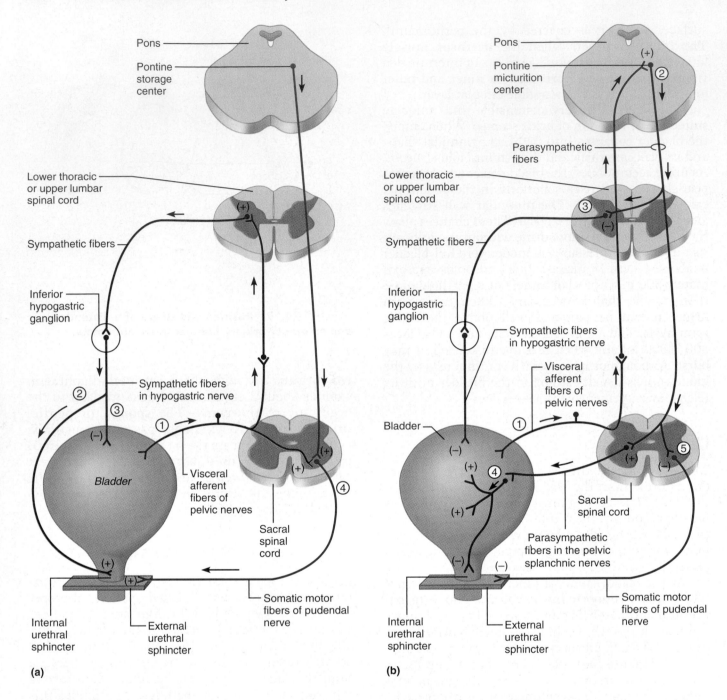

(a)

(b)

FIGURE 24.20   Neural circuits controlling continence and micturition. (a) Storage reflexes: As the bladder fills with urine, distension of the bladder wall stimulates visceral afferent fibers ①. These initiate spinal reflexes stimulate contraction of the internal and external sphincters and inhibit contraction of the detrusor muscle of the bladder, e.g., ②–④. **(b)** Voiding reflexes: Upon activation by visceral afferents ①, parasympathetic outflow via the pontine micturition center ② stimulates detrusor muscle contraction, inhibits contraction of the internal sphincter ④, and inhibits sympathetic ③ and somatic (pudendal) outflow to the bladder and external sphincter ⑤. Consequently, the bladder empties.

Micturition

Micturition (mik″tu-rish′un; *mictur* = urinate), also called **voiding** or **urination,** is the act of emptying the bladder. As urine accumulates, distension of the bladder walls activates stretch receptors there. Impulses from the activated receptors travel via visceral afferent fibers to the sacral region of the spinal cord, setting up spinal reflexes that (1) increase sympathetic inhibition of the bladder detrusor muscle and internal sphincter (temporarily) and (2) stimulate contraction of the external urethral sphincter by activating pudendal motor fibers (Figure 24.20a).

When about 200 ml of urine has accumulated, afferent impulses are transmitted to the brain, creating the urge to void. Bladder contractions become more frequent and more urgent and if it is convenient to empty the bladder (a decision made by the cerebral cortex), voiding reflexes are initiated. Visceral afferent impulses activate the *micturition center* of the dorsolateral pons. Acting as an on/off switch for micturition, this center signals the parasympathetic neurons that stimulate contraction of the detrusor muscle and relaxation of the internal and external sphincters, expelling urine (Figure 24.20b).

When one chooses not to void, reflex bladder contractions subside within a minute or so and urine continues to accumulate. Because the external sphincter is voluntarily controlled, we can choose to keep it closed and postpone bladder emptying temporarily. After another 200–300 ml or so has collected, the micturition reflex occurs again and, if urination is delayed again, is damped once more. The urge to void eventually becomes irresistible and micturition occurs when urine volume exceeds 500–600 ml, whether one wills it or not. After normal micturition, only about 10 ml of urine remains in the bladder.

🅗 HOMEOSTATIC IMBALANCE

After the toddler years, **incontinence** is usually a result of emotional problems, physical pressure during pregnancy, or nervous system problems. In *stress incontinence*, a sudden increase in intra-abdominal pressure (during laughing and coughing) forces urine through the external sphincter. This condition is common during pregnancy when the heavy uterus stretches the muscles of the pelvic floor and the urogenital diaphragm that support the external sphincter. In *overflow incontinence*, urine dribbles from the urethra whenever the bladder overfills.

In **urinary retention,** the bladder is unable to expel its contained urine. Urinary retention is normal after general anesthesia (it seems that it takes a little time for the detrusor muscle to regain its activity). Urinary retention in men often reflects hypertrophy of the prostate gland, which narrows the urethra, making it difficult to void. When urinary retention is prolonged, a slender rubber drainage tube called a **catheter** (kath'ĕ-ter) must be inserted through the urethra to drain the urine and prevent bladder trauma from excessive stretching. ●

Review Questions

Multiple Choice/Matching

(Some questions have more than one correct answer. Select the best answer or answers from the choices given.)

1. The lowest blood concentration of nitrogenous waste occurs in the (a) hepatic vein, (b) inferior vena cava, (c) renal artery, (d) renal vein.

2. The glomerular capillaries differ from other capillary networks in the body because they (a) have a larger area of anastomosis, (b) are derived from and drain into arterioles, (c) are not made of endothelium, (d) are sites of filtrate formation.

3. Damage to the renal medulla would interfere *first* with the functioning of the (a) glomerular capsules, (b) distal convoluted tubules, (c) collecting ducts, (d) proximal convoluted tubules.

4. Which is reabsorbed by the proximal convoluted tubule cells? (a) Na^+, (b) K^+, (c) amino acids, (d) all of the above.

5. Glucose is not normally found in the urine because it (a) does not pass through the walls of the glomerulus, (b) is kept in the blood by colloid osmotic pressure, (c) is reabsorbed by the tubule cells, (d) is removed by the body cells before the blood reaches the kidney.

6. Filtration at the glomerulus is directly related to (a) water reabsorption, (b) arterial blood pressure, (c) capsular hydrostatic pressure, (d) acidity of the urine.

7. Renal reabsorption (a) of glucose and many other substances is a T_m-limited active transport process, (b) of chloride is always linked to the passive transport of Na^+, (c) is the movement of substances from the blood into the nephron, (d) of sodium occurs only in the proximal tubule.

8. If a freshly voided urine sample contains excessive amounts of urochrome, it has (a) an ammonia-like odor, (b) a pH below normal, (c) a dark yellow color, (d) a pH above normal.

9. Conditions such as diabetes mellitus, starvation, and low-carbohydrate diets are closely linked to (a) ketosis, (b) pyuria, (c) albuminuria, (d) hematuria.

Short Answer Essay Questions

10. What is the importance of the adipose capsule that surrounds the kidney?

11. Trace the pathway a creatinine molecule takes from a glomerulus to the urethra. Name every microscopic or gross structure it passes through on its journey.

12. Explain the important differences between blood plasma and renal filtrate, and relate the differences to the structure of the filtration membrane.

13. Describe the mechanisms that contribute to renal autoregulation.

14. Describe what is involved in active and passive tubular reabsorption.

15. Explain how the peritubular capillaries are adapted for receiving reabsorbed substances.

16. Explain the process and purpose of tubular secretion.

17. How does aldosterone modify the chemical composition of urine?

18. Explain why the filtrate becomes hypotonic as it flows through the ascending limb of the loop of Henle. Also explain why the filtrate at the tip of the loop of Henle (and the interstitial fluid of the deep portions of the medulla) is hypertonic.

19. How does urinary bladder anatomy support its storage function?

20. Define micturition and describe the micturition reflex.

25

FLUID, ELECTROLYTE, AND ACID-BASE BALANCE

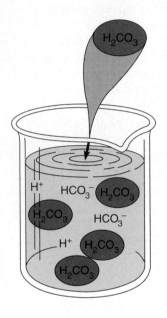

Body Fluids (pp. 883–885)

1. List the factors that determine body water content and describe the effect of each factor.

2. Indicate the relative fluid volume and solute composition of the fluid compartments of the body.

3. Contrast the overall osmotic effects of electrolytes and nonelectrolytes.

4. Describe factors that determine fluid shifts in the body.

Water Balance and ECF Osmolality (pp. 885–890)

5. List the routes by which water enters and leaves the body.

6. Describe feedback mechanisms that regulate water intake and hormonal controls of water output in urine.

7. Explain the importance of obligatory water losses.

8. Describe possible causes and consequences of dehydration, hypotonic hydration, and edema.

Electrolyte Balance (pp. 890–897)

9. Indicate the routes of electrolyte entry and loss from the body.

10. Describe the importance of ionic sodium in fluid and electrolyte balance of the body, and indicate its relationship to normal cardiovascular system functioning.

11. Describe mechanisms involved in regulating sodium and water balance.

12. Explain how potassium, calcium, and anion balance of plasma is regulated.

Acid-Base Balance (pp. 897–904)

13. List important sources of acids in the body.

14. List the three major chemical buffer systems of the body and describe how they resist pH changes.

15. Describe the influence of the respiratory system on acid-base balance.

16. Describe how the kidneys regulate hydrogen and bicarbonate ion concentrations in the blood.

17. Distinguish between acidosis and alkalosis resulting from respiratory and metabolic factors. Describe the importance of respiratory and renal compensations to acid-base balance.

Have you ever wondered why on certain days you don't urinate for hours at a time, while on others you void every few minutes? Or why on occasion you cannot seem to quench your thirst? These situations and many others reflect one of the body's most important functions: maintaining fluid, electrolyte, and acid-base balance.

Cell function depends not only on a continuous supply of nutrients and removal of metabolic wastes, but also on the physical and chemical homeostasis of the surrounding fluids. This was recognized with style in 1857 by the French physiologist Claude Bernard, who said, "It is the fixity of the internal environment which is the condition of free and independent life." In this chapter, we first examine the composition and distribution of fluids in the internal environment and then consider the roles of various body organs and functions in establishing, regulating, and altering this balance.

Body Fluids

Body Water Content

If you are a healthy young adult, water probably accounts for about half your body mass. However, not all bodies contain the same amount of water. Total body water is a function not only of age and body mass, but also of sex and the relative amount of body fat. Because of their low body fat and low bone mass, infants are 73% or more water (this high level of hydration accounts for their "dewy" skin, like that of a freshly picked peach). After infancy total body water declines throughout life, accounting for only about 45% of body mass in old age. A healthy young man is about 60% water; a healthy young woman about 50%. This difference between the sexes reflects the fact that females have relatively more body fat and relatively less skeletal muscle than males. Of all body tissues, adipose tissue is *least* hydrated (containing up to 20% water); even bone contains more water than does fat. By contrast, skeletal muscle is about 65% water. Thus, people with greater muscle mass have proportionately more body water.

Fluid Compartments

Water occupies two main **fluid compartments** within the body (Figure 25.1). A little less than two-thirds by volume is in the **intracellular fluid (ICF) compartment,** which actually consists of trillions of tiny individual "compartments": the cells. In an adult male of average size (70 kg, or 154 pounds), ICF accounts for about 25 L of the 40 L of body water. The remaining one-third or so of body water is outside cells, in the **extracellular fluid (ECF)**

compartment. The ECF constitutes the body's "internal environment" referred to by Claude Bernard and is the external environment of each cell. The ECF compartment is divisible into two *subcompartments:* (1) **plasma,** the fluid portion of blood, and (2) **interstitial fluid (IF),** the fluid in the microscopic spaces between tissue cells. There are numerous other examples of ECF that are distinct from both plasma and interstitial fluid—lymph, cerebrospinal fluid, humors of the eye, synovial fluid, serous fluid, secretions of the gastrointestinal tract—but most of these are similar to IF and are usually considered part of it.

Composition of Body Fluids

Electrolytes and Nonelectrolytes

Water serves as the *universal solvent* in which a variety of solutes are dissolved. Solutes may be classified broadly as *electrolytes* and *nonelectrolytes.* **Nonelectrolytes** have bonds (usually covalent bonds) that prevent them from dissociating in solution; therefore, no electrically charged species are created when nonelectrolytes dissolve in water. Most nonelectrolytes are organic molecules—glucose, lipids, creatinine, and urea, for example. In contrast, **electrolytes** are chemical compounds that *do* dissociate into ions in water. (See Chapter 2 if necessary to review these concepts of chemistry.) Because ions are charged particles, they can conduct an electrical current—hence the name *electrolyte.* Typically, electrolytes include inorganic salts, both inorganic and organic acids and bases, and some proteins.

Although all dissolved solutes contribute to the osmotic activity of a fluid, electrolytes have much greater osmotic power than nonelectrolytes because each electrolyte molecule dissociates into at least two ions. For example, a molecule of sodium chloride (NaCl) contributes twice as many solute particles as glucose (which remains undissociated), and a molecule of magnesium chloride ($MgCl_2$) contributes three times as many:

$$NaCl \rightarrow Na^+ + Cl^-$$ (electrolyte; two particles)

$$MgCl_2 \rightarrow Mg^{2+} + 2Cl^-$$ (electrolyte; three particles)

$$glucose \rightarrow glucose$$ (nonelectrolyte; one particle)

Regardless of the type of solute particle, water moves according to osmotic gradients—from an area of lesser osmolality to an area of greater osmolality. Thus, electrolytes have the greatest ability to cause fluid shifts.

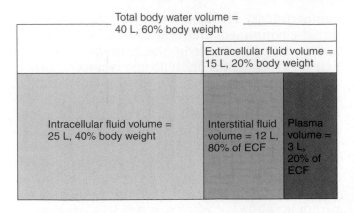

Total body water volume = 40 L, 60% body weight

Extracellular fluid volume = 15 L, 20% body weight

Intracellular fluid volume = 25 L, 40% body weight

Interstitial fluid volume = 12 L, 80% of ECF

Plasma volume = 3 L, 20% of ECF

FIGURE 25.1   The major fluid compartments of the body. [Values are for a 70-kg (154-pound) male.]

Electrolyte concentrations of body fluids are usually expressed in **milliequivalents per liter (mEq/L)**, a measure of the number of electrical charges in 1 liter of solution. The concentration of any ion in solution can be computed using the equation

$$mEq/L = \frac{ion\ concentration\ (mg/L)}{atomic\ weight\ of\ ion} \times \begin{array}{c} no.\ of \\ electrical \\ charges\ on \\ one\ ion \end{array}$$

Thus, to compute the mEq/L of sodium or calcium ions in solution in plasma, we would determine the normal concentration of these ions in plasma, look up their atomic weights in the periodic table (see Appendix D), and plug these values into the equation:

$$Na^+: \frac{3300\ mg/L}{23} \times 1 = 143\ mEq/L$$

$$Ca^{2+}: \frac{100\ mg/L}{40} \times 2 = 5\ mEq/L$$

Notice that for ions with a single charge, 1 mEq is equal to 1 mOsm, whereas 1 mEq of bivalent ions (those with a double charge like calcium) is equal to 1/2 mOsm. In either case, 1 mEq provides the same degree of reactivity.

Comparison of Extracellular and Intracellular Fluids

A quick glance at the bar graphs in Figure 25.2 reveals that each fluid compartment has a distinctive pattern of electrolytes. Except for the relatively high protein content in plasma, however, the extracellular fluids are very similar. Their chief cation is sodium, and their major anion is chloride. However, plasma contains somewhat fewer chloride ions than interstitial fluid, because the nonpenetrating plasma proteins are normally anions and plasma is electrically neutral.

In contrast to extracellular fluids, the ICF contains only small amounts of Na^+ and Cl^-. Its most abundant cation is potassium, and its major anion is HPO_4^{2-}. Cells also contain substantial quantities of soluble proteins (about three times the amount found in plasma).

Notice that sodium and potassium ion concentrations in ECF and ICF are nearly opposite (Figure 25.2). The characteristic distribution of these ions on the two sides of cellular membranes reflects the activity of cellular ATP-dependent sodium-potassium pumps, which keep intracellular Na^+ concentrations low and K^+ concentrations high. Renal mechanisms can reinforce these ion distributions by secreting K^+ into the filtrate as Na^+ is reabsorbed from the filtrate.

Electrolytes are the most abundant solutes in body fluids and determine most of their chemical and physical reactions, but they do not constitute the *bulk* of dissolved solutes in these fluids. Proteins and some of the nonelectrolytes (phospholipids, cholesterol, and neutral fats) found in the ECF are large molecules. They account for about 90% of the mass of dissolved solutes in plasma, 60% in the IF, and 97% in the ICF.

Fluid Movement Among Compartments

The continuous exchange and mixing of body fluids are regulated by osmotic and hydrostatic pressures. Although water moves freely between the compartments along osmotic gradients, solutes are unequally distributed because of their size, electrical charge, or dependence on active transport. Anything that changes the solute concentration in any compartment leads to net water flows.

Exchanges between plasma and IF occur across capillary membranes. The pressures driving these fluid movements are described in detail in Chapter 18 on pp. 632–635. Here we will simply review the outcome of these mechanisms. Nearly protein-free plasma is forced out of the blood into the interstitial space by the hydrostatic pressure of blood. This filtered fluid is then almost completely reabsorbed into the bloodstream in response to the colloid osmotic (oncotic) pressure of plasma proteins. Under normal circumstances, the small net leakage that remains behind in the interstitial space is picked up by lymphatic vessels and returned to the blood.

Exchanges between the IF and ICF are more complex because of the selective permeability of cell membranes. As a general rule, two-way osmotic flow of water is substantial. But ion fluxes are restricted and, in most cases, ions move selectively by active transport. Movements of nutrients, respiratory

? *What is the major cation in ECF? In ICF? What is the intracellular anion counterpart of ECF's chloride ions?*

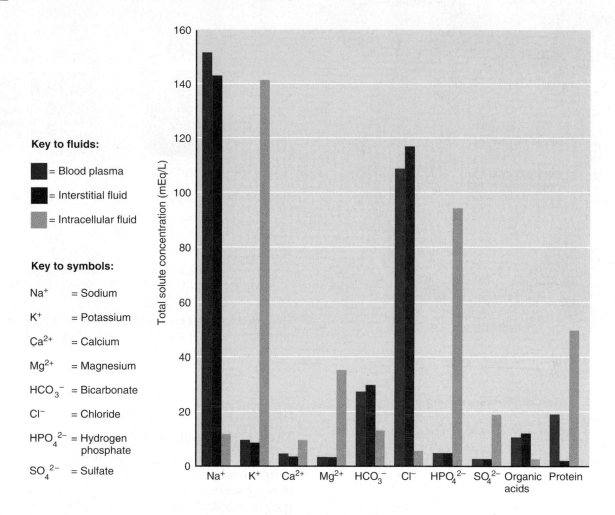

Key to fluids:

■ = Blood plasma

■ = Interstitial fluid

■ = Intracellular fluid

Key to symbols:

Na⁺   = Sodium

K⁺   = Potassium

Ca²⁺   = Calcium

Mg²⁺   = Magnesium

HCO₃⁻ = Bicarbonate

Cl⁻   = Chloride

HPO₄²⁻ = Hydrogen phosphate

SO₄²⁻ = Sulfate

FIGURE 25.2   Electrolyte composition of blood plasma, interstitial fluid, and intracellular fluid.

gases, and wastes are typically unidirectional. For example, glucose and oxygen move into the cells and metabolic wastes move out.

Plasma circulates throughout the body and links the external and internal environments (Figure 25.3). Exchanges occur almost continuously in the lungs, gastrointestinal tract, and kidneys. Although these exchanges alter plasma composition and volume, compensating adjustments in the other two fluid compartments follow quickly so that balance is restored.

Many factors can change ECF and ICF volumes. Because water moves freely between compartments, however, the osmolalities of all body fluids are equal (except during the first few minutes after a change in one of the fluids occurs). Increasing the ECF solute content (mainly the NaCl concentration) can be expected to cause osmotic and volume changes in the ICF—namely, a shift of water out of the cells. Conversely, decreasing ECF osmolality causes water to move into the cells. Thus, the ICF volume is determined by the ECF solute concentration. These concepts underlie all events that control fluid balance in the body and should be understood thoroughly.

Water Balance and ECF Osmolality

For the body to remain properly hydrated, water intake must equal water output. *Water intake* varies widely from person to person and is strongly

■ *Na⁺, K⁺, HPO₄²⁻, protein anions.*

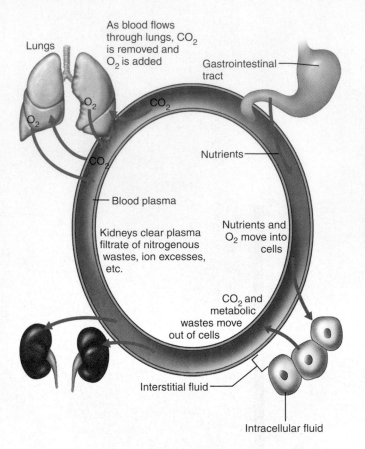

FIGURE 25.3 The continuous mixing of body fluids. Blood plasma is the communicating medium between cells' external and internal environments. Exchanges between the plasma and tissue cells (intracellular fluid) are made through the interstitial space.

influenced by habit, but it is typically about 2500 ml a day in adults (Figure 25.4). Most water enters the body through ingested liquids and solid foods. Body water produced by cellular metabolism is called **metabolic water** or **water of oxidation.**

Water output occurs by several routes. Water that vaporizes out of the lungs in expired air or diffuses directly through the skin is called **insensible water loss.** Some is lost in obvious perspiration and in feces. The balance (about 60%) is excreted by the kidneys in urine.

Healthy people have a remarkable ability to maintain the tonicity of their body fluids within very narrow limits (285–300 mOsm/L). A rise in plasma osmolality triggers (1) thirst, which prompts us to drink water, and (2) release of antidiuretic hormone (ADH), which causes the kidneys to conserve water and excrete concentrated urine. On the other hand, a decline in osmolality inhibits both thirst and ADH release, the latter followed by output of large volumes of dilute urine.

? *How would the values shown here be affected by (1) drinking a six-pack of beer? (2) a fast in which only water is ingested?*

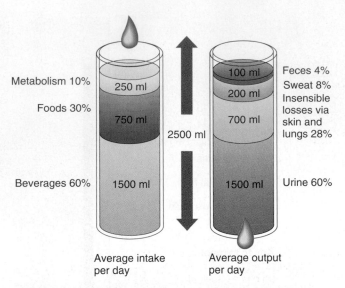

FIGURE 25.4 Major sources of water intake and output. When intake and output are in balance, the body is adequately hydrated.

Regulation of Water Intake
The Thirst Mechanism

Thirst is the driving force for water intake, but the **thirst mechanism** is poorly understood. It appears that an increase in plasma osmolality of only 2–3% results in a dry mouth and excites the hypothalamic *thirst center.* A dry mouth occurs because the rise in plasma oncotic pressure causes less fluid to leave the bloodstream. Because the salivary glands obtain the water they require from the blood, less saliva is produced. The same response is produced by a decline in blood volume (or pressure). However, because a substantial decrease (10–15%) is required, this is the less potent stimulus.

The hypothalamic thirst center is stimulated when its *osmoreceptors* lose water by osmosis to the hypertonic ECF, or are excited by baroreceptor inputs, angiotensin II, or other stimuli. Collectively, these events cause a subjective sensation of thirst,

(1) Much more water intake from beverages; much greater water output in urine (and perhaps in vomit, if too much alcohol was imbibed, an output route not illustrated). (2) Beverage intake would represent a higher percentage of water intake; water intake from foods and metabolism would drop drastically. Less urine would be made and excreted by the kidneys.

which motivates us to get a drink (Figure 25.5). This mechanism helps explain why it is that some cocktail lounges and bars provide free *salty* snacks to their patrons.

Curiously, thirst is quenched almost as soon as we begin drinking water, even though the water has yet to be absorbed into the blood. The damping of thirst begins as the mucosa of the mouth and throat is moistened and continues as stretch receptors in the stomach and intestine are activated, providing feedback signals that inhibit the thirst center. This premature quenching of thirst prevents us from drinking more than we need and overdiluting our body fluids, and allows time for the osmotic changes to come into play as regulatory factors.

As effective as thirst is, it is not always a reliable indicator of need. This is particularly true during athletic events, when thirst can be satisfied long before sufficient liquids have been drunk to maintain the body in top form. Additionally, elderly or confused people may not recognize or heed thirst signals, and fluid-overloaded renal or cardiac patients may feel thirsty despite their condition.

Regulation of Water Output

Output of certain amounts of water are unavoidable. Such **obligatory water losses** help to explain why we cannot survive for long without drinking. Even the most heroic conservation efforts by the kidneys cannot compensate for zero water intake. Obligatory water loss includes the *insensible water losses* described above, water that accompanies undigested food residues in feces, and a minimum daily **sensible water loss** of 500 ml in urine. Obligatory water loss in urine reflects the facts that (1) when we eat an adequate diet, our kidneys must excrete 900–1200 mOsm of solutes to maintain blood homeostasis, and (2) human kidneys must flush urine solutes (end products of metabolism and so forth) out of the body in water.

Beyond obligatory water loss, the solute concentration and volume of urine excreted depend on fluid intake, diet, and water loss via other avenues. For example, if you perspire profusely on a hot day, much less urine than usual has to be excreted to maintain water balance. Normally, the kidneys begin to eliminate excess water about 30 minutes after it is ingested. This delay reflects the time required to inhibit ADH release. Diuresis reaches a peak in 1 hour after drinking and then declines to its lowest level after 3 hours.

Because pretzels are salty, blood osmolality would increase, which would depress salivation, thus increasing thirst. ■

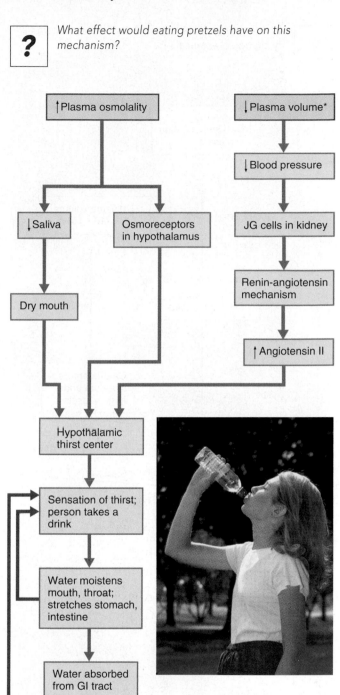

? What effect would eating pretzels have on this mechanism?

(*Minor stimulus)

Key:
- Increases, stimulates
- Reduces, inhibits
- Initial stimulus
- Physiological response
- Result

FIGURE 25.5 The thirst mechanism for regulating water intake. The major stimulus is reduced osmolality of blood plasma. (Not all effects of ADH and angiotensin II are depicted.)

At what point in this flowchart would plasma osmolality change dramatically?

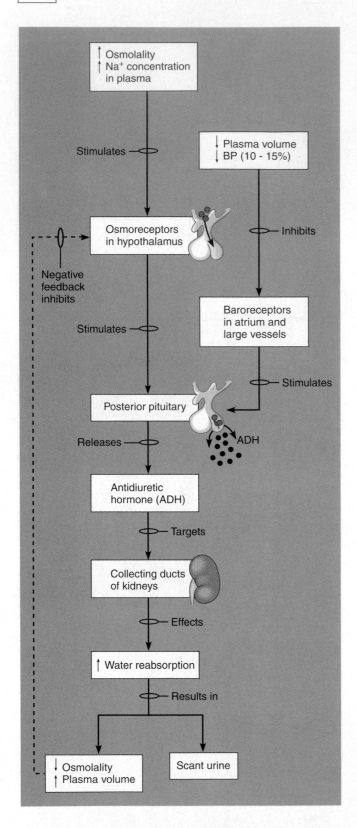

FIGURE 25.6   **Mechanisms and consequences of ADH release.**

■ At increased water absorption.

The body's water volume is closely tied to a powerful water "magnet," ionic sodium. Moreover, our ability to maintain water balance through urinary output is really a problem of sodium *and* water balance because the two are always regulated in tandem by mechanisms that serve cardiovascular function and blood pressure. However, before dealing with Na^+ issues, we will recap ADH's effect on water output.

Influence of ADH

The amount of water reabsorbed in the renal collecting ducts is proportional to ADH release. When ADH levels are low, most of the water reaching the collecting ducts is not reabsorbed but simply allowed to pass through because the number of aquaporins in the luminal membranes of the principal cells is at a minimum. The result is dilute urine and a reduced volume of body fluids. When ADH levels are high, aquaporins are inserted in the principal cell luminal membranes and nearly all of the filtered water is reabsorbed; a small volume of concentrated urine is excreted.

Osmoreceptors of the hypothalamus sense the ECF solute concentration and trigger or inhibit ADH release from the posterior pituitary accordingly (Figure 25.6). A decrease in ECF osmolality inhibits ADH release and allows more water to be excreted in urine, restoring normal Na^+ levels in the blood. An increase in ECF osmolality stimulates ADH release both directly by stimulating the hypothalamic osmoreceptors and indirectly via the renin-angiotensin mechanism. (The latter stimulus is not illustrated in Figure 25.6 but is shown in Figure 25.9.)

ADH secretion is also influenced by *large* changes in blood volume or blood pressure. Under these conditions, baroreceptors in the atria and various blood vessels sense a decrease in BP and reflexively trigger an increase in ADH secretion from the posterior pituitary. The key word here is "large" because changes in ECF osmolality are much more important as stimulatory or inhibitory factors. Factors that trigger ADH release by reducing blood volume include prolonged fever; excessive sweating, vomiting, or diarrhea; severe blood loss; and traumatic burns. Figure 25.9 summarizes how renal mechanisms involving aldosterone, angiotensin II, and ADH tie into overall controls of blood volume and blood pressure.

Disorders of Water Balance

Few people really appreciate the importance of water in keeping the body's "machinery" working at peak efficiency. The principal abnormalities of water balance are dehydration, hypotonic hydration, and edema, and each of these conditions presents a special set of problems for its victims.

Dehydration

When water output exceeds intake over a period of time and the body is in negative fluid balance, the result is **dehydration.** Dehydration is a common sequel to hemorrhage, severe burns, prolonged vomiting or diarrhea, profuse sweating, water deprivation, and diuretic abuse. Dehydration may also be caused by endocrine disturbances, such as diabetes mellitus or diabetes insipidus (see Chapter 15). Early signs and symptoms of dehydration include a "cottony" or sticky oral mucosa, thirst, dry flushed skin, and decreased urine output (*oliguria*). If prolonged, dehydration may lead to weight loss, fever, and mental confusion. Another serious consequence of water loss from plasma is inadequate blood volume to maintain normal circulation and ensuing *hypovolemic shock*.

In all these situations, water is lost from the ECF (Figure 25.7a). This is followed by the osmotic movement of water from the cells into the ECF, which equalizes the osmolality of the extracellular and intracellular fluids even though the total fluid volume has been reduced. Though the overall effect is called dehydration, it rarely involves only a water deficit, because most often electrolytes are lost as well.

Hypotonic Hydration

When the ECF osmolality starts to drop (usually this reflects a deficit of Na^+), several compensatory mechanisms are set into motion. ADH release is inhibited, and as a result, less water is reabsorbed and excess water is quickly flushed from the body in urine. But, when there is renal insufficiency or when an extraordinary amount of water is drunk very quickly, a type of cellular *overhydration* called **hypotonic hydration** may occur. In either case, the ECF is diluted—its sodium content is normal, but excess water is present. Thus, the hallmark of this condition is **hyponatremia** (low ECF Na^+), which promotes net osmosis into the tissue cells, causing them to swell as they become abnormally hydrated (Figure 25.7b). These events must be reversed quickly—for example, by intravenous administration of hypertonic mannitol, which reverses the osmotic gradient and "pulls" water out of the cells. Otherwise, the resulting electrolyte dilution leads to severe metabolic disturbances evidenced by nausea, vomiting, muscular cramping, and cerebral edema. Hypotonic hydration is particularly damaging to neurons. Uncorrected cerebral edema quickly leads to disorientation, convulsions, coma, and death.

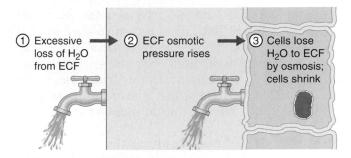

① Excessive loss of H_2O from ECF → ② ECF osmotic pressure rises → ③ Cells lose H_2O to ECF by osmosis; cells shrink

(a) Mechanism of dehydration

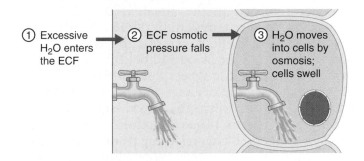

① Excessive H_2O enters the ECF → ② ECF osmotic pressure falls → ③ H_2O moves into cells by osmosis; cells swell

(b) Mechanism of hypotonic hydration

FIGURE 25.7 **Disturbances in water balance.**

Edema

Edema (ĕ-de′mah; "a swelling") is an atypical accumulation of fluid in the interstitial space, leading to tissue swelling. Edema may be caused by any event that steps up the flow of fluid out of the blood or hinders its return.

Factors that accelerate fluid loss from the blood include increased blood pressure and capillary permeability. Increased capillary hydrostatic pressure can result from incompetent venous valves, localized blood vessel blockage, congestive heart failure, or high blood volume. Whatever the cause, the abnormally high capillary hydrostatic pressure intensifies filtration at the capillary beds.

Increased capillary permeability is usually due to an ongoing inflammatory response. Recall from p. 675 that inflammatory chemicals cause local capillaries to become very porous, allowing large amounts of exudate (containing not only clotting proteins but also other plasma proteins, nutrients, and immune elements) to form.

Edema caused by hindered fluid return to the blood usually reflects an imbalance in the colloid osmotic pressures on the two sides of the capillary membranes. For example, **hypoproteinemia** (hi″po-pro″te-ĭ-ne′me-ah), a condition of unusually low levels of plasma proteins, results in tissue edema because protein-deficient plasma has an abnormally low colloid osmotic pressure. Fluids are forced out of the capillary beds at the arterial ends by blood pressure as usual, but fail to return to the blood at the venous ends. Thus, the interstitial spaces become congested with fluid. Hypoproteinemia may result from protein malnutrition, liver disease, or *glomerulonephritis* (in which plasma proteins pass through "leaky" renal filtration membranes and are lost in urine).

Although the cause differs, the result is the same when lymphatic vessels are blocked or have been surgically removed. The small amounts of plasma proteins that seep out of the bloodstream are not returned to the blood as usual. As the leaked proteins accumulate in the IF, they exert an ever-increasing colloid osmotic pressure, which draws fluid from the blood and holds it in the interstitial space. Because excess fluid in the interstitial space increases the distance nutrients and oxygen must diffuse between the blood and the cells, edema can impair tissue function. However, the most serious problems resulting from edema affect the cardiovascular system. When fluid leaves the bloodstream and accumulates in the interstitial space, both blood volume and blood pressure decline and the efficiency of the circulation is severely impaired.

Electrolyte Balance

Electrolytes include salts, acids, and bases, but the term **electrolyte balance** usually refers to the salt balance in the body. Salts are important in controlling fluid movements and provide minerals essential for excitability, secretory activity, and membrane permeability. Although many electrolytes are crucial for cellular activity, here we will specifically examine the regulation of sodium, potassium, and calcium. Acids and bases, which are more intimately involved in determining the pH of body fluids, are considered in the next section.

Salts enter the body in foods and fluids, and small amounts are generated during metabolic activity. For example, phosphates are liberated during catabolism of nucleic acids and bone matrix. Obtaining enough electrolytes is usually not a problem. Indeed, most of us have a far greater taste than need for salt. We shake table salt (NaCl) on our food even though natural foods contain ample amounts and processed foods contain exorbitant quantities. The taste for very salty foods is learned, but some liking for salt may be innate to ensure adequate intake of these two vital ions.

Salts are lost from the body in perspiration, feces, and urine. When we are salt-depleted, our perspiration is more dilute. Even so, on a hot day much more salt than usual can be lost in sweat, and gastrointestinal disorders can lead to large salt losses in feces or vomitus. Thus, the flexibility of renal mechanisms that regulate the electrolyte balance of the blood is a critical asset. Some causes and consequences of electrolyte imbalances are summarized in Table 25.1.

H   HOMEOSTATIC IMBALANCE

Severe electrolyte deficiencies prompt a craving for salty foods and often exotic foods, such as smoked meats or pickled eggs. This is common in those with *Addison's disease,* a disorder entailing deficient mineralocorticoid hormone production by the adrenal cortex. When electrolytes other than NaCl are deficient, a person may even eat substances not usually considered foods, like chalk, clay, starch, and burnt match tips. This appetite for abnormal substances is called *pica.* ●

The Central Role of Sodium in Fluid and Electrolyte Balance

Sodium holds a central position in fluid and electrolyte balance and overall body homeostasis. Indeed, regulating the balance between sodium input and output is one of the most important renal functions. The salts $NaHCO_3$ and $NaCl$ account for 90–95% of all solutes in the ECF, and they contribute about 280 mOsm of the total ECF solute concentration (300 mOsm). At its normal plasma concentration of about 142 mEq/L, Na^+ is the single most abundant cation in the ECF and the only one exerting *significant* osmotic pressure. Additionally, cellular plasma membranes are relatively impermeable to Na^+, but some does manage to diffuse in and must be pumped out against its electrochemical gradient. These two qualities give sodium the primary role in controlling ECF volume and water distribution in the body.

It is important to understand that while the sodium content of the body may change, *its ECF concentration normally remains stable* because of immediate adjustments in water volume. Remember, *water follows salt.* Furthermore, because all body fluids are in osmotic equilibrium, a change in plasma Na^+ levels affects not only plasma volume and blood pressure, but also the ICF and IF volumes. In addition, sodium ions continuously move back and forth between the ECF and body secretions. For example, about 8 L of Na^+-containing secretions (gastric, intestinal, and pancreatic juice, saliva, bile) are spewed into the digestive tract daily, only to be almost completely reabsorbed. Finally, renal

TABLE 25.1 Causes and Consequences of Electrolyte Imbalances

Ion	Abnormality (Serum Value)	Possible Causes	Consequences
Sodium	Hypernatremia (Na^+ excess in ECF: > 145 mEq/L)	Dehydration; uncommon in healthy individuals; may occur in infants or the confused aged (individuals unable to indicate thirst) or may be a result of excessive intravenous NaCl administration	Thirst: CNS dehydration leads to confusion and lethargy progressing to coma; increased neuromuscular irritability evidenced by twitching and convulsions
	Hyponatremia (Na^+ deficit in ECF: <130 mEq/L)	Solute loss, water retention, or both (e.g., excessive Na^+ loss through burned skin, excessive sweating, vomiting, diarrhea, tubal drainage of stomach, and as a result of excessive use of diuretics); deficiency of aldosterone (Addison's disease); renal disease; excess ADH release	Most common signs are those of neurologic dysfunction due to brain swelling. If sodium amounts are actually normal but water is excessive, the symptoms are the same as those of water excess: mental confusion; giddiness; coma if development occurs slowly; muscular twitching, irritability, and convulsions if the condition develops rapidly; systemic edema; congestive heart failure in cardiac patients. In hyponatremia accompanied by water loss, the main signs are decreased blood volume and blood pressure (circulatory shock)
Potassium	Hyperkalemia (K^+ excess in ECF: >5.5 mEq/L)	Renal failure; deficit of aldosterone; rapid intravenous infusion of KCl; burns or severe tissue injuries which cause K^+ to leave cells	Nausea, vomiting, diarrhea; bradycardia; cardiac arrhythmias, depression, and arrest; skeletal muscle weakness; flaccid paralysis
	Hypokalemia (K^+ deficit in ECF: <3.5 mEq/L)	Gastrointestinal tract disturbances (vomiting, diarrhea), gastrointestinal suction; chronic stress; Cushing's disease; inadequate dietary intake (starvation); hyperaldosteronism; diuretic therapy	Cardiac arrhythmias, flattened T wave; muscular weakness; alkalosis; hypoventilation; mental confusion
Phosphate	Hyperphosphatemia (HPO_4^{2-} excess in ECF: >6 mEq/L)	Increased intestinal absorption; decreased urinary loss due to renal failure; hypoparathyroidism; major tissue trauma	No direct clinical symptoms because an excess or deficit in phosphate is usually accompanied by a decrease or increase in Ca^{2+} levels
	Hypophosphatemia (HPO_4^{2-} deficit in ECF: <1 mEq/L)	Decreased intestinal absorption; increased urinary output; hyperthyroidism	
Chloride	Hyperchloremia (Cl^- excess in ECF: >105 mEq/L)	Dehydration; increased retention or intake; hyperkalemia	Metabolic acidosis due to enhanced loss of bicarbonate; weakness; stupor; rapid, deep breathing; unconsciousness
	Hypochloremia (Cl^- deficit in ECF: <95 mEq/L)	Vomiting; overhydration; hypokalemia; excessive ingestion of alkaline substances; aldosterone deficiency	Metabolic alkalosis due to bicarbonate retention
Calcium	Hypercalcemia (Ca^{2+} excess in ECF: >5.8 mEq/L or 11 mg%)*	Hyperparathyroidism; excessive vitamin D; prolonged immobilization; renal disease (decreased excretion); malignancy; Paget's disease; Cushing's disease accompanied by osteoporosis	Bone wasting, pathological fractures; flank and deep thigh pain; kidney stones, nausea and vomiting, cardiac arrhythmias and arrest; depressed respiration, coma
	Hypocalcemia (Ca^{2+} deficit in ECF: <4.5 mEq/L or 9 mg%)*	Burns (calcium trapped in damaged tissues); increased renal excretion in response to stress and increased protein intake; diarrhea; vitamin D deficiency; alkalosis	Tingling of fingers, tremors, tetany, convulsions; depressed excitability of the heart, bleeder's disease
Magnesium	Hypermagnesemia (Mg^{2+} excess in ECF: >6 mEq/L)	Rare (occurs when Mg is not excreted normally); deficiency of aldosterone; excessive ingestion of Mg^{2+}-containing antacids	Lethargy; impaired CNS functioning, coma, respiratory depression
	Hypomagnesemia (Mg^{2+} deficit in ECF: <1.4 mEq/L)	Alcoholism; loss of intestinal contents, severe malnutrition; diuretic therapy	Tremors, increased neuromuscular excitability, convulsions

* 1 mg% = 1 mg/100 ml

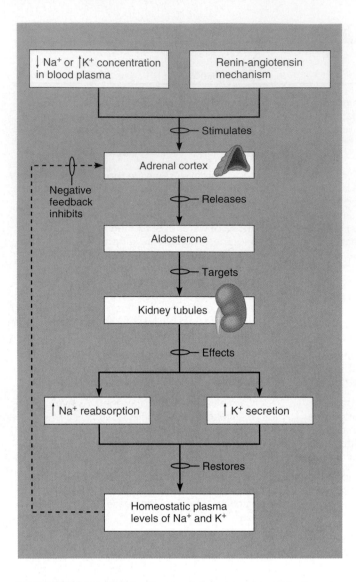

FIGURE 25.8 Mechanisms and consequences of aldosterone release.

acid-base control mechanisms (discussed shortly) are coupled to Na$^+$ transport.

Regulation of Sodium Balance

Despite the crucial importance of sodium, receptors that specifically monitor Na$^+$ levels in body fluids have yet to be found. Regulation of the Na$^+$-water balance is inseparably linked to blood pressure and volume, and involves a variety of neural and hormonal controls. Reabsorption of Na$^+$ does *not* exhibit a transport maximum, and in healthy individuals nearly all Na$^+$ in the urinary filtrate is reabsorbed. We will begin our coverage of sodium balance by reviewing the regulatory effect of aldosterone. Then we will examine various feedback loops that interact to regulate sodium and water balance and blood pressure.

Influence of Aldosterone

The hormone **aldosterone** "has the most to say" about renal regulation of sodium ion concentrations in the ECF. But whether aldosterone is present or not, some 65% of the Na$^+$ in the renal filtrate is reabsorbed in the proximal tubules of the kidneys and another 25% is reclaimed in the loops of Henle (see Chapter 24).

When aldosterone concentrations are high, essentially all the remaining filtered Na$^+$ is actively reabsorbed in the distal convoluted tubules and collecting ducts. Water follows if it can, that is, if the tubule permeability has been increased by ADH. Thus, aldosterone usually promotes both sodium and water retention. When aldosterone release is inhibited, virtually no Na$^+$ reabsorption occurs beyond the distal tubule. So, although urinary excretion of large amounts of Na$^+$ *always* results in the excretion of large amounts of water as well, the reverse is *not* true. Substantial amounts of nearly sodium-free urine can be eliminated as needed to achieve water balance.

The most important trigger for aldosterone release from the adrenal cortex is the renin-angiotensin mechanism mediated by the juxtaglomerular apparatus of the renal tubules (see Figures 25.8 and 25.9). When the juxtaglomerular (JG) apparatus responds to (1) sympathetic stimulation, (2) decreased filtrate osmolality, or (3) decreased stretch (due to decreased blood pressure), the JG cells release renin. Renin catalyzes the series of reactions that produce angiotensin II, which prompts aldosterone release. Conversely, high renal blood pressure and high filtrate osmolality depress release of renin, angiotensin II, and aldosterone. The adrenal cortical cells are also directly stimulated to release aldosterone by elevated K$^+$ levels in the ECF (Figure 25.8).

Aldosterone brings about its effects slowly, over a period of hours to days. The principal effects of aldosterone are to diminish urinary output and increase blood volume. However, before these factors can change by more than a few percent, feedback mechanisms for blood volume control come into play.

HOMEOSTATIC IMBALANCE

People with Addison's disease (hypoaldosteronism) lose tremendous amounts of NaCl and water to urine. As long as they ingest adequate amounts of salt and fluids, people with this condition can avoid problems, but they are perpetually teetering on the brink of hypovolemia and dehydration. ●

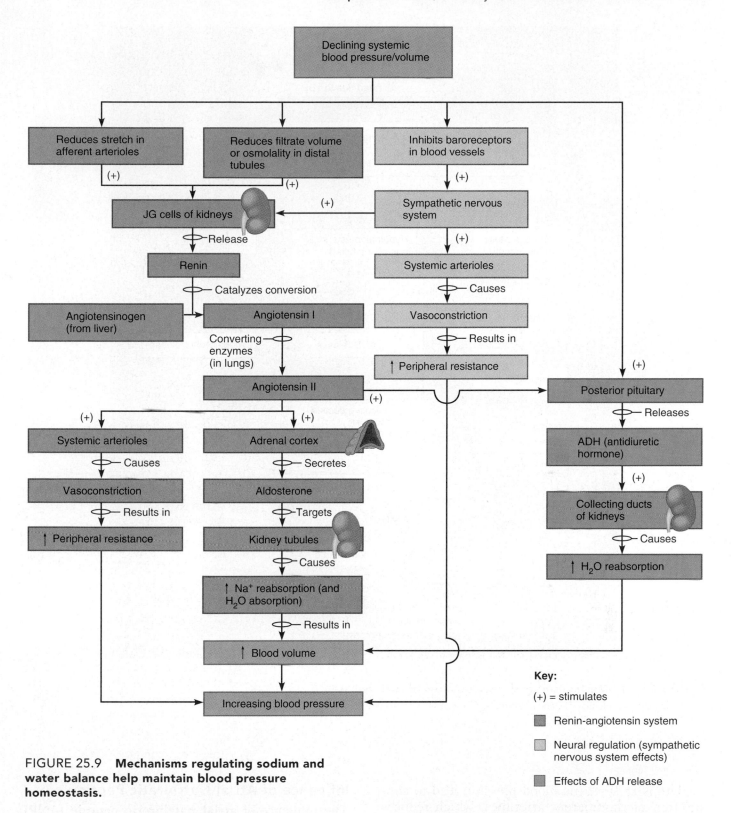

FIGURE 25.9 **Mechanisms regulating sodium and water balance help maintain blood pressure homeostasis.**

Cardiovascular Baroreceptors

Blood volume is carefully monitored and regulated to maintain blood pressure and cardiovascular function. As blood volume (hence pressure) rises, baroreceptors in the heart and in the large vessels of the neck and thorax (carotid arteries and aorta) alert the cardiovascular centers in the brain stem. Shortly after, sympathetic nervous system impulses to the kidneys decline, allowing the afferent arterioles to dilate. As the glomerular filtration rate rises, Na^+ output and water output increase. This phenomenon, called *pressure diuresis,* reduces blood volume and blood pressure.

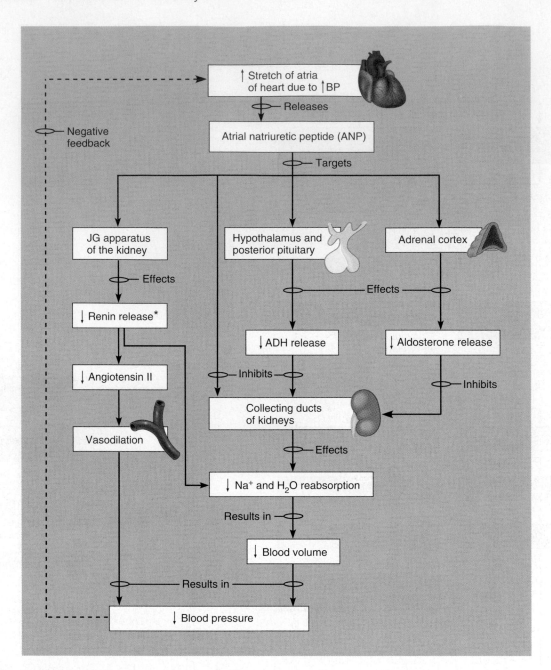

FIGURE 25.10   Mechanisms and consequences of ANP release.

*↓ renin release also inhibits ADH and aldosterone release and hence the effects of those hormones.

Drops in systemic blood pressure lead to constriction of the afferent arterioles, which reduces filtrate formation and urinary output and increases systemic blood pressure (see Figure 25.9). Thus, the baroreceptors provide information on the "fullness" or volume of the circulation that is critical for maintaining cardiovascular homeostasis. Because Na^+ concentration determines fluid volume, the baroreceptors might be regarded as "sodium receptors."

Influence of Atrial Natriuretic Peptide

The influence of **atrial natriuretic peptide (ANP)** can be summarized in one sentence: It reduces blood pressure and blood volume by inhibiting nearly all events that promote vasoconstriction and Na^+ and water retention (Figure 25.10). A hormone that is released by certain cells of the heart atria when they are stretched by the effects of elevated blood pressure, ANP has potent diuretic and natriuretic (salt-excreting) effects. It promotes excretion of Na^+ and water by the kidneys, presumably by inhibiting the ability of the collecting duct to reabsorb Na^+ and by suppressing the release of ADH, renin, and aldosterone. Additionally, ANP acts both directly and indirectly (by inhibiting renin-induced generation of angiotensin II) to relax vascular smooth muscle; thus it causes vasodilation. Collectively, these effects reduce blood pressure.

Influence of Other Hormones

Female Sex Hormones   The **estrogens** are chemically similar to aldosterone and, like aldosterone, enhance NaCl reabsorption by the renal tubules. Because water follows, many women retain fluid as their estrogen levels rise during the menstrual cycle. The edema experienced by many pregnant women is also largely due to the effect of estrogens. **Progesterone** appears to decrease Na^+ reabsorption by blocking the effect aldosterone has on the renal tubules. Thus, progesterone has a diuretic-like effect and promotes Na^+ and water loss.

Glucocorticoids   The usual effect of **glucocorticoids**, such as cortisol and hydrocortisol, is to enhance tubular reabsorption of Na^+, but they also promote an increased glomerular filtration rate that may mask their effects on the tubules. However, when their plasma levels are high, the glucocorticoids exhibit potent aldosterone-like effects and promote edema.

Regulation of Potassium Balance

Potassium, the chief intracellular cation, is required for normal neuromuscular functioning as well as for several essential metabolic activities. Nevertheless, it can be extremely dangerous. Because the relative ICF-ECF potassium concentration directly affects a cell's resting membrane potential, even slight changes in K^+ concentration in the ECF have profound effects on neurons and muscle fibers. K^+ excess in the ECF decreases their membrane potential, causing depolarization, often followed by reduced excitability. Too little K^+ in the ECF causes hyperpolarization and nonresponsiveness. The heart is particularly sensitive to K^+ levels. Both too much and too little K^+ (hyperkalemia and hypokalemia respectively) can disrupt electrical conduction in the heart, leading to sudden death (Table 25.1).

Potassium is also part of the body's buffer system, which resists changes in the pH of body fluids. Shifts of hydrogen ions (H^+) into and out of cells induce corresponding shifts of K^+ in the opposite direction to maintain cation balance. Thus, ECF potassium levels rise with acidosis, as K^+ leaves and H^+ enters the cells, and fall with alkalosis, as K^+ moves into the cells and H^+ leaves them to enter the ECF. Although these pH-driven shifts do not change the total amount of K^+ in the body, they can seriously interfere with the activity of excitable cells.

Regulatory Site: The Cortical Collecting Duct

Like Na^+ balance, K^+ balance is maintained chiefly by renal mechanisms. However, there are important differences in the way this balance is achieved. The amount of Na^+ reabsorbed in the tubules is precisely tailored to need, and Na^+ is *never* secreted into the filtrate.

In contrast, the proximal tubules predictably reabsorb about 55% of the filtered K^+, and the thick ascending limb of Henle's loop absorbs another 30% or so, leaving less than 15% to be lost in urine regardless of need. The responsibility for K^+ balance falls chiefly on the cortical collecting ducts, and is accomplished mainly by changing the amount of K^+ *secreted* into the filtrate.

As a rule, K^+ levels in the ECF are excessive, and the rate at which the principal cells of the cortical collecting ducts secrete K^+ into the filtrate is accelerated over basal levels. (At times, the amount of K^+ excreted may actually exceed the amount filtered.) When ECF potassium concentrations are abnormally low, K^+ moves from the tissue cells into the ECF and the renal principal cells conserve K^+ by reducing its secretion and excretion to a minimum. Note that these principal cells are the same cells that mediate aldosterone-induced reabsorption of Na^+ and the ADH-stimulated reabsorption of water. Additionally, *type A intercalated cells,* a unique population of collecting duct cells, can reabsorb some of the K^+ left in the filtrate (in conjunction with active secretion of H^+), thus helping to reestablish K^+ (and pH) balance. However, keep in mind that the main thrust of renal regulation of K^+ is to *excrete* it. Because the kidneys have a limited ability to retain K^+, it may be lost in urine even in the face of a deficiency. Consequently, failure to ingest potassium-rich substances eventually results in a severe deficiency.

Influence of Plasma Potassium Concentration

The single most important factor influencing K^+ secretion is the K^+ concentration in blood plasma. A high-potassium diet increases the K^+ content of the ECF. This favors entry of K^+ into the principal cells of the cortical collecting duct and prompts them to secrete K^+ into the filtrate so that more of it is excreted. Conversely, a low-potassium diet or accelerated K^+ loss depresses its secretion (and promotes its limited reabsorption) by the collecting ducts.

Influence of Aldosterone

The second factor influencing K^+ secretion into the filtrate is aldosterone. As it stimulates the principal cells to reabsorb Na^+, aldosterone simultaneously enhances K^+ secretion (see Figure 25.8). Thus, as plasma Na^+ levels rise, K^+ levels fall proportionately.

Adrenal cortical cells are *directly* sensitive to the K^+ content of the ECF bathing them. When it increases even slightly, the adrenal cortex is strongly stimulated to release aldosterone, which increases K^+ secretion by the exchange process just described. Thus, K^+ controls its own concentrations in the ECF via feedback regulation of aldosterone release. Aldosterone is also secreted in response to the renin-angiotensin mechanism previously described.

Ⓗ HOMEOSTATIC IMBALANCE

In an attempt to reduce NaCl intake, many people have turned to salt substitutes, which are high in potassium. However, heavy consumption of these substitutes is safe only when aldosterone release in the body is normal. In the absence of aldosterone, hyperkalemia is swift and lethal regardless of K^+ intake (Table 25.1). Conversely, when a person has an adrenocortical tumor that pumps out tremendous amounts of aldosterone, ECF potassium levels fall so low that neurons all over the body hyperpolarize and paralysis occurs. ●

Regulation of Calcium and Phosphate Balance

About 99% of the body's calcium is found in bones in the form of calcium phosphate salts, which provide strength and rigidity to the skeleton. Ionic calcium in the ECF is important for normal blood clotting, cell membrane permeability, and secretory behavior. Like Na^+ and K^+, ionic calcium has a potent effect on neuromuscular excitability. Hypocalcemia increases excitability and causes muscle tetany. Hypercalcemia is equally dangerous because it inhibits neurons and muscle cells and may cause life-threatening cardiac arrhythmias (Table 25.1).

Calcium ion levels are closely regulated, rarely deviating from normal limits. Under normal circumstances about 98% of the filtered Ca^{2+} is reabsorbed owing to the interaction of two hormones—parathyroid hormone and calcitonin. The bony skeleton provides a dynamic reservoir from which calcium and phosphate can be withdrawn or deposited to maintain the balance of these electrolytes.

Influence of Parathyroid Hormone

The most important controls of Ca^{2+} homeostasis are exerted by **parathyroid hormone (PTH)**, released by the tiny parathyroid glands located on the posterior aspect of the thyroid gland in the pharynx. Declining plasma levels of Ca^{2+} directly stimulate the parathyroid glands to release PTH, which promotes an increase in calcium levels by targeting the following organs:

1. **Bones.** PTH activates osteoclasts (bone-digesting cells), which break down the bone matrix, resulting in the release of Ca^{2+} and HPO_4^{2-} to the blood.
2. **Small intestine.** PTH enhances intestinal absorption of Ca^{2+} indirectly by stimulating the kidneys to transform vitamin D to its active form, which is a necessary cofactor for Ca^{2+} absorption by the small intestine.
3. **Kidneys.** PTH increases Ca^{2+} reabsorption by the renal tubules while decreasing phosphate ion reabsorption. Thus, calcium conservation and phosphate excretion go hand in hand. Hence, the *product* of Ca^{2+} and HPO_4^{2-} concentrations in the ECF remains constant, preventing calcium-salt deposit in bones or soft body tissues.

Some Ca^{2+} is reabsorbed passively in the PCT via diffusion through the paracellular route (a process driven by the electrochemical gradient). However, PTH-dependent Ca^{2+} reabsorption occurs mainly in the DCT and is driven by a Ca^{2+}-ATPase pump.

As a rule, 75% of the filtered phosphate ions (including $H_2PO_4^-$, HPO_4^{2-}, and PO_4^{3-}) are reabsorbed in the PCT by active transport. Phosphate reabsorption is regulated by its transport maximum. Amounts present in excess of that maximum simply flow out in urine. PTH inhibits active transport of phosphate by decreasing its T_m.

When ECF calcium levels are within normal limits (9–11 mg/100 ml blood) or high, PTH secretion is inhibited. Consequently, release of Ca^{2+} from bone is inhibited, larger amounts of calcium are lost in feces and urine, and more phosphate is retained. Hormones other than PTH alter phosphate reabsorption. For example, insulin increases it while glucagon decreases it.

Influence of Calcitonin

Produced by the parafollicular cells of the thyroid gland, the hormone **calcitonin** is released in response to rising blood Ca^{2+} levels. Calcitonin targets bone, where it encourages deposit of calcium salts and inhibits bone reabsorption. Calcitonin is an antagonist of PTH, but its contribution to calcium and phosphate homeostasis is minor to negligible compared to that of PTH.

Regulation of Anions

Chloride is the major anion accompanying Na^+ in the ECF and, like sodium, Cl^- helps maintain the osmotic pressure of the blood. When blood pH is within normal limits or slightly alkaline, about 99% of filtered Cl^- is reabsorbed. In the PCT, it moves passively and simply follows sodium ions out of the filtrate and into the peritubular capillary blood. In most other tubule segments, Na^+ and Cl^- transport are coupled.

When acidosis occurs, less Cl^- accompanies Na^+ because HCO_3^- reabsorption is stepped up to restore blood pH to its normal range. Thus, the choice between Cl^- and HCO_3^- serves acid-base regulation. Most other anions, such as sulfates and nitrates, have transport maximums, and when their concentrations in the filtrate exceed their renal thresholds, excesses spill into urine.

Acid-Base Balance

Because of their abundant hydrogen bonds, all functional proteins (enzymes, hemoglobin, cytochromes, and others) are influenced by H^+ concentration. It follows then that nearly all biochemical reactions are influenced by the pH of their fluid environment, and the acid-base balance of body fluids is closely regulated. (For a review of the basic principles of acid-base reactions and pH, see Chapter 2.)

Optimal pH varies from one body fluid to another, but not by much. The normal pH of arterial blood is 7.4, that of venous blood and IF is 7.35, and that of ICF averages 7.0. The lower pH in cells and venous blood reflects their greater amounts of acidic metabolites and carbon dioxide, which combines with water to form carbonic acid, H_2CO_3.

Whenever the pH of arterial blood rises above 7.45, a person is said to have **alkalosis** (al"kah-lo'sis) or **alkalemia.** A drop in arterial pH to below 7.35 results in **acidosis** (as"ĭ-do'sis) or **acidemia.** Because pH 7.0 is neutral, chemically speaking 7.35 is not acidic. However, it is a higher-than-optimal H^+ concentration for most cells, so any arterial pH between 7.35 and 7.0 is called **physiological acidosis.**

Although small amounts of acidic substances enter the body via ingested foods, most hydrogen ions originate as metabolic by-products or end products. For example, (1) breakdown of phosphorus-containing proteins releases *phosphoric acid* into the ECF, (2) anaerobic respiration of glucose produces *lactic acid*, (3) fat metabolism yields other organic acids, such as fatty acids and *ketone bodies*, and (4) the loading and transport of carbon dioxide in the blood as HCO_3^- liberates hydrogen ions. Finally, although hydrochloric acid produced by the stomach is not really *in* the body, it is a source of H^+ that must be buffered if digestion is to occur normally in the small intestine.

The H^+ concentration in blood is regulated sequentially by (1) chemical buffers, (2) the brain stem respiratory center, and (3) renal mechanisms. Chemical buffers act within a fraction of a second to resist pH changes and are the first line of defense. Within 1–3 minutes, changes in respiratory rate and depth are occurring to compensate for acidosis or alkalosis. The kidneys, the body's most potent acid-base regulatory system, ordinarily require hours to a day or more to effect changes in blood pH.

Chemical Buffer Systems

Recall that acids are *proton donors*, and that the acidity of a solution reflects only the *free* hydrogen ions, not those bound to anions. *Strong acids*, which dissociate completely and liberate all their H^+ in water (Figure 25.11a), can dramatically change a solution's pH. By contrast, *weak acids* dissociate only partially (Figure 25.11b), and so have a much slighter effect on pH. However, weak acids are efficient at preventing pH changes, and this feature allows them to play important roles in chemical buffer systems.

Bases are *proton acceptors*. Strong bases are those that dissociate easily in water and quickly tie up H^+. Conversely, weak bases are slower to accept protons.

A **chemical buffer** is a system of one or two molecules that acts to resist changes in pH when a strong acid or base is added. They do this by binding to H^+ whenever the pH drops and releasing them when pH rises. The three major chemical buffer systems in the body are the *bicarbonate, phosphate,* and *protein buffer systems.* Anything that causes a shift in H^+ concentration in one fluid compartment simultaneously causes a change in the others. Thus, the buffer systems actually buffer one another, so that any drifts in pH are resisted by the *entire* buffer system.

Bicarbonate Buffer System

The **bicarbonate buffer system** is a mixture of carbonic acid (H_2CO_3) and its salt, sodium bicarbonate ($NaHCO_3$), in the same solution. Although it also buffers the ICF, it is the *only* important *ECF* buffer.

Carbonic acid, a weak acid, does not dissociate to any great extent in neutral or acidic solutions. Thus, when a strong acid such as HCl is added to this buffer system, most of the existing carbonic acid remains intact. However, the bicarbonate ions of the salt act as weak bases to tie up the H^+ released by the stronger acid (HCl), forming *more* carbonic acid:

$$HCl + NaHCO_3 \rightarrow H_2CO_3 + NaCl$$
strong acid   weak base   weak acid   salt

To prevent the pH shift that occurs in situation (a), would it be better to add a strong base or a weak base? Why?

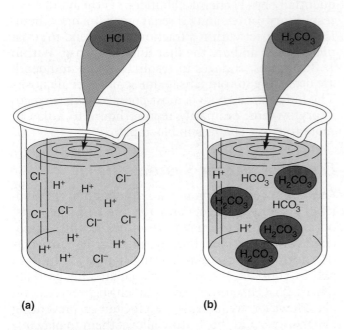

(a)   **(b)**

FIGURE 25.11   Dissociation of strong and weak acids. (a) When added to water, the strong acid HCl dissociates completely into its ions (H^+ and Cl^-). **(b)** By contrast, dissociation of H_2CO_3, a weak acid, is very incomplete, and some molecules of H_2CO_3 remain undissociated (symbols shown in green circles) in solution.

Because it is converted to the weak acid H_2CO_3, HCl lowers the pH of the solution only slightly.

When a strong base such as sodium hydroxide (NaOH) is added to the same buffer solution, a weak base such as sodium bicarbonate ($NaHCO_3$) does not dissociate under the alkaline conditions and so does not contribute to the rise in pH. However, the added base forces the carbonic acid to dissociate further, donating more H^+ to tie up the OH^- released by the strong base:

$$NaOH \quad + \quad H_2CO_3 \quad \rightarrow \quad NaHCO_3 \quad + \quad H_2O$$
strong base   weak acid   weak base   water

The net result is replacement of a strong base ($NaOH$) by a weak one ($NaHCO_3$), so that the pH of the solution rises very little.

Although the bicarbonate salt in the example is sodium bicarbonate, other bicarbonate salts function

in the same way because the HCO_3^- is the important ion, not the cation it is paired with. In cells, where little Na^+ is present, potassium and magnesium bicarbonates are part of the bicarbonate buffer system.

The buffering power of this type of system is directly related to the concentrations of the buffering substances. Thus, if acids enter the blood at such a rate that all the available HCO_3^-, often referred to as the **alkaline reserve,** are tied up, the buffer system becomes ineffective and blood pH changes. The bicarbonate ion concentration in the ECF is normally around 25 mEq/L and is closely regulated by the kidneys. The concentration of H_2CO_3 is just over 1 mEq/L but the supply of carbonic acid (which comes from the CO_2 released during cellular respiration) is almost limitless, so obtaining that member of the buffer pair is usually not a problem. The carbonic acid content of the blood is subject to respiratory controls.

Phosphate Buffer System

The operation of the **phosphate buffer system** is nearly identical to that of the bicarbonate buffer. The components of the phosphate system are the sodium salts of dihydrogen phosphate ($H_2PO_4^-$) and monohydrogen phosphate (HPO_4^{2-}). NaH_2PO_4 acts as a weak acid. Na_2HPO_4, with one less hydrogen atom, acts as a weak base.

Again, H^+ released by strong acids is tied up in weak acids:

$$HCl \quad + \quad Na_2HPO_4 \quad \rightarrow \quad NaH_2PO_4 \quad + \quad NaCl$$
strong acid   weak base   weak acid   salt

and strong bases are converted to weak bases:

$$NaOH \quad + \quad NaH_2PO_4 \quad \rightarrow \quad Na_2HPO_4 \quad + \quad H_2O$$
strong base   weak acid   weak base   water

Because the phosphate buffer system is present in low concentrations in the ECF (approximately one-sixth that of the bicarbonate buffer system), it is relatively unimportant for buffering blood plasma. However, it is a very effective buffer in urine and in ICF, where phosphate concentrations are usually higher.

Protein Buffer System

Proteins in plasma and in cells are the body's most plentiful and powerful source of buffers, and constitute the **protein buffer system.** In fact, at least three-quarters of all the buffering power of body fluids resides in cells, and most of this reflects the buffering activity of intracellular proteins.

As described in Chapter 2, proteins are polymers of amino acids. Some of the linked amino acids have exposed groups of atoms called *organic acid*

A weak base, because it would act as a buffer by tying up H^+. ■

(carboxyl) groups (—COOH), which dissociate to release H^+ when the pH begins to rise:

$$R^\star—COOH \rightarrow R—COO^- + H^+$$

Other amino acids have exposed groups that can act as bases and accept H^+. For example, an exposed —NH_2 group can bind with hydrogen ions, becoming —NH_3^+:

$$R—NH_2 + H^+ \rightarrow R—NH_3^+$$

Because this removes free hydrogen ions from the solution, it prevents the solution from becoming too acidic. Consequently, a single protein molecule can function reversibly as either an acid or a base depending on the pH of its environment. Molecules with this ability are called **amphoteric molecules** (am"fo-ter'ik).

Hemoglobin of red blood cells is an excellent example of a protein that functions as an intracellular buffer. As explained earlier, CO_2 released from the tissues forms H_2CO_3, which dissociates to liberate H^+ and HCO_3^- in the blood. Meanwhile, hemoglobin is unloading oxygen, becoming reduced hemoglobin, which carries a negative charge. Because H^+ rapidly binds to the hemoglobin anions, pH changes are minimized. In this case, carbonic acid, a weak acid, is buffered by an even weaker acid, hemoglobin.

Respiratory Regulation of H^+

Respiratory system regulation of acid-base balance provides a *physiological buffering system*. Although such a buffer system acts more slowly than a chemical buffer system, it has one to two times the buffering power of all the body's chemical buffers combined. As described in Chapter 21, the respiratory system eliminates CO_2 from the blood while replenishing its supply of O_2. Carbon dioxide generated by cellular respiration enters erythrocytes in the circulation and is converted to bicarbonate ions for transport in the plasma:

$$CO_2 + H_2O \overset{\text{carbonic}}{\underset{\text{anhydrase}}{\rightleftharpoons}} \underset{\substack{\text{carbonic}\\\text{acid}}}{H_2CO_3} \rightleftharpoons \underset{\substack{\text{bicarbonate}\\\text{ion}}}{H^+ + HCO_3^-}$$

The first set of double arrows indicates a reversible equilibrium between dissolved carbon dioxide and water on the left and carbonic acid on the right. The second set indicates a reversible equilibrium between carbonic acid on the left and hydrogen and bicarbonate ions on the right. Because of these equilibria, an increase in any of these chemical species pushes the reaction in the opposite direction. Notice also that the right side of the equation is equivalent to the bicarbonate buffer system.

In healthy individuals, CO_2 is expelled from the lungs at the same rate it is formed in the tissues. During carbon dioxide unloading, the reaction shifts to the left, and H^+ generated from carbonic acid is reincorporated into water. Because of the protein buffer system, H^+ produced by CO_2 transport is not allowed to accumulate and has little or no effect on blood pH. However, when hypercapnia occurs, it activates medullary chemoreceptors (via cerebrospinal fluid acidosis promoted by excessive accumulation of CO_2) that respond by increasing respiratory rate and depth. Additionally, a rising plasma H^+ concentration resulting from any metabolic process excites the respiratory center indirectly (via peripheral chemoreceptors) to stimulate deeper, more rapid respiration. As ventilation increases, more CO_2 is removed from the blood, pushing the reaction to the left and reducing the H^+ concentration.

When blood pH rises, the respiratory center is depressed. As respiratory rate drops and respiration becomes shallower, CO_2 accumulates; the equilibrium is pushed to the right, causing the H^+ concentration to increase. Again blood pH is restored to the normal range. These respiratory system–mediated corrections of blood pH are accomplished within a minute or so.

Changes in alveolar ventilation can produce dramatic changes in blood pH—far more than is needed. For example, a doubling or halving of alveolar ventilation can raise or lower blood pH by about 0.2 pH unit. Because normal arterial pH is 7.4, a change of 0.2 pH unit yields a blood pH of 7.6 or 7.2—both well beyond the normal limits of blood pH. Actually, alveolar ventilation can be increased about 15-fold or reduced to zero. Thus, respiratory controls of blood pH have a tremendous reserve capacity.

Anything that impairs respiratory system functioning causes acid-base imbalances. For example, net carbon dioxide retention leads to acidosis; hyperventilation, which causes net elimination of CO_2, can cause alkalosis. When the cause of the pH imbalance is respiratory system problems, the resulting condition is either **respiratory acidosis** or **respiratory alkalosis** (see Table 25.2 on p. 903).

Renal Mechanisms of Acid-Base Balance

Chemical buffers can tie up excess acids or bases temporarily, but they cannot eliminate them from the body. And while the lungs can dispose of the **volatile**

---
* Note that R indicates the rest of the organic molecule, which contains many atoms.

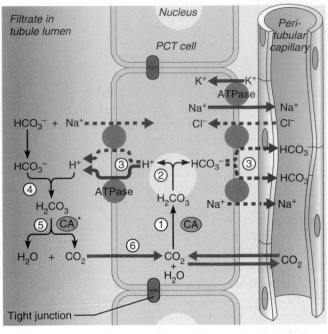

FIGURE 25.12   **Reabsorption of filtered HCO_3^- is coupled to H^+ secretion.** (Numbers by reaction lines indicate the sequence of events.)

① CO_2 combines with water within the tubule cell, forming H_2CO_3.

② H_2CO_3 is quickly split, forming H^+ and HCO_3^-.

③ For each H^+ secreted into the filtrate, a bicarbonate ion (HCO_3^-) enters the peritubular capillary blood by cotransport (either via symport with Na^+ or via antiport with Cl^-).

④ Secreted H^+ can combine with HCO_3^- present in the tubular filtrate, forming carbonic acid (H_2CO_3). Thus, HCO_3^- disappears from the filtrate at the same rate that HCO_3^- (formed within the tubule cell) is entering the peritubular capillary blood.

⑤ The H_2CO_3 formed in the filtrate dissociates to release carbon dioxide and water.

⑥ Carbon dioxide then diffuses into the tubule cell, where it acts to trigger further H^+ secretion.

Primary active transport processes are indicated by solid red reaction arrows; secondary active transport is indicated by a dashed red arrow; passive processes (simple diffusion and facilitated diffusion) are indicated by blue arrows.

*The breakdown of H_2CO_3 to CO_2 and H_2O in the tubule *lumen* is catalyzed by carbonic anhydrase *only* in the PCT.

Key:

➡ = Primary active transport

▪▪▶ = Secondary active transport

(CA) = Carbonic anhydrase

➡ = Passive transport (diffusion)

● = Protein carrier

---

acid carbonic acid by eliminating CO_2, only the kidneys can rid the body of other acids generated by cellular metabolism: phosphoric, uric, and lactic acids, and ketone bodies. These acids are sometimes referred to as **metabolic (fixed) acids,** but this terminology is both unfortunate and incorrect because CO_2 and hence carbonic acid are also products of metabolism. Additionally, only the kidneys can regulate blood levels of alkaline substances and renew chemical buffers that are used up in regulating H^+ levels in the ECF. (HCO_3^-, which helps regulate H^+ levels, is lost from the body when CO_2 flushes from the lungs.) Thus, the ultimate acid-base regulatory organs are the kidneys, which act slowly but surely to compensate for acid-base imbalances resulting from variations in diet or metabolism or from disease.

The most important renal mechanisms for regulating acid-base balance of the blood involve (1) conserving (reabsorbing) or generating new HCO_3^-, and (2) excreting HCO_3^-. If we look back at the equation for the operation of the carbonic acid–bicarbonate buffer system of the blood, it is obvious that losing a HCO_3^- from the body produces the same net effect as gaining a H^+, because it pushes the equation to the right, increasing the H^+ level. By the same token, generating or reabsorbing HCO_3^- is the same as losing H^+ because it pushes

the equation to the left, decreasing the H^+ level. Therefore, to reabsorb bicarbonate, H^+ has to be secreted, and when we excrete excess HCO_3^-, H^+ is retained (not secreted).

Because the mechanisms for regulating acid-base balance depend on H^+ being secreted into the filtrate, we consider that process first. Secretion of H^+ occurs mainly in the PCT and in type A intercalated cells of the collecting duct. The H^+ secreted is obtained from the dissociation of carbonic acid, created from the combination of CO_2 and water within the tubule cells (Figure 25.12), a reaction catalyzed by *carbonic anhydrase*. For each H^+ secreted into the tubule lumen, one Na^+ is reabsorbed from the filtrate, maintaining the electrochemical balance.

The rate of H^+ secretion rises and falls with CO_2 levels in the ECF. The more CO_2 in the peritubular capillary blood, the faster the rate of H^+ secretion. Because blood CO_2 levels directly relate to blood pH, this system can respond to both rising and falling H^+ concentrations. Notice that secreted H^+ can combine with HCO_3^- in the filtrate, generating CO_2 and water. In this case, H^+ is bound in water. The rising concentration of CO_2 in the filtrate creates a steep diffusion gradient for its entry into the tubule cell, where it promotes still more H^+ secretion.

Conserving Filtered Bicarbonate Ions: Bicarbonate Reabsorption

Bicarbonate ions (HCO_3^-) are an important part of the bicarbonate buffer system, the most important inorganic blood buffer. If this reservoir of base, or *alkaline reserve*, is to be maintained, the kidneys must do more than just eliminate enough hydrogen ions to counter rising blood H^+ levels. Depleted stores of HCO_3^- have to be replenished. This is more complex than it seems because the tubule cells are almost completely impermeable to the HCO_3^- in the filtrate—they cannot reabsorb them. However, the kidneys can conserve filtered HCO_3^- in a rather roundabout way, also illustrated in Figure 25.12. As you can see, dissociation of carbonic acid liberates HCO_3^- as well as H^+. While the tubule cells cannot reclaim HCO_3^- directly from the filtrate, they can and do shunt HCO_3^- generated within them as a result of H_2CO_3 splitting into the peritubular capillary blood. Accompanying HCO_3^- into the blood is the Na^+ that enters the tubule cell as H^+ is pumped out. Thus, reabsorption of HCO_3^- depends on the active secretion of H^+ via an H^+-ATPase and more importantly by a Na^+-H^+ antiporter, which uses the lumen–to–tubule cell Na^+ gradient to drive the transport process. In the filtrate H^+ combines with filtered HCO_3^-. For each filtered HCO_3^- that "disappears," a HCO_3^- generated within the tubule cells enters the blood—a one-for-one exchange. When large amounts of H^+ are secreted, correspondingly large amounts of HCO_3^- enter the peritubular blood, and the net effect is that HCO_3^- is almost completely removed from the filtrate.

Generating New Bicarbonate Ions

Two renal mechanisms commonly carried out by the type A intercalated cells of the collecting ducts generate *new* HCO_3^- that can be added to plasma. Both mechanisms involve renal excretion of acid, *via secretion and excretion* of either H^+ or ammonium ions in urine. Let's examine how these mechanisms differ.

Via Excretion of Buffered H⁺   As long as *filtered bicarbonate* is reclaimed (Figure 25.12), the secreted H^+ is *not excreted or lost* from the body in urine. Instead, the H^+ is buffered by HCO_3^- in the filtrate and ultimately becomes part of water molecules (most of which are reabsorbed).

However, once the filtered HCO_3^- is "used up" (this has usually been accomplished by the time the collecting ducts are reached), any additional H^+ secreted is excreted in urine. More often than not, this is the case.

Reclaiming filtered HCO_3^- simply restores the bicarbonate concentration of plasma that exists at the time. However, a normal diet introduces new H^+ into the body, and this additional H^+ must be coun-

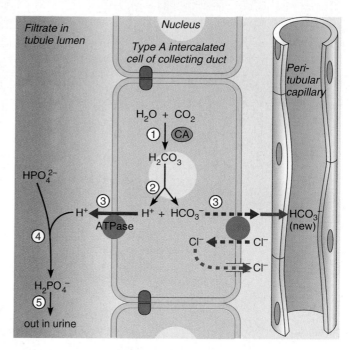

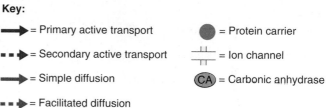

Key:

➤ = Primary active transport

▪▪▶ = Secondary active transport

➤ = Simple diffusion

▪▪▶ = Facilitated diffusion

● = Protein carrier

╫ = Ion channel

(CA) = Carbonic anhydrase

FIGURE 25.13   Generation of new HCO_3^- via buffering of excreted H^+ by monohydrogen phosphate (HPO_4^{2-}). H^+ from the dissociation of H_2CO_3 is actively secreted by a H^+-ATPase pump and combines with HPO_4^{2-} in the lumen. HCO_3^- generated at the same time leaves the basolateral membrane via an antiport carrier in a HCO_3^--Cl^- exchange process which maintains electroneutrality in the duct cells.

teracted by the generation of new HCO_3^- (as opposed to filtered HCO_3^-) which moves into the blood to counteract acidosis. This process of alkalinizing the blood is the way the kidneys compensate for acidosis. The excreted H^+ also must bind with buffers in the filtrate. Otherwise, a urine pH incompatible with life would result. (H^+ secretion ceases when urine pH falls to 4.5.) The most important urine buffer is the *phosphate buffer system*, specifically its weak base *monohydrogen phosphate* (HPO_4^{2-}).

The components of the phosphate buffer system filter freely into the tubules, and about 75% of the filtered phosphate is reabsorbed. However, their absorption is inhibited during acidosis; therefore, the buffer pair becomes more and more concentrated as the filtrate moves through the renal tubules. As shown in Figure 25.13, the type A intercalated cells secrete H^+ actively via a H^+-ATPase pump and via a

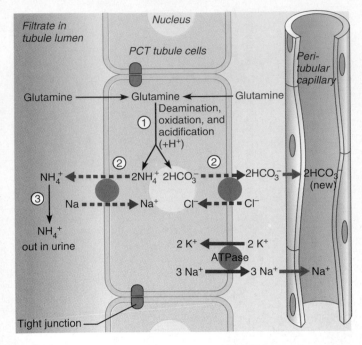

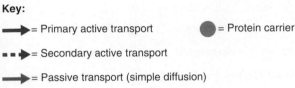

Key:

➡ = Primary active transport   ● = Protein carrier

▪▪▶ = Secondary active transport

➡ = Passive transport (simple diffusion)

FIGURE 25.14   Generation of new HCO₃⁻ via glutamine metabolism and NH₄⁺ secretion. PCT cells metabolize glutamine to NH_4^+ and HCO_3^-. NH_4^+ (a weak acid) is then actively secreted by occupying the H^+ site on the H^+-Na^+ cotransporter. The new HCO_3^- generated moves into the peritubular capillary blood.

K^+-H^+ antiporter (not illustrated). The secreted H^+ combines with HPO_4^{2-}, forming $H_2PO_4^-$ which then flows out in urine. Bicarbonate ions generated in the cells during the same reaction move into the interstitial space via a HCO_3^--Cl^- antiport process and then move passively into the peritubular capillary blood. Notice again that when H^+ is being excreted, "brand new" bicarbonate ions are added to the blood—over and above those reclaimed from the filtrate. Thus, in response to acidosis, the kidneys generate new HCO_3^- and add it to the blood (alkalinizing the blood) while adding an equal amount of H^+ to the filtrate (acidifying the urine).

Via NH₄⁺ Excretion
The second and more important mechanism for excreting acid uses the ammonium ion (NH_4^+) produced by glutamine metabolism in the PCT cells (Figure 25.14). For each glutamine metabolized (deaminated, oxidized, and acidified by combination with H^+), two NH_4^+ and

two HCO_3^- result. The HCO_3^- moves through the basolateral membrane into the blood. Ammonium ions are weak acids that donate few H^+ at physiological pH. These, in turn, are excreted and lost in urine. As with the phosphate buffer system, this buffering mechanism replenishes the alkaline reserve of the blood, because as NH_4^+ is secreted the newly made sodium bicarbonate enters the blood.

Bicarbonate Ion Secretion
When the body is in alkalosis, another population of intercalated cells (type B) in the collecting ducts exhibit net HCO_3^- secretion (rather than net HCO_3^- reabsorption) while reclaiming H^+ to acidify the blood. Overall the type B cells can be thought of as "flipped" type A cells and the HCO_3^- secretion process can be visualized as the exact opposite of the HCO_3^- reabsorption process illustrated in Figure 25.12. However, the predominant process in the nephrons and collecting ducts is HCO_3^- reabsorption, and even during alkalosis, the amount of HCO_3^- excreted is much less than the amount conserved.

Abnormalities of Acid-Base Balance
All cases of acidosis and alkalosis can be classed according to cause as *respiratory* or *metabolic* (Table 25.2).

Respiratory Acidosis and Alkalosis
Respiratory pH imbalances result from some failure of the respiratory system to perform its normal pH-balancing role. The partial pressure of carbon dioxide (P_{CO_2}) is the single most important indicator of the adequacy of respiratory function. When respiratory function is normal, the P_{CO_2} fluctuates between 35 and 45 mm Hg. Generally speaking, higher values indicate respiratory acidosis, and lower values indicate respiratory alkalosis.

Respiratory acidosis is the most common cause of acid-base imbalance. It most often occurs when a person breathes shallowly or when gas exchange is hampered by diseases such as pneumonia, cystic fibrosis, or emphysema. Under such conditions, CO_2 accumulates in the blood. Thus, respiratory acidosis is characterized by falling blood pH and rising P_{CO_2}.

Respiratory alkalosis results when carbon dioxide is eliminated from the body faster than it is produced, causing the blood to become more alkaline. This is a common result of hyperventilation (see p. 745). Respiratory acidosis is frequently associated with pathology, but respiratory alkalosis rarely is.

TABLE 25.2   Causes and Consequences of Acid-Base Imbalances

Condition and Hallmark	Possible Causes; Comments
Metabolic acidosis uncompensated (uncorrected) (HCO_3^- <22 mEq/L; pH <7.4)	**Severe diarrhea:** bicarbonate-rich intestinal (and pancreatic) secretions rushed through digestive tract before their solutes can be reabsorbed; bicarbonate ions are replaced by renal mechanisms that generate new bicarbonate ions

Renal disease: failure of kidneys to rid body of acids formed by normal metabolic processes

Untreated diabetes mellitus: lack of insulin or inability of tissue cells to respond to insulin, resulting in inability to use glucose; fats are used as primary energy fuel, and ketoacidosis occurs

Starvation: lack of dietary nutrients for cellular fuels; body proteins and fat reserves are used for energy—both yield acidic metabolites as they are broken down for energy

Excess alcohol ingestion: results in excess acids in blood

High ECF potassium concentrations: potassium ions compete with H^+ for secretion in renal tubules; when ECF levels of K^+ are high, H^+ secretion is inhibited |
| **Metabolic alkalosis** uncompensated (HCO_3^- >26 mEq/L; pH >7.4) | **Vomiting or gastric suctioning:** loss of stomach HCl requires that H^+ be withdrawn from blood to replace stomach acid; thus H^+ decreases and HCO_3^- increases proportionally

Selected diuretics: cause K^+ depletion and H_2O loss. Low K^+ directly stimulates the tubule cells to secrete H^+. Reduced blood volume elicits the renin-angiotensin mechanism, which stimulates Na^+ reabsorption and H^+ secretion

Ingestion of excessive sodium bicarbonate (antacid): bicarbonate moves easily into ECF, where it enhances natural alkaline reserve

Constipation: prolonged retention of feces, resulting in increased amounts of HCO_3^- being reabsorbed

Excess aldosterone (e.g., adrenal tumors): promotes excessive reabsorption of Na^+, which pulls increased amount of H^+ into urine. Hypovolemia promotes the same relative effect because aldosterone secretion is increased to enhance Na^+ (and H_2O) reabsorption |
| **Respiratory acidosis** uncompensated (P_{CO_2}>45 mm Hg; pH <7.4) | **Impaired gas exchange or lung ventilation** (e.g., in chronic bronchitis, cystic fibrosis, emphysema): increased airway resistance and decreased expiratory air flow, leading to retention of carbon dioxide

Rapid, shallow breathing: tidal volume markedly reduced

Narcotic or barbiturate overdose or injury to brain stem: depression of respiratory centers, resulting in hypoventilation and respiratory arrest |
| **Respiratory alkalosis** uncompensated (P_{CO_2} <35 mm Hg; pH >7.4) | **Direct cause is always hyperventilation:** hyperventilation in pain/anxiety, asthma, pneumonia, and at high altitude represents effort to raise P_{O_2} at the expense of excessive carbon dioxide excretion

Brain tumor or injury: abnormality of respiratory controls |

Metabolic Acidosis and Alkalosis

Metabolic pH imbalances include all abnormalities of acid-base imbalance *except* those caused by too much or too little carbon dioxide in the blood. Bicarbonate ion levels below or above the normal range of 22–26 mEq/L indicate a metabolic acid-base imbalance.

The second most common cause of acid-base imbalance, **metabolic acidosis,** is recognized by low blood pH and HCO_3^- levels. Typical causes of metabolic acidosis are ingestion of too much alcohol (which is metabolized to acetic acid) and excessive loss of HCO_3^-, as might result from persistent diarrhea. Other causes are accumulation of lactic acid during exercise or shock, the ketosis that occurs in diabetic crisis or starvation, and, infrequently, kidney failure.

Metabolic alkalosis, indicated by rising blood pH and HCO_3^- levels, is much less common than metabolic acidosis. Typical causes are vomiting of the acidic contents of the stomach (or loss of those secretions through gastrointestinal suctioning), intake of excess base (antacids, for example), and constipation, in which more than the usual amount of HCO_3^- is reabsorbed by the colon.

Effects of Acidosis and Alkalosis

The absolute blood pH limits for life are a low of 7.0 and a high of 7.8. When blood pH falls below 7.0, the central nervous system is so depressed that the person goes into coma and death soon follows. When blood pH rises above 7.8, the nervous system is overexcited, leading to such characteristic signs as muscle tetany, extreme nervousness, and convulsions. Death often results from respiratory arrest.

Respiratory and Renal Compensations

When an acid-base imbalance is due to inadequate functioning of one of the physiological buffer systems (the lungs or kidneys), the other system tries to compensate. The respiratory system attempts to compensate for metabolic acid-base imbalances, and the kidneys (although much slower) work to correct imbalances caused by respiratory disease. These **respiratory** and **renal compensations** can be recognized by changes in plasma P_{CO_2} and bicarbonate ion concentrations. Because the compensations act to restore normal blood pH, the pH may actually be in the normal range when the patient has a significant medical problem.

Respiratory Compensations As a rule, changes in respiratory rate and depth are evident when the respiratory system is attempting to compensate for metabolic acid-base imbalances. In metabolic acidosis, the respiratory rate and depth are usually elevated—an indication that the respiratory centers are stimulated by the high H^+ levels. The blood pH is low (below 7.35) and the HCO_3^- level is below 22 mEq/L. As the respiratory system "blows off" CO_2 to rid the blood of excess acid, the P_{CO_2} falls below 35 mm Hg. By contrast, in respiratory acidosis, the respiratory rate is often depressed and *is the immediate cause of the acidosis* (with some exceptions such as COPD).

Respiratory compensation for metabolic alkalosis involves slow, shallow breathing, which allows CO_2 to accumulate in the blood. A metabolic alkalosis being compensated by respiratory mechanisms is revealed by a pH over 7.45 at least initially, elevated bicarbonate levels (over 26 mEq/L), and a P_{CO_2} above 45 mm Hg.

Renal Compensations When an acid-base imbalance is of respiratory origin, renal mechanisms are stepped up to correct the imbalance. For example, a hypoventilating individual will exhibit acidosis. When renal compensation is occurring, both the P_{CO_2} and the HCO_3^- levels are high. The high P_{CO_2} is the cause of the acidosis, and the rising HCO_3^- level indicates that the kidneys are retaining bicarbonate to offset the acidosis. Conversely, a person with renal-compensated respiratory alkalosis will have a high blood pH and a low P_{CO_2}. Bicarbonate ion levels begin to fall as the kidneys eliminate more HCO_3^- from the body by failing to reclaim it or by actively secreting it. Note that the kidneys cannot compensate for alkalosis or acidosis if that condition reflects a *renal* problem.

Review Questions

Multiple Choice/Matching

(Some questions have more than one correct answer. Select the best answer or answers from the choices given.)

1. Body water content is greatest in (a) infants, (b) young adults, (c) elderly adults.

2. Potassium, magnesium, and phosphate ions are the predominant electrolytes in (a) plasma, (b) interstitial fluid, (c) intracellular fluid.

3. Sodium balance is regulated primarily by control of amount(s) (a) ingested, (b) excreted in urine, (c) lost in perspiration, (d) lost in feces.

4. Water balance is regulated primarily by control of amount(s) (use choices in question 3).

5 through 10. Answer questions 5 through 10 by choosing a response from the following:

(**a**) ammonium ions   (**e**) hydrogen ions   (**h**) potassium
(**b**) bicarbonate   (**f**) magnesium   (**i**) sodium
(**c**) calcium   (**g**) phosphate   (**j**) water
(**d**) chloride

5. Two substances regulated (at least in part) by the influence of aldosterone on the kidney tubules.

6. Two substances regulated by parathyroid hormone.

7. Two substances secreted into the proximal convoluted tubules in exchange for sodium ions.

8. Part of an important chemical buffer system in plasma.

9. Two ions produced during catabolism of glutamine.

10. Substance regulated by ADH's effects on the renal tubules.

11. Which of the following factors will enhance ADH release? (a) increase in ECF volume, (b) decrease in ECF volume, (c) decrease in ECF osmolality, (d) increase in ECF osmolality.

12. The pH of blood varies directly with (a) HCO_3^-, (b) P_{CO_2}, (c) H^+, (d) none of the above.

13. In an individual with metabolic acidosis, a clue that the respiratory system is compensating is provided by (a) high blood bicarbonate levels, (b) low blood bicarbonate levels, (c) rapid, deep breathing, (d) slow, shallow breathing.

Short Answer Essay Questions

14. Name the body fluid compartments, noting their locations and the approximate fluid volume in each.

15. Describe the thirst mechanism, indicating how it is triggered and terminated.

16. Explain why and how sodium and water balance are jointly regulated.

17. Describe the role of the respiratory system in controlling acid-base balance.

18. Explain how the chemical buffer systems resist changes in pH.

19. Explain the relationship of the following to renal secretion and excretion of hydrogen ions: (a) plasma carbon dioxide levels, (b) phosphate, and (c) sodium bicarbonate reabsorption.

26

THE REPRODUCTIVE SYSTEM

1. Describe the common function of the male and female reproductive systems.

Anatomy of the Male Reproductive System (pp. 906–912)

2. Describe the structure and function of the testes, and explain the importance of their location in the scrotum.

3. Describe the structure of the penis, and indicate its role in the reproductive process.

4. Describe the location, structure, and function of the accessory reproductive organs of the male.

5. Discuss the sources and functions of semen.

Physiology of the Male Reproductive System (pp. 912–920)

6. Describe the phases of the male sexual response.

7. Define meiosis. Compare and contrast it to mitosis.

8. Outline events of spermatogenesis.

9. Discuss hormonal regulation of testicular function and the physiological effects of testosterone on male reproductive anatomy.

Anatomy of the Female Reproductive System (pp. 920–929)

10. Describe the location, structure, and function of the ovaries.

11. Describe the location, structure, and function of each of the organs of the female reproductive duct system.

12. Describe the anatomy of the female external genitalia.

13. Discuss the structure and function of the mammary glands.

Physiology of the Female Reproductive System (pp. 929–937)

14. Describe the process of oogenesis and compare it to spermatogenesis.

15. Describe ovarian cycle phases, and relate them to events of oogenesis.

16. Describe the regulation of the ovarian and uterine cycles.

17. Discuss the physiological effects of estrogens and progesterone.

18. Describe the phases of female sexual response.

Sexually Transmitted Diseases (pp. 937–938)

19. Indicate the infectious agents and modes of transmission of gonorrhea, syphilis, chlamydia, genital warts, and genital herpes.

Most organ systems of the body function almost continuously to maintain the well-being of the individual. The **reproductive system,** however, appears to "slumber" until puberty. The **primary sex organs,** or **gonads** (go'nadz; "seeds"), are the *testes* in males and the *ovaries* in females. The gonads produce sex cells, or **gametes** (gam'ēts; "spouses") and secrete **sex hormones.** The remaining reproductive structures—ducts, glands, and external genitalia (jen-ĭ-ta'le-ah)—are referred to as **accessory reproductive organs.** Although male and female reproductive organs are quite different, their common purpose is to produce offspring.

The male's reproductive role is to manufacture male gametes called *sperm* and deliver them to the female reproductive tract, where fertilization can occur. The complementary role of the female is to produce female gametes, called *ova* or *eggs.* When these events are appropriately timed, a sperm and egg fuse to form a fertilized egg, the first cell of the new individual, from which all body cells will arise. The male and female reproductive systems are equal partners in events leading up to fertilization, but once fertilization has occurred, the female uterus provides a protective environment in which the embryo develops until birth.

Sex hormones—androgens in males and estrogens and progesterone in females—play vital roles both in the development and function of the reproductive organs and in sexual behavior and drives. These hormones also influence the growth and development of many other organs and tissues of the body.

Anatomy of the Male Reproductive System

Males are males because they possess testes. The sperm-producing **testes,** or male gonads, lie within the *scrotum.* From the testes, the sperm are delivered to the body exterior through a system of ducts including (in order) the *epididymis,* the *ductus deferens,* the *ejaculatory duct,* and finally the *urethra,* which opens to the outside at the tip of the *penis.* The accessory sex glands, which empty their secretions into the ducts during ejaculation, are the *seminal vesicles, prostate gland,* and *bulbourethral glands.* Take a moment to trace the duct system in Figure 26.1, and identify the testis and accessory glands before continuing.

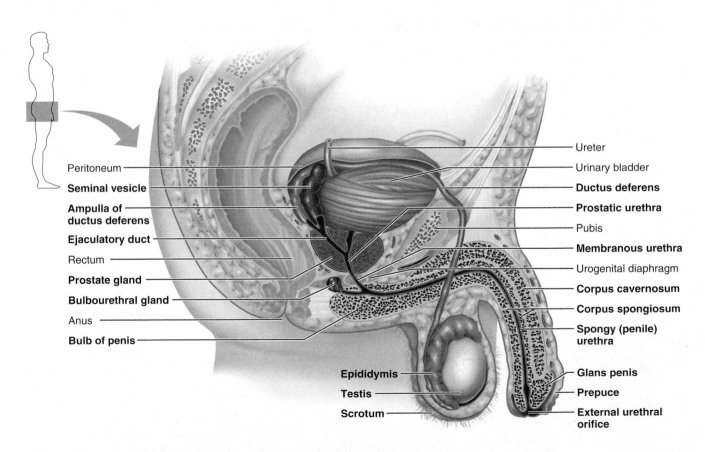

FIGURE 26.1   Reproductive organs of the male, sagittal view. A portion of the pubis of the coxal bone has been left to show the relationship of the ductus deferens to the bony pelvis. (See *A Brief Atlas of the Human Body,* Figures 57 and 58.)

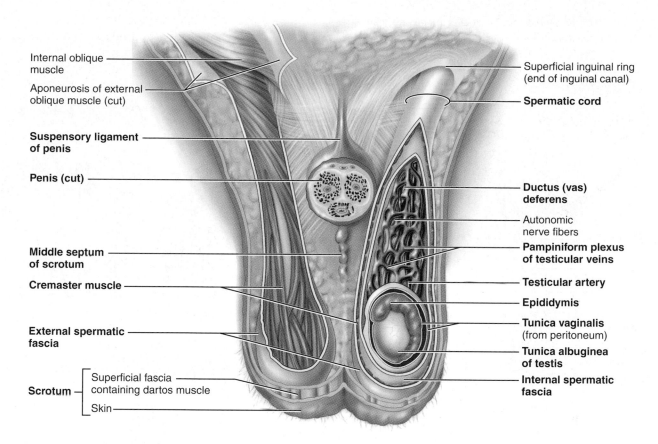

Internal oblique
muscle

Aponeurosis of external
oblique muscle (cut)

Suspensory ligament
of penis

Penis (cut)

Middle septum
of scrotum

Cremaster muscle

External spermatic
fascia

Scrotum ⎰ Superficial fascia
containing dartos muscle
⎱ Skin

Superficial inguinal ring
(end of inguinal canal)

Spermatic cord

Ductus (vas)
deferens

Autonomic
nerve fibers

Pampiniform plexus
of testicular veins

Testicular artery

Epididymis

Tunica vaginalis
(from peritoneum)

Tunica albuginea
of testis

Internal spermatic
fascia

FIGURE 26.2 **Relationships of the testis to the scrotum and spermatic cord.** The scrotum has been opened and its anterior portion removed.

The Scrotum

The **scrotum** (skro'tum; "pouch") is a sac of skin and superficial fascia that hangs outside the abdominopelvic cavity at the root of the penis (Figure 26.2). Covered with sparse hairs, the scrotal skin is more heavily pigmented than that elsewhere on the body. The paired oval testes lie suspended in the scrotum, separated by a midline septum which provides one compartment for each testis. This seems a rather vulnerable location for a man's testes, which contain his entire genetic heritage. However, viable sperm cannot be produced at core body temperature (36.2°C). Hence the superficial location of the scrotum, which provides a temperature about 3°C lower, is an essential adaption. Furthermore, the scrotum responds to temperature changes. When it is cold, the testes are pulled closer to the warmth of the body wall, and the scrotum becomes shorter and heavily wrinkled to reduce heat loss. When it is warm, the scrotal skin is flaccid and loose to increase the surface area for cooling, and the testes hang lower. These changes in scrotal surface area help maintain a fairly constant intrascrotal temperature and reflect the activity of two sets of muscles. The **dartos muscle** (dar'tos; "skinned"), a layer of smooth muscle in the superficial fascia, wrinkles the scrotal skin. The **cremaster muscles** (kre-mas'ter; "a suspender"), bands of skeletal muscle that arise from the internal oblique muscles of the trunk, elevate the testes.

The Testes

Each plum-sized testis is approximately 4 cm (1.5 inches) long and 2.5 cm (1 inch) in diameter and is surrounded by two tunics. The outer tunic is the two-layered **tunica vaginalis** (vaj"ĭ-nal'is), derived from the peritoneum (see Figures 26.2 and 26.3). Deep to this is the **tunica albuginea** (al"bu-jin'e-ah; "white coat"), the fibrous capsule of the testis. Septa extending from the tunica albuginea divide the testis into 250 to 300 wedge-shaped *lobules* (Figure 26.3), each containing one to four tightly coiled **seminiferous tubules** (sem"ĭ-nif'er-us; "sperm-carrying"), the actual "sperm factories."

The seminiferous tubules of each lobule converge to form a **tubulus rectus,** a straight tubule that conveys sperm into the **rete testis** (re'te), a tubular network on the posterior side of the testis. From the rete testis, sperm leave the testis through the *efferent ductules* and enter the *epididymis* (ep"ĭ-did'ĭ-mis), which hugs the external testis surface.

Which of the tubular structures shown here are the "sperm factories"?

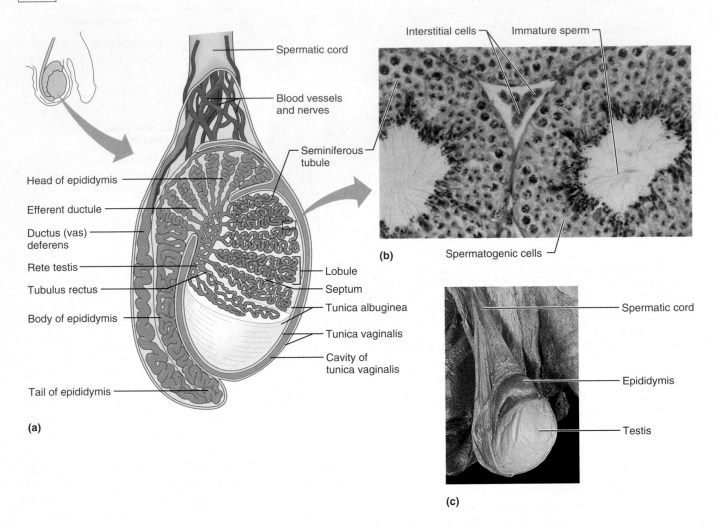

FIGURE 26.3   Structure of the testis. (a) Partial sagittal section of the testis and associated epididymis. **(b)** Cross-sectional view of portions of the seminiferous tubules, showing the spermatogenic (sperm-forming) cells making up the epithelium of the tubule walls and the interstitial cells in the loose connective tissue between the tubules (400×). **(c)** External view of a testis from a cadaver; same orientation as in (a).

Lying in the soft connective tissue surrounding the seminiferous tubules are the **interstitial cells,** (Figure 26.3). These cells produce androgens (most importantly *testosterone*), which they secrete into the surrounding interstitial fluid. Thus, the sperm-producing and hormone-producing functions of the testis are carried out by completely different cell populations.

The long **testicular arteries,** which branch from the abdominal aorta (see *gonadal arteries* in Figure 18.22c, p. 647) supply the testes. The **testicular veins** draining the testes arise from a vinelike network called the **pampiniform plexus** (pam-pin'ĭ-form; "tendril-shaped") that surrounds the testicular artery (see Figure 26.2). This plexus absorbs heat from the arterial blood, cooling it before it enters the testes. Thus, this plexus provides an additional way to maintain the testes at their cool homeostatic temperature.

The testes are served by both divisions of the autonomic nervous system. Associated sensory nerves transmit impulses that result in agonizing pain and nausea when the testes are hit forcefully. The nerve fibers are enclosed, along with the blood vessels and lymphatics, in a connective tissue sheath called the **spermatic cord** (see Figure 26.2).

■ *Seminiferous tubules.*

Although *testicular cancer* is relatively rare (affecting 1 of every 20,000 males), it is the most common cancer in young men (ages 15 to 35). A history of mumps or orchitis (inflammation of the testis) increases the risk, but the most important risk factor for this cancer is *cryptorchidism* (nondescent of the testes). Because the most common sign of testicular cancer is a painless solid mass in the testis, self-examination should be practiced by every male. If detected early, the cure rate is impressive. Over 90% of testicular cancers are cured by surgical removal of the cancerous testis (*orchiectomy*) followed by radiation therapy and chemotherapy. ●

The Penis

The **penis** ("tail") is a copulatory organ, designed to deliver sperm into the female reproductive tract (Figure 26.4). The penis and scrotum, which hang suspended from the perineum, make up the external reproductive structures, or **external genitalia,** of the male. The male **perineum** (per"ĭ-ne'um; "around the anus") is the diamond-shaped region located between the pubic symphysis anteriorly, the coccyx posteriorly, and the ischial tuberosities laterally. The floor of the perineum is formed by muscles described in Chapter 10 (pp. 306–307).

The penis consists of an attached *root* and a free *shaft* or *body* that ends in an enlarged tip, the **glans penis.** The skin covering the penis is loose, and it slides distally to form a cuff of skin called the **prepuce** (pre'pūs), or **foreskin,** around the glans. Frequently, the foreskin is surgically removed shortly after birth, a procedure called *circumcision* ("cutting around").

Internally, the penis contains the spongy urethra and three long cylindrical bodies (*corpora*) of erectile tissue. *Erectile tissue* is a spongy network of connective tissue and smooth muscle riddled with vascular spaces. During sexual excitement, the vascular spaces fill with blood, causing the penis to enlarge and become rigid. This event, called *erection*, enables the penis to serve as a penetrating organ. The midventral erectile body, the **corpus spongiosum** (spon"je-o'sum; "spongy body") surrounds the urethra. It expands distally to form the glans and proximally to form the part of the root called the **bulb of the penis.** The bulb is covered externally by the sheetlike bulbospongiosus muscle and is secured to the urogenital diaphragm. The paired dorsal erectile bodies, called the **corpora cavernosa** (kor'por-ah kă-ver-no'sah; "cavernous bodies"), make up most of the penis and are bound by fibrous tunica albuginea. Their proximal ends form the **crura of the penis** (kroo'rah; "legs"; singular, crus). Each crus is sur-

rounded by an ischiocavernosus muscle and anchors to the pubic arch of the bony pelvis.

The Male Duct System

As mentioned earlier, sperm travel from the testes to the body exterior through a system of ducts. In order (proximal to distal), the **accessory ducts** are the epididymis, the ductus deferens, the ejaculatory duct, and the urethra.

The Epididymis

The comma-shaped **epididymis** (*epi* = beside; *didym* = the testes) is about 3.8 cm (1.5 inches) long (see Figures 26.1 and 26.3a, c). Its *head*, which joins the efferent ductules, caps the superior aspect of the testis. Its *body* and *tail* are on the posterolateral area of the testis. Most of the epididymis consists of the highly coiled *duct of the epididymis* with a true uncoiled length of about 6 m (20 feet). Some pseudostratified epithelium cells of the duct mucosa exhibit long, nonmotile microvilli (*stereocilia*), which absorb excess testicular fluid and pass nutrients to the sperm in the lumen.

The immature, nearly nonmotile sperm that leave the testis are moved slowly through the duct of the epididymis. As they move along its tortuous course (a trip that takes about 20 days), the sperm gain the ability to swim. When a male is sexually stimulated and ejaculates, the smooth muscle in the epididymis walls contracts, expelling sperm into the next segment of the duct system, the *ductus deferens*. Sperm can be stored in the epididymis for several months, but if held longer, they are eventually phagocytized by epithelial cells of the epididymis.

The Ductus Deferens and Ejaculatory Duct

The **ductus deferens** (duk'tus def'er-ens; "carrying away"), or **vas deferens,** is about 45 cm (18 inches) long. It runs upward as part of the spermatic cord from the epididymis through the inguinal canal into the pelvic cavity (see Figure 26.1). It is easily palpated as it passes anterior to the pubic bone. It then loops medially over the ureter and descends along the posterior bladder wall. Its terminus expands to form the **ampulla** and then joins with the duct of the seminal vesicle (a gland) to form the short **ejaculatory duct.** Each ejaculatory duct enters the prostate gland; there it empties into the urethra.

The ductus deferens propels live sperm from their storage sites, the epididymis and distal part of the ductus deferens, into the urethra. At the moment of ejaculation, the thick layers of smooth muscle in its walls create strong peristaltic waves that rapidly squeeze the sperm forward.

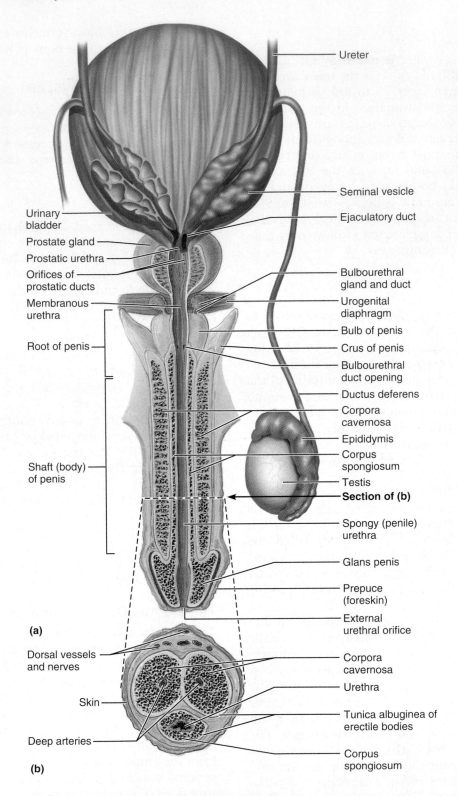

Ureter

Seminal vesicle

Ejaculatory duct

Urinary bladder

Prostate gland

Prostatic urethra

Orifices of prostatic ducts

Bulbourethral gland and duct

Membranous urethra

Urogenital diaphragm

Bulb of penis

Root of penis

Crus of penis

Bulbourethral duct opening

Ductus deferens

Corpora cavernosa

Epididymis

Corpus spongiosum

Shaft (body) of penis

Testis

Section of (b)

Spongy (penile) urethra

Glans penis

Prepuce (foreskin)

External urethral orifice

(a)

Dorsal vessels and nerves

Corpora cavernosa

Urethra

Skin

Tunica albuginea of erectile bodies

Deep arteries

Corpus spongiosum

(b)

FIGURE 26.4 **Male reproductive structures. (a)** Posterior view showing longitudinal (coronal) section of the penis. **(b)** Transverse section of the penis.

As Figure 26.3 illustrates, part of the ductus deferens lies in the scrotal sac. Some men opt to take full responsibility for birth control by having a **vasectomy** (vah-sek'to-me; "cutting the vas"). In this relatively minor operation, the physician makes a small incision into the scrotum and then cuts through and ligates (ties off) the ductus deferens. Sperm are still produced for the next several years, but they can no longer reach the body exterior. Eventually, they deteriorate and are phagocytized.

Vasectomy is simple and provides highly effective birth control (close to 100%).

The Urethra

The **urethra** is the terminal portion of the male duct system (see Figures 26.1 and 26.4). It conveys both urine and semen (at different times), so it serves both the urinary and reproductive systems. Its three regions are (1) the *prostatic urethra*, the portion surrounded by the prostate gland; (2) the *membranous urethra* in the urogenital diaphragm; and (3) the *spongy (penile) urethra*, which runs through the penis and opens to the outside at the *external urethral orifice*. The spongy urethra is about 15 cm (6 inches) long and accounts for 75% of urethral length.

Accessory Glands

The **accessory glands** include the paired seminal vesicles and bulbourethral glands and the single prostate gland (Figures 26.1 and 26.4). These glands produce the bulk of *semen* (sperm plus accessory gland secretions).

The Seminal Vesicles

The **seminal vesicles** (sem'ĭ-nul) lie on the posterior bladder wall. These glands are fairly large, each about the shape and length (5 to 7 cm) of a little finger. Their secretion, accounting for about 60% of the volume of semen, is a yellowish viscous alkaline fluid containing fructose sugar, ascorbic acid, a coagulating enzyme *(vesiculase)*, and prostaglandins. As noted, the duct of each seminal vesicle joins that of the ductus deferens on the same side to form the ejaculatory duct. Sperm and seminal fluid mix in the ejaculatory duct and enter the prostatic urethra together during ejaculation.

The Prostate Gland

The **prostate gland** (pros'tāt) is a single doughnut-shaped gland about the size of a chestnut (see Figures 26.1 and 26.4). It encircles the urethra just inferior to the bladder. Enclosed by a thick connective tissue capsule, it is made up of 20 to 30 compound tubular-alveolar glands embedded in a mass (stroma) of smooth muscle and dense connective tissue. The prostatic gland secretion plays a role in activating sperm and accounts for up to one-third of the semen volume. It is a milky, slightly acid fluid that contains citrate (a nutrient source), several enzymes (fibrinolysin, hyaluronidase, acid phosphatase), and prostate-specific antigen (PSA). It enters the prostatic urethra via several ducts when prostatic smooth muscle contracts during ejaculation.

HOMEOSTATIC IMBALANCE

The prostate gland has a reputation as a health destroyer (perhaps reflected in the common mispronunciation "prostrate"). Hypertrophy of the prostate gland, which affects nearly every elderly male, distorts the urethra. The more the man strains to urinate, the more the valvelike prostatic mass blocks the opening, enhancing the risk of bladder infections (cystitis) and kidney damage. Traditional treatment has been surgical, but some newer options are gaining popularity, such as using microwaves or drugs to shrink the prostate; and inserting and inflating a small balloon to compress the prostate tissue away from the prostatic urethra. Another option is a procedure in which a catheter containing a tiny needle is inserted and burstlike releases of radiofrequency radiation are used to incinerate excess prostate tissue.

Prostatitis (pros"tah-ti'tis), inflammation of the prostate, is the most common reason for a man to consult a urologist, and prostate cancer is the third most prevalent cancer in men. Screening for prostate cancer typically involves digital examination of the prostate through the anterior rectal wall and assay of PSA levels in the blood. PSA, although a normal component of semen, is a tumor marker and its rising serum level follows the clinical disease course of prostate cancer. When possible, prostate cancer is treated surgically. Alternative therapies for metastasized cancers involve castration or therapy with drugs (flutamide) that block androgen receptors, or LHRH* analogues (such as leuprolide), which inhibit gonadotropin release. Deprived of the stimulatory effects of androgens, the prostatic tissue regresses and urinary symptoms typically decline. ●

The Bulbourethral Glands

The **bulbourethral glands** (bul"bo-u-re'thral), also called **Cowper's glands,** are pea-sized glands inferior to the prostate gland (Figures 26.1 and 26.4). They produce a thick, clear mucus prior to ejaculation that drains into the spongy urethra and neutralizes traces of acidic urine in the urethra.

Semen

Semen (se'men) is a milky white, somewhat sticky mixture of sperm and accessory gland secretions. The liquid provides a transport medium and nutrients and contains chemicals that protect and activate the sperm and facilitate their movement.

---

* LHRH (luteinizing hormone–releasing hormone) is analogous to GnRH (gonadotropin-releasing hormone). See discussion of GnRH's effects on p. 531.

Mature sperm cells are streamlined cellular "missiles" containing little cytoplasm or stored nutrients. The fructose in seminal vesicle secretion provides nearly all their fuel. Prostaglandins in semen decrease the viscosity of mucus guarding the entry (cervix) of the uterus and stimulate reverse peristalsis in the uterus, facilitating sperm movement through the female reproductive tract. The presence of the hormone *relaxin* and certain enzymes in semen enhance sperm motility. The relative alkalinity of semen as a whole (pH 7.2–7.6) helps neutralize the acid environment of the male's urethra and the female's vagina, thereby protecting the delicate sperm and enhancing their motility. Sperm are very sluggish under acidic conditions (below pH 6). Semen also contains an antibiotic chemical called **seminalplasmin,** which destroys certain bacteria. Clotting factors found in semen coagulate it just after it is ejaculated. Then, its contained fibrinolysin liquefies the sticky mass, enabling the sperm to swim out and begin their journey through the female duct system.

The amount of semen propelled out of the male duct system during ejaculation is relatively small, only 2–5 ml, but there are between 50 and 130 million sperm per milliliter.

Physiology of the Male Reproductive System

Male Sexual Response

Although there is more to it, the chief phases of the male sexual response are (1) *erection* of the penis, which allows it to penetrate the female vagina, and (2) *ejaculation,* which expels semen into the vagina.

Erection

Erection, enlargement and stiffening of the penis, results from engorgement of the erectile bodies with blood. When a man is not sexually aroused, arterioles supplying the erectile tissue are constricted and the penis is flaccid. However, during sexual excitement a parasympathetic reflex is triggered that promotes release of nitric oxide locally. NO relaxes vascular smooth muscle, causing these arterioles to dilate, which allows the erectile bodies to fill with blood. Expansion of the corpora cavernosa of the penis compresses their drainage veins, retarding blood outflow and maintaining engorgement. The corpus spongiosum expands but not nearly as much as the cavernosa; its main job is to keep the urethra open during ejaculation. Erection of the penis is one of the rare examples of parasympathetic control of arterioles. Another parasympathetic effect is stimulation of the bulbourethral glands, which causes lubrication of the glans penis.

Erection is initiated by a variety of sexual stimuli, such as touching the genital skin, mechanical stimulation of the pressure receptors in the penis, and erotic sights, sounds, and smells. The CNS responds to such stimuli by activating parasympathetic neurons that innervate the internal pudendal arteries serving the penis. Sometimes erection is induced solely by emotional or higher mental activity (the thought of a sexual encounter). Emotions and thoughts can also inhibit erection, causing vasoconstriction and resumption of the flaccid penile state.

Ejaculation

Ejaculation (*ejac* = to shoot forth) is the propulsion of semen from the male duct system. While erection is under parasympathetic control, *ejaculation* is under sympathetic control. When impulses provoking erection reach a certain critical level, a spinal reflex is initiated, and a massive discharge of nerve impulses occurs over the sympathetic nerves serving the genital organs (largely at the level of L_1 and L_2). As a result,

1. The reproductive ducts and accessory glands contract, emptying their contents into the urethra.

2. The bladder sphincter muscle constricts, preventing expulsion of urine or reflux of semen into the bladder.

3. The bulbospongiosus muscles of the penis undergo a rapid series of contractions, propelling semen at a speed of up to 500 cm/s (200 inches/s) from the urethra. These rhythmic contractions are accompanied by intense pleasure and many systemic changes, such as generalized muscle contraction, rapid heartbeat, and elevated blood pressure.

The entire ejaculatory event is referred to as **climax** or **orgasm.** Orgasm is quickly followed by muscular and psychological relaxation and vasoconstriction of the penile arterioles, which allows the penis to become flaccid once again. After ejaculation, there is a latent period, ranging in time from minutes to hours, during which a man is unable to achieve another orgasm. The latent period lengthens with age.

Spermatogenesis

Spermatogenesis (sper"mah-to-jen'ĕ-sis; "sperm formation") is the sequence of events in the seminiferous tubules of the testes that produces male gametes—*sperm* or *spermatozoa.* The process begins around the age of 14 years in males, and continues throughout life. Every day, a healthy adult male makes about 400 million sperm. It seems that nature has made sure that the human species will not be endangered for lack of sperm.

Before we describe the process of spermatogenesis, let's define some terms. First, having two sets of

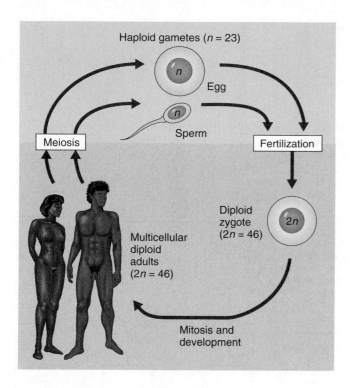

FIGURE 26.5   **The human life cycle.**

most body cells divide) distributes replicated chromosomes equally between the two daughter cells. Consequently, each daughter cell receives a set of chromosomes identical to that of the mother cell. Meiosis, on the other hand, consists of two consecutive nuclear divisions, and its product is four daughter cells instead of two, each with *half* as many chromosomes as typical body cells. Thus, meiosis reduces the chromosomal number by half (from $2n$ to n) in gametes. Mitosis and meiosis are compared in Figure 26.6.

Meiosis

The two nuclear divisions of meiosis, called *meiosis I* and *meiosis II*, are divided into phases for convenience. Although these phases are given the same names as those of mitosis (prophase, metaphase, anaphase, and telophase), events of meiosis I are quite different from those of mitosis, as detailed next.

Recall that prior to mitosis all the chromosomes are replicated. Then the identical copies remain together as *sister chromatids* connected by a centromere throughout prophase and during metaphase. At anaphase, the centromeres split and the sister chromatids separate from each other so that each daughter cell inherits a copy of *every* chromosome possessed by the mother cell (see Figure 26.6). Let's look at how meiosis differs.

Meiosis I   As in mitosis, chromosomes replicate before meiosis begins. But in prophase of meiosis I, an event never seen in mitosis (nor in meiosis II for that matter) occurs. The replicated chromosomes seek out their homologous partners and pair up with them along their entire length (Figure 26.7). This alignment takes place at discrete spots along the length of the homologues—more like buttoning together than zipping. As a result of this process, called **synapsis,** little groups of four chromatids called **tetrads** are formed. During synapsis, a second unique event called crossover occurs. **Crossovers** are formed within each tetrad as the free ends of one maternal and one paternal chromatid wrap around each other at one or more points. Crossover allows an exchange of genetic material between the paired maternal and paternal chromosomes.

During metaphase I, the tetrads line up randomly at the spindle equator; that is, either the paternal or maternal chromosome can be on a given side of the equator. During anaphase I, the sister chromatids representing each homologue behave as a unit—almost as if replication had not occurred—and the *homologous chromosomes* (each still composed of two joined sister chromatids) are distributed to opposite ends of the cell.

Thus, when meiosis I is completed, the following conditions exist: Each daughter cell has (1) *two*

chromosomes, one from each parent, is a key factor in the human life cycle (Figure 26.5). The normal chromosome number in most body cells is referred to as the **diploid chromosomal number** (dip'loid) of the organism, symbolized as **$2n$.** In humans, this number is 46, and such diploid cells contain 23 pairs of similar chromosomes called **homologous chromosomes** (ho-mol′ŏ-gus) or **homologues.** One member of each pair is from the male parent (the *paternal chromosome*); the other is from the female parent (the *maternal chromosome*). Generally speaking, the two homologues of each chromosome pair look alike and carry genes that code for the same traits, though not necessarily for identical expression of those traits. (Consider, for example, the homologous genes controlling the expression of freckles; the paternal gene might code for an ample sprinkling of freckles and the maternal gene for no freckles.)

The number of chromosomes present in human gametes is 23, referred to as the **haploid chromosomal number** (hap′loid), or **n;** gametes contain only one member of each homologous pair. When sperm and egg fuse, they form a fertilized egg that reestablishes the typical diploid chromosomal number of human cells.

Gamete formation in both sexes involves **meiosis** (mi-o′sis; "a lessening"), a unique kind of nuclear division that, for the most part, occurs only in the gonads. Recall that *mitosis* (the process by which

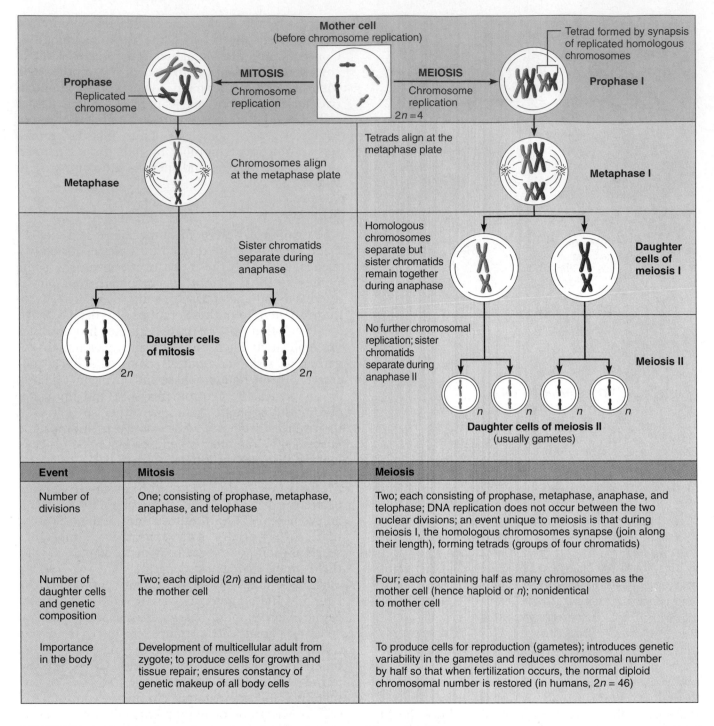

FIGURE 26.6   **Comparison of mitosis and meiosis in a mother cell with a diploid number (2n) of 4.** Mitosis is shown on the left, meiosis on the right. (Not all the phases of mitosis and meiosis are shown.)

copies of one member of each homologous pair (either the paternal or maternal) and none of the other, and (2) a diploid amount of DNA but a *haploid* chromosomal number because the still-united sister chromatids are considered to be a single chromosome. Since meiosis I reduces the chromosome number from 2n to n, it is sometimes called the **reduction division of meiosis.**

Meiosis II   The second meiotic division, meiosis II, mirrors mitosis in every way, except that the chromosomes are *not* replicated before it begins. Instead, the chromatids in the two daughter cells of meiosis I are simply parceled out among four cells. Because the chromatids are distributed equally to the daughter cells (as in mitosis), meiosis II is sometimes referred to as the **equational division of meiosis** (Figure 26.7).

FIGURE 26.7 **Meiosis.** The series of events in meiotic cell division for an animal cell with a diploid number (2*n*) of 4. The behavior of the chromosomes is emphasized.

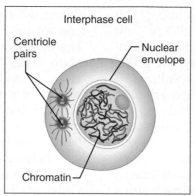

Interphase cell

Centriole pairs

Nuclear envelope

Chromatin

Interphase events
As in mitosis, meiosis is preceded by events occurring during interphase that lead to DNA replication and other preparations needed for the cell division process. Just before meiosis begins, the replicated chromatids, held together by centromeres, are ready and waiting.

Meiosis I

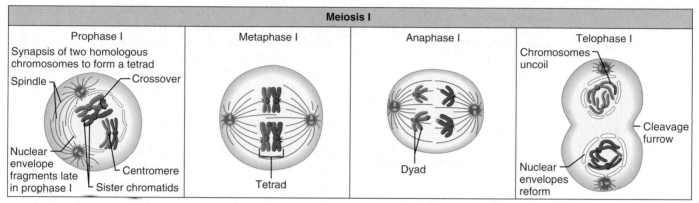

Prophase I

Synapsis of two homologous chromosomes to form a tetrad

Spindle — Crossover

Nuclear envelope fragments late in prophase I — Centromere — Sister chromatids

Metaphase I

Tetrad

Anaphase I

Dyad

Telophase I

Chromosomes uncoil

Cleavage furrow

Nuclear envelopes reform

Prophase I
As in prophase of mitosis, the chromosomes coil and condense, the nuclear membrane and nucleolus break down and disappear, and the spindle is formed. However, a unique event not seen in mitosis, called synapsis, occurs in prophase I of meiosis. Synapsis involves the coming together of homologous chromosomes to form tetrads, little packets of four chromatids. While in synapsis, the "arms" of adjacent homologous chromatids become wrapped around each other, forming several points of crossover or chiasmata. Generally speaking, the longer the chromatids, the more chiasmata are formed. (This process of crossover is shown in one tetrad of the prophase I view, and the result of that one event of crossing over is followed through meiosis II.) Prophase I is the longest period of meiosis, accounting for about 90% of the total period. By its end, the tetrads have attached to the spindle and are moving toward the spindle equator, and the sister chromatids have exchanged parts at points of crossover.

Metaphase I
During metaphase I, the tetrads align on the spindle equator in preparation for anaphase.

Anaphase I
Unlike the anaphase events of mitosis, the centromeres do not break during anaphase I of meiosis, and so the sister chromatids (dyads) remain firmly attached. However, the homologous chromosomes do separate from each other and the dyads are moved toward opposite poles of the cell.

Telophase I
The nuclear membranes re-form around the chromosomal masses, the spindle breaks down, and the chromatin reappears as telophase and cytokinesis are completed, forming two daughter cells. The daughter cells (now haploid) enter a second interphase-like period, called interkinesis, before meiosis II occurs. There is no second replication of DNA before meiosis II.

Meiosis II

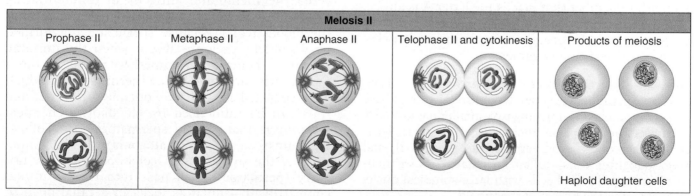

Prophase II

Metaphase II

Anaphase II

Telophase II and cytokinesis

Products of meiosis

Haploid daughter cells

Meiosis II begins with the products of meiosis I (two haploid daughter cells) and undergoes a mitosis-like nuclear division process referred to as the equational division of meiosis. After progressing through prophase, metaphase, anaphase, and telophase, followed by cytokinesis, the product is four haploid daughter cells each genetically different from the original mother cell. During human spermatogenesis the daughter cells remain interconnected by cytoplasmic extensions.

Meiosis accomplishes two important tasks: (1) It reduces the chromosomal number by half and (2) it introduces genetic variability. The random alignment of the homologous pairs during meiosis I provides tremendous variability in the resulting gametes by scrambling genetic characteristics of the two parents in different combinations. Variability is increased further by crossover—during late prophase I, the homologues break at crossover points and exchange chromosomal segments (see Figure 26.7). As a result, it is likely that no two gametes are exactly alike, and all are different from the original mother cells.

Spermatogenesis: Summary of Events in the Seminiferous Tubules

A histological section of an adult testis shows that most cells making up the epithelial walls of the seminiferous tubules are in various stages of cell division (Figure 26.8). These cells, collectively called **spermatogenic cells** (*spermatogenic* = sperm forming), give rise to sperm in the following series of cellular divisions and transformations.

Mitosis of Spermatogonia: Forming Spermatocytes
The outermost tubule cells, which are in direct contact with the epithelial basal lamina, are stem cells called **spermatogonia** (sper″mah-to-go′ne-ah; "sperm seed"). The spermatogonia divide more or less continuously by *mitosis* and, until puberty, all their daughter cells become spermatogonia.

Spermatogenesis begins during puberty, and from then on, each mitotic division of a spermatogonium results in two distinctive daughter cells—*types A* and *B*. The **type A daughter cell** remains at the basement membrane to maintain the germ cell line. The **type B cell** gets pushed toward the lumen, where it becomes a **primary spermatocyte** destined to produce four sperm. (To keep these cell types straight, remember that just as the letter A is always at the beginning of our alphabet, a type A cell is always at the tubule basement membrane ready to begin a new generation of gametes.)

Meiosis: Spermatocytes to Spermatids
Each primary spermatocyte generated during the first phase undergoes meiosis I, forming two smaller haploid cells called **secondary spermatocytes.** The secondary spermatocytes continue on rapidly into meiosis II, and their daughter cells, called **spermatids** (sper′mah-tidz), are small round cells with large spherical nuclei seen closer to the lumen of the tubule.

Spermiogenesis: Spermatids to Sperm
Each spermatid has the correct chromosomal number for fertilization *(n)*, but is nonmotile. It still must undergo a streamlining process called **spermiogenesis,** during which it sheds its excess cytoplasmic baggage and forms a tail. Details of this process appear in Figure 26.9a. The resulting **sperm,** or **spermatozoon** (sper″mah-to-zo′on; "animal seed"), has a head, a midpiece, and a tail, which correspond roughly to *genetic, metabolic,* and *locomotor regions,* respectively. The **head** of the sperm consists almost entirely of its flattened nucleus, which contains compacted DNA. Adhering to the top of the nucleus is a helmetlike **acrosome** (ak′ro-sōm; "tip piece"). The lysosome-like acrosome is produced by the Golgi apparatus and contains hydrolytic enzymes that enable the sperm to penetrate and enter an egg. The sperm **midpiece** contains mitochondria spiraled tightly around the contractile filaments of the tail. The **tail** is a typical flagellum produced by a centriole. The mitochondria provide the metabolic energy (ATP) needed for the whiplike movements of the tail that will propel the sperm along its way in the female reproductive tract.

Role of the Sustentacular Cells
Throughout spermatogenesis, descendants of the same spermatogonium remain closely attached to one another by cytoplasmic bridges (see Figure 26.8). They are also surrounded by and connected to supporting cells of a special type, called **sustentacular cells** or **Sertoli cells,** which extend from the basal lamina to the lumen of the tubule. The sustentacular cells, bound to each other by tight junctions, divide the seminiferous tubule into two compartments. The **basal compartment** extends from the basal lamina to their tight junctions and contains spermatogonia and the earliest primary spermatocytes. The **adluminal compartment** lies internal to the tight junctions and includes the meiotically active cells and the tubule lumen.

The tight junctions between the sustentacular cells form the **blood-testis barrier.** This barrier prevents the membrane antigens of differentiating sperm from escaping through the basal lamina into the bloodstream. Because sperm are not formed until puberty, they are absent when the immune system is being programmed to recognize one's own tissues early in life. The spermatogonia, which are recognized as "self," are outside the barrier and thus can be influenced by bloodborne chemical messengers that prompt spermatogenesis. Following mitosis of the spermatogonia, the tight junctions of the sustentacular cells open to allow primary spermatocytes to pass into the adluminal compartment—much as locks in a canal open to allow a boat to pass.

In the adluminal compartment, spermatocytes and spermatids are nearly enclosed in recesses in the sustentacular cells (see Figure 26.8), anchored there by a particular glycoprotein on the spermatogenic

What cell type results from meiosis?

?

FIGURE 26.8 **Spermatogenesis.** **(a)** Scanning electron micrograph of a cross-sectional view of a seminiferous tubule (350×). (© R. G. Kessel and R. H. Kardon, *Tissues and Organs: A Text-Atlas of Scanning Electron Microscopy,* W. H. Freeman and Company, 1979, all rights reserved.) **(b)** Events of spermatogenesis, showing the relative position of various spermatogenic cells. **(c)** Enlarged view of a portion of the wall of the seminiferous tubule, showing the spermatogenic cells surrounded by sustentacular cells (colored gold). (*Note:* The process of spermatogenesis encompasses the phases through the production of sperm. Conversion of the haploid spermatids to sperm cells is specifically called spermiogenesis.)

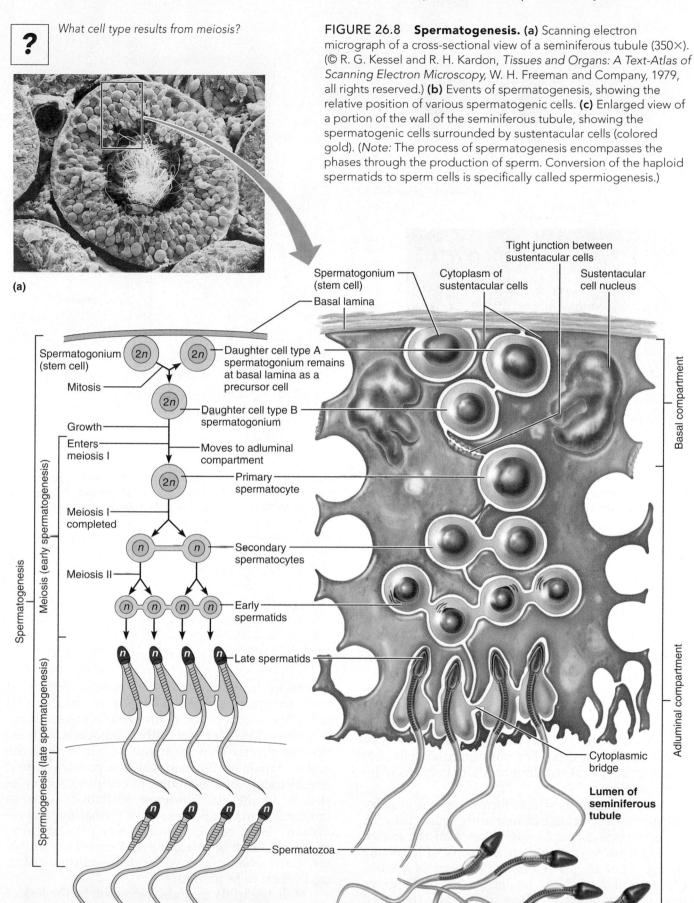

(a)

(b)

(c)

■ *spermatids.*

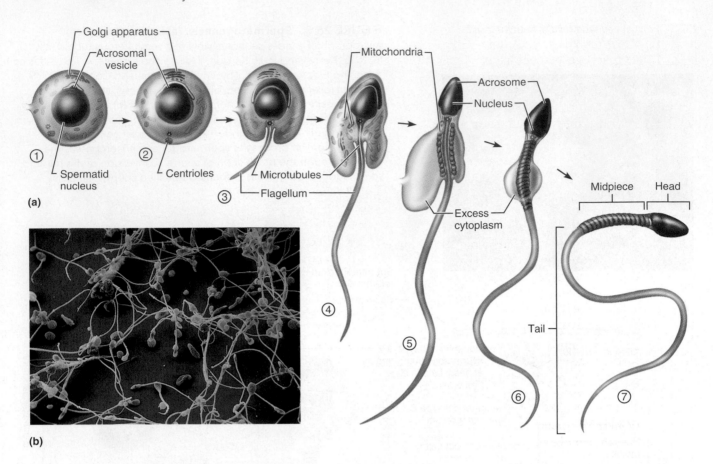

Golgi apparatus

Acrosomal vesicle

Mitochondria

Acrosome

Nucleus

Spermatid nucleus

Centrioles

Microtubules

Flagellum

Excess cytoplasm

Midpiece   Head

Tail

(a)

(b)

FIGURE 26.9 Spermiogenesis: transformation of a spermatid into a functional sperm. (a) The stepwise process of spermiogenesis consists of ① packaging of the acrosomal enzymes by the Golgi apparatus, ② forming the acrosome at the anterior end of the nucleus and positioning the centrioles at the opposite end of the nucleus, ③ elaboration of microtubules to form the flagellum, ④ mitochondrial multiplication and their positioning around the proximal portion of the flagellum, and ⑤ sloughing off excess cytoplasm. ⑥ Structure of an immature sperm that has just been released from a sustensxtacular cell. ⑦ Structure of a fully mature sperm. **(b)** Scanning electron micrograph of mature sperm (600×).

cell's surface. The sustentacular cells deliver nutrients to the dividing cells, move them along to the lumen, secrete **testicular fluid** (rich in androgens and metabolic acids) that provides the transport medium for sperm in the lumen, and dispose of the excess cytoplasm sloughed off the spermatids as they transform into sperm. The sustentacular cells also produce chemical mediators that help regulate spermatogenesis.

Spermatogenesis—from formation of a primary spermatocyte to release of immature sperm into the lumen—takes 64 to 72 days. Sperm in the lumen are unable to "swim" and are incapable of fertilizing an egg. They are pushed by the pressure of the testicular fluid through the tabular system of the testes into the epididymis, where they gain increased motility and fertilizing power.

HOMEOSTATIC IMBALANCE

According to some studies, a gradual decline in male fertility has been occurring in the past 50 years. Some believe the main cause is environmental toxins, PVCs, or especially compounds with estrogenic effects. These compounds, which block the action of male sex hormones as they program sexual development, are now found in our meat supply as well as in the air. Common antibiotics such as tetracycline may suppress sperm formation; and radiation, lead, certain components of pesticides, marijuana, lack of selenium, and excessive alcohol can cause abnormal (two-headed, multiple-tailed, etc.) sperm to be produced.

Male infertility may also be caused by the lack of a specific type of Ca^{2+} channel (Ca^{2+} is needed for normal sperm motility), anatomical obstructions, and hormonal imbalances. A low sperm count accompanied by a high percentage of immature sperm may hint that a man has a *varicocele* (var′ĭ-ko-sēl″), a condition that hinders drainage of the testicular vein, resulting in an elevated temperature in the scrotum that interferes with normal sperm development. ●

Hormonal Regulation of Male Reproductive Function

The Brain-Testicular Axis

Hormonal regulation of spermatogenesis and testicular androgen production involves interactions between the hypothalamus, anterior pituitary gland, and testes, a relationship sometimes called the **brain-testicular axis.** The sequence of regulatory events, shown schematically in Figure 26.10, is as follows:

① The hypothalamus releases **gonadotropin-releasing hormone (GnRH),** which controls the release of the anterior pituitary gonadotropins, **follicle-stimulating hormone (FSH)** and **luteinizing hormone (LH).** (Both FSH and LH were named for their effects on the female gonad.) GnRH reaches the anterior pituitary cells via the blood of the hypophyseal portal system.

② Binding of GnRH to pituitary cells (gonadotrophs) prompts them to secrete FSH and LH into the blood.

③ FSH stimulates spermatogenesis indirectly by stimulating the sustentacular cells to release **androgen-binding protein (ABP).** ABP prompts the spermatogenic cells to bind and concentrate **testosterone,** which in turn stimulates spermatogenesis. Thus, FSH makes the cells receptive to testosterone's stimulatory effects.

④ LH binds to the interstitial cells, prodding them to secrete testosterone (and a small amount of estrogen). LH is therefore sometimes called **interstitial cell–stimulating hormone (ICSH)** in males. Locally, testosterone serves as the final trigger for spermatogenesis. Testosterone entering the bloodstream exerts a number of effects at other body sites.

⑤ Testosterone inhibits hypothalamic release of GnRH and acts directly on the anterior pituitary to inhibit gonadotropin release. **Inhibin** (in-hib′in), a protein hormone produced by the sustentacular cells, serves as a barometer of the normalcy of spermatogenesis. When the sperm count is high, inhibin release increases and it inhibits anterior pituitary release of FSH and hypothalamic release of GnRH. When sperm count falls below 20 million/ml, inhibin secretion declines steeply.

As you can see, the amount of testosterone and sperm produced by the testes reflects a balance among three sets of hormones: (1) GnRH, which

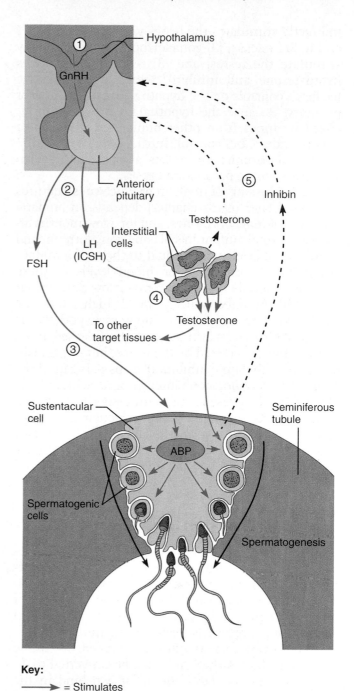

Key:

——▶ = Stimulates

----▶ = Inhibits

FIGURE 26.10 Hormonal regulation of testicular function, the brain-testicular axis. ① The hypothalamus releases gonadotropin-releasing hormone (GnRH). ② GnRH stimulates the anterior pituitary to release gonadotropins— follicle-stimulating hormone (FSH) and luteinizing hormone (LH), also called interstitial cell–stimulating hormone (ICSH). ③ FSH causes the sustentacular cells to release androgen-binding protein (ABP). ④ LH acts on the interstitial cells, stimulating their release of testosterone. ABP binding of testosterone then enhances spermatogenesis. ⑤ Rising levels of testosterone and inhibin (released by the sustentacular cells) exert feedback inhibition on the hypothalamus and pituitary.

indirectly stimulates the testes via its effect on FSH and ICSH release; (2) gonadotropins, which directly stimulate the testes; and (3) testicular hormones (testosterone and inhibin), which exert negative feedback controls on the hypothalamus and anterior pituitary. Because the hypothalamus is also influenced by input from other brain areas, the whole axis is under CNS control. In the absence of GnRH and gonadotropins, the testes atrophy, and sperm and testosterone production ceases.

Development of male reproductive structures (discussed later in the chapter) depends on prenatal secretion of male hormones, and for a few months after birth, a male infant has plasma gonadotropin and testosterone levels nearly equal to those of a midpubertal boy. Soon thereafter, blood levels of these hormones recede and they remain low throughout childhood. As puberty nears, much higher levels of testosterone are required to suppress hypothalamic release of GnRH. So as more GnRH is released, more testosterone is secreted by the testes, but the threshold for hypothalamic inhibition keeps rising until the adult pattern of hormone interaction is achieved, evidenced by the presence of mature sperm in the semen. Maturation of the brain-testicular axis takes about three years, and once established, the balance between the interacting hormones remains relatively constant. Consequently, an adult male's sperm and testosterone production also remains fairly stable.

Mechanism and Effects of Testosterone Activity

Testosterone, like all steroid hormones, is synthesized from cholesterol. It exerts its effects by activating specific genes to transcribe messenger RNA molecules, which results in enhanced synthesis of certain proteins in the target cells. (See Chapter 15.)

In some target cells, testosterone must be transformed into another steroid to exert its effects. In the prostate gland testosterone must be converted to *dihydrotestosterone (DHT)* before it can bind in the nucleus, and in certain neurons of the brain, testosterone is converted to *estrogen* (es'tro-jen) to bring about its stimulatory effects. Thus, in that case at least, a "male" hormone is transformed into a "female" hormone to exert its masculinizing effects.

As puberty ensues, testosterone not only prompts spermatogenesis but has multiple anabolic effects throughout the body (see Table 26.1, p. 936). It targets accessory reproductive organs—ducts, glands, and the penis—causing them to grow and assume adult size and function. In adult males, normal plasma levels of testosterone maintain these organs: When the hormone is deficient or absent, all accessory organs atrophy, semen volume declines markedly, and erection and ejaculation are impaired. Thus, a man becomes both sterile and impotent. This situation is easily remedied by testosterone replacement therapy.

Male secondary sex characteristics—that is, features induced in the *nonreproductive* organs by the male sex hormones (mainly testosterone)—make their appearance at puberty. These include the appearance of pubic, axillary, and facial hair, enhanced hair growth on the chest or other body areas in some men, and a deepening of the voice as the larynx enlarges. The skin thickens and becomes oilier (which predisposes young men to acne), bones grow and increase in density, and skeletal muscles increase in size and mass. The last two effects are often referred to as the *somatic effects* of testosterone (*soma* = body).

Testosterone also boosts basal metabolic rate and influences behavior. It is the basis of the sex drive (libido) in both males and females, be they heterosexual or homosexual. Thus, although testosterone is called a "male" sex hormone, it should not be specifically tagged as a promoter of male sexual activity. In embryos, the presence of testosterone masculinizes the brain.

The testes are not the only source of androgens; the adrenal glands of both sexes also release androgens. However, the relatively small amounts of adrenal androgens are unable to support normal testosterone-mediated functions when the testes fail to produce androgens, so we can assume that it is the testosterone production by the testes that supports male reproductive function.

Anatomy of the Female Reproductive System

The reproductive role of the female is far more complex than that of a male. Not only must she produce gametes, but her body must prepare to nurture a developing embryo for a period of approximately nine months. **Ovaries,** the **female gonads,** are the primary reproductive organs of a female, and like the male testes, ovaries serve a dual purpose: They produce the female gametes (ova) and sex hormones, the **estrogens*** and **progesterone** (pro-ges'tě-rōn). The accessory ducts (uterine tubes, uterus, and vagina) transport or otherwise serve the needs of the reproductive cells and a developing fetus.

As illustrated in Figure 26.11, the ovaries and duct system, collectively known as the **internal genitalia,** are mostly located in the pelvic cavity. The female's accessory ducts, from the vicinity of

---

* The ovaries produce several different estrogens (*estradiol, estrone,* and *estriol*). Estradiol is the most abundant and is most responsible for estrogenic effects.

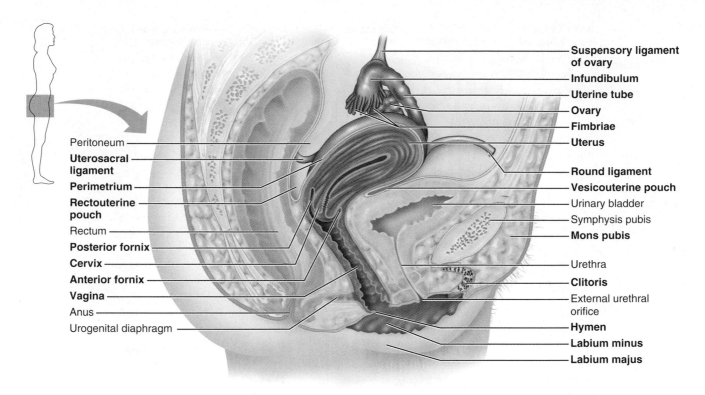

Peritoneum

Uterosacral ligament

Perimetrium

Rectouterine pouch

Rectum

Posterior fornix

Cervix

Anterior fornix

Vagina

Anus

Urogenital diaphragm

Suspensory ligament of ovary

Infundibulum

Uterine tube

Ovary

Fimbriae

Uterus

Round ligament

Vesicouterine pouch

Urinary bladder

Symphysis pubis

Mons pubis

Urethra

Clitoris

External urethral orifice

Hymen

Labium minus

Labium majus

FIGURE 26.11 **Internal organs of the female reproductive system, midsagittal section.** (See *A Brief Atlas of the Human Body,* Figure 59.)

the ovary to the body exterior, are the *uterine tubes,* the *uterus,* and the *vagina.* The external sex organs of females are referred to as the **external genitalia.**

The Ovaries

The paired ovaries flank the uterus on each side (see Figure 26.11). Shaped like almonds and about twice as large, each ovary is held in place within the peritoneal cavity by several ligaments. The **ovarian ligament** anchors the ovary medially to the uterus; the **suspensory ligament** anchors it laterally to the pelvic wall; and the **mesovarium** (mez"o-va´re-um) suspends it in between (see Figures 26.11 and 26.14a). The suspensory ligament and the mesovarium are part of the **broad ligament,** a peritoneal fold that "tents" over the uterus and supports the uterine tubes, uterus, and vagina. The ovarian ligaments are enclosed by the broad ligament.

The ovaries are served by the **ovarian arteries,** branches of the abdominal aorta (see Figure 18.22c), and by the *ovarian branch of the uterine arteries.* The ovarian blood vessels reach the ovaries by traveling through the suspensory ligaments and mesovaria.

Like a testis, an ovary is surrounded externally by a fibrous **tunica albuginea** (Figure 26.12), which is in turn covered externally by a layer of cuboidal epithelial cells called the *germinal epithelium,* which is continu-

ous with the peritoneum. The term *germinal epithelium* is a misnomer because this layer does not give rise to ova. The ovary has an outer cortex, which houses the forming gametes, and an inner medullary region, containing the largest blood vessels and nerves, but the relative extent of each region is poorly defined.

Embedded in the highly vascular connective tissue of the ovary cortex are many tiny saclike structures called **ovarian follicles.** Each follicle consists of an immature egg, called an **oocyte** (o´o-sīt; *oo* = egg), encased by one or more layers of very different cells. The surrounding cells are called **follicle cells** if a single layer is present, and **granulosa cells** when more than one layer is present. Follicles at different stages of maturation are distinguished by their structure. In a **primordial follicle,** one layer of squamouslike follicle cells surrounds the oocyte. A **primary follicle** has two or more layers of cuboidal or columnar-type granulosa cells enclosing the oocyte; it becomes a **secondary follicle** when fluid-filled spaces appear between the granulosa cells and then coalesce to form a central fluid-filled cavity called an *antrum.* At the mature **vesicular follicle,** or **Graafian follicle** (graf´e-an), stage, the follicle bulges from the surface of the ovary. The oocyte of the vesicular follicle "sits" on a stalk of granulosa cells at one side of the antrum. Each month in adult women, one of the ripening follicles ejects its oocyte from the ovary, an

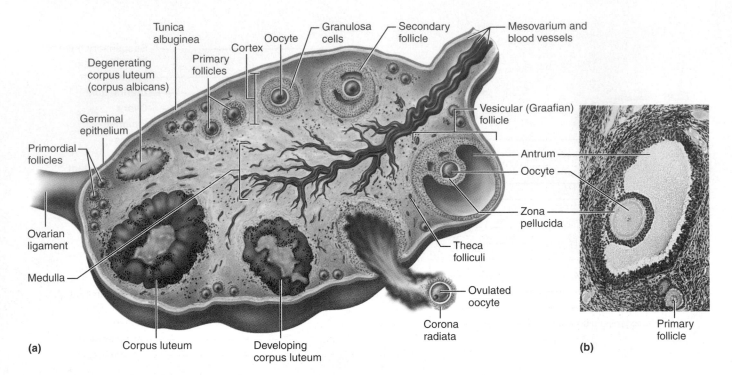

FIGURE 26.12   **Structure of an ovary.** **(a)** The ovary has been sectioned to reveal the follicles in its interior. Note that not all of these structures would appear in the ovary at the same time. **(b)** Photomicrograph of a mature vesicular (Graafian) follicle (60×).

event called **ovulation** (Figure 26.13). After ovulation, the ruptured follicle is transformed into a very different looking glandular structure called the **corpus luteum** (lu'te-um; plural, corpora lutea), which eventually degenerates. As a rule, most of these structures can be seen within the same ovary. In older women, the surfaces of the ovaries are scarred and pitted, revealing that many oocytes have been released.

The Female Duct System

The Uterine Tubes

The **uterine tubes** (u'ter-in), also called **fallopian tubes** and **oviducts,** form the initial part of the female duct system (Figures 26.11 and 26.14a). They receive the ovulated oocyte and provide a site where fertilization can occur. Each uterine tube is about 10 cm (4 inches) long and extends medially from the region of an ovary to empty into the superolateral region of the uterus via a constricted region called the **isthmus** (is'mus). The distal end of each uterine tube expands as it curves around the ovary, forming the **ampulla;** fertilization usually occurs in this region. The ampulla ends in the **infundibulum** (in"fun-dib'u-lum), an open, funnel-shaped structure bearing ciliated, fingerlike projections called **fimbriae** (fim'bre-e; "fringe") that drape over the ovary. Unlike the male duct system, which is continuous with the tubules of the testes, the uterine tubes have little

or no actual contact with the ovaries. An ovulated oocyte is cast into the peritoneal cavity, and many oocytes are lost there. However, the uterine tube performs complex movements to capture oocytes—it bends to drape over the ovary while the fimbriae

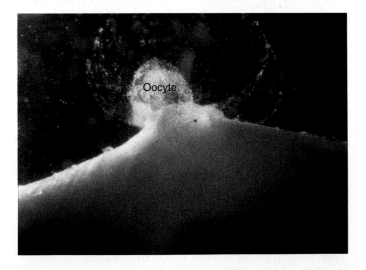

FIGURE 26.13   **Ovulation.** A secondary oocyte is released from a follicle at the surface of the ovary (30×). (The orange mass below the ejected oocyte is part of the ovary. Primary and secondary oocytes are explained in Figure 26.19.)

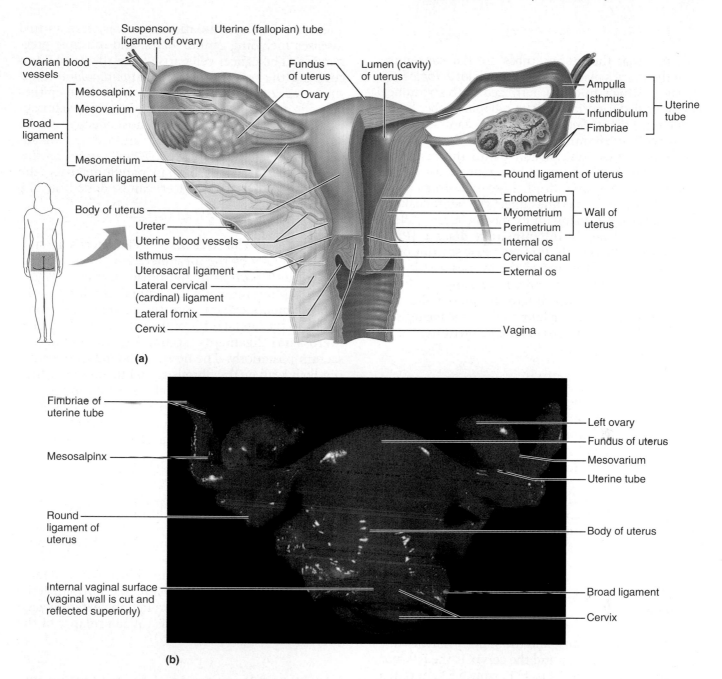

FIGURE 26.14 Internal reproductive organs of a female. (a) Posterior view of the female reproductive organs. The posterior walls of the vagina, uterus, and uterine tubes and the broad ligament have been removed on the right side to reveal the shape of the lumen of these organs. **(b)** Anterior view of the reproductive organs of a female cadaver. (See *A Brief Atlas of the Human Body*, Figure 60.)

stiffen and sweep the ovarian surface. The beating cilia on the fimbriae then create currents in the peritoneal fluid that tend to carry the oocyte into the uterine tube, where it begins its journey toward the uterus.

The uterine tube contains sheets of smooth muscle, and its thick, highly folded mucosa contains both ciliated and nonciliated cells. The oocyte is carried toward the uterus by a combination of muscular peristalsis and the beating of the cilia. Nonciliated

cells of the mucosa have dense microvilli and produce a secretion that keeps the oocyte (and sperm, if present) moist and nourished.

Externally, the uterine tubes are covered by visceral peritoneum and supported along their length by a short mesentery (part of the broad ligament) called the **mesosalpinx** (mez″o-sal′pinks; "mesentery of the trumpet"; *salpin* = trumpet), a reference to the trumpet-shaped uterine tube it supports (see Figure 26.14a).

HOMEOSTATIC IMBALANCE

The fact that the uterine tubes are not continuous with the ovaries places women at risk for *ectopic pregnancy* in which an ovum, fertilized in the peritoneal cavity or distal portion of the fallopian tube, begins developing there. Such pregnancies naturally abort, often with substantial bleeding. Another potential problem is infection spreading into the peritoneal cavity from other parts of the reproductive tract. Gonorrhea bacteria and other sexually transmitted microorganisms sometimes infect the peritoneal cavity in this way, causing an extremely severe inflammation called **pelvic inflammatory disease (PID)**. Unless treated promptly, PID can cause scarring of the narrow uterine tubes and of the ovaries and lead to sterility, and even death. Indeed, scarring and closure of the uterine tubes, which have an internal diameter as small as the width of a human hair in some regions, is one of the major causes of female infertility. ●

The Uterus

The **uterus** (Latin for "womb") is located in the pelvis, anterior to the rectum and posterosuperior to the bladder (see Figures 26.11 and 26.14). It is a hollow, thick-walled muscular organ that functions to receive, retain, and nourish a fertilized ovum. In a premenopausal woman who has never been pregnant, the uterus is about the size and shape of an inverted pear, but it is usually somewhat larger in women who have borne children. Normally, the uterus flexes anteriorly where it joins the vagina, causing the uterus as a whole to be inclined forward or *anteverted*. However, the organ is frequently turned backward, or *retroverted*, in older women.

The major portion of the uterus is referred to as the **body** (see Figure 26.11 and Figure 26.14). The rounded region superior to the entrance of the uterine tubes is the **fundus**, and the slightly narrowed region between the body and the cervix is the *isthmus*. The **cervix** of the uterus is its narrow neck, or outlet, which projects into the vagina inferiorly. The cavity of the cervix, called the **cervical canal**, communicates with the vagina via the *external os* (*os* = mouth) and with the cavity of the uterine body via the *internal os*. The mucosa of the cervical canal contains *cervical glands* that secrete a mucus that fills the cervical canal and covers the external os, presumably to block the spread of bacteria from the vagina into the uterus. Cervical mucus also blocks the entry of sperm, except at midcycle, when it becomes less viscous and allows sperm to pass through.

 HOMEOSTATIC IMBALANCE

Cancer of the cervix is common among women between the ages of 30 and 50. Risk factors include fre-quent cervical inflammations, sexually transmitted diseases including genital warts, and multiple pregnancies. The cancer cells arise from the epithelium covering the cervical tip. In a *Papanicolaou (Pap) smear*, or cervical smear test, some of these epithelial cells are scraped away and then examined for abnormalities. A Pap smear is the most effective way to detect this slow-growing cancer, and women are advised to have one every year. When results are inconclusive, a test for human papillomavirus (the main cause of cervical cancer) can be done from the same Pap sample. ●

Supports of the Uterus   The uterus is supported laterally by the **mesometrium** ("mesentery of the uterus") portion of the broad ligament (Figure 26.14a). More inferiorly, the **lateral cervical (cardinal) ligaments** extend from the cervix and superior vagina to the lateral walls of the pelvis, and the paired **uterosacral ligaments** secure the uterus to the sacrum posteriorly. The uterus is bound to the anterior body wall by the fibrous **round ligaments**, which run through the inguinal canals to anchor in the subcutaneous tissue of the labia majora. These ligaments allow the uterus a good deal of mobility, and its position changes as the rectum and bladder fill and empty.

 HOMEOSTATIC IMBALANCE

Despite the many anchoring ligaments, the principal support of the uterus is provided by the muscles of the pelvic floor, namely the muscles of the urogenital and pelvic diaphragms (Table 10.7). These muscles are sometimes torn during childbirth. Subsequently, the unsupported uterus may sink inferiorly, until the tip of the cervix protrudes through the external vaginal opening. This condition is called **prolapse of the uterus.** ●

The undulating course of the peritoneum produces several cul-de-sacs, or blind-ended peritoneal pouches. The most important of these are the *vesicouterine pouch* (ves"i-ko-u'ter-in) between the bladder and uterus, and the *rectouterine pouch* between the rectum and uterus (see Figure 26.11).

The Uterine Wall   The wall of the uterus is composed of three layers (Figure 26.14a). The **perimetrium,** the outermost serous layer, is the visceral peritoneum. The **myometrium** (mi"o-me'tre-um; "muscle of the uterus") is the bulky middle layer, composed of interlacing bundles of smooth muscle, that contracts rhythmically during childbirth to expel the baby from the mother's body. The **endometrium** is the mucosal lining of the uterine cavity (Figure 26.15); it is a simple columnar epithe-

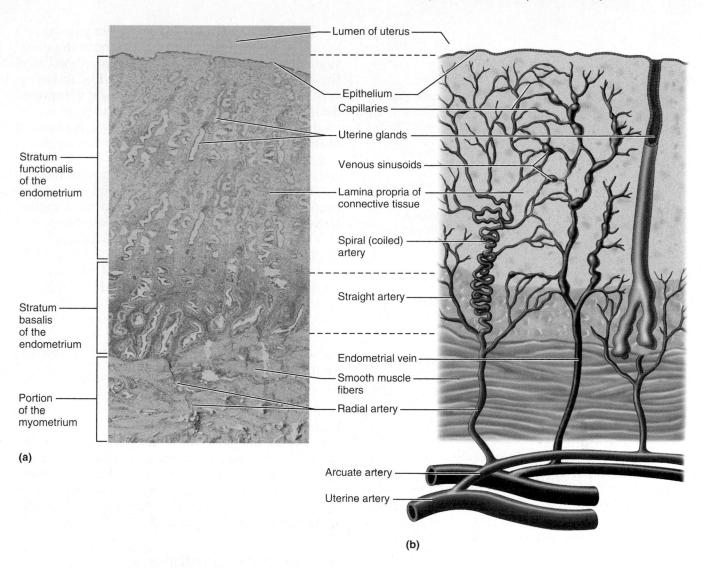

FIGURE 26.15   **The endometrium and its blood supply. (a)** Photomicrograph of the endometrium, longitudinal section, showing its functionalis and basalis regions (35×). **(b)** Diagrammatic view of the endometrium, showing the straight arteries that serve the stratum basalis and the spiral arteries that serve the stratum functionalis. The thin-walled veins and venous sinusoids are also illustrated.

lium underlain by a thick lamina propria. If fertilization occurs, the young embryo burrows into the endometrium (implants) and resides there for the rest of its development.

The endometrium has two chief *strata* (layers). The **stratum functionalis** (fungk-shun-a′lis), or **functional layer,** undergoes cyclic changes in response to blood levels of ovarian hormones and is shed during menstruation (approximately every 28 days). The thinner, deeper **stratum basalis** (ba-să′lis), or **basal layer,** forms a new functionalis after menstruation ends. It is unresponsive to ovarian hormones. The endometrium has numerous *uterine glands* that change in length as endometrial thickness changes.

To understand the cyclic changes of the uterine endometrium (discussed later in the chapter), it is essential to understand the vascular supply of the uterus. The **uterine arteries** arise from the *internal iliacs* in the pelvis, ascend along the sides of the uterus, and send branches into the uterine wall (Figure 26.14a). These branches break up into several **arcuate arteries** (ar′ku-āt) within the myometrium. The arcuate arteries send **radial branches** into the endometrium, where they in turn give off **straight arteries** to the stratum basalis and **spiral (coiled) arteries** to the stratum functionalis. The spiral arteries repeatedly degenerate and regenerate, and it is their spasms that actually cause the functionalis layer to be shed during menstruation. Veins in the endometrium are thin-walled and form an extensive network with occasional sinusoidal enlargements.

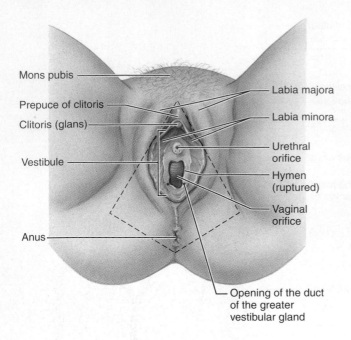

Mons pubis

Prepuce of clitoris

Clitoris (glans)

Vestibule

Anus

Labia majora

Labia minora

Urethral orifice

Hymen (ruptured)

Vaginal orifice

Opening of the duct of the greater vestibular gland

FIGURE 26.16   The external genitalia (vulva) of the female. The region enclosed by the dashed lines is the perineum.

The Vagina

The **vagina** ("sheath") is a thin-walled tube, 8–10 cm (3–4 inches) long. It lies between the bladder and the rectum and extends from the cervix to the body exterior (see Figure 26.11). The urethra is embedded in its anterior wall. Often called the *birth canal,* the vagina provides a passageway for delivery of an infant and for menstrual flow. Since it receives the penis (and semen) during sexual intercourse, it is the *female organ of copulation.*

The distensible wall of the vagina consists of three coats: an outer fibroelastic *adventitia,* a smooth muscle *muscularis,* and a *mucosa* marked by transverse ridges or *rugae,* which stimulate the penis during intercourse. The mucosa is a stratified squamous epithelium adapted to stand up to friction. Certain mucosal cells *(dendritic cells)* act as antigen-presenting cells and are thought to provide the route of HIV transmission from an infected male to the female during sexual intercourse. (AIDS, the immune deficiency disease caused by HIV, is described in Chapter 20.) The vaginal mucosa has no glands; it is lubricated by the cervical mucous glands. Its epithelial cells release large amounts of glycogen, which is anaerobically metabolized to lactic acid by resident bacteria. Consequently, the pH of a woman's vagina is normally quite acidic. This acidity helps keep the vagina healthy and free of infection, but it is also hostile to sperm. Although vaginal fluid of *adult* females is acidic, it tends to be alkaline in adolescents, predisposing sexually active teenagers to sexually transmitted diseases.

In virgins (females who have never participated in sexual intercourse), the mucosa near the distal **vaginal orifice** forms an incomplete partition called the **hymen** (hi'men) (Figure 26.16). The hymen is very vascular and tends to bleed when it is ruptured during the first coitus (sexual intercourse). However, its durability varies. In some females, it is ruptured during a sports activity, tampon insertion, or pelvic examination. Occasionally, it is so tough that it must be breached surgically if intercourse is to occur.

The upper end of the vaginal canal loosely surrounds the cervix of the uterus, producing a vaginal recess called the **vaginal fornix.** The posterior part of this recess, the *posterior fornix,* is much deeper than the *lateral* and *anterior fornices* (see Figures 26.11 and 26.14a). Generally, the lumen of the vagina is quite small and, except where it is held open by the cervix, its posterior and anterior walls are in contact with one another. The vagina stretches considerably during copulation and childbirth, but its lateral distension is limited by the ischial spines and the sacrospinous ligaments.

HOMEOSTATIC IMBALANCE

The uterus tilts away from the vagina. Hence, attempts by untrained persons to induce an abortion by entering the uterus with a surgical instrument may result in puncturing of the posterior wall of the vagina, followed by hemorrhage and—if the instrument is unsterile—peritonitis.   ●

The External Genitalia

As mentioned earlier, the female reproductive structures that lie external to the vagina are called the *external genitalia* (see Figure 26.16). The external genitalia, also called the **vulva** (vul'vah; "covering") or **pudendum** ("shameful"), include the mons pubis, labia, clitoris, and structures associated with the vestibule.

The **mons pubis** (mons pu'bis; "mountain on the pubis") is a fatty, rounded area overlying the pubic symphysis. After puberty, this area is covered with pubic hair. Running posteriorly from the mons pubis are two elongated, hair-covered fatty skin folds, the **labia majora** (la'be-ah mah-jor'ah; "larger lips"). These are the female counterpart, or *homologue,* of the male scrotum (that is, they derive from the same embryonic tissue). The labia majora enclose the **labia minora** (mi-nor'ah; "smaller"), two thin, hair-free skin folds, homologous to the ventral penis. The labia minora enclose a recess called the **vestibule,** which contains the external opening of the urethra more anteriorly as well as that of the vagina. Flanking the vaginal opening are the pea-size **greater vestibular glands** (not illustrated), homolo-

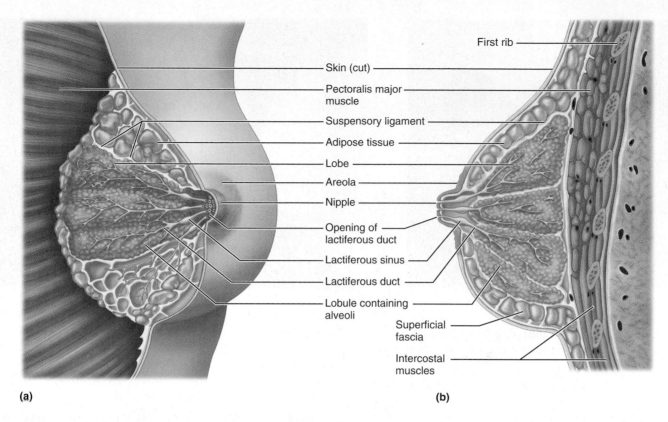

Labels (left to right, part a and b):

First rib

Skin (cut)

Pectoralis major muscle

Suspensory ligament

Adipose tissue

Lobe

Areola

Nipple

Opening of lactiferous duct

Lactiferous sinus

Lactiferous duct

Lobule containing alveoli

Superficial fascia

Intercostal muscles

(a)   (b)

FIGURE 26.17 **Structure of lactating mammary glands. (a)** Anterior view of a partially dissected breast. **(b)** Sagittal section of a breast.

gous to the bulbourethral glands of males. These glands release mucus into the vestibule and help to keep it moist and lubricated, facilitating intercourse.

Just anterior to the vestibule is the **clitoris** (klit'o-ris; "hill"), a small, protruding structure, composed largely of erectile tissue, that is homologous to the penis of the male. Its exposed portion is called the glans. It is hooded by a skin fold called the **prepuce of the clitoris,** formed by the junction of the labia minora folds. The clitoris is richly innervated with sensory nerve endings sensitive to touch, and it becomes swollen with blood and erect during tactile stimulation, contributing to a female's sexual arousal. Like the penis, the clitoris has dorsal erectile columns (corpora cavernosa), but it lacks a corpus spongiosum. In males, the urethra carries both urine and semen and runs through the penis, but the female urinary and reproductive tracts are completely separate, and neither runs through the clitoris.

The female **perineum** is a diamond-shaped region located between the pubic arch anteriorly, the coccyx posteriorly, and the ischial tuberosities laterally. The soft tissues of the perineum overlie the muscles of the pelvic outlet, and the posterior ends of the labia majora overlie the *central tendon,* into which most muscles supporting the pelvic floor insert (see Table 10.7).

The Mammary Glands

The **mammary glands** are present in both sexes, but they normally function only in females (Figure 26.17). The biological role of the mammary glands is to produce milk to nourish a newborn baby, so they are important only when reproduction has already been accomplished.

Developmentally, mammary glands are modified sweat glands that are really part of the *skin,* or *integumentary system.* Each mammary gland is contained within a rounded skin-covered breast within the superficial fascia, anterior to the pectoral muscles of the thorax. Slightly below the center of each breast is a ring of pigmented skin, the **areola** (ah-re'o-lah), which surrounds a central protruding **nipple.** Large sebaceous glands in the areola make it slightly bumpy and produce sebum that reduces chapping and cracking of the skin of the nipple. Autonomic nervous system controls of smooth muscle fibers in the areola and nipple cause the nipple to become erect when stimulated by tactile or sexual stimuli and when exposed to cold.

Internally, each mammary gland consists of 15 to 25 **lobes** that radiate around and open at the nipple. The lobes are padded and separated from each other by fibrous connective tissue and fat. The interlobar connective tissue forms **suspensory ligaments**

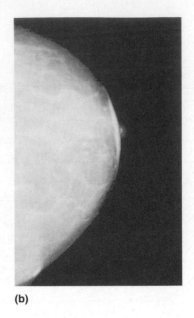

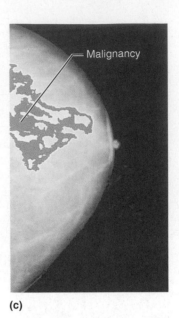

Malignancy

(a)　　　　　　　　　　　　　　(b)　　　　　　　　　　　　　　(c)

FIGURE 26.18　**Mammograms. (a)** Photograph of a woman undergoing mammography. **(b)** Normal breast. **(c)** Breast with tumor.

that attach the breast to the underlying muscle fascia and to the overlying dermis. As suggested by their name, the suspensory ligaments provide natural support for the breasts, like a built-in brassiere. Within the lobes are smaller units called **lobules,** which contain glandular **alveoli** that produce milk when a woman is lactating. These compound alveolar glands pass the milk into the **lactiferous ducts** (lak-tif′er-us), which open to the outside at the nipple. Just deep to the areola, each duct has a dilated region called a **lactiferous sinus** where milk accumulates during nursing.

The description of mammary glands given here applies only to nursing women or women in the last trimester of pregnancy. In nonpregnant women, the glandular structure of the breast is largely undeveloped and the duct system is rudimentary; hence breast size is largely due to the amount of fat deposits.

Breast Cancer

Invasive breast cancer, the most common malignancy of U.S. women, strikes nearly 200,000 American women each year. One woman in eight will develop this condition. Breast cancer usually arises from the epithelial cells of the ducts, not from the alveoli. A small cluster of cancer cells grows into a lump in the breast from which cells eventually metastasize.

Known risk factors for developing breast cancer include (1) early onset menses and late menopause; (2) no pregnancies or first pregnancy later in life; (3) previous history of breast cancer; and (4) family history of breast cancer (especially in a sister or mother). Some 10% of breast cancers stem from hereditary defects and half of these can be traced to dangerous mutations in a pair of genes, dubbed *BRCA1* (breast cancer 1) and *BRCA2*, which virtually guarantee that the carriers will develop breast cancer. However, more than 70% of women who develop breast cancer have no known risk factors for the disease.

Breast cancer is often signaled by a change in skin texture, puckering, or leakage from the nipple. Early detection by breast self-examination and mammography is unquestionably the best way to increase one's chances of surviving breast cancer. Since most breast lumps are discovered by women themselves in routine monthly breast exams, this simple examination should be a health maintenance priority in every woman's life. The American Cancer Society had recommended scheduling **mammography**—X-ray examination that detects breast cancers too small to feel (less than 1 cm)—every two years for women between 40 and 49 years old and yearly thereafter (Figure 26.18); however, some authorities are suggesting that yearly is too frequent.

Once diagnosed, breast cancer is treated in various ways depending on specific characteristics of the lesion. Current therapies include (1) radiation therapy, (2) chemotherapy, and (3) surgery, often followed by irradiation or chemotherapy to destroy stray cancer cells. Newer on the therapy scene are Herceptin, a drug containing bioengineered antibodies that jam estrogen receptors that control aggressive growth in breast cancer cells; and tamoxifen, an antiestrogen compound that significantly improves the outcome for those with early- or late-stage breast cancer. Until the 1970s, the standard treatment was **radical mastectomy** (mas-tek'to-me; "breast cutting"), removal of the entire affected breast, plus all underlying muscles, fascia, and associated lymph nodes. Most physicians now recommend **lumpectomy,** less extensive surgery in which only the cancerous part (lump) is excised, or **simple mastectomy,** removal of the breast tissue only (and perhaps some of the axillary lymph nodes).

Many mastectomy patients opt for breast reconstruction to replace the excised tissue. Currently tissue "flaps," containing muscle, fat, and skin taken from the patient's abdomen or back, are providing acceptable alternatives for "sculpting" a natural-looking breast.

Physiology of the Female Reproductive System

Oogenesis

Gamete production in males begins at puberty and continues throughout life, but the situation is quite different in females. A female's total supply of eggs is already determined by the time she is born, and the time span during which she releases them extends from puberty to menopause (about the age of 50).

Meiosis, the specialized nuclear division that occurs in the testes to produce sperm, also occurs in the ovaries. In this case, female sex cells are produced, and the process is called **oogenesis** (o"o-gen'e-sis; "the beginning of an egg"). The process of oogenesis (Figure 26.19) takes years to complete. First, in the fetal period the **oogonia,** the diploid stem cells of the ovaries multiply rapidly by mitosis and, then enter a growth phase and lay in nutrient reserves. Gradually, *primordial follicles* begin to appear as the oogonia are transformed into **primary oocytes** and become surrounded by a single layer of flattened follicle cells. The primary oocytes begin the first meiotic division, but become "stalled" late in prophase I and do not complete it. By birth, a female has her lifetime supply of primary oocytes; of the original 7 million oocytes approximately 2 million of them escape programmed death and are already in place in the cortical region of the immature ovary. Since they remain in their state of suspended animation all through childhood, the wait is a long one—10 to 14 years at the very least!

At puberty, perhaps 400,000 oocytes remain and beginning at this time a small number of primary oocytes are activated each month. However, only one is "selected" each time to continue meiosis I, ultimately producing two haploid cells (each with 23 replicated chromosomes) that are quite dissimilar in size. The smaller cell is called the **first polar body.** The larger cell, which contains nearly all the cytoplasm of the primary oocyte, is the **secondary oocyte.** The events of this first maturation division are interesting. A spindle forms at the very edge of the oocyte (see Figure 26.19, left), and a little "nipple," into which the polar body chromosomes will be cast, appears at that edge. This sets up the polarity of the oocyte and ensures that the polar body receives almost no cytoplasm or organelles.

The first polar body may continue its development and undergo meiosis II, producing two even smaller polar bodies. However, in humans, the secondary oocyte arrests in metaphase II and it is this cell (not a functional ovum) that is ovulated. If an ovulated secondary oocyte is not penetrated by a sperm, it simply deteriorates. But, if sperm penetration does occur, it quickly completes meiosis II, yielding one large **ovum** and a tiny **second polar body** (see Figure 26.19). The union of the egg and sperm nuclei constitutes fertilization. What you should realize now is that the potential end products of oogenesis are three tiny polar bodies, nearly devoid of cytoplasm, and one large ovum. All of these cells are haploid, but only the ovum is a *functional gamete.* This is quite different from spermatogenesis, where the product is four viable gametes—spermatozoa.

The unequal cytoplasmic divisions that occur during oogenesis ensure that a fertilized egg has ample nutrients for its seven-day journey to the uterus. Without nutrient-containing cytoplasm the polar bodies degenerate and die. Since the reproductive life of a female is at best about 40 years (from the age of 11 to approximately 51) and typically only one ovulation occurs each month, fewer than 500 oocytes out of her estimated pubertal potential of 400,000 are released during a woman's lifetime. Again, nature has provided us with a generous oversupply of sex cells.

The Ovarian Cycle

The monthly series of events associated with the maturation of an egg is called the **ovarian cycle.** The ovarian cycle is best described in terms of two consecutive phases. The **follicular phase** is the period of follicle growth, typically indicated as lasting from the first to the fourteenth day of the cycle. The **luteal phase** is the period of corpus luteum activity, days 14–28. The "typical" ovarian cycle repeats at intervals of 28 days, with *ovulation* occurring mid-cycle. However, cycles as long as 40 days or as short as 21 days are fairly common. In such cases, the length of the follicular phase and timing of ovulation

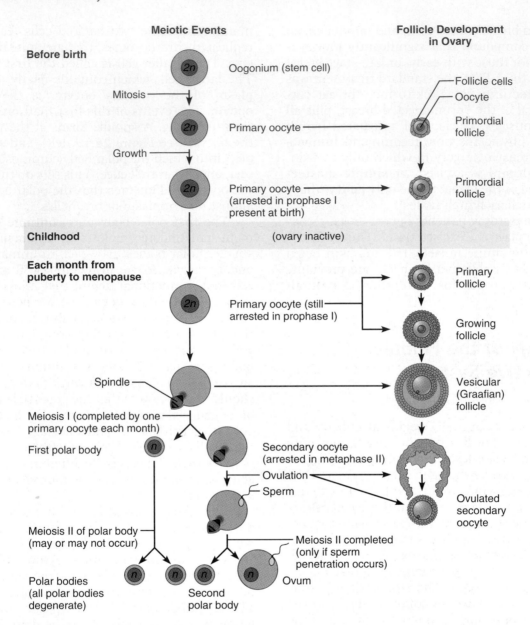

FIGURE 26.19 **Events of oogenesis.** Left, flowchart of meiotic events. Right, correlation with follicle development and ovulation in the ovary.

vary, but the luteal phase remains constant: It is 14 days between the time of ovulation and the end of the cycle.

The Follicular Phase

Maturation of a primordial follicle to the mature state occupies the first half of the cycle and involves several events as shown in Figure 26.20, steps 1–6.

A Primordial Follicle Becomes a Primary Follicle

When the primordial follicles ① are activated (a process directed by the oocyte), the squamouslike cells surrounding the primary oocyte grow, becoming cuboidal cells, and the oocyte enlarges. The follicle is now called a primary follicle ②.

A Primary Follicle Becomes a Secondary Follicle

Next, follicular cells proliferate, forming a stratified epithelium around the oocyte ③. As soon as more than one cell layer is present, the follicle cells take on the name *granulosa cells*. The granulosa cells are connected to the developing oocyte by gap junctions, through which ions, metabolites, and signaling molecules can pass and from this point on, bidirectional "conversations" occur between the oocyte and granulosa cells, so they guide one another's development. One of the signals passing from the granulosa cells to the oocyte "tells" the oocyte to grow.

In the next stage ④, a layer of connective tissue condenses around the follicle, forming the **theca fol-**

FIGURE 26.20 **Schematic view of the ovarian cycle: development and fate of ovarian follicles.** The numbers on the diagram indicate the sequence of events in follicle development, *not* the movement of a developing follicle within the ovary. ① Primordial follicle containing a primary oocyte surrounded by flattened cells. ② A primary follicle containing a primary oocyte surrounded by cuboidal follicle cells. ③ and ④ The growing primary follicle, which is secreting estrogen as it continues to mature. ⑤ The secondary follicle with its forming antrum. ⑥ The mature vesicular follicle, ready to be ovulated. Meiosis I, producing the secondary oocyte and first polar body, occurs in the mature follicle. ⑦ The ruptured follicle and ovulated secondary oocyte surrounded by its corona radiata of granulosa cells. ⑧ The corpus luteum, formed from the ruptured follicle under the influence of LH, produces progesterone (and estrogens). ⑨ The scarlike corpus albicans.

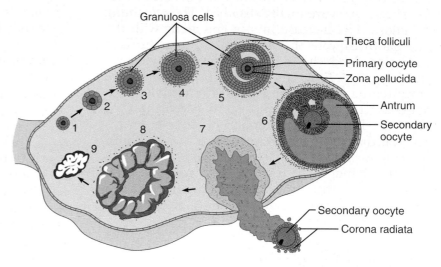

liculi (the'kah fah-lik'u-li; "box around the follicle"). As the follicle grows, the thecal and granulosa cells cooperate to produce estrogens (the inner thecal cells produce androgens, which the granulosa cells convert to estrogens). At the same time, the granulosa cells secrete a glycoprotein-rich substance that forms a thick transparent membrane, called the **zona pellucida** (pĕ-lu'sid-ah), around the oocyte.

In phase ⑤, clear liquid accumulates between the granulosa cells and eventually coalesces to form a fluid-filled cavity called the **antrum** ("cave"). The presence of an antrum distinguishes the new secondary follicle from the primary follicle.

A Secondary Follicle Becomes a Vesicular Follicle

The antrum continues to expand with fluid until it isolates the oocyte, along with its surrounding capsule of granulosa cells called a **corona radiata** ("radiating crown"), on a stalk on one side of the follicle. When a follicle is full size (about 2.5 cm, or 1 inch, in diameter), it becomes a vesicular follicle ⑥ and bulges from the external ovarian surface like an "angry boil." This usually occurs by day 14.

As one of the final events of follicle maturation, the primary oocyte completes meiosis I to form the secondary oocyte and first polar body (see Figure 26.19). Once this has occurred (⑥ in Figure 26.20), the stage is set for ovulation. At this point, the gran-

ulosa cells send another important signal to the oocyte that says, in effect, "Wait, do not complete meiosis yet!"

Ovulation

Ovulation occurs when the ballooning ovary wall ruptures and expels the secondary oocyte (still surrounded by its *corona radiata*) into the peritoneal cavity ⑦. Some women experience a twinge of pain in the lower abdomen when ovulation occurs. This episode, called *mittelschmerz* (mit'el-shmārts; German for "middle pain"), is caused by the intense stretching of the ovarian wall during ovulation.

In the ovaries of adult females, there are always several follicles at different stages of maturation. As a rule, one follicle outstrips the others to become the *dominant follicle* and is at the peak stage of maturation when the hormonal (LH) stimulus is given for ovulation. How this follicle is selected, or selects itself, is still uncertain, but it is probably the one that attains the greatest FSH sensitivity the quickest. The others degenerate and are reabsorbed.

In 1–2% of all ovulations, more than one oocyte is ovulated. This phenomenon, which increases with age, can result in multiple births. Since different oocytes are fertilized by different sperm, the siblings are *fraternal*, or nonidentical, twins. Identical twins result from the fertilization of a single oocyte by a single sperm, followed by separation of the fertilized egg's daughter cells in early development.

■ *1–5 contain primary oocytes; 6 contains a secondary oocyte.*

The Luteal Phase

After ovulation, the ruptured follicle collapses, and the antrum fills with clotted blood. This *corpus hemorrhagicum* is eventually absorbed. The remaining granulosa cells increase in size and along with the internal thecal cells they form a new, quite different endocrine gland, the *corpus luteum* ("yellow body") (see Figure 26.20, step (8)), that begins to secrete progesterone and some estrogen. If pregnancy does not occur, the corpus luteum starts degenerating in about ten days and its hormonal output ends. In this case, all that ultimately remains is a scar (9) called the *corpus albicans* (al'bĭ-kans; "white body"). On the other hand, if the oocyte is fertilized and pregnancy ensues, the corpus luteum persists until the placenta is ready to take over its hormone-producing duties in about three months.

Hormonal Regulation of the Ovarian Cycle

Ovarian events are much more complicated than those occurring in the testes, but the hormonal controls set into motion at puberty are similar in the two sexes. Gonadotropin-releasing hormone (GnRH), the pituitary gonadotropins, and, in this case, ovarian estrogen and progesterone interact to produce the cyclic events occurring in the ovaries.

Establishing the Ovarian Cycle

During childhood, the ovaries grow and continuously secrete small amounts of estrogens which inhibit hypothalamic release of GnRH. As puberty nears, the hypothalamus becomes less sensitive to estrogen and begins to release GnRH in a rhythmic pulselike manner. GnRH, in turn, stimulates the anterior pituitary to release FSH and LH, which act on the ovaries.

Gonadotropin levels continue increasing for about four years and, during this time, pubertal girls are still not ovulating and thus are incapable of getting pregnant. Eventually, the adult cyclic pattern is achieved, and hormonal interactions stabilize. These events are heralded by the young woman's first menstrual period, referred to as **menarche** (mĕ-nar'ke; *men* = month, *arche* = first). Usually, it is not until the third year postmenarche that the cycles become regular and all are ovulatory.

Hormonal Interactions During the Ovarian Cycle

The waxing and waning of anterior pituitary gonadotropins (FSH and LH) and ovarian hormones and the positive and negative feedback interactions that regulate ovarian function are as described next.

Notice that events 1–8 in the following discussion directly correspond to the same numbered steps in Figure 26.21. A 28-day cycle is assumed.

(1) On day 1 of the cycle, rising levels of GnRH from the hypothalamus stimulate increased production and release of follicle-stimulating hormone (FSH) and luteinizing hormone (LH) by the anterior pituitary.

(2) FSH and LH stimulate follicle growth and maturation and estrogen secretion. FSH exerts its main effects on the follicle cells, whereas LH (at least initially) targets the thecal cells. (Why only *some* follicles respond to these hormonal stimuli is still a mystery. However, there is little doubt that enhanced responsiveness is due to formation of more gonadotropin receptors.) As the follicles enlarge, LH prods the thecal cells to produce androgens. These diffuse through the basement membrane, where they are converted to estrogens by the granulosa cells. Only tiny amounts of ovarian androgens enter the blood, because they are almost completely converted to estrogens within the ovaries.

(3) The rising estrogen levels in the plasma exert *negative feedback* on the anterior pituitary, inhibiting its release of FSH and LH, while simultaneously prodding it to synthesize and accumulate these gonadotropins. Within the ovary, estrogen increases estrogen output by intensifying the effect of FSH on follicle maturation. *Inhibin*, released by the granulosa cells, also exerts negative feedback controls on FSH release during this period.

(4) Although the initial small rise in estrogen blood levels inhibits the hypothalamic-pituitary axis, high estrogen levels have the opposite effect. Once estrogen reaches a critical blood concentration, it exerts *positive feedback* on the brain and anterior pituitary.

(5) High estrogen levels set a cascade of events into motion. There is a sudden burstlike release of accumulated LH (and, to a lesser extent, FSH) by the anterior pituitary about midcycle (also see Figure 26.22a).

(6) The LH surge stimulates the primary oocyte of the dominant follicle to complete the first meiotic division, forming a secondary oocyte that continues on to metaphase II. LH also triggers ovulation at or around day 14. Whatever the mechanism, blood stops flowing through the protruding part of the follicle wall and within 5 minutes, that region of the follicle wall bulges out, thins, and then ruptures. The role (if any) of FSH in this process is unknown. Shortly after ovulation, estrogen levels decline. This probably reflects the damage to the dominant estrogen-secreting follicle during ovulation.

(7) The LH surge also transforms the ruptured follicle into a corpus luteum (hence the name "luteinizing" hormone), and stimulates the newly formed endocrine gland to produce progesterone and estrogen almost immediately after it is formed.

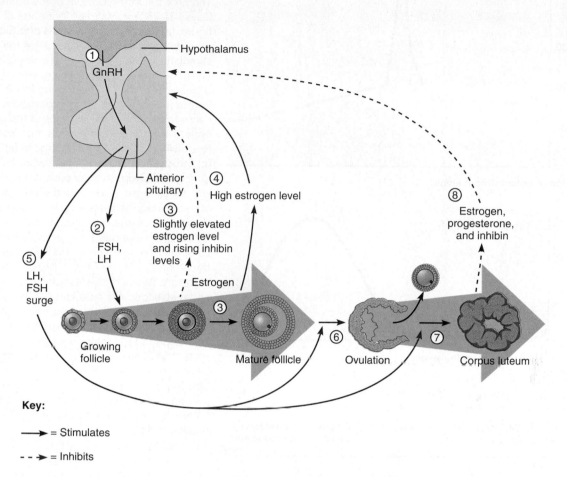

FIGURE 26.21 Feedback interactions in the regulation of ovarian function. Numbers refer to events listed in the text. Events that follow step 8 (negative feedback inhibition of the hypothalamus and anterior pituitary by progesterone and estrogens) are not depicted, but involve a gradual deterioration of the corpus luteum and, therefore, a decline in ovarian hormone production. Ovarian hormones reach their lowest blood levels around day 28.

⑧ Rising progesterone and estrogen blood levels exert a powerful negative feedback effect on anterior pituitary release of LH and FSH. Corpus luteum release of inhibin enhances this inhibitory effect. Declining gonadotropin levels inhibit the development of new follicles and prevent additional LH surges that might cause additional oocytes to be ovulated.

As LH blood levels fall, the stimulus for luteal activity ends, and the corpus luteum degenerates. As goes the corpus luteum, so go the levels of ovarian hormones, and blood estrogen and progesterone levels drop sharply.

The marked decline in ovarian hormones at the end of the cycle (days 26–28) ends their blockade of FSH and LH secretion, and the cycle starts anew.

The Uterine (Menstrual) Cycle

Although the uterus is where the young embryo implants and develops, it is receptive to implantation only for a very short period each month. Not surprisingly, this brief interval is exactly the time when a developing embryo would normally begin implanting, about seven days after ovulation. The **uterine**, or **menstrual** (men'stroo-al), **cycle** is a series of cyclic changes that the uterine endometrium goes through each month as it responds to changing levels of ovarian hormones in the blood. These

Stimulation of LH (and FSH) surges by rising estrogen levels, steps 4 and 5. ■

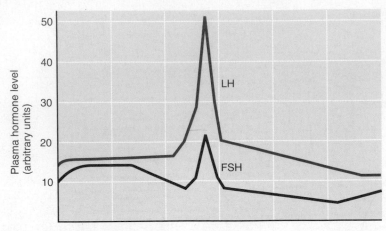

(a) Fluctuation of gonadotropin levels

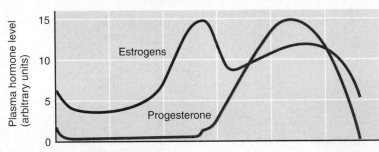

(b) Fluctuation of ovarian hormone levels

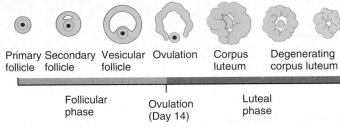

(c) Ovarian cycle

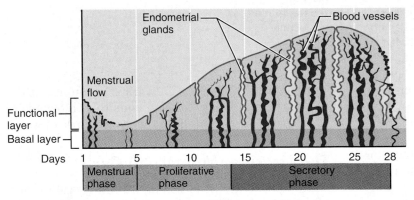

(d) Uterine cycle

FIGURE 26.22 **Correlation of anterior pituitary and ovarian hormones with structural changes of the ovary and uterus.** The time bar at the bottom of the figure, reading Days 1 to 28, applies to all four parts of this figure. **(a)** The fluctuating levels of pituitary gonadotropins in the blood (FSH = follicle-stimulating hormone; LH = luteinizing hormone). These hormones regulate the events of the ovarian cycle. **(b)** The fluctuating levels of ovarian hormones (estrogens and progesterone) that cause the endometrial changes of the uterine cycle. The high estrogen levels are also responsible for the LH/FSH surge in (a). **(c)** Structural changes in the ovarian follicles during the 28-day ovarian cycle are correlated with (d) changes in the endometrium of the uterus during the uterine cycle. Though not shown here, some evidence indicates that follicular development may actually *begin* on day 25 of the preceding cycle. **(d)** The three phases of the uterine cycle: menstrual, proliferative, and secretory. Basically, the first phase is a shedding and the second is a rebuilding of the functional zone of the endometrium. Both of these events occur before ovulation and together they correspond to the follicular phase of the ovarian cycle. The third phase, the secretory phase which begins immediately after ovulation, enriches the blood supply and provides the nutrients that prepare the endometrium to receive an embryo. It corresponds in time to the luteal phase of the ovarian cycle.

endometrial changes are coordinated with the phases of the ovarian cycle, which are dictated by gonadotropins released by the anterior pituitary.

The events of the uterine cycle, depicted in Figure 26.22d, are as follows:

1. Days 1–5: Menstrual phase. In this phase, **menstruation** (men"stroo-a'shun), the uterus sheds all but the deepest part of its endometrium. (At the beginning of this stage, ovarian hormones are at their lowest normal levels and gonadotropins are beginning to rise. Then FSH levels begin to rise.) The thick functional layer of the endometrium detaches from the uterine wall, a process that is accompanied by bleeding for 3–5 days. The detached tissue and blood pass out through the vagina as the menstrual flow. By day 5, the growing ovarian follicles are starting to produce more estrogen (see Figure 26.22b).

2. Days 6–14: Proliferative (preovulatory) phase. In this phase, the endometrium rebuilds itself: Under the influence of rising blood levels of estrogen, the basal layer of the endometrium generates a new functional layer. As this new layer thickens, its glands enlarge and its spiral arteries increase in number (also see Figure 26.15). Consequently, the endometrium once again becomes velvety, thick, and well vascularized. During this phase, estrogens also induce synthesis of progesterone receptors in the endometrial cells, readying them for interaction with progesterone.

Normally, cervical mucus is thick and sticky, but rising estrogen levels cause it to thin and become crystalline, forming channels that facilitate the passage of sperm into the uterus.

Ovulation, which takes less than five minutes, occurs in the ovary at the end of the proliferative stage (day 14) in response to the sudden release of LH from the anterior pituitary. LH also converts the ruptured follicle to a corpus luteum.

3. Days 15–28: Secretory (postovulatory) phase. This 14-day phase is the most constant timewise. During the secretory phase the endometrium prepares for implantation of an embryo. Rising levels of progesterone from the corpus luteum act on the estrogen-primed endometrium, causing the spiral arteries to elaborate and converting the functional layer to a secretory mucosa. The uterine glands enlarge, coil, and begin secreting nutritious glycogen into the uterine cavity. These nutrients sustain the embryo until it has implanted in the blood-rich endometrial lining.

Increasing progesterone levels also cause the cervical mucus to become viscous again, forming the *cervical plug*, which blocks sperm entry and plays an important role in keeping the uterus "private" in the event an embryo has begun to implant. Rising progesterone (and estrogen) levels inhibit LH release by the anterior pituitary. If fertilization has not occurred, the corpus luteum begins to degenerate toward the end of the secretory phase as LH blood levels decline. Progesterone levels fall, depriving the endometrium of hormonal support, and the spiral arteries kink and go into spasms. Denied oxygen and nutrients, the lysosomes of the ischemic endometrial cells rupture, and the functional layer begins to self-digest, setting the stage for menstruation to begin on day 28. The spiral arteries constrict one final time and then suddenly relax and open wide. As blood gushes into the weakened capillary beds, they fragment, causing the functional layer to slough off. The menstrual cycle starts over again on this first day of menstrual flow.

Figure 26.22 also illustrates how the ovarian and uterine cycles fit together. Notice that the menstrual and proliferative phases overlap the follicular stage and ovulation in the ovarian cycle, and that the uterine secretory phase corresponds to the ovarian luteal phase.

Extremely strenuous activity can delay menarche in girls and can disrupt the normal menstrual cycle in adult women, even causing *amenorrhea* (a-men"o-re'ah), cessation of menses. Female athletes have little body fat, and fat deposits help convert adrenal androgens to estrogens. In addition, hypothalamic controls are blocked in some way by severe physical regimens. These effects are usually totally reversible when the athletic training is discontinued, but a worrisome consequence of amenorrhea in young, healthy adult women is that they suffer dramatic losses in bone mass normally seen only in old age. Once estrogen levels drop and the menstrual cycle stops (regardless of cause), bone loss begins.

Although menses has traditionally been viewed as a somewhat messy way of discarding a uterine lining "fattened" in anticipation of a baby that was never conceived, its adaptive value and expense to the female body in terms of tissue, blood, and nutrient (particularly iron) loss have recently been questioned. Why not just keep the prepared endometrium for the next cycle? A controversial view of menses proposed by Maggie Profit of the University of California at Berkeley may give you something to think about and debate. The uterus is a hospitable receptacle for bacteria and viruses delivered via the male penis and piggy-backing in semen. This being so, she theorizes that menses is an aggressive way of cleaning house. Bleeding rids the body of the uterine lining where pathogens are likely to linger, and menstrual blood is loaded with macrophages which provide active protection.

Extrauterine Effects of Estrogens and Progesterone

With a name meaning "generators of sexual activity," estrogens are analogous to testosterone, the male steroid. As estrogen levels rise during puberty, they (1) promote oogenesis and follicle growth in the ovary and (2) exert anabolic effects on the female reproductive tract (see Table 26.1). Consequently, the uterine tubes, uterus, and vagina grow larger and become functional—ready to support a pregnancy. The uterine tubes and uterus exhibit enhanced motility; the vaginal mucosa thickens; and the external genitalia mature.

Estrogens also support the growth spurt at puberty that makes girls grow much more quickly than boys during the ages of 12 and 13. But this growth is short-lived because rising estrogen levels also

TABLE 26.1   Summary of Hormonal Effects of Gonadal Estrogens, Progesterone, and Testosterone

Source, Stimulus, Effects	Estrogens	Progesterone	Testosterone
Major source	Ovary: developing follicles and corpus luteum	Ovary: mainly the corpus luteum	Testes: interstitial cells
Stimulus for release	FSH (and LH)	LH	ICSH and declining levels of inhibin produced by the sustentacular cells
Feedback effects exerted	Both negative and positive feedback exerted on anterior pituitary release of gonadotropins	Negative feedback exerted on anterior pituitary release of gonadotropins	Negative feedback suppresses release of ICSH by the anterior pituitary (and perhaps release of GnRH by the hypothalamus)
Effects on reproductive organs	Stimulate growth and maturation of reproductive organs and breasts at puberty and maintain their adult size and function. Promote the proliferative phase of the uterine cycle; stimulate production of watery (crystalline) cervical mucus and activity of fimbriae and uterine tube cilia. Promote oogenesis and ovulation by stimulating formation of FSH and LH receptors on follicle cells. Stimulate capacitation of sperm in the female reproductive tract by affecting vaginal and uterine secretions. During pregnancy stimulate growth of the uterus and enlargement of the external genitalia and mammary glands.	Cooperates with estrogen in stimulating growth of breasts and in regulating the uterine cycle (promotes the secretory phase of the uterine cycle); stimulates production of viscous cervical mucus. During pregnancy, quiets the myometrium and acts with estrogen to cause mammary glands to achieve their mature milk-producing state.	Stimulates formation of male reproductive ducts, glands, and external genitalia. Promotes descent of the testes. Stimulates growth and maturation of the internal and external genitalia at puberty; maintains their adult size and function. Required for normal spermatogenesis via effects promoted by its binding to ABP on spermatogenic cells; suppresses mammary gland development.
Promotion of secondary sex characteristics and somatic effects	Promote lengthening of long bones and feminization of the skeleton (particularly the pelvis); inhibit bone reabsorption and then stimulate epiphyseal closure; promote hydration of the skin; stimulate female pattern of fat deposit, and appearance of axillary and pubic hair. During pregnancy act with relaxin (placental hormone) to induce softening and relaxation of the pelvic ligaments and pubic symphysis.		Stimulates the growth spurt at puberty; promotes increased skeletal mass and then epiphyseal closure at the end of adolescence; promotes growth of the larynx and vocal cords and deepening of the voice; enhances sebum secretion and hair growth, especially on the face, axillae, genital region, and chest.
Metabolic effects	Generally anabolic; stimulate Na$^+$ reabsorption by the renal tubules, hence inhibit diuresis; enhance HDL (and reduce LDL) blood levels (cardiovascular sparing effect)	Promotes diuresis (antiestrogenic effect); increases body temperature	Generally anabolic; stimulates hematopoiesis; enhances the basal metabolic rate
Neural effects	Feminize the brain		Responsible for sex drive (libido) in both sexes; masculinizes the brain; promotes aggressiveness

cause the epiphyses of long bones to close sooner, and females reach their full height between the ages of 15 and 17 years. In contrast, the aggressive growth of males continues until the age of 19 to 21 years.

The estrogen-induced secondary sex characteristics of females include (1) growth of the breasts; (2) increased deposit of subcutaneous fat, especially in the hips and breasts; (3) widening and lightening of the pelvis (adaptations for childbirth); (4) growth of axillary and pubic hair; and (5) several metabolic effects, including maintaining low total blood cholesterol levels (and high HDL levels) and facilitating calcium uptake, which helps sustain the density of the skeleton. (These metabolic effects, though initiated under estrogen's influence during puberty, are not true secondary sex characteristics.)

Progesterone works with estrogen to establish and then help regulate the uterine cycle and promotes changes in cervical mucus (see Table 26.1). Its other effects are exhibited largely during pregnancy, when it inhibits uterine motility and takes up where estrogen leaves off in preparing the breasts for lactation. Indeed, progesterone is named for these important roles (*pro* = for; *gestation* = pregnancy). However, the source of progesterone and estrogen during most of pregnancy is the placenta, not the ovaries.

Female Sexual Response

The **female sexual response** is similar to that of males in most respects. During sexual excitement, the clitoris, vaginal mucosa, and breasts engorge with blood; the nipples erect; and increased activity of the vestibular glands lubricates the vestibule and facilitates entry of the penis. These events, though more widespread, are analogous to the *erection* phase in men. Sexual excitement is promoted by touch and psychological stimuli and is mediated along the same autonomic nerve pathways as in males.

The final phase of the female sexual response, *orgasm,* is not accompanied by ejaculation, but muscle tension increases throughout the body, pulse rate and blood pressure rise, and the uterus begins to contract rhythmically. As in males, orgasm is accompanied by a sensation of intense pleasure and followed by relaxation. Orgasm in females is not followed by a refractory period, so females may experience multiple orgasms during a single sexual experience. A man must achieve orgasm and ejaculate if fertilization is to occur, but female orgasm is not required for conception. Indeed, some women never experience orgasm, yet are perfectly able to conceive.

Sexually Transmitted Diseases

Sexually transmitted diseases (STDs), also called *venereal diseases (VDs),* are infectious diseases spread through sexual contact. The United States has the highest rates of infection among developed countries. Over 12 million people in the United States, a quarter of them adolescents, get STDs each year.

As a group, STDs are the single most important cause of reproductive disorders. Until recently, the bacterial infections gonorrhea and syphilis were the most common STDs but, within the past decade, viral diseases such as genital herpes and AIDS (mostly transmitted during sexual intercourse) have taken center stage. AIDS, caused by HIV, the virus that cripples the immune system, is described in Chapter 20. The other important bacterial and viral STDs are described here. Condoms are effective in helping to prevent the spread of STDs, and their use is strongly urged, particularly since the advent of AIDS.

Gonorrhea

The causative agent of **gonorrhea** (gon"o-re'ah) is *Neisseria gonorrhoeae,* which invades the mucosae of the reproductive and urinary tracts. These bacteria are spread by contact with genital, anal, and pharyngeal mucosal surfaces. Most cases occur in adolescents and young adults.

Commonly called "the clap," the most frequent symptom of gonorrhea in males is *urethritis,* accompanied by painful urination and discharge of pus from the penis (penile "drip"). Symptoms vary in women, ranging from none (about 20% of cases) to abdominal discomfort, vaginal discharge, abnormal uterine bleeding, and occasionally, urethral symptoms similar to those seen in males.

Untreated gonorrhea can cause urethral constriction and inflammation of the entire male duct system. In women, it causes pelvic inflammatory disease and sterility. These consequences declined with the advent in the 1950s of penicillin, tetracycline, and certain other antibiotics. However, strains resistant to those antibiotics are becoming increasingly prevalent.

Syphilis

Syphilis (sif'ĭ-lis), caused by *Treponema pallidum,* a corkscrew-shaped bacterium, is usually transmitted sexually. However, it can be contracted congenitally from an infected mother. Indeed, an increasing number of women trading sex for "crack" (a cocaine derivative) are passing syphilis on to their fetuses. Fetuses infected with syphilis are usually stillborn or die shortly after birth. The bacterium easily penetrates intact mucosae and abraded skin and enters the local lymphatics and the bloodstream. Within a few hours of exposure, an asymptomatic bodywide infection is in progress. The incubation period is typically two to three weeks, at the end of which a red, painless primary lesion called a *chancre* (shang'ker) appears at the site of bacterial invasion. In males, this is typically the penis, but in females the lesion often goes undetected within the vagina or on the cervix. The primary lesion ulcerates and becomes crusty; then it heals spontaneously and disappears after one to a few weeks.

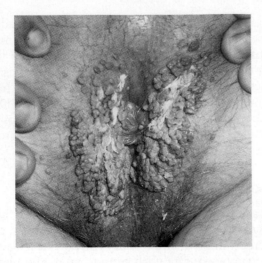

FIGURE 26.23   **Genital warts on the vulva.**

If syphilis is untreated, its secondary signs appear several weeks later. A pink skin rash all over the body is one of the first symptoms. Fever and joint pain are common. These signs and symptoms disappear spontaneously in three to twelve weeks. Then the disease enters the *latent period* and is detectable only by a blood test. The latent stage may last a person's lifetime (or the bacteria may be killed by the immune system), or it may be followed by the signs of *tertiary syphilis*. Tertiary syphilis is characterized by *gummas* (gum'ahs), destructive lesions of the CNS, blood vessels, bones, and skin. Penicillin, which interferes with the ability of dividing bacteria to synthesize new cell walls, is still the treatment of choice for all stages of syphilis.

Chlamydia

Chlamydia (klah-mid'e-ah; *chlamys* = cloak) is a largely unrecognized, silent epidemic that infects perhaps 4 to 5 million people yearly, making it the most common sexually transmitted disease in the United States. Chlamydia is responsible for 25–50% of all diagnosed cases of pelvic inflammatory disease (and, consequently, at least a 1 in 4 chance of ectopic pregnancy), and each year more than 150,000 infants are born to infected mothers. About 20% of men and 30% of women infected with gonorrhea are also infected by *Chlamydia trachomatis*, the causative agent of chlamydia.

Chlamydia is a bacterium with a viruslike dependence on host cells. Its incubation period within the body cells is about one week. Symptoms include urethritis (involving painful, frequent urination and a thick penile discharge); vaginal discharge; abdominal, rectal, or testicular pain; painful intercourse; and irregular menses. In men, it can cause arthritis as well as widespread urogenital tract infection. In women, 80% of whom suffer *no* symptoms from the infection, its most severe consequence is sterility. Newborns infected in the birth canal tend to develop conjunctivitis and respiratory tract inflammations including pneumonia. The disease can be diagnosed by cell culture techniques and is easily treated with tetracycline.

Genital Warts

The *human papillomavirus (HPV)*—actually a group of about 60 viruses—is responsible for the sexual transmission of **genital warts** (Figure 26.23). About a million Americans develop genital warts each year, and it appears that HPV infection increases the risk for certain cancers (penile, vaginal, cervical, and anal). Indeed, the virus is linked to 80% of all cases of invasive cervical cancer.

Treatment is difficult and controversial. Some prefer to leave the warts untreated unless they become widespread, whereas other clinicians recommend their removal by cryosurgery or laser therapy.

Genital Herpes

Human herpesviruses (*herpes simplex, Epstein-Barr virus*, and others) are among the most difficult human pathogens to control. They remain silent for weeks or years and then suddenly flare up, causing a burst of blisterlike lesions. Direct transmission of herpes simplex virus type 2, the common cause of **genital herpes** (her'pēz), is via infectious secretions. The painful lesions that appear on the reproductive organs of infected adults are usually more of a nuisance than a threat. However, congenital herpes infections can cause severe malformations of a fetus, and herpes infections have been implicated as a cause of cervical cancer. Most people who have genital herpes do not know it, and it has been estimated that one-quarter to one-half of all adult Americans harbor the type 2 herpes simplex virus. The antiviral *acyclovir*, which speeds healing of the lesions and reduces the frequency of flare-ups, is the drug of choice for treatment. Inter Vir-A, an antiviral ointment, provides some relief from the itching and pain that accompany the lesions.

Review Questions

Multiple Choice/Matching

(Some questions have more than one correct answer. Select the best answer or answers from the choices given.)

1. The structures that draw an ovulated oocyte into the female duct system are (a) cilia, (b) fimbriae, (c) microvilli, (d) stereocilia.

2. The usual site of embryo implantation is (a) the uterine tube, (b) the peritoneal cavity, (c) the vagina, (d) the uterus.

3. The male homologue of the female clitoris is (a) the penis, (b) the scrotum, (c) the penile urethra, (d) the testis.

4. Which of the following is correct relative to female anatomy? (a) The vaginal orifice is the most dorsal of the three openings in the perineum, (b) the urethra is between the vaginal orifice and the anus, (c) the anus is between the vaginal orifice and the urethra, (d) the urethra is the more ventral of the two orifices in the vulva.

5. Secondary sex characteristics are (a) present in the embryo, (b) a result of male or female sex hormones increasing in amount at puberty, (c) the testis in the male and the ovary in the female, (d) not subject to withdrawal once established.

6. Which of the following produces the male sex hormones? (a) seminal vesicles, (b) corpus luteum, (c) developing follicles of the testes, (d) interstitial cells.

7. Which will occur as a result of nondescent of the testes? (a) male sex hormones will not be circulated in the body, (b) sperm will have no means of exit from the body, (c) inadequate blood supply will retard the development of the testes, (d) viable sperm will not be produced.

8. The normal diploid number of human chromosomes is (a) 48, (b) 47, (c) 46, (d) 23, (e) 24.

9. Relative to differences between mitosis and meiosis, choose the statements that apply *only* to events of meiosis. (a) tetrads present, (b) produces two daughter cells, (c) produces four daughter cells, (d) occurs throughout life, (e) reduces the chromosomal number by half, (f) synapsis and crossover of homologues occur.

10. Match the key choices with the descriptive phrases below.

Key: **(a)** Androgen-binding   **(e)** Inhibin
               protein   **(f)** LH
           (b) Estrogens   **(g)** Progesterone
           (c) FSH   **(h)** Testosterone
           (d) GnRh

___, ___ **(1)** Hormones that directly regulate the ovarian cycle

___, ___ **(2)** Chemicals in males that inhibit the pituitary-testicular axis

___ **(3)** Hormone that makes the cervical mucus viscous

___ **(4)** Potentiates the activity of testosterone on spermatogenic cells

___, ___ **(5)** In females, exerts feedback inhibition on the hypothalamus and anterior pituitary

___ **(6)** Stimulates the secretion of testosterone

11. The menstrual cycle can be divided into three continuous phases. Starting from the first day of the cycle, their consecutive order is (a) menstrual, proliferative, secretory, (b) menstrual, secretory, proliferative, (c) secretory, menstrual, proliferative, (d) proliferative, menstrual, secretory, (e) secretory, proliferative, menstrual.

12. Spermatozoa are to seminiferous tubules as oocytes are to (a) fimbriae, (b) corpus albicans, (c) ovarian follicles, (d) corpora lutea.

13. Which of the following does not add a secretion that makes a major contribution to semen? (a) prostate, (b) bulbourethral glands, (c) testes, (d) vas deferens.

14. The corpus luteum is formed at the site of (a) fertilization, (b) ovulation, (c) menstruation, (d) implantation.

15. The sex of a child is determined by (a) the sex chromosome contained in the sperm, (b) the sex chromosome contained in the oocyte, (c) the number of sperm fertilizing the oocyte, (d) the position of the fetus in the uterus.

16. FSH is to estrogen as estrogen is to (a) progesterone, (b) LH, (c) FSH, (d) testosterone.

17. A drug that "reminds the pituitary" to produce gonadotropins might be useful as (a) a contraceptive, (b) a diuretic, (c) a fertility drug, (d) an abortion stimulant.

Short Answer Essay Questions

18. Why is the term *urogenital system* more applicable to males than to females?

19. The spermatid is haploid, but it is not a functional gamete. Name and describe the process during which a spermatid is converted to a motile sperm, and describe the major structural (and functional) regions of a sperm.

20. Oogenesis in the female results in one functional gamete—the egg, or ovum. What other cells are produced? What is the significance of this rather wasteful type of gamete production—that is, production of a single functional gamete instead of four, as seen in males?

21. List three secondary sex characteristics of females.

22. Describe the events and possible consequences of menopause.

23. Define menarche. What does it indicate?

24. Trace the pathway of a sperm from the male testes to the uterine tube of a female.

25. In menstruation, the stratum functionalis is shed from the endometrium. Explain the hormonal and physical factors responsible for this shedding. (Hint: See Figure 26.22.)

26. Both the epithelium of the vagina and the cervical glands of the uterus help prevent the invasion and spread of vaginal pathogens. Explain how each of these mechanisms works.

27. Some anatomy students were saying that the bulbourethral glands (and the urethral glands) of males act like city workers who come around and clear parked cars from the street before a parade. What did they mean by this analogy?

28. A man swam in a cold lake for an hour and then noticed that his scrotum was shrunken and wrinkled. His first thought was that he had lost his testicles. What had really happened?

Appendix A
Two Important Metabolic Pathways

Step 1 Two-carbon acetyl CoA is combined with oxaloacetic acid, a 4-carbon compound. The unstable bond between the acetyl group and CoA is broken as oxaloacetic acid binds and CoA is freed to prime another 2-carbon fragment derived from pyruvic acid. The product is the 6-carbon citric acid, for which the cycle is named.

Step 2 A molecule of water is removed, and another is added back. The net result is the conversion of citric acid to its isomer, isocitric acid.

Step 3 The substrate loses a CO_2 molecule, and the remaining 5-carbon compound is oxidized, forming a α-ketoglutaric acid and reducing NAD^+.

Step 4 This step is catalyzed by a multi-enzyme complex very similar to the one that converts pyruvic acid to acetyl CoA. CO_2 is lost; the remaining 4-carbon compound is oxidized by the transfer of electrons to NAD^+ to form $NADH+H^+$ and is then attached to CoA by an unstable bond. The product is succinyl CoA.

Step 5 Substrate-level phosphorylation occurs in this step. CoA is displaced by a phosphate group, which is then transferred to GDP to form guanosine-triphosphate (GTP). GTP is similar to ATP, which is formed when GTP donates a phosphate group to ADP. The products of this step are succinic acid and ATP.

Step 6 In another oxidative step, two hydrogens are removed from succinic acid, forming fumaric acid, and transferred to FAD to form $FADH_2$. The function of this coenzyme is similar to that of $NADH+H^+$, but $FADH_2$ stores less energy. The enzyme that catalyzes this oxidation-reduction reaction is the only enzyme of the cycle that is embedded in the mitochondrial membrane. All other enzymes of the citric acid cycle are dissolved in the mitochondrial matrix.

Step 7 Bonds in the substrate are rearranged in this step by the addition of a water molecule. The product is malic acid.

Step 8 The last oxidative step reduces another NAD^+ and regenerates oxaloacetic acid, which accepts a 2-carbon fragment from acetyl CoA for another turn of the cycle.

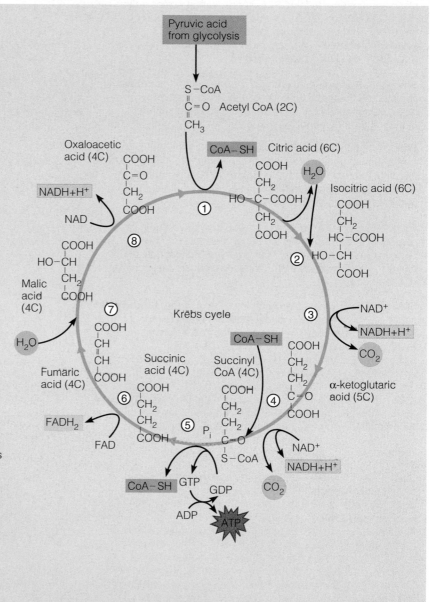

Krebs Cycle (Citric Acid Cycle) All but one of the steps (step 6) occur in the mitochondrial matrix. The preparation of pyruvic acid (by oxidation, decarboxylation, and reaction with coenzyme A) to enter the cycle as acetyl CoA is shown above the cycle. Acetyl CoA is picked up by oxaloacetic acid to form citric acid; and as it passes through the cycle, it is oxidized four more times [forming three molecules of reduced NAD ($NADH + H^+$) and one of reduced FAD ($FADH_2$)] and decarboxylated twice (releasing 2 CO_2). Energy is captured in the bonds of GTP, which then acts in a coupled reaction with ADP to generate one molecule of ATP by substrate-level phosphorylation.

Step 1   Glucose enters the cell and is phosphorylated by the enzyme hexokinase, which catalyzes the transfer of a phospate group, indicated as P_i, from ATP to the number six carbon of the sugar, producing glucose-6-phosphate. The electrical charge of the phosphate group traps the sugar in the cell because of the impermeability of the plasma membrane to ions. Phosphorylation of glucose also makes the molecule more chemically reactive. Although glycolysis is supposed to *produce* ATP, in step 1 ATP is actually consumed—an energy investment that will be repaid with dividends later in glycolysis.

Step 2   Glucose-6-phosphate is rearranged and converted to its isomer, fructose-6-phosphate. Isomers, remember, have the same number and types of atoms but in different structural arrangements.

Step 3   In this step, still another molecule of ATP is used to add a second phosphate group to the sugar, producing fructose-1,6-bisphosphate. So far, the ATP ledger shows a debit of –2. With phosphate groups on its opposite ends, the sugar is now ready to be split in half.

Step 4   This is the reaction from which glycolysis gets its name. An enzyme cleaves the sugar molecule into two different 3-carbon sugars: glyceraldehyde 3-phosphate and dihydroxyacetone phosphate. These two sugars are isomers of one another.

Step 5   The isomerase enzyme interconverts the 3-carbon sugars, and if left alone in a test tube, the reaction reaches equilibrium. This does not happen in the cell, however, because the next enzyme in glycolysis uses only glyceraldehyde phosphate as its substrate and is unreceptive to dihydroxyacetone phosphate. This pulls the equilibrium between the two 3-carbon sugars in the direction of glyceraldehyde phosphate, which is removed as fast as it forms. Thus, the net result of steps 4 and 5 is cleavage of a 6-carbon sugar into two molecules of glyceraldehyde phosphate; each will progress through the remaining steps of glycolysis.

Step 6   An enzyme now catalyzes two sequential reactions while it holds glyceraldehyde phosphate in its active site. First, the sugar is oxidized by the transfer of H from the number one carbon of the sugar to NAD, forming $NADH + H^+$. Here we see in metabolic context the oxidation-reduction reaction described in Chapter 24. This reaction releases substantial amounts of energy, and the enzyme capitalizes on this by coupling the reaction to the creation of a high-energy phosphate bond at the number one carbon of the oxidized substrate. The source of the phosphate is inorganic phosphate (P_i) always present in the cytosol. As products, the enzyme releases $NADH + H^+$ and 1,3-bisphosphoglyceric acid. Notice in the figure that the new phosphate bond is symbolized with a squiggle (~), which indicates that the bond is at least as energetic as the high-energy phosphate bonds of ATP.

THE TEN STEPS OF GLYCOLYSIS   Each of the ten steps of glycolysis is catalyzed by a specific enzyme found dissolved in the cytoplasm. All steps are reversible. An abbreviated version of the three major phases of glycolysis appears in the lower right-hand corner of the second page.

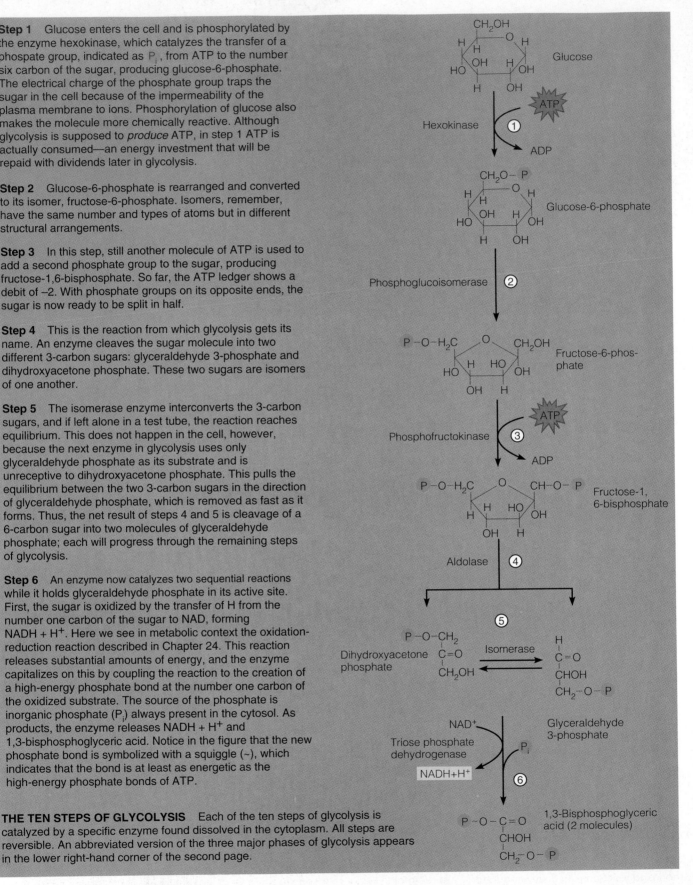

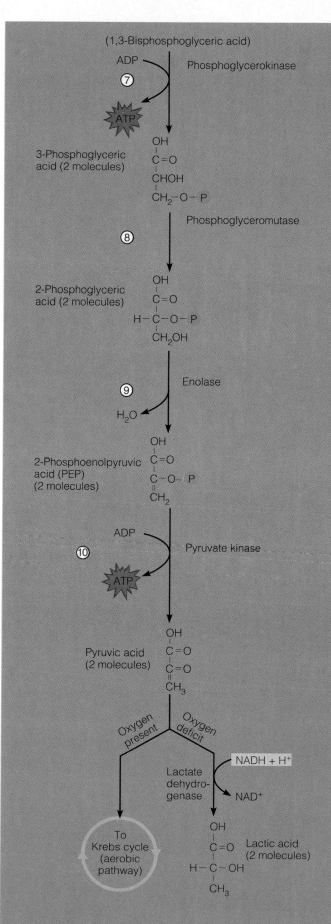

(1,3-Bisphosphoglyceric acid)

ADP → Phosphoglycerokinase
(7)
ATP

3-Phosphoglyceric acid (2 molecules)

OH
|
C=O
|
CHOH
|
CH₂–O– P

Phosphoglyceromutase
(8)

2-Phosphoglyceric acid (2 molecules)

OH
|
C=O
|
H–C–O– P
|
CH₂OH

Enolase
(9)
H₂O

2-Phosphoenolpyruvic acid (PEP) (2 molecules)

OH
|
C=O
|
C–O– P
||
CH₂

ADP → Pyruvate kinase
(10)
ATP

Pyruvic acid (2 molecules)

OH
|
C=O
|
C=O
||
CH₃

Oxygen present   Oxygen deficit

Lactate dehydrogenase   NADH + H⁺ → NAD⁺

To Krebs cycle (aerobic pathway)

Lactic acid (2 molecules)

OH
|
C=O
|
H– C– OH
|
CH₃

Step 7 Finally, glycolysis produces ATP. The phosphate group, with its high-energy bond, is transferred from 1,3-bisphosphoglyceric acid to ADP. For each glucose molecule that began glycolysis, step 7 produces two molecules of ATP, because every product after the sugar-splitting step (step 4) is doubled. Of course, two ATPs were invested to get sugar ready for splitting. The ATP ledger now stands at zero. By the end of step 7, glucose has been converted to two molecules of 3-phosphoglyceric acid. This compound is not a sugar. The sugar was oxidized to an organic acid back in step 6, and now the energy made available by that oxidation has been used to make ATP.

Step 8 Next, an enzyme relocates the remaining phosphate group of 3-phosphoglyceric acid to form 2-phosphoglyceric acid. This prepares the substrate for the next reaction.

Step 9 An enzyme forms a double bond in the substrate by removing a water molecule from 2-phosphoglyceric acid to form phosphoenolpyruvic acid, or PEP. This results in the electrons of the substrate being rearranged in such a way that the remaining phosphate bond becomes very unstable; it has been upgraded to high-energy status.

Step 10 The last reaction of glycolysis produces another molecule of ATP by transferring the phosphate group from PEP to ADP. Because this step occurs twice for each glucose molecule, the ATP ledger now shows a net gain of two ATPs. Steps 7 and 10 each produce two ATPs for a total credit of four, but a debt of two ATPs was incurred from steps 1 and 3. Glycolysis has repaid the ATP investment with 100% interest. In the meantime, glucose has been broken down and oxidized to two molecules of pyruvic acid, the compound produced from PEP in step 10.

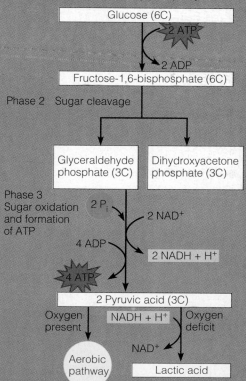

Phase 1   Sugar activates by phosphorylation

Glucose (6C)
2 ATP
2 ADP
Fructose-1,6-bisphosphate (6C)

Phase 2   Sugar cleavage

Glyceraldehyde phosphate (3C)   Dihydroxyacetone phosphate (3C)

Phase 3 Sugar oxidation and formation of ATP
2 Pᵢ
2 NAD⁺
4 ADP
2 NADH + H⁺
4 ATP

2 Pyruvic acid (3C)

Oxygen present   NADH + H⁺   Oxygen deficit
NAD⁺
Aerobic pathway   Lactic acid

Appendix B
Answers to Multiple Choice and Matching Questions

Chapter 1
1. c; 2. a; 3. e; 4. a, d; 5. (a) wrist (b) hip bone (c) nose (d) toes (e) scalp; 6. neither c nor d would be visible in the median sagittal section; 7. (a) dorsal (b) ventral (c) dorsal (d) ventral; 8. b; 9. b

Chapter 2
1. b, d; 2. d; 3. b; 4. a; 5. b; 6. a; 7. a; 8. b; 9. d; 10. a; 11. b; 12. a, c; 13. (1)a, (2)c; 14. c; 15. d; 16. e; 17. d; 18. d; 19. a; 20. b; 21. b; 22. c

Chapter 3
1. d; 2. a, c; 3. b; 4. d; 5. b; 6. e; 7. c; 8. d; 9. a; 10. a; 11. b; 12. d; 13. c; 14. b; 15. d; 16. a; 17. b; 18. d

Chapter 4
1. a, c, d, b; 2. c, e; 3. b, f, a, d, g, d; 4. b; 5. c; 6. b

Chapter 5
1. a; 2. c; 3. d; 4. d; 5. b; 6. b; 7. c; 8. c; 9. b; 10. b; 11. a; 12. d; 13. b

Chapter 6
1. e; 2. b; 3. c; 4. d; 5. e; 6. b; 7. c; 8. b; 9. e; 10. c; 11. b; 12. c; 13. a; 14. b

Chapter 7
1. (1)b, g; (2)h; (3)d; (4)d, f; (5)e; (6)c; (7)a, b, d, h; (8)i; 2. (1)g, (2)f, (3)b, (4)a, (5)b, (6)c, (7)d, (8)e; 3. (1)b, (2)c, (3)e, (4)a, (5)h, (6)e, (7)f

Chapter 8
1. (1)c, (2)a, (3)a, (4)b, (5)c, (6)b, (7)b, (8)a, (9)c; 2. b; 3. d; 4. d; 5. b; 6. d; 7. d

Chapter 9
1. c; 2. b; 3. (1)b, (2)a, (3)b, (4)a, (5)b, (6)a; 4. c; 5. a; 6. a; 7. d; 8. a; 9. (1)a, (2)a, c, (3)b, (4)c, (5)b, (6)b; 10. a; 11. c; 12. c; 13. c; 14. b

Chapter 10
1. c; 2. c; 3. (1)e, (2)c, (3)g, (4)f, (5)d; 4. a; 5. c; 6. d; 7. c; 8. c; 9. b; 10. d; 11. b; 12. a; 13. c; 14. d; 15. a; 16. a

Chapter 11
1. b; 2. (1)d, (2)b, (3)f, (4)c, (5)a; 3. b; 4. c; 5. a; 6. c; 7. b; 8. d; 9. c; 10. c; 11. a; 12. (1)d, (2)b, (3)a, (4)c

Chapter 12
1. a; 2. d; 3. c; 4. a; 5. (1)d, (2)f, (3)e, (4)g, (5)b, (6)f, (7)i, (8)a; 6. b; 7. c; 8. a; 9. (1)a, (2)b, (3)a, (4)a, (5)b, (6)a, (7)b, (8)b, (9)a; 10. d; 11. (1)d, (2)e, (3)d, (4)a

Chapter 13
1. b; 2. c; 3. c; 4. (1)d, (2)c, (3)f, 4(b), (5)e, (6)a; 5. (1)f, (2)i, (3)b, (4)g, h, l, (5)e, (6)i, (7)c, (8)k, (9)l, (10)c, d, f, k; 6. (1)b 6; (2)d 1, 8; (3)c 2; (4)c 5; (5)a 4; (6)a 3, 9; (7)a 7; (8)a 7; (9)d 1; (10)a 3, 4, 7, 9; 7.(1) a, 1, 5; (2)a, 3, 5; (3)a, 4; (4)a, 2; (5)c, 2; 8. d; 9. a; 10. d; 11. c; 12. b; 13. c; 14. d; 15. a; 16. b; 17. c; 18. d; 19. b; 20. a; 21. b; 22. b; 23. d; 24. b; 25. d; 26. b; 27. b; 28. e; 29. b; 30. c; 31. c; 32.b

Chapter 14
1. d; 2. (1)S, (2)P, (3)P, (4)S, (5)S, (6)P, (7)P, (8)S, (9)P, (10)S, (11)P, (12)S; 3. b; 4. c; 5. a

Chapter 15
1. b; 2. a; 3. c; 4. d; 5. (1)c, (2)b, (3)f, (4)d, (5)e, (6)g, (7)a, (8)h, (9)b and e, (10)a; 6. d; 7. c; 8. b; 9. d; 10. b; 11. d; 12. b; 13. c; 14. d

Chapter 16
1. c; 2. c; 3. d; 4. b; 5. d; 6. a; 7. a; 8. b; 9. c; 10. d

Chapter 17
1. a; 2. c; 3. b; 4. c; 5. b; 6. b; 7. c; 8. d; 9. b

Chapter 18
1. d; 2. b; 3. d; 4. c; 5. e; 6. d; 7. c; 8. b; 9. b; 10. a; 11. b; 12. c; 13. b

Chapter 19
1. c; 2. c; 3. a, d; 4. c; 5. a; 6. b; 7. a; 8. b; 9. d

Chapter 20
1. c; 2. a; 3. d; 4. d, e; 5. a; 6. d; 7. b; 8. c; 9. d

Chapter 21
1. b; 2. a; 3. c; 4. c; 5. b; 6. d; 7. d; 8. b; 9. c, d; 10. c; 11. b; 12. b; 13. b; 14. c; 15. b; 16. b

Chapter 22
1. c; 2. d; 3. d; 4. b; 5. b; 6. a; 7. d; 8. d; 9. b; 10. c; 11. c; 12. a; 13. d; 14. d; 15. b; 16. c; 17. a

Chapter 23
1. a; 2. c; 3. c; 4. d; 5. b; 6. c; 7. a; 8. d; 9. d; 10. a; 11. b; 12. d; 13. c; 14. d; 15. a

Chapter 24
1. d; 2. b; 3. c; 4. d; 5. c; 6. b; 7. a; 8. c; 9. a

Chapter 25
1. a, 2. c; 3. b; 4. b; 5. h, i, j; 6. c, g; 7. e, h; 8. b; 9. a, e; 10. j; 11. b, d; 12. a; 13. c

Chapter 26
1. a, b; 2. d; 3. a; 4. d; 5. b; 6. d; 7. d; 8. c; 9. a, c, e, f; 10. (1)c, f; (2)e, h; (3)g; (4)a; (5)b, g, e; (6)f; 11. a; 12. c; 13. d; 14. b; 15. a; 16. b; 17. c

Glossary

Pronunciation Key

' = Primary accent
" = Secondary accent

Pronounce:

a, fa, āt	as in	fate	o, no, ōt	as in	note
ă, hă, at		hat	ŏ, frŏ, ŏg		frog
ah		father	oo		soon
ar		tar	or		for
e, stre, ēt		street	ow		plow
ĕ, hĕ, en		hen	oy		boy
er		her	sh		she
ew		new	u, mu, ūt		mute
g		go	ŭ, sŭ, un		sun
i, bi, īt		bite	z		zebra
ĭ, hĭ, im		him	zh		measure
ng		ring			

Abdomen (ab'do-men) Portion of the body between the diaphragm and the pelvis.

Abduct (ab-dukt') To move away from the midline of the body.

Abscess (ab'ses) Localized accumulation of pus and disintegrating tissue.

Absolute refractory period Period following stimulation during which no additional action potential can be evoked.

Absorption Process by which the products of digestion pass through the alimentary tube mucosa into the blood or lymph.

Accessory digestive organs Organs that contribute to the digestive process but are not part of the alimentary canal; include the tongue, teeth, salivary glands, pancreas, liver.

Accommodation The process of increasing the refractive power of the lens of the eye; focusing.

Acetabulum (as"ĕ-tab'u-lum) Cuplike cavity on lateral surface of the hip bone that receives the femur.

Acetylcholine (ACh) (as"ĕ-til-ko'lēn) Chemical transmitter substance released by some nerve endings.

Acetylcholinesterase (AChE) (as"ĕ-til-ko"lin-es'ter-ās) Enzyme present at the neuromuscular junction that prevents continued muscle contraction in the absence of additional stimulation.

Achilles tendon. *See* Calcaneal tendon.

Acid A substance that releases hydrogen ions when in solution (compare with Base); a proton donor.

Acid-base balance Situation in which the pH of the blood is maintained between 7.35 and 7.45.

Acidosis (as"ĭ-do'sis) State of abnormally high hydrogen ion concentration in the extracellular fluid.

Actin (ak'tin) A contractile protein of muscle.

Action potential A large transient depolarization event, including polarity reversal, that is conducted along the membrane of a muscle cell or a nerve fiber.

Activation energy The amount of energy required to push a reactant to the level necessary for action.

Active immunity Immunity produced by an encounter with an antigen; provides immunological memory.

Active site Region on the surface of functional (globular) proteins that fit and interact chemically with other molecules of complementary shape and charge.

Active transport Membrane transport processes for which ATP is provided, e.g., solute pumping and endocytosis.

Adaptation (1) Any change in structure or response to suit a new environment; (2) decline in the transmission of a sensory nerve when a receptor is stimulated continuously and without change in stimulus strength.

Adduct (a-dukt') To move toward the midline of the body.

Adenine (A) (ad'ĕ-nēn) One of the two major purines found in both RNA and DNA; also found in various free nucleotides of importance to the body, such as ATP.

Adenohypophysis (ad"ĕ-no-hi-pof'ĭ-sis) Anterior pituitary; the glandular part of the pituitary gland.

Adenoids (ad'en-noids) Pharyngeal tonsil.

Adenosine triphosphate (ATP) (ah-den'o-sēn tri"fos'fāt) Organic molecule that stores and releases chemical energy for use in body cells.

Adipocyte (ad'ĭ-po-sīt) An adipose, or fat cell.

Adipose tissue (ad'ĭ-pōs) Areolar connective tissue modified to store nutrients; a connective tissue consisting chiefly of fat cells.

Adrenal glands (uh-drē'nul) Hormone-producing glands located superior to the kidneys; each consists of medulla and cortex areas.

Adrenergic fibers (ad"ren-er'jik) Nerve fibers that release norepinephrine.

Adrenocorticotropic hormone (ACTH) (ah-dre'no-kor"tĭ-ko-trō'pik) Anterior pituitary hormone that influences the activity of the adrenal cortex.

Adventitia (ad"ven-tish'e-ah) Outermost layer or covering of an organ.

Aerobic (a'er-ōb"ik) Oxygen-requiring.

Aerobic endurance The length of time a muscle can continue to contract using aerobic pathways.

Aerobic respiration Respiration in which oxygen is consumed and glucose is broken down entirely; water, carbon dioxide, and large amounts of ATP are the final products.

Afferent (af'er-ent) Carrying to or toward a center.

Afferent (sensory) nerve Nerve that contains processes of sensory neurons and carries nerve impulses to the central nervous system.

Afferent (sensory) neuron Nerve cell that carries impulses toward the central nervous system; initiates nerve impulses following receptor stimulation.

Agglutination (ah-gloo"tĭ-na'shun) Clumping of (foreign) cells; induced by cross-linking of antigen-antibody complexes.

Agonist (ag'o-nist) Muscle that bears the major responsibility for effecting a particular movement; a prime mover.

AIDS Acquired immune deficiency syndrome; caused by human immunodeficiency virus (HIV); symptoms include severe weight loss, night sweats, swollen lymph nodes, opportunistic infections.

Albumin (al-bu'min) The most abundant plasma protein.

Aldosterone (al-dos'ter-ōn) Hormone produced by the adrenal cortex that regulates sodium ion reabsorption.

Alimentary canal (al"ĭ-men'tar-e) The continuous hollow tube extending from the mouth to the anus; its walls are constructed by the oral cavity, pharynx, esophagus, stomach, and small and large intestines.

Alkalosis (al"kah-lo'sis) State of abnormally low hydrogen ion concentration in the extracellular fluid.

Allantois (ah"lan'to-is) Embryonic membrane; its blood vessels develop into blood vessels of the umbilical cord.

Alleles Genes coding for the same trait and found at the same locus on homologous chromosomes.

Allergy (hypersensitivity) Overzealous immune response to an otherwise harmless antigen.

Alopecia (al"o-pe'she-ah) Baldness.

Alpha (α) helix The most common type of secondary structure of the amino acid chain in proteins; resembles the coils of a telephone cord.

Alveolar (acinar) gland (al-ve'o-lar) A gland whose secretory cells form small, flasklike sacs.

Alveolar ventilation rate (AVR) An index of respiratory efficiency; measures volume of air wasted and flow of fresh gases in and out of alveoli.

Alveolus (al-ve′o-lus) (1) One of the microscopic air sacs of the lungs; (2) tiny milk-producing glandular sac in the breast.

Alzheimer's disease (altz′hi-merz) Degenerative brain disease resulting in progressive loss of memory and motor control, and increasing dementia.

Amino acid (ah-me′no) Organic compound containing nitrogen, carbon, hydrogen, and oxygen; building block of protein.

Ammonia (NH_3) Common waste product of protein breakdown in the body; a colorless volatile gas, very soluble in water and capable of forming a weak base; a proton acceptor.

Amniocentesis A common form of fetal testing in which a small sample of fluid is removed from the amniotic cavity.

Amnion (am′ne-on) Fetal membrane that forms a fluid-filled sac around the embryo.

Amoeboid motion (ah-me′boyd) The flowing movement of the cytoplasm of a phagocyte.

Amphiarthrosis (am″fe-ar-thro′sis) A slightly movable joint.

Ampulla (am-pul′lah) A localized dilation of a canal or duct.

Amylase Digestive system enzyme that breaks down starchy foods.

Anabolism (ah-nab′o-lizm) Energy-requiring building phase of metabolism in which simpler substances are combined to form more complex substances.

Anaerobic (an-a′er-ōb-ik) Not requiring oxygen.

Anaerobic glycolysis (gli-kol′ĭ-sis) Energy-yielding conversion of glucose to lactic acid in various tissues, notably muscle, when sufficient oxygen is not available.

Anaerobic threshold The point at which muscle metabolism converts to anaerobic glycolysis.

Anaphase Third stage of mitosis in which a full set of daughter chromosomes move toward each pole of a cell.

Anastomosis (ah-nas″to-mo′sis) A union or joining of nerves, blood vessels, or lymphatics.

Anatomy Study of the structure of living organisms.

Androgen (an′dro-jen) A hormone that controls male secondary sex characteristics, such as testosterone.

Anemia (ah-ne′me-ah) Reduced oxygen-carrying ability of blood resulting from too few erythrocytes or abnormal hemoglobin.

Aneurysm (an′u-rizm) Blood-filled sac in an artery wall caused by dilation or weakening of the wall.

Angina pectoris (an-ji′nah, an″jĭ-nah pek′tor-is) Severe suffocating chest pain caused by brief lack of oxygen supply to heart muscle.

Angiotensin II (an″je-o-ten′sin) A potent vasoconstrictor activated by renin; also triggers release of aldosterone.

Anion (an′i-on) An ion carrying one or more negative charges and therefore attracted to a positive pole.

Anoxia (ah-nŏk′se-ah) Deficiency of oxygen.

Antagonist (an-tag′o-nist) Muscle that reverses, or opposes, the action of another muscle.

Anterior pituitary. *See* Adenohypophysis.

Antibody A protein molecule that is released by a plasma cell (a daughter cell of an activated B lymphocyte) and that binds specifically to an antigen; an immunoglobulin.

Anticodon (an″ti-ko′don) The three-base sequence complementary to the messenger RNA (mRNA) codon.

Antidiuretic hormone (ADH) (an″tĭ-dī″yer-eh′tik) Hormone produced by the hypothalamus and released by the posterior pituitary; stimulates the kidneys to reabsorb more water, reducing urine volume.

Antigen (Ag) (an′tĭ-jen) A substance or part of a substance (living or nonliving) that is recognized as foreign by the immune system, activates the immune system, and reacts with immune cells or their products.

Anucleate cell (a-nu′kle-āt) A cell without a nucleus.

Anus (a′nus) Distal end of digestive tract; outlet of rectum.

Aorta (a-or′tah) Major systemic artery; arises from the left ventricle of the heart.

Aortic body Receptor in the aortic arch sensitive to changing oxygen, carbon dioxide, and pH levels of the blood.

Apgar score Evaluation of an infant's physical status at 1 and 5 minutes after birth by assessing five criteria: heart rate, respiration, color, muscle tone, and reflexes.

Apnea Breathing cessation.

Apocrine sweat gland (ap′o-krin) The less numerous type of sweat gland; produces a secretion containing water, salts, proteins, and fatty acids.

Apoenzyme (ap′ō-en-zī m) The protein portion of an enzyme.

Aponeurosis (ap″o-nu-ro-sis) Fibrous or membranous sheet connecting a muscle and the part it moves.

Appendicitis (ă-pen′dĭ-si′tis) Inflammation of the appendix (wormlike sac attached to the cecum of the large intestine).

Appendicular Relating to the limbs; one of the two major divisions of the body.

Appositional growth Growth accomplished by the addition of new layers onto those previously formed.

Aqueous humor (a′kwe-us) Watery fluid in the anterior chambers of the eye.

Arachnoid (ah-rak-noid) Weblike; specifically, the weblike middle layer of the three meninges.

Areola (ah-re′o-lah) Circular, pigmented area surrounding the nipple; any small space in a tissue.

Areolar connective tissue A type of loose connective tissue.

Arrector pili (ah-rek′tor pi′li) Tiny, smooth muscles attached to hair follicles; cause the hair to stand upright when activated.

Arrhythmia (a-rith′me-ah) Irregular heart rhythm caused by defects in the intrinsic conduction system.

Arteries Blood vessels that conduct blood away from the heart and into the circulation.

Arteriole (ar-tē r′e-ōl) A minute artery.

Arteriosclerosis (ar-tē r′e-o-skler-o′sis) Any of a number of proliferative and degenerative changes in the arteries leading to their decreased elasticity.

Arthritis Inflammation of the joints.

Arthroscopic surgery (ar-thro-skop′ik) Procedure enabling a surgeon to repair the interior of a joint through a small incision.

Articular capsule Double-layered capsule composed of an outer fibrous capsule lined by synovial membrane; encloses the joint cavity of a synovial joint.

Articular cartilage Hyaline cartilage covering bone ends at movable joints.

Articulation (joint) The junction of two or more bones.

Association areas Functional areas of the cerebral cortex that act mainly to integrate diverse information for purposeful action.

Association neuron (interneuron) Nerve cell located between motor and sensory neurons that shuttle signals through CNS pathways where integration occurs.

Astigmatism (ah-stig′mah-tizm) A condition in which unequal curvatures in different parts of the lens (or cornea) of the eye lead to blurred vision.

Astrocyte (as′tro-sī t) A type of CNS supporting cell; assists in exchanges between blood capillaries and neurons.

Ataxia (ah-tak′se-ah) Disruption of muscle coordination resulting in inaccurate movements.

Atelectasis (at″ĕ -lik′tah-sis) Lung collapse.

Atherosclerosis (a″ther-o″skler-o′sis) Changes in the walls of large arteries consisting of lipid deposits on the artery walls; the early stage of arteriosclerosis.

Atmospheric pressure Force that air exerts on the surface of the body.

Atom Smallest particle of an elemental substance that exhibits the properties of that element; composed of protons, neutrons, and electrons.

Atomic mass number Sum of the number of protons and neutrons in the nucleus of an atom.

Atomic number The number of protons in an atom.

Atomic symbol The one- or two-letter symbol used to indicate an element; usually the first letter(s) of the element's name.

Atomic weight   The average of the mass numbers of all the isotopes of an element.

ATP (adenosine triphosphate) (ah-den′o-sēn tri″fos′fā t)   Organic molecule that stores and releases chemical energy for use in body cells.

Atria (a′tre-ah)   The two superior receiving chambers of the heart.

Atrial natriuretic peptide (ANP) (a′tre-al na″tre-u-ret′ik)   A hormone released by certain cells of the heart atria that reduces blood pressure and blood volume by inhibiting nearly all events that promote vasoconstriction and Na$^+$ and water retention.

Atrioventricular (AV) bundle (a″tre-o-ven-tri′kyoo-ler)   Bundle of specialized fibers that conduct impulses from the AV node to the right and left ventricles; also called bundle of His.

Atrioventricular (AV) node   Specialized mass of conducting cells located at the atrioventricular junction in the heart.

Atrioventricular (AV) valve   Valve that prevents backflow into the atria when the ventricles are contracting.

Atrophy (at′ro-fe)   Reduction in size or wasting away of an organ or cell resulting from disease or lack of use.

Auditory ossicles (ah′sih-kulz)   The three tiny bones serving as transmitters of vibrations and located within the middle ear: the malleus, incus, and stapes.

Auditory tube   *See* Pharyngotympanic tube.

Autoimmunity   Production of antibodies or effector T cells that attack a person's own tissue.

Autolysis (aw″tol′ĭ-sis)   Process of autodigestion (self-digestion) of cells, especially dead or degenerate cells.

Autonomic nervous system (ANS)   Efferent division of the peripheral nervous system that innervates cardiac and smooth muscles and glands; also called the involuntary or visceral motor system.

Autonomic (visceral) reflexes   Reflexes that activate smooth or cardiac muscle and/or glands.

Autoregulation   The automatic adjustment of blood flow to a particular body area in response to its current requirements.

Autosomes   Chromosomes number 1 to 22; do not include the sex chromosomes.

Avogadro's number (av″o-gad′rōz)   The number of molecules in one mole of any substance, 6.02×10^{23}.

Axial   Relating to the head, neck, and trunk; one of the two major divisions of the body.

Axolemma (ak″so-lem′ah)   The plasma membrane of an axon.

Axon   Neuron process that carries impulses away from the nerve cell body; efferent process; the conducting portion of a nerve cell.

B cells   Also called B lymphocytes; oversee humoral immunity; their descendants differentiate into antibody-producing plasma cells.

Baroreceptor (bayr″o-re-sep-tor)   Pressoreceptor; receptor that is stimulated by pressure changes.

Basal body (ba′sal)   The elongated part of a cell centriole that forms the bases of cilia and flagella.

Basal ganglia   *See* Basal nuclei.

Basal lamina (lam′ĭ-nah)   Noncellular, adhesive supporting sheet consisting largely of glycoproteins secreted by epithelial cells.

Basal metabolic rate (BMR)   Rate at which energy is expended (heat produced) by the body per unit time under controlled (basal) conditions: 12 hours after a meal, at rest.

Basal nuclei (basal ganglia)   Specific gray matter areas located deep within the white matter of the cerebral hemispheres.

Basal surface   The surface near the base or interior of a structure; nearest the lower side or bottom of a structure.

Base   A substance capable of binding with hydrogen ions; a proton acceptor.

Basement membrane   Extracellular material consisting of a basal lamina secreted by epithelial cells and a reticular lamina secreted by underlying connective tissue cells.

Basophil (ba′zo-fil)   White blood cell whose granules stain deep blue with basic dye; has a relatively pale nucleus.

Benign (be-nīn′)   Not malignant.

Biceps (bi′seps)   Two-headed, especially applied to certain muscles.

Bicuspid (mitral) valve (mi′tral)   The left atrioventricular valve.

Bile   Greenish-yellow or brownish fluid produced in and secreted by the liver, stored in the gallbladder, and released into the small intestine.

Bilirubin (bil′i-roo′bin)   Red pigment of bile.

Biofeedback   Training that provides an awareness of visceral activities; enables an element of voluntary control over autonomic body functions.

Biogenic amines (bi″o-jen′ik am′inz, ah′mē nz)   Class of neurotransmitters, including catecholamines and indolamines.

Bipolar neuron   Neuron with axon and dendrite that extend from opposite sides of the cell body.

Blastocyst (blas′to-sist)   Stage of early embryonic development; the product of cleavage.

Blood pressure (BP)   Force exerted by blood against a unit area of the blood vessel walls; differences in blood pressure between different areas of the circulation provide the driving force for blood circulation.

Blood-brain barrier   Mechanism that inhibits passage of materials from the blood into brain tissues; reflects relative impermeability of brain capillaries.

Bolus (bo′lus)   A rounded mass of food prepared by the mouth for swallowing; any soft round mass.

Bone (osseous) tissue (os′e-us)   A connective tissue that forms the bony skeleton.

Bone remodeling   Process involving bone formation and destruction in response to hormonal and mechanical factors.

Bone resorption   The removal of osseous tissue; part of the continuous bone remodeling process.

Bony thorax (thoracic cage)   Bones that form the framework of the thorax; includes sternum, ribs, and thoracic vertebrae.

Bowman's capsule (bo-manz).   *See* Glomerular capsule.

Boyle's law   States that when the temperature is constant, the pressure of a gas varies inversely with its volume.

Bradycardia (brad″e-kar′deah)   A heart rate below 60 beats per minute.

Brain death   State of irreversible coma, even though life-support measures may have restored other body organs.

Brain stem   Collectively the midbrain, pons, and medulla of the brain.

Brain ventricle   Fluid-filled cavity of the brain.

Branchial groove (brang′ke-al)   An indentation of the surface ectoderm in the embryo; the external auditory canals develop from these.

Bronchioles   The branching air passageways inside the lungs.

Bronchus (brong′kus)   One of the two large branches of the trachea that leads to the lungs.

Buffer   Chemical substance or system that minimizes changes in pH by releasing or binding hydrogen ions.

Bulk (vesicular) transport   The movement of large particles and macromolecules across a plasma membrane.

Burn   Tissue damage inflicted by intense heat, electricity, radiation, or certain chemicals, all of which denature cell proteins and cause cell death in the affected areas.

Bursa (ber′sa)   A fibrous sac lined with synovial membrane and containing synovial fluid; occurs between bones and muscle tendons (or other structures), where it acts to decrease friction during movement.

Bursitis   Inflammation of a bursa.

Calcaneal tendon (kal-ka′ne-al)   Tendon that attaches the calf muscles to the heelbone (calcaneus); also called the Achilles tendon.

Calcitonin (kal″sih-to′nin)   Hormone released by the thyroid that promotes a decrease in calcium levels of the blood; also called thyrocalcitonin.

Calculus (kal′ku-lus)   A stone formed within various body parts.

Callus (kal′lus)   (1) Localized thickening of skin epidermis resulting from physical trauma; (2) repair tissue (fibrous or bony) formed at a fracture site.

Calorie (cal) Amount of energy needed to raise the temperature of 1 gram of water 1° Celsius. Energy exchanges associated with biochemical reactions are usually reported in kilocalories (1 kcal = 1000 cal) or large calories (Cal).

Calyx (ka′liks) A cuplike extension of the pelvis of the kidney.

Canaliculus (kan″ah-lik′u-lus) Extremely small tubular passage or channel.

Cancer A malignant, invasive cellular neoplasm that has the capability of spreading throughout the body or body parts.

Capillaries (kap′il-layr″ēs) The smallest of the blood vessels and the sites of exchange between the blood and tissue cells.

Carbohydrate (kar″bo-hi-drāt) Organic compound composed of carbon, hydrogen, and oxygen; includes starches, sugars, cellulose.

Carbonic acid–bicarbonate buffer system Chemical system that helps maintain pH homeostasis of the blood.

Carbonic anhydrase (kar-bon′ik an-hi′dras) Enzyme that facilitates the combination of carbon dioxide with water to form carbonic acid.

Carcinogen (kar″sĭ′no-jin) Cancer-causing agent.

Cardiac cycle Sequence of events encompassing one complete contraction and relaxation of the atria and ventricles of the heart.

Cardiac muscle Specialized muscle of the heart.

Cardiac output (CO) Amount of blood pumped out of a ventricle in one minute.

Cardiac reserve The difference between resting and maximal cardiac output.

Cardiogenic shock Pump failure; the heart is so inefficient that it cannot sustain adequate circulation.

Cardiovascular system Organ system that distributes the blood to deliver nutrients and remove wastes.

Carotene (kar′o-tēn) Yellow to orange pigment that accumulates in the stratum corneum epidermal layer and in fatty tissue of the hypodermis.

Carotid body (kar-rot′id) A receptor in the common carotid artery sensitive to changing oxygen, carbon dioxide, and pH levels of the blood.

Carotid sinus (si-nus) A dilation of a common carotid artery; involved in regulation of systemic blood pressure.

Cartilage (kar′tĭ-lij) White, semiopaque connective tissue.

Cartilage bone (endochondral bone) Bone formed by the calcification of hyaline cartilage structures.

Cartilaginous joints (kar″tĭ-laj′ĭ-nus) Bones united by cartilage; no joint cavity is present.

Catabolism (kat-tab′o-lizm) Process in which living cells break down substances into simpler substances.

Catalyst (kat′ah-list) Substance that increases the rate of a chemical reaction without itself becoming chemically changed or part of the product.

Cataract Clouding of the eye's lens; often congenital or age-related.

Catecholamines (kat″ĕ-kol′ah-menz) Epinephrine and norepinephrine.

Cation (kat′i-on) An ion with a positive charge.

Caudal (kaw′dul) Literally, toward the tail; in humans, the inferior portion of the anatomy.

Cecum (se′kum) The blind-end pouch at the beginning of the large intestine.

Cell differentiation The development of specific and distinctive features in cells, from a single cell (the fertilized egg) to all the specialized cells of adulthood.

Cell division period [mitotic (M) phase] One of two major periods in the cell life cycle; involves the division of the nucleus (mitosis) and the division of the cytoplasm (cytokinesis).

Cell life cycle Series of changes a cell goes through from the time it is formed until it reproduces itself.

Cell-mediated immunity Immunity conferred by activated T cells, which directly lyse infected or cancerous body cells or cells of foreign grafts and release chemicals that regulate the immune response.

Cell membrane. See Plasma membrane.

Cellular respiration Metabolic processes in which ATP is produced.

Cellulose (sel′u-lōs) A fibrous carbohydrate that is the main structural component of plant tissues.

Central (Haversian) canal (hah-ver′zhan) The canal in the center of each osteon that contains minute blood vessels and nerve fibers that serve the needs of the osteocytes.

Central nervous system (CNS) Brain and spinal cord.

Centriole (sen′tre-ol) Minute body found near the nucleus of the cell; active in cell division.

Centrosome (cell center) A region near the nucleus which contains paired organelles called centrioles.

Cerebellum (ser″ĕ-bel′um) Brain region most involved in producing smooth, coordinated skeletal muscle activity.

Cerebral aqueduct (ser′ĕ-bral, sĕ-re′bral) The slender cavity of the midbrain that connects the third and fourth ventricles; also called the aqueduct of Sylvius.

Cerebral cortex The outer gray matter region of the cerebral hemispheres.

Cerebral dominance Designates the hemisphere that is dominant for language.

Cerebral palsy Neuromuscular disability in which voluntary muscles are poorly controlled or paralyzed as a result of brain damage.

Cerebral white matter Consists largely of myelinated fibers bundled into large tracts; provides for communication between cerebral areas and lower CNS centers.

Cerebrospinal fluid (CSF) (ser″ĕ-bro-spi′nal) Plasmalike fluid that fills the cavities of the CNS and surrounds the CNS externally; protects the brain and spinal cord.

Cerebrovascular accident (CVA) (ser″ĕ-bro-vas′ku-lar). Condition in which brain tissue is deprived of a blood supply, as in blockage of a cerebral blood vessel; a stroke.

Cerebrum (ser′ĕ-brum) The cerebral hemispheres and the structures of the diencephalon.

Cervical vertebrae The seven vertebrae of the vertebral column located in the neck.

Cervix Lower outlet of the uterus extending into the vagina.

Chemical bond An energy relationship holding atoms together; involves the interaction of electrons.

Chemical energy Energy stored in the bonds of chemical substances.

Chemical equilibrium A state of apparent repose created by two reactions proceeding in opposite directions at equal speed.

Chemical reaction Process in which molecules are formed, changed, or broken down.

Chemoreceptor (ke″mo-re-sep′ter) Receptors sensitive to various chemicals in solution.

Chemotaxis (ke″mo-tak′sis) Movement of a cell, organism, or part of an organism toward or away from a chemical substance.

Cholecystokinin (CCK) (ko″le-sis″to-ki′nin) An intestinal hormone that stimulates gallbladder contraction and pancreatic juice release.

Cholesterol (ko-les′ter-ol″) Steroid found in animal fats as well as in most body tissues; made by the liver.

Cholinergic fibers (ko″lin-er′jik) Nerve endings that, upon stimulation, release acetylcholine.

Chondroblast (kon′dro-blast) Actively mitotic cell form of cartilage.

Chondrocyte (kon′dro-sīt) Mature cell form of cartilage.

Chorion (kor′e-on) Outermost fetal membrane; helps form the placenta.

Chorionic villi sampling (ko″re-on′ik vil′i) Fetal testing procedure in which bits of the chorionic villi from the placenta are snipped off and the cells karyotyped. This procedure can be done as early as 8 weeks into the pregnancy.

Choroid (ko′roid) The vascular middle tunic of the eye.

Choroid plexus (ko′roid plex′sus) A capillary knot that protrudes into a brain ventricle; involved in forming cerebrospinal fluid.

Chromatin (kro′mah-tin)   Structures in the nucleus that carry the hereditary factors (genes).

Chromosomes (kro′mo-somz)   Barlike bodies of tightly coiled chromatin; visible during cell division.

Chronic obstructive pulmonary disease (COPD)   Collective term for progressive, obstructive respiratory disorders; includes emphysema, chronic bronchitis.

Chyme (kīm)   Semifluid, creamy mass consisting of partially digested food and gastric juice.

Cilia (sil′e-ah)   Tiny, hairlike projections on cell surfaces that move in a wavelike manner.

Circle of Willis   An arterial anastomosis at the base of the brain.

Circumduction (ser″kum-duk′shun)   Movement of a body part so that it outlines a cone in space.

Cirrhosis (sĭ-ro′sis)   Chronic disease of the liver, characterized by an overgrowth of connective tissue or fibrosis.

Cisterna chyli (sis-ter′nah ki′li)   An enlarged sac at the base of the thoracic duct; the origin of the thoracic duct.

Cisternae (sis-ter′ne)   Any cavity or enclosed space serving as a reservoir.

Cleavage   An early embryonic phase consisting of rapid mitotic cell divisions without intervening growth periods; product is a blastocyst.

Clonal selection (klo′nul)   Process during which a B cell or T cell becomes sensitized through binding contact with an antigen.

Clone   Descendants of a single cell.

Coagulation   Process in which blood is transformed from a liquid to a gel; blood clotting.

Cochlea (kok′le-ah)   Snail-shaped chamber of the bony labyrinth that houses the receptor for hearing (the organ of Corti).

Codon (ko′don)   The three-base sequence on a messenger RNA molecule that provides the genetic information used in protein synthesis.

Coenzyme (ko-en′zīm)   Nonprotein substance associated with and activating an enzyme, typically a vitamin.

Cofactor   Metal ion or organic molecule that is required for enzyme activity.

Collagen fiber   The most abundant of the three fibers found in the matrix of connective tissue.

Colloid (kol′oid)   A mixture in which the solute particles do not settle out readily and do not pass through natural membranes.

Colloid osmotic pressure (kol′oid ahz-mah′tik)   Pressure created in a fluid by large nondiffusible molecules, such as plasma proteins that are prevented from moving through a (capillary) membrane. Such substances tend to draw water to them.

Colon   Regions of the large intestine; includes ascending, transverse, descending, and sigmoid portions.

Combination (synthesis) reaction   Chemical reaction in which larger, more complex atoms or molecules are formed from simpler ones.

Complement   A group of bloodborne proteins, which, when activated, enhance the inflammatory and immune responses and may lead to cell lysis.

Complementarity of structure and function   The relationship between a structure and its function; i.e., structure determines function.

Complementary base   The degree of base pairing between two sequences of DNA and/or RNA molecules.

Complete blood count (CBC)   Clinical test that includes a hematocrit, counts of all formed elements and clotting factors, and other indicators of normal blood function.

Compound   Substance composed of two or more different elements, the atoms of which are chemically united.

Concentration gradient   The difference in the concentration of a particular substance between two different areas.

Conducting zone   Includes all respiratory passageways that provide conduits for air to reach the sites of gas exchange (the respiratory zone).

Conductivity   Ability to transmit an electrical impulse.

Cones   One of the two types of photoreceptor cells in the retina of the eye; provide for color vision.

Congenital (kun-jeh′nih-tul)   Existing at birth.

Congestive heart failure (CHF)   Condition in which the pumping efficiency of the heart is depressed so that circulation is inadequate to meet tissue needs.

Conjunctiva (kon″junk-ti′vah)   Thin, protective mucous membrane lining the eyelids and covering the anterior surface of the eye itself.

Connective tissue   A primary tissue; form and function vary extensively. Functions include support, storage, and protection.

Connexons (kŏ-nek′sonz)   Hollow cylinders made of transmembrane proteins that connect adjacent cells at gap junctions, allowing chemical substances to pass through.

Consciousness   The ability to perceive, communicate, remember, understand, appreciate, and initiate voluntary movements.

Contraception   The prevention of conception; birth control.

Contractility   Muscle cell's ability to move by shortening.

Contraction   To shorten or develop tension, an ability highly developed in muscle cells.

Contralateral   Relating to the opposite side.

Control center   One of three interdependent components of homeostatic control mechanisms; determines the set point.

Cornea (kor′ne-ah)   Transparent anterior portion of the eyeball; part of the fibrous tunic.

Corona radiata (kor-o′nah ra-de-ah′tah)   (1) Arrangement of elongated follicle cells around a mature ovum; (2) crownlike arrangement of nerve fibers radiating from the internal capsule of the brain to every part of the cerebral cortex.

Coronary circulation   The functional blood supply of the heart; shortest circulation in the body.

Cortex (kor′teks)   Outer surface layer of an organ.

Corticosteroids (kor″tĭ-ko-stĕ′roidz)   Steroid hormones released by the adrenal cortex.

Cortisol (hydrocortisone) (kor′tih-sol)   Glucocorticoid produced by the adrenal cortex.

Covalent bond (ko-va′lent)   Chemical bond created by electron sharing between atoms.

Cranial nerves   The 12 nerve pairs that arise from the brain.

Craniosacral division   Another name for the parasympathetic division of the autonomic nervous system.

Cranium (cranial bones) (kra′ne-um)   Bony protective encasement of the brain and organs of hearing and equilibrium; also called the skull.

Creatine kinase (kre′ah-tin)   Enzyme that catalyzes the transfer of phosphate from phosphocreatine to ADP, forming creatine and ATP; important in muscle contraction.

Creatine phosphate (CP) (fos-fāt)   Compound that serves as an alternative energy source for muscle tissue.

Creatinine (kre-at′ĭ-nin)   A nitrogenous waste molecule which is not reabsorbed by the kidney; this characteristic makes it useful for measurement of the GFR and glomerular function.

Crista ampullaris   Sensory receptor organ within the ampulla of each semicircular canal of the inner ear; dynamic equilibrium receptor.

Cross section   A cut running horizontally from right to left, dividing the body into superior and inferior parts.

Crystals   Large arrays of cations and anions held together by ionic bonds; formed when an element or compound solidifies or is in a dry state.

Cutaneous (ku-ta′ne-us)   Pertaining to the skin.

Cutaneous sensory receptors   Receptors located throughout the skin that respond to stimuli arising outside the body; part of the nervous system.

Cyclic AMP   Important intracellular second messenger that mediates hormonal effects; formed from ATP by the action of adenylate cyclase, an enzyme associated with the plasma membrane.

Cystic fibrosis (CF)   Genetic disorder in which oversecretion of mucus clogs the respiratory passages, predisposes to fatal respiratory infections.

Cytochromes (si'to-krōmz) Brightly colored iron-containing proteins that form part of the inner mitochondrial membrane and function as electron carriers in oxidative phosphorylation.

Cytokines Chemical mediators involved in cellular immunity; include lymphokines, and monokines.

Cytokinesis (si"to-ki-ne'sis) The division of cytoplasm that occurs after the cell nucleus has divided.

Cytoplasm (si'to-plazm) The cellular material surrounding the nucleus and enclosed by the plasma membrane.

Cytosine (C) (si'to-sēn) Nitrogen-containing base that is part of a nucleotide structure.

Cytoskeleton Literally, cell skeleton. An elaborate series of rods running through the cystol, supporting cellular structures and providing the machinery to generate various cell movements.

Cytosol Viscous, semitransparent fluid substance of cytoplasm in which other elements are suspended.

Cytotoxic T cell Effector T cell that directly kills (lyses) foreign cells, cancer cells, or virus-infected body cells. Also called a killer T cell.

Deamination (de'am"ih-na'shun) Removal of an amine group from an organic compound.

Decomposition reaction Chemical reaction in which a molecule is broken down into smaller molecules or its constituent atoms.

Defecation (def"ih-ka'shun) Elimination of the contents of the bowels (feces).

Deglutition (deg"loo-tish'un) Swallowing.

Dehydration (de"hi-dra'shun) Condition of excessive water loss.

Dehydration synthesis Process by which a large molecule is synthesized by covalently bonding smaller molecules together.

Dendrite (den'drīt) Branching neuron process that serves as a receptive, or input, region; transmits the nerve impulse toward the cell body.

Depolarization (de-po"ler-ah-za'shun) Loss of a state of polarity; loss or reduction of negative membrane potential.

Dermatome (der'mah-tōm) Portion of somite mesoderm that forms the dermis of the skin; also the area of skin innervated by the cutaneous branches of a single spinal nerve.

Dermis Layer of skin deep to the epidermis; composed of dense irregular tissue.

Desmosome (dez'muh-sōm) Cell junction composed of thickened plasma membranes joined by filaments.

Diabetes insipidus (di"ah-be'tez in-sih'pih-dus) Disease characterized by passage of a large quantity of dilute urine plus intense thirst and dehydration caused by inadequate release of antidiuretic hormone (ADH).

Diabetes mellitus (DM) (meh-li'tus) Disease caused by deficient insulin release, leading to inability of the body cells to use carbohydrates.

Dialysis (di-al'ah-sis) Diffusion of solute(s) through a semipermeable membrane.

Diapedesis (di"ah-pě-de'sis) Passage of blood cells through intact vessel walls into tissue.

Diaphragm (di'ah-fram) (1) Any partition or wall separating one area from another; (2) a muscle that separates the thoracic cavity from the lower abdominopelvic cavity.

Diaphysis (di-af'ĭ-sis) Elongated shaft of a long bone.

Diarthrosis (di"ar-thro'sis) Freely movable joint.

Diastole (di-as'to-le) Period of the cardiac cycle when either the ventricles or the atria are relaxing.

Diastolic pressure (di-as-tah'lik) Arterial blood pressure reached during or as a result of diastole; lowest level of any given ventricular cycle.

Diencephalon (interbrain) (di"en-seh'fuh-lon) That part of the forebrain between the cerebral hemispheres and the midbrain including the thalamus, the third ventricle, and the hypothalamus.

Differential white blood cell count Diagnostic test to determine relative proportion of individual leukocyte types.

Diffusion (dĭ-fu'zhun) The spreading of particles in a gas or solution with a movement toward uniform distribution of particles.

Digestion Chemical or mechanical process of breaking down foodstuffs to substances that can be absorbed.

Digestive system System that processes food into absorbable units and eliminates indigestible wastes.

Dipeptide A combination of two amino acids united by means of a peptide bond.

Diploë (dip'lo-e) The internal layer of spongy bone in flat bones.

Diploid chromosomal number The chromosomal number characteristic of an organism, symbolized as $2n$; twice the chromosomal number (n) of the gamete; in humans, $2n = 46$.

Diplopia (dĭ-plo'pe-ah) Double vision.

Dipole (polar molecule) Nonsymmetrical molecules that contain electrically unbalanced atoms.

Disaccharide (di-sak'ah-rīd", di-sak'ah-rid) Literally, double sugar; e.g., sucrose, lactose.

Dislocation (luxation) Occurs when bones are forced out of their normal alignment at a joint.

Displacement (exchange) reaction Chemical reaction in which bonds are both made and broken; atoms become combined with different atoms.

Distal (dis'tul) Away from the attached end of a limb or the origin of a structure.

Diuretics (di"u-ret'iks) Chemicals that enhance urinary output.

Diverticulum (di"ver-tik'u-lum) A pouch or sac in the walls of a hollow organ or structure.

DNA (deoxyribonucleic acid) (de-ok"sĭ-ri"bo-nu-kla'ik) A nucleic acid found in all living cells; it carries the organism's hereditary information.

DNA replication Process that occurs before cell division; ensures that all daughter cells have identical genes.

Dominant traits Occurs when one allele masks or suppresses the expression of its partner.

Dominant-recessive inheritance Reflects the interaction of dominant and recessive alleles.

Dorsal (dor'sul) Pertaining to the back; posterior.

Double helix The secondary structure assumed by two strands of DNA, held together throughout their length by hydrogen bonds between bases on opposite strands.

Down-regulation Decreased response to hormonal stimulation; involves a loss in the number of receptors and effectively prevents the target cells from overreacting to persistently high hormone levels.

Duct (dukt) A canal or passageway; a tubular structure that provides an exit for the secretions of a gland, or for conducting any fluid.

Ductus (vas) deferens Extends from the epididymis to the urethra; propels sperm into the urethra by peristalsis during ejaculation.

Duodenum (du"o-de'num) First part of the small intestine.

Dura mater (du'rah ma'ter) Outermost and toughest of the three membranes (meninges) covering the brain and spinal cord.

Dynamic equilibrium Sense that reports on angular or rotatory movements of the head in space.

Dyskinesia (dis-kĭ-ne'ze-ah) Disorders of muscle tone, posture, or involuntary movements.

Dyspnea (disp-ne'ah) Difficult or labored breathing; air hunger.

Eccrine glands (ek'rin) Sweat glands abundant on the palms, soles of feet, and the forehead.

Ectoderm (ek'to-derm) Embryonic germ layer; forms the epidermis of the skin and its derivatives, and nervous tissues.

Edema (ě-de'mah) Abnormal accumulation of fluid in body parts or tissues; causes swelling.

Effector (ef-ek'ter) Organ, gland, or muscle capable of being activated by nerve endings.

Efferent (ef'er-ent) Carrying away or away from, especially a nerve fiber that carries impulses away from the central nervous system.

Elastic cartilage Cartilage with abundant elastic fibers; more flexible than hyaline cartilage.

Elastic fiber Fiber formed from the protein elastin, which gives a rubbery and resilient quality to the matrix of connective tissue.

Electrical energy   Energy formed by the movement of charged particles across cell membranes.

Electrocardiogram (ECG or EKG) (e-lek″tro-car′de-o-gram″)   Graphic record of the electrical activity of the heart.

Electrochemical gradient   The distribution of ions involving both a chemical and an electrical gradient interacting to determine the direction of diffusion.

Electroencephalogram (EEG) (e-lek″tro-en-sef′ah-lo-gram″)   Graphic record of the electrical activity of nerve cells in the brain.

Electrolyte (e-lek′tro-līt)   Chemical substances, such as salts, acids, and bases, that ionize and dissociate in water and are capable of conducting an electrical current.

Electrolyte balance   Refers to the balance between input and output of salts (sodium, potassium, calcium, magnesium) in the body.

Electromagnetic radiation   Energy form that travels in waves.

Electron   Negatively charged subatomic particle; orbits the atom's nucleus.

Electron shells (energy levels)   Regions of space that consecutively surround the nucleus of an atom.

Element   One of a limited number of unique varieties of matter that composes substances of all kinds; e.g., carbon, hydrogen, oxygen.

Embolism (em′bo-lizm)   Obstruction of a blood vessel by an embolus (blood clot, fatty mass, bubble of air, or other debris) floating in the blood.

Embryo (em′bre-o)   Developmental stage extending from gastrulation to the end of the eighth week.

Emesis   Reflexive emptying of the stomach through the esophagus and pharynx; also known as vomiting.

Encephalitis (en″seh-fuh-lī′tis)   Inflammation of the brain.

Endergonic reaction   Chemical reaction that absorbs energy, e.g., an anabolic reaction.

Endocardium (en″do-kar′de-um)   Endothelial membrane that lines the interior of the heart.

Endochondral ossification (en″do-kon′dral)   Embryonic formation of bone by the replacement of calcified cartilage; most skeletal bones formed by this process.

Endocrine glands (en′do-krin)   Ductless glands that empty their hormonal products directly into the blood.

Endocrine system   Body system that includes internal organs that secrete hormones.

Endocytosis (en″do-si-to′sis)   Means by which fairly large extracellular molecules or particles enter cells, e.g., phagocytosis, pinocytosis, receptor-mediated endocytosis.

Endoderm (en′do-derm)   Embryonic germ layer; forms the lining of the digestive tube and its associated structures.

Endogenous (en-doj′ĕ-nŭs)   Originating or produced within the organism or one of its parts.

Endometrium (en″do-me′tre-um)   Mucous membrane lining of the uterus.

Endomysium (en″do-mis′e-um)   Thin connective tissue surrounding each muscle cell.

Endoplasmic reticulum (en″do-plaz′mik rĕ-tik′u-lum)   Membranous network of tubular or saclike channels in the cytoplasm of a cell.

Endosteum (en-dos′te-um)   Connective tissue membrane covering internal bone surfaces.

Endothelium (en″do-the′le-um)   Single layer of simple squamous cells that line the walls of the heart, blood vessels, and lymphatic vessels.

Energy   The capacity to do work; may be stored (potential energy) or in action (kinetic energy).

Energy intake   Energy liberated during food oxidation.

Energy output   Sum of energy lost as heat, as work, and as fat or glycogen storage.

Enzyme (en′zīm)   A protein that acts as a biological catalyst to speed up a chemical reaction.

Eosinophil (e″o-sin′o-fil)   Granular white blood cell whose granules readily take up a stain called eosin.

Ependymal cell (ĕ-pen′dĭ-mul)   A type of CNS supporting cell; lines the central cavities of the brain and spinal cord.

Epidermis (ep″ĭ-der′mis)   Superficial layer of the skin; composed of keratinized stratified squamous epithelium.

Epididymis (ep″ĭ-dĭ′dĭ-mis)   That portion of the male duct system in which sperm mature. Empties into the ductus (or vas) deferens.

Epidural space   Area between the bony vertebrae and the dura mater of the spinal cord.

Epiglottis (eh″puh-glah′tis)   Elastic cartilage at the back of the throat; covers the opening of the larynx during swallowing.

Epileptic seizures   Abnormal electrical discharges of groups of brain neurons, during which no other messages can get through.

Epimysium (ep″ĭ-mis′e-um)   Sheath of fibrous connective tissue surrounding a muscle.

Epinephrine (ep″ĭ-nef′rin)   Chief hormone produced by the adrenal medulla. Also called adrenalin.

Epiphyseal plate (e″pĭ-fis′e-ul)   Plate of hyaline cartilage at the junction of the diaphysis and epiphysis that provides for growth in length of a long bone.

Epiphysis (e-pif′ĭ-sis)   The end of a long bone, attached to the shaft.

Epithalamus   Most dorsal portion of the diencephalon; forms the roof of the third ventricle with the pineal gland extending from its posterior border.

Epithelium (epithelial tissue) (ep″ĭ-the′le-ul)   Pertaining to a primary tissue that covers the body surface, lines its internal cavities, and forms glands.

Erythrocytes (e-rith′ro-sīts)   Red blood cells.

Erythropoiesis (ĕ-rith″ro-poi-e′sis)   Process of erythrocyte formation.

Erythropoietin (EPO) (ĕ-rith″ro-poi′ĕ-tin)   Hormone that stimulates production of red blood cells.

Esophagus (ĕ-sof′ah-gus)   Muscular tube extending from the laryngopharynx through the diaphragm to join the stomach; collapses when not involved in food propulsion.

Estrogens (es′tro-jenz)   Hormones that stimulate female secondary sex characteristics; female sex hormones.

Eupnea (ūp-ne′ah)   Normal respiratory rate and rhythm.

Eustachian tube (u-sta′shan).   *See* Pharyngotympanic tube.

Exchange (displacement) reaction   Chemical reaction in which bonds are both made and broken; atoms become combined with different atoms.

Excitation-contraction (E-C) coupling   Sequence of events by which transmission of an action potential along the sarcolemma leads to the sliding of myofilaments.

Excretion (ek-skre′shun)   Elimination of waste products from the body.

Exergonic reaction   Chemical reaction that releases energy, e.g., a catabolic or oxidative reaction.

Exocrine glands (ek′so-krin)   Glands that have ducts through which their secretions are carried to a particular site.

Exocytosis (ek″so-si-to′sis)   Mechanism by which substances are moved from the cell interior to the extracellular space as a secretory vesicle fuses with the plasma membrane.

Exogenous (ek-sah′jeh-nus)   Developing or originating outside an organ or part.

Exons   Amino acid–specifying informational sequences (separated by introns) in the genes of higher organisms.

Extension   Movement that increases the angle of a joint, e.g., straightening a flexed knee.

Exteroceptor (ek″ster-o-sep′tor)   Sensory end organ that responds to stimuli from the external world.

Extracellular fluid (ECF)   Internal fluid located outside cells; includes plasma and interstitial fluid.

Extracellular materials   Substances found outside the cell; include interstitial fluid, blood plasma, and cerebrospinal fluid.

Extracellular matrix   Nonliving material in connective tissue consisting of ground substance and fibers that separates the living cells.

Extrasystole (ek″strah-sis′to-le)   Premature heart contraction.

Extrinsic (ek-strin′sik)   Of external origin.

Extrinsic eye muscles   The six skeletal muscles which attach to and move each eye.

Exudate (ek'syoo-dāt)   Material including fluid, pus, or cells that has escaped from blood vessels and been deposited in tissues.

Facilitated diffusion   Passive transport process used by certain molecules, e.g., glucose and other simple sugars too large to pass through plasma membrane pores. Involves movement through channels or movement facilitated by a membrane carrier.

Fallopian tube (fah-lō'pē-un).   *See* Uterine tube.

Fascia (fash'e-ah)   Layers of fibrous tissue covering and separating muscle.

Fascicle (fas'ĭ-kl)   Bundle of nerve or muscle fibers bound together by connective tissues.

Fatty acids   Linear chains of carbon and hydrogen atoms (hydrocarbon chains) with an organic acid group at one end. A constituent of fat.

Feces (fe'sēz)   Material discharged from the bowel; composed of food residue, secretions, bacteria.

Fenestrated (fen'es-tra-tid)   Pierced with one or more small openings.

Fertilization   Fusion of the sperm and egg nuclei.

Fetus   Developmental stage extending from the ninth week of development to birth.

Fiber   A slender threadlike structure or filament.

Fibrillation   Condition of rapid and irregular or out-of-phase heart contractions.

Fibrin (fi'brin)   Fibrous insoluble protein formed during blood clotting.

Fibrinogen (fi-brin'o-jin)   A blood protein that is converted to fibrin during blood clotting.

Fibrinolysis   Process that removes unneeded blood clots when healing has occurred.

Fibroblast (fi'bro-blast)   Young, actively mitotic cell that forms the fibers of connective tissue.

Fibrocartilage   The most compressible type of cartilage; resistant to stretch. Forms vertebral discs and knee joint cartilages.

Fibrocyte (fi'bro-sīt)   Mature fibroblast; maintains the matrix of fibrous types of connective tissue.

Fibrosis   Proliferation of fibrous connective tissue called scar tissue.

Fibrous joints   Bones joined by fibrous tissue; no joint cavity is present.

Filtrate   A plasma-derived fluid that is processed by the renal tubules to form urine.

Filtration   Passage of a solvent and dissolved substances through a membrane or filter.

First-degree burn   A burn in which only the epidermis is damaged.

Fissure (fih'zher)   (1) A groove or cleft; (2) the deepest depressions or inward folds on the brain.

Fixator (fix'a-ter)   Muscle that immobilizes one or more bones, allowing other muscles to act from a stable base.

Flagellum (flah-jel'lum)   Long, whiplike extension of the plasma membrane of some bacteria and of sperm; propels the cell.

Flexion (flek'shun)   Movement that decreases the angle of the joint, e.g., bending the knee from a straight to an angled position.

Flexor (withdrawal) reflex   Reflex initiated by a painful stimulus (actual or perceived); causes automatic withdrawal of the threatened body part from the stimulus.

Fluid mosaic model   A depiction of the structure of the membranes of a cell as phospholipid bilayers in which proteins are dispersed.

Follicle (fah'lih-kul)   (1) Ovarian structure consisting of a developing egg surrounded by one or more layers of follicle cells; (2) colloid-containing structure of the thyroid gland.

Follicle-stimulating hormone (FSH)   Hormone produced by the anterior pituitary that stimulates ovarian follicle production in females and sperm production in males.

Fontanels (fon"tah-nelz')   Fibrous membranes at the angles of cranial bones that accommodate brain growth in the fetus and infant.

Foramen (fo-ra'men)   Hole or opening in a bone or between body cavities.

Forebrain (prosencephalon)   Anterior portion of the brain consisting of the telencephalon and the diencephalon.

Formed elements   Cellular portion of blood.

Fossa (fos'ah)   A depression, often an articular surface.

Fovea (fo've-ah)   A pit.

Fracture   A break in a bone.

Free radicals   Highly reactive chemicals with unpaired electrons that can scramble the structure of proteins, lipids, and nucleic acids.

Frontal (coronal) plane   Longitudinal (vertical) plane that divides the body into anterior and posterior parts.

Fulcrum   The fixed point on which a lever moves when a force is applied.

Fundus (fun'dus)   Base of an organ; part farthest from the opening of the organ.

Gallbladder   Sac beneath the right lobe of the liver used for bile storage.

Gallstones (biliary calculi)   Crystallized cholesterol that obstructs the flow of bile from the gallbladder.

Gamete (gam'ēt)   Sex or germ cell.

Gametogenesis (gam"eh-to-jen'eh-sis)   Formation of gametes.

Ganglion (gang'gle-on)   Collection of nerve cell bodies outside the CNS.

Ganglionic neuron (gan"gle-ah'nik)   Autonomic motor neuron that has its cell body in a peripheral ganglion and projects its (postganglionic) axon to an effector.

Gap junction   A passageway between two adjacent cells; formed by transmembrane proteins called connexions.

Gastrin   Hormone secreted in the stomach; regulates gastric juice secretion by stimulating HCl production.

Gastroenteritis   Inflammation of the gastrointestinal tract.

Gastrulation (gas"troo-la'shun)   Developmental process that produces the three primary germ layers (ectoderm, mesod xzs, and endoderm).

Gene   One of the biological units of heredity located in chromatin; transmits hereditary information.

Genetic code   Refers to the rules by which the base sequence of a DNA gene is translated into protein structures (amino acid sequences).

Genitalia (jen"i-ta'le-ă)   The internal and external reproductive organs.

Genome   The complete set of chromosomes derived from one parent (the haploid genome); or the two sets of chromosomes, i.e., one set from the egg, the other from the sperm (the diploid genome).

Genotype (jen'o-tīp)   One's genetic makeup or genes.

Germ layers   Three cellular layers (ectoderm, mesoderm, and endoderm) that represent the initial specialization of cells in the embryonic body and from which all body tissues arise.

Gestation period (jes-ta'shun)   The period of pregnancy; about 280 days for humans.

Gland   Organ specialized to secrete or excrete substances for further use in the body or for elimination.

Glaucoma (glaw-ko'mah)   Condition in which intraocular pressure increases to levels that cause compression of the retina and optic nerve; results in blindness unless detected early.

Glial cells (gle'al)   The supporting cells in the central nervous system.

Glomerular capsule (glō-mer'yoo-ler)   Double-walled cup at end of a renal tubule; encloses a glomerulus. Also called Bowman's capsule.

Glomerular filtration rate (GFR)   Rate of filtrate formation by the kidneys.

Glomerulus (glo-mer'u-lus)   Cluster of capillaries forming part of the nephron; forms filtrate.

Glottis (glah'tis)   Opening between the vocal cords in the larynx.

Glucagon (gloo'kah-gon)   Hormone formed by alpha cells of pancreatic islets; raises the glucose level of blood.

Glucocorticoids (gloo"ko-kor'tĭ-koidz)   Adrenal cortex hormones that increase blood glucose levels and aid the body in resisting long-term stressors.

Gluconeogenesis (gloo"ko-ne"o-jen'ĕ-sis) Formation of glucose from noncarbohydrate molecules.

Glucose (gloo'kōs) Principal blood sugar; a hexose.

Glycerol (glis'er-ol) A modified simple sugar (a sugar alcohol).

Glycocalyx (cell coat) (gli"ko-kal'iks) A layer of externally facing glycoproteins on a cell's plasma membrane that determines blood type; involved in the cellular interactions of fertilization, embryonic development, and immunity, and acts as an adhesive between cells.

Glycogen (gli'ko-jin) Main carbohydrate stored in animal cells; a polysaccharide.

Glycogenesis (gli"ko-jen'ĕ-sis) Formation of glycogen from glucose.

Glycogenolysis (gli"ko-jĕ-nol'ĭ-sis) Breakdown of glycogen to glucose.

Glycolipid (gli"ko-lip'id) A lipid with one or more covalently attached sugars.

Glycolysis (gli-kol'ĭ-sis) Breakdown of glucose to pyruvic acid—an anaerobic process.

Goblet cells Individual cells (unicellular glands) that produce mucus.

Golgi apparatus (gol'je) Membranous system close to the cell nucleus that packages protein secretions for export, packages enzymes into lysosomes for cellular use, and modifies proteins destined to become part of cellular membranes.

Golgi tendon organs Proprioceptors located in tendons, close to the point of skeletal muscle insertion; important to smooth onset and termination of muscle contraction.

Gonad (go'nad) Primary reproductive organ; i.e., the testis of the male or the ovary of the female.

Gonadocorticoids (gon"ah-do-kor'tĭ-koidz) Sex hormones, primarily androgens, secreted by the adrenal cortex.

Gonadotropins (gon"ah-do-trōp'inz) Gonad-stimulating hormones produced by the anterior pituitary.

Graafian follicle (graf'e-an). See Vesicular follicle.

Graded muscle responses Variations in the degree of muscle contraction by changing either the frequency or strength of the stimulus.

Graded potential A local change in membrane potential that varies directly with the strength of the stimulus, declines with distance.

Graves' disease Disorder resulting from hyperactive thyroid gland.

Gray matter Gray area of the central nervous system; contains cell bodies and unmyelinated fibers of neurons.

Growth hormone (GH) Hormone that stimulates growth in general; produced in the anterior pituitary; also called somatotropin (STH).

Guanine (G) (gwan'ēn) One of two major purines occurring in all nucleic acids.

Gustation (gus-ta'shun) Taste.

Gut-brain peptides Neuropeptides produced by nonneural tissue that are widespread in the gastrointestinal tract.

Gyrus (ji'rus) An outward fold of the surface of the cerebral cortex.

Hair follicle Structure with outer and inner root sheaths extending from the epidermal surface into the dermis and from which new hair develops.

Hapten (hap'ten) An incomplete antigen; has reactivity but not immunogenicity.

Haversian system (hah-ver'zhen). See Osteon.

Heart attack (coronary). See Myocardial infarction.

Heart block Impaired transmission of impulses from atrium to ventricle resulting in dysrhythmia.

Heart murmur Abnormal heart sound (usually resulting from valve problems).

Heimlich maneuver Procedure in which the air in a person's own lungs is used to expel an obstructing piece of food.

Helper T cell Type of T lymphocyte that orchestrates cellular immunity by direct contact with other immune cells and by releasing chemicals called lymphokines; also helps to mediate the humoral response by interacting with B cells.

Hematocrit (he-mat'o-krit) The percentage of erythrocytes to total blood volume.

Hematoma (he"mah-to'mah) Mass of clotted blood that forms at an injured site.

Hematopoiesis (hem"ah-to-poi-e'sis) Blood cell formation; hemopoiesis.

Heme (hēm) Iron-containing pigment that is essential to oxygen transport by hemoglobin.

Hemocytoblast (he"mo-si'to-blast) Bone marrow cell that gives rise to all the formed elements of blood.

Hemoglobin (he'muh-glo-bin) Oxygen-transporting component of erythrocytes.

Hemolysis (he-mah'lĕ-sis) Rupture of erythrocytes.

Hemophilia (he"mo-fil'e-ah) A term loosely applied to several different hereditary bleeding disorders that exhibit similar signs and symptoms.

Hemopoiesis (he"mo-poi-e'sis) Hematopoiesis.

Hemorrhage (hem'or-ij) Loss of blood from the vessels by flow through ruptured walls; bleeding.

Hemostasis (he"mo-sta'sis) Stoppage of bleeding.

Heparin Natural anticoagulant secreted into blood plasma.

Hepatic portal system (hĕ-pat'ik) Circulation in which the hepatic portal vein carries dissolved nutrients to the liver tissues for processing.

Hepatitis (hep"ah-ti'tis) Inflammation of the liver.

Hernia (her'ne-ah) Abnormal protrusion of an organ or a body part through the containing wall of its cavity.

Heterozygous (het"er-o-zi'gus) Having different allelic genes at one locus or (by extension) many loci.

Hilton's law Any nerve serving a muscle producing movement at a joint; also innervates the joint itself and the skin over the joint.

Hilus (hi'lus) The indented region of an organ from which blood and/or lymph vessels and nerves enter and exit.

Hippocampus Limbic system structure that plays a role in converting new information into long-term memories.

Histamine (his'tuh-mēn) Substance that causes vasodilation and increased vascular permeability.

Histology (his-tol'o-je) Branch of anatomy dealing with the microscopic structure of tissues.

HIV (human immunodeficiency virus) Virus that destroys helper T cells, thus depressing cell-mediated immunity; symptomatic AIDS gradually appears when lymph nodes can no longer contain the virus.

Holocrine glands (hol'o-krin) Glands that accumulate their secretions within their cells; secretions are discharged only upon rupture and death of the cell.

Homeostasis (ho"me-o-sta'sis) A state of body equilibrium or stable internal environment of the body.

Homologous (ho-mol'ŏ-gus) Parts or organs corresponding in structure but not necessarily in function.

Homozygous (ho-mo-zi'gus) Having identical genes at one or more loci.

Hormones Steroidal or amino acid–based molecules released to the blood that act as chemical messengers to regulate specific body functions.

Humoral immunity (hu'mer-ul) Immunity conferred by antibodies present in blood plasma and other body fluids.

Huntington's disease Hereditary disorder leading to degeneration of the basal nuclei and the cerebral cortex.

Hyaline cartilage (hi'ah-lĭn) The most abundant cartilage type in the body; provides firm support with some pliability.

Hydrochloric acid (HCl) (hi"dro-klor'ik) Acid that aids protein digestion in the stomach; produced by parietal cells.

Hydrogen bond Weak bond in which a hydrogen atom forms a bridge between two electron-hungry atoms. An important intramolecular bond.

Hydrogen ion (H⁺) A hydrogen atom minus its electron and therefore carrying a positive charge (i.e., a proton).

Hydrolysis (hi"drah'lă-sis) Process in which water is used to split a substance into smaller particles.

Hydrophilic (hi″dro-fil′ik)   Refers to molecules, or portions of molecules, that interact with water and charged particles.

Hydrophobic (hi″dro-fo′bik)   Refers to molecules, or portions of molecules, that interact only with nonpolar molecules.

Hydrostatic pressure (hi″dro-stă′tik)   Pressure of fluid in a system.

Hydroxyl ion (OH⁻) (hi-drok′sil)   An ion liberated when a hydroxide (a common inorganic base) is dissolved in water.

Hyperalgesia   Pain amplification.

Hypercapnia (hi″per-kap′ne-ah)   High carbon dioxide levels in the blood.

Hyperemia   An increase in blood flow into a tissue or organ; congested with blood.

Hyperglycemic (hi″per-gli-se′mik)   Term used to describe hormones such as glucagon that elevate blood glucose level.

Hyperopia (hi″per-o′pe-ah)   A condition in which visual images are routinely focused behind the retina; commonly known as farsightedness.

Hyperplasia (hi″per-pla′ze-ah)   Accelerated growth, e.g., in anemia, the bone marrow produces red blood cells at a faster rate.

Hyperpnea (hi″perp-ne′ah)   Deeper and more vigorous breathing, but with unchanged respiratory rate, as during exercise.

Hypersensitivity (allergy)   Overzealous immune response to an otherwise harmless antigen.

Hypertension (hi″per-ten′shun)   High blood pressure.

Hypertonic (hi″per-ton′ik)   Excessive, above normal, tone or tension.

Hypertonic solution   A solution that has a higher concentration of nonpenetrating solutes than the reference cell; having greater osmotic pressure than the reference solution (blood plasma or interstitial fluid).

Hypertrophy (hi-per′trah-fe)   Increase in size of a tissue or organ independent of the body's general growth.

Hyperventilation   Increased depth and rate of breathing.

Hypocapnia   Low carbon dioxide levels in the blood.

Hypodermis (superficial fascia)   Subcutaneous tissue just deep to the skin; consists of adipose plus some areolar connective tissue.

Hypoglycemic (hi″po-gli-se′mik)   Term used to describe hormones such as insulin that decrease blood glucose level.

Hyponatremia   Abnormally low concentrations of sodium ions in extracellular fluid.

Hypoproteinemia (hi″po-pro″te-ĭ-ne′me-ah)   A condition of unusually low levels of plasma proteins causing a reduction in colloid osmotic pressure; results in tissue edema.

Hypotension   Low blood pressure.

Hypothalamic-hypophyseal tract (hi″po-thah-lam′ik–hi″po-fiz′e-al)   Nerve bundles that run through the infundibulum and connect the neurohypophysis and the hypothalamus.

Hypothalamus (hi″po-thal′ah-mus)   Region of the diencephalon forming the floor of the third ventricle of the brain.

Hypotonic (hi″po-ton′ik)   Below normal tone or tension.

Hypotonic solution   A solution that is more dilute (containing fewer nonpenetrating solutes) than the reference cell. Cells placed in hypotonic solutions plump up rapidly as water rushes into them.

Hypoventilation   Decreased depth and rate of breathing.

Hypovolemic shock (hi″po-vo-le′mik)   Most common form of shock; results from extreme blood loss.

Hypoxia (hi-pok′se-ah)   Condition in which inadequate oxygen is available to tissues.

Ileocecal valve (il″e-o-se′kal)   Site where the small intestine joins the large intestine.

Ileum (il′e-um)   Terminal part of the small intestine; between the jejunum and the cecum of the large intestine.

Immune system   A functional system whose components attack foreign substances or prevent their entry into the body.

Immunity (im′ūn′ĭ-te)   Ability of the body to resist many agents (both living and nonliving) that can cause disease; resistance to disease.

Immunocompetence   Ability of the body's immune cells to recognize (by binding) specific antigens; reflects the presence of plasma membrane-bound receptors.

Immunodeficiency   Any congenital or acquired condition causing a deficiency in the production or function of immune cells or certain molecules (complement, antibodies, etc.) required for normal immunity.

In vitro (in ve′tro)   In a test tube, glass, or artificial environment.

In vivo (in ve′vo)   In the living body.

Incompetent valve   Valve which does not close properly.

Incontinence   Inability to control micturition voluntarily.

Indolamines   Biogenic amine neurotransmitters; include serotonin and histamine.

Infarct (in′farkt)   Region of dead, deteriorating tissue resulting from a lack of blood supply.

Infectious mononucleosis   Highly contagious viral disease; marked by excessive agranulocytes.

Inferior (caudal)   Pertaining to a position near the tail end of the long axis of the body.

Inferior vena cava   Vein that returns blood from body areas below the diaphragm.

Inflammation (in″flah-ma′shun)   A nonspecific defensive response of the body to tissue injury; includes dilation of blood vessels and an increase in vessel permeability; indicated by redness, heat, swelling, and pain.

Infundibulum (in″fun-dib′u-lum)   (1) A stalk of tissue that connects the pituitary gland to the hypothalamus; (2) the distal end of the fallopian (uterine) tube.

Inguinal (ing′wĭ-nal)   Pertaining to the groin region.

Inner cell mass   Accumulation of cells in the blastocyst from which the embryo develops.

Innervation (in″er-va′shun)   Supply of nerves to a body part.

Inorganic compound   Chemical substances that do not contain carbon, including water, salts, and many acids and bases.

Insertion   Movable attachment of a muscle.

Insulin   A hormone that enhances the carrier-mediated diffusion of glucose into tissue cells, thus lowering blood glucose levels.

Integration   The process by which the nervous system processes and interprets sensory input and makes decisions about what should be done at each moment.

Integumentary system (in-teg″u-men′tar-e)   Skin and its derivatives; provides the external protective covering of the body.

Intercalated discs (in-ter′kah-la″ted)   Gap junctions connecting muscle cells of the myocardium.

Interferon (IFN) (in-ter-fēr′on)   Chemical that is able to provide some protection against virus invasion of the body; inhibits viral growth.

Internal capsule   Band of projection fibers that runs between the basal nuclei and the thalamus.

Internal respiration   Exchange of gases between blood and tissue fluid and between tissue fluid and cells.

Interoceptor (in″ter-o-sep′tor)   Sensory receptor nerve ending in the viscera, which is sensitive to changes and stimuli within the body's internal environment; also called visceroceptor.

Interphase   One of two major periods in the cell life cycle; includes the period from cell formation to cell division.

Interstitial cells   Cells located in the loose connective tissue surrounding the seminiferous tubules; they produce androgens (most importantly testosterone), which are secreted into the surrounding interstitial fluid.

Interstitial fluid (IF) (in″ter-stish′al)   Fluid between the cells.

Interstitial lamellae   Incomplete lamellae that lie between intact osteons, filling the gaps between forming osteons, or representing the remnants of an osteon that has been cut through by bone remodeling.

Intervertebral discs (in″ter-ver′teh-brul)   Discs of fibrocartilage between vertebrae.

Intracapsular ligament   Ligament located within and separate from the articular capsule of a synovial joint.

Intracellular fluid (ICF) (in″trah-sel′u-ler)   Fluid within a cell.

Intrinsic factor   Substance produced by the stomach that is required for vitamin B_{12} absorption.

Introns Noncoding segment or portion of DNA that ranges from 60 to 100,000 nucleotides long.

Involuntary muscle Muscles that cannot ordinarily be controlled voluntarily.

Involuntary nervous system The autonomic nervous system.

Ion (ī'on) Atom with a positive or negative electric charge.

Ionic bond (ī-ah'nik) Chemical bond formed by electron transfer between atoms.

Ipsilateral (ip"sih-la'ter-ul) Situated on the same side.

Irritability Ability to respond to stimuli.

Ischemia (is-ke'me-ah) Local decrease in blood supply.

Isograft Tissue graft donated by an identical twin.

Isomer (i'so-mer) One of two or more substances that has the same molecular formula but with its atoms arranged differently.

Isometric contraction (i"so-me'trik) Contraction in which the muscle does not shorten (the load is too heavy) but its internal tension increases.

Isotonic contraction (i"so-tah'nik) Contraction in which muscle tension remains constant at a given joint angle and load, and the muscle shortens.

Isotonic solution A solution with a concentration of nonpenetrating solutes equal to that found in the reference cell.

Isotopes (i'so-tōps) Different atomic forms of the same element, vary only in the number of neutrons they contain; the heavier species tend to be radioactive.

Jejunum (jĕ-joo'num) The part of the small intestine between the duodenum and the ileum.

Joint (articulation) The junction of two or more bones.

Joint kinesthetic receptor (kin"es-thet'ik) Receptor that provides information on joint position and motion.

Juxtaglomerular apparatus (JGA) (juks"tah-glo-mer'ular) Cells of the distal tubule and afferent arteriole located close to the glomerulus that play a role in blood pressure regulation by releasing the enzyme renin.

Karyotype (kar'e-o-tīp) The diploid chromosomal complement, typically shown as homologous chromosome pairs arranged from longest to shortest (X and Y are arranged by size rather than paired).

Keratin (ker'ah-tin) Fibrous protein found in the epidermis, hair, and nails that makes those structures hard and water-repellent; precursor is keratohyaline.

Ketones (ketone bodies) (ke'tōnz) Fatty acid metabolites; strong organic acids.

Ketosis (kē-tō'sis) Abnormal condition during which an excess of ketone bodies is produced.

Killer T cell *See* Cytotoxic T cell.

Kilocalories (kcal) *See* Calorie.

Kinetic energy (ki-net'ik) The energy of motion or movement, e.g., the constant movement of atoms, or the push given to a swinging door that sets it into motion.

Krebs cycle Aerobic metabolic pathway occurring within mitochondria, in which food metabolites are oxidized and CO_2 is liberated, and coenzymes are reduced.

Labia (la'be-ah) Lips; singular, labium.

Labor Collective term for the series of events that expel the infant from the uterus.

Labyrinth (lab'ĭ-rinth") Bony cavities and membranes of the inner ear.

Lacrimal (lak'ri-mal) Pertaining to tears.

Lactation (lak-ta'shun) Production and secretion of milk.

Lacteal (lak'te-al) Special lymphatic capillaries of the small intestine that take up lipids.

Lactic acid (lak'tik) Product of anaerobic metabolism, especially in muscle.

Lacunae (lah-ku'ne) A small space, cavity, or depression; lacunae in bone or cartilage are occupied by cells.

Lamella (lah-mel'ah) A layer, such as of bone matrix in an osteon of compact bone.

Lamina (lam'ĭ-nah) (1) A thin layer or flat plate; (2) the portion of a vertebra between the transverse process and the spinous process.

Large intestine Portion of the digestive tract extending from the ileocecal valve to the anus; includes the cecum, appendix, colon, rectum, and anal canal.

Larynx (lar'ingks) Cartilaginous organ located between the trachea and the pharynx; voice box.

Latent period Period of time between stimulation and the onset of muscle contraction.

Lateral Away from the midline of the body.

Leukemia Refers to a group of cancerous conditions of white blood cells.

Leukocytes (loo'ko-sīts) White blood cells; formed elements involved in body protection that take part in inflammatory and immune responses.

Leukocytosis An increase in the number of leukocytes (white blood cells); usually the result of a microbiological attack on the body.

Leukopenia (loo"ko-pe'ne-ah) Abnormally low white blood cell count.

Leukopoiesis The production of white blood cells.

Lever system Consists of a lever (bone), effort (muscle action), resistance (weight of object to be moved), and fulcrum (joint).

Ligament (lig'ah-ment) Band of regular fibrous tissue that connects bones.

Ligands Signaling chemicals that bind specifically to membrane receptors.

Limbic system (lim'bik) Functional brain system involved in emotional response.

Lipid (lih'pid) Organic compound formed of carbon, hydrogen, and oxygen; examples are fats and cholesterol.

Lipolysis (lī-pol'ĭ-sis) The breakdown of stored fats into glycerol and fatty acids.

Liver Lobed accessory organ that overlies the stomach; produces bile to help digest fat, and serves other metabolic and regulatory functions.

Lumbar (lum'bar) Portion of the back between the thorax and the pelvis.

Lumbar vertebrae The five vertebrae of the lumbar region of the vertebral column, commonly called the small of the back.

Lumen (loo'min) Cavity inside a tube, blood vessel, or hollow organ.

Luteinizing hormone (LH) (lu'te-in-ī z"ing) Anterior pituitary hormone that aids maturation of cells in the ovary and triggers ovulation in females. In males, causes the interstitial cells of the testis to produce testosterone.

Lymph (limf) Protein-containing fluid transported by lymphatic vessels.

Lymph node Small lymphatic organ that filters lymph; contains macrophages and lymphocytes.

Lymphatic system (lim-fat'ik) System consisting of lymphatic vessels, lymph nodes, and other lymphoid organs and tissues; drains excess tissue fluid from the extracellular space and provides a site for immune surveillance.

Lymphatics General term used to designate the lymphatic vessels that collect and transport lymph.

Lymphocyte Agranular white blood cell that arises from bone marrow and becomes functionally mature in the lymphoid organs of the body.

Lymphokines (lim'fo-kīnz) Proteins involved in cell-mediated immune responses that enhance immune and inflammatory responses.

Lysosomes (li'so-sōmz) Organelles that originate from the Golgi apparatus and contain strong digestive enzymes.

Lysozyme (li'so-zī m) Enzyme in sweat, saliva, and tears that is capable of destroying certain kinds of bacteria.

Macromolecules Large, complex molecules containing from 100 to over 10,000 amino acids.

Macrophage (mak'ro-fāj") Protective cell type common in connective tissue, lymphatic tissue, and certain body organs that phagocytizes tissue cells, bacteria, and other foreign debris; important as an antigen presenter to T cells and B cells in the immune response.

Macula (mak′u-lah) (1) Static equilibrium receptor within the vestibule of the inner ear; (2) a colored area or spot.

Malignant (muh-lig′nent) Life threatening; pertains to neoplasms that spread and lead to death, such as cancer.

Malignant melanoma (mel″ah-no′mah) Cancer of the melanocytes; can begin wherever there is pigment.

Mammary glands (mam′mer-e) Milk-producing glands of the breast.

Mandible (man′dĭ-bl) Lower jawbone; U-shaped, largest bone of the face.

Mast cells Immune cells that function to detect foreign substances in the tissue spaces and initiate local inflammatory responses against them; typically found clustered deep to an epithelium or along blood vessels.

Mastication (mas″tĭ-ka′shun) Chewing.

Meatus (me-a′tus) External opening of a canal.

Mechanical advantage (power lever) Condition that occurs when the load is close to the fulcrum and the effort is applied far from the fulcrum; allows a small effort exerted over a relatively large distance to move a large load over a small distance.

Mechanical disadvantage (speed lever) Condition that occurs when the load is far from the fulcrum and the effort is applied near the fulcrum; the effort applied must be greater than the load to be moved.

Mechanical energy The energy directly involved in moving matter; e.g., in bicycle riding, the legs provide the mechanical energy that moves the pedals.

Mechanoreceptor (meh″kĕ-no-rē-sep′ter) Receptor sensitive to mechanical pressure such as touch, sound, or exerted by muscle contraction.

Medial (me′de-ahl) Toward the midline of the body.

Medial lemniscal system (lem-nis′kul) The pathway to the cerebral cortex for discriminative touch, pressure, vibration, and conscious proprioception.

Median (midsagittal) plane Specific sagittal plane that lies exactly in the midline.

Mediastinum (me″de-ah-sti′num) A subdivision of the thoracic cavity containing the pericardial cavity.

Medulla (mĕ-dul′ah) Central portion of certain organs.

Medulla oblongata (mĕ-dul′ah ob″long-gah′tah) Inferiormost part of the brain stem.

Meiosis (mi-o′sis) Nuclear division process that reduces the chromosomal number by half and results in the formation of four haploid (*n*) cells; occurs only in certain reproductive organs.

Melanin (mel′ah-nin) Dark pigment formed by cells called melanocytes; imparts color to skin and hair.

Melatonin (mel″ah-to′nin) A hormone secreted by the pineal gland that inhibits secretion of GnRH by the hypothalamus; secretion peaks at night.

Membrane potential Voltage across the plasma membrane.

Membrane receptors A large, diverse group of integral proteins and glycoproteins that serve as binding sites.

Memory cells Members of T cell and B cell clones that provide for immunological memory.

Menarche (mĕ-nar′ke) Establishment of menstrual function; the first menstrual period.

Meninges (mĕ-nin′jēz) Protective coverings of the central nervous system; from the most external to the most internal, the dura mater, arachnoid, and pia mater.

Meningitis (mĕ-nin-ji′tis) Inflammation of the meninges.

Menopause Period of life when, prompted by hormonal changes, ovulation and menstruation cease.

Menstruation (men″stroo-a′shun) The periodic, cyclic discharge of blood, secretions, tissue, and mucus from the mature female uterus in the absence of pregnancy.

Merocrine glands (mer′o-krin) Glands that produce secretions intermittently; secretions do not accumulate in the gland.

Mesencephalon (mes″en-sef′ah-lon) One of the three primary vesicles of the developing brain; becomes the midbrain.

Mesenchyme (meh′zin-kī m) Common embryonic tissue from which all connective tissues arise.

Mesenteries (mes″en-ter′ēz) Double-layered extensions of the peritoneum that support most organs in the abdominal cavity.

Mesoderm (mez′o-derm) Primary germ layer that forms the skeleton and muscles of the body.

Mesothelium (mez″o-the′le-um) The epithelium found in serous membranes lining the ventral body cavity and covering its organs.

Messenger RNA (mRNA) Long nucleotide strands that reflect the exact nucleotide sequences of the genetically active DNA and carry the message of the latter.

Metabolic (fixed) acid Acid generated by cellular metabolism that must be eliminated by the kidneys.

Metabolic rate (mĕt″ah-bol′ik) Energy expended by the body per unit time.

Metabolic water (water of oxidation) Water produced from cellular metabolism (about 10% of our body's water).

Metabolism (mĕ-tab′o-lizm) Sum total of the chemical reactions occurring in the body cells.

Metaphase Second stage of mitosis.

Metastasis (mĕ-tas′tah-sis) The spread of cancer from one body part or organ into another not directly connected to it.

Metencephalon (afterbrain) A secondary brain vesicle; anterior portion of the rhombencephalon of the developing brain; becomes the pons and the cerebellum.

Microfilaments (mi″kro-fil′ah-ments) Thin strands of the contractile protein actin.

Microglia (mi-kro′gle-ah) A type of CNS supporting cell; can transform into phagocytes in areas of neural damage or inflammation.

Microtubules (mi″kro-tu′bū lz) One of three types of rods in the cytoskeleton of a cell; hollow tubes made of spherical protein that determine the cell shape as well as the distribution of cellular organelles.

Microvilli (mi″kro-vil′i) Tiny projections on the free surfaces of some epithelial cells; increase surface area for absorption.

Micturition (mik″tu-rish′un) Urination, or voiding; emptying the bladder.

Midbrain (mesencephalon) Region of the brain stem between the diencephalon and the pons.

Midsagittal (median) plane Specific sagittal plane that lies exactly in the midline.

Milliequivalents per liter (mEq/L) The units used to measure electrolyte concentrations of body fluids; a measure of the number of electrical charges in 1 liter of solution.

Mineralocorticoid (min″er-al′ō-kor′tih-koyd) Steroid hormone of the adrenal cortex that regulates mineral metabolism and fluid balance.

Minerals Inorganic chemical compounds found in nature; salts.

Mitochondria (mi″to-kon′dre-ah) Cytoplasmic organelles responsible for ATP generation for cellular activities.

Mitosis Process during which the chromosomes are redistributed to two daughter nuclei; nuclear division. Consists of prophase, metaphase, anaphase, and telophase.

Mixed nerves Nerves containing the processes of motor and sensory neurons; their impulses travel to and from the central nervous system.

Molar (mo′lar) A solution concentration determined by mass of solute—one liter of solution contains an amount of solute equal to its molecular weight in grams.

Molarity (mo-lar′ĭ-te) A way to express the concentration of a solution; moles per liter of solution.

Mole (mōl) A mole of any element or compound is equal to its atomic weight or its molecular weight (sum of atomic weights) measured in grams.

Molecule Particle consisting of two or more atoms joined together by chemical bonds.

Monoclonal antibodies (mon″o-klo′nal)   Pure preparations of identical antibodies that exhibit specificity for a single antigen.

Monocyte (mon′o-sīt)   Large single-nucleus white blood cell; agranular leukocyte.

Monokines   Chemical mediators that enhance the immune response; secreted by macrophages.

Monosaccharide (mon″o-sak′ah-rīd)   Literally, one sugar; building block of carbohydrates; e.g., glucose.

Morula (mor′u-lah)   The mulberry-like solid mass of blastomeres resulting from cleavage in the early conceptus.

Motor areas   Functional areas in the cerebral cortex that control voluntary motor functions.

Motor end plate   Troughlike part of a muscle fiber's sarcolemma that helps form the neuromuscular junction.

Motor nerves   Nerves that carry impulses leaving the brain and spinal cord, and destined for effectors.

Motor unit   A motor neuron and all the muscle cells it stimulates.

Mucous membranes   Membranes that form the linings of body cavities open to the exterior (digestive, respiratory, urinary, and reproductive tracts).

Mucus (myoo′kus)   A sticky, thick fluid secreted by mucous glands and mucous membranes; keeps the free surface of membranes moist.

Multinucleate cell (mul″tĭ-nu′kle-āt)   Cell with more than one nucleus, e.g., skeletal muscle cells, liver cells.

Multiple sclerosis (MS)   Demyelinating disorder of the CNS; causes patches of sclerosis in the brain and spinal cord.

Multipolar neurons   Neurons with three or more processes; most common neuron type in the CNS.

Muscarinic receptors (mus″kah-rin′ik)   Acetylcholine-binding receptors of the autonomic nervous system's target organs; named for activation by the mushroom poison muscarine.

Muscle fiber   A muscle cell.

Muscle spindle (neuromuscular spindle)   Encapsulated receptor found in skeletal muscle that is sensitive to stretch.

Muscle tension   The force exerted by a contracting muscle on some object.

Muscle tone   Sustained partial contraction of a muscle in response to stretch receptor inputs; keeps the muscle healthy and ready to act.

Muscle twitch   The response of a muscle to a single brief threshold stimulus.

Muscular dystrophy   A group of inherited muscle-destroying diseases.

Muscular system   The organ system consisting of the skeletal muscles of the body and their connective tissue attachments.

Myelencephalon (spinal brain)   A secondary brain vesicle; lower part of the developing hindbrain, especially the medulla oblongata.

Myelin sheath (mi′ĕ-lin)   Fatty insulating sheath that surrounds all but the smallest nerve fibers.

Myoblasts   Embryonic mesoderm cells from which all muscle fiber develops.

Myocardial infarction (MI) (mi″o-kar′de-al in-fark′shun)   Condition characterized by dead tissue areas in the myocardium; caused by interruption of blood supply to the area.

Myocardium (mi″o-kar′de-um)   Layer of the heart wall composed of cardiac muscle.

Myofibril (mi″o-fi′bril)   Rodlike bundle of contractile filaments (myofilaments) found in muscle cells.

Myofilament (mi″o-fil′ah-ment)   Filament that constitutes myofibrils. Of two types: actin and myosin.

Myoglobin (mi″o-glo′bin)   Oxygen-binding pigment in muscle.

Myogram   A graphic recording of mechanical contractile activity produced by an apparatus that measures muscle contraction.

Myometrium (mi″o-me′tre-um)   Thick uterine musculature.

Myopia (mi-o′pe-ah)   A condition in which visual images are focused in front of rather than on the retina; nearsightedness.

Myosin (mi′o-sin)   One of the principal contractile proteins found in muscle.

Myxedema (mik″sĕ-de′mah)   Condition resulting from underactive thyroid gland.

Nares (na′rez)   The nostrils and nasal portion of the pharynx.

Natural killer (NK) cells   Defensive cells that can lyse and kill cancer cells and virus-infected body cells before the adaptive immune system is activated.

Necrosis (nĕ-kro′sis)   Death or disintegration of a cell or tissues caused by disease or injury.

Negative feedback mechanisms   The most common homeostatic control mechanism. The net effect is that the output of the system shuts off the original stimulus or reduces its intensity.

Neonatal period   The four-week period immediately after birth.

Neoplasm (ne′o-plazm)   An abnormal mass of proliferating cells. Benign neoplasms remain localized; malignant neoplasms are cancers, which can spread to other organs.

Nephron (nef′ron)   Structural and functional unit of the kidney; consists of the glomerulus and renal tubule.

Nerve cell   A neuron.

Nerve fiber   Axon of a neuron.

Nerve growth factor (NGF)   Protein that controls the development of sympathetic ganglionic neurons; secreted by the target cells of postganglionic axons.

Nerve impulse   A self-propagating wave of depolarization; also called an action potential.

Nerve plexuses   Interlacing nerve networks that occur in the cervical, brachial, lumbar, and sacral regions and primarily serve the limbs.

Nervous system   Fast-acting control system that triggers muscle contraction or gland secretion.

Neural tube   Fetal tissue which gives rise to the brain, spinal cord, and associated neural structures; formed from ectoderm by day 23 of embryonic development.

Neuroglia (nu-rog′le-ah)   Nonexcitable cells of neural tissue that support, protect, and insulate the neurons; glial cells.

Neurohypophysis (nu″ro-hi-pof′ĭ-sis)   Posterior pituitary plus infundibulum; portion of the pituitary gland derived from the brain.

Neuromuscular junction   Region where a motor neuron comes into close contact with a skeletal muscle cell.

Neuron (nu′ron)   Cell of the nervous system specialized to generate and transmit nerve impulses.

Neuron cell body   The biosynthetic center of a neuron; also called the perikaryon, or soma.

Neuronal pools   Functional groups of neurons that process and integrate information.

Neuropeptides (nu″ro-pep′tĭds)   A class of neurotransmitters including beta endorphins and enkephalins (which act as euphorics and reduce perception of pain) and gut-brain peptides.

Neurotransmitter   Chemical released by neurons that may, upon binding to receptors of neurons or effector cells, stimulate or inhibit them.

Neutral fats   Consist of fatty acid chains and glycerol; also called triglycerides or triacylglycerols. Commonly known as oils when liquid.

Neutralization reaction   Displacement reaction in which mixing an acid and a base forms water and a salt.

Neutron (nu′tron)   Uncharged subatomic particle; found in the atomic nucleus.

Neutrophil (nu′tro-fil)   Most abundant type of white blood cell.

Nicotinic receptors (nik″o-tin′ik)   Acetylcholine-binding receptors of all autonomic ganglionic neurons and skeletal muscle neuromuscular junctions; named for activation by nicotine.

Nociceptor (no″se-sep′tor)   Receptor sensitive to potentially damaging stimuli that result in pain.

Nondisjunction   Failure of sister chromatids to separate during mitosis or failure of homologous pairs to separate during meiosis; results in abnormal numbers of chromosomes in the resulting daughter cells.

Nonpolar molecules   Electrically balanced molecules.

Norepinephrine (NE) (nor″ep-ĭ-nef′rin)   A catecholamine (biogenic amine) neurotransmitter and adrenal medullary hormone, associated with sympathetic nervous system activation.

Nuclear envelope   The double membrane barrier of a cell nucleus.

Nucleic acid (nu-kle′ik)   Class of organic molecules that includes DNA and RNA.

Nucleoli (nu-kle′o-li)   Dense spherical bodies in the cell nucleus involved with ribosomal subunit synthesis and storage.

Nucleoplasm (nu′kle-o-plazm″)   A colloidal fluid, enclosed by the nuclear membrane.

Nucleosome (nu′kle-o-sōm)   Fundamental unit of chromatin; consists of a strand of DNA wound around a cluster of eight histone proteins.

Nucleotide (nu′kle-o-tīd)   Building block of nucleic acids; consists of a sugar, a nitrogen-containing base, and a phosphate group.

Nucleus (nu′kle-is)   (1) Control center of a cell; contains genetic material; (2) clusters of nerve cell bodies in the CNS.

Nutrients   Chemical substances taken in via the diet that are used for energy and cell building.

Oblique section   A cut made diagonally between the horizontal and vertical plane of the body or an organ.

Occlusion (ah-kloo′zhun)   Closure or obstruction.

Octet rule (rule of eights) (ok-tet′)   The tendency of atoms to interact in such a way that they have eight electrons in their valence shell.

Olfaction (ol-fak′shun)   Smell.

Oligodendrocyte (ol′ĭ-go-den′dro-sīt)   A type of CNS supporting cell that composes myelin sheaths.

Oocyte (o′o-sīt)   Immature female gamete.

Oogenesis (o″o-jen′ĕ-sis)   Process of ovum (female gamete) formation.

Ophthalmic (op-thal′mik)   Pertaining to the eye.

Optic (op′tik)   Pertaining to the eye or vision.

Optic chiasma (op′tik ki-az′muh)   The partial crossover of fibers of the optic nerves.

Organ   A part of the body formed of two or more tissues and adapted to carry out a specific function; e.g., the stomach.

Organ system   A group of organs that work together to perform a vital body function; e.g., the nervous system.

Organelles (or″gah-nelz′)   Small cellular structures (ribosomes, mitochondria, and others) that perform specific metabolic functions for the cell as a whole.

Organic compound   Any compound composed of atoms (some of which are carbon) held together by covalent (shared electron) bonds.

Organic   Pertaining to carbon-containing molecules, such as proteins, fats, and carbohydrates.

Organism   The living animal (or plant), which represents the sum total of all its organ systems working together to maintain life.

Origin   Attachment of a muscle that remains relatively fixed during muscular contraction.

Osmolality   The number of solute particles dissolved in one liter (1000 g) of water; reflects the solution's ability to cause osmosis.

Osmolarity (oz″mo-lar′ĭ-te)   The total concentration of all solute particles in a solution.

Osmoreceptor (oz″mo-re-sep′tor)   Structure sensitive to osmotic pressure or concentration of a solution.

Osmosis (oz-mo′sis)   Diffusion of a solvent through a membrane from a dilute solution into a more concentrated one.

Osmotic pressure   The pressure applied to a solution to prevent the passage of solvent into it; the tendency to resist further (net) water entry.

Ossicles (ah′sih-kulz).   *See* Auditory ossicles.

Ossification (os″ĭ-fi-ka′shun).   *See* Osteogenesis.

Osteoblasts (os′te-o-blasts)   Bone-forming cells.

Osteoclasts (os′te-o-klasts)   Large cells that resorb or break down bone matrix.

Osteocyte (os′te-o-sīt)   Mature bone cell.

Osteogenesis (os″te-o-jen′e-sis)   The process of bone formation; also called ossification.

Osteoid (os′te-oid)   Unmineralized bone matrix.

Osteomalacia (os″te-o-mah-la′she-ah)   Disorder in which bones are inadequately mineralized; soft bones.

Osteon (os′te-on)   System of interconnecting canals in the microscopic structure of adult compact bone; unit of bone; also called Haversian system.

Osteoporosis (os″te-o-po-ro′sis)   Increased softening of the bone resulting from a gradual decrease in rate of bone formation.

Ovarian cycle (o-vayr′e-an)   Monthly cycle of follicle development, ovulation, and corpus luteum formation in an ovary.

Ovary (o′var-e)   Female reproductive organ in which ova (eggs) are produced; female gonad.

Ovulation (ov″u-la′shun)   Ejection of an immature egg (oocyte) from the ovary.

Ovum (o′vum)   Female gamete; egg.

Oxidases   Enzymes that catalyze the transfer of oxygen in oxidation-reduction reactions.

Oxidation (oks′ĭ-da″shun)   Process of substances combining with oxygen or the removal of hydrogen.

Oxidation-reduction (redox) reaction   A reaction that couples the oxidation (loss of electrons) of one substance with the reduction (gain of electrons) of another substance.

Oxidative phosphorylation (ok′sĭ-da″tiv fos″for-ĭ-la′shun)   Process of ATP synthesis during which an inorganic phosphate group is attached to ADP; occurs via the electron transport chain within the mitochondria.

Oxygen debt   The volume of oxygen required after exercise to oxidize the lactic acid formed during exercise.

Oxyhemoglobin (ok″sĭ-he″mo-glo′bin)   Oxygen-bound form of hemoglobin.

Oxytocin (ok″sĭ-to′sin)   Hormone secreted by hypothalamic cells that stimulates contraction of the uterus during childbirth and the ejection of milk during nursing.

Paget's disease (paj′ets)   Disorder characterized by excessive bone breakdown and abnormal bone formation.

Palate (pal′at)   Roof of the mouth.

Pancreas (pan′kre-us)   Gland located behind the stomach, between the spleen and the duodenum; produces both endocrine and exocrine secretions.

Pancreatic juice (pan″kre-at′ik)   Bicarbonate-rich secretion of the pancreas containing enzymes for digestion of all food categories.

Papilla (pah-pil′ah)   Small, nipplelike projection; e.g., dermal papillae are projections of dermal tissue into the epidermis.

Parasagittal planes   All sagittal planes offset from the midline.

Parasympathetic division   The division of the autonomic nervous system that oversees digestion, elimination, and glandular function; the resting and digesting subdivision.

Parasympathetic tone   State of parasympathetic effects; e.g., unnecessary heart accelerations; normal activity levels of digestive and urinary tracts.

Parathyroid glands (par″ah-thi′roid)   Small endocrine glands located on the posterior aspect of the thyroid gland.

Parathyroid hormone (PTH)   Hormone released by the parathyroid glands that regulates blood calcium level.

Parietal (pah-ri′ĕ-tal)   Pertaining to the walls of a cavity.

Parietal serosa   The part of the double-layered membrane that lines the walls of the ventral body cavity.

Parkinson's disease   Neurodegenerative disorder of the basal nuclei involving abnormalities of the neurotransmitter dopamine; symptoms include persistent tremor and rigid movement.

Partial pressure   The pressure exerted by a single component of a mixture of gases.

Parturition (par″tu-rish′un)   Culmination of pregnancy; giving birth.

Passive immunity   Short-lived immunity resulting from the introduction of "borrowed antibodies" obtained from an immune animal or human donor; immunological memory is not established.

Passive (transport) processes Membrane transport processes that do not require cellular energy (ATP), e.g., diffusion, which is driven by kinetic energy.

Pathogen (path′o-jen) Disease-causing microorganism.

Pectoral (pek′tor-al) Pertaining to the chest.

Pectoral (shoulder) girdle Bones that attach the upper limbs to the axial skeleton; includes the clavicle and scapula.

Pedigree Traces a particular genetic trait through several generations and helps predict the genotype of future offspring.

Pelvic girdle (hip girdle) Consists of the paired coxal bones that attach the lower limbs to the axial skeleton.

Pelvis (pel′vis) (1) Basin-shaped bony structure composed of the pelvic girdle, sacrum, and coccyx; (2) expanded proximal portion of the ureter within the kidney.

Penis (pe′nis) Male organ of copulation and urination.

Pepsin Enzyme capable of digesting proteins in an acid pH.

Peptide bond (pep′tid) Bond joining the amine group of one amino acid to the acid carboxyl group of a second amino acid with the loss of a water molecule.

Perforating canals Canals that run at right angles to the long axis of the bone, connecting the vascular and nerve supplies of the periosteum to those of the central canals and medullary cavity; also called Volkmann's canals.

Pericardium (per″ĭ-kar′de-um) Double-layered serosa enclosing the heart and forming its superficial layer.

Perichondrium (per″ĭ-kon′dre-um) Fibrous, connective-tissue membrane covering the external surface of cartilaginous structures.

Perimysium (per″ĭ-mis′e-um) Connective tissue enveloping bundles of muscle fibers.

Perineum (per″ĭ-ne′um) That region of the body spanning the region between the ischial tuberosities and extending from the anus to the scrotum in males and from the anus to the vulva in females.

Periosteum (per″e-os′te-um) Double-layered connective tissue that covers and nourishes the bone.

Peripheral congestion Condition caused by failure of the right side of the heart; results in edema in the extremities.

Peripheral nervous system (PNS) Portion of the nervous system consisting of nerves and ganglia that lie outside of the brain and spinal cord.

Peripheral resistance A measure of the amount of friction encountered by blood as it flows through the blood vessels.

Peristalsis (per″i-stal′sis) Progressive, wavelike contractions that move foodstuffs through the alimentary tube organs (or that move other substances through other hollow body organs).

Peritoneum (per″ĭ-to-ne′um) Serous membrane lining the interior of the abdominal cavity and covering the surfaces of abdominal organs.

Peritonitis (per″ĭ-to-ni′tis) Inflammation of the peritoneum.

Permeability That property of membranes that permits passage of molecules and ions.

Peroxisomes (pĕ-roks′ĭ-sōmz) Membranous sacs in cytoplasm containing powerful oxidase enzymes that use molecular oxygen to detoxify harmful or toxic substances, such as free radicals.

Petechiae (pe-te′ke-ah) Small, purplish skin blotches caused by widespread hemorrhage due to thrombocytopenia.

Peyer's patches (pi′erz) Lymphoid organs located in the small intestine.

pH unit (pe-āch) The measure of the relative acidity or alkalinity of a solution.

Phagocytosis (fag″o-si-to′sis) Engulfing of foreign solids by (phagocytic) cells.

Phagosome (fag′o-sōm) Vesicle formed as a result of phagocytosis; formed when an endocytic vesicle fuses with a lysosome.

Pharmacological dose A drug dose that is dramatically higher than normal levels of that substance (e.g., hormone) in the body.

Pharyngotympanic tube Tube that connects the middle ear and the pharynx. Also called auditory tube, eustachian tube.

Pharynx (fayr′inks) Muscular tube extending from the region posterior to the nasal cavities to the esophagus.

Phenotype (fe′no-tīp) Observable expression of the genotype.

Phospholipid (fos″fo-lip′id) Modified lipid, contains phosphorus.

Phosphorylation The mechanism of ATP synthesis; includes substrate-level and oxidative phosphorylation.

Photoreceptor (fo″to-re-sep′tor) Specialized receptor cells that respond to light energy.

Physiological acidosis (as″ĭ-do′sis) Arterial pH lower than 7.35 resulting from any cause.

Physiological dose A drug dose that replicates normal levels of that substance (e.g., hormone) in the body.

Physiology (fiz″e-ol′o-je) Study of the function of living organisms.

Pineal gland (body) (pin′e-al) A hormone-secreting part of the diencephalon of the brain thought to be involved in setting the biological clock and influencing reproductive function.

Pinocytosis (pe″no-si-to′sis) Engulfing of extracellular fluid by cells.

Pituitary gland (pĭ-tu′ih-tayr″e) Neuroendocrine gland located beneath the brain that serves a variety of functions including regulation of gonads, thyroid, adrenal cortex, lactation, and water balance.

Placenta (plah-sen′tah) Temporary organ formed from both fetal and maternal tissues that provides nutrients and oxygen to the developing fetus, carries away fetal metabolic wastes, and produces the hormones of pregnancy.

Plasma (plaz′mah) The nonliving fluid component of blood within which formed elements and various solutes are suspended and circulated.

Plasma cells Members of a B cell clone; specialized to produce and release antibodies.

Plasma membrane Membrane, composed of phospholipids, cholesterol, and proteins that encloses cell contents; outer limiting cell membrane.

Platelet (plāt′let) Cell fragment found in blood; involved in clotting.

Pleura (ploo′rah) Two-layered serous membrane that lines the thoracic cavity and covers the external surface of the lung.

Pleural cavities (ploo′ral) A subdivision of the thoracic cavity; each houses a lung.

Plexus (plek′sus) A network of converging and diverging nerve fibers, blood vessels, or lymphatics.

Polar molecules Nonsymmetrical molecules that contain electrically unbalanced atoms.

Polarized State of a plasma membrane of an unstimulated neuron or muscle cell in which the inside of the cell is relatively negative in comparison to the outside; the resting state.

Polycythemia (pol″e-si-the′me-ah) An excessive or abnormal increase in the number of erythrocytes.

Polymer A substance of high molecular weight with long, chainlike molecules consisting of many similar (repeated) units.

Polypeptide (pol″e-pep′tĭd) A chain of amino acids.

Polyps Benign mucosal tumors.

Polysaccharide (pol″e-sak′ah-rīd) Literally, many sugars, a polymer of linked monosaccharides; e.g., starch, glycogen.

Pons (1) Any bridgelike structure or part; (2) the part of the brain stem connecting the medulla with the midbrain, providing linkage between upper and lower levels of the central nervous system.

Pore The surface opening of the duct of a sweat gland.

Positive feedback mechanisms Feedback that tends to cause the level of a variable to change in the same direction as an initial change.

Posterior pituitary. *See* Neurohypophysis.

Postganglionic axon (fiber) (post″gang-gle-ah′nik) Axon of a ganglionic neuron, an autonomic motor neuron that has its cell body in a peripheral ganglion; the axon projects to an effector.

Potential energy Stored or inactive energy.

Preganglionic neuron Autonomic motor neuron that has its cell body in the central nervous system and projects its axon to a peripheral ganglion.

Presbyopia (pres″be-o′pe-ah) A condition that results in the loss of near focusing ability; typical onset is around age 40.

Pressoreceptor A nerve ending in the wall of the carotid sinus and aortic arch sensitive to vessel stretching.

Pressure gradient Difference in hydrostatic pressure that drives filtration.

Primary active transport A type of active transport in which the energy needed to drive the transport process is provided directly by hydrolysis of ATP.

Prime mover Muscle that bears the major responsibility for effecting a particular movement; an agonist.

Process (1) Prominence or projection; (2) series of actions for a specific purpose.

Progesterone (pro-jes′ter-ōn) Hormone partly responsible for preparing the uterus for the fertilized ovum.

Prolactin (PRL) (pro-lak′tin) Adenohypophyseal hormone that stimulates the breasts to produce milk.

Pronation (pro-na′shun) Inward rotation of the forearm causing the radius to cross diagonally over the ulna—palms face posteriorly.

Prophase The first stage of mitosis, consisting of coiling of the chromosomes accompanied by migration of the two daughter centrioles toward the poles of the cell, and nuclear membrane breakdown.

Proprioceptor (pro″pre-o-sep′tor) Receptor located in a joint, muscle, or tendon; concerned with locomotion, posture, and muscle tone.

Prostaglandin (PG) (pros″tah-glan′din) A lipid-based membrane-associated chemical messenger synthesized by most tissue cells that acts locally as a hormonelike substance.

Prostate gland Accessory reproductive gland; produces one-third of semen volume, including fluids that activate sperm.

Protein (pro′tēn) Complex substance containing carbon, oxygen, hydrogen, and nitrogen; composes 10–30% of cell mass.

Prothrombin time Diagnostic test to determine status of hemostasis system.

Proton (pro′ton) Subatomic particle that bears a positive charge; located in the atomic nucleus.

Proton acceptor A substance that takes up hydrogen ions in detectable amounts. Commonly referred to as a base.

Proton donor A substance that releases hydrogen ions in detectable amounts; an acid.

Proximal (prok′si-mul) Toward the attached end of a limb or the origin of a structure.

Pseudounipolar neuron (soo″do-u″nĭ-po′lar) Another term for unipolar neuron.

Psychosomatic illnesses Emotion-induced illnesses.

Puberty Period of life when reproductive maturity is achieved.

Pulmonary (pul′muh-nayr-e) Pertaining to the lungs.

Pulmonary arteries Vessels that deliver blood to the lungs to be oxygenated.

Pulmonary circuit System of blood vessels that serves gas exchange in the lungs; i.e., pulmonary arteries, capillaries, and veins.

Pulmonary edema (ĕ-de′muh) Leakage of fluid into the air sacs and tissue of the lungs.

Pulmonary veins Vessels that deliver freshly oxygenated blood from the respiratory zones of the lungs to the heart.

Pulmonary ventilation Breathing; consists of inspiration and expiration.

Pulse Rhythmic expansion and recoil of arteries resulting from heart contraction; can be felt from outside the body.

Pupil Opening in the center of the iris through which light enters the eye.

Purkinje fibers (pur-kin′je) Modified cardiac muscle fibers of the conduction system of the heart.

Pus Fluid product of inflammation composed of white blood cells, the debris of dead cells, and a thin fluid.

Pyloric sphincter (pi-lor′ik sfink′ter) Valve of the distal end of the stomach that controls food entry into the duodenum.

Pyramidal (corticospinal) tracts Major motor pathways concerned with voluntary movement; descend from the frontal lobes of each cerebral hemisphere.

Pyruvic acid An intermediate compound in the metabolism of carbohydrates.

Radioactivity The process of spontaneous decay seen in some of the heavier isotopes, during which particles or energy is emitted from the atomic nucleus; results in the atom becoming more stable.

Radioisotope (ra″de-o-i′so-tōp) Isotope that exhibits radioactive behavior.

Ramus (ra′mus) Branch of a nerve, artery, vein, or bone.

Rapid eye movement (REM) sleep Stage of sleep in which rapid eye movements, an alert EEG pattern, and dreaming occur; also called paradoxical sleep.

Reactant A substance taking part in a chemical reaction.

Receptor (re-sep′tor) (1) A cell or nerve ending of a sensory neuron specialized to respond to particular types of stimuli; (2) molecule that binds specifically with other molecules, e.g., neurotransmitters, hormones, antigens.

Receptor-mediated endocytosis One of three types of endocytosis in which engulfed particles attach to receptors before endocytosis occurs.

Receptor potential A graded potential that occurs at a sensory receptor membrane.

Recessive traits A trait due to a particular allele that does not manifest itself in the presence of other alleles that generate traits dominant to it; must be present in double dose in order to be expressed.

Reduction Chemical reaction in which electrons and energy are gained by a molecule (often accompanied by gain of hydrogen ions) or oxygen is lost.

Referred pain Pain felt at a site other than the area of origin.

Reflex Automatic reaction to stimuli.

Refraction The bending of a light ray when it meets a different surface at an oblique rather than right angle.

Regeneration Replacement of destroyed tissue with the same kind of tissue.

Relative refractory period Follows the absolute refractory period; interval when a threshold stimulus is unable to trigger an action potential unless the stimulus is particularly strong.

Renal (re′nal) Pertaining to the kidney.

Renal autoregulation Process the kidney uses to maintain a nearly constant glomerular filtration rate despite fluctuations in systemic blood pressure.

Renal clearance The volume flow rate (ml/min) at which the kidneys clear the plasma of a particular solute; provides information about renal function.

Renin (re′nin) Substance released by the kidneys that is involved with raising blood pressure.

Rennin Stomach-secreted enzyme that acts on milk protein; not produced in adults.

Repolarization Movement of the membrane potential to the initial resting (polarized) state.

Reproductive system Organ system that functions to produce offspring.

Resistance exercise High-intensity exercise in which the muscles are pitted against high resistance or immovable forces and, as a result, muscle cells increase in size.

Respiration The processes involved in supplying the body with oxygen and disposing of carbon dioxide.

Respiratory system Organ system that carries out gas exchange; includes the nose, pharynx, larynx, trachea, bronchi, lungs.

Resting membrane potential The voltage that exists across the plasma membrane during the resting state of an excitable cell; ranges from −50 to −200 millivolts depending on cell type.

Reticular fibers (ret-tik′u-lar) Fine network of connective tissue fibers that form the internal supporting framework of lymphoid organs.

Reticular formation   Functional system that spans the brain stem; involved in regulating sensory input to the cerebral cortex, cortical arousal, and control of motor behavior.

Reticular lamina   A layer of extracellular material containing a fine network of collagen protein fibers; together with the basal lamina it is a major component of the basement membrane.

Reticulocyte (rĕ-tik′u-lo-sīt)   Immature erythrocyte.

Retina (ret′ĭ-nah)   Neural tunic of the eyeball; contains photoreceptors (rods, cones).

Rhombencephalon (hindbrain) (romb″en-sef′ah-lon)   Caudal portion of the developing brain; constricts to form the metencephalon and myelencephalon; includes the pons, cerebellum, and medulla oblongata.

Ribosomal RNA (rRNA)   A constituent of ribosome; exists within the ribosomes of cytoplasm and assists in protein synthesis.

Ribosomes (ri′bo-sōmz)   Cytoplasmic organelles at which proteins are synthesized.

RNA (ribonucleic acid) (ri′bo-nu-kle′ik)   Nucleic acid that contains ribose and the bases A, G, C, and U. Carries out DNA's instructions for protein synthesis.

Rods   One of the two types of photosensitive cells in the retina.

Rotation   The turning of a bone around its own long axis.

Rugae (ru′ge)   Elevations or ridges, as in stomach mucosa.

Rule of nines   Method of computing the extent of burns by dividing the body into a number of areas, each accounting for 9% (or a multiple thereof) of the total body area.

S (synthetic) phase   The part of the interphase period of the cell life cycle in which DNA replicates itself, ensuring that the two future cells will receive identical copies of genetic material.

Sagittal plane (sa′jih-tul)   A longitudinal (vertical) plane that divides the body or any of its parts into right and left portions.

Saliva   Secretion of the salivary glands; cleanses and moistens the mouth and begins chemical digestion of starchy foods.

Saltatory conduction   Transmission of an action potential along a myelinated fiber in which the nerve impulse appears to leap from node to node.

Sarcolemma   The plasma membrane surface of a muscle fiber.

Sarcomere (sar′ko-mēr)   The smallest contractile unit of muscle; extends from one Z disc to the next.

Sarcoplasm   The nonfibrillar cytoplasm of a muscle fiber.

Sarcoplasmic reticulum (SR) (sar″ko-plaz′mik rĕ-tik′u-lum)   Specialized endoplasmic reticulum of muscle cells.

Satellite cell   A type of supporting cell in the PNS; surrounds neuron cell bodies within ganglia.

Schwann cell   A type of supporting cell in the PNS; forms myelin sheaths and is vital to peripheral nerve fiber regeneration.

Sclera (skle′rah)   White opaque portion of the fibrous tunic of the eyeball.

Scrotum (skro′tum)   External sac enclosing the testes.

Sebaceous glands (oil glands) (se-ba′shus)   Epidermal glands that produce an oily secretion called sebum.

Sebum (se′bum)   Oily secretion of sebaceous glands.

Second-degree burn   A burn in which the epidermis and the upper region of the dermis are damaged.

Second messenger   Intracellular molecule generated by the binding of a chemical (hormone or neurotransmitter) to a plasma membrane receptor; mediates intracellular responses to the chemical messenger.

Secondary sex characteristics   Anatomic features, not directly involved in the reproductive process, that develop under the influence of sex hormones, e.g., male or female pattern of muscle development, bone growth, body hair distribution, etc.

Secretion (se-kre′shun)   (1) The passage of material formed by a cell to its exterior; (2) cell product that is transported to the exterior of a cell.

Secretory vesicles (granules)   Vesicles containing proteins that migrate to the plasma membrane of a cell and discharge their contents from the cell by exocytosis.

Section   A cut through the body (or an organ) that is made along a particular plane; a thin slice of tissue prepared for microscopic study.

Segregation   During meiosis, the distribution of the members of the allele pair to different gametes.

Selectively permeable membrane   A membrane that allows certain substances to pass while restricting the movement of others; also called differentially permeable membrane.

Semen (se′men)   Fluid mixture containing sperm and secretions of the male accessory reproductive glands.

Semilunar valves (sĕ″me-loo′ner)   Valves that prevent blood return to the ventricles after contraction.

Seminiferous tubules (sem″ĭ-nif′er-us)   Highly convoluted tubes within the testes; form sperm.

Sense organs   Localized collections of many types of cells working together to accomplish a specific receptive process.

Sensory (afferent) nerves   Nerves that contain processes of sensory neurons and carry impulses to the central nervous system.

Sensory areas   Functional areas of the cerebral cortex that provide for conscious awareness of sensation.

Sensory receptor   Dendritic end organs, or parts of other cell types, specialized to respond to a stimulus.

Serosa (serous membrane) (se-ro′sah)   The moist membrane found in closed ventral body cavities.

Serous fluid (sē r′us)   Clear, watery fluid secreted by cells of a serous membrane.

Serum (sē r′um)   Amber-colored fluid that exudes from clotted blood as the clot shrinks and then no longer contains fibrinogen.

Sesamoid bones (ses′ah-moid)   Short bones embedded in tendons, variable in size and number, many of which influence the action of muscles; largest is the patella (kneecap).

Severe combined immunodeficiency syndromes (SCIDs)   Congenital conditions resulting in little or no protection against disease-causing organisms of any type.

Sex chromosomes   The chromosomes, X and Y, that determine genetic sex (XX = female; XY = male); the 23rd pair of chromosomes.

Sex-linked inheritance   Inherited traits determined by genes on the sex chromosomes, e.g., X-linked genes are passed from mother to son, Y-linked genes are passed from father to son.

Sexually transmitted disease (STD)   Infectious disease spread through sexual contact.

Signal sequence   A short peptide segment present in a protein being synthesized that causes the associated ribosome to attach to the membrane of rough ER.

Simple diffusion   The unassisted transport across a plasma membrane of a lipid-soluble or very small particle.

Sinoatrial (SA) node (si″no-a′tre-al)   Specialized myocardial cells in the wall of the right atrium; pacemaker of the heart.

Sinus (si′nus)   (1) Mucous-membrane-lined, air-filled cavity in certain cranial bones; (2) dilated channel for the passage of blood or lymph.

Skeletal cartilage   Composes most of the skeleton in early fetal life; articular cartilage, nasal cartilage in the adult skeleton.

Skeletal muscle   Muscle composed of cylindrical multinucleate cells with obvious striations; the muscle(s) attached to the body's skeleton; voluntary muscle.

Skeletal system   System of protection and support composed primarily of bone and cartilage.

Skull   Bony protective encasement of the brain and the organs of hearing and equilibrium; includes the facial bones. Also called the cranium.

Small intestine   Convoluted tube extending from the pyloric sphincter to the ileocecal valve where it joins the large intestine; the site where digestion is completed and virtually all absorption occurs.

Smooth muscle   Spindle-shaped cells with one centrally located nucleus and no externally visible striations (bands). Found mainly in the walls of hollow organs.

Sodium-potassium (Na⁺-K⁺) pump   A primary active transport system that simultaneously drives Na^+ out of the cell against a steep gradient and pumps K^+ back in.

Sol-gel transformation   Reversible change of a colloid from a fluid (sol) to a more solid (gel) state.

Solute (sol′yoot)   The substance that is dissolved in a solution.

Solute pump   Enzyme-like protein carrier that mediates active transport of solutes such as amino acids and ions uphill against their concentration gradients.

Somatic nervous system (so-ma′tik)   Division of the peripheral nervous system that provides the motor innervation of skeletal muscles; also called the voluntary nervous system.

Somatic reflexes   Reflexes that activate skeletal muscle.

Somatosensory system   That part of the sensory system dealing with reception in the body wall and limbs; receives inputs from exteroceptors, proprioceptors, and interoceptors.

Somite (so′mīt)   A mesodermal segment of the body of an embryo that contributes to the formation of skeletal muscles, vertebrae, and dermis of skin.

Spatial discrimination   The ability of neurons to identify the site or pattern of stimulation.

Special senses   The senses of taste, smell, vision, hearing, and equilibrium.

Specific gravity   Term used to compare the weight of a substance to the weight of an equal volume of distilled water.

Sperm (spermatozoa)   Male gamete.

Spermatogenesis (sper″mah-to-jen′ĕ-sis)   The process of sperm (male gamete) formation; involves meiosis.

Sphincter (sfink′ter)   A circular muscle surrounding an opening; acts as a valve.

Spinal cord   The bundle of nervous tissue that runs from the brain to the first to third lumbar vertebrae and provides a conduction pathway to and from the brain.

Spinal nerves   The 31 nerve pairs that arise from the spinal cord.

Splanchnic circulation (splangk′nik)   The blood vessels serving the digestive system.

Spleen   Largest lymphoid organ; provides for lymphocyte proliferation, immune surveillance and response, and blood-cleansing functions.

Spongy bone   Internal layer of skeletal bone. Also called cancellous bone.

Sprain   Ligaments reinforcing a joint are stretched or torn.

Static equilibrium   Sense of head position in space with respect to gravity.

Stenosis (stě-no′sis)   Abnormal constriction or narrowing.

Steroids (stě′roidz)   Group of chemical substances including certain hormones and cholesterol; they are fat soluble and contain little oxygen.

Stimulus (stim′u-lus)   An excitant or irritant; a change in the environment that evokes a response.

Stomach   Temporary reservoir in the gastrointestinal tract where chemical breakdown of proteins begins and food is converted into chyme.

Stressor   Any stimulus that directly or indirectly causes the hypothalamus to initiate stress-reducing responses, such as the fight-or-flight response.

Stroke.   *See* Cerebrovascular accident.

Stroke volume (SV)   Amount of blood pumped out of a ventricle during one contraction.

Stroma (stro′mah)   The basic internal structural framework of an organ.

Structural (fibrous) proteins   Consist of extended, strandlike polypeptide chains forming a strong, ropelike structure that is linear, insoluble in water, and very stable; e.g., collagen.

Subcutaneous (sub″kyu-ta′ne-us)   Beneath the skin.

Substrate   A reactant on which an enzyme acts to cause a chemical action to proceed.

Sudoriferous gland (su″do-rif′er-us)   Epidermal gland that produces sweat.

Sulcus (sul′kus)   A furrow on the brain, less deep than a fissure.

Summation   Accumulation of effects, especially those of muscular, sensory, or mental stimuli.

Superficial   Located close to or on the body surface.

Superior   Refers to the head or upper body regions.

Superior vena cava   Vein that returns blood from body regions superior to the diaphragm.

Supination (soo″pĭ-na′shun)   The outward rotation of the forearm causing palms to face anteriorly.

Suppressor T cells   Regulatory T lymphocytes that suppress the immune response.

Surfactant (ser-fak′tant)   Secretion produced by certain cells of the alveoli that reduces the surface tension of water molecules, thus preventing the collapse of the alveoli after each expiration.

Suspension   Heterogeneous mixtures with large, often visible solutes that tend to settle out.

Suture (soo′cher)   An immovable fibrous joint; with one exception, all bones of the skull are united by sutures.

Sweat gland.   *See* Sudoriferous gland.

Sympathetic division   The division of the autonomic nervous system that activates the body to cope with some stressor (danger, excitement, etc.); the fight, fright, and flight subdivision.

Sympathetic (vasomotor) tone   State of partial vasoconstriction of the blood vessels maintained by sympathetic fibers.

Symphysis (sim′fih-sis)   A joint in which the bones are connected by fibrocartilage.

Synapse (sin′aps)   Functional junction or point of close contact between two neurons or between a neuron and an effector cell.

Synapsis (sĭ-nap′sis)   Pairing of homologous chromosomes during the first meiotic division.

Synaptic cleft (si-nap′tik)   Fluid-filled space at a synapse.

Synaptic delay   Time required for an impulse to cross a synapse between two neurons.

Synaptic knobs (boutons)   Axonal terminals; the bulbous distal endings of the terminal branches of an axon.

Synaptic vesicles   Small membranous sacs containing neurotransmitter.

Synarthrosis (sin″ar-thro′sis)   Immovable joint.

Synchondrosis (sin″kon-dro′sis)   A joint in which the bones are united by hyaline cartilage.

Syndesmosis (sin″des-mo′sis)   A joint in which the bones are united by a ligament or a sheet of fibrous tissue.

Synergist (sin′er-jist)   Muscle that aids the action of a prime mover by effecting the same movement or by stabilizing joints across which the prime mover acts to prevent undesirable movements.

Synostosis (sin″os-to′sis)   A completely ossified joint; a fused joint.

Synovial fluid   Fluid secreted by the synovial membrane; lubricates joint surfaces and nourishes articular cartilages.

Synovial joint   Freely movable joint exhibiting a joint cavity; also called a diarthrosis.

Synthesis (combination) reaction   A chemical reaction in which larger, more complex atoms or molecules are formed from simpler ones.

Systemic (sis-tem′ik)   Pertaining to the whole body.

Systemic circuit   System of blood vessels that serves gas exchange in the body tissues.

Systole (sis′to-le)   Period when either the ventricles or the atria are contracting.

Systolic pressure (sis-tah′lik)   Pressure exerted by blood on the blood vessel walls during ventricular contractions.

T cells   Lymphocytes that mediate cellular immunity; include helper, killer, suppressor, and memory cells. Also called T lymphocytes.

T tubule (transverse tubule) Extension of the muscle cell plasma membrane (sarcolemma) that protrudes deeply into the muscle cell.

Tachycardia (tak″e-kar′de-ah) A heart rate over 100 beats per minute.

Target cell A cell that is capable of responding to a hormone because it bears receptors to which the hormone can bind.

Taste buds Sensory receptor organs that house gustatory cells, which respond to dissolved food chemicals.

Telencephalon (endbrain) (tel″en-seh′fuh-lon) Anterior subdivision of the primary forebrain that develops into olfactory lobes, cerebral cortex, and corporal striata.

Telodendria The terminal branches of an axon.

Telophase The final phase of mitosis; begins when migration of chromosomes to the poles of the cell has been completed and ends with the complete separation of the two daughter cells.

Tendon (ten′dun) Cord of dense fibrous tissue attaching muscle to bone.

Tendonitis Inflammation of tendon sheaths, typically caused by overuse.

Testis (tes′tis) Male primary reproductive organ that produces sperm; male gonad.

Testosterone (tes-tos′tĕ-rōn) Male sex hormone produced by the testes; during puberty promotes virilization, and is necessary for normal sperm production.

Tetanus (tet′ah-nus) (1) A smooth, sustained muscle contraction resulting from high-frequency stimulation; (2) an infectious disease caused by an anaerobic bacterium.

Thalamus (tha′luh-mis) A mass of gray matter in the diencephalon of the brain.

Thermogenesis (ther″mo-jen′ĕ-sis) Heat production.

Thermoreceptor (ther″mo-re-sep′ter) Receptor sensitive to temperature changes.

Third-degree burn A burn that involves the entire thickness of the skin; also called a full-thickness burn. Usually requires skin grafting.

Thoracic duct Large duct that receives lymph drained from the entire lower body, the left upper extremity, and the left side of the head and thorax.

Thorax (tho′raks) That portion of the body trunk above the diaphragm and below the neck.

Threshold stimulus Weakest stimulus capable of producing a response in an irritable tissue.

Thrombin (throm′bin) Enzyme that induces clotting by converting fibrinogen to fibrin.

Thrombocyte (throm′bo-sīt) Platelets; cell fragments that participate in blood coagulation.

Thrombocytopenia (throm″bo-si″to-pe′ne-ah) A reduction in the number of platelets circulating in the blood.

Thrombus (throm′bus) A clot that develops and persists in an unbroken blood vessel.

Thymine (T) (thi′mēn) Single-ring base (a pyrimidine) in DNA.

Thymus gland (thi′mus) Endocrine gland active in immune response.

Thyrocalcitonin (thi″ro-kal″sĭ-to′nin). *See* Calcitonin.

Thyroid gland (thi′roid) One of the largest of the body's endocrine glands; straddles the anterior trachea.

Thyroid hormone (TH) The major hormone secreted by thyroid follicles; stimulates enzymes concerned with glucose oxidation.

Thyroid-stimulating hormone (TSH) Adenohypophyseal hormone that regulates secretion of thyroid hormones.

Thyroxine (T₄) (thi-rok′sin) Iodine-containing hormone secreted by the thyroid gland; accelerates cellular metabolic rate in most body tissues.

Tight junction Area where plasma membranes of adjacent cells are fused.

Tissue A group of similar cells and their intercellular substance specialized to perform a specific function; primary tissue types of the body are epithelial, connective, muscle, and nervous tissue.

Tissue perfusion Blood flow through body tissues or organs.

Tonicity (to-nis′ĭ-te) A measure of the ability of a solution to cause a change in cell shape or tone by promoting osmotic flows of water.

Tonsils A ring of lymphocyte tissue around the entrance to the pharynx.

Trabecula (trah-bek′u-lah) (1) Any of the fibrous bands extending from the capsule into the interior of an organ; (2) strut or thin plate of bone in spongy bone.

Trachea (tra′ke-ah) Windpipe; cartilage-reinforced tube extending from larynx to bronchi.

Tract A collection of nerve fibers in the central nervous system having the same origin, termination, and function.

Transcription One of the two major steps in the transfer of genetic code information involving the transfer of information from a DNA gene's base sequence to the complementary base sequence of an mRNA molecule.

Transduction (trans-duk′shun) The conversion of the energy of a stimulus into an electrical event.

Transepithelial transport (trans-ep″ĭ-the′le-al) Movement of substances through, rather than between, adjacent epithelial cells connected by tight junctions, such as absorption of nutrients in the small intestine.

Transfer RNA (tRNA) Short-chain RNA molecules that transfer amino acids to the ribosome.

Transfusion reaction Occurs when the donor's red blood cells are attacked by the recipient's plasma agglutinins.

Translation One of the two major steps in the transfer of genetic code information, in which the information carried by mRNA is decoded and used to assemble polypeptides.

Transverse (horizontal) plane A plane running from right to left, dividing the body into superior and inferior parts.

Tricuspid valve (tri-kus′pid) The right atrioventricular valve.

Triglycerides (tri-glis′er-īdz) Fats and oils composed of fatty acids and glycerol; are the body's most concentrated source of energy fuel; also known as neutral fats.

Triiodothyronine (T₃) (tri″i-o″do-thi′ro-nēn) Secretion and function similar to those of thyroxine.

Tripeptide A combination of three amino acids united by means of peptide bonds.

Trophoblast (tro′fo-blast) Outer sphere of cells of the blastocyst.

Tropic hormone (trōp′ik) A hormone that regulates the function of another endocrine organ.

Trypsin Proteolytic enzyme secreted by the pancreas.

Tubular reabsorption The movement of filtrate components from the renal tubules into the blood.

Tubular secretion The movement of undesirable substances (such as drugs, urea, excess ions) from blood into filtrate.

Tumor An abnormal growth of cells; a swelling; cancerous at times.

Tunica (too′nĭ-kah) A covering or tissue coat; membrane layer.

Tympanic membrane (tim-pan′ik) Eardrum.

Ulcer (ul′ser) Lesion or erosion of the mucous membrane, such as gastric ulcer of stomach.

Umbilical cord (um-bĭ′lĭ-kul) Structure bearing arteries and veins connecting the placenta and the fetus.

Umbilicus (um-bĭ′lĭ-kus) Navel; marks site where umbilical cord was attached in fetal stage.

Unipolar neuron Neuron in which embryological fusion of the two processes leaves only one process extending from the cell body.

Unmyelinated fibers (un-mi′ĕ-lĭ-nāt″ed) Axons lacking a myelin sheath and therefore conducting impulses quite slowly.

Up-regulation Increased target cell formation of receptors in response to increasingly higher levels of the hormones to which they respond.

Uracil (U) (u′rah-sil) A smaller, single-ring base (a pyrimidine) found in RNA.

Urea (u-re′ah) Main nitrogen-containing waste excreted in urine.

Ureter (u-re′ter)   Tube that carries urine from kidney to bladder.

Urethra (u-re′thrah)   Canal through which urine passes from the bladder to outside the body.

Uric acid   The nitrogenous waste product of nucleic acid metabolism; component of urine.

Urinary bladder   A smooth, collapsible, muscular sac that stores urine temporarily.

Urinary system   System primarily responsible for water, electrolyte, and acid-base balance and removal of nitrogenous wastes.

Uterine tube (u′ter-in)   Tube through which the ovum is transported to the uterus. Also called fallopian tube.

Uterus (u′ter-us)   Hollow, thick-walled organ that receives, retains, and nourishes fertilized egg; site where embryo/fetus develops.

Uvula (u′vu-lah)   Tissue tag hanging from soft palate.

Vaccine   Preparation that provides artificially acquired active immunity.

Vagina   Thin-walled tube extending from the cervix to the body exterior; often called the birth canal.

Valence shell (va′lens)   Outermost electron shell (energy level) of an atom that contains electrons.

Varicosities   Knoblike neuronal swellings containing mitochondria and synaptic vesicles.

Vas (vaz)   A duct; vessel.

Vasa recta (va′sah rek′tah)   Capillary branches that supply loops of Henle in the medulla region of the kidney.

Vascular   Pertaining to blood vessels or richly supplied with blood vessels.

Vascular spasm   Immediate response to blood vessel injury; results in constriction.

Vasoconstriction (vas″o-kon-strik′shun)   Narrowing of blood vessels.

Vasodilation (vas″o-di-la′shun)   Relaxation of the smooth muscles of the blood vessels producing dilation.

Vasomotion (vas″o-mo′shun)   Intermittent contraction or relaxation of the precapillary sphincter beds resulting in a staggered blood flow when tissue needs are not extreme.

Vasomotor center (vas″o-mo′ter)   Brain area concerned with regulation of blood vessel resistance.

Vasomotor fibers   Sympathetic nerve fibers that cause the contraction of smooth muscle in the walls of blood vessels, thereby regulating blood vessel diameter.

Veins (vānz)   Blood vessels that return blood toward the heart from the circulation.

Ventral   Pertaining to the front; anterior.

Ventricle   (1) Paired, inferiorly located heart chambers that function as the major blood pumps; (2) cavities in the brain.

Venule (ven′ūl)   A small vein.

Vertebral column (spine) (ver′ti-brul)   Formed of a number of individual bones called vertebrae and two composite bones (sacrum and coccyx).

Vesicle (vě′sĭ-kul)   A small liquid-filled sac or bladder.

Vesicular follicle   Mature ovarian follicle.

Vestibule   An enlarged area at the beginning of a canal, i.e., inner ear, nose, larynx.

Villus (vil′us)   Fingerlike projections of the small intestinal mucosa that tremendously increase its surface area for absorption.

Visceral (vis′er-al)   Pertaining to an internal organ of the body or the inner part of a structure.

Visceral muscle   Type of smooth muscle; its cells contract as a unit and rhythmically, are electrically coupled by gap junctions, and often exhibit spontaneous action potentials.

Visceral organs (viscera)   A group of internal organs housed in the ventral body cavity.

Visceral serosa (se-ro′sah)   The part of the double-layered membrane that lines the outer surfaces of organs within the ventral body cavity.

Viscosity (vis′kos′ĭ-te)   State of being sticky or thick.

Visual field   The field of view seen when the head is still.

Vital capacity (VC)   The volume of air that can be expelled from the lungs by forcible expiration after the deepest inspiration; total exchangeable air.

Vital signs   Includes pulse, blood pressure, respiratory rate, and body temperature measurements.

Vitamins   Organic compounds required by the body in minute amounts.

Vocal folds   Mucosal folds that function in voice production (speech); also called the true vocal cords.

Volatile acid   An acid that can be eliminated by the lungs; carbonic acid is converted to CO_2, which diffuses into the alveoli.

Volkmann's canals   *See* Perforating canals.

Voluntary muscle   Muscle under strict nervous control; skeletal muscle.

Voluntary nervous system   The somatic nervous system.

Vulva (vul′vuh)   Female external genitalia.

Wallerian degeneration (wal-er′ē-an)   A process of disintegration of an axon that occurs when it is crushed or severed and cannot receive nutrients from the cell body.

White matter   White substance of the central nervous system; myelinated nerve fibers.

Xenograft   Tissue graft taken from another animal species.

Yolk sac (yōk)   Endodermal sac that serves as the source for primordial germ cells.

Zygote (zi′gōt)   Fertilized egg.

Photo and Illustration Credits

Photo Credits

Chapter 1
1.1: Photodisc.
1.8, 1: Jenny Thomas/Addison Wesley Longman.
1.8a: Simon Fraser/Royal Victoria Infirmary, Newcastle Upon Tyne/Science Photo Library/Photo Researchers.
1.8b: Science Photo Library/Photo Researchers.
1.8c: CNRI/Science Photo Library/Photo Researchers.

Chapter 2
2.21c: Computer Graphics Laboratory, University of California, San Francisco.

Chapter 3
3.9a–c: David M. Philips/Visuals Unlimited.
3.12b: Birgit H. Satir, Dept. of Anatomy and Structural Biology, Albert Einstein College of Medicine.
3.14: John Heuser, Heuser Lab.
3.17b: Don Fawcett/Visuals Unlimited.
3.18b: Barry King, University of California, Davis School of Medicine.
3.20b: Don Fawcett/Visuals Unlimited.
3.22: K. G. Murti/Visuals Unlimited.
3.24a: Mary Osborn, Max Planck Institute.
3.24b: Frank Solomon and J. Dinsmore, Massachusetts Institute of Technology.
3.24c: Mark S. Ladinsky and J. Richard McIntosh, University of Colorado.
3.26c: David M. Philips/Visuals Unlimited.
3.28b,c: Don W. Fawcett/Visuals Unlimited.
3.29a: Ada Olins/Biological Photo Services.
3.29b: GF Bahr, Armed Forces Institute of Pathology.
3.32: Conly Rieder.
3.38b: E. Kiselva/D. Fawcett/Visuals Unlimited.

Chapter 4
4.2a: G. W. Willis/Visuals Unlimited.
4.2b: Allen Bell, University of New England/Benjamin Cummings.
4.2c: Cabisco/Visuals Unlimited.
4.2d: R. Calentine/Visuals Unlimited.
4.2e,f: Allen Bell, University of New England/Benjamin Cummings.
4.6a: J. A. Buckwalter and L. Rosenberg. From *Collagen Relat. Res.* (1983) 3:489–504, Amsterdam: North Holland, 1975.
4.8a,b: Ed Reschke.
4.8c: Carolina Biological/Visuals Unlimited.

4.8d: Allen Bell, University of New England/Benjamin Cummings.
4.8e: Ed Reschke/Peter Arnold.
4.8f,g: Ed Reschke.
4.8h,i: Allen Bell, University of New England/Benjamin Cummings.
4.8j,k: Ed Reschke.
4.10: Biophoto Associates/Photo Researchers.
4.11a: Eric Graves/Photo Researchers.
4.11b: Ed Reschke.
4.11c: Allen Bell, University of New England/Benjamin Cummings.

Chapter 5
5.2b: Dennis Strete/Fundamental Photographs.
5.3a: Cabisco/Visuals Unlimited.
5.3b: John D. Cunningham/Visuals Unlimited.
5.5: CNRI/Science Photo Library/Photo Researchers.
5.6b: Carolina Biological Supply/Phototake.
5.6d: Manfred Kage/Peter Arnold.
5.7a: Bart's Medical Library/Phototake.
5.7b: P. Marazzi/SPL/Photo Researchers.
5.7c: Zeva Oelbaum/Peter Arnold.
5.8b: Myrleen Ferguson/PhotoEdit.
5.8c: P. Marazzi/Science Photo Library/Photo Researchers.
5.8d: John Meyer/Custom Medical Stock Photography.

Chapter 6
6.6c: Allen Bell, University of New England/Benjamin Cummings.
6.9: Ed Reschke.
6.15a,b: P. Motta, Department of Anatomy, University "La Sapienza," Rome/Science Photo Library/Photo Researchers.
Table 6.2, 1: Lester Bergman/Corbis.
2: ISM/Phototake.
3: SIU/Peter Arnold.
4: SIU/Visuals Unlimited.
5,6: University of Iowa's Virtual Hospital.

Chapter 7
7.5, 3: R.T. Hutchings.
7.6c: Dave Roberts/Science Photo Library/Photo Researchers.
7.7, 3: R.T. Hutchings.
7.9a: R.T. Hutchings.
7.24c: Lester Bergman/Corbis.
7.25, 1–6: R.T. Hutchings.
7.26b: Department of Anatomy and Histology, University of California, San Francisco.
7.28a: From *A Stereoscopic Atlas of Human Anatomy* by David L. Bassett.
7.29b: Biophoto Associates/Science Source/Photo Researchers.
7.30, 1–4: R.T. Hutchings.

Table 7.4, 3–6: From *A Stereoscopic Atlas of Human Anatomy* by David L. Bassett.

Chapter 8
8.3b: Mark Neilsen, University of Utah/Addison Wesley Longman.
8.5a–g: John Wilson White/Addison Wesley Longman.
8.6a–e: John Wilson White/Addison Wesley Longman.
8.8d: L. Bassett/Visuals Unlimited.
8.10c: VideoSurgery/Photo Researchers.
8.11b: From *A Stereoscopic Atlas of Human Anatomy* by David L. Bassett.
8.12c: From *A Stereoscopic Atlas of Human Anatomy* by David L. Bassett.
8.13: Elaine Marieb.
8.14: CNRI/Science Photo Library/Photo Researchers.

Chapter 9
9.1: Thomas S. Leeson.
9.2b: John D. Cunningham/Visuals Unlimited.
9.3a: Marian Rice.
9.4e: John Heuser.
9.6, 1–3: H. E. Huxley, Brandeis University, Waltham, MA.
9.12b: Eric Graves/Photo Researchers.
Table 9.4, 4: Eric Graves/Photo Researchers.
5: Marian Rice.
6: Department of Anatomy and Histology, University of California, San Francisco.

Chapter 10
10.4a: Paul Waring/BioMed Arts Associates/Benjamin Cummings.
10.5a: Paul Waring/BioMed Arts Associates/Benjamin Cummings.
10.9c: From *A Stereoscopic Atlas of Human Anatomy* by David L. Bassett.
10.10c: From *A Stereoscopic Atlas of Human Anatomy* by David L. Bassett.
10.13b, 1: From *A Stereoscopic Atlas of Human Anatomy* by David L. Bassett.
10.16c: L. Bassett/Visuals Unlimited.
10.25c: Charles Thomas, Kansas University Medical Center.

Chapter 11
11.4a: Manfred Kage/Peter Arnold.
11.5d: Don W. Fawcett/ Photo Researchers.
11.17b: Oliver Meckes/Ottawa/Photo Researchers.
11.18: John E. Rash, Department of Biomedical Sciences, Colorado State University.

Chapter 12
12.6b: Robert A. Chase.
12.6c: A. Glauberman/Photo Researchers.
12.7: Volker Steger/Peter Arnold.

12.10a: Peter J. Ocello, Office of Intellectual Properties, Michigan State University.
12.11b, 1: Pat Lynch/Photo Researchers.
12.14: Leonard Lessin/Peter Arnold.
12.17a,c: L. Bassett/Visuals Unlimited.
12.17d: Biophoto Associates/Photo Researchers.
12.20a: Alexander Tsiara/Photo Researchers.
12.23b: From *A Stereoscopic Atlas of Human Anatomy* by David L. Bassett.
12.26b,c,d: L. Bassett/Visuals Unlimited.

Chapter 13
13.4a: Richard Tauber/Benjamin Cummings.
13.7b: From *A Stereoscopic Atlas of Human Anatomy* by David L. Bassett.
13.9c: Ed Reschke/Peter Arnold.
13.10: A. L. Blum/Visuals Unlimited.
13.18b: Stephen Spector/Benjamin Cummings.
13.21d: P. Motta/Department of Anatomy/University "La Sapienza," Rome/Science Photo Library/Photo Researchers.
13.25c: I. M. Hunter-Duvar, Department of Otolaryngology, The Hospital for Sick Children, Toronto.
13.26a: From *Tissues and Organs: A Text Atlas of Scanning Electron Microscopy* by R. Kessel and R. Kardon, © W. H. Freeman, 1979.
Table 13.2c: William Thompson/Addison Wesley Longman.

Chapter 15
15.7b: Ed Reschke.
15.9a,b: Photo Researchers.
15.10b: From *Color Atlas of Histology* by Leslie P. Garner and James L. Hiatt, © Williams and Wilkins, 1990.
15.12b: Ed Reschke.
15.14a,b: Charles B. Wilson, Neurological Surgery, University of California Medical Center, San Francisco.
15.16: Carolina Biological Supply/Phototake.

Chapter 16
16.2: Ed Reshke/Peter Arnold.
16.8a,b: Stan Flegler/Visuals Unlimited.
16.10a: J. O. Ballard.
16.10b: Ed Reschke/Peter Arnold.
16.10c,d: J. O. Ballard.
16.10e: Carden Jennings Publishing Ltd.
16.14: Meckes/Ottawa/Photo Researchers.
16.15, 1–4: Jack Scanlon, Holyoke Community College.

Chapter 17
17.4a: A. & F. Michler/Peter Arnold.
17.4c: L. Bassett/Visuals Unlimited.
17.4f: A. & F. Michler/Peter Arnold.
17.8b: From *A Stereoscopic Atlas of Human Anatomy* by David L. Bassett.
17.8c: Lennart Nilsson, *The Body Victorious*, New York: Dell, © Boehringer Ingelheim International GmbH.
17.8d: L. Bassett/Visuals Unlimited.
17.11a: Ed Reschke.
17.19: Phil Jude/Science Photo Library/Photo Researchers.
17.23: Robert Brenner/PhotoEdit.

Chapter 18
18.1a: Gladden Willis/Visuals Unlimited.
18.20c: Science Photo Library/Photo Researchers.

Chapter 19
19.3: Francis Leroy, Biocosmos/Science Photo Library/Photo Researchers.
19.4b: Biophoto Associates/Photo Researchers.
19.6c: Mark Neilsen/Benjamin Cummings.
19.6d: LUMEN Histology, Loyola University Medical Education Network, http://www.lumen.luc.edu/lumen/MedEd/Histo/frames/histo_frames.html.
19.7: Michael Abbey/Photo Researchers.
19.8: John Cunningham/Visuals Unlimited.
19.9: Biophoto Associates/Science Source/Photo Researchers.

Chapter 20
20.1a: Lennart Nilsson, *The Body Victorious*, New York: Dell, © Boehringer Ingelheim International GmbH.
20.12b: Arthur J. Olsen, The Scripps Research Institute.
20.18b: Andrejs Liepins/Science Photo Library/Photo Researchers.

Chapter 21
21.2a: John White Wilson/Addison Wesley Longman.
21.3a: From *A Stereoscopic Atlas of Human Anatomy* by David L. Bassett.
21.4c,d: From *A Stereoscopic Atlas of Human Anatomy* by David L. Bassett.
21.6b: University of San Francisco.
21.6c: Science Photo Library/Photo Researchers.
21.8b: Carolina Biological Supply/Phototake.
21.9a: R. Kessel/Visuals Unlimited.
21.10a: Richard Tauber/Benjamin Cummings.
21.10b: From *A Stereoscopic Atlas of Human Anatomy* by David L. Bassett.
21.11: Science Photo Library/Photo Researchers.

Chapter 22
22.8, 2,3: From *Tissues and Organs* by R. Kessel and R. Kardon, © W. H. Freeman, 1979.
22.9b: Science Photo Library/Photo Researchers.
22.12a: Biophoto Associates/Photo Researchers.
22.12b: From *Color Atlas of Histology* by Leslie P. Garner and James L. Hiatt, © Williams and Wilkins, 1990.
22.14b: From *Color Atlas of Histology* by Leslie P. Garner and James L. Hiatt, © Williams and Wilkins, 1990.
22.21d: LUMEN Histology, Loyola University Medical Education Network, http://www.lumen.luc.edu/lumen/MedEd/Histo/frames/histo_frames.html.
22.22a: P. Motta/Department of Anatomy/University "La Sapienza," Rome/Science Photo Library/Photo Researchers.
22.22b: Secchi-Lecaque-Roussel-UCLAF/CNRI/Science Photo Library/Photo Researchers.
22.23a,b: From *A Stereoscopic Atlas of Human Anatomy* by David L. Bassett.
22.24b: From *A Stereoscopic Atlas of Human Anatomy* by David L. Bassett.
22.26b: Victor Eroschenko, University of Idaho/Benjamin Cummings.
22.30a: From *A Stereoscopic Atlas of Human Anatomy* by David L. Bassett.
22.31: Ed Reschke/Peter Arnold.

Chapter 23
23.24: Laurent/B. Hop Ame/Photo Researchers.
23.26: Omni-Photo.

Chapter 24
24.1b: L. Bassett/Visuals Unlimited.
24.2b: Richard Tauber/Benjamin Cummings.
24.3a: From *A Stereoscopic Atlas of Human Anatomy* by David L. Bassett.
24.4c: Biophoto Associates/Photo Researchers.
24.5c: From *Tissues and Organs* by R. Kessel and R. Kardon, © W. H. Freeman, 1979.
24.7b: P. Motta and M. Castellucci/Science Photo Library/Photo Researchers.
24.17: Biophoto Associates/Photo Researchers.

Chapter 25
25.5: Spencer Grant/Stock Boston.

Chapter 26
26.3b,c: Ed Reschke.
26.8a: From *Tissues and Organs* by R. Kessel and R. Kardon, © W. H. Freeman, 1979.
26.9b: Manfred Kage/Peter Arnold.
26.12b: Ed Reschke.

26.13: C. Edelman/La Vilette/Photo Researchers.
26.14b: From *A Stereoscopic Atlas of Human Anatomy* by David L. Bassett.
26.15a: Victor Eroschenko, University of Idaho/Benjamin Cummings.
26.18a: Spencer Grant/Stock Boston.
26.18b,c: Richard D'Amico/Custom Medical Stock Photography.
26.23: Kenneth Greer/Visuals Unlimited.

Illustration Credits

All illustrations by Imagineering STA Media Services unless otherwise noted.

Chapter 1
1.1: Jeanne Koelling.
1.2: Adapted from Campbell, Reece, and Mitchell, *Biology*, 5e, F40.10a, © Benjamin Cummings, 1999.
1.3: Vincent Perez/Wendy Hiller Gee.
1.10a: Adapted from Seeley, Stephens, and Tate, *Anatomy & Physiology*, 4e, F1.15a, New York: WCB/McGraw-Hill, 1998.
1.10b: Adapted from Marieb and Mallatt, *Human Anatomy*, 3e, F1.9, © Benjamin Cummings, 2003.
1.13: Adapted from Marieb and Mallatt, *Human Anatomy*, 3e, F1.10, © Benjamin Cummings, 2003.

Chapter 2
2.5b: Adapted from Campbell, Reece, and Mitchell, *Biology*, 5e, F2.13, © Benjamin Cummings, 1999.

Chapter 3
3.2: Tomo Narashima.
3.5: Tomo Narashima.
3.5b,c: Adapted from Lodish, et al., *Molecular Cell Biology*, 3e, F24-35, F24-38, New York: Freeman, © Scientific American Books, 1995.
3.17a: Tomo Narashima.
3.18a,c: Tomo Narashima.
3.20: Tomo Narashima.
3.21: Tomo Narashima.
3.23: Adapted from Campbell, Reece, and Mitchell, *Biology*, 5e, F7.16, © Benjamin Cummings, 1999.
3.24: Tomo Narashima.
3.25: Adapted from Campbell, *Biology*, 4e, F7.21, © Benjamin Cummings, 1996.
3.26: Tomo Narashima.
3.28a: Tomo Narashima.
3.32: Adapted from Campbell, Reece, and Mitchell, *Biology*, 5e, F12.5, © Benjamin Cummings, 1999.
3.33: Adapted from Campbell, Reece, and Mitchell, *Biology*, 5e,

F12.14, © Benjamin Cummings, 1999.

Chapter 4
4.6: From Mathews, Van Holde, and Ahern, *Biochemistry*, 3e, F9.24, © Benjamin Cummings, 2000.

Chapter 5
5.1: Tomo Narashima.

Chapter 6
6.2: Adapted from Tortora and Grabowski, *Principles of Anatomy and Physiology*, 9e, F7.2, New York: Wiley, © Biological Sciences Textbooks and Sandra Reynolds Grabowski, 2000.

Chapter 7
7.1: Nadine Sokol.
7.2a,b: Nadine Sokol.
7.3a,b: Nadine Sokol.
7.4a–c: Nadine Sokol.
7.5: Nadine Sokol.
7.6a,b: Nadine Sokol.
7.7: Nadine Sokol.
7.8: Nadine Sokol.
7.9b: Nadine Sokol.
7.10: Nadine Sokol.
7.12: Nadine Sokol.
7.13: Kristin Otwell.
7.14: Laurie O'Keefe.
7.15: Laurie O'Keefe.
7.16: Laurie O'Keefe.
7.17: Laurie O'Keefe.
7.20: Laurie O'Keefe.
7.21: Nadine Sokol.
7.22: Laurie O'Keefe.
7.23: Laurie O'Keefe.
7.24a,b: Laurie O'Keefe.
7.26: Laurie O'Keefe.
7.27: Laurie O'Keefe.
7.28b: Laurie O'Keefe.
7.29a: Laurie O'Keefe.
7.31: Kristin Otwell.
Table 7.2: Kristin Otwell.
Table 7.3: Kristin Otwell.
Table 7.4: Laurie O'Keefe.

Chapter 8
8.2: Barbara Cousins.
8.4: Barbara Cousins.
8.8a–c,e: Barbara Cousins.
8.10a,b: Barbara Cousins.
8.11a,c,d: Barbara Cousins.
8.12a,b,d: Barbara Cousins.

Chapter 9
9.2a: Raychel Ciemma.
9.3: Adapted from Marieb and Mallatt, *Human Anatomy*, 3e, F10.4, © Benjamin Cummings, 2003.
9.17: Adapted from Martini, *Fundamentals of Anatomy & Physiology*, 4e, F10.14, Upper Saddle River, NJ: Prentice Hall, © Frederic H. Martini, 1998.
9.24: Adapted from Marieb and Mallatt, *Human Anatomy*, 3e, F10.12, © Benjamin Cummings, 2003.

Chapter 10
10.1: Adapted from Martini, *Fundamentals of Anatomy & Physiology*, 4e, F11.1, Upper Saddle River, NJ: Prentice Hall, © Frederic H. Martini, 1998.
10.4b: Raychel Ciemma.
10.5b: Raychel Ciemma.
10.6: Raychel Ciemma.
10.7: Raychel Ciemma.
10.8: Raychel Ciemma.
10.9a,b,d,e: Raychel Ciemma.
10.10a,b: Raychel Ciemma.
10.11: Raychel Ciemma.
10.12: Raychel Ciemma.
10.13: Raychel Ciemma.
10.14: Raychel Ciemma.
10.15: Raychel Ciemma.
10.16a,b: Raychel Ciemma.
10.18: Laurie O'Keefe.
10.19: Raychel Ciemma.
10.20: Wendy Hiller Gee.
10.21: Raychel Ciemma.
10.22: Raychel Ciemma.
10.23: Raychel Ciemma.
10.25a,b,d–f: Raychel Ciemma.

Chapter 12
12.8: Adapted from Marieb and Mallatt, *Human Anatomy*, 3e, F13.10, © Benjamin Cummings, 2003.
12.15: Adapted from Marieb and Mallatt, *Human Anatomy*, 3e, F13.20, © Benjamin Cummings, 2003.
12.19: Adapted from Marieb and Mallatt, *Human Anatomy*, 3e, F13.24, © Benjamin Cummings, 2003.

12.28: Stephanie McCann.
12.32: Adapted from Marieb and Mallatt, *Human Anatomy*, 3e, F13.34, © Benjamin Cummings, 2003.

Chapter 13
13.2c,d: Adapted from Marieb and Mallatt, *Human Anatomy*, 3e, F16.1, © Benjamin Cummings, 2003.
13.5: Charles W. Hoffman.
13.19: Adapted from Marieb and Mallatt, *Human Anatomy*, 3e, F16.17, © Benjamin Cummings, 2003.
13.25: Charles W. Hoffman.
13.26b: Charles W. Hoffman.
13.30: Stephanie McCann.

Chapter 15
15.1: Adapted from Marieb, *Essentials of Human Anatomy & Physiology*, 7e, F9.3, © Benjamin Cummings, 2003.

Chapter 16
16.4: Nadine Sokol.

Chapter 17
17.1: Wendy Hiller Gee.
17.2: Barbara Cousins.
17.3: Barbara Cousins.
17.4b,d,e: Barbara Cousins.
17.6: Barbara Cousins.
17.8a: Barbara Cousins.
17.9: Barbara Cousins.
17.10: Barbara Cousins.
17.14a: Barbara Cousins.

Chapter 18
18.1: Adapted from Tortora and Grabowski, *Principles of Anatomy and Physiology*, 9e, F21.1, New York: Wiley, © Biological Sciences Textbooks and Sandra Reynolds Grabowski, 2000.
18.17b: Barbara Cousins.
18.19b: Barbara Cousins.
18.20b,d: Barbara Cousins.
18.21b: Barbara Cousins.
18.22b–d: Barbara Cousins.
18.23b,c: Barbara Cousins.
18.24b: Barbara Cousins.
18.25b,c: Barbara Cousins.
18.26b: Barbara Cousins.
18.27b,c: Barbara Cousins.
18.28b,c: Barbara Cousins.

Chapter 19
19.1, 1: Adapted from Marieb and Mallatt, *Human Anatomy*, 3e, F20.1, © Benjamin Cummings, 2003.

Chapter 20
20.14: Adapted from Johnson, *Human Biology*, 2e, F9.13, © Benjamin Cummings, 2003.

Chapter 21
21.5: Adapted from Marieb and Mallatt, *Human Anatomy*, 3e, F21.6, © Benjamin Cummings, 2003.
21.13, 1,3: Adapted from Marieb and Mallatt, *Human Anatomy*, 3e, F21.15d, © Benjamin Cummings, 2003.

21.23: Adapted from Berne and Levy, *Physiology*, 3e, F36.9, St. Louis, MO: Mosby-Year Book, © Mosby-Year Book Europe Ltd., 1993.
21.27: Adapted from Marieb and Mallatt, *Human Anatomy*, 3e, F21.16, © Benjamin Cummings, 2003.

Chapter 22
22.6: Adapted from Seeley, Stephens, and Tate, *Anatomy & Physiology*, 4e, F24.2, New York: WCB/McGraw-Hill, © McGraw-Hill, 1998.
22.7: Cyndie Wooley.
22.14a: Kristin Otwell.

Chapter 24
24.1a: Linda McVay.
24.5a,b: Adapted from Marieb and Mallatt, *Human Anatomy*, 3e, F23.8, © Benjamin Cummings, 2003.
24.18: Linda McVay.

Chapter 26
26.16: Carla Simmons.

Index

Note: Page numbers in **boldface** indicate a definition. A *t* following a page number indicates tabular material, and an *f* indicates an illustration.

continues

pep-, peps-, pept- *digest* pepsin, a digestive enzyme of the stomach; peptic ulcer

per-, permea- *through* permeate; permeable

peri- *around* perianal, situated around the anus

phago- *eat* phagocyte, a cell that engulfs and digests particles or cells

pheno- *show, appear* phenotype, the physical appearance of an individual

phleb- *vein* phlebitis, inflammation of the veins

pia *tender* pia mater, delicate inner membrane around the brain and spinal cord

pili *hair* arrector pili muscles of the skin, which make the hairs stand erect

pin-, pino- *drink* pinocytosis, the process of a cell in small particles

platy- *flat, broad* platysma, broad, flat muscle of the neck

pleur- *side, rib* pleural serosa, the membrane that lines the thoracic cavity and covers the lungs

plex-, plexus *net, network* brachial plexus, the network of nerves that supplies the arm

pneumo *air, wind* pneumothorax, air in the thoracic cavity

pod- *foot* podiatry, the treatment of foot disorders

poly- *multiple,* polymorphism, multiple forms

post- *after, behind* posterior, places behind (a specific) part

pre-, pro- *before, ahead of* prenatal, before birth

procto- *rectum, anus* proctoscope, an instrument for examining the rectum

pron- *bent forward* prone; pronate

propri- *one's own* proprioception, awareness of body parts and movement

pseudo- *false* pseudotumor, a false tumor

psycho- *mind, psyche* psychogram, a chart of personality traits

ptos- *fall* renal ptosis, a condition in which the kidneys drift below their normal position

pub- *of the pubis* puberty

pulmo- *lung* pulmonary artery, which brings blood to the lungs

pyo- *pus* pyocyst, a cyst that contains pus

pyro- *fire* pyrogen, a substance that induces fever

quad-, quadr- *four-sided* quadratus lumborum, a muscle with a square shape

re- *back, again* reinfect

rect- *straight* rectus abdominis, rectum

ren- *kidney* renal, renin, an enzyme secreted by the kidney

retin, retic- *net, network* endoplasmic reticulum, a network of membranous sacs within a cell

retro- *backward, behind* retrogression, to move backward in development

rheum- *watery flow, change, or flux* rheumatoid arthritis, rheumatic fever

rhin-, rhino- *nose* rhinitis, inflammation of the nose

ruga- *fold, wrinkle* rugae, the folds of the stomach, gallbladder, and urinary bladder

sagitt- *arrow* sagittal (directional term)

salta- *leap* saltatory conduction, the rapid conduction of impulses along myelinated neurons

sanguin- *blood* consanguineous, indicative of a genetic relationship between individuals

sarco- *flesh* sarcomere, unit of contraction in skeletal muscle

saphen- *visible, clear* great saphenous vein, superficial vein of the thigh and leg

sclero- *hard* sclerodermatitis, inflammatory thickening and hardening of the skin

seb- *grease* sebum, the oil of the skin

semen *seed, sperm* semen, the discharge of the male reproductive system

semi- *half* semicircular, having the form of half a circle

sens- *feeling* sensation; sensory

septi- *rotten* sepsis, infection; antiseptic

septum *fence* nasal septum

sero- *serum* serological tests, which assess blood conditions

serrat- *saw* serratus anterior, a muscle of the chest wall that has a jagged edge

sin-, sino- *a hollow* sinuses of the skull

soma- *body* somatic nervous system

somnus *sleep* insomnia, inability to sleep

sphin- *squeeze* sphincter

splanchn- *organ* splanchnic nerve, autonomic supply to abdominal viscera

spondyl- *vertebra* ankylosing spondylitis, rheumatoid arthritis affecting the spine

squam- *scale, flat* squamous epithelium, squamous suture of the skull

steno- *narrow* stenocoriasis, narrowing of the pupil

strat- *layer* strata of the epidermis, stratified epithelium

stria- *furrow, streak* striations of skeletal and cardiac muscle tissue

stroma *spread out* stroma, the connective tissue framework of some organs

sub- *beneath, under* sublingual, beneath the tongue

sucr- *sweet* sucrose, table sugar

sudor- *sweat* sudoriferous glands, the sweat glands

super- *above, upon* superior, quality or state of being above others or a part

supra- *above, upon* supracondylar, above a condyle

sym-, syn- *together, with* synapse, the region of communication between two neurons

synerg- *work together* synergism

systol- *contraction* systole, contraction of the heart

tachy- *rapid* tachycardia, abnormally rapid heartbeat

tact- *touch* tactile sense

telo- *the end* telophase, the end of mitosis

templ-, tempo- *time* temporal summation of nerve impulses

tens- *stretched* muscle tension

tertius *third* peroneus tertius, one of three peroneus muscles

tetan- *rigid, tense* tetanus of muscles

therm- *heat* thermometer, an instrument used to measure heat

thromb- *clot* thrombocyte; thrombus

thyro- *a shield* thyroid gland

tissu- *woven* tissue

tono- *tension* tonicity; hypertonic

tox- *poison* antitoxic, effective against poison

trab- *beam, timber* trabeculae, spicules of bone in spongy bone tissue

trans- *across, through* transpleural, through the pleura

trapez- *table* trapezius, the four-sided muscle of the upper back

tri- *three* trifurcation, division into three branches

trop- *turn, change* tropic hormones, whose targets are endocrine glands

troph- *nourish* trophoblast, from which develops the fetal portion of the placenta

tuber- *swelling* tuberosity, a bump on a bone

tunic- *covering* tunica albuginea, the covering of the testis

tympan- *drum* tympanic membrane, the eardrum

ultra- *beyond* ultraviolet radiation, beyond the band of visible light

vacc- *cow* vaccine

vagin- *a sheath* vagina

vagus *wanderer* the vagus nerve, which starts at the brain and travels into the abdominopelvic cavity

valen- *strength* valence shells of atoms

venter, ventr- *hollow cavity, belly* ventral (directional term); ventricle

ventus *the wind* pulmonary ventilation

vert- *turn* vertebral column

vestibul- *a porch* vestibule, the anterior entryway to the mouth and nose

vibr- *shake, quiver* vibrissae, hairs of the nasal vestibule

villus *shaggy hair* microvilli, which have the appearance of hair in light microscopy

viscero- *organ, viscera* visceroinhibitory, inhibiting the movements of the viscera

viscos- *sticky* viscosity, resistance to flow
vita- *life* vitamin
vitre- *glass* vitreous humor, the clear jelly of the eye
viv- *live* in vivo
vulv- *a covering* vulva, the female external genitalia
zyg- *a yoke, twin* zygote

Suffixes

-able *able to, capable of* viable, ability to live or exist
-ac *referring to* cardiac, referring to the heart
-algia *pain in a certain part* neuralgia, pain along the course of a nerve
-apsi *juncture* synapse, where two neurons connect
-ary *associated with, relating to* coronary, associated with the heart
-asthen *weakness* myasthenia gravis, a disease involving paralysis
-atomos *indivisible* anatomy, which involves dissection
-bryo *swollen* embryo
-cide *destroy or kill* germicide, an agent that kills germs
-cipit *head* occipital
-clast *break* osteoclast, a cell that dissolves bone matrix
-crine *separate* endocrine organs, which secrete hormones into the blood
-dips *thirst, dry* polydipsia, excessive thirst associated with diabetes
-ectomy *cutting out, surgical removal* appendectomy, cutting out of the appendix
-ell, -elle *small* organelle
-emia *condition of the blood* anemia, deficiency of red blood cells
-esthesi *sensation* anesthesia, lack of sensation
-ferent *carry* efferent nerves, nerves carrying impulses away from the CNS
-form, -forma *shape* cribriform plate of the ethmoid bone
-fuge *driving out* vermifuge, a substance that expels worms of the intestine
-gen *an agent that initiates* pathogen, any agent that produces disease
-glea, -glia *glue* neuroglia, the connective tissue of the nervous system
-gram *data that are systematically recorded, a record* electrocardiogram, a recording showing action of the heart
-graph *an instrument used for recording data or writing* electrocardiograph, an instrument used to make an electrocardiogram
-ia *condition* insomnia, condition of not being able to sleep
-iatrics *medical specialty* geriatrics, the branch of medicine dealing with disease associated with old age
-ism *condition* hyperthyroidism
-itis *inflammation* gastritis, inflammation of the stomach
-lemma *sheath, husk* sarcolemma, the plasma membrane of a muscle cell

-logy *the study of* pathology, the study of changes in structure and function brought on by disease
-lysis *loosening or breaking down* hydrolysis, chemical decomposition of a compound into other compounds as a result of taking up water
-malacia *soft* osteomalacia, a process leading to bone softening
-mania *obsession, compulsion* erotomania, exaggeration of the sexual passions
-nata *birth* prenatal development
-nom *govern* autonomic nervous system
-odyn *pain* coccygodynia, pain in the region of the coccyx
-oid *like, resembling* cuboid, shaped as a cube
-oma *tumor* lymphoma, a tumor of the lymphatic tissues
-opia *defect of the eye* myopia, nearsightedness
-ory *referring to, of* auditory, referring to hearing
-pathy *disease* osteopathy, any disease of the bone
-phasia *speech* aphasia, lack of ability to speak
-phil, -philo *like, love* hydrophilic, water-attracting molecules
-phobia *fear* acrophobia, fear of heights
-phragm *partition* diaphragm, which separates the thoracic and abdominal cavities
-phylax *guard, preserve* anaphylaxis, prophylactic
-plas *grow* neoplasia, an abnormal growth
-plasm *form, shape* cytoplasm
-plasty *reconstruction of a part, plastic surgery* rhinoplasty, reconstruction of the nose through surgery
-plegia *paralysis* paraplegia, paralysis of the lower half of the body or limbs
-rrhagia *abnormal or excessive discharge* metrorrhagia, uterine hemorrhage
-rrhea *flow or discharge* diarrhea, abnormal emptying of the bowels
-scope *instrument used for examination* stethoscope, instrument used to listen to sounds of parts of the body
-some *body* chromosome
-sorb *suck in* absorb
-stalsis *compression* peristalsis, muscular contractions that propel food along the digestive tract
-stasis *arrest, fixation* hemostasis, arrest of bleeding
-stitia *come to stand* interstitial fluid, between the cells
-stomy *establishment of an artificial opening* enterostomy, the formation of an artificial opening into the intestine through the abdominal wall
-tegm *cover* integument
-tomy *to cut* appendectomy, surgical removal of the appendix
-trud *thrust* protrude, detrusor muscle
-ty *condition of, state* immunity, condition of being resistant to infection or disease
-uria *urine* polyuria, passage of an excessive amount of urine
-zyme *ferment* enzyme